T4-ALG-969

Developing and Sustaining World Fisheries Resources

The State of Science and Management

2nd *World Fisheries* CONGRESS

Editors: DA Hancock, DC Smith, A Grant, JP Beumer

CSIRO
AUSTRALIA

National Library of Australia Cataloguing-in-Publication entry

World Fisheries Congress (2nd : 1996 : Brisbane, Qld.).
Developing and sustaining world fisheries resources : the state of science and management : 2nd World Fisheries Congress proceedings.

Bibliography
ISBN 0 643 05985 7

1. Fisheries – Congresses
2. Fishery management – Congresses
I. Hancock, D.A. (Donald Alexander), Smith, D.C., Grant, A. and Beumer, J.P.
II. CSIRO
III. Title

338.3727

This book is available from:

CSIRO Publishing
PO Box 1139 (150 Oxford Street)
Collingwood, VIC 3066
Australia

Tel: (03) 9662 7666 Int: +(613) 9662 7666
Fax: (03) 9662 7555 Int: + (613) 9662 7555
Email: sales@publish.csiro.au
http://www.publish.csiro.au

Editorial and production manager: Marta Veroni

Editors

D.A. Hancock
29 Woodlands Way
Quindalup
Western Australia 6281, Australia.

A. Grant
Marine and Freshwater Research
PO Box 1139
Collingwood, Victoria 3066, Australia.

D.C. Smith
Marine and Freshwater Resources Institute
PO Box 114
Queenscliff, Victoria 3225, Australia.

J.P. Beumer
Fisheries Group
GPO Box 3129
Brisbane, Queensland 4001, Australia.

Text Editors

D. Mahon
Marine and Freshwater Resources Institute
PO Box 114
Queenscliff, Victoria 3225, Australia.

M. Veroni
CSIRO Publications
PO Box 1139
Collingwood, Victoria 3066, Australia.

Scientific Editors

We gratefully acknowledge the time and effort of the following individuals:

N.J. Bax
D. Campbell
M. Cappo
R.W. Day
P.I. Dixon
G.J. Gooley
D.L. Grey
R.K. Griffin
L. Gunthorpe
P.D. Jackson
I.A. Knuckey
R.C.J. Lenanton
J.M. Lyle
C.M. MacDonald
A.K. Morison
G. Newman
C. O'Brien
J.G. Pepperell
B.F. Phillips
A.E. Punt
R. Reichelt
P.C. Rothlisberg
K.R. Rowlings
A.D. M. Smith
D.J. Staples
I.M. Suthers
R.E. Thresher
T.I. Walker

Organizing Committees

International Steering Committee

Members of Fisheries Societies worldwide

National Organizing Committee

John Glaister	New South Wales Fisheries (*Congress Chair*)
Nicholas Bax	CSIRO Division of Fisheries, Tasmania
John Beumer	Queensland Department of Primary Industries
Patricia Dixon	Centre for Marine Science, University of New South Wales
Darryl Grey	Northern Territory Department of Primary Industry & Fisheries
Don Hancock	Australian Society for Fish Biology, Western Australia
Martine Kinloch	South Australian Research and Development Institute
Ted Loveday	Queensland Commercial Fishermen's Organisation
Alexander Morison	Marine and Freshwater Resources Institute, Victoria (*Congress Treasurer*)
Julian Pepperell	Pepperell Research and Consulting, New South Wales
Bruce Phillips	Curtin University of Technology, Western Australia
Peter Rothlisberg	CSIRO Division of Fisheries, Queensland
Andrew Sanger	Inland Fisheries Commission, Tasmania
David Smith	Marine and Freshwater Resources Institute, Victoria
Peter Young	CSIRO Division of Fisheries, Tasmania

Program Committee

Bruce Phillips	Curtin University of Technology, Western Australia (*Program Chair*) (Themes 3 & 6 convenor)
Patricia Dixon	Centre for Marine Science, University of New South Wales (Theme 5 convenor)
Darryl Grey	Northern Territory Department of Primary Industry & Fisheries (Theme 4 convenor)
Robert Kearney	University of Canberra, Australian Capital Territory (Theme 2 convenor)
Robert Lewis	South Australian Research and Development Institute (Theme 2 convenor)
Jeremy Lyle	Department of Primary Industry and Fisheries, Tasmania (Theme 6 convenor)
Julian Pepperell	Pepperell Research and Consulting, New South Wales (Theme 3 convenor)
James Penn	Fisheries Department of Western Australia (Theme 1 convenor)
Derek Staples	Bureau of Resource Sciences, Australian Capital Territory (Themes 1 & 5 convenor)

Publications Committee

David Smith	Marine and Freshwater Resources Institute, Victoria (*Publications Chair*)
John Beumer	Queensland Department of Primary Industries
Ann Grant	Marine and Freshwater Research (*journal*)
Donald Hancock	Australian Society for Fish Biology, Western Australia
Alexander Morison	Marine and Freshwater Resources Institute, Victoria

Congress Sponsors

Principle Sponsor

Major Sponsors

Sponsors

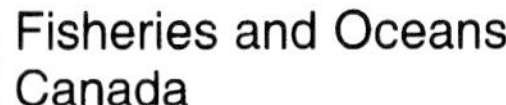

AMSC

Australasian Marine Science Consortium

Contents

CONTENTS

Contents

Contents

Contents

Preface

The Second World Fisheries Congress held in Brisbane, Australia, in July–August 1996 was a defining moment for fisheries science and management. It allowed world leaders in science, industry and fisheries management to come together and examine the status of our world resources.

We heard from Dr Pamela Mace, our keynote speaker, that supply from marine capture fisheries and inland capture fisheries has peaked and that the only likely source of increased production is through aquaculture or better use of existing resources. Fleet overcapacity, environmentally sound technologies and habitat protection are seen as world wide issues. Advances have occurred in fisheries science, and management and industry is increasingly assuming a greater leadership role. Global conservation standards, long-term strategic planning at a national level and a willingness to share responsibility are all themes that emerged from the Congress.

These proceedings encapsulate the majority of the discussions and ideas arising from the Congress. They are a unique record of this landmark meeting, and as such, the Commonwealth Scientific and Industrial Research Organisation is proud to have sponsored their preparation and publication. The Australian Society of Fish Biology, which hosted the Congress largely through generous sponsorship from the Fisheries Research and Development Corporation, is similarly proud to be associated with this important publication.

On behalf of all who took part, including the large group of organisers, sponsors and others who dedicated time, resources and funding to support the Congress and this publication, we commend this publication to you as a record of this landmark congress.

J.P. Glaister
Congress Chair

P.C. Young
CSIRO

P. Dixon
Vice President ASFB

Keynote Presentation

Developing and Sustaining World Fisheries Resources: The State of the Science and Management

Pamela M. Mace [1]

National Marine Fisheries Service, Northeast Fisheries Science Centre, 166 Water Street, Woods Hole, MA 02543, USA.

Dedication

This paper is dedicated to the memory of Peter A. Larkin, one of the great pioneers of modern fisheries science; a superb scientist, an outstanding educator, and a friend and mentor to me during the course of my graduate studies at the University of British Columbia. His humour and wisdom will always be remembered.

Summary

The popular characterization of the world's fisheries is that they are on the brink of disaster. FAO data suggest that the supply from marine capture fisheries (about 83 million tonnes (t) in total, but only about 54 million t available for direct human consumption) and inland capture fisheries (about 6.5 million t) may be approaching a limit, with almost 70% of marine capture fisheries classified as fully or overexploited. World fishing fleets are operating at an estimated net loss of about $US54 billion ($10^9$), presumably offset by government subsidies. Each year, commercial marine fisheries result in about 28.7 million t of bycatch and 27 million t of discards, equivalent to about one-third of landings from marine capture fisheries. Aquaculture has exhibited phenomenal growth, but also spectacular failures in some sectors. Debate exists about whether aquaculture production (18.5 million t of animal products and 6.9 million t of plant products in 1994) can expand sufficiently to meet projected increasing demand for food and income that probably cannot be provided by natural systems. Poor science, poor management, poor data, inadequate institutions, inadequate policy, and a host of other factors have been implicated in this 'crisis'. However, a small set of fundamental problems surpasses all others in requiring immediate and focussed remedial action; indeed, they are the key to solving many of the other problems. For capture fisheries, overcapacity is the single most important factor threatening the long-term viability of exploited fish stocks and the fisheries that depend on them. Global fleet capacity must be reduced substantially, perhaps by as much as 50%, to levels commensurate with sustainable resource productivity. For aquaculture, the most pressing need is to solve the dilemma of promoting expansion, while at the same time demanding the development of environmentally sound technologies and farming practices.

But perhaps the state of the world's fisheries is not as bad as the popular portrayal. There have been considerable advances in fisheries science, particularly computer modelling, methods for quantitative analysis, resource survey techniques and biotechnology; a number of innovative management systems are now being tested; several depleted stocks have been rebuilt; there is considerable room for expansion in aquaculture; scientists,

[1] The views expressed are those of the author, not necessarily those of the National Marine Fisheries Service.

environmentalists and fishers themselves are beginning to have a major positive impact on management objectives and practices that will enhance prospects for sustainability; and several revolutionary international agreements have recently been concluded. The challenge for the future is to build on existing successes to effect a transition to environmentally and economically sound fisheries on a global and local basis. This is a formidable task. The transition will have a high price in monetary terms, and an even higher price in social terms. Therefore, the transition will be resisted and there will be pressure to preserve the *status quo*. Unfortunately, except in rare cases, the *status quo* is no longer a viable option. The groundrules that have brought fisheries to a crossroad must be abandoned in favour of new paradigms that emphasize long-term sustainability over short-term gain. The model for the future must incorporate global conservation standards, formulation of holistic national policies focussing on long-term objectives, and greater involvement of fishers and other stakeholders in the management process. All players need to be united in their determination to implement effective management, and to be responsible and accountable for their decisions and actions.

The transition has already begun.

INTRODUCTION: FISHERIES AT A CRITICAL JUNCTURE

Fisheries are important producers of food and income. Fish account for 19% of the total human consumption of animal protein (FAO 1993*a*), with about 1×10^9 people relying on fish as their primary source of animal protein (Williams 1994). Fisheries provide direct or indirect incomes to about 200 million people (Garcia and Newton in press). One-third of the world catches is exchanged through international trade, the volume of which doubled between 1980 and 1990 (Garcia and Newton in press). The total value of fish trade in 1993 was about $US40 billion, of which developing countries accounted for about 48% (FAO 1995*a*).

However, the popular characterization of the world's fisheries is that they are on the brink of disaster. As a whole, capture fisheries are being exploited at or beyond the maximum sustainable yield (MSY) (Garcia and Newton in press). Overall, there is little or no further room for expansion. Many stocks are overexploited. Many fishing fleets are overcapitalized. Globally, fishing revenues do not cover costs (FAO 1993*a*). Many fisheries would probably not be viable without government grants and subsidies. Even in non-capital-intensive fisheries, destructive fishing practices such as dynamiting and cyanide poisoning have decimated coastal stocks (Williams 1994). Over the past five years, dozens of articles have been written in newspapers, popular magazines and journals highlighting these and related problems (e.g. Parfit and Kendrick 1995; Safina 1995).

Several recent spectacular examples of overfishing have accelerated the debate. After years of overfishing, the Grand Banks cod fishery collapsed dramatically in 1992 resulting in the loss of about 40 000 jobs in Newfoundland. In 1994, in response to severe overfishing leading to severely depleted stocks, the US New England Fishery Management Council initiated plans to reduce fishing effort on groundfish by 50% and indefinitely closed more than 3000 square miles of Georges Bank, which historically had supported one of the most diverse and prolific marine ecosystems in the world (Collins 1994). In the 1990s, the Black Sea has been characterized as the marine ecological catastrophe of the century (Garcia and Newton in press), due to a combination of uncontrolled heavy fishing, nutrient inputs and the introduction of exotic species (Caddy and Griffiths 1990).

Overfishing and overcapitalization have fuelled increasing numbers of violent conflicts. In 1995, Canadian officials arrested a Spanish trawler on the high seas, creating an international uproar. During 1993–96, traditional fishers in India burned trawlers, held trawler operators hostage, and organized strikes to protest the presence of foreign vessels inside their 200-mile zone. In 1996, fishers in Atlantic Canada threw stones through the windows of the New Brunswick Premier's house and office, while fishers in Pacific Canada occupied the offices of the Department of Fisheries and Oceans to protest a proposed plan to reduce fleet capacity.

This paper examines the causes of this 'crisis', whether a crisis really exists, impediments to progress, and future prospects and challenges. The focus is primarily on marine capture fisheries and secondarily on aquaculture, with lesser emphasis on inland capture fisheries (which suffer many of the same problems as marine capture fisheries), and subsistence and recreational fisheries (for which relatively few comprehensive data are available).

BACKGROUND: GLOBAL STATISTICS

World total fish production

Global production from fisheries began to accelerate dramatically in the 1950s (Fig. 1). The annual growth rate in production from marine fisheries (which account for about 85% of world fish yields) averaged about 6.8% in the 1950s, 7.4% in the 1960s, 1.7% in the 1970s (declining due to the collapse of the Peruvian anchoveta), 3.6% in the 1980s, and 0.5% in the first three years of the 1990s, at which time FAO speculated that world fisheries production appeared to be reaching a plateau (FAO 1995*a*). Total world fish production (marine and inland, capture and culture, but excluding cultivated plants) surpassed the 100 million t mark for the first time in 1989. During the early 1990s, total production declined slightly, but subsequently increased to new highs of 102.3 million t in 1993 and 109.6 million t in 1994, the most recent years for which FAO statistics are

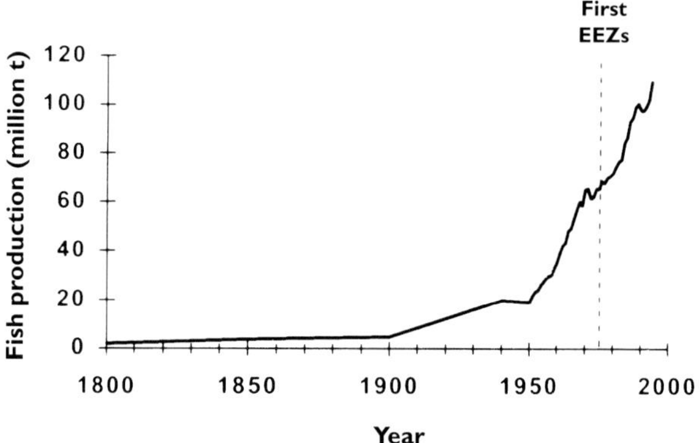

Fig. 1. World total fisheries production. Source: Hilborn (1990), FAO (1995*a*), FAO (unpublished).

available. The increase in production between 1993 and 1994 represents a record increase (7.3 million t) from one year to the next.

Marine capture fisheries

According to FAO statistics, total world landings from marine capture fisheries increased five-fold from 1950 to 1990. Over this period, the annual production of marine fish grew somewhat more rapidly than the human population. In 1993, the supply from marine capture fisheries was about 78 million t, but only about 50 million t was available for direct human consumption with the remainder used for industrial products such as feed for agriculture including aquaculture. In 1994, the supply was about 83 million t, with 54 million t for direct human consumption. FAO estimates that about 44% of the world's marine stocks are heavily to fully exploited, 25% overexploited, depleted or recovering and 32% with potential for expansion; thus nearly 70% of exploited species are fully exploited, overexploited, or rebuilding (Garcia and Newton in press).

Alverson *et al.* (1994) have estimated that commercial marine fisheries result in about 28.7 million t of bycatch and 27 million t of discards, equivalent to almost one-third of landings from marine capture fisheries. Shrimp fisheries account for the largest share (37.2%) of world discards, with discard ratios up to 15 times the landed catch. Actual removals of marine fishery resources are likely to be even higher than the sum of recorded landings and estimated discards, due to unreported landings, fish lost to spoilage before or after landing, subsurface mortality that occurs when fish encounter fishing gear but are not captured, and ghost fishing mortality which is caused by lost gear that continues to fish.

Inland capture fisheries

The rate of increase in yields from inland capture fisheries has not been as rapid as that from marine capture fisheries, increasing from about 5 million t in 1970 to 6.5 million t in 1990, with subsequent landings near this level (FAO 1995*a*, 1995*b*; FAO unpublished; Fig. 2). In 1992, the most important producers of the world inland catch were Asia (54%), Africa (25%), republics of the former USSR (7%), North America (4.5%) and South America (4.5%) (FAO 1995a). Virtually all inland fish caughtare used for human consumption. Recreational and subsistence fisheries are widespread but are not generally incorporated in the statistics reported to FAO. Commercial fisheries are also locally important, with some export markets for high-value fish and for ornamental species (FAO 1995*a*). According to FAO (1995*b*), most major inland fisheries are fully exploited or overexploited and there are few if any large inland fisheries with a confirmed potential for significant expansion. Thus, present yields probably cannot be substantially increased through traditional capture fisheries; however, yields could potentially be increased by enhancement activities such as fish stocking and habitat rehabilitation (FAO 1995*b*).

Subsistence and recreational fisheries

FAO statistics do not distinguish between fish captured commercially and recreationally. In fact, most of the recreational

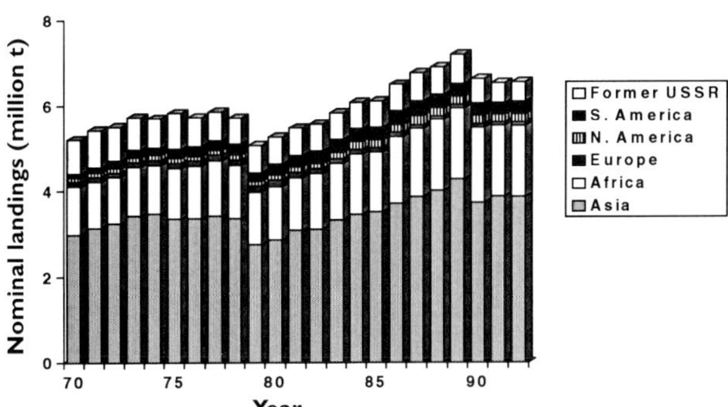

Fig. 2. Nominal landings from inland capture fisheries by continent. Source: FAO (1995*a*).

and subsistence catch is probably not included in national statistics. Therefore, no comprehensive global statistics are available for subsistence and recreational fisheries. However, subsistence fisheries may be the most important fisheries of all in many areas, and are obviously extremely important contributors to the food security of developing nations (e.g. Bailey and Zerner 1992; Williams 1994; Parfit and Kendrick 1995; Pinkerton and Weinstein 1995). Recreational fisheries are already the most important components of a few fisheries, particularly in developed countries, but also increasingly in some developing countries. In the United States, it has been estimated that in 1995, 14 million marine anglers made 65.6 million fishing trips and caught a total of 339.1 million finfish, releasing half alive, and retaining about 100 000 t; in fact, this is a substantial underestimate since it excludes fishing in Alaska, Hawaii, Washington and Texas and fishing for Pacific coast salmon (NMFS 1996*b*). In the US Atlantic, blue marlin, white marlin, sailfish and spearfish have been reserved exclusively for recreational use since 1988.

Aquaculture

Aquaculture production has been expanding rapidly (Fig. 3). Total production in 1993 was about 16.5 million t (excluding plants), a 2.4-fold increase over the 1984 production level of 6.9 million t (FAO 1995*a*). In 1994, production rose another 12.6% to 18.6 million t. In addition, production of cultured plants is significant (about 6.9 million t in 1994). By 1993, aquaculture had increased in importance to 16% of the world food fish supply (compared with 12% in 1984), with about 60% of aquaculture production from inland waters (FAO 1995*a*). The bulk of production is from developing nations, with Asia producing nearly 90% of the global total in 1994. China alone produced about 60% of the global total, followed by India (6.3%) and Japan (5.6%) (Fig. 4). In 1992, finfish, crustaceans and molluscs accounted for 68%, 24% and 7% respectively of total production (excluding plants). In the same year, aquaculture accounted for 60% of the world supply of freshwater finfish, 40% of the supply of molluscs, 30% of marine shrimp supplies, and 43% of salmon supplies, but only 5% of the world supply of marine fish (FAO 1995*a*).

However, the recent expansion of aquaculture has been accompanied by serious environmental concerns and disease

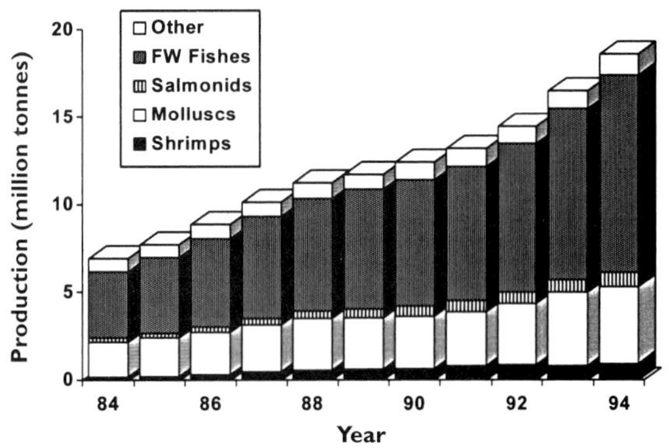

Fig. 3. Trends in aquaculture production by species groups. Source: FAO (1995*a*), FAO (unpublished).

outbreaks. Mangrove forests and other coastal habitats are being destroyed at an alarming rate to accommodate aquaculture projects. The shrimp culture industries in Taiwan and China have exhibited spectacular declines within the past 8 years, due to disease problems and environmental degradation.

The key problems

A large number of factors have been implicated in the currently perceived 'crisis' in fisheries. Although all may be relevant to a greater or lesser degree (and some will be examined further below), I contend that a small set of fundamental problems surpasses all others in requiring immediate and focussed remedial action. For capture fisheries, overcapacity (including both amounts of gear and numbers of participants) is the single most important factor threatening the long-term viability of exploited fish stocks and the fisheries that depend on them. For aquaculture, the most pressing need is to solve the dilemma of promoting expansion, while at the same time demanding the development of environmentally sound technologies and farming practices.

Overcapacity: the most important problem for capture fisheries

I use the term 'overcapacity' to mean either excessive amounts of capital in the form of fishing vessels and gear (i.e. overcapitalization), or excessive numbers of participants, or both. Many coastal and subsistence fisheries with relatively little capital investment also suffer from 'overcapacity', in the form of too many participants or too many dependents for the size or condition of the resource. Massive rural poverty and landlessness may cause 'Malthusian overfishing', which can result in the use of destructive fishing practices (e.g. dynamiting and cyanide poisoning) to overexploit local resources. In this section, I focus on fishing fleet capacity in developed nations, since few estimates of the extent of overcapitalization or 'overparticipation' are available for small fisheries.

Trends in fleet capacity

Between 1970 and 1992, the world's decked vessels increased in number from 580 980 to 1 178 160 and in gross registered tonnage (GRT) from 13.6 billion GRT to 26 billion GRT; over the same period, the world's undecked vessels increased in number from 1.5 million to 2.3 million (Fig. 5; FAO 1995*a*). This rate of growth in world fishing capacity is way out of proportion to the rate of growth of landings. Over the period 1970–89, the global industrial fleet size was increasing at a relative rate double that shown by global landings (FAO 1993*b*). But the problem is greatly magnified by the rate of increase in fishing power due to technology improvements. Using 1980 as the base year, Fitzpatrick (1995) estimated average technology coefficients of 0.54 in 1965, 1.0 in 1980 and 2.0 in 1995; i.e. fishing power is estimated to have increased about four-fold since 1965. This suggests that the rate of growth in catching capacity may have been as high as eight times the rate of growth in landings!

Today the industrial fishing fleet represents 30% of the world's total shipping over 100 GRT (FAO 1995*a*). Fishing fleets are grossly overcapitalized on a global basis and, with some exceptions, on a local basis. FAO (1993*a*) estimated that the total annual operating cost of the global fishing fleet for 1989 was $US92 200 million, without including returns to capital or allowances for debt servicing, which was estimated to add about $US32 000 million to fishing costs. Assuming estimated gross revenues of about $US70 000 million in 1989 implies a net loss of $US54 000 million. Some of this deficit may be covered by subsidies in the form of grants, loans, tax preferences, price support programmes, general unemployment and welfare programmes, and other direct and indirect forms of financial

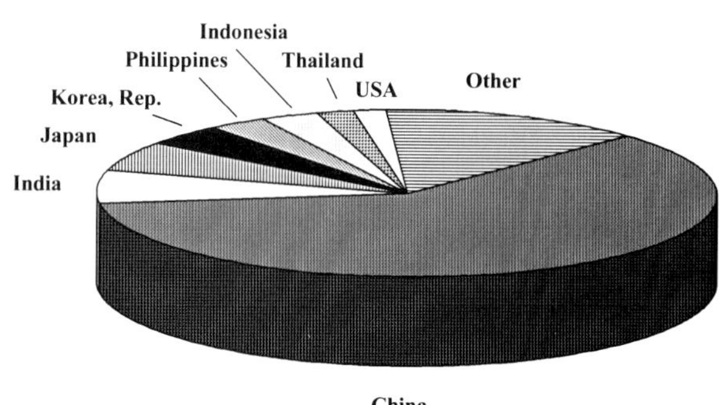

Fig. 4. Aquaculture production by principal producers in 1994. Source: FAO (unpublished).

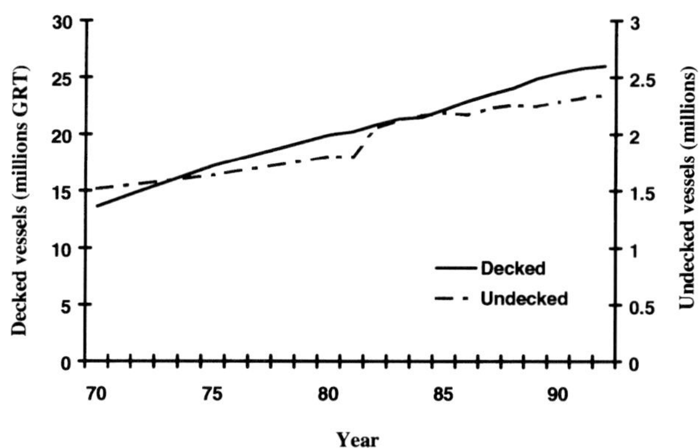

Fig. 5. Trends in gross registered tonnage (GRT) of decked fishing vessels and numbers of undecked fishing vessels. Source FAO (1995*a*).

aid; but the deficit is probably overestimated due to under-reporting of landings and incomplete or inaccurate data on costs and earnings. Also, the FAO estimate of a net loss of $US54 000 million (a figure that is widely cited) was based on 1989 data and may not be so relevant today, due to the substantial scaling down of fishing operations conducted by the former Soviet Union.

Estimates of the extent of overcapacity

Interpretation of indicators of overcapacity such as declining catch-per-vessel or vessel unit (e.g. GRT) and dramatically reduced fishing seasons is complex. For example, in the Alaskan Pacific halibut fishery the length of the fishing season decreased from 47 days in 1977 to 2 days in 1994 prior to the implementation of individual transferable quotas (ITQs), even though quotas and landings increased through much of this period (Fig. 6). However, most of these vessels also participated in alternative fisheries, making it difficult to quantify the extent of overcapacity associated specifically with the Pacific halibut fishery. Although overcapacity (as inferred from the severely compressed fishing season) appears to have been extremely high in the Pacific halibut fishery, the magnitude of overcapacity is likely to be much less if all regional fisheries are taken into consideration.

The inherent difficulty of estimating overcapacity is probably one of the reasons why there are relatively few quantitative estimates of the extent of fleet overcapacity. In some cases, bio-economic models have been used to estimate the amount of overcapacity, but most estimates are actually guesses based on some historical reference year when the fleet was assumed to be operating at a much higher level of economic efficiency. Table 1 provides a summary of examples of both types of estimates. For example, Garcia and Newton (in press) produced quantitative global estimates by fitting a production model to global landings data and gross tonnage data adjusted for fishing power increases, using Fitzpatrick's (1995) coefficients. Their analysis indicates a need for world fishing capacity to be reduced by 25% for revenues to cover operating costs, or 53% for revenues to cover total costs. Although estimates vary from fishery to fishery

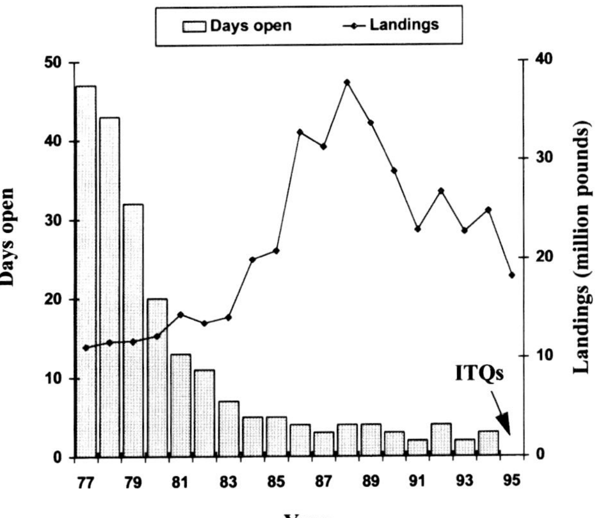

Fig. 6. Trends in season length and landings in the Alaskan Pacific halibut fishery. Source: International Pacific Halibut Commission Annual Reports.

Table 1. Estimates of overcapacity

Global
For revenues to cover operating costs, world capacity should be reduced by 25%; for revenues to cover total costs, the world fleet should be reduced by 53%[1]
World fishing overcapacity estimated to be about 30%[2]
International and National
European Union — in 1996, the EU has determined that EU fishing fleet capacity needs to be reduced by as much as 40% over the next 6 years
Russian Federation — 2/3 reduction required
United States — depending on the fishery, estimates of overcapacity vary from 0 to 75%[3]
Australian fisheries — many have a larger catching capacity than needed to take the catch efficiently[4]
Individual fisheries
Bristol Bay red salmon — Amount of 'unnecessary gear' = 33% during 1934–39, 83% during 1955–59[5]
British Columbia salmon and roe herring — in 1982, fleets needed to be reduced by 50%[6]
British Columbia salmon — in 1996, a plan to reduce the number of licences by 50% is being implemented[7]
British Columbia rockfish fishery — 20–30 vessels in 1980–85, compared with more than 100 vessels in 1995 for similar total landings[8]
Canadian Atlantic groundfish — in 1986, the licensed capacity of the Nova Scotia inshore fleet was about 4 times that needed to catch the quota (estimate precedes recent stock collapses)[9]
New England groundfish — net economic value could be maximized by a 70% reduction in fishing effort[10]
New England lobster — about one-half of the present fishing effort would be needed to achieve economic efficiency or marginal cost pricing[11]
Gulf of Mexico shrimp — recent harvests could be taken by one-third of fleet; i.e. fleet overcapacity about 67%[12]
Bering Sea pollock — catching capacity appears to be double or more the annual quota[13]
US Atlantic surf clam — after ITQs were implemented in 1990, the fleet was reduced by 54% within 2 years[14]
Southeast US wreckfish fishery — in the 3 years following the introduction of ITQs, the number of permits decreased from 91 to 21 and the number of vessels reporting landings decreased from 44 to 11[14]
US Pacific sablefish — within a year of implementation of the ITQ programme in 1994, the number of vessels that participated in the hook and line fishery decreased from 1139 to 690, despite measures implemented to discourage consolidation[15]
New Zealand inshore fisheries — estimated amount of overcapitalization $NZ28 million in 1983, about 20%.[16]
Australian Northern Prawn Fishery — recommended effort in 1993 was 50% 1986 level[4]

[1]Garcia and Newton in press; [2]House of Lords 1996; [3]Sissenwine and Swartz 1992, NMFS 1996*a*; [4]McLoughlin *et al.* 1993; [5]Crutchfield and Pontecervo 1969; [6]Pearse 1982; [7]Department of Fisheries and Oceans Canada (DFO); [8]Laura Richards, DFO, pers. comm.; [9]DFO 1988, Halliday *et al.* 1992; [10]Edwards and Murawski 1993; [11]Bell 1972; [12]Ward and Sutinen 1994; [13]Miller *et al.* 1994, NMFS 1996*a*; [14]NMFS 1996*a*; [15]Joe Terry, NMFS, pers. comm.; [16]Anon. 1984; Sissenwine and Mace 1992.

(Table 1), overall reductions of the order of at least 50% seem to be required for many fisheries.

Other examples in Table 1 show that overcapacity is not a new problem. For example, the Bristol Bay red salmon fishery was already substantially overcapitalized in the 1930s. Similarly, the British Columbia salmon fleet, which became subject to a major fleet reduction plan in 1996, was already double its presumed optimal size in 1982 (Pearse 1982).

The overcapacity problem is exacerbated by latent effort: the situation where the number of licences in a fishery far exceeds the number of active participants in any given season. For example, less than half of the licensed vessels in both the mobile and fixed gear Atlantic Canada groundfish fleets are active (Greg Peacock, Department of Fisheries and Oceans Canada, pers. comm.). The number of permits in the US Atlantic shark fishery is about 2500, whereas only about 50 vessels are substantially dependent on the fishery. The existence of latent effort means that, even in limited-access fisheries, any improvement in stock size or prices can quickly be dissipated by redeployment of previously inactive vessels.

Consequences of overcapacity

I contend that unless the overcapacity problem is solved, attempts to address many other important fisheries problems may largely be wasted. In overcapitalized fisheries, reductions in fleet capacity are a precondition to the success of management measures designed to mitigate overfishing, solve bycatch issues, eliminate environmentally destructive fishing practices, reduce under-reporting, and improve government–industry relations.

From an economic perspective, overcapacity is equated with an excessive quantity of vessels or fishing gear that are not fully utilized (overcapitalization). From a conservation (and social) perspective, the most important manifestation of the problem is too many people dependent on fisheries for their livelihoods. This leads to a situation where the average fishing enterprise is economically marginal (by definition for open access fisheries at equilibrium) or close to it (the empirical outcome even in many limited access fisheries, such as those limited entry licensing fisheries that have failed to match fleet capacity to resource productivity). Thus, when environmental conditions deteriorate or fish prices decline, or both (as is happening in Canadian and mainland US Pacific salmon fisheries at present), fishers cannot absorb the downturn, no matter how ephemeral. Economically marginal fishers are concerned about making this month's boat payment, and usually cannot afford to sacrifice today for the unguaranteed possibility of increased catches or profits at some unspecified time in the future. Economically marginal fishers generally cannot absorb the additional costs associated with seeking out fishing grounds where bycatch and discards may be minimized; they may fish on the grounds with the highest net revenues, regardless of bycatch and other factors that may have little direct influence on profits. Economically marginal fishers often cannot afford to spend time carefully deploying and retrieving gear in ways that will minimize bycatch or maximize survival of fish released. Economically marginal fishers may not even want to expend the time and expenses necessary to recover lost gear, even though such gear may continue to 'ghost fish' for many years.

When a high proportion of fishers are economically marginal, the net result is likely to be, (i) increased pressure on scientists to conduct 'optimistic' assessments and increased challenges of the validity of the science, (ii) increased pressure on managers to select total allowable catches (TACs) from the upper, risk-prone confidence intervals of projected catch distributions, (iii) increased pressure on governments to provide financial aid (i.e. subsidies) to prop up failing businesses, and (iv) increased incentive to circumvent fishing regulations, including under-reporting of landings and use of destructive fishing practices. In addition, highliners who are doing well may not want to change the *status quo*. Even those operating at the margin who recognize the overcapacity and overfishing problems may nevertheless want to protect a 'way of life', hoping that the status of the resource or the market will improve in the future.

Environmental degradation: the most important problem for aquaculture

Until a couple of decades ago, when technologies were less intensive, aquaculture was generally considered an environmentally sound technology (Pillay 1992). However, recent dramatic crashes of some aquaculture operations, most notably shrimp, have drawn increased attention to the consequences of uncontrolled expansion that has not paid due regard to environmental impacts. Extensive shrimp operations (those with low stocking densities and no supplemental feed or fertilizer inputs) have resulted in the destruction of large areas of coastal marshes, including mangrove swamps. Total mangrove area in the Philippines, estimated to have been 450 000 ha in the early 1900s, had been reduced to about 149 400 ha by 1988 (Reyes *et al.* 1994). About 50% of the mangrove forests in the Philippines have been developed into brackishwater fish ponds (Saclauso 1989). The area converted in Thailand is estimated to be about 27% and in Ecuador about 13–14% (Pillay 1992). An estimated 300 000 ha of brackish ponds are in use in Indonesia, and the majority of these have been constructed in coastal mangrove areas, of which there are roughly 3.6 million ha (Chamberlain 1991). Some experts have suggested that the production of extensive shrimp ponds constructed in mangrove wetlands is less than that of an intact mangrove ecosystem (references cited in Aiken and Sinclair 1995). On the other hand, intensive shrimp farming operations (those with high stocking densities, reliance on feed inputs and large discharges of wastes) have resulted in disease problems and localized pollution. Diseases and environmental degradation resulted in spectacular crashes of the shrimp culture industries in Taiwan in 1988 and in China in 1993.

Pollution and disease are also problems in intensively-stocked pen and cage farms in marine or enclosed freshwater areas. Contamination from pesticides and antibiotics used in intensive operations is another concern. In some cases, overfishing of marine stocks to provide brood stock and fish feed may be occurring. For example, an estimated 70% of the trawl catch from the Gulf of Thailand is classified as 'trash fish', primarily used to produce fish meal for aquaculture (World Bank 1991). Other problems include land subsistence and salination when water is withdrawn for use in aquaculture projects, and genetic consequences of escapes of farm-bred fish.

Synthesis: a common problem

What overcapacity in capture fisheries and environmental degradation in aquaculture operations have in common is that the fishing industry is not paying the full costs of doing business; the former due to subsidies and government grants that mask net operating losses and to lack of consideration of the future costs of rebuilding depleted stocks, and the latter due to failure to mitigate environmental impacts. Governments are subsidizing the demise of capture fisheries, and not paying due attention to the effects of current aquaculture practices on the environment. The social costs of the degradation of ecosystems and the environment have yet to be paid by future generations.

OTHER SHORTCOMINGS AND FAILURES

Overcapacity and environmental degradation are the symptoms, the result of systems gone awry. What has gone wrong? Poor science, poor management, poor data, inadequate institutions, poor policy, and a host of other factors have often been blamed for the crisis situation that is currently perceived to exist. As a fisheries scientist who has delved increasingly into the fisheries management arena, the ordering of these 'shortcomings' reflects the frequency of the complaints I tend to hear the most. However, as I will try to elaborate below, I believe the reverse order is more appropriate in terms of the relative magnitude of the inadequacies.

A. Inadequate science

The foundations of modern fisheries science were laid down by the late 1950s by numerous works including those of Thompson and Bell (1934) on yield-per-recruit, Graham (1935) on the 'law of fishing' (the expansion of fishing power as stocks become depleted), Schaefer (1954, 1957) on stock production models, Ricker (1954) on stock and recruitment, Gordon (1954) on the economic theory of common property resources, and Beverton and Holt (1957) on dynamic pool models, stock and recruitment, and many other topics. Smith (1994) provides a comprehensive account of these studies and earlier work that was influential in the development of the scientific basis of fisheries management. Subsequent developments include stock assessment methods and standardized resource surveys. Cohort analysis and virtual population analysis (VPA) were developed during the mid-1960s to early 1970s. Nowadays, there are several stock assessment organizations throughout the world that routinely use VPA and other assessment models to estimate historical trends in fishable biomass and fishing mortality. Many extensions of the original VPA methodology have been developed to improve estimation techniques and provide statistical diagnostics. These include various *ad hoc* methods, separable VPA and ADAPT (Gavaris 1988). Alternative stock assessment methods that are widely used include size-based methods (reviewed by Gallucci *et al.* 1996), non-equilibrium stock production models (e.g. Prager 1994) and stock synthesis (Methot 1990).

Assessing the condition of fish stocks and fisheries is difficult due to the complexity of the marine environment and because of the long and expensive time series of data required to apply most stock assessment models. Complexity and lack of data are also responsible for the fact that most methods for assessing stock status are based on a single species approach (an exception is the multi-species VPA methodology developed by the International Council for the Exploration of the Sea (ICES); Sparre 1991 provides an explanation of the method). A vast array of multi-species models has been developed for marine fisheries, particularly predator-prey models (reviewed by Collie 1995), most of which involve only two or three species, but these have rarely been useful in a stock assessment or fisheries management context because of the large number of alternative hypotheses that can potentially explain the observations, the lack of incorporation of all key biological and environmental interactions, or the lack of data on the nature and extent of these interactions. Multi-species mathematical models are capable of producing an almost unlimited diversity of combinations of multiple stable and unstable equilibria, regular and chaotic fluctuations, and extinctions.

Throughout the world it is certainly true that there is a need for considerably more scientific research to provide a sound basis for managing fish stocks and fisheries; however, I contend that it is not so much the state of the art of the science that is limiting, it is simply that most fisheries systems have not yet been adequately researched. Limitations of the current science usually pale in comparison to the inability to implement stock assessment recommendations and to control fleet capacity. In most cases where scientific stock assessments have been conducted, the shortcomings of science have not generally been the primary cause of fisheries failures. However, there are notable exceptions; some due to a lack of adequate science and some to a lack of sufficient science (where adequacy refers to quality, and sufficiency refers to quantity). Two examples of stock assessments that were, in retrospect, grossly inaccurate are Atlantic cod off Newfoundland and Labrador and orange roughy off New Zealand. It has been argued that the first is an example of inadequate science (overly-optimistic assessments) while the second is an example of insufficient science (lack of knowledge of basic life history parameters). However, alternative interpretations exist.

The spawning biomass of northern cod off Newfoundland and Labrador declined from an estimated 1.6 million t in 1962 to 22 000 t in 1992 (Fig. 7). This has resulted in the loss of up to

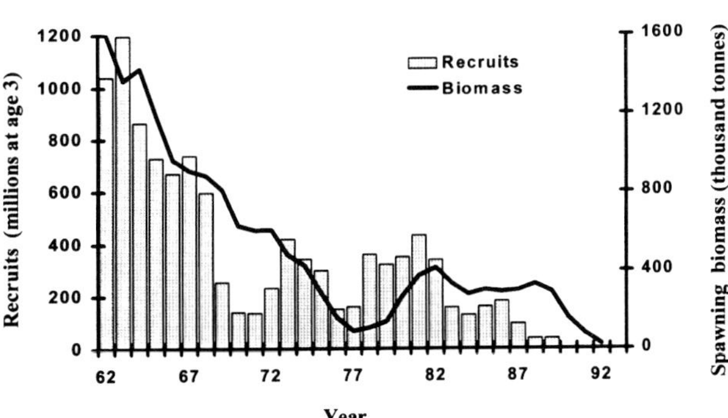

Fig. 7. Trends in spawning biomass and recruitment for northern cod. Source: Bishop *et al.* 1993; Ram Myers, Department of Fisheries and Oceans, Canada, pers. comm.

40 000 jobs and a five-year aid package (1992–97) amounting to more than $CAN1.5 billion in unemployment benefits and other assistance. Several factors may have contributed to the collapse of this fishery, but a number of authors (e.g. Hutchings and Myers 1994; Hutchings 1996; Walters and Maguire 1996) have concluded that overly-optimistic stock assessments were a major culprit. Hutchings and Myers (1994) provided a comprehensive review of the evidence that environmental effects precipitated the collapse. They concluded that harvest rates less than 17% per year would have permitted the stock to sustain itself. Beginning in 1977, the management target ranged from 80 to 100% $F_{0.1}$, where $F_{0.1}$ was estimated to be 18%. However, harvest rates frequently exceeded the target, especially in the 1980s. Hutchings and Myers (1994) attributed most of the blame to overestimation of stock size coupled with high uncertainty in research survey abundance estimates. However, it is also generally true that the official total allowable catch was set above the level based on scientific advice and was in turn exceeded by reported landings (Aiken and Sinclair 1995). Other factors that have been implicated in the stock collapse include foreign overfishing, under-reporting and discarding, industry opposition to TAC reductions, government subsidies leading to excessive fishing capacity, predation by marine mammals, and environmental effects such as water temperature changes that may have affected recruitment, distribution and mortality (Walters and Maguire 1996). Debate about the relative importance of unpredictable environmental factors and whether there was sufficient scientific information to predict or prevent the collapse continues (Hutchings 1996).

In contrast to the long history of research on northern cod, little was known about the population dynamics of orange roughy when a deepwater (800–1200 m) fishery for this species began to develop off the coast of New Zealand in the late 1970s. Scientists, managers, and the fishing industry were all misled by the huge spawning aggregations being discovered, and the associated high catch rates. Large biomass was implicitly associated with high productivity. Since demographic parameters were completely unknown, it was considered conservative to assume that they were similar to those of a 'typical' or average temperate-water teleost (i.e. natural mortality and Brody growth coefficient both equal to 0.2, and ages of maturity and recruitment both about 5 years), and stock assessments and yield projections were based on these assumptions. However, subsequent studies (Mace *et al.* 1990; Francis 1995) showed that natural mortality is probably less than 0.05, the age of maturity may exceed 20–25 years, and maximum age may be in excess of 150 years. As a consequence, estimates of long-term sustainable yield were reduced to about 20% of initial estimates. In the meantime, fleet capacity, although regulated by individual quotas almost from the beginning, quickly built to levels in excess of that needed to harvest the long-term sustainable yield.

Lack of sufficient science is also problematic in the aquaculture industry where research has focussed on improving existing technologies or developing new ones for increased production, rather than evaluating the effects of farming on the external environment (Pillay 1992). As is the case for capture fisheries, much more research is needed, but the research base needs to diversify to focus more on environmental problems. Scientific and engineering solutions are required to develop productive, disease-resistant strains of species for cultivation, develop low cost feeds with minimal input of wild fish stocks, use water and other inputs more efficiently, treat or mitigate the effects of biological wastes, and develop environmentally friendly systems that minimize destruction of coastal habitats.

B. Inadequate or inappropriate management goals

The development of the theory and practice of fisheries management has lagged developments in fisheries science. Simple restrictions such as time and area closures, gear restrictions and various forms of exclusive access date back several centuries, but such measures were usually restricted to lakes and rivers and nearshore areas. Until recently, the expansion of coastal and offshore fishing fleets was largely uncontrolled, although some nations, New Zealand for example, have had limited entry off and on since the turn of this century.

To a large extent, management has been guided by the 'paradigm' of inexhaustibility of marine resources dating from the 19th century. For example, Thomas Huxley stated in 1883: 'I believe that the cod fishery, the herring fishery, the pilchard fishery, the mackerel fishery and probably all the great sea-fisheries are inexhaustible; that is to say, nothing we can do seriously affects the number of fish' (Smith 1994). The paradigm of inexhaustibility has been extremely influential in guiding fisheries management, at least until many of 'the great sea-fisheries' collapsed. Even today, there are many who do not believe that the limits of capture fisheries have been reached, and others — those whose livelihoods depend on fishing — who do not want to recognize the limits for the particular fisheries in which they are involved.

Thus, the focus of management has been on development and expansion, rather than control and restraint. Even today, most fisheries are open access fisheries, and most limited access fisheries suffer from overcapacity. And while it is widely recognized that open access is an anachronism, there is widespread disagreement about the alternatives. Many economists argue in favour of individual fishing share systems (often called quasi-property rights systems) such as individual transferable quotas (ITQs) and related systems. Sociologists argue for community-based management or co-management, with most decisions being made at the local level and little if any control by a central government. The environmental community appears to be divided on the issue. For example, while many environmental groups recognize the benefits of fishing share systems, Greenpeace has mounted an aggressive anti-ITQ campaign in New Zealand and the United States. Similarly, a recent publication from the World Wildlife Fund (Kemf *et al.* 1996) recommends that management programmes should be focussed on limiting effort and restricting access to fisheries; yet it states in the same paragraph that limited-access programmes 'should not be allowed to create private property rights in any fishery'. This is contradictory. In effect, any system that is not open access is a form of property right. The main difference between a limited entry system (with fleet-wide quota) and an individual quota system is that in the former case a pre-set number of vessels with exclusive rights to a particular fishery are each individually permitted to try to take as many fish as possible before the quota is exhausted, whereas in the latter case a limited but variable number of vessels (which is likely to be more closely matched to overall quotas and market conditions) is permitted to take a fixed share of the overall quota — and to take this share in a manner calculated to fulfill other objectives such as maximizing quality and price, and minimizing costs and wasteful bycatch. In addition, community-based management, which is more widely advocated by the environmental community and a variety of other stakeholders, is obviously a form of property right that may exclude participation by those outside the community.

With few exceptions, TACs, limited entry and indirect controls such as time and area closures and gear restrictions have failed to promote biologically and economically sustainable fisheries. One of the reasons is the lack of comprehensive management goals. Often, biological goals may be stated explicitly (e.g. to maintain fishing mortality at or below some target fishing mortality rate such as F_{msy} or $F_{0.1}$), while economic and social goals are not at all well articulated. Unstated objectives such as minimization of interference with the *status quo* or maximization of employment often seem to take precedence. This results in a failure to inhibit investments during periods of growth in markets or stock abundance, and a failure to promote disinvestment during periods of declining stock abundance or profitability.

However, it would be unfair to blame fisheries managers entirely for this predicament. Fisheries managers, who are usually government employees, often have extremely limited power to implement the necessary controls, due to the lack of strong national policies, or the lack of political will to implement strong policies even if such exist on paper. It is difficult to develop and implement management strategies to satisfy long-term objectives when those objectives have not been specified precisely and may be subject to the whim of politicians.

C. Inadequate data and statistics

As fisheries management strategies become increasingly complex, and the long-term sustainability of fisheries becomes increasingly challenged, the need for accurate and complete data assumes increasing importance. Depending on the models employed, stock assessment estimation methods and management alternatives may require data on basic demographic parameters (e.g. growth rates and natural mortality), commercial landings, discards, age or length composition of landings, commercial catch-per-unit-effort, and standardized fishery-independent resource surveys. In general, long time series of data are required for fisheries science and management, but relatively few exist.

Globally, the quality of commercial fisheries data is poor and may be diminishing. For example, ICES has not published official statistics on commercial catch and effort since 1992 (and then only covering the period up to 1988), due to absence of data. The Northwest Atlantic Fisheries Organisation (NAFO) has also experienced delays in publication of official statistics. Similarly, the most recent publication of fisheries statistics of the United States (NMFS 1996*b*) is much less complete than previous annual reports. In New Zealand, the conversion to an ITQ management system in 1986 was accompanied by a substantial deterioration in the quality and quantity of commercial effort and catch-per-unit-effort statistics. In the Philippines, a re-organization of government has resulted in a considerably reduced ability to collect data for stock assessments and management (World Bank 1991).

Data collection systems for commercial fisheries still rely largely on paper log-book records compiled by fishers who often perceive that landings data they provide will ultimately be used to curtail their activities. Misreporting, non-reporting and careless reporting are commonplace. Increasingly, conscientious stakeholders in developed nations are questioning why fisheries agencies do not make more use of modern electronic recording and tracking systems. Yet there seems to be a general reluctance to impose such systems on individual fishing enterprises. In addition, electronic recording systems would not solve all reporting problems and may not be the best mechanism for obtaining data on bycatch, discards and cryptic mortality. Observer programmes can be invaluable for the acquisition of accurate commercial fisheries data, but they are expensive and require that fishing vessels be large enough to accommodate one or two extra people.

Even if fishers complied completely with all reporting requirements, the resources devoted to compiling, auditing, archiving and managing such databases are generally inadequate. It is common for fisheries landings and effort statistics to be 1–3 years in arrears. In addition, whenever systems change (e.g. due to a new management regime, new database management software, or reorganization of government departments), a year or two of data may have to be excluded temporarily, and sometimes permanently due to the inability to ever catch up with the backlog.

Other data limitations include a lack of unbiased samples for the size or age frequency of the catch, lack of knowledge of the extent of adoption and the relative effectiveness of changes in fishing technology (which is essential for interpreting trends in commercial catch-per-unit-effort), and lack of fishery-independent data, particularly fishery-independent estimators of trends in stock abundance.

Lack of credibility in the fundamental data used to assess and manage fisheries fuels attacks on the credibility of the science and the validity of management actions. The fishing industry is aware that the data on which assessments are based are shaky at best: fishers know that landings data are incomplete or inaccurate because they or their compatriots provided the data, they know that discarding and other wasteful fishing practices are common but poorly reported, and they may be aware that knowledge of demographic parameters and the complexity of ecosystem interactions is poor. Generally, the quality of scientific assessments is not the real issue, at least not in terms of methodology; the quality of the data on which the assessments are based is much more problematic.

D. Inadequate institutions and other mechanisms for involving stakeholders

Existing institutions have failed to check undisciplined fishing behaviour that has led to overcapacity and overexploitation. In fact, existing institutions appear to have created an atmosphere of distrust between scientists, managers, politicians, small-scale commercial fishers, large-scale commercial fishers, recreationists, subsistence and traditional fishers, environmentalists and other stakeholders. User conflicts are becoming increasingly more intense and violent. Each group uses the others as scapegoats. Members of the fishing industry accuse stock assessment scientists of being overly-conservative and calls for peer reviews of stock assessments are becoming increasingly frequent in many countries, including the United States. In addition, there is widespread distrust by the fishing industry of management conducted by government authorities. Decision systems tend to be cumbersome and bureaucratic. Response times to resolve resource or fishery problems are slow. Current institutions also typically lack the flexibility to respond to fisheries problems in creative ways. Poor communication is also an issue.

Another problem that limits the proficiency of institutions is the lack of skilled experts. Even in industrialized, relatively affluent countries, there is a lack of experts in the fields of fish stock assessments, fisheries socio-economics and fisheries management. When there is a need for peer review of the scientific validity of stock assessments, there is a relatively small pool of qualified individuals outside of government management agencies. In addition, by attempting to micromanage for maximum biological or economic yields and failing to control fleet capacity, the governments of many industrialized countries may have created impossible jobs for management authorities. Throughout much of the world, government scientists and managers are expected to produce better science and management with progressively fewer resources (More with Less), while fishers are expected to divide fewer fish amongst a larger number of participants (Less with More). Institutions need to recognize and resolve these mismatches.

Although there is strong agreement on the need for institutional reform, there is no such agreement on the type of reform that is needed. This is mainly because of the wide diversity of opinions about the role of fisheries as generators of food, income, employment, recreation and existence value. There is no general agreement on the goals for fisheries, particularly social goals, but also economic and biological goals. The appropriate balance between top-down and bottom-up management will vary between fisheries and countries. What is the role of government, what is the role of fishers' cooperatives, communities and other stakeholders; who should be responsible and who should be accountable? It is now widely recognized that regulations imposed by governments can be readily circumvented if the fishing industry does not support them; i.e. effective resource management cannot be achieved without the active involvement of those affected. But the other extreme of community-based management is also prone to failure, particularly when there is potential for takeovers by private sector or local government elites (see Bailey and Zerner 1992 for examples from Indonesia), or when the effectiveness of such systems is undermined by the absence of some key features such as relatively low human population density, homogeneity of kin- or territorially-based communities, the use of relatively simple extractive technologies (Bailey and Zerner 1992), and lack of major external markets.

In general, the most effective institutional arrangement will probably involve shared management of fisheries resources, with the balance between government and user control varying from fishery to fishery.

E. Inadequate national policies and international standards

To date, fisheries management has suffered from a lack of strategic and long-term planning at the local, national, regional and international levels. National policies and international conservation standards and guidelines, which have been urgently needed to reverse the trends of the last few decades, are only now beginning to gain importance. Current national policies, stated or implicit, may include provision of employment, contributing to food supplies and food security and earning foreign exchange, but almost invariably the perspective is short-term. The resources allocated to the collection of commercial, subsistence and recreational fisheries data and to the conduct of scientific research are generally inadequate. Even when new and 'revolutionary' management regimes are put in place they are rarely effectively evaluated, monitored, or enforced. Long-term and holistic objectives have been well-articulated in numerous recent international agreements, but in most cases have yet to be adhered to in practice.

Aside from failing to develop comprehensive, operational national fisheries strategic plans, politicians have in fact exacerbated the problem by intervening on an *ad hoc* basis to overturn management decisions or bail out failing fishing enterprises. Fisheries are still treated as the employer of last resort. In addition, national government policies relevant to fisheries are often unco-ordinated and conflicting, with some government agencies taking action to restrict fishing activities while others take actions to promote it. This fragmented authority needs to be replaced by an all-encompassing regulatory framework that considers biological, social and economic concerns and formulates clear, consistent, strong and unambiguous objectives for the conservation and management of natural marine resources, including integrated coastal zone management. National governments need to ratify current international agreements and, beyond that, to actually implement the agreements domestically. In addition, there is an urgent need for national aquaculture policies.

Synthesis: a common problem

Whether one is referring to developed or developing nations, nations with or without comprehensive fisheries policies, or nations with or without strong science, management and data collection programmes; when fisheries fail, it is largely the result of a lack of *effective* controls. Groundfish fisheries in Atlantic Canada and the northeastern United States (New England) make an interesting comparison. Until recently, all New England fisheries were open access, with only indirect controls on harvest (e.g. time-area closures and mesh size regulations). By comparison, Atlantic Canada appears to have had much stricter controls, with explicit limits on participation (through limited entry licensing) and extensive regulation of harvesting (including quotas). Yet both arrived at the same end-point at roughly the same time; that end-point being collapsed groundfish stocks, and fleet capacities around four times that required to exploit the resources at target levels or to maximize net economic value (Table 1; Halliday *et al.* 1992; Edwards and Murawski 1993).

Of the five classes of factors identified above, which have been the most important causes of fisheries failures? I contend that, to date, lack of national policies and institutional failures have been more limiting than science, management or data. Sound national and international policy and effective institutions are essential for providing the necessary environment to foster good science, management and data collection programmes. As stated by World Bank (1991), in both developed and developing nations, the most pressing problems facing improved management and development appear to be of an institutional, economic and social nature rather than a technical one. However, improvements in all five areas will be required to ensure the future viability of the world's fisheries. There is no question that much more scientific research is required, in addition to much stronger management, much more and better data (basic biology, population dynamics, commercial landings, bycatch, discards, costs and earnings, research surveys), improved institutions that better define roles, responsibilities and accountabilities of all players, and strengthened national policies.

Successes

But perhaps the condition of the world's natural fisheries resources is not as bad as the popular portrayal. Garcia and Newton's (in press) analysis indicates that when all species are considered together, the world's fisheries resources appear to be exploited at about the MSY (Maximum Sustainable Yield) level. There is still room for expansion in some individual fisheries. And many previously depleted stocks are recovering; for example, Atlantic US striped bass (Fig. 8A), herring in Norway, the Gulf of Maine and Georges Bank (Fig. 8B), northwest Atlantic mackerel, US Pacific sardine, some stocks of Alaska salmon, finfish in the North Sea during the two world wars, sperm whales, gray whales (e.g. eastern Pacific gray whales; Fig. 8C), bowhead whales (e.g. western Arctic bowhead whales; Fig. 8D), other whales and many pinniped stocks. Marine teleosts, in particular, seem to have remarkable resilience, with few known cases of extinction of marine populations by fishing (one case being the Irish Sea common skate, *Raia batis*; Brander 1981), and no known cases of extinction of entire marine species through fishing. In most cases, marine fishery resources are likely to be able to recover provided fishing pressure is alleviated sufficiently (Myers *et al.* 1995*a*).

In addition, many wild stocks have been well-managed (at least from a conservation perspective) for long periods of time. Examples from large-scale fisheries include Falkland Islands squid (Basson and Beddington 1993), Pacific halibut (with respect to quotas, not fleet capacity; Fig. 6), US Pacific groundfish stocks such as Pacific whiting and pollack, Canadian Georges Bank scallops, New Zealand hoki stocks, and Western Australian rock lobsters (Brown and Phillips 1994). Also, many small-scale community-based fisheries management systems have endured for decades or even centuries; e.g. fisheries co-operative associations in Japan (Short 1992; Yamamoto and Short 1992), small villages on the shores of Lake Titicaca in Peru (Pinkerton and Weinstein 1995), local fisheries management systems in Indonesia (Bailey and Zerner 1992), and related systems elsewhere in the Asia-Pacific region, the Caribbean, South America, Africa and the Middle East.

The scientific basis for managing fish stocks is also much more solid than the popular portrayal. Since the mid-1950s, the level of sophistication and variety of analytical stock assessment methods, optimization procedures, computer simulation models and technological advancements has escalated. Scientists have had the capability to build comprehensive multi-component ecosystem models for at least two decades (e.g. Anderson and Ursin 1977). Although the lack of data on life history characteristics and species and environmental interactions may limit the utility of such approaches, knowledge about fish population dynamics and basic demographic parameters has increased to the extent that global syntheses of important population processes can be attempted (e.g. Myers *et al.* 1995*b*). Such work is facilitated by the development of computerized databases that incorporate a wide range of statistics and parameter estimates from fish stocks throughout the world. One of the most important of these is ICLARM's FishBase (Froese and Pauly 1996), a fisheries database supported primarily by the European Commission since its inception in 1989, which currently includes FAO catch and aquaculture statistics, Myers *et al.* (1995*b*) spawner-recruit data, museum collections, yield-per-recruit analyses and other valuable information. Enhanced data collection programmes have also enabled the application of some multi-species assessment techniques, such as multi-species VPA (Virtual Population Analysis) which has been extensively applied by ICES. Other important developments include the increased application and sophistication of risk assessment methodologies, adaptive management experiments (Sainsbury 1991), acoustics, other survey techniques, biotechnology, genetics, food technology and food safety. Techniques for collecting commercial fisheries data are gradually being brought into the electronic age with, for example, satellite vessel monitoring systems and electronic logbooks for reporting landings (although their use is currently not very extensive).

A wide array of innovative approaches to fisheries management and the associated decision-making process are also being tested. These include (1) ITQ systems in New Zealand, Australia, Canada, Iceland, the United States, Chile, South Africa and elsewhere, (2) other individual fishing share programmes such as the days-at-sea programmes designed by the US New England Fishery Management Council, (3) restructuring of the management hierarchy to empower fishing communities and other stakeholders to play a more active role in decision making, but also to be accountable for those decisions (for example, the Australian model, which incorporates the Australian Fisheries Management Authority (AFMA), Management Advisory Committees (MACs) and

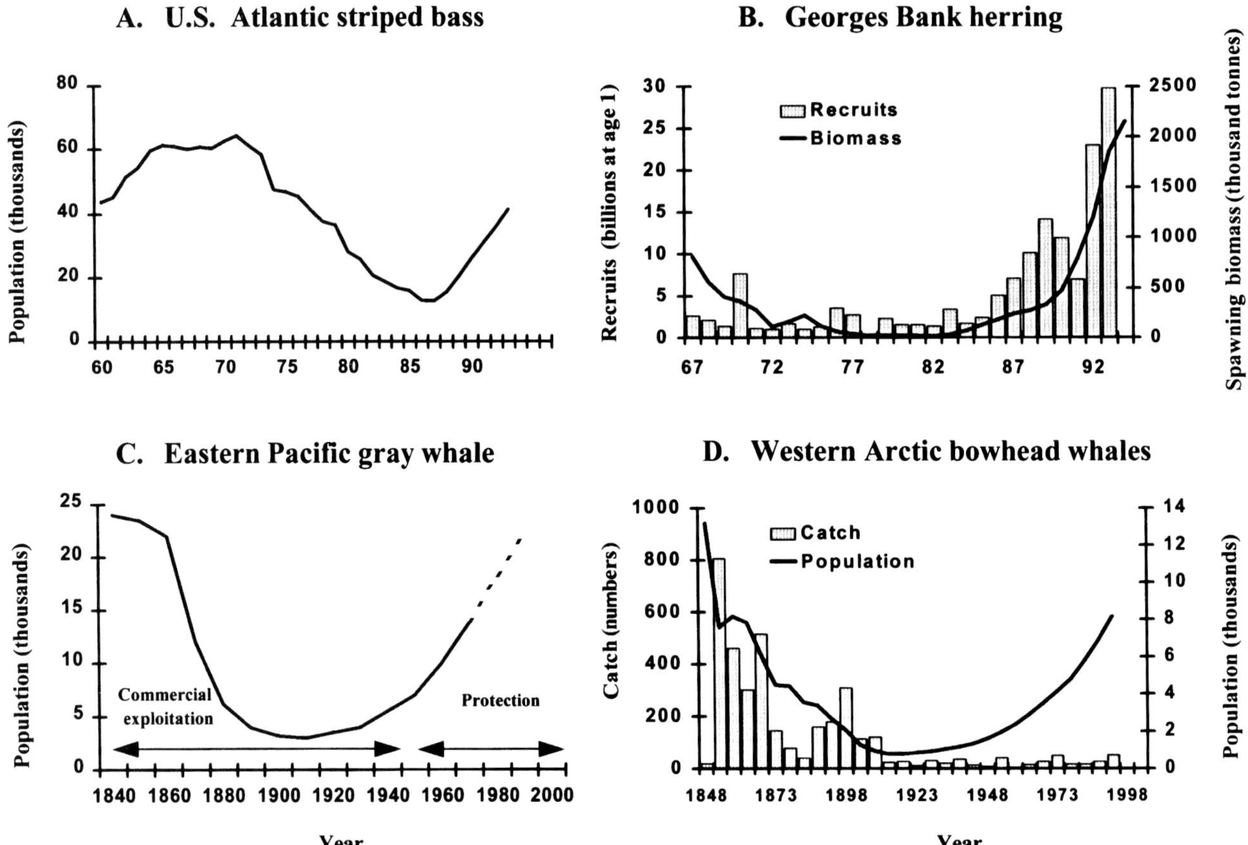

Fig. 8. Examples of depleted stocks that have recovered. Sources: (A) Rugolo *et al.* (1994), (B) NMFS (1996c), (C and D) Steve Swartz, NMFS, pers. comm.

specialist subcommittees that together attempt to ensure that managers, scientists, industry and other stakeholders all work together in an atmosphere of cooperation (AFMA 1995), (4) community development quotas (CDQs) developed in western Alaska in the early 1990s, (5) bottom-up co-ordinated local action to restore fish populations and associated habitat (see Pinkerton and Weinstein 1995 for several examples), and (6) unique methods for dealing with bycatch and discards such as the ban on discards recently implemented in Norway (House of Lords 1996). Several nations are accelerating attempts to deal with fleet overcapacity (e.g. the European Union, Canada, Australia and the Philippines), and non-selective or destructive fishing gears and practices are gradually being eliminated (e.g. large scale high seas drift-netting was banned by United Nations Resolution 46/215 in 1991). The jury is still out on the likely long-term success of many of these measures.

Overall, the environmental community is playing an increasingly positive role. More environmental organizations are becoming involved in fisheries issues, and they are gradually educating and awakening the consciences of the public, managers and fishers alike. Rather than focussing their efforts on attacking governments and the fishing industry, environmental groups are increasingly adopting constructive approaches, such as embarking on special projects designed to help communities devise and implement sustainable fishing practices (Kemf *et al.* 1996).

The cause for optimism has been considerably enhanced by the successful completion of several promising new and recent international agreements pertaining to fisheries: for example, the 1982 United Nations Convention on the Law of the Sea (UNCLOS), the High Seas Compliance agreement, the 1993 International Convention on Biological Diversity, the United Nations Conference on Straddling Fish Stocks and Highly Migratory Fish Stocks, and the FAO Code of Conduct for Responsible Fisheries (FAO 1995*d*). As a result of UNCLOS, many distant water fleets have already been forced to scale down. For example, the Japanese distant water fleet, which was landing about 4 million t of fish in the early 1970s (Royce 1988), now takes less than 1 million t (FAO 1995*a*). The Russian distant water fleet is currently undergoing substantial downsizing following the breakup of the former Soviet Union.

One advance that needs to be singled out from the rest is the development of the precautionary approach, as embodied in the Straddling Stocks agreement and the FAO Code of Conduct, and detailed in FAO (1995*c*) and FAO (1995*d*). Key elements of the precautionary approach include the application of prudent foresight, consideration of the needs of future generations, avoidance of overexploitation and overcapacity, immediate implementation of corrective measures, use of the best scientific evidence available and appropriate placement of the burden of proof (Table 2; FAO 1995*c*). Examples of precautionary measures for capture fisheries include early implementation of controlled

Table 2. Elements of the precautionary approach for capture fisheries (summarized from FAO 1995c)

The precautionary approach recognizes that changes in fisheries systems are only slowly reversible, difficult to control, not well understood, and subject to changing environment and human values.

The precautionary approach involves the application of prudent foresight. Among other things, it requires:

- consideration of the needs of future generations and avoidance of changes that are not potentially reversible;
- explicit consideration of undesirable or unacceptable outcomes, including overexploitation of resources, overdevelopment of harvesting capacity, loss of biodiversity, major physical disturbances of sensitive biotopes, or social or economic dislocations;
- implementation of necessary corrective measures without delay, to achieve their purpose on a timescale not to exceed two or three decades;
- use of the best scientific evidence available, along with programmes to improve data collection and statistics, enhance research on the stock and fisheries, and incorporate uncertainty and risk assessments into the analyses;
- harvesting and processing capacity commensurate with estimated sustainable levels of the resource, with increases in capacity further constrained when resource productivity is highly uncertain;
- all fishing activities having prior management authorization and being subject to periodic review;
- an established legal and institutional framework for fishery management; and
- appropriate placement of the burden of proof.

The precautionary approach to fishery management is applicable even with very limited information.

Table 3. Examples of precautionary measures (summarized from FAO 1995c)

Always control access to the fishery early, before problems appear. An open access fishery is not precautionary.

Place a cap on both fishing capacity and total fishing mortality rate.

Develop operational target reference points (management goals) and limit reference points (e.g. minimum acceptable biomass or maximum acceptable fishing mortality).

If limit reference points are exceeded, implement recovery plans immediately to restore the stock.

Encourage responsible fishing through, for example, some form of tenure of fishing rights.

Encourage development of fisheries that are economically viable without long-term subsidies.

Establish data collection and reporting systems.

Avoid harvesting immature fish, unless there is strong protection of the spawning stock.

Use area closures to limit risks to the resource and environment by providing refuges for stocks and to protect habitat.

Develop management plans cooperatively with stakeholders.

access, development of operational target reference points (management goals) and limit reference points, development of recovery plans to restore stock status, encouragement of responsible fishing practices and cooperative development of management plans (Table 3; FAO 1995*c*). A precautionary approach has also been developed for aquaculture (FAO 1995*c*). The precautionary approach, if fully adopted, may well be one of the most important paradigms for the future viability of fisheries.

Optimism is also warranted by the considerable room for expansion in aquaculture, and projections that indicate that aquaculture will probably be able to compensate for anticipated static supplies from capture fisheries to meet increased world demand for fish products at least until the year 2010, provided, of course, that development proceeds in an environmentally sustainable manner. Overall, aquaculture is a success story despite recent, widely publicized failures in some segments of the industry. Sorgeloos and Sweetman (1993) suggest that examples of successful aquaculture initiatives include the 100 000 t Atlantic salmon farming industry in Norway, which pioneered genetic research, selective breeding, the development of high quality feeds, and the introduction of salmon farming in other nations; the sea-cage culture of groupers, snappers, sea bass and jacks in Asia; the 65 million sea bass and sea bream fry produced for grow-out in cage systems in the Mediterranean; the development of penaeid shrimp hatcheries in Asia; and the rapid expansion of seaweed farming in Asia. The Chinese system of integrated crop-livestock-fish farming (Pillay 1992) is another example. Some authors (e.g. Csavas 1994) have also characterized shrimp farming as one of the most outstanding commercial success stories in the history of aquaculture, having increased from about 0.1 million t in 1982 to 0.7 million t in 1992. Shrimp culture has become a major source of export earnings in many developing countries (Aiken and Sinclair 1995), the downside being that in most cases development has proceeded at the expense of the environment. In 1994, the fish, shellfish and seaweeds produced by farming were estimated to be worth nearly $US40 billion (Fish Farming International 1996).

Impediments to progress

It is important to build on these successes. However, there are a large number of obstacles to overcome.

Human population growth

Pressures of population growth and increased wealth will result in increased demand for fish products in the future. FAO (1995*a*) estimates that, relative to the 1993 production of food fish of 72.3 million t (from marine, inland and aquaculture sources), the demand for food fish may increase by 19 million t (26%) by the year 2010. FAO considers such an increase in the production of food fish to be possible if aquaculture production can be substantially increased, if capture fisheries can be managed more rationally, and if bycatch can be used more effectively. The downside is that *unless* fisheries management does actually improve and fish stocks do actually rebuild, *per capita* fish consumption levels may decline substantially, particularly in developing nations where protein deficiencies may already exist. Expanding population and expanding international and national markets are already eroding the ability of communities to manage local resources.

Lack of political will

In the context of national priorities, the long-term sustainability of fisheries is often not considered an issue of paramount importance to politicians. Often, it may be more expedient to make concessions to vocal constituents, rather than to confront the difficult and potentially unpopular decisions necessary to ensure the future viability of natural resources and the communities that depend on them. As a result, the governments of many countries are still subsidizing fleet expansion rather than subsidizing fleet reduction and redeployment or retraining of fishers. Unless fleet size can be reduced to levels commensurate with resource productivity, both globally and locally, there is little hope for meaningful progress towards ensuring sustainability of marine capture fisheries.

Ingrained beliefs and practices that impede sustainability

A major threat to fisheries sustainability is that there are a number of inappropriate 'groundrules' or mind-sets that have formed guiding principles for fisheries management to date. Most, if not all of these need to be abandoned. Overcapacity, and the belief systems and mind-sets that impede reductions in fleet capacity, are the fundamental problems requiring immediate and focussed remedial action to ensure the sustainability of marine resources and associated ecosystems. Other 'groundrules' that have contributed towards the current situation of overcapacity and overfishing also need to be abandoned and replaced with appropriate alternatives that embrace precautionary, risk-averse approaches. An extraordinary culture change—that is, a complete change in the way most people perceive fisheries—is required before the new paradigms of precautionary approaches (Tables 2 and 3) and risk-averse decision making can become the norm. This section examines some of the obstacles to effecting the necessary culture change.

The legacy of the inexhaustibility paradigm

Fishing is still perceived by some to be the 'last frontier'. While it is true that new fishing grounds are still being discovered (although at a diminishing rate), and new markets and new product lines are being developed, thus facilitating the utilization of previously-undesirable species, the available global statistics suggest that sustainable landings from capture fisheries are unlikely to increase much over recent levels. Some groups may recognize that the inexhaustibility paradigm no longer applies, but nevertheless they act as if it does. For example, in a recent report, the Scottish Fishermen's Federation agreed that fleet downsizing was needed to reduce fishing effort but stated that they did not really want to reduce their effort because 'there is still a fair amount of fish in the sea at the moment and fishing is our way of life. Fishing is all there is ... there is no alternative or anything else to do' (House of Lords 1996).

Belief in the status quo

There seems to be a general belief that the *status quo* has inherent merit, particularly if the *status quo* involves large numbers of small-scale fishers. Many people still hold onto, and revere, the image of the small-scale fisher as someone who is generally poor, struggles to make ends meet, and works under adverse and often unsafe conditions, battling the elements, conquering Nature, risking his life. The image of a fisher as a wealthy, successful, professional businessperson seems abhorrent to many. But a *status quo* that involves fleet overcapacity, marginal or negative profits, and an often-high fatality or injury rate should not be vigorously protected. Belief in the *status quo* also translates into resistance to new management initiatives. However, it is important to recognize that management initiatives are usually formulated in response to a failure of the *status quo*. Often, the *status quo* is not a viable option for the future, regardless of the management system.

Fishing as a birthright

Many fishers and those who aspire to engage in fishing in the future perceive fishing as a right rather than a privilege. This belief affects attitudes about the need for responsible fishing practices, record keeping, and supplying data to management authorities. It also results in the onus of the 'burden of proof' falling on the management authorities rather than the fishing industry. Scientists and managers must prove that the resources are in jeopardy before implementing restrictive management measures. Reversal of the burden of proof requires that the fishing industry show that its activities are not seriously jeopardizing the long-term sustainability of the resource and the associated marine community. There is also a need to change the ethic of fishing to one where the privilege of earning a livelihood from fishing is accompanied by responsibilities in the form of codes of conduct, duties of care, and obligations to record and supply governments or other authorities with the data needed to assess and manage fishery resources.

Greed is rewarded

To date, those who have fished the hardest have generally obtained the largest shares when quotas are allocated between individuals, gears, regions or nations. To an extent, this practice is logical in that individuals or corporations who have invested the most in prosecuting particular fisheries should not be unduly penalized. However, the practice has often been extended to situations where fleets have violated existing regulations that were intended to restrict increases in fishing effort. Illegal, irresponsible and unsustainable fishing practices have generally resulted in few if any repercussions for the countries and fishing fleets that have exhibited such practices. In addition, civil disobedience is being increasingly used as a measure to argue for special allocation privileges. Gains from sustainable management must accrue to those who have behaved responsibly by not overexploiting, by taking care to minimize bycatch and discards, and by allowing depleted fish stocks to recover.

Conflicting perspectives

As fishing pressure increases and stocks decline, there will be some who recognize that a problem is developing, and others who deny it. It is even possible to have severely depleted stocks and substantial overcapacity, yet to have no consensus on the existence of a problem, let alone its causes, consequences, and solutions. Perceptions about the status of the stock, the status of management, and the solutions required to improve the fishery generally vary among the players involved. Stock assessments may indicate declining catch-per-unit-effort, declining average size of harvested fish, and increasing reliance on newly recruited year classes; yet fisheries managers may want to give the benefit of the doubt to those whose livelihoods depend on revenues from the resource; commercial fishers may claim that they see no evidence of declines in abundance; recreational fishers may claim that the resource is in jeopardy because of commercial fishing practices, but that recreational fishing has relatively little impact and so should be allowed to continue; and politicians may be swayed by the arguments of their most vocal constituents.

The US fishery for Atlantic bluefin tuna provides a good example of the extent to which perspectives about stock status and management options can diverge. Management of this open access fishery is a nightmare. More than 27 000 permit holders (commercial and recreational combined) compete for a national quota of only 1311 t (corresponding to an average of about one half of a fish per permit holder per year). In order to ensure that all participants have the opportunity to catch their 'fair share' of this valuable resource (an individual bluefin tuna can be worth as much as $US20 000), the National Marine Fisheries Service (NMFS) divides the quota between gear types, geographic areas and months or weeks, and devotes substantial staff resources to continually adjust these allocations by in-season openings and closings to try to make quota available for each fishing sub-fleet as the bluefin tuna migrate along a semi-predictable route adjacent to the coastlines of about 13 US states. Scientific stock assessments indicate that the bluefin tuna spawning stock is currently around 13% of the levels estimated for the mid-1970s (Fig. 9; International Commission for the Conservation of Atlantic Tunas 1996). Yet commercial fishers claim that the stock has increased substantially in recent years and commercial fishing representatives have been pressuring politicians to substantially increase the national quota. In addition, most commercial sub-fleets (categorized by gear type and fishing area), as well as the recreational sector, have attempted to enlist political support to each obtain a larger share of the national quota. On the other hand, the environmental community has attempted to obtain a

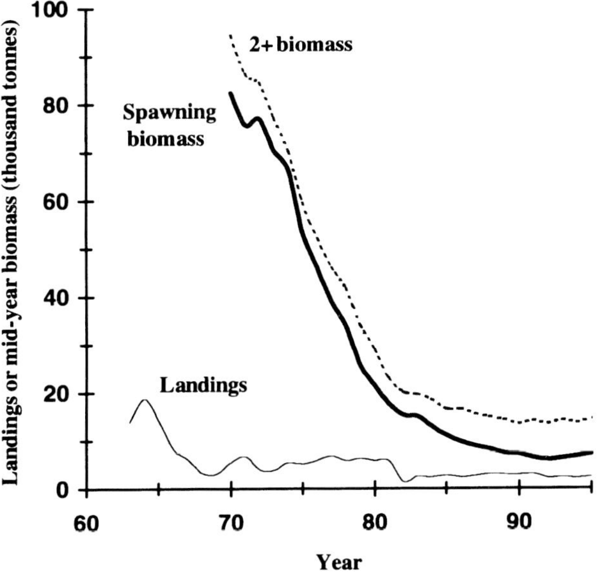

Fig. 9. Estimated stock biomass and landings of Atlantic bluefin tuna. Source: International Commission for the Conservation of Atlantic Tunas (1996).

Commission for International Trade of Endangered Species (CITES) listing for Atlantic bluefin tuna, and has suggested that the commercial fisheries should be severely curtailed. Other groups have suggested that the recreational fishery, which targets juveniles, should be abolished until stocks rebuild. Recreational fishery representatives have countered that bluefin tuna fishing is a significant seasonal industry for coastal communities and that recreational fishing is more important to the national economy than commercial fishing. Similar debates routinely take place in fisheries for other species.

Use of inappropriate scapegoats

Attacks on the quality or validity of scientific stock assessments are becoming commonplace but, with few exceptions, the quality of the science is usually an inappropriate scapegoat. Stock assessments may be imprecise, but there is little evidence to support the widely-held belief (particularly by members of the commercial fishing industry) that scientists (particularly government scientists) tend to produce overly-conservative assessments. Challenges to scientific stock assessments are becoming an effective means of delaying proposed restrictions. In the United States, there is an increasing tendency to ask the courts to determine the validity of the science. Also, openness about uncertainty in stock assessments has tended to provide an excuse for political compromise and is frequently used as an argument for maintaining *status quo* landings. The fact is that stock assessments will probably always be imprecise, but the appropriate response to imprecise assessments is to manage conservatively.

The environment is another inappropriate scapegoat. Participants in fisheries often claim that fish stocks are inherently variable, and that fishers have little effect on stock condition relative to the physical environment or to the effects of other predators such as seals, other marine mammals, birds and piscivorous fish. When stocks collapse, there are often protracted arguments about whether it was 'environment' or 'overfishing'. Obviously, both are important. High fishing mortality accelerates and prolongs stock declines when environmental conditions become unfavourable.

MSY is also an example of an inappropriate scapegoat. It has often been claimed that MSY is an inappropriate management goal because fisheries managed for MSY have often been overexploited. However, by definition, MSY is sustainable. In fact, many fisheries have supposedly been 'managed' using MSY on paper, but few have embraced the concept of MSY in practice. For example, the International Convention for the Conservation of Atlantic Tunas (ICCAT) explicitly specifies MSY as the management objective, yet in practice management measures have not been implemented at all until stock assessments have shown that fisheries have progressed well beyond MSY, and even then management recommendations have been more appropriate to preserving the *status quo* rather than rapidly rebuilding stocks to MSY levels. If all fisheries had been managed at MSY and, in particular, if fleet capacities had been commensurate with MSY (even on a single-species basis), fisheries would be much better off today.

Oversimplified objective functions

On the other hand, simple objectives such as maximization of yields may not be sufficient management goals, by themselves, for the future. The fisheries literature is replete with theoretical models that use sophisticated analytical and simulation techniques to find optimal solutions for various biological and socio-economic objective functions, but the most common objective is to maximize single quantities such as MSY or MEY (Maximum Economic Yield). In reality, given the uncertainty of estimates of stock size and fishing mortality, the unpredictability of future environmental effects, multi-species interactions, and inherent difficulties in the practical implementation of management strategies, maximization of yields may be an impossible ideal. Even if maximum sustainable yields could be estimated exactly, the combined effect of overcapitalized and economically marginal fishing fleets that need to land large quantities of fish in order to be profitable, imperfect monitoring systems and wasteful fishing practices will almost invariably result in target catches being overshot. Thus, if the objective was to achieve the maximum, it will be exceeded, and therefore goals that emphasize maximization of yield may ultimately sacrifice sustainability. The combination of uncertainties about stock status and the gold rush mentality of fisheries led Ludwig *et al.* (1993) to question the whole concept of sustainable use of fisheries; however, Rosenberg *et al.* (1993) correctly countered that the real problem is the difficulty of estimating and achieving optimal yields, particularly when optimum is equated with maximum.

Rather than focussing on *maximization* of simple objective functions, fisheries management models should be based on *optimization* of multi-attribute objectives. Fisheries objective functions need to incorporate all relevant socio-economic objectives, biological constraints, uncertainty in estimates of current and future stock status and, perhaps most importantly, limitations on the ability to implement or achieve management targets. In fact, these uncertainties and limitations suggest that it may be better to aim for a moderate level of exploitation in capture fisheries, rather than attempting to exploit the entire surplus production and therefore overshooting and over-exploiting. Although 'suboptimal' levels of exploitation would ultimately sacrifice potential yields, benefits include reduced risk of recruitment overfishing of target species, reduced bycatch and discarding, less disruption of the structure and function of marine biological communities, less frequent need for assessments and adjustments to total allowable catches (thus allowing scientists and managers the luxury of time to develop long-term research programmes and management strategies), and less need to micromanage.

Short-term economic gain versus long-term sustainability

In fisheries management, it is common for short-term economic gain to take priority over long-term sustainability. This is understandable given that fishing mortalities far higher than the fishing mortality associated with maximum sustainable yield (F_{msy}) can be sustained in the short-term. The fishing mortality that maximizes cumulative yield (F_{mcy}) can even exceed the fishing mortality associated with extinction (F_T) (Mace 1994; Mace 1996), at least over the short term. Non-linear cost functions can also lead to situations where the fishing mortality that maximizes economic yields (F_{mey}) exceeds F_{msy} and even F_T, contrary to the classic bio-economic model (Mace 1996).

Short-term perspectives are characteristic of open access fisheries, whether the fisheries are in an expanding phase (in which case there is strong resistance to restrictive regulations) or in an overcapitalized and overfished state (in which case there may be strong resistance to quota reductions and strong incentive to circumvent regulations). However, a short-term perspective is also common in limited access fisheries, including fisheries regulated by ITQs. An ITQ quota holder who is able to sneak in a little extra catch benefits directly in the short term, while the long-term consequences (a somewhat reduced stock size due to a somewhat increased fishing mortality) are spread across all quota holders. In addition, fishers often tend to trade off the certainty of the present against the uncertainty of the future (meaning that they have high discount rates). Such uncertainties may be large; for example, uncertainties about future abundance and distribution of the resource, future stock assessment findings, real or anticipated changes in government leaders or government policy, concern about continued physical ability to participate in fisheries, and fear of possible downturns in market demands and prices (Mace 1993).

As stated in a recent report (House of Lords 1996): all over the world, the long-term sustainability of fish stocks is being sacrificed in favour of the short-term protection of employment in the fishing industry. But the lesson to be learned from the collapse of the Grand Banks cod fishery is that ignoring the warning signs year after year can end in the loss of thousands rather than hundreds of jobs.

Synthesis: a common problem

In many fisheries, stakeholders cannot even agree on the problem, let alone the solutions. Defining and reconciling objectives pertaining to all dimensions of fisheries (biological, economic and social) is such a difficult task that it is rarely addressed in a comprehensive and transparent manner.

Future prospects and challenges

The world's fisheries are already in the process of transition, but without appropriate intervention the outcome may not be positive. To quote an old proverb: 'If you don't change direction, you'll get where you're going. But before you can change direction, you must first accept where you're going, and then decide that you don't want to get there'. We have enough science, enough technology, enough eloquent, forward-thinking words on paper, enough appreciation of the problems, enough spectacular examples of fishery failures — and enough successes — to know what needs to be done. However, the old saying that recognizing the problem is half the solution does not apply to fisheries. Inertia comes from lack of political will, lack of monetary commitment, and pleas from fish harvesters and others to maintain a way of life that, unfortunately, in many cases is rapidly becoming an anachronism.

The challenge for the future is to discard outmoded beliefs and practices that impede sustainability, and to build on existing successes to effect a transition to environmentally and economically sound fisheries both at global and local levels. The key criteria are environmental sustainability (which will require an overall reduction in fishing mortality rates on wild stocks and greater concern for environmental impacts of aquaculture projects), economic and social viability (which will require a much reduced fleet capacity and a much-reduced dependence on wild stock fisheries for employment) and cooperative management (which will require a realigning of the roles and responsibilities of governments, participants and other stakeholders). There are five major challenges that need to be met.

1. Reduce fleet capacity and reduce dependence on wild fish stocks

A substantial reduction in fishing effort is urgently needed throughout fisheries world-wide. Global fishing capacity should be reduced by at least 50%. Unfortunately, historical attempts to freeze or reduce fleet capacity have been notoriously unsuccessful due to factors such as lack of mobility of capital, prospects of short-term economic gain, lack of alternative employment opportunities, and local values and ideals. Canada implemented some of the earliest limited entry licence systems (e.g. Pacific salmon in 1969 and roe-herring in 1974). However, these systems have failed to control expansions in fleet capacity. While the number of vessels in the British Columbia salmon fleet has steadily declined since 1969, the capacity of the fleet has grown substantially. Fourteen years ago, Pearse (1982) recommended that the salmon and roe-herring fleets should be reduced to half their present size over a 10-year period. This year (1996), the Canadian government is in the process of implementing an aggressive plan to reduce the number of licences in the British Columbia salmon fleet by 50% (Department of Fisheries and Oceans Canada 1996).

Currently, government-sponsored buy-out programmes are in progress or recently completed in the European Union, northeast and northwest United States, Pacific Canada, Japan, Taiwan, Australia, New Zealand and most other major fishing nations. However, their effectiveness at reducing fleet capacity over the long term remains to be seen. To date, government-funded buy-out programmes have almost invariably failed to achieve the objective of permanently reducing fleet capacity. Experience suggests that privatization may be a much more effective mechanism for bringing fleet capacity in line with resource productivity; essentially, implementation of property rights promotes an industry-funded 'buy-out' of less efficient fishing units (see ITQ examples in Table 1).

2. Define and implement workable access rights systems

One of the greatest challenges of all is to resolve the issue of the role of 'property rights' in the sustainability of fisheries. Property-rights systems, including ITQs and related systems and community-based management or collective private property, are the appropriate response to the recognition that fisheries can no longer be considered the 'employer of last resort'. In addition, the establishment of use rights is important for protecting artisanal fishers from unequal competition with industrial vessels. I agree with van der Elst (1997), that: 'there simply is no alternative but to harvest natural resources according to an acceptable system of access rights'. But these systems are not panaceas. When stocks decline unpredictably, ITQ quota holders and small communities may be just as economically marginal as open access participants. When coastal communities continue to rely heavily on fisheries for income and employment, without attempting to diversify the economic base, short-term social and economic considerations may continue to override conservation and sustainability. Thus, it is essential to match fleet capacity and participation to resource productivity in order to make access rights systems workable. In addition, several important decisions need to be made when designing access control systems; the most important of these include initial allocations, the extent to which transferability of rights and consolidation should be permitted and the appropriate distribution of the costs and benefits of restricted access (including user-pays principles and the distribution of resource rents).

3. Implement the precautionary approach

A precautionary approach (Tables 2 and 3) is vital for ensuring the long-term viability of the world's fisheries. However, precautionary approaches and related concepts such as risk-averse decision making may be difficult to sell due to pressures of population growth and associated demands for fish protein, the legacy of inexhaustibility, short-term objectives taking precedence over the long-term, and lack of political will to make the tough but necessary decisions. We need the courage and the means to implement effective management, even to close down fisheries in trouble or to suspend aquaculture projects that are environmentally destructive. However, having the means to implement effective management equates to having sufficient financial and other resources to mitigate negative social impacts of fisheries closures or downscaling of fleet capacity. Before precautionary approaches can become the norm, we need to minimize or eliminate overcapacity and excessive reliance on capture fisheries for food and employment.

A potential net result of reducing fleet capacity, reducing reliance on capture fisheries, implementing access controls and adopting the precautionary approach is that fish stocks will be exploited at levels well below maximum sustainable yields. However, the negative effect of reduced yields may be outweighed by benefits such as reduced risk of recruitment overfishing of target species, reduced bycatch and discarding, less ghost fishing by lost or abandoned gear, less disruption of the structure and function of marine biological communities, improved human safety, higher quality products for human consumption, increased profitability of the harvesting sector (leading to reduced pressure on scientists to make optimistic stock assessments, reduced pressure on managers to make risk-prone management decisions, reduced incentive to circumvent regulations, and reduced requests for government subsidies to bail out failing businesses), less frequent need for assessments and changes in total allowable catches, and less need to micromanage.

4. Expand aquaculture without destroying natural environments

FAO (1995*a*) estimates that if supplies from capture fisheries do not increase significantly, aquaculture production will need to produce approximately 31 million t by the year 2010 in order to maintain current *per capita* levels of consumption. This will require that the estimated 1993 production be doubled in a period of 17 years, a challenge that is believed to be feasible considering recent rates of expansion, technical knowledge and availability of financing. Nevertheless, proper planning, environmental considerations, proper system management and disease control will have to play a more important role than at present if crashes in production are to be avoided (FAO 1995*a*).

Priorities for aquaculture research are identifying substitutes for fish meal and reducing the volume of fish required as feed, improved husbandry, disease control, waste treatment and other means of mitigating environmental degradation. The need to replace extensive and intensive shrimp culture systems with semi-intensive systems is becoming widely recognized (Bailey 1992; Csavas 1994). Ultimately, aquaculture operations that destroy coastal habitats, use wild fish to produce cultured fish, and release untreated waste products into the environment must be phased out and replaced with environmentally sound practices. There are at least three promising possibilities, each of which provides challenges to both scientists and engineers: (1) small-scale diversified aquaculture operations including polyculture and integrated aquaculture (animal husbandry with aquaculture or agriculture with aquaculture), which have long been a Chinese tradition (Lin 1994), (2) inland-based recirculating systems with complete capture of wastes, and (3) open ocean cage culture.

5. Develop effective policies and institutions to achieve the preceding objectives

The changes required to achieve environmental sustainability, economic viability and cooperative management will be extremely costly and controversial. Monetary costs are only one factor; social costs may be much more important and problematic overall. The structure and function of institutions will need to be modified to achieve an appropriate balance between top-down (government-based) and bottom-up (community-based) management. The model for the future must incorporate global conservation standards and guidelines, holistic national policies focussing on long-term objectives for fisheries management and aquaculture, strengthened fisheries management plans for all fisheries, and increased active involvement of all stakeholders at all levels (Fig. 10). The importance of maintaining sustainable fisheries needs to be elevated on national agendas, but there is also a need to reduce micro-level political interference. Funding for biological, economic and social data as bases for decision making needs to be increased. Subsidies need to be redirected from fleet expansion to social adjustment programmes, and from capture fisheries to aquaculture. Fishing decisions should be devolved to the lowest appropriate level, but governments should not abdicate responsibility for stewardship of national natural resources. Community development programmes should be encouraged, but governments should insist on environmental assessments as prerequisites for all major developmental programmes. Fishers, environmentalists, scientists, managers, politicians and other stakeholders all need to be united in their determination to implement effective management, and to be responsible and accountable for their decisions and actions. A commitment to cooperative management and cooperative problem solving is imperative, now and into the future.

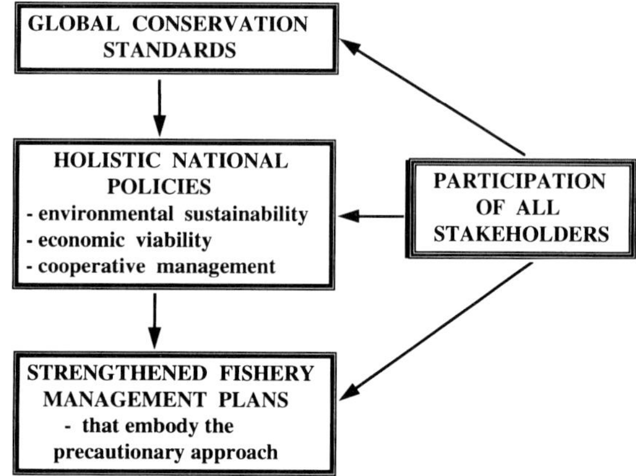

Fig. 10. Key elements for successful management for the future.

Acknowledgments

My sincere thanks to the Congress programme committee for the honour of being chosen as keynote speaker for the Second World Fisheries Congress. I am also very grateful to the Australian Fisheries Management Authority (AFMA) for sponsoring my participation in the Congress. Many colleagues provided data, verification of case studies used as examples, comments on earlier versions of this manuscript, or photographic materials for the oral presentation based on this paper. I am much indebted to Sandy Argue, Leo Brander, Karyl Brewster-Geisz, Bill Clark, Steve Edwards, Charlie Ess, Brenda Figuerido, Bill Folsom, Kevin Friedland, Amy Gautam, Jay Ginter, John Glaister, Susan Hanna, Dick Harvey, Anne Hollowed, Rebecca Lent, Doreen Liew, Paul Macgillivray, Sarah McLaughlin, Jim McVey, Ben Muse, Chris Newton, Bob O'Boyle, Andrea Oliver, Greg Peacock, Bruce Phillips, Julie Porter, Laura Richards, Don Robertson, Becky Rootes, Tony Smith, Margo Schulze, Gary Shepherd, Michael Sissenwine, Linda Stathoplos, Max Stocker, Steve Swartz, Joe Terry, Robin Tuttle, Dennis Weidner and Mark Wildman.

References

AFMA (1995). AFMA Corporate Plan 1995–2000. Australian Government Printing Service, Canberra.

Aiken, D, and Sinclair, M (1995). From capture to culture: exploring the limits of marine productivity. *World Aquaculture* **26**(3), 21–34.

Alverson, D L, Freeburg, M H, Murawski, S A, and Pope, J G (1994). A global assessment of fisheries bycatch and discards. *FAO Fisheries Technical Paper* 339. 235 pp.

Anderson, K P, and Ursin, E (1977). A multispecies extension to the Beverton and Holt theory of fishing, with accounts of phosphorus circulation and primary production. *Meddr. Danm. Fisk.-og Havunders. N.S.* **7**, 319–435.

Anon. (1984). Inshore finfish fisheries: proposed policy for future fisheries management. NZ Ministry of Agriculture and Fisheries, Wellington. 31 pp.

Bailey, C (1992). Coastal aquaculture development in Indonesia. In 'Contributions to Fishery Development Policy in Indonesia'. (Eds R B Pollnac, C Bailey and A Poernomo) pp. 57–72. Central Research Institute for Fisheries, Ministry of Agriculture, Indonesia.

Bailey, C, and Zerner, C, (1992). Local management of fisheries resources in Indonesia: opportunities and constraints. In 'Contributions to Fishery Development Policy in Indonesia'. (Eds R B Pollnac, C Bailey and A Poernomo) pp. 38–56. Central Research Institute for Fisheries, Ministry of Agriculture, Indonesia.

Basson, M, and Beddington, J R (1993). Risks and uncertainties in the management of a single-cohort squid fishery: the Falkland Islands *Illex* fishery as an example. In 'Risk evaluation and biological reference points for fisheries management'. (Eds S J Smith, J J Hunt and D Rivard) pp. 253–60. *Canadian Journal of Fisheries and Aquatic Sciences Special Publication* 120.

Bell, F W (1972). Technological externalities and common-property resources: an empirical study of the US northern lobster fishery. *Journal of Political Economy* **80**, 148–58.

Beverton, R J H, and Holt, S J (1957). On the dynamics of exploited fish populations. Ministry of Agriculture, Fisheries and Food (UK), *Fishery Investigation, Series* 2, **19**. 533 pp.

Bishop, C A., Murphy, E F, Davis, M B, Baird, J W, and Rose, G A (1993). An assessment of the cod stock in NAFO Divisions 2J+3KL. *NAFO Scientific Council Research Document* 93/86, Series No. N2271. 51 pp.

Brander, K (1981). Disappearance of common skate, *Raia batis*, from Irish Sea. *Nature* **290**, 48–9.

Brown, R S, and Phillips, B F (1994). The current status of Australia's rock lobster fisheries. In 'Spiny Lobster Management'. (Eds B F Phillips, J S, Cobb, and J Kittaka) pp. 33–63 (Fishing News Books: London).

Caddy, J F, and Griffiths, R C (1990). A perspective on recent fishery-related events in the Black Sea. *GFCM Studies and Reviews* **63**, 43–71.

Chamberlain, G W (1991). Shrimp farming in Indonesia: 1-Growout techniques. *World Aquaculture* **22**(2), 12–27.

Collie, J S (1995). Predator-prey models in marine fisheries. Source document for the FAO world conference on the security of fish supplies. Kyoto, Japan, December 1995.

Collins, C H (1994). Beyond denial: the northeastern fisheries crisis — causes, ramifications and choices for the future. Unpublished report, Boston, Massachusetts, USA.

Crutchfield, J A, and Pontecervo, G (1969). The Pacific Salmon Fisheries: a Study of Irrational Conservation. John Hopkins Press. 220 pp.

Csavas, I (1994). Important factors in the success of shrimp farming. *World Aquaculture* **25**(1), 34–56.

Department of Fisheries and Oceans Canada (1988). Report of the Scotia-Fundy Groundfish Industry Capacity Advisory Committee. Department of Fisheries and Oceans (DFO), Hallifax, Nova Scotia, Canada. 14 pp.

Department of Fisheries and Oceans Canada (1996). Pacific salmon fishery revitilization program. Department of Fisheries and Oceans (DFO), Ottawa, Canada.

Edwards, S F, and Murawski, S A (1993). Potential economic benefits from efficient harvest of New England groundfish. *North American Journal of Fisheries Management* **13**(3), 437–49.

FAO (1993*a*). Marine fisheries and the law of the sea: a decade of change. Special chapter (revised) of The State of Food and Agriculture 1992. *FAO Fisheries Circular* No. 853. FAO, Rome. 65 pp.

FAO (1993*b*). Review of the state of world marine fishery resources. *FAO Technical Paper* No. 335. FAO, Rome. 136 pp.

FAO (1995a). The state of world fisheries and aquaculture. FAO, Rome. 57 pp.

FAO (1995*b*). Review of the state of world fishery resources:inland capture fisheries. *FAO Fisheries Circular* No. 885. FAO, Rome. 63 pp.

FAO (1995c). Precautionary approach to fisheries. Part 1: Guidelines on the precautionary approach to capture fisheries and species introductions. *FAO Fisheries Technical Paper* 350/1. 52 pp.

FAO (1995*d*). Code of Conduct for Responsible Fisheries. FAO, Rome. 41 pp.

Fish Farming International (1996). World farm production is boosted again by China. *Fish Farming International* **23**(7), 5.

Fitzpatrick, J (1995). Technology and fisheries legislation. Paper presented to the International Technical Consultation on the Precautionary Approach to Capture Fisheries. Lysekil, Sweden, 6–13 June 1995.

Francis, R I C C (1995). The longevity of orange roughy: a reinterpretation of the radiometric data. New Zealand Fisheries Research Document 95/2. 13 pp.

Froese, R and Pauly, D (Eds) (1996). FishBase 96: concepts, design and data sources. ICLARM, Manila, Philippines. 179 pp.

Gallucci, V F, Amjoun, B, Hedgepeth, J, and Lai, H L (1996). Size-based methods of stock assessment of small-scale fisheries. In 'Stock Assessment: Quantitative Methods and Applications for Small-Scale Fisheries'. (Eds V F Gallucci, S B Saila, D J Gustafson and B J Rothschild) pp. 9–81 (CRC Press: New York).

Garcia, S, and Newton, C (in press). Current situation, trends and prospects in world capture fisheries. In 'Global Trends in Fisheries Management'. (Eds E Pikitch, D Huppert and M Sissenwine). *American Fisheries Society Monograph Series.*

Gavaris, S (1988). An adaptive framework for the estimation of population size. Canadian Atlantic Fisheries Scientific Advisory Committee (CAFSAC) Research Document 88/29. 12 pp.

Gordon, H S (1954). The economic theory of common property resources: the fishery. *Journal of Political Economy* **62**, 124–42.

Graham, M (1935). Modern theory of exploiting a fishery, and its application to North Sea trawling. *Journal du Conseil International pour l'Exploration de la Mer* **13**, 264–74.

Halliday, R G, Peacock, F G, and Burke, D L (1992). Development of management measures for the groundfish fishery in Atlantic Canada: a case study of the Nova Scotia inshore fleet. *Marine Policy* (1992),411–26.

Hilborn, R. (1990). Marine biota. In 'The earth as transformed by human action: global and regional changes in the biosphere over the past 300 years'. (Eds B L Turner II, W C Clark, R W Kates, J F Richards, J T Matthews and W B Meyer) pp. 371–85.

House of Lords (1996). Fish stock conservation and management. Report of the House of Lords Select Committee on Science and Technology. HL Paper 25. London: HMSO. 77 pp.

Hutchings, J A (1996). Spatial and temporal variation in the density of northern cod and a review of hypotheses for the stock's collapse. *Canadian Journal of Fisheries and Aquatic Sciences* **53**, 943–62.

Hutchings, J A, and Myers, R A (1994). What can be learned from the collapse of a renewable resource? Atlantic cod, *Gadus morhua*, of Newfoundland and Labrador. *Canadian Journal of Fisheries and Aquatic Sciences* **51**, 2126–46.

International Commission for the Conservation of Atlantic Tunas (1996). Report of the Standing Committee on Research and Statistics (SCRS). ICCAT. Madrid, Spain.

Kemf, E, Sutton, M, and Wilson, A (1996). Wanted Alive: Marine Fishes in the Wild. World Wildlife Fund for Nature, Gland, Switzerland.

Lin, J (1994). Aquaculture in China. *World Aquaculture* **25**(4), 39–40.

Ludwig, D, Hilborn, R, and Walters, C (1993). Uncertainty, resource exploitation, and conservation: lessons learned from history. *Science* **260**, 17.

Mace, P M (1993). Will private owners practice prudent resource management? *Fisheries* **18**(9), 29–31.

Mace, P M (1994). Relationships between common biological reference points used as thresholds and targets of fisheries management strategies. *Canadian Journal of Fisheries and Aquatic Sciences* **51**, 110–122.

Mace, P M (1996). The bioeconomic consequences of risk-averse management strategies. (unpublished ms).

Mace, P M, Fenaughty, J M, Coburn, R P, and Doonan, I J (1990). Growth and productivity of orange roughy (*Hoplostethus atlanticus*) on the north Chatham Rise. *New Zealand Journal of Marine and Freshwater Research* **24**, 105–19.

McLoughlin, K, Staples, D, and Maliel, M (1993). Fishery status reports 1993. Australian Government Publishing Services, Canberra.

Methot, R D (1990). Synthesis model: an adaptable framework for the analysis of diverse stock assessment data. In 'Proceedings of the symposium on application of stock assessment techniques to gadids'. (Ed Loh-Lee Low). *International North Pacific Fisheries Commission* **50**, 259–77.

Miller, M, Lipton, D, and Hooker, P (1994). Profile of change: a review of offshore factory trawler operations in the Bering Sea/Aleutian Islands pollock fishery. National Marine Fisheries Service (NMFS), United States Department of Commerce, Washington, DC.

Myers, R A, Barrowman, J, Hutchings, N J A, and Rosenberg, A A (1995*a*). Population dynamics of exploited fish stocks at low population levels. *Science* **269**, 1106–8.

Myers, R A, Bridson, J, and Barrowman, N J (1995*b*). Summary of worldwide spawner and recruitment data. *Canadian Technical Report of Fisheries and Aquatic Sciences* 2024. 303 pp.

NMFS (1996*a*). Our Living Oceans '95. NOAA Technical Memorandum NMFS-F/SPO-xx. 165 pp.

NMFS (1996*b*). Fisheries of the United States, 1995. Current Fishery Statistics No. 9500. National Marine Fisheries Service (NMFS), United States Department of Commerce. Washington, D.C. 126 pp.

NMFS (1996*c*). 21st Northeast Regional Stock Assessment Workshop (21st SAW), Stock Assessment Review Committee (SARC). NEFSC Reference Document 96-05d, Woods Hole, MA, USA.

Parfit, M, and Kendrick, R (1995). Diminishing returns. *National Geographic* **88**(5), 2–37.

Pearse, P H (1982). Turning the tide: a new policy for Canada's Pacific fisheries. Report of the Commission on Pacific Fisheries Policy, Department of Fisheries and Oceans, Canada.

Pillay, T V R (1992). Aquaculture and the Environment. John Wiley and Sons, Inc., New York. 189 pp.

Pinkerton, E, and Weinstein, M (1995). Fisheries that work: sustainability through community-based management. The David Suzuki Foundation, Vancouver, British Columbia, Canada. 199 pp.

Prager, M H (1994). A suite of extensions to a non-equilibrium surplus production model. *Fishery Bulletin* **92**, 374–89.

Reyes, A B, De Sagun, R B, and Munoz, J C (1994). Fishing effort reduction policies: the Philippine situation. In 'Indo-Pacific Fishery Commission, Proceedings of the Symposium on Socio-economic Issues in Coastal Fisheries Management'. pp. 191–205. Bangkok, Thailand, November 1993. *RAPA Publication* 1994/8. 442 pp.

Ricker, W E (1954). Stock and recruitment. *Journal of the Fisheries Research Board of Canada* **11**, 559–623.

Rosenberg, A A, Fogarty, M J, Sissenwine, M P, Beddington, J R, and Shepherd, J G (1993). Achieving sustainable use of renewable resources. *Science* **262**, 828–9.

Royce, W F (1988). Comments on Japanese fishery management. In 'Fishery Science and Management: objectives and limitations'. (Ed W S Wooster) pp. 243–6. *Lecture Notes on Coastal and Estuarine Studies* 28. (Springer-Verlag: New York).

Rugolo, L J, Crecco, V A, and Gibson, M R (1994). Modeling stock status and the effectiveness of alternative management strategies for Atlantic coast striped bass. Summary Report to the Striped Bass Management Board, Atlantic States Marine Fisheries Commission, Washington, D.C. 30 pp.

Saclauso, C A (1989). Brackishwater aquaculture: threat to the environment? *Naga*, July, 1989. 6–9.

Sainsbury, K J (1991). Application of an experimental approach to management of a tropical multispecies fishery with highly uncertain dynamics. *ICES Marine Science Symposium* **193**, 301–19.

Safina, C (1995). The world's imperiled fish. *Scientific American* **273**(5), 46–53.

Schaefer, M (1954). Fisheries dynamics and the concept of maximum equilibrium catch. Sixth Annual proceedings of the Gulf and Caribbean Fisheries Institute, 1953. 1–11.

Schaefer, M (1957). Some considerations of the population dynamics and economics in relation to management of commercial marine fisheries. *Journal of the Fisheries Research Board of Canada* **14**, 669–81.

Short, K (1992). The Japanese coastal fisheries management system based on exclusive fishing rights. In 'International Perspectives on Fisheries Management'. (Eds T Yamamoto and K Short) pp. 43–66. Tokyo: ZENGYOREN (National Federation of Fisheries Cooperative Associations). 527 pp.

Sissenwine, M P, and Mace, P M (1992). ITQs in New Zealand: the era of fixed quotas in perpetuity. *Fishery Bulletin US* 90, 147–60.

Sissenwine, M P, and Swartz, S (1992). Analysis of the potential economic benefits from rebuilding US fisheries. National Marine Fisheries Service (NMFS), United States Department of Commerce. Washington, DC.

Smith, T (1994). Scaling Fisheries. Cambridge University Press, Cambridge, UK. 392 pp.

Sorgeloos, P, and Sweetman, J (1993). Aquaculture success stories. *World Aquaculture* **24**(1), 4–14.

Sparre, P (1991). Introduction to multispecies virtual population analysis. *ICES Marine Science Symposium* **193**, 12–21.

Thompson, W F, and Bell, F H (1934). Biological statistics of the Pacific halibut fishery. 2. Effects of changes in intensity upon total yield and yield per unit gear. Report of the International Fisheries Commission 8.

van der Elst, R, Branch, G, Butterworth, D, Wickens, P and Cochrane, K (1996). How can fisheries resources be allocated; who owns the fish? In 'Developing and Sustaining World Fisheries Resources: The State of Science and Management. Second World Fisheries Congress, Brisbane 1996'. (Eds D A Hancock, D C Smith, A Grant and J P Beumer) pp. 307–14 (CSIRO Publishing: Melbourne).

Walters, C J, and Maguire, J J (1996). Lessons for stock assessment from the northern cod collapse. *Reviews in Fish Biology and Fisheries* **6**, 125–37.

Ward, J, and Sutinen, J G (1994). Vessel entry-exit behaviour in the Gulf of Mexico shrimp fishery. *American Journal of Agricultural Economics* **76**(4), 916–23.

Williams, M (1994). Transition in the contribution of living aquatic resources to sustainable food security. Unpublished *ICLARM Report.* 67 pp.

World Bank (1991). Fisheries and aquaculture research capabilities and needs in Asia: studies of India, Thailand, Malaysia, Indonesia, the Philippines, and the ASEAN region. *World Bank Technical Paper* No. 147. Fisheries Series.

Yamamoto, T, and Short, K (Eds) (1992). International Perspectives on Fisheries Management. Tokyo: ZENGYOREN (National Federation of Fisheries Cooperative Associations). 527 pp.

Why do some fisheries survive while others collapse?

Theme 1

Why Do Some Fisheries Survive and Others Collapse?

M. Sinclair,[A] *R. O'Boyle,*[A] *D.L. Burke*[B] *and G. Peacock*[B]

[A]Marine Fish Division, Fisheries and Oceans Canada, PO Box 1006, Dartmouth, NS, B2Y 4A2, Canada.
[B]Fisheries and Oceans Canada, PO Box 550, Halifax, NS, B3J 2S7, Canada.

Summary

The concepts underlying fisheries management are introduced by linking the Gordon-Schaefer bio-economic model to recruitment overfishing. Activities which lower the cost curve are high risk to the degree that the open access equilibrium point coincides with higher effort levels and lower spawning stock biomass. Collapse is defined to occur in those cases when fishing effort is not restricted and spawning stock biomass is reduced to levels below that required for moderate to good recruitment. Changes in the ecosystem can alter the spawning stock biomass which is required for moderate recruitment, and thus the effort levels which will generate collapse. The activities of fisheries management are introduced in a second model. Within the framework of these models two cod stocks on the Scotian Shelf off Nova Scotia, Canada, are described with the aim of illustrating the principles underlying survival and collapse. The shelf is characterized by strong gradients in bottom temperature, generally less than 4°C on the eastern Shelf. Grey seal abundance is considerably higher within the eastern stock area. Recruitment dropped sharply in the mid-1980s for the eastern Scotian Shelf stock and biomass has declined to very low levels. The fishery was closed in 1993. Recruitment to the western Scotian Shelf stock has not declined, and the spawning stock biomass is at moderate levels. One stock has collapsed and the other has survived. The fisheries management activities, changing environmental conditions, and seal predation changes on juvenile cod, are described. For both management units, fishing effort was not restricted to target levels since the introduction of quota management in 1977. As a result, there has been growth overfishing. It is concluded, however, that differences in management and fishing activities between the two areas cannot account for the differential stock responses. Colder environmental conditions and increases in juvenile natural mortality due to seal predation are concluded to be contributing factors to the decline in stock production and recruitment for the eastern stock. The reasons for the overfishing of both stocks under quota management are analysed using the fisheries management model. The issues illustrated by the cod examples are then used to draw conclusions on the management activities that are necessary to prevent the collapse of fisheries. Three key processes described within the Gordon-Schaefer model are used to structure the conclusions. These are: changes in the slope of the cost curve; control of fishing effort; and the dynamic relationship between the slope of the cost curve, the open access equilibrium point and minimum spawning stock biomass.

Introduction

We will describe two cod (*Gadus morhua*) fisheries off Nova Scotia, Canada, to introduce the theme of why some fisheries survive and others collapse. The landed values for the marine fisheries of the maritime provinces of Atlantic Canada from 1984 to 1995 are shown in Figure 1. Although the overall landings have been at near record levels during recent years, the groundfish component has been declining steadily. It is this component that we will focus on. The Gordon-Schaefer bio-economic model (Fig. 2) is used to illustrate the ideas underlying fisheries management and to define survival and collapse of fisheries. At relatively low levels of fishing effort there is a maximum in the net revenue (i.e. the difference between the

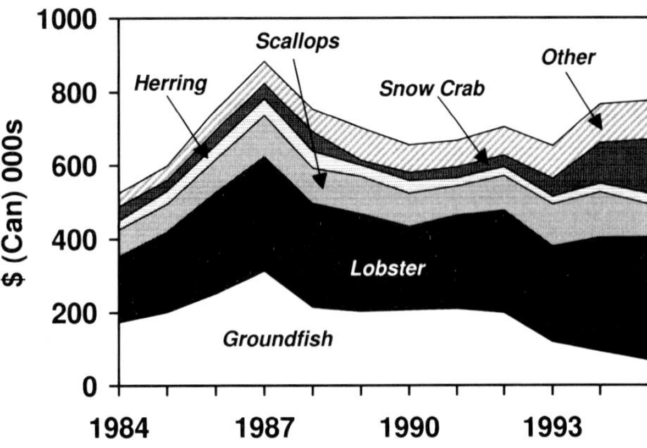

Fig. 1. Trends in the total landed value in thousands of dollars from 1984 to 1995 for marine fisheries of the Maritime provinces of Atlantic Canada (Nova Scotia, New Brunswick, and Prince Edward Island).

revenue curve and the cost curve is greatest at the fishing mortality level defined as F_{MEY}). Without effective controls on fishing effort, the fishing industry reaches exploitation levels at or beyond the point at which costs exceed revenues — the so-called open access equilibrium point (OAE). However, high effort levels that result in low economic benefits may or may not coincide with biological collapse. In this context it is useful to consider two categories of overfishing. 'Growth overfishing' occurs when annual fishing mortality extracts from the target population a sufficiently high proportion of the resource such that the maximal growth potential is exceeded (i.e. effort levels beyond F_{MSY}). 'Recruitment overfishing' occurs when fishing practices impair the reproductive potential of the resource. This can occur by elimination of spawning components within a management unit comprising a complex of spawning populations, and/or reduction of the spawning stock biomass to such low levels that the generation of new recruits to the fishery is reduced to lower levels than occur with moderate to high levels of adult abundance. The 'minimum spawning stock biomass' (MSB) can be defined as the adult abundance level required to prevent recruitment overfishing.

For the purposes of this introductory presentation to Theme 1, we define a fishery collapse to occur when effort levels generate stock biomass levels below MSB. The lower the cost curve, the more severe the reduction in spawning stock biomass if effort is not effectively restricted. In contrast, if the OAE point coincides with an abundance level above MSB, even though growth overfishing is occurring and no net revenue is being generated, biological collapse has not occurred. Under this latter situation, a reduction in fishing effort leads to rapid rebuilding of the resource. In contrast, if recruitment overfishing has occurred and the abundance level is below MSB, the stock may not rebuild for many years when effort is reduced or the fishery closed. Growth overfishing is relatively benign, whereas recruitment overfishing generates biological collapse and socio-economic dislocation.

The processes illustrated in Figure 2 can be used to structure a discussion on the reasons why some fisheries survive and others collapse. These are:

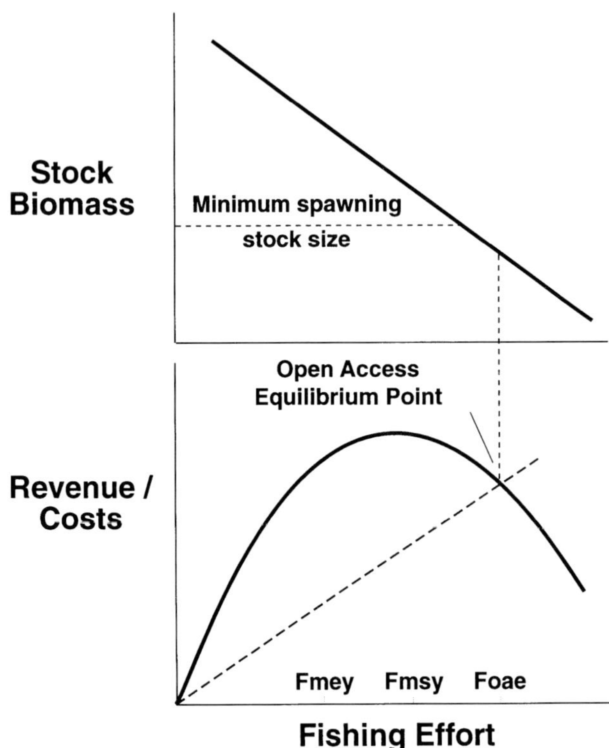

Fig. 2. The relationship between minimum spawning stock size (MSB), fishing effort and the open access equilibrium point is illustrated using the Gordon (1954) and Schaefer (1954) bio-economic model.

1. activities which alter the slope of the cost curve (such as government subsidization and increase in technological efficiencies),
2. the control of fishing effort, and
3. the relationship between fishing effort, OAE and MSB.

The conceptual model is an abstraction of a complex set of biological and economic processes, and as such is an oversimplification. What has been left out is the importance of environmental variability and ecosystem structure on resource production. MSB is obviously not a fixed point, being influenced by environmental conditions. To include temporal variability of the environmental conditions in Figure 2, we could consider a range of production curves corresponding to 'good' and 'poor' time periods.

In addition to the above underlying bio-economic concepts, it is helpful in a discussion of survival and collapse to identify the fisheries management functions. These are illustrated in Figure 3. Strategic planning, usually at the level of national or state governments for overall fisheries policy, addresses relatively long time scales, and is supported by both business and resource analysis activities. The management planning occurs on annual or multi-year time scales in relation to short-term objectives for a particular fishery within the overall strategic framework. Catch and effort monitoring, fishing entitlements, and enforcement are activities required to implement the management plan. The entitlements box involves activities which control access to the

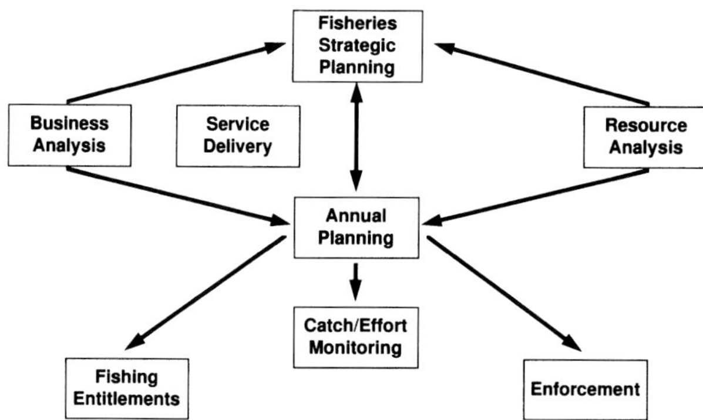

Fig. 3. The activities of fisheries management structured within eight boxes of a fisheries management model.

resource, including the transfer of licences or the buying and selling of quotas (Individual Transferable Quotas). The service delivery box involves activities, usually within governments, that provide resources for the business line and an analytical function on how well the fisheries management business line of a department is performing. We have found the model a useful structure for the evaluation of the effectiveness of different fisheries management approaches. In summary, Figure 2 illustrates in a simple way the ideas underlying fisheries management and provides a definition for collapse, whereas Figure 3 identifies the range of activities required by a management authority to generate sustainable use.

The two cod management units off Nova Scotia (4VsW cod and 4X cod) (Fig. 4) have several features which make them interesting case histories. The continental shelf is well within the 200-mile limit and thus the two cod management units do not require international management. There have been no directed foreign fisheries on these stocks since extension of jurisdiction. Marine fisheries in Canada are a federal responsibility, thus only one level of government has responsibility for management. In principle, this should simplify institutional structures. Canada and the United States have invested heavily over several decades in scientific research on both the oceanographic environment and the fisheries resources of the Scotian Shelf and Gulf of Maine area. As a result, these shelf seas are amongst the best understood from an ecosystem perspective in the world's oceans. There are strong environmental gradients along the shelf, and differences in predation on cod by seals between 4VsW and 4X. A sophisticated enforcement capability including vessels, aeroplanes, helicopters, and dockside enforcement by well trained fisheries officers has been in place since 1977. A management team has developed annual management plans based on elaborate consultation with the fishing industry and representatives of the provincial governments of Nova Scotia and New Brunswick. Canada has tried to achieve multiple objectives for fisheries since the extension of jurisdiction. Underachievement of fisheries management objectives has not been due to a lack of funding or commitment. The Department of Fisheries and Oceans' costs of groundfish management for the Scotia-Fundy area are discussed in this volume by O'Boyle and Zwanenburg (1997). In this sense, the survival and collapse of diverse fisheries under comprehensive management regimes provide important case histories.

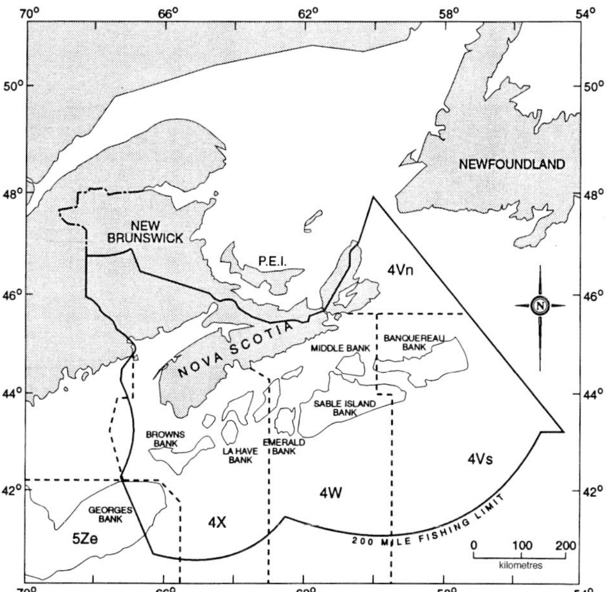

Fig. 4. Map of the Atlantic Provinces of Canada showing the boundaries of the Scotia-Fundy area of the Department of Fisheries and Oceans, and the NAFO Divisions (4V, 4W, 4X and 5Ze).

The trends in spawning stock biomass and recruitment from 1958 to the present for the two cod stocks are shown in Figures 5 and 6. Recruitment levels have been very low for 4VsW cod since the mid-1980s, and spawning stock biomass had declined in 1994 to the lowest level estimated. There is anecdotal evidence that several spawning components in 4VsW have been overfished (Younger *et al.* 1996), and egg and larval surveys indicate that the spring spawning component on Western Bank has been lost (Frank *et al.* 1994). The low level of spawning stock biomass, and the evidence of consistently low recruitment, led to a closure of the 4VsW cod fishery in September 1993. On the western shelf (4X), spawning stock biomass is at moderate levels in recent years, and recruitment during the past decade has been similar to the 1958 to 1985 levels. The 4X cod fishery is

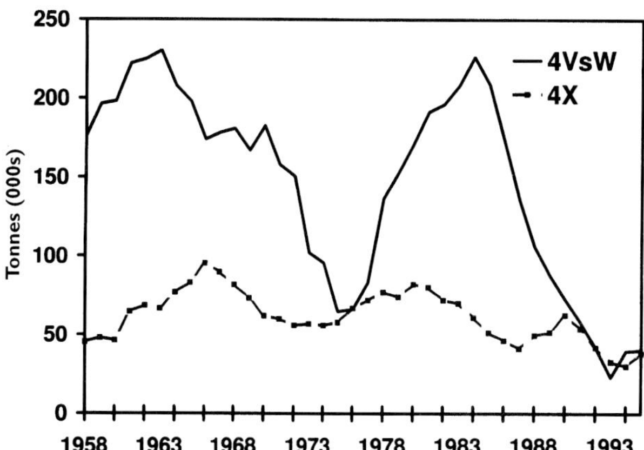

Fig. 5. Trends in estimates of spawning stock biomass for 4VsW cod (age 4+) and 4X cod (ages 3+) from 1958 to 1995.

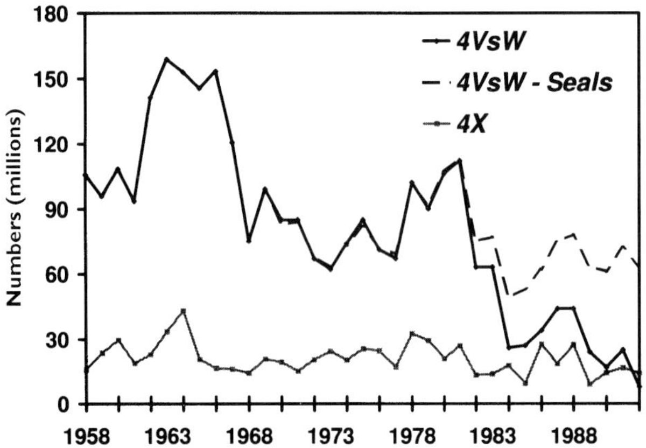

Fig. 6. Trends in estimates of recruitment (abundance at age 1) for 4VsW cod and 4X cod from 1958 to 1992. The dashed line shows estimates of 4VsW recruitment without seal consumption. The feeding model assumes that seals consume a constant proportion of their diet as cod and that natural mortality other than that due to seal predation is 0.2.

ongoing. The trends in weights-at-age from 1970 to 1995 are shown in Figure 7. Age 7 cod in 4VsW were considerably higher in the 1990s than during the previous two decades. In contrast, there is no trend in the weight-at-age of cod in 4X. It is also noted that cod growth is more rapid on the western shelf (4X) compared to the eastern shelf (4VsW). The observations shown in Figures 5 and 6 illustrate that the 4VsW cod fishery has collapsed, whereas that in 4X has survived. Why?

Fisheries management

The conservation objectives for groundfish fisheries off Atlantic Canada were established following a 1976 policy discussion, in anticipation of extension of jurisdiction and the implementation of a Canadian management regime (see Angel *et al.* 1994). The objectives have been to prevent both growth overfishing and recruitment overfishing. The strategy has been to fish at a constant level of effort corresponding to fishing mortality of $F_{0.1}$ (close to F_{MEY} in Fig. 2). The core tactic or tool has been single species annual quotas by management areas. The linkage between the conservation objectives and the tactic of quota management is established under item 8 of the basic principles of the annual groundfish management plan. This item states the rules for setting the total allowable catch (TAC) for groundfish management units off Atlantic Canada:

8.1 If the stock assessment provides evidence of levels of spawning stock biomass likely to endanger recruitment, fishing effort in the coming year will be reduced to allow immediate growth in spawning stock biomass.

8.2 Where the $F_{0.1}$ level for the next year differs 10% or 10 000 t from the current year TAC, the following formula would apply:
a) 50% rule;
b) for larger reductions — twice $F_{0.1}$ rule.

There has been a difference in the use of closed areas between the two cod management units. In 4X, Browns Bank (Fig. 4) has been closed during the spring spawning period for cod and haddock since 1970. Since 1987, there has been a closed area to dragging on Western Bank (4W) in order to reduce discarding of juvenile cod and haddock. The area has been closed to all groundfish gears since 1993.

The relationship between the scientific advice, the TACs, and the reported landings for 4VsW and 4X cod between 1977 and 1995 are shown in Figures 8 and 9. For both management units the discrepancies are similar. The TACs frequently exceeded the advice, particularly when stocks were declining. Also, in 4X the reported landings often exceeded the TACs. This was partially due to the practice prior to 1993 of allowing fleets to continue fishing on a bycatch basis after their quota had been reached. Finally, there were years when the TAC was not limiting. The dotted line illustrates what the $F_{0.1}$ advice should have been, given our present estimation of stock trends. The advice was overly optimistic during most of the time period. In sum, Figures 8 and 9 indicate similar problems with the implementation of quota management for both cod management units between 1977 and 1992.

Additional implementation problems, misreporting practices, and constraints to enforcement of quota regulations are analysed in Angel *et al.* (1994). They conclude that misreporting was a

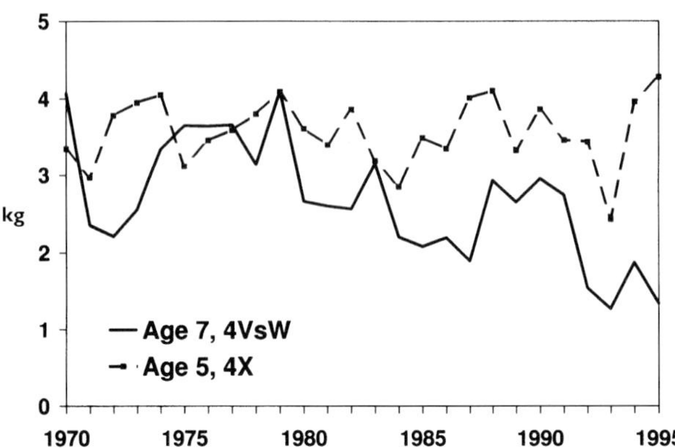

Fig. 7. July weights-at-age in kilograms for 4VsW (age 7) and 4X (age 5) cod from 1970 to 1995.

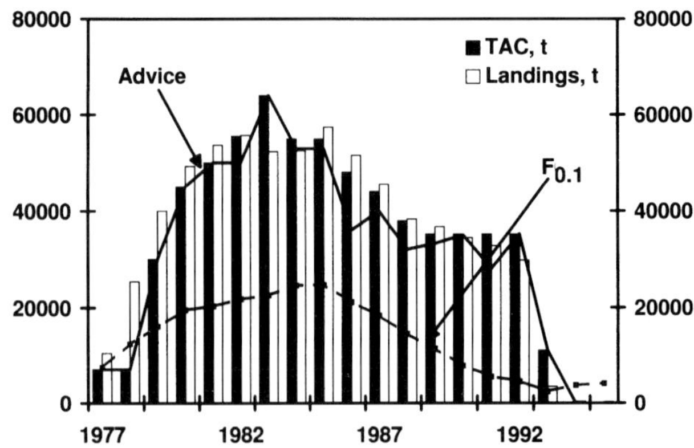

Fig. 8. Trends in past scientific advice, present perception of correct $F_{0.1}$ advice, total allowable catches (TACs) set by the Ministers of DFO, and reported landings (all in tonnes) for 4VsW cod from 1977 to 1995.

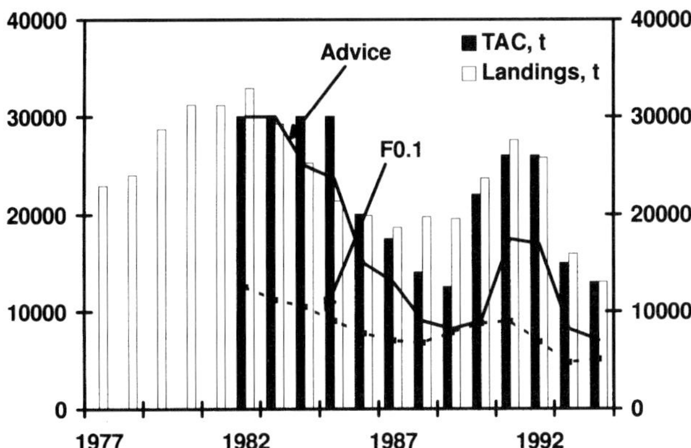

Fig. 9. Trends in scientific advice, total allowable catches (TACs) set by the Ministers of DFO, and reported landings (all in tonnes) for 4X cod from 1977 to 1995. There were no quotas for this management unit until 1982.

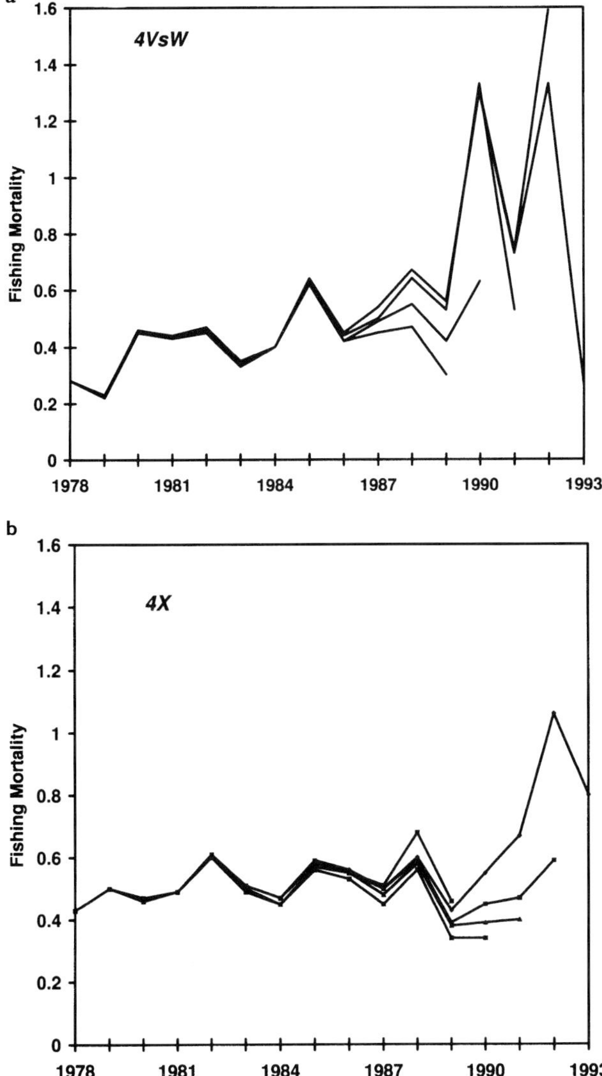

Fig. 10. Trends in estimates of fishing mortality using the sequential population analysis model (SPA) for (**a**) 4VsW and (**b**) 4X cod. The underestimates of fishing mortality illustrate the so-called retrospective problem for these two management units.

major issue between 1977 and 1992 for both management units, even though the degree could not be estimated with confidence. As a result, the landings of cod have been significantly underestimated and discarding of cod juveniles has been high during certain years. It was concluded that most of the regulations dealing with quotas were unenforceable at sea. The report did not find major differences in misreporting practices between 4VsW and 4X.

The poor quality of the data on catches of cod has compromized the ability to provide accurate estimates of stock abundance. For both management units, there has been a so-called retrospective problem. Each assessment indicates that the previous year's estimates of abundance have been overly optimistic (Fig. 10a, b). The problem is a complex one with different causes for different management units. It is partially due to poor data on catches. It is noted that the retrospective pattern for both management units has been a problem, with the 4VsW unit being particularly severe.

The aggregate trends in fishing effort by three fleet sectors in 4VsW and 4X are shown in Figure 11a, b. In both areas there is a steady increase in days fishing from 1977 to the late 1980s. Since 1992, days fishing has declined dramatically in 4VsW due to the moratorium on cod in this area and the bycatch nature of the haddock fishery. In 4X the declines in annual days fishing by the aggregate fleet have also been marked, in this case due to stricter implementation of quota management for cod, haddock, and pollock. The descriptions of fishing effort trends in 4VsW and 4X indicate that quota management did not generate a

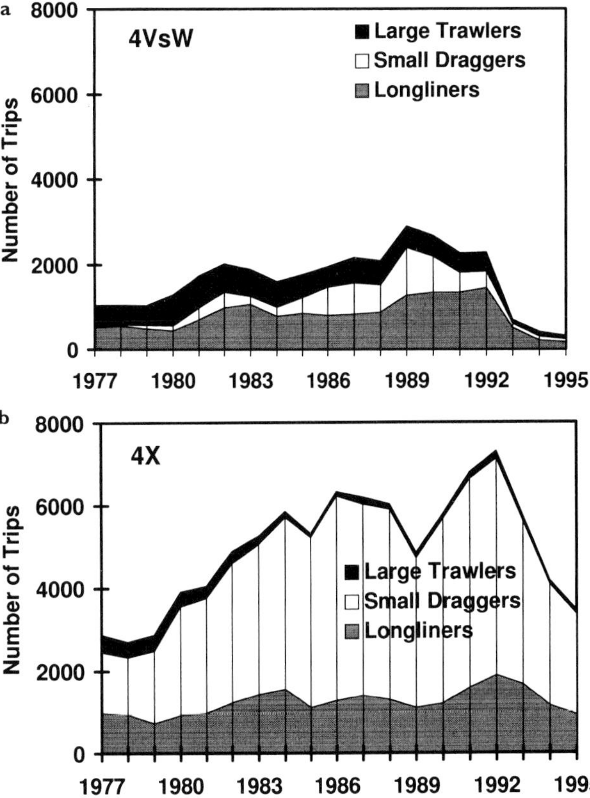

Fig. 11. Trends in numbers of groundfish trips between 1977 and 1995 for large trawlers, small draggers, and longliners in (**a**) 4VsW and (**b**) 4X.

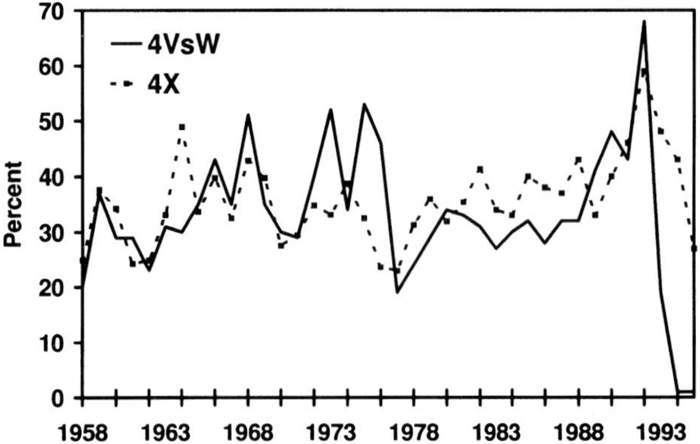

Fig, 12. Trends in estimates of exploitation rate (%) for 4VsW and 4X cod between 1958 and 1994. The $F_{0.1}$ target for these two management units is about 20%.

constant level of fishing effort. There have been more fishing trips annually in 4X by the small draggers and longliners. However, large trawler effort has been higher in 4VsW.

The estimates of trends in exploitation rate, based on the assumption that natural mortality has been constant, are shown in Figure 12. The temporal patterns are similar for the two management units, but the increases in the early 1990s in 4VsW and the subsequent decline in 1993 following the moratorium are more marked. The sharp increases in 4VsW between 1989 and 1992 are not consistent with the effort trends (Fig. 11). With respect to the strategic target of $F_{0.1}$ (an exploitation rate of about 20%), the management actions until 1994 have not been successful. From 1977 to 1993, exploitation rates moved further from the target for both management units, and growth overfishing occurred between 1980 and 1993.

It is difficult to account for the different stock trends (Figs. 5 and 6) of the cod population complexes in respectively 4VsW and 4X due to management actions and fishing practices. Fishing effort and exploitation levels appear to have been higher in 4X than in 4VsW, and the implementation of quota management similar in both areas. Yet 4VsW cod has collapsed and 4X cod has survived.

Environmental and Ecosystem Change

The Scotian Shelf oceanographic conditions are influenced by several large scale circulation features and the topography of the continental shelf. The Labrador Current and the outflow from the Gulf of St. Lawrence result in southward flow in the surface circulation of cool, relatively fresh water. The remnants of the Labrador current tend to track the continental slope with diversions onto the shelf, whereas part of the outflow from the Gulf of St. Lawrence generates a coastal current along the eastern and southern shores of Nova Scotia. Due to the complex topographic structure of the Scotian Shelf, there is recirculation over banks and coastal upwelling of deeper water. Off southwest Nova Scotia and in the Bay of Fundy, strong tidal mixing contributes to the circulation. The Gulf Stream, which flows in a northeasterly direction offshore from the shelf, sheds rings which result in a cross-shelf transport of warm saline slope water, both at the surface and through channels into the basins. This combination of circulation features generates strong latitudinal gradients in both surface and bottom water. Large areas of the bottom off Labrador, Newfoundland, and in the southern Gulf of St. Lawrence are characterized by sub-zero temperatures. The eastern Scotian Shelf (4VW) is also an area of cold average bottom conditions (predominantly less than 4°C). In contrast, the western Scotian Shelf (4X) bottom conditions are considerably warmer (mostly greater than 4°C). The along-shelf gradients in bottom temperature are marked (see Loder *et al.* 1996*b*).

Decadal scale variability in oceanographic conditions for the Scotian Shelf have been described by Loder *et al.* (1996*a*). The 1960s were a cold period, as have been the 1990s. The differences in July bottom water temperature between 1990 to 1994 and 1975 to 1989 are shown in Figure 13. The striking feature is the increase in below-2°C bottom conditions, particularly in 4VsW. Bottom temperatures in 4X have also cooled, but most of the area is still above 4°C (Table 1). About 18 000 km^2 of cod habitat have changed from above 4°C to below 4°C in 4VsW, with 60% of the area below 4°C during recent years. In contrast, 11 000 km^2 of cod habitat have changed to below 4°C in 4X, with 25% of the area below this temperature level. About 40% of 4VsW has been below 2°C in this decade. In sum, both management units have experienced cooler bottom water conditions during recent years, but the cooling has been more pronounced in 4VsW. The colder conditions probably do not cause an increase in natural mortality for cod, but they reduce stock production (Campana *et al.* 1995) and restrict the area of distribution. The latter process may have led to increases in catchability (i.e. higher fishing mortality per unit of effort).

Table 1. Comparison of Bottom Temperatures in 4VsW and 4X during 1975–89 and 1990–94

	4VsW 1975–89		4X 1975–89	
°C	**%**	**Area (sq km)**	**%**	**Area (sq km)**
<0				
0–2	13.77	14 961	2.23	1695.4
2–4	25.03	27 207	5.53	4205.4
4–6	22.56	24 522	18.05	13718.3
6–8	19.88	21 603	40.99	31162.4
8–10	17.11	18 590	31.71	24105.2
10–12	1.53	1 661	1.47	1120.1
12≤	0.12	133	0.02	15.7
sum		108 677		76022.5
	4VsW 1990–94		**4X 1990–94**	
°C	**%**	**Area (sq km)**	**%**	**Area (sq km)**
<0	0.51	501.2		
0–2	39.48	39028.9	13.76	8898.5
2–4	20.83	20588.2	12.4	8018.7
4–6	18.18	17968.2	21.57	13951.4
6–8	9.8	9684.9	22.52	14563.5
8–10	10.22	10108	26.41	17083.9
10–12	0.91	903.1	3.35	2166.9
12≤	0.08	77.2		
sum		98859.7		64682.9

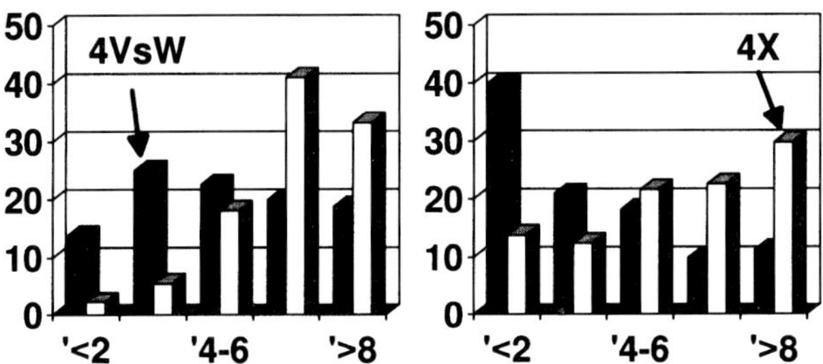

Fig. 13. The percentage of the areas within 4VsW and 4X that is characterized by different July temperature conditions in 1990 to 1994 compared with 1975 to 1989.

Mahon and Sandeman (1985) have described a biogeographic discontinuity in fish communities across the Scotian Shelf close to the line separating 4X and 4W. The southern limit of distribution of several cold water species occurs on the eastern Scotian Shelf (i.e. capelin, Greenland halibut), and the northern limit of distribution of several warm water species occurs on the western shelf (i.e. butterfish, ocean pout). The recent colder conditions in 4VsW have resulted in increases in abundance of north temperate fish species such as capelin, with a concomitant change in fish community structure in 4VsW (Frank *et al.* 1996). In 4X, the fish community composition has been relatively stable.

CHANGES IN NATURAL MORTALITY OF COD

Grey seal abundance has been increasing in abundance on the Scotian Shelf during the past two decades by about 12% annually (Fig. 14). Reproduction occurs in two separate locations: within the Gulf of St. Lawrence on the ice, and on Sable Island. It is interpreted that the separate reproduction areas support two populations with limited mixing. However, individuals migrate over large areas during their annual feeding migrations. Stobo *et al.* (1990) have described the geographic and seasonal distribution of grey seals on the Scotian Shelf. If the tag returns from coastal locations reflect the offshore distribution, the seal abundance in 4X is about 25% that estimated in 4VsW. The diet of grey seals on Sable Island has been monitored annually since 1989. Although variable, about 5–15% of the diet by wet weight has been observed to be juvenile cod. Mohn and Bowen (in press) have estimated the annual consumption of cod by seals in 4VsW in both numbers and weight (Fig. 15a, b). Since the mid-1980s the seals are estimated to have consumed more cod (by number) than are landed by the fishery. The average predation mortality on age 1–3 cod by seals has been estimated to be as high as 0.35 between 1989 to 1993, the years of the 4VsW cod collapse. However, there is considerable uncertainty on whether the overall 4VsW cod natural mortality has increased over time due to the increasing consumption of cod by seals. There are no data on the consumption of juvenile cod by other predators (including cannibalism), and it has been argued that other predators may be reducing their consumption of cod in parallel with the increasing consumption by grey seals (Mohn and Bowen in press).

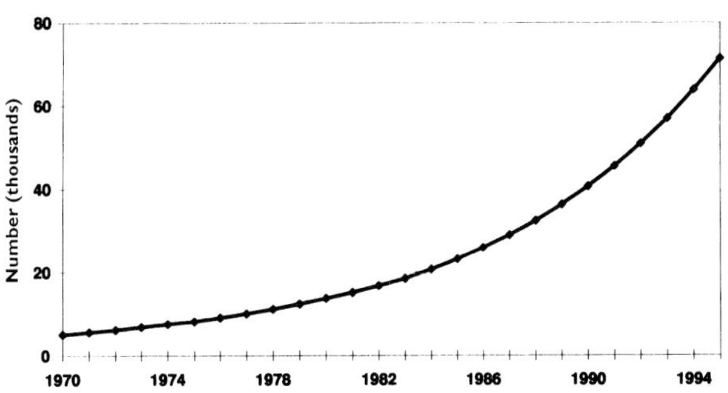

Fig. 14. Trends in the number of grey seals of the Sable Island populations (4VsW stock area) from 1970 to 1993 (from Mohn and Bowen 1996).

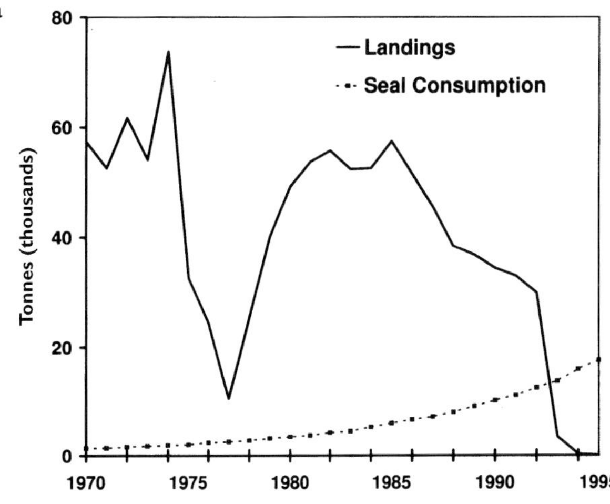

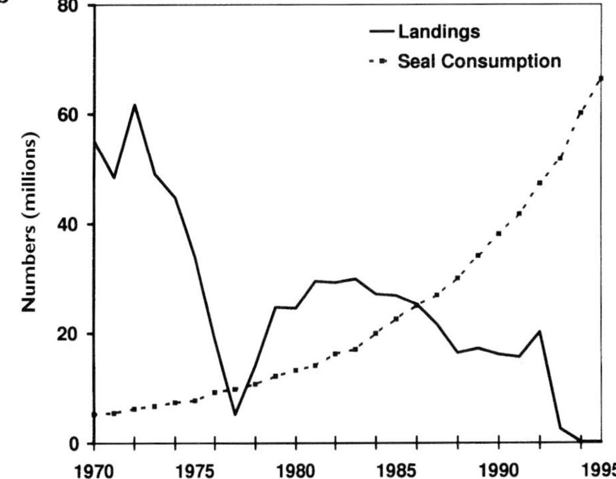

Fig. 15a, b. Comparison of the annual consumption of 4VsW cod by grey seals with the landings from the fishery from 1970 to 1995 (from Mohn and Bowen in press).

Total juvenile mortality rates (Z), i.e. natural (M) plus discarding mortalities due to fishing (F), can be estimated from the annual research vessel groundfish surveys on the Scotian Shelf (Fig. 16). Survivorship ratios for each age and year were calculated along cohorts. The Zs for ages 1–3 were averaged for each year, and a three-year moving average filter applied. No corrections were made for variations in the catchability-at-age of the survey gear. The peaks in the Z values for 4VsW juvenile cod in 1974 and 1984 may be due to high discarding by, respectively, the distant water draggers (mid-1970s) and the Canadian fleet (mid-1980s). The rise in Z values since 1988 is not thought to be due to

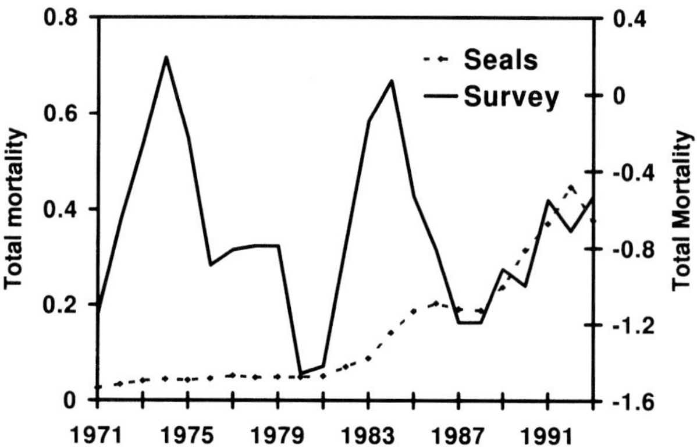

Fig. 16. Trends in estimates of 4VsW cod juvenile mortality due to seals (from Mohn and Bowen in press) and of total mortality (Z) from the July groundfish survey. The seal predation model estimates assume that cod are a constant proportion of the diet and that natural mortality from other causes is 0.2 for ages 1, 2 and 3.

discarding, because of the use of observers on some 'offshore' vessels and the change in mesh sizes. The recent increases in Z parallel the cod mortality estimates due to seals.

The research vessel survey observations and the model results of Mohn and Bowen (in press) support the inference that juvenile cod natural mortality has increased in 4VsW during recent years due to seal predation, and that a comparable increase has not occurred in 4X. Given the lack of precision of the estimates (due to the constraints of the survey design and methodology on estimation of juvenile abundance), firm conclusions cannot yet be made (Sinclair *et al.* unpublished).

Causes of the cod fishery collapses

How do the descriptions of the 4VsW and 4X cod fisheries help in resolving the present controversy in Atlantic Canada on the relative roles of overfishing, environmental change, and seal predation on the cod collapses? The Scotian Shelf observations support the conclusion that all three factors have been important. Table 2 summarizes the similarities and differences. With low levels of seal predation and good environmental conditions, the 4X cod population complex has been resilient to the high levels of fishing effort. In contrast, coincident with higher levels of seal predation and a reduction in the proportion of the cod habitat above 2°C, the cod population complex in 4VsW has collapsed under similar levels of fishing effort and essentially the same management system. The Scotian Shelf observations suggest that the diagnosis by Hutchings and Myers (1994), Hutchings (1996), and Myers *et al.* (in press) is unbalanced. If the eastern Scotian Shelf cod is representative of cod management units to the north and east, there is evidence to infer that poor environmental conditions and increases in seal predation on juvenile cod have contributed to the cod collapses off Atlantic Canada.

It is difficult to assess the relative importance of fishing practices, environmental change, and seal predation to the sharp decline in recruitment to the 4VsW cod population complex in the mid-1980s. If the high estimates of seal predation by Mohn and Bowen (in press) are assumed to be representative, the recruitment at age 1 did not drop as sharply in 1983 (Fig. 6). It is also to be noted that the estimates of recruitment do not take into account discarding. The Z trends shown in Figure 16 suggest that discarding of juvenile cod was high in the mid-1970s. The anecdotal information collected from fishers also

Table 2. Summary of similarities and differences between 4VsW and 4X cod

	4VsW Cod	4X Cod
Bottom Temperature	40% of area below 2°C during 1990s	14% of area below 2°C during 1990s
Fish Growth	Declining weights-at-age	Stable weights-at-age
Fish Community	Increase in 'northern' species in 1990s	No change, except for increase in dogfish
Predators	Predation by seals increasing, some evidence that M has increased for juveniles	Low predation by seals, M on juveniles shows no trend
Conservation Objectives	Prevent recruitment and growth overfishing	Prevent recruitment and growth overfishing
Strategy	Constant effort at $F_{0.1}$ target	Constant effort at $F_{0.1}$
Core Tactics	Single species quotas	Single species quotas
Ancillary Tactics	Western Bank juvenile closure	Browns Bank spawning closure
Misreporting	High	High
'Retrospective' Problem	Important	Important
Fishing Effort	Increased till late 1980s	Increased till early 1990s
Fleet Sector Composition	Landings dominated by offshore draggers	Landings partitioned more evenly between several fleet sectors
Unemployment Insurance Incentives	Common policy	Common policy
Fishing Mortality	Exceeded F_{MAX}, the growth overfishing target, since 1980	Exceeded F_{MAX}, the growth overfishing target, since 1980
Recruitment	Recruitment declined since mid-1980s	Recruitment stable
Spawning Components	Loss of spring spawners, and some of the smaller spawning areas	No evidence of change

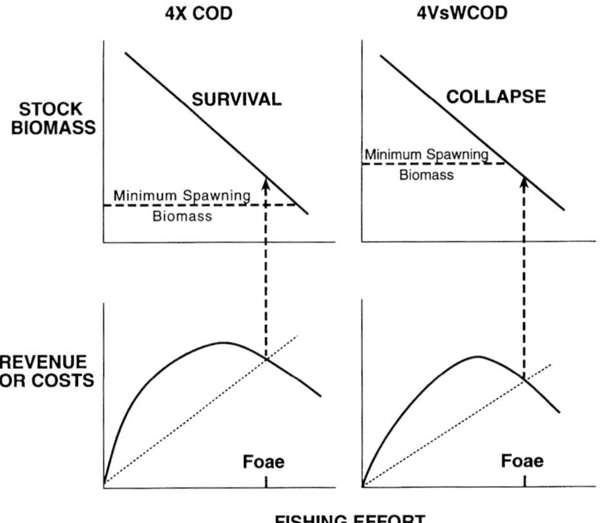

Fig. 17. Schematic representation of the 'survival' of 4X cod and the 'collapse' of 4VsW cod at high levels of fishing effort due to differences in the relationship between minimum spawning stock biomass and the open access equilibrium point.

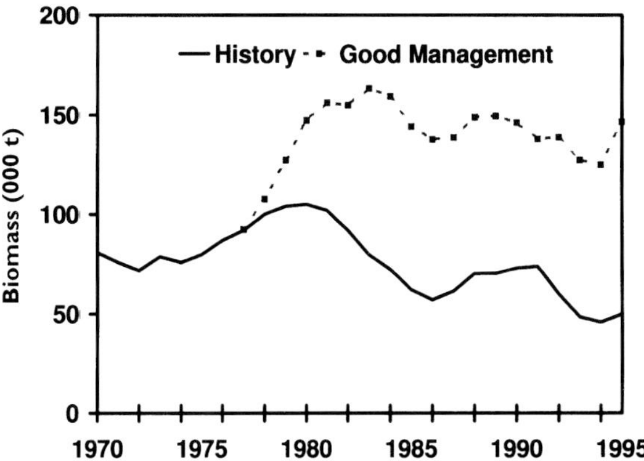

Fig. 18. Impacts of fishing beyond the $F_{0.1}$ target on 4X cod biomass. The lower line illustrates the estimates of stock biomass (age 1+) at historical levels of exploitation. The upper line estimates the stock biomass that would have occurred if fishing had been restricted at the $F_{0.1}$ level assuming the recruitment trends in Figure 6.

suggests that the larger dragger fleet eliminated some of the cod spawning components in 4VsW (Younger *et al.* 1996).

The differential responses of 4X and 4VsW cod to overfishing are schematically represented in Figure 17 using the Gordon-Schaefer bio-economic model. There was overfishing for both management units as fishing effort was not sufficiently constrained by quota management (Fig. 11). At the high effort and low biomass levels, the fleets could still make a profit. For 4X cod, the spawning stock biomass has been sufficient to generate reasonable recruitment. As a result, there has been sustainable overfishing. In contrast for 4VsW, recruitment has declined sharply and with sustained high levels of fishing effort the fishery has collapsed. To date there is little evidence that management actions taken since 1993 are generating stock rebuilding. The collapse (in spite of a fishery closure) may be of a long-term nature.

The costs of overfishing due to shortfalls in the quota management regime are illustrated in Figures 18 and 19. If the $F_{0.1}$ strategy had been fully implemented from 1977 to the present, assuming the recruitment patterns estimated in Figure 6, the 4X cod biomass would have been relatively stable at about 150 000 t (two to three times the observed stock biomass). Also at these higher biomass levels, it is possible that recruitment levels could have been higher. For 4VsW the analysis is more complex. With perfect implementation of the $F_{0.1}$ strategy, the stock biomass would have initially increased three-fold until the mid-1980s. Subsequently, the stock would have declined dramatically even under good management (due to the poor recruitment observed since 1983). If the high estimates of seal predation are representative, much of the loss of cod production in 4VsW has been due to changes in juvenile natural mortality. However, if fishing practices are partly responsible for the declines in recruitment, the impacts of overfishing are underestimated in Figure 19. The analysis for both 4VsW and 4X cod suggests that the $F_{0.1}$ strategy is a reasonable one if recruitment is not impaired.

The cod examples illustrate the complexity of fisheries collapses. Fishing practices, environmental variability, and ecosystem change can all play a role. The only factor that can be controlled for harvest fisheries is fishing effort. Why was fishing effort not held constant at the target $F_{0.1}$ level between 1977 and 1993?

MANAGEMENT SHORTFALLS

The second model described in the introduction addresses the components of a fisheries management system (Fig. 3). We will use this model to summarize why fishing effort was not held constant at the appropriate target, and to draw some management related conclusions on why some fisheries survive and others collapse. Angel *et al.* (1994) and Burke *et al.*. (1996)

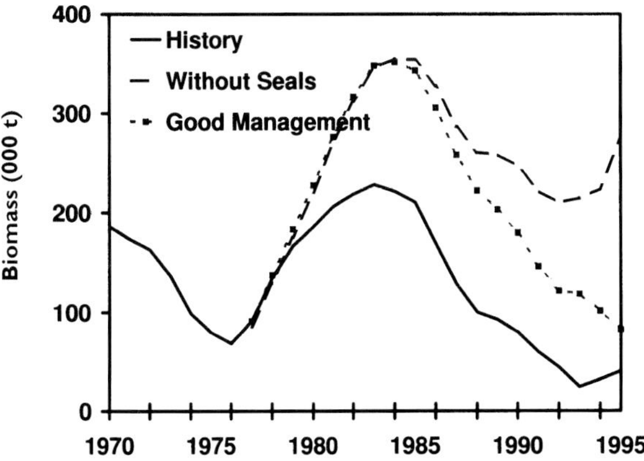

Fig.19. Impacts of fishing beyond the $F_{0.1}$ target and of seal predation on 4VsW cod stock biomass (age 3+). The lower line illustrates the estimates of stock biomass at historical levels of exploitation. The upper lines estimate trends in stock biomass with fishing at the $F_{0.1}$ level with and without seal predation on cod juveniles (i.e. assuming the recruitment trends for 4VsW cod shown in Figure 6).

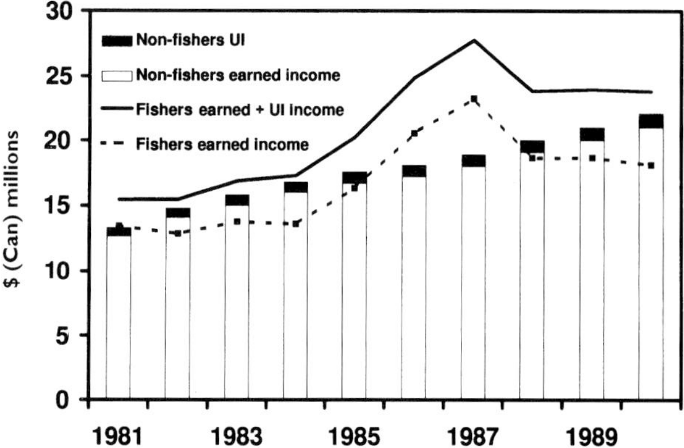

Fig. 20. Trends in annual income and unemployment insurance benefits for Nova Scotia workers and fishers between 1981 and 1990.

have analysed groundfish management in this geographic area, and we will use these studies to illustrate concepts.

An analysis of the *Fisheries Strategic Planning* activities for Atlantic Canada groundfish between 1977 and the present suggest that a key problem has been conflicting strategic objectives at the level of the federal government. The economy of the Maritimes Provinces and Newfoundland are characterized by high levels of unemployment and the fishery has been used for regional development initiatives. The income support programmes have encouraged participation in the fishery beyond levels justified on the basis of sound business principles (Fig. 20). Even though the Department of Fisheries and Oceans (DFO) has clear responsibilities under the Fisheries Act for the sustainable use of fisheries resources, other departments at the federal level (and provincial governments) have a major impact on the fisheries sector. This point is illustrated by the government expenditures on fisheries management in Atlantic Canada (Fig. 21). A relatively small proportion is controlled by DFO. Under multiple objectives, the relative roles of economic efficiency and distribution of wealth vary by fishery, area, and time. The focus on conservation objectives is often lost. A second major problem at the strategic planning level relates to the access policies and principles. Canada has not decided upon

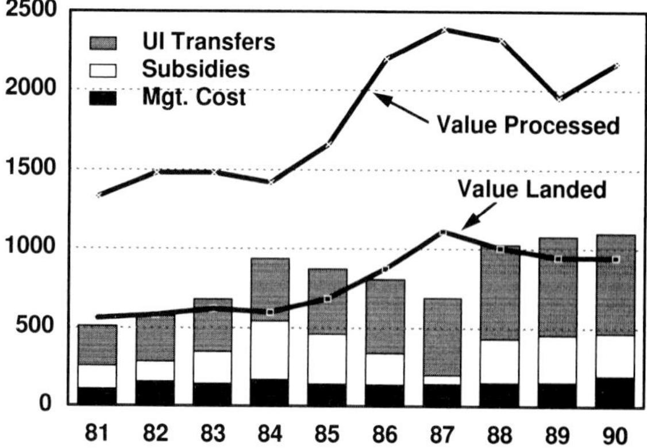

Fig. 21. Trends in processed value and landed value of marine fisheries in Atlantic Canada, as well as government expenditures in support of fisheries. The costs of fishing and processing by industry are not included.

the principles to be used to provide secure access to marine renewable resources. Access rules and resource shares have been adjusted arbitrarily by government. This fundamental lack of security of access has generated negative incentives for conservation by the 'tenants,' and has undermined the management process. These two strategic planning issues, multiple (and conflicting) objectives, and uncertainty in the principles underlying security of access, continue to constrain the fisheries management system in Canada.

The *Business Analysis* activity for groundfish in the Scotia-Fundy Region has been relatively underfunded. This has resulted in an imbalance. There has been limited capacity for analysis of economic and social impacts of strategic options. Few business and social parameters are being monitored. The costs, benefits, and effectiveness of many regulatory changes are not being adequately evaluated.

By contrast, the Scotia-Fundy *Resource Analysis* activities have been well funded during the 1977 to 1995 period. A shortfall up until about 1990 was weak cooperation and communication with the fishing industry. Also, the oceanographic conditions and species interactions were not well discussed during the peer review of the status of the stocks. The misreporting problem and survey catchability trends have contributed to the so-called retrospective problem illustrated in Figure 10a, b. The above issues have been addressed during the past several years. The peer review process has been decentralized and representatives of the fishing industry involved at each step of the process. Joint industry/science surveys have been initiated, and a society formed between fishers and scientists to promote conservation (the 'Fishermen Scientist Research Society'). Formal committee structures have been struck to incorporate scientific advice on changing oceanographic conditions and species interactions into the stock status reports. There is an awareness of the retrospective problem, and approaches have been developed to adjust estimates of abundance and fishing mortality. One issue that is still unresolved is the inability to reach consensus by science on stock status levels that would trigger implementation of rule 8, point 1. There have not been operational definitions of minimum spawning stock levels below which there is concern about recruitment overfishing.

The *Service Delivery* function should be able to provide analyses on the costs of the diverse management activities, and thus allow redistribution of departmental resources and adjustment of organizational constraints that impede delivery of the groundfish management business line. The management structure of the department has not facilitated this activity, and the financial system has made it difficult to analyse costs of fisheries management activities. The recent departmental reductions of about 30–40%, as well as the transfer of costs of some services to the fishing industry, have provided a focus on deficiencies in the analytical and managerial aspects of service delivery function (i.e. the ability of the organization to easily adjust activities and expenditures in response to changing needs of a departmental business line). The function itself has not been well recognized due to lack of definition of business lines within the department until the past several years.

The *Annual Planning* activity (Fig. 3) has been a central focus of departmental activity from 1977 to the present. The shortfalls until recently have been the lack of specificity on how the annual groundfish management plans would achieve specific fishing capacity changes and effort level targets. The linkages between management actions and the expected changes in fishing practices have not in general been made explicit. As a result, the appropriate parameters have not always been monitored in order that the efficacy of the management action could be evaluated. The new focus on formal legal partnerships between the department and different fleet sectors of the fishing industry has led to a redefinition of the annual planning process under the title 'integrated harvesting plans'. As a result, the deficiencies noted above are being addressed. A peer review process for the management plans, parallel to that for stock assessments, is envisioned. The core regulatory tool for the achievement of the conservation objectives of groundfish has been single species quota by restricted geographic areas. As a result, the focus on real-time monitoring has been on quota. Descriptions of trends in fishing effort and technology, as well as changes in fishing practices, have not been monitored with the same timeliness. This has led to a lack of understanding by management and science of the growth of fishing power, and the inability to evaluate within season quota/effort disparities. The management planning function for groundfish is in the process of expanding its use of regulatory tools for the control of fishing mortality (quota, days fishing, seasonal closed areas for protection of spawning components, closed areas for the protection of juveniles, and more effective use of gear restrictions). In parallel, a wider range of approaches is being used in the plans to regulate fishing capacity (EAs, ITQs, licence stacking, licence buy-backs).

The *Catch and Effort Monitoring* function for groundfish has improved significantly since the introduction of an industry-funded dockside monitoring programme (DMP), for the ITQ fleet (1991), and for the EA fleet (1993). Under DMP, a third party weighs out all landings at a limited number of ports. The data are forwarded directly to the department within several days and the accuracy of the information has been evaluated to be high. The remaining fleet sectors under competitive quotas (the longliners, handliners, and gill-netters) are in the process of migrating from a so-called catch monitoring system (weekly reports by vessel essentially on the 'honour system') to DMP. As noted above under the planning box, to date (even though information on effort is recorded), the monitoring system does not generate real-time outputs on accumulated days fishing by vessel in relation to accumulated landings. Furthermore, the information management system is not user-friendly, resulting in a lack of 'real-time' analysis by industry and science of harvesting activities in relation to the annual plan. These information accessibility issues are presently being addressed.

The *Fishing Entitlements* activity is perhaps the most contentious aspect of the overall structure in the conceptual model. Due to the deficiencies in strategic planning summarized above, the security of access to the groundfish resource, or the very nature of the entitlements defined under licences and quota shares, is fragile. The Fisheries Act empowers the Minister with 'absolute authority' over resource access. The quota sharing arrangements are still fluid, and new licences have been added and others removed over the years. The exchange of 'property' (licences and quota) is not regulated under a formal structure similar to a stock exchange or registry of deeds. This uncertainty in the permanence of the entitlements has led to fleet sector bickering for shares, civil disobedience over departmental actions, and a lack of a long-term view by participants. As a result, it has been difficult to focus on conservation concerns for sustainable use. The new licensing policy, involving 'core' fishers, and the migration towards property rights and community allocations within the fixed gear sector are recent steps towards improved management of fishing entitlements.

The *Enforcement* activities review indicated that there was good success in applying regulation on foreign fishing vessels operating in Scotia-Fundy area waters and in controlling fishing along the Canada/US boundary in the Gulf of Maine. The success in controlling closed areas was also considered to be effective. Enforcement at-sea of regulations against discarding, trans-shipping, and misreporting by area was least successful. DMP has improved compliance at landing points, but enforcement at-sea of conservation rules requires attitude changes on the part of fishers. There is some evidence that these can be fostered through a combination of severe penalties, a tighter control over fishing practices, enhanced real-time monitoring and reporting, and maturation of ITQ/EA systems which provide a longer-term ownership perspective on resources. This improvement is reflected in Figures 11 and 12 by the movement of the fishing mortality towards the $F_{0.1}$ target for 4X cod in recent years.

The brief summaries of Scotia-Fundy groundfish activities in each of the boxes (Fig. 3) of the model illustrate the complex nature of fisheries management. In this case history, there are multiple causes for the prognosis of survival of the 4X cod fishery and the clear collapse of the 4VsW cod fishery. The 1995 workshop, with representatives of the fishing industry, led to 32 recommendations amongst the diverse components of the model to improve the management system. It is of interest that the group identified deficiencies in every box of the model.

Why, then, did the Canadian management regime not achieve the stated conservation objectives? The strategy of constant effort at the $F_{0.1}$ level is considered to be appropriate, with the caveat that there needs to be additional specific actions to ensure the protection of spawning components for those management units comprising a number of spawning components. The tactic of quota management, however, has been fraught with problems. The retrospective study infers that shortfalls have been due to a systems failure in the implementation of the chosen core regulatory tactic, rather than the result of a failure in a particular management function. Given the indications of reductions in fishing effort and exploitation rate for 4X cod under quota management, it is tentatively concluded that this tactic can be made to work within the Canadian context. The next few years are critical ones.

Conclusions

The two cod management units off Nova Scotia have been used to illustrate the issues underlying why some fisheries survive and others collapse. Can we draw some general conclusions from these examples? We will return to the three key processes underlying fisheries management.

A lowering of the cost curve due to technological efficiencies and capacity reduction can clearly be a positive development in a fishery if: (i) the real costs are borne by the participants; and (ii) effort can be restricted at the target level. Artificial lowering of the cost curve, by often-hidden government subsidies, can be a negative development for a fishery, particularly if the management regime cannot control fishing effort. Lowering the cost curve moves the OAE point to higher levels of fishing effort coincident with lower stock biomass levels that may result in recruitment overfishing. The alteration of the cost curve involves the management of capacity. This function is best left to the fishing industry and market forces, with minimal government intervention. To allow efficient capacity adjustment, the following points are important:

- security of access is a prerequisite;
- real costs of fishing need to be internalized;
- subsidies minimized;
- a capacity reduction mechanism needs to be in place (such as community quotas, licence stacking, TURFS, ITQs).

Effort has to be controlled at a prescribed level. This function frequently is best carried out by a government agency. The following are essential requirements:

- adequate information system on catch and effort;
- meaningful sanctions that act as a deterrent to poor fishing practices;
- enforcement internalized within the industry so that the regulatory activities are dealing with a small fraction of the participants.

The final process focusses on the uncertainty of fisheries management that arises from the lack of understanding of the controls of fisheries production. Given this fundamental constraint, a precautionary approach is needed, and a rapid response to negative signals is essential. To err in the direction of short-term socio-economic objectives at the cost of the conservation objectives can lead to serious long-term social costs. Important points are:

- the protection of the reproductive capacity of the target species;
- a precautionary approach to prevent recruitment overfishing;
- rapid adjustment of effort downwards in response to ecosystem change suggesting lower production.

Acknowledgments

We acknowledge and appreciate the generous contribution of data and analyses provided by Bob Mohn, Paul Fanning, Don Clark, Kees Zwanenburg, Doreen Liew, and Bob Branton; as well as the help with the illustrations by Art Cosgrove and manuscript preparation by Darlene Guilcher.

References

Angel, J R, Burke, D L, O'Boyle, R N, Peacock, F G, Sinclair, M, and Zwanenburg, K C T (1994). Report of the workshop on Scotia-Fundy groundfish management from 1977 to 1993. *Canadian Technical Report of Fisheries and Aquatic Sciences* 1979, vi + 175pp.

Burke, D L, O'Boyle, R N, Partington, P, and Sinclair, M (1996). Report of the second workshop on Scotia-Fundy groundfish management. *Canadian Technical Report of Fisheries and Aquatic Science* 2100, vii + 247 pp.

Campana, S E, Mohn, R K, Smith, S J, and Chouinard, G A (1995). Spatial implications of a temperature-based growth model for Atlantic cod (*Gadus morhua*) off the eastern coast of Canada. *Canadian Journal of Fisheries and Aquatic Sciences* **52**, 2445–56.

Frank, K T, Drinkwater, K F, and Page, F H (1994). Possible causes of recent trends and fluctuations in Scotian Shelf/Gulf of Maine cod stocks. *ICES Marine Sciences Symposium* **198**, 110–20.

Frank, K T, Carscadden, J E, and Simon, J E (in press). Recent excursions of capelin (*Mallotus villosus*) to the Scotian Shelf and Flemish Cap during anomalous hydrographic conditions. *Canadian Journal of Fisheries and Aquatic Sciences.*

Gordon, H S (1954). The economic theory of a common property: The fishery. *Journal of Political Economy* **62**(2), 124–42.

Hutchings, J A (in press). A spatial model for the decline of northern cod and a review of hypotheses for the stock's collapse. *Canadian Journal of Fisheries and Aquatic Sciences.*

Hutchings, J A, and Myers, R A (1994). What can be learned from the collapse of a renewable resource? Atlantic cod, *Gadus morhua*, of Newfoundland and Labrador. *Canadian Journal of Fisheries and Aquatic Sciences* **51**, 2126–46.

Loder, J W, Hannah, C G, and Petrie, B (1996*a*). Seasonal and decadal variability of hydrography and circulation in the Georges Bank region. EOS **76**(3), OS197.

Loder, J W, Han, G, Hannah, C G, Greenberg, D A, and Smith, P C (1996*b*). Hydrography and baroclinic circulation in the Scotian Shelf region: Winter *vs* summer. *Canadian Journal of Fisheries and Aquatic Sciences Supplement* (in press).

Mahon, R, and Sandeman, E J (1985). Fish distributional patterns on the continental shelf and slope from Cape Hatteras to the Hudson Strait - a trawl's eye view. In 'Towards the Inclusion of Fishery Interactions in Management Advice'. (Ed R Mahon) pp. 137–152. *Canadian Technical Report of Fisheries and Aquatic Sciences* No. 1347.

Mohn, R, and Bowen, W D (in press). Grey seal predation on the eastern Scotian Shelf: modelling the impact on Atlantic cod. *Canadian Journal of Fisheries and Aquatic Sciences.*

Myers, R A, Hutchings, J A, and Barrowman, N J (in press). Hypotheses for the decline of cod in the North Atlantic. *Marine Ecology Progress Series.*

O'Boyle, R, and Zwanenburg, K (1997). A comparison of the benefits and costs of quota *versus* effort-based fisheries management. In 'Developing and Sustaining World Fisheries Resources: The State of Science and Management. Second World Fisheries Congress, Brisbane 1996'. (Eds D A Hancock, D C Smith, A Grant and J P Beumer) pp. 283–90 (CSIRO Publishing: Melbourne).

Schaefer, M B (1954). Some aspects of the dynamics of populations important to the management of the commercial marine fisheries. *Bulletin International American Tropical Tuna Commission* **1**, 26–56.

Sinclair, A F, Myers, R A, and Hutchings, J A (unpublished). Seal fisheries interactions in eastern Canada: can we detect increased mortality of juvenile cod? (Under review).

Stobo, W T, Beck, B, and Horne, J K (1990). Seasonal movements of grey seals (*Halichoerus grypus*) in the Northwest Atlantic. In 'Population Biology of Sealworm (*Pseudoterranova decipiens*) in Relation to its Intermediate and Seal Hosts'. (Ed W D Bowen) pp. 199–213. *Canadian Bulletin of Fisheries and Aquatic Sciences* 222.

Younger, A, Healey, K, Sinclair, M, and Trippel, E (1996). Fisheries knowledge of spawning locations on the Scotian Shelf. In 'Report of the Second Workshop on Groundfish Management'. (Eds D L Burke, R N O'Boyle, P Partington, M Sinclair) pp. 147–56. *Canadian Technical Report of Fisheries and Aquatic Sciences* 2100.

The Frequency and Severity of Fish Stock Declines and Increases

Ray Hilborn

School of Fisheries Box 357980, University of Washington, Seattle, WA 98195-7980 USA.

Summary

Trends in spawning stock biomass, recruitment and catch were examined for 129 stocks of fish for which over 20 years of data were available. These data show that dramatic declines and increases in abundance are common. 10% of all stocks examined showed a 10-fold decline in spawning stock biomass and recruitment using a 5-year running average. 6% of stocks showed a 10-fold increase in both spawning stock biomass and recruitment. 19% of stocks exhibited a 5-fold decline and 14% showed a 5-fold increase. Of 31 stocks of Clupiformes, 23% showed a 10-fold decline in spawning stock biomass, while 16% showed a 10-fold increase. 43 stocks of Gadiformes were examined: 12% showed a 10-fold decline and none showed a 10-fold increase. Of the 15 pleuronectiform stocks examined, none showed a 10-fold decline or increase. Of 52 salmoniform stocks examined, 8% showed a 10-fold decline while 7% showed a 10-fold increase. These data illustrate that large scale fluctuations in fish stocks are widespread, and that in general increases are not quite as common as declines. The results by taxonomic group indicate that Clupiformes are the most variable, while Pleuronectiformes seem to be the most stable.

Introduction

There is widespread and growing concern about fish stock collapses. The closure of the northern cod fishery in eastern Canada and the depressed state of New England groundfish stocks are only two of a large number of fish stocks that have received attention in the general media. Other stocks that have attracted significant attention from environmental groups include the Atlantic bluefin tuna, Pacific salmon on the Columbia and Sacramento rivers, orange roughy in New Zealand, and southern bluefin tuna in the Indian Ocean and western Pacific. In some cases these species have been proposed or listed as endangered species, they may have been mentioned as candidates for CITES (Convention for International Trade Endangered Species) listing, or environmental groups may have brought suits to reduce levels of exploitation. In addition to these fish species, numerous species of marine mammals and reptiles (sea turtles, stellar sea lions, several species of porpoise) have attracted similar attention because of their potential collapse in association with commercial fisheries. The coverage of these issues in the general scientific press provides the impression that (1) stock collapses are quite common, (2) stock collapses are almost always the result of overfishing, and (3) professional fisheries managers are frequently unable to responsibly manage the resources they control. Within the professional fisheries community there is also widespread concern, accentuated by the fact that world fish catches have been declining for several years, and that reviews of the status of stocks in several countries indicate that a significant proportion of fish stocks is overexploited. Fisheries managers need to face up to very real failures and to revise fisheries management procedures to make sure that we maximize the potential returns from the fish resources.

However, before we can revise our management process in light of lessons about fish stock collapses, we need to understand what they are, what causes them, and how best to respond to the potential danger of collapse. The role of natural variability needs to be determined before we examine the frequency and concern about fish stock collapses. We need to understand what is known about the medium- and long-term behaviour of fish stocks in both their harvested and unharvested conditions. It is not widely appreciated that many, if not most, fish stocks exhibit large fluctuations in abundance in the absence of fishing. Indeed many stocks have been documented to collapse, by our definition, long before fishing power was a significant factor in determining fish stock abundance. Cushing (1982), has documented the periodic rise and collapse of the Norwegian and Swedish herring stocks in the 17th, 18th and 19th centuries. Soutar and Isaacs (1969) have shown that the California sardine collapsed regularly over the last 1000 years. Caddy and Gulland (1983) classified fish stocks into four groups of behaviour, of which three showed very strong periodic changes in abundance that would probably fall into the 'collapsed' category at their low points. How common, then, are fish stock collapses? The northern cod fishery in the 1990s meets our definition of collapse, as does the California sardine fishery, where harvesting ceased in the 1950s and the stock has only in recent years rebuilt to significant numbers. There are numerous other examples but, clearly, examining the history of well-documented stocks is not a reliable method to understand the frequency of stock collapse, because many stocks that are collapsed will not have been studied. With a longer time horizon we can see that most Atlantic salmon (*Salmo salar*) stocks in the eastern US have collapsed, as have some baleen whale stocks. Other whale stocks, such as the gray whale, probably never collapsed, and simply recovered from a period of severe overexploitation.

There are at least three mechanisms that can lead to stock collapse. Most obvious is habitat destruction. The demise of Atlantic salmon in the eastern US and much of Europe is clearly caused by loss of habitat, building of dams and weirs that prevent access, and changes in water quality. Habitat destruction is also clearly a dominant factor in many marine species that utilize wetlands. Secondly many stocks collapse from a change in environmental conditions. The Norwegian and Swedish herring fisheries discussed by Cushing (1982) are good examples. Most of the long-term collapses of marine species that have been documented for epochs prior to the mid 20th century were probably caused by environmental change. The third cause of collapse is low stock abundance, caused or aggravated by intensive fishing pressure. How does intensive fishing lead to stock collapse? A population may collapse either because of changes in adult survival or recruitment. The two traditional explanations for stock collapse at low stock densities are either depensatory predation, or failure of individuals at low densities to mate as successfully as at higher densities. Almost all evidence is that changes in recruitment are the cause of the stock collapse - populations fail to increase when fishing is stopped because the number of fish recruiting to the population does not exceed the mortality of the recruited population.

MATERIALS AND METHODS

Myers *et al.* (1990, 1995) have assembled histories of spawning stock biomass, recruitment, catch and fishing mortality rate for several hundred stocks. At the time of the analysis, 129 of these stocks had a history of 20 or more years of spawning stock biomass and recruitment data. I calculated a 5-year running average for spawning stock biomass and recruitment as follows:

$$\overline{R}_y = \frac{R_{y-2} + R_{y-1} + R_y + R_{y+1} + R_{y+2}}{5}$$

$$\overline{S}_y = \frac{S_{y-2} + S_{y-1} + S_y + S_{y+1} + S_{y+2}}{5}$$

where $\overline{R}_y$ is the recruitment in year y, is the running average recruitment, $\overline{S}_y$ is the spawning stock biomass in year *y* and is the running average spawning stock biomass. I chose three levels of stock decline and increase, 10-fold, 5-fold and 3-fold. If was at any point 0.1 of a previous value then the stock was classified has having undergone a 10-fold decline in spawning stock biomass. Similarly if at any point the value was 10 times higher than a previous value, the stock was classified as having undergone a 10-fold increase. This classification procedure was used for spawning stock biomass and recruitment at the 10-fold, 5-fold and 3-fold levels.

RESULTS

Table 1 shows the results for spawning stock biomass, and Table 2 shows the results for recruitment. Perhaps the most striking feature is the frequency of 10-fold declines and increases: almost

Table 1. Frequency of declines and increases in spawning stock biomass

		Decline			Increase		
Taxonomic group	**n**	**10-fold**	**5-fold**	**3-fold**	**10-fold**	**5-fold**	**3-fold**
Clupeiformes	31	23%	39%	65%	16%	35%	58%
Gadiformes	43	12%	26%	40%	0%	2%	21%
Perciformes	4	0%	25%	50%	0%	25%	25%
Pleuronectiformes	15	0%	13%	33%	0%	0%	0%
Salmoniformes	52	8%	10%	27%	10%	13%	35%
Total	145	11%	21%	40%	7%	14%	32%

Table 2. Frequency of declines and increases in recruitment

Taxonomic group	n	Decline 10-fold	Decline 5-fold	Decline 3-fold	Increase 10-fold	Increase 5-fold	Increase 3-fold
Clupeiformes	31	16%	39%	65%	6%	32%	52%
Gadiformes	48	17%	25%	48%	4%	10%	25%
Perciformes	9	22%	33%	44%	22%	22%	44%
Pleuronectiformes	16	0%	13%	19%	0%	0%	6%
Salmoniformes	45	7%	13%	33%	13%	20%	42%
Total	149	12%	23%	44%	8%	17%	35%

10% of stocks in this database showed changes of this order. Increases are not quite as common as declines, but nearly so. There is also considerable difference between taxonomic groups, with Clupeiformes and Salmoniformes showing the most dramatic changes, and Salmoniformes showing more increases than declines. Gadiformes show the fewest recoveries.

Discussion

This analysis shows that declines and increases are quite common among marine fish stocks and that major increases in spawning stock biomass are nearly as common as major declines. I have made no attempt in this analysis to determine if these fluctuations were due to the impacts of fishing, or to natural changes; indeed such determination is quite difficult and will undoubtedly require a case-by-case analysis. These will hopefully move many observers of fish populations to recognize the amount of variability common in fish populations. It seems highly likely that much of the variation in abundance seen in these stocks is due to natural variation in recruitment rather than being due to the effects of fishing alone. It seems probable that significant declines in abundance may be common in natural populations, and that fisheries management regimes should be adapted to such large-scale variation. Certainly the frequency of increases is encouraging; we tend to focus on the many documented declines and think that most fish populations go down, but do not come up. Finally, we must be aware that the data sets examined are unlikely to be a random sample of fish populations and the selection bias probably favours species that have not collapsed; if they had collapsed, the data series might have been terminated or never been put into a database. Thus the analysis described here is a beginning of a programme to describe the variability of fish populations. A systematic effort to collect the histories of as many stocks as possible in as unbiased a method as possible would be of great utility.

Acknowledgments

This study would not have been possible without the extensive work of Ram Myers and his colleagues in assembling this data set. Christina Tonitto performed much of the data analysis.

References

Caddy, J F, and Gulland, J A (1983). Historical patterns of fish stocks. *Marine Policy* **7**, 267–78.

Cushing, D (1982). 'Climate and fisheries.' (Academic Press: London).

Myers, R A, Blanchard, W, and Thompson, K R (1990). Summary of north Atlantic fish recruitment 1942–1987. *Canadian Technical Report of Fisheries and Aquatic Sciences* No 1743.

Myers, R A, Bridson, J, and Barrowman, N J (1995). Summary of worldwide stock and recruitment data. *Canadian Technical Report of Fisheries and Aquatic Sciences* No 2024.

Soutar, A, and Isaacs, J D (1969). A history of fish populations inferred from fish scales in anaerobic sediments off California. *California Marine Research Committee* CalCOFI **13**, 63–70.

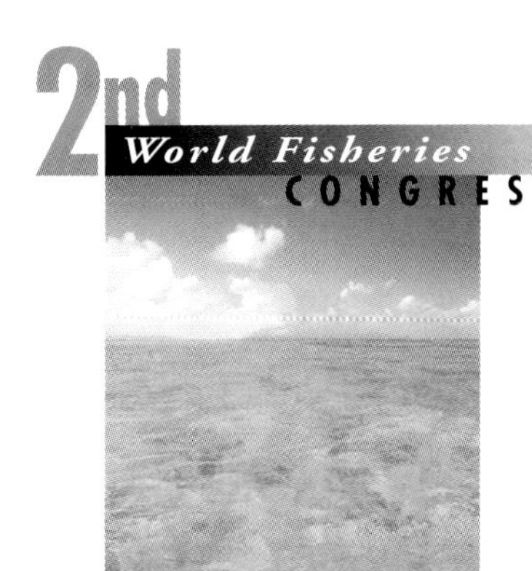

A Critical Review of Tuna Stocks and Fisheries Trends World-wide, and Why Most Tuna Stocks are not yet Overexploited

Alain Fonteneau

Orstom Scientist, Inter-American Tropical Tuna Commission, 8604 La Jolla Shores Drive, La Jolla CA 92037 USA.

Summary

A two-year programme compared the status of the tuna species and stocks exploited world-wide. Most tuna species, especially the tropical species, are still showing increasing catches world-wide. Very large oceanic habitat is the major cause for this. Most tunas undertake extensive seasonal migrations, for instance between their feeding and spawning zones; the thermoregulation capabilities of the tunas, especially efficient for the adults of the largest species, allow them to migrate and feed in 'a wide range of water temperatures'. Furthermore, most tuna stocks appear to have significant fractions that are still not available to the fisheries, being located in a remote zone, too scattered or too deep. Those fractions of tuna stocks mostly unavailable to the fisheries may often act as natural refuges because of their incomplete mixing with the fully-exploited fractions of stocks. These conditions probably well explain why most tuna stocks have in the past shown very few cases of recruitment overfishing.

However, the positive conclusion that most tuna stocks are not yet overexploited, probably cannot be extrapolated to the future. Various bluefin stocks are already in bad shape. The bad condition of these temperate tuna stocks can readily be explained by various factors, among others a very high value on the sashimi market (allowing a sustainable fishery, even at very low catch rates). Various tuna stocks can also suffer local overexploitation. For most tuna species and stocks, the exploitation rate has been constantly increasing as shown by the trends of their catches. However, most tuna stock assessments have been quite inefficient in estimating the real maximum sustainable yield of those stocks. The biological specificities of the highly migratory stocks also account for the major difficulties faced by scientists and international commissions involved with stock assessments. Various tuna stocks may soon join the large group of already-overfished resources, because very few unexploited fishing zones remain for the major tuna stocks.

Introduction

Trends of the tuna fisheries and peculiarities of tuna stocks

This paper reviews global trends in tuna fisheries. It then reviews stock assessment and management problems specific to the various tuna stocks, in relation to their biology. Finally it discusses the prospects for rational management and conservation of tuna stocks.

Until recently the tuna resources world-wide have been providing increasing and large catches (over 4 million t, see Fig. 1).

A surprising observation on the tuna fisheries is that '**tuna management problems**' are **still relatively minor for most tuna species**, (with the critical exception of bluefin tuna stocks) at least in comparison with most other valuable fisheries. This quite good status of most tuna stocks is not the result of a better stock assessment; nor is it a result of active and efficient management, for very few management actions have been implemented on tuna stocks (again with the exception of the bluefin tuna, Western Atlantic and Southern bluefin stocks).

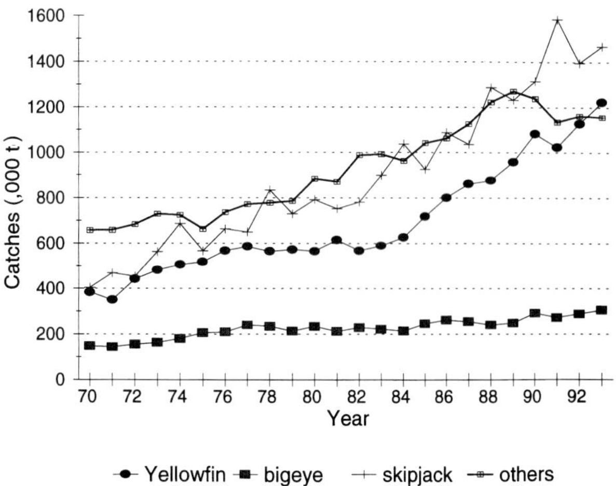

Fig. 1. Yearly global catches of tunas by species.

A SUMMARIZED WORLD-WIDE OVERVIEW OF THE TUNA FISHERIES

A world-wide database on tuna fisheries

The following overview of the tuna fisheries is based on the analysis of a world-wide database recently developed by the authors. It combines most of the catch and effort data, at a 5° square aggregation, collected on tuna fisheries since 1952 for the longline fisheries, and 1969 for the surface fisheries. A scientific atlas of the tuna fisheries and stocks will soon be published (jointly by Inter-American Tropical Tuna Commission (IATTC) and Institute Francais de Recherche Scientifique pour le Development en Cooperation (ORSTOM)) and will review the data used, and the methods and hypotheses used in their preparation (for instance conversion from numbers to weight, raising of log-books, species composition, etc.).

The completeness and reliability of this database remain often questionable; however, it will be assured that these data indicate without major bias the global patterns and trends of the major tuna fisheries.

Trends of catches by species and oceans

The trends of the yearly catches (taken from IATTC, IPTP, ICCAT and FAO statistical bulletins) are given by ocean for the major tuna species and stocks in Figure 2 .

Those catch trends are quite interesting to analyse and they can well be classified in three categories:

1. Species like yellowfin and skipjack (Fig. 2 a, b and e) are typically tropical species taken predominantly by surface gears. They show, for most stocks, high levels of catches and a permanently increasing trend. Those trends (and the present scientific analyses) indicate that most of those stocks are not yet overexploited or even fully exploited.

2. Species like bigeye and albacore (Figs 2c and 2d) are more temperate tunas, showing yearly catches which are (a) taken primarily by the longline gear, (b) at a lower absolute level and (c) quite stable over time (albacore) or increasing slowly (bigeye). Extensive stock assessments are necessary to explain in detail the status of each of those stocks, but it will appear that those tuna stocks have always lower biomasses and lower productivities, and they may often be reaching already a status of full exploitation. Recruitment overfishing was never observed for any of these stocks.

3. The yearly catches of bluefin tuna, the largest and most temperate tuna, often show a variable and decreasing trend (Figs. 2f and 2g), the catches being always at a quite low level (in comparison with the catches of tropical tunas) and often limited by strict regimes of quotas. Those bluefin stocks are easily overexploited, the two best examples being the southern bluefin or the west Atlantic bluefin.

AN OVERVIEW OF THE TUNA FISHING ZONES FOR YELLOWFIN, SKIPJACK AND BIGEYE BY OCEAN

Longline fisheries

An important factor to take into account in the analysis of tuna fisheries and stocks is the wide distribution of those stocks and fisheries (Fig. 3). It is quite spectacular to observe that the entire range of most tuna stocks was already fished in every ocean by the longliners (primarily Japanese) during the mid-1950s and 1960s. This large extension of the longline fisheries indicates well the distribution of the adults, as the longline gear catches mainly large individuals. Figure 4 shows well for yellowfin and bigeye this relative stability of the sizes of the areas fished by the longliners. However, it is quite striking to notice, when comparing the historical and present fishing zones of the longliners, that important changes in the targeted species were observed during their histories in the intertropical area. Whereas the historical longline fisheries were primarily targeting yellowfin and albacore, they are now primarily targeting species such as bigeye and bluefin, for the 'sashimi' market in specific fishing zones (at specific depths, for instance deeper for bigeye). The Eastern Atlantic Ocean longline fishery provides a good example of this spectacular change of target species (Fig. 5).

SURFACE FISHERIES

The pole and line bait boats used the major fishing gear for tunas world-wide until the late 1960s, but this changed in the early 1970s, and at present the surface fisheries are dominated by purse seine catches. The economic efficiency of purse seiners and their ability to catch large quantities of tunas, independently of bait availability, are the key factors explaining the spectacular development of the purse seine tuna fisheries world-wide. The first historical major fishing zone for the tuna purse seiners was the eastern Pacific Ocean (early 1960s), followed by the Eastern Atlantic Ocean (mid 1960s) and Western Pacific and Indian Oceans (early 1980s). The tuna fisheries have shown recently (since the early 1980s) a spectacular increase in those two last areas. The average fishing map of the surface fisheries shows the area presently exploited in each ocean for yellowfin, bigeye and skipjack (Fig. 6). Figure 7 gives an estimation of the change in the sizes of the areas significantly exploited by the surface fisheries and shows at a world-wide scale this increase of exploited areas. It can be

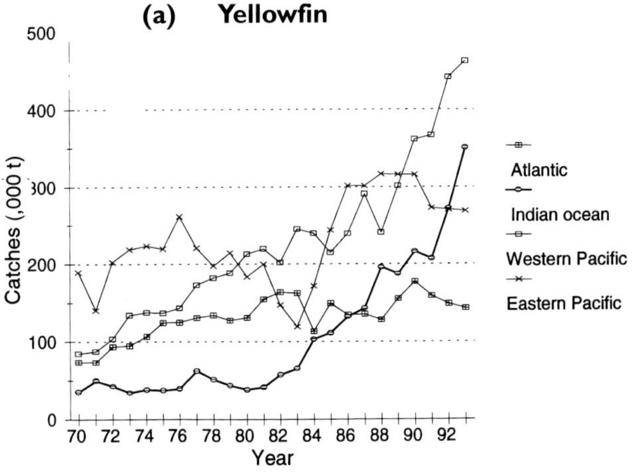

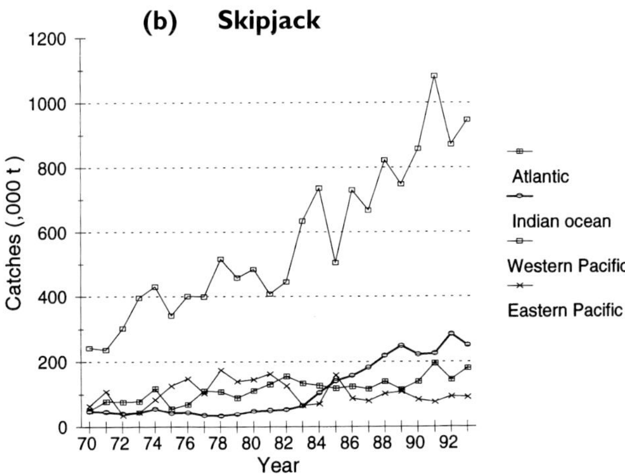

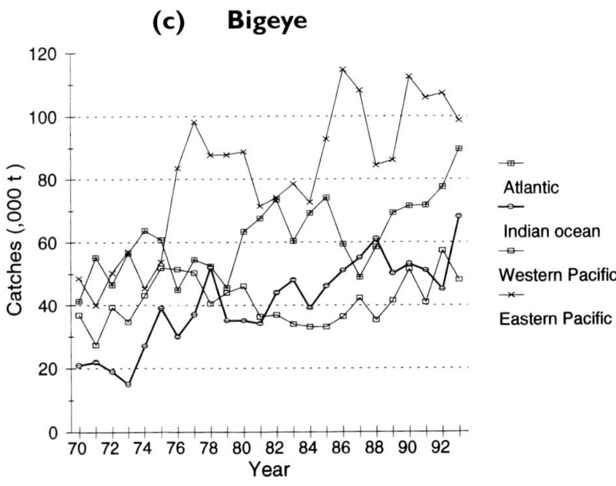

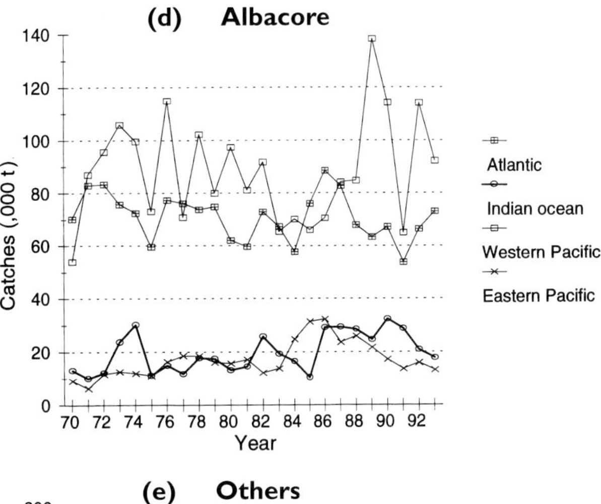

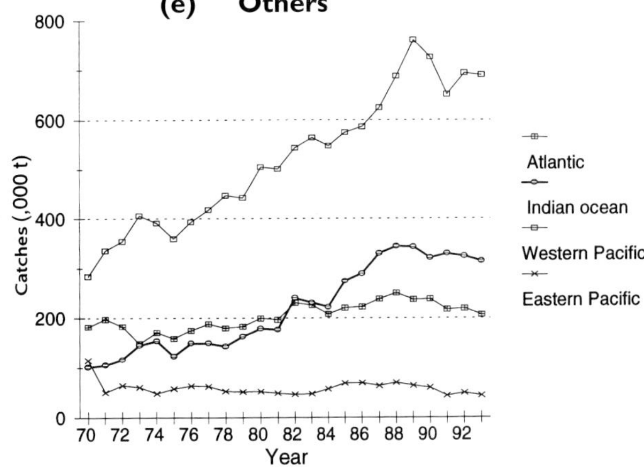

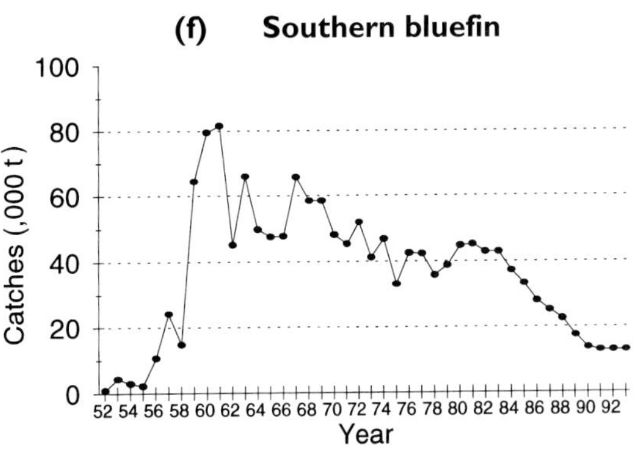

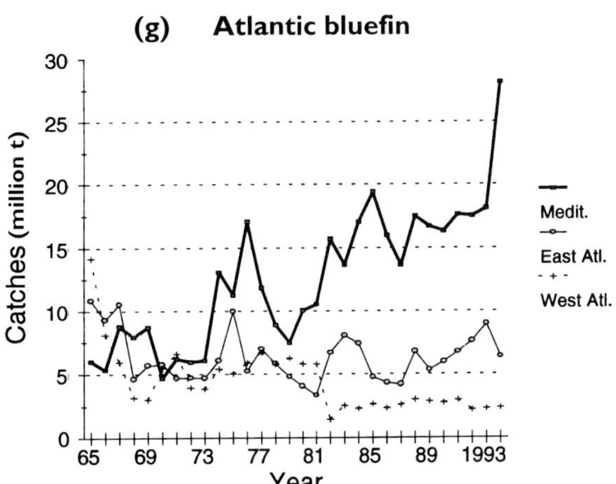

Fig. 2. Yearly catches of tunas and species and ocean for (a) Yellowfin, (b) Skipjack, (c) Bigeye, (d) Albacore, (e) tunas and billfishes other than yellowfin, skipjack, bigeye and albacore, (f) Southern bluefin tuna and (g) West Atlantic bluefin tuna.

(a) Yellowfin

(b) Bigeye

O 500 t

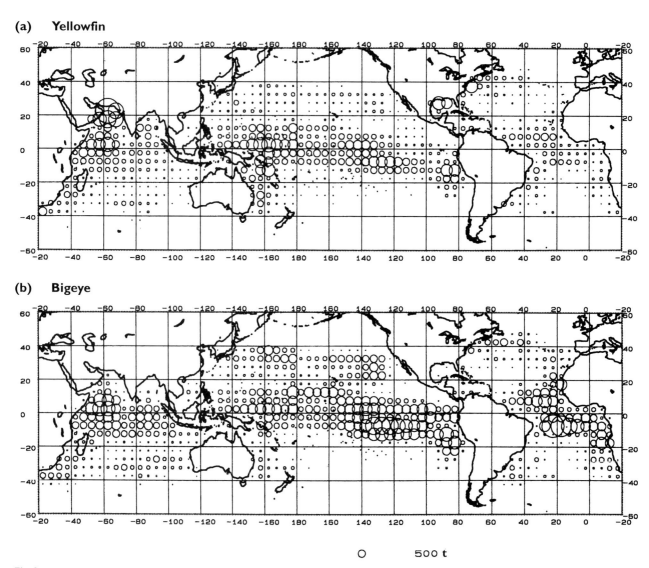

Fig. 3. Average 5 degree square fishing maps of (a) yellowfin and (b) bigeye tunas by the longline fisheries during recent years, period 1989–1993.

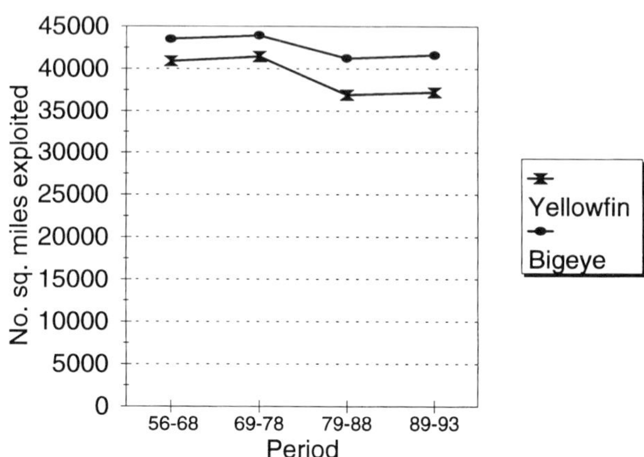

Fig. 4. Estimated changes in the size of the area exploited world-wide by the longline fisheries: areas with (a) yellowfin and (b) bigeye catches greater than 10 t yearly per 5 degree squares.

noticed that this increasing trend is more spectacular for yellowfin than for skipjack.

Longline and surface fishing zones

It is quite interesting to compare the major fishing zones of the longline and surface gears for the tuna species heavily exploited by both gears, such as yellowfin and bigeye.

The yellowfin fisheries are dominated quantitatively by the surface fisheries as shown by Figure 8 (about 90% and 64% of the total yellowfin catches in the Atlantic and Indian Oceans respectively during the period 1989 to 1993); most of those catches by the surface fisheries are taken between 20°N and 10°S (world-wide average = 94%). The area exploited by the longline fisheries, catching the adult yellowfin, extends to 40°N and 40°S, but the majority of the catches is from the central 20°N to 10°S area (world-wide average during the period 1989–1993 = 66%).

The world-wide bigeye fisheries are dominated quantitatively by the longline fisheries as shown by Figure 9, catching respectively

YFT
BET
1500 t
YFT
BET
1500 t

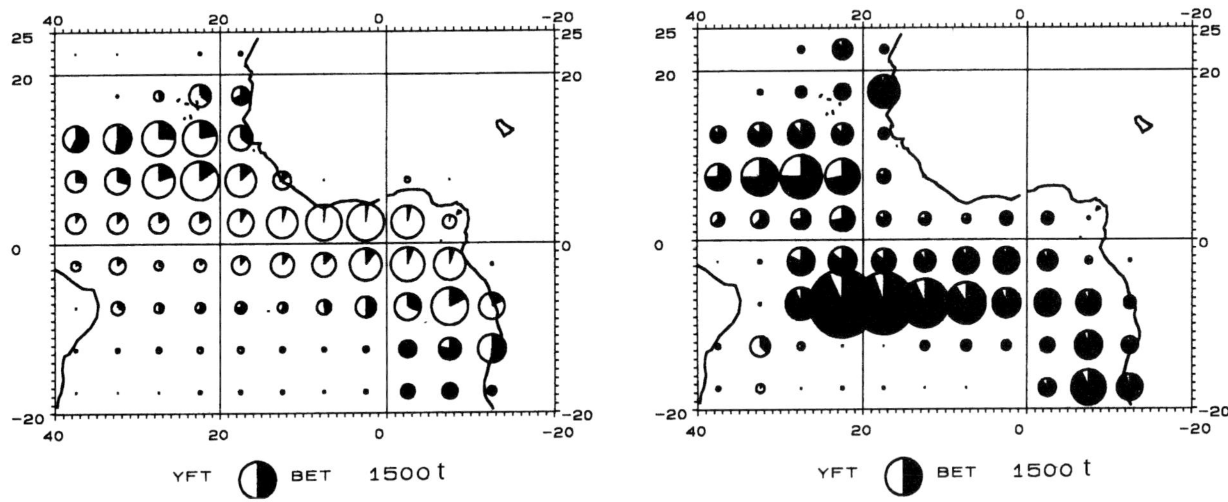

Fig. 5. Average fishing maps of yellowfin and bigeye tunas by the longline fisheries during the historical period (1956–68), and during recent years (1989–93) in the Gulf of Guinea.

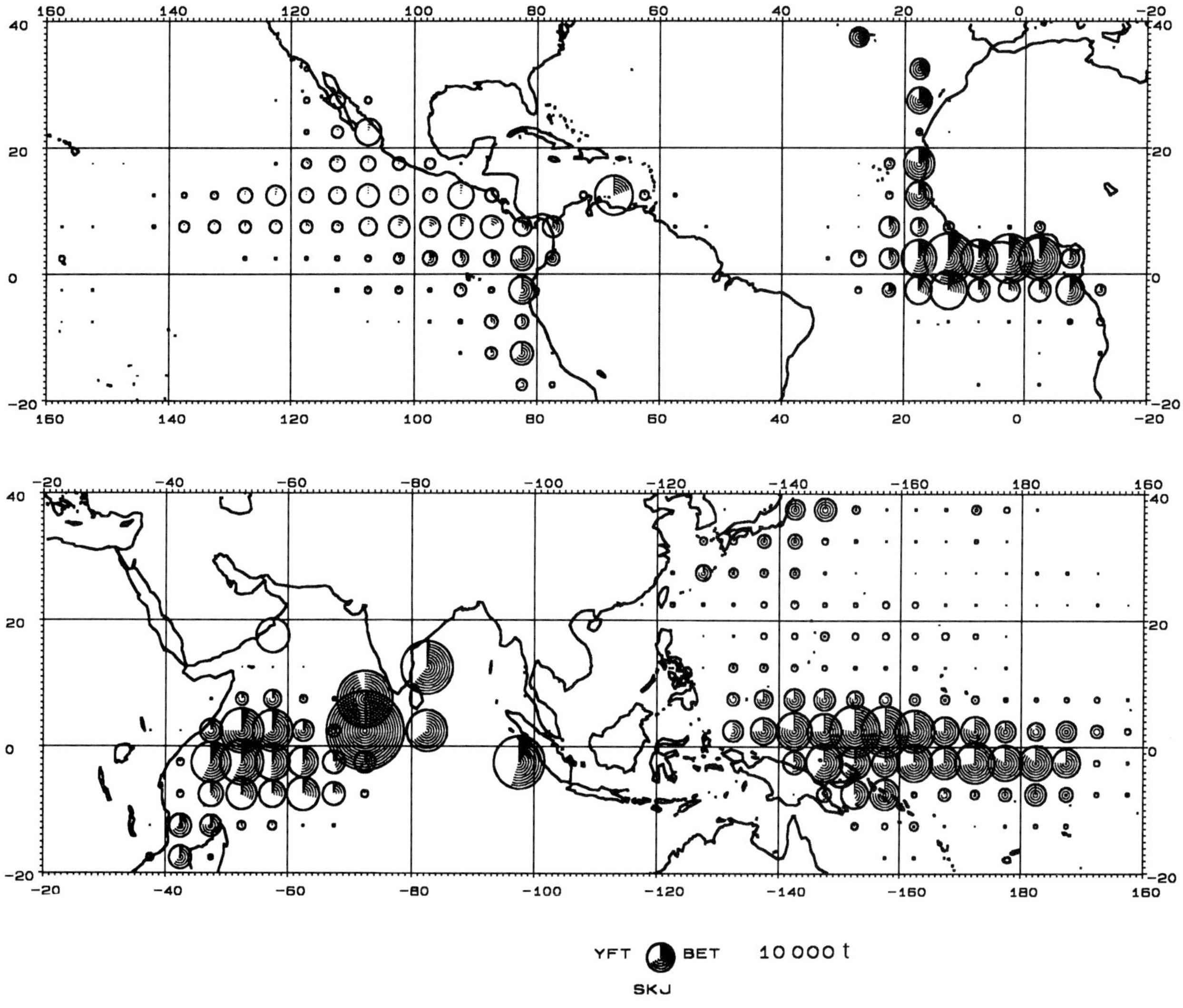

YFT
BET
10 000 t
SKJ

Fig. 6. Average fishing maps of yellowfin, bigeye and skipjack tunas by surface fisheries during recent years (1989–1993).

about 57% and 82% of the Indian and Atlantic Oceans catches during the period 1989 to 1993 (the proportion is unknown in the Pacific Ocean, because the small bigeye taken in the Western Pacific Ocean are classified as yellowfin in the statistics). The surface fisheries are catching mostly small individuals (in the equatorial areas). The longline fisheries are exploiting large geographical areas between 40°N and 40°S. However, the longline bigeye catches are predominantly taken in the 15°N to 15°S area (average 1989–1993 = 81%).

Tuna stock assessment: a difficult task

Tuna stock assessment and management may be considered a very difficult task. The results may appear to be quite successful, if judged by the trends of catches for the major stocks. However, this quite positive result was not connected in most cases to very efficient research and management of the stocks. Stock assessment was often a very difficult task for various combined reasons.

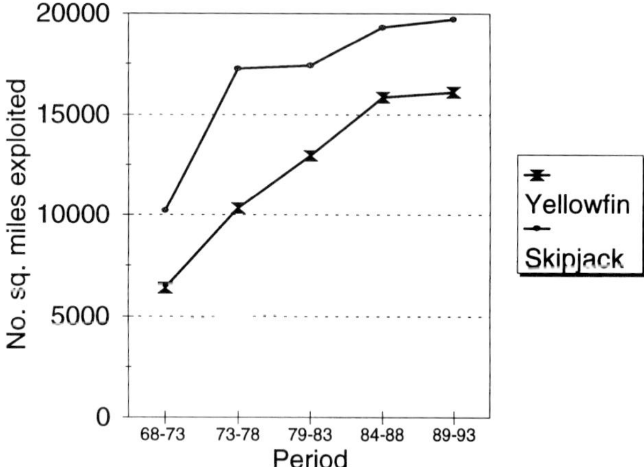

Fig. 7. Estimated changes in the sizes of the areas exploited world-wide by the surface fisheries: areas with (a) yellowfin and (b) skipjack catches greater than 100 t yearly per 5 degree squares.

Problems with the catch, effort and size data

Catch and effort statistics, associated with a good sampling of the sizes taken, are not always collected; and often they are not made available to the scientists or to the *ad hoc* tuna commissions, for various reasons. This is an old problem faced for many exploited stocks. However, the offshore distribution of most tuna stocks outside any EEZ, often increases those statistical problems, which are exacerbated in various areas, such as in large parts of the Pacific Ocean, by the lack of an efficient international tuna commission responsible to collect, and process and disseminate those scientific statistics.

Geographical stock structure, tuna migrations and cryptic biomass

Most tunas belong to the 'migratory species' category. Their stocks are distributed over large oceanic basins. They can be described, following MacCall (1990), as a distribution with central '... areas of high basic suitability with correspondingly a high local density of individuals, and ... areas of lower basic habitat with a lower density of individuals'. Those areas of lower densities are most often located in the periphery. This irregular basin well described and analysed by MacCall, will show a variable shape as in Figure 10. Furthermore, the shape of this basin will be variable for most tuna stocks, seasonally (Fig. 11) and interannually (during the El Niño events). In general, the time and area strata with high densities and large biomasses are predominantly exploited by the fisheries, for obvious economic reasons. Those areas of high densities are predominantly located in the central basins. Some exceptions can be noted for temperate tunas such as bluefin, a species which is showing most often discontinuous basins in its distribution, often with large catches in the periphery of its distribution.

It will then be fundamental to evaluate more precisely the movement patterns of the tunas inside this basin: following the typology given by MacCall (1990) and fairly well applied to tunas, those movements can be classified in two categories:

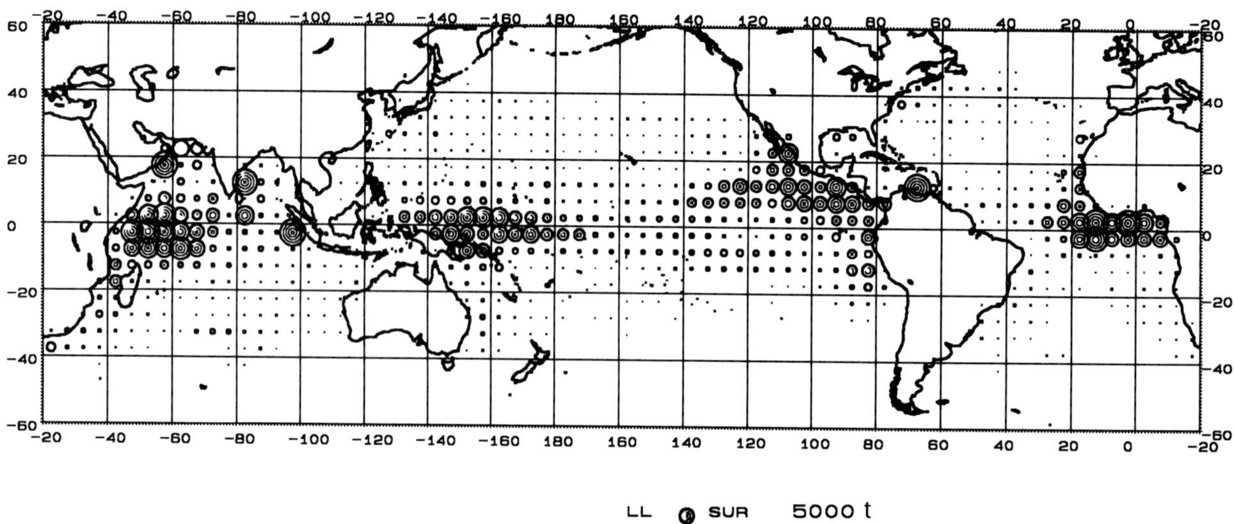

Fig. 8. Average fishing map of yellowfin (by 5 degree square, 1989–1993) by longline and surface gears.

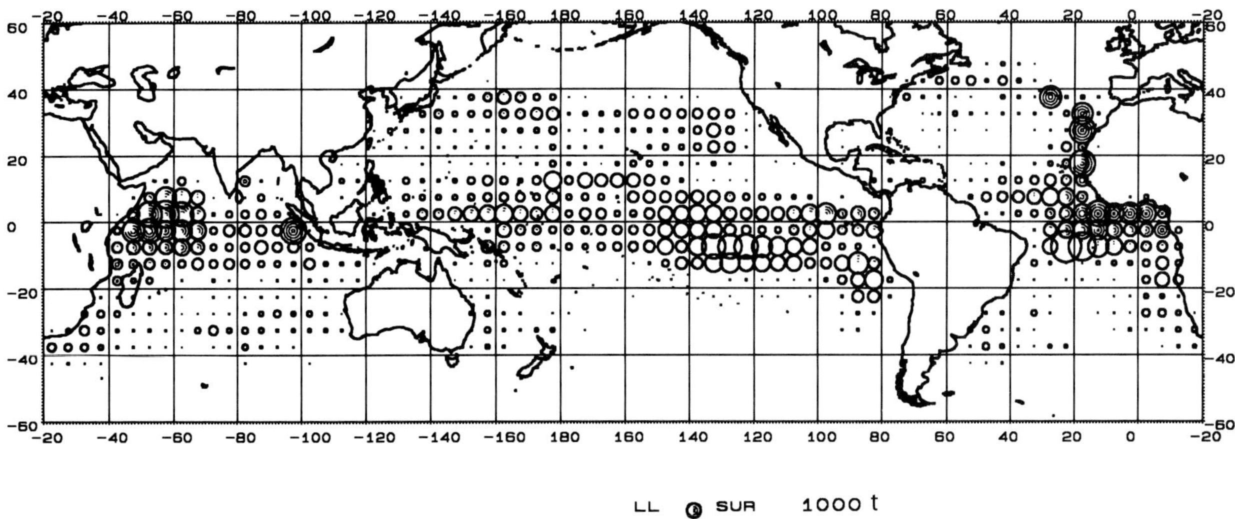

Fig. 9. Average fishing map of bigeye (by 5 degree squares, 1989–1993) by longline and surface gears.

- In his ***low viscosity*** pattern (this pattern could well be called a 'fluid stock'), movement of individuals is rapid with respect to growth and harvesting rates, so that harvests have a roughly equivalent effect on the population wherever they are taken.
- In the ***high viscosity*** pattern, losses due to harvesting are replaced mainly by *in situ* population growth, with immigration proceeding at a negligible rate.

Tunas were classified by the Caracas lawyers in the early 1980s as being 'highly migratory species'. A general thinking at this time (at least for the lawyers) was probably that most (or all?) tuna stocks should be classified in the ***low viscosity*** category, with rapid and permanent movement. However, even if there is no doubt that most tuna species can undertake extensive migrations, there is now an increasing tendency among the tuna scientists world-wide to consider that the low viscosity hypothesis is not a correct one for most tuna stocks (Hilborn and Sibert 1986). Most tuna stocks probably belong to categories with an '***intermediate viscosity***', somewhere between those two extremes of low and high viscosity. In this type of situation, local overfishing may occur, without reducing significantly the overall total biomass of the total stock distributed in the entire basin. This '***intermediate viscosity***' of tunas will be different in each tuna species and stock. It will be a key factor to study in order to make any realistic stock assessment and management of every tuna stock. Viscosity is probably quite low for the temperate tuna species undertaking **advective migrations** (such as bluefin tunas, Fig. 12c), but quite high for the equatorial species like skipjack tuna, undertaking primarily **diffusive movements** at a smaller scale (Fig. 12a). This concept of viscosity could be a key parameter in the understanding of the interactions between various tuna fisheries catching the same stock in different areas or/and at different ages Shomura *et al.* (1994). It could then also play a key role in the rational management of tuna stocks: the hypothesis of an intermediate viscosity for various tuna stocks could explain well the relatively good shape of most tuna stocks and fisheries, significant fractions of tuna stocks remaining potentially '**cryptic**' because of the very large sizes of most tuna basins (with only the areas of high densities being exploited heavily), creating

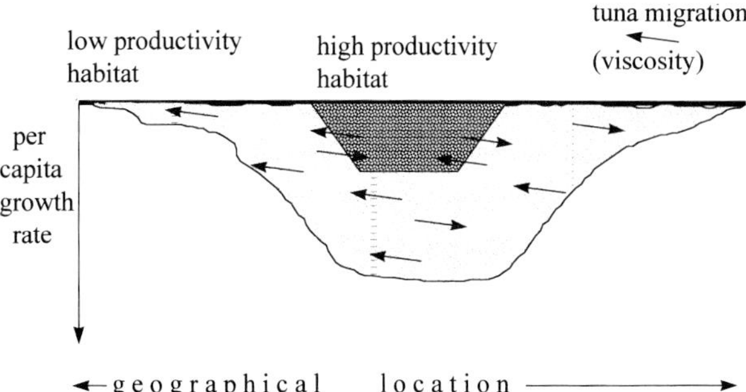

Fig. 10. The basin typology as proposed by MacCall (1990): geographical heterogeneity of the productivity of a migratory stock in an oceanic basin.

unintentionally '**natural reserves**' of tuna biomasses. Most of those fractions of stocks are cryptic due to economic factors: primarily value of the tuna species, but also cost to operate the fisheries, and catch rates.

Production models and changes of the exploited areas

It is quite clear that past performances of tuna stock assessment have often been quite limited: a global error arising from the use of global production models (modelling trends in catch and effort relationships) was to conclude that the stocks were already overexploited when their real exploitation rates were quite low (hopefully this type of erroneous conclusion was not really dangerous for stock conservation). It was the consequence of various combined circumstances:

- In the case of yellowfin tuna in the eastern Atlantic and Pacific Oceans, for example, productivity was estimated by production models. In each case a purse seine fishery exploited a **coastal fraction** of a stock in a large basin with a tuna population showing intermediate viscosity. The real productivity of the stocks was underestimated when the fisheries were too coastal. The Maximum Sustainable Yield (MSY) presently estimated may be more realistic as most of the distribution areas are now explored and fished.

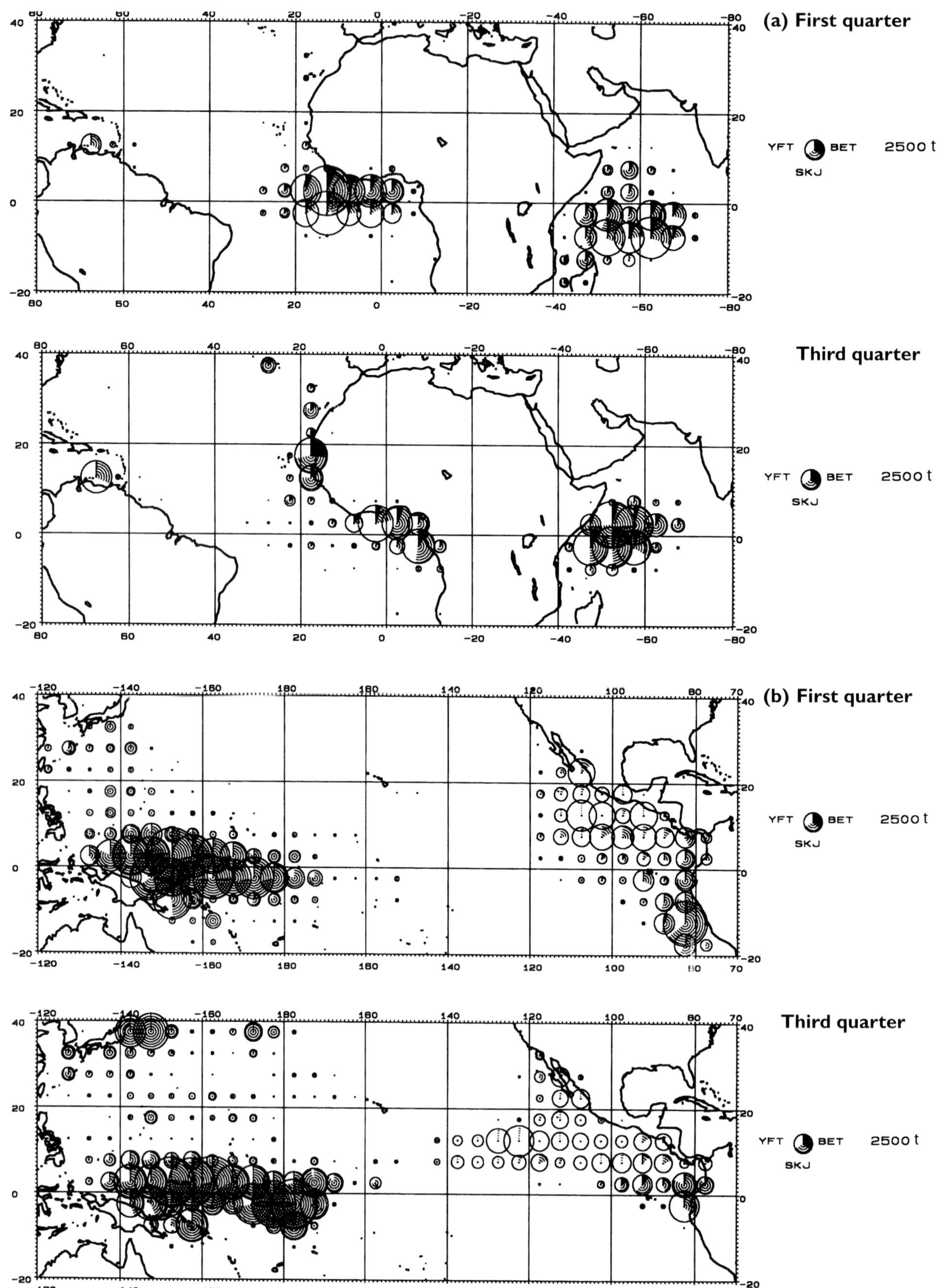

Fig. 11. The seasonality of the tuna fisheries shown by the average quarterly catches (surface fisheries) during the 1st and 3rd quarters of the average period 1989–1993. (a) Atlantic and Indian oceans, and (b) Pacific Ocean.

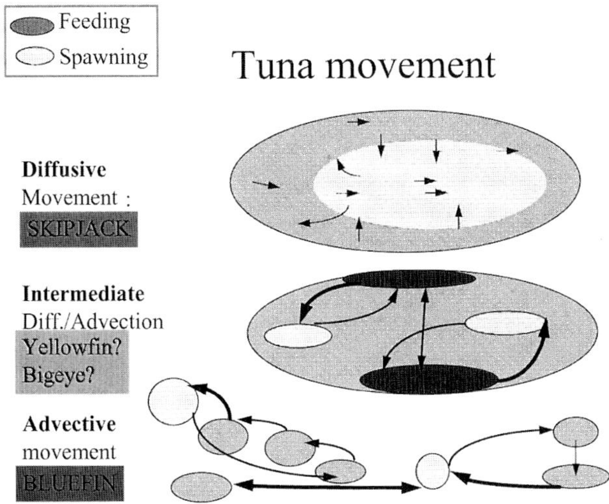

Fig. 12. A schematical overview of the various types of tuna migrations between their feeding and spawning areas.

- When the MSY of yellowfin tuna was estimated from the **longline fisheries** [in the Atlantic Ocean by production models (see Le Guen and Wise 1967), or in the Indian Ocean by Virtual Population Analysis (VPA) type models (see Wang and Tanaka 1988)], the MSY or recruitment levels were dramatically underestimated, as the potential catches of large yellowfin by the longliners are always quite small in comparison with the real potential catches (by the purse seiners).

ANALYTICAL MODELS

Various tuna stocks are permanently monitored by VPA analysis (analysing the catch-at-age matrix) of various types, at least when the available data allow this type of analysis (e.g. for Atlantic bluefin, Eastern Pacific bluefin and southern bluefin tuna). Those results have been very interesting and conclusive in a few cases (see the annual reports of IATTC, the Commission for the Conservation of Southern Bluefin Tuna, and of the International Commission for the Conservation of Atlantic Tuna).

However, the results obtained by this type of analysis were too often quite uncertain and variable over time for many tuna stocks under study.

The major difficulties in the 'tuna VPA' were probably due to a combination of various reasons such as:

- The uncertainties in the **age-specific natural mortality rate.** It is quite obvious that natural mortality rate is variable with age for most tuna species, especially for the large tunas undertaking extensive migrations between their equatorial nurseries and their temperate feeding zone where the adults are concentrated. The hypothetical natural mortality rate used by scientists for stock assessment is most often assumed to be constant. The uncertainties are major, and may have a serious impact on all the subsequent yield-per-recruit analyses, especially those studying the yield-per-recruit interactions between fisheries for juveniles and fisheries for adults (as shown by Lapointe *et al.* 1989). If the natural mortality rate of juvenile tuna is high, the real numbers of those fishes could be greater than the number estimated by the VPA.
- The **impossibility to age the catches** routinely. Most VPAs are based only upon the size structure of the catches and a subsequent estimation of their age, based upon a statistical average growth. This problem is worsened by the differential sexual growth often observed for tunas.
- The difficulty to obtain **reliable tuning indices** for most migrating stocks. Most tuning indices are obtained from the fisheries themselves, and they are often biased. Those biases can be of various types: (1) they often underestimate the real decrease of the stocks (for instance because of the increased fishing powers of the fleets); or (2) they overestimate the real decrease of the total biomass, because they are taken primarily in the fished area, wherein local overfishing (and high viscosity) often produces **locally depleted fractions of stocks.** Most tuning indices used in tuna stock assessment were often too pessimistic in their decreasing trends, because they reflected local overfishing rather than the real trend of the total biomass of the stock. A good VPA analysis should seek to estimate N, the total number of fish at age swimming in the basin, not the local number of fish (n) in the exploited fraction of the basin. N and n may be very different when the viscosity is high, or identical when the viscosity is low (a rare case for tuna stocks?). A subsequent problem from this VPA bias was that often, the recruitments calculated by the scientists were artificially increasing when the fisheries expanded geographically in the basin (to explain the increased catches). The North Atlantic swordfish fishery (ICCAT 1994) may be a good example of this potential bias, with levels of the estimated recruitment increasing over time in proportion to the increased catches and increased fishing zones.

One interesting fact to keep in mind is that this bias is quite safe in terms of management, as it will always produce an underestimation of the potential productivity of the stocks, leading unwillingly but *de facto* to a **precautionary management of the stock.**

Increased fishing efficiency

Various other biases often found in the analysis of the tuna stock status may be more dangerous. The major one is probably the **permanent and spectacular increase of fishing power** observed world-wide for both the surface and longline fisheries. This dangerous bias was also a major one for demersal fisheries (for example in the Canadian cod collapse), but is most of the time very difficult to correct. Various efforts have been made by scientists to estimate those changes and the extent to which they haved improved fleet efficiencies (Punsly and Deriso 1991; Marsac 1992; Gascuel *et al.* 1993).

For example potential sources of increase in purse seiners' fishing power might include:

- Development of bird radar, a new device allowing the remote location of tuna schools associated with bird aggregations;
- Improvement in purse seine design, allowing the nets to close faster (less time lost, operation more efficient) and deeper

(allowing to fish with a deep thermocline, such as in the Indian Ocean and Western Pacific Ocean). Those changes are well documented for the Atlantic and Indian Oceans;

- World-wide development of fishing on artificial logs and sea mounts;
- Spectacular technological improvement in sonar and echosounding devices;
- Generalization and improvement of the satellite imagery transmitted in real time to the boats, allowing more efficient research; and
- Probably a multitude of other factors secretly developed by fishermen to increase their fishing efficiency and their catch rates.

However, it is nearly impossible for scientists to be sure that the increase of fishing efficiencies has been well estimated and without bias: this potential bias may be critical for various stocks and areas, such as the yellowfin stocks in the eastern Pacific, Atlantic and Indian Oceans.

Conclusion

Prospects for the rational management and conservation of the tuna resources world-wide

The positive conclusion, that most tuna stocks are not yet overexploited, probably cannot be extrapolated to the next millennium.

For most tuna species and stocks, the overall exploitation rate has been constantly increasing during recent years, as shown by the increasing trends of their catches. Various tuna stocks can also suffer local overexploitation and locally-depleted biomass, which is not very satisfactory for the concerned countries, even when the global stocks may still be in relatively good shape. Furthermore, very large numbers of small juvenile tunas are often taken in increasing numbers and proportion (for instance yellowfin and bigeye, in association with the increasing fisheries on floating objects), reducing too often the yields-per-recruit of those stocks. The real MSY of those underexploited tuna stocks remain most of the time difficult to evaluate for various reasons, among others the biological specificity of the highly migratory stocks and the institutional problems faced by the tuna scientists. The major difficulties presently faced by most tuna commissions world-wide to manage efficiently their tuna stocks are quite obvious. In the next 10 or 20 years, there is an increasing and possibly critical danger of overexploitation, as very few potential fishing zones remain now unexploited for the major tuna stocks, especially for all tuna species with a high value on the market. The most serious potential management problems will probably be faced in the Pacific Ocean, despite the co-ordinated effects of the South Pacific Commission (SPC) and the IATTC where there is an obvious and potentially dangerous lack of a tuna commission internationally responsible for the tuna statistics, research and management in the entire Pacific Ocean.

Acknowledgments

This paper was primarily based on a world-wide database recently built, at a 5° square/month level, combining more or less all the existing catch and effort statistics on tunas. I take this opportunity to give my deep and sincere thanks to the various bodies and scientists who have provided generously their help and their data to this project. Special thanks to the various IATTC scientists involved in the collection and processing of the eastern Pacific data, especially Michael Hinton. The ICCAT, SPC and IPTP staff have also provided a valuable input to this work, namely Papa Kebe for the ICCAT, David Ardill and Alejandro Anganuzzi for the IPTP, and Tony Lewis, John Hampton and Tim Lawson for the SPC. The national statistical offices from Japan, Korea, Taiwan and USA have also offered their full cooperation and their data. Special thanks should be given to the Shimizu and Tohoku scientists in Japan, because they have collected and released the most complete tuna data set. Ziro Suzuki was very helpful in obtaining this data set and I want to address to him my special thanks. Furthermore the ORSTOM team of scientists in the Indian Ocean, namely R. Pianet and F. Marsac, also provided valuable information on the Indian Ocean surface fisheries.

References

International Commission for the Conservation of Atlantic Tuna (1994). Report for biennal period, 1992–93, part II (1993).

Inter-American Tropical Tuna Commission (1994). Annual report of the Inter-American Tropical Tuna Commission.

Gascuel, D, Fonteneau, A, and Foucher, F (1993). Analyse de l'évolution des puissances de pêche par l'analyse des cohortes: application aux senneurs exploitant l'albacore (*Thunnus albacares*) dans l'Atlantique est. *Aquatic Living Resources* **6**, 15–30.

Hilborn, R, and Sibert, J (1986). Is international management of tunas necessary? *South Pacific Commission Newsletter* **38**, 31–40.

Laloe, F (1989). Un modèle global avec quantité de biomasse inaccessible dépendant de la surface de pêche. Application aux données de la pêche d'albacores (*Thunnus albacares*) de l'Atlantique est. *Aquatic Living Resources* **2**, 231–9.

Lapointe, M F, Peterman, R M, and MacCall, A D (1989). Trends in fishing mortality rate along with errors in natural mortality rate can cause spurious time trends in fish stock abundances estimated by virtual population analysis (VPA). *Canadian Journal of Fisheries and Aquatic Sciences* **46**, 2129–139.

Le Guen, J C, and Wise, J (1967). Méthode nouvelle d'application du modèle de Schaefer aux populations exploités d'albacores (*Thunnus albacares*) dans l'Atlantique. Cahiers. O.R.S.T.O.M, Séries. Océanographie **5**, 79–93.

MacCall, A D (1990). 'Dynamic geography of marine fish populations.' University of Washington Press. 153 pp.

Marsac, F (1992). Etude des relations entre l'hydroclimat et la pêche thonière hauturière tropicale dans l'océan Indien occidental. Thèse de doctorat, Université de Bretagne occidentale, Brest. 353 pp.

Punsly, R G, and Deriso, R B (1991). Estimation of the abundance of yellowfin tuna, *Thunnus albacares*, by age groups and regions within the eastern Pacific ocean. *Inter-American Tropical Tuna Commission Bulletin* 20, (2).

Shomura, R S, Majkowski, I J, and Langi, S (Eds) (1994). Interactions of Pacific tuna Fisheries. *FAO Fisheries Technical Paper 336*. Vol. 1: Summary report and papers on interactions. 326 pp.

Wang, C H, and Tanaka, S (1988). Development of a multicohort analysis method and its application to the Indian Ocean yellowfin tuna length composition. *Far Seas Fisheries Research Laboratory Bulletin* **25**, 1–72.

Successes and Failures in the Management of Atlantic Herring Fisheries: Do We Know Why Some have Collapsed and Others Survived

Robert L. Stephenson

Department of Fisheries and Oceans, Biological Station, St. Andrews, New Brunswick, Canada E0G 2X0.

Summary

Four major Atlantic herring stocks have recently recovered from spectacular collapses resulting from a combination of heavy fishing pressure, recruitment failure and lack of effective management. Others have narrowly avoided collapse through fortuitous recruitment and management action to restrict fishing. Few if any stocks have escaped major fluctuations in spawning stock biomass. Collapse has most often been attributed to a complex rather than a single cause, making it difficult to anticipate and to arrest. Major unresolved issues which are relevant to prevention of future stock collapse include: care of individual spawning areas (sub-stocks) within management areas, improved abundance and recruitment indices, improved understanding of ecosystem and multi-species interactions, development of appropriate management thresholds and targets, and effective control of overcapacity and improved technology. There is a need for further development of a management approach which considers multiple factors, and which can react quickly to complex and changing conditions.

Introduction

Atlantic herring are evaluated and managed in approximately 30 units around the North Atlantic (Stephenson 1991, 1992). The fisheries for herring in these areas are diverse. The suite includes several that are very old and well-studied, and some that have been the subject of innovative management (including early experiments in limited entry, catch restrictions and quotas, and individual transferable quotas, ITQs). Together, they provide a very useful case study for comparisons in fisheries management.

In spite of early and often innovative management, and considerable recent regulation, there have been several instances of failure of herring fisheries. These include collapses of the notable Norwegian spring-spawning, North Sea, Georges Bank and Icelandic summer-spawning, herring fisheries. Indeed, the collapses of the major herring stocks in the Northwest Atlantic in the 1960s and early 1970s have been termed 'the most striking phenomena in the history of the European fisheries' (Jakobsson 1985).

A major symposium held in Aberdeen in July 1978 addressed the issue of problems in assessment and management of pelagic stocks at a time when several major herring and other pelagic fisheries had just collapsed (Saville 1980). It has been almost 20 years since that major conference, and there have been considerable advances in management. The collapsed stocks have rebounded, and are now supporting commercial fisheries again. Although a number of case studies on the performance of herring fisheries and aspects of management have been summarized in recent years — particularly as a result of recent symposia on herring and pelagic fisheries (e.g. Brett 1985; Kawasaki *et al.* 1990; Wespestad *et al.* 1991) — it is appropriate to re-investigate the issue of success and failure in the management of Atlantic herring fisheries in general, and in the context of the theme of this Congress, address the question of what we now know about why some have some collapsed and others have survived.

The record of stock collapse

While one could argue about the definition of collapse — and there is certainly the need to differentiate between fishery collapse and stock collapse — there are, by any definition, dramatic examples of both stock and fishery collapse in herring.

Norwegian spring-spawning (Atlanto-Scandian) herring (Jakobsson 1985), (Fig. 1a). The stock of herring originating in spawning areas along the coast of Norway is estimated to have exceeded 11 million t (spawning stock biomass, SSB) in 1956, but decreased rapidly to collapse in about 1970. SSB was estimated to have been only 9000 t in 1972. The pre-collapse period was marked by a large increase in fishing effort and new technology. Landings which had been in the order of 750 000 t to 1.5 million t for more than 15 years peaked at 1.7 million t in 1966, just prior to the collapse. Fishing mortality (F) was estimated to have been less than 0.2 throughout the period 1950 to 1963, but rose quickly to values as high as 2.25 around the time of the collapse. Recruitment, which is pulsed in this northern stock, failed when the SSB was below 2.5 million t.

North Sea herring, (Saville and Bailey 1980; Burd 1985; Bailey and Steele 1992), (Fig. 1b). The SSB of autumn-spawning herring in the North Sea decreased from 5 million t in 1947 to an estimated 45 000 t by the mid-1970s. The three main spawning components which make up the North Sea complex (Downs herring spawning in the southern North Sea and English Channel, Bank or central North Sea, and Orkney-Shetland area spawners) failed at different times. Annual landings, which fluctuated between 500 000 t and 700 000 t throughout the period 1947–1963, increased sharply to about 1.2 million t in 1965 before decreasing over a decade to collapse. Fishing mortality, which had been less than 0.5 through the 1950s, rose sharply in the late 1960s to values between 1.0 and 1.5 until the fishery was closed in 1977.

Icelandic summer-spawning herring (Jakobsson 1985), (Fig. 1c). The stock of summer-spawning herring from the coast of Iceland is estimated to have grown during the 1954–1963 period of consistent, high recruitment and low exploitation to a level of about 300 000 t (in 1962) but then decreased to a depleted state (minimum SSB of 11 000 t) within about six years. Landings which were less than 50 000 t y^{-1} over the period 1950–1960 increased to 125 000 t in 1963, just prior to stock collapse. Fishing mortality which had been less than 0.5 from 1950 to 1960 rose sharply to values of 0.75–1.5 during the collapse period. Recruitment was at its lowest level on record during collapse.

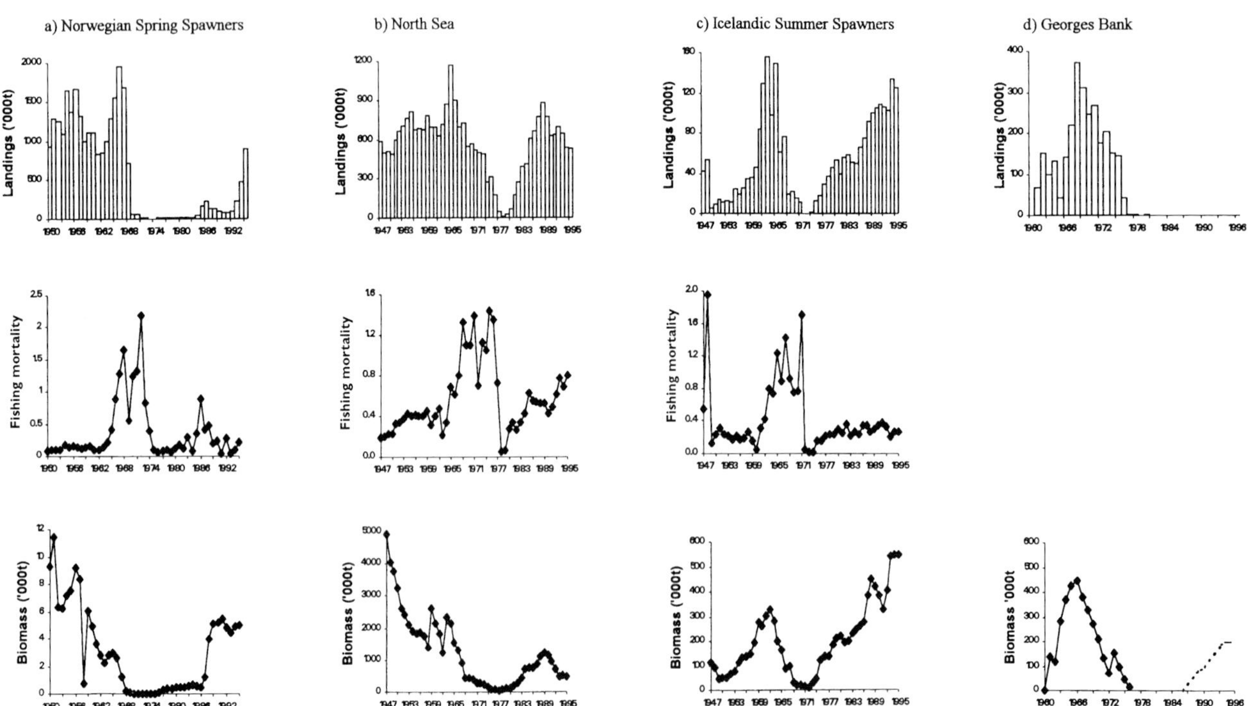

Fig. 1. Landings (top), fishing mortality (middle) and spawning stock biomass (lower) through stock collapse and recovery of (a) Norwegian spring-spawning; (b) North Sea; (c) Icelandic summer-spawning; and (d) Georges Bank Atlantic herring stocks. Data from Anon. 1995*a* (a–c) and Melvin *et al.* 1996.

Georges Bank herring (Anthony and Waring 1980), (Fig. 1d). A fishery for autumn-spawning herring on Georges Bank in the western Atlantic began in 1961, grew to over 374 000 t by 1968, then declined rapidly to collapse by 1977. The population which is estimated to have exceeded 1.3 million t in 1966 had virtually disappeared by 1977. As in the other examples of collapsed herring stocks, F increased substantially and recruitment decreased during the collapse period.

COMMON CHARACTERISTICS OF COLLAPSE

These fisheries had a number of similarities preceding and during stock collapse. All were subjected to record high landings just prior to collapse, by fleets which were increasing effort and technological capability. This was reflected in a sharp rise in F, to values greater than 1 in all cases and to a value greater than 2 in the case of Norwegian spring-spawners. All show evidence of low recruitment as collapse began, or recruitment failure in the final stages of collapse, or both.

In all cases, the stock collapse was rapid. The time from peak landings to fishery closure (or failure) ranged from 3–5 years (Norwegian spring-spawners) to about a decade for the other examples. In all cases the spawning stock biomass was reduced to very low levels, and in the case of Georges Bank to the point that herring were undetectable in surveys. The events are correctly termed stock collapse rather than just fishery collapse.

REASONS FOR COLLAPSE

A number of factors have been put forward as contributing to these examples of stock collapse. Prime among these have been heavy fishing pressure, reduced recruitment, high exploitation of juveniles, and lack of effective management restriction.

The large and rapid increases in landings in each area have been attributed to increasing fishing effort and improvements in fishing technology. Jakobsson (1985) shows a coincidence between very high levels of F leading to collapse in the Icelandic fishery and the introduction of sonar-fishing and the power-block. Increased effort in the North Sea was, in large part, due to an influx of additional large purse seiners following collapse of the Norwegian spring-spawning herring (Bailey and Steele 1992). Even the short development of the Georges Bank fishery shows the rapid technological progression from a gill-net fishery by vessels of a single nation (1961–63) through bottom-trawl to high numbers of large, specialized purse seiners and pelagic trawlers of distant water fleets from many nations (Anthony and Waring 1980). Saville (1980), in summarizing conclusions from the 1978 Aberdeen meeting, makes the point that there were at that time few pelagic stocks which had been subject to modern fishing technology whose state did not give cause for alarm.

Depressed recruitment has been implicated in each case. Low recruitment in years leading up to collapse, combined with high fishing effort, was considered to have initiated collapse in Icelandic summer-spawning and Norwegian spring-spawning herring stocks (Jakobsson 1985). A similar situation is true for Georges Bank (Anthony and Waring 1980). In the Norwegian spring-spawning and North Sea examples, recruitment failure after spawning stock biomass had been reduced has been interpreted as indicating recruitment overfishing below critical SSB levels of 2.5 million t (Norwegian spring-spawners) and 800 000 t (North Sea) (Anon. 1995*a*). However, Corten (1986) and Bailey (1990) suggest that recruitment in the North Sea was also depressed during these years by environmental conditions.

A high exploitation rate on young herring is considered to have been a major factor of special importance in the decline of the Norwegian spring-spawning herring. A fishery directed at juvenile herring, (mainly 0–2 y) took as much as 500 000 t in 1967 and 1968, far outnumbering the catch of adult herring, and fishing-out entire year-classes (e.g. 1963, 1964 and 1966) prior to maturation (Jakobsson 1985). Dragesund *et al.* (1980) suggest that protection of 0 and 1 y groups of herring in the 1960s would have been the only measure needed to prevent depletion of this stock.

Lack of restriction on fishing by management, and failure to take management action, has been cited as a contributing factor in all cases. 'Managers and scientists failed to take action to prevent the [North Sea] collapse...because of the problem of quantifying the catches of that stock' (Burd 1985). In all cases agreement and implementation of management was confounded by multi-national jurisdiction. Burd (1985) points out that the first management action in the North Sea was not taken until 15 years after the need was first discussed. Saetersdal (1980) and Jakobsson (1985) record that the fishery on Atlanto-Scandian herring collapsed without any restriction having been imposed on it. Anthony and Waring (1980) state with respect to Georges Bank that 'although the managers of ICNAF[1] accepted the proper issues...and set constraints to reach these goals, the TACs (total allowable catches) were set more for economic than for conservation reasons. The 1972 TAC was two to three times that recommended by the scientists.'

EXAMPLES OF NEAR COLLAPSE

West of Scotland herring

This fishery experienced a large increase in catch, with resulting high F and decreasing SSB, similar to adjacent fisheries in the early 1970s. Fishery closure for two years (1979, 1980) arrested the decline in SSB at 62 000 t and with improved recruitment the stock quickly rebounded (Anon. 1995*a*).

Southern Gulf St. Lawrence (4T) herring

The introduction of a purse seine fishery to Canada's Gulf of St. Lawrence in the mid-1960s led to a rapid increase in landings from approximately 34 000 t taken by gill-nets between 1935 and 1966 to an average of 166 000 t per year from 1967–1972. The stock, which is estimated to have been over one million t in 1969, declined rapidly to below 20 000 t before severe restriction on TAC and a series of years of good recruitment reversed the decline (Iles 1993; Claytor *et al.*. 1995).

[1] ICNAF: International Commission for Northwest Atlantic Fisheries

Bay of Fundy (4WX) herring

This population of herring from spawning areas off southwest Nova Scotia was at a low level in the mid 1970s, and the fishery was threatened by a sharp drop in the fishmeal market — the mainstay of the industry. In this case, a novel, early co-management structure involving elimination of the meal market, implementation of an individual boat-quota system, and marketing improvement, allowed rational rebuilding of both the stock and the fishery (Iles 1993; Stephenson *et al.* 1993).

Recovery from collapse or near collapse

While recovery has taken place in all of these stocks, the rate has differed. Predictably, those in which SSB decline was arrested early recovered faster than those in which the SSB was reduced to extremely low levels. In the cases of the North Sea, Norwegian spring-spawning herring and Georges Bank herring, SSB appears to have been reduced to well below 1% of its peak, and recovery was slow. In these cases, there seems to have been recruitment failure which took longer to overcome. Management measures implemented in re-opening of the European fisheries included minimum size-at-first-capture, seasonal closures, area closures, TACs, and effort restrictions (Jakobsson 1985). In the case of Georges Bank, extended jurisdiction placed it under control of the USA and Canada, who left it closed to fishing until satisfied with the degree of recovery. Recovery, once it began, was quite rapid — especially in the absence of fishing.

Stocks which have 'survived'

Stocks which have not collapsed (survived) have not, it seems, been spared major fluctuations in population size. The major Baltic herring stock (Subdivisions 25–29 and 32), the largest stock which has been free of recent signs of collapse, has exhibited SSB fluctuations of almost 50% in the past two decades with a relatively stable fishery as a result of recruitment fluctuation (Anon. 1995*a*). Smaller, especially more northerly, stocks exhibit even greater fluctuations of the same sort.

Of considerable concern, even in stocks which have not shown signs of collapse, is the potential erosion of stock components within apparently appropriate overall levels of fishing mortality. The collapse of the North Sea stock is recognized as having occurred in a progressive way through different sub-units. In the case of the Bay of Fundy herring stocks on the east coast of Canada (NAFO 4WX), the focus of fishing on individual spawning grounds has resulted in progressive erosion (stock collapse) of individual sub-units of the stock complex, even when the fishery has operated within an apparently suitable overall TAC. In order to protect the diversity of spawning units, management has evolved towards a regime based on spawning area.

Current issues related to stock collapse and survival

The Atlantic herring fisheries span considerable ranges in most major elements of management: underlying biological stock structure (from single stocks to complexes with a number of spawning units); degree of biological understanding; fishery type and intensity (low effort by passive gear through intense effort by sophisticated vessels); assessment forum and structure (ICES,[2] Canada and USA); management context (single-nation control through trans-boundary situations to complex jurisdictions); management structure, and suite of management measures.

Management deficiencies have been attributed, in whole or in part, to most of these elements, as well as to natural dynamics such as environmental or ecosystem changes. The following section highlights current issues which are considered to be critical to prevention of future stock collapse.

a) Biological basis for assessment and estimation of stock size

Stephenson (1991, 1992), in reviewing aspects of the assessment of Atlantic herring concluded that there was a number of difficulties and shortcomings related to biological bases for management. Three of these were:

i. *Lack of understanding of stock structure and the need for increased emphasis on care of sub-stocks.* Stock structure remains a major issue in many management units. Few assessment units are made up of a single population. Most are complexes (e.g. the North Sea) which contain a number of spawning areas. Many of these spawning areas seem to be discrete on time scales that are relevant to management — and thus require greater consideration in management.

   Stock assessments and management assume discreteness of stock management areas. However, catch statistics, and even survey indices, are confounded by extensive migration and mixing of herring — particularly at the juvenile and adult stages. Recent assessments of US Atlantic herring, for example, have been on an aggregate stock complex (5Y+5Z) because of the inability to partition catch and survey information (Anon. 1995*b*). This further complicates the care of stock components.

ii. *Weak/poor abundance indices and lack of recruitment indices.* Herring assessments generally suffer from weak independent tuning surveys. Abundance indices based on larvae are not age-structured, and are therefore of limited analytical use. Trawl indices, of the type commonly used for groundfish, have been successfully applied in only a few cases. The most promising methodology appears to be acoustic survey - if survey design and concurrent sampling of species composition and age structure are sufficient. Herring enter fisheries at a young age, and there is generally a lack of recruitment information.

iii. *Deficiencies in the estimation of stock size.* Retrospective analyses indicate some deficiencies in estimation of stock size during the major collapses cited above (e.g. Saville and Bailey 1980; Sinclair *et al.* 1985). These are not considered to have been major elements of collapse and substantial methodological improvement has been made in the decades since. Recent advancements in presentations of uncertainty and risk will improve the use of stock size estimates in effective management.

[2] ICES: International Council for the Exploration of the Sea

b) Environmental and ecosystem effects

The long history of herring fisheries prior to recent mechanization demonstrates that these stocks are prone to major fluctuations. Jakobsson (1985) records that historically 'great herring periods alternated with periods when no herring fishing took place'. This periodicity is demonstrated well by fluctuations in the long catch histories of the Atlanto-Scandian (Devold 1963; Jakobsson 1985) and Bay of Fundy (weir) fisheries. In some cases environmental changes may have played a part in the major collapses — but generally their influence appears to have been minor compared with the effects of the fisheries (Saville 1980). There are undoubtedly several mechanisms by which environmental influences are acting — at various stages in the life history — but these have been difficult to demonstrate. In some cases there has been a high degree of coincidence between adjacent stocks, indicating the influence of broad environmental factors.

Increasing discussion recently has focussed on the role of herring within multi-species systems. Multi-species interactions, the position of herring as a major forage species, and changes in natural mortality may have been involved in stock abundance fluctuations to a greater degree than was appreciated in the past — and will certainly become more dominant issues in management in future.

c) Continued increase in mechanization

Increasing mechanization was perhaps the most significant contributing factor to the major collapses outlined above. Jacobsson wrote in 1985: 'There is no doubt that the existing modern and efficient fleets could fish the TACs allocated to them many times each year', and that situation remains. The improvements in sonar, navigation and capture technology continue to make herring more vulnerable to fishing. In the words of a respected captain in the Bay of Fundy herring purse seine fishery, 'we are all highliners now'. The schooling behaviour of herring makes it possible for modern fleets to maintain very high catch rates until the stock is depleted (Jakobsson 1985).

A potentially positive aspect of increasing mechanization is the improved opportunity to make use of technological observation from the fleet on distribution and relative abundance, as industry surveys of populations for management purposes.

d) Management and regulation

i. *Negative fishery behaviour.* Herring fisheries, as is the case with most others, have suffered from violations of regulations and detrimental fishing practices. There has been overshooting of TACs, erroneous catch recording, dumping at sea, and species mis-identification (e.g. juvenile herring as sprat in the mixed industrial fishery of the North Sea). The 4WX herring fishery, for example, has had a persistent problem with erroneous reporting and recording of landings, which precluded an analytical assessment for a few years and resulted in two major catch reconstructions covering most of the time series of the modern fishery.

ii. *Thresholds and targets.* The Aberdeen conference concluded that pelagic stocks seem to be more susceptible to collapse than other (groundfish) stocks, and require their own management reference points, with a lower rate of fishing mortality. While there has been some progress in this regard (including establishment of minimum thresholds for some ICES stocks), there is a great need for continued development of biological thresholds and targets of use to management for herring.

iii. *Pro-active management.* Management of herring stocks has failed in in the past in large part 'because one has usually done too little too late' (Jakobsson 1985). Improved management of the herring stocks demands a more effective dialogue among all vested interests, and greater readiness on the part of management authorities to react quickly, particularly in situations of adverse and unpredictable changes. The move to co-management increases the likelihood of such progress (Stephenson and Lane 1995; Lane and Stephenson, 1997).

Collapse has most often been considered to be the result of a complex rather than a single cause, making it difficult to anticipate and to arrest. This review shows the need to continue to work toward a management approach which considers multiple factors, and which can react quickly to complex and changing conditions. This would be facilitated by a management system which included structured decision making, on the basis of all available data, in a climate of uncertainty and with explicit consideration of risk.

REFERENCES

Anon. (1995*a*). Report of the ICES Advisory Committee on Fishery Management, 1994. *ICES Cooperative Research Report* No. 210. International Council for the Exploration of the Sea: Copenhagen.

Anon. (1995*b*). Environmental assessment and preliminary fishery management plan for the Atlantic herring fishery of the north-western Atlantic. (US Department of Commerce, National Marine Fisheries Service, Northeast Fisheries Center, Woods Hole, Mass).

Anthony, V C, and Waring, G (1980). The assessment and management of the Georges Bank herring fishery. *Rapports et proces-verbaux des Reunions, Conseil International pour L'Exploration de la Mer* **177**, 72–111.

Bailey, R S (1990). Changes in the North Sea herring population over a cycle of collapse and recovery. In 'Long-term variability of pelagic fish populations and their environment'. Proceedings of the International Symposium, Sendai, Japan, 14–18 November 1989. (Eds T Kawasaki, S Tanaka, Y Toba and A Taniguchi) pp. 191–98 (Pergamon Press).

Bailey, R S, and Steele, J H (1992). North Sea herring fluctuations. In 'Climate variability, climate change and fisheries'. (Ed M H Glantz) pp. 213–30 (Cambridge University Press: Cambridge).

Brett, J R (Ed) (1985). Proceedings of the Symposium on the Biological Characteristics of herring and Their Implications for Management. *Canadian Journal of Fisheries and Aquatic Sciences* **4** (Suppl. 1), 1–278.

Burd, A C (1985). Recent changes in the central and southern North Sea herring stocks. *Canadian Journal of Fisheries and Aquatic Sciences* **42**(Suppl. 1), 192–206.

Claytor, R, Dupuis, H, Mowbray, F, Nielson, G, LeBlanc, C, Poulin, L, Bourque, C, and MacDougall, C (1995). Assessment of the NAFO Division 4T Southern Gulf of St. Lawrence herring stock, 1994. *Department of Fisheries and Oceans Atlantic Fisheries Research Document* 95/69. DFO, Canada.

Corten, A (1986). On the causes of the recruitment failure of herring in the central and northern North Sea in the years 1972–1978. *Journal du Conseil. Conseil International pour l'Exploration de la Mer* **42**, 281–94.

Devold, F (1963). The history of the Atlanto-Scandian herring. *Rapports et procès-verbaux des Reunions, Conseil International pour L'Exploration de la Mer* **154**, 98–108.

Dragesund, O, Hamre, J, and Ulltang, O (1980). Biology and population dynamics of the Norwegian spring-spawning herring. *Rapports et proces-verbaux des Reunions, Conseil International pour L'Exploration de la Mer* **177**, 43–71.

Iles, T D (1993). The management of Canadian Atlantic herring fisheries. In 'Perspectives on Canadian marine fisheries management'. (Eds L S Parsons and W H Lear) pp. 123–50. *Canadian Bulletin of Fisheries and Aquatic Sciences* **226**. National Research Council, Canada.

Jakobsson, J (1985). Monitoring and management of the Northeast Atlantic herring stocks. *Canadian Journal of Fisheries and Aquatic Sciences* **42**(Supp 1), 207–21.

Kawasaki, T, Tanaka, S, Toba, Y, and Taniguchi, A (1990). 'Long-term Variability of Pelagic Fish Populations and their Environment.' Proceedings of the International Symposium, Sendai, Japan, 14–18 November 1989 (Pergamon Press).

Lane, D, and Stephenson, R L (1997). Fisheries Management Science: Integrating the roles of science, economics, sociology and politics in effective fisheries management. In 'Developing and Sustaining World Fisheries Resources: The State of Science and Management. Second World Fisheries Congress, Brisbane 1996'. (Eds D A Hancock, D C Smith, A Grant and J P Beumer) pp. 177–82 (CSIRO Publishing: Melbourne).

Melvin, G D, Fife, F J, Power, M J, and Stephenson, R L (1996). The 1996 Review of Georges Bank (5) herring. *DFO Atlantic Fisheries Research Document* 96/29.

Saetersdal, G (1980). A review of past management of some pelagic stocks and its effectiveness. *Rapports et proces-verbaux des Reunions, Conseil International pour L'Exploration de la Mer* **177**, 505–12.

Saville, A (1980). Discussion and conclusions of the Symposium on the Biological Basis of Pelagic Fish Stock Management. *Rapports et proces-verbaux des Reunions, Conseil International pour L'Exploration de la Mer* **177**, 513–7.

Saville, A, and Bailey, R (1980). The assessment and management of the herring stocks in the North Sea and to the west of Scotland. *Rapports et proces-verbaux des Reunions, Conseil International pour L'Exploration de la Mer* **177**, 112–42.

Sinclair, M, Anthony, V C, Iles, T D, and O'Boyle, R N (1985). Stock assessment problems in Atlantic herring (*Clupea harengus*) in the northwest Atlantic. *Canadian Journal of Fisheries and Aquatic Sciences* **42**(5), 888–98.

Stephenson, R L (1991). Comparisons of tuning methods used in herring stock assessments in the northeast and northwest Atlantic. *International Council for the Exploration of the Sea C.M.* 1991/H:39. 8 pp.

Stephenson, R L (1992). An overview of herring assessments in the northeast and northwest Atlantic. *International Council for the Exploration of the Sea C.M.* 1992/H:26. 7 pp.

Stephenson, R L, and Lane, D E (1995). Fisheries management Science: a plea for conceptual change. *Canadian Journal of Fisheries and Aquatic Sciences* **52**, 2051–6.

Stephenson, R L, Lane, D E, Aldous, D G, and Nowak, R (1993). Management of the 4WX Atlantic herring (*Clupea harengus*) fishery: An evaluation of recent events. *Canadian Journal of Fisheries and Aquatic Sciences* **50**, 2742–57.

Wespestad, V, Collie, J, and Collie, E (Eds) (1991). 'Proceedings of the International Herring Symposium, Anchorage, Alaska, October 23–25, 1990 (9th Lowell Wakefield Fisheries Symposium).' (University of Alaska: Fairbanks).

THE BLUEFIN TUNA (*THUNNUS THYNNUS*) FISHERY IN THE CANTABRIAN SEA (NORTHEAST ATLANTIC)

J. L. Cort and V. O. de Zárate

Instituto Español de Oceanografia, Apartado 240, Santander 39080, Spain.

Summary

The artisanal baitboat fleet targeting bluefin tuna in the Cantabrian Sea (northeast Atlantic) has been in operation since the 1940s. This fishery mainly exploits young fish of the eastern Atlantic bluefin tuna stock migrating towards this area. The historical series of catch, effort and abundance indices are examined for a period of 20–30 years, together with biological, ecological and tag-recapture data. Furthermore, the current yield of the fishery is analysed taking as a reference the general state of the bluefin tuna stock assessed by the scientific committee of the International Commission for the Conservation of Atlantic Tunas (ICCAT).

INTRODUCTION

North Atlantic bluefin tuna, *Thunnus thynnus*, has for differing reasons, become one of the most controversial species fished in recent years. These reasons include the high commercial value of its meat, particularly on the Japanese market, and the precarious state of this fish resource in the western part of the Atlantic Ocean since the end of the 1970s. Therefore, the International Commission for the Conservation of Atlantic Tunas (ICCAT) adopted restrictive fishing measures in 1981 for the countries (Canada, Japan and the United States of America) that catch this species in the western Atlantic, and implemented minimum size regulations. In spite of these measures, the expected results for bluefin stock recovery have not been attained. Consequently, in 1991 and 1993, the restrictive measures on bluefin tuna fishing were further intensified for those countries.

The pessimistic outlook for the resource in the western part of the Atlantic Ocean contrasts with the apparently healthy state of the stocks on the eastern side of the Atlantic, including that of the Mediterranean, where catches have increased year after year and which, in recent years, have recorded their highest values ever.

This paper discusses a bluefin baitboat fishery that is one of the most traditional fisheries in the eastern Atlantic and, one which has been carried out for almost 50 years by artisanal fishers of northern Spain and southern France.

MATERIALS AND METHODS

This paper uses official statistical data published in ICCAT Statistical Bulletins. These data correspond to the total catches, to the number of vessels and to the fishing effort that is expended by the French and Spanish fleets in the Cantabrian Sea and the Bay of Biscay.

Information relative to stock structure is taken from size data used by the ICCAT bluefin tuna stock assessment groups. Size/age keys were obtained from the reading of fin ray spines from tuna samples from the Cantabrian Sea fishery and applied to these data.

The monitoring of fishing effort on age 2 fish was carried out using nominal data, as days at sea, from the baitboats that direct theirs fishing activities exclusively at bluefin tuna. The series includes the most reliable data.

For the development of abundance indices for the age 2 group, nominal catch data, as numbers of fish, were used, taking into account that fishing effort was stable during the 23 years considered in the data series, and that the area fished remained constant.

Tagging/recovery data have been published in numerous research works on this species. In this paper more emphasis has been placed on those data that correspond to the Spanish tagging cruises initiated 20 years ago.

BLUEFIN TUNA IN THE CANTABRIAN SEA AND THE BAY OF BISCAY

Area of study

The area included in this study is located in the eastern Cantabrian Sea, known as the Bay of Biscay, between approximately 43°N–47°N and 1°W–5°W.

Fishing and the fishing fleet

The first experiments on bluefin tuna fishing by baitboat were carried out in the early 1950s, and since then fishing by troll, the traditional fishing method up to that time, has gradually diminished.

The development of the fleet over the past 30 years shows (Table 1) that very distinct variations can be observed between the fleets of the countries involved in this fishery. Although activity by the French (St. Jean de Luz) fleet has declined, and is currently almost non-existent, the Spanish (Fuenterrabia) bluefin fishery has continued.

Table 1. The bluefin tuna fleet of Fuenterrabía (Spain) and St. Jean de Luz (France)

	St. Jean de Luz			Fuenterrabía		
Year	Gross t (40–79)	Gross t (80–120)	TOTAL	Gross t (40–79)	Gross t (80–120)	TOTAL
1965	26	26	52	–	–	30
1966	33	31	64	–	–	28
1967	34	12	46	–	–	29
1968	25	9	34	–	–	30
1969	25	0	25	26	6	32
1970	34	0	34	22	6	28
1971	33	0	33	23	7	30
1972	28	0	28	24	10	34
1973	30	0	30	27	10	37
1974	25	0	25	25	9	34
1975	30	0	30	23	8	31
1976	20	0	20	18	5	23
1977	18	0	18	19	6	25
1978	17	0	17	18	7	25
1979	10	0	10	13	9	22
1980	8	0	8	13	9	22
1981	6	0	6	13	8	21
1982	5	0	5	11	7	18
1983	9	0	9	13	5	18
1984	8	0	8	13	8	21
1985	9	0	9	13	12	25
1986	9	0	9	13	7	20
1987	8	0	8	12	8	20
1988	8	0	8	12	7	19
1989	8	0	8	9	10	19
1990	7	0	7	7	10	17
1991	7	0	7	9	7	16
1992	7	0	7	6	8	14
1993	7	0	7	7	18	25
1994	6	0	6	7	10	17
1995	6	0	6	7	19	26

In addition to these vessels, numerous vessels comprising the fleet that targets albacore, but which catch bluefin tuna, primarily of age 1 fish (4–7 kg), also need to be be taken into consideration.

Fishing seasons

The Cantabrian Sea is an area where juvenile bluefin tuna congregate during the spring, summer and autumn. The appearance of the first bluefin tuna schools takes place in May when the water temperature barely reaches 15°C. The most abundant group, that of age 2 fish, whose weight is from 7 to 8 kg, is the first to arrive in the area. Meanwhile, within the water masses between 17 and 20°C, the remainder of the juvenile groups, i.e. up to age 5 that weigh from 40 to 50 kg, and small adults that can weigh more than 100 kg, start arriving in the Bay of Biscay from west of the Iberian Peninsula. Their final recruitment to the area concludes in August, when the water temperature exceeds 20°C. The area where fish congregate remained constant for all the years of observation (Cort 1990). Then, starting in autumn, when the water temperatures start to cool, bluefin tuna leave the Cantabrian Sea and emigrate towards the south.

Tagging and migration

Tagging cruises carried out by the Spanish Institute of Oceanography (IEO) in the Bay of Biscay started in 1976 and ended in 1991. Using baitboat vessels, some 5663 bluefin tuna were tagged, mainly juveniles of ages 1–3 (60–119 cm). Recaptures total 361 (up to December, 1995), of which 17 were transAtlantic migrations (recovered in the fisheries off the eastern coast of the United States), seven were recovered in the Mediterranean Sea, and six in the eastern Atlantic, far from the Bay of Biscay. The remainder, up to the total of 361, were recovered in the Cantabrian Sea and the Bay of Biscay in various years. From the time-area analysis of tag recaptures the conclusion was reached that juvenile bluefin tuna return to the Bay of Biscay at least until they reach age 5. They arrive from the west and north-west of the Cantabrian Sea. Their dispersal within the Bay of Biscay is wide and their movements are of minor magnitude at first, but increase as the fishing season advances towards autumn.. The return migration to the wintering areas takes place towards the west and north-west.

From tag recaptures obtained (Fig. 1), interaction is shown between the Bay of Biscay fishery and the juvenile summer fishery in the north-western Mediterranean (Gulf of Lion) and the juvenile summer fishery in the eastern and western Atlantic (USA fishery).

During the change from the juvenile phase to the spawning stage, further interaction is shown between the Bay of Biscay fishery and the winter fishery of the Canarian-Saharan area; the spawning fishery of the area close to the Strait of Gibraltar; the spawning fishery in the Tyrrhenian Sea; the western Mediterranean (Balearic Islands area) and the Gulf of Lion, and the Japanese winter longline fishery in the central area of the Atlantic Ocean (Cort 1994).

A summary (Table 2) of the bluefin tagging cruises in the eastern Atlantic and Mediterranean, since the beginning of such research activities, shows that Spain, with 12 129 bluefin tuna tagged in the Gulf of Cadiz, Bay of Biscay and western

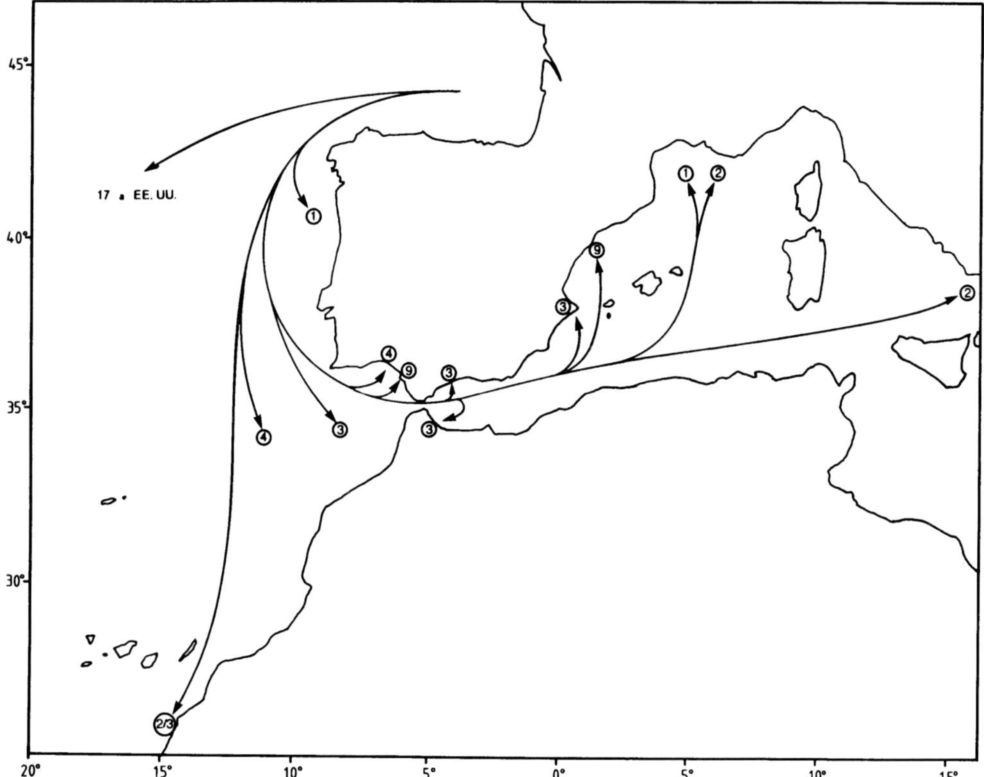

Fig. 1. Recaptures and migrations of bluefin tuna tagged in the Bay of Biscay. Numbers in circles indicate years at liberty.

Table 2. Bluefin tagging experiments in the eastern Atlantic and Mediterranean

Country	No. tagged	Recaptures Total	(%)	Atl.→Med.	Med.→Atl.	Trans-Atl.	Ref. (*)
Italy (1911–1912)	30	–	–	–	–	–	(1)
Portugal (1931–1935)	107	–	–	–	–	–	(2)
Norway (1957–1962)	237	25	(10.5)	–	–	–	(3)
Spain (Atl.) (1960–1967)	312	21	(6.7)	6	–	–	(4)
Portugal (1960)	50	–	–	–	–	–	(5)
Italy (1963–1968)	296	6	(2.0)	–	–	–	(6)
France–Portugal (1967–1972)	34	5	(14.7)	–	–	2	(7)
CIESM(**) (1972)	8	–	–	–	–	–	(8)
Morocco (Atl.) (1972–1978)	195	21	10.8	–	–	–	(9)
Spain (Atl.) (1976–1991)	5 663	361	(6.4)	11	–	17	(10)
Spain (Med.) (1977–1994)	5 173	167	(3.2)	–	14	–	(11)
France (Med.) (1991)	8	–	–	–	–	–	(12)
Italy (1991)	16	–	–	–	–	–	(12)
TOTAL	12 129	606	(5.0)	17	14	19	

* (1) Heldt 1927; (2) Rodríguez-Roda 1980; (3) Hamre 1963, 1964; (4) Rodríguez-Roda 1969, 1980; (5) Vilela 1960; (6) Arena and Li Greci 1970; (7) Aloncle 1973; Mather III *et al.* .1973; (8) CIESM 1972; (9) Lamboeuf 1975; Brêthes 1978, 1979; Brêthes and Mason 1979; (10) Cort 1990; (11) Rey and Cort 1986; (12) Cort and de la Serna 1993.

** International Commission for the Scientific Exploration of the Mediterranean Sea.

Mediterranean, has provided 92% of the total sample and 90.6% of the recaptures. In addition, it can be seen that 97% of the recaptures occur in the same part of the Ocean where the tagged fish were released (eastern Atlantic or Mediterranean). Of the 606 bluefin recoveries, 19 (3.1%) were recovered in the area close to the eastern coast of the United States.

STATUS OF THE BLUEFIN TUNA STOCKS IN THE ATLANTIC OCEAN

Bluefin tuna catches in the Bay of Biscay fishery

The historical series of catches, from 1960 up to the present, show (Table 3 and Fig. 2) significant stability in the catches over the past 35 years.

The age distribution (Table 4) of bluefin tuna catches in this area between 1966 and 1995 shows 5 299 211 bluefin tuna were caught. Summarizing Table 4 and assigning an average weight for each age group (Cort 1990), a general view is obtained of how the catch is distributed, with the total weight reaching 61 521 t (Table 5).

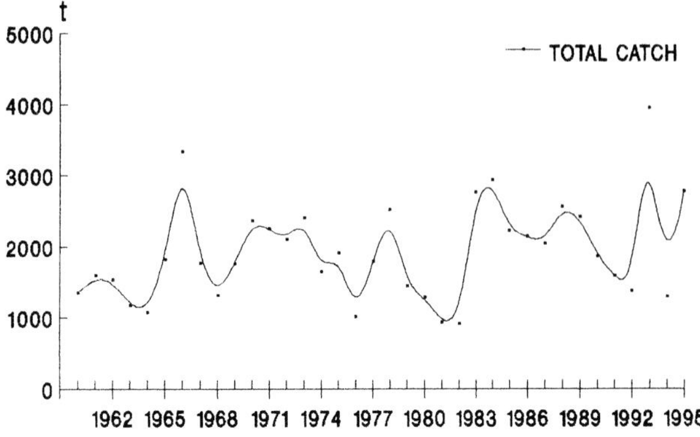

Fig. 2. Total annual bluefin tuna catches (t) by baitboats in the Bay of Biscay.

Table 3. Bluefin tuna catch and effort (baitboat)

Year	Catch (t)	Catch (Age 2) number	Effort (Age 2) days at sea
1960	1 352	–	–
1961	1 598	–	–
1962	1 537	–	–
1963	1 178	–	–
1964	1 079	–	–
1965	1 820	–	–
1966	3 335	88 389	–
1967	1 771	28 730	–
1968	1 314	25 083	–
1969	1 761	90 327	–
1970	2 362	115 412	–
1971	2 254	74 172	–
1972	2 103	53 319	–
1973	2 410	87 114	1 700
1974	1 648	67 129	1 730
1975	1 911	12 8804	1 585
1976	1 013	56 327	1 050
1977	1 791	79 082	1 350
1978	2 522	66 970	2 030
1979	1 448	13 318	1 280
1980	1 286	32 255	1 430
1981	938	38 247	1 130
1982	914	33 033	1 050
1983	2 758	53 040	1 200
1984	2 931	147 017	1 050
1985	2 228	97 035	1 430
1986	2 147	52 905	1 090
1987	2 045	124 700	1 450
1988	2 561	66 370	1 110
1989	2 417	97 311	1 540
1990	1 860	42 107	1 250
1991	1 589	74 699	1 170
1992	1 376	74 367	1 150
1993	3 940	198 733	1 390
1994	1 294	24 330	760
1995	2 772	83 133	1 710

Table 4. Age structure of bluefin tuna (numbers of fish) in the Cantabrian Sea (1966–1995) (baitboats)

	Number/Age cohort								
Year	**1**	**2**	**3**	**4**	**5**	**6**	**7**	**8**	**Total**
1966	9 357	88 389	88 157	19 190	772	298	0	16	206 179
1967	3 536	28 730	56 641	6 973	371	0	68	280	96 599
1968	10 463	25 083	15 880	15 543	2 294	24	14	32	69 333
1969	34 554	90 327	16 640	5 074	1 211	0	0	0	147 806
1970	52 283	115 412	29 075	8 590	4 364	1773	876	0	212 373
1971	5 953	74 172	12 349	5 963	8 784	3 468	1 628	471	112 788
1972	516	53 319	18 975	2 828	6 802	3 151	1 728	1 035	88 354
1973	20 044	87 114	6 145	2 489	5 929	4 592	2 767	2 185	131 265
1974	1 513	67 129	28 786	3 380	1 948	542	278	92	103 668
1975	40 005	128 804	13 008	6 581	1 958	550	59	3	190 968
1976	1 237	56 327	12 880	2 323	1 398	408	134	19	74 726
1977	30 744	79 082	18 056	5 431	333	167	96	7	133 916
1978	165 820	66 970	9 813	14 252	3 736	380	49	34	261 054
1979	38 234	13 318	15 798	14 676	3 036	628	148	18	85 856
1980	41 962	32 255	10 649	3 908	3 116	1 218	472	29	93 609
1981	80 078	38 247	3 902	1 707	386	109	82	36	124 547
1982	23 728	33 033	9 389	3 249	871	367	0	0	706 37
1983	357 517	53 040	18 346	1 468	166	45	0	0	430 582
1984	56 331	147 317	40 968	5 144	1 180	222	97	35	251 294
1985	17 738	97 035	47 611	7 821	1 262	252	11	0	171 730
1986	202909	52 905	7 341	7 755	772	185	24	1	271 892
1987	47 237	118 488	8 289	4 430	2 007	204	60	12	180 727
1988	284 444	66 370	10 383	2 265	855	448	88	20	364 873
1989	180 988	97 311	9 177	3 315	296	140	39	9	291 275
1990	111 618	42 107	20 781	6 816	1 425	306	10	1	183 064
1991	52 230	74 699	10 339	5 534	1 457	210	62	8	144 539
1992	45 229	74 367	14 022	1 547	676	39	46	0	135 926
1993	41 993	191 900	59 016	16 145	2 498	258	7	0	311 817
1994	24 782	24 330	17 453	8 745	505	10	4	1	75 830
1995	165 366	83 133	30 636	2 645	150	51	0	3	281 984
1966–95	2 148 409	2 200 713	660 505	195 787	60 558	20 045	8 847	4347	5 299 211

The development (Table 3 and Fig. 3) of fishing effort, as days at sea, is directed at the age 2 group by the tuna fleet that exclusively fishes bluefin tuna in the Cantabrian Sea. Fishing power increased since 1977, when sonar was installed on the tuna fishing vessels, and the effect was noted immediately (1978). In spite of that, the trend in effort is one of general stability (at about 1300 days at sea/year). In 1994 and 1995 variations occurred with respect to the rest of the series. There was a decline in effort in 1994 due to the vessels leaving the fishery because of the lack of fish, whereas in 1995 the opposite was the case.

An index of abundance (Fig. 4), based on the catches of age 2 fish (Table 3), shows an increasing trend, particularly since 1979, the year in which the lowest catch of the series was recorded. It should be noted that this abundance is affected by

Table 5. Bluefin tuna general age structure, by mean weight (W), for 1966–1995

Age	W	Fish	Catch (kg)	% Fish	%W
1	5	2 148 409	10 742 045	41	17
2	11	2 200 713	24 207 843	42	39
3	20	660 505	13 210 100	12	21
4	37	195 787	7 244 119	4	12
5	55	60 558	3 330 690	1	5
6	73	20 045	1 463 285	0	2
7	94	8 847	831 618	0	1
8	113	4 347	491 211	0	1
Total		5 299 211	61 520 911	100	100

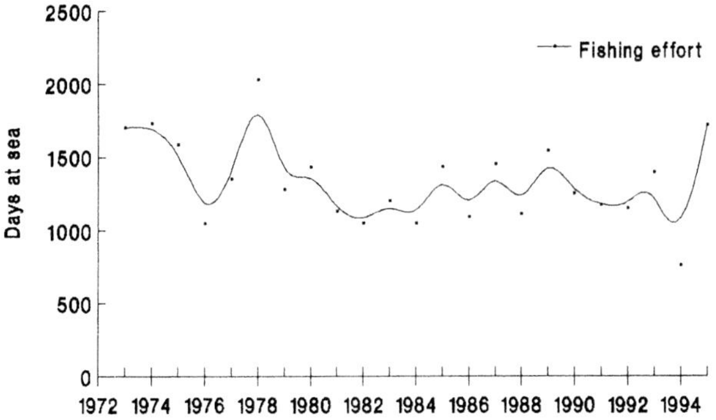

Fig. 3. Fishing effort (days at sea) in the bluefin tuna fishery for the Age 2 cohort in the Bay of Biscay.

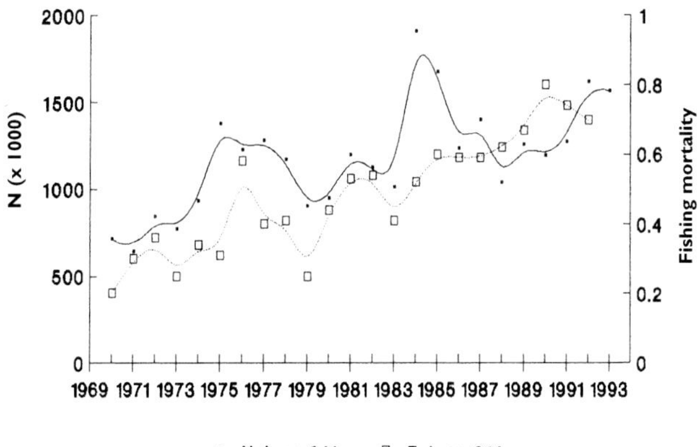

Fig. 5. Eastern Atlantic and Mediterranean bluefin tuna stock size (N) and fishing mortality (F) (ICCAT 1995).

recruitment from the Mediterranean and by the proportion of bluefin tuna that cross the Strait of Gibraltar. Therefore, the maximum catch values observed, with nine-year intervals (1975, 1984 and 1993), indicate that two years prior to each of these, there were important bluefin yields in the Mediterranean, and that in those years the rates of juvenile bluefin crossing the Strait of Gibraltar to join the young fish fisheries in the eastern Atlantic, were high. The trend obtained using catch-per-unit-effort (CPUE) as an index of abundance is the same, although in the early years of the series the fishing effort is not documented.

Using data available from the Spanish tagging cruises in the Mediterranean, it was estimated that 60–70% of the bluefin recruitments from the spawning areas in the western Mediterranean, crossed the Strait of Gibraltar to join the surface fisheries of the eastern Atlantic, mainly the baitboat fishery in the Bay of Biscay (Cort and de la Serna 1993).

Results of ICCAT bluefin tuna stock assessments (eastern stock, including the Mediterranean)

The last year in which a scientific committee meeting was held to assess the state of the eastern Atlantic bluefin tuna stocks was 1994 (ICCAT 1995). Based on the results of the committee , the following observations can be noted:

(i) Abundance of age groups 2–4 (8–40 kg) has been constantly increasing since the early 1970s. However, fishing mortality exerted on these age groups is very high (Fig. 5). In this sense, the abundance estimated for age-class 2 in the Bay of Biscay (which corresponds to a part of the stock) showed an increasing trend, which was not as marked as that obtained in the general context.

(ii) The most abundant cohort in the last 20 years was the 1982 age-class (without considering 1991, apparently more abundant). Of the 2 392 000 bluefin tuna born in that year and recruiting to the fisheries in 1993, only 7% survived fishing and natural mortality four years hence. About 75% of this mortality was exerted in the Mediterranean; the rest in the Bay of Biscay fishery.

(iii) The assessment shows a very marked decline in the abundance of age groups 4+, 5–7 and 8+. In particular, it is estimated that the numbers of the age 8+ group have decreased approximately 87% between 1970 and 1993, and about 83% between 1983 and 1993.

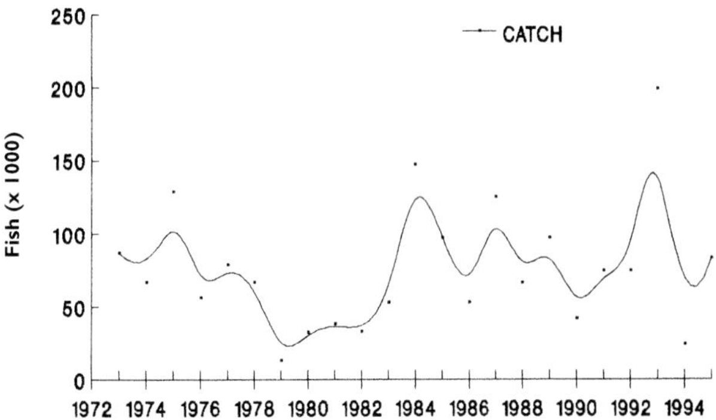

Fig. 4. Abundance of Age 2 fish for the bluefin tuna fishery in the Bay of Biscay.

Conclusions and recommendations of the scientific committee

The committee concluded that in the assessments carried out by ICCAT there are considerable discrepancies in the information used as a basis for such work. This is due for the most part to the fact that many of the countries, such as Italy, Tunisia and Turkey, to cite the most important, that fish in the Mediterranean are not members of this International Commission. In some cases there are no basic data to carry out the stock assessments. Another, more recent problem, is the appearance of the so-called uncontrollable fleets (Oriental longliners that fly flags of convenience). These factors have an effect on the quality of data that are available and used. Hence, the results should be carefully evaluated. In spite of this, the state of the eastern stock, which includes the Mediterranean, does not seem for the moment to be endangered.

The ICCAT stock assessment group recommended compliance with the bluefin tuna minimum size limit (6.4 kg). The group also recommended avoiding the catch of fish less than 0.8 kg, which during certain periods of the years are very abundant in the Mediterranean and which are targeted by the sport and commercial artisanal fisheries.

In the scientists' report there is mention of the increase in the level of bluefin tuna fishing mortality in the eastern Atlantic stock, a level which should be maintained at that of 1975 (the year that the bluefin management measures came into force). This measure has not been complied with in the Mediterranean fisheries, where there have been constant increases since then.

References

Aloncle, H (1973). Marquage de thons rouges dans le Golfe de Gascogne. *ICCAT Collective Volume of Scientific Papers* I, 445–58.

Arena, P, and Li Greci, F (1970). Marquage de thonides en Mer Tyrrhenienne. *Journées ichtyologiques*, 115–19. Rome, CIESM.

Brêthes, J C (1978). Campagne de marquage de jeunes thons rouges au large des côtes du Maroc. *ICCAT Collective Volume of Scientific Papers*, VII (2), 313–7.

Brêthes, J C (1979). Sur les premières recupérations des thons rouges marqués en juillet 1977 au large du Maroc. *ICCAT Collective Volume of Scientific Papers* VIII (2), 367–9.

Brêthes, J C, and Mason, J M Jr. (1979). Bluefin tuna tagging off the Atlantic coast of Morocco in 1978. *ICCAT Collective Volume of Scientific Papers* VIII (2), 329–32.

CIESM (1972). CIESM Rapport d'activités.

Cort, J L (1990). Biología y pesca del atún rojo, *Thunnus thynnus* (L.) del Mar Cantábrico. Publicaciones Especiales, *Instituto Español de Oceanografía*, No. 4. 272 pp.

Cort, J L (1994) Cimarrón II. *(Ediciones Servicio Central de Publicaciones del Gobierno Vasco)*. 75 pp.

Cort, J L, and de la Serna, J M (1993). Revisión de los datos de marcado/recaptura de atún rojo (*Thunnus thynnus*, L.) en el Atlántico este y mediterráneo. *ICCAT, Doc. SCRS/93/81*, 8 pp.

Hamre, J (1963). Tuna tagging experiment in Norwegian waters. *FAO Fisheries Report* (6)3, 1125–32.

Hamre, J (1964). Observations on the depth range of tagged bluefin tuna based on pressure marks on the Lea tags. *ICES, Scombriform Fish Committee* (151). 5 pp.

Heldt, H (1927). Le thon rouge (*Thunnus thynnus*, L.). Mise à jour de nos connaissances sur le sujet. *Bulletin Statitistique Oceanographique Salammbô* **7**. 24 pp.

ICCAT (1995). *Informe del período bienal 1994-95. I parte (1994)*, ICCAT Vol. 2. 305 pp.

Lamboeuf, M (1975). Contribution à la connaissance des migrations des jeunes thons rouges a partir du Maroc. *ICCAT Collective Volume of Scientific Papers* IV, 141–4.

Mather, F. J III, Mason, J, and Jones, A C (1973). Distribution of fisheries and life history data relevant to identification of Atlantic bluefin tuna stocks. *ICCAT Collective Volume of Scientific Papers* II, 234–58.

Rey, J C, and Cort, J L (1986). The tagging of the bluefin tuna (*Thunnus thynnus*) in the Mediterranean: history and analysis. *CIESM*. 2 pp.

Rodriguez-Roda, J (1969). Resultados de nuestras marcaciones de atunes en el Golfo de Cádiz durante los años 1960 a 1967. *Publicaciones Tecnicas Junta Estudios de Pesca* **8**, 153–8.

Rodriguez-Roda, J (1980). Nuevas rutas en las migraciones transatlánticas del atún. *IBERICA, Actualidad científica* **207**, 8–10.

Vilela, H (1960). Estudos sobre a biologia dos atuns do Algarve. *Boletin de Pesca* **69**, 11–34.

Are Southern Australian Shark Fisheries Sustainable?

J. D. Stevens,[A] *T. I. Walker*[B] *and C. A. Simpfendorfer*[C]

[A] CSIRO Division of Fisheries, GPO Box 1538, Hobart, Tasmania, Australia.
[B] Marine and Freshwater Resources Institute, PO Box 114, Queenscliff, Victoria, Australia.
[C] Western Australian Marine Research Laboratories, PO Box 20, North Beach, Western Australia, Australia.

Summary

Historically, shark fisheries have been characterized as 'boom or bust' operations with the specialized life history strategies of sharks making this group particularly vulnerable to overexploitation. Australia is one of only several countries with integrated research and management plans for its shark fisheries.

A south-eastern fishery targets school and gummy sharks. School sharks have a low productivity and current catches are not considered sustainable. Assessments put the present biomass at about 20–59% of virgin levels. On the other hand, gummy sharks are relatively productive and current catches are considered sustainable. A recent management measure is aimed at reducing mesh sizes and targeting a restricted range of young year classes while allowing non-entrapment of the larger, more fecund adult breeding stock.

A Western Australian fishery for dusky sharks is based on new-born fish; its sustainability will not be apparent for a few more years until the first-fished age classes recruit into the breeding population. Western Australian whiskery shark stocks are currently at 25% of virgin levels and management measures to reduce effort are in place.

Introduction

Historically, fisheries directed at individual shark species have been characterized as 'boom and bust' operations and this has been attributed to the specialized life-history strategy of sharks making them particularly vulnerable to overexploitation. Often-cited examples are the Californian soupfin shark (*Galeorhinus galeus*) and North Atlantic porbeagle (*Lamna nasus*) fisheries and several stocks of spiny dogfish (*Squalus acanthias*) and basking sharks (*Cetorhinus maximus*) (Anderson 1990; Compagno 1990; Bonfil 1994). In the case of some spiny dogfish and basking shark stocks, economic and market factors were also involved in the collapse of the fishery and it is difficult to disentangle these effects from biological factors. For basking sharks, which have an enormous range, localized depletion rather than stock collapse may provide a better explanation. Where the stock is more isolated, as in the Californian school shark, and where the fishery is intensive and expands rapidly, this is more likely to reflect a true stock collapse. In 1973, Holden published a paper questioning whether long-term sustainable fisheries for elasmobranchs were possible. While citing the collapse or marked declines of several shark fisheries he reported that in other instances, notably the ray fisheries of the north east Atlantic, catch rates had remained relatively stable over a 20-year period before starting to decline. More recent examples of declining shark fisheries such as the thresher (*Alopias vulpinus*) and angel shark (*Squatina californica*) fisheries on the West coast of the USA (Cailliet *et al.* 1993) and the school shark fishery in Brazil and Argentina (G Chiaramonte, Museo Argentino de Ciencias Naturales, Buenos Aires, Argentina, personal communication) show that the lessons of history have not been learnt. Current concerns over global overfishing of shark stocks caused largely by massive increases in the shark fin trade have highlighted the requirement for rational management of elasmobranch stocks.

The susceptibility of elasmobranchs to stock collapse is attributed to their life-history strategy with slow growth, late attainment of sexual maturity, low fecundity, low natural mortality and relatively direct relationship between the size of the breeding stock and recruitment. Teleost stocks characteristically show considerable short-term variability in recruitment due primarily to environmentally induced changes in survival of eggs and larvae. Because of their high fecundity–high mortality strategy, their capacity to compensate for population change through density-dependent mechanisms is greater than for elasmobranchs. The capacity of elasmobranch species to sustain fishing pressure will depend on the overall productivity of the species and its capacity for density-dependent change. Holden (1974) felt that these compensatory mechanisms were probably manifested through changes in fecundity while Wood *et al.* (1979) suggested changes in natural mortality were more likely.

Only three countries, Australia, New Zealand and USA, have management plans for elasmobranch fisheries. This is mainly a consequence of the lower importance (volume and price) of this group of fishes. While shark fisheries in southern Australia are only worth some $A22 million dollars, shark meat is an important relatively high-priced product on the domestic market and the fishery has a long history and is socially complex. Consequently Australia has invested in an integrated research and management plan for its principal southern shark fisheries; these comprise a south-eastern fishery for school (*Galeorhinus galeus*) and gummy (*Mustelus antarcticus*) sharks, and a south-western fishery for dusky(*Carcharhinus obscurus*), whiskery (*Furgaleus mackii)* and gummy sharks (Fig. 1).

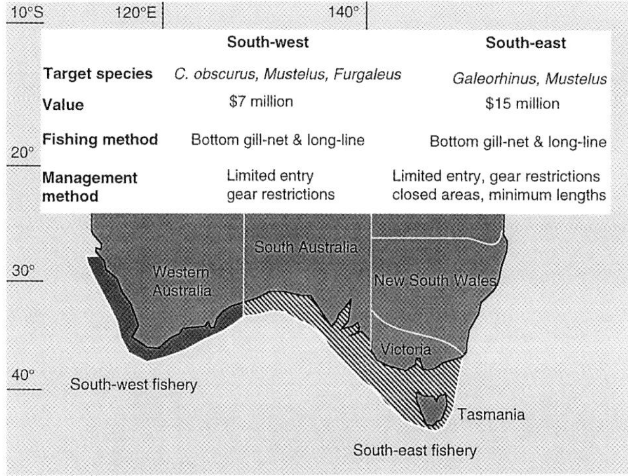

Fig. 1. Management areas and fishery synopsis for the southern Australian shark fishery.

SOUTHERN AUSTRALIA'S SHARK FISHERIES

In south-east Australia, exploitation of school sharks began in the 1920s using longlines but increased dramatically during the war years with the market for shark liver oil; catches peaked at 1307 t carcass weight (carcass weight × 1.5 = live weight) in 1949 (Fig. 2A) and the fishery spread from inshore to offshore waters (Walker *et al.* 1995). With the collapse of the liver market and re-establishment of the shark meat market, there was no further growth of the fishery until the 1960s. In 1964, gill-nets were introduced and school shark production rose rapidly, peaking in 1969 at 2105 t (Fig. 2A). Although the proportion of gummy shark in the catch was increasing (Fig. 2B), school shark still comprised the bulk of the catch. Following the ban on the sale of large school shark in 1972 because of high mercury levels, gummy shark took over as the principal species in the fishery. With relaxation of the mercury regulations, catches of both species increased, reaching a new peak in 1986 at 3779 t (Fig. 2A and B). Concerns on overfishing of school shark had been expressed as early as 1945, but escalating catches which reached

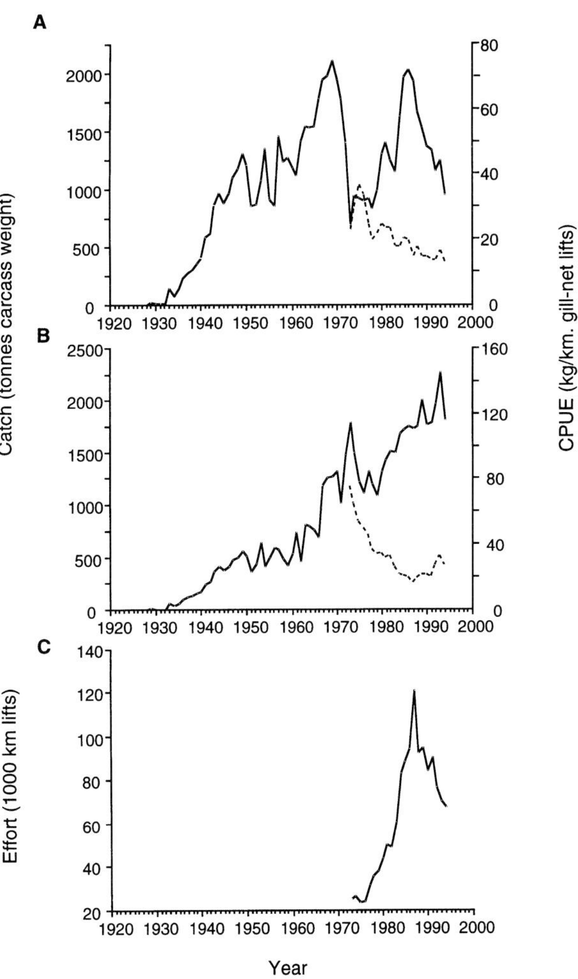

Fig. 2. South-east shark fishery. Catch (t carcass weight), total equivalent gill-net effort (1000 km, gill-net lifts) and CPUE (kg/km. gill-net lifts) by year: **A.** School shark (solid line catch, dashed line CPUE). **B.** Gummy shark (solid line catch, dashed line CPUE). **C.** Fishing effort. (Data from Southern Shark Fishery Monitoring Database, 6 June 1996).

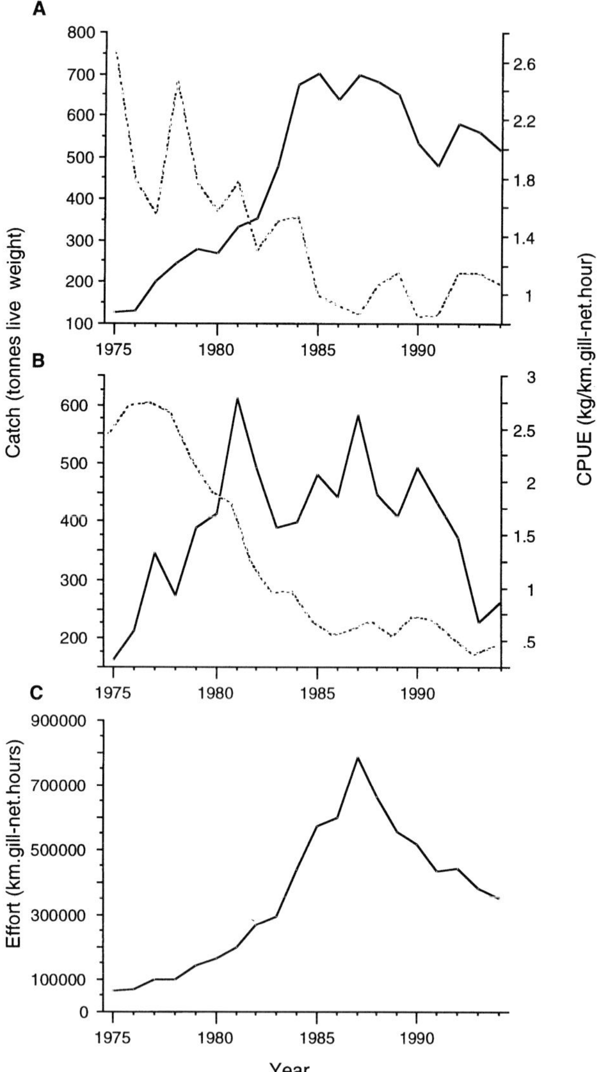

Fig. 3. South-west shark fishery. Catch (t live weight), total equivalent gill-net effort (km. gill-net. hours), and CPUE (kg/km. gill-net. hour) by year: **A.** Dusky shark (solid line catch, dashed line CPUE). **B.** Whiskery shark (solid line catch, dashed line CPUE). **C.** Fishing effort.
(Data from Fisheries Department of Western Australia Catch and Effort Database).

a second peak in 1987 (Fig. 2A), led to introduction of the management plan in 1988. Management comprises limited entry, minimum legal lengths, gear controls restricting effort in the net and hook sectors, closure of nursery areas and some inshore waters, recommended mesh sizes, and plans for a government buy-back to further reduce effort in the fishery. Research on school and gummy sharks has been carried out since the 1940s (Olsen 1954, 1984; Walker 1992, 1994). Currently a five-year strategic research plan covering further biological and ecological studies and periodic stock assessments has been developed by various government/industry shark committees and is linked to the management plan.

A shark fishery in southern Western Australia using longlines started in the 1940s but expanded rapidly during the late 1970s and early 1980s with the introduction of gill-nets (Heald 1987) (Fig. 3). The catch of dusky sharks increased from about 100 t in 1975 to about 700 t between 1985 and 1990 (Fig. 3A) and the catch of whiskery sharks increased from about 150 t in 1975 to over 600 t in 1981 (Fig. 3B). Concerns over rapidly-increasing effort and declining catch rates (Fig. 3) in the mid-1980s led to the implementation of a management plan, comprising limited entry and gear restrictions (Lenanton *et al.* 1989)

Are catches of gummy, school, dusky and whiskery sharks in southern Australia sustainable?

Gummy shark research has shown that this species is one of the more productive elasmobranchs. Sexual maturity is attained as early as three years (more usually 4–5), growth is relatively fast and maximum longevity is around 16 years (Table 1). Gummy sharks are fairly fecund, and although the average litter size is 14, fecundity increases exponentially with maternal size so that the largest sharks have litters of 30–40. Females breed each year off South and Western Australia (Lenanton *et al.* 1990), and every second year in Bass Strait. The species is demersal on the continental shelf and upper slope of southern Australia south of 30°S, ranging from the intertidal to about 300 m depth. Genetically the stock appears homogeneous across southern Australia from Bunbury, WA, to Eden, NSW, (R Ward and M Gardner, CSIRO Fisheries, Hobart), and while large individual migrations have been demonstrated, the average movement rate is low suggesting regional sub-stocks. Birth is not restricted to specific nursery areas. Stock assessments, using catch rates and deterministic age-structured models, suggest that the population has been reduced to between 40–55% of virgin biomass and that current catches are sustainable (Walker *et al.* 1995). Gummy sharks are mainly taken by gill-net and the smaller (15 cm) mesh sizes used in Bass Strait are now being encouraged elsewhere. This additional management measure is aimed at targeting a restricted range of young year classes while allowing non-entrapment of the larger, more fecund, adult breeding stock.

School sharks are much less productive than gummy sharks. They mature later at eight years for males and ten years for females. Growth is slow and the species lives to about 60 years of age. While average litter sizes are high (averaging 30), litter size increases only slightly with maternal size and females only breed every 2–3 years (Table 1). School sharks have restricted nursery areas, many of which are in inshore estuarine embayments where they are more vulnerable to human-induced mortality. Current stock assessments based on catch rates and stochastic age-structured models put the biomass at 20–59% of virgin levels (Punt and Walker 1996). However, despite concerns expressed since the 1940s, the stock has not collapsed. A number of factors, some fortuitous, may have aided the school shark. School sharks are known to spend some of their time (presently unquantifiable) in the upper water column where they are not vulnerable to the demersal gear employed in the fishery. In addition, they spend some time in oceanic waters beyond the continental slope; although not truly oceanic a number of recent tag returns of New Zealand sharks demonstrate trans-Tasman migrations suggesting a more complex stock structure. School

Table 1. Biological parameters for target species taken in the southern Australian shark fishery

Parameter	Gummy	School	Dusky	Whiskery
Age at maturity (years)	4 (male) 5 (female)	8 (male) 10 (female)	20	5
Longevity (years)	16	60	50	15–20
Litter size (number)	Average 14 Max 38	Average 30 Max 43	Average 14 Max 19	Average 19 Max 28
Size at birth (cm)	30	33.5	90	25
Breeding frequency (years)	1–2	2–3	2	2
Distribution	Southern Australia	Southern Australia/ New Zealand	*Circum*-Australia	Mainly south Western Australia
Max. depth range (m)	300	800	400	330

sharks have also received some respite during the mercury controls, and exploitation of the gummy shark has also taken some of the pressure off school shark stocks. However, current catches of school shark are not considered to be sustainable.

Dusky sharks are slow growing and do not mature until about age 20. Although females produce only 10–19 (average 14) pups probably every second year, the large birth size (Table 1)suggests that natural mortality is low. Dusky sharks are found throughout Australian waters and pupping is also known to occur on the east coast but little is known about stock structure. Fishing mortality is restricted to the 0+ and 1+ age classes in the south-west and is estimated at 30%. The adult stock is not subject to any significant fishing pressure. The sustainability of Western Australian dusky shark catches will not be apparent for a few more years until the first-fished age classes recruit into the breeding population (Simpfendorfer *et al.* 1995).

Whiskery sharks occur in temperate waters of western and southern Australia on or near the bottom at depths down to 330 m. They mature at about five years and live for 15–20 years. An average of 19 pups are produced every second year (Table 1). Assessments based on analysis of catch rates and deterministic age-structured models indicate that stocks of whiskery shark are presently overexploited and currently at about 25% of virgin levels (Simpfendorfer *et al.* 1995). The models suggest the sustainable catch is about 250–300 t.

CONCLUSIONS

Adequate stock assessment and management of shark fisheries require monitoring of catch and fishing effort, development of models incorporating the peculiarities of shark reproduction, survival, complex movement and distribution patterns, and the selectivity characteristics of gill-nets and longlines. With careful management, sharks can provide particularly stable fisheries as this group is not subject to the same recruitment variability as teleosts and invertebrates. For sharks, natural mortality is probably highest in the first few year-classes and, as fecundity usually increases with size, the largest females contribute most in terms of reproductive output. Modelling of southern shark stocks suggests there are advantages in harvesting a restricted number of younger cohorts while protecting the breeding stock. This can be achieved by adopting maximum legal lengths and by using mesh sizes which select young sharks while allowing non-entrapment of larger individuals.

REFERENCES

Anderson, E D (1990). Fishery models as applied to elasmobranch fisheries. In 'Elasmobranchs as living resources: Advances in the biology, ecology, systematics, and the status of the fisheries'. Proceedings of the second United States-Japan workshop East-West Center, Honolulu, Hawaii 9–14 December 1987. (Eds H L Pratt, S H Gruber and T Taniuchi). *NOAA Technical Report NMFS* 90. 518 pp.

Bonfil, R (1994). Overview of world elasmobranch fisheries. *FAO Fisheries Technical Paper* 341. 119 pp.

Cailliet, G M, Holts, D B, and Bedford, D (1993). A review of the commercial fisheries for sharks on the west coast of the United States. In 'Shark conservation'. Proceedings of an international workshop on the conservation of elasmobranchs held at Taronga Zoo, Sydney, Australia, 24 February 1991. (Eds J Pepperell, J West and P Woon). Zoological Parks Board of New South Wales. 147 pp.

Compagno, L J V (1990). Shark exploitation and conservation. In 'Elasmobranchs as living resources: Advances in the biology, ecology, systematics, and the status of the fisheries'. Proceedings of the second United States–Japan workshop East-West Center, Honolulu., Hawaii 9–14 December 1987 (Eds H L Pratt, S H Gruber and T Taniuchi). *NOAA Technical Report NMFS* 90. 518 pp.

Heald, D I (1987).The commercial shark fishery in temperate waters of Western Australia. *Fisheries Report 75* (WA Fisheries Department). 71 pp.

Holden, M J (1973). Are long-term sustainable fisheries for elasmobranchs possible? In 'Fish stocks and recruitment'. (Ed B B Parish). *Rapports et Proces-Verbaux des Reunions Conseil International pour l'Exploration de la Mer.* **164**, 360–67.

Holden, M J (1974). Problems in the rational exploitation of elasmobranch populations and some suggested solutions. In 'Sea Fisheries Research'. (Ed F R Harden-Jones) chapter 7, pp. 117–37 (Paul Elek: London).

Lenanton, R, Millington, P, and Smyth, C (1989). Shark and chips. Research and management into southern WA's edible shark fishery. *Western Fisheries* **1**(3), 17–23.

Lenanton, R C J, Heald, D I, Platell, M, Cliff, M, and Shaw, J (1990). Aspects of the reproductive biology of the gummy shark, *Mustelus antarcticus* Gunther, from waters off the south coast of Western Australia. *Australian Journal of Marine and Freshwater Research* **41**, 807–22.

Olsen, A M (1954). The biology, migration, and growth rate of the school shark, *Galeorhinus australis* (Macleay) (Carcharhinidae) in south-eastern Australian waters. *Australian Journal of Marine and Freshwater Research* **5**, 353–410.

Olsen, A M (1984). Synopsis of biological data on the school shark *Galeorhinus australis* (Macleay 1881). *FAO Fisheries Synopsis* 139, Rome. 42 pp.

Punt, A E, and Walker, T I (1996). Stock assessment and risk analysis for 1996 for the school shark *Galeorhinus galeus* (Linnaeus) off southern Australia using a spatially aggregated age-structured population dynamics model. Shark FAG document SS/96/D9.49 pp.

Simpfendorfer, C, Hall, N, Lenanton, R, and Donohue, K (1995). Fisheries status and stock assessment for the southern and west coast demersal gillnet and demersal longline fisheries. Unpublished report, Fisheries Department of Western Australia, (3), November 1995.

Walker, T I (1992). A fishery simulation model for sharks applied to the gummy shark, *Mustelus antarcticus* Günther, from southern Australian waters. *Australian Journal of Marine and Freshwater Research* **43**, 195–212.

Walker, T I (1994). Fishery model of gummy shark, *Mustelus antarcticus,* for Bass Strait. In 'Proceedings of Resource Technology '94. New Opportunities Best Practice'. University of Melbourne, Melbourne, 26–30 September 1994. (Ed I Bishop). pp 422–38 (The Centre for Geographic Information Systems and Modelling: The University of Melbourne).

Walker, T I, Stone, T, Battaglene, T, and McLoughlin, K (1995). Fishery assessment report: the southern shark fishery 1994. (Australian Fisheries Management Authority: Canberra).

Wood, C C, Ketchen, K S, and Beamish, R J (1979). Population dynamics of spiny dogfish (*Squalus acanthias*) in British Columbia waters. *Journal of the Fisheries Research Board of Canada* **36**, 647–55.

Prediction of Snapper (*Pagrus auratus*) Recruitment from Sea Surface Temperature

M. P. Francis, A. D. Langley[A] *and D. J. Gilbert*

National Institute of Water and Atmospheric Research, PO Box 14-901, Kilbirnie, Wellington, New Zealand.
[A] Present address: Sanford Ltd, PO Box 443, Auckland, New Zealand.

Summary

Ten trawl surveys were conducted in the Hauraki Gulf, New Zealand, to estimate 1+ snapper year-class strength (YCS), and to develop a recruitment prediction model. After correcting for changes in catchability, 96% of the variation in YCS was explained by mean sea surface temperature during February–June of the 0+ year. YCS predictions based on temperature for the 1981–1988 year-classes were strongly correlated with YCS estimates for recruited snapper derived from commercial longline age-frequency data, thus validating the prediction model. Snapper recruit to the adult population at 3–5 years, enabling prediction of YCS 3–5 years before recruitment. The 1993 and 1994 year-classes were predicted to be weak or below average in strength, and the 1995 and 1996 year-classes were predicted to be above average. Forward projection of the snapper population model using predicted recruitment suggests that biomass will increase slowly at current catch levels.

Introduction

The number of fish recruiting to an exploited population may vary from year to year. The effect of such recruitment variation on the size of a population depends on the magnitude of that variation, and the relative abundances of recruiting and recruited fish. If recruitment variation is large enough to have a significant impact on population size, estimates of recruitment variation and a means for predicting future recruitment are essential for forecasting stock sizes.

The New Zealand snapper, *Pagrus auratus* (Bloch and Schneider 1801), is most abundant over mud and sand in less than 50 m depth. The snapper population along the north-east North Island coast (SNA 1) supports New Zealand's most valuable inshore demersal fishery. Commercial catches (5000–6000 t per year) have been constrained since 1986 by Individual Transferable Quotas, and the recreational catch was estimated to be 2800 t in 1994 (Gilbert *et al.* 1996). In the absence of major changes in fishing mortality, natural mortality and growth rates, recruitment variability is likely to be the main factor affecting future population sizes. For snapper, recruitment occurs at about the minimum legal size (which varies from 25 to 27 cm fork length) at an age of 3–5 years (Gilbert *et al.* 1996).

To quantify snapper recruitment variation, a series of 10 spring trawl surveys was carried out in the Hauraki Gulf (36° 30' S,

175°E), the main SNA 1 nursery area. Using the data from the first seven surveys, Francis (1993) found a very high correlation ($r = 0.97$) between 1+ snapper year-class strength (YCS) and mean April–June sea surface temperature (SST) during the 0+ year. The present study aimed to update the temperature-recruitment model proposed by Francis (1993), to re-evaluate the magnitude of recruitment variability, and to predict future recruitment for use in SNA 1 population forecasting models. A summary of the main differences between the study of Francis (1993) and the present study is given in Table 1.

Methods

Ten stratified random trawl surveys were carried out in the Hauraki Gulf by the 28 m research vessel *Kaharoa* in October–November of consecutive years from 1984 to 1994, with the exception of 1991. The aim was to estimate the YCS of 1+ snapper. The survey encompassed an area of 11 720 km^2 between depths of 10 m and 150 m. About 30% of this area consists of untrawlable foul ground and an exclusion zone surrounding a submarine cable. Francis (1993) scaled up his estimates of snapper YCS to account for areas of untrawlable bottom by assuming that the density of 1+ snapper was the same over untrawlable bottom as over trawlable bottom in an adjacent stratum. In the present study, no scaling was done because the assumption of equal densities could not be tested, and only a relative index of abundance was required.

The surveys were designed to optimize the estimation of snapper abundance by using a two-phase strategy (Francis 1984). Trawl stations were randomly placed within strata, subject to the restriction that tows should be a minimum of 1 nautical mile (1.85 km) apart. All of the tows analysed were made during daylight hours. The number of trawl tows ranged from 47 to 85 per survey (mean 69).

Table 1. Differences between the snapper temperature-recruitment model reported by Francis (1993) and the revised model reported in the present study

	Francis (1993)	Present study
Number of trawl surveys	7	10
Survey area	Scaled for untrawlable ground	Not scaled for untrawlable ground
Trawl door-spread	Constant	Varies with warp length
Estimation of numbers of 1+ snapper	Mode truncation	Age–length key
Year-class strength estimation	Trawl survey index	Trawl survey index adjusted for catchability
Months used to predict YCS from SST	April–June	February–June
Prediction model validation	Qualitative comparison with age-frequency data	Quantitative correlation with estimates from age-structured model

A high-opening otter trawl with a theoretical headline height of 5–6 m and codend mesh of 30 or 40 mm was used for all surveys. The towing speed was 3.0–3.5 knots (5.6–6.5 km.h^{-1}) and the distance towed along the bottom was measured by Doppler log or Global Positioning System (GPS). Distance towed was standardized at 0.7 nautical miles (1.30 km) for all surveys except for the first, during which three tow lengths were used: 0.2, 0.5 and 1.0 nm (0.37, 0.93, and 1.85 km) depending on water depth (shorter tows were used in shallow water). Headline height was routinely measured with a headline transducer on 8 of the 10 surveys. Door-spread was not measured routinely. Instead, it was estimated from the amount of warp deployed based on gear trials (Francis *et al.* 1995). Mean survey door-spread ranged from 65–83 m. Further details of the survey methods were provided by Francis *et al.* (1995).

Most snapper caught were measured to the centimetre below fork length (FL), but for large catches only random subsamples were measured. Otoliths were removed from random sub-samples of snapper, and fish ages in years were determined as described by Davies and Walsh (1995). The theoretical birthday was taken as 1 January, after Paul (1976), and year-classes were numbered after their first full calendar year. For example, snapper spawned during the 1982–83 spawning season were defined as the 1983 year-class.

The relative number of snapper of a given age-class present in the survey area was estimated by applying an age-length key to the weighted length-frequency distribution for each survey. This procedure differed from that of Francis (1993) who defined age-classes by truncating length-frequency modes at the troughs between modes.

Sea surface temperature (SST) was measured daily at 0930 h from 1967 to 1996 at the University of Auckland Leigh Marine Laboratory climate station on the north-western side of the Hauraki Gulf. Correlation coefficients were calculated between snapper YCS and monthly mean SST for each month in the period between the onset of spawning (September) and December of the following year. To protect against Type I error inflation when calculating multiple correlation coefficients, significance levels were adjusted by the Dunn-Sidak method (Sokal and Rohlf 1981; Day and Quinn 1989). Recruitment data are usually log-normally distributed (Hennemuth *et al.* 1980), so YCS was $\log_e$ transformed before analysis.

Results

In most surveys, there were clear length-frequency modes at 8–12 cm and 14–18 cm (Francis *et al.* 1995), corresponding with the 0+ and 1+ age-classes respectively (Paul 1976; Francis 1994). In some surveys however, the 0+ and 1+ modes were almost non-existent. Most of the snapper caught were less than 5 years old, and there was high variability in the relative abundance of the age-classes within each survey (Francis *et al.* 1995).

Estimates of 1+ snapper relative YCS in the survey area are shown in Table 2. Log_e(YCS) was unrelated to Leigh SST measured at the time of the surveys (mean of the October and November monthly means) ($r = 0.32$, $P > 0.05$) indicating that variations in YCS could not be explained by temperature-induced variations

Table 2. Estimated 1+ snapper year-class strength (YCS), and year-class strength adjusted for catchability (YCS_{adj}), in the Hauraki Gulf survey area. No survey was conducted of the 1990 year-class

Year-class	YCS	YCS_{adj}
1983	0.68	1.24
1984	2.00	3.64
1985	2.79	5.08
1986	3.17	5.78
1987	1.43	2.61
1988	3.92	3.92
1989	10.04	10.04
1990	–	–
1991	3.47	3.47
1992	1.22	1.22
1993	1.39	1.39

in catchability. Correlation coefficients between $\log_e$(YCS) and mean monthly SST are shown in Figure 1. Coefficients were significant at the Dunn-Sidak adjusted probability level (P = 0.0032) for March, May and June, and the February correlation was marginally non-significant. For the present study, the 5-month period February–June during the 0+ year was chosen as a predictor of 1+ YCS, compared with April–June used by Francis (1993).

A significant positive correlation (P < 0.01) between $\log_e$(YCS) and mean February–June SST explained 79% of the variance in YCS (Fig. 2):

$$\text{Log}_e(\text{YCS}) = -16.839 + 0.969 \text{ SST} \quad (1)$$

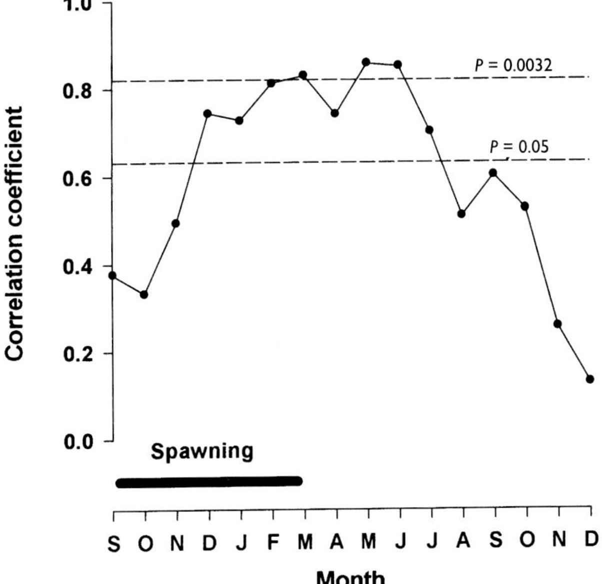

Fig. 1. Plot of correlation coefficients between $\log_e$(snapper YCS) and mean monthly sea surface temperature for the period between the onset of spawning and the following December. Each point has a sample size of 10. Dashed lines indicate significance at P = 0.05 and P = 0.0032 (the Dunn-Sidak adjusted significance level).

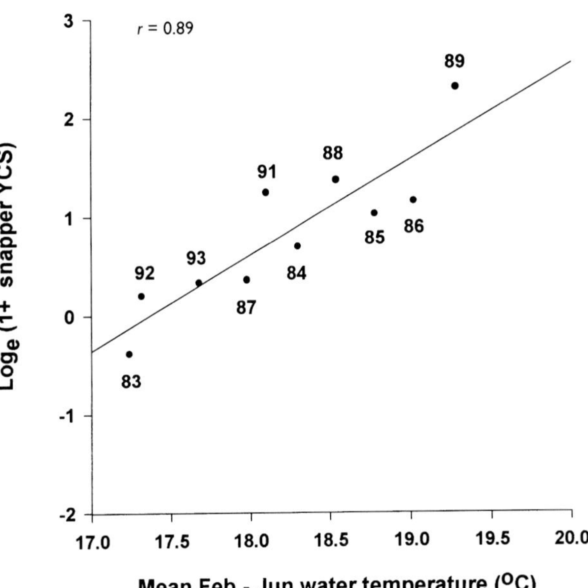

Fig. 2. Relationship between $\log_e$(snapper YCS) and mean February–June sea surface temperature during the 0+ year. Numbers indicate year-classes.

Inspection of the residuals from this model revealed a strong pattern: the first five surveys had negative residuals, and the last five surveys had positive residuals (Fig. 3, YCS model). The slopes of regressions fitted separately to the first five and last five surveys were not significantly different (P = 0.17). An analysis of covariance showed that, after adjustment for SST, YCS estimates for the last five surveys were greater than YCS estimates for the first five surveys by a factor of 1.82, i.e. the catchability of 1+ snapper appears to have increased by 82% half-way through the survey series. This is assumed to have been caused by alterations to the net between the 1988 and 1990 surveys which involved increases in the weight of the groundrope and the flotation of the headline (Francis *et al.* 1995). Much of the pattern in the residuals of equation (1) could be explained statistically by variations in mean headline height among surveys (r^2 = 0.64 for the 8 surveys for which headline height was measured). However, the magnitude of the change in headline height (an increase of 18% between the means for the first five and the last five surveys) does not seem sufficient to account for an 82% mean increase in catch rates of 1+ fish. Furthermore, the increased weight of the groundrope may have improved the seabed contact of the net, and reduced escapement of 1+ snapper underneath the gear. Because the reasons for the increase in catchability are uncertain, and because headline height measurements are available for only 8 of the 10 surveys, headline height has not been incorporated as an independent variable in the recruitment model.

Instead, the YCS estimates for the first five surveys were adjusted (YCS_{adj}) for low catchability by multiplying them by 1.82 (Table 2). A linear regression of $\log_e(YCS_{adj})$ against February–June SST explained 96% of the variation in YCS_{adj}:

$$\text{Log}_e(\text{YCS}_{adj}) = -17.045 + 0.997 \text{ SST} \quad (2)$$

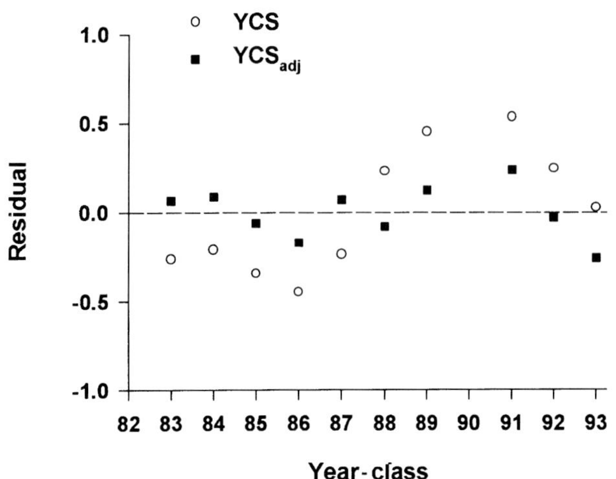

Fig. 3. Residuals from equation (1) (YCS) and equation (2) (YCS_{adj}) *versus* year-class.

The residuals for equation 2 are much smaller than for equation 1, and are randomly distributed around zero (Fig. 3).

Mean February–June SST ranged from 17.24 to 19.72°C during 1967–95. Predicted YCS for the 1967–1995 year-classes was calculated using equation (2) (Fig. 4). Mean recruitment, measured as mean $\log_e(YCS_{adj})$, was 1.47 ± 0.69. Predicted YCS showed a 3–5 year cycle which is probably related in part to the El Niño–Southern Oscillation (ENSO) cycle. The Southern Oscillation Index explains 41% of the variability in Leigh SST (Francis and Evans 1993).

The temperature-recruitment model predicts strong year-classes (more than one s.d. above the mean) in 1970, 1971, 1974, 1978, 1981 and 1989, and weak year-classes (more than one s.d. below the mean) in 1969, 1983, 1992 and 1993 (Fig. 4). The 1994 year-class is predicted to be below average in strength, and the 1995 and 1996 year-classes are predicted to be above average.

Discussion

Observed snapper YCS_{adj} varied by a factor of 8.2 times over an 11-year period (Table 2). This is half the variability reported by Francis (1993), who based his estimate on YCS rather than YCS_{adj}. The 8.2-fold variation from the present study is similar to the 7-fold variation over 3 years, and 10-fold variation over 5 years, reported in two Japanese studies of *P. auratus* (= *P. major* in Matsumiya *et al.* 1980 and Kojima 1981).

Hauraki Gulf YCS estimates for the 1981–1989 year-classes after they had recruited to the adult population were obtained by Maunder and Starr (in press) from an age-structured model that used catch-at-age data from commercial longline samples collected during 1989–94. The 1981 year-class was consistently strong as predicted by the temperature-recruitment relationship. Similarly, the 1983 year-class was consistently weak, as expected from the trawl survey estimate of 1+ YCS. After excluding the 1989 year-class (because it had not fully recruited to the commercial fishery), there was a highly significant correlation between the present YCS_{adj} predictions and Maunder and Starr's estimates of recruited YCS (r = 0.88, N = 8, P < 0.01). This validates the use of the Hauraki Gulf trawl survey estimates of YCS_{adj}, and the relationship between YCS_{adj} and SST, for predicting the strength of recruited year-classes.

Some caution is required when making predictions, because the regression model in equation (2) is 'over-fitted' relative to the precision of the input data. The amount of variation in YCS explained by the model is improbably high considering that the individual YCS estimates have coefficients of variation in the range 15–30% (Francis 1993). Nevertheless, predictions of

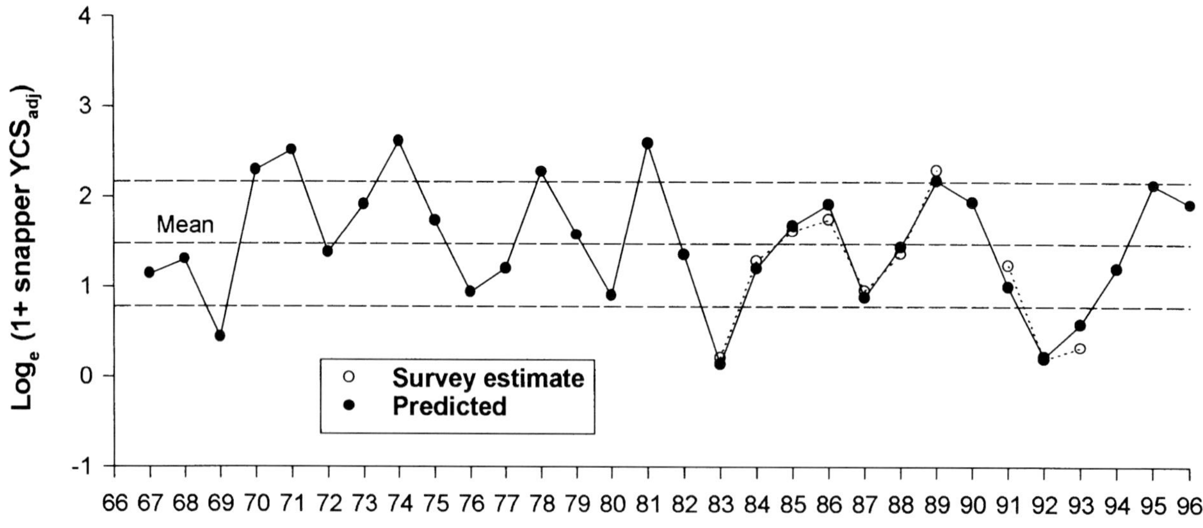

Fig. 4. Predicted and observed $\log_e$(snapper YCS_{adj}) for the period 1967–96. Predicted values were calculated using equation (2). Observed values were the trawl survey estimates of YCS_{adj} after scaling to the same mean as the equivalent predicted values. Horizontal lines indicate the mean ± 1 s.d. of the predicted $\log_e(YCS_{adj})$ estimates.

future relative YCS from both equations (1) and (2) are very similar because the SST regression coefficients for equation (1) (0.969) and for equation (2) (0.997) are nearly the same.

Trawl surveys enable estimation of 1+ snapper YCS 2–4 years before they recruit to the fishery at age 3+ to 5+. A regression model relating YCS to SST in the 0+ year extends the lead time to 3–5 years. Extreme summer and autumn SSTs usually result from strong ENSO events, which can be monitored using the Southern Oscillation Index. In some years it may be possible to forecast the likelihood, but not the magnitude, of very weak or very strong year-classes several months before the year-class has been spawned.

SST explained nearly all of the observed variation in snapper YCS. However, SNA 1 stock biomass was constant over the period 1984-1993 (Gilbert *et al.* 1996), so our results provide no information on whether YCS is also affected by stock size.

Age-structured models of the SNA 1 population have been developed and applied during the last few years (Annala and Sullivan 1996; Gilbert *et al.* 1996). Those models incorporate predicted recruitment indices for the 1993–1996 year-classes based on the temperature-recruitment relationship reported here, and stochastic recruitment based on the time series of indices shown in Figure 4 for subsequent year-classes. The SNA 1 population is projected to increase slowly over the next 10 years at both the current and proposed commercial catch quotas (4938 t and 3000 t respectively) (Annala and Sullivan 1996).

ACKNOWLEDGMENTS

We thank the Masters and crew of *Kaharoa* for their assistance with the trawl surveys, and J. Drury, K. George, J. Hadfield, B. Hartill, P. Horn, L. Paul, C. Walsh, and other Ministry of Fisheries and NIWA staff for their help in the collection and analysis of data and discussions on snapper recruitment. We also thank members of the Snapper Fishery Assessment Working Group and Ian Doonan for useful comments on this manuscript.

REFERENCES

Annala, J H, and Sullivan, K J (Compilers and Editors) (1996). Report from the Fishery Assessment Plenary, April–May 1996: stock assessments and yield estimates. (Report held at National Institute of Water and Atmospheric Research, Wellington, New Zealand).

Davies, N M, and Walsh, C (1995). Length and age composition of commercial snapper landings in the Auckland Fishery Management Area, 1988–94. *New Zealand Fisheries Data Report* **58**. 86 pp.

Day, R W, and Quinn, G P (1989). Comparisons of treatments after an analysis of variance in ecology. *Ecological Monographs* **59**, 433–63.

Francis, M P (1993). Does water temperature determine year class strength in New Zealand snapper (*Pagrus auratus*, Sparidae)? *Fisheries Oceanography* **2**, 65–72.

Francis, M P (1994). Growth of juvenile snapper, *Pagrus auratus*. *New Zealand Journal of Marine and Freshwater Research* **28**, 201–18.

Francis, M P, and Evans, J (1993). Immigration of subtropical and tropical animals into north-eastern New Zealand. In 'Proceedings of the Second International Temperate Reef Symposium'. (Eds C N Battershill, D R Schiel, G P Jones, R G Creese and A B MacDiarmid) pp. 131–136 (NIWA Marine: Wellington).

Francis, M P, Langley, A D, and Gilbert, D J (1995). Snapper recruitment in the Hauraki Gulf. *New Zealand Fisheries Assessment Research Document* 95/17. 27 pp.

Francis, R I C C (1984). An adaptive strategy for stratified random trawl surveys. *New Zealand Journal of Marine and Freshwater Research* **18**, 59–71.

Gilbert, D J, Sullivan, K J, Davies, N M, McKenzie, J R, Francis, M P, and Starr, P J (1996). Population modelling of the SNA 1 stock for the 1995–96 fishing year. *New Zealand Fisheries Assessment Research Document* 96/15. 39 pp.

Hennemuth, R C, Palmer, J E, and Brown, B E (1980). A statistical description of recruitment in eighteen selected fish stocks. *Journal of Northwest Atlantic Fisheries Science* **1**, 101–11.

Kojima, K (1981). Growth of the red sea bream (*Pagrus major*) in young stages in Yuya Bay, the Japan Sea. *Bulletin of the Seikai Regional Fisheries Research Laboratory* **56**, 55–70.

Matsumiya, Y, Endo, Y, and Azeta, M (1980). Estimation of the abundance of 0-group red sea bream in Shijiki Bay. *Bulletin of the Seikai Regional Fisheries Research Laboratory* **54**, 315–20.

Maunder, M N, and Starr, P J (in press). Validating the Hauraki Gulf snapper pre-recruit trawl surveys and temperature recruitment relationship using catch at age analysis with auxiliary information. *New Zealand Fisheries Assessment Research Document.*

Paul, L J (1976). A study on age, growth, and population structure of the snapper, *Chrysophrys auratus* (Forster), in the Hauraki Gulf, New Zealand. *Fisheries Research Bulletin* 13. 62 pp.

Sokal, R R, and Rohlf, F J (1981). 'Biometry.' 2nd edn. (Freeman: New York).

Role of Fishing and Sea Lamprey-induced Mortality in the Rehabilitation of Lake Trout in the US waters of Lake Superior

C. P. Ferreri[A] *and W. W. Taylor*[B]

[A] The Pennsylvania State University, School of Forest Resources, University Park, Pennsylvania 16802, USA.
[B] Michigan State University, Department of Fisheries and Wildlife, East Lansing, Michigan 48824, USA.

Summary

Efforts to rehabilitate lake trout populations, which historically supported an important commercial fishery in Lake Superior, have met with limited success due to continuing excessive mortality by sea lamprey predation and commercial fishing. The role of these mortalities in lake trout population dynamics was evaluated using a Leslie projection matrix to calculate the finite rate of population growth (λ). Elasticity analysis indicated that survival, particularly during the pre-reproductive ages, made the greatest contribution to the lake trout population growth rate. Using sensitivity analysis, it was found that reducing fishing mortality had a greater effect on the population growth rate than reducing sea lamprey-induced mortality by an equal percentage. Fishery managers can reduce fishing mortality of pre-reproductive lake trout by increasing the legal size limit thereby increasing their population growth rate in Lake Superior and facilitating their rehabilitation to historic levels of abundance.

Introduction

Lake trout (*Salvelinus namaycush*) populations in the United States waters of Lake Superior currently support an important tribal commercial fishery and are often referred to as the 'bread and butter' of the recreational fisheries (Hansen 1994). Historically, lake trout in Lake Superior played an important ecological role as a top predator (Christie 1974; Ryder *et al.* 1981) and supported a lucrative commercial fishery averaging two million kg annually until 1949 (Hile *et al.* 1951). Lake trout commercial yields declined precipitously during the 1950s as a result of overexploitation and sea lamprey (*Petromyzon marinus*) predation (Pycha and King 1975). In response to the loss of this valuable commercial fishery, the governments of the United States of America and Canada initiated several efforts to rehabilitate lake trout populations in Lake Superior. In the US waters of Lake Superior, stocking of hatchery-reared lake trout began in the early 1950s in an effort to supplement stocks that were declining (Lawrie and Rahrer 1973). In 1958, the Great Lakes Fishery Commission initiated a chemical control programme for sea lampreys in the tributaries to Lake Superior (Smith and Tibbles 1980). Finally, severe harvest restrictions were placed on the commercial fishery in 1962 in an effort to further reduce lake trout

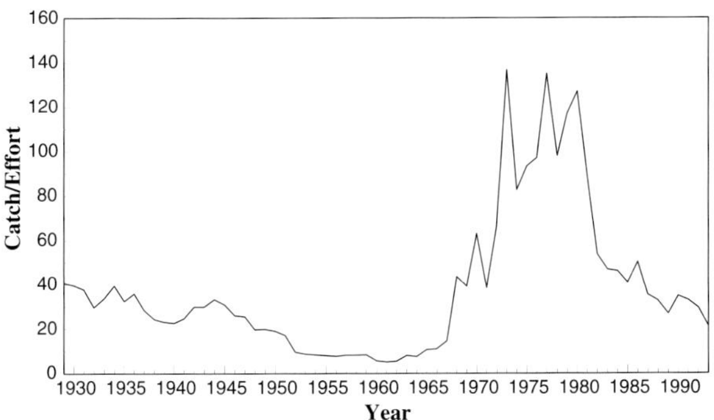

Fig. 1. Catch-per-unit-effort (1929–1993) of lake trout in the US waters of Lake Superior. (Assessment of catch/effort of lake trout adapted from Hansen et *al.* (1995)).

mortality in this lake (Krueger *et al.* 1986). Although progress toward rehabilitation of lake trout populations has been slow, abundance of wild lake trout in the US waters of Lake Superior has generally increased from 1970 to 1993 (Fig. 1; Hansen *et al.* 1995). However, lake trout population abundance in these waters still has not reached historical levels.

Researchers have hypothesized that excessive mortality due to sea lamprey predation and commercial fishing continue to hinder lake trout populations from attaining population growth rates necessary for reaching historical levels of abundance (Hansen *et al.* 1995). Total annual mortality of lake trout in the US waters of Lake Superior has often been estimated to exceed the maximum permissible level of 45% (Hansen 1994). Although it is generally agreed that lake trout total annual mortality must be decreased to achieve rehabilitation goals, it has been difficult for fishery managers to agree on the magnitude of the reduction necessary. Additionally, there has been considerable debate as to whether it would be more beneficial, in terms of lake trout population growth, to further reduce commercial fishing or sea lamprey-induced mortality. The goal of this study was to evaluate the impact of commercial fishing and sea lamprey-induced mortality on lake trout rehabilitation efforts in the US waters of Lake Superior. In particular, we were interested in determining at what ages survival was most important in determining the population growth rate of lake trout populations in Lake Superior and how variations in the survival schedule affected this growth rate.

One way to assess the relative contribution of survival at different ages to the overall population growth rate is to perform an elasticity analysis on the parameters of a Leslie matrix (Crouse *et al.* 1987; Caswell 1989). The Leslie matrix integrates information on growth, survival and reproduction into a single measure of population growth rate, λ (lambda). As such, the Leslie matrix allows for direct comparison of population growth or decline under a wide variety of conditions (Hayes and Taylor 1990). An age-based Leslie matrix was used to represent the dynamics of the current lake trout population in the US waters of Lake Superior and to evaluate the impact of various management strategies aimed at reducing lake trout mortality on its population growth rate.

METHODS

The population dynamics of lake trout in the area of Lake Superior overlapping Lake Superior Management Zones MI-4, MI-5, and MI-6 (Fig. 2; Hansen *et al.* 1995) was evaluated during the period from 1985 to 1993. This study site and time period were chosen because wild lake trout have dominated the populations in this area since 1985 and lake trout populations have been routinely monitored by the National Biological Service which provided information regarding age-specific survival (Hansen *et al.* 1995).

Leslie matrix

The Leslie matrix (Caswell 1989) parameters, age-specific survival and fecundity rates, were estimated for wild lake trout in the US waters of Lake Superior using published literature and reports. The age-specific fecundity of lake trout was determined by combining information on length at age (Ebener 1990) with information on weight-specific fecundity (Peck 1988) as described by Ferreri and Taylor (1996).

Age-specific mortality was determined as the sum of the instantaneous rates of natural, sea lamprey-induced and commercial fishing mortalities. It was assumed that natural mortality remained at the same level as in historical periods ($M = 0.2$, Sakagawa and Pycha 1971). Total mortality was averaged from 1986-1988 in the statistical district MI-4 which included most of the waters where fish had been collected ($Z = 0.88$, Ebener *et al.* 1989). The age-specific sea lamprey-induced mortality calculated by Ebener (1990) was used for that component of the total mortality. Finally, fishing mortality, which began to significantly impact the population when the fish reached 51 cm in length (Pycha 1980), was determined by subtraction of natural and sea lamprey-induced mortality from the total mortality. The length, weight, fecundity and survival-at-age used to derive the Leslie matrix which represented current lake trout populations in the US waters of Lake Superior are listed in Table 1. The finite rate of population growth, λ, was calculated as the principle eigenvalue of the Leslie matrix (Pielou 1977; Caswell 1989). A value of λ greater than 1 indicates an increasing population, equal to 1 indicates a stable population and less than 1 indicates a decreasing population (Pielou 1977). One way to interpret λ is in terms of a percent increase or decrease per year. For example, when $\lambda = 1$, there is no change per year. If $\lambda = 1.2$,

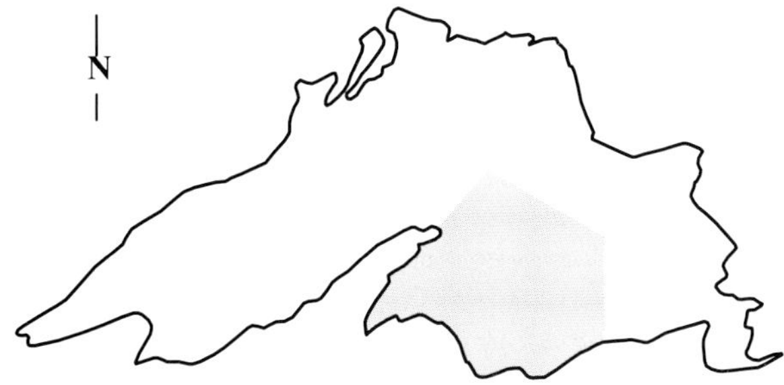

Fig. 2. Lake Superior and the location of the study site (shaded area) within the US waters.

Table 1. Parameters used to construct Leslie matrices for the US waters of Lake Superior. Length is in cm, weight is in kg, P(x) is age specific survival and F(x) is age-specific fecundity

age	cm	kg	P(x)	F(x)
1	3.58	0.00	0.82	0.00
2	23.66	0.11	0.82	0.00
3	35.40	0.37	0.82	0.00
4	43.74	0.72	0.77	0.00
5	50.20	1.10	0.76	0.00
6	57.00	1.62	0.41	8.41
7	61.00	2.00	0.41	68.78
8	63.10	2.22	0.41	174.22
9	66.50	2.62	0.41	499.21
10	69.40	2.98	0.41	811.51
11	71.10	3.22	0.41	929.34
12	74.40	3.70	0.41	1175.46
13	78.10	4.30	0.41	1479.92
14	79.60	4.56	0.41	1612.29
15	83.10	5.21	0.41	1942.04
16	85.20	5.63	0.41	2154.38
17	86.30	5.86	0.41	2270.07
18	87.31	6.07	0.41	2378.68
19	88.87	6.41	0.41	2552.95
20	90.36	6.75	0.41	2724.33
21	91.77	7.08	0.41	2892.91
22	93.12	7.41	0.41	3058.78
23	94.41	7.73	0.41	3222.04
24	95.64	8.05	0.41	3382.78

the population is increasing at a rate of 20% per year, while if $\lambda = 0.8$, the population is decreasing at a rate of 20% per year.

Relative contribution of age-specific survival

Once the Leslie matrices were parameterized, the sensitivity of the population growth rate to changes in age-specific survival and fecundity was determined using elasticity analysis, a type of sensitivity analysis that allows one to determine the proportional change in the population growth rate in response to a proportional change in age-specific survival or fecundity (Caswell 1989). The elasticity (e_{ij}) of λ with respect to the matrix element a_{ij} was calculated as:

$$e_{ij} = \frac{a_{ij}}{\lambda}\left(\frac{v_i w_j}{<v,w>}\right)$$

where w_j was the right eigenvector corresponding to λ, v_i was the left eigenvector corresponding to λ, and $<v,w>$ was the scalar product of the two vectors (deKroon *et al.* 1986; Caswell 1989). The calculated elasticities summed to 1 so that the relative contribution of matrix elements, age-specific survival and fecundity, to the population growth rate can be directly compared (deKroon *et al.* 1986; Caswell 1989). Thus, matrix elements associated with higher elasticities make a higher relative contribution to population growth rate than matrix elements associated with lower elasticity values. Comparison of elasticities allowed determination of which factors were most important in determining the population growth.

Managing fishing and sea lamprey-induced mortality

The effect of different management scenarios aimed at controlling total mortality of current lake trout populations in the US waters of Lake Superior was evaluated by changing either fishing or sea lamprey-induced mortality to a specified percentage of the current value in the matrix, and recalculating λ. Fishing mortality was allowed to vary from 0% to 200% of its current value, and the effect of sea lamprey-induced mortality was evaluated at 0%, 50%, 100%, and 200% of its current rate.

Additionally, the effects of changing the minimum legal size limit regulations on the population growth rate of lake trout were determined by eliminating fishing mortality below the hypothesized minimum legal size and recalculating λ. Currently, the minimum legal size limit for commercially caught lake trout in Lake Superior is 43.2 cm, but net efficiency (114 mm stretch mesh gill-nets) selects for lake trout 51 cm and larger in size (Pycha 1980). Thus, the effect of raising the minimum legal size limit from 55.8 cm to 76.2 cm, by intervals of approximately 5 cm, on lake trout population growth rate was evaluated.

RESULTS

Relative contribution of age-specific survival

Lake trout populations in the US waters of Lake Superior were declining by about 7% per year between 1985 and 1993. Survival of lake trout during the pre-reproductive ages (younger than age 6) made the greatest relative contribution to population growth, and survival was more important than fecundity in determining population growth rate until age 11 (Fig. 3). The relative contribution of survival and fecundity of the older lake trout in the populations was nearly identical, although it was much less than the relative contribution of either survival or fecundity at earlier ages.

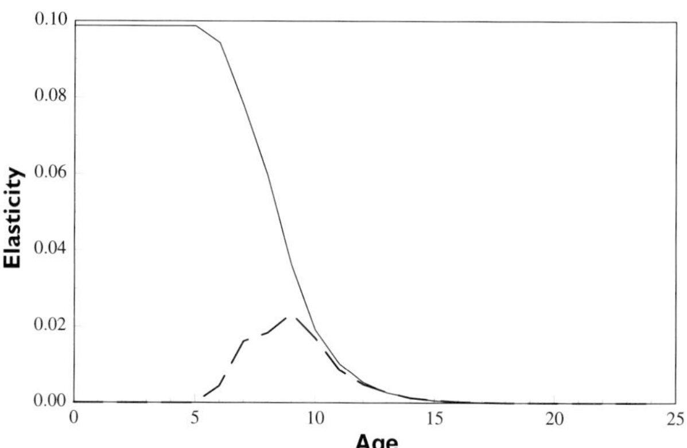

Fig. 3. Elasticities (relative contribution) of age specific survival (—) and fecundity (– –) rates to lake trout population growth in the US waters of Lake Superior.

Managing fishing and sea lamprey-induced mortality

During the period from 1985 to 1993, lake trout populations in the US waters of Lake Superior were declining at a rate of 7% per year (λ = 0.930) given 100% of the current fishing mortality and 100% of the current sea lamprey-induced mortality (Fig. 4). If commercial fishing mortality was allowed to increase further, given the current level of sea lamprey-induced mortality, the current lake trout population in the US waters would decline at even faster rates. In fact, fishing mortality would need to be reduced by approximately 30% to achieve a stable population in the US waters. In order to achieve a growing population in these waters, fishing would need to be reduced even further given the current rates of sea lamprey-induced mortality. However, if sea lamprey-induced mortality could be reduced to one-half of its current level, fishing would only need to be reduced by 15% in order to attain a stable population. Further, if sea lamprey-induced mortality could be eliminated completely, fishing mortality would only need to be reduced by approximately 1% in order to maintain a stable population. On the other hand, given current fishing mortality levels, if sea lamprey-induced mortality doubled, the population would decrease at approximately 13% per year, a rate similar to the rate of decline experienced by lake trout populations during the 1950s before sea lamprey control was effective (Ferreri and Taylor 1996). In order to achieve a stable population given 200% sea lamprey-induced mortality, fishing mortality would have to be reduced by over 60%.

Given the current individual growth and fecundity schedule of lake trout populations in the US waters of Lake Superior, reducing fishing mortality would have a greater effect on the population growth rate than reducing sea lamprey-induced mortality. For example, if the current fishing mortality rate was reduced by 20% given the current sea lamprey-induced mortality, λ would increase to 0.973. On the other hand, if the current sea lamprey-induced mortality was reduced by 20% given the current level of fishing mortality, λ only increases to 0.943; a total of 3% per year less than when fishing was reduced by the same percentage.

One management strategy to reduce fishing mortality is to raise the minimum legal size limit. In our current model, commercial fishing mortality begins to impact the population when lake trout reach a size of 56 cm. Raising the minimum legal size limit from 56 cm to 76 cm allows the population growth rate to change from decreasing at 7% annually to increasing at approximately 17% annually (Table 2). In order to achieve a growing population, the minimum legal size limit would need to be approximately 66 cm. If fishing was eliminated completely, the lake trout population in the US waters of Lake Superior has the potential to grow at approximately 18% annually given the current level of sea lamprey-induced mortality.

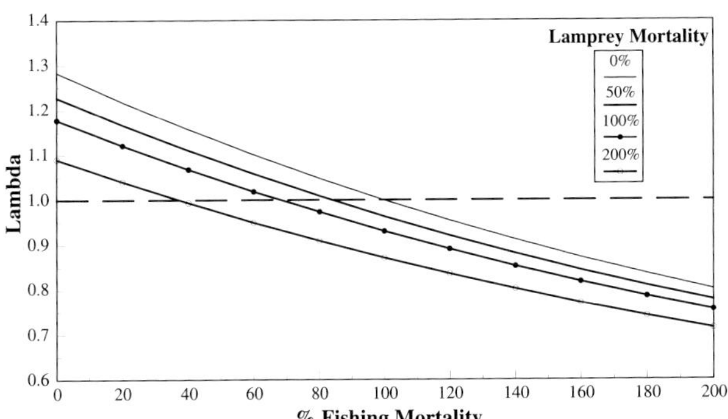

Fig. 4. Trade-off between commercial fishing and sea lamprey-induced mortalities (as percentage of current value) for lake trout populations in the US waters of Lake Superior.

DISCUSSION

Evaluation of the relative contribution of age-specific fecundity and survival rates to the population growth rate can help guide the concentration and direction of management efforts (Crouse *et al.* 1987). In general, the present analysis showed that lake trout total annual mortality needs to be dramatically reduced in order to allow the population to grow to reach rehabilitation goals. Further, elasticity analysis indicates that survival, especially during the pre-reproductive ages, was the most important factor contributing to the population growth rate of lake trout populations in the US waters of Lake Superior. Fishery managers have basically two choices by which to control lake trout mortality: they can manage fishing-induced mortality through fishing regulations, or they can manage sea lamprey-induced mortality by altering the sea lamprey control programme. Trade-off analysis has shown that greater gains, in terms of lake trout population growth, will be made by reducing fishing mortality than by reducing sea lamprey-induced mortality by equal percentages. This finding is most likely due to the fact that fishing mortality is greater than sea lamprey-induced mortality during the years right before maturity. Historical and current records indicate that lake trout tend to reach maturity at about 60 cm (Eschmeyer 1955; Ebener *et al.* 1989). Currently, lake trout are fished using 114 mm stretch mesh gill-nets which target lake trout that are approximately 56 cm in length (Pycha 1980). Sea lamprey, on the other hand, are size-selective predators that target the largest prey item available (Swink 1991). Thus, fishing mortality is greatest during several of the pre-reproductive years when survival is most important in determining the population growth rate. As a result, reducing fishing mortality under the current scenario will have a greater impact on the population growth rate than reducing sea lamprey-induced mortality.

Table 2. Change in population growth rate of lake trout populations in relationship to changes in the minimum legal size limit (cm)

Size (cm)	λ
56	0.9302
61	0.9870
66	1.0882
71	1.1600
76	1.1672
no fishing	1.1770

Given the current levels of commercial fishing mortality in the US waters, sea lamprey-induced mortality would have to be completely eliminated for lake trout to achieve stable population growth. After nearly 40 years of sea lamprey control efforts, it seems unlikely that eliminating sea lamprey from the Great Lakes is possible. In fact, sea lamprey populations in the Great Lakes are likely to increase rather than decrease. The Clean Water Act has improved water quality in many tributaries to the Great Lakes allowing sea lamprey to use more habitat within each tributary to spawn as well as colonize new streams (Moore and Lychwick 1980; Ferreri *et al.* 1995; Zint *et al.* 1995). In addition, current funding policies regarding the Great Lakes Fishery Commission's chemical control programme has made it difficult to maintain an adequate sea lamprey control programme and clearly will not allow enhancement of the programme in response to increased sea lamprey abundances (Great Lakes Fishery Commission 1993). If sea lamprey abundance increases due to improved water quality in the Great Lakes, lake trout mortality can be expected to increase as well (Ferreri *et al.* 1995). In this case, commercial fishing mortality will have to be reduced even further if lake trout populations are to achieve stable population growth.

One strategy to reduce fishing mortality is to increase the minimum legal size limit. Elasticity analysis showed that the population growth rate of lake trout in the US waters of Lake Superior was most sensitive to changes in the survival of lake trout to maturity. Although lake trout in these areas reach maturity at approximately 60 cm (Ebener *et al.* 1989), they can be legally harvested at 43.2 cm (Hansen *et al.* 1995). The selectivity of the commercial gill-nets used in these waters peaks at 56 cm (Pycha 1980). With the current size limit and gear, many of the lake trout caught by the fishery are immature. One way to limit the impact of the fishery is to raise the minimum size limit of legally-harvested lake trout to over 60 cm. This action would protect lake trout during their critical juvenile years until they reached maturity and spawned. Analysis of the effects of raising the minimum legal size limit on the lake trout population growth rate indicated that a minimum size limit of 66 cm resulted in a growing population. However, for this strategy to be effective, gear modifications would need to occur so that lake trout smaller than 66cm would not be captured.

In summary, survival, particularly during the pre-reproductive ages, makes the greatest contribution to the population growth rate of lake trout in the US waters of Lake Superior. As such, Great Lakes fishery managers should focus on developing strategies to enhance the survival of lake trout of pre-reproductive age. Reducing fishing mortality, either by increasing the minimum legal size limit to above 61 cm or by reducing effort, will have a greater effect on the population growth rate than reducing sea lamprey-induced mortality by an equal percentage.

References

Caswell, H (1989). 'Matrix population models.' (Sinauer Associates, Inc.: Massachusetts).

Christie, W J (1974). Changes in the fish species composition of the Great Lakes. *Journal Fisheries Research Board Canada* **31**, 827–54.

Crouse, D T, Crowder, L B, and Caswell, H (1987). A stage-based population model for loggerhead sea turtles and implications for conservation. *Ecology* **68**, 1412–23.

deKroon, H, Plaiser, A, Groenendael, J, and Caswell, H (1986). Elasticity: the relative contribution of demographic parameters to population growth rate. *Ecology* **67**, 1427–31.

Ebener, M P (1990). Assessment and mark-recapture of lake trout spawning stocks around the Keweenaw Peninsula area of Lake Superior. *Great Lakes Indian Fish and Wildlife Commission, Biological Services Division Administrative Report* 90-8.

Ebener, M P, Selgeby, J, Gallinat, M, and Donoforio, M (1989). Methods for determining total allowable catch of lake trout in the 1842 treaty-ceded area within the Michigan waters of Lake Superior, 1990–1994. *Great Lakes Indian Fish and Wildlife Commission, Biological Services Division Administrative Report* 89-11.

Eschmeyer, P H (1955). The reproduction of lake trout in southern Lake Superior. *Transactions of the American Fisheries Society* **84**, 47–74.

Ferreri, C P, and Taylor, W W (1996). Compensation in individual growth rates and its influence on lake trout population dynamics in the Michigan waters of Lake Superior. *Journal of Fish Biology* **49**, 763–77.

Ferreri, C P, Taylor, W W, and Koonce, J F (1995). The effects of improved water quality and stream treatment rotation on sea lamprey abundance: implications for lake trout rehabilitation in the Great Lakes. *Journal of Great Lakes Research* **21**, 176–84.

Great Lakes Fishery Commission (1993). Annual Report. (Great Lakes Fishery Commission: Michigan).

Hansen, M J (Ed) (1994). The state of Lake Superior in 1992. Great Lakes Fishery Commission Special Publication 94–1. (Great Lakes Fishery Commission: Michigan).

Hansen, M J, Peck, J W, Schorfhaar, R G, Selgeby, J H, Schreiner, D R, Schram, S T, Swanson, S T, MacCallum, W R, Burnham-Curtis, M K, Curtis, G L, Heinrich, J W, and Young, R J (1995). Lake trout (*Salvelinus namaycush*) restoration in Lake Superior, 1959–1993. *Journal of Great Lakes Research* **21**, 152–75.

Hayes, D B, and Taylor, W W (1990). Reproductive strategy in yellow perch (*Perca flavescens*): effects of diet ontogeny, mortality, and survival costs. *Canadian Journal of Fisheries and Aquatic Sciences* **47**, 921–7.

Hile, R, Eschmeyer, P H, and Lunger, G F (1951). Status of the lake trout fishery in Lake Superior. *Transactions of the American Fisheries Society* **80**, 278–312.

Krueger, C C, Swanson, B L, and Selgeby, J H (1986). Evaluation of hatchery-reared lake trout for reestablishment of populations in the Apostle Islands region of Lake Superior, 1960–1984. In 'Fish Culture in Fisheries Management'. (Ed R H Stroud) pp. 93–107 (American Fisheries Society: Maryland).

Lawrie, A H, and Rahrer, J F (1973). Lake Superior: a case history of the lake and its fisheries. *Great Lakes Fishery Commission Technical Report No 19*.

Moore, J D, and Lychwick, T J (1980). Changes in mortality of lake trout (*Salvelinus namaycush*) in relation to increased sea lamprey (*Petromyzon marinus*) abundance in Green Bay, 1974–78. *Canadian Journal of Fisheries and Aquatic Sciences* **37**, 2052–6.

Peck, J W (1988). Fecundity of hatchery and wild lake trout in Lake Superior. *Journal of Great Lakes Research* **14**, 9–13

Pielou, E C (1977). 'Mathematical ecology.' (John Wiley and Sons: New York).

Pycha, R L (1980). Changes in mortality of lake trout (*Salvelinus namaycush*) in Michigan waters of Lake Superior in relation to sea lamprey (*Petromyzon marinus*) predation, 1968-1978. *Canadian Journal of Fisheries and Aquatic Sciences* **37**, 2063–73.

Pycha, R L, and King, G R (1975). Changes in the lake trout population of southern Lake Superior in relation to the fishery, the sea lamprey, and stocking, 1950-1970. *Great Lakes Fishery Commission, Technical Report No.* 28.

Ryder, R A, Kerr, S R, Taylor, W W, and Larkin, P A (1981). Community consequences of fish stock diversity. *Canadian Journal of Fisheries and Aquatic Sciences* **38**, 1856–66.

Sakagawa, G T, and Pycha, R L (1971). Population biology of lake trout (*Salvelinus namaycush*) of Lake Superior before 1950. *Journal of the Fisheries Research Board Canada* **28**, 65–71.

Smith, B R, and Tibbles, J J (1980). Sea lamprey (*Petromyzon marinus*) in Lakes Huron, Michigan, and Superior: history of invasion and control, 1936–78. *Canadian Journal of Fisheries and Aquatic Sciences* **37**, 1780–801.

Swink, W D (1991). Host-size selection by parasitic sea lampreys. *Transactions of the American Fisheries Society* **120**, 637–43.

Zint, M T, Taylor, W W, Carl, L, Edsall, C C, Heinrich, J, Sippel, A, Lavis, D, and Schaner, T (1995). Do toxic substances pose a threat to rehabilitation of lake trout in the Great Lakes? A review of the literature. *Journal of Great Lakes Research* **21**, 530–46.

Lower Danube Fisheries Collapse and Prediction

Nicolae Bacalbaşa-Dobrovici[A] *and Radu Suciu*[B]

[A] University of Galaţi, str. Domnească 47, 6200 Galaţi, România.
[B] Danube Delta Institute, str. Babadag 165, 8800 Tulcea, România.

Summary

The lower Danube Romanian fisheries produced 20 000–30 000 t fish yearly during the middle of this century. In recent years the quantities dropped to 5000–8000 t yearly.

During the last 30 years, due to anthropomorphic factors, the lower Danube fisheries regressed. The semi-migratory species (*Cyprinus carpio, Leuciscus idus*, etc) have lost their previous importance. Sturgeon (*Huso huso, Acipenser gueldenstaedti* and *A. ruthenus*) are endangered. After the 1989 political and economical changes, overfishing and poaching have increased and there is a lack of an infrastucture capable of undertaking monitoring and surveillance.

Overfishing occurred in Romania a hundred years ago. Through political will, adequate legislation and with reduced material means, the lower Danube fisheries were restored by Dr Grigore Antipa. Presently it is necessary to use past experience. The improvement in water quality and the restoration of some of the ~half million hectares dammed in floodplain is strongly needed.

The Danube River

Location and hydrology

The Danube River is situated in Central and Southeast Europe, in a catchment basin extending between 50° and 42°N (Fig. 1). Its catchment basin of 807 000 km^2 is 1630 km long, from west to east, and is situated largely in natural forest zones except for the Panonian lowland and the Bărăgan east plains.

Climatically, over two-thirds of the Danube catchment basin lies within the *Cfb* zone, one-fifth in the *Dfb* and about one-twentieth in the *Cfa* zone; the remainder is characterized by high montane climate (Strahler 1969). The Danube is the second longest river in Europe (2857 km) and has a variable flow between 1610 and 15 540 m^3 s^{-1} at Ismail Ceatal (the first bifurcation) (Stančik *et al.* 1988).

The Danube is usually divided into three regions: the upper Danube from the source to Vienna (890 km), the middle region between Vienna and the Iron Gates, and the lower Danube down to the Black Sea (Fig. 1). More than one-third of its basin (36%) and 1075 km (38%) of its length are situated in Romania.

The lower portion (942 km) (Fig. 2) has 3 subdivisions: the Iron Gates II reservoir; from its barrage to the Prut; and then to the Black Sea. The Iron Gates II Dam is situated 80 km downstream

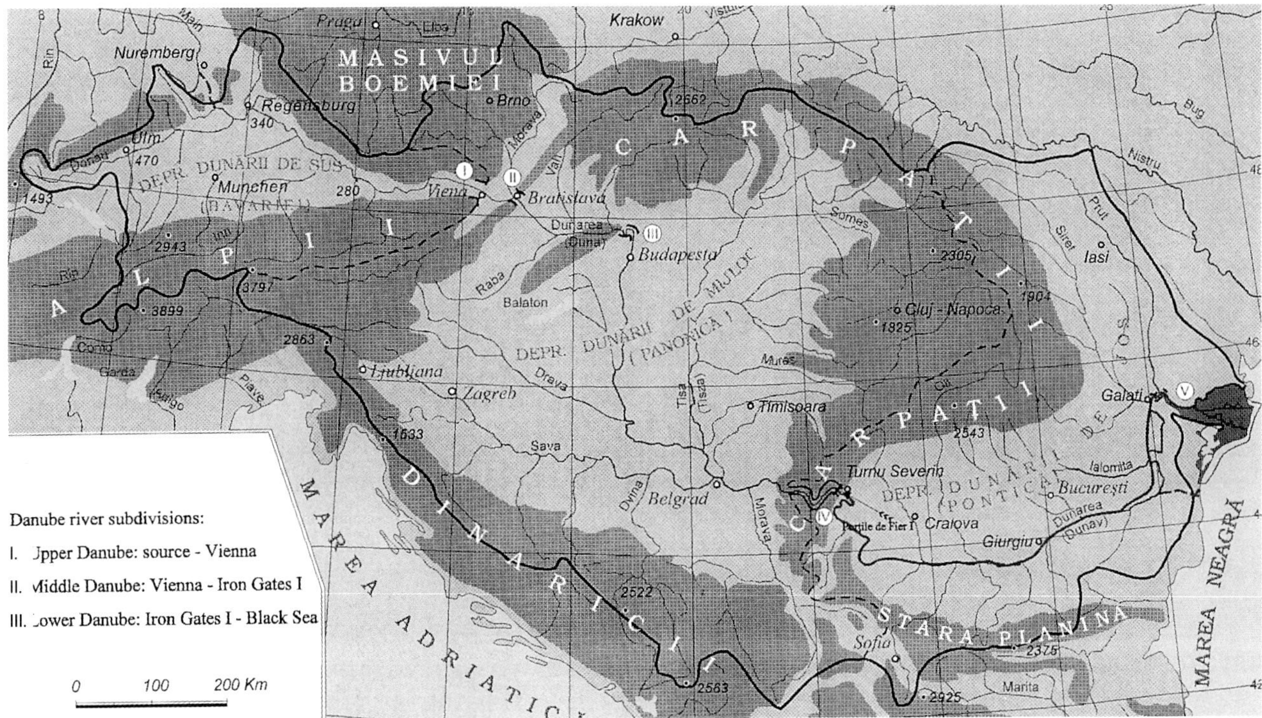

Fig. 1. Catchment basin of the Danube river. Danube river subdivisions. I. Upper Danube: source to Vienna. II. Middle Danube: Vienna to Iron Gates I. III. Lower Danube: Iron Gates I to Black Sea.

of Iron Gates I and since 1984 has cut off all fish migration. Between this dam and the delta, the river receives several tributaries including the Olt, Siret and Prut. All of these have storage reservoirs on them and as a result alluvial transport has decreased considerably in recent years (Miron 1983). The flood plain here previously exceeded 500 000 ha, but only about 10% of this area is still seasonally inundated.

The early state of the lowest reaches, the delta, when over 400 000 ha were flooded, was described in the monograph of Banu (1967). Now the Danube Delta (460 000 ha) has been transformed. Some western parts of the riverine-delta have been dammed. The lakes of the northern delta and the Razim–Sinoe complex (70 000 ha) to the south have controlled aquatic regimes. Tens of thousands of hectares of ponds have been built in the central and lower delta.

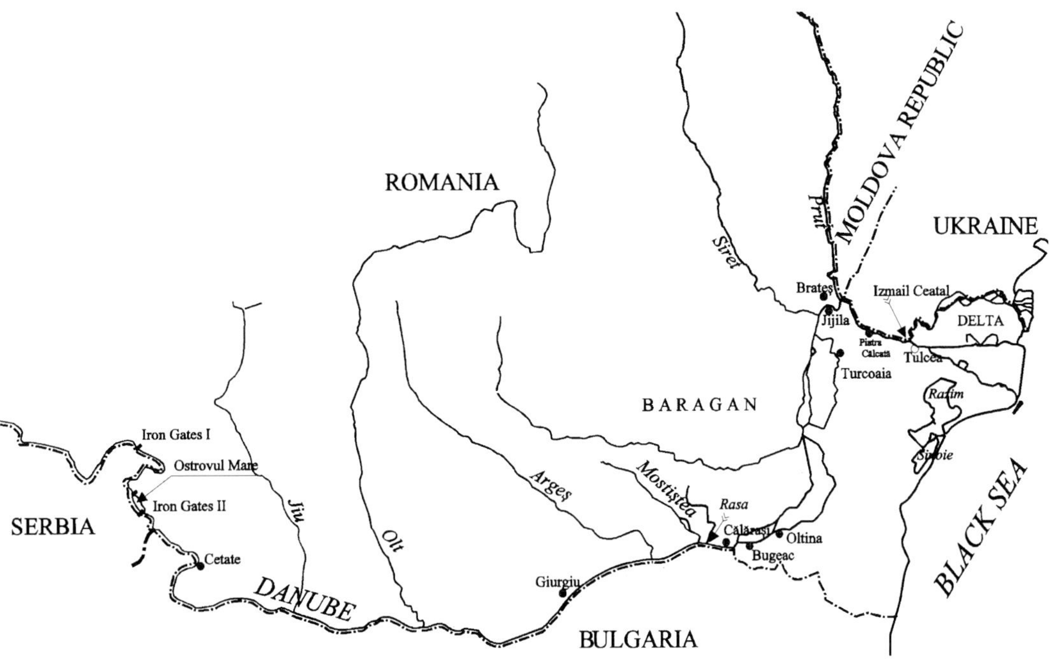

Fig. 2. Lower Danube. ● - Fish ponds.

Chemistry of the lower Danube

The Danube belongs to the hydro-carbonate series of rivers. There is evidence that a slow increase in dissolved salts is occurring from year to year (Ivanov 1982; Strainer *et al.* 1982). The conductivity is now 379–455 µS cm^{-1} (Anon.1993). The dissolved oxygen (DO_2) maintains values at the 73% limit. A severe shortage of DO_2 occurs in all shallow stagnant waters under ice cover.

The nutrients are higher than in the previous decades: the ammonium ions (0.28–0.43 mg N L^{-1}) have recorded increases, while nitrates (1.93–1.94 mg N L^{-1}), orthophosphates (0.10–0.15 mg P. L^{-1}) and total phosphorus (0.16–0.22 mg P L^{-1}) have been maintained at a relatively constant level compared with previous years (Anon.1993). In the same period, petroleum hydrocarbons and PCBs show a significant impact along the lower Danube section, at certain points along the river from 700 km up river to the delta; chromium and cadmium show higher values downstream of the Iron Gates (Equipe Cousteau *et al.* 1993). Concentrations significantly exceeding the Romanian and European Union Standards were measured for the heavy metals Fe, As, Zn, Se and Hg along the Romanian sector of the lower Danube river (Anon. 1995).

Almost all of the Danube may be considered beta mesosaprobic (Marcoci and Cure 1981). The saprobic index (2.05–2.13) is stationary. Saprobic quality : IInd class (Anon.1993). According to the pH, oxidation–reduction capacity and redox potential values and to their correlation diagram, the Danube waters are included in the 'septic water' category (Anon. 1995).

FISH AND FISHERIES OF THE LOWER DANUBE

Fish species

The Danube, with over 100 species, is faunistically the richest river of the entire zoogeographical area (Bănărescu 1964). When marine species that penetrate into the Danube discharge zone are taken into account, there are 24 fish families (15 freshwater, 2 introduced, 3 migratory and 4 brackishwater) and Petromyzontidae. About one-third of the families have direct economic importance. In the first quarter of this century, *Carassius auratus gibelio* appeared in the river, and in the 1970s became one of the dominant forms. The majority of new fish species, however, was introduced by man, including *Lepomis gibbosus,* and, in the past 30 years, the eastern Asian *Ctenopharyngodon idella, Hypophthalmichthys molitrix, Aristichthys nobilis, Mylopharingodon piceus* and the incidentally introduced *Pseudorasbora parva.* Some indigenous Danubian migratory species are no longer found (*Acipenser sturio* and *A. nudiventris*).

The number of species in the lower reaches some ten years ago was 73 (Bacalbaşa-Dobrovici *et al.* 1984). A new species is presently being described, but some marine species no longer enter the isolated lagoons.

There are three principal groupings of fish (Bacalbaşa-Dobrovici 1989):

- species living only in the river channel (*Acipenser ruthenus, Barbus barbus, Zingel streber, Z. zingel*);
- species living in the permanent water of the floodplain and the inundated zones (*Tinca tinca, Carassius carassius, Misgurnus fossilis,* etc.) presently sharply declining; and
- semi-migratory species, feeding in the marine fringes of the delta, in the delta itself or in the lower reaches of the river and performing anadromous reproductive migrations. These same species are also present in the middle reaches of the Danube, but their migrations to the freshly flooded zones are shorter, e.g. *Cyprinus carpio, Leuciscus idus, Stizostedion lucioperca* and *Silurus glanis.* After spawning, adult fish tend to return to the channel with the first movement of water off the floodplains. Fish remaining in permanent pools and backwaters are intensively fished and in shallower waters are eaten by fish-eating birds or, during severe winters, die from anoxia. Fisheries on the floodplain are based mainly on Cyprinidae and Percidae.

In the maritime zone of the delta, changes in water level, frequently associated with variations in salinity, produce changes in the fish population. Fish enter or leave the river or maritime delta or foredelta. The penetration of freshwater species into the foredelta may be hazardous. Mortalities of fish follow after onshore winds produce rapid changes in salinity.

Migrations

For the lower Danube, including the Delta, Antipa (1916) distinguished typical migrations and aperiodical movements, caused by deterioration of environmental conditions. Besides cyclical migrations of floodplain species (river–floodplain–river), in spring some upstream spawning migrations occur with semi-migratory species of the delta. During heavy floods, at the end of Spring and the beginning of Summer, fish of the delta move to the foredelta for feeding. Aperiodical migrations consist of movements to avoid anoxic conditions and in response to the changing salinities in the foredelta.

Migrations of typical migratory species are for spawning (migratory sturgeon and Clupeidae) and feeding (Mugilidae). Although migrations have remained basically the same, distances and intensities have been drastically reduced, especially for migrations upstream of the delta, where, instead of great quantities of *Cyprinus carpio,* there are smaller runs of *Carassius auratus gibelio.* In the last decade we observed in June a new migration of the eastern Asian species, especially *Hypophthalmichthys molitrix* and to a lesser extent *Aristichthys nobilis,* to the Iron Gate II dam and in the tributaries Jiu and Olt.

Anadromous species from the Black Sea (*Salmo trutta labrax, Alosa pontica* and the sturgeons *Huso huso, Acipenser gueldenstaedti and A. stellatus*) now spawn between Brăila and Ostrovul Mare. The Danube shad was not affected by river works. By contrast, the Danube sturgeon fishery has declined since the beginning of the century. Peak migrations of *Huso huso* occur during February–April and September–November. *Acipenser güldenstaedti* also has two migration peaks, February–May and August–October, and *A. stellatus* peaks in March or April–May and then usually August–September or October. The migratory clupeids *Alosa pontica* and *A. caspia tanaica* occur in the river during the Spring flood. Fry of the

Danube shad drift downstream at the water surface. The drift of the fry formerly lasted four months, but that period is now much reduced because the river is now contained by dykes. When fry reach the sea they are 60 mm or more in length and are dispersed throughout the foredelta (Lyashenko 1953). The smallest Danubian clupeid, *Clupeonella c. cultriventris*, is presently drastically reduced.

In the lower Danube, transverse migrations across the floodplain are now restricted to the small island of Brăila and the Delta system, which are not dammed. Even here migrations have been reduced, especially through a marked decrease in number of semi-migratory fish from the delta.

Productivity

The shallow nature of the floodplain in Romania permitted estimates to be made of floodplain standing stocks (Antipa 1910). These estimates remain valid even now. The quantity of fish concentrated in the lake of a closed flood-plain depression represents the whole production of that depression, and cases have been recorded of total catches of between 2000 and 3000 kg ha^{-1} in such waters. The fish production under unconstrained flood regimes depends on flood height, timing and duration, climatic conditions (especially during the spawning period), and the species structure of the community. The major subdivisions of the lower Danube floodplain had an average production of 80–110 kg ha^{-1}. Methods to determine standing stock in the main channels are less reliable.

Commercial fisheries were always important in the lower Danube. Through modified ecological conditions in the lower Danube, the creation of the Iron Gates I and II reservoirs and the decrease of fish stocks, many fishing methods are now disappearing and catch rates for most fishing gears have been reduced compared with 30 years ago.

Aquaculture

Lakes used for irrigation and fish exploitation, or for fish rearing, exist both in the Romanian Danube zone Rasa, (Mostiştea Bugeac, Oltina in its middle sector and Razim in the delta) and in the Ukrainian zone. The best fish production is obtained from the lakes exclusively designed for that purpose.

There are many pond farms in the presently dammed floodplain (Cetate, Giurgiu, Călăraşi, Turcoaia, Jijila, Brateş, Piatra Călcăta). Such farms with hatcheries provide the bulk of the fish production in the middle sector of the lower Danube zone (Bacalbaşa-Dobrovici 1985). The largest pond farms are constructed in the delta. Fish farm construction began in the 1960s; between 1970 and 1989, the area of fish polders attained 53 007 ha but, despite the commitment, aquaculture was not successful in the delta, and many polders were abandonned. Fish farming takes place in custom-built polders which occupy about 40 600 ha. They are operated by ten state-owned or joint-venture companies. The polders are state-owned and controlled by the regional council Tulcea (Baboianu and Goriup. 1995). Normal fish culture has been developed in only some 3000–5000 ha.

Fish species mentioned in the text are listed in Table 1.

Table 1. Fish species mentioned in text

Acipenseridae
Acipenser nudiventris (Lovetzky)
Acipenser ruthenus L.
Acipenser stellatus (Pallas)
Acipenser sturio L.
Huso huso L.
Clupeidae
Alosa caspia tanaica (Pavlov)
Alosa pontica (Eichwald)
Clupeonella cultriventris (Nord.)
Salmonidae
Salmo trutta labrax (Pallas)
Esocidae
Esox lucius L.
Cyprinidae
Aristichthys nobilis (Rich.)
Barbus barbus L.
Carassius carassius (L.)
Carassius auratus gibelio (Bloch)
Ctenopharyngodon idella (Val.)
Cyprinus carpio L.
Hypophthalmichthys molitrix (Val.)
Leuciscus idus L.
Mylopharingodon piceus (Rich.)
Pseudorasbora parva (Schlegel)
Tinca tinca L.
Cobitidae
Misgurnus fossilis L.
Siluridae
Silurus glanis L.
Centrarchidae
Lepomis gibbosus L.
Percidae
Stizostedion lucioperca L.
Zingel streber (Siebold)
Zingel zingel L.

HISTORY OF THE LOWER-DANUBE FISHERIES

Records of the lower Danube fisheries date back to 335 BC (Giurescu 1964). During the Middle Ages, complex regulations concerning fishing rights for the Danube floodplain lakes included provisions for their maintenance in good condition.

The first collapse of the lower Danube fisheries occurred during the end of the nineteenth century owing to the lack of legislation, and overfishing. The activity of Grigore Antipa improved the bad situation. During the first 15 years of the twentieth century, the

Danube fisheries became an example of good exploitation for a great river in its lower reaches (about 1000 km).

Deterioration of the environment became significant only after the mid-twentieth century. Damming and drainage works were extended to include about 0.5 million ha of the lower Danube. Water turbidity, increased through the intensification of sand and gravel extraction and maintained by the action of propellers of vessels, is detrimental to the fry of sturgeon as well as to the utilization by fish of strips of vegetation near the river shore. The frequent waves raised by some of the larger ships also have a negative effect on the fry.

In Romania about 3 million ha are equipped for irrigation. Irrigation is the greatest consumer of Danube water; pumping stations, near the river bank, pump the water from May to July, when some spawn and fry are also present. There is no known way to avoid pumping pelagic spawn and small fry which then die on the irrigated fields.

The environmental problems of the lower Danube are worsened through the economic activity in the whole catchment area, where the human population is roughly 86 million and many major cities and small towns along the river lack sewage treatment facilities for domestic and industrial waste. Concomitantly hydrologic modifications along the course of the river have compromised its capacity for self-purification. Presently more than 35 dams span the Danube river before it reaches the Black Sea. Pollutant loadings in the Danube river are high, but its capacity to retain nutrients and heavy metals in the floodplains is diminished. This has resulted in eutrophication of lakes in the delta with the loss of the submerged rooted plants (Pringle *et al.* 1993). Also, the north-western reaches of the Black Sea are subject to blooms of algae with partial hypoxia and even anoxia near the bottom, an important trophic region for sturgeons.

Collapse

The environmental issues have contributed to a slow and continual decline of the lower Danube fisheries. Rising investment in fisheries resulted an increase in total production until the late 1980s. Since 1989 production has fallen sharply (Fig. 3). The collapse of the active fishery in the Atlantic Ocean and Black Sea is complete. Inland fisheries are also in decline. It is difficult to give correct figures for the lower Danube catches at present. There are many errors; the old system of data collection is unable to cope with the changing situation. Some of the production obtained in large natural lakes is included in the aquaculture statistics (Cowx and Pritchard 1994). The illegal/unreported fishing, which takes regionally about half of the total catch, does not appear in the statistics, especially for the valuable species.

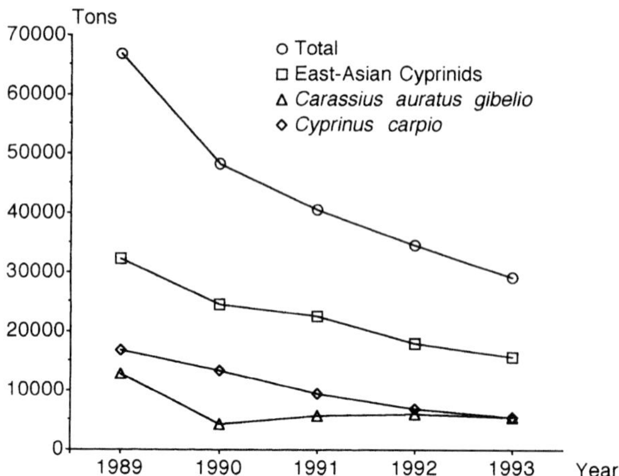

Fig. 3. Decrease in nominal catches in the inland waters of Romania (1989–1993).

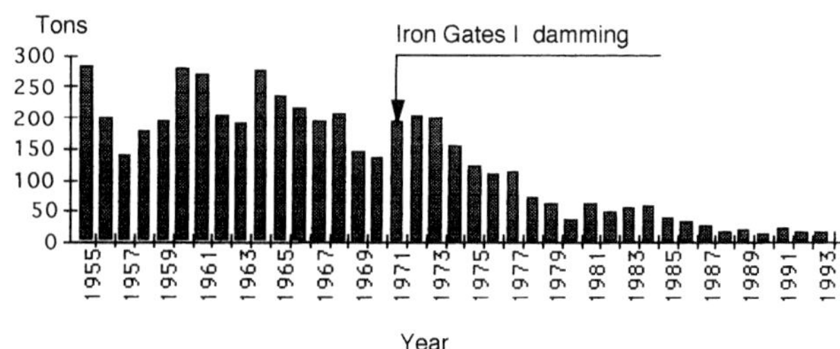

Fig. 4. Decreasing sturgeon catches at the Romanian mouths of the Danube river (1955–1993).

Although the Romanian fish statistics are very approximate, we can state about the commercial fish species that:

- sturgeon, especially *Huso huso*, *Acipenser gueldenstaedti* and *Acipenser stellatus* are endangered (Bacalbaşa-Dobrovici 1991*a*, 1991*b*, 1993, 1994) (Fig. 4);
- the runs of semi-migratory species (*Cyprinus carpio*, *Leuciscus idus*, *Silurus glanis*) are diminished (Bacalbaşa-Dobrovici 1985; 1989; Bacalbaşa-Dobrovici *et al.* 1990) (Figs 5 and 6);
- while the indigenous Danubian cyprinids give low yields, the newly-introduced eastern Asian cyprinids are caught each year in increasing quantities (Fig. 5);
- the size of spawners is decreasing (Staraş *et al.* 1994);
- the catch-per-effort is smaller;
- although for some species catches appear to be increasing, this is due to overfishing (Fig. 6).

The sharp decline of the lower Danube fisheries after 1989 is due to many factors:

- the confusing and conflicting legislation relating to fisheries;
- the splitting of the fisheries sector under the jurisdiction of several ministries;
- the lack of co-ordination of the dispersed management of fisheries;
- there is no infrastructure to undertake monitoring and surveillance except the Danube Delta Biosphere Reserve Authority;
- the conflicting interests of different groups involved in the exploitation of Danube fish;
- the dispersed but heavy poaching of spawners and fry.

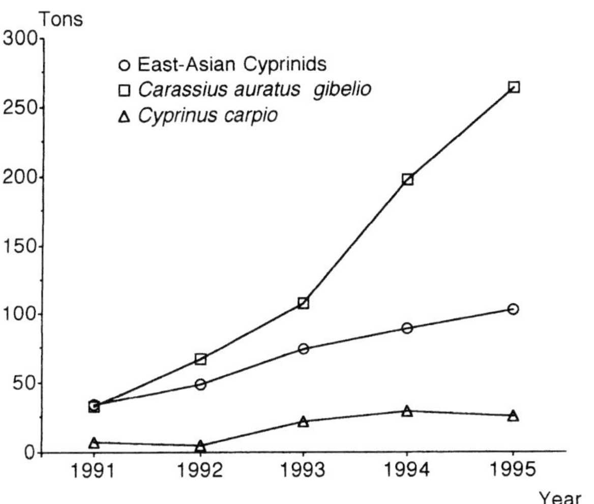

Fig. 5. Variation of catches in the most important cyprinids in the lower Danube (1991–1995).

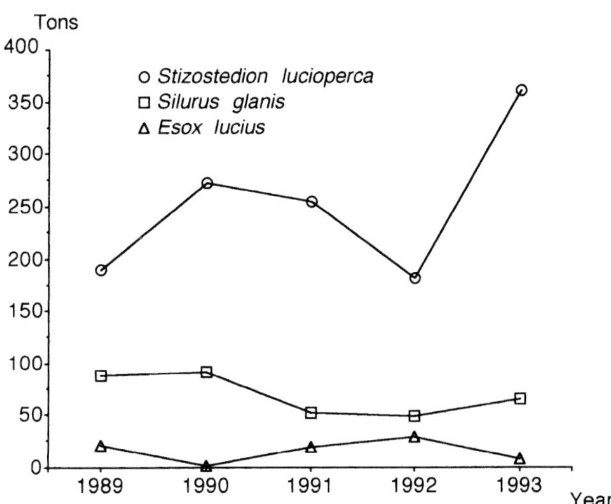

Fig. 6. Variation of nominal catches in overfishing regime in the inland waters of Romania (1989–1993).

PREDICTION

The present lower Danube fisheries collapse has two components. The first is the gradual environmental degradation that began in the 1960s. The second, beginning in the same period but accentuated in the 1990s is produced ultimately by heavy overfishing.

Environmental reconstruction is feasible through studies, international collaboration and considerable investments. But the economic situation of the country does not permit large investments for some years.

Overfishing — near collapse — occurred in Romania a hundred years ago. Through political will, adequate legislation and with reduced material means, the lower Danube fisheries were restored through the activity of Dr Grigore Antipa. However, it is true that the environmental problems did not exist at that time.

Presently it is necessary to utilize past experience of the restoration of the Danube fisheries. That is:

- a legislative framework that reflects the need to manage fisheries as part of an integrated strategy for the sustainable use of natural resources;
- the control of the fisheries through a special agency;
- a clear capture-fisheries strategy;
- controlled access to capture fishery resources by a system of licences and franchises, which include the responsibility for its management;
- surveillance and monitoring of the fisheries;
- organization and control of the fish market.

This strategy was utilized with success one hundred years ago by Dr Antipa. Now it could improve the Romanian lower Danube fisheries to the level of the late 1980s, before the collapse.

Romania has a good corp of specialists in fishery and aquaculture. We have contacted the Government and have discussed the restoration of the fisheries with the President of the country. There are moves to develop new legislation and organization of the fisheries, but with modest financial means.

Without improvement in water quality and restoration of some floodplain habitat to enhance juvenile output and semi-migratory species growth, there is little possibility that the Danube fish productions of the middle of this century will again be attained.

ACKNOWLEDGMENT

The authors wish to thank colleague Julian Nichersu of the Danube Delta Institute Tulcea, for his contribution in computerized cartography.

REFERENCES

Anon. (1993). Report regarding the Danube water quality characteristics between I - XII 1991. *Mediul înnconjurător*, Bucureşti **4**, 4–20.

Anon. (1995). EROS 2000 Project. Romanian Center of Marine Geology and Geoecology, Bucharest.

Antipa, G (1910). 'Regiunea inundabilă a Dunării.' (Carol Goebl: Bucureşti).

Antipa, G (1916). 'Pescăria şi pescuitul în România.' (Academia Română: Bucureşti, Fond Adamachi 8(46)).

Baboianu, G, and Goriup, P (1995). Management Objectives for Biodiversity Conservation and Sustainable Development in the Danube Delta Biosphere Reserve, Romania. Draft (Information Press: Oxford).

Bacalbaşa-Dobrovici, N (1985). The effects on fisheries of non-biotic modification of the environment in the East Danube river area. In 'Habitat modifications and freshwater fisheries'. (Ed J Alabaster) pp. 13–27 (FAO by Butterworths: London).

Bacalbaşa-Dobrovici, N (1989). The Danube River and its Fisheries. In 'Proceedings of the International Large River Symposium'. (Ed D P Dodge). *Canadian Special Publication on Fisheries and Aquatic Sciences* **106**, 455–68.

Bacalbaşa-Dobrovici, N (1991*a*). Die Rettung der Donauwanderstöre. *Fischer und Teichwirt* **6**, 206–7.

Bacalbaşa-Dobrovici, N (1991*b*). Statut des différantes espèces d' ésturgeons dans le Danube roumain: problèmes liés a leur maintenance. In 'Acipenser'. (Ed P Williot) pp. 185–192 (Centre National de machinisme agricole du génie rural des eaux et forêts: Bordeaux).

Bacalbaşa-Dobrovici, N (1993). The Romanian sturgeon salvation needs a consequent strategy and tactics. *Scientific Annals of the Danube Delta Institute* **II**, 221–6.

Bacalbaşa-Dobrovici, N (1994). Evolution of critical elements for the lower Danube sturgeon maintainance. *Scientific Annals of the Danube Delta Institute* **III**/1, 191–4.

Bacalbaşa-Dobrovici, N, Bănărescu, P, Nolčik, J. Janisch, R, Janović, D, Keiz, C and Weber, E (1984). Das Vorkommen einzelner Fischarten im Donaustrom und Überschwemmungsgebiet im Jahre 1983. Wissenschaftliche Kurzreferate, 24. Arbeitstagung der I.A.D., Szentendre/Ungarn, **II**, 149–56.

Bacalbaşa-Dobrovici, N, Nicolau, C, and Niţu, M (1990). Fisheries management and the hydraulic regime in the Danube Delta. In 'Management of freshwater fisheries'. Proceedings of a symposium organized by the European Inland Fisheries Advisory Commission (Eds W L T van Densen, B Steinmetz and R H Hughes) pp. 447–61 (Pudoc: Wageningen).

Bănărescu, P (1964). 'Pisces — Osteichthyes. Fauna R.P.R. XIII.' (Editura Academiei RPR: Bucureşti).

Banu, A C (Coord) (1967). 'Limnologia sectorului românesc al Dunării.' (Academia RSR: Bucureşti).

Cowx, I G, and Pritchard, M P (1994). Capture fisheries environment and resource management in Romania. (manuscript).

Equipe Cousteau, International Atomic Agency, Marine Environment Laboratory Monaco and Water Resources Research Centre (VITUKI) Institute for Water Pollution Control, Budapest, Hungary (1993). The Danube for whom and for what? (Report on Pollution Data). *Mediul înconjurător* **4**(2), 21–32.

Giurescu C (1964). 'Istoria pescuitului (i a pisciculturii în România.' (Editura Academiei RPR: Bucureşti).

Ivanov, K (1982). Über chemische und termische Belastungen der Donau und ihrer Nebenflüße. *Schweizerische Zeitschrift für Hydrologie* **44**(2), 181–93.

Lyashenko, A F (1953). Juvenile Black Sea shad biology and their quantification, (Russian). In 'Black Sea shad and biological basis of their fishery'. *Trudy Instituta Gidrobiologii* **28** pp. 85–229 (Academiya Nauk Ukrainskoj SSR: Kiev).

Marcoci, S, and Cure, V (1981). Die Dynamik der saproben Struktur des Planktons im rumänischen Donauabschnitt. 22. Arbeitstagung der I.A.D. Basel, 209–11.

Miron, I (1983). 'Lacul de acumulare Izvorul Muntelui-Bicaz.' (Editura Academiei RSR: Bucureşti).

Pringle, C, Vellidis, G, Heliotis, F, Bandacu, D, and, Cristofor, S (1993).Environmental problems of the Danube Delta. *American Scientist* **81**, 350–61.

Stančik, A *et al.* (1988). 'Hydrology of the River Danube' (Priroda: Bratislava).

Staraş, M, N`vodaru, I, and Cernişencu, I (1994). Remarks on fish stocks and fisheries status in the Danube Delta Biosphere Reserve. *Scientific Annals of the Danube Delta Institute* **III** / 1, 227–32.

Strahler, A N (1969). 'Physical geography' 3rd edn. (J Wiley and Sons, Inc.: New York).

Strainer, M, Diaconu, K, and Bradatan, M (1982). Dynamik einiger chemischer Indikatoren, erforscht im Wasser des Donaustromes. 23. Arbeitstagung der IAD, Wien. Wissenschaftliche Kurzreferate, 21–3.

Short- and Long-term Fluctuations in Catches of Elvers of the Japanese Eel, *Anguilla japonica*, in Taiwan

Wann-Nian Tzeng

Department of Zoology, College of Science, National Taiwan University, Taipei, Taiwan 10617, Republic of China.

Summary

Elvers of the Japanese eel, *Anguilla japonica*, in Taiwan have been insufficient to meet the demand for farming since the eel culture industry was developed in 1965. Annual catches of elvers were independent of demand and revealed an approximately nine-year cycle in fluctuation, with good catches in 1969–70, 1978–79 and 1990–91. Glass eels migrate along with the cold China Coastal Current in winter, and thus elvers were more abundant on the west coast than the east coast. Catches increased with decreasing temperatures (r = –0.75). Japan has the highest catch of elvers within the distribution range of the Japanese eel. Its catch revealed an approximately 11-year cycle with a peak in 1979 and then drastically declining in recent years. The rise and fall of catches of elvers from *A. japonica* correspond to those of the American (*A. rostrata)* and European (*A. anguilla)* eels. Overfishing and habitat degradation were probably the main causes of the recent declines.

Introduction

Japanese eel, *Anguilla japonica* Temminck and Schlegel, is a catadromous fish, widely distributed in the western Pacific, south from the Philippines, through Taiwan, mainland China, Korea and north to Japan (Tesch 1977). It is the most abundant of the four species of anguillid eels in Taiwan (Tzeng 1982, 1983*a*; Tzeng and Tabeta 1983). The Japanese eel spawns in the middle Pacific Ocean, approximately 14–16°N, 134–143°E to the west of the Mariana Islands (Tsukamoto 1992), in June–July (Tsukamoto 1990; Tzeng 1990). The eel larvae (leptocephali) passively migrate with the North Equatorial and Kuroshio Currents to the coastal waters of these Asian countries, metamorphose to glass eels and then become elvers in estuaries. The migration from the spawning ground to the estuaries takes approximately 5–6 months (Tzeng 1990; Tzeng and Tsai 1992). In Taiwan, elvers are harvested in the estuaries for pond culture (Tzeng 1985). Due to overfishing of elvers, construction of hydroelectric dams and pollution from domestic sewage and industrial wastes, eels have become very rare in the rivers (Tzeng *et al.* 1994, 1995). Most of the eels marketed are supplied from pond culture.

A number of studies on the biology of elvers in estuaries of Taiwan has been conducted, including on age and birthdate (Tabeta *et al.* 1987; Tzeng 1990; Umezawa and Tsukamoto

1990), stock identification and larval migration (Tzeng and Tsai 1992, 1994; Sang *et al.* 1994; Tzeng *et al.* 1994; Liu and Liao 1995), timing of estuarine immigration in relation to environmental conditions (Tzeng 1984*a*, 1985; Chen-Lee *et al.* 1994), fishing exploitation rates (Tzeng 1984*b*), and otolith microchemistry and migratory history (Tzeng 1994, 1995; Tzeng and Tsai 1994; Cheng and Tzeng 1996). However, little is known of the causes of long-term fluctuations in catches of the Japanese eel elvers (Tzeng 1983*b*, 1986). This paper attempts to elucidate the causes of fluctuations in catches of the Japanese eel elvers, particularly those in Taiwan.

DATA SOURCE AND ANALYSIS

Daily catches of Japanese eel elvers in the Shuang-chi River estuary in northeastern Taiwan were collected from 1980 to 1995. The number of elvers collected was recorded by species, site, fishing gear, fishing date and hours. Rainfall and runoff were also recorded. Water temperature has been measured since 1991 and its relation to the catch-per-unit of fishing effort of elvers (CPUE, number of elvers per day) was analysed. The fishers divide the elvers into white- and black- types according to the degree of pigmentation on the tail. The white was *Anguilla japonica*, the black included three species, *A. marmorata*, *A. celebecensis* and *A. bicolor pacifica* (Tzeng 1983*a*, 1983*b*). *A. japonica* was dominant in the catches (Tzeng 1983*a*) and its catch data were used in this study.

In addition, the annual catches of elvers in the coastal waters of Taiwan in the fishing season from November to March (Tzeng 1983*b*), 1965 to 1995, were calculated from the monthly catch data of the Taiwan Fisheries Yearbook (Anon 1965–1995). These data were compared with the demand of elvers for eel farming. The demand was estimated by taking the total area of eel ponds in Taiwan and multiplying by the mean stocking density (12 000 elvers/ha) (Tzeng 1983*b*). The annual catch during the period from 1965 to 1995 was also analysed with spectral analysis to identify cyclic fluctuations.

Catches of elvers in Japan, Korea, mainland China and Taiwan, and the amount of elvers stocked in culture ponds in these countries and in Malaysia, were also analysed from 1972 to 1995. Japanese eel elvers do not occur in Malaysia, but are imported for farming. The data on catch and stocking of elvers in these countries were obtained from the report of the Japan Eel Importers' Association (Anon 1996). The long-term catch data of the elvers in Japan were analysed with spectral analysis and also compared with those of American and European eels in the Atlantic Ocean to determine if the fluctuations in catches of elvers were a global phenomenon.

RESULTS

Distribution of elvers in relation to coastal current

The oceanic current systems are different between the east and west coasts of Taiwan. Catches of elvers were more abundant in north and west coastal waters than in east coastal waters (Fig. 1). This indicates that glass eels migrate with the cold China Coastal Current to Taiwan. This has been validated from the immigration of elvers in the coastal waters reaching a peak during the period of the lowest winter water temperatures (Tzeng 1985), and the daily ages of elvers arriving at estuaries along the west coast being older in the south than in the north (Cheng and Tzeng 1996).

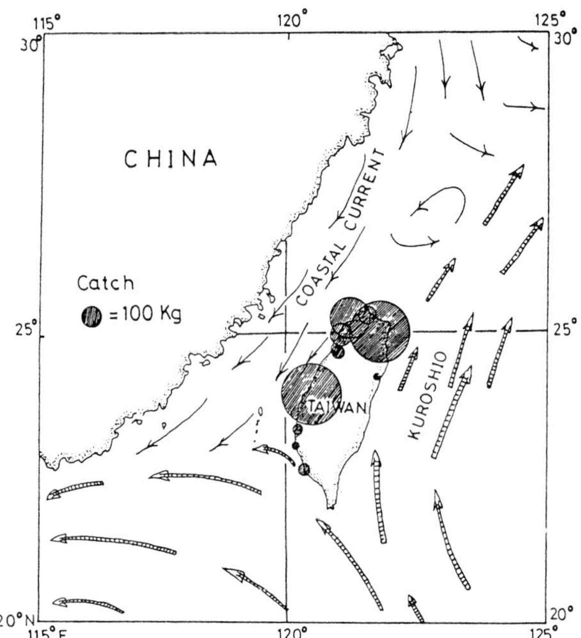

Fig. 1. Relationship between oceanic currents and catches of elvers in Taiwan in the month of January (after Chu 1963 and Chen 1975) (Solid circles, catch of elvers).

Site-specific relationship between elver catch and temperature

The CPUE and water temperature were negatively correlated in the Shuang-chi River estuary of north-eastern Taiwan ($r = -0.75$, Fig. 2). However, these factors were positively correlated in Japan (Doi 1972). Apparently the effect of temperature on elver catch was opposite between the southern-most and northern-most distribution ranges of the Japanese eel.

Catch and demand of elvers

The demand for elvers for eel farming in Taiwan has increased year by year since the eel culture industry was developed in 1965 (Fig. 3). Regression coefficient (*b*) of demand (*y*) on year (*t*) by

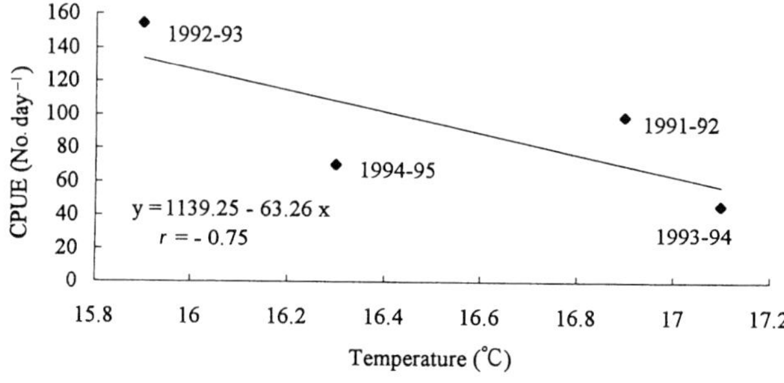

Fig. 2. Relationship between temperature and CPUE (No./day) of elvers in the Shuang-chi River estuary, north-eastern Taiwan, 1991–95.

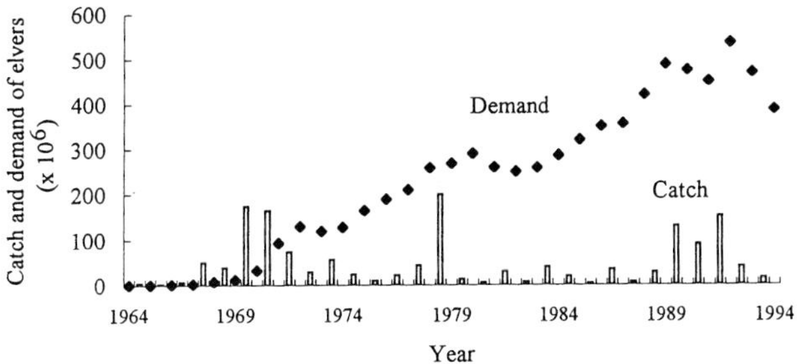

Fig. 3. Catch and demand in number of elvers for pond culture in Taiwan, 1964–1994.

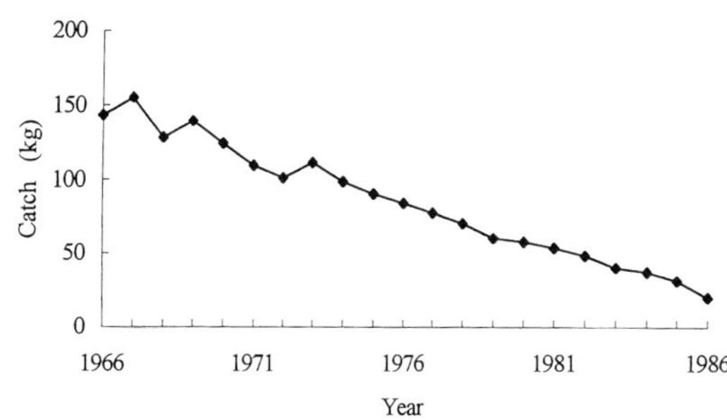

Fig. 5. Annual catch of the adult river eel, *Anguilla marmorata,* by a set-net in Penlin Town, Taipei Country, 1966–1986. (C.S. Tzeng unpublished data.)

the trend analysis of $y = a + bt$ was highly significant ($P < 0.0001$) but that of catch on year was not significant ($P > 0.8$). The correlation coefficient (r) between catch and demand was also not significant ($r = 0.04$, $F = 0.045$, $P > 0.05$). These indicated that the fluctuations in catches of elvers are a natural phenomenon and not related to the increasing demand. The local supply of elvers for farming has fallen far short of meeting the demand since 1971 (Fig. 3).

On the other hand, catches in Taiwan showed periodic fluctuations during the period from 1964 to 1994 with peak catches in 1969–70, 1978–79 and 1990–91 (Fig. 3). The periodogram shows that catch fluctuated in cycles of approximately 9.3 and 7.0 years (Fig. 4). The period of 28-year cycle was negligible because the time series was only 30 years.

The decline in downstream-migrating adult eels

Catches of adult river eel, *Anguilla marmorata,* in the upstream of Tanshui River in northern Taiwan, collected by a set-net during the downstream migration in autumn, were recorded from 1966 to 1985 (Fig. 5). Trend analysis indicated that the catches significantly decreased year by year since 1966 ($P < 0.0001$). The decline was probably due to the recruitment failure resulting from overfishing of elvers in the river mouth and the construction of hydroelectric dams obstructing the upstream migration of the elvers.

Differences in catch of elvers within the distribution range

The annual catches of the Japanese eel elvers in Taiwan, mainland China, Korea and Japan during the period from 1972 to 1995 are shown (Fig. 6a) with catches in Japan the highest; reaching a peak of approximately 130 t in 1979 and then drastically decreasing to approximately 50 t. Catches in China have increased since 1972 and reached peaks of approximately 40 t in 1982, in 1990 and in the last two years. Catches in Taiwan and Korea were very low, generally less than 10 t.

The time-series of the stocking amounts of elvers for pond culture in Japan correspond fairly well with the catch (Fig. 6b). However, the stocking amounts in Taiwan were much higher than the catch, suggesting that lots of elvers were imported for culture from other countries. Recently, the eel culture industry has been established in China and Malaysia, and many Taiwanese eel farmers have emigrated to these countries, resulting in a significant decrease in stocking amounts of elvers in Taiwan in recent years (Fig. 6b).

Japan has a long history of eel culture, stable fishing effort and an abundance of elver supply. Its catch data were considered to be appropriate for evaluating the long term fluctuations.

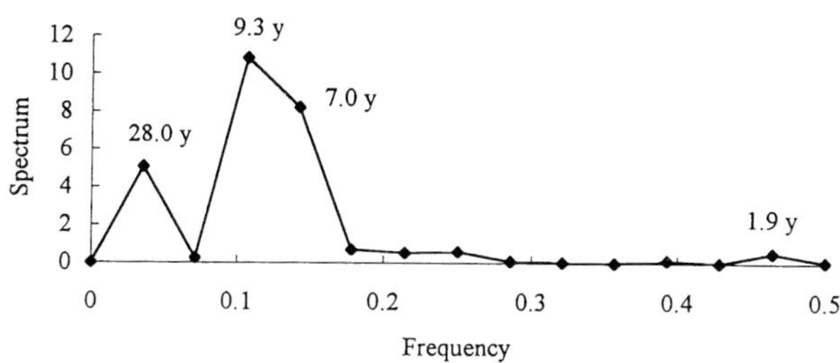

Fig. 4. Periodogram of the catch-time series of elvers in Taiwan.

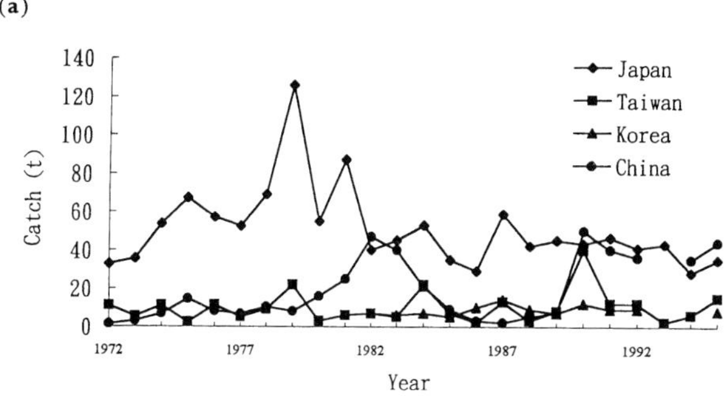

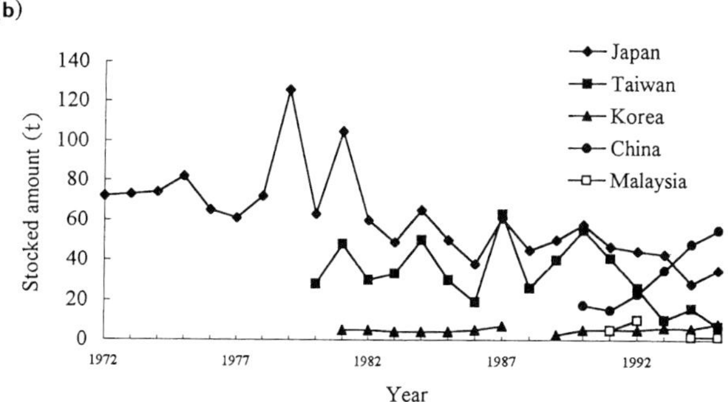

Fig. 6. Catch (A) and stocked amount (B) of Japanese eel elvers in Japan, Taiwan, Korea, China and Malaysia, 1972–1995. 1 kg = 5000–5500 elvers.

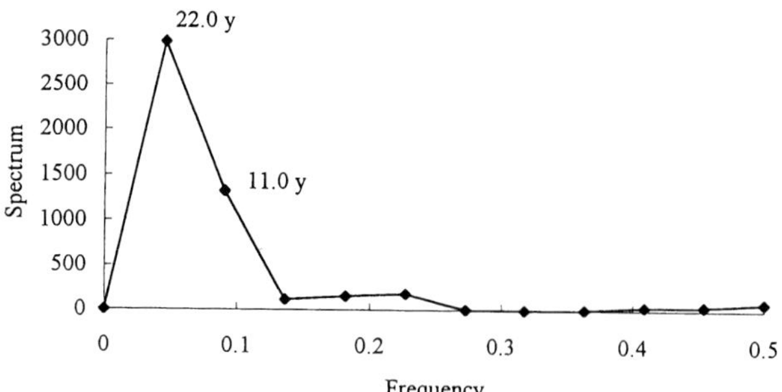

Fig. 7. Periodogram of the catch-time series of elvers in Japan.

According to its time-series from 1972 to 1995, recruitment of Japanese eel seemed to reach a peak in 1979 and fluctuated in a cycle of approximately 11.0 years, but decreased continuously in recent years (Figs 6 and 7). The period of 22-year cycle was negligible because the time series was only 24 years. This phenomenon of decline in recruitment was also found in the American eel and the European eel (Castonguay *et al.* 1994*b*); and the rise and fall of the catches were synchronized across the three species.

Discussion and conclusions

On the basis of the pond areas for eel culture, the demand of elvers for farming in Taiwan is approximately 50 t per year. However, the annual catch of elvers in Taiwan averages approximately 10 t (Fig. 3). Catches are unable to satisfy the demand because the exploitation rate of elvers in Taiwan has been very high and was estimated to be 47.15% in 1983 (Tzeng 1984*b*). To meet the demand, elvers have to be imported from other countries, but catches have also decreased in these countries. At the same time, China and Malaysia compete in the eel market, and thus the importation of elvers has become more difficult for Taiwan, resulting in a decrease in the stocking amounts in recent years (Fig. 6).

Due to increasing demand and high prices, elvers were harvested with many overlapping cross-river set-nets at the river mouths in Taiwan. Also, as a result of construction of hydroelectric dams and pollution from domestic and industrial wastes, adult eel populations in the rivers have decreased dramatically (Fig. 5). Nowadays, it is very rare to find an eel larger than 40 cm in Taiwan (Tzeng *et al.* 1994) and China (Guan *et al.* 1994). To increase local elver recruitment, hormone-induced mature eels have been releasing to enhance the spawning stock of the Japanese eel since 1976 by the Taiwan Fisheries Research Institute (Liao *et al.* 1994). However, the stocking rates have been small and their effects are unclear.

Catches of fish are dependent on population size, environmental carrying-capacity, availability and fishing effort. The CPUE of elvers in north-eastern Taiwan was inversely correlated with temperature, irrespective of the inter-annual (Fig. 2) and daily catch data (Tzeng 1985; Chen-Lee *et al.* 1994). Meanwhile, the catch along the western coast of Taiwan affected by the cold China Coastal Current was higher than along the eastern coast, well-influenced by the warm Kuroshio Current (Fig. 1). However, the Kuroshio Current greatly contributes to the transportation of elvers in the northern range of their distribution, with higher temperatures resulting in greater catches in Japan (Doi 1972). The above evidence suggests that the environmental effects on short-term variation in catches of elvers are site-specific.

Catches of elvers in Taiwan show a nine-year cycle with peaks in 1969–70, 1978–79 and 1990–91, (Figs 3 and 4). The level of catch was similar in these three periods. Although the demand in Taiwan increased to a very high level in recent years, the catch did not show a corresponding increase, indicating the catch in Taiwan has reached to its maximum. The cycle in catches of elvers in Japan was approximately eleven-years with the peak in 1979, which coincided with that of Taiwan (Figs 6 and 7). The synchronous occurrence in peak catches suggests that the elvers of Taiwan and Japan are from the same population as indicated by Sang *et al.* (1994). However, catches of elvers in Taiwan have not shown a decline in recent years as has occurred in Japan (Figs 3 and 6). This may indicate that there are different fluctuation mechanisms influencing catches of elvers between Taiwan and Japan.

Catches of elvers in Japan were the most abundant of all Asian countries. These reached a peak in 1979 and have decreased drastically since then (Fig. 6). The timing of rises and falls of catches corresponded with those of the American and European eels (Castonguay *et al.* 1994*b*). Decreases in catches of elvers of American and European eels were hypothesized to be the result of a weak Gulf Stream that transports the larvae from the spawning ground in the Sargasso Sea to the Atlantic coasts of North America and Europe (Castonguay *et al.* 1994*a*, *b*). The Kuroshio Current that transports Japanese eel larvae in the western Pacific was reported to fluctuate at a cycle of five years (Sugimoto *et al.* 1995). The long-term change of catches in Japan did not match the cyclic change of the Kuroshio (Figs 6 and 7). It appears that the decrease in catches of elvers in Japan could not be completely explained by the changes in the Kuroshio Current alone. As mentioned above, overfishing of elvers and habitat degradation, by the construction of hydro-electric dams, that may cause the decrease of parent stock and obstruct upstream migration, may be the causal factors of the long-term variations in catches of Japanese eel elvers.

Acknowledgments

This study was financially supported by the National Science Council of Republic of China (Contract No. NSC 85-2311-B002-032). The author is grateful to Mr S. G. Yue for collecting the catch data, Ms C. E. Wu for preparing the manuscript and Dr T. F. Tsai , Dr J. Beumer and Dr D. C. Smith for reviewing the manuscript.

References

Anon (1965-1995). Fisheries yearbook of Taiwan area (1965-1995). Taiwan Fisheries Bureau, Department of Agriculture and Forestry, Provincial Government of Taiwan.

Anon (1996). Catches and pond-culture amounts of juvenile eel in the Asian countries. *Monthly Report of Japan Eel Importer's Association* '96/1, No. 210, p. 10.

Castonguay, M, Hodson, P V, Couillard, C M, Eckersley, M J, Dutil, J D, and Verreault, G (1994*a*). Why is recruitment of the American eel declining in the St. Lawrence River and Gulf ? *Canadian Journal of Fisheries and Aquatic Sciences* **42**, 899–908.

Castonguay, M, Hodson, P V, Moriarty, C, Drinkwater, K F, and Jessop, B M (1994*b*). Is there a role of ocean environment in American and European eel decline? *Fisheries Oceanography* **3**(3), 197–203.

Chen, T S (1975). Preliminary study of the elver of *Anguilla japonica* in Taiwan. *China Fisheries Monthly* **268**, 11-19.

Chen-Lee, Y L, Chen, H Y, and Tzeng, W N (1994). Reappraisal of the importance of rainfall in affecting catches of *Anguilla japonica* elvers in Taiwan. *Australian Journal of Marine and Freshwater Research* **45**, 185–90.

Cheng, P W, and Tzeng, W N (1996). Timing of metamorphosis and estuarine arrival across the dispersal range of the Japanese eel *Anguilla japonica*. *Marine Ecology Progressive Series* **131**, 87–96.

Chu, T Y (1963). The oceanography of the surrounding waters of Taiwan. *Report of the Institute of Fisheries Biology, Ministry of Economic Affairs and National Taiwan University* **1**(4), 29-44.

Doi, T (1972). How to predict future catches in theoretical treatments. *Japan Fisheries Resources Conservation Association, Tokyo, Japan*. 60 pp.

Guan, R, Wang, X, and Ke, G (1994). Age and growth of eels *Anguilla japonica* in a Chinese river. *Journal of Fish Biology* **45**, 653–60.

Liao, I C, Kuo, C L, Yu, T C, and Tzeng, W N (1994). Release and recovery of Japanese eel, *Anguilla japonica*, in Taiwan. *Journal of Taiwan Fisheries Research* **2**(1), 1–6.

Liu, L L, and Liao, H H (1995). Allozyme variation of the Japanese eel *Anguilla japonica* among collections from three different locations in Taiwan. *Journal of the Fisheries Society of Taiwan* **22**(3), 247–53.

Sang, T K, Chang, H Y, and Chen, C T (1994). Population structure of the Japanese eel, *Anguilla japonica*. *Molecular Biology and Evolution* **11**, 250–60.

Sugimoto, T, Tadokoro, K, and Furushima, Y (1995). Climate and weather effects on the chlorophyll concentration in the north western North Pacific. In 'Biogeochemical Processes and Ocean Flux in the Western Pacific'. (Eds H Sakai and Y Nozaki) pp. 575–92 (Terra Scientific Publishing Company (TERRAPUB), Tokyo).

Tabeta, O, Tanaka, K, Yamada, J, and Tzeng, W N (1987). Aspects of the early life history of the Japanese eel *Anguilla japonica* determined from otolith microstructure. *Nippon Suisan Gakkaishi* **53**, 1727–34.

Tesch, F W (1977). 'The Eel: Biology and Management of Anguillid Eels.' (Chapman and Hall: London). 434 pp.

Tsukamoto, K (1990). Recruitment mechanism of the eel *Anguilla japonica* to the Japanese coast. *Journal of Fish Biology* **36**, 659–71.

Tsukamoto, K (1992). Discovery of the spawning area for Japanese eel. *Nature* **356**(6372), 789–91.

Tzeng, W N (1982). New record of the elver, *Anguilla celebesensis* Kaup, from Taiwan. *Bioscience* **19**, 57–66.

Tzeng, W N (1983*a*). Species identification and commercial catch of the Anguillid elvers from Taiwan. *China Fisheries Monthly* **366**, 16–23.

Tzeng, W N (1983*b*). Seasonal and long-term changes of the catch of Anguillid elvers in Taiwan. *Journal of the Fisheries Society of Taiwan* **10**, 1–7.

Tzeng, W N (1984*a*). Dispersion and upstream migration of marked Anguillid eel, *Anguilla japonica*, elvers in the estuary of the Shuang-Chi River, Taiwan. *Bulletin of Japanese Society of Fisheries Oceanography* **45**, 10–20.

Tzeng, W N (1984*b*). An estimate of the exploitation rate of *Anguilla japonica* elvers immigrating into the coastal waters of Shuang-Chi River, Taiwan. *Bulletin of the Institute of Zoology Academia Sinica* **23**, 173–80.

Tzeng, W N (1985). Immigration timing and activity rhythms of the eel, *Anguilla japonica*, elvers in the estuary of northern Taiwan, with emphasis on environmental influences. *Bulletin of the Japanese Society of Fisheries Oceanography* No. **47–48**, 11–28.

Tzeng, W N (1986). Resources and ecology of the Japanese eel *Anguilla japonica* elvers in the coastal waters of Taiwan. *China Fisheries Monthly* **404**, 19–24.

Tzeng, W N (1990). Relationship between growth rate and age at recruitment of *Anguilla japonica* elvers in a Taiwan estuary as inferred from otolith growth increments. *Marine Biology* **107**, 75–81.

Tzeng, W N (1994). Temperature effects on the incorporation of strontium in otolith of Japanese eel, *Anguilla japonica* Temminck and Schlegel. *Journal of Fish Biology* **45**, 1055–66.

Tzeng, W N (1995). Migratory history recorded in otoliths of the Japanese eel, *Anguilla japonica*, elvers as revealed from SEM and WDS analyses. *Zoological Studies*, **34** (Supplement 1), 234–6.

Tzeng, W N, and Tabeta, O (1983). First record of the short-finned eel *Anguilla bicolor pacifica* from Taiwan. *Bulletin of Japanese Society of Scientific Fisheries* **49**, 27–32.

Tzeng, W N, and Tsai, Y C (1992). Otolith microstructure and daily age of *Anguilla japonica* Temminck & Schlegel elvers from the estuaries of Taiwan with reference to unit stock and larval migration. *Journal of Fish Biology* **40**, 845–57.

Tzeng, W N, and Tsai, Y C (1994). Changes in otolith microchemistry of the Japanese eel, *Anguilla japonica*, during its migration from the ocean to the rivers of Taiwan. *Journal of Fish Biology* **45**, 671–83.

Tzeng, W N, Cheng, P W, and Lin, F Y (1994). Relative abundance, sex ratio and population structure of the Japanese eel *Anguilla japonica* in the Tanshui River system of northern Taiwan. *Journal of Fish Biology* **46**, 183–201.

Tzeng, W N, Hsiao, J J, Shen, H P, Chern, Y T, Wang, Y T, and Wu, J Y (1995). Feeding habit of the Japanese eel, *Anguilla japonica*, in the streams of northern Taiwan. *Journal of the Fisheries Society of Taiwan* **22**(4), 279–302.

Umezawa, A, and Tsukamoto, K (1990). Age and birth date of the glass eel, *Anguilla japonica*, collected in Taiwan. *Nippon Suisan Gakkaishi* **56**, 1199–201.

Fluctuations in the Fisheries of Kenya's Rift Valley Lakes: Causes and Prospects for the Future

S. Mucai Muchiri

Department of Fisheries, Moi University, PO Box 3900, Eldoret, Kenya.

Summary

Lakes Turkana, Baringo and Naivasha are lakes in the eastern arm of the Great Rift Valley that provide important commercial fisheries, and to a small extent provide recreation. The fishery of Lake Turkana is based on 12 main species of fish out of a total of 48 species. Tilapias, *Lates* spp., *Citharinus, Distichodus, Labeo* and *Hydrocynus* are the most important. In Lake Baringo three species (*Oreochromis niloticus, Clarias gariepinus* and *Protopterus aethiopicus*) are the most important of the seven species present in the lake. Lake Naivasha hosts a total of five species of which *Oreochromis leucostictus, Tilapia zillii* and *Micropterus salmoides* are of commercial importance. The three fisheries support an estimated 10 000 fishers plus people engaged in fish industry related services. Yet these fisheries depict wide fluctuations in fish landings.

Habitat variability has been identified as one of the most important factors influencing the fisheries of the three lakes. Lake level fluctuations are shown to be closely followed by similar fluctuations in fish catches. The observed fluctuations in lake levels are as a result of climatic factors combined with human activities which include damming of rivers and abstraction of water for irrigation. Variability in submerged vegetation cover has also been important in the three lakes.

Other anthropogenic influences on the fishery of Lake Naivasha take the form of fishing pressure and species introductions while in Lake Baringo, catchment degradation leading to excessive silt loading in the lake has played a more important role.

The persistence of these fisheries in the face of besetting environmental and anthropogenic factors is attributable, for the greater part, to the resilience of the tilapias that constitute the most important catch in each of the three Rift Valley lakes. It is suggested that an integrated approach to catchment management is necessary for the achievement of maximum sustainable fisheries in the rift valley lakes.

Introduction

Fish populations in nature are continually subject to changes in size structure and distribution, being influenced by both intrinsic and extrinsic environmental variables. These natural variables may be physical or biotic. Human involvement in certain instances enhances the speed and scale of change or deviates the direction of change in fish populations. The magnitude of change in exploited fish populations is therefore dependent upon the combined effects of natural and human disturbances and the capacity of such populations to absorb these effects. Some of the disturbances arise from farther afield and become manifested through linking systems. For example, the ecology of lake or floodplain fish populations will invariably be influenced by activities occurring in the whole catchment as well as variations occurring within the ecosystems. Management of exploited fish populations may, therefore, ultimately depend on how accurately we can predict the responses of such populations to variations in their environments.

Analyses of the fisheries of three lakes (Turkana, Baringo and Naivasha) lying within Kenya's portion of the Great Rift Valley (Fig. 1) help to illustrate the foregoing. The fisheries of these three lakes support an estimated total of 10 000 fishers plus people engaged in fishing industry-related activities. These fisheries, however, exhibit wide fluctuations in landings (Fig. 2). The reasons for these fluctuations and factors that have allowed the persistence of the fisheries of Lakes Turkana, Baringo and Naivasha are the subject of this paper.

THE LAKES AND THEIR FISH COMMUNITIES

Lake Turkana

The recent status of Lake Turkana ecosystem has been described in greater detail by Hopson (1982) and Kolding (1989). Lake Turkana is a large (265 km long, 30 km wide and up to 115 m deep) endorheic lake lying at between 2°27'N and 4°40'N and at an altitude of 375 m a.s.l. (metres above sea level). Surface water temperature varies slightly at between 27.2 and 29.4°C while the lake bottom temperature ranges between 25.4 and 26.4°C (Kolding 1989). The lake is surrounded by land falling in class VI of Pratt *et al.* (1966), who described it as very arid and of very low ecological potential. Lake Turkana receives the largest proportion (80–90%) of its water from the perennial River Omo, which drains the highlands of south eastern Ethiopia and enters the lake through a marshy delta in the north. Other smaller rivers enter the lake seasonally. The Turkwell flows from the slopes of Mt. Elgon on the border of Kenya and Uganda in the west and is seasonal in its lower reaches. With the recent construction of a hydroelectric dam across its gorge, the river has practically ceased flowing into the lake, except during exceptionally heavy rains. The Kerio river collects its water from the Tugen and Marakwet Hills in the south of Lake Turkana.

The fish community of Lake Turkana is the richest of the Rift Valley lakes. It comprises 18 families of fish with a total of 48 species, 7 of which are endemic to the lake. The present fish community of Lake Turkana is typically nilotic and there is zoogeographical evidence that suggests past connections of the Omo-Turkana basin with the Nile system (Roger 1944; Butzer 1971). The cichlids, cyprinids and characids are the more dominant species having 7, 7, and 9 species, respectively; 12 of the 48 species feature frequently in the fishery of Lake Turkana. Tilapias, *Lates* spp. *Citharinus, Distichodus, Labeo* and *Hydrocynus* are the more important.

Lake Baringo

Lake Baringo is a shallow (average 3.5 m in depth) freshwater lake on the floor of the Rift Valley at an altitude of 975 m a.s.l. and about 150 km south of Lake Turkana (*cf.* Fig. 1). The lake has an average surface area of 130 km^2 but is subject to wide variations in size, which are dependent on rainfall on the highland areas of its catchment. Lake Baringo has no surface outlet but may have a subterranean outlet as suggested by Beadle (1932), which may explain the freshness of the lake water. The principal rivers that flow into Lake Baringo are the Perkerra, Molo and Endao, all of which flow from the highlands south of the lake. The other smaller rivers include Tangulbei, Mukutan and Ol Arabel, which drain the eastern scarp of the Baringo section of the Rift Valley and flow generally southward into Lake Baringo. The first three rivers have been perennial in the past, albeit with greatly varying volumes of water. These rivers flow through very dry land that is severely eroded by flood water during periods of high rainfall upon which huge torrents bring with them heavy loads of silt and debris.

Silt that enters the lake has very low organic content and therefore tends to remain in suspension aided by wind which is more prevalent in the afternoons. The suspended silt gives Lake Baringo a brown colour resulting in very low transparency. The average secchi depth readings for March 1996 was 3.5 cm. This is one of the lowest records of secchi depth transparency recorded in Lake Baringo. Kallqvist (1987) recorded secchi readings of 5–7 cm in 1976–77 and 20 cm in 1979. The 1979 figure had also been observed by Beadle (1932) in 1931. The present low transparency is as a result of the shallow sediments (average 2 m deep of lake water) being kept in suspension rather than from high productivity.

Plankton primary production is based mainly on the blue-green alga *Microcystis aeruginosa* which dominates the phytoplankton community at times forming blooms on the water surface. The macrophyte community of lake Baringo is sparse and confined mainly on the southern shoreline at river entry points. *Cyperus* spp. dominate the large emergent plant community, while *Potamogeton* spp. are the more common submerged macrophytes.

The fish fauna of Lake Baringo comprises six naturally occurring species and one introduced species. This is atypical of tropical lakes, which normally support many species of fish (Lowe-McConnell 1987). This may suggest a recent drying up of the lake and only being replenished by riverine species. Based on previous studies (Okoric 1974) and available preliminary data (unpublished), the trophic ecology of Lake Baringo fishes is comprised of short food chains arising from:

- **detritus** where *Clarias gariepinus* feeds both on detrital aggregates associated with the bottom and predates on the other detritivores, *Oreochromis niloticus* and *Labeo gregorii. Protopterus aethiopicus* is also a detritivore but there is no evidence of it falling prey to *Clarias.*
- **plankton** provides forage for *Oreochromis* and *Barbus* spp. *Barbus* is mainly a zooplankton feeder but is also a facultative insectivore.

Oreochromis niloticus, C. gariepinus, P. aethiopicus and *B. gregorii* form the basis of a commercial fishery at Lake Baringo.

Lake Naivasha

Like Baringo, Lake Naivasha is a freshwater lake, covering an area of approximately 150 km^2. The lake is the highest of the

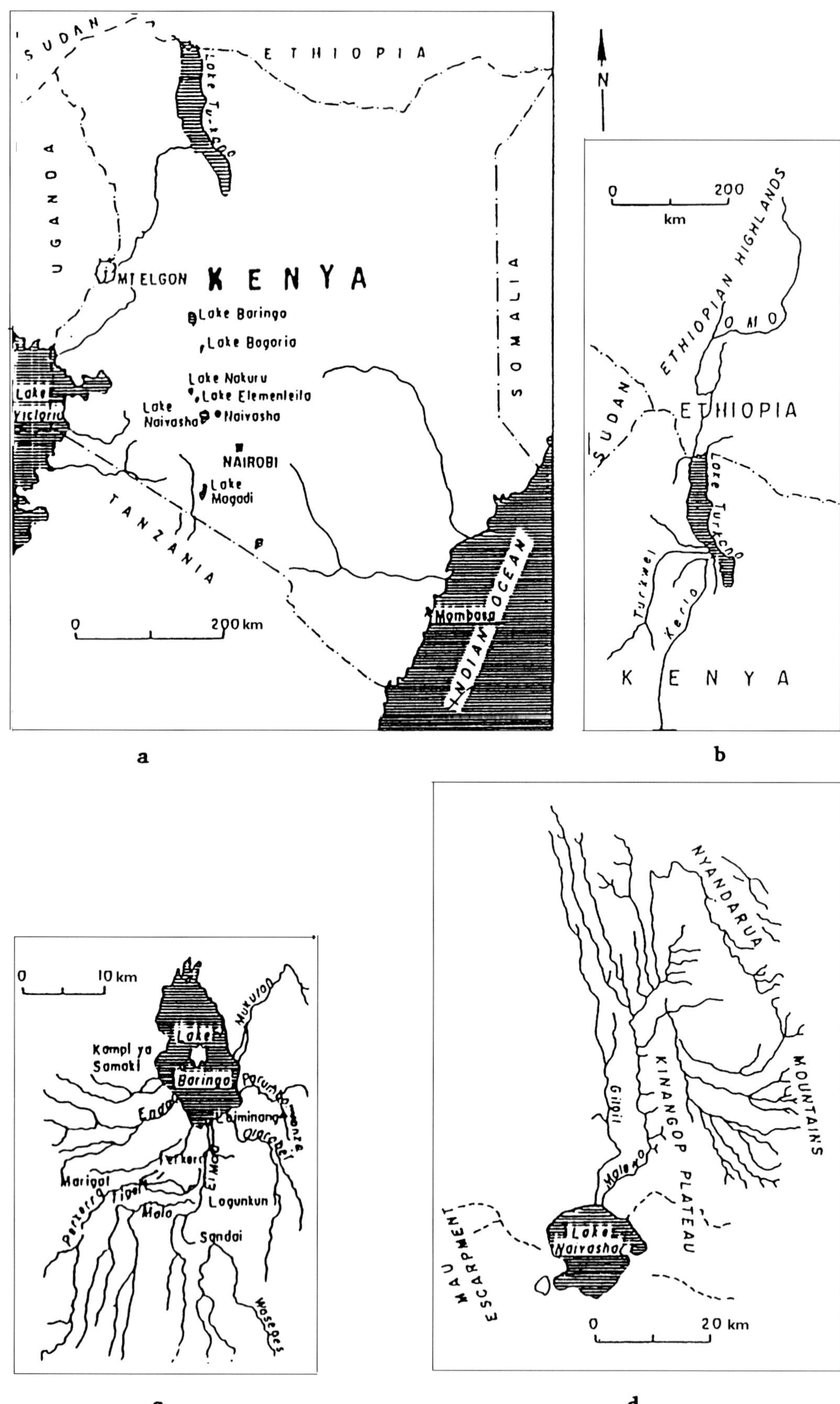

Fig. 1. Location map of (a) the rift valley lakes; (b) Lake Turkana and its catchment area; (c) Lake Baringo and its catchment area; (d) Lake Naivasha and its principal rivers.

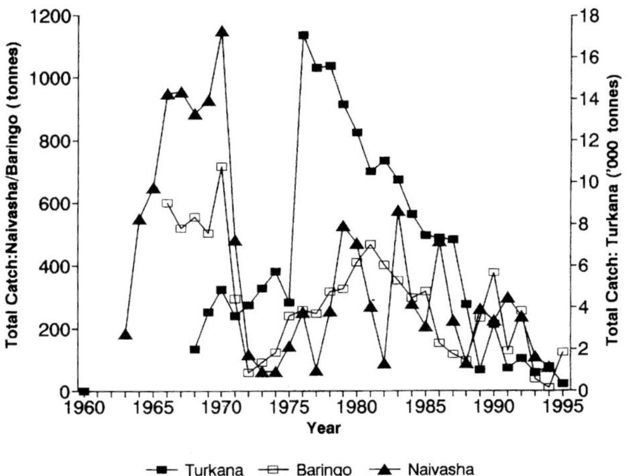

Fig. 2. Changes in total fish catches from the rift valley lakes (Data obtained from the Department of Fisheries, Nairobi).

Rift Valley lakes at an altitude of 1890 m a.s.l. It lies in a closed basin and receives 90% of its water from the perennial River Malewa (1533 km^2 drainage area) and the rest from one ephemeral stream (River Gilgil, drainage area = 151 km^2), direct rainfall and ground seepage. The lake is very shallow, being, for the most part, between 4 and 6 m in depth, and exhibits considerable fluctuations in water levels with a history of complete drying out in the last century. Water temperatures are generally in the range of 20–25°C. Vegetation cover has varied considerably in the last few years. Belts of *Cyperus papyrus* dominate the margins, but these are often reduced by harvesting and burning. Gaudet (1977) recorded 25 species of shallow-water macrophyte community, 10 submerged, 8 floating and 7 associated with floating vegetation. Between 1975 and 1983 submerged and floating-leaved macrophytes were absent from the lake. The reasons for this variation are discussed by Harper *et al.* (1990) and Harper (1992).

Lake Naivasha hosts a total of five species of fish which are a remnant community from two native species and several introductions (Muchiri and Hickley 1991). Muchiri *et al.* (1995) have found an average persistence time to be 36.9 to 29.7 years for the species whose start and end of their presence in the lake can be deduced with any certainty. The food and feeding habits of the fish species of Lake Naivasha have been described by Hickley *et al.* (1994), Muchiri *et al.* (1994) and Muchiri *et al.* (1995). The tilapias were shown to consume large amounts of detrital aggregates comprising fine silt, decaying vegetable debris and leaf litter together with associated animal communities. *O. leucostictus* also takes small proportions of green algae and diatoms. The submerged macrophytes are consumed by *T. zillii* together with the dedicated invertebrate macroherbivore, crayfish, *Procambarus clarkii*, (Harper *et al.* 1990). *T. zillii* also feeds on *Micronecta* and other insects when not consuming detritus and macrophytes. The largemouth bass has a mixed diet starting with zooplankton and as it grows larger consumes macroinvertebrates (especially *Micronecta* and crayfish), fish and frogs. *Barbus* and the guppy, *Lebistes reticulata,* feed on insects and zooplankton. Muchiri *et al.* (1994) identified several unexploited or underexploited fish feeding niches at Lake Naivasha and consequently suggested consideration of careful introduction of more species to diversify the ecosystem and enhance the fishery potential. Presently, the two tilapias (*O. leucostictus* and *T. zillii*), the largemouth bass (*M. salmoides*) and the invertebrate *P. clarkii* support an important canoe-based commercial fishery. The bass also provides a sport fishery.

CAUSES OF FISHERIES FLUCTUATIONS

Water level fluctuations

Historical changes in water levels of the Rift Valley lakes have been reviewed by Hopson (1982). He suggested that lakes in the eastern arm of the Rift Valley, in Kenya, show similar trends. Fluctuations in the levels of Lake Turkana since 1888 parallel those of Lake Naivasha (Butzer 1971; Richardson and Richardson 1972). On the other hand total fish landings from lakes Turkana, Baringo and Naivasha appear to correspond fairly closely with water level changes (Fig. 3). Regression of fish catches against lake levels of Turkana and Naivasha (whose water level data are more or less complete) show significant positive relationships (at $P < 0.05$).

1. Turkana

   Total catch (t) = 8774 + 1202 water level ($r = 0.47$)

2. Naivasha

   Total catch (t) = 216 + 82.6 water level ($r = 0.40$)

In Lake Turkana, the wide recession of water has led to the loss of large littoral areas including the tilapia-rich Ferguson's Gulf. Due to the shallow depth of Lake Baringo and Naivasha, any change in water level is reflected in wide recession or flooding of large areas of lake margins. These wide changes in horizontal movement have important ecological implications for fish habitats particularly for those important fish communities that occur in the shallow littoral zones. Lowe-McConnell (1982) listed habitat drying and flooding as some of the important factors that control tilapia numbers in fish communities. Tilapias were also shown by Fryer and Iles (1972) to find shelter from predation and fishing pressure in flooded marginal terrestrial vegetation on the shores of Lake Victoria during rising water.

The offshore species of Lake Turkana are not so much directly affected by fluctuations of lake water levels as by the discharge of the affluent rivers flowing into the lake. Beadle (1981) suggests that the River Omo, having a large catchment well covered with forests and other vegetation provides a rich source of nutrients that form the basis of the lake's productivity. Harbott (1982) and Talling (1986) demonstrated that the production of *Microcystis aeruginosa*, the dominant alga in Lake Turkana, is influenced by the seasonality of the River Omo flow regime. This production then supports secondary production based on zooplankton and other planktivores (Ferguson 1982) which would be reflected in the pelagic fish production. Although Melack (1976) did not include Lake Turkana in the determination of a relationship between fish yields and primary productivity, he demonstrated that the two parameters have a good positive correlation.

Aquatic vegetation

No data are available for the Rift Valley lakes to categorically link fish yields with variability of aquatic vegetation. However, it has

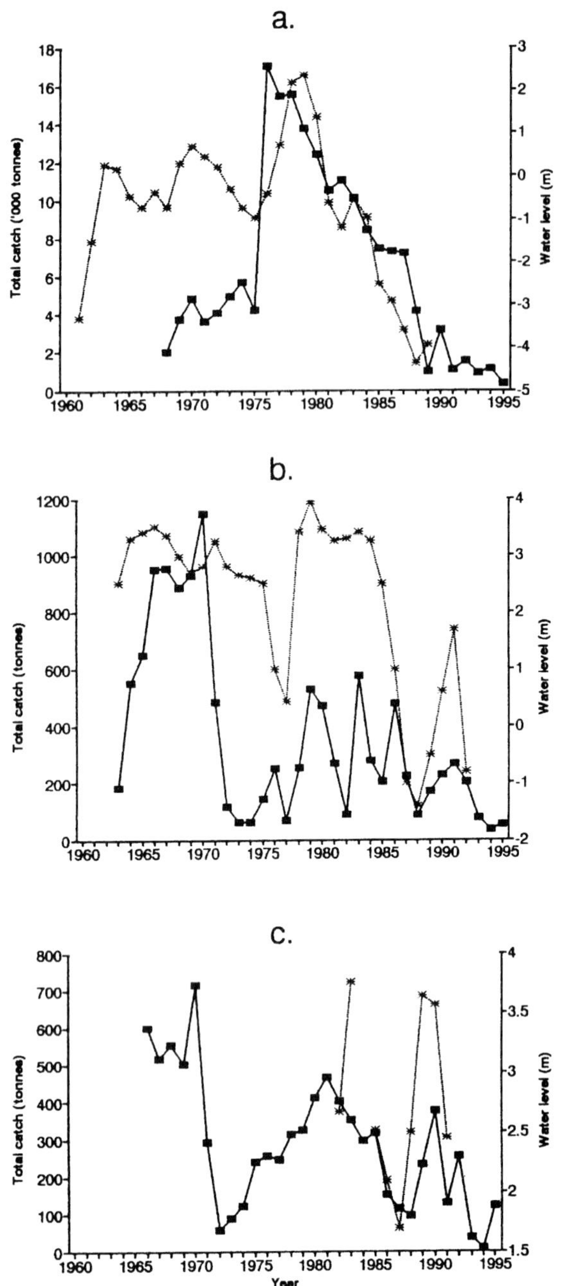

Fig. 3. Total catch and water level changes in the Rift Valley lakes. (a) Lake Turkana - 375 m a.s.l is taken as 0 metres; (b) Lake Naivasha - 1890 m a.s.l. taken as 0 metres; (c) Lake Baringo - 975 m a.s.l is taken as 0 metres. (Lake level data obtained from the Ministry of Water Development, Nairobi).

been possible to adduce evidence of such a relationship based on direct observation and results of studies carried out elsewhere. Of the three Rift Valley lakes, Lake Naivasha has proportionately greater abundance of macrophytes. But, as discussed above, the macrophyte community at Naivasha has had mixed fortunes over the years. The role of aquatic plants in the success of fish populations through feeding and breeding has been shown in Lake Naivasha for tilapia (Muchiri and Hickley 1991; Muchiri *et al.* 1995) and for largemouth bass (Hickley *et al.* 1994). It has been shown that during periods of total absence of submerged macrophytes catches generally declined but as the macrophytes recovered there was evidence of upward trends in fish abundance.

A similar relationship was also demonstrated from 30 Texas reservoirs (Durocher *et al.* 1984) where 'a highly significant positive relationship ($P < 0.01$) was found between percent submerged vegetation (up to 20%) and both the standing crop of largemouth bass and numbers being recruited to harvestable size.' It is desirable, however, to consider the total trophic status of the water body. This is particularly important because extensive aquatic macrophytes have the effect of reducing chlorophyll a and total phosphorus concentrations and increase in secchi disc transparencies which may yield underestimated trophic status (Canfield *et al.* 1983; Harper 1992).

Influences of aquatic macrophytes on fish populations through feeding and breeding can either be direct or indirect. Direct macrophyte consumers are likely to be influenced by variation in macrophyte abundance. In the Rift Valley lakes, only *T. zillii* consumes macrophytes. However, as demonstrated in Naivasha, *T. zillii* is not an absolute macrophyte feeder and is able to consume other food items even in the presence of an abundance of aquatic plants. Indirectly, macrophytes provide a source of detritus for the detritivores. In Naivasha largemouth bass was found to feed in areas more closely associated with macrophytes but did not consume the plants. Aquatic macrophytes play a more crucial role in providing nesting grounds for fish, and refugia for fish threatened by predators. Since the macrophyte feeders are able to take alternative prey items, it appears that the role of macrophytes in aquatic ecosystems in influencing fish populations is more important to reproduction than feeding.

Human impacts

Human impacts on fisheries yields come in various forms. In the Rift Valley lakes the more important impacts are overexploitation of the fisheries resources, introductions of alien species and catchment degradation in the wider catchment areas. The Lake Turkana Fisheries Project (Hopson 1982) indicated some degree of overfishing for certain stocks (especially those of *Citharinus, Distichodus, Lates* and the tilapias). While Kolding (1989) acknowledges overexploitation of tilapias, he does not accept the whole concept of overfishing in Lake Turkana and instead has put more weight on the effects of variability in water levels as the major influencing factor of fish yields. In Naivasha, and Baringo there are indications of overfishing. Muchiri and Hickley (1991) have shown this for Naivasha, while in Baringo overfishing has been cited several times in the *Annual Statistical Bulletins* of Kenya's Fisheries Department. As the number of fish to catch decreases, the fishers respond by reducing the mesh sizes of the gill-nets used. This results in recruitment overfishing where a large proportion of fish is caught before reaching spawning size or having spawned only once (Siddiqui 1977). Muchiri *et al.* (1994) have demonstrated that overfishing in Naivasha only occurs because of the presence of a small number of target fish whereas more than 60% of the lake ecosystem is not utilized by fish (hence the suggestion of a careful introduction of more fish species to enhance the fishery).

The impacts of species introductions are well documented (see for example Pitcher and Hart 1995). Most introductions have resulted in negative impacts as many authors have suggested for the Nile perch situation in Lake Victoria. In the Rift Valley lakes, results of introductions can be viewed as positive from the fisheries point of view. The introduction of *Protopterus aethiopicus* into Lake Baringo has enhanced the fishery in that lake, while in Naivasha the entire fishery is dependent on introduced species. There have not been any species introductions into Lake Turkana.

Poor land uses in agriculture and pastoralism in areas that are steeply sloping and in areas that already have sparse vegetation cover have resulted in extensive soil erosion. Erosional effects in Baringo area are especially noteworthy. The clearing of extensive areas of papyrus at Naivasha has substantially reduced the buffer zone that helps trap silt and nutrients from agricultural runoff. More and more land that constitutes the catchment area of Lake Naivasha is being cleared for agriculture. This is also true for Lake Baringo area where more land is being put under irrigation.

Damming of rivers for irrigation and hydroelectric power generation is a feature common to the catchments of the three Rift Valley lakes. The result of this damming is a reduction of water that reaches the lakes. With the present extensive use of the River Omo water for irrigation in Ethiopia and the damming of the Turkwell River for hydroelectric power generation, the important sources of nutrients for the open water of Lake Turkana may be severely affected. The remaining recourse for the open water productivity in this lake is the frequent mixing of water, caused by the prevailing south-east winds (Hopson 1982), that brings up nutrients occurring at the bottom.

REASONS FOR PERSISTENCE OF THE FISHERIES

The fisheries of the three Rift Valley lakes have at all times been predominantly inshore, at Lake Turkana within water less than 15 m deep. Yet it is the littoral zones that are affected most by agents of habitat change. To survive in this kind of habitat, the fish species have to be adaptable to such changes. Across the Rift Valley lakes, tilapias appear to be the most successful in occupying this rapidly changing environment and hence supporting the fisheries (Fig. 4). The key to their success lies in their flexibility in feeding and breeding. *Oreochromis niloticus* (the more important species in Turkana and the only tilapia in Baringo), has been reported to take varying food types in different water bodies or at different times in the same lakes (e.g. Moriarty and Moriarty 1973; Moriarty *et al.* 1973; Balirwa 1990). *O. niloticus* is able to consume phytoplankton, detritus and epiphytic algae. Similarly, *O. leucostictus* (the species that dominates the fishery of Lake Naivasha), consumes a wide variety of food items which include phytoplankton, detritus and macroinvertebrates. In Lake Naivasha, as well as Albert and George, in lagoons fringing Lake Victoria and in ponds and dams where the species has been introduced, the importance of bottom deposits has been stressed (Fish 1955; Lowe-McConnell 1958; Welcomme 1970; Muchiri *et al.* 1995). The ability to consume detritus is an important adaptation that allows dependence on a rich and almost unlimited food source (Bowen 1980; Mann 1988) which dampens out effects of food limitation in aquatic ecosystems (Muchiri *et al.* 1995).

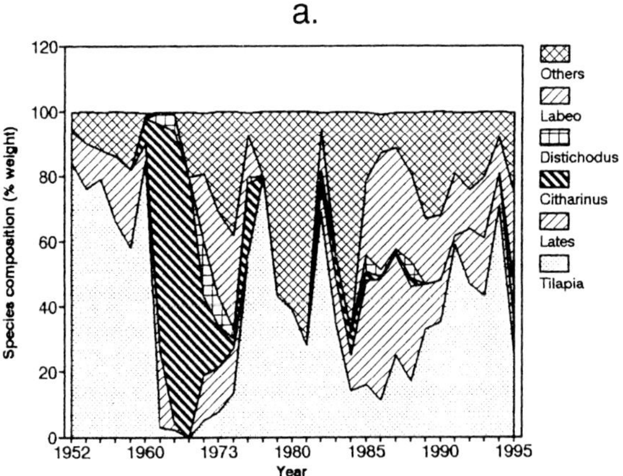

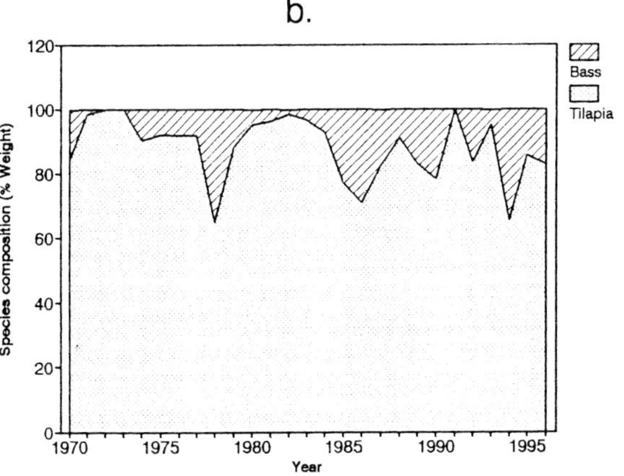

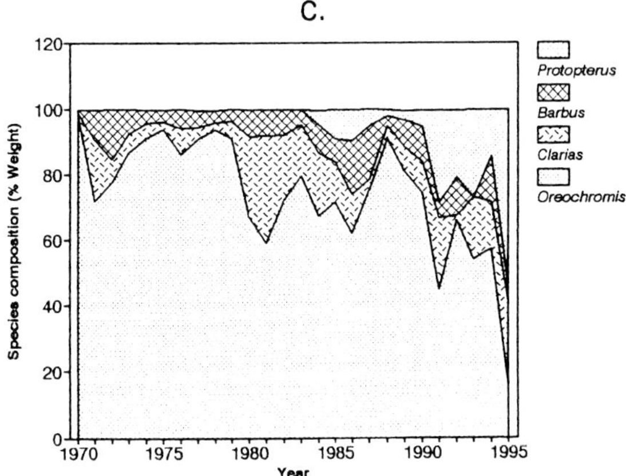

Fig. 4. Species composition from the fishery of the rift valley lakes, Kenya; (a) Lake Turkana, (b) Lake Naivasha, (c) Lake Baringo (Data obtained from the Department of Fisheries, Nairobi. For Lake Turkana between 1952 and 1970 data after Kolding, 1989).

A common feature of breeding in cichlids is that of parental care (Fryer and Iles 1972). This feature allows rapid reproduction. Mouthbrooders, particularly of the genus *Oreochromis* have been more successful in the Rift Valley lakes because of their ability to reproduce even with receding water levels, thus overcoming variations in their habitats. Presence of macrophytes appears to be more crucial in breeding of substrate spawners such as *Tilapia zillii* (Fishelson 1966; Spataru 1978) which also occurs in Lakes Turkana and Naivasha.

The future

Overfishing notwithstanding, fluctuating water level has been demonstrated as a major factor in the performance of the fisheries of the Rift Valley lakes. The effects of water fluctuations are more pronounced in shallow aquatic ecosystems such as the gently sloping shore of Lake Turkana and whole lake areas of Lake Baringo and Naivasha. As the Rift Valley lakes depend almost entirely on water received in the wider catchment areas, it is imperative that management of lake communities (including fish) be linked to whole catchment management. With regard to the Rift Valley lakes, attention should be paid to choice of appropriate land uses in the catchment areas. The need to provide suitable land for agriculture is in direct conflict with conservation needs of these areas. It is the forested highland areas that are suitable for agriculture because of their good soils and rainfall, and whereas the areas immediately in the vicinity of the lakes have rich soils (especially around Naivasha and Baringo), the soils are too dry to support any crops except under irrigation. Water used in agriculture in these regions is often indeterminate (although in certain cases official abstraction quotas exist) so that even with accurate hydrological data (which are scarce) it is difficult to accurately predict amounts of water that reach the lakes. In Naivasha some of the water that eventually reaches the lake is also abstracted for irrigation of large export-oriented floricultural farms.

The major obstacle to integrated catchment management in Kenya's Rift Valley basins is that several groups have vested interests in the resources, yet there is a lack of co-ordination in their use and management. The River Omo catchment is entirely within the boundaries of Ethiopia which means that international goodwill is important in the management of the Lake Turkana ecosystem. For those falling within the borders of Kenya (as it is with Baringo and Naivasha catchments), several government bodies and private groups have interests in the resources. The ministries in charge of water, agriculture, forestry, fisheries and rural development are all stakeholders with individual priorities. At Naivasha, there is the influential Lake Naivasha Riparian Owners Association which has on several occasions in the past influenced the direction of management for the Lake Naivasha ecosystem.

Sound management proposals have been made in the past (e.g. Harper *et al.* 1990 for Lake Naivasha and its catchment areas, and Bryan 1994 for the Baringo basin). However, in the long-term, management of the Rift Valley fisheries, and by extension whole lake ecosystems, will depend on the formulation of mechanisms that will allow co-ordination of activities of the various interest groups. One parastatal body exists in Baringo that is meant to be an umbrella management body. However, the Kerio Valley Development Authority (KVDA), like the other regional development authorities in Kenya, is not an effective co-ordinating body. Also the KVDA appears to lay more emphasis on agricultural food, and hydroelectric and fuelwood energy production with less attention to other issues. It appears then that a more powerful body with wider interests is desirable. An inter-ministerial commission would in this case play a more meaningful role of coordinating the activities of various interest groups. The Inter-ministerial Commission on Marine Resources in Brazil provides a good example as it has played a more effective role in co-ordinating national and state activities that manage marine aquatic resources (Dugan 1990). The private sector should also be co-opted in such a commission to represent the greater majority of the resource users. The Omo-Turkana basin requires an inter-governmental commission similar to one created recently for Lake Victoria.

Acknowledgments

I am grateful to the Director of Fisheries for access to the fish catch data. Drs Barasa Wangila and Hussein Isaack read the manuscript and made useful comments. Funds to cover my expenses to the Second World Fisheries Congress were provided by the Canadian International Development Agency (CIDA) through a Memoral University (Canada) — Moi University (Kenya) linkship programme.

References

Balirwa, J S (1990). The effects of ecological changes in Lake Victoria on present trophic characteristics of *Oreochromis niloticus* in relation to species role as a stabilizing factor of biomanipulation. *International Agricultural Centre, Wageningen, Occasional Paper* No. 3, 58–66.

Beadle, L C (1932). Scientific results of the Cambridge Expedition to East African Lakes 1930 — 1. 4. The waters of some East African lakes in relation to their fauna and flora. *Journal of the Linnean Society (Zoology)* **38**, 157–211.

Beadle, L C (1981). 'The Inland Waters of Tropical Africa — An Introduction to Tropical Limnology.' 2nd edition, (Longman: London and New York). 475 pp.

Bowen, S H (1980). Detrital non-amino acids are the key to rapid growth in *Tilapia mossambica* in Lake Valencia, Venezuela. *Science* **207**, 1216–18.

Bryan, R B (1994). Land degradation and the development of land use policies in a transitional semi-arid region. In 'Soil Erosion, Land Degradation and Social Transition - Geological analysis of a semi-arid tropical region, Kenya'. (Ed R B Bryan) (Catena Verlag: Cremlingen-Destedt). 248 pp.

Butzer, K M (1971). 'Recent History of an Ethiopian Delta — The Omo River and the Level of Lake Rudolf.' *The University of Chicago, Department of Geography Research Paper* No. 136. 184 pp.

Canfield, D E Jr., Langland, K A, Maceina, M J, Haller, W T, Shireman, J V, and Jones, J R (1983). Trophic state classification of lakes with aquatic macrophytes. *Canadian Journal of Fisheries and Aquaculture Sciences* **40**, 1813–19.

Dugan, P J (Ed) (1990). 'Wetlands Conservation: a Review of Current Issues and Required Action.' (IUNC: Gland, Switzerland). 96 pp.

Durocher, P P, Provine, W C, and Kraai, J E (1984). Relationship between abundance of largemouth bass and submerged vegetation in Texas reservoirs. *North American Journal of Fisheries Management* **4**, 84–8.

Ferguson, A J D (1982). Studies on the zooplankton of Lake Turkana. In 'Lake Turkana. A report on the findings of the Lake Turkana Project 1972–1975'. Overseas Development Administration, London **1**, 1334–55.

Fish, G R (1955). The food of *Tilapia* in East Africa. *Uganda Journal* **19**, 85–9.

Fishelson, L (1966). Cichlidae of the genus *Tilapia* in Israel. *Bamindgen* **18**, 67–80.

Fryer, G, and Iles, T D (1972). 'The Cichlid Fishes of the Great Lakes of Africa: Their Biology and Evolution.' (TFH Publications: Neptune City, New Jersey).

Gaudet, J M (1977). Natural drawdown on Lake Naivasha, Kenya, and the formation of papyrus swamps. *Aquatic Botany* **3**, 1–47.

Harbott, B J (1982). Studies on algal dynamics and primary productivity in Lake Turkana. In 'Lake Turkana. A report on the findings of the Lake Turkana Project 1972–1975'. Overseas Development Administration, London, **1**, 1331–55.

Harper, D M (1992). The ecological relationships of aquatic plants at Lake Naivasha, Kenya. *Hydrobiologia* **232**, 65–71.

Harper, D M, Adams, C and Mavuti, K M (1992). The aquatic plant communities of the Lake Naivasha wetland, Kenya: patterns, dynamics and conservation IV. *International Wetlands Conference*, Ohio, 1992.

Harper, D M, Mavuti, K M, and Muchiri, S M (1990). Ecology and management of Lake Naivasha, Kenya in relation to climatic change, alien species introductions and agricultural development. *Environmental Conservation* **17**, 228–36.

Hickley, P, North, R, Muchiri, S M, and Harper, D M (1994). The diet of largemouth bass, *Micropterus salmoides*, in Lake Naivasha, Kenya. *Jounal of Fish Biology* **44**, 607–19.

Hopson, A J (Ed) (1982). 'Lake Turkana. A Report on the Findings of the Lake Turkana Project 1972–1975'. Overseas Development Administration, London, **1–6**. 1614 pp.

Kallqvist, T (1987). 'Primary production and phytoplankton in Lake Baringo and Naivasha.' *Norwegian Institute for Water Research (NIVA) Report No.* E-8041905. 59 pp.

Kolding, J (1989). 'The fish resources of Lake Turkana and their environment'. Candidate Scientific thesis, University of Bergen and Final Report of Project KEN 043, NORAD, Oslo. 262 pp.

Lowe-McConnell, R H (1958). Observations on the biology of *Tilapia nilotica* in East African waters. *Reviews Zoology and Botany of Africa* **57**, 129–70.

Lowe-McConnell, R H (1982). Tilapias in fish communities. In 'The Biology and Culture of Tilapias'. (Eds R S V Pullin and R H Lowe-McConnell), ICLARM Conference Proceedings 7, 83–113.

Lowe-McConnell, R H (1987). 'Ecological Studies in Tropical Fish Communities.' (Cambridge University Press). 382 pp.

Mann, K H (1988). Production and use of detritus in freshwater, estuarine and coastal marine ecosystems. *Limnology and Oceanography* **33** (4, part 2), 910–30.

Melack, J M (1976). Primary production and fish yields in trical lakes. *Transactions of the American Fisheries Society* **105**, 575–80.

Moriarty, D J W, and Moriarty, C M (1973). The assimilation of carbon from phytoplankton by two herbivorous fishes: *Tilapia nilotica* and *Haplochromis nigripinis*. *Journal of Zoology* (London) **171**, 41–55.

Moriarty, D J W, Darlington, J P E C, Dunn, L C, Moriarty, C M, and Terlin, M P (1973). Feeding and grazing in Lake George. Proceedings of the Royal Society of London, Series B, **184**, 299–319.

Muchiri, S M, and Hickley, P (1991). The fishery of Lake Naivasha, Kenya. In 'Catch Effort Sampling Strategies: Their Application in Freshwater Fisheries Management'. (Ed I G Cowx) pp. 382–92 (Blackwell: Oxford).

Muchiri, S M, Hickley, P, Harper, D M, and North, R (1994). The potential for enhancing the fishery of Lake Naivasha, Kenya. In 'Rehabilitation of Freshwater Fisheries'. (Ed I G Cowx) pp. 348–58 (Blackwell: Oxford).

Muchiri, S M, Hart, P J B, and Harper, D M (1995). The persistence of two introduced tilapia species in Lake Naivasha, Kenya, in the face of environmental variability and fishing pressure. In 'The Impact of Species Changes in African Lakes'. (Eds T J Pitcher and P J B Hart) pp. 299–319 (Chapman and Hall: London).

Okoric, O O (1974). On the ecology and exploitation of the fisheries of an East African Rift Valley Lake. Part 1. On the bionomics and population structure of *Tilapia nilotica*, Linnaeus, 1757 in Lake Baringo, Kenya. Unpublished report of the East African Freshwater and Fisheries Research Organisation, Jinja. 24 pp + 26 tables + 7 figures.

Pratt, D J, Greenway, P J, and Gwynne, M D (1966). A classification of East African rangeland. *Journal of Applied Ecology* **3**, 369–82.

Pitcher, T J, and Hart, P J B (1995). 'The Impact of Species Changes in African Lakes.' (Chapman and Hall: London). 601 pp.

Richardson, J L, and Richardson, A E (1972). History of an African rift valley lake and its climatic implications. *Ecological Monographs* **42**, 499–534.

Roger, J (1944). Mollusques fossiles et subfossiles du basin du lac Rudolphe. In 'Mission Scientifique de l'Omo, 1932–33'. (Ed C Armboug) *Bulletin of Museum of Natural History* (Paris) 2, 60–230.

Siddiqui, A Q (1979). Changes in fish composition in Lake Naivasha, Kenya. *Hydrobiologia* **64**(2), 131–8.

Spataru, P (1978). Food and feeding habits of *Tilapia zillii* (Gervais) (Cichlidae) in Lake Kinneret (Israel). *Aquaculture* **14**, 327–38.

Talling, J F (1986). The seasonality of phytoplankton in Africian lakes. *Hydrobiologia* **138**, 139–60.

Welcomme, R L (1970). Studies of the effects of abnormally high water levels on the ecology of fish in certain shallow regions of Lake Victoria. *Journal of Zoology* (London) **4**, 39–55.

Bycatch in the Gulf of Mexico Shrimp Fishery

James M. Nance and Elizabeth Scott-Denton

National Marine Fisheries Service, Galveston Laboratory, 4700 Avenue U, Galveston, Texas 77551 USA.

Summary

Over the past five years a total of 3653 observer days have been secured by shrimp bycatch observers in the Gulf of Mexico and along the east coast of the United States of America. Analysis revealed that on average about 27 kg of organisms per hour are taken during trawling operations in the Gulf of Mexico. Examination of the composition of the organisms revealed that about 68% of the catch by weight is composed of finfish (mostly groundfish), 16% by commercial shrimp species, 13% by non-commercial shrimp crustaceans and 3% by non-crustacean invertebrates.

Although groundfish species make up the majority of the bycatch taken in shrimp trawls, three species (king mackerel, *Scomberomorus cavalla*, Spanish mackerel, *S. maculatus*, and red snapper, *Lutjanus campechanus*) have received a great deal of attention because of their commercial and recreational importance and the potential for significant impacts on their population abundance through shrimp trawling activities. Average catch of these three species is generally below 0.5 kg per hour

Introduction

The incidental harvest of non-target species or bycatch by the US Gulf of Mexico shrimp fishery is a controversial and volatile issue. The principal gear used in this fishery is the otter-trawl, a relatively non-selective bottom-trawl that catches a variety of finfish and invertebrate species. Many of these species are released dead, injured or stressed. Conservation agencies and environmental organizations generally view shrimping as a destructive or wasteful fishery that negatively impacts many other living marine resources (Fowle and Bierce 1992).

Bycatch in the shrimp fishery has long been a topic of concern (Lindner 1936). Information on species composition of shrimp fleet discards in the Gulf of Mexico has been documented by several studies (Hildebrand 1954; Bullis and Carpenter 1968; Moore *et al.* 1970; Bryan and Cody 1975; Chittenden and McEachran 1975; Drummond 1976; Pavella 1977; Warren 1981; Bryan *et al.* 1982; Nichols *et al.* 1987; Powers *et al.* 1987). Although shrimp trawl bycatch has been identified as a potential problem for over sixty years, little has been accomplished in reducing or eliminating bycatch. Despite numerous attempts to utilize bycatch, shrimp trawlers continue to discard several t of finfish each year.

In recent years, the Gulf of Mexico shrimp fishery has experienced increased scrutiny regarding the impacts of trawl bycatch on natural resources. Shrimp fishery bycatch discards in the Gulf of Mexico have been estimated at several billion fish per year, with most of the catch composed of groundfish such as croaker, seatrout, porgies and spot (Nichols *et al.* 1990). In addition to the once abundant stocks of groundfish that have suffered reduction over the past few decades, many other species are transient or temporary residents on the shrimp grounds and are captured in shrimp trawls at certain times of the year or during certain life stages. Although these species may represent only a minor component of the total bycatch, their losses may have significant adverse effects on their population abundances. Excessive bycatch in shrimp trawls has been suggested as the primary cause for declines in stocks of some commercially important finfish, endangered sea turtles and other living resources in the Gulf of Mexico (Henwood and Stuntz 1987; Goodyear and Phares 1990; Caillouet *et al.* 1991; Goodyear 1991; Alverson *et al.* 1994).

As one of the largest and most valuable US fisheries (National Marine Fisheries Service Fisheries Statistics Division 1994), the shrimp industry could be severely altered and impacted if effective solutions to the bycatch problem are not found. Additionally, major long-term harvest reductions could be imposed on other important commercial and recreational fisheries targeting finfish species affected by shrimp trawler bycatch. While these actions would be required to protect and in some cases rebuild affected fish populations, their implementation would not be without economic and social disruptions to those fisheries.

In February 1992, a joint commercial/government research programme was initiated between the National Marine Fisheries Service (NMFS) Southeast Fisheries Science Center (SEFSC) and the Gulf and South Atlantic Fisheries Development Foundation, Inc. (Foundation) to collect species-specific bycatch data to characterize catch rates by number and weight taken by the shrimp fishery during commercial operations in the US Gulf of Mexico. The goals of this joint research programme were to: (1) update bycatch estimates temporally and spatially, (2) manage and maintain bycatch characterization data sets, (3) analyse bycatch characterization data on the temporal and spatial catch rates of finfish and shrimp, and (4) provide data to estimate total bycatch of selected species for stock assessment analysis. This research effort provides essential data to Southeast Regional Office (SERO), Gulf and South Atlantic Fishery Management Councils, fish and wildlife departments of Gulf and south Atlantic states, associations of commercial shrimpers and recreational fishermen, and legislators and elected officials at all levels of government. This paper provides a general overview of the results from this ongoing research programme.

MATERIALS AND METHODS

The sampling design used in this research effort follows the guidelines as set forth in the Research Plan Addressing Finfish Bycatch in the Gulf of Mexico and South Atlantic Shrimp Fisheries, prepared by the Foundation, under the direction of a Steering Committee composed of individuals representing industry, environmental, state and Federal interests (Hoar *et al.* 1992). The intent of the sampling design was to survey the shrimp fishery during commercial operations and not to simply establish a research survey study of the bycatch or the finfish populations. The sampling universe consisted of all tows from all vessels shrimping in the Gulf of Mexico. Parameters of interest were the catch totals and size distributions of bycatch species of finfish and invertebrates incidentally taken by the shrimp fleet.

The quantity and type of bycatch differ with fishing location, season, depth, trawl type and turtle excluder device (TED) type. Stratification by each of these variables will tend to minimize the variances of catch estimates. The use of trawl and TED types as stratification variables, however, was impossible since this information was not available '*a priori*'. Thus, only location, season and depth can effectively be used as stratification variables. Twenty-four stratification strata were identified using three seasons (pre-summer: January through April, summer: May through August, and post-summer: September through December), four locations (Florida, Alabama-Mississippi, Louisiana, and Texas), and two depth zones (nearshore, ≤10 fm; and offshore, >10 fm). The sample unit consisted of a single subsample from a trawl haul.

NMFS-trained observers were used to collect the trawl haul subsamples and record the data from the fishery. A subsample was obtained from a randomly-selected net after each tow. The data collected consisted of tow weight, species composition, species abundance, species weight and length measurements by species groups. Preliminary research tows indicated that a subsample of 13 kg per towing hour was adequate to ensure that most species taken in the catch were adequately represented in the subsample. A detailed description of the on-board sampling procedures is contained in the NMFS Bycatch Characterization Sampling Protocol (Galveston Laboratory 1992). All observers, whether funded by NMFS or through the Foundation, were required to collect data following this protocol. To further standardize the data collection methods, observers from the various programmes were required to successfully complete a five-day training workshop established during the first year of the bycatch research programme.

Statistical analysis of the data from the characterization trips has been accomplished using a variety of statistical methods including, but not be limited to, ANOVA, linear and multiple regression, t-tests and spatial statistics. The purpose of these analyses was to detect significant differences in the catch-per-unit-effort (CPUE) of selected bycatch species by season, location and depth. No statistical analysis of the data is presented in this initial paper, and all general comparisons will be only in weight values.

RESULTS AND DISCUSSION

Over the past five years (February 1992 through October 1995) a total of 3653 sea-days of sampling effort has been achieved by NMFS observers (1359 days) and non-NMFS observers (2294 days) in the Gulf of Mexico and along the east coast of the United States of America. Most of the effort occurred in waters

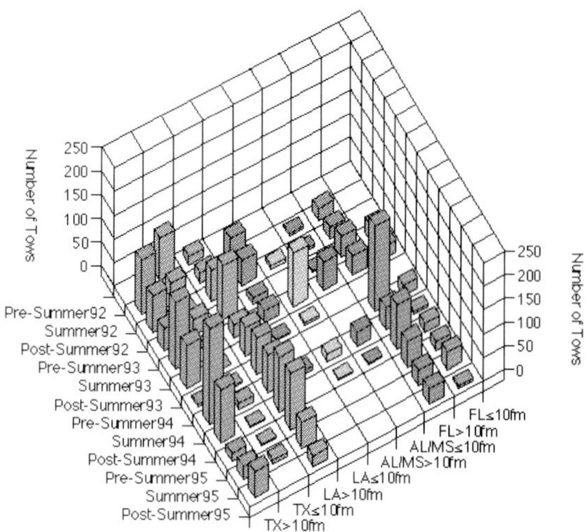

Fig. 1. Number of observed tows depicted by season, location and depth. (TX - Texas, LA - Louisiana, AL/MS - Alabama / Mississippi, FL - Florida).

off Texas (1090 days) and Louisiana (943 days), followed by Florida (686 days) and Alabama–Mississippi (160 days). These sea-days were accomplished during 544 trips, varying in length from 1 to 54 days. From these sea days, bycatch data have been collected from 5045 individual tows, with several hundred different species being documented from the trawls. The majority of tows sampled occurred in the offshore waters (>10 fm) off Louisiana and Texas (Fig. 1). The Alabama–Mississippi area had the fewest sampled tows. This pattern, however, is indicative of the commercial shrimping effort in the areas, since most of the effort and catch is also from the Louisiana and Texas offshore areas.

When all tows collected in the Gulf of Mexico were combined for analysis, the statistics revealed that on the average about 27 kg of organisms per hour are taken during trawling operations. Analysis of the composition of the organisms showed that about 68% of the catch by weight is composed of finfish, 16% by commercial shrimp species (brown shrimp, *Penaeus aztecus*, white shrimp, *P. setiferus*, pink shrimp, *P. duorarum*, seabobs, *Xiphopenaeus kroyeri*, sugar shrimp, *Trachypenaeus constrictus*, and rock shrimp, *Sicyonia brevirostris*), 13% by non-commercial shrimp crustaceans and 3% by non-crustacean invertebrate species. The top 10 species caught in shrimp trawls in the Gulf of Mexico by weight were longspine porgy (15%, *Stenotomus caprinus*), brown shrimp (9%), Atlantic croaker (9%, *Micropogonias undulatus*), inshore lizardfish (6%, *Synodus foetens*), pink shrimp (3%), Gulf butterfish (3%, *Peprilus burti*), lesser blue crab (2%, *Callinectes similis*), white shrimp (2%), longspine swimming crab (2%, *Portunus spinicarpus*) and rock shrimp (2%). The other 47% of the catch not accounted for by these ten species was composed of several hundred finfish and invertebrate species.

Shrimp trawl catch-per-hour averages for each location and depth zone are presented in Figure 2. In each case the offshore zone had a lower catch rate than that of the nearshore area for the same location. The Alabama–Mississippi area had the highest overall combined (nearshore + offshore) catch rate, while

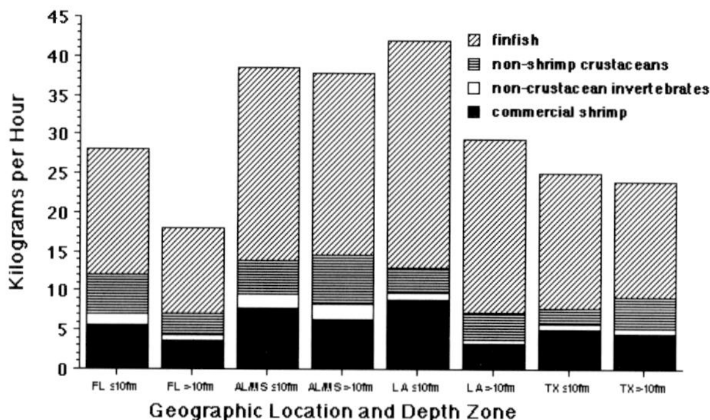

Fig. 2. Average kilograms of organisms per hour by location and depth. (TX - Texas, LA - Louisiana, AL/MS - Alabama / Mississippi, FL - Florida).

the Louisiana nearshore area had the greatest catch rate for a single area. Offshore Florida had the lowest single area catch rate. In each area, the lowest catch rates were for non-crustacean invertebrates, followed by non-shrimp crustaceans and commercial shrimp. Finfish had the highest catch rates in all locations and depths. On the average, about four kg of finfish are taken as bycatch for every one kg of commercial shrimp harvested in the trawl. This is less than the 10:1 finfish to shrimp weight ratio reported in the early 1980s (Pellegrin *et al.* 1981). The use of TEDs in shrimp trawls can probably explain some of the decrease (Renaud *et al.* 1990), while overall reduction in some finfish populations in the Gulf of Mexico may also be contributing to the change in the ratio (Chittenden and McEachran 1975; Nichols *et al.* 1990).

In the Florida nearshore area the iridescent swimming crab (*Portunus gibbesii*) comprised 11% of the average weight in a typical trawl, followed by the leopard searobin (*Prionotus scitulus*) at 6% and the sand perch (*Diplectrum formosum*) at 6%. In the Florida offshore area the blotched swimming crab (*Portunus spinimanus*) constituted the greatest average weight at 9%, followed by the shoal flounder (*Syacium gunteri*) at 7% and the inshore lizardfish at 5%. In the Alabama–Mississippi nearshore area the Atlantic croaker made up the greatest percentage of the catch in a typical trawl at 32%, followed by the sand seatrout (*Cynoscion arenarius*) and the lesser blue crab each at 7%. In the Alabama–Mississippi offshore area the longspine porgy represented the species with the greatest overall average weight percentage in the catch at 14%, followed by the inshore lizardfish and the Atlantic croaker, each at 5%.

In Louisiana, the Atlantic croaker represented the greatest percentage of the nearshore catch at 18%, with Gulf menhaden (*Brevoortia patronus*) at 9% and longspine porgy at 8%. In the offshore waters, longspine porgy was greatest at 20%, followed by the inshore lizardfish and the Atlantic croaker, each at 10%. Catch composition in the Texas area was very similar to that for Louisiana. The nearshore catch in Texas was dominated by Atlantic croaker at 15%, followed by Gulf butterfish at 14% and longspine porgy at 6%. In the offshore area the longspine porgy represented the greatest component of the catch at 25%, followed by the Atlantic croaker at 6% and the inshore lizardfish at 6%. It can be observed from the data that on average the Atlantic croaker was the bycatch species that was represented

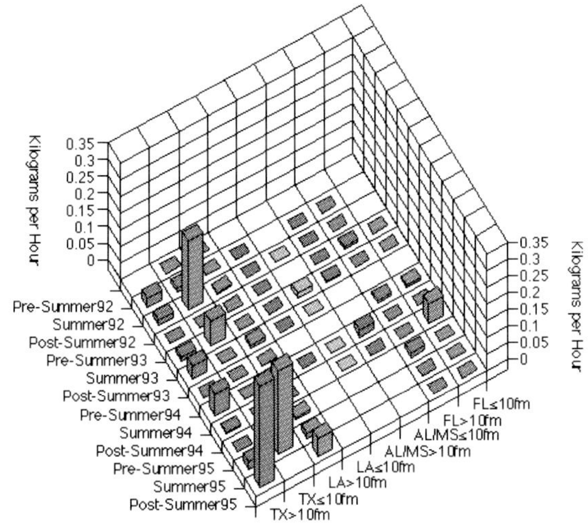

Fig. 3. Average kilograms of king mackerel (*Scomberomorus cavalla*) per hour by season, location and depth. (TX - Texas, LA - Louisiana, AL/MS - Alabama / Mississippi, FL - Florida).

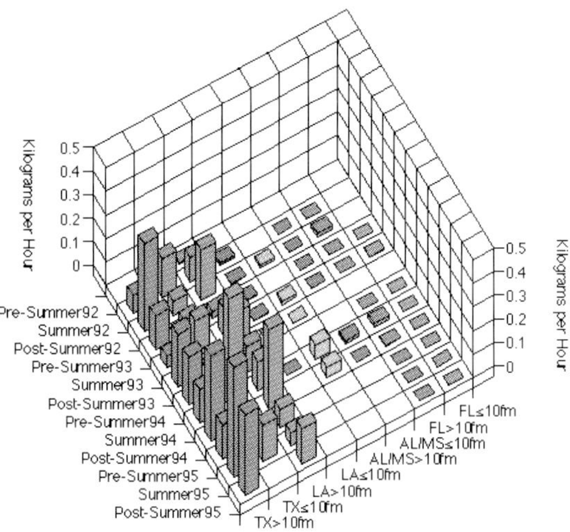

Fig. 5. Average kilograms of red snapper (*Lutjanus campechanus*) per hour by season, location and depth. (TX - Texas, LA - Louisiana, AL/MS - Alabama / Mississippi, FL - Florida).

most frequently in the nearshore trawls for most areas, while the longspine porgy typically dominated the offshore trawls.

Although groundfish species make up the majority of the bycatch taken in shrimp trawls, several species that represent only minor components of the total bycatch have received a great deal of attention because of their commercial and recreational importance, and the potential for significant impacts on their population abundance. Three of these species are king mackerel (*Scomberomorus cavalla*), Spanish mackerel (*S. maculatus*) and red snapper (*Lutjanus campechanus*). Catches of these three species are far less than those observed for some of the common groundfish species. The highest catch rates for king mackerel have been recorded in the Texas area, typically during the summer seasons nearshore and the post-summer seasons offshore

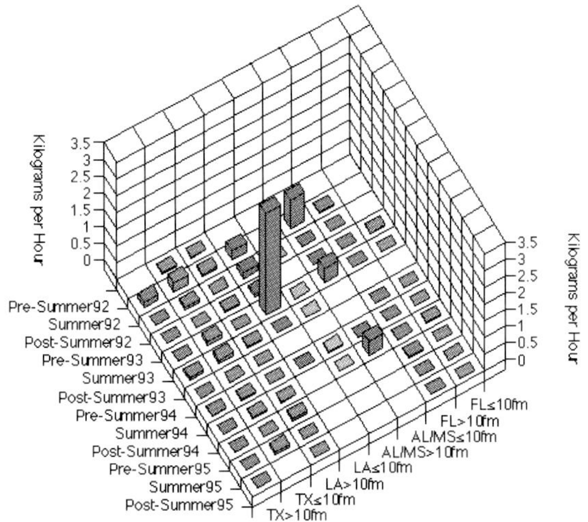

Fig. 4. Average kilograms of Spanish mackerel (*Scomberomorus maculatus*) per hour by season, location and depth. (TX - Texas, LA - Louisiana, AL/MS - Alabama / Mississippi, FL - Florida).

(Fig. 3), with none of the CPUE values over 0.35 kg per hour. On average, king mackerel had a median length of 232 mm. The highest catch rates for Spanish mackerel occurred in the nearshore waters off Texas, Louisiana and Alabama–Mississippi (Fig. 4). A very high catch rate (for Spanish mackerel) was noted during the pre-summer season in 1993 off Louisiana, but most values are below 0.5 kg per hour. On the average, Spanish mackerel had a median length of 179.5 mm. Red snapper are rarely caught in the Florida and Alabama–Mississippi areas (Fig. 5). Highest catch rates are experienced in the offshore waters of Texas and Louisiana during summer and post-summer seasons. None of the catch rates exceeded 0.5 kg per hour. On the average, red snapper had a median length of 109 mm.

Overall the shrimp trawl bycatch characterization research programme has yielded a great deal of important information on catch rates for various invertebrate and finfish species in the US Gulf of Mexico. Estimates of the magnitude of the bycatch for a particular species can be determined by extrapolating these average catch rates with average hours of shrimping effort in the same depth, location and season. However, these estimated values must be used in the context of a stock assessment to determine the actual impacts that shrimp trawling activities have on a particular population.

REFERENCES

Alverson, D L, Freeberg, M H, Murswski, S A, and Pope, J G (1994). A global assessment of fisheries bycatch and discards. *FAO Fisheries Technical Paper* No. 339. *Rome, FAO*, 1994. 233 pp.

Bryan, C E, and Cody, T J (1975). A study of commercial shrimp, rock shrimp, and potentially commercial finfish 1973–75. Part III. Discarding of shrimp and associated organisms in the Texas brown shrimp (*Penaeus aztecus* Ives) grounds. *Texas Parks and Wildlife, Coastal Fisheries Branch Publication*, 1975. 38 pp.

Bryan, C E, Cody, T J, and Matlock, G C (1982). Organisms captured by the commercial shrimp fleet on the Texas brown shrimp (*Penaeus aztecus* Ives) grounds. *Texas Parks and Wildlife Department Technical Series* No. 31. 26 pp.

Bullis, H R Jr., and Carpenter, J S (1968). Latent fishery resources of the central west Atlantic region. In 'The future of the fishing industry of the United States'. *University of Washington Publication in Fisheries, New Series Volume IV*: 61–4.

Caillouet, C W, Duronslet, M J, Landry, A M, Revera, D B, Shaver, D J, Stanley, K M, Heinly R W, and Stabenau, E K (1991). Sea turtle strandings and shrimp fishing effort in the northwestern Gulf of Mexico, 1986–1989. *Fishery Bulletin* **89**, 712–18.

Chittenden, M E, Jr., and McEachran, J D (1975). Fisheries on the white and brown shrimp grounds in the northwestern Gulf of Mexico. *Proceedings of the American Fisheries Society, 105th Annual Meeting, Las Vegas, Nevada, September 13, 1975*. 5 pp.

Drummond, S B (1976). Shrimp fleet discard survey. *Oceanic Resource Surveys and Assessment Task Status Report, NOAA, National Marine Fisheries Service, Southeast Fisheries Science Center,* October 1976. pp. 19–27.

Fowle, S, and Bierce, R (Eds) (1992). Proceedings of the shrimp trawl bycatch workshop, November 22–23, 1991. St. Petersburg, FL. Center for Marine Conservation Publication. 183 pp.

Galveston Laboratory (1992). Shrimp Trawl Bycatch Characterization Sampling Protocol Manual for Data Collection. *National Marine Fisheries Service, Southeast Fisheries Science Center, Galveston Laboratory, Galveston, TX. Laboratory Report*. 62 pp.

Goodyear, C P (1991). Sensitivity of shrimp-bycatch mortality estimates to natural mortality for red snapper. *National Marine Fisheries Service, Southeast Fisheries Science Center, Miami Laboratory, Miami, FL. Laboratory Report, Contribution, MIA 90/91–28*. 6 pp.

Goodyear, C P, and Phares, P (1990). Status of red snapper stocks of the Gulf of Mexico - Report for 1990. *National Marine Fisheries Service, Southeast Fisheries Science Center, Miami Laboratory, Miami, FL. Laboratory Report, Contribution* # CRD-89/90–05.

Henwood, T A, and Stuntz, W E (1987). Analysis of sea turtle captures and mortalities during commercial shrimp trawling. *Fishery Bulletin* **85**, 813–17.

Hildebrand, H H (1954). A study of the fauna of the brown shrimp (*Penaeus aztecus* Ives) grounds in the western Gulf of Mexico. *Publication of the Institute of Marine Sciences, University of Texas* **3**(2), 234–366.

Hoar, P, Hoey, J, Nance, J, and Nelson, C (1992). A research plan addressing finfish bycatch in the Gulf of Mexico and south Atlantic shrimp fisheries. *Gulf and South Atlantic Fisheries Development Foundation, Inc. Publication*. 114 pp.

Lindner, M J (1936). A discussion of the shrimp trawl-fish problem. *Louisiana Conservation Review, October 1936*, 12–17, 51.

Moore, D H, Brusher, H A, and Trent, L (1970). Relative abundance, seasonal distribution, and species composition of demersal fishes off Louisiana and Texas, 1962–64. *Contribution in Marine Science* **15**, 45–70.

National Marine Fisheries Service Fisheries Statistics Division (1994). Fisheries of the United States, 1994. Current Fishery Statistics No. 9400. 113 pp.

Nichols, S, Shah, A, Pellegrin, G, Jr., and Mullin, K (1987). Estimates of annual shrimp fleet bycatch for thirteen finfish species in the offshore waters of the Gulf of Mexico. *NMFS Southeast Fisheries Science Center, Pascagoula Laboratory, Pascagoula, MS. Laboratory Report, October 1987*. 25 pp.

Nichols, S, Shah, A, Pellegrin, G, Jr., and Mullin, K. (1990). Updated estimates of annual shrimp fleet bycatch in the offshore waters of the Gulf of Mexico. *NMFS Southeast Fisheries Science Center, Pascagoula Laboratory, Pascagoula, MS. Laboratory Report, August 1990*. 25 pp.

Pavella, J S (1977). An analysis of the finfish discards resulting from commercial shrimp trawling in the north central Gulf of Mexico. *M.S. Thesis, University of South Mississippi*. 143 pp.

Pellegrin, G J, Jr., Drummond, S B, and Ford, R S, Jr. (1981). The incidental catch of fish by the northern Gulf of Mexico shrimp fleet. *NMFS Southeast Fisheries Science Center, Pascagoula Laboratory, Pascagoula, MS. Laboratory Report*. 49 pp.

Powers, J E, Goodyear, C P, and Scott, G P (1987). The potential effect of shrimp fleet bycatch on fisheries production of selected fish stocks in the Gulf of Mexico. *National Marine Fisheries Service, Southeast Fisheries Science Center, Miami Laboratory, Miami, FL. Laboratory Report, Contribution* # CRD-87/88-06. 7 pp.

Renaud, M, Gitschlag, G, Klima, E, Shah, A, Nance, J, Caillouet, C, Zein-Eldin, Z, Koi, D, and Patella, F (1990). Evaluation of the impact of turtle excluder devices (TEDs) on shrimp catch rates in the Gulf of Mexico and south Atlantic, March 1988 through July 1989. *NOAA Technical Memorandum, NMFS-SEFC-254*. 165 pp.

Warren, J R (1981). Population analysis of the juvenile groundfish on the traditional shrimp grounds in Mississippi Sound before and after the opening of the shrimp season. *M.S. Thesis, University of South Mississippi*. 94 pp.

Exploitation of Norwegian Spring-spawning Herring (*Clupea harengus* L.) Before and After the Stock Decline: Towards a Size-Selective Fishery

Aril Slotte[A] *and Arne Johannessen*[B]

[A] Institute of Marine Research, PO Box 1870, N-5024 Bergen, Norway.
[B] Department of Fisheries and Marine Biology, University of Bergen, Bergen High Technology Centre, N-5020 Bergen, Norway.

Summary

The fishery of Norwegian spring-spawning herring was unregulated before the stock collapse in the late 1960s after improved fishing technology between the 1940s and 1960s had increased fishing mortality substantially. The major part of the catch was processed to meal and oil, and the size of the fish was unimportant. Thus, there was an extensive fishery on the immature herring which had a serious impact on the stock decline.

Recently the stock has recovered and reached a biomass level almost comparable to that of the 1950s, but the fishing pattern has been modified. Today the herring fishery is size-selective, and influenced by three major factors. Firstly, there is a minimum landing size of 25 cm. Secondly, migrations of the adult herring are size-dependent and thirdly, the large herring is of higher economic value compared to medium- and small-sized herring.

Introduction

The Norwegian spring spawning herring is a classical example of a fish stock which has undergone stock depletion and successful rebuilding (Fig. 1). This stock peaked at a total biomass level of about 16 million t in 1954 and by the end of the 1960s it decreased to below 100 000 t (Anon. 1996). Prior to the stock collapse the fishery was unregulated, and thus the fishing effort made serious impact on the state of the stock. The introduction of new technology, such as echo-sounder in the late 1940s, sonar in the late 1950s and the power block in the early 1960s, increased the catch potential of the stock substantially (Figs 2 and 3). The individual size of the herring was unimportant, and the major part of the catch was processed to meal and oil (Fig. 4). The fishery on immature herring was quite extensive in the 1960s (Fig. 2) and this had a serious influence on the collapse of the stock (Dragesund *et al.* 1980). From 1970 onwards the fishery on immature herring was banned and regulative control of the fishery on adult herring was introduced. Recruitment to the stock was poor throughout the 1970s, and growth of the spawning stock in recent years is almost entirely attributed to the 1983 year-class. In 1991, this year-class contributed with 80–90% of the spawning stock in terms of numbers (Anon. 1991). A total stock biomass of about 9 million t in 1995 is attributed mainly to recruitment of the 1991 and 1992 year-classes (Anon. 1996).

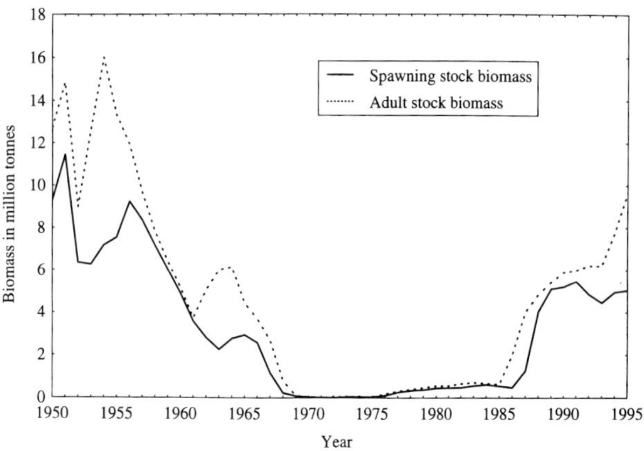

Fig. 1. Stock biomass of Norwegian spring-spawning herring 1950–1995 (Anon. 1996). Spawning stock biomass includes all mature herring, and adult stock includes all herring ≥3 years old (including immature herring).

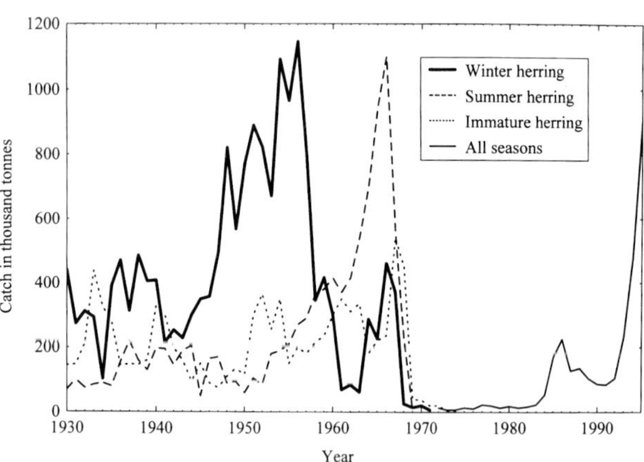

Fig. 2. Catches of Norwegian spring-spawning herring 1930–1995. The winter herring fishery includes Norwegian catches during the pre-spawning and spawning season in January–April. The summer herring fishery includes international catches (dominated by Iceland and Russia) in the Norwegian Sea in May–August. Immature herring was dominated by 'small' (0–1 year olds) and to some extent 'fat' herring (1–4 year olds). Data from 1930–1971 are adapted from Dragesund *et al.* (1980), and data from 1972–1995 are taken from Anon. (1996).

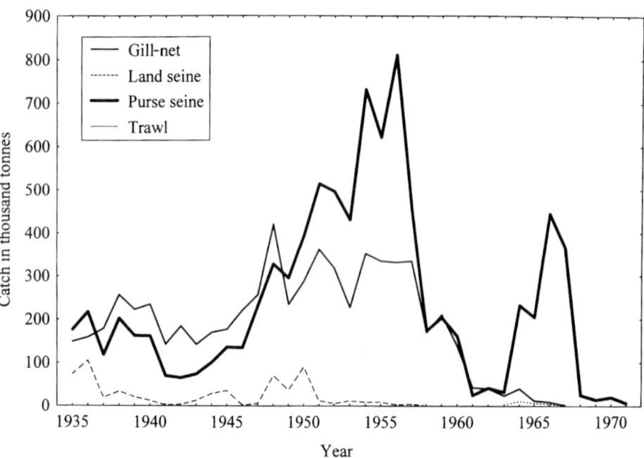

Fig. 3. The Norwegian winter herring fishery taken by different gear categories in 1935–1971. Data from statistical year books of the Norwegian fishery.

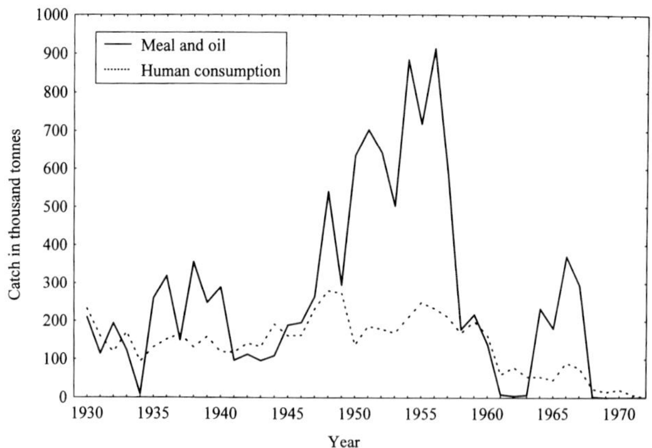

Fig. 4. The Norwegian winter herring fishery by utilization in 1930–1971. Data source: Statistical year books of the Norwegian fishery.

The purpose of this paper is to describe the present fishery on the Norwegian spring-spawning herring and evaluate the impact on the population.

Materials and methods

Data

Areas and locations referred to in this paper are given in Figure 5. The areas correspond to the statistical areas applied by the Institute of Marine Research. The year 1995 is applied in this paper as an example of the present exploitation of the

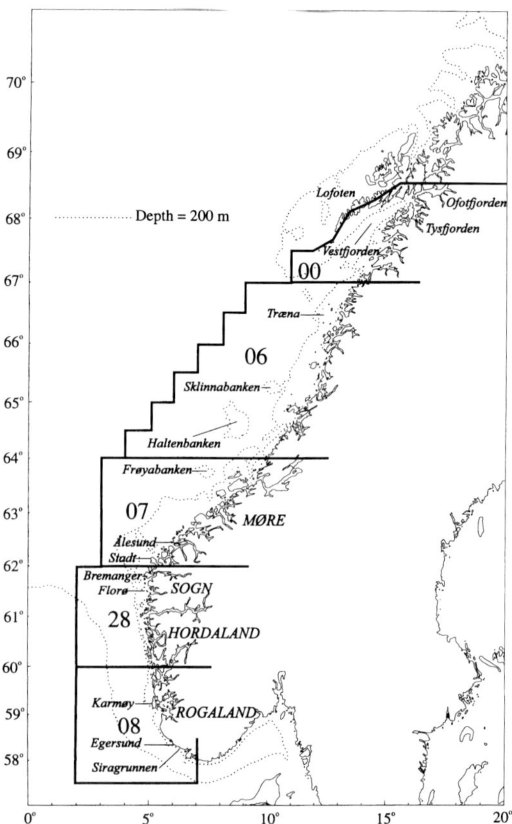

Fig. 5. Statistical areas and locations referred to in the paper.

Norwegian spring-spawning herring stock. Commercial catch data were provided by *The Norwegian fishermen's sales organisation for pelagic fish.* All herring catches landed in Norway are recorded with the following: catch size in kilograms, catch position in terms of statistical area and location, and date of capture. The catches processed for human consumption are divided into five size groups by individual weights as follows: group 1 (>333 g), group 2 (200–333 g), group 3 (125–200 g), group 4 (83–125 g) and group 5 (<83 g). The proportion of the total catch (kg) is calculated for each size group by taking subsamples of the catch at landing.

Estimation of year-class composition

The year-class composition in the catch was estimated by a new method combining data from biological samples with catch by size group from the fishery. The analysis is described in a working document presented at the ICES Northern Pelagic and Blue Whiting Working Group 1996 (Slotte and Røttingen 1996).

The year-class composition of the adult stock during the pre-spawning and spawning period (January–April) in 1995 was represented by relative abundance estimates obtained during a research survey applying acoustic methodology and trawling by the R/V *Michael Sars* in February–March 1995, whereas estimates obtained in February-March 1996 were considered representative for the age composition in the wintering period (August–December) in 1995 (Anon. 1996).

RESULTS

Fishery

Several nations participated in the catching of Norwegian spring-spawning herring in 1995, and Norway caught the largest portion (59%) of a total catch of about 900 000 t (Table 1). Norwegian and Russian catches were taken in the wintering area in Vestfjorden (area 00) and in the spawning areas along the Norwegian coast (mainly areas 06 and 07) (Fig. 5), whereas Icelandic and Faroese catches were taken mainly within Faroese waters and to some extent in international waters. Other nations took all their catches in international waters. The mean individual weight of the herring was high in the catches taken by all nations (Table 1). This indicates a size-selective fishery, which is based on three main groups of underlying factors: regulative, biological and economic.

Regulative factors

Regulative measures were introduced when the stock collapsed. In 1970 the fishery for small herring was banned, and a minimum landing size of 20 cm was introduced. This limit was increased to 25 cm in 1975. If catches of herring below 25 cm are taken in any area, then the fishery in these areas is closed until the size composition has changed.

Biological factors

Co-ordinated surveys with research vessels from the Faroes, Iceland, Norway and Russia recorded the distribution of Norwegian spring-spawning herring in the Norwegian sea during the summer 1995 (Anon. 1995). Size-dependent migration of the herring make large individuals more available in some areas. Larger individual mean size during the summer was recorded for the herring distributed in international and Faroese waters than for those in Norwegian waters. Thus the high mean weight during this period is not a result of size-selective fishing, rather than a result of size-dependent migration. However, in the wintering area (area 00) and spawning areas (mainly 06 and 07) the fishery was size-selective. This is confirmed by comparing the year-class composition in catches with that of the stock (age ≥ 3) during the pre-spawning and spawning season (January–April) (Fig. 6) and during the winter season (August–December) (Fig. 7). This fishery depended on size-dependent migration of the herring as described below.

The spawning migration started in early January and by the end of the month the stock had left the wintering area in Ofotfjorden, Tysfjorden and inner Vestfjorden. Figure 8 shows that there was little variation in the size-group composition during this period. The fishery was still size-selective and occurred in those parts of the fjords inhabited by herring of larger size.

Table 1. Total catch (t) and mean individual weight (gram) of Norwegian spring-spawning herring 1995 by nation (Anon. 1996)

Nation	Total catch (t)	Mean weight (g)
Norway	529 838	306
Russia	100 000	281
Iceland	173 418	313
Faroes	57 084	264
Netherlands	7 969	355
Denmark	30 131	313
UK (Scotland)	230	313
Germany	556	355
Greenland	3 000	313
Total	902 226	302

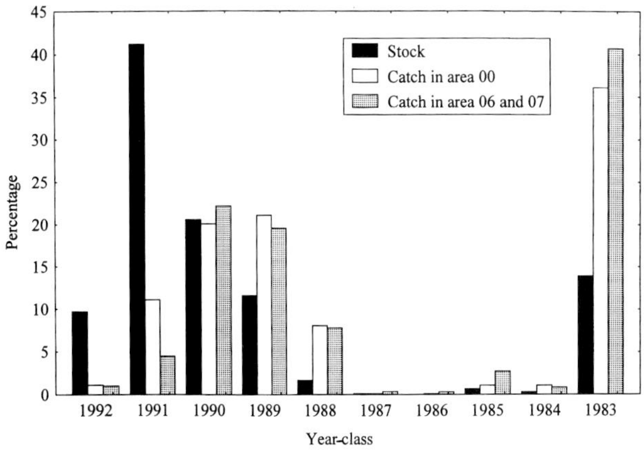

Fig. 6. Year-class composition of the stock, and in commercial catches of the Norwegian spring-spawning herring during the prespawning and spawning period January–April 1995 in area 00 and 06+07 (see Fig. 5).

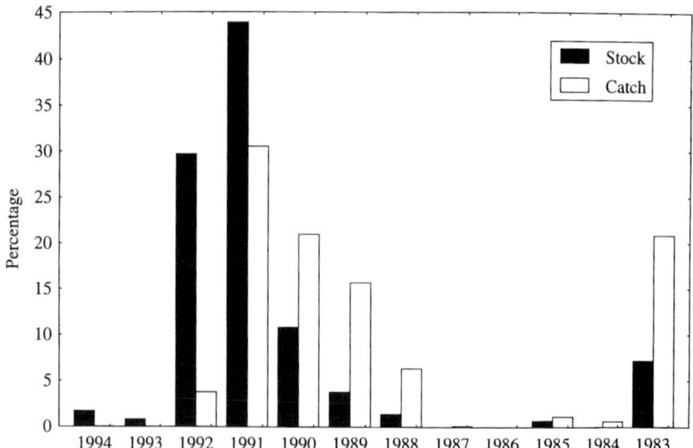

Fig. 7. Year-class composition of the stock, and in commercial catches of the Norwegian spring-spawning herring during the wintering period 1995 in Vestfjorden (area 00).

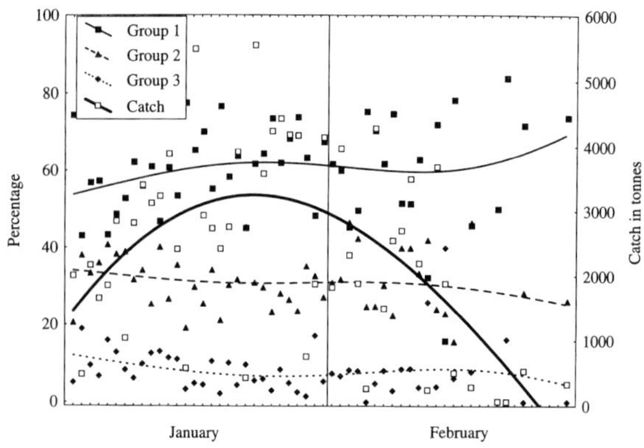

Fig. 8. Variations in relative proportion of size-groups in commercial catches of Norwegian spring-spawning herring in area 00 during the pre-spawning period in January–February 1995 (catches of herring in size group 4–5 were rare).

Spawning migration along the coast was size-dependent, with the largest herring arriving at the spawning grounds off Møre before the smaller ones. The fishery was also most intense during the period with the largest herring present (Figs 9 and 10). In late January a small fishery developed south of Vestfjorden (off Træna in area 06), which was based on the first portion of the southward-migrating herring. During February the fishery was more widespread in area 06, and catches of herring migrating southwards from Vestfjorden were still recorded off Træna and also farther south around Sklinnabanken and Haltenbanken. During the period February-April the main fishery was located off Møre (area 07), south of Frøyabanken, with important catching localities off Ålesund.

Only a minor part of the stock migrated south of Stadt and in February-March a small fishery (below 2000 t) was located off the Sogn district (area 28), mainly off Bremanger and Florø. Some herring migrated farther south to the Hordaland (area 28) and Rogaland (area 08) districts, aiming for the spawning grounds off Karmøy, Egersund and Siragrunnen, and a small catch of only 540 t was taken in this area.

Maturation of recruit spawners (3–4 year olds) was somewhat delayed (unpublished data) compared to repeat spawners (5 years and older). The recruit spawners tended to spawn closer to the wintering area, and the northernmost spawning area was recorded off Lofoten. Thus the migration distance and energy expenditure for migration was reduced.

An opposite trend in the migration pattern was observed during the fall when the herring arrived in Vestfjorden for wintering, and the fishery started. The fish size then increased from August and stabilized in October–November, when the fishery also peaked (Fig. 11). As in January, herring of larger size occurred at some locations of the fjord, allowing a size-selective fishery to take place.

Economic factors

The herring is of higher economic value when processed for human consumption compared to oil and meal, and the

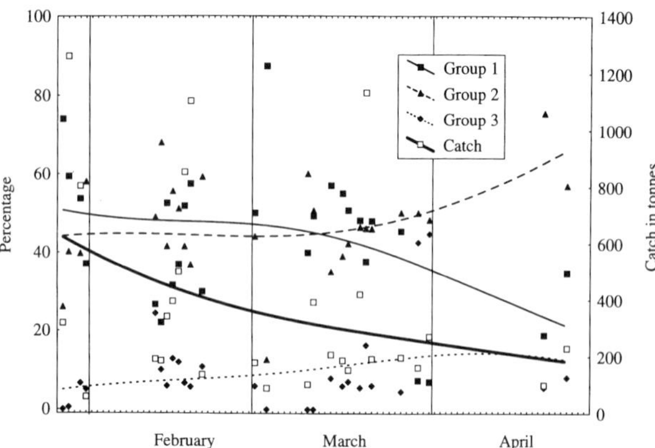

Fig. 9. Variations in relative proportion of size-groups in commercial catches of Norwegian spring-spawning herring in area 06 during the pre-spawning and spawning period in January–April 1995 (catches of herring in size group 4–5 were rare).

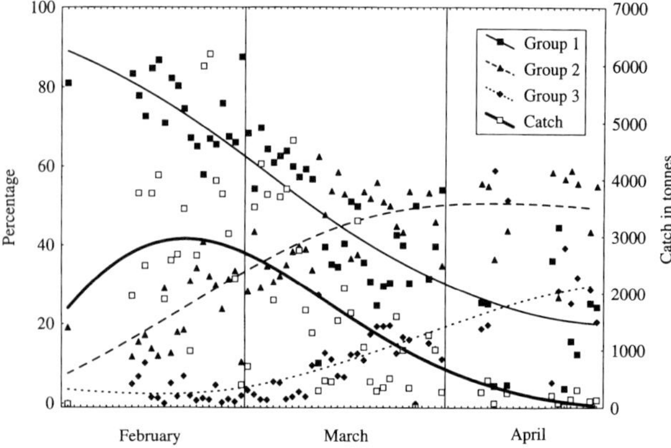

Fig. 10. Variations in relative proportion of size-groups in commercial catches of Norwegian spring-spawning herring in area 07 during the pre-spawning and spawning period in February–April 1995 (catches of herring in size group 4–5 were rare).

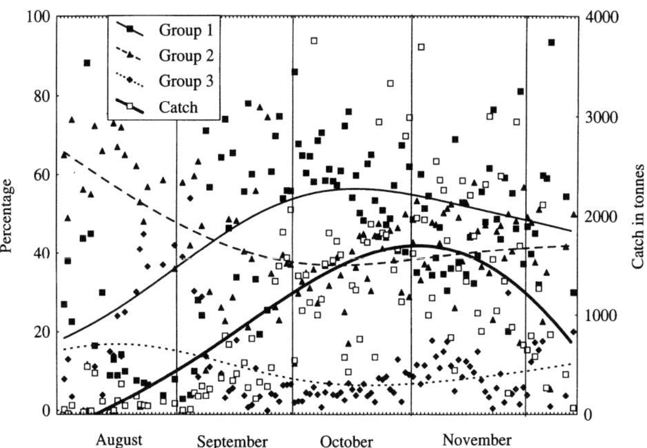

Fig. 11. Variations in relative proportion of size-groups in commercial catches of Norwegian spring-spawning herring in area 00 during the wintering period in August–December 1995 (catches of herring in size group 4–5 were rare).

fishermen are on average paid about twice the price when landing catches for consumption. In 1995, 74% of the total catch was processed for consumption, and catches reduced to oil and meal production was most common in January and during August-September (Fig. 12). The value of a catch depends on size and quality of the herring. Fish in size groups 1 and 2 are of higher economic value than those in size group 3 (the fishermen are on average paid 1.5 times higher price for size groups 1 and 2), whereas herring in size groups 4 and 5 are considered unsuitable for human consumption. As a rule, trawl-caught herring as well as purse seine catches which are dominated by size groups 3–5 are processed for meal and oil, whereas purse seine catches dominated by larger herring (size groups 1 and 2) are used for human consumption. In January the fishermen are allowed new annual quotas and many trawlers take their herring catch quotas early in the year in order to be able to participate in other fisheries later in the year. This may explain the amount of herring processed for meal and oil during this period. In August–September, the majority of herring available are of small size and not preferred for human consumption.

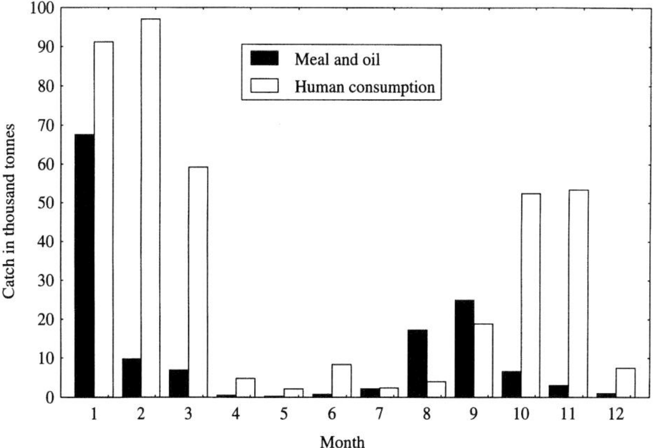

Fig. 12. Monthly catches of Norwegian spring-spawning herring processed for meal and oil, or directed for human consumption in 1995.

Discussion

Size-dependent migration

Why are migrations of Norwegian spring-spawning herring size-dependent? During the winter period in the Vestfjorden area, catches taken in some parts of the fjord comprise herring of larger size than in other parts. This indicates that schooling of herring is size-dependent. Several studies propose that it may be more beneficial to the individual fish in a school if there is little variation in size and state of the fish. Pitcher *et al.* (1996) observed that splitting and joining of individual herring in schools in the open sea can occur frequently. This implies that segregation by size may be a mechanism to explain why catches with fish of different size are segregated by area. Similarly, hydrodynamics may provide advantages for herring in a school when swimming close to neighbours of similar size (Pitcher *et al.* 1985), and Nøttestad *et al.* (1996) observed that school dynamics depend closely on the state of the herring.

During spawning migration however, other considerations have to be made. Herring are known to arrive at spawning grounds in waves, with the largest fish first and the smaller ones behind, and this is believed to be a result of delay in maturation rate from large to smaller herring (Lambert 1987, 1990; Ware and Tanasichuk 1989). In recent years, maturation of recruit spawners of Norwegian spring-spawning herring was somewhat delayed compared to older conspecifics (unpublished data). Recruit spawners tended to spawn closer to the wintering area, which reduced their migration distance and energy costs. A delay in maturation with decreasing size does not seem to be a fact among the repeat spawners (unpublished data) .

Studies of Blaxter (1969), and Ware (1975, 1978) support the hypothesis that migration speed is related to fish size. Ware (1975, 1978) found that optimal cruising speed, defined as the velocity at which the total energy expenditure per unit distance travelled is minimal, increases with the size of pelagic fish. It seems therefore likely that herring which migrate towards spawning grounds without foraging, travel at a speed which will minimize the energy expended and optimize the reproductive output. This may explain why the large herring arrive at the spawning grounds earlier than the smaller ones, and the smallest individuals prefer to spawn closer to their wintering area. The size-dependent feeding migration as observed in the Norwegian Sea (Anon. 1995) is probably also a result of increased optimal swimming speed with fish size.

Effect of size-selective fishery

Hamre and Toresen (1986) studied effects of minimum landing size regulation in the fishery of Norwegian spring-spawning herring. They found that by increasing the minimum landing size from 20 cm to 27 cm, little gain in yield-per-recruit was obtained, but the effect on the spawning stock biomass was considerable. Similar effects on the stock are expected when the landing size is further increased as a result of size-selective fishery. However, there may also be negative effects of a size-selective fishery. Intense fishing pressure on the frontal part of herring under migration may reduce the likelihood that a considerable portion of the stock will reach the southern spawning grounds, which were also regarded as the most important ones in the 1930s (Runnstrøm 1941). Catch rates off Møre in February 1995 were up to 6000 t/day. When comparing catches at Møre with abundance estimates of the portion of the stock spawning at Karmøy (10 000–30 000 t) in 1990–1994 (Johannessen *et al.* 1995), it is not unreasonable that this fishery reduces the portion of the spawning stock migrating southwards. This is supported by the fact that the herring reaching Karmøy in late February and March has a similar size and year-class composition to the herring caught off Møre in mid February (Johannessen *et al.* 1995).

Reducing the proportion of the stock spawning at the southern grounds may also have negative effects on recruitment of the herring. Dragesund (1970) suggested that the more important factors determining recruitment to the stock were: (1) the size and the geographical spread of the spawning area; (2) duration of the spawning period; (3) rate of dispersion of larvae from the hatching grounds; and (4) the match or mismatch between available food and the hatching of herring larvae. The spawning period off Karmøy is longer than off Møre (Johannessen *et al.* 1995) and larvae from more southern grounds may be dispersed over a larger area (Svendsen *et al.* 1995). Thus increased spawning in this area would increase the likelihood of good recruitment.

Although increased fishing pressure on the larger fish has no obvious advantage for the recruitment potential of herring from the southern spawning grounds, the overall effect for the stock seems positive under the present migration regime of Norwegian spring-spawning herring.

Acknowledgments

This study was funded by the Norwegian Research Council. The authors would like to thank *The Norwegian fishermen's sales organisation for pelagic fish* for making data on catch by size group available for statistical analysis.

References

Anon. (1991). Report of the Atlanto-Scandian Herring and Capelin Working Group. *ICES CM*. Assess **17**, 1–62.

Anon. (1995). Report on surveys of the distribution and migrations of the Norwegian spring spawning herring and the environment of the Norwegian Sea and adjacent waters during the spring and summer of 1995. (ICES Working Group Document).

Anon. (1996). Report of the Northern Pelagic and Blue Whiting Fisheries Working Group. *ICES CM*. Assess **14**, 1–33.

Blaxter, J H S (1969). Swimming speed of fish. *FAO Fisheries Reports* **62**, 69–100.

Dragesund, O (1970). Factors influencing year-class strength of Norwegian spring spawning herring (*Clupea harengus* L.). *Fiskeridirektoratets Skrifter Serie Havundersøkelser* **15**, 381–450.

Dragesund, O, Hamre J, and Ulltang, Ø (1980). Biology and population dynamics of the Norwegian spring spawning herring. *Rapports et Procès-Verbaux des Réunions du Conseil International pour l'Exploration de la Mer* **177**, 43–71.

Hamre, J, and Toresen, R (1986). The effects of minimum landing size regulation in the fishery of Norwegian spring spawning herring. *ICES C.M.* 1986/H:51.

Johannessen, A, Slotte, A, Bergstad, O A, Dragesund, O and Røttingen, I (1995). Reappearence of Norwegian spring spawning herring (*Clupea harengus* L.) at spawning grounds off southwestern Norway. In 'Ecology of Fjords and Coastal Waters'. (Eds H R Skjoldal, C Hopkins, K E Erikstad and H P Leinaas) pp. 347–63.

Lambert, T C (1987). Duration and intensity of spawning in herring (*Clupea harengus*) as related to the age structure of the mature population. *Marine Ecology Progess Series* **39**, 209–20.

Lambert, T C (1990). The effect of population structure on recruitment in herring. *Journal du Conseil International pour l'Exploration de la Mer* **47**, 249–55.

Nøttestad, L, Aksland, M, Beltestad, A, Fernø, A, Johannessen, A, and Misund, O A (1996). Schooling dynamics of Norwegian spring spawning herring (*Clupea harengus* L.) in a coastal spawning area. *Sarsia* **80**, 277–84.

Pitcher, T J, Magurran, A E, and Edwards, J I (1985). Schooling mackerel and herring choose neighbours of similar size. *Marine Biology* **86**, 319–22.

Pitcher, T J, Misund, O A, Fernø, A, Totland, B, and Melle, V (1996). Adaptive behaviour of herring schools in the Norwegian Sea as revealed by high-resolution sonar. *ICES Journal of Marine Science* **53**, 449–52.

Runnstrøm, S (1941). Quantitative investigations on herring spawning and its yearly fluctuations at the west coast of Norway. *Fiskeridirektoratets Skrifter Serie Havundersøkelser* **6**(8), 5–71.

Slotte, A, and Røttingen, I (1996). The Norwegian fishery on Norwegian spring spawning herring in 1995. Working document in the ICES Northern Pelagic and Blue Whiting Working Group, Bergen, 23–29 April 1996.

Svendsen, E, Fossum, P, Skogen, M, Eriksrød, G, Bjørke, H, Nedreaas, K, and Johannessen, A (1995). Variability of the drift patterns of spring spawned herring larvae and the transport of water along the Norwegian shelf. *ICES C.M* /Q:25.

Ware, D M (1975). Growth, metabolism, and optimal swimming speed of a pelagic fish. *Journal of Fisheries and Research Board of Canada* **32**, 33–41.

Ware, D M (1978). Bioenergetics of pelagic fish: theoretical change in swimming speed and ration with body size. *Journal of Fisheries Research Board of Canada* **35**, 220–8.

Ware, D M, and Tanasichuk, R V (1989). Biological basis of maturation and spawning waves in Pacific herring (*Clupea harengus pallasi*). *Canadian Journal of Fisheries and Aquatic Sciences* **46**, 1776–84.

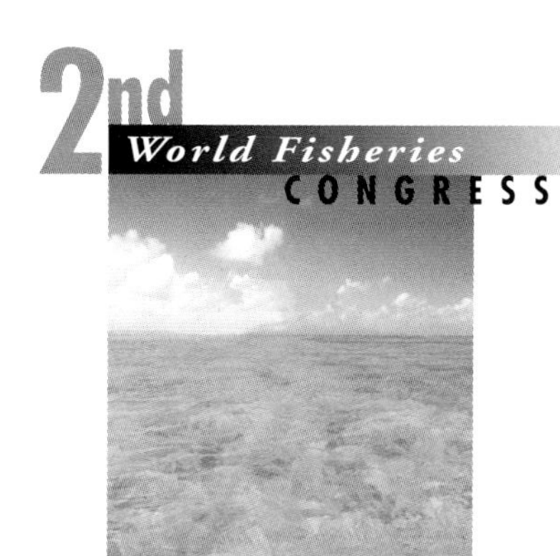

Fisheries and Climate Change: the IPCC Second Assessment

John T. Everett

NOAA, National Marine Fisheries Service, 1315 East-West Highway, Silver Spring, MD 20910, USA.

Summary

'Fisheries. Climate-change effects interact with those of pervasive overfishing, diminishing nursery areas, and extensive inshore and coastal pollution. Globally, marine fisheries production is expected to remain about the same; high-latitude freshwater and aquaculture production are likely to increase, assuming that natural climate variability and the structure and strength of ocean currents remain about the same. The principal impacts will be felt at the national and local levels as species mix and centers of production shift. The positive effects of climate change, such as longer growing seasons, lower natural winter mortality, and faster growth rates in higher latitudes, may be offset by negative factors such as changes in established reproductive patterns, migration routes, and ecosystem relationships.' These words and others from the IPCC second assessment, and some background, are presented and discussed.

Introduction

The second Scientific Assessment of Climate Change of the Intergovernmental Panel on Climate Change (IPCC) is in distribution (IPCC 1995*a*, *b*). As a member of the IPCC team, the author led an international team of authors, contributors and reviewers to assess the impacts of climate change on world freshwater and marine fisheries and on aquaculture, and to suggest adaptation measures. Dr. H. Suzanne Bolton provided enormous amounts of scientific support to the team.

In the text that follows, the author presents material from several IPCC documents in a manner that captures the IPCC mission and the findings of interest to the fisheries community. The same team helped in preparing or reviewing these other documents. In some cases, the author has elaborated on the IPCC narrative to provide amplification or rationale.

The IPCC was established in 1988 to provide an authoritative statement of scientific opinion on climate change. Several hundred scientific experts served on three Working Groups. Their work was broadly peer-reviewed and subjected to full governmental reviews. Working Group I deals with the science of climate change itself, Working Group II deals with the impacts and response strategies, and Working Group III with broad socio-economic issues such as the costs and benefits of global mitigation efforts in energy, forestry and agriculture. All the significant fisheries materials are included in Working Group II reports.

Nature of the issue

Human activities are increasing the atmospheric concentrations of greenhouse gases, which tend to warm the atmosphere, and, in some regions, aerosols, which tend to cool the atmosphere. These changes in greenhouse gases and aerosols, taken together, are projected to lead to regional and global changes in climate and climate-related parameters such as temperature, precipitation, soil moisture, and sea level. Based on the range of sensitivities of climate to increases in greenhouse gas concentrations and plausible ranges of emissions, climate models, taking into account greenhouse gases and aerosols, project an increase in global mean surface temperature of about 1 to 3.5°C by the year 2100, and an associated increase in sea level of about 15 to 95 cm. The reliability of regional-scale predictions is still low, and the degree to which climate variability may change is uncertain. However, potentially serious changes have been identified, including an increase in some regions in extreme high-temperature events, floods, and droughts. Anticipated consequences include effects on fires, pest outbreaks, and composition, structure, and functioning of the ecosystems, including primary productivity.

Working Group I summary of predicted effects of climate change (IPCC 1995*a*)

Working Group I evaluated results from Global Circulation Models (GCMs), other simulation models, observed trends, palaeo-ecological data and other information. This was provided to Working Group II with information to apply to physical, chemical, geological, biological and socio-economic systems of significance today and in the future. Working Group II then assessed impacts and developed possible response strategies.

Working Group I emphasized that 'the role of the ocean, including changes in sea-ice, in modulating future climate change' is an important factor limiting the accuracy of prediction and detection of climate change by existing models. The very limitation in understanding physical, chemical and biological systems in the ocean and their interactions is one of the key factors limiting the ability of existing models to provide Working Group II scientists with more precise guidance.

Owing to the coarse resolutions of GCMs (needed to make the calculations feasible with today's computers) the outputs are too broad to support good estimates of impacts on populations whose territories and behaviour are far more constrained. The models are quite limited in their ability to make local or, to a large extent, even regional projections in time scales relevant to many biological processes. While it is easy to dismiss the significance of the projections encompassed in the resulting documents, the limits of models simply have to be acknowledged and proper caveats applied. Even with the limitations, the model projections provide bounds that can be considered in addressing potential living marine resource responses. The models are improving, assisted by greater interplay between modellers and their user community. A workshop across many disciplines of modellers is planned for September 1996 and should be of great assistance in improving the relevance of GCM outputs.

Many of the physical and chemical changes in which Working Group I has the greatest confidence focus around the greenhouse effects, the human contribution of aerosols, and large areal and temporal-scale events. Many of the important processes in the oceans, wetlands, lakes and streams are inadequately described within the GCMs so that the predictability of climate impacts diminishes as geographical scale is reduced.

The following sections summarize the key findings of relevance to fisheries.

Temperature

Global average temperature is expected to increase by 1 to 3.5°C by 2100. High northern latitudes should warm more than the average. Night-time and winters should warm more than the average. Changes in oceanic conditions will lag behind changes on the continents perhaps by about ten years and rise as much or nearly so as the land. Exceptions might occur in a belt around Antarctica and in the high-latitude North Atlantic. The modelled sea surface temperature (SST) around Antarctica increases more slowly because the effective oceanic thermal inertia is very large due to the deep vertical mixing of water in regions of upwelling and formation of Antarctic Bottom Water. The predicted temperature increase in the high-latitude North Atlantic is also slower due to thermal inertia from the continued operation of the thermohaline-driven current system 'conveyor belt' and the formation of a deep water mass, the North Atlantic Deep Water. GCMs predict a gradual weakening of the thermohaline circulation in the North Atlantic. However, if the thermohaline circulation weakens sufficiently, due to increased precipitation and resultant freshwater capping of the ocean surface layer, the operation of the conveyor belt may change suddenly rather than gradually. Over the past hundred years, the temperature of the land and ocean surface has increased by about 0.3 to 0.6°C. There is no consistent, global pattern of change in variability. The temperatures of the last hundred years may be as warm as any sustained period of several centuries in the past 10 000 years. There is strong evidence that the North Atlantic warmed many times during the last glacial period by as much as 5 to 7°C over a few decades then cooled over centuries. There is weaker evidence that similar rapid changes may have occurred in the previous inter-glacial period as well. There is no information available that indicates similar rapid temperature changes in other regions. The last 10 000 years have been very stable and it is unlikely that mean global temperatures have varied by more than 1°C in one century.

Precipitation

There should be a few percent increase in precipitation, but it may be more concentrated in time. Increases are expected in low and high latitudes, with greater evaporation in mid-latitudes. This can affect regional ocean salinity, watershed flows, turbidity, and related factors.

Sea level rise

Sea level is forecast to rise 15 to 95 cm by 2100. Changes will occur from thermal expansion and melting of ice, with regional variations due to dynamic effects of wind and atmospheric pressure patterns, regional ocean density differences, land motion and oceanic circulation. Tide gauge and other information show a 10–25 cm rise in sea level during this century. No acceleration in the rate of rise has been detected in this period.

Pollutants

Changes in the magnitude and temporal pattern of pollutant loading in the coastal ocean, lakes, and rivers will occur as a result of changes in sea level, precipitation and runoff.

Stability

Once greenhouse gas emissions stabilize, climate stability will not be reached for centuries, largely because of the thermal mass of the ocean.

Currents and upwelling

Freshwater influx from the movements and melting of sea ice or ice sheets may lead to a weakening of the global thermohaline circulation, leading to cooling in the North Atlantic and possibly causing unpredictable instabilities in the climate system. There are competing arguments as to whether oceanic and coastal upwelling would increase or decrease. In any case, there is no ability to make reliable forecasts at the regional scale as to which governs the upwelling systems. Forces leading the oceans to vary in temperature and currents on decadal scales of natural variability are not well understood and are not well modelled.

Storms

It is uncertain whether the frequency and severity of tropical cyclones will increase due to climate change. There are no definite trends observed over the last 50 years. Similarly, while there is some evidence of regional changes in non-tropical storminess, there is no evidence of any uniform increase. Also, there are no clear trends in variability of extreme events at the global scale, although there are some regional trends in both directions.

El Niño

It is uncertain whether the intensity or frequency of El Niño Southern Oscillation events in the Pacific might change as a result of global warming. The instrumental record of the last 120 years indicates that the post-1989 period of ENSO activity is unusual, but some important information is not available. There is some pre-instrument information that indicates that such behaviour has happened before.

Sea-ice

Considerable reductions in sea-ice are expected. The Northwest Passage and Northern Sea Route of Russia will probably be opened up for routine shipping. GCM experiments predict large reductions in sea-ice extent but produce widely varying results and do not portray extent and seasonal changes of sea-ice for the current climate very well. One model estimates that with a doubling of greenhouse gases, sea-ice would cover only about 50% of its present area. Another projects a 43% reduction for the Southern Hemisphere and a 33% reduction for the Northern Hemisphere. The global area of sea-ice is projected to shrink by up to 17×10^6 km^2. One prediction is that in the Northwest Passage and Northern Sea Route, a century of warming would lead to a decline in winter fast-ice thickness from 1.8 to 2.5 m at present to 1.4 to 1.8 m and a shipping season of 41–100 days. A possible feedback with snow thickness may alter these relationships and actually lead to an increase in ice thickness. Given the complexity of many interactions and feedbacks, it is not at all clear what will be the overall effect of an air temperature increase on the Arctic ice cover and upper ocean. In the Antarctic, where the sea-ice cover is divergent and where land boundaries are less important, it is more reasonable to suppose that the main effect of global warming will be a simple retreat of the ice edge southward. Even here, a complex set of feedback mechanisms comes into play when the air temperature changes. The balance of open water areas, upper ocean structure and pycnocline depth adjusts itself to minimize the impact of changes, tending to preserve an ice cover even though it may be thinner and more diffuse. There has been no trend in sea-ice extent at either pole since 1973, when satellite data became available. Measurements of total sea-ice mass in the Arctic show no trends since 1979 when monitoring began.

UV-B

The IPCC does not study ozone depletion and resultant increases in UV-B radiation. Other groups have this role. The following is taken from the latest (1994) WMO/UNEP science and impact assessments. (WMO 1995; WMO/UNEP 1994).

(a) If there is general compliance with the Montreal Protocol to limit ozone-depleting chemicals, stratospheric chlorine and bromine levels (the primary chemicals) are expected to peak around the turn of the century. The ozone layer will be most affected by human-influenced perturbations and susceptible to natural variations (e.g. major volcanic eruptions or a very cold northern winter) during this period. After about the turn of the century, as stratospheric chlorine and bromine decline, ozone losses will diminish and the ozone layer may return to normal about the middle of the next century, assuming other global changes are not significant.

(b) In clear oceanic waters, penetration to several tens of metres has been shown. However, the transparency of the water strongly depends on the water type. In coastal waters with particulates and dissolved organics, concentrations of UV-B may penetrate less than 1 metre to the 1% level.

(c) Generally, the size of the Antarctic ozone hole and its intensity has increased since its onset in the late 1970s. The areal extent of the ozone hole was approaching the size of the Antarctic continent in the mid-1980s and now is greater than the continent of Antarctica. At its maximum, the ozone hole has exceeded 20 million km^2 for each of the years since 1989. Ozone losses are increasing in mid-latitude areas and the Arctic as well.

(d) The latest science assessment notes that the growth rates of several major ozone-depleting substances have slowed, demonstrating the expected impact of the Montreal Protocol and its Amendments and Adjustments. Subsequent to the assessment, the abundance of methyl chloroform (which is entirely of anthropogenic origin) was actually shown to be declining (Ravishankara and Albritton 1995). Methyl chloroform is the first substance regulated under the Protocol that has shown a distinct decrease in atmospheric abundance, not just a decrease in its rate of growth.

(e) The successful recovery of the ozone layer will depend upon the adherence of nations to the Montreal Protocol and the 1992 Copenhagen Amendments. Increasing concentrations of halogens, such as bromine, are still of concern.

Working Group II Overview (IPCC 1995*b*)

Working Group II of the IPCC was charged with reviewing the state of knowledge concerning the impacts of climate change on physical and ecological systems, human health, and socio-economic sectors. Working Group II also was charged with reviewing available information on the technical and economic feasibility of a range of potential adaptation and mitigation strategies. The assessment provides scientific, technical, and economic information that can be used, *inter alia*, in evaluating whether the projected range of plausible impacts constitutes 'dangerous anthropogenic interference with the climate system,' as referred to in Article 2 of the United Nations Framework Convention on Climate Change (UNFCCC), and in evaluating adaptation and mitigation options that could be used in progressing towards the ultimate objective of the UNFCCC.

Human health, terrestrial and aquatic ecological systems, and socio-economic systems (e.g. agriculture, forestry, fisheries, and water resources) are all vital to human development and well-being and are all sensitive to changes in climate. Whereas many regions are likely to experience the adverse effects of climate change, some of which are potentially irreversible, some effects of climate change are likely to be beneficial. Hence, different segments of society can expect to confront different changes and the need to adapt to them.

Policymakers are faced with responding to the risks posed by anthropogenic emissions of greenhouse gases in the face of significant scientific uncertainties. It is appropriate to consider these uncertainties in the context of information indicating that climate-induced environmental changes cannot be reversed quickly, if at all, due to the long time scales associated with the climate system. Decisions taken during the next few years may limit the range of possible policy options in the future because high near-term emissions would require deeper reductions in the future to meet any given target concentration. Delaying action might reduce the overall costs of mitigation because of potential technological advances but could increase both the rate and the eventual magnitude of climate change, hence the adaptation and damage costs.

Policymakers will have to decide to what degree they want to take precautionary measures by mitigating greenhouse gas emissions and enhancing the resilience of vulnerable systems by means of adaptation. Uncertainty does not mean that a nation or the world community cannot position itself better to cope with the broad range of possible climate changes or protect against potentially costly future outcomes. Delaying such measures may leave a nation or the world poorly prepared to deal with adverse changes and may increase the possibility of irreversible or very costly consequences. Options for adapting to change or mitigating change that can be justified for other reasons today (e.g. abatement of air and water pollution) and which will make society more flexible or resilient to anticipated adverse effects of climate change appear particularly desirable.

Terrestrial and aquatic ecosystems

Ecosystems contain the Earth's entire reservoir of genetic and species diversity and provide many goods and services critical to individuals and societies. These goods and services include (i) providing food, fibre, medicines, and energy; (ii) processing and storing carbon and other nutrients; (iii) assimilating wastes, purifying water, regulating water runoff, and controlling floods, soil degradation, and beach erosion; and (iv) providing opportunities for recreation and tourism. These systems and the functions they provide are sensitive to the rate and extent of changes in climate. Mean annual temperature and mean annual precipitation can be correlated with the distribution of the worlds major biomes.

The composition and geographic distribution of many ecosystems will shift as individual species respond to changes in climate; there will likely be reductions in biological diversity and in the goods and services that ecosystems provide for society. Some ecological systems may not reach a new equilibrium for several centuries after the climate achieves a new balance.

Lakes, streams, and wetland

Inland aquatic ecosystems will be influenced by climate change through altered water temperatures, flow regimes, and water levels. In lakes and streams, warming would have the greatest biological effects at high latitudes, where biological productivity would increase, and at the low-latitude boundaries of cold- and cool-water species ranges, where extinctions would be greatest. Warming of larger and deeper temperate zone lakes would increase their productivity; although in some shallow lakes and in streams, warming could increase the likelihood of anoxic conditions. Increases in flow variability, particularly the frequency and duration of large floods and droughts, would tend to reduce water quality and biological productivity and habitat in streams. Water-level declines will be most severe in lakes and streams in dry evaporative drainages and in basins with small catchments. The geographical distribution of wetlands is likely to shift with changes in temperature and precipitation. There will be an impact of climate change on greenhouse gas release from non-tidal wetlands, but there is uncertainty regarding the exact effects from site to site.

Coastal systems

Coastal systems are economically and ecologically important and are expected to vary widely in their response to changes in climate and sea level. Climate change and a rise in sea level or changes in storms or storm surges could result in the erosion of shores and associated habitat, increased salinity of estuaries and freshwater aquifers, altered tidal ranges in rivers and bays, changes in sediment and nutrient transport, a change in the pattern of chemical and microbiological contamination in coastal areas, and increased coastal flooding. Some coastal ecosystems are particularly at risk, including saltwater marshes, mangrove ecosystems, coastal wetlands, coral reefs, coral atolls, and river deltas. Changes in these ecosystems would have major negative effects on tourism, freshwater supplies, fisheries, and biodiversity. Such impacts would add to modifications in the functioning of coastal oceans and inland waters that already have resulted from pollution, physical modification, and material inputs due to human activities.

Oceans

Climate change will lead to changes in sea level, increasing it on average, and also could lead to altered ocean circulation, vertical mixing, wave climate, and reductions in sea-ice cover. As a result, nutrient availability, biological productivity, the structure and functions of marine ecosystems, and heat and carbon storage capacity may be affected, with important feedbacks to the climate system. These changes would have implications for coastal regions, fisheries, tourism and recreation, transport, off-shore structures, and communication. Palaeo-climatic data and model experiments suggest that abrupt climatic changes can occur if freshwater influx from the movement and melting of sea-ice or ice sheets significantly weakens global thermohaline circulation.

IPCC Second Assessment Report. Chapter 16 — Fisheries (Everett *et al.* 1995)

The 15 000 word report, Chapter 16 — Fisheries, was distilled to the 115 words in the opening summary of this paper, and then approved by the IPCC for presentation to 'Policy Makers'. At times, it hardly seems worth the years of effort that went into it, but in the context that 'Fisheries' received the same level of treatment and page-count as other sectors, such as agriculture, energy, forestry, water supply, and coastal zone, we in the fisheries sector sat at the table with everyone else and we had our voices heard. Fisheries scientists from around the world participated in the drafting and reviewing of the document. Without this tremendous support, we would not have taken the major step that this chapter represents.

The Fisheries Lead Authors are:

Dr John T. Everett, Convening Lead Author, NOAA, NMFS, USA, assisted by Dr H. Suzanne Bolton, also of NMFS

Dr Jean-Paul Troadec
ORSTOM
Inst. Français de Recherche pour le Développement en Coopération
Brest, France

Dr Ezekiel Okemwa
Kenya Marine and Fisheries Research Institute
Mombasa, Kenya

Dr Henry A. Regier
Institute for Environmental Studies, University of Toronto
Toronto, Canada

Dr Daniel Lluch Belda
Centro de Investigaciones Biologica de Baja California
La Paz, BCS, Mexico

Dr Andrei Krovnin
Russian Federal Research Institute of Fisheries and Oceanography (VNIRO)
Moscow, Russia

Chapter 16 is a major accomplishment. This synthesis document, combined with the rigorous nature of the IPCC review process, will make this a valuable contribution for years to come. The authors' initial task was to use only what was printed in the peer-reviewed literature to make an assessment of the global impact of climate change on fisheries. The immediate problem was that the literature generally deals with regional-specific findings that pertain to one or a few species. There is no global synthesis in the literature. The Fisheries Lead Authors reasoned that if we hypothesized regional and global impacts on the fish themselves and on the human users, and suggested adaptation measures, and the world fisheries community agreed with these findings, then we would have moved the understanding forward one giant step. We stuck our necks out, very uncomfortably far for some of us, called our work a draft, and sent it out for peer review. We made some revisions and then it went through two formal IPCC peer and country (government) reviews, about one year apart. We do know of the five of us Lead Authors, and the 26 Contributing Authors, and the 19 IPCC Official Peer Reviewers, but there are probably hundreds of fisheries experts whom we will never know within the United Nations member countries who received the documents for review as part of each nation's clearance process. I doubt that many fisheries documents in the literature have had such a complete review.

I believe we went as far as our knowledge of the physical environment and its linkage to biota would support. Future global circulation models will probably provide better information to biologists than that available now. Regional specificity is not now good enough to support detailed hypotheses, and there are enough regional differences among models to suggest, at least to this author, that agreements among models at the regional scale, may be more due to coincidence than to certainty.

It is useful to quote verbatim the 'Fisheries ' Executive Summary, as it was crafted and reviewed by many fisheries experts:

> Any effect of climate change on fisheries will occur in a sector that is already characterized, on a global scale, by full utilization, massive over-capacity, and sharp conflicts between fleets and other competing uses of aquatic ecosystems. Climate change impacts are likely to exacerbate existing stresses on fish stocks, notably overfishing, diminishing wetlands and nursery areas, pollution, and UV-B radiation. The effectiveness of actions to reduce the decline of fisheries depends on our capacity to distinguish among these stresses and other causes of change. This capacity is insufficient and, although the effects of environmental variability are increasingly recognized, the contribution of climate change to such variability is not yet clear.
>
> While overfishing has a greater effect on fish stocks than climate change today, progress is being made on the overfishing problem. Overfishing results from an institutional failure to adjust harvesting ability to finite and varying fish yields. Conventional management paradigms, practices, and institutions — inherited from the period when fish stocks were plentiful — are not appropriate for the new situation of generally full exploitation, especially of

important fish stocks. Although the Law of the Sea represents an important step in the proper direction, only a few countries have adopted the institutional arrangements needed to regulate the access of fishing fleets to fishery resources. The UN Conference on Straddling and Highly Migratory Fish Stocks, and the FAO Code of Conduct for responsible fisheries, are likely to accelerate the adoption and effective implementation of regulatory mechanisms. Should climate change develop according to the IPCC scenarios, it may become more important than overfishing over the 50 to 100 year period of this 1995 climate assessment.

- Globally, under the IPCC scenarios, saltwater fisheries production is hypothesized to be about the same, or significantly higher if management deficiencies are corrected. Also, globally, freshwater fisheries and aquaculture at mid to higher latitudes could benefit from climate change. These conclusions are dependent on the assumption that natural climate variability and the structure and strength of wind fields and ocean currents will remain about the same. If either changes, there would be significant impacts on the distribution of major fish stocks, though not on the global production [medium confidence].
- Even without major change in atmospheric and oceanic circulation, local shifts in centres of production and mixes of species in marine and fresh waters are expected as ecosystems are displaced geographically and changed internally. The relocation of populations will depend on properties being present in the changing environments to shelter all stages of the life cycle of a species.
- While the complex biological relationships among fisheries and other aquatic biota and physiological responses to environmental change are not well understood, positive effects, such as longer growing seasons, lower natural winter mortality, and faster growth rates in higher latitudes, may be offset by negative factors such as a changing climate that alters established reproductive patterns, migration routes, and ecosystem relationships [high confidence]. Changes in abundance are likely to be more pronounced near major ecosystem boundaries. The rate of climate change may prove a major determinant of abundance and distribution of new populations. Rapid change due to physical forcing, will usually favour production of smaller, low priced, opportunistic species that discharge large numbers of eggs over long periods [high confidence]. However, there are no compelling data to suggest a confluence of climate change impacts that would affect global production in either direction, particularly because relevant fish population processes take place at regional or smaller scales for which GCMs are insufficiently reliable.
- Regionally, fresh water gains or losses will depend on changes in amount and timing of precipitation, on temperatures, and on species tolerances. For example, increased rainfall during a shorter period in winter could still lead to reduced levels in summer in river flows, lakes, wetlands, and thus, in freshwater fisheries. Marine stocks that reproduce in freshwater (e.g. salmon), or require reduced estuarine salinities, will be similarly affected [high confidence].
- Where ecosystem dominances are changing, economic values can be expected to fall until long-term stability (i.e. at about present amounts of variability) is reached [medium confidence]. National fisheries will suffer if institutional mechanisms are not in place that enable fishers to move within and across national boundaries [high confidence]. Subsistence and other small scale fishers, lacking mobility and alternatives, are often most dependent on specific fisheries and will suffer disproportionately from changes [medium confidence].
- Because natural variability is so great relative to global change, and the time horizon on capital replacement (ships and plant) is so short, impacts on fisheries can be easily overstated and there will likely be relatively small economic and food supply consequences [medium confidence].
- An impact ranking can be constructed. The following categories are listed in descending order of sensitivity, positive or negative, to climate change [medium confidence]:
  1. Freshwater fisheries in small rivers and lakes, in regions with larger temperature and precipitation change;
  2. Fisheries within Exclusive Economic Zones, particularly where access regulation mechanisms artificially reduce the mobility of fishing groups and fleets and their capacity to adjust to fluctuations in stock distribution and abundance;
  3. Fisheries in large rivers and lakes;
  4. Fisheries in estuaries, particularly where there are species without migration or spawn dispersal paths, or estuaries impacted by sea level rise or by decreased river flow;
  5. High seas fisheries.

Adaptation options providing large benefits irrespective of climate change follow [medium confidence]:

- Design and implement national and international fishery management institutions that recognize shifting species ranges, accessibility, and abundances and that balance species conservation with local needs for economic efficiency and stability;
- Support innovation by research on management systems and aquatic ecosystems;
- Expand aquaculture to increase and stabilize seafood supplies, to help stabilize employment, and carefully, to augment wild stocks;
- In coastal areas, integrate the management of fisheries with other uses of coastal zones;
- Monitor health problems (e.g. red tides, ciguatera, cholera) that could increase under climate change and harm fish stocks and consumers.

Recommendation

Now that the IPCC documents are being distributed, it is time to take action. For this Congress, I would argue for addressing only the first option. I believe it is the most important. This option, seeks to build into resource management regimes the concept that fish abundances and distributions will change over time as a consequence of climate change and also that resource productivity and industrial capacity should be in balance. THIS APPROACH IS NEEDED NOW! Perhaps climate change awareness can serve as a vehicle to help bring this essential message to the attention of political leaders and fishery managers. The natural variability in regional climate, that sometimes takes places over decades, represents greater (or as great) changes as will come with climate change. The only difference may be the permanence of the changes. If we can adapt our institutions to deal with climate change, we will have addressed the decadal scale natural variability issues that plague many existing institutions and the industries they regulate.

REFERENCES

Everett, J T, Okemwa, E, Regier, H A,Troadec, J-P, Krovnin, A, and Lluch-Belda, D (1995). Fisheries. In 'The IPCC Second Assessment Report, Volume 2: Scientific-Technical Analyses of Impacts, Adaptations, and Mitigation of Climate Change'. (Eds R T Watson, M C. Zinyowera and R H Moss.) (Cambridge University Press: Cambridge and New York). 31 pp.

IPCC (1995*a*). Climate Change—1995: The IPCC Second Assessment Report, Volume 1:The Science of Climate Change. (Eds J J Houghton, L G Meiro Filho, B A Callander, N Harris, A Kattenberg and K Maskell). (Cambridge University Press: Cambridge and New York).

IPCC (1995*b*). Climate Change—1995: The IPCC Second Assessment Report, Volume 2: Scientific–Technical Analyses of Impacts, Adaptations, and Mitigation of Climate Change. (Eds R T Watson, M C Zinyowera and R H Moss). (Cambridge University Press: Cambridge and New York).

Ravishankara, A R, and D L Albritton (1995). Methyl chloroform and the atmosphere. *Science* **269**, 183–4.

WMO (1995). 'Scientific Assessment of Ozone Depletion: 1994.' World Meteorological Organization Global Ozone Research and Monitoring Project—Report No. 37, Geneva.

WMO/UNEP (1994). 'Environmental Effects of Ozone Depletion: 1994.' Assessment. United Nations Environment Programme, Nairobi.

Squid Recruitment in the Genus *Illex*

R. K. O'Dor,[A] *P. G. Rodhouse*[B] *and E. G. Dawe*[C]

[A] Department of Biology, Dalhousie University, Halifax, Nova Scotia, Canada B3H 4J1.
[B] British Antarctic Survey, Natural Environment Research Council, High Cross, Madingley Road, Cambridge CB3 0ET, UK.
[C] Science Branch, Department of Fisheries and Oceans, PO Box 5667, St. John's, Nfld., Canada A1X 5X1.

Summary

Squid are among the short-lived species replacing declining traditional fisheries. Higher global market values than for many fish, mean that larger, more valuable catches are possible for cephalopods, which grow rapidly with high production to biomass ratios. However, annual cephalopods, with no reserves once a year class is overfished, require an understanding of stock structure and highly variable recruitment dynamics to protect directed fisheries. Increment analysis for age and growth and stomach content analysis indicate flexible seasonality and feeding strategies allowing a wide range of habitats. *Illex coindetii* occupies the Mediterranean, Caribbean and eastern Atlantic margin from 55°N to 20°S but never sustains major fisheries. On western boundaries, *I. illecebrosus* collapsed after intense, widely distributed fishing, while *I. argentinus* has survived much longer as the world's largest squid fishery, apparently stabilized by a complex of stocks, breeding widely dispersed in space and time. Selective gears in feeding areas are probably self-limiting, allowing adequate escapement, provided squid are not pursued by fishing into breeding areas. Recruitment is limited by food production and by the stability of interactions between adult migrations and ocean dynamics, which strongly influence early growth and survival.

Introduction

This paper summarizes new observations, perspectives and conclusions from the FAO Fisheries Technical Paper, 'Squid Recruitment Dynamics' (in press), to help fishers and fishery managers avoid or minimize risks in squid fisheries. The IOC/FAO's International Recruitment Program (IREP) began encouraging such studies at its 1981 meeting in Halifax, and in 1987 the FAO commissioned the Cephalopod International Advisory Council (CIAC) to produce a volume on squid recruitment. Authors gathered at the ICES Shellfish Life Histories Symposium in 1990 and began organizing information on three commercial species in the same genus: *Illex coindetii, I. illecebrosus* and *I. argentinus.* The concept was to focus on a genus of closely-related commercial squid that are widely distributed in the eastern and western boundary currents of both the North and South Atlantic and then to look for generalizations. Twenty-five authors from a dozen countries assembled data from unpublished or grey literature sources, and then provided synthetic reviews. The volume demonstrates advances over the last decade that provide the necessary basis for a comprehensive understanding of recruitment in cephalopods, making them a prime target for advancing knowledge of marine recruitment processes in general.

Although the volume focusses on *Illex* species as examples of the ommastrephids that sustain the world's largest commercial squid fisheries, it also provides some comparisons and contrasts with *Loligo* species which may help to manage these and similar, primarily localized and/or artisanal fisheries. Chapters reviewing individual species suggest that, while stocks in eastern boundary current systems rarely reach abundances capable of sustaining directed fisheries, stocks from western boundary currents like the Gulf Stream and Brazil Current can build up to very high, but also highly variable, biomasses. Thus eastern boundary stocks are, perhaps, under less threat than western. Large fleets that build up when abundances are high are prone to overfish as stocks decline, to avoid short-term financial difficulties. The consequences appear often to be long-lasting stock collapses leading to both biological and economic disasters.

REVIEWS

The reviews of two of the three species included meetings of principal researchers to ensure complete coverage. This provided the first comprehensive review of *I. coindetii* over its full range and an integration of records of *I. argentinus* over its range. Surviving *I. illecebrosus* researchers provided an update of events since the collapse of the fishery and a re-interpretation of published information in light of new observations on age, growth and migrations.

Perhaps the most interesting result of the reviews was the recognition of the remarkable similarity between the populations of all species and all ranges. All species appear to have nearly year-round hatching with seasonal variations in abundance likely to be attributable to larval survival rather than egg production, since fecundity is high compared with *Loligo* species and usually exceeds 100 000 eggs per female. Growth rates (Table 1) also vary seasonally, typically increasing progressively as the exposure temperatures of hatchlings increase through the season. However, high mobility and opportunistic feeding allow squid to survive and thrive under a wide range of conditions. The key question for recruitment prediction and fishery stabilization seems to be whether this adaptability arises from phenotypic plasticity (i.e. any individual squid can do well in any environment) or from genotypic variation (i.e. selective breeding produces races adapted to particular areas, oceanographic features or seasons).

Evidence is accumulating that squid are not homogeneous and that races exist which need to be managed as stocks to achieve stability and predictability. This paper presents a brief overview of the physical factors that squid require to thrive and of the biological factors that limit their survival and growth.

PHYSICAL FACTORS

The life cycles and recruitment success of squids, for example *I. illecebrosus* (Fig. 1) (O'Dor and Dawe in press), are closely linked to environmental features, particularly oceanic currents (O'Dor 1992). Generally, eggs and paralarvae are transported passively and adults must migrate back 'upstream' to spawn. The year-round warm waters of the Gulf Stream provide for rapid development of eggs, which are packaged by spawning females in large floating egg balloons (O'Dor and Balch 1985) that

Table 1. Linear regression statistics by sex and month of hatching for the genus *Illex*

Species / Study / Month of hatching	Slope: Male	Slope: Female	Intercept: Male	Intercept: Female
Illex coindetii				
Sanchez (1995)				
Oct.–Mar.	0.36	0.35	11.2	37.7
Apr.–Sep.	0.42	0.38	1.9	31.1
Illex illecebrosus				
Dawe and Beck (in press)				
Mar.	0.53	0.71	108.2	84.0
Apr.	0.60	1.03	105.6	46.3
May	0.75	1.29	110.8	26.5
Illex argentinus				
Rodhouse and Hatfield (1990)				
May	0.35	0.94	137.0	-7.6
Jun.	0.70	0.99	46.1	-13.2
Jul.	0.81	1.03	26.6	-10.7
Aug.	1.00	1.30	-9.8	-7.3

transport them north as the eggs develop, delivering them to the highly productive slope-water mixing zones. A close coupling exists between biological and physical properties (Fig. 2). The gelatinous balloons sink until reaching water of equal density and then move north at whatever speed this water provides, which may range up to 250 km day^{-1}. Development can take as little as 6 days at 26°C or as much as 26 days at 13°C (O'Dor *et al.* 1986), so currents could easily move eggs thousands of kilometres. The influences of adult spawning times and of sites on the places where recruits occur are certainly profound and possibly unpredictable. Typical upward swimming by hatchlings makes it likely that they will be entrained in the downwelling zone at the edge of the current. This assures slower transport and access to production which is concentrated here. As they grow, migration occurs into cooler waters and eventually onto the continental shelf and into the fishery.

The situation for *I. argentinus* is much more complex. The range of this species (Fig. 3) includes one of the world's widest continental shelves, with the Brazil Current and two branches of the Falkland (Malvinas) Currents meeting at the freshwater lens created by the Rio de la Plata. This allows a tremendous range of strategies to close the life cycle. In austral winter and spring, paralarvae are found to the northern limits of sampling, 33°S off Brazil. In summer, paralarvae have been found as far as 45°S off Argentina, and it has been suggested that eggs of the South Patagonic stock are spawned in the cold waters of the Falkland (Malvinas) Current, which would require that they be carried north to warmer waters for development (Kronkiewicz 1986), a possible alternative to adult migration. The critical management question is whether the several identified stocks are biologically 'tuned' to particular physical patterns by intense selection, giving rise to distinct races, or simply the result of a common pool of paralarvae becoming part of various stocks as a result of haphazard distribution by the physical processes.

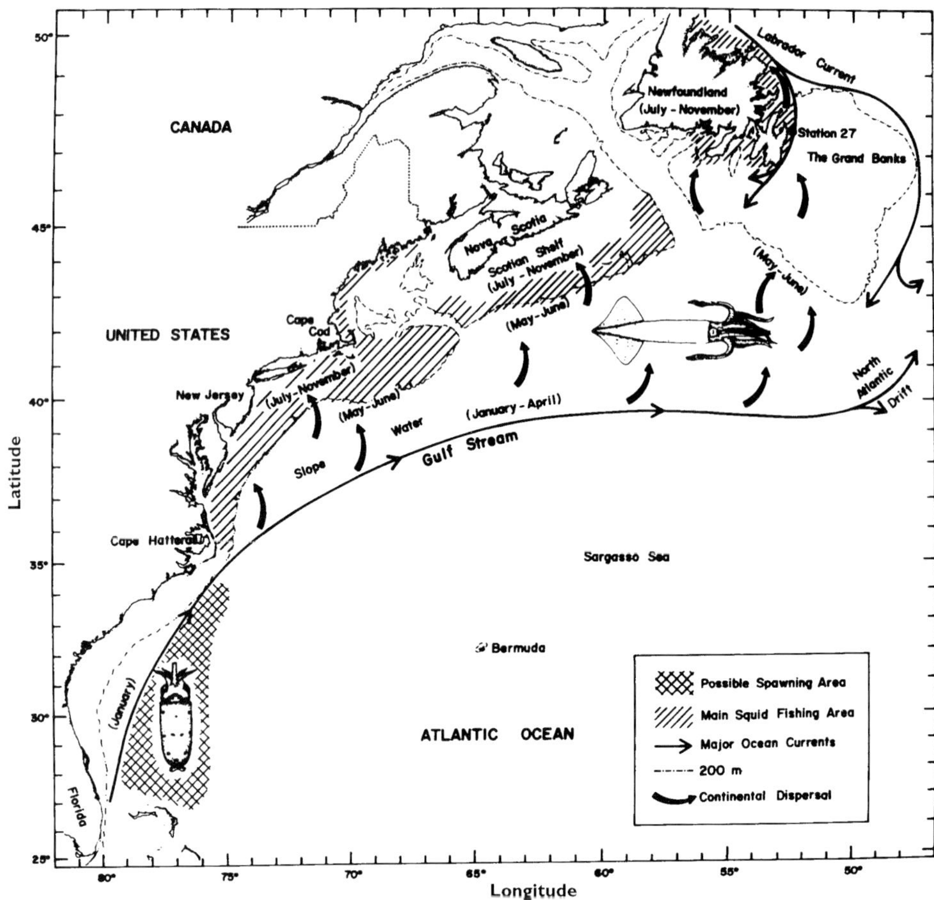

Fig. 1. Representation of the life cycle of *Illex illecebrosus* in relation to the main Northwest Atlantic Ocean currents and sampling sites (Dawe and Warren 1993).

The absence of major, directed fisheries for *I. coindetii* means there is less information available to develop specific models linking ocean dynamics to biology, but this species does occupy the entire eastern boundary of the Atlantic Ocean from Norway to Angola, as well as the Mediterranean and Black Seas (Jereb and Ragonese 1995; Roper and Mangold in press; Sanchez *et al.* in press). Although current systems are less powerful here than those on the western boundary, much of this area is dominated by coastal upwelling systems that can be highly productive (Parrish *et al.* 1983). However, the offshore winds that induce coastal upwelling also disperse pelagic-spawning coastal organisms (Parrish *et al.* 1981) and may disperse food concentrations required for paralarvae (Lasker 1978). The lack of stable, large-scale current systems may be related to failure to develop highly specialized stocks or races that could give rise to directed fisheries and may, as well, explain the large range of this single species. This suggests that biological specialization is an important element in stabilizing the large commercial stocks in the western Atlantic.

Biological factors

Superimposed on the physical variation that influences recruitment are biological factors that are at least partially uncoupled from ocean dynamics. These biological factors can be extrinsic (having to do with changes in the ecosystem) or intrinsic (having to do with characteristics of the squid). The authors of the FAO volume explore both types, ranging from trophic relationships to genetic relationships and how they may be altered by the fishery.

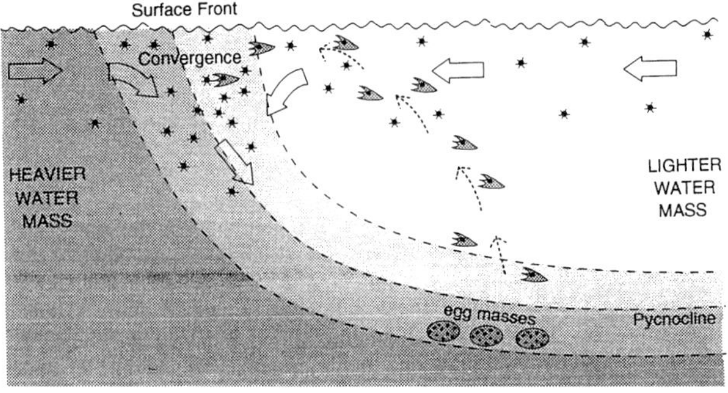

Fig. 2. Within the pycnocline, squid egg masses find a neutrally buoyant zone where they are suspended at mid-depth at suitable development temperatures and under conditions of reduced predation. Upon hatching, squid paralarvae may rise to the surface layer, to be carried into the convergent frontal zone by the density-driven flow. Dashed lines, density isopleths; dotted arrow, path of paralarvae; open arrows, density driven flow; * larval food (Bakun and Csirke in press).

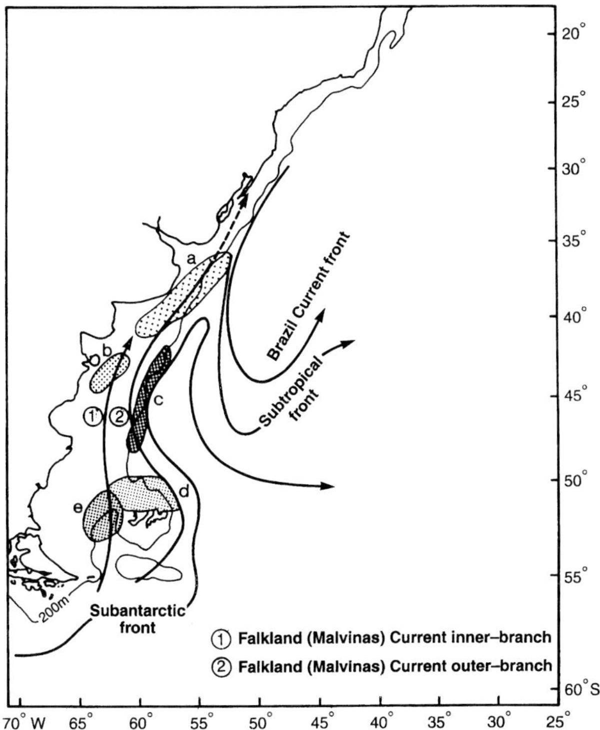

Fig. 3. Current patterns in the distribution range of the Argentinean squid *Illex argentinus* in the Southwestern Atlantic (modified from Peterson 1992) and its main fishing grounds (based on Brunetti 1990; Csirke 1987). (a) Winter and spring fishing grounds of the Argentinean and Uruguayan fleets, (b) Summer fishing grounds of the Argentinean fleet, (c) Autumn and winter fishing grounds of the international fleet, (d) Autumn and winter fishing by the international fleet off the Falkland (Malvinas) Islands, (e) Recent fishing for sub-adults (Haimovici *et al.* in press).

Dawe and Brodziak (in press) investigate such trophic relationships with data sets consisting of yearly estimates of abundances for *Illex* and those fish species which interact with *Illex* as predators, prey or competitors. Simple correlation analysis is applied to each data set to determine significant interactions. Results suggest that, on the continental shelf, recruitment of *Illex* may be affected by the abundance of predators and competitors. The relatively simple system is illustrated in Newfoundland waters (Fig. 4), where recruitment of *I. illecebrosus* to a coastal fishery area near the limit of its distribution may be related to abundance of fish prey.

Another approach to such interactions uses life-history reconstruction from statoliths, (the squid equivalent to fish otoliths), and gladii (squid pens) (Arkhipkin and Perez in press). Statoliths in this genus show daily rings, providing good age and hatching dates when specially preserved. However, daily growth increments on pens provide a unique, direct record of linear growth which is preserved in normal, formalin-fixed museum specimens. For example, the records on squid pens can be used as an index of environmental conditions for seasons and areas that are not subject to intense sampling, such as the growth rates (Fig. 5) for late season juvenile *I. illecebrosus* caught at the edge of the continental shelf after leaving the Gulf Stream. In contrast to earlier data (Table 1), the growth rates here are decreasing as the season progresses.

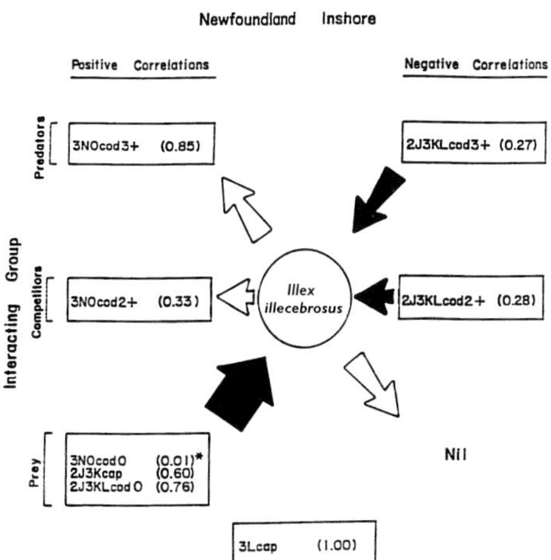

Fig. 4. The relative importance of three types of trophic interactions on *Illex illecebrosus* recruitment in Newfoundland coastal waters, based on the occurrence of positive *versus* negative correlations with relevant cod and capelin stocks and age groups (*p*-values marked by an asterisk were judged to be statistically significant). Thickness of the dark arrows represents relative importance of interactions which could affect *Illex illecebrosus* recruitment (Dawe and Brodziak in press).

Forsythe (1993) has proposed that seasonally increasing growth rates result from increasing temperatures based on size-at-age data and *ad libitum* feeding in laboratory reared squids, but the results (Fig. 5) suggest that juveniles may be food-limited once the peak of the spring bloom has passed. The mobility of juvenile squid, of course, allows selection of various temperature and prey regimes by vertical and horizontal migration, so resolving the question of oceanic dynamics *versus* food is complex. Positions (Fig. 6) of a series of squid samples recorded in relation to physical structures at the edge of the Gulf Stream, and growth history analysis of these and similar samples indicates that juvenile growth actually increases as squid move inshore from the warmer waters, probably in pursuit of fish prey as suggested by the predator-prey correlations.

Understanding population dynamics is critical for managing and predicting the abundance of short-lived species like squids,

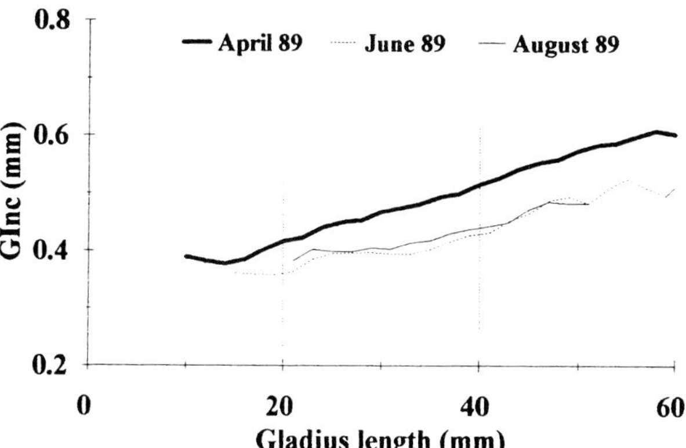

Fig. 5. Seasonal variation in juvenile squid growth rates. Mean size-specific variation of gladius growth increments (GInc) for 60 *Illex illecebrosus* juveniles collected in the indicated months of 1989 near 42°N, 61°W (Perez 1995).

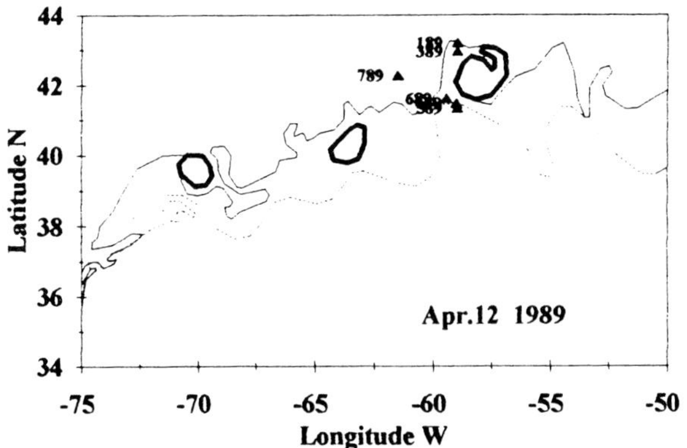

Fig. 6. Satellite images of the dynamic changes in the northern boundary of the Gulf Stream (dotted line), the shelf/slope front (solid line) and the warm-core rings (thick line) during a sampling period in April, 1989. Samples used for gladius growth rate analysis, indicated by triangles, ranged from 189 (4 April 1989) to 789 (17 April 1989) (Perez 1995).

because there are no pre-recruit year classes to project in virtual population analyses. Early season abundance estimates based on acoustic surveys (Starr and Thorne in press) of subadult 'microcohorts' (Caddy 1991) may be useful. Acoustic surveys may also give insights into school structure and stability, which have important implications for population genetics. Carvalho and Nigmatullin (in press) point out that squid, in general, have low levels of mean observed heterozygosity per locus (0.00–0.07, compared to an average value of 0.15 for marine invertebrates) and proportions of polymorphic loci (0.17–0.28 *vs* 0.59). Such low values are typically indicative of population bottlenecks and founder events and/or recent population expansions. In contrast, high values for Nei's mean genetic identities occur in reproductively isolated stocks. This suggests that fished stocks include many specialized breeding units, perhaps at the level of microcohorts or even schools, that increase and decrease rapidly in numbers and need to be considered independently for management.

Adult behaviours that place some egg balloons in optimal conditions, while others become vagrants (*sensu* Sinclair 1988), may result in high survival rates for sibling paralarvae, which could produce the patterns of inbreeding seen (O'Dor in press). This would require very different reproductive and life-history strategies among ommastrephids, like *Illex*, than those of *Loligo*, which involve multiple matings over massive communal egg beds that tend to stabilize gene pools. Rodhouse *et al.* (in press) model the potential impact of selective mortality on an *Illex* fishery and show, for example, that seasonal fishing restrictions have the potential to select for early maturing squid that could quickly reduce average body size and influence catch rates and markets.

Conclusions

Unlike loliginid squids, often fished on spawning grounds, ommastrephids are typically fished on feeding grounds. In contrast to *I. illecebrosus*, the *I. argentinus* fishery appears to have been stabilized by a complex mix of stocks which breed widely dispersed in space and time. As these stocks move rapidly through feeding areas to breeding areas, fishing with selective gears in feeding areas is probably self-limiting, allowing adequate escapement, provided squid are not pursued into breeding areas. These stocks appear now to be under increasing threat (Pearce 1996). Total recruitment is limited both by production (e.g. *I. coindetii*) and by stability of the interactions between adult migrations and ocean dynamics, which strongly influence early growth and survival (e.g. *I. illecebrosus*).

References

Arkhipkin, A, and Perez, J A A (in press). Life-history reconstruction. In 'Squid Recruitment Dynamics'. (Eds P G Rodhouse, E G Dawe and R K O'Dor.) *FAO Fisheries Technical Paper* (FAO: Rome).

Bakun, A, and Csirke, J (in press). Environmental processes and recruitment variability. In 'Squid Recruitment Dynamics'. (*loc. cit.*).

Brunetti, N E (1990). The evolution of the *Illex argentinus* (Castellanos 1960) fishery. *Informes Técnicos de Investigación Pesquera, Consejo Superior de Investigaciones Barcelona* **155**. 19 pp. [in Spanish].

Caddy, J (1991). Daily rings on squid statoliths: An opportunity to test standard population models? In 'Squid Age Determination Using Statoliths'. Proceedings of the International Workshop, 9–14 October 1989, Instituto di Tecnologia della Pesca e del Pescato. (Eds P Jereb, S Ragonese and S V Boletzky). pp. 53–66 (N.T.R.–I.T.P.P. Special Publication No. 1, Mazara del Vallo, Sicily, Italy.

Carvalho, G R, and Nigmatullin, Ch M (in press). Stock structure analysis and species identification. In 'Squid Recruitment Dynamics'. (*loc. cit.*).

Csirke, J (1987). The Patagonian Fishery Resources and the Offshore Fisheries in the South-West Atlantic. *FAO Fisheries Technical Paper* No. 289. 75 pp. (FAO: Rome).

Dawe, E G, and Beck, P C (in press). Population structure, growth and sexual maturation of short-finned squid (*Illex illecebrosus*) at Newfoundland. *Canadian Journal of Fisheries and Aquatic Sciences*.

Dawe, E G, and Brodziak, J K T (1996). Trophic relationships, ecosystem variability and recruitment. In 'Squid Recruitment Dynamics'. (*loc. cit.*).

Dawe, E G, and Warren, W (1993). Recruitment of short-finned squid in the Northwest Atlantic Ocean and some environmental relationships. *Journal of Cephalopod Biology* **2**(2), 1–21.

Forsythe, J W (1993). A working hypothesis of how seasonal temperature change may impact the field growth of young cephalopods. In 'Recent Advances in Cephalopod Fisheries Biology'. (Eds T Okutani, R K O'Dor and T Kubodera). pp. 133–143 (Tokai University Press: Tokyo).

Haimovici, M, Brunetti, N E, Rodhouse, P G, Csirke, J, and Leta, R H (in press). *Illex argentinus*. In 'Squid Recruitment Dynamics'. (*loc. cit.*).

Jereb, P, and Ragonese, S (1995). An outline of the biology of the squid *Illex coindetii* in the Sicilian channel, Central Mediterranean. *Journal of the Marine Biological Association of United Kingdom* **75**, 373–90.

Kronkiewicz, A (1986). Growth and life cycle of squid, *Illex argentinus*, from Patagonian and Falkland Shelf and Polish fishery of squid for this region; 1978–1985. *ICES C.M. 1986/K*: **27**. 25 pp.

Lasker, R (1978). The relation between oceanographic conditions and larval anchovy food in the California Current: Identification of the factors leading to recruitment failure. *Rapport et Procès-Verbaux des Rèunions Conseil International pour l'Exploration de la Mer* **173**, 212–30.

O'Dor, R K (1992). Big squid in big currents. *South African Journal of Marine Science* **12**, 225–35.

O'Dor, R K (in press). Squid life history strategies. In 'Squid Recruitment Dynamics'. (*loc. cit.*).

O'Dor, R K, and Balch, N (1985). Properties of *Illex illecebrosus* egg masses potentially influencing larval oceanographic distribution. *Northwest Atlantic Fisheries Organization Science Council Studies* **9**, 69–76.

O'Dor, R K, and Dawe, E G (in press). *Illex illecebrosus*. In 'Squid Recruitment Dynamics'. (*loc. cit.*).

O'Dor, R K, Foy, E. A, Helm, P L, and Balch, N (1986). The locomotion and energetics of hatchling squid, *Illex illecebrosus*. *American Malacological Bulletin* **4**, 55–60.

Parrish, R H, Nelson, C S, and Bakun, A (1981). Transport mechanisms and reproductive success of fishes in the California Current. *Biological Oceanography* **1**, 175–203.

Parrish, R H, Bakun, A Husby, D M, and Nelson, C S (1983). Comparative climatology of selected environmental processes in relation to eastern boundary current pelagic fish production. In 'Proceedings of the expert consultation to examine changes in abundance and species composition of neritic fish resources', San José, Costa Rica, 18–29 April 1983 (Eds G D Sharp and J Csirke). pp. 731–77. *FAO Fisheries Report* No. 291 (3), (FAO Rome).

Pearce, F (1996). After the Falklands bonanza. *New Scientist* **149** (2017), 32–5.

Perez, J A A (1995). The early life history of the short-finned squid, *Illex illecebrosus* (Cephalopoda: Ommastrephidae) as reconstructed from the gladius structure. Ph.D Thesis, Dalhousie University. Halifax, Canada. 150 pp.

Peterson, R G (1992). The boundary currents in the western Argentine Basin. *Deep-Sea Research* **39**, 623–44.

Rodhouse, P G, and Hatfield, E M C (1990). Dynamics of growth and maturation in the cephalopod *Illex argentinus* de Castellanos, 1960 (Teuthoidea: Ommastrephidae). *Philosophical Transactions of the Royal Society of London* **B 329**, 229–41.

Rodhouse, P G, Murphy, E J, and Coelho, M L (in press). Impact of fishing on squid life histories. In 'Squid Recruitment Dynamics'. (*loc. cit.*).

Roper, C F E, and Mangold, K (in press). Systematic and distributional relationships of *Illex coindetii* to the genus *Illex*. In 'Squid Recruitment Dynamics'. (*loc. cit.*).

Sánchez, P (1995). Age and growth of *Illex coindetii*. *ICES Marine Science Symposia* **199**, 441–4.

Sanchez, P, González, A F, Jereb, P, Laptikhovsky, V V, Mangold, K, Nigmatullin, Ch, and Ragonese, S (in press). *Illex coindetii*. In 'Squid Recruitment Dynamics'. (*loc. cit.*).

Sinclair, M (1988). 'Marine Populations. An Essay on Population Regulation and Speciation.' 252 pp. (University of Washington Press: Seattle).

Starr, R M, and Thorne, R E (in press). Acoustic assessment of squid stocks. In 'Squid Recruitment Dynamics'. (*loc. cit.*).

Protecting Vulnerable Stocks in Multi-species Prawn Fisheries

J. W. Penn, R. A. Watson, N. Caputi and N. Hall

Western Australian Marine Research Laboratories, PO Box 20, North Beach, WA 6020, Australia.

Summary

Information from the unusual collapse and rebuilding of individual penaeid stocks within the Western Australian multi-species trawl fisheries assists in developing population models for penaeids and identifies potentially vulnerable prawn stocks. The Exmouth Gulf stock of *Penaeus merguiensis* collapsed in the 1960s and has not recovered. The *P. esculentus* stocks in Shark Bay and Exmouth Gulf suffered recruitment overfishing during the early 1980s, but significant reductions in the fishing effort directed at this species have resulted in increased breeding stock and improved catches; spawning stock–recruitment relationships have been developed for the two *P. esculentus* stocks. No such relationship is clear for the *Penaeus latisulcatus* stock in Shark Bay, but the data suggest that recruitment may be influenced by an environmental effect at the time of recruitment. The ability of a fishery to exert high levels of pre-spawning fishing mortality is a common factor in penaeid fisheries showing recruitment overfishing. Local geography, the position of the fishery within a species range, the presence of other species in a fishery, and the catchability, appear to affect susceptibility to overfishing.

Introduction

The penaeid prawn stocks, which support major tropical and subtropical fisheries throughout the world, have historically been regarded as resilient to recruitment overfishing. Until the early 1980s, there were no documented spawning stock–recruitment relationships (SRRs) for penaeids, and it was believed that the well-documented environmental effects on recruitment were more likely to be the cause of the typically large catch variations from penaeid fisheries (see Garcia 1983 for review). In addition to environmentally driven variations in recruitment obscuring any underlying SRR, data limitations have also contributed to the difficulties in developing reliable recruitment and, particularly, spawning stock indices for the study of SRRs for penaeids (Gulland 1984).

Events in the multi-species penaeid fisheries of Western Australia (WA) during the early 1980s (Penn and Caputi 1985), however, led to renewed scientific debate about the potential impact of fishing on such stocks. Significant reductions in recruitment of tiger prawn stocks in Shark Bay and Exmouth Gulf, and evidence of a discernible SRR (Penn and Caputi 1986), have led to wider acceptance that specific management action to maintain breeding stock levels will be required for some penaeid fisheries.

The purpose of this paper is to review and update the spawning stock and recruitment relationships developed for Western Australian penaeid stocks, and to use these data with information from other recorded cases of penaeid recruitment overfishing to identify stocks in situations in which precautionary management strategies should be adopted.

WESTERN AUSTRALIAN CASE STUDIES

Industrial fishing for penaeid prawns began along the WA coastline during the early 1960s and has been subject to a staged development under a system of limited-entry fishing regulations (Bowen and Hancock 1984). In the early 1960s two major WA fisheries developed along the desert coastline at Shark Bay (26°S) and at Exmouth Gulf (22°S). Smaller discrete fisheries operating off the coast at Onslow, Nickol Bay, and in the Kimberley Bays (Fig. 1) developed in the late 1960s to early 1970s, and a new fishery began off Broome in the mid 1980s. Limited entry management arrangements for each of these fisheries were introduced during 1980–90.

The dominant species in these fisheries, in order of importance, are western king prawns (*P. latisulcatus*), brown tiger prawns (*P. esculentus*), banana prawns (*P. merguiensis*) and endeavour prawns (*Metapenaeus endeavouri*), with total catches in the order of 3500 t having been taken by approximately 90 vessels in 1994 (Anon. 1996). Scientific literature covering the biological and fishery data used to manage the most abundant stocks in the two major fisheries has been summarized by Penn *et al.* (1989).

EXMOUTH GULF FISHERY

The trawl fishery in Exmouth Gulf exploits four species: banana prawns, tiger prawns, western king prawns, and endeavour prawns, with banana prawn and tiger prawn stocks showing a high degree of annual variability (Fig. 2). The fleet was developed incrementally under limited-entry management to a maximum of 23 vessels in 1979–82; however, this number has subsequently been reduced to 19 (1984) and 16 (1990) through industry-funded buy-back arrangements.

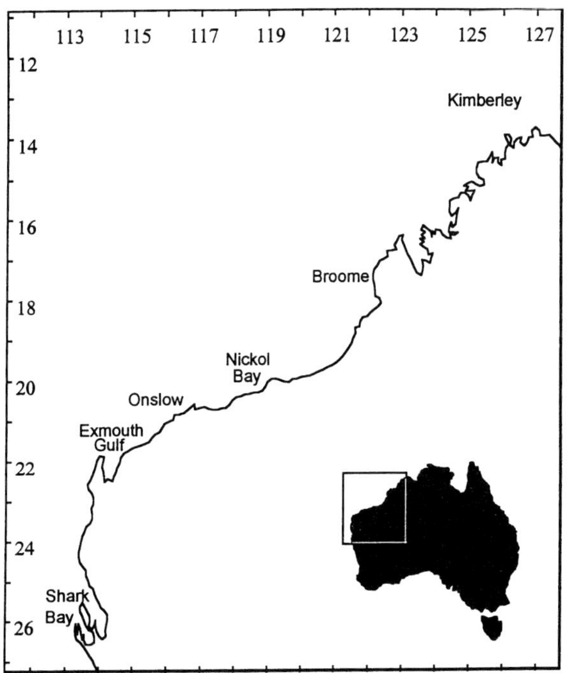

Fig. 1. Outline of the Western Australian coastline showing the areas of the major prawn trawl fisheries.

Banana prawn stock

Trawling for prawns in Exmouth Gulf began in 1963 with a small fleet that fished in daylight for schools of banana prawns (52 t taken). However, after three years of fishing (1964, 60 t; 1965, 57 t; 1966, 39 t), the catch declined to 22 t in 1967, and by 1968 was an insignificant proportion of the fishery. During 1965 and 1966, the fleet switched to night fishing on the more abundant but less catchable tiger and king prawns. Although no longer targeted by the fleet, small catches of banana prawns (less than 1 t) have continued to be recorded in most subsequent years, indicating that the species is still present but at a consistently low level. On three occasions since 1967, schools of banana prawns have occurred associated with years of high summer rainfall (1976, 17 t; 1990, 5 t; 1995, 9 t); however, on each occasion catches have not been sustained.

Noting that this banana prawn stock is at the southern limit of the species range, the decline in recruitment during the 1960s could be attributed to either a change in an unknown environmental factor that significantly reduced recruit survival, or to an adverse effect of fishing on spawner abundance that led to lower numbers of recruits, or to both factors. The consistent occurrence of fishable quantities (schools) during surveys prior to the development of the fishery and then for three years at low levels of fishing effort (Penn 1984) supports the hypothesis that the effect of fishing pressure has been a significant contributing factor. Similarly, the failure of the stock to rebuild to previous production levels, i.e. above 50 t, despite the occurrence over a 29-year period of the high-rainfall conditions usually favourable to this species, adds further support to the hypothesis that fishing has had a negative impact.

Tiger prawn stock

Night trawling targeting tiger prawns had become the major fishing activity in Exmouth Gulf by 1966. Tiger prawns dominated the catch through to 1980, but with significant year-to-year variations in production (Fig. 2). These annual catch variations from 200 to 1200 t appear to be related to

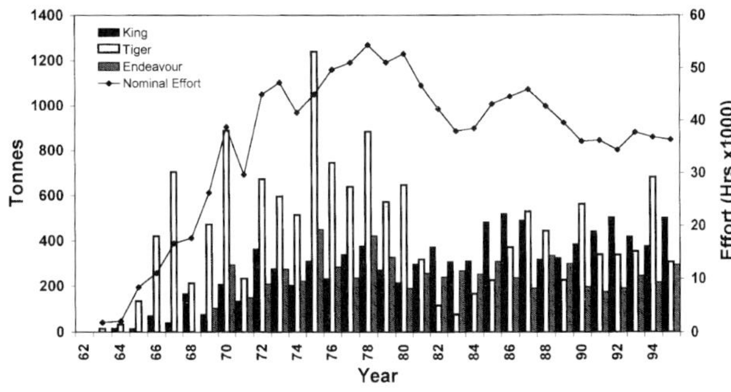

Fig. 2. Catches of the major prawn species taken in the Exmouth Gulf fishery (1963–95) and nominal hours of trawling effort.

cyclone events, although not all cyclones produced good catches (Penn and Caputi 1986). In 1981 and 1982 a significant decline in recruitment, following an escalation in effective fishing effort through vessel replacements and targeted fishing on new recruits, raised concerns that recruitment overfishing had occurred (Penn and Caputi 1985). From the very detailed database for this stock, a strong relationship between spawning stock levels in Spring, and subsequent Autumn recruitment to the fishery and cyclonic activity measured as rainfall was developed (Penn and Caputi 1986). This first recorded SRR for a penaeid stock has been subsequently updated (Penn *et al.* 1995; Caputi *et al.* 1997), and with current years added is shown in Figure 3. Although the basic SRR remains evident (r = 0.64), the initial hypothesis to explain the positive and negative impacts of cyclones (Penn and Caputi 1986) has not been supported by recent data. However, exclusion of the two years with extreme cyclone events (1971 and 1975) from the analysis shows a continuing strong relationship between spawning stock and recruitment in typical environmental circumstances (r = 0.74). Further assessment of the effects of cyclones on the SRR is constrained by the limited number of cyclone events occurring.

The ability to demonstrate a SRR for this stock exists because of a number of unusual factors. There is relatively little environmental variability in Exmouth Gulf, and those variations that do occur are significant and discrete 'events'. Secondly, there is a detailed database for the fishery which includes the necessary 'contrast' in the spawning-stock levels to allow recruitment to be assessed at both high and low spawning-stock levels. Similarly, the database contains sufficient contrast in levels of effort to allow the impact of fishing on survival to spawning to be assessed. The decline in effort through the early 1980s came about through radical management action including roster fishing by only one-third of the fleet during the recruitment period (March–April) and the closure of the whole tiger prawn area before and during spawning (August to November) in several years.

The impact of fishing as the major cause of the decline in spawning stock was demonstrated by the within-season relationship between an index of recruitment, effective fishing effort, and subsequent spawner index (abundance) (Penn and Caputi 1986). In an update by Penn *et al.* (1995) this relationship continues to show a significant multiple correlation (r = 0.95). The effect of fishing is confirmed by the response of the breeding stock, and subsequent recruitment, to management changes that reduce fishing effort and mortality.

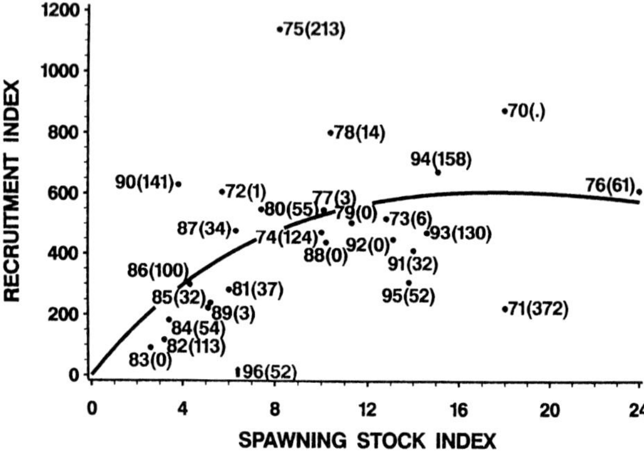

Fig. 3. Relationship between spawning stock and recruitment for tiger prawns (*P. esculentus*) in Exmouth Gulf. Recruitment year is shown, with summer rainfall (mm) in parenthesis.

When the recruitment-to-spawner relationship (RSR) and the SRR are combined (Penn and Caputi 1986), the SRR and RSR intersect (i.e. the spawning stock and recruitment are in equilibrium) at about 40 000 h effective fishing effort. This level of fishing effort has been shown historically to result in the maximum sustainable catch of just over 400 t (Caputi 1992); however, the occurrence of catches below this level in more recent years may be due to targeting of larger prawns for market reasons.

Following the collapse of the tiger prawn stock in 1982, management arrangements were introduced to reduce pre-spawning effort on tiger prawns (Penn *et al.* 1989). The resulting variable closure of the tiger prawn grounds is designed to allow a constant escapement of tiger prawns sufficient to provide an optimal level of spawning stock irrespective of annual recruit strength.

King prawn stock

The king prawn stock is most abundant in the northernmost oceanic sector of the Gulf. Production of this species lagged behind that from the tiger prawn stock in the early 1960s before becoming a regular component of the fishery by 1970 (Fig. 2). The subsequent annual production reflects the overall effort in the fishery, as well as the level of targeting of king prawn areas by the fleet. The level of targeting has generally been a function of the annual abundance of king prawns relative to the tiger prawns that occur in the more protected southern sector of the Gulf. No specific SRR analysis has yet been undertaken on this stock, but the production from the stock since 1982 (Fig. 2) does allow some SRR issues to be examined.

With the decline in tiger prawns (1982), management measures were introduced in 1983 to close the southern half of the Gulf and refocus all of the effort onto the more northern king prawn grounds where spawning by king prawns occurs (Penn 1980). This significant additional effort was permitted on the basis that the stock would be resilient to fishing (Penn 1984). The annual production from this stock (after 1983) now ranges from about 300 t to 500 t (mean 415 t) compared with 200–400 t (mean 291 t) in the previous decade, a 40% increase.

This increased targeted effort on king prawns has clearly resulted in higher catch levels and therefore can be assumed to have decreased the abundance of spawning king prawns over an extended period. Despite these changes there has been no downward trend in king prawn catch and the typical cycle of variation, presumably due to some environmental effects on recruit survival, has continued. With the effective closure of the tiger prawn fishery, levels of effort similar to those that had collapsed the tiger prawn stock were transferred to the king prawn stock; hence, it can be inferred that king prawns are less likely to be susceptible to recruitment overfishing in this fishery at least.

Endeavour prawn stock

The endeavour prawn stock tends to overlap the distribution of the tiger prawns in the southern sector of the Gulf and, to some extent, the king prawns in the north (Penn *et al.* 1989). It is a lower-value species, and effort is generally not targeted towards its capture, thus making the interpretation of the catch and catch-per-unit-effort (CPUE) data more complex. No SRR analysis has been undertaken on this stock; however, production did not decline in parallel with the tiger prawn stock despite the application of similar fishing pressure during the late 1970s and early 1980s. From this catch history it appears that the endeavour prawn stock in Exmouth Gulf is less likely to be affected by fishing than the tiger prawn stock.

SHARK BAY FISHERY

The fishery for prawns in Shark Bay began in 1962, following exploratory fishing in the late 1950s. The major species taken are western king prawns and tiger prawns, with endeavour prawns and a variety of smaller species (coral prawns) taken as a minor bycatch. Since the early 1980s, the dynamics of the fishery have also been affected by the availability of a scallop species (*Amusium balloti*), which is taken as a bycatch. This actively-swimming scallop provides a highly variable trawl catch ranging between 500 t and 20 000 t live weight annually, with a 'typical' year producing approximately 3000 t. The scallops are taken as a bycatch by the dedicated prawn trawl fleet (27 vessels) and as a target catch by a dedicated fleet of 14 full-time scallop trawlers. The availability of scallops can have a highly variable impact on the prawn fishery and cause significant bias in the prawn catch rates in some years. The scallop stock largely overlaps the distribution of the mature spawning segment of the king prawn stock (Penn 1980) and therefore causes the greatest bias in the king prawn catch–effort information. On some occasions (Fig. 4), when the scallops have been very abundant, e.g. 1991–93, their presence has effectively reduced the effort on prawns by reducing the maximum duration of prawn trawl shots from approximately 1.5 h to a few minutes, after which the catches became too heavy to bring onboard.

King prawn stock

Trawling for king prawns began in the southern sheltered waters of the Bay in 1962. By 1970 king prawn catches had increased to approximately 1000 t and all parts of the stock were being exploited. Management of the fishery was based on a limited-entry system (Bowen and Hancock 1984), which allowed controlled expansion of the fleet. Management was, however, fairly conservative, because a severe short-term decline in king prawn catches in the 1966 and 1967 seasons had raised concerns about the sustainability of the stock and/or the habitat. Both catch and fleet size were expanded through to the end of the 1970s, before the decline in the alternative tiger prawn stock caused a reassessment of the optimum fishing effort. Subsequently (1990), an industry-funded buy-back scheme was introduced, which reduced the fleet from 35 to 27.

Assessment of the status of the king prawn stock has traditionally focussed on stock production models (Hall and Penn 1979; Penn *et al.* 1989), to set target levels of effort. More recently, Caputi *et al.* (1997) investigated the relationship between stock and recruitment and found no significant relationship up to present levels of effort. However, a relationship between recruitment and an environmental factor (Caputi *et al.* 1996; Lenanton *et al.* 1991) has been found. This positive correlation ($r = 0.51$) was between recruitment and sea level (sea level reflects the strength of the Leeuwin Current, which brings warm tropical water to the Bay in autumn–winter each year). The hypothesis proposed to explain this effect is that temperature affects catchability (Penn 1984; Joll and Penn 1990) and/or growth of king prawns at recruitment time (April–May). Examination of the effects of spawning stock and environment (sea level) simultaneously (Fig. 5) confirms that spawning stock does not have a significant influence on recruitment.

The king prawn stock appears to be fairly resilient to fishing pressure in Shark Bay; with the decline in the alternative species (tiger prawns) in the 1980s, additional targeting of effort towards king prawns occurred, which appears to have resulted in increased catch levels from the same level of recruits but no longer-term reduction in recruitment. The lower catches since 1989 appear to be the result of lower effort levels resulting from

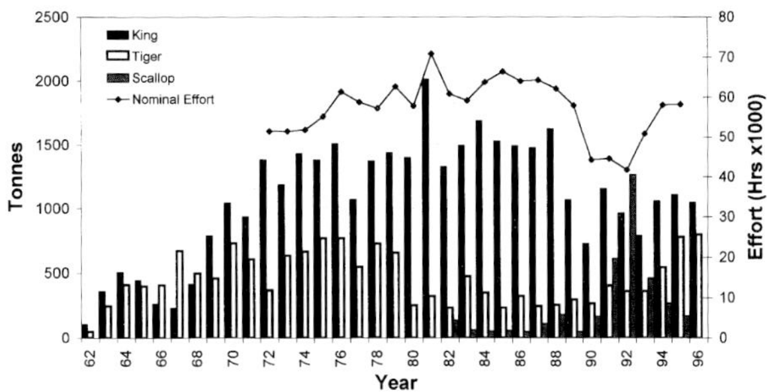

Fig. 4. Catches of the two major prawn species and scallops (meat weight) taken from the Shark Bay fishery (1962–95) and nominal hours of trawling by the prawn fleet.

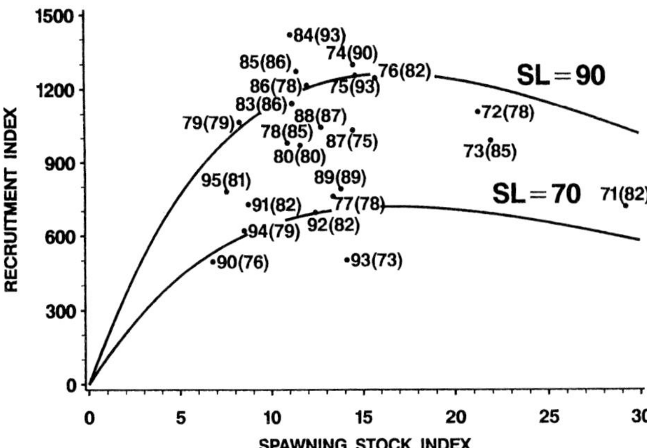

Fig. 5. Relationship between spawning stock, recruitment and sea level (reflecting the influence of the Leeuwin Current) for the king prawn stock in Shark Bay, derived from the indices described in Caputi et *al.* (1997). Recruitment year is shown, with sea level in parenthesis.

the buy-back scheme and an unusually weak Leeuwin Current in the early 1990s due to an extended El Niño Southern Oscillation (ENSO) event (Caputi *et al.* 1996), and a corresponding increase in abundance of scallops. This unusually high biomass of scallops both biased the king prawn CPUE data, on which the recruitment index is based, and reduced the effort on king prawns, thus resulting in lower catch. The interactions of these factors are still being evaluated, but the return of catches to expected levels of 1000 to 1200 t with the end of the 3–4-year ENSO event, and higher effort levels, suggest that catch downturn was driven by the environment rather than by spawning stock abundance.

Tiger prawn stock

Fishing for tiger prawns developed as an activity secondary to fishing for king prawns but became a major part of the catch in the mid 1960s. During the 1970s, tiger prawn catches increased to an average of about 650 t per year (Fig. 4) and tended to have a reciprocal relationship with king prawn catches each year owing to targeting by the vessels. Following an increase in effective effort on tiger prawns in the late 1970s, resulting from vessel replacements and improved targeting on localized areas of high tiger prawn abundance using radar technology, the catch declined to an average of 300 t during the 1980s (Fig. 4). Assessment of this decline indicated that recruitment of tiger prawns was strongly correlated with spawning stock levels, and that fishing effort had been responsible for reduced spawning stock levels (Penn *et al.* 1989). No significant environmental effect on recruitment to explain this decline could be identified.

Examination of the SRR (Fig. 6) updated from Penn *et al.* (1995) indicates that the original assessment that recruitment declined as a result of reduced breeding stock levels continues to be supported ($r = 0.64$) by the data from additional years (Caputi *et al.* 1997). Because tiger prawns form only a minor proportion of the catch from Shark Bay, management action to improve levels of tiger prawn breeding stocks was not introduced until the late 1980s. This lack of management action when compared with the situation in Exmouth Gulf has

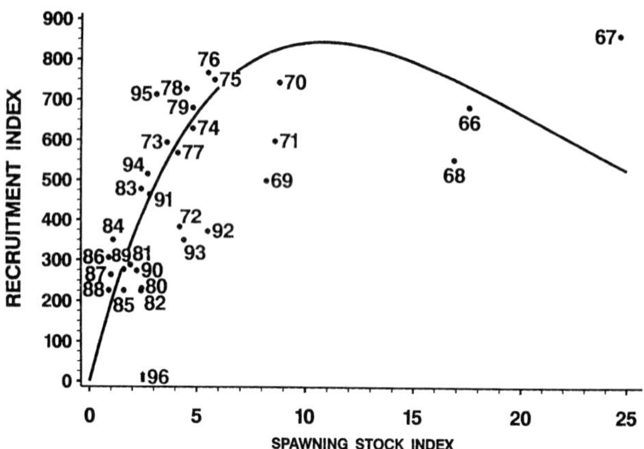

Fig. 6. Relationship between spawning stock and recruitment for tiger prawns in Shark Bay (1966–95). Data points show the year of recruitment.

enabled the two fisheries to be treated as a fortuitous case of experimental management, with the Shark Bay stock being a 'control' for Exmouth Gulf for a period of ten years (Penn *et al.* 1995). A further implication flowing from the Shark Bay assessment is that the tiger prawn stock in this fishery appears to have reached a recruitment equilibrium at about 300 t production, equivalent to a recruitment index of 250–350 in the 1980s (Fig. 6). The increase in tiger prawn recruitment since 1991 appears to be due to an increase in spawning stock (Caputi *et al.* 1997), as a result of an overall reduction in effort due to a buy-back scheme and because most prawn trawlers focussed on scallop and king prawn areas, away from tiger prawn areas.

Control of exploitation in the Shark Bay prawn fishery is achieved by a system of both temporal and spatial closures primarily aimed at optimizing the yield of the more abundant king prawns. Owing to the largely overlapping distribution of the king and tiger prawns, specific management to protect breeding stocks of the more susceptible tiger prawns has been restricted to overall controls on effort through the seasonal closure system (Penn *et al.* 1989), and to a general buy-back scheme in 1990. This buy-back scheme was designed to provide effort levels corresponding to equilibrium catches of about 1100 t of king prawns and 500 to 600 t of tiger prawns, while maintaining tiger prawn spawner indices of approximately 5 kg h^{-1}. In 1996, however, a new specific closure to protect tiger prawn spawning areas before and during the August to October spawning season (Penn *et al.* 1995) has been proposed. This system is being modelled on the Exmouth Gulf pre-spawning closure system, which is designed to leave an optimal (safe) level of spawning stock each year.

Review of penaeid recruitment overfishing

In reviewing data from a number of penaeid fisheries prior to 1981, Penn (1984) hypothesized that more-catchable species should be first to show evidence of recruitment overfishing. Using data from observations in aquaria, Penn (1984) categorized penaeid species into three broad behavioural–catchability types, of which the more catchable type-2 and type-3 schooling species were expected to be most susceptible to fishing. Subsequent field experiments (Penn, unpublished) using the methods of Joll and Penn (1990) have confirmed the differences in catchability in the field between the type-1 (lowest catchability) and type-2 (medium catchability) species which provide the majority of the Western Australian case studies. In reviewing the hypothesis of Penn (1984) and the subsequent fate of the various species in the Western Australian fisheries it is apparent that these particular stocks generally fit the 1984 hypothesis. However, a number of other stocks (Table 1) now considered to have suffered from recruitment overfishing are not all type-2 or type-3 species. Conversely, a number of type-3 stocks identified as being at high risk by Penn (1984), e.g. *P. merguiensis* (Gulf of Carpentaria, Staples *et al.* 1995) and *P. setiferus* (Gulf of Mexico, Nance 1995), have continued without significant change since 1981. Similarly, the heavily fished type-2 stock of *P. aztecus* in the Gulf of Mexico (Nance 1995) has shown a slight increasing trend in recruitment since the 1960s.

Table 1. Penaeid prawn stocks considered to have suffered from recruitment overfishing, their behaviour catchability type (after Penn 1984), and reference

Type 1(low)	*P. latisulcatus* (South Australia)	Gulf St Vincent	Morgan (1995)
	P. latisulcatus (South Australia)	Venus Bay	Anon. (1978)
Type 2(medium)	*P. esculentus*	Exmouth Gulf (Western Australia)	Penn *et al.* (1995)
	P. esculentus	Shark Bay (Western Australia)	Penn *et al.* (1995)
	P. esculentus[A]	Gulf of Carpentaria (Northern Australia)	Somers (1994)
	P. semisulcatus	Saudi Arabia (Arabian Gulf)	Morgan and Garcia (1982)
Type 3(high)	*P. merguiensis*	Exmouth Gulf (Western Australia)	Penn *et al.* (1989) this paper
	P. setiferus	Western Gulf of Mexico (Mexico)	Gracia (1996)

[A]Recruitment overfishing has not been formally reported for this stock; however, a significant reduction in pre-spawning fishing effort has been legislated, and catches have returned to about half of the historical peak production level.

DISCUSSION

Subsequent to the reviews of the possibilities for recruitment overfishing in penaeid fisheries (Garcia 1983; Penn 1984), two clearly documented cases of recruitment overfishing have now been reported for *P. esculentus* stocks in WA (Penn and Caputi 1986; Penn *et al.* 1995). Data on *P. esculentus* in the Gulf of Carpentaria (Somers 1994) also suggest that recruitment overfishing may have contributed to a significant reduction in catch from that fishery. Recruitment overfishing is also apparent in two of the three stocks of *P. latisulcatus* in South Australia (J. Keesing, pers. comm.; Morgan 1995) and has occurred with *P. setiferus* in Mexico (Gracia 1991). The historically depleted stock of *P. merguiensis* in Exmouth Gulf (Penn 1984) has also failed to recover over a 25-year period. Similarly the previously significant fishery for schooling *P. semisulcatus* in Arabian Gulf waters has apparently not recovered to its levels of the 1950s and 1960s, although some sectors have staged short-lived recoveries, e.g. off Saudi Arabia (Morgan and Garcia 1982), following significant reductions in fishing activity.

Although the number of stock failures in which recruitment overfishing has been implicated have increased during the 1980s, the majority of penaeid stocks still continue to show considerable resistance to fishing pressure. In most of these resilient penaeid stocks, variation in recruitment is more often related to environmental influences on recruitment (Staples *et al.* 1995) rather than to breeding stock abundance, although these data are rarely available. Such correlations between recruitment and environmental factors have mostly been shown in temperate and subtropical stocks where such environmental variation is probably greater. Conversely, recruitment variability should be less in tropical fisheries where most penaeid stocks live; however, recruitment in these areas has received little research attention to date (Staples *et al.* 1995).

There are also many penaeid fisheries where significant multi-year cycles in catches occur and an environmental influence on recruitment is assumed but cannot be identified. In these situations there is a possibility, yet to be evaluated, that heavy but intermittent fishing pressure on spawning stocks could generate such cycles in recruitment. For example the SRR/RSR model developed for the Exmouth Gulf tiger prawn stock (Penn and Caputi 1986) suggests that if fishing severely depletes a stock and the fleet moves away when recruitment fails, the stock would recover fully in two seasons from an almost total collapse. That is, with a relatively mobile fleet focussing heavy fishing pressure on a series of stocks, there is the potential to generate three- or four-year cycles in recruit abundance, which would give the appearance of being environmentally driven.

Penn (1984) hypothesized that more-catchable species would be the first to show the effects of fishing on spawning stocks and exhibit recruitment overfishing. The present review indicates that there are now examples from all three catchability types in which recruitment overfishing has been implicated. Further examination of the three depleted Western Australian stocks (Penn and Caputi 1985, 1986; Penn *et al.* 1995) indicates a number of factors in addition to catchability that appear to have contributed to the decline in spawning stocks and recruitment. Firstly, both *P. esculentus* stocks aggregate or concentrate at time of spawning, owing to their geographic situation in embayments and their preference for discrete and spatially restricted habitats. Secondly, both stocks are exploited in multi-species situations, where there is sufficient overlap to allow the alternative species (*P. latisulcatus*) to effectively subsidize the cost of fishing. This has allowed fishing to continue to reduce spawner biomass to levels normally uneconomic to trawl. Another factor common to these stocks has been the more restricted spawning season for the tiger prawns near the southern limit of their range (lat. 22–26°S); this has allowed a high level of pre-spawning effort. In both tiger prawn stocks, the reduction in spawning biomass has been to approximately 20% of virgin levels before recruitment reductions became evident in the catch. In contrast, the alternative low-catchability type-1 species in these fisheries, *P. latisulcatus*, does not aggregate at spawning, is not habitat specific, and is in the middle of its geographic range (Penn 1980); in these stocks, spawning is protracted and a low proportion of the fishing effort is pre-spawning.

Examination of the other cases of possible recruitment overfishing suggests that the position of stock within the species geographic range could be a common factor. For example, the two *P. latisulcatus* stocks in South Australia are at the extreme end of the range of the species (lat. 35°S) where they can be expected (Penn 1980) to have a particularly short spawning season and do not start spawning until their second year of life (Morgan 1995). In South Australia the majority of fishing effort and hence mortality can therefore occur prior to spawning. Both Gulf St Vincent and Venus Bay stocks also occur in embayments, which are likely to cause stock aggregation at spawning time. The *P. semisulcatus* stocks in the Arabian Gulf are in a similar situation, being at the northern limit of their range (30°N), and they sometimes exhibit schooling (aggregation behaviour) at high stock sizes (Morgan and Garcia 1982; Penn 1984), thereby

Table 2. Frequency of some risk factors contributing to identified cases of recruitment overfishing. Abbreviations in parentheses refer to localities identified in Table 1

		Risk Factors			
		Embayment Location	Range Position	Catchability Type	Multi/Single Species
P. latisulcatus	(GSV)	Yes	Edge	Low	Single
P. latisulcatus	(VB)	Yes	Edge	Low	Single
P. esculentus	(EG)	Yes	Edge	Medium	Multi
P. esculentus	(SB)	Yes	Edge	Medium	Multi
P. esculentus	(GOC)	No	Middle	Medium	Multi
P. semisulcatus	(AG)	No	Edge	Medium	Multi
P. merguiensis	(EG)	Yes	Edge	High	Multi
P. setiferus	(Mex)	No	Middle	High	Single

increasing the potential impact of fishing. Similarly, *P. merguiensis* in Exmouth Gulf (Penn and Caputi 1986) is a (highly aggregating) schooling species at the southern limit of its range (lat. 22°S), where historic fishing would have occurred mostly in the pre-spawning period.

In contrast, the tiger prawn stock (*P. esculentus*) in the Gulf of Carpentaria, which has shown signs of recruitment depression (Somers 1994) is closer to the centre of its geographic range, and is not confined in an embayment. However, in one of the major fishing areas in the northwestern Gulf the stock does tend to be concentrated in relatively discrete areas, owing to the presence of islands where a large highly sophisticated fleet congregates each year.

In summary, this review of instances of recruitment overfishing identified during the past decade suggests that a combination of factors has usually contributed to the cases in which a reduction of spawning stock appears to have led to recruitment failure. These examples suggest that the stocks at higher risk are those where the circumstances of the stock and associated fishery enable the generation of significant pre-spawning fishing mortality. This is often most likely in situations where the species is at the edge of its geographic range so that spawning season and hence egg production is more limited (Penn 1980). These fisheries also tend to be in localities where direct environmental effects on recruit variability are likely to be greatest, and where low stock sizes can also occur by chance because of environmental effects but are then maintained low if heavy fishing effort continues. Similarly, a variety of factors, e.g. behaviour (after Penn 1984) and habitat preference, can contribute to the increased catchability and exploitation of a stock. Equally, the situation of minor species in a multiple-stock fishery can contribute to unusually high levels of fishing mortality. The frequency of occurrence of these factors for the eight stocks identified in Table 1 shows that embayment and geographic range factors occur in five and six cases, that behaviour type is spread across all three types, and that multi-species fisheries occur in five instances (Table 2).

Identification of high-risk penaeid stocks using this array of factors is suggested as a measure of directing research towards fisheries where it is most likely to provide relevant information to support management needs. In these fisheries where the stocks are at higher risk, there also needs to be an increased focus on the development and maintenance of long-term fishery databases designed to enable both the effects of fishing effort on spawning stocks and the effects of environment and spawning stock on recruitment to be determined (Caputi 1992). Penaeid stocks assessed to be in the high-risk category should also be subject to more conservative management under the precautionary principle advocated by Garcia (1996). Given that the increasing use of electronic technology by fishermen to identify areas of localized stock abundance has the potential to significantly increase exploitation rates in all penaeid fisheries, monitoring of spawning stock levels should be given increased research priority in future.

References

Anon. (1978). Prawn tagging results in the Gulf of St Vincent and Investigator Strait. *South Australian Fishing Industry Council* **2**(2), 10–12.

Anon. (1996). State of the Fisheries 1994–95. (Fisheries Department Western Australia). 134 pp.

Bowen, B K, and Hancock, D A (1984). The limited entry prawn fisheries of Western Australia: research and management. In 'Penaeid Shrimps, their Biology and Management'. (Eds J A Gulland and B J Rothschild) pp. 272–89 (Fishing News Books: Surrey, England).

Caputi, N (1992). Spawning stock–recruitment relationships. In 'Recruitment Processes, Australian Society for Fish Biology Workshop (held in Hobart, August 1991)'. (Ed D A Hancock) pp. 142–9.

Caputi, N W, Fletcher, R, Pearce, A, and Chubb, C (1996). Effect of the Leeuwin current on the recruitment of fish and invertebrates along the Western Australian coast. *Marine and Freshwater Research* **47**, 147–55.

Caputi, N, Penn, J W, Joll, L M, and Chubb, C F (1997). Stock-recruitment–environment relationships for invertebrate species of Western Australia. *Canadian Fisheries and Aquatic Sciences Special Publication* 125.

Garcia, S (1983). The stock–recruitment relationship in shrimps: reality or artefacts and misinterpretations? *Oceanographie Tropicale* **18**, 25–48.

Garcia, S M (1996). Stock–recruitment relationships and the precautionary approach to management of tropical shrimp fisheries. *Marine and Freshwater Research* **47**, 59–65.

Gracia, A (1991). Spawning stock–recruitment relationships of white shrimp in the southwestern Gulf of Mexico. *Transactions of the American Fisheries Society* **120**, 519–27.

Gracia, A (1996). White shrimp (*Penaeus setiferus*) recruitment overfishing. *Marine and Freshwater Research* **47**, 59–65.

Gulland, J A (1984). Introductory guidelines to shrimp management: some further thoughts. In 'Penaeid Shrimps, their Biology and Management'. (Eds J A Gulland and B J Rothschild) pp. 290–9 (Fishing News Books: Surrey, England).

Hall, N G, and Penn, J W (1979). Preliminary assessment of effective effort in a two species trawl fishery for penaeid prawns in Shark Bay, Western Australia. *Rapports et Procès-verbaux des Réunions, Conseil international pour l'Exploration de la Mer* **175**, 147–54.

Joll, L M, and Penn, J W (1990). The application of high resolution navigation systems to Leslie–Delury depletion experiments for the measurement of trawl efficiency under open sea conditions. *Fisheries Research* **9**, 41–55.

Lenanton, R C, Joll, L M, Penn, J W, and Jones, K (1991). The influence of the Leeuwin Current on coastal fisheries in Western Australia. In 'The Leeuwin Current: an Influence on the Coastal Climate and Marine Life of Western Australia'. (Eds A F Pearce and D I Walker). *Journal of the Royal Society of Western Australia* **74**, 101–14.

Morgan, G R (1995). Assessment, management and research support for the Gulf St Vincent prawn fishery. *South Australian Fisheries Management Series* **12**. 35 pp.

Morgan, G R, and Garcia, S (1982). The relationship between stock and recruitment in the shrimp stocks of Kuwait and Saudi Arabia. *Oceanographie Tropicale* **17**(2), 133–7.

Nance, J M (1995). Stock assessment report 1994 Gulf of Mexico shrimp fishery. (NMFS. SEFSC Galveston Laboratory: United States). 15 pp.

Penn, J W (1980). Spawning and fecundity of the western king prawn, *Penaeus latisulcatus* Kishinouye, in Western Australian waters. *Australian Journal Marine and Freshwater Research* **31**, 21–35.

Penn, J W (1984). The behaviour and catchability of some commercially exploited penaeids and their relationship to stock and recruitment. In 'Penaeid Shrimps, their Biology and Management'. (Eds J A Gulland and B J Rothschild) pp.115–40 (Fishing News Books: Surrey, England).

Penn, J W and Caputi, N (1985). Stock recruitment relationships for the tiger prawn (*Penaeus esculentus*) fishery in Exmouth Gulf, Western Australia and their implications for management. In 'Second Australian National Prawn Seminar'. (Eds P C Rothlisberg, B J Hill and D J Staples) pp. 165–73 (Cleveland: Australia).

Penn, J W, and Caputi, N (1986). Spawning stock–recruitment relationships and environmental influences on the tiger prawn (*Penaeus esculentus*) fishery in Exmouth Gulf, Western Australia. *Australian Journal of Marine and Freshwater Research* **37**, 491–505.

Penn, J W, Hall, N G, and Caputi, N (1989). Resource assessment and management perspectives of the penaeid prawn fisheries of Western Australia. In 'Marine Invertebrate Fisheries: their Assessment and Management'. (Ed J F Caddy) pp. 115–40.

Penn, J W, Caputi, N, and Hall, N G (1995). Spawner–recruit relationships for the tiger prawn (*Penaeus esculentus*) stocks in Western Australia. *ICES Marine Science Symposia (Actes du symposium)* **199**, 320–33.

Somers, I F (1994). Species composition and distribution of commercial penaeid prawn catches in the Gulf of Carpentaria, Australia, in relation to depth and sediment type. *Australian Journal of Marine and Freshwater Research* **45**, 317–36.

Staples, D J, Vance D J, and Loneragan, N R (1995). Penaeid prawn recruitment variability: effect of the environment. In 'Proceedings of the Workshop on Spawning Stock–Recruitment Relationships (SRRs) in Australian Crustacean Fisheries'. (Eds A J Courtney and M G Cosgrove) pp. 41–8 (Department of Primary Industries: Queensland).

Chilean Resources of Benthic Invertebrates: Fishery, Collapses, Stock Rebuilding and the Role of Coastal Management Areas and National Parks

Juan C. Castilla

Estación Costera de Investigaciones Marinas, Las Cruces; and Facultad de Ciencias Biológicas, Pontificia Universidad Católica de Chile, Casilla 114–D, Santiago, Chile.

Summary

The fishery for benthic invertebrates (about 60 species) in Chile is based exclusively on small-scale (=artisanal) diving operations. The fleet is composed of about 11 000 small vessels and wooden boats (each type being less than 15 t of gross register), and there are approximately 12 700 registered divers. The annual shellfish landings range between 140 000 and 160 000 t, or 2% of the total landing. Their annual export value ($US140 million) represented 11% of the total fishery revenue in 1994. The foreign market demand for Chilean shellfish, particularly the muricid gastropod *Concholepas concholepas* ('loco'), have resulted in stock collapses. Year-round total closures of fisheries (e.g. loco) have been enforced. This paper describes the fishery collapses and stock rebuilding. Further, the institutionalization of new management tools for benthic resources and the use of adaptive management procedures and co-management strategies are described. Organized communities of fishers ('Caletas') are part of coastal management plans experimenting with common-property rights through the allocation of small (less than 1–2 km^2) Management and Exploitation Areas (MEAs). The future role of the MEAs and those of Coastal Marine Preserves and Parks are discussed. These are foreseen as comprising a national network serving conservation and management purposes.

Introduction: Fishery Components and Problems

The Chilean marine and freshwater fisheries are based on large-scale (industrial) fisheries, small-scale (artisanal) fisheries, and mariculture. In the past 15 years the three have experienced spectacular growth in landings and export revenues(Fig. 1A), but at the same time have faced problems. The small-scale fishery, with small vessels and boats below 15 t of gross register, increased landings from approximately 0.1 million t (1974) to 0.9 million t (1994). The inshore fishes and gastropods contributed most of the increase. The mariculture and industrial and artisanal fisheries have positioned Chile as one of the four leading fishery countries, accounting for approximately 10% of the world total fish production (FAO 1995).

Fishing in Chile has become an important commodity in trade. The total net export value of the production steadily increased from $US100.6 million in 1976 to $US1366.4 million in 1994 (Fig. 1B).

Small-scale fisheries also show some of the problems originating from open-access policies. Nevertheless, they present a series of characteristics that facilitate Hardin's (1968) solution to the 'Tragedy of the Commons' paradigm (Berkes 1985; Castilla *et al.* in press). These are: (a) the comparatively reduced scale of fishery operation; (b) the existence of community 'fishery and cultural traditions' in connection with the resource, environ ment and sustainability; (c) the existence of a multi-species fishery; (d) the existence of possibilities to resolve property-right issues and develop co-management and community-based management schemes; (e) the possibility of increasing the 'transparency' of the fishery activities such as stock assessment, landings, harvesting and commercialization; (f) the use of the fish extracted for food consumption and direct local commercialization; (g) the individual character of the fishery and lack of bycatch problems.

This paper reviews the official Chilean fishery landing and export revenue statistics for 1965–94 (Bustamante and Castilla 1987), mainly focussing on the shellfish resources. It also integrates small-scale fishery characteristics for benthic invertebrates as well as the role played by new variables such as the increase in foreign shellfish demand by first-world countries. Finally, it incorporates the new small-scale fishery management rules institutionalized in the 1991 Fishery and Aquaculture Law (Castilla 1994), and discusses the actual role played by coastal Management and Exploitation Areas (MEAs) and the future role foreseen for Marine Preserves and Parks.

STATISTICS: LANDING AND EXPORT REVENUE

The information on fishery landings and export revenues was extracted from the Servicio Nacional de Pesca (SERNAP) annual reports and from the Boletín Estadístico Pesquero, Sistema de Información Pesquera of the Instituto de Fomento Pesquero (IFOP) Chile. Information on the small-scale fishery fleet is seldom reported and data on their fishing effort do not exist.

Hancock (1969), Castilla and Becerra (1975) and Bustamante and Castilla (1987) analysed the landing patterns (1960–85) for the small-scale invertebrate wild stocks and cultured species landed by the artisanal fleet in Chile. According to the past and present Chilean legislation these resources (approximately 60 species) can be fished exclusively by the small-scale sector. The extractive activity is based on diving operations (Castilla 1988, 1994; Aranda *et al.* 1989; SERNAP 1994; Castilla *et al.* in press). Along the coastline there also exists a poorly studied subsistence fishery based on rocky intertidal harvesting of algae and shellfish by food-gathering fisherfolk (Durán *et al.* 1987). SERNAP (1994) estimated their number at over 6000 people. Accordingly, the interpretation of the Chilean artisanal fishery patterns is limited. In spite of these factors, efforts have been made to describe the artisanal fisheries more comprehensively (Bustamante and Castilla 1987). According to SERNAP (1994) there are 10 864 small artisanal vessels and 12 732 registered divers. It is estimated that this sector provides direct employment to over 70 000 fishers connected with sea-going, land-based and commercial activities. Trends in the landings of this artisanal fishery show two phases: a first representing almost exclusively domestic consumption, with landings between 70 093 t (1968) and 128 558 t (1978); and a second, with an increase of exports, where landings increased to

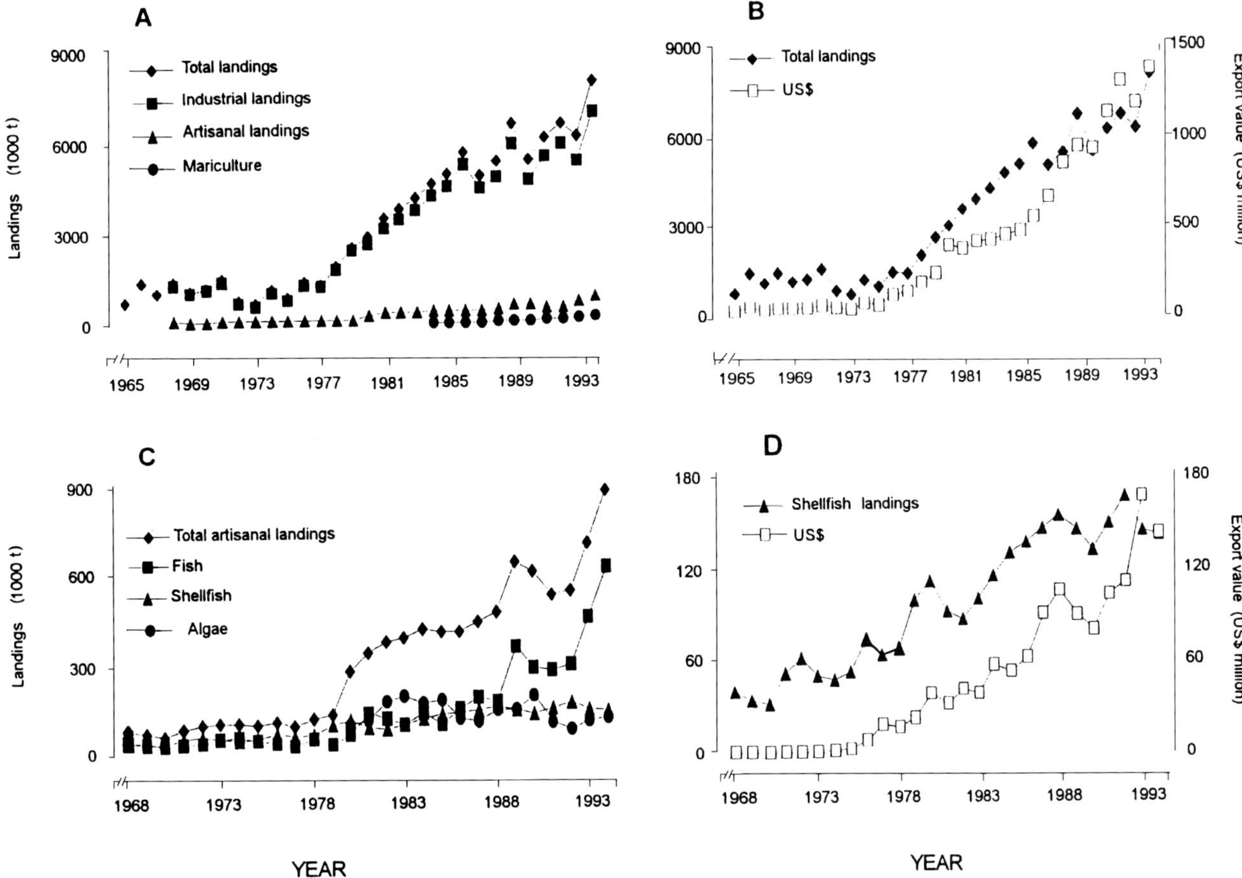

Fig. 1. Landings and export revenue of the Chilean industrial, artisanal and mariculture fisheries between 1965 and 1994.

274 352 t (1979) and reached a peak of 878 913 t in 1994 (Fig. 1C). In fact, during the early 1980s the artisanal sector suddenly faced the opportunity to substantially increase small-scale fishery activities and revenues through the penetration of foreign markets, by competing with international shellfish products (Castilla 1990). This development occurred under a policy of open access in Chile.

There are three main wild stock fisheries in the Chilean artisanal activity: inshore finfish, shellfish and algae (Fig. 1C). The artisanal finfish products, with few exceptions such as the swordfish, *Xiphias gladius*, are for domestic consumption (it is difficult to extract export revenues from this commodity from present statistical reports). Shellfish are also heavily consumed in the country, but there also exist detailed statistics about their export revenues. The same is true for the algae. Figure 1D shows that shellfish landings ranged between about 100 000 t (1979), at the start of the boom in exports, and 140 000–165 000 t (1993–94). The revenues from the export trade experienced a steady increase during this period from approximately $US22 million (1979) to $US140–165 million in 1993–94(Fig. 1D).

Within the shellfish products there are three main wild stock categories: molluscs, crustaceans, and other species such as sea urchins. In 1994 the sea urchin *Loxechinus albus* accounted for about 93% of 'other species'. The mollusc landings showed the most spectacular increase from approximately 28 000 t (1968) to over 88 000 t (1994) (Fig. 2A). Among the molluscs the muricid gastropod 'loco', *Concholepas concholepas*, is by far the most significant and valuable species. Its landings and fishery management (Castilla 1995) had a great impact on the total mollusc landings in the country and their export revenues (Fig. 2B). Revenues from the mollusc wild stocks in 1993 (mostly due to the elevated international price of the loco) was $US119 million.

The Chilean marine gastropod fishery is one of the richest in the world. Four muricid or muricid-related species are extracted: the loco, the 'trumulco' *Chorus giganteus*, the 'locate' *Thais chocolata* and the 'caracol rubio' *Xanthochorus cassidiformis*. Furthermore, there are 10 species of large key-hole limpets of the genus *Fissurella* (Oliva and Castilla 1992), and snails such as *Tegula* spp. and *Trophon* spp. Between 1968 and 1994, catches have been dominated by the loco (Fig. 2C, D). Four phases can be seen in the landings (Fig. 2C). First, before 1978, the gastropod fishery relied exclusively on locos, and consumption was mostly domestic. Second, the sudden increase in landings during the early 1980s was due to the opening of foreign markets in Asia, lasting for about 10 years and boosting the export revenue for gastropods. In this phase there was a diversification of the gastropod fisheries, with the locate replacing the loco in foreign markets. Third, between 1989 and 1992 fishing for locos was

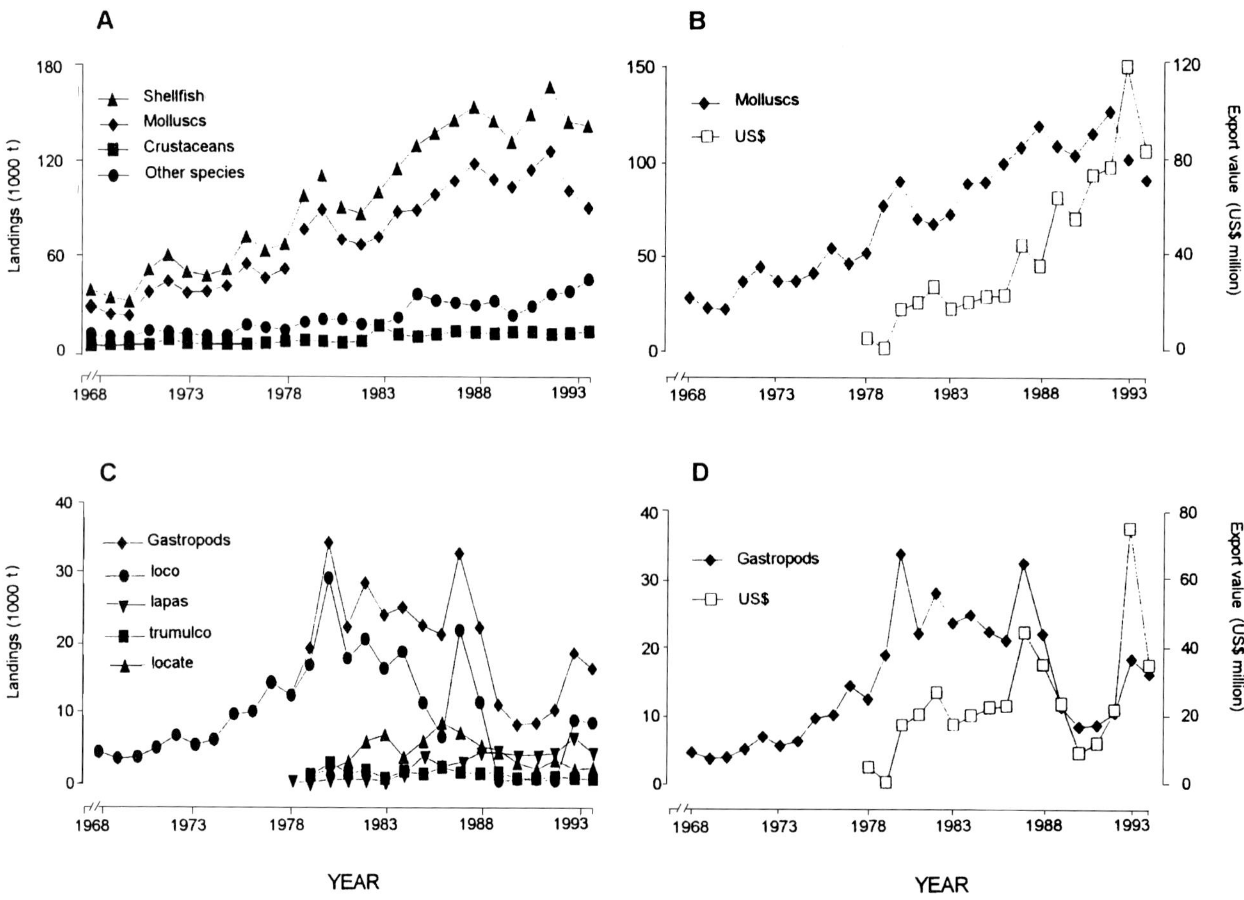

Fig. 2. Landings and export revenue of the Chilean shellfisheries between 1968 and 1994.

completely banned in Chile. Finally, new management regulations were implemented in 1993 (Castilla 1994) and the international price of locos in 1993 (following the 4-year fishery ban) increased markedly, yielding an enormous $US64 million revenue (Fig. 2D).

Lessons from the loco fishery

Castilla (1995) and Castilla and Fernández (1996) analysed the fishery for locos. Five phases were observed between 1960 and 1994 (Fig. 3): 1960–75, exclusive domestic consumption; 1976–82, initial export phase; 1983–88, overexploitation; 1989–92, total closure of the fishery; and 1993 to the present, implementation of new fishery management measures in tune with the 1991 Law. Part of the regulations are based on a Total Allowable Catch (TAC) that is divided into individual quotas, as well as on co-management and community-based strategies (Castilla *et al.* in press). The main lessons derived from the loco fishery in Chile can be grouped into 'business as usual' and 'novelties'.

Business as usual

(1) The opening of external markets and lack of appropriate management practices in the country (a 'race for the loco'), lack of stock assessments, and open access policies produced an overexploitation of the species in the early 1980s.

(2) Attempts to regulate fishing activities were taken too late (the middle and late 1980s).

(3) Most probably, the 'ratchet effect' (Ludwig *et al.* 1993) was present in the early 1980s 'loco fishery scenario'; in 1982 the Chilean coast was affected by the strongest El Niño –Southern Oscillation event of the century, and according to Castilla and Camus (1992) the increase in water temperature could have affected the recruitment of locos or caused natural declines of loco populations already overfished, particularly in northern Chile.

(4) Economic and social problems derived from the overall decline triggered the harsh management policy of a total ban on loco fishing for four years. This resulted in an increase in poaching. According to Castilla (1995), around 4000–6000 t of locos per year were taken illegally during this period.

Novelties

(1) The 1989–90 'loco fishery scenario' was the catalyst for the implementation of new management tools for the rational use of benthic marine resources in the 1991 Chilean Fishery and Aquaculture Law (Castilla 1994). The loco played a crucial role because of the revenues involved and the economic and social significance of its fishery.

(2) Management regulations were implemented for 'fully-exploited benthic resources'. They were based on the establishment of TAC and individual quotas (Castilla 1994).

(3) Regular stock assessment programmes were implemented.

(4) A novel approach of co-management and community ownership rights over reduced extensions of coastal segments were included in the 1991 Law. A system of Management and Exploitation Areas for benthic resources was implemented — usually of less than 1–2 km^2 of sea bottom, exclusively dedicated to small-scale fisher communities (Castilla 1994). Preliminary results of natural re-stocking of locos, sea urchins and key-hole limpets (Castilla and Fernández 1996) in coves or caletas formed by organized fisher communities (particularly in central Chile) are promising and offer hopes for the future sustainable use of benthic resources (Payne and Castilla 1994; Minn and Castilla 1995; Pino and Castilla 1995; Castilla and Fernández 1996; Castilla and Pino 1996; Castilla *et al.*

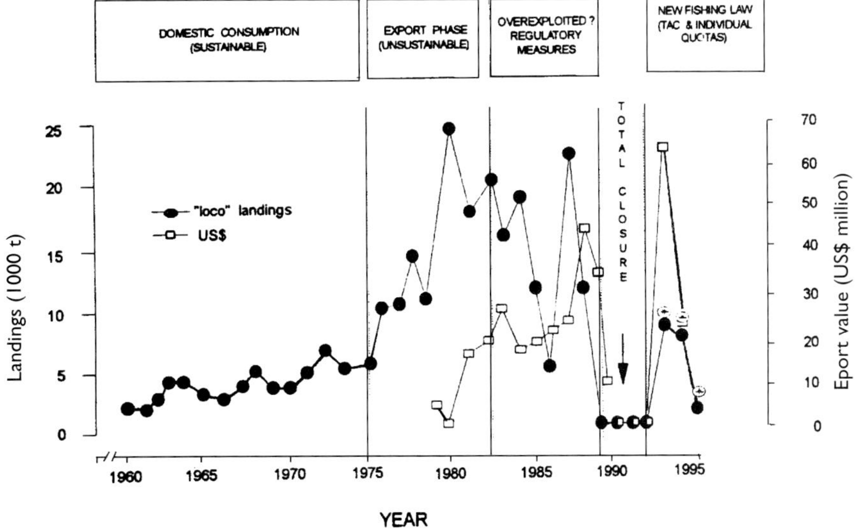

Fig. 3. 'Loco', *C. concholepas*, landings and revenue values between 1960 and 1994. The arrow indicates that during the four years of total fishery closure high levels of poaching occurred (Castilla 1995). Circles with asterisks show the official TAC in tonnes.

in press). Presently, small-scale fisher communities are participating in the processes and are eager to obtain MEAs to promote the sustainable use of benthic resources within them.

(5) The above measures have significantly improved the social organization of fisher coastal communities and promoted the formulation of internal community rules (varying from community to community) to cope with the new challenges of enhancing and using 'their benthic resources' in a sustainable manner within the MEAs.

(6) A major result (e. g. at Caleta El Quisco in central Chile) has been the realization by the fisher communities that rebuilding of invertebrate stocks occurs 'naturally' if the MEAs are subjected to reasonable periods (2–4 years) of 'no-take' or 'resting'. Joint teams of scientists and divers from the communities are conducting stock assessments, adding transparency to the process, and later validating the results through commercial extraction. Catch-per-unit-effort (CPUE) of benthic resources, such as loco and sea urchins, has greatly increased within the MEAs as compared with common access fishing grounds (Table 1). Furthermore, under the new set of rules the commercialization of products is becoming more controlled by the communal organization.

THE TIP OF THE ICEBERG

Although the loco fishery plays a critical role in the Chilean small-scale fishing activity, and in 1993 represented 50% of the total revenue from shellfish, there are other unique invertebrate species in urgent need of more rational management. The sea urchin *Loxechinus albus* is one of them. The species has a very high demand and price on the foreign market. *L. albus* landings have increased 20-fold from around 2000 t (1975) to nearly 40 000 t (1994). Key-hole limpets and the stonecrab *Homalaspis plana* are also vulnerable to overexploitation during the coming years. Fortunately, the present Chilean Fishery and Aquaculture Law contains novel regulations and tools to confront the 'business as usual scenario'. In Chile, the more promising solutions to stop the overexploitation of shellfishes are seen through the implementation of these tools, rather than through aquaculture. This is particularly the case for the most valuable benthic, non-filter-feeding invertebrate species.

THE COMMONS: COMMUNAL–INDIVIDUAL DILEMMA

Hardin (1968) illustrated brilliantly the tragedy of the commons and ways to solve it. In the Chilean artisanal fisheries this tragedy is a reality, and open-access fishery policies of the past exacerbated the issue. On the other hand, community-based management schemes address only one part of the dilemma's equation. It is my view that the Chilean coastal realm, owing to its long extension and macro-configuration, possesses unique geographical characteristics that permit experimentation with mixed management and conservation approaches. For instance, the small MEAs, representing the interests of the community and individuals, could be latitudinal combined with a system of Marine Coastal Preserves and National Parks representing the interests of the commons. Both could act as stock-rebuilding centres and at the same time as seeding grounds for other fished areas. Conservation and the rational use of resources could be served simultaneously (Castilla 1996).

DISCUSSION

Bustamante and Castilla (1987) drew attention to the rich Chilean fauna of invertebrates that are fished for by diving, and warned of the risk of overexploitation faced by several species. Approximately 10 years later, some of the predictions have come to pass. Nevertheless, positive lessons were learned. The present benthic invertebrate management tools, institutionalized in the 1991 Chilean Fishery and Aquaculture Law, are rewarding, particularly because most of them were derived from progress in basic ecological and fishery research made in the country (Castilla and Durán 1985; Castilla 1988; Castilla 1994; Castilla *et al.* 1994; Castilla *et al.* in press). As stressed in this paper the small-scale fishery activity in Chile has a 'societal dimension'. The legislator implemented a set of management procedures balancing 'the commons, communal and individual rights', thereby accounting for the existence of 'real fisher communities' and traditions. The scheme used in this Law is radically different from the Individual Transferable Quotas (ITQ) scheme, for instance in practice in New Zealand (Dewees 1996). ITQs used as the only tool to solve fishery problems within traditional small-scale fishery societies can create 'virtual communities'; they can, eventually, work against the final goal of sustainable use of the resources incorporating the users. ITQs are quasi-property rights depending on economic and political realities and industry goals (Dewees 1996). Most importantly, they are rights that, if not used adequately by the recipients, can destroy traditional fishery societies. Small-scale fishery co-management and communal-rights (viewed as alternative approaches) in Chile are in their infancy, but preliminary results are promising. In any case, the key to success must always be to understand not only

Table 1. Comparison of size and CPUE (per hour) between El Quisco MEA and open access areas for two dived - benthic species heavily exploited in Central Chile. Mean values and standard deviations, or range of mean values, are provided. s = loco's summer ban lifting; w = loco's winter ban lifting, 1993 (partially taken from Castilla and Fernández 1996)

	Management and Exploitation Area (MEA)		Open Access Diving Areas	
Species	Size (cm)	CPUE	Size (cm)	CPUE
Loco	110–117	280–540[s]		
(*C. concholepas*)	107–118	91–186[W]	103–108	15–143
Sea urchin	102 (7.0)		78.8 (5.5)	
	97.1 (8.1)	409 (219)	87.6 (8.1)	129 (77.9)

how the fish behave, but also how the fishers behave. Rational fisheries do not exist if both components are not well tuned.

The Chilean MEAs, co-management- and community-based schemes, jointly with the development of a national policy to establish large Marine Preserves and Parks, can be seen as the most promising approach to combining conservation and sustainable use of benthic resources in the country (Castilla 1996).

Acknowledgments

I acknowledge the financial support from Projects Fondecyt No. 193/0684, the European Economic Community, Contract Nº CI1* CT93 and Minera Escondida Ltd. During the completion of the paper I benefited from a PEW Fellowship. I sincerely thank C. Pino who collected the statistics and helped with the figures; J. Alvarado and M. Fernández and an anonymous referee suggested important modifications to an early draft.

References

Aranda, E Bustos, E, and Chomaly, J (1989). Estado de situación del sector pesquero artesanal en Chile. *Revista de la Comisión Permanente del Pacífico Sur* **18**, 7–33.

Berkes, F (1985). Fishermen and 'the tragedy of the commons'. *Environmental Conservation* **12**, 199–206.

Bustamante, R H, and Castilla, J C (1987). The shellfishery in Chile: an analysis of 26 years of landings (1960–1985). *Biología Pesquera* **16**, 79–97.

Castilla, J C (1988). La problemática de la repoblación de mariscos en Chile: diagnóstico, estrategias y ejemplos. *Investigaciones Pesqueras (Chile)* **35**, 41–8.

Castilla, J C (1990). Clase Magistral: importancia y proyección de la investigación en Ciencias del Mar en Chile. *Revista de Biología Marina (Montemar), Valparaíso* **25**, 1–18.

Castilla, J C (1994). The Chilean small-scale benthic shellfisheries and the institutionalization of new management practices. *Ecology International Bulletin* **21**, 47–63.

Castilla, J C (1995). The sustainability of natural resources as viewed by an ecologist and exemplified by the fishery of the mollusk *Concholepas concholepas* in Chile. In 'Defining and Measuring Sustainability'. (Eds M Musaninghe and W Shearer). pp. 153–9 (The International Bank for Reconstruction and Development/The World Bank: Washington, DC.)

Castilla, J C (1996). La futura Red Chilena de Parques y Reservas Marinas y los conceptos de conservación, preservación y manejo en la legislación nacional. *Revista Chilena de Historia Natural* **69**, 253–70.

Castilla, J C, and Becerra, R M (1975). The shellfishery in Chile: an analysis of the statistics 1960–1973. In 'International Symposium on Coastal Upwelling Proceedings'. (Ed J E Valle.) pp. 61–90 (Universidad del Norte: Coquimbo, Chile.)

Castilla, J C, and Camus, P A (1992). The Humboldt–El Niño scenario: coastal benthic resources and anthropogenic influences, with particular reference to the 1982–83 ENSO. In 'Benguela Trophic Functioning'. (Eds A I L Payne, K H Brink, K H Mann and R Hilborn). *South African Journal of Marine Sciences* **12**, 703–12.

Castilla, J C, and Durán, R (1985). Human exclusion from the rocky intertidal zone of central Chile: the effects on *Concholepas concholepas* (Gastropoda). *Oikos* **45**, 391–9.

Castilla, J C, and Fernández, M (1996). Small-scale benthic fisheries in Chile: a lesson on co-management and sustainable use of benthic invertebrates. National Academy of Science International Conference on Ecosystem Management for Sustainable Marine Fisheries, National Research Council, Ocean Studies Board (OSB), February 19–24, Monterey, California, 1–17.

Castilla, J C, and Pino, C (1996). The small-scale fishery of the red sea urchin, *Loxechinus albus* (Molina 1782), in Chile and the Management and Exploitation Area of Caleta El Quisco. *Out of the Shell* **5**, 5–8.

Castilla, J C, Branch, G M, and Barkai, A (1994). Exploitation of two critical predators: the gastropod *Concholepas concholepas* and the rock lobster *Jasus lalandii*. In 'Rocky Shores: Exploitation in Chile and South Africa'. (Ed W R Siegfried). *Ecological Studies* **10**, 101–30. (Springer: Berlin.)

Castilla, J C, Manriquez, P, Alvarado, J, Rosson, A, Pino, C, Espoz, C, Soto, R, Oliva, D, and Defeo, O (in press). Artisanal 'Caletas' as units of production and co-managers of benthic invertebrates in Chile. *Canadian Journal of Fishery and Aquatic Sciences* (Special Publication).

Dewees, C M (1996). Summary of individual quota systems and their effects on New Zealand and British Columbia fisheries. National Academy of Science International Conference on Ecosystem management for Sustainable Marine Fisheries. National Research Council, Ocean Studies Board (OSB), February 19–24, Monterey, California, pp. 1–8.

Durán, L R, Castilla, J C, and Oliva, D (1987). Intensity of human predation on rocky shores at Las Cruces, Central Chile. *Environmental Conservation* **14**, 143–9.

FAO (1995). The state of world fisheries and aquaculture. Food and Agriculture Department of the United Nations. Rome. 57 pp.

Hancock, D A (1969). La pesquería de mariscos en Chile. Instituto de Fomento Pesquero (IFOP), Publicación No. **5**, 1–94.

Hardin, G (1968). The tragedy of the commons. *Science* **162**, 1243–8.

Ludwig, D, Hilborn, R, and Walters, C (1993). Uncertainty, resource exploitation, and conservation: lessons from history. *Science* **260**, 17–36.

Minn, I, and Castilla, J C (1995). Small-scale artisanal fishing and benthic invertebrate management in Caleta Las Cruces, central Chile. *Out of the Shell* **5**, 11–15.

Oliva, D, and Castilla, J C (1992). Guía para el reconocimiento y morfometría de diez especies del género *Fissurella* Bruguière, 1789 (Mollusca:Gastropoda) comunes en la pesquería y conchales indígenas de Chile central y sur. *Gayana, Zool.* **56**, 77–108.

Payne, H E, and Castilla, J C (1994). Socio-biological assessment of Common Property Resource Management: small-scale Fishing Unions in Central Chile. *Out of the Shell* **4**, 10–14.

Pino, C, and Castilla, J C (1995). The key-hole limpets (*Fissurella* spp.) in the Chilean artisanal fishery. *Out of the Shell* **5**, 8–10.

SERNAP (1994). Anuario Estadístico de Pesca, Servicio Nacional de Pesca, Chile.

Development, Perspectives and Management of Lobster and Abalone Fisheries off Northwest Mexico, under a Limited Access System

A. Vega,[A] *D. Lluch-Belda,*[B] *M. Muciño,*[A] *G. León,*[A] *S. Hernández,*[B] *D. Lluch-Cota,*[B] *M. Ramade*[C] *and G. Espinoza*[A]

[A] Instituto Nacional de la Pesca/CRIP La Paz. Km. 1 Carr. Pichilingue, La Paz B.C.S., Mexico.
[B] Centro de Investigaciones Biológicas del Noroeste. PO. Box 128, La Paz B.C.S. Mexico.
[C] Fed. Reg. Soc. Coop. Industria Pesquera. 'Baja California', Ensenada, B.C. Mexico.

Summary

The role of a limited access system, as well as the influence of environmental variability, are analysed with regard to the development of abalone and lobster fisheries off northwest Mexico. In both fisheries, regulatory controls have followed traditional approaches; mainly through minimum legal sizes, closed seasons, and total catch and effort limitations. In spite of some bio-ecological and management similarities, each fishery evolved differently after 1972/73. While lobster landings have been relatively stable, with quasi-oscillatory fluctuations, the abalone fishery collapsed, presumably due to the combined effects of inadequate policies and environmental factors. Results suggest that management should be strengthened through further cooperation with industries (co-management) and multi-disciplinary approaches, and that climate variability must be considered in management strategies, both in the short term and long term.

Introduction

Abalone and lobster fisheries sustain many fishing communities along the west coast of the Baja California Peninsula. Fluctuations in the catches of both fisheries have large social and economic impacts. Moreover, both are important with regard to Mexican exports of fish products: abalone revenues rank third and lobster fourth ($US33 and $U21 million, respectively). Biologically, both fisheries rely on long lived, slow growing and and relatively sedentary groups of species. Five species form abalone catches, but green *(Haliotis fulgens)* and yellow (*H. corrugata*) abalone account for about 95% of total landings. The remainder of the catch is red (*H. rufescens*), black (*H. cracherodii*) and the chine or white abalone (*H. sorenseni*) (De Buen, 1960; Guzmán 1992; Ortíz and León 1988; Vega *et al.* 1994). Three lobster species are caught, but the subtropical-temperate red rock lobster (*Panulirus interruptus*) constitutes 95–97% of the catch, the remainder being comprised of two tropical species: the blue (*P. inflatus*) and green (*P. gracilis*) lobster (Vega *et al.* 1996).

Despite some early studies, systematic monitoring of catches did not begin until early 1970s. Since then, most of the research has focused on the reproductive biology, structure (size, sex) and abundance of the population, in order to enable basic fishing regulations. Recent research is attemting to bring together the effects of fishing and the impact of the regional environment (Guzmán *et al.* 1991, Vega and Lluch-Cota 1992). Regional environment changes have been shown to have a large impact on small pelagics (Lluch-Belda *et al.* 1986, 1991) but has not been stated for benthic long-living resources. For abalone, Cox (1962) had suggested some environmental effects on abundance, but the idea of a causal relationship between climate and abundance for these species has been re-examined only recently (Guzmán 1994; Vega *et al.* 1994).

A possible relationship between abundance of the lobster species and climate fluctuations in Baja California was suggested from the cyclic pattern of catches during the last five decades (Vega and Lluch-Cota 1992). This is coincident with the hypothesis previously suggested by Johnson (1960, 1971) and Pringle (1986), in the sense that anomalous ocean currents, such as El Niño, affect its pattern of abundance and concentration of larvae and subsequently catches. Similar environmental impacts have been suggested for Hawaiian lobster (Polovina and Mitchum 1992), and, in Western Australia, Phillips *et al.* (1994) reported a close relationship between recruitment and catches in several spiny lobster fisheries and ENSO variability.

Usually fishing effort has been considered by many as the main factor affecting abundance variations and little attention has been paid on the potential effects of climate. So, in this paper we analyse the development of the two fisheries, in the context of a limited access system and from the perspective of the fishing regime evolution and environmental variability.

METHODS AND SOURCES OF INFORMATION

Our main sources of information were the research programmes carried out by the Instituto Nacional de la Pesca (INP) through its Regional Fisheries Centers at La Paz, Bahía Tortugas and Ensenada. Fishermen's organizations provided catch data and descriptions of their internal polices and socio-economic function. The effectiveness of management was inferred from indicators such as trends in total catches and effort, development of industrial infrastructure and performance of fishermen's organizations.

The relationship between abundance and climate was explored using the SST observations contained in the COADS data set (Roy and Mendelssohn 1994). Monthly anomalies were first computed from one-degree squares along the coastal zone between the 28° and 26°N latitude. Since no significant differences were found between squares, all series were averaged into a single regional index.

Present study is focused in the central region because it is the most productive for both fisheries, and historically has contributed approximately 76% and 85% of the total landings for lobster and abalone, respectively .

RESULTS AND DISCUSSION

Historic background and performance of fishermen's cooperatives

Although they are the oldest remaining fisheries in Mexico, continuous catch records are available only since 1929 for lobster and since 1940 for abalone. Foreign enterprises began harvesting these resources in third quarter of 19th century and remained in Mexico until the mid 1930s (Cox 1962; Mateus 1986; Vega *et al.* 1996), when national laws stated that these species were to be exploited exclusively by Mexican fishermen's organizations, through fishing cooperatives. This provided a unique scheme of limited entry. Most of cooperatives were established between 1938–1954 (Table 1). The fishing zone for each one is indicated in Figure 1 by the respective block number. Thus, there are 10 cooperatives in the central zone, which account for 53% of the lobster traps and 60% of abalone diving boats. The northern zone has 28% and 20%, respectively, whereas the southern has 19% and 20%, respectively.

Together with the nationalization of these fisheries, industrial development played a major role in its evolution. Before 1930 production was limited to two processing plants in northern Baja California (Cox 1962; Mateus 1986), but between 1930 and 1965 nine additional privately owned plants were established and production expanded. As profits increased, cooperatives installed their own plants and private owners were gradually replaced. Currently, cooperatives own most of the industrial capacity (12 plants for processing frozen and live lobster and canning of abalone). Most cooperatives are economic units, organized as social enterprises, and manage harvesting and industrialization as a vertically integrated process. The cooperatives have achieved the ability both to adjust to regulatory polices and to explore different management alternatives under the legal frame established.

Management system and fishing regime

By law, both resources are managed at the federal level by the Ministry of Environment, Natural Resources and Fisheries. However, in practice, management is carried out under three strategies: restricted fishing rights, official regulatory mechanisms and community-based decisions. Fishing rights were first allocated according to the 1936/1938 legislation (which resulted in a limited entry regime); but since 1992 the law

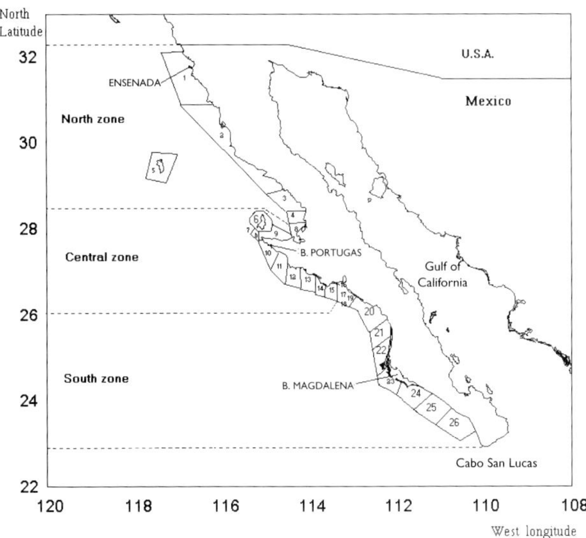

Fig. 1. Zones of study and approximated fishing area by cooperative along west Baja California

Table 1. Fishing cooperatives fishing abalone and lobster in Baja California (average number of boats and traps during 1993–1995)

N°	Cooperatives per zones	Year of origen	Lobster Fishery # boats	# traps/boats	# traps/Coop.	Abalone Fishery # diving boats
	Northern zone:					
1	Litoral de Baja California	1983	16	75	1200	15
2	Ensenada	1936	56	92	5152	20
3	Rafael Ortega Cruz	1945	15	69	1035	8
4	Siempre viva	1984	5	50	250	
5	Abuloneros y langosteros	1955	7	40	280	7
	Subtotal = 5 Cooperatives		**99**	**65**	**7917**	**50**
	Central zone:					
6	Pescadores Nacionales de Abulón	1942	20	70	1400	22
7	Buzos y Pescadores	1942	21	65	1365	16
8	Luis Gomez Z.	1974	8	40	320	
9	La Purisima	1944	30	70	2100	11
10	Bahia Tortugas	1944	22	60	1320	15
11	Emancipacion	1939	25	60	1500	22
12	California San Ignacio	1936	16	60	960	20
13	Leyes de Reforma	1974	19	55	1045	12
14	Progreso	1944	40	65	2600	25
15	Punta Abreojos	1943	45	60	2700	10
	Subtotal = 10 Cooperatives		**246**	**61**	**15 310**	**153**
	Southern zone:					
16	Laguna de San Ignacio	1969	8	25	200	2
17	19 de Septiembre	1975	8	25	200	2
18	Ejidal Cadeje	1974	10	30	300	2
19	Ejidal San Jose de Gracia	1979	8	30	240	2
20	Puerto Chale	1958	55	35	1925	15
21	La Poza	1957	10	30	300	6
22	Pescadores de Puerto San Carlos	1969	15	30	450	5
23	Bahia Magdalena	1953	38	35	1330	15
24	General Meliton Albañez	1959	10	30	300	3
25	Ejidal Punta Lobos	1975	6	25	150	
26	Todos Santos	1967	6	25	150	
	Subtotal = 11 Cooperatives	**1968**	**174**	**32**	**5545**	**52**
	TOTAL		519		28 772	255

allows rights to be transferable (Secretaria de Pesca, 1992). Fishing rights for each cooperative only apply to a fishing area or 'economic zone', which is strictly respected by other cooperatives and defended against outside fishermen (Table 1 and Fig. 1). Federal regulations have followed traditional approaches, mainly through minimum legal sizes, closed seasons and limited fishing effort (Vega *et al.* 1994, 1996). In addition and unlike Southern California (Cox 1962; Duffy 1973), sport fishing of these resources is prohibited in Mexico. Despite this and other differences, most California regulations were applied in Baja California until specific studies were carried out.

Usually any given regulatory mechanism in Mexican fisheries has remained constant for very long periods. In the case of abalone, the most significant change occurred until 1982-83, when a latitudinal scheme of closed seasons and specific minimum legal sizes by zones was established (Fig. 2); in spite of it had been reccomended since 1980 (Inst. Nac. de Pesca,

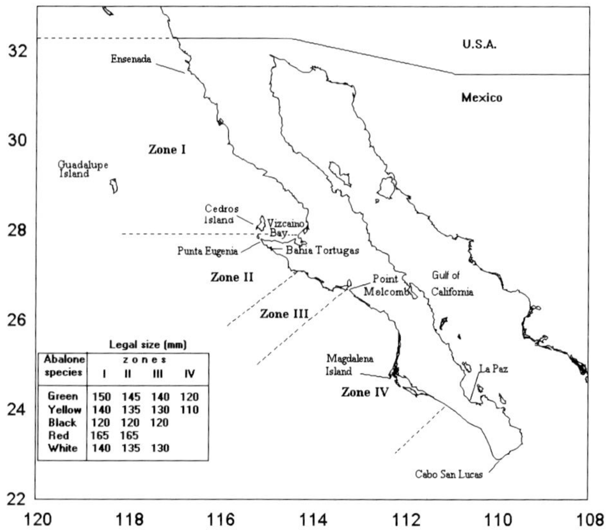

Fig. 2. Regime of the closed seasons and legal size for the abalone fishery in Baja California.

1980). Other changes were the reduction of the fishing season (from 10 to 7 months), the enforced fulfillment of minimum legal sizes and the mandatory use of log-books for catch and effort recording. A quota system was proposed in 1973 (Lluch-Belda *et al.* 1973), but not effectively introduced until 1989–90, as a total allowable catch (TACC) for each co-operative. Annual TACCs are determined on the basis of direct censuses, through joint assessments by governmental researchers and the cooperative's technical staff. TACCs are then distributed among the diving units, in accordance with internal policies of each cooperative.

Lobster regulations include protection of ovigerous females, closed seasons, minimun legal sizes, restriction to fishing by trapping, and a limited number of boats and traps are authorized per cooperative. Until 1992 the closed season was the same for all fishing areas, but recent research has permitted the management through delayed-stepped closures (Fig. 3), that better reflect the time–latitudinal lags in the reproductive cycle (Vega *et al.* 1996). The minimun legal size was set at a total length of 265 mm until 1960, but since then a carapace length (CL) of 82.5 mm has been used (Vega and Lluch-Cota 1992; Vega *et al.* 1996)

However, the main support for management tools is the combination of limited entry and strict delineation of cooperatives' fishing areas. This, as well as their control over harvesting and processing, has allowed long periods of stabilized effort levels. Moreover, cooperatives have been evolving towards community-based self-regulatory management, with self control of effort and other internal fishing decisions (Table 2) aimed at strengthening the federal regulations. Close collaboration with the federal ministry has meant that information requirements are being fulfilled and co-management strategies are now applied, for example, joint stock assessment for determination of the abalone quotas, and gradual adjustment of the lobster zonal closed season. Indeed, for both fisheries there is a clear tendency towards active co-management.

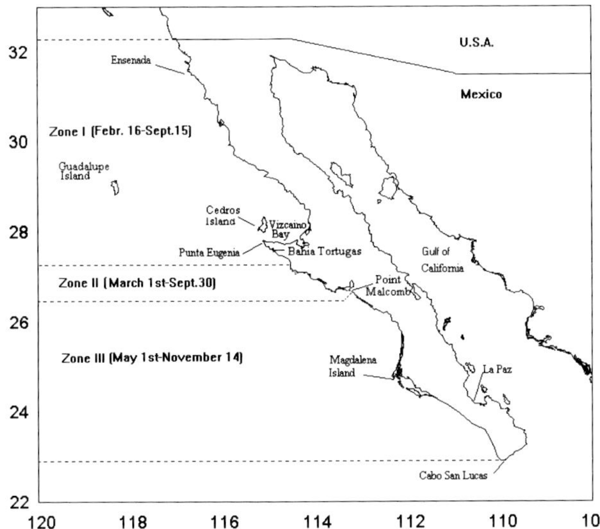

Fig. 3. Regime of the closed seasons for the lobster fishery in Baja California.

Catch trends as related to fishing regime and environmental changes

Figure 4 shows the historical development of both fisheries with respect to catch levels as well as organizational and technological stages. The period from the early 1900s to the mid-1940s corresponds to an under exploited stage for both fisheries, with open access and little regulation. Most cooperatives and plants were installed between 1938 and 1950, thus fisheries expansion occurred after the end of the second world war and until 1958. During this second period, lobster production increased to average almost 1000 t, while abalone catches reached a historical maximum around 5000 t; partially because of the use of new fishing grounds and improved fishing equipment. From 1959 to 1973 sustained yields were achieved under the autonomy of recently restructured cooperatives. During this period, catch stability is evident for both resources, with only moderate fluctuations until 1974–75, but after that each follows different trends despite similarities in habitat and management.

Figures 5 and 6 show catches and effort trends and SST variations for the central west region. It is evident that lobster landings

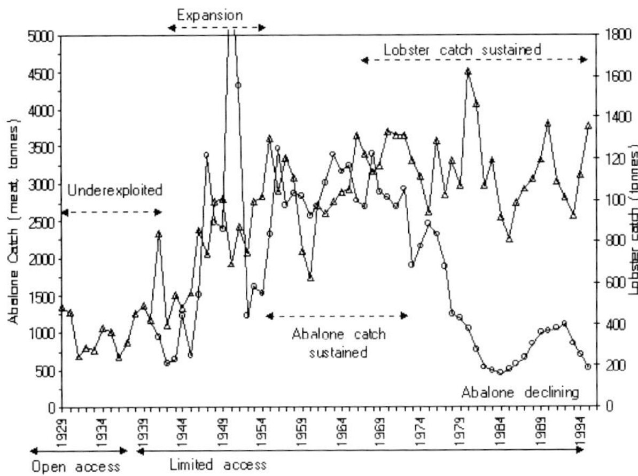

Fig. 4. Stages in the history of lobster and abalone fisheries in Baja California.

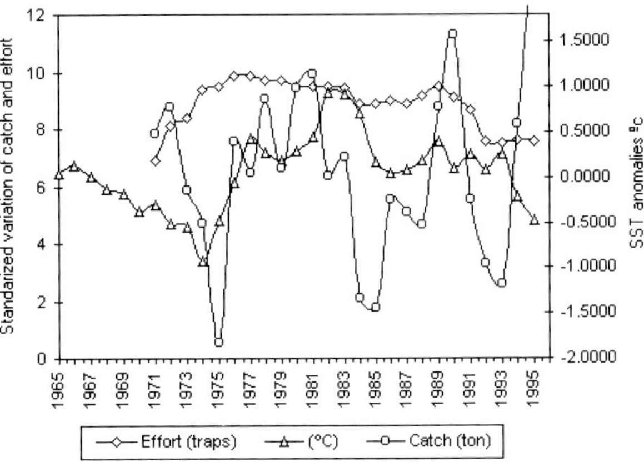

Fig. 5. Variability of lobster catch and effort versus SST anomalies. Variation of catch and effort is measured as (annual value-average)/(standard deviation).

Table 2. Internal regulatory measures of the Cooperatives, in addition to official regulations.

Regulation	Abalone	Lobster
Minimum legal size	Abalone is fished 5–10 mm larger than official size Fishermen are obliged to carry on board a graduated ruler to verify legal size. If sublegal abalone is detected, internal sanctions, including temporary suspensions, are applied.	Fishermen carry a metallic gauge to verify legal size. All sublegal lobster are returned immediately to sea. If fishermen retain sublegal lobster, internal sanctions, includingtemporary suspensions, are applied. Only sublegal lobster that are damaged or injured by fishes may be landed.
Rotation of beds and closures of areas	Temporal rotation of abalone beds and/or closures of some areas for 2 or 4 years, aimed at enhancing recruitment. It is prohibited to harvest or pick up abalone in rocky areas during low tides.	Not applicable
Closed seasons	Through quota system, the fishing season has been reduced. This is because of quotas that allow cooperatives to delay the beginning of fishing or hasten its closure.	Most cooperatives agreed to protect berried females, hasten season closures by 1–2 months.
Quotas	An annual total allowable quota per cooperative is distributed equally among divers. A quota, per day or by season, is assigned for the number of abalones. This helps harvest larger animals and improve yield.	No catch quota is applied but there is an effective quota effort: only a fixed quantity of traps per boat is permitted. This number ranges from 30 to 70 depending on the Cooperative.
Limited entry	The crew of a diving boats is very stable. There is self-protection among the members and it is not easy to admit new members, except those replacing retiring ones.	Similar to abalone. It is customary to move old fishermen into abalone crews, which used to be replaced by young people. In some cooperatives fishermen have agreed to share both resources, at least in most of the central zone.
Ecological protection and enhancement	For protection of juvenile abalone in its cryptic habitat, it is not permitted to remove chitons for lobster bait; or octopuses with diluted chlorine from the caves or under stones in the . intertidal zone	Not applicable
	Cooperatives are obligated to restock and enhance heavily exploited banks; as well as to encourage abalone culture and juvenile transplanting (still in pilot and experimental phases).	
Surveillance and compliance	All cooperatives have a compliance council, which polices internal and government policies as well as protecting resources from poachers or free riders. Several honorary inspectors, who work together with official inspectors, are authorized for each cooperative. Cooperatives spent considerable amounts of money in these activities .	Same. If some fisherman is detected violating any internal or officials rules he is severely sanctioned, even removed from the Cooperative.
Fishing gears	Divers must use a graduated harvesting iron to verify legal size, in order to prevent the catch of sublegal sizes. No type of sport fishing gear is permitted.	Is prohibited to set gillnets on rocky bottoms where lobster inhabit. Selectivity experiments are ongoing for escape vents introduction in traps.
Other measures	Domestic consumption has been drastically reduced to 6–8 legal abalone per month for each fisherman, and these are unloaded from the internal crew's quota. Some cooperatives have eliminated the domestic consumption.	Domestic consumption controlled as in the case of abalone: just 6–8 legal lobster per month for each fisherman are authorized.

continued the upward trend until the fishery reached its greatest development at the end of the 1960s and mid 1970s. Since then catches show a cyclic pattern, ranging from 900 to 1800 t, but fluctuating around an annual average of 1300 t. Sustained yield was probably reached by the end of the 1970s, without further possibilities of expansion, since all available fishing grounds were allocated to the established cooperatives (Fig. 1). In fact, this is the main reason that effort tended to remain relatively stable, with small or moderate variations. Despite this, recent catch trends show greater variability (Fig. 5) on an interannual and decadal scale than those stages before late 1950s (Fig. 4) and a quasi-oscillatory pattern probably associated with abundance changes due to environmental effects.

For abalone, the period of stable catches ended by 1972/73, and was followed by a sharp decline towards a historical minimum in 1983/84, to about 15% of the average in such period. This has been considered as the collapsing phase of the fishery (Guzmán *et al.* 1991; Guzmán, 1992; 1994). Fortunately prices increased sharply after 1980 so the revenues did not collapse dramatically and the activity has maintained its social and economic importance. Among relevant adjustments enforced after 1982/83, cooperatives reduced their number of diving boats by 50% and effectively applied self-regulatory measures (Table 2).

These, and probably also the favourable environmental conditions during 1984–1990 (mainly the cooling of sea water and recovery of the seaweed beds), allowed the abalone harvest to increase to about 1100 t by 1991 (Fig. 6). However, after 1991 there was again a significant decline, coincident with the 1992/93 warm period associated with El Niño anomalies. Cox (1962) mentioned similar, negative weather effects on the

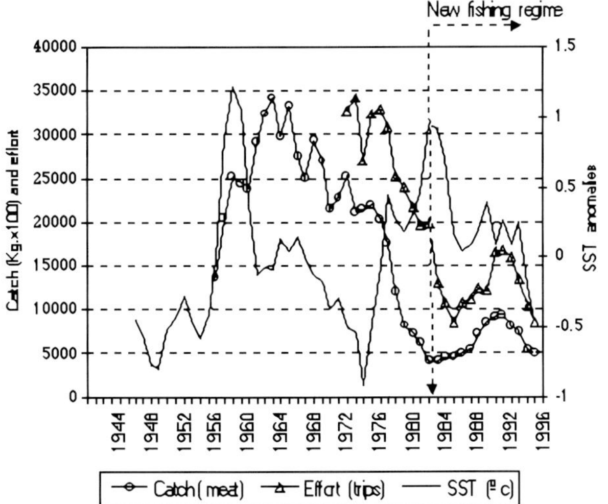

Fig. 6. Abalone Catch and fishing effort and SST anomalies, in Baja California Mexico.

California fishery, and downward catch trends have ocurred since 1970 in both California (CalCOFI Rep. 1993, 1994) and Baja California, coinciding with the strongest warming trend from the mid 1970s to 1983, despite management differences. The hypothesis of a causal relationship between oceanographic–climatic changes and abalone abundance variations has been suggested by some authors (Tegner 1989; Davis *et al.* 1992; Guzmán *et al.* 1991, Guzmán 1994; Vega *et al.* 1994). For example, Tegner (1989) suggested that violent storms and El Niño events are negative factors affecting the food supply and survival. Some authors suggest that El Niño has some responsibility in its depletion of black abalone (Tissot 1990).

Finally, and still poorly studied, there are other ecological factors related to warm sea temperatures, which affect the depletion of abalone stocks, for example, the increase of potential competitors and diseases (Miller and Miller 1993; Lafferty and Kuris 1993; Richards and Davis 1993). Among these, the 'withering syndrome' may be the cause of massive mortality of abalone off Southern California (Lafferty and Kuris 1993) and Baja California (Instituto Nacional de Pesca, unpublished data). In Mexico it was detected in 1990 in the central zone of Baja California, affecting first the black and later the green abalone. The problem was particularly severe from 1991/92 until 1994, coinciding with the recent decline in total catch. Despite research efforts to elucidate the deseases, causes still are poorly understood, but fortunately abalone with the 'withering syndrome' have been disappearing naturally.

CONCLUSIONS AND RECOMMENDATIONS

It appears that the objective of a sustainable fishery has been more or less achieved for lobster, although there is evidence that the fishing regime does not fully explain the changes in abundance. If so, greater account of the climatic–oceanographic effects could provide a more predictive approach, enabling the development of catch forecast models useful for management purposes.

While doing so may lead to improve management of the lobster fishery, abalone administration clearly demands a multi disciplinary approach (Sharp 1995) since traditional strategies have failed. Today this fishery is based upon a severely diminished stock, and is maintained only because of its high monetary value. However, the decline of catches after the stabilized period is probably the result of several factors, including difficulties in stock assessment, failures to effectively enforce regulation and the lack of knowledge of environmental effects, among others. In a way, the current status is reflecting the accumulated effects of inadequate polices and low compliance before 1981/82, which may have been boosted by unfavourable conditions, such as the warm anomalies during El Niño events and long-term warming of sea water from 1974 to the early 1990s (Figs 4 and 6). Special attention should be paid to measuring the levels of non-compliance (mainly by poachers) and its role in depletion of abalone stocks in some areas.

Future research must identify and explain the factors determining stock fluctuations, as well as the causal mechanisms behind the spatial and temporal changes in distribution and abundance of these benthic resources. For this, both the fishing regime and the effects of oceanic variability should be taken into account. Both fisheries have been fully developed, with no possibilities of further expansion. Therefore, limited access must be consolidated as a management tool (Water. 1991), preferably under an active co-management scheme between government, cooperatives (industry) and the scientific community. This could be formalized rapidly through a regional management council, for each fishery or both together.

ACKNOWLEDGMENTS

This work was partially funded by the grant SIMAC94-CM02, in a joint research effort by INP/CRIP, CIBNOR and the 'Baja California' Fishing Cooperatives Federation. Thanks are expressed to those Institutions. Also we thank Bruce Phillips, Nick Caputi, Neil Andrews and Chris Chubb for reviewing the manuscript and for their helpful suggestions.

REFERENCES

CalCOFI (1993): Fisheries review 1992, CalCOFI Rep. **34**, 16–18.

CalCOFI (1994): Fisheries review 1993, CalCOFI Rep. **35**, 14–16.

Cox, K. W. (1962). California abalones, family Haliotidae. *California Fish and Game. Fish Bulletin* 118; 133 pp.

Davis G. E., Richads, D. , Haaker, P. L., and Parker, D.O. (1992). Abalone population decline and fishery management in Southern California. In 'Abalone of the World: Biology, Fisheries and Culture'. (Eds S A Shepherd, M J Tegner, and S A Guzman del Próo) pp. 237–249 (Fishing News Books: Oxford). 608 pp.

De Buen, F. (1960). Abulones de Baja California. Moluscos del genero *Haliotis*. Universidad de Chile. *Revista de Biología Marina* **10**, 201–7.

Duffy, J.M. (1973). The status of the California spiny lobster resource. *Marine Research Technical Report* No. 10, 15 pp.

Guzmán Del Proo, S A, Lluch-Belda, D, Lluch-Cota, D B, Hernandez, S, and Salinas Zavala., C A (1991). Efecto de cambios climáticos en la abundancia del abulón en la costa Pacífica de la Península de Baja California. Taller México-Australia sobre reclutamiento de recursos marinos bentónicos de la Península de Baja California (unpublished MS).

Guzmán Del Proo, S A (1992). A review of the biology of Abalone and its Fishery in Mexico. In 'Abalone of the World: Biology, Fisheries and Culture'. (Eds S A Shepherd, M J Tegner, and S A Guzman del Próo) pp. 341–60 (Fishing News Books: Oxford, London). 608 pp.

Guzmán Del Proo, S A (1994). Biología, ecología y dinámica de población del abulón (*Haliotis spp*) de Baja California, México. Tesis Doctoral. Inst. Politecnico Nacional/ Esc. Nal. Cienc. Biologicas.

Johnson, M W (1960). The offshore drift of larvae of the California spiny lobster *Panulirus interruptus*. CalCOFI Rep. **7**, 147-61 pp.

Johnson, M. W. (1971). The Palinurid and Scyllarid lobster larvae of the tropical Eastern Pacific and their distribution as related to the pervailing hydrography. *Bulletin Scripps Institution of Oceanography* **19**, 1–36 pp.

Instituto Nacional de la Pesca (1980). Análisis de la Pesquería de Abulón en Baja California. Fundamentos biológicos para un nuevo régimen de explotación del recurso (Diagnósis). Departamento de pesca/Inst. Nal. de la Pesca. pp. 1–86.

Lafferty, K D and Kuris, A M (1993). Mass mortality of abalone *Haliotis cracherodii* on the California Channel Islands: Tests of epidemiological hypothesis. *Marine Ecology Progress Series* **96**, 239–48.

Lluch-Belda, D, Guzman, S A, Marin, V, and Ortiz Q M (1973). La Pesquería de Abulón en Baja California. Un Análisis de su desarrollo y perspectivas futuras S.I.C.-Sbría. de Pesca/Inst. Nal.Pesca. INP/51, 16, 26 pp.

Lluch-Belda, D., Magallón, F.J. and Schwartzlose, R.A. (1986). Large fluctuations in the sardine fishery in the Gulf of California: Possible causes. CalCOFI Rep. XXVII, 136–40.

Lluch-Belda, D, Hernández-Vazques, S, and Schwartzlose, R A (1991). A hipothetical model for the fluctuations of the California sardine population (*Sardinops sagax caerulea*). In 'Long Term Variability of Pelagic Fish Populations and their Environment' (Eds Kawasaky *et al.*) pp. 293–9. Proceedings of the International Symposium, Sendai, Japan, 14–18 November 1989 (Pergamon Press).

Mateus, H. (1986). Semblanza de la Pesca en Baja California. Historia y desarrollo. Sec. Pesca, 1a. ed., México, D.F.

Miller, A.C., and Lawrenz-Miller, S.E. (1993). Long-term trends in black abalone, *Haliotis cracherodii*, Leach 1814, populations along the Palos Verdes Peninsula, California. *Journal of Shellfish Research* **12**(2), 195–200.

Ortíz, Q M, and León C G (1988). Recursos pesqueros de México y sus perspectivas. Recurso Abulón (*Haliotis* spp.) In 'Recursos Pesqueros del Pais'. Sría. Pesca/Inst.Nal. de la Pesca, pp. 11–51.

Phillips, B P, Pearce, A F, Lichtield, R, and Guzman Del Proo, S A (1994). Spiny lobster catches and the ocean environment. In 'Spiny Lobster Managment' (Eds B F, Phillips, J S, Cobb, and J Kittaka). pp 250–61. (Fishing News Books: London).

Polovina, J J and Mitchum, G T (1992). Variabilty in spiny lobster *Panulirus marginatus* recruitment and sea level in the Northwestern Hawaiian Islands. *Fiseriesh Bulletin U.S.* **90**, 483–93.

Pringle, J D (1986). California spiny lobster (*Panulirus interruptus*) larval retention and recruitment: a review and synthesis. *Canadian Journal Fisheries and Aquatic Sciences* **43**, 2142–52.

Richards D V, and Davis, G E (1993). Early warnings of modern populations collapse in black abalone, *Haliotis cracherodii*, Leach 1814, at the California Channel Islands. *Journal of Shellfish Research* **12**(2), 189–94

Roy, C and Mendelssohn R (Eds) (1994). COADS on CD-ROM. Volume 5: Eastern Pacific Ocean. ORMSTON/NOAA. Comaparative Eastern Ocean System Program.

Secretaria de Pesca (1992). Ley Federal de Pesca y su Reglamento.

Sharp, G D (1995). It's about time: new beginnings and old ideas in fisheries science. *Fisheries Oceanography* **4**(4), 324–41.

Tegner, M. J. (1989). The California abalone fishery: Production, ecological interactions, and prospects for the future. In 'marine Invertebrates Fisheries: Their Assessment and Management' (Ed J F Caddy), pp. 401–20 (Wiley: New York).

Tissot, B N (1990). El Niño responsible for decline of black abalone off Southern California. *Hawaii Shell News* **38**(6), 3–4.

Vega V A. and Lluch-Cota D B (1992). Análisis de las fluctuaciones en los volumenes de langostas (*Panulirus* spp.), del litoral oeste de la Península de Baja California, en relación con el desarrollo histórico de la pesquería y la variabilidad del marco ambiental. In 'Mem. Taller Intern. México–Australia sobre reclutamiento de recursos marinos bentónicos de Baja California': pp 191 212.

Vega V A, Leon Carballo, G and Muciño Diaz, M (1994). Sinopsis de informacion biologica, pesquera y acuacultural de los abulones (*Haliotis* spp.) de la Peninsula de Baja California, Mexico. (convenio SEPESCA/CIBNOR). 117 pp.

Vega V A, Espinoza Castro, G, and Gomez Rojo, C (1996). Pesqueria de langosta (*Panulirus* spp.). In 'Estudio del potencial Pesquero y Acuicola de Baja California Sur' (Eds. M. Casas-Valdez and G. Ponce-Díaz), Vol 1, pp. 227–61.

Waters, J R (1991). Restricted access vs. open access methods of management toward more effective regulation of fishing effort. *Marine Fisheries* **53**(3), 1–10.

Fish Shoaling Behaviour as a Key Factor in the Resilience of Fisheries: Shoaling Behaviour Alone can Generate Range Collapse in Fisheries

Tony J. Pitcher

Fisheries Centre, University of British Columbia, Vancouver, Canada.

Summary

Range collapse, a progressive and dangerous reduction in spatial range, often exacerbates stock collapse in fisheries by making fish easy to locate and concentrating the harvesting power of the fishing fleet. Whereas previous models emphasize environmental determinants of range, such as habitat quality, this paper shows how shoaling behaviour alone can cause range collapse independent of habitat. Shoaling behaviour is therefore a critical element of the resilience of fisheries to human harvest. Shoaling behaviour is already known to accelerate stock collapse in overharvested fisheries; the new model presented here shows that simple behavioural rules can rapidly generate a spatial pathology from which recovery may be slow. The model enhances our insight of the fragility of shoaling fish species to human harvest. Diagnostics of spatial pathology presaging impending collapse are expensive to gather: this work suggests that data for preventative management might be gathered cost-effectively by observing the dynamics among and within fish shoals.

Introduction

When fish populations collapse, two linked phenomena generally occur; *stock collapse* and *range collapse.* Stock collapse has long been the concern of fishery scientists but range collapse, a progressive and dangerous reduction in spatial range of the stock, has received relatively little attention.

Stock collapse is defined as a rapid reduction in stock abundance, and is distinguished from short-term natural fluctuations. Although one school of thought seeks environmental correlates of collapse (e. g. Sinclair *et al.* 1985; Mann 1992; Hansen *et al.* 1994), sufficiently powerful mechanisms driving stock collapse can be generated by the impact of harvest on fish population dynamics and fish behaviour without the need to invoke environmental change. The stock collapse problem has generated a considerable amount of research, including a search for depensatory mechanisms embedded in the recruitment process (Myers 1993; Myers *et al.* 1995), and for management policies that reduce the risk of becoming trapped in a downward spiral of stock abundance. But such analyses are not sufficient. Recent disastrous fishery collapses show that this 'classical' fishery science is unhelpful without spatial modelling (Walters and Maguire 1996).

Range collapse makes a stock collapse more serious because the progressive concentration of remaining fish into a reducing area makes fish easier to locate and concentrates the fishing power of a fleet that has been built with profits from a previous era of higher abundance. Range collapse is often represented as being driven by environmental forces through competition among fish for optimal habitat. For example, one of the most powerful theories underlying range collapse, known as the 'basin' model, has been put forward by MacCall (1990). From the practical management perspective, recent work has sought to describe the spatial pathology that may be diagnostic of impending catastrophe (Hutchings 1996). But obtaining such spatial information is expensive because it derives from survey data that must be widespread and independent of catch rates from the fishery.

This paper puts forward a new model for range collapse that is based on the shoaling behaviour of fish. Unlike previous models, it suggests that spatial range dynamics, including range collapse, may be generated in the absence of significant environmental gradients in space and time. This approach implies that shoaling behaviour is a critical feature of the resilience of fisheries to human harvest and raises the prospect of obtaining cost-effective, early warning of impending problems by monitoring behavioural parameters of shoaling fish.

Shoaling in fishes

Shoaling behaviour is defined as fish choosing to remain together for social reasons (Pitcher 1983). Substantial experimental evidence shows that shoaling is driven by the second-to-second evaluation by individual fish of the cost and benefits of behavioural decisions to join, leave or stay with other fish (review: Pitcher and Parrish 1993). Costs and benefits shift dynamically with shoal size and, outside the spawning season, are largely determined by defence against predators and foraging for food. Hence, shoal size is fluid and responds rapidly to local conditions of food and predators (Schneider 1989). Experiments show that shoaling fish make quantitative risk-balancing trade-offs between foraging rewards and perceived predation risk (Godin and Smith 1988; Pitcher *et al.* 1988). The size of the group is one important elective adjustment that is made to adjust trade-offs to the current regime. In the absence of behaviour that excludes newcomers from the group, a behaviour not described for fishes, shoal size rapidly tracks zero net benefit to individuals in the local area (Pitcher and Parrish 1993).

Adjustment of shoal size to the prevailing food/predation regime is possible only if fish shoals both split and meet so that they have the opportunity to merge and exchange members. These processes occur at a meso-scale that has traditionally been difficult to observe and there is consequently very little information. Recently, modern side-scan high-resolution sonar has made it easier to gather data about fish shoals at sea (Misund *et al.* 1995, 1996). Field measurements of shoaling herring (*Clupea harengus*) revealed a dynamic regime of joining and splitting among schools in the Norwegian Sea (Pitcher *et al.* 1996), behaviour that is a prerequisite for the model described in this paper.

Stock collapse and range collapse in shoaling fishes

A pernicious mismatch between specific adaptations that evolved to fit them to the pelagic niche and the operation of human fisheries renders shoaling fish especially vulnerable to overexploitation (Pitcher 1995). Ecological, physiological and behavioural adaptations in foraging, spawning, migration and shoaling may contribute to the fragility of these resources. Shoaling is an especially important factor as it may lead to a depensatory stock collapse through fish catchability increasing as stock abundance declines (Clark 1974). The impact of shoaling on stock collapse conspires with that of the fishers themselves, who attempt to maintain catch rates as abundance declines by employing sophisticated technology to locate and capture fish shoals. Pitcher (1995) provides a comparative analysis of models that set out to describe this process. A surplus production model in which catchability increases as stock size declines (Csirke 1988), and one in which catch rate is constant irrespective of stock abundance (Pitcher 1995), were compared with the classic Schaefer fishery model, which assumes that catch rates are exactly proportional to abundance, and catchability (the slope of the latter relationship) is constant. The conclusion was that models that incorporate shoaling effects exhibited higher fragility and lower resilience in the face of harvest. Bio-economic analysis of these models presents an even more dismaying scenario. In the constant catch rate fishery models, investment deriving from high profits increases the catching power of vessels to speed a precipitous fishery collapse that is unheralded by biological or economic indicators (Mackinson, Sumaila and Pitcher in press).

Although they provide a convincing, and indeed alarming, analysis of a fishery collapse that can occur in the absence of environmental forcing, these surplus production models pool stock parameters and are unable to track the spatial elements of range collapse. Spatial analysis is essential for better insight that might lead to forecasting and the detection of measurable early-warning symptoms that might avert a fishery collapse. For example, retrospective spatial analysis of trawl surveys of the northern cod (*Gadus morhua*) stock in Newfoundland has revealed such symptoms: habitat occupied by the fish became more patchy and concentrated during to the fishery collapse (Hutchings 1996). Early warning symptoms in the spatial abundance data that might have warned of the collapse in sufficient time to avert catastrophe required sampling throughout the cod's range and would have needed to have been more extensive the earlier the warning was required. Models that explicitly take range collapse into account will provide more robust analytical tools and will suggest what features might best be measured.

The MacCall model

MacCall's basin model (1990) is such an explicitly spatial analysis of range collapse. Habitat suitability is assumed to be highest in a core area for the species. This core area is termed the optimal habitat. Range contraction (and expansion) is driven in this model by competition among fish for occupancy of this optimal habitat. As abundance increases, competition forces individuals to move outside this area into progressively sub-optimal habitats. Population growth, and expansion of range, occurs until '*r*' (the intrinsic rate of natural increase) falls to zero, while the fitness of individual fish, as reflected in growth and mortality, remains the same throughout the range because the density in sub-optimal habitats is lower to compensate. As abundance reduces, the converse occurs and individuals gain the opportunity to move back towards the optimal core area as competition lessens. (The speed with which the range expansion and contraction occurs is determined by a 'viscosity' parameter in MacCall's model.) The basin is illustrated in Figure 1A. Implicit in the basin model are underlying environmental factors driving the contraction and expansion of range though changes in the quality of the core habitat, reflected in population growth '*r*'. Kawasaki (1993) considers the Japanese sardine (*Sardinops sagax*) in the Kuroshio Current to expand and collapse under forces similar those applying in the basin model.

The basin model predicts that fish outside the core area will be found at lower density to achieve equal fitness, a testable prediction that has not yet been validated. But a more important aspect of the basin model is that it regards changes in environmental factors, in the form of habitat suitability, as the principal template upon which range dynamics are structured.

A ranking of habitat quality among fishery locations could provide early warning of spatial problems under the MacCall model. The ranking might be used to predict the order in which local stocks would be expected to disappear at the start of a range collapse. Fish vanishing from low ranking sites could be used as an indicator.

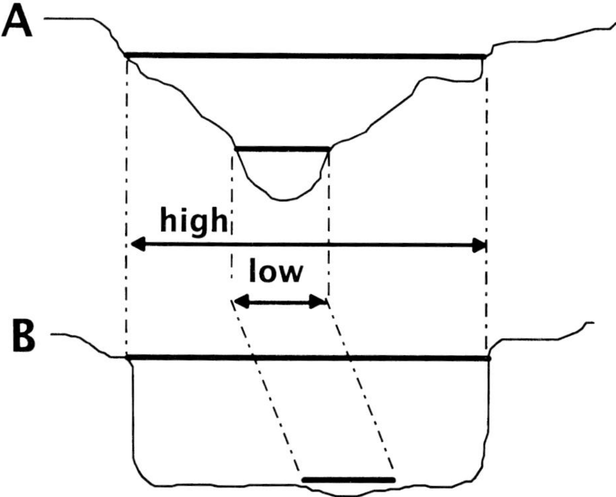

Fig. 1. Conceptual basis of alternative range collapse models. Vertical axis and profile of basins indicate relative habitat quality: deeper basin is better habitat. Spatial range of the fish stock at high and low abundance are indicated by arrows. **A** (top), MacCall (1990) basin model structured on habitat quality; **B** (lower), Shoal model described in this paper. Range at low abundance is as for MacCall basin model, but is not constrained to any particular location (see text for details).

The shoal behaviour model

An alternative model for range collapse, presented in this paper, is summarized in the same terms as the MacCall basin model in Figure 1B. Habitat quality is assumed to be constant geographically across a wide area of the species' range, producing a 'flat-bottomed basin'. At the lip of the basin, fitness declines as environmental parameters determine the edge of the animals' range. MacCall-style basin effects could operate at this edge. But the new model operates entirely within the flat-bottomed area.

The model consists of a simulation that tracks the locations of moving fish shoals in progression of time steps as their numbers become reduced by mortality or increased by population growth. Mortality is assumed to occur in discrete shoals — this is an over-simplification when abundance is reduced by natural factors, but is not unrealistic for purse seine or mid-water pair-trawl fisheries. The model assumes that shoal size is constant, and that adjustments in shoal numbers are made by splitting or by combining when shoals meet and may exchange members.

Movement of each shoal in the model is determined by distance to the nearest neighbour shoal. In the real world, information about this distance could be gained by the frequency with which shoals meet each other. How the model shoals move is determined by three cases, probably simplifying a continuous gradient in the real world. First, if the distance to the nearest neighbour shoal (NNS) is less than a threshold distance away (d'), the shoal moves one step in a random direction. Secondly, if the NNS is greater than d' but less than d'' away, the reference shoal moves towards NNS with a velocity related to the strength of a shoal attraction parameter, α. The parameter α simulates a shoaling tendency that differs among species (Pitcher 1986). A control for the model sets α to zero. Finally, if NNS is farther away than d'', the reference shoal moves at random.

Figure 2 illustrates the conceptual regime in this model, Appendix A provides the mathematical details of an algorithm that implements this concept, and Figure 3 shows the graphical output of one typical run and control run of the relevant computer program. The model is clearly sufficient to generate

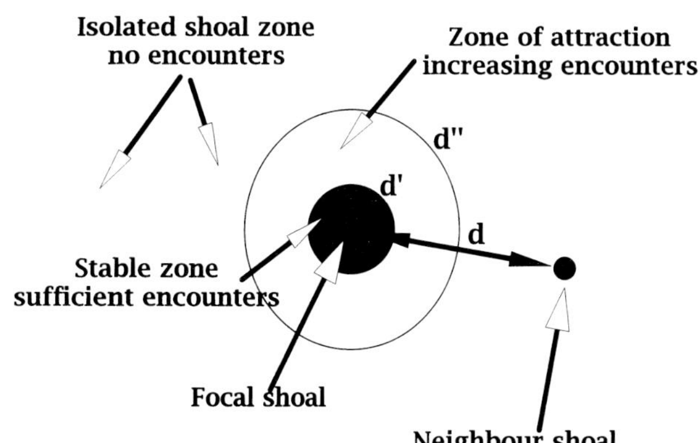

Fig. 2. Conceptual diagram of the nearest neighbour shoal algorithm used to generate range collapse. Mathematical details are given in Appendix A.

range collapse. It generates range expansion when shoal numbers increase rather than decrease.

In runs of this model, swimming speed (*s*) is not critical to obtaining range collapse, but when swimming speeds are higher (*qs*) for shoals that are alone (NNS > *d″*), the result is achieved more rapidly under a wider range of parameter values. (In the version of the model presented here, $q = 2$.)

The version of the model described here is the simplest that will generate range collapse, but alternatives to basing movement on the distance of one nearest neighbour are being sought. For example, an algorithm based on the local density of shoals is probably a more realistic reflection of shoal encounter rate. An alternative model drives movement behaviour by actually tracking shoal meetings, but achieves essentially the same effect.

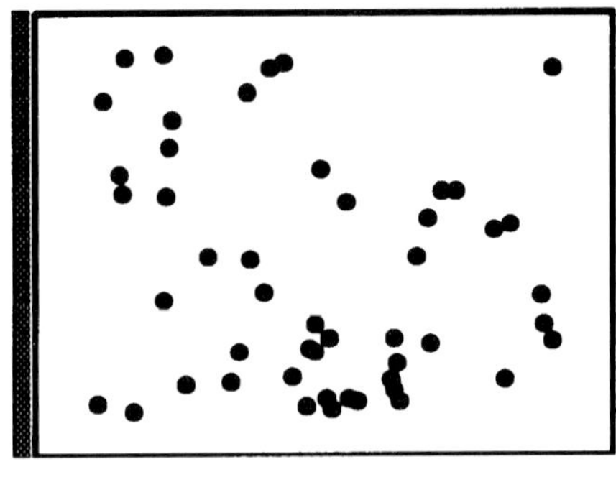

A. Shows random location of shoals in arena representing their range at start of the simulation. Grey bar at left indicates number of shoals alive. Shoal size is assumed constant in this model, so that mortality of shoals is proportional to mortality of fish. Details of the algorithm are given in the text and in Appendix A.

B. Location of shoals when approximately half have died. A range collapse towards the bottom left hand corner of the arena is in progress.

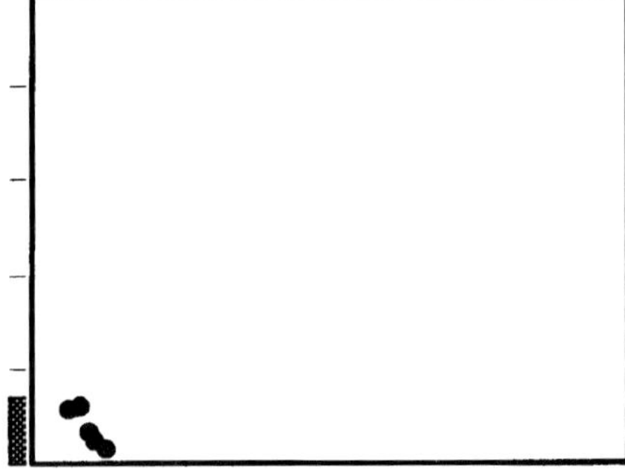

C. Location of shoals when approximately nine-tenths have died. Range collapse is complete.

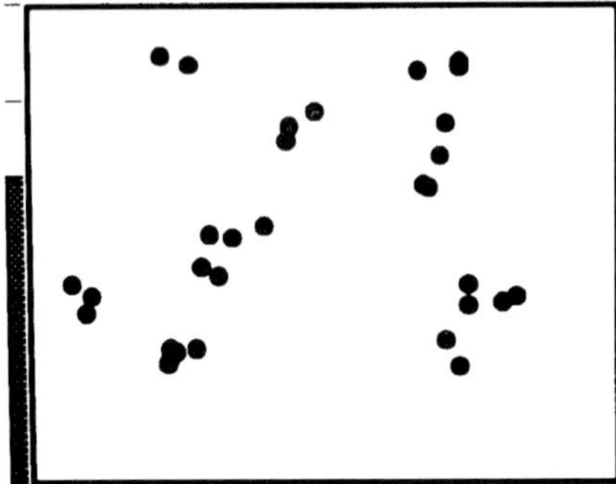

D. Half-completed control run in which shoal attraction parameter, α, is set to zero. Shoals are distributed at random in clusters generated by the movement dynamics.

Fig. 3. Some Results of the Range Collapse Simulation Model.

DISCUSSION

This paper suggests that shoaling behaviour alone can cause range collapse and is therefore a critical element of the resilience of fisheries to human harvest. Not all stock collapses are accompanied by range collapse. For example, hake (*Merluccius* spp.) stocks appear to be quite resilient to harvest and relatively fast to rebuild. This resilience can also be ascribed to behavioural factors, such as cannibalism and other hake traits that lead to spatial dynamics that differ from those of shoaling fish (Pitcher and Alheit 1994). The relative number of refugia from which stocks that have suffered a collapse can rebuild is a determinant of resilience: ways in which shoal behaviour can mitigate collapse and facilitate rebuilding are currently under investigation.

MacCall's basin model might be augmented by the inclusion of fish social factors because the two mechanisms of range dynamics are not mutually incompatible. Both models are similar in that they conflate the changes in individual fitness that likely drive the spatial dynamics of fish populations in the ocean. MacCall's model is predicated on population growth rather than individuals whereas the shoal model is structured on shoals as proxy for individuals. But the two models produce contrasting spatial predictions of what happens during a range collapse. The basin model has the fish disappearing from fishery locations in reverse order of habitat suitability. The shoaling model on the other hand, predicts that fish will vanish from sites at random. A spatial diagnostic of collapse would be clumped groups of sites vanishing simultaneously, while fish remain at the pre-collapse density in only one or a few clumped sites.

The model presented here is based on units of whole fish shoals, but a desirable next step would be an individually-based model (IBM), that can track individual fitness in relation to trade-offs among food, predators and social proximity (Tyler and Rose 1994).

ACKNOWLEDGMENTS

I would like to thank Anders Ferno, Mark Mangel, Ole Arve Misund, Daniel Pauly and Chuck Hollingworth for discussion of the ideas in this paper and comments on the draft manuscript.

REFERENCES

Clark, C (1974). Possible effects of schooling on the dynamics of exploited fish populations. *Journal du Conseil International pour L'Exploration de la Mer* **36**, 7–14.

Csirke, J (1988). Small shoaling pelagic fish stocks. In 'Fish Population Dynamics: 2nd edition'. (Ed J A Gulland) pp. 271–303 (John Wiley: New York).

Godin, J G, and Smith, S A (1988). A fitness cost of foraging in the guppy. *Nature* **333**, 69–71.

Hansen, B, Gaard, E, and Reinert, J (1994). Physical effects on recruitment of Faroe Plateau cod. *ICES Marine Science Symposium* **198**, 521–8.

Hutchings (1996). Spatial and temporal variation in the density of northern cod and a review of hypotheses for the stock's collapse. *Canadian Journal of Fisheries and Aquatic Sciences* **53**, 943–62.

Kawasaki, T (1993). Recovery and collapse of the Far Eastern sardine. *Fisheries Oceanography* **2**, 244–53.

MacCall, A D (1990). 'The Dynamic Geography of Marine Fish Populations.' (University of Washington Press: Seattle). 153 pp.

Mackinson, S, Sumaila, R, and Pitcher, T J (in press). Bioeconomics and catchability: fish and fishers behaviour during stock collapse. *Fisheries Research.*

Mangel, M (1990). Resource Divisibility, Predation and Group Formation. *Animal Behaviour* **39**, 1163–72.

Mann, K H (1992). Physical influences on biological processes: how important are they? *South African Journal of Marine Science* **12**, 107–21.

Misund, O A, Aglen, A, and Frónas, E (1995). Mapping the shape, size and density of fish schools by echo integration and a high-resolution sonar. *ICES Journal of Marine Science* **52**, 11–20.

Misund, O A, Ferno, A, Pitcher, T J, and Totland, B (1996). Variation in horizontal area and relative echo intensity of herring schools as recorded by a high-resolution sonar. (submitted).

Myers, R A (1993). Depensatory recruitment and the collapse of fisheries. 6th Conference on Natural Resource Modelling and Analysis, St. Johns, Canada.

Myers, R A, Barrowman N J, Hutchings, J A, and Rosenberg, A A (1995). Population dynamics of exploited fish stocks at low population levels. *Science* **269**, 1106–8.

Pitcher, T J (1983). Heuristic definitions of shoaling behaviour. *Animal Behaviour* **31**, 611–13.

Pitcher, T J (1986). The functions of shoaling behaviour. In 'The Behaviour of Teleost Fishes'. (Ed T J Pitcher) pp. 294–337 (Croom Helm: London). 553 pp.

Pitcher, T J (1995). The Impact of Pelagic Fish Behaviour on Fisheries. *Scientia Marina* **59**, 295–306.

Pitcher, T J, and Alheit, J (1994). What Makes A Hake? A Review of the Critical Biological Features that Sustain Global Hake Fisheries. In 'Hake: Fisheries, Ecology and Markets'. (Eds T J Pitcher and J Alheit) pp. 1–15 (Chapman and Hall: London). 703 pp.

Pitcher, T J, and Parrish, J (1993). The Functions of Shoaling Behaviour. In 'The Behaviour of Teleost Fishes: 2nd Edition'. (Ed T J Pitcher) pp. 363–439 (Chapman and Hall: London). 715 pp.

Pitcher, T J, Lang, S H, and Turner, J R (1988). A risk-balancing trade-off between foraging rewards and predation hazard in shoaling fish. *Behavioral Ecology and Sociobiology* **22**, 225–8.

Pitcher, T J , Misund, O A, Ferno, A, Totland, B, and Melle, V (1996). Adaptive behaviour of herring schools in the Norwegian Sea as revealed by high-resolution sonar. *ICES Journal of Marine Science* **53**, 449–52.

Schnieder, D C (1989). Identifying the spatial scale of density dependent interaction of predators with schooling fish in the southern Labrador Current. *Journal of Fish Biology* **35** (suppl. A), 107–15.

Sinclair, M, Tremblay, M J, and Bernal, P (1985). El Niño events and variability in a Pacific mackerel (*Scomber japonicus*) survival index: support for Hjort's second hypothesis. *Canadian Journal of Fisheries and Aquatic Sciences* **42**, 602–8.

Tyler, J A, and Rose, K A (1994). Individual variability and spatial heterogeneity in fish population models. *Reviews in Fish Biology and Fisheries* **4**, 91–123.

Walters, C J, and Maguire, J J (1996). What can we learn from the northern cod collapse? *Reviews in Fish Biology and Fisheries* **6**, 125–37.

Appendix A. Shoal Attraction Algorithm

M shoals are located at time *t* at coordinates

$$L(x,y)_{i,t}$$

Each shoal contains a constant number of individual fish,

$$N$$

Shoals suffer mortality at rate *Z*,

$$M_t = M_0 \exp(-Z_t)$$

The nearest neighbour, *j*, of each shoal, *i*, is located distance *d* away,

$$d = \min\{L(x,y)_{i,t} - L(x,y)_{k,t}\} \quad k \neq i$$

Each shoal changes location between time *t* and *t*+1

$$L(x,y)_{i,t} + s_{i,t}$$

where $s_{i,t}$ – movement speed of shoal *i* at time *t*

Speed depends on three zones defined by distances to nearest shoal *j*,

d' and d''

$s_{i,t} = s$ for $d < d'$

$s_{i,t} = qs$ for $d > d''$

$s_{i,t} = s_{i,t}'$ for $d'' < d > d'$

$s_{i,t}' = (L(x,y)_{i,t} - L(x,y)_{j,t})/a$

where $q > 1$, speed multiplier for isolated shoals

where $a = i/(1 - \alpha)$; $1 > \alpha > 0$

Shoal attraction factor α varies from 0 to 1

The model is implemented by plotting the locations of simulated shoals for *t* time steps.

(A Visual Basic module for Excel is available from the author.)

Sources of Unaccounted Mortality in Fish Capture Technologies

F. Chopin,[A] *D. L. Alverson,*[B] *Y. Inoue,*[C] *T. Arimoto,*[D] *P. He,*[E] *P. Suuronen,*[F] *and G. I. Sangster*[G]

[A] Frank Chopin, National Research Institute of Fisheries Engineering, Hasaki, Japan.
[B] Dayton L. Alverson, National Resource Consultants, Seattle, Washington, USA.
[C] Yoshihiro Inoue, Nat. Res. Inst. Fisheries Engineering, Hasaki, Japan.
[D] Takafumi Arimoto, Tokyo University of Fisheries, Tokyo, Japan.
[E] Pingguo He, Fisheries and Marine Institute, Newfoundland, Canada.
[F] Petri Suuronen, Finnish Fisheries and Game Institute, Helsinki, Finland.
[G] Graham Sangster, Marine Laboratory, Aberdeen, Scotland.

Summary

With recent trends in fisheries management towards low risk or precautionary fishery management strategies, the need to identify significant sources of biological waste associated with commercial capture technologies is becoming increasingly important. In capture fisheries, reported catches are generally considered the only source of fishing-induced mortality. However, in addition to commercial reported catches, illegal, recreational and artisinal landings may be non-reported components of the landed catch. There are also a number of discrete unaccounted technical sources of fishing mortality that can occur during the capture–escape process including; discards, ghost fishing, escape, avoidance and drop-out mortality. Collectively, they can form the basis of a catch mortality model where:

$$F = [F_{CL} + F_{AL} + F_{RL}] + F_B + F_D + F_E + F_G + F_0 + F_A + F_P + F_H$$

Research to date has shown that for some gear types, the probability of mortality is a function of stresses and injuries received during the capture and/or escape process. This paper reviews sources of unaccounted fishing mortality and suggests that identifying, quantifying and reducing sources of biological waste should be an integral component of precautionary management requiring detailed investigation of fish–fishing gear interactions.

Introduction

In the mid 19th century two schools of thought had emerged with respect to the impact of fishing on marine resources; those who believed in the inexhaustible nature of the sea fisheries (Huxley 1884) and those who felt that the bottom fisheries were declining annually and catches were steadily diminishing (Garstang 1903). However, with the emergence of fishery science as a discipline in the early 1900s and evidence of the effects of fishing on the decline of fish catches, the fishing industry entered the 20th century with the knowledge that the inexhaustible nature of the sea was a myth and that some fisheries were already being depleted. This change in direction represented a major transformation in philosophical thought in the approach to harvesting of marine resources and resulted in governments setting high priorities nationally and internationally for investigating the impacts man could have on commercial food fishes (Brown-Goode 1886; Anon. 1903). However, it was not until 1948 that FAO started collecting global marine fish catch statistics and only as recently as 1993 that FAO recognized the general failure of most management systems to adequately contain the pressure from improved technology (FAO 1993).

Most fish resources are now considered to be either fully fished or overexploited. Investigations into the impacts of fishing and the development and implementation of policies aimed at achieving sustainability have not been successful as is evidenced by the level of biological waste that remains in capture fisheries. While some forms of biological waste such as discards have been identified, the full range of fishing-induced mortalities for each capture technology and fishery remains unclear. Moreover, some of the basic technological strategies for reducing biological waste, such as improving the size and species-selective characteristics of the fishing gear, do not take into account the fate of fish and other animals encountering the gear but eluding capture. This paper, reviews the most recent efforts to identify the various types of unaccounted fishing mortalities associated with the capture process and suggests that establishing fisheries on a sustainable basis requires a paradigm shift in how biological waste and capture technologies are managed.

The fish capture process

Survival and mortality are two terms used to describe the fate of fish encountering fishing gears and suggest that the process of survival or mortality is a simple binary issue. These terms represent two broad categories of physiological condition but provide little information on the likelihood of becoming a statistic in either category. An alternative approach is to consider fish as having a range of conditions that, prior to capture, are governed by various environmental and biological factors. Prior capture conditions may be modified as a result of interaction with the fishing gear, and can alter the probability of capture injury and death.

The wide range of fish capture methods, the variety of environmental situations in which they are used and the variable pre-capture conditions of fish suggest that the magnitude of injury and stress fish will be subjected to during capture and escape may vary significantly over time and space. For example, fish entering an otter trawl generally undergo some degree of forced swimming, confinement, overcrowding and damage due to contact with the fishing gear, debris or other fish as the net is fishing , being retrieved or during sorting of the catch. The amount of time the fish are in the trawl, the quantity of other fish and debris in the cod-end, the towing speed and the depth the gear is retrieved from, and the retrieval rate, may all affect the degree of injury and level of stress response. Fish caught by other fishing gears may experience a different set of stressors and injuries. Nikonorov (1975) made some early attempts at defining different stages of the capture process and suggested that fish capture involves fish passing through zones of influence, action and retention. Thus the range of influence of the gear is not only where the fish are retained e.g. hook, cod-end, etc. but also includes parts of the fishing equipment that guide, herd, alarm or scare fish. The probability of survival for fish not retained is most likely to be a function of the cumulative injuries and stresses received during the complete fishing process.

Fish injuries

There have been several attempts made to quantify the level of external injuries associated with fish capture including assessment of 'life state' charts that record the degree of physical damage, body deformation, blood loss and body movement (Hoag 1975; Rogers *et al.* 1986; Main and Sangster 1990). Main and Sangster (1988, 1990); Soldal *et al.* (1991); and Soldal and Isaksen (1993) investigated the occurrence of scale removal in gadoids escaping from a demersal trawl; Engas *et al.* (1990) attempted to simulate injuries to cod, haddock and saithe from netting in a tank experiment, and Suuronen *et al.* (1995) studied trawl-induced injuries in Baltic cod. The occurrence of more traumatic and ultimately fatal injuries has yet to be investigated in detail, but the following have been observed in cod-end escapees: contusions — on the snout and forehead, the flanks, the operculum edges and caudal fin; eye damage (cataracts); fin rot — most frequent on the caudal fin (occasionally severe with the whole of the caudal fin degenerating and exposing the end of the spine); hernia (including intestinal protrusion from the anal orifice). Damage to the skin, depending on its severity and extent, could result in the loss of osmoregulatory control, mechanical protection, protection from pathogen invasion and sensory reception. The impact of damage to fish eyes will be dependent on the importance of eyesight to the species but may include a reduction in the foraging capability or increased susceptibility to predators. Severe fin rot may result in lost swimming ability and susceptibility to pathogens. Punctured abdomen can result from contact with netting, debris and other fish during fishing, gear retrieval and sorting prior to discarding. Mortality is expected to be high due to the initial trauma and associated internal damage as well as making the fish highly susceptible to pathogen/parasitic invasion. Anal hernia is indicative of external pressures on the internal organs of fish and may result from forces applied in the gear (cod-end, seine bunt etc.) prior to escape or release, or incurred during retrieval or sorting of the catch

Fish stress

Stress is a response mechanism which enables the fish to avoid or overcome potentially threatening, noxious or harmful situations (Pickering 1993). There are several parts of the fish endocrine system that are sensitive to stress but the predominant endocrine response is an elevation in circulating levels of corticosteroids and catecholamines. A variety of stressors associated with capture and escape have been recorded including; fatigue, damage, confinement, overcrowding and barotrauma. How a fish reacts to one or more stressors will depend upon the fish species, fish condition, the magnitude of the stressors which are a function of gear type (trawl, seine, hook, gill-net etc.), and mode of operation (depth, towing speed, retrieval rate etc.).

The most commonly used tests to measure chronic stress associated with capture and handling are plasma cortisol and plasma glucose, both of which have limitations. Plasma cortisol levels are commonly used in aquaculture as an index of fish stress (Donaldson 1981) and recovery from stress (Pickering *et al.* 1993) but have only been used on a few fishing gears (see Chopin *et al.* 1995). Analysing cortisol samples taken from fish at sea is difficult due to the problems in identifying the specific stressors to which the fish have been exposed and obtaining control group fish whose stress accurately reflect the natural condition of the school. Laboratory experiments in which fish were exposed to specific capture suggest that stress is a function of gear type and capture with some gear allowing fish to survive by adapting to the capture condition (Chopin and Inoue 1996; Chopin *et al.* in press).

Sources of accounted and unaccounted fishing mortality

Unaccounted fishing mortalities derive their name from a series of fishing induced mortalities that are not included in estimations of fishing mortality (F) (Ricker 1976). In a review of literature on fish condition after escape from fishing gears, Chopin and Arimoto (1995) and Chopin *et al.* (1996) developed a general catch mortality model to describe a variety of discrete types of unaccounted fishing mortality associated with fish capture. This model was later expanded by members of the International Council for the Exploration of the Sea (Anon. 1995) Sub Group on fishing mortality (Fig. 1). Fishing mortality was defined as: 'The sum of all fishing induced mortalities occurring directly as a result of catch or indirectly as a result of coming into contact with fishing gears'. These were expressed by a general fishing mortality equation:

$$F = [F_{CL} + F_{AL} + F_{RL}] + F_B + F_D + F_E + F_G + F_O + F_A + F_P + F_H$$

Where: $[F_{CL} + F_{AL} + F_{RL}]$ represent fishing mortality associated with reported commercial, artisanal and recreational fish landings respectively.

F_B Illegal and misreported landings

F_D Mortality associated with discards

F_E Mortality associated with fish after escape from fishing gear

F_G Mortality associated with ghost fishing

F_O Mortality associated with fish passively dropping off or out of fishing gears

F_A Mortality associated with fish avoiding the fishing gear

F_P Mortality associated with predation after escape

F_H Mortality due to changes in habitat associated with fishing.

To date, information on the full range of fishing-induced mortalities for different species and gear types is not available. The following provides a summary of data collected to date:

Discard mortality estimates

Research directed at understanding the consequences of fish mortality associated with the capture and discard of undersized target species has a long history (Holt 1895; Allen 1896; Garstang 1903; Russel and Edser 1926; Baranov 1977). However, it was not until the latter half of the 20th century that discard studies gained momentum as managers focussed their efforts on maximizing yield-per-recruit and achieving it to some extent by improving the size-selective properties of the fishing gear. In their development of regional and global bycatch and discard estimates, Alverson *et al.* (1994) estimated that the mean weight of global discards was 27×10^6 t based on a catch of 83×10^6 t (Table 1).

Most of the data collected on discard rates has focussed on determining only the actual number or weights of animals discarded with little effort made to record the level of mortality. Factors contributing to discard mortality include the type of

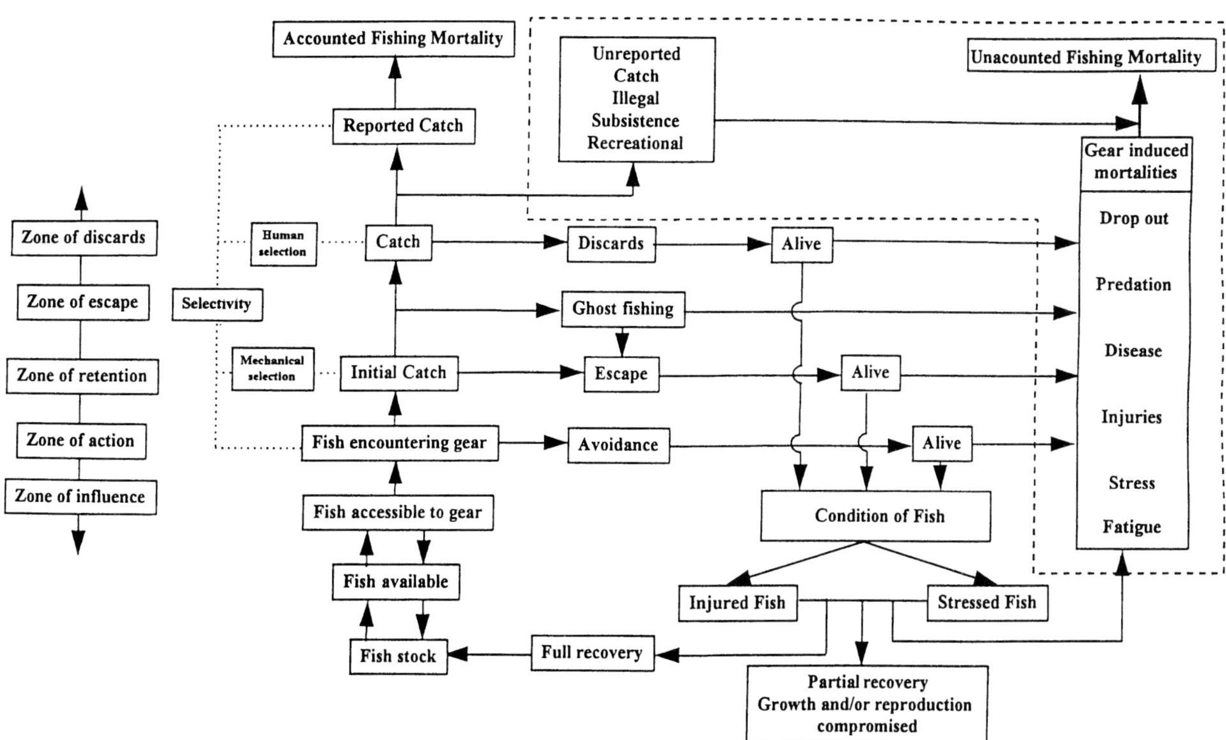

Fig. 1. Unaccounted mortality associated with different stages of the fishing process. (Chopin *et al.* 1996).

Table 1. Estimated discard weight by major world region per year (Source: Alverson *et al.* 1994)

Area	Estimated Discard t	Area	Estimated Discard t
Northwest Pacific	9 131 752	Eastern Central Pacific	767 444
Northeast Atlantic	2 671 346	Northwest Atlantic	685 949
West Central Pacific	2 776 726	East Central Atlantic	595 232
Southeast Pacific	2 601 640	Mediterranean and Black Sea	564 613
West Central Atlantic	1 600 897	Southwest Pacific	293 394
Western indian Ocean	1 471 274	Atlantic Antarctic	35 119
Southwest Atlantic	802 884	Indian Ocean Antarctic	10 018
Eastern Indian Ocean	802 189	Pacific Antarctic	109
TOTAL t			27 012 099

gear deployed, its mode of operation and retrieval, the amount and composition of the catch, as well as the sorting methods and duration of sorting the catch and environmental conditions. Of the data collected to date, there is a large range in the discard mortality rates for particular species and gear types, reflecting combinations of factors including variations in pre-capture condition, experimental methodologies and fishing practices.

The potential for errors in estimating global discards is enormous in light of non-standardized reporting procedures in which experimental methodologies vary between regions and fisheries, as well as confusion over what constitutes bycatch and discards. Additionally, much of the data on discards originates from the Northern temperate fisheries with large gaps existing for other regions of the world. While significant efforts have been made to quantify the level of discards, there is a distinct lack of data on the fate of discards once they are returned to the sea.

Escape mortality

Mortalities occurring after escape have been estimated from laboratory experiments using wild and cultured fish, during non-commercial fishing trials and by direct observation of commercial fishing gears. Post-release observation periods have ranged from 12 h for discard mortalities to in excess of 40 days for otter trawl escape mortalities. Chopin and Arimoto (1995) reviewed the level of mortality of fish escaping or released from fishing gears and noted that mortalities could be either immediate or delayed and that fish escaping from fishing gears may die as a direct result of physical damage and stress, or indirectly due to a reduced capacity to escape predators or resist disease.

Experimental methodologies

The methodology of full-scale survival experiments for towed fishing gears consists of three main steps: (a) collection, (b) transportation, and (c) holding and monitoring of the fish. Usually, escapees are collected by a cover/cage mounted around the cod-end or any selective device that is under investigation. Escapees are usually transferred to a remote cage site because of the difficulty of monitoring on the fishing grounds.

The most extensive mortality experiments on groundfish escaping from otter trawls to date have been reported by Sangster and Lehmann (1993), and Sangster *et al.* (1996). Using commercial trawlers, haddock (*Melanogrammus aeglifinus L.)* and whiting (*Merlangus merlangus L.*) were studied after escape from 70, 90, 100 and 110 mm diamond mesh cod-ends. The mortality rates of haddock and whiting controls were 0%. The mortality rates for the haddock and whiting experimental groups were 36–52% and 40–48% (70 mm cod-end), 18–21% and 22–27% (90 mm cod-end), 17–27% and 23–33% (100 mm cod-end) and 11–15% and 14–17% (110 mm cod-end), respectively.

Full-scale field experiments of fish survival are complex, and many unaccounted or uncontrollable factors may bias results (e. g. Main and Sangster 1990; Anon. 1994; Sangster and Lehmann 1994; Chopin and Arimoto 1995; Suuronen 1995). Recent underwater observations suggest that the small-meshed, hooped cover that is often used to collect escapees may seriously affect the geometry and movements of the cod-end and thus fish escape behaviour and mortality (Suuronen 1995). Further, there is always some uncertainty about the cause of death of the treatment fish (escapees) even though control group fish held in cages for several weeks have low mortalities. Other problems that have yet to be resolved include: the duration of caging in survival experiments (Anon. 1994); determining the level of stress associated with physical contact of fish with the cover (Soldal *et al.* 1993; Suuronen *et al.* 1996), and the extremely short time period after starting fishing when fish can be used in the experiment.

To date, escape mortality research has tended to focus only on short-term individual effects in the absence of long-term ecological factors such as increased predation risk or impaired growth or reproductive capacity. The inability to collect escapees from long tows and large cod-end catches (i. e. under commercial fishing conditions) is probably one of the most serious problems with all present techniques. Other priority areas which require further development are (a) flow-free transportation of escapees to holding cages, (b) capture of adequate numbers and size-classes of non-disturbed control fish, and (c) methodology for a quick and reliable assessment of skin-damages in fish. Since methodological approaches for estimating the survival of cod-end escapees are still at the development stage, some caution is still needed when interpreting these data.

Ghost fishing mortality

There is no clear definition of ghost fishing and it is difficult in many cases to distinguish between fishing mortalities associated with lost fishing gears, discarded fishing gears and fishing gear debris jettisoned from fishing vessels. For the purpose of this paper, ghost fishing mortality is defined as the number of fish and other animals killed directly or indirectly as a result of injuries and stress incurred due to encountering complete or partial fishing gears that have been lost during the process of setting, fishing or retrieving or cannot be physically retrieved. They do not include mortalities arising from the dumping or discharging of garbage such as scraps of netting and twine. This distinction is made to isolate specific fishing-related mortalities from those associated with waste disposal practices, the former being directly related to the fishing operation while the latter are related to the practice of dumping and discharging marine debris by the operator. For a review of the impacts of marine debris see Laist (in press). To date, although ghost fishing is known to occur with some gear types, notably passive fishing gears such as pots, gill-nets and trammel nets, there has been little systematic research carried out to quantify the full extent of ghost fishing mortality of various fishing gears and there are serious gaps in our knowledge of the short- and long-term unaccounted fishing mortalities associated with ghost fishing.

Recording the number of ghost fishing gears

Ideally, the best way of determining the number of ghost fishing gears would be for fishers to voluntarily report the location and nature of lost, discarded or irretrievable fishing gears. However, to date, most information on the number of lost, irretrievable or discarded fishing gears is collected indirectly by fishery research institutes, government agencies or other non-fishery research groups. Indirect surveys, while collecting some valuable information, tend not to allow a thorough investigation of the various types of losses (complete/partial loss during setting and, retrieving; gear which cannot be located etc.) or the conditions under which the loss occurred (fishing on wrecks, rocky ground, rough sea conditions, ice, deep water, high/low energy coasts, lakes etc.) and most importantly, their exact or estimated position. To date the number of lost, irretrievable or discarded fishing gears have been estimated for only a few fisheries and fishing gear types (Table 2). Survey estimates are derived from a variety of sources including shore-based surveys, counting the number of gears observed in a fixed area using manned and remote underwater vehicles and counting the number of gears snagged by a grapnel towed behind a vessel. Although surveys by observation or retrieval have some benefits, there has seldom been an evaluation of the efficiency of any survey method employed.

Quantifying ghost fishing mortality

Having determined that fishing gears are left in the sea for a variety of reasons, the next task is to determine the number of fish and animals killed as a result of encountering the gear and how this varies temporally and spatially. Matsuoka (in press) proposed a general ghost fishing gear model that describes a variety of categories of losses and suggested that ghost fishing mortality is time-dependent due to physical degradation of the fishing gear. For fishing fleets that set fishing gears in a variety of locations (high energy coasts, high current, shallow water, etc.) it is quite likely that fishing gears lost in different regions will be likely to have different long-term fishing potential. Surprisingly, while it is generally recognized that fishing gears such as pots and traps can, when left in the sea continue to keep fishing, there has been only a limited effort to quantify the full extent of the problem. Table 3 summarizes the work to date in measuring the unaccounted fishing mortality of pots and traps left in the sea.

Eliminating or reducing ghost fishing mortality

Although the amount of ghost fishing has not been quantified for most fisheries, it is generally perceived as a significant source of unaccounted fishing mortality for some gear types in some fisheries. Efforts to reduce the numbers of fish and other animals killed by gears left in the sea include retrieval of lost gill-nets by fishers at the end of the fishing season and use of degradable panels of netting in pots and nets. However, the efficacy of either of these methods to date has not been ascertained because no quantitative analysis of ghost fishing gears has been made. A key to improving research in this field would be to develop standardized survey methodologies for identifying and quantifying the short- and long-term impacts of ghost fishing for target and non-target species.

Table 2. Estimated number of lost fishing gears per annum

Species	Area	Estimated no. lost gears Pots (P), Gill-nets (GN)	Reference
Dungeness Crab	Fraser River	2834 (11%) – P	Breen (1987)
Dungeness Crab	Columbia River	6577 (18%) – P	Muir *et al.* (1984)
King crab	Alaska	10% – P	High and Worland (1979)
King Crab	Bering Sea	10 000–20 000 (10–20%) – P	Stevens *et al.* (1993)
Snow Crab	NB Canada	2466 – P	Mallet *et al.* (1988)
Snow Crab	NF Canada	8.3% – P	Miller (1977)
Lobster	New England	$42 - 63 \times 10^3$(20–30%) – P	Smolowitz (1978)
Lobster	New England	93 000–187 000 – P	Breen (1990)
Cod	NF Canada	5000 – GN	Fosnaes (1975)
Groundfish	Atlantic Canada	8000 (2%) – GN	Vienneau and Moriyasu (1994)
Groundfish	New England	39 per sq. nm	Carr *et al.* (1985)

Table 3. Estimated fishing mortality or duration of ghost fishing for pots and gill-nets

Species	Gear type	Estimated no. lost gears	Reference
Snow crab	Pot	44.3 crabs per pot = 100 t	Stevens *et al.* (1993)
Snow crab	Pot	0.5% landed catch	Mallet *et al.* (1988)
Lobster	Pot	670 t	Miller (1977)
Sablefish	Pot	326 t	Scarbrook *et al.* (1988)
Groundfish	Gill-net	3600 t fish per annum	Smolowitz (1978)
Groundfish	Gill-net	74 days: 25 fish + 48 crabs/net	Vienneau and Moriyasu (1994)
Herring	Gill-net	Nets ghost fish for 7 years	Carr *et al.* (1985)
Dungeness Crab	Pot	10 per pot per year	Breen (1990)
Salmon	Gill-net	2 years (fish) and 6 years (crabs)	High (1985)

FUTURE STRATEGIES IN CONSERVATION HARVESTING TECHNOLOGY

Fisheries management reforms and technological change over the last thirty years have had some important cosmetic effects in terms of reducing the level of non-target species and sizes discarded in some fisheries. However, most of these efforts have been without any direct measure of the reduction in numbers of fish killed and mostly focussed on only one category of fishing-induced mortality, i.e. discards. We have tried to show that the fish capture process can result in a number of discrete types of fishing mortality all of which have the potential for being quantified by gear type and fishery. Clearly, there are significant gaps in our understanding of the full extent of fishing-induced mortalities for most gear types and species of fish and with respect to experimental methodologies. Only data on discards have been thoroughly reviewed and their impact, estimated at 27 million t, is approximately 35% of the reported global catch. However, the errors associated with scaling-up information from vessel trip discards to annual fleet discards are enormous. Discard rates from vessels are subject to large fluctuations from vessel to vessel, as well as spatially and temporally, as fleets and fish move over the fishing grounds. Despite these shortcomings, the best guess of 27 million t with a range of 17.9–39.5 million t should be more than cause for concern and a reason for improving the precision of estimates.

While discard levels in some fisheries have been researched extensively, discard mortalities as well as all other types of fishing-induced mortality have only recently been a topic for investigation. What is common to each category is a lack of standardized experimental methodologies that make it difficult to make valid comparisons between mortalities measured in different fisheries. These problems are compounded by a lack of time-series data in all fisheries making it difficult to assess whether the level of biological waste is increasing or decreasing and for what specific technical, biological, economic or social reason.

There are several critical steps necessary to mitigate the problem of biological waste in fisheries. Firstly, all fishing-induced mortalities and not just reported catches need to be measured. Historically, these other mortalities have been an unmeasured component of natural mortality (M) making it impossible to determine which specific capture technologies and fishing practices result in the highest levels of biological waste. Consequently, there remains great confusion among industry and the public as to which fishing gears and fishing practices should be promoted and which ones give cause for concern. A greater effort is required to identify and quantify the types of unaccounted mortalities such as discards, ghost fishing , escape mortality, etc. by gear type and by fishery and how they may be incorporated into fishing mortality (F). Secondly, reliable estimates of biological waste must be more than a single event and collecting time-series data from standardized experiments repeated in problem fisheries on a annual basis is necessary. In this way, a barometer of biological waste through sentinel monitoring programmes can be used to measure the effects of technological change. Thirdly, specific technologies aimed at mitigation of biological wastes such as discards and ghost fishing should be encouraged and carried out with an assessment of their impact in relation to a quantitative assessment of the reduction of fishing-induced mortalities.

To date much of the research on discard mortality has focussed on improving mechanical selection of the catch in parts of the gear where the fish accumulate in large numbers (cod-end, bag net, seine bunt) rather than investigating a broader range of options where selection can take place. There has only been a limited amount of research on targeting species and size assemblages spatially and temporally on the fishing grounds. The advent of scientific sounders and improved designs of fishing gears may make it possible to adjust fishing practices dynamically rather than current practices of setting the gear for a fixed number of hours or days before retrieval. In towed fishing gears such as seines and trawls, there have been few attempts to select fish in the early stages of capture (zones of influence and action) with most efforts being made to release fish in the after-part of the gear. Additionally, most devices are passive in operation relying on sieving and direct physical contact in order to separate the catch. Further research is necessary to investigate non-contact methods of separating the catch including the use of artificial stimuli such as light, sound and water flow to modify encounter behaviour and to extend investigations to include zones of influence and action.

Long-term impacts of ghost fishing probably started to increase soon after the use of low-cost synthetic fibres used in fishing gears during the 1950s and 1960s. The lack of research into how ghost fishing efficiency changes with time and environmental conditions as well as reliable estimates of the number of fishing gears left in the sea prohibit any detailed evaluation. Despite these unknowns, ghost fishing is already known to occur with pots and gill-nets and various efforts to reduce ghost fishing such as making parts of the gear biodegradable or easier to locate have been considered. However, while some advances have been made with pots, no effective solutions have yet been found to reduce ghost fishing of gill-nets.

Despite a long history of research into gear selectivity as a tool for reducing the capture of non-target sizes of commercial food fishes, much of this work has been based on the assumption that fish escaping from fishing gears survive. It is only recently that research efforts have been initiated to validate this assumption and most of this has been directed towards short-term investigations and relatively large-sized fish. Topics for further investigation include establishing a reliable index of fish condition prior to capture and to determine how this may affect mortality estimates, long-term effects of injuries and stress on escapees. Currently, no research has been carried out to investigate the mortality of very small fish (O and 1+ group fish) that pass through the fishing gears.

Other types of fishing mortality such as drop-out mortality, avoidance mortality and the impacts of habitat modification on fish stocks are seldom investigated and need further investigation. Likewise, fishing mortalities such as non-reported, misreported and illegal catches are also difficult to estimate but there is anecdotal information to suggest that they may be significant in some fisheries. Consequently, although we now recognize a variety of different types of fishing-induced mortality and perceive some of these to be significant in some fisheries and gear types, they remain to date as an unmeasured component of natural mortality.

On a global level, the issue of world food security and biological waste in fisheries require a different approach to the same problem. While the work of Alverson *et al.* (1994) highlights the importance of biological waste, what is now required is a reliable barometer of the ability to reduce unaccounted fishing mortalities. A United Nations-based field monitoring team to carry out standard sampling programmes in key problem fisheries on a continuing basis could have a significant impact on improving the quality of time-series data for a variety of fishing-induced mortalities. Since developing appropriate methodologies is a necessity and likely to be developed iteratively as data are collected, free access to data is likely to accelerate the process of development. For this reason, the process of developing analytical methods might best be carried out by allowing researchers around the world to access UN data sets stored on the Internet. The requirement to carry out analysis for monitoring biological waste would still rest with the UN but the complex task of researching methodologies could be shared with a broader group of fishery scientists.

In much the same way that contemporary thinking about the impacts of man on harvesting marine resources underwent a complete overhaul at the start of the 20th century, a similar approach is now needed to identify, quantify and reduce the level of biological waste that currently exists in capture fisheries globally. Currently, the amount of money and research applied to measuring reported catches seems disproportionate to the amount spent on measuring other fishing-induced mortalities which may account for over 32.5% of the reported catch. Simply stated, a paradigm shift to measuring the full range of fishing-induced mortalities must be the starting point towards achieving sustainability.

REFERENCES

Allen, E J (1896). The regulations of the local sea fisheries committees in England and Wales. *Journal of the Marine Biological Association of the United Kingdom* **4**, 386–95.

Alverson, D L, Freeberg, M H, Pope, J G, and Murawski, S A (1994). A global assessment of fisheries bycatch and discards: A summary overview. *FAO Fisheries Technical paper* No. 339. Rome, FAO 1994. 233 pp.

Anon. (1903). The second international conference for the exploration of the sea, Christiana, 1901. *Journal of the Marine Biological Association* **6**, 389–412.

Anon. (1994). Report of the sub-group on methodology of fish survival experiments. *ICES C.M.* 1994/B:8. 46 pp.

Anon. (1995). Report of the sub-group on unaccounted mortality in fisheries. *ICES CM* 1995/B:1.

Baranov, F I (1977). Selected works on fishing gear Vol. III Theory of fishing. Israel Program for Scientific Translations. Jerusalem 1977. Jerusalem ISBN 0 7065 1575 7.

Breen, P A (1987). Mortality of Dungeness crabs caused by lost traps in the Fraser River Estuary, British Columbia. *North American Journal of Fisheries Management* **7**, 429–35.

Breen, P A (1990). Report of the working group on ghost fishing. *Proceedings of the Second International Conference on Marine Debris.* II, 1216–25.

Brown-Goode (1886). Inquiry into the decrease of food-fishes. B. The propagation of food-fishes in the waters of the United States. *United States Commission of Fish and Fisheries.* Washington, Government printing office. 1886. 1203 pp.

Carr, H A, Amaral, E H, Hulbert, A W, and Cooper, R (1985). Underwater survey of simulated lost demersal and lost commercial gillnets off New England. *Proceedings of the Workshop on the Fate and Impact of Marine Debris*, 26–29 November 1984, Honolulu, Hawaii. NOAA TM NMFS SWFC **54**, 438–47.

Chopin, F S M, and Arimoto, T, (1995). The condition of fish escaping from fishing gears—a review. *Fisheries Research* **21**, 315–27.

Chopin, F S M, and Inoue, Y (1996). Stress and survival in the capture process. In 'Behavioural Physiology of fish in the capture process'. (Eds T Arimoto and K Nanba) (Japanese Society for Scientific Fisheries, Series) **108**, 116–28. [In Japanese].

Chopin, F S M, Arimoto, T, Okamoto, N, Inoue, Y, and Tsunoda, A (1995). The use of plasma cortisol kits for measuring the stress response in fish due to handling and capture. *Journal of Tokyo University of Fisheries* **82**, 79–90.

Chopin, F, Inoue, Y, and Arimoto, T (1996). Development of a catch mortality model. *Fisheries Research* **25**, 377–82.

Chopin, F S M, Inoue, Y, and Arimoto, T (1996). A comparison of the stress response and mortality of sea bream *Pagrus major* captured by hook and line and trammel net. *Fisheries Research* **28**, 277–89.

Donaldson, E M (1981). The pituitary-interrenal axis as an indicator of stress in fish. In 'Stress in Fish'. (Ed. A D Pickering) pp. 11–47 (Academic Press: New York/London).

Engas, A, Isaksen, B, and Soldal, A V (1990). Simulated gear injuries on cod and haddock, a tank experiment. *ICES Working Group on Fishing Technology and Fish Behaviour*, Rostock 23–25 April 1990. 9 pp.

FAO (1993). Review of the state of world marine fishery resources. *FAO Fisheries technical paper 335*. Rome, FAO. 136 pp.

Fosnaes, T (1975). Newfoundland cod war over use of gill-nets. *Fishing News International*. June, 40–3.

Garstang, W (1903). The impoverishment of the sea. A critical summary of the experimental and statistical evidence bearing upon the alleged depletion of the trawling grounds. *Journal of the Marine Biological Association of the United Kingdom* **6**, 1–69.

High, W L (1985). Some consequences of lost fishing gear. Proceedings of the Workshop on the Fate and Impact of Marine Debris, 26–29 November 1984, Honolulu, Hawaii. NOAA TM NMFS SWFC **54**, 430–7.

High, W L, and Worlund, D D (1979). Escape of King Crab, *Paralithodes camtschatica*, from derelict pots. *NOAA Technical Report*. NMFS SSRF–734. 13 pp.

Hoag, S H (1975). Survival of halibut released after capture by trawls. *International Halibut Commission, Scientific Report* No. 57. 18 pp.

Holt, E W L (1895). An examination of the present state of the Grimsby trawl fishery with especial reference to the destruction of immature fish. *Journal of the Marine Biological Association of the United Kingdom* **3**, 339–448.

Huxley, T H (1884). Inaugural Address. *Fisheries Exhibition Literature* **4**, 1–22.

Laist, D W (in press). Entanglement of marine life in marine debris including a comprehensive list of species with entanglement and ingestion records. In 'Seeking Global Solutions.' (Eds J M Coe, and D B Rogers) Proceedings of the Third International conference on marine debris, 8–13 May 1994, Miami, Florida. (Springer-Verlag, New York).

Main, J, and Sangster, G I (1988). A report on an investigation to assess the scale damage and survival of young fish escaping from a demersal trawl. *Scottish Fisheries Working Paper* No. 3/88.

Main, J, and Sangster, G I (1990). An assessment of the scale damage to, and survival rates of young gadoid fish escaping from the cod-end of a demersal trawl. *Scottish Fisheries Research Report* No. 46. 28 pp.

Mallet, P Y, Chiasson, Y, and Moriyasu, M (1988). A Review of catch, fishing effort and biological trends for the 1987 southwestern Gulf of St. Lawrence snow crab, *Chionoecetes opilio* fishery. *CAFSAC Research Document*. 88/32. 39 pp.

Matsuoka, T (in press). Ghost fishing efficiency. In Proceedings of the Fourth Asian Fisheries Forum, October 12–16 1996. Beijing, China.

Miller, R J (1977). Resource under utilization in a Spider Crab industry. *Fisheries* **2**(3), 9–13.

Muir, W D, Durkin, J T, Coley, T C, and McCabe Jr, G T (1984). Escape of captured Dungeness crabs from crab pots in the Columbia river estuary. *Journal of North American Fisheries Management* **4**, 552–5.

Nikonorov, I V (1975). Interactions of fishing gear with fish aggregations. Israeli Program for Scientific Translations, Jerusalem 1975. 216 pp.

Pickering, A D (1993). Endocrine-induced pathology in stressed salmonid fish. *Fisheries Research* **17**, 35–50.

Ricker, W E (1976). Review of the rate of growth and mortality of Pacific salmon in salt water, and noncatch mortality caused by fishing. *Journal of the Fisheries Research Board of Canada* **33**, 1483–524.

Rogers, S G, Langston, H T, and Targett, T E (1986). Anatomical trauma to sponge-coral reef fishes captured by trawling and angling. *Fisheries Bulletin* **84**, 697–704.

Russel, E S, and Edser, T (1926). The relation between cod-end mesh and size of fish caught. Preliminary experiments with the trouser trawl. *Journal du Conseil* **1**, 39–54.

Sangster, G I, and Lehmann, K (1993). Assessment of the survival of fish escaping from fishing gears. *ICES C.M.* 1993/B:2.

Sangster, G I, and Lehmann, K (1994). Commercial fishing experiments to assess the scale damage and survival of haddock and whiting after escape from four sizes of diamond mesh codends. *ICES C.M.* 1994/B:38. 24 pp.

Sangster, G I, Lehmann, K and Breen, P A (1996). Commercial fishing experiments to assess the survival of haddock and whiting after escape from four sizes of diamond mesh cod-ends. *Fisheries Research* **25**, 323–46.

Scarsbrook, J R, McFarlane, G A, and Shaw, W (1988). Effectiveness of experimental escape mechanisms in Sablefish traps. *North American Journal of Fisheries Management* **8**, 158–61.

Smolowitz, R J (1978). Trap design and ghost fishing: Discussion. *Marine Fisheries Review* **40**, 59–67.

Soldal, A V, and Isaksen, B (1993). Survival of cod and haddock after escape from a Danish seine. *ICES Fish Capture Committee C.M.* 1993/B:2.

Soldal, A V, Isaksen, B, Marteinsson, J E, and Engas, A (1991). Scale damage and survival of cod and haddock escaping from a demersal trawl. *ICES Fish Capture Committee C.M.* 1991/B:44.

Soldal, A V, Isaksen, B, Marteinsson, J E, and Engas, A (1993). Survival of gadoids that escape from a demersal trawl. *ICES Marine Science Symposia* **196**, 122–7.

Stevens, B G, Haaga, J A, and Donaldson, W E (1993). Underwater observations on behavior of King Crabs escaping from crab pots. *AFSC Processed Report* 93–06.

Suuronen, P (1995). Conservation of young fish by management of trawl selectivity. *Finnish Fisheries Research* **15**, 97–116.

Suuronen, P, Erickson, D, and Orrensalo, A (1996). Mortality of herring escaping from pelagic trawl codends. *Fisheries Research* **3**(4), 305–21.

Suuronen, P, Lehtonen, E, Tschernij, V, and Larsson, P O (1995). Skin injury and mortality of Baltic cod escaping from trawl cod-ends equipped with exit windows. In 'Conservation of young fish by management of trawl selectivity' Document VIII, Academic Dissertation, Helsinki, Dec 1995.

Suuronen, P, Perez-Comas, J A., Lehtonen, E, and Tschernij, V (1996). Size related mortality of herring (*Clupea harengus L.*) escaping through a rigid sorting grid and trawl codend meshes. *ICES Journal of Marine Science*, **53**, 691–700.

Vienneau, R., and Moriyasu, M (1994). Study of the impact of ghost fishing on snow crab, *Chionoecetes opilio*, by conventional conical traps. *Canadian Technical Report of Fisheries and Aquatic Sciences* No.1984. 9 pp.

Factors Contributing to Collapse yet Maintenance of a native Fish Community in the Desert Southwest (USA)

John N. Rinne[A] *and Jerome A. Stefferud*[B]

[A] USDA Forest Service, Rocky Mountain Station, The Southwest Forest Science Complex 2500 S. Pineknoll Drive, Flagstaff, Arizona 86001, USA.
[B] USDA Forest Service, Tonto National Forest 2324 E. McDowell Road Phoenix, Arizona 85010, USA.

Summary

Data presented indicate that the ratio of population numbers of native to non-native fish species in the Verde River, Arizona fish community is variable but most consistently 85:15. In part, the relative numbers of each component are dependent on river location and species of fish. Characteristic stream hydrography, in terms of location, season, and severity and duration of flood events, also appears to be a contributing factor for maintaining this ratio. Although a 7-year flood event collapsed populations of both native and non-native species, the proportions of the two groups changed only slightly (82:18 to 85:15). By comparison, a 50+ year flood event resulted in native species increasing by 11% (86–97%). Reproduction and recruitment of natives is enhanced post-flood, apparently as a result of cleaning and rejuvenation of stream substrates. Based on a decade of data (1986 to 1996), the natural hydrography of the upper Verde River, as opposed to biological or native/non-native species interactions, is suggested as the controlling factor in sustaining native fishes in this large, low desert river.

Introduction

Patterns of stream discharge and the presence of introduced fishes have been suggested to interact to influence native fishes and their sustainability in streams of the arid American southwest (Minckley and Meffe 1987; Rinnee and Minckley 1991; Rinne 1993, 1995a; and Stefferud and Rinne 1995). Commencing with the water reclamation era in the early 1900s, dams have been ubiquitously imposed on most rivers in this region (Rinne 1991, 1993, 1995b), drastically and irreversibly altering aquatic habitats (Miller 1960; Rinne and Minckley 1991). Non-native fish introductions commenced before 1900 and have steadily escalated (Rinne 1991, 1993, 1995a). The combination of these two impacts has markedly reduced both the range and numbers of all native species of fishes, rendering some extinct (Minckley and Douglas 1991), yet native fishes persist in most drainage basins in the Southwest. However, where both altered, unnatural stream hydrography and non-native fish exist, native fish generally are either reduced from historic levels or are absent (Minckley and Deacon 1968).

In all but a few streams, these two factors, when combined, have eliminated native fish species and modified community structure and dynamics, often within a few years after the action.

Streams where the native fauna remains predominant are a rarity in the Southwest. The exact mechanism that sustains native species in the presence of non-native species is not understood, but is hypothesized to be related to cycles of extreme droughts and floods that characterize streams in this region. Because of the ever-increasing demand for water for a growing populace, and the generally imperilled status of the southwestern native fish fauna (Rinne and Minckley 1991), clarification of this question is important at this time.

The upper Verde River, central Arizona, from its source at Sullivan Lake to the mouth of Sycamore Creek 60 km downstream is a rarity among southwestern rivers. It is the only reach of large (1.3–7.0 m^3 sec^{-1} annual mean discharge) perennial desert river in Arizona where surface flows are not altered by dams or diversions. Further, this reach of river retains a native fish fauna that is largely intact. Six of the eight native species (desert sucker, *Catostomus clarki*; Sonora sucker, *Catostomus insignis*; roundtail chub, *Gila robusta*; longfin dace, *Agosia chrysogaster*; spikedace, *Meda fulgida* and speckled dace, *Rhinichthys osculus*) are present in viable numbers (Stefferud and Rinne 1995); all are either state or federally-listed species. Two species (Colorado squawfish, *Ptychocheilus lucius* and razorback sucker, *Xyrauchen texanus*), which have been documented in archaeological sites along the Verde (Minckley and Alger 1968) but are no longer to be found, are currently objects of re-introduction and restoration efforts. Nevertheless, an assemblage of non-native fish occurs in this reach of river, including red shiner (*Cyprinella lutrensis*), fathead minnow (*Pimephales promelas*), common carp (*Cyprinis carpio*), green sunfish (*Lepomis cyanellus*), smallmouth bass (*Micropterus dolomieui*), yellow bullhead (*Amerius natalis*), and channel catfish (*Ictalurus punctuatus*). Reasons for complete retention of the native fish are unclear. Whether the native species are maintaining their populations in the presence of non-native species, or whether the non-native species are unable to become dominant due to hydrologic conditions (water quantity and quality) and stream gradient is not known. Few published data are available to document either the history of encroachment by non-native species or trends in the dynamics of southwestern fish communities.

Although a number of studies on fishes and their habitats have been undertaken in the Verde River, none have investigated the long-term interrelationships of the native and non-native fish communities relative to flood disturbance events and to each other. The relationship of base and flood flows to sustainability of native fishes in the upper Verde River needs to be determined in order to provide input to water management plans for the system. The headwater Big Chino Valley aquifer is a major source of baseflow for the Verde River above Sycamore Creek; and there are proposals to extract its water to supply municipal needs to towns in the watershed. Depletion of discharge from the aquifer could alter baseflow in the river and affect sustainability of native fish. Other proposals to divert water directly from the Verde River and construction of mainstream dams have been made. However,. the unresolved question of their effects on the native fish has deferred approval by resource agencies. Continued growth in the region will probably increase pressure for water development and diversion from the upper Verde. Resource management agencies need to have reliable data in time and space to chart a course of action that will conserve native fish in the river.

In Spring 1994, we initiated a study to define the roles and relative influence of physical (hydrology, geomorphology) and biological (non-native fish population and community dynamics) factors on the sustainability of native fishes in time (10 years) and space (seven localities over 60 km of river) in this reach of free-flowing river. Initiation of the study immediately followed a 50+ year flood event in this reach of river. Although only in the third year of study, the intent of this paper is to provide baseline data and to discuss the role of these two factors in the collapse, re-establishment, and sustainability of the native fish community in a large, perennially-flowing desert river.

Methods

Study sites were located primarily on reasonable access and linear disposition over the 60 km reach. Specific location was based on channel morphology and types of habitat present (Stefferud and Rinne 1995; Rinne and Stefferud 1996). Seven sites were chosen commencing at a point 7.6 km east of Sullivan Lake and ending at the confluence of Sycamore Creek (Fig. 1). Sites were linearly separated by distances of 3.3–16.2 km.

Initially, we proposed to sample 250–300 m of stream at each site; however, exact length was dependent on habitat complexity at the site. In general, a riffle-pool sequence was selected, then the study site was expanded to include other significant habitat types present (Rinne and Stefferud in press). Habitat types were sampled for fish with backpack DC electro-fish units, seines (3.2 mm mesh), dip-nets, and trammel nets progressing from downstream to upstream. All fish captured within a habitat type were identified to species, enumerated and returned alive to the stream. All sites were sampled in Spring; however, only the upper- (Burnt Ranch) and lower-most (Sycamore) sites were sampled in Autumn. Reduced (seven to two) sampling protocol

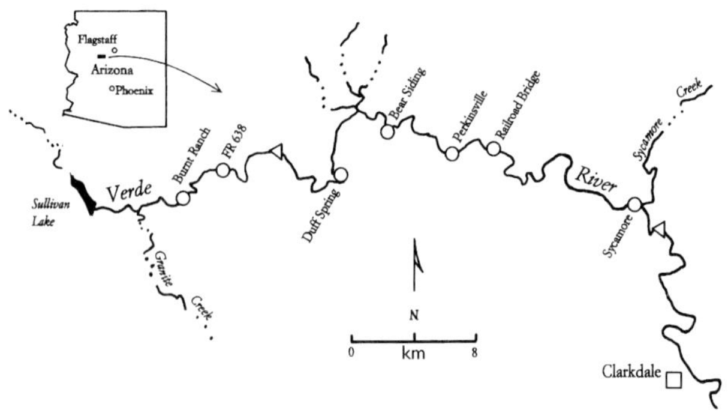

Fig. 1. The upper Verde River indicating the seven sites sampled between 1993 and 1996. Triangles indicate location of the Paulden and Clarkdale stream gauges.

Table 1. Fish community dynamics in the upper Verde River, Spring 1994–96. Values in ( ) and [ ] are percent change from previous Spring, and bold numbers are native : non-native proportions. Species abbreviations are shown and are the same for Tables 2 and 3

Species	Spring 1994	Spring 1995 (%)	Spring 1996 (%)
Native species			
Desert sucker (CACL)	2644	247 (–90)	471 (+90)
Sonoran sucker (CAIN)	1810	322 (–82)	654 (+103)
Roundtail chub (GIRO)	776	341 (–56)	259 (–24)
Longfin dace (AGCH)	1319	12 (–99)	282 (+2250)
Spikedace (MEFU)	428	72 (–83)	140 (+94)
Speckled dace (RHOS)	171	25 (–85)	68 (+172)
Total (%) [%change]	7148 **(82)**	1019 **(85)** [–86]	1874 **(85)** [+714]
Introduced Species			
Red shiner (CYLU)	1473	97 (–93)	275 (+183)
Common carp (CYCA)	23	6 (–74)	13 (+117)
Smallmouth bass (MIDO)	14	10 (–28)	32 (+220)
Green sunfish (LECY)	5	29 (+480)	6 (–79)
Yellow bullhead (AMNA)	36	32 (–11)	6 (–84)
Fathead minnow (PIPR)	6	0 (–100)	0 (—)
Total (%) [% change]	1558 **(18)**	174 **(15)** [–89]	332 **(15)** [+71]

was based on both consideration of impact to fishes and to address the potential relative influences of summer monsoon *versus* winter flood events.

RESULTS

Numbers of most fish species decreased markedly between Spring 1994 and 1995 sampling; only green sunfish increased (Table 1). Native fish decreased by between 56 and 99% (mean, 83%) non-natives by 11 to 100%. Yellow bullhead was reduced only 11% and smallmouth bass 28%. Red shiner decreased by 93%, and no fathead minnows were found in Spring 1995.

Bi-annual sampling at the upper- and lower-most sites in the river offers a more detailed picture of population dynamics (Table 2). All native species, except roundtail chub, declined from Spring 1994 to Autumn 1994 at the Burnt Ranch site, decreased further by Spring 1995 before increasing in Autumn 1995 and Spring 1996. Most non-native species varied little between samples. The introduced red shiner population at Burnt Ranch held stable between Spring and Autumn 1994 sampling and then declined markedly (>80%) by Spring 1995 only to then increase markedly (95%) by Autumn 1995. By comparison, at the Sycamore site, red shiner increased over 100-fold between Spring and Autumn 1994 only to collapse (-99%) to very low population levels in 1995 and 1996. At Sycamore, common carp numbers were low from Spring 1994 to Spring 1996. Similarly, smallmouth bass numbers remained low during all sample periods, disappeared at Sycamore in Spring 1995, only to increase slightly by Autumn 1995 at this site. Green sunfish were absent at these two sites in 1994, increasing at the Burnt Ranch site by Autumn 1995. Numbers of yellow bullhead were generally low and declined steadily from Spring 1994 to Spring 1995 before increasing in Autumn 1995. Fathead minnows were essentially absent from all sampling.

At all seven sites, native fishes predominated (82 to 85%) in numbers over the non-native species (15 to 18%) (Table 1; Fig. 2). However, such predominance of native species was variable, comprising 54% of the fish community in Autumn 1994 and increasing to 85% over the entire study area by Spring 1995. By Autumn 1995 the native fish community had increased to comprise 88% of the total fish community. Fish population data collected in an independent study of the Verde River in 1986 (USFWS 1989) indicated that native species comprised 88% of the fish community (Table 3). Sampling by the Arizona Game and Fish Department immediately after the 50+year flood event in Spring 1993 (Fig. 3) indicated natives were the dominant component (96%) of the fish community by Autumn 1993 (Table 3).

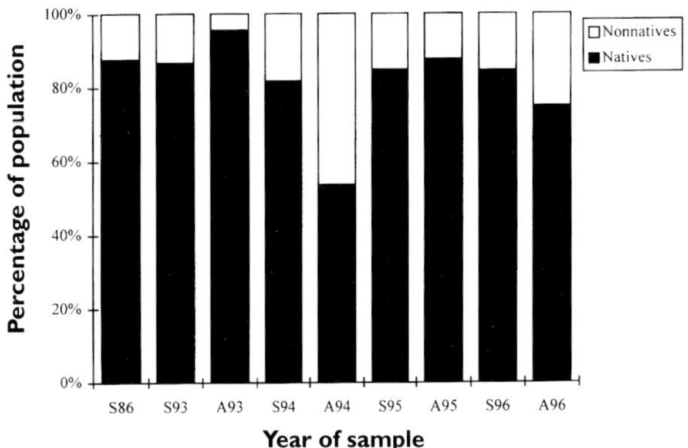

Fig. 2. Proportional numbers of native to non-native species of fishes in the upper Verde River, 1986–1996. Ratios for 1993 through 1995 are for Spring (S) and autumn (A), respectively.

Table 2. Fish community dynamics based on Spring 1994, Autumn 1994, Spring 1995, Autumn 1995, and Spring 1996 sampling at Burnt Ranch (BR) and Sycamore Creek (SC) study sites. The total number at the two sites is indicated. Order of species as in Table 1

Species	Spring 94			Autumn 94			Spring 95		
	BR	SC	Total	BR	SC	Total	BR	SC	Total
Native species									
CACL	339	379	718	31	93	124	15	29	44
CAIN	278	223	551	214	25	239	60	37	97
GIRO	15	165	180	50	17	67	3	104	107
AGCH	1072	1	1073	94	0	94	0	0	0
MEFU	257	92	349	93	0	93	33	17	50
RHOS	0	19	19	0	1	1	0	0	0
Introduced species									
CYLU	39	3	42	50	395	445	7	5	12
CYCA	1	4	5	67	1	68	0	2	2
MIDO	2	3	5	2	3	5	3	0	3
LECY	0	0	0	0	0	0	0	2	2
AMNA	2	10	12	1	6	7	2	1	3
PIPR	0	0	0	5	0	5	0	0	0

Species	Autumn 95			Spring 96		
	BR	SC	Total	BC	SC	Total
Native species						
CACL	44	77	121	79	38	117
CAIN	103	93	196	92	41	133
GIRO	40	6	46	23	25	48
AGCH	397	0	397	91	1	92
MEFU	290	0	290	33	51	84
RHOS	0	12	12	0	0	0
Introduced species						
CYLU	151	7	158	88	9	97
CYCA	3	0	3	1	1	2
MIDO	0	9	9	5	0	5
LECY	7	0	7	1	0	1
AMNA	5	8	13	1	0	1
PIPR	0	0	0	0	0	0

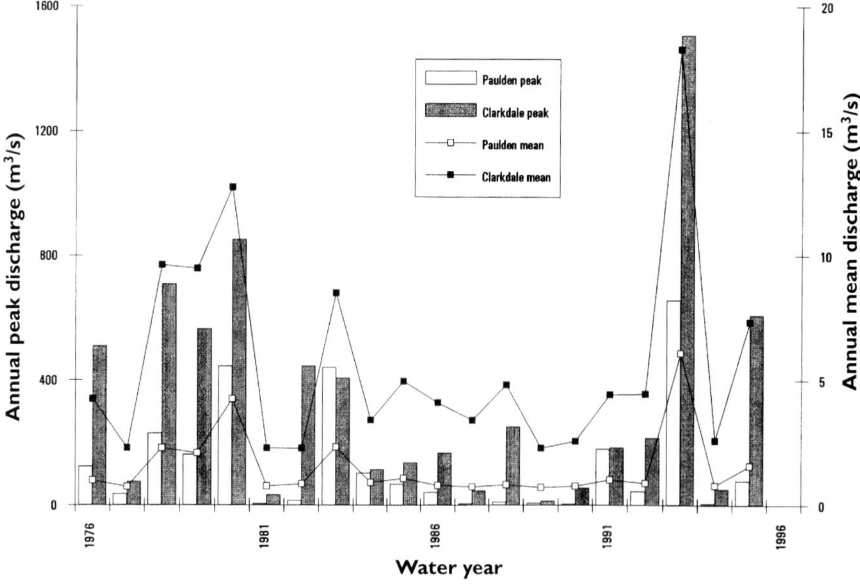

Fig. 3. Annual peak and mean annual discharge at Paulden and Clarkdale gauges during water years 1976 to 1995 (US Geological Survey records).

Table 3. Relative percentages of the fish community collected in the Verde River 1986–96. Species designation as in previous tables except for ICPU, channel catfish; species ratios (%) are indicated by season

Species	Spr 86	Spr 93	Aut 93	Spr 94	Aut 94	Spr 95	Aut 95	Spr 96
Native species								
CACL	18.7	26.2	20.9	30.4	10.8	25.7	14.7	21.3
CAIN	13.0	17.9	27.8	20.8	20.8	25.3	16.8	29.6
GIRO	11.0	1.8	1.6	8.9	5.8	26.8	4.3	11.7
AGCH	25.1	25.5	34.4	15.2	8.2	0.9	26.6	12.7
MEFU	19.1	3.7	2.5	4.9	8.1	5.7	20.1	6.3
RHOS	0.8	10.0	8.6	2.0	0.1	2.0	4.7	3.0
%	88.0	85.0	96.0	82.0	54.0	85.0	88.0	85.0
Non-native Species								
CYLU	6.6	7.4	2.9	16.9	38.8	7.6	10.6	12.4
CYCA	0.1	1.3	0.4	0.3	5.9	0.5	0.2	0.6
MIDO	0.6	1.5	0.0	0.0	0.0	2.3	0.5	0.3
AMNA	3.3	1.1	0.6	0.4	0.6	2.3	1.0	0.3
ICPU	0.4	0.4	0.0	0.1	0.0	0.2	0.0	0.0
PIPR	0.0	3.1	0.1	0.1	0.4	0.0	0.0	0.0
%	12.0	15.0	4.0	18.0	46.0	15.0	12.0	15.0

DISCUSSION

Minckley and Meffe (1987) contrasted differences in flood hydrology between southwestern arid lands and those of lowland mesic regions of the central and eastern USA. They reported that most of the annual water yield in southwestern systems was produced during high discharges in brief periods of time, whereas low discharges produced a far greater proportion of total yield from mesic systems. They suggested the differential effects of these patterns on fishes was significant to the native : non-native mix of species now present in the Southwest. Minckley and Meffe (1987) concluded that native southwestern fishes were better adapted to withstand the effects of large floods than non-native fishes. During severe floods, non-native species were either displaced or killed, whereas native species maintained position in or adjacent to channel habitats, persisted in micro-refuges, or rapidly recolonized if displaced. Based on our data from the Verde River following 50+ and 7-year flood events, it appears that floods of this magnitude negatively impact and dramatically reduce both native and non-native fish communities. The 7-year event in Winter (March) 1995 reduced the native fishes in the upper Verde by 86% and the non-native community by 89%. However, the native: non-native community composition ratio remained almost identical between years, respectively (82:18 and 85:15). Further, by Spring 1996 the native fish component had increased, on average, over 700%; non-natives, despite marked increases in red shiner and smallmouth bass, increased by an average of only 71%.

Data from work previous to our study indicated that following almost a decade (1984-92) of drought and low flows, non-natives only comprised 12% of the total fish community (Marty Jackle, US Bureau of Reclamation, pers. comm.; USFWS 1989). Following the massive flooding in 1993, non-natives were reduced to only 4% of the total fish community. Therefore, Minckley and Meffe's hypotheses on the mechanism of interactions and control of the native and non-native fish community in the Southwest may be partially correct. First, not only the non-natives are impacted by flooding, but the natives as well. Immediately post-flood, native fish rebound rapidly; non-natives also increase but more slowly, and some ratio of the two is established. The ratio appears, in part, to be dependent on the subsequent annual hydrographs, but has been consistently 85:15, native to non-native species. However, the fact that non-natives comprised only 12% of the total fish community in the late 1980s (see above) after a four-year period of low flow suggests other mechanisms may influence the native : non-native ratio in the Verde River.

Green sunfish was the only species to increase in absolute numbers in the Verde River fish community between 1994 and 1995 (5–29). We suggest it was a result of a differential magnitude of flooding over the entire study area. The flood in 1995, in contrast to that in 1993, affected mostly the reach of river downstream from Sycamore creek (i.e. the Clarkdale gauge). The majority (26–29 individuals) of the increase in green sunfish numbers occurred in the upper reaches of the river (sites 2–5). The peak discharge at these sites, as indicated by the Paulden gauge, suggests that flows were either below the threshold to negatively impact this species, or of the proper magnitude to overflow ponds on the watershed and introduce sunfish into the river, or a combination of both. Similarly, we suggest that the input of fathead minnow from watershed stock tanks may have been the reason for the appearance of this species immediately post-flood in the Spring 1993 sample at Burnt Ranch. By Autumn 1993, fatheads had disappeared and the native fish component reached 96% of the fish community.

Precipitation and stream hydrographs are stochastic and unpredictable in the Southwest and floods of significance (i.e. >400 m^3 sec^{-1}) in the upper Verde River appear to occur randomly. Cycles of flood and drought, however, are evident, and ensuing years of low flow are now probable following the 1993 and 1995 floods. Accordingly, based on the historic hydrograph, the probability for a period of low flows from now to the year 2000 increases. Consequently, an excellent opportunity exists to study in detail native *versus* non-native fish interactions during reduced flooding and low flow conditions.

Typical of desert streams and rivers, flooding can be significant in the Verde River. Larger floods, estimated to have recurrence intervals of 50–60 years (US Geological Survey 1992), appear to 'reset' the biological and geomorphological reference baseline. That is, riparian vegetational succession was set back to a base level, channel morphology was modified, and stream-bed materials were sorted and rearranged. Flood events in 1993 (50+ year) and 1995 (7-year) altered channel morphology and aquatic habitats by eroding stream banks and restructuring and 'invigorating' substrate materials.

After each flood, total abundance of fish was reduced and the population structure of the various species altered. Based on data collected in Winter 1993 and 1995 flood events, both 50+ and 7-year return intervals reduced both native and non-native species. Nevertheless, natives still comprised 85% or more of the fish community. Following these events the natives rebound quickly in numbers in response to the restructuring or 'invigorating' of substrate materials (Mueller 1984). Spawning success and recruitment probably contribute to the quick re-establishment of native fish populations. Although non-native species begin to increase at the same time (Table 2,) their lowered numbers also would favour a rapid increase in native fish numbers because of the lack of competition and more probably predation (Minckley 1983; Rinne and Alexander 1995: Rinne 1995*b*). Further, recovery rates of non-native fish populations were variable. Red shiner were markedly reduced by the 7-year flood event, common carp were less reduced, and smallmouth bass sustained their numbers.

Long-term (>5 years) studies of native fish populations in low, desert rivers of similar or larger-size are non-existent. Several long-term studies in smaller streams are ongoing: Aravaipa Creek has a 25-year record of fish community dynamics (W L Minckley pers. comm.), and there are 10-year records of native fishes and habitat associations for several streams in the upper San Francisco and Gila River drainages in New Mexico (D L Propst pers. comm.). Whereas these existing data sets may be useful for comparing effects of floods and droughts on native species, none of the studies are in stream reaches that have a significant non-native fish component. Thus the Verde River project is unique in the Southwest and offers an excellent opportunity to test and refine current hypotheses relative to native : non-native fish interactions as influenced by drought and flooding over the long term.

References

Miller, R R (1960). Man and the changing fish fauna of the American southwest. *Papers of the Michigan Academy of Science, Arts, and Letters* **46**, 365–404.

Minckley, W L (1983). Status of the razorback sucker, *Xyrauchen texanus* (Abbott), in the lower Colorado River basin. *The Southwestern Naturalist* **28**, 165–87.

Minckley, W L, and Alger, N T (1968). Fish remains from an archaeological site along the Verde River Yavapai County, Arizona. *Plateau* **40**, 91–7.

Minckley, W L, and Deacon, J E (1968). Southwestern fishes and the enigma of 'Endangered species'. *Science* **159**, 1424–32.

Minckley, W L, and Douglas, M E (1991). Discovery and extinction of western fishes: a blink of the eye in geologic time. In 'Battle Against Extinction'. (Eds W L Minckley and J E Deacon) pp. 7–17 (University of Arizona Press: Tucson).

Minckley, W L, and Meffe, G K (1987). Differential selection by flooding in stream fish communities of the arid American southwest. In 'Community and Evolutionary Ecology of North American Stream Fishes'. (Eds W J Matthews and D C Heins) pp. 93–104 (University of Oklahoma Press: Norman).

Mueller, G (1984). Spawning by *Rhinichthys osculus* (Cyprinidae) in the San Francisco River, New Mexico. *The Southwestern Naturalist* **29**, 354–6.

Rinne, J N (1991). An approach to management and conservation of a declining regional fish fauna: southwestern USA. *Proceedings Symposium on Wildlife Conservation, 5th International Conference Zoology* **5**, 56–60.

Rinne, J N (1993). Declining southwestern Aquatic habitats and fishes: Are they sustainable? Sustainability symposium. USDA Forest Service General Technical Report RM-247, 256–65.

Rinne, J N (1995*a*). The effects of introduced fishes on native fishes: Arizona, southwestern United States. In 'Proceedings of the World Fisheries Congress, Theme 3'. (Ed D P Phillips) pp. 149–59 (Oxford and IBH Publishing Company: New Delhi).

Rinne, J N (1995*b*). Interactions of predation and hydrology on native southwestern fishes: Little Colorado spindace in Nutrioso Creek, Arizona. In 'Hydrology and Water Resources in Arizona and the Southwest'. Vols 22–25. pp. 33–38. Proceedings of the 1995 meeting of the Arizona Section, American Water Resource Association and the Hydrology Section, Arizona-Nevada Academy of Sciences, April 22, 1995, Flagstaff, Arizona. 114 pp.

Rinne, J N, and Alexander, M A (1995). Non-native salmonid predation on two threatened native species: Preliminary results of field and laboratory studies. *Proceedings Desert Fishes Council* **26**, 114–16.

Rinne, J N, and Minckley, W L (1991). Native fishes of arid lands: A dwindling resource of the desert southwest. USDA Forest Service General Technical Report RM–206, 1–45.

Rinne, J N, and Stefferud, J A (1996). Relationships of native fishes and aquatic macrohabitats in the Verde River, Arizona. In 'Hydrology and Water Resources in Arizona and the Southwest'. Vol. **26**, pp. 13–22 (American Water Resources Association and the Hydrology Section Arizona-Nevada Academy of Sciences: Flagstaff).

Stefferud, J A, and Rinne, J N (1995). Preliminary observations on the sustainability of fishes in a desert river: The roles of streamflow and introduced fishes. In 'Hydrology and Water Resources in Arizona and the Southwest'. Vols 26–32, pp. 22–25, (American Water Resources Association and the Hydrology Section, Arizona-Nevada Academy of Sciences: Flagstaff).

US Fish and Wildlife Service (1989). Fish and wildlife coordination act substantiating report: Central Arizona Project, Verde and East Verde River water diversions, Yavapai and Gila counties, Arizona.

US Geological Survey (1992). Basin characteristics and streamflow statistics in Arizona as of 1989. US Geological Survey, Water-Resources Investigations Report 91–4041.

Issues and Outcomes for Theme 1

Why do Some Fisheries Survive While Others Collapse?

M. Sinclair

Marine Fish Division, Fisheries and Oceans Canada Bedford Institute of Oceanography, PO Box 1006, Dartmouth, NS, B2Y 4A2, CAnada

Introduction

In the introductory talk for this Theme, I defined 'survival' and 'collapse' of fisheries from the perspective of the Gordon-Schaefer bio-economic model. Collapse is considered to occur at high levels of fishing effort such that the spawning stock biomass is below the minimum required for moderate to good recruitment. It was pointed out that management and industry activities which lower the cost curve generate high risk. Under this situation, the fishing industry can continue to make a profit at very low levels of stock biomass at high levels of fishing effort. If the open-access equilibrium point occurs at effort levels which correspond to spawning stock biomass below the minimum required for reasonable levels of recruitment, the fishery will collapse if there is ineffective effort control. The model illustrates the role of cost curve changes through technological efficiencies and subsidies on long-term sustainability. The characteristics of 'collapse' are that recruitment is impaired, and thus stock biomass does not respond to management actions on time scales of interest to the fishing industry and fishing communities. In contrast, if the open-access equilibrium point occurs at effort levels at which spawning stock biomass is above the minimum level for reasonable recruitment, economic and biological overfishing through ineffective effort control does not generate fishery 'collapse'. The above definition of collapse is used to summarize the diverse studies presented in Theme 1.

Ray Hilborn provided a similar definition of collapse, and introduced the term sustainable overfishing which perhaps sounds like an oxymoron but is not. Many fisheries around the world are in this category. They are characterized by growth overfishing and loss of economic potential. However, if recruitment is not compromised the fisheries can be sustainable at suboptimal levels. There were several examples of such fisheries provided in the theme presentations: tropical tunas; Baja California spiny lobster; Nova Scotia American lobster; and Western Australian prawn stocks. Perhaps the most striking example is the Mediterranean bluefin tuna stock, which has generated about 20 000 t annually for 500 years at high levels of fishing effort.

A key issue for Theme 1 is the identification of biological and technological characteristics of these fisheries which have been observed to be sustainable at very high levels of fishing effort and/or ineffective management. In what sense are they different from fisheries which have been observed to collapse due to recruitment failure? Two characteristics were discussed. Several fisheries in the so-called sustainable overfishing category have refugia to present technology. For example, Alain Fonteneau concluded that there are large parts of the distributional range for species of tropical tunas that are not presently economical to fish. The same situation was proposed for horse mackerel off Chile by Patricio Barria. There were also examples of collapses of fisheries due to the lack of refugia. Aril Slotte described the collapse of Norwegian spring spawning herring at a time when all areas of the annual cycle of adult migration were fished, as well as the juvenile distributional area. Juan Carlos Castilla summarized the collapse of the fishery on 'loco' (the gastropod muricid *Concholepas concholepas*) off Chile by small scale artisanal diving operations. There appear to be no refuges from fishing pressure. A second characteristic of sustainable overfishing is a passive and selective harvesting technology for species with a relative lack of aggregation during various parts of the life history. Spiny lobster off Baja California and American lobster off Nova Scotia are tentative examples. I will return to this issue of characteristics of fisheries that appear to be sustainable at high levels of fishing effort at the end of this summary.

Causes of collapse of fisheries

Based on the Theme 1 presentations, a classification scheme for collapsed fisheries is provided. Collapse, as stated in the introduction, involves recruitment failure for an extended period (a decade or more), and slow response to management actions. All of the studies describing collapse of fisheries involved fishing effort levels higher than the target of the management plan, or alternatively there were no limitations on fishing effort. There were no examples of fisheries which have collapsed under a management regime which was meeting the target fishing mortality levels. Also, there are examples of collapse of low technology artisanal fisheries, as well as high technology industrial fisheries.

The first category of cause of collapse is entitled 'predominantly overfishing'. Examples are:

- Norwegian spring spawning herring — Aril Slotte;
- Baja California abalone — Armando Vega and Daniel Lluch Belda;
- Chilean 'loco' — Juan Carlos Castilla;
- Gulf of Trieste clams — Donatella Del Piero;
- Northern and southern bluefin tuna — Alain Fonteneau.

There are several biological phenomena occurring in these examples, although there is frequently uncertainty in the interpretation of the life history aspects of overfishing. For the herring examples (notably southwest Nova Scotia herring which is not yet considered to be in the collapse category), there is evidence of a loss of spawning components of management units comprised of stock complexes. For the sedentary benthic mollusc fisheries which have collapsed, one interpretation is that under high fishing effort the low density of molluscs (which are broadcast spawners) prevents successful recruitment. Thirdly, some examples suggest that the extreme reduction in biomass of a single stock management unit results in impaired recruitment. The discussions highlighted the issue that there is considerable uncertainty concerning what reproductive processes are being compromised when overfishing generates collapse.

The second category is collapse due to bycatch mortality. There was only a single example, red snapper in the Gulf of Mexico. Recruitment of this species to the fishery is low due to the high bycatch of juvenile snapper by the shrimp fishery. James Nance inferred that bycatch is preventing a recovery of this high value fishery. It may not, however, have been the cause of the collapse.

The third category is collapse due to changes in natural mortality associated with heavy fishing. Paola Ferreri and W.W. Taylor described the decline in the Lake trout populations of Lake Superior caused by increases in natural mortality due to sea lamprey in the 1950s. Although the interpretation is controversial, I argued that there is considerable evidence that the cod stock on the eastern Scotian Shelf off Nova Scotia has collapsed in part due to increases in juvenile natural mortality caused by increases in grey seal predation.

The fourth category involves environmental change, and in some cases evidence of so-called regime shifts. Examples from the Theme presentations are:

- Chilean anchovy — N. Ehrhardt, P. Barria and R. Serra;
- Eastern Australian gemfish — K. Rowling;
- Japanese eel — W.-N. Tzeng.

The best documented examples of the role of regime shifts on ecosystem structure were provided by S. R. Hare and co-authors for the northeast Pacific, and by D. Lluch-Belda for eastern boundary currents.

The fifth category is collapse due to habitat loss, primarily for freshwater fisheries. Mucai Muchiri provided a number of examples of severe habitat loss for fisheries of the rift valley lakes of Kenya. Water levels have declined in some lakes and turbidity has increased dramatically in others due to forestry and agriculture activities within the drainage systems. R.A. Englund and R. Filbert described declines in fish abundance in the streams of Oahu due to diversion and channelization. D.D. Reid, J.H Harris and G.A. White summarized the declines of dominant species (Murray cod, golden and silver perch) due to river regulation and barriers on the flood-plain river system of New South Wales, Australia, since the 1950s. A marine example was provided by T. Kadiri-Jan, whose poster described the impact of changes in the coastal zone environment on habitat requirements for juvenile coconut crabs, in New Caledonia.

A sixth category is collapse due to pollution. As is the case for habitat loss, this cause is particularly severe in freshwater and estuarine environments. N. Bacalbaşa-Dobrovici described the loss of production of migratory fish in the lower Danube. During the past 30 years due to eutrophication, the fisheries on migratory species have collapsed. K.H. Chu and co-authors documented the decline in fishery production of commercially important crustaceans in the Zhujiang (Pearl River) estuary of China. The industrialization and urbanization in this delta area have resulted in heavy pollution.

The final category is where there is no explanation for the fishery collapse. The only example of this type is the decline of the squid fishery off Nova Scotia and Newfoundland (*Illex illecebrosis*) which was identified by R. O'Dor and co-authors. The fishery collapsed in the early 1980s after a few seasons of intense, widely distributed fishing. Other squid fisheries around the world have remained stable over much longer time periods. Perhaps this example should be in category 1 (i.e. 'primarily overfishing'), but the declines are difficult to interpret due to fishing, given the broad geographic range of the species and the relatively short duration of the offshore fishery.

REASONS FOR SURVIVAL AND PREVENTION OF COLLAPSE

A large number of papers presented in Theme 1 addressed diverse reasons for the survival of specific fisheries, as well as ideas on management approaches to prevent collapse. The approaches are grouped into several categories.

The first category is entitled fortuitous. As stated in the introduction, in hindsight those fisheries involving refuges where fishing is either impossible given present technology or presently uneconomical have survived relatively uncontrolled fishing effort. The gummy shark fishery off southern Australia described by J. Stevens appears to fit this category. A second type of fortuitous survival results from constraints of the harvesting technology in relation to the life history characteristics. J. Penn and co-authors considered the hypothesis that fisheries on burrowing prawns should be more sustainable than those fisheries on prawn species that cannot avoid the gear to the same degree. The concept is an interesting one, but the empirical observations were mixed.

A second category for prevention is the reduction in natural mortality due to control of predation. The single example provided was the use of chemical poisons to reduce the parasite phase of sea lamprey in the Great Lakes. The approach reduced predation by 90% and contributed to rebuilding of the Lake trout populations.

A third category is the prevention of fishing on juveniles. Aril Slotte argued that the elimination of the juvenile fishery on the Norwegian spring spawning fishery has been a key element in the apparent sustainability of the present fishery since the rebuilding of this stock.

J. Penn and co-authors described how controlling days-at-sea resulted in higher yields from several prawn stocks off Western Australia. These species, however, appear to be resilient to recruitment overfishing. This was the only example where direct control of days-at-sea was identified as a tool to prevent fisheries collapse.

Surprisingly, there were few papers presented in the Theme session addressing the degree to which annual quotas can prevent fisheries collapse. I presented evidence that the stricter implementation of quota management for cod on the western Scotian Shelf off Nova Scotia appears to be reducing fishing effort to the conservation target. However, this fishery is not one which had collapsed and it is not argued that quota management prevented a collapse during the late 1980s and early 1990s when fishing effort was very high. J.A. Koslow and co-authors described how quota management has resulted in a major resource decline for orange roughy off southeastern Australia. Similar conclusions were drawn for the role of quota management in the decline of snapper along the northeast North Island coast of New Zealand by M.P. Francis and co-authors, and for Atlantic herring off Nova Scotia by R. Stephenson. Thus, in principle, quota management is a tool to prevent collapse of fisheries, but most of the examples presented addressed the degree to which stocks had declined to low levels under this tactic. However, it may well be that the stocks would have declined to even lower levels under alternative tactics of effort control.

The explicit uses of refuges and closed areas (in contrast to the accidental or fortuitous refuges identified under the first category) is the sixth category of approaches to prevent fisheries collapse. J.C. Castilla provided an interesting example of the use of protected areas to enhance the recovery of the 'loco' fishery off Chile. The cooperatives are involved, and initial experiments are well received by these artisanal fishers. Within Theme 3, D. Pollard provided an overview of the success of harvest refugia as a tool for inshore fish population enhancement off eastern Australia. During discussions, there was considerable support for the use of this tool to achieve conservation objectives.

The seventh category of approaches involves restrictions on harvesting technology. E. Pikitch and co-authors described changes in fishing practices for Pacific halibut that should result in a reduction in juvenile bycatch mortality. J. Nance and E. Scott-Denton presented results of modification of shrimp trawls that reduce bycatch of juvenile red snapper substantially. N.J.F. Rawlinson and D.T. Brewer described the survival rate of diverse fish species escaping square mesh cod-ends during trawling.

An eighth category was recommended in a presentation on herring fisheries by R. Stephenson. He described the use of within-season cooperative management between industry and government in order to ensure that all spawning components are being sustained for a quota management regime within a management unit based on a stock complex.

The ninth category of approaches to prevent collapse is the improved use of recruitment forecasts in order to adjust effort levels and to reduce juvenile discarding when strong year-classes are entering the fishery. M. Francis and co-authors described the surprising accuracy in prediction of snapper recruitment off North Island, New Zealand, using a temperature index. The paper also suggested that the use of an ENSO–Southern Oscillation index may provide earlier predictions. B. Megrey and co-authors described a modelling approach to the prediction of Alaska Walleye pollock and the use of the prediction in management decisions.

The final issue addressed under the heading of prevention of collapse is the need in some fishery situations for improved bilateral or regional institutions for both assessment and management. A. Fonteneau made this point for some species of tropical tunas. M. Muchiri made a similar recommendation for the management of fisheries in east African lakes for which several countries share the drainage system.

In conclusion, with respect to the prevention of collapse of fisheries, the papers presented within Theme 1 infer that a package of tools is frequently required to effectively control fishing effort at levels consistent with conservation objectives. If quotas are the core regulatory tool chosen, it may prove useful to associate fleet sector and even individual quotas with an estimate of the number of sea days and the hours of gear usage that are required to catch that quota. With real-time monitoring of the effort and reported landings, discrepancies between the two can be identified in a timely manner. If the data are inconsistent, it is likely that there is either misreporting or that the quota is incorrectly set with respect to actual biomass. Quota management regimes need to monitor effective effort with equal timeliness and accuracy as they do for catch. Quota and days-at-sea can in some cases be complemented by spawning area closures when a management unit comprises several spawning components of uncertain relative abundance levels. Finally, gear restrictions that minimize bycatch mortalities and habitat alteration provide a fourth complementary tool.

Concluding points

During the Congress, several of the introductory speakers have stressed that capacity reduction is the number one priority for sustainable fisheries. While I do not quibble with that statement, it needs qualification. Capacity reduction does not by itself result in effort control at the target level required to meet the conservation objectives. Also, it is too early to conclude whether capacity reduction schemes enhance the conservation ethic of the remaining participants. Thus, parallel with capacity reduction, there has to be more effective implementation of the diverse effort control regimes that are available. Also, what can be done if fishing capacity is not reduced, which undoubtedly will be the case for many fisheries around the world for some time. Do we have to wait for capacity reduction to have occurred before we can have sustainable fisheries? It would appear from the papers presented in Theme 1 that, in situations where capacity cannot be reduced for whatever reasons, greater attention needs to be paid to the effort control package.

Hopefully we can learn some lessons from the fortuitous examples of sustainable overfishing. In the introduction to this summary, it was stated that extensive refuges and constraints to harvesting technology tend to be associated with fisheries that have survived in spite of poor or no management. In the terms of the Gordon-Schaefer bio-economic model, a fallback position under conditions of excess capacity which cannot be reduced, is to adopt management measures which raise the cost curve, such as extensive marine protected areas (or broodstock closed areas) and constraints to harvesting technology. The economists find this language distasteful and reminiscent of Schmacher's 'small is beautiful' approach to development. Such measures which effectively raise the cost curve reduce the potential rent that can be derived from a resource, but they may allow sustainable overexploitation rather than collapse. This may be all that can be achieved for certain fisheries for which capacity cannot be reduced. If fishing overcapacity is a necessary evil, it is essential to have robust tools for effort control in order to ensure the protection of reproductive capacity.

What are the roles of science, economics, sociology and politics in fisheries management?

Theme 2

What are the Roles of Science, Economics, Sociology and Politics in Fisheries Management?

B. K. Bowen

Associate - Western Australian Fisheries Department, St George's Terrace, Perth, WA 6000 Australia.

Summary

Fisheries management has historically focussed on the biological aspects, being concerned primarily with the protection and conservation of fish stocks. However, fisheries management has now progressed to another level of complication, requiring greater consideration of a variety of elements such as advances in fishing technology, stricter controls on catches, requirement for access rights, and the effects of fishing on the seabed and on biodiversity. This, in turn, has led to the need for a clearer definition of the role of the manager, and the roles of science, economics, sociology and politics in fisheries management. The directions are set through the political process, although the manager is the focal point and the driving force. However, a process has to be established to bring together the array of disciplines needed to provide the specialist advice. The specialists have an essential role in the provision of information and advice on technical aspects of the decision-making process, such as fisheries modelling, the need for environmental policies, the use of the precautionary principle, and maximizing economic efficiency in the exploitation of fisheries resources.

Introduction

The title given to me for my talk poses a question about the roles of science, economics, sociology and politics in fisheries management. The first thing that I need to explain is that I have no specific expertise in any of these disciplines. However, I have been involved in fisheries management for most of my working life. This has been with the Western Australian Fisheries Department, until 1991, but other associations with fisheries work in past years have included being Chairman of the National Fisheries Research and Development Council and a Director on the Board of the Australian Fisheries Management Authority. Currently, I am predominantly involved in environmental assessment in Western Australia. The thoughts expressed in this paper reflect my Australian experience.

In considering what I might say to the Congress, I am reminded of a discussion segment from Jonathan Lynn and Antony Jay's whimsical book *Yes Minister*. Sir Humphrey Appleby, as Head of the Department of Administrative Affairs, is having lunch with Sir Desmond Glazebrook, a senior banker of many years' standing, who is seeking to be appointed to a Government Authority on his retirement. Sir Humphrey, after listing the many and various Authorities available, including that of the White Fish Authority, asked Sir Desmond if he knew anything about their activities. The answer was in the negative, so Sir

Humphrey then asked 'what *did* he know about?' Sir Desmond explained that there was nothing that he knew — 'after all', he said, 'I'm a banker. It's not required'.

I felt much the same as Sir Desmond when I was asked to speak about the role of science, economics, sociology and politics in fisheries management. Does a fisheries manager actually need to know anything about these subjects or, like the banker, is it not required?

Fisheries management has changed dramatically over the last 50 years, and it is still changing at a rapid pace. Dawson (1980) reported that:

> Historically, fishery management has focused primarily on the biological aspects of the fisheries, concerned primarily with the protection and conservation of fish stocks. It is now clear that if fisheries are to make their full contribution to society, social, economic, political and environmental factors must be incorporated into the management process. Greater consideration of the people involved in or affected by the fishery — from boat-builders and fishermen to fish merchants and consumers — is necessary. Fishery management is necessarily concerned with the application of restrictive (regulatory) measures, but is not only concerned with these. Its scope is wider and includes a strong role in planning and executing fishery development.

Fisheries management in 1996 has progressed to another level of complication, at least in some parts of the world, requiring greater consideration of a variety of elements such as advances in technology, stricter controls on catches, requirement for access rights and increased concern about the effects of fishing on the seabed and on biodiversity. This, in turn, has led to a need for a clearer definition of the roles of science, economics, sociology and politics in fisheries management.

This theme talk explores some of those roles, commencing with politics, within the context of the changing scene. However, in doing so, I have found it necessary also to discuss the integrating process and the fisheries manager.

The politician

Irrespective of the form of government, there will be a person within the political process who is responsible for setting the policy directions for the fisheries management and formulating the administrative arrangements under which management will be conducted. In Australia, that person is the relevant Minister, who is part of the elected government. The Minister will, of course, receive an array of advice from government officers, members of the fishing industry and the public generally.

The starting point for sound fisheries management is the political will to recognize fisheries as an important element in the nation's economy, and to provide the philosophy, policy and structure within which the industry and the fishery specialists can work. The Minister responsible for Fisheries at the Commonwealth level, as well as those in the States (Provinces), have been very active in this regard during the past five years or so, and have established new arrangements for fisheries management. There have been a number of reasons for this activity:

a. Technological advances have resulted in higher fishing pressure, and for some fisheries this has resulted in their collapse.

b. There has been a nation-wide examination of the concept of ecologically sustainable development (ESD).

c. Community concern about the effects of fishing on the fish stocks, including the bycatch, and on the seabed has increased.

d. The fishing bycatch can include mammals and reptiles which are in need of special protection.

e. The governments have needed to clarify their role as custodians of the fish resources on behalf of the general community.

f. The cost of managing the exploitation of the fish stocks by the fishing industry has increased for a number of reasons, including the growth in the complexity of the regulations.

g. Economic efficiency has become more focussed as a factor in fisheries management.

h. The philosophy of continuous improvement in management includes a review process.

Fisheries Ministers throughout Australia have been very active in their commissioning of reviews and introducing new fisheries legislation. The review process commenced in 1989 when the then Commonwealth Minister, the Hon. John Kerin, presented a government policy statement entitled 'New Directions for Commonwealth Fisheries Management in the 1990s'. His opening words were:

> This policy statement presents a blueprint for the future management of those fisheries which are under the control of the Commonwealth. The document is therefore important because it explains why the Commonwealth Government is making certain changes and how things will be done in the future.

One of the changes made was the establishment of the Australian Fisheries Management Authority. The Chairman of that Authority, Mr Jim McColl, will be speaking at this Congress on its role, organizational structure and management arrangements.

Two other references to Ministerial policy-setting in recent time will suffice to demonstrate the very necessary role of government in setting the managerial style and philosophy for fisheries management. In February 1995, the then Tasmanian Minister for Fisheries, the Hon. Robin Gray (1995), wrote an article entitled 'The role of Government in fisheries management' setting out the reasons for the introduction of new legislation. He wrote:

> The role for governments, as custodians of the resource on behalf of the general community, is to ensure that fisheries resources are used in an ecologically sustainable manner and as efficiently as possible.
>
> Governments must also ensure a return to the community — the owners of the resource.

> In doing so, governments have the responsibility of ensuring that the basis for the sharing of the resource among all users is clearly understood and is acceptable as equitable and that the allocation of fisheries resources and their level of utilisation is consistent with the needs of both present and future generations.
>
> It is, therefore, vital that the legislative basis for fisheries management sets out clearly appropriate objectives and provides the necessary tools for achieving these objectives.

In Western Australian, the Minister for Fisheries, the Hon. Monty House, released a package of documents in June 1995 following a review which involved input from more than 1000 individuals, from the commercial, recreational and aquaculture sectors, and from government agencies with an interest in the long-term future of fish and fisheries. In December 1994, Mr House said:

> The review clearly identified new challenges in:
>
> - developing agency management structures and services which are able to meet the demands and challenges of an increasing complex fisheries sector;
> - defining an acceptable funding base to meet these demands;
> - dealing with increasing public interest in the aquatic environment and its conservation;
> - encouraging and assisting new industry development.

Mr House went on to say that it was the Government's role to:

> - protect the community resource for the maximum benefit and enjoyment of all;
> - enable maximum sustainable commercial gain from that resource; and
> - provide opportunity and encouragement for appropriate growth and development in the commercial and recreational sectors, aquaculture and other areas such as tourism and Aboriginal interests.

The political process is an essential element in the integration of sound fisheries management. It is through this process that there will be established:

a. The legislative framework within which the common property fisheries resource will be managed and administered, and including the objectives to be pursued.

b. The philosophical approach to resource sharing, and involvement by industry and the community generally in decision making.

c. The funding arrangements for management, with special emphasis on the nature of the costs to be paid by industry and those which will be provided by government from the public purse.

The objectives set down in the government legislation describe, in broad terms, the outcomes to be achieved. The Australian Fisheries Management Act 1991 provides one example, and includes (i) efficient and cost-effective fisheries management; (ii) exploitation of fisheries resources in an ecologically sustainable manner, and in particular having regard to the impact of fishing on non-target species and on the marine environment; (iii) maximizing economic efficiency in the exploitation of fisheries resources; and (iv) achieving government targets in relation to the recovery of costs from the fishing industry.

These objectives clearly establish the framework within which fisheries management has to operate. However, the outcomes expected to be achieved, in accord with those objectives, also provide a guide to the nature of the technical expertise required. The fishing industry also needs to be involved, and its advice combined with the expertise of the specialists through an integrating process.

THE INTEGRATING PROCESS

Fisheries management, in all of its complexities, has developed as a result of there being a commercial fishing industry, together with the ever-expanding number of recreational fishermen. In many countries, the fishing industries have been encouraged by their governments to establish efficient fleets to explore and exploit the fishing grounds, both within their domestic waters as well as in international waters. Fishing has been recognized as an important part of their country's primary industries, producing seafood products for a variety of markets.

Recreational fishing is equally important and, for a growing number of fisheries, the pressure exerted by this segment is greater than that exerted by the commercial sector. The thoughts expressed in this Theme talk relate mostly to the commercial sector, but they are equally relevant to the recreational activities where the process of management through corporate decision making can be achieved.

The fishing industry world-wide has been successful in its endeavours to exploit an array of fish stocks, but technological advances have resulted in many of these stocks being overfished.

From my experience, there is now an increasing desire by the industry to operate at a level of exploitation which makes best use of the resource in terms of sustainability, financial returns per unit of fishing, and returns to the industry generally. The imperative for this approach is, of course, now embodied within the fisheries objectives set down in the Acts of Parliament. However, the objectives are achieved more readily if the industry and the managing authority are in agreement about the outcomes being sought, and this is being achieved through an integrating process of management advisory committees.

In Australia, these Management Advisory Committees, or MACs as they have been termed, have been established by legislation to bring the fishing industry into a formal process of discussion and decision making. Perhaps the earliest MAC formed in Australia was the Western Australian Rock Lobster Industry Advisory Committee, which was in general use in the 1950s and then prescribed in legislation in 1966. The MACs typically have a majority of their members selected from the industry, together with managers and scientists, but they can also include others from economics, sociology, recreational fishing and the broad environmental movement. Nowadays, the chairman is selected to be independent of the specific areas of interest represented on the MAC.

The process of MAC discussions, and the requirement to prepare public management plans, provides a substantial focus for a consideration of the roles of the members of those MACs in relation to the information needed for fisheries management. Here, I am making the assumption that there is a commitment to continuous improvement in the quality of management advice. This also assumes that each member of a MAC, and the instrumentalities and associations supporting those members, are committed to the same philosophy of continuous improvement in the outcomes to be achieved. Incorporated in this philosophy is the discipline of setting objectives, performance criteria and strategies, and establishing a system to audit the performance. The establishment of the MAC system has highlighted the importance of that group of people broadly referred to as fisheries managers.

The fisheries manager

The fisheries manager is the focal point and the driving force of successful fisheries management. The reporting mechanisms can range from the traditional Departmental structure under Ministerial control to that of a Commission or Authority whereby most of the decisions are taken by an expert group specifically selected by the government to undertake that task. Whatever the arrangements, the overall function of the professional fisheries manager will remain the same, i.e. to develop management proposals and oversee the implementation of sound management in relation to the specific fishery or fisheries allocated to his or her responsibility.

The elements of management and the method of its implementation will be varied and complex, depending on the type of fishery and the nature of the external influences, but the principles of establishing specific objectives and giving attention to the process will be common to all. The formal discipline in which the manager is trained is not the most important factor. What is important is that the manager is able to drive the process, has a sound understanding of the fishery and understands the specialist advice which is necessary and available.

There is a temptation, at times, for the manager to take on the role of the specialist, and attempt to provide advice at a level of detail which more properly belongs to that specialist. Whilst, under some circumstances, the manager may be able to operate at the specialist level, in general, it is a dangerous path to follow. Fisheries management will be most successful when there is a clear process for decision making and a mutual understanding of all of the participants, each operating at his or her professional level to produce an integrated best-practice outcome. Management advisory committees play an essential role in this process.

The roles of the scientist and economist in fisheries management need to be considered within the context of the work of the MACs because it is through this process that successful fisheries management is achieved. Of course, the specialists have a much wider brief within their respective disciplines, and their input to the MACs is only one element.

I am conscious that the title given to me for this talk also included sociology, and that the social aspects of resource sharing, both within the commercial and recreational sectors and between those sectors, is very important, as is the effect of management decisions on isolated communities or communities having specific requirements. These matters have usually been considered within the context of the political decisions or as part of a socio-economic analysis rather than through the professional input of advice from the social scientist. However, the role of social scientists is increasing, not only in the area of resource-use conflict but also in the non-market evaluation of environmental and conservation values.

The scientist

The framework within which the fisheries scientist operates is established by the objectives of the employing research institute, and these can be wide ranging. The Fisheries Division of the Australian Commonwealth Scientific and Industrial Rresearch Organisation (CSIRO) has set out its objective as:

> To develop a sound scientific basis for the use and conservation of Australia's marine living resources and environment; to provide scientific advice to environmental, industry and resource managers to ensure the economic sustainability of these resources and enhance the competitiveness of dependent industries.

The strategies established to achieve this objective include the undertaking of research in support of the fishing industries, and communicating and applying the results of the research. Much of the communication and application of that research is undertaken through the process of the management advisory committees.

The requirement for information about a resource being fished can be defined in simple terms, although the gathering of that information is often difficult, time-consuming and costly. The four overriding requirements are:

a. The resource must be maintained at or above a level such that the average recruitment does not fall as a result of the reduced abundance brought about by fishing.

b. The effect of environmental variability on the level of recruitment needs to be understood and considered in the context of resource abundance.

c. The size at which fish are allowed to be caught needs to take into account the combination of growth and mortality.

d. The effect of the fishing gear on the abundance of both the target and the non-target species, and also on the seabed, needs to be understood.

Each fisheries scientist will contribute data to the management process in accordance with his or her particular discipline. This is essential for fisheries management to progress. However, the success of the management process depends not only on the quality of the data provided by the scientists but also on the corporate relationship developed between the scientist, the manager and the fishing industry. In Australia, this has been achieved through the structure of the MAC. Some of the elements of fisheries management where the scientist has an essential role, and which is also important to the success of the corporate process, are discussed below. Each element is related one with the other, but can be separated for the purpose of the discussion.

Information provided by the industry

One of the principal sources of data needed by the scientist is the fishermen's records of the catch and the effort expended in taking that catch. However, fishermen spend most of their working time at sea, and they can be a very valuable source of additional data if time is taken to explore that avenue. Most fishermen are very keen observers, and those observations can form the basis of a hypothesis to be tested. From my experience, there are at least three elements of stock assessment about which most fishermen have a very clear general understanding. These are:

a. Breeding animals need to be adequately protected.

b. The number of small animals entering the fishery changes from year to year, and information of a qualitative nature on this change can often be obtained from discussions with fishermen.

c. The weather and sea conditions change, and the fishermen develop a good understanding of the part these elements play in the catchability of the target species, and thus in the catch-per-unit-effort.

A research programme can be enhanced considerably if the scientist spends time at sea with fishermen exploring the knowledge gained by the industry. This not only allows the scientist to understand better the information being provided by the industry, including catch and effort data, but also allows the fishermen to understand and have confidence in the data requirements of the scientist. In addition, the interaction between scientists and the industry is essential to the process of industry understanding the possible relationships between fishing effort, breeding stock abundance, recruitment success and the environmental influences.

Fisheries management will be successful if all parties to the advisory and decision-making process have confidence that the estimates of abundance of the resource being fished have been based on the best information available. Problems will occur if the scientist presents an estimate of abundance which is at variance with the general observations of the industry. That, of course, is not to say that the scientist is wrong, but perceptions play a major role in the process of fisheries management. If there are differences between the estimates provided by the scientist and the observations of the industry, there is a need for the scientists to discuss with industry members the basis of their observations so that they can understand the abundance estimates within the context of those observations.

Stock assessment

Each stock assessment report to the management advisory committees, and to the fishing industry generally, is subject to intense questioning, and often there are requests for peer reviews. The scientist has an essential role in presenting the stock assessment reports to the MACs and guiding the discussions in relation to their management implications. However, there is still some way to go in the development of the process linking the work of the scientists in the research institutes, the preparation of the stock assessment reports, including their content, and the functions of the management advisory committees.

Fisheries modelling

Models used to assist in the fisheries management decision making can be of two types. Allen (1993) wrote: 'A broad distinction can be made between use of models for estimation, in which estimated values of parameters or variables are obtained by analytical or numerical solution of implicit equations, and simulation, in which the behaviour of a variable, such as a fish population, is explored in its response to changes in external factors. Both models are widely used in fisheries studies.'

The use of models has been essential in the development of the array of projects forming a research programme, and to bring into management focus the results from those projects. Discussions of models have mostly been undertaken within the research laboratories. However, in recent times, when the effects of heavy fishing pressure have become more evident, and the objectives of fisheries management have become clearer in terms of maintaining the resource through the introduction of catch limits or the rigorous introduction of gear controls, the estimation and simulation models have been increasingly presented at management discussions. The basic theory of the models has not changed greatly since the 1960s, but computers have provided the power for an array of estimates to be calculated quickly and for simulation models to be used to provide estimates of future catches under a variety of management strategies.

The use of models by scientists in fisheries management will increase, but the scientists have a real challenge to bring to the fisheries management discussions the added degree of clarity and integration of thought which can be provided through the use of models, but at the same time ensure that the limitations of the data are recognized and explained. As described by Allen (1993), 'the greatest problem which will face fisheries modellers will be to obtain the ever increasing amounts and kinds of data which they will need to run their models in the real world. Progress will thus be data limited.'

The precautionary principle

The Rio Declaration has set down a number of principles, and one of these is the precautionary approach. Ray Hilborn (1996) provided a graphic description in relation to the application of the precautionary approach to fisheries management. 'This ship (the precautionary approach) is fuelled by high-powered public sentiment that arose from the publicity surrounding a series of widely publicised fish stock collapses. The precautionary approach won't go away, and you really have two options: turn sharply, get out of the way and be left half-swamped bobbing in the wake, or get aboard, go up to the wheelhouse, put your hand on the wheel and help direct its course.' It is, of course, the latter advice which will win the day, but the scientist has a vital role to play in assisting fisheries management in defining precautionary fishing.

There is a tendency for some people, perhaps even scientists, to pursue a particular course of action, in relation to proposed fishing activities, and cite the precautionary principle as the basis for their argument. The debate on its use requires much more than the mere introduction of the words, and this has been achieved through the recent publication of the FAO Fisheries Technical Papers on the precautionary approach to fisheries, parts 1 and 2.

The precautionary approach has two essential elements: the probability of an event happening and the consequences if the event occurs. This moves the debate into an understanding of risk analysis and risk assessment so that judgements can be made about the most appropriate action. Here, I am defining risk analysis as the calculation of risk, and risk assessment as the overall process which includes risk analysis, risk management, risk perception and risk communication.

Fisheries management is about making judgements within the bounds defined by the objectives established. The scientist has an essential role to play in this process by providing information to the system of fisheries management on the broad subject of risk assessment as well as the more technical aspects of probabilities and consequences. It is important that each management advisory committee discusses and decides upon the meaning of precautionary fishing in relation to the fishery of its interest. As stated by Ray Hilborn (1996), 'Fishing industries worldwide need to get involved in the process of defining precautionary fishing, because if you don't define it, someone else will'.

The development of environmental policies

Fishing has an impact not only on the target species but also on other species taken in the fishing gear. Additionally, the fishing gear can affect the seabed. These impacts have resulted in community criticism of some aspects of the activities of the fishing industry. Critical newspaper articles appear, approaches are made for action to be taken at the political level, and applications are lodged under endangered species legislation. A recent example in Australia, which is of international importance, is the application to have prawn trawling in the northern waters of Australia listed, under the Commonwealth Endangered Species Protection Act 1992, as a key threatening process to turtles and two species of fish.

This concern expressed by segments of the community about the impacts of fishing is relatively new in terms of the time allocated to the subject on the agendas of meetings on fisheries management. This should not viewed as a problem, but rather as a challenge to the fisheries managers to pursue further their nation's responsibilities under the broad requirements of the Rio Declaration On Environment and Development and within the concept of Ecologically Sustainable Development.

Responsible fisheries management incorporates sound principles which relate not only to the sustainable use of the target species, but also to the other elements of the environment. This is not to say that fishing will not have some impact on the environment. To suggest that trawling will not affect the seabed in some manner is akin to suggesting that farming will not change the nature of the farmland selected and approved for that purpose. If trawling is approved through the proper process of government, the community must accept that some changes will result.

Fisheries management has a responsibility to address the broad concept of sustainability through the development of environmental policies. Only by this means will the fishing industry be in a position to place before the community a statement which ensures the long-term future of wild-stock fisheries.

This leads me to a brief reference to quality assurance certification. Many readers will be familiar with the ISO 9000 series of documents about business quality management produced by the International Standards Organisation (ISO). In Australia, there is usually a requirement that a business seeking a government contract has to have certification to the ISO 9000 standard of quality assurance. There is now an international draft ISO 14000 series of documents on environmental management systems. These documents provide guidelines for continuous improvement in environmental management. There is an increasing number of businesses and government agencies operating within the management principles of the ISO 9000 series, and so too will the ISO 14000 principles of environmental management become important for all manner of organizations, including the fishing industry.

The role of the scientist is pivotal to the development of environmental policies by those responsible for fisheries management so that there will be continuous improvement in environmental understanding and action.

The economist

The economist plays a major role in fisheries management through the provision of information and advice on an array of topics, including cost benefit analysis, economic rent, economic performance, fleet adjustment schemes, and product values under different management strategies. Within Australia, this work has been undertaken under the broad objective of the national economic research organization ABARE (Australian Bureau of Agricultural and Resource Economics), whose role is to enhance the economic performance of the fishing industry and increase efficiency in the use of marine resources, through economic research and policy analysis, market and resource assessment, and surveys and communication. However, at least within Australia, the role of economics in fisheries management has not been as clearly identified as it has for fisheries science, perhaps because each operator in the industry has his or her own idea about the most appropriate strategy for their individual economic success.

Economists will continue to have a significant role in the provision of information on aspects of the economic performance of the fishing fleets. However, the objective of 'maximising the economic efficiency in the exploitation of fisheries resources', as set out in the Australian Fisheries Management Act, presents a challenge to the economists to provide information to assist the discussions about proposed adjustments to reduce the number of boats in the fishing fleets. There have been a number of adjustment schemes implemented, such as those for the prawn fishery in the Gulf of Carpentaria, and the economic considerations have been essential to the discussions leading to the adjustment decisions.

One of the most recent analyses of fleet size has been that undertaken by Lindner (1994) in relation to the rock lobster fishery off Western Australia. In the 1970s there were 836 boats in this fishery. However, this number has been reduced to 638 as at 30 June 1994 through a process which permitted the trading of rock lobster pots. For a number of social reasons, the management committee proposed that the fleet size should not be permitted to be reduced below 600 boats. The paper by Lindner demonstrated, *inter alia*, that a decision on the minimum fleet size had to be a trade-off between economic efficiency and social and other benefits which may be perceived to operate, such as regional development or employment opportunities. Lindner showed that, in terms of economic efficiency, the introduction of a minimum fleet size rule would reduce the total industry profits by $A15 million to $A20 million per annum.

CONCLUDING COMMENTS

By way of providing some concluding comments, I have decided to do two things: firstly, to provide a brief example of the fisheries management integrating process by reference to the Western Australian rock lobster fishery; and secondly, to summarize the elements necessary for successful fisheries management.

The rock lobster fishery

In 1992, the Minister for Fisheries, as well as a number of people engaged in the Western Australian fishing industry, had sought information on the benefits, or otherwise, of moving the management of the rock lobster fishery from an input-control limited entry system, based on pot numbers, which had been extant since 1963, to an output TAC/ITQ method of management. The matter was referred to the relevant MAC, the Rock Lobster Industry Advisory Committee (RLIAC), which is comprised principally of industry members from the rock lobster fishing and processing sectors, but also with members from the Fisheries Department and a member with recreational fishing interests.

The RLIAC appointed a steering committee which decided upon an examination framework and arranged for the preparation of specialist papers on economics, marketing and law enforcement, as well as an overview paper to integrate the specialist information with that already available on the population status of the rock lobster and the management implications of moving from a relatively sophisticated limited entry input system to one using TACs and ITQs.

Each of the authors of the papers collaborated on an ongoing basis with the steering committee and with each other.

The papers have formed the basis for wide-ranging industry discussions and advice about the future management directions for the rock lobster fishery, which is the most valuable fishery in Australia. For decades the industry had caught rock lobster under the rules of a minimum legal size of 76 mm carapace length and a season from 15 November to the following 30 June. The papers challenged the managers and industry generally to think about the advantages and disadvantages of changing both the times at which rock lobsters were caught and the minimum legal length. They also, for the first time, had information on the economic performance of the fishing sector as a whole, and this added a degree of economic focus to the discussions which had not hitherto been available.

The input-control system of management has remained in place, but the discussions have resulted in a better understanding of the important management issues and the fine-tuning of the management package.

Summary of the elements necessary for successful fisheries management

The following points, in summary form, have been developed from my experience in fisheries management and my work with specialists, but I believe that the summary will apply to fisheries management generally.

a. Provision of appropriate legislation which sets the policy directions and establishes the broad objectives.

b. Recognition of the fishing industry as part of corporate management, and definition of the role of industry in providing management advice.

c. Establishment of an appropriate administrative structure for fisheries management and determination of the arrangements for the provision of operational support, such as law enforcement, and specialist information from research institutions.

d. Establishment of appropriate arrangements for an integrating process, such as the appointment of management advisory committees, to bring together the relevant expertise for discussions and the preparation of advice to the decision-making authority.

e. Commitment by all participants to continuous improvement in the fisheries management process, including regular performance reviews.

f. Ongoing examination, by all participants involved in fisheries management, of their roles with a view to continuous improvement in their contribution to the corporate decision-making process.

g. Ongoing examination of the agenda required for sound fisheries management, including the concepts of ecologically sustainable development and environmental policy.

h. The preparation and implementation of fisheries management plans formulated in accordance with the elements summarized in (a) to (g) above.

REFERENCES

Allen, K R (1993). A personal retrospect of the history of fisheries modelling. In 'Population dynamics for fisheries management'. (Ed D A Hancock) pp. 21–28. Australian Society for Fish Biology Workshop Proceedings, Perth 24–25 August 1993.

Dawson, C L (1980). Glossary of terms and concepts used in fishery management. ACMRR Working Party on the scientific basis of determining management measures. *FAO Fisheries Report* **236**, 115–30.

FAO (1995). Precautionary approach to fisheries. Parts 1 and 2. FAO Technical Paper 350/1 and 350/2.

Gray, R (1995). The role of Government in fisheries management. In 'Fishing Today 8(1)'. (Ed T Walker) pp. 13–14 (Turtle Press: Hobart Australia).

Hilborn, R (1996). The implications of a precautionary approach for commercial fisheries. In 'Fishing Today 9(1)'. (Ed T Walker) pp. 24–25 (Turtle Press: Hobart Australia)

Lindner, B (1994). Long term management strategies for the western rock lobster fishery. Vol. 2. Economic efficiency of alternative input and output based management systems. Fisheries Department of Western Australia, Fisheries Management Paper No. 68.

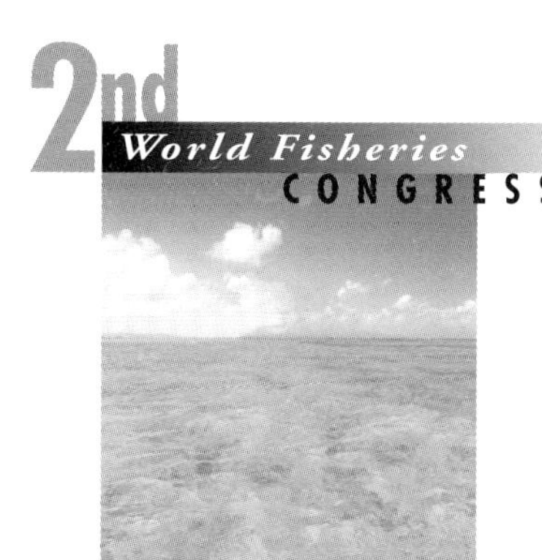

Fisheries Management Science: Integrating the Roles of Science, Economics, Sociology and Politics in Effective Fisheries Management

Daniel E. Lane[A] *and Robert L. Stephenson*[B]

[A] Faculty of Administration, University of Ottawa, Ottawa, Ont. Canada K1N 6N5.
[B] Department of Fisheries and Oceans, Biological Station, St. Andrews, N.B. Canada E0G 2X0.

Summary

Recent fishery failures, combined with changing views on government regulation, point to the critical and urgent need for a new approach in fisheries management. Future management must focus on integrating biological, economic, social and political considerations, rather than solely on the state of fish populations. We have proposed integration of the traditional fields of fisheries science and fisheries management, with the scientific problem-solving approaches of management science, to form 'Fisheries Management Science'. Fisheries Management Science provides a framework for structured decision making employing the techniques of operations research/management science. A case study, based on the Bay of Fundy herring fishery, moves toward the conceptual and practical aspects of Fisheries Management Science — including development of a co-management structure characterized by industry involvement, interdisciplinary support teams, modelling of spatial/temporal elements of the fishery for in-season management, explicit consideration of risk, and decision-performance monitoring.

Introduction

The problems associated with managing fisheries have been well-documented in the literature (Wooster 1988; Hilborn and Walters 1992; Ludwig *et al.* 1993; Smith *et al.* 1993). Difficulties in effective decision making in fisheries arise from the inherent variability of marine ecosystems, the unobservable nature of aspects of the natural dynamics of fish stocks, multiple and conflicting objectives, and management bureaucracies that are not responsive to required rapid change. There are few examples of integrated management approaches to problem-solving in fisheries.

Stephenson and Lane (1995) presented a critique of the current state of fisheries science and fisheries management and noted the lack of an appropriate context for the management of commercial fisheries exploiting marine fish stocks. They proposed a framework for developing strategic management alternatives and for evaluating these relative to scientific, economic, sociological, and political considerations using the structured techniques of decision analysis from the field of management science. The term Fisheries Management Science (FMS) was coined to denote the interdisciplinary roles of fisheries management, fisheries science, and management science in dealing with fisheries problems.

This paper elaborates on the structural and operational components of FMS and presents the experience to date of applying the paradigms of the FMS framework to the commercial herring fishery in the Bay of Fundy and the Scotian Shelf area of Canada's Atlantic coast (NAFO divisions 4WX). This case study reveals that greater involvement of participants in the management process, and enhanced responsibility in decision making, provides the means by which interdisciplinary considerations may be integrated into fisheries policy. This progress toward effective 'co-management' is improved through the direct involvement of fisheries managers, fisheries policy analysts, and fisheries scientists in the spatial and temporal activities of the fishery while acting as an interdisciplinary support team (rather than as the 'regulator') toward achieving mutually acceptable goals for the fishery.

The following section outlines the dynamics and requirements for the development of a context-based FMS system for a commercial fishery. This discussion identifies the steps and tools essential to the evolution of an appropriate interdisciplinary management programme. In addition, the herring case study and the structures and dynamics that are evolving within that system toward successful realization of FMS, are presented.

Fisheries management decision making

Management systems may be described by (1) the strategic plan of the organization, (2) the participants and how they are organized, (3) the management process that identifies how and by whom decisions are taken, and (4) the tools required to assist in the decision-making process.

Strategic plan

The mission statement of the management organization reflects the strategic plan of the fishery. It articulates the business of the fishery, its operating environment, its clients, expectations, competitive advantage, and scope of operations and provides the ultimate context and guidelines for management decision making at the strategic and operational levels. The mission statement defines the multiple objectives of the fishery (including providing an operational definition of stock conservation, and the desirable level of commercial activity and socio-economic viability), the relative importance of each objective and how they are to be realized in the management action plan.

Fisheries have generally been negligent in developing and applying strategic planning statements for their management system (MacKenzie 1974). Where such statements exist, they often prescribe non-specific doctrine such as 'commitments to stock conservation', 'economic viability of the industry' and 'a strong seasonal employment sector'. Consequently, management of most fisheries is without a real operational context to guide decision making.

Organization

To deal with the complexities of fisheries problems, the management organization must be flexible and balanced to consider the wide range of policy impacts on its participants. Fisheries systems have participants that include the fishing industry, government central agency representatives, and other community stakeholders. The diverse constituencies of these groups must contribute to, and be represented in, policy setting at the strategic (longer term) as well as at the operational (intra-seasonal) level. Finally, the organization must be empowered to make decisions on behalf of its constituents (Jentoft 1989).

Management process

The management process is the means by which the legitimate organization is empowered to deal with the problems of the fishery. The process must treat complex problems in a structured way by detailing (i) the definition of the problem, decision alternatives, specific objectives, and constraining factors, (ii) the generation of alternatives for problem resolution, (iii) the evaluation of the potential effectiveness of alternative decisions in relation to the stated mission of the organization, and (iv) tracking and feedback response to the actual *versus* anticipated impacts of the implemented decisions (Lane and Stephenson 1995).

Fisheries management problems have multiple and conflicting objectives: (a) economic objectives — measuring the performance of commercial fishery sectors toward achieving a minimum desirable level of return; (b) social objectives — measuring the public benefits derived from ownership and consumption of the resource including maintaining target levels of seasonal employment in fish harvesting and processing; (c) biological objectives — measuring stock attributes (e.g. a desirable age composition) over and above the establishment of minimum abundance criteria. All objectives of fisheries problems are measurable valuations used directly to compare the expected performance of alternative policy options. Expectations of these measures provide the means of ranking the potential effectiveness of alternative strategies for decision making.

Tools for decision making

The management process requires tools to assist in the development and evaluation of decision alternatives. The most useful tools take the form of appropriately conceived quantitative models of the fishery system used for anticipating, projecting, and estimating multiple measures of performance. Analytical models must also take into account the links among biological, economic, and social considerations as well as system uncertainty and errors in observation. Making decisions with knowledge of the range of output possibilities (risk assessment) will provide the basis for risk management. Finally, monitoring and the ongoing accountability of past decisions, based on desirable performance over the planning period, determine the effectiveness of the decision-making strategy. Such ongoing monitoring of decision performance, and continuous improvement over time are the ideas behind 'total quality management' and 'management by objectives'. When actual observations vary significantly from expected results, this 'signal' causes an adjustment in decision making and operational strategy to take effect. Feedback from systems which are out of control are not behaving as expected. This feedback initiates continuous adjustment in an effective decision process.

It is envisaged that a management system grounded in FMS would involve the establishment of a well-defined strategic plan, a representative and multi-disciplinary organization empowered by a management process that includes all participants in a decision-making process, supported by tools to measure the expected performance and feedback of policy options for the fishery. The move toward making FMS operational requires a change in long-established approaches in existing fisheries. The question of how we evolve from *status quo* to FMS is dependent on the starting point of the *status quo* position. In the following section, the experience of the Scotia-Fundy commercial herring fishery in moving toward a new management framework based on FMS is presented as a case study.

THE SCOTIA-FUNDY HERRING FISHERY

Fishery background and management context

The 4WX herring fishery is the largest herring fishery in the western Atlantic, with annual landings in the order of 100 000 t. The commercial fishery involves a variety of gear types including fixed gears (weirs, shutoffs, and gill-nets), and a dominant mobile gear sector fleet of approximately 35 purse-seine vessels that take over 90% of the annual catch. The commercial fishery has survived major changes in market emphasis and demand, and it has been dominated at different times by sardine, fish meal, fillet and roe markets.

This herring fishery has been at the forefront of innovative fisheries management (Stephenson *et al.* 1993). It was one of the first fisheries to be managed under limited entry (since 1970), was among the first to come under nationally allocated Total Allowable Catches (TACs) (1972), and was the first modern commercial marine fishery to implement individual vessel quotas shares (1976). There is a history of involvement of the fishing industry in the management process through the Atlantic Herring Management Committee (AHMC, established 1972) and, its successor, the Scotia-Fundy Herring Advisory Committee (SFHAC, since 1981). Management planning has been by annual management plans developed by the federal government Department of Fisheries and Oceans (DFO) in collaboration with the SFHAC, but a general continuity on elements such as gear sector sub-allocation, and Individual Transferable Quota (ITQ) transfers has been imposed by a longer term '10-year' plan established in 1983 and rolled over in recent years (after the end of the original 10-year plan in 1993). Although the industry has been discussing the form of long-term planning to replace the 1983 plan for several years, progress has been slow. In 1994, a major change in stock status prompted development of an in-season management system.

In-season management of 1995

Problems in the fishery in 1994 (lack of market-sized large fish, poor physiological condition, and unusual fish behaviour) led to a postponement of the 1995 management plan pending a revised biological evaluation. The updated biological assessment (spring 1995) summarized the problems in the 1994 fishery and documented low larval survey abundance from 1994 spawning. The stock assessment indicated that the spawning stock had declined from about 600 000 t in the late 1980s to perhaps as low as 200 000 t. Consequently, the Department of Fisheries and Oceans of Canada insisted on a cautious approach in the management of the 4WX stock complex and set a reduced TAC for 1994–1995 at 80 000 t (reduced from 150 000 t). Moreover, this TAC was set only on an interim basis to be reviewed throughout the course of the fishery.

The resulting 4WX herring management plan stressed the importance of monitoring progress and signals in the 1995 summer fishery, particularly related to the spawning grounds, and required an in-season re-evaluation of the fishery. To meet these requirements, the Scotia-Fundy herring purse seine monitoring working group (MWG), a sub-committee of the SFHAC, was established to evaluate information from the fishery on an ongoing basis. The committee was comprised of representatives from industry (the purse-seine fleet, and the processing sector), and the federal government (fisheries operations/management, and fisheries scientists).

The importance of timely and effective decisions to be made by the MWG during this season necessitated new information and structured approaches to dealing with it. The information included (i) joint industry and DFO monitoring of stock size in fishing areas; (ii) rapid compilation of data for dissemination to the MWG; (iii) analysis and use of the data in a form appropriate for consensus decision making. The MWG was provided with the mandate and empowered to make decisions on areal fishing limits for the remainder of the summer purse-seine fishery. The committee met routinely in person or by conference call to review new information and to decide on a course of action. Considerable progress was made on obtaining appropriate information on which to base decisions. Information collected from the summer fishery included:

- *Statistics* — records of all vessels' searching activity and catch locations were available on a daily basis and were summarized and plotted weekly.
- *Sampling* — very thorough coverage of all aspects of the summer fishery for size and biological characteristics resulted from the increased presence of biologists on the fishing grounds, observers, and from sampling done by members of the industry (vessels and plants). Length-frequency observations by fishing ground and week were made available and discussed while the fishery was in progress.
- *Surveys* — a series of stock surveys was undertaken of major spawning areas using commercial vessels. Sonars and sounders were used to document the number, location and approximate size of herring schools. In most successful surveys, special vessels worked together to provide rigorous coverage of the target areas.

Since the primary concern was for the status of the stock (especially abundance of fish on individual spawning grounds), biological observations formed much of the in-season information brought to the MWG for discussion. It was essential to the MWG to develop criteria against which these observations could be compared. To this end, tables of expected biological feedback measures (size and age composition, relative local abundance, location and distribution, fat and feed content) including an expected range of values for each were constructed. This 'checklist' was used as the basis for decision making from the biological observations on individual spawning components and discrete fisheries. It took considerable effort on the part of the MWG participants, involving weekly conference calls and meetings which became more frequent during the fishing period. Increased sampling and more rapid summary and dissemination of information were also required to enable estimates of stock availability prior to fishing.

Negative observations or a high degree of uncertainty led to the imposition of further restrictions on the industry by the MWG in the form of closures of individual spawning grounds. In addition to the biological checklist, the MWG found it natural to include multi-disciplinary aspects (equitable distribution among the fleet, monitoring and enforcement, prices, markets) in the discussions. The recommended solutions often took these aspects into account implicitly.

Extension of the in-season MWG management process into the subsequent winter fishery was complicated by controversy arising from stakeholders external to the decision-making committee. In this instance, there was concern on the part of other stakeholders that access of purse seiners to particular local concentrations of herring stocks should not be permitted. There was fundamental disagreement regarding the MWG's authority to manage fish in this area. After a successful lobby of local and regional politicians by the stakeholders, access was restricted contrary to the wishes of the industry members of the MWG. This event has since jeopardized the coordinated workings of the MWG.

Future directions

The experience of the Scotia-Fundy herring fishery with in-season management in 1995 represents a major step toward realizing the integrated approach of FMS. The empowered MWG, comprised of the main participants in the fishery system and the consensus-based decision process using real-time quantitative measures and tools as decision aids, marks a radical change away from disciplinary, hierarchical and government-controlled management, toward the inclusion of inter-disciplinary aspects in decision making.

The following analyses the in-season management experience of the MWG and addresses its needs in the evolution toward decision making through FMS.

Need for a strategic plan

The MWG was conceived in an atmosphere of crisis surrounding the perceived declining status of herring stock abundance and the potential for overexploitation by the dominant purse-seine fleet. As such, the immediate concern and direction provided to the MWG was that of monitoring and reacting appropriately to signals about the biological status of the stock. The short-term urgency of this situation precluded further discussion on other longer-term issues and considerations. Although decisions taken by the MWG were nearly always rationalized on the basis of industry economic viability and the equitable distribution of fishing rights, these issues were only considered implicitly in the decision-making process.

The specification of a strategic plan would entrench explicitly what has already been accounted for in decision making. Moreover, the formal statement of these considerations in a mission statement would provide future committees with guidelines on how to deal with differing problems not simply related to urgent biological concerns. As such, the organization would be obligated to record their complete justification of actions as a means of sustaining the process. It is recommended that a strategic planning exercise take place in this fishery in preparation for upcoming management committees and the issues to follow. The rigour of this process would help the organization to focus on how it will make future decisions and would serve as documentation to future decision makers.

Organization

The four main groups active in the management of the herring fishery include (i) the independent fishermen's associations representing the different gear sectors (weirs, gill-nets, and seiners) and geographically defined local groups; (ii) active shore-based processing firms specializing in herring; (iii) the governments of Canada through the Department of Fisheries and Oceans and the provincial governments of Nova Scotia and New Brunswick; and (iv) the diverse group of community members, fishermen in other fisheries, special interest and lobby groups who are impacted by the activities of the Scotia-Fundy herring fishery. The MWG was comprised of representatives from a portion of the fishing industry (harvesting and processing) as well as government officials. The industry representatives were all individuals who were empowered by their respective associations to act on their behalf. However, they represented only the dominant purse-seine gear sector of the harvesting sector at the exclusion of fixed-gear fishermen. There were no members of the MWG to specifically represent the wider 'stakeholders' community.

The exclusion of the latter group diminished the scope and decision-making ability of the MWG as was evidenced by the protests of the lobby against the purse seiners' access to some winter 1995 herring aggregations. The problem of effective representation of this diverse stakeholders' group notwithstanding, it would be incumbent on future MWG organizations to establish more formal information links to its peripheral community. More importantly, the exclusion of fixed gear representatives from herring decision-making bodies risks creating further animosity between these different harvesting groups. Inasmuch as the fixed-gear groups have parallel associations to the seiners, representatives from these groups should be included in the direct management process to the extent that they will be affected by decisions taken in the fishery.

Management process

The key development in the management process has been its 'scale' of application. Historically, the management decisions were made based on annual stock assessments as inputs to twice-yearly meetings which fixed the annual management plan. In-season interventions were reserved to enforce occasional variance orders and to monitor the aggregate exploitation of the fishery. In contrast, the in-season decision-making process tracks vessel activity from fishing area to fishing area. This in-season perspective not only provided the means needed to manage exploitation of the individual spawning grounds, but also required a direct, ongoing communications link among all the participants. Consequently, decision making moved from an annual standardized procedure, to ongoing multiple and directed decision interventions in consultation with participants. The industry's link to the ongoing decision process lies in its acquisition of real-time data and the use of industry observation. When data are required at this level, their reliability and usefulness are clearly enhanced.

Initially, the roles and responsibilities of the participants in the in-season management process were discussed and debated. They have since become more comprehensive, but as yet not fully structured or empowered by legislation to consider the multi-objective issues of the fishery. Nevertheless, a better understanding and appreciation has been realized for the contributions of each participant group toward accomplishing the organization's mutual goals. Harvesters and processors have become more responsible for providing timely and accurate information from the observations of their own constituents, for maintaining the confidence of their membership in the process, and for developing and supporting the consensus approach to decision making. Government researchers have become more committed to providing practical, relevant information directly complementary to in-season decision making. Government administrators have become more responsible for the orderly prosecution and implementation of the decisions by providing logistical and communications support to the fishing industry and all participants. Finally, other stakeholders have participated in the consensus-building process by sensitizing the decision-making body to the ramifications of their decision alternatives.

Formalized and enhanced in-season roles of the participants define a Management Team approach to making decisions. The Team would ultimately be held accountable for decisions made and therefore be responsible for tracking the performance of past decisions by comparing the actual results of decision making with anticipated results. This feedback should be incorporated into adjustments in future decisions toward achieving pre-defined strategic goals.

Tools for decision making

The experience of in-season management in this fishery resulted in development of a tool (in the form of a biological checklist) to assist in the interpretation of observations from the fishery. In support of management decisions for the herring fishery, other specific tools are being developed for structured decision making. The suite of descriptive and analytical tools includes the following:

- Stock database — a spatial–temporal database of in-season data, including vessel logbook entries, and real-time stock survey information. This database is compiled through a geographical information system (GIS) to provide summary descriptive analysis of data in graphical and map formats. A database and GIS (using the MAPINFO software package) for the herring fishery was prepared based on historical log-book data for 1990–92. Analysis is continuing with regard to the form and content of descriptive information useful to in-season decision making.
- Industry database — fishery operating data used to report a snapshot (e.g. monthly or quarterly) of the economic performance of the sectors of the fishery. This tool also provides seasonal projections of the socio-economic performance of the industry. This information has been used to analyse the intra-seasonal and inter-seasonal impacts of market fluctuations for herring products (e.g. roe price changes), changes in herring catches, and adjustments in quota allocations to gear sectors (Lane and Stephenson 1996).
- Simulation analysis — in-season stock abundance estimates throughout the fishery. This analysis is based on a probabilistic model of herring stock dynamics and the relative strengths of spawning stocks. Current data observations are used to update stock size estimates calculated using Bayesian statistical analysis. This tool describes stock status in the form of probability distributions. These estimates are used to simulate the expected impacts of in-season exploitation policies on stock size as well as on the economic performance of the industry (Lane and Stephenson 1993).
- Risk management — quantitative analysis of management alternatives. The probabilistic analysis described above describes the range of anticipated outcomes' measures and their possibility of occurrence for alternative management policies. It remains to be determined how non-dominated alternative policies can be compared. The risk management tool identifies tradeoffs among the multiple objectives and performance measures of the fishery and analyses these trade-offs to provide a resulting ranking of alternative policy options.
- Aggregate stock assessment — estimation of the size of the herring stock. This analytical tool provides a static estimate of the aggregate abundance of the stock. Assessment methods use Virtual Population Analysis (VPA, least square parameter estimation) methods to calculate point estimates of cohort strength and total population size (Stephenson *et al.* 1995). These traditional analyses can be compared to the independently derived in-season stock estimates to improve stock estimation reliability.
- Feedback analysis — this tool compares the actual multi-objective impacts of an implemented decision policy with the recorded and anticipated performance of the decision taken. The gap between the actual and expected decision performance is the measure of the ability to manage and control the system. A wide gap requires a review of the uncertainty of decision evaluation and implies a more conservative approach to decision making (Lane and Stephenson 1995).

The suite of tools to support management decision making represent a well-established set of methods for decision analysis from the field of operations research. As such, there is substantial formal and applied evidence of the usefulness of such systems (Lane 1992).

CONCLUSIONS

An effective fisheries management system can be developed by providing the appropriate representation, scale, and decision-making responsibility to participants in the fishery. In reaction to a crisis situation, the Scotia-Fundy herring fishery is evolving toward a more direct, participative co-management system. By the actions of its decision-making committee, it has demonstrated the integration of the roles and responsibilities of fisheries science and socio-economics. It is important to follow-up on the goodwill that exists since the 1995 experience — and the large investment in time and effort of the original participants — by entrenching and empowering, through legislation, the roles and responsibilities of the participants in the management process.

References

Hilborn, R, and Walters, C J (1992). 'Quantitative fisheries stock assessment: choice, dynamics, and uncertainty'. (Chapman and Hall: New York).

Jentoft, S (1989). Fisheries co-management: delegating government responsibility to fisheries organizations. *Marine Policy*, April, 137–54.

Lane, D E (1992). Management Science in the Control and Management of Fisheries: An Annotated Bibliography. *American Journal of Mathematical and Management Sciences* **12**(2,3), 101 52.

Lane, D, and Stephenson, R (1993). A decision making framework for providing catch advice in fisheries. University of Ottawa, Faculty of Administration, Program of Research in Inter disciplinary System Management (PRISM) Working Paper 93–62.

Lane, D E, and Stephenson, R L (1995). A decision making framework for the development of management plans. *Atlantic Fisheries Research Document* 95/80.

Lane, D E, and Stephenson, R L (1996). SATURN: A Framework for Integrated Analysis in Fisheries Management. *Special Issue of INFOR* **34**(3), 156–80.

Ludwig, D, Hilborn, R, and Walters, C J (1993). Uncertainty, resource exploitation, and conservation: lessons from history. *Science* **260**, 36–7.

Mackenzie, W C (1974). Conceptual aspects of strategic planning for fishery management and development. *Journal of the Fisheries Research Board of Canada* **31**, 1705–12.

Smith, S J, Hunt, J J, and Rivard, D (Eds) (1993). Risk Evaluation and Biological Reference Points for Fisheries Management. *Canadian Special Publication of Fisheries and Aquatic Sciences* 120.

Stephenson, R L, and Lane, D E (1995). Fisheries science in fisheries management: a plea for conceptual change. *Canadian Journal of Fisheries and Aquatic Sciences* **52**(9), 2051–6.

Stephenson, R L, Lane, D E, Aldous, D, and Nowak, R (1993). Management of the 4WX Atlantic Herring (*Clupea harengus*) Fishery: An Evaluation of Recent Events. *Canadian Journal of Fisheries and Aquatic Sciences* **50**, 2742–57.

Stephenson, R L, Power, M J, Sochasky, J B, Fife, F J, Melvin, G D, Gavaris, S, Iles, T D, and Page, F (1995). Evaluation of the stock status of 4WX herring. *DFO Atlantic Fisheries Research Document* 95/83.

Wooster, W S (Ed) (1988). 'Fishery Science and Management: Objectives and Limitations'. Lecture Notes on Coastal and Estuarine Studies, **28**. (Springer-Verlag: New York).

THE STAKEHOLDER SATISFACTION TRIANGLE: A MODEL FOR SUCCESSFUL MANAGEMENT

L. A. Nielsen,[A] *B. A. Knuth,*[B] *C. P. Ferreri,*[A] *S. L. McMullin,*[C] *R. Bruch,*[D] *C. E. Glotfelty,*[A] *W. W. Taylor*[E] *and D. A. Schenborn*[D]

[A] Penn State School of Forest Resources, University Park, PA USA.
[B] Cornell University Department of Natural Resources, Ithaca, NY USA.
[C] Virginia Tech Department of Fisheries and Wildlife Sciences, Blacksburg, VA USA.
[D] Wisconsin Department of Natural Resources, Madison, WI USA.
[E] Michigan State University Department of Fisheries and Wildlife, East Lansing, MI USA.

Summary

Fisheries management has emphasized scientific substance and institutional processes as the basis for making decisions. However, because most fisheries decisions require the support — both informal and formal — of the stakeholders in the decisions, effective fisheries management also requires emphasis on the relationships between stakeholders and professional managers. As shown in fisheries management cases involving trout fishing restoration (Bighorn River, Montana), salmon harvest policy (Lake Ontario, New York), anadromous fish restoration (Susquehanna River, Pennsylvania), comprehensive watershed planning (Lake Winnebago, Wisconsin), and sea lamprey control (Laurentian Great Lakes), building stakeholder relationships can overcome substantial impediments and present new opportunities. Such relationships are built on personal, local, sustained, broad, and cooperative work with local people, politicians, businesses, and organizations; the result is mutual understanding and trust. Adding 'relationship" to substance and process completes the stakeholder satisfaction triangle, improving the chances for successful management.

INTRODUCTION

Interest in effectiveness has broken out like a feverish sweat on the brow of fisheries management. Fuelled partly by budget cuts and partly by citizen dissatisfaction with government, the drive towards effectiveness has infected strategic plans, organizational visions, and accountability measures (Nielsen 1993). Coupled with a growing recognition of the diversity of stakeholders (Decker *et al.* 1996), the emphasis on effectiveness reflects two fundamental issues for today's resource management institutions and professionals — why have we accomplished less than we could, and how can we do better? This paper emphasizes the latter question, suggesting that we embed management within the concept of the stakeholder satisfaction triangle.

Before we explore that question, some reflection on the first question is needed — why have we done less than we could? We contend that fisheries professionals have envisioned fisheries management as governed by biological and ecological facts and principles — with an occasional reference to economic returns — and placed themselves in the role of interpreters of that material into management philosophies, laws, regulations, policies, and programmes. This attitude reveals itself explicitly in descriptive statements of many institutions, including, for example, the American Fisheries Society, which states that it 'promotes scientific research and enlightened management of aquatic resources for optimum use and enjoyment by the public.'

No one can question the importance of scientific knowledge and objective data for decision-making, but such knowledge fits the idiom of being 'necessary but not sufficient.' As fisheries management decisions have become more central to the lives of citizens and their communities and as more fishery stocks have become marginalized, our science has been called into question for its quality (i.e. repeatability, predictability, and generality) and for its subjectivity (Decker *et al.* 1991). Professional reaction to these challenges has typically been confrontational, based on positional bargaining that established a biological/ecological position opposite a utilization position (utilization varying from recreational to commercial to habitat-altering). The upshot has been the development of explicit procedures for institutions to divulge and gather information and opinions. Rather than satisfying either agencies or the public, these procedures tend to pass on the conflict to political appointees, judges, and legislators. The result has been delay, wasted funds, decisions based on demonstrable power — and declining fisheries.

The Stakeholder Satisfaction Triangle

This situation can be improved *via* a widely accepted model for conflict reduction — the stakeholder satisfaction triangle. The model asserts that satisfactory decisions are based on three major axes — substance, process, and relationship. For any situation, all three aspects are important, but importance will vary according to the specific situation, the stakeholders involved and the stage of decision making.

Substance refers to the technical and factual content of the situation. This is the realm of the scientist and technician, for whom data collection and analysis are most comfortable and fulfilling. Throughout this century, but especially since World War II, the explosion of scientific inquiry has catapulted substance to a dominant role in societal affairs (what professional field has resisted the trend to be called a science?). Because most fisheries professionals are zoologists, ecologists, limnologists, oceanographers, or economists, most fisheries management has been conceptualized as a substance-driven activity. Fisheries profesionals love substance.

Process refers to the explicit steps that are followed in a management decision. This is the realm of the administrator, lawyer, and watch-dog group, who are interested in ensuring that they have a chance to be heard on every decision. Process is institutionalized into the democratic/representative political system, *via* requirements for public comment and advertising of all pending decisions. The United States of America, for example, has many environmental consideration laws, which do not mandate particular actions or evaluatory criteria, but only require that all decisions be publicly reviewed and that public comments be explicitly addressed by the implementing authority (e.g. the environmental impact assessment process under the National Environmental Policy Act). Fisheries professionals hate process.

Relationship refers to the development of positive networks among individuals with direct or indirect interest in or influence over a management decision. This is the realm of the politician, journalist, entrepreneur, and civic leader, who wish to know, trust, and have access to the decision maker. Relationship is important to people who want to be understood and who demand confidence that their values and needs will be considered in the decision. Relationship occurs outside the formal procedures that collect official comments; it is built with frequent, informal, and non-specific communication and with a broad commitment to a community, whether localized or dispersed. Fisheries professionals avoid relationship.

Through time, fisheries management has emphasized substance and has added process as a reality of modern government. For the future, however, the addition of relationship and the synthesis of all three elements will be necessary for success. A recent study of state agencies revealed eight characteristics of effectiveness (McMullin 1993). Whereas a few characteristics related to substance (e.g. credibility based on a bottom line of keeping the resource first, strong development of agency personnel) and process (e.g. open, equitable decision-making process responsive to public), most of the characteristics clearly emphasized relationship-building or the integration of all three elements. These included (1) being pro-active in anticipating future issues; (2) listening to the public, understanding their desires, and involving them in decisions; (3) providing employees wide latitude to do their jobs their way and take risks; (4) decentralized structure and participative decision making; and (5) strong public support that can be mobilized when needed.

In the case studies that follow, we explore examples of fisheries management that demonstrate the importance of relationship and the integration of substance, process, and relationship in the successful implementation of fisheries decisions.

Shortcutting on relationship: gridlock on Pacific salmon harvest policy in New York

Lake Ontario forms a long border between the US State of New York and the Canadian Province of Ontario. A significant portion of Lake Ontario and its fishery falls to New York jurisdiction, managed by the New York State Department of Environmental Conservation (NYSDEC). Like all of the Laurentian Great Lakes, the fishery of Lake Ontario has passed through rapidly changing conditions over the past century. Beginning in the late 1960s, stocking of Pacific salmon and other species created new recreational fisheries of enormous popularity. These new fisheries have also created a new set of stakeholders and issues (information in this section comes from Connelly *et al.* 1992).

As salmon fishing developed in the 1970s, the life history of Pacific salmon created both a problem and opportunity for fisheries managers. Adult Pacific salmon began migrating up suitable streams, where they would eventually die. Managers wished to use the adult salmon most effectively, both to create a fishery and to avoid the streambank accumulation of rotting salmon carcasses. Therefore, believing that migrating salmon would not strike a lure, they allowed snagging — the use of heavy, unbaited hooks dragged across the streams in the hope of encountering and impaling a salmon anywhere on its body. Salmon fishing in general, and snagging as a specific style, became very popular. The Salmon River in eastern Lake Ontario experienced a 6-fold increase in fishing effort, 920-fold increase in angling expenditures, and a 30-fold increase in salmon catch from 1975 to 1989; about 20% of this total effort was attributable to snagging.

The popularity of salmon angling produced significant negative effects as well. These included increased traffic and parking congestion, smelly and ugly carcasses, littering and trespassing, conflicts among anglers and citizens, and increased demand for local services. NYSDEC also became concerned that they had under-estimated the negative impacts of snagging — increased fishing regulation violations, unrealistic demands for salmon harvest, focus on harvest as the most important part of angling, transfer of the snagging approach to other fisheries, and the ethical repugnance of snagging compared with traditional angling techniques. Other Great Lakes' states voiced similar misgivings about snagging and began to eliminate snagging.

Furthermore, studies of Pacific salmon behaviour conducted during the 1980s demonstrated that migrating fish would strike at brightly coloured baits, provided that angling density was not so high that it disturbed the fish excessively. This new information fit well with NYSDEC's growing concern for the ethics and general negative impacts of snagging. Consequently, in 1988, NYSDEC concluded that it should eliminate snagging.

NYSDEC instituted an event to convince their stakeholders that snagging should be eliminated. They avoided the typical public hearing, however, in favor of a 'Salmon Summit' at which invited participants discussed issues and alternatives for their resolution. The summit involved local business leaders and local government officials in drafting a 5-year fishery management plan for the Salmon River. The summit concluded that snagging should be phased out over a three-year period.

One stakeholder group, Pacific Salmon Unlimited, objected to the snagging ban because of an expected economic loss. They challenged the legal authority of NYSDEC to ban snagging, as an attempt to interfere in the legislature's sole authority to create public policy. On the advice of legal counsel, NYSDEC backed away from the agreed plan and returned to the legislatively created State Environmental Quality Review, similar to federal environmental review processes under NEPA. Under this formal process, NYSDEC decided to ban snagging by 1994, subject to further administrative and judicial review.

The problems in implementing this change result from the absence of effective relationships. A predilection, based on substance, to change the harvest policy was rushed through the relationship-building process, leading to a procedural and legal gridlock. The cases below demonstrate how patience in relationship building and integration of all three triangle elements can improve outcomes.

Bringing the community together: Winnebago Comprehensive Management Plan

Lake Winnebago is located in east-central Wisconsin, part of the Fox River drainage that comprises 17% of the state's inland waters. Winnebago is a multi-purpose resource, serving as water supply for industrial and domestic use, waste disposal for the same users, recreation for anglers, boaters, and swimmers, hydropower supply, commercial fishing, and commercial wild celery/sago pondweed harvest. Through time, the lake and its surrounding watershed have been heavily exploited and, consequently, have declined in quality (information in this section comes from Anon. 1989).

Management had been hampered by the sheer size of the watershed, the complexity of ecological and use-related factors, and conflicting interests among users. The system is managed by the Wisconsin Departments of Natural Resources (WDNR) and Agriculture, the US Army Corps of Engineers, and by many local, state, and federal agencies with jurisdiction over some aspect of land, water, air, waste, and biological resources. In the past, these groups had each focussed primarily on a single issue or resource.

To help overcome this problem, the WDNR initiated a comprehensive management planning process in 1986. This process would eventually produce the Winnebago Comprehensive Management Plan, a basin-wide approach to identifying issues, considering alternative solutions, and implementing chosen alternatives.

The Plan builds on the enormous substantive background accumulated by the WDNR and other agencies during decades of monitoring and research. The availability of this information, produced by an agency widely recognized as among the nation's best, certainly contributed to the success. Furthermore, the agency's proclivity to serious comprehensive planning and strategic thinking also created a working structure in which a plan could be contemplated and facilitated. This process, however, was not legally required or mandated in the usual sense, but was one that allowed stakeholders to think together and agree on what needed to be done. From the beginning, the Plan focussed on inclusiveness, communication, and mutual decision making. Most importantly, perhaps, it was long-term and intuition driven. The leaders knew that building credibility with the participants required their constant attention and a personal understanding that would allow best judgement of what was possible and timely.

Prior to the start of the formal planning process, the co-ordinator conducted an extensive appraisal of individuals, agencies, and organizations with a stake in the management or use of the Winnebago system. This developed an understanding of individuals and groups who were potential planning committee members; administrators from the WDNR and other agencies who should be kept informed; and private citizens and citizen's organizations whose support would be necessary.

The first formal step was a management and research workshop on the Winnebago system and several subsequent issue analysis sessions in the fall of 1986, to identify priority problems. These problems were categorized and eventually assigned to three planning committees (Biota and Habitat, Nutrient and Eutrophication, and User). The Biota and Habitat Committee included mostly WDNR biologists and other professionals with technical expertise on fisheries and wildlife management. The Nutrient and Eutrophication Committee included a wide variety of state and local water, waste, soil, and land use professionals. The User Committee included the broadest representation, including both professionals and a wide variety of business, civic, angling, and environmental leaders. In total, the committees met 19 times in one year, involving more than 2500 person-hours.

The products of these committees entered a more expansive public participation process at each stage in the Plan development. The public participation process involved a wide range of citizens from the region and beyond. Opportunities to learn about the planning effort and communicate with the committees and staff were ensured by (1) mailing information regularly to more than 1000 addressees; (2) conducting dozens of interviews with a wide range of people; (3) holding 19 'Public Information Exchange Meetings' to discuss planning products as these were developed; (4) making dozens of presentations to civic, sporting, and business organizations; (5) displaying a poster series at various events, and (6) working closely with the media to maintain regional and state-wide exposure and understanding.

The planning process eventually led to a comprehensive Plan that passed through two drafts before being finalized in September, 1989, by the Secretary of the Wisconsin Department of Natural Resources. The WDNR accepted a primary role in Plan implementation, allocating two full-time professionals to oversee actions. The User Committee has remained in place as a representative of stakeholder interests and to ensure the continuous building of relationships. A continuing involvement process has been used since 1989 to monitor accomplishments and deal with new issues. The Plan included 120 recommendations for action; to date 73 have been completed or are in progress. Citizens formed several new groups to implement the plan, including one that now raises $US75 000 annually for Lake Winnebago walleye management.

The project co-ordinator attributes the implementation success to the deep ownership held by the public in the Winnebago plan as 'their plan,' developed through long-term public involvement. The basic foundation is that the public trusts that WDNR will never skirt issues the public thinks important, will always be honest, and will involve them in making the decisions that affect them. The active and continuing involvement of the stakeholders assures agency follow-through on the plan.

Evolving a long-term strategy: American shad restoration on the Susquehanna River

American shad have a history in the Susquehanna River that is familiar in the stories of anadromous species across the world. The Susquehanna River, which drains most of eastern Pennsylvania and smaller portions of New York and Maryland, is the largest tributary of the Chesapeake Bay. A host of diadromous fishes, including four *Alosa* species, striped bass, and American eels, inhabited the Susquehanna at remarkable densities in earlier centuries. Through time, however, their abundances have declined, due to overfishing, water pollution, and damming to serve the Pennsylvania canal system. So alarming was the decline of American shad in the Susquehanna in the 19th Century that it was the stimulus for the creation of the Pennsylvania Fish Commission in 1866, charged specifically with reversing the trend (information in this section comes from St. Pierre 1992; and Stranahan 1993).

American shad continued to decline, however, and worsened with the creation of a series of hydropower dams between 1900 and 1930. The largest and most downstream, the 95-ft high Conowingo Dam, virtually closed the Susquehanna as a habitat for American shad. Since then, power companies and state and federal agencies have battled to balance power production and cost with fisheries restoration. They have argued over the substance of fish restoration, a most difficult and uncertain field of knowledge; they have battled through the processes of permits and court proceedings; and, finally, they have built the relationships necessary for a win-win strategy.

Beginning in the early 1900s, a series of studies and technical opinions debated whether or not American shad and other alosids would use fish passage devices. Earliest studies concluded that shad would not use ladders, but experience from western states in the 1940s showed that shad had ascended ladders there. Consequently, while new studies in the 1950s and 1960s agreed that shad passage was possible, they failed, however, to recommend constructing ladders, because the river upstream of the dams was too polluted for shad survival. Finally in 1969, a coalition of state and federal agencies succeeded in convincing the Philadelphia Electric Company (PECO) to work cooperatively with them to restock the upper river with shad eggs and to construct an experimental fish lift at Conowingo Dam. The Conowingo fish lift, built by PECO for over $US1 million, began operating in 1972. Fish collected at the life were transported by truck around the three additional upstream hydroelectric dams. Fish eggs were collected annually from other rivers and placed in spawning boxes in the Susquehanna, but the results were poor. In 1976, therefore, the Pennsylvania Fish Commission dedicated the world's only shad hatchery to grow fingerlings for stocking in the river.

Throughout this time, federal, state, and utility officials were working together *via* a series of formal committees devoted first to shad restoration and later to restoration of all anadromous species. The membership of the committees regularly expanded to include more utilities and agencies, becoming more representative of the full set of responsible authorities with potential to influence restoration. In 1979, the US Fish and Wildlife Service established a full-time co-ordinator to develop a 'Strategic Plan for Restoration of Migratory Fishes to the Susquehanna River.'

Cooperation and confrontation were alternately displayed in the formal processes that occurred beginning in the 1970s. Across that period, relicensing of the hydropower dams was under consideration by the Federal Energy Regulatory Commission (FERC). From 1976 to 1980, agencies and utilities engaged in open battle in formal reviews and hearings. With no clear resolution in sight, the presiding FERC judge awarded new licences to the utility companies, but required that the question of shad restoration be handled in subsequent mandatory hearings. Again, these hearings produced a gridlock. Eventually, the power utilities agreed to fund additional studies from 1981–1985. In 1985, a similar agreement extended funding for shad restoration activities and studies another decade. Part of this agreement was a commitment that operators of three upstream dams would create fish passages on their dams when PECO installed sufficient permanent fish passage devices at Conowingo Dam to ensure a large enough population of spawning fish to reach their dams. Throughout this period, more shad were reaching the base of Conowingo and being transported upstream, and more shad were being hatched and stocked in the river.

Through another series of formal hearings and appeals, PECO sought to void its responsibility for passing more adult shad. During these proceedings, however, a PECO nuclear plant was ordered to be closed by federal regulators as a safety hazard. The resulting public dissatisfaction led to a major turnover in the company's leadership and a complete reversal of policy and strategy. Based on the decades of joint work and relationship building through its membership on the formal committees, PECO concluded that since it benefitted from the river, it also needed to invest in its sustainability. The company withdrew its objections and implemented a series of shad restoration programmes, including minimum flow regimes, changes in turbine operation, and the construction of a permanent fish passage facility capable of handling 750 000 shad and 5 million other alosids annually. The facility was built for $US12.5 million and began passing shad in 1991. The other utilities have lived up to their end of the bargain and have begun building fish passage devices on their dams. By Spring, 1999, the entire 350-mile length of the river in Pennsylvania will be open for fish passage. This series of commitments for fish passage construction comprise the largest effort of its type ever undertaken for American shad. It has paved the way for an ambitious plan to remove hundreds of smaller blockages on tributaries throughout the Chesapeake Bay basin.

Involving all parties: Bighorn River, Montana

The Bighorn River was once a warm, silty stream flowing leisurely near Billings, Montana. An upstream dam built in 1967, however, changed the river, cooling and clearing the water and producing a brown/rainbow trout fishery that soon became world-renowned. For a short time in the late 1970s, the Bighorn was closed to public fishing because of claims by the local native American tribe that it controlled the river. The temporary halt in fishing allowed fish density to rise and trophy fish to grow, so that when the river was again opened in 1981, fishing was better than ever. The Bighorn's reputation continued to grow, along with fishing pressure, guiding services, and user conflicts. By 1986, public and private dissatisfaction had reached the point that something needed to be done (information in this section comes from McMullin and Nielsen 1991).

Management of the Bighorn fishery falls to the Montana Department of Fish, Wildlife and Parks (MDFWP). In previous attempts to prescribe management changes on other Montana rivers, MDFWP had been badly scarred by their failure to address the relationship side of the stakeholder triangle. For the Bighorn River, they adopted a different style that emphasized the public as a partner.

Public opinion was centred on the need for regulatory changes to reduce fishing pressure and harvest. MDFWP professionals, however, also believed that habitat degradation, associated with poor water quality, insufficient flow, and low food supplies for trout, were also major impediments to a satisfactory fishery. They concluded that an educational programme would be important to building the substantive base for management decisions.

MDFWP devised a multi-step public involvement process, outside the formal hearing process. The first step was to engage local communities in determining the management goals for the fishery. Public meetings were held in two riverside towns, where most of the guides and interested citizens lived. The meetings were broadly advertised through the media, resulting in attendance about three times higher than typical MDFWP hearings. Each meeting included briefings by MDFWP professionals, followed by a facilitated workshop to determine goals. The meetings yielded a single unanimous conclusion — the Bighorn should be managed to produce trophy trout.

The next step was drafting of a management plan by MDFWP biologists, designed to accomplish the publicly-endorsed goal. The draft plan was written in lay terms, targeted at local citizens, guides, outfitters, anglers, local legislators, and civic leaders. Hundreds of copies of the plan were sent out, and ten presentations about the plan were made to civic and sporting groups over a 90-day comment period.

The third step was gathering reactions to the draft plan. Each of the distributed copies of the plan included a pre-addressed survey that the reader could return to MDFWP. Response to the plan was overwhelmingly positive, ranging from 69 to 83% for the stated objectives. Moreover, more than half of respondents admitted to changing their opinion on Bighorn management after reviewing the plan.

After making modifications to the plan that were suggested by the public comments, MDFWP presented the Plan to its policy-setting Commission, which formally adopted it for implementation. The strength of building this plan based on broad community participation became obvious when one interest group offered a rival plan that focussed only on restrictive regulations. The Commission noted that a similar option had been considered and rejected during the public involvement process and, therefore, they dismissed the rival plan immediately. Moreover, the Commission ordered MDFWP to use a similar public involvement process to develop management plans for the state's ten most important fisheries.

Developing effective partnerships: sea lamprey control in the Laurentian Great Lakes

The least welcome immigrant to the Laurentian Great Lakes has been the sea lamprey, a primitive, jawless fish that parasitizes valuable commercial and game fishes. Beginning with the modification of the Welland Canal between Lakes Ontario and Erie in 1919, sea lampreys have found their way into Lakes Erie, Michigan, Huron, and Superior. Along with overfishing, sea lamprey invasion proved disastrous to Great Lakes fisheries, and the long story of sea lamprey control provides a major example of an evolving stakeholder satisfaction triangle on an international scale (information in this section comes from Smith and Tibbles 1980, and Ferreri *et al.* 1995).

The collapse of important fisheries prompted the United States and Canadian governments to sign the Convention on Great Lakes Fisheries in 1955, establishing the Great Lakes Fishery Commission. High on the new commission's agenda was control of sea lampreys. The Commission approached this task by building the needed substance. They sponsored extensive studies of the life history of the sea lamprey and experiments on various control techniques, including screening more than 6000 potential lamprey-killing chemicals. Scientists quickly identified an effective chemical (TFM, or 3-trifluoromethyl-4-nitrophenol), and the Commission launched a TFM-based control programme in Lake Superior tributaries in 1958. The control programme was eventually expanded to all the Great Lakes and has been responsible for significantly reducing sea lamprey abundance, thereby creating an opportunity for overall rehabilitation of Great Lakes fisheries.

Because the commission is an independent agency apart from state, provincial, and tribal authorities that manage Great Lakes fisheries, many of the commission's early actions were taken without consulting those authorities. Similarly, the fisheries management authorities often did not consult each other about their actions, including extensive stocking of Pacific salmon throughout the 1970s. Unfortunately, fish do not recognize political boundaries, and it soon became clear that co-ordinated management was essential. The Great Lakes Fishery Commission, building on its successful sea lamprey control programme, provided a focus for informal co-ordination of management efforts. Gradually, state and provincial authorities organized more formally, first *via* a series of Lakes' Committees, and then, in 1980, *via* the Strategic Great Lakes Fishery Management Plan (tribal authorities signed on in 1993). The plan was a major breakthrough in which separate fisheries agencies agreed to discuss any management action that may have effects beyond their jurisdictional boundaries and to act only after consensus had been reached among all relevant parties.

The plan formalized the decision-making process for the Great Lakes fisheries. Fisheries goals and objectives are set by lake committees composed of state, provincial, and tribal authorities; lake committees are advised by technical committees composed of scientists. Lake committees meet yearly to review progress, evaluate current objectives, and revise their management plans. These formal groups, which evolved over the years based on strong working relationships, form the core of long-standing bonds upon which management has succeeded.

A test of those relationships occurred in the early 1990s. Each year, the cost of TFM and other programmes continued to increase, but funding from US and Canadian governments remained flat. Furthermore, improving water quality in the Great Lakes increased the potential for sea lampreys to thrive in more tributaries, possibly necessitating expansion of the control programme. Participants realized that funding shortfalls threatened the future of the control programme. Because of their strong relationships, committee members were able to join forces quickly to mount a powerful and successful campaign to increase the Commission's budget, particularly to support sea lamprey control. Within a political atmosphere focussed on budget cutting, the Commission's voice was heard and echoed through two nations.

Building relationships

These examples demonstrate the importance of building relationships among stakeholders as part of any natural resource decision. An important aspect of such relationships is that their value comes into play at different times in different situations. For Montana's Bighorn River, the relationships allowed a competing but inadequate idea to be discarded quickly in formal hearings. For American shad on the Susquehanna River, long-standing relationships among agencies and power companies provided the context for a new industry executive team to make rapid and popular commitments to fish restoration. For Wisconsin's Lake Winnebago, broad relationships provided ways for neighbours — people and institutions — to link their individual interests and commitments into a larger whole. For Great Lakes sea lamprey control, an evolving network allowed diverse agencies to rally behind a common need. It is the nature of relationships to be fluid and unique, as opposed to either substance or process, which are by nature determinate and prescriptive.

Consequently, we cannot write a prescription for where, when, or how relationships will be useful and important. Relationships are the true 'human dimension' in resource management — and, therefore, are infinitely surprising and interesting. We can, however, provide some guidelines for effectively building relationships among stakeholders.

Relationships are about people. People build relationships by getting to know and understand other people. If culture can be defined as a web of talk, then relationships are the strands of that web, built by people talking with one another about the many things that affect their lives. The oldest political axiom is that it isn't what you know, but whom you know that matters. And all of those 'whoms' are people.

Relationships are about lives. We develop relationships by spending time with people regularly as part of our daily lives. We learn to understand other people's interests, beliefs, fears, and aspirations. Through scores of small activities and interactions, we come to understand, rely on and trust each other. Then, when a major issue arises, we can believe in the goodwill of others to help us. For example, when asked how we could improve media relations, journalists always reply that we must talk with them often and informally — not just when a crisis erupts.

Relationships are about neighbourhoods. Building relationships requires an orientation to neighbours and communities. For fisheries managers, this means that the local managers must work with landowners, business owners, teachers, religious leaders, civic groups, and politicians. This is a particularly difficult concept for fisheries professionals, who, as ecologists, tend to think at large scales — ecosystem management is the perfect example of our desire to expand the playing field. But at least as important as a global view is the ability to get something done at home. As the bumper-stickers say, 'Think globally, act locally.'

Relationships are two-way streets. Relationships require fisheries professionals to commit their 'time, talent, and treasure' to actions beyond their immediate concern. And these may be actions that are inconsistent with optimality principles, such as using the hatchery tank-truck and crew to water a new municipal garden. Relationships may require supporting another project or viewpoint now, because it is a higher overall priority than a fishery project, with the anticipation of reciprocity later.

Relationships are self-selecting. Building relationships often requires opening oneself to people whom you might ordinarily avoid. Perhaps the most rewarding part of relationship-building is the chance to learn more about people, to make friends where one never expected. Being open to relationships means welcoming people on their terms, as they become friends and partners in the business of building community — and resource sustainability.

In summary, we offer the process of building relationships as the essential third leg in the triangle of management success. Building relationships is long-term and individual; it requires investing in a great number of people and activities that do not have immediate pay-off. For ecologists, however, the knowledge that 'everything is connected to everything else' should be easily transferable to our relations with people. The connections we make today with our neighbours and colleagues will eventually turn out to be connected to our sustainable management of fisheries resources.

References

Anon. (1989). 'Management of the Lake Winnebago System'. (Wisconsin Department of Natural Resources: Madison).

Connelly, N A, Knuth, B A, and Dawson, C P (1992). The failure of success in natural resource policy: Pacific salmon harvest policy in New York state. *Policy Studies Review* **11**(2), 24–36.

Decker, D J, Krueger, C C, Baer Jr, R A, Knuth, B A, and Richmond, M E (1996). From clients to stakeholders: A philosophical shift for fish and wildlife management. *Human Dimensions of Wildlife* **1**(1), 70–86.

Decker, D J, Shanks, R E, Nielsen, L A, and Parsons, G R (1991). Ethical and scientific judgments in management: Beware of blurred distinctions. *Wildlife Society Bulletin* **19**, 523–7.

Ferreri, C P, Taylor, W W, and Koonce, J F (1995). Effects of improved water quality and stream treatment rotation on sea lamprey abundance: Implications for lake trout rehabilitation in the Great Lakes. *Journal of Great Lakes Research* **21**, 176–84.

McMullin, S L (1993). Characteristics and strategies of effective state fish and wildlife agencies. *Transactions of the North American Wildlife and Natural Resources Conference* **58**, 206–10.

McMullin, S L, and Nielsen, L A (1991). Resolution of natural resource allocation conflicts through effective public involvement. *Policy Studies Journal* **19**, 553–9.

Nielsen, L A (1993). Sharing success: The rationale for management effectiveness research. *Transactions of the North American Wildlife and Natural Resources Conference* **58**, 201–5.

St. Pierre, R (1992). 'History of the American shad restoration program on the Susquehana River'. (United States Fish and Wildlife Service: Harrisburg, Pennsylvania).

Smith, B R, and Tibbles, J J (1980). Sea lamprey (*Petromyzon marinus*) in Lakes Huron, Michigan, and Superior: History of invasion and control, 1936–1978. *Canadian Journal of Fisheries and Aquatic Sciences* **37**, 1780–801.

Stranahan, S Q (1993). 'Susquehanna: River of dreams'. (Johns Hopkins: Baltimore).

Can Fishery Catch Data Supplement Research Cruise Data? A Geographical Comparison of Research and Commercial Catch Data

Richard M. Starr[A] *and David S. Fox*[B]

[A] University of California Sea Grant Extension Program, PO Box 440, Moss Landing, California 95039, USA.
[B] Oregon Department of Fish and Wildlife, 2040 SE Marine Science Drive, Newport, Oregon 97365,USA.

Summary

Log-books maintained by participants in the US west coast groundfish trawl fishery provide a detailed set of catch and effort data with broad temporal and spatial coverage. We developed a geographical information system (GIS) to compare 10 years of data from the Oregon commercial trawl fishery with data from US National Marine Fisheries Service research cruises conducted at the same time in the same area. We compared log-book and research catch locations by overlaying catch-per-unit-effort maps and evaluating the geographic co-occurrence of the polygons defined by isopleths. We also compared biomass estimates produced by the two data types. Our results indicate that commercial fishery log-books provide data about fish distribution and abundance that are comparable to research surveys. We believe that log-books can be used to augment research studies and improve estimates of the distribution and abundance of selected species.

Introduction

As the number and size of fishing vessels increased in the 1960s and 1970s, there was a corresponding increase in the world harvest of marine fish species. For the past 20 years, however, catches have been declining in many world fisheries. During that time, the technology available to locate and capture fishes outpaced the availability of information with which to manage fisheries. Presently, fishermen have a much greater capability to locate and harvest fish than scientists have of assessing fish stocks.

One way to increase information available for fishery management is to use information collected by harvesters. The west coast of the United States is home to a well-developed commercial trawl fishery. Over 377 vessels operated in the Oregon, Washington, and California groundfish trawl fishery in 1991 (PFMC 1991); they are all required to maintain log-book records. These fishery log-books provide a detailed set of catch and effort data with broad temporal and spatial coverage. With the availability of efficient geographical information systems (GIS), tools now are available to display and analyse large spatial data bases.

We developed and used a GIS to compare data from the Oregon commercial trawl fishery with data from US National Marine Fisheries Service (NMFS) research cruises conducted at the same time in the same area (Fig. 1). In this paper, we suggest that log-

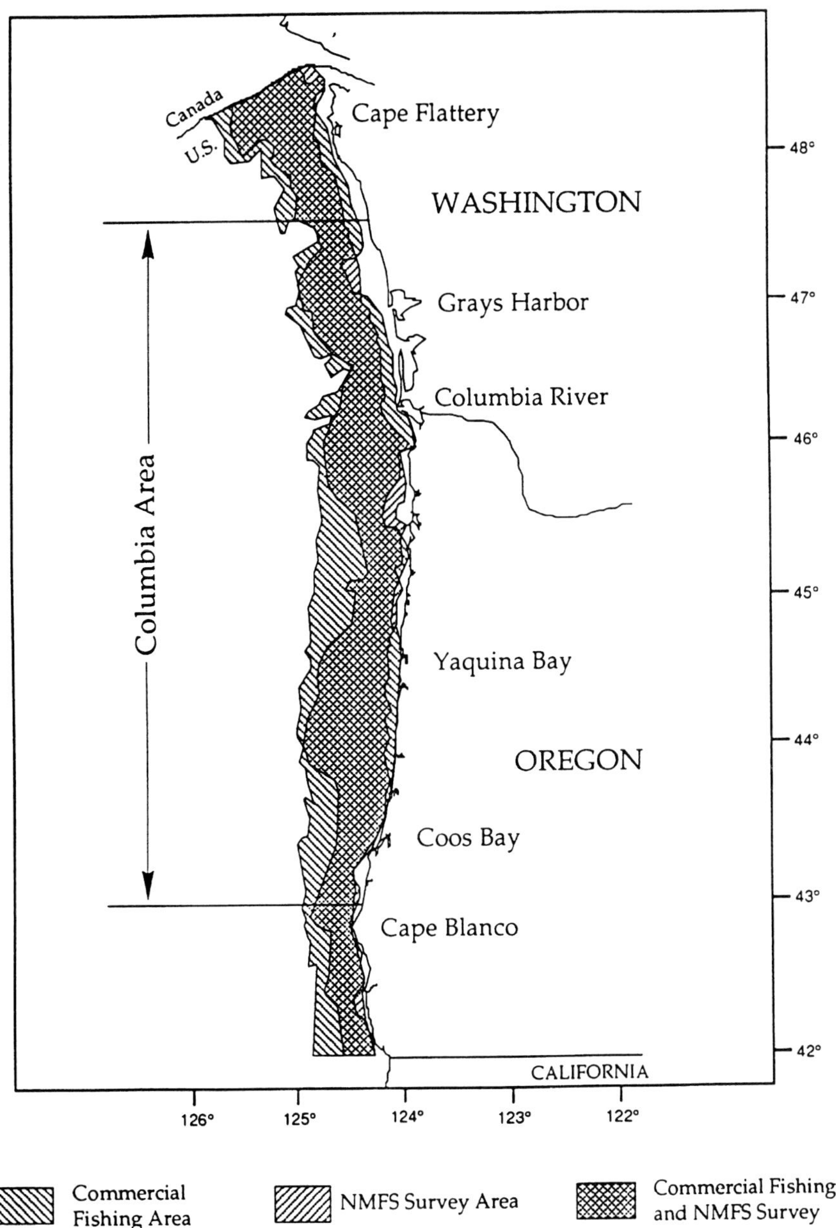

Fig. 1. Commercial fishing and NMFS survey areas in the Columbia Management Area along the Oregon and Washington coast.

book data can complement research data and be used to improve estimates of the distribution and abundance of selected species.

METHODS

The research catch data used in this study originated from a series of NMFS Pacific west coast bottom-trawl surveys of groundfish resources conducted in 1980, 1983, 1986, and 1989. The surveys, often referred to as triennial trawl surveys, contained excellent spatial and temporal overlap with the commercial fishery data set. The NMFS triennial trawl survey methods are described by Gunderson and Sample (1980); Weinberg *et al.* (1984); Coleman (1986); Coleman (1988); Weinberg (1994) and Weinberg *et al.* (1994).

Commercial fishing data used in these analyses were compiled from groundfish bottom trawl fishery log-books collected by the Oregon Department of Fish and Wildlife (ODFW) from 1980–1989. This ten-year log-book data base contains over 130 000 individual records, each representing a single trawl tow. Information for each record includes vessel number, gear type, port of landing, date, latitude and longitude, effort in trawl hours, and catch in pounds for each species or market category reported.

Our analysis included five species: Dover sole (*Microstomus pacificus*), English sole (*Pleuronectes vetulus*), sablefish (*Anoplopoma fimbria*), yellowtail rockfish (*Sebastes flavidus*), and shortspine thornyhead (*Sebastolobus alascanus*). We computed

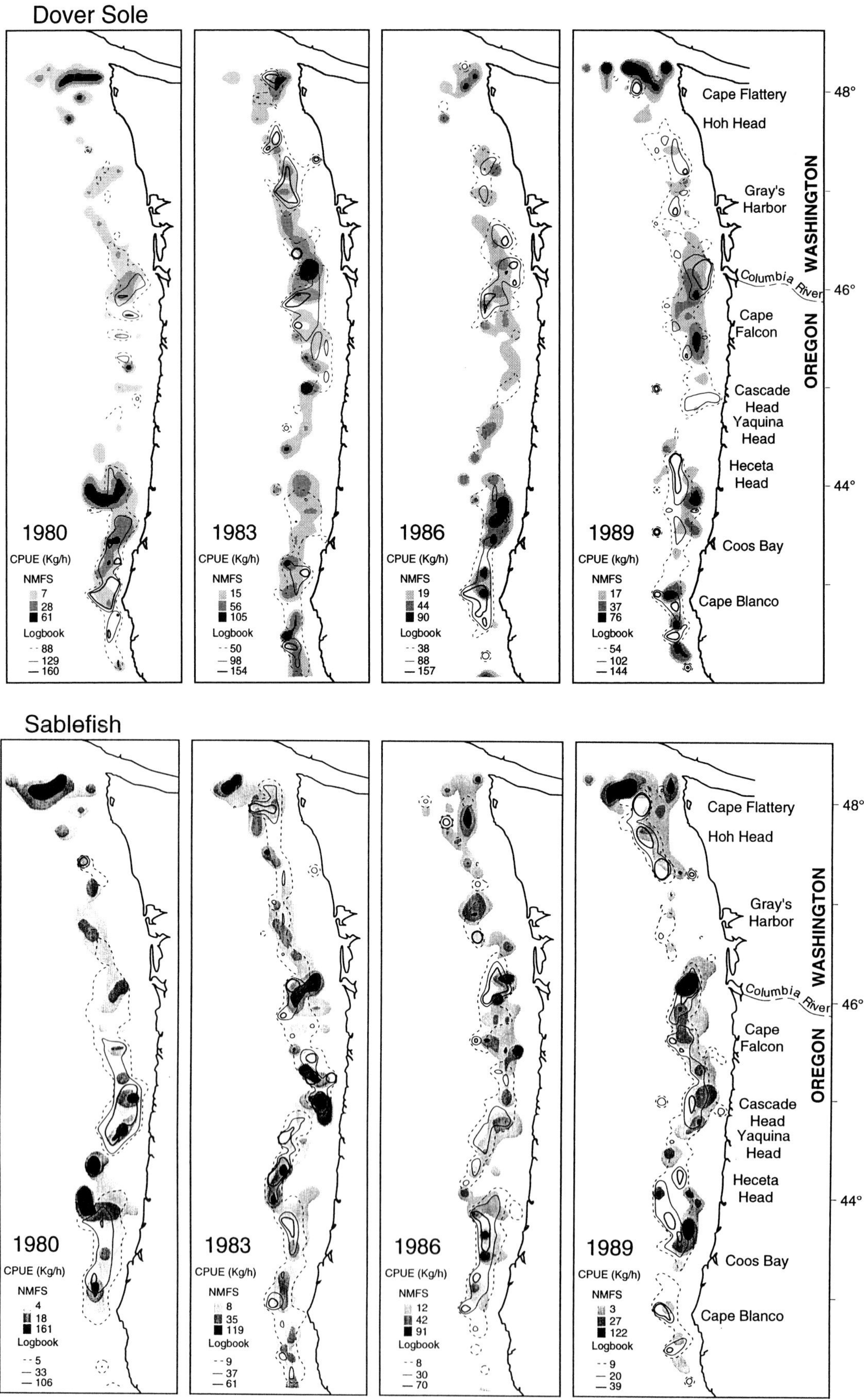

Fig. 2. Isopleths of the 50th, 75th, and 90th percentile of CPUE for commercial fishery and NMFS triennial trawl cruises for a) Dover sole, b) sablefish,

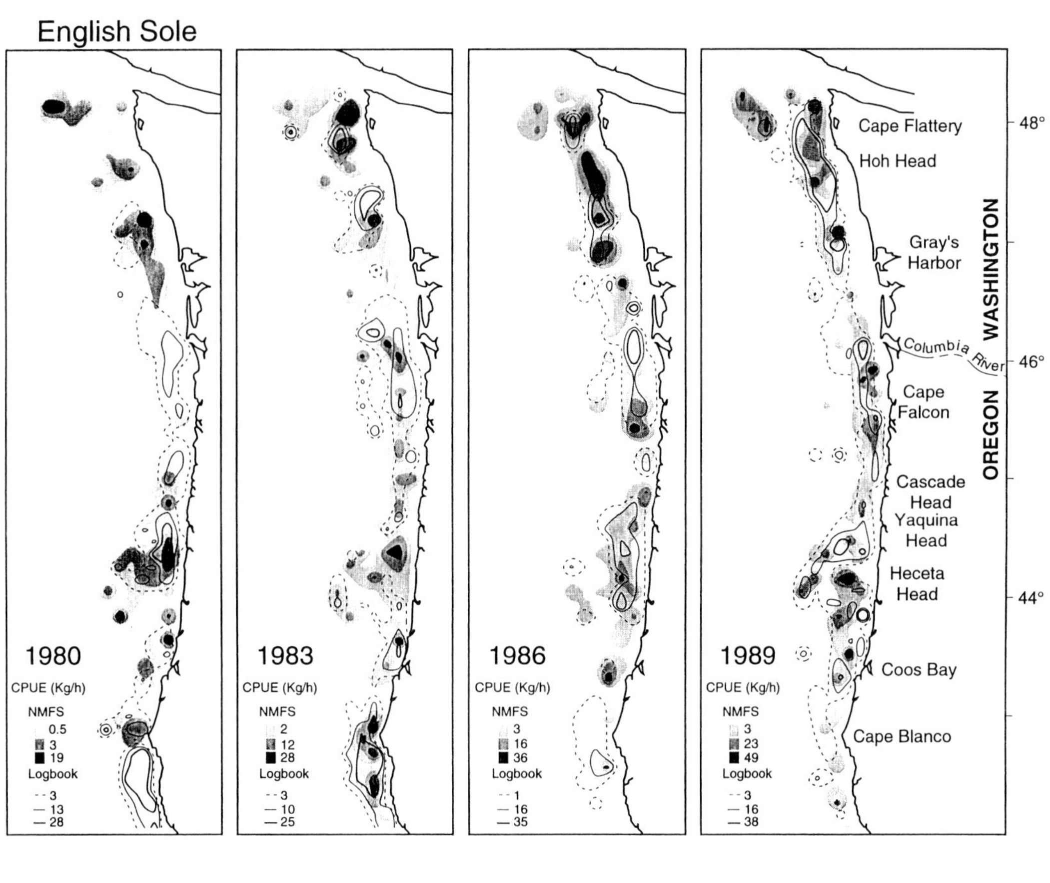

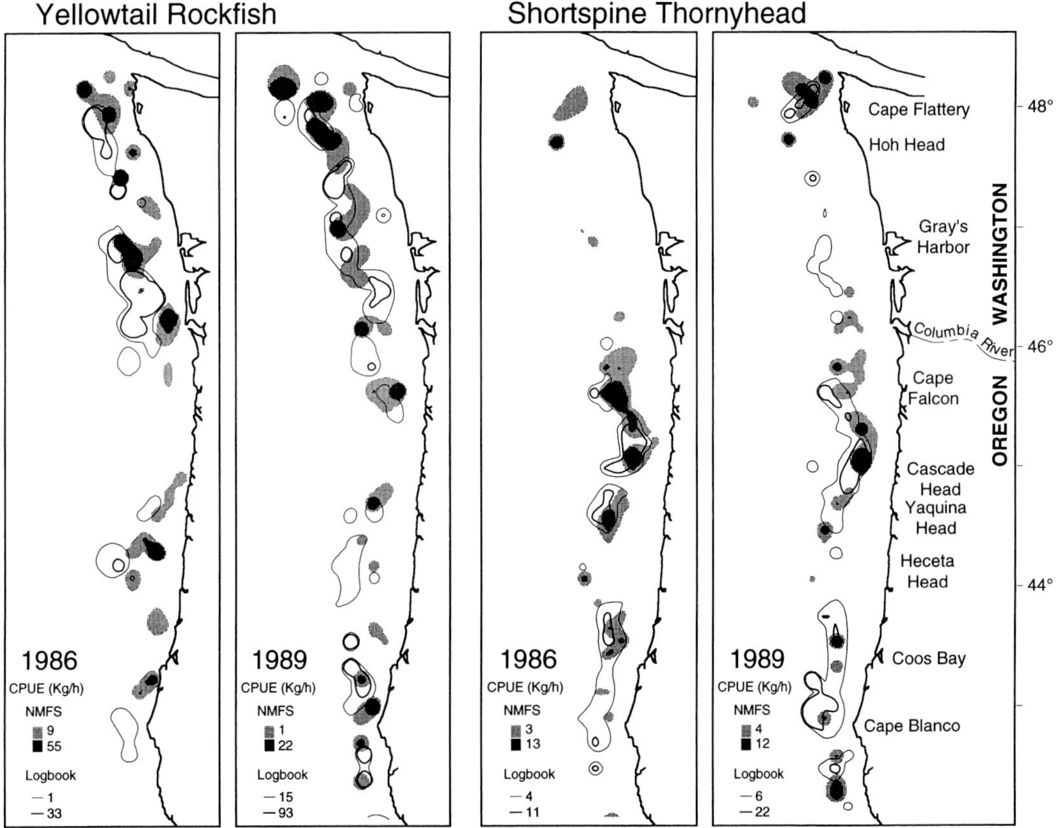

Fig. 2. Isopleths of the 50th, 75th, and 90th percentile of CPUE for commercial fishery and NMFS triennial trawl cruises for c) English sole, d) yellowtail rockfish, and e) shortspine thornyhead.

catch-per-unit-effort (CPUE), expressed in kg/h, for each species in each tow in both the research and commercial fishery databases.

Isopleths representing the 50th, 75th, and 90th percentile levels of CPUE were computed and plotted for each species, year, and data type, using commercial GIS and contouring software packages. For each species, log-book and research catch locations were compared by overlaying the CPUE maps and evaluating the geographic co-occurrence of the polygons defined by the isopleths. We computed the total surface area of polygons and the total area of overlap between the two data types. We also compared biomass estimates from the log-book data with similar values generated from NMFS data. We estimated biomass from the log-book data using an area swept methodology similar to that employed by NMFS (Gunderson and Sample 1980).

Results

For all species, the 50th, 75th, and 90th percentile levels of CPUE were of similar magnitude and spatially co-occurred (Fig. 2). In each of the data sets, areas of high catch rates occurred in similar locations from year to year. In a few cases, one of the data sets did not identify a high catch location. In those cases, the discrepancy was caused by lack of either commercial or research sampling effort.

In terms of percent overlap between data types, more than 66% of high log-book catch locations (defined by isopleths of 50th percentile of CPUE) occurred in the same area as high research catches for Dover sole, English sole, and sablefish (Table 1). Overlap ranged from 60–81% for the 50th percentile polygons, from 23–63% for the 75th percentile polygons, and from 13–47% for the 90th percentile polygons (Table 1). Overlap of surface area for shortspine thornyhead CPUE isopleths averaged 58% and 57%, for the 50th and 75th percentile of CPUE respectively. Overlap of surface area for yellowtail rockfish CPUE isopleths averaged 59% and 19%, for the 50th and 75th percentile of CPUE respectively.

Log-book biomass estimates for all species were similar in magnitude to those derived from research data, with the exception of log-book biomass estimates for Dover Sole which were three to six times greater than those derived from research catches (Fig. 3). Biomass estimates for English sole, sablefish, and yellowtail rockfish exhibited overlapping 95% confidence intervals for all years compared.

One of the primary advantages to using commercial catch data for understanding stock dynamics is that the fishery data provide a larger data set with which to interpret trends. In this study, log-book data showed trends different from those of the research data. Log-book biomass estimates for Dover sole exhibited a general decreasing trend from 1980 through 1985, and the trend appeared to be flat or slightly increasing from 1986 through 1989. The biomass trend derived from research catches was flat for Dover sole for the entire time period.

Log-book biomass estimates for sablefish exhibited a general decreasing trend with slight increases in 1986 and 1987. Biomass trends derived from research catches fluctuated out-of-phase with log-book estimates. Log-book biomass estimates for English sole exhibited a flat trend from 1980–1985, then a sharply increasing trend from 1985–1989. Biomass trends derived from research catches were similar. Log-book biomass estimates for yellowtail rockfish exhibited an increasing trend from 1985–1989. The two research data points resulted in a flat trend in biomass. Log-book biomass estimates for shortspine thornyhead rockfish exhibited a flat trend from 1984–1988. The biomass estimate for 1989 was twice that of previous estimates. The two research data points resulted in a decreasing trend in biomass.

Table 1. Summary of spatial overlap between NMFS and commercial catch areas. Values are percent overlap of surface areas enclosed by the 50th, 75th, and 90th percentile CPUE isopleths of each data type. Data reflect only areas that included sampling effort for both NMFS and log-book data sets

Species	Year	Percentiles		
		50th	75th	90th
Dover sole	1980	60.3	47.1	15.4
	1983	74.4	44.1	25.0
	1986	75.9	41.4	33.3
	1989	60.8	22.6	12.5
Average		67.9	38.8	21.6
Sablefish	1980	65.0	44.8	23.1
	1983	80.5	62.9	46.7
	1986	76.1	41.7	44.4
	1989	72.6	51.9	25.0
Average		73.6	50.3	34.8
English sole	1980	66.7	54.5	40.0
	1983	65.0	47.2	33.3
	1986	66.7	48.3	30.8
	1989	66.7	50.0	22.7
Average		66.3	50.0	31.7
Yellowtail	1986		41.4	16.7
rockfish	1989		38.5	21.4
Average			40.0	19.1
Shortspine	1986		64.7	80.0
thornyhead	1989		51.7	33.3
Average			58.2	56.7

Discussion

In the 1980s, fishery log-books were used by Gabriel and Tyler (1980), Gabriel (1982), and Tyler *et al.* (1984) to describe the distribution of commercial fishing effort off Oregon and Washington. Since that time, west coast fishery log-book data have received only limited use in stock assessments because scientists and managers have assumed that log-book CPUE data do not provide an accurate index of fish abundance. This assumption is based upon the concept that research catches reflect actual fish distribution and abundance, but commercial

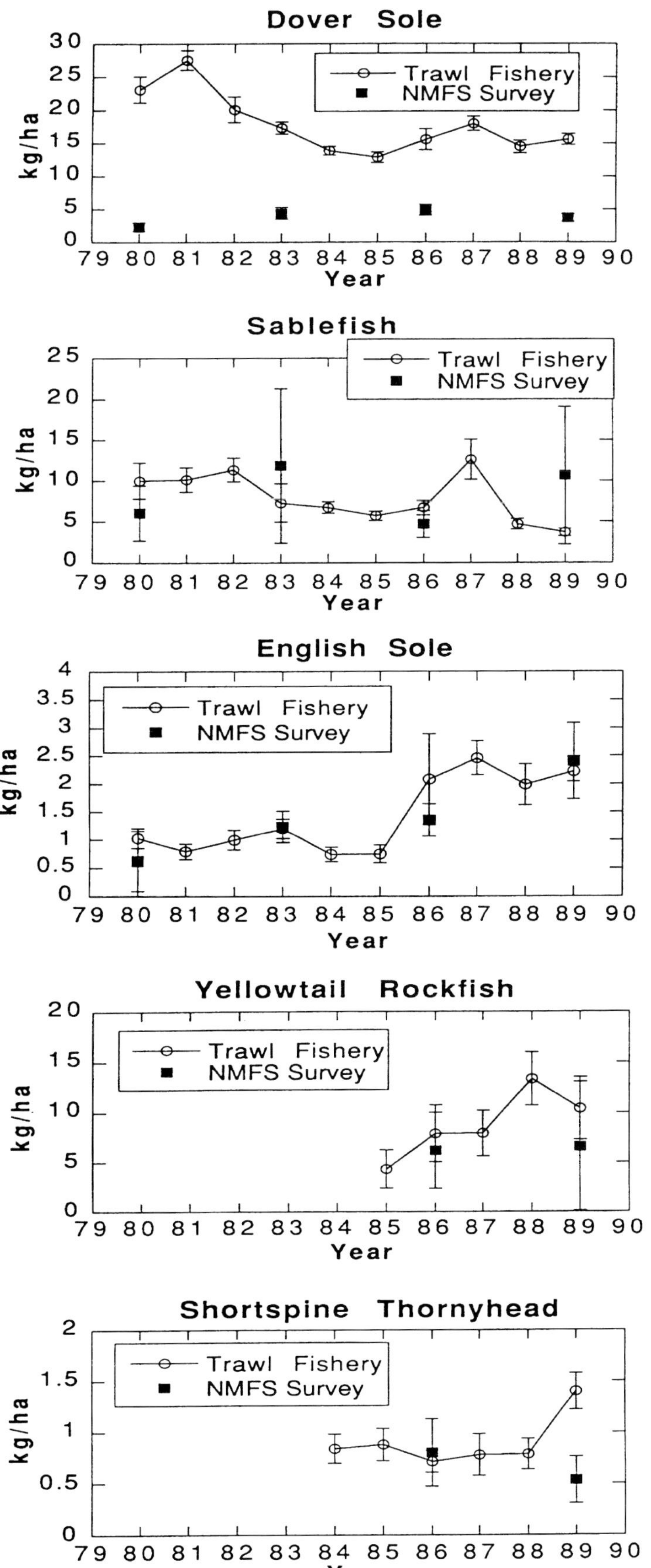

Fig. 3. Biomass estimates and 95% CI derived from trawls conducted from July to September in the commercial fishery and NMFS triennial trawl cruises.

fishing patterns and catches are more greatly influenced by market conditions, regulations, weather, and proximity to port.

Our analyses, however, indicated a reasonable correspondence between log-book and research data with a few exceptions. Both log-book and research data sets clearly showed specific areas that consistently produced high catch rates. The relationship between research and log-book biomass estimates varied from good to poor, depending on species. The relationship between data sets was stronger for species that were broadly distributed and for which fishing techniques were similar between research and commercial vessels.

The increased number of biomass estimates generated by log-book data provided a different view of biomass trends for several of the species, and also provided a greater degree of certainty in evaluating biomass trends. Both research and log-book estimates of biomass for yellowtail rockfish were highly variable. The high variance in the data indicates that neither research trawl nor commercial trawl gear adequately estimate the relative abundance of yellowtail rockfish. This suggests that, for some species, methods of estimating abundance other than trawl surveys are necessary.

Factors affecting the use of commercial fishery data

Cooperation from a large number of commercial fishermen is obviously critical for log-books to be useful. Proper collection, preparation, and screening of log-book data are also essential to maximizing the usefulness of the information. The procedures we used for collecting and processing log-book data included catch recording by fishermen, log-book collection, log-book screening, computer data entry, and error checking. The numerous fishing vessels and trips provided a large quantity of data, so we were selective of information included in the analyses. Only those log-books that were complete, legible, and had location information for each tow were analysed. We used only those log-books that had closely corresponding fish landing records, enabling the use of weights recorded at the dock. Computer error checking procedures included automated and manual review checks to detect tows that had unreasonable locations with respect to depth or distance from a previous tow. The application of these quality assurance steps resulted in retention of about 50% of the tows executed by the fishery.

The discard of fishes in the commercial fishery provides a potentially major discrepancy between log-book and research estimates of fish abundance. The problem is especially difficult to address if discard occurs at differential rates in different locations. If discard rates are known and are consistent in time and space, however, biomass estimates derived from log-books can be scaled for comparison with research estimates.

A critical assumption in the use of any trawl surveys is the fishing efficiency of the gear and resulting catchability of fishes. For estimating biomass, NMFS assumed a catchability of 1.0 for their surveys, indicating that all fish in the path of a trawl would be caught, and there is no herding effect (Methot *et al.* 1994). We made the same assumption for commercial trawls. For short periods, constant catchability may be a valid assumption. Over longer periods, such as a decade or more, the assumption of constant catchability is probably invalid. The technological advances in navigational electronics and fishing gear that have occurred in the last decade suggest that fishing efficiency and catchability have probably changed. Studies to determine catchability should be conducted periodically to ensure that CPUE estimates are comparable from year to year.

Use of log-book data to augment research

Our results suggest that commercial fishery log-book data provide estimates of fish distribution and abundance that are similar to research data. It follows, then, that commercial fishery log-book data may be used to evaluate or augment research data. Designing large research efforts typically involves compromises in terms of costs, temporal and spatial coverage, and sampling intensity within the study area. The commercial fishery samples every month of every year, in all trawlable areas, and at many thousand sites. Log-book data, then, may provide information to help identify sampling error, increase sample size, or fill spatial and temporal gaps in research data.

The additional information provided by log-books may be useful in tuning population models and helping managers to identify the degree of risk and uncertainty in stock assessments; thus aiding in the development of risk-based fishery models, such as proposed by Hilborn *et al.* (1993). Additionally, log-books may become increasingly important if area-specific management techniques become more common. Log-books may become the only cost-effective way to increase the amount of information available in specific areas.

Acknowledgments

We thank Ken Weinberg of NMFS for providing the triennial trawl data for our use. This paper was funded in part by a grant from the National Sea Grant College Program, National Oceanic and Atmospheric Administration, U.S. Department of Commerce, under grant number NA 89AA-D-SG138, through the California Sea Grant College. The views expressed herein are those of the authors and do not necessarily reflect the views of NOAA or any of its sub-agencies.

References

Coleman, B A (1986). The 1980 Pacific west coast bottom trawl survey of groundfish resources: estimates of distribution, abundance, age and length composition. *NOAA Technical Memorandum* NMFS-F/NWC-100. Seattle, WA. 181 pp.

Coleman, B A (988). The 1986 Pacific west coast bottom trawl survey of groundfish resources: estimates of distribution, abundance, age and length composition. *NOAA Technical Memorandum* NMFS-F/NWC-152. Seattle, WA. 145 pp.

Gabriel, W L (1982). Structure and dynamics of northeastern Pacific demersal fish assemblages. PhD thesis, Oregon State University, Corvallis, OR. 298 pp.

Gabriel, W L, and Tyler, A V (1980). Preliminary analysis of Pacific coast demersal fish assemblages. *Marine Fisheries Review* **42**(3–4), 83–8.

Gunderson, D R, and Sample, T M (1980). Distribution and abundance of rockfish off Washington, Oregon, and California during 1977. *Marine Fisheries Review* **42**(3–4), 2–16.

Hilborn, R, Pikitch, E K, and Francis R C (1993). Current trends in including risk and uncertainty in stock assessment and harvest decisions. *Canadian Journal of Fisheries and Aquatic Sciences* **50**, 874–80.

Methot, R, Lauth, R Shaw, F, and Wilkins M (1994). Assessment of the west coast sablefish stock in 1994. pp. B-1-B-87. In 'Appendices to the status of the Pacific coast groundfish fishery through 1994 and recommended acceptable biological catches for 1995'. Pacific Fishery Management Council, Metro Center, Suite 420, 2000 SW First Avenue, Portland, OR 97201.

PFMC (1991). Pacific Fishery Management Council groundfish PacFIN report. *Pacific States Marine Fisheries Commission*. 45 SE 82nd Drive, Suite 100, Gladstone, OR 92027.

Tyler, A V,. Beals, E L, and Smith, C L (1984). Analysis of logbooks for recurrent multi-species effort strategies. *International North Pacific Fisheries Commission Bulletin* **42**, 39–46.

Weinberg, K L, Wilkins, M E, and Dark, T A (1984). The 1983 Pacific west coast bottom trawl survey of groundfish resources: estimates of distribution, abundance, age and length composition. *NOAA Technical Memorandum*. NMFS-F/NWC-70. Seattle, WA. 376 pp.

Weinberg, K L (1994). Rockfish assemblages of the middle shelf and upper slope off Oregon and Washington. *US Fisheries Bulletin* **92**, 620–32.

Weinberg, K L, Wilkins, M E, Lauth, R R, and Raymore, P A Jr. (1994). The 1989 Pacific west coast bottom trawl survey of groundfish resources: estimates of distribution, abundance, age and length composition. *NOAA Technical Memorandum* NMFS-AFSC-33. Seattle, WA. 168 pp.

Cost-effective Recovery of Endangered Snake River Salmon

Brian Garber-Yonts and R. Bruce Rettig

Agricultural and Resource Economics, Oregon State University, Corvallis, Oregon 97331, USA.

Summary

Anadromous fish are affected by a diverse array of human activities. Managing declining stocks is complicated by the multitude of potential actions for improving survival. Managers must decide what subset of potential actions will achieve stock rebuilding objectives with the least harm to the regional economy. Cost-effectiveness analysis combines quantitative analysis of recovery actions using biological simulation models with economic estimates of the opportunity costs of recovery measures. Monte Carlo simulation estimates the probability of stock recovery under alternative scenarios. Linking recovery probabilities with the costs of associated scenarios allows the depiction of a cost-effectiveness frontier, which identifies the least cost alternatives achieving increasing levels of recovery probability. Results suggest that improvement over current survival levels may be achieved at costs much lower than those of current management. While results have important policy implications, neither equity concerns nor impacts on other biological populations are included in the analysis.

Introduction

Conserving anadromous fish in the US Pacific Northwest is one of the region's most strongly supported environmental policies. Over the past century and a half, economic development of the Pacific coast of North America has taken a severe toll on anadromous salmon populations (*Oncorhynchus tshawytscha, O. kisutch, O. nerka, O. keta, O. gorbuscha, O. mykiss, O. clarki*; chinook, coho, sockeye, chum, and pink salmon, steelhead and sea-run cutthroat trout, respectively) — the same species that define the region's history, culture and identity. Concern for the decline in many stocks, disillusionment with hatcheries and other measures to protect anadromous fish, and a rising social appreciation for protection of genetic diversity have contributed to a public commitment to conserve the remaining wild stocks. But will this support continue in the face of increasingly higher prices in the form of taxes, utility bills, and job losses?

Although concerns for these and other anadromous stocks, for which petitions are now pending, have long been shared by scientists, a careful review by Nehlson *et al.* (1991) sparked wide discussion and commitment to action. On 20 November 1991, the National Marine Fisheries Service (NMFS) listed Snake River sockeye salmon as endangered under the Endangered Species Act of 1973, and on 22 April 1992, Snake River spring and summer chinook (identified as a single stock in the NMFS listing) and fall chinook salmon were listed as threatened. Two years later, as stocks continued to decline, NMFS reclassified the Snake River chinook stocks as endangered.

On the one hand, realists understand that the consequences of more than a century of fishing, logging, mining, agriculture, urban growth, and construction of dams for flood control, navigation, irrigation, and hydropower cannot be undone. On the other hand, broad public support exists for maintaining diverse salmonid populations distributed broadly through as much of their historic range as possible. However, the growing costs have created opposition and caution even among salmon recovery proponents. For example, the US Congress recently capped expenditure for the Bonneville Power Administration on fish and wildlife programmes in the Columbia River Basin at $US435 million per year (with possible supplementation in years of poor water conditions or if required by order of the judicial system). From another perspective, a recent study (NMFS 1995*b)* placed the cost to carry out the Snake River Endangered Salmon Recovery Plan (NMFS 1995*a*) over the next seven years at $US388 million per year. This cost does not reflect costs borne by non-federal sources such as state and local governments and private parties. Measures to protect salmon species in the Sacramento River Basin, the Puget Sound, and coastal river systems, also present large and rising costs.

What recovery actions are proposed? The broad range of management actions discussed as alternatives for improving survival of Snake River salmon stocks reflects the fact that a salmon is subjected to many sources of human-induced mortality at each phase of its life cycle. The result for managers is that the range of possible management actions is very large and the selection of the set of measures to implement is a complex analytical task. Management alternatives analysed in this research include structural and managerial changes in the system of reservoirs and dams on the Snake and Columbia Rivers, including alterations in flow regimes and reservoir elevations during the smolt migration season, structural alteration of the dams to facilitate these and to improve passage at the dams by either spilling water over the dams (fish spill) or routing smolts through turbine bypass channels. Other alternatives are reduction of smolt predators, up-river collection of smolts and transport *via* barges to a release point below the most downstream dam, changes in the in-river and coastal fisheries that harvest Snake River salmon, structural changes to improve upstream passage at dams, and changes in irrigation withdrawals and summertime flows in the spawning and rearing habitats to reduce associated mortality. Individual actions were combined to comprise management scenarios for simulation modelling, described below. Many other management alternatives have been proposed but were not included in this research due to lack of sufficient data to allow quantitative analysis.

The biological models

How can we possibly grasp the implications of so many simultaneous policy changes in an ecosystem driven by wide climatic variation and continuing human interventions? 'Ecosystem models are always wrong, in the sense that reality conforms to their numerical projections only very rarely. Models are indispensable because without them human misunderstanding persists, unaware of its errors' (Lee 1993 p. 62). Among the simulation models to which Lee was referring are two used in our analysis. The Columbia River Salmon Passage (CRiSP) model developed by the Center for Quantitative Science (CQS) at the University of Washington traced salmon smolts from spawning habitat or hatchery to the ocean. Model outputs from CRiSP were inputs to the Salmon Life Cycle Model developed by researchers at the US Forest Service and Resources for the Future.

The Columbia River Salmon Passage Model (Center for Quantitative Science 1993) simulates the migration of smolts from an identified point of release (hatchery releases and wild stocks are not treated differently in CRiSP, but may be identified and modelled separately) through the hydropower system on the Snake and Columbia River and tributaries, to the Columbia River estuary. The model incorporates several submodels, including travel time, reservoir mortality (predation), dam passage, nitrogen supersaturation, and a flow/velocity relationship which uses a hydroregulation model (SAM, HYSSR, and HYDROSIM, see CQS 1993, pp. 7–16 for discussion) output to indicate daily flow levels. CRiSP1.4 can be run in a mode that simulates passage under a single set of conditions and is thus effectively a deterministic model, or in Monte Carlo mode. The latter, which is most relevant to this paper, involves running the model iteratively, allowing each stochastic variable to vary within the user-defined range to generate a probability distribution for survival to the estuary.

Calibration of the model involves extensive geographic, hydrologic and biological data sets and is an ongoing process under the *aegis* of the Center for Quantitative Science at the University of Washington. Since the beginning of the modelling reported in this paper, both the model and calibration have been superseded. Further analysis should include the most current model and data. However, given the large range of uncertainties inherent in modelling a system as complex as that of the Columbia River and the life-cycle of anadromous fishes, caution should always be taken in interpreting output that allows relative comparisons at best and should not be regarded as definitive.

The Stochastic Life-Cycle Model (SLCM) developed by Lee and Hyman (1992) simulates the entire life cycle of anadromous fishes and incorporates several stochastic processes in both fish biology and habitat. SLCM uses output from CRiSP (mean percent smolt survival and coefficient of variation) to represent survival in the juvenile stage of the salmon life cycle. The model calculates survival through the successive life stages from an initial specified escapement of spawners into a given sub-basin, through egg, fry, smolt, and adult stages in the ocean and river, and finally to the number of adult females returning to spawn. Unlike CRiSP (when run in Monte Carlo mode; SLCM performs only Monte Carlo simulations), which initiates each run of the model with the same number of smolts released and generates a probability distribution of percent passage, SLCM is run for successive generations and may be run for as long a period, e.g. 50 years (approximately 10 generations), as the user specifies. Thus, the SLCM simulation results are in the form of frequency distributions for the population size of each life stage after the specified number of years. The frequency distribution of sub-basin escapements of spawning adults can thus be interpreted to determine the probability of the adult population maintaining the target population sizes for the respective stocks.

The output from CRiSP, specifically the mean and coefficient of variation of smolt passage to the estuary, is used in the SLCM both in the calibration of the model to historical conditions, and as parameter settings for forward projections. Output from the SLCM is then used to scale alternatives on the Y-axis of the cost-effectiveness frontier. The cumulative percentage of simulations resulting in populations above the respective target population level identified by NMFS represents, for the purpose of this analysis, the probability of survival of a given endangered stock.

COSTS OF ALTERNATIVES

The costs of the recovery actions were the results of a long-term planning project undertaken by the US Corps of Engineers and associated agencies called the System Operation Review, original analysis by Daniel Huppert and David Fluharty of the University of Washington and other material produced by the NMFS-convened Economic Technical Committee, co-ordinated by Huppert and Fluharty. In this section, we review those research efforts, identify major categories of recovery costs, and describe the order of magnitude of some of them.

Many federal, state, and local agencies share responsibility for managing the wide variety of services of the Columbia River Basin. The Columbia River System consists of 30 major dams built by the federal government and both federal and non-federal facilities providing power production, irrigation, navigation, flood control, recreation, fish and wildlife habitat, and municipal and industrial water supplies. There is a need to co-ordinate a system operating strategy for managing the multiple uses of the systems into the 21st century which will (a) provide interested parties with a continuing and increased long-term role in system planning, (b) renegotiate and (c) renew a contractual arrangement among the region's major hydro-electric-generating utilities and affected federal agencies, and renew or develop new contracts that divide Canada's share of Columbia River Treaty downstream power benefits and obligations among three participating public utility districts and the Bonneville Power Administration (US Army Corps of Engineers *et al.* 1995). The System Operation Review (SOR) began in 1990 and was completed late in 1995. This planning exercise drew on a set of economists in the three agencies, consultants to the agencies, and sister federal agencies (especially the Forest Service and the National Marine Fisheries Service). Operating under the public consultation requirements of the National Environmental Policy Act, much additional information and review of this study were obtained from state and local agencies and from affected interest groups.

As the SOR went on, policy analysts found it difficult to separate the charge of their study from other changes taking place, many driven by concern for declining salmon stocks, and from studies being carried out by other groups under related, but different laws. As part of the Endangered Species Act processes, NMFS contracted with researchers at the University of Washington to develop estimates of the economic impacts of designating critical habitat for Snake River salmon stocks, and then continued this arrangement with estimates of costs of recovery. The subsequent reports (Huppert *et al.* 1992; Huppert and Fluharty 1995) were based on original analysis and information from consultation with an Economic Technical Committee. This committee consisted of representatives of various federal agencies (especially those developing the SOR), the Northwest Power Planning Council (which had its own charge of identifying changes in river operations to protect fish and wildlife in the Columbia River Basin), states, Indian tribes, groups making heavy use of the river such as irrigators and major power users, and the general public.

Salmon recovery costs associated with hydropower dominate other categories. Most policy analyses, including the SOR, assume that foregone hydropower will lead to the purchase of alternative sources of power generation, that this will increase power prices, and that the quantity of power sold would decline with higher prices. Alternatives that inhibit the use of river flows for electricity generation during peak demand periods yield large cost estimates. The baseline for fish spill levels is an agreement between the Northwest Power Planning Council and the Corps of Engineers to try to achieve 70% fish passage efficiency (the fraction of smolts routed downstream through routes other than through hydro-power turbines) for yearling migrants and 50% fish passage efficiency for sub-yearling migrants. If this fish spill were to stop, hydroelectricity costs would decline by approximately $US30 million per year. On the other hand, if efficiency rates were to rise from 70/50 to 80/70, costs would rise by $9 million per year.

The SOR identified many options for modifying flows in the river. If current flow requirements were changed to follow patterns found in the decade before Snake River wild salmon were listed under the Endangered Species Act, hydropower costs would decline by $45 million per year. Drawing the Snake River down to 'natural river' level for 4.5 months would increase power costs by $85 million. A less expensive natural river drawdown for 2 months, which appeared in an earlier draft of the SOR and was included in our analysis, is no longer under consideration, while a year-around drawdown, which was added recently and was not included in our analysis, would raise power costs above baseline by $167 million per year.

Next to hydropower costs in size are the public expenditures to modify dams so they can operate at lower river levels. The amortized value of these expenses ranges from $89 million per year for a drawdown to a natural river level for 4.5 months per year to $7.8 million to draw down only Lower Granite dam to a fixed level below the minimum operating pool. The drawdown scenarios also impose other costs. Costs to irrigators to move pumps and lift water higher would range from $1.4 million to $4.5 million per year depending on the depth of drawdown and how long the drawdown took place. Natural river conditions would stop barge traffic on the lower Snake River, resulting in higher costs to their customers ranging from $2.1 million per year for a part-time one-dam drawdown to $37.5 million per year to maintain the lower Snake at natural river levels year round. Reduced recreation, mostly associated with reduced access to the river, would add costs ranging from $90 million per year for the natural river operation to $40 million for a more limited drawdown. The value of recreationally and commercially caught salmon would decline by somewhere between $6 million and $11 million per year depending on the drawdown option selected. Drawdown would also add between $3 million and $5 million per year to costs of municipal and industrial water use.

The costs of flow changes and reservoir drawdown overwhelm other, but important cost categories. Predator control, especially expenditures to reduce squawfish which congregate below dams and prey on migrating smolts, is an essential element in recovery options. Beginning with the 1991 expenditure of a half million dollars per year, our model runs suggest that an additional 90% increase in control effectiveness could cost $23.8 million per year, but would be a 'best buy' in recovery strategies, assuming that high levels of predator control are biologically advisable. Compensating fishermen for reduced harvests for a sum of just over $6 million per year and improvements to adult passage costing about $2.5 million are other investments with high returns in fish survival.

Finally, much debate is taking place in the Pacific Northwest over costs to maintain and improve riparian habitats critical to spawning and rearing habitat. While detailed analysis of this phase of the cycle was not included in the research, we follow the lead of other analysts who argue that provision of adequate salmon habitat in these areas is mandated by laws other than protection of Snake River endangered species. Since these habitat costs are not attributed to salmon recovery, they become part of the baseline for all alternatives and do not appear in this analysis.

TRACING THE COST-EFFECTIVENESS FRONTIER

Cost-effective recovery alternatives are those combinations of actions that attain a specific probability of species recovery at less cost than any other alternatives with the same biological effectiveness; alternatively, they are those combinations that maximize the probability of recovery for any given level of expenditure. The set of cost-effective recovery alternatives generate an envelope which is called the cost-effectiveness frontier. Because of the controversial nature of some biological modelling assumptions and the incompleteness and imprecision of cost estimates, the combinations discussed in this section reflect only crudely the idea of cost-effectivness frontiers.

Figure 1 displays two cost-effectiveness frontiers. In Fig. 1A, points marked with ▲ indicate the most cost-effective recovery strategies. Why do so many of these indicate high probabilities of salmon recovery while spending less money than the baseline case suggests? Although the scenarios are complex, the principal reason is that baseline hydrosystem management includes spills and flow mixes that are very expensive and may be less biologically effective than expanded predator control programmes, tighter restrictions on harvest, expanded smolt-transportation programmes, and additional adult passage improvements. However, the biological scientific community is sharply divided on the effectiveness of barging smolts past the dams. Although the data suggest that survival of smolts to release points downriver is improved with smolt-transportation, many argue that disappointing adult survival statistics suggest that smolt transportation creates stresses that result in increased ocean mortality.

To address this controversial issue, a second cost-effectiveness frontier, with points marked with ■, is traced for options that exclude use of smolt transportation. All options that would present recovery probabilities above 50% include major commitments to predator control and elimination of all commercial and recreational fisheries that encounter endangered salmon stocks. To achieve recovery probabilities above 90% also requires drawing the river down to its natural level.

In Fig. 1B, alternatives are identified that would recover all or nearly all endangered spring/summer runs and 90% or more of the endangered fall chinook runs. This Figure suggests that some strongly supported measures, such as increased fish spill and drawdown, would cost much more than other options, while providing almost no gain in species recovery. This dramatic relationship can also be used to identify weaknesses in the work

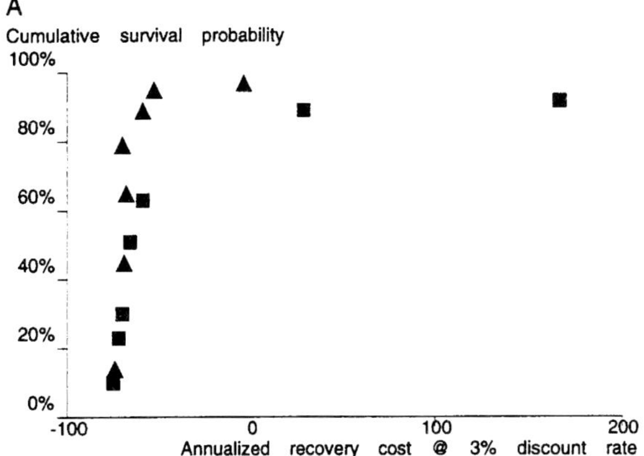

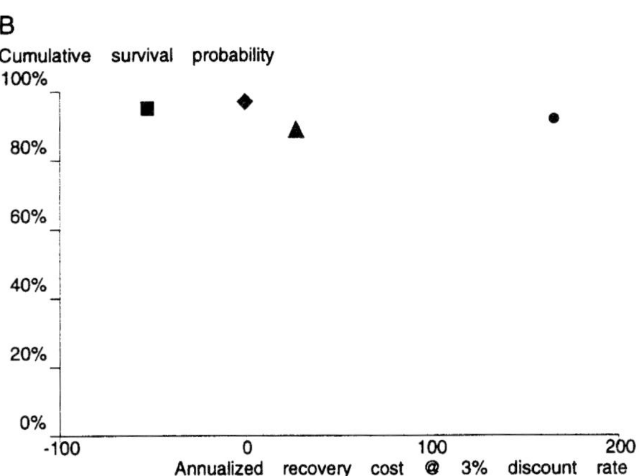

Fig. 1. Cost-effectiveness frontier for recovery of endangered wild Snake River chinook salmon. Costs (($ millions) are annual payments amortized at 3%. **A.** Fall run recovery options. Symbols: ▲, least-cost alternatives if smolt transportation is restricted; ■, alternatives without restricting the use of smolt transportation. **B.** Alternatives to recover both fall and summer/spring runs (all include elimination of harvest). Symbols: ■, baseline predator control, no fish spill, 5-dam smolt transport, pre-baseline flow, baseline upstream passage; ◆, 90% predator reduction, increased fish spill, 4-dam smolt transport, pre-baseline flow, baseline upstream passage; ▲, 90% predator reduction, baseline fish spill, no smolt transport, baseline flow, baseline upstream passage; ●, 90% predator reduction, no fish spill, no smolt transport, 4.5-month drawdown to natural river, adult passage improvements.

we and other cost-effectiveness analysts have done. Some of these concerns can be addressed with this methodology and some fall outside its purview.

Addressing uncertainties with sensitivity analysis

The scientific community in the Pacific Northwest is sharply divided about the biological effects of the major policies discussed in this paper: smolt transportation, drawdown of reservoirs, and shaping of water flows. For example, there is consensus that stress-induced mortality results from nitrogen supersaturation associated with fish spill. Also, everybody agrees that concentrating smolts within barges may create stress, facilitate the exchange of disease organisms, and have other harmful effects. Whether the positive gains in accelerating the speed of downstream migration and protection from other sources of morbidity and mortality offset these losses cannot be readily settled with limited data and scientific information. However, the relative importance of uncertainties with regard to model results can be addressed with sensitivity analysis.

The structure of both the SLCM and CRiSP models facilitates the comparison of model dynamics under different assumptions about functional forms, parameter distributions and point values. Comparison of model results under different assumptions allows identification of key areas of uncertainty.

Sensitivity analysis for a single management scenario is, however, very time consuming and is only tractable for a small set of alternatives. The approach taken in this analysis of comparing a broad range of management scenarios precluded the analysis of uncertainty of results. The next phase of the research should focus on a sensitivity analysis of the alternatives identified in this research as cost-effective.

To keep the scope of the research reported in this paper manageable, we concentrated on politically plausible options. One extension that could be pursued might be to select recovery policies that conformed to models of social equity. For example, the political process includes many statements to the effect that all parties that contributed to the decline in salmon should share in recovery. Because we focussed on cost-effective recovery, we did not wish to exclude the possibility that some groups might make greater sacrifices than others. Just such large losses occur with the sharp reduction in commercial and recreational fishery harvest in all the cost-effective strategies identified. If social considerations required equal sacrifices, one could translate this into selections of strategies to be modelled. For example, harvest strategies could have been limited to no more than a 50% (or 30% or 10%) reduction in fishing rates.

Recognizing missing economic information

Every policy analyst who attempts to identify the costs of fish and wildlife protection in the Columbia River Basin notes that the categories of costs identified are incomplete and that most cost estimates are crude. Even with this degree of care most omit a critical category, which we also have had to exclude for lack of information. This is the category that economists call transaction costs.

What are transaction costs and what is their relevance to fishery decision making? In general terms, they are the costs associated with negotiation and enforcement of an agreement between two or more parties. Consider one difficult example, which relates to our estimate of the cost of reducing the size of fishing fleets.

In our analysis, we estimated the decline in net revenues received by fishermen due to greater harvest restrictions. For example, if regulations require that harvests be cut in half, the cost of harvest restriction would be half of the baseline level of net revenues earned by fishermen. However, negotiation and enforcement of harvest restrictions increases transaction costs (which we overlook). Rather than buying out fishermen, seasons and other regulations are set that will lead to a reduced level of fishing with the greatest reductions coming in areas with higher incidences of endangered fish. Before these regulations are imposed, much research takes place. Landings records are collected and analysed by state and federal agencies. Streams are sampled to learn the number of returning spawners. Records are kept at hatcheries and analysed. Tagging experiments are carried out so that agencies can estimate the time and area that wild fish are most likely to be harvested.

Collection and analysis of data is only the beginning. Next, results are distributed to affected industries, other agencies, and the interested public. Town hall meetings are held to discuss implications of the results with fish harvesters. State fishery commissions and regional fishery management councils meet to develop initial recommendations, and then public hearings are held, with careful documentation of what is said at hearings and what is mailed into agencies. After regulations are set, several agencies must enforce the regulations and monitor activities to learn how well regulations are enforced. Violators either pay uncontested fines or they and the public pay additional costs for adjudication. Throughout the process, fishermen incur additional costs to prove they are in compliance and they may also use costly strategies to evade detection of illegal activities.

Thus, part of recovery costs is the difference between what management costs would be without endangered species protection and the costs with additional protection. If public agencies decide to purchase fishing licences, and possibly also vessels or gear, from fishermen, additional transaction costs are created. These costs may be sound investments if a smaller fleet ensures greater profitability and viability of the fishing fleet and lower future regulatory costs and if these payments address social equity considerations by avoiding placing an undue fraction of recovery costs on a few people.

Failure to consider transaction costs means that we and other analysts may be sharply underestimating costs of some options such as harvest reduction. Our analysis concludes that a very cost-effective alternative is to reduce the ocean harvest by British Columbia trollers on stocks that include Snake River fish. However, these reductions are part and parcel of difficult negotiations between the United States and Canada over stock interception (Huppert 1995). Our simplified assumption that costs can be approximated by the lost net value of reduced Canadian harvests ignores the costly, difficult, and protracted international negotiation process; it also ignores the expenses within Canada if the Canadian government had to enforce reductions in the ocean troll fishery.

Conclusions

Cost-effectiveness analysis cannot identify the best approaches and cannot exclude specific approaches. However, it can play a helpful role in fishery policy by identifying costly and problematic recovery options. Although not formally part of the analysis, equity considerations can be addressed by compensation to those bearing large shares of the cost (Berry and Rettig 1994) and can help guide choice of feasible and practical alternatives. Finally, cost-effectiveness can be highly complementary to biological analysis by identifying the value of missing information and encouraging adaptive management approaches to learning more about biological systems while taking timely actions to conserve our valuable endangered species.

References

Berry, H, and Rettig, R B (1994). Who should pay for salmon recovery? Pacific Northwest Extension Publication, PNW 470.

Center for Quantitative Science (1993). Columbia River Salmon Passage Model, CRiSP1: Documentation for Version 4. (Center for Quantitative Science, University of Washington: Seattle).

Huppert, D D (1995). US/Canada Salmon Wars: Why the Pacific Salmon Treaty has not brought peace, New Directions in Marine Affairs, Report 1 (School of Marine Affairs, University of Washington: Seattle).

Huppert, D D, and Fluharty, D L (1995). Economics of Snake River salmon recovery: A report to the National Marine Fisheries Service. (School of Marine Affairs, University of Washington: Seattle).

Huppert, D D, Fluharty, D L, and Kenney, E S (1992). Economic effects of management measures within the range of potential critical habitat for Snake River endangered and threatened salmon species. A report to the National Marine Fisheries Service. (School of Marine Affairs, University of Washington: Seattle).

Lee, D C, and Hyman J B (1992). The Stochastic Life-Cycle Model (SLCM): simulating the population dynamics of anadromous salmonids. USFS Intermountain Research Station, Research Paper INT-459 (US Forest Service: Boise, Idaho).

Lee, K (1993). 'Compass and Gyroscope: Integrating Science and Politics for the Environment'. (Island Press: Washington, DC).

Nehlson, W, Williams, J, and Lichatowich, J A (1991). Pacific salmon at the crossroads: stocks at risk from California, Oregon, Idaho, and Washington. *Fisheries* **16**(2), 4–21.

NMFS (1995*a*). NMFS Proposed Recovery Plan for Snake River Salmon (US National Marine Fisheries Service: Portland, Oregon).

NMFS (1995*b*). NMFS Proposed Recovery Plan, Chapter VI, Addendum, Agency budget costs and implementation schedule for proposed recovery tasks (US National Marine Fisheries Service: Portland, Oregon).

US Army Corps of Engineers (1995). US Department of Energy Bonneville Power Administration, and US Bureau of Reclamation (1995). Columbia River System Operation Review: final environmental impact statement, Appendix O, Economic and Social Impact (US Army Corps of Engineers: Portland, Oregon).

Fuzzy Control Theory Applied to American Lobster Management

Saul B. Saila

University of Rhode Island, Graduate School of Oceanography, Narragansett, Rhode Island 02882-1197 USA.

Summary

Quota-based management of an offshore fishery for the American lobster *Homarus americanus* in the northwest Atlantic region was investigated as an adaptive control system which involved fuzzy control theory. The system model consisted of two major modules. First, a neural network module provided a forecast of lobster stock abundance two years ahead, based on a time series of trawl survey data. Results of this forecast, as well as temperature data, were then combined in a fuzzy rule-based expert system to provide final quota estimates. Simulation studies with this system indicated that it was robust and accurate relative to more traditional approaches. Fuzzy control theory allows control of complex processes which cannot be effectively treated using conventional control theory, because the precise nature of the many factors and their interactions are unknown.

Introduction

Adaptive management of natural resources based on modern control theory was thoroughly explored by Walters (1986). However, applications of this methodology into subsequent fisheries management-related activities have been few. There may be several reasons for this, but one of the most important is believed to be that marine fisheries are such complex processes that they cannot be satisfactorily controlled by the results of conventional control theory. Suitable methods for the effective implementation of adaptive control for complex non-linear systems with high levels of uncertainty have become available fairly recently, but apparently they have not been adequately recognized by fisheries scientists.

Conventional control theory has had tremendous success in uses where the system is very well-defined, such as in missile and space vehicle guidance. However, this theory has not been very effectively applied to control complex systems such as commercial fisheries, because the precise structure of the system is virtually unknown. Nevertheless, there have been some applications. For example, simulation studies in which linear feedback controllers were coupled with mathematical models of populations have been reported by Tanaka (1980) and Muratori *et al.* (1989). Fishery systems were considered to be too non-linear to be well controlled by such linear feedback controllers (Horwood *et al.* 1990; Jacobs *et al.* 1991). They identified fishery management as a complex non-linear, as well as stochastic control problem, and suggested that this class of problems could be solved suboptionally by mathematical simplification. The above and other techniques that have been applied to adaptive control with varying success seem to have one important common feature. It is that they rely heavily upon a valid and accurate model of the process to be controlled. If a process cannot be usefully modelled within the framework assumed by the theory, then satisfactory control cannot be achieved.

Many complex processes (including commercial fisheries), may have a serious model deficiency. That is, although models used to simulate fisheries are often mathematically elegant, they are not realistically validated and may contain parameters which are not time-invariant and which also have high variances. In addition, some fisheries are characterized by a large amount of *a priori* information available only in qualitative form by having performance criteria which are only specified linguistically. A simple example of the latter might be recruitment success characterized as very high, high, moderate, low, and very low.

What seems to be needed for situations such as the above, is a technique for handling qualitative information and uncertainty in a rigorous way. This technique will necessitate making some assumptions about the way that inexactness is described, but in doing so, it must be powerful enough to handle the many ways this arises. Furthermore, its implementation for control purposes must be straightforward and not too computationally intensive. Fuzzy control engineering is believed to be such a technique. Two recent volumes which address fuzzy control theory and engineering with applications are Pedrycz (1995) and Bardossy and Duckstein (1995). This brief paper in no way explains fuzzy adaptive control, and the interested (or sceptical) fishery scientist is referred to the above references for details. Suffice it to say that fuzzy set theory has evolved into many important application areas, including adaptive control of complex non-linear systems containing considerable uncertainties. Fuzzy sets are a generalization of ordinary set theory, which allows imprecise and qualitative information to be expressed in an exact way. The notion of a fuzzy set, when combined with fuzzy implication statements and fuzzy composition, produces a fuzzy control system. This system is a collection of implication statements which causally link input and output fuzzy sets to replace the conventional mathematical model of control theory. The composition operation allows the model to be used in the same way that matrix multiplication allows a state space model to be used. It is, therefore, suggested that fuzzy control theory may offer a rigorous and practical new twist to addressing complex adaptive control problems in fisheries which include uncertainties and critical information expressed in linguistic form. A recent issue of *Ecological Modelling* dealt exclusively with fuzzy logic in which Tuma *et al.* (1996) reached essentially the same conclusion for complex industrial production processes.

Some fuzzy control concepts

Fuzzy set theory, a generalization of conventional set theory, was introduced by Zadeh (1965). Industrial applications of fuzzy controllers ranging from cameras to trains have resulted in a multi-billion dollar industry in Japan. However, applications of these concepts to ecology and fisheries have been very limited to date. These few applications to ecological modeling include Bosserman and Ragade (1982) and Salski (1992). Saila (1992) and Sakuramato (1995) provide some initial applications to fisheries. The following material attempts to briefly describe the concept of fuzziness in a form which seems appropriate for control problems in fisheries science. Both this section, as well as the control systems model to be described, utilize information related to the American lobster fishery. This section is largely derived from Pedrycz (1995) and Tong (1977).

Let us specify linguistic measures of American lobster relative recruitment success on the closed interval (0.10, 0.50). Now suppose that one of the measures of relative recruitment (R) is about 0.25. An ordinary set which describes this can be expressed in terms of a membership function μ, which can take values of either 0 or 1. If $\mu(R) = 0$, then the relative recruitment value R is not a member of the set. If $\mu(R) = 1$, then it is a member of the set. Graphically, this might be represented by the rectangular function shown in Figure 1a. A fuzzy set which represents the same concept is shown in Figure 1b. In this case, it is evident that the membership function takes on all values between 0 and 1.

The ordinary set of Figure 1a is very precise in its meaning, by having a sharp transition from membership to non-membership. The fuzzy set (Fig. 1b) allows the qualitativeness of the measure to be reflected in a gradual membership transition. Using this idea, qualitative information can be represented mathematically and handled rigorously.

Fuzzy sets are combined in a manner similar to ordinary sets by means of some simple definitions. For details well beyond those

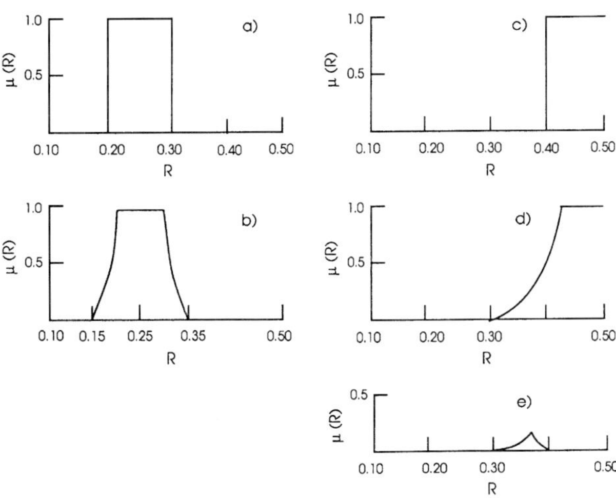

Fig. 1. **a.** Illustration of a non-fuzzy set of relative recruitment indices for the American lobster derived from trawl survey data. This figure illustrates a value of about 0.25. **b.** Illustration of a fuzzy set of relative recruitment indices similar to Figure 1a. This figure illustrates a value of about 0.25. **c.** A non-fuzzy set of relative recruitment indices. A value of relative recruitment much greater than 0.25 is shown. **d.** A fuzzy set of relative recruitment indices similar to that produced in Figure 1c. **e.** Illustration of the intersection of two fuzzy sets of Figure 1c and d.

introduced here, see Li and Yen (1995), for example. The union of two sets (A+B) is described by the membership function:

$$\mu A{+}B\ (x) = \max\ [\mu A\ (x),\ \mu B\ (x)]$$

The intersection of two sets (A.B) is defined by

$$\mu A.B\ (x) = \min\ [\mu A\ (x),\ \mu B\ (x)]$$

Negation (A) is defined by

$$\mu A\ (x) = [1 - \mu A\ (x)]$$

Now suppose that there is another estimate of relative recruitment much greater than 0.25. A non-fuzzy set to express this might be as shown in Figure 1c with a similar fuzzy set as shown in Figure 1d. Then the intersection of these two measures produces the null set in the non-fuzzy case, and the set shown in Figure 1e for the fuzzy case. This is a very basic property of the fuzzy concept. That is, linguistic measures are not necessarily mutually exclusive, being able to provide information in what seems to be disjoint situations.

We next consider the concept of a fuzzy implication statement. In terms of an extension to the example provided, a linguistic functional relationship between a temperature anomaly (T) and relative recruitment might be: IF {temperature anomaly is large negative} THEN {relative recruitment value is much greater than 0.20}.

If a fuzzy set of temperature anomalies is provided, the implication is a two-dimensional fuzzy set *P* in the product space of temperature anomalies and relative recruitment values. This is given by

$$\mu_P = \mu A.B = \min\ [\mu A\ (T),\ \mu B\ (R)]$$

In general, a useful description of the behavior of a relationship will require several fuzzy implication statements (rules), each of which will produce a set $P^{(i)}$. The individual $P^{(i)}$s are combined to provide an overall *P* by calculating the union of them all such that

$$P = P^{(1)} + P^{(2)} + \ldots + P^{(N)}$$

In this case, $P^{(i)}$ is the two-dimensional set produced by the ith rule, and N is the number of rules.

The above collection of rules is considered to be a fuzzy algorithm. It is a very powerful concept since it is the fuzzy equivalent of the system i.e. a fuzzy model. The dual of this is a fuzzy algorithm that describes the behavior of a controller.

Fuzzy composition is a required concept for exploiting the basic ideas of fuzzy sets. Basically, it allows an algorithm to convey information about situations which are not normally encountered. However, it is difficult to explain briefly, and the interested reader is referred to Chapter 3 of deSilva (1995) for an effective and detailed explanation of fuzzy composition.

The basic idea of a fuzzy set and the two extensions to it, namely implication and composition, are thought to be the essential theory behind the control system to be described herein.

In many areas of fisheries science, it is believed that the concept of fuzzy sets better reflects the inherent uncertainty of the data than point estimates, such as are usually applied. Indeed, although seemingly less precise than so-called crisp data, fuzzy data are, in fact, a more adequate representation of reality under uncertain conditions (Klir and Folger 1988). Therefore, it is suggested that somewhat subjectively-defined membership functions should not be judged as weak model parts. Instead, they should be judged as the best available technique for expressing uncertainty, which seems to be ubiquitous in biological systems.

Method

An integrated software platform for design and implementation of adaptive and expert systems termed O'INCA™ was utilized for developing the quota-based American lobster management system. The O'INCA design framework allows for integration of neural networks and fuzzy logic, as well as combining a graphical user interface (GUI), design validation, simulation/debugging, C code generation and design documentation into a unified environment. In spite of the sophistication of this software, it does not relieve the user of important system design problems, such as the appropriateness of measurements for inputs and determining the primary fuzzy sets to be used together with the rules which will form the control algorithm. Another basic problem is concerned with the numerical descriptions of linguistic variables, their range, and quantification. Although the capabilities of the software platform far exceed the requirements for this application, they serve effectively for the application.

System design

It is generally recognized that a quota-based management system is appropriate for single-species fisheries, especially for invertebrates with relatively long-life histories, such as the American lobster (Squires *et al.* 1995). The following describes what is believed to be a novel approach to such an undertaking. The background information and data used for this study were obtained primarily from Anon. (1993), A Report of the 16th Northeast Regional Stock Assessment Workshop (16th SAW).

The model framework for the quota-based management system based on the Georges Bank and south offshore fishery for the lobster is briefly described in this section. The National Marine Fisheries Service (NMFS) has conducted annual surveys in this region since 1970. The management system utilizes annual standardized trawl surveys which provide relative abundance indices for juveniles and fully recruited females, as well as surface temperature anomalies in the Georges Bank region as inputs. Based on this information, the system outputs a quota projection for two years beyond the current date. This time frame is considered necessary in order to accommodate any regulatory changes for new quota alternatives through the management agencies.

Table 1 provides definitions and ranges for the variables used in the system design. A diagram of the management system is illustrated in Figure 2a to indicate the modules and variables which were used. One neural network module and one fuzzy logic were required. Based on the input variables shown in Table 1 and Figure 2a, the neural network is used to develop a projection of the survey index value for fully recruited adults two years into

Table 1. Definition of variables used in the construction of the lobster management system

Input variable	Data range	Output variable	Data range
Neural Network Module			
GBFR	0.10–0.50	GBFR + 2	0.10–0.50
GBFR-1	0.10–0.50		
GBPR	0.02–0.25		
GBPR-1	0.02–0.25		
Fuzzy Logic Module			
Input variable	**Data range**	**Output variable**	**Data range**
PFR	0.10–0.50	QUOTA	2000–10 000
T-ANOM	−1.00-+2.50		

Note: GBFR refers to survey index numbers per standard tow of fully recruited female lobsters from the Georges Bank and South region. GBPR refers to pre-recruits of both sexes using the same survey index. The −1 refers to a one-year lagged value for the index and +2 refers to a two-year ahead prediction. PFR refers to predicted full recruit derived from the neural net, and T-ANOM refers to Georges Bank surface temperature anomalies lagged by five years.

the future. From this projection, as well as from sea surface temperatures anomalies lagged by five years, the fuzzy logic module provides an estimate of the projected quota two years in advance.

In this application, the neural network module (Fig. 2b) utilizes four inputs and produces one output. The neural network architecture is a typical three-layer backpropagation network with six nodes in the hidden layer. The supervised learning algorithm which was used is based on a gradient approach to minimizing residual error. Training parameters for this application included the learning rate (set to 0.1), gain (set to 1.0), and momentum (set to 0.5). The training file consisted of a 20-year time series of input variables, and known outputs were developed from available data. A specified tolerance of <0.20 was used as a terminating condition for training.

From the fuzzy logic module (Fig. 2c), it is clear that two inputs and one output were utilized. A fuzzy set that relates a domain interval to its degree of membership in a particular set describes a membership function of the corresponding variable. PFR input and the QUOTA output membership functions consisted of five triangular-shouldered labels which partitioned the universe of discourse (data range) into five overlapping categories, each occupying about 40 % of the total interval. It is suggested that much of the uncertainty resulting from significant measurement errors in the survey indices and catch data can be effectively treated in this manner.

The temperature anomaly data were initially regressed against pre-recruit index values lagged by stepwise annual increments. The best fit was found to be with a five-year lag. However, the fit of the regression model, although statistically significant, only had a coefficient of determination (*R*-square) value of 0.22. Therefore, only three labels for this membership function were used.

One rulebase consisting of 15 rules was developed from the three membership functions. After membership functions are defined over each variable's universe of discourse, relations between input

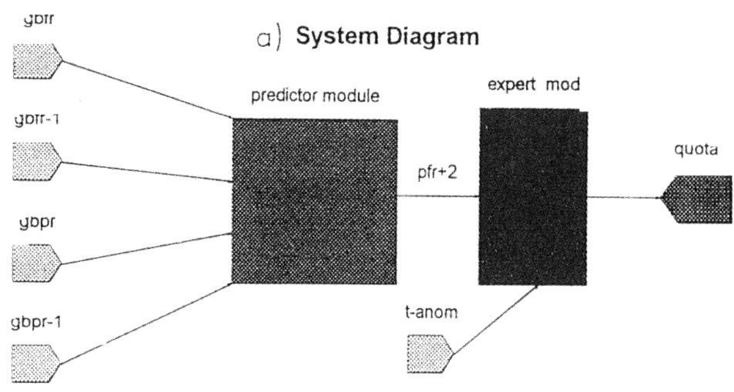

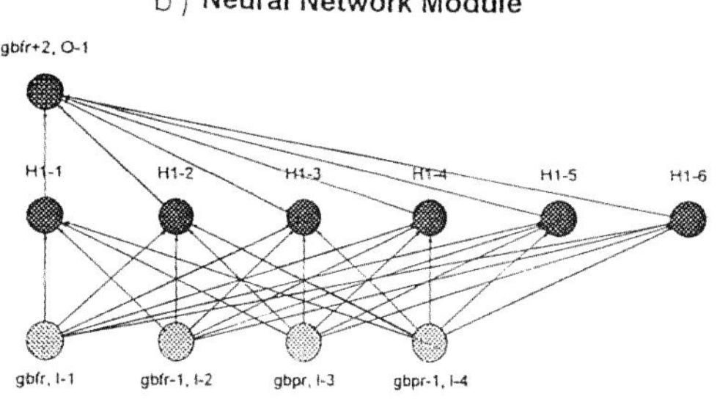

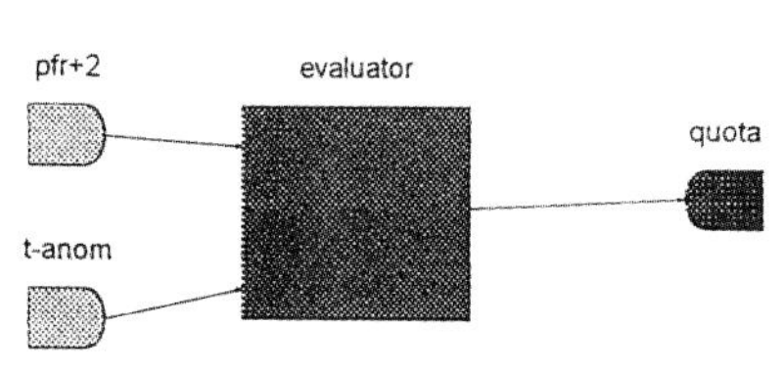

Fig. 2. (**a**) System diagram of the proposed quota-based American lobster management system, (**b**) neural network module and (**c**) fuzzy logic module.

and output fuzzy sets can be defined through a set of rules of the form: IF {condition} THEN {action} . This implementation of a fuzzy control system is clearly in the form of a knowledge-based expert system. The input objects to the fuzzy logic module accept crisp (real) numbers and produce fuzzified data that are propagated to the rulebase. The output object, QUOTA, in the system accepts data in the form of a fuzzy set from the rulebase and produces output in the form of a predicted quota. The defuzzification method utilized in this case was the centroid method, sometimes termed the centre of gravity method. Many different strategies exist for rule evaluation. However, the max-min inference method was utilized. In this method, the min operation $[\mu_A(x) \wedge \mu_B(y)]$ is used for the AND conjunction, and the max operation $[\mu_A(x) \vee \mu_B(y)]$ is used for the OR disjunction for evaluating the grade of the antecedent class in each rule.

RESULTS

The results obtained by using the neural network module in a simulation mode are illustrated in Table 2. It is evident from this table that all the projections from the neural net-model were

Table 2. Results derived from the neural network module showing predicted index values for fully-recruited female lobsters based on the training set

Year	Observed index two years ahead	Predicted value two years ahead
1971	0.447	0.407
1972	0.192	0.231
1973	0.284	0.274
1974	0.280	0.295
1975	0.451	0.406
1976	0.318	0.373
1977	0.330	0.300
1978	0.317	0.284
1979	0.340	0.339
1980	0.359	0.347
1981	0.285	0.318
1982	0.322	0.326
1983	0.249	0.300
1984	0.312	0.332
1985	0.212	0.280
1986	0.322	0.334
1987	0.330	0.300
1988	0.347	0.357
1989	0.406	0.343
1990	0.339	0.309

within the error tolerance specified by the maximum training error of 0.20. The results shown in Table 3 were derived from a simulation run using the entire system model. The year for each projection is shown in Column 1. Column 2 lists the projected quota in tonnes for each year. Columns 3–7 indicate the degree of membership (belief value) within a domain interval for each of the five labels. In this approach, functions overlap to indicate that a value may belong to different sets at the same time but with different degrees of membership.

Discussion

The example illustrated in this report is a very simple application of the O'INCA design framework methodology to produce a fuzzy control system for lobster management in a given area. Much more complex systems employing several neural networks, fuzzy logic, and user-defined modules are possible with this design tool. However, the example attempts to demonstrate the practicality of this approach.

From this example, it was demonstrated that the neural network could provide a relatively precise and robust procedure for time-series prediction. Saila (1996) has already indicated the overall superiority of neural nets over conventional time-series procedures. However, it is generally recognized that a large amount of measurement error may be associated with fisheries data, which suggests that good precision does not necessarily mean good accuracy. Fuzzification of data in the fuzzy logic model effectively acknowledges and accommodates this uncertainty which is usually ignored in many fishery measurement models.

Table 3. Results from the system model showing the projected quotas (t)and the relative weights calculated for each of the rulebase variables in the output

Year	QUOTA (t)	Database weights				
		Very low	Low	No increase	Medium increase	High increase
1973	8065	0.00	0.00	0.00	0.65	0.42
1974	5128	0.25	0.51	0.71	0.00	0.00
1975	5557	0.04	0.20	0.52	0.00	0.00
1976	5846	0.00	0.62	0.92	0.00	0.00
1977	7688	0.00	0.00	0.08	0.54	0.29
1978	7798	0.00	0.00	0.23	0.71	0.77
1979	5955	0.04	0.00	0.92	0.01	0.00
1980	5566	0.00	0.20	0.55	0.00	0.00
1981	6898	0.00	0.00	0.51	0.49	0.25
1982	6925	0.00	0.00	0.60	0.40	0.36
1983	6415	0.00	0.00	0.52	0.24	0.06
1984	6605	0.00	0.00	0.52	0.33	0.11
1985	5948	0.00	0.00	0.52	0.01	0.00
1986	6694	0.06	0.00	0.60	0.41	0.17
1987	5495	0.00	0.25	0.60	0.00	0.00
1988	6752	0.00	0.00	0.57	0.49	0.19
1989	5950	0.00	0.00	0.68	0.01	0.00
1990	7512	0.00	0.00	0.28	0.72	0.52
1991	6231	0.00	0.00	0.84	0.17	0.03
1992	6107	0.00	0.00	0.88	0.10	0.02
1993	6808	0.00	0.00	0.55	0.76	0.21
1994	6350	0.00	0.00	0.63	0.01	0.00

Although the lagged temperature anomaly was also used as input to the fuzzy logic module, its influence on the projected quota was relatively small. However, its inclusion illustrates the versatility of the development used. Examination of the data in Columns 3–7 of Table 3 suggests that the simulation run with the fuzzy control system resulted in a relatively stable situation with only five indicated changes in the existing quota over the 20-year time frame. Of these five changes, four suggested a modest increase in the quota and one indicated a decrease.

In summary, complex resources, such as commercial fisheries, have been found to be very difficult to control by means of conventional control theory, primarily due to inadequate knowledge of their behaviour. Control theory is defined in this case as the formalization and generalization of actions that may be taken. Traditionally, system control has been based on engineering control theory. A control problem is usually described by two types of variables, namely the control and the state variables. The control variables govern the evaluation of the system from one step to the next, and the state variables describe the behaviour of the system in any step. The control problem has been addressed as a mathematical programming problem. Attempts to apply stochastic dynamic programming to such problems have been plagued by dimensionality problems. It is suggested that fuzzy control may be a reasonable alternative.

Some suggested advantages of the lobster management model based on fuzzy control theory are:

1) The neural network module provides a precise and adaptive forecasting tool for short-term forecasts.

2) The fuzzy logic module seems to accommodate uncertainty and unrealistic precision, as well as complex interactions to provide a robust control.

3) The fuzzy rulebase is transparent and relatively easy to understand.

4) The resulting management model seems to be robust. The model output does not change much by small changes in the input values.

REFERENCES

Anon. (1993). Report of the 16th Northeast Regional Stock Assessment Workshop (16th SAW). (Northeast Fisheries Science Center Reference Document 93–16: Woods Hole, Massachusetts 02543).

Bardossy, A, and Duckstein, L (1995). 'Fuzzy rule-based modeling with applications to geophysical, biological, and engineering systems.' (CRC Press: Boca Raton, Florida).

Bosserman, R W, and Ragade, R K (1982). Ecosystem analysis using fuzzy set theory. *Ecological Modelling* **16**, 191–208.

deSilva, C W (1995). 'Intelligent control fuzzy logic application.' (CRC Press: Boca Raton, Florida).

Horwood, J W, Jacobs, O L R, and Ballance, J H (1990). A feedback control law to stabilize fisheries. *Journal du Conseil International pour l'Exploration de la Mer* **42**, 57–64.

Jacobs, O L R, Ballance, J D, and Horwood, J W (1991). Fishery management as a problem in feedback control. *Automatica* **27**(4), 627–39.

Klir, G J, and Folger, T S (1988). *Fuzzy Sets, Uncertainty, and Information.* (Prentice-Hall: London).

Li, H X, and Yen, V C (1995). 'Fuzzy sets and fuzzy decision making.' (CRC Press: Boca Raton, Florida).

Muratori, S, Rinaldi, S, and Trincher, B (1989). Performance evaluation of positive regulators for population control. *Modeling Identification and Control* **10**, 125–34.

Pedrycz, W (1995). 'Fuzzy sets engineering.' (CRC Press: Boca Raton, Florida).

Saila, S B (1992). Application of fuzzy graph theory to successional analyses of a multispecies trawl fishery. *Transactions of the American Fisheries Society* **121**, 211–23.

Saila, S B (1996). Guide to some computerized artificial intelligence methods. In 'Computers in Fisheries Research'. (Eds B A Megrey and E Moksness) pp. 8–403 (Chapman and Hall: London).

Sakuramato, K (1995). A method to estimate relative recruitment from catch-at-age data using fuzzy control theory. *Fisheries Science* **61**(3), 401–5.

Salski, A (1992). Fuzzy knowledge-based models in ecological research. *Ecological Modelling* **63**, 103–12.

Squires, D, Kirkley, J, and Tisdall, C A (1995). Industrial transferable quotas as a fisheries management tool. *Fisheries Science* **3**(2), 141–69.

Tanaka, S (1980). A theoretical consideration on the management of a stock-fishery system by catch quota and its dynamical properties. *Bulletin of the Japanese Society for Scientific Fisheries* **46**, 1477–83.

Tong, R M (1977). A Control Engineering Review of Fuzzy Systems. *Automatica* **13**, 559–69.

Tuma, A H, Hassis, D, and Rentz, O (1996). A comparison of fuzzy expert systems, neural networks, and neuro-fuzzy approaches. *Ecological Modelling* **85**, 93–8.

Walters, C (1986). '*Adaptive Management of Renewable Resources.*' (Macmillan: New York).

Zadeh, L A (1965). Fuzzy sets. *Information and Control* **8**, 338–53.

The Collapse of the Eastern Australian Gemfish Stock — Issues for Management and the Role of Fisheries Science

K. R. Rowling

NSW Fisheries Research Institute, PO Box 21, Cronulla NSW 2230 Australia.

Summary

In the 1970s the New South Wales demersal trawl fishery began exploiting continental slope resources occurring at depths of 200 to 600 m. Large quantities of gemfish *Rexea solandri* were caught during a winter pre-spawning migration at depths of 350 to 450 m, and landings peaked at over 5000 tonnes in 1980. Following declines in catch rate and mean fish size, a Total Allowable Catch (TAC) of 3000 t was imposed on the fishery in 1988, despite significant opposition from the catching sector. In the following years it became apparent that the gemfish stock had been subject to a series of poor recruitments, and the TAC was reduced in response to stock assessments indicating a significant decline in spawning biomass. In 1993 and subsequent years the TAC was set to zero, and gemfish catchers have become involved in research surveys to test the validity of the biological data on which the stock assessment is based, and to estimate the current level of abundance of mature gemfish.

Introduction

In the 1970s the New South Wales (NSW) demersal trawl fishery began exploiting continental slope resources occurring at depths of 200–600 m. A number of deepwater species were targeted, the most prominent being gemfish *Rexea solandri*, which were mainly caught during a winter pre-spawning migration at depths of 350–450 m. Gemfish landings increased rapidly, and peaked at over 5000 t in 1980. However a succession of poor cohorts, spawned during the late 1980s, led to a collapse in the eastern gemfish stock, and the virtual closure of the fishery in 1993. The cause of the stock collapse remains unknown, although variability of the oceanographic conditions in the spawning area is thought to be a significant factor.

In this paper, I want to deal with some of the implications for management resulting from the collapse of the gemfish resource, and touch on the role that fisheries science has played in underpinning the management regime for the fishery.

The fish

The gemfish is a large predatory member of the family Gempylidae (snake mackerels) occurring at depths of 100–600 m off southern Australia and New Zealand (Nakamura and Parin 1993). They attain a maximum age of about 16 years, which corresponds to a length to caudal fork (LCF) of 116 cm and a weight of about 15 kg. Female gemfish mature at 4–6 years of age (60–75 cm LCF) when individual fish weigh 2–4 kg. The majority of fish in the breeding population are between 5 and 8 years of age, and 3–6 kg in weight (Rowling 1990).

Biochemical research (Paxton and Colgan 1993) has identified two separate breeding stocks of gemfish off southern Australia. The fishery information presented in this paper relates to the eastern stock, which occurs from Tasmania to northern New South Wales (Fig. 1). Eastern gemfish spawn during a short period from late July to mid August each year, on areas of the upper continental slope between latitudes 32° and 33°S. Prior to spawning, mature gemfish form aggregations in depths of 350–450 m. In most years, these aggregations are first contacted by the fishery in eastern Bass Strait waters (39–40°S) in early June, after which the aggregations appear to migrate northward towards the spawning grounds, usually arriving off Sydney (34°S) by early to mid July (Fig. 2). The majority of the commercial catch is taken during this short pre-spawning migration.

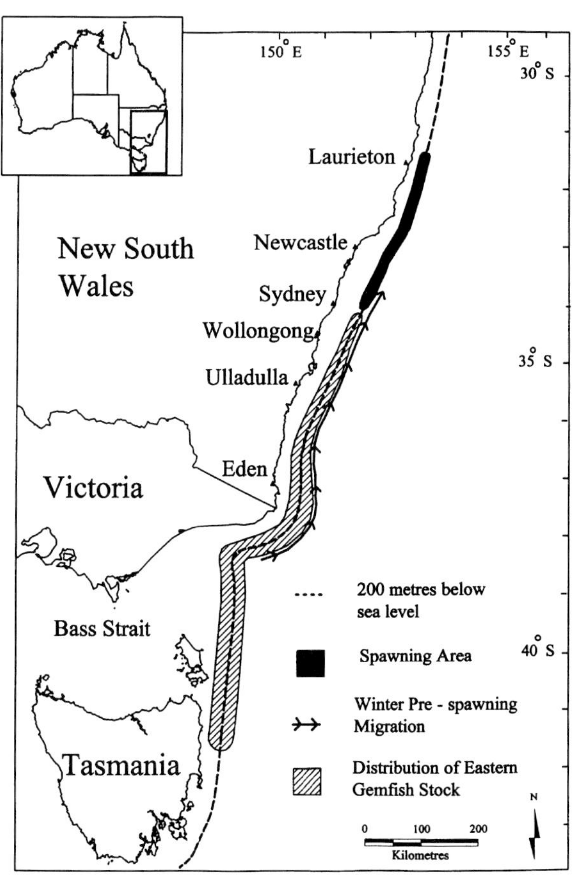

Fig 1. Distribution of the eastern gemfish stock off south eastern Australia.

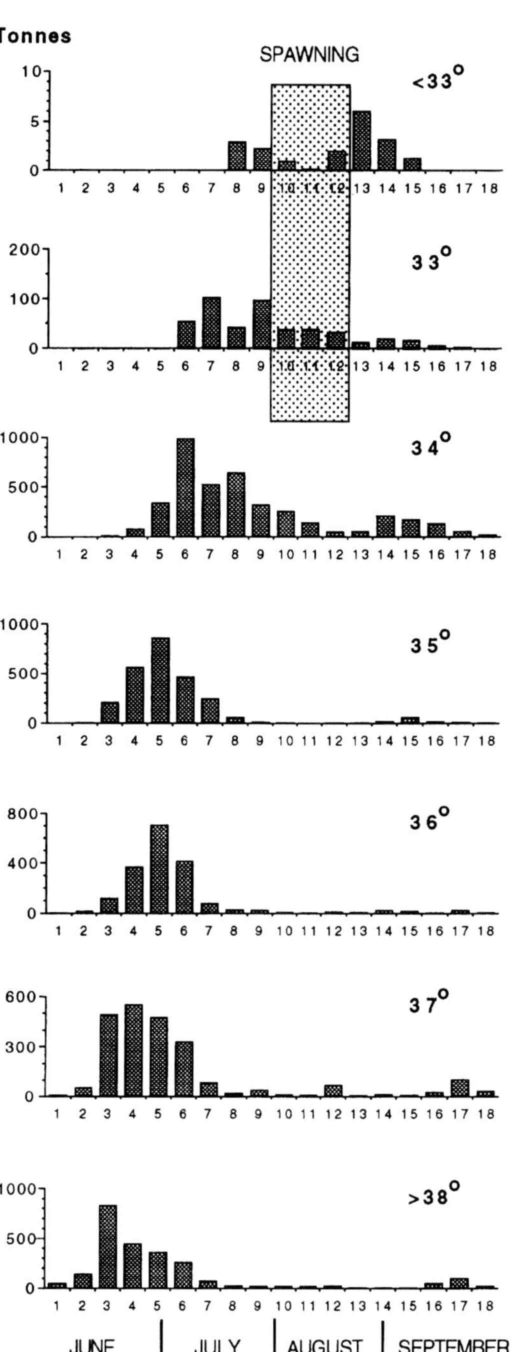

Fig. 2. Distribution of eastern gemfish catches weekly by latitude and time of season from logbook data for the years 1986–1993.

THE FISHERY

Gemfish are caught mainly by demersal trawling, although small quantities are also taken by droplines. In the early 1970s, the NSW coastal trawl fishery, which had previously operated solely on continental shelf grounds, began fishing deeper waters along the adjacent continental slope. A number of species were found to be seasonally abundant on these grounds, with the highest catch rates occurring during the winter months. The pre-spawning aggregations of gemfish were easily targeted with demersal trawls,

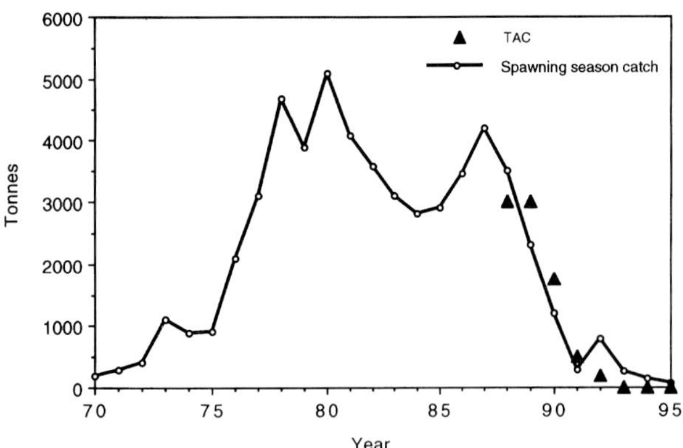

Fig. 3. Landings of eastern gemfish during the winter seasons from 1970 to 1995 and TAC for 1988–1995.

and catches of 5–10 t for individual trawl shots were common, with maximum catches in excess of 30 t per shot.

Landings of gemfish increased rapidly during the late 1970s, and peaked at over 5000 t in the 1980 season (Fig. 3). At this time, the fishery involved between 60 and 80 trawlers, operating out of the main coastal ports in eastern Victoria and southern NSW. These boats were generally 20–25 m in length, and many had been purpose built for catching gemfish. After the initial peak, annual landings slowly declined to around 3000 t before reaching a secondary peak of 4300 t in 1987.

Original stock assessment

A research programme was established in 1975 to monitor catch rates in the NSW trawl fishery, and the size and age composition of the main species caught. The results of this research showed a significant fall in standardized catch rates of gemfish during the early 1980s, in parallel with a decline in the average size of fish in the catch (Fig. 4). These results were released to industry and the fishery managers in 1987, together with a recommendation that catches of gemfish not be allowed to exceed 3000 t per season (Rowling 1987).

In 1988, gemfish became the first species in the South East Fishery to be subject to quota management when a 3000 t Total Allowable Catch (TAC) was implemented. This figure was considered to approximate the sustainable yield for the eastern gemfish stock. In 1988, the TAC was set as a 'global' catch limit for the whole fleet, and landing of gemfish was banned after the TAC had been caught by early July. Following significant public protests from the fishing industry, the fishery was re-opened in early August and a further 500 t of gemfish were caught before the end of the 1988 season. In subsequent seasons, the TAC was apportioned to individual vessels on the basis of their gemfish catch history.

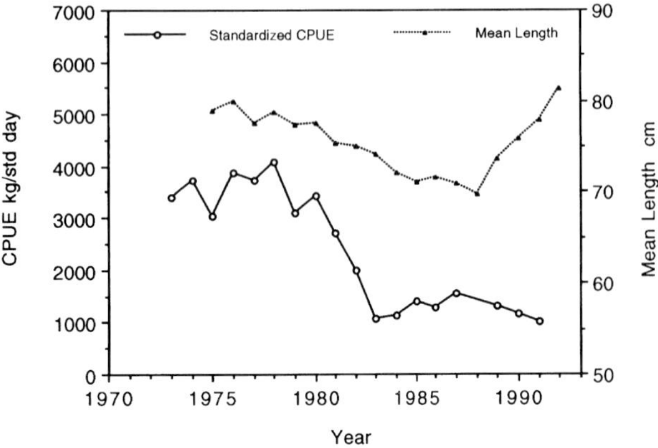

Fig. 4. Standardized CPUE and mean size of fish from spawning run catches of eastern gemfish.

Industry reaction

There was a very strong negative reaction by industry to the imposition of restrictions on the gemfish catch — the general opinion was that restrictions were unnecessary, as gemfish were only vulnerable to capture during the short period prior to spawning. There was also considerable criticism of the validity of the research findings, mainly to do with the perception that measurements made ashore would not accurately represent the true size composition because of sorting of the catch carried out aboard fishing vessels.

I believe this reaction was largely due to the fact that industry could not easily observe at first hand the changes in catch rate and mean fish size which had been found during the research work. Because gemfish form aggregations, peak catch rates remained high despite substantial declines in catch rates averaged over whole seasons. Catches of pre-spawning gemfish continued to be comprised of large, mature fish between 60 and 110 cm in length, and the actual distribution of fish sizes in a given catch could not be ascertained from casual observation. There were also considerable variations in these indices both within a season and geographically between seasons, which helped to mask the changes detected by the research programme.

An additional concern raised by fishers was that the research work had been carried out in the fishers' cooperatives and at the central markets, and did not directly involve members of the catching sector. In response to the concerns expressed by industry about the validity of measurements made during the shore-based research programme, a team of observers was deployed during the 1989 season to measure samples from catches as they were landed aboard trawlers. The results of this research (Rowling *et al.* 1990) confirmed that measurements made ashore were representative of the size composition of catches from the pre-spawning aggregations, and the trends observed in the research data could fairly be considered to reflect changes in the gemfish population.

Recruitment collapse

Size composition data collected during 1989 also provided the first indication of a problem with recruitment to the gemfish stock. The data clearly showed a significant reduction in the relative number of gemfish around 60 cm LCF (4-year old fish, spawned in 1985) in the 1989 pre-spawning run. The TAC for 1990 was reduced to 1750 t in recognition that the 1985 cohort might be poorly represented as 5 year olds in that season, and close monitoring of the catches from the pre-spawning run confirmed that both the 1985 and 1986 cohorts were poorly recruited to the 1990

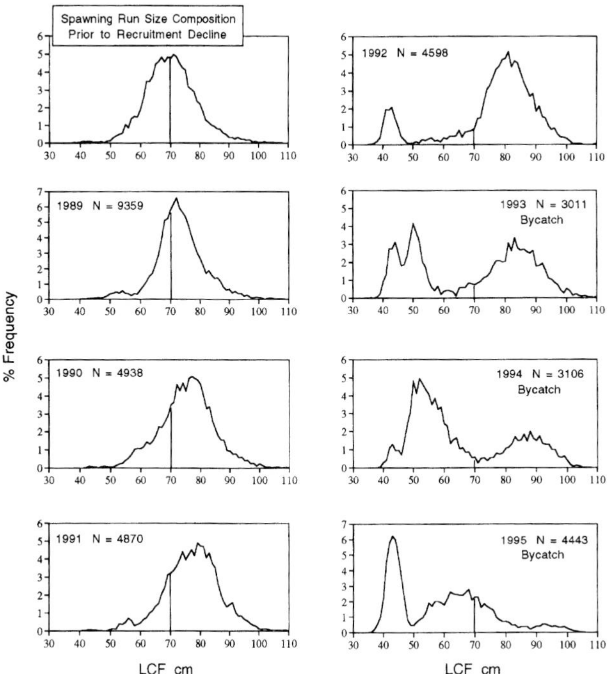

Fig. 5. Size composition of eastern gemfish from spawning season catches, 1989–1995.

spawning population. The fishery landed just 1200 t during the 1990 season, well short of the reduced TAC.

In response to evidence of continued poor recruitment of all cohorts spawned from 1985 to 1989 (Fig. 5), which suggested a significant decline in the abundance of mature fish, the TAC for eastern gemfish was progressively reduced to 500 t in 1991 and just 200 t in 1992. This dramatic reduction in the TAC provoked considerable public debate about the stock assessment for eastern gemfish, and in 1992 some vessels significantly over-caught their allocated quotas. Because gemfish aggregate prior to spawning, relatively high catch rates could still be attained when aggregations were encountered.

In 1993 and subsequent seasons, eastern gemfish have been subject to a zero TAC, as the poorly recruited cohorts have reached those age classes which previously dominated the mature population. There is no doubt that the implementation of the zero TAC significantly reduced the impact of fishing on the reduced stock. However, during the period of the zero TAC, fishers have been allowed to land a limited bycatch of gemfish taken while targeting other species which occur in the same depth range. These arrangements have resulted in the discarding at sea of large incidental catches of gemfish, which have been mostly unreported. The result has been inaccurate estimates of total catch for recent seasons, but more importantly these management arrangements have caused resentment amongst industry members who do not agree with having to discard large, quite valuable catches, given that the majority of the gemfish in such catches are killed by the trauma of being hauled to the surface in a trawl net.

There have also been variations between seasons in the quantity of gemfish allowed to be landed under the daily trip limit arrangements, and different trip limits have applied to different methods of capture and under different jurisdictions throughout the period of the zero TAC. This has resulted in a considerable degree of uncertainty about the applicable regulations, and has increased the difficulty of effectively enforcing the trip limit regulations.

POPULATION MODELLING

The availability of an extensive time series of data on changes in the composition of gemfish catches led to the development of age-structured population models, which are useful for describing the history and current status of the gemfish resource.

In 1988, Dr K.R. Allen developed a cohort analysis which explicitly accounted for the characteristics of the gemfish stock, especially differences in growth and mortality rates between the sexes. When this analysis was published (Allen 1989) it formed the basis for the first quantitative description of the dynamics of the eastern gemfish stock. The cohort analysis was updated as further data became available through to 1992, after which the zero TAC prevented estimation of a valid CPUE index. Trends in biomass and recruitment from the most recent cohort analysis (Allen 1994) are shown in Fig. 6. The analysis indicates a significant decline in mature biomass in the early 1980s, to about 45% of the pre-exploitation level, followed by a period of stable or slightly increasing biomass to 1988. Mature biomass then declined sharply as the poor cohorts spawned in the mid 1980s progressively entered the mature population.

Recently, dynamic population models have been developed using Bayesian and maximum likelihood methods, and the results of these models have been carried forward to simulate the effects of various future harvest strategies for the gemfish stock (Smith 1996). While the trend in mature biomass is similar to that resulting from the cohort analysis, the estimates of original biomass are larger and the initial stock declines are greater in the dynamic models, to between 25% and 35% of pre-exploitation levels (Fig. 7). The Bayesian analyses show different trends in recent biomass depending on the CPUE series used in the model;, however, none of the dynamic population model outputs suggest any abrupt change in biomass during the 1989–92 period, when the poor cohorts became the dominant

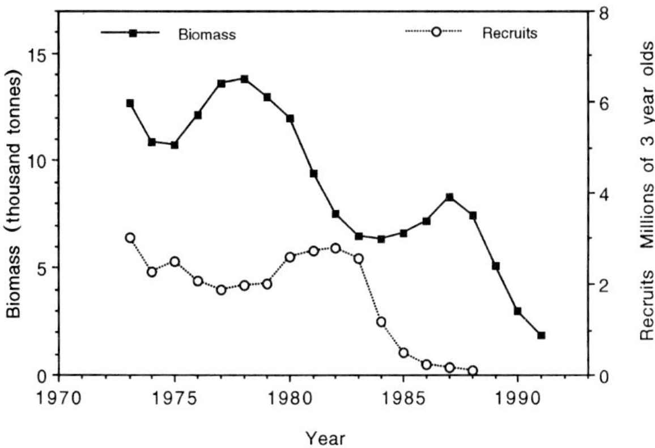

Fig. 6. Trends in mature biomass and recruitment from a cohort analysis of the eastern gemfish stock (Allen 1994).

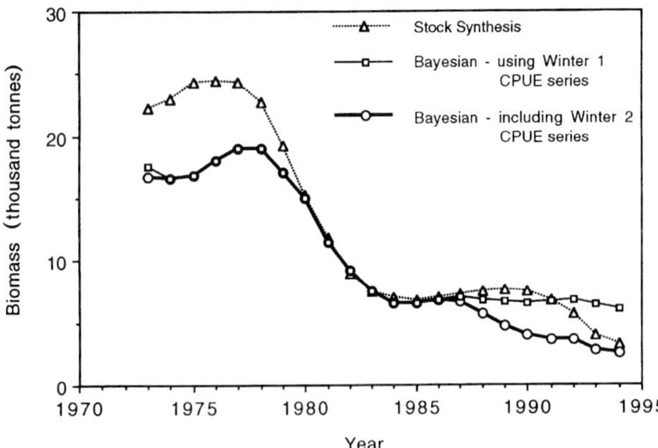

Fig. 7. Estimates of mature biomass of eastern gemfish from dynamic population models (after Smith, 1996).

components of the mature population. The accuracy of the biomass estimates from these models for recent years, and the assumptions in the models concerning selectivity of the fishery, are still the subject of debate.

Current situation

In the period since the TAC was set to zero in 1993, there has been an increased involvement by the catching sector in research designed to assess the validity of the stock assessment for eastern gemfish. Initially, the focus of the industry-based research was on possible bias in the size composition data collected from pre-spawning catches. The results of this work (Prince and Wright 1994) confirmed that the stock assessment had succeeded in describing the main trends in the gemfish population, and that there had been a period of extremely low recruitment to the stock, which justified the cautious management approach. I believe that this work represented the beginning of a process of education of those directly involved in the fishery about the methods used in fisheries assessment and population modelling, and the importance of ensuring objectivity in the data collection and assessment processes.

During the 1996 winter season, some members of the gemfish catching sector have been involved in a series of trawl surveys aimed at determining the relative strengths of the 1990 and 1991 cohorts, on which any recovery of the gemfish stock will be dependent. The results of these surveys are yet to be discussed, and there remains much work to do before the survey results can be incorporated into the stock assessment for eastern gemfish. However, it is through developments such as this that the controversy surrounding the stock assessment for eastern gemfish will be reduced.

Conclusions

The collapse of the eastern gemfish stock, and the issues subsequently raised for management of the fishery targeting that stock, reinforce the absolute need for appropriate long-term research programmes aimed at understanding the dynamics of exploited populations. I am sure that management of the gemfish fishery through this difficult period would not have been as successful if a suitable research programme had not been in place for the fishery prior to the collapse of the stock.

The expressions of industry discontent with the assessment and management of the fishery for eastern gemfish show it is also highly desirable to involve industry from the outset in both the planning and operational phases of research on commercial fisheries. This, I believe, is one of the best ways of reducing the controversy that always seems to surround stock assessments — however, there is often little enthusiasm for being involved in monitoring-type research programmes when fisheries are stable or developing.

Finally, the collapse of the eastern gemfish stock, and the very significant economic ramifications for the fishery, emphasize the need to take special care when dealing with aggregating fish stocks, which are more susceptible not only to overfishing but also to unfavourable environmental conditions.

References

Allen, K R (1989). A cohort analysis of the gemfish stock of southeast Australia. Unisearch Ltd, University of New South Wales, Kensington.

Allen, K R (1994). Further comparison of cohort analysis results. Unpublished manuscript, March 1994.

Nakamura, I, and Parin, N V (1993). FAO Species Catalogue, Vol. 15, Snake Mackerels and Cutlassfishes of the World. FAO Fisheries Synopsis No. 125, Volume 15. FAO (UN), Rome.

Paxton, J, and Colgan, D (1993). Biochemical genetics and stock assessment of common gemfish and ocean perch. Final Report, Fisheries Research and Development Corporation Project 91/35. Australian Museum, Sydney.

Prince, J D, and Wright, G (1994). A description of the size structure of the 1993 eastern Australian winter gemfish aggregations and a synthesis of industry's perspective on the existing stock assessment. Final Report, Fisheries Research and Development Corporation Project 93/057 and Fisheries Resources Research Fund Project 19/12, Biospherics Pty Ltd, Leederville.

Rowling, K R (1987). The need for catch controls in the gemfish fishery. NSW Fisheries Research Institute Internal Report No. 26, June 1987.

Rowling, K R (1990). Changes in the stock composition and abundance of spawning gemfish *Rexea solandri* (Cuvier), Gempylidae, in south-eastern Australian waters. *Australian Journal of Marine and Freshwater Research* **41**, 145–63.

Rowling, K R, Smith, D C, Laurenson, F A, Brown, L, and Broadhurst, M (1990). Onboard observation of the 1989 gemfish fishery. Final Report, Fishing Industry Research and Development Committee Project 88/126. NSW Fisheries Research Institute, Cronulla.

Smith, A D M (1996). Evaluation of harvest strategies for Australian fisheries at different levels of risk from economic collapse. Final Report, Fisheries Research and Development Corporation Project T93/238. CSIRO Division of Fisheries, Hobart

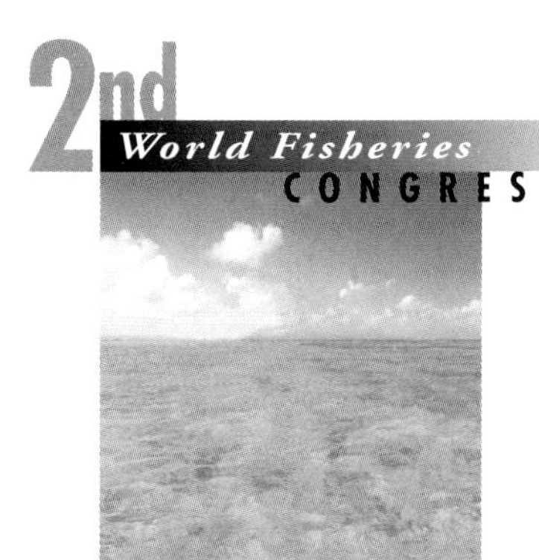

Your Work is of Value, Prove it or Perish

D. Baker

South Australian Research and Development Institute, PO Box 120, Henley Beach, South Australia 5022, Australia.

Summary

Fisheries scientists and managers 'know' that their work is critical and beneficial to the natural resource, but to outsiders evidence of this value may not be obvious. Ongoing financial pressures mean that both public and industry funding is becoming increasingly outcome driven. The performance of fisheries, fisheries management and fisheries research must increasingly be benchmarked with readily understood performance indicators if they are to retain the support of policy makers and the public. Few such quantitative measures are used in the fisheries arena. As a result the work may be overlooked and undervalued in the eyes of society.

Performance indicators must be targeted to the needs of all clients of the service provided, be they industry, recreational fishers, tourists, the general public or politicians. To be an effective tool, a suite of indicators must be established, indicative of the interests of all clients. The use of multiple indicators will also reduce the risk of a single incorrect measurement providing a false impression. Information should be compiled and reported on appropriate periodic basis. Variations in the data between periods or from targets should be explained to increase comprehension by laypersons.

This paper will illustrate how this approach can be applied and what the benefits and disbenefits to fisheries scientists and managers could be.

Introduction

World-wide, the seemingly ever-increasing financial pressures facing government and the private sector are impacting upon the resources made available to fisheries research and management. Funding is becoming more difficult to secure and is increasingly outcome-driven. Those providing the finance want to know that they are getting value from their investment. The fisheries sector is not unique in this. Most areas, be they private or public sector, are increasingly required to demonstrate accountability in a formal manner, by providing both quantitative and qualitative evidence of service provision, and valuation of outcomes.

In the case of fisheries management (enforcement, compliance and administration) and particularly fisheries research, financial data do not adequately represent all aspects of the service provided. While the outcomes of fisheries management and research may be in part financial this does not represent the true value of the work. In some cases there may be little or no obvious financial gain from a project, yet there is a high probability of important and productive outcomes.

Fisheries scientists and managers 'know' that their work is critical and beneficial to the natural resources. But do the external stakeholders? Most importantly, do those providing the funding understand this? Great emphasis is placed on the 'word' of the service provider and the acknowledgment of professionals working in the field (i.e. 'Our clients know we are doing a good job because we tell them we are, and we are respected in our field'). Increasingly this is not enough. 'Experts' may not be as respected and trusted by the public as they/we would like to believe. The acceptance of the merit of the work by peers is not sufficient, as they are not a client. The publication of articles and books may be an indicator of success to fellow-professionals but it does not necessarily follow that it is valued by stakeholders and clients.

If the fisheries management and research sectors are to retain the financial support of policy makers and fund providers, they need to provide evidence of their relative value to the stakeholders. There need to be quantitative, qualitative, financial and non-financial data which illustrate the benefits to the stakeholders of investment in these sectors, to the stakeholders. Many other sectors (e.g. private companies and other public sector arenas) are much better placed to provide relevant evidence of the benefits of their work as there are many built-in measures involved (i.e. profit, sales, production). The fisheries management and research sectors haves a comparative disadvantage which they must rapidly overcome, when competing for increasingly scarce funding. Without adequate benchmarking of fisheries performance, management and research, the sector will lose investment in the long run to those other sectors, that are able to demonstrate an apparent higher return on investment. The consequence of this will be a lower funding base, leading to less research, fewer managers, fewer scientists and a potentially catastrophic impact upon the natural resource.

Performance indicators

A key element of the movement towards increased accountability is the incorporation of performance indicators in the reporting process. The use of formal performance indicators in government had its origins in the USA in the 1960s and came into widespread use in Britain in the 1980s (Carter *et al.* 1992). Their use has become widespread in recent years and they can be found in areas such as strategic management planning, programme performance budgeting, state of the environment reporting (Department of the Environment, Sport and Territories 1994), funds allocation in universities (Hattie *et al.* 1991), and quality assessment (Goedegebuure *et al.* 1990).

The use of performance indicators is growing and it is inevitable that they will have an increased role in fisheries. The issue for managers and scientists to consider is whether they practically embrace the concept and play the leading role in the definition of the performance indicators, or reactively wait for a regime to be imposed upon them. The great danger in the latter is that if the indicators are not correctly defined the true value of the work may be understated, the competitiveness of managers and scientists would be reduced and consequently investment may decline. Despite producing valuable outcomes, managers and scientists may find themselves out of work and the resource inadequately stewarded.

Benefits of performance indicators to fisheries

The requirement to report upon and/or be answerable to performance indicators may seem to be an undesirable bureaucratic intrusion on the 'real' work of fisheries scientists and managers, competing for their scarce time. If viewed reactively, this can be the case, but if practically embraced they can become a valuable tool to achieve advantageous results for the natural resource, by increasing competitiveness to win 'a bigger piece of the funding pie'. A well-devised set of indicators produced periodically can be beneficial through the provision of evidence of:

- **effectiveness.** The investment has been made for a purpose. Performance indicators could be used to demonstrate just how successful the work is at achieving the desired outcomes by simply measuring the outputs. For example, funding may have been provided to increase the health of fish stocks in a particular fishery. Indicators should be chosen to illustrate the impact upon fish health resulting from the work, as well as measuring the benefits of healthier fish stocks. Healthier stocks lead to a larger commercial catch and increased sales and/or price. The quality and quantity of recreational fishing and overall environmental health should also improve as a consequence.
- **efficiency.** There is always pressure to do more with less resources, implying inefficiency. Performance indicators can show how efficiently the inputs of investment are utilized. The measures would tend to take the form of input relative to output. In the above example this could be the dollar investment relative to increase in stock numbers and/or stock value (i.e. it cost $x to obtain each additional fish/$).
- **need for increased investment.** Performance indicators may show that current procedures and strategies are not achieving the client's desired outcome and that additional investment would add value. In the above example, the performance indicators may show that fish health and numbers are improving, but increased funding may be necessary to produce agreed results or be warranted to maximize return on investment.
- **allocation of funding.** Performance indicators can be used to provide evidence of the marginal benefits of work (i.e. work in some areas is more effective and/or efficient than in others). Greater overall benefit would be achieved by giving priority to those tasks with the higher marginal benefits.
- **trends in service provision.** Over-time performance indicators can be used to indicate to clients the trends in outcomes, identifying the relationships between investment and outcomes (e.g. every dollar invested yields $x dollars benefit to the fisheries sector; efficiency is increasing at rate of y).
- **resource and sustainability.** Performance indicators can provide evidence of the condition of the resource and its sustainability, highlighting threats. Although similar procedures may already be in place, formal reporting may be less ambiguous and be more likely to receive prompt attention due to the acceptance of the process by all stakeholders.

- **credibility and reliability.** In the medium to long-term, performance indicators can be used as proof of the success and benefits of previous work when new investment is being sought (i.e. when applying for new or additional funding, a research team which has a proven and documented track record in efficiently and effectively carrying out projects of a related nature, has a competitive advantage over teams without such evidence).

- **value to non-commercial clients.** Historically, quantified evidence has tended to focus on commercial interests, particularly on those directly providing the investment. Performance indicators provide the opportunity to include all interests (e.g. recreational users, organizations and individuals with conservation and biodiversity concerns etc.). Spin-off benefits to these non-commercial interests of work can thus be formally captured and added into the benefits equation. Performance indicators can thus increase the opportunities to elicit moral and possibly financial support from these interests in the future.

PUTTING THEORY INTO PRACTICE

The first step in the process of establishing performance indicators is to identify the target audience. Who are the stakeholders/clients, who represents them, and therefore who do we need to convince that the work is beneficial? The goals and desired outcomes of the user groups dictate the form of the performance indicators required. It is of little use to provide evidence which is difficult for a client to understand or has no relevance to them. It is logical, therefore, to formulate the indicators with consultation and agreement with the user groups. The client profile will vary between organizations; however, the types of user groups that need to be considered include:

- **industry.** Commercial fishers and processors. In many cases this will be focussed through representative bodies, but in most cases consultation with 'grass roots' will provide more useful information and ensure client satisfaction.

- **recreational users.** Recreational fishers are the most obvious non-commercial client group, but consideration should be given to other groups (i.e. boat users, water skiers, divers, campers, bird watchers, sightseers etc.). Official bodies are likely to exist for many of these, but perhaps even more so than in the case of industry, the question of how representative they are arises. Particularly as some of the relevant bodies may only represent a small percentage of participants, 'grass roots' consultation may be warranted.

- **politicians.** As representatives of the people and as individuals who have a significant influence on policy (and the purse-strings), politicians are important stakeholders. As politicians are not always directly accessible, involvement of their advisers could be considered.

- **government departments.** Much of the work in the fisheries arena is not done in isolation and may involve collaboration with government departments or impact upon areas for which they have responsibilities.

- **other governments.** The work may impact upon areas under the jurisdiction of other governments or levels of government (i.e. local, state, federal, international) or be subject to inter-governmental agreements.

- **community interest groups.** Ranging from local community action groups to international non-government organizations (e.g. World Wide Fund for Nature and Greenpeace) who may have an interest in the area.

- **indigenous population.** The work may impact upon an area in which an indigenous population has significant interest and possibly ownership.

- **general public.** This is potentially the most important client group, which is often overlooked in the planning process. Only some members of society are represented in the above categories; however, many others, while not members of organizations and/or community groups, may still derive value from the work. This value might be relatively small on an individual scale yet be considerable when looked at on a cumulative basis (Baker and Pierce 1994). It also is relevant to remember that it is the general public who pay the taxes that fund public sector work, who vote for politicians and buy the products and services of the private sector. Not including the needs and values of the general public could result in an underestimation of the value of the work.

This list is not intended to be absolutely definitive and it is logical that there will be some cross-membership between the groups.

The next step in implementing performance indicators is to consider what measures best quantify the merit of the work. These can take various forms including:

- **scientific and environmental data.** There are many potential measures, such as creel surveys, catch rates, harvest value, fish health, size, breeding rates, stock and recruitment indices, biodiversity, water quality parameters, etc. It can be argued that much of these data are already being collected, but there is a need to formalize the process and present it in a standardized and agreed format, packaged for clients on a regular basis. Scientific data are also certain to form the basis of many of the indicators that can be used to assess the state of the environment, although these may need to be purpose-designed, and non-scientific data should not be overlooked. The use of environmental indicators is still in the formative stages; however the global trend towards implementation of State of the Environment reporting (OECD 1991) will hasten this.

- **application of the accounting concepts of stocks and flows to scientific data.** The premise of treating natural resources as capital (stock) and increases in resources as a build-up and decreases as a rundown in natural capital could be useful. This would present the data in a form already familiar to, and accepted by, many key players. In particular it would fit into the 'business' nature of commercial clients (Pearce *et al.* 1989).

- **economic indicators.** There are already many economic measures in common use for many areas of management and planning. The important task is to select those appropriate to the task at hand. There are ways of measuring industry performance (profit, production value, economic impact, employment level, labour costs, capital costs, pollution clean-up costs, management costs, tourism impacts, project costs, etc.). Once again, an advantage with these is that they are in a form familiar to many clients.

- **social indicators.** Much fisheries work should be seeking to address the needs of its obvious clients and society in general. Therefore, it is desirable to obtain an indication of client/public perceptions, their values and measure their behaviour. Are the public aware of the work? What are their needs in relation to the work? Are their needs being met? Do they think that their needs are being met? Surveys are one obvious way of doing this, for example, the US Forest Service have conducted a study of the opinions of the major clients of their Fisheries programme (Forsgren and Loftus 1993). A more qualitative measure is keeping a record of correspondence with the public, copies of relevant community interest groups' reports and newsletters etc. Anecdotal evidence should not be overlooked.
- **visual measures.** Often overlooked is the use of visual evidence. Some benefits are obvious to the eye (e.g. a rehabilitated area) but this relies on having seen the 'before' to compare with the 'after'. This can be overcome by keeping a photographic or video record as a performance indicator.

Once potential indicators have been identified another round of consultation with the stakeholders/clients should take place. This process should perhaps be screened with peak bodies, with 'grass roots' input also desirable. This is to produce an agreed suite of indicators that provides meaningful information for the clients and addresses their needs and priorities in the area. Thus an agreed level of performance is set for success or failure to be judged against. This process is summarized in Figure 1.

STEP 1: **Consult with clients**

Identify and consult with ALL clients, as to their needs.

STEP 2: **Consider performance indicators**

Identify potential performance indicators which will meet the needs of the various clients.

STEP 3: **Consult with clients**

Consult further with clients, produce an agreed suite of performance measures. Agree on outcomes.

STEP 4: **Implement process**

Produce indicators and report to clients on a regular agreed basis.

STEP 5: **Feedback**

Consult with clients as to the effectiveness of suite and amend as necessary.

Fig. 1. Implementation Process.

Effective implementation

As stated earlier, performance indicators can be seen as a hindrance rather than a help. Many organizations keep statistics which require considerable time to collect, yet the majority of employees can see no significant benefit. How can a regime of performance indicators be applied so that they are a useful tool to scientists and managers alike and not simply another repository of useless data? This can be achieved by optimizing the following issues:

- The scale at which performance indicators are implemented
- The number of performance indicators to be used
- The reporting process
- Cost of the process.

Scale

There is a tendency for performance indicators to be implemented on a macro-scale (i.e. whole of company/ departments or possibly on a national level). This is understandable, and it may be desirable to be able to measure how the company or department as a whole is performing on the relevant issue. With issues of a primarily financial nature this is relatively easy and effective to implement as there is a common measuring unit, money. This allows data to be gathered on a fine scale, be aggregated at the various levels and provide meaningful information at all levels. This however, is not necessary feasible or desirable with non-monetary data.

If non-monetary indicators are established on a 'global' level, the regime must be implemented with a bottom-up approach. That is, the performance indicators should be identified and established at the micro-level. It is at this fine scale that the policy decisions of management are put into practice, where the marginal value decisions are made and where the practical understanding lies. The statistical needs of management farther up the line in relation to performance indicators can be taken into account by effectively treating management as a 'client'. This is already done with much financial and workforce information. If the regime is established on a top-down basis there is a possibility that the indicators set will only meet the requirements of senior management and not effectively reflect the value of the work done.

A further advantage of a fine-scale approach is accountability. The smaller the scale the more evident it becomes where responsibility lies and credit is due. Using the more global approach, the good work done in one area may be hidden because of failures in another.

In the fisheries arena, the appropriate micro-level for implementation of performance indicators would be on a project or site-specific basis (i.e. a distinct fishery or part thereof). Data collected at the micro-level can easily be summarized to a higher level as required by clients.

Number of indicators

How many indicators are required? The simple answer is — as many as are needed; however this must be qualified. If every single possible indicator is used, the result could be hundreds, the process would become unwieldy, confusing, costly and of use to no-one. Conversely, using a single indicator except in the simplest of cases is not really going to present a full picture, particularly when faced with a diverse client profile. What is required is a suite of performance indicators representative of the needs of all of the relevant client groups. It is desirable that the indicators are complementary and meet the needs of more than one client. Basically the number and make-up of the suite will be determined on a case-by-case basis. It will be a product of the consultation process, with clients identifying which performance indicators explain most of what they want to know. An optimal suite of indicators should be a mix of those indicators that address the prioritized information needs of more than one client and those indicators that address the variation in the needs of the various clients.

A further advantage of having multiple indicators is that of error reduction. It is a fact of life that errors occur. With complementary indicators it is more likely that errors will be more apparent and trends will not be overlooked.

Reporting

Performance indicators must be regularly monitored and made available in an appropriate format to the relevant parties. The whole process is next to useless if the data are not maintained and reported. Clients need to know what is happening with the work. It should be considered as important as financial reporting and treated as a requirement. The timing of reporting would be a case-by-case decision, but common sense should be used. There are not many cases where monthly reporting would be warranted and a five-yearly reporting period would be likely to have relatively few benefits. Simply to conform with existing custom (financial and organizational reports) annual reporting has some appeal, although some indicators may require more/less frequent measurement. First and foremost, whatever reporting cycle adopted, it should have the agreement of the clients.

Once the reporting regime is established, part of the regular reporting should involve comment on variations in the data between periods or from targets. This will make the information more readily understandable to the layperson.

Cost

Perhaps the most important consideration of all. No one wants or needs a process which requires a large portion of available funds simply to report on what has been done with those funds (or what could have been done with those funds if time and money wasn't been wasted reporting on them). Once again common sense must play a part. If some of the data have already been collected (internally or externally), then this should be used or adapted if necessary. If the best potential indicator is prohibitively expensive look to second best indicators for alternatives. Take into account the cost-effectiveness of options as this may have an influence upon the number of indicators used.

The magnitude of the project is a key factor in how much should be spent. The larger the amount of resources involved in the project the more it is likely that a higher degree of accountability will be required and the justification for increased reporting is greater. One approach to this issue is the establishment of a monitoring budget within each project, allocating a specified percentage of funding for the process. This mirrors the manner in which the funding for financial accounting is allocated in most projects.

SUMMARY

Current trends point to an increased requirement for accountability, and performance indicators have a role to play in this. Fisheries scientists and managers are not immune to this trend and eventually a performance indicator regime in some form will be imposed upon them whether they like it or not. If the process is embraced at an early stage there is potential to shape the agenda in a manner which will create a tool which will prove invaluable in continuing and expanding their work. A well-designed suite of performance indicators tailored to the needs of the relevant clients will provide greater evidence and understanding of the value of the work undertaken by scientists and managers than in the current situation. Failure to adapt to this changing environment could have serious adverse impacts upon fisheries managers, scientists and most importantly the natural resources.

REFERENCES

Baker, D L, and Pierce, B E (1994). If Sustainability Fails, Who Loses Most? In 'Recreational Fishing: What's the Catch?' (Ed D A Hancock) pp. 159–166 (Australian Government Publishing Service: Canberra).

Carter, N, Klien, R, and Day, P (1992). 'How Organisations Measure Success: The Use of Performance Indicators in Government.' pp. 5–51, 165–183 (Routledge: London).

Department of the Environment, Sport and Territories (1994). 'State of the Environment Reporting: Framework for Australia.' pp. 1–42(Australian Government Publishing Service: Canberra).

Forsgren, H, and Loftus, A J (1993). Rising to a Greater Future: Forest Service Fisheries Program Accountability. *Fisheries*, **18**, No.5, 15–20.

Goedegebuure, L C J, Maassen, P A M, and Westerheijden, D F (1990). Quality Assessment in Higher Education. In 'Peer Review and Performance Indicators: Quality Assessment in British and Dutch Higher Education'. (Eds L C J Goedegebuure, P A M Maassen and D F Westerheijden) pp. 15–36 (Uitgeverij Lemma: Utrecht).

Hattie, J, Tognolini, J, Adams, K, and Curtis, P (1991). 'An Evaluation of a Model for Allocating Research Funds Across Departments within a University Using Selected Indicators of Performance.' (Australian Government Publishing Service: Canberra).

OECD (1991). 'OECD Environmental Indicators: a Preliminary Set.' (OECD: Paris).

Pearce, D, Markandya, A, and Barbier, E. (1989). 'Blueprint for a Green Economy'. (Earthscan: London).

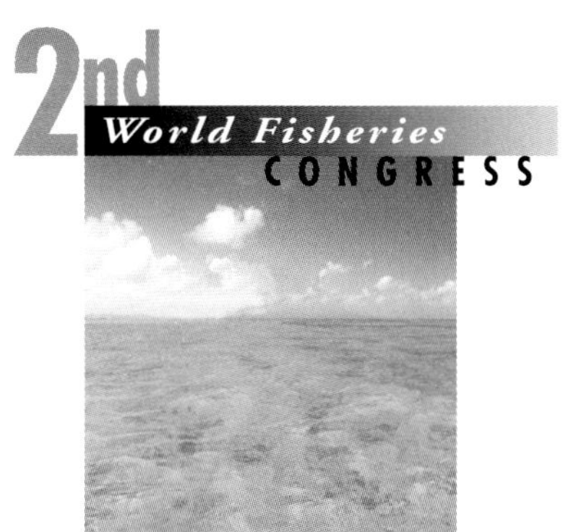

The Socio-economic Effects and Implications of Fishermen's Community Trusts in South Africa

Pieter G. du Plessis[A] *and De Wet Schutte*[B]

[A] Department of Business Management, University of Stellenbosch, South Africa.
[B] Human Sciences Research Council, Cape Town, South Africa.

Summary

There is great concern in many circles regarding the socio-economic conditions of the fishing communities along the South African coast. The allocation of area-specific fishing quotas to the registered Fishermen's Community Trusts (FCTs) was introduced in an effort to improve the socio-economic conditions of these communities. The primary objective with this paper is to give an overview of the investigation into the establishment and the functioning of these FCTs in South Africa. Although the final outcome of the investigation shows that the FCTs are contributing to the relief of the problem, some form of intervention in the system is necessary to place their operations on a more sound footing. The recommendation is for a 'Mother Trust' to be created to act as an intermediary between the Quota Board and the FCTs, in such a manner that the autonomy of the individual FCTs is not violated. As a consequence of the development of a new fisheries policy for South Africa, this recommendation has not yet been implemented. It is, however, expected that this, or something similar, will be formalized in 1997.

Introduction and background to fishermen's community trusts

This paper deals with a brief overview of the nature and activities of Fishermen's Community Trusts (FCTs) in South Africa. It only touches on the outstanding features and events as these occurred over time, and points out the major problems and recommendations for the improvement of the system. It portrays the efforts initiated by the Quota Board and the Chief Directorate of Sea Fisheries to improve the socio-economic conditions and general standard of living of fishermen's communities along the South African coast.

At a meeting of the Quota Board on 14 May 1992, an inquiry into the socio-economic conditions prevailing in fishing communities along the West Coast of South Africa was announced. Concern was expressed regarding the existing socio-economic conditions and general standard of living in these communities and that something had to be done to improve the situation. At subsequent meetings, the possibility of awarding area-specific fishing quotas to these communities was discussed and approved in principle. On 9 December 1992, the Chief Directorate of Sea Fisheries was requested to initiate procedures for the establishment of these community development trusts.

This paper is mainly based on the Report of the Committee of Inquiry into Fishermen's Community Trusts, tabled on 31 December 1994 with the Minister of Environmental Affairs and Tourism, Cape Town.

The Directorate had to develop guidelines for a proforma constitution (Deed) for the FCTs and for procedures and conditions for their establishment. The Quota Board indicated that the following interest groups were to be represented on the Board of Trustees of each FCT:

- the local authorities
- the organized fishing industry
- fishermen and fishing industry workers
- community leaders such as teachers, lawyers and clergy, as well as groups such as women's committees and spiritual groups
- nominations by the Chief Directorate of Sea Fisheries.

It had to be emphasized at all times that the creation of the FCTs was intended only for those members of communities who were dependent on fishing and not for entire communities. These Trusts were also not to encroach on the jurisdiction of other agencies such as the government departments of Health, Welfare and Education. FCTs were to provide support in areas where there was none available from the State or other established sources.

At the meeting of the Quota Board on 8 June 1993, disappointment was expressed with regard to the slow progress being made with the establishment of the FCTs. It was reported that the matter was much more complicated than originally expected. For example, the community quota of 3000 t of hake was considered to be totally inadequate. The Quota Board announced that the fishing industry had to accept, in principle, that the FCTs formed an integral part of the industry and as such had a right to a quota. In future the community quotas would be considered in the same light as those for all other quota holders.

By 1 November 1993, 15 FCTs had been established. The criteria for quota allocation were also announced:

- number of fishermen/factory workers in an area
- existing infrastructure
- alternative employment opportunities in an area
- average income earned by fishermen/factory workers in an area
- assistance available from quota holders/employers in an area.

Following voluntary contributions by the existing quota holders, the total community quota available was increased to 5783 t. The Quota Board was asked to award quotas only to those FCTs where the Board of Trustees had been elected by an open meeting to which the 'whole community' had been invited.

Matters giving rise to heated debate and incidents during this establishment phase of the FCTs were, *inter alia*:

- rumours of cash disbursement to fishermen through the FCT
- defining a fishermen's community and fishermen
- onset of the first general democratic election in South Africa.

During this period there was unrest, violence, protests and even incidents of hostages being held because of fishermen's communities and other groups, demonstrating their dissatisfaction with the development of the FCTs at that time.

The Minister of Environmental Affairs thus appointed a Committee of Inquiry (of which both authors were members) into FCTs on 28 September 1994 to report within 40 days.

COMMITTEE OF INQUIRY INTO FISHERMEN'S COMMUNITY TRUSTS

The terms of reference of this Committee of Inquiry into FCTs were to advise the Minister on the following matters:

- the advisability of the allocation of quotas and/or other fishing rights to fishing communities (groups of people within coastal communities whose income depends mainly on the commercial fishing sector)
- should both fishermen and fish factory workers benefit from the Trusts?
- should money received by the Trusts be paid out as cash?
- is it legally permissible to allocate community quotas?
- should allocations be made geographically or by industry sector?
- what alternative system could replace the present Community Trust system?

The Committee conducted its investigation by means of open meetings in all the registered FCTs; by personal interviews with the Board of Trustees and with identified individuals in each FCT; and by inviting written or oral presentations from any interested party or group. Significant and representative responses were achieved through these methods.

OPERATIONAL ASPECTS OF FCTS

The following is a brief overview of some of the outstanding findings regarding operational aspects of the FCTs.

The Board of Trustees

Residents in most of the communities indicated that they had first heard of the concept and formation of FCTs at the meetings which served as a forum for the election of pilot committees. These pilot committees were to become the first trustees for the individual FCTs and to compile and register the deeds of trust for their FCTs. Relatively few residents were interested in these initial activities and consequently did not participate in the election of trustees.

Once the FCTs were registered and rumours began circulating that money could be forthcoming to the communities, interest and participation in activities increased dramatically. Individuals and groups in many communities now complained of never being consulted during the election process and consequently started opposing and criticizing the work performed by the trustees. These differences placed much stress on the trustees and in some instances developed into sessions of verbal abuse and even physical skirmishes.

Despite the opposition and problems caused by dissatisfied community members, the trustees continued to perform their tasks relatively successfully in terms of the deeds of trust.

General administration and financial management

It was not expected of the Committee to conduct comprehensive administrative and financial audits. From its investigation it became clear that there are relatively few FCTs experiencing problems in this regard. All the trustees indicated an awareness that the systems were legally subject to independent audit.

Initiating community projects

It quickly became apparent that few of the FCTs had the expertise, creativity and initiative for the implementation of projects that would be in the interest of the communities. All the FCTs with successful projects running had stated clearly that there would be no 'cash handouts.'

Some trustees indicated that they were hesitant to initiate projects as most caused dissent and unrest in their communities. They felt that their communities first had to be educated with regard to the principle of community service and benefits that do not provide immediate cash in hand.

Project management

Although the FCTs had been in operation for less than two years, several had already launched successful projects. Most of these are well-managed and contributing to the overall improvement of the community. There are also a few examples where the projects were not managed according to acceptable standards and where supporting documentation could not be produced.

Few of these projects can yet be labelled as being major successes in line with the initial intention of creating long-term benefits, job opportunities and infrastructure that can render positive long-term returns to the fishing community. Most are of a short-term nature and aimed at merely relieving immediate needs. This area requires the cultivation of initiative, long-term planning and vision.

The following is a selected list of successful and unsuccessful projects embarked upon by FCTs:

- school fees and scholarships to family members
- covering the cost of funerals for members of the community
- settling overdue accounts
- contributing to educational tours for school children in the community
- financing stock for small home shops
- donations to schools for educational purposes
- buying a tractor for moving boats in and out of the water
- buying chairs for a community hall
- cash handouts
- transportation cost for community members to participate in protest action in Cape Town
- loans for purchasing fishing gear, equipment and housing
- purchase of building materials for housing
- freezing facilities for fish caught
- cemetery renovation
- food parcels.

Acceptance of the FCTs concept in the communities

Even though several individuals felt that much more should be done for the enhancement of their communities, no-one wanted the FCTs to be abolished. FCTs are considered a step in the right direction and may be the beginning of something better.

Given the major differences between the needs and circumstances of many of the communities, it must be expected that these differences will lead to different approaches and applications of the available funds.

General evaluation of the functioning of FCTs

Considering that the main aim with the FCTs is the social, economic and educational enhancement of the fishing communities and the general improvement of their quality of life, it must be concluded that this has not yet been achieved. However, taking into consideration that there was nothing before and that this is only the beginning of what is hopefully to become a permanent institution, it can only improve.

The following are some of the major reasons for the slow start of FCTs:

- the concept of FCTs is new and there were no precedents
- for most of the trustees, this is a new experience, most never having served on similar bodies
- the hasty implementation of the concept, without sufficient training and preparation, placed undue pressure on the trustees with too much being expected of them
- in an attempt to win the favour of the community, some FCTs embarked on cash handouts and food parcels. They all agreed that these were the projects that created the most problems and would not be repeated.

Specific problems and reasons for conflict in some fishing communities

The following are some of the outstanding causes of problems encountered in the system and in some FCTs. Most of these problems can be attributed to inexperience and a lack of education with the concept of FCTs. They can easily be remedied and should not be repeated in future:

Inadequate infrastructure and preparation time

The Chief Directorate of Sea Fisheries was expected to launch a major new concept, involving the lives of thousands of fisher folk, right from scratch and in a very short period of time. At that point there was insufficient manpower, no experience on the proposal, an overload of normal duties and insufficient funds to properly manage the process. The communities to be reached also had no exposure to such a concept and did not know what was expected. They had to be convinced to cooperate to make the system work to their own long-term advantage. Even though the Directorate made a valiant effort with its limited means, it was not enough to educate and prepare the communities for what was to come. Much more time was needed to convince the communities that their cooperation was a prerequisite for the concept of FCTs to reach its objectives.

The period between the decision to undertake this major project and the actual launching of the FCTs in the communities, was less than one year. The announcement of the launch and the actual election of trustees occurred simultaneously in communities totally unprepared for it. Even though these events caused many problems and delays to get the project off the ground, many have been overcome, but special effort is needed to correct the wrong impressions and damage caused in the initial process.

The addition of new FCTs

Although the original Quota Board model for FCTs was aimed at enhancing rural, relatively isolated fishing communities, several urban groups demanded registration as FCTs, once the potential benefits became known. The groups were mainly from Cape Town, Port Elizabeth and False Bay. Since these fisher folk do not live in single geographical locations, it would have been impossible to launch community development projects through the FCTs concept. These groups then became aggressive and assertive in their demands to be registered as FCTs and caused incidents where the assistance of the police was required to get the situation under control.

Given the geographical circumstances of these groups and the limited budget available, the guidelines for qualifying FCTs must clearly delineate the principles and requirements for future registration. This matter has yet to be formalized.

Who are the real fishermen and fishing communities who must benefit?

Fishermen at grassroots level in the rural areas are as a rule not organized, which, combined with their geographically undefined locations, caused several problems during the inception stages. Answers to the questions as to: 'Who and what is a fisherman?' and 'What is a fishing community?' had not yet been defined although the answers were crucial for the registration of the respective communities. An independent investigation by the Human Science Research Council (HSRC) concluded, contrary to general belief, that there were only three communities along the West Coast and South Coast where more than half of the inhabitants were directly involved in the fishing industry.

Within the so-called fishing communities, the trustees had further problems in identifying the intended beneficiaries of the FCTs. Line-fishermen consider themselves the only 'real' fishermen as they claim to suffer most from the unpredictability of the sea and the resource. In spite of this, all the boards of trustees agreed on including all these fishermen and fish factory workers as beneficiaries of the FCTs. Boat owners and persons identified as factory management were excluded and in some instances also skippers of fishing vessels. The questions as to where relatives, widows and children of fishermen fit has not yet been resolved.

The ultimate solution would be for the trustees to implement a generally agreed-upon system of registration for those fishermen who qualify in their communities. It is not yet clear whether the trustees can handle this sensitive matter without external assistance.

Naming of the individual FCTs and the composition of the Board of Trustees

The names selected by some of the FCTs caused uncertainty and dissent in several communities. Referring only to the 'ABC Community Trust' created expectations that the entire community was included in the ambit of the benefits and the operations of the trust. Tensions ran high, especially where there were rumours of possible cash handouts. In those communities where the name of the trust refers specifically to 'ABC Fishermen's Trust', there was much less tension.

A factor also contributing to uncertainty was the set of guidelines for the composition of the boards of trustees. The fact that 'outsiders' such as teachers, lawyers, clergymen and others had to be included created the impression that more than just the fisher folk were to benefit. The original reason for this composition was to add expertise not normally found amongst fisher folk, on the board.

Party politics

The launching of the concept of FCTs coincided with the running-up to the first democratic general election in South Africa. In several communities, the persons aspiring to become trustees in the FCTs, were also those with an interest and involvement in the political elections. In some of the communities where funding had already been channelled to the FCT, several trustees applied the funding for, *inter alia*, food parcels, announcing that it came from their respective political parties. Trust funding was in some instances apparently also used for more direct political campaign expenditures such as organizing rallies and transportation to meetings and events. The FCT activities in these communities were frequently marked by wild political rhetoric which seriously marred the acceptability of the trusts.

In the FCTs where the motives were honourable, the trustees were strongly outspoken against the use of FCT infrastructure and funds for any political manipulation or purposes.

Cash handouts

Experience showed that cash handouts to the community do not contribute to meeting the objectives of the FCTs. Handouts caused major upsets and dissatisfaction since there was not enough cash available to satisfy even a small portion of the community. Nonetheless, many of those that considered themselves to be the 'real' fishermen exerted a great deal of pressure on the trustees for cash handouts. They argued that because of chronic poverty, these funds rightfully belonged to them and should be handed over to relieve their situation, before there could be any thought of community projects. It is important that the circle of poverty and misery must be broken, but this is more likely to be achieved with long-term enhancement through community projects, rather than with one-off cash handouts.

The trust deeds of all the FCTs do not state clearly that cash handouts are not permitted. When the often more militant members of such communities become frustrated with their failure to get cash, they seem to lose total interest in the FCTs and see no further use for them.

Geographical areas served by FCTs

A sense of dissatisfaction was experienced in some FCTs because of the geographical area to be served with the available funds. This matter will have to be addressed in future to prevent hostilities between groups or communities in the same geographical areas.

Allocation of quotas to the individual FCTs

FCTs indicated, with only one exception, that they were not satisfied with the amount of the total quota that was allocated. They stated clearly that the 'dissatisfaction' should not be confused with 'ungratefulness.' It was often put, 'We had nothing before, at least now we have something.'

This is a very sensitive matter which must be resolved, preferably by all the FCTs.

Communication between the boards of trustees and their communities

Ignorance and confusion as regards the concept of FCTs prevails amongst the constituents in most communities, mainly because the elected trustees failed to communicate the necessary information. No instance could be traced where the contents of the specific deeds of trust had been discussed with the communities. Most members of the communities questioned did not even know what such a deed was, or that their FCT had one.

This lack of communication can to a large extent be blamed on the inexperience of the trustees who should be educated in the responsibilities of trustees and of their accountability to the community.

Competence of trustees

It is important for the success of every FCT that at least one of the trustees must have vision, initiative, competence and a thorough knowledge of the community, and above all, is accepted as a leader. In most instances where such a person was found, he or she was not a fisherman, but came from the ranks of the clergy, teachers or business. In many instances a gap existed between the 'needy' fishermen and the 'well off' trustees, making it difficult to constitute a board which is generally acceptable and competent.

Relatively few boards of trustees could fill the requirements of legal, financial and management skills from within their own ranks. In many instances these services were provided by outside consultants on a contract basis.

Differences in the deeds of trust

Every FCT determines the details of its deed of trust within the guidelines provided. In many instances the members of different communities exchanged information as to the activities of their respective trusts. This amounted to unfair pressure on trustees, especially in instances where loans and cash handouts are not made, and made it extremely difficult for trustees to keep the peace and still adhere to their deeds of trust. Because of this, several trustees have indicated their unwillingness to continue as trustees, or their intention of not being available for re-election.

Various combinations and permutations of the above-mentioned issues were observed in all fishermen's communities and can be seen as the root cause of the unstable conditions which developed. Many of these matters will be resolved through time and experience. Others will have to be dealt with specifically through training, assistance and education to ensure that all FCTs are run in accordance with the purpose for which they were established.

Conclusions as regards FCTs

The following conclusions on matters of principle for the functioning of FCTs are important for their future acceptability and ability to achieve the aims for which they were established.

Should the concept of FCTs be continued?

Although many problems are being experienced with the operation of the FCTs, the greater majority of the communities and of the trustees indicated that they should be continued to become a permanent feature of the infrastructure. The main reasoning was that although there had never before been anything by which fisher folk could be assisted, at least now there was something.

The Committee of Inquiry was of the opinion that the FCTs should not be allowed to continue in exactly the same format. However, there were sufficient positive features which could be enhanced and expanded to create an infrastructure that was simpler and easier to steer towards its original aims.

Who should benefit from the FCTs?

There is no general agreement in the communities as to whom should benefit from the activities of the FCTs. There are potentially three main groups which could be considered for benefits:

- fishermen who actually go out to sea to catch fish
- workers in fish factories and processing facilities at sea and on land
- the broader community of which the fishing community forms an integral part.

Further permutations of these groups could include retired persons, widows and their dependants.

The first two categories listed can be subdivided into permanent employees of fishing companies, earning a fixed salary with benefits, and the fishermen and workers that do not have permanent employment, salaries or benefits. They are dependent on the weather, the seasons and the availability of the resource at any point in time.

It is concluded, in consensus with the original objective of the Quota Board, that the beneficiaries of the trusts ought to be primarily those fishermen and fish factory workers who do not have a permanent or fixed income and benefits. However, because of the complexity and features of each individual community, it is matter which must be resolved within each community.

Should cash handouts be allowed?

There is consensus that as a rule, no direct payment of trust funds to the community should be allowed. Cash handouts were singled out as the most important cause of problems in the operations of FCTs. In all the FCTs where handouts were made, conflict and hostilities erupted.

However, due to the special circumstances prevailing in many communities the trustees should be allowed to decide on matters equivalent to handouts, for example:

- emergency aid in the form of food parcels in deserving cases
- payment of outstanding rent
- payment of outstanding utility accounts
- assistance when a disaster such as storm or fire strikes.

Such assistance should be approached with the greatest discretion and transparency. No direct provision of cash should occur in any situation.

LEGAL STATUS OF FCTS

The Appellate Division of the Supreme Court distinctly expressed the view that the ownership of trust property does not vest in the trust but in the beneficiaries of the trust, therefore a trust cannot be a legal person [CIR *v* Mac Neillie's Estate 1961(3) SA 833 (A) on 840 and confirmed in Braun *v* Blann and Botha 1984(2)SA 850 (A)].

In terms of Clause A1 of the Quota Board's guidelines approved by the Minister on 10 March 1993, a quota allocation shall only be made to a legal or a natural person who is a South African citizen, which includes a registered SA company, a partnership, a trust, closed corporation or cooperative. Although a trust is not a legal or natural person, it is clearly the intention of the Quota Board to include the trust as a qualifying entity in respect of the allocation of a quota.

In a subsequent judgement on September 27, 1995, a Supreme Court judge ruled that quotas allocated to FCTs were illegal. Before this judgement, the trustees of the individual FCTs had met on 27 September 1994, to form the South African Association for FCTs (SAAFCTs). Their objective was to jointly manage matters of common interest to all the FCTs. After thorough consultation, they decided on 26 January 1996, to register a cooperative with the FCTs as the individual shareholders. The cooperative would now be the recipient of the annual quota allocated by the Quota Board, which would then be paid out to the individual FCTs according to the ratio applying at the inception of the system.

This event came as a natural evolution due to circumstances and is in line with the major recommendation of the Committee of Inquiry into FCTs which was made in December 1994, as outlined below:

Can the quota allocation to the FCTs be withdrawn?

According to the provisions in the Sea Fisheries Act, No 12 of 1988, the Quota Board can, under certain conditions, suspend, withdraw or reduce the quotas allocated. However, the Quota Board had already indicated that they intend the granting of quotas to fishing communities to become a permanent part of the system and that there is no reason to consider the withdrawal of the quota. Pending the outcome and recommendations of the Inquiry into the FCTs, it could have been considered an alternative.

The source of funding for the FCTs

There were suggestions that a levy on all quota species could be considered as an alternative source of funding for the FCTs. Preliminary calculations showed that in order to maintain the level of financial support for FCTs, the levy on all quota species would at least have to be doubled. Such a ruling would also deny the individual FCTs the right to decide on catching their own portion of the quota. Such a levy system would in practice be much simpler and easier to administer than the quota system and would spread the burden more evenly over the entire industry. However, it would be difficult to convince the formal fishing industry of the merit of such an approach after it had already rejected any suggestions along these lines.

ALTERNATIVE SOLUTIONS

The Committee of Inquiry considered the following alternative solutions to the FCTs:

Abolition of the concept of FCTs, without any replacement

The total abolition of the Community Quota System was seriously considered for the following reasons:

- fishermen in certain (particularly large) communities were mobilized to oppose anything but direct individual benefits. Some were not prepared to consider any other alternative than cash handouts
- certain groups had spread the word that, because of the imminent political changes at that time, the entire Quota Board system was going to be changed in the near future and that individual fishermen would be allocated their own quota to deal with as they wished
- it would be difficult to create an official administrative infrastructure which would not require significant powers of control to administer the quota system effectively. The Chief Directorate was not supportive and did not have the capacity to become involved to this extent

- the deep-sea and pelagic trawling industry is being penalized because of the loss of a part of their potential TAC being used to finance community development and enhancement. This is in essence a government responsibility to be financed through normal income tax contributions
- initially the community quota system operated as a dispenser of charity, resulting in dependence rather than enhancement and self-development
- the FCT system had resulted in bitter dissent, division and unrest in several communities. There were indications and threats of bloodshed, arson and revolution in those communities.

Alternative approaches to upliftment and development of communities

This approach implies that communities should be assisted through the implementation of special enhancement and development programmes. The following are examples:

- that FCTs be incorporated in the government's Reconstruction and Development Plan (RDP)
- that private enterprise in the fishing industry be called upon through social responsibility to become actively involved in the enhancement of the people in their industry
- that quotas be given directly to individual fishermen
- that existing government departments should shoulder the responsibilities for these communities.

After considering these alternatives, the Committee of Inquiry forwarded the following recommendations for dealing with the concept of community quotas.

An intermediary between the Quota Board and the FCTs

These recommendations were forwarded on the date of the report being handed over to the Minister. In the time that had elapsed, several developments have caused a number of the recommendations to become outdated or to be inappropriate. However, the basic principle contained in the recommendations is still valid and is expected to be implemented during 1996 or early 1997. The following is a brief outline of the more important issues at stake.

The final recommendation of the Committee is based on the assumption that:

- the principle of the FCTs must continue
- individual trusts must continue to exist independently
- individual FCTs have a legal claim to their portion of the quota.

The Mother Trust (MT)

The investigation showed that some form of intervention was needed in order to place the FCTs on a sounder footing. The creation of a Mother Trust (MT) with its own trust deed, to act as an intermediary between the Quota Board and the FCTs, would provide a way of facilitating without violating the autonomy of the individual trusts. The chairmen of the individual FCTs would form the management board of the MT, under the chairmanship of a person appointed by them.

Functions of the MT

The MT would be the legal recipient of the total community quota and administer its allocation. No transfer of funds to the FCTs would be considered unless a formal project proposition is submitted by the FCT to the board of the MT. Proposed projects must be considered and approved by the board of the MT. Once a project had been approved and funds forwarded, it would become the responsibility of the board of the MT to monitor progress. The board would create the infrastructure to perform these functions and shoulder ultimate responsibility.

Specific problems to be dealt with by the MT

The following are the main issues that normally caused problems in the functioning of the FCTs, which would be dealt with by the MT:

- unallocated funds to be kept out of the hands of the trustees of the FCTs, thus relieving the pressure for cash handouts or other unapproved payments. The FCTs will only have funds allocated to approved projects
- the MT to have the authority to determine the criteria for acceptable projects. This authority to be vested and exercised through the joint decision making of the chairmen of the individual FCTs
- If a justified project is not forwarded or initiated by the individual FCTs, no funding would be received
- this procedure would force the trustees of the individual FCTs to get closer to their constituencies in order to generate proposals which will contribute positively to the enhancement of the communities
- should a FCT not qualify for its portion of the total quota in a specific year, this portion would be kept in reserve for future use
- uniform standards can be set and applied in a transparent system with all FCTs receiving the same treatment according to their merit. This would be especially true in the case of:
  - administrative expenses forming part of project plans
  - creation of the infrastructure for supporting services and assistance in the form of a creativity centre, project management, administrative assistance and many more
- the system would become self-regulating with the joint board of chairmen of the individual FCTs acting as watchdogs and decision makers for their own affairs
- responsibility for the allocation of funds would be indirectly placed in the hands of the communities that are to benefit
- the MT would now decide on the admission of new FCTs applying for registration.

The creation of such a Mother Trust changes the process of funding around so that funds would not be paid over to the individual FCTs, thereby placing them in a dilemma over what should be done with these funds. The funding would now be placed with the MT and it is up to the individual FCTs to come forward with justified proposals as to how the funding should be applied in their communities.

CONCLUSION

As of July 1996, the MT had not yet been registered. The recommendations of the Committee of Inquiry were received favourably in the office of the Minister. However, since the Government of National Unity came to an end early in 1996, there has been a cabinet reshuffle and a new Minister has taken over the responsibility for Sea Fisheries.

At the present time a new Fisheries Policy is in the process of being developed for South Africa. It is unlikely that major decisions will now be taken that could become invalid once the new policy is in place.

It is generally expected that the cooperative registered by the SAAFCTs in January 1996 will ultimately perform the functions of the proposed MT and that its board of directors will thus become responsible for performing the functions as envisaged for the MT.

Even though there are many uncertainties as to what is to happen in the near future, it seems likely that the FCTs, which in the meantime continue to operate as before, will remain a permanent part of the infrastructure and that the main principles contained in the Report of the Committee of Inquiry into FCTs will be applied.

Post-Apartheid Fisheries Management Policy in South Africa: The Need for a Change in Management Philosophy

Trevor Hutton,[A] *Kevern L. Cochrane*[B] *and Tony J. Pitcher*[A]

[A] Fisheries Centre, University of British Columbia, Vancouver, BC, Canada, V6T 1Z4.
[B] Fishery Resources Division, FAO, via delle Terme di Caracalle 00100, Rome, Italy.

Summary

The transition to democracy in South Africa provides an opportunity to explore alternative fisheries management policies. A brief summary is presented of past fisheries management policy, which was based on central control, and the current policy-development framework. Co-management is being evaluated as an option that shares management responsibility between government and user groups in order to foster cooperation. This is in line with the aims of the new government which are to increase user-participation and set up democratic structures in the fisheries. As an example, a pilot project was designed to consider the potential role that co-management could play in a specific community. There are recently established representative groups in this community; however, formal structures need to be created and attention given to education and capacity-building. A general will to establish a new cooperative philosophy is essential to create a sustainable and equitable harvest of marine resources to the overall benefit of all in South Africa.

Introduction

The previous dispensation in South Africa was based on a system of institutionalized racism where an undemocratic government relied on central control to administer Apartheid policy. The first all-party multi-racial elections took place in April 1994 and there is presently an interim government negotiated on the basis of national unity. The onus is now on the new government to address the inequalities of the past. Furthermore, South Africa has a rapidly increasing population and a consequent growing need for housing, health care, education and food. It is within this context that one needs to consider future fisheries management policy in South Africa.

The aim of this paper is to provide an overview of problems and issues in the development of fisheries management policy in South Africa, using a pilot study on a small fishing community in the Western Cape as an example. The central theme of the paper is the need for a paradigm shift from management based on central control by the state to shared democratic management. This reflects common problems facing the governance of fisheries world-wide.

Fisheries management policy in South Africa

At present, South Africa does not have a comprehensive fisheries management policy, rather, various laws and regulations are *de facto* statements of policy. The country has an array of complex and highly integrated fisheries, including commercial, recreational and subsistence fisheries. Over 25 000 people are employed in the formal fisheries sector. Within the commercial sector, catches are dominated by the demersal and the pelagic fisheries, e.g. these two fisheries accounted for 88–95% of the reported catch for the period from 1975 to 1992 (Boonstra 1993).

Since 1948, a series of Sea Fisheries Acts have limited entry by imposing a broad array of statutory controls and restrictions on fishing including licensing requirements and quota allocations. Indeed over the last few decades the central government has played the major role in assuming responsibility for management of marine resources. At present, the Department of Environmental Affairs and Tourism administers government policy through the Sea Fisheries Act (1988). The Chief Directorate: Sea Fisheries is the primary fisheries management agency within this ministry. Research is undertaken by the Sea Fisheries Research Institute (SFRI) of the Chief Directorate, and administration through Exploitation Control and Administration.

Two new processes have been fostered to increase user involvement, extend consultation and improve equity. First, the Sea Fisheries Research Institute has established scientific working groups that include independent scientists, and discussion groups with various sectors of the fishing industry. Inputs from all these sources are fed to the Sea Fisheries Advisory Committee (SFAC), which provides final advice on resource management to the Minister. SFAC itself is made up of participants from different sectors of the industry and other interest groups. Secondly, allocation of quota is now undertaken by an independent body (the Quota Board) that became fully operational in 1991 and whose remit is to ensure stability in the industry, encourage the social development of fishing communities and create jobs.

Nevertheless, these essentially bureaucratic moves have failed to be widely perceived as effectively addressing past inequities. Often, the legitimacy of existing regulations is not recognized, and participation and consent is proving difficult to establish. There is considerable resentment of the system, user participation in management is still considered by many to be limited, and only a few new entrants to the fisheries have been accommodated.

RECONSTRUCTION AND DEVELOPMENT

The present national fisheries management system still reflects aspects of the previous government's policy. Despite the existence of the SFAC and the independent Quota Board, there is the perception of top-down imposition of technical measures by the central government regulatory body. With the political emancipation of South Africa this approach is being challenged by many of those who claim they have previously been denied access both to fishing opportunities and participation in the management process.

This is resulting in a thorough re-examination of the existing system with a view to the development of an equitable approach consistent with the aims of the government's Reconstruction and Development Programme (RDP) (African National Congress 1994). Key objectives in this programme include, e.g. increasing employment, sustainable utilization, increasing user participation in management, earning foreign exchange, economic efficiency, and equity in the distribution of benefits. The major limiting factor in terms of the contribution of fisheries to the objectives of the RDP is the production potential of the living marine resources upon which South Africa's fisheries depend. Almost all of these resources are currently being fully exploited and, in some cases, overexploited. There is therefore not the capacity simply to allow all of those who wish to participate in a fishery to do so.

Can the structures, most of which were established prior to the RDP, meet these new objectives? J. Issel (in Hirshon 1995) argues that Apartheid laws forced people who sought a living from fisheries to take employment with the large fishing companies. At the same time, a commonly held belief is that the fishing companies, generally seen as 'big business', have made huge profits by overexploiting common resources while failing to provide job security for their employees, or creating tangible benefits for the rest of the community (Manuel and Glazewski 1991). These issues ensure that access to South Africa's living marine resources is politically a highly sensitive topic.

One experiment that failed provides an example. Fishermen's Community Trusts were established in late 1992 with the primary objective of development of the coastal fishing communities. In order to fund these Trusts, a certain share of the total allowable catch (TAC), predominantly of the hake fishery, was set aside, e.g. in 1995 a total of 6504 t of the hake TAC of 138 166 t was set aside. Thirty-three Community Trusts had registered by November 1995. As part of the agreement the quotas were sold to the established companies in the industry. In practice it was found that aspects such as identification of the beneficiaries in the communities, led to conflict, including violence on occasions. A heated debate as to the merits of the Trusts culminated in a Supreme Court decision at the end of 1995 where it was decided that under the present Act the Trusts could not receive community quotas.

A number of other options that address RDP objectives are being considered. For example, a South African marine lawyer, J. Glazewski (pers. comm.) believes that fishers' rights should be strengthened to make future investment secure. In order to satisfy the calls for redistribution, existing stakeholders should make a sacrifice in favour of new entrants. Following a one-off redistribution, security would be provided by guaranteeing fishery access for a substantial period of time.

CURRENT FISHERIES POLICY DEVELOPMENT

Following South Africa's first democratic elections in April 1994 it was deemed appropriate for a comprehensive new fisheries policy to be developed in an open and transparent manner. The Fisheries Policy Development Committee (FPDC) was set up by the Minister of Environmental Affairs and Tourism in April 1995. The Committee consists of representatives from industry, the coastal provinces (the regional fishing fora, described below), the provincial governments, labour, the environmental sector and the recreational sector. A draft FDPC policy document covers institutional structures, policy objectives, principles and fisheries research. The focus of fisheries research in South Africa, with a long-standing international reputation in the traditional hard sciences of oceanography, ecology and fishery assessment, will, as in many other countries, need to include social and economic perspectives, and may include user participation in research and management.

Regional fishing fora, which aim to represent fisheries interest groups at 'grassroots' level, are organizations which have grown out of the Fishing Forum originally set up in the Western Cape by members of the African National Congress and other interested parties. There are now regional fora in all four maritime provinces, each having representation on the FPDC with some of the provinces having more than one regional forum.

It is possible that the regional fishing fora will become part of the management process. There are some clauses in the draft FDPC policy document which imply that management authority could under certain circumstance be delegated to lower levels (regional and local) and the fishing fora could be responsible for users having greater access to decision-making bodies.

Co-management and user participation

The present global crisis in fisheries management practices is often attributed to inappropriate institutional arrangements for policy formulation, implementation and enforcement and to some degree, the lack of legitimacy of management regimes (Jentoft 1989). Under co-management systems, legitimacy should be improved by transferring more responsibility to user groups and including them in the management process, leading to a system of co-management. Co-management and derivatives thereof have been presented as beneficial alternatives to the central government control management systems which currently exist in many countries (Pinkerton 1989). Negotiated formal or informal agreements between participants (user-groups and organisations) and state regulatory institutions would contribute to meeting the RDP objective of democratization of management to maximize user participation in decision making.

User participation in the form of co-management arrangements is unlikely to solve all the problems in South African fisheries. However, to design an appropriate fisheries management policy and relevant legislature for the future, the cooperative participation of those who will be most directly affected by it is essential. Experience elsewhere (Pinkerton 1989) suggests that co-management can lead to resolution of conflict and a re-direction of effort toward common interests and goals.

An illustrative example: the Arniston fishing community

This study considers the line-fishery on the South East Cape Coast focussing on Arniston (Fig. 1), where a pilot research project considered the potential role of co-management. The project required a preliminary assessment of the biological characteristics of the resource, such as distribution in time and space, and the socio-economic and political environment, that is the role of local organizations and recent policy initiatives by the

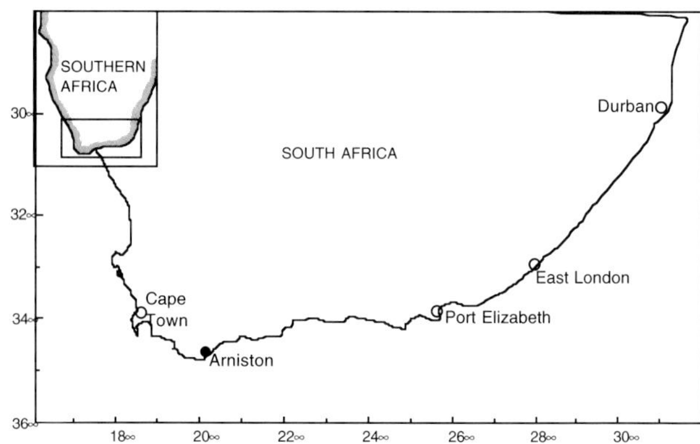

Fig. 1. Arniston: location for pilot project.

central government, the FPDC and the regional fishing fora. The aim of the pilot project was to collect information that could be compared with information from successful experiences with co-management in other countries (Jentoft and McCay 1995), and hence to evaluate the potential of such an approach in South Africa.

The multi-species linefishery in South Africa dates back to the 1600s. Despite its fairly long catch history, detailed biological research aimed at the management of the target species only began in the early 1980s. The number of species important to the linefishery is in the order of 95 (Van Der Elst and Adkin 1991). On a national level, linefishing is pursued by a fleet of ~3250 commercial vessels, between 57 000 recreational ski-boats and an estimated 750 000 shore anglers (Van Der Elst and Adkin 1991).

In 1985, a new national linefish management plan was established to manage the resource through a system of licensing based on classification as commercial (A permit) or semi-commercial (B permit) (Stander 1995). To minimize competition between the sectors, in October 1992 new fish categories reflecting conservation status were introduced: these comprise critical, restricted, exploitable, recreational and bait, species. The South African Marine Linefish Management Association (SAMLMA), an independent association which includes representatives of all the fishery sectors (but not the communities), is recognized by the Chief Directorate of Sea Fisheries and has a formal role in management.

Arniston was chosen for the pilot study primarily because considerable information was already available as it was one of thirteen 'coloured' fishing communities surveyed by Schutte (1993) and it is one of the few fishing communities exclusively involved in a single fishery. Arniston is a relatively closed community with little influx of labour and few alternative sources of employment apart from fishing (Schutte 1993 pers. comm.).

An initial study was undertaken using a Rapid Rural Appraisal approach (Rhoades 1987). Information was obtained from interviews with 'key informants', by semi-structured interviews (SSIs, see Pido and Chua 1992), secondary information (Schutte's socio-economic study conducted in 1993), and authors' personal observations. For the pilot study a total of seventeen interviews were conducted (there are approximately more than 100 full-time fishers in Arniston).

The main species in the annual catch of the community in weight and in monetary returns are yellowtail (*Seriola lalandi*: Carangidae), geelbek (*Atractoscion aequidens*: Sciaenidae) and kob (*Argyrosomus* spp. Sciaenidae). These are migratory species which are seasonally available and are fished by the rest of the commercial linefishery, the semi-commercial sector and to a degree by the recreational sector. Hence local community-based management is not a viable option.

As in Schutte's (1993) social survey, current size restrictions and bag limits are still a major issue in the community. There are feelings of mistrust amongst fishers as some feel their concerns were not included in the formation of regulations. In general, fishers do not understand the ecological reasons underlying regulations such as closed seasons, size restrictions and marine reserves.

Any co-management agreements would have to be fostered between government and some local organization. One of the key factors was the role of the local Arniston Fishermen's Community Trust. The Trust had a major influence on Arniston community by facilitating the evolution of a local organization concerned with local fisheries management issues. The agenda at meetings held to discuss distribution of trust money also focussed on issues relating to management of the fishery. The chairman of the Trust thus became the formal representative of the community in terms of local management concerns and issues.

Recent events in the community are leading directly toward providing the opportunity for user participation in fisheries management issues. A local Fishermen's Forum was set up in early 1996 as a result of developments at the more senior level (the FPDC) and to increase the bargaining power of the local community in the policy-development process. The aim of the Forum is act as a local organization which represents the fishers in the community and was set up by the leadership of the Community Trust which is still a legal entity. Whereas the direct role of the Trust was to administer the funds the Fishermen's Forum is solely concerned with fisheries management issues. It is too early to judge the success of the local Forum as it has only recently been established. However, it has the potential to play a critical role in facilitating the process of local user participation in the management process.

The linefishery is a multi-sector, multi-species fishery and as such, effective management would require that contested and shared resources be regulated through a single umbrella authority, able to follow a consistent overall plan and to co-ordinate all the user groups. The present management system administered by central government possesses a competent administrative facility, an advanced scientific capability and, most essentially, the regulatory capacity. However, judging by the grievances detected among the Arniston community, it lacks support from, and communication with, these stakeholders.

Considering the opposing interests of the user groups (commercial, semi-commercial, recreational), it seems essential for an umbrella authority to retain responsibility for management of the widely distributed or migratory resources and to strive to minimize conflict and obtain consensus on management issues. At present, the Chief Directorate: Sea Fisheries is the obvious agency, with the necessary capacity for this task. However, given the strong propensity for controversy and confrontation within the fishery there is clearly a need for the establishment of a co-management process between government and user groups, in an effort to foster cooperation and understanding.

Co-management can only be established by formalizing arrangements that define the hierarchical organizational structure and specific responsibilities of all parties in the process (Lane and Stephenson 1995). The lines of authority, reporting structure, information links, cost recovery and exchange of data correspondence among all parties need to be determined. For Arniston, the most obvious recommendation that can be made is that the local Fishermen's Forum should represent the problems being experienced by the community by being formally incorporated into the regional forum as an integral part of the new fisheries management policy. Formal agreements between local forums, regional forums and more senior government structures should be put in place.

Discussion

The value of user involvement in fisheries management was alluded to in the discussion on co-management. As Jentoft and McCay (1995) note, the question is not so much if and why user groups should be involved, but how. Generally when such a suggestion is made the trend is to include individual fishers as active participants in fisheries decision-making bodies. In South Africa some elements of user participation are present in the form of advisory channels of communication, but this falls short of sharing management responsibility.

A more dynamic partnership between government and interest groups could be fostered in South Africa, using the capabilities of the interest groups, complemented by the ability of the fisheries administration and scientists, to provide enabling legislation and administrative and technical assistance. This would also require the enhancement of skills in both groups and an educational thrust on the part of the government including increasing awareness of the management process.

One of the most important aspects at the local level is representation. The onus is on the users to establish or facilitate the formation of truly representative bodies which will be responsible for local decisions. In some cases this is a lot to ask of communities who have previously been denied these opportunities, but in Arniston the local Fishermen's Forum has attempted to begin the process. Another issue which is a constraint to local-level management, in the case of Arniston and other communities dependent on the linefishery, is the shared nature of the marine resources.

An initial task will be to create mechanisms to increase participation in consultation regarding the content of regulations with respect to these shared resources. This has, as already described, been happening in some sectors in South Africa. The most critical issue which is plaguing the success of this process is what Jentoft (1989) referred to as distributional effects. The central debates have not been about how to manage the resource, and what structures or institutions are needed, but who has a quota or a licence. There is an urgent need to resolve the issue of allocation through negotiation and conflict resolution. A widely accepted system of access is essential to improving the management of the resource and the fisheries.

Once this has been accomplished there is the potential for the delegation of some part of the management responsibility and decision making to user groups, providing that the capacity is there to ensure responsibilities are carried out competently. To ensure that communities have suitable management capacity, appropriate new consultative and decision-making bodies and structures for different fisheries and areas will need to be created.

Conclusion

A change in management philosophy which promotes user participation in future fisheries management has been presented as an alternative to past fisheries management policy in South Africa. This shift is essential in order to fulfil the basic principles and objectives of the RDP. The implementation of regulations will fail if they are not perceived as being legitimate by the majority of the users. It is possible that co-management could increase the legitimacy of these regulations through the process of promoting cooperation and shared management responsibility. Co-management has the potential to contribute to effective resource management if formal structures are created and attention given to education and capacity building. Thus, it is important that the concept of co-management is recognized as an essential component of fisheries management policy in South Africa.

Acknowledgments

This work was funded by the South African Foundation for Research Development. We thank Dr De Wet Schutte, Ms Maria Hauck, Dr P. Wickens, Mr Andrew Penney and Mr Barry M. Clark for helpful comments.

References

African National Congress (1994). 'The Reconstruction and Development Program.' (Umanyano Publications: Johannesburg).

Boonstra, H G van D (Ed) (1993). South African Commercial Fisheries Review 1991 (**1**). (Chief Directorate: Sea Fisheries: Cape Town).

Hirshon, G (1995). Angling for more equity: but allocating more quotas to small fishermen may cause problems. *Financial Mail*, February 17th, 28–30.

Jentoft, S (1989). Fisheries co-management: delegating government responsibility to fishermen's organisations. *Marine Policy* **13**(2), 137–54.

Jentoft, S, and McCay, B (1995). User participation in fisheries management: lessons drawn from international experiences. *Marine Policy* **19**(3), 227–46.

Lane, D E, and Stephenson, R L (1995). Matching technical measures with multiple objectives through comanagement. *International Council for the Exploration of the Sea*, C.M. 1995/S:11.

Manuel, F, and Glazewski, J (1991). The Oceans: Our common heritage. In 'Going green: people, politics and the environment in South Africa'. (Eds J Cock and E Koch) (Oxford University Press: Cape Town).

Pido, M D, and Chua, T E (1992). A framework for Rapid Rural Appraisal of coastal environments. In *International Centre for Living Aquatic Resource Management Conference Proceedings* No. **37**, 133–48.

Pinkerton, E (1989). Introduction: attaining better fisheries management through co-management — prospects, problems, and propositions. In 'Co-operative management of local fisheries: new directions for improved management and community development'. (Ed E Pinkerton) pp. 3–33 (University of British Columbia Press: Vancouver).

Roades, R E (1987). Basic field techniques for Rapid Rural Appraisal. In *Proceedings of the 1985 International Conference on Rapid Rural Appraisal.* Systems Research and Farming Systems Research Projects (Khon Kaen, Thailand: University of Khon Kaen).

Schutte, D (1993). 'n Ontleding van die ontwikkelingspotentiaal van geselekteerde vissersgemeenskappe aan die Wes-en Suidkus. R.G.N.: Kaap.

Stander, G (1995). Aspects of the development and regulation of South Africa fisheries. In 'Review of international experiences in access rights and their implications for fisheries management in South Africa'. Access Rights and Resource Implications Task Group. pp. 51–71 (Sea Fisheries Research Institute: Cape Town, South Africa).

Van Der Elst, R P, and Adkin, F (Eds) (1991). Marine linefish. Priority species and research objectives in southern Africa. *Special Publication of the Oceanographic Research Insitute, South Africa* **1**. 132 pp.

Converting Research Results into Improved Management using Bayesian Statistics: the Example of Pacific Salmon Gauntlet Fisheries

J. P. Scandol[A] *and C. J. Walters*

UBC Fisheries Centre, 2204 Main Mall, Vancouver BC V6T 1Z4, Canada.
[A] Present address: 22 Theodore St, Balmain NSW 2041, Australia

Summary

Converting the results of scientific research into improved management can be a difficult process. Fisheries managers require quantitative methods which allow scientific results to be embedded within management procedures. Bayesian statistics provides a methodology for combining prior information with data to generate post observation (posterior) statistical inferences. We suggest that in some circumstances Bayesian statistics provides an excellent mechanism for absorbing scientific results. The fishery we use to illustrate this process is the gauntlet-style fishery on sockeye salmon. Using a variety of research methods, scientists have and continue to develop methods which allow pre-season estimates of salmon run sizes, run timing anomalies and run diversion (distribution between alternative migration routes). None of these methods are completely reliable but they do present sources of information that could be quantitatively integrated into assessment procedures. Such information can be interpreted as informative prior probability distributions within a Bayesian inference scheme, to be combined with each fishing season's survey and exploitation rate data. The resulting running estimates can provide the basis for an in-season adaptive harvesting procedure.

Introduction

Science and management

Stephenson and Lane (1995) discussed the inter-relationships between fisheries management, fisheries science and management science. This study considers two of these components by presenting a formal method for using elements of decision theory to facilitate communication between fisheries management and fisheries science for the gauntlet fisheries for Pacific salmon in the North Pacific.

Scientists continue to take on the challenge of trying to elucidate the underlying causes of variability in fish stocks, so that less emphasis has to be placed upon inductive methods. What many scientists don't realise is that fisheries management is a complex, uncertain process even if fish population dynamics are excluded. New scientific ideas are not always integrated into fisheries management, and should not be. However, many researchers would like to think that their work is relevant to some appropriate goal (such as sustainable development). We therefore need to design mechanisms that enable new information to be blended into the management processes such that several goals are met. Such mechanisms should: (1) include the ability to revert to results that would have been obtained without adding additional information; (2) present a fixed and interpretable target to which scientists can work; (3) enable decision makers to adjust degrees of belief in the new information; (4) indicate when new observations contradict the new scientific input, and discard the scientific model if required. We suggest that Bayesian statistical inference meets all of these objectives, and we present a case of how such methodology might be used for the management of Fraser River sockeye salmon (*Onchorhynchus nerka*).

Subjective probability

Several interpretations of probability exist (Efron 1986; Hamming 1991). The most common interpretation is a frequentist one where the probability of a proposition is the expected long-term frequency that such a proposition is true. An alternative interpretation is termed 'subjective' probability which is far less restrictive. Here probability is interpreted as an individual's degree of belief in a proposition. Apart from some fascinating historical and philosophical considerations (Gigerenzer *et al.* 1989) there were practical reasons why the application of Bayesian statistical inference in the sciences has been constrained. Utilization of many Bayesian inference algorithms requires complex analytical operations which can now be solved numerically. The example presented in this paper is one such case. Application of Bayesian methods to fisheries problems is not new. Recent papers include (Fried and Hilborn 1988; Hilborn *et al.* 1994; Walters and Ludwig 1994). Hilborn and Walters 1992 give a detailed exposition of the application of Bayesian methods for fisheries problems.

Bayes' Theorem in statistical inference is best illustrated with the simple example of a coin toss. This quintessential random process illustrates the benefits and responsibilities of Bayesian methods. Most people, when asked what the probability of a head will be, will suggest 0.5. However, this is a subjective judgment because it requires assumptions about the credibility of the person who supplied the coin, etc. True frequentists would reply that they do not know what the probability is, they need to collect some data first! Bayesians can incorporate the subjective information they know about a system into the estimation of the probability using Bayes' theorem:

$$P(H|D)=\frac{P(D|H).P(H)}{\sum[P(D|H).P(H)} \qquad (1)$$

This equation reads: the probability of a hypothesis given the data is equal to the probability of the data given the hypothesis multiplied by the probability of the hypothesis divided by the total probability of the data.

Salmon runs in space and time

Figure 1 illustrates our basic conception of how salmon migrate from the high-seas to their natal streams *via* the intensive coastal US and Canadian fisheries. Three components of this diagram require emphasis. Firstly, there is considerable uncertainty associated with the actual number of fish that return, (estimates are made using highly variable stock-recruitment and environment-recruitment relationships). Secondly, the time of the run peak — often interpreted as the timing anomoly from the mean — is also uncertain. The run peak is the date that the maximum number of fish pass a particular point. Finally, the percentage of fish that migrate to the Fraser River *via* the north of Vancouver Island (as opposed to the south), or the northern diversion rate changes between years. These are the three core variables that can make Fraser River sockeye stock assessment error-prone and lead to management errors. For example, when the peak run occurs early it may mislead managers into thinking a larger run is yet to come.

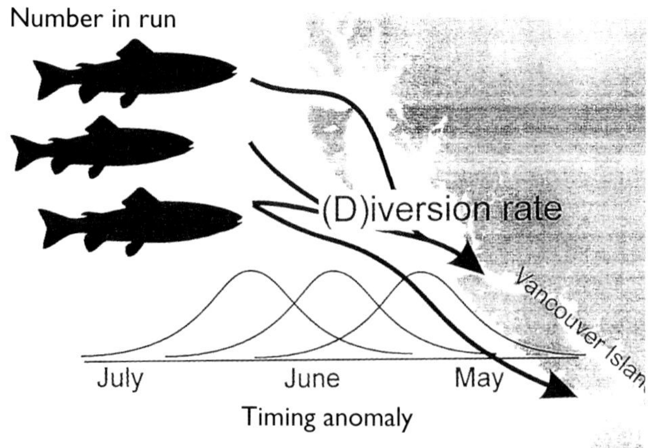

Fig. 1. Schematic representation of salmon runs returning to the Fraser River in British Columbia, Canada. Uncertainty is present in: the total run size (N); the timing anomaly (T); and the diversion rate (D) around Vancouver Island.

Management objectives are based upon firstly reaching escapement goals and then secondly meeting complex (and frequently disputed) allocation goals between native, Canadian and US fishers (Woodey 1987). To achieve these objectives the Pacific Salmon Commission has developed methods and fostered partnerships that collect and interpret catch and effort data, test-fishing information and acoustic information. These sources allow for real time updating of daily data such that in-season adjustments can be made to pre-season plans. The fishing fleet consists of a highly effective troll, seine and gill-net fleet (which can take around 70% of the run) as well as active native and sport fisheries. In 1994 it became evident that existing management methods were not sufficient to safely sustain the resource (Fraser 1995). Failure to take into account uncertainties in catch, effort, diversion rate, pre-season runsize, in-river survival and exploitation rates resulted in the 'disappearance' of over one million sockeye salmon in 1994 (Fraser 1995). A key recommendation of the resultant Fraser report was to adopt a risk-averse management strategy. Such strategies require estimates of the probability distributions of outcomes and the consequent costs associated with those outcomes (Morgan and Henrion 1990).

At the University of British Columbia we have undertaken a large multi-disciplinary modelling project which aimed at generating pre-season forecasts of: run size, run timing and diversion rate (Healey *et al.* 1994). It was believed that by utilizing the extensive physical datasets available for the north Pacific and constructing well-calibrated bioenergetic models of sockeye salmon, we could generate unique forecasts. We used a

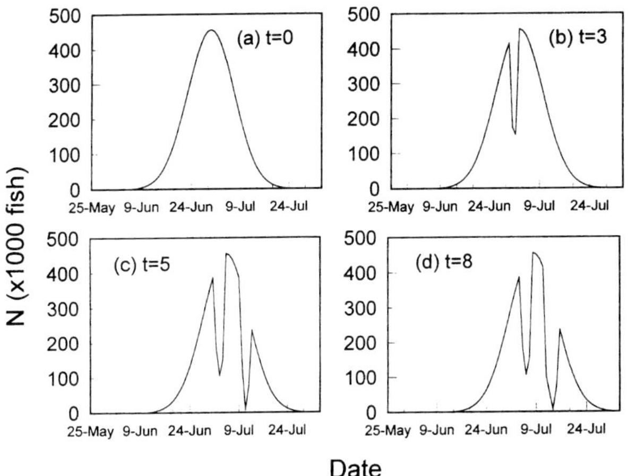

Fig. 2. Snapshots in time of a simulated salmon run being harvested by commercial fisheries. Time (t) is the number of days since the start of the simulated run.

range of methods for these forecasts including regression analysis (Cox and Hinch, in press), individual-based models (Walter *et al.* in press) and dynamic optimization models. Given that all these forecasts would include considerable uncertainty, and that the in-season assessment procedures were so well established (Woodey 1987) it seemed appropriate to seek assessment procedures that could extend existing methods and enable the potential blending of new types of forecasts into the calculations. This paper summarizes some initial analyses of that method.

MATERIALS AND METHODS

Fraser River sockeye salmon runs can be modelled as a one-dimensional bell-shaped spatial distribution of fish moving onto and along the coast in time (Cave and Gazey 1994). Figure 2a illustrates a 'run-timing curve' which is interpreted as the number of fish an observer would see over time at a single point in space. The total area under the curve represents the total run size, and width specifies the run spread. Most runs have been observed to have a spread (interpreted in our model as one standard deviation of the run-timing curve) of around 7 days. An additional complication is that the salmon runs are split between the north and south routes around Vancouver Island. We discretized the model such that our salmon run is represented as (using similar notation to Cave and Gazey 1994):

$$N_{i,t,r} = \frac{D \cdot N}{\alpha\sqrt{2\pi}} \cdot \exp\left\{-\frac{1}{2}\cdot\left(\frac{t-T-t_{\max}}{\sigma}\right)\right\} \qquad (2)$$

Here $N_{i,t,r}$ represents the initial (before harvest) number of salmon in spatial element i, of day t using route r. The spread of the run is 7 days, t_{max} is the day of the run peak, T is the timing anomaly, D is the diversion rate, and N is the total number of salmon in the run. For this analysis we considered run size (N) of 8 million fish, no timing anomaly ($T = 0$), a peak day of June 30 and a 90% diversion rate (D). The salmon run is propagated in space and time by defining the size of the spatial elements such that each daily timestep sees the run move forward one spatial element. Thus:

$$N_{i+1,t+1,r} = N_{i,t,r} - \sum_f C^f_{i,t,r} \qquad (3)$$

The term $C^f_{i,t,r}$ represents the catch taken in element i at time t on route r by fishery f. We sum this over all fisheries to calculate the total elemental harvest. Commercial fisheries can be specified to be operating within a particular domain of the space-time array $\delta_{i,j}$ = 0 fishery inactive, 1 if fishery active). Assuming that we have an estimate of the harvest rate (u_f) for each fishery we can predict the catch for fishery f from the model using:

$$C^f = \sum_{i,t,r} N_{i,t,r} \cdot u_f \cdot \delta_{i,j} \qquad (4)$$

Figures 2b, 2c and 2d illustrate the impact on the run of a series of commercial fisheries which impact time-space elements of the run. These five simulated fisheries were specified to typify the high exploitation rates that seine and gill-net fisheries generate. The four sub-plots within Figure 2 illustrate how 'holes' are cut into the salmon run as the season progresses. Due to the different swim speeds of the fish these holes tend to be refilled with time, but this complication requires additional model parameterization which is beyond the scope of this paper.

Using data collected (or simulated results) from each fishery (C_o), we can calculate the value of the likelihood function for the model $M(N,D,T)$ (which generated estimated catch data C_m) using:

$$L\left(C|M(N,D,T)\right) = \exp\left\{-\frac{1}{2}\cdot\sum_f \frac{\ln\left(C^f_o / C^f_m\right)}{V^f}\right\} \qquad (5)$$

Here V^f is the variance of the log-scaled catch from fishery f (obtained from historical records). Using numerical optimization (in this case complete enumeration) we can determine the parameter values of N, D and T (which are used to calculate C_m) which are most likely to generate dataset C_o, hence specify the maximum likelihood estimate for N, D and T. The Bayesian approach is affected by including prior probability values for each value of N, D and T that we consider during the optimization process. This modifies the above equation (5) so that:

$$P\left(M(N,D,T)|C\right)\alpha Pp(N)\cdot Pp(D)\cdot Pp(T)\cdot L(\ldots) \qquad (6)$$

We can calculate the actual posterior probability distribution of the run size by integrating out (summing over) the nuisance parameters (D and T) and then re-normalizing the product of the likelihood function and the priors.

RESULTS

Using the base case salmon run illustrated in Figure 2 we determined the impact of a range of prior probability distributions for N, D and T on the maximum posterior value of N and the 90% credibility intervals. These calculations also provide us with point estimates on the most probable values of D and T. Initially we used uninformative prior probability distributions for N, D and T. Uninformative priors are constant for the entire range of parameter

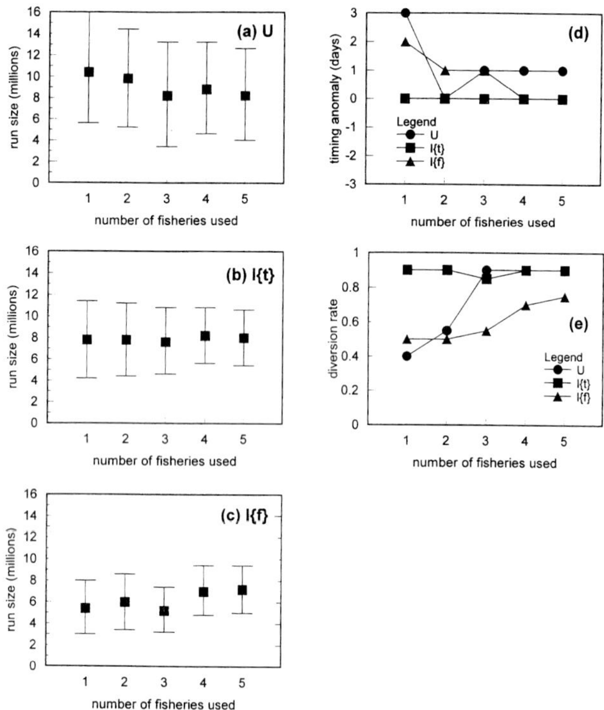

Fig. 3. Plots illustrating the impact of additional fisheries data on: (a) variation in estimated run size using uninformative priors (U); (b) variation in estimated run size using correct informative priors (I{t}); (c) variation in estimated run size using incorrect informative priors (I{f}); (d) most probable value of timing anomaly for U, I{t} and I{f}; (e) most probable value of the diversion rate for U, I{t} and I{f}.

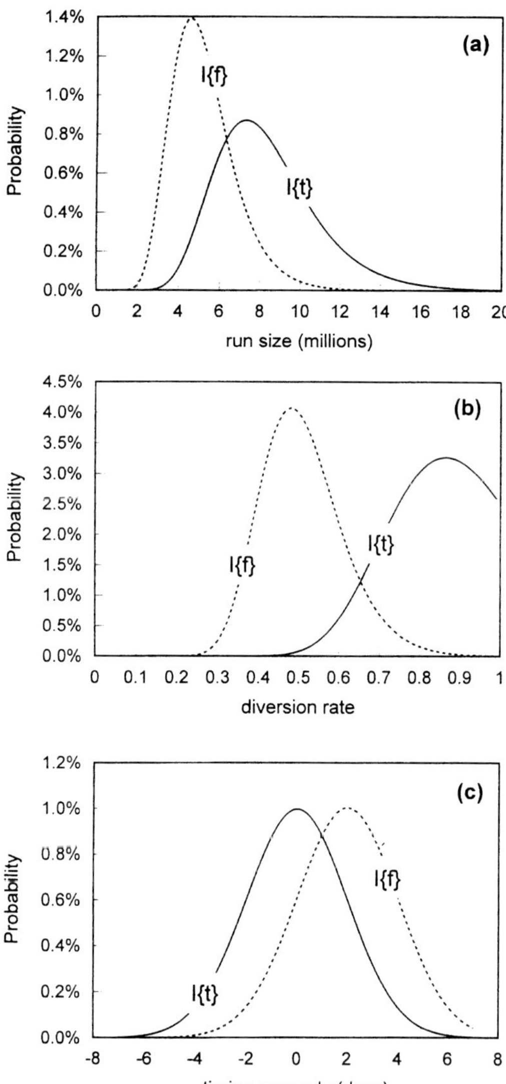

Fig. 4. Prior probability distributions used in the assessment calculations illustrated in Fig. 3. (a) Priors representing the correct (I{t}) and incorrect (I{f}) run size *N*; (b) Priors representing the correct (I{t}) and incorrect (I{f}) diversion rate *D*; (c) Priors representing the correct (I{t}) and incorrect (I{f}) timing anomaly *T*.

values evaluated, so apart from the nominal information contained in specifying the parameter space bounds these priors do not add additional information to the calculations. Figure 3a illustrates the estimated run size (actual value 8 million fish) as we add additional catch information from fisheries. As expected, using a small amount of data, the credibility bounds are wide, but as we add data the most probable value slowly converges upon actual run size. Note that the 90% credibility intervals remain wide for this estimation procedure. Figure 3d and 3e (series U) illustrate the estimated value of timing anomaly *T* (actual value 0) and diversion rate *D* (actual value 90%) as we add fishery data. Both of these parameters eventually converge to their true values.

The second test of this procedure used the prior probability distributions as illustrated in Figure 4. Two sets of priors were tested; the first set was biased towards the true value of the parameters (series I{t}), whilst the second was incorrectly biased (series I{f}). We suggest that the results of many scientific studies, including those from the salmon project we described in the Introduction, could be interpreted as Bayes prior probability distributions. Figure 3b indicates the consequence of using correct prior probability distribution. For small amounts of data, the assessment calculations are more accurate and suggest less variability than when using an uninformative prior. As we add information, the most probable value remains relatively constant but the credibility intervals become smaller. In contrast, using an incorrect prior probability distribution (Fig. 3c), a downward bias for expected run size is caused when small amounts of data are available; however, as information is augmented to the calculations the assessment begins to converge to the true value. Note that for virtually all calculations the true value (8 million fish) is contained within the 90% credibility intervals even when the most probable value of the prior is 5 million fish. A similar trend exists for the parameters *D* and *T*. For *T* (Fig. 3d) a correct prior value reproduced the exact timing anomaly and a biased prior *T* does not appear to affect the accuracy of the calculations for long. Similarly, the diversion rate *D* (Fig. 3d) is corrected more slowly with a biased prior than an unbiased one, but the correction eventually does take place.

Discussion

Bayesian methods enable us to blend prior information of a process with data collected within inference procedures. If the prior information is correct then the inference converges more rapidly to the true result yet if the prior information is incorrect this initial bias is corrected as more data are collected. The benefits of this process are that prior information is not wasted and that a well-defined mechanism exists for using secondary information within decision making. This mechanism may encourage scientists to consider the application of their work and this may provide greater focus for fisheries science.

Two recommendations in implementing such a procedure are suggested: (1) The inference calculations should be constructed such that the models only influence the calculations when very little data are present (e.g. the prior probability distributions should have a high variance). (2) Should the model forecasts be incorrect, then the data need to be able to correct the bias in the assessment as soon as strong evidence is available. If a risk-averse management strategy has been specified then the expected value of the prior probability estimate of run size should be biased downwards by at least one standard deviation from the pre-season estimate.

The lack of acceptance of Bayesian methods in applied science is surprising, but it appears a rift in the frequentist and Bayesian schools of thought at the turn of the nineteenth century is responsible (Gigerenzer *et al.* 1989). It is important to note that there is a convi.ncing argument that Bayesian methods are an excellent model for the actual process of scientific induction. For example (Howson and Urbach 1989) have suggested that the Bayesian approach gives a superior insight into the nature of scientific reasoning than rival methods. This puts those who question the application of Bayes' Theorem within the sciences in a more defensive position.

A worrisome aspect of our results is the influence of fishing itself on the quality of the assessments. We see no obvious way to obtain accurate run size estimates within a salmon fishing season without fishing hard enough to potentially risk depletion of the run. This 'dual effect' of control needs to be studied more carefully in harvest management situations.

ACKNOWLEDGMENTS

We would like to thank Jim Cave and Jim Woody (Pacific Salmon Commission) for their encouragement of this project. This research was funded by an NSERC strategic grant to study the biophysical controls of salmon migration and production (awarded to Mike Healey, Paul LeBlond and Carl Walters).

REFERENCES

Cave, J D, and Gazey W J (1994). A preseason simulation model for fisheries on Fraser River sockeye salmon (*Oncorhynchus nerka*). *Canadian Journal of Fisheries and Aquatic Sciences* **51**, 1535–49.

Cox, S P, and Hinch, S P (in press). Changes in size at maturity of Fraser River sockeye salmon (1952–1993) and associations with temperature. *Canadian Journal of Fisheries and Aquatic Sciences.*

Efron, B (1986). Why isn't every one a Bayesian? *American Statisician* **40**, 1–11.

Fraser, J A (1995). Fraser River Sockeye 1994: Problems and Discrepancies. Fraser River Sockeye Public Review Board, Vancouver.

Fried, S M, and Hilborn, R (1988). Inseason forecasting of Bristol Bay, Alaska, sockeye salmon (*Oncorhynchus nerka*) abundance using Bayesian probability theory. *Canadian Journal of Fisheries and Aquatic Sciences* **45**, 850–5.

Gigerenzer, G, Swijtink, Z, Porter, T, Daston, L, Beatty, J, and Krüger, L (1989). 'The Empire of Chance: How probability changed science and everyday life.' (Cambridge University Press: Cambridge).

Hamming, R W (1991). 'The Art of Probability for Scientists and Engineers'. (Addison-Wesley: Redwood City).

Healey, M C, Thomson, K, Hinch, S, Leblond, P, Scandol, J, Walters, C, Ingraham, J, Huato, L, and Jardine, I (1994). Biophysical controls of salmon migration and production in the northeast pacific. *ICES Mini-symposium on fish migration* **1**, 22.

Hilborn, R, and Walters, C J (1992). 'Quantitative Fisheries Stock Assessment: Choice, Dynamics and Uncertainty.' (Chapman and Hall: London).

Hilborn, R, Pikitch, E K, and McAllister, M K (1994). A Bayesian estimation and decision analysis for an age-structured model using biomass survey data. *Fisheries Research* **19**, 17–30.

Howson, C, and Urbach, P (1989). 'Scientific Reasoning: The Bayesian Approach.' (Open Court: La Salle).

Morgan, M G, and Henrion, M (1990). 'Uncertainty: A Guide to Dealing with Uncertainty in Quantitative Risk and Policy Analysis'. (Cambridge University Press: Cambridge).

Stephenson, R L, and Lane, D E (1995). Fisheries management science: a plea for conceptual change. *Canadian Journal of Fisheries and Aquatic Sciences* **52**, 2051–6.

Walter, E E, Scandol, J P, and Healy, M C (in press). A reappraisal of the ocean migration patterns of sockeye salmon by individual based modelling. *Canadian Journal of Fisheries and Aquatic Sciences.*

Walters, C J, and Ludwig, D (1994). Calculation of Bayes posterior probability distributions for key population parameters. *Canadian Journal of Fisheries and Aquatic Sciences* **51**, 713–22.

Woodey, J C (1987). In-season management of Fraser River sockeye salmon (*Oncorhynchus nerka*): meeting multiple objectives. In 'Sockeye salmon (*Oncorhynchus nerka*) population biology and future management'. (Eds H D Smith, L Margolis and C C Wood) pp. 367–74. *Canadian Special Publications in Fisheries and Aquatic Sciences* 96.

MONITORING FISHERIES EFFORT AND CATCH USING A GEOGRAPHICAL INFORMATION SYSTEM AND A GLOBAL POSITIONING SYSTEM

G. J. Meaden[A] *and Z. Kemp*[B]

[A] Canterbury Christ Church College, North Holmes Road, Canterbury, Kent, CT1 1QU, UK.
[B] Computing Laboratory, University of Kent, Canterbury, Kent, CT2 7NF, UK.

Summary

At the present time most of the world's commercial fishing vessels are using manual data-logging systems, many of which incorporate crude geo-referencing of fishery activities. Additionally, many fishery management problems are concerned with variations or inequities in the spatial domain. To overcome these problems, an integrated Fisheries Computer Aided Monitoring system (FISHCAM) has been designed and is illustrated here. The system presently integrates a Geographical Information System (GIS) and a Global Positioning System (GPS) with a relational database which may be housed in an on-board computer. The main functionality is in producing a range of graphical, textual and mapping output. Since the database is linked to a proprietary GIS, it has the capacity for integration with a wide range of external data sets, i.e. for any spatial and temporal analyses which fisheries managers might require. Observations are also made on the system's important implementation considerations.

INTRODUCTION AND OBJECTIVES

Lessons learned from recent fish stock collapses make it clear that effective fisheries management is a prerequisite for fisheries of all scales, and in all marine areas. It is widely agreed that 'better management' would, amongst other factors, be a direct function of the quality and quantity of the information available. The present availability of catch statistics for most fisheries is extremely variable and often unreliable (Holden 1994). For instance, most catch statistics are landing statistics, with the difference between them being made up by 'black' fish (fish that are illegally landed and not recorded) and/or by discards. Although discard estimates are made in some fisheries, it is always desirable to have reliable total catch statistics as an accurate informational source for management purposes. Area misreporting of catches also occurs and this results from fishermen not wishing to reveal good fishing grounds, though it may be done through ignorance or a lack of appreciation of the necessity for accurate data.

The importance of precise locational catch and effort data may be for reasons directly related to the management of spatially imposed quotas, or indirectly for a range of associated reasons which will be outlined below. At present the reported locational accuracy of fishery-related data is often very poor, e.g. in UK waters the current log-book requires only that catch location be recorded from within an ICES statistical rectangle, which may

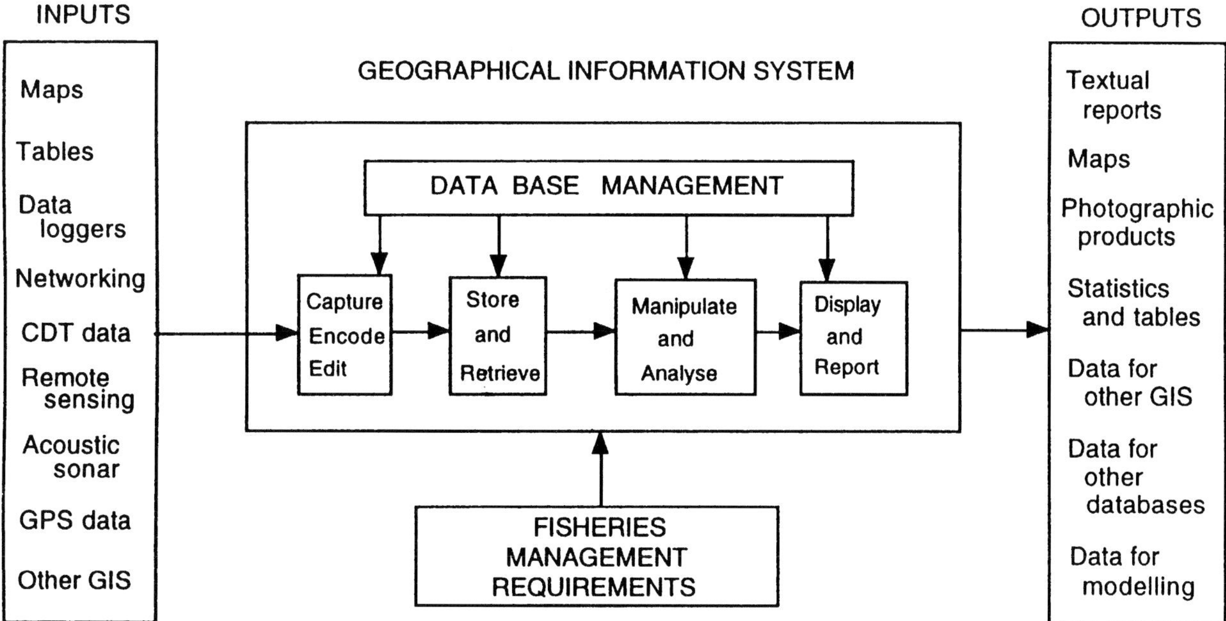

Fig. 1. The major characteristics of a marine resources GIS.

cover over 4000 km^2. Precise locational data are required to take account of irregular temporal and spatial resource distributions of fish stocks, variations in the potential fisheries effort, habitat distribution variations, and inequities in socio-economic aspects of a fishery (Meaden and Do Chi 1996). Given this concern with location-based problems, it is essential that any Information Technology (IT) based management tool is capable of considerable analytical functionality in the spatial domain (Hilborn and Walters 1992).

An IT tool which is specifically adapted to handling spatially-related problems, is the Geographical Information System (GIS). This is basically a system of integrated hardware and software components, which allow the input, storage, manipulation, analysis and output of spatially associated data (Fig. 1). GIS has been extensively adopted for management purposes across a range of terrestrial-based activities, and it has been amongst the fastest growing sectors of the IT market (Rhind 1993; Anon. 1994). Although a few attempts have been made to introduce GIS into marine fisheries situations, its implementation has been slow, for reasons which have been discussed by Meaden and Do Chi (1996).

A further technology directed at the spatial domain is the Global Positioning System (GPS). This satellite-based system allows highly accurate, geo-referenced positional data to be obtained, for any static or moving object almost anywhere on Earth. GPSs range from small hand-held receivers which give geo-referenced user locations to the nearest 100 m, to differential systems capable of accuracies within a few cm. These latter systems may continuously plot locational data at pre-set time intervals; they are fully programmable; they can be utilized in all-weather situations and can be linked with other systems to form fully integrated surveying, measuring and data collection units (Gilbert 1994). The potential for GPS use in fisheries management has already been realized, with vessels from certain countries being equipped with GPS as a means of monitoring the location of their activities. Future use of GPS will increase since the technology has a direct benefit when used as an aid to navigation, and since the monitoring of vessels in international waters will be more important in the future, i.e. following the United Nations Convention on the Law of the Sea agreement on the management of migratory and straddling stocks.

The aims of this paper are to show how the technologies of database management systems, GISs and GPSs can be brought together to create a Fisheries Computer Aided Monitoring system (FISHCAM). This system could be applied to all the main fishing methods, and it could be introduced aboard commercial fishing vessels of any size. The overall objectives of FISHCAM are to enable fishing activities to be analysed in a flexible and extensible manner. In this paper details are given on the individual components of FISHCAM, how the system would be used on board a fishing vessel, and how the geo-referenced data collected on fishing activities, when linked to other data sets through the utility of an existing GIS, can provide the basis for a range of modelling and analyses of other fisheries-related activities. Since the range of data collected through FISHCAM will be similar to, or in advance of, that which is presently collected through the manual log-book system, then a major reason for adoption of FISHCAM is that it would replace the present fisheries log-books.

THE FISHCAM SYSTEM'S DESIGN

FISHCAM's functional design must take cognizance of several factors including:

(a) The need to provide essential, comprehensible data for fisheries management;

(b) The ability to generate data which are in a format capable of being integrated with the mapping, visualization and spatial analysis capabilities normally associated with GIS;

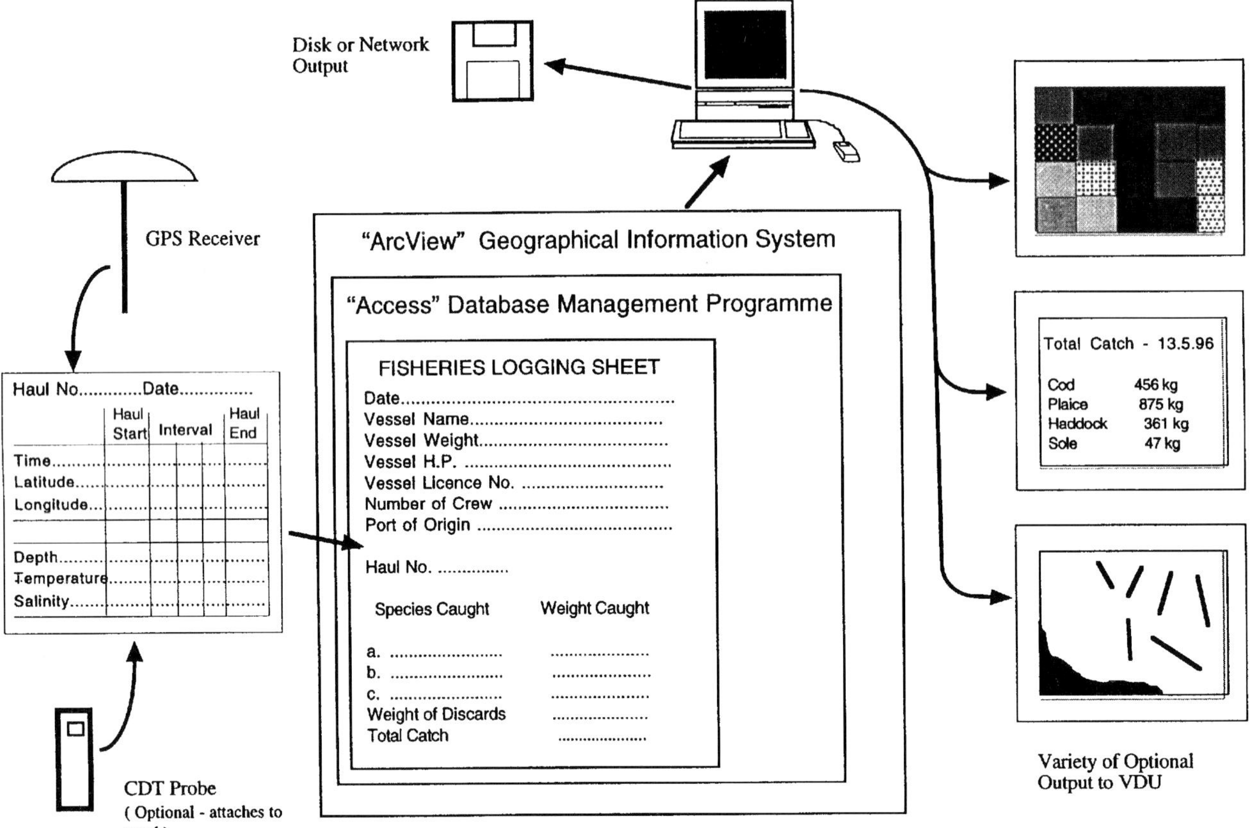

Fig. 2. Schematic illustration of the FISHCAM system.

(c) The need to be easily learned and able to minimize crewing responsibility and effort;

(d) The need to be housed so that it is effective in withstanding a range of marine conditions.

FISHCAM's basic configuration is shown in Figure 2. It consists of a number of components, each of which could take a variety of brand and functional forms, and which could be integrated so as to satisfy a range of expenditure limitations and/or functional requirements. FISHCAM's operations and its utility are initially described as three separate working components, and then in terms of its requisite database.

(i) *GPS Recording.* Because the need for very precise locational catch and effort related data is not presently justified, then a GPS operating to an accuracy of perhaps 10 to 30 m would be sufficient. This equipment could be purchased, complete with antenna, for less than $US2000. The primary requisite for the receiver would be its ability to be linked to a data storage device.

(ii) *Data Logging.* Although some GPS have integrated programmable data loggers, and there is a range of custom-built loggers into which a GPS could be integrated, it is advised that GPS input is into a lap-top or other computer processing unit, *via* an external port. The computer would need to be loaded with the FISHCAM database program.

(iii) *Other Environmental Recordings.* FISHCAM has been designed with the capability of integration with a range of electronic measuring or recording devices. Initially, conductivity, depth and temperature (CDT) equipment might be attached to the trawl tow line, but a range of other devices could be integrated and temporally synchronized to the GPS locational data. Such devices may record variables such as current speed, chlorophyll levels and plankton counts, with more sophisticated acoustic sonar devices producing data on bottom type, plus estimations of fish biomass and/or distributions.

It will be important that FISHCAM is correctly installed and the responsible management authorities would need to set out minimum material criteria for the complete system. Once FISHCAM was operational then routinely it would be necessary to set up a new log, perhaps to record each day's activities. Upon commencement of the fishing trip, static data, referring to trip details, date, port of departure, crewing, etc. would be appropriately entered. Once the fishing grounds were reached the responsible crew member would key in the time of the first trawl 'shot', and this would be automatically registered *via* the integrated GPS to a geo-referenced position. For other types of fishing, the time when the seine-or drift-nets, or longlines, first entered the water would be similarly recorded. Upon tow completion, associated time and locational data would also be recorded. Once the nets had been emptied and fish sorting was completed, the configuration of FISHCAM would allow for the weight of all species caught and recorded to be assigned to the area defined by the trajectory of the trawl sweep. In some instances it might be acceptable to record a unit number of boxes filled or a volumetric proportion of bins filled, or alternatively estimates of catch volumes from a known cod-end carrying capacity could be made. A problem might be in determining the weight of discards, though theoretically we see no reason why their weight should not be estimated, if necessary from quantities temporarily stored.

Turning to the FISHCAM database structure, it was evident during the early stages of the system's design, that a very flexible data model would be needed to underpin the system. Due to the requirements to flexibly amalgamate the spatial, aspatial and temporal data that may arise, a limited geo-spatial model such as that underlying many proprietary GISs would not do. For this reason, FISHCAM integrates a relational database with a GPS system at one end and a GIS at the other (see Fig. 2). Alternative configurations are possible, but for the initial implementation prototype, it was considered that the optimal solution would be to use the GPS to ensure accurate spatial attributes for the fishing activities, the GIS for its mapping, display and spatial analytical functions, and the database for providing the open-ended query and retrieval capabilities. It is worth emphasizing that the two fundamental themes of space and time provide a common thread that runs through the various perspectives that relate to the system. These perspectives can best be viewed as the short-term, the medium-term and the longer-term goals of the system.

From the **short-term** perspective the system provides the functionality for the capture, retrieval and analysis of all data on fishing activities as and when they occur. The data recorded are comprehensive enough to encompass a range of queries that arise with respect to marine fishing — 'What?', 'How much?', 'Where?', 'When?' and 'By whom?'. In other words, input includes details of species fished, quantities thereof, spatial location of fishing (perhaps related to specific management areas), dates or times when the activities take place, and the ownership or registration number of boats that are participating in the fishery. These core data can be used to provide monitoring information on a day-to-day basis.

From the **medium-term** perspective, FISHCAM can be used to retrieve summary information in a variety of ways, e.g. queries may pertain to individual boats, trips, dates, ports, countries of origin, etc. The design of the system also provides for the inclusion of data on fishing quotas so enabling management bodies to maintain control of this important regulatory aspect. It is envisaged that the flexible data model will provide management bodies (such as Ministry of Agriculture, Fisheries and Food in the UK) with sufficient data to support a range of policy and decision-making processes which could be related to space or time. The following give an indication of the typical querying capabilities:

(a) Total weight of species s_1, by boat owner B_1

(b) Total weight of species s_2, by all UK registered boats

(c) Summary of the weight of all species fished between time T_1 and T_2

(d) Change in the total catch weight of species s_3, between time T_1 and T_2

(e) Total catch per unit effort in user specified marine areas

(f) Total catch weight of species s_4 in year Y_1, compared with EU fishing quotas.

From the **long-term** perspective, FISHCAM can be used to integrate fishing data with environmental data. The current version of the system is being designed with the potential to include physical parameters relating to CDT measurements of the water at geo-referenced locations along trawl transects. As mentioned earlier, future versions may include the capability of integrating remotely sensed marine data with the fishing activity data, to enable long-term modelling and spatio-temporal analyses of the marine environment. Some details of possible extensions to the management capabilities of the system are presented in the next section.

The user interface with the system is illustrated in Figure 3 (a and b), which presents views of some of the data capture

a

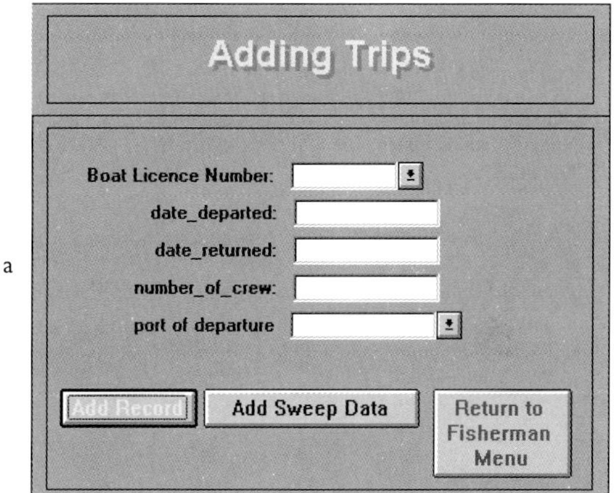

b

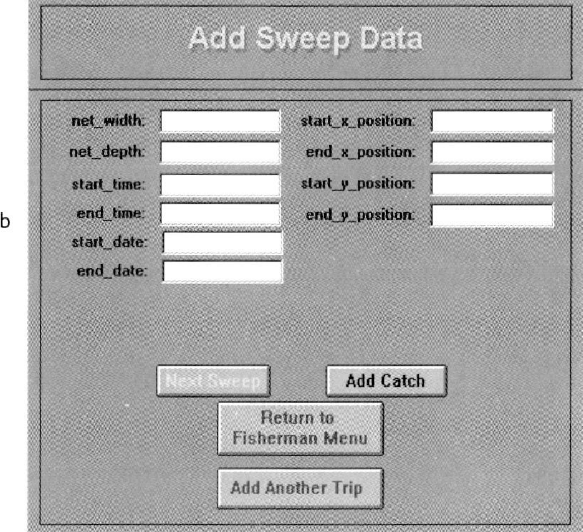

Fig. 3. The user interfaces allowing data to be entered to the FISHCAM system showing (a) Fishing vessel trip data (b) Trawl sweep data.

Fig. 4. The user interface illustrating the access to 'Help' facilities (Optionally *via* the 'Internet').

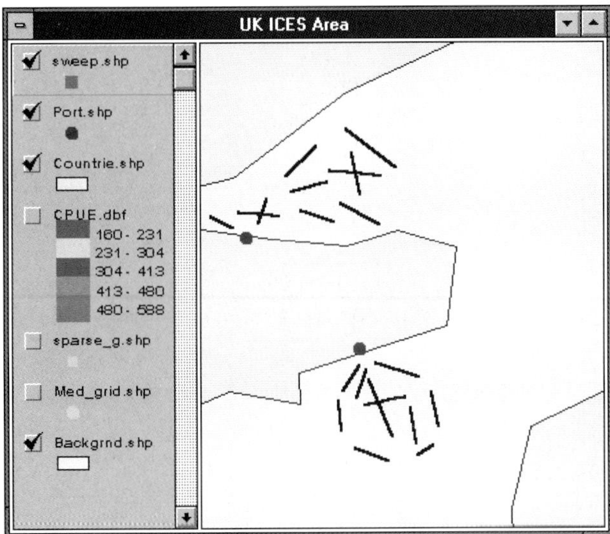

Fig. 5. Visualization of trawl sweep locations achieved through the integration of FISHCAM to a proprietary GIS.

facilities. Allied to this there is a comprehensive 'help' system (widely accessible *via* a standard Internet browser), as illustrated in Figure 4. The hierarchical design of the user interface naturally reflects the short–medium and long-term perspectives of the system. The functionality available for on-board data capture is presented to users *via* traversal of a defined and intuitive path through the interface. These users need not be concerned with the retrieval and querying capabilities, which are intended for overall monitoring and decision making by fisheries managers.

FISHCAM'S OUTPUT AND VISUALIZATION CAPABILITIES

In keeping with the philosophy of FISHCAM, the system's output capabilities are also designed to be flexible. The type of output and its style of presentation varies depending on the information retrieved and the form in which it is to be presented. Output can be in a textual form; this form applies to *ad hoc* queries to the database, generally involving aspatial attributes, as well as tables and statistical summaries of one or more variables. Graphical output can either take the form of graphs or a range of thematic maps. Graphs can be in a variety of formats which visually highlight numerical ranges or distributions of variables in space, or numerical changes over time. Mapping output is clearly referenced to the spatial attributes applicable to fishing activity, as well as any marine environmental data held in the database.

It is in the provision of this mapping capability that the decision to include a GIS in the FISHCAM configuration comes into its own. While it is possible to provide cartographic display capabilities using custom-built coding, for reasons of efficiency, and a disinclination to 're-invent the wheel', a proprietary GIS was used. This capability enables map displays of any relevant variable, e.g. catch-per-unit-effort, catch-per-species in requisite areas, the trajectory of fishing sweeps, etc. and mapping can be assigned to any physical area over which a management interest exists. The scale of mapping is a function of both the data-input resolution and the purposes for which the mapping is required.

The GIS not only enables the mapping of existing database variables, but it also provides powerful spatial analysis functions, such as overlay and buffering. This enables the visualization of variables pertinent to fishing activities at any point in time or over different times. This capability is extremely important in a system such as FISHCAM which is intended to support the decision-making process. The capacity to visualize fishing effort in the physical context in which it occurs is a very potent tool and is cognitively more effective than equivalent alphanumeric output. Figures 5 and 6 illustrate how FISHCAM's query retrieval and map display capabilities can be used to visualize the application's domain space.

At its present developmental stage, FISHCAM has yet to be linked to other marine data sets, but since this capability exists, then the possibilities offered for investigative management are almost infinite. An illustrative listing of these includes:

(a) Establishing the 'normal' range of a fish stock and the measurable fluctuations in this range;

(b) Measuring the degree to which catches are matched to a range of local physical environmental indicators;

(c) Establishing the relationship between fish concentrations and spatial and temporal variations in food supply and type;

(d) Matching indices of fish protein availability to local and/or international needs;

(e) Measurement of disturbance to the sea-bottom ecology by trawling activities;

(f) Studying the spatial variability in fish community structures;

(g) Selecting locations for fish stock enhancement programmes using the knowledge gained of optimum fish biomass environments;

(h) Monitoring of spatially related experimental fishing programmes.

a

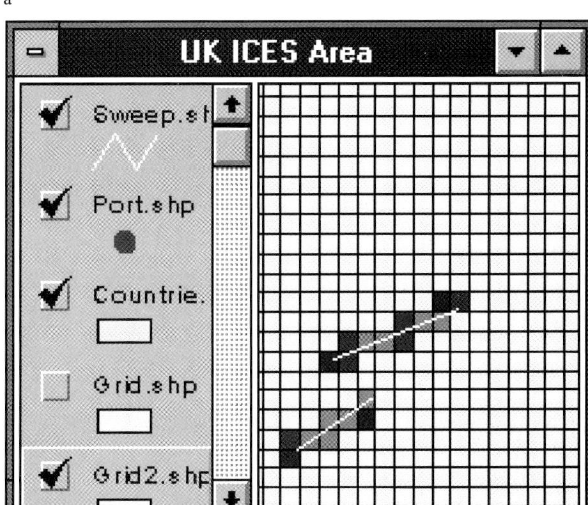

b

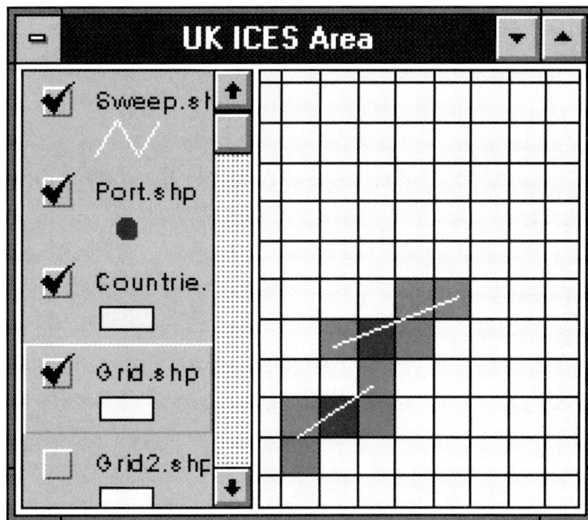

Fig. 6. Visualizations of catch-per-unit-effort as allocated to grid cells along trawl sweep lines, and achieved through the integration of FISHCAM to a proprietary GIS (a) At a small resolution (b) At a larger resolution.

It is envisaged that a subsequent version of the system will further extend these output facilities. An important pre-requisite of efficient fisheries management is adequate information about change, i.e. both in fishing activities and the environment in which they occur. By linking changes over time with the display functionality, it would be possible to display a chronologically ordered series of maps relevant to a user-specified temporal interval. This serial map display could be presented in the form of an animation or 'fly-through' in time. This means of displaying either absolute change, or the rate of change, in relevant variables will highlight important trends and could well prove to be an effective tool in the management of an activity which operates in a dynamic 3-D space.

SOME FISHCAM IMPLEMENTATION CONSIDERATIONS

Although a system such as we suggest might help in overcoming some of the shortcomings of the present fisheries monitoring and data-logging systems, and notwithstanding the fact it will prove to be a valuable spatial analytical tool, there are problems and considerations which are important to acknowledge:

(a) Given the sheer number of commercial fishing vessels within a fishery, FISHCAM might take a long while to implement. We suggest that implementation should proceed sequentially, either on the basis of all vessels exploiting particular fisheries or on the basis of geographical areas of need. For various reasons it would seem advisable to initiate the system in a deep sea pelagic fishery.

(b) Clearly any legal obstacles to implementation would need to be overcome and exact systems parameters need to be established.

(c) We suggest that the FISHCAM database software should be provided free of charge, i.e. since data gained will be advantageous to management and since it will replace the present paper-based logging system. For some vessel classes, or in other specified instances, the whole system, i.e. including hardware, might need to be freely supplied.

(d) Although there might be some resentment that the system constitutes 'extra work', we envisage that, apart from minimal system training needs, FISHCAM operation will entail little extra work or responsibility which is beyond the present logging requirements.

(e) From a management perspective there would be some concern that the system could be abused, i.e. mainly since fishermen might be loath to reveal good fishing locations. We can make several comments here. Clearly it is our aim to demonstrate the system, and we must assume that its introduction would be based on law and a democratic decision. It would also appear that, although there might be initial concern, once the system was operative it should soon become acceptable, i.e. on the basis that its purposeful long-term design is to promote better fisheries potential. Most fishermen must be well aware that catches of this open-access resource are, on balance, rapidly diminishing, and, if they are not aware, then perhaps an educative process is necessary in order to instil a conservation ethos.

(f) A final consideration relates to marine environmental data availability. Although the range of data sets to which the information collected by FISHCAM could theoretically be matched in a GIS environment, is very large, the actual availability of any sets may be restricted in terms not only of their physical existence, but also in cost terms, data format and/or structure terms, in the scale of their input parameters and in terms of their 'timeliness'. To obtain appropriate data sets, the potential user would need to have good access to data-search facilities. We do not underestimate the present difficulties which may be faced in assembling requisite data. Indeed, it was with this consideration in mind that one of the design functionalities of FISHCAM included the facility for integrating a range of data collection devices.

Acknowledgments

We would like to acknowledge the help given by Richard Fisher, Dave Hoare and Craig Mold in the preparation of the FISHCAM system.

References

Anon. (1994). 'World Geographical Information Systems Software and Services Markets'. (Frost and Sullivan: New York).

Gilbert, C (1994). Integrating other measuring devices with GPS positions. *Mapping Awareness*, **8**(5), 44–7.

Hilborn, R, and Walters, C J (1992). 'Quantitative Fisheries Stock Assessment: Choice, Dynamics and Uncertainty.' (Chapman and Hall: New York).

Holden, M (1994). 'The Common Fisheries Policy.' (Fishing News Books: Oxford).

Meaden, G J, and Do Chi, T (1996). Geographical Information Systems: Applications to Marine Fisheries. *FAO Fisheries Technical Paper* No. 356. (FAO: Rome).

Rhind, D (1993). Maps, information and geography: A new relationship. *Geography*, **78**(339), 156–9.

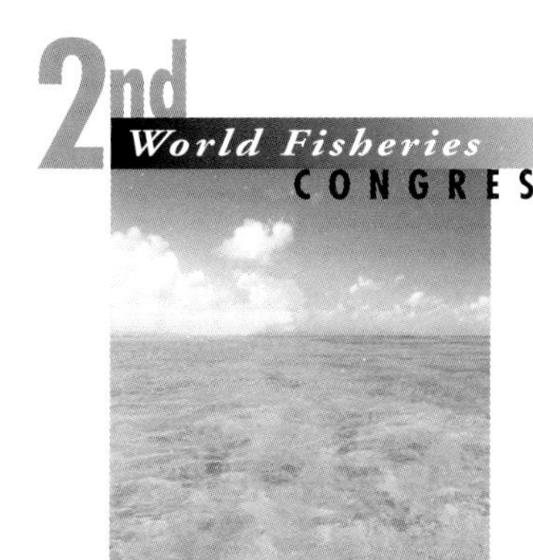

New Technology for Fisheries Management — Implementing a Satellite-Based Vessel Monitoring System

Philip Marshall

Australian Fisheries Management Authority (AFMA), PO Box 7051, Canberra Mail Centre, ACT 2610, Australia.

Summary

The requirements for successful implementation of satellite-based vessel monitoring systems (VMS) are outlined with reference to the experience of the Australian Fisheries Management Authority (AFMA) with this technology. VMS has an important part to play in fisheries management and its potential will be achieved on a global scale with further development and standardization.

Introduction

In the past 40 years the fishing industry has been very successful in applying new technology to the process of harvesting fish, effectively increasing fishing effort. Technologies developed for warfare such as radar, sonar and satellite navigation have industrialized fishing. The sophistication of current fishing techniques has further increased through the use of aircraft for fish spotting, satellite weather and sea-surface monitoring and a myriad of new communication and electronic devices. When coupled with the increases in fishing vessel size and numbers we now have the situation where it is generally accepted that the fishing industry can harvest fish faster than the fish can reproduce.

In this industrialized fishing world it is obviously critical to ensure effective management of the fish resources. The AFMA approach to fisheries management is documented and presented elsewhere (McColl and Stevens 1997). This includes a particular management philosophy with a range of legislative and consultative measures. Also playing its part is the use of technology, in many cases the same technologies used for harvesting by the fishing industry. It is important to note that technology is not a panacea and it must be part of a comprehensive management framework. Neither should the potential of technology be underestimated in fisheries management. Appropriate use of technology can enable changes in management practices and provide new opportunities just as it has in fish harvesting.

Since its inception, AFMA has adopted a strategy of maximizing its administrative efficiency through the use of information technology (IT). AFMA has applied IT to systems in the licensing, quota monitoring, catch-and-effort data, compliance, and financial areas. Perhaps the most novel use of technology has been the development and implementation of the Vessel Monitoring System (VMS). It is AFMA's aim to integrate all systems, including VMS, into a single information base.

WHAT IS A VMS?

In concept, a state of the art VMS consists of three components:

On-board equipment

VMS is currently a cooperative system where only participating vessels are monitored. Each participating vessel must carry an operating transmitter or transceiver (sometimes incorrectly referred to as a transponder) which is capable of sending position and other data to a satellite in space. A transceiver is preferred as it will be capable of receiving messages from shore or from other vessels. The transmitter or transceiver must have an integrated means of fixing a position and hence calculating speed and course. The Global Positioning System (GPS) used so successfully by the fishing industry, is the generally preferred method because of its high level of accuracy. In the South Pacific, the on-board equipment is often collectively referred to as the Automatic Location Communicator (ALC).

Transmission medium

A means of moving data between the ALCs on vessels and the monitoring agency is required. This will involve the use of at least one satellite and an earth station which is capable of handling the communications traffic of all ALCs and forwarding that traffic to the monitoring agency *via* a secure public data network such as X25.

Monitoring station

Within a monitoring agency such as AFMA, there must be a means of collecting data from the earth station, storing those data for subsequent review, analysing the data to detect exceptional conditions of interest to monitoring officers, and displaying those data in a meaningful way, typically against a background map. A specialized Geographical Information System (GIS) is also a highly desirable element of the monitoring station particularly for historical and statistical analysis of both position and catch data.

It should be noted that a VMS need not necessarily be satellite-based and could use other transmission media such as High Frequency radio. However, when all of the requirements of a VMS are considered, particularly coverage of the range of modern fishing vessels, satellite is clearly the most effective.

WHY IMPLEMENT A VMS?

There are two primary business areas for which a VMS is relevant. These are compliance with management rules and data collection.

Compliance with fisheries management rules

Ideally, fisheries management rules are designed to achieve sustainable, equitable, and profitable fishing. Fisheries management controls typically include a mix of limiting vessel numbers through licences, limiting vessel access to particular areas, restrictions on gear types, and quotas on the amounts of particular species which may be taken. VMS assists in maintaining the effectiveness of these management rules through providing reports to monitoring agencies of the positions of vessels. Reports are sent from equipment on-board licenced vessels automatically at relatively frequent time intervals so that conclusions may be made about the activities of those vessels.

While this requires some cooperation from the vessel operator in terms of maintaining the equipment on board, there are a number of technical requirements for the VMS so that the integrity of the data is assured to a high level. It must be understood that as a consequence of reported data, legal investigations and prosecutions may result if the activities of vessel operators contravene the management rules.

Data collection

Catch and effort data are often a primary source of information relating to the status of fisheries. There are considerable benefits in collecting catch-and-effort data *via* VMS. Benefits are derived from faster delivery of data to the monitoring agency, reductions in cost of data entry and increases in accuracy through both minimizing data handling and the direct interaction between the vessel operator and the data entry/editing program.

Catch data and other fishing activity data such as reports about a vessel's intentions, may also have a compliance-related function. For example, catch reports may be used to monitor quota in an output-controlled fishery.

AFMA's VMS

Domestic trawl

On 1 January 1994, AFMA officially implemented its first VMS in a deep sea trawl fishery with approximately 30 vessels. The fishery concerned is managed by an Individual Transferable Quota (ITQ) system operating across several geographical zones. Vessels fishing for the species *Hoplostethus atlanticus*, known as orange roughy in Australia and New Zealand, are required by permit condition to carry an operational ALC which has been approved by AFMA.

The satellite services most commonly used for fisheries monitoring are Inmarsat C or the Argos system. Within the Pacific region, New Zealand, the USA, French Polynesia and New Caledonia have active VMS which use either Inmarsat C or Argos. AFMA's system is compatible with that of Australia's near neighbour, New Zealand, and will be compatible with the regional system currently being planned by the South Pacific

Forum Fisheries Agency (FFA). This facilitates the movement of vessels between the Exclusive Economic Zones (EEZs) of each country. It also means reduced cost and increased ease of movement for vessel operators through use of commonly accepted equipment.

Vessel operators must purchase and install an ALC. Choice of ALC type is limited to a list of equipment which has been approved by AFMA as meeting the AFMA specification. AFMA also approves commercial installers and maintainers of the equipment against another set of requirements.

At this time only Inmarsat C ALCs are approved. Inmarsat C is used because its design specification includes a monitoring capability. This capability takes the form of software to support automated periodic reporting of a variety of position and status information as well as allowing a degree of customization for particular user requirements. This means that transceivers built to the Inmarsat C standard can be used off-the-shelf to meet AFMA's functional requirements with a minimum of customization. It is therefore effective as a compliance tool and has a relatively low cost. Importantly, Inmarsat C is acceptable to the fishing industry for its ancillary features which provide the vessel operator with communications and sea safety capabilities.

Within AFMA there is a communications link to the Inmarsat Land Earth Station (LES) operated by Telstra in Perth, Western Australia. Position reports are sent from the vessel at programmed time intervals using the Inmarsat data-reporting mode. These reports are retrieved from the AFMA 'mailbox' at the LES and are stored in the monitoring base station in Canberra. A Windows PC-based package, the Terravision Fleet Tracking System, supplied by Terravision Ltd of Perth, Australia, has been installed and further refined to meet AFMA's monitoring requirements.

Basically, Terravision stores the position reports in a database and allows the display of vessel positions and tracks on a map background. Most importantly Terravision performs some analysis of the reports with the aim of identifying events which may be of interest to the monitoring officers.

Within AFMA we believe that the ability of the base station to perform monitoring tasks in a quasi expert system manner will be the main determinant of VMS cost-effectiveness. VMS can certainly be effective but it does generate a large amount of data and the personnel resources required to maintain and monitor it could be excessive if the computer system cannot identify events of interest, and in many cases respond to those events automatically. Further, aside from real time analysis, there must be an historical and statistical analysis capability to identify trends and to cross check VMS data against data from other sources such as log-books, sightings information, and observer reports. It is in this area where AFMA believes that a GIS will play a major role as part of an integrated information systems strategy.

WHY DID AFMA IMPLEMENT VMS?

In the domestic trawl fishery the use of VMS was intended to increase the effectiveness of the ITQ system and hence address overfishing of the orange roughy resource. It was suspected that there was a high degree of non-compliance with the zone-based quotas, primarily through misreporting the location of the catch. In one particular zone the cost of catching is relatively low and the cost of quota is relatively high because of large spawning aggregations in this zone. As a result, there was a temptation to report catch from this zone against another quota zone or outside of the EEZ altogether.

It is interesting to note that many operators in this fishery placed considerable pressure on AFMA to raise the level of compliance as they wanted their investment in the fishery to be secured. This has become a continuing trend with many operators in other fisheries seeing similar benefits and also proposing innovative ways of applying VMS to achieve improvements in both compliance and management arrangements.

That the Australian fishing industry is to a large extent cooperating and even proposing use of VMS is probably testimony to the consultative approach to fisheries management. This experience is not common to VMS implementations in all other countries. However, both the New Zealand and Hawaiian experiences are the same as that of Australia, and consultation is a significant factor in the success of all three VMS implementations. The consultative approach does have one drawback. It can mean lengthy delays before consensus is reached on implementing VMS and this can have consequences for the fish stocks and achieving effective management.

AFMA'S RESULTS

The introduction of VMS has been very successful. The realization of its positive effect has grown within AFMA, other fisheries management organizations and the fishing industry. Interest in use of the system in other fisheries is widespread and is growing. In the orange roughy fishery a distinct change has occurred in the catch profile by zone such that catch reports are now considered to be more realistic and now conform to the quota system intentions. In the more than two years of operating its VMS, AFMA has not brought a prosecution based on information from the VMS. The main effect of the system has been as a deterrent and this has been confirmed through discussions with various vessel operators and crew. This should be viewed as a positive result for all concerned, as all parties are 'winners'. Also on a positive note, verbal reports from some operators have indicated that the aggregation of fish in the primary, limited access, spawning ground has increased compared to the two previous years, possibly because of reduced fishing pressure.

Australia has a large fishing zone but AFMA has only a limited budget for compliance activities such as surveillance. One of the main benefits of the VMS is its use as a tool in identifying suspect behaviour and just as importantly, in confirming legitimate behaviour. With the assistance of VMS, AFMA now has a high level of confidence that 90% of the vessels are behaving legitimately allowing it to target its limited resources on the 10% of vessels which have a high threat potential. VMS can therefore achieve more efficient compliance activity and, importantly, it can assist in identifying the actual level of compliance.

Monitoring and compliance officers have also used the system to good effect in a number of targeted surveillance operations. The system has allowed AFMA to detect potential breaches of fishing conditions and to coordinate boardings and inspections. Further, the communications capability provided by Inmarsat C has allowed AFMA to quickly and reliably contact the vessel operator at sea and either question or provide direction about activity which has come to attention *via* VMS monitoring. In this context VMS is used in a preventative rather than a curative manner and adds to the deterrent effect of VMS.

One of the most positive aspects is the very favourable view of the VMS from the fishing industry. Operators have found the communications capabilities of the Inmarsat transceivers to be of great value in addition to providing more effective management of the fishery.

There have been a number of negatives in the first two years of VMS usage. These were generally in the category of technical problems which are to be expected with the introduction of relatively new technology. Most of the problems AFMA experienced are now unlikely to occur provided monitoring agencies are aware of them and the latest software is used. A high degree of reliability can now be expected with Inmarsat C based on more recent experience.

Some limitations of the system have also been realized. Position data provide good information on where a vessel may be fishing but they do not provide sufficient evidence to satisfy a court that the vessel was actually fishing. VMS data may be used as supporting evidence but further conventional compliance measures must be used in conjunction to enable successful prosecutions.

Many of the problems experienced were operational and demonstrated the need for appropriate installation, maintenance and diagnostic procedures. These must also be backed up with adequate staffing resources who can attend to the operational aspects of the system as well as the monitoring function itself.

Japanese tuna vessel reporting

AFMA manages a second fishery where it has developed a VMS. Through a bilateral agreement with Japan, up to 100 tuna long-line vessels operate within the Australian EEZ annually. At the time of writing this paper, there has been no agreement on bilateral access for 1996. However, should such an agreement be reached, it would be a requirement that these vessels report both position and catch data to AFMA *via* VMS. This requirement applies to any licenced foreign vessel operating within the Australian EEZ.

The reporting system for the Japanese longline vessels was operational on five vessels during 1995. The development of this system was a collaborative effort between AFMA and the Japanese Fisheries Agency. The system differs from that used by AFMA in the domestic trawl fishery in that the use of an additional technology, Inmarsat A, is permitted and it includes, as its major objective, catch-and-effort data reporting.

The scale of the use of VMS with the Japanese vessels has been too small to draw many definite conclusions. The development of the system was time-consuming and highlighted the problems involved in developing customized processes. The Inmarsat A system in particular required a purpose-built communications protocol. The system development issues and operational difficulties associated with use of the on-board data entry and transmission equipment, highlight the need for a standard approach to the design of such systems so that development costs are minimized and life is made simpler for the vessel operator regardless of the EEZ in which the vessel operates.

Despite the problems caused by lack of standards, the catch-reporting function shows considerable promise. Data have been successfully received and stored and indications are that, with further refinement, the data collection process will be simpler, more accurate, less expensive and more timely than the current system of log-books and radio reports. The refinements required include improvements to documentation and procedures, and software changes in the communications protocol to ensure positive end-to-end acknowledgment of report transmission.

Aside from the conclusions relating to catch reporting, evaluation of the Japanese reporting system has led AFMA to conclude that the Inmarsat A variant is unsuitable for a compliance role as it does not adequately satisfy all of the technical requirements.

Requirements for an effective VMS

System requirements

To meet the typical business needs of a VMS, AFMA considers that a VMS should satisfy the following nine generic criteria:

1. *Tamper proof operation.* Neither the vessel operator nor any other person should be able to interfere with, or eavesdrop on, the transmissions between the on-board equipment and the monitoring agency. The position fix must be authentic and it must be authenticated as originating from the vessel or at least the VMS equipment on the vessel.
2. *Transmission of appropriate vessel activity (position) data.* The content of the position reports must be sufficient to adequately interpret the operations of the vessel. As a minimum, AFMA considers that latitude, longitude, speed, course, date and time are required. Status information relating to the operation of the VMS equipment is highly desirable.
3. *Transmission of appropriate catch and effort data.* Catch and effort data reported must be sufficient for the required scientific and management functions to be carried out. The timing of transmission of catch-and-effort reports must relate to an appropriate, management-defined, reporting period. The transmission must be auditable and not allow the sender or receiver to fraudulently claim or deny transmission of any catch report message.
4. *Reliable continuous operation, 24 hours/365 days per year.* The whole system must be operational at all times and in all environmental conditions where fishing activities may potentially be carried out.
5. *Position information independent of vessel operator control.* The system must be fully-automatic and self-initializing.

The vessel operator must not be able to affect the intended monitoring function or avoid detection of improper activities through control of the timing of position reports, generating false positions or timing of fishing activities.

6. *Positional accuracy to an appropriate level for the monitoring requirement.* The accuracy and reliability of the position reports should be such that required borders can be enforced and that the credibility of positions is accepted by courts of law.
7. *Capable of tracking subject vessels throughout their range.* The system must be capable of tracking vessels throughout the area being monitored and preferably throughout the range of the vessels being monitored so that vessels cannot avoid continuous monitoring and potentially avoid detection of improper activities.
8. *Timeliness of transmission appropriate to monitoring requirement.* Reports must reach the monitoring agency within a time sufficient to enable an appropriate response to be undertaken.
9. *Messaging capability to and from the vessel.* A messaging capability **from** the vessel is mandatory for catch reporting. A messaging capability **to** the vessel is highly desirable as it permits actions to be queried and 'reminds' vessel operators of the monitoring function (preventative monitoring).

For the catch-reporting function, AFMA's experience suggests the need to make two-way communication mandatory so that the vessel operator can receive a positive 'end-to-end' confirmation that the exact contents of what was sent were received. This prevents the operator from falsely claiming transmission of a message and provides the operator with proof of all legitimate transmissions. An electronic signature as used in electronic commerce may be a useful addition or alternative.

Legal framework

Not all vessel operators will view VMS as desirable. There must be a sound legal framework which requires all operators to meet a uniform standard at the same time. Such a framework must include penalties which are of sufficient size that non-participation, misuse or tampering are effectively discouraged. AFMA has a two-level legal requirement to ensure continuous VMS usage. To allow for genuine equipment failures a vessel may report *via* radio as an interim measure but ultimately AFMA can require the vessel to return to port.

While every attempt should be made in the system's technical infrastructure to maintain its integrity, no technical system is infallible. Much of the delay in implementing VMS in the past was due to costly attempts to develop a system which was completely secure. The general approach in many countries including Australia, is to accept a degree of risk in the system integrity and make sure that any breaches are detected and the consequences do not make the breach worthwhile.

Integrated management arrangements

VMS is not by itself a solution for all of the industrialized fishing problems of the world. VMS is no substitute for a well thought out and pragmatically-implemented management plan. However, VMS can be an effective part of such a plan. For VMS to be effective it must be applied to a set of compliance management scenarios where it can truly add value. VMS must also be supported by an operational framework including personnel resources for monitoring and maintaining the system, and conventional compliance measures such as aerial surveillance, at-sea boarding, catch documentation, landing inspections etc.

Fishing industry cooperation

Current VMS are participative systems by nature of the need for an ALC to be carried on board the fishing vessel. Mandating use through legislative instruments is required to ensure conformity. However, there is a higher chance of successfully achieving system aims if the operators are not hostile to its introduction. Hostility can be avoided if the majority of operators can see clear benefits in the system for themselves and there are no significant costs or disadvantages. It is difficult to avoid some costs. In Australia and New Zealand the user-pays principle has been applied and the fishing industry has been required to fund the purchase of the ALC directly and pay AFMA's costs of running the system through annual levies.

There are two issues where negative impact on the fishing industry must be minimized in the implementation of the VMS. Firstly, the security of both position and catch information must be maintained. This information can be highly commercially sensitive and VMS provides it more accurately and speedily than through previous reporting methods. Secondly, the system must not be too onerous for the use of the vessel operators. It must not unduly interfere with their normal activities and must be relatively easy to use.

Benefits for the fishing industry

The benefits are often portrayed as only the ancillary capabilities of the on-board equipment for sea safety and private communications between the vessel and shore, and other vessels. These benefits should not be understated and AFMA's experience is that they are highly valued by operators. Indeed as more innovative applications are developed to make use of the satellite communications systems, the value to operators may significantly increase.

The primary benefit of VMS for operators is sometimes lost in consideration of the ancillary benefits of various systems. VMS can improve the management of fisheries. It can make management arrangements more effective. If a higher level of compliance with management arrangements is achieved then those arrangements should lead to a sustainable fishery which profits all stakeholders. This has been well demonstrated in the orange roughy fishery.

A further consequence of more effective compliance is cost containment. AFMA's experience with VMS to date does not indicate a reduction in compliance costs. In fact, in the early stages of implementation there are additional costs to industry for the purchase of ALCs. However, there is clear evidence that more effective compliance can be achieved while containing costs to pre-VMS levels. There is potential for a reduction in compliance costs in the longer-term as experience and confidence builds in the system itself and its ability to monitor the level of compliance.

VMS, like all good technologies, should also enable changes to management arrangements which may produce benefits in productivity and flexibility and which may not have been practical previously. This has been demonstrated in the Southern Ocean where AFMA has been able to permit a vessel to trawl for *Dissostichus eleginoides* (Patagonian toothfish) near a sensitive environmental area, due in part to AFMA's ability, to monitor its activities *via* the VMS. Considerable potential exists in other fisheries where VMS can achieve advantages in various ways. For example, near real time management of a fishery may be possible based on the ready availability of catch information.

Commercial fishing is coming under greater scrutiny from both the environmental and recreational fishing sectors. VMS has a significant role to play in satisfying these sectors that commercial fishing is not performed in sensitive areas. In Hawaii, the US National Marine and Fisheries Service implemented a VMS for these reasons. There are a number of fisheries in Australia where the commercial sector could benefit from this type of VMS application.

Future developments

VMS is still in its infancy. It is being developed and implemented in a variety of fisheries throughout the world. In Australia, AFMA has a number of fisheries where VMS is considered viable and valuable and it is expected that implementation will proceed over the next few years. Within AFMA new uses are still being found for the system in the compliance role and further benefits are expected to accrue from the system in this role. The major enhancement to the system's use will be in extending its function to the data collection role. This will primarily cover standard catch-and-effort data but there are a variety of other data which can be captured by the system automatically through the attachment of appropriate sensors. Sea surface temperature and other environmental measurements are examples of this type of data.

Outside of Australia a similar pattern is being followed. New Zealand and other Forum Fisheries Agency (FFA) countries, Japan, the USA and the European Union countries are very active in developing and implementing VMS. While there are some exceptions, most of the applications have concentrated on compliance with management arrangements within EEZs. Use of these systems for data collection will grow substantially in the short- to medium-term in the same way as with AFMA's plans in Australia.

The use of VMS for regional monitoring of fishing vessels is now happening in the European Union and the FFA. The use of VMS has been foreshadowed in the United Nations agreement relating to the conservation and management of straddling and highly migratory fish stocks. These developments will ultimately see VMS monitoring extended to the high seas. The nature of such monitoring will evolve through the political process and may simply be monitoring of vessels by their flag state. It is conceivable that United Nations agreements will ultimately require all countries to monitor all of their own flagged fishing vessels operating on the high seas. Cooperative arrangements between countries for the exchange of data will also be required.

The vessel operators will also reap further benefits through the use of VMS. National VMS results have already identified the potential of VMS as a significant aid in successful fisheries management. Hopefully, better management arrangements in which VMS plays a part, will produce more stable, sustainable and profitable fisheries on a global scale. Vessel operators will also find new ways of using the VMS equipment for their own purposes. The VMS equipment is already being used very successfully for fleet management purposes by both small and large fishing companies. In other industries, technology is playing a major role in terms of electronic commerce and it is here that significant potential exists for the fishing industry. For example, the VMS equipment will enable more reliable and direct communications between the fishing vessel and fish processing and marketing agencies which could lead to positive changes in fish marketing practices.

Success Factors

Many of the factors relevant to the success of any given VMS have been covered above. Solutions for most of these issues have been identified through the experiences of those agencies who have implemented systems. There are some issues which require further effort to enable VMS to achieve its full potential on a global scale.

There is a need for off-the-shelf, easy to implement, low cost, commonly-accepted standard systems. Where VMS has been implemented to date, fisheries agencies have played a pioneering role, by necessity involving themselves deeply in many of the technical issues associated with the development of appropriate system software, the use of the satellite services and particularly in the selection of the on-board equipment. Many other fisheries agencies do not currently have the required expertise or funds to successfully undertake these technical tasks.

Software suppliers and satellite service providers have made good progress in supplying more usable, off-the-shelf solutions at a lower cost and a process of refinement will continue as in most technological fields. Further efforts are required particularly in the area of on-board equipment and data reporting. Fisheries administrations must play a more active role in establishing standard requirements so that the technology providers can develop appropriate solutions and a degree of uniformity and simplicity can be introduced globally. Such standards should address both the equipment and the content and format of data to be reported. An internationally recognized agency, such as the Food and Agriculture Organisation (FAO), could play an invaluable role in establishing standards and gaining their acceptance by all countries.

Fishing is now a global and industrialized business which must be managed on a more global basis and this has been foreshadowed in the United Nations. It must be possible for all countries to implement technological solutions such as VMS and those solutions should allow the fishing industry to move between jurisdictions in a controlled manner but with relative ease and minimal cost.

The political will to implement effective fisheries management is the major factor in the future use of VMS. Technology issues have been addressed and improvements will continue. Bringing together industry, scientific, management and political interests in a common purpose is the only way to resolve the fishing problems created by technology and allow technology to be used as an effective part of the solution.

Reference

McColl, J, and Stevens, R (1997). Australian Fisheries Management Authority Organizational Structures and Management Philosophy. In 'Developing and Sustaining World Fisheries Resources: The State of Science and Management. Second World Fisheries Congress, Brisbane 1996'. (Eds D A Hancock, D C Smith, A Grant and J P Beumer) pp. 655–60 (CSIRO Publishing: Melbourne).

Industry Must be Part of the Fisheries Management System

M. R. France

14 Neil Street, Osborne Park, W A 6016, Australia.

Summary

With the support and cooperation of the fishing industry any fisheries management regime will be more efficient and effective. Biologists will often find that the missing piece to the puzzle of this imprecise science can be provided by fishermen. Economists can prove many of their assumptions, theories and models by reviewing financial and practical information provided by the fishing industry. Environmentalists will learn that fishermen are the original conservationists and are concerned about ecologically sustainable development. Fishermen are out there every day working with the environment. The fishing industry often creates essential infrastructure in remote and developed areas and becomes an integral part of the social fabric of rural and urban communities. When the fishing industry unites it can and does apply considerable pressure on the political system. It is important to involve the fishing industry in the decision-making processes affecting fisheries management. They have much to contribute.

Introduction

What does the statement *industry must be part of the fisheries management system* have to do with this theme, you may ask? The answer is — *everything.* If you are not convinced, ask yourself the following questions:

- Where would fisheries science be if fishermen didn't discover the various fish species and develop the different resources, assist in the research projects and provide a substantial amount of research dollars?
- How would the economists come up with their Gross Values of Production (GVPs) and test their economic theories for fisheries management regimes without the benefit of the financial and practical information provided by fishermen?
- What would replace the social and economic infrastructure which has been developed by the fishing industry leading to fish becoming the fourth largest export primary industry in Australia?
- Who would keep fisheries managers honest if the political machine known as the fishing industry did not exist?

Radical Views? — Some may think so, but having spent my life in the fishing industry and a good part of it being involved in the management process, I believe I know. The role of industry is an integral part of the science, economics, sociology and politics of fisheries management.

Unfortunately, many of the traditional bureaucrats who manage fisheries, as well as a large number of scientists, economists and politicians, would regard the suggestion that industry should play a vital role in fisheries management as being akin to the well-worn statement of allowing the inmates to run the asylum. Those of us in the industry asylum are not too sure at times just who the inmates are.

Fortunately times are changing and, in this country at least, many enlightened thinkers now realize it is absurd to try to successfully manage fisheries without the active participation of the fishing industry in the process. This has led to the establishment of Management Authorities consulting with Industry Advisory Committees to manage some of our fisheries. However old habits die hard and industry is still faced with the tendency for some government bureaucrats to revert to form from time to time. Governments too often consult in a token patronizing manner with little intention of taking serious note of the industry input. This usually brings about a poor result, culminating in the implementation of ineffective management regimes, dissatisfied fishermen and costly, avoidable court action. In such cases, the cynics among us watch the changes occurring in fisheries management philosophies around the world, and wonder whether the promises by fisheries managers of partnerships, consultation and quality management are nothing more than rhetoric.

Notwithstanding, the evolution of this industry has seen many changes in fisheries management. Managers these days are no longer mere licensing clerks. The responsibility of a fisherman goes much further than just being a catcher of fish. The experience of the past where fisheries were overexploited as a result of governments issuing too many licences when there was too little knowledge of fish stocks, and the subsequent need for industry support and funding to bring about rationalization in fisheries, has resulted in the need to involve fishermen in the management process and for fishermen to address the broader issues. Today, individual fishermen, industry groups and management advisory committees are continually faced with making decisions on conservation, resource sharing, research priorities, management costs and native title. A far cry from the good old days when all the average fisherman did was go fishing.

Today's concerns about the environment and sustaining resources assume a considerably higher priority than in past years. For the traditional fisherman, it has been a huge adjustment to come to terms with such issues as biodiversity, ecologically sustainable development and biological reference points. Without doubt we have a large number of confused fishermen out there.

Add in the 'Precautionary Principle', which seems so widely embraced by so many as justification in some instances for doing nothing, and you have an industry that becomes more frustrated and unsettled. We all recognise that buzz words and catch phrases come and go. The value they have when current, why they come and why they go, are matters about which linguists and sociologists (and fishermen) can speculate, but in technical fields such terms come as bright ideas to be discarded when the errors they represent and the damage they can do are recognized. This, in my view, is especially the case, and a matter of some concern, when the term 'precautionary approach' is used.

Much time and a great deal of writing has been devoted to this term by the United Nations, the FAO and various national fishery bodies. Accounts of what it is thought to mean are given at length in two papers from FAO, one of which was perhaps prepared for, while the other followed from, a meeting in Lysekil, Sweden, in June 1995, of which FAO published a report.

We are told in these documents that the precautionary approach *involves the application of prudent foresight; — it involves developing, within management strategies and plans, explicit consideration of precautionary actions that will be taken to avoid specific outcomes; — it involves explicit consideration of undesirable and potentially unacceptable outcomes and provides contingency and other plans to avoid or mitigate such outcome.* In effect the precautionary approach is to carry on with what we are doing and to do so, as most people do, with caution.

It is probably fair to assume that the word 'precautionary' was chosen for its supposed incantatory magical power: words from whose utterances follow miraculous changes: *abracadabra* and the stocks are saved! By the words precautionary approach we are to understand that the problems of fisheries will be resolved by something more and other than the established practices. I don't believe there are new methods of identifying risks or of measuring probabilities: there are no new rules to be observed in coming to decisions except that things are to be done under an exhortation to be cautious.

In the *Australian Intergovernmental Agreement on the Environment* (May 1992) section 3.5.1 states words to the effect that ... *In the application of the Precautionary Principle, public and private opinions should be guided by:*

1. Careful evaluation to avoid where possible serious irreversible damage to the environment, and
2. Assessment of the risk weighted consequences of the various options ...

This all should mean the consequences any environmental or precautionary approach decisions have on the industry must be measured and the risk weighed against the benefits. I wonder how many of the champions of the environmental issues or those who tout the precautionary approach have even considered the industry position. It is likely they never will unless the industry is involved in the decision-making process!

After the UN Conference in August 1995 in New York relating to the Conservation and Management of Straddling Fish Stocks and Highly Migratory Fish Stocks, the Fisheries Minister of Canada, Mr Brian Tobin, was quoted in the New York times as saying ... *We have made it clear that the desire to harvest fish must take a back seat to the need to sustain the fisheries ...* It is doubtful if the fishing industry totally accepts this sentiment. Many of us believe harvesting and sustaining resources should rank equally, rather than one being more important than the other.

Because of their heavy financial investment in the fishing industry, fishermen are the original conservationists. They are generally not the short-sighted rapists and pillagers they are often accused of being. Fishermen are interested in tomorrow — their futures are entirely dependent on the long-term biological sustainability of the resource. But as with most things the media prefer to sensationalize and talk about rape and slaughter of fish stocks and ecosystems, instead of the sound and successful management which is evident in many of our fisheries. Tragically the industry is generally too busy going about the business of fishing to respond to the sensationalist newspaper articles which typify journalism in this country. All too often the image and reputation of the fishing industry portrayed by the media is completely wrong.

Good fishermen are by nature fiercely competitive and independent and therefore do not easily respect excessive constraint. They are surprisingly innovative, and can and often will circumvent surveillance and enforcement regulations that are dreamt up by theorists if the incentive to do so exists. It is therefore ridiculous to try to implement regulations that have not first been thoroughly discussed with the industry. It is also important to recognize that much of the success of surveillance and enforcement regulations is dependent on the effectiveness of the management system in place. We all know the problems of enforcement in a fishery managed by quotas. For instance, if you try to manage and police quotas in a remote high-value fishery such as the Northern Prawn Fishery in Australia (as proposed by one misguided economist in the early 1990s) you are clearly facing an almost impossible mission. Clearly the incentive for fishermen to circumvent the rules can be minimized if good fisheries management regimes are implemented in consultation with industry, and strong property rights are in place.

The often-quoted and seriously misguided philosophy that fisheries are a common property resource is, in my view, the root cause of many of the problems. Under such circumstances who is accountable? Governments continue to recover costs of managing fisheries from industry. Research costs are often levied back to the industry. Recreational fishermen believe they have a rightful share of fishing resources, yet the professional section of the industry carries the brunt of the cost and criticism if things go wrong. I believe Government should have a strong influence in the administration of the fisheries resource on behalf of the owners, but the industry should have a heavy hand in the management as they are the major investors and have the greatest incentive to protect their investment, i.e. the biological and economic well-being of the fisheries resource. Whilst fisheries managers continue to insist on preserving what they deem as their divine right to suspend or cancel a fishing boat licence if a fisheries regulation is breached, the fish resource will be in jeopardy. Cancelling a fishing licence is akin to cancelling the registration of a vehicle if the driver is convicted of a traffic offence. The difference is that the fishing boat licence can in many cases be worth millions of dollars. It is regrettably a sign of the old tired mentality which does not recognize the knowledge, expertise and financial commitment on which this industry is based.

There is little doubt that fishermen often know as much or more about the resource than many scientists, and the industry definitely benefits when fishermen and biologists work together. Regrettably this is not a regular occurrence due to myopic attitudes on the part of some fishermen and scientists. Many fishermen are not good communicators and are often intimidated and do not speak up at organized workshops and conferences. When the bold soul attempts to speak, he is often ridiculed by some scientist who is more articulate and uses scientific jargon the fishermen often do not understand. The tragedy is that such scientists deny themselves the opportunity of hearing the views from the people in the field with a wealth of practical experience because the fishermen usually clam up if they feel embarrassed. Unfortunately the most knowledgeable fishermen are often the least communicative. They have a tendency to guard their knowledge jealously. So when scientists and industry get together they should listen to each other's point of view. Beneath the ramblings of many fishermen there is often that valuable bit of information which can make this business of fisheries science a little more precise.

Fishermen must also acknowledge that the scientists make a valuable contribution. They must realise that the scientists rarely can prove their conclusions before the damage has been done and therefore they are usually advancing opinions. Of course they can be wrong and scientists must not believe they are infallible. But scientific opinion joined with industry knowledge and intuition can be a powerful and successful combination. The compulsory financial contribution by the fishing industry to the Fisheries Research and Development Corporation, and the inclusion of eminent scientists on management advisory committees has forced closer cooperation between fishermen and scientists, but there is still need for improvement in many fisheries. Mutual respect and trust must continue to be established between the fishing industry and scientists. Both have much to contribute and both have a considerable investment in this industry.

The Australian Northern Prawn Fishery (value about $A150 million/year) is one interesting case study of all the things we should and should not do in managing a fishery. During the early 1960s the Government virtually ignored the industry and established management regimes that resulted in the usual over capitalization and effort blow out. The fishery was spread over thousands of miles in very remote areas with no real infrastructure. The Government used the perceived opportunity to force fishermen, as conditions of licences, to develop communities and establish shore-based facilities. The combination of too many boats, too little scientific knowledge and an arrogant bureaucracy resulted in a severely depleted resource, a fragmented industry and huge financial losses.

The fishing industry was not entirely blameless. As with every community there can be a few rotten apples in every barrel. This was the case during the early development years in the Northern Prawn Fishery. There were some instances of fishermen and companies not playing the game when catches exceeded the capacity of shore facilities to cope with the volume. Others blatantly defied management rules by fishing during closed seasons and in permanently closed areas.

Thankfully this situation does not apply today. The vast majority of the owners and skippers of the 128 trawlers operating in the fishery acknowledge the need for management to preserve the prawn resources and scrupulously observe the rules. The number of breaches detected these days during a season are minimal and generally are of a minor nature and do not warrant prosecution.

Desperate times often bring out the best in people. When the fishing industry unites it can apply enormous pressure on the management and political system. To overcome the problems in the Northern Prawn Fishery, industry got together and, working with Government and scientists, embarked on a massive re-structuring programme which turned the Northern Prawn Fishery into one of the pacesetters in fisheries management in Australia and around the world today. The industry now has a formal place in the management process.

The Federal Government has provided a difficult task by setting the combined objectives of biological sustainability and economic efficiency as a basis for managing Commonwealth fisheries. Examination of the initiatives taken to provide for a sustainable Northern Prawn Fishery (and a glimpse into potential problems in the future) demonstrates that industry must be an essential element in the management decision-making process if the delicate balance of achieving these objectives is going to be met. We should learn from past experiences when planning for the future, but I often wonder if we do.

The Australian Orange Roughy Fishery: A Process for Better Resource Management

Darby M. Ross[A] *and David C. Smith*[B]

[A] 8/32 Myers St, Geelong, Victoria 3220, Australia.
[B] Marine and Freshwater Resources Institute, PO Box 114, Queenscliff, Victoria 3225 Australia.

Summary

The Australian orange roughy (*Hoplostethus atlanticus*) fishery is an important case study for new or developing fisheries. The fishery developed rapidly in the late 1980s, from less than 200 t in 1985 to about 28000 t in 1989. Industry's expectations were high: catch rates of 100 t per two-minute bottom time were recorded. Parallels were drawn with a larger orange roughy fishery in New Zealand waters. Total allowable catches (TACs) were imposed in the fishery in 1989/90. There were two significant contributions to the assessment and management of this fishery. First, in 1987, the research effort in southern Australia was apportioned amongst research institutions rather than the previous open competition for funds. Secondly, in 1989 at industry's instigation, an industry–technical liaison committee was established. This committee developed a stock protection strategy, determined research needs, and evaluated and funded research, through a voluntary industry levy. As a result of the close working relationships between industry, scientists and managers, the overall reduction in catch from a peak of 41 000 t in 1990 to current landings of less than 10000 t proceeded in a relatively ordered manner.

There are two broad principles which can be drawn from this fishery. First, industry must be organized to address concerns within its ranks before negotiations. Second, scientists must interact with industry both formally and informally. Common ownership of problems, solutions and outcomes is a critical component of successful fishery management.

Introduction

The management of new or developing fisheries is often extremely difficult. Industry is optimistic due to high catch rates. Scientists do not have a time-series of fishery data upon which to base assessments and, in most cases, the responses of the population to exploitation are not yet apparent. Resource managers receive conflicting advice. Too often, too little is done too late.

The Australian orange roughy (*Hoplostethus atlanticus*) fishery is an important case study. The fishery has had a short history, developing rapidly in the late 1980s, from less than 200 t in 1985 to about 28 000 t in 1989 (Lyle 1994). Catch rates were extremely high: up to 100 t per two-minute bottom time were recorded and large marks were seen on echo sounders. Industry's expectations were high and parallels were drawn with a larger orange roughy fishery in New Zealand. Little was known of the biology of the species in Australian waters; productivity and stock size were unknown. Consequently, scientific advice at the commencement of the fishery was limited, and in some cases conflicting, and drew heavily on the 'New Zealand experience'. Additionally, several research institutes commenced major research projects simultaneously but initially with little co-ordination.

It was against this background that the Government Industry Technical Liaison Committee (GITLC) was formed. This was the first such committee in south-eastern Australia to provide a forum for industry, scientists and managers to address scientific issues and requirements. In this paper the role of GITLC is described and evaluated. It is argued that this committee provided a framework for avoiding or at the very least reducing, many of the problems which confronted the fishery. First, however, the history of the fishery is briefly reviewed, and the research background and management during the early part of the fishery are discussed.

HISTORY OF THE AUSTRALIAN ORANGE ROUGHY FISHERY

Orange roughy are widely distributed on the mid-slope regions of temperate waters (Gomon *et al.* 1994). In Australian waters they are found at depths from 700 to 1400 m but the bulk of the catch is taken between 800 and 1200 m. The species was first recorded in Australian waters off New South Wales in 1972 (Lyle *et al.* 1989). Although widely distributed at low densities, orange roughy form dense spawning and non-spawning aggregations from which large catches can be taken.

The bulk of orange roughy have been taken as part of the South East Fishery (SEF; formerly known as the South East Trawl Fishery). This is one of Australia's oldest fisheries commencing in 1914. Until the 1970s, the SEF was based on shelf resources, primarily off NSW and eastern Victoria. During the 1970s, the fishery expanded to western Victoria and South Australia, Tasmania, and to waters deeper than 200 m (Fig. 1).

Orange roughy were first taken commercially off Tasmania in 1982; early catches were taken mostly as an adjunct to shelf and upper-slope trawling operations (Lyle 1994). Catches remained at relatively low levels (less than 400 t) until late 1986 when a non-spawning aggregation off Sandy Cape, western Tasmania was discovered (Fig. 1). Over the next two years other non-spawning aggregations were found off Port Davey and Beachport and, together with catches from dispersed fish, annual landings ranged from 4000 to 8500 t (Chesson 1996). These aggregations did not recur annually. Orange roughy became the most important species in the SEF, resulting in a complete restructuring of the fishery with considerable fishing effort moving from the traditional area on the east coast to fish waters adjacent to western Bass Strait and Tasmania.

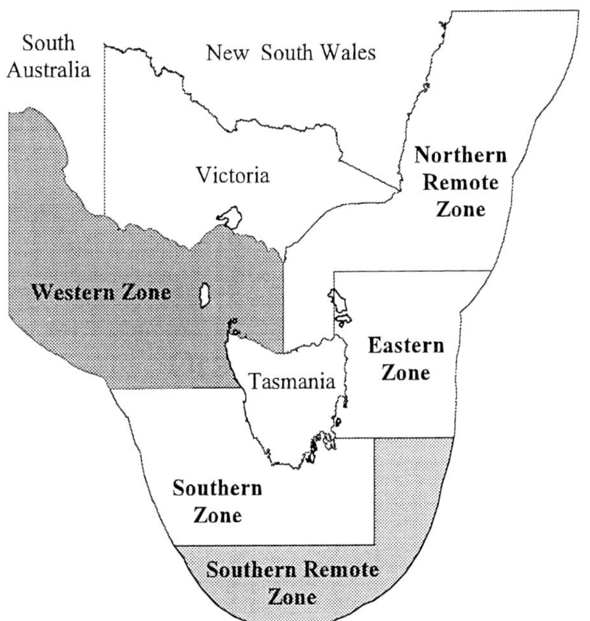

Fig. 1. Orange roughy management zones in the South East Fishery.

Landings increased dramatically in 1989 when a major spawning aggregation was discovered off the east coast off Tasmania along with non-spawning aggregations off southern Tasmania. Landings peaked at about 41 000 t in 1990. With the introduction of management measures (see below), landings have subsequently declined, to about 7000 t in 1995 (Chesson 1996).

Substantial quantities of orange roughy were taken in the adjoining Great Australian Bight trawl fishery during the late 1980s. Non-spawning aggregations were fished in 1988 and 1989, with landings of almost 3000 and 4000 t, respectively. Catches declined quickly and have remained under 1000 t since. Orange roughy are also taken in the Remote Zones but annual landings have been variable ranging from 40 t to almost 2000 t between 1989 and 1994 (Chesson 1996).

MANAGEMENT AND ASSESSMENT

The SEF management boundaries are from Barranjoey Pt, just north of Sydney NSW, to Cape Jervis, South Australia including waters around Tasmania. Jurisdiction is from 3 nautical miles (nm) to the 200 nm limit of the Australian Fishing Zone. Coastal waters out to 3 nm are under State jurisdiction except where varied by the Offshore Constitutional Settlement (OCS).

The SEF, a Commonwealth fishery, is managed by the Australian Fisheries Management Authority (AFMA), (a statutory authority). AFMA replaced the Australian Fisheries Service (AFS) (a division of the Department of Primary Industries) in 1992. The South East Trawl Management Advisory Committee (SETMAC) (with participants from industry, management and science) is a liaison body between AFMA and industry and provides advice regarding the trawl component of the fishery on a variety of issues including research. In December 1991, management of the SEF changed from primarily input controls to output control in the form of an individual transferable quota (ITQ) system. Different TACs apply to each zone in the fishery (Fig. 1).

Research and monitoring of the fishery has been, and continues to be, undertaken by several research agencies. These include State Government research institutes, CSIRO, and other Commonwealth agencies, notably the Bureau of Resource Sciences (BRS). In latter years, industry scientists have also played an important role. With so many agencies involved, and with direct competition for funds, it was inevitable that the research effort was fragmented and, in some cases, duplicated.

The Demersal and Pelagic Fish Research Group (DPFRG), formed in 1973 as a forum for review and co-ordination of fisheries research conducted in south eastern Australia, played an increasing role in the SEF and by the mid 1980s was providing advice to AFS on the status of SEF fish stocks. DPFRG comprised scientists only: industry and resource managers were not directly involved. Despite the limitations discussed below, DPFRG, in the absence of other arrangements, performed a useful function and this was acknowledged by the Chair of the group being made a member of SETMAC.

The rapid increase in orange roughy landings during the late 1980s saw a major research effort directed at the fishery. The nature of the fishery, particularly the extreme depths at which orange roughy are taken, meant that funding requirements were very high. A workshop was held in 1987 to co-ordinate, and to avoid duplication of research (Williams 1989). The positive outcome was that the research effort in southern Australia was apportioned amongst research institutions according to the expertise and location of that particular institution, rather than the previous open competition for funds.

Industry arrangements

Until the advent of the orange roughy fishery, industry in the South East Trawl (SET) had not been able to form effective fora for discussion of common interests and problems. There were several distinct industry sectors within the fishery, and operators in each sector had little interest in others.

The orange roughy fishery generated a mass movement into the South West (SW) Sector (equivalent to the Eastern, Southern and Western Zones in Fig. 1), as it was then called. In the late 1980s operators met and formed the South East Trawl Fishing Industry Association (SETFIA). It was noted at the inaugural meeting that the only Commonwealth fishery industry believed it had an effective input into the management of was the Northern Prawn Fishery, where operators had been meeting in formal associations since soon after the introduction of the first management plan. The orange roughy situation gave the necessary focus and common interest for this to occur in the SW Sector.

It was obvious to industry members that the massive catch rates of the time would see a rapid management response, and most operators recognized the value of a united front, whence meaningful negotiations could be fostered. In a short time the association had a membership representing well over 80% of the catch capacity in the SW Sector, and worked quickly and effectively to formulate agendas and protocols to ensure that the representations it made were ratified before presentation.

At this time there was little trust between industry, scientists and managers. Industry was concerned that managers would draw information from the scientific community, interpret this with little consultation and then make decisions unilaterally. This focussed industry attention on the need to enter that process as far upstream as possible. The risk identified was that of uncertainty, one of the most dangerous components of commercial decision making. The concept of GITLC was thus proposed, and on the simple premise that paying an entry fee would ensure a good seat at the theatre, a voluntary levy was proposed and endorsed.

These commercially oriented motivations in no way denigrate the formation of GITLC, and in fact add to its integrity: too frequently, industry stands back from the information-gathering and decision-making process, then resorts to political spilling tactics to frustrate management initiatives. In short, the GITLC approach is healthier and more honest.

The SETFIA strategy was set in place quite early, and followed successfully throughout. This was simply:

- that all criticisms of scientific work would be made through GITLC
- that once the liaison process had occurred, negotiation for TACs would not be based on ambit claims, but would be pitched within the boundaries defined by the assessment.

By and large, through SETFIA, industry was in pretty good shape for the challenge of managing orange roughy.

The Government Industry Technical Liaison Committee

Until the formation of GITLC there was no formal mechanism for consideration of SEF research issues by industry, scientists and managers. The outcome of this was that liaison between industry and scientists was poor; misunderstanding and mistrust were common. For example in 1988, industry questioned the value of research in the fishery (estimated at over $2 million) particularly the apparent lack of results on orange roughy. Industry was also concerned at DPFRG's advice that the sustainable yield for orange roughy could be less than 1000 t. This advice was based on limited data; primarily biomass estimates derived from trawl surveys and preliminary results on productivity. Such results did not accord with fishermen's own experience and from the perceived similarity with the much larger New Zealand fishery. This advice also conflicted with earlier authoritative scientific comment that suggested a very large biomass. Consequently, industry, initially, had little confidence in the research conducted in the fishery.

Following the discovery of the major spawning aggregation off Tasmania in 1989, catches rose dramatically. Concern was expressed, by all sectors, at wastage due to burst nets and dumping. The perceived need to reduce catches became more urgent. Considerable progress had been made by scientists but the early biomass estimates (20000 to 50000 t) were obviously wrong and stratified random trawl surveys were inappropriate for a species from which the bulk of the catch came from targeting aggregations.

At industry's instigation SETMAC established GITLC. This committee was 'to assist in the development of a program of scientific evaluation from a management perspective to interact with DPFRG as appropriate'. SETMAC also agreed that priority be given to supporting the development of a strategy for reacting to management needs and be funded through the imposition of a specific levy on orange roughy fishermen.

Membership of the committee was flexible and initially, AFS provided the convenor and secretariat. The first meeting was held in August 1989. After the current status of research was described, GITLC developed a stock protection strategy which included real-time catch monitoring and voluntary closures to allow scientific assessment (biomass estimation and biological parameters). Future research needs were also identified.

It can be seen from the committee's terms of reference listed below that GITLC had a broad brief, providing considerable advice on the fishery:

1. to assist in the development of a program of scientific evaluation from a management perspective
2. to seek advice on, and to consider, applications for industry funds for research into orange roughy and to make recommendations to SETMAC on these applications
3. to make recommendations for research needed on orange roughy and seek to arrange for the conduct of additional research required
4. to receive annual reports from organisations or persons who have received funding and to comment to SETMAC on these reports
5. to make recommendations to the Minister on the stock protection strategy as endorsed by SETMAC
6. to monitor the effectiveness of the stock protection strategy and to make recommendations as required
7. to recommend to the SETMAC Sub-committee on Finance and Surveillance, proposed levy rates to apply to SW Sector endorsement holders, and expenditures for the operation of the stock protection strategy.

It is not proposed here to give a detailed description of GITLCs subsequent meetings and recommendations. However, GITLC was closely involved in orange roughy research and provided advice on the phased reduction of the orange roughy TAC until 1993. Through GITLC, a biomass assessment and catch monitoring programme was funded. The biomass assessment component dealt specifically with trialling acoustic and egg surveys for orange roughy. The committee also provided the framework for choosing industry vessels for surveys and was involved in developing a code of practice for the fishery. With so little known of stock structure in the fishery, GITLC supported an 'adaptive management' approach to management of the fishery. Later it took responsibility for other major SW Sector species such as blue grenadier (*Macruronus novaezelandiae*). The committee also played an extremely important role in disseminating research results.

GITLC was replaced, in 1993, by the South East Fishery Assessment Group (SEFAG).

DISCUSSION AND CONCLUSIONS

So, how successful was GITLC in providing a forum for industry, scientists and managers and in meeting its objectives? It would be disingenuous to argue that committees like GITLC solve all the problems in communication between groups. However, by all measures GITLC was successful. It provided an active and vigorous forum. There was not always agreement, and debate was often heated, but there was common ownership and mutual respect.

GITLC forced all parties to communicate more effectively. It was certainly difficult for some scientists to expose the results of their research in this forum prior to formal review. Conversely, industry had to learn that the results may change between meetings as more analyses were undertaken. It could be argued that SEFAG is more rigorous and its role more clearly defined. Yet, despite this (or perhaps because of it) most participants would agree that GITLC was the more dynamic forum.

The most important outcome of GITLC was that the stock protection strategy was generally accepted. This was despite incomplete information and difficult-to-believe results such as orange roughy living to greater than 100 years of age. As a result of the close working relationships between industry and scientists, within GITLC, the overall reduction in catch from the peak in 1990 to current landings of less than 10 000 t proceeded in a relatively ordered manner.

An interesting comparison can be drawn with the fishery for gemfish (*Rexea solandri*) in the SEF. Catch histories of the two species are shown in Fig. 2. The fishery for gemfish commenced in the mid 1970s and was the subject of a major monitoring project directed by Kevin Rowling of NSW Fisheries Research Institute. By the early 1980s, key stock indicators such as size, age and catch rates were showing that fishing was impacting the stock (Rowling 1990, 1994, 1997). By the late 1980s, all indications were of a recruitment collapse and there was no or little recruitment for 4 years. The situation was so severe that a zero TAC was applied in 1993 after several years during which the TAC 'chased' catches downwards. Despite these clear indicators, sections of industry were still arguing that all was well with the fishery and the 'scientists' had it all wrong. Some very unpleasant debate followed. There is no doubt that a 'GITLC' for gemfish would have ensured a more reasoned debate.

There are two broad principles which can be drawn from the orange roughy fishery. First industry must be organized to address

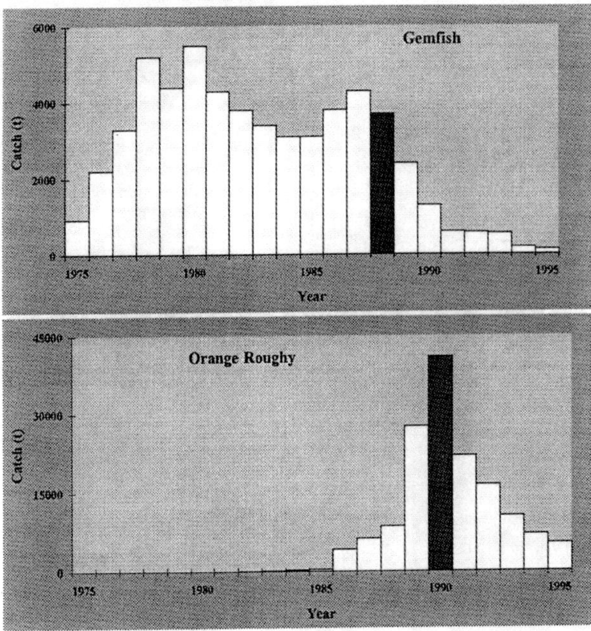

Fig. 2. Catch histories for gemfish and orange roughy. The dark bars represents the years in which TACs were introduced.

concerns in its ranks before negotiations. Second, scientists must interact with industry both formally and informally. Common ownership of problems, solutions and outcomes is a critical component of successful fishery management.

References

Chesson, J (Ed) (1996). The South East Fishery 1995. Fishery Assessment Report compiled by the South East Fishery Assessment Group. Australian Fisheries Management Authority, Canberra.

Gomon, M F, Glover, J C M, and Kuiter, R H (1994). The Fishes of Australia's South Coast. The Flora and Fauna of South Australia Handbooks Committee, State Print, Adelaide.

Lyle, J M (1994). Orange roughy (*Hoplostethus atlanticus*). In 'A Scientific Review of the South East Fishery with Particular Reference to Quota Management'. (Ed R D J Tilzey). Bureau of Resource Sciences Bulletin. (Australian Government Publishing Service: Canberra).

Lyle, J M, Evans, K R, and Wilson, M A (1989). A summary of Orange Roughy Biological Information: 1981–1986. Department of Sea Fisheries, Tasmania. Technical Report 39. 47pp.

Rowling, K R (1990). Changes in the Stock Composition and Abundance of Spawning Gemfish *Rexea solandri* (Cuvier), Gempylidae, in South Eastern Australian Waters. *Australian Journal of Marine and Freshwater Research* **41**, 145–63.

Rowling, K R (1994). Gemfish (*Rexea solandri*). In 'A Scientific Review of the South East Fishery with Particular Reference to Quota Management'. (Ed R D J Tilzey). Bureau of Resource Sciences Bulletin (Australian Government Publishing Service: Canberra).

Rowling, K R (1997). The collapse of the eastern Australian gemfish stock — Issues for management and the role of fisheries science. In 'Developing and Sustaining World Fisheries Resources: The State of Science and Management. Second World Fisheries Congress, Brisbane 1996'. (Eds D A Hancock, D C Smith, A Grant and J P Beumer) pp. 210–14 (CSIRO Publishing: Melbourne).

Williams, M (1989). Orange Roughy Research in Australia: A Case Study for Research Co-ordination. *Search* **20**, 130–4.

Sustainably Managing Sustainable Management

G. M. Kailis

M G Kailis Group of Companies, 50 Mews Road, Fremantle WA 6160, Australia.

Summary

Sustainable management processes require us to treat fisheries management as a set of human interactions to be managed, and not biological indicators to be applied. The process must involve user groups, researchers and fishery managers so as to minimize conflict and maximize commitment to sustainable management. If users identify and accept management processes, the implementation of outcomes will rely less on expensive and ineffectual coercive measures. For fishery managers, changes to the process will lead to more satisfying roles by broadening decision-making responsibility, encouraging independent resolution of conflict and clarifying core responsibilities on behalf of the community.

Introduction

Fisheries management occurs in a changing political, social, economic and biological environment. A fisheries management system must be able to deal with political agenda, technological change, varying user demand and fluctuations in biological systems which cause variability in recruitment. In addition, there is increasing community interest in *how* fisheries managers achieve fish stock management objectives and concern at the costs of management. Management systems must not only be effective in achieving biological outcomes but must be cost-efficient, cost-effective as well as having regard for those in the community who are affected by them.

Traditional bureaucratic control systems based on centralized control run a grave risk of failing under the weight of expectations outlined above. The resultant fallout will not only lead to undesirable resource outcomes it will also wreak a human toll on fisheries managers. A well-crafted management process should create a decision-making environment which encourages users of marine resources to participate in management to provide sufficient flexibility to respond to internal and external challenges. The need to engage users in the management of fisheries has been recognized internationally with the editors of the *Economist* noting in 1995 that: 'Governments that regulate their domestic fleets most successfully work with fishermen rather than against them.' (Anon. 1995).

This paper considers the reasons for involving users in management and some points on how to manage, and maximize, the benefits of that involvement.

BACKGROUND

The perspective outlined below is informed and influenced by my training in law and management, and by my practical experience in relation to private industry and public sector fisheries management. I have had experience in a wide range of industry and statutory bodies whose tasks have included advising on appropriate bureaucratic and legal structures for Western Australian fisheries, sustainable development of Australian fisheries and National fisheries research priorities. My involvement, as a Director of the Australian Fisheries Research and Development Corporation, has been especially useful in observing a wide range of industry and government attitudes to research and management in the eight separate Australian fisheries jurisdictions.

Based on these experiences, the following key issues are identified:

1. We will almost never have enough information about fish stocks (and fishing activities) to determine and maintain the appropriate level of effort.

   "We thought that modern fisheries science could become adept at assessing the abundance of fish in the sea and their productivity but we have failed miserably in that." (Walters, 1996).

2. There is no 'right way' to manage a fishery.

Increases in expectations of fisheries management have made traditional objectives such as a sustainable breeding stock only a small part of fisheries management. There is increasing emphasis on broader concepts such as community benefit and increased concern for the costs of management. The Western Australian Fish Resources Management Act (Fish Resources Management Act 1994, Western Australia) contains eight separate specific and somewhat conflicting heads of objectives. Included in the legislative objectives for the Australian Fisheries Management Authority (Australian Fisheries Management Authority 1995) are those of 'cost-effective management' and 'ensuring accountability'.

3. All fisheries management requires individuals to forego immediate economic or social benefits for future benefits to themselves and society.

Where users are generally supportive of management programmes, the involvement of users in developing management plans will result in plans that are more effective and less costly to run. Fishery management plans which do not have the support of users rely on coercion for their effectiveness. Coercion without community or user group support is almost always expensive and ineffective. Where valuable rights are involved there is also the potential for corruption of public officials. The net benefits of regulation to the community will depend on the costs of implementation, monitoring and enforcement of fishery management measures as well as increasing value of output and reduced fishing costs (Anderson 1992).

4. Conflict over resources by user groups can poison the management process and increase pressures for over-exploitation as each group tries to shift the burdens of restrictions to other user groups.

The potential for user conflict and competition has increased due to factors including increased urban development, changes in land use, rapid expansion of tourism and increasing recreational use (Resource Assessment Commission 1993). Australia and other countries also have to make provision in the management process for users who had previously been marginalized, such as indigenous peoples, and new uses of marine resources such as mariculture.

5. Adequate research funding must be sourced to properly resource fishery management if we are to ensure the sustainable exploitation of our marine resources.

In addition to improving our poor understanding of how most biological ecosystems work and how disturbance affects them (FRDC in press), we must cover new areas of community and scientific interest, as in implementing ecological sustainable development.

The Fisheries Research and Development Corporation of Australia which supports fishery research in Australia recently re-organized its programmes to reflect increasing interest in, and demand for, research into areas such as ecosystem protection.

A sustainable management process must address the above issues. Achieving sustainable development of fishery resources requires a framework in which environmental and equity issues can be integrated with resource decision-making (Commonwealth of Australia 1991).

THE CORE OF A SUSTAINABLE PROCESS — IDENTIFICATION AND ACCEPTANCE

In developing a management framework, an indispensable ingredient for success is that participants (including fishery managers and researchers) must identify with and support the process. This is analogous to a socialization process with users becoming incorporated into the system and identifying positively with it. Failure in the management process to foster acceptance of, and identification with, the fishery management system means that coercion, with all its inherent failings, is the prime mechanism for control.

A good example of incorporation of a user group within the system can be found in Western Australia. There has been considerable success with the recruitment of Voluntary Fisheries Liaison Officers (VFLOs). The VFLO programme recruits recreational fishers to assist in education and maintenance of the recreational fisheries management system. VFLOs have no statutory powers but have special identifying clothing and patrol important centres of recreational activity advising recreational fishers of management rules and monitoring compliance. In addition to potentially reducing the costs of education and management, the use of VFLOs has the effect of incorporating within the system one of the user groups and co-opting members of that group into maintaining the system.

Coercion merely results in compliance. Where coercion fails and participants in fishery management do not identify with the system, the result is action that is either indifferent to, or intentionally contrary to, management plans and can significantly undermine the success of management.

Indifference by a key participant can be as damaging as active dissent. For example, the common risk of failing to include researchers in the management process is highlighted by Table 1. Where there is no coercion or identification, researchers may act independently of the needs of management, and research programmes may fail to generate information relevant to management issues. Fishery managers are frequently left with the difficult task of reconciling industry demands and scientific advice (McKinnon 1993), a task which is made more difficult if extensive scientific research is not useful for the management of the fishery.

COMMENCING A SUSTAINABLE MANAGEMENT PROCESS

The sheer variety, diversity and character of fisheries is such that a simple check list for management processes designed to be sustainable could be impractical and misleading. In Western Australia, a wide range of consultative mechanisms is used, ranging from formal meetings of statute-based management advisory committees to more informal processes such as licence-holder meetings. The resulting interaction is often described in Western Australia as Co-management (Fisheries Department of Western Australia 1994). Based on my experience, the following have been useful in a wide range of situations:

1. Establish a research committee

Research committees encourage participants to focus on long-term outcomes of fishery management. Where dialogue between user groups, fisheries managers and researchers has broken down over management issues, a research committee is a less conflict-laden environment to encourage recommencement of a positive dialogue. In addition research committees are the key to binding researchers into the fishery management

Table 1. Interaction between identification and coercion (adapted from Hogg and Abrams 1988)

ATTITUDES OF PARTICIPANTS→ ACTIONS BY AUTHORITY↓	Identification Support	Non-identification	Dis-satisfaction
Coercion	Compliance	Compliance	Compliance
No Coercion	Conformity	Independence	Anti-Conformity/ Counter-Conformity

process. For example, the CSIRO Fisheries Division is a participant in the Research Committee of the Northern Prawn Management Advisory Committee.

2. User rights — confirmation not diminution

A sustainable management process cannot be built where users' interests are being constantly threatened. Threats undermine any incentive for a user to give up immediate benefits for future gains. In the process leading to the formation of the Australian Fisheries Management Authority (AFMA), a critical success factor was an explicit undertaking that existing rights would not be re-allocated through auctions or competitive bidding, and legislation would formally recognize the ongoing nature of rights in existing developed fisheries (Department of Primary Industries and Energy Publications Unit 1989). The ongoing success of AFMA can in large part be attributed to adherence by the independent board to these principles.

Any management process which contemplates, either implicitly or explicitly, confiscation and re-allocation of users' rights, or which encourages threats to confiscate those rights, will lead to enduring bitterness and an alienation of users from the management system. As Machiavelli put it in his advice to the Prince:

> but above all things he *(the Prince)* must keep his hands off the property of others because men more quickly forget the death of their father than the loss of their patrimony. (Machiavelli 1513, p.131).

An important advantage of confirmation of existing users' rights is a reduction in rent-seeking behaviour by user groups with conflicting objectives, e.g. recreational *versus* commercial. Users are encouraged to participate in a management process to maximize joint utilization of the resource and to negotiate over allocation of the resource rather than to seek external political intervention. A typical example of rent-seeking behaviour was reported in the journal 'The Queensland Fisherman' where recreational groups were resisting bag limits and, rather than accepting mutual responsibility for controlling effort, were lobbying government for additional restrictions to first be placed on commercial users (Anon. 1996).

3. Adequate resourcing of the management process and industry–user consultation

As the management process now includes a wide variety of participants, adequate resourcing of participation in the process is required. Effective consultation requires time and effort by all participants. In Western Australia, the Fish Resources Management Act 1994 provides for a portion of statutory fees and levies to flow to statutory advisory committees and industry associations. These bodies organize users into coherent cohesive groups to participate in management processes, help to provide a consistency of thought and approach across fisheries, and communicate back to user groups on management information and advice.

At a federal level, although similar levies funding the consultation process are not directed to industry associations, independent executive officers of the various statutory Management Advisory Committees are funded by industry levies to organize and assist industry participation.

User groups may not form spontaneously without direct encouragement by fishery managers. It would not be unusual for there to be strong antipathy amongst members or factions of a user group, whether for historical or personal reasons. Fishery managers should be prepared to act positively and assist user groups to form. Once user groups have been formed, managers must give regard to advice that emanates from those groups over advice received through individual consultations. A failure to respect advice received from those groups will undermine incentives for users to cooperate and compromise in coming to a corporate response. In 1987 the Pearl Producers' Association in Western Australia was formed largely as a result of strong encouragement by the then Executive Director of the Fisheries Department, Mr B. K. Bowen. Prior to its formation, industry–government interaction was based on individual consultations and largely ineffective in either generating usable advice or engendering committment to the management system. Since 1987, the Pearl Producers' Association has proved to be an effective industry body and now undertakes significant roles in marketing and safety, in addition to assisting industry involvement in the management process.

The use of statutory powers to direct part of compulsorily levied fees to provide assistance to user groups does raise some difficult ethical issues. In Australia the persistent non payment of levies usually leads to a loss of fishing entitlements and so to the loss of livelihood for a defaulting payer. The levy structure should not infringe the fundamental human right of freedom of association. The levies and fees should be seen as part of a user's contribution to funding an orderly management process in an effective and cost-efficient manner. There must be careful public scrutiny to ensure that groups which are publicly funded do effectively represent their users and are recognized by the users they purport to represent. The uses to which that money is put by user groups must also be carefully monitored to prevent diversion of compulsorily levied funds to other causes.

4. Respect Management Advisory Committee (MAC) advice

In preparing this paper, I am aware that other papers are likely to cover MACs and their role in Australia. It is well recognized that MACs are a mechanism for consultation and for generating management action. They also have an important function in fostering the identification by users with the management system and the acceptance of outcomes from the management process. It must be recognized that as a consequence of this dual role of MACs a failure to accept MAC advice not only risks undermining the management process and hindering effective fishery outcomes but also risks alienating users from the system.

In order to avoid breakdown of the system as a whole, where MAC advice is not implemented, policies outlining the general circumstances where MAC advice may be rejected should be settled up front. In Western Australia in 1992, a decision by the then Minister of Fisheries, contrary to advice generated through the well-established MAC process for the Rock Lobster fishery, led to a perception of political interference with fishery management. It should be noted that the course taken by the Minister was widely regarded as less restrictive than that advocated by the MAC. Commercial fishers responded to the Ministerial action by taking political action outside the normal fishery management process with a vote of censure at the Annual General Meeting of the WA Fishing Industry Peak Body. Individual fishers and fisher groups took further direct action by campaigning against the Minister in a subsequent election. This incident in Australia's longest established fishery advisory committee highlights the damage that can be done to a MAC process by actual (or perceived) political interference, combined with a failure to clearly outline for all parties the circumstances when MAC advice may not be followed.

My recommendation to those who have the responsibility to act on MAC advice is to establish that they will only disregard that advice where:

i) the effect of taking the advice would be to dis-enfranchise a minority group on the MAC or a group not involved in the MAC process;

ii) the advice would not lead to an ecologically sustainable outcome, or

iii) there is some strong overwhelming public interest which conflicts with the MAC advice.

The latter consideration of 'the public interest' should generally only be judged by a politician responsible to the broader community and not by appointed government officers. The decision and the reasons for it should become public knowledge.

5. User charges

Users should contribute to the cost of management regardless of whether they fish for pleasure or profit (Industry Commission 1992). User charges are an equitable way to ensure that those who benefit from management contribute to the system, increasing resources available for research and management.

User charges can have an additional benefit of encouraging users to minimize the costs of management. Significant efficiency benefits are only likely when users are fully engaged in improving the system and are prepared to modify their behaviour in order to reduce costs and raise efficiency. There should be a clear link between reduced costs and reduced user charges. Using user charges merely as a tool to generate public revenue will break the link between user activities and the costs of management of users. Incentives for user involvement will be reduced and potential benefits would be lost. Any decision to raise public revenues for resource access should therefore be separated from the issue of issues of user charges.

Cultural changes

Success in reforming management processes within a sustainable management framework will require significant cultural change for those in fishery management who are used to a command-and-control system. The role of government officers with statutory responsibilities will move away from direct control and decision making to ensuring that the community interest is served through the process and that the process works effectively and efficiently. There will be resistance by traditional fishery managers to such changes.

In turn, many users will object to accepting the responsibilities flowing from recognition of their rights and increased participation in management. As stated above, users should contribute to the cost of management regardless of whether they fish for pleasure or profit. In many cases users will see a significant and immediate rise in user charges, and this has certainly been the general experience in Australia. Our own company has seen user charges for fishery operations in Exmouth Gulf more than double due to recent moves to user charges in Western Australia.

In Australia, these necessary cultural changes are assisted by wide consultation before the introduction of changes and the liberal use of expert reports and broadly based working groups. There is often an accompanying re-organization of the fishery management bureaucracy with all senior and middle management positions being declared vacant, re-advertized and new appointees sought. New statutory arrangements for management and consultation are adopted as a result of the reforms, including mandatory consultation in certain cases. Reforms that have increased funding by users also promise greater accountability to those users. Based on my personal experience with the reform process in Western Australia, consistent pressure over a number of years will be required before the process is completed.

Conclusion

Sustainable management processes require us to treat fisheries management as a set of human interactions to be managed and not as biological indicators to be applied. The process must involve user groups, researchers and fishery managers in order to minimize conflict and maximize commitment to sustainable management. If users identify and accept management processes, the implementation of outcomes will rely less on expensive and ineffectual coercive measures. For fishery managers, changes to the process will lead to more satisfying roles by broadening decision-making responsibility, encouraging independent conflict resolution and by clarifying core responsibilities on behalf of the community.

As a final note for those government officials and politicians who may still be concerned at losing power and authority:

> Never let any Government imagine that it can choose perfectly safe courses; rather let it expect to take very doubtful ones because it is found in ordinary affairs that one never seeks to avoid one trouble without running into another but prudence consists in knowing how to distinguish the character of troubles, and for choice to take the lesser evil. (Machiavelli 1513, p.177).

References

Anderson, L G (1992). Enforcement issues in selecting fisheries management policy. *Marine Resource Economics* **6**, 261–77.

Anon. (1995). No place like home. *The Economist*, April 22, 1995.

Anon. (1996). Bag Limits Reveal 'creed of greed' by Sunfish Lobby. *The Queensland Fisherman*, January 1996, 5–7.

Australian Fisheries Management Authority (1995). Protecting our Fishing Future. Corporate Plan 1995–2000. (Australian Fisheries Management Authority: Canberra).

Commonwealth of Australia (1991). Ecologically Sustainable Development Working Groups Final Report: Fisheries. (Australian Government Publishing Service: Canberra).

Department of Primary Industries and Energy Publications Unit (1989). New Directions for Commonwealth Fisheries Management in the 1990s. (Australian Government Publishing Service: Canberra).

Fisheries Department of Western Australia (1994). Report of the Fisheries Portfolio Review to Minister for Fisheries November 1994.

FRDC (in press). Fisheries Research and Development Corporation Research and Development Plan 1996–2001, Investing for Tomorrow's Catch 1996–2001.

Hogg, M A, and Abrams, D (1988). 'Social Identifications.' (Routledge: London).

Industry Commission (1992). Cost Recovery for Managing Fisheries. Report No. 17. (Australian Government. Publishing Service: Canberra).

Machiavelli, N (1513). 'The Prince.' (Wordsworth Editions Ltd: Ware).

McKinnon, K R (1993). Review of Marine Research Organisations. (State Government Press: Canberra).

Resource Assessment Commission (1993). 'Coastal Zone Inquiry Final Report' (Australian Government Publishing Service: Canberra).

Walters, C (1996). In 'Fishing on thin ice'. (Ed C Anderson) *Western Fisheries*, January 1996, 41–44.

Ownership and Management of Traditional *Trochus* Fisheries at West Nggela, Solomon Islands

S. Foale

Department of Zoology, University of Melbourne, Parkville, Victoria 3052, Australia.

Summary

The *Trochus* management system at West Nggela (Florida Group, Central Province), Solomon Islands, was studied to determine the importance of Customary Marine Tenure (CMT), economic factors and Traditional Ecological Knowledge (TEK) in the management of *Trochus* stocks on a number of reefs in the area. *Trochus* populations were estimated using mark-recapture. The high price of *Trochus*, relative to the effort involved in collecting and selling them, appeared to motivate regular poaching on reefs where ownership was disputed or that were remote from surveillance. *Trochus* populations on all these reefs appeared to be overfished as a consequence. Undisputed reefs in front of villages were diligently managed, but even these were depleted compared with well-managed *Trochus* fisheries elsewhere in the Pacific. TEK played a relatively minor role in both the harvesting and conservation of stocks. Various axioms of TEK are discussed in the paper.

Introduction

At West Nggela in the Solomon Islands (Fig. 1), *Trochus niloticus* populations on reefs are managed by a system of serial prohibitions, or 'tambus', during which fishing for *Trochus* is banned for a period, usually between three months and a year. The tambu is announced to the community, usually at a church service (most Nggela people belong to the Anglican denomination), and is usually, though not always, advertised by a stick erected on the reef crest, which may have a leaf or a shell tied to the top. Tambus can be declared either by a member of the owning family ('custom' tambu) or by a Priest ('Mama') or Brother ('TaSiu') of the Anglican church ('church' tambu), on behalf of that member. The declaration of the tambu may involve some ritual, the form of which varies among individuals. These days, church tambus apparently command more respect from the population than custom tambus, and are widely regarded as the most effective way to protect a reef from poaching. At the end of the tambu, the owning family, usually *via* the church leader, announces that the tambu is over, after which the member of the owning family, who declared the tambu, commences the harvest, usually with the help of two or three members of the extended family.

Trochus are collected by diving with small plastic goggles, as most *Trochus* habitat is subtidal at West Nggela. Harvests may

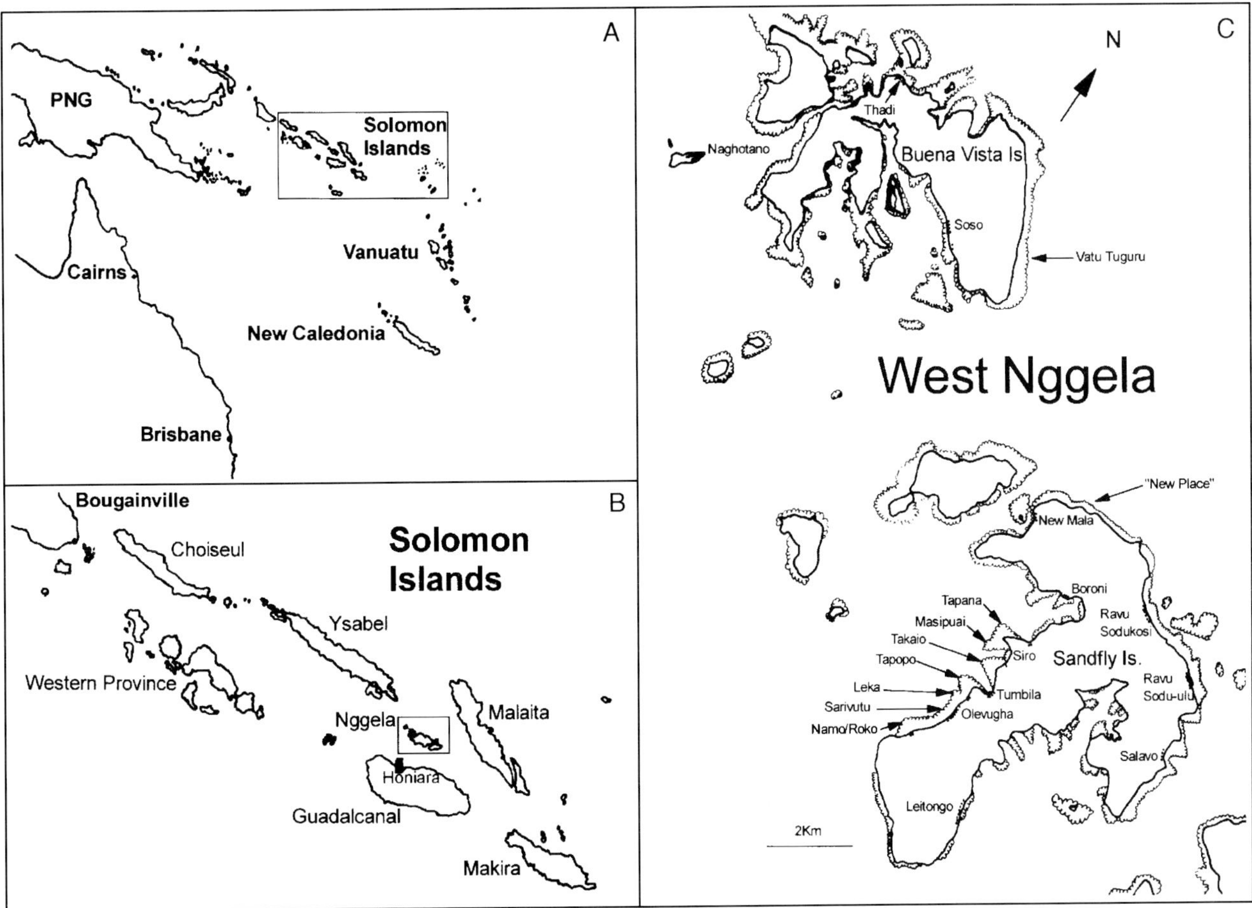

Fig. 1. Map of West Nggela (C), showing all major villages, and reefs included in the study. A: Location of Solomon Islands; B: Location of Nggela within the Solomons.

last from three days to a week, during which time the person who instigated the tambu has exclusive access to the stock. When that person is satisfied with the harvest, the reef is opened to the rest of the community for anything from a week to several months, before the next tambu is declared and the reef is again closed to fishing for *Trochus*.

This is the ideal scenario, but often the management system is compromised. This paper looks at the effectiveness of the tambu system in managing *Trochus* stocks at West Nggela, and explores factors which result in its breakdown and the depletion of *Trochus* stocks on some reefs.

MATERIALS AND METHODS

1. Stock assessments

Trochus stock assessments were attempted, using the Peterson mark-recapture technique (Seber, 1982; Nash, 1985), on ten reefs in the Sandfly / Buena Vista area of West Nggela (Fig. 1). Up to four local divers, with expertise in diving for *Trochus*, assisted the author in the collection and marking of *Trochus* within the area to be harvested. The team attempted to mark between 50 and 100 shells in the size range typically harvested (>6 cm). All marked animals were also measured (maximum basal diameter). Marking ranged across as much of the area demarcated for harvesting as possible, to ensure that marked and unmarked individuals had the same chance of being captured in the harvest. Marking usually required two days of diving. Marks were made with pencil on the nacre just inside the aperture of the shell, following the method of Nash *et al.* (1995). Marked *Trochus* were usually returned to where they were found. Given that the date for harvesting a reef was usually set in advance, marking was timed to precede the harvest by about one to two weeks. In cases where the harvest was delayed to more than four weeks after marking, the marking exercise was repeated to avoid the risk of loss of the first mark from overgrowth by nacre.

Most diving was done on breath-hold. Subsequent to the harvest by the reef owners, all *Trochus* in the harvest were measured and inspected for marks (boiling of the shells and removal of the animal for domestic consumption had no effect on the mark).

The area of each reef was plotted on a digitizer from high resolution black and white aerial photographs. Distances between two or more recognizable points on each reef were measured while in the field using a handheld Global Positioning System receiver, to calibrate the scale. The reef areas were used to calculate the densities of *Trochus*.

Table 1. Estimates of population sizes and densities of *Trochus* on ten reefs at West Nggela

Reef	Tambu length	Harvest size >8 cm/All	Count (>8 cm)	Count (All)	Habitat Area (ha)	Density (>8 cm/All) (*Trochus*/ha)	Poaching?	Disputed?	Remote?
New Place	0		Too few marked		ND	NA	Yes	No	Yes
Takaio	8		Too few marked		13.6	NA	Yes	Yes	No
Masipuai	9		Too few marked		13.8	NA	Yes	Yes	No
Tapana	9		25 marked, 13 harvested		5.3	NA	Maybe	No	No
Namo/Roko	9		Too few marked		ND	NA	Yes	No	Semi
Vatu Tuguru	9	15/52	ND	741 (278; 2214)P	18.7	ND/39.5	Maybe	No	Yes
Tapopo	11	91/160	261 (209; 313) B	ND	9.5	27.6/ND	No	No	No
Leka	11	46/88	87 (69; 104) B	175 (160; 190) B	6.4	13.5/32.3	No	No	No
Salavo	12	207/295	436 (303; 569) B	789 (487; 1254) P	10.3	42.5/76.9	No	No	No

ND = No data; NA = not applicable

All refers to the fishable *Trochus* only

Population estimates include all animals removed in the harvest and at any other time after marking

Numbers in brackets are the 95% confidence limits

B = Binomial Distribution, P = Poisson Distribution

2. Ownership and management of reefs

Details about ownership and boundaries of each reef were obtained by interviews and informal discussions (in the Nggela language) with people in each village. Family trees were constructed for several of the groups to facilitate an understanding of the way in which power over reef resources is formally distributed and rotated within an ownership group. Ownership disputes were discussed with many fishers and reef owners to clarify the effects on management of *Trochus* and other reef resources. Problems with enforcement of tambus on disputed and remote reefs were also covered. The author also attended two land/reef court cases. Subsequent discussions with leaders from each party, as well as with one of the judges for a case, helped to elucidate issues relevant to reef resource management. Interviews, observation and participation in some harvests provided information about harvesting systems, including division of labour and sharing of the catch.

3. Economic importance of *Trochus*

Reef owners and fishers were interviewed to determine:

a. the relative importance of income from the sale of *Trochus* in their overall annual income,

b. the cash return-per-unit-effort compared with other cash-earning activities, and

c. ways in which money earned from the sale of *Trochus* was spent, and how different types of expenditure were prioritized within an ownership group.

4. Traditional ecological knowledge

Fishers, including reef owners, were interviewed on the following subjects:

a. *Trochus* emergence patterns — diel, lunar, and seasonal,

b. natural mortality, including loss of shells due to boring polychaetes ('rotten tops'), predation by hermit crabs, and any other factors, and

c. reasons for recent decreases in catches.

Sufficient fishers were interviewed, separately, to obtain consensus between at least five people on each point. In the case of emergence patterns, concurrent opinions were provided by at least 40 people over the course of the study.

The positions of ten wild *Trochus* were marked with subtidal floats and the location and behaviour of each animal was recorded on 21 separate dives (8 during the day, and 13 at night, between 6.50 p.m. [dusk] and 9 p.m.), which spanned five weeks, including two full moons and one new moon. These observations were designed to gain an impression of spawning timing, diurnal emergence and feeding patterns, and extent of movement over the reef within that time frame. When an individual *Trochus* could no longer be found, a new one was sought and its position marked.

The number of 'rotten tops' (i.e. shells rendered unsaleable due to boring polychaete worms) was noted for all harvests inspected and divers were asked to recall the number of rotten tops they left on the reef during each harvest.

The number of hermit crabs occupying *Trochus* shells was noted for each marking dive, and fishers were asked the number of hermits in *Trochus* shells they encountered during their harvests.

Results

Stocks and management

Population estimates from mark-recaptures are presented in Table 1. Marking was attempted for an additional three reefs, but these were abandoned due to densities being too low for the mark-recapture technique to be used. Reefs which were undisputed and located in front of villages were relatively free from poaching and supported harvestable quantities of *Trochus*. Even the highest densities, however (Salavo), are low compared with many fisheries elsewhere in the Pacific (Adams *et al.* 1992).

Figure 2 shows the size-frequencies of *Trochus* harvests, some of which were not part of a mark-recapture exercise. Despite the

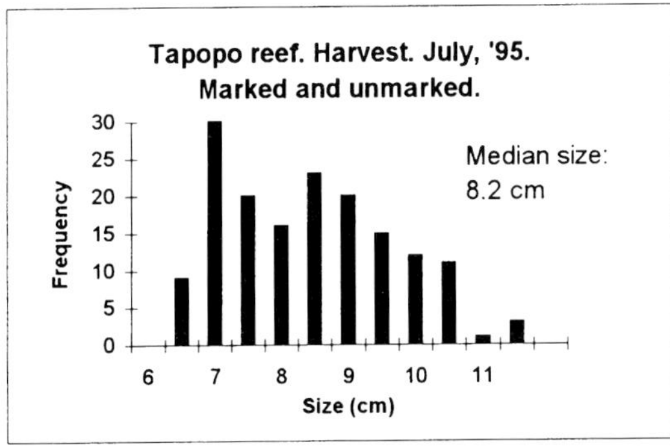

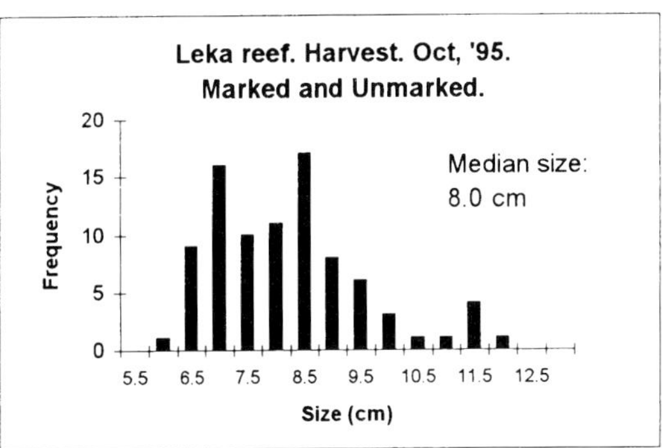

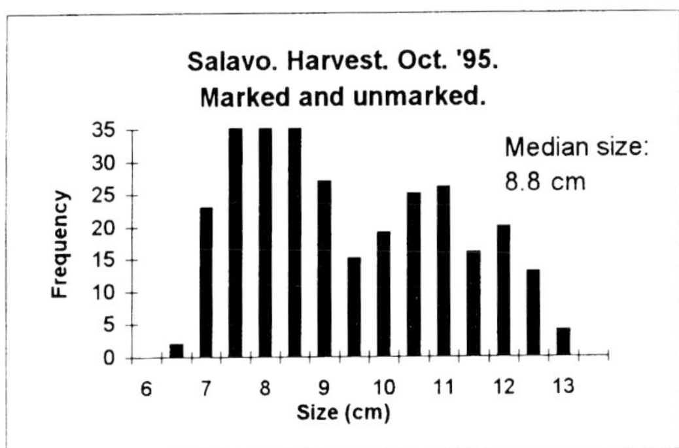

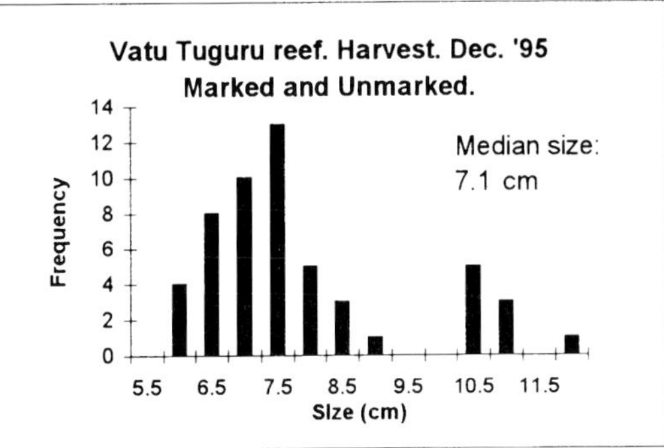

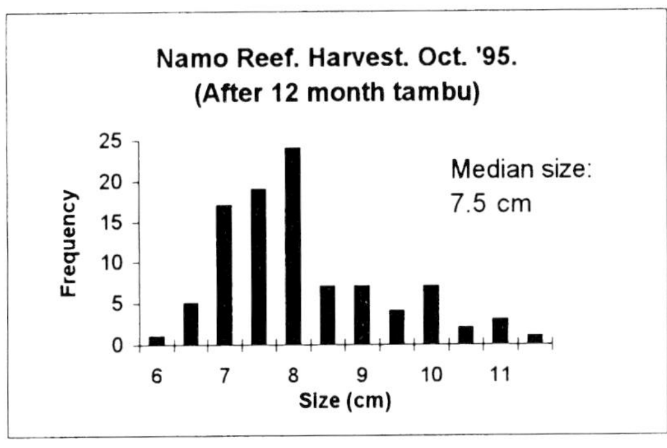

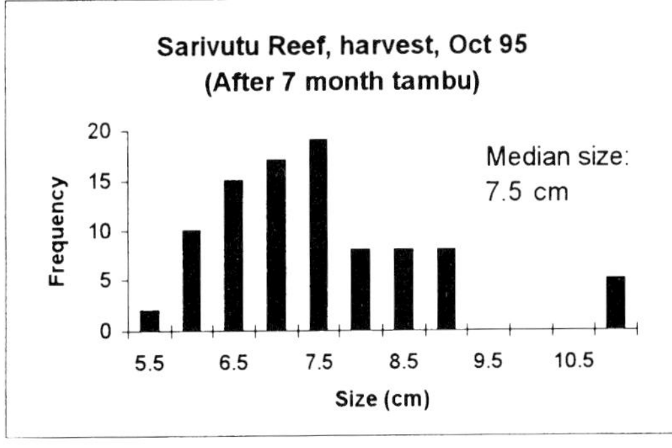

Fig. 2. Size-frequencies of *Trochus* harvests from various reefs around Sandfly/Buena Vista, 1995, Namo and Sarivutu are small reefs near Olevugha Village.

introduction of minimum and maximum size limits of 8 cm and 12 cm respectively by Solomon Islands Fisheries Division in 1993, it is clear that *Trochus* smaller than 8 cm and larger than 12 cm are still being harvested at West Nggela. Villagers were always able to sell the entire harvest in Honiara. Assuming *Trochus* grow from approximately 6 cm to 8 cm in their third year (Nash 1993), the size-frequencies indicate that on most reefs a large proportion of the population are not surviving past their first year in the fishery.

On Nggela, as in many parts of Melanesia, reefs are regarded as simply an extension of the land with respect to ownership. Sedentary and sessile reef resources, principally *Trochus* and bêche-de-mer, are subject to relatively strict fishing prohibitions compared with fish, which are accessible to non-reef owners. Land and reef is inherited matrilinealy and marriage patterns are predominantly virilocal (the groom's family pays a bride-price, which gives them rights over the woman's labour, and usually requires her to live in the husband's village). This results in women tending to move away from the land and reef that their children will inherit. Although a woman's brother has equal rights over their mother's land, his children will relinquish all primary rights when he dies, while her children will inherit primary rights. If the woman and her children no longer reside on that land, but the brother and his children do, an arrangement is usually made to allow those children to stay on after their father's death (i.e. they are granted usufructory rights), as long as they do not proceed with any major income-generating project on that land without consulting the primary rights holder first.

The Nggela people have a traditional institution for purchasing land (and reef, where contiguous), which is commonly referred to as 'Huihui'. This takes the form of a public ceremony which must be attended by all of the chiefs and heads of families related to each of the parties involved in the transaction, as well as the Paramount Chief of the district. In most cases the purchase price includes one or more pigs, some ceremonial baskets of food, and sometimes a small amount of cash (custom money in the past). While the total cash value of this package is small compared with the modern freehold value of the land, acquisition of land by Huihui is not open to everyone. It is usually restricted to members of a family who do not stand to inherit the land matrilinealy. In practice, the most common case is when a man or woman wishes to acquire some of his or her father's land. However, there have been cases where land was purchased by non-relatives, but a discussion of these is beyond the scope of this paper. In many cases a number of people contribute to the total price in a Huihui transaction, and as a result, each will acquire rights to the land purchased. These rights are also only heritable matrilinealy and subsequent patrilineal bequeathals must be arranged through further Huihuis or similar ceremonies.

Disputes over ownership often arise as a result of disagreement between two parties as to whether or not, or how much, a maternal ancestor contributed to a Huihui which might have occurred so long ago that there are no longer any living witnesses and there is no written record of the event. Cases of violations of land and marine tenure abound, but are usually tolerated unless a large amount of money comes into play, such as in the event of a development of some sort.

Both Masipuai and Takaio reefs were under dispute at the time of the study (see Table 1), as a result of an attempt by one of two parties, each claiming ownership of the adjacent land, to close down the school there and sell the land to a resort developer. Consequently the *Trochus* stocks had been subject to heavy fishing pressure despite the prior proclamation of *Trochus* tambus by both parties claiming ownership of the reefs. Several people asserted that these two reefs had been highly productive in the past and would be so again if the tambus were respected and they were allowed to recover. This of course would not happen until the dispute was settled.

Vatu Tuguru reef was not under dispute, but the reef was harvested before the tambu was lifted. The first, unofficial, harvest was made a week before the scheduled harvest, by a member of a different branch of the family from that of the person who declared the tambu. This explains the low numbers in the official harvest (Table 1). After careful investigation of the family tree in this particular case, it became clear that the person who declared the tambu (and harvested second) in fact had a somewhat dubious claim to ownership of the reef, based upon an alleged contribution to a Huihui by one of his maternal ancestors about four generations back. His claim was categorically denied by the other branch of the family. This is a good example of a case where a violation of CMT is tolerated (no-one objected to him declaring the tambu), because relatively little is at stake. Vatu Tuguru is also far from any settlement, and there were claims of poaching occurring prior to both of these harvests.

'New Place', while not under dispute, was in a remote location, well out of sight of the nearest village. The reef was subject to constant and heavy fishing pressure by the primary rights holders as well as by poachers. The reason for not placing any tambu on the reef was that it would not be respected due to the fact that it could not be enforced, so the owners simply fished it down in competition with poachers.

Table 2. Primary sources of cash income for a sample of villagers at West Nggela, 1995

Source of income (and market)	No. of people
Trochus (exported *via* Honiara)	4
Artisanal fin-fishery (Honiara)	19
Aquarium fish (exported *via* Honiara)	11
Wildlife (exported *via* Honiara)	2
Lime (Honiara)	3
Bêche-de-mer (exported via Honiara)	4
Remittances/employment	12
Garden produce, including betelnut	6
Other	2

Leka, Tapopo and Salavo, by contrast, are all undisputed reefs, which are located in front of villages. Namo and Roko are undisputed and in front of a small, relatively isolated hamlet, but were poached when the owner went away to another island for two weeks to tend his gardens. There may be a number of reasons for the low number of *Trochus* marked at Tapana, including heavy fishing in the past, the small size of the reef, and relative scarcity of good *Trochus* habitat. The reef was in sight of the owner's hamlet, but may have been poached at some time during the tambu.

Economic importance of *Trochus*

Sixty three people (four women and 59 men) from West Nggela were asked to name their most important source of cash income for 1995. The results are summarized in Table 2. Most of these people engaged in more than one of the listed activities, often as part of a group.

Of all these activities, diving for *Trochus* is unanimously regarded as having the highest cash return per unit effort, even though the total return per individual over the course of a year may be less than other slower earning activities such as artisanal fishing or marketing of garden produce. Unlike fresh fish and other primary products, *Trochus* shell needs no processing (apart from removing the meat, which is really a subsistence bonus), can be stored indefinitely at no cost, sold quickly and in any quantity, and is relatively easy and cheap to transport. For most of 1994 and 1995, villagers were receiving between SBD 12 and SBD15[1] (approximately US$US3.50–US$4.35) per kg for whole *Trochus* shell in Honiara. Most villagers in Nggela regard *Trochus* as 'pure cash just sitting on the reef'.

Of the 55 men who listed something other than *Trochus* as their major cash-earner, 32 derived some income from *Trochus* during 1995. The largest harvest, 183 kg (1420 shells) was taken from a 31.6 ha area of reef in front of Ravu Sodu-ulu Village, after an 11-month tambu. Most harvests, however would be roughly an order of magnitude smaller than this.

Systems of power-sharing within an ownership group varied widely amongst groups, but where ownership was shared by a number of siblings, the senior male often made executive decisions as to who would be allocated the right to make a

[1]SDB – Solomon Island Dollars

tambu on the reef in any one year. Siblings usually took turns to make tambus, and priority was often allocated on the basis of worthiness of the intended expenditure. For example, a tambu to pay for the cementing of a grave or to meet school fees would take priority over one which was earmarked to pay for a cassette player.

Traditional ecological knowledge (TEK)

There were relatively few axioms of TEK which were relevant to *Trochus* management and harvesting. They included: 1) that *Trochus* are easiest to find two or three days after full moon, 2) that if reefs are closed to fishing for longer than a year, too many *Trochus* are lost to hermit crabs and shell borers ('rotten top'), and 3) that Cyclone Ida in 1972 was the main reason *Trochus* are relatively scarce at Nggela today.

Investigation of *Trochus* behaviour revealed that, although some *Trochus* did appear to spawn one day after full moon (after sunset), there was no evidence that they were more easily found on the following days. Borer damage was very rare, with only one *Trochus* affected out of all the harvests examined during the study. This was at Salavo, where a relatively larger proportion of the harvest exceeded 12 cm than for the other sites (Fig. 2). Borer damage rarely affects *Trochus* smaller than 12 cm (Nash 1993). Of 11 divers interviewed, 10 reported seeing no borer damage during their harvests. However, it appears that a proportion of *Trochus* are taken over by hermit crabs (Table 3).

DISCUSSION

The evidence presented here shows that at West Nggela, *Trochus* stocks are not able to be managed unless they are (a) under unambiguous ownership, and (b) accessible to continuous surveillance by the owner. It is also clear that, even where poaching can be prevented, stocks are heavily fished and relatively impoverished compared with other fisheries in the Pacific. According to Adams *et al.* (1992), a healthy but well-fished reef supports around 100 shells per ha of suitable habitat, and reefs which are fished relatively lightly support up to 300 shells per ha. At Aitutaki (Cook Islands) the Island Council will not allow a harvest to proceed until densities reach approximately 600 shells/ha. The highest density found in this study was at Salavo: 42.5 shells/ha (>8 cm) and 76.9 shells/ha (>6 cm). However, the large reef at Ravu Sodu-ulu, from which the harvest of 183 kg was taken, (see results) may have supported a higher density than Salavo. If one assumes a similar fraction of the population was removed at Ravu Sodu-ulu as at Leka, Tapopo and Salavo (i.e. 1/3 to 1/2), then it is likely that the unfished densities of *Trochus* (>6 cm) were between 90 and 135 shells per ha. Ravu Sodu-ulu, like Leka, Tapopo and Salavo, was undisputed and situated in front of a large village.

Table 3. Number of hermit crabs occupying *Trochus* shells found during marking dives in 1995

Reef	Day/Night	Hermits	*Trochus*
Tapopo	Night	11	46
Leka	Night	2	30
Leka	Day	1	81
Salavo	Day	1	55
Vatu Tuguru	Day	4	83

The economic data presented shows that, while *Trochus* have a high value and are highly sought after, they are not the most important source of income for most people (see also Turner 1994). This may explain why poaching appears to be tolerated in many cases. However, given the high cash return for *Trochus* relative to effort, at the time of the study, it is clear that the incentive for poaching is high, as is the incentive for adventurous ownership claims over reefs. It should be noted that the export of whole *Trochus* shells from Solomon Islands was banned from 1 January 1996, and this had the effect of depressing the buying price in Honiara by up to one- third. The ban was designed to relieve some of the pressure on stocks and to limit exports to processed shell only (i.e. button blanks made in Honiara).

The TEK relating to emergence and predation/shell damage appeared to have relatively little importance to harvesting and management of *Trochus* stocks at the time of this study. However, the TEK axiom relating to emergence is in my opinion worthy of further study. Similar TEK is also reported for other parts of the Pacific (Nash 1993). Previous studies indicate that an increase in emergence, and/or aggregative behaviour, is related to reproduction (Nash 1985, 1993). Given that *Trochus* spawn year round at low latitudes and most commonly on a lunar cycle (Nash 1993), it seems likely that only a fraction of the population is spawning at any one time (see also Nash 1985). However it is possible that a proportion of the population may reside for much of the time (including when feeding at night) in deep holes and crevices in the reef, and these individuals, after spawning, may, for a short time at least, hide in shallower recesses at the top of the reef where they are easier to find. It should also be noted that reef owners at West Nggela were never observed to time their harvests to take advantage of the increase in availability of *Trochus* predicted by the TEK.

Since rotten top tends to be restricted to shells larger than 12 cm basal diameter (Nash 1993), this condition is likely to affect a greater proportion of the population if fishing pressure is reduced, since more individuals would reach an old age. Virgin populations may have very high proportions of worm damage, as reported for Pohnpei (FSM) in 1946 (Clarke and Ianelli 1995). Compliance with the maximum size limit of 12 cm would turn such a problem to the advantage of the reef owners, by increasing gamete production and potential recruitment.

Hermit crabs, on the other hand, select a wide range of shell sizes (Nash 1985). Although a reduction in fishing pressure may not affect the proportion of *Trochus* taken by hermits, it is likely that more frequent harvesting would reduce shell damage from pagurization, thus a greater proportion of hermit-occupied shells would still be saleable.

The assertion by a great many fishers at West Nggela that Cyclone Ida in 1972 is to blame for low *Trochus* abundances, even today, is worthy of comment. Even allowing for exaggeration, the differences between today's catches and the pre-cyclone catches reported by fishers, are very large. The massive habitat destruction caused by the cyclone would obviously have drastically depleted the *Trochus* population at the time. Given a growth rate of up to 10 cm per year for the plate coral, *Acropora hyacinthus* (Veron 1986), it is hard to believe there has not been substantial recovery of suitable shelter for *Trochus* in the ensuing 23 years. This and the abundant evidence presented here, of constant and heavy fishing pressure on *Trochus* at Nggela, lead me to suggest that the latter is likely to be the more significant factor contributing to the impoverished state of *Trochus* populations at West Nggela today. Indeed, the apparent unwillingness of many West Nggela fishers and reef owners to view their own activities as having an influence on the diminished densities of *Trochus* at the moment could be regarded as an obstacle (though not an insurmountable one) to any attempts to improve management of *Trochus* with outside assistance.

Conclusion

In conclusion, it appears that the complexities of the ownership system, in combination with burgeoning economic pressures, have led to very weak management of *Trochus* resources at West Nggela. Johannes (1994) has advocated programmes of information exchange, between fisheries officers and fishers, ideally culminating in a synergy of fisheries science and local knowledge which would facilitate rapid improvements in stock management, and I suspect this approach has considerable potential at West Nggela. However it appears that problems of enforcement are unlikely to be resolved easily. Some improvements may arise through a systematic registration of land and reefs (this is already being attempted in other parts of the Solomons), and perhaps a codification of the rules of tenure and inheritance, though there are still problems with these approaches (Graham, 1994; Turner, 1994) and a discussion of either is beyond the scope of this paper.

References

Adams, TJH, Leqata, J, Ramohia, P, Amos, M, and Lokani, P (1992). Pilot survey of *Trochus* and bêche-de-mer resources in the Western Province of Solomon Islands with options for management. Draft Report, South Pacific Commission, Noumea.

Clarke, RP, and Ianelli, JN (1995). Current paradigms in *Trochus* management and opportunities to broaden perspectives. South Pacific Commission *Trochus* Information Bulletin No. **4**, 3–28.

Graham, T (1994). Flexibility and the codification of traditional fisheries management systems. South Pacific Commission Traditional Marine Resource Management and Knowledge Information Bulletin No. **3**, 2–6.

Johannes, RE (1994). Government-supported, village-based management of marine resources in Vanuatu. . FFA Report # 94/2. Forum Fisheries Agency, Honiara. 27 pp.

Nash, W (1985). Aspects of the biology of *Trochus niloticus* and its fishery in the Great Barrier Reef region. Northern Fisheries Research Centre, Cairns, Australia.

Nash, W (1993). *Trochus*. In 'Nearshore Marine Resources of the South Pacific'. (Eds A Wright and L Hill) pp. 451–96 (Institute of Pacific Studies: Suva).

Nash, W, Tuara, P, Terekia, O, Munro, D, Amos, M, Leqata, J, Mataiti, N, Teopa, M, Whitford, J, and Adams, T (1995). The Aitutaki *Trochus* fishery: A case study. Inshore Fisheries Research Project Technical Document No. 9. (South Pacific Commission:, Noumea, New Caledonia).

Seber, GAF (1982). 'The Estimation of Animal Abundance and Related Parameters.' 2nd edn. (Griffin and Co: London).

Turner, JW (1994). Sea change: Adapting Customary Marine Tenure to commercial fishing. The case of Papua New Guinea. In 'Traditional Marine Tenure and Sustainable Management of Marine Resources in Asia and the Pacific'. (Eds G R South, D Goulet, S Tuqiri, and M Church) pp. 141–154 (International Ocean Institute — South Pacific: Suva).

Veron, J E N (1986). 'Corals of Australia and the Indo-Pacific.' (Angus and Robertson: Australia).

BIOLOGICAL ADVICE ON FISHERY MANAGEMENT IN THE NORTH ATLANTIC 1970–1995

Hans Lassen[A] *and Ralph Halliday*[B]

[A] Danish Institute for Fisheries Research, Charlottenlund Slot, DK 2920 Charlottenlund, Denmark.
[B] Marine Fish Division, Bedford Institute of Oceanography, PO Box 1006, Dartmouth, Nova Scotia, Canada, B2Y 4A2.

Summary

Fisheries management in the North Atlantic is based on assessments of the status of the fish stocks as input to political discussion and decision on appropriate management measures. This process is repeated annually.

The biological advice is provided through international scientific cooperation within the International Council for the Exploration of the Sea (ICES), within the Northwest Atlantic Fisheries Organization (NAFO prior to 1979 International Commission for Northwest Atlantic Fisheries, ICNAF) and after 1976 nationally by coastal states for stocks within the Extended Fishing Zones (Canada and USA).

The paper uses a sample of data for five cod stocks from across the North Atlantic. These stocks have been exploited at high mortality levels and proved susceptible to recruitment failures leading to the collapse of the fisheries. The fishing mortalities are not correlated between stocks and do not depend on two key economic determinants, the prices of the fish and the price of oil, the latter as a proxy for the costs of fishing. This suggests that local mechanisms are important for the development of fishing pressure.

The paper discusses how better results may be obtained in the future. The authors suggest that the institutional framework of fisheries management and a new fisheries science will both be required.

INTRODUCTION

The fisheries in the North Atlantic (FAO areas 21 and 27, see Fig. 1) yield 10–15 million t of fish or about 15% of the world total marine catch of approximately 90 million t (FAO Fisheries Statistics: Catches and Landings 1993). The most important species are the groundfish (demersal species) Atlantic cod, haddock and saithe — called pollock in the Northwest Atlantic — together with the pelagic species Atlantic herring, Atlantic mackerel and capelin. These species account for about 50% of the catch tonnage. The fisheries in the Northeast Atlantic are substantially larger in total than in the Northwest Atlantic and catch trends are rather different in the two areas.

The most important participants in the North Atlantic fisheries are the member states of the European Union (EU) which harvest about 35% of the total yield, including Norway (about 15%), and Iceland, USA and Canada (about 10% each). Other important exploiters are former members of the Soviet Union and the Faeroe Islands.

The North Atlantic states have a long history in the progressive development of international and more recently, domestic institutions also for the management of fisheries. Scientists in these countries have been among the leaders in the development of population dynamics and its applications to the regulation of fishing.

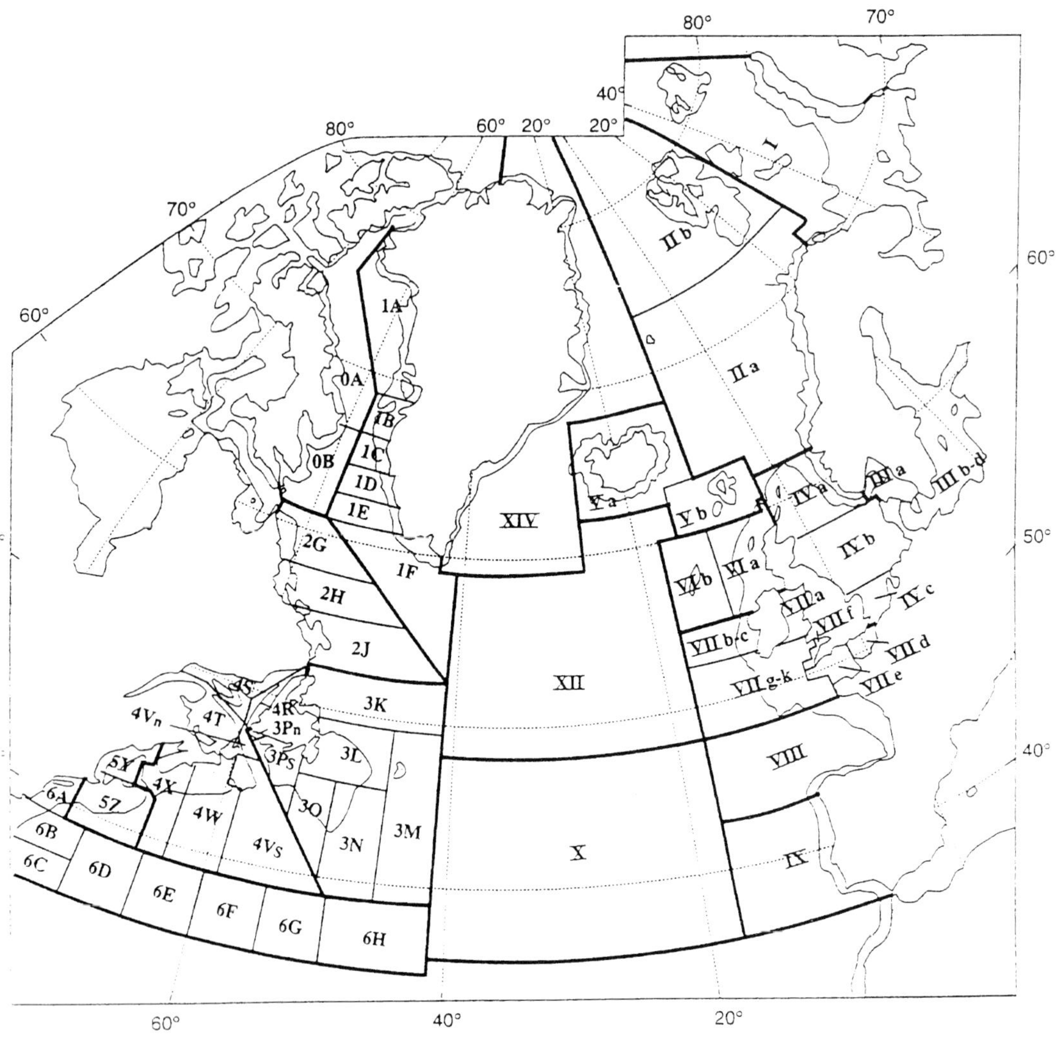

Fig. 1. Map of the North Atlantic showing Statistical Areas as defined by ICES and ICNAF for the Northeast and Northwest Atlantic (east and west of 42°W, respectively). Heavy lines are Subarea and Statistical Area boundaries, light lines are Division and Subdivision boundaries.

The fisheries management model used throughout the North Atlantic is to regulate the fisheries based on biological assessment of the fish stocks. This paper reviews the effectiveness of the advisory process in influencing the exploitation level of fish stocks. The limitations of the North Atlantic fishery management model are identified and potentially more effective approaches are discussed.

Administrative frameworks for management

Two commissions, the Northeast Atlantic Fisheries Commission (NEAFC) in the east, and the International Commission for the Northwest Atlantic Fisheries (ICNAF) in the west, established around 1950 the framework for regulation of North Atlantic Fisheries, many elements of which apply to this day. Their objective was to maximize the physical yield from the resources on an ongoing basis, through rational exploitation. The advice required thus came from biologists and concerned the level of productivity of fish stocks. When limits on the level of exploitation of fish stocks became a necessity in the 1970s, the Commissions were the vehicles used to negotiate sharing arrangements.

Dissatisfaction with the performance of international commissions, in the North Atlantic and elsewhere, and pressure from coastal states with extensive continental shelves, led to widespread acceptance at the Third United Nations Law of the Sea Conference of extensions of fishery jurisdiction up to 200 nautical miles (nm) offshore. By about 1977, all nations adjacent to North Atlantic fishing grounds had declared such limits.

Most of the continental shelves lie within the jurisdictions of the EU member countries, Norway, Faeroe Islands, Iceland, Greenland, Canada and the USA. It is the EU, rather than member states, which is responsible for fishery policy, thus the outer bounds of member state jurisdictional claims define the boundary of the EU regulatory area. In addition, there are expansive areas of ocean still outside national jurisdictions. Two new international conventions were negotiated to regulate fishing in these areas. These established a new NEAFC commission for the Northeast Atlantic, and a North Atlantic Fisheries Organization (NAFO) for the Northwest Atlantic.

It is unfortunate that the new lines on the oceans have little meaning for the fish. A large proportion of the yield from North

Atlantic fish stocks comes from stocks which are not distributed exclusively in one jurisdiction. This is particularly so in the Northeast Atlantic, where Sætersdal (1984) estimated that at least 80% of the long-term yield was expected to come from shared stocks. Although the problem is less severe in the Northwest Atlantic, at least 50% of the overall harvest from the North Atlantic comes from shared resources. Thus, there is still a large international element to North Atlantic fisheries management which has led to negotiations of many bilateral and multilateral agreements, to numerous political conflicts, and even some physical confrontations at sea.

The model for administration of domestic fisheries continued after 1977 to be that of a government agency responsible for decision making. This agency was aided by biological advisory bodies advising on the state of the stocks, and by advisory bodies representing the interests of fishery participants. Finally there was support by an enforcement agency to implement and enforce regulations. The greatest deviations occurred in the US system, where regional councils were established to draw up management plans, and in the EU, where enforcement authority remained at the national level of government. It is outside the scope of this paper to describe the function of domestic systems in detail, but accounts are given for Canada by Parsons (1993); the EU by Holden (1994); Iceland by Arnason (1995); the USA by Finch (1985); Kelly (1978) and Miller *et al.* (1990), and a comparison of all domestic and international systems in the North Atlantic is provided by Halliday and Pinhorn (1996).

From the present perspective, the important point is that biological advice was a central element of the managerial process of all these agencies, both international and domestic, and all made provision to obtain it in a systematic way. Northeast Atlantic management agencies, including the new NEAFC depend almost entirely on the International Council for the Exploration of the Sea (ICES) for biological advice, as did the old NEAFC and the Permanent Commission before it. In the Northwest Atlantic, ICNAF maintained a scientific committee of its own (the Standing Committee on Research and Statistics — STACRES) which died with its parent body, but was replaced by a Scientific Council as part of NAFO. Although the Scientific Council has the scope to advise member states on their domestic stocks if requested, and has done so for some Greenlandic and Canadian stocks, Canadian authorities have preferred to maintain their own scientific advisory system for most stocks that fall entirely within domestic jurisdiction. The USA, which was not a member of NAFO until 1995, of necessity also developed a domestic system. It is the purpose of these scientific bodies to co-ordinate the input of data to provide a thorough review of analyses to ensure their validity, and to develop a consensus of opinion on the status of stocks, and on potential yields, which can be passed on to the recipients of advice.

BIOLOGICAL ADVICE FOR FISHERY MANAGEMENT

The development of fish stock assessment techniques and management approaches has been constantly intermingled. The period up until around 1970 focussed on regulating the size at first capture or in general the exploitation pattern, and the corresponding assessment analysis was yield-per-recruit (Beverton and Holt 1957). The period after 1970 focussed on restricting the removals from the stocks and the corresponding type of analysis was Virtual Population Analysis (VPA), called SPA (Sequential Population Analysis) in the Northwest Atlantic. For a review of these techniques see Megrey (1989). This was originally an analysis of catch data only but was extended to include both catch and survey data. In the late 1980s, on the northwest side of the Atlantic the ADAPT framework (Gavaris 1989) was introduced and accepted as a general approach to VPA. In the Northeast Atlantic area the XSA method (Darby and Flatman 1995), which is modelled after the ADAPT framework, became the standard assessment tool in ICES. The development since 1950 has been summarized in Table 1.

Table 1. Development of management advice since 1950

Period	Theoretical basis	Management objective on which the biological advice was based	Main Emphasis of sampling strategy	Management Advice
1950–1970	Yield/Recruit (Beverton and Holt)	Optimize yield on a long-term basis	Average length/age compositions from stock and fisheries	Regulate mesh size in trawl fisheries
1970–1980	VPA	*Status quo* fishing mortality/Optimal fishing mortality F_{max}, $F_{0.1}$	Annual age compositions from catches supplemented by recruitment indices (and whole stock estimates in the Northwest Atlantic) from R/V surveys	Regulate total removals from the stock, e.g. by setting Total Allowable Catch (TAC) limits annually
1980–1996	Tuned VPA/ADAPT	Maintaining stock within 'safe biological limits', or above MBAL (Minimum Biological Allowable often interpreted as Minimum Spawning Stock Biomass)	Annual age compositions from catches and stocks. Emphasis on better utilization of R/V data. In the Northeast Atlantic, shift from recruitment surveys to surveys covering the entire length range in the population	Unchanged except that TAC advice sometimes supplemented/ replaced by advice on effort reduction
1996–	Unchanged	Precautionary Approach	Unchanged	Unchanged

Biological reference points

In order to provide management advice there was a need for well-defined reference points. These reference points were originally targets, e.g. fishing mortality calculated to *maximize* the yield. More recently reference points have been established as upper limits of fishing mortalities or lower limits on spawning stock biomass (SSB). These reference points were established on *sustainability* considerations. These two approaches must be considered as hierarchical : the *sustainability* of the fish resources must be ensured before objectives concerning output *maximization* can be pursued. This hierarchy is implicit in the precautionary approach (FAO 1995*a*). Application of the precautionary approach to fisheries management would as a minimum require that management should ensure that the direct effects of fishing do not result in the productive potential of stocks being eliminated.

Two examples of biological reference points relating to *sustainability* are presented below. These levels must be considered as limits which should not be exceeded, and not as targets:

- F_{MED} — If fishing mortality is kept at or below F_{MED} the spawning stock biomass (SSB) will on average be kept at or above a level at which historically, the stock has been able to reproduce itself. F_{MED} can be overestimated, and therefore be misleading, in cases where exploitation has been high and the recruitment of the stock low for a large part of the time series used for the calculation of F_{MED} . For a more detailed discussion, see Sissenwine and Shepherd (1987).

- MBAL — Minimum Biological Acceptable Limit. MBAL is defined by the level of a spawning stock below which the probability of poor recruitment increases as spawning stock size decreases. This is built on the hypothesis that recruitment apart from a possible stock-recruitment relationship, is white noise, (see Serchuk and Grainger 1992).

Fish stocks exhibit variation in recruitment between years but also shifts in productivity levels which are longer lasting, e.g. the gadoid outburst in the North Sea (Cushing 1984) or the corresponding event in the Baltic Sea cod in 1975-1984. It may therefore be most appropriate to define reference levels relative to sustainability in terms of the upper limit of the fishing mortality which can be allowed on a fish stock instead of a minimum spawning stock biomass.

The above reference points relate to a stock–recruitment relationship. Such relationships are usually poorly known if they exist at all except for stocks at extremely low SSB levels. There are other approaches to conservation principles for fisheries (see Olver *et al.* 1995).

Two reference fishing mortalities relating to *yield maximization* have often been used in formulating management objectives. These reference levels may be considered as targets provided the fishing mortality is below the level indicated by the reference levels relating to sustainability:

- F_{max} is the fishing mortality corresponding to the maximum yield in a steady state situation (MSY) (Beverton and Holt 1957).

- $F_{0.1}$ is the fishing mortality for which the increase in steady state yield from increasing the fishing mortality is one-tenth of the yield increase when applying fishing mortality to the unfished stock (Gulland and Boerema 1973). $F_{0.1}$ is used because F_{max} often is ill-defined due to the flat-topped nature of the yield curve. $F_{0.1}$ has furthermore been demonstrated to be a reasonable reference point for the exploitation of pelagic stocks, which historically have been subject to collapse when fished at levels above $F_{0.1}$. $F_{0.1}$ has also be adopted, e.g. by Canada and by ICNAF/NAFO, to provide a safeguard against overexploitation due to possible errors of estimation of stock size or to compensate for enforcement deficiencies.

For a detailed discussion of biological reference points, see Caddy and Mahon (1995).

The form of biological advice

Throughout the century various scientific committees have served as advisory bodies to North Atlantic management authorities both international and domestic. However, there has always been interchange of principles and methods as well as of scientific personnel among those scientific committees and thus a substantial degree of standardization in the form of advice given. ICES is the oldest of these advisory bodies and has had the most widespread influence and thus is used here as an example of how the form of advice has evolved.

Until the mid-1970s, ICES had no formalized principles or objectives for providing advice, but the increasing demand for advice on TAC levels then required a more rigorous approach. TAC advice was then based on F_{max} although, if spawning and fishable stock sizes were below acceptable levels, a stepwise rebuilding programme was to be designed (ICES 1977).

In 1978 the Advisory Committee on Fishery Management (ACFM) replaced the previous Liaison Committee. The form of advice remained the same but, as most stocks were fished far above F_{max} or $F_{0.1}$, ACFM was required to recommend gradual reductions in F. Initially, reductions recommended were 10% per year but, as little progress was being made, larger reductions in F were recommended from 1980. There were stocks that were suffering from recruitment failure or which were being fished in excess of the levels indicated by biological reference points. The concept of 'safe biological limits' was introduced at the beginning of the 1980s, see e.g. Hoydal (1983) and the F_{MED} family of reference points was introduced in 1987.

Although management agencies did not provide ICES with clear statements of their objectives, ACFM was nonetheless criticized for assuming responsibilities for the selection of management objectives and for determining the rates at which objectives should be reached. In response ACFM in 1991, formulated its own objective: 'To provide the advice necessary to maintain viable fisheries within sustainable ecosystems' (ACFM 1992). The ICES advisory approach based on the MBAL reference point is described in more detail by Serchuk and Grainger (1992) and is summarized in Table 2.

Table 2. The ICES advisory approach

Assessment category	Form of ACFM advice
SSB at or below the 'minimum biologically acceptable level' (MBAL) or expected to become so in the near future at current levels of fishing mortality rate	Recommendation aiming at rebuilding the stock above MBAL, the severity of the advice depending on the precise state of the stock
SSB above MBAL and not in any immediate danger of falling below that level	A range of options together with short-term impact statements corresponding to those options, but no recommendation
Data inadequate to carry out a qualitative assessment.	Normally a precautionary TAC based on recent catch levels

COD IN THE NORTH ATLANTIC

Cod (*Gadus morhua)* is the most important groundfish in the North Atlantic fisheries. Garrod and Schumacher (1994) reviewed the fisheries and the stock structure of this species in the North Atlantic. These stocks have generally been exposed to increased fishing mortality during the 1970–1995 period. Several cod stocks are under moratoria at present, e.g. Northern Cod (NAFO Divisions 2J+3K-L), Southern Grand Banks Cod (NAFO Division 3N-O) and Eastern Scotian Shelf Cod (NAFO Divisions 4Vs-W) (Fig. 1). The yield of cod has been decreasing throughout the period 1970–1995 both in the Northwest and the Northeast Atlantic.

The present study uses five cod stocks from the North Atlantic to demonstrate trends in stock development. The examples are chosen so that each form of management organization found in the North Atlantic is represented: nationally managed stocks [Eastern Scotian Shelf cod by Canada (Fanning *et al.* 1995), and Icelandic cod by Iceland (ICES 1996*a*)]; managed bilaterally [Northeast Arctic cod — by Norway and Russia (ICES 1996*b*)]; multi-nationally [North Sea cod — by EU and Norway (ICES 1996*c*)] or managed through an international fisheries organization [Southern Grand Bank cod 3N-O by NAFO (Stansbury *et al.* (1995)].

There is a substantial amount of data available for all five stocks. All major fisheries on these stocks supply catch and effort statistics. The fisheries are sampled and the catch-at-age by year matrix is available for the entire period which we consider. There are for all five stocks well-established bottom trawl survey series available. There has also been invested a considerable amount of research on understanding the biology of cod, see e.g. the ICES Cod and Climate Symposium 1992, Reykjavik. Full analytical assessments are available for all five stocks and are considered to be fairly reliable.

All five stocks have been managed with the objective, among several, to keep fishing mortality at a constant level. This level varies among the stocks between the biological reference points $F_{0.1}$, F_{max}, and maintaining the fishing mortality constant at a higher level. All stocks are or have recently been at a very low biomass level; Eastern Scotian Shelf cod and Southern Bank cod are currently under moratoria because of very low spawning stock biomass estimates, Icelandic cod was at a low level around 1990, Northeast Arctic cod in the first half of the 1980s, and the North Sea cod is presently at the lowest level seen in the available time series (since the mid-1960s).

Table 3 shows the annual average yield for 1970–1994, and Figure 2 shows the fishing mortality for the fully exploited age groups for this period. These levels should be related to a biological reference level, e.g. F_{MED}. This is done in Table 3 where the average 1980–1992 fishing mortalities for the fully-exploited age groups are compared to three biological reference points $F_{0.1}$, F_{max} and F_{MED}. The 1980–1992 time period was chosen because of the introduction of the 200 miles EEZ towards the end of the 1970s. The actual fishing mortalities were well above $F_{0.1}$ and F_{max} and at or even above the F_{MED} reference point for all stocks. Hence the resulting fishing mortalities are far from what was intended, and furthermore two of the stocks have actually collapsed.

The quality of the assessments is often judged based on "retrospective" analysis. This analysis is a comparison of the population and mortality estimates of the more recent years based on truncated time series, e.g. truncate the data series at 1990 to estimate the fishing mortality in 1990, then truncate it in 1991 to again estimate the fishing mortality for 1990, etc. This analysis is limited in as much as misapprehensions of the biology at the time of the assessment, but later remedied, will not be included. Figure 3 shows retrospective plots of the estimated fishing mortalities. The general conclusion is that the assessments are subject to substantial uncertainty. There are quite large differences between the five stocks represented, with both over- and under-estimates. The pattern suggests particular problems with the Northeast Arctic and the Eastern Scotian Shelf Cod.

Table 3. Average annual yield (1970–1995) and comparison of the average fishing mortalities for fully exploited age-groups 1980–1992 with three biological reference points. These biological reference points were recalculated by the present authors based on data available in the assessment documents

Stock	Average (1970–94) annual yield '000 t	Fully-exploited Age groups	Average F (1980–1992)	$F_{0.1}$	F_{max}	F_{MED}
North Sea Cod (IV)	198	2–8	0.88	0.14	0.24	0.81
Northeast Arctic Cod (I and II)	559	5–10	0.69	0.11	0.21	0.46
Icelandic Cod (Va)	361	5–10	0.77	0.21	0.37	0.45
Southern Bank Cod (3N-O)	42	5–10	0.42	0.10	0.16	0.40
Eastern Scotian Shelf Cod (4Vs-W)	40	7–9	0.75	0.10	0.15	0.41

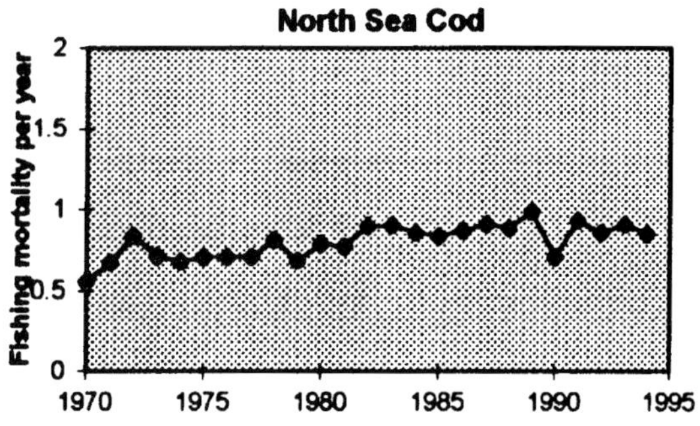

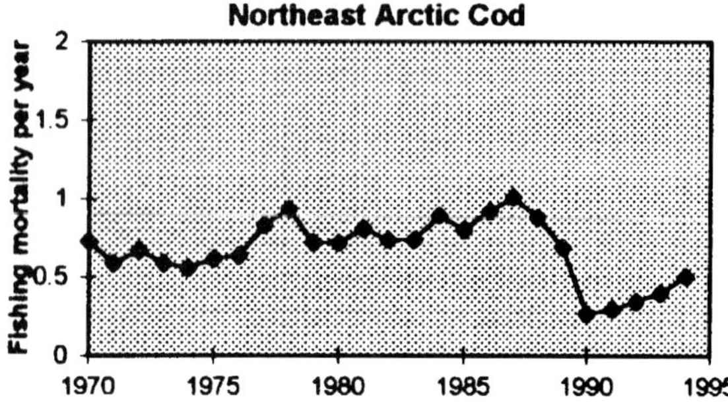

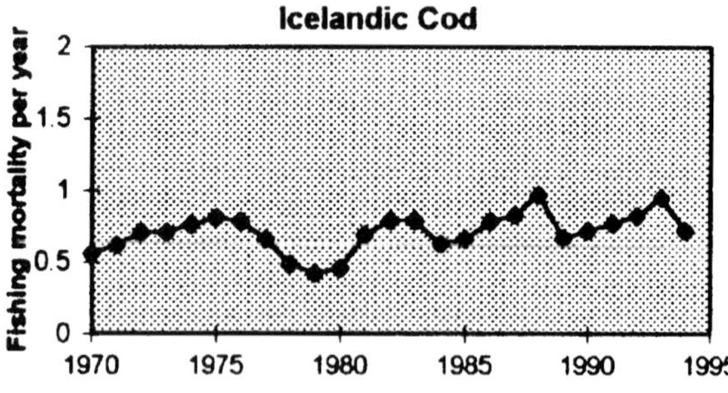

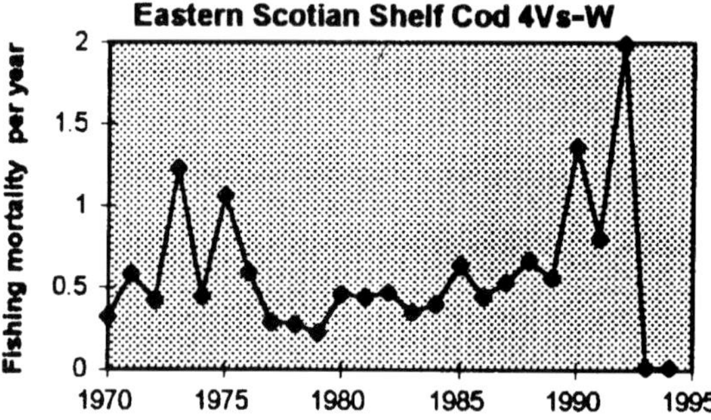

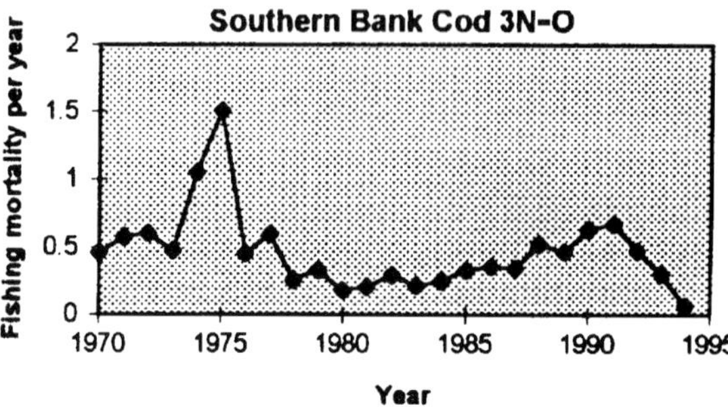

Fig. 2. Fishing mortalities for fully recruited age groups for five cod stocks from the North Atlantic.

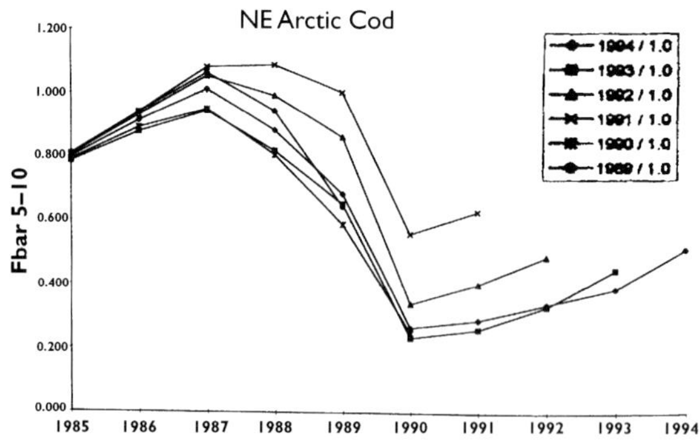

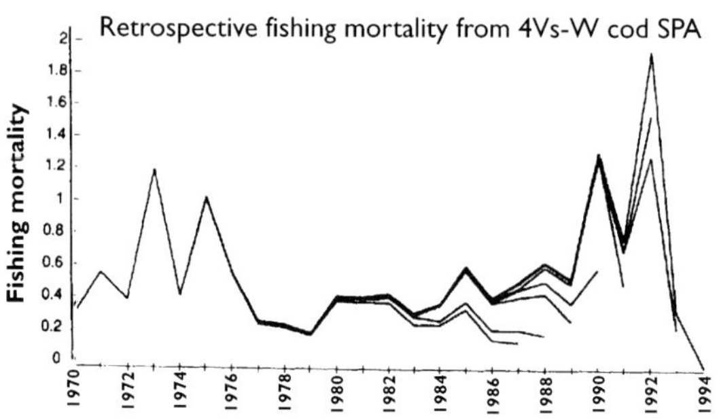

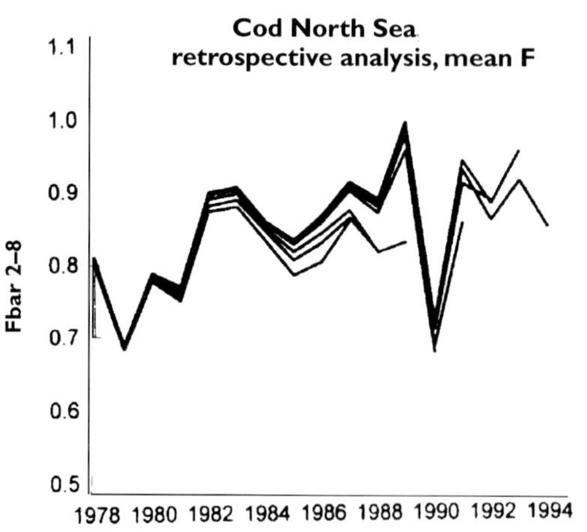

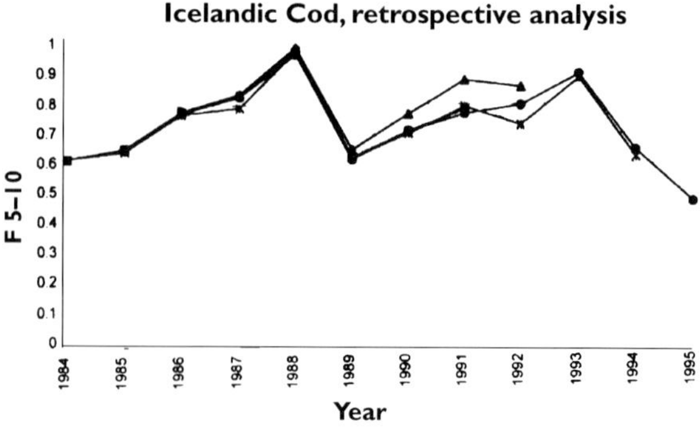

Fig. 3. Retrospective pattern for mean fishing mortality for fully recruited age groups for five cod stocks from the North Atlantic. After ICES (1996*a–c*) with kind permission of the General Secretary of ICES, and Fanning *et al.* (1995).

The biological advice has been the same for decades: fish stocks are exploited above biological reference points related to optimal use of the growth potential of the resource, i.e. MSY, F_{max}, $F_{0.1}$. For example ICES (1979) stated for the roundfish in the North Sea "... that these stocks (cod, haddock and whiting) are severely overexploited, that fishing effort should be considerably reduced and effective action taken to improve the exploitation pattern....". And again in 1979 a 20% reduction in fishing mortality were recommended (ICES 1980). Advisory documents are full of warnings of the possibility of a stock collapse if a series of bad year-classes should occur. The biological assessment shows that the fisheries are susceptible to recruitment failure and that the effects will be immediate because the fisheries depend on 1–2 year-classes only.

The biological advisory system has however, not provided predictions of stock collapses, i.e. recruitment failures. The biological advice traces decline in SSB but is unable to establish when recruitment fails.

What controls the level of fishing mortality?

To better achieve biologically-based management objectives, understanding of the fisheries system as a whole is required. Such better understanding allows deliberate changes in the overall system which lead to a more rational exploitation of the stocks.

Fishing is an economic enterprise and, as such, should react to changes in the economic conditions for the fishery, i.e. fish prices and costs of fishing (Hanneson 1993). The traditional analysis of an open access fishery will only be partly valid since, from the early 1950s, there have been deliberate management efforts to control fishing mortality, first through technical regulation of the fisheries and later by introducing TACs and quotas. In spite of these management efforts the fishing mortality for each of the five North Atlantic cod stocks examined has been, or is, at a very high level.

The development of the fishing mortalities in the five stocks (Table 4) are not well correlated. The multiple correlation coefficients for the fishing mortalities for the fully-exploited age groups for the period 1970–1992 (Table 5) suggest that the

Table 4. Annual mean fishing mortalities by stock and annual indices of fish product prices (Fresh, chilled or frozen) and oil prices together with the OECD GNP index with 1980 = 1

Year	Fishing mortalities					North America import prices $US/kg year prices	Crude oil prices ($US) per barrel) (Light Dubai year prices)	OECD GNP index
	North East Arctic	Icelandic	North Sea	NAFO 3N-O	NAFO 4Vs-W			(1980 = 1)
1970	0.73	0.56	0.56	0.47	0.33	0.62		0.37
1971	0.59	0.62	0.67	0.58	0.59	0.68		0.40
1972	0.67	0.71	0.84	0.60	0.42	0.78	1.90	0.44
1973	0.59	0.71	0.71	0.48	1.23	0.90	2.83	0.50
1974	0.56	0.76	0.68	1.05	0.45	0.93	10.41	0.55
1975	0.62	0.81	0.71	1.51	1.07	0.96	10.70	0.62
1976	0.64	0.78	0.71	0.46	0.60	1.16	11.63	0.65
1977	0.83	0.66	0.71	0.60	0.29	1.38	12.38	0.71
1978	0.93	0.49	0.81	0.26	0.28	1.43	13.03	0.83
1979	0.72	0.43	0.69	0.33	0.23	1.58	29.75	0.92
1980	0.72	0.46	0.79	0.18	0.46	1.71	35.69	1.00
1981	0.81	0.69	0.77	0.21	0.44	1.86	34.32	0.99
1982	0.74	0.79	0.90	0.29	0.47	1.88	31.80	0.98
1983	0.74	0.79	0.91	0.22	0.35	1.85	28.78	0.98
1984	0.89	0.63	0.86	0.26	0.40	1.96	28.07	0.98
1985	0.8	0.66	0.83	0.33	0.64	2.00	27.63	1.00
1986	0.92	0.78	0.87	0.36	0.45	2.27	12.97	1.19
1987	1.01	0.83	0.91	0.35	0.54	2.64	16.92	1.38
1988	0.88	0.97	0.89	0.53	0.67	2.77	13.22	1.48
1989	0.69	0.67	0.99	0.47	0.56	2.84	15.69	1.49
1990	0.27	0.71	0.71	0.63	1.37	3.09	20.50	1.63
1991	0.3	0.77	0.93	0.68	0.81	3.23	16.56	1.69
1992	0.35	0.8	0.86	0.48	1.99	3.17	17.21	1.79
1993	0.4	0.9	0.91	0.31	0.02	2.97	14.90	1.79
1994	0.51	0.64	0.85	0.07	0.02		14.76	
1995		0.47					16.09	

Table 5. Multiple correlation coefficients of the fishing mortalities for five cod stocks in the North Atlantic for the period 1970-1992

Stock	Southern Bank Cod (3N-O)	Eastern Scotian Shelf Cod 4Vs-W	Northeast Arctic Cod (I and II)	Icelandic Cod (Va)	North Sea Cod (IV)
Southern Bank Cod (3N-O)	1				
Eastern Scotian Shelf Cod (4Vs-W)	0.31	1			
Northeast Arctic Cod (I and II)	–0.42	-0.68	1		
Icelandic Cod (Va)	0.33	0.41	-0.12	1	
North Sea Cod (IV)	–0.31	0.04	0.17	0.40	1

general driving forces have at least different weight in the different fisheries.

Indices of fish prices and fishing costs pertinent to North Atlantic fisheries as a whole were developed to see if the effect of general economic conditions could be detected in mortality rate trends. As an indicator of fish prices we use the annual average prices for import to the North American market of fish: fresh, chilled or frozen (FAO Fisheries Statistics: Commercial values and products 1975, 1983, 1993). These prices have been deflated by the ratio of OECD-Gross National Products (GNP) in fixed 1980 prices and actual prices ($US). It is difficult to find a simple economic indicator which reflects the actual costs of fishing. Oil is the single most dominating variable-cost factor for a fishing vessel, accounting for approximately 50% of total costs of running a fishing vessel. Labour costs are in most North Atlantic fisheries based on a system of sharing the income from the landings. Here we have chosen, as a first proxy, spot crude oil prices. Spot crude prices are available from 1972 for Arabian light/Dubai oil (BP Statistical Review of World Energy 1996). As for the prices on fish we have deflated the oil prices using the ratio of OECD-GNPs in fixed 1980 prices and running year prices ($US).

The relationship between fishing mortality and prices ($US 1980 level) was investigated using multiple regression analysis with the coefficients " and $:

Fishing mortality = intercept + " *oil price + $ *fish price

However, generally for the five cod stocks we did not find this relationship to be statistically significant at the 5% level (Table 6). This suggests that mechanisms other than general economic factors are of importance in understanding the level of fishing mortality and how this develops.

In the Northeast Atlantic it will be fish product prices in an European currency which may be relevant. This price level was high in the mid 1980s because of the currency exchange rate and may during that period have fuelled the tendency to increase the fishing mortalities. Furthermore, the actual costs seen by the fishermen are influenced by local conditions. There are numerous cases throughout the North Atlantic of government subsidies to the fishing industry through, for example, exemption of fuel oil from taxes, grants and cheap loans for fleet renewal, and income support. In any case, it appears that local factors predominate the development of fishing mortality trends, rather than general economic conditions.

THE FUTURE OF BIOLOGICAL ADVICE FOR FISHERIES MANAGEMENT

The above analysis indicates that the fish stocks in the North Atlantic are neither effectively utilized nor exploited on a sustainable basis. The consequences are fishery collapses, economic hardship among participants, and social disruption in their communities. The fishing industry world-wide, and specifically in the North Atlantic, have gone through a number of crises (FAO 1995*b*). The collapse of the Faeroese fishing industry around 1990 (Mørkøre 1991) and the Breton crisis (Delbos and Prémel 1996) may serve as examples.

Robust mechanisms for the development and delivery of biological advice on management of fish stocks have long been in place in the North Atlantic, and much of this advice has been accepted by management agencies and translated into, sometimes very complex, regulatory frameworks for the control of fishing. Although present examples show that regulatory actions occasionally succeed in reducing fishing mortality, this occurs only when recruitment failure triggers an economic and social crisis which demands drastic action at a political level; fishing mortality was not reduced (or prevented from increasing) through effective, ongoing, regulation.

It is not likely that some increase in the accuracy of the biological advice provided, or a simple change in management tactics, would remedy this situation. The fundamental problem is over-investment in catching capacity stemming from the common property status of the resource (Gordon 1954). This motivates

Table 6. Multiple regression analysis of the mean average fishing mortality for fully-exploited age-groups (see Table 1) with oil and fish prices in $US 1980 level as independent explanatory variables

Stock	Probability of H_0: model not significant (F-test)	Multiple *r*-squared	Intercept	Coefficient oil price	Coefficient fish price	Residual standard error
Southern Bank Cod (3N-O)	0.027	0.32	2.77	–0.0088	–1.1642	0.26
Eastern Scotian Shelf Cod (4Vs-W)	0.294	0.12	0.92	–0.0151	-0.0101	0.12
Northeast Arctic Cod (I and II)	0.159	0.18	–0.30	0.0059	0.4797	0.20
Icelandic Cod (Va)	0.058	0.26	0.60	–0.0066	0.1303	0.12
North Sea Cod (IV)	0.087	0.23	0.23	–0.0022	0.3453	0.086

fishermen to maximize present yields to pay today's bills, and places them in fundamental disagreement with the principle underlying fish stock management, which is to forego yield now to reap greater yields (or ensure continuing yields) in future. As the role of fishermen in the standard North Atlantic management model is that of lobbyists, it is not surprising that the history of management is one of conflict between fishermen's representatives and fisheries bureaucracies, with hapless biologists being disparaged by both sides. The lack of acceptance of regulations by fishermen leads to a high level of violations, and enforcement has proven generally ineffective in deterring such illegal behaviour. For example, an estimated 30–40 000 t of cod per year were removed from Flemish Cap (in the NAFO Regulatory Area) during a moratorium on fishing in the late 1980s (NAFO Redbooks 1990–91).

It is a truism that securing the productivity of the fish stocks is a prerequisite for the long-term well-being of a fishing industry. However, fishing is an economic activity, and the objective of fishery management must surely be to encourage an economically-viable fishing industry which makes a net contribution to the well-being of society. The traditional management model, with is preoccupation with the conservation of fish stocks and its heavy dependence on biological advice, neglects the management of fisheries. Indeed, the failure to view fishery management as fundamentally an economic and social issue might well be considered the root of the problems which made conservation objectives virtually impossible to achieve. Thus, the management model must change. Mechanisms which introduce quasi-property rights in fisheries, such as Individual Transferable Quotas (ITQs) and others which delegate management responsibilities to participants, offer ways through which the fundamental motivations driving investment decisions by fishermen may be altered to be consonant with conservation. In the context of this paper, however, the question is what this diagnosis implies for the future role of biological advice.

Emphasis should be placed by biologists on the precautionary principle and on the sustainability of stocks. The ICES ACFM (Serchuk and Grainger 1992) has established, in their form of advice, the concept of the Minimum Biological Acceptable Limit formulated in spawning stock biomass units. This is a precautionary measure which guards against the possibility that spawning stock might be reduced to a point where recruitment is adversely affected. This is a sensible, even if arbitrary, approach but we propose that such a limit be formulated as a restriction of the mortality allowed because of inherent variability in fish stock productivity. A key issue is how to institutionalize such an absolute upper limit. We advocate that this be done in the legal frameworks for fisheries management. Such limits would be based on sustainability considerations, e.g. F_{MED}. The precautionary principle should be used to establish limits where there are no assessments - this would lead to such stocks being only very lightly exploited. An example in the north Atlantic, where this at present may be appropriate, is the expanding fisheries for deep-water species.

Missing from biological advice at the moment is any realistic prognosis of resource productivity on a time-scale longer than one or two years. There is substantial variability in the productivity of the fish stocks with longer frequencies than yearly as used in most simulation models. This variability is best documented for pelagic fish stocks, see e.g. Kawasaki and Omori (1988); Baumgartner *et al.* (1992) but is also seen in cod (Cushing 1984; Buch and Hansen 1988; Larreñeta 1988). Also, flatfish seem to show a similar variability in productivity, e.g. American plaice (Bowering *et al.* in press; Bowering, Morgan and Brodie unpublished data). It is however not only the recruitment which varies in a fish stock; growth shows some variability, see Brander (1995) for a discussion of growth variability of Atlantic cod. We are however, constrained in our efforts since we only have 20–30 years of data on which to base the biological advice. To include the desired longer time span in our considerations we must rely on theoretical considerations on how best to extrapolate the existing data to this longer time perspective. These models have mostly been built on equilibrium considerations not taking the inherent variability in stock productivity into account. Most importantly, it is essential that research be conducted which broadens our understanding of these longer-term variations in fish production and which may result in some predictive capability.

Other than defining the bounds of sustainable harvest levels in short- and longer-term contexts, it can be asked what else biologists have to offer to the management of fisheries. Management measures are political answers to technical, economic and social questions. The relevant analyses and advice should come from practitioners of these disciplines, not form biologists. The biologists will have to step down from their position as sole scientific advisors on fisheries management to make more room for socio-economic anaylsis. However, it would equally be a mistake to push them aside completely. There is a role that biologists must play in the integration of their analyses in broader models of fisheries systems, and they have an essential contribution to make to the debate on strategies for compatibility in the management of fish stocks and fisheries.

REFERENCES

Arnason, R (1995). The Icelandic fisheries: evolution and management of a fishing industry. (Fishing News Books: Oxford). 177 pp.

Baumgartner, T R, Soutar, A, and Ferreira-Bartrina, V (1992). Reconstruction of the history of Pacific Sardine and Northern Anchovy Populations over the past two millennia from sediments of the Santa Barbara Basin, California. *Cooperative Oceanic Fishery Investigations CALCOFI Report* 33, 24–40.

Beverton, R J H, and Holt, S J (1957). On the Dynamics of Exploited Fish Populations. Ministry of Agriculture, Fisheries and Food Fishery Investigations Series II, Volume XIX.

Bowering, W R, Brodie, W B, and Morgan, M J (in press). Changes in the Abundances and certain population parameters of American Plaice (*Hippoglossoides platessoides*) on St. Pierre Bank (NAFO Subdivision 3Ps) during 1972–94 with implications for fisheries management. *North American Journal of Fisheries Management.*

Brander, K M (1995). The effect of temperature on growth of Atlantic Cod (*Gadus Morhua L.*). *ICES Journal of Marine Science* V52, N1 (Feb), 1–10.

Buch, E, and Hansen, H H (1988). Climate and Cod Fishery at West Greenland. In 'Long term changes in Marine Fish Populations'. (Eds T Wyatt and M G Larrañeta) (Vigo).

Caddy, J F, and Mahon, R (1995). Reference points for Fishery management. *FAO Technical Paper* No. 347.

Cushing, D (1984). The Gadoid outburst in the North Sea. *Journal du Conseil International pour l'Exploration de la Mer* **41**(2), 159–66.

Darby, C D, and Flatman, S (1995). Manual on XSA. MAFF, Lowestoft, UK.

Delbos, G, and Prémel, G (1996). The Breton Fishing Crisis in the 1900s: Local Society in the Throes of Enforced Change. In 'Fisheries Management in Crisis'. (Eds K Crean and D Symes) (Fishing News Books: Oxford).

Fanning, L P, Mohn, R K, and MacEachern, W J (1995). DFO Atlantic Fisheries Research Document 95/73.

FAO (1995*a*). Precautionary Approach to fisheries. Part 1: Guidelines on the precautionary approach to capture fisheries and species introductions. *FAO Technical Paper* No. 350 Part 1. Rome. FAO. 52 pp.

FAO (1995*b*). Review of the State of the World Fishery Resources: Marine fisheries. *FAO Fisheries Circular* No 884 Rome FAO. 105 pp.

Finch, R (1985). Fishery management under the Magnuson Act. *Marine Policy* **9**, 170–9.

Garrod, D J, and Schumacher, A (1994). North Atlantic Cod: The broad canvas. In 'Cod and Climate Change Proceedings of a symposium'. (Eds J Jakobsson, O S Astthorsson, R J H Beverton, B Bjoernsson, N Daan, KT Frank, J Meinke, B Rothschild, S Sundby and S Tilseth) pp. 59–76. International Council Exploration of the Sea Vol. 198.

Gavaris, S (1989). An adaptive framework for the estimation of population size. CAFSAC Research Document 88/29. 12 pp.

Gordon, H S (1954). The economic theory of a common property resource: the fishery. *Journal of Political Economy* **62**, 124–42.

Gulland, J A, and Boerema, L K (1973). Scientific advice on catch levels. *Fishery Bulletin* **71**(2), 325–35.

Halliday, R G, and Pinhorn, A T (1996). North Atlantic Fishery Management Systems: A comparison of Management Methods and Resource Trends. *Journal of Northwest Atlantic Fisheries Sciences* **20**, 9–135.

Hannesson, R (1993). Bioeconomic Analysis of Fisheries. FAO (Fishing News Books: Oxford).

Holden, M (1994). The Common Fisheries Policy: Origin, Evaluation and Future. (Fishing News Books: Oxford).

Hoydal, K (1983). ICES procedures in formulating management advice. FAO FIIP/R289 suppl 3 No. 289(3), 215–30.

ICES (1977). Form of Advice provided by the ACFM of ICES. *Cooperative Research Report* No. 62 (ICES: Copenhagen).

ICES (1979). Reports of the ICES Advisory Committee on Fishery Management. *Cooperative Research Report* No 85. ICES February 1979, Copenhagen.

ICES (1980). Reports of the ICES Advisory Committee on Fishery Management. *Cooperative Research Report* No 93. ICES February 1980. Copenhagen.

ICES (1996*a*). Report of the North Western Working Group. ICES CM 1996/Assess:15.

ICES (1996*b*). Report of the Arctic Fisheries Working Group. ICES CM 1996/Assess:4.

ICES (1996*c*). Report of the Working Group on the Assessment of Demersal Stocks in the North Sea and Skagerrak. ICES CM 1996/Assess:6.

Kawasaki, T, and Omori, M (1988). Fluctuations in three major sardine stocks in the Pacific and the global trend in temperature. In 'Long term changes in Marine Fish Populations'. (Eds T Wyatt and M G Larrañeta) (Vigo 1988).

Kelly, J E (1978). The Fishery Conservation and Management Act of 1976. *Marine Policy* **2**, 30–6.

Larrañeta, M G (1988). Fish Recruitment and environment. In 'Long term changes in Marine Fish Populations'. (Eds T Wyatt and M G Larrañeta) (Vigo 1988).

Megrey, B A (1989). Review and comparison of age-structured stock assessment models from theoretical and applied points of view. *American Fisheries Society, Symposium* No. 6, 8–48.

Miller, M M, Hooker, P J, and Fricke, P H (1990). Impressions of ocean fisheries management under the Magnuson Act. *Ocean Development and International Law* **21**, 263–87.

Mørkøre, J (1991). Et korporativt forvaltningsregimes sammenbrud — erfaringer fra det færøske fiskeri i nationalt farvand. In 'Fiskerireguleringer': Nordiske Seminar og Arbejdsrapporter No. 516, Nordisk Ministerråd, Copenhagen (in Danish) (The collapse of a cooporative management system — experiences from the Faroese fishery in domestic waters).

NAFO (1995). NAFO Redbooks. NAFO, NAFO Headquarter, Wyse Road 192, Dartmouth, NS, Canada.

Olver, C H, Shuter, B J, and Minns, C K (1995). Towards a definition of conservation principles for fisheries management. *Canadian Journal of Fisheries Aquatic Sciences* **52**, 184–94.

Parsons, L S (1993). Management of marine fisheries in Canada. *Canadian Bulletin of Fisheries and Aquatic Sciences* 2225. 784 pp.

Sætersdal, G (1984). Shared stocks and fishery management in the Northwest Atlantic under the EEZ regime. In 'OECD. Experiences in the management of national fishing zones'. pp. 123–32 (Organization for Economic Cooperation and Development: Paris).

Serchuk, F, and Grainger, R (1992). Form of ACFM Advice. ICES CM; 1992/Assess:20.

Sissenwine, M P, and Shepherd, J G (1987). An alternative perspective on recruitment overfishing and biological reference points. *Canadian Journal of Fisheries and Aquatic Sciences* **44**(4), 913–18.

Stansbury, D, Bishop, C A, Murphy, E F, and Davis, M B (1995). An assessment of the Cod Stock in NAFO Div. 3NO. NAFO SCR Document 95/70 Serial No. N2585.

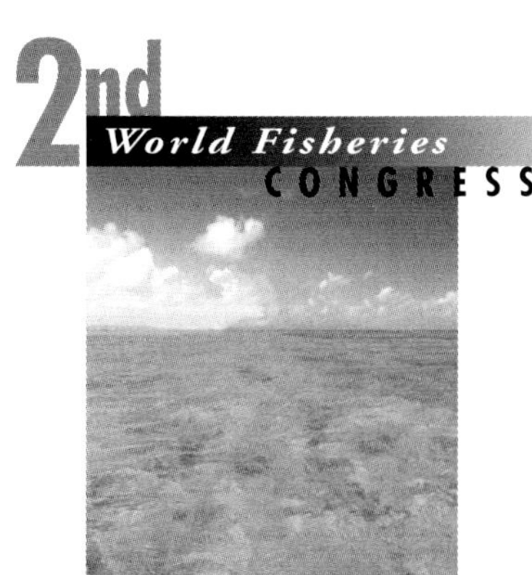

A Comparison of the Benefits and Costs of Quota *versus* Effort-based Fisheries Management*

R. O'Boyle and K. C. T. Zwanenburg

Marine Fish Division, Bedford Institute of Oceanography, PO Box 1006, Dartmouth, Nova Scotia, Canada, B2Y 4A2.

Summary

Prior to 1970, the regulation of Canada's East Coast fisheries was achieved through effort controls but since then, catch controls or quotas have become the backbone of fisheries management. The recent failure of the Atlantic coast fishery is being blamed by some on deficiencies in quota management; and effort controls, which appear to have been successfully implemented in some fisheries, are being proposed as an alternative. In this paper, we define the elements of quota- and effort-based management, and examine how these have been implemented in six fisheries off Nova Scotia: groundfish (multi-species, quota-based), herring (single species, quota-based), offshore scallop and lobster (single species, quota-based), and inshore scallop and lobster (single species, effort-based). For each fishery, the costs of management are evaluated in relation to the benefits, the latter being defined by attainment of some conservation objective. While two of the four quota-based fisheries did not attain the conservation objective, it was not possible to judge the success of effort-based management, as the two fisheries considered had no stated objective. Management costs for all six fisheries were comparable and less than ten percent of the fisheries' processed value, indicating that any of the management system examined could be cost effective. There were, however, differences in how these costs were distributed, with the successful systems having a slightly higher proportion of enforcement costs. As quota management has been effectively implemented in some fisheries, it is not possible, or even desirable, to choose categorically between the two management approaches. Rather, it is essential to ensure that whatever system is implemented has the appropriate balance of clearly stated objectives, strategies, regulations, enforcement and assessment. Recommendations are presented which may improve the implementation of groundfish fisheries management off the coast of Nova Scotia.

Introduction

Prior to 1970, the groundfish fisheries of Canada's Atlantic coast were regulated by effort controls. Difficulties in controlling fishing effort led to the implementation of quotas in 1970 and since then, these have been used as the principal means of controlling harvesting. The collapse of the groundfish fisheries in the early 1990s has been blamed by some on deficiencies in catch or quota management and has prompted consideration of an effort-based system, similar to that used in the current, relatively successful, lobster fishery. There is little analysis to support this contention and, as the pre-1970 experience in Canada has shown, effort-based management is not without its problems. Each management approach is likely to have its strengths and weaknesses and is best suited to the particular circumstances of each fishery. It is perhaps more appropriate to ask how much each system costs, relative to the benefits that it generates.

In Nova Scotia, quotas are used by the Canadian Department of Fisheries and Oceans (DFO) in the management of the groundfish, herring, offshore lobster and offshore scallop fisheries

* An earlier version of this paper was presented at the 1995 Second Groundfish Workshop of the Scotia-Fundy Region, Department of Fisheries and Oceans, and was published in the Workshop Proceedings (Burke *et al.* 1996).

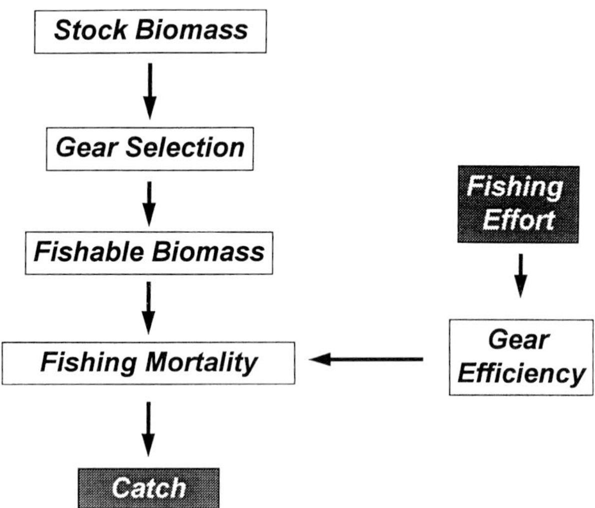

Fig. 1. Conceptual model of a fishery.

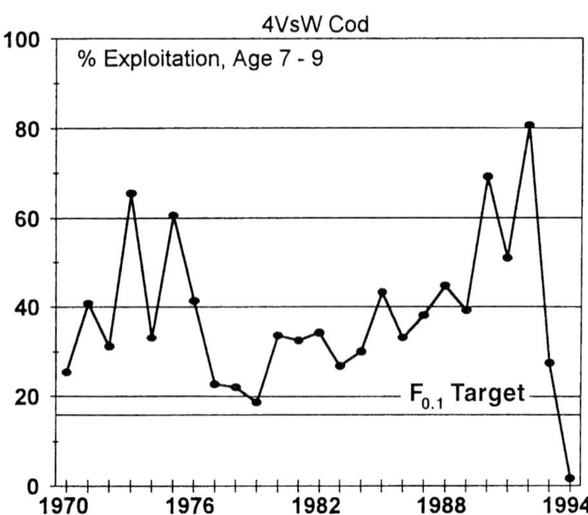

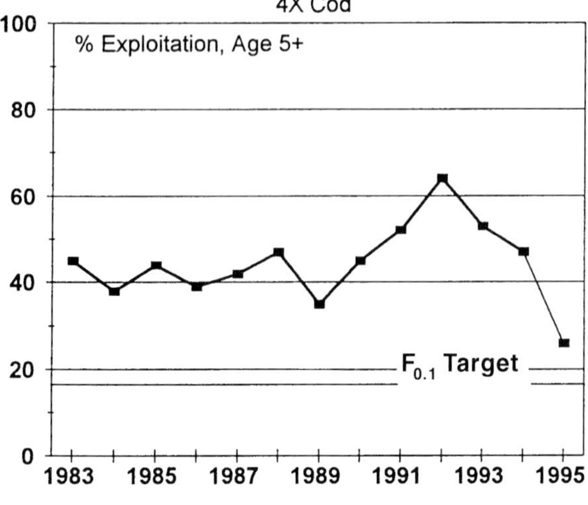

Fig. 2. Trends in exploitation rates for two Scotian Shelf cod stocks.

and are not used in the inshore scallop and lobster fisheries. This situation has allowed us to undertake a comparative analysis of the utility of catch *versus* effort controls. These fisheries are first examined to determine the degree to which they can be classified as being catch- or effort-controlled. Then, a comparison is made of their success in achieving stated goals, relative to the cost of management. The distribution of these costs among the various elements of each system is examined to determine the key features that lead to management success. Based on this, recommendations are made to improve the management of the groundfish resources off Nova Scotia.

A DEFINITION OF CATCH- AND EFFORT-BASED MANAGEMENT

In order to classify a fishery as being catch- or effort-controlled, it is essential to define what constitutes each. Of the total stock biomass, only a fraction, the fishable biomass, can be harvested, due to the regulation of a gear's selectivity properties (Fig. 1). These are intended to limit exploitation on particular life history stages, such as juveniles or spawning adults. They include regulations on mesh, hook and vent sizes, restrictions on fish size retained, and closed areas and/or seasons.

Fishing effort takes a fraction of the fishable biomass. This is termed the fishing mortality (F). Fishing mortality can be regulated either directly through limiting catch, or indirectly through limiting fishing effort. It is indirect in the case of effort controls because the efficiency of fishing must also be regulated. Regulations on gear efficiency involve restrictions on the size of boats and/or gear. Catch regulations involve gear selection and catch controls, while effort regulations involve gear selection, effort and gear efficiency controls (Table1).

The regulations used in the 1995 groundfish, herring, inshore and offshore lobster, and inshore and offshore scallop fisheries off Nova Scotia were classified as per Table 1 based on interviews with the lead biologist for each fishery (Table 2). As expected, almost all fisheries regulate gear selection. The

Table 1. Regulations employed in catch- and effort-controlled fisheries

Catch Regulations	Effort Regulations
Gear Selection	Gear Selection
• mesh/hook/vent size	• mesh/hook/vent size
• animal size/berried females (lobsters)	• animal size/berried females (lobsters)
• nursing/spawning areas	• nursery/spawning areas
Catch	Effort
• TAC	• number of licences
• ITQ/EA	• gear quantity
• Dockside monitoring	• area closures
• Discard/bycatch	• season closures
	• trip limits
	Gear Efficiency
	• boat size
	• gear size

Table 2. Classification of the Scotian Shelf fisheries. Dashes indicate presence of regulation

System	Regulation	Groundfish	Herring	Offshore Scallop	Offshore Lobster	Inshore Scallop	Inshore Scallop
	Mesh/Hook/Vent Size	–			–	–	–
	Fish Size			–	–	–	–
	Nursery/Spawning Area	–			–		–
Catch							
	TAC	–	–	–	–		
	ITQ	–	–	–	–		
	DMP	–	–	–	–		
	Discard/Bycatch	–					
	Mesh/Hook/Vent Size	–			–	–	–
	Fish Size			–	–	–	–
	Nursery/Spawning Area	–			–		–
Effort	No of Licences	–	–	–	–	–	–
	Gear Quantity						–
	Area Closure	–			–	–	–
	Season Closure	–			–	–	–
	Trip Limits	–					
	Boat Size	–		–		–	–
	Gear Size				–		–

groundfish, herring, offshore scallop and offshore lobster fisheries all regulate catch and are classified as catch-controlled fisheries. However, these fisheries also employ some regulations of effort. The inshore scallop and lobster fisheries employ regulations of effort but do not directly control catch, and are thus classified as effort-controlled fisheries. In practice, therefore, catch-controlled fisheries employ a combination of catch and effort regulations while effort-controlled fisheries employ only effort regulations.

MANAGEMENT SUCCESS UNDER CATCH- AND EFFORT-CONTROL

We wish to determine whether or not management using either catch or effort controls has been successful in achieving the stated objectives of the fishery. Since 1977, the multi-species groundfish fishery has been managed to limit fishing mortality on each stock to the level of $F_{0.1}$ ($F_{0.1}$ is the level of fishing mortality for which the marginal increase in yield-per-recruit due to a small increase in fishing mortality is 1/10th of the marginal increase in yield at very low levels of fishing mortality). This management objective was designed to not only conserve the resource but also to ensure high catch rates and thus generate economic benefits (Gulland and Boerema 1973). However, fishing mortality has consistently exceeded the $F_{0.1}$ target (Fig. 2), often by a considerable amount. Thus catch-controlled groundfish fisheries have been unable to attain their stated conservation objective.

There has been some speculation that catch controls cannot be effectively employed in multi-species fisheries like groundfish on the Scotian Shelf. However, even the single-species, catch-controlled, herring fishery has not limited fishing mortality to the stated target of $F_{0.1}$ (Fig. 3; Anon. 1996).

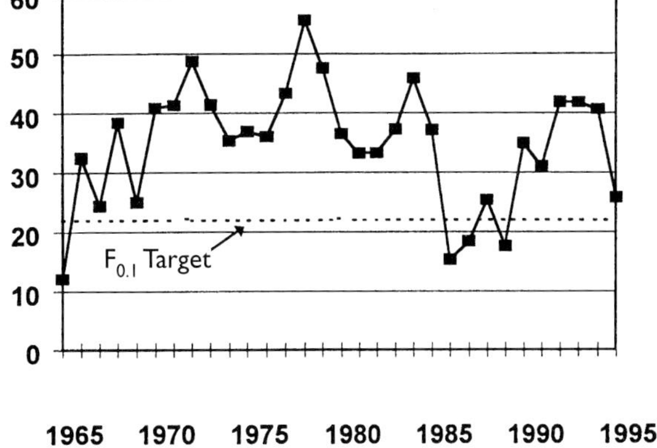

Fig. 3. Trend in exploitation rate for the 4WX herring stock.

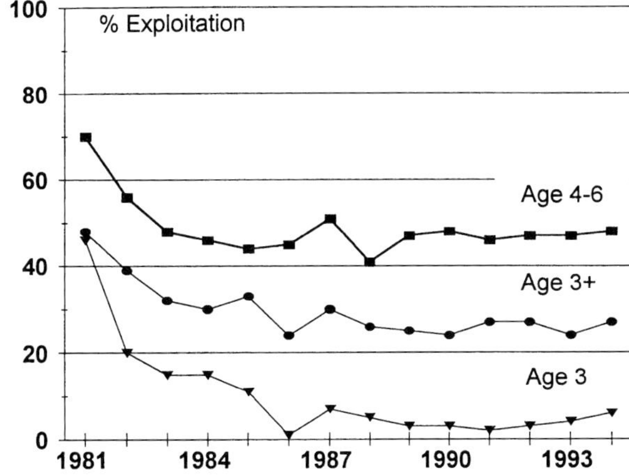

Fig. 4. Trend in Exploitation Rate for the Georges Bank Offshore Scallop Stock.

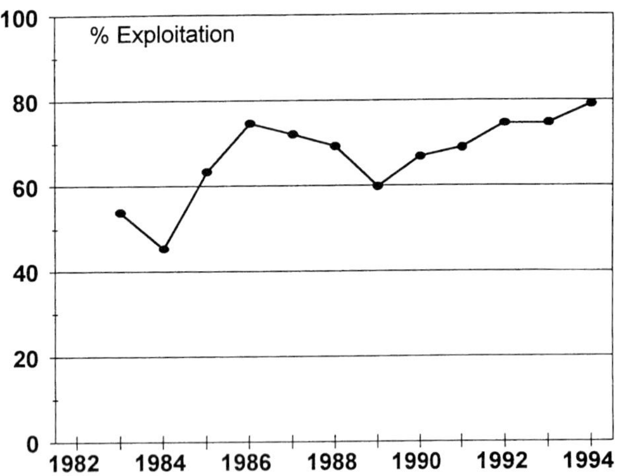

Fig. 5. Trend in exploitation rate for the southwest Nova Scotia inshore lobster stock.

In contrast to groundfish and herring, the results of the offshore scallop (Fig. 4) and offshore lobster fisheries are quite different (Anon. 1995). In the case of scallops, the $F_{0.1}$ conservation target has been achieved. For lobster, while a conservation target like $F_{0.1}$ has not been established, an annual quota of 720 t has been set as a cap to exploitation and has not been exceeded. For these fisheries then, catch controls have been successful in achieving the stated goals.

The situation is less clear for the effort-controlled inshore scallop and lobster (Fig. 5) fisheries since these fisheries have no explicitly stated conservation objectives. It is therefore not possible to state whether or not management targets have been achieved. Notwithstanding this, the inshore scallop fishery has experienced declining catches and increasing effort in recent years and there are concerns that the resource is in serious decline (Anon. 1995). While inshore lobster landings remain at historical highs, exploitation rates are exceptionally high, with 80% of the resource harvested each year. Recent assessments (Anon. 1995) of this resource have raised concerns that the apparent high productivity cannot continue in the face of such high exploitation rates.

Based on these observations, it is not possible to categorically state that catch-controlled fisheries are inherently less successful than effort-controlled fisheries in attaining conservation objectives. While quotas have certainly not been successful in managing groundfish and herring fisheries, they do appear to have been successful for offshore scallop and lobster. Similarly, it cannot be said that effort controls are the answer to management success. There is evidence that at least one of the two effort-controlled fisheries examined was experiencing resource decline.

It is evident then that both catch and effort controls can work in some situations, but not in others. Some understanding of the differences among these systems — and thus perhaps the reasons for success — can be obtained from an analysis of the costs to manage each fishery and how these are distributed among the various components of the management system.

THE COSTS OF MANAGING THE SCOTIAN SHELF FISHERIES

To allow us to compare the six fisheries, DFO management costs in 1993 and 1994 were expressed as a percentage of both the landed and processed value for each resource (Table 3). Management costs include all aspects of the management system, but do not include overheads from non-regional sources (costs of management activities in DFO, Ottawa).

In 1993 and 1994, the landed value of the groundfish fisheries was \$Can127 million and \$Can93 million respectively (Fig. 6). Processed value was about 2.5 times landed value each year. DFO management costs were \$23 million and \$18 million in 1993 and 1994, respectively, or ~18–20% and 7–8% of the landed and processed value, respectively in each year. In comparison, the herring fishery generated relatively low landed value (\$13–16 million annually) but processing increased this more than seven-fold. Management costs in this fishery were 11–18% and 2% of the landed and processed value, respectively. For scallops, processing increased the landed value of \$106–115 million by 1.4 times. Management costs were 3–6% of the landed/processed value. Finally, the most lucrative fishery, lobster, generated a landed value of \$136–163 million, with management costs being 5–6% of this — and even lower (4–5%) when considering the processed value.

Table 3. Management cost and revenue by Scotian Shelf fishery

Year	Fishery	Landed value 10^6\$Can	Processed value as percent of landed value	Management cost 10^6\$Can	Cost as Percent of landed value	Cost as percent of landed value
	Groundfish	126.6	262	22.9	18	7
1993	Herring	16.4	717	1.8	11	2
	Scallop	105.5	140	6	6	4
	Lobster	135.6	128	8.3	6	5
	Groundfish	92.7	248	18.2	20	8
1994	Herring	12.8	753	2.3	18	2
	Scallop	115.4	140	5.1	4	3
	Lobster	163.4	134	8.4	5	4

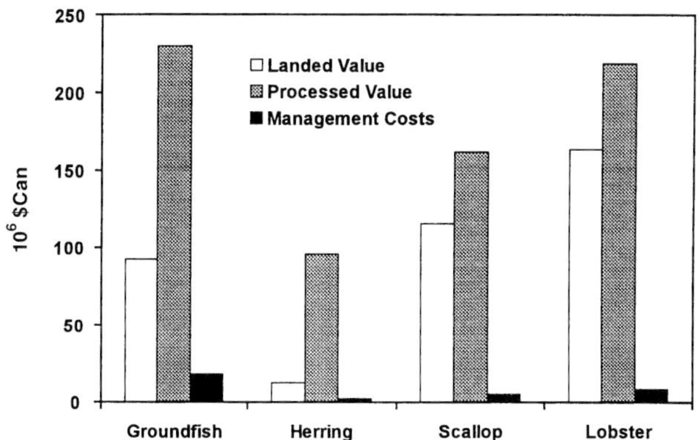

Fig. 6. Comparison of 1994 landed and processed value to the management costs of each fishery.

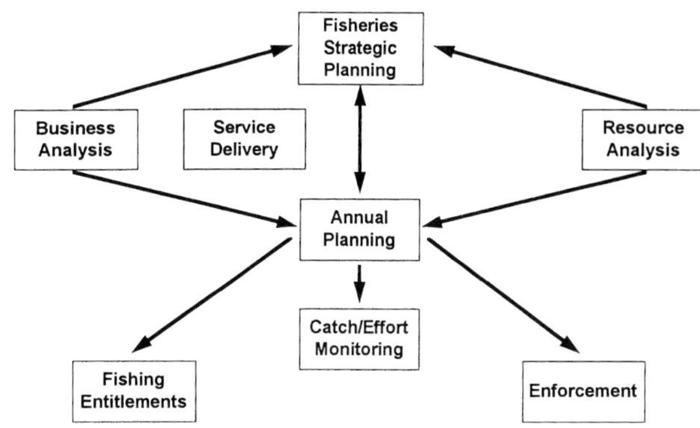

Fig. 8. Elements of the Burke (1995) model of a fisheries management organization.

It is sometimes stated that fisheries generate less wealth than the cost of managing them. These results, however, show clearly that the management costs for these fisheries were only a fraction of either the landed or processed values of the resource. Management costs were less than 10% of the processed value of the fisheries considered, and varied little among these fisheries. These costs do not include other forms of government support, such as unemployment insurance and vessel loans, which can significantly increase the overall cost to government of the fishery. These funds often lower the individual fisherman's costs and can influence involvement in the fishery. It can be argued that without these funds, the overall participation rate in the fishery would decline, and so too would the management costs. Thus the costs presented above are likely to be on the high side.

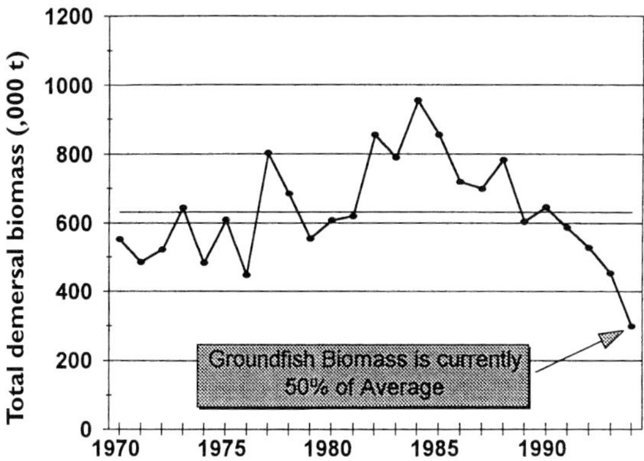

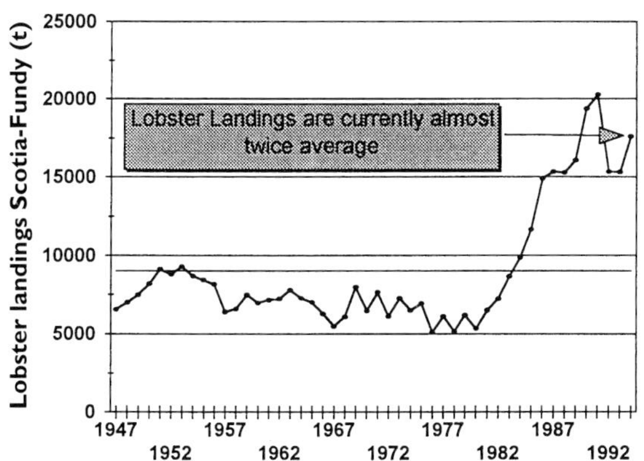

Fig. 7. Trends in long-term biomass (groundfish) and yield (lobster) of resources off Nova Scotia.

These comparisons do not take into account the current state of these resources. In the case of groundfish, resource biomass, and thus yield, is about 50% below the 1970–1995 average. The decline has been particularly precipitous since 1990 and has not allowed time for the management organization to adjust its costs. In contrast, lobster landings have recently been almost twice the long-term. A better measure of the comparison among these fisheries is to consider not the current yield, but the resource's long-term potential (Fig. 7). Taking these observations into account, the relative management costs among these fisheries would be similar.

To understand how these costs are distributed among the components of management, it is necessary to have a model of the management system. The model used in the present study (Fig. 8; Burke 1995) was developed as part of internal DFO discussions on organizational changes. Its intent is to ensure not only that all elements of the management system are present but also that they function together effectively. This model is therefore ideally suited for use in the comparison of the Scotian Shelf fisheries.

According to this model, an effective management system consists of:

- **Strategic Planning**, which creates the management institutions and their support, including the legislation for all elements of the system. It also defines the objectives and strategies of fisheries management.
- **Annual Planning**, which develops the annual harvest plans, including details of the regulatory package.
- **Business Analysis**, which provides Strategic and Annual Planning with advice and information on all aspects of the economic, social and business issues relevant to the fishery.
- **Resource Analysis**, which undertakes the biological stock assessment and research on the impacts of harvesting.

- **Service Delivery**, which provides ongoing monitoring of the costs and functioning of the management system.
- **Catch and Effort Monitoring**, which provides timely, accurate data related to the use of fishing entitlements as part of Annual Planning.
- **Fishing Entitlements**, which distributes licences and co-ordinates and administers fishing entitlements, such as ITQs. And finally,
- **Enforcement**, which is responsible for monitoring compliance with the regulations and takes action against violators.

To examine the distribution of management costs in the fisheries, the detailed expenditure information for 1993–94 and 1994–95 for each fishery was classified according to each element of the model. The expenditures were not initially recorded with the Burke model in mind. This made assigning the costs to each specific management element difficult. As a consequence, some of this was necessarily subjective. However, in the case of resource analysis and enforcement, the two main elements in all these fisheries, the classification is relatively accurate because the costing for them is readily identifiable. The problem in expenditure classification as outlined here underlines the need for an effectively designed service delivery module. Following this classification process, the percent distribution of these costs were averaged over the two fiscal years to provide one estimate per element of each fishery.

When comparing the distribution of management costs for groundfish with those of herring (Tables 4 and 5, and Figs 9 and 10), it is clear that herring has the highest relative expenditure on resource analysis of all the fisheries examined. In contrast, enforcement costs for herring were lower than for the other fisheries. This is consistent with the relatively lean package of regulations employed in this fishery (Table 2).

The distribution of the inshore and offshore lobster fishery management costs are given in Table 5 and Figure 10. Resource analysis costs are 15% of the total for the inshore resource and 9% of the total for the offshore. Enforcement costs appear to be relatively high, being 58% for the inshore and 67% for the offshore resource. In comparing the resource analysis costs for the lobster fishery with those of groundfish and herring, one could conclude that effort-based management is less costly. However, this is not a fair comparison. It is more appropriate to compare the resource analysis costs between the inshore (effort-controlled) and offshore (catch-controlled) fisheries. When this is done, it appears that relatively more costs are incurred in the effort-based fishery, contrary to the notion that catch-based systems require relatively more funding for resource analysis. The relatively high enforcement costs for offshore lobsters are in

Table 4. Percent distribution of groundfish and herring management costs by element of the Burke (1995) model

Element	Groundfish	Herring
Strategic Planning	3.37	1.37
Annual Planning	1.92	0.90
Service Delivery	12.99	12.24
Business Analysis	3.56	6.24
Resource Analysis	30.96	56.45
Fishing Entitlements	0.11	0.16
Catch/Effort Monitoring	2.52	2.81
Enforcement	44.58	19.84

Table 5. Percent distribution of lobster management costs by element of the Burke (1995) model

Element	Inshore Lobster	Offshore Lobster
Strategic Planning	3.26	3.46
Annual Planning	2.70	2.80
Service Delivery	12.57	12.60
Business Analysis	1.23	3.89
Resource Analysis	14.68	8.73
Fishing Entitlements	0.64	0.01
Catch/Effort Monitoring	7.20	1.21
Enforcement	57.73	67.30

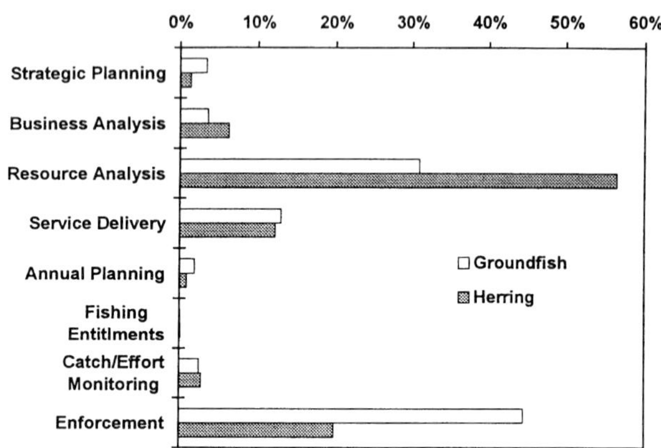

Fig. 9. Distribution of groundfish and herring management costs by element of the Burke (1995) model.

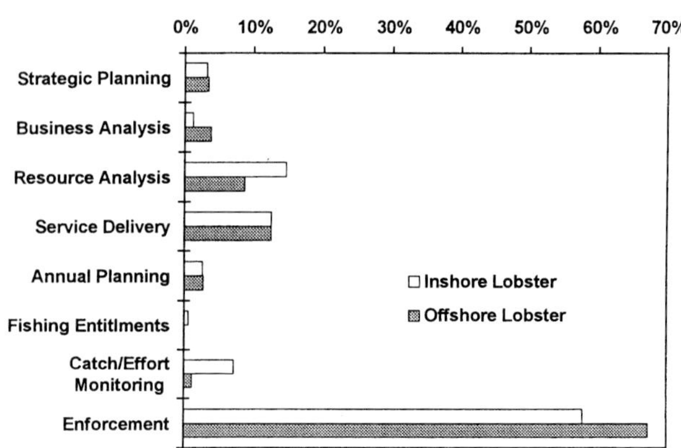

Fig. 10. Distribution of lobster management costs by element of the Burke (1995) model.

Table 6. Percent distribution of scallop management costs by element of the Burke (1995) model

Element	Inshore Scallop	Offshore Scallop
Strategic Planning	2.52	3.10
Annual Planning	2.08	2.55
Service Delivery	14.03	12.52
Business Analysis	2.93	1.70
Resource Analysis	27.61	18.03
Fishing Entitlements	0.27	0.04
Catch/Effort Monitoring	4.02	2.01
Enforcement	46.53	60.05

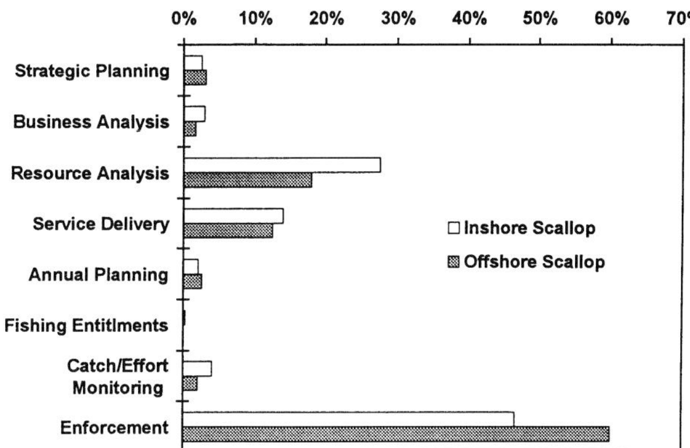

Fig. 11. Distribution of scallop management costs by element of the Burke (1995) model.

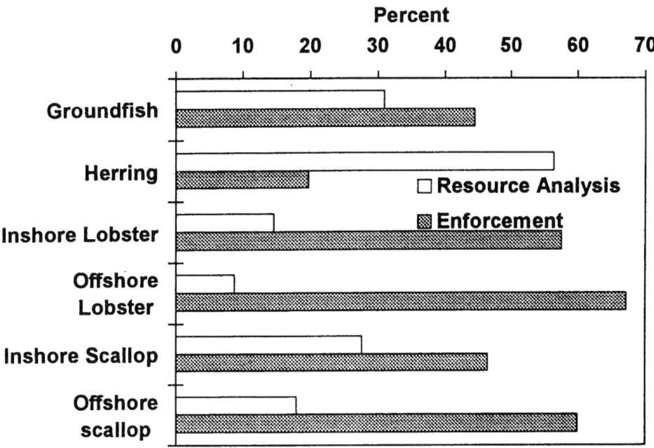

Fig. 12. Comparison of resource analysis and enforcement costs by fishery as a percent of total management costs.

line with expectations, although those for inshore lobster exceed those of groundfish on a proportional basis.

In the scallop fishery (Table 6 and Fig. 11), the resource analysis costs for the effort-controlled inshore fishery are higher than for the catch-controlled offshore fishery. This is again contrary to expectations. In the offshore fishery, a relatively high percentage of funding is directed towards enforcement. Overall, the distribution of costs in the inshore fishery is not dissimilar to

Table 7. Summary of recommendations made at the Second Workshop on Scotia-Fundy Groundfish Management (Burke *et al.* 1996)

Model Element	Recommendation
Strategic Planning	• Delegate decision making closest to those concerned. • Delegate decision making closest to those concerned. • Continue quota management as core regulatory measure, but complement with effective effort controls. • Undertake real-time monitoring of effort as well as catch.
Annual Planning	• Maintain existing spawning areas and define new ones. • Explore the implementation of quotas by year-class.
Business Analysis	• Develop a price monitoring system. • Undertake an analysis of the costs and effectiveness of enforcement. • Evaluate the degree of capacity reduction under current property rights fisheries. • Evaluate the costs of the new catch and effort monitoring system.
Resource Analysis	• Evaluate most appropriate time of annual planning cycle to conduct most up-to-date assessment. • Provide technical input to new regulatory initiatives. • Provide estimates of minimum spawning biomass per stock. • Provide annual estimates of both catch and effort per stock.
Service Delivery	• Define new planning and financial systems to allow linking of the management actions to their costs and benefits.
Catch and Effort Monitoring	• Institute full, industry-funded, dockside monitoring of landings. • Develop indices of at-sea discarding to count against quotas. • Implement real-time, effort monitoring system. • Develop user-friendly decision-support system for managers and industry.
Fishing Entitlements	• Respect existing fleet share arrangements. • Set criteria for inter-annual carryover allocations per licence.
Enforcement	• Develop industry-led sanction system. • Establish mechanism for industry to advise DFO on enforcement improvements. • Implement targeted observer coverage. • Implement easily enforced effort regulations. • Develop code of responsible fishing practices.

those in groundfish, with apparently the same net result in resource management.

In summary, the distribution of resource analysis costs between comparable catch- and effort-controlled fisheries was not in accordance with expectations, while those for enforcement were (Fig. 12). Indeed, for those fisheries that have been managed successfully, a relatively high proportion of funds has been expended on enforcement.

Concluding Remarks

The two fisheries that have been relatively successful in attaining their stated objectives (offshore lobster and scallop) have been under catch control; however, neither the catch-controlled groundfish or herring fisheries were able to meet stated objectives. It was not possible to judge the success of the effort-controlled fisheries as these had no explicitly stated objectives. Therefore, one cannot say categorically that effort controls are superior to catch controls. The costs of managing either system were in general less than 10% of the resource's processed value, indicating that either approach can be cost-effective. There was a slight tendency for resource analysis costs to be higher under effort-based management. This was unexpected and further examination is necessary to validate this observation. While there is a hint of a direct relationship between enforcement costs and management success, it is by no means certain. Effort-based management appeared to be less costly to enforce than the comparable catch-based system; however, catch-based management appeared to be achieving its management objectives, at least in some cases, while this was not obvious in the case of effort-based management. Again, further investigation of the information is required to be more categorical about these conclusions.

While it appears that catch-based management can be a cost-effective means of managing some fisheries, data limitations precluded a more in-depth analysis. However, what became most evident from this analysis is the need to strike the right balance among the interacting elements of the management system, and to ensure that each element is working properly. This was a theme of the Second Workshop on Scotia-Fundy groundfish management, held in October, 1995 (Burke *et al.* 1996). After extensive discussion on the merits of catch and effort controls, it was concluded that rather than adopt a single approach, it was more important to ensure that the management system as a whole was functioning properly. Within this framework, regulations of catch or effort, or some combination of both, can be potentially effective. With this in mind, a series of recommendations (Table 7) was made, using the Burke (1995) model. In conclusion, before any management approach is accepted or rejected, it is essential to conduct an evaluation, such as was done here, in the context of a management model and, where possible, improve those elements of the system that need improving. Only by evaluating each of these elements against their ability to achieve stated objectives will it be possible to either improve existing management schemes or design effective new ones.

Acknowledgments

We wish to thank Doug Pezzack and Ginette Robert for providing the information on the lobster and scallop fisheries. Special thanks go to Phillip Stirling, who had the tenacity to wade through volumes of financial records.

References

Anon. (1995). Scotia-Fundy Spring 1995 Stock Status Report for Pelagics, Invertebrates, and Marine Mammals. *DFO Atlantic Fisheries Scotia-Fundy Regional Stock Status Report* 95/1.

Anon. (1996). 4WX Herring. *DFO Atlantic Fisheries Stock Status Report.* 96/18.

Burke, D L (1995). A Business and Information Model for the Fisheries Management Business Line in the Department of Fisheries and Oceans. Unpublished Manuscript.

Burke, D L, O'Boyle, R N, Partington, P, and Sinclair, M (1996). Report of the Second Workshop on Scotia-Fundy Groundfish Management. *Canadian Technical Report of Fisheries and Aquatic Sciences* 2100: vii + 247 pp.

Gulland, J A, and Boerema, L K (1973). Scientific Advice on Catch Levels. *Fisheries Bulletin (US)* **71**, 325–35.

Quantification of Objectives, Strategies and Performance Criteria for Fishery Management Plans — an Australian Perspective

Anthony D. M. Smith

CSIRO Division of Marine Research, Hobart, Tasmania, Australia.

Summary

There are at least eight separate management jurisdictions in Australia. During the 1990s there has been some convergence between jurisdictions in their approach to development of management plans and the use of quantitative performance criteria as an important element of such plans. Performance criteria are increasingly used to monitor the success of management plans (and management agencies) in meeting their stated or legislated objectives. Over and above their use in monitoring performance, these quantitative measures have also been used to help develop and evaluate alternative harvest strategies for some managed stocks. Three examples are described, involving fisheries for orange roughy, school shark and gemfish. These examples illustrate the complex interactions between objectives, strategies and performance measures.

Introduction

Objectives, strategies and performance criteria are closely related. Performance criteria should measure the success of management strategies in achieving specific management objectives. In this sense they function as a means of auditing the success of management. For biological or conservation objectives, such as ecological sustainability, performance criteria often make use of biological reference points. These have been used for many years in fisheries management but have been the focus of further research and debate recently (FAO 1993; Smith *et al.* 1993).

It is useful to distinguish two ways in which performance criteria can be used. First, they can be used in a monitoring or audit role to measure the performance of current management strategies relative to management objectives. Second, they can be used to help develop and evaluate management strategy options, and in particular to evaluate future harvest strategies (e.g. Kirkwood and Smith 1996).

This paper investigates some current trends in fishery management in Australia, with particular reference to the development of formal management plans and the use within them of quantitative performance criteria and indicators. Three case studies are explored to further investigate and illustrate the relationship between objectives, management strategies and performance measures.

Some current trends in fisheries management in Australia

Fisheries in Australia are managed under at least eight separate jurisdictions. Each of the six States as well as the Northern Territory has jurisdiction over fish resources in adjacent waters, and some resources are managed federally (that is, by the Commonwealth of Australia). Additionally, some fisheries are managed under joint jurisdiction, including some internationally shared resources.

There have been major changes in fisheries management in Australia in the last decade. Some of the more evident trends include:

- greater participation of the fishing industry in decision making
- moves to recover costs of management from industry
- a gradual shift to greater use of output controls — Total Allowable Catches (TACs), Individually Transferable Quotas (ITQs)
- greater public concern with broader environmental issues such as effect of fishing, bycatch, and loss of habitat
- increasing involvement of non-fisheries government agencies in decisions concerning fisheries management
- a more formal requirement for development of management plans
- requirements for specifying objectives and performance criteria for management, and formally reporting against them
- more frequent use of quantitative stock assessment, better documentation of scientific advice, and development of processes to provide annual status reports
- increasing, although still limited, use of biological reference points, trigger points, and pre-agreed management responses
- increasing use of methods of stock assessment which take explicit account of uncertainty
- a move towards framing stock assessment advice as an evaluation of the consequences of alternative management strategies or decisions.

One other important aspect of fisheries management in Australia is that most jurisdictions are now operating under new or recently updated fishery Acts (see Table 1). Almost all of these Acts now specify the objectives of fisheries management and almost all refer in one way or another to ecologically sustainable development (ESD) as a specific objective. However, there are also some interesting and important differences between jurisdictions in the structures and processes for fisheries management. For example both the Federal Government and Queensland have set up statutory authorities for fisheries management, removing day-to-day management from the direct control of government, although both authorities are bound by their respective Acts and report to the relevant Minister.

Table 1. Fisheries legislation in Australian jurisdictions

Jurisdiction	Year of Act	ESD Objective in Act	Management Plans in Act	Performance Criteria in Act
Federal	1991	Yes	Yes	Yes
Queensland	1994	Yes	Yes	No
New South Wales	1994	Yes	Yes	Yes
Victoria	1995	Yes	Yes	Yes
Tasmania	1995	Yes	Yes	No
South Australia	1982 (1991)	No	No	No
Western Australia	1994	Yes	Yes	No
Northern Territory	1988 (1994)	No	Yes	No

The year given is the year of the Act. The year in brackets refers to more recent amendments to the Act. The last three columns indicate whether there is explicit reference within the Act to an ESD objective, to management plans, and to the use of performance criteria within management plans.

Some of the more recent fisheries Acts have included fairly detailed specifications on the form and content of management plans. For example the Victorian Fisheries Act states that a management plan must, among other things, specify the objectives of the plan, the management tools and measures to be used to achieve those objectives, and performance indicators, targets and monitoring methods. Similar obligations are specified in other jurisdictions (see Table 1).

Even where these requirements are not formally part of the legislation, they tend to be implemented in practice. For example, in Western Australia there is an annual report to Parliament which tables a range of performance indicators relating to Departmental objectives for fisheries management (Anon. 1995). Performance indicators relating to the objective of ecological sustainability include the proportion of fisheries for which annual stock assessments have been completed, whether breeding stock levels are increasing, decreasing or adequate, and a comparison between predicted and actual catches.

Three Australian case studies

The examples explored in this section are all from Federal (or joint Federal/State) managed fisheries. To this extent they are not representative of fishery management practices throughout all Australian management jurisdictions. They do, however, serve to illustrate some important relationships between the development of management objectives and strategies, and the use of quantitative performance criteria.

Orange roughy

Fishing for orange roughy (*Hoplostethus atlanticus*) in Australia occurs mainly in waters adjacent to Tasmania. Orange roughy is the most important component of the South East Fishery, which is managed under Federal jurisdiction. Orange roughy comprises one of fourteen quota species in the fishery. Although there is not yet a formal management plan for the whole South East Fishery, the Australian Fisheries Management Authority (AFMA) has developed management objectives, strategies, and performance criteria for orange roughy (along with most of the other quota species). Much of the following is summarized in Bax (1996).

AFMA's overall objective is to ensure that the orange roughy resource is utilized in a manner consistent with the principles of ecologically sustainable development. This objective is further defined as maintaining the long term productivity of each stock in the fishery. Currently three stocks are recognized, each with its own management zone.

AFMA's current strategy to achieve these objectives is as follows:

Maintain the spawning biomass of each orange roughy stock above 30% of the spawning biomass at the onset of significant commercial fishing (1988).

Where there is a greater than a 50% chance that a stock is below 30% of the 1988 spawning biomass, then future TACs will be set such that the biomass reaches 30% of unfished biomass (B_0) by 2004.

Where there is greater than a 50% probability that the stock is below 20% of the 1988 spawning biomass, then the TAC will be zero and remain at zero until that probability exceeds 50%.

Where the stock is at such a low level that the recovery to 30% B_0 cannot be attained by 2004, then years of zero TACs may be set to ensure stock recovery.

The 'strategy' outlined above is a type of decision rule which makes use of an assessment of stock depletion in conjunction with two biological reference points, a target (30% B_0) and a threshold (20% B_0). In practice, the assessment of stock depletion is obtained using a Bayesian stock reduction analysis (similar to that described in Francis 1992). Recent assessments have evaluated the consequences of alternative future TAC trajectories in relation to the probability of exceeding the specified target and threshold. On the basis of such an assessment in 1994, a three-year strategy to stabilize catches in the eastern zone and reduce catches in the southern zone was agreed and implemented.

The orange roughy fishery is one of the earliest examples of the formal use of biological reference point targets and thresholds in Australian fisheries management. The evolution of these performance criteria is shown in Table 2.

Table 2. Evolution of biological reference points for orange roughy in the South East Fishery

Year	Target	Threshold	East	South	Combined
1990	50	–	67	–	–
1991	50	–	60	–	45
1992	30–50	20	40	43	24
1993	30	20	38	41	40
1994	30	20	28	35	21

Target and threshold reference points used in framing management advice, together with assessed status of Eastern, Southern and Combined stocks, by year. All figures are percentages of unexploited biomass.

Southern shark

This case illustrates some aspects of the interplay between science and management (including a variety of user interests) in developing and framing objectives, strategies and performance criteria.

The Southern Shark fishery is managed under joint federal and State arrangements, although the day-to-day management is undertaken by AFMA. Two species, school shark and gummy shark, comprise the bulk of the catch and the fishery is currently managed using input controls (mainly effort and gear restrictions). The assessment of the stocks has been contentious for a number of years, but it is generally accepted that school shark are more depleted than gummy shark. It is not possible to reliably target the two species separately in those areas and depths where they occur concurrently.

While the fishery is a managed fishery, there is currently no formal management plan in place (the previous management plan having lapsed in late 1992), and the assessment advice is framed against objectives and performance criteria developed by the management advisory committee for the fishery (SharkMAC) and by AFMA. In 1995, and within an overall objective of sustainable use, the immediate objective was to halt the decline in school shark stocks, and to attempt to rebuild to at least 40% of initial biomass over a 10–15 year period (Walker *et al.* 1995). The aim with the gummy shark stocks was to stabilize them at 1994 levels.

The assessment of the status of the school shark stocks has been revised recently by the Southern Shark Fisheries Assessment Group (Anon. 1996*a*). This assessment group includes representatives of the fishing industry and the management agencies as well as scientists. SharkFAG's advice to SharkMAC in May 1996 included the following information:

- the 1995 mature biomass is between 13 and 43% of the corresponding unexploited equilibrium biomass;
- despite the uncertainty in the estimates, recent catches are substantially above estimates of Maximum Sustainable Yield;
- there is a high probability that current fishing effort will lead to further reductions in the stock. An immediate reduction of 20–30% in effort directed at school shark would be needed to stabilize the resource at its current level in 15 years with a 50% probability.

The ranges in these estimates reflect some of the underlying uncertainty in aspects of the assessment. Part of the assessment report also included an evaluation of probabilities of exceeding current biomass in 15 years under different effort reduction strategies. These probabilities were evaluated across a range of sensitivity tests. Further analyses combined the results across sensitivity tests by giving each an equal weighting.

The response of SharkMAC to this advice was in part to better define and refine the management objectives and performance criteria. There was clarification of the basis for reference points, with mature biomass being selected. There was agreement that the school shark stocks were too low and needed rebuilding. Rather than set a target for rebuilding (as a % of B_0), the decision was made to specify a probability that the mature biomass would be above its current (1996) level after 15 years, and to use this to select the level of effort reduction that would be required. After considerable debate, and making use of a graph which showed probability of increase against level of effort reduction, SharkMAC decided that there should be an 80% probability that, at the start of 2011, school shark stocks would be no lower than at present. This choice of probability corresponded to a required reduction in the effective effort directed at school shark of 42%. This corresponds to a reduction in catch to 518 t p.a. from 1998 (the 1995 catch was 959 t). The Committee also noted that 'from a risk aversion viewpoint a 90% probability was desirable, however the cost benefit of achieving this probability was very high' (Anon. 1996*b*). The management advisory committee and fishery assessment group are now in the process of evaluating specific combinations of management measures which would achieve the requisite reduction in catch to 518 t per year.

Eastern gemfish

Like orange roughy, eastern gemfish (*Rexea solandri*) has been an important component of the South East Fishery (SEF). A sequence of poor year-classes in the late 1980s (Rowling 1994) resulted in successive reductions in TAC for this species until a zero TAC was set in 1993. The TAC has remained at zero since 1993, although a trip limit has been established to take account of bycatch from targeting of other species in the SEF. The current management objective for eastern gemfish is to allow recovery of the stock to 40% of the spawning biomass in a reference year (1979).

A recent Bayesian assessment of the status of the stock (Punt 1996) has allowed an evaluation of a range of feedback harvest strategies for this species (Smith *et al.* 1996). Some of the results of this evaluation are shown in Table 3. The harvest strategies are limited to constant fishing mortality or constant catch strategies. A range of performance indicators is presented including the 'sustainability' indicator — the probability that the strategy will result, after 25 years, in the spawning stock (measured as spawning potential or egg production) exceeding the performance target ($P(E_{fin}>0.4E_{79})$) (Table 3). It is evident from Table 3 that there is a tradeoff between the objectives of maximizing catch, minimizing variation in catch, and achieving sustainability targets. The results also suggest that the biological target may have been set too high.

Table 3. Evaluation of harvest strategies for eastern gemfish

Harvest strategy	Continuing catch (t)	Average catch (t)	Catch variability (%)	$P(E_{fin}>0.4E_{79})$
$F_{0.1}$	1450	1930	26	0.22
$F_{0.2}$	1330	1693	27	0.30
$F_{0.3}$	1060	1302	26	0.41
C = 1000 t	1000	1000	3	0.48
C = 2000 t	1910	1947	5	0.28
C = 3000 t	1710	2439	10	0.16

Performance of six harvest strategies over 25 years evaluated *via* simulation. Performance indicators include continuing ('sustainable') catch, average catch over the entire period, variation in catch from year-to-year, and probability that the final spawning potential will exceed the target level. For details, see Smith *et al.* (1996).

Discussion

The preceding examples have illustrated some of the interactions between management objectives, performance criteria and harvest strategies. In each case a general management objective (resource sustainability) has been given operational definition through use of a specific biological reference point, whether target or threshold. As it happens, in each case the performance criterion has been expressed as a probability of exceeding a reference biomass or spawning level (most often represented as a depletion level). In Australia, it seems, biomass reference points are more often used than fishing mortality reference points.

The examples also illustrate the use of performance criteria as an explicit component of a harvest strategy. In the orange roughy case, the performance criteria (expressed as probabilities of exceeding targets or thresholds) are used as triggers to guide decisions on future TACs. This approach presupposes a stock assessment method which allows estimation of the success of a particular decision or strategy in achieving the performance criterion. An important aspect of such methods is the ability to capture and deal with uncertainty in the assessment and evaluation. Bayesian methods are particularly suitable in this respect and feature in two of the three case studies mentioned.

Finally, the examples illustrate an approach to fisheries assessment and management which has been described as 'management strategy evaluation' (Smith 1994). Two of the examples (orange roughy and shark) involve evaluation of non-adaptive harvest strategies (prespecified TAC levels and effort reduction scenarios respectively), while the gemfish example explicitly explores feedback harvest strategies. The aim in each case, though, is to explore the tradeoffs between alternative objectives (represented by performance criteria or indices) across a range of possible management strategies.

The examples described in this paper represent a very small and biased subset of fishery assessment and management plans and strategies in Australia. Most of the jurisdictions are moving towards the development of formal management plans which include the use of performance indices and criteria as explicit measures of management success. Some are also linking such measures to pre-agreed management actions. One of the challenges for scientists will be to find ways of helping to develop and evaluate harvest strategies for fisheries where formal quantitative stock assessment is difficult, and where performance indices and management strategies are based on the limited types of data available in so many fisheries.

Acknowledgements

I am very grateful for the time and help given to me by senior fisheries managers and scientists in each of the State and federal jurisdictions. They are too numerous to mention individually, and most will probably feel that little of our discussions and their input is reflected in this review. For this I am solely to blame. I am grateful for the help given to me on the three case examples by Trysh Stone, André Punt, John Wallace and Nic Bax.

REFERENCES

Anon. (1995). Annual Report 1994–1995. Fisheries Department of Western Australia, Perth.

Anon. (1996*a*). School shark stock assessment April 1996. Report to the Southern Shark Fishery Management Advisory Committee (SharkMAC) from the Southern Shark Fishery Assessment Group (SharkFAG). Australian Fisheries Management Authority, Canberra.

Anon. (1996*b*). Minutes of the twenty sixth meeting of the Southern Shark Fishery Management Advisory Committee. Australian Fisheries Management Authority, Canberra.

Bax, N J (Comp.) (1996). Orange roughy (*Hoplostethus atlanticus*). Stock Assessment Report for the South East Fishery Assessment Group. Australian Fisheries Management Authority, Canberra.

FAO (1993). Reference points for fishery management: their potential application to straddling and highly migratory resources. *FAO Fisheries Circular* No. 864.

Francis, R I C C (1992). The use of risk analysis to assess fishery management strategies: a case study using orange roughy (*Hoplostethus atlanticus*) on the Chatham Rise, New Zealand. *Canadian Journal of Fisheries and Aquatic Sciences* **49**, 922–30.

Kirkwood, G P, and Smith, A D M (1996). Assessing the precautionary nature of fishery management strategies. In 'Precautionary Approach to Fisheries'. pp. 141–58. Part 2: scientific papers. *FAO Fisheries Technical Paper.* No. 350, Part 2. (Rome, FAO).

Punt, A E (1996). Preliminary stock assessments of eastern gemfish (*Rexea solandri*) using Bayesian and maximum likelihood methods. In 'Evaluation of harvest strategies for Australian fisheries at different levels of risk from economic collapse'. (Ed A D M Smith) pp. 6–67. Final Report to the Fisheries Research and Development Corporation, CSIRO Division of Fisheries, January 1996. 164 pp.

Rowling, K R (1994). Assessment of the gemfish (*Rexea solandri*) stock using a 'customised' age structured analysis. In 'Population Dynamics for Fisheries Management'. (Ed D A Hancock) pp. 168–72. *Australian Society for Fish Biology Workshop Proceedings*, Perth, 24–25 August 1993 (Australian Society for Fish Biology, Perth).

Smith, A D M (1994). Management strategy evaluation — the light on the hill. In 'Population Dynamics for Fisheries Management'. (Ed D A Hancock) pp. 249–53. *Australian Society for Fish Biology Workshop Proceedings*, Perth, 24–25 August 1993. (Australian Society for Fish Biology, Perth).

Smith, S J, Hunt, J J, and Rivard, D (1993). Risk evaluation and biological reference points for fisheries management. *Canadian Special Publication of Fisheries and Aquatic Sciences* **120**. viii+442 pp.

Smith, A D M, Punt, A E, Wayte, S E, and Klaer, N L (1996). Evaluation of harvest strategies for eastern gemfish (*Rexea solandri*) using Monte Carlo simulation. In 'Evaluation of harvest strategies for Australian fisheries at different levels of risk from economic collapse'. (Ed A D M Smith) pp. 120–160. Final Report to the Fisheries Research and Development Corporation, CSIRO Division of Fisheries, January 1996. 164 pp.

Walker, T, Stone, T, Battaglene, T, and McLoughlin, K (1995). The Southern Shark Fishery 1994. *Fisheries Assessment Report*, Southern Shark Fishery Assessment Group. Australian Fisheries Management Authority, Canberra.

Toward Entrepreneurial Wild Capture Fisheries[1]

B. Pierce

Inland Waters Research and Development Program, South Australian Research and Development Institute - Aquatic Sciences, PO Box 120, Henley Beach, South Australia, 5022, Australia.

Summary

After a century of rapid growth, world capture fisheries face a crisis of vision sparked by perceived plateauing production, questioning of scientific capabilities, and a burden of over-investment. To survive and break through to a new era of enhanced performance, core weaknesses of western institutionalized fisheries management must be confronted with urgency and an expectation of dramatic improvement. Characteristics such as slow responsiveness, lack of focus, a 'satisficing' approach, and a failure to link manager motivation to stakeholder benefits can be overcome through adoption of an entrepreneurial business model of management. A culture shift in the areas of client focus, benefit/outcome driving, and a bias for action offer, as evidenced by case studies, underexploited opportunities for the fish, stakeholders and managers. The challenge for 21st century fisheries managers will be to exceed stakeholders' dreams for tomorrow by constantly pursuing absolute excellence in their pathfinding, decision making and implementation roles today.

Introduction

Over the past century, world capture fisheries have undergone dramatic and sustained growth in production, scientific knowledge, and technical capacity. Not unlike the rights of passage inherent in coming of age, capture fisheries are currently facing a crisis of vision sparked by perceived stabilized global production (Csavas 1994), serious questioning of scientific capacities and directions (Finlayson 1994), and acknowledgement of excessive and misdirected technological investment (Alverson *et al.* 1994; Mace 1997). Bringing capture fisheries through to maturity requires a new vision of confidence in their future — and daring to confront core management flaws rather than simply focus on operational shortcomings.

During the 'mining' phase on unexploited stocks, production and economic 'success' are seldom a reflection of management success, but globally, most virgin stocks are a thing of the past.

[1] The original contribution titled *Integrated fisheries management: structured solutions to managing political, economic and scientific chaos in degraded wild stock fisheries* linked business and strategic planning with a structured approach to quantifying fisheries management outcomes by treating individual fisheries as experiments. After submission of this abstract, *FAO Fisheries Report* (519) became available which details a materially similar system; therefore, this further advancement towards pro-active capture fisheries management was pursued.

Technological improvements and innovation can extend initial success by continuing to provide incremental benefits which are largely management-independent, but which still rest upon and are limited by the health of the natural production-base. Whether you view the production-base of wild capture fisheries world-wide as containing too many collapsed, collapsing or overexploited stocks (Ludwig *et al.* 1993), or feel the picture is brighter, 'better' management has become the central hub on which efficiency, stock health, quality and most other operational benefits depend now and in the future.

Fisheries management has consistently lagged behind and reacted to racing industry development (here including both commercial and recreational harvest sectors), but seems to have been largely saved the scrutiny of failings (and celebration of successes) by a combination of factors such as overall industry expansion, appeals to the fluctuating nature of the resource (Hjort 1914) and, in this author's opinion, by the dedication and massive overwork of the people who self-select into this field. Fisheries management, taken here to include research, enforcement, policy and administrative components contributing to the overall system, was created by society to resolve *all* the problems it perceived it had created in particular fisheries (e.g. overharvest), and magically inherited a host of 'problems' it may not have caused (e.g. species dominance shifts, habitat degradation) or didn't cause (e.g. highly variable stock abundance). Given considerable evidence that large western governmental institutions rarely are able to achieve such complex social objectives (Drucker 1980; Naisbitt 1982), it is perhaps surprising that the capabilities and achievements of institutionalized fisheries management have only recently begun to be seriously questioned (Ludwig *et al.* 1993; Anon. 1994; Finlayson 1994; Crean and Symes 1996; Baker and Pierce 1997).

The objective here is *not* to pre-empt the vision resetting process necessary for the myriad different capture fisheries world-wide. Rather, it is to challenge each person contributing to the management systems of such fisheries to believe that they are able to dramatically and sustainably improve the wild capture fishery they work with—and that taking a truly entrepreneurial approach to management is the first place to start. Evidence will first be presented in support of sustained improvement as the only survival option for our future fisheries. Three fisheries arenas where dramatic results can be expected will be proposed (client focus, benefit/outcome focus, and bias for action). Primary emphasis has been placed on transferring demonstrated techniques from highly motivated people-management fields such as business, management science, and sociology. While emphasis will be on western management systems, many opportunities identified should be cross-cultural, or can assist in preventing the transfer of suboptimal management technology to developing countries (e.g. under the strategy for international fisheries research, World Bank *et al.* 1992).

THE CHALLENGE: IMPROVE TO SURVIVE

Commercial terrestrial hunting is largely extinct as an occupation globally, being replaced by captive culture production. The aquatic equivalent, wild capture fisheries, face an analagous if delayed progression down the same pathway unless they remain sufficiently economically viable (based on successfully managed stocks) to retain control over the habitat production system on which they are absolutely dependent. As with any living system/organism, the immediate and ongoing challenge to fisheries is to respond and adapt to change. A few such changes include:

- responding to the demand not just for stable, but for ever-increasing development opportunities to achieve Sustainable Development (Perrings 1994). Successful adaptation will accrue economic and social benefits and support; failure to adapt, the inverse result.
- responding to an inability to meet and dominate world market demand (Csavas 1994). Successful adaptation means retention of market share, increased profitability, and a basis for re-investment in the future; failure to adapt could accentuate privatization of habitat by competing aquaculture interests (e.g. South Australia has recently dedicated 3400 ha of coastal waters to aquaculture which must adversely impact wild fish production capacity therein).
- responding to increased societal perception that world fish stocks are overexploited and unable to sustain harvest (e.g. Ludwig *et al.* 1993; Finlayson 1994) in the face of increasing multiple-use/multiple-objective conflicts (McGlade 1989). Successful adaptation is essential to survival of both the management system and the commercial harvest sector (Jeffries 1995; Loveday 1995).
- meeting world community demands to reduce bycatch wastage (Alverson *et al.* 1994). Success could yield a stronger production base, a contribution to further increased production (e.g. at least a significant part of the estimated 27 million t of annual world bycatch; Alverson *et al.* 1994) and potential product diversification; failure could result in further societal restriction on access to harvestable stocks.

CAN INSTITUTIONALIZED FISHERIES MANAGEMENT SYSTEMS DELIVER THESE IMPROVEMENTS?

Consistent elements of a generalized fisheries management system (Fig. 1) include a chain of responsibility by management inputs through managers to a political authority who is

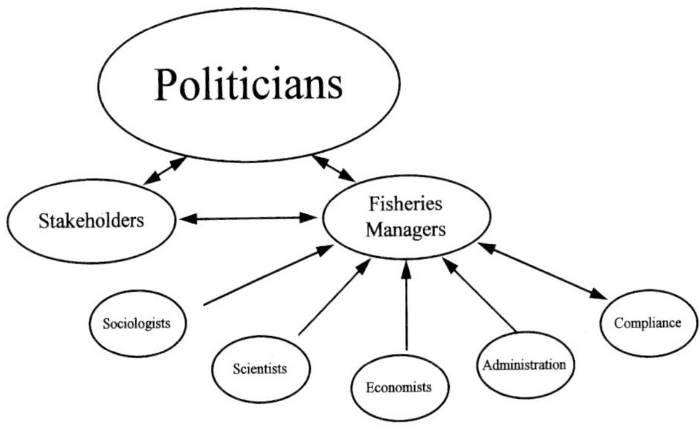

Fig. 1. A generalized model of western fisheries management with arrows implying significant communication pathways. Emphasis is on hierarchy and centralized authority.

responsible for implementation of actions (and to stakeholders, amongst others). Except for data acquisition, contact by lower level advisors with stakeholders is seldom rewarded. Managers and advisors are relatively independent of stakeholder performance feedback. Cost of implementation of such systems is seldom reported. Australian fisheries management costs in the vicinity of 10% of gross landed value (1992/3; FRDC 1992, together with relevant State and Territory annual reports) to in excess of 25% of landed value in some depressed fisheries (O'Boyle and Zwanenburg 1997).

If the status of major capture fisheries is a legitimate performance indicator for such management systems, few could be defined as highly successful. The traditional western (fisheries) management hierarchy suffers from what may be tragic flaws on optimization, motivation, responsiveness, and authority criteria:

Optimization: Where do we optimize fisheries? Virtually everywhere we 'satisfice' fisheries, yet the management concept that was worth a Nobel prize in 1979 is not used in any ASFA (Aquatic Sciences and Fisheries Abstracts) reference in aquatic sciences in the past decade. Does it sound familiar that 'In a political system there are far too many constituencies to optimize; one must try to determine the one area in which optimization is required. But in all other areas — their number in a political system is always large — one tries to satisfice, that is, to find the solution in which enough of the constituencies can acquiesce. One tries to find a solution that will not create opposition, rather than one that will generate support. Satisficing is what politicians mean when they talk of an 'acceptable' compromise (Drucker 1980).' How often have we satisficed our way to stock collapse?

Motivation: Maslow's well documented hierarchy of human needs/motives (basic physiological needs, safety needs, social needs, self-esteem needs and self-actualization needs) are rarely harnessed to get management and stakeholders working congruently and positively towards societal objectives for fisheries. For example, while managers are seldom fired for fisheries failures, commercial sector stakeholders constantly face the marketplace equivalent even when they recognize that stocks are unacceptably depressed. While recreational and conservation sector stakeholders are more likely to be motivated by higher order, less fundamental needs (e.g. self-actualization), their numeric dominance heavily weights their directives regardless of how these relate to management direction(s). Place this 'one hand tied behind your back' disadvantage for managers within a flexible political/social structure and the medium-term results under conditions of conflict are predictable, if often suboptimal.

Responsiveness: Studies of large organizations demonstrate that they change (read 'adapt') very slowly (Peters and Waterman 1982; Peters 1992). Perhaps the multi-decadal cycle in ascendency and descendency of the concept of maximum sustainable yeild (MSY) provides a fair indicator that fisheries institutions are not immune to this lethargy. On the other hand, it is widely accepted that the rate of change is accelerating in many areas impinging on humans and the environment (Naisbitt 1982; Peters 1992). Certainly fish stocks, markets, and technology can vary dramatically over a daily to annual basis.

Authority: Under western democratic regimes, authority is granted by consent of the governed and operates through a relatively complex bureaucracy. Is it surprising that upper management, despite assurances they 'know what to do' invariably desire more power, more resources, and more support (both internal and external). Given world trends towards decentralization, reduced funding and increased 'user-pays' accountability to stakeholders (Naisbitt 1982, 1994), this wish list seems increasingly improbable.

Such weaknesses are only important if there is evidence that superior management systems exist. Peters and Waterman (1982) have found seven key characteristics associated with successful (their 'excellent') large business organizations:

- bias toward action (*vs* bias toward inaction)
- client focussed (*vs* client independent)
- entrepreneurship and autonomy (*vs* bureaucracy)
- hands-on and value driven (*vs* hands-off and inertia driven)
- single-minded (*vs* responsible for everything)
- simple and lean (*vs* complicated)
- mistake-tolerant within core values (*vs* mistake-penalizing, regardless).

How many fisheries management entities would rate as excellent against these criteria?

Underexploited entrepreneurial opportunities

Pre-industrialization, fish stocks managed themselves as evidenced by virgin stock levels and the continued survival of those stocks through time. It follows that fisheries management is inescapably people management (Charles 1995; Clay and McGoodwin 1995). In exploring options for dramatic improvement of capture fisheries outcomes, exploitation of alternative management models therefore falls out as a logical priority. The entrepreneurial business model, redirected towards desired fisheries goals, may better harness rapid responsiveness, motivation, focussed objective achievement and efficiency capabilities. Three components integral to entrepreneurship and parallelling the pathfinding, decision making and implementation functions within management (Peters and Waterman 1982) are initially considered for exploitation to reap enhanced fisheries benefits from capture fisheries: client focus, benefit/outcome focus, and bias for action.

Client focus (Sleeping with the enemy!)

Protected by the mantle of science (Finlayson 1994) and reinforced with an ever-increasing armoury (remember MSY, hatchery-based enhancement, to newer fads like ITQs and adaptive management), managers of the world's fisheries often seem to view themselves as going forth to battle the greedy exploiters (and ignorant recreationals) who, let loose upon society's common property/open access resources, would rationally destroy them (Ostrom 1990). The traditional management structure (Fig. 1) essentially guarantees an adversarial relationship between stakeholders and management.

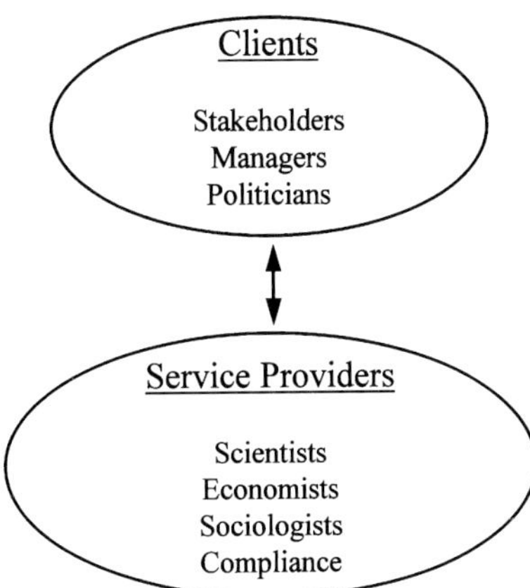

Fig. 2. A simplified entrepreneurial model of fisheries management relationships. Emphasis is on mutually beneficial client/service provider interactions.

Reframing this relationship, in keeping with the trend towards 'user pays, user says,' into a win/win partnership (Fig. 2) obviously removes conflict, but may seem like sleeping with the enemy.

In a people-management framework, stakeholders are the people we must manage the fish through. Potential benefits to enlisting them on the same side include increased observation and feedback from people working daily with the resource, their numeric and political power, as well as potential for their increased investment in management. Diverse options exist to implement such a pathfinding system including:

- Privatization/community-ization (e.g. Japanese coastal fishery territorial use/ownership rights; Ruddle 1987).
- Authority status (e.g. the Australian Fisheries Management Authority, AFMA) with stakeholder funding and, in part, accountability, linked to arms-length government responsibility.
- Co-management/cooperative management (Rettig *et al.* 1989) which tends to have longer arms-length accountability/responsibility links to stakeholders and greater governmental linkage.
- joint venture strategic and operational partnerships with individual stakeholders (e.g. Loch *et al.* 1995).
- Individual contact (i.e. getting up from behind the desk and getting out with stakeholders on a personal and individual basis).

Case Study: In 1993, management of inland water fisheries in South Australia shifted from agency-based (Department of Fisheries) to co-management-based (Scalefish Management Committee). Research advice became an obligate component of the committee structure at the same time as facing increased financial accountability to industry whose research co-funding responsibility increased towards full cost-recovery. This forced greatly increased contact between researchers and grass-roots industry which produced unanticipated benefits, including:

- discovery that industry representatives consistently supported *more* conservative management measures than research indicated were required for sustainability (e.g. industry demanded more restrictive size limits and seasonal closures for Murray cod, *Maccullochella peeli*, as well as a complete moratorium on all harvest of silver perch, *Bidyanus bidyanus*, and freshwater catfish, *Tandanus tandanus*).
- access to a wealth of observational data concerning the resource, as well as positive questioning of current scientific thinking, which has spurred and directed research (e.g. previous thinking was that epizootics in wild stocks were rare events, while with industry assistance, it has now been demonstrated that they are in fact regular, pervasive and water quality-driven impacts on the lower River Murray fishery).
- cooperative rather than antagonistic approaches to overcoming problems and achieving improvements in the performance of the fishery (e.g. research proposals are now regularly co-sponsored by industry and often include extensive industry involvement in project implementation and development).

Benefit/outcome focus (Adding real mega-value!)

Reading the annual reports emanating from most western institutionalized fisheries management agencies would lead you to believe that these fisheries were all being managed successfully and that each year was better than the last. Annual reports written by stakeholders might present a more realistic picture as well as highlight the critical fishery-specific benefits/ performance indicators. Few such benefits accrue from a reactive management approach; rather, pro-active, planned and milestoned strategies to yield additional benefits from fisheries can successfully add new value:

- by providing a measure against which decision making (and necessary reactive action) can be made.
- by spreading investment risks across multiple opportunities, multiple partners, and through time.
- by providing agreed and exact targets and performance indicators (see Baker 1997).
- and by providing the basis, within the fishery, for sustained growth in value on which to provide and attract funds for necessary re-investment in the future of the fishery.

Case Study: Central Australia hosts a unique native fish fauna which persists in a primarily ephemeral water system spread over 1/8th of Australia. Illegally introduced common carp, *Cyprinus carpio*, feral in a large dam within the Cooper Creek drainage posed a threat to this fishery and ecosystem. Prior institutionalized management sought eradication as the primary benefit with introduction of one angling species as a secondary benefit (Hall 1988). A second exercise under the later co-management structure expanded the benefit focus dramatically to include:

- commercial fishers, with researchers and recreational fishers, harvesting at least 75% of extant saleable stocks pre-treatment, thus adding value and preventing resource waste in the eyes of the community.
- recreational fishers, with commercial and research assistance, harvesting significant numbers of adult native sport fish

which were moved to metropolitan sites to cost-effectively enhance existing, but limited, angling opportunities there.

- restocking of the post-treatment water to support a recreational angling opportunity in the desert!
- restocking of the post-treatment impoundment with a full complement of native Cooper Creek fishes to attempt to create a significant genetic refugia (and pro-active conservation safeguard) should introduced species or habitat threats impact the core wild stock in future.
- education of South Australians through all partners jointly approaching diverse media to convey the message that introduced species like carp provide major adverse impacts while ecosystem-based management using native species results in far greater commercial, recreational and conservation values (see Coombs 1996).
- full cost recovery by the dam owner, but at a reduced cost due to the many volunteers contributing time and other resources.

Bias for action (Nike's™: Just do it!)

Great directions and great decisions are nothing without great implementation. Institutionalized management stifles action through the number of human links in the chain which must be successfully passed for ideas to become reality. Hurdles which must be overcome for individuals and organizations to embrace a culture biased for action may include:

- 'negative inertia' inherited from existing for years in a narrowly defined hierarchical niche lacking emphasis on action.
- lack of examples/models to emulate.
- lack of (perceived) opportunities to act due to lack of client focus and benefit focus.
- lack of rewards (or presence of disincentives) for undertaking new activities that yield additional benefits.
- failure to see such activity as part of your role (e.g. 'I'm a scientist, its not my place to ...').
- fear of failure.

Case Study*:* Feral exotic species have had major impacts on terrestrial and aquatic environments in Australia. Regulations under the South Australian Fisheries Act (1982) sought to prevent additional adverse introductions by significantly restricting those exotic fish species which could be kept and traded within the aquarium industry. A significant backlash within the pet industry and hobby resulted in relatively entrenched government/industry positions. Scientists and compliance officers were essentially dragged into this 'lion's den' and forced to actively interact with, in particular, enthusiastic hobbyists. The result over a period of three years was a process of co-education with mutual benefits, including:

- greatly increased understanding within government of the hobby and of the characteristics of specific exotic fish.
- increased appreciation within the hobby of unique Australian native fishes to the point where the South Australian Native Fish Association was founded and is now the largest aquarist society in South Australia.
- both 'positions' shifted significantly (e.g. previously vociferous opponents of this legislation eventually actively supported measures to increase protection of local aquatic habitat and native fish species).
- numerous hobbyists, previously focussed on keeping exotic fishes, became active volunteers assisting field and laboratory research projects on wild fisheries, including one who gained sufficient skills to win employment within government in that capacity.

References

Alverson, D L, Freeberg, M H, Murawski, S A, and Pope, J G (1994). A global assessment of fisheries bycatch and discards. *FAO Fisheries Technical Paper* 339, 1-233.

Anon. (1994). 'Review of fisheries in OECD member countries.' (Organisation for Economic Co-operation and Development: Paris).

Baker, D (1997). Your work is of value, prove it or perish. In 'Developing and Sustaining World Fisheries Resources: The State of Science and Management. Second World Fisheries Congress, Brisbane 1996'. (Eds D A Hancock, D C Smith, A Grant and J P Beumer) pp. 215–19 (CSIRO Publishing: Melbourne).

Baker, D L, and Pierce, B E (1997). Does fisheries management reflect societal values?: Contingent valuation evidence for the River Murray. *Fisheries Management and Ecology* **4**, 1–15.

Charles, A T (1995). Fishery science: the study of fishery systems. *Aquatic Living Resources* **8**, 233–9.

Clay, P M, and McGoodwin, J R (1995). Utilising social sciences in fisheries management. *Aquatic Living Resources* **8**, 203–7.

Coombs, R (1996). Native fish replace European carp at Leigh Creek. *Outback* (35), 7.

Crean, K, and Symes, D (Eds) (1996). 'Fisheries management in crisis.' (Fishing News Books: Oxford).

Csavas, I (1994). World aquaculture status and outlook. *INFOFISH International* (5/94), 47–54.

Drucker, P F (1980). 'Managing in turbulent times.' (Harper and Row: New York).

FAO (1995). Report of the expert consultation on guidelines for responsible fisheries management. Wellington, New Zealand, 23–27 January 1995. *FAO Fisheries Report* 519, 1–100.

Finlayson, A C (1994). 'Fishing for truth.' (Institute of Social and Economic Research, Memorial University of Newfoundland: St John's).

FRDC (Fisheries Research and Development Council) (1992). 'Research and development plan, 1992–3 to 1996–7.' (FRDC: Canberra).

Hall, D A (1988). The eradication of European carp and goldfish from the Leigh Creek Retention Dam. *SAFISH* **12**, 15–16.

Hjort, J (1914). Fluctuations in the great fisheries of Northern Europe. *Rapports et Proces-Verbaux des Reunions, Conseil international pour l'Exploration de la Mer* **20**, 1–228.

Jeffriess, B (1995). Fisheries management planning: a national perspective. In 'Fisheries management planning. Proceedings of the Second National Fisheries Managers Workshop, Bribie Island, Qld'.(Ed N Taylor-Moore) pp. 27–34 (Queensland Fisheries Management Authority: Brisbane).

Loch, J S, Moriyasu, M, and Jones, J B (1995). An improved link between industry, management and science: review of case history of the Southwestern Gulf of St. Lawrence snow crab fishery. *Aquatic Living Resources* **8**, 253–65.

Loveday, T (1995). How successful have fisheries managers been? In 'Fisheries management planning. Proceedings of the Second National Fisheries Managers Workshop, Bribie Island, Qld'. (Ed N Taylor-Moore) pp. 37–44 (Queensland Fisheries Management Authority: Brisbane).

Ludwig, D, Hilborn, R, and Walters, C (1993). Uncertainty, resource exploitation, and conservation: Lessons from history. *Science-Washington DC* **260**(5104), 17, 36.

Mace, P M (1997). Developing and sustaining world fisheries resources: The state of the science and management. In 'Developing and Sustaining World Fisheries Resources: The State of Science and Management. Second World Fisheries Congress, Brisbane 1996'. (Eds D A Hancock, D C Smith, A Grant and J P Beumer) pp. 1–20. (CSIRO Publishing: Melbourne).

McGlade, J M (1989). Integrated fisheries management models: Understanding the limits to marine resource exploitation. *American Fisheries Society Symposium* **6**, 139–65.

Naisbitt, J (1982). 'Megatrends: ten new directions transforming our lives.' (Warner Books: New York).

Naisbitt, J (1994). 'Global paradox.' (Allen and Unwin: St. Leonards, Australia).

O'Boyle, R, and Zwanenburg, C T (1997). A comparision of the benefits and costs of quota *versus* effort-based fisheries management. In 'Developing and Sustaining World Fisheries Resources: The State of Science and Management. Second World Fisheries Congress, Brisbane 1996'. (Eds D A Hancock, D C Smith, A Grant and J P Beumer) pp. 283–90. (CSIRO Publishing: Melbourne).

Ostrom, E (1990). 'Governing the commons.' (Cambridge University Press: Cambridge).

Perrings, C (1994). Biotic diversity, sustainable development, and natural capital. In 'Investing in natural capital'. (Eds A Jansson, M Hammer, C Folke and R Costanza) pp. 92–112 (Island Press: Washington DC).

Peters, T (1992). 'Liberation management.' (MacMillan: London).

Peters, T, and Waterman, R H (1982). 'In search of excellence'. (Harper and Row: New York).

Rettig, R B, Berkes, F, and Pinkerton, E (1989). The future of fisheries co-management: A multi-disciplinary assessment. In 'Co-operative management of local fisheries'. (Ed E Pinkerton) pp. 273–89 (University of British Columbia Press: Vancouver).

Ruddle, K (1987). Administration and conflict management in Japanese coastal fisheries. *FAO Fisheries Technical Paper* 273, 1–94.

World Bank, United Nations Development Programme, Commission of the European Communities, and Food and Agriculture Organization of the United Nations (1992). A study of international fisheries research. *Policy and Research Series* **19**, 1–104.

Issues and Outcomes for Theme 2

What are the Roles of Science, Economics, Sociology and Politics in Fisheries Management

B.K. Bowen

Associate — Western Australian Fisheries Department, St George's Terrace, Perth, Australia.

The Organizing Committee had decided that the poster sessions would provide the principal focus for the World Fisheries Congress, and the Committee was not disappointed.

There were 44 excellent posters providing information within the Theme 2 topic of the role of science, economics, sociology and politics in fisheries management. The information ranged from examples of the place of science in fisheries management, through an array of technical descriptions of scientific endeavour to examples of co-management in the management process.

The socio-economic aspects of fisheries management also attracted attention, together with information on the state of the environment and a consideration of biodiversity in the marine environment.

The posters provided an excellent demonstration of the array of elements involved in the process of fisheries management; and all those who were involved are to be congratulated on the thought given and time taken to prepare and display them.

All participants to the Congress had ample opportunity in a relaxed manner to inspect the posters and discuss them with the authors. Time does not permit me to say more about the posters, other than to repeat my thanks, on behalf of all who attended the Congress, for the outstanding displays.

There were four concurrent sessions of Congress talks, and while many of us would have liked to have attended a number of those concurrent sessions, this was not possible. For the benefit of those members of Congress who were not able to attend the Theme 2 talks, I will endeavour to provide a summary of the highlights.

If I were asked to summarize the thrust of the papers, it would be that there is an increasing desire to have continuous improvement in fisheries management, but also the process is becoming increasingly complex. Progress will be achieved by improving the understanding, by all of the interested parties, of the issues involved. This can be achieved through both the formal and the informal process of co-management.

Dan Lane set the scene for the Theme 2 discussions by exploring elements of the integrating process in fisheries management. The tasks are complicated, but the coming together of fisheries management and fisheries science was an essential starting point. He termed this integrating process as management science, incorporating problem-solving, accountability and feedback mechanisms, interdisciplinary methodologies and strategic planning.

Integration often comes about as a result of crisis situations, but there was a challenge for integration to be undertaken before a crisis arose. Dan drew attention to the increasing importance of the designated fisheries manager in the management process, but also said that there was very little training available for this essential task.

Larry Nielson drew attention to the role of the fisheries manager by emphazising the importance of building relationships to provide for effective outcomes in fisheries management. His description of the stakeholders satisfaction triangle reminded us that whilst there were the important elements of substance and process, there was also a third element of equal importance, that of relationships.

Relationships are about people rather than institutions; they are about peoples' lives, about a two-way communication process and about understanding the needs of all manner of people rather than only those one chooses to select. A great example of the need for scientists to understand those engaged in a fishery, perhaps even as an overriding consideration of the successful outcome of fisheries management, was given by Simon Foale. His work was on the trochus fishery of Nest Gala in the Solomon Islands, where an understanding of the rules for establishing the rights of ownership, which were passed down through the family system, were essential to the management of the fishery.

Kevin Rowlings used the example of the eastern gemfish stock, which went into crisis overfishing, to explore the different attitudes of people where science and industry were not in a relationship or co-partnership mode. Even though the reduction in the gemfish fishery was such that the total allowable catch was now set at zero, the work recently undertaken to bring the scientific understanding to the industry and to involve the industry in the 'ownership' of the data had reduced considerably the management problems. A basis was being built on which the integration of the process could proceed, and this should bring about a better management outcome when the stocks improved to provide for the possibility of a TAC being set once again.

Bryan Pierce told us, from his experiences in South Australia, about the need to focus on the array of people required to be brought into the discussions on fisheries management. These included both the commercial and recreational sectors of a fishery as well as people from the broad environmental movement and the community generally. By expanding the areas of discussions, opportunities for management action may arise which could not have been envisaged originally.

David Baker challenged us to focus at all times on the goals to be achieved and the indicators by which our performance can be measured. Are we doing what we said we would do? and are we continuously seeking to improve our effectiveness towards the goal of successful fisheries management?

The role of the scientist was also considered by Kathryn Matthews who discussed the dilemma posed for the scientist when management action was judged to be significantly at variance with the advice provided by the scientist. The means by which scientists make their views known to the public needs to be fully explored and carefully considered.

Having heard a number of speakers with a background in fisheries management, the Theme 2 audiences were then addressed by some of the leaders in the Australian fishing industry. Murray France and Darby Ross are both from the fishing industry with years of experience at sea and in fleet management. They provided a clear industry perspective on co-management which called for a recognition of the part industry played, not only as the producer of seafood products but also as a provider of some of the data required for analyses of the state of the stocks. They spoke about the benefits which flow from an involvement of industry in the decision-making process on fisheries management. They stressed that there was a need for ongoing discussions leading to an understanding of concepts such as the precautionary principle and ecologically sustainable development so that the manifestation of these principles was meaningful to the practicalities of each fishery.

George Kailis, a manager in the fishing industry and with legal and administrative training, demonstrated that for sustainable fisheries management to be achieved, the process adopted had also to be sustainable. He established some guiding principles in relation to co-management, and reminded us that a central point was the clear establishment of the rights of individuals to access the resource. If this were not the case the industry would lack confidence in the process and that process would not be sustainable.

Tony Smith, an Australian fisheries scientist, drew together an array of information, from the Australian experience, about the fisheries objectives described in a number of new Fisheries Acts recently proclaimed. They demonstrated that as well as the traditional stock maintenance objectives there were now objectives about economic performance, ecologically sustainable development and cost recovery. George Kailis also drew this to our attention, and reminded us that their implementation needed discussion because some of the objectives could be judged to be contradictory, one with the other.

Tony Smith went on to discuss the changing scene for scientists in their involvement in assisting with the achievement of the objectives; an example being a change from the position of clearly establishing what the total allowable catch should be to one of providing information about the effect on the stocks if various levels of catches were allowed to be taken. Also, the scientist was now giving greater attention to the establishment of reference points for the stocks being exploited. He used the orange roughy fishery to provide an example of the role of the scientist.

Turning now to the South African experience under the new regime, Pieter Du Plessis and Trevor Hutton, in separate papers, described the opportunity to explore new ways of managing the fisheries, including greater user participation. The discussion on fisheries management now being undertaken in South Africa also provided an opportunity for further consideration of questions such as who should control the quotas and what are the most appropriate means of sharing the benefits.

The papers identified above were about the broad approaches to fisheries management. However, there were also papers which provided valuable technical information. Bruce Rettig spoke about the costs of recovery strategies to assist in our management understanding of the achievement of recovery goals with the least economic sacrifice. The example used focussed on the Snake River chinook and sockeye salmon.

Other technical papers, adding to our understanding of the data required for fisheries management, were given by:

(a) Rick Starr on the reliability of log-book information compared with research catch data;

(b) Saul Saila on the value of exploring the use of Fuzzy Control Theory in fisheries analysis;

(c) Robert O'Boyle who compared the outcomes using catch controls and gear unit controls, considering both the fishing mortalities generated by their use and the costs involved in their management;

(d) James Scandol who reminded us about the need for timely data using Bayesian statistics whereby the information from within-season activities were added to the pre-season forecasts thus increasing the validity of the data;

(e) Geoffery Meadon who enticed us to consider the use of GIS to add substantially to the precision of the data provided by industry; and

(f) Phillip Marshall on a satellite-based information system.

The last speaker in Theme 2, Hans Lassen, brought together many of the matters of importance raised by previous speakers when he described the nature of the biological advice provided to fisheries management for the North Atlantic fisheries from 1970 through to 1995. Many of the complexities and concerns seen in other fisheries were evident in the North Atlantic, ranging from the institutional structure, overcapacity, fishing pressure higher than the objectives set, and generally a pessimistic outlook to the possibility of continuous improvement in fisheries management.

As Serge Garcia said, even though we know the words to describe sustainable fishing, the practice of sound fisheries management is much more difficult to achieve. Continuous examination of the roles of all of those involved in fisheries management and continuous improvement in the process will remain a challenge between now and the Third World Fisheries Congress.

How can fisheries resources be allocated? Who owns the fish?

Theme 3

How Can Fisheries Resources Be Allocated … Who Owns The Fish?

Rudy van der Elst,[A] *George Branch,*[B] *Doug Butterworth,*[B] *Pattie Wickens*[B] *and Kevern Cochrane*[C]

[A] Oceanographic Research Institute, PO Box 10712, Marine Parade, 4056, KwaZulu-Natal, South Africa.
[B] Zoology Department, University of Cape Town, PO Rondebosch, Western Cape, South Africa.
[C] Food and Agricultural Organization, Viale delle, Terme di Caracalla, 00100, Rome, Italy.

Summary

This paper reviews the basic elements of property rights, discusses the steps that need to be considered in developing an access system and identifies various allocation procedures. In particular a possible model is proposed for re-allocating access to a previously marginalized fisher community, as in South Africa, with the costs completely internalized. Examples are drawn from a wide range of management systems in different countries. Progress with the development of a new fisheries policy in South Africa highlights the importance of resolving issues of access. In an attempt to simplify the access allocations, different types of fisheries are grouped into four categories, each of which has different biological, management and investment characteristics.

Introduction

As the earth's natural resources dwindle in the face of growing human demand, so the competition for these resources becomes more intense. Central to this problem is the question of ownership and by implication who can have access. Nowhere is this more evident than when resources are perceived to be common property as in the case of fish in the sea. Rights of access have become one of the most pressing and vexing issues is fisheries management world-wide, not only because of the potential effects on sustainability of resources but also because of the growing legitimization of people's democratic rights.

Many nations have been wrestling with the issue of access rights for decades or, in some cases, centuries. Many have gained valuable experience in the successes and failures of different approaches. Conversely, many other nations have had little exposure to the developing theory and practice of access rights. It is thus important that the knowledge and experience already gained internationally be considered by those countries embarking on developing policies on access rights.

How then to practically develop and implement a system of access rights that will achieve social acceptance, sustainable usage of resources and optimal economic returns … if indeed this is possible?

The authors of this paper have recently all been directly and practically involved in seeking solutions to some real problems concerning access. Several expert groups in South Africa have been engaged in drawing up position documents on related topics (Cochrane *et al.* 1995; Branch *et al.* 1996; van der Elst *et al.* 1996) and we report on some of their deliberations here. These technical documents were instigated by the South African Department of Sea Fisheries and the South African National Fisheries Policy Development Committee. However, we do not speak on their behalf.

There are three major issues considered in this paper:

- Basic property rights
- Steps to be considered in developing access systems
- Recent South African developments.

Basic property rights

It is an undeniable fact that, for centuries, people have considered fish in the sea to be a common property: natural resources which belonged to those who took possession by catching them. 'Many of today's fishing problems stem from the traditional right of free access to the stocks' (Pearce 1994). Just like the air we breathe, the birds that fly and the rivers that flow, there has been an attitude that no single user has exclusive rights to these resources and no one can prevent others from sharing in their exploitation. Why set fish apart from other natural resources such as wildlife and land? Much can be attributed to the high cost of defending exclusive rights, invariably higher than the returns which harvesting the resource may bring (Christy and Scott 1965). Many fish species are virtually impossible to defend against exploitation by others while the wildlife in gameparks or a hectare of land is readily protected. Quite where to draw the line is often unclear. Does a fish that has been caught, tagged and released remain the property of the original fisher who merely releases it to grow bigger and fatter? Does a fish cultured in a fish farm and released into the wild have a specific owner ? What is clear is that common property resources are at risk from overexploitation. Lack of ownership and lack of control will not only dissipate profits but could result in collapse of the resource. There simply is no alternative but to harvest natural resources within the limits of an acceptable system of access rights.

The need for developing a system of sharing resources becomes more urgent as the resources dwindle and the demand rises. Competition for the declining resource may eventually become physical — in more ways than one. The fact that total global marine landings appear to have passed their peak is mirrored in the increasing number of scientific papers dealing with issues of access and user conflict (Fig. 1).

The debate surrounding the true nature of property and property rights will continue, irrespective of theories and policies. What is more important is to identify different property regimes that may be socio-economically acceptable and biologically sustainable. Ownership and access are not confined to the individual but include a wide spectrum of different levels of rights ranging from people and their communities to national and international claims of ownership. Four basic categories of property rights were proposed by Berkes and Kislalioglu (1991), and a fifth is added here:

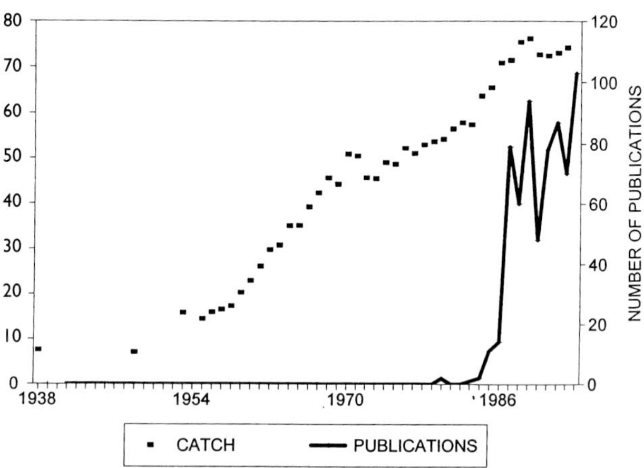

Fig. 1. The historic relationship between the annual global catch (millions of tons) of marine fish and publications dealing with issues of access to marine resources. Catch data sourced from FAO and publications from a key word search in Aquatic Sciences Extracts.

(i) Common property: Here there is no ownership and there are no property rights. Hence there is no form of restriction over entry to the resource and harvesting is unregulated.

(ii) Private property: Here exclusive rights to the resource have been allocated to individuals or companies to create private property rights. Since the resource is held in private, this includes the right to trade in the resource and the setting of catch and effort levels by the owners. The role of government is limited to the establishment and enforcement of such property rights.

(iii) Communal property: Here the resource is held and controlled by an identifiable community of users. The resource is then managed by that community of users, possibly in association with government to a greater or lesser degree.

(iv) State or national property: Here the resources are exclusively managed by the government of a state or country. Under such a system the government has sole jurisdiction over the use and allocation of the resource with management decisions made by government personnel.

(v) Global or international property: Although resources that fall outside the territorial or EEZ waters of nations may well be considered common property, there are increasingly more cases where nations have signed international conventions that effectively determine the shared ownership and access of such resources. Many examples exist, e.g. International Commission for the Conservation of Atlantic Tunas (ICCAT), Commission for the Conservation of Antarctic Marine Living Resources (CCAMLR).

These definitions are in fact extremes and most fishery examples around the world are a mixture of these property regimes. What is more important is to determine to what extent it is desirable for all or some fisheries to move to private ownership with the state being responsible for the protection of such rights. Alternatively, to determine to what extent it is desirable that the state retains control over initial and future allocation of resources.

It is an inescapable and well documented fact, that common property fisheries, where there is complete open access, tend to become depleted. Often such fisheries lead to excessive capital investment as stocks dwindle when fishers attempt to maximize their landings with improved gear and labour. Hence, not only is the stock depleted but the profitability of individual fishers is eroded. In the absence of sole or limited ownership, the profits are shared by all fishers, hence there is little incentive for them to leave the fishery to permit stock rebuilding and recovery. Economic efficiency can thus be improved by preventing excessive entry to the fishery but possibly linked to distributing the profits — or rent — more widely. For this reason, common property regimes cannot be recommended as a means of management for any stocks that are in demand for human use.

STEPS IN DEVELOPING ACCESS SYSTEMS

There seems to be consensus (if that is possible in fisheries!) that the fish resources of a country in fact belong to the people of that country. It is seen as the task of government to allocate the rights to these resources on an equitable, economically viable and biologically sustainable basis. This will involve one or more systems of access rights. Some of the steps to be taken in developing such systems are discussed.

Choosing effort or catch as the mechanism of limiting access

There is a variety of controls that can assist in establishing a system of access rights. Obviously, none will be effective without information on the stock, including information on its population dynamics and size of the sustainable catch. However, each mechanism has different implications. The mechanisms employed are of three types, namely 'input', 'output' and 'technical' controls. Input controls place limits on the amount of effort that can be put into the capture of the resource. For example, a limit may be placed on the number of boats or the number of nets that may be employed. Output controls place limits on the amount of the resource that may be harvested. Usually this takes the form of setting a Total Allowable Catch (TAC) which limits the overall catch. In most cases the TAC is set as a tonnage which is then divided proportionally between those people or companies who have been given the rights to harvest that resource. Technical measures may influence both input and output controls and include the establishment of protected zones, closed seasons and size limits.

Limiting effort by means of controlling entry, seasons or type of gear is common and has the advantage that the most efficient and skilled fishers should prosper most. It may also be possible to calculate levels of effort that will reduce the need for annual assessment of stocks. However, in multi-species fisheries it is possible that limiting effort on one species may result in under- or overutilization of other species. Furthermore, constant improvements in technology, and hence efficiency, need to be continually checked if the effort allocations are not to escalate beyond sustainability. In turn, fishers are likely to invest more in technological aids which may lead to overcapitalization. Australian fisheries, for example, have largely shifted from an effort limitation to one of catch control for the above reasons.

The nature of the fishery also needs to be considered. There are some fisheries where effort limitations are unlikely to be effective: for example, when the catch-per-unit-effort is not proportional to abundance, as in the case of shoaling species such as anchovy and tuna. In the case of fisheries that are based on recreation, subsistence or cultural practices, effort limitations are invariably unacceptable.

Limiting catch by way of setting TACs is perhaps the most commonly used method of controlling access to large fisheries around the world. Australia, Canada, Iceland, New Zealand, Holland and South Africa are among many nations which limit total catch (Wilder 1994). In Australia, federal government policy requires fishery managers to demonstrate that other systems are superior to transferable quotas before they will be considered (Anon. 1986).

Although limiting the catch overcomes the problem of 'creeping efficiency' in effort, there is a greater potential to cheat and hence greater compliance costs. In the case of New Zealand, with its transferable quota system, much of the control is managed through tightly monitored landing and audit records, creating a 'paper trail' that is said to have improved compliance control (Crothers 1988). Overall, the New Zealand experience appears to have been successful, with improved economic performance and better stock status (Annala 1996). However, and not surprisingly, it has not been a universal roaring success with the informal fishers such as those in the recreational and subsistence sectors. In many fisheries of developing countries, catch controls must fail simply through lack of capacity and infrastructure to undertake reliable monitoring.

Catch controls in multi-species fisheries can also lead to wastage when fish are dumped because their quotas have already been filled. A system of 'over-runs' and 'under-catches' that allow a proportion (say 10%) of the quota to be carried forward to the following year partly eases this problem (Annala 1996). Clearly, there are strengths and weaknesses in both effort and catch control. It is equally evident that the access to some fisheries is better controlled by one or the other method, or indeed a combination of both input and output controls.

How to allocate access rights

Once the basic method of controlling access has been defined, it remains for rights to be allocated and a number of factors need to be considered here. Four are suggested by Matthiasson (1992):

- who should obtain rights?;
- whether such rights are transferable and divisible;
- the time period for which rights should be granted;
- the process by which the initial allocation is made.

Who should be given access?

Essentially, access rights can be granted to individuals, companies, vessels or groups of people in communities or cooperatives. It is argued by Matthiasson (1992) that the nature of the recipients can have an impact on the economic efficiency of the fishery. Granting rights to vessels and communities rather than individuals may well improve efficiency as more capable fishers will be able to participate in the fishery. This in effect is achieved through transfer of rights.

Many nations, such as Australia, Canada, Chile, Holland, New Zealand, South Africa and the USA, have adopted quota schemes that are granted to individuals, companies or communities. The European Union has allocated communal rights to member nations to fish in the EEZ of each other. Iceland, on the other hand, has adopted a policy whereby access is linked to specific vessels (Arnason 1996).

Examples exist of access rights granted to communities. Japan and Fiji provide good examples of such systems. In Japan, the ancient system of clan ownership of local marine areas has been entrenched in law. Local cooperatives have jurisdiction over fishing grounds, although this still occurs within guidelines and regulations set by the national government (Kawaguchi and Naruko 1993; Pinkerton 1993). Such rights cannot be loaned, rented or transferred. There are interesting parallels here with the communal usage of wildlife in parts of Kenya where small landowners jointly share the benefits of hunting migratory species that occur on their communal lands.

Customary marine tenure is also in place in many parts of the South Pacific, apparently with success, often leading to Territorial Use Rights (TURFs) such as in Fiji. These are legally recognized and in many cases appear to be well co-managed (Christy 1982; Ruddle *et al.* 1992).

Other examples of granting access to groups or cooperatives exist, not necessarily belonging to the same community. In Turkey and Mexico examples exist of successful cooperatives where access to fish certain fishing grounds is vested in those which have open membership. Most appear to result in a high level of co-management.

Whichever system is developed it must be legitimate and enjoy the support of all participants. Whoever receives the rights of access, their overall catch still needs to be constrained, possibly by means of a TAC; whether this is internally or externally imposed depends on the level of co-management that has been achieved.

Transferability and divisibility of rights

Once rights have been allocated, a decision needs to be made whether such rights are restricted to the original holder or whether they can be freely transferred or divided. Transferability of rights can lead to greater economic efficiency as the most efficient fishers are likely to have the biggest share. It can also offer opportunities for equity and broadening access in cases where rights have been entrenched by a few operators, such as in the case of South Africa. In New Zealand, some rights were repurchased by the government and allocated for redistribution to Maori people. However, facilitating broader access or new entrants invariably calls for some measure of government intervention, possibly at the expense of economic efficiency.

It is clear that the nature of the fishery needs to be considered when deciding on the transferability and hence the concentration of rights. In some fisheries, where capital investment is great, it may be necessary to concentrate rights to allow for international competitiveness.

A special case of transferability of rights lies in the individual transferable quota (ITQ) system pioneered in Iceland (1976) and New Zealand (1986). Here, quotas as a proportion of TAC, are traded on free market principles with only limited government control. There are many reported cases of improved efficiency, greater employment and better resource management in regions where ITQs have been implemented. However, problems still remain and the final chapter on its success remains to be written (Edwards 1994; Annala 1996).

Tenure of rights

Most important is the consideration of the time period for which rights are allocated. This can vary from a short period to perpetuity. It is argued that longer periods are generally preferable in that they engender ownership and responsibility towards the resource amongst the right holders. Short-term rights may lead to 'mining' and virtual open-access if guarantees for renewed access are not given. However, this may also arise towards the end of a long-term tenure.

Long-term rights may be difficult to reverse, especially if errors have been made in allocations. Hence there is a greater emphasis on the initial correct allocation of rights. In Chile this is addressed by decreasing the annual quota by a small amount annually and allowing this to be auctioned off. This means new entrants can be accommodated and the quota holders are less likely to plunder stocks towards the end of their tenure.

New Zealand has opted for allocating rights in perpetuity on the basis that this will assist with capital investment and ensure maximum economic efficiency. However, such indefinite tenure is unlikely to find social acceptance in many fisheries, especially amongst those who have no, or only limited, access. Once again the nature of the fishery needs to be a significant consideration in deciding on tenure.

Initial allocation process

Most important is the initial system of allocation. Matthiasson (1992) suggests that there are five basic systems:

(i) Public enterprise: Here the state has sole ownership, does all the fishing and distributes the profits to the general public. This solves the problem of ownership conflict but invariably leads to inefficient use of resources, especially as the profit motive is absent. Moreover, the government must allocate, harvest and monitor — i.e. be player and referee at one time!

(ii) Harvesting under government contract: Essentially similar to (i) but harvesting is contracted out — probably resulting in less benefit for the people but greater overall economic efficiency.

(iii) Granting rights: This process needs to be socially acceptable and fair. In most cases (such as New Zealand and Iceland) access is initially granted on the basis of historic performance, whereafter free market forces take over. In instances where certain groups of people have been historically excluded (as happened in South Africa), specific steps must be considered to

restore a fair system, such as re-allocating some of the past rights to previously marginalized fishers.

(iv) Renting or selling rights: Here a fixed rental or purchase price is set and considerable funds are generated. These can cover costs of management and administration and the public sector can draw significant benefits. In Namibia, a 'resource rental' system has been introduced which provides tangible income to the state from the use of living resources and is structured to encourage participation by nationals.

(v) Auction: Though similar to renting, it is possible that an auction or tender system may well result in a higher financial return for the state, but is likely to favour those who have sufficient finances to outcompete others at an auction. Thus financial standing is the sole criterion deciding who earns the right to fish. In many instances other factors may be considered more important, including historical performance and labour employment.

The South African experience

Armed with all these diverse and glowing experiences around the world, how does one practically proceed with developing and implementing a new fishery allocation process. Such opportunities seldom present themselves, but have arisen recently in the case of South Africa with its restored democracy. Here we have a new country with a new and popular President, a new anthem, a new flag and new policies to be developed. An official Reconstruction and Development Programme (RDP) has created huge expectations in the country, not least amongst its many fishers who expect to benefit from 'broader allocation of resources' (RDP 1994). All that has not been restored are the fish resources!

Access to South African fish resources has in the past not been equitable and a real challenge now exists to accommodate these previously marginalized fishers. However, it is clear that most fisheries are already fully subscribed and only modest opportunities for increased harvests are to be found.

In order to resolve the issue a number of steps have been taken. First, it was decided to adopt a scheme not unlike the New Zealand Treaty of Waitangi which had settled Maori claims. Hence, a substantial quota of hake, which was part of an increased TAC, was technically allocated to disadvantaged fishers through a system of 'community trusts'. However, most of these fishers lacked capacity to access the offshore capital-intensive hake resources and effectively these quotas became 'paper quotas' that were then later sold back to the large companies. These 'paper quotas' clearly had considerable value but unfortunately a combination of poor implementation, inadequate benefits to satisfy everyone's expectations, and unbridled demand for 'free fish,' led to a fiasco that not only ended in riots but further tarnished the entire fisheries management process.

Next, a most eventful and stormy public meeting was held by the Minister of Environment Affairs and Tourism (responsible for fisheries) with several hundred irate fishers. Here he announced the creation of a National Fisheries Policy Development Committee (FPDC) to devise a new fisheries policy for the nation. This has been a most painful two-year process. Hindsight is an exact science but the chairperson, Mr Mandla Gxanyana, and his large committee faced a daunting task. How do you cut up a cake that is already being consumed? Certainly not everyone can be satisfied. Politically correct, but practically impossible was the rule that all decisions were to be by consensus. This has meant that many potentially good ideas were scrapped by the committee because one sector (or one person on the committee) objected. The FPDC comprised various users and regional representatives, most with only moderate experience in the wider principles of fisheries management, but each representing the views of a particular sector and, hence, with vested interests in the outcome. Scientists were purposely excluded, which retarded the development of possible models. Instead, much of the debate was clouded with labour issues and inter-provincial (state) claims to stocks which further slowed progress. Nevertheless, the committee served one very important function. For the first time in the history of fishery management in South Africa, it brought together all the main players with a stake in fisheries, including recreationists, conservationists, managers, industry, small business and even representatives of previously marginalized communities who had turned to poaching. After initially slow progress, the committee appointed several specialist task teams to make progress and this proved vital to the development of the policy. Amongst these task teams one was appointed to deal with possible options for controlling access and another to deal with the question of subsistence fishers (Branch *et al.* 1996; van der Elst *et al.* 1996).

Table 1. Classification of South African Fisheries

Fishery Types			
Offshore	**Demersal**	**Inshore**	**Intertidal**
Offshore pelagic Deepwater trawling Demersal offshore trawling Midwater trawling Tuna longlining and poling Natal & south coast lobster Offshore prawn trawling Demersal longlining Boat linefish	Demersal longlining Demersal inshore trawling Boat linefish Inshore pelagic Squid West coast lobster traps	West coast lobster hoopnets Seals Inshore linefish Abalone Inshore nets (drift, set, trek nets) Whelks, octopus Kelp Specimen collection Biologically active components	Limpets Rock mussels Sand mussels Seaweeds Winkles Bait organisms Red bait

Feedback between these specialist task groups and the committee led to active interactions. Particularly useful were a series of workshops. These interactions allowed a draft policy to be hammered out. In June 1996 the Draft Fisheries Policy was handed to the Minister of Environmental Affairs and Tourism, and it is intended that this be released as a 'white paper' for public comment before it can be implemented. It remains to be seen what finally emerges and how it will be implemented.

What has become very clear is the fact that the nature of a particular fishery is a primary determinant in considering who should have access to them. Consequently, the Access Rights Task Team based many of their deliberations on this fact and devised a system of classifying the different fisheries in South Africa. Four basic groups were identified, depending on the amount of technology and finance involved in harvesting, the accessibility and mobility of stocks and the ease of control over harvesting (see Table 1). The main features of these four types can be gauged from Figure 2. It is clear that there are great differences between the four types in terms of the biology of stocks, the nature of the fisheries, the characteristics of users, as well as the types and levels of control that can realistically be applied to each.

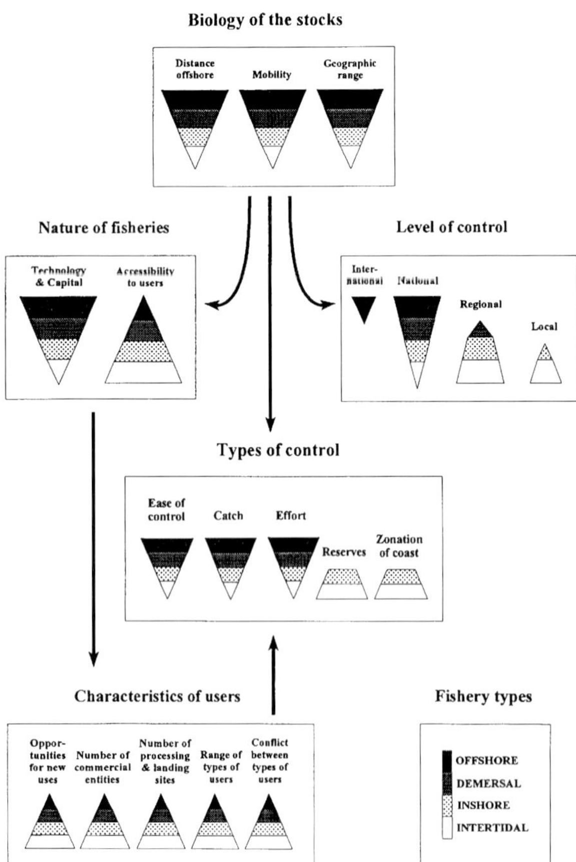

Fig. 2. Summary of the ways in which the biology of the stocks influence the nature of the fishery, the types of control, the appropriate regional level of control and the characteristics of the users.

Some of the major conclusions

Predictably there was no universal acceptance of all the various proposals and suggestions although there was much more consensus on more issues than anticipated. A number of suggestions were proposed:

In general:

- Open access was rejected.
- Allocation should be fair, equitable and transparent, and recognize past inequalities.
- Foreigners should be excluded from the EEZ unless South African companies lack capacity to harvest a particular resource.
- Control systems should be a combination of catch, effort and technical controls, tailor-made to ensure sustainability of each stock.
- Rights should be transferable by sale.
- Once sold, sellers cannot reclaim rights.
- Duration of rights should be long-term but can be withdrawn if abused, or bought back if needed for re-allocation.
- There should be payment for rights. This could include a purchase price and/or an annual rental.
- Rights should initially be allocated according to a set of 'broadly accepted' criteria. Criteria would include historical involvement, capacity to access the resource, and proven past discrimination. Contravention of the rules and regulations should result in forfeiting of rights. 'Paper quotas' should not be permitted.

For recreational users:

- Recreationals would also be expected to pay for access in the form of a levy.
- There should be open entry for all persons although limits would be placed on catches.
- Certain species should be allocated for primarily recreational use and custodianship.
- Recreationals should not be allowed to sell their catch.

For subsistence users:

- These should be specifically identified and licensed.
- Different regulations (such as bag limits) may be applied to subsistence and recreational users.
- They should be allowed to sell part of their catch, but only locally.
- Subsistence users may be given preferential rights along specific zoned sections of coast similar to TURFs.

How to allocate rights to new entrants:

The re-allocation process

In the South African context there is a special predicament. Not only do resources have to be fairly allocated to new entrants, but the disparities of the past necessitate a special case of re-allocation. Given that most fisheries are already fully used, this creates big problems. In some cases it will be necessary to reduce allocation from existing right-holders and give to those who can demonstrate past discrimination. However, this should be done in a manner which does not place the fishery at risk because threats to any fishery imply a threat to the established labour forces in that fishery. When taking a decision on which stocks can most readily be transferred in this way and with these goals, the nature of the fishery plays an important role. So, for example, nearshore resources which are readily accessible, comparatively sedentary and can be harvested relatively inexpensively, may be suitable candidates for allocation to new entrants into the fishery. On the other hand, offshore resources that are very mobile and require high technology and high capital investment may best be left in the hands of the established industry.

Among all participants concerned with the development of policy there was consensus (for once!) that the fish belong to the people of South Africa and furthermore that the fish have a value. Hence, any system of re-allocation should also be economically sound and rights should be obtained at a price. However, as current holders did not pay for their rights in the first place, it may seem unfair to expect new entrants (especially those previously disadvantaged) to now have to pay.

A model has thus been proposed which includes a 'transition period' of redistribution (Fig. 3). During this time, (possibly 5–10 years) an annual percentage of some stocks could be redistributed from current holders to previously disadvantaged users. The nature and size of such transfers would be carefully determined to avoid jeopardizing the viability of the fishery and its labour force. The rate of attrition would depend on the nature of the stock and the relative efficiency with which it can be harvested by different economic sectors. For instance, in the case of capital-intensive offshore fisheries, the attrition is unlikely to exceed 1%–2% per year. The current holders would be compensated for this attrition, in recognition of the fact that their investment developed the fishery in the first place. There might also have to be compensation paid for any retrenchments that would have to be made. New entrants would be expected to pay for their rights and should not be able to sell them during this transitional period. Those who had been politically disadvantaged could be eligible for government support in the form of soft loans, expertise, etc. The current holders cannot be expected to pay outright for the rights that they previously obtained for free, as this would be prohibitive. However, they would be expected to pay a 'Special Transition Levy' during this period. This levy, in addition to the normal management levy, would be imposed on all right-holders in all fisheries and would be held in trust. This Trust would then provide the source of funds for assisting disadvantaged entrants as well as providing compensation to be paid to those who would lose a proportion of their rights.

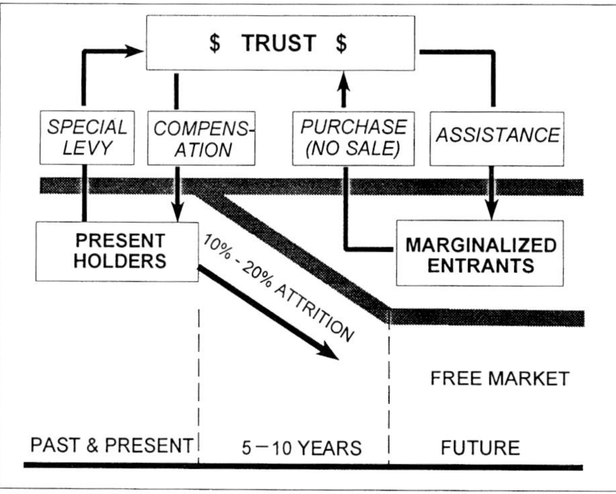

Fig. 3. Possible re-allocation model.

After the Transition period had lapsed, all rights would be entrenched on a long-term basis and the entire system would be market driven thereafter. Only macro-government policy would subsequently be needed such as possible buy-back of rights should this be necessary for management or further re-allocations.

One further advantage is that the entire process will be internalized within the fishery and not dependent on already overstretched national budgets. The model should not interfere with bio-economic equilibria nor should it result in subsidized fishing that could unfairly influence international trade pricing.

This model is not unlike aspects of the New Zealand ITQ system; just packaged differently.

Conclusion

No matter how important we consider scientific information and economic efficiency as inputs to decisions about fisheries management, they are only complementary to resolving the issue of access to resources. Without popular support, good science will remain a junior partner and decisions by 'politicians' will prevail.

Unless you manage fisheries through the barrel of a gun (which does happen of course!) there is no alternative to setting up structures that will facilitate discussion and transfer of information. Above all there must be (and be seen to be) fair play. In South Africa, the development of a new fisheries policy and associated models for access are yet to be accepted and implemented but the discussions and deliberations that have taken place at dozens of fishing fora purposely set up throughout the country (not infrequently under a tree in the bush!) have been a major positive development.

Fisheries scientists should play a much greater role in selling their product. There is huge potential — and need — for educating fishers, managers and politicians. Considering the advancing state of multi-media technology one cannot help but be reminded of the initiatives taken by the late Philip Sluczanowski in his endeavours to break the technology gap (Sluczanowski *et al.* 1993; Sluczanowski and Prince 1994).

Despite the many nations represented at this Congress, many more are not present. Many of these have no, or only poorly, developed methods of controlling access and many are primarily small-scale tropical fisheries that continue to be marginalized in a global sense (Pauly 1994). This means that a significant part of the global catch remains subject to open access and potentially at risk, especially in developing countries where food security is paramount. Ultimately, open access must lead to uncontrolled harvest, the 'tragedy of the commons' and collapse of economically viable stocks.

There is no universal model for allocating fish. It is also not possible to satisfy all demands and expectations. Social norms of nations and peculiarities of each resource and its users will have to influence access consideration. However, this should not preclude or indeed delay the development of sustainable access systems in all fisheries.

Acknowledgments

Many individuals have contributed to the discussions reported here. In particular the following are acknowledged: M. Gxanyana, R. Ball, P. Zulu, M. Sowman and D. Baird.

References

Annala, J H (1996). New Zealand's ITQ system: have the first eight years been a success or failure? *Reviews in Fish Biology and Fisheries* **6**, 43–62.

Anon. (1986). New directions for Commonwealth fisheries management in the 1990s: A government policy statement. Australian Government Printing Service, Canberra.

Arnason, R (1996). On the IRQ fisheries management system in Iceland. *Reviews in Fish Biology and Fisheries* **6**, 63–90.

Berkes, F, and Kislalioglu, M (1991). Community-based management and sustainable development. In: 'La recherche face a la peche artisanale'. Symposium International (Ed J R Durand *et al.*) pp. 567–74 (Orstom-Ifremer, Montpellier, France).

Branch, G, Baird, D, Cochrane, K, Moola, Z, Butterworth, D, and Sowman, M (1996). Review of access rights options for South Africa. Final report of the Access Rights Technical Committee appointed by the Fisheries Policy Development Working Committee. 70 pp.

Christy, F T (1982). Territorial use rights in marine fisheries: definitions and conditions. *FAO Fisheries Technical Paper*. No 227, Rome.

Christy, F T, and Scott, A (1965). The common wealth in ocean fisheries. (Johns Hopkins University: Baltimore). 281 pp.

Cochrane, K. Bross, R, Butterworth, D, Hutton, T, Laan, R, Jackson, L, Martin, R, Payne, A, Penney, A, Roelofse, A, and Shannon, L V (1995). Review of international experiences in access rights and their implications for fisheries management in South Africa. Sea Fisheries Research Institute Access Rights and Resource Implications Task Group. 89 pp.

Crothers, S (1988). Individual transposable quotas: the New Zealand experience. *Fisheries* **13**(1), 10–12.

Edwards, S F (1994). Ownership of renewable ocean resources. *Marine Resource Economics* **9**, 253–73.

Kawaguchi, K, and Naruko, T (1993). An overview of the coastal fisheries management system in Japan. *FAO Fisheries Report* **474**(2), 417–25. Kobe.

Matthiasson, T (1992). Principles for distribution of rent from a 'commons'. *Marine Policy* **16**(3), 210–31.

Pauly, D (1994). Small-scale fisheries in the tropics: marginality, marginalization and some implications for fisheries management. *ICLARM contribution* No 1070.

Pearce, P H (1994). Fishing rights and policy: the development of property rights as instruments of fisheries management. In 'The state of the world's fisheries resources'. Proceedings of the First World Fisheries Congress Plenary Session. (International Science Publisher: Lebanon). pp. 76–91.

Pinkerton, E (1993). Local fisheries co-management. A review of international experiences and their implications for salmon management in British Columbia. *Internal Report*. 20 pp.

RPD (1994). The reconstruction and development programme, African National Congress, Umanyano, Johannesburg. 1–147.

Ruddle, K, Hviding, E, and Johannes, R (1992). Marine resource management in the context of customary marine tenure. *Marine Resource Economics* **7**, 249–73.

Sluczanowski, P R W, and Prince, J D (1994). User interface adds value to fisheries model — ABASIM. *Agricultural Systems and Information Technology* **6**(1), 44–6.

Sluczanowski, P R W, Walker, T I, Stankovic, H, Forbes, S, Tonkin, J, Schenk, N H, Prince, J D, and Pickering, R (1993). SharkSim: A computer graphics model of a shark fishery. In 'Shark Conservation'. Proceedings of an International Workshop on the Conservation of Elasmobranchs held at Taronga Zoo, Sydney Australia. 24 February 1991. (Eds J Pepperell, J West and P Woon) pp. 45–7.

van der Elst, R P, Butterworth, D, Hecht, T, de Wet Schutte, D, and Salo, K (1996). Relief measures for marine subsistence fisherfolk in South Africa. Report of the Technical Relief Measures Task Team appointed by the Fisheries Policy Development Committee. 19 pp.

Wilder, M (1994). Quota systems in internationl wildlife and fisheries regimes. Environmental policy: from regulation to economic instruments. (Centre for Studies and Research in International Law and International Relations, Hague Academy of International Law.) 56 pp.

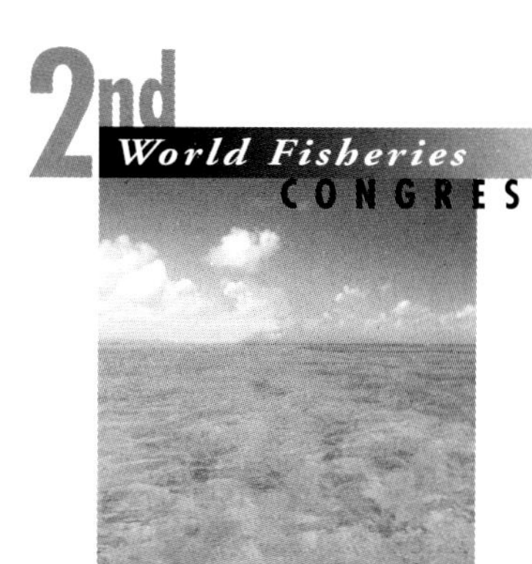

Importance of Fish and Fisheries to Aboriginal Communities

Danny Chapman

NSW Aboriginal Land Council, South East Coast Branch, Shop 2 Chapman House, 34D Orient Street, Batemans Bay, NSW 2536, Australia.

Summary

This paper examines the importance of fisheries to the Indigenous people of Australia. It explores the history of the fishing industry from pre-colonization to the present day.

Evidence of Aboriginal fishing activities can be traced back 7000 years and this can be found in middens around estuaries and the ocean shore-line. After colonization and due to the competition for fertile land Aboriginal people were 'rounded up' and moved onto the less fertile land which was generally found around the shore-line. In order to survive, the Aboriginal people were forced to catch fish and to rely on the ocean to sustain them.

From colonization to the middle of the 1980s, fishing generally remained unregulated and overexploited. Around this time the Government started to recognize that the industry would not survive without further tightening of the regulations. Historically when there is conflict between black and white races, usually the white prevails and this could not be better illustrated in the fishing industry. The first people forced out of the industry were the Aboriginal people, and the prosecutions still happen today.

In 1993 the High Court of Australia recognized, in MABO, that Aboriginal people had Native Title. The NSW Court of Appeal also found that Native Title extended to include fishing rights. However, in the Mason *v* Tritton case there was not sufficient evidence found to support the claim of Mason that he was fishing under his Native Title.

Introduction

This paper will underline the importance of fish primarily to the coastal Aboriginal people in south eastern NSW (hereafter the 'South-east coast') and to a lesser extent the inland Aboriginal people, and will describe events from pre-colonization to the present day.

I propose to cover the following points:

- The role of fishing in the lives of Aboriginal people for sustenance and as a traditional practice;
- How the importance of fishing to Aboriginal people has endured and thrived since colonization, at times with the cooperation of the NSW Government (hereafter 'the Government'), and at other times despite the actions of the Government;
- How government regulation of the fishing industry has affected Aboriginal people; and
- Discussion on the effect of the recognition of native title rights by the High Court of Australia in ***Mabo No.2*** ((1992)

175 CLR 1) on both Aboriginal fishing and the fishing industry in general.

It will not be possible to explore all the areas in the depth that is required to fully appreciate the position that Aboriginal people find themselves in today. It should be pointed out at this early stage that this paper is presented from an Aboriginal point of view.

Aboriginal fishing before the European invasion

It is without doubt that Aboriginal people occupied the Australian continent first, long before the Europeans settled our land. In fact, some archaeologists have found evidence that we have been in Australia for at least 40 000 years. During that time Aboriginal people enjoyed all the resources the land and waters had to offer, including fish and shellfish (both referred to hereafter as 'fish').

On the coastal fringe of Australia, in particular the south-east coast on which I will be concentrating, there is much archaeological evidence of Aboriginal activity found along the shore-line and estuaries dating back 7000 years. Evidence of coastal Aboriginal activity before 7000 years is now under the sea due to the rising of the seas following the end of the last ice age.

The main focus of this evidence is on areas called middens where the catch was consumed. Middens typically contain fish shells and bones and the bones of land animals as well. There are many large middens along the south-east coast and indeed along the whole of the NSW coast. Middens continued to be used in the post-contact period.

The importance of fish to Aboriginal people on the south-east coast was extensive. Fish meant sustenance either as a food source or for trade purposes. The Aboriginal people of the south-east coast moved up and down the coastal area between what is now La Perouse and the Victorian border and fishing was important in supporting this.

Although fishing by Aboriginal people has been described as an economic tradition by anthropologists and the like, this does not convey the full picture of its importance. It is certainly true that fishing is primarily an economic activity, but it is also a communal activity. A large percentage of the Aboriginal people from the south-east coast currently fish and have always done so. Knowledge, customs and techniques related to fishing are passed down from generation to generation.

Even if a time came when fishing is no longer necessary for the survival of our people we would continue to fish. It is a practice that is entrenched in our culture that you will see has endured the many changes to the lives of our people over the last 200 years or so and it will continue to endure.

Feasting on fish was also an important part of ceremonies and still plays a central role when our people get together.

Fishing after the invasion

While the importance of fishing to the Aboriginal people of the south-east coast continues to this day its practical applications have changed along with the changes in the circumstances of Aboriginal people since colonization. The constancy of its role in the lives of my people and its adaptability to the changes in our lives stamp it as a truly living tradition. As such, the other impacts of European settlement provide the context for the changes to its applications to the lives of Aboriginal people.

I will now attempt to show how several factors and the effect of economic downturns have not only intensified our reliance on fishing but have also improved our ability to catch fish. Some of these factors include: improvements in technology, the availability of a wider trading market, the establishment of Aboriginal reserves by the Government and the restrictions of access to land resources, and the attitude of the Government toward Aboriginal fishing.

The first white settlers moved into the south-east coastal area in 1819, looking primarily for fertile land for agricultural needs. The foresters, sealers and gold miners soon followed. Along with this intrusion came conflicts and competition for land which led to massacres and the poisoning of Aboriginal people at Eden.

On the south-east coast, Aboriginal people soon used their ability to catch fish to their advantage following European settlement. Trade developed from the outset, often as much to the advantage of the settlers as to the Aboriginal people. For example, the town of Broulee, which is between Moruya and Batemans Bay, was often short of food in the 1830s because the supply boats came infrequently. It is well known that the Aboriginal people saved the settlement from starvation several times by supplying fish.

Aboriginal people were also employed in the fishing industry very soon after settlement. For example, in 1839 the two whaling boats in Eden were crewed exclusively by Aboriginal people. During the second half of the 19th century and up to the First World War, Aboriginal people used fishing as a means of self sufficiency. We were being driven off the fertile land and into reserves by Government acting on the demands of farmers. Most of these reserves were located near the coast on the sandy land unsuitable for agriculture and soon became residential bases for fishing activities.

As our capacity to live from the fertile land was diminished we came to rely more on fish. In recognition of this trend and to reduce the need to supply rations, the Government provided some fishing boats. The Government's systematic dispossession of Aboriginal people from their lands on the south-east coast was actually facilitated by Aboriginal fishing traditions. As fish was one of the tradeable commodities to which Aboriginal people had access, fish were traded for other necessities.

During the years between the two World Wars the reliance on fishing increased even further. Commercial fishing by Aboriginal people also increased, as evidenced by reports of non-Aboriginal people making official complaints to the Fisheries Department about the threat of Aboriginal fishing to their livelihoods. By 1926, about 75% of the Aboriginal reserves on the south-east coast had been revoked to make way for urban expansion and more agriculture. This further decreased the ability of Aboriginal people to live from the land and increased their dependency on fish.

The Depression was devastating; as the fish markets collapsed and opportunities for self-sufficiency evaporated. Aboriginal people were denied the unemployment benefits and often the rations which were being made available to non-Aboriginal people. This was despite Aboriginal people in those times being able to hold fishing licences and pay the appropriate fees to the Government.

Since the Second World War conditions have become even more difficult. The amount of 'public' land available for use has been further reduced, as has the availability of employment to Aboriginal people. There has been more non-Aboriginal fishing activity, which has led to depletions of fishing resources and increased regulation of the fishing industry. Aboriginal people are now the most economically-disadvantaged group on the south-east coast.

Aboriginal people continue to rely heavily on the sea for sustenance. As economic conditions become more difficult, opportunities close off and we head back to the sea. This is one reason why fishing is now such a widespread practice among Aboriginal people on the south-east coast.

In the post-war years, not only has the dispossession of land continued but the Government has also sought to start restricting our access to fish, especially to trade. This has occurred through legislation and regulation. The continual reduction in our access to land and sea resources makes the goal of self-sufficiency more and more difficult to achieve.

Nevertheless, our people continue to fish both to eat and to trade as we have always done. We will continue to take the economic opportunities which present themselves in order to survive. Fishing has always provided those opportunities and even though they are constantly under attack, I believe fishing will continue to provide us with opportunities.

GOVERNMENT REGULATION OF FISHING

Just as we have relied more on the sea when resources become scarce, so do non-Aboriginal people. With the continual rise in the population of Australia there is increased demand for fish. The opening up of international markets and improvements in refrigeration technology and transport have further increased the demand for fish. Some fish, such as abalone have become an industry almost totally supported by demand from overseas.

The increase in demand for fish led to many more people becoming involved in commercial fishing. The increase in commercial fishing activity has resulted in the fish resources being depleted. In the case of some species the supplies have decreased markedly.

It is not, however, just the number of people involved in commercial fishing and the amount of fish caught that has produced shortages and endangered fish supplies. Commercial fishing typically involves hauling whatever fish are there and keeping the most commercially-valuable species. The immediate demands of the market do not necessarily encourage consideration of factors such as the continued existence of any given species.

By contrast, Aboriginal people employ a 'circular' method of fishing where we fish for whatever is in season at the time. For example, there are specific times of the year when prawns are plentiful and so at that time of year we target prawns. It is the same with other species of fish such as mullet and so on. This method ensures sustainability of fish resources because, by catching the species that are most plentiful at any given time, no species become endangered.

The Government controls fishing by issuing licences and imposing bag limits. Initially the Government issued general fishing licences under which any type of fish could be caught. While there are some Aboriginal people with fishing licences this is more the exception than the rule. There are now less than ten Aboriginal people with general fishing licences on the south-east coast. The communal and supposedly 'irregular' (which we dispute) way Aboriginal people generally conduct fishing activities does not necessarily fit well with a licensing scheme based on the allocation of fish quotas to individuals. The general fishing licence limitation scheme failed to prevent serious depletion of some species of fish.

ABALONE FISHING

One example of stock depletion is blacklip abalone (*Halitois rubra*). The arrival in the late 1960s of non-Aboriginal professional divers using hookah equipment, enabling them to stay submerged for long periods, resulted in depleted abalone stocks on the south-east coast by the mid 1970s. Unable to obtain licences in other Australian States, many of these divers came to New South Wales where they could operate with few restrictions. Most, of the commercial catch of abalone is exported as Australians are not prepared to pay the high prices abalone command. The professional divers rarely eat abalone either.

Aboriginal people on the south-east coast on the other hand have been eating abalone for at least the last 3000 years. Abalone shells are found in significant numbers in middens all along the south-east coast. Abalone were taken for food and for trade, and this continues today. In contrast to non-Aboriginal divers, Aboriginal divers do not use hookah equipment. The only equipment aids used, if any, are wetsuits, flippers and snorkels. This restricts abalone diving to areas near the shore and this practice protects the species.

There are accounts of Aboriginal people selling abalone both domestically and for export in the late 1950s and early 1960s, to South-East Asian people. Even though the non-Aboriginal abalone divers only arrived on the south-east coast in the late 1960s, by the mid-1970s the rampant taking of abalone for commercial purposes had led to a crisis.

In 1980 the NSW Government responded to abalone stock depletion by introducing a scheme which restricted the right to take abalone for commercial purposes to those with special licences. In 1984 abalone was declared a restricted fishery in New South Wales. In deciding who was to be granted special abalone fishing licences, the Government set out certain criteria, the most important of which was the amount of abalone a diver had already taken. The use of hookah equipment meant that it was the non-Aboriginal divers who had taken the most abalone and so they were granted most of the licences.

Most Aboriginal divers did not meet the entry criterion of catches of 200 kg of abalone per month for six months in each year over a three year period. Only two out of the 59 licences issued went to Aboriginal people. The use of sustainable fishing practices counted for very little.

There are other reasons why Aboriginal people did not meet the Government's entry criteria for abalone fishing licences. A large number of Aboriginal people from any given community engaged in fishing, share the work around. Therefore, the amount any individual caught was inevitably less than the prescribed catch.

The criteria applied by the Government in issuing licences to fish were not sensible. Licences were issued to those who caught the most fish, yet the reason for the licensing system was to ensure that fish resources were not fished out. Therefore, the people who were responsible for the depletion of fish resources were the ones rewarded with licences. There would have been no need for a licensing system if these people had been fishing with the sustainability of the resources in mind.

In the space of approximately 15 years, Aboriginal abalone divers went from a situation where we were the only people fishing for abalone to having our traditional abalone fishing practices outlawed. Our current status under the regulatory scheme is the same as any recreational diver. We are limited to 10 abalone per diver per day while the average licence holder is entitled to 9 t per diver per year.

The restricted licence scheme has had little success in preserving abalone stocks. Abalone are becoming more and more difficult to find each year. Over the next fifteen years, we have found that NSW Fisheries is prepared to go to any length to prosecute us for diving for abalone. NSW Fisheries now even deploys an aeroplane to control the coast to check whether Aboriginal people are diving. The fisheries agency has taken a confrontationist approach which has only increased tensions. For example, these days NSW Fisheries officers only have to see Aboriginal people near the ocean and they pull them up and search their vehicles.

Ongoing relations with NSW Fisheries

The relationship between NSW Fisheries and Aboriginal communities on the south-east coast has been very strained for some time. Successive governments have met with us but there has been no action. We have told those governments that we will continue to exercise our traditional rights even if they do nothing. The fact that fishing continues despite the constant and threatening presence of NSW Fisheries officers only underlines the strength not only of the sustenance aspect of Aboriginal fishing but particularly of the tradition.

The unwillingness of governments to address this problem will be detrimental to all in the long run. We are concerned that abalone fishing is the thin end of the wedge and that the same thing will happen in other restricted fisheries in the future. The new legislation in New South Wales covering commercial fishing, the ***Fisheries Management Act 1994***, has compounded this concern. Also, the High Court's decision in ***Mabo No.2*** has left little doubt in our minds that the common law will recognize the existence of native title fishing rights for Aboriginal people in Australia.

Both these factors will have a significant impact on the management of fishing resources in New South Wales in the future and bear much deeper analysis than I have time for here. I do however, want to make some comments about both of them. I will begin with the ***Fisheries Management Act 1994***.

Fisheries Management Act 1994 (NSW) (hereafter 'the new Act')

The new Act expressly does not affect native title rights. The Act created a system which can be loosely described as granting property rights in fish to existing licence holders. Licence holders were able to apply to the Minister for Fisheries for 'shares' in a fishery. Applications could be made based on catch history either by an individual or under a boat licence. Commercial fishers are still required to be licensed.

If a person without a licence wishes to gain access to a fishery, he or she must purchase shares from an existing shareholder in a recognized fishing operation. A person will not be issued with a fishing licence unless he or she holds shares in a recognized fishing operation. Individual fishing boats must be licensed also. Boat and licence packages sell for between $A70 000 and $A500 000 depending on the size of the boat. Hence the cost of buying into the industry is prohibitive for most Aboriginal people.

The new Act has put further pressure on Aboriginal fishing. The Government is currently restricted by the Act from redressing past injustices by issuing fresh licences to Aboriginal people. The Act also provides NSW Fisheries Officers with wide powers to search vehicles, boats and premises and to seize equipment.

Again, the new Act's operation is at odds with its supposed intent of ensuring sustainable fishing. Shares and quotas were allocated to existing licence holders on the basis of catch history which favours those who have been taking the most fish already. While there is no requirement that fishers catch a certain amount in the future, the workings of a quota system encourage licence holders to fish up to their quota even if the fish resources are being severely stretched.

Negotiations were commenced last year with a view to introducing a special type of licence for Aboriginal people. However, this proposal has yet to be implemented.

Native title and Aboriginal fishing

The common law of Australia has recognized traditional rights since the decision of the High Court in ***Mabo No.2*** in 1993. Justice Kirby, in his judgement on the NSW Court of Appeal case in ***Mason v. Tritton*** ((1994) 34 NSWLR 572), stated that he believed that the law in Australia recognizes a form of native title or traditional right of indigenous people in fish. The claim for native title in this case failed however, because the standard of proof required was not satisfied.

That the traditions of Aboriginal fishing continue to exist to this day is undoubted. It was acknowledged in ***Mabo No.2*** that a traditional right is capable of modernization. Therefore, for the law to be in any way a genuine attempt to recognize these traditions, native title rights in fishing must be recognized. The scope of the 'rights' which are to be recognized by the law will be a measure of Australia's willingness to accept Aboriginal traditions.

Although the courts have yet to recognize native title fishing rights in any particular case, these rights will continue to be asserted. There are at least two native title claims in NSW which include waters which have already been registered. These are at Wellington which is inland and Byron Bay which is coastal. There will be more to follow. It is merely a matter of time before there is a successful claim.

It would be difficult for the NSW Government to sustain an approach which attempts to legislate to extinguish these rights or to argue that they have already been extinguished. Even if it succeeds in arguing that native title fishing rights have been extinguished or impaired there may well be large amounts of compensation payable to those whose rights were extinguished.

Firstly, it was presumed before ***Mabo No.2*** that no such rights existed. It would therefore be difficult to argue that any legislation prior to ***Mabo No.2*** intended to extinguish those rights.

Secondly, NSW law only allowed the declaration of restricted fisheries in 1979. This is significant because the argument that declaring a restricted fishery exhibits an intention to extinguish an Aboriginal fishing right may be put (although I do not agree with it). Even if such an argument was to succeed, the Aboriginal claimant would have the opportunity of claiming for compensation under the Native Title Legislation due to the extinguishment or impairment of native title rights inconsistent with the ***Racial Discrimination Act 1975 (Commonwealth).***

Thirdly, the States only acquired the right to legislate in relation to the territorial sea from the Commonwealth in 1980 through the ***Coastal Waters (State Powers) Act 1980 (Commonwealth)*** and ***Coastal Waters (State Title) Act 1980 (Commonwealth).*** Significantly, any right or title to the property in the seabed beneath the coastal waters was excluded from the handover. Native title rights would fall into this category.

Fourthly, any future attempts by the Government to extinguish native title fishing rights in NSW by legislation would be inconsistent with the ***Native Title Act 1993 (Commonwealth)*** and the legislation would be invalid to the extent of this inconsistency. In Australia, Commonwealth laws prevail over State laws to the extent of any inconsistency.

CONCLUSION

The current 'stand off' between NSW Fisheries and Aboriginal people continuing to practice their fishing traditions is unlikely to be resolved by litigation. In any case the sustainability of fishing resources is a continuing concern for all.

There need to be negotiated agreements and ongoing co-operation between the Government and those with native title fishing rights. The first step is for the Government to recognize the existence of these rights so that our people who continue to exercise them do not continue to feel under siege.

Agreements relating to co-management of fishing resources, as have happened in other countries, between the Government and Aboriginal groups are just one way in which cooperation can occur. Other options are also available.

Already negotiations involving the Government and Aboriginal groups have commenced with a view to setting up a viable aquaculture industry in NSW. Establishing and operating abalone farms is a way in which coastal Aboriginal communities can head towards self-sufficiency. The re-seeding of the coast with abalone from hatcheries should also address the problem of dwindling abalone stocks.

How Many Fish Can A Sparrow Eat? — Litigating Aboriginal Claims To Fisheries — The Canadian Experience

Hugh Braker

PO Box 1178, Port Alberni, British Columbia, Canada V9Y7M1.

Summary

Aboriginal rights and treaty rights over the resources of the seas and rivers are increasingly disputed between aboriginal people, the State and other fishing interests. In Canada, by vigorously defending prosecution for fishing offences in the criminal courts, aboriginal people are promoting their rights and challenging the economic and social policies of the State. A recent spate of litigation for aboriginal claims to fish resources highlights the complexity of current jurisdictional arrangements. Management of Fisheries and Oceans and of Indians and Lands Reserved for Indians falls under Federal jurisdiction, whereas management of processing of fish and fish products falls under the jurisdiction of the Provinces. Management of inland fisheries is implemented through various agreements between the Federal and Province governments.

The seminal case for litigation began in May 1984 when Ronald Edward Sparrow of the Musqueam First Nation Indian Band was prosecuted for fishing with excess net in the Fraser River at Canoe Pass, Vancouver. Despite the innocuous offence, the *Sparrow Case* ended with appeals in the Supreme Court and is pivotal to the development of aboriginal rights law in Canada. Since then, not only have there been numerous other aboriginal litigants, but non-aboriginal commercial fishing interests have sought to contest aboriginal rights. As the *Sparrow Case* is further interpreted, various resource allocation issues, such as the validity of existing commercial fishing licence conditions, are raised. The issues of aboriginal rights to sell fish, the authority of the State over conservation of fish stocks and protection of fish habitat, and the obligations of fisheries agencies to consult with their constituencies need clearer definition.

Introduction

It is inevitable with growing demand for the resources of the seas and rivers that there is increasing conflict between the nation states and those who hold aboriginal rights or aboriginal treaty rights. In Canada that conflict is manifest in the courts. Canada is not the only jurisdiction in which aboriginal people have turned to the courts to protect their right to the resources of the seas and rivers. In Washington State, for example, aboriginal people rely on the courts to settle conflicting claims to the resources; see cases *United States v Washington* (1974); *Puvallup Tribe v Washington Department of Game* (1968); *Sohappy v Smith* (1969); and *United States v Washington* (1980).

Unlike Washington State and almost every other jurisdiction, until recently the consistent pattern of aboriginal fishing rights litigation in Canada has been in the criminal courts. As a result of attempts by the State to suppress aboriginal fishing practices through the use of criminal or quasi-criminal legislation, aboriginal people have found themselves arguing aboriginal rights or even title in defence of charges in court.

As the Supreme Court of Canada stated in the case of *R v Sparrow* (1990):

> the trial for a violation of penal prohibition may not be the most appropriate setting in which to determine the existence of an aboriginal right.

Not only is the burden of proof different in a criminal trial but the rules of discovery, evidence and procedure are different. Nonetheless, given the competition for the resources and the attempts by the state to strictly define aboriginal fishing rights, the criminal courts in Canada have become the forum for the development of aboriginal fishing rights law.

Aboriginal Canadians share a profound belief in their sovereignty and ownership of the resources. This partly explains the effort and determination with which First Nations defend their members charged with offences. Such efforts can cost in excess of $Can15 million when a case proceeds to the Court of Appeal, particularly when the defence to a fisheries charge is aboriginal rights.

Canadian courts may have been hesitant to use criminal trials to develop aboriginal rights law because of the potential effects of the decisions. As will be illustrated in this paper, the courts in Canada have given aboriginal people not just a bare recognition of their rights but a basis to challenge the economic and social policy of the State itself. It is that aspect of the litigation that this paper will expand upon. While there are many other aspects of the aboriginal fishing rights litigation in Canada deserving of scrutiny, it is the challenge to economic and social policy and government control that will be examined here.

Any examination of the litigation in Canada must begin with a statement on the question of jurisdiction in Canada. Fisheries and Oceans as well as Indians and Lands Reserved for Indians fall within the jurisdiction of the Federal (national) Government. Businesses connected with the processing of fish and fish products fall within the purview of the Provinces (states). In addition, most Provinces have reached agreements with the Federal Government on how to manage most inland fisheries. Further complicating the division of powers between the Provinces and the Federal Government is Section 35 (1) of the Canadian Constitution which states:

> The existing Aboriginal and treaty rights of the Aboriginal peoples of Canada are hereby recognized and affirmed.

THE SPARROW CASE

On a warm May 25th evening in 1994, Ronald Edward Sparrow, a member of the Musqueam First Nation (Indian Band) got into his boat and headed for Canoe Pass on the south arm of the Fraser River delta near Vancouver, British Columbia. The Fraser River is famed for its salmon runs. The fishing permit for his First Nation and issued under the authority of the regulations to the Fisheries Act, RSC 1970, c. F-14, permitted Sparrow to fish with a net no longer than 25 fathoms. Sparrow took a 45 fathom net with him that evening and in so doing set the stage for the seminal case on aboriginal fishing rights law in Canada. Sparrow was stopped and charged by officers from the Department of Fisheries and Oceans and convicted at trial. Eventually his appeals ended in the Supreme Court of Canada. The Sparrow case is now a test case referred to by courts throughout the Commonwealth despite the innocuous facts of the case. Marvin Storrow, Q C, counsel for Sparrow, subsequently revealed that no one envisaged that the Sparrow case would become so important to the development of aboriginal rights law in Canada.

After finding that the Musqueum people have an aboriginal right to fish, and finding that the priority of the right of the Musqueum people is second only to conservation, the Court noted that there will be controversy and conflict. The Court said, in quoting *R v Sparrow* (1990, p. 173) from its earlier decision in the case *Jack v R* (1979):

> The federal Regulations become increasingly strict in regard to the Indian fishery over time, as first the Commercial fishery developed and then sport fishing became common. What we can see is an increasing subjection of the Indian fishery to regulatory control.

and also said it recognized the possibility:

> of conflict between aboriginal fishing and the competitive commercial fishery.

In the *R v Sparrow* (1990, p. 176) case the Court stated:

> We recognize the existence of this conflict and the probability of its intensification as fish availability drops, demand rises and tensions increase.

Here the court was accurately foreshadowing events to come. The greatest conflict over fishing rights in recent years has not been between the government and aboriginal people, but between aboriginal people and the commercial and sport fishing industries. That is not to deny that there is conflict between the government and aboriginal people over fishing rights. However, because the government has changed its strategy to one of engaging aboriginal people in negotiations for fishery agreements, conflict between government and aboriginal people has decreased. Through negotiations the aboriginal people have attempted to redefine the rules with respect to resource allocation and to challenge the existing social and economic policy objectives of the government.

In *R v Sparrow* (1990, p. 181) the Supreme Court of Canada said:

> By giving aboriginal rights constitutional status and priority, Parliament and the provinces have sanctioned challenges to social and economic policy objectives embodied in legislation to the extent that aboriginal rights are affected.

The Court further stated that the Constitutional provision in Canada recognizing and affirming the aboriginal rights gives a strong check on legislative power and a measure of control over government conduct (*R v Sparrow* 1990, p. 181).

These words of the Supreme Court have, in part, led to increased conflict and provided scope for aboriginal litigation in Canada.

CLAIMING ABORIGINAL RIGHTS IN THE COURTS

A group of cases worthy of note are those of *R v Jack, John and John* (1996); *R v Little* (1996) and *R v Sampson and Elliott*

(1996). In these cases the salient facts were similar and involved salmon stocks from Vancouver Island. It was shown that the salmon stocks, when migrating to their natal rivers to spawn, first pass Alaska, northern British Columbia and the east and west coasts of Vancouver Island. While passing through these areas the stocks are fished by other user groups such as commercial and sport fishers. By the time the salmon reached their natal rivers, where the aboriginal river fishers are seeking access to them, there are only enough left for spawning purposes and none for harvest.

In *R v Jack, John and John* (1996) the British Columbia Court of Appeal acquitted the accused aboriginals as the Court found that the fisheries scheme did not afford the aboriginal fishery the priority that had been mandated by *R v Sparrow* (1990). The Court quoted from the trial judgement where the trial judge said:

> The end result of all of this is that barely enough of the chinook indigenous to the Leiner and Tahsis Rivers reach their destinations to perpetuate the stock. And that means that the heaviest burden of conservation is borne by the Indian, the person who has a constitutionally protected right to first priority after only valid conservation allotments.

In none of the cases did the aboriginals say that they were sufficiently consulted by the Department of Fisheries and Oceans. It is what the Court had to say on the lack of consultation by the Department of Fisheries and Oceans that the case has been most cited for. The Court said the alleged consultations by the Department of Fisheries did not meet the 'criteria indicated in Sparrow' (*R v Sparrow* 1990, pp. 132–133) as:

> They did not cover all of the conservation measures which were implemented and which affected the availability of chinook at the mouth of the Leiner River.

and

> There was a duty on the DFO (Department of Fisheries and Oceans) to ensure that the Indian Band was provided with full information on the conservation measures and their effect on the Indians and other user groups.

and

> The DFO had a duty to fully inform itself of the fishing practices of the aboriginal group and their views of the conservation measures.

and

> It was the duty of the DFO to inform the Band of the conservation measures being implemented before they were implemented.

This was a very strong judgement on the issue of the need for the Department of Fisheries and Oceans to consult in a meaningful way with aboriginal people. The words of the Court of Appeal were echoed in the companion judgement of *R v Sampson and Elliott* (1996, p. 208) where the Court said:

> the requirement of consultation, as set out in *Sparrow*, is not fulfilled by the DFO merely waiting for a Band to raise the question of its Indian food fish requirements, discussing those requirements, and attempting to fulfill those requirements. Consultation embraces more than the foregoing. It includes being informed of the conservation measures being implemented.

In my view, the *R v Jack, John and John* (1996) case and the *R v Sampson and Elliott* (1996) case place a heavy duty on the government to consult in a meaningful way with aboriginal people. There will have to be more disclosure by the government to aboriginal groups. Greater demands will be placed on fisheries biologists and managers to explain their decisions and recommendations. Aboriginal people will now be better able to question the basis for fisheries programmes. Both cases have given the aboriginal people greater power to challenge social and economic policy objectives and to exercise some control over government conduct. Other fishery groups such as commercial and sport groups do not enjoy that power. As aboriginal people seek to challenge social and economic policy using the information that the government must now disclose to them, and as the aboriginal people seek to exercise greater control over government conduct armed with the same information, it is perhaps without doubt that there will be further conflict between aboriginal fishers and other user groups.

In support of this were the words of the Court of Appeal in *R v Little* (1996), the facts of which were similar to *R v Jack, John and John* (1996). The Court said government policy which:

> had the effect of accommodating the interests of the commercial fishery and the economic benefits brought to the Province by the sports fishery, at the expense of both conservation and the aboriginal food fishery

and the allocation of priorities in this case did not comply with the constitutional rights and treaty rights of the aboriginal fisher.

Challenging aboriginal rights in the courts

In October of 1995 John Cummins, a Reform Party member of Parliament, commenced an action in British Columbia Supreme Court seeking, *inter alia*, to challenge the power of the Minister of Fisheries to enter into agreements with First Nations allowing the First Nations to sell portions or all of their quota from First Nations communal fishing licences issued by the Department of Fisheries and Oceans. Cummins, also a commercial fisher, belongs to a party that some in Canada would call right wing and which is the third largest in Canada.

In June 1996, Cummins sought interlocutory relief in Federal Court to stop the sale of fish caught in the Somass River by aboriginal fishers pursuant to the their communal licences issued by the Department of Fisheries and Oceans. Cummins also sought in that action to limit the numbers of fish caught by the aboriginal fishers. His applications were rejected by the Federal Court as the Court felt that he had not met the test necessary for interlocutory relief.

Not finished in his attempts to stop the aboriginal fisheries, in July 1996 Cummins argued yet another case in Federal Court. In this action Cummins sought a court order to stop all fishing of Fraser River sockeye salmon stock until one million, six hundred thousand mature sockeye reached the spawning grounds in the headwaters of the Fraser River. As an alternative,

Cummins sought a court order to stop the Minister of Fisheries from allowing aboriginal fishers from catching Fraser River sockeye in such numbers that the total number of mature Fraser sockeye escaping to the spawning grounds would be less than one million, six hundred thousand. As another alternative, Cummins sought a court order to prohibit the Minister of Fisheries from allowing the sale of fish caught pursuant aboriginal communal fishing licences which impact on Fraser sockeye stock.

In the case, *Cummins et al.* v *The Minister of Fisheries and Oceans et al.* (1996) all three applications by Cummins were unsuccessful. Justice Campbell of the Federal Court of Canada ruled that he did not have jurisdiction to hear the applications by Cummins. Justice Campbell said that Cummins sought to have the:

> Court act as a regulatory authority exercising power paramount to that of the Minister (of Fisheries).
>
> The relief requested on the Notice of Motion is wrapped up in the essential question of the number of fish which should reach the spawning ground this year. In my opinion, this question is not suitable for judicial solution. I agree with the Respondents' argument that to attempt a decision of this sort is simply a 'second guess' of the decisions that the Minister has already made on expert advice.

Cummins had argued that an injunction was needed in order to preserve the sockeye until a judicial review application could be heard. Cummins cited the decision of the British Columbia Court of Appeal in *MacMillan Bloedel Ltd* v *Mullin et al.* (1985). In that case aboriginal groups claimed aboriginal title to an island where MacMillan Bloedel Ltd sought to fell trees for timber and were successful in getting an injunction to stop the logging. Cummins argued that the MacMillan Bloedel Ltd case applied in his application. Justice Campbell found that Cummins did not have a constitutional aboriginal right to maintain whereas in the MacMillan Bloedel Ltd case the aboriginal people did.

The *Cummins* cases are reflective of several factors. Firstly, there is a sense of helplessness among commercial fishers as they look for someone to blame for the depletion of the fisheries resource. Commercial fishers failed to acknowledge that in British Columbia First Nations, communal licensed fishers only harvest 5% to 10% of the available catch of salmon whereas the commercial fishers harvest over 80% of the available stock (based on Department of Fisheries and Oceans figures with sport fishers accounting for the balance).

Secondly, there is a sense of betrayal felt by users of the resource, other than aboriginals. The sense of betrayal is fostered by a belief that 'special rights' have been 'given' to First Nations. Inherent in this belief is a misunderstanding of the nature of aboriginal rights and a misconception that First Nations citizens are not 'worthy' of these 'special rights'.

Lastly, there is a lack of perspective by commercial fishers of the problems in the fishing industry. In particular, commercial fishers fail to acknowledge the other causes of problems in the industry and the causes for the depletion of the resources.

As can be seen from the above cases, aboriginal people will use the words of the Supreme Court of Canada to define their aboriginal rights to challenge the social and economic policies embedded in legislation. The aboriginal people, by ensuring that the government lives up to the requirements of the term 'consultation', as is used in the *Sparrow* case, will seek to even the field of knowledge and thus exercise more control over government conduct and the power of the legislatures and Parliament. This will result in further friction with other fishery resource user groups.

UNRESOLVED ISSUES

As the *R v Sparrow* (1990) case is further interpreted, other large questions will be placed before the courts. Among the most difficult for the courts to resolve is the question of whether aboriginal people have the right to sell the fish they catch pursuant to their aboriginal right. The Supreme Court of Canada is currently considering three cases it has reserved decision on. *R v NTC Smokehouse Ltd* (1993), *R v Vanderpeet* (1993) and *R v Gladstone* (1993), heard by the Supreme Court of Canada in the first week of December 1995, all raise the issue of whether aboriginal people have the right to sell their fish. The Supreme Court of Canada has already made an important determination in *R v Nikalj* (1996) unreported, *SCC Reg no. 23804*, by reversing its 1993 decisions. This determination found that while requiring aboriginal people to hold a licence is not in itself an infringement of their aboriginal right, terms of a licence which infringe aboriginal rights and which are not severable from a licence may however, result in a licence being invalid.

A more difficult question for the courts to consider will be the question of what the Supreme Court of Canada meant by the term conservation in the *Sparrow* case. Non-aboriginals have argued that conservation includes the concept of enhancing a stock. Others have argued that conservation means preventing extinction, while still others have argued that conservation means maintaining a sufficient pool of genetically diverse stock.

A third difficult question for the courts will be to what extent the aboriginal people can challenge environmental decisions based on their aboriginal rights to fish. Aboriginal people claim that they must have the right to ask the courts to protect the spawning areas of the salmon otherwise the aboriginal right will become an empty right.

Lastly, no court has yet examined the thorny issue of what numbers of fish, if any, should be attached to the aboriginal catch. Part of the answer to that issue may be determined by the Supreme Court of Canada's decisions when hearing the cases on whether aboriginal people have the right to sell fish.

When hearing the courts address these issues, only one sure outcome can be predicted and that is that there will continue to be conflict between aboriginal and other fishers.

LEGAL CASES

Cummins et al. v The Minister of Fisheries and Oceans et al. (1996). (July 20, 1996) unreported, FCCTD #T-1474–96 (Vancouver Registry).

Jack v R (1979). 1 CNLR 25.

MacMillan Bloedel Ltd v Mullin et al. (1985). 3WWR 577.

Pulvallup Tribe v Washington Department of Game (1968). 391 US 392 (known as Puyallup 1).

R v Gladstone (1993). 4 CNLR 75 (BCCA).

R v Jack, John and John (1996). 2 CNLR 113 (BCCA).

R v Little (1996). 2 CNLR 136 (BCCA).

R v Nikalj (1996). Case unreported.

R v NTC Smokehouse Ltd (1993). 4 CNLR 75 (BCCA).

R v Sampson and Elliott (1996). 2 CNLR 184 (BCCA).

R v Sparrow (1990). 3 CNLR 160 at 172.

R v Sparrow (1990). p. 181.

R v Vanderpeet (1993). 4 CNLR 221 (BCCA).

SCC Reg no. 23804 (1993). 4 CNLR 117 (BCCA).

Sohappy v Smith (1969). 302 F. Supplement 899 (D. Ore).

United States v Washington (1974). 384 F. Supplement 312 (D Wash); aff'd (1975), 520 F. 2d 676 (9th Cir.); Certiorari denied, (1976) 96 S. Ct.877 (known as Boldt 1).

United States v Washington (1980). 506F. Supplement (known as Boldt 2).

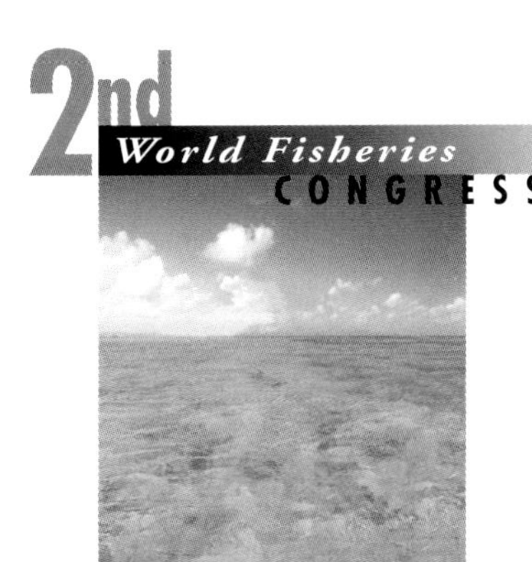

Maori Fisheries Rights and the Quota Management System

Sir Tipene O'Regan

Treaty of Waitangi Fisheries Commission, New Zealand.

Summary

Ten years ago the Quota Management System (QMS) provided the framework for managing the main commercial fish species in New Zealand. The provision of property rights enabled New Zealand's age-old struggle over Treaty of Waitangi fishing rights to be settled, with considerably less pain than might have been expected. The Treaty of Waitangi Fisheries Commission now manages considerable fisheries, assets on behalf of Maori, giving Maori the pre-eminent position in the industry. The paper describes the means for achieving this, and expected trends for the future, including greater direct involvement by quota owners in the purchase and provision of management services for their fisheries, together with the improved ecological understanding of fisheries needed for setting total allowable catches (TACs).

Introduction

It is now ten years since the Quota Management System (QMS) was established as the framework for managing the main commercial fish species in New Zealand. Such a milestone provides a useful opportunity to reflect upon this quite extraordinary and revolutionary change in fisheries and fishing rights and the way we conduct our affairs and our business in New Zealand. Most reflection and debate centres around what the experience with the QMS teaches us about the way to manage fisheries. However, the events of 1986 are important for another reason, because in many ways the introduction of the quota management system itself, provided a mechanism by which the Treaty fishing rights of our tribes could finally be given effect.

This outcome represents a very great irony. Before 1986 a battle had been going on which began shortly after New Zealand's founding in 1840. That battle had been waged in the courts and by Parliamentary petition, over the nature and extent of Maori fishing rights, secured and guaranteed by the Treaty of Waitangi. Another major question was the Crown's duty to protect those rights, by what the Courts have called the Crown's duty of 'active protection' arising from the Treaty.

The long history of argument from the Maori side about the destruction of their Treaty rights in fisheries had not been able to be met, other than by pious Parliamentary reference, while the practical reality was that the government assumed the rights in fish unto itself and granted access to whomsoever it thought fit. The Maori tribal Treaty partner was not a class generally 'thought fit'.

When we had the 'Commons' out there, everyone who wished to fish went out fishing and the fish belonged to no one until they were caught. The whole model of the 'Commons', from our point of view, was a Western device, driven by the power culture, for preventing particular Treaty property rights of Maori being given effect. There are interesting parallels with other 'public' assets similarly defined as belonging to 'all New Zealanders'. Restricting the use of the 'Commons' by licensing and a host of government regulations, accumulated since the Oyster Fisheries Act 1866, merely confirmed the practical dispossession of Tribes from their ancient fishing rights which had supposedly been secured and guaranteed to them by the Treaty of Waitangi.

Establishment of Maori fishing rights

The great irony to which I referred was revealed when New Zealand moved towards a model of property rights in fisheries under a quota management system. Although it was a model driven by a whole lot of notions, principally sustainability, it was very much centred on ideas of property as far as the fishers were concerned. The irony was that the model provided a forum, for the first time since 1840, in which Treaty of Waitangi rights in fisheries could actually be transferred and given effect. The right had become, in both legal and economic senses, tangible.

It was intended that the rights would be delivered as a form of property to the predominantly non-Maori fishing industry. Maori simply said to the Crown and to the courts: 'there are already extant property rights in these fisheries and it is therefore not possible for the Crown to allocate Individual Transferable Quotas (ITQs) without violating those existing property rights'. 'Sovereignty is not ownership. It's not yours to give away.' In 1986 the Courts agreed with that view, at least on an interim basis, and a whole new cycle of litigation and negotiation began.

Its important to note here that there had always been statutory protection of Maori rights but the rights were 'formless' in legal terms whilst they inhabited a context of a 'Commons' property right. They could not be given effect to. They were at best, defensive, as the Te Weehi case demonstrated. In the Te Weehi case the High Court found, on appeal, that the exercise of customary or non-commercial rights was exempted from the relevant fisheries laws as long as particular circumstances were present (Te Weehi *v* Reginal Fisheries Officer (1986) 1 NZLR 680).

Without waiting for, or risking direction from, the Courts on this matter, the government stopped the process of allocating ITQs (after ITQs had been allocated to 29 species on the basis of catch history) and commenced a process of negotiation with Maori. These initial negotiations proceeded through 1987 and 1988, culminating in the Maori Fisheries Act 1989.

The Maori property right of which I speak is a collective right belonging to Tribes. They are the Treaty 'partners'.

The concept that there can be a collective private property in fisheries is not a foreign idea to Maori any more than it is to Westerners in the form of a Company. To Maori though, it represents a component of *Rangatiratanga*, or chiefly dominion as part of the collective property of a Tribe. However, the nature and extent of those rights had not been authoritatively described or established by the Courts, and Parliament had deliberately avoided it - fearful of total defeat of the position it had assumed. It preferred to maintain its dignity by partial acknowledgment.

The Waitangi Tribunal embarked upon a process of investigation in the 1980s and produced two reports on fisheries claims (Muriwhenua Fishing Report (WAI 22) June 1988, and Ngai Tahu Sea Fisheries Report (WAI 27) 1992). These reports provided some common basis for negotiations between Maori and the government, although only the first of these reports was available during the early stages of negotiation.

The outcome of the negotiation process was an interim settlement (driven by an impending General Election) under which the government promised to enter the quota market and re-purchase, on a willing buyer/willing seller basis, 10% of ITQ for every species in every fisheries management area and transfer that ITQ to a Maori Fisheries Commission. The Commission, in conjunction with *Iwi* (Tribes), began the development of options for the permanent allocation of those assets to *Iwi*.

Meanwhile, High Court proceedings were being prepared by Maori to determine the 'nature and extent' of Maori fishing rights as a prelude to further claims against the Crown.However, these proceedings were forestalled and then superseded by further complex negotiations which led to the so-called Sealord settlement of 1992.

Under this Deed of Settlement, the Crown agreed to fund Maori into a 50/50 joint venture with Brierley Investments Limited to bid for Sealord Products Limited — New Zealand's largest fishing company and the holder of some 28% of the total New Zealand total allowable commercial catch. In return, Maori agreed that all their current and future fishing claims in respect of commercial fishing rights would be fully satisfied and discharged.

The Deed of Settlement also promised Maori 20% of quota for all species not yet in the Quota Management System, which is a system of fisheries management given explicit support by Maori under that Deed. In addition, special regulations, which would protect and control Maori customary non-commercial fishing rights, were to be produced. These and other key elements of the Deed of Settlement were reflected in the Treaty of Waitangi (Fisheries Claims) Settlement Act 1992. Upon its passage, the Maori Fisheries Commission was reconstituted as the Treaty of Waitangi Fisheries Commission, which has a wider range of statutory functions than its predecessor.

The Waitangi Tribunal had found that Maori possessed exclusive fishing rights out to the 12-mile territorial limit and a development right which extended beyond that to the edge of the Economic Zone. Some Maori groups considered that the Interim Settlement of 1989 and the 'Sealord deal' did not adequately reflect the strength and value of these rights.

Partly as a consequence of the belief that the agreement was short on quantum, the Deed of Settlement expressly allowed for Maori to continue to strengthen their position in the fishing industry through use of their own resources. Consequently, the Commission has developed something of a tradition of using its own resources to achieve strategic improvements to the overall Maori position in the fishing industry by further acquisition.

The result of this activity is that the Commission now manages the following assets on behalf of Maori:

- 10% of all ITQ in all quota species (approximately 57 000 t);
- 50% of Sealord Products Ltd. (which owns, processes and exports approximately 24% of New Zealand's quota), and in which Maori hold a pre-emptive right to acquire the interests of Brierley Investments Ltd should they wish to sell;
- 74% of Moana Pacific Ltd.;
- 100% of Prepared Foods Processing Ltd. (New Zealand's largest processor of abalone);
- 100% of Chathams Processing Ltd. (operator of the main processing facilities on the Chatham Islands);
- 100% of Pacific Marine Farms (producer, processor and exporter of Pacific oysters with projected 1996 production of 1 million dozen oysters);
- Cash reserves of approximately $NZ71 million.

As well, a number of Tribes have acquired significant volumes of quota on the market in their own right, largely by the use of bank finance.

Even though these assets have not yet been allocated to *Iwi*, there has been a dramatic growth in Maori presence in the 'business and activity of fishing' in recent years. Since 1990, 55 new *Iwi* fishing companies have been established and nearly 20% of *Iwi* own or operate their own processing and/or marketing capacity. Much of this activity has been facilitated by Commission policies of making the 10% of quota it owns available to *Iwi* through discounted annual leases (Bevin and Boyd 1996*a*, *b*). The accumulated value of these quota lease discounts since 1993/94 is said to be in the order of $NZ51 million. In addition the Commission funds approximately $NZ1 million per annum for training and scholarships to assist Maori gain employment in, or improve technical or management skills applicable in, any aspect of the fishing industry.

All of this change has been unsettling for others in the New Zealand industry. The introduction of the QMS followed by six years of negotiations with Maori provided both licence and quota holders with considerable uncertainty over the quality of their rights. Not surprisingly perhaps, licence holders feared that a government which had admitted some responsibility for the expropriation of Maori rights might seek to remedy that situation by expropriating some of theirs.

So far those fears have proved groundless and the pre-eminent position of Maori in the industry, now it has been achieved, is one of the major sources of security in the legal framework of property rights represented by the QMS. The transfer has taken place at no cost to other industry players and a number have used the opportunity to exit the fisheries sector on terms which would not have been possible without Individual Transferable Quota.

The use of ITQ as the means of settling Treaty claims to commercial fisheries was critical to the removal of resource rents from ITQ in 1994. Those rents were designed by the Crown to remove so called 'super profits'. The debilitating effect of those prospective rents arose from the fact that it was not at all easy to distinguish super profits from normal profits which had been generated from investments in the fishing production chain extending from fisheries management and enhancement, through processing to marketing. As a result, the theoretical incentives on quota owners to invest in projects which benefitted their fishery (and thereby increased the value of ITQ) would have been undermined. The defeat of the proposal by Maori on Treaty rights grounds has been an important factor in reshaping wider Industry attitudes to the Maori presence.

FUTURE NEEDS

The situation today is that the first chapter in the story of the QMS is drawing to a close. That chapter has seen the establishment of the system and its integration with Maori commercial interests. Now, the more interesting issue is how that system is going to evolve. I suggest that we will see two trends. The first will involve greater direct involvement by quota owners in the purchase and provision of management services for their fisheries.

The second will be continual refinement in the ecological understanding of fisheries which will be reflected in the setting of total allowable catches (TACs). The latter is particularly consistent with Maori cultural perspectives on natural resources and fisheries in particular.

The QMS has reached a stage in its development where quota owners have sufficient maturity and strong incentives to take more direct responsibility for the management of commercial fisheries. New Zealand quota owners pay the full costs of fisheries management services carried out by government but have little influence over the nature of these services or the efficiency with which they are delivered.

Consequently, there is considerable potential to deliver more value from the $NZ50 million or so spent each year on fisheries research, administration and enforcement. If that potential is to be realized, there must be a closer relationship between the funders of these services (quota owners) and the deliverers of those services who, where possible, should be exposed to the disciplines of a contestable market.

Critical standards and specifications can still be set, audited and enforced by government but quota owners should be able to consider spending on fisheries services to be a normal investment rather than a form of taxation.

For this to occur, a dramatic re-organization is required across the fishing industry. That re-organization is already being manifest in the establishment of fisheries management companies or quota-owner organizations. These companies can be expected to have four primary functions:

1. the collection of funds to finance fisheries management activities;

2. the development and imposition of management rules affecting shareholders;
3. the representation of shareholders in TAC and other fisheries fora; and
4. the protection of harvesting rights and promotion of management rights by quota owners.

Currently these companies draw heavily upon consensus as their normal mode of operation. Consensus about fishing, of course, is never easy and that has confronted the industry with the uncomfortable fact that consensus now means obtaining substantial agreement with Maori.

However, I am confident that these new arrangements will continue to evolve and will ultimately flourish because the effect of these debates is to clarify the common interests of all quota owners, Maori or non-Maori, in the future health of the fisheries that we share and in the economic performance of the export industry to which we belong.

Summary

To summarize the first evolutionary trend then, I would say that we will move from a position where Maori and non-Maori share the harvesting rights in commercial fisheries resources to the point where we share real responsibility for the management of those resources. The incentives unleashed by the improved security of ITQ property rights in the post-settlement era provide the foundation for new roles for both government and quota owners in the management of fisheries.

The second evolutionary direction I predict is in the priority which must be given to fisheries research to operate the QMS better. In its present form, a number of important linkages in the food chain, in the inter-relationships of species and a range of biological and geographical factors have been to a large extent, excluded from consideration. In my view, that exclusion must be rectified.

Such factors need to be included in the process of refinement so the system becomes much more reflective of the actual biological, geographical and physical characteristics of the resource and its environmental context. There are plainly serious questions to be asked of a uniform management model imposed on species as diverse as squid, hoki, abalone, flounders and crayfish. The system can be said to work but it remains a crude tool, indeed, in respect of the diversity of species and their inter-relationships.

A simple example is the relationship between kina (sea urchins), paua (abalone) and the seaweeds on which the latter live, or for which the two former compete for food, and the species that predate upon the kina. There is a fundamental relationship in the ecological balance between those resources. If you don't manage them as a set then you tend to destroy the productive balance of those resources, particularly the paua. Unfortunately, there seems to be some perverse law of nature that if we disrupt that balance, the most valuable species appear to be the ones which suffer most.

That is just one example, there are plenty of others. An explosion of paddle crabs as a result of the predating species being overfished has the potential to damage the treasured bivalve shellfish on which they, in turn, predate. Similarly, we need to better understand the relationship between juvenile hoki (whiptail or blue grenadier) and hapuka (groper). We don't manage these inter-relationships of species very well. In fact, we eventually ignore their existence in a system sense. I think it is an area with a need to develop much more refinement — and promptly.

In many ways, a lot of the Maori criticism of the QMS has been based on a traditional cultural view of species relationships and their management, which has given rise to criticism and cynicism about the model itself. The alleged single-species fixation of the QMS undermines its perceived status as a sustainability weapon. This is perhaps an ill-conceived limitation because there is no reason why TAC-setting procedures cannot make better use of the more sophisticated Maori ecological understanding. Indeed, a more rigorous scientific perspective demands it.

At the end of the day, we still need to control the level and pattern of harvest of all species. What I suggest is that we will see the emergence of both better ecological understanding and a more sophisticated approach to TAC setting than a slavish adherence to the objective of maximum sustainable yield in single species.

Conclusion

To conclude, I return to the nature of the property right. The very fact of having that right, the right of access to fish in the form of transferable property, has meant that New Zealand's age-old struggle over Treaty fishing rights has been able to be settled.

One of the most interesting points to be made is that, on the whole, this very large transfer of assets to Maori which has taken place as a result of the quota management system being brought in, has really been accomplished without any great pain. Certainly, that is, compared to the pain that *could* have been created.

That is a very considerable political achievement. It is also a very considerable economic achievement. There is, though, a long way to go. The system itself, needs to be refined and will continue to evolve. We have righted an historic wrong, or have gone a very substantial way to doing so. We have created a new economic base for Maori in this sector and we have at the same time, established a world-class model for fisheries management. That is, in national terms, perhaps even in international terms, a considerable achievement.

Even though the quota management system was not invented, developed and conceived to give effect to Treaty rights, it provided, conceptually, the key to that historic resolution. In New Zealand history that may yet stand as its greatest triumph.

References

Bevin, J, and Boyd, R (1996*a*). Review of the 1995/1996 Wetfish/Paua Lease Distribution. Report prepared for Te Ohu Kai Moana, June 1996.

Bevin, J, and Boyd, R (1996*b*). Survey of the Status of Iwi Fishing Operations. Report prepared for Te Ohu Kai Moana, June 1996.

A Novel Approach for Rebuilding Fish Stocks and Ecological Balance to Help Small Scale and Traditional Fisheries

D. S. Sheshappa

University of Agricultural Sciences, College of Fisheries, Mangalore - 575001, India.

Summary

The inshore fishing zones of India up to 10 km from the shore have been fully exploited if not overexploited by both the mechanized and the traditional sectors. Most of the coastal regions have adopted mechanization. This has either resulted in the complete abolition of the small-scale and traditional fishing activities or affected the socio-economic structure of the artisanal sectors. Government should evolve policies regarding modern fishing technology and technologies for rebuilding stock and maintaining ecological balance, and should ensure employment for the coastal people with sustained fishing. New approaches to be implemented are: implementing strict fisheries regulations, arranging awareness programmes for fishermen, encouraging the mechanized sectors to venture into offshore and deep-sea regions, and developing artificial fish habitats through the installation of artificial reefs.

Introduction

India is the seventh largest fish-producing country in the world. Of the total marine fish production of 2.27 million t in 1993–94, 90% is from inshore waters up to 50 m depth. This is almost equivalent to the potential catch as estimated from the biomass of 2.28 million t, indicating that the inshore region is fully exploited.

Fishing was the main source of livelihood of the coastal people until the 1950s, their fishing operations extending to about 10 miles from the shore. Although some of the coastal districts of the country are continuing their traditional fishing activities, most districts introduced mechanized fishing in the 1960s. The Government of the coastal states encouraged mechanized fishing both by providing training through the training centres and also by financial subsidies. This has resulted in severe competition between the mechanized sector and the traditional sector, for harvesting the same resources from the same fishing grounds.

Present Status of Small-Scale and Traditional Fisheries

Mechanized fishing has been introduced on a large scale along the south-west coast of the country: bottom-trawling for shrimps in the 1960s and purse seining for pelagic resources of oil sardine (*Sardinella longiceps*) and Indian mackerel (*Rastrelliger kanagurta*), white sardine (*Kowala coval*), anchovies (*Stolepherus spp.*), etc. in the 1970s were introduced without any control of fishing effort. This radically changed the pattern of marine landings; the modern gears became so efficient that bottom-trawls caught everything in their path, and the purse seines, in addition to their normal pelagic catches, could catch shrimps when they form small schools near the surface in September, immediately after the monsoon season. High prices were paid for the shrimp, so the revenue was substantial. The traditional fishery has almost completely ceased. For example, in the coastal state of Karnataka, the 'Rampani' was one of the most primitive methods of fishing for the capture of sardine and mackerel schools in large quantities. Each operational unit involved more than one hundred fishermen and thus was a popular group activity of the village. This was almost wiped out in the early 1980s as a result of the introduction of purse seines in the mid 1970s, both the fishing methods being aimed at the same resources. Thus, the coastal fishermen were forced to abandon their profession and become workers on board the mechanized boats owned by a few affluent families and other fish traders. The government also encouraged 'Rampani' people to buy purse-seine boats by providing subsidies.

During the 1980s, the traditional fishermen started protesting against the operation of the mechanized boats in the inshore waters, and the matter was resolved through the intervention of the government by means of some consensus agreements between the two sectors and also by the enforcement of fisheries regulations.

Present status of fish resources and ecology of inshore waters

Modern fishing methods have not only been responsible for the maximum exploitation or overexploitation of some of the valuable resources of the inshore waters but have also caught large quantities of bycatch including the juveniles of many commercially important species. Bottom-trawling and purse seining have damaged the catfish resources of the south-west coast of India through large-scale destruction of the eggs (H. P. C. Shetty, pers. comm.). This has exacerbated the deterioration of the socio-economic status of the coastal small-scale fishermen.

The coastal inshore waters, which form the breeding ground for many of the commercially-valuable fish species, became environmentally degraded in terms of fish habitat because of:

* increased fishing effort by the bottom-trawlers and purse seiners in the region;
* discharge of sewage and industrial effluents;
* inflow of agricultural chemicals and pesticides; and
* coastal erosion due to reclamation of mangrove areas etc. for brackish-water aquaculture.

The traditional sectors are also causing considerable damage to the coastal resources because of the operation of highly-efficient fishing craft, gears and methods in the monsoon season; the mechanized fishery is banned during this period because it is considered to be the breeding season for many of the fish resources of the region. In the traditional fishery, the development of large canoes made of fibreglass-reinforced plastic and propelled by powerful outboard engines, the use of monofilament for gill-nets and the operation of encircling nets of the purse-seine type for catching breeders in large quantities, have been found to be detrimental to resource conservation. These methods appear to be as efficient as the mechanized boats and hence should not be allowed to operate during the monsoon period. Thus, the coastal zone is becoming a barren water body that needs attention and actions to make it productive again.

Protection of the inshore zone

There is an urgent need to protect the coastal zones from indiscriminate fishing and to ban misuse of the water bodies so that the fisheries activities may be sustained. Technologies have to be evolved and implemented for rebuilding stock and maintaining the ecological balance of the region, and employment for the coastal people must be assured by the adoption of uniform policies by the respective governments.

Important approaches in this direction are:

(1) Strict implementation of fisheries regulations as a first step in fisheries resource management by both the state and central governments.

(2) Arranging an awareness programme among fishermen with regard to their responsibilities and rights in protecting the coastal water bodies.

(3) Encouraging the mechanized fishing fleet operators to step into hitherto unexploited or underexploited offshore and deep-sea resources.

(4) Strict control of the quality of the effluents discharged into the sea by industry.

(5) Development of artificial fish habitats all along the coastal zone.

First approach: fisheries regulation

(a) The water bodies should be zoned between the traditional small-scale sector, small mechanized sector and offshore sector for efficient fisheries management and control. In 1978, with modifications in 1980, the Government of India fixed certain fisheries regulations to be implemented by the respective governments of the maritime states. They are:

* the area up to 10 km from the shore is reserved for the exclusive use of the traditional and small-scale fishermen and non-mechanized craft operators.
* the small mechanized boats are to operate beyond 10 km from the shore.
* the vessels of 20 m length and more to operate beyond 23 km from the shore.

(b) Restrictions should be imposed on the type of fishing crafts and fishing methods and on the fishing season in each region. Only selective fishing methods, especially passive ones, should be permitted. All types of dragged and encircling gears, which are non-selective, should be completely banned in the region.

Second approach: fishermen awareness programme

The fishermen of the region should be educated about their rights and responsibilities in safeguarding the water bodies and also about acceptance of the rules and regulations to be implemented by the Government. The education programmes should be on the following topics.

* The type of fish species available in the area, their breeding seasons, breeding grounds, feeding habits, age and size at maturity, natural mortality, etc.
* The need for the restricted zone and the regulation of fishing methods, mesh-construction and fishing season.
* The effects of indiscriminate exploitation of juveniles, of catches exceeding maximum sustainable yield and of bycatch. The maximum fishing effort required for the given resources, the type of fishing methods and their catching efficiency and selectivity in fishing in order to achieve responsible exploitation.
* The profitable way of marketing to make the small-scale fisheries sector an attractive and adaptable resource for the community.
* Energy conservation by utilizing sail technology on the boats.
* Training in alternative technologies such as the development of artificial fish habitat, pearl oyster farming for pearl production, mussel farming, clam culture, mud crab culture, etc.

Third approach: encouraging offshore and deep-sea fishing

The mechanized fleets, especially the trawlers, should be encouraged to venture into offshore and deeper-water regions beyond the traditional zone to reduce their deleterious fishing activity in the inshore zone. This could be done by provision of information and training in the following: the various offshore resources available, their grounds and the type of fishing methods adopted; location of the fishing grounds by echo-sounder; basic navigation and use of Global Positioning System (GPS) navigation; radio communication using VHF; vessel safety, fire fighting, survival at sea, rules of the road, and navigation signals during fishing and cruising; handling of fish on board the vessel and preservation etc. Further, the respective governments should subsidize the acquisition of the additional equipment necessary for deep-sea fishing. Purchase of fuel also should be subsidized, because the fuel cost represents the major operating expenditure for the deep-sea fishing vessels.

Fourth approach: pollution-free effluent discharge from the industries

Many industries, such as fertilizers, petroleum, thermal power plants, chemical plants etc. are being established, or will be established, along the coasts. They discharge large quantities of effluents into the sea. Unless these discharges are pollution free, they will affect the biomass of the sea and hence the ecology of the region. These should be strictly regulated to protect the environment.

Fifth approach: development of artificial fish habitat

This is one of the most important approaches to be implemented for rebuilding fish stocks and maintaining the ecological balance, and it is one of the novel approaches for the development of small-scale and traditional fisheries in such regions.

Fish habitats normally arise through biological or morphological processes occurring in the sea bottom (Fernandez 1996). However, any external object or stable structure may be placed in the sea as an artificial fish habitat to attract, to aggregate and to regenerate pelagic, demersal, migratory and residential fishes.

Artificial Reef (AR) technology is being practised all over the world. It is reported to increase fish catches by 20% to 4000%, to prevent overfishing and to increase aquatic fish habitat (Grove and Sona 1991). Traditional artificial fish habitats on the Coromondal coast were reported by Hornell (1924). Bergstrom (1983) reviewed fish-aggregating devices in India and South-East Asia. These habitats have been created by the traditional fishermen of the south-west coast of India by dumping rock fastened with coconut fronds onto the sea bottom to attract fish closer to the shore; the fish thus aggregated were caught by beach-seine operation (Fernandez 1996). Sanjeeva Raj (1996) reported that concrete ring reefs erected to the south of Madras were dominated by a variety of species such as rabbit fish (*Siganus* spp.), parrot fish (*Scarus* spp.), large croakers (*Epinephelus* spp.) etc. Deposition of egg masses of cuttlefishes and squids and accumulation of algae, bivalves and barnacles were also reported.

Marine life such as bio-fouling organisms, fish fry, fingerlings and adult fish colonize the reef for shelter, feeding and breeding. New food chains are formed at the reef. Thus, the ARs are eco-friendly technologies, and Indian coastal waters, which are so much devastated by the mechanized trawlers, can be revitalized through use of ARs.

Selection of sites for the ARs is important; they should not be damaged by the waves and traffic at sea, and they should be located in places where the coastal fishermen had their traditional activities before these were supplanted by mechanized fishing.

Various materials can be used for the construction of artificial reefs and fish-aggregating devices, such as concrete rings tied with old tyres, iron barrels, stones packed inside coir or rope nets, concrete wastes from demolished structures, high-density polyethylene (HDPE) materials etc. Raja (1996) developed ARs from HDPE material; the fishermen themselves can easily construct these structures (Fig. 1) and erect them at the pre-determined location (Figs 2 and 3), and can install them for demarcation of the zone for the exclusive use of traditional fishermen (Fig. 4). The ARs can be installed off each village, preferably between 5 and 10 km from the shore, all along the coast. The structures should be looked after by the entire community of the village as a joint responsibility to avoid poaching and theft by intruders.

Benefits of artificial reefs

(1) The coastal fishery stocks can be restored to their original level, thereby providing the coastal fishermen with economic stability through sustained fishing.

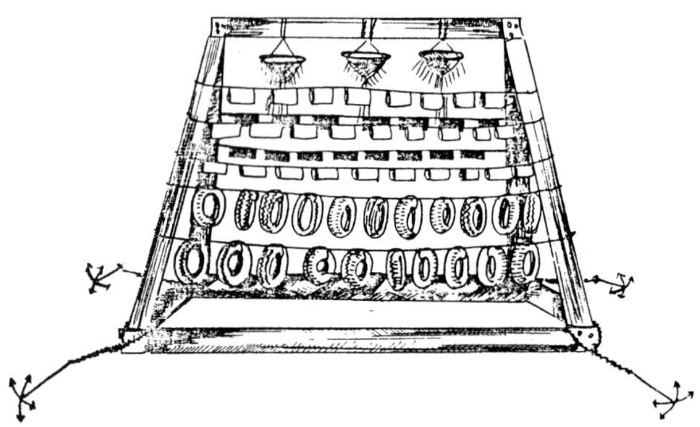

Fig. 1. Artificial reef structure made of HDPE material.

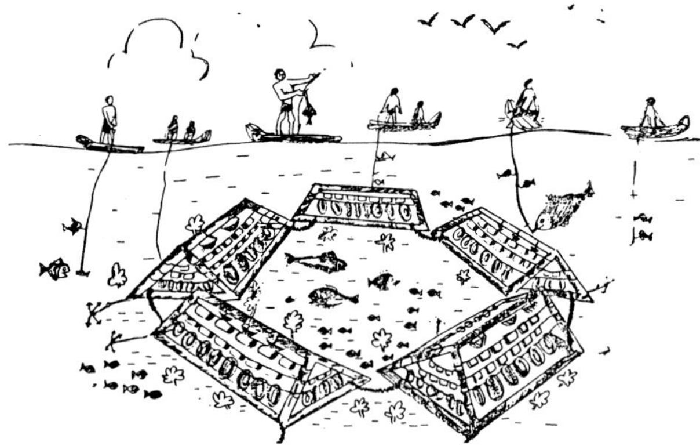

Fig. 2. Artificial reef in a pentagon shape for effective fish aggregation and fishing.

(2) ARs are low-cost eco-friendly technologies for the artisanal fishermen.

(3) The reefs can be constructed and erected by the fishermen themselves after adequate training, and the expenditure involved can be shared through the local fishermen's cooperative societies in which they are the shareholders.

(4) The economic activity of each coastal village will be increased because of distribution of food production throughout the coastal region, and allied business activities such as ice production, marketing etc.

(5) There are employment opportunities for women in marketing the catches.

(6) There will be overall improvement in the living standard of the fishermen's community.

(7) Demarcation of the coastal inshore zone for the exclusive use of the traditional sector can be achieved automatically since the ARs obstruct the passage of the bottom-trawlers.

(8) The most selective methods of fishing, such as hook and line, traps, gill-nets etc. can be employed profitably.

(9) Since the fishing grounds are very close to the shore, the fishermen can reach the ground by rowing or under sail, thereby saving fuel and achieving a high cost–benefit ratio.

(10) Certain ARs could be developed to attract tourists for game fishing with hook and line, SCUBA diving, underwater photography etc. ARs could also be used for developing communities of reef ornamental fishes.

Constraints of artificial reefs

(1) The ARs may be buried by the silting of the substratum, and monsoon waves may overturn the structures and cause shifting.

(2) Transportation of AR structures and erection in the exact location is very difficult. Overturning of the structure should be avoided in order to create a more natural habitat on the sea bottom.

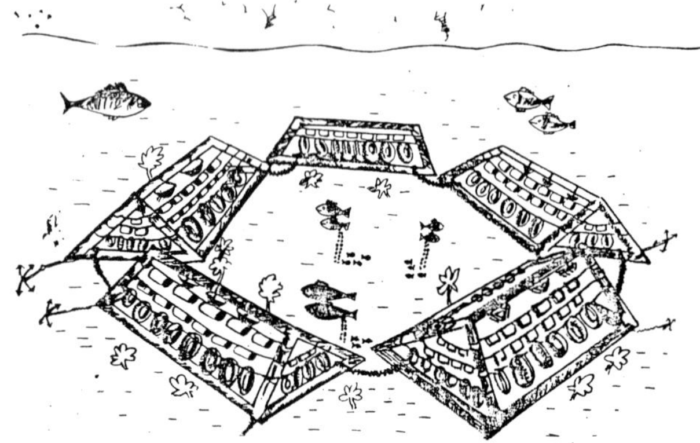

Fig. 3. Fish breeding at artificial reef.

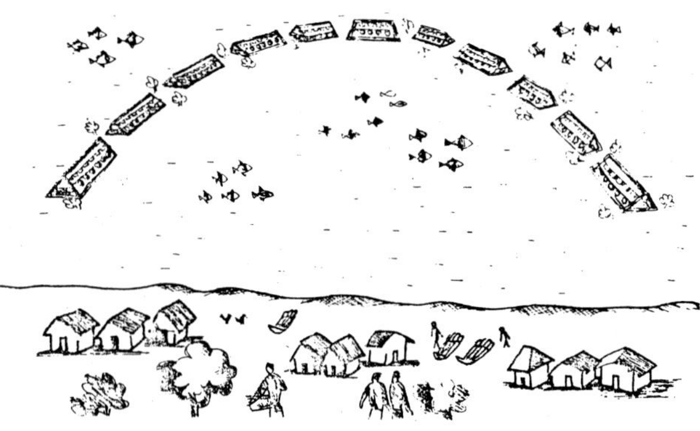

Fig. 4. Artificial Reef for demarcation of territorial fishing grounds for the traditional fishermen.

(3) The maintenance and guarding of the ARs can be achieved only if there is good cooperation among the families of the community in the village.

REFERENCES

Bergstrom, M (1983). Review of experiences with and present knowledge about fish aggregating devices. Bay of Bengal Programme, BOBP/WP/23, Madras.

Fernandez, J (1996). Artificial fish habitat — a community programme for bio-diversity conservation. *Bulletin of the Central Marine Research Institue* **48**, 42–56.

Grove, R S, and Sona, C J (1991). Artificial habitat technology in the world — today and tomorrow. In 'Recent Advances in Aquatic Habitat Technology'. Proceedings of the Japan–US Symposium on Aquatic Habitat for Fisheries, Tokyo, Japan, 3–9.

Hornell, J (1924). The fishing methods of the Madras Presidency. *Madras Fish Bulletin* **18**, 60–6.

Raja, G (1996). Fish aggregation devices and artificial reefs. *Bulletin of the Central Marine Research Institute* **48**, 8–10.

Sanjeeva Raj, P J (1996). Artificial reefs for a substantial coastal ecosystem in India involving fisherfolk participation. *Bulletin of the Central Marine Research Institute* **48**, 1–3.

Towards Management of Lake Tanganyika's Fisheries

George Hanek,[A] *G.V. Everett,*[B] *O.V. Lindqvist*[C] *and Hannu Mölsä*[C]

[A] FAO, B.P. 1250, Bujumbura, Burundi.
[B] FAO, Viale delle Terme di Caracalla, 00100 Rome, Italy.
[C] University of Kuopio, PO Box 1627, 70211 Kuopio, Finland.

Summary

The characteristics of Lake Tanganyika, its fisheries and existing fisheries management practices are outlined. The 'Research for the Management of the Fisheries on Lake Tanganyika' (LTR), a regional, multi-disciplinary research project is described. The project is funded by the Finnish International Development Agency (FINNIDA) and executed by the Food and Agriculture Organization of the United Nations (FAO) in close cooperation with the University of Kuopio.

The LTR's ultimate objective is to formulate a coherent lake-wide fisheries management policy based on the 'best scientific evidence'. The details of three years of sampling are discussed, together with an outline of complementary actions required to establish the Lake Tanganyika Fisheries Organization.

Introduction

Lake Tanganyika is shared by Burundi, Tanzania, Zaire and Zambia. The lake has been geographically isolated for about 20 million years; its fish species and fisheries resources therefore have special features not found elsewhere. The fisheries of Lake Tanganyika mostly target the abundant pelagic resources consisting of two clupeids and their four endemic predators (Family Centropomidae).

The unique pelagic fishery with its seasonal and inter-annual fluctuations in catches and catch composition, defies standard fisheries planning and management. Studies undertaken by FAO during the 1960s and 1970s in each of the four countries concluded that simultaneous lakewide research was needed to understand high fish production in the apparently nutrient-poor waters (Coulter 1981). In 1977 the four countries asked FAO's Committee for Inland Fisheries of Africa (CIFA) to create a Sub-Committee for Lake Tanganyika. This body was specifically requested to formulate and find funding for a regional fisheries research project. The (FINNIDA) Finnish International Development Agency provided the funding and a FAO-executed regional project of 5 years duration commenced in 1992.

THE CHARACTERISTICS OF LAKE TANGANYIKA AND ITS FISHERIES

Lake characteristics

Lake Tanganyika (3°20'–8°48'S and 29°03'–31°12'E) lies at 773 m above mean sea level; it is 673 km long, has a surface area of 32 900 km^2 and a maximum width of 48 km. The maximum depth is 1470 m, making it the second deepest lake in the world. The average depth is 570 m and the volume is 18 800 km^3 (Coulter 1966). The percentage surface areas under jurisdiction of the four riparian States are Burundi (8%), Tanzania (41%), Zaire (46%) and Zambia 6% (Coenen *et al.* 1993).

Fisheries resources

The lake is known internationally for the spectacular variety of its endemic cichlid fish fauna (Coulter 1991). It is composed of a remarkable and genetically-diverse benthic community with a contrastingly simple pelagic community. The benthic community consists of almost 300 fish species of which over two-thirds are endemic (Poll 1986). The pelagic fish community is composed of six endemic, non-cichlid species: two schooling clupeids, *Limnothrissa miodon* (Boulenger 1906) and *Stolothrissa tanganicae* Regan, 1917, and their major predators, four members of the genus *Lates* (Centropomidae), *L. stappersii, L. angustifrons, L. mariae* and *L. microlepis.*

The potential yield of Lake Tanganyika, is estimated to be between 295 000 and 460 000 t, based on 90 kg/ha/year (Mikkola and Lindqvist 1989) and 140 kg/ha/year (Coulter 1981), respectively.

Fisheries characteristics

Most fishing is done at night as virtually all fishing methods (e.g. purse seines, lift-nets, beach seines and scoop-nets) rely on clupeids being attracted to light. Fishing activities, therefore, practically cease every month during the full moon.

There are three recognizable types of fisheries on Lake Tanganyika; industrial, artisanal and traditional. The industrial fishery was started in 1954, when Greek fishermen introduced the purse seine. A typical industrial fishing unit consists of a 16 to 20 m long steel vessel, a purse seine and auxiliary steel boat, 5 lamp boats and a total crew of 30–40 fishermen. Presently, there are 13 units (which are semi-active) in Burundi, one in Tanzania (Kigoma), 16 in Zambia (Mpulungu) and 21 in Zaire (4 in Moba and 17 in Kalemie). The number of these units is decreasing in Burundi where 23 units were active in 1976 (Bellemans 1991). It has remained almost constant in the other countries, although recently a number of Kalemie-based units moved to Zambia.

The artisanal fishery in the northern part of the lake uses mainly catamarans and to a lesser extent trimarans, although the latter have totally disappeared from Burundi (Bellemans 1991). A typical catamaran fishing unit consists of two (three for trimarans) 6–7 m long mainly wooden plank hulls, a lift-net (55–65 m circumference), 6–7 lamps and an average of 4.7 fishermen. There were respectively 604 and 67 active catamaran and 'Apollos' (a larger catamaran: 7–9 m long canoe, lift-net of up to 100 m of opening circumference, 14–19 lamps and an average of 8–11 fishermen) in Burundi in 1992 (Coenen 1994). There were 739 catamarans in the Kigoma and Rukwa Regions of Tanzania (Chakraborty *et al.* 1992) and 45 catamarans in the Uvira and Fizi zones of Zaire (Maes *et al.* 1991). There are very few catamarans in the southern part of the lake (e.g. five in Zambia in 1994; Mwape 1994). The majority of fishing units in the south are beach seines operating at night, with lights, mainly to catch clupeids (Hoekstra and Lupikisha 1992).

The traditional subsistence fishery uses many different fishing gears (gill-nets, hook and line, scoop-net, longlines, traps, mosquito-nets, etc.). Although all are generally inefficient, many people are involved in their use around the lake.

The extent of lakewide fishing pressure was estimated from the density of fishing craft observed during aerial surveys (Hanek *et al.* 1993). In total 13 976 canoes were counted. The highest density of canoes per km of shoreline was recorded in Burundi (11.3 canoes/km or a total of 1802 canoes), followed by Zaire (10.3 canoes/km or 7570 canoes), Tanzania (6.0 canoes/km or 3839 canoes) and Zambia (3.6 canoes/km or 765 canoes). During the same study (*ibid*) 459 fish landing sites were recorded, 34 in Burundi, 127 in Tanzania, 257 in Zaire and 41 in Zambia. The greatest number of fish landing sites around the lake (192 or 41.8%) were classified into Category II (having 11–30 canoes). There were fewer Category I (1–10 canoes) landing sites (147 or 33.3%). The largest sites (>81 canoes) were recorded at 8 sites in Burundi.

Fishing effort, total catch, value, social and economic importance, processing and marketing

According to the October 1993 Frame Survey (FS) the fishing fleet in Burundi was composed of 13 purse seiners, 671 lift-net units (604 catamarans and 67 Apollo units) and 298 traditional fishing units. A major decline of the catch-per-unit-effort (CPUE) has been recorded over the last ten years for the industrial as well as the traditional fisheries for Burundi (Coenen and Nikomeze 1994). For example, the average CPUE/night for the industrial fishery in Burundi decreased from 1173 kg/night/unit in 1983 to 150 kg/night/unit in 1993 and now appears to be unprofitable. However, the artisanal lift-net fishery, due to the use of bigger nets, better fishing lamps and the choice of more productive fishing grounds, manages to maintain its CPUE at a profitable level. For example, the CPUE for 'Apollos' was 300 kg/night/unit in 1993 (Coenen and Nikomeze 1994).

Fluctuations in catch composition have also been determined. In 1993, the clupeid catch in Burundi accounted for 67% of the total catch, *L. stappersii* for 31.6%, other *Lates* spp. for 0.2%. Clupeids are generally the most abundant species, although there is often an inverse relationship in catch numbers between clupeids and *L. stappersii.* The total fish catch for Lake Tanganyika for 1992 was estimated at 167 000 t, shared as follows: Burundi 24 000 t, Tanzania 80 000 t, Zaire 50 000 t and Zambia 13 000 t (Coenen 1994). The value of the catch has been estimated at approximately $US26 million in 1992 (*ibid*).

Over one million people are dependent on the Lake Tanganyika fisheries including some 40 000 fishermen and their families and those involved in fish processing and marketing (*ibid*).

Considerable differences exist in the level of fisheries development around the lake. There is an extensive and costly infrastructure (cold stores, processing plants, refrigerated trucks, etc.) in Mpulungu, Zambia, and to a lesser extent in Kalemie, Zaire. No such facilities are available elsewhere, notably in Bujumbura and Kigoma. While a detailed inventory of the fisheries infrastructures of Lake Tanganyika is made regularly(Hanek 1993, 1995), its overall value has not been determined.

Fish processing is not well developed; clupeids are either sold fresh or sun-dried. An improved method of washing them in brine and then drying them on racks has been introduced but is rarely used. Only in Mpulungu and in Kalemie do cleaning, brining, freezing and sometimes smoking (particularly of *L. stappersii*) take place. Recently, the canning of clupeids and *L. stappersii* was developed in Zambia (Hanek 1994). External marketing of catches in excess of local needs is difficult and complex due to transportation problems. With the exception of the very north of the lake most roads are tangential. The shores are steep and few roads link the populations around the edges of the lake, particularly the extensive shorelines of Zaire and Tanzania. Fish, particularly clupeids, are thus traded along the coast by 'water-taxis' or by the ferries *M/V Liemba* and *M/V Mwongozo* at ports between Mpulungu and Bujumbura. Major outlets for dry fish are the 'Copperbelt' complex of large towns in Zambia, the Zairian cities of Lubumbashi, Bukavu and Goma, and in Rwanda (Hanek 1994).

Present fisheries management status

At various times, the four riparian countries have introduced their own fisheries regulations to control fishing effort. These have never been introduced on a 'lake-wide' basis and, due to inadequate formulation and very limited enforcement capability, they have been ineffective. The authorities of each country generally agree that the fisheries of the lake must be managed on a regional basis. Regional cooperation among riparian countries exists in the context of regional, intergovernmental organizations such as the Southern African Development Community (SADC), Preferential Trade Area (PTA) and Communauté Economique des Pays des Grands Lacs (CEPGL), as well as in the context of joint bilateral commissions. However, these institutions are generally concerned more with macro-economic and political aspects and are not structured to co-ordinate specific fisheries management activities (Gréboval 1990).

The lack of a regional organization to manage the fisheries resources was discussed at the Third Meeting of the Committee for Inland Fisheries of Africa (CIFA) which consequently established, in 1977, the Sub-Committee for Lake Tanganyika. The main objective of the sub-committee was to formulate a regional research project. Subsequent CIFA meetings debated a number of issues including harmonization of fisheries legislation, collection and analysis of fisheries statistics, and the conditions of access to the fisheries. However, no real progress has been made to date. There are three main obstacles to progress. First there are legal differences: Tanzania and Zambia possess an Anglo-Saxon legal framework, whereas Burundi and Zaire operate within the Roman civil law system. Second, there is a lack of biological knowledge of the fisheries resources and there are differences in fishing practices and development in the four riparian countries. Finally, Lake Tanganyika is not only a fishing ground but also a border and a place of communication and trade. It is difficult therefore for any agreement not to have wide-reaching implications (Bonnucci 1990).

There is a great diversity among the riparian countries with regard to fishing categories. Burundi classifies its fisheries into the categories used in this paper (industrial, artisanal and traditional). Zaire recognizes recreational and individual fishing, taking into account the type of craft and gear used. There is no comparable classification in Tanzania or Zambia, although the licence fee exacted in Tanzania depends on the type of craft employed. Due to these differences a considerable disparity in licence fees exists, particularly for the industrial fishery. There is a considerable variation in the regulations to protect the fisheries, for example, in minimum mesh size and prohibited gear. In Burundi and Zaire, mesh sizes less than 4mm and beach seines are prohibited. In Zambia the minimum mesh size for monofilament nets is 120 mm and 10 mm for all other types of nets but the use of beach seines is legal. In Tanzania the use of monofilament nets is prohibited in all national inland waters (Bonnucci 1990).

Towards the management of Lake Tanganyika fisheries resources

Role of Lake Tanganyika research

The objective of the LTR is to provide the data to formulate a fisheries management policy. The LTR Project Document details two phases, a 1.5 year preparatory phase and a 3.5 year execution phase. The preparatory phase, which started in January 1992, was successfully completed on schedule. The execution phase involves the coordination and execution of a complex multi-disciplinary research programme (SSP). Its design was fully tested and has been subsequently modified and refined. Its demanding execution started, as scheduled, in July 1993. It has six major components:

(i) *Hydrodynamics*. This involves studies of the upwelling and downwelling phenomena of the nutrient rich waters and their effects on pelagic biological production. It includes water current measurements and the collection of meteorological data.

(ii) *Remote sensing*. Studies are undertaken of the spatial and seasonal distribution of the upwelling phenomena in relation to surface water temperature.

(iii) *Fish and zooplankton biology*. Many variables are measured for the six targeted pelagic fish species. Intensive zooplankton sampling is carried out down to 300 m.

(iv) *Genetic structure of pelagic fish*. Studies are undertaken to determine the possible genetic discreteness of pelagic fish stocks.

(v) *Limnology and carbon/energy budget*. Most of the significant parameters are measured at regular depth intervals down to 300 m.

(vi) Fisheries data. These are collected and analysed, and reporting of annual Catch Assessment Surveys and Frame Surveys standardized.

The activities outlined above are complemented by hydroacoustic studies and integrated sampling surveys for other components using the project's research vessel *R/V Tanganyika Explorer*. A series of lake-wide hydroacoustic cruises continues to be conducted in order to estimate the biomass of the different target fish species.

Regular sampling and data entry/analysis occurs simultaneously in three main LTR stations at Bujumbura (Burundi), Kigoma (Tanzania) and Mpulungu (Zambia). Less detailed and less frequent sampling is also done at six LTR substations located in Karonda (Burundi), Kipili (Tanzania), Nsumbu (Zambia) and in Uvira, Kalemie and Moba (all in Zaire). The sampling comprises: regular weekly sampling for hydrodynamics, limnology, zooplankton, fisheries biology and statistics; intensive sampling (over a 24 h cycle) every 6 weeks for limnology and zooplankton; seasonal sampling for limnology and fish genetics; regular surveys are organized for fisheries statistics; and lakewide research cruises (8–9/year) using *R/V Tanganyika Explorer*.

Considerable human resources are required to execute the SSP and over 120 people are involved in data collection, analysis and reporting. The amount of data collected by LTR automatic meteorological and hydrological instruments and by LTR staff is considerable. Over four megabytes of data/information are received every month. Consequently, the management of LTR's data banks is a complex task.

LTR completed three full years of sampling in July 1996. By the end of 1996 the final science report will be presented to all four participating countries. It will present the summaries of the results and conclusions of each research component, detail both the qualitative and quantitative trophic relationships (links between components), provide the application of dynamic pool, empirical and bio-economic models for the pelagic fisheries and propose initial fisheries management measures. Proposed management measures may include fishing effort (intensity, gear type, by area), closed seasons,and protected areas.

In addition LTR has initiated and provided a forum to discuss a host of relevant issues for the authorities of the four participating countries through the annual joint meetings of the LTR's Coordination and International Scientific Committees. It has proposed standardized reporting of fisheries statistical data which has now been adopted by all four countries. It has introduced the principles of fisheries management which were highlighted during the LTR Workshop on the Management of Lake Tanganyika Pelagic Stocks. Both the authorities and the representatives of artisanal and industrial fishermen of all four participating countries were invited to the workshop. The LTR has also prepared the Lake Tanganyika Fisheries Directory which is updated annually. This provides detailed listing of all participants in the fisheries including Government, fisheries training, education and/or research institutions, the private sector, suppliers of fishing gear and material, projects and fisheries cooperatives and associations.

Biological parameters must be combined with the monitoring of economic parameters so as to assess social and economic effects contributing to optimal management of the resources. This has been proposed for the project's second phase which will last three years and will consolidate the information gained during the first phase. Management of the main fish stocks will be stressed and directed towards establishing a legal framework for the management and development of the main fisheries. Uniform legislation will be formulated together with the corresponding regulations that should be adopted by each country. Not only will national management authorities and their legislative ability need to be strengthened but a viable lakewide management organization will have to be established.

Role of governments

The implementation of a fisheries management plan will demand a number of complementary measures ranging from agreeing and establishing the required institutional and legal framework to approving the eventual management entity. They will all require additional financial resources and all are clearly outside LTR's Terms of Reference.

It was recognized at both the last session of CIFA's Sub-Committee for Lake Tanganyika and at the Fourth Joint Meeting of the LTR's Committees (Hanek and Craig 1995), that the major constraint to the management of the lake's fisheries is the lack of a legislative governing authority. It was recommended therefore, that the riparian countries establish the Lake Tanganyika Fisheries Organization (LTFO) with a mandate to undertake research programmes, co-ordinate and implement fishery management measures, as well as to co-ordinate and enforce fisheries regulations. Many valuable lessons were learned during the establishment of the Lake Victoria Fisheries Organization (LVFO) which should be considered when forming a LTFO. Principally there must be early and careful consensus on expectations and functions of the fisheries organization. International technical, legal and financial support is essential to the overall framework of such an organization. Full-time rather than part-time participation throughout the region will hasten the process for Lake Tanganyika as it has done for Lake Victoria.

At the end of LTR involvement it should be possible to implement regional management of Lake Tanganyika's pelagic fisheries on sound biological, economic and social criteria. This will benefit the people of the area as well as the conservation of the resource. Indeed, the conservation of the lake and its resources is inseparable from its role as a provider to the people living on its shores.

REFERENCES

Bellemans, M S (1991). Historique des pêcheries artisanales et coutumières au Burundi de 1952 à 1991. PNUD/FAO BDI/90/002, Rapport de terrain No. 5. 54 pp.

Bonnucci, N (1990). The outlook for the harmonization of Lake Tanganyika Fisheries Legislation. In 'Principles of Fisheries Management and Legislation of Relevance to the Great Lakes of East Africa: Introduction and Case Studies'. (Ed D Gréboval) UNDP/FAO Regional Project for Inland Fisheries Planning (IFIP), RAF/87/099-TD/05/90 (En). 41 pp.

Chakraborty, D, Lyimo, E O, and Shila, N C (1992). Fisheries inventories in Lake Tanganyika. URT/87/016, Field Document No. 10. 28 pp.

Coenen, E J (1994). Frame Survey results for Lake Tanganyika, Burundi (28–31.10.1992) and comparison with past surveys. FAO/FINNIDA Research for the Management of the Fisheries on Lake Tanganyika. GCP/RAF/271/FIN-TD/18 (En). 26 pp.

Coenen, E J, and Nikomeze, E (1994). Lake Tanganyika, Burundi, results of the 1992–1993 Catch Assessment Surveys. FAO/FINNIDA Research for the Management of the Fisheries on Lake Tanganyika. GCP/RAF/271/FIN-TD/24 (En). 27 pp.

Coenen, E J, Hanek, G, and Kotilainen, P (1993). Shoreline classification of Lake Tanganyika based on the results of an aerial frame survey (29.09.–03.10.1992). FAO/FINNIDA Research for the Management of the Fisheries on Lake Tanganyika.GCP/RAF/271/FIN-TD/10 (En). 11 pp.

Coulter, G W (1966). Hydrobiological processes and the deep water fish community in Lake Tanganyika, Vols 1 and 2. PhD. thesis. Queens University, Belfast.

Coulter, G W (1981). Biomass, production and potential yield of the Lake Tanganyika pelagic fish community. *Transactions of the American Fisheries Society* **110**, 325–35.

Coulter, G W (1991). 'Lake Tanganyika and its life.' (Oxford University Press).

Gréboval, D (Ed) (1990). Principles of Fisheries Management and Legislation of Relevance to the Great Lakes of East Africa: Introduction and Case Studies. UNDP/FAO Regional Project for Inland Fisheries Planning (IFIP), RAF/87/099-TD/05/90 (En). 41 pp.

Hanek, G (Ed) (1993). 1993 Lake Tanganyika Fisheries Directory/ Répertoire des pêches du lac Tanganyika 1993. FAO/FINNIDA Research for the Management of the Fisheries on Lake Tanganyika. GCP/RAF/271/FIN-TD/14 (En & Fr). 69 pp.

Hanek, G (1994). Management of Lake Tanganyika Fisheries. FAO/FINNIDA Research for the Management of the Fisheries on Lake Tanganyika. GCP/RAF/271/FIN-TD/25 (En). 21pp.

Hanek, G (Ed) (1995). 1995 Lake Tanganyika Fisheries Directory/Répertoire des pêches du lac Tanganyika 1995. FAO/ FINNIDA Research for the Management of the Fisheries on Lake Tanganyika. GCP/RAF/271/FIN-TD/33 (En and Fr). 87 pp.

Hanek, G, and Craig, J F (Eds) (1995). Report of the Fourth Joint Meeting of the LTR's Coordination and International Scientific Committees. FAO/FINNIDA Research for the Management of the Fisheries on Lake Tanganyika. GCP/RAF/271/FIN-TD/42 (En). 53 pp.

Hanek, G, Coenen, E J, and Kotilainen, P (1993). Aerial frame survey of Lake Tanganyika Fisheries. FAO/FINNIDA Research for the Management of the Fisheries on Lake Tanganyika. GCP/RAF/271/FIN-TD/09 (En). 29 pp.

Hoekstra, T M, and Lupikisha, J (1992). The artisanal capture fisheries of Lake Tanganyika, Zambia: major socioeconomic characteristics of its fishermen and their fishing units. UNDP/FAO Regional Project for Inland Fisheries Planning (IFIP), RAF/87/099-TD/41/92 (En). 93 pp.

Maes, M, Leendertse, K, and Mambona Wa Bazolana (1991). Recensement des unités de pêche zaïroise dans la partie nord du lac Tanganyika. Projet Régional PNUD/FAO pour la Planification des Pêches Continentales (PPEC). RAF/87/099–WP/09/91 (Fr).` 61 pp.

Mikkola, H, and Lindqvist, O V (1989). Report on a Project Mobilization Mission. Preparatory phase GCP/RAF/271/FIN, FAO, Rome, July 1989, 104 pp.

Mwape, L M (1994). Frame survey Lake Tanganyika (Zambian waters). FAO/FINNIDA Research for the Management of the Fisheries on Lake Tanganyika. Field Report, October 1994. 10 pp.

Poll, M (1986). Classification des Cichlidae du lac Tanganyika: tribus, gendres et espèces. Académie de Belgique, *Mémoires de la classe des sciences* **45**, 1–163.

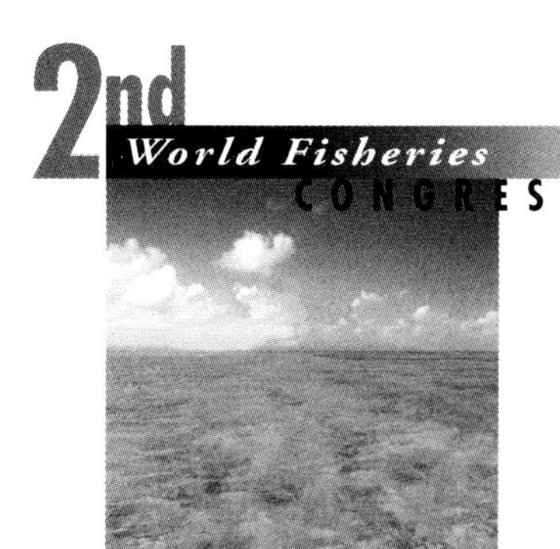

International Bio-Political Fisheries Management: Sashimi on the Cutting Edge

John Mark Dean

Marine Science Program and Center for Environmental Policy, University of South Carolina, Columbia, SC. 29208 USA.

Summary

Any attempt to present the issues of the management of highly migratory species of fishes must make significant compromises in coverage. There is one issue that drives all other considerations and that is 'allocation', which is the division of the available fish among all parties that desire a share of the fishery resource. The allocation conflict is a result of two basic facts: the fisheries generally have open access and they take place in international waters. Historical, cultural, economic, and political dimensions of the Atlantic bluefin tuna are used as a case study. The allocation issue is presented from the perspectives of policy development, management procedures and practices, implementation, enforcement and dispute resolution. Lack of information about the basic biology of the fishes limits the resolution of the allocation issue and the paper concludes with comments and concerns about the future management of highly migratory fishes.

Introduction

How do you allocate shares in a fishery when a single fish is worth $US10 000? I address the issue of allocation of a highly migratory species, the bluefin tuna of the western Atlantic Ocean, as a case study. There is no presumption that it is the only or best example. It is a most contentious allocation issue, it is very important to the United States fishing community, it receives a great deal of media and political attention, and there is an abundance of data.

It is worthwhile to have an overview of how this fish surfaced as a resource that commands this much attention. Details are available in the 1994 National Research Council report, 'An Assessment of Atlantic Bluefin Tuna', (NRC Report), the Final Environmental Impact Statement for a regulatory amendment by the U.S. National Marine Fisheries Service (1995) and the annual reports of the Standing Committee on Research and Statistics (SCRS) of the International Commission for the Conservation of Atlantic Tunas (ICCAT). I will present only a superficial overview of the biology of the fish for those who are not immersed in issues of this fishery. There is an abundance of published and grey literature with extensive information in SCRS reports.

The northern bluefin tuna (*Thunnus thynnus*) is a highly migratory scombrid found in the western Atlantic from

Labrador and Newfoundland into the Gulf of Mexico and Caribbean Sea and as far south as Brazil. In the eastern Atlantic, its historical range was from the coast of Norway to 40 degrees south latitude of the coast of Africa. It is among the largest teleosts of pelagic waters and reaches 320 cm fork length, 680 kg in weight and is the largest of the tunas. Bluefin are very opportunistic, aggressive, high-level carnivores and forage, depending upon seasonal availability of prey, on many types of fishes, crustaceans and cephalopods. Tuna form schools that are usually made up of individuals of similar size which can be mixed with schools of other tunas such as albacore, yellowfin, bigeye and skipjack. It is believed that the females in the western Atlantic become sexually mature at about 196 cm fork length (148 kg) and that fish of that size are eight years of age. The assumption is that fish greater than that size are sexually mature, and therefore the spawning population is made up of all fish eight years or older, which is a critical point in stock assessment determinations. Spawning of the Atlantic population is believed to take place in two areas, the Gulf of Mexico and the western Mediterranean Sea. The two spawning areas apparently have slightly different periods of peak spawning activity, the Gulf group spawns in mid-April through mid-June while the Mediterranean group spawning peak is later, in June and July. Their growth rate varies somewhat and the age of sexual maturation is presumed to be slightly different, with the Mediterranean group maturing at a smaller size.

Statement and problem

What is the origin of the problem and is it unique? After all, there are other big fish in the ocean and there are many different kinds of tunas, some of which are much more numerous than bluefin. The SCRS estimates that the western Atlantic bluefin tuna stock is 20% of 1970 levels and there are increasing numbers of fishers who want to harvest this high-value fish each year. On the basis of the knowledge of the reproductive biology and some limited tagging results, ICCAT separated the northern Atlantic bluefin into two stocks for management purposes in 1981. The line of longitude dividing the eastern and western stock was originally set at 40 degrees, which was later changed to 45 degrees. The two-stock working hypothesis has been used for stock assessment purposes to establish total allowable catch quotas since that time. The countries most affected by such management decisions are the United States, Japan and Canada.

Analysis

The most fundamental aspect of the problem is that of dealing with a common pool or common property resource (CPR). It is well understood by fishers and biologists that highly migratory species of fishes (HMS) do not understand or respect political jurisdictions. As long as nations had limited seaward authority (5–10 km), distant water fishing nations did not feel constrained and tuna fishing nations operated in international waters under well-recognized concepts of freedom of the seas. However, after the virtually universal adoption of the 200 mile exclusive economic zone (EEZ) that followed the UN Law of the Sea Convention, tuna now were within EEZs and became subject to domestic national jurisdictional regimes. Fisheries that had been outside the coastal state purview now became of direct economic interest to those nations. Although residence of fish within their EEZ afforded the coastal states the authority to manage the fish, including regulation and allocation, there were clearly HMS management problems that could not be solved solely with coastal state domestic jurisdiction. This was particularly difficult for the United States because its position and policy was based on the claim that tunas alone did not come under coastal state jurisdiction. Thus, management of tunas could, and should, be handled by international agreement alone. It is worth noting the inconsistency of that position with the US's very rigorous domestic management of billfish (marlins and sailfish) and swordfish, all species that are also regulated by the Atlantic Tunas Convention Act and the ICCAT treaty. That position was dramatically altered by the 1990 amendments to the Magnuson Fishery Conservation and Management Act which brought the tunas under domestic management. The most recent international action, the United Nations Treaty on Straddling Stocks, is just now on the table for adoption. This will be a most important element for possible transformation of management of HMS. But it is simply too early to realize the impact it might have upon adoption and implementation.

The theory of management of CPRs has been most recognized, and to a degree popularized, by Hardin's 1968 paper 'The Tragedy of the Commons'. There is some concern that this point of view has precluded the discussion of successful CPR management regimes and those are well considered by Ostrom (1990). It is not my intention to discuss different HMS management regimes. There are several excellent collections of articles in symposia (Joseph and Greenough 1979; Stroud 1990, 1994; Kruse *et al.* 1993), and an excellent discussion of different management regimes and the national interests of such regimes in Parsons (1993).

Because the western bluefin stock assessment has shown a significant decline in spawning stock biomass since the early 1970s (NRC 1994), there has been a severe harvest quota since 1981 on the western stock which is used for scientific monitoring. Even with sustained quotas in the 1980s, the stock assessment showed continued reduction with some flattening of the curve in the mid-1990s. The quota for 1995 and 1996 is 2200 t each year. Fishers are very critical of the two-stock hypothesis because it leads to reduced catches for the US, Canada and Japan, but not for nations fishing the eastern stock. They note that with sustained reductions in quota for the west over a 15 year period, there have been increased catches in the east. The fishers who harvest the western stock are additionally upset because the eastern stock has only one real ICCAT conservation measure, *via* minimum size, which is routinely ignored by those countries fishing that stock.

The western Atlantic Ocean fishery was historically a recreational fishery and there was not a viable commercial fishery until the early 1970s. What events transformed the western bluefin with a value of $US0.55/kg and used for catfood, to a fish worth as much as $US88/kg and the centrepiece at Japanese sashimi restaurants? A major technological advancement made it feasible to capture the large north Atlantic bluefin in prime condition and market them in Japan for a premium price. In addition, the economy of Japan moved into high gear and it became one of the seven major economies in the world with a high level of disposable income. Japan's culture celebrates high quality fish and conducts a great deal of business and entertainment in sashimi restaurants. In Japan it is traditional that bluefin sets the price standard for all tunas and also for yellowtail, squid, octopus, clams, sea urchin roe, eel, salmon and the other sashimi items. In January 1970, Boeing introduced its new large passenger plane, the 747, into commercial service, and Japan Air Lines (JAL) took delivery of their first planes in July of 1970. The 747 carries from 360–490 passengers, depending upon seating configuration and range needed. When fully loaded with passengers it also has the capacity to carry large amounts of freight over great distances without refuelling.

One of JAL's first destinations for their 747s to and from Tokyo was New York city, and other international carriers quickly followed. A bluefin of excellent quality, with high fat content and colour, caught on the coast of Massachusetts in the afternoon of June 1 can be on the Tsukiji auction market floor in Tokyo at 4:00 am on June 3 or 4. Such fish can bring more than $US88/kg. It is estimated that there is an investment of about $US9/kg in a bluefin as it arrives at Tsukiji auction. The auction price has an annual fluctuation of $US22–110/kg, with an average of about $US33/kg although prices were lower in 1995 than in previous years.

In other fisheries, we presume that market economics work, with a reduction of effort and harvest when populations get low. That is not true in this fishery, as the cost of capture is relatively low, since bluefin can be harvested very close to shore by vessels with low operating expenses. Many of the anglers for bluefin are part-time non-professionals and the value per fish is very high. That is, the value to the boat relative to the expenses of capture and marketing is so high that it is presently worth the effort to catch the last fish. It is not in the fisher's self interest to let someone else land it or let it go.

We must consider who the players are in this dramatic high-value fishery. It is they who are affected by the allocation issue. It is very important to understand that the individuals engaged in resource utilization (appropriation) are not the same individuals who make, monitor, and enforce the rules. This is especially true of international regimes which manage HMS. Fishers' activities and agendas are differentiated from the diplomats, Non-Government Organization (NGO) and international civil servants who sit at the negotiating table or occupy the seats against the wall and the lounges in the lobbies. Also, the scientists from government agencies, corporations, academia, and NGOs are significant players and their roles must be recognized. When issues become politicized, science and public policy become indistinguishable. Management scientists cannot be removed from the realm of politics and advocacy.

ICCAT STRUCTURE AND PARTICIPATION

This elaborate tuna ballet is performed annually in the theatre of the Secretariat of The International Commission for the Conservation of Atlantic Tunas, which was signed and implemented in 1969. The organization consists of 22 signatory members with authority derived from Article 64 of the Draft Convention of the Law of the Sea, 1980. Its charter is to conserve and manage tuna and related species throughout their ranges in a manner that achieves the maximum sustainable catch.

United States participation is authorized by the Atlantic Tunas Convention Act of 1975 (ATCA), which was re-authorized in 1995. The Act stipulates that there will be no more than three commissioners appointed by the President, to serve no more than three two-year terms. One of the three can be a US government employee. The other two, from the private sector, must 'be knowledgeable and experienced with regard to fishing in the Atlantic Ocean, Gulf of Mexico or the Caribbean Sea' and represent the commercial and recreational fishing interests. The commissioners are assisted by an Advisory Committee, called for in ATCA, of 20 individuals appointed by the commissioners and five from the regional fishery management councils. They must be knowledgeable and represent the various fisheries that are governed by the convention. The Advisory Committee has the opportunity to offer comments on all proposed research programmes and initiatives, reports, recommendations for action and policy, and regulations of the Commission. The International Division of NOAA's NMFS provides administrative and technical staff support to the Advisory Committee and the Commissioners. The US delegation to the Commission meeting is made up of the Commissioners, six members of the Advisory Committee, NOAA and State Department staff, and US congressional representation. The process of policy development for the US position is relatively transparent, public and well-reported. This is a significantly different process than that followed by other ICCAT nations. One result of the openness and multiplicity of interests that must be represented in the US position is that the image of US national representation at ICCAT is one of sustained and continued bargaining rather than unitary national interest.

ICCAT functions with a structure that consists of the Commission (composed of not more than three delegates from any one member), the Council which is an elected body made up of the chairman, vice-chairman, and representatives from four to eight member nations that performs functions assigned to it by the convention or the Commission, and the Executive Secretary who is responsible for the budget and co-ordination of ICCAT programmes and activities. This is a particularly critical role as the Secretariat is responsible for the collection and analysis of data to accomplish the Commission mission. The Standing Committee on Research and Statistics (SCRS) provides advice and makes recommendations to the subject area panels and the Commission. The panels which review the species under their purview, such as bluefin tuna, collect scientific and other information, propose recommendations for joint actions and recommend studies by member nations.

The Commission is responsible for the development and formulation of regulatory proposals and actions. Those that are approved by the Commission are submitted to member governments for approval. If there are no objections in six months, each party to the convention is then responsible for implementing and enforcing the regulations that have been approved by ICCAT. This is particularly important because the

UN Straddling Stocks Treaty will delegate the management of straddling stock fisheries to those international management regimes that are already in place. Therefore, ICCAT will be the responsible party for implementation of management of HMS in the Atlantic Ocean.

Discussion

Fishers using the western bluefin tuna resource, whether as recreational charter skippers or clients, or in the general angling or commercial categories, think that the SCRS assessment is wrong. They are convinced that there is a plot in NMFS and the conservation community to put them out of business to protect the fish for the 'tree huggers', 'porpoise petters' and 'recreational anglers', and the fishing fleets of the eastern Atlantic. These terms are used with the same passion and pointedness as 'communist' and 'pinko' were used during the cold war era and with the same agenda to discredit opponents and gain political advantage. User groups work aggressively and effectively to affect the outcome by influencing their politicians. Public policy formulation is one of the clearest demonstrations of the participatory democratic political process. If a developed and implemented public policy, whether domestic or international, is counter to personal objectives,an attack can be made on the position of an opponent rather than focussing on the interests and substance of the issue.

Presently, the polar positions are based upon accepting or rejecting the stock assessment, which is dependent upon perceptions of, and confidence in, the quality of the data used in the development of the stock assessments. In addition, there is great concern and criticism of whether the assumptions of the population dynamics models are being met by the modellers. The recreational and commercial sectors argued for years for a review of the data and processes of the stock assessment conducted by the SCRS, with major participation by US scientists. At present, the fishing community's faith in these outcomes ranks lower than confidence ratings of politicians. In preparation for the 1994 ICCAT meeting, the US Department of Commerce's NOAA acquiesced to the demands of the user groups and the advice of the Advisory Committee and requested that the NRC conduct an independent review of the scientific basis for management of the Atlantic bluefin tuna. It is commonly observed that domestic groups pursue their interests by pressuring the government to adopt favourable policies, and politicians seek power by constructing coalitions or serving local constituents' needs at the national level. At the international level, national governments seek to maximize their own ability to satisfy domestic pressures, while minimizing the adverse consequences of foreign developments. The outcome of the international negotiations is that policies and practices that could be rational and advantageous for the players at the domestic level may be impolitic and disadvantageous for those players at the international level.

Analyses of ICCAT proceedings and the discussions of the US ICCAT Advisory Committee show that most nations do not go to the Commission meeting with negotiating positions and national policies built upon strong technical information. Rather, they do their negotiation for position and outcome based primarily on political persuasion and a perceived short term domestic political and economic advantage. That is precisely the point at which, in my opinion, science can and should enter the allocation process. The records are very clear that the allocation of the total allowable catch of bluefin tuna, often settled at the eleventh hour of the Commission meeting, has been disputed on the basis of the outcome of the separate stock assessments by the SCRS of eastern and western stocks since 1981. The result is that the fishers harvesting the western stock will always argue that the assessment is flawed, and there is therefore no resolution or closure of the issue.

There have been requests for fundamental research to address this issue since the 1981 decision. The NRC Review clearly shows that no coherent research programme has been initiated by the management agencies in that period, so the US position at the ICCAT table is not strongly supported by independent research and analyses. Critical assumptions and data needs for the numerical models are not met and research programmes to resolve those problems are seriously lacking.

Research needs

The most fundamental research need is for data that will help resolve the issue of stock resolution. Does the bluefin tuna population consist of a single genetic pool? Are the two spawning units from the same genetic unit? Do the two spawning areas contribute different amounts to the fishery? Is there mixing of the offspring and the adult spawners on the fishing grounds? Is there home fidelity to the spawning sites? Can bluefin spawn at either site? Are there any other spawning sites in the Atlantic?

Specific examples of fundamental research which directly address the management issues are: 1) Genetics: There had been no US work on the molecular genetics of bluefin until 1994. Studies are currently underway at several academic institutions in the US utilizing the now routine techniques of molecular genetics to test the hypothesis that there is no difference in the spatial population structure of Atlantic bluefin tuna: 2) Age and Growth: the assumption of the VPA model is that the age of each individual fish is known by direct measurement of length. The NRC Report was critical of the lack of specific information on the age and growth of bluefin tuna. In practice, the VPA model is run using the 'age slicing' technique, whereby the size of the fish is used as a surrogate for age. It is well known that many fish show indeterminate growth, and growth is age-dependent. Therefore, larger fish are generally older fish. In fact, there is a great deal of variability in size-at-age for pelagic fishes. There has been no sustained research programme on the estimation of age of bluefin tuna to date. There have been significant advances in the techniques of age estimation with major advances in methodologies in recent years: 3) Movements: There has been no comprehensive or coherent research project on the tracking of movement or migration of bluefin with tagging studies since the early 1970s. The NRC Report clearly states that tagging studies have not been carried out in a manner that would permit resolution of fish movement, abundance estimates or stock mixing. This is not to say tagging is not occurring. What is not in place is a project with testable hypotheses designed to be conducted over a long period of time: 4) Reproduction: There is very little known about the reproductive biology of bluefin. The NRC Report expressed serious reservations about the quality and quantity of the data used for the establishment of the age of first reproduction which is very important in the VPA. The only current US project addressing the reproductive characteristics of bluefin is an extremely modest one initiated in 1992 at the New England Aquarium in Boston, Massachusetts: 5) Abundance: There is a small project in place for a fishery-independent stock abundance estimate, by aerial survey, initiated with private financial support in 1993.

The projects now underway that address these issues: population genetics, reproductive biology and the fishery-independent aerial survey, were proposed outside the NMFS and NOAA. The government did not originate the studies necessary to develop information that would lead to resolution of the issues. Rather, their identification, initiation and the political pressure necessary to get the required funding to carry out the research came from the ICCAT Advisory Committee and the fishery user groups. If these research projects are properly funded and carried out for the necessary length of time, the results will speak directly to the issues that currently remain unresolved. Thus, the issue of conducting the VPA on a one- or two-stock model, which is used to set the total allowable catch for each nation, would not have to remain as a sticking point of negotiation, act as a distraction to the Commissioners and be a most divisive and contentious issue for the domestic fishery. It does not mean that the allocation would not be difficult, only that it would be much clearer how the allocation to each nation is made. It is feasible to remove such issues from the table while an added benefit is that the results will permit the management scientists to focus on cleaning up the landings data and other technical matters that will provide better quality stock assessments.

A VIEW OF THE FUTURE?

What are the features of institutions for management of international fisheries that have led to apparently flawed policies and undesirable outcomes? I think a significant one is the inability of the Commissions to develop and implement rigorous independent research programmes that provide data and analyses that are beyond question. It is always possible, probable and appropriate that there will be different interpretations of experimental results. However, if the design is valid and the data are collected in a proper manner, then the dispute can be largely resolved.

Can organizations manage large-scale CPR or are the characteristics of smaller, or local, institutions more appropriate as management units? Small international organization management structures of certain CPRs seem to be workable. These include well-documented efforts on the halibut in the North Pacific, and plaice and herring in the North Sea. Scale might be the issue, as boundaries of large marine ecosystems are diffuse. The state of our basic knowledge of functional relationships of such systems and HMS is very limited.

I suggest that the most desirable outcomes that we should expect from such an institution are the systematic collection, review and dissemination of data, the co-ordination of research and periodic assessment of the relevant science. They do provide an ongoing forum for negotiation. A fundamental issue is whether the institution should have a resident scientific staff or simply collect and analyse data from the domestic scientists of the member states, as is done by ICCAT. Does this compromise scientific principles? Does this confound the data and analyses? Is the output suspect because of the potential for bias and falsification with this structure? Such questions are not currently addressed. The result has been that the research priorities are those of the national ministries or scientists and do not always address the needs of the Commission. National results are viewed with great scepticism and cynicism by the fishing community of each nation. Several management models exist that should be examined to specifically address the issue of which structure will deliver the highest quality of applied science. They include, but are not limited to, FAO, the Inter-American Tropical Tuna Commission, and the North Atlantic Salmon Commission. We should carefully analyse the issues, determine which of them can appropriately be addressed with scientific research and then develop a plan to directly address the limitations in the science data base. If we do not do so, all those with an interest in fishery resources can, will, and should continue to challenge management plans on the basis that the science is flawed. Domestic user groups and nations that do not receive the allocation they seek will then use the political process to support their personal position, no matter how extreme. Allocation of resource shares to nations is one of the most fundamental roles of international management regimes. If we want and expect fishery policy to be firmly based upon science, we must conduct excellent independent science and incorporate the results into the process.

REFERENCES

Final Environmental Impact Statement for a Regulatory Amendment for the Western Atlantic Bluefin Tuna Fishery (1995). Silver Springs, MD. USA: NOAA–NMFS. Highly Migratory Species Management Division.

Hardin, G (1968). The tragedy of the commons. *Science* **162**, 1243–8.

Joseph, J, and Greenough, J (1979). 'International Management of Tuna, Porpoise, and Billfish.' (University of Washington Press: Seattle, WA).

Kruse, G, Eggers, D M, Marasco, R J, Pautzke, C, and Quinn II, T J (1993). 'Proceedings of the International Symposium on Management Strategies for Exploited Fish Populations.' Report No. 93–02. (Alaska Sea Grant College Program: University of Alaska Fairbanks).

National Research Council (1994). 'An Assessment of Atlantic Bluefin Tuna.' (National Academy of Sciences: Washington, DC. USA).

Ostrom, E (1990). 'Governing the Commons: The Evolution of Institutions for Collective Action.' (Cambridge University Press: New York).

Parsons, L S (1993). Management of marine fisheries in Canada. *Canadian Bulletin of Fisheries and Aquatic Sciences.* **221**. 763 pp.

Stroud, R (1990). 'Planning the Future of Billfishes.' Part 1. (National Coalition for Marine Conservation: Savannah, GA. USA).

Stroud, R (1994). 'Conserving America's Fisheries.' (National Coalition for Marine Conservation: Savannah, GA. USA).

Management of New Zealand's Snapper Fishery: Allocation of a Limited Resource Between Commercial and Non-Commercial Users

Kevin Sullivan

Ministry of Fisheries, PO Box 1020 Wellington, New Zealand.

Summary

The snapper fishery is the most valuable coastal finfish fishery in New Zealand, earning over $NZ50 million in exports in 1995. Most of this catch comes from the Hauraki Gulf. The snapper fishery in the Hauraki Gulf is also the most important recreational fishery in New Zealand. In this study, the effects of a range of management measures were modelled to determine the long-term allocation between the different sector groups as the stocks rebuild. Stronger recreational controls and reduced Total Allowable Commercial Catch (TACC) limits have been imposed on the fishery recently to rebuild the stock towards B_{MSY}. Projections of the model show that reductions in TACC to allow for rebuilding result in an increased non-commercial yield. As the stock increases in abundance, the non-commercial share could be up to 60% of the total yield. As no overall limit is defined for non-commercial use, their total catch will inevitably increase.

Introduction

In 1986 a Quota Management System (QMS) of Individual Transferable Quotas (ITQs) was introduced into most of the fisheries in New Zealand. At this time the inshore fisheries appeared to be overexploited, in particular the stocks of snapper appeared to be at very low levels of biomass. The landings of snapper in New Zealand had declined from a peak catch in 1978 of nearly 18 000 t to only 9000 t in 1985. With the introduction of the QMS, Total Allowable Commercial Catches (TACCs) for snapper were set below the prevailing catch levels to allow rebuilding.

Under the Fisheries Act, the Total Allowable Catch (TAC) is based on the principle of maximum sustainable yield, as qualified by economic, environmental and social factors. In setting the TACC for commercial fisheries, the Minister must allow first for recreational, traditional, Maori and other non-commercial interests.

The snapper fishery is the most valuable coastal finfish fishery in New Zealand, earning over $NZ50 million in exports in 1995; most of this catch comes from the Hauraki Gulf. The snapper fishery in the Hauraki Gulf is also the most important recreational fishery in New Zealand.

In this study the effect of a range of management measures was modelled to determine the long-term allocation between the different sector groups as the stocks rebuild. Stronger recreational controls and reduced TACC limits have been imposed on the fishery recently to rebuild the stock towards B_{MSY} (Biomass at maximum sustainable yields). Projections of the model show that reductions in commercial TACC to allow for rebuilding could result in an increased non-commercial yield unless additional controls are introduced; the non-commercial share could be up to 60% of the total yield by the year 2005.

The current situation highlights the problem of allocation in fisheries where recreational fishers are important. Quota holders in the commercial fishery were given an expectation that a share of the resource was allocated to them under the policy of the QMS. However, as no overall limit is defined for non-commercial use, their total catch could increase above the current level. To manage the snapper fishery in the future, a policy of allocation is required which will be sustainable for the snapper stocks and also equitable to those involved in both the commercial and non-commercial fisheries.

BIOLOGY OF NEW ZEALAND SNAPPER

The New Zealand snapper (*Pagrus auratus*) belong to the family Sparidae (sea breams). They are abundant in the warmer waters of New Zealand and similar species also occur in Australian and Japanese waters. The distribution of snapper in New Zealand is probably limited by water temperatures and available shelf area. Snapper have been studied for many years in New Zealand and most aspects of the life history are well understood (Paul 1976).

Snapper are demersal fish found in depths down to about 200 m, but are most abundant in 10–60 m. They are a dominant fish in the inshore coastal community and occupy a wide range of habitats including rocky reefs and sand and mud bottom. Small snapper are found mainly in shallow bays, estuaries and harbours. They show seasonal movements from inshore to deeper waters to 60 m. Adult fish also make seasonal migrations to spawning areas. Tagging studies have shown long-distance movements by individual fish, but generally movement is localized. These small-scale movements are likely to be associated with feeding behaviour and responses to weather and tides.

Snapper are serial spawners, releasing many batches of eggs over an extended season during spring and summer. The major spawning grounds which have been identified correspond closely with the nursery grounds of young snapper. The short planktonic phase of snapper larvae means that they are not subjected to large-scale drift by currents. Young fish school in shallow water and sheltered areas and move out to deeper waters in winter. The fish disperse more widely as they grow older. They reach maturity at 25–30 cm in length. Large schools of snapper congregate before spawning and move onto the spawning grounds usually in November. The spawning season may extend to January-February in some areas before the fish move off to feeding grounds. The winter grounds are thought to be more widespread in deeper waters, when commercial catch rates of snapper are lowest (June-August).

Paul (1976) showed that year-class strength in the snapper stock was closely related to water and air temperatures during the summer months, warmer years resulting in stronger year-classes. Francis (1993) was able to quantify the relationship between year-class strength and water temperature using data from random trawl surveys of juvenile snapper.

The growth rate of snapper varies from area to area. There is some evidence that growth rate has increased in recent years, possibly as a density-dependent effect of lower stock sizes. Snapper have a marked seasonal growth pattern with rapid growth from November to May and then a slowing or cessation of growth from June to September (when water temperatures are lowest). Snapper may live up to 60 years or more and have a very low rate of natural mortality. They are generalist feeders, eating whatever is most abundant, but diet changes with increasing size.

Based on biological characteristics (growth rates, genetic differences) and geographical separation (spawning grounds) there are thought to be seven major stocks of snapper in New Zealand (Sullivan 1985). Results of numerous tagging programmes support this separation of stocks and show movement is generally limited between areas. The north-east coast of New Zealand has three snapper stocks, the largest of which is centred on the Hauraki Gulf.

COMMERCIAL FISHERY FOR SNAPPER IN THE HAURAKI GULF

The commercial fishery for snapper up to 1971 was summarized by Paul (1977). The commercial fishery developed in the late nineteenth century, although the Maori inhabitants had fished snapper for food and trade for at least the previous 500 years, based on archaeological records. The main methods were initially handlining (replaced by longlines), set-netting and beach seining. As early as 1908 there was concern about the local depletion of snapper in areas of the Hauraki Gulf. Steam trawlers dominated the fishery from about 1915, and by 1920 there was more concern about depleted stocks. Danish seining was introduced in the Hauraki Gulf in late 1923 and proved very successful on fishing grounds which were previously unprofitable. Conflict between the various methods led to area closures for Danish seining and trawling, measures which continue in force to the present day.

The economic depression in the thirties saw a decrease in steam trawling and a large increase in part-time and subsistence fishers. An export market to Australia developed but in an uncontrolled manner due to the fragmented nature of the industry. Snapper was competing in the same markets with tarakihi (*Nemadactycus macopterus*), the other prime inshore finfish species in New Zealand. Following the war, motor trawlers became dominant and many seiners were also converted to trawling.

Although the abundance of snapper seemed to be declining in the forties, the fifties saw increased abundance and higher catch rates by seiners. At this time a trawl fishery for snapper developed on the west coast of the North Island which resulted in reduced fishing effort in the Hauraki Gulf.

Following a period when catch limits restricted the landings of snapper, catches by Danish seine increased dramatically in 1968. This rise coincided with the introduction of a larger and higher-opening seine net. However, it appears that the abundance of snapper in the Hauraki Gulf increased at this time as other methods also showed increased catches. Paul (1977) suggested that strong year-classes entered the Hauraki Gulf snapper stock at this time. It appears that the catches of trawl and seine vessels were always dominated by small fish and therefore trends in the catch follow trends in recruitment.

Since the early 1980s the longline fishery has increased in importance as the Japanese market for spiked fish (*iki jime*) was developed. The fish are killed by spiking the brain immediately on capture. The fish are then chilled in an ice slurry to enhance the bright skin colour and preserve the quality. Fish are landed within 24 hours, packed in ice and air-freighted daily to the market (Harvie 1982); up to 95% of the snapper catch is exported. There are seasonal fluctuations in price reflecting the volume of supply. A more recent development in the snapper fishery has been the export of live fish to the Japanese market. Although the size of this market is small and the highest quality must be achieved, the returns are much greater per kilogram of fish.

Following poor catches in 1982 and 1983, snapper stocks in New Zealand were considered to be depleted. Reported catches from the Hauraki Gulf had fallen from 10 128 t in 1978 to only 6256 t in 1983. However, 1982 and 1983 corresponded with the occurrence of unusually cold water temperatures around New Zealand and poor fishing success for recreational fishers as well. Catches improved slightly in 1984 and 1985 and from 1 October 1986 were subject to TACC constraints within the QMS.

Quota management system

The Quota Management System of Individual Transferable Quotas was introduced on 1 October 1986, despite objections by Maori over ownership of fisheries resources (see Appendix A). For the 26 species within the scheme a TACC was set by stock (or area). The initial TACCs were either based on catch levels in the fishery in 1983 or on the limited stock assessment advice available. Quota holders were allocated a specified tonnage (to the nearest 100 kg) which was their individual share of the TACC, which they could use, sell or lease as they wished. A limit of 20% was set on the amount of quota in any stock which could be held by an individual quota holder. The allocations were based on the catch history of the fishers in the previous few years and were given free to current holders of fishing permits.

Fishers were asked to nominate which two of the years from 1982 to 1984 they wished to be used to determine their average annual catch history. New entrants to the fishery were given a nominal history based either on their most recent catch or on their financial commitment (investment in vessels, gear etc.).

A regional Objections Committee and a national Quota Appeal Authority (QAA) were instituted to review appeals by individuals concerning their catch histories recorded in the fisheries statistics. The QAA spent over three years resolving these objections, arbitrating on revised catch histories for individual fishers and companies and recommending large increases in some cases. These increases have been additional to the original TACC levels.

The sum of catch histories in most inshore fisheries exceeded the TACCs for the stocks. Before any administrative cuts were made to individual shares, a competitive tender (buy-back) was used to provide the opportunity for individuals to sell out of the fishery voluntarily. All fishers were paid at the same rate per tonne, equal to the highest bid accepted, for each species. The Government spent $45 million in buying out 15 800 t of catch history (three-quarters of the target reductions). However, for some species *pro-rata* cuts were still necessary to reduce the sum of catch history entitlements to the TACC level. A condition of these *pro-rata* cuts was that any future increase in TACCs would first be used to reinstate the amounts cut. This condition applied only for the original quota holders to whom the allocations were made, i.e. was not transferable.

The initial intention was that Government would adjust TACCs in future years by entering the market and buying and selling quota, using the funds built up from quota sales by competitive tender (TACC increases) and resource rentals. However, a fund was never established, although over $70 million was raised from increased TACCs after 1986. (Note: this tender was for mainly deep-water species and raised more than the initial cost of the buy-back in the inshore fishery).

The original scheme of fixed tonnages of ITQ did not last long and following prolonged consultations with industry, the QMS was changed to a proportional scheme whereby each ITQ became a proportion of the total TACC. In hindsight this seemed a more realistic management system for fish stocks, but it also shifted the financial risk from Government to the quota holders. It meant that future increases in TACCs could not be tendered out by the Government, but would automatically be owned by quota holders. Conversely, quota holders would also bear the full consequences of decreased TACCs.

Management of the Hauraki Gulf commercial fishery for snapper

The Hauraki Gulf snapper fishery has a long history of management measures which have been implemented to reduce conflict between the various fishing methods, reduce juvenile mortality and improve the yield-per-recruit (YPR) from the stock. Separate lines define the areas closed to trawl, Danish seine and pair trawl; mesh size is controlled by a 125 mm minimum, and the minimum legal size (MLS) for snapper is 25 cm. The original intention of area closures was to protect the inner Hauraki Gulf waters from the more efficient fishing methods. They also helped to reduce the conflict between fishers using different methods.

In 1978 the Hauraki Gulf Advisory Council was formed, which concluded that a reduction in the number of vessels was needed. In 1982 a Controlled Fishery of limited entry licensing was established for Hauraki Gulf wetfish. Although this excluded some vessels from the Hauraki Gulf, it did not reduce the number of vessels to an appropriate level. It also displaced some effort into adjacent areas where fishing effort was already excessive.

Since 1986, catch has been limited by the TACC for fishstock SNA 1, which includes snapper from the Hauraki Gulf and adjacent areas. All the other input controls remain in force but the fishery is no longer controlled by licences. Access to the fishery is on the basis of quota ownership.

With the introduction of the QMS, the TACC for SNA 1 was set at 4710 t, based on the results of the 1983 and 1984 tagging programmes (Sullivan *et al.* 1988). This TACC was only 57% of the sum of the catch histories. Consequently, 45% of the total cost of the buy-back was spent on reductions in snapper holdings prior to the introduction of the QMS. However, even after two tender rounds, *pro-rata* cuts of 20% were still required in the SNA 1 fishstock to reduce the sum of ITQ entitlements to the TACC level. Quota holders who took the cut in preference to the compensation offered, had an expectation that future TACC increases would allow them higher ITQs (when the cut amounts of snapper were reinstated).

In the following years, decisions by the QAA gradually increased the TACC, as ITQs were granted or increased as a result of the appeal process. The TACC for the SNA 1 fishstock reached 6300 t by 1990–91, higher than the reported landings in the last year before the QMS was introduced and causing most of the benefit of the original reductions to be lost.

As a result of an updated stock assessment in 1992 using CPUE indices (Gilbert and Sullivan 1994), the TACC was cut to 4900 t. This reduction in TACC was intended to speed up the rate of rebuilding and improve the recreational access to the resource. However, a large-scale tagging programme which was carried out in 1993 throughout the north-east snapper fishery, showed that the Hauraki Gulf snapper stock was still only at half B_{MSY}.

The Minister of Fisheries moved to cut the TACC from 1 October 1995 to 3000 t, within an overall TAC of 5600 t (2300 t was set aside for recreational fishers and 300 t was made a specific allowance for Maori customary take). However, an injunction was granted to the fishing industry by the court which to date has prevented this catch level taking effect. The commercial fishery is still operating under the previous TACC of 4938 t.

NON-COMMERCIAL FISHERY FOR SNAPPER

The non-commercial use includes traditional (subsistence) and Maori use as well as recreational catch. Although freshwater fishers are required to buy a licence, none is required for marine fishing. The only restrictions are on the type and size of fishing gear, minimum fish sizes and bag limits which are set for most prime species. This means that there is no real upper limit set to the total non-commercial catch, which could increase with greater participation. Education rather than enforcement is seen as the key to compliance in these fisheries, as the effort is widespread and highly seasonal, peaking during the summer holiday period.

RECREATIONAL FISHING

Fishing has always been popular in New Zealand as the climate, extensive coastline and the wide variety and abundance of fish combine to provide excellent opportunities for fishers. However, with the increased exploitation of inshore stocks after 1970, the quality of recreational fishing declined. The failure of management to protect recreational fisheries resulted in localized depletion, low catch rates and small average size of fish.

The national policy on marine recreational fishing which was released in 1989 sets out the philosophy used to manage recreational fishing (MAF 1989). The main objective of the policy is to ensure recreational users have access to a reasonable share of fishery resources, while daily catch limits are used to ensure the catch is shared more equitably amongst users. Where the resources are insufficient to meet the needs of all potential users, recreational fishers may be given precedence.

Although input controls are used to manage the non-commercial fishery, the QMS is the major management tool in the commercial fishery for snapper. TACCs are set for the commercial fishery after allowance has been made for non-commercial users. This allowance provides for free access to a share of the harvest.

In recent years there have been several changes in the rules governing the snapper fishery. The recreational bag limit was reduced to 15 fish in 1994 and to 9 fish from 1 October 1995. The minimum legal size was increased to 27 cm (from 25 cm which is still the minimum size limit for commercial fishers) on 1 December 1994. These measures are likely to have reduced the non-commercial catch of snapper. Based on the size distribution of fish landed at boat ramps in 1994, the change in size limit was estimated to reduce the weight of the annual catch by 10%. Based on the distribution of bag sizes in the same survey, the recent change in bag limit to 9 fish is estimated to reduce the annual catch by another 8%.

ESTIMATES OF NON-COMMERCIAL CATCH

Estimates of catch are not available for all the various non-commercial users. Therefore the catch estimated for the recreational fishery has been taken here to represent the total non-commercial catch.

The National Marine Recreational Fishing Survey in 1987 found that about 24% of the population in the north of New Zealand had participated in marine recreational fishing in the previous 12 months. The survey did not attempt to quantify the catch but identified the most important species and areas of recreational catch and the methods used. Snapper was the most commonly taken fish. Line fishing from shore and from small boats were the most common fishing methods.

In the 1983 and 1984 snapper tagging programmes up to 30% of tag recaptures were returned by recreational fishers in some areas (Sullivan *et al.* 1988). External loop tags were used in this study, which required voluntary return of recaptures by fishers. The annual non-commercial catch was estimated to be about 1200 t (Hauraki Gulf/Bay of Plenty), based on the catches recorded by the commercial fishery and the comparable recovery of tags. This was about 17% of the total catch of 7100 t (including under-reporting) from the stocks.

In 1994, a combined telephone and diary survey was carried out in the northern areas of New Zealand to estimate the catch of the major species taken by the non-commercial fishery. The snapper catch was estimated at 2100 t from the Hauraki Gulf/Bay of Plenty stocks. Based on this figure the recreational share of the catch had increased to 33% of the total estimated catch (6300 t). It is likely that the non-commercial catch had increased over this period. However, there is evidence to suggest that the 1994 estimate reflects a good fishing season in that year (because of warmer autumn temperatures and settled weather) and may have been higher than the average catch in recent years. A comparison of CPUE from boat ramp surveys showed a 50% increase from 1991 to 1994 (Sylvester 1995), a period when stock size was unlikely to be changing. It appears that the non-commercial catch of snapper may be variable and depend on factors such as predominant weather patterns, water temperatures and overall effort (which may be related to localized fishing success), as well as on the abundance of fish. With improved access of recreational fishers to boats and more efficient fishing gear, effective fishing effort may have increased. However, the catch of individual fishers has been disappointing. The ramp surveys showed that 65% of fishers did not catch any snapper in trips surveyed in 1991, and 48% were unsuccessful in 1994.

In 1996, a further telephone and diary survey is being carried out to estimate the recreational catch of snapper and other species; results will be available in early 1997.

Stock assessment of Hauraki Gulf snapper

1993 tagging programme results

In 1993, a tag and release experiment was carried out along the north-east coast of the North Island to estimate the stock size of snapper (Gilbert *et al.* 1996). Fish were caught by trawl and longline and tagged with internal coded-wire tags. Commercial catches were examined for tags throughout the following year and the length frequency of catches was measured for a number of fishing methods. Using tag recaptures over the year and catch records from the fishery, the stock size of snapper was estimated. Allowance was made for the estimated rates of initial mortality, tag loss and detection of tags which were recovered using electronic wanding equipment by researchers.

As the tag recovery data showed a high level of mixing between the Bay of Plenty and the Hauraki Gulf, these stocks were assessed together (Annala and Sullivan 1996). Under a range of assumptions the stock size of the Hauraki Gulf/Bay of Plenty snapper in December 1993 was estimated at 28 000 to 34 000 t using Petersen estimates which were corrected for growth throughout the year. These biomass estimates were very similar to the sum of the estimates from two previous tagging programmes in 1983 (Bay of Plenty) and 1984 (Hauraki Gulf) of 34 400 t.

Modelling

To assess the current stock status and to predict the effect of alternative management measures it is necessary to model the dynamics of the snapper stock. Two alternative models were compared, the first assuming the stock was in equilibrium at the time of the tagging programme in 1993, and the second modelling the total catch history of the stock from 1850 (Gilbert 1994). The stock status in both cases was similar: the 1993 biomass was below half B_{MSY} (Table 1). B_{MSY} was estimated to be 25.3% B_0 using the growth, mortality and selectivity parameters assumed in the model.

Equilibrium model

In this model the assumption was made that the sub-stocks were in equilibrium at the time of the 1993 tagging programme and that the total catch in the 1993–94 year would maintain the stock at the same biomass. The 1993–94 catch by the non-commercial fisheries was assumed to be 2100 t. With a 10% allowance for under-reporting the commercial catch was assumed to be 4275 t.

At each level of fishing mortality a yield-per-recruit analysis may be used to predict a corresponding equilibrium level of mid-year biomass. Using the range of stock size above, the observed catch/biomass ratio in 1993–94 was 22.3% to 18.7%. At equilibrium for these exploitation rates, the stock was predicted to be at 8.7% to 11.3% B_0 respectively.

Total catch history model

Gilbert (1994) modelled the history of the stocks from 1850 to the present using a stock reduction technique. This model was fitted to the estimates of biomass from both the 1993 and 1983–84 tagging programmes (Annala and Sullivan 1996). The estimate of virgin biomass was 292 700 t, which gives the stock status in 1993–94 at 9.6% to 11.6% B_0 (B_0 = virgin biomass), a very similar result to the equilibrium model. Gilbert found that this model was not sensitive to catch history in the early years but was sensitive to the period over which mean recruitment was assumed: higher and lower mean recruitment resulted from using periods of warmer and colder temperatures.

Projections

In modelling the future projections of the snapper stock there were three major uncertainties to consider: current stock biomass, mean recruitment and recreational catch. The current

Table 1. Status of the Hauraki Gulf/Bay of Plenty snapper stocks for two alternative models of the population, and two biomass estimates from the 1993 tagging programme. The 1994 biomass is the start of year biomass for the fishing year 1993–94 (R = recruitment of 4-year-old snapper, B = biomass)

	Mean R (million)	B_{1994}/B_{MSY}	B_{1994}/B_0
Equilibrium model			
Lower 1994 biomass (28 000 t)	10.1	0.34	0.087
Higher 1994 biomass (34 000 t)	9.5	0.45	0.113
Total catch history model			
Lower 1994 biomass (28 000 t)	9.2	0.38	0.096
Higher 1994 biomass (34 000 t)	9.2	0.46	0.116

biomass was assumed to be 32 000 t (range 28 000 to 34 000 t) based on the estimates from the 1993 tagging programme (Annala and Sullivan 1996). The mean recruitment was assumed to be 9 million four-year-old fish (range 8–11 million) based on the recent modelling results with the total catch history model of D. J. Gilbert (pers. comm.). The recreational catch was modelled by assuming that the fishing mortality by this sector increased by 2% a year (range 1 to 3%). The effect of the change in the recreational bag limit to 9 fish on 1 October 1995 was estimated to result in a drop of 8% in the fishing mortality of the non-commercial sector. This was assumed to continue for all years after 1995–96. The size limit change to 27 cm (1 December 1994) was included by adjusting the catch of the four-year-old year-class (survival of four-year-olds returned to the water was assumed to be 80%).

The model was projected forward assuming a constant commercial catch from the Hauraki Gulf and Bay of Plenty of 80% of the SNA 1 TACC plus 10% allowance for under-reporting. The TACC was assumed to be 3000 t from 1996–97. Recruitment was assumed to be at the mean level for all year-classes after 1996, the last year for which an index is available.

The results of the baseline analysis and sensitivity to the major uncertainties are shown in Table 2. With the reduced TACC of 3000 t (from 1996–97) the stock increases in abundance by about 50% by the year 2005 to reach 69% of B_{MSY}. With a deterministic model, the stock does not reach B_{MSY} within the next 10 years. However, when using stochastic simulations of future recruitment the stock reaches B_{MSY} by 2005 in a very small proportion of the runs.

Until the year 1999 the recreational catch is modelled to remain below 2000 t per year as poor recruitment is predicted to enter the fishery. However, Table 2 shows that as the stock size rebuilds after this, the recreational share of the total catch increases to be above 50% by the year 2005. Projections of the model for longer periods are not considered to be particularly realistic, as further management changes are likely to occur. However, when the recreational catch is increased by 1% or more per year, longer model runs indicated that the stock is not able to reach B_{MSY} but reaches a maximum and then declines.

Table 2. Recreational share of total snapper catch in the year 2005 and stock status of the Hauraki Gulf/Bay of Plenty snapper stocks in 2005 from a projection model with TACC of 3000 t from 1996–97 (B = biomass, F = fishing mortality)

	Recreational share (%)	B_{2005}/B_{MSY}
Basecase	53.7	0.69
Lower 1994 biomass (28000 t)	53.6	0.73
Higher 1994 biomass (34000 t)	53.9	0.59
1% increase in recreational F	51.4	0.71
3% increase in recreational F	55.9	0.67
Lower mean recruitment (8 million)	50.3	0.68
Higher mean recruitment (11 million)	59.3	0.55

INCREASING MAXIMUM SUSTAINABLE YIELD

There is potential for increasing the yield from snapper stocks by improvements in yield-per-recruit (YPR) and reduction in pre-recruit mortality. The following measures were modelled to quantify the increase in MSY:

(a) Increase MLS for the commercial longline fishery from 25 cm to 30 cm.

YPR increased by 3.4%, while the increase in incidental mortality was estimated to be 1.2%; a net increase of 2.4% in MSY results from this measure.

(b) Increase MLS for recreational fishers from 27 cm to 30 cm.

YPR increased by 2.1%, while the increase in incidental mortality was estimated to be 0.8%; a net increase of 1.3% in MSY results from this measure.

(c) Remove the MLS of snapper for trawl and Danish seine catches.

Catches of snapper below the MLS of 25 cm must be discarded with the current fisheries regulations. However, most of the undersized fish discarded by trawl and Danish seine vessels are thought to die. Therefore the removal of the MLS for these methods would make these small fish count against the quota and reduce total mortality by a small fraction. The added advantage of this measure is that vessels would have a much greater incentive to avoid areas of high juvenile abundance. This measure would increase the rate of rebuilding of the stock, but was not quantified.

(d) Area closures and mesh size changes.

Closing additional areas to commercial fishing by trawl and Danish seine vessels could reduce the pre-recruit mortality of fish below 4 years, and provide a proportional increase in MSY. A 10% decline in pre-recruit mortality was estimated to increase MSY by about 2%, under the assumption of current pre-recruit mortality of 20% per year. Increased mesh sizes would also improve YPR and reduce pre-recruit mortality, but the impact is more difficult to quantify, as the likely survival rates of fish passing through the meshes is unknown.

DISCUSSION

Since the QMS was introduced, the question over who owns the resources has been hotly debated. The initial allocation based on catch history gave quota holders privileged access, for which they were expected to pay a resource rental. The scheme has changed in response to the various problems experienced. ITQs which were originally denominated in tonnes are now proportional shares of the TACC. The resource rentals have been abolished and replaced by cost recovery. The nature and extent of Maori fishing rights were never resolved, but a compromise was reached which involved compensation within the new management system (Appendix A).

There have also been calls to define recreational fishing rights more closely, setting tonnage limits for each stock, which could be traded in the same way as commercial ITQs (Ackroyd *et al.* 1990; Pearse 1991; Anon. 1992; Horton 1994). However, this is neither practical nor acceptable politically and reflects a lack of understanding of the nature of the fishing right. The current generation do not own the recreational right but are merely guardians of the right. As future generations must also have the same rights as current users, the rights cannot be sold off or transferred to other sector groups.

Under the Fisheries Act, the TACs are based on the principle of maximum sustainable yield, as qualified by economic, environmental and social factors. In setting the TACC for commercial fisheries the Minister must allow for recreational, traditional, Maori and other non-commercial interests. This is not necessarily a priority right.

It is apparent that the yield curve for snapper is very flat, so that a large increase in stock biomass will not increase yields substantially. At the current stock level of about half B_{MSY}, the stock is producing 90% of the MSY. However, at such low biomass the recreational fishery is not considered to have adequate catch rates. The recent TACC reduction is designed to meet the requirements of the Fisheries Act, to rebuild the stock to the level that will support the MSY. In achieving this goal, a small increase in total yield will result. Improvements in the recreational fishery are also expected to occur.

The critical question remains about the appropriate allocation to non-commercial and commercial sectors. The current legislation does not specify how the non-commercial share should be determined. The fishing industry is unhappy that any rebuilding of the stock will result in a greater proportion of the total yield being taken by non-commercial users. They regard their property right of ITQs as a basis for a fixed share of the available yield based on the relative shares in 1985 when the QMS was introduced. However, it is important to note that the non-commercial share in 1985 was probably based on the minimum historic level of biomass.

In the short-term the recreational catch is not expected to increase because poor recruitment is predicted to enter the fishery until 1998. However, rebuilding of the stock will inevitably result in greater recreational catch (over 50% of the total yield in the model projections), unless further restrictions are introduced. Stronger management measures on individual fishers will only temporarily control the recreational catch level. To manage the snapper fishery in the longer term, a policy of allocation is required which will be sustainable for the snapper stocks and also equitable to those involved in both the commercial and non-commercial fisheries.

References

Ackroyd, P, Hide, R P, and Sharp, B M H (1990). New Zealand's ITQ system: prospects for the evolution of sole ownership corporations. Report to MAF Fisheries, Wellington. 86 pp.

Annala, J H, and Sullivan, K J (Comps.) (1996). Report from the Fishery Assessment Plenary, April-May 1996: stock assessments and yield estimates. (Unpublished report held in NIWA library, Wellington). 308 pp.

Anon. (1992). Sustainable fisheries (*Tiakina nga taonga a tangaroa*): Report of the Fisheries Task Force to the Minister of Fisheries on the review of fisheries legislation. New Zealand Ministry of Agriculture and Fisheries. 75 pp.

Francis, M P (1993). Does water temperature determine year class strength in New Zealand snapper (*Pagrus auratus*, Sparidae)? *Fisheries Oceanography* **2**(2), 65–72.

Gilbert, D J (1994). A total catch history model for SNA 1. *New Zealand Fisheries Assessment Research Document* 94/24. 16 pp.

Gilbert, D J, and Sullivan, K J (1994). Stock assessment of snapper for the 1992–93 fishing year. *New Zealand Fisheries Assessment Research Document* 94/3. 37 pp.

Gilbert, D J, Sullivan, K J, Davies, N M, McKenzie, J R, Francis, M P, and Starr, P J (1996). Population modelling of the SNA 1 stock for the 1995–96 fishing year. New Zealand Fisheries Assessment Research Document 96.

Harvie, R (1982). A code of practice for air- freight chilled fish. New Zealand Fishing Industry Board. 20 pp.

Horton, C T (1994). Report to the Fisheries *ad-hoc* Ministerial committee. (Internal report by the Chairman of the Fisheries Legislative Review Committee). 15 pp.

Levine, H B (1989). Maori fishing rights: ideological developments and practical impacts. *Maritime Anthropological Studies* **2**(1), 21–33.

MAF (1989). National policy for marine recreational fisheries. New Zealand Ministry of Agriculture and Fisheries. 9 pp.

Paul, L J (1976). A study on age, growth and population structure of the snapper, *Chrysophrys auratus* (Forster), in the Hauraki Gulf, New Zealand. *Fisheries Research Bulletin* **13**. 62 pp.

Paul, L J (1977). The commercial fishery for snapper, *Chrysophrys auratus* (Forster), in the Auckland region, New Zealand, from 1900 to 1971. *Fisheries Research Bulletin* **15**. 84 pp.

Pearse, P H (1991). Building on progress: fisheries policy development in New Zealand. (A report prepared for the Minister of Fisheries). New Zealand Ministry of Agriculture and Fisheries. 28 pp.

Sullivan, K J (1985). Snapper. In 'Background papers for the 1985 Total Allowable Catch recommendations'. (Comps and Eds J A Colman, J L McKoy and G G Baird) pp. 187–214 (Unpublished report, held in NIWA (Fisheries) Greta Point library, Wellington).

Sullivan, K J, Hore, A J, and Wilkinson, V H (1988). Snapper. In 'Papers from the workshop to review fish stock assessments for the 1987–88 New Zealand fishing year'.(Comps and Eds G G Baird and J L McKoy) pp. 251–75. (Unpublished report, held in NIWA (Fisheries) Greta Point library, Wellington).

Sylvester, T (1995). Initial results of the Northern boat ramp survey. *Seafood New Zealand* **3**(1), 11–13.

Waitangi Tribunal (1988). Report of the Waitangi Tribunal on the Muriwhenua fishing claim. New Zealand Waitangi Tribunal. 370 pp.

Waitangi Tribunal (1992). The Ngai Tahu sea fisheries report 1992. New Zealand Waitangi Tribunal. 409 pp.

APPENDIX A: MAORI FISHING

Fishing was an integral part of the life of coastal Maori tribes from the first occupation of New Zealand. The provision of food and trade with inland tribes were traditional practices both before and after the arrival of European settlers. In 1840, when the Maori tribes signed the Treaty of Waitangi with the Queen of England, the treaty guaranteed the 'full, exclusive and undisturbed possession of their lands, forests, fisheries and other properties which they may collectively or individually possess for so long as they might wish and desire to retain the same in their possession' (Waitangi Tribunal 1988). The important feature of this agreement was that tribal *rangatiratanga* (authority) to manage the resources would be preserved.

However, the Treaty was not upheld in practice and over time, in dealing with Maori fishing rights, it was ignored. In 1866 the Oyster Fishery Act allowed Maori the use of the resource for food only, not for sale, thereby excluding them from the commercial fishery which had developed. Similarly, other legislation failed to protect Maori fishing rights, despite continual protests. By 1908, when comprehensive fisheries management measures were first introduced in New Zealand, Maori fishing rights were merely defined in terms of a few oyster reserves, reserved exclusively for local use.

Maori grievances concerning land, fish and other *taonga* (prized possessions) have continued through to the present. In 1975, the Waitangi Tribunal was established by Government to hear these grievances. In 1985 the jurisdiction of the Tribunal was extended back to 1840, which allowed formal Maori claims on fisheries to be addressed. At this time, the QMS was being developed and the scheme was implemented in 1986 despite objections by Maori.

Maori objected to the introduction of the QMS as a matter of principle, for the property right created by ITQs would deprive them of ownership guaranteed by the Treaty. Despite their vigorous protests, the legislation was introduced in the Fisheries Amendment Act 1986, without any specific provision for Maori. Of further concern was that many Maori commercial fishers received no allocation of ITQ under the QMS because of their prior removal from the fishery administratively in 1983 as part-time fishers. In October 1987 four Maori parties went to the High Court which brought down an injunction which prevented the inclusion of any additional species in the quota system.

In this period the courts also set precedence with decisions on a number of individual cases involving Maori fishers. In 1986 the courts confirmed the traditional right of Maori to take shellfish, and in 1988, upheld that the Fisheries Act 1983 did not apply to Maori commercial fishers who were fishing in accord with Maori fishing rights (Levine 1989).

In 1988, the Muriwhenua claim to fish resources of northern New Zealand was reported back to Government, recommending that the Crown negotiate a settlement with Maori (Waitangi Tribunal 1988). In 1992 the Tribunal reported on the Ngai Tahu fisheries claim on fish resources around the South Island (Waitangi Tribunal 1992). Once more the injustices to Maori concerning fisheries were made explicit. However, by this time the Crown had already begun negotiations to compensate Maori for the loss of their fishing rights.

Levine (1989) points out that the implementation of the QMS in 1986 strengthened Maori claims to fishing rights. The 1986 amendment gave away property rights in the fisheries that the Crown had already guaranteed to Maori (Waitangi Tribunal 1988). The creation of ITQs was based on the assumption of Crown ownership of the resources. This was obviously at odds with the guarantees of the Treaty, which had once more been given official recognition when the Waitangi Tribunal was established in 1975.

A joint working party of the Crown and Maori representatives was established in 1988 to determine how Maori fishing rights should be given effect. Following much debate and many changes to the original proposal, the Maori Fisheries Act 1989 was passed. This provided for 10% of all quota in the QMS to be given to Maori over a four-year period ($NZ10 million was also provided for the purpose of fisheries development) and the establishment of a Maori Fisheries Commission to manage the allocation and control of these assets. This was seen as an interim settlement, and enabled Maori participation in the commercial fishery to develop while further negotiations with the Crown continued.

A final settlement was reached with the 1992 Treaty of Waitangi (Fisheries Claims) Settlement Act. The Crown agreed to provide $150 million over three years to allow Maori to purchase 50% of the largest fishing company in New Zealand. In addition, Maori would be allocated 20% of all the quota of new species brought into the QMS. Non-commercial Maori fishing rights are still subject to the Treaty and give obligations to the Crown. These have resulted in the enactment of provisions to allow *mataitai* reserves and *taiapure* (local fisheries) where tribal management can be established in areas traditionally important to Maori.

Although the nature and extent of Maori fishing rights were never resolved by the courts, a compromise was finally reached which recognized the Maori claims to commercial fisheries resources, while retaining the new management system in place. The Commission now faces the difficult task of sharing the benefits of these assets among the Maori tribes.

The Allocation of Inshore Marine And Estuarine Fish Resources in Australia: The Need for a Precautionary Decision-Making Paradigm?

Noel G. Taylor-Moore

Marine Fisheries, Queensland Department of Primary Industries, GPO Box 2454, Brisbane, Queensland, Australia 4001.

Summary

The key issue facing inshore marine and estuarine fisheries management in Australia is the complex nature of resource allocation. The frequent inability of fisheries management planning to produce acceptable allocation outcomes has forced a fundamental rethink of fisheries management throughout Australia. The aim of this paper is to propose a fisheries resource allocation policy framework, based on a precautionary approach, as a means of achieving these outcomes.

Introduction

Pressures on inshore marine and estuarine fisheries resources have been increasing, both through greater access to fish stocks by commercial, recreational and indigenous fishers to meet food, recreational and cultural needs, and through impacts on fish habitat as access by other users of the marine wetlands increases to meet the needs of population growth. Management strategies must be holistic as society seeks the sustainable use of local, regional, national and global fisheries resources.

The aim of this paper is to propose a policy framework for a fisheries resource allocation paradigm, based on a precautionary approach, which includes a multi-dimensional matrix as part of an holistic decision-making process.

Australian context

Australia is a small fishing nation by world standards, accounting for about 0.2% of world fish production. Production of fisheries resources is comparatively low because Australia is the driest continent on earth with a coastline of 36 733 km, containing 783 estuaries and enclosed marine waters, of which 415 are tropical, 170 subtropical and 198 temperate. The majority of these marine areas are intertidal flats, mangroves, seagrass beds and saltmarsh, 70% of which are believed to have moderate to high fisheries and conservation values. However, a third of these marine areas are under real threat from population pressures. Moreover, Australia is one of the most highly urbanized countries in the world with 85% of its 18 million population concentrated adjacent to marine wetlands areas upon which inshore fisheries resources depend. Areas threatened include Botany Bay (New South Wales (NSW)), Port Phillip Bay (Victoria (Vic)) and Trinity Inlet (Queensland (Qld)) (Saenger 1989). Other areas such as Moreton Bay (Qld), have potential threats and need careful management.

Australia's finfish fisheries, excluding tuna species, are a small proportion of the gross value of production (GVP) of Australia's fisheries but a high proportion of the total volume of production — 21% of \$A1.75 x 10^{12} and 59% of 218 000 t respectively (ABARE 1995). The estuarine and inshore marine fisheries, although low as a proportion of GVP, are high in the socio-political and ecological sensitivities associated with their management — not uncommon in global terms. Examples of such fisheries are barramundi (*Lates calcarifer*), snapper (*Pagrus auratus*), tailor *(Pomatomus saltatrix*), Australian salmon and/or Australian herring (*Arripis* spp.), and whiting (*Sillago* spp. and *Sillaginodes punctata*), which are briefly described below to highlight some of these sensitivities as a basis for proposing a new fisheries resource allocation paradigm. These multi-jurisdictional and multi-user group fisheries have a GVP of \$A54 x 10^6 and a commercial catch of 14 250 t (ABARE 1995).

Barramundi is an important tropical species in four jurisdictional areas of Australia: Northern Territory (NT), Western Australia (WA), Torres Strait and Qld. The fishery has an estimated commercial catch of 1000 t with a GVP of \$A3.4 x 10^6 (ABARE 1995). The recreational catch of barramundi is estimated to be about the same as the commercial fishery in the NT and about half that of the Qld commercial fishery. Policy issues include: sustainability of the Qld east coast fishery given human impacts, such as lack of fishways and ponded pastures, on barramundi; Local Authorities seeking tourism based on recreational fishing; designation of marine parks; and Aboriginal community access to barramundi stocks for commercial purposes.

Snapper is a key fishery in most Australian States with a commercial catch of about 4300 t and a GVP of \$A17 x 10^6 (ABARE 1995). The estimated significance of the recreational catch varies from ten times that of the commercial fishery in Qld to about half in NSW, and a third in South Australia (SA). Policy issues include: concerns about sustainability of the fishery as a result of habitat degradation and overfishing; impact of trawling on juveniles; impacts of longlines and fish aggregation devices on adult stocks; sharing of snapper stocks, at different phases of the life cycle, between inshore and offshore recreational fishers; and conflicts associated with the use of different gear by commercial and recreational fishers.

Tailor is a small commercial fishery of some 200 t with a GVP of \$A0.5 x 10^6 (ABARE 1995) but a significant recreational fishery in Qld, WA, NSW, and Vic. Policy issues include: concerns that tailor stocks may be declining given the downturn in recreational catches; the variability and measurement of stock abundance; various disturbances to schooling stocks and nursery areas by other user groups; and that overfishing may be occurring because of their schooling nature and spawning aggregations, at places such as Fraser Island, Qld.

Australian salmon and Australian herring support key commercial fisheries in the southern States of Australia, mainly WA, with a commercial catch of 5000 t and a GVP of \$A3.2 x 10^6 (ABARE 1995). Policy issues include: juvenile susceptibility to disturbance from marine activities and urbanization; technology creep through the use of spotter planes; allocation conflicts between anglers and commercial operators; and the growth of Australian salmon/herring-based tourism.

Whiting is a valuable commercial and recreational species. The commercial take is about 3750 t, including trawled whiting, with a GVP of \$A29 x 10^6. King George whiting (KGW) is the premium whiting species for Australia's southern States, including Tasmania (Tas), with the recreational sector taking less than the commercial sector. The recreational fishery for other whiting species (*Sillago* spp.) is unknown in most States but anecdotal evidence suggests it is significant. Policy issues concerning the KGW fishery include: the low levels of spawning; intense recreational fishing of juveniles and sub-adults leaving nursery areas; very high levels of recreational fishing effort in waters adjacent to metropolitan areas; impacts of commercial netting on juveniles in nursery areas; conflicts over use of haul nets *versus* handlines by commercial operators; and perceptions about the effects of netting on angler catches.

In summary, there is an array of fisheries allocation challenges relating to Australian inshore marine and estuarine fisheries. The nature of these challenges, the allocation process and the resultant outcomes are multi-dimensional, but are not unique to Australia.

THE RESOURCE ALLOCATION PROBLEM PARADIGM?

The current resource allocation paradigm cannot meet these challenges because it is production-oriented, efficiency- and equity-based and uses mainly input controls and regulatory frameworks. But a paradigm shift is occurring in fisheries management in Australia (Taylor-Moore 1995) as a result of factors such as multiple-use management of common property resources, ecological risks and uncertainties, cultural mores and perceptions, and legal and institutional arrangements. However, this paper will focus on the implication of ecological risks and uncertainties for fisheries resource allocation. In this context the basis of this shift has been a growing recognition of three main policy areas: ecologically sustainable development (ESD), the precautionary principle and the requirements of international environmental instruments. These policy areas will shape fisheries legislation, statutory fishery management plans and therefore the behaviour of fishers.

Firstly, sustainability (ESD), although integral to any interventionist policy and decision-making process, is only one of many interrelated objectives. Policy relating to sustainability differs comprehensively from other policy as it focusses on irreversibility, ecological risks and thus limits to human behaviour, complex interactions with other problems, spatial and temporal complexities, pervasive uncertainty, lack of defined property rights and stakeholder responsibilities (Dovers

1995). Such factors distort allocation signals and thus the productive use of fisheries resources.

Secondly, the incorporation of ESD into the objectives of fisheries legislation automatically introduces the precautionary principle (scientific uncertainty and its associated threats) into the core of the allocation process. The precautionary principle is '*a decision making process for applying caution*' (Harding and Fisher 1993) which is transparent, focusses on the nature and level of risks and uncertainties, is transdisciplinary, and highlights diverse and high levels of stakeholder participation. Harding (1994) also suggests that the '*precautionary principle institutionalises caution*' and is therefore the basis of a dramatic shift in stakeholder and decision makers' perceptions because it focusses on '*shifting the burden of proof*' (Cameron 1993) from the decision maker to the fisher or parties affecting the environment. Garcia (1994) suggests that an unfettered application of the precautionary principle could lead to serious unplanned social and economic repercussions.

Thirdly, international environmental instruments can affect fisheries resources in two main ways. One way is by legally-binding treaties or conventions, such as the UN Convention on the Law of the Sea (1982), which directly affects fisheries through the 'optimum use of fisheries resources', and the Convention on Biological Diversity (1992) which has indirect effects on fisheries through seeking broader ecological constraints. A second way is through 'soft laws' which are not legally binding but which relate to more fisheries-specific constraints and management methods, such as Agenda 21, United Nations Conference on Environment and Development (1992) which incorporates the precautionary principle (Tsamenyi and McIlgorm 1995). They also suggest that the major threats facing the fishing industry from formally accepting these requirements are greater attention to bycatch, endangered species, area closures and limitations on fishing methods.

These three main policy issues lead to a precautionary approach to resource allocation which places production within an acceptable level of ecological risk and uncertainty, with resource failure as the bottom line.

Critical allocation policy questions arise from these three areas. For example:

- Who owns the fish, how and when should the Government intervene in the allocation process?
- How does the need for sustainable fisheries and the precautionary principle impact on allocation decisions?
- What information is required to ensure ecological, economic and socio-political risks and uncertainties are accommodated?
- What forms of consultation mechanisms and negotiating frameworks are needed to ensure stakeholder empowerment in the decision-making process?
- What are the threats and opportunities from international environmental instruments?
- Have fisheries agencies, industry and other stakeholders learnt from previous decisions?
- What integrative processes are needed to link fisheries management to other natural resource management approaches such as integrated catchment management?
- What are the best forms of inter-agency co-ordination for multiple-use management?

A new paradigm is needed to integrate these policy issues into the allocation process.

Resource allocation under a precautionary approach

A precautionary fisheries resource allocation paradigm is a decision-making framework which attempts to reconcile: the complex needs, values and perceptions of a range of stakeholders, multiple objectives of society and marine resource users, and transdisciplinary sustainability problems and threats. However, this reconciliation process has constraints, including: constitutional and legislative limitations, greater intrusion of social justice into the allocation process, market imperfections and the 'level playing field', institutional failure to act, uncertainties and risks of imperfect information, communication breakdown and over-consultation with stakeholders, and significantly, the need for production.

The following policy principles, not in any order of priority, are suggested as a means of meeting the allocation challenges facing Australian inshore marine and estuarine fisheries and as a model for other fisheries and jurisdictions:

1. *formal statutory planning processes* with specific objectives, measurable outcomes and management plans as the basis of allocation in the context of natural resource management and multiple-use management;
2. *risk management decision making* based on a precautionary approach to uncertainties and threats, performance indicators, reference points and best available information;
3. *clarification of resource security, access rights and cost recovery* for all users of marine wetlands;
4. *intra- and inter-generational equity and social justice* for all affected parties;
5. *formal and systematic consultation* to promote 'ownership' of allocation decisions through greater civic participation and inter-agency co-ordination using formal negotiating frameworks taking into account different social and cultural mores;
6. *community knowledge and empowerment* through extension and education awareness;
7. *flexible decision-making organizational arrangements* incorporating options such as self-governance, market-driven models, expert policy groups, co-management committees, management advisory committees and zonal advisory committees;
8. *formal learning processes* for fisheries agencies and stakeholders through case studies and other learning strategies; and
9. *a multi-dimensional framework* for assessing the impacts and outcomes of allocation decisions.

Together, these nine principles provide the basis of a multi-dimensional fisheries resource allocation paradigm, but it is Principle 9 which provides the formal integrative basis of the policy and thus the paradigm.

Resource allocation is an holistic process involving the determination of fishery-specific objectives and the application of allocation measures to achieve a range of multi-dimensional outcomes. These outcomes are the result of a complex set of resource redistributions.

Principle 9 is an holistic framework which portrays allocation outcomes as interactions between the measures available for allocating fisheries resources and the multi-dimensional context within which these decisions are made. This principle is applied in this paper as an outcomes matrix centred on Moreton Bay (Table 1) which is a significant Australian subtropical marine wetland area. Moreton Bay (Latitude 27°15' S, Longitude 153°15' E), is a marine park which stretches some 120 km and is bounded by a series of large sand islands, servicing some two million people and a range of inshore marine and estuarine fisheries. It is the location of a major fishery by Australian standards with some 27% of Queensland's commercial estuarine finfish production and a significant recreational fishery for whiting, snapper, tailor, bream, flathead, crabs and a range of less common species, providing some 1.5 million angling days each year (Quinn 1992).

The major dimensions and examples of associated objectives relevant to Moreton Bay are: *biological dimension* (sustainable fisheries, fisheries habitat protection); *ecological dimension* (sustainable ecosystems, bycatch reduction, threatening processes modified, ecological risks reduced, marine plants protected); *industrial dimension* (changes in fleet and processing sector characteristics); *economic dimension* (viable fisheries and fishers, seafood supply, compensation for fishery adjustment, resource security and planning certainty); *social dimension* (increased fishing opportunities, fair access, *ex gratia* payments for fishery adjustment, successful consultation); *political dimension* (reduced conflicts, acceptance of decisions); *cultural dimension* (maintaining lifestyles, totem species and recognition of sacred sites and cultural values, self/co-management). How can these fishery-specific objectives be achieved for Moreton Bay?

The range of allocation measures which could be applied in Moreton Bay include: *access controls* (recreational-only fishing areas, beach access, fishing zones, marine and national parks); *tenure controls* (licences and permits, fishery shares, their transferability and longevity); *input controls* (gear, fisher numbers and vessels); *output controls* (bag, trip and possession limits, quotas and their longevity and transferability); *temporal and spatial controls* (seasonal and area closures, refugia, declared habitat areas, proclaimed fishing zones); *species controls* (protected and threatened species, spawning and nursery areas, size limits, protected marine plants, translocations, totem species and regulated species); and *financial controls* (access fees, quota levies, cost recovery, economic rent, peak body support, licence and permit fees). The application of an allocation control causes a redistribution in the use of fisheries resources. Outcomes are the result of these redistributions. What are the outcomes of the application of the above controls in the context of Moreton Bay?

Outcomes for the estuarine and inshore marine fisheries of Moreton Bay (Table 1) are the result of a redistribution of:

- *wealth and political power* to commercial fishers who gain limited entry licences and remain after adjustments, e.g. limited entry licences for a Moreton Bay fishery; to indigenous fishers as a result of native title allocations; or to recreational fishers gaining greater access to fisheries resources through recreational-only fishing areas such as Pumicestone Passage in the northern section of the Bay;
- *decision-making power* shifting to the community as Management Advisory Committees, Co-management Committees, Zonal Advisory Committees, panels of experts, etc. broaden the debate to a multiple-use and natural resource management perspective;
- *ownership of fisheries resources* from the community to users as resource security, access and property rights are more defined with a likely result of monopolistic competition or concentration of ownership in the more lucrative fisheries, financial security for fishers and greater planning certainty for all user groups;
- *access to fish stocks* traded between and within the indigenous, commercial and recreational sectors through resource partitioning such as weekend commercial fishing closures;
- *fish stocks for ecosystem maintenance* away from fishers through marine park closures, refugia, protected species, seasonal closures, size limits and spawning closures;
- *access to fisheries habitat* between the fishing sectors and other users through the use of refugia, fish sanctuaries, Local Authorities limiting beach access, marine park zoning and other coastal management planning processes;
- *regional economic development* as the various fishing sectors grow through population shifts, demands for more recreational and tourism opportunities as Local Authorities seek to close rivers, and for more fisheries infrastructure such as boat ramps, marinas with seafood outlets; and
- *access to fishing areas* by non-commercial fishers as the commercial sector is adjusted or restructured through a Moreton Bay fishing zone and/or a buy-back scheme based on sustainability, commercial viability and social conflict reduction to accommodate the demands of recreational anglers and tourism.

In the context of Moreton Bay, each *row* of the matrix (Table 1) provides options to achieve fishery-specific allocation objectives. For example, ecological objectives can be achieved through a range of allocation measures — limiting bycatch through access controls like marine park conservation zones, changing fisher attitudes through tenure controls such as transferable licences, limiting habitat damage by input controls such as gear modifications, fish movements enhanced by spatial controls such as fishways, and protecting spawners by temporal controls such as seasonal closures. Specific ecological objectives, such as sustainable ecosystems, can be met through a range of allocation measures.

Allocation Dimensions	Fisheries Resource Allocation Measures						
	Access Controls	Tenure Controls	Input Controls	Output Controls	Temporal, Spatial Controls	Species Controls	Financial Controls
Biological	• catch and effort changes and distributions • sustainability of target species	• temporal effort changes • user group catch size and distribution	• latent effort realized • size composition • catch perceptions • fleet behaviour • gear effects	• sustainable limits • catch shares • stock size uncertainty	• marine park zoning effects • effects on catch and effort	• regulated species • health and disease • spawning closure effects • size limit impacts	• changes in effort • information costs • research levies
Ecological	• bycatch reduction • fish movements • habitat loss by development • conservation effects of marine park zoning	• stewardship attitudes	• bycatch changes • habitat modification • key threatening processes	• wastage • high grading • bycatch • protected species	• impacts of refugia • success of fishways	• translocations • changes in take of protected species • disease transmission	• amenity fees • polluter pays • internalizing of externalities • information
Industrial	• fleet infrastructure changes	• fleet redeployment	• boat replacement • technology & safety	• operational planning	• motherships • fleet attributes and maintenance	• boat and processing plant design	• innovations • buy-back schemes
Economic	• loss of income • regional growth • tourism gains/losses • compensation • resource security	• windfall gains • *ex gratia* payments • superannuation • finance stability • buy-back schemes	• investment warnings • financial viability of fisher • inefficiencies of fishing gear	• seafood supply • concentration of ownership	• seafood supply • closures affecting tourism • seasonal incomes	• commercial - only species • fishing gear	• cost recovery • economic rent • adjustment incentives • licence renewal
Social	• family dislocation • increased amenity • perceived access rights • code of practice	• dislocation • continued conflict • recreational sector purchase	• competitive fishing gear • different sector rules	• bag limit effects • angling licences • changing style of operations	• generational equity • sectoral expectations • special areas	• recreational species • creed of greed	• gear sales tax • consultation costs • arbitrary social cost levy
Political	• community support • fisher expectations • community expectations • creed of greed • *ex gratia* payments	• conflict reduction • community perceptions • self-adjustment	• netting in sensitive areas • Local authority limitations • inshore fish kills	• recreational TAC effects • bag limit reactions and demands	• Local Government area closures • resource partitioning	• 'cuddly' species	• recreational licence fee • peak organization support
Cultural	• indigenous rights • sacred site access	• commercial fisheries • artisanal fisheries	• traditional hunting gear	• indigenous TAC	• indigenous fishing areas	• totem species	• levels of charge

Table 1. A multi-dimensional matrix of the outcomes of resource allocation in Moreton Bay, Queensland

Each *column* of Table 1 highlights the multi-dimensional nature of the potential outcomes resulting from the application of a specific allocation measure. For example, the application of access controls: changing catch compositions through recreational-only fishing areas — the *social* dimension; conservation of protected species through marine park zoning — the *ecological* dimension; fleet infrastructure changes through fishery zoning — the *industrial* dimension; resource security through access rights — the *economic* dimension; and loss of access to sacred sites through fishery zoning — the *cultural* dimension.

The multi-dimensional perspectives provided by the matrix highlight the impacts allocation decisions would have on the fishers and the community at large, possible allocation options and decision-making frameworks under a precautionary allocation paradigm. It is an evaluation *schema* which may indicate improvements in the decision-making process, hence reducing the level of unplanned outcomes.

The policy framework, and specifically the allocation matrix, has a range of applications to other fisheries and jurisdictions. The degree of maturity of the fishery will determine the specificity of the matrix; for example an offshore developmental fishery may not require the full matrix, whereas a restructuring of the barramundi fishery would. The development of fishery management plans could incorporate the matrix as part of the review of a fishery so that all outcomes, including potential unplanned effects, can be identified and assessed.

The current approach to fisheries resource allocation only emphasizes some of the allocation dimensions. However, the matrix forces decision makers to consider all ramifications, internalizes the new paradigm and is the basis of a precautionary approach. The most powerful uses of such a matrix are its application in developing more holistic fisheries policies and legislation, monitoring, auditing and evaluation of allocation processes, as a diagnostic and training tool for managers and industry, and as a framework for formalized learning through case studies.

REFERENCES

ABARE (Australian Bureau of Agriculture and Resource Economics) (1995). Australian Fisheries Statistics 1995.

Cameron, J (1993). The precautionary principle — Core meaning, constitutional framework and procedures for implementation. In 'The Precautionary Principle' Conference Proceedings, Institute of Environmental Studies, University of New South Wales, 20–21 September.

Dovers, S (1995). Sustainability, policy and ecological economics: hubris or humility. In 'Ecological Economics'. Conference, Coffs Harbour, November 19–23. pp. 383–94 (Department of Primary Industries and Energy).

Garcia, S M (1994). The precautionary approach to fisheries with reference to straddling fish stocks and highly migratory fish stocks. Food and Agriculture Organisation of the United Nations, Rome. *Fisheries Circular* 871. 76 pp.

Harding, R (1994). Sustainability:Background Paper Interpretation of the principles. Fenner Conference on the Environment. Australian and New Zealand Environment and Conservation Council.

Harding, R, and Fisher, L (1993). A transdisciplinary approach to the precautionary principle. In 'The Precautionary Principle'. Conference Proceedings, Institute of Environmental Studies, University of New South Wales, 20–21 September.

Quinn, R H (1992). Fisheries resources of the Moreton Bay region. Queensland Fish Management Authority.

Saenger, P (1989). An inventory of Australian estuaries and enclosed marine waters. The Australian Recreational and Sport Fishing Confederation.

Taylor-Moore, N G (1995). Fisheries Management Planning in the 1990s. In 'Proceedings of the Second National Fisheries Managers Workshop'. pp. 1–14. Bribie Island, Queensland. Queensland Fisheries Management Authority.

Tsamenyi, M, and McIlgorm, A (1995). International environmental instruments: their effect on the fishing industry (FRDC Project). University of Wollongong and Australian Maritime College.

An Economic Assessment of Reallocating Salmon and Herring Stocks from the Commercial Sector to the Recreational Sector in Western Australia

M. S. van Bueren,[A] *R. K. Lindner*[A] *and P. B. McLeod*[B]

[A] Faculty of Agriculture , University of Western Australia, Nedlands, Western Australia, 6907.
[B] Department of Economics, University of Western Australia, Nedlands, Western Australia, 6907.

Summary

This study quantified the economic gains and losses from reallocating fish stocks between recreational and commercial users in the Australian salmon (*Arripis truttaceus*) and Australian herring (*A. georgianus*) fisheries of Western Australia. Under current management the commercial sector was estimated to generate short-run profits of between $A260 000 and $780 000 per annum. A comparative measure of benefits for the recreational sector was obtained using the contingent valuation and travel cost methods. Based on a survey of 97 anglers, economic surplus from recreational fishing ranged between $A15 and $33 per day, amounting to between $A1.6 and $3.4 million in aggregate benefits. It is difficult to determine how sensitive these benefits are to changes in the availability of salmon or herring but in the case of salmon, anglers were willing to pay, on average, $5.55 for an additional fish. This far exceeds the net commercial value of salmon, estimated to be $0.84/fish. The results for herring are less conclusive, as the survey was unable to obtain a reliable estimate of recreational benefits at the margin for this species.

Introduction

Competing demands by various user groups for coastal and fishery resources in Australia are becoming a major issue in fisheries management. From an economic viewpoint, stocks should be allocated to those groups which gain the greatest marginal value from using the resource, as this is consistent with the goal of maximizing economic benefits to society. While economic efficiency is an important benchmark, it is usually disregarded when formulating policy because of the inherent difficulties in measuring non-market benefits. Only a handful of Australian studies have measured the net economic benefits of recreational fishing (e.g. Collins 1991; Dragun 1991; Staniford and Siggins 1992). A more comprehensive suite of studies has been undertaken in the United States, but it is generally recognized that further research is necessary to improve our methods of measuring benefits from recreational angling.

The objective of this study was to refine survey instruments commonly used to value non-market goods, with a view to making some statement about their suitability for inclusion in future assessments of fisheries allocation. The salmon and herring fisheries in Western Australia are examined, two fisheries for which there is considerable community pressure to revise allocation of stocks between recreational and commercial fishers. The fisheries are located on the south and lower west coasts of

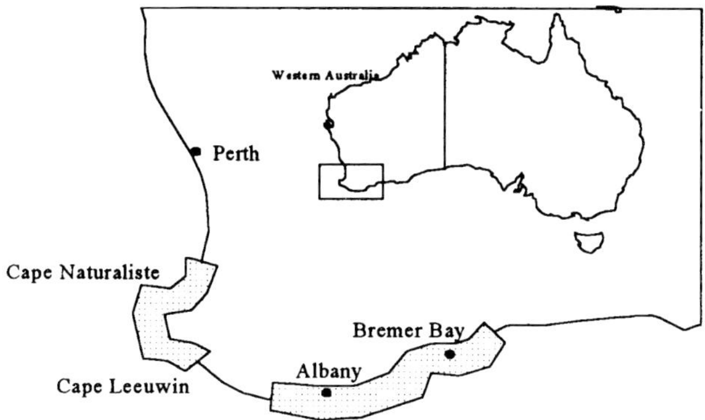

Fig. 1. Location of the West Australian commercial salmon and herring fisheries, shown by the shaded areas on this map.

Western Australia (Fig. 1). Relative to other Western Australian fisheries the commercial sector is small, comprising only 34 licensed operators. In comparison, the recreational sector involves over 100000 active anglers, many of whom target salmon and herring. Other features that characterize each sector are summarized in Table 1.

Although competition for access to salmon and herring stocks has been present for many years, the problem is becoming more acute as the number of anglers continues to grow. There is no scientific evidence that commercial fishing is having an adverse impact upon recreational catch or *vice versa.* Rather, it is considered that stocks are more greatly influenced by environmental factors such as the Leeuwin Current (Lenanton 1994). The competition between anglers and commercial operators therefore appears to arise mainly out of prejudices or misconceptions about each other's activities rather than true competition for catch.

Methods

The economic performance of commercial salmon and herring fishers was evaluated based on short-run profits generated in the 1992 season. Producers' surplus was not estimated because supply and demand functions could not be derived from the data available. On the supply side, the biology of salmon and herring fish stocks is not fully understood and the relationships between catch and effort are not well defined. This made it difficult to estimate parameters of the production function that are necessary for estimating marginal costs. Demand was assumed to be perfectly elastic, which meant that no benefits were attributed to consumers of commercially-caught fish. Similarly, no value was placed on producers' surplus which might accrue in processing markets. Both these assumptions are thought to be reasonable because there are numerous substitutes for salmon and herring.

No effort was made to quantify the non-market 'lifestyle benefits' that are associated with professional fishing. However, some account was taken of these benefits by not including operators' own labour costs in the profit calculation. While it is acknowledged that lifestyle benefits are substantial, it is argued that operators may be able to obtain similar benefits by continuing to fish for salmon and herring on a non-commercial scale.

The recreational sector was evaluated by quantifying the economic benefits enjoyed by anglers, net of what they spend to go fishing. This value is termed economic surplus, and is a valid measure of net benefits to society. Anderson (1983) provided a theoretical base for analysing the size of economic surplus from recreational fishing. He postulated that an individual's marginal value for a day of fishing is a decreasing function of the number of days fished over the course of a year. Economic surplus is

Table 1. A summary of the Western Australian salmon and herring fisheries, describing the features of the recreational and commercial sectors

Commercial sector	Recreational sector
Annual catch:	**1994 catch [B]:**
• 1500–2500 t salmon	• 220 t salmon (10% share of total)
• 800–1500 t herring	• 358 t herring (26% share of total)
• season extends Feb. through to April	
• first stage processing by 11 local firms	
Effort:	**Effort [C]:**
• 34 licensed fishers	• more than 100000 active fishers
• 125 people directly employed in fishery [A]	• 12% targeting salmon
	• 39% targeting herring
Regulations:	**Regulations:**
• limited entry	• open access
• area licensing; fishers restricted to beaches or zones	• bag limits apply
• gear restricted to beach-seine nets or trap nets.	• netting banned
• season restrictions apply	

[A] Western Australian Fisheries Department 1992/93 Annual Report.

[B] Western Australian Fisheries Department 1994 Salmon and Herring Creel Survey

[C] Australian Bureau of Statistics (1989).

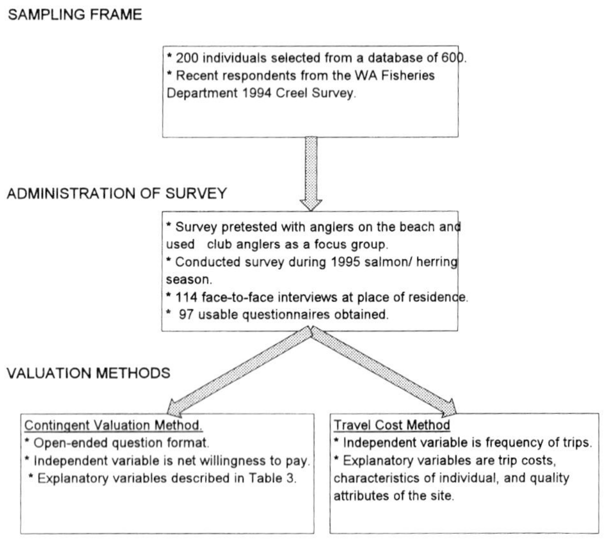

SAMPLING FRAME

* 200 individuals selected from a database of 600.
* Recent respondents from the WA Fisheries Department 1994 Creel Survey.

ADMINISTRATION OF SURVEY

* Survey pretested with anglers on the beach and used club anglers as a focus group.
* Conducted survey during 1995 salmon/ herring season.
* 114 face-to-face interviews at place of residence.
* 97 usable questionnaires obtained.

VALUATION METHODS

Contingent Valuation Method.
* Open-ended question format.
* Independent variable is net willingness to pay.
* Explanatory variables described in Table 3.

Travel Cost Method
* Independent variable is frequency of trips.
* Explanatory variables are trip costs, characteristics of individual, and quality attributes of the site.

Table 2. Flow chart showing methods used to survey recreational anglers

equal to the area under this demand curve, less the costs incurred to go fishing.

In the absence of a market for recreational fishing, and hence prices, we need to resort to survey techniques for eliciting individuals' demand curves. The flow chart in Table 2 outlines how the sampling frame was selected and how the survey was administered. Open-ended contingent valuation was the primary method used to value benefits. This entailed asking respondents to firstly estimate their total expenses for a specific fishing trip known to both angler and researcher. They were then asked to nominate how much the cost of a trip could rise before deciding that the particular trip in question was not worth the expense. Individual responses were analysed statistically by building a linear regression model that included a variety of variables to explain the variation in net willingness to pay (Table 3). The travel-cost method was used to provide supplementary estimates of angler benefits so the validity of estimates derived from contingent valuation could be checked.

For the purposes of resource allocation, policy makers are particularly interested in the recreational value of incremental changes in catch rate. The marginal value of catch was elicited in two ways, both of which utilized contingent valuation. The first method involved asking anglers to state their preferred daily catch of each species and the extra trip costs that they would be willing to pay if their preferred catch was realized. The marginal value of catch was then calculated by dividing the extra willingness to pay figure by the difference in actual to preferred catch levels. The second method involved partially differentiating the net willingness to pay function with respect to salmon and herring catch to derive a demand curve for each species.

Results

The mean income and costs of commercial salmon and herring fishing operations in Western Australia are summarized by Table 4. Survey data indicated that, on average, the short-run profit for commercial operators in 1991/92 was $24 239 which equates to a whole-sector profit of $630 227 (both estimates in 1996 $A). Allowing for the inter-seasonal variation in prices and catch, it is expected that profits are likely to range between $260 000 to $780 000. The survey does not provide any information on the marginal net benefit of commercially-caught salmon and herring, but if we assume that half of the profits are attributable to each species, then the average marginal value of salmon is $0.84/fish and $0.06/fish for herring. These figures are calculated using average weights for salmon and herring of 3.2kg and 0.25kg, respectively.

Table 3. Description of explanatory variables in the contingent valuation model

Variable name	Expected sign	Description
INCOME	+ve	Annual income of angler
EMPLOY	+ve	Dummy variable for employment status of angler
SALM	+ve	Angler's mean daily catch of salmon
HERR	+ve	Angler's mean daily catch of herring
ALLFISH	+ve	Angler's mean daily catch of all fish
GEAR	+ve	Total value of all fishing gear owned by the angler
ACTIVITY	+ve	Dummy variable for whether or not fishing was the primary activity undertaken on the trip
CLUB	+ve	Dummy variable for whether or not angler is a member of a fishing club
SATISFY	-ve	An index measuring the extent to which catch is essential for producing a satisfying fishing trip
HRSFISH	+ve	Angler's mean number of hours spent fishing at the site
WEATHER	+ve	An index measuring the angler's perception of the suitability of weather and sea conditions for fishing
CONGEST	-ve	Total number of anglers on the beach
PROBS	-ve	An index derived from the summation of five responses, each of which measured the extent to which the angler perceived there to be a problem with his/her fishing trip. The five problems relate to access roads, facilities, pollution, crowding, and commercial fishers
TRIPCOST	-ve	Travel, accommodation, and fishing expenses incurred by the angler

The survey of recreational anglers found that on average, individuals spend $21 per day, or alternatively $902 per year, on fishing. This provides a measure of the minimum value that

Table 4. Economic performance of West Australian commercial salmon and herring fishers in 1991/92. Data from McLeod and McGinley (1994). Income and costs of both operations were combined. Means are based on a sample of seven operators. Aggregate values assume the sector comprises 26 active operators. All values expressed in 1996 $A

	Mean per active licence ($)	Aggregate for sector ($)
Gross Income	58 136	1 511 542
Short-run costs:		
Operating costs	15 935	414 315
Vehicle overheads	5 936	154 339
Office overheads	1 600	41 596
Capital costs	10 426	271 066
Total short-run costs	33 897	881 315
Short-run profit	24 239	630 227

recreational fishers place on their sport. These estimates compare favourably with an earlier study undertaken by Lindner and McLeod 1991, who found that anglers spent between $575 to $1008 per annum (1996 $A) on gear, travel, and other fishing inputs. In addition to expenditures, anglers obtained a positive amount of economic surplus. Table 5 lists four comparable estimates of economic surplus which range between $15 and $33 per day. The lower bound value is possibly the more accurate figure because a number of high bids lifted the mean willingness to pay. The contingent valuation estimate of $18 needs to be interpreted with caution as the regression model had a very poor adjusted r^2 value of 0.128. It is not unusual for social survey techniques to produce low r^2 statistics. For example, in one of the Australian studies mentioned earlier, Staniford and Siggins (1992) obtained an r^2 value of 0.24. Mitchell and Carson (1989) suggest 0.15 as a minimum.

Table 5. Estimates of net economic benefits from recreational fishing in southern Western Australia Individual benefits were aggregated using data from a creel survey conducted by the Western Australian Fisheries Department that estimated total angler effort in 1994 to be 104 000 angler days. Both valuation models were estimated by Ordinary Least Squares using the computer package Microfit V3.0 (Pesaran and Pesaran 1991).

Source of estimate	Economic surplus estimates: Individual ($/day)	Aggregate ($mill/yr)
Sample mean willingness to pay *	27.36 $^{\Delta}$	2.85
Sample median willingness to pay *	15.00	1.56
Contingent valuation model	17.97	1.87
Travel cost model	32.77	3.41

* Of the 97 people interviewed, 13 were unable to offer a bid and a further 10 gave zero bids. The zero bids were included in the sample mean and median calculations.

$^{\Delta}$ Mean willingness to pay has a high standard error of $48.96

The most likely explanation for the low explanatory power is that the chosen variables did not capture all the important attitudinal and lifestyle characteristics of the individual, as only TRIP COST and SATISFY were statistically significant. Two other factors may have contributed to the poor model fit. Firstly, the use of trip costs as a payment vehicle may have biased the bids upwards by encouraging respondents to offer a bid which was some arbitrary proportion of their trip costs. This theory is supported by the positive sign on the TRIP COST coefficient. Secondly, respondents did not seem to confine their bids to a specific trip but rather made an *ex ante* valuation of fishing in general. This was despite repeated reminders that we were only interested in an *ex post* value for the trip. These two factors could also be the reason why site-quality and catch-rate variables were not significant.

The marginal value of salmon and herring could not be estimated from the contingent valuation regression because the catch variables were not significant. Instead, a mean value for salmon and herring at the margin was elicited by asking respondents directly about the value they placed on catching a preferred number of fish. Including zero and non-zero bids, the mean marginal value of salmon was calculated to be $5.55/fish. This estimate has a high standard error of $13.02 and is therefore not a very reliable measure of salmon's marginal value. In the case of herring, only one respondent was willing to pay for a preferred increase in catch rate. The most probable reason for such a large number of zero bids for herring is that anglers are reasonably satisfied with their current catches of this species and further increases in catch may only marginally improve their satisfactions from fishing.

In general, only a small proportion of the total sample were prepared to pay more for their preferred number of fish. This could be explained if we believe that respondents made an *ex ante* valuation of fishing instead of an *ex post* valuation of their actual trip and, in the process of doing so, included the possibility of catching more salmon or herring when formulating their initial bid. In light of this observation, the use of contingent valuation for eliciting the marginal value of catch is questionable, at least one based on the format adopted by this survey.

DISCUSSION AND CONCLUSION

This study has shown that the welfare generated by recreational fishing is significant, being in the order of $A1.5 million to $A3.4 million annually. As these benefits apply to the general fishing experience, and not just to salmon and herring fishing, they cannot be compared directly to profits in the commercial sector. It is difficult to determine how sensitive the recreational benefits are to changes in the availability of salmon or herring, but the results do provide some economic justification for shifting the allocation of salmon away from the commercial sector and towards the recreational sector. This conclusion is reached by comparing the average marginal value of salmon as a sport fish, estimated to be $5.55/fish, compared to the average net value of $0.84/fish in its commercial use. The results for herring are less conclusive, as the survey was unable to elicit marginal values for this species.

It is acknowledged that due to the asymmetry of recreational and commercial fishing, the task of making valid comparisons of economic value at the margin is extremely difficult. As such, the numerical values presented in this study need to be interpreted with caution. While the results adequately demonstrate the direction in which salmon and herring stocks should be allocated to improve economic efficiency, the models developed are not capable of determining an optimal allocation, or of simulating the economic consequences of various management strategies available for reallocating fishery resources. This is because a continuous demand function for recreationally-caught fish could not be derived using the chosen survey techniques. Discrete choice modelling is an alternative approach being considered by the authors. In future work there is scope for modelling allocation as a dynamic process and for analysing catch rate as an expectation function for each individual angler.

References

Anderson, L G (1983). The demand curve for recreational fishing with an application to stock enhancement activities. *Land Economics* **59**(3), 279–86.

Australian Bureau of Statistics (1989). Recreational fishing in Western Australia, July 1987.

Collins, P B (1991). Estimating the marginal value of a fish to recreational anglers in South Australian Fisheries: An appraisal of contingent valuation and the travel cost model. (Flinders University of South Australia).

Dragun, A K (1991). A comparative review of commercial and recreational fishing in Port Phillip Bay and Western Port. (Victorian Department of Conservation and Environment).

Lenanton, R C (1994). Western Fisheries Research Committee Annual Report 1993/94. Fisheries Department of Western Australia.

Lindner, R K and Mcleod, P B (1991). The economic impact of recreational fishing in Western Australia. *Fisheries Management Paper* No. 38. Fisheries Department of Western Australia.

McLeod, P B, and McGinley, C(1994). Economic Impact Study: Commercial Fishing in Western Australia. *Fisheries Management Paper* No. 61. Fisheries Department of Western Australia.

Mitchell, R C, and Carson, R T(1989). Using surveys to value public goods: The contingent valuation method. Washington, DC.

Pesaran, M H, and Pesaran, B(1991). 'Microfit Version 3 User Manual.' (Oxford University Press, Oxford, UK).

Staniford, A J, and Siggins, S K(1992). Recreational fishing in Coffin Bay: Interactions with the commercial fishery. *Fisheries Research Report* No. 32 Fisheries Department of South Australia.

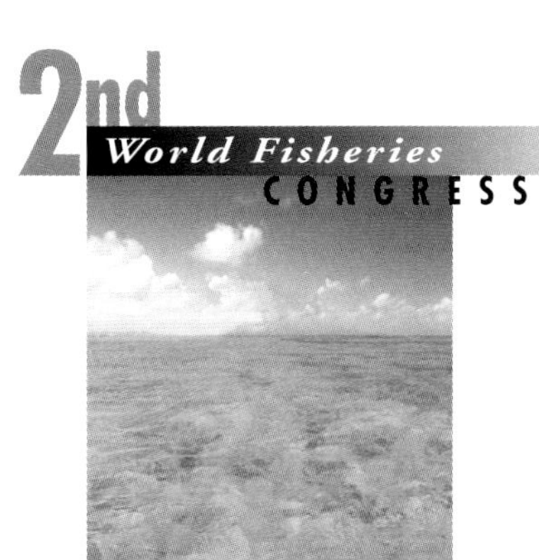

Optimal Allocation Of Fisheries Quotas

G. R. Morgan

WA Marine Research Laboratories, PO Box 20, North Beach 6020, Western Australia.

Summary

A model shows that the long-term incentive for reducing operational costs in a quota-managed fishery is independent of the total allowable catch (TAC). However, the short-term affordability of cost-reduction processes is maximized at the original Maximum Economic Yield. If TACs are set near the Maximum Sustainable Yield, variable quotas provide reduced biological risk, give greater long-run catches and generate larger economic rents than fixed quotas. However, at TACs set at lower levels, fixed quotas generally produce greater catch and economic benefits.

Allocation of the TAC involves both the allocation of economic rent between the fishery participants and the wider community, and the allocation of the TAC between fishery participants; only when both issues are addressed together can economic rent allocation be optimized. In Australia, the separation of TAC allocation from the allocation of economic rent (through a 'cost recovery' process) has led to sub-optimal rent allocation, reduced incentives for cost reduction by the regulatory authority, misconceptions on resource ownership and a weakened regulatory authority. Suggestions for alleviating some of these problems include the sharing of resource rent generated through reductions in 'recoverable costs' between the fishery participants and the regulatory authority.

Introduction

The theoretical bases of quota management (particularly transferable quotas) have not been adequately established and, in many fisheries, quotas are determined without a clear understanding of either the potential impact on the resource or the implications of quota allocation decisions on the operations and economic viability of the participants in the fishery. Often, such as in the quota-managed fisheries for abalone in Western Australia, the quota-setting decisions are based on historical catches and monitoring of annual catch levels, and allocation decisions are based solely on past participation in the fishery; returns to the community are limited to cost recovery of expenditure on Government administration, compliance and research. There are usually no theoretical bases for establishing at what level the quota, or total allowable catch (TAC) should be set, the impact on participants' operations of different levels of quota or how the quota should be allocated to achieve desired economic and social goals.

This paper addresses some of the theoretical issues involved, both in the determination of fisheries quotas (or TACs) and in the allocation of these quotas in an economically efficient manner. Many of the results have general practical application in quota-managed fisheries and, it is to be hoped, will lead to a better process of identifying those fisheries that are suitable for quota management, determining appropriate TACs and allocating the TAC in an economically efficient and equitable manner.

Bases for establishing quotas

Why quotas?

The move to a system of transferable quota management in fisheries was prompted by initiatives in the 1970s and 1980s (Christy 1973, 1982; Clark 1982; Copes 1986) that recognized that the then biologist's view of fisheries management by input controls (usually to a target of Maximum Sustainable Yield, MSY) was economically inefficient and required ever-increasing restrictions to counteract the inevitable move to more efficient operations. Such increases in operational efficiency ran the risk of overexploitation of the resource and reduced returns to individual operators.

The prime motive behind quota management is therefore to maximize the economic rent generated from a resource by providing incentives for maximizing operational efficiency (equivalent to minimizing operational costs) to take a fixed catch rather than to constrain operational efficiency to a fixed level of input. An assumed reduction in management costs also assists in this process. Quota management therefore encourages the maximization of resource rent **at any given level of catch quota**. The total resource rent generated from a resource will depend not only on the extent to which operational costs are minimized but also on the level of quota established.

The relationship between TAC, stock sustainability and economic performance — where should TACs be set in relation to the production curve?

In establishing TACs, the prime objective of moving to quota management of inducing operational cost minimization (within the constraints of stock sustainability) should be considered in the context of what is an appropriate TAC. A simple Excel spreadsheet model established theoretical production and cost curves (Fig. 1) for a hypothetical fishery under quota management with a fixed number of operators (100). The cost curve assumed a fixed cost for each operator and a variable cost that was linearly related to the level of fishing effort. Operational cost reduction was undertaken in 10 steps, each a reduction of 10% of the original cost. The cost of achieving each step of operational cost reduction was assumed to be an exponential relationship in which initial cost reductions were easier (and cheaper through technological innovation) to achieve than subsequent cost reductions.

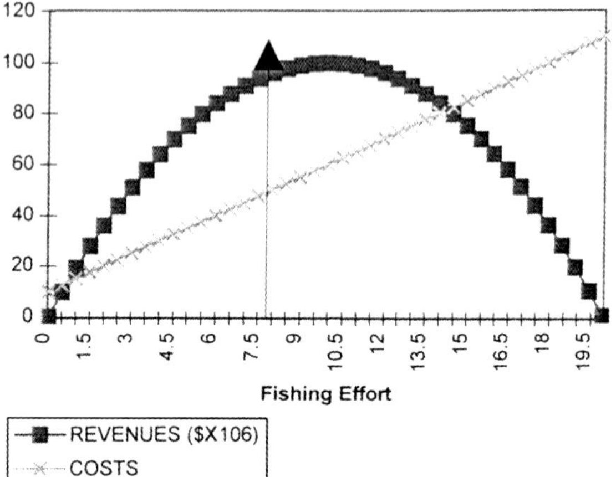

Fig. 1. Revenue and operational cost functions used in the modelling exercise

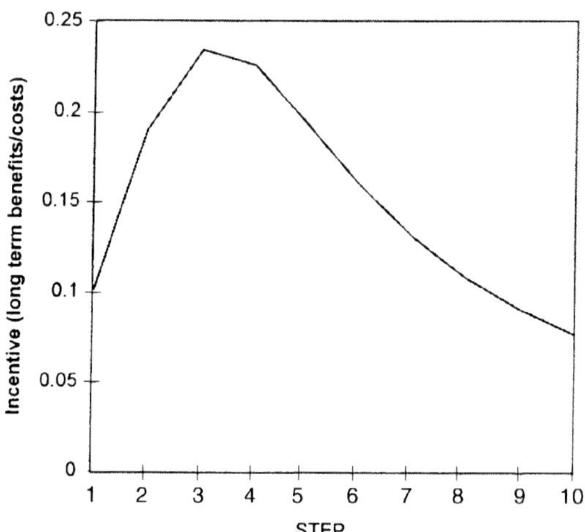

Fig. 2. Changes in the long-term incentive for cost reduction with each cost reduction step achieved. Each step represents a 10% cost reduction.

The incentive for achieving cost reductions was measured in two ways. First, a long-term benefit/cost ratio was established in which the long-term, undiscounted resource rent generated per operator was compared with the immediate costs of achieving each step of operational cost reduction. Second, the short-term 'affordability' of the cost-reduction process was determined by calculating the relation that the cost of each step of operational cost reduction bore to the overall resource rent generated per operator.

The principle conclusions were as follows.

- The incentive for each step of operational cost reduction (measured as the long-term benefit/cost ratio) was independent of the TAC and of the precise relationship between operational cost and fishing effort.
- The incentive for operational cost reduction increased with each step in the cost-reduction process, reached a maximum and then declined (Fig. 2). This implies that operational cost reduction in a quota-managed fishery will be an increasingly attractive process until the costs of technology of achieving further reductions make those reductions increasingly unattractive.
- The short-term affordability of cost reduction was greatest at the original (i.e. before the cost reduction) Maximum Economic Yield (MEY) and declined with distance away from that point (Fig. 3).
- For the case studied, the short-term affordability was relatively flat over a wide range encompassing both MEY and MSY. This indicates that TACs set anywhere within this range will have little effect on short-term affordability of operational cost reduction.

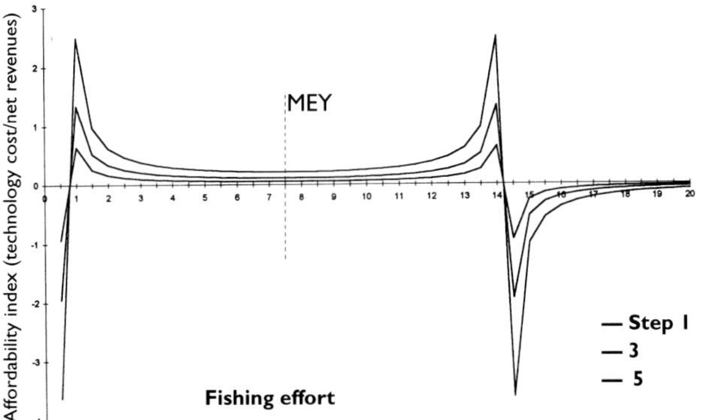

Fig. 3. Affordability of cost reduction at different levels of fishing effort.

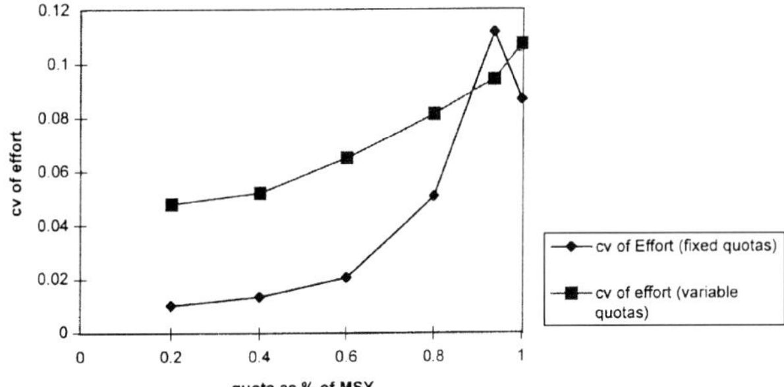

Fig. 4. Coefficient of variation of effort required to take fixed or variable quota under conditions of variable recruitment.

- The short-term 'affordability' of operational cost reduction decreased rapidly as the (original) point of zero economic rent generated is approached, and a discontinuity existed at the point of zero economic rent (Fig. 3).
- If recruitment was variable, year-to-year variation in fishing effort required to take a fixed quota increased rapidly as the TAC level approached MSY (see below and Fig. 4), although at MSY this variability declined as quota could not be taken in all years.

Hence, if quotas are to achieve their aim of maximizing economic efficiency, there needs to be a long-term incentive for achieving cost reductions and those cost reductions need to be affordable in the short term. Long-term incentives for cost reduction have been shown to be independent of levels of fishing effort and affordability of those reductions is maximized at initial MEY. Year-to-year variability in fishing effort required to take a quota is high at MEY and probably higher than at the point of MSY. Therefore, given that biological risk of stock reduction is also greater at MSY than at lower levels, quota levels might best be set near MEY and be adjusted upwards to new MEYs as costs are reduced. It is unlikely that optimal quota levels will ever be at MSY.

The effects of variable recruitment and fixed *v.* variable TACs

In any real fishery situation, recruitment to the fishery will be expected to vary over time both in response to environmental influences and as a result of exploitation. The extent of variation will depend on the species in question, and the effect of such year-to-year variation on the biomass of the exploited stock will depend on the exploitation rate and the extent to which older year classes contribute to the fishery.

The question of the most appropriate strategy for a quota-managed fishery under a scenario of variable recruitment was examined by constructing a stochastic spreadsheet model in which recruitment to a fishery with five exploited year classes was allowed to vary randomly (recruitment was normally distributed with a mean of 1 and standard deviation of 1) over an 8-year period. The mean production function and cost function was identical to Figure 1. The effects of fixed and variable quotas were then examined at various levels of TAC, with the variable quota being adjusted each 'year' in response to changes in stock biomass as a result of recruitment variation. The fishing effort required to take a fixed quota varied in accordance with the variability in recruitment. The extent of such variation (measured as the coefficient of variation of effort) increased rapidly with increasing TAC (measured as the percentage of MSY) before declining as the MSY level was approached (Fig. 4); this decline was partly a result of the inability to take these relatively high quotas in years when recruitment was low. At fixed TACs set at a level less than MSY, quota levels were reached in almost all years, although the year-to-year variation in fishing effort needed to take the quota was high.

With variable recruitment, it might be expected that adjusting the TAC each year in response to recruitment (and stock biomass) levels would not only produce higher average catches but also generate higher economic rents from the fishery. Within the modelling structure described above, total catches over the 8-year period were 2.1% higher at MSY with a variable TAC than with a fixed TAC. At TAC levels less than about 0.8 of the long-run MSY, variable TACs actually produced a smaller total catch than fixed TACs (Table 1). Also, under a variable quota-management system, the variation in fishing effort needed to take the TAC was significantly larger than the equivalent fixed-quota scenario (Fig. 4), except at levels a little less than MSY, where variability of effort for fixed quotas was higher.

The discounted economic rent generated under a fixed and variable quota system was also examined by the above modelling technique and with a discount rate of 7%. In general (Fig. 5), fixed quotas produce higher levels of economic rent from the fishery at all levels of TAC except at a point a little less than MSY (in this case, at MEY) where variable quotas produced higher

Table 1. Catches over an 8-year period under fixed and variable quota management systems with variable recruitment

Quota Level (% of MSY)	Catch (Fixed quota)	Catch (Variable quota)
0.2	160	144
0.4	320	311
0.6	480	474
0.75 (MEY)	600	592
0.8	640	638
1.0 (MSY)	767	790

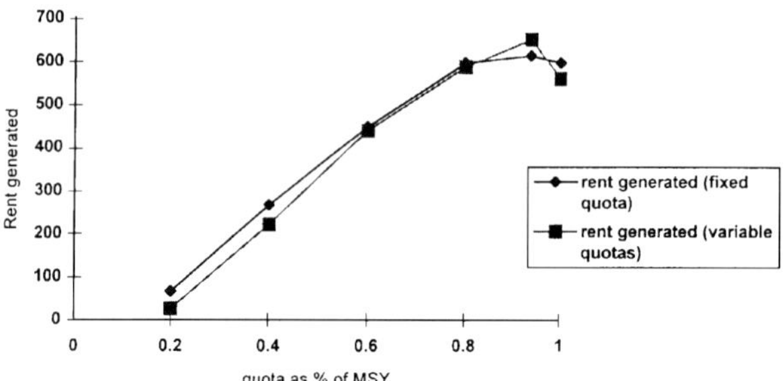

Fig. 5. Discounted economic rent generated under variable and fixed quotas at various levels of quota.

economic rents. This difference is related to the increased variability of fishing effort required to take a variable quota, which results in uncertainty over long-term licence valuations and hence a long-term reduction in discounted rent generated.

Hence, variable quotas adjusted in accordance with random variations in recruitment result in significantly higher levels of year-to-year variation in fishing effort required to take the quota than is the case with fixed quotas, and the total long-term catch is generally less; the increased protection to the stock may be important if the quotas are set near MSY or if stock abundance is low. The long-term economic rent generated from the modelled fishery is usually greater with fixed than with variable quotas except at points approaching MSY. Since the modelling techniques used in this analysis were generalized production and cost functions applicable to many fisheries, these results probably have general application. However, similar analyses need to be undertaken for specific fisheries.

The choice between fixed and variable quotas depends on the extent of stock (particularly breeding stock) protection that is considered necessary and the objectives of maximizing economic rent and minimizing effort variability. If TACs are set at levels near MSY, the biological risk of a stock decline may be significant and hence management by variable quotas will usually be the preferred option. However, if TACs are set at lower levels (particularly near MEY), fixed quotas will usually produce greater long-term economic rents while minimizing the variability in fishing effort required to take the quota.

Competition for quota, quota turnover and vertical integration

If operational cost reduction is to be funded from retained profits from the industry, these need to be sufficient also to generate acceptable returns on capital invested. The affordability of cost-reduction initiatives (Fig. 3) decreases the further away from MEY the quota is set. This has significant implications if quota is freely transferable through an efficient market process and potential buyers of quota are not constrained by funding through retained profits from the quota-managed fishery. In such cases, it would be expected (J. Penn, pers. comm.) that the rate of transfer of quota units would follow the same relationship as that shown in Figure 3, with an increase in turnover of quota units being expected the further away from MEY the quota was set. New entrants bringing new capital to the fishery would not be so constrained by decreased affordability of operational cost reduction and could therefore benefit from the increased profits such reduction in costs would generate. Established quota holders within the fishery would be subject to such funding constraints and increasingly, as quotas were established further away from MEY, would be unable to fund further cost reductions. Turnover in quota would therefore be expected to increase.

Likewise, quota turnover might be expected to increase as the relatively inexpensive cost reductions are made and further operational cost reductions become not only less affordable to existing quota holders (Fig. 3) but also less attractive as long-term benefit:cost ratios decline (Fig. 2). This process is obviously also important in vertical integration within the industry and in the commonly observed trend (as in New Zealand, Clark 1993) towards quota purchase by processors etc.

Research requirements for establishing TACs

Prioritization of research needs to take into account the type of quota system established, including the relative benefits of fixed and variable TACs and the level at which TACs are to be set. It is assumed here that the basic information on the stock is available to set TACs initially at some known position on the production function.

If TACs are to be set near the long-run MSY, variable quotas may be the most appropriate strategy. Year-to-year variation in fishing effort will be large, with higher fishing effort levels being applied in years when stock biomass (including the breeding stock biomass) is highest. However, if quotas are fixed, with TAC levels near the long-run MSY, the highest fishing effort will be applied in years when stock biomass (including the breeding stock biomass) is lowest and the biological risk of stock reduction increases.

Hence, under a fixed-quota regime, an understanding of the breeding stock–recruitment relationship (and its variability) is critical. Close monitoring of overall levels of stock biomass (particularly breeding stock biomass) must be given high priority. Setting TACs at this level also reduces the short-term incentive for operational cost reduction through reduced affordability. As a result, frequent updated information on operational costs and profits are needed to maintain quotas at levels that will continue to provide incentives for cost reductions. An index of the affordability of further cost reductions may be through the turnover rate of quota units in the fishery, although any market constraints need to be taken into account.

If TACs are to be fixed and set at levels less than the MSY (perhaps nearer MEY) then biological research priorities can be less focussed on breeding stock-recruitment relationships and more orientated towards stock biomass monitoring, including aspects such as regional differences in growth and mortality. Again, frequent updated information on operational costs and profits are needed to maintain quotas at levels that will continue to provide incentives for cost reductions.

If the TAC is set near the long-run MSY, research costs will be higher than if it is set lower (and long-run economic rents generated will be lower); research costs will be higher for a fixed quota management regime than a variable quota one (Table 2). If the TAC is set lower than the MSY, research costs will be higher with variable quotas than with a fixed quota (Table 2).

THE CENTRAL PROBLEM — ALLOCATION BETWEEN THE REGULATOR AND THE REGULATED

Allocation of the resource rent between the participants and the community for which the regulating agency acts, requires a knowledge of what resource rent is being generated, and often this is not available. The practice has been for the government to impose an initial minimal resource rent levy (in the form of a management fee or tax) with a commitment to increase such fees over time so that the government is eventually the major beneficiary of the resource rent generated. For example, the 'management fee' for the Australian bluefin tuna fishery was initially set to recover 38% of attributable management costs (Wesney 1989) but with a commitment to move towards full management and partial research-cost recovery (Anon. 1994). Without a detailed knowledge of the resource rent being generated (or how it was changing over time in response to quota management), an incremental approach was adopted, which presumably could be halted when it was seen that the allocation process was optimal. During such a process, economic rent from the fishery might be expected to increase as the incentives for operational cost reduction (see above) take effect. As a result, an incremental approach to resource-rent allocation will constantly re-adjust itself to changing circumstances without a knowledge of how that rent was generated. Rent generated by use of the community resource will be treated in the same way as rent generated through operational cost reduction by the fishermen. Such an approach will therefore have significant impact on the 'affordability' of future cost reductions (see above) and may either hinder the process of cost reduction or lead to increased market turnover of quota.

In most fisheries to date, the initial quota fee or price has been set so low that quota holders have profited. This may be useful in initially encouraging the acceptance of the concepts of quota management but it has also had unfortunate consequences such as encouraging speculation and, where a secondary market exits, of dissipating resource rent in subsequent transaction fees. In addition, initially low fees can attract and retain inefficient firms within the industry since the windfall profits are capitalized into the value of the original quota holder company.

In all quota-managed fisheries to date, the allocation of the resource rent generated from the fishery between the participants and the government has been done by administrative decision. This is economically inefficient (Morgan 1995) since it is highly unlikely that the regulatory authority would have sufficient detail of the demand structure for quotas to optimize the quota price and the rent accruing to the regulatory authority. In the communications, airlines and financial industries, both theory and practice have shown (Morgan 1995) that allocation of resource rents by auction or tender is the only method that has maximized economic efficiency in the allocation process. Allocation by a suitably designed auction or tender process has addressed a wide range of both economic and public policy objectives and brought out the real costs of non-economic policy goals. This is an important point; the usual view of the auction process is that it addresses revenue goals only which, in the case of natural resource allocation, is generally not the primary goal.

Morgan (1995) suggested that resource-rent allocation in fisheries would eventually follow the same path once the need to provide incentives for fishermen to adopt the concepts of quota management had diminished. Since that time, Western Australia has moved to examine a process of tendering for quota in some of its quota-managed fisheries although, to date, no such process has been implemented. However, the ability to adopt a tender process in quota allocation exists in most fisheries legislation in Australia.

Resource ownership issues

The concept of a resource rent has been introduced into many managed fisheries (particularly in Australia and New Zealand) through the rationale of 'cost recovery' for management and support services to the fishery. Those costs have been calculated and a formula has been agreed with industry by which costs are recovered. But is 'cost recovery' significantly different in concept from a resource rent where users of the resource pay the community an access fee? The question of resource ownership has been raised repeatedly by the fishing industry in Australia: if the industry is paying for a service to manage 'their' fishery, then this implies some form of industry ownership of the resource; under a resource rent understanding, the ownership of the resource is clearly recognized as being the wider community, for which the Government acts as a custodian and manager.

Allocation between participants

If the allocation of TAC between participants in a fishery is done together with the allocation of economic rent between participants and the regulatory authority, the regulatory authority becomes one more stakeholder in the allocation process and an auction or tender process will not only allocate resource rent in an economically efficient way (with revenues being generated which can offset the costs of management, research and development of the fishery, as well as providing, if needed, a

Table 2: Prioritization of aspects of biological and economic research under various quota management scenarios

Quota Scenario	Stock-Recruitment	Stock Biomass	Growth, mortality	Economic data
Fixed TAC near MSY	Critical	High	Low	High
Variable TAC near MSY	High	High	Low	High
Fixed TAC near MEY	Low	Medium	Medium	High
Variable TAC near MEY	Medium	High	Medium	High

return to the wider community) but will also identify those potential users of the resource with the highest use-values for the fishery in question (Morgan 1995). The auction or tender process can be designed to address non-economic goals such as the protection of existing rights in a fishery, matching quota to capacity, and avoiding collusive practices and monopoly situations. The extensive theory of auctions and tenders that has been developed in other industries (e.g. Riley and Samuelson 1981; McAfee and McMillan 1987) suggests designs appropriate for a fisheries context. For example, providing designated bidders with a price preference in an otherwise conventional auction design will increase the quota won by the designated bidders (who could be pioneers or existing operators in the fishery) while at the same time increasing revenues generated from the auction. Price preferences are therefore an efficient means for protecting existing or implied rights to the resource to whatever degree is considered appropriate. Similarly, quota can be matched to capacity in an economically efficient way by incorporating simultaneous bidding on small quota units together with multiple rounds of sealed bids (Morgan 1995).

If the allocation of TAC between participants is made separately from the allocation of economic rent between participants and the regulatory authority, this implies that administrative decisions regarding the sharing of resource rent between the regulator and the regulated would have been carried out first (since it is difficult to envisage the allocation of resource rent by an auction or tender process with two bidders, the regulator and the group of potential participants in the fishery). Although this process has been the norm in quota-managed fisheries, particularly in Australia, with resource rent being allocated through a process of 'cost recovery' for specified management, compliance and research costs related to the fishery. Such 'cost recovery' principles have also been applied to non-quota-managed fisheries as a means of allocation of economic rent. In the fishery for the western rock lobster, which is Australia's largest fishery, management by effort limitation (input controls) results in significant dissipation of economic rent (Lindner 1994). Moreover, under the present management arrangements for that fishery, economic rent is unlikely to exceed 69% of the rent potentially achievable under alternative (i.e. transferable quota) management.

TAC has been allocated between participants in quota-managed fisheries with ingenuity to take into account previous history in the fishery. Perhaps the most successful is the 'Adjusted Preferred Method' pioneered in South Australia, which essentially gives participants the choice of how they want previous catch history to be taken into account (average catch over the past *x* years, catch in the best *y* of *x* years, catch in the previous season etc.) and then adjusts the total back to the established TAC, resource-rent allocation having been previously and separately addressed through a 'cost recovery' process. At the other extreme, in fisheries such as the Western Australian abalone fishery, allocation between participants is based simply on an equal share of the TAC for all participants with, again, resource-rent allocation being separately addressed through a 'cost recovery' process.

One of the significant practical problems in dealing with allocation issues as two separate processes (apart from the resource ownership perceptions that such an approach engenders) is that the focus of allocation between the regulated and the regulator shifts to a debate of what are 'recoverable costs' for a particular fishery. This debate has prompted significant Government expenditure in establishing new accounting and financial management procedures to identify expenditure for individual fisheries. It has also been a driving force (initiated by the fishing industry) behind the minimization of those recoverable costs and in the general move to the provision of more efficient management, compliance and research services by Government. In New Zealand, this process has continued through to Government deregulation and the competitive tendering of these services.

The 'cost recovery' concept of quota management is a solution (albeit imperfect) to only one part of the allocation process — that of allocating the resource rent between the regulator and the regulated. Because such a concept avoids completely the question of the real value of quota units, it is inevitably economically inefficient and leads to significant dissipation of resource rent.

Under a cost-recovery system, the benefits of improvements in cost efficiency of providing management and regulatory services flow back to the industry through a reduction in the 'recoverable costs', rather than to the broader community and Government regulator; hence the incentive for achieving such efficiencies is driven from outside the regulatory authority, and the relationship between regulator and regulated may be reversed, with a vocal industry driving the process of change within the regulatory authority. A possible solution is for any additional economic rent generated through a process of reduction in 'recoverable costs' to be shared between the participants in the fishery and the regulatory authority. Under such a scheme, incentives for improved efficiency and cost reduction would be partially retained within the regulatory authority, providing an internal impetus and a common goal with industry for cost reduction.

The result of a weakening of the power of the regulatory authority may have serious consequences for long-term stock sustainability. Under a 'cost recovery' process for allocation of economic rent, the incentive is always for a reduction in recoverable costs. Unless a strong regulatory authority or an enlightened industry exists, this leads to a focus on short-term objectives. The operation of the Australian Fisheries Management Authority is one of several examples in Australia in which research and compliance activities have diminished partly in response to cost-recovery processes, with those functions increasingly being devolved to Management Advisory Committees for specific fisheries. In Western Australia, the problem of ensuring that long-term stock sustainability continues to be addressed has been approached by defining particular 'non-negotiable' activities (such as monitoring catch, fishing effort and breeding stock levels and minimum compliance activities) whose costs are recovered. This effectively defines a minimum economic rent accruing to the regulatory authority. However, it falls far short of optimizing the economic rent from the fishery in question or in addressing the allocation of that rent in an equitable manner.

When the processes of allocation are separate, economic rent must be allocated by administrative decision although it is conceivable that allocation of TAC between participants in the fishery may be done by a tender or auction process. If overall allocation of resource rent is to be done in the most economically efficient and equitable manner, the two components of the allocation process need to be addressed together, preferably through a well designed auction or tender process.

Maximizing economic efficiency in the allocation process

Apart from having a suitably designed auction or tender process that can address both economic and non-economic objectives, economic efficiency in allocation can be enhanced by measures that reduce or eliminate resource-rent dissipation. Such dissipation can result from excessive transfer costs incurred by the regulatory authority or the inefficient operation of a secondary market for quota. In a fisheries context, an inefficient primary quota allocation process may protract industry restructuring (Squires *et al.* 1994) and lead to monopoly market failures (Anderson 1991). Efficient secondary markets for quotas cannot quickly correct the deficiencies of initial quota allocation (Morgan 1995).

Brokers operate in secondary markets for quotas and licences in most countries where quota management is practised although their operations have generally not received adequate attention in a fisheries context. By matching sellers and buyers, they assist in minimizing the dissipation of economic rent and hence achieving economic efficiency. However, few, if any, regulatory authorities have examined ways in which these brokers' activities can be assisted to achieve greater efficiency and, often, they are viewed as adding to economic rent dissipation through their fee structure.

In conclusion, the full implications of moving to quota management have often not been fully appreciated in fisheries where this process has taken place, and the quota-management techniques applied have often failed to address the allocation and equity issues in an efficient manner. A significant factor in this is often the lack of information on both the biological and economic status of the fishery in question. These data are essential for establishing and monitoring the success of quota-managed fisheries.

References

Anderson, L G (1991). A note on market power in ITQ fisheries. *Journal of Environmental Economics and Management* **21**, 291–6.

Anon. (1994). Australia advances ITQ discourse in Seattle. Australian Fisheries, September 1994, 6–8.

Christy, F T (1973). Fisherman quotas: a tentative suggestion for domestic management. *Occasional paper* 19, Law of the Sea Institute, Rhode Island.

Christy, F T (1982). Territorial use rights in marine fisheries:definitions and conditions. *FAO Fisheries Technical Paper* 227, FIPP/T277.

Clark, C W (1982). Towards a predictive model for the economic regulation of commercial fisheries. *Canadian Journal of Fisheries and Aquatic Sciences* **37**, 1111–29.

Clark, I (1993). Individual transferable quotas: the New Zealand experience. *Marine Policy*, September 1993, 340–2.

Copes, P (1986). A critical review of the individual quota as a device in fisheries management. *Land Economics* **62**(3), 278–91.

Lindner, B (1994). Economic efficiency of alternative input and output based management systems in the western rock lobster fishery. Fisheries Department of Western Australia, *Fisheries Management Paper* 68, unpag.

McAfee, R P, and McMillan, J (1987). Auctions and bidding. *Journal of Economic Literature* **25**, 699–738.

Morgan, G (1995). Optimal fisheries quota allocation under a transferable (TQ) management system. *Marine Policy* **19**(5), 379–90.

Riley, J, and Samuelson, W (1981). Optimal auctions. *American Economic Review* **71**, 381–92.

Squires, S, Alauddin, D M, and Kirkley, J (1994). Individual transferable quota markets and investment decisions in the fixed gear sablefish industry. *Journal of Environmental Economics and Management* **27**, 185–204.

Wesney, D (1989). Applied fisheries management plans: individual transferable quotas and input controls. In 'Rights Based Fishing'. (Eds P A Neher, R Arnason and N. Mollett) pp. 267–88 (NATO ASI Series, Series E, Applied Sciences).

Institutional Reforms: Realigning the Roles and Responsibilities of Fishermen and the Government

Steven Wright[A] *and Ted Gale*[B]

[A] Department of Fisheries and Oceans, Vancouver, British Columbia, Canada.
[B] Department of Fisheries and Oceans, Ottawa, Ontario, Canada.

Summary

Self-management has become a popular addition to the lexicon of public resource managers, particularly among fishery managers. But can all management decisions in the fishery be devolved to the fishermen themselves? Does self-management necessarily mean the substitution of private decisions for public decisions or can a larger degree of self-direction be accommodated within traditional models of public administration?

In Canada's Pacific commercial fisheries, the agency responsible for fisheries management, the Department of Fisheries and Oceans (DFO), has launched an initiative aimed at divorcing the distributional (licensing and allocation) function from the operational side of fisheries management. This is not a self-management initiative in the usual sense. Decisions about who will be given access to the resource and what their allocation should be will remain the responsibility of government.

There are many arguments supporting the splitting of these decisions from the other aspects of fisheries management. The Economic Council of Canada, in a study of fisheries policy, concluded 'fishery officials should be as insulated as much as possible from decisions about who is to participate, so as to depersonalize and depoliticize the choice of gear and fishermen'.

Licensing and allocation decisions involve social consequences that extend well beyond the community of commercial fishermen. In other words, fishermen's private interests do not necessarily equate to the public interest. For example, commercial licensing and allocation decisions can affect processing operations, employment and the viability of coastal communities. While some may argue that access and allocations should be left to the marketplace, where individuals could 'vote with their dollars' for commercial access to the resource, it is unlikely that the general public would accept such an unfettered approach to the distribution of fishing privileges. The public could also be concerned about the possibility of excessive concentration of commercial licences in a few hands, particularly the domination of the fishery by corporations able to outbid individuals. For these reasons, the government has decided that in assigning fishing privileges and setting allocations, government rather than industry or the market should be the decision maker. While fishermen will not be given a direct *role*, they will be given a direct *voice* in decision making.

This paper provides a detailed description of the proposal to establish an arm's length, quasi-judicial Board to make commercial licensing and allocation decisions and apply sanctions in Canada's Pacific commercial fisheries.

INTRODUCTION

Canada's Pacific fishery resources sustain valuable aboriginal, recreational and commercial fisheries. Pacific commercial fisheries generate landed values of about \$Can500 million dollars and wholesale values of almost \$Can1.0 × 10^9. The industry provides full and part-time employment for about 25 000 people on vessels and in processing plants. Almost 400 000 anglers participate in the recreational fishery, and sportfishery-related tourism generates about \$180 million in revenues annually. Almost 90 000 aboriginals, living primarily on lands adjacent to water, depend on fish for food, social and ceremonial purposes.

Management of Canada's Pacific fisheries is complicated by a variety of biological, social and economic factors. This paper will outline how Canada proposes to address some of these complexities, focussing on allocation conflicts, conservation concerns and problems with the consultative process and the enforcement system.

Allocation conflicts between and within the major fishery sectors have been exacerbated by the cyclical nature of stock abundance in many species, most notably salmon, increased competition and a growing demand by competing users of the fish resource for a greater share. The cyclical nature of stock abundance in salmon can be demonstrated by the fact that recorded catches dropped from a level of 35 million pieces in 1983 to only 19 million in 1984 and rebounded to 42 million in 1985. Obviously, years of low catch place a significant financial burden on a commercial fishing fleet geared to harvest all fish available in a productive year. While competing commercial gear types are sensitive to changes in allocations to gear types at any time, allocation decisions in years of low catches are especially contentious.

On Canada's Pacific coast, salmon fisheries account for the largest share of total landed value of commercial catches. This fishery has, however, experienced increasing competition in world markets and a growing world supply, most notably from aquaculture operations. Consequently, salmon prices, which peaked in 1988, fell by about 50% within three years. Salmon prices are expected to remain low. The combination of low catches with low prices has increased allocation conflicts between commercial gear types.

Despite the profound changes affecting these commercial fisheries, the way many fundamental fisheries management decisions are made has remained the same: the Minister of Fisheries and Oceans and departmental staff make the critical decisions about who gets to fish (the licensing decision), and how much (the allocation decision), behind closed doors. These decisions are made 3000 miles away from the Pacific coast in Ottawa. Lack of information about how and on what basis these decisions are made sometimes makes them seem arbitrary and unfair.

Recognizing the need for change, an elaborate system of advisory committees has evolved to consult and offer advice on licences, allocations and general fishery management. This consultative system has been designed to give the industry more systematic influence in fisheries management. Yet it is still not known how and why final decisions are made. Many industry members suspect that the system is somehow failing them and that some groups have gained unfair advantage over others.

Efforts to have commercial groups resolve allocation conflicts have been largely unsuccessful. When industry members have been unable to resolve allocation disputes, they turn to the Minister to act as arbiter. Invariably, some interest groups in the industry believe they can be more effective by opting out of industry negotiations and appealing directly to the Minister. Consequently, managerial attention is diverted from fundamental management activities, and the Minister is deflected from larger policy issues, toward resolving allocation disputes.

Many participants in the industry also argue that the system for enforcing fisheries rules has let them down. Handling fisheries violations — breaches of the Fisheries Act, its regulations and licence terms and conditions — through the criminal courts has proved slow, time-consuming and expensive. The courts know little about the fisheries and tend to underrate the seriousness of fisheries violations. The low fines handed down by the courts generally do little to discourage illegal fishing — in fact, the fines are often viewed as little more than a cost of doing business.

Moreover, the system today gives industry only a small role in enforcement. Many contend that fishermen need more say in setting the penalties for failing to respect the rules that govern their fishery. They also need assurances that illegal fishing will be dealt with swiftly and fairly and that penalties handed down will reflect the severity of the violation.

A PACIFIC FISHERIES BOARD

To address these issues, Canada is proposing to establish an independent, quasi-judicial Board to make critical management decisions in the Pacific commercial fisheries. This Board will be based in the Pacific region rather than in Ottawa, to increase both the sensitivity and the accessibility of decision makers.

The new Board would be delegated some key powers currently exercised by the Minister and DFO under the *Fisheries Act*, namely, licensing and allocation access in the marine commercial fisheries. The Board would also take over from the criminal courts the power to impose sanctions for fisheries violations by commercial licence holders.

The Minister of Fisheries and Oceans would remain responsible for conservation and would set the overall levels of harvest for the marine commercial fisheries. The Department would retain full responsibility for recreational and international fisheries. The Department would also continue its role of managing the aboriginal food fishery and resolving land claims.

Within the marine commercial fisheries, the Minister would still set the broad policy framework, with input from industry and other levels of government. But the Minister would permanently give up the power to decide *individual* licensing and resource allocation cases. The Board, operating under ministerial policy and conservation directions, and within the limits of the overall harvest, would decide exactly who gets the fish and how much. The Board would also assess penalties for breaking the rules. The Board's decisions could not be appealed to the Minister. A judicial review of Board decisions by the Federal Court would always be possible, although such a review would not look at the substance of the Board decisions: it would consider whether a Board had exceeded its jurisdiction or ignored some fundamental principle of natural justice (e.g. procedural fairness).

The Pacific Board would have no more than five members — and possibly fewer. Members of the Board would be knowledgeable about the fisheries but would have no direct or indirect financial stake in it. The Minister of Fisheries and Oceans would be responsible for recommending to the Governor in Council (in effect, the Cabinet) the appointment or reappointment of Board members. Members would be appointed for a fixed period of five years on a full-time basis. These five members of the Board would constitute an Executive Board. Provision would also be made for appointing additional members to the Board; they would be appointed by the Governor in Council for a period of up to three years. The purpose of having additional members would be to assist the Executive Board in coping with its workload of allocation, licensing and sanctions hearings. These additional members would not be decision makers for allocation and licensing. They could merely hold hearings and make recommendations on allocation and licensing to the Executive Board. The additional members would be decision makers on licence appeals or sanctions cases.

A formal policy framework is needed to serve as the context for Board decisions. Although there are currently many specific departmental rules for licence holders, e.g. vessel replacement rules, there are few clear statements of framework policies for the marine commercial fisheries. Before the Board could be established, DFO must articulate and codify its policies on marine commercial fisheries. In many instances, particularly in the area of licensing, this could entail developing general statements of policy where none exists today. Industry would be involved in the exercise of codifying the policy framework.

The Minister would have the power to issue general policy directives, to set or alter the management framework, e.g. licence transferability. The Minister could direct more specifically that the number of licences in a fishery be increased or that an exploratory fishery be opened. These policy directives would be binding on the Board. But the legislation would restrict the Minster's ability to intervene by policy directives in Board decisions. The Minister could not issue directions so specific as to determine individual cases or intervene on cases already before the Board for decision.

The Minister would also have the power to issue binding conservation directives to the Board, to ensure the conservation and protection of fisheries resources. For example, it may become necessary to change an allocation after the Board has ordered it because of a decrease in the resource. The Board would be compelled to make this change.

The new Board would address the shortcomings of the criminal justice system. Appropriate penalties, providing an effective deterrent to illegal fishing, would be handed down swiftly and fairly by the Board acting as a knowledgeable 'fisheries court'. In particular, retention of a licence would now be linked with the individual fisherman's willingness to abide by the rules of the fishery. Industry could help develop penalty schedules.

The *Fisheries Act* would be amended so that breaches of conditions of licence and regulations by licensed commercial fishermen would be taken out of the criminal courts and handled as administrative matters. Obstruction, habitat destruction and Criminal Code offences (e.g. fraud, assault) would still be dealt with through the courts.

The Board would apply a range of administrative sanctions. These would include one or more of: forfeiture of fish, gear and vessel used in the violation; quota reduction; licence suspension; non-renewal or cancellation; or financial penalties of up to $15 000. Minor violations would be made ticketable violations subject to a maximum financial penalty of $2000.

The burden of proof would be lower than in criminal cases. In criminal cases, proof must be established beyond a reasonable doubt. Under the Board, proof would be established on the balance of probabilities.

Benefits to the Department of Fisheries and Oceans

The concept of an arm's length Board offers many advantages to the Department of Fisheries and Oceans. A Pacific Board provides an expert, independent body able to make the difficult commercial licensing and allocation decisions. In consequence, the Minister and Department could focus on policy-setting, conservation and fisheries operations.

The Board's public hearings process could be used to receive public input on contentious policy issues such as licence fees.

Since the Board will determine who is eligible to appear before it in its public hearings, this will force industry organizations to become more accountable (i.e. to demonstrate who they are speaking for and to confirm their mandate). Since input to the Board will be public, industry advisors will be accountable for the advice they give.

An administrative tribunal, comprising persons expert in fisheries matters, will provide a speedy and effective sanctioning process. The ticketing provisions in particular would accelerate the sanctioning process and reduce the time spent by fishery officers in court proceedings. A Pacific Board will, therefore, promote more effective and efficient use of enforcement resources.

Benefits to industry

There are also significant benefits to industry in this proposal. The public hearings process ensures an open process with direct industry input into licensing and allocation decisions. Fishermen will have direct access to the decision-making process and the process will be open to public view. There will be no room for political interference. Decisions will be taken locally. Licensing, allocation and sanction decisions will be made-in-Pacific decisions, responsive to local needs.

Since most of the Department's existing advisory groups are licence-based, a board would provide a forum for currently disenfranchised voices to be heard (e.g. coastal communities dependent on fish processing for employment).

Political accountability is maintained. The Minister continues to set the policy framework but can not intervene in individual cases.

Creation of the Board will allow industry to make business decisions with more confidence. There will be less risk of surprise and consequently greater security, continuity and predictability. At the same time the system will be flexible. The Board structure and processes can evolve over time, in line with a changing industry. It will allow time for a large degree of industry self-management. It will support industry's evolution rather than inhibit it.

ISSUES

In moving forward with this proposal, industry has raised concerns at the development stage and several challenges are anticipated with implementation once the legislation is passed.

Development

At the development stage, there have been quite understandable and predictable problems in garnering the support of some fishermen's groups for the concept

Of the specific problems identified by industry groups, three are key. First, some industry advisory bodies believe that the Board's allocation function undermines their role in allocations. These groups also contend that public hearings would give the views of parties with only a minor interest in fisheries issues (e.g. environmental groups) as much significance as the views of licence holders. A contrast with those who seek even further devolution of authority to industry are those who oppose the 'depoliticization' of the decision-making process. These argue that allocation decisions should remain the responsibility of a politician who is directly accountable to the voters. Obviously, lobby groups and their lobbyists are comfortable with, and believe they can be more successful in, closed-door negotiations with a Minister.

Second, Board membership is contentious. While the legislation requires Board members to be knowledgeable about and have experience in fisheries, there was considerable concern about the possibility that the government might appoint political friends or, worse still, well-known academics or public figures with a personal policy agenda.

Finally, the issue of Board costs grew in significance over the consultation period as major budget reductions and increased user fees were announced. Industry was concerned about the additional costs associated with operating an independent Board and believed these costs would be passed on to licence holders in the form of higher licence fees.

In response to these concerns, a number of modifications to the legislation have been proposed. First, it has been proposed to strengthen industry's role in allocations by allowing industry groups representing a majority of licence-holders to submit allocation proposals to the Board for ratification. The Board would then merely ratify industry's allocation agreement, provided the agreement was in the public interest and consistent with every applicable policy and conservation directive.

Second, the Minister agreed to refer candidates for appointment to the Board to a parliamentary Standing Committee for review. These committees comprise members of all recognized political parties in the House. The candidate and the Minister himself could be called to appear before the Committee for up to two weeks of hearings, at which industry could also testify.

Finally, several proposals for minimizing Board costs were suggested. For example, Executive Board membership may be reduced and the sanctioning process has been further streamlined to ensure that the use of oral hearings is minimized in cases of minor infractions.

Implementation

The operational linkages between fisheries managers and the Board will be complex. It will be extremely important to ensure an effective working relationship develops between the Board and the Department. Any confusion or error resulting from an inability to deliver sound decisions in a timely manner could severely undermine the Board's credibility with industry (and the Department).

The quality of appointees to the Board will be a key determinant of the Board's success. Board processes and decisions must also be designed to ensure public confidence and affirm legitimacy with core constituents. Processes should, to the extent possible, induce industry itself to develop its own consensus among various interests.

CONCLUSION

The Pacific Fisheries Board represents a new approach for making crucial management decisions in Canada's Pacific commercial fisheries. The public hearings process provides for transparent decision making and provides a forum for fishermen to have a direct voice in decision making. If the Department of Fisheries and Oceans is successful in implementing this initiative we believe there will be significant benefits to industry in terms of better decisions and that fishermen's confidence in the ability to effectively and fairly manage the fisheries will be enhanced.

The proposed changes reflect an evolution in the relationship between fishermen and fishery managers. As with any evolutionary change this initiative will survive so long as it is better than current approaches.

Fisheries Co-Management: A Comparative Analysis

Sevaly Sen and Jesper Raakjaer Nielsen

Institute for Fisheries Management & Coastal Community Development, The North Sea Centre, 9850 Hirtshals, Denmark.

Reprinted from Marine Policy, Vol. 20, Sevaly Sen and Jesper Raakjaer Nielsen, Fisheries Co-management: A Comparative Analysis, No. 5, pp. 405–18, 1996, with kind permission from Elsevier Science Ltd., The Boulevard, Langford Lane, Kidlington OX5 1GB, UK.

Summary

This paper is based on a review of 22 case studies on fisheries co-management in small-scale, semi-industrial and industrial fisheries in developing and developed countries in Africa, Asia, the Caribbean, Europe, North America and the Pacific. Case studies are classified according to a typology of co-management arrangements. The typology is based on the nature of the decision-making arrangements between governments and users. Decision-making arrangements refer to the roles of governments and user groups, the management tasks and the stages in the management process.

Eleven case studies are analysed in detail. The analysis shows that co-management covers a wide variety of collaborative arrangements between governments and users. On the basis of the information from these case studies, a number of observations is made concerning the determinants of the type of co-management regime in place. Determinants include the capabilities and aspirations of user groups, the type of approach, the difficulty of the decision to be taken, the type of management tasks, the stage in the management process, boundaries, types of user groups and political culture and social norms. The paper concludes with the issues requiring further research.

Introduction

This paper describes a comparative analysis of case studies on fisheries co-management arrangements found in the fisheries management literature. The research forms part of the work carried out under the Fisheries Co-management Research Project, a collaborative exercise between the Institute for Fisheries Management and Coastal Community Development (IFM), the International Centre for Living Aquatic Resource Management (ICLARM) and collaborating national research institutions in Asia and Africa. Through literature reviews[1] and case study field research in selected countries, the project aims to develop a set of globally or regionally applicable fisheries co-management models. The project has drafted a research framework which is an adapted version of the Institutional Analysis and Development framework for common property resources, developed by the Workshop in Political Theory and Policy Analysis (Ostrom 1990; Oakerson 1992; Ostrom *et al.,* 1994). The research framework

1 Literature was sourced from a variety of places and included both published and unpublished reports. Martin's (1989, 1992) bibliography of Common Pool Resources and Collective Action provided valuable guidance on relevant documentation. Other sources of literature included the ICLARM and North Sea Centre library and the Food and Agriculture Organisation. Articles were also obtained from researchers, academics and development workers who were or have been involved with fisheries co-management and/or community-based management. A bibliography of fisheries co-management has now been compiled by the Project (IFM/ICLARM, in prep).

uses an institutional approach to understand decision-making arrangements and was intended as a tool to enable systematic and comparative analysis of the literature review and the field research. Whilst analysis of the field research using the framework is feasible, it was found that insufficient information was available from case studies in the literature review to enable the research framework to be applied. However, the literature review did provide many interesting examples of different types of fisheries management arrangements known as 'co-management' and enabled both a typology of co-management to be constructed as well as comparative analysis to be carried out between cases. A number of well-known case studies (e.g. Lofoten fishery, Atlantic Surf Clam Fishery) had been reviewed in detail by other authors so it was decided that the analysis should focus on case studies which had not received the same exposure. The purpose of this analysis was to distil similarities and differences in co-management regimes between and within types and the effects of the different arrangements on sustainability, efficiency and equity. Finally, elements of success and failure as well as future research are identified.

This paper aims to complement and supplement the comparative research on user participation and co-management undertaken by Jentoft and McCay (1995). This covered a number of well documented co-management cases in developed countries (e.g. Canada, USA, Denmark, Japan, Norway).

WHAT IS CO-MANAGEMENT?

The effectiveness of existing fisheries management regimes in maintaining or achieving sustainable resource utilization is constantly debated, as fisheries in many parts of the world continue to be under pressure or in crisis. In recent years there has been growing recognition that user groups have to become more actively involved in fisheries management, if the regime is to be both effective and legitimate (Jentoft 1989; Raakjaer Nielsen and Vedsmand 1995). In this analysis, fisheries co-management is defined as an arrangement where responsibility for resource management is shared between the government and user groups and is considered to be one solution to the growing problems of resource overexploitation (Hanna 1992; Pinkerton 1989; Berkes 1989; Jentoft 1989;).

Based on the above definition, co-management is considered to be different from community-based resource management (CBRM) because government is also involved in the decision-making process concerning the management of the fishery. This traditional marine tenure system, traditional fisheries management systems and community-based resource management are not considered to be co-management because government is not involved in the decision-making process. This clear delineation can be awkward in practice. It could be argued that CBRM regimes are co-management if they are recognized in national legislation or they form part of sectoral development policies. However, for the purposes of this analysis, CBRM regimes have been excluded.

The exclusion of community-based management regimes from the analysis reduced the number of case studies available for review. Much of the literature which purports to describe

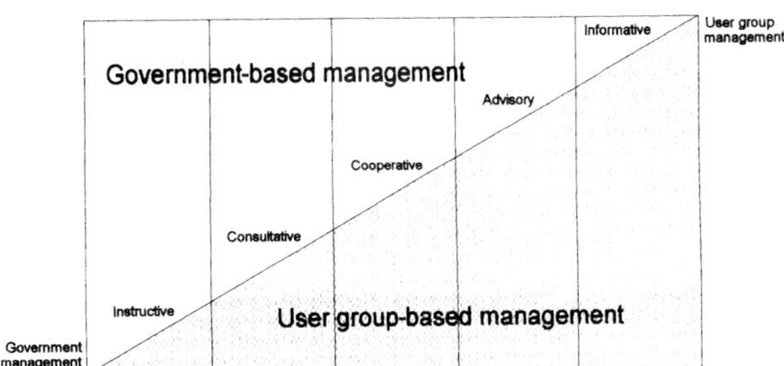

Fig. 1. Spectrum of co-management arrangements.
Adapted from McCay 1993 and Berkes 1994.

'fisheries co-management' case studies, in fact described community-based management arrangements. In addition to the reviews of case studies, articles on theoretical or general country descriptions of co-management arrangements were reviewed. On the basis of this theoretical and empirical literature review (IFM/ICLARM, in prep), fisheries co-management arrangements can be classified into five broad types according to the role government and users play as shown in Figure 1.

Type A: *Instructive* — There is only minimal exchange of information between government and users. This type of co-management regime is only different from centralized management in the sense that the mechanisms exist for dialogue with users, but the process itself tends to be government informing users on the decisions they plan to make.

Type B: *Consultative* — Mechanisms exists for government to consult with users but all decisions are taken by government.

Type C: *Cooperative* — This type of co-management is where government and users cooperate together as equal partners in decision-making.

Type D: *Advisory* — Users advise government of decisions to be taken and government endorses these decisions.

Type E: *Information* — Government has delegated authority to make decisions to user groups who are responsible for informing government of these decisions.

However, this typology is a simplification of a very complex situation. There is a multitude of tasks which can be co-managed under each type at different stages in the management process. Thus co-management covers a broad spectrum of possible collaborative decision-making between government and user groups encompassing:

- the roles of government and user groups in decision making;
- the types of management tasks which can and want to be co-managed by user groups and government and;
- the stage in the management process when co-management is introduced (i.e. planning, implementation, evaluation).

Roles of government and user groups

In an idealized co-management scenario, both government and user groups cooperate as equal partners for all management tasks and at all stages in the management process. However, in most co-management arrangements, the role of government and user groups vary. Jentoft and McCay (1995) note that the role user groups play in the decision-making process depends on who and how they are represented. They describe two types of representation for users: functional (based on gear types) and territorial (based on geography). In many developing countries and some developed countries, there may be other types of user-group representation based on socio-cultural variables such as ethnicity, gender or religion. The role of user groups in the decision-making process will depend on their relative negotiating capabilities, knowledge and strengths *vis a vis* each other and with government. Some groups may feel alienated or poorly represented and decide to boycott the decision-making process. There may also be other stakeholders who have a legitimate right to be represented in the co-management process, such as scientists and social scientists, and those representing the public interest, such as environmentalists. The type of representation is often determined by the political culture of the country and whether participatory or representative democracy is encouraged or discouraged.

Jentoft and McCay (1995) and Raakjaer Nielsen and Vedsmand (1995) emphasize the importance of level, and related to this, scale. 'Level' refers to the level at which decision making should and does take place, namely local, regional, national or supra-national. 'Scale' refers to the fisheries resource system and the management tasks to be undertaken. If the system and/or tasks are large and complex then decision making at a local level may not be effective or sufficient. In these cases, some management decisions may have to be made at the national level. Increasing the scale implies that there will be more diverse membership and representation so that direct democracy might become difficult. There might also be greater inequalities.

Management tasks

Depending on the particular institutional and organizational set-up, different management tasks may be suitable for different forms of co-management decision making. There are a number of management tasks which include policy formulation, resource estimation, access rights, harvesting regulations, market regulations, monitoring, control and enforcement.

For some tasks, for example policy formulation, it may be desirable to have full and equal decision-making by government and stakeholders. For others, such as access rights, it may be more appropriate that the decision is made by government, based on consultation with user groups. Different management tasks might be subject to different types of co-management between government and users. With regard to the type of management tasks which can be co-managed, there is a need to differentiate between decentralization (i.e. moving responsibilities to a lower level of government) and delegation, which might mean transfer of responsibilities from government to a user-group organization (national or local). Depending on the management function itself, as well as the political and social context, some may be appropriate for decentralization and others for delegation. For example, it might not be appropriate to delegate to user groups management functions which are also in the public interest. Both decentralization and delegation require that capabilities and aspirations exist at another level of government and within user groups to carry out these functions. In addition, governments may be reluctant to relinquish some or all of their authority for all management functions (Pomeroy 1993).

Stage in the management process

Another dimension to co-management is the stage in the process at which users become involved: planning, implementation or evaluation. Under an ideal co-management regime, user groups should be involved at all stages of the co-management process, but what actually occurs might be quite different. Hanna (1996) points out that management processes are established to achieve particular objectives so the cost-effectiveness of the process has to be compared to other possible processes. A centralized approach at the planning stage will tend to have lower design costs than a cooperative approach as it is likely to take less time to reach decisions. However, implementation, monitoring and enforcement of the programme might be more costly because the regime is not considered legitimate by users who have had very little say in its design. Conversely, if there is a lack of information to manage the fishery, the co-management approach might lead to lower transaction costs at the planning and implementation phase because fishers can provide information on fishing patterns, catches and the status of the resource.

Evaluating fisheries co-management

Evaluating fisheries co-management can relate to the meeting of management objectives or its impact on the resource and its users. Evaluation does not necessarily entail quantifying these outcomes, but assessing whether co-management has had a positive or negative effect on them. The three main types of outcomes considered most relevant for evaluating a co-management arrangement are sustainability, efficiency and equity. According to Hanna (1996) these are defined as follows:

- *Sustainability* can be divided into stewardship and resilience. Stewardship is the tendency for resource users to maintain productivity and ecological characteristics of the resource. Resilience is the ability of the system to absorb and deal with changes and shocks.
- *Efficiency* refers to the cost-effectiveness of the arrangement, in particular whether it has reduced transactions costs or improved the net returns to the fishery.
- *Equity* is divided into representation, process clarity, homogeneous expectations and distributive effects. Representation refers to the extent to which users and stakeholders are represented. Process clarity concerns the transparency of the management process. Homogeneous expectations is the extent to which participants have similar expectations concerning the management process and its objectives. Distributive effects concern the extent to which the management process has led to a more or less equitable distribution of benefits.

Table 1. Case Studies Reviewed

Typology	Number of case studies
Instructive	2
Consultative	5
Cooperative	6
Advisory	4
Informative	5

CASE STUDIES FROM THE LITERATURE REVIEW

The 22 case studies in the literature review were classified into a broad typology of co-management arrangements according to the co-management spectrum and based on an *overall* assessment of the type of co-management regime in place. The limited information available in the literature made it possible only to classify the co-management regime in a general sense. A breakdown by management tasks and stages in the process was not possible. In general, case studies referred to particular fisheries rather than countries, so it was possible to have more than one type of fisheries co-management regime operating within the same country. Table 1 gives the number of case studies reviewed by type.

This section provides a brief description of eleven selected case studies from the literature review, according to co-management type. Case studies providing the most information about the co-management arrangement were selected. The classification of case studies (Fig. 1 and Table 1) has been based on the information available in the literature review only. It is therefore acknowledged that there are limitations to such a classification as greater knowledge about each case study might result in a change of classification.

TYPE A: INSTRUCTIONAL

Inland Waters, Bangladesh (Ahmed *et al.* 1995)

In 1986, the government of Bangladesh initiated a New Fisheries Management Policy aimed at improving and sustaining open inland water fisheries as well as providing greater equity in the distribution of benefits of the fishery. The main policy instrument was to discontinue the leasing of public water bodies to middlemen and replace this system with direct access rights to fishers. The government hoped for a partnership between themselves and fisher communities. One of the most notable features of the new fisheries management regimes which came into operation following the introduction of this new policy was the active participation of Non Governmental Organizations (NGOs) as intermediaries between fishers and government. In some of the models described in the paper, the role of government (and NGOs) is *instructive* rather than cooperative. The NGOs take a central role in organizing and representing fishers, until such time the time they are sufficiently organized to represent themselves. One NGO took over the management of over 800 waterbodies, obtaining long-term leases and entering into a cooperative arrangement with landless people to fish the waterbodies. However, the NGO makes all financial and management decisions and group members (fishers) are only responsible for labour inputs.

Lake Kariba, Zambia (Malasha 1996)

Lake Kariba is a man-made lake shared between two countries, Zambia and Zimbabwe. In 1994, the Zambian government decided to implement a fisheries co-management regime for the open access artisanal gill-net fishery on the Zambian side of the lake. This was thought to be a solution to the problems of falling catches, limited resources for enforcement (for fisheries and for smuggling) and poor living conditions of fishers and their families as many were based in temporary settlements. The lakeshore was divided into four zones under the jurisdiction of zonal fisheries management committees which are comprised of traditional authorities, users, NGOs, entrepreneurs and village committee representatives. The zones were further sub-divided into designated fishing villages. All fishers and their families were compelled to move from their fishing camps or their own villages to these designated villages. Dual residency of villages was prohibited. Within each village, fishing village management committees were formed which report to the zonal committees. Both committees have only monitoring roles in the fishery and decide where development funds should be spent. It is envisaged that a number of new conflicts will arise as a result of the movement of fisher families to these designated villages. In addition, village management committees appear to be making arbitrary decisions, hold meetings irregularly and lack basic organizational and management skills. At the zonal committee level, the interests of small-scale fishers are poorly represented. However, the process is still in an early stage.

This is a case of *instructive* co-management as all decisions concerning the fishery are taken by the government although users have been involved in the process.

TYPE B: CONSULTATIVE

Lake Malombe, Malawi (Bland and Donda 1995; Donda 1995)

Lake Malombe, one of the smaller lakes in Malawi, is relatively shallow and is showing signs of serious overexploitation. In 1993, the government of Malawi initiated a participatory fisheries management programme aimed at achieving greater participation of fishing communities around the lake. Committees were established to represent the interests of fishing communities. Government fisheries extension workers trained committee members on the biological basis of fisheries regulations and were involved in general institutional capacity building. Fisheries regulations were developed through negotiation between the two groups. Enforcement is carried out by the Fisheries Department. The process appears to be *consultative* especially with regard to the determination of harvesting rules, although there are also some instructional aspects to the regime especially with regard to government decisions on the policy and ways to achieve objectives for Lake Malombe.

San Miguel Bay, The Philippines (Pomeroy and Pido 1995)

In 1991, the government decentralized the management of nearshore fisheries to municipalities and local fishing communities. The fishery in San Miguel Bay is a multi-gear open access fishery. By 1993, it was evident that government management was not working. The fishery was overexploited and there were increasing incidences of conflict between user groups. In the same year, the San Miguel Bay Management Council was established to design and implement a management plan for the Bay. The Council comprises representatives from user groups, local government, NGOs, peoples' organizations, academics and the police. It is advised and supported by a number of advisory and administrative committees and task forces which comprise representatives from different administrative levels (i.e. municipal, district, province). The majority of posts is held by the government. The main tasks of the Council are to provide day-to-day policy guidance and administration, to co-ordinate plans and legislation of local governments and external authorities, and to act as an advocate to national government on matters requiring legislation and support to implement the plan.

This case study is classified as *consultative* as decisions are taken by government after consulting with users.

TYPE C: COOPERATIVE

Pacific Fishery Management Council (PFMC), USA (Hanna 1992)

The Pacific Fishery Management Council is the focus of decision authority in fishery management. Regulations are formally enacted by the Council, advised by three advisory committees which represent users. Processors and consumers are represented in one committee and scientists and economists in the other two. During the planning process to develop a new licence limitation programme for the West Coast groundfish fishery, users had an active role throughout the design and development phase. The programme is classified as *cooperative* co-management, where government and users are in partnership.

Customary Fishing Rights Areas, Fiji (Cooke and Moce 1995; Ruddle 1995)

The fisheries co-management regime in the fishing rights areas of Fiji is, in general, a cooperative effort between national government and users. The fishing rights areas (*qoliqoli*), officially termed Customary Fishing Rights Areas, are under the control of clan chiefs and recognized by government. Management of the subsistence fishery is decided upon and controlled by the traditional authorities but responsibility for the management of the small-scale commercial fishery is shared between the traditional clan chiefs and government in a complex arrangement. Licences for commercial fishing are issued by the Fisheries Division, but before applying a fisher must first obtain a permit from the social unit in whose area he or she intends to operate. This is issued by the District Commissioner, with the consent of the tribal group. Thus, the traditional authority determines whether commercial fishing can occur and the conditions on the licensee concerning target species, permitted gear, areas exclusion and conservation rules. There is not, however, a uniformity of approach. Management strategies depend on the individual Fisheries Officer and the chief involved. Whilst this arrangement is classified as *cooperative* co-management, in some areas there remains confusion concerning the ownership of fishing rights and management rights which has resulted in conflicts between government and traditional authorities.

TYPE D: ADVISORY

Days at Sea Regulation in the Kattegat, Denmark (Raakjaer Nielsen and Vedsmand 1995)

The days at sea regulation in the Kattegat is an experiment aimed at solving the problems of discard and mis-reporting created by the quota system in the sole and *Nephrops* fishery and the conflict between fishers and scientists concerning resource estimation. Prompted by fisher protests, negotiations between the government and the industry concerning the management of the fishery led to the establishment of a working group. Representatives from the government, the fishers' association and a research institute meet monthly as a working group and make decisions concerning the number of days at sea to be allocated and to review progress. These decisions are communicated to the Danish Regulation Advisory Board which is a consultative board advising the Minister. However, all proposals put forward by the working group have so far been accepted. This co-management arrangement is therefore considered *advisory*.

TYPE E: INFORMATIVE

Producer Organization Management in the Dutch Flatfish Fishery (Hoefnagel and Smit 1995)

In 1993, a co-management regime for the flatfish fishery was implemented in the Netherlands following quota overfishing of flatfish and increasingly poor relations between fishers and government. Responsibility for managing individual fishers' quotas has been devolved to groups of fishers who pool their individual quotas. All group members have to be members of the same Producer Organization, but group membership is not compulsory. These groups are responsible for implementing and enforcing regulations, imposing sanctions and organizing intra-group quota exchanges. The government retains responsibility for controlling the national quota (allocated by the European Union) and all tasks relating to the implementation of the Common Fisheries Policy. Fishers were encouraged to join groups through an incentive scheme which gave them slightly higher days at sea and the possibility to rent quota during the whole year. This arrangement is classified as *informative* as the Producer Organization informs government of the regulations it has implemented.

Groups are quite homogeneous which aids cooperation. Only 17% of vessels have chosen to be outside the system. Overall, the system seems to have been successful as national quotas have not been exceeded, levels of inputs and outputs have been stabilized and fishers are satisfied with the group system. However, as part of this success might be attributable to relatively high quotas and low catches, assessment of whether the co-management programme achieved its goals is difficult to determine.

The Matjes Herring Fishery, Denmark (Raakjaer Nielsen and Vedsmand 1995)

The matjes herring is a special quality North Sea/Skagerrak herring sold in the Netherlands as a snack. The season lasts only three months a year. The fishery, undertaken by purse seiners from Denmark, Sweden and Norway, is co-ordinated and managed by the Matjes Committee. The committee holds regular meetings with representatives from the fishers' organizations, buyers and processors. Decisions on regulations and quota distribution, as well as monitoring and enforcement are undertaken by producers' organizations in each country. The decision-making process itself is considered transparent as information is distributed to all participants in the fishery. This is thought to be made possible by a relatively small number of participants using the same gear. This arrangement is considered *informative* as the committee has received the approval of the Danish government and informs the government on the decisions they have taken.

Mechanized Beach Seine Fishery, Inhassoro, Mozambique (Kristiansen *et al.*1995)

In 1981, the beach-seine fishers were organized into a Fishers Association. This enabled the District Administrator to discuss and negotiate with one group. The Association regulates fishing activities, calls meetings, which local government officials attend, and makes decisions concerning the opening and closure of the fishery. The government officials do not interfere in the decisions which are taken. A letter (also signed by the District Administration) is then sent to Provincial Authorities informing them of the decisions taken. This is submitted in proposal form to central government for authorization. As the administrative approval process takes a long time, many of the decisions proceed without formal government authorization. In effect, the authorization procedure *informs* government of the decisions taken by the Association.

Corporate Wetfisheries Management, Faroe Islands (Moerkoere 1992)

Between 1975 to 1990, the wetfish fishery around the Faroe Islands was regulated by the Raw Fish Fund. The regulatory mechanism was a subsidy/tax regime together with minimum price fixing. The Fund was established according to Faroe Island legislation. The aim of the Fund was to ensure an acceptable income for fishers and to use it as an instrument to conserve fish stocks as no quota system was in place. It was intended that the Fund would be self-financing. In good years it would generate a surplus which could be used in bad years. It was foreseen that there would be no government subsidies. The Fund was administered by a Board whose members included industry representatives (catching and processing) and government. The industry were in the majority.

Pressure from interest groups represented on the boards led to high subsidy levels and taxation seldom used. The Fund was not self-financing and the government had to intervene with additional subsidies. High subsidy levels and distorted prices led to overcapitalization of the fishery and a collapse of fish stocks. The result was that the country was on the brink of bankruptcy.

While the Fund was in operation, it *informed* government of decisions taken. This is a case of a co-management arrangement which failed. This can be attributed to irresponsible behaviour by the Fund's Board members due to their inability to withstand pressure from various interest groups in the fisheries sector.

DISCUSSION

This literature review covered a variety of different co-management arrangements in five regions: Africa, Asia, Europe, North America and the Pacific. They included artisanal, semi-industrial and industrial fisheries in both freshwater and marine habitats. Information on the co-management arrangement varied. Although most of the case studies provided a general overview, there was limited information on the details of the arrangement. With these constraints in mind, the following discussion focusses on whether there are any commonalities and differences between and within types and the effects of the arrangement on outcomes. Finally, elements of success and failure as well as future research issues are identified.

Type of co-management arrangement

As described earlier, the type of co-management regime is determined by the role of users and governments in decision making. Based on the literature review some form of co-management seems always to be appropriate. However, this does not imply that delegated self-management is more appropriate than informative or consultative management arrangements. The proper design principles depend upon the context and conditions under which the co-management arrangement has to work.

These will often evolve gradually (Jentoft and McCay 1995) through a process of 'muddling through'. North (1990) observed that institutional change often occurs as marginal adjustments of old structures rather than radical innovations or total reorganization. The process is always dynamic. Thus, the type of co-management arrangements described in the case studies may change. However, a number of observations can be made which determine the type of co-management regime in place.

(1) ***Capabilities and aspirations of user groups.*** The way governments decentralize or delegate management authority has an effect on the type of co-management regime in place. Although the aim of government might be cooperative co-management, this can only be achieved if users are also willing and capable of taking on shared responsibilities. Cooperative, advisory and informative co-management occurred in situations where user groups were able and willing to take up the responsibility (Matjes, PO Management and PFMC). Unorganized or poorly represented user groups, low levels of education, and lack of empowerment all hindered a more equal participation in the decision-making process. The review indicated that developing countries trying to initiate co-management may be working with communities where there is no existing

organization of user groups so that these have to be introduced. Here the type of co-management arrangement is more likely to be instructive or consultative until user groups are organized and capable of cooperating more equally in the management process (Bangladesh inland waters and Lake Malombe). Thus, existing organizations of user-groups are not a pre-requisite to co-management *per se*, but the nature of user group organizations does play an important role in determining the type of co-management regime.

(2) ***Top-down or bottom-up approaches.*** The type of approach influences the type and nature of the user-group participation in decision making. It is more likely that the type of co-management approach is instructional or consultative when it is top-down, and advisory or informative when it is bottom-up. In some of the case studies, governments have actively pursued a policy to promote co-management (Bangladesh inland waters, Lake Kariba, Lake Malombe, San Miguel Bay, PFMC and Faroes). In these cases the type of co-management tended to be instructive or consultative. Whilst the type of arrangement is also affected by the capabilities of users, in some of the cases the implication is that governments want to play the dominant role in the management process. In the Matjes fishery and the beach seine fishery in Inhassoro, co-management was 'bottom-up', emanating from well-organized user groups. This is reflected in the type of arrangement which is informative.

(3) ***Difficult decisions.*** Greater user participation in co-management also occurs when governments are unwilling to deal with the political, social or economic responsibility of taking hard decisions, preferring to let the user groups deal with these problems (Dutch PO Management and Faroes).

(4) ***Management tasks.*** The type of co-management arrangement to be implemented also depends on the management tasks to be undertaken. In general, although the cases studies are not explicit about management tasks, there is evidence that the more specific the tasks are (harvesting and market regulation), the lower the level at which decisions are taken. In these situations, it more likely that the type of co-management arrangement will be advisory or informative (Dutch PO Management, Matjes, Inhassoro and Kattegat). Very little information was available on the policy formulation process, but there are some indications that, where it does take place, it tends to be instructive or consultative. This observation is supported in the general co-management literature (Jentoft and McCay 1995).

(5) ***Stage in the management process.*** In general., the information from the case studies indicates that co-management arrangements, whatever the type, occur during implementation and only occur to a minor extent in planning (Lake Kariba and PFMC). There is no clear evidence from the case studies on user participation in evaluation. However, in the case of the Dutch POs, the Matjes and Kattegat fisheries, there are indications that the implementation process is being continually evaluated by government and user groups.

(6) ***Boundaries.*** The importance of boundaries in fisheries co-management has been thoroughly discussed in the literature (Ostrom 1990; Pinkerton 1989). In general the literature indicates that the more clearly defined the boundaries, the greater the role of users in the decision-making process. However, the boundaries issue is very complex as in any fishery there are many boundaries (physical, social, technical, economic, political). The case studies demonstrate that many kinds of boundaries are in place, physical (Lake Malombe, PFMC and Faroes), residence (Kariba and San Miguel Bay), organizational (Dutch PO) and socio-cultural (Fiji), lack of land ownership (Bangladesh) and resource (Danish Matjes fishery and Kattegat). Although these are the 'clearer' boundaries indicated, there is often a mixture of boundaries which determines (who, where and how) the type of co-management arrangement.

(7) ***Types of user groups.*** In most cases user groups were homogeneous either functionally, territorially or socio-culturally. This contributed to group cohesion. Socio-cultural homogeneity was also important for collaboration between user groups (PFMC and Kattegat). Conversely, where there was socio-cultural heterogeneity in multi-user group situations, co-management was more difficult with government taking a bigger role in decision making (Lake Kariba).

(8) ***Political culture and social norms.*** The political culture and social norms of the country and/or society also affect the type of co-management arrangement. Societies not familiar with political empowerment may find it difficult to participate on an equal basis with government (Bangladesh inland waters). The political (modern and traditional) structure in the country may also exclude certain types of co-management arrangements and/or encourage others (Fiji, Lake Kariba).

Anticipated outcomes

Although many of the co-management arrangements described in the cases studies are still at an early stage of implementation, some overall observations can be made concerning outcomes. Outcomes do not refer to actual outcomes, as in most cases it is too early to assess them, but to the outcomes anticipated by the co-managers.

In practically all of the cases, the main rationale for introducing a co-management arrangement was the fact that the fishery was nearing overexploitation or was already overexploited. In this respect, co-management was a form of crisis management, seen as a way to impose stewardship over the fish resources (Lake Kariba, Lake Malombe and Inhassoro). In other cases, co-management was implemented in order to prevent or resolve conflicts among user groups or between user groups and government. In some cases this was in addition to the problem of overexploitation (San Miguel Bay and Dutch PO management) or where there was no overexploitation (Fiji and Kattegat). With regard to conflict management, co-management was introduced to make the management process more resilient. Co-management was also seen as one way to increase the resiliency of the system to changes in the market (Matjes) or in the system (Faroes).

With regard to the four components of equity, information from the case studies was sparse. Greater representation in the co-management process was clearly a goal but it was not clear how well stakeholders and users were represented. However, in all except one of the cases, it appears that users and/or stakeholders were better represented than before. It seems evident, that process clarity is great as a result of co-management. There seems to be no difference whether the decision-making arrangement is a Village committee (Bangladesh inland waters, Lake Kariba and Lake Malombe), a management council (San Miguel Bay, PFMC, Kattegat and Dutch POs), a fishing committee (the Matjes committee), or has no formal structure (Fiji).

CONCLUSIONS

The analysis of the literature on co-management has covered artisanal, semi-industrial and industrial fisheries in both developed and developing countries. It has been restricted to cases which fit within the broad definition of co-management as some form of collaborative arrangement between government and user groups. Many arrangements which have been described as co-management have therefore been excluded as they describe some form of community-based management (traditional or modern) which is not, according to the definition used in this analysis, co-management.

The typology used in the analysis provides a simplified way to classify the variety of exiting co-management arrangements. This should be viewed as a guide only, as the situation is likely to be much more complex. Different management tasks may be subject to different decision-making arrangements, and it is possible that in a particular fishery, the whole spectrum of different co-management arrangements is present. However, as so little information was available in the literature review on decision-making arrangements, it was only possible to carry out a broad classification. Furthermore, the classification is not static. A co-management arrangement which is classified as consultative today may be cooperative in the future. This will depend on the factors that determine the type of co-management arrangement. Based on the review of twenty-two case studies, of which eleven were reviewed in detail, some key determinants were identified.

The capabilities and aspirations of user-groups are clearly an important determinant of the type of co-management arrangement. In fisheries where user group organizations are weak or non-existent, the government is more likely to have a more dominant role. The case studies from some of the developing countries verify this. The type of approach is also important. It is more likely that the role of government is greater when the co-management approach is top-down. When the approach is bottom-up, the role of user groups is greater. With regard to management tasks, the more specific the tasks, the lower the level at which decisions are taken, and the greater the involvement of user groups. There was no evidence from the case studies, that users had any significant influence on policy formulation. In the case of developing countries, decisions on whether there should be a co-management approach appears to be taken entirely by government.

Other determinants of the type of co-management arrangement include the stage in the management process when co-management is introduced, the type of boundaries, the types of user groups and the political culture and social norms of the country or society in question.

With the exception of a few case studies, it seems clear that co-management is a form of crisis management. Governments, observing the failure of their own management regime take the decision to bring users into the management process. The anticipated outcome is that sustainability, efficiency and equity of the resource and its users will be improved. In all the cases reviewed, it has not been possible to evaluate whether these anticipated outcomes have been achieved.

This review should be regarded as an early contribution to the process of understanding the different types of co-management arrangements that exist. There remain significant gaps in the information concerning these arrangements. If these are filled, a better understanding of the general circumstances under which different co-management arrangements work and the factors which influence the movement from one type of co-management arrangement to another could be achieved. Of particular importance are details concerning the types of decision-making arrangements in place and the key factors which determine and affect these arrangements. Little is known of the effect of co-management arrangements on outcomes (sustainability, equity and efficiency) both in terms of the process itself and the impact on the resource and its users. In order to evaluate outcomes, more information is required on user/stakeholder representation, capabilities and participation in the decision-making processes and on whether compliance and ability to enforce management rules have increased as a result of co-management.

The case studies currently being carried out by the IFM/ICLARM Co-management Project in Africa and Asia should contribute to these information requirements. A research framework based on the Institutional Analysis and Development Framework will enable both systematic and comparative analysis of information of co-management arrangements in developing countries for a variety of fisheries. The comparative analysis should enable distillation of both the necessary pre-conditions for a particular type of co-management arrangement as well as identification of the key factors which affect the arrangement and its outcome.

ACKNOWLEDGMENT

This paper is the result of a literature review undertaken by the project entitled: *Fisheries Co-management: A Worldwide Collaborative Research Project,* funded by the Danish International Development Agency. The paper has benefited from the helpful comments given by Fikret Berkes, Susan Hanna, Svein Jentoft, Sten Sverdrup-Jensen and Tomas Vedsmand.

REFERENCES

Ahmed, M, Capistrano, D, and Hossain, M (1995). Fisheries Co-management in Bangladesh — Experiences with GO-NGO-Fishery Partnership Models. Presented at the Fifth Conference of the International Association for the Study of Common Property, Bodo, Norway, 24–28 May 1995.

Berkes, F, (1989).Co-Management and the James Bay Agreement. In 'Co-Operative Management of Local Fisheries: New Directions for Improved Management and Community Development'. (Ed E Pinkerton) pp. 189–208.

Berkes, F, (1994). Co-Management: bridging the two solitudes. *Northern Perspectives* **22**(2–3), 18–20.

Bland, S J R, and Donda, S J (1995). Common Property and Poverty: Fisheries Co-management in Malawi. Presented at the Fifth Conference of the International Association for the Study of Common Property, Bodo, Norway, 24–28 May 1995.

Cooke, A, and Moce, K (1995). Current trends in the management of *qoliqoli* in Fiji. *South Pacific Commission Traditional Marine Resource Management and Knowledge Information Bulletin.* No. 5. April 1995. Noumea, New Caledonia.

Donda, S J (1995). Fisheries Co-management in Malawi. Presented at the Workshop on Fisheries Co-management, Hirtshals, Denmark 29–31 May 1995.

Hanna, S A (1992). Creating User Group Vested Interest in Fishery Management Outcomes: A Case Study of the Pacific Fishery Management Council. Presented at The World Fisheries Congress, Athens, Greece, May 3–8, 1992.

Hanna, S A (1996). User participation and fishery management performance within the Pacific Fishery Management Council. *Ocean and Coastal Management* **26**, 23–44.

Hoefnagel, E, and Smit, W (1995). Experiences in Dutch Co-management on Marine Fish Resources. *Agricultural Economics Research Institute LEI-DLO*, The Hague, The Netherlands.

Jentoft, S (1989). Fisheries Co-Management: delegating government responsibility to fishermen's organisations. *Marine Policy* **13**, 137–54.

Jentoft, S, and McCay, B J (1995). User participation in fisheries management. Lessons drawn from international experiences. *Marine Policy* **19**(3), 227–46.

Kristiansen, A, Poiosse, E, Machava, M, Santana, P, and Meisfjord, J (1995). Co-Management of Fisheries in Inhassoro, Inhambane Province, Mocambique: A case study. Ministerio de Agricultura e Pesca, Instituto de Desenvolvimento da pesca de Pequena Escala (IDPPE).

Malasha, I (1996). In search of a New Management Regime on the Northern Shores of Lake Kariba. *CASS Occasional Paper.* University of Zimbabwe.

Martin, F (1989). Common Pool Resource and Collective Action. A Bibliography Vol. 1. Bloomington, Indiana University, Workshop in Political Theory and Policy Analysis.

Martin, F (1992). Common Pool Resource and Collective Action. A Bibliography Vol. 2. Bloomington, Indiana University, Workshop in Political Theory and Policy Analysis.

McCay, B J (1993). Management Regimes. Property Rights and The Performance of Natural Resource Systems. Background Paper prepared for the September 1993 Workshop, The Beijer International Institute of Ecological Economics.

Moerkoere, J (1992). The Collapse of a Corporate Management System: experiences from the Faroese fishing industry. Paper presented at the World Fisheries Conference Athens, Greece 3–8 May 1992.

North, D (1990). 'Institutional Change and Economic Performance'. (Cambridge University Press: Cambridge.)

Oakerson, R J (1992). Analyzing the Commons: A Framework. In 'Making the Commons Work: Theory, Practice and Policy'. (Ed D W Bromley) pp. 41–59 (Institute for Contemporary Studies Press: San Francisco).

Ostrom, E (1990). 'Governing the Commons: the Evolution of Institutions for Collective Action.' (Cambridge University Press: Cambridge.)

Ostrom, E, Gardner, R, and Walker, J (1994). 'Rules, Games and Common Pool Resources.' University of Michigan Press, Ann Arbor.

Pinkerton, E (Ed) (1989). 'Co-operative Management of Local Fisheries.' (University of British Columbia Press: Vancouver.)

Pomeroy , R S (1993). A Research Framework for Coastal Fisheries Co-management Institutions. *NAGA, The ICLARM Quarterley* **16**(1).

Pomeroy, R S, and Pido, M (1995). Initiatives Towards Fisheries Co-management in the Philippines. The case of San Miguel Bay. *Marine Policy* **19**(3), 213–226.

Raakjaer Nielsen, J, and Vedsmand, T (1995). Fisheries Co-management: An alternative strategy in fisheries — cases from Denmark OECD, in press.

Ruddle, K (1995). A guide to the literature on traditional community-based fishery management in Fiji. *South Pacific Commission Traditional Marine Resource Management and Knowledge Information Bulletin* No. 5. April 1995. Noumea, New Caledonia.

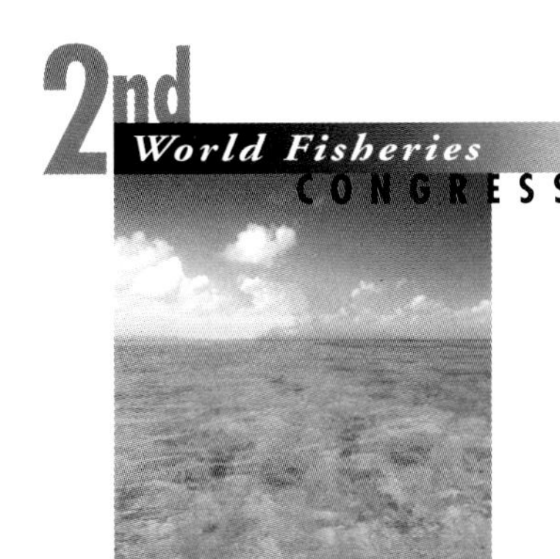

Franchising Fisheries Resources, an Alternative Model for Defining Access Rights in Western Australian Fisheries

J. W. Penn, G. R. Morgan and P. J. Millington

Western Australian Marine Research Laboratories, PO Box 20, North Beach, Western Australia, 6020.

Summary

The limited entry management arrangement and system of individually transferable effort units (ITEs) first developed in the 1960s and now extended to cover all major Western Australian commercial fisheries, has a conceptual basis which closely parallels the successful franchising model being used extensively in the wider business community.

The contractual relationships between franchisees, master franchisors and their shareholders, is similar to the relationship between commercial fishermen and the Government representing other community stakeholders, and could be used as the basis for future legislation to manage fisheries.

Western Australian commercial fishermen, like other franchisees, are in most major fisheries provided with a defined operating territory, which limits uneconomic competition between fishing units, and gear/vessel limits which balance capital inputs with potential turnover (catch), while allowing each fisherman to operate as an independent entrepreneur. Similarly, Government as a master franchisor has responsibility for keeping the number and distribution of a variety of franchisees in balance with the stocks' capacity to reproduce, requires franchisees to provide information for strategic planning, while charging fees for both services provided and a return to shareholders (the community).

The use of the franchise model for understanding fisheries management concepts and possibly for fisheries legislation is therefore proposed as a commercially-tested model. This model provides a basis for establishing clearly-defined reciprocal rights, obligations and franchise fee structures for resource users, while allowing Government conservation obligations and resource allocation issues to be dealt with in a commercially rational contract framework.

Introduction

Commercial fisheries management arrangements developed in Western Australia to conserve stocks have by definition restricted access to fisheries resources. These restrictions in access have led to relatively high goodwill values for those commercial vessels with access rights in each limited entry fishery (now defined as a 'managed fishery' under the new Fish Resources Management Act 1994). A market in those access rights has developed, and reflects both the annual value of production from those vessels holding licences and also the expected future sustainable yield from the fishery concerned. Increasing value of these transferable access rights, both in Western Australia (current market value at about $A2.5 x 10^{12}$) and elsewhere in Australia has led to an increasing debate on the nature and duration of these rights. Recently the new Fish Resources Management Act, proclaimed by the Western Australian Government in October 1995, has added a degree of security to these rights. However, this legislation tends to focus primarily on the obligations of the access licence holder in the context of resource conservation.

The increased value of these commercial fishing access rights, and demands for an increasing share of the fisheries resources by recreational fishermen, who also have been given non-renewable access rights to some fisheries, has resulted in increasing political debate about the ownership of common property fish resources. These resource-sharing debates are becoming the major issue facing fisheries management in the 1990s. Similarly, leases in Western Australia over 'crown' waters for aquaculture, particularly pearl farming, also create goodwill values but these leases may conflict with other potential public sector users, e.g. recreational boating for fishing or yachting.

The purpose of this paper is to record the evolution of fisheries access rights, both commercial and recreational, in the very successful Western Australian fisheries. It examines the parallels between this development and that of the business franchising industry with which there are many similarities, but which has a more commonly understood legal basis in terms of contractual obligations. Secondly, the paper examines the franchise model to determine whether it can provide a more commercially understandable framework which could better define the rights attached to all fishing and aquaculture licences, and thus provide a more logical framework for resolving resource sharing issues.

History of development of the Western Australian fisheries management system

Active management of WA fisheries resources began in the early 1960s with the purpose of controlling exploitation and creating an orderly system to allow development of the State's newly discovered shellfish fisheries (Bowen and Hancock 1982). This initiative was taken to avoid 'the tragedy of the commons' syndrome (Hardin 1968) and was based on the stock assessment and theoretical management strategies being developed by scientists (Beverton, Holt and Gulland) working on the European North Sea fisheries at that time.

The approach involved controlling fishing effort using a system of limited entry licences created under the Fisheries Act of Western Australia 1905, with matching federal legislation being involved where stocks were fished in both State and Commonwealth waters. Early experience with limited entry, using vessel numbers alone, led to a realization that in the competitive environment of fishing, fishermen would automatically seek to maximize their catch by increasing their quantity or size of fishing gear, which was not controlled, and hence increase exploitation rates. This led to a refocussing of management onto additional direct controls on catching power using gear units, such as a specific licence on pot (trap) numbers in relation to vessel size in the Western Rock Lobster Fishery. A vessel replacement policy was also introduced which prohibited increases in vessel size. Similar controls were also placed on the two major prawn trawl fisheries in Shark Bay and Exmouth Gulf, which developed in 1962 and 1963 respectively (Penn *et al.* 1989). These measures focussed primarily on net length and vessel size. The regulations in these fisheries were designed to directly control the catching power of the whole fleet for conservation reasons, but also had the secondary impact of significantly reducing the trend towards individual vessels using increased capital inputs to gain a competitive edge within the fleet and a greater overall share of the available catch. Although there was still a variety of uncontrolled factors particularly in relation to electronic technology, e.g. radar and global positioning systems (GPS), which have resulted in increased effectiveness of the fishing effort, by and large these have not resulted in a high degree of overcapitalization. Secondly, the early implementation of a system of unitization of fishing gear in the lobster fishery and ability to both trade and accumulate gear units to allow larger vessels to be built under the vessel replacement regulations, has had the added benefit of allowing industry restructuring. As a result the rock lobster fleet numbers have automatically declined as owners have sought to increase their vessel size and accumulate additional gear units per boat (Fig. 1). Gear unitization linked to vessel capacity (also unitized) has been implemented in the major prawn trawl fisheries in Shark Bay and Exmouth Gulf.

During the 1980s limited entry management arrangements incorporating both individually transferable effort (ITE) units

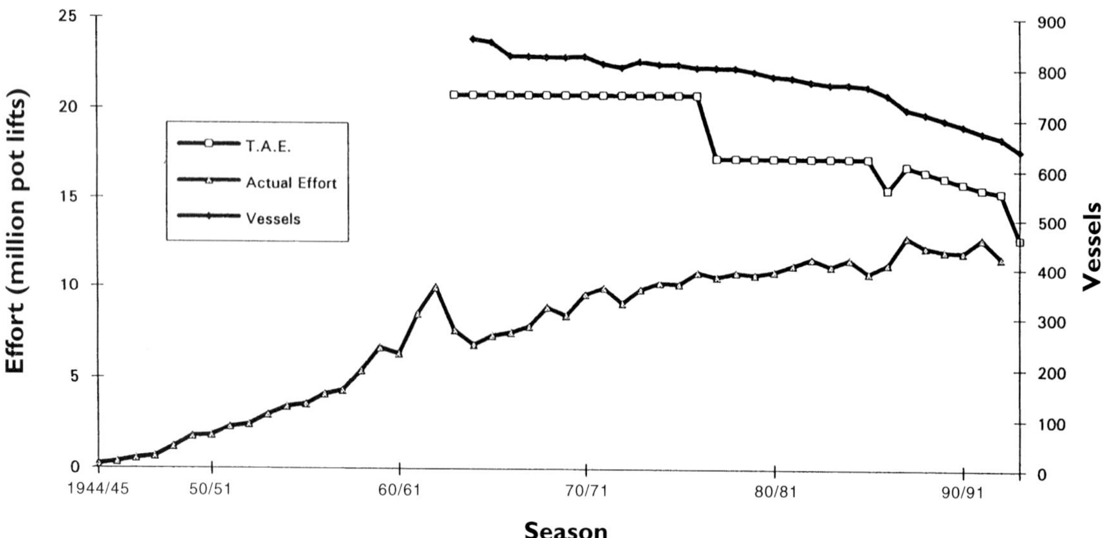

Fig. 1. Trends in catch, actual fishing effort (in pot lifts), total allowable effort (TAE), and fleet number in the Western Australian rock lobster fishery.

and individually transferable quota (ITQ) systems, were extended to essentially all significant commercial fisheries in Western Australia. In most of those fisheries managed using input controls, the fishing has been unitized, often incorporating time gear units, so that proportional adjustments to gear usage can be undertaken annually. This system allows for automatic fleet restructuring by individual vessel licensees purchasing additional gear units, thus phasing some vessels out of each fleet, and making the remaining vessels more viable. The evolution of this system into annually variable gear 'quotas' to match incoming recruitment levels has now been introduced for the rock lobster fishery (1993/94), where an 18% temporary pot reduction was used to reduce a catch peak, improve survival to the breeding stock, and redistribute the catch both throughout the year and between years for market purposes (Anon. 1993). The 18% reduction in pot usage indirectly led to a significant reduction in fleet size, and an increase in licence goodwill values (Fig. 2).

Management of recreational fishing in Western Australia parallelled the introduction of commercial management, that was first applied to the major high-value commercial fisheries during the 1960s. In addition to minimum size limits and bag limits, specific licences were introduced for the high value rock lobster and freshwater crayfish (recreationally fished only) stocks and general recreational fishing using nets. In most cases the recreational licence regulations were introduced to complement the major commercially-fished species and maintained some balance in access rights between the recreational and commercial sectors.

In the early 1990s, in recognition of the increasing impact of recreational fishing, a comprehensive review of that sector was undertaken (Anon. 1991). As a result, a wide range of additional input controls was introduced including gear limitations and minimum legal sizes. The existing bag limits were also considerably tightened and extended to cover most edible species, to educate users in both the conservation ethos and the sharing of resources. Controls on the total numbers of recreational licences were not however considered necessary at that time.

In summary, the full range of input tools is used for management of wild stock fisheries. For commercial fishing these include direct constraints on effort through boat size limits, controls on the amount of gear which can be deployed (usually effected through the system of unitization), closed areas and closed seasons. Recreational fishermen are similarly subject to a range of input controls such as gear limits, closed areas and seasons. In addition management also uses, where appropriate, output controls, through individually transferable quotas in the case of commercial fishermen and bag limits for recreational fishermen. Pragmatically, to contain enforcement costs, guarantee sustainability and control overcapitalization, some input controls have also been retained whenever output quotas have been introduced.

In addition to wild stock fisheries, Western Australia also has a long history of pearl oyster fishing and pearl culture, with leases and licences being granted over crown waters for that purpose through the Pearling Act 1990. The value of cultured pearl production was over $A180 million in 1995. Trading of pearl shell quotas and associated farming leases is facilitated through the commercial licence transfer register administered by the Fisheries Department of Western Australia. These leases have been granted in past years through a simple planning and public comment process. Due to the remoteness of pearl oyster fishing and culture sites, these leases have previously had little impact on other users of coastal waters. However, expansion of the pearling industry into more populated areas in recent years has

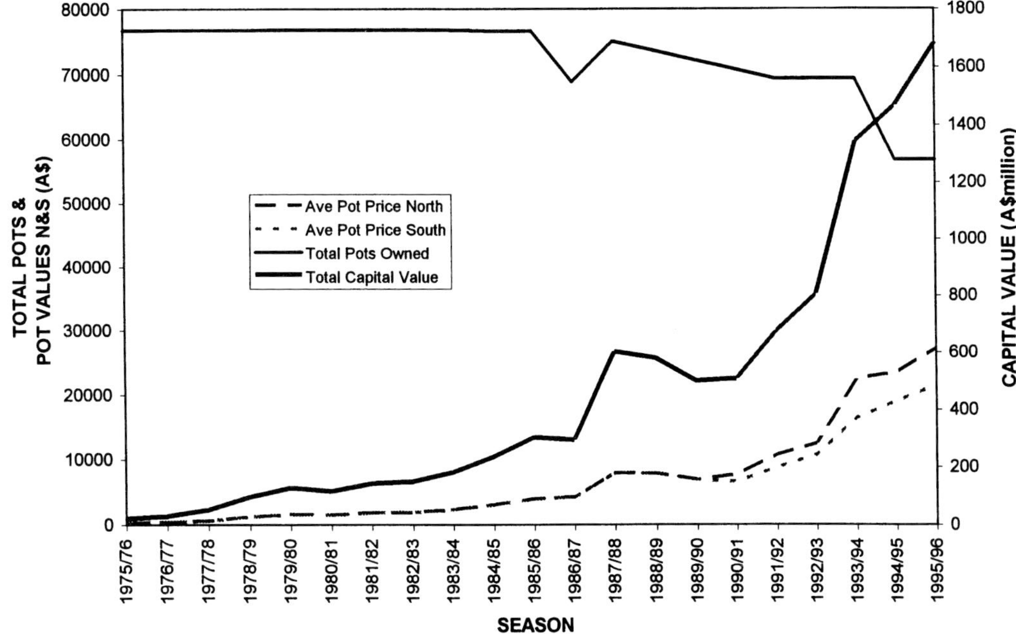

Fig. 2. Aggregate 'goodwill' values for the Western Australian rock lobster pot licences and pot numbers permitted to be used in the fishery for the period 1975–1996.

resulted in conflict with other recreational users seeking to exercise their access rights to crown waters, and again raised the issue of the 'nature' of pearling leases and licences.

More recently, the Western Australian Government has been encouraging the development of general aquaculture, including cage and longline culture in bays and estuaries, usually through licences and water leases issued through the Fish Resources Management Act. These industries are at a very early stage of development, but are already beginning to compete for space with other users of coastal waters.

The other major aquatic resource user group, the passive users, are catered for through protected fish observation areas and Fish Habitat Protection areas under the Fish Resources Management Act, and a two- (soon to be three-) tier system of marine reserves administered under the Conservation and Land Management Act 1984. There are complicated public processes required before the declaration and management of such reserves. However, the exploitation of fish resources within them is generally subject to the Fish Resources Management Act.

Current access to the Western Australian aquatic resources

Since the permanent freeze on the issue of new WA commercial fishing licences in 1983, access to most of the 60 or so unit commercial fisheries can only be achieved through purchase of existing licences. This purchase and transfer is effected through the licence transfer system and register operated under the Fish Resource Management Act. The register of licences now permits the recording of other interests on licences to enhance security for lenders, thus improving access to development finance.

For a small number of fisheries, mostly coastal and estuarine, licences are generally not transferable. However, access is, in some cases, permanently permitted through 'grandfather' clauses which allow family members to continue historic fishing activities.

A similar situation to transferable commercial fishing licences exists for entry to the pearl oyster fishing and culture industry. Here access to pearl oyster quotas and associated water leases can be gained through market mechanisms facilitated by the Government licence register. For the emerging mariculture industry, access to suitable areas of crown water is still possible through a difficult cross-agency allocation process. In time, however, it is expected that access to scarce prime sites will increasingly occur through market mechanisms and the licence and water lease register.

Recreational fishing licences, in contrast to commercial licences, are all of limited duration, mostly purchased annually. There is no overall restriction on access, except indirectly through the cost of licences. Where resources are becoming limited, e.g. abalone, recreational access to resources is limited by extensive closed seasons and tight bag limits.

However, the resources to which the commercial fishermen have traditionally had most access, both biological and in space and time, are also increasingly being sought by other user groups, e.g. recreational fishing, aquaculture and non-extractive users (boating, recreational diving, etc.). These trends now require active mechanisms to adjust access shares to fish resources and crown waters which are both equitable and acceptable to both the competing sectors and the Government representing the community at large. To adjust access shares however, requires the valuation of licences, which again raises the questions of the nature of licences and leases issued by the Government.

This need for an equitable reallocation process and its potential costs to the community, coupled with tighter budgetary regimes within Government has also focussed attention on the costs involved in managing fish resources and the potential for a direct return to the community for access to obviously valuable resources. This focus on Government expenditure reduction has recently (1995/96) led to the imposition of full cost-recovery through additional licence fees on the more profitable commercial fisheries in Western Australia.

The Franchise model

Franchising as a business management tool began in the nineteenth century in the USA (Singer Sewing Machine Co) and Britain (breweries — public houses, Dres (1991)), and is now a widely-used and successful system applied throughout the Western world. Franchising combines the strengths of large business with the motivation and entrepreneurship of small business.

The essential element of modern franchising is an organizational structure where a franchisee is granted a long term 'right' under contractual terms to engage in a business using a system prescribed by a franchisor or 'owner' of the system. The contractual relationship must have significant economic benefits to both parties and provide a less risky business environment for the franchisee. Additional benefits are that the franchisee has a degree of independence, access to reliable management advice, an improved ability to obtain finance for the operation and standard operating procedures, etc. In return for these benefits, however, the franchisee accepts a variety of restrictions on business activity, which may include territorial limits, controls on production capacity (capitalization), enforced conformity to standard methods of operation, etc.

In successful franchises, the franchisor attempts to ensure the ongoing viability of the franchisees by setting, for example, appropriate territories, promoting a positive image of the franchisees' business and providing strategic management for ongoing development of all franchise operations. In return for this service, franchisors extract fees including:

i. An initial franchisee fee (usually small to cover training and recruitment).

ii. An annual franchise service fee or royalty (profit margin based on turnover).

iii. An advertising levy (to promote franchise products, etc).

iv. An accounting or administrative fee (to cover cost of services provided).

These fees are negotiated and set at levels which are economically viable for both parties.

In assessing the franchise 'model' as a possible legislative framework for defining licences and leases in the WA fisheries sector, it is useful to compare the existing roles of the fisheries participants with their franchise business counterparts (Fig. 3). In this context the franchisor or Government fisheries agency has the added complexity of operating multiple and frequently competitive franchises. The equivalent roles of Government in a Westminster System, and fish resource users in the franchise model, are outlined below. (In other political systems a 'Minister' may be an equivalent member of a Cabinet (e.g. Secretary)).

Government sector (master and multiple franchisor)

(a) Minister for Fisheries — Chairman of the master franchise company is responsible to the company shareholders (or voting public) for achieving an equitable return on assets, i.e. fish stocks, and associated crown waters.

(b) Fisheries Agency (Master Franchise Administrator)

(i) Corporative Executive — represents the Board of the master franchisor company and has overall responsibility for sustaining, managing and avoiding conflicts between the variety of franchises issued.

(ii) Management and Licensing Division — develops and registers franchise contracts (management plans) and facilitates franchise contract transfers and fee collections (licensing branch). The Division also deals with dispute resolution between franchise groups.

(iii) Research Division — works directly with franchisees to compile information on production from whole sector, undertakes independent research to forecast future trends and advises on strategic directions to maintain long term viability of each sector.

(iv) Enforcement Division — ensures compliance with franchise rules (management plans) by both franchisees and non franchise holders (general public) throughout the state.

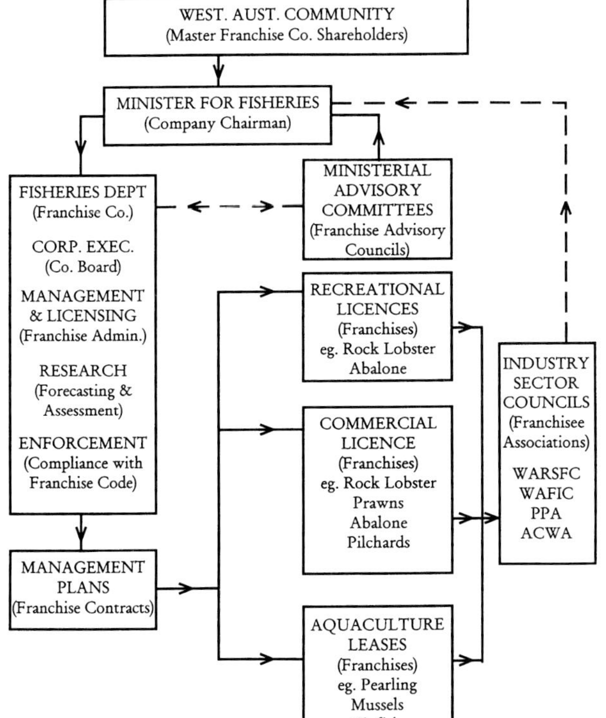

Fig. 3. Proposed franchising model framework for the management of the Western Australian fisheries sector.

Management Advisory Committees (Franchise Advisory Councils) — are made up of experts covering *all* aspects of the particular franchise sector (including a relevant Government representative) and provide strategic advice to the Minister (Master Franchisor) on methods to enhance the viability of their respective franchisees. The most successful franchise operators are selected to sit on these committees.

Fish utilization sector (franchisees)

This group represents a wide range of resource users who have been provided with a right to use fish resources:

(a) Commercial users — with transferable and ongoing rights (franchises) to take specific groups of fish species for commercial gain, e.g. rock lobsters, prawns, abalone, scallops, snapper and tropical reef fish.

(b) Recreational users — with annual franchise rights to take certain species for personal use, e.g. rock lobsters, abalone and trout.

(c) Aquaculturists — with ongoing lease rights (franchises) to areas of 'crown' water for the commercial production of fish products, e.g. pearl and mussel farmers.

(d) Fish processors — some fish processors which have exclusive licences to process certain categories of fish, e.g. rock lobsters, freshwater crayfish, etc. have in effect ongoing franchise rights to particular sectors.

(e) Additional user groups not specifically covered by existing legislation but which would probably be covered under a franchising model are recreational charter fishing operators, regional aboriginal communities and passive (non exploitive) users who seek long-term rights to crown waters or some resources.

Industry sector councils (franchisee associations)

Equivalent groups covered here are the WA Fishing Industry Council (WAFIC), the Pearl Producers Association (PPA), the Aquaculture Council of WA (ACWA), and the WA Recreational and Sport Fishing Council (WARSFC). These bodies represent their respective fishing sectors (franchisees) in contract negotiations regarding fee levels with the Minister (Master Franchisor), promote the value of fishing licences (franchises), and represent licence or lease holders (franchisees) in disputes between sectors.

DISCUSSION

Over the past decade the obligations of the licensees (franchisees) in Western Australia have been progressively better defined through a series of legislative changes, while the obligations of the Master Franchisor (the Government and its administration) to more clearly define the 'rights' being issued to the different users has received less attention. The WA Government initiative to apply cost-recovery to its services to the more profitable commercial fishing and pearl culture industries in 1995/96, generated considerable debate by the various resource users as to their ongoing rights and responsibilities in managing the State's fisheries. The franchise model described here provides a considerable number of analogies to the system of licensing which has evolved in Western Australia since the 1960s.

By allowing the development of a market in access rights and addressing the issue of overcapitalization, the Western Australian management systems have both increased the success of WA fisheries and controlled exploitation levels to achieve sustainability for most of its managed fisheries. This contrasts with many other jurisdictions in which fish stocks are either not being sustained, or the regulators have chosen (by action or inaction) to allow overcapitalization or unlimited access, thus dissipating the resource rent which can be obtained from a sustainably-managed and profitable fishery.

Obviously in some jurisdictions there has been a conscious decision to dissipate the resource rent and allow fishing and/or harvesting to become a component of the social safety net. However, to allow uncontrolled harvesting makes sustainability of resources difficult to ensure. To guarantee sustainability, it is necessary and usually critical to have some controls on access to resources and, if economically-efficient harvesting is to be achieved, controls are also needed on the levels of capitalization. In the case of WA fisheries the level of capitalization in the fleet has been kept to reasonable levels by the linkage of vessel size to catching capacity through the system of vessel and gear unitization.

Working from this basis of managing fish resources to ensure both sustainability and economic viability, the Government in Western Australia has in effect adopted the role of the Master Franchisor. It has developed management plans (contracts) with its franchisees to maintain a degree of exclusivity from competition, thus promoting the sustainable use of fish stocks, cash flow and generally, profits. The Master Franchisor guarantees, for example, that there will be no other franchises licensed by it within a certain catchment area. The WA Government, in granting limited access rights to fishermen, has provided an equivalent form of contract. Similarly, the issue of pearling leases, with exclusion zones, provides for a degree of commercial protection for pearl farming operators.

Should the size of the market within the catchment area grow, normal franchise contracts have provision for the franchise area to be split, as long as the total turnover for the original franchise is not adversely affected. The exact details obviously vary with the contract. This mechanism is similar to the method adopted in the development of Western Australia's prawn fisheries (Bowen and Hancock 1982) where additional vessels were incorporated within the fishery as the catch expanded under test fishing. These vessels paid an entry fee or relinquished another valuable fishing licence (franchise) to gain entry.

Within the common fast-food franchise agreements there are also controls on the production capacity of the franchise, e.g. kitchen equipment, which defines maximum possible throughput . This is analogous to gear controls linked to vessel size/capacity regulations which control the catching power of licence holders under most WA management plans.

With respect to the costs and the flow of profits, further franchising analogies can be made. In most franchises there are two types of fees payable to the Master Franchisor. Firstly there is the franchise service fee which is directly comparable to recently-imposed cost recovery fees, and includes the fees for services such as advertising contract administration and legal protection of the franchise operation against unauthorized competing fishing activities. Secondly, there is a fee related to the turnover which represents the profit margin to the Master Franchisor. This is analogous to the return to the community usually expressed as a resource rent charge. In the WA commercial fisheries this return to the community is represented by a 'better interests of the fishing industry' fee which is presently returned to the fisheries sector to address the myriad issues inherent in sustainable resource management.

The major return to the community owners of the resources is however, through the income tax benefits which flow from a profitable commercial sector. Similarly in the case of the Western Australian recreational sector their licence or franchise fees provide a return to the community which is used to enhance the sustainability of the exploited resources. Secondary benefits also flow from this sector's activities through economic multipliers such as employment in supporting non-fishing industries.

What the commercial Master Franchisor cannot guarantee is that the creation of franchises for a similar but not directly-related product in the same catchment area will not affect the profitability of that franchise. In the Government situation a degree of control is possible, although with the multi-faceted nature of Government, the ability of the Fishery Agency to control all such developments is often limited. The structures of Government in Western Australia are still developing to cope with these 'whole of Government' marine resource allocation problems. For example there may be a decision to create a marine park with new commercial benefits to the State. However, the planning process for the marine park may involve a decision to exclude certain historical fishing activities. Such actions lead to demands for compensation for the alienation of access. Where such an 'access' right is not well defined, either in law or conceptually, there is inevitably considerable conflict between the various fishing and/or aquaculture sectors and the Government representing the community, especially when the loss of access needs to be valued.

Using the franchise model, it should be easier to conceptualize such a scenario as an exercise by the Master Franchisor to start a new type of franchise in an existing franchise catchment area and in doing so creating a different flow of benefits. For example it is not uncommon for a Master Franchisor to franchise out two different but related fast-food outlets (e.g. chicken and hamburgers) on the same site. Whether antagonisms or synergies are generated would be the subject of detailed economic evaluation before a decision is made. In natural resource allocation issues, similar planning is also required prior to any decision which adversely affects an existing access holder. This in turn should trigger some change in the 'contract' between Government (the Master Franchisor) and the fishing sector (the franchisee), which has had the prime historical access.

As shown in the preceding examples, the franchise model provides a conceptual framework which can guide the wide variety of people now involved in decision making for fisheries (e.g. fishermen, lawyers, investors, public servants and politicians) to understand and anticipate the often conflicting responses to the management planning process. The use of a conceptual model can also help to put a higher degree of rigour into the decision making and 'contracting' process. At a minimum the franchise model should be used to provide a checklist of matters that must be considered by the Master Franchisor (Government) and the franchisees (resource users) when decisions are being made on access changes.

CONCLUSIONS

The paper has shown that considering contemporary fisheries management directions in Australia within the framework of an already well established commercial model such as franchising can have a number of advantages.

Firstly, such a franchising model clearly establishes ownership of the resource as lying with the community at large, for whom the Minister and the Regulatory Authority act to administer exploitation of the resource under 'contracts' with various 'franchisees' which have ongoing access to it. Costs for access to the resource (the 'franchise fee') may consist of a cost-recovery component, which assists in offsetting administrative costs of management, research etc, and a resource rent charge which provides a return to the wider community as shareholders. However, these costs must be economically realistic and acceptable to both parties to the contract.

Secondly, a franchising model clearly defines not only the rights and obligations of the participants in the fishery or aquaculture venture (the franchisees) but also the rights and most importantly obligations of the Government as the Master Franchisor. These Government obligations to its various franchisees have usually not been made explicit in fisheries management arrangements up to the present time, apart from the general obligation to conserve the resource.

This definition of Government obligations to its franchisees has important implications in Government decisions on equity issues since, under a franchise contract arrangement, the Government would need to consider the economic impact of any equity decisions on franchises already in existence and, presumably, buy back franchises or adjust access fees accordingly in order to reflect the impact of those equity decisions on existing franchisees. Again, this is rarely the situation in current resource management arrangements where Government often inadvertently issues additional 'franchises' (for example, in the form of additional recreational fishing licences) without explicitly considering impacts on existing franchise holders (for example, commercial fishermen) or adjusting access fees to maintain profitability of those existing franchise holders.

The management plans under which many commercial fisheries in Australia already operate could be the precursor of a formal 'franchise' agreement or contract between the Government as the Master Franchisor and participants in the fishery as franchisees. However, for such management plans to be consistent with the proposed franchise model, additional elements would need to be added to address a number of issues which are presently not commonly incorporated into fisheries management plans. Also, the present management plans would need to be expanded to include contractual arrangements with all users of a fisheries resource or habitat.

The additional elements which would need to be incorporated into management plans are as follows:

1. Obligations of the Government as the Master Franchisor. These obligations would include maintenance of the resource, maintenance (but not guarantee) of a level of profitability of franchisees, orderly control of competition, systems for renewal of franchise contracts, maintenance of franchise transfer mechanisms, etc.
2. Obligations of the franchisees. As well as the usual resource-related items such as adhering to time or area restrictions, size limits, gear restrictions etc, these would include issues such as requirements to pay for services from the franchisor (e.g. compliance and research services), payments for access to the resource, the period of the franchise arrangement, arrangements for sub-franchisees, etc.
3. Rules for expanding or decreasing the numbers of franchises, such as recreational licences, commercial fishing licences, and compensating for imposition of competing activities, e.g. conservation areas, aquaculture leases, etc.
4. Rules for setting fee structures and an independent arbitration system for dispute resolution.

The parallels pointed out in this paper between the evolving system of fisheries management in Australia and the commercial model of franchising is, in the authors' opinion, not a coincidence. The similarities arise from a common goal of optimizing the value of assets to the owners of those assets (whether the assets are fish stocks owned by the community or fast-food market share 'owned' by a hamburger franchisor) through a process of controlled access, orderly operations and maintenance of consistent standards. In the case of management of fish stocks, an added dimension is the need to consider non-economic social goals although, as Morgan (1995) has pointed out, appropriate procedures for equity allocation can at least bring out the real cost of any non-economic goals.

In Australia in recent years, there has been a commitment between the Federal and State Governments towards a National Competition Policy. Inherent in this policy is a commitment to the removal of statutory barriers to competition, on the assumption that economic efficiencies will be realized. These barriers could include limitations on entry to fisheries and other input controls. The appropriateness of such an approach for natural resource management regimes, where the fundamental objective is sustainability, have yet to be argued in the appropriate forum. The concept of a franchise model in natural resource management may contribute to a more informed debate.

For controlled access fisheries management systems to continue to address the question of optimizing the value of the resource, the relationship between the owners of the resource (the community) and those who have access to that resource needs to be made more explicit and transparent. This paper has presented a model, based on established commercial franchising, which addresses that need and offers a conceptual and possible legal framework for the Government and fish resource users to share the common goal of maximizing the benefits derived from fishing while ensuring sustainable management of fisheries resources and aquatic ecosystems.

References

Anon. (1991). The future for Recreational Fishing. Final report of the Recreational Fishing Advisory Committee. *Fisheries Department Western Australian Fisheries Management Paper* 41. 73 pp.

Anon. (1993). Rock Lobster Industry Advisory Committee discussion paper: Management proposals for 1993/94 and 1994/95 western rock lobster season. Fisheries Department, *Western Australian Fisheries Management Paper* 54. 25 pp.

Bowen, B K, and Hancock, D A (1982). The Limited Entry Prawn Fisheries of Western Australia: Research and Management. *Fisheries Research Bulletin Western Australia* 27, 1–20.

Dres, A W (1991). The economic analysis of franchising and its regulation. In 'Franchising and the law: theoretical and comparative approaches in Europe and the United States'. (Ed C Joerges) pp. 133–142. Nomos 1991.

Hardin, G (1968). The tragedy of the commons. *Science* **162**, 1243–8.

Morgan, G (1995). Optimal fisheries quota allocation under a transferable (TQ) management system. *Marine Policy* **19**(5), 379–90.

Penn, J W, Hall, N G, and Caputi, N (1989). Resource assessment and management perspectives of the prawn fisheries of Western Australia. In 'Marine Invertebrate fisheries: their assessment and management'. (Ed J F Caddy) pp. 115–140 (J Wiley and Sons Inc. New York).

Concepts and Practice of Individual Transferable Quotas for the Management of Fisheries — An Overview

B. D. Shallard

Bruce Shallard and Associates, PO Box 27409, Wellington, New Zealand.

Summary

This paper describes the implementation and key features of management of fisheries by Individual Transferable Quota (ITQ). A description of the problems of conventional fisheries management approaches is given together with the steps taken to move to quota based systems. The benefits of ITQ are outlined as are the implementation steps. The process used in New Zealand involved defining management policy and goals, setting Total Allowable Catch (TAC) and the Total Allowable Commercial Catch (TACC), before allocating quotas to individuals.

The features of the process are the provision of quota trading, the enforcement approach and access charges. The paper concludes with the results of the New Zealand experience and the benefit to the New Zealand fishery, fishing industry and the economy.

Introduction

This paper describes the implementation and key features of management of fisheries by means of a Quota Management System (QMS) based on Individual Transferable Quotas (ITQs). The paper explains some of the key features of The Quota Management System of fisheries management under the following categories:

- How does a Quota Management System operate?
- What are the problems of conventional fisheries management?
- What can be gained by moving to a property rights system?
- The New Zealand experience.
- QMS implementation issues.
- The results of the New Zealand experience.
- Conclusion.

How does a quota management system operate?

In essence, a Quota Management System provides for management of fisheries in the following manner:

- The sustainable part of the fish stock, by species, by area, is determined by scientific stock assessment.

- Based on this sustainable concept, Total Allowable Catches (TACs) are set on an annual basis for each fish stock by area.
- Allowance within the TAC, is made for traditional (artisanal) and/or recreational fishing.
- The balance, the Total Allowable Commercial Catch (TACC), is allocated to individual commercial quota holders as ITQ.
- ITQs once allocated to quota holders, are fully tradable or leasable as a property right.
- The Government maintains an efficient monitoring system to allow it to keep track of catch against quota.
- All or most other 'conventional' input controls, such as bag limits, mesh sizes, closed seasons and closed areas can be eliminated or reduced.
- Quota holders have the right to catch up to their quota at any time during the fishing year thus removing the 'race for fish'.

What are the problems for conventional fisheries management?

From the middle of the twentieth century onwards, many Governments have encouraged commercial fishing through a range of mechanisms for a variety of reasons. Many of these involved subsidies — some of which continue to encourage production to feed rapidly-growing populations.

This, coupled with the vastly expanding options for catching and processing fish, has meant that the sustainability of some fish stocks in some cases is threatened. While knowledge of the biological profiles of species has been poor, the clear indications of overfishing have led to the imposition of input controls as a method of management.

Input controls came in many forms, including limits imposed on the size and capacity of vessels, or sizes of nets, coupled with closed areas and/or seasons. The problems of input controls are soon obvious:

- They are very expensive, and in many cases difficult to adequately police.
- The effort and thus costs of fishing activity increase considerably as fisheries move to maintain production levels despite the controls imposed.
- The yields based on catch-per-unit-effort decline.
- The industry becomes overcapitalized and uncompetitive.
- Over-regulation leads to inefficiencies and excessive costs both for the industry and for fisheries management.

The manager's response

The conventional response from fishery managers when faced with this situation has been to consider, in addition, some form of output control.

The first step in moving to output controls has been to place limits on the total amount of fish which can be extracted on a seasonal or annual basis. The main problems with such blanket limits have been:

- The 'race for fish' which has serious economic ramifications. This simply means that each fisher must catch his or her share before his competitor. It leads in the extreme, to a very short season of one day in some cases with attendant inefficiency
- The difficulties and costs of policing.

A second method of restricting output has involved making permanent, non-transferable allocations to individuals or separate entities, but this entails:

- Restricting national economic behaviour as each entity must attempt to maximize its portion.
- Restricting efficiencies of a competitive industry with no market for the trading of rights to take fish.

What can be Gained by Moving to a Quota Management System?

A Quota Management System brings in the concept of property rights based on ITQ as a tool.

The QMS facilitates the conservation of threatened or depleted stocks by:

- Allowing for a direct limit to be placed on the output from the fishery by the setting of effective Total Allowable Catches.
- Encouraging, indeed requiring, the fisher to 'farm' the resource, rather than have the attitude of the hunter/gatherer.
- Reducing the need for the 'race for fish' which is prevalent under an input control system. Overcapitalization is reduced as fishers do not need to have larger, better and faster vessels to catch fish before others, as they have a guaranteed right of access to the fishery over the year.
- Instilling attitudes of compliance with rules that serve the future interest of industry participants themselves. The high level of self-monitoring and the concept of ownership of fishing rights achieve this.

The QMS encourages economic and efficient behaviour to benefit both industry and country by:

- The reduction of costs to the advantage of the fisher and the economic well-being of the country.
- Allowing fishers to adjust their role in the fishery in terms of whether they wish to maintain current levels of activity, reduced to reflect a lifestyle or an economic requirement, or indeed to expand rapidly into a particular fishery whilst reducing in others.
- Encouraging capital inflow into the commercial fishery from other sectors of the economy, as the property rights approach produces an asset worth investing in, thus broadening the financial base of the commercial fishery and allowing its diversified development.
- Encouraging the inefficient operator to leave the fishery, particularly, if the State concludes that assistance is desirable to encourage these inefficient operators to leave.

- Providing a means of assessing fishery value and an asset for raising capital to underpin fishing ventures. In particular to allow for more exploration of hitherto underutilized resources or use of more efficient and technically appropriate fishing methods.

The QMS reduces the need for State involvement in:

- Setting rules or making arbitrary allocations between individuals or groups of participants.
- Maintaining a large and costly policing force, as effective operation of the QMS allows a much higher level of industry self-policing to protect the asset base.

THE NEW ZEALAND EXPERIENCE

Having now considered conventional fisheries management problems, and some of potential gains from a Quota Management System based on ITQ, it is helpful to look at the situation in New Zealand, and how New Zealand approached a Quota Management system.

The Quota Management System was introduced into New Zealand fisheries for 27 species on 1 October 1986. There had also been, for a limited range of species, a quota management scheme in existence in New Zealand since 1 April 1982. These species were limited to the deepwater, to relatively unexploited stocks that came within New Zealand's management purview as the result of the declaration of a 200-mile Exclusive Economic Zone by New Zealand in 1978, and had limited seasonal transferability.

In line with its overall economic policy, it had been the New Zealand Government's declared intention in this area to bring market forces to bear in the management of fisheries resources. The QMS has been used as the vehicle for this.

NEW ZEALAND'S ECONOMIC ZONE

New Zealand is a small island nation, with a temperate climate, situation in the South Pacific Ocean between 30 and 50° S and at ~ 170° E.

The country includes a number of small outlying islands and, as a consequence, its 200–mile exclusive economic zone, is 1 300 000 square nautical miles, more than 15 times its land mass.

Although New Zealand's waters are substantial, there is very little continental shelf and much of the area is not biologically productive. 72% of the zone is deeper than 1000 metres, 22% between 200 and 1000 and only 6% less than 200 metres.

In the late 1970s/early 1980s, New Zealand's inshore fisheries came under severe biological and economic pressure with the problems outlined earlier in this paper. As a result of this the New Zealand Government considered a new *'Management Philosophy'* which said:

- Fisheries are a public resource;
- The Government is resource trustee and manager;
- The people, as resource owners, have free and open access to the resource, subject to conservation conditions;
- Those people who derive direct economic benefit from the resource pay for the cost of management.

MANAGEMENT POLICY AND GOALS

From management philosophy, management policy was developed. The key components were conservation and allocation. In relation to conservation, the goal was the wise use of renewable resources by limiting catches to maximize sustainable production.

In terms of allocation it was decided that recreational fishers should get the *'first cut'* of the TAC, subject to daily bag limits such as 30 per person per day for main finfish species and less for lobster (6) and abalone (10), and trading restrictions prohibiting recreational fishers from selling their catch.

Commercial fishers were then allocated the remainder of the TAC with the goal of maximizing the net economic return to the nation.

INDIVIDUAL TRANSFERABLE QUOTA (ITQ)

Having looked at the management philosophy and policy, a definition of ITQ was needed as follows:

> *'An Individual Transferable Quota system allocates to individuals the transferable or tradable* ***right to harvest*** *a specific sustainable quantity of the surplus fish stock production'.*

QMS IMPLEMENTATION

There were a number of important steps to be taken before the QMS could commence operation. The individual fisheries had to be identified and formalized, the TAC and TACC established, and individual quotas allocated.

SPECIES

The species included in the quota system from 1 October 1986 totalled 26 finfish species and abalone. Jack mackerel, squid, rock lobster and scallops have been added since that date.

QUOTA AREAS

The quota areas were established relating to the fishstock boundaries for each species, with the result that some species, being the one stock for the whole NZ zone, are managed within the one area, while other species have a number of areas. It is important to remember that a quota is a quantity of one species in one area.

TOTAL ALLOWABLE CATCHES — TOTAL ALLOWABLE COMMERCIAL CATCHES

The TAC and TACC for each species was established within each quota management area, and given Legal determination by Government.

ALLOCATION OF INDIVIDUAL QUOTAS

Individual Transferable Quotas were allocated on the basis of each person's historical participation in each fishery. Fishers were advised of their catch by species by area over a three-year period prior to the introduction of the quota system. They were able to select two out of the three years of catch history, the average forming the basis of their quota.

In most fisheries, the sum of all the individual catch histories exceeded the Total Allowable Catch levels. To reduce the indicative quotas to the appropriate level, Government instituted a scheme to buy out the excess.

Government buy-back

The Government buy-back was the major preliminary step to operating the QMS. The objective was to buy-back some potential quota holders catch history, to match the level of fishing activity with the TACC.

The rationale for this was to facilitate industry rationalization, increase industry acceptance of the policy, improve compliance, and to reduce the effects of any initial administrative cuts that would be required to reduce historical catch levels to the TACC.

Principal features of the scheme

A. Quota Trading

Quota trading is the key to success. It is a continuous adjustment process that allows for economic rationalization and Industry fine-tuning (matching catch with quota).

Quota may be freely traded subject to a limit of foreign ownership of 24.9% and an 'aggregation limit' of 20% maximum holding of inshore species by area. Trades may be made in lots of 100kg or more and must be registered with the Government. Quota can be traded in perpetuity or by lease or sublease. Trades must be notified to the Government with details of the seller, buyer, amount of quota and the price.

The reason for foreign ownership limits was to remove concerns about New Zealand fisheries being dominated by the other nations that fished in New Zealand waters, particularly Japan, South Korea and Taiwan. The reason for the aggregation limits was to protect small-scale fishing operators from large companies totally dominating the industry.

B. Enforcement

Prior to the introduction of the QMS the enforcement approach in New Zealand's fisheries management policies was of the traditional *'game warden'* type. The emphasis in conventional enforcement activity is the physical apprehending of wrong-doers. New Zealand had a poor record of convictions, inefficient use of enforcement officers and a concentration on minor offending.

It was the intention with the introduction of the QMS, to change the nature and focus of the enforcement activity.

The QMS monitors the flow of product from approved landing places through to final sale. The enforcement focus is now *'on land'* rather than the conventional at sea surveillance and policing role. The monitoring of this flow is the new QMS enforcement task and leads to investigation and prosecution of those who defraud the process.

The features of the new organization are catch, product and accounting systems surveillance, monitoring and management of information, strategic and technical intelligence analysis and targeted investigations/audits.

To effect this product flow control New Zealand has legislated for the following:

- Fish landed in New Zealand are deemed to have been caught in the EEZ;
- Catch, effort and processing at sea are recorded (by monthly returns);
- Landing in New Zealand is at specified landing points;
- Landing/destination/purchase invoices are recorded (by monthly returns);
- Fishers sell only to licensed fish receivers (LFR);
- LFR keep product and financial records (and make monthly returns);
- Non LFR keep product and financial records;
- Quota owners have a record of portfolio status (monthly);
- Fish exports are documented.

C. Access Charges (Cost of Management)

The third feature of the scheme is access charges. Resource rentals have been used as access charges to cover the cost of gaining preferential access rights to a public resource. Resource rentals have now been replaced in New Zealand by a charge to the commercial industry of the full cost to Government of management that relates to the commercial fishery.

QMS implementation issues

While many benefits of the QMS are immediate, some facets of implementing such a system require time before the benefits become apparent. Implementing QMS requires:

- Increased sophistication on the part of both the Government managers and the industry in areas of general management, information systems and reporting.
- A higher level of scientific analysis of fisheries capability and sustainability, requiring in-depth scientific stock assessment processes to be developed.
- Retraining and redirection of the enforcement role. It needs to be noted that it will be some time before this function is fully operational.

The precise nature of the QMS requires careful consideration. Some of the major issues are:

- The imposition of some financial risk for the Government and the industry if they are effectively guaranteed or not guaranteed respectively. The risks of both options need to be considered.
- Countries that introduced quotas without tradability have invariably found that they fail. Tradability of quotas allows people to enter and leave the industry as they wish. It allows

industry rationalization driven by the market and provides a means for signalling the 'value' of the fisheries both collectively and individually. It also encourages efficiency and innovation which must go hand-in-hand with an effective fishery.

- The QMS should be able to deal with variability in stocks. It is quite clear that stocks of many species vary greatly from year-to-year or over a cycle of five, ten or fifteen years. Features, such as under- and over-fishing rights, the possibility of seasonal adjustments, and some form of trade-off between the species should be considered.
- Consideration of whether quotas are expressed in precise tonnage terms or whether they are in proportional terms (whether they are firm or non-firm rights). New Zealand started with quotas that were precisely defined in tonnage terms, and has subsequently changed to a system of proportional quotas. The latter arrangement has worked more effectively for New Zealand, but it is possible in some fisheries to have a mixture of firm and non-firm rights.

THE RESULTS OF THE NEW ZEALAND EXPERIENCE

The gains from the QMS in light of the New Zealand experience are many:

- New Zealand's stressed inshore fisheries which were, in some cases, near commercial collapse in the mid 1980s, are now showing healthy signs of recovery.
- New Zealand's major rock lobster fishery brought into the ITQ scheme in 1990 is now showing healthy growth.
- The New Zealand commercial fishing industry has prospered under the QMS with quota as assets on company balance sheets, allowing companies to expand. New Zealand's major export fish products being deepwater fish, such as hoki and orange roughy, have been major contributors to this expansion.
- The system has brought the development of a much more united fishing industry taking responsibility for its share of the management of New Zealand's fisheries.
- A cooperative approach between Government and industry has been engendered by all participating in the benefits gained by implementation.
- The system has allowed New Zealand's indigenous Maori people to become involved in the business and activity of fishing both at the local traditional level and as part of the larger commercial industry.
- New Zealand has refined its Fisheries Legislation over this period but has maintained as a central core the QMS and is now in the process of bringing into the scheme the remainder of the commercial species caught in New Zealand waters over and above the 31 species that are currently covered.
- The QMS has allowed Government to cease intervening directly in management of individual fishing operations. The Government role is to monitor and ensure that the rules are applied fairly and equitably and enforcement is applied where necessary.
- Benefits and responsibility are placed with the fishing industry.

CONCLUSION

In conclusion the New Zealand experience would confirm that the introduction of a QMS has been of benefit to the fishery, the fishing community, and the economy:

- The scheme has assisted the New Zealand fishing industry in enhancing its status within the New Zealand economy and in the international market place.
- Fish stocks have been protected and enhanced under the scheme.
- There is never a better time to introduce the QMS than when stocks and the fishing industry are at a level of concern about the future.

Professor Ray Hilborn of Washington University, Seattle, USA, returned to New Zealand in March 1995 and concluded that 'on economic performance New Zealand fisheries and fishing industry rated at 'A' in the top 5% in the world', that 'New Zealand was way ahead of most countries in terms of allocating fishing resources through the commercial quota system', and that 'New Zealand was one of the few countries which had turned its fishery into a profitable and sustainable system using quota'.

Results of the system have been favourable for New Zealand in the ten-year period since 1986. The arguments for introducing a Quota Management System using ITQs are compelling, and the difficulties of conventional fishery management will continue unless world fisheries move to some form of quota management system.

Critical Tests for Variation Indicate mtDNA Characters are Powerful for Mixed Stock Analysis

B. L. Brown,[A] *J. M. Epifanio,*[B] *C. J. Kobak*[C] *and P. E. Smouse*[D]

[A] Ecological Genetics Laboratory, Virginia Commonwealth University, Richmond, VA, 23284.USA.
[B] Department of Fisheries and Wildlife, Michigan State University, East Lansing, MI 48224, USA.
[C] Xybion Medical Systems Corporation, 240 Cedar Knolls Road, Cedar Knolls, NJ 07927, USA.
[D] Center for Theoretical and Applied Genetics, Rutgers University, New Brunswick, NJ 08903, USA.

Summary

Because American shad, *Alosa sapidissima,* are now harvested primarily from the open ocean, rather than from rivers of origin, there is a growing potential for disputes over stock ownership. To assess the origin of shad from coastal mixtures, genetic mixed stock analysis (MSA) was applied using mitochondrial (mt) DNA variation as a discriminating character array. Sampling efficiency and geographical and temporal stability of mtDNA data were examined before estimating the contribution of broadly distributed river populations to the mixed ocean fisheries.

From the coastal mixed fishery, 1888 individuals from 19 rivers (17 stocks) and 520 individuals from the coastal mixed stock assemblage were assayed. Restriction fragment analysis of mtDNA furnished sufficient information to delineate stocks geographically. Because temporal variance between yearly stock samples was not significant, samples across years for a river were pooled. Two quadratic programming approaches were used to estimate the contributing stock composition of ocean mixtures in two years which proved equally effective. MSA results indicated that composition of the coastal harvests is dynamic and variable from year to year suggesting that the population assemblages migrating through the coastal harvests are more dynamic than previously thought.

Introduction

Alosa sapidissima, the American shad, is an anadromous species in the herring family (Clupeidae) whose natural range extends along the eastern coast of North America from southern Florida to South Aulatsivik in northern Labrador (Scott and Scott 1988). Its range was artificially expanded to the western coast of North America from California to Alaska following introduction into the Columbia and Sacramento Rivers in the 1880s (Scott and Scott 1988). Shad migrate extensively along the North American east coast (from south to north in the spring and from north to south in the fall) and are generally believed to be highly philopatric (Melvin *et al.* 1986, 1992) resulting in accumulation of genetic differences among river populations over time. Consequently, river stocks are characterized by different assemblages of genotypes. Yet few genetic data have been collected which could be used to evaluate the fidelity of shad homing to their natal rivers, aside from evidence by Nolan *et al.* (1991) supporting accurate region-specific homing of female shad. As they illustrated, mitochondrial DNA analysis can be informative when other genetic techniques, such as allozyme analysis or meristics, do not show adequate differentiation to successfully delineate fish populations.

Currently, shad are extensively harvested along the coasts of Virginia, Maryland, Delaware, North Carolina and New Jersey during their annual pre-spawning migration to natal tributaries. However, shad populations in many east coast drainages are now mere vestiges of their former abundances. Therefore, to responsibly and effectively manage stocks of shad, knowledge of the composition of the coastal harvest is necessary. Pursuant to the Chesapeake Bay Agreement, governors of the three Bay states adopted and signed in 1989 the Chesapeake Bay Alosid Management Plan which assigned high priority to research into developing stock identification procedures which would enable specific river stock identification in mixed-stock alosid intercept fisheries (CEC 1989; ASMFC 1989, 1990, 1991). Such techniques allow estimation of the impact of Atlantic coast fisheries on regional restoration efforts. The current study was designed to exhaustively explore the potential effect of the coastal intercept fisheries using genetic mixed stock analyses. We used genetic stock identification procedures that permitted identification of specific river stocks in mixed-stock alosid intercept fisheries and estimation of the portions of stocks represented in coastal harvests of two specific states, Virginia and Maryland. This was accomplished by comparing the genetic stock composition of shad harvested in Atlantic Ocean waters off Virginia and Maryland with genotypes of shad from major populations in eastern North American tributaries. Estimates of the proportion of shad harvested that originated from each tributary examined were then used to evaluate the impact of the coastal shad fisheries on restoration efforts and to provide suggestions for regulatory measures.

METHODS

Field sampling

In 1992 and 1993, field personnel from cooperating state, provincial, and federal agencies collected approximately 3000 individual shad from 18 east coast drainages and one west coast drainage which traditionally support large runs (Fig. 1). Sample size from each river in each year was approximately 100 fish. Rivers were selected to provide the broadest possible geographic coverage both within the mid-Atlantic region and along the North American east coast, within the confines of the project budget and duration. Sampling was not attempted in other east coast tributaries which did not support abundant American shad populations. Every effort was made to sample only spawning aggregations or individuals committed to upriver migration to maximize the chance that collections represented discrete breeding populations. In several instances, notably the Pamunkey and James Rivers of Virginia, shad were harvested over a protracted period to ensure complete representation of the entire spawning run and to avoid 'snapshot' sampling. American shad were also sampled from commercial fisheries at coastal Virginia landing locations (Rudee Inlet, Wachapreague, and Chincoteague) and from Ocean City, Maryland during 1992 and 1993 (Fig. 1).

Columbia River, Washington was included because these shad were used to rehabilitate the Susquehanna River run. Columbia River shad were originally introduced from Sacramento River, California in the 1890s and the Sacramento River population

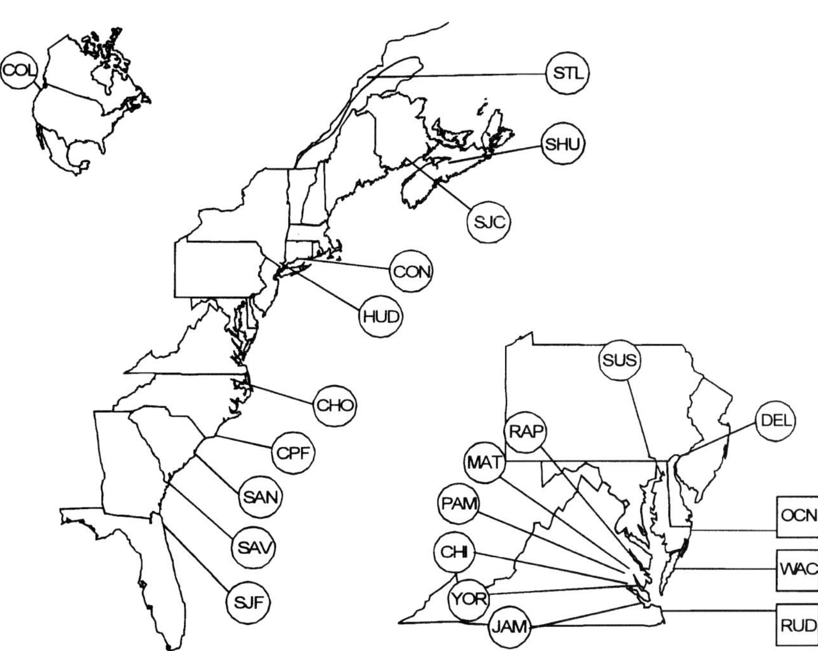

Fig. 1. Sampling localities in North America for American shad (*Alosa sapidissima*) collected in 1992 and 1993. Sites where baseline stocks were sampled are circled. Commercial landing locations along the mid-Atlantic coast of North America where shad samples were collected are enclosed by a square. Abbreviations are: COL: Columbia River, Washington; STL: Saint Lawrence River, Quebec; SHU: Shubenacadie River, Nova Scotia; SJC: Saint John River, New Brunswick; CON: Connecticut River, Connecticut; HUD: Hudson River, New York; DEL: Delaware River, New Jersey; SUS: Susquehanna River, Maryland; RAP: Rappahannock River, Virginia; MAT: Mattaponi River, Virginia; PAM: Pamunkey River, Virginia; CHI: Chickahominy River, Virginia; YOR: York River, Virginia; JAM: James River, Virginia; CHO: Chowan River, North Carolina; CPF: Cape Fear River, North Carolina; SAN: Santee River, South Carolina; SAV: Savannah River, Georgia; SJF: Saint John's River, Florida; OCN: Ocean City, Maryland; WAC: Wachapreague and Chincoteague, Virginia; and RUD: Rudee Inlet, Virginia.

was itself established by an introduction of Hudson River shad in 1870 (Scott and Scott 1988). Today, American shad occur from southern California to the Bering Sea and Kamchatka. Approximately 57 million hatchery-reared Columbia River fry have been released into Susquehanna River over the past decade as part of the shad restoration programme (Richard St. Pierre, US Fish and Wildlife Service, Harrisburg, Pennsylvania, pers. comm.).

Laboratory protocols

Genetic characterization of shad stocks was conducted as described by Epifanio *et al.* (1995) and Brown *et al.* (1996) isolating the molecule known as mitochondrial DNA (mtDNA) and then characterizing it with each of 14 restriction endonucleases (see Table 1). The digested mtDNA was electrophoresed in 0.8–1.0% agarose horizontal gels overnight, stained with ethidium bromide, and fragment patterns were visualized under UV light. In most cases, DNA was then transferred from each gel onto a nylon filter (Southern 1975), probed with caesium chloride-purified shad mtDNA, and the fragments visualized by a non-radioactive colorimetric protocol (BluGene Detection Kit, GIBCO BRL, Gaithersburg, MD 20898, USA).

Data analysis

Restriction fragment patterns were given alphabetic codes and the fragment pattern code from all digests for an individual were concatenated to form a fish's haplotype. Shad are heteroplasmic for mtDNA size and/or site variation, i.e. individual shad sometimes have mtDNA molecules of different size and/or nucleotide sequence (Bentzen *et al.* 1988). This phenomenon has been reported for invertebrates (Bermingham 1990; Brown and Paynter 1991; Zouros *et al.* 1992) and several fishes, including shad (Bentzen *et al.* 1988; Epifanio *et al.* 1995). Because regular inheritance for size variation has not been adequately demonstrated in shad or other taxa, length heteroplasmy was suppressed for the present analyses. Restriction site heteroplasmy, however, was scored because stable transmission has been demonstrated in other taxa (Wilkinson and Chapman 1991; Yang and Griffiths 1993; Skibinski *et al.* 1994). Site heteroplasmy required careful data coding, because mtDNA is presumed to be non-recombinant and haploid in all individuals. This mixing of two haplotypes in an individual can create some new haplotypic classes, but not for every occurrence. In each case, the individual sites were evaluated and if the string of sites was identical to either previously observed root pattern, that was the genotype code assigned. In cases where the composite coding was novel and did not correspond to either of the two root genotypes, a new letter was given to designate the heteroplasmic genotype. This method of 'dominant' data entry was selected over a double coding form ('codominant' 2, 1, 0) because MSA results do not depend on the choice of coding in the case of mtDNA data (Epifanio *et al.* 1995). Using the simpler 'dominant' coding ultimately collapsed the number of source population haplotypes from 221 to 127.

Genetic differences among shad samples from east coast rivers occur primarily as variation in the frequencies of shared mtDNA haplotypes, some ubiquitous and some common among only a few populations, punctuated by unique haplotypes limited to particular river systems (Bentzen *et al.* 1988, 1989; Brown and Chapman 1991; Nolan *et al.* 1991). Shad populations have likely diverged recently, since the most recent North American glaciation. Low levels of gene flow due to intermigration and supplementation from hatcheries may also cause many anadromous species not to demonstrate fixed (or 'polarized') mtDNA differences among populations. Therefore, as with other methods of genetic stock identification, it is not usually possible to assign with great confidence *individual* fish to their natal rivers (Brown *et al.* 1996; Epifanio *et al.* 1995; Hilborn and Walters 1992; Smouse *et al.* 1982). In instances such as this, the standard method developed to determine relative contributions of different source stocks to a mixed population is maximum likelihood analysis (reviewed by Pella and Milner 1987; Smouse *et al.* 1990; Xu *et al.* 1994). This method of estimating stock composition has been widely employed and has

Table 1. Correct allocation percentages for placing individual shad into their true river based on different sets of restriction enzymes. The numbers in the 'Enzymes Used' column refer to restriction enzymes 1) *Aat* I; 2) *Apa* I; 3) *Bcl* I; 4) *Bgl* I, 5) *Dra* I; 6) *Eco* RI, 7) *Eco* RV, 8) *Hind* III; 9) *Kpn* II; 10) *Pst* I; 11) *Pvu* II; 12) *Sal* I; 13) *Sma* I; 14) *Sst* II. Codominant and Dominant refer to the treatment of site heteroplasmy as described in the text. For a detailed description of other metrics tested by this procedure, see Epifanio *et al.* (1995)

Enzymes Used	# Haplo.	Codominant	# Haplo.	Dominant
Absence of Genetic Data	0	6.67	0	6.67
(2,12)	16	15.79	10	15.38
(9,12)	17	16.80	11	15.89
(6,9,12)	26	19.64	18	18.72
(4,6,9,12)	38	21.36	28	19.84
(4,6,9,10,12)	46	22.67	33	20.75
(3,4,6,9,10,12)	55	23.38	40	21.26
(1,3,4,6,9,10,12)	64	23.68	48	21.26
(1,3,4,6,7,9,10,12)	84	26.11	60	23.18
(1–12,14)	112	28.44	85	23.18
(1–14)	116	27.23	89	23.89

demonstrated utility for evaluating salmon and other mixed-stock fisheries along the western North American coast (Pella and Milner 1987) heretofore using solely allozyme data sets. The assumptions of conditional maximum likelihood analysis were reviewed by Utter and Ryman (1993) and can be briefly summarized as (1) baseline stocks which potentially contribute to the mixture are genetically distinguishable, i.e. overlapping but distinct genetic arrays, (2) sampling of baseline stocks is sufficiently precise to identify a significant portion of genetic diversity within each, (3) all source populations represented in the mixture are part of the baseline data set, and (4) a sufficiently large random sample is obtained from the mixed-stock fishery. In addition, the procedure requires that separate characters used to identify populations (e.g. allelic isozymes, parasite occurrence, or mtDNA) be mutually independent.

For American shad mtDNA data, validity of these assumptions was addressed in various manners. The first assumption, identifiability of baseline stocks, was explored by performing a *chi*-square analysis developed by Roff and Bentzen (1989) to evaluate the significance of mtDNA genetic difference among populations. This analysis has several advantages over traditional contingency tests including greater sensitivity to geographic and temporal genetic variation as well as maintenance of high significance levels despite small sample sizes.

Analysis of molecular variance, AMOVA (Excoffier *et al.* 1992) was used to gauge the overall divergence among baseline populations and to partition genetic variation into components representing divergence among regional sets of rivers, among rivers within a region, and within single source rivers. The substrate for AMOVA was a matrix whose elements were genetic distance coefficients between all pairs of individuals. Because it was not known, *a priori*, whether genetic variation in American shad was geographically organized, we explored several biologically plausible hypotheses for regional grouping. The first hypothesis separated primarily semelparous populations (drainages south of Chesapeake Bay) from iteroparous populations (Chesapeake Bay and northward).

The second assumption, precision of baseline sampling, was addressed with an *a posteriori* analysis of sampling intensity. This analysis, a modification of the combinatorial approaches of Hebert *et al.* (1988) and Bernatchez *et al.* (1989), provided a relative indication of the impact of sample sizes on the recovery of rarer haplotypes. The relationship between haplotype diversity and sample size was evaluated by selecting random samples of 10, 20, 30, etc. (with replacement) from each population. This process was repeated 1000 times for each population to obtain the average number of haplotypes detected at each level of sampling intensity. This analysis facilitated determination of the intensity of haplotype sampling within populations. The second assumption was also evaluated using the analysis of allocation efficacy described in Epifanio *et al.* (1995) and expanded to include multiple-year collections as described by Brown *et al.* (1996).

The third assumption, that all source populations potentially contributing to the coastal harvest were included in the baseline data set, was addressed as outlined in Xu *et al.* (1994) and Brown *et al.* (1996) using a computer program called SHADRACQ developed specifically for mtDNA mixed stock analysis. A consideration of the magnitude and stability of Q calculated for each source population and for the composition estimates of the mixed harvest gave an indication of whether the available sources were sufficient to model the mixture sample. Q is the 'residual sum of squares,' a measure of the failure of the model to fit the data. According to Xu *et al.* (1994) a large change between the Q calculated for conditional and unconditional indicates that, due to poor fit, the algorithm has 'adjusted' the source haplotype frequencies substantially, in an attempt to create a better fit for the mixture. Because the mixture sample can contain a non-trivial number of 'singletons' (haplotypes recorded in single fish), not seen in any of the archived source populations, we must consider two possibilities: (a) additional stocks are contributing to the mix that have not previously been sampled for our archival set of baselines; or (b) the singletons are in fact from the archived stocks but they represent rare variants that have simply not been recovered in previous samples. Source populations with very small sample sizes are particularly vulnerable to being 'adjusted' by SHADRACQ, because the algorithm allows for the fact that their genetic frequencies were poorly estimated. When Q values exceed 0.05 for individual source populations, either the gene frequencies were poorly estimated or there may have been an unidentified source population contributing to the mixture.

The fourth assumption of maximum likelihood analysis, that an unbiased sample was collected from the admixture population, was addressed by evaluating the statistical probabilities that haplotypes unique to one of the source populations would be overlooked in various size samples taken from the mixtures as outlined by Grewe *et al.* (1993). Conservatively assuming that each of the baseline populations contributed equally, the average percentage of individuals with haplotypes unique to their river of origin was used to calculate the probabilities that some genotypes would be missed in mixture samples of different sizes.

An analysis of the ability to allocate individual shad to their natal streams was also employed to assess both the anticipated effectiveness of mixed-stock analysis with mtDNA data and the minimum number of restriction enzymes necessary to obtain acceptable allocation results. Methods of optimizing stock resolution from restriction site variation were investigated by comparing four distance metrics (measures of genetic distance among individuals) that addressed two considerations; first, how best to treat site heteroplasmy: as a 'codominant' character or as a 'dominant' character; and second, whether to include information on genetic similarity between haplotypes or to simply treat all haplotypes as equally different. Obviously, allocation of individual fish is not the intent of mixed stock analysis; however, the degree to which allocation exceeds random chance (for example, 1/17 = 0.059, or 5.9%, if 17 populations are present in the baseline) was a useful tool to assess the anticipated effectiveness of mixture analyses with the data at hand during each year of the study. The specific statistical methods developed are described in detail by Epifanio *et al.* (1995). This method was also used by Brown *et al.* (1996) to evaluate the potential effects of combining temporal samples from the same stock in two different ways: an 'allocate then pool' strategy and a 'pool then allocate' strategy. Here, individuals were allocated to the genetically closest source population by computing the genetic distance between each individual and each population mean calculated without that individual. Individuals were then placed in populations that were least distant. Finally, the fractions of correctly allocated individuals were tallied as a measure of success. A success rate significantly higher than chance (i.e. 6.2% for 'allocate then pool' and 5.9% for 'pool then allocate') indicated strong genetic stock divergence and considerable statistical power and utility for subsequent mixed-stock analysis; a low success rate implied minimal genetic stock divergence and low statistical power for mixed-stock analysis. This analysis was useful for confirming which baseline samples to pool and for determining whether temporal subsamples should be pooled.

The final requirement of maximum likelihood analysis, character independence, was addressed through choice and usage of a numerical algorithm employed to estimate the proportion of ocean-intercept shad destined for each east coast tributary. Since all restriction sites on the mitochondrial genome are linked, restriction fragment patterns are theoretically not independent of one another. Programs were selected that allowed mtDNA haplotypes to be entered as 'states' of a single character rather than as genetic characters inherited under Mendelian laws. Entering haplotypes as states of a single character was equivalent to treating them as 'multiple alleles at a single locus' and satisfied the final requirement of the analysis. One program, GIRLSEM, was particularly useful for analysis of the present data (Pella 1986) because it used a combination of two procedures to rapidly converge on maximum likelihood estimates of composition and provided means for calculating standard deviation for the composition estimates. A second program, SHADRACQ, was also used to estimate the contributing stock composition of the open ocean collections. SHADRACQ was developed by Xu *et al.* (1994) specifically for finding the correct allocation fractions for shad when the fish were categorized using mtDNA frequency data, and addressed the need to assume that baseline frequencies were estimates.

Results and discussion

American shad mtDNA

From approximately 3000 American shad collected and assayed in 1992 and 1993, we recovered complete testable haplotypic data sets for 1888 baseline individuals (n = 988 shad in 1992 and n = 900 shad in 1993) and 520 individuals from the coastal mixed stock assemblage. Our assay suite of 14 restriction endonucleases recognized 100 polymorphic cleavage sites (approximately 3.5% of the entire mtDNA genome). The variation in cleavage sites revealed 127 baseline haplotypes (using dominant coding) among all baseline collection sites over the two years. Results of this study complemented results from previous studies (Bentzen *et al.* 1988, 1989; Nolan *et al.* 1991). All studies found identical common haplotypes and similar levels of within and among population genetic variation.

Selecting the optimal set of restriction enzymes

As a first approximation, including additional genetic information should improve allocation success because population samples differ in the frequencies of most haplotypes. The natural predilection is thus to add as many characters as possible to provide increasing resolution (e.g. Lynch and Crease 1990). Theoretically, if enough characters are added, each population can be characterized as a non-overlapping collection of unique haplotypes. Beyond a certain point, however, it was determined that sampling variation (noise) inherent in characterizing large numbers of characters on small samples of individuals overwhelmed the variation due to stock divergence (signal). Additional resolution was lost because restriction sites on the mtDNA do not segregate independently, i.e. many restriction sites are informationally redundant. Our strategy was to assay mixtures using a subsample of restriction sites selected to optimize variation due to divergence *versus* individual sampling variation. The strategic question became 'How much variation is enough?' To answer that question, the allocation exercise described in Epifanio *et al.* (1995) was repeated with different subsets of the 14 restriction enzymes, allocating each individual to that source population (out of 15 populations sampled in 1992 and included in the optimization analysis) for which the genetic distance was smallest.

While it was not possible to explore all (2^{14} - 1) combinations of the 14 restriction enzyme sets, it was possible to evaluate allocation success when using all single-enzymes, all two-enzyme sets, all three-enzyme sets, and stepwise from that point on (Table 1). A 'natural' cascade of enzymes emerged that was reasonably robust. The same pattern occurred regardless of whether haplotypes were coded as dominant or codominant. The best cascade of enzymes began with *Sal* I, followed by *Sal* I + *Apa* I, and then *Eco* RI + *Kpn* I + *Sal* I. Beyond this point informational redundancies became important. We found that adding the less polymorphic restriction enzymes, rather than additional redundant yet greatly polymorphic restriction enzymes, was most useful. Stepwise exploration suggested the best quartet (21.4% correct allocation), quintet (22.7% correct allocation) and sextet (23.4% correct allocation). Some modest additional gains were available by selecting different metrics which peaked at 28.4% for 13 restriction enzymes but declined to 27.2% for 14 restriction enzymes. It is notable that 28% correct allocation of individual shad is almost twice the rate achievable for west coast salmon using allozyme data (salmon have more donor stocks than shad).

These gains were each a considerable improvement over the success rate achievable by random allocation (i.e. 0.059 for 17 potential contributing populations) but confirmed that mtDNA markers did not permit assignment of an individual fish to its source with a high degree of confidence. Yet, although correct allocation fractions were much less than the ideal 100% expected from populations with non-overlapping distributions of fixed characters, the best allocation fraction (28.4%) was a marked improvement over the random allocation fraction (5.9%). The reader is reminded that this was not the mixed-stock analysis, merely a test of the ability to allocate individual fish to their natal tributary, used as an indication of how many of the 14 restriction enzymes used were actually necessary to obtain the most accurate and precise allocation with the least expenditure of time and funds.

It was this result that may be most important to fishery managers: only small to modest improvements in allocation success (and thus genetic resolution when a mixed-stock analysis is undertaken) were made by using more than 6 restriction enzymes, due to the informational redundancies inherent in a set of rigidly-linked and non-recombining genetic markers. The lesson is that pursuing additional variable mtDNA sites for American shad beyond what has been done by this study is not an economically promising strategy. The likelihood of uncovering a variably structured stretch of mtDNA diagnostic for rivers is low considering the distributions already uncovered. Additional restriction enzymes would merely represent an incremental increase in expense, while adding minimal additional (or even reducing) resolution.

Baseline stocks

Differences among east coast shad populations occurred primarily as variation in the frequencies of the common and shared mtDNA haplotypes. Some haplotypes were ubiquitous, others were common to only a few populations, none were both frequent and unique to a particular river system. Variation in shad mtDNA was characterized by the single most common haplotype (AAAAAAAAAAAAAA), 13 widely distributed haplotypes, and 113 additional haplotypes that occurred in five or fewer populations (Complete set of tabular data may be purchased from: The Depository of Unpublished Data, Document Delivery, CISTI, National Research Council Canada, Ottawa, Canada K1A 0S2). Many of the latter occurred in a single population and were represented by a single individual, henceforth referred to as 'singletons.' Fifty-seven percent of individuals had haplotypes found only in their natal river, although none occurred at a large enough frequency to be useful as a river-specific signature or 'private polymorphism' as described by Neel (1978). The absence of polarized mtDNA differences is presumed to be due to the geologically recent separation of shad populations and possibly due to low levels of straying. Permutational *chi*-squares between each pair of populations indicated significant stock differentiation in all comparisons except James / Chickahominy Rivers and Pamunkey / Mattaponi Rivers. Both of these tributary pairs were subsequently pooled reducing the number of potential stocks from 19 to 17 because both met three criteria: (i) the distributions were not significantly different *via chi*-square analyses, (ii) genetic distances between samples were small (Epifanio *et al.* 1995), and (iii) the sources were geographically proximal (from the same watershed).

AMOVA including all populations did not convey a detailed geographic pattern of inter-stock differences (results not shown). Several biogeographical groupings that might be useful for inter-jurisdictional management purposes failed to produce significant among-region components of variance. However, an AMOVA conducted for each possible pair of populations separately permitted between-population variance components to become a measure of genetic distance which was useful in the determination of the optimum suite of enzymes to employ for MSA.

The adequacy of baseline stock population sampling (*a posteriori* analysis of sampling intensity) indicated that our sample sizes generally exceeded the minimum for establishing precise mean frequencies. The average sample sizes necessary to detect 95% of the haplotypes already observed in shad ranged from n = 30 to n = 90, stabilizing near n = 70 for most populations. As predicted, the number of rare variants increased with increased sample size. Notably, the relative frequencies of the more polymorphic haplotypes within populations stabilized as sample sizes approached 40. In fact, frequencies of the 14-enzyme polymorphic mtDNA haplotypes, and even the cumulative frequency of singletons, stabilized at sample sizes of approximately 50–70.

Coastal admixtures

Chi-square analyses to evaluate the significance of mtDNA genetic difference among coastal harvest samples collected in 1992 and 1993 indicated that coastal harvests were in general significantly different from one another excepting Wachapreague and Ocean City samples collected in 1992 which were not significantly different from one another (Table 2). The similarity between Wachapreague and Ocean City paralleled the findings of an earlier study (Brown and Chapman 1991) where geographically proximate landing sites along the Delmarva Peninsula had similar catch compositions.

Estimates of the necessary mixed-assemblage sample sizes to detect a significant portion, 95%, of source population variation

Table 2. ***Chi*-square comparisons among samples from mid-Atlantic coastal shad harvests collected in 1992 and 1993. Values were calculated by the method of Roff and Bentzen (1989). Entries below the diagonal are the observed *chi*-square values; entries above the diagonal are the probabilities that the observed values were due to chance. R, Rudee Inlet, Virginia; W, Wachapreague (includes Chincoteague and Quinby) Virginia; OC, Ocean City, Maryland**

Location	R 1992 (n = 99)	W 1992 (n = 64)	OC 1992 (n = 87)	R 1993 (n = 151)	W 1993 (n = 119)
R 1992		0.015	0.006	0.002	0.034
W 1992	59.464		0.174	0.005	<0.001
OC 1992	53.429	37.509		0.212	0.374
R 1993	62.157	59.289	31.528		<0.001
W 1993	47.915	49.140	26.904	45.279	

ranged from n = 58–298 depending on the frequencies of rare haplotypes in the reference populations. Xu *et al.* (1994) recommended that n = 100 was a minimum size for precise analysis of fishery mixtures. The 1992 and 1993 mixture sample sizes indicated in Table 2 appeared to have been adequate as evidenced by the small number of coastal-intercept fish with unique haplotypes (only 2–5%) with the exception of the sample collected in 1992 from Wachapreague for which 14% of shad had genotypes not previously recorded. However, the fact that any components of the mixture samples could not be classified by the baseline data set suggests that either the unique haplotypes actually existed in the reference populations but were missed in the samples collected or that coastal-intercept fish with unique haplotypes were derived from populations not included in the present baseline data set.

Sixteen of the seventeen baseline stocks were included in the analysis of coastal harvest composition; Columbia River was excluded as a source because it could not *a priori* have been a contributor. Conditional composition estimates using SHADRACQ and GIRLSEM were similar for all stocks that made substantial contributions to a harvest. When model fit was good, the less abundant stock estimates also coincided. Frequency estimates for the five mixture samples are presented in Table 3. One general observation was that shad from St. John River, New Brunswick, Canada were a frequent and sizeable component of the coastal harvest. Appearance of all other stocks was sporadic and variable with respect to magnitude. In general, Rudee Inlet landings consistently had contributions from Chesapeake Bay and southern stocks while Wachapreague landings consistently indicated contributions from northern and Canadian stocks. However, while there was evidence of regionally-targeted stocks, different stocks were harvested within the regions.

The question of whether all source populations potentially contributing to the coastal harvest were included in the baseline data set was further addressed through considering the magnitude and stability of Q, calculated by the SHADRACQ algorithm for each source population and for the composition estimates of the mixture. We do, in fact, see an indication that the existing source population data were not adequate to model some mixture samples (Table 3). This was most apparent for Rudee Inlet 1992 and Wachapreague 1992 coastal harvests. For the Rudee Inlet 1992 analysis, Q-values for source populations were large, implying that the sampled source populations did not provide an adequate description of the mixture. In the case of Wachapreague 1992, there was a large difference between the conditional and unconditional Q-values indicating that a large adjustment in the haplotype frequencies of one or more source populations was necessary to achieve a close fit for the mixture. One of the prime candidates for a missing stock is the population of the Potomac River. While repeated attempts to sample the spawning run have failed to yield enough shad to characterize the population, information from local fisherpeople indicates a population of substantial size. It could also be that Susquehanna River or some other 'inconstant' population is represented in these mixtures. Standard deviations for conditional contribution estimates, calculated by GIRLSEM, yielded a similar message. When the source populations included in the analysis did not yield a good fit to the mixture, standard deviations were large. This was an additional indication that we were missing, or had poorly characterized, one or more source populations.

Conclusions

This study investigated coastal shad fisheries in 1992 and 1993 along the east coast of North America. It detected representatives of Canadian, northeast, mid-Atlantic and southeastern stocks and documented that severely depleted lower Chesapeake Bay stocks were also harvested, sometimes in significant numbers. Based on the outcome of the present genetic analysis, further management of the coastal intercept fishery is warranted to promote the rebuilding of individual shad stocks. The study implicates coastal fishing pressure as a possible contributor but not the sole cause of the reduced numbers of American shad in east coast rivers.

Although it would be beneficial to fishery management, coastal admixtures were not found to be static from year to year. While some stocks were consistently detected in the coastal harvests, presence of other stocks varied widely by location and year. According to the genetic estimates of stock composition for coastal harvests, three management strategies would be consistent with the goal of promoting recovery of shad to fishable abundance. First, to avoid undue fishing pressure on depressed and recovering shad stocks, coastal harvest of shad should cease. As a less drastic measure, managers could use composition estimates, the poundage harvested in the ocean fishery and the average size of intercept shad to estimate the numbers of shad harvested for each tributary and compare the estimated level of fishing pressure with acceptable F_{max} levels. Coastal fishery quotas could then be introduced that maintain total annual harvest from each tributary below the recommended F_{max} levels. Alternatively, and at the very least, future management efforts should be directed toward annual estimation of the composition of coastal admixture harvests so that managers could accurately evaluate the benefits and disadvantages of prosecuting coastal shad fisheries. While such 'real time fisheries management' may be more acceptable to users of the resource, its cost may outweigh the economic benefit to the state or province of allowing the fishery to persist.

The presence in Virginia and Maryland's coastal intercept fisheries of shad destined for tributaries in other states has coastwide implications and will therefore impact management decisions for other east coast shad stocks as well. The yearly analysis of genetic variation in baseline shad populations from Connecticut to South Carolina provides managers along the east coast with valuable information on stock assessment and the timing of return of various cohorts to their natal tributaries. Perhaps most important, the dynamic nature of coastal harvest composition indicates that it would be difficult or impossible to predict from one year to the next exactly which stocks would be harvested and therefore difficult to implement measures to responsibly manage coastal fisheries for the conservation of any single state or province's spawning stocks in the absence of some form of real-time MSA.

Table 3. Estimated stock composition of American shad harvests in coastal intercept fisheries during 1992 and 1993. Unconditional estimates and Q-values were generated by SHADRACQ. 'Conditional' refers to conditional estimates obtained with either SHADRACQ or GIRLSEM (both programs generated equivalent values). Population abbreviations are as described for Fig. 1

	Ocean City, MD 1992			Rudee Inlet, VA 1992			Wachapreague, VA 1992		
Population	**Cond.**	**Uncond.**	**Q**	**Cond.**	**Uncond.**	**Q**	**Cond.**	**Uncond.**	**Q**
1 SHU							0.1945	0.0000	0.0000
2 SJC	0.4327	0.4153	0.000678	0.1916	0.1916	0.1711			
3 CON									
4 DEL	0.5293	0.5609	0.002003						
5 RAP									
6 YOR				0.1744	0.1740	0.2032	0.1520	0.0000	0.0000
7 PAM							0.2061	0.0000	0.0000
8 JAM				0.0539	0.0542	0.0368			
9 CPF	0.0125	0.0000	0.000000	0.0366	0.0368	0.0040			
10 SAN				0.0639	0.0639	0.0422			
11 SAV							0.0324	0.0000	0.0000
12 SJF							0.1122	0.1452	0.0001
13 STL				0.3213	0.3211	0.3681	0.1211	0.1480	0.0001
14 HUD							0.1794	0.7068	0.0503
15 SUS				0.1583	0.1584	0.1745			
16 CHO	0.0256	0.0238	0.001008				0.0022	0.0000	0.0000
Cond. Q	3.398			6.814			9.119		
Uncond. Q	2.599	D_{st} =	0.1931	4.379	D_{st} =	0.0683	3.978	D_{st} =	0.1240
Q for mix	2.184	M_{st} =	0.2005	3.349	M_{st} =	0.0828	1.942	M_{st} =	0.0408

	Ocean City, MD 1993			Wachapreague, VA 1992		
Population	**Cond.**	**Uncond.**	**Q**	**Cond.**	**Uncond.**	**Q**
1 SHU	0.0851	0.0368	0.0000	0.0096	0.0000	0.0000
2 SJC	0.1694	0.1100	0.0001	0.3362	0.3305	0.0003
3 CON				0.1228	0.0000	0.0000
4 DEL	0.2039	0.2486	0.0010	0.0174	0.0124	0.0000
5 RAP						
6 YOR	0.0467	0.0312	0.0000			
7 PAM	0.2246	0.0838	0.0002			
8 JAM				0.2392	0.5373	0.0055
9 CPF	0.0006	0.0000	0.0000	0.0181	0.0000	0.0000
10 SAN	0.1364	0.2935	0.0070			
11 SAV	0.1262	0.1961	0.0029	0.1554	0.0707	0.0001
12 SJF						
13 STL						
14 HUD						
15 SUS	0.0071	0.0000	0.0000	0.0821	0.0374	0.0000
16 CHO	0.0002	0.0000	0.0000	0.0192	0.0117	0.0002
Cond. Q	5.936			2.308		
Uncond. Q	4.141	D_{st} =	0.0568	1.837	D_{st} =	0.0854
Q for mix	3.124	M_{st} =	0.0490	1.313	M_{st} =	0.0767

ACKNOWLEDGMENTS

This study was a result of the generous and diligent efforts of more than 40 collaborators at US and Canadian agencies who collected, preserved, and shipped, often at their own expense, samples of shad from each of the rivers and coastal fisheries investigated. Peter Grewe and John Graves provided helpful comments and instruction on some laboratory techniques. Many graduate and undergraduate students at Virginia Commonwealth University assisted with laboratory processing of samples and with computer support without monetary compensation. We would also like to thank Arthur J. Butt, David C. Taylor and Summer S. W. Schultz for helpful comments on the manuscript. JME and BLB were funded by the Virginia Marine Resources Commission (VMRC–F-110–R) and the Chesapeake Scientific Investigations Foundation, Inc. (CSIF–VCU–92–010, CSIF-VCU–93–012). PES and CJK were funded by the New Jersey Agricultural Experiment Station (NJAES–32102/USDA).

References

ASMFC (1989). Chesapeake Bay Alosid Management Plan. (Atlantic States Marine Fisheries Commission: Washington, District of Columbia).

ASMFC (1990). Shad and river herring workshop, 1990. (Atlantic States Marine Fisheries Commission: Washington, District of Columbia).

ASMFC (1991). Fishery management plan review for shad and river herring. (Atlantic States Marine Fisheries Commission: Washington, District of Columbia).

Bentzen, P, Leggett, W C, and Brown, G G (1988). Length and restriction site heteroplasmy in the mitochondrial DNA of American shad (*Alosa sapidissima*). *Genetics* **118**, 509–18.

Bentzen, P, Brown, G G, and Leggett, W C (1989). Mitochondrial DNA polymorphism, population structure, and life history variation in American shad (*Alosa sapidissima*). *Canadian Journal of Fisheries and Aquatic Sciences* **46**, 1446–54.

Bermingham, E (1990). Mitochondrial DNA and the analysis of fish population structure. In 'Electrophoretic and isoelectric focusing techniques in fishery management'. (Ed D H Whitmore) pp. 197–221 (CRC Press: Boca Raton).

Bernatchez, L, Dodson, J J, and Boivin, S (1989). Population bottlenecks: influence on mitochondrial DNA diversity and its effect in coregonine stock discrimination. *Journal of Fish Biology* **35A**, 233–44.

Brown, B L, and Chapman, R (1991). Genetic analysis of American shad entering Chesapeake Bay. Report to Chesapeake Bay Commission (Chesapeake Bay Commission:Annapolis).

Brown, B L, and Paynter, K T (1991). Mitochondrial DNA analysis of natural variation and selectively inbred Chesapeake Bay oysters, *Crassostrea virginica*. *Marine Biology* **110**, 343–52.

Brown, B L, Epifanio, J M, Smouse, P E, and Kobak, C J (1996). Temporal stability of mtDNA haplotype frequencies in American shad stocks: to pool or not to pool across years? *Canadian Journal of Fisheries and Aquatic Sciences* **53**, 2274–83.

CEC (1989). Chesapeake Bay Alosid Management Plan: an agreement commitment report to the Chesapeake Executive Council. (Chesapeake Executive Council: Annapolis).

Epifanio, J M, Smouse, P E, Kobak, C J, and Brown, B L (1995). Measuring mitochondrial DNA divergence among populations of American shad (*Alosa sapidissima*). I. How much variation is enough? *Canadian Journal of Fisheries and Aquatic Sciences* **52**, 1688–702.

Excoffier, L, Smouse, P E, and Quattro, J M (1992). Analysis of molecular variance inferred from metric distances among DNA haplotypes: Application to human mitochondrial DNA restriction data. *Genetics* **131**, 479–91.

Grewe, P M, Krueger, C C, Aquadro, C F, Bermingham, E, Kincaid, H L, and May, B P (1993). Mitochondrial DNA variation among lake trout strains stocked into Lake Ontario. *Canadian Journal of Fisheries and Aquatic Sciences* **50**, 2397–403.

Hebert, P D N, Ward, R D, and Weider, L J (1988). Clonal diversity patterns and breeding-system variation in *Daphnia pulex*, an asexual-sexual complex. *Evolution* **42**, 147–59.

Hilborn, R, and Walters, C J (1992). Quantitative fisheries stock assessment: Choice, dynamics and uncertainty. (Routledge, Chapman and Hall, Inc: New York).

Lynch, M, and Crease, T J (1990). The analysis of population survey data on DNA sequence variation. *Molecular Biology and Evolution* **7**, 377–94.

Melvin, G D, Dadswell, M J, and Martin, J D (1986). Fidelity of American shad, *Alosa sapidissima* (Clupeidae), to its river of previous spawning. *Canadian Journal of Fisheries and Aquatic Sciences* **43**, 640–6.

Melvin, G D, Dadswell, M J, and McKenzie, J A (1992). Usefulness of meristic and morphometric characters in discriminating populations of American shad *Alosa sapidissima* Osteichthyes Clupeidae inhabiting a marine environment. *Canadian Journal of Fisheries and Aquatic Sciences* **49**, 266–80.

Neel, J V (1978). Rare variants, private polymorphisms, and locus heterozygosity in Amerindian populations. *American Journal of Human Genetics* **30**, 465–90.

Nolan, K, Wirgin, I I, and Grossfield, J (1991). Discrimination among Atlantic coast populations of American shad (*Alosa sapidissima*) using mitochondrial DNA. *Canadian Journal of Fisheries and Aquatic Sciences* **48**, 1724–34.

Pella, J J (1986). The method of fitting expectations applied to computation of conditional maximum likelihood estimates of stock composition from genetic marks Report. No. 171 (US DOC/NOAA/NMFS/NWAFC Auke Bay Laboratory: Auke Bay).

Pella, J J, and Milner, G B (1987). Use of genetic marks in stock composition analysis. In 'Population genetics and fishery management'. (Eds N Ryman and F Utter) pp. 247–76 (University of Washington Press: Seattle).

Roff, D A, and Bentzen, P (1989). The statistical analysis of mitochondrial DNA polymorphisms: *Chi*-square and the problem of small samples. *Molecular Biology and Evolution* **6**, 539–45.

Scott, M G, and Scott, W B (1988). Atlantic fishes of Canada. *Canadian Journal of Fisheries and Aquatic Sciences* **219**, 1–731.

Skibinski, D O, Gallagher, C, and Beynon, C M (1994). Sex-limited mitochondrial DNA transmission in the marine mussel *Mytilus edulis*. *Genetics* **138**, 801–9.

Smouse, P E, Spielman, R, and Park, M (1982). Multiple-locus allocation of individuals to groups as a function of the genetic variation within and differences among human populations. *American Journal of Human Genetics* **119**, 445–63.

Smouse, P E, Waples, R S, and Tworek, J A (1990). A genetic mixture analysis for use with incomplete source population data. *Canadian Journal of Fisheries and Aquatic Sciences* **47**, 620–34.

Southern, E M (1975). Detection of specific sequences among DNA fragments separated by gel electrophoresis. *Journal of Molecular Biology* **98**, 503–17.

Utter, F, and Ryman, N (1993). Genetic markers and mixed stock fisheries. *Fisheries* **18**, 11–21.

Wilkinson, G S, and Chapman, A M (1991). Length and sequence variation in evening bat D-loop mitochondrial DNA. *Genetics* **128**, 607–18.

Xu, S, Kobak, C J, and Smouse, P E (1994). Constrained least squares estimation of mixed population stock composition from mtDNA haplotype frequency data. *Canadian Journal of Fisheries and Aquatic Sciences* **51**, 417–25.

Yang, X, and Griffiths, A J (1993). Male transmission of linear plasmids and mitochondrial DNA in the fungus *Neurospora*. *Genetics* **134**, 1055–62.

Zouros, E, Freeman, K R, Ball, A O, and Pogson, G H (1992). Direct evidence for extensive paternal mitochondrial DNA inheritance in the marine mussel *Mytilus*. *Nature* **359**, 412–14.

Computer Simulation of Fisheries Closures

R. A. Watson[A] *and C. T. Turnbull*[B]

[A] Western Australian Marine Research Laboratories, Fisheries Department of Western Australia, PO Box 20, North Beach, Western Australia 6020.
[B] Northern Fisheries Centre, Queensland Dept of Primary Industries, PO Box 5396, Cairns, Qld 4870, Australia.

Summary

Simulation modelling was used to examine the benefits of seasonal and spatial closures of two prawn fisheries which differed in their recruitment patterns; one had a single annual recruitment pulse (annual) while the other had two each year (biannual). An optimization procedure was used to assign monthly fishing effort (within realistic constraints) which would maximize annual catch value. Reductions in values resulting from uncertainty in recruitment timing were examined.

An 'ideal' pattern of monthly fishing effort for each fishery was fitted using an exhaustive search method because other methods could not find the global optimum. Catch values resulting were used as a basis of comparison within each fishery with closure results. For the annual fishery, the best seasonal closure produced 98% of the value of the 'ideal', spatial closures 102%, and combined seasonal and spatial closures 104%. Relative values for the biannual fishery were similar except for combined closures which produced 116%. Generally, however, spatial closures outperformed combined and seasonal closures when recruitment timing was uncertain. Egg production was generally 30–40% of an unfished stock and was highest for combined closures.

Introduction

Modification of fishing effort levels and patterns through seasonal and spatial closures is common in the management of many fisheries. Closures are used to protect breeding stocks (Morgan 1984; Penn *et al.* 1997), and in annual fisheries (those with a single recruitment pulse) they are widely used to maximize the value of landings by delaying the harvest of individuals until they are of a preferred size (Morgan 1984; Somers 1985).

Before closures are introduced as part of the management of a fishery it is usually necessary to project the likely impact on fish stocks, the economic performance of the fishery and even the effect on individual fishers. Accurate predictions are difficult because of the complexity of these systems, the annual variation due to environmental effects, and the necessity to predict the way in which the pattern of fishing effort will be altered. This requirement to predict the response of fishers to the proposed changes is a major determinant in the success of any management strategy, and yet may be the least predictable part of the system.

Simulation modelling offers an approach to exploring the possible consequences of fisheries closures and has been used to investigate a range of fisheries closures (Nichols 1982; Gribble and Dredge 1994; Watson and Restrepo 1994). Penaeid fisheries are particularly amenable to this approach as most species are short-lived, recruitment patterns are generally simple, and the commercial value of individuals often increases rapidly with size. Moreover, many of these fisheries are annual or nearly so in nature so that closures can be developed to prevent both recruitment and growth overfishing.

Computer models of tropical penaeid fisheries have been used to investigate seasonal (Watson *et al.* 1993*b*) and spatial closures (Die and Watson 1992*a*). They have also been used to assess the importance of critical habitat to the value of penaeid fisheries (Watson *et al.* 1993*a*). Critical to these models is the basis for determining and representing recruitment patterns (Carothers and Grant 1987; Watson *et al.* 1996), rates of migration and growth (Cohen and Fishman 1980; Watson and Turnbull 1993), and the response of fishers to closures (Allen and McGlade 1986; Watson *et al.* 1993*b*).

To consolidate this work we propose to model two tropical fisheries, one with an annual, and the other with a biannual recruitment pattern (two pulses each year), and to examine the relative advantages of seasonal, spatial, and combined seasonal-spatial closures on yields, catch values, and egg production. Annual recruitment is often assumed in models of prawn fisheries; however, in tropical fisheries such as northern Australia, biannual recruitment also occurs. The performance of closures will also be examined under circumstances where the timing of the recruitment pattern or offshore migration is stochastic as in nature.

Model

Model equations

Equations for the processes included in the simulation model (such as growth and natural mortality) have been published elsewhere (Restrepo and Watson 1991; Die and Watson 1992*a*; Watson *et al.* 1993*b*) and space restriction do not allow their duplication here.

Model parameters

Parameters specific to the prawn species and fishery being simulated were drawn from descriptions of the fishery for *Penaeus esculentus*, the brown tiger prawn, in the Torres Strait of northern Queensland, Australia (Watson *et al.* 1993*b*). Size-specific prices represent those for the same species in the northern fisheries of Western Australia (WA Dept of Fisheries unpublished data). There was no information available on the seasonal variation of prices, therefore we used the same price structure for all months of the simulation but acknowledge that variations by season do occur and that production levels also influence prices paid. The maximum monthly (11 000 hrs) and maximum annual (108 000 hrs) fishing efforts used were taken from the prawn fishery of Shark Bay, Western Australia.

Annual recruitment fishery

In this scenario all recruits (1.6 x 10^8) entered the fishery in the month of January at age 0+ months (within their first month) (Fig. 1).

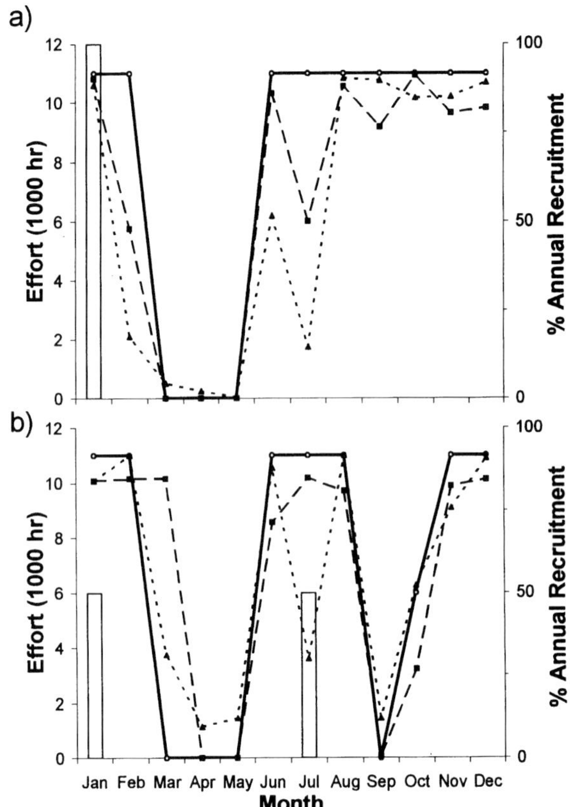

Fig. 1. Recruitment and monthly fishing effort patterns used in the simulation of the (a) annual recruitment and (b) biannual recruitment prawn fisheries. Bars are recruitment numbers, solid lines are 'ideal' (exhaustive search) monthly efforts producing maximum annual catch value, dotted lines are monthly efforts optimized (simplex method) to maximize value without a closure, and dashed lines are monthly efforts optimized (simplex method) to maximize value in the presence of a seasonal and/or spatial closure.

Biannual recruitment fishery

As in the annual recruitment scenario, 1.6 x 10^8 recruits entered the fishery during their first month of age except there were two recruitments of prawns each year, half entering in January and half in July (Fig. 1).

Stochastic recruitment timing

In order to evaluate the effects of the variation in recruitment timing, often observed with wild prawn fisheries, on the performance of closure strategies, the timing of recruitment used was stochastic in nature. For these evaluations 200 trials were completed during which the recruitment patterns described above were shifted by a random factor drawn from a normal distribution (mean=0 months, s.d.=0.23 month). The approximate 95% confidence limits of recruitment timing are two weeks in advance and two weeks after the mean shown in Figure 1.

Fishing effort optimization and ideal value fishery

An exhaustive search was used to determine the 'ideal' monthly pattern of fishing effort to maximize commercial value in the absence of any seasonal or spatial closures (Fig. 1 solid lines).

This will be referred to as the 'ideal value' for each of the annual and biannual recruitment fisheries. Note: as the lifespan of these prawns is about 18 months there is residual stock from the previous year available to the fishery. The annual yields, values, and egg productions resulting from this pattern of fishing effort in the annual and biannual recruitment fisheries were used as the basis for comparison with subsequent investigations of seasonal and spatial closures.

The exhaustive method of investigating the best distribution of monthly fishing effort was not practical for assessing a wide range of possible seasonal and/or spatial closures; therefore we used a modification of the downhill simplex method of Nelder and Mead (Sprott 1991). Parameter estimates of monthly fishing effort were constrained to 11 000 hours (the maximum for the modelled fishery) and similarly to an annual maximum of 108 000 hours. This was achieved by a combination of constraining individual monthly effort parameter searches within the simplex program and through the use of a penalty function when the sum of monthly fishing effort exceeded 108 000 hours in total. The optimum pattern of fishing efforts resulting were similar to, but not identical to, those resulting from exhaustive searches (Fig. 1); therefore the results of closure evaluations are expressed as a percentage of the maximum produced by the ideal fishery described above. During the evaluation of a closure the closed months were assigned no fishing effort and were not included in the optimization procedure.

RESULTS

The 'ideal' value fishery

The ideal monthly pattern for the biannual recruitment fishery had two periods of no fishing compared to the single period for the annual recruitment fishery (Fig. 1). The maximum value from the biannual recruitment fishery using the exhaustive search was 88% of the annual recruitment fishery. As expected the ideal pattern of monthly fishing effort for the biannual recruitment fishery included two periods without fishing; however, these periods were expected to be symmetrical. The first no-fishing period, corresponding to the January recruitment, was, as expected, identical to that for the annual recruitment fishery which also had a January recruitment. The second no-fishing period, corresponding to the July recruitment, was expected to be of a similar duration but to occur six months later. This was not the case (Fig. 1b), and in addition, the ideal fishing effort for October in the biannual recruitment fishery was 6000 hours which was different from that for all other months, which were either at the monthly maximum of 11 000 hours or had no fishing (closed). There are two explanations for this lack of symmetry. Two months after the July recruitment, when individuals are becoming vulnerable to fishing gear (through gear selectivity), the fishery should be closed for three months to prevent the harvest of these individuals until they increase in value; however, valuable individuals from the January recruitment are still numerous and this moderates the duration of the ideal closure. It would also appear that monthly time steps are too long or coarse to allow symmetry in the closure periods. Smaller time steps such as weeks might have allowed the fishery to be closed for the first half of October and open (fished at maximum rate of 11 000 hours month^{-1}) for the second half.

Using different values for the initial guess, the optimization program converged to solutions with noticeably different parameter (monthly effort) estimates and with similar resulting commercial values. This behaviour suggests that this is a global rather than a local optimization problem, and that appropriate global optimization methods such as simulated annealing (Kirkpatrick *et al.* 1983) or genetic algorithms (Gallagher and Sambridge 1994) should be used.

Seasonal closure

Value

A plot of the response surface of value to closure starting month and duration reveals that some closures approached to within 5% of that produced by the ideal value fishery (black shaded areas of Figure 2). The optimum seasonal closure (with no accompanying spatial closure) for the annual recruitment fishery was a three-month closure starting in March (two months after recruitment) (Figure 2a marked with white cross). This closure produced values that were 98% of the ideal value fishery (which had a very similar distribution of fishing effort — Figure 1). Adding uncertainty to the recruitment timing in conjunction with this closure reduced fishery value by a small margin to 97%.

The best closure for the biannual recruitment fishery was a two-month closure starting in April (Figure 2b — marked with white cross). This closure produced a value 98% of the ideal value fishery. Biannual recruitment caused the value response surface

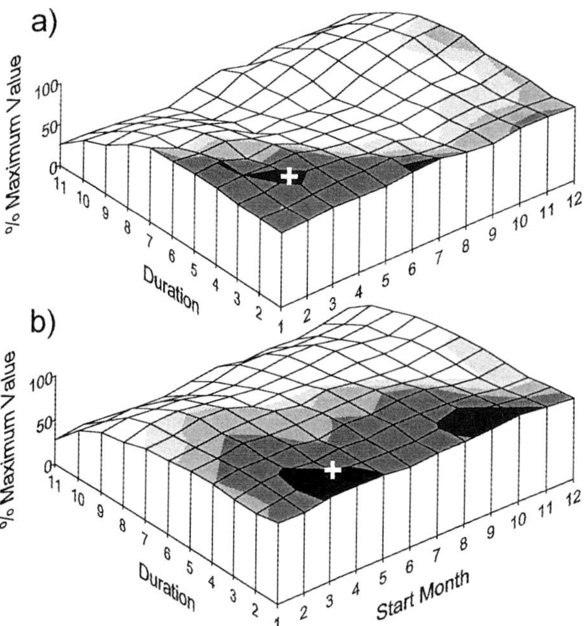

Fig. 2. Surface plot of percent maximum effort for seasonal closures of different durations and starting months for the (a) annual recruitment and (b) biannual recruitment fisheries. Black areas are those combinations resulting in annual catch values within 5% of the maximum from an ideal fishery and a white cross marks the best combination of duration and starting month.

to be smoother than for annual recruitment causing a wider range of seasonal closures to produce similar catch values (Fig. 2b — note the extent of shaded surfaces indicating closures producing values within 20% of the maximum produced by the ideal value fishery). Uncertainty in the timing of recruitment reduced value from this closure to 96%.

Yield

Although effort allocation was optimized to produce the maximum value not yield, the results from seasonal closure combinations are still instructive. For the annual recruitment fishery the maximum annual yield resulted from a two-month closure starting in March. This yield was 97% of that produced by the ideal value fishery and 3% more than the yield produced by the closure producing the highest value.

For the biannual recruitment fishery the maximum yield was produced by a three-month closure starting in March. This yield was 3% higher than that produced by the ideal value fishery and 6% higher than the closure which produced the highest value.

Eggs

As expected, any closure of the fishery generally improved egg production compared to a simulated fishery that operated year round. The three-month closure which produced the maximum value for the annual recruitment fishery increased annual egg production by 20% over that of the ideal value fishery and allowed 33% of the eggs production of an unfished stock.

The two-month closure maximizing value for the biannual recruitment fishery increased annual egg production by 7% over that of the ideal value fishery and allowed 35% of the egg production of an unfished stock.

Spatial closure

Value

Using a spatial closure width of 8 km, the value of the annual recruitment fishery reached 102% of that of the ideal value fishery, outperforming the best seasonal closure (Fig. 3). Changes to value introduced by uncertainty in recruitment timing were small, indicating that this variability has little or no impact on spatial closure performance.

The best spatial closure width for the biannual recruitment fishery was 7 km offshore and produced a value 104% of the ideal value fishery. Uncertainty in recruitment timing reduced value to 96%.

Yield

The spatial closure of the annual recruitment fishery which produced the maximum value (8 km offshore) produced the same yield as that of the best seasonal closure. The best spatial closure of the biannual recruitment fishery (7 km) produced 96% of the yield of the best seasonal closure.

Eggs

Egg production resulting from the best spatial closure of the annual recruitment fishery for value (8 km offshore) was 33% of the unfished egg production and 120% of the egg production

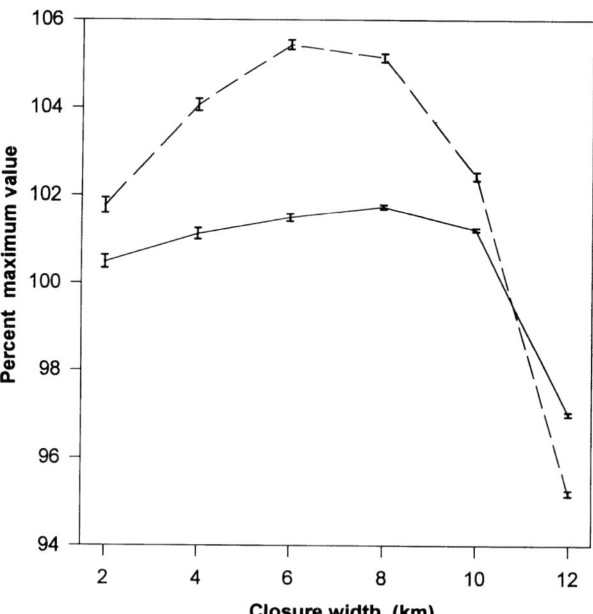

Fig. 3. Percent of maximum annual landed value resulting from simulated spatial closures extending to varying distances offshore for the annual recruitment (solid line) and biannual recruitment fisheries (dashed line) with 95% confidence limits. No concurrent seasonal closures were simulated.

associated with the ideal value fishery. For the biannual recruitment fishery there was 39% of the unfished egg production and 120% of that of the ideal value fishery.

Seasonal-spatial closure

Value

The best combination of seasonal and spatial closures of the annual recruitment fishery was a closure from April to July (inclusive) with a spatial closure width of 7 km offshore which produced 104% of the value of the ideal value fishery. This was the best single closure of this fishery and outperformed either the solely seasonal or spatial closures in the absence of uncertainty in recruitment timing. With the addition of uncertainty in the recruitment timing the value was reduced to 94% of that produced by the ideal no closure fishery which was less than that accomplished by spatial closures alone.

The best combined seasonal and spatial closure for the biannual recruitment fishery was a closure of June and July with a spatial closure width of 9 km which produced 116% of the ideal value fishery. This closure outperformed any single seasonal or spatial closure of this fishery. Uncertainty in recruitment timing reduced this considerably to 85%.

Yield

The yield for the best combined seasonal and spatial closure of the annual recruitment fishery was 86% of that of the ideal value fishery, while for the biannual recruitment fishery this was 96%.

Eggs

The egg production of the best combined seasonal and spatial closure of the annual recruitment fishery was 44% of the unfished level and 162% of the ideal value fishery. The best combined closure of the biannual recruitment fishery was 43% of the unfished level and 132% of the egg production from the ideal value fishery.

Closure strategies

A decision tree can be drawn up outlining the process of choosing a closure management strategy (Fig. 4). Ability to make these decisions would rely on accurate information and on suitable motivation. If fishers work in cooperation to achieve the global goal of maximizing the annual value of the catch (such as our ideal value fishery) there would clearly be no need to consider a closure. As this is seldom the case it becomes important to know whether stock migration or recruitment patterns allow effort to be targeted on individuals of the optimal size to maximize value of the fishery. Compulsory changes in fishing effort patterns might gain the benefits that would have been achieved through cooperation.

Migrating prawns can be managed through spatial closures; however, if prawns do not continue to migrate until they are at least the optimal harvest size then these spatial closures should be augmented by a seasonal closure. If the animals do not migrate then it becomes important to know whether recruitment patterns are simple, that is consisting of animals of a single size and value. If this is the case then seasonal closures can be effective in maximizing value, otherwise it may be difficult improve the value of the fishery through seasonal or spatial closures.

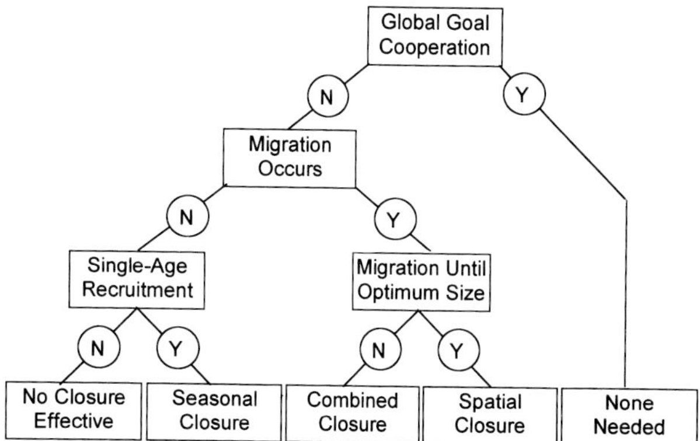

Fig. 4. Decision tree for choosing closure strategies to maximize annual catch value from a prawn fishery.

DISCUSSION

Seasonal closures were equally effective in maximizing harvest value in our two simulated fisheries, the annual and biannual recruitment fisheries, and were similar in their starting month and duration. It is likely that constraints on the maximum fishing effort which the optimization program could assign to any one month (11 000 hours) may have reduced the performance of seasonal closures particularly in the case of the annual recruitment fishery where all individuals reach the optimal size in the same month and can not all be harvested using the fishing effort allowed.

Spatial closures did better and outperformed the ideal value fishery which used the optimal distribution of monthly fishing effort but did not use a spatial closure. For the annual recruitment fishery the benefits of spatial closures were not eroded by uncertainty in recruitment timing like they were for seasonal closures but our investigations indicate that they are vulnerable to the equivalent, that is variation in migration timing and rates. Unlike the annual recruitment fishery, spatial closures of the biannual recruitment fishery were reduced by uncertainty in recruitment timing and this is likely to have been a consequence of the more complex recruitment pattern.

Combined seasonal and spatial closures outperformed any separate seasonal or spatial closure for both the annual and biannual recruitment fisheries. Combined closures produced higher catch values and higher egg production than either seasonal or spatial closures alone. Because combined closures include seasonal closures, their benefits were reduced by uncertainty in recruitment timing, particularly that of the biannual recruitment fishery. Improvements with the use of a combined closure *versus* either a seasonal or spatial closure alone were most evident for the biannual recruitment fishery. It should be noted, however, that the ideal value used as a benchmark for the biannual fishery was 88% of that for the annual recruitment fishery. Nevertheless it demonstrates how an appropriate closure can maximize the value produced by a fishery even if complex recruitment patterns make this difficult and restrict the maximum value obtainable.

Improvements to the process of optimizing the distribution of monthly fishing effort to maximize the landed value produced by our simulated closures must await work on new global optimization algorithms. These may include modifications of the simulated annealing method to better accommodate parameter constraints and achieve a true global optimum. With more efficient algorithms it may be possible to attempt a nested approach whereby the simulation program attempts to optimize fishery value by investigating different closures (combinations of starting month, duration, and distance offshore), and within each closure the monthly distribution of fishing effort is also optimized. Since the parameter values for closure starting month and duration are integer values, this would require a different approach to optimization than we have employed here.

Closures can be used by fisheries managers for many purposes. When the object is to maximize the value of annual landings then closures are used to modify the distribution of fishing effort in time and/or space in order to alter the mortality rates of selected components of the stock. These components typically are lower in their current landed value than the rest of the stock. In some fisheries it is easy to protect individuals until they reach an optimal value but in others it is not. This depends on the degree of spatial separation possible between stock components or cohorts of different values.

Changes to patterns of fishing effort in response to closures are not always predictable but they are the greatest determinant of their success as a management measure. Work by Allen and McGlade (1986); Die and Watson (1992b), and others has shown how differing strategies amongst fisheries can significantly influence the effects of management measures.

Many other factors may also influence closure effectiveness. Spatial closures are weakened by variable migration timing and rates, while seasonal closures are undermined by variable recruitment timing, multi-size or value recruitment, and by the persistence of individuals from one fishing season to the next. In Western Australia many prawn fisheries start the year by harvesting high value individuals which have survived from the previous year. These 'residuals' are soon harvested — often before 'new' recruits from the current year reach the optimal harvest size to maximize the value of the fishery. In these fisheries spatial closures must augment seasonal closures in order to prevent growth and recruitment overfishing.

Acknowledgment

The authors would like to acknowledge the support of the Fisheries Department of Western Australia. We are grateful for the assistance of Neil Sumner, Norm Hall, Gary Morgan and Ian Towers.

References

Allen, P, and McGlade, J M (1986). Dynamics of discovery and exploitation: the case of the Scotian shelf groundfish fisheries. *Canadian Journal of Fisheries and Aquatic Sciences* **43**, 1187–200.

Carothers, P E, and Grant, W E (1987). Fishery management implications of recruitment seasonality: simulation of the Texas fishery for the brown shrimp, *Penaeus aztecus*. *Ecological Modelling* **36**, 239–68.

Cohen, M D, and Fishman, G S (1980). Modeling growth-time and weight and length-time relationships in a single year-class fishery with examples for North Carolina pink and brown shrimp. *Canadian Journal of Fisheries and Aquatic Sciences* **37**, 1000–11.

Die, J D, and Watson, R A(1992*a*). A per-recruit simulation model for evaluating spatial closures in an Australian penaeid fishery. *Aquatic Living Resources* **5**, 145–53.

Die, J D, and Watson, R A(1992*b*). Dissipation of spatial closure benefits as a result of non-compliance. *Mathematics and Computers in Simulation* **33**, 451–6.

Gallagher, K, and Sambridge, S (1994). Genetic algorithms: a powerful tool for large-scale nonlinear optimization problems. *Computers and Geosciences* **20**, 1229–36.

Gribble, N A, and Dredge, M C L (1994). Mixed-species yield-per-recruit simulation of the effect of seasonal closure of a central Queensland coastal prawn trawling ground. *Canadian Journal of Aquatic Sciences* **51**, 998–1011.

Kirkpatrick, S, Gelatt, D D, and Vecchi, M P (1983). Optimization by simulated annealing. *Science* **220**, 671–80.

Morgan, G R (1984). Recruitment and closed seasons for *Penaeus semisulcatus* in Kuwait. In: Final Report, Proceedings Third Shrimp and Fin Fisheries Management Workshop, Fin Fisheries Session, 4–5 Dec, 1982. pp. 173–80 (Ed C P Mathews).

Nichols, S (1982). Impacts on shrimp yields of the 1981 fishery conservation zone closure off Texas. *US National Marine Fisheries Service Marine Fisheries Review* **44**, 31–7.

Penn, J W, Watson, R A, Caputi, N, and Hall, N (1997). Protecting vulnerable stocks in a multi-species prawn fisheries. In 'Developing and Sustaining World Fisheries Resources: The State of Science and Management. Second World Fisheries Congress, Brisbane 1996'. (Eds D A Hancock, D C Smith, A Grant and J P Beumer) pp. 383–90 (CSIRO Publishing: Melbourne).

Restrepo, V R, and Watson, R A (1991). An approach to modeling crustacean egg-bearing fractions as a function of size and season. *Canadian Journal of Aquatic Sciences* **48**, 1431–6.

Somers, I F (1985). Maximising value per recruit in the fishery for banana prawns, *Penaeus merguiensis*, in the Gulf of Carpentaria. In: Second Australian National Prawn Seminar. (Eds P C Rothlisberg, B J Hill and D J Staples) pp. 185–91. Cleveland, Australia.

Sprott, J C (1991). 'Numerical recipes — routines and examples in Basic.' Cambridge University Press, Sydney. 398 pp.

Watson, R A, and Restrepo, V R (1994). Evaluating closed season options with simulation for a tropical shrimp fishery. *ICES Marine Science Symposia* **199**, 391–8.

Watson, R A, and Turnbull, C T (1993). Migration and growth of two tropical penaeid shrimps within Torres Strait, Northern Australia. *Fisheries Research* **17**, 353–68.

Watson, R A, Coles, R G, and Lee Long, W J (1993a). Simulation estimates of annual yield and landed value for commercial penaeid prawns from a tropical seagrass habitat, northern Queensland, Australia. *Australian Journal of Marine and Freshwater Research* **44**, 211–9.

Watson R A, Die, D J, and Restrepo, V R (1993*b*). Closed seasons and tropical penaeid fisheries: a simulation including fleet dynamics and uncertainty. *North American Journal of Fisheries Management* **13**, 326–36.

Watson, R A, Turnbull, C T, and Derbyshire, K J (1996). Identifying tropical penaeid recruitment patterns. *Australian Journal of Marine and Freshwater Research* **47**, 77–85.

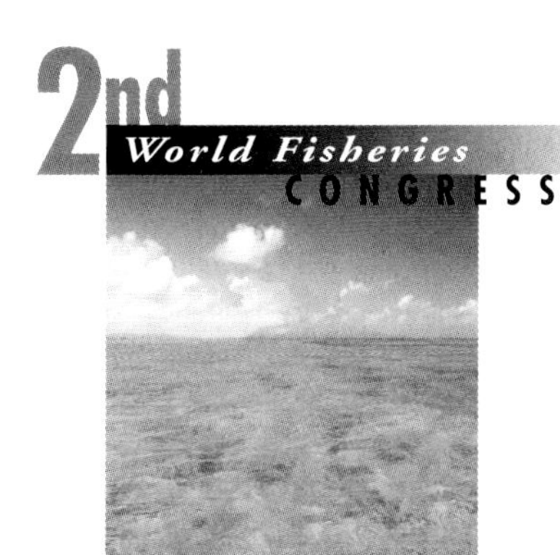

Controlling Illegal Fishing in Closed Areas: The Case of Mackerel off Norway

P. J. B. Hart

Department of Zoology, University of Leicester, University Road, Leicester LE1 7RH, UK.

Summary

Closed areas are used in a number of fisheries to conserve stocks and limit effort. A particular case is the closure in the early autumn of the Norwegian coastal area to European Union (EU) purse seiners fishing for mackerel. EU vessels caught fishing in the Norwegian zone are penalized by large fines. A stochastic dynamic model of a fishing trip examines the combination of detection probability by the Coast Guard and fine size, that excludes EU boats from the Norwegian zone. The influence of variation in the probability of a catch and catch size is examined. As the Norwegian zone becomes more attractive, the probability of detection and fine size have to be larger to force total exclusion. Also, it is optimal for vessels to become risk-prone as the fishing trip comes to an end and the ship's hold is still less than half full.

Introduction

A commonly-used fisheries management measure to protect fish stocks is to close sea areas. Often the closure applies to a subset of the fleet fishing in the areas leaving some vessels fishing around the edge of the closed zone. Exclusion from the zone creates the temptation to cheat the system and to enter for a quick catch. This temptation is particularly strong if the catch rates inside the zone are assessed as much higher than outside. An example of such a system is the closure of Norwegian coastal waters to EU vessels in the late autumn mackerel (*Scomber scombrus*) fishery (Hempel 1995), and another is the so-called 'Irish Box' to the west of Ireland from which, until recently, Spanish vessels were excluded. In this paper the problem is analysed in terms of the Norwegian mackerel fishery.

Creating a closed area must be accompanied by a method of deterring would–be intruders. The Norwegian Coast Guard has adopted a policy of inflicting very high fines, in the order of 850 000–5 000 000 NOK or £85 000–500 000, on EU vessels caught fishing for mackerel in the Norwegian zone. This policy will work if the rate of detection is sufficient to make fishermen believe that there is a good chance of being caught. A very low detection rate will make the fine a remote possibility so nullifying its effect. For the authorities, maintaining a given detection rate is costly as it requires ships at sea or aeroplanes overflying the area. This paper shows that the successful application of the Norwegian policy must be designed around the cost-benefit equation used by fishermen when making decisions. This can only be done by examining the strategy fishermen might take in the face of a particular management measure.

This paper presents a model of a vessel's fishing trip in the region of a boundary between an open and a closed area. It can be used to determine the optimal combination of detection probability and fine size to keep fishermen out of the closed area. The model has also been used to show how the deterrent effect of fines can be modified by properties of the resource such as catch probability and size. A similar problem was tackled by Sutinen and Andersen (1985).

The model

The fishing trip is divided into twenty days and it is assumed that the fisherman decides once a day whether to fish in the open or the closed area. Each area has its own probability of making a successful catch and catch size. Should the fisherman decide to fish in the closed zone, he will have to cope with a particular probability of being detected by the Coast Guard and, if detected, a fine of a given amount of money. These last two factors, detection probability and fine size, are the variables to be manipulated to force exclusion.

It is assumed that the fisherman is aiming to end the trip with a full hold which will ensure maximum income from the trip. It is assumed further, that at any point in the trip the fullness of the hold will influence the decision the skipper makes about where to fish. Hold fullness is the state variable of the model and decisions made are contingent upon it. The full list of variables and their symbols is given in Table 1.

The decisions possible at the beginning of each day are shown in Figure 1. The system is modelled using stochastic dynamic programming (Mangel and Clark 1988). This makes it possible to calculate the optimal sequence of zone choice which will maximize the hold fullness over the 20 day trip. At the start of each day the computer program implementing the model evaluates the expected income at time t, given that the state is x and the time horizon is T, using the following equation.

$$\begin{aligned}\text{Income}(x, t, T) = \max\{&\chi(1)\text{Inc}(x_1, t{+}1, T) + (1-\chi(1)\text{Inc}(x_2, t{+}1, T);\\ &\delta(2)[\chi(2)\text{Inc}(x_3, t{+}1, T) - \phi(2)] + (1-\delta(2))\\ &(\chi(2)\text{Inc}(x_3, t{+}1, T)] + \delta(2)[1-\chi(2)\\ &\text{Inc}(x_4, t{+}1, T) - \phi(2)] + (1-\delta(2))[(1-\chi(2))\\ &\text{Inc}(x_4, t{+}1, T)]\}\,. \qquad (1)\end{aligned}$$

The values of the state variable, x_1 to x_4 are given by the following difference equations.

$$x_1 = x + C(1) \qquad (2)$$

$$x_2 = x_4 = x \qquad (3)$$

$$x_3 = x + C(2) \qquad (4)$$

Final income was calculated from

$$\text{Inc}(x, T, T) = px\,100 - \alpha T \qquad (5)$$

Table 1. A list of the variables used in the model, the symbols representing them, and the units and values used in the simulations

Variable	Symbol	Units	Values
Trip length	T	days	20
Unit of decision	t	days	–
Fishing zone	z	–	European (1) Norwegian (2)
Price of fish	p	£UK	£100 t^{-1}
Catch size per set	$C(z)$	tonnes	Variable
Hold fullness	x	tonnes	Variable
Max hold fullness	X	tonnes	1000 t
Ship running costs	α	£UK	£1000 day^{-1}
Fine imposed detection^{-1}	ϕ	£UK	£0–80000
Probability of detection by coast guard	δ	–	$0 \le \delta \le 1$
Probability of making a catch in zone z	$\chi(z)$	–	$0 \le \chi(z) \le 1$
Income from trip given x at time t	$Inc(x,t,T)$	GBP	Variable

Multiplying by one hundred is necessary to convert x into tonnes, as in the program one unit of x is equal to 100 tonnes.

The program output for each time unit is the expected income at T for each value of the state variable and the optimal zone choice to achieve maximum expected income. An example of the optimal choices is shown in Figure 2A.

Values for the variables such as hold capacity and catch size, are loosely based on data gathered from the horse mackerel (*Trachurus murphyi*) fishery off Chile (Hancock *et al.* 1995). Ship running costs and prices are arbitrary. This is not important as the purpose of the present study is to show the general behaviour of the fishing boat in relation to fine size and detection rate. The model is applicable in its present form to a pelagic fishery.

Analysis of the model

The optimal policy is analysed at two levels. The first looks at how the optimal policy changes within a fishing trip as hold

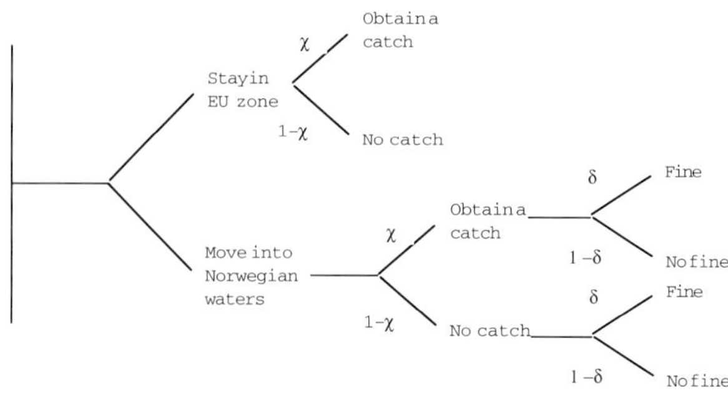

Fig. 1. The possible decisions that could occur in each time unit making up a fishing trip. Probabilities of events would not change from one time unit to another. See Table 1 for the meaning of symbols.

fullness and time change. At the second level, successful exclusion from the closed zone is determined as a function of the probability of detection and fine size. Complete exclusion is also changed by the probability of making a catch and catch size in the two zones. The effects of these are analysed.

The optimal policy within a fishing trip

Optimal policy within a fishing trip, as a function of hold fullness and time, is shown in Figure 2A for an environment where the probabilities of making a catch and catch size in the European zone are both less than in the Norwegian zone. In this situation the optimal policy is to become more risk-prone when hold fullness is low and the trip is near its end. In this situation, it is worth taking the risk of obtaining a good-sized catch from the Norwegian zone despite the possibility of being fined. The alternative is to stay in the EU zone, take a smaller catch with less probability and end up with fewer fish in the hold by the end of the trip.

The pattern of Figure 2A is dependent on the fact that $\chi(1)<\chi(2)$ and $C(1)<C(2)$. That this is the case is demonstrated by Figure 2B which results from a run of the model when $\chi(1)<\chi(2)$ but $C(1)>C(2)$. In this case the probability of finding a school of fish is less in the EU zone, but when found it yields a bigger catch. The model predicts that skippers should stay in the EU zone except for the middle part of the trip, the exact timing of which depends on hold fullness.

The response to changing detection probability and fine size

The model was run with fixed values of $\chi(1) = 0.2$ and $\chi(2) = 0.4$ and five values each of $C(1)$ and $C(2)$ but always with $C(1)<C(2)$. For each run $\delta(2)$ was changed in 0.1 steps from 0 to 1.0 while the fine was changed from £0 to £80 000 in steps of £10 000. Each policy matrix was similar to Figure 2A if a mixed strategy was optimal, and for each the percentage of the 190 possible states that dictated intrusion into the Norwegian zone was calculated and sample output is shown in Figure 3A. As the ratio between $C(1)$ and $C(2)$ increased from 1:2, 1:3, 1:4, to 1:5 the range increased of the values of $\delta(2)$ and $\phi(2)$ that were associated with a mixed policy (Fig. 3B–D). The higher the ratio between $C(1)$ and $C(2)$, the more risk-prone the fishing skipper would be expected to be.

An interesting property of the result in Figure 3 is the symmetry shown by the level of intrusion into the Norwegian zone. The plane of symmetry is the top left/bottom right diagonal and it means that given levels of detection and fine are equivalent in their effects on the optimal policy to be adopted by the fisherman. A fine of £70 000 with a detection probability of 0.1 has the same effect as a detection probability of 0.7 and a fine of £10 000. From the Coast Guard's point of view the equivalence disappears as it would cost much more to produce a detection probability of 0.7 than of 0.1. From their perspective, increasing the fine would be the best policy. This asymmetry could have implications for the stability of the system, as the low detection rate preferred by the Coast Guard could provide an incentive to the fisherman to invest

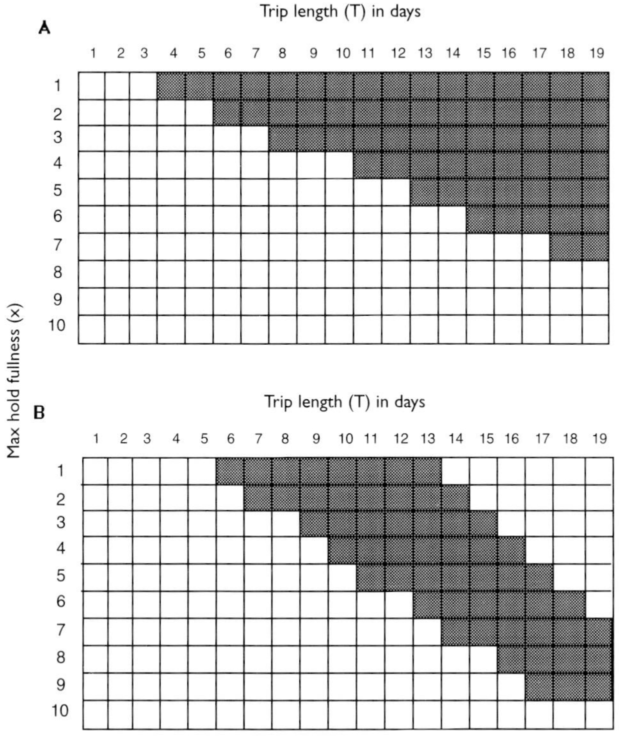

Fig. 2. The optimal policy for fishermen under different catch conditions. A. The policy for a trip when $\delta = 0.2$, $\phi = £20000$, $\chi(1) = 0.2$, $\chi(2) = 0.4$, $C(1) = 1$ and $C(2) = 4$. B. The policy when $\delta = 0.1$, $\phi = £10000$, $\chi(1) = 0.1$, $\chi(2) = 0.6$, $C(1) = 4$ and $C(2) = 1$. White signifies the European zone as optimal and grey the Norwegian.

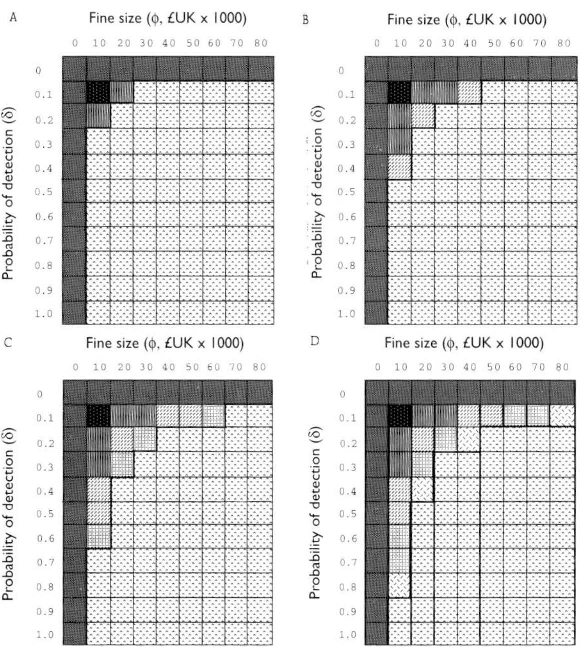

Fig. 3. Conditions under which one of three policies are optimal; to be in the Norwegian zone all the time, to intrude into the Norwegian zone some of the time or never to intrude. The shading corresponds to the percentage of the 190 X–T combinations (Figure 2) for which intrusion into the Norwegian zone is optimal. For each panel, $\chi(1) = 0.2$, $\chi(2) = 0.4$ and $C(1) = 1$. A. $C(2) = 2$. B. $C(2) = 3$. C. $C(2) = 4$. D. $C(2) = 5$.

in devices that reduced personal probability of detection. This would be made easier when δ is low.

It will not always be true that the probability of obtaining a catch in the EU zone will be less than in the Norwegian zone. Setting $C(1) = 1$ and $C(2) = 4$, the model was run for all combinations of $\delta(2)$ and $\phi(2)$ with $\chi(1) = 0.4$ and $\chi((2) = 0.2$. The result was that the combinations of $\delta(2)$ and $\phi(2)$ that dictated a mixed strategy were reduced from 13 to 3. The frequency of the mixed strategy increased to 10 when $\chi(1) = \chi(2) = 0.4$.

A further possibility is that although the catch probability in the EU zone could be less than in the Norwegian zone, the size of catch could be greater. This would simulate a situation where the EU zone contained fewer schools of fish but they were larger, so producing a larger catch. Setting $C(1) = 4$ and $C(2) = 1$, and $\chi(1) = 0.1$ coupled with $\chi(2) = 0.2$, 0.4 or 0.6, the model was run with the full range of $\delta(2)$ and $\phi(2)$. The difference between $\chi(1)$ and $\chi(2)$ had to be greater than 0.3 to have an effect, with the incidence of mixed strategies rising from 0 to 2 at $\chi(1) = 0.1$ and $\chi(2) = 0.6$. Under these conditions, intrusion into the Norwegian zone was only worthwhile when $\chi(1) = 0.1$ and $\chi(2)$ = £10 000. The model predicts that skippers with an empty hold should be risk-prone in the middle of the trip, but the incidence gets later in the trip as the hold fills (Fig. 2B).

Discussion

The principal result from the model is that the fisherman might be expected to respond variably to a regime of fines imposed to exclude them from specific areas. The probability of detection by the policing authority and the fine size will determine whether or not zone intrusion is worth the risk. The response will be further modified by the density and distribution of fish schools in the two zones. Because of this, a fixed policy of fines would not guarantee the exclusion of unwanted vessels. A flexible policy would be more workable and cheaper to implement. There is a precedent for this in Norway where flexible closed areas to protect juvenile fish are already used as a management policy (Anon. 1996). Catches are monitored to determine the proportion of juveniles. When this gets too high, the area is closed. It is then monitored further to determine when the proportion of juveniles has fallen below the desired level, after which the area is opened again.

In a pelagic fishery such as that for mackerel, it would pay the authorities to monitor catch rates of individual vessels, as they are more likely to cheat when they have made poor catches and are near the end of their time at sea. This will be limited by fuel capacity, a factor not yet included in the model. The availability of the resource could also be monitored to determine the relative quality of the open and closed zones. In the case of mackerel, catch rates from EU and Norwegian vessels outside and inside the Norwegian zone, would provide information to the Coast Guard on the likelihood of intrusion. Combined with a model, such as the one described here, monitoring could make it possible to decide on levels of surveillance required to provide a sufficient deterrent. In this way surveillance could be reduced when the EU zone was more productive, so saving on the cost of surveillance boats at sea or aeroplanes in the air. The benefit of this would depend on the cost of monitoring resource availability. This might not be great in that most of the data could be gathered by radio direct from the vessels at sea or through satellite monitoring.

The model described is very simple and would need to be elaborated considerably before it could be used by a policing agency. Variables that need to be included are the cost and time of travelling from one zone to another, the fuel capacity of the vessel, the time it takes to pump a catch on board and how changes in it might alter the probability of being detected. A further important variable would be the time spent searching for a school which can vary considerably (Hancock *et al.* 1995).

The most important message conveyed by this study is that devising management measures is a waste of time unless consideration is given to how fishermen will respond. This can only be determined by studying directly the decision-making process used by fishermen to choose what to do and devising models of the sort described in this paper that are based on the fishermens' actions.

Acknowledgments

I thank Drs Kjell and Anne Gro Salvanes for reading an early draft of the manuscript and making valuable comments.

References

Anon. (1996). Fish stock conservation and management. House of Lords, Session 1995–96, 2nd Report of the Select Committee on Science and Technology, (Her Majesty's Stationary Office: London). HL Paper 25, 1–77.

Hancock, J, Hart, P J B, and Antezana, T (1995). Searching behaviour and catch of horse mackerel (*Trachurus murphyi*) by industrial purse–seiners off south–central Chile. *ICES Journal of Marine Science* **52**, 991–1004.

Hempel, E (1995). Why Norwegian Coast Guard is getting tougher. *Fishing News International* **34**(1), 4–5.

Mangel, M, and Clark, C W (1988). 'Dynamic modeling in behavioural ecology.' (Princeton University Press: Princeton).

Sutinen, J G, and Anderson, P (1985). The economics of fisheries law enforcement. *Land Economics* **61**(4), 387–97.

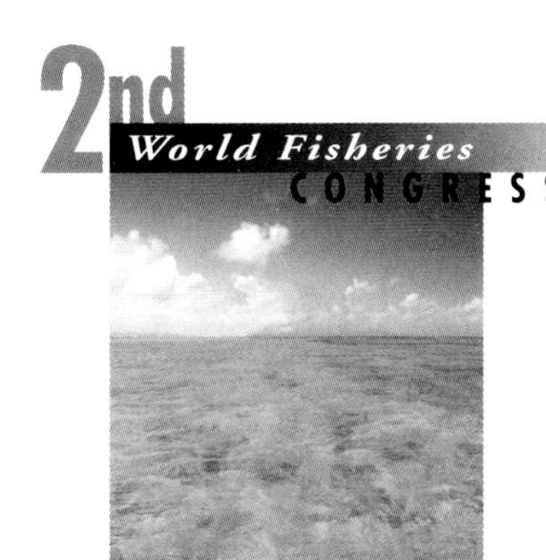

Florida Keys National Marine Sanctuary: Retrospective (1979–1995) Reef Fish Assessment and a Case for Protected Marine Areas

Jerald S. Ault,[A] *James A. Bohnsack*[B] *and Geoffrey A. Meester*[A]

[A] University of Miami, Rosenstiel School of Marine and Atmospheric Science, 4600 Rickenbacker Causeway, Miami, Florida 33149, USA.

[B] National Marine Fisheries Service, Southeast Fisheries Science Center, 75 Virginia Beach Drive, Miami, Florida 33149, USA.

Summary

Designating certain marine areas as 'protected' — usually by officially naming them 'sanctuaries' or 'reserves' — is a revolutionary resource-management tool used in mitigating habitat degradation and overfishing. In reserves, critical habitats are protected in an effort to conserve biodiversity and environmental quality, and to sustain resource usage. Despite widespread interest in this new concept, few quantitative paradigms or management models are available for implementing or developing strategic plans for reserves in terms of their overall design, number, total area, and proximity to physical features. The Florida Keys National Marine Sanctuary (FKNMS) is a national treasure with rich subtropical multi-species reef fisheries, a multi-billion dollar tourist economy, and unique aesthetic qualities. The increase in the protection of this valuable habitat and its resources brought about by its status as a sanctuary underscores the need for an adaptive management strategy that defines the structure and function of reserves within the FKNMS. In support of reef fish management, in our preliminary assessments we employed a systems approach using a nexus of advanced visualization, data assimilation, and quantitative analysis techniques to develop: a spatially-explicit model that links relatively sparse survey estimates of reef fish densities and sizes to key physical factors; and, a multi-species assessment index that uses the metabolic variable 'average size' as a biological indicator of stock status or health. These approaches can be very helpful in making fishery management decisions concerning the FKNMS resources, and can help to define the evolving role of marine protected areas in fishery management.

Introduction

The Florida Keys are a national treasure of the United States that support rich tropical multi-species coral reef fisheries, a multi-billion dollar tourist economy, and unique aesthetic qualities. The coral reef tract supports exceptional species diversity and many valuable fishery resources (e.g. groupers, snappers, lobsters, and shrimps). Tropical reef ecosystems are well-known among the most complex on earth, with extremely complicated biological and physical interactions, making decision making difficult by conventional fishery management methods. The Florida Keys reef fishery is a prominent component of an important 'ecosystem-at-risk', as it is one of the nation's most significant and most stressed marine resources under NOAA management (NOAA 1995). While public uses, demands, and conflicts for these resources have continued to increase sharply over time, the status and biological dynamics of reef fishery resources are not well understood. Harvesting has long been known to deleteriously alter the statistics of population 'health' (Ricker 1963). In general, traditional fishery management approaches fail for reef fishes because they do not effectively control fishing effort.

Heightened concerns about cumulative impacts from human disturbances and the proximity to the growing Miami urban centre resulted in the 1990 establishment of the 2800 nm^2 Florida Keys

National Marine Sanctuary (FKNMS). In the FKNMS, a total of 24 areas are set off as marine ecological reserves where access is restricted in an effort to conserve biodiversity and environmental quality, and to sustain fishery resources. Designating certain marine areas as 'protected' — usually by officially naming them 'sanctuaries' or 'reserves' — is a revolutionary resource-management tool used in mitigating habitat degradation and overfishing. Reef fishes with specific home ranges make excellent candidates as spatial access management indicators, thus Marine Fishery Reserves (MFR) are an essential element of the 1995 FKNMS draft management plan (DOC 1995). Because of their relatively high site fidelity, strategic use of MFRs for reef fish could protect the most valuable segments of spawning stocks, build stock biomass, and augment yields in adjacent exploited areas *via* increased adult fluxes of larger fish (Polacheck 1990; Russ *et al.* 1992; DeMartini 1993; Russ and Alcala 1996). Despite widespread interest in this 'new' concept, few quantitative paradigms or management models are available for implementing or developing strategic plans for reserves in terms of their overall design, number, total area, and location (Man *et al.* 1995).

In this paper we conduct a preliminary baseline status assessment of the reef fish resources in the FKNMS retrospectively over the last two decades, the period for which there are detailed fishery-independent survey records. This is done prior to implementation of marine 'replenishment' reserves in the FKNMS to provide the basis for determination of future changes, particularly those brought about by introduction of a national marine sanctuary.

Methods

The Florida Keys coral reef ecosystem sits parallel to the Florida current and Florida Bay, stretching southward 200 km from Key Biscayne to the Dry Tortugas. The subtropical marine environment encompasses many varied habitats including estuaries, lagoons, coral islands, seagrass beds, and an extensive coral reef system (Fig. 1). Florida Bay and a network of coastal estuaries situated proximally to the Keys may serve a 'nursery' function for many juvenile fishes and lobsters that occupy the reefs as adults. These areas may ultimately be affected by substantial changes in freshwater outflows from Everglades restoration efforts (Harwell *et al.* 1996). The clear water and high diversity of reef fish of the Florida Keys provide a unique environment to assess the status and trends of multi-species fisheries. Here we use a systems approach to link reef fish abundance, biomass and community distribution with human impacts, environmental variability, and resource health through a comprehensive reef fish monitoring and assessment programme in the FKNMS. The systems approach integrates data assimilation, sampling theoretics, statistical analysis, mathematical modelling and management activities to facilitate and improve performance of the fishery management institution (Ault and Fox 1989; Rothschild *et al.* 1996).

Reef fish statistical surveys

Fishery-independent quantitative visual survey estimates of reef fish population abundance and size were taken continuously from 1979 to 1995 along the entire reef track of the Florida Keys ranging from Biscayne National Park to Key West out to the Dry Tortugas. The samples covered a variety of depths at approximately 70 reef sites (Fig. 2 and Table 1). The visual survey method follows Bohnsack and Bannerot (1986). The visual survey method is a reliable, non-invasive, non-destructive method of quantitatively sampling coral reef fish communities that provides information critical to stock assessment, such as quantitative observations on the presence-absence, abundance, and average length data. A database management system was set up and utilized to manage and analyse information collected through visual surveys of reef fish from 1979 to the present. The extensive visual survey database contains 4147 visual survey samples (Table 1) taken at approximately 70 reef sites during 1979 to 1995, including information on 226 species of reef fishes (Table 2), contained in 42 different biological and physical variables. This database, coupled with information from local, state and federal agencies, was used to determine impacts on the Florida Keys reef fish community due to

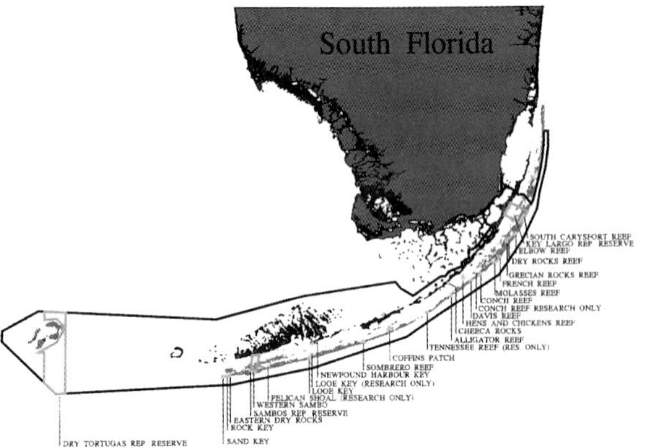

Fig. 1. Map showing South Florida regional ecosystem including Florida Bay, the coral reef tract with its spatial relationship to the Florida Keys National Marine Sanctuary (dark polygon) and the Miami urban centre. Replenishment reserves are shown as the paler polygons. Sanctuary Preservation Areas (SPAs) are named. Boundaries of the Dry Tortugas ecological reserve have not yet been determined.

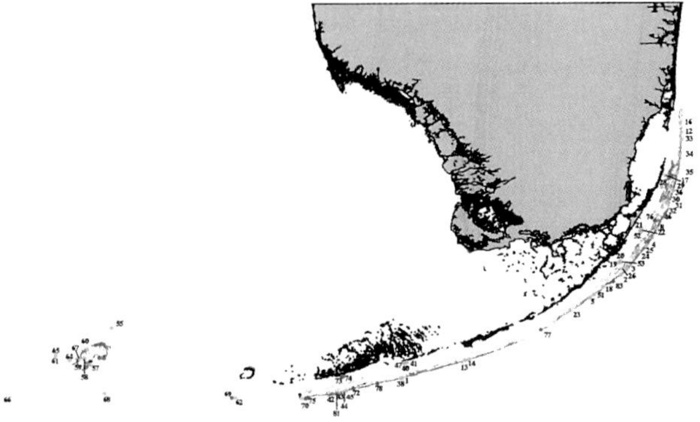

Fig. 2. The Florida Keys National Marine Sanctuary showing the Replenishment Reserves and Sanctuary Preservation Areas and locations of the 67 reef sites where 4147 visual reef fish samples have been taken from 1979 to 1995.

Table 1. Fishery-independent visual survey sampling effort in the Florida Keys reef tract from 1979 to 1995 by reef zone habitat and depth intervals. The offshore zone is exposed to the Florida Current and Gulf Stream. Depth class units are feet

	Reef Zone Habitat				
Depth Class	Artificial Reef	Inshore	Offshore	Total	Percent
$x < 10'$	0	791	147	938	22.62
$10' \leq x < 20'$	2	735	923	1660	40.03
$20' \leq x < 30'$	76	383	635	1094	26.38
$30' \leq x < 40'$	28	21	128	177	4.27
$40' \leq x < 50'$	9	8	127	144	3.47
$50' \leq x < 60'$	0	6	95	101	2.44
$60' \leq x < 70'$	0	1	32	33	0.79
Totals	**115**	**1945**	**2087**	**4147**	
Percent	**2.77**	**46.90**	**50.33**		**100.0**

Table 2. Fishes seen by visual survey divers in the Florida Keys from 1979 to 1995

Class	Order	Family	Genus	Species
Sharks				
Elasmobranchiomorphi	Lamniformes	Carcharhinidae	Carcharhinus	1
		Rhincodontidae	Ginglymostoma	1
		Sphyrnidae	Sphyrna	2
	Rajiformes	Dasyatidae	Dasyatis	1
		Mobulidae	Manta	1
		Myliobatidae	Aetobatis	1
		Urolophidae	Urolophus	1
Bony Fishes				
Osteichthyes	Anguilliformes	Muraenidae	Gymnothorax	3
	Atheriniformes	Atherinidae	Atherinomorus	1
			Hypoatherina	1
		Belonidae	Strongylura	1
			Tylosurus	1
		Exocoetidae	Hemiramphus	1
	Aulopiformes	Synodontidae	Synodus	2
	Beryciformes	Holocentridae	Holocentrus	5
			Myripristis	1
	Clupeiformes	Clupeidae	Harengula	1
			Jenkinsia	1
	Elopiformes	Elopidae	Megalops	1
	Gasterosteiformes	Aulostomidae	Aulostomus	1
		Fistulariidae	Fistularia	1
	Lophiiformes	Ogcocephalidae	Ogcocephalus	1
	Perciformes	Acanthuridae	Acanthurus	3
		Apogonidae	Apogon	1
		Blenniidae	Hypleurochilus	1
			Ophioblennius	1
			Scartella	1
		Callionymidae	Paradiplogrammus	1
		Carangidae	Alectis	1
			Caranx	5
			Decapterus	2
			Elagatis	1
			Seriola	1
			Trachinotus	1
		Centropomidae	Centropomus	1
		Chaetodontidae	Chaetodon	4
		Cirrhitidae	Amblycirrhitus	1
		Clinidae	Acanthemblemaria	2
			Emblemaria	1
			Hemiemblemaria	1
			Malacoctenus	3
			Paraclinus	2
		Echeneidae	Echeneis	1
		Ephippidae	Chaetodipterus	1
		Gerreidae	Eucinostomus	1

Table 2. Fishes seen by visual survey divers in the Florida Keys from 1979 to 1995 (continued)

Class	Order	Family	Genus	Species
			Gerres	1
		Gobiidae	Coryphopterus	4
			Gnatholepis	1
			Gobionellus	1
			Gobiosoma	2
			Ioglossus	2
			Microgobius	2
		Grammatidae	Gramma	1
		Haemulidae	Anisotremus	2
			Haemulon	11
		Inermiidae	Inermia	1
		Kyphosidae	Kyphosus	1
		Labridae	Bodianus	1
			Clepticus	1
			Doratonotus	1
			Halichoeres	6
			Hemipteronotus	3
			Lachnolaimus	1
			Thalassoma	1
		Lutjanidae	Lutjanus	9
		Malacanthidae	Malacanthus	1
		Mullidae	Mulloidichthys	1
			Pseudupeneus	1
		Opistognathidae	Opistognathus	2
		Pempheridae	Pempheris	1
		Pomacanthidae	Centropyge	1
			Holacanthus	3
		Pomacentridae	Pomacanthus	2
			Abudefduf	1
			Chromis	5
			Microspathodon	1
			Pomacentrus	6
		Priacanthidae	Priacanthus	2
		Scaridae	Cryptotomus	1
			Nicholsina	1
			Scarus	5
			Sparisoma	7
		Sciaenidae	Equetus	4
			Odontoscion	1
		Scombridae	Scomberomorus	3
		Serranidae	Diplectrum	1
			Epinephelus	6
			Hypoplectrus	8
			Liopropoma	1
			Mycteroperca	5
			Paranthias	1
			Rypticus	1
			Serranus	4
		Sparidae	Archosargus	1
			Calamus	4
			Diplodus	1
		Sphyraenidae	Sphyraena	2
	Pleuronectiformes	Bothidae	Bothus	2
	Scorpaeniformes	Scorpaenidae	Scorpaena	1
	Tetraodontiformes	Balistidae	Aluterus	2
			Balistes	2
			Cantherhines	2
			Canthidermis	1
			Melichthys	1
			Monacanthus	2
		Ostraciidae	Lactophrys	5
		Tetraodontidae	Canthigaster	1
			Chilomycterus	1
			Diodon	2
			Sphoeroides	1
			Total Species	**226**

increased pressure and disturbance from human activities, particularly fishing.

The visual survey data collection involved 12 divers competent in fish identification over the 17 year period. To fully utilize the database containing visual census survey estimates, it was necessary to determine diver efficiency, variability, and intercomparability. To accomplish the estimation of relative fishing power for specific survey divers, we used the 'fishing power' model of Robson (1966) as the principal statistical method implemented through multivariable regression analysis (Neter *et al.* 1985). Data from the Dry Tortugas were used for this purpose since the nine-day sampling trip involved a controlled experiment to test diver inter-comparability. We standardized the survey data relative to the most experienced diver represented in the database. This relative fishing power coefficient can then be used to standardize the entire database so that each visual 'catch-per-unit-effort' measurement is comparable over time and space. Comparable standard estimates of stock size and condition were also generated using the headboat catch and effort data from Bohnsack *et al.* (1994) by computing estimates of average size in the exploitable phase of the population for the period 1981–1995. Average weight was converted to average length using the parameters of Bohnsack and Harper (1988). Spatial trends in abundance were analysed using scientific visualization software (Advanced Visualization System, AVS, and Interactive Data Language, IDL) and Arc/Info GIS (Geographic Information System) to plot abundance by reef site throughout the Florida Keys over time.

COMPUTER AND ANALYTICAL MODELS

To assess the status of the multi-species fishery, we developed an object-oriented computer model, REEFS (reef-fish equilibrium exploitation fishery simulator), and parameterized it for the Florida Keys. The computer model incorporates the following theoretical concepts. Maximum age of the stock (or life span) and the natural mortality rate were normalized to make the survivorship at maximum age 5% of the unexploited initial cohort size. The population model was the von Foerster-McKendrick form of the conservation equation (Ault and Olson 1996), with the intent to express the equation for population in terms of the average numbers at an average size. The continuous partial differential equation can be modified to a stochastic age-independent model for ensemble number at a given length for the entire population (Ault and Rothschild 1991) as

$$\tilde{N}_{\gamma}\left(L_{\gamma}\right)=\int_{t_r}^{t_\lambda} R(\tau-a)s(a)\Theta(a)P\left(L_{\gamma}|a\right)da$$

where $R(\tau\text{-}a)$ is recruitment, $S(a)$ is survivorship at age a, $\Theta(a)$ is sex ratio at age, and $P(L_\gamma|a)$ is the probability of being length L given the fish is age a. Population dynamics parameters for each fish stock were gleaned from the summaries in Claro (1994) and additional taxa-specific literature. The simulation model REEFS was used to calibrate all models and estimations. While the growth, mortality and recruitment dynamics of tropical coral reef fishes are similar in composition to higher latitude fishes, the timing of key events in reproduction and ageing may be quite unlike those of higher latitude fishes (Ault and Fox 1990; DeMartini 1993). The relative 'health' of a reef fish community is a sensitive indicator of direct and indirect stress on the entire marine ecosystem and is best described using a physiological variable like average size in some phase of the stock's life history. To monitor the 'health' of the *s* stocks over time we use a metabolic variable, average size, in the exploitable phase of the stock as an indicator variable for a given species.

$$\bar{L}(t)=\frac{F(t)\int_{t'}^{t_\lambda}N(a,t)L(a,t)da}{F(t)\int_{t'}^{t_\lambda}N(a,t)da}$$

A simple power relationship exists between length $L(a)$ and weight $W(a)$ at age. Equation (2) implies that the fraction in the fishery catch should be equivalent to the average size of the exploitable phase of the population remaining in the sea. To estimate the total mortality of reef fish, we used a length-based method (Ault and Ehrhardt 1991; Ehrhardt and Ault 1992; Ault *et al.* 1996; Gayanilo *et al.* 1996).

$$\left[\frac{L_\infty-L_\lambda(t)}{L_\infty-L'(t)}\right]^{\frac{Z(t)}{K}}=\frac{Z(t)\left(L'(t)-\bar{L}(t)\right)+K\left(L_\infty-\bar{L}(t)\right)}{Z(t)\left(L_\lambda-\bar{L}(t)\right)+K\left(L_\infty-\bar{L}(t)\right)}$$

In general, average length of reef fish in the exploitable phase of the stock ($L_\lambda - L'$) is strongly correlated with population size in both numbers and biomass. Average size can be used as *an* indicator variable of population health because the statistic is robust to population estimation measure (i.e. visual or headboat survey data). Iterative application of the method over data for each species provides time-series information on reef fish mortality rates, and thus abundance, for all the species in the survey. The spawning potential ratio (SPR) is defined as the current over the unexploited spawning stock biomass (Mace *et al.* 1996).

RESULTS

Trends in recreational, commercial, and headboat effort since 1965 are shown (Fig. 3). Note that recreational effort has increased sharply during this period. Since 1981, the largest increase in fishing effort has clearly come from the recreational sector. This component continues to show a dramatic, unchecked increase; while commercial and headboat nominal fishing effort have been relatively stable over the same period.

Comparative analysis of average length in the exploitable phase of the stock for three species, one within each of three taxa of economically important reef fish (i.e. groupers, snappers, grunts), are shown in Figure 4. Average length estimates are derived from headboat catch and effort statistics from 1981–1995, while the visual survey data are for the period 1979–1995. Headboat data are used in the comparative analysis since they provide consistent data both in the effort record and catch statistics. The comparative estimates of

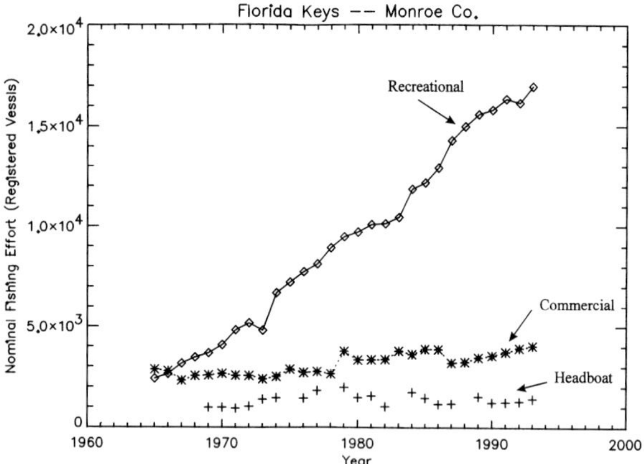

Fig. 3. Nominal fishing effort by recreational, commercial, and headboat vessels during the period 1965–1995.

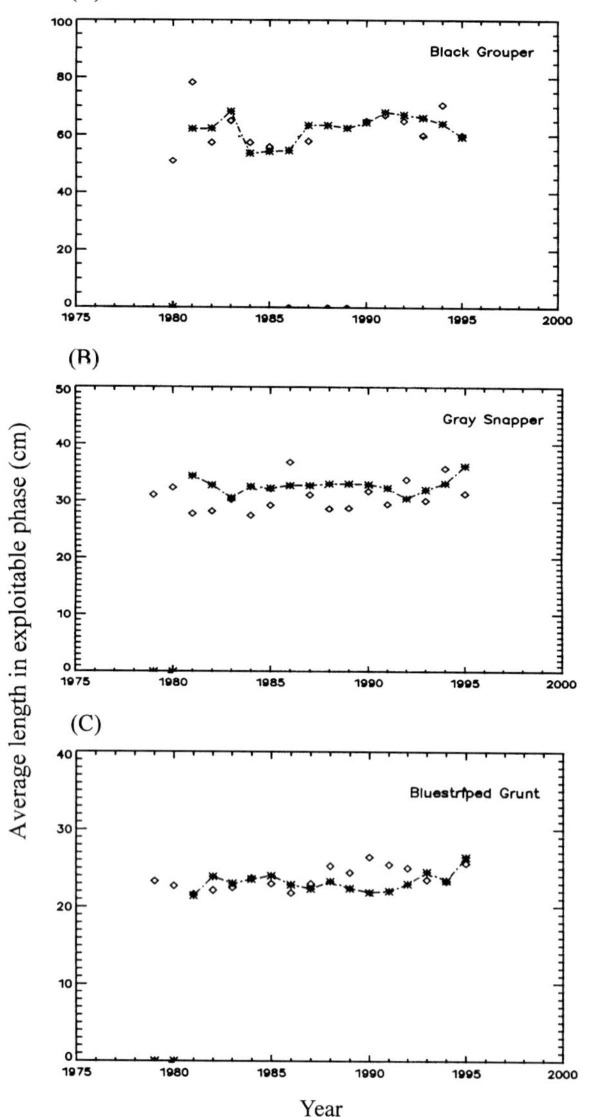

Fig. 4. Estimates of average length in the exploitable phase of the stock within the FKNMS from 1979 to 1995 using headboat (∗) and visual survey (◊) data for: (A) black grouper; (B) gray snapper; and, (C) bluestriped grunt.

Table 3. Model variables and life history parameters for Florida Keys reef fish

Symbol	Definition	Units
s	Species of fish	
γ	Year class	
a	Cohort age class	
t_λ	Oldest age class	Years
t_r	Age of recruitment	Years
t_m	Minimum age of maturity	Years
t'	Minimum age of capture	Years
L'	Minimum size of capture	mm
L_λ	Oldest size class	mm
$W(a,t)$	Initial weight at age	g
$N(a,t)$	Initial number at age	Fish
$M(a)$	Natural mortality	Year^{-1}
$F(a,t)$	Fishing mortality	Year^{-1}
$\Theta(a)$	Sex ratio at age	dimensionless
W_∞	Ultimate weight	kg
L_∞	Ultimate length	mm
K	Brody growth coefficient	Year^{-1}
t_0	Age at which size equals 0	Years
α_{WL}	Weight-length scalar	dimensionless
β_{WL}	Weight-length power	dimensionless
$S(a)$	Survivorship to age a	dimensionless

average length in the exploitable phase of the stock from the two data sources are in fairly good agreement. In some cases, the visual observations scattered like normal residuals around the headboat estimates. In others, particularly the grunts, the headboat estimates shows a flat trend, while the visual estimates show increasing trends.

Using average length from headboat and visual survey data, we estimated the annual fishing mortality rates (F) for several families of exploited reef fishes using population dynamics parameters from the fishery (Tables 3 and 4). Parameter sets for the three families of important fishery species cluster or group out in fairly distinct patterns, suggesting characteristic population-dynamic signatures within taxa. Table 5 provides estimates of annual fishing mortality rates derived from average length statistics for representative members of three taxa each for visual and headboat data. The F's derived by sampling were calibrated with analytical estimates from simulation. In general, the current average sizes of the exploitable phase for many of the economically important fishery populations are marginally above the minimum size of capture regulated by fishery management. The F's indicate that the fishing pressure is significant enough that very few fish survive very long after reaching the length of the minimum size. Another indication of the significant changes occurring within the reef community of the FKNMS is the quasi 'logistic' growth of the barracuda population over the period covered by the visual survey. Not only has the barracuda stock increased in abundance, but the average size of the individuals has also increased significantly, subsequently leading to an increase in biomass for the entire population.

Table 4. Florida Keys reef fish population dynamics parameters used in mortality estimation and fishery simulations

Species	Population Parameters											
	M	t_λ	L_∞	W_∞	K	t_0	t_m	L'	t'	α_{WL}	β_{WL}	L_λ
Groupers												
Black grouper	0.15	20	1200	31.587	0.16	–0.300	4.0	508	39	4.27e-6	3.2051	1153.07
Mycteroperca bonaci												
Red grouper	0.18	17	938	11.865	0.153	–0.099	4.0	508	61	1.1314e-5	3.0350	869.01
Epinephelus morio												
Graysby	0.20	15	415	1.1397	0.13	–0.940	3.0	203.2	52	1.2238e-5	3.0439	362.46
Epinephelus cruentatus												
Snappers												
Gray snapper	0.300	10	722.32	5.2458	0.1358	–0.863	2.0	254.0	29	3.0486e-5	2.8809	556.16
Lutjanus griseus												
Yellowtail snapper	0.214	14	454.69	1.2968	0.2088	–0.712	2.0	304.8	56	7.7482e-5	2.7180	433.44
Ocyurus chrysurus												
Mutton snapper	0.214	14	938.74	14.058	0.1290	–0.738	2.0	304.8	29	1.574e-5	3.0112	797.75
Lutjanus analis												
Grunts												
White grunt	0.375	8	511.85	3.062	0.1859	–0.776	1.5	203.2	24	8.354e-6	3.1612	410.25
Haemulon plumieri												
Bluestriped grunt	0.500	6	289.64	0.4712	0.4841	–0.0106	1.0	203.2	31	1.9436e-5	2.9996	273.54
Haemulon sciurus												
Tomtate	0.333	9	441.56	1.889	0.0909	–2.0949	2.0	203.2	57	6.193e-6	3.2077	279.89
Haemulon aurolineatum												
Barracuda, Great	0.20	15	1238.3	14.034	0.1724	–0.4610	3.0	619.2	44	4.107e-6	3.0825	1151.54
Sphyraena barracuda												

Table 5. Fishing mortality rate coefficients estimated from analysis of average size in the exploitable phase of the stock from 1979–1995 in the Florida Keys for representatives of three economically important taxa. The symbol n(v) is the number of annual visual survey samples, HB is headboat survey, VS is visual survey, and SPR% is percent spawning potential ratio

Species		Black Grouper		Gray Snapper		Bluestriped Grunt	
Year	n(v)	HB	VS	HB	VS	HB	VS
1979	13				0.693		0.288
1980	145		55.06		0.476		0.718
1981	213	0.754	0.081	0.243	2.263	2.775	2.207
1982	189	0.646	1.343	0.417	1.860	0.000	1.311
1983	507	0.323	0.470	0.807	0.880	0.491	0.935
1984	228	3.456	1.343	0.449	2.656	0.172	0.065
1985	124	2.743	1.819	0.489	1.219	0.000	0.514
1986	33	2.517		0.425	0.056	0.614	1.832
1987	73	0.562	1.228	0.424	0.686	1.093	0.506
1988	263	0.566		0.388	1.547	0.298	0.000
1989	318	0.634		0.387	1.469	1.103	0.000
1990	282	0.501	0.470	0.402	0.574	1.729	0.000
1991	280	0.331	0.373	0.479	1.139	1.439	0.000
1992	256	0.365	0.470	0.821	0.303	0.534	0.000
1993	196	0.412	0.894	0.523	0.929	0.000	0.172
1994	340	0.512	0.247	0.377	0.133	0.239	0.296
1995	687	0.961	0.894	0.100	0.642	0.000	0.000
Total	**4147**	**Median 0.566**	**0.894**	**0.424**	**0.880**	**0.491**	**0.288**
SPR%		6.94	3.16	32.05	16.99	67.19	77.16

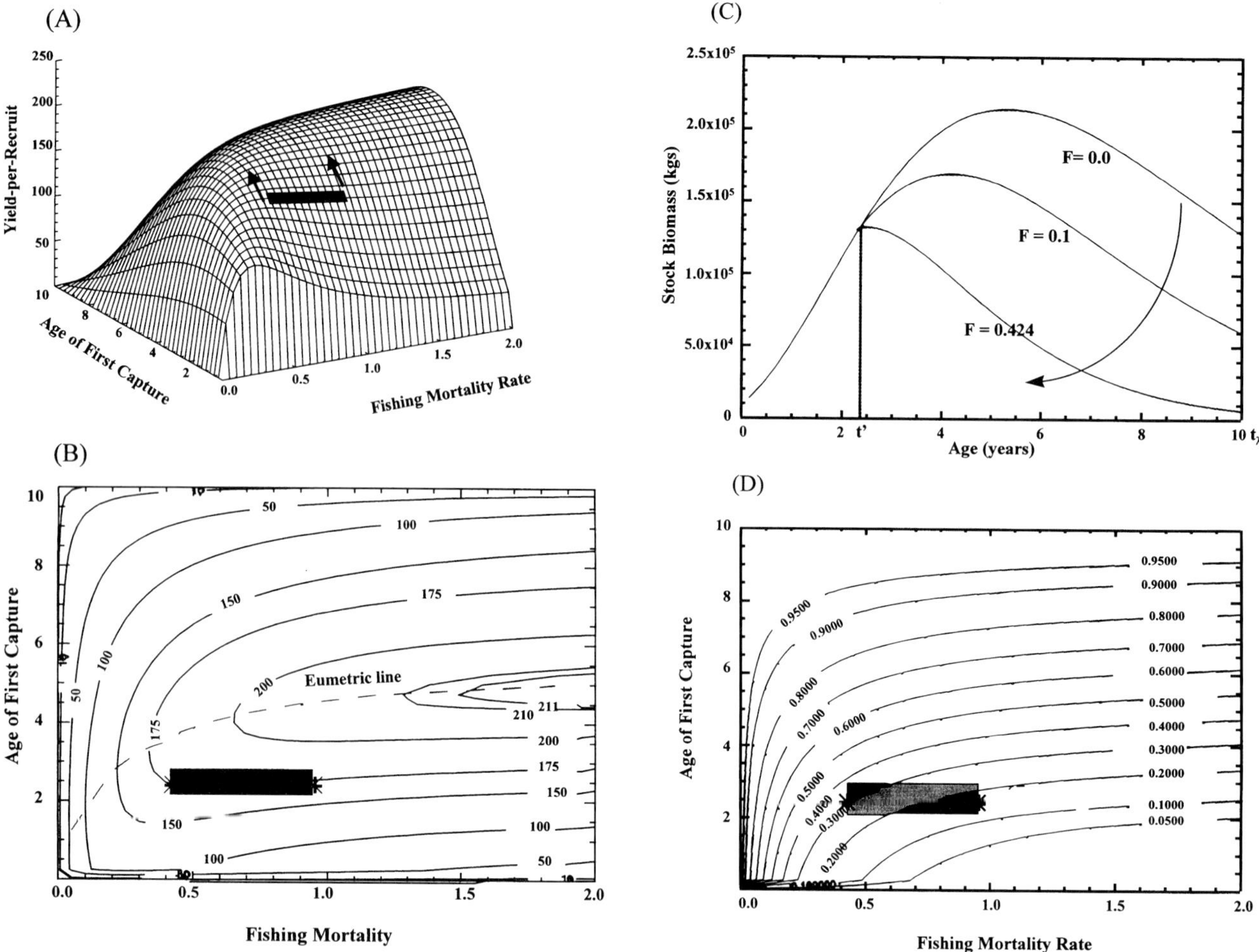

Fig. 5. Example of analytical yield analyses for gray snapper (*Lutjanus griseus*): (A) yield-per-recruit surface: (B) yield-per-recruit isopleth; (C) spawning stock biomass on age; and (D) spawning potential ratio. Parameter values for the REEFS model are from Table 4. Rectangular areas in panels B and D show range of F and t' estimates. Area in rectangle of panel D below 0.30 SPR indicates range of estimates that exceed the Federal definition of overfishing following Mace *et al.* (1996).

As an example, we conducted a management analysis for gray snapper (*Lutjanus griseus*) with the fishery parameters derived from our analyses. Median *F*s for both data types were used as conservative measures to bound the range of feasible rates operable in the fishery, and to provide uncertainty bounds for the overall problem. Figure 5 shows the results from yield-per-recruit (YPR) and spawning potential ratio (SPR) analyses. The YPR analysis suggests that the current size of first capture L' may be resulting in 'growth overfishing', and should be increased to maximize the YPR (Fig. 5A–B). How to move the fishery over the yield surface is of paramount concern to managers. Figure 5C shows the progressive reductions of stock biomass from an unexploited state with no fishing to the lower bound *F* estimate. The current range of *F* is likely also to be contributing to 'recruitment overfishing', pushing the SPR below Federal minimum standards. The minimum estimated *F* still reduces the SPR to about 32% of the unexploited state. If the upper bound is used, then the SPR is reduced to about 17% of the unexploited spawning stock biomass (Fig. 5D). When compared across the three taxa, we find that most of the groupers, and many of the snappers are apparently overfished (Fig. 6).

Discussion

Assessments based on fishery-independent data are reliable measures of reef fish abundance, population dynamics and community composition (Gunderson 1993). But, to date, the scope of applications in tropical coral reef ecosystems has been limited. Through population growth, 'average size' is a metabolic variable that embeds mortality and reproductive processes. As a result, its use as an indicator variable of population status or health has it roots in traditional fisheries management.

Over *at least* the last two decades, the period for which there are spatially-explicit fishery-independent survey records, economically and ecologically important Florida Keys reef fish populations appear to have been heavily fished. Simple correlation of nominal fishing effort with the average size in the exploitable phase of the population for different reef fish species indicates that increased fishing pressure is diminishing the size and biomass of desirable fish populations (e.g. groupers and snappers) in the Florida Keys. According to the National Marine Fisheries Service's definitions of overfishing in US Fishery management plans (Mace *et al.* 1996), the exploitation rates

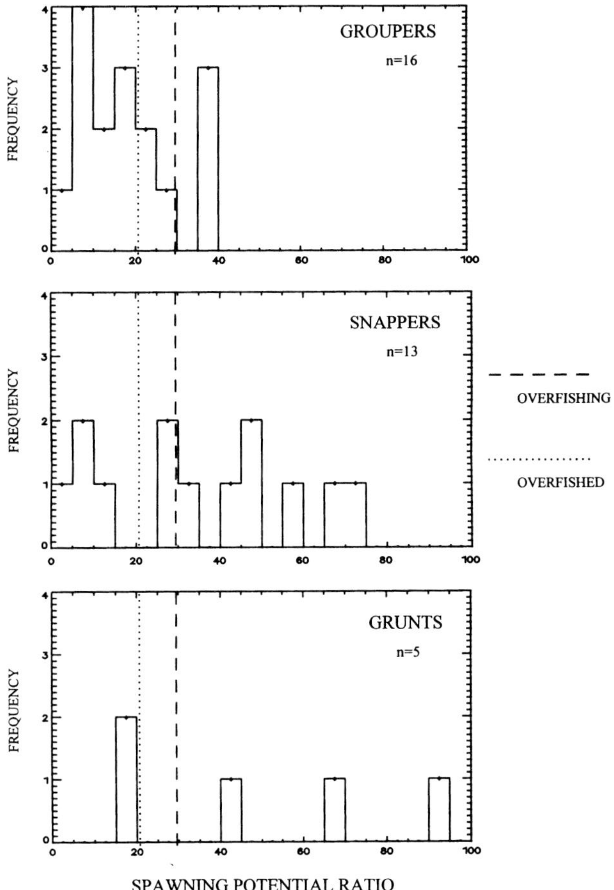

Fig. 6. Histograms showing the numbers of species within 3 taxa (i.e. groupers, snappers, and grunts) and the current estimated spawning potential ratios.

estimated herein suggest many of these stocks are presently 'overfished'. 'Juvenesence', or the process of making the stock young by diminishing the abundance of older size classes over time, has long been known to result from heavy fishing (e.g. Beverton and Holt 1957; Ricker 1963). In the Keys, the effect of recreational fishing effort is an enigma since, like an artisanal fishery, it is distributed heterogeneously, has diffuse areas of operation, and the fishery is generally not well sampled. Escalating exploitation levels appear to be proportional to human population growth along the South Florida coastal fringe. Since South Florida population size is projected to increase four-fold over the next 25 years, it seems likely to continue to increase, as will habitat degradation, suggesting a much longer-term potential to significantly alter reef fish community structure and dynamics in the Keys.

Significant changes in abundance amongst key reef fish stocks have affected community dynamics and composition. It is interesting to note that there is currently little directed fishing effort for barracuda. This stems from the fact that barracuda have been implicated as a fish affected by 'ciguatera', a paralytic nerve toxin that originates in algae and is subsequently passed up the food chain, accumulating in top piscivore predators. Human consumption can result in convulsions and death. It may be surprising then that when compared to all other species on the reef in year by year rankings of stock biomass, barracuda have now apparently become the dominant source of biomass within the Florida Keys reef fish community. It is interesting to speculate as to why such changes occurred. Most prominent are the apparent reductions in competition by other top predators, such as groupers, snappers, and sharks that have been fished intensively over long periods of time. As a consequence, barracuda have practically no other principal predators, or their predators are of limited abundance. But on the other hand, barracudas have a large and increasing prey base (e.g. grunts). Claro (1991) hypothesized a similar process in the Golfo de Batabano, Cuba, where chronic overharvesting of snappers resulted in abundance and ultimately community shifts in favour of grunts. Such dramatic increases in the abundance, average size, and biomass of a top predator within the FKNMS reef fish community will obviously have significant impacts on the dynamics of the entire community, particularly on key prey species (including many sport fishing favourites like snappers and grunts). In addition, excessive predation from particularly dominant predators like barracuda may counteract reductions in fishing mortality F on certain species brought about by traditional fishery management policy.

Fishing effort on multi-species stocks in the Keys is highly variable in both time and space, but it is generally very intense. Since the number of recreational anglers using the Florida Keys resources will continue to increase, without some form of effective management intervention such as establishing marine reserves, reef fish stocks in the Florida Keys will continue to decline and perhaps collapse. Generally, those stocks in trouble are by and large those with low spawning potential ratios SPR. Looking broadly, critical population (minimum size at first capture L') and fishery (F) parameters should be adjusted to mitigate the 'growth overfishing' and 'recruitment overfishing' syndromes already prevalent in the fishery. This should be done in a multi-species context to optimize the biotic and fishery potential of the reef fish assemblage as a whole. The most compelling reason for implementing a spatial protection approach is that other traditional approaches (e.g. control of F *via* vessel effort and gear restrictions, or a size-based control) habitually fail because they do not effectively control effective fishing effort. Improved fishing (better vessel design, hydroacoustics, hydraulics) and navigation (Global Positioning System GPS, Loran C) technologies have made the average fishermen more efficient over time and space. These improved technologies and growing populations of fishermen, particularly among recreational anglers even with bag limits, can do great damage. Fishermen have always been great at increasing their effort. Bycatch mortality and high fishing effort from the expanding fleets limit the effectiveness of size limits. In theory, every fish can be caught once it reaches minimum legal size. In multiple species fisheries, the more vulnerable species continue to be caught causing serial overfishing (Bohnsack and Ault 1996). In addition, the depths of capture for some species limit the effectiveness of minimum size regulations. Even low hooking mortality does not work because of the long lives (some measured in decades) of many economically important species. Each fish can potentially be caught many times during its residence in the exploitable phase of the population, eventually turning a low probability mortality event into a certitude.

Strategic use of spatial closures can potentially control nominal fishing effort more effectively under the pressures and conditions prevalent in the Florida Keys, particularly for resident fishes with limited home ranges. None of the traditional management approaches effectively deals with the issues of genetic selection. Spatial closures allow wild genotypes to persist and perhaps proliferate. Thus, it is imperative that Florida Keys reef fish populations receive further protection to obtain production levels commensurate with achievement of long-term goals for marine biodiversity and sustainable fisheries. The former productivity levels of the Key reef fishery may be recoverable; but, this will require innovative approaches to controlling the fishing mortality rate *F*. Coupled with vigilant enforcement, marine reserves are certain to reduce the intensity, and *de facto* reduce the variability, of *F* to levels near zero within reserves. Marine reserves may also reduce the risk of recruitment overfishing by maintaining a critical spawning stock biomass (Bohnsack and Ault 1996).

The preliminary results of our 17-year retrospective multi-species assessments are encouraging; and they provide a baseline for reference against change after implementation of a spatial management regime. But, these can realistically only be considered first order estimates of stock status. The intrinsic error in population estimates derived from both the fishery-independent surveys and fishery-dependent data efforts ultimately propagates as these are used to compute average size and mortality rates for each stock. However, these statistics can be improved through a number of steps. Use of a class of mechanistic models of population density distributions like diffusion-taxis are warranted, as these models help to explore the degree to which spatial distributions can be described solely as consequences of accumulated individual responses to local biotic and abiotic features (Murray 1989). A better understanding of these relationships would allow researchers to build on the associations of animals and habitat developed from relatively sparse sampling in large spatial domains to effectively 'fill-in' maps with population estimates for those areas that were not physically sampled. Secondly, to increase survey efficiency, the relative fishing powers of survey divers and headboat gears require intercalibration. In the water, divers in the visual survey see what headboat fishermen will ultimately bring in over the gunnel as catch. In theory, fishing mortality should be strictly proportional to true population abundance by a scale factor, the catchability coefficient. In adopting these two recommendations, programme benefits will include cost-effective reef fish sampling designs and assessments critical to evaluation of sustainable fisheries, as perhaps an indicator of restoration success. High precision monitoring data will also guide development of analytical 'structured' population and 'trophic-stack' community models, and allow accurate forecasts and hindcasts of multi-species reef fish populations. Such efforts are central to the protection of marine biodiversity, to promote sustainable fisheries, and to further the science of ecosystem management.

Finally, we recommend that future synoptic sampling designs should combine innovative optical (e.g. green band lasers, stereoscopic camera arrays) and hydroacoustic methods in multi-stage sampling designs to improve estimates of biomass and abundance for species groups over more area and deeper waters. One hypothesis that we are currently exploring is that the Florida Keys consist of two zones, the upper and lower Keys, that are distinctly different in species composition and abundance. Use of scientific data visualization techniques, in conjunction with new statistical methods and analytical models applied to the database, will allow us to test several biological hypotheses. Because the dynamics of multi-species fish assemblages appear to be regulated by trophodynamics at individual, population and community levels, it is important to provide assessments of the entire reef fish community in the context of the physical environment. We believe these baseline analyses are key to the use of marine reserves to build sustainable fisheries and maintain marine biodiversity.

Acknowledgments

We thank Guillermo Diaz, Doug Harper, Ken Lindeman, Jiangang Luo, Elizabeth Maddox and Joe Serafy for technical assistance. This research was sponsored by NOAA and the United States Man and the Biosphere Marine and Coastal Ecosystem Directorate Grant No. 4710-142-L3-B, and the NOAA Coastal Ocean Program Grant No. NA37RJO200.

References

Ault, J S, and Ehrhardt, N M (1991). Correction to the Beverton and Holt Z-estimator for truncated catch length frequency distributions. *Illarm Fishbyte* **9**(1),37–9.

Ault, J S, and Fox Jr, W W (1989). FINMAN: simulated decision analysis with multiple objectives. In 'Mathematical Analysis of Fish Stock Dynamics'. (Eds E F Edwards and B A Megrey). *American Fisheries Society Symposium* **6**, 166–79.

Ault, J S, and Fox Jr, W W (1990). Simulation of the effects of spawning and recruitment patterns in tropical and subtropical fish stocks on traditional management assessments. *Gulf and Caribbean Fisheries Institute* **39**, 361–88.

Ault, J S, and Olson, D B (1996). A multicohort stock production model. *Transactions of the American Fisheries Society* **125**(3), 343–63.

Ault, J S, and Rothschild, B J (1991). A length-based numerical model for simulating resource decision dynamics. *International Council for the Exploration of the Sea C.M. 1991/D*:24, Statistics Committee.

Ault, J S, McGarvey, R N, Rothschild, B J, and Chavarria, J (1996). Stock assessment computer algorithms. In 'Stock Assessment: Quantitative Methods and Applications for Small Scale Fisheries'. (Eds V F Gallucci, S Saila, D Gustafson and B J Rothschild) pp. 501–15 (Lewis Publishers (Division of CRC Press). Chelsea, Michigan).

Beverton, R JH, and Holt, S J (1957). On the dynamics of exploited fish populations. Ministry of Agriculture, Fisheries and Food, Fishery Investigations, Series II 19. 533 pp.

Bohnsack, J A, and Ault, J S (1996). Management strategies to conserve marine biodiversity. *Oceanography* **9**(1), 73–82.

Bohnsack, J A, and Bannerot, S P (1986). A stationary visual census technique for quantitatively assessing community structure of coral reef fishes. *NOAA Technical Report NMFS* 41. 35 pp.

Bohnsack, J A, and Harper, D E (1988). Length-weight relationships of selected marine reef fishes from the southeastern United States and the Caribbean. *NOAA Technichal Memorandum NMFS-SEFC*-215. 31 pp.

Bohnsack, J A, Harper, D E, and McCellan, D B (1994). Fisheries trends from Monroe County, Florida. *Bulletin of Marine Science* **54**(3), 982–1018.

Claro, R (1991). Changes in fish assemblages structure by the effect of intense fisheries activity. *Tropical Ecology* 32(1), 36–46.

Claro, R (1994). Ecología de los peces marinos de Cuba. Centro de Investigaciones de Quintana Roo, Mexico.

DeMartini, E E (1993). Modeling the potential of fishery reserves for managing Pacific coral reef fishes. *Fishery Bulletin* **91**, 414–27.

Department of Commerce, US (DOC) (1995). Florida Keys National Marine Sanctuary Draft Management Plan/Environmental Impact Statement. US Department of Commerce, NOAA, Washington, DC.

Ehrhardt, N M, and Ault, J S (1992). Analysis of two length-based mortality models applied to bounded catch length frequencies. *Transactions of the American Fisheries Society* **121**(1),115–22.

Gayanilo Jr, F C, Sparre, P, and Pauly, D (1996). FAO-ICLARM stock assessment tools. *FAO Computerized Information Series Fisheries.* 126 pp.

Gunderson, D R (1993). Surveys of fisheries resources. John Wiley and Sons New York.

Harwell, M A, Long, J F, Bartuska, A M, Gentile, J H, Harwell, C C, Myers, V, and Ogden, J C (1996). Ecosystem management to achieve ecological sustainability: the case of south Florida. *Environmental Management* **20**(4),497–521.

Mace, P, Botsford, L, Collie, J, Gabriel, W, Goodyear, P, Powers, J, Restrepo, V, Rosenberg, A, Sissenwine, M, Thompson, G, and Witzig, J (1996). Scientific review of definitions of overfishing in US fishery management plans. National Marine Fisheries Service.

Man, A, Law, R, and Polunin, N V C (1995). Role of marine reserves in recruitment to reef fisheries: a metapopulation model. *Biological Conservation* **71**, 197–204.

Murray, J D (1989). 'Mathematical biology.' Springer-Verlag. 767 pp.

Neter, J, Wasserman, W, and Kutner, M H (1985). Applied linear statistical models. Second Edition. Richard D. Irwin. Homewood, Illinois.

NOAA (1995). Our living oceans. NOAA/NMFS, Department of Commerce. 160 pp.

Polacheck, T (1990). Year around closed areas as a management tool. *Natural Resource Modeling* **4**(3), 327–54.

Ricker, W E (1963). Big effects from small causes: two examples from fish population dynamics. *Journal of Fisheries Research Board of Canada* **20**, 257–64.

Robson, D S (1966). Estimation of relative fishing power of individual ships. *ICNAF Research Bulletin* **3**, 5–14.

Rothschild, B J, Ault, J S, and Smith, S G (1996). A systems science approach to fisheries stock assessment and management. In 'Stock Assessment: Quantitative Methods and Applications for Small Scale Fisheries'. (Eds V F Gallucci , S Saila, D Gustafson and B J Rothschild) pp. 473–92 (Lewis Publishers (Division of CRC Press). Chelsea, Michigan).

Russ, G R, and Alcala, A C. (1996). Do marine reserves export adult fish biomass? Evidence from Apo Island, central Philippines. *Marine Ecology Progress Series* **132**, 1–9.

Russ, G R, Alcala, A C, and Cabanban, A S (1992). Marine reserves and fisheries management on coral reefs with preliminary modelling of the effects on yield per recruit. Proceedings of the Seventh International Coral Reef Symposium, Guam **2**, 978–85.

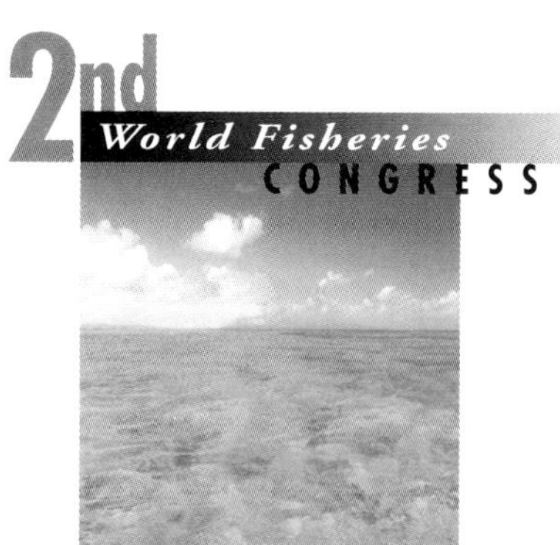

Issues and Outcomes for Theme 3

How Can Fisheries Resources be Allocated...Who Owns the Fish ?

Rudy van der Elst

Oceanographic Research Institute, PO Box 10712 Marine Parade, 4056 KwaZulu-Natal, South Africa.

If anyone arrived at this Congress in any doubt about the importance of allocating resources they will have changed their minds by now. Throughout this four-day Congress the theme of resource allocation emerged as a major issue — often an impediment — to successful fisheries management. Allocating fisheries resources as a problem is remarkably independent of the nature of the fishery, ranging from small-scale traditional fisheries to high-seas industrial operations. Those who attended the presentations by Sir Tipany O'Regan, Danny Chapman, Hugh Braker and Steven Wright will have been exposed to the real meaning of access to, and ownership of, fish.

There were about 280 posters on display at the Congress, yet only 20 of these (7%) could comfortably be accommodated in this Theme. This was disappointing because the problems associated with access and ownership were actually raised in so many papers and posters. Presumably, one can take this low percentage to be an indication of the difficult and thorny nature of this topic.

Clearly there are different levels of scale involved in the question of access, broadly divisible into macro, meso and micro issues.

Macro issues

At the macro level of allocation one would include the international rights and access to resources, especially those that fall outside the EEZ of nations or those that are shared. There was rather little attention given to this aspect at the Congress for the simple reason that good progress has already been made in this regard over the past decade. The development of national economic zones and the Law of the Sea convention (UNCLOS), has to a large extent ameliorated problems here, evident from the presentation by Serge Garcia. However, the paper by John Dean on bluefin tuna allocation and management reminded us how complex this process still remains.

Meso issues

Within this level can be included the allocation of national resources of a country to its people by some appropriate system. Clearly this remains an area of great concern and also one where much progress still needs to be made. Various examples emerged from both the poster and paper presentations, indicating varying levels of sophistication, depending on the situation and country concerned.

In India a system of zonation was proposed, intended to assist in broad allocation of resources by restricting access to specific zones for certain gear types. Thus, the inshore 10 nm of coast would be open only to small-scale, non-mechanized fishers while further offshore trawling and purse seining would be promoted. The inshore zone would be protected and managed to protect spawning grounds and substrate while the two zones would be separated by a necklace of fish aggregating devices (FADS) maintained by local village communities.

There were increasingly optimistic reports about the transferable QMS system of New Zealand and the desire by others to implement similar systems.

There were other innovations. A 'Franchise' model was described by Peter Millington for the Western Australian rock lobster fishery that could provide a mechanism for facilitating the development of appropriate allocation systems — even if the basic ideas date back a century to the Singer Sewing Machine Company!

Then there was the case of South Africa where re-allocation is a unique challenge. A 'model-of-change' was proposed that would broaden access to those previously denied entry. A fishery-wide transfer levy would offset compensation and new-entry costs until the system was entirely market driven, hence internalizing the system and not influencing either the bio-economic equilibrium or international pricing.

Despite the progress made in many countries with developing systems of access, it would appear that most of the world's fishing nations have yet to devise and implement acceptable systems that will ensure sustainable use of resources. This is especially true in developing nations. A poster dealing with the plight of marginalized fishers in Bangladesh and their call for common property rights and cheap credit highlighted this vividly.

MICRO ISSUES

Included here are the much smaller-scale and localized fisheries. While they may be small in scale, the questions of access and ownership at this level are often also the most difficult to resolve. The Congress may well have demonstrated progress with international allocation arrangements but locally we see continued conflict, especially in multi-user, shared and dwindling resources. Numerous papers and posters highlighted this problem but rather few offered solutions. It would appear that many situations exist where questions of local access can be readily resolved but that it may be a question of administrative inertia or lack of political will that prevents this. For example, the competition for King George whiting in Gulf St Vincent (South Australia) as well as a variety of species in Port Phillip Bay (Victoria), could surely be resolved through some form of administrative zonation? Perhaps the recreational anglers should fish closer to the urban centres with commercial rights allocated beyond.

Several examples of successful arrangements to reduce conflict were presented. For example, the prawn trawler bycatch off KwaZulu-Natal comprised species of value to linefishers. By an administrative change in permit condition which shortened the season, a very modest drop in prawn harvest can now result in a significant increase in linefish catches, benefitting numerous small-scale operators. What was once a dilemma of granting access to one or other user is now resolved to accommodate both. Similarly, a combination of input controls on defined fishing grounds, such as gear type, were recommended by K. J. Sullivan to assist with the division of access to snapper stocks in the New Zealand commercial and non-commercial fisheries. Noel Taylor-Moore proposed a decision-making matrix which would incorporate a suite of relevant factors, based on the precautionary principle, to assist in allocating specific estuarine and inshore fish species.

Clearly the reasons why access and ownership in the smaller-scale fisheries remain largely unresolved lies in the fact that it involves more people at a more personal, and hence political, interface.

There is a real need to consider the plight of artisanal and subsistence fishers, a point raised from the floor by Amanda Vincent. These fishers are indeed not well represented at this Congress, yet they are also the most numerous and needy of support because the risks to them do not simply lie in economic efficiency but rather in food security and simply survival. Pamela Mace noted this too when she concluded that the social costs of a collapsed fishery may well be higher than the economic costs. A greater input from social scientists at this Congress would have been justified.

One topic that should have enjoyed more attention was the question of allocating resources to stock rebuilding and especially to marine refugia. There was a convincing poster by Dave Pollard on this topic which demonstrated potential benefits, including value-enhancement through improved non-consumptive use. In many regions and countries around the world where the capacity to introduce complex and economically-based systems of access may be limited, the use of carefully planned and sited marine refugia could significantly enhance sustainability of resources in an otherwise open access system.

There appears to be a major, but solvable, impediment to the allocating of many fish resources, marine and fresh water. This lies in our apparent inability to adequately demonstrate comparable values, and, by implication, benefits, of different allocation strategies. Is recreational angling more valuable than commercial harvesting of similar stocks or indeed the non-consumptive use of such stocks? Do we in fact have adequate information on the respective landings made by the respective sectors? Apparently not and this dilemma was particularly well highlighted in the posters by Tony Pitcher *et al.* and Richard Tilzey, suggesting that this represents an important future field of study.

There is no universal model for allocating fish resources and the system of allocation is dependent not only on the sustainability and maximizing economic rent of the resource but also on its social acceptability in the country concerned. In particular, the question of how much control a government wants to retain or is prepared to devolve to its people and market forces needs to be decided. Great challenges are facing the fishery authorities of many nations over the next few years in developing acceptable systems of access, especially considering the almost universal desire to reduce input to most fisheries and develop better output controls.

Finally, allocation of resources will never be successful without the support of people. Fishers, managers and politicians will need to make decisions but they can only do so if they are informed and have opportunities to participate. This calls for education. I see this as a most important supporting activity over the next few years in the urgent need to develop acceptable systems of access for many fisheries around the world, especially amongst those that remain open access.

How can aquaculture help sustain world fisheries?

Theme 4

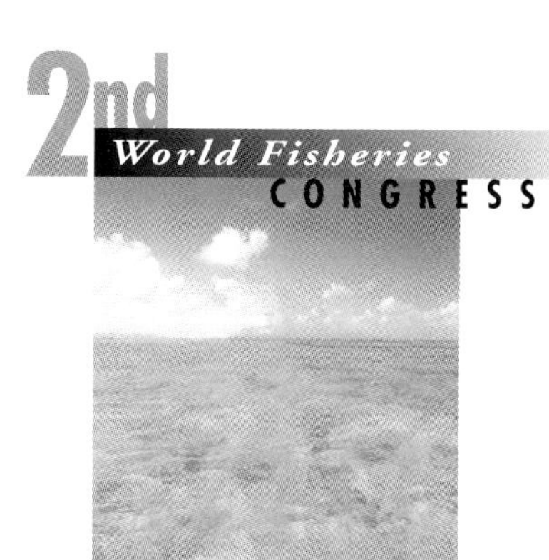

How Can Aquaculture Help Sustain World Fisheries?

I Chiu Liao

Taiwan Fisheries Research Institute, Keelung, Taiwan 202.

Summary

The Earth is three-quarters water, containing a probable inexhaustible seafood resource for human needs. However, in recent years, world capture fisheries production has levelled off or declined, reaching a peak in 1989 at 88.67 million t. This suggests that the capture fisheries are being harvested near to their maximum sustainable yield. With the burgeoning world population, coupled with a growing demand for seafood, aquaculture has been increasingly tapped to fill the widening supply gap created by the stagnating capture fisheries production.

While aquaculture may be seen as the answer to the shortcomings of capture fisheries, the prospects are not all that encouraging. As the aquaculture industry grows, so will conflicts with other users over such resources as water and land, particularly along the coasts. Four suggested major issues are expected to shed light on how aquaculture will be able to sustain world fisheries. These are resource partitioning, coastal zone management, stock enhancement, and new technology. The challenges for aquaculture are clear and the trends in aquaculture development will be examined to give some answers, referencing in some cases the Asian aquaculture experience, where a majority of the world's aquaculture products are produced.

Introduction

The ocean is the Earth's last great frontier, a vast repository of aquatic products, with still yet unexplored resources and opportunities. For thousand of years, it has been seen as an inexhaustible seafood resource, sustaining many coastal communities. World fisheries production has been expanding since the 1970s, with help from technical and technological innovations and huge investments. In 1989, world fisheries production exceeded the 'magic' figure of 100 million t and peaked in 1993[1] with 101.4 million t (FAO 1995*a*) (Fig. 1). However, breaking down these figures between capture fisheries and aquaculture shows a different picture (Fig. 2). Capture fisheries production peaked in 1989 at 88.67 million t and has since declined or fluctuated near this level, indicating that world capture fisheries are being harvested near their maximum sustainable yield. Overfishing, however, is not the only threat to the world's fisheries, with the destruction of fish habitat and pollution also impacting on and reducing stocks.

[1] Online data from FAO (1996) puts the production in 1993 at 102.3 million t. FAO (1996) shows world fisheries production in 1994 at a record 109.6 million t, with most of the increase coming from marine production (mainly attributable to anchoveta catches in the Southeast Pacific). Since these data are only for 1994 and the other available data are either incomplete or not yet updated, this Theme paper used the data from FAO (1995a).

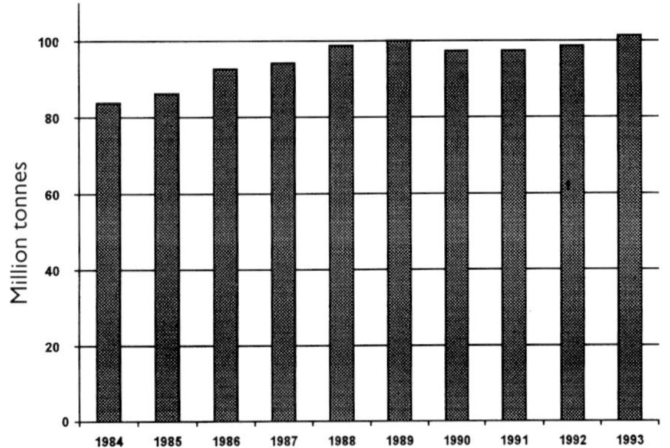

Fig. 1. World fisheries production, 1984–1993. Data from FAO (1995*a*).

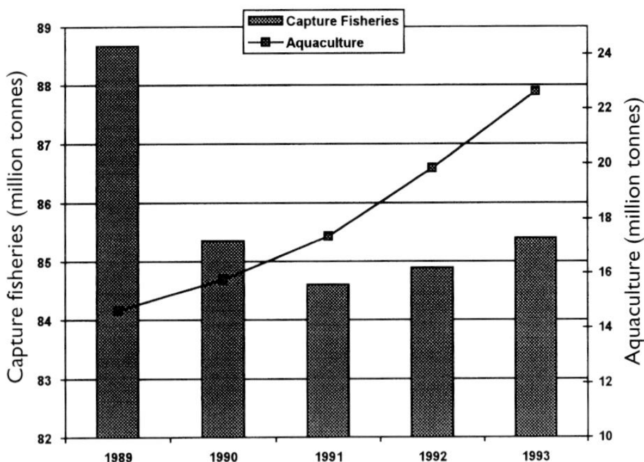

Fig. 2. World capture fisheries and aquaculture production, 1989–1993. Data from FAO (1995*a*, 1995*b*).

At the same time, overall aquaculture production has been increasing. From 15% in 1989, the share of aquaculture reached 22% of the total production in 1993. However, much of this production is concentrated in one region — Asia, which accounts for 80–85% of the world's aquaculture production.

With the fluctuations in capture fisheries production, aquaculture is expected to play an increasingly important role in world fisheries, especially with the increasing population and the increasing demand for medium- and high-grade fishery products.

How much of a role aquaculture will play will depend primarily on filling the supply gap created by the stagnating capture fisheries production. New (1991) predicted that aquaculture should fill a gap of 19.6 million t in the year 2000, 37.5 million t in 2010, and 62.4 million t in 2025. In making this prediction, New (1991) assumed that world population will reach 8.5×10^9 in 2025, as predicted by a United Nations Population Study in 1991, production of capture fisheries will level off at 100 million t (excluding seaweeds and marine mammals) by 2000 and demand will not increase beyond its 1989 level, that is, availability of fish and seafood will have to stay at 19.1 kg/capita/year (excluding again seaweeds and marine mammals). Reassessing and updating New's (1991) assumptions, Csavas (1995) predicted that aquaculture should produce 24.5 million t in 2000, 37.2 million t in 2010 and 54.8 million t in 2020, out of the forecast world fisheries production of 84.5 million t, 97.2 million t and 114.8 million t, respectively. From these figures, the role of aquaculture towards and beyond the 21st century is evident.

But would aquaculture be able to fill this demand? How would it fill this demand? Can it help sustain world fisheries? How can it help sustain world fisheries? These are the questions that have to be answered and the challenges that aquaculture has to face. The directions with which aquaculture has developed in recent years can give us some answers, with the experience in Asia, where most of the world's aquaculture is practised, tapped in some cases.

Constraints and Strategies

Aquaculture is entering a new era, an era in which it will be expected to be more compatible with its environment and other users of resources. In other words, aquaculture has to become more sustainable. The changes in world fisheries that are taking place are certain to continue to impact aquaculture dramatically.

For the purpose of this Theme, four major issues, identified by the Congress organizers, where aquaculture has impacted or will significantly impact world fisheries are briefly reviewed and discussed. These issues include *resource partitioning*, e.g. access to stocking materials; *coastal zone management*, e.g. regulation of resource use; *stock enhancement*, e.g. economic viability; and *new technology*, e.g. biotechnology and transgenism. The first three issues are related, as they deal mainly with the environment. The last issue is distinctive, though not any less important or relevant to the sustainability of aquaculture.

Resource partitioning

Most of the stocking materials used in aquaculture are still collected from natural sources. In recent years, however, the supply has become depleted due to overexploitation, or has been unpredictable due to the limited access to broodstock and seedstock. Artificial propagation has been seen as a solution.

In the aquaculture industry, the method most often employed for inducing maturation and spawning is the injecting or implanting of reproductive hormones. However, this has been found to inflict physiological stress and injuries to the fish, making them susceptible to disease, sometimes even causing death (Liao *et al.* 1972). Future research on induced maturation and spawning should thus explore physiological, nutritional, ecological and endocrinological approaches.

Despite the established mass propagation and culture techniques of many aquaculture species, some major species still need further consideration. In eel culture, one difficulty met by the industry is the seed supply. Eel culture has been a major industry for several decades in Japan and Taiwan. In recent years, some countries have also become major players in the industry, notably China and Malaysia, compounding the problem of seedstock access. The industry is totally dependent on natural sources for its supply of glass eel. However, the supply is limited and very unpredictable, often causing difficulties in the production process. The industry is thus placing its hopes on artificial propagation. In Taiwan, Japan and China, some breakthroughs have already been achieved. Tanaka *et al.* (1995) reported recently their success in the feeding of rotifers, which are not natural food items of eel larvae, to Japanese eel, *Anguilla japonica*, larvae and growing them through 18 days. This is significant because rotifers are the most common first food in marine fish larviculture. In Taiwan, Japanese eel larvae had been successfully reared through their 31st day (Yu *et al.* 1993; Yu and Tsai 1994). However, considerable efforts have still to be made before eel larviculture techniques can be completely established.

Coastal zone management

Sustainability has long been considered to be a much more acute problem in the coastal areas, with habitat destruction and pollution as major concerns.

Illegal logging activities have destroyed a lot of marsh lands and mangroves, which are the nursery grounds for juvenile fish, prawns, and other organisms. As a result, the numbers of fingerlings and juveniles of many species are significantly reduced. On the other hand, considerable catches of broodstocks of fish and prawns for artificial propagation also reduce the natural sources.

Pollution caused by industry wastewaters and outlet waters brought about by urban life has seriously affected the quality of coastal waters and thus, the fisheries resource. As the population increases, land, air, and water pollution will become even more severe. There is a need to improve methods for waste treatment and utilization.

The continuing increase in the spread and diversity of culture species and technologies, and intensification of aquaculture practices, have contributed greatly to the increase in aquaculture production. This development has, however, come with some cost to the environment, such as deterioration in water quality, habitat destruction and loss of biodiversity. Attention should be given to environmental externalities, which could mean designing systems that encompass a relatively large and varied system of multiple products (Bardach 1995).

Competition for water and land among several industries has adversely affected the sustainability of aquaculture. For example, there is a regulation in Taiwan limiting the expansion of aquaculture beyond its current level, and in some cases, a reduction in the area devoted to aquaculture.

In many aquaculture countries, dependence on the land has been decreasing, especially with opposition from concerned groups against new aquaculture sites. Thus aquaculture has recognized the need to use more of the ocean, once an exclusive domain of capture fisheries, going farther offshore.

Stock enhancement

Stock enhancement, in its earliest form, began in Japan during the Meiji Period (1867–1912) and much earlier in the West. By means of artificial propagation and aquaculture technology, millions and even billions of seeds can be released to enhance stocks. Artificial reefs can be built and cast to make more habitats for aquatic animals. Such stock enhancement can help restock the natural resource, but it requires a preliminary pollution-free environment. This is also a complex activity. Fortunately, knowledge on stock enhancement has been accumulating. In Taiwan, the stock enhancement programme developed in late 1987 included the identification of suitable species, mass production of seeds, appropriate release sites, regulation of fishing season and fishing methods, tracking research, and result assessment. The fisherman's or aquafarmer's enthusiasm and concern for resource protection and the government's policies on resource management have a great impact on the success of the stock enhancement programme (Liao 1996).

One of the difficulties in the implementation of stock enhancement programmes is determining the optimum stocking rate. A different optimum stocking rate may be estimated, depending on the viewpoint used, e.g. oceanography, ecology and economics (Ida and Hayashizaki 1994). In Japan, fish prices tend to decrease when the production of a fish species exceeds 100 000 t. The success in the stock enhancement of chum salmon, *Oncorhynchus keta*, illustrates this point. With the aggressive release of chum salmon, production increased dramatically resulting in a decrease in its unit price. This caused a serious adverse effect on the income of chum salmon fishermen. There is need, therefore, to determine the optimum stocking rate when considering stock enhancement to make it economically viable. Tables 1 and 2 show the species and numbers released by Taiwan and Japan.

New technology

Despite the success of traditional methods, such as selective breeding, in increasing production and species diversity, there is still need for more new technologies. These new technologies should be able to respond to such challenges as overcoming the high feed costs, disease outbreaks, and limitations of the culture facilities and environment.

One of the increasingly adopted new technologies in aquaculture is biotechnology. Using conventional means of selective breeding may take several generations to accomplish. With the old method, a specific character of one species is impossible to transfer to another species using interspecific crosses because of difference in chromosomes. Transgenic technology or transgenesis breaks through these barriers, allowing transfer of a specified gene from one species to another species (Zhu *et al.* 1986; Chong and Vielkind 1989; Chen *et al.* 1991). This technology can be used to produce 'superfishes' —

Table 1. Fingerlings, subadults and adults released in Taiwan (1976–1995) (Data from Liao 1996)

Category	No. of species	Quantity (x1000)	Species (common name)
Finfish	7	6 447	Black sea bream, red sea bream, goldlined sea bream, thornfish. gray snapper, Japanese eel, marbled eel
Crustacean	6	58 727	Grass prawn, kuruma prawn, bear prawn, sand shrimp, redtail prawn, swimming crab
Mollusc	4	5 631	Small abalone, short-necked clam, purple clam, *Tapes* spp.

Table 2. Fingerlings produced and released in Japan (1994) (Data from Liao 1996)

Category	No. of species	Fingerling production (x1000)	Fingerlings released (x1000)	Species (common name)
Finfish	35	84 970	69 034	Herring, cod, blanquillo, striped jack, yellow amberjack, yellowtail, horse mackerel, rainbow fish, blue emperor, Japanese parrotfish, rock porgy, Japanese sea bass, chicken grunt, Japanese croaker, red spotted grouper, blue spotted grouper, black sea bream, red sea bream, sandfish, bluespotted mud hopper, fat cod, bar tailed flathead, black rock fish, jacopever, rock fish, devil stinger, Japanese flounder, roundnose flounder, dab, marbled sole, slime flounder, barfin flounder, spotted halibut, tiger puffer
Crustacean	14	597 037	342 047	Kuruma prawn, green tiger prawn, yellow sea prawn, speckled shrimp, northern prawn, pink shrimp, Japanese spiny lobster, Hanasaki king crab, horsehair crab, mangrove crab, green mangrove crab, swimming crab, hair crab
Mollusc	27	2 998 040	13 964 773	Ear shell, abalone, button shell, top shell, Japanese pearl oyster, marine snail, ark shell, native mussel, scallop, ezo giant scallop, freshwater clam, cockle, smooth giant clam, common hard clam, shield clam, short-necked clam, hen clam, northern clam, gaper, sand clam, razor shell, octopus
Echinoderm	7	74 228	75 425	Sea urchin, sea cucumber

healthy, tasty, and fast-growing — and ensure year-round production (Liao 1994). It improves market competition of aquacultural products due to the lower costs. However, transgenic technology has encountered some environmental and ethical questions, and there have been calls for its judicious use.

Any expansion of aquaculture to help sustain world fisheries should be complementary; it should not be grounded on competition. An answer would be stock enhancement, whereby aquaculture helps increase the stock for capture fisheries. Another collaboration would be cage aquaculture. An improvement of aquaculture engineering and introduction of modern management techniques can contribute to the sustainable development of aquaculture. For example, advances in cage aquaculture technology in Norway provided a much-needed boost in its aquaculture production. From a low production in the early 1990s, Norway recovered in 1994, when it produced 207 000 t of Atlantic salmon *Salmo salar*. With its accumulated cage aquaculture knowledge, Norway is expected to more than triple its production by the end of this century.

Competition for scarce natural resources, such as groundwater, land, and energy, has prompted the development of superintensive aquaculture. Superintensive aquaculture allows increased control over the biotic and abiotic components of the system and the maximization of the production potential. Compared with conventional intensive aquaculture and based on 50 t/year of eel production in Taiwan, superintensive recirculating aquaculture, at a stocking density of 80 kg/m^2 (against intensive aquaculture's 2 kg/m^2), requires 93% less water per kg, 96% less space, and 80% less labour (Table 3). Aquaculture practices that are wasteful of resources will not be sustainable in the face of these highly valued and tightly regulated resources (Chamberlain and Rosenthal 1995). These systems, however, generate a large amount of waste. Improvements in mechanical and biotechnical, notably, microbial, means for using the waste as fertilizer should be considered (Bardach 1995).

Other Issues

The negative consequences of the aquaculture industry have been attributed to inadequate planning, management and implementation or enforcement of aquaculture policy (Liao *et al.* 1995). On the other hand, the success of aquaculture derives from its development as a source of income or profit, rather than for subsistence or food. There is a need to change the attitude from one of profit-mindedness to providing more food. Because of the profit motivation, the trend in aquaculture is towards the culture of high-value species, which many of the world's population who really need animal protein cannot easily afford.

One of the critical areas in the aquaculture industry is its marketing system. Thus in Taiwan, fishermen's associations receive subsidies from the central and local governments to assist

Table 3. Comparison of superintensive recirculating aquaculture and conventional intensive aquaculture in Taiwan, based on 50 t/year eel production

	Superintensive recirculating aquaculture	Conventional intensive aquaculture
Average stocking density	80 kg/m^2	2 kg/m^2
Culture area	1000 m^2	25 000 m^2
Water requirement per kg of eel produced	0.7 t	10 t
Average labour requirement	2 persons	10 persons
Water quality	More stable	Less stable
Contamination from outside	Less likely	More likely
Chemical used	Much less	40 times more
Wastewater	Less	10 times more

fishermen and aquafarmers. One of the association's goals is to shorten the marketing channels through production-marketing cooperation among members. In 1993, Taiwan's total aquaculture production was 285 275 t. About 27%, or 70 210 t of the total aquaculture production, was sold on the wholesale market for consumption and at local production areas (Department of Agriculture and Forestry 1993). However, only 9% or 23 117 t of the total production went through the cooperative production-marketing programme that allows products to be auctioned directly on the wholesale market. Therefore, it is unlikely that the present cooperative system will be able to provide a satisfactory service to aquafarmers (Liao *et al.* 1995).

Market demand for traditional aquaculture products has not kept pace with increases in production, and the oversupply has resulted in lower prices (Liao *et al.* 1995). Since the size of aquaculture farms is usually relatively small, depressed prices greatly affect aquafarmers, particularly those culturing traditional species, such as tilapia *Oreochromis* spp. and milkfish *Chanos chanos*. There is also a need to utilize processing technology, to increase added values, to produce environment-friendly feeds, to establish quality control, and to adopt effective policies and marketing systems.

Mass mortality caused by disease outbreaks has been experienced in cultured prawns in Taiwan and Southeast Asian countries. In Taiwan, diseases plummeted the prawn culture industry into a crisis (Liao 1989*a*, *b*, 1990). Almost a decade has passed and the prawn culture industry is still trying to recover from the deleterious effects. One preventive measure against diseases is the development of vaccines. Disease prevention through the use of an improved diet, more efficient management of the culture environment, and a highly reliable system of sanitation should be further explored.

There is also a growing interest and concern about interactions between aquaculture and the environment. Aquaculture development in many nations has been hampered by the limitation of available land. Aside from this, competition with other sectors or interests, such as recreational, navigational, traditional and aesthetic, have posed major constraints on the development of aquaculture. Direct competition has also occurred with existing wild fisheries. On the other hand, in Taiwan, the recent overproduction of milkfish fry has resulted in a fall of wild fry collection.

Getting the government's attention for resource management, information dissemination and personnel training is also important. These personnel should be appropriate and conscientious, those who can devote their time and effort toward the practice of aquaculture. This will be a problem since many of the younger generation prefer the urban areas for work. The interest in a seagoing career has dwindled because of the improvement in Taiwan's standard of living. There is thus a need for more innovative recruitment schemes.

In Taiwan, unrealistic and arbitrary proposals from the government without the benefit of research or consultations with trained aquaculture specialists have impeded the growth of aquaculture. These policies include a change in the use of aquacultural areas, shifting from 40% marine culture and 60% inland culture, to 60% marine and 40% inland culture. (Liao *et al.* 1995). The regulatory framework should be restructured through a wide range of legislation, taking into consideration the fast pace aquaculture is developing. An example of this policy support is in agriculture. Agriculture flourished because the irrigation infrastructure was already in place when its development was promoted. In Taiwan, discussions on exclusive aquaculture zones are in place but legislation on this has yet to find fruition.

In some countries, opposition against new aquaculture sites is gaining ground, with the public objecting to the negative environmental impact of aquaculture. One of the demands is the cessation of further development of the mangrove environment for aquaculture purposes. The use of mangroves for aquaculture, aside from its impact on the environment, such as habitat destruction, resulted in a reduction in the natural source of fry.

There is also a need for promoting information transfer to help determine the direction of aquaculture. Each aquaculture developing nation can then train others through information exchange and technology transfer.

Conclusion

In recent years, the success of aquaculture has been attributed, not to its development for subsistence or food, but to its development as a source of income or profit. It is time, at this stage of its development, to think otherwise. The world-wide expansion of aquaculture is making an increasingly important contribution to world fisheries production. This is especially so with the declining capture fisheries catch, which is traditionally the main source of aquatic products. Aquaculture should not only meet the increasing demand for medium- and high-grade fishery products, but also for traditional products.

One of the promising avenues by which aquaculture can meet the demand for high-value fisheries is by extending itself beyond its traditional land-based system (Liao and Shyu 1992). Cage aquaculture is one such possibility. In Asia, where there is a great potential for this new technology, the Japanese have already made advanced efforts in sea farming and ranching (Nakahara 1992). Asia and other countries, where limited and increasing costs of land and poor coastal zone management are problems, have great potential for cage aquaculture and can benefit from some of the advances in other countries, such as Norway and Scotland (Dahle 1995).

In Japan, a group of fishermen, recognizing the harm that denudation of a nearby forest has done to their means of livelihood and the fisheries ecosystem, recently took steps to correct it. With shovels and seedlings, they climbed up the mountains to replenish the barren land in the sincere hope that the seedlings that they planted will grow to become forests. In nature, the forests yield fallen leaves, which will eventually decay to provide important nutrients for plankton, commencing a complex food web cycle that will eventually trickle down to the fishermen and back and so on. Without the forests, the plankton would have less nutrients to subsist, thus breaking the cycle that includes shrimps and small and big fishes. This thoughtful gesture by the fishermen to restore the forest ecosystem will eventually redound toward helping sustain the fisheries.

Unlike aquaculture, the benefits and inputs in capture fisheries cannot be planned or controlled. Fisheries rely on nature for their inputs. If this ability is misused, disaster can result as experiences in aquaculture have shown. A concrete example is the prawn culture industry crisis that occurred in Taiwan and Southeast Asia. Proper planning is, therefore, important to avoid past pitfalls.

Aquaculture is certainly up to the challenges that were posed at the beginning of this paper. Looking at the past and present directions as has been shown will help to anticipate the needs of future generations. With such technologies as stock enhancement, cage aquaculture, and superintensive aquaculture for food production, a pollution-free environment, and appropriate and resourceful people, aquaculture can indeed help sustain world fisheries.

Acknowledgments

I would like to thank the Australian Institute of Marine Science and the R.O.C. National Science Council for providing support for my participation in this Congress. Mr Darryl Grey, Theme 4 Convenor, and Dr John Glaister, Congress Chair, were instrumental in inviting me to address the congress as the Theme 4 speaker for which this paper was submitted. My special and sincere gratitude to them. Thanks are also due Dr Herman T. Weng and Mr Jonathan VA A. Nuñez for their help in the preparation of the manuscript.

References

Bardach, J (1995). Aquaculture and sustainability. *World Aquaculture* **26**(10), 2.

Chamberlain, G, and Rosenthal, H (1995). Aquaculture in the next century: Opportunities for growth challenges of sustainability. *World Aquaculture* **26**(1), 21–5.

Chen, T T, Lin, C M, and Knight, K (1991). Application of transgenic fish technology in aquaculture. *Bulletin of the Institute of Zoology, Academia Sinica Monograph* **16**, 375–86.

Chong, S C, and Vielkind, J R (1989). Expression and fate of CAT reporter gene microinjected into fertilized medaka (*Oryzias latipes*) eggs in the form of plasmid DNA, recombinant phage particles and its DNA. *Theoretical and Applied Genetics* **78**, 369–80.

Csavas, I (1995). The status and outlook of world aquaculture, with special reference to Asia. In 'Aquaculture Towards the 21st Century: Proceedings of the INFOFISH-AQUATECH '94 Conference'. (Eds K P P Nambiar and T Singh) pp. 1–13 (INFOFISH: Kuala Lumpur, Malaysia).

Dahle, L A (1995). Offshore fish farming — Recent developments. In 'Aquaculture Towards the 21st Century: Proceedings of the INFOFISH-AQUATECH '94 Conference'. (Eds K P P Nambiar and T Singh) pp. 169–84 (INFOFISH: Kuala Lumpur, Malaysia).

Department of Agriculture and Forestry (1993). 'Department of Agriculture and Forestry, 1993. Taiwan Area Agricultural Products Wholesale Market Yearbook.' 1993 Edition. (Department of Agriculture and Forestry, Provincial Government of Taiwan: Taiwan, ROC) (In Chinese).

FAO (1995*a*). 'FAO Yearbook, Fishery Statistics: Commodities, Vol. 77.' *FAO Fisheries Series* No. 45, FAO Statistic Series No. 124. (FAO Fisheries Department, Food and Agriculture Organization of the United Nations: Rome, Italy). 428 pp.

FAO (1995*b*). Aquaculture Production Statistics, 1984–1993. FAO Fisheries Circular No. 815, Revision 7. (FAO Fisheries Department, Food and Agriculture Organization of the United Nations: Rome, Italy). 186 pp.

FAO (1996). Global Fishery Production in 1994. (Fishery Information, Data and Statistics Unit, Food and Agriculture Organization of the United Nations: Rome, Italy). (Available online at URL: http://www.fao.org/waicent/faoinfo/fishery/catch/catch94a.htm).

Ida, H, and Hayashizaki, K (1994). Stock enhancement and density dependence in chum salmon, *Oncorhynchus keta*. *The Iden* **48**(5), 37–43. (In Japanese).

Liao, I C (1989*a*). *Penaeus monodon* culture in Taiwan: Through two decades of growth. *International Journal of Aquaculture and Fisheries Technology* **1**(1), 16–24.

Liao, I C (1989*b*). Taiwanese shrimp culture: A molting industry. In 'Proceedings: Shrimp World IV'. (Eds K Chauvin, P Menesses, W Chauvin and A Cuccia) pp. 55–83 (Shrimp World, Inc.: New Orleans, Louisiana, USA).

Liao, I C (1990). Aquaculture in Taiwan. In 'Aquaculture in Asia'. (Ed M M Joseph) pp. 345–69 (Asian Fisheries Society, Indian Branch: Mangalore, Karnataka, India).

Liao, I C (1994). Role of biotechnology in aquaculture in Taiwan. In 'Socioeconomics of Aquaculture'. Tungkang Marine Laboratory Conference Proceedings 4. (Eds Y C Shang, P S Leung, C S Lee, M S Su, and I C Liao) pp. 15–36 (Tungkang Marine Laboratory, Taiwan Fisheries Research Institute: Tungkang, Pingtung, Taiwan).

Liao, I C (1996). Status, problems, and prospects of stock enhancement in Taiwan. Paper presented at the Asia-Pacific Conference on Science and Management of Coastal Environment, 25–28 June 1996, Hong Kong.

Liao, I C, and Shyu, C Z (1992). Evaluation of aquaculture in Taiwan: Status and constraints. In 'Resources and Environment in Asia's Marine Sector'. (Ed J B Marsh.) pp. 185–97 (Taylor and Francis: Washington, DC, USA).

Liao, I C, Tseng, L C, and Cheng, C S (1972). Preliminary report on induced natural fertilization of grey mullet, *Mugil cephalus* Linnaeus. *Aquaculture* **2**(1), 17–21. (In Chinese, English abstract).

Liao, I C, Lee, W C, and Hsu, Y K (1995). Aquaculture in Taiwan: Toward a sustainable industry. In 'International Cooperation for Fisheries and Aquaculture Development: Proceedings of the 7th Biennial International Conference of the International Institute of Fisheries Economics and Trade, Vol. 2'. (Ed D S Liao) pp. 1–13 (National Taiwan Ocean University: Keelung, Taiwan and International Institute of Fisheries Economics and Trade: Oregon, USA).

Nakahara, H (1992). Japanese efforts in marine ranching development. In 'Resources and Environment in Asia's Marine Sector'. (Ed J B Marsh) pp. 199–216 (Taylor and Francis: Washington, DC, USA).

New, M B (1991). Turn of the millennium aquaculture: Navigating troubled waters or riding the crest of the wave? *World Aquaculture* **22**(3), 28–49.

Tanaka, H, Kagawa, H, Ohta, H, Okuzawa, K, and Hirose, K (1995). The first report of eel larvae ingesting rotifers. *Fisheries Science* **61**(1), 171–2.

Yu, T C, and Tsai, C L (1994). The artificial propagation of *Anguilla japonica*. *Taiwan Fisheries Research Institute Newsletter* **61**, 5–8. (In Chinese).

Yu, T C, Tsai, C L, Tsai, Y, and Lai, J Y (1993). Induced breeding of Japanese eels, *Anguilla japonica*. *Journal of Taiwan Fisheries Research* **1**(1), 27–34.

Zhu, Z, Xu, K, Li, G, Xie, Y, and He, L (1986). Biological effects of human growth hormone gene microinjected into the fertilized eggs of loach, *Misgurnus anguillicaudatus* (Canter). *Kexue Tongbao* **31**, 988–90.

The Myth of Sustainability in Managed Wild Fisheries: The Economic Case for Aquaculture Subsidies

Trellis G. Green[A] *and James R. Kahn*[B]

[A] Department of Economics and International Business, University of Southern Mississippi, PO Box 5072, Hattiesburg, Mississippi 39406-5072 USA.
[B] Department of Economics, Stokeley Management Center, University of Tennessee Knoxville, Tennessee 37996-0550 USA, and Oak Ridge National Laboratory, Oak Ridge, Tennessee 37831-6205, USA.

Summary

This study recasts the role of aquaculture as an alternative, *demand-side* policy to supplement traditional, *supply-side* fisheries management. This works by expanding the availability of cultured fish substitutes, which lowers the demand for wild fish. It is shown that traditional policy cannot guarantee sustainability when fish demand accelerates. The key finding is that markets fail in both aquaculture and the wild fishery so as to impede the optimal substitution of aquaculture by private investors. The primary reason is because the market price of fish lies below its true scarcity value. A methodology is devised for measuring aquaculture's beneficial impact on wild fisheries in order to assess the feasibility of promoting aquaculture (subsidies) when the market fails to provide the socially optimum level. The study concludes that increased aquaculture subsidization is justified when social net benefits of aquaculture exceed full social costs, including potential deleterious effects on the wild fishery.

Depletion of ocean fisheries-policy failure

Despite 'traditional' regulatory policy, ocean fisheries began to decline, particularly in developed countries, after world landings peaked in 1989 (Hargis *et al.* 1986). The United Nations Food and Agriculture Organization has reported that 13 of 17 major ocean fisheries are in serious trouble. Paradoxically, the demand for ocean resources, for many reasons, began to accelerate (Edwards 1992) just as the overall pattern of depletion appeared to be worsening (Chandler and Turnbull 1990).

One reason cited in the past for this policy failure was that policies were not implemented, but this is no longer a valid explanation. Today, fisheries without any policies to mitigate the overexploitive impact of open-access externalities are the exception rather than the rule. Even collapsing fisheries feature some degree of traditional fisheries management.

When policy did fail it was usually attributed to improper selection and implementation, rather than the nature of policy itself (Milon 1993). This is a well-known, contributory factor when open-access type regulations, for example, are selected to limit effort rather than entry. Thus, restrictions on gear, season, fishing location, and minimum size have been implemented to reduce fishing efficiency. However, the potential for depletion under open-access remains strong because entry is not directly controlled and rent is dissipated (Anderson 1986).

Experts reached a consensus that the 'proper' policy was to efficiently harvest stocks by limiting access through taxation, licensing, or the individual transferable quota (ITQ). However, taxation proved to be politically infeasible and practically unworkable. As for licensing, evolving technologies like sonar, spotter planes, and Global Positioning Systems (GPS), still resulted in higher catch-per-unit-effort by limited entrants, which can shrink fish populations. This is known as 'capital stuffing' (Copes 1986).

Gradually, the ITQ became the preferred limited-access tool because it theoretically circumvents capital stuffing, but ITQs also have problems (Muse and Schelle 1989). First, ITQ incentives encourage fishers to maximize return on the individual quota. This does not control incidental catch nor other actions ('high grading') which can adversely affect targeted and non-targeted stocks (Copes 1986). Second, ITQs are difficult to enforce, which can lead to 'quota busting' by individual fishers (Hide and Ackroyd 1990).

The problem with all traditional policies, both open-access and limited-access, is that they focus exclusively on the 'supply side' by restricting inputs (effort and technology) and outputs (harvest). Their inherent weaknesses are exacerbated by accelerating global demand, which increases the returns to harvest. Thus, the incentives are increased for capital stuffing under a licensing regime, for quota busting with ITQs, and for entry under open-access regulations.

This paper argues that supply-side policies, which emphasize how to produce the harvest, cannot sustain a fishery forever when subjected to continually increasing demand. Our goal is to recast the role of aquaculture as an alternative, 'demand-side' policy to supplement traditional policy. Another goal is to devise a methodology to assess the feasibility of subsidizing aquaculture when the market fails to provide the socially optimal level.

The myth of sustainable yield

Even an inefficiently regulated, open-access fishery is said to be characterized by sustainability, in the sense that fish stocks remain stable whenever harvest equals natural growth. Since every harvest level which is less than or equal to maximum natural growth can be sustainable, the economic literature has centred around choosing the optimal level of harvest or effort in order to determine social benefits of limited-access policy. This concept of sustainable yield is predicated on a static demand curve, held constant at some initial position.

However, human population and income do not remain constant, tending to rise over time. Likewise, the societal demand for fish would be expected to increase. Edwards (1992) found that seafood demand is rising because of a shift in preferences, independent of population growth. If demand pressure is sufficiently high, a stock can be driven to extinction, unless it remains viable as it approaches zero and the cost of catching the last remnants of the stock is exceedingly high (Anderson 1986). In the absence of limited-access, higher market prices create temporary rents, which attract more effort. Moreover, if the fish stock is characterized by a critically *depensated* growth function, the fish stock could be driven to zero even if the cost of catching the last remnants is exceedingly high.

Because real world-demand for fish is continually increasing (and the productivity of the ecosystem is generally constant,[1] *the regulated open-access fishery is inherently unsustainable.* Limited-access may lessen the impact of increasing demand, but in and of itself it cannot prevent accelerating demand pressure from reducing per capita consumption and depleting some fishery stocks in the long-run. In addition to increasing demand, factors such as environmental degradation and improper management might also be important.

Demand pressure is an important problem in developing countries, where population is growing rapidly and agricultural productivity is diminishing from environmental degradation. For example, rapid population growth in urban areas of Brazil, coupled with the conversion from locally-managed common property fisheries to open-access commercial fisheries, is amplifying the ongoing depletion of fishery resources in the Amazon system.

Similarly, in the USA, mammoth increases in demand for red drum (*Sciaenops ocellatus*), associated with a Cajun food fad called 'blackened redfish', nearly drove the stock to extinction (Rockland 1988). Cod, salmon, striped bass, giant bluefin, and other fish and shellfish have also exhibited declining yields.

Welfare implications of aquaculture/open-access fishery interactions

In an open-access fishery, which dissipates rent, social benefits consist entirely of consumers' and producers' surpluses. Open-access equilibrium can be modelled by specifying a different supply curve for each level of fish stocks (Kahn and Kemp 1985). This is shown in Figure 1a, where *S0* is the low cost supply function associated with maximum stock (carrying capacity), and *S9* is a higher cost supply function near stock depletion. Line *D1* represents the existing demand for fish.

Since there is a unique equilibrium catch associated with each level of stock, there is only one point on each economic supply curve that can be characterized as a biological equilibrium. This is obtained by mapping the equilibrium catch function onto the family of supply curves. The locus of biological equilibria in Figure 1a represents this mapping and illustrates the set of biological equilibrium points. An economic equilibrium occurs at point *b* where the locus of biological equilibria intersects fish demand *D1* at catch Q_7 and price P_7.

The economic net benefits associated with this open-access equilibrium are shown in Figure 1b. It should be noted this analysis assumes open-access, which is reasonable because of its *de facto* prevalence and the political and industry resistance to

[1] Ecosytem productivity might not be constant if other anthropogenic activities affect the ecosystem, e.g. pollution could reduce it while pollution abatement could increase it. However, ecosystems do have a maximum productivity associated with a given state of the world.

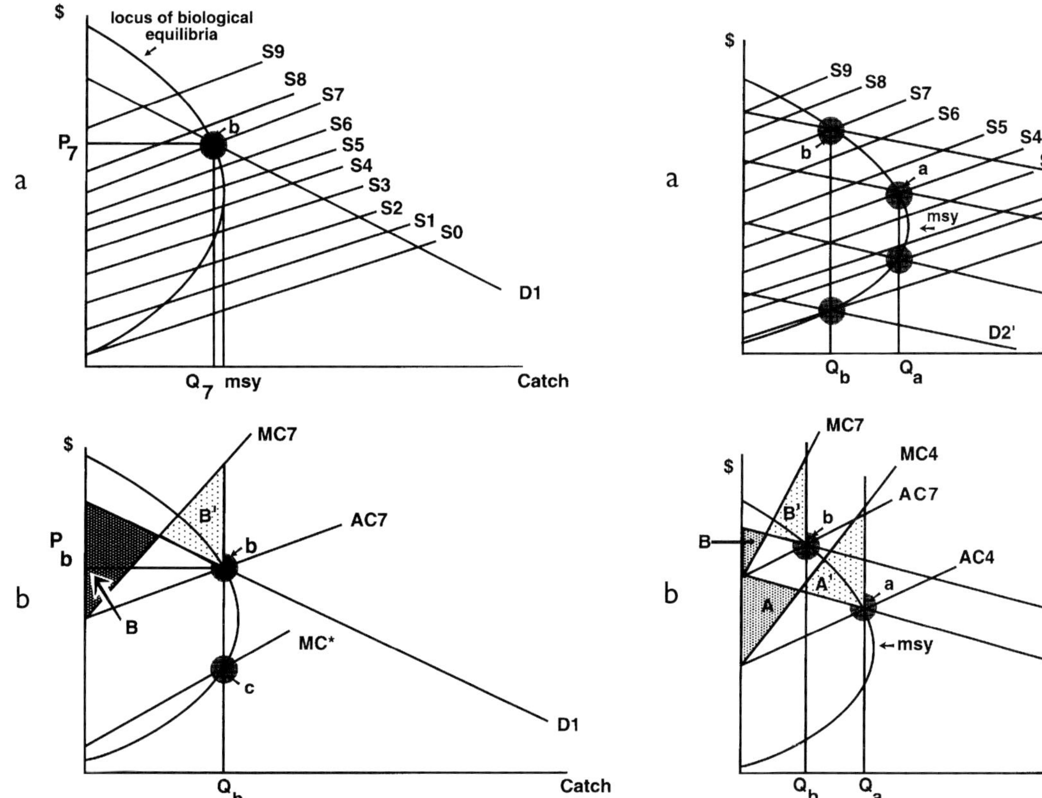

Fig. 1. Open-access bioeconomic model showing (**a**) equilibrium and (**b**) economic net benefits (B-B').

Fig. 2. Effects of aquaculture on the wild fishery (**a**) open-access equilibrium (**b**) economic net benefits [(A-A')-(B-B')].

limited-access historically. The open-access assumption does not limit the generality of these results because, in the limited-access case, one only needs to consider fishery rent in addition to consumers' and producers' surpluses, and the results are similar.

Since open-access implies that the inverse of the supply function *S7* is the average cost function (*AC7*) and not the marginal cost function, the marginal cost function is identified at *MC7*. Net social benefit is the shaded area between marginal cost function *MC7* and demand function *D1*. Note that there is a positive area where demand is greater than marginal cost (area *B*) and a negative area where marginal cost is greater than demand (area *B*').

Thus, net social benefits associated with an open-access equilibrium at point *b* are equal to area *B* minus area *B*'. The negative area *B*' is the open-access externality associated with overexploitation, since the same Q_b catch could also be generated with far less effort at point *c* on *MC**. This efficiency savings, generated by fewer vessels and greater fish abundance, represents the foregone *rent* that could have been earned from the fishery.

The welfare impact of aquaculture on the open-access fishery will be examined through its effects on the demand for wild fish. Cultured fish are a substitute for wild fish, so the creation of this substitute will lower the demand for wild fish. In Fig. 2a, if the initial equilibrium on demand *D1* lies at point *b* above the maximum sustainable yield (msy), a reduction in demand from *D1* to *D1*' will increase wild catch from Q_b to Q_a. Conversely, if the initial equilibrium lies below msy on demand *D2*, then a reduction in demand to *D2*' will reduce wild catch from Q_a to Q_b. In both cases the equilibrium moves clockwise along the locus of biological equilibria, indicating an increase in the stocks of wild fish. Thus, one unambiguous outcome of aquaculture promotion from demand reduction (*ceteris paribus*[2]) is an increase in the population of wild fish (Anderson 1985).

The direct welfare consequence of this aquaculture-induced reduction in fish demand is ambiguous, however. As shown in Figure 2b, the shift in demand changes wild fishery net benefits from area (*B-B*') to area (*A-A*'). Depending on the slopes of wild fish demand and supply functions, and where they intersect the locus of biological equilibria, the change in net benefits from the introduction of aquaculture could be positive or negative. This is an empirical matter that will vary across fisheries.

Because wild stocks increase, there would be additional beneficial effects of aquaculture to factor into the benefit-cost calculus, regardless of whether benefits are positive or negative in the commercial wild fishery. First, there is the increase in recreational fishing benefits from enhanced wild stocks. Second, there is a spectrum of passive use values associated with higher stocks, including those directly related to the stocks and those related to the increased productivity of the ecosystem in general.

[2] Other things being equal

Thus, even if the *direct*, open-access benefits of aquaculture in Figure 2b [(*A-A*')-(*B-B*')] for a given fishery were small or negative, the *indirect* recreational and ecosystem benefits could still be large enough to make society better off. This is particularly true for species, like red drum, which have a relatively high recreational value (Green 1994).

Does the market provide the right amount of aquaculture?

Aquaculture's share of world seafood output is no more than 20%, yet this is not necessarily indicative of too little aquaculture. Aquaculture industries range from salmon ranches in Norway and oyster beds in North America, to prawn and carp farms in developing countries like Thailand. Optimal aquaculture substitution requires that markets function so that fish price fully reflects scarcity cost, i.e. markets must not fail.

Because aquaculture is a private good, one might expect the market to generate optimal increases in the quantity of cultured output, when scarcity in depleted wild fisheries increases prices in the long-run. In fact, the literature suggests that this type of *substitution* is the most feasible scarcity-mitigating response mechanism when scarcity drives up resource prices (Smith 1979). New discovery and technological progress are two other mitigating responses to scarcity. However, new discoveries are unlikely to mitigate scarcity in fisheries since fishery reserves are largely identified and exploited. Technological progress for wild fisheries actually increases scarcity by intensifying fishing pressure. Moreover, this process can be exacerbated by biological lags (Clark 1990), which mask the crossing of critical population thresholds.

The market might not be capable of generating optimal aquaculture substitution, however, because the wild fishery is subject to price distorting market failures. First, market price might not incorporate the right signals since the price of wild fish does not embody foregone recreational or passive use benefits. Price will lie below its true scarcity cost. Second, in the presence of open-access externalities in the wild fishery, price is based on average cost rather than marginal cost (Anderson 1986). Furthermore, since rent eventually disappears under open-access, scarcity rent cannot properly signal current or impending increases in scarcity.

Within the market for cultured fish, market failure also occurs because pioneering research and development (R&D) of a highly risky nature is necessary to discover efficient culturing techniques. Initial capital investment can be large (Huffman *et al.* 1992). The aquaculture firm must bear all the risks of the R&D investment, yet non-risk-takers are not excluded from sharing the benefits of pioneering information. These *public goods* benefits of R&D cannot be factored into the *private goods* decision making of firms when they decide upon their optimal level of R&D investment. Additionally, other conventional R&D arguments, such as the importance of risk pooling, would be applicable to aquaculture (Cohen and Noll 1991). Second, if aquaculture is characterized by declining average costs (Adams and Pomeroy 1992), where increased scale lowers the marginal costs of marketing and distribution, then subsidization may be required to *incubate* this industry.

Demand-side fisheries policy: aquaculture promotion

The scale of aquaculture activities depends on its profitability, which in turn is a function of the price of the cultured fish and the cost of producing it. Conversely, the price of the cultured fish depends on the price of the wild fish. Thus, policies that either raise the price of wild fish or reduce the cost of cultured fish will promote aquaculture substitution.

This can be reduced to a single metric by looking at the *substitution ratio*, which is the ratio of the price of wild fish to the unit cost of producing cultured fish. Aquaculture becomes more profitable, *ceteris paribus*, as the substitution ratio increases, either from increasing wild price or reducing culture cost. Likewise, aquaculture becomes less profitable if the substitution ratio falls, and production decreases. However, one of the reasons we do not have more aquaculture is because the prevailing substitution ratio for many fisheries, especially those with a high recreational value, lies below the socially optimum ratio. The market fails, of course, because the market price of wild fish is not high enough to reflect the social cost of harvest, which is the true scarcity price in terms of foregone recreational, ecosystem, and passive use values.

Obviously, correcting this market failure by eliminating the discrepancy between the price of wild fish and their social cost should be a policy objective. To this end, all traditional supply-side policies have been geared to raise the numerator of the substitution ratio (market price). Unfortunately, they are doomed to fall short because of the unrealistic prospect of internalizing the opportunity costs of harvest (value of fish in recreational or ecosystem use) into market price. This, coupled with the practical difficulties of implementing limited-access, makes removal of this price distortion unlikely.

Consequently, aquaculture substitution is likely to be sub-optimal because the observed market price of wild fish remains relatively low in comparison to the costs of cultured fish. If only policy could eliminate the price discrepancy created by the divergence of the true social cost of wild harvest (scarcity price) from its observed private cost (market price), then the substitution ratio could rise and aquaculture could expand to a more optimal level. Supply-side policy cannot do this by merely limiting fishing inputs and outputs, especially when demand is increasing.

This study advocates the consideration of policies that stimulate aquaculture production by lowering the cost of culture in the substitution ratio's denominator. This is a 'demand-side' approach to wild fishery management because policy makers control the availability of wild fish substitutes (demand), rather than wild harvest inputs and outputs (supply). Increased production of substitutes reduces the demand for wild fish as shown in Figure 2b.

Production and/or R&D subsidies are classic policy tools which can be used to lower culture costs and expand aquaculture output. If aquaculture is characterized by declining average costs

(Adams and Pomeroy 1992), then any initial, subsidy-driven expansion in scale could move aquaculture down the long-term average cost curve and reduce future costs. More importantly, as a scarcity-mitigating response, the substitution approach can confront the problem of increasing demand, while traditional policies can be used to remove open-access externalities.

R&D activities, which have been shown to suffer from market failure, can be expanded in a variety of ways. First, R&D activities in the culture industry can be promoted by direct subsidies or guaranteed loans to private investors. Second, R&D can be encouraged by government-sponsored R&D at universities, cooperative extensions, and government research laboratories.

Aquaculture is not the only demand-side approach. The National Marine Fisheries Service in the USA has sponsored the *Marine Fisheries Initiative* (MARFIN), which seeks to protect vulnerable stocks by developing 'underutilized' wild stocks like butterfish. However, the resultant pressure has pushed other previously underutilized species toward decline, e.g. the shark fishery. Thus, promotion of underutilized stocks might be only a temporary solution.

In many areas of the world, increased fishery pressure is in large part due to a breakdown in the productivity of terrestrial agriculture. Soil depletion, desertification, and deforestation have reduced terrestrial food productivity, resulting in an increased demand for marine food sources. Even though aquaculture can play an important role in relieving demand pressure in these circumstances, the policy priority should be in reversing the decline in terrestrial productivity.

CONCLUSIONS

In response to the decline of major ocean estuarine, and inland fisheries, we examine the need to broaden the fisheries policy mix to include demand-side approaches. The objective is to expand aquaculture output as a commodity substitute for overexploited wild fish, thereby reducing wild fish demand. Traditional fisheries policies, such as limited-access, are supply-side in nature. They possess shortcomings exacerbated by increasing global demand, which make sustainability difficult to achieve, even under limited-access.

A simple model identifies three components of aquaculture's social net benefits derived from the wild fish stock. These include benefits of direct wild fishery use, indirect recreational use, and indirect passive use (including ecosystem services). These social net benefits, in addition to private benefits received by aquaculturists (profit), can be estimated for use in benefit-cost analysis.

The principal finding is that market failures in the wild fishery and aquaculture impede the optimal substitution of fish farms by the private market. The underlying reason is because the price of wild fish lies below its true social cost. Thus, fish price cannot properly signal aquaculture as a scarcity-mitigating response in all fisheries.

As a result, the substitution ratio of wild fish price to aquaculture cost lies below the desired social optimum. Traditional policy cannot raise the numerator of this ratio enough because it cannot fully internalize externalities of wild harvest on recreational, passive, and ecosystem uses.

Demand-side policy, such as increased aquaculture subsidization, can raise the substitution ratio by reducing the cost of culture. Potential policies include direct production subsidies, guaranteed R&D loans, or R&D development grants to universities and cooperative extensions. It could also be incorporated into co-management schemes (Pinkerton 1994).

Multi-disciplinary research is needed to identify viable candidate fisheries, which are characterized by chronic depletion and culture feasibility. This would involve biological studies, sociological studies, market risk analysis, and industry structure models. Market and non-market valuation techniques are needed to quantify aquaculture net benefits and adverse effects.

Subsidy policy is justified if the present value of net social benefits from aquaculture exceeds the applicable social costs. Applicable social costs should include the opportunity cost of programme funds plus any adverse effects of aquaculture on wild fish genetics, biodiversity, ecosystem habitat, and disease. Research in the area of adverse effects has just begun (Herk and Rogers 1989).

Although our understanding of the social costs and benefits of aquaculture is at a developmental stage, there appear to be significant advantages to pursuing aquaculture as a policy for protecting the continued viability of wild stocks. These demand-side policies can certainly enhance the sustainability and non-harvest benefits associated with the wild fishery, and will lead to a more socially-optimal aquaculture scale. Such policies could also engender a more efficient harvesting of the wild fishery.

REFERENCES

Adams, C M, and Pomeroy, R S (1992). Economies of size and integration in commercial hard clam culture in the southeastern United States. *Journal of Shellfish Research* **11**, 169–76.

Anderson, J L (1985). Market interactions between aquaculture and the common-property commercial fishery. *Marine Resource Economics* **2**, 1–24.

Anderson, L G (1986). 'The Economics of Fisheries Management.' (Johns Hopkins: Baltimore).

Chandler, W J, and Turnbull, A (1990). Needs assessment of the National Marine Fisheries Service. National Fish and Wildlife Foundation, Washington D.C.

Cohen, L R, and Noll, R G (1991). 'The Technology Pork Barrel.' (Brookings Institute: Washington D.C.).

Clark, C W (1990). 'Mathematical Bioeconomics.' (Wiley and Sons: New York).

Copes, P (1986). The individual quota in fisheries management. *Land Economics* **62**, 278–91.

Edwards, S F (1992). Are consumer preferences for seafood strengthening? tests and some implications. *Marine Resource Economics* **7**, 141–51.

Green, T G (1994). Allocation between commercial and recreational sectors in stressed marine fisheries. *Society and Natural Resources* **7**, 39–55.

Hargis, W J Jr., Baker, R J, Bemiss, F, Cato, J C, Harville, J P, Haynie, A W, Lyman, H, Mehos, J A, Peterson, J G, Smith, C L, and Towell, W E (1986). NOAA fishery management study. Unpublished report: National Marine Fisheries Service, Silver Spring, Maryland, USA.

Herk, W H, and Rogers, B D (1989). Threats to coastal fisheries. In 'Marsh management in coastal Louisiana: effects and issues-proceedings of a symposium'. Biological Report 89-22 (Eds W G Duffy and D Clark) pp. 196–212. US Fish and Wildlife Service, Washington, D.C.

Hide, R P, and Ackroyd, P (1990). Depoliticizing fisheries management: Chatham Islands' Pata (Abalone) as a case study. Center for Resource Management, Lincoln University, New Zealand.

Huffman, D C, Baldridge, T R, and Dellenbarger, L E (1992). Study indicates redfish culture requires $750 000 investment. *Water Farming Journal* **7**, 10.

Kahn, J R, and Kemp W M (1985). Economic losses associated with the decline of submerged aquatic vegetation in the Chesapeake Bay. *Journal of Environmental Economics and Management* **12**, 245–63.

Milon, J W (1993). US fisheries management and economic analysis: implications of the Alaskan groundfish controversy. *American Journal of Agricultural Economics* **75**, 1177–82.

Muse, B, and Schelle, K (1989). Individual fisherman's quotas: a preliminary review of some recent programs. *CFEC Report* No. 89–1, Alaska Commercial Fisheries Entry Commission, Juneau, Alaska.

Pinkerton, E (1994). Economic and management benefits from the coordination of capture and culture fisheries: the case of Prince William Sound pink salmon. *North American Journal of Fisheries Management* **14**, 262–77.

Rockland, D B (1988). User conflicts and 'blackened' economic analysis: a case history of the red drum fishery in the Gulf of Mexico. In 'Marine Fishery Allocations and Economic Analysis'. (Ed J W Milon) pp. 18–23. *Southern Natural Resources Economics Committee* No. 26, Mississippi State, MS.

Smith, V. K (1979). 'Scarcity and Growth Reconsidered.' (Resources for the Future: Washington D.C.).

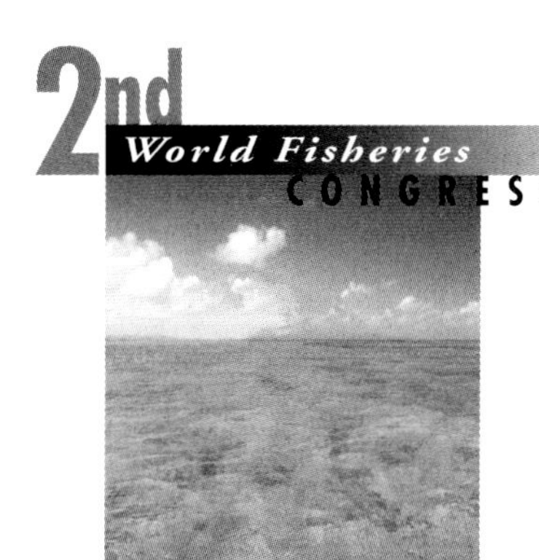

World Inland Fisheries and Aquaculture — Changing Attitudes to Management

R. L. Welcomme

FAO, Via delle Terme di Caracalla, Rome, Italy.

Summary

The exponentially increasing human population is affecting fisheries. Statistics indicate a static and, in some cases, declining inland capture fisheries sector which can be traced to environmental impacts and overfishing, but in some areas results from a shift from commercial to recreational fisheries. There is divergence between the temperate zones, which emphasize recreational and aesthetic values of inland waters, and the tropics, which prioritize food production. Both systems rely increasingly on stocking and enhancement to maintain preferred species. Capture fisheries are being converted into aquaculture systems through increased control over the biotic and abiotic components of the system. Intensification of inland fisheries calls for substantial inputs of resources, manpower and funding which require adequate guarantees for the protection of the investment. There is an increasing trend to assign responsibility for management to local groups and to otherwise limit access to the fishery. The need for investment is leading to growing external involvement in the fishery and loss of autonomy by the fishermen.

Introduction

Current policies for food and agriculture aim at securing the sustainability of the world's resources in the interests of long-term food security. This aim finds expression in a growing concern for the long-term stability of the world's ecosystems as reflected by such documents as Agenda 21 of the United Nations Conference on Environment and Development, and The Convention on Biological Diversity. Conservation and sustainability are being pursued against increasing pressures on natural resources which are changing in the way in which they are used. These changes are particularly evident in inland waters which serve as collectors for all processes occurring in their basins. At the same time the living aquatic resources of river and lake basins are called on to provide protein-rich food for human populations. The fisheries sector world-wide has entered a critical period and the ways in which the aquatic environment and living aquatic resources are viewed are changing rapidly. Documentation of these trends and their extrapolation to the future is essential in the medium term for the allocation of funds and formulation of policies for inland water fisheries and aquaculture.

Context

Inland fisheries and aquaculture are strongly conditioned by the context in which they are being pursued. The changes now occurring are responses to the demographic, economic and political evolution of the world at the present time. Several factors may be singled out from many as particularly important in determining trends in resource use:

Human population

The United States (UN 1993) proposes three scenarios for human population growth into the next century, the medium one of which predicts a population of around 10×10^9 people by 2040. The growing population places increasing demands on natural resources, not only as a life support system through the services that a healthy ecosystem can provide, but also as a source of food and recreation. Changes in human distribution also alter the direction of resource use. The major trends of the past decades have been towards urbanization, and shifts to live along lakes, rivers and the marine fringes of the continents. There is a move away from subsistence agriculture at the rural level, and a growing demand for transfer of food through commerce to urban markets. These tend to be more varied than those of rural environments and require a corresponding diversification of species and products. Urbanization implies a greater environmental pressure on the resource as the amounts of waste grow, and other population-induced effects concentrate more around the available water.

Water supply

Water is now becoming one of the primary limiting factors to the growth of societies as supplies struggle to keep up with steadily increasing demand. By the middle of next century the medium population growth prediction indicates that nearly half the world's population will live in 58 countries experiencing severely restricted access to water. In other parts of the world *per capita* water supply will be substantially decreased. Against this background of increasing pressure on a fundamental resource, it is clear that the competing consumers of that resource, including fisheries and aquaculture, will have to become more efficient and focused in the way in which they use water. Furthermore, the need to conserve water quality for human consumption means that intensive fisheries and aquaculture practices will either have to produce less pollution and eutrophicating nutrients or will have to be associated with downstream water-treatment systems.

Feed and other resources

Most aquacultural production is based on low-input systems relying on low protein agricultural by-products and pond fertilizers. Such resources are also used by agriculture, animal husbandry and poultry rearers. As the general demand for food rises, the intensification of agricultural activities in parallel with aquaculture will lead to competition, and less resources may be available to the fisheries sector. Production of about 1.26 million t of finfish and 0.8 million t of shrimp in 1996 was totally dependent on fish meal. This source of fish feed is finite and its limited supply means that the culture of the carnivorous species that depend on it can only be expanded if a substitute high protein feed is found. In the meantime, if efforts to market the small pelagics that are used for fish meal directly for human consumption are successful, the availability of high protein feeds may decline. Both these trends, together with the rapid increases in other farm operating costs, point to the need for improvements in the efficiency of aquaculture and intensively managed systems generally.

Land availability

As populations rise, the proportion of land needed to produce food will increase, although much of the land already under culture may deteriorate through bad practices. Some agricultural land will be obtained by the draining of swamps and the isolation of floodplains, thereby diminishing the area of water available for capture and culture fisheries. As available land areas decline, there will be increased pressure on land for all uses. Indeed in many Asian countries land-use policies prohibit the conversion of arable land to aquaculture. The resulting competition means that fisheries and aquaculture must use the available areas more efficiently. It also implies limits to the spread of traditional freshwater aquaculture systems which will eventually direct future expansion to the marine environment, to the closer integration of aquaculture and agricultural systems, and to highly efficient water-recycling systems. The search for efficiency is reinforced by the current conservation-oriented attitude which places a premium on maintaining some ecosystems intact, rehabilitating degraded systems and by limiting practices, such as introductions and stocking, which negatively affect the gene pool.

Political and economic climate

Recent years have seen a shift in political and economic orientation from socialist models, involving open access to fisheries and subsidies to the fishing and aquaculture sectors, to capitalist or open-market ones which imply private ownership and financial self-sufficiency of capture and culture fisheries. At the same time there has been a growth in concern about the future of natural resources that encourages the development of sustainable approaches to management which are sometimes inconsistent with the pure, short-term, market approach.

Current situation

Analysis of the statistics submitted by the Member Nations to FAO (FAO 1995 *a* and *b*) can be used to derive time series for review of the various fisheries sectors. For the purposes of this analysis, there are three main problems. Firstly, inland capture fisheries statistics are not reported as a separate category but have to be derived by the subtraction of the aquaculture statistics from the total reported inland production. As the two categories are not always reported by species groups in the same manner, this may lead to some discrepancies. Secondly, the definition of aquaculture currently in use encompasses many practices intermediate between capture fisheries and intensive aquaculture. As practices such as culture-based fisheries are at the centre of the changing approach to management, present reporting systems are not sufficiently sensitive to permit documentation of these trends. Thirdly, recreational fisheries are not fully accounted for and frequently go unreported. Analysis of the statistics, therefore, has to be founded on an extensive knowledge base which enables a fuller interpretation of trends.

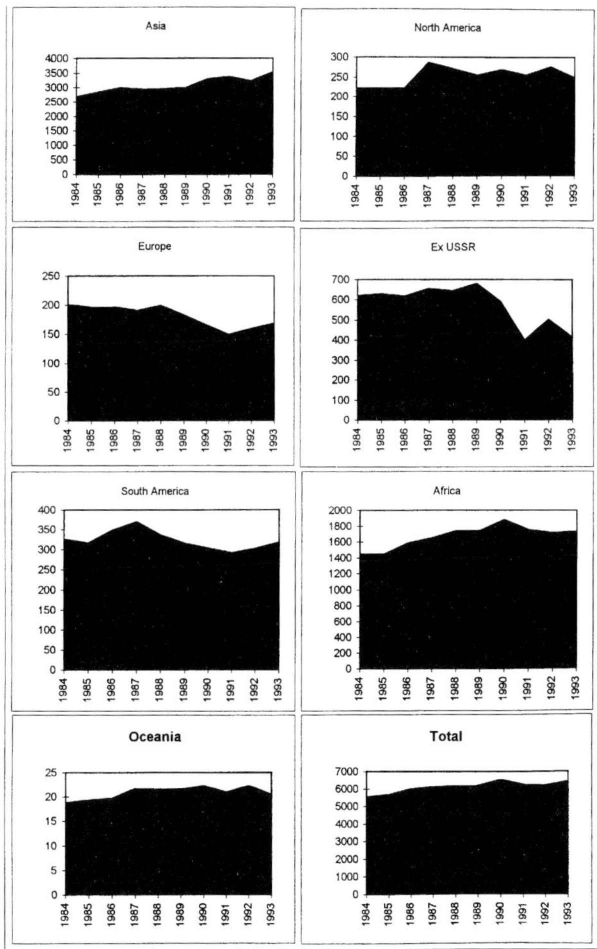

Fig 1. Time series of inland fisheries catch by continent. *x* axis = year; *y* axis = catch in thousand t.

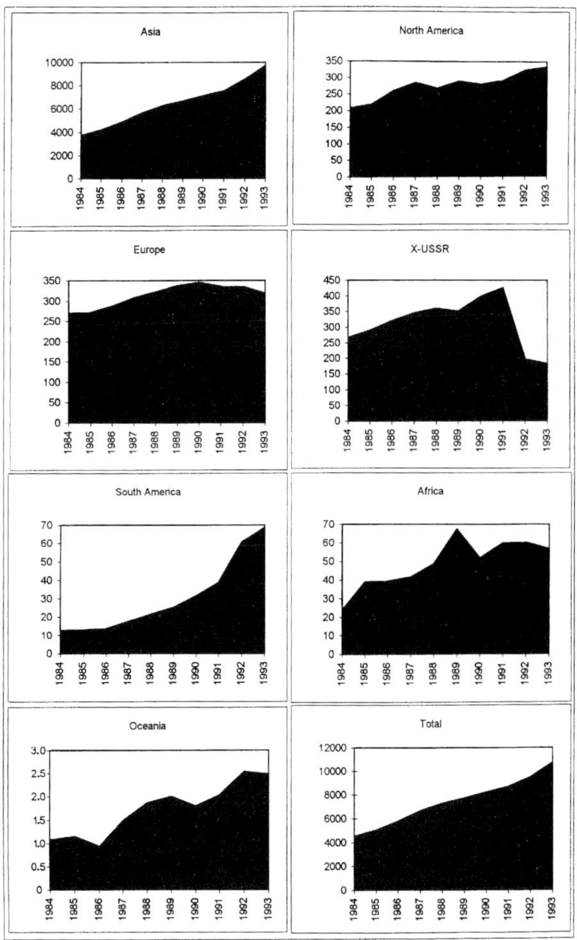

Fig. 2. Time series of aquaculture production by continent. *x* axis = year; *y* axis = production in thousand t.

Subject to the above caveats, the statistics for inland and diadromous fisheries and aquaculture indicate the following trends:

Capture fisheries

After several decades of growth, catches stabilized on most continents after 1984 (Fig. 1). In many cases there have been declines in catch although in the last two years several continents have reported slight increases. The main continent registering an increase is Asia, which accounts for 50% of the catch from world inland capture fisheries. The overall trends only indicate total catch and do not provide important information about changes in the exploited assemblages. It is well known that the multi-species fisheries of most inland waters undergo a series of changes in response to fishing pressures which force the assemblage towards smaller species and sizes (Welcomme 1995). Overfishing at the assemblage level is therefore characterized by a loss of larger species and a reduction in mean size of catch. Using these criteria as applied to numerous reports, as well as the formal fishery statistics, FAO (1995*c*) concluded that most major inland fisheries are fully exploited or in some cases overexploited and, with very few exceptions, there are no large inland fisheries with confirmed potential for significant expansion.

The poor state of exploited fish stocks is aggravated by the equally poor state of the environment of many of the rivers and lakes (see for example Dudgeon 1992 for Asian rivers, or Lu 1994 for Chinese rivers). Long-term pollution and eutrophication arising from industrial, domestic and agricultural waste waters have degraded water quality. Massive dam building, channelization, water diversion and water abstraction have regulated and modified rivers. It is known that environmental degradation tends to induce many of the same changes in assemblage structure as does fishing (Rapport *et al.* 1985). It also produces some effects that are unique, such as the promotion of invasion by exotics, the removal of entire assemblages, or the sterilization of adult fish. In most of the world's inland waters the present degraded condition is due to an interaction of both sources of stress.

Aquaculture

The yield from aquaculture has increased steadily over the last decade and is now the only aquatic resource sector to show consistent gains (Fig. 2). The global rise in the tonnage of aquaculture products varies with area. Most, some 91%, comes from Asia and of that, 66% is from China. In fact China accounts for 60% of all inland aquaculture production, which means that the global statistics are extremely sensitive to those reported from that country. Aside from Asia there is a tendency for increases in North and South America, and Oceania. European production has fallen slightly since 1990 and yield from the countries of the former USSR suffered a dramatic decline after 1991. In both cases these are mainly due to the collapse of centrally planned aquaculture due to the withdrawal of subsidies under the financial restructuring of East European economies.

Despite the well-publicized successes of luxury aquaculture products such as crustaceans (shrimp) and salmon which contributed 5.7% and 4.3% of world production respectively in 1993, the majority of world aquaculture products is still cyprinids, mostly Chinese and Indian carps which contributed 47% of total production in 1993. Even tilapias which show promise of becoming the new wave of expansion in fish farming only comprise 3% of current yields. Aquaculture is therefore, still largely dependent on relatively low food chain freshwater species raised in semi-intensive polyculture systems, and further expansion of culture of the higher valued carnivorous species is limited by the availability of high protein feeds. Furthermore there is competition within the aquaculture sector for high protein feeds. Most high value species are marine, and the greater availability of coastal areas suitable for aquaculture may lead to the development of production from such areas in preference to inland waters.

Aquaculture is expanding world-wide and is becoming accepted in areas outside its traditional confines in Asia and Europe. The growing success of aquaculture in new areas derives from its development as a source of income rather than for subsistence, and its incorporation into local agriculture practice as a means of diversification of income and diet. As a result, more flexible integrated culture systems which include fish are being adopted in many regions but a considerable potential for further expansion remains. It is highly probable that a similar situation with regard to non-reported yields applies in aquaculture as in inland fisheries. Much non-market-oriented production and production from mixed systems such as rice-fish culture certainly escapes registration.

Recreational fisheries

Recreational fisheries are extremely important in some parts of the world and dominate management of inland water resources particularly in the North and South temperate zones where the number of anglers runs into millions. For example, The European Inland Fisheries Advisory Commission (EIFAC) estimated the number of anglers active in 20 of 31 European countries to be 20.2 million for a population of 417 million or some 4.8% of the population (O'Grady 1995). Angler numbers were even higher in North America where 17.8% of the Canadian population and 11.9% of the United States population held recreational fishing licences in 1975 and 1978 respectively (Grover 1980). Recreational fisheries are not confined to the affluent industrialized countries. They have a significant but largely unassessed impact on inland fisheries management policies in South America, particularly Argentina, Brazil, Bolivia and Chile.

Recreational fisheries sometimes produce a considerable amount of fish for domestic consumption much of which goes unrecorded in the official fishery statistics. For example nine European countries estimate that about 165 000 t are caught by the recreational fishery. These same countries reported 94 877 t of inland fishery catch to FAO in 1993, some of which originated in the recreational sector but most came from commercial fisheries. Catch rates from recreational fisheries of 30–94 kg/ha (EIFAC 1996) for Europe, which are comparable with the yield of intensively fished tropical floodplains and lakes, indicate that most modern recreational fisheries are intensive rather than extensive in nature.

Subsistence fisheries

The products of subsistence fisheries are also rarely included in the records. Subsistence fisheries are generally spatially diffuse and carried out by numerous unlicensed individuals. Fishing sites range from small rivers and water bodies to the casual offtake of rice-fish farming. The quantities caught are probably impressive. For example, Welcomme (1976) estimated that the official catch from African rivers could probably be doubled if the yield from the smaller rivers and streams were to be included in the sum. It is probable that similar situations exist throughout most of the world and that the real catch from inland waters is considerably higher, perhaps even double, the catch as derived from official reports.

Current management practices

Current management of inland fisheries is delimited into capture and culture by the current definition of aquaculture used by FAO for statistical purposes. This definition states that:

> Aquaculture is the farming of aquatic organisms including fish, molluscs, crustaceans and aquatic plants. Farming implies some sort of intervention in the rearing process to enhance production, such as regular stocking, feeding, protection from predators, etc. Farming also implies individual or corporate ownership of the stock being cultivated. For statistical purposes, aquatic organisms which are harvested by an individual or corporate body which has owned them throughout their rearing period contribute to aquaculture while aquatic organisms which are exploitable by the public as a common property resource, with or without appropriate licences, are the harvest of fisheries.

This definition, however, has been a source of some confusion as it fails to classify the intermediate practices that fall within the general terms 'fisheries enhancement' or 'culture-based' fisheries (New and Crispoldi-Hotta 1992). As it stands, many managed inland fisheries including recreational fisheries now fall under the aquaculture category. To avoid this, an amendment is proposed whereby the definition of farming is recast as follows: 'farming implies rearing organisms during most or all of their life cycle' thus excluding from aquaculture those fisheries maintained only by stocking. This definition is reinforced by a more precise categorization of the various types of fisheries practice into either aquaculture or capture fisheries (Table 1).

Fisheries management seeks to influence the fishery through actions on (i) the fishery; (ii) the fish stock and (iii) the environment.

Table 1. Proposed classification of various fisheries practices into capture fisheries and aquaculture

Production from	Designation Aquaculture	Fisheries
Hatcheries	*	
Ponds	*	
Tanks	*	
Raceways	*	
Cages	*	
Pens	*	*
Barrages	*	
Stocked lakes and reservoirs		
• with other enhancement (predator control and/or fertilization, habitat modifications)	*	
• no other intervention		*
Unstocked lakes and reservoirs		
• with enhancement (fertilization and/ or predator control, habitat modifications)	*	
• no enhancement		*
Ranching of anadromous fish		*
Fish and crustaceans caught in open waters		*
Privately owned recreational fisheries		*
Fish and other animals harvested from brush parks		
• managed over time and with other enhancement	*	
• harvested on an install and harvest basis		*
Fish and other animals harvested from		
• fish aggregating devices		*
• artificial reefs		*
Molluscs		
• subject to open fishery		*
• from owned and managed grow-out site	*	
Enhanced marine fisheries		*
Harvest of natural seaweed beds		*
Harvest of planted and suspended seaweed	*	
Rice-fish culture		
• from stocked rice-paddy	*	
• from unstocked rice-paddy		*
Lagoon (including vallicoltura) production	*	
Private, tidal ponds (tambaks)	*	

Figure 3 illustrates the major factors influencing an exploitable stock of fish. The various management practices seek to influence these factors in various ways:

Capture fisheries seek to influence f (fishery mortality) by regulating the rate of extraction of fish from the natural population. Such regulation is usually attempted through

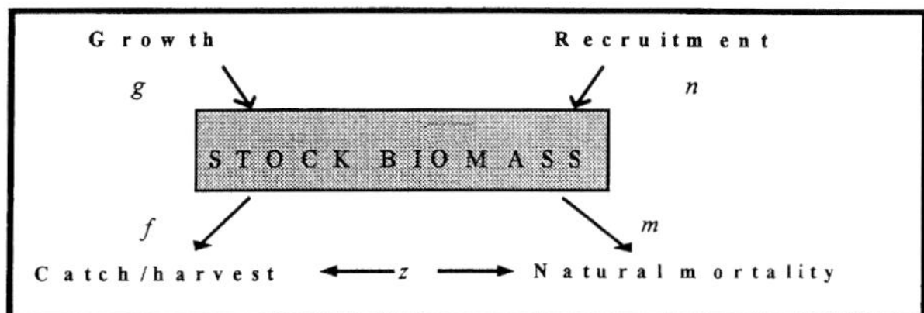

Fig. 3. Major factors influencing exploitable stocks of fish. (*g*, growth; *n*, number; *f*, fishing mortality; *m*, natural mortality; *z*, total mortality).

'traditional' controls over the fishery such as mesh size regulations, gear prohibitions, closed seasons, catch quotas and other limitations on access. The steady degradation in both the quantity and quality of the catch over the last five years indicates that such traditional systems of management are largely ineffective. There are several reasons for this. In the first place regulations attempt to apply single-species criteria to multi-species situations. Secondly, many of the regulations are unenforceable on spatially diffuse fishing communities. Thirdly, the open access nature of many fisheries and subsidies from government have encouraged fishermen numbers in excess of the support capacity of the fishery. Given the degraded state of the fishery at current levels of exploitation it appears that increases in yield can not be expected from classically managed capture fisheries world-wide.

Aquaculture tends to ignore f as it tends to harvest all of the biomass at a single fishing at a predetermined time. Rather the culturist seeks to control all other processes — n (numbers) through stocking with young organisms produced by artificial reproduction or extracted from nature, g (growth) through feeding and m (mortality) through full control of the pond, raceway or cage environment.

Intensification of fisheries. There is a range of management practices intermediate between capture fisheries and aquaculture which generally fall under the term *culture-based* or *enhanced fisheries* and which together contribute to a process that could be termed intensification of the fishery. These practices are usually adopted in a stepwise manner leading to a progressive increase in fishery production per unit area of water through increasing controls on number (n), growth (g) and mortality (m) as follows:

i) **Stocking** natural waters to improve recruitment (n), biases fish assemblage structure to favoured species, or maintain productive species that would not breed naturally in the system;

ii) **Introduction** of new species to exploit underutilized parts of the food chain or habitats not colonized by the resident fauna;

iii) **Fertilization** to raise the general level of productivity (g);

iv) **Elimination** of unwanted species that either compete with (g), or (m) predate upon desired species;

v) **Constituting** an artificial fauna of selected species (g);

vi) **Engineering** of the environment to improve levels of reproduction (n), shelter and vital habitat (m);

vii) **Cage culture** and parallel intensification of effort of the capture fishery;

viii) **Modification** of the water body to cut off bays and arms to serve for intensive fish ponds;

ix) **Aquaculture** through management of the whole system as an intensive fish pond;

x) **Genetic modification** to increase growth, production, disease resistance and thermal tolerance of the stocked or cultured material.

TRENDS IN RESOURCE USE

There is an overall tendency to increase the degree of human control over the process of fish production from freshwater. Few inland fishery systems are now exempt from some degree of intervention and simple capture fisheries managed by classical controls on the fishery alone are probably now limited to larger lakes and rivers. In most cases the impacts of other human uses, as well as greatly intensified fisheries, have necessitated continuing involvement in management. The intensification of a fishery is associated with a number of other social, institutional and financial changes which are a precondition for the successful transition from capture to culture.

Contrasting strategies in developed and developing countries

There are two major strategies for management of inland fisheries which depend largely on the affluence and the population pressures of the societies concerned. In the less affluent tropical regions, where the populations are increasing, the emphasis is on production of fish for food and for export to developed countries. This strategy has been described by Zweig (1985) as management for survival. In more affluent societies, where populations are relatively stable, resources are used more for recreation and the waters themselves are subject to increasing pressures for conservation and aesthetic rehabilitation. In this region there have been long-term trends to suppress commercial in favour of recreational fisheries in natural waters while using intensive aquaculture to produce selected inland species for profit. This polarization, aggravated by the purchasing power of the temperate zone countries, has resulted in a tendency to move products of aquaculture and capture fisheries from the developing to the developed world.

Ownership

Increased human control of the fishery and its environment requires investment of time, funding and other resources which have to be protected. There is a parallel trend therefore, to transfer the rights to the fish from the public to the private domain so that the new owners can protect their interests and are also empowered to negotiate with other users of the resource. In the past most fisheries were subject to traditions which regulated access, gear types, the sharing of resources and, in general, the way in which the fishery was integrated into the general fabric of the society. As such, fisheries represented a type of managed commons that avoided the 'tragedy' that classically confronts unmanaged systems of this type. It is only comparatively recently that the opening of access to aquatic resources in search of social equity, and as a solution to unemployment and landlessness, created the anarchic conditions under which degradation and collapse rapidly ensue. There has since been increasing tendency on the part of many governments to reverse the move away from local management. The core of these policies in both developed and developing countries has been to charge users with the management and conservation of the resource in exchange for the right to enjoy the benefits of the resource and the protection of any investments made. This effectively means an assignment of ownership or rights to individuals or groups of individuals which corresponds to the enclosures of common land. Inevitably this has meant that, while the majority of the population is excluded from the resource and often from the political process controlling it, there is a *quid pro quo* in the higher sustainable yield available for consumption that results from better managed systems.

The trend to 'privatization' of the resource is observable in many sectors. The recreational fisheries of Europe are now largely managed by groups of angling associations who are assigned the rights to certain waters. These groups are responsible for stocking the waters and for maintaining the environment in a condition suitable for fish. At the same time they can and do sue other users for damages resulting to the fishery through pollution and environmental degradation. In most cases licences are only issued to members of the group subject to the successful completion of a course of instruction. Fishermen operating in tropical lake, reservoir and river systems are increasingly being charged with the management of the resources they fish. In many cases this involves a return to traditional management systems which very often require the qualification of fishermen through rituals. Ideally, such groups are protected by the state from incursions by others seeking to use the resource. However, after the breakdown in traditional management over the last few decades it is not always easy to simply return to a system whose justification and sanctions have been eroded. Under another system for the assignment of fishing rights, the lot fisheries that are so common in Southeast Asia, the term of rental has been extended from one to several years thereby giving protection to the lessees and encouraging them to improve the fishery.

Assigning the rights to manage and exploit dams and reservoirs is more complicated. Such water bodies being man-made rarely fall within tradition and are not usually located on private land. They generally belong to some entity outside the fishery, a power-generating company for example or to a corporate entity such as a village or irrigation authority. This means that the rights to manage and draw benefit from a fishery have to be negotiated with third parties. In some areas of the world such as Southeast Asia this has not proved a problem, but in others such as Africa it has posed a serious limitation to expansion of this type of management.

Finance

Financing of the fishery is undergoing radical changes on several fronts. The transition from centrally planned to free market economies has had far reaching consequences for inland fisheries and aquaculture in many parts of the world. In centrally planned systems, fisheries activities were generally organized on the basis of cooperatives which were subsidized by the state. Enhancement of fisheries through stocking was based on state-financed hatcheries which rarely recuperated the cost of the stocking material issued to the fishery enterprises, even though there were occasional attempts to make the fishery pay for itself by supporting the hatcheries. It is not by accident that the most successful of the intensified fisheries in terms of increases in yield have been in countries such as China, Cuba, the former USSR or Poland whose fisheries were centrally controlled. More recently the transition to free market economies has removed subsidies causing the collapse of many of the fisheries and, incidentally, of many of the hatcheries supporting them. This has been attributed to failure to adapt to the new circumstances by both the economic and institutional sectors, and the re-establishment of intensively managed systems in these countries presupposes that they are cost-effective. One problem has been that the transition occurred without any funding being made available to the former fisheries enterprises. Positive signs are, however, forthcoming in both China and Cuba where the transition to economically self-sustaining fisheries is well under way.

Private funding of stocking has also shifted from the public to the private sector in many recreational fisheries. The stocking of rivers and lakes with sport-fish species used to be offered as a public facility in part paid for by licences whereas the angler associations are now expected to pay in full for this service.

The trend to economic bondage by fishermen to external financing agencies has increased in recent years. For example, on Lake Victoria, the need to finance deep-water craft and heavy duty nets to catch the introduced Nile perch, together with the need to sell the catch to processing and export firms has robbed the fishermen of their independence. Similar problems have occurred wherever the intensification of the capture fishery has required investment in gear and boats. The intensification of fisheries involves costs far above those normally incurred by simple capture fisheries. In most cases the fishermen's communities are unable to support this additional cost and funding is sought from third parties from outside the fishery. The greater predictability, more controlled harvesting season and more concentrated harvest attracts investment from businessmen who then take the major part of the profit. This means that the intensification process is frequently accompanied by a loss of independence by the fishermen and a growing dependence on credit or other external forms of funding

Education

Intensive management of a resource implies a greater knowledge of its functioning than that needed for a simple extractive activity. In some countries the development of the new intensified systems of management has been accompanied by the emergence of research, often empirical, on the most appropriate ways to manage the systems and there is already a wealth of information in countries such as China, Cuba, India or Poland on stocking, feeding and fertilization ratios. Research alone is not enough, because the extension of the results and the application of the formula by the fishermen also requires a higher level of preparation. This trend reinforces the drift from simple fishermen to a more complex organization of the fishery.

Greater diversity and higher value species

At present the majority of the world's inland fisheries and aquaculture production has been of low value species for local consumption. This will probably continue to be the situation for some time to come. However, growing urbanization, possibilities to profit from urban and export markets and the growing affluence of many countries implies that future markets will become increasingly oriented towards more expensive species. This has already been seen to be the case in aquaculture in Asia where the relative proportion of the traditional carps is falling in favour of more valuable fishes. In aquaculture there has been a trend to diversification in fin-fish and mollusc production over the years (Table 2). Interestingly crustacean production halted the trend to diversification in the late 1980s and now appears to be concentrating on fewer species with the bulk of production (90%) being from only five species. This points to a trend to diversify the number of species under culture but to concentrate the bulk of the production in comparatively few species.

CONCLUSIONS

The demand by expanding human populations for food is driving a shift from low cost, extensive, rural activities oriented to the supply of local communities, to high cost, intensive activities oriented to supplying urban communities and foreign markets. Capture fisheries in waters which depend on the complex trophic webs of natural systems for their sustainability have generally reached the limits of their productive capacity. Unmodified systems are now rare and reconstructed faunas

Table 2. Number of species reported as being cultured in various environments

	1984	1985	1986	1987	1988	1989	1990	1991	1992	1993
Inland and brackish										
Crustacea	13	14	14	15	16	17	17	17	16	15
fish	70	70	80	79	83	86	91	93	95	95
molluscs	2	2	2	2	2	2	2	2	2	2
Marine										
Crustacea	12	16	16	18	18	17	15	15	14	14
fish	22	28	34	38	36	35	37	38	38	42
molluscs	38	40	42	45	48	50	52	52	51	51

aimed at survival within modified systems are becoming more common. Fisheries are maintained increasingly by stocking, and managers frequently aim to increase productivity by fertilization, ecosystem manipulation and the elimination of competitors and predators. Further gains in production are made by incorporating cage and cove-culture systems into the environment. Changes are also evident at the social, economic and political levels, whereby inland fisheries are becoming more capital-intensive and controlled by financiers external to the fishery. Many of the world's waterbodies are heading towards management by extensive, semi-intensive or intensive techniques for standard assemblages of fish which correspond best to market demands. This intensification of the fishery is going on at different rates in different parts of the world and in different systems. It is already completed in farm ponds, is far advanced in many small reservoirs and has already begun in larger water bodies and in the marine coastal environment.

Acknowledgments

This synoptic view of the trends in inland fisheries and aquaculture emerged from discussions with the members of the Inland Water Resources and Aquaculture Service of FAO as well as such consultants as Malcom Beveridge and Ian Dunn who assisted in the framing of the definition of aquaculture and in position papers for the Kyoto Conference. As such, the author is expressing a larger consensus and wishes to acknowledge the role played by his colleagues in formulating these ideas.

References

Dudgeon, D (1992). Endangered ecosystems: a review of the conservation status of tropical Asian rivers. *Hydrobiologia* **248**, 167–91.

EIFAC (1996). Report of Workshop on Recreational Fishery Planning and Management Stategies in Eastern Europe, Zilina, Slovakia, 22–25 August 1995. *EIFAC Occasional Paper* 32.

FAO (1995*a*). FAO Yearbook of fishery statistics, catches and landings. *FAO Fisheries Series* **44** vol 76. 687 pp.

FAO (1995*b*). Aquaculture production statistics 1984-1993. *FAO Fisheries Circular* 815 rev. 7. 186 pp.

FAO (1995*c*). Review of the state of world fishery resources: Inland capture fisheries. *FAO Fisheries Circular* 885. 63 pp.

Grover, J H (Ed) (1980). Allocation of Fishery Resources. Proceedings of the Technical consultation on Allocation of Fishery Resources, Vichy, France, 20–23 April, 1980. FAO, Rome. 623 pp.

Lu, X (1994). A review of river fisheries of China. *FAO Fisheries Circular* 862. 47 pp.

New, M, and Crispoldi-Hotta, A (1992). Problem in the application of the FAO definition of aquaculture. *FAO Aquaculture Newsletter* **1**, 5–8.

O'Grady, K (Ed) (1995). Review of inland fisheries and aquaculture in the EIFAC area by sub-region and sub-sector. Sub-regional and sub-sectoral reports presented at the EIFAC Consultation on Management Strategies for European Inland Fisheries and Aquaculture for the 21st Century during the European Inland Fisheries Advisory Commission eighteenth session, Rome, Italy, 17–25 May 1994. *FAO Fisheries Report* 509 Supplement 1. 79 pp.

Rapport, D J, Regier, H A, and Hutchinson, T C (1985). Ecosystem behaviour under stress. *American Naturalist* **125**, 617–40.

UN (1993). World population prospects. The 1992 revision. UN, New York, 135. 677 pp.

Welcomme, R L (1976). Some general and theoretical considerations on the fish yield of African rivers. *Journal of Fisheries Biology* **8**, 351–64.

Welcomme, R L (1995). Relationship between fisheries and the integrity of river systems. *Regulated Rivers: Research and Management* **11**, 121–36.

Zweig, R D (1985). Freshwater aquaculture in management for survival. *Ambio* **14**(20), 66–74.

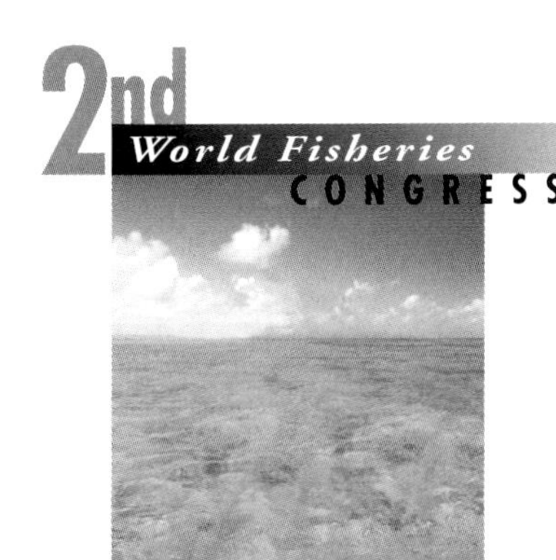

The Fisheries–Aquaculture Relationship in New Zealand; is it Competitive or Complementary?

Robert W. Hickman

Aquaculture Research Centre, National Institute of Water and Atmosphere, P O Box 14-901, Wellington, New Zealand.

Summary

The 30-year history of aquaculture in New Zealand has been one of competition for resources. The three export-producing aquaculture activities (Pacific oysters, mussels and salmon) have had to compete with numerous alternative users for the coastal water. Existing capture fisheries are in direct competition with aquaculture. Expansion into more open waters and new aquaculture ventures will lead to further competition with capture fisheries. Fishery-aquaculture collaboration has also occurred, an outstanding example being the enhancement of the Tasman Bay scallop fishery. Aquaculture techniques are also being used for rejuvenating the dredge oyster fishery in Foveaux Strait. Aquaculture was developed by individuals and specialist companies, but traditional fishing companies are now heavily involved. Expansion of New Zealand's aquaculture will remain low volume, high quality, export-orientated. The market promotion necessary for the quality farmed products fully complements the value-added promotion of New Zealand's capture fishery exports.

Introduction

The total world catch of fish, shellfish and seaweeds, is currently just over 100 million t. The proportion of the total that comes from aquaculture is increasing (Anon. 1994). This is the result of stagnation of the harvest from capture fisheries which has drifted around 80 million t since 1989, combined with rapid expansion of aquaculture production, which went up from 7 to 14 million t during 1984–1992 and jumped to 16 million t in 1993 (FAO 1994). It is projected that demand for aquaculture product will be around 25 million t by the year 2000. If this demand is met, it would raise the aquaculture proportion to over 20% of world fisheries production, because 'the capture fisheries of the world are (already) being harvested at close to their maximum sustainable yield' (Chamberlain and Rosenthal 1995).

New Zealand Fisheries and Aquaculture

In New Zealand both capture fisheries and aquaculture are still expanding. The fishing industry has set itself a target figure of $NZ2000 million for the export value of its fish products by the year 2000; currently New Zealand's seafood exports are worth just over $NZ1200 million. It is anticipated that increased aquaculture production will contribute significantly towards

Table 1. Landings of the New Zealand coastal and deepwater commercial fisheries at ten year intervals from 1960 to 1990 (Derived from data in Paul 1986)

	Landings in t (to the nearest thousand)			
	1960	1970	1980	1990
Coastal	32 000	47 000	91 000	135 000
Deepwater	—	—	177 000	334 000
Total	32 000	47 000	268 000	469 000

meeting the 2000 target but aquaculture will still comprise less than 10%, by either volume or value, of the country's total seafood exports.

New Zealand's coastal fisheries were de-licensed in 1964 and subsequently 'expanded at an accelerating rate' (Watkinson and Smith 1972). Deep-water fisheries, which were 'discovered in the late 1970s during the international rush to claim and explore 200-mile fishing zones', have maintained the acceleration (Paul 1986). Commercial oyster farming was also legalized in 1964, but aquaculture only started to develop after the 1971 Marine Farming Act came into force (Anon. 1989). The past three decades have thus been a period of rapid expansion for both the fishing and the aquaculture sectors of the industry (Tables 1 and 2).

Fishing and aquaculture in competition

New Zealand has 5000 km of coastline, most of it washed by, in the words of the New Zealand Mussel Industry Council, 'the cool clean oceans of the South Pacific ... which meet the highest water quality criteria'. Nevertheless, the 30-year history of aquaculture in New Zealand has been one of competition for the industry's most essential resource, namely the coastal water. As in other countries, there are many competing interests for the shallow, sheltered, unpolluted stretches of coastline that are the preferred sites for aquaculture activity. Recreational, navigational, traditional and aesthetic interests have posed major constraints on the development of aquaculture in this country with its strong tradition of public access to all coastal waters (Hickman 1991). New Zealand now has three established export-producing aquaculture activities; namely oyster, mussel and salmon farming. Each of them has had to compete with the multitude of alternative users for access to their most essential resource. One such alternative use for potential marine farming water is fishing, and on occasions direct competition has occurred between aquaculture and traditional fishery interests.

Pacific oyster (*Crassostrea gigas*) farming, which is established in harbours throughout the northern half of the North Island and is spreading into the top of the South Island, has alienated intertidal areas for set-netting. Longline farming of the greenshell mussel (*Perna canaliculus*), which is based in the Marlborough Sounds and at Coromandel and Stewart Island, and sea-cage farming of quinnat salmon (*Oncorhynchus tshawytscha*), which is the main method for farming salmon at Stewart Island and in the Marlborough Sounds, are incompatible with line, trawl and dredge fishing. Salmon farming by sea-run ranching from sites around the South Island has on occasions found itself in competition with both the recreational fishers on the river banks and the commercial trawlers off the river mouths.

Table 2. Exports of New Zealand aquaculture products over the period 1985–1995 (Derived from NZ Fishing Industry Board statistics)

	Exports (to the nearest t)		
	Mussels	Oysters	Salmon
1985	1 849	764	—
1986	2 707	972	—
1987	4 125	778	—
1988	5 794	736	345
1989	6 917	845	1 186
1990	8 570	832	1 874
1991	10 658	953	2 389
1992	12 783	988	3 030
1993	14 893	1 123	3 336
1994	17 825	1 295	3 756
1995	21 300	1 435	5 061

Although there is as yet no significant export of farmed product, the techniques for farming paua, or New Zealand abalone, are well established (Tong *et al.* 1992) and several land-based farms are currently being developed. There is interest in barrel culture on longlines as a method of farming paua which would put it into the same competitive situation as mussel farming. There is also considerable interest in the possibilities for ranching and enhancement of paua (Schiel 1992; Schiel and Mercer 1992). These activities have potential for competition with both the commercial and the recreational fisheries. Ranching would require rights over the areas of seabed on which the juveniles had been seeded to grow to harvest size. Enhancement of natural populations raises questions of ownership, fishing rights, size limits and how to distinguish the enhanced paua from the naturally settled, wild ones.

Alienation of available sea surface and/or seabed by the physical presence of the marine farming structures is the main point of contention between wild and culture fisheries. Conflict occurs despite the fact that marine farming licences confer no rights over the seabed, and the siting of marine farms has hitherto generally been restricted to locations considered unimportant for commercial fishing. There is, however, continuing pressure from within the industry for further expansion of the existing marine farming activities. For example, mussel farmers in the Coromandel region of the North Island are looking to move into more open waters outside the established farming areas, and to do so on a large scale, by taking out licences covering more than 1000 ha, as compared to the traditional 3 ha mussel farm licence. This scale of development, together with new ventures such as cage farming of marine finfish, will inevitably lead to further competition with the wild fisheries. The competition may not be just for space. The proposed scale of development has raised questions, and prompted research into whether the very intensive methods used for longline farming of mussels are sustainable in open water, that is probably less productive than

Table 3. Annual landings of dredged mussels and annual production of farmed mussels in New Zealand at 5-year intervals, 1965–1995 (Derived from data in Watkinson and Smith 1972; Anon. 1989; and MAF Fisheries statistics)

	Landings/production (in t)	
	Dredged mussels	Farmed mussels
1965	1 619	—
1970	1 023	—
1975	1 629	—
1980	1 555	3 050
1985	11	10 750
1990	—	29 400
1995	—	60 000*

* estimated value.

in the near-shore embayment locations traditionally used for mussel farming. On the positive side mussel farms provide sheltered habitat for a number of fish species and could thus benefit some wild fisheries.

Fishery *versus* aquaculture competition has also occurred at the product end of the industry. The rapid rise in production of farmed greenshell mussels over the years following the first significant harvest in 1977 was matched by the complete demise of the wild dredge fishery (Table 3) which gradually found that there was no market for its large, tough, slow-grown, wild mussels, when matched against the smaller, succulent, fast-grown, farmed product (Hickman 1991). Mussel farming has, in fact, gone on to expand the domestic market to a level well in excess of that achieved by the wild fishery, at the same time as developing major new export markets.

FISHING AND AQUACULTURE IN COLLABORATION

Just as there has been competition between capture and culture fisheries, there has also been collaboration. The enhancement of New Zealand's southern scallop fishery provides an outstanding example of the benefits of collaboration. Aquaculture provided the technology — bags of monofilament nylon suspended from longlines — for catching the scallop spat in prodigious quantities and for on-growing them to a size suitable for reseeding directly onto the seabed. Standard dredge fishing methods are used for harvesting the crop after 2–3 years, and the fishery now operates on the basis of rotational seeding and cropping of extensive areas of seabed. This highly successful fusion of complementary fishery and aquaculture techniques has rejuvenated, expanded and stabilized the Tasman Bay/Golden Bay fishery, and perhaps provided the model by which to enhance other fish and shellfish resources (Anon. 1995).

Indeed, the application of aquaculture techniques for reseeding is currently being tried as a potential means of rejuvenating the wild fishery for the New Zealand dredge oyster (*Tiostrea lutaria*) that used to exist in Foveaux Strait at the bottom of the South Island (Street 1995). This large (10 000 t/year) natural fishery was decimated by the protozoan parasite *Bonamia* in the late 1980s and was closed to commercial dredging in 1992 (Doonan *et al.* 1994). At the other end of the South Island, hatchery techniques for production of dredge oyster spat are being considered as a means of supplying seed for enhancement of the small dredge fishery that coexists with the scallop fishery in the Tasman Bay/Golden Bay region. These techniques were developed primarily to establish dredge oyster farming as a possible diversification for mussel farmers (Hickman 1992).

Marine farming in New Zealand was developed by individuals and specialist aquaculture companies with assistance from government agencies such as the New Zealand Fishing Industry Board, the Ministry of Agriculture and Fisheries, and the Marlborough Harbour Board. It was not until about the mid-1980s that traditional fishing companies got involved. They brought into mussel farming their expertise in areas such as bulk handling and processing. They also increased the stability and profitability of mussel farming by providing vertical integration throughout the growing, harvesting, processing and marketing sectors. Fishing companies are now major participants in the Pacific oyster, mussel and salmon farming industries. They are involved in commercial development of abalone (*Haliotis iris*, *H. australis*) farming, and they are showing interest in aquaculture of rock lobster (*Jasus edwardsii*), snapper (*Pagrus auratus*), turbot and brill (*Colistium nudipinnus*, *C. guntheri*), and several surf clam species (*Mactra* spp., *Spisula* spp.).

Lobster farming, by on-growing juveniles through to marketable size, has already raised the spectre of competition with the rock lobster fishery for the puerulus or settlement stage, and steps are being taken to resolve this resource supply problem. A possible solution may be the surrender of a set amount of fishing quota in return for permission to collect a specified number of puerulus for farming. The quota trade arrangements would vary for different fishing areas, but one current proposal is to surrender 1 t of quota for 25 000 puerulus stage juveniles. Complementary rock lobster fishing and farming activities would seem to be a logical development, particularly in collaboration with a mussel farming interest, since mussels have been shown to be an excellent food for lobsters in captivity (James 1995). The complementary interests of fishing and aquaculture are also exemplified by the fact that the Tasman Bay scallop enhancement company has funded research into snapper rearing, aimed at the production of juveniles for either enhancement of wild stocks or sea-cage farming (Tait 1995).

The ability to produce juveniles in large numbers, which generally results from the development of aquaculture techniques, can facilitate research into aspects of fish and shellfish biology, such as natural recruitment and juvenile growth. These are essential to enhancement and farming projects but can also bring better understanding of the wild fisheries.

THE FISHERIES–AQUACULTURE RELATIONSHIP IN THE FUTURE

Despite New Zealand's extensive natural potential for aquaculture, any expansion will of necessity remain low volume, high quality, export orientated. The current 60 000 t production of greenshell mussels and their low price (retailing at less than $NZ2/kg) on the domestic market might suggest otherwise, but on the world market, which is the outlet for the majority of the crop, NZ mussels occupy only a very small, high value, specialist niche. Farmed New Zealand salmon exporters likewise target the high value end of the world market by taking advantage of the different harvesting season. Interest in rock lobster and paua farming is fuelled by their high value on Asian markets.

Seafood is 'a highly traded commodity' and New Zealand seafood is exported at a much higher level than the '38% of world production (that) was traded internationally in 1990' (ICLARM 1995). The New Zealand seafood industry has put much effort into overseas marketing, placing great emphasis on quality and added-value, and has been very successful (Anon. 1989). Complementary marketing of its products can benefit both the aquaculture and the capture sectors of the industry. Traditional fish sales will benefit from the clean green image which has been established for New Zealand's aquaculture products. Marketing of new farmed-species and products will be easier through the existing outlets for wild seafood.

The history of aquaculture in New Zealand has been marked by periods of over- or under-supply and consequent low or high market realization. We fortunately do not seem to have met the forecast that 'for stand-alone fish farms, a farmer can expect a total or near-total crop failure at least once in 10 years and probably, on average, twice in 10 years (from disease, equipment failure, adverse climatic conditions, red tides, theft, etc.)' (ICLARM 1995). However, integration of diversified culture and capture enterprises may help to reduce some of the risks involved in both these seafood industry activities.

The unanimity in the aims of both fishery and aquaculture sectors to develop their industry towards the year 2000 (and the $2000 million target!) and beyond, overshadows the differences that have arisen between them during the years of rapid expansion in both sectors.

References

Anon. (1989). Directions in: Foreign Exchange Earnings. The New Zealand Aquaculture Industry. Report prepared for the NZ Trade Development Board, May 1989. 55 pp.

Anon. (1994). Farming heads for quarter of world aquatic harvest. *Fish Farming International* **21**(12), 10–11.

Anon. (1995). Model fishery setting the pace. The southern scallop enhancement success story. *Seafood New Zealand* **3**(4), 8–11.

Chamberlain, G, and Rosenthal, H (1995). Aquaculture in the next century; opportunities for growth, challenges of sustainability. *World Aquaculture* **26**(1), 21–5.

Doonan, I J, Cranfield, H J, and Michael, K P (1994). Catastrophic reduction of the oyster, *Tiostrea chilensis* (Bivalvia: Ostreidae), in Foveaux Strait, New Zealand, due to infestation by the protistan *Bonamia* sp. *New Zealand Journal of Marine and Freshwater Research* **28**, 335–44.

FAO (1994). Aquaculture Production 1986–1992. *FAO Fisheries Circular No. 815 Revision 6 (FIDI/C815 Rev.6)*. 216 pp.

Hickman, R W (1991). *Perna canaliculus* (Gmelin) in New Zealand. In 'Estuarine and Marine Bivalve Mollusk Culture'. (Ed W Menzel) pp. 325–34 (CRC Press: Boca Raton, USA).

Hickman, R W (1992). Spat production. In 'Farming the Dredge Oyster: Proceedings of a Workshop, July 1992'. (Eds P J Smith, G G Baird, and M F Beardsell) pp. 7–11. *New Zealand Fisheries Occasional Publication No 7.*

ICLARM (1995). 'From Hunting to Farming Fish.' Report by Consultative Group on International Agricultural Research (ICLARM: Manila, Philippines). 20 pp.

James, P (1995). How do rock lobsters feed? *Aquaculture Update* **11**, 8–9.

Paul, L J (1986). 'New Zealand Fishes. An Identification Guide.' (Reed Methuen: Auckland). 184 pp.

Schiel, D R (1992). The enhancement of paua (*Haliotis iris* Martyn) populations in New Zealand. In 'Abalone of the World; Biology, Fisheries and Culture'. (Eds S A Shepherd, M J Tegner, and S A Guzman del Proo) pp. 474–84 (Blackwell: Oxford).

Schiel, D R, and Mercer, S F. (1992). Commercial-scale seeding of paua, *Haliotis iris*, into natural habitats at the Chatham Islands. Report to The Chatham Islands Shellfish Reseeding Association and MAF Fisheries, September 1992. 29 pp.

Street, R J (1995). Oyster enhancement trials in Foveaux Strait. *Seafood New Zealand* **3**(9), 32–3.

Tait, M (1995). Snapper enhancement research at Mahanga Bay. *Aquaculture Update* **11**, 5.

Tong, L J, Moss, G A, Redfearn, P, and Illingworth, J (1992). A manual of techniques for culturing paua, *Haliotis iris*, through to the early juvenile stage. *New Zealand Fisheries Technical Report* No 31. 21 pp.

Watkinson, J G, and Smith, R (comps) (1972). 'New Zealand Fisheries.' (New Zealand Marine Department: Wellington). 91 pp.

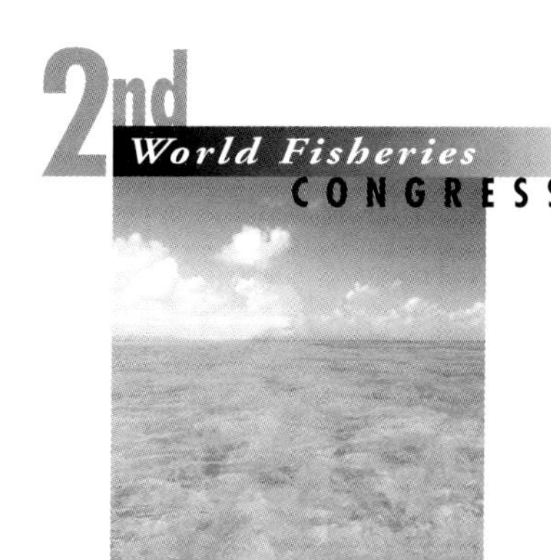

Coastal Planning and Aquaculture: Marine Farming Development Plans for Tasmania

Richard McLoughlin

Marine Resources Division, Department of Primary Industry and Fisheries, GPO Box 192B, Hobart, Tasmania 7001, Australia.

Summary

Marine aquaculture in Tasmania began to reach a critical stage in the early 1990s with industry demand for growth outstripping supply of readily available coastal sites. A relatively new feature of this situation was that applications for new marine farm sites began to meet with occasionally vigorous opposition from local residents and community groups. It became clear that existing legislation was inadequate to properly manage industry growth as well as account for other coastal zone uses and users.

In response, the Tasmanian Government has enacted a new legislative package concurrent with the development of regional coastal aquaculture plans encompassing management and expansion of the industry. In these plans, areas of water are set aside for marine farming — called zones — which are similar in legal status and planning focus to those on land (e.g. residential or light industrial zones). It has become clear that preparation of these management systems has encompassed a comprehensive coastal planning exercise. In this paper the background to the new legislative framework and initial results of regional coastal aquaculture planning in Tasmania are outlined.

Introduction

Tasmania has recently introduced new legislation to control coastal marine farming (mariculture). The legislation it replaced — an amended Fisheries Act — was only 13 years old, but was already quite radically insufficient for a modern regional planning system involving a fast-growing seafood production sector. The new legislation is called the *Marine Farming Planning Act* and, as its title indicates, it involves a planning-based emphasis.

Examination of the general coastal management literature indicates that it is a new system, and it may indicate a way forward for other jurisdictions that have to balance competing demands as they seek to make the most of their coastal marine farming potential. Obviously the extent to which this system may provide a helpful example in other situations depends on the specific planning frameworks already existing in these situations. However, the general principles of local community and coastal user-group consultation, expertise-based decision-making and focus on ecologically and financially sustainable production, should be generally applicable.

Tasmania lies off the south-east coast of mainland Australia, comprising a small group of islands with a population of around 490 000. The climate is cool temperate, with coastal waters dominated by the surrounding Southern Ocean, resulting in a cool, well-flushed marine environment (Fig. 1). Sections of the coast are rugged with many inlets, bays and minor estuaries providing ideal environments for aquaculture. Clean and unpolluted waters characterize the marine environment surrounding the State, and general interest in aquaculture is high.

Marine farming is not new in Tasmania, but until well into the 1980s it was only a small-scale industry. The legislation and planning framework is, therefore, the State's first attempt to comprehensively manage an industry that clearly involves large and extensive investment, and that can realistically be expected to provide a substantial proportion of the seafood production of the State for the long-term future. Despite relatively small size, Tasmania was Australia's second largest producer of seafood in 1995. Total value in 1994/95 was in the order of $A215 million, of which aquaculture production accounted for approximately $110 million. Much of this production value is provided by salmonid culture, which grew rapidly in the late 1980s and early 1990s. A view of the growth of Tasmanian aquaculture is shown in Figure 2.

Until the new direction represented by salmonid culture in the 1980s, marine farming in Tasmania basically meant the culture of Pacific oysters (*Crassostrea gigas*) on racks in intertidal areas. However, the potential of finfish farming was on the minds of politicians and fisheries managers in Tasmania — as it was elsewhere in the world — by the early 1970s. However, the process of technological development was slow. Thus, when the Fisheries Act received a major amendment in 1982 to provide an expanded set of controls for marine farming, the practical

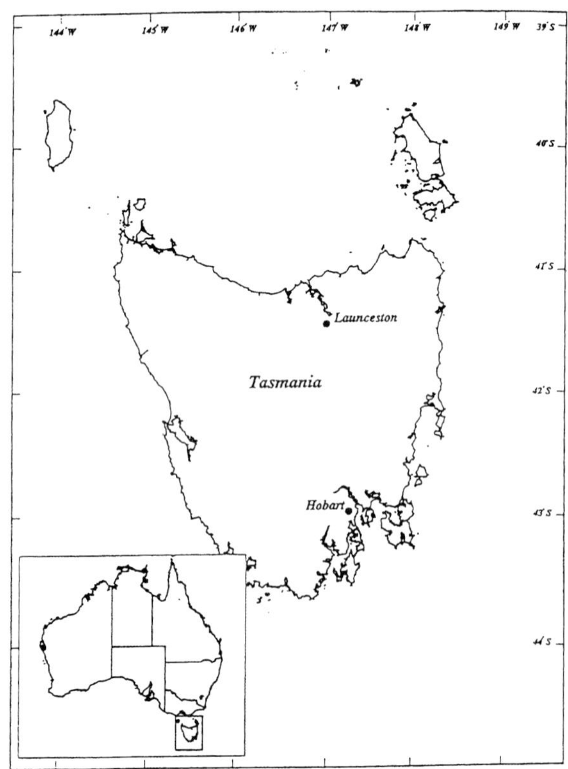

Fig. 1. Map of Tasmania.

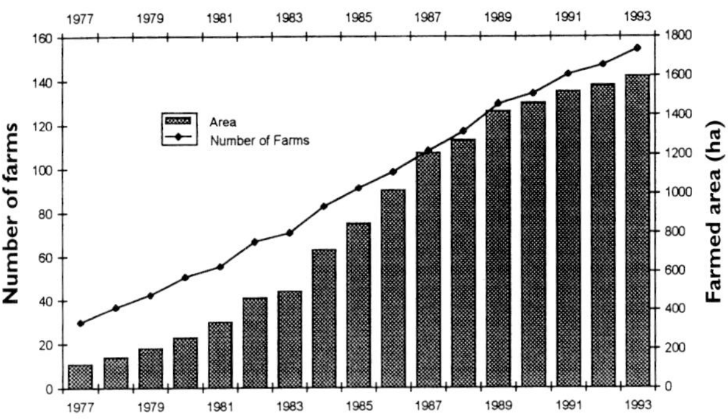

Fig. 2. Histogram of growth in marine farming in Tasmania 1977–1993.

implications of extensive finfish farming still hadn't been seen. The new legal framework involved an attempt to predict future problems. It is therefore not surprising that it turned out not to be ideal, but at least the new opportunities that opened up after 1982 had a framework sufficient to allow the industry to become established.

The 1982 system and its problems

As noted, marine farming activities in Tasmania were controlled by a 1982 amendment of the *Fisheries Act 1959*. The main characteristics of the legislative framework that was established then for regulating marine farming were as follows:

- an application was made to the Department of Primary Industry and Fisheries (DPIF) for the area intended to be used for marine farming;
- there was a limited right for people to object to the granting of approvals for marine farming activities (appeals were limited to persons directly affected);
- if all legal appeals (if any) against the marine farm were rejected, a lease or permit was granted which provided exclusive use to the holder, usually for a period of 20 years; and
- a licence that was valid for 12 months was issued; which outlined specific operational conditions that must be complied with by the lessee.

There was of course a range of other legislative controls, not specific to marine farming, which also impacted on marine farming activities. They included, for example, legislation controlling employee industrial health and safety, and standards for processing and handling products for human consumption. However, in addition there were more directly relevant controls, such as those on pollution and land-use planning, none of which was integrated with the framework of marine farm regulation.

From a resource management perspective, there were a number of deficiencies in the 1982 regulatory framework. These included that:

- there were no guidelines by which applicants and applications for new marine farms could be assessed (e.g. site and species suitability, financial resources, etc.)
- although a statutory mechanism existed under the *Land Use and Planning Approvals Act* 1993 to prepare Maritime

Planning Schemes that could potentially address the interface between the marine farm users, on-land users and the broader issues of protection of coastal environment, no such schemes were actually prepared;

- there was no explicit provision in the Act for the development and implementation of controls which govern marine farm activities;
- there was only limited opportunity to modify the boundaries of a marine farm once granted;
- there was a very limited framework for consideration of the environmental consequences of marine farm proposals;
- rights of appeal against the issue of marine farm approvals were very restrictive and differed according to whether an application was deemed to be for a lease or a permit; and
- appeals against allocation decisions were expensive for both applicants and appellants and were frequently heard in an adversarial court system.

The demand for new marine farm sites grew rapidly in the late 1980s and early 1990s, and applications began to generate increasing numbers of appeals, sometimes from people who were directly affected by the proposed farm site and sometimes from people not affected at all. At the same time, the Government was moving towards a new and integrated land-use planning system and there was a growing recognition that the existing fisheries legislation was inappropriate for dealing with the range of issues that needed to be considered when planning for marine farming.

Therefore with appeals by then slowing down most applications in the courts, in late 1993 a moratorium was placed on the granting of new marine farms. This resulted in a high level of (predictable) frustration from the aquaculture sector, which was by then starting to enjoy the benefits of increasing demand for products, linked to a 'clean and green' image associated with Tasmanian seafoods. Allied with this was a recognition in government that aquaculture represented one of the few agri-business sectors in the Tasmanian economy with real prospects for significant growth in value and production. Additionally, the location of these businesses in mainly small regional economies clearly provided substantial benefits in terms of employment and infrastructure investment.

By this time the context created by the State's new land-use planning regime was a definite spur to action. It was becoming evident that there was also a need for a new regulatory framework for marine farming, and that this would need to:

(**a**) conform to a new resource management and land-use planning framework — which is based explicitly on Ecologically Sustainable Development (ESD) principles — and, particularly, to a general State Coastal policy;

(**b**) be consistent with the long-awaited revision of the State's broader fisheries management legislation (embodied in a new *Living Marine Resources Management Act*); and

(**c**) aim to achieve the following specific ends:

- the formulation of a clear basis for facilitating the future development and management of the marine farming industry in Tasmania;
- statutory processes and other mechanisms for the recognition and resolution, where possible, of concerns involving marine farming;
- better environmental management programmes
- improved mechanisms for the general management and control of marine farming operations; and
- an end to the moratorium on new marine farms.

In developing this response to predicted industry growth, the Government recognized the need to identify and consider the full range of community interests and views in relation to the use of a *public asset*, the Tasmanian coastline. It has also given unequivocal recognition to the importance of the industry and the need to ensure its sustainable development. The keystone of the policy development undertaken to provide the new systems was the provision of high levels of security on all sides; for the community regarding sustainable and acceptable use of its coastal environments, and for the industry in undertaking its financial planning and investments in a relatively stable regulatory environment.

An obvious strategy was to emulate existing systems and programmes applied elsewhere, but very few legislative-based coastal planning schemes with a focus on facilitation of aquaculture appear to have been devised and/or implemented. Local government management schemes devised in response to perceived (and real) conflict between traditional users and new marine farms have been implemented in Scotland and to a lesser extent in Ireland (T. Pearson, pers. comm.). An aquaculture development strategy developed for the State of Maine (Anon. 1990), and followed by a comprehensive set of additional documents relating to regulations and environmental monitoring and assessment, identified the necessary requirements for creating a suitable investment climate for aquaculture, without explicitly recommending new specific planning legislation.

Comprehensive biophysical and aquaculture opportunity studies conducted for the State of British Columbia have provided a solid footing on which to base aquaculture planning and development, although up to seven government jurisdictions are involved in the process (Ricker *et al.* 1989). In New Zealand, regional coastal strategies (e.g. Marlborough Regional Policy Statement, Anon. 1993) have attempted to provide a holistic view of coastal uses and influences and provide sets of general principles to guide development in the region, mainly for local government authorities. Similar strategies, based primarily on local government planning and development of coastal assets, are common in Australia and elsewhere. Most, however, appear to treat aquaculture as a coastal activity *impacting* on more traditional uses, and therefore requiring specific controls and quite often multiple approval processes.

The Tasmanian legislative controls

The critical elements of the new Marine Farming Planning Act, together with the relevant controls on marine farming, are as follows:

- the requirement for there to be Marine Farming Development Plans (MFDPs) for marine farms. These will cover whole districts or regions (such as bays and estuaries) as well as individual farms, thus superseding the need to undertake *ad hoc* and site-by-site assessment;
- the identification in these MFDPs of Marine Farming Zones (MFZs), very carefully selected and defined areas in which farming itself can occur;
- the need to monitor and manage the environmental impact of all marine farming proposals;
- the opportunity for broad community input into the development plans; and
- the expansion and strengthening of provisions covering licensing, management controls and economic return to the community.

The legislation to achieve these goals:

- outlines processes (through the MFDPs) for identifying areas of State waters to be set aside for marine farming. This relies largely on an expertise-based planning review panel that is separate from the 'planning authority' (in this case, the Department of Primary Industry and Fisheries);
- outlines processes to be followed in allocating areas of water to particular persons. In a major departure from past practice, this is to be achieved through an expertise-based board separate from *both* the planning authority and the panel of planning experts;
- outlines processes to be followed if there is a desire to increase the size of a lease area, when the allowable area for marine farming in a zone has not been fully developed;
- creates offences for environmental infringements and for operational matters such as not keeping equipment within a lease area; and
- provides a mechanism for enforcement of the various management controls included in the MFDP.

The Marine Farming Planning Act is separated into three distinct functions:

1. The preparation of MFDPs
2. The allocation of leases in the marine farm zones
3. The administration of marine farms under the new Act.

The focus of this paper is on the planning aspects of the new processes. Nonetheless the other functions are briefly described in order to indicate the integrated nature of the whole system.

The preparation of Marine Farming Development Plans

These Plans are the most innovative feature of the new system, and constitute its basic mechanism. These are described first in an outline of their main aims and processes; and secondly as a four-part run-through of the actual process of developing a Plan.

Overview

The new system is seen as part of the wider State planning system. A very simple indication of the main aspects of this system, and of how the new fisheries legislation fits into it, is shown in Figure 3. Consistent with the policy directions set by the State Coastal Policy and other maritime planning legislation, the objectives for the Marine Farming Development Plans are:

- the orderly management and development of marine farming;
- to promote the sustainable development of the State's living marine resources and the maintenance of ecological processes and genetic diversity;
- to provide for the fair, orderly and sustainable use and development of the State's living marine resources;
- to encourage industry and public involvement in marine farming planning; and
- to facilitate economic development in accordance with the objectives set out above.

In allocating areas of water as suitable for future marine farming, the development plans use the zoning concept, more familiar in terrestrial settings. Zones have traditionally been used in planning schemes and land management plans as a way of providing a clear indication of the different controls applying to different areas of a plan. The use of zoning in MFDPs provides continuity with adjacent terrestrial zoning, and allows for better integration of planning for the management of the coastal areas in the future.

The identification of marine farming zones in the Development Plans (which generally involve under 3% of all the waters in the

Tasmanian State Coastal Policy
prepared under the *State Policies and Projects Act 1993*
to guide all use and development in the Coastal Zone

Land Use Planning Schemes and Interim Orders prepared under the *Land Use and Approvals Act 1993*	**Marine Farming Development Plans** prepared under the *Marine Farming Planning Act 1995*	**Fishery Management Plans** and Controls prepared under the *Living Marine Resources Act 1995*	**State of the Environment Reports** prepared under the *State Policies and Projects Act 1993*
To provide strategic planning and coordinated action to achieve the objectives of the resource management and planning system of Tasmania	To provide sustainable management of marine resources for marine farming use	To provide sustainable management of the State's wild fisheries resources	The biophysical environment, socio-economic features and coastal economic activities would be addressed for the Coastal Zone
Local councils and Marine Boards are planning authorities with responsibilities to prepare land and maritime planning schemes	DPIF has responsibility to prepare MFDPs	DPIF has the responsibility to prepare Management Plans and administer controls	The Tasmanian Sustainable Development Advisory Council to establish a Coastal, Estuarine and Marine Reference Group

Fig. 3. Legislative hierarchy for coastal resource management in Tasmania.

plan area) is considered to provide the best and clearest approach. The zoning strategy aims to give all users of the coastal waters a straightforward indication of which areas are available for different uses, and help minimize the conflict between competing users of these waters.

A range of planning principles and selection criteria is used when setting aside areas of water for marine farming. These amount to guidelines for the establishment of the zones, and in turn provide a basis for the calculation of the proportion of the area within each zone which can actually be used for marine farming (known as the 'maximum leasable area', or MLA). These principles include (among others):

- minimizing environmental and ecological impact;
- maximizing sustainable fish productivity;
- maintaining navigable passages for recreational and commercial vessels;
- preserving other major water uses; and
- maintaining a reasonable distance between marine farming activities for sound stock husbandry purposes.

When drawing up the marine farming zones these principles are applied, and potential areas of conflict are examined and considered prior to the public release of the draft plans. Potential for conflict might involve such factors as proximity to marine reserves and national parks, or to shoreline areas planned for tourism or urban development, or the frequent use of the area by recreational vessels. It is clearly essential to minimize the perceived and real impact of marine farming on other users of the coastal waters and on the marine environment.

THE PROCESS

First draft

This is the phase during which the proposed zones are first identified (Fig. 4). It begins with preliminary identification of MFZs by:

- cataloguing applications for, and expressions of interest in, farming for that area;
- collecting and assessing available environmental data;
- consulting with the marine farming industry; and
- identifying other users and potential restrictions on zone boundaries.

Preliminary boundaries of marine farm zones are then drawn. Once there are lines on the map to discuss, the first consultation gets under way. Government agencies are consulted, as well as interest groups. There is a series of public meetings in the relevant area, as well as meetings for specific groups. From the issues that have arisen, necessary alterations are made. At this stage there is thus a first draft ready to be taken to the Minister, who can approve its release for the second and more formal round of consultation, review and revision.

Preparation of a Draft Marine Farming Development Plan (MFDP)	
1.	Consultation with farmers.
2.	Review of marine farm files and overseas literature.
3.	Collection and collation of environmental data.
4.	Identification of other users and zone boundary restrictions.
5.	Initial outline of zones.
6.	Consultation with Government agencies and interest groups.
7.	Community meetings and consultation.
8.	Review of zone boundaries.
9.	Preparation of draft Plan.
10.	Approval by Minister for Primary Industry and Fisheries for initial review by industry.

Initial Review	
1.	Internal review by Tasmanian Aquaculture Council.
2.	Review and comment on draft by Marine Farming Planning Advisory Committee.
3.	Review and recommendation on release of draft by Marine Farming Planning Review Panel.
4.	Draft MFDP submitted to the Minister for decision on public release.

Public Exhibition	
1.	Sixty (60) day period for public comment and representations.

Resolution of Concerns	
1.	Comments, concerns and representations are collated and considered by the Department for Primary Industry and Fisheries (DPIF) with the preparation of a report to the Marine Farming Planning Review Panel.
2.	The Panel will consider the representations, and where appropriate, conduct a hearing in relation to representations made.
3.	Draft Plan modified as necessary by the Panel.

Final Plan	
1.	Marine Farming Planning Review Panel submits plan to Minister for Primary Industry and Fisheries for approval.
2.	If approved it is subsequently gazetted.

Implementation	
1.	Implementation of MFDP.

Review	
1.	The MFDP will be under regular and periodic review to ensure primary objectives are met, and to allow for changing circumstances that may be relevant. A statutory process for alterations to the MFDP is outlined in the legislation.

Fig. 4. Planning process for marine farming.

Final Draft

The draft is now examined by three bodies external to the Department itself. The first is the Tasmanian Aquaculture Council, for the industry. It then goes to a body that incorporates the relevant user group interests in the Plan area, and it is constituted individually for each Plan. In other words, this Marine Farming Planning Advisory Committee (MFPAC) is not expert-based, but explicitly seeks to represent local interests. Thus a typical MFPAC will have between 12 and 20 members (usually around 15), and include representatives of local councils, farming groups, progress associations, recreational boaters, divers and fishers, professional fishers and local conservation and other community groups. Changes to this early draft can, and often are, made at this stage.

Finally, the Plan goes for the first time to the Marine Farming Planning Review Panel (MFPRP). This body is set up under the legislation, and its constitution has been the subject of lengthy discussion and negotiation, both inside Government and with industry and others in the community. Almost every important interest group lobbied to have representation on the Panel. But the Government remained firm to the fundamental principle that this body should not be representative, but expert, and that any breach in this foundation would fatally weaken the long-term credibility of the Panel.

The result is a body that has generally been recognized in the community as providing a balanced expert reviewing capacity.

The membership is appointed by the Governor — i.e. by the Government — as follows:

1. a *chairperson*;
2. a person with *expertise in planning* nominated by the Chairperson of the Land Use Planning and Review Panel (one of the major entities under the State's new planning system);
3. the *Director of Environmental Management* (a senior statutory position in the State's Department of Environment and Land Management);
4. a person with ability in *marine resource management*;
5. a person with ability to assess *boating, recreational and navigational issues*;
6. a person with experience in *marine farming*; and
7. a person *nominated by the Minister*.

The Panel is thus able to assess the draft professionally, from planning and environmental viewpoints in particular. Its assessment includes one or more on-site inspections of all proposed zones, from the water and/or from land. By the time the MFPRP has finished its considerations, the Plan is ready to go to the Minister, who may now approve its statutory release for public exhibition.

Public exhibition

The Act specifies that the Draft Plan should then be put out for public exhibition for 60 days. This is obviously a most important part of the process, since the public and industry now have — unlike at the earlier public meetings — a detailed document to examine. The reasons for decisions are laid out, and evidence described; therefore objections can be properly argued and detailed.

The descriptions of the MFZs follow a standard format, allowing readers to see quickly the basic information on the history (if any) of marine farming in the area, the environmental conditions, the other uses in the area, and what marine farming activities are proposed.

Review and resolution

Once the 60 days have elapsed, the comments, concerns and representations are assessed by the Department (in its capacity as the planning authority). It is charged with responsibility for preparing a report on the consultation for the MFPRP. This report is then considered by the Panel. It is able, where it feels the issue has been satisfactorily clarified, to amend the Plan. On the other hand, if it feels there are issues on which it cannot confidently make a final determination, it may call and conduct formal public hearings, to allow the interested parties a final opportunity to argue their cases.

The Panel puts the finishing touches to the Plan, and presents it for the third time to the Minister. In the last analysis the Minister has the right to approve or to send it back for further work. If, however, the Minister approves the Plan it is gazetted and comes into operation. The Plans must, by law, be reviewed every 10 years and may be reviewed on demand at two-yearly intervals. But there are significant restrictions on such on-demand reviews, and even the 10-year reviews are intended to maintain a high degree of security for farms operating leases under the Plans. The process makes no apologies, therefore, for allowing every opportunity for public input and debate, but once a planning decision has been made, the Marine Farm Zones have a high degree of security for marine farmers.

The allocation of Marine Farm Zones

The issue of coastal resource allocation, as opposed to planning, was regarded as a critical area needing specialist attention, given the situation in 1993 of marine farm applications outstripping supply of available sites. When considering the issue of how applications for new marine farms should actually be assessed, and who would make the recommendations on individual grants of marine farm leases, the Government decided from the start that an independent body would be appropriate. This was deemed essential where there were competing applications for the same site. Thus, although the Minister retains the final decision, there is an important element of arm's length dealing as far as the Government and its managing Department is concerned.

Again, the make-up of the chosen body — the Board of Advice and Reference, or BAR — was the subject of much fruitless lobbying by groups seeking a place. But the functions of the BAR are relatively narrow, though weighty. Despite the discretion allowed for under the Act, the BAR's job is essentially two-fold:

1. to recommend the appropriate method for allocation of authorizations in particular MFZs (such as tender, direct grant, etc); and
2. to assess the individual applicants, and make recommendations to the Minister.

Therefore the Government again stuck to the principle of expertise-based decision making, and indeed the final composition is actually the one originally proposed. It has only three members:

1. a person who is a *qualified legal practitioner*;
2. a person with experience and knowledge in *marine farming and the seafood industry*; and
3. a person with experience in *business and commerce*.

The BAR also recommends on the appropriate means of allocating leases in particular cases (which may include competitive tenders and auctions), and on how people holding 'certificates of preference' will participate in the process. These certificates may be issued by the Minister to people who had previously submitted applications under the old Fisheries Act for, or who have held permits for relevant experimental work in, the area covered by the MFZ in question. They thus recognize significant prior commitment.

THE ADMINISTRATION OF MARINE FARMS UNDER THE NEW ACT

The legislation integrates the administration and planning roles required by Government in that the farms will be controlled under two instruments: their leases and their licences. The former will provide the basic authorization to occupy and use the waters concerned, and will include only very general conditions. But the majority of the operating conditions are set out in the annual licence, which will thus provide the main control on how farms actually conduct their businesses. An important aspect of the controls exercised over leases will be the environmental conditions, and these are embodied in individual quality-assurance requirements, tailored for each lease and reflected in licence conditions.

EARLY RESULTS OF THE PLANNING PROCESS, AND A BRIEF CONCLUSION

The MFDP process to date has proved to be remarkably successful in identifying both new areas available for marine farming, and in providing substantial area increases to existing farms. Most importantly, this has been done in a climate of active community involvement and consultation. On the whole there has been a low level of community concern, but even where concern has been expressed it seems clear that the existence of the new processes greatly eases the tensions, and provides transparent and acceptable fora for working through these concerns.

This is reflected in results from the first few of the new Plans where, for example, it is proposed that the total leasable marine farm area in the Huon River estuary be increased from 125 to 385 ha (a 300% increase), in the D'Entrecasteaux Channel from 355 to 893 ha (a 251% increase), and in the Tasman Peninsula and Norfolk Bay region from 235 to 540 ha (a 229% increase). That the process has substantially worked is evident from the fact that for the Huon River Estuary Plan, the 300% increase in area was achieved with only two objections, both of which were resolved by the Review Panel *via* the public hearing process. Substantial increases in production, income and employment in the region are therefore forecast in the next few years.

However it has been inevitable and important in terms of upholding the integrity of the system, that the planning process itself also established that in many apparently promising areas there are fragile and ecologically significant sites that are not suitable for aquaculture — beds of seagrass are an example. As a result, development may in many areas be quite constrained. This has been a disappointment in some respects, but it is also right and proper in view of the overall objectives quoted above, with their strong emphasis on ecological sustainability.

Obviously, these large incremental increases in available water should result, over the next decade or so, in corresponding increases in production as the farms are taken up and production begins. Values for the industry at farm gate, presently nearing $A110 million annually, should reflect this increase in production. The investment in the legislation and its corresponding requirement for skilled staff and financial resources appears to have been well worth the investment even though it is only in its second year of operation.

Finally, it is acknowledged that all this is creating virtually a new class of property, based on the value of the water itself. Developed marine farm sites with existing leases are transferable, and high prices have occasionally been paid for good sites. The Government is sensitive to the fact that, by making available large new areas of marine farm sites from what is now regarded as a 'free access' public resource, a great deal of capital value is created and potentially passed into the 'private' economic sphere of activity.

Maintaining a reasonable return to the community for the exclusive commercial use of these community-owned assets — the coastal waterways of Tasmania — will doubtless be the subject of further debate (as well as the subject of deliberation by the BAR, as it considers the appropriate means of allocating leases). It is also a reminder that, as on land, any planning system may have significant economic and financial consequences.

Therefore, while a planning system seems indisputably sensible in this context, that does not mean that it has been particularly easy to implement. It has solved many problems, but by their very nature coastal resource use, planning and allocation will remain as hard decisions, perhaps akin to forestry resource use and management. While relatively new, it does appear that the present Tasmanian system does incorporate appropriate and effective structural safeguards, and that the interests of all parties — the industry, the Government, and the various community stakeholders — can be satisfied within this framework.

REFERENCES

Anon. (1990). An aquaculture production strategy for the State of Maine (includes associated documents). *Maine Department of Marine Resources, Technical report, March 1990.* 103 pp.

Anon. (1993). Marlborough Regional Policy Statement. Marlborough District Council, New Zealand. 78 pp.

Ricker, K E, McDonald, J W, and de Lang Boom, B (1989). Biophysical suitability of the western Johnstone Strait, Queen Charlotte Strait and West Vancouver Island regions for salmonid farming in net cages. *Ministry of Agriculture and Fisheries, Province of British Columbia.* 126 pp.

Aquaculture and its Environment: A Case for Collaboration

U. C. Barg, D. M. Bartley, A. G. J. Tacon and R. L. Welcomme

Food and Agriculture Organization of the United Nations, Rome, Italy.

Summary

Aquaculture is currently one of the fastest growing food production systems in the world with production increasing at an average rate of 9.4% per year over the past decade. Most of global aquaculture output is produced in developing countries, and, significantly, in low-income food-deficit countries. There is considerable diversity of aquaculture practices with regard to species, environments and systems utilized, with very distinct resource use patterns involved and basic differences existing between systems managed for survival and those managed for profit. The aquaculture community is increasingly confronted with environmental issues such as environmental impacts of and on aquaculture practice as well as environmental interactions between aquafarms. Much of the current controversy is centred around environmental degradation resulting in some cases from inadequate co-ordination and management of development, as well as from irresponsible practices by some entrepreneurs engaged in shrimp or salmonid culture which risk bringing the whole sector into disrepute. In addition, aquaculture is to face increasing competition for limited resources, such as water, land and feed inputs, lack of recognition as a legitimate resource user, lack of institutional and legal support, over-regulation and bad publicity. Responsibilities for sustainable aquaculture development need to be shared among aquafarmers, manufacturers and suppliers of aquaculture inputs, processors and traders of aquaculture products, financing institutions, government authorities, researchers, special interest and pressure groups, professional associations, and non-governmental organizations. It is now generally accepted that increasing efficiency in resource use and minimizing adverse environmental interactions will be major goals for the next decade; this will require commitment and willingness to collaborate by all those involved, either directly or indirectly, in aquaculture development. This paper calls for constructive dialogues among partners in aquaculture development. The FAO Code of Conduct for Responsible Fisheries and associated guidelines is presented as one possible framework for such dialogues.

Introduction

'To watch: Farmed fish will become a major source of protein. The World Bank will encourage its development and hopes it will meet 40% of world demand for fish by 2010'. The Economist *(1995): The World in 1996.*

Aquaculture is increasingly being recognized as a globally important food-producing sector. With stagnating yields from capture fisheries and increasing demand for fish and fishery products, expectations for aquaculture to increase its contribution to the world's production of aquatic food are very high. Throughout its long history, aquaculture has contributed significantly to food supply and rural development, particularly among resource-poor farmers in Asia. Moreover, with more than 80% of the world's population expected to be living in developing countries by 2010 (Alexandratos 1995; FAO 1995*a*), there is hope that aquaculture will continue to strengthen its role in contributing to food security and poverty alleviation in many developing countries.

Aquaculture, in common with many other sectors, uses natural resources and interacts with the environment. Food producing activities have been the greatest shapers of landscape and environment throughout the world. While some societies are prepared to accept further modifications to their environment in the interests of more diversified or efficient food production, others place a higher value on the conservation of the existing landscape and seek to return to past states. Aquaculture, as a producer of much needed food, is increasingly confronted with issues of environmental protection. It is now generally accepted that increasing efficiency in resource use and minimizing adverse environmental interactions will be major goals in aquaculture development for the next decade, which will require commitment and willingness to collaborate by all those involved, either directly or indirectly.

The purpose of this paper is to highlight the very diverse nature of aquaculture and the ways in which this diversity relates to environmental and developmental issues. This paper calls for constructive dialogues among partners in aquaculture development, and the FAO Code of Conduct for Responsible Fisheries and associated guidelines is also presented as one possible framework for such dialogues.

PRODUCTION, ENVIRONMENT AND DIVERSITY OF AQUACULTURE

Aquaculture is growing fast

Aquaculture is currently one of the fastest growing food production systems in the world with production increasing at an average rate of 9.4% per year over the past decade; total world aquaculture production more than doubled by weight from 10.4 to 25.5 million t and tripled by value from \$US13.1 to 39.8×10^9 between 1984 and 1994, respectively (FAO 1996; Tacon 1996). In 1994, aquaculture contributed 21.7% of total world fisheries landings, providing a total of 12.1 million t of high quality animal food (figure excludes gutting and shelling, bait fish and aquatic plants) for direct human consumption, which was composed of 93.7% fish, 4.0% molluscs and 2.3% crustaceans.

It is the Asian aquafarmers who currently produce more than 90% of the world's aquaculture output. Over 86% of the 1994 production total was grown in developing countries; 75% of global aquaculture tonnage was produced by farmers in low-income food-deficit countries (LIFDCs).[1] In particular, over 80% of finfish cultured globally was produced within LIFDCs, with 95% being lower value (in marketing terms) inland freshwater fish species (i.e. mainly Chinese and Indian carps feeding low on the aquatic food chain) and produced as a polyculture of complementary fish species at low stocking densities within semi-intensive or extensive farming systems. In contrast, over 60% of total finfish production within developed countries is based on the monoculture of high-value carnivorous fish species within intensive farming systems, with fish being reared within ponds, tanks, pens or net-cages at high stocking densities and usually fed on high-quality manufactured *aquafeeds* or natural feeds (Tacon *et al.* 1995). Aquaculture production within LIFDCs has been growing over 5-times faster (i.e. 13%/y) than within developed countries (i.e. 2.2%/y) between 1984 and 1994.

[1] LIFDCs including all food-deficit countries with a *per capita* income below the level used by the World Bank to determine eligibility for International Development Association assistance, *ca.* \$US1345 in 1993.

Environment and resources: challenges to aquaculture

In its transition to become a major food-producing sector, aquaculture faces significant challenges, some of which have been related to the ways aquaculture interacts with the environment. Since aquaculture depends heavily on natural resources such as water, land, seed and feed, it is essentially the use of these resources by aquafarmers, as well as their access to appropriate quantity and quality of these resources, which will determine the scale of impact of these environmental interactions. There is a wide range of interactions between aquaculture and the environment, including environmental impacts of aquaculture, environmental interactions between aquafarms, and external environmental impacts on aquaculture. These interactions have already been reviewed in the literature (e.g. Rosenthal *et al.* 1988; Chua *et al.* 1989; Pullin 1989; GESAMP 1991; Iwama 1991; Barg 1992; Munday *et al.* 1992; Pillay 1992; Pullin 1993; Beveridge *et al.* 1994; Campton 1995; FAO/NACA 1995; Phillips 1995*a*, 1995*b*).

Environmental impacts ***of*** aquaculture have been associated mainly with *high-input high-output intensive systems* (e.g. culture of salmonids in raceways and cages), the effects of which included nutrient and organic enrichment of recipient waters resulting in build-up of anoxic sediments, changes in benthic communities and the eutrophication of lakes. *Large-scale shrimp culture* has resulted in degradation of wetlands, salination of agricultural and drinking water supplies, and land subsidence due to groundwater abstraction. However, misapplication of chemicals, collection of seed from the wild and use of fishery resources as feed inputs, are also causing concern. Mollusc culture has been held responsible for local anoxia of bottom sedimentation and increased siltation.

Aquaculture is the principle reason for the introduction of freshwater fishes (Welcomme 1988; Bartley and Subasinghe, unpublished data) and experience has shown that the introduced species will eventually enter the natural ecosystem (either through purposeful release or accidental escape). Thus, non-native species in culture can adversely impact local resources through hybridization, predation and competition, transmission of disease, and changes in habitat, e.g. burrowing, plant removal, sediment mobilization and turbidity.

Environmental interactions ***between*** farms, include self-pollution and transmission of diseases, and occur in areas where the high density of farms forces use of water contaminated by neighbouring installations, with significant losses of farmed stocks and financial returns. Effects can also occur at a distance with interchange of living material between farms and a consequent spread of disease.

External environmental impacts ***on*** aquaculture have resulted in some areas due to changes in water quality. These impacts include water releases which affect water supply to farms, as well as physical degradation of aquatic habitats and aquatic pollution by other developments which cause mass mortalities, disease outbreaks, and product contamination. Such impacts are in addition to the adverse effects on composition, abundance and distribution of wild fish populations, which affect capture fisheries. Where non-native species or genetically improved species, e.g. products of selective breeding, monosex populations etc. are utilized in culture, there is the risk that 'wild' fish will enter the production ponds and cause unwanted breeding that will undo the progress of controlled breeding and domestication. There is also the risk of disease introduction from the wild fish.

Some types of aquaculture have produced social conflicts within local communities, e.g. the introduction and rapid expansion of shrimp culture has in some areas resulted in marginalization of people (Bailey 1988). Cage culture installations sometimes have raised aesthetic concerns (EC 1995) and conflict with tourism and traditional fisheries, and the enclosure of some public waters for culture has deprived other users of access.

Aquaculture will face competition for finite resources, especially water and land. Pressure on resources continues to increase due to population growth, coupled with unequal distribution of food, other goods and services, and increasing degradation of aquatic environments (e.g. Chua and Pauly 1989; Chua *et al.* 1989; Dudgeon 1992; Burbridge 1994; Petr and Morris 1995; Muir 1995; Barg *et al.* in press). Regulatory measures to reduce environmental degradation by major offenders are sometimes lacking or not enforced. In contrast, aquaculture, a relatively minor offender (if it is an offender at all), is often not recognized by government authorities, other resource users, major polluters and the public in general as a legitimate resource user. This has produced a climate under which there is little protection of the aquafarmers' needs for adequate quality and quantity of resources (Van Houtte 1994; FAO/NACA 1995). Long-term consequences of uncoordinated resource utilization in river basins or coastal zones are sometimes ignored or neglected, particularly with regard to the social and economic value of goods and services provided by aquatic ecosystems, their genetic resources and biodiversity. Feed resources for some aquaculture practices are also limited. This applies principally to ingredients for the manufacture of compound feeds for high-value carnivorous fish and shrimp species. Increased demand for fishmeal and fish oil may push up prices in a situation of finite supply from marine capture fisheries (Tacon 1995).

The pressure to use resources more efficiently, to increase competitiveness and to respond to market forces in some areas is resulting in trends toward intensification of aquaculture production. These are associated with more sophisticated farm management, shift to monoculture of high-value species, and the targeting of more affluent consumers. There is an increased risk that such trends to intensification will increase environmental impacts if inappropriate planning and management of such farming systems and, in particular, the inefficient use of resources and inputs such as equipment and chemicals, are not avoided.

However, the current debate on environmental impacts of aquaculture is due mainly to irresponsible practices by relatively few aquaculture entrepreneurs, who risk bringing the whole sector into disrepute. In some cases this is leading to attempts to control coastal aquaculture while neglecting the need to support and develop sustainable inland culture systems producing food affordable to resource-poor consumers (Bailey and Skladany 1991). There is much scope for the aquaculture community to improve the public image of the sector, as well as to seek collaboration with other interest groups to reverse trends of continued degradation of aquatic environments and resources.

Diversity in aquaculture: issues changing perceptions

The impressive growth rate of aquaculture production, together with the call for sustainable development of aquaculture, show positive impacts. Above all, there is an increasing awareness, both within and outside the aquaculture community, of the very diverse nature of aquaculture. To date aquaculture has generally been viewed as a single activity only, but nowadays it is clear that it actually represents the aquatic counterpart of all terrestrial agriculture under the one name. Such a broadened view, which is not new to many experts particularly in countries with major aquaculture production, is facilitating a more detailed consideration of technological, economic, social, and environmental issues as well as of institutional and legal factors governing aquaculture development. Since some of the current controversy on aquaculture and the environment has been biased by lack of insight into the complex nature of aquaculture, it is convenient to highlight some of the basic differences and issues in aquaculture:

Diversity in species and systems. Aquaculture organisms under culture include 141 species and other taxa of finfish, 31 taxa of crustacea, and 59 taxa of molluscs, in addition to seaweeds, other aquatic plants, frogs, turtles, crocodiles, sea-squirts, and sponges (Garibaldi 1996). Specific biological characteristics require diverse farming systems within different environments. Farming systems are very different as to the level of control and input of resources by the farmer. Most global production is based predominantly on semi-intensive and extensive systems, and on aquaculture-based fisheries, producing affordable finfish for domestic rural markets and subsistence. Intensive systems often produce moderate to high value species for export or domestic urban markets. However, shrimp, usually grown as a cash crop, can be reared in either extensive, semi-intensive or intensive systems.

Diversity in farmers. Significant differences exist between farming approaches, resource use patterns, the farmers themselves, and the needs for food security and rural development. Generally, the farming approach employed by aquafarmers in developing countries (and in particular LIFDCs) has been targeted more toward management for survival (Zweig 1985), whereas the farming approach used in developed countries has been targeted more toward management for profit (Tacon 1996). Traditional aquaculture in developing countries of the South (Pillay 1996), evolved as small-scale peasant enterprises to suit rural economies. Major characteristics of such aquaculture include (i) family ownership, (ii) polyculture, (iii) integration with crop and animal farming activities, (iv) waste recycling and beneficial use of farm wastes, (v) diversification of food production and spreading of farmers' risks among different commodities, (vi) provision of off-season work for farmers and wage-earners, and (vii) a general means of improving nutrition and incomes.

Diversity in impacts and environments. The potential for environmental impact from aquaculture is related to a wide range of interrelated factors including availability, amount and quality of resources utilized, species cultured, farming systems management, and environmental characteristics of the location of the aquafarm, e.g. little significant pollution effects result from small-scale low-input aquafarms. Impacts from intensive farms in open coastal areas with good water flow are usually minimal. While the risk of eutrophication by aquaculture may be significant in some inland water bodies, there is little related evidence for most of coastal aquaculture. Effects resulting from aquaculture waste discharges might be minimal because of the 'environmental carrying capacity' of many recipient water bodies (GESAMP 1986, 1991) has not been exceeded. Many types of aquaculture actually contribute to environmental improvement (Phillips 1995a), e.g. by reducing use of pesticides in rice fish culture, by utilizing nutrients in wastewater-based culture, or by absorbing nutrients in seaweed or mollusc farming thereby reducing risks of coastal eutrophication, etc. In programmes for the restoration and recovery of endangered fish species and stocks, hatcheries and culture systems have been used to provide a temporary sanctuary and to increase numbers of individuals for re-introduction into the wild (Johnson and Jensen 1991; Hedgecock *et al.* 1994).

AQUACULTURE DEVELOPMENT AND THE CODE OF CONDUCT FOR RESPONSIBLE FISHERIES

Origin and scope of the Code

During the last decade, issues such as sustainable development, environmental interactions and long-term sustainability of aquaculture received increasing attention at local, national and international levels (Chamberlain and Rosenthal 1995; PACON 1995; Pullin 1995; Reinertsen and Haaland 1995; Creswell 1996). Special attention is also given to these issues as relevant to developing countries, particularly in Asia (Pullin 1993; FAO/NACA 1995; Bagarinao and Flores 1995; ADB/NACA 1996; NACA 1996; SEAFDEC/FAO in prep.). Following the 1992 Cancún Conference on Responsible Fishing,and the United Nations Conference on Environment and Development, FAO was requested by its member countries to draft an International Code of Conduct for Responsible Fisheries. Many experts and representatives from governments, intergovernmental and non-governmental organizations (NGOs) participated in the formulation of the Code during several technical consultations and sessions of the FAO Committee on Fisheries. The Code (see Appendix I) was adopted by the FAO Conference in November 1995 (FAO 1995*b*).

The Code is designed to follow the provisions of UNCED Agenda 21 and the Convention on Biological Diversity, in addition to other international agreements and legal instruments relevant to fisheries. The Code sets out principles and international standards of behaviour for responsible practices with a view to ensuring the effective conservation, management and development of living aquatic resources, while recognizing the nutritional, economic, social, environmental and cultural importance of fisheries and aquaculture, and the interests of all those involved in these sectors. The essence of the Code is contained in Chapters on General Principles, Fisheries Management, Fishing Operations, Aquaculture Development, Integration of Fisheries into Coastal Area Management, Post Harvest Practices and Trade, and Fishery Research. The Code is voluntary, i.e. legally non-binding (except for the Compliance Agreement) though it is expected that States and international organizations, including NGOs, will promote the Code and its effective implementation.

The Code's Article 9 on Aquaculture Development contains provisions relating to aquaculture, including culture-based fisheries and their responsible development in areas of national jurisdiction and within transboundary aquatic ecosystems, to conservation of genetic diversity and ecosystems, and to responsible practices at the production level. The Article calls for appropriate legal and administrative frameworks, and the production and update of management plans for sustainable use of resources shared by aquaculture and other activities. Emphasis is given to the need to assure that aquaculture development does not compromise the livelihood of and access to resources by local communities. International cooperation can be furthered by the recommendation for States to establish means to collect, share and disseminate information such as databases and networks dealing with aspects of aquaculture development.

The Article requires advance evaluation and monitoring of impacts, acknowledges that farmed stocks may adversely affect natural ones, and states that efforts should be made to minimize this risk. Especially important will be the appropriate use of alien species and genetically modified organisms. The Article stresses international cooperation and advisement when such organisms are placed in or near transboundary water bodies. The role of aquaculture in the recovery of endangered species is also part of the Code.

In relation to aquaculture production, the Article addresses the role of rural communities, producer organizations and fish farmers. Recommendations relate, in particular, to appropriate use of feeds, feed additives and fertilizers, disease control and other chemicals, to safe disposal of wastes that may adversely affect human health and environment, and to assurance of food safety of aquaculture products.

Given the complex issues, the different interests and the wide-ranging problems fisheries and aquaculture are facing throughout the world, it is indeed remarkable that governments, experts, and other interested parties from both developing and developed countries, succeeded, through collaboration and negotiation, in the preparation and adoption of the Code. This consensus can be regarded as a strong call for commitment to self-regulation and monitoring, to more precautionary approaches, as well as to sustainable development in fisheries and aquaculture. The tasks ahead in most cases are very challenging and are likely to be fulfilled only through collaboration.

Implementation of the Code: tasks, roles and responsibilities

The Code stipulates actions to be taken by States. However, it is also meant to address persons, interest groups or institutions, public or private, who are involved in or concerned with aquaculture. A major target here is to stimulate and facilitate the development of guidelines of responsible practice containing norms and standards, agreed upon by farm owners or operators, and other businesses and activities which support or are otherwise associated with aquatic farming. Given the diversity of aquaculture practices, and of the political, social and economic conditions in which they take place, such guidelines may well be formulated in such ways as to meet specific requirements at local, national, regional and international levels. Major benefits which can be derived from the development of such voluntary codes of practice and guidelines are:

- public image of the aquaculture sector can be improved through adherence to established norms and adequate self regulation;
- aquaculturists, associated by agreement on common standards and objectives, are in a better position to defend their interests, and to negotiate for rights and privileges against competing interests;
- there will be greater common understanding and agreement on specific measures which can or should be implemented to ensure sustainable development of aquaculture;
- roles and responsibilities of persons, interest groups or institutions, public or private, can be identified and negotiated, with a view to assure and confirm their commitment and contribution to sustainable development of aquaculture; and
- as part of integrated area management, responsible aquaculture acknowledges its interaction with other sectors in the conservation and efficient use of resources, and therefore, can request that those sectors do not compromise the availability of resources of adequate quantity and quality required by aquaculture and fisheries.

Collaboration for responsible aquaculture development

Collaboration by many will be required to further promote responsible aquaculture development. Examples are presented here of major areas for possible collaboration among government authorities, researchers, aquafarmers, manufacturers and suppliers of aquaculture inputs, and other interest groups.

Government authorities in many countries will continue to play a major role in promoting and regulating aquaculture development. In many countries, for example, existing administrative and legal frameworks may need to be adjusted to address the specific characteristics and needs of the sector. Collaboration may be required in designating or establishing a competent authority in charge of aquaculture or in establishing appropriate linkages with other authorities such as those concerned with agriculture, rural development, water resources, environment, health, education and training, etc. Likewise, legal provisions and regulatory measures may need to be streamlined so as to clearly set forth the responsibilities and privileges of aquaculturists. Aquaculture is still not being recognized as the aquatic equivalent to agriculture, and there is scope for increasing awareness of both public institutions and the general public about aquaculture and its similarities with agriculture. This may be achieved through collaborative efforts by aquafarmers, authorities, media and non-governmental initiatives.

In many countries there is a continued need to produce and regularly update development plans for supporting, regulating and reporting on the aquaculture sector. Involvement of local communities and authorities, aquafarmers, and resource management experts in aquaculture development planning, can help in assuring that aquafarms are sited in locations which prevent social conflicts, unacceptable environmental degradation or exposure of aquafarms to external environmental threats. Zoning and site regulations may be specified to conform with the requirements of plans for regional development and river basin or coastal area management plans. Government authorities, in collaboration with aquafarmers, may evaluate the possible benefits and consequences of the introduction of new or different aquaculture products, methods or technologies to ascertain whether they are likely to contribute to food supply and rural development as well as to the economy and to the welfare of their citizens generally. Alternatively, they may result in significant public problems such as abandoned capital investment, requirements for subsidy, or excessive demands on scarce or critical resources (feedstuffs, land, water) needed for more important products or activities.

Obviously, environmental assessment is an important area for collaboration by authorities, researchers and aquafarmers. Consultations among all concerned should ensure that procedures for environmental impact assessment and monitoring are sufficiently flexible, taking into account that the scale and cost of such efforts may well have to be adjusted to the scale of the perceived impact of a given aquaculture operation. Contributions by scientists to this area of collaboration can be crucial. For example, for advance evaluation (i.e. pre-impact assessment) to have any meaning, there must be predetermined standards, i.e. acceptable limits of impacts. However, our knowledge of many ecosystems and their genetic diversity is often incomplete, especially in many developing countries and tropical regions. Setting standards for allowable genetic 'effluent' is still very difficult, because of the scarcity of information on the effects of aquaculture/wild animal interaction, survival of aquaculture escapees, and their impact on ecosystems. Available models for the assessment and prediction of ecological impacts of aquaculture wastes are usually only applicable in temperate regions, and cost-effective and rapid assessment methods are needed which can be applied easily in tropical environments and developing countries. Given the current debate, environmental scientists can also help by clearly distinguishing between actual and hypothetical environmental hazards resulting from aquaculture practice.

Manufacturers and suppliers of inputs such as equipment, feeds and chemicals, also can contribute significantly to responsible aquaculture, by providing inputs appropriate and specific to systems and species, and by advising farmers on use and limitations of such inputs. For example, feed manufacturers and suppliers have a responsibility to provide quality feeds and to advise on effective feeding methods. Manufacturers, suppliers and aquafarmers can collaborate on the safe and effective use of drugs, pesticides, hormones, etc. by adjusting the quantities and rates of delivery of these products to obtain the desired effects with minimum wastage, as well as by paying close attention to withdrawal periods to ensure contaminant-free products. Aquafarmers, hatchery operators and scientists can collaborate on appropriate procedures for the production of seed and the development of improved stocks, in particular to avoid inbreeding and outbreeding depression, and to reduce risks of genetic, disease or ecological impacts by escaped or released fish.

Resource-poor farmers with no access to information may need to be assisted through local capacity-building activities. Local communities and authorities as well as non-governmental initiatives may find it useful to include aquaculture components into agricultural extension programmes in their areas. Rural development experts may advise on formation of aquafarmers' associations. Media may contribute to initiatives for sustainable aquaculture development and to more balanced public perceptions by showing efforts and living conditions of rural aquafarmers, particularly in developing countries.

CURRENT EFFORTS AND OUTLOOK

Aquaculture has provided increased food and income to many parts of the world with relatively minor adverse environmental effects. The majority of developing country aquaculture involves low-input environmentally benign fish culture; developed country aquaculture is often highly regulated with environmental protection policies. It is the few highly publicized examples of severe habitat degradation that threaten aquaculture's continued growth. However, the trends within many developing countries toward the use of more intensive aquaculture systems and more higher-value species often in sensitive coastal areas will greatly increase the potential for environmental damage and may put additional stress on the socio-economic structure of local communities. Therefore, it is essential that the aquaculture industry and all the stakeholders involved adopt a strong commitment for cooperation and self regulation.

The international community, through FAO, is developing international guidelines in support of the implementation of the Code of Conduct for Responsible Fisheries. Some of these guidelines are well advanced while others are in draft form. Steps and mechanisms for further elaboration of these guidelines and implementation of the Code are being derived from FAO's governing bodies through input to the biennial meeting of the Committee on Fisheries (COFI). FAO appreciates the interest of the international community in implementing the Code. Participation is encouraged, where appropriate, in COFI and other appropriate international fora. During first meetings of the Subsidiary Body on Scientific, Technical and Technological Advice (SBSTTA), it was recommended that relevant articles of the Convention on Biological Diversity are reflected in the guidelines for implementation of the Code of Conduct.

In addition to FAO's efforts, there are numerous local, national, regional and international initiatives promoting sustainable development of aquaculture. FAO encourages collaboration in this regard, as well as in relation to the implementation of the Code's principles. It is hoped that collaboration in the preparation and implementation of specific guidelines for responsible aquaculture will also contribute to greater recognition of aquaculture, particularly in terms of its benefits and of the diversity of practices and people involved.

The potential for aquaculture to contribute further to enhanced food supplies and incomes, particularly in developing countries, has been reiterated at the recent International Conference on the Sustainable Contribution of Fisheries to Food Security (FAO/Japan 1995). However, this potential may be reduced in view of trends of increasing pressure on aquatic environments and competition for finite resources by non-aquaculturists, and partly also by negative perceptions resulting from relatively few cases of severe environmental degradation and social disruption caused by certain types of aquaculture.

Many will continue to be concerned with the sustainability of the aquaculture industry, and with environmental issues in aquaculture. Many will also be concerned with issues of sustainable development and aquaculture, and its possible contribution to poverty alleviation and assurance of sustained food supply in both rural and urban areas of developing countries. In the meantime, many aquafarmers, like most of their terrestrial counterparts, will have to continue to attempt solving problems on their farms while struggling with constraints such as inadequate access to resources, natural and financial, lack of institutional and legal support, or unavailability of appropriate information.

Given the diversity in aquaculture and the sometimes very different perceptions of 'sustainability', more balanced and informed approaches are required to effectively address developmental and environmental issues at any given location. Constructive dialogues among responsible partners can help to integrate these requirements. Obviously, participation of farmers and their communities in such dialogues is essential. However, when negotiating responsibilities for sustainable development in local areas, obligations do not rest only with farmers and their communities. Responsibilities beyond the farm level need to be shared by many stakeholders. Providing an 'enabling environment' for sustainable development in aquaculture, as in agriculture, is the responsibility of people in governments and their institutions, the media, financial institutions, pressure groups, associations, non-governmental organizations, as well as of social and natural scientists, manufacturers and suppliers of inputs, processors and traders of products. Fairness and responsible attitudes in consultations and negotiations between countries or regions may also help sustainable aquaculture development.

As for aquaculture, agreements on responsible action, or codes of practice, may be considered in a positive or pro-active sense in that their aim is to ensure the continued sustainable growth of the aquaculture sector and to protect and assist it in collaboration with all other interested parties. The Code of

Conduct for Responsible Fisheries provides a voluntary framework for reference to initiate or strengthen constructive dialogues and collaboration between responsible partners, for sustainable development of fisheries and aquaculture.

References

ADB/NACA (1996). Aquaculture sustainability action plan. Regional study and workshop on aquaculture sustainability and the environment (RETA 5534). Asian Development Bank and Network of Aquaculture Centres in Asia-Pacific. NACA, Bangkok, Thailand. 21 pp.

Alexandratos, N (Ed) (1995). 'World Agriculture: Towards 2010. An FAO Study.' FAO, Rome and John Wileys, Chichester. 488 pp.

Bagarinao, T U, and Flores, E E C (Eds) (1995). Towards Sustainable Aquaculture in Southeast Asia and Japan. Proceedings of the Seminar-Workshop on Aquaculture Development in Southeast Asia, held 26-28 July 1994 in Iloilo City, Philippines. SEAFDEC Aquaculture Department, Iloilo, Philippines. 254 pp.

Bailey, C (1988). The social consequences of tropical shrimp mariculture development. *Ocean Shoreline Management* **11**, 31–44.

Bailey, C, and Skladany, M (1991). Aquacultural development in tropical Asia: A re-evaluation. *Natural Resource Forum* **15**(1), 66–71.

Barg, U C (1992). Guidelines for the promotion of environmental management of coastal aquaculture development. FAO *Fisheries Technical Paper* 328 122 pp.

Barg, U, Dunn, I, Petr, T, and Welcomme, R L. (in press). Inland fisheries and water management. In 'Handbook of Water Resources and the Environment'. (Ed A K Biswas) (International Journal of Water Resources Development).

Beveridge, M C M, Ross, L G, and Kelly, L K (1994). Aquaculture and biodiversity. *Ambio* **23**(8),497–502.

Burbridge, P R (1994). Integrated planning and management of freshwater habitats, including wetlands. *Hydrobiologia* **285**, 311–22.

Campton, D E (1995). Genetic effects of hatchery fish on wild populations of Pacific salmon and steelhead: what do we really know? *American Fisheries Society Symposium* **15**, 337–53.

Chamberlain, G, and Rosenthal, H (1995). Aquaculture in the next century. Opportunities for growth - challenges of sustainability. *World Aquaculture* **26**(1), 21–5.

Chua, T E, and Pauly, D (Eds) (1989). Coastal area management in South East Asia: policies, management strategies and case studies. *ICLARM Conference Proceedings* 19. 254 pp.

Chua, T E, Paw, J N, and Guarin, F Y (1989). The environmental impact of aquaculture and the effects of pollution on coastal aquaculture development in Southeast Asia. *Marine Pollution Bulletin* **20**(7), 335–43.

Creswell, R L (Ed) (1996). Book of abstracts of 'World Aquaculture '96', 29/01–2/02 1996, Bangkok, Thailand. World Aquaculture Society, Baton Rouge, USA. 498 pp.

Dudgeon, D (1992). Endangered ecosystems: a review of the conservation of tropical Asian rivers. *Hydrobiologia* **248**, 167–219.

EC (1995). Aquaculture and the environment in the European Community. Luxembourg: Office for Official Publications of the European Communities. 89 pp.

FAO (1995a). Food, agriculture and food security: the global dimension. Information paper prepared for the 1996 World Food Summit. WFS96/TECH/1. 35 pp.

FAO (1995b). Code of Conduct for Responsible Fisheries. Rome, FAO. 48 pp.

FAO (1996). Aquaculture — A source of hope. Fact sheets prepared for the 1996 World Food Summit. Rome, FAO.

FAO/Japan (1995). Safeguarding future fish supplies: key policy issues and measures. Main Document contributed to the International Conference on Sustainable Contribution of Fisheries to Food Security, Kyoto, Japan, 4-9 December 1995, organized by the Government of Japan, in collaboration with the Food and Agriculture Organization of the United Nations (FAO). KC/FI/95/1. 50 pp.

FAO/NACA. (1995). Regional Study and Workshop on the Environmental Assessment and Management of Aquaculture Development. FAO and Network of Aquaculture Centres in Asia-Pacific. Bangkok, Thailand. NACA Environment and Aquaculture Development Series No. 1. NACA, Bangkok, Thailand. 492 pp.

Garibaldi, L (1996). List of animal species used in aquaculture. FAO Fisheries Circular 914 (FAO: Rome). 38 pp.

GESAMP (IMO/FAO/Unesco/WMO/WHO/IAEA/UN/UNEP Joint Group of Experts on the Scientific Aspects of Marine Pollution) (1986). Environmental Capacity. An approach to marine pollution prevention. *Report Studies GESAMP* 30, 49.

GESAMP (IMO/FAO/Unesco/WMO/WHO/IAEA/UN/UNEP Joint Group of Experts on the Scientific Aspects of Marine environmental Protection) (1991). Reducing Environmental Impacts of Coastal Aquaculture. *Report Studies GESAMP* 47, 35.

Hedgecock, D, Siri, P, and Strong, D R (1994). Conservation biology of endangered Pacific salmonids: introductory remarks. *Conservation Biology* **8**, 863–4.

Iwama, G K (1991). Interactions between aquaculture and the environment. *Critical Reviews Environmental Control* **21**(2), 177–216.

Johnson, J E, and Jensen, B L (1991). Hatcheries for endangered freshwater species. In 'Battle Against Extinction'. (Eds W L Minckley and J E Deacon) pp. 199–217 (University of Arizona Press: Tucson).

Muir, J F (1995). Aquaculture development trends: perspectives for food security. Contribution to the International Conference on Sustainable Contribution of Fisheries to Food Security, Kyoto, Japan, 4–9 December 1995, organized by the Government of Japan, in collaboration with the Food and Agriculture Organization of the United Nations (FAO). KC/FI/95/TECH/4. 133 pp.

Munday, B, Eleftheriou, A, Kentouri, M, and Divanach, P. (1992). The interactions of aquaculture and the environment: A bibliographical review. Report prepared for the Commission of European Communities; Directorate General for Fisheries. 325 pp.

NACA (1996). The second five year programme of the Network of Aquaculture Centres in Asia-Pacific (NACA), 1996–2000. Theme: Aquaculture Sustainability. NACA, Bangkok, Thailand. 27pp.

PACON (1995). Proceedings of the PACON conference on sustainable aquaculture 95. Honolulu, Hawaii, USA, 11–14 June 1995. 441 pp.

Petr, T, and Morris, M (Eds) (1995). Indo-Pacific Fishery Commission. Papers contributed to the Regional Symposium on Sustainable Development of Inland Fisheries under Environmental Constraints. Bangkok, Thailand, 19–21 October 1994, and Country reports presented at the sixth session of the IPFC Working Party of Experts on Inland Fisheries. Bangkok, Thailand, 17–21 October 1994. *FAO Fisheries Report* No. 512 (Supplement). Rome, FAO. 262 pp.

Phillips, M J (1995a). Aquaculture and the environment - striking a balance. In 'Aquaculture towards the 21st century'. (Eds K P P Nambiar and T Singh) pp. 26–55. Proceedings of INFOFISH-AQUATECH'94 International Conference on Aquaculture, Colombo, Sri Lanka, 29–31 August 1994. Organized by INFOFISH and the Sri Lanka Export Development Board.

Phillips, M J (1995b). Shrimp culture and the environment. In 'Towards Sustainable Aquaculture in Southeast Asia and Japan'.(Eds T U Bagarinao and E E C Flores) pp. 37–62 (SEAFDEC Aquaculture Department: Iloilo, Philippines).

Pillay, T V R (1992). Aquaculture and the Environment. Oxford, Fishing News Books, Blackwell. 189 pp.

Pillay, T V R (1996). The challenges of sustainable aquaculture. *World Aquaculture* **27**(2), 7–9.

Pullin, R S V (1989). Third World aquaculture and the environment. *NAGA ICLARM Quarterly* **12**(1), 10–13.

Pullin, R S V (1993). An overview of environmental issues in developing country aquaculture, In 'Environment and aquaculture in developing countries'. (Eds R S V Pullin, H Rosenthal and J L MacLean) pp. 1–19. *ICLARM Conference Proceedings* 31. 359 pp.

Pullin, R S V (1995). Growth and aquaculture sustainability. NAGA/ICLARM Quarterly (July 1995), 19–20..

Reinertsen, H, and Haaland, H (Eds) (1995). Sustainable fish farming. Proceedings of the first international symposium on sustainable fish farming, Oslo, Norway, 28–31 August 1994. Baalkema, Rotterdam. 307pp.

Rosenthal, H, Weston, D, Gowen, R, and Black, E (1988). Report of the *ad hoc* study group on environmental impact of mariculture. Copenhagen, *ICES Cooperative Research Report* 154. 83 pp.

SEAFDEC/FAO (in preparation). Report and proceedings of SEAFDEC/FAO Expert Meeting on the Use of Chemicals in Aquaculture in Asia, held 20–22 May 1996, at the Aquaculture Department of the Southeast Asian Fisheries Development Center in Tigbauan, Iloilo, Philippines.

Tacon, A G J (1995). Aquaculture feeds and feeding in the next millennium: major challenges and issues. *FAO Aquaculture Newsletter* **10**, 2–8.

Tacon, A G J (1996). Trends in aquaculture production, with particular reference to low-income food-deficit countries 1984–1993. *FAO Aquaculture Newsletter* **11**, 6–9.

Tacon, A G J, Phillips, M J, and Barg, U C (1995). Aquaculture feeds and the environment: the Asian experience. *Water Science and Technology* **31**(10), 41–59.

The Economist (1995). The World in 1996. The world in figures: industries. p. 93.

Van Houtte, A (1994). The legal regime of aquaculture. FAO *Aquaculture Newsletter* **7**, 10–15.

Welcomme, R L (1988). International introductions of inland aquatic species. FAO Fisheries Technical Paper 294.

Zweig, R (1985). Freshwater aquaculture in management for survival. *Ambio* **14**(22), 66–74.

APPENDIX I. CODE OF CONDUCT FOR RESPONSIBLE FISHERIES

ARTICLE 9 — AQUACULTURE DEVELOPMENT

9.1 Responsible development of aquaculture, including culture-based fisheries, in areas under national jurisdiction

9.1.1 States should establish, maintain and develop an appropriate legal and administrative framework which facilitates the development of responsible aquaculture.

9.1.2 States should promote responsible development and management of aquaculture, including an advance evaluation of the effects of aquaculture development on genetic diversity and ecosystem integrity, based on the best available scientific information.

9.1.3 States should produce and regularly update aquaculture development strategies and plans, as required, to ensure that aquaculture development is ecologically sustainable and to allow the rational use of resources shared by aquaculture and other activities.

9.1.4 States should ensure that the livelihoods of local communities, and their access to fishing grounds, are not negatively affected by aquaculture developments.

9.1.5 States should establish effective procedures specific to aquaculture to undertake appropriate environmental assessment and monitoring with the aim of minimizing adverse ecological changes and related economic and social consequences resulting from water extraction, land use, discharge of effluents, use of drugs and chemicals, and other aquaculture activities.

9.2 Responsible development of aquaculture including culture-based fisheries within transboundary aquatic ecosystems

9.2.1 States should protect transboundary aquatic ecosystems by supporting responsible aquaculture practices within their national jurisdiction and by cooperation in the promotion of sustainable aquaculture practices.

9.2.2 States should, with due respect to their neighbouring States, and in accordance with international law, ensure responsible choice of species, siting and management of aquaculture activities which could affect transboundary aquatic ecosystems.

9.2.3 States should consult with their neighbouring States, as appropriate, before introducing non-indigenous species into transboundary aquatic ecosystems.

9.2.4 States should establish appropriate mechanisms, such as databases and information networks to collect, share and disseminate data related to their aquaculture activities to facilitate cooperation on planning for aquaculture development at the national, subregional, regional and global level.

9.2.5 States should cooperate in the development of appropriate mechanisms, when required, to monitor the impacts of inputs used in aquaculture.

9.3 Use of aquatic genetic resources for the purposes of aquaculture including culture-based fisheries

9.3.1 States should conserve genetic diversity and maintain integrity of aquatic communities and ecosystems by appropriate management. In particular, efforts should be undertaken to minimize the harmful effects of introducing non-native species or genetically altered stocks used for aquaculture including culture-based fisheries into waters, especially where there is a significant potential for the spread of such non-native species or genetically altered stocks into waters under the jurisdiction of other States as well as waters under the jurisdiction of the State of origin. States should, whenever possible, promote steps to minimize adverse genetic, disease and other effects of escaped farmed fish on wild stocks.

9.3.2 States should cooperate in the elaboration, adoption and implementation of international codes of practice and procedures for introductions and transfers of aquatic organisms.

9.3.3 States should, in order to minimize risks of disease transfer and other adverse effects on wild and cultured stocks, encourage adoption of appropriate practices in the genetic improvement of broodstocks, the introduction of non-native species, and in the production, sale and transport of eggs, larvae or fry, broodstock or other live materials. States should facilitate the preparation and implementation of appropriate national codes of practice and procedures to this effect.

9.3.4 States should promote the use of appropriate procedures for the selection of broodstock and the production of eggs, larvae and fry.

9.3.5 States should, where appropriate, promote research and, when feasible, the development of culture techniques for endangered species to protect, rehabilitate and enhance their stocks, taking into account the critical need to conserve genetic diversity of endangered species.

9.4 Responsible aquaculture at the production level

9.4.1 States should promote responsible aquaculture practices in support of rural communities, producer organizations and fish farmers.

9.4.2 States should promote active participation of fish farmers and their communities in the development of responsible aquaculture management practices.

9.4.3 States should promote efforts which improve selection and use of appropriate feeds, feed additives and fertilizers, including manures.

9.4.4 States should promote effective farm and fish health management practices favouring hygienic measures and vaccines. Safe, effective and minimal use of therapeutants, hormones and drugs, antibiotics and other disease control chemicals should be ensured.

9.4.5 States should regulate the use of chemical inputs in aquaculture which are hazardous to human health and the environment.

9.4.6 States should require that the disposal of wastes such as offal, sludge, dead or diseased fish, excess veterinary drugs and other hazardous chemical inputs does not constitute a hazard to human health and the environment.

9.4.7 States should ensure the food safety of aquaculture products and promote efforts which maintain product quality and improve their value through particular care before and during harvesting and on-site processing and in storage and transport of the products.

Environmentally Sustainable Aquaculture Production — an Australian Perspective

Nigel Preston,[A] *Ian Macleod, Peter Rothlisberg and Brian Long*

CSIRO Division of Fisheries, PO Box 120 Cleveland, Qld 4163 Australia.
[A] and the Cooperative Research Centre for Aquaculture

Summary

Australia has a diverse aquaculture industry which includes tropical, subtropical and temperate sectors. Over the last three years production has increased by 40% in tonnage and 80% in value. Most of the production is from the coastal zone where the most valuable species are pearl oysters, Atlantic salmon, edible oysters, tuna and prawns. Low production levels in relation to coastal area, and low levels of water pollution generally, permit a high standard of aquaculture products. However, there have already been significant production losses due to adverse environmental impacts on and from aquaculture; these must be prevented if the industry is to advance. The intensive cage-culture of fish requires particularly conservative management and continual monitoring. The edible oyster industry has been adversely affected by pollution and this must be more carefully considered in the process of site selection. The more remotely located pearl oyster industry is less affected by pollution but needs to prevent the introduction of pest species. The expanding prawn farming industry needs to develop viable alternative strategies for pond and effluent nutrient management to permit environmentally sustainable development. The effectiveness of environmental monitoring programmes could be considerably improved with the use of automated data-loggers, which could also provide early warning of adverse environmental impacts. Direct nutrient tracer experiments are needed to validate environmental impact or carrying-capacity models. In general, for site selection and coastal zone management purposes, Geographic Information Systems and decision-support systems are considered to offer a powerful aquaculture planning tool.

Introduction

Australia has a diverse aquaculture industry which includes tropical, subtropical and temperate sectors. Over the past three years (1992 to 1995) aquaculture production in Australia increased by 40% in tonnage and 80% in value, and now comprises 26% of the value of total fisheries production (Table 1). In the same period total capture fisheries production tonnage declined by 11%. Most of the production is from the coastal zone with only about 10% of total production value coming from freshwater species (ABARE 1995).

Of the species farmed in Australia, pearl oysters contribute approximately half the estimated value of the total aquaculture production (Table 2). The other species farmed in the coastal zone are produced for seafood markets, the main species being edible oysters, salmon, tuna and prawns. Rainbow trout is currently the most valuable inland species, with freshwater crayfish, barramundi and silver perch emerging as new industries (Treadwell *et al.* 1992).

Australia has lower aquaculture production levels in relation to coastal areas than other countries in the region, including New Zealand (Csavas 1993). In 1990 Australia produced 0.01 kg of aquaculture production per 1000 ha of coastal land, compared with 1.29 kg for the equivalent area in New Zealand and

Table 1. Summary of the production and value of Australian aquaculture and fisheries statistics for 1992 and 1995 (source ABARE 1995)

Sector	1992	1995	% change
Aquaculture production (t.10^3)	16.8	23.6	+40%
Fisheries production (t.10^3)	245.8	218.2	−11%
Aquaculture value ($A.$10^6$)	255.3	458.5	+80%
Fisheries value ($A.$10^6$)	1492.7	1744.5	+16%

33.75 kg in Japan. This low level of production and comparatively low levels of coastal water pollution permit a very high health standard for aquaculture products. In order to maintain this standard it is imperative to minimize or prevent adverse environmental impacts on and from aquaculture.

Since the mid-1980s there has been a significant shift in the approach used for aquaculture planning, from an emphasis on estimating potential production levels for each target species and region (e.g. FAO 1984), to an increasing focus on sustainable production levels (Csavas 1993). To achieve sustainable production, aquaculture will have to accommodate the rapid increase in human population and urbanization of coastal areas, the associated decline in natural resources (e.g. broodstock for aquaculture) and the need to improve methods of aquaculture waste disposal.

Australian aquaculture is relatively new and, because of increased community awareness of environmental issues, the industry is being subjected to closer scrutiny than that traditionally reserved for agriculture. Aquaculture producers, particularly those directly affected by nutrient run-off from agriculture, feel that they are unfairly treated in relation to permitted water quality parameters of effluent which limit production levels, and in bearing the costs of upstream and downstream monitoring programmes. However, both industry and government recognize that environmental impacts pose a real threat to sustainable production and, as outlined below, these concerns are species-specific and often site-specific. The Australian aquaculture industry is currently being reviewed, with management plans being developed and implemented for most States. These plans generally adopt a conservative approach, and recognize the need to collect and analyse quantitative information on environmental impacts on and from aquaculture. In the interim, relatively strict environmental controls are generally considered to be constraining rapid or extensive development.

In this paper the status of Australian aquaculture is examined in relation to environmentally sustainable production of the main coastal zone species. The principal focus is on the physical, chemical and biological environment and how best to improve the associated planning and monitoring procedures. Clearly this should not be done in isolation from other major issues such as disease (e.g. Lightner 1992), nutrition (e.g. New 1995) and market demand (e.g. FAO 1984); however this paper makes no attempt to address these additional issues in any detail.

Table 2. Summary of Australian aquaculture statistics (ABARE 1995). Table shows tonnage and value for each species in 1995. 'Others' include silver perch, redclaw crayfish and barramundi.

Species	Production (t) 1995	Value ($A.$10^3$) 1995	Value % 1995
Pearl oysters	n/a	206 215	45.0
Atlantic salmon	6 000	67 000	14.6
Tuna	2 250	64 000	14.0
Edible oysters	8 434	42 204	9.2
Trout	4 426	38 592	8.4
Prawns	1 648	27 736	6.0
Others	925	12 761	2.8
Total	23 683	458 508	100

Fish farming

Intensive cage-culture of fish in multiple-use water bodies probably presents a more demanding challenge to achieving environmentally sustainable production than any other form of aquaculture. Current cage-culture practices result in a high input of nutrients to the water column and sediments in the form of fish faeces and uneaten food (Gowan and Bradbury 1987). Reducing the amount of feed wasted by improving feed formulations (e.g. high energy, low nutrient feeds) and improving feeding regimes is a high priority for intensive cage-culture of fish and other forms of aquaculture that use pelleted feeds (OECD 1989). In shallow waters the adverse effects of enriched sediments beneath fish farms can have a negative effect on production, thus forcing farmers to find alternative sites (Gowen and Rosenthal 1993). Other major concerns are the use of antibiotics and disinfectants, and release of these chemicals to the natural environment (Austin 1993).

Salmon and tuna are the two main species farmed in cages in Australia. Atlantic salmon, *Salmo salar,* are farmed in net cages in inshore waters in the Huon estuary in south-east Tasmania (Fig. 1). The cage-culture technology is well developed and is based on techniques developed in Europe and North America.

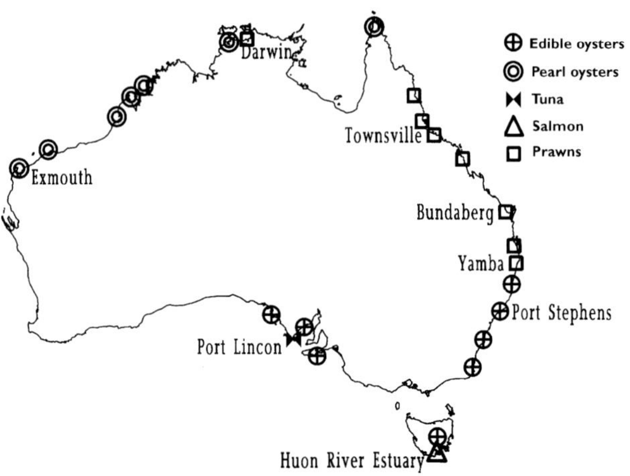

Fig. 1. Aquaculture production sites in Australia.

Production of Atlantic salmon has expanded rapidly from the first commercial harvest in 1987 to 6000 t in 1995. The potential expansion of salmon farming is considered to be partly limited by a shortage of sites unless the technology for offshore cultivation is developed (Treadwell *et al.* 1992).

In response to the rapid growth of aquaculture in Tasmania, the State Government has implemented a regional coastal aquaculture planning strategy and enacted legislation to control development (see McLoughlin 1997). An important component of the new Marine Farming Planning Act is the need to monitor and manage the environmental impact of all marine farming. The effects of salmon farming on the environment of the Huon estuary have been examined (Woodward *et al.* 1992) and key environmental monitoring variables have been identified. However, in common with other aquaculture sectors, information about optimal sampling design and cost-effectiveness is lacking.

The Tasmanian management plans for siting salmon farms take into account the water flow and exchange rates, water depth and distance between farms. Approximate limits on carrying capacity have been estimated in relation to stocking density, the ratio of open area to stocked cage area and the minimum depth of water under the cages. Other important factors included in the plan are the preservation of areas for other water uses, such as recreation and shipping, buffers between farms and shore access by farmers.

Tuna farming is a more recent development than salmon farming in Australia. Wild-caught juvenile southern bluefin tuna (*Thunnus maccoyii*) are stocked in cages in the coastal waters of Port Lincoln, South Australia. The fish are on-grown to increase weight and improve condition before marketing. Initial production of 26 t in 1991 increased to 2250 t in 1995. A comprehensive aquaculture management plan has been developed for the tuna farming region with a proposed review of the plan in 1998 (Bond 1993). Limits have been placed on the maximum volume of sea-cages and the number of hectares occupied by cages in response to concerns about long-term nutrient levels in the region.

The tuna management plan considers that, at the current scale and stocking densities, tuna farming will result in relatively minor localized impact on the water quality in the immediate vicinity (<50 m) of a sea-cage operation. The management plan does identify concerns about marked changes to sediment chemistry and biota in the immediate vicinity of sea-cages. The changes to the sediment biota are considered to be reversible and periodic shifting of sea-cages ('fallowing') has been recommended to ameliorate long-term impacts. An initial period of two years was recommended between fallowing but optimization of the duration of cage deployment at individual sites has not been attempted. In April 1996 Australian tuna farmers suffered losses of up to 75% of total production which was attributed to asphyxiation of fish by sediments re-suspended by a severe storm.

Preliminary estimates of the environmental holding capacity of salmon- or tuna-cage culture have relied on modelling nutrient outputs calculated from mass balance equations and flushing regimes of receiving waters (e.g. Gowan and Bradbury 1987). This approach has provided a first approximation for cage-culture in Australia, but such models have yet to be validated by direct measurement of changes in nutrient levels and they give no indication of the levels of temporal and spatial variation in nutrient loads (see Beveridge and Phillips 1993).

Studies of macrobenthic communities (Ritz *et al.* 1989) and stable carbon isotope measurements (Ye *et al.* 1991) of sediments beneath and adjacent to Australian salmon sea-cages have provided indirect estimates of the distance that organic enrichment from fish feeds extends from the cages. Direct tracing of nutrients with enriched stable isotopes (e.g. Kling 1994) could provide more precise determination of the origin of nutrients in aquaculture waterways with multiple sources of nutrient inputs. Nutrient tracing would also permit quantification of the rate of accumulation and recovery of sediments. This would assist in determining the carrying capacity of farm sites and in optimizing the frequency and duration of fallowing. Such studies should simultaneously determine the physical properties of each region, such as the rate of water removal and vertical stratification, which play an important role in the response of the environment to the nutrient load (Aure and Stigebrandt 1990). This information will enhance the design of farm site monitoring programmes and the deployment of moored data-loggers. Continuous monitoring of critical variables, such as dissolved oxygen, could provide early warning of adverse environmental impacts and negative effects on farm production.

OYSTERS

In contrast to the cage-culture of fish, the cultivation of oysters on racks or rafts depends entirely on the availability of planktonic food. It has been argued that the farming of bivalves results in the net-removal of nutrients from the environment at the time of harvest (Folke and Kautsky 1989). However, oysters and other edible bivalves selectively accumulate some bacterial and chemical contaminants, including heavy metals, that can pose a serious health threat to consumers. For this reason, together with consumer awareness of the potential problems, the production of cultured edible bivalves has recently levelled off or declined in many countries (Pullin 1993).

Farming edible oysters is one of Australia's oldest aquaculture industries, having commenced in the 1870s (Nell 1993). Historically the Sydney rock oyster (*Saccostrea commercialis*) has been the most important oyster species. Sydney rock oysters are grown in the intertidal region on sticks or trays. The peak of production was in 1976/77 (14.6 million dozen). Recently the production of this species has declined (11 million dozen in 1988 to 6 million dozen in 1992). Reasons for this include the introduction of Pacific oysters (*Crassostrea gigas*), pollution, toxic algal blooms and disease (Treadwell *et al.* 1992; Nell 1993). As the production of Sydney rock oysters has declined there has been a significant increase in Pacific oyster production, mainly in Tasmania and South Australia, which is expected to continue.

As in other countries, pollution from agricultural or urban waste in Australia poses a threat to sustaining current oyster farm sites and developing new ones. In the past the use of antifouling paints containing tributyl tin oxide (TBTO) affected the NSW Sydney rock oyster industry (Batley *et al.* 1989). Although TBTO has now been banned, threats from other chemical pollutants still remain. In many areas future lease allocations are unlikely because of conflict with the increasing water areas demanded by recreational water-users (Nell 1993). The granting of new leases is also hampered by concerns about environmental carrying-capacity. Attempts are now being made to develop predictive models of carrying-capacities for oyster farms based on an assessment of primary productivity and production-linked nutrient cycles.

Pearl oysters

Pearl oyster farming is Australia's most valuable aquaculture industry and the technology is well developed. The industry depends principally on the collection of pearl oysters from the wild, but hatcheries are also used. Pearl oysters are farmed in racks, on lines or cages hung from rafts or resting on the sea floor. The level of production from wild stocks is relatively stable and is not expected to increase substantially. In future the successful development of hatcheries coupled with optimal selection of new grow-out sites could result in the expansion of the industry.

The pearl and edible oyster industries in Australia provide contrasting examples of factors affecting environmentally sustainable production. The pearl oyster industry is predominantly situated in remote locations relatively free from adverse threats of pollution from agricultural or urban waste. Pearls and pearl shell are non-perishable and thus unaffected by bacterial contamination that can result from poor handling of edible molluscs (FAO 1988). Adverse environmental impacts on the industry could stem from poor water quality in hatcheries located too close to sources of chemical or biological contaminants. A potential threat to the grow-out phase is the introduction of exotic pests, including toxic microalgae or predators, *via* ballast water in ships; a threat shared by many other aquaculture industries (Hallegraeff *et al.* 1990; Maclean 1993).

Prawn farming

Prawn farming in Australia has developed only recently and relatively slowly compared with most other prawn farming regions. From 1984 to 1995 production rose from 15 t to 1600 t. Approximately 350 ha of ponds are currently in production from farms distributed over 2000 km of coastline, ranging from tropical (Darwin) to temperate (Yamba) conditions (Fig. 1). Striking contrast in rates of industry development are provided by countries such as Thailand where production rose from 20 000 t in 1984 to 160 000 t in 1994 (Lin 1995). Based on current production levels and future predictions, a reasonable scenario is that the Australian industry will continue moderate growth, which could result in an increase of the total pond production area to 1100 ha producing 3500–6000 t by the year 2000 (Macarthur Consulting 1995).

Despite the relatively moderate projected growth, there are justifiable concerns about potential adverse environmental impacts of prawn farming. Some locations suitable for pond aquaculture are adjacent to unique and environmentally sensitive areas such as the Great Barrier Reef and other marine parks. The major criticisms levelled at prawn farming practices elsewhere must be taken into account in developing a sustainable industry in Australia.

The construction and operation of earth ponds for aquaculture in the coastal zone for prawn farming has disadvantages and advantages in relation to environmentally sustainable production. The disadvantages are well illustrated by the frequently cited adverse effects of prawn farming which include the destruction of coastal vegetation, salinization of land and pollution of waterways (Phillips *et al.* 1993; Primavera 1993). On the positive side, the use of water treatment reservoirs and partial or total recirculating systems (Lin 1995; Sandifer and Hopkins 1996) could provide far greater control of influent and effluent water quality for land based aquaculture systems than can be achieved in open waterways.

In parts of Asia and South America prawn farming has resulted in the mass destruction of mangroves and alteration of the hydrological characteristics of adjacent areas (Phillips *et al.* 1993; Primavera 1993). In Australia the clearing of mangroves is strictly controlled. Indeed all marine plants including mangroves are protected and may only be cleared under permit. It is now well-recognized that mangroves are sub-optimal sites for prawn farming because of acid sulfate soils, high pond construction costs and loss of important natural fisheries habitats (Csavas 1993). Studies of the habitat requirements of prawn species that sustain commercial fisheries in Australia have further emphasized the importance of mangroves (Staples *et al.* 1985).

Current prawn farming practices have also been criticized for the release of pond effluent containing elevated levels of dissolved and particulate nutrients derived mainly from the input of feed and fertilizers (Phillips *et al.* 1993). Failure to develop viable alternative strategies for pond and effluent nutrient management and treatment are considered to be a major constraint to environmentally sustainable development, particularly in areas where unrestricted development has occurred (Phillips *et al.* 1993; Primavera 1993).

Currently the recommended or permitted levels for nutrients and suspended solids discharged from Australian prawn farms vary from State to State (Macarthur Consulting 1995). Prawn farmers are concerned about being forced to comply with increasingly stringent effluent discharge levels. However, there is a lack of quantitative water quality data as the basis for legislation. An important issue to resolve is the relationship between influent and effluent water quality. For example, major rivers on the east Australian coast, such as the Clarence, Logan or Burdekin, can have very high nutrient and suspended solid loads at peak flow. This may have major effects on farm water quality and result in elevated levels of suspended solids in effluents being wrongly attributed to farm activities alone.

The potential for prawn farmers to improve influent or effluent water quality is well recognized. A number of options have been trialled or proposed, including improved farm design, sedimentation ponds, polyculture with filter feeders, recirculating systems and the timing of discharge to spread effluent loading (Lin 1995). However, quantitative information on the efficiency or cost-effectiveness of these techniques has not been determined for any Australian prawn farm, although trials have commenced (Jones and Preston 1996).

SITE SELECTION AND COASTAL ZONE MANAGEMENT

Many of the problems associated with low production and adverse environmental impacts on or from aquaculture could be avoided by better site selection (Phillips *et al.* 1993; Preston *et al.* 1995). There is increasing recognition that aquaculture site selection in the coastal zone is a complex task. Coastal aquaculture activities can have impacts on or from a broad range of activities, including agriculture, urbanization, tourism, coastal zoning and environmental restrictions, boating, shipping and capture fisheries. For land-based aquaculture, site selection also requires decisions about soil types, vegetation, topography, distance from waterways, protected habitats, potential sites of upstream pollution and permitted effluent discharge in receiving waters.

Given the complexity of site selection we have started to investigate the potential of Geographical Information Systems (GIS) in aquaculture planning. In this paper we provide an example of the application of GIS in the preliminary screening of potentially suitable sites for prawn farming. The area selected was approximately 35 km of coastline in south-east Queensland bisected by the Logan river (Fig. 2). There are already five prawn farms in the area with a combined total of 50 ha of ponds. The total land area examined was 85 870 ha. Two classes of land were quantified: optimal, and suitable but sub-optimal.

Using the TOPO250K digital spatial data supplied by Australian Land Information Group (AUSLIG), four constraint layers were built. These were elevation and slope, distance-to-waterfront, distance-to-urban development, and wetland occurrence (Table 3). By adding successive constraints the area suitable for aquaculture is reduced. The area of interest was initially restricted to all mainland less than 2 km from coastal or estuarine waterfront, a total of 22 987 ha. Topographic constraints

Table 3. Application of Geographical Information System to Logan River, southeast Queensland. Prescribed areas and geographic constraints

Prescribed area	Abbreviation	Area (ha)	% area
Total mainland area	MLAND	85 870	100
No wetland	WTLND	81 455	95
Area >3 km from urban development	RBND	32 576	38
Area <2 km from waterfront	DIST2	22 987	27
Area <1 km from waterfront	DIST1	13 687	16
Elevation <15 m or slope <5%	TOPO	18 723	22

Additive constraints	DIST1	%	DIST2	%
Area near waterfront	13 687	16	22 987	27
+TOPO	10 180	12	13 986	16
+TOPO +WTLND	6 776	8	9 602	11
+TOPO +WTLND + RBND	1 702	2	3 245	4

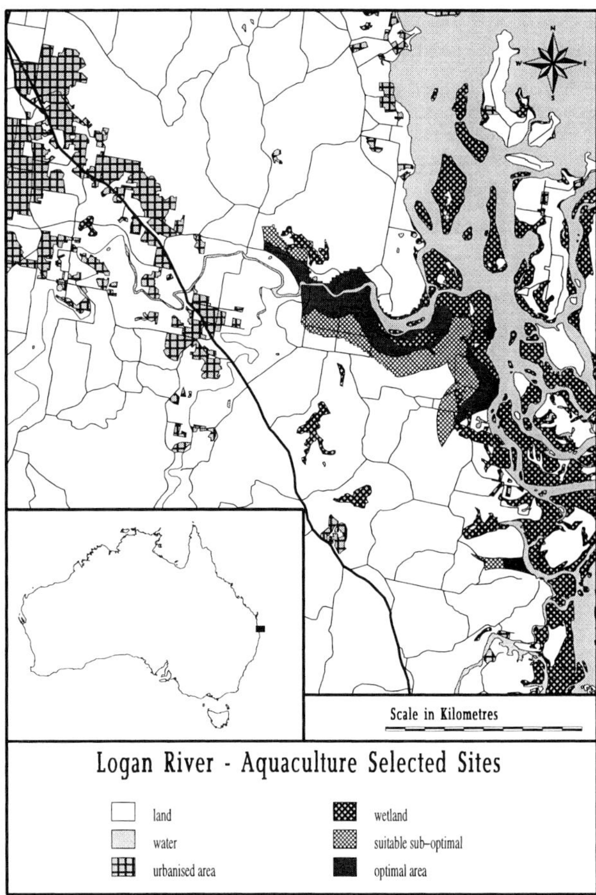

Fig.2. Logan River, southeast Queensland — application of Geographical Information System technology to aquaculture site suitability.

excluded all areas over 15 m in elevation or 5% slope, which reduced the area to 13 986 ha. Land classed as wetland was then excluded, leaving an area of 9602 ha. This was further constrained by excluding all areas within 3 km of urbanized development. This left an area of interest of 3245 ha which was sub-classed by distance-to-waterfront into optimal areas (less than 1 km from waterfront) and sub-optimal (less than 2 km from waterfront). The final result of this example was an area of 1702 ha identified as being optimal and an additional 1543 ha identified as sub-optimal, giving a total area of 3245 ha being suitable (Fig. 2). Further investigation with additional regional data, such as soil surveys, water quality analysis, ownership and availability would further refine the classification.

This example does not imply that the defined areas are considered as optimal for prawn farming. In fact the seasonal temperatures at this latitude restrict the grow-out season for one of the farmed species (*Penaeus monodon)* to a few months, compared with year-round production at lower latitudes (Preston *et al.* 1995). The example does, however, demonstrate that GIS, combined with information from environmental monitoring programmes, could provide a powerful tool to assist in planning the future development of Australian aquaculture. The GIS permits multiple layers of additional restrictions to be added. These could include the planning zones for all other coastal zone activities, proximity to industrial, agricultural or domestic waste discharges, power supplies, urban centres and airports.

Conclusion

In summary, despite the comparatively low level of aquaculture production in relation to coastal area, Australia faces significant challenges to achieve environmentally sustainable production. The nature of the challenge varies according to the species and management techniques. Intensive cage-culture of fish requires particularly conservative management and continual monitoring in order to minimize adverse environmental impacts and to avoid negative impacts on production.

Urbanization and resultant water pollution, coupled with strict government regulations, could result in declining aquaculture production of edible oysters in the affected areas. By comparison, the relatively remote location and non-perishable nature of pearl oysters reduces the level of concern about urbanization and contaminants. However, in common with most other species, sustainable production could be threatened by the introduction of exotic pests, including toxic microalgae or predators. Prawn farming in Australia has the opportunity to learn from mistakes made in other countries. Failure to develop viable alternative strategies for pond and effluent nutrient management and treatment are a constraint to environmentally sustainable development.

The generally conservative nature of industry and government aquaculture management plans should provide the opportunity to gather sufficient data from existing sites to considerably improve future site selection. It is suggested that the precision of many of the existing environmental monitoring programmes could be considerably enhanced by the use of continual recording, automated data-loggers. Such loggers could also provide early warning and more accurate modelling of adverse environmental impacts. Direct nutrient tracer experiments are also needed to validate environment impact or carrying-capacity models.

Finally, GIS and associated decision support systems are considered to offer a powerful aquaculture planning tool. They facilitate the increasingly complex tasks associated with integrating aquaculture planning into multi-sectorial coastal zone planning and management, and aid prospective growers in optimizing site selection.

Acknowledgments

We would like to thank Richard McLoughlin, Colin Shelly, Glen Schipp, Michael Gunzburg, John Nell and Brian Jeffries for supplying information and for their comments on drafts.

References

ABARE (1995). Australian aquaculture production. Australian fisheries statistics. Australian Bureau of Agricultural and Resource Economics, Canberra, Australia. 52 pp.

Aure, J, and Stigebrandt, A (1990). Quantitative estimates of the eutrophication effects of fish farming on fjords. *Aquaculture* **90**, 135–56.

Austin, B (1993). Environmental issues in the control of bacterial diseases of farmed fish. In 'Environment and Aquaculture in Developing Countries'. (Eds R S V Pullin, H Rosenthal, and J L Maclean) pp. 237–51. ICLARM Conference Proceedings 31. 359 pp.

Batley, G E, Fuhua, C, Brockbank, C I, and Flegg, K J (1989). Accumulation of tributyltin by the Sydney Rock Oyster, *Saccostrea commercialis. Australian Journal Marine Freshwater Research* **40**, 49–54.

Beveridge, M C M, and Phillips, M J (1993). Environmental impact of tropical inland aquaculture. In 'Environment and Aquaculture in Developing Countries'. (Eds R S V Pullin, H Rosenthal and J L Maclean) pp. 213–36. ICLARM Conference Proceedings 31. 359 pp.

Bond, T (1993). Port Lincoln Aquaculture Management Plan. Resource Management Division, Department of Environment and Land Management, Adelaide, South Australia.

Csavas, I (1993). Aquaculture development and environmental issues in the developing countries of Asia. In 'Environment and Aquaculture in Developing Countries'. (Eds R S V Pullin, H Rosenthal and J L Maclean) pp. 74–101. ICLARM Conference Proceedings 31. 359 pp.

FAO (1984). A study of methodologies for forecasting aquaculture development. *FAO Fisheries Technical Paper* 248. 47 pp.

FAO (1988). Report of the Seventh Session of the Indo Pacific Fishery Commission (IPFC) Working Party of Experts on Aquaculture, 1-6 August, 1988, Bangkok. *FAO Fisheries Report* 411. 37 pp.

Folke, C, and Kautsky, N (1989). The role of ecosystems for a sustainable development of aquaculture. *Ambio* **18**, 234–43.

Gowan, R, and Bradbury, N (1987). The ecological impact of salmonid farming in coastal waters: a review. *Oceanography and Marine Biology Annual Review* **25**, 563–75.

Gowan, R J, and Rosenthal, H (1993). The environmental consequences of intensive coastal aquaculture in developed countries: what lessons can be learnt. In 'Environment and Aquaculture in Developing Countries'. (Eds R S V Pullin, H Rosenthal and J L Maclean) pp. 102–15. ICLARM Conference Proceedings 31. 359 pp.

Hallegraeff, G M, Bolch, C J, Bryan, J, and Koerbin, B (1990). Microalgal spores in ship's ballast water: a danger to aquaculture. In 'Toxic Marine Phytoplankton'. (Eds E Graneli, B Sundstrom, L Edler and D M Anderson) pp. 475–80 (Elsevier: New York).

Jones, A B, and Preston, N P (1996). Biofiltration of shrimp pond effluent by oysters in a raceway system. World Aquaculture 96. Abstracts 188 pp.

Kling, G W (1994). Ecosystem-scale experiments. The use of stable isotopes in fresh waters. In 'Environmental Chemistry of Lakes and Reservoirs'. (Ed L A Baker) pp. 1084–92 (American Chemical Society).

Lightner, D V (1992). Shrimp virus disease: diagnosis, distribution and management. In 'Proceedings of the special session on shrimp farming'.(Ed J Wyban) pp. 238–53 (World Aquaculture Society: Baton Rouge, Louisiana).

Lin, C K (1995). Progression of intensive marine shrimp culture in Thailand. In 'Swimming Through Troubled Waters: Proceedings of the Special Session on Shrimp Farming, Aquaculture 95'. (Eds C L Bowdy and J S Hopkins) pp. 13–23. (World Aquaculture Society: Baton Rouge, Louisiana, USA).

Macarthur Consulting (1995). Australian prawn farming industry research and development plan. Fisheries Research and Development Corporation, Canberra, Australia. 73 pp.

McLoughlin, R (1997). Coastal planning and aquaculture: Marine farming development plans for Tasmania. In 'Developing and Sustaining World Fisheries Resources: The State of Science and Management. Second World Fisheries Congress, Brisbane 1996'. (Eds D A Hancock, D C Smith, A Grant and J P Beumer) pp. 455–61 (CSIRO Publishing: Melbourne).

Maclean, J L (1993). Developing-country aquaculture and harmful algal blooms. In 'Environment and Aquaculture in Developing Countries'. (Eds R S V Pullin, H Rosenthal and J L Maclean) pp. 252–84. ICLARM Conference Proceedings 31. 359 pp.

Nell, J A (1993). Farming the Sydney Rock Oyster (*Saccostrea commercialis*) in Australia. *Reviews in Fisheries Science* **1**(2), 97–120.

New, M B (1995). Aquafeeds for sustainable aquaculture. In 'PACON Sustainable Aquaculture '95, Hawaii'. 281 pp.

OECD (1989). Aquaculture, developing a new industry. OECD publications, Paris, France. 126 pp.

Phillips, M J, Kwei Lin, C K, and Beveridge, M C M (1993). Shrimp culture and the environment: lessons from the world's most rapidly expanding warmwater aquaculture sector. In 'Environment and Aquaculture in Developing Countries'. (Eds R S V Pullin, H Rosenthal and J L Maclean) ICLARM Conference Proceedings 31. 359 pp.

Preston, N P, Burford, M A, Jackson, C J, and Crocos, P C (1995). Sustainable shrimp farming in Australia - prospects and constraints. In 'PACON Sustainable Aquaculture' 95, Hawaii'. pp. 308–16.

Primavera, J H (1993). A critical review of shrimp pond culture in the Philippines. *Reviews in Fisheries Science* **1**(2), 151–201.

Pullin, R S V (1993). Discussion and recommendations on aquaculture and the environment in developing countries In 'Environment and Aquaculture in Developing Countries'. (Eds R S V Pullin, H Rosenthal and J L Maclean) pp. 312–38. ICLARM Conference Proceedings 31. 359 pp.

Ritz, D A, Lewis, M E, and Ma Shen (1989). Response to organic enrichment of infaunal macrobenthic communities under salmonid cages. *Marine Biology* **103**, 211–4.

Sandifer, P A, and Hopkins, J S (1996). Conceptual design of a sustainable pond-based shrimp culture system. *Aquaculture Engineering* **15**, 41–52.

Staples, D J, Vance, D J, and Heales, D S (1985). Habitat requirements of juvenile penaeid prawns. In 'Second National Prawn Seminar'. (Eds P C Rothlisberg, B J Hill and D J Staples) pp. 47–54 (NPS2, Cleveland: Australia).

Treadwell, R, McKelvie, L, and Maguire, G B (1992). Potential for Australian Aquaculture. Australian Bureau of Agricultural and Resource Economics, Canberra, Australia. 82 pp.

Woodward, I O, Gallagher, J B, Rushton, M J, Machin, P J, and Mihialenko, S (1992). Salmon Farming and the Environment of the Huon Estuary, Tasmania. Technical Report.

Ye, Li-Xun, Ritz, D, Fenton, G, and Lewis, M (1991). Tracing the influence on sediments of organic waste from a salmonid farm using stable isotope analysis. *Journal of Experimental Marine Biology and Ecology* **145**, 161–74.

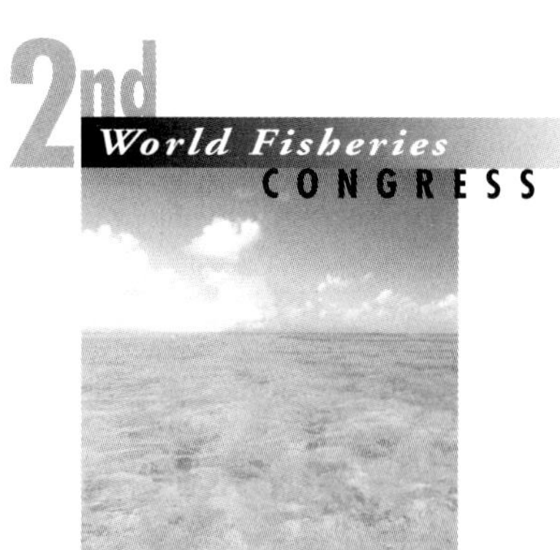

Conservation of Genetic Resources for Aquaculture[1]

Roger S. V. Pullin

International Center for Living Aquatic Resources Management (ICLARM), MCPO Box 2631, 0718 Makati, Metro Manila, Philippines.

Summary

It seems likely that aquaculture will continue to comprise a wide range of production systems, species and products, requiring a wide range of breeds for good results. This will depend upon the conservation of aquatic genetic resources for their sustainable use. Over the last decade, research on genetics in aquaculture has focussed more on genetic manipulations and quantitative genetics than on genetic resources characterization and conservation. Some genetic resources, especially in freshwater, are being depleted before they have been characterized and evaluated for aquaculture potential. However, documentation and conservation efforts are increasing and the importance of intraspecific genetic variation is becoming more widely recognized. The main requirement for achieving progress in the conservation and sustainable use of genetic resources for aquaculture is *information*, especially in terms of accurate genetic characterization based upon molecular genetics data. Public awareness, education and far-sighted policies are essential to ensure that genetic resources for aquaculture are conserved for present and future use.

Genetic resources for what kind of aquaculture?

The conservation of genetic resources for aquaculture begs the question, *for what kind of aquaculture?* There is an intense debate in progress on how best to feed the world (e.g. Anon. 1996; Ross 1996) and it is difficult to forecast who will be the farmers of the future and what systems they will use. Integration of aquaculture with other enterprises (e.g. agriculture, forestry, horticulture, wastewater re-use) is possible throughout a wide range of scale and intensity of operations, and it has been argued that integrated farms are both less risky than stand-alone fish farms and provide a more reliable ecological basis for sustainable aquaculture (Pullin 1995). Martinez i Prat (1995) has warned against intensive monoculture and has cited the boom-and-bust cycles of some shrimp farming operations as examples stating:

> *Genetic diversity is suffering as well ... breeding has focused mostly on short term yield concerns ... Only in recent years — through the International Center for Living Aquatic Resources Management (ICLARM) — has an aquacultural development programme taken into account the role of genetic resources, both to obtain improved performance and to avoid genetic erosion. In most other cases, experimental genetic management technologies such as hybridization, ploidy, gynogenesis and gene transfer have been used as shortcuts to obtain high yielding fish.*

[1] ICLARM Contribution No. 1301.

Conversely, the *Economist* (Anon. 1996) has suggested that most public concerns about modern, high technology aquaculture are baseless and has described selective breeding as *'an old-fashioned and long-winded form of genetic manipulation'* compared to *'new science, which allows direct manipulation of the genetic make-up of plants and animals ... After all, no one can fully predict how organisms crafted in a laboratory will interact with existing plants and animals'*.

This assertion unwittingly supported the rationale for a broad definition of genetically modified organisms (GMOs), including not only the products of 'genetic engineering' but also selectively-bred strains, hybrids etc. (Pullin 1994).

With such diverse viewpoints and possibilities it is probably safe to assume that aquaculture will comprise a wide range of production systems, including small- to large-scale, non-intensive to intensive, non-integrated to fully integrated, and involving, as is presently the case, diverse climates, water composition and structures (e.g. cages, pens, ponds, raceways and tanks). Moreover, hatchery, nursery and grow-out operations will usually be separate. Consumer preferences (e.g. for 'ecocertification' of production methods, and for a wide range of high quality produce) are likely to be increasingly influential. It is therefore probable that aquaculture will need a wide range of breeds tailored to specific production systems and products. Therefore, a wide range of genetic resources, in wild and farmed populations, will be needed to maximize near- and long-term options.

Table 1 lists some of the descriptors used for aquatic genetic resources and comments on their use and limitations. Aquatic biodiversity and aquatic genetic resources for aquaculture have been extensively reviewed recently (McAndrew *et al.* 1993; Beveridge *et al.* 1994; Maclean and Jones 1995; Pullin 1996). The diversity of exploited aquatic organisms is already very high: of the 24 600 known species of finfish, over 5000 are used by humans. Other exploited aquatic animals total several hundred more species, and possibly thousands of aquatic species have attributes through which they could become resources for humans (Brummett and Katambalika 1996; Pullin 1996).

Much of aquaculture also depends upon organisms that comprise aquatic food webs which also contribute to the maintenance of water quality, e.g. phytoplankton and micro-organisms as biological aerators and metabolite processors in fishponds. Conservation strategies for aquatic species and populations must address intraspecific diversity if the future 'evolvability' of aquatic species is to be preserved for both natural and farm environments (see reviews in Nielsen and Powers 1995). This is a basic issue for implementing the US Endangered Species Act (ESA), which recognizes the need to protect Evolutionarily Significant Units (ESUs) (Waples 1991, 1995; Moritz *et al.* 1995).

APPLICATION OF GENETICS IN AQUACULTURE

With the exception of the common carp (*Cyprinus carpio*), some salmon, trout and catfish species and a few others, captive breeding of farmed finfishes and purposeful genetic selection began only a few decades ago. Most farmed aquatic organisms probably still resemble wildtypes genetically. Some may be worse than wildtypes in terms of their performance because of accumulated inbreeding (e.g. Eknath and Doyle 1990) or inadvertent selection for undesirable traits. Farmers' and

Table 1. Some descriptors of genetic resources for aquaculture, excluding the products of genetic manipulation and molecular genetics (e.g. homozygous clones, hybrids, polyploids, transgenes, individual genes)

Descriptor	Origin(s): users	Comments: example(s)
Species (sp.)	Classical taxonomy (Linnaeus 1758) conservationists, legislators, resource managers, scientists, statisticians.	Not often used by farmers or the public: e.g. *Oreochromis niloticus* Linnaeus, *Cyprinus carpio* (Linnaeus), natural and artifical hybridization and introgression confound this descriptor e.g. for *Oncorhynchus* spp. (Bartley *et al.* 1990).
Subspecies (ssp.)	Classical taxonomy and modern revisions: taxonomists, few fish breeders, few legislators.	Not widely used but important for some aquaculture species: e.g. *Oreochromis niloticus niloticus* (Linnaeus), *O. n. cancellatus* (Nichols), *O. n. filoa* subsp.n. *O. n. vulcani* (Trewavas), and *O. n. baringoensis*, *C. carpio* var. *nudus*.
Variety (var.)	Copied from plant taxonomy: aquarists, few fish breeders, scientists.	Not widely used: e.g. common carp varieties *C. carpio* var. *communis*, *C. carpio* var. *specularis*, *C. carpio* var. *flavipinnis*, *C. carpio* var. *nudus*.
Strain/Breed	Copied from plant and livestock breeding: fish breeders, scientists, seed suppliers.	Not well defined and not often based on verifiable breeding histories: e.g. see strains Table of FishBase (Froese and Pauly 1996); the GIFT strain Nile tilapia has a well-defined history (Eknath 1995).
Evolutionarily Significant Unit (ESU)/Distinct population segment	United States Endangered Species Act 1978 and policy revisions (for discussion, see Waples 1995): exploratory use by conservationists, legislators and scientists.	Utility, scientific basis and legislative applicability still being explored; main test cases are distinct riverine populations of Pacific salmon (*Oncorhynchus* spp.).
Common name	Local knowledge (traditional and modern): conservationists, economists; farmers, fishers; the public; resource managers; scientists; statisticians.	Very widely used; 'all that most people know about most fish'. e.g. Common Names Table of FishBase, with 50 000 entries (Froese and Pauly 1996).

Table 2. **Classification by major subject of papers presented at the five International Symposia on Genetics in Aquaculture (ISGA) held from 1982–1994. For ISGA II-IV, poster paper abstracts were included in this analysis. AF = application of genetic research on farms; CS = chromosome manipulation and sex control; DB = research for domestication and breeding programmes; EI = environmental impacts; GA = general reviews on genetics in aquaculture; MT = molecular genetics and gene transfer; PG = population genetics and genetic resources, including molecular genetic techniques in these areas; QG = quantitative genetics, including hybridization, inbreeding etc.**

ISGA	AF		CS		DB		EI		GA		MT		PG		QG			
No.	No.	(%)	No.	(%)	No.	(%)	No.	(%)	No.	(%)	No.	(%)	No.	(%)	No.	(%)	Total	References
I	–	(–)	7	(18)	4	(11)	1	(3)	4	(11)	–	(–)	3	(8)	19	(50)	38	Wilkins and Gosling (1983)
II	2	(3)	12	(18)	11	(16)	–	(–)	6	(9)	–	(–)	19	(28)	17	(25)	67	Gall and Busack (1986)
III	–	(–)	9	(17)	11	(20)	1	(2)	3	(6)	6	(11)	10	(19)	14	(26)	54	Gjedrem (1990)
IV	–	(–)	24	(27)	14	(16)	–	(–)	4	(4)	23	(26)	15	(17)	10	(11)	90	Gall and Chen (1993)
V	5	(3)	43	(24)	22	(15)	6	(4)	2	(1)	15	(10)	28	(19)	26	(18)	147	IAGA-MGPL (1994)
Totals	7	(2)	95	(24)	62	(16)	8	(2)	19	(5)	44	(11)	75	(19)	86	(22)	396	

breeders' rights and intellectual property instruments for their protection are therefore not yet as pressing concerns in aquaculture as in agriculture, where they are highly developed (e.g. Crucible Group 1994). However, there is a paucity of data comparing the genetic characteristics of wild and farmed populations of aquatic species.

Table 2 groups, by major topic, the contributions (about 400) to the five International Symposia on Genetics in Aquaculture (ISGA) held so far. The proportion of contributions on population genetics and genetic resources for aquaculture is only 19%, compared to 35% for genetic manipulations (chromosomal, sex control and gene transfer) and 38% for quantitative genetics, domestication and breeding combined. Contributions describing the application of genetic research on fish farms total a mere 2%, probably reflecting that these symposia have attracted mainly on-campus and laboratory research contributions. There have also been very few contributions on the environmental impacts of the application of genetics in aquaculture.

Loss of genetic resources for aquaculture

Aquatic genetic resources are being rapidly lost. During this century, it is conservatively estimated that about 980 fish species are known to have become threatened and about 160 of these are considered endangered. Almost all are freshwater species, which are adversely affected by human activities and particularly vulnerable because their habitats are restricted, with limited opportunities for recolonization. Freshwater finfish are the most threatened of all significantly exploited vertebrates (Bruton 1995) because of the rapid growth of human populations and the consequent increased demands for freshwater and for watershed lands to accommodate the needs of agriculture, aquaculture, domestic water supply, forestry, human settlements, industry, power generation, recreation, transportation, and waste treatment and disposal. Marine species, such as those associated with coral reefs and mangroves, are less threatened with species extinction, but local populations, and thus at least some (ESUs), face overexploitation and habitat destruction, especially from capture fisheries.

Aquaculture can have large negative impacts through water abstraction, harmful effluents, disease transmission and clearance or fragmentation of habitats (e.g. mangroves). In addition to the purposeful releases of fish to enhance fisheries, farmed fish often escape from aquaculture installations. They may contribute to the depletion or loss of wild fish stocks (e.g. by predation, competition for food or territory or by spreading diseases), to changes in natural aquatic habitats (e.g. clearance of vegetation or increased turbidity) and to genetic change. Such serious threats to aquatic biota and their habitats are becoming more widely recognized (e.g. Wager and Jackson 1993; Maitland 1994). However, water resources developments policies and projects often ignore the economic importance of fish for food and of fish genetic resources (e.g. FAO 1995 *a, b*).

Tilapia genetic resources for aquaculture — an example of the value of genetic resources

In order to plan strategic research for the genetic improvement of Nile tilapia (*Oreochromis niloticus*) as a model species for tropical aquaculture, ICLARM sought in 1986 the help of a Norwegian team, led by Dr Trygve Gjedrem (NORAGRIC/AKVAFORSK), that had pioneered the genetic improvement of farmed Atlantic salmon (*Salmo salar*). A proposal was made to apply the same approach to tilapia as for salmon, i.e. the characterization of genetic resources, followed by their evaluation and selective breeding (Gjedrem and Pullin 1986).

An international workshop (Pullin 1988) confirmed that stocks of *O. niloticus* in Asia were derived from small founder populations and had suffered from inbreeding and introgression of genes from other species. Hence, decisions were taken to collect and to evaluate some wild *O. niloticus* populations from

Africa and some Asian farmed strains. The Genetic Improvement of Farmed Tilapia (GIFT) project was then launched.

Breeders (150–160) or fingerlings (200–800) from open waters in Egypt, Ghana, Kenya and Sénégal became, after rigorous quarantine procedures, the founder populations of four wildtypes, to be compared with four strains farmed in Asia: 'Israel', 'Singapore', 'Taiwan' and 'Thailand'. In comparative trials, these strains were grown in eleven test environments. The overall result was that the wild strains, with the exception of the strain from Ghana, grew as well or better than the farmed strains. The wild strains from Egypt and Kenya were the best performers and the ranking of the strains, with some variations, were broadly consistent among environments (Eknath *et al.* 1993).

These results were interpreted as indicating low genotype × environment interaction and, following a subsequent 8 × 8 diallele crossing experiment that indicated low heterosis, the GIFT project team chose a pure breeding strategy, i.e. the development of a synthetic strain, containing appropriate genetic contributions from all strains, upon which selective breeding would then be performed for successive generations. This proved highly successful and has since led to a Philippine National Tilapia Breeding Program, with a national foundation to sustain it. In on-farm trials in the Philippines, GIFT strain fish have been found to have increased growth by about 60% and to have almost 50% better survival, when compared to widely farmed commercial strains (Eknath 1995).

In general, the results showed that there were significant differences in growth performance among the different strains of this species, including differences among wild populations of the same subspecies *O. niloticus niloticus* from Egypt, Ghana and Sénégal (the wild Kenyan strain was *O. niloticus vulcani*).

Biochemical characterization of these strains also showed differences (Macaranas *et al.* 1995). Superimposing the strain performance rankings of Eknath *et al.* (1993) on the dendrogram of genetic distances shows that genetic similarity bears little relation to culture performance (Fig. 1). The Thailand strain came originally from Egypt (*via* Japan), so its genetic proximity to the Egypt strain here makes sense. The Ghana strain, (consistently the worst performer in these Philippine trials) is also genetically close to these two. Moreover, the Sénégal strain (an 'average' performer in both trials) is genetically farther from this cluster than the lower-ranked Singapore strain and, in one trial, the Israel strain. The Kenya strain ranked highly in performance with the Egypt strains, but it appears as a more distant subspecies on the dendogram.

This mismatch of genetic identity and performance data will not surprise geneticists. The phenotypic differences observed may be the result of one or more genetic differences other than those upon which the genetic identity data are based. However, aquaculturists can note from this example the need to document breeding material thoroughly and to test the performance of strains and breeds thoroughly, rather than uncritically accept reports on the origin and genetic identity of broodstock or indeed to assume that such information can predict performance.

Through a project called Dissemination and Evaluation of Genetically Improved Tilapia in Asia (DEGITA), GIFT strain Nile tilapia are currently under investigation in Bangladesh, China, the Philippines, Thailand and Vietnam to compare their performance with currently farmed strains. In Bangladesh, where Nile tilapia culture has a short history with very limited broodstock management, the GIFT strain appears to be superior to local fish on-station (in ponds and cages) and on-farms (village ponds). In China, Thailand and Vietnam, where there are both longer histories of tilapia farming and greater climatic variation, and therefore the possibility of both natural and artificial selection of local strains to their environments, the GIFT strain is not always superior to local strains. For example, in the north of Vietnam where the low temperature limits of tilapia farming are approached, a Vietnamese strain gave about 90% survival compared to the GIFT strain's 64% during a particularly cool growing period in 1995; whereas survival of both strains appeared similar in a warmer year (1994) (M.M. Dey, pers. comm.).

The full analysis of these DEGITA and other on-station and on-farm trials should reveal to what extent locally available resources of this exotic species in Asia, even where no artificial selection has occurred, may be valuable in breeding programmes. In African locations, where wild and farmed tilapia strains exist, introduction of the products of breeding programmes located elsewhere is probably unwise, unless their potential impacts upon nationally and locally important resources have been thoroughly evaluated and they pose no significant risks.

Important lessons for the conservation and use of tilapia genetic resources are emerging from the GIFT and DEGITA projects. Firstly, the wide divergence in performance between different populations of the same subspecies (e.g. *O. n. niloticus*, Egypt *vs.* Ghana) could not have been foreseen from any available published data. Secondly, although the GIFT synthetic strain can be assumed to contain a useful mixture of genes, contributed in differing proportions from the eight strains tested, nothing can be predicted about what might have resulted if yet more wild strains of *O. n. niloticus* or of other *O. niloticus* sspp., as recognized by Trewavas (1983) and Seyoum and Kornfield (1992), had been

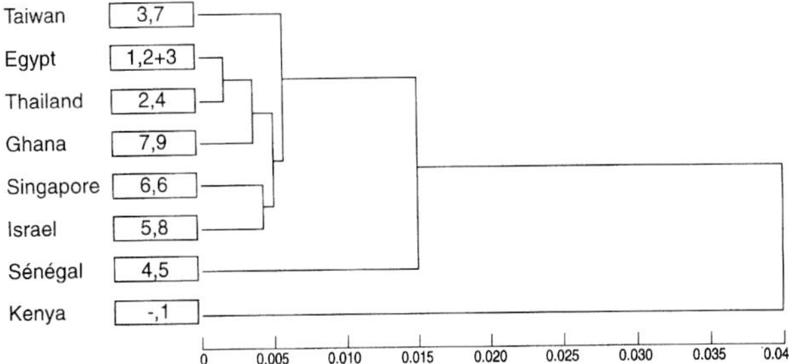

Fig. 1. Genetic relationships among strains of Nile tilapia (*Oreochromis niloticus*) based on Nei's genetic distance (after Macaranas *et al.* 1995) with corresponding rank orders for the performance of these stains in two comparative trials (Eknath *et al.* 1993). The first figure gives the rank order among seven strains from the first trial (Kenya not available) and the second the rank order among nine strains (including Kenya and with the progeny of two introductions from Egypt). The dendogram refers to the lowest ranked entry for the Egypt strains, though both are considered genetically close.

available and included. It is probable that these are also valuable sources of genetic variance for future breeding programmes and perhaps, (though from the GIFT project results this seems less likely for *O. niloticus*) for heterosis in hybridization.

This leads to a major question — what should be conserved for aquaculture from the total aquatic genetic resources in the world? In an ideal world, the answer would be 'everything', because the value of a genetic resource cannot be forecast until it is thoroughly investigated. Some fish genetic resources are, however, diminishing before they have been described. For example, most of the spring runs of Chinook salmon (*Oncorhynchus tshawytscha*) in California disappeared before it was possible to document their genetics, and fall-run fish are now venturing less far inland (D.M. Bartley, pers. comm.).

Information needs

The information required for characterizing, evaluating and conserving aquatic genetic resources is often not available. For example, plans to privatize water resources (as envisaged in a growing number of countries) can only take proper account of the implications for fish conservation and use if the status of the resident fish populations is documented in a format accessible to planners and developers.

Information on the world's aquatic genetic resources for their conservation and use is held by many individuals and organizations at the local, national, regional and international levels, e.g. clubs, community organizations, government departments, museums, non-governmental organizations, universities, UN bodies, etc. These sources are generally held independently of each other and in a multiplicity of formats ranging from local knowledge and amateur records (published or unpublished) to professional publications and databases. They are also recorded in a large number of local, national and international languages, and with different parameters and different descriptors for location, time repeatability, etc. Such scattered and unstructured information is usually *de facto* inaccessible.

Information in the form of molecular genetic markers will be increasingly needed for breeding programmes (e.g. see Garcia and Benzie 1995). However, as fish breeders search for genes to improve complex commercial traits such as survival and growth, they will find in the relevant literature data that mainly refer to neutral markers, i.e. not able to be related to actual or potential aquaculture performance unless allied to performance trials. For fish, there are relatively few clear-cut examples of enzyme polymorphisms that are related to commercial traits; most concern disease resistance associated with specific genotypes (e.g. Withler and Evelyn 1990; Ibarra *et. al.* 1994).

Genetic resources conservation *in situ* and *ex situ*

When used for aquatic genetic resources, the term '*in situ*' normally means wild or feral populations, and does not include their relatives on farms, as is sometimes the case in agriculture. Aquatic genetic resources in open waters are usually subject to very different ownership and access (public and open), compared to those on farms (private and restricted). McAndrew *et al.* (1993) have reviewed the issues that pertain to *in situ* and *ex situ* conservation of genetic resources for aquaculture, emphasizing *ex situ* methods, such as genetic manipulations combined with cryopreservation, and their prospects for further development. *Ex situ* genetic resources for aquaculture, on farms and in genebanks or germplasm collections, are largely under private ownership, though this may be community-based (e.g. village ponds) and not only restricted to individuals, families or corporations. There are also some *ex situ* collections (broodstocks, juveniles and cryopreserved sperm) held for research, conservation or breeding purposes, or combinations of all of these, by publicly funded institutions such as government fish farms and hatcheries, public aquaria and universities (Pullin 1993; Harvey 1996).

For aquaculture, conservation of genetic resources *in situ* is of paramount importance because such resources greatly exceed, in genetic diversity and abundance, those that can be conserved *ex situ*. Therefore it is difficult at this stage of the development of aquaculture to envisage a global plan of action or multilateral framework for access to aquatic genetic resources similar to those being developed for plant genetic resources, the latter of which is largely reliant on both *ex situ* collections comprising hundreds of thousands of accessions as well as on limited *in situ* resources. More realistically, an interim period of increased support for the characterization and evaluation of aquatic genetic resources, *in situ* and *ex situ*, is probable, i.e. to document more fully their genetic diversity and performance and to make this information widely available. This would assist both the exercising of national rights over genetic resources of actual or potential value, and the fulfilment of national obligations under the Convention on Biological Diversity (CBD). Moreover, intellectual property regimes for aquatic genetic resources, consistent with the CBD, GATT/TRIPS, and the rights of indigenous peoples, breeders and farmers, could then be developed equitably, based on accurate scientific information. The major constraints to such a scenario are time, financial resources and, especially in the developing world, facilities and skilled personnel. *In situ* genetic resources are particularly at risk and can become threatened with serious depletion or total loss in the absence of data on their status and value.

Conservation decisions and mechanisms

The US Endangered Species Act (ESA) defines a species as '*any distinct population segment of any species of vertebrate fish or wildlife which interbreeds in nature*' (Waples 1991). Hence, genetic distinctiveness (for example, from enzyme or DNA data) is not an explicit requirement for protection under this Act; though some would maintain that this is implied. Extensive studies and commentaries on *Oncorhynchus* spp. (e.g. Bartley *et al.* 1990, 1992*a*, *b*; Hedrik 1994; Hedgecock *et al.* 1994) show the complexities of the genetic information that is facing those who will operate the ESA. Some populations are genetically similar and from the same river but have different life histories. This raises the issue of habitat and life history; a population, its habitat and life history combined as an *in situ* resource worthy of protection. For example, winter run Chinook salmon (*O. tshawytscha*) in the Sacramento River have unique properties such as fat content (D.M. Bartley, pers. comm.) but cannot be distinguished by allozymes from other runs in the same river (Bartley *et al.* 1992*b*). As Hedgecock *et al.* (1994) pointed out, this greatly complicates the question of 'what is an ESU?'.

Solutions to these problems require balanced policies, backed by appropriate national and international legislation, that are consistent with the CBD. 'Balance' in conservation and development policies is always difficult to achieve. It should, however, be possible to establish risk assessment procedures to weigh potential benefits against environmental costs. A risk-benefit approach on whether to conserve or to permit the loss of aquatic ESUs would have to consider the interests of present and future generations of humans. Moritz *et al.* (1995) have provided some useful examples of how this might be approached with respect to listing under the ESA.

Cost may be the most important criterion. Livestock breeders have long had to face this issue in choosing which breeds and species to conserve. Cunningham (1995) has implied that it should relate to use in present or foreseeable farming systems. In aquaculture, where farming systems are still evolving rapidly and few have been well proven over a long period, and where domestication and breeding programmes have not progressed far, this approach would not yet be widely applicable. However, Cunningham (1995) has also explored ultimate return, i.e. cost ratios for conserving livestock breeds as small populations and has suggested that, given discount rates of under 5%, expected benefits of 100–200 times annual costs could repay a 50-year conservation effort for a breed. A similar approach could be explored for both *in situ* and *ex situ* conservation of genetic resources for aquaculture. At this early stage in the development of aquaculture, as domestication proceeds, most breeding programmes and genetic resources conservation and use could include the long-term objective of minimizing the loss of genetic diversity as well as more immediate objectives of improving traits.

POLICY AND PUBLIC SUPPORT

The conservation of genetic resources for aquaculture will depend as much on voluntary actions and responsible stewardship of the natural environment by the general public and the private sector, as on legislation, protocols and intellectual property systems. Public awareness and education to explain the importance of conservation measures for aquatic genetic resources are essential to protect them and their habitats. *In situ* conservation of aquatic genetic resources can be combined with other conservation efforts (for example, gameparks and coastal nature reserves) and with recreational activities in tourist-valued habitats, such as coral reefs and sport fisheries areas (Pullin 1990).

It will be a tragedy if the world proceeds to forfeit much of its genetic resources for aquaculture and has to face, as for agriculture, the need to base future breeding programmes on a diminishing number of domesticated strains of a few species and small populations of wild relatives. This would constrain the variety of species and breeds available to farmers and the variety of produce to consumers. To avert such a scenario requires political will and adequate resources. The limited resources available for genetic research in aquaculture will need to be used not only for new technologies for genetic improvement, but also for the characterization, evaluation and conservation of genetic resources. Preferably, these areas of work should be combined so that characterization, conservation and breeding goals are all approached *together.*

In order to generate the accurate and up-to-date information that is vital to conservation efforts and sustainable use of aquatic genetic resources, reliable and cost-effective field and laboratory methods are needed for their characterization. The further development and application of molecular genetic techniques is therefore critical. The recent reviews in Carvalho and Pitcher (1995) show the range of techniques that is emerging. The most widely used techniques in the near future will probably be microsatellites and random amplified polymorphic DNA (RAPD) (K.A. Naish, pers. comm.).

It is a cause for concern that in some legal and other instruments for the conservation of genetic resources, international conventions are framed using definitions mainly at the species level. As Ryman *et al.* (1995) have shown, exploited fishes require conservation measures based on intraspecific diversity. Moreover, the 'governance' approach to natural resources management (Kooiman 1993) has yet to be applied significantly to genetic resources *per se.* Clearly, if the conservation of genetic resources for use in aquaculture is to succeed, such instruments and perspectives must be applied at all three levels of biodiversity: genes, species (and ESUs) and ecosystems.

ACKNOWLEDGMENTS

I wish to thank Dr Devin Bartley of FAO, Dr Kerry-Anne Naish of the University of Wales, Swansea, and numerous ICLARM colleagues for their help towards finalization of this paper.

REFERENCES

Anon. (1996). Science and Technology. Growing pains. *Economist* **339**, 7962 (April 20, 1996) 81–3.

Bartley, D M, Gall, G A E, and Bentley, B (1990). Biochemical genetic detection of natural and artificial hybridization of chinook and coho salmon in Northern California. *Transactions of the American Fisheries Society* **119**, 431–7.

Bartley, D M, Bentley, B, Olin, B, Gall, P G, and Gall, G A E (1992*a*). Population genetic structure of coho salmon *(Oncorhynchus kisutch)* in California. *California Fish and Game* **78**, 88–104.

Bartley, D M, Bentley, B, Brodziak, J, Gomulkiewicz, R, Mangel, M, and Gall, G A E (1992b). Geographic variation in population genetic structure of chinook salmon from California and Oregon. *Fishery Bulletin* **90**, 77–100 [authorship amended per errata, *Fishery Bulletin* **90**(3), iii].

Beveridge, M C M, Ross, L G, and Kelly, L A (1994). Aquaculture and biodiversity. *Ambio* **23**, 497–502.

Brummett, R E, and Katambalika, K (1996). Protocols for the development of indigenous species under Malawian smallholder conditions. *Aquaculture Research* **27**, 225–33.

Bruton, M N (1995). Have fishes had their chips? The dilemma of threatened fishes. *Environmental Biology of Fishes* **43**, 1–27.

Carvalho, G R, and Pitcher, T J (Eds) (1995). 'Molecular genetics in fisheries.' (Chapman and Hall: London). 141 pp.

Crucible Group (1994). People, patents and plants: the impact of intellectual property on biodiversity, conservation, trade and rural society. International Development Research Centre, Ottawa.

Cunningham, P (1995). Genetic diversity in domestic animals: strategies for conservation and development. In 'Beltsville Symposia in Agricultural Research XX: Biotechnology's role in the genetic improvement of farm animals'. (eds R H Miller, V G Pursel and H D Norman) pp. 13–23 (American Society of Animal Science, Savoy, IL).

Eknath, A (1995). The Nile tilapia. In 'Conservation of fish and shellfish resources: managing diversity'. (Eds J Thorpe, G Gall, J Lannan and C. Nash) pp. 177–191. (Academic Press: London).

Eknath, A E, and Doyle, R W (1990). Effective population size and rate of inbreeding in aquaculture of Indian major carps. *Aquaculture* **85**, 293–305.

Eknath, A E, Tayamen, M M, Palada-de Vera, M S, Danting, J C, Reyes, R A, Dionisio, E E, Capili, J B, Bolivar, H L, Circa, A V, Bentsen, H B, Gjerde, B, Gjedrem, T, and Pullin, R S V (1993). Genetic improvement of farmed tilapias: the growth performance of eight strains of *Oreochromis niloticus* tested in different farm environments. *Aquaculture* **111**, 171–88.

FAO (1995*a*). Water development for food security. FAO/WFS96/Tech 12. Advance Unedited Version. 39 pp.

FAO (1995*b*). Preparations of an international expert consultation on the management of protected areas and sustainable rural development. Working Party on Wildlife Management and National Parks, African Forestry and Wildlife Commission. FAO/AFWC/WL/95b.

Froese, R, and Pauly, D (Eds) (1996). FishBase. A biological database on fish. International Center for Living Aquatic Resources Management, Manila, Philippines.

Gall, G A E, and Busack, C A (1986). 'Genetics in aquaculture II.' (Elsevier Science Publishers: Amsterdam). 386 pp.

Gall, G A E, and Chen, Hong Xi (1993). 'Genetics in aquaculture IV'. (Elsevier Science Publishers: Amsterdam). 332 pp.

Garcia, D K, and Benzie, J A H (1995). RAPD markers of potential use in penaeid prawn *(Penaeus monodon)* breeding programs. *Aquaculture* **130**, 137–44.

Gjedrem, T (1990). 'Genetics in aquaculture III.' (Elsevier Science Publishers: Amsterdam). 346 pp.

Gjedrem, T, and Pullin, R S V (1986). A new breeding program for the development of tilapia culture in development countries. Report to the Rockefeller Foundation. International Center for Living Aquatic Resources Management, Manila, Philipppines. 15 pp.

Harvey, B (1996). Banking fish genetic resources: the art of the possible. In 'Biodiversity, science and development: towards a new partnership'. (Eds F di Castri and T Younès) pp. 439–45 (CAB International: Wallingford, UK).

Hedgecock, D, Siri, P, and Strong, D R (1994). Conservation biology of endangered Pacific salmonids: introductory remarks. *Conservation Biology* **8**, 863–4.

Hedrik, R P (Ed) (1994). Endangered Pacific salmonids. Special Section. *Conservation Biology* **8**, 863–4.

IAGA-MGPL (1994). Fifth International Symposium on Genetics in Aquaculture. Program and Book of Abstracts. International Association for Genetics in Aquaculture at the Marine Gene Probe Laboratory of Dalhousie University, Halifax, Nova Scotia, Canada. 164 pp.

Ibarra, A M, Hedrick, R P and Gall, G A E (1994). Genetic analysis of rainbow trout susceptibility to the myxosporean, *Ceratomyxa shasta*. *Aquaculture* **120**, 239–62.

Kooiman, J, (Ed) (1993). 'Modern governance.' (SAGE Publications: London). 280 pp.

Linnaeus, C (1758). '*Systema Naturae per Regna Tria Naturae secundum Classes, Ordinus, Genera, Species cum Characteriibus, Differentiis Synonymis, Locis*.' 10th ed., Vol. 1. Holmiae Salvii. 824 pp.

Macaranas, J M, Agustin, L Q, Ablan, M C A, Pante, M J R, Eknath, A E, and Pullin, R S V (1995). Genetic improvement of farmed tilapias: biochemical characterization of strain differences in Nile tilapia. *Aquaculture International* **3**, 43–54.

Maclean, R H, and Jones, R W (1995). Aquatic biodiversity conservation: a review of current issues and efforts. Strategy for International Fisheries Research, Ottawa, Canada. 59 pp.

Maitland, P S (1994). Conservation of freshwater fish in Europe. *Nature and Environment* **66**. (Council of Europe Press, Strasbourg). 50 pp.

Martinez i Prat, A R (1995). Fishing out aquatic diversity. *Seedling, Quarterly Newsletter of Genetic Resources Action International* **12**(2), 2–13.

McAndrew, B J, Rana, K J, and Penman, D J (1993). Conservation and preservation of genetic variation in aquatic organisms. In. 'Recent advances in aquaculture'. (Eds J F Muir and R J Roberts) pp. 236–95 (Blackwell Scientific Publications: Oxford).

Moritz, C, Lavery, S, and Slade, R (1995). Using allele frequency to define units for conservation and management. *American Fisheries Society Symposium* **17**, 244–62.

Nielsen, J, and Powers, D B (Eds) (1995). Evolution and the aquatic ecosystem: defining unique units in population conservations. *American Fisheries Society Symposium* **17**. 435 pp.

Pullin, R S V (Ed) (1988). Tilapia Genetic Resources for Aquaculture. *ICLARM Conference Proceedings* 16. 108 pp.

Pullin, R S V (1990). Down-to-earth thoughts on conserving aquatic genetic diversity. *Naga, the ICLARM Quarterly* **14** (2), 3–6.

Pullin, R S V (1993). *Ex-situ* conservation of the germplasm of aquatic animals. *Naga, the ICLARM Quarterly* **16**(2/3), 11–13.

Pullin, R S V (1994). Exotic species and genetically modified organisms in aquaculture and enhanced fisheries: ICLARM's position. *Naga, the ICLARM Quarterly* **17**(4), 19–24.

Pullin, R S V (1995). Growth and sustainability of aquaculture. *Naga, the ICLARM Quarterly* **18**(3), 19–20.

Pullin, R S V (1996). Biodiversity and aquaculture. In 'Biodiversity, science and development: towards a new partnership'. (Eds F di Castri and T Younès) pp. 409–23. (CAB International: Wallingford, UK).

Ross, E B (1996). Malthusianism and agricultural development: false premises, false promises. *Biotechnology and Development Monitor* **26**, 24.

Ryman, N, Utter, E, and Laikre, L (1995). Protection of intraspecific biodiversity of exploited fishes. *Review in Fish Biology and Fisheries* **5**, 417–46.

Seyoum, S, and Kornfield, I (1992). Taxonomic notes on the *Oreochromis niloticus* subspecies-complex (Pisces: Cichlidae), with a description of a new subspecies. *Canadian Journal of Zoology* **70**, 2161–5.

Trewavas, E (1983). Tilapiine fishes of the genera *Sarotherodon, Oreochromis* and *Danakilia*. British Museum (Natural History), London, UK.

Wager, R, and Jackson, P (1993). The action plan for Australian freshwater fishes. Australian Nature Conservation Agency, Canberra. 122 pp.

Waples, R S (1991). Pacific salmon, *Oncorhynchus* spp., and the definition of 'species' under the Endangered Species Act. *Marine Fisheries Review* **53**, 11–22.

Waples, R S (1995). Evolutionary significant units and the conservation of biological diversity under the Endangered Species Act. *American Fisheries Society Symposium* **17**, 8–27.

Wilkins, N P, and Gosling, E M (1983). 'Genetics in aquaculture.' (Elsevier Science Publishers: Amsterdam).

Withler, R E, and Evelyn, T P T (1990). Genetic variation and resistance to bacterial kidney disease within and between two strains of coho salmon from British Columbia. *Transactions of the American Fisheries Society* **119**, 1003–9.

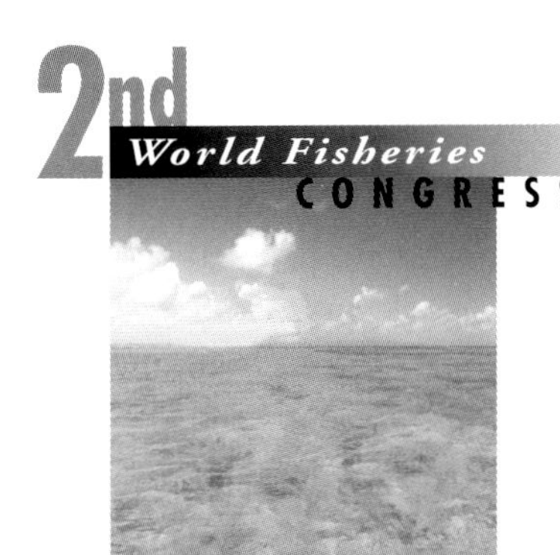

A Responsible Approach to Marine Stock Enhancement[1]

H. L. Blankenship[A] *and K. M. Leber*[B]

[A] Washington Department of Fish and Wildlife, 600 Capitol Way North, Olympia, Washington 98501-1091, USA.
[B] The Oceanic Institute, Makapuu Point, Waimanalo, Hawaii 96795, USA.

Summary

A 'Responsible Approach' concept with several key components is described. Each component is considered essential to control and optimize enhancement. The components include the need to: (1) prioritize and select target species for enhancement; (2) develop a species management plan that identifies harvest opportunity, stock rebuilding goals, and genetic objectives; (3) define quantitative measures of success; (4) use genetic resource management to avoid deleterious genetic effects; (5) use disease and health management; (6) consider ecological, biological and life-history patterns when forming enhancement objectives and tactics; (7) identify released hatchery fish and assess stocking impacts; (8) use an empirical process for defining optimum release strategies; (9) identify economic and policy guidelines; and (10) use adaptive management. Developing case studies with cod *Gadus morhua*, red drum *Sciaenops ocellatus*, striped mullet *Mugil cephalus*, and white seabass *Atractoscion nobilis* are used to verify that the responsible approach to marine stock enhancement is practicable.

Introduction

Marine fish populations are declining world-wide. In the United States, present abundance trends are known for only 15 of the most important marine stocks; about half of them are declining (Anon. 1991*a*, 1992). Current harvest rates on most declining stocks are far in excess of exploitation levels needed to maintain the high long-term average yields that could be achieved through contemporary fishery management practices. Projected increases in human population size world-wide suggest this trend will continue into the future (Anon. 1991*b*).

[1] Subject matter in this article has been reprinted with permission from *American Fisheries Society Symposium* **15**, 167–75.

Three principal tactics are available to fishery managers to replenish depleted stocks and manage fishery yields: regulating fishing effort; restoring degraded nursery and spawning habitats; and increasing recruitment through propagation and release (stock enhancement). The first two methods form the basis for the present federal approach to managing marine fisheries in the United States. The potential of the third method is only starting to be documented with marine fishes.

Marine stock enhancement is not a new concept. In fact, hatchery-based stock enhancement was the principal technique used in an attempt to restore marine fisheries during the late 1800s and early decades of the twentieth century. However, stock enhancement fell out of favour among fishery biologists after a half-century of hatchery releases had produced no evidence of an increased yield. Cod *Gadus morhua*, haddock *Melanogrammus aeglefinus*, pollack *Pollachius virens*, winter flounder *Pseudopleuronectes americanus* and Atlantic mackerel *Scomber scombrus* were stocked. Regrettably, when the last of the early marine hatcheries in the United States closed in 1948, after 50 years of stocking marine fishes, the technology had progressed no further than the stocking of unmarked, newly hatched fry. This was partly a result of the early approach to assessment, in which the success of hatchery programmes was judged by numbers of fry stocked rather than by numbers of adults surviving to enter the fishery (Richards and Edwards 1986).

A NEW AND RESPONSIBLE APPROACH

Two general problems have restricted development of marine stock-enhancement technology this century. Lack of an evaluation capability to determine whether hatchery releases were successful has been a major obstacle. Before the development of modern marking methods, fish tagging systems were not applicable to the small size of the early life-history stages released by hatcheries. The other impediment to development of marine enhancement has been the inability to culture marine fishes beyond early larval stages to the juvenile stage (fingerlings and larger sizes).

A new approach to marine stock enhancement is long overdue. Faced with declining stocks and an expanding world population, managers around the globe are looking at marine enhancement with renewed interest. To develop and evaluate stock enhancement's full potential, a process is needed for designing and refining stock-enhancement tactics based on the combined effects of managing the resource (i.e. the interactive effects of hatchery practices, release strategies, harvest regulations and habitat restoration on the condition of the managed stock).

Recent advances in both tagging technology and marine fish culture provide basic tools for a new approach to marine enhancement. The technology is now available for benign tagging of fish from juvenile through adult life stages (Bergman *et al.* 1992). Such tagging provides the basis for a quantitative assessment of the success of stock enhancement. Several marine fishes can be cultured to provide a wide range of life stages for release (e.g. McVey 1991; Honma 1993). Together, these tools allow an empirical evaluation of survival of cultured fish in the wild, and feedback on the effects of hatchery release can be used to refine enhancement strategies. Impacts of release on wild stocks, and the fisheries based on them, can be quantified and evaluated. Survival can be examined over a range of hatchery practices and release variables (such as culture practices, fish size at release, release magnitude, release site and season) to identify optimum combinations of hatchery and release strategies.

These new tools provide the basis for significantly increasing wild stock abundances. To ensure their successful use and avoid repeating the past mistakes experienced in both marine and freshwater enhancement, development of marine stock-enhancement programmes must be approached carefully. The expression 'a responsible approach to marine stock enhancement' embraces a logical and conscientious strategy for applying aquaculture technology to help conserve and expand natural resources. This approach prescribes several key components as integral parts in developing, evaluating, and managing marine stock-enhancement programmes. Each component is considered essential to control and to optimize the results of enhancement. The components include the need to (1) prioritize and select target species for enhancement; (2) develop a species management plan that identifies harvest opportunity, stock rebuilding goals and genetic objectives; (3) define quantitative measures of success; (4) use genetic resource management to avoid deleterious genetic effects; (5) use disease and health management; (6) consider ecological, biological, and life-history patterns when forming enhancement objectives and tactics; (7) identify released hatchery fish and assess stocking impacts; (8) use an empirical process for defining optimum release strategies; (9) identify economic and policy guidelines; and (10) use adaptive management. Combining new marine fish culture and tagging technologies with these ten principles is gaining support as a *responsible approach* to marine stock enhancement.

Empirical data suitable for accurately assessing the impact of hatchery releases on wild populations are often lacking. Partly because of this uncertainty, there is an increasing division of conservationists into two camps — one adamantly favouring increased fishing regulations and habitat protection and restoration in preference to hatchery releases, the other supporting propagation and release as an additional tool to manage fisheries and restore declining stocks. This split must be reconciled. Is stock enhancement of marine fishes a powerful, yet undeveloped technology for rebuilding depleted wild stocks and increasing fishery yields? Or are emerging marine enhancement programmes merely futile attempts at recovering precious resources, thus diverting money and attention away from habitat restoration and the regulations needed to control overfishing? To realize the full potential of marine enhancement for the conservation and rapid replenishment of declining marine stocks, this technology must be developed to supplement and replenish marine stocks responsibly and quickly.

We must act now to assess the potential of marine stock enhancement through carefully planned research programmes. Using strong inference (Platt 1964), which is essentially the scientific method, and addressing all of the components of the responsible-approach concept, research programmes will either document the value of marine enhancement or reveal that enhancement is not a useful concept. Without determined and careful attention to the 10 points listed above, marine hatchery releases in the 1990s may serve only to fuel divisiveness between the two conservationist camps, with little or no positive impact on natural resources.

Applying the responsible-approach concept to new stock-enhancement initiatives is straightforward. Existing enhancement programmes may find it useful to review the 10 components discussed here. Incorporating those components expanded upon below, that are not already part of ongoing enhancement programmes, should provide a measurable increase in the realized effectiveness of replenishment efforts.

Prioritize and select target species for enhancement

In the absence of a candid and straightforward method, targeting species for stock enhancement can become a difficult and biased process. Unless attention is focussed on the full spectrum of criteria that can be used to prioritize species, consideration of an immediate need by an advocacy group or simply the availability of aquaculture technology can become the driving factors in species selection. Commercial and recreational demand are obviously important criteria, but should they take precedence over other factors?

To reduce the bias inherent in selecting species, a semi-quantitative approach was developed in Hawaii to identify selection criteria and prioritize species for stock enhancement research (Leber 1994). This approach involved four phases: an initial workshop, where selection criteria were defined and ranked in order of importance; a community survey, which was used to solicit opinions on the selection criteria and generate a list of possible species for stock-enhancement research; interviews with local experts to rank each candidate species with regard to each selection criterion; and a second workshop, in which the results of the quantitative species-selection process were discussed and a consensus was sought. This decision-making process focussed discussions, stimulated questions and quantified participants' responses. Panellists' strong endorsement of the ranking results and selection process used in Hawaii demonstrates the potential for applying formal decision making to species selection in other regions.

A critical step in removing bias from the species-selection process lies in the type of numerical analysis used. The relative importance of the various criteria can be used in the analysis by factoring the degree to which each fish meets each criterion by the criterion weight. This produces a score for each species. This same concept is used to determine dominance in ecological studies of species assemblages (i.e. relative abundance multiplied by frequency of occurrence in samples). Use of a trained facilitator to conduct the workshops also reduces bias by focussing activities on achieving results and by encouraging participation by all present.

Formal decision-making tools have been used effectively to prepare comprehensive plans for fisheries research (Bain 1987). Mackett *et al.* (1983) discuss the interactive management system for the Southwest Fisheries Center of the National Marine Fisheries Service. Similar processes have been used for research on North Pacific pelagic fisheries, in strategic planning for Hawaii's commercial fishery for skipjack tuna *Katsuwonus pelamis* (Boggs and Pooley 1987), and for a 5-year scientific investigation of marine resources of the main Hawaiian Islands (Pooley 1988).

Develop a species management plan

A management plan identifies the context into which enhancement fits into the total strategy for managing stocks. The goals and objectives of stock-enhancement programmes should be clearly defined and understood prior to implementation. The genetic structure of wild stocks targeted for enhancement should be identified and managed according to objectives of the enhancement programme. What is the population being enhanced? Can it be geographically defined? Clearly, in the interest of both production aquaculture and conservation, effort must be made to maintain genetic diversity (Kapuscinski and Jacobson 1987; Shaklee *et al.* 1993*a*, *b*).

Assumptions and expectations about the performance and operation of the enhancement programme necessary to make it successful should be identified (such as post-release survival, interactions with wild stocks, long-term fitness and disease). Critical uncertainties about the basic assumptions that would affect the choice of production and management strategies should likewise be identified and prioritized. Evaluation of these uncertainties should be an integral part of the species management plan, and a feedback loop to evaluate and change production and management objectives should be included.

Define quantitative measures of success

Without a definition of success, how do you know if or when you have it? Explicit indicators of success are clearly needed to evaluate stock-enhancement programmes. The objectives of enhancement programmes need to be stated in terms of testable hypotheses. To be testable, a hypothesis must be falsifiable (Popper 1965). Depending on enhancement objectives, multiple indicators of success may be needed. These could include statements such as 'Hatchery releases will provide at least a 20% increase in annual landings of *Polydactylus sexfilis* in the Kahana Bay recreational fishery by the third year of the project', or 'Monitoring will show less than 3% change in the frequency of rare alleles (frequency less than 0.05) after five years of hatchery releases (this assumes that a control for the effects of environmentally induced change in allele frequencies is possible).'

Numerous indicators should be identified to track progress over time. Although simplistic, indicators like the two examples above could be linked to success and would provide a basis for evaluating enhancement efforts during the initial period of full-scale releases. Clearly, to examine such indicators requires a reliable, quantitative marking and assessment system for tracking hatchery fish.

Use genetic resource management

The need for genetic resource management in stock-enhancement programmes is currently the subject of intense public debate, and its importance cannot be overrated. Responsible guidelines are now becoming available to aid resource managers in revitalizing stocks without the loss of genetic fitness that could follow from inbreeding in the hatchery and subsequent outbreeding depression in the wild (Kapuscinski and Jacobson 1987; Shaklee *et al.* 1993*a*, 1993*b*). Once the genetic status of the target stock and the genetic goals of the enhancement programme are identified, the approach for managing genetic resources is similar to the approach for managing other enhancement objectives (e.g. controlling the level of impact of stocked fish on abundances of the target population). This approach includes (1) identifying the genetic risks and consequences of enhancement; (2) defining an enhancement strategy; (3) implementing genetic controls in the hatchery and a monitoring and evaluation programme for wild stocks; (4) outlining research needs and objectives; and

(5) developing a feedback mechanism. These points are discussed in detail by Kapuscinski and Jacobson (1987) and Shaklee *et al.* (1993*a*, 1993*b*).

A genetic resource management plan should encompass genetic monitoring prior to, during and after enhancement, as well as proper use of a sufficiently large and representative broodstock population and spawning protocols, to maintain adequate effective broodstock population size. Prior to enhancement, a comprehensive genetic baseline evaluation of the wild population should be developed to describe the level and distribution of genetic diversity. This baseline should at least include the geographical range of the particular stock targeted for enhancement. This monitoring should take place over a long enough period of time to capture short-term fluctuation or long-term change. The baseline can be used as a basis to determine an effective population or broodstock size to minimize the undesirable genetic effects of inbreeding, changes in allele frequencies, and loss of alleles. Genetic monitoring of the broodstock and its released progeny should be undertaken to measure success. Long-term genetic monitoring of the wild stock after enhancement should also occur to measure possible loss of genetic diversity, which might be attributed to enhancement efforts.

Maintenance and proper use of a sufficient broodstock population may be one of the toughest and most expensive components of marine stock enhancement. It is also one of the most important. The typically high fecundity rate of marine fish provides the opportunity for a greatly reduced effective population size in a hatchery environment because relatively few adults could potentially contribute a large number of eggs. Fortunately, however, marine fish are genetically more homogeneous than freshwater and anadromous species on a relative scale, and genetic studies show relatively little stock separation due to geographic, clinal or temporal factors (Gyllensten 1985; Waples 1987; Bartley and Kent 1990; King *et al.* 1995). In vagile marine species gene flow is often sufficient to homogenize the genetic structures over broad areas. Regardless, sufficient numbers of broodstock must be utilized so that the genetic diversity (including rare alleles) of the fish being released is the same over time as in their wild counterparts.

Hubbs-Sea World Research Institute (Hubbs) has been an early promoter of a responsible genetic management plan. Hubbs leads a consortium of Californian researchers who are evaluating the feasibility of enhancement of white seabass *Atractoscion nobilis*. Although the genetic profile of progeny from an individual spawn may differ from wild spawns, use of multiple hatchery spawns can approximate the genetic variability observed in the wild. Bartley and Kent (1990) successfully used this concept with white seabass and showed that over 98% of the genetic variability observed in the wild could be maintained with an effective population of 100 brood fish.

The red drum (*Sciaenops ocellatus*) enhancement programme of Texas Parks and Wildlife Department provides a good example of the maintenance of a large broodstock with yearly replenishment (McEachron *et al.* 1995). Texas has 140–170 adult broodstock for its programme, with an annual replacement of at least 25%. In Norway, studies of allele frequencies are being used to compare broodstock and their progeny with the wild populations of Atlantic cod (Svasand *et al.* 1990).

Use disease and health management

Disease and health guidelines are important to both the survival of the fish being released and the wild populations of the same species or other species with which they interact. Florida Department of Environmental Protection has developed an aggressive and responsible approach in this area in association with its red drum enhancement project (Landsberg *et al.* 1991). Florida's policy requires that all groups of fish pass a certified inspection for bacterial and viral infections and parasites prior to release. Maximum acceptable levels of infection and parasites in the hatchery populations are established on the basis of the results of screening healthy wild populations.

Form enhancement objectives and tactics

During the design phase of enhancement programmes, ecological factors that can contribute to the success or failure of hatchery releases should be considered. Predators, food availability, accessibility of critical habitat, competition over food and space, environmental carrying capacity and abiotic factors, such as temperature and salinity, are all key variables that can affect survival, growth, dispersal, and reproduction of cultured fish in the wild. Predatory losses and food availability have long been thought to be among the principal variables that mediate recruitment success in wild populations (Houde 1987; Lasker 1987).

Habitat degradation in marine environments can also affect recruitment success. For example, seagrass meadows are important nursery habitats for fishes and crustaceans (Kikuchi 1974). In vegetated aquatic environments, habitat availability and habitat quality (e.g. structural complexity) have been shown to mediate survival from predators (Crowder and Cooper 1982; Stoner 1982; Main 1987). In some cases, habitat degradation in marine environments may be so complete that certain habitats are unsuitable for stock enhancement (Stoner 1994). To enhance fisheries in some locales, restoration of coastal habitat may be the first priority.

Marine stock enhancement should never be used as mitigation to justify loss of habitat. However, enhancement efforts with cultured fishes can fill a void where critically important habitats such as coastal wetlands and estuaries, which provide nurseries for early life stages, are irretrievably lost or degraded.

In addition to ecological factors, there may be physiological and behavioural deficits in hatchery-reared fish that strongly reduce survival in the wild (e.g. swimming ability, feeding, predator avoidance, agonism, schooling and habitat selection). In Japan, Tsukamoto (1993) has evaluated the effect of behaviour on survival of cultured red sea bream, *Pagrus major*, released into the sea. Tsukamoto's results indicate that a predator-avoidance behaviour, (tilting), in which wild fish lie flat against the substratum, may be reduced or absent in cultured fish during the first few days after release into the sea. Abnormal tilting behaviour was directly correlated with mortality rate. For certain learned behaviours, exposure of wild fish in hatchery microcosms to behavioural cues and responses may be needed to overcome behavioural deficits (Olla and Davis 1988).

A solid understanding of the ecological and biological mechanisms mediating target species abundances can require exhaustive field studies for each species considered for enhancement. Whole careers have been dedicated to understanding mechanisms behind animal distributions and abundance; it does not seem practical to hold off on stock-enhancement research until the ecological mechanisms are completely understood. However, failure to consider such factors can result in poor performance of released fish at best and perhaps negative impacts on natural stocks (Murphy and Kelso 1986).

Our viewpoint is that preliminary, pilot-scale experimental releases with subsequent monitoring of cultured fish afford a direct method for evaluating assumptions about the impacts of uncontrolled environmental factors. For example, assumptions about carrying capacity in particular release habitats can and should be evaluated through pilot releases conducted prior to full-scale enhancement at those sites (Leber *et al.* 1995*b*). This approach is elaborated below.

Identify released hatchery fish and assess stocking impacts

One of the most critical components of any enhancement effort is the ability to quantify success or failure. Without some form of assessment, one has no idea to what degree the enhancement was effective or, more critically, which approaches were totally successful, partially successful, or a downright failure. Natural fluctuations in marine stock abundance can mask successes and failures. Maximization of benefits cannot be realized without the proper monitoring and evaluation system.

Tagging or marking systems that are benign and satisfy the basic assumption that identified fish are representative of untagged counterparts are essential, but were not available until relatively recently. The detrimental effects of external tags are well documented (Isaksson and Bergman 1978; Hansen 1988; McFarlane and Beamish 1990), and few fishery managers or researchers defend their use today, especially with juvenile fish. Useful information retrieved from external tags is usually restricted to migration and growth rates of relatively large fish (Scott *et al.* 1990; Trumble *et al.* 1990).

In recent years, a few identification systems (e.g. coded wire tags, passive integrated transponder tags, genetic markers and otolith marks) have been developed, that meet the requirements that identified fish are representative of the species with regard to behaviour, biological functions and mortality factors, and thus provide unbiased data (Buckley and Blankenship 1990). The story of the development and now widespread use of the coded wire tag (Jefferts *et al.* 1963) is well known, and it is fair to say that it has revolutionized the approach to stock enhancement (Soloman 1990).

With an unbiased tag or mark, quantitative assessment of release impact is possible. In developing enhancement programmes, evaluation of hatchery contributions can be partitioned into at least four distinct stages: initial survival, survival through the nursery stage, survival to adulthood (entry into the fishery), and successful contribution to the breeding pool. In Hawaii, the proportion of hatchery fish in field samples taken after pilot releases of striped mullet, *Mugil cephalus*, has been as high as 80% in initial collections, 50% in some nursery habitats through the tenth month after release, and (in a recreational fishery in Hilo, Hawaii) as high as 20% of the catch (Leber 1995; Leber *et al.* 1995*a*, *b*). In Norway, genetic markers are beginning to show that released Atlantic cod produce viable offspring in the wild (Jorstad 1994).

Assessment of the effects of release should go further than evaluation of survival and contribution rates of hatchery fish. Evaluation of interactions between hatchery fish and wild stocks is also critical. Clearly, evaluation of genetic impact is important. It is equally important to understand whether hatchery releases increase abundances in the wild or simply displace the wild stocks targeted for enhancement. At least one experimental study in Hawaii has documented that released hatchery fish can indeed increase abundances in a principal nursery habitat without displacing wild individuals (Leber *et al.* 1995*b*).

Use an empirical process to define optimum release strategies

Just as preliminary releases can be used to evaluate ecological assumptions, pilot release experiments afford a means of quantifying and controlling the effects of release variables and their influence on the performance of cultured fish in coastal environments (Tsukamoto *et al.* 1989; Svasand and Kristiansen 1990; Leber 1995; Willis *et al.* 1995).

Experiments to evaluate fish size at release, release season, release habitat, and release magnitude should always be conducted prior to launching full-scale enhancement programmes. These experiments are a critical step in identifying enhancement capabilities and limitations and in determining release strategy. They also provide the empirical data needed to plan enhancement objectives, test assumptions about survival and cost-effectiveness, and model enhancement potential. The lack of monitoring to assess survival of the fish released by marine enhancement programmes in the first half of this century was the single greatest reason for the failure of those programmes to increase stock abundances and fishery yields (Richards and Edwards 1986).

On the basis of results of pilot experiments by The Oceanic Institute in Hawaii, hatchery-release variables were steadily refined to maximize striped mullet enhancement potential. This resulted in an increase in recapture rates by at least 400% over a 3-year period (Leber *et al.* 1995*a*, *b*). During the third year of pilot studies in Kaneohe Bay, hatchery fish provided at least 50% of the striped mullet in net samples during the entire 10-month collection period after releases. An understanding of how fish size-at-release and release habitat affected survival were the primary factors needed to increase recapture rates. However, understanding the interaction of release season with size-at-release and release habitat also had significant impact on refinement of release strategies. The apparent doubling effect on abundances in the third year was achieved with a release of only 80 000 juveniles into the principal striped mullet nursery habitat in Kaneohe Bay, the largest estuary in Hawaii. A subsequent study documented that mullet releases did not displace wild juveniles from that nursery habitat (Leber *et al.* 1995*b*). Thus, hatchery releases in Kaneohe Bay appear to be increasing population size in the primary nursery habitat. Clearly, these pilot experiments are crucial for managing enhancement impact.

Identify economic and policy objectives

Initially, costs and benefits can be estimated and economic models developed to predict the value of enhancement. This information can be used to generate funding support through reprioritization, legislation or user fees. The information can contribute to an explicit understanding with policy makers and the general public on the time frame that is needed for components such as adaptation of culture technology and pilot release experiments before full-scale releases can begin. The education of the public and policy makers on the need for, and benefits of, a responsible approach is also important. In Florida, pressure is mounting to drop the responsible-approach concept involving pilot-scale releases and instead plant millions of red drum fry as a neighbouring state has done (Wickstrom 1993), on the assumption that the bigger the numbers planted, the better.

Use adaptive management

Adaptive management is a continuing assessment process that allows improvement over time. The key to this improvement lies in having a process for changing both production and management objectives (and strategies) to control the effects of enhancement. Essentially, adaptive management is the continued use of the nine key components above to ensure an efficient and wise use of a natural resource. The use of adaptive management is central to the successful application of the approach outlined above. Some minimum level of ongoing assessment is needed, superimposed over a moderate research framework that provides a constant source of new information. New ideas for refining enhancement are thus constantly considered and integrated into the management process.

Conclusion

The need for marine stock enhancement has been identified, and we must learn from mistakes made in the past. The necessity for and benefit from following a responsible approach in implementing enhancement cannot be overemphasized. Organizations that have subscribed to this new approach are starting to see the benefits. Without pilot experiments, Hawaii researchers would not have increased survival rate by over 400% in Kaneohe Bay in just three years nor provided a 20% contribution to the catch in the recreational fishery in Hilo Bay. What is needed now is a concerted effort by the managers of new and existing enhancement programmes to use, evaluate and refine the approach described here.

Given the world-wide decline in fisheries catch rates, bold new initiatives are needed to revitalize fisheries. We need to take care, though, to preserve existing stocks as we work to restore and increase the harvest levels of those stocks by using cultured fishes.

Acknowledgments

We thank Devin Bartley, Don Kent, Rich Lincoln, Stan Moberly and Scott Willis, who have greatly contributed to the development of the ideas expressed here. We also thank Maala Allen, Paul Bienfang, Churchill Grimes, Gary Sakagawa, Kimberly Smith and Dave Sterritt for insightful comments on the manuscript. Order of authorship was determined by the flip of a coin.

References

Anon. (1991*a*). Our living oceans: first annual report on the status of U.S. living marine resources. NOAA Technical Memorandum, NMFS/SPO-1.

Anon. (1991*b*). *Food and Agriculture Organization of the United Nations* yearbook 70 (1990), fishery statistics.

Anon. (1992). Our living oceans: first annual report on the status of U.S. living marine resources. NOAA Technical Memorandum, NMFS/SPO-1.

Bain, M B (1987). Structured decision making in fisheries management: trout fishing regulation on the Au Sable River, Michigan. *North American Journal of Fisheries Management* **7**, 475–81.

Bartley, D M, and Kent, D B (1990). Genetic structure of white seabass populations from the southern California Bight Region: applications to hatchery enhancement. *California Cooperative Oceanic Fisheries Investigations Report* **31**, 97–105.

Bergman, P K, Haw, F, Blankenship, H L, and Buckley, R M (1992). Perspectives on design, use, and misuse of fish tags. *Fisheries* **17**(4), 20–4.

Boggs, C H, and Pooley, S G (Eds) (1987). Strategic planning for Hawaii's aku industry. *Southwest Fisheries Center Administrative Report*. H-87–1, 22, Honolulu, Hawaii.

Buckley, R M, and Blankenship, H L (1990). Internal extrinsic identification systems: Overview of implanted wire tags, otolith marks and parasites. *American Fisheries Society Symposium* **7**, 173–82.

Crowder, L B, and Cooper, W E (1982). Habitat structural complexity and interaction between bluegills and their prey. *Ecology* **63**, 1802–13.

Gyllensten, U (1985). The genetic structure of fish: differences in the intraspecific distribution of biochemical genetic variation between marine, anadromous and freshwater species. *Journal of Fisheries Biology* **26**, 691–9.

Hansen, L P (1988). Effects of Carlin tagging and fin clipping on survival of Atlantic salmon (*Salmo salar L.*) released as smolts. *Aquaculture* (Netherlands) **70**, 391–4.

Honma, A (1993). *Aquaculture in Japan*. Japan FAO Association. Chiyoda-Ku, Tokyo.

Houde, E D (1987). Fish early life dynamics and recruitment variability. *American Fisheries Society Symposium* **2**, 17–29.

Isaksson, A, and Bergman, P K (1978). An evaluation of two tagging methods and survival rates of different age and treatment groups of hatchery-reared Atlantic salmon smolts. *Journal of Agricultural Research in Iceland* **10**(1), 74–99.

Jefferts, K B, Bergman, P K, and Fiscus, H F (1963). A coded-wire identification system for macro-organisms. *Nature* (London) **198**, 460–2.

Jorstad, K E (1994). Cod stock enhancement studies in Norway — genetic aspects and the use of genetic tagging. *World Aquaculture Society*. New Orleans, Louisiana.

Kapuscinski, A R, and Jacobson, L D (1987). Genetic guidelines for fisheries management. University of Minnesota, Minnesota Sea Grant College Program, *Sea Grant Research Report* 17, Duluth.

Kikuchi, T (1974). Japanese contributions on consumer ecology in eelgrass (*Zostera marina*) beds, with special reference to trophic relationships and resources in inshore fisheries. *Aquaculture* **4**, 145–160.

King, T L, Ward, R,. Blandon, I R, Colura, R L, and Gold, J R (1995). Using genetics in the design of red drum and spotted seatrout stocking programs in Texas: a review. *American Fisheries Society Symposium* **15**, 499–502.

Landsberg, J H, Vermeer, G K, Richards, S A, and Perry, N (1991). Control of the parasitic copepod *Caligus elongatus* on pond-reared red drum. *Journal of Aquatic Animal Health* **3**, 206–9.

Lasker, R (1987). Use of fish eggs and larvae in probing some major problems in fisheries and aquaculture. *American Fisheries Society Symposium* **2**, 1–16.

Leber, K M (1994). Prioritization of marine fishes for stock enhancement in Hawaii. The Oceanic Institute. Honolulu, Hawaii.

Leber, K M (1995). Significance of fish size-at-release on enhancement of striped mullet fisheries in Hawaii. *Journal World Aquaculture Society* **26**(2), 143–53.

Leber, K M, Sterritt, D A, Cantrell, R N, and Nishimoto, R T (1995*a*). Contribution of hatchery-released striped mullet, *Mugil cephalus*, to the recreational fishery in Hilo Bay, Hawaii. Hawaii Department of Land and Natural Resources, Division of Aquatic Resources, Technical Report 94–03.

Leber, K M, Brennan, N P, and Arce, S. M (1995*b*). Marine enhancement with striped mullet: are hatchery releases replenishing or displacing wild stocks? *American Fisheries Society Symposium* **15**, 367–87.

Mackett, D J, Christakis, A N, and Christakis, M P (1983). Designing and installing an interactive management system for the southwest fisheries center. In 'Coastal Zone 83'. (Eds O T Magoon and H Converse) pp. 518–27 (American Society of Civil Engineers: New York).

Main, K L (1987). Predator avoidance in seagrass meadows: prey behavior, microhabitat selection and cryptic coloration. *Ecology* **68**(1), 170–80.

McEachron, L W, McCarty, C E, and Vega, R (1995). Beneficial uses of marine fish hatcheries, Enhancement of red drum in Texas coastal waters. *American Fisheries Society Symposium* **15**, 161–6.

McFarlane, G A, and Beamish, R J (1990). Effect of an external tag on growth of sablefish and consequences to mortality and age at maturity. *Canadian Journal of Fisheries and Aquatic Sciences* **47**, 1551–7.

McVey, J P (Ed) (1991). *Handbook of Mariculture*, Volume II Finfish Aquaculture. CRC Pres Inc., Boco Raton, Florida.

Murphy, B R, and Kelso, W E (1986). Strategies for evaluating freshwater stocking programs: past practices and future needs. In 'Fish culture in Fisheries Management'. (Ed R H Stroud) pp. 303–16 (American Fisheries Society. Bethesda, Maryland).

Olla, B L, and Davis, M (1988). To eat or not be eaten. Do hatchery reared salmon need to learn survival skills? *Underwater Naturalist* **17**(3), 16–18.

Platt, J R (1964). Strong inference. *Science*, **146**(3642), 347–53.

Pooley, S G (Ed) (1988). Recommendations for a five-year scientific investigation on the marine resources and environment of the main Hawaiian Islands. Southwest Fisheries Center Administrative Report H-88–2, Honolulu, Hawaii.

Popper, K R (1965). Conjectures and refutations, the growth of scientific knowledge. Harper and Row, New York.

Richards, W J, and Edwards, R E (1986). Stocking to restore or enhance marine fisheries. In 'Fish Culture in Fisheries Management'. (Ed R H Stroud) pp. 75–80 (American Fisheries Society, Bethesda, Maryland).

Scott, E L, Prince, E D, and Goodyear, C D (1990). History of the cooperative game fish tagging program in the Atlantic Ocean, Gulf of Mexico, and Caribbean Sea, 1954–1987. *American Fisheries Society Symposium* **7**, 841–53.

Shaklee, J B, Busack, C A, and Hopley, Jr. C W (1993*a*). Conservation genetics programs for Pacific salmon at the Washington Department of Fisheries: living with and learning from the past, looking to the future. In 'Selective breeding of fisheries in Asia and the United States'. (Eds K L Main and E Reynolds) pp. 110–41 (The Oceanic Institute. Honolulu, Hawaii).

Shaklee, J B, Salini, J, and Garrett, R. N (1993*b*). Electrophoretic characterization of multiple genetic stocks of barramundi perch in Queensland, Australia. *Transactions of the American Fisheries Society*. **122**, 685–701.

Soloman, D J (1990). Development of stocks: Strategies for the rehabilitation of salmon rivers. In 'Strategies for the rehabilitation of salmon rivers'. (Ed D Mills) pp.35–44 (Linnean Society of London, London).

Stoner, A W (1982). The influence of benthic macrophytes on foraging behavior of pinfish, *Lagodon rhomboides*. *Journal of Experimental Marine Biology and Ecology* **104**, 249–74.

Stoner, A W (1994). Significance of habitat and stock pre-testing for enhancement of natural fisheries: experimental analyses with queen conch *Strombus gigas*. *Journal of the World Aquaculture Society* **25**, 155–65.

Svasand, T, and Kristiansen, T S (1990). Enhancement studies of coastal cod in western Norway, part IV. Mortality of reared cod after release. *Journal du Conseil International pour l'Exploration de la Mer* **47**, 30–39.

Svasand, T, Jorstad, K E, and Kristiansen, T S (1990). Enhancement studies of coastal cod in western Norway, part I. Recruitment of wild and reared cod to a local spawning stock. *Journal du Conseil International pour l'Exploration de la Mer* **47**, 5–12.

Trumble, R J, McGregor, I R, St-Pierre, G, McCaughran, D A, and Hoag, S H (1990). Sixty years of tagging Pacific halibut: A case study. *American Fisheries Society Symposium* **7**, 831–40.

Tsukamoto, K (1993). Marine fisheries enhancement in Japan and the quality of fish for release. European Aquaculture Society Special Publication 19. 556 pp.

Tsukamoto, K, Kuwada, H, Hirokawa, J, Oya, M, Sekiya, S, Fujimoto, H, and Imaizumi, K (1989). Size-dependent mortality of red sea bream, *Pagrus major*, juveniles released with fluorescent otolith-tags in News Bay, Japan. *Journal of Fish Biology* **35** (Supplement A), 59–69.

Waples, R S (1987). Multispecies approach to the analysis of gene flow in marine shore fishes. *Evolution* **41**, 385–400.

Wickstrom, K (1993). Biscayne redfish: yes! *Florida Sportsman*, (March), 90–91.

Willis, S A, Falls, W W, Dennis, C W, Roberts, D E, and Whitchurch, P G (1995). Assessment of season-of-release and size-at-release on recapture rates of hatchery-reared red drum in a marine stock enhancement program in Florida. *American Fisheries Society Symposium* **15**, 354–65.

Evaluation of the Relative Importance of Hatchery-Reared and Wild Fish in the Restoration of Lake Superior Lake Trout

Michael J. Hansen,[A] *James R. Bence,*[B] *James W. Peck*[C] *and William W. Taylor*[B]

[A] University of Wisconsin-Stevens Point, College of Natural Resources, Stevens Point, Wisconsin 54481, USA.
[B] Michigan State University, Department of Fisheries and Wildlife, East Lansing, Michigan 48824, USA.
[C] Michigan Department of Natural Resources, 488 Cherry Creek Road, Marquette, Michigan 49855, USA.

Summary

Juvenile, hatchery-reared lake trout *Salvelinus namaycush* were stocked, in conjunction with controls on sea lampreys *Petromyzon marinus* and commercial fisheries, to restore lake trout into Lake Superior following their collapse in the 1950s. We evaluated the contribution of wild and stocked lake trout to recruitment to spawning populations in Michigan's waters of Lake Superior using Ricker stock-recruitment models. Data were from large-mesh gill-net assessment fisheries prosecuted each Spring in 1959–1993. Estimates of recruitment rates at low stock sizes (the density-independent term of the Ricker model) were near replacement rates for stocked fish, but near zero for wild fish. We conclude from this that stocked lake trout fuelled population recoveries in Michigan's waters of Lake Superior. The poor performance of wild fish may be caused by confounding factors such as gill-net fishing effort, which reduced survival to recruitment at the same time that wild fish became abundant.

Introduction

Lake Superior is the largest of the Laurentian Great Lakes and the second largest (by volume) freshwater lake in the world. Historically, lake trout was a dominant top predator in this lake and supported a substantial commercial fishery with an average harvest of 2 million kg from 1913 to 1950 (Hansen *et al.* 1995). Excessive mortality caused by a combination of commercial fisheries and predation by the exotic sea lamprey (*Petromyzon marinus*) is generally believed to have caused the collapse of lake trout populations in the 1950s. Following the collapse, juvenile, hatchery-reared lake trout *Salvelinus namaycush* were stocked, in conjunction with controls on sea lampreys and fisheries, to restore lake trout into Lake Superior (Lawrie and Rahrer 1972, 1973; Pycha and King 1975).

The contribution of stocked lake trout to population recovery in Lake Superior has been the subject of considerable debate. Several authors have noted that stocked lake trout generally attempted to spawn near inshore release sites, rather than on offshore reefs where spawning historically occurred (e.g. Dryer and King 1968; Pycha and King 1975). A view emerged that stocked lake trout were impaired in their ability to find suitable spawning grounds, and therefore spawned on sites that were inappropriate for incubation of eggs (Eshenroder *et al.* 1983). Supporting this view, Krueger *et al.* (1986) found that wild spawners contributed significantly more young lake trout than did stocked spawners in the Apostle Islands area of western Lake Superior. In contrast, Peck (1979) and Peck and Schorfhaar (1991) found that inshore plantings of lake trout contributed substantially to spawning aggregations on offshore reefs in Michigan that were historically important for reproduction.

One problem in determining the importance of stocked lake trout to recruitment in Lake Superior was that residual fish from wild stocks were thought to be present in most areas (Lawrie 1978), and hence Eshenroder *et al.* (1983) concluded that stock restoration in Lake Superior was not an appropriate example for generalizing to the other Great Lakes where wild stocks were believed to be extinct.

The contribution of stocked lake trout to recruitment from a particular spawning stock has only been tested for two spawning stocks in eastern Wisconsin waters of Lake Superior (Krueger *et al.* 1986; Schram *et al.* 1995). A similar analysis across broader areas was used to determine whether stocked lake trout contributed to recruitment to the spawning population (Hansen *et al.* 1995). We argue below that the stock-recruitment models used in all these analyses were not plausible depictions of the stock-recruitment process. Our objective here was to develop a plausible model for quantifying the importance of stocked and wild lake trout on population recoveries in Michigan's waters of Lake Superior. We use this model to evaluate whether stocked lake trout contributed substantially to the recruitment of wild lake trout in an area one generation later.

METHODS

Stock assessment of lake trout in Lake Superior is carried out by defined management areas (Hansen 1996) (Fig. 1). Here we evaluate stock-recruitment for Michigan areas, but exclude MI1, MI2 and MI8 because too few years of data were available.

Trends in relative abundance of lake trout were monitored with 'large mesh' gill-nets fished in each lake trout management area in 1959–1993. These nets were of 114-mm stretched-mesh, 210/2 multi-filament nylon twine, 18 meshes deep, hung on the H basis, and fished from late April through early June. Nets were of non-uniform length, so catch-per-unit-effort (CPUE) was defined as the number of fish caught per km of net. Sets were also of non-uniform duration, particularly in 1959–1969, so

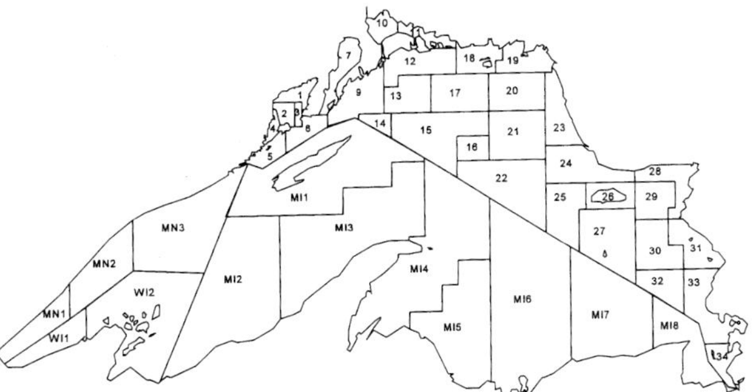

Fig. 1. Lake Superior lake trout management areas. US management areas are denoted by state: MI, Michigan; MN, Minnesota; WI, Wisconsin. Areas marked by numbers only are in Canadian waters.

CPUE was standardized to net-nights using corrections derived from an experiment in 1970: net-nights = 1.00, 1.52, and 1.80 for sets of one, two, and three or more nights' duration, respectively (Curtis *et al.* in press).

Hatchery-reared lake trout were all marked by removal of a fin (clipping) before stocking, so the CPUE of unclipped lake trout was assumed to be of wild fish and the CPUE of clipped lake trout was assumed to be of stocked fish. Means of CPUE (over lifts within a year) were calculated for each area as indices of abundance. These indices of abundance were used as recruitment and spawner-stock size values because age-specific catches were not available across all areas and years. We paired the CPUE of each parental stock with CPUE of wild fish eight years later, which were assumed to be the recruits produced by the parental stock, given a generation time of eight years (Krueger *et al.* 1986; Hansen *et al.* 1995).

Previous analyses of the contributions by wild and stocked parents to recruitment of lake trout in Lake Superior were based on a linear model (Krueger *et al.* 1986; Hansen *et al.* 1995; Schram *et al.* 1995), in which:

$$R_i = \beta_0 + \beta_w W_i + \beta_s S_i = \varepsilon_I \tag{1}$$

where R is an index of wild recruits, W is an index of wild of wild parental stock, S is an index of stocked parental stock, ε is residual variance, and I indicates the year. However, two of these previous analyses (Hansen *et al.* 1995; Schram *et al.* 1995) used the logarithms of the CPUEs, which implies an underlying model in which the progeny of wild and stocked lake trout are multiplied together to produce the mixture of recruits.

Plausible models should add together the progeny produced by the two parental stocks, instead of multiplying them. The analysis could then estimate the relative contribution of each parental stock to the mixture of recruits. Equation 1 (using data on the arithmetic scale) defines such a model (Krueger *et al.* 1986), but includes density-dependence only indirectly through the intercept, and when fit by ordinary least-squares assumes additive and homogenous errors. Here we investigate a variety of models both with and without density-dependence. When density-dependence was included, we did so using a variant of the Ricker stock-recruitment model (Ricker 1975). In general terms these models can be written as:

$$R_i = f(W_I, S_I) \times \varepsilon_I \tag{2}$$

where f(.) defines the deterministic part of the stock-recruitment equation.

We assume that stock-recruitment errors can be approximated by a log-normal distribution (e.g. Peterman 1981; Hilborn and Walters 1992), and fit the model by ordinary least squares after log transformation of both sides of equation 2:

$$\begin{gathered} \log_e(R_i) + \log_e(f(S_i, W_i)) + \log(\varepsilon_i) \\ \log(\varepsilon_i) \sim N(0, \sigma^2) \end{gathered} \tag{3}$$

Table 1. Candidate models

Model	Equation	Description
1	$R_i = \alpha(W_i + S_i)$	Recruitment directly proportional to total stock size, no density-dependence and stocked and wild fish equivalent.
2	$R_i = \alpha(W_i + S_i)e^{-\beta(W_i + S_i)}$	Ricker stock-recruitment function based on total stock size, includes density-dependence and stocked and wild fish equivalent.
3	$R_i = \alpha_W W_i + \alpha_S S_i$	Recruitment directly proportional to stock size, no density-dependence, different proportionality constant for stocked and wild fish.
4	$R_i = \alpha(kW_i + S_i)e^{-\beta(kW_i + S_i)}$	Ricker stock-recruitment function, includes density-dependence. Both density-dependent and density-independent terms apply to stocked fish equivalents. 'k' converts wild fish into stocked equivalents.
5	$R_i = (\alpha_W W_i + \alpha_S S_i)e^{-\beta_W W_i - \beta_S S_i}$	'Multi-species' Ricker stock recruitment function. Stocked and wild fish have different density-dependent and density-independent terms.

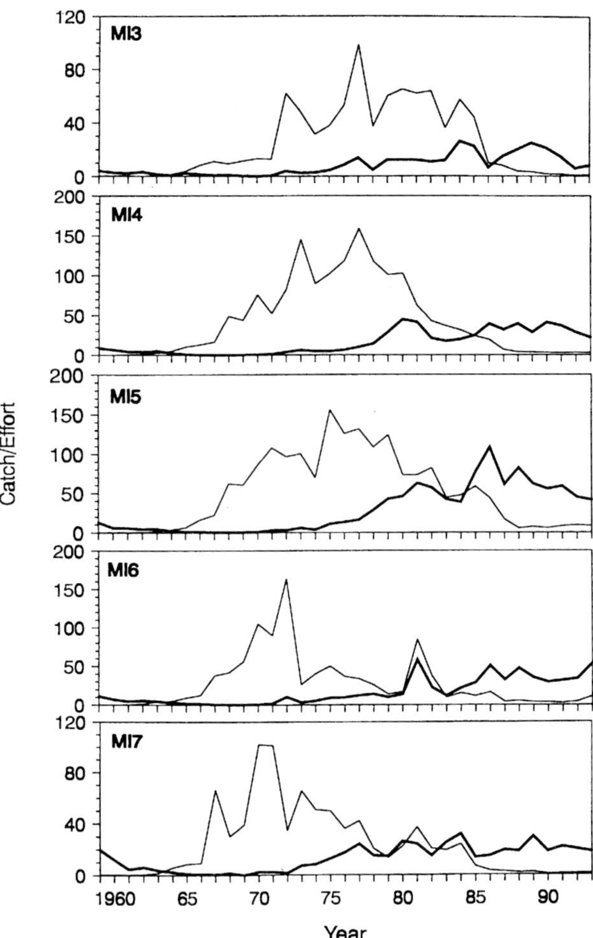

Fig. 2. Catch-per-unit-effort (CPUE, fish km^{-1} net-nights^{-1}) of wild and stocked lake trout from 1959 through 1993 for Michigan management units on Lake Superior. Bolder lines denote wild CPUE and lighter lines stocked CPUE.

Parameter estimates and asymptotic standard errors were obtained numerically using a quasi-Newton method, and checked by considering a variety of starting values and Simplex as an alternative search method.

We evaluated five different stock-recruitment models ranging from one (model 1) to four parameters (model 5), when fitted to single areas (Table 1). Model 1 treated all fish equally and assumed recruitment was directly proportional to spawner density. Model 5 allowed different reproductive success for stocked and wild fish and for different density-dependence from stocked and wild fish. Models 2–4 were of intermediate complexity. Each time we increased the number of parameters we tested whether the additional parameter led to a significantly better fit by *F*-tests based on 'extra sums of squares' (Bates and Watts 1988).

We also evaluated models designed to achieve parsimony by assuming that parameters were either identical across areas (All Areas Combined) or linked in some fashion across areas (All Areas Linked). We considered linked parameters for models 3–5, which contained analogous parameters for wild and stocked fish. For a pair of such parameters $\theta_{S,A}$ and $\theta_{W,A}$ for area *A* (which could be either $\alpha_{S,A}$, $\alpha_{W,A}$ or $\beta_{S,A}$, $\beta_{W,A}$), $\theta_{W,A}$ was replaced by $\theta_{S,A} \times k_\theta$ in our model formulation. This is equivalent to assuming that the relative values of $\theta_{S,A}$ and $\theta_{W,A}$ remain constant over areas, and allows us to estimate a single k_θ instead of a θ_W, for each area. To apply this linked approach, we fitted a single model that described each of the areas simultaneously. This required distinguishing among areas through the use of indicator variables, so that area-specific parameters would be applied to the correct observations. For the model we selected as best in general, we used 'extra sum of squares' *F*-tests to test whether All Areas Linked improved the fit over All Areas Combined, and whether All Areas Summed (i.e. the model fitted separately for each area and then sums of squares and degrees of freedom summed) improved the fit over All Areas Linked.

Results

Catch-per-unit-effort (CPUE) showed large fluctuations over time. In general wild fish were scarce early in the time series and became more abundant later (Fig. 2). Stocked fish were initially scarce or absent, increased to a peak, and then declined. The contrast present in these data appears sufficient to allow estimation of stock-recruitment relationships, but we note that there is one period of increasing numbers of wild fish at which time stocked fish declined in abundance.

Table 2. Tests for significance of extra parameters

F-tests (one degree of freedom) used to compare each model with models containing one fewer parameter.

Comparison		Area MI3	Area MI4	Area MI5	Area MI6	Area MI7	All Areas Combined	All Areas Linked	All Areas Summed
Model 5 *vs* 4	*F*	1.99	8.92	6.87	8.48	0.89	23.46	16.58	5.71
	P	0.17	<0.01	0.015	<0.01	0.36	<0.01	<0.01	<0.01
Model 4 *vs* 3	*F*	19.36	9.01	3.45	19.29	31.29	42.52	18.04	16.81
	P	<0.01	<0.01	0.076	0.076	<0.01	<0.01	<0.01	<0.01
Model 4 *vs* 2	*F*	35.86	45.87	35.12	20.76	35.72	106.67	—	31.30
	P	<0.01	<0.01	<0.01	<0.01	<0.01	<0.01		<0.01
Model 3 *vs* 1	*F*	11.37	29.78	39.51	2.24	3.33	48.70	—	11.30
	P	<0.01	<0.01	<0.01	0.15	0.081	<0.01		<0.01
Model 2 *vs* 1	*F*	1.34	0.88	4.95	1.34	1.35	0.06	—	1.84
	P	0.26	0.36	0.036	0.26	0.26	0.81		0.11

Significantly better fits ($P < 0.05$) were generally achieved as additional parameters were added to the models (Table 2). We selected the four parameter model (5) because it produced significantly better fits than the three parameter model (4) in three of five areas and for All Areas Combined, All Areas Linked and All Areas Summed (Table 2). We selected the All Areas Linked version of Model 5 as our model of choice because it produced a significant improvement over All Areas Combined ($F_{8,120} = 5.42$, $P < 0.01$), but was not improved on by All Areas Summed ($F_{8,112} = 1.48$, $P = 0.17$).

Estimated recruitment rates at low stock sizes indicated that stocked parents (α_s) made significant contributions to recruitment and that wild parents ($\alpha_w = k_\alpha \times \alpha_s$) appeared to make relatively little contribution (Table 3). Stocked parents apparently reproduced at near-replacement levels when population sizes were low.

Positive values for β_s in all areas indicated that the presence of stocked fish appeared to inhibit recruitment (Table 3). Plots of stock-recruitment functions show that density-dependence imposed by stocked fish apparently limits the ability of stocked fish to replace themselves (Fig. 3).

Density-dependence for wild fish is represented by the product $k_\alpha \times \beta_s$. Since k_β is negative (Table 3), increases in density of wild fish increases the production of recruits per spawner. The density-independent term ($\alpha_w = k_\alpha \times \alpha_s$) is quite small, however, and over the ranges of observed CPUEs, the stock-recruitment model predicted that wild fish produced at 10% or less of replacement levels.

Table 3. Parameter estimates for Model 5, All Areas Linked

Parameter	Area	Estimate	Asymptotic standard error
α_s	MI3	0.725	0.117
α_s	MI4	0.574	0.0866
α_s	MI5	0.820	0.126
α_s	MI6	1.12	0.167
α_s	MI7	0.965	0.152
β_s	MI3	0.0204	0.00475
β_s	MI4	0.0.00878	0.00235
β_s	MI5	0.00699	0.00228
β_s	MI6	0.0224	0.00234
β_s	MI7	0.0200	0.00368
k_α	All	0.0112	0.00778
k_α	All	−1.18	0.290

DISCUSSION

Our analysis of stock-recruitment relationships in Michigan management areas indicated that stocked lake trout were important contributors to the recruitment of wild lake trout. The importance of stocked lake trout in restoration of wild stocks is supported by evidence that the individual growth curves for wild fish now resemble those for fish from source populations (Ferreri and Taylor 1996) and that wild fish in central Michigan's waters are genetically less diverse than, and

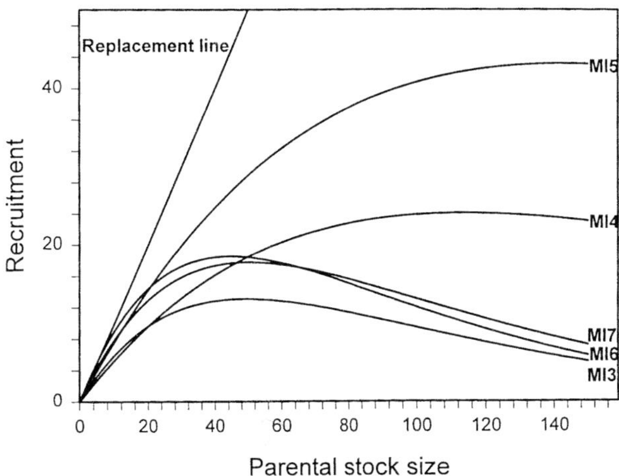

Fig. 3. Estimated relationships between recruitment and parental stock for stocked spawners (assuming no wild spawning stock). Plotted functions based on parameters in Table 3, units are fish km^{-1} net-nights^{-1}.

lack rare alleles present in wild, stocks that received little or no stocked fish from hatcheries (Burnham Curtis 1993). The production of extant wild populations through stocking of hatchery fish in Lake Superior has broader implications. Restoration of lake trout in the other Laurentian Great Lakes through stocking of hatchery fish now seems feasible, even though these lakes lack residual wild populations of lake trout.

Our results agree with those of Hansen *et al.* (1995) on the role of hatchery fish, but unlike that linear analysis, the present nonlinear analysis indicated that wild lake trout were unimportant to the recruitment of wild progeny. The models used here are more plausible than those used previously because progeny are added rather than multiplied. Our results contrast with earlier reports that CPUE of wild lake trout explained more variation in recruitment than stocked lake trout at two spawning areas in Wisconsin's waters of Lake Superior, Sand Cut Reef (Krueger *et al.* 1986) and Gull Island Shoal (Schram *et al.* 1995). This does not appear to be an artifact of using different models. We fitted model 5 to the Gull Island Shoal data in Schram *et al.* (1995) and found results that also indicated that wild parents explained more of the variation in recruitment than abundance of stocked parents for that data set.

We believe that the poor performance of wild fish in our analysis was likely to have been caused by the fact that conditions for survival of their progeny deteriorated as they became abundant. Hansen *et al.* (1994) reported declining survival of stocked fish in Michigan's waters of Lake Superior during the 1970s and early 1980s, and this decline may well apply to survival of naturally produced fish. While stock sizes can explain part of the decline in survival, Hansen *et al.* (1996) found that survival was most strongly related to commercial gill-net fishing effort. This was the same period in which wild CPUE was increasing (Fig. 2).

Recent results suggest that the future of lake trout restoration may be more promising, at least in the short term, than is suggested by our stock-recruitment model results for wild fish. In particular, gill-net fishing effort declined by 45% during 1990–1993 in Michigan's waters of Lake Superior due to conversion to trap nets, movement of fishers to other Great Lakes, and poor markets (Hansen *et al.* 1995). Recruitment of lake trout to spawning populations should be promoted if this lower fishing mortality can be maintained. It is encouraging that CPUE of young wild fish caught in small mesh gill-nets (fish too small to be vulnerable to the large mesh nets) has not declined as hatchery spawning stock has declined in the past few years (Michigan Department of Natural Resources, unpublished data). The recent cessation of stocking of hatchery fish in Michigan should allow a test of whether wild fish can sustain themselves under current conditions. To conclude, we showed here that this opportunity for restoration of self-sustaining lake trout stocks was created by the earlier stocking of hatchery fish.

Acknowledgments

We thank all past and present members of the Lake Superior Technical Committee for their input and efforts. Wayne R. MacCallum provided the map of Lake Superior lake trout management areas. US Federal Aid in Sport Fish Restoration Act funds provided partial support for this work. Some of the research reported herein was completed while MJH was employed by the National Biological Service, Great Lakes Science Center.

References

Bates, D M, and Watts, D G (1988). 'Nonlinear regression analysis and its applications.' (John Wiley and Sons: New York).

Burnham Curtis, M K (1993). Intralacustrine speciation of *Salvelinus namaycush* in Lake Superior: An investigation of genetic and morphological variation and evolution of lake trout in the Great Lakes. Ph.D. dissertation, University of Michigan, Ann Arbor Michigan. 268 pp.

Curtis, G L, Selgeby, J H, and Schorfhaar, R G (in press). Decline and recovery of lake trout populations near Isle Royale, Lake Superior, 1929–1990. *Transactions of the American Fisheries Society.*

Dryer, W R, and King, G R (1968). Rehabilitation of lake trout in the Apostle Islands region of Lake Superior. *Journal of the Fisheries Research Board of Canada* **25**, 1377–403.

Eshenroder, R L, Poe, T P, and Olver, C H (1983). Strategies for rehabilitation of lake trout in the Great Lakes: proceedings of a conference on lake trout research. Great Lakes Fishery Commission Technical Report 40.

Ferreri, C P, and Taylor, W W (1996). Compensation and individual growth rate and its influence on lake trout population dynamics in the Michigan waters of Lake Superior. *Journal of Fish Biology* **49**, 763–77.

Hansen, M J (Ed) (1996). A lake trout restoration plan for Lake Superior. Great Lakes Fishery Commission. 34 pp.

Hansen, M J, Ebener, M P, Schorfhaar, R G, Schram, S T, Schreiner, D R, and Selgeby, J H (1994). Declining survival of lake trout stocked during 1963–1986 in US waters of Lake Superior. *North American Journal of Fisheries Management* **14**, 395–402.

Hansen, M J, Peck, J W, Schorfhaar, R G, Selgeby, J H, Schreiner, D R, Schram, S T, Swanson, S T, MacCallum, W R, Burnham-Curtis, M K, Curtis, G L, Heinrich, J W and Young, R J (1995). Lake trout (*Salvelinus namaycush*) populations in Lake Superior and their restoration in 1959–1993. *Journal of Great Lakes Research* **21**(Supplement 1), 152–175.

Hansen, M J, Ebener, M P, Schorfhaar, R G, Schram, S T, Schreiner, D R, Selgeby, J H, and Taylor W W (1996). Causes of declining survival of lake trout stocked in US waters of Lake Superior in 1963–1986. *Transactions of the American Fisheries Society* **125**, 831–43.

Hilborn, R, and Walters, C J (1992). 'Quantitative fisheries stock assessment: choice, dynamics and uncertainty.' (Chapman and Hall: New York).

Krueger, C C, Swanson, B L, and Selgeby, J H (1986). Evaluation of hatchery-reared lake trout for reestablishment of populations in the Apostle Islands region of Lake Superior, 1960–84. In 'Fish culture in fisheries management'. (Ed R H Stroud) pp. 93–107 (American Fisheries Society: Bethesda, Maryland).

Lawrie, A H (1978). The fish community of Lake Superior. *Journal of Great Lakes Research* **4**, 513–49.

Lawrie, A H, and Rahrer, J F (1972). Lake Superior: effects of exploitation and introductions on the salmonid community. *Journal of the Fisheries Research Board of Canada* **29**, 765–76.

Lawrie, A H, and Rahrer, J F (1973). Lake Superior: a case history of the lake and its fisheries. Great Lakes Fishery Commission Technical Report 19.

Peck, J W (1979). Utilization of traditional spawning reefs by hatchery lake trout in the upper Great Lakes. Michigan Department of Natural Resources Fisheries Research Report 1871.

Peck, J W, and Schorfhaar, R G (1991). Assessment and management of lake trout stocks in Michigan waters of Lake Superior. *Michigan Department of Natural Resources Fisheries Research Report* 1956.

Peterman, R M (1981). Form of random variation in salmon smolt-to-adult relations and its influence on production estimates. *Canadian Journal of Fisheries and Aquatic Sciences* **38**, 1113–9.

Pycha, R L, and King, G R (1975). Changes in the lake trout population of southern Lake Superior in relation to the fishery, the sea lamprey, and stocking, 1950–70. *Great Lakes Fishery Commission Technical Report* 28.

Ricker, W E (1975). 'Computation and interpretation of biological statistics of fish populations.' *Fisheries Research Board of Canada Bulletin* **191**.

Schram, S T, Selgeby, J H, Bronte, C R, and Swanson, B L (1995). Stock size, egg deposition, and recruitment of the Gull Island Shoal lake trout population, 1972–1992. *Journal of Great Lakes Research* **21**(Supplement 1), 225–32.

Assessment of Stock Enhancement of Barramundi *Lates calcarifer* (Bloch) in a Coastal River System in Far Northern Queensland, Australia

D. J. Russell and M. A. Rimmer

Queensland Dept of Primary Industries, Northern Fisheries Centre, PO Box 5396, Cairns QLD 4870, Australia.

Summary

As part of an experiment to investigate the efficacy and cost-benefits of stock enhancement, nearly 69 000 barramundi fingerlings were released into the Johnstone River in northern Queensland from 1993 to 1996. All stocked fish were marked with coded wire tags to allow their discrimination from wild stocks and to determine size, site and date of release. The initial experimental design used two size-classes of fish (30–40 mm and 50–60 mm total length) which were released into freshwater, estuarine and upper-tidal habitats. The stocked fish took about three years to reach the minimum legal size of 580 mm total length and comprised between 15 and 19% of barramundi sampled. No significant difference was found in the numbers of fish returned from the two different stocking size-classes. While some stocked fish did make intra-riverine movements, the majority were recaptured within 3 km of their original release site. Angler record cards and commercial catch data are being used in an effort to detect measurable increases in catch-per-unit-effort.

Introduction

Barramundi, *Lates calcarifer* (Bloch), is a large, catadromous centropomid found throughout most of the Indo-west Pacific including the coastal rivers of tropical Australia (Greenwood 1976). In north-eastern Australia, barramundi spawn in coastal waters and estuaries just before or during the wet season from October to February (Dunstan 1959; Russell and Garrett 1985). In a tagging study to determine the movements of juvenile barramundi in north Queensland tidal creeks, Russell and Garrett (1988) found that most barramundi were recaptured at their original release location.

Barramundi are a popular and highly sought-after recreational and commercial food and sport-fish and in recent years fishers have expressed concern over perceived declines in stock numbers. There is a paucity of information on the health of barramundi stocks but available statistics from the east Queensland commercial fishery suggest there has been a historical decline in stock size (Russell 1987). More recent data from the Gulf of Carpentaria suggest that commercial catch rates have fallen by 30% since 1981 (R. Garrett pers. comm.). The reasons for the decline of the fishery are contentious but habitat degradation and overexploitation appear to be major factors (Russell 1987). Since 1981 fisheries managers have attempted to respond to this situation and have progressively introduced a range of management initiatives directed at effort reduction (Russell 1988). Further restrictions are likely to be introduced (R. Garrett pers. comm). Major components of the management strategy are: a closed season between November and January inclusive; area closures; a recreational angling bag limit of five fish; commercial gear restrictions and; a minimum legal size of 580 mm and maximum legal length of 1200 mm.

To assist in the recovery of exploited fish stocks, Leber *et al.* (1995) suggested that restrictive management measures could be augmented by an additional management strategy–stock enhancement. The development in the late 1980s and early 1990s of efficient and cost-effective technology for producing barramundi fingerlings (Rutledge and Rimmer 1991) has made stock enhancement a viable management option in northern Australia. In Queensland, there is wide support for stock-enhancement programmes, particularly amongst recreational fishers. Numerous community-based stocking groups have been formed to promote stock enhancement in coastal streams. In 1990, the Russell-Mulgrave became the first open river system to be stocked with hatchery-reared barramundi and since then, fish have been released into many coastal streams in eastern Queensland and in the Gulf of Carpentaria. Government agencies are now actively encouraging a responsible approach to stocking (Blankenship and Leber 1995), but little is known about appropriate techniques and methodologies for using stock enhancement as a tool for assisting the recovery of barramundi populations in coastal waters.

This paper outlines the preliminary results of a study to develop appropriate stocking strategies and to assess their impacts on the barramundi fishery of the Johnstone River in northern Queensland and to document the cost-benefits of this stocking programme.

STUDY SITE

The Johnstone River rises on the Atherton Tableland and flows into the Coral Sea near the township of Innisfail (17°32'S,146°02'E) in north Queensland. The river bifurcates five km from its mouth into the North and South Johnstone Rivers. It has a small catchment of about 1630 km^2 and sugar cane growing is the dominant land use on the coastal plain (Russell and Hales 1993*a*). The river has a narrow coastal plain less than 30 km wide and an escarpment prevents the upstream movement of fish from coastal areas to the upper tableland. Agricultural activities have impacted on wetland habitat within the catchment, with an overall reduction of about 60% over the past fifty years (Russell and Hales 1993*a*). The river supports a multi-species recreational line fishery and a seasonal commercial gill-net fishery which is restricted to the lower estuary. Five part-time commercial fishers operate in the catchment. The river has not been previously stocked with barramundi.

METHODS

Barramundi fingerlings used for stocking were spawned from broodstock held at the Northern Fisheries Centre (NFC) in Cairns and grown-out to between 25 and 35 mm total length (TL) using extensive larval rearing techniques (Rutledge and Rimmer 1991). After harvesting from the ponds, fish were held at the NFC hatchery prior to their release and on-grown as necessary. The original experimental design required that equal numbers of two size-classes of fish, one small (30–40 mm TL), and one large (50–60 mm TL), be stocked in three different habitat types (lower estuarine, upper-tidal and freshwater) in the Johnstone River. This design was undertaken annually for three years beginning in 1992/93, but was modified in an effort to obtain more information on the suitability of other secondary stocking locations. The first of these changes, starting in 1993/94, involved the stocking of a freshwater swamp system with large fish and later, in 1995/96, an extra two estuarine sites were stocked with large size-class fish. To distinguish the hatchery fish from wild fish, all barramundi were marked with coded wire tags prior to release. Russell and Hales (1992) successfully implanted coded wire tags into the cheeks of barramundi as small as 30 mm TL and obtained high survival and retention rates, and found that tagging had no significant effect on long-term survival or growth. They were able to achieve tagging rates of between 250 and 270 fish/hour.

Initially, monitoring involved using a boat-mounted electro-fisher (Smith-Root 7.5GPP) to catch barramundi which were subsequently scanned with a wand detector (Northwest Marine Technology). A small mesh (2 mm) beam-trawl was also used to sample fish in the lower estuary. All fish found to contain tags were retained and the tags were extracted and decoded. Anglers were asked to retain the heads of all legal size barramundi captured and return them to a central repository where they were subsequently checked for the presence of tags. As an incentive, a small reward was offered to anglers who returned tagged fish. Regular inspections were also made of the commercial catches from the Johnstone River.

Catch and effort data on the recreational fishery in the Johnstone River were collected using voluntary angler record cards. These cards requested details of the size and number of the species caught, the number of anglers in the party and the time spent fishing. Anglers were requested to complete the cards after each fishing trip and return them using a post-free mailing address or conveniently located drop boxes. Cards were widely available and a field liaison officer was available to provide support and assistance where necessary.

Data were stored on a relational database, and size data were analysed using a Generalized Linear Model (GLM) with a logit link and binomial error distribution. In analyses of the proportions of stocked fish in the population, only size cohorts which were likely to contain stocked fish (i.e. <650 mm TL) were considered. As stocked fish have only recently begun to enter the commercial fishery, no assessment of changes in commercial catch has been undertaken.

RESULTS

Stocking

Since 1992/3 a total of 68 770 tagged hatchery-bred barramundi have been released into the Johnstone River at six

Table 1. Numbers of fish stocked at locations in the Johnstone catchment. Primary release sites are marked with an asterisk and n.s. indicates sites which were not stocked in that year

Location	1992/93	1993/94	1994/95	1995/96	Total
North Johnstone R. (upper tidal)*	2072	6111	6667	2899	17 749
North Johnstone R. (freshwater)*	2107	6224	6345	2828	17 504
Ninds Creek (estuary)*	2096	6308	6424	3073	17 901
Bulguru freshwater swamp	n.s.	2064	3466	3383	8913
Coconuts (lower estuary)	n.s.	n.s.	n.s.	3321	3321
Town Reach (estuary)	n.s.	n.s.	n.s.	3382	3382
All Locations	6275	20 707	22 902	18 886	68 770

separate sites (Table 1). In the first three years, approximately equal numbers of the two size-classes were released at the primary stocking sites listed in Table 1. In 1995/96, only large fish (50–60 mm TL) were released.

Tag returns

By the middle of March 1996, 85 microtagged barramundi from three age classes had been recaptured. Of these, one was from a commercial catch, six were from the recreational fishery and the remainder were from irregular research sampling. In the research samples, stocked barramundi constituted about 19% of the catch. Catches from release locations (Table 1) were not considered in these analyses to avoid any biases associated with stocked fish remaining resident at these sites.

Since the opening of the 1996 barramundi season (1 February 1996), 62 fish from the recreational fishery and 19 from the commercial catch have been scanned for tags. Of those, seven were found to be tagged and they ranged in size from 590 mm to 615 mm TL. While this number is small, it can be expected to increase as fish from successive stockings are recruited into the fishery. Tagged fish comprised about 15% of all barramundi of the same size class (<650 mm TL) which have been returned by fishers. It is thought that the number of tagged fish which have been caught is actually much higher than those recorded because, for various reasons, some anglers have chosen not to return the heads for scanning.

Recreational fishery catch and effort

From December 1993 to March 1996, 1142 angler record cards had been received detailing the fishing activities of 2267 anglers over 2950 hours. Data were supplied on more than 2800 fish from over 40 freshwater, estuarine and marine species. Most common were, the fresh water sooty grunter (*Hephaestus fuliginosus*, n = 992), mangrove jack (*Lutjanus argentimaculatus*, n = 340) and barramundi (n = 298). The majority of the remaining fish were predominantly marine species.

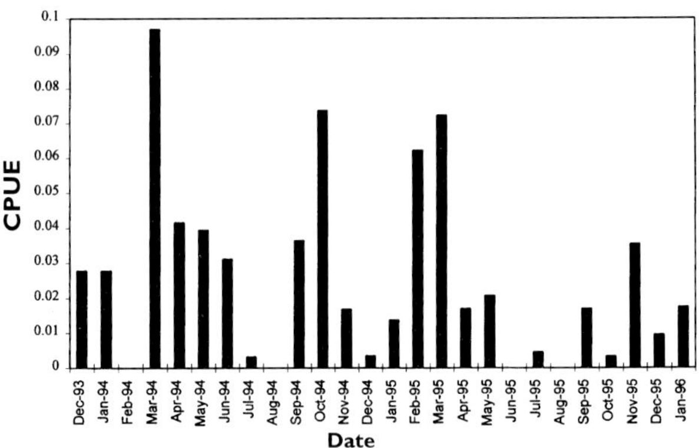

Fig. 1. CPUE (number of fish caught per angler hour) for barramundi from the Johnstone River.

Monthly catch-per-unit-effort (CPUE) is less that 0.1 fish/angler hour and was highly variable within and between years (Fig. 1). Monthly CPUE appears to be lower during the cooler winter months with peaks before the start of the closed season (October) and immediately after the opening of the fishery in February.

Size at stocking

Table 2 shows the number of fish of the two size-classes (large and small fish) released at the primary locations within the catchment and the number of subsequent recaptures. There was no significant difference ($p > 0.10$) in the numbers of large and small size fish that have been recaptured to date with predicted means of 0.16% (se: 0.028) for small fish and 0.19% (se: 0.028) for large fish.

Release locations

Some stocked fish from each of the three primary release locations shown in Table 2 have been recaptured. The initial sampling strategy, which involved extensive electro-fishing at only one of the release locations (North Johnstone R. upper-tidal), has resulted in a high number of recaptures from that site, so it is difficult to come to any conclusions about a preferred stocking location. However, the effects of release site may become clearer as more information becomes available from the recreational and commercial fishery. Some 8913 fish were released into the Bulguru freshwater swamp at the headwaters of

Table 2. Total number of fish of the two size-classes (small 30–40 mm TL and large 50–60 mm TL) released between 1992/93 and 1994/95 and the number of subsequent recaptures to 1 March 1996

Location	Small		Large	
	Released	Recaptured	Released	Recaptured
North Johnstone R. (upper tidal)	6969	25	7881	33
North Johnstone R. (freshwater)	6785	8	7891	6
Ninds Creek (estuary)	7054	1	7774	5

Table 3. Numbers of barramundi which have moved more than 3 km or stayed resident at stocking location. Some movements data are unavailable

Location	No Movement	Movement
North Johnstone R. (upper tidal)	45	16
North Johnstone R.(freshwater)	2	11
Ninds Creek (estuary)	4	4
All sites	51 (62%)	31 (38%)

Ninds Creek between 1993/94 and 1995/96, but none of them has yet been recaptured. The reasons for this are unclear but may be related to periodic declines in the water quality of swamp waters. A low dissolved oxygen saturation (6.6%) and depressed pH (5.3) were measured subsequent to a release in February 1996, and these were thought to have resulted in the newly stocked fish exhibiting distressed behaviour.

Movements

Most (61%) of the fish that were recaptured were within 3 km of their release site (Table 3). The remainder had undertaken intra-riverine movements of up to 37 km. There did not appear to be a discernible pattern to those movements. Fish stocked at the freshwater site in the North Johnstone moved downstream to the upper tidal areas (20 km downstream) and into the South Johnstone River. Fish released in the tidal area of Ninds Creek moved upstream into upper tidal freshwater areas of the North Johnstone and South Johnstone Rivers. No movements to coastal foreshores or into other river systems have yet been recorded.

Growth

The growth of tagged fish recaptured during this study is shown in Figure 2. The first batch of fish released in 1992/93 took about three years to reach the legal size of 580 mm TL. Growth is rapid during the first twelve months, but then slows considerably. There also appears to be a relationship between water temperature in the Johnstone River and growth of barramundi, with slower growth during the cooler mid-year months. Growth rates increase again in the summer with the onset of warmer water temperatures (Fig. 2).

Cost-benefits of stocking

If the 69 000 fish stocked during this current study were all purchased from a commercial hatchery (at 25 cents each) then the total cost would amount to $A17 250. Previous studies (Russell 1988) have suggested that barramundi will be caught in the recreational and commercial fisheries of north-eastern Queensland in a ratio of around 1:3. The value of barramundi to the commercial and recreational sectors is also different. Rutledge *et al.* (1990) estimated that the direct cost incurred by a recreational fisher in north-east Queensland to catch a barramundi was approximately $50. To a commercial fisher, the average 2.5 kg barramundi at $10/kg is worth $25. Using these data, only about 550 of the fish released (<1%) would have to be subsequently caught to recover the purchase price of the fingerlings.

DISCUSSION

Barramundi stocked into the Johnstone River over three consecutive breeding seasons since 1992/93 have become successfully established and began to enter the fishery in early 1996. Estuarine, freshwater and upper tidal habitats all appear to be suitable locations for stocking, although not enough data are yet available to determine if any one site is superior. A number of factors, including water quality, can make some locations unsuitable for stocking and these need to be carefully

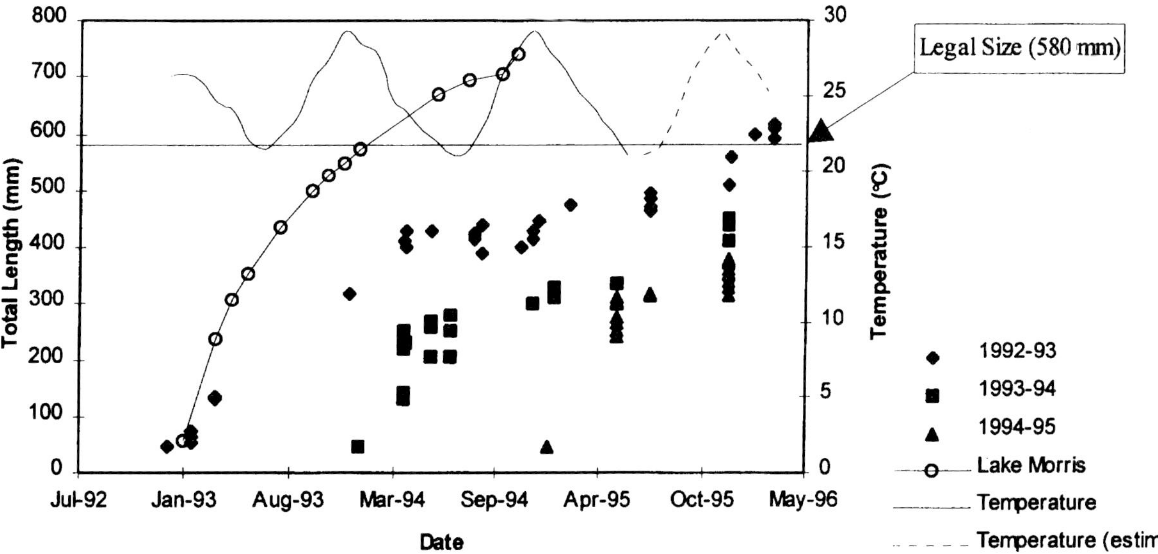

Fig. 2. Size of recaptured fish stocked from 1992/93 to 1994/95. Fish stocked in different years are represented by different markers. Growth of barramundi stocked in Lake Morris is shown for comparison. The solid line shows the running monthly average for water temperatures in the North Johnstone River from January 1993 to June 1995 while the broken line is an estimate of water temperatures from August 1995 to March 1996.

evaluated prior to the release of fish. For example, there is no indication of survival of any of the fish released into the secondary stocking site in the Bulguru freshwater swamp. Water quality measurements found low dissolved oxygen levels after the last stocking in February 1996, which were consistent with the distressed behaviour exhibited by the fish immediately after release. Low dissolved oxygen in streams after periods of heavy rain can result from influx or re-suspension of oxygen-demanding materials as a result of storm water input (Graczyk and Sonzogni 1991). Bishop (1980) recorded natural fish kills in the Northern Territory occurring as a result of oxygen deprivation due to exposure to anoxic bottom waters disturbed by flood rains. While coastal swamps and lagoons are natural nursery habitats for juvenile barramundi (Moore 1982; Russell and Garrett 1983,1985), these areas can be quite volatile. Such habitats should be selected as stocking sites only after a rigorous assessment of their suitability.

Other studies have shown that for some species, size-at-release is an important determinant of later survival. For example, Leber (1995) found that size-at-release of mullet (*Mugil cephalus*) influenced survival and that the critical release size for this fish was 70 mm TL. The results of this study indicate that the size of fish at stocking (30–40 mm TL or 50–60 mm TL) did not affect their survival. However, these are preliminary results and further examination of the survival of different sizes classes needs to be undertaken to determine the optimal (in cost-benefit terms) size-at-release.

Stocked fish in the Johnstone River took approximately three years to reach the minimum legal size of 580 mm TL. This is consistent with the growth rates of natural stocks of barramundi in Papua New Guinea (Reynolds and Moore 1982) and the Gulf of Carpentaria (Davis and Kirkwood 1984). However, barramundi stocked into artificial impoundments have substantially higher growth rates. MacKinnon and Cooper (1987) found growth of barramundi in Lake Tinaroo to exceed that estimated for wild stocks in Papua New Guinea and in the Gulf of Carpentaria. Substantially faster growth rates were also found by A. Hogan (pers. comm.) for barramundi stocked into Lake Morris, a small, tropical impoundment in north Queensland (Fig. 2). Hogan's data show that the fish in Lake Morris reached the minimum legal size in just over a year, compared to with three years for fish stocked into coastal rivers. The reasons for this extraordinary disparity may be related to lack of competition and an abundance of prey in impoundments.

Recaptures of stocked fish have shown that some barramundi do make substantial intra-riverine movements, both upstream and downstream, although most fish (62%) were recaptured within 3 km of their release site. There is no evidence, as yet, of coastal or inter-riverine movements although evidence from earlier Australian studies suggests that some limited movements do occur (Davis 1986; Russell and Garrett 1988).

The Johnstone River supports a substantial multi-species recreational fishery of which barramundi is an important component. There is also a seasonal, multi-species gill-net fishery in the estuary, of which barramundi is also a major component. There also appears to be strong partitioning of the barramundi resource based on spatial and age distribution. The smaller fish, which are more likely to be found in the freshwater or upper-tidal areas, appear more likely to be caught by the recreational fishery (Fig. 3). The larger fish, greater than about 700 mm TL, make up a large proportion of the commercial catch (Fig. 3). This is the result of active gill-net selection of larger fish (Russell 1988) and spawning movements of mature adults from freshwater to commercial fishing grounds in estuarine and coastal waters (Davis 1986).

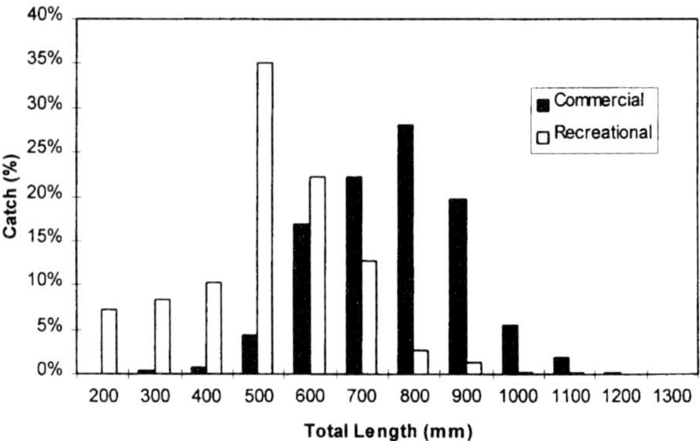

Fig. 3. Size distribution of barramundi caught in the recreational (N = 963) and commercial (N = 590) fisheries in north-eastern Queensland. (Data source: Russell and Hales 1993*b*; Russell 1988).

Catch rates of barramundi in the recreational fishery prior to the recruitment of stocked fish are low (Fig. 1), but comparable to those in similar wet tropic coast streams (Russell *et al.* 1996). The impact that stocked fish will have on these catch rates is yet to be determined, but the high proportions of stocked fish in both research sampling (19%) and commercial and recreational catches (15%) suggest that the contribution may be substantial. In addition, the amount of catch (550 fish) needed to provide nett economic benefit for the local community should be easily achievable.

Addressing the issue of falling catches of some fisheries purely through restrictions on gear and effort may not be sufficient to arrest the decline, and fisheries managers may have to look at other measures. Stock enhancement is one fisheries management tool which, if used judiciously, has the potential to provide significant long-term benefits to a fishery. Before the implementation of stocking programmes, it is advantageous to first determine the best release strategies to ensure maximum survival on a least cost basis and to ascertain the impacts that these releases could have on existing wild stocks. The use of such information in the planning of a stock-enhancement programme, together with the application of the responsible approach principles (Blankenship and Leber 1995), should ensure a cost-effective programme with a high likelihood of success.

Acknowledgments

We wish to thank Mr David Reid who provided valuable advice on the statistical analyses, and Messrs P Hales, G. Vallance and Ms S. Helmke for technical assistance. Mr A. Hogan kindly gave us access to growth data on Lake Morris barramundi.

REFERENCES

Bishop, K A (1980). Fish kills in relation to physical and chemical changes in Magela Creek (East Alligator River system, Northern Territory) at the beginning of the tropical wet season. *Australian Zoologist* **20**, 485–500.

Blankenship, H L, and Leber, K M (1995). A responsible approach to marine stock enhancement. *American Fisheries Society Symposium* **15**, 167–75.

Davis, T L O (1986). Migration patterns of barramundi, *Lates calcarifer* (Bloch), in van Diemen Gulf, Australia, with estimates of fishing mortality in certain areas. *Fisheries Research* **4**, 243–58.

Davis, T L O, and Kirkwood, G P (1984). Age and growth studies on barramundi, *Lates calcarifer* (Bloch) in northern Australia. *Australian Journal of Marine and Freshwater Research* **35**, 673–90.

Dunstan, D J (1959). The barramundi *Lates calcarifer* (Bloch) in Queensland waters. CSIRO Division of Fisheries and Oceanography, Technical Paper Number 5. 22 pp.

Graczyk, D J, and Sonzogni, W C (1991). Reduction of dissolved oxygen concentration in Wisconsin streams during summer runoff. *Journal of Environmental Quality* **20**, 445–51.

Greenwood, P H (1976). A review of the family Centropomidae (Pisces: Perciformes). *Bulletin of the British Museum (Natural History) Zoology* **29**, 4–81.

Leber, K M (1995). Significance of fish size-at-release on enhancement of striped mullet fisheries in Hawaii. *Journal of the World Aquaculture Society* **26**, 143–53.

Leber, K M, Brennan, N P, and Arce, S M (1995). Marine enhancement with striped mullet: are hatchery releases replenishing or displacing wild stocks? *American Fisheries Society Symposium* **15**, 376–87.

MacKinnon, M R, and Cooper, P R (1987). Reservoir stocking of barramundi for enhancement of the recreational fishery. *Australian Fisheries* **46**(7), 34–7.

Moore, R (1982). Spawning and early life history of barramundi, *Lates calcarifer* (Bloch) in Papua New Guinea. *Australian Journal of Marine and Freshwater Research* **33**, 647–61.

Reynolds, L F, and Moore, R (1982). Growth rates of barramundi, *Lates calcarifer* (Bloch) in Papua New Guinea. *Australian Journal of Marine and Freshwater Research* **33**, 663–70.

Russell, D J (1987). Review of juvenile barramundi (*Lates calcarifer*) wildstocks in Australia. In 'Management of Wild and Cultured seabass/barramundi (*Lates calcarifer*)'. (Eds J W Copland and D L Grey) pp. 44–9. Proceedings of an International Workshop, Darwin, Australia. Australian Centre for International Agricultural Research, Proceedings Number 20.

Russell, D J (1988). An assessment of the east Queensland inshore gill net fishery. Report Number QI88024 (Queensland Department of Primary Industries, Brisbane). 57 pp.

Russell, D J, and Garrett, R N (1983). Use by juvenile barramundi, *Lates calcarifer* (Bloch), and other fishes of temporary supralittoral habitats in a tropical estuary in northern Australia. *Australian Journal of Marine and Freshwater Research* **34**, 805–11.

Russell, D J, and Garrett, R N (1985). Eary life history of barramundi, *Lates calcarifer* (Bloch), in north-eastern Queensland. *Australian Journal of Marine and Freshwater Research* **36**, 191–201.

Russell, D J, and Garrett, R N (1988). Movements of juvenile barramundi, *Lates calcarifer* (Bloch), in north-eastern Queensland. *Australian Journal of Marine and Freshwater Research* **39**, 117–23.

Russell, D J, and Hales, P W (1992). Evaluation of techniques for marking juvenile barramundi *Lates calcarifer* (Bloch), for stocking. *Aquaculture and Fisheries Management* **23**, 691–9.

Russell, D J, and Hales, P W (1993*a*). Stream Habitat and Fisheries Resources of the Johnstone River Catchment. Report No.QI93056. QI96008 (Queensland Department of Primary Industries, Brisbane). 59 pp.

Russell, D J, and Hales, P W (1993*b*). A survey of the Princess Charlotte Bay recreational barramundi fishery. Report No. QI93049. (Queensland Department of Primary Industries, Brisbane). 18 pp.

Russell, D J, Hales, P W, and Helmke, S A (1996). Stream habitat and fish resources in the Russell and Mulgrave Rivers Catchment. Report Number QI96008 (Queensland Department of Primary Industries, Brisbane). 52 pp.

Rutledge, W P, and Rimmer, M A (1991). Culture of larval sea bass, *Lates calcarifer* (Bloch), in saltwater rearing ponds in Queensland, Australia. *Asian Fisheries Science* **4**, 345–55.

Rutledge, W, Rimmer, M, Russell, D J, Garrett, R, and Barlow, C (1990). Cost benefit of hatchery-reared barramundi, *Lates calcarifer* (Bloch), in Queensland. *Aquaculture and Fisheries Management* **21**, 443–8.

Demand for Recreational Fishing and Stock-Enhancement Programmes in the Northeast Region of Taiwan

David S. Liao

Institute of Fisheries Economics, National Taiwan Ocean University, 2 Pei-Ning Road, Keelung, Taiwan, ROC.

Summary

This paper analyses the demand for recreational fishing off the north-eastern coast of Taiwan and discusses some implications for stock-enhancement programmes that have been undertaken by fishery agencies to improve sustainability of fishing stocks in the region. Survey data from coastal sport fishermen and charter boat fishermen were used in an empirical demand model. Catch rates, trip costs and fishing experience significantly affected the demand for recreational fishing. Catch rate had a significant effect on the number of fishing trips taken, suggesting that catch rate should be an important determinant of the demand for recreational fishing. Fishermen are responsive to changes in fishing success. Catch rate is therefore critical to good management for recreational fisheries. Catch rates can be improved by the stock-enhancement programmes.

Introduction

Recreational fishing has become a popular outdoor activity in the north-eastern region of Taiwan. Increased participation in recreational fishing has created problems for fisheries management. Assessing the demands of recreational and commercial fishing on fisheries resources is one of the most difficult tasks confronting fisheries managers. Historically, emphasis has been placed on the commercial harvest of fisheries resources. As a result, many of the available data concerning the biological and economic impacts of fisheries utilization pertain to commercial fishing industries and there is very little information relating to recreational fishing. Information about the relationship between recreational fishing demand and stock-enhancement programmes has been limited. Increases in catch due to stock-enhancement programmes could increase economic benefits for recreational anglers. Furthermore, stock-enhancement programmes could improve sustainability of fisheries resources for recreational and commercial fishing. Thus, the purpose of this analysis is to provide some basic economic information on recreational fishing and stock-enhancement programmes.

A regional analysis of recreational fishing involved data from surveys of three groups: charter boat anglers, charter boat operators, and shore fishermen. The survey of charter boat

Table 1. Average fishing trips of charter boats in north-eastern Taiwan, 1993

Item	Taipei County and Keelung City	Ilan County	Total
January	5.30	7.40	5.97
February	5.30	8.50	6.32
March	11.00	10.02	10.69
April	18.80	16.60	18.10
May	18.80	15.80	17.30
June	19.00	16.60	18.24
July	20.00	17.80	19.30
August	18.73	17.50	18.34
September	18.63	16.00	17.79
October	14.42	13.00	14.24
November	7.00	10.50	8.11
December	5.80	8.80	6.73
Total Fishing Trips	162.36	158.52	161.13

operators was conducted by personal interview; 44 out of 90 operators (48.9%) were interviewed. The survey of charter boat anglers was designed to obtain information on anglers' socio-economic characteristics, numbers of fishing trips, catch, etc., and 83 anglers were interviewed. In addition, 58 shore fishermen were surveyed to provide data for conducting the empirical demand model estimation.

DEMAND FOR CHARTER BOAT RECREATIONAL FISHING

Generally, the charter boats in north-eastern Taiwan operate all year round (Table 1). In 1993, the charter boats in Taipei county and Keelung City averaged 162 fishing trips per boat for the year and those in Ilan county had 159 fishing trips per boat. Thus, there was no marked difference between the two geographical areas in the number of fishing trips in 1993. Extrapolation from the data on the average number of fishing trips per boat suggested that the total number of fishing trips by all charter boat fishermen in the north-eastern region was 14 536 (Table 2). The total number of charter boat fishermen by location was also estimated. There were 133 224 charter boat anglers during the study period, of whom 108 083 (81%) fished in the coastal waters of Taipei county and Keelung city and 25 141 in the offshore waters of Ilan county.

The demand model for charter boat fishing uses the standard assumption that the cost of accessing the fishing ground works like a price in its effect on recreational fishing demand. The underlying theory and several practical concerns in estimating travel cost demand models have been thoroughly reviewed (McConnell 1975; Gum and Martin 1977; McConnell and Strand 1981; Bockstael *et al.* 1987). The travel cost model has been extended to include fishing quality as well as demand determinants such as price, income and socio-economic variables. The semilog function for the model was chosen on the basis of past empirical experience (Ziemer *et al.* 1980; Vaughan and Russell 1982; Strong 1983; Smith *et al.* 1983; Smith 1988; Liao 1992), and because it explains fishing quality in a reasonable way. The semilog function ensures that a negative price coefficent and positive quality coefficient infer increasing welfare with increasing quality. Thus, the travel cost demand model for charter boat fishing is specified as follows:

$TRIPS = Exp(B_0+B_1\ EXP+B_2\ CAT+B_3\ INC+B_4\ TTC+B_5\ OTC+B_6\ CAR+B_7\ OC_1+B_8\ OC_2+B_9\ OC_3+B_{10}\ OC_4+B_{11}\ EDU)$, where TRIPS = number of charter boat fishing trips taken by anglers; EXP = number of years of charter boat fishing experience; CAT = fish catch per trip; INC = estimated annual income; TTC = total trip costs including travel costs, charter fees and time costs; OTC = total costs for other types of fishing trips; CAR = car ownership (= 1 if car owner, = 0 otherwise); OC_1 = if employed in the manufacturing sector, = 0 otherwise; OC_2 = 1 if employed in the construction sector, = 0 otherwise; OC_3 = 1 if employed in the scale sector, = 0 otherwise; OC_4 = 1 if employed in the other sector, = 0 otherwise; EDU = educational level (=1 none, =2 primary school, =3 secondary school =4 high school, =5 college or higher); B_0, B_1, B_2, B_3, B_4, B_5, B_6, B_7, B_8 B_9, B_{10}, B_{11} are regression parameters to be estimated.

On the basis of the survey data, the above model was estimated by using ordinary least-squares method. The results are shown in Table 3. Years of charter boat fishing experience by the anglers have a significant effect on the number of trips taken. Catch per trip is statistically significant; this is consistent with economic theory and previous studies and suggests that catch rate should be an important determinant of the demand for charter recreational fishing, i.e. that fishermen are responsive to changes in fishing success. The total trip cost variable accounts for both trip costs and time costs. Time cost is a measure of the opportunity cost of time in terms of lost wages. In the estimated equation, the expected relationship between the number of trips and total trip costs is seen. The estimated coefficient of trip cost is significant and negative. An increase in trip cost results in a decrease in the number of trips. The estimates of the car ownership variable suggest that car ownership by charter boat fishermen does lead to an increase in the number trips.

Table 2. Estimated total number of anglers and fishing trips by charter boats in north-eastern Taiwan, 1993

	No. trips per boat	×	No. charter boats	=	Total No. fishing trips	×	No. anglers per trip	=	Total No. anglers
Taipei County & Keelung City	162.36		70		11 365.20		9.51		108 083.05
Ilan County	158.52		20		3170.40		7.93		25 141.27
Total					14 535.60				133 224.32

Table 3. Estimated demand function for charter boat recreational fishing in north-eastern Taiwan, 1993

$r^2 = 0.5799$

Variable	Coefficients	t-value	P
Constant	1.9528100	2.143	0.0356
EXP	0.1165340	3.964	0.0002
CAT	0.0170780	2.806	0.0065
INC	0.0000001	1.166	0.2474
TTC	–0.0002910	–4.048	0.0001
OTC	0.0003260	1.201	0.2336
CAR	1.4558480	2.734	0.0079
OC_1	–0.8441550	–0.687	0.4941
OC_2	–0.5890600	–0.819	0.4154
OC_3	–0.0585200	–0.079	0.9369
OC_4	–0.4602950	–0.687	0.4946
EDU	–0.1814330	–1.503	0.1373

Table 4. Evaluation by charter boat anglers of the catch per trip, 1993 (1, very unsatisfactory; 2, unsatisfactory; 3, average; 4, satisfactory; 5, very satisfactory)

	1	2	3	4	5
Taipei country & Keelung city	2%	11%	32%	43%	13%
Ilan country	0%	19%	22%	47%	12%
Combined	1%	14%	28%	46%	12%

The catch rate of 27.34 kg per trip for Ilan county fishermen was higher than that reported from Taipei county and Keelung city (24.65 kg). Fishermen were asked to evaluate their average catch per trip. Region-wide, 12% of the respondents indicated that it was very satisfactory and 46% that it was satisfactory (Table 4).

Respondents were asked to identify the principal target species of their fishing trips in 1993. Charter boat operators fished all year round for yellow porgy, hairtail, spotted mackerel, red bullseye, small scale blackfish, marbled rockfish, grouper, porgy, and red horsehead. Operators fished primarily for yellowfin sea bream from June to February and for sea perches, snapper, and formosan squid during April to July. Striped bonito was targeted primarily during February to August.

Demand for shore recreational fishing

Shore recreational fishing has become a popular outdoor activity in the north-eastern region. A survey of 58 shore anglers indicated that the mean one-way distance fishermen travelled from home to fishing sites was 61.1 km. Fishing party size averaged 3.2 persons. The length of fishing time per trip by shore anglers averaged 6.1 h. The average recreational shore fisherman had made approximately 13.3 trips in the past 6 months. This is equivalent to 2.2 trips per month per anglers.

The survey data from shore fishermen were also used to estimate demand function for shore recreational fishing. Trip demand for shore recreational fishermen is specified as a semilog arrhythmic equation:

$TRIPS = Exp(B_0+B_1\ EXP+B_2\ CAT+B_3\ INC+B_4\ TTC+B_5\ OTC+B_6\ FEC+B_7\ FPS+B_8\ OC_1+B_9\ OC_2+B_{10}\ OC_3+B_{11}\ OC_4+B_{12}\ EDU+B_{13}\ AGE)$, where TRIPS = number of fishing trips taken by anglers; EXP = the number of years of shore fishing experience; CAT = fish catch per trip; INC = the estimated annual income; TTC = total trip costs; OTC = total costs for other types of fishing trips; FEC = Fixed equipment costs; FPS = Fishing party size; OC_1 = 1, if employed in the manufacturing sector; =0 otherwise; OC_2 = 1, if employed in the scale sector; =0 otherwise; OC_3 = 1, if employed in the government sector; =0 otherwise; OC_4 = 1, if employed in the other sector; =0 otherwise; EDU = 1, high school or college education; =0 otherwise; AGE = age of fishermen; and B_0, B_1, B_2, B_3, B_4, B_5, B_6, B_7, B_8, B_9, B_{10}, B_{11}, B_{12}, B_{13}, are regression parameters and to be estimated.

Coefficient estimates for the shore demand function are reported in Table 5. Considering the trip cost variable, the coefficient was negative and statistically significant in the demand function. This relationship was expected, since trips should decrease as the costs of access increase. The same relationship was found in the demand for charter boat recreational fishing analysis. Again, family income failed to exert any significant influence on the dependent variable. As for the charter boat fishing, years of shore fishing experience exerted a significant influence. The catch rate variable was positive in sign and significant, and it exerted its positive impact upon additional fishing trips.

Catch by anglers is an important determinant of trip frequency. However, about 54% of shore fishermen in the north-eastern region classed their catch per trip as unsatisfactory and 24% as very unsatisfactory, with only 17% classing it as 'average', 5% as 'satisfactory' and some as 'very satisfactory'. This suggests that stock-enhancement programmes are important for development and management of recreational fisheries in the region.

Table 5. Estimated demand function for shore recreational fishing in north-eastern Taiwan, 1994

$r^2 = 0.5024$

Variable	Coefficients	t-value	P
Constant	2.2609860	1.976	0.0545
EXP	0.0286570	2.552	0.0143
CAT	0.0492980	4.162	0.0001
INC	0.0000002	0.729	0.4696
TTC	–0.0003150	–2.456	0.0181
OTC	0.0000277	1.213	0.2317
FEC	0.0000025	0.444	0.6595
FPS	0.0178240	0.438	0.6632
OC_1	–0.1588990	–0.184	0.8550
OC_2	–0.0348770	–0.041	0.9679
OC_3	0.0883540	0.097	0.9228
OC_4	0.1078390	0.123	0.9026
EDU	0.2376070	1.105	0.2753
AGE	–0.0184320	–1.505	0.1395

Table 6. Fingerlings or seeds released in north-eastern Taiwan, 1988–1992

Year	Species	Taipei Country	Keelung City	Ilan Country	Total
1988	Black sea bream		16 750		16 750
	Thornfish			10 000	10 000
	Small abalone	72 000	72 000	74 300	218 300
1989	Black sea bream		27 817		27 817
	Red sea bream		5313		5313
	Small abalone	75 000		80 000	155 000
1990	Goldlined sea bream			89 010	89 010
	Thornfish	20 000			20 000
	Small abalone	100 000			100 000
1991	Black sea bream		62 500		62 500
	Goldlined sea bream			113 637	113 637
	Small abalone	100 000			100 000
1992	Goldlined Sea bream	50 000	75 000	65 000	190 000
	Small abalone	100 000			100 000

Source: Taiwan Fisheries Bureau

STOCK-ENHANCEMENT PROGRAMMES IN THE NORTH-EASTERN REGION

In addition to building artificial reefs and establishing resource protection zones to create fishing grounds, the present stock-enhancement programme aims at stocking fry or seeds. The prefecture governments and Taiwan Fisheries Research Institute (TFRI) have been releasing fingerlings and seeds in the north-eastern waters (Table 6).

The stock-enhancement programme has been evaluated by the Taiwan Fisheries Research Institute (Wu and Kuo 1994). Tag–recapture data from northern Taiwan during 1989–93 (Table 7) shows considerable variation in recapture rates among different areas. Survival rate was impressive. Thus, the catch rate of recreational fishing can be improved by the contribution of hatchery-released fish to the ocean. The cost of production for fingerlings released is usually low. At present, the amount of finfish fingerlings released in the region is too limited to have a significant impact on stocks. To accelerate stock-enhancement programmes, more attention should be paid to enhance fishery regulations, prevent water pollution, augment basic studies on fingerlings released, and set up sea farming centres (Liao 1996). In Taiwan and many other countries, the objective of stock-enhancement programmes has been to rebuild fishing stocks for commercial fisheries. There is a need to establish cooperation between biologists and economists for ultimate success of the stock-enhancement programmes. The biologists must develop the applicable aquaculture technologies for improving commercial

Table 7. Records of tagged black sea bream in the north-eastern waters of Taiwan during 1989–93 projects by Taiwan Fisheries Research Institute (TFRI)

Release data	Hatchery	Method of growing	No. released (x 1000)	Size of fish (mm)	Average size (mm)	Tag types	Release locality	Recapture rate (%)
Sep. 30, 1989	TFRI Tainan Branch	Cage	9.4	56–148	97	Anchor tag	Ta-Wu-Lun	0.20
Aug. 18, 1990	TFRI Tainan Branch	Pond	9.9	55–163	89	Anchor tag	Ho-Pin-Tao	1.42
Nov. 9, 1991	TFRI Tainan Branch	Pond	4.4	67–230	118	Anchor tag	Wai-Mo-Shan	0.11
Sep. 26, 1992	Breeder (Fisherman)	Pond	11.6	53–229	87	Anchor tag	Chang-Tan-Li (outer)	0.12
Sep. 9, 1993	Breeder (Fisherman)	Pond	11.6	55–157	92	Anchor tag	Chang-Tan-Li (inner)	0.70
			6.2	44–110	71	Ventral fin clipped		0.70

Source: Wu and Kuo (1994)

and recreational fishery resources, and economists should assess the benefits of such technologies and enhancement activities.

References

Bockstael, N E, Strand, I E, and Hanemann, W M (1987). Time and the recreational demand model. *American Journal of Agricultural Economics* **69**, 293–302.

Gum, R L, and Martin, W E (1977). Structure of demand for outdoor recreation. *Land Economics* **53**, 43–55.

Liao, D S (1992). Economic Analysis of the 1991 South Carolina Shrimp Baiting Fishery. *South Carolina Marine Resources Centre Technical Report* No. 81.

Liao, I C (1996). Status, Problems, and Prospects of stock enhancement in Taiwan. Prepared for the 'Asia-Pacific Conference on Science and Management of Coastal Environment', 25–28 June 1996, Hong Kong.

McConnell, K E (1975). Some Problems in Estimating the Demand for Outdoor Recreation. *American Journal of Agricultural Economics* **57**, 330–4.

McConnell, K E, and Strand, I E (1981). Measuring the cost of time in recreational demand analysis: An Application to Sport Fishing. *American Journal of Agricultural Economics* **63**, 153–6.

Smith, V K (1988). Selection and recreation demand. *American Journal of Agricultural Economics* **70**, 29–36.

Smith, V K, Desvousges, W H, and McGivney, M P (1983). The opportunity cost of travel time in recreation deman models. *Land Economics* **59**, 259–77.

Strong, E J (1983). A note on the functional form of travel cost models with zones of unequal populations. *Land Economics* **59**, 64–72.

Vaughan, W J, and Russell, C R (1982). Valuing a fishing day. An application of a symmetric varying parameter model. *Land Economics* **58**, 450–63.

Wu, C C, and Kuo, C L (1994). Movement and recapture of black porgy (*Acanthopagrus schlegeli*) released in northern waters off Taiwan. *Journal Taiwan Fisheries Research Institute* **2**, 1–13.

Ziemer, R F, Musser, W N, and Hill, R C (1980). Recreational demand equations: Functional form and consumer surplus. *American Journal of Agriculture Economics* **62**, 136–41.

Can Aquaculture Help Restore and Sustain Production of Giant Clams?

J. D. Bell, A. M. Hart, T. P. Foyle, M. Gervis and I. Lane

International Centre for Living Aquatic Resources Management (ICLARM), Coastal Aquaculture Centre, PO Box 438, Honiara, Solomon Islands.

Summary

Giant clams (Tridacnidae) have been depleted in many developing countries in the Indo-Pacific, mainly to supply the market for adductor muscle. Aquaculture has not contributed to restoring production for this market because economic returns appear to be marginal and farmers have been unwilling to grow giant clams for the 7–10 years needed to reach an acceptable size. Aquaculture has, however, developed two alternative markets for cultured giant clams. Specimens of 50–100 mm shell length (SL) are popular in the tropical marine aquarium industry, and clams of 120–150 mm SL are suitable as sashimi (live seafood). In the Solomon Islands, simple farming methods have been developed for small-scale village farmers to supply giant clams to these markets. Several species of giant clams have been reared by villagers for the aquarium market, after 8–10 months grow-out, at relatively good profits. One species, *Tridacna derasa*, has proved particularly suitable for the live seafood trade and can be grown to 150 mm SL in 18–24 months with a survival of 85%. Establishment of village farms has paved the way for cost-effective restoration of giant clams in the wild. Village farmers maintain a proportion of their 'seed' clams in protective cages until large enough to escape predation, then they are placed on reefs. The success of such restocking programmes will depend on understanding how to distribute the clams in ways that maximize their survival and reproductive success.

Introduction

Wild stocks of several species of giant clams (Tridacnidae) have been depleted, sometimes to the point of extinction, in recent decades throughout much of the Indo-Pacific (Munro 1993*a*; Lucas 1994). In the case of the larger species (*Tridacna gigas* and *T. derasa*), the decline was due mainly to heavy fishing pressure for the lucrative market in Taiwan for adductor muscle (Dawson 1985). In several of the island nations of the Pacific, however, stocks of several species have also decreased through continued local harvesting by the expanding populations (Munro 1993*a*). As a result of these declines, giant clams were listed under the Convention on International Trade in Endangered Species (CITES) (Wells *et al.* 1983). Good natural stocks of the larger species now exist only on the Great Barrier Reef (GBR) of Australia, and in parts of Micronesia (Lucas 1994).

In this paper, we summarize various efforts to restore and sustain the production of giant clams through aquaculture. We focus mainly on the recent progress in the establishment of small-scale farming systems intended to benefit coastal villagers in developing countries. We also outline how the farming of giant clams by villagers has paved the way for cost-effective enhancement of wild stocks.

INITIAL ATTEMPTS TO FARM *TRIDACNA GIGAS*

After the life cycle of giant clams was closed in the mid 1970s by La Barbera (1975) and Jameson (1976), the apparent economic advantages of supplying the market for adductor muscle led to a flurry of research projects in the Asia-Pacific region. This research was done initially by the University of Papua New Guinea and the Micronesian Mariculture Demonstration Centre (MMDC) in Palau in 1976 (see Munro 1993*a*, 1993*b*), and later through the Australian Centre for International Agricultural Research (ACIAR) and the International Centre for Living Aquatic Resources Management (ICLARM). As a result of this concerted effort, reliable methods now exist for the mass production of juveniles (Braley 1992; Calumpong 1992; Heslinga *et al.* 1990; Crawford *et al.* 1986). Much has also been published about the predators of giant clams (e.g. Govan 1994, 1995), and their anatomy and histology (Norton and Jones 1992).

Despite this knowledge, production of large giant clams for the adductor muscle trade has not been implemented. So far, it has not proved to be practical; economically or socially. Using data from the GBR, Tisdell *et al.* (1993) estimated that the optimal grow-out period needed to maximize the net present value, or capitalized value of a farm, was 10 years. In some habitats, e.g. intertidal reefs on the GBR, good survival during such a long grow-out period is technically feasible (Lucas 1994), but this is not the general situation in the Pacific. In the Solomon Islands, for example, Hambrey and Gervis (1993) showed that survival of *T. gigas* was likely to vary greatly over the grow-out period needed for the adductor muscle to reach a suitable size, making economic returns marginal, and consequently coastal villagers are not prepared to provide the necessary husbandry for the estimated minimum grow-out period of seven years.

Although the farming of giant clams to supply the market for adductor muscle has not proved viable, two other forms of farming; i.e. rearing smaller giant clams for the aquarium trade and sashimi markets, promise to provide an attractive source of income for coastal villagers.

DEVELOPMENT OF THE AQUARIUM MARKET

Marketing of giant clams for the aquarium trade began in 1987 with *Tridacna derasa* (Heslinga *et al.* 1988). Since then the market for cultured marine organisms in the tropical marine aquarium trade has been expanding steadily (J. Walch pers. comm.). This is due in part to increased awareness of the role that aquaculture can play in the conservation of coral reef organisms. Cultured giant clams have proved especially popular because they have attractive mantles, do not need to be fed due to their symbiotic relationship with zooxanthellae, and because they assist with the assimilation of ammonium waste products. The greatest demand is for individuals of 50–100 mm shell length (SL).

To test whether grow-out for the aquarium market would be viable for coastal villagers, we established a series of 14 experimental farms spread across 500 km of the Solomon Islands. The objectives of our project were to: (1) develop appropriate farming techniques for production of small clams by coastal villagers, (2) quantify rates of growth and survival during grow-out to market size and

Table 1. Mean survival and growth of *Tridacna derasa*, *T. gigas*, and *T. squamosa* after 8 months grow-out at village farms in the Solomon Islands

	T. derasa	*T. gigas**	*T. squamosa*
No. of experimental farms	14	30	10
Mean growth ±SD (mm month^{-1})	7.9 ± 1.1	3.7 ± 1.3	5.3 ± 1.0
Mean survival ±SD (%)	93.0 ± 11.0	54.0 ± 18.6	66.6 ± 15.5

* From earlier village trials, usually without replication of cages at a site, in which grow-out time was 9.5 months.

(3) identify environmental factors affecting growth and survival to aid in the selection of optimum grow-out sites.

Replicated grow-out trials are in progress at these experimental village farms for several species of giant clams: *Tridacna gigas, T. derasa, T. crocea, T. maxima, T. squamosa* and *Hippopus hippopus.* To date, several strong trends have emerged. First, juvenile clams stocked at a size of 20–25 mm SL can be grown-out effectively in a simple production unit of 4 × 0.36 m^2 galvanized wire mesh (19 mm) cages placed on a trestle in the shallow subtidal zone (Fig. 1). For the first two months of grow-out, a 'settlement ring' of 5 mm mesh is inserted in the cage to prevent escapement (Fig. 1). Second, most species can be reared to market size by villagers within 8–10 months (Table 1). So far, the fastest growing and most robust species has proved to be *T. derasa,* followed by *T. squamosa* and *T. gigas.* Third, water flow, temperature and the number of predators (mainly ranellid gastropods) in the experimental cages can affect growth rates significantly (Foyle *et al.* in press; Bell *et al.* in press).

To assess the economic viability of farming giant clams for the aquarium trade at an appropriate commercial scale, we established a total of 26 farms in late 1994. In 1995, ~25 000 juvenile giant clams were sold to aquarium markets in the USA on behalf of the village farmers. The clams had a total value of $US76 000 which represented an average net profit of $US1300 per annum per farmer. This is equivalent to the annual wage of a full-time skilled labourer in the Solomon Islands, yet required only one full day's work per week. This aspect of the small-scale farming system is particularly attractive to villagers as it allows

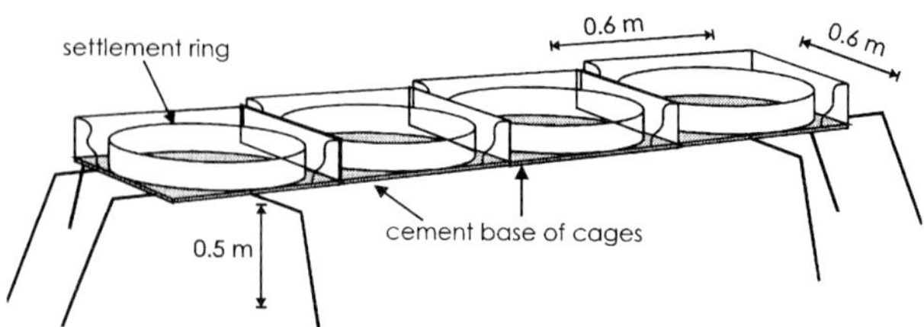

Fig. 1. Basic production unit used by village clam farmers in the Solomon Islands. Each cage has sides of 19 mm mesh, the four cages are placed on a trestle made of 12 mm steel rods. Settlement rings are used for the first two months of grow-out (see text).

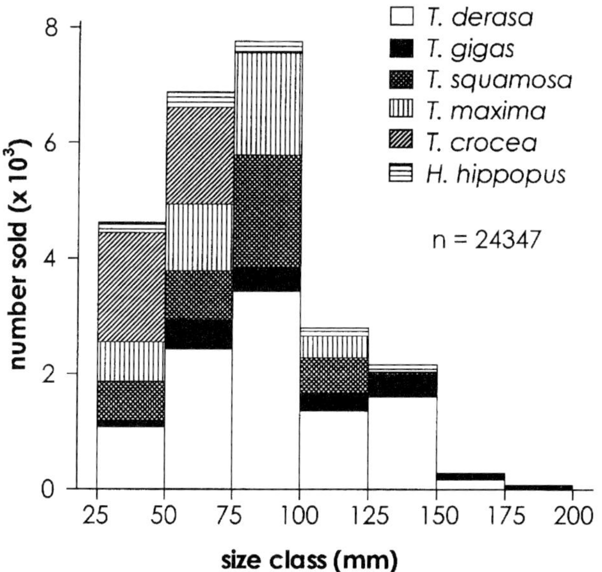

Fig. 2. Size-frequency distribution of giant clams sold to the aquarium trade in 1995 by village farmers in the Solomon Islands.

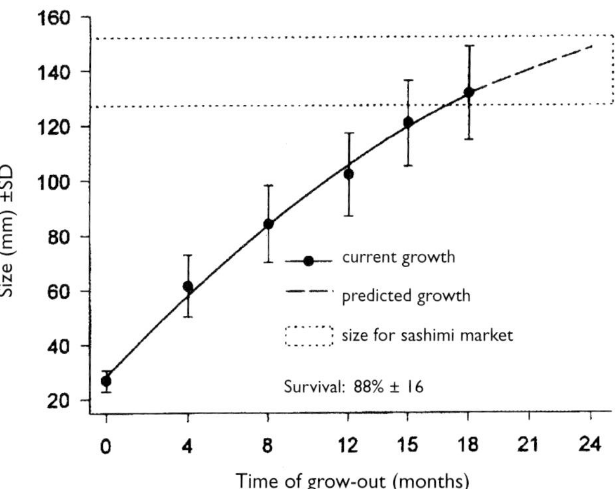

Fig. 3. Mean size of *Tridacna derasa* during 18 months grow-out at 14 village farms in the Solomon Islands.

them time for their traditional activities, e.g. fishing, gardening and producing cocoa and copra.

The initial sales of clams to the aquarium market indicated that the greatest demand was for individuals in the size ranges of 75–100 mm SL (Fig. 2). *Tridacna derasa* accounted for 41% of sales, *T. squamosa, T. maxima* and *T. crocea* accounted for 17%, 16% and 14% respectively, while *T. gigas* and *H. hippopus* together accounted for <13% of total sales (Fig. 2). However, these results reflect the composition of stocks held by village farmers in the Solomon Islands. Recent information from distributors of tropical marine aquarium animals is that the greatest demand and highest prices are for *T. crocea* and *T. maxima* with iridescent blue or green mantles. Hatcheries which can produce such seed will obviously have a competitive edge in supplying the aquarium trade.

Although there is undoubtedly some room for expansion, the aquarium market for giant clams is unlikely to be large enough to support substantial numbers of village farmers from a variety of developing countries. So far, the major benefit of farming giant clams for the aquarium trade has been to lay the foundation for the expansion of small-scale aquaculture operations. In the Solomon Islands, village farmers have now been trained in husbandry of clams, maintenance of accounts and inventories and in transporting their clams to distributors. These skills have prepared the farmers for the development of a larger industry to supply giant clams to live seafood markets in southeast Asia.

POTENTIAL OF *TRIDACNA DERASA* FOR SUPPLYING THE LIVE SEAFOOD MARKET

The increased demand for live seafood in southeast Asia, and the traditional appeal of giant clams to consumers in Taiwan and southern Japan, indicates that there may be good potential for supplying giant clams of 120–150 mm SL to live seafood markets on a relatively large scale. *Tridacna derasa* appears to be the most suitable species for such markets. There are two main reasons for this. First, taste tests have demonstrated that *T. derasa* is well accepted as sashimi, and that it has higher ratings than *T. gigas* and *H. hippopus* (Peavy and Riley 1993*a, b*). Second, predictions based on mean growth and survival estimates from village farms in the Solomon Islands indicate that *T. derasa* has a faster rate of growth than those of other species (Table 1). Our data indicate that it can be grown out to the size required for the sashimi market in ≤2 years (Fig. 3), with 85% survival. At the best sites, *T. derasa* attains 150 mm SL in 18 months, with 98% survival

We are now testing village farming of *T. derasa* for the sashimi market at a pilot commercial scale. Using site selection criteria developed from earlier trials, an additional 26 village farms have been set up and supplied with 3000 'seed' clams of 20–25 mm SL. Production of clams of 150 mm SL from these farms is projected to average >4 tonnes per month between September 1997 and September 1998. This production will provide the opportunity to make a realistic economic evaluation of farming *T. derasa* as live seafood. To facilitate the marketing of live *T. derasa* of 150 mm SL, we will be developing and evaluating (a) domestic and international transport systems for distributing large quantities of live clams, (b) systems for holding clams in transit and at market destinations (these should be no more complicated than those already used for the marketing of live coral reef fish), and (c) the demand for smaller live clams by the traditional consumers of giant clam adductor muscle.

A MODEL FOR GIANT CLAM FARMING IN DEVELOPING COUNTRIES

Gervis *et al.* (1995) have developed a model for village farming of giant clams based on the holistic approach advocated by Naegel (1995). It has the following components: (a) a centralized, commercially operated hatchery which sells seed to village farmers at a profit, (b) small-scale village farmers who grow-out the seed to market size, (c) a distributor who purchases the seed from the village growers and sells it to an exporter, and (d) an exporter who supplies the overseas markets. In most cases, the production of seed and collection of clams from the village growers for delivery to the exporter would be undertaken by the one enterprise. Under the model proposed by Gervis *et al.* (1995), village farmers do not have to meet the relatively expensive costs of establishing and operating a hatchery, and the hatchery owners do not have to secure their own access to numerous good sites for growing-out the clams. The model also has the advantage of distributing benefits to several small-scale operators in a variety of coastal villagers.

Restocking of giant clams to restore wild populations

Development of village-based giant clam farming is only one way of increasing the production of giant clams on a sustainable basis. Restoration of stocks of wild giant clams will provide the potential for increased harvests from subsistence, recreational, and commercial fisheries. The restoration of depleted stocks was also a major reason for the development of aquaculture methods for giant clams (Copland and Lucas 1988).

Although the production of giant clam seed is now routine, restocking these animals has posed problems not usually encountered for most other marine fisheries species; individuals are virtually sessile and remain vulnerable to predation by carnivorous gastropods and fish for at least three years (Calumpong and Solis-Duran 1993). The only way to overcome this problem is to 'farm' the juveniles until they are large enough to escape predation. One way of defraying the high cost of doing this is to combine restocking programmes with the establishment of small-scale farms. We have initiated this scheme in the Solomon Islands. During the harvest of clams for market, each village farmer is required, under a policy agreement, to set aside 20 individuals at random from each cohort for restocking.

We will be recording the growth and survival of clams dedicated to restocking during each of three phases: (a) the first 9–18 months of grow-out for market in cages at village farms — these data are already being collected from the experimental farms, (b) a second period of grow-out in cages, at reduced density, for a further 18–27 months until the clams are robust enough to be 'released' onto reefs, and (c) the period following placement onto coral reefs until the clams reach sexual maturity. During the last phase, individuals will be tagged and placed into microhabitats within the reef selected to maximize growth and minimize predation. This strategy will vary among species; for example, *T. crocea* needs to be placed in cavities in dead *Porites* colonies (Isa 1995), whereas *T. gigas* performs best in relatively open spaces.

Apart from the obvious questions about whether the rates of survival and growth of 'released' clams are cost-effective, important considerations for restocking of giant clams are: a) what stocking density maximizes fertilization success? b) which locations should be stocked to maximise successful settlement of juveniles? and (c) what is the contribution of propagules of restocked clams to the total population? As the female generation time of giant clams is as long as nine years for some species (Lucas 1994), these questions can only be answered by substantial, long-term restocking programmes.

Although restocking is an attractive means of re-establishing fisheries for giant clams, there are risks involved. Release of hatchery-reared juveniles has the potential to alter the gene frequencies of wild stock, and introduce diseases to conspecifics and other species, unless it is done in a responsible way (Blankenship and Leber 1995; Munro and Bell in press). Indeed, Benzie (1993) has shown that there are major differences in gene frequencies among populations of giant clams within their range, and among wild and cultured stocks from the same location. The genetic diversity of giant clams can be maintained by avoiding the introduction of conspecifics to areas with sufficient wild broodstock, and ensuring that seed for aquaculture is derived from multiple cohorts (see review by Munro and Bell in press for more details). In cases where it is necessary to introduce individuals to re-establish populations, giant clams should be produced from stocks with similar genetic structure to the relict population (see Benzie 1993), and crossbred with any remaining wild stock. Munro (1993*b*) describes protocols to minimise the risk of introducing pathogens to the wild during re-establishment of giant clams.

In addition to the work in Solomon Islands, re-establishment of giant clams is underway in the Philippines (Calumpong and Solis-Duran 1993; Mingoa-Licuanan 1993), Fiji (E. Ledua pers comm.), Micronesia (S. Lindsay pers comm.), and Tonga (K. Udagawa pers. comm.). Continued funding and sound monitoring programmes are needed to evaluate the success of these initiatives.

Conclusions

Despite initial setbacks to farming of giant clams to supply the market for adductor muscle, viable small-scale ventures producing giant clams for the aquarium trade are emerging in Pacific island nations. These village farms are poised to produce larger quantities of clams for the live seafood trade in southeast Asia. The development of village farms has also provided a means of re-stocking giant clams in the wild, and demonstrated that farming and restocking will need to be complementary activities: re-establishment of giant clams is unlikely to be cost-effective in the absence of farming, and aquaculture will be impeded without adequate broodstock in the wild. We conclude that methods for farming and restocking giant clams are now developed to the point where they hold much promise as tools for restoring and sustaining production of giant clams.

Acknowledgments

Much of the research on giant clams in Solomon Islands summarized in this paper was supported by the Australian Centre for International Agricultural Research (ACIAR), and the European Union's STABEX program. We thank Jay Maclean and John Munro for their comments on the draft manuscript.

References

Bell, J D, Lane, I, Gervis, M, Soule S, and Tafea, H (in press). Village-based farming of the giant clam, *Tridacna gigas* (L), for the aquarium market: initial trials in the Solomon Islands. *Aquaculture Research*..

Benzie, J A H (1993). Review of the population genetics of giant clams. In 'Genetic aspects of conservation and cultivation of giant clams'. (Ed P Munro) pp. 1–6. *ICLARM Conference Proceedings* No. **39**, International Center for Living Aquatic Resources Management, Manila, Philippines.

Blankenship, H L, and Leber, K M (1995). A responsible approach to marine stock enhancement. *American Fisheries Society Symposium* **15**, 165–75.

Braley, R D (Ed) (1992). 'The giant clam: hatchery and nursery culture manual.' *ACIAR Monograph* **15**, Australian Centre for International Agricultural Research, Canberra. 144 pp.

Calumpong, H P (Ed) (1992). 'The giant clam: an ocean culture manual.' *ACIAR Monograph* **16**, Australian Centre for International Agricultural Research, Canberra. 68 pp.

Calumpong, H P, and Solis-Duran, E (1993). Constraints in restocking Philippine reefs with giant clams. In 'Biology and Mariculture of giant clams'. (Ed W K Fitt) pp. 94–8. *ACIAR Proceedings* No. **47**. Australian Centre for International Agricultural Research, Canberra.

Crawford, C M, Nash, W J, and Lucas, J S (1986). Spawning induction, and larval and juvenile rearing of the giant clam, *Tridacna gigas*. *Aquaculture* **58**, 281–95.

Copland, J W, and Lucas, J S (Eds) (1988). 'Giant clams in Asia and the Pacific.' *ACIAR Monograph* No. **9**, Australian Centre for International Agricultural Research, Canberra. 274 pp.

Dawson, B (1985). Taiwanese clam boat fishing in Australian waters. *Australia-Asia Papers* No. 33, School of Modern Asian Studies, Griffith University, Brisbane, Australia.

Foyle, T P, Bell, J D, Gervis, M, and Lane, I (in press). Survival and growth of juvenile fluted giant clams, *Tridacna squamosa*, in large-scale village grow-out trials in the Solomon Islands. *Aquaculture*.

Gervis, M, Bell, J D, Foyle, T P, Lane, I, and Oengpepa, C (1995). Giant clam farming in the South Pacific, past experience and future prospects: An ICLARM perspective. Paper presented to the workshop on present status and potential of aquaculture research and development in Pacific Island countries, Nuku'alofa, Tonga, November 1995. 14 pp.

Govan, H (1994). Predators of maricultured tridacnid clams. *Proceedings of the 7th International Coral Reef Symposium* **2**, 749–53.

Govan, H (1995). *Cymatium muricinum* and other ranellid gastropods: major predators of cultured tridacnid clams. ICLARM Technical Reports No. **49**, International Center for Living Aquatic Resources Management, Manila, Philippines. 136 pp.

Hambrey, J, and Gervis, M (1993). The economic potential of village based farming of giant clams (*Tridacna gigas*) in Solomon Islands. In 'Biology and Mariculture of giant clams'. (Ed W K Fitt) pp. 138–46. *ACIAR Proceedings* No. **47**. Australian Centre for International Agricultural Research, Canberra.

Heslinga, G A, Watson, T C, and Isamu, T (1988). Giant clam research and development in Palau. In 'Giant clams in Asia and the Pacific'. ACIAR Monograph No. 9, pp. 49–50. Australian Centre for International Agricultural Research, Canberra.

Heslinga, G A, Watson, T C, and Isamu, T (1990). 'Giant clam farming.' Pacific Fisheries Development Foundation (NMFS/NOAA), Honolulu, Hawaii. 179 pp.

Isa, J (1995). Fisheries resource management development by stock enhancement in Okinawa. Paper presented to the Joint FFA/SPC Workshop on the Management of South Pacific Inshore Fisheries, Noumea, New Caledonia, 26 June–7 July 1995. 6 pp.

Jameson, S C (1976). Early life history of the giant clams *Tridacna crocea*, *T. maxima*, and *Hippopus hippopus*. *Pacific Science* **30**, 219–33.

La Barbera, M (1975). Larval and post-larval development of the giant clams, *Tridacna maxima* and *Tridacna squamosa* (Bivalvia: Tridacnidae). *Malacologia* **15**, 69–79.

Lucas, J S (1994). The biology, exploitation, and mariculture of giant clams (Tridacnidae). *Reviews in Fisheries Science* **2**, 181–223.

Mingoa-Licuanan, S (1993). Country report: Philippines. In 'Genetic aspects of conservation and cultivation of giant clams'. (Ed P Munro) pp. 40–3. *ICLARM Conference Proceedings* No. **39**, International Center for Living Aquatic Resources Management, Manila, Philippines.

Munro, J L (1993*a*). Giant clams. In 'Nearshore Marine Resources of the South Pacific'. (Eds A Wright and L. Hill) pp. 431–49. Forum Fisheries Agency, Honiara, Institute of Pacific Studies, Suva.

Munro, P E (1993*b*). 'Genetic aspects of conservation and cultivation of giant clams.' ICLARM Conference Proceedings No. **39**, International Center for Living Aquatic Resources Management, Manila, Philippines.

Munro, J L, and Bell, J D (in press). Enhancement of marine fisheries resources. *Reviews in Fisheries Science*.

Naegel, L C A (1995). Research with a farming systems perspective needed for the development of small-scale aquaculture in non-industrialized countries. *Aquaculture International* **3**, 277–91.

Norton, J H, and Jones, G W (1992). 'The giant clam: An Anatomical and Histological Atlas.' *ACIAR Monograph* **14**, Australian Centre for International Agricultural Research, Canberra. 142 pp.

Peavy, S, and Riley, J (1993*a*). Consumer accceptance of giant clams in the local foodservice industry. Research Report No. 93–03, National Aquaculture Center, Kosrae, Federated States of Micronesia. 22 pp.

Peavy, S, and Riley, J (1993*b*). Sensory evaluation of three giant clam species: Research Report No. 93–05, National Aquaculture Center, Kosrae, Federated States of Micronesia. 14 pp.

Tisdell, C A, Tacconi, L, Barker, J R, and Lucas, J S (1993). Economics of ocean culture of giant clams, *Tridacna gigas*: internal rate of return analysis. *Aquaculture* **110**, 13–26.

Wells, S M, Pyle, R M, and Collins, N M (1983). Giant Clams. In 'The IUCN Invertebrate Red Data Book'. pp. 97–107. Gland, Switzerland.

Immunoindicators of Environmental Pollution/Stress and of Disease Outbreak in Aquaculture

R. Dinakaran Michael

Fish Immunology Laboratory, Postgraduate and Research Department of Zoology, The American College, Madurai 625 002, India.

Summary

The objectives of this paper were to present the immunotoxic effect of some of the aquatic pollutants (heavy metals and pesticides) on the immunity in the tilapia, *Oreochromis mossambicus* (Peters) and to develop a model for using the possible immunomodulatory effect of these pollutants as an indicator of environmental pollution/stress for predicting microbial infection and the consequent disease outbreak in aquaculture. When groups of *O. mossambicus* were exposed to very small concentrations (0.1 to 10% of the LD/LC_{50} concentrations) of heavy metals such as Chromium, Mercury, Nickel, Lead or Aluminium and pesticides such as Monocrotophos, Endosulfan or Quinolphos, there was significant reduction in the magnitude of antibody response and the number of leucocytes/lymphocytes. In the field test, when fish were maintained in cages in polluted waters, the magnitude of immune response was reduced. There is a broad spectrum of immunological assays which can be used for indicating environmental pollution/stress. If these tests indicate that the immune system of the fish is compromised, microbial disease outbreak can be predicted and steps can be taken to save the fish.

Introduction

Aquaculture productivity constitutes a significant portion of national income in many countries in Asia. Large scale mortalities of fish often occur in ponds and lakes due to environmental pollution/stress followed by microbial infection. Tripathi *et al.* (1978) reported substantial loss due to mortality and cessation of growth of fish in ponds in West Bengal. India as a result of epidemic infections. He also reported that infection depends on the quality of the water and the crowding of fish. In India seven bacterial genera have so far been identified as pathogens of freshwater fishes (Lipton and Lakshmanan 1986). The recent outbreak of Epizootic Ulcerative Syndrome (EUS) which caused mass mortalities among wild as well as cultured fishes (Kumar and Dey 1988; Jhingran 1990) brought great loss in aquaculture productivity. Outbreak of diseases of such a dimension in India and elsewhere clearly indicates the need for developing an indicator or a warning system for prediction of disease outbreak so that prompt prophylactic measures can be taken to save the fish.

Fish immune system

Though fish were the first vertebrates to develop a specific adaptive immune system, they are quite able to protect themselves from microbial infection. The lymphoid organs of bony fishes (Manning 1994) like tilapia are the thymus, teh

Table 1. Stress/pollution-induced susceptibility to diseases

Sl. No	Fish	Infection/Diseases	Reference
	STRESS		
1.	Atlantic salmon	Infection (*Vibrio salmonicida*)	Wiik *et al.* 1989
2.	Coho salmon	Infection	Maule *et al.* 1987
3.	Brown trout	Saprolegnia infection and furunculosis	Pickering and Duston 1983 Pickering and Pottinger 1985
4.	Winter flounder	Finrot disease	Mahoney *et al.* 1973
5.	Rainbow trout	Infection	
6.	Atlantic salmon Rainbow trout	Infection	Fevolden *et al.* 1991
	POLLUTION		
7.	Carp and brown trout	Viral infection	Anderson *et al.* 1984
8.	Coho salmon, eel, carp, brown trout	Vibriosis	Zeeman and Brindley 1981
9.	Blue gourami	Bacterial and viral infection	O'Neill 1981*b*
10.	Zebra fish	Bacterial infection	Anderson *et al.* 1984

spleen and the head-kidney which are involved in the differentiation, maturation and proliferation of T- and B-cells of the specific immunity. In addition, a variety of non-specific immune mechanisms are also found in fishes. Fishes strikingly differ from higher vertebrates in the absence of lymphnodes, tonsils and bursa/bone marrow and in the presence of head kidney. Both specific humoral and cell mediated immune responses have been shown in fish.

STRESS/POLLUTION-INDUCED IMMUNOSUPPRESSION

The physiology of fish is continuously challenged by different kinds of stressors in its environment and the primary response to stressors is the release of corticosteroids (Wedemeyer *et al.* 1984). Elevated cortisol (the main corticosteroid) level has been shown to result in decreased immune response (Anderson *et al.* 1982; MacArthur *et al.* 1984; Stave and Roberson 1985) and increased susceptibility to infection and disease (Table 1).

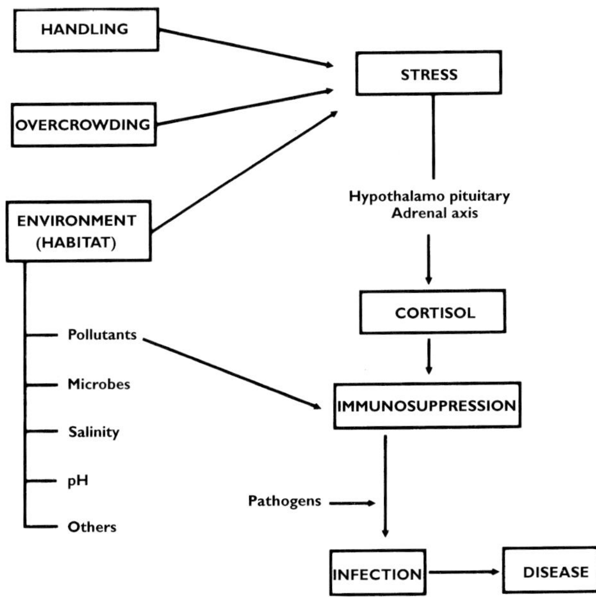

Fig. 1. Stress/pollution-induced susceptibility to infection.

The stressors which can affect fish immunity are either chemical or non-chemical in nature (Fig. 1). The former includes pollutants such as heavy metals, pesticides and farm fertilizers. The non-chemical stressors are the physical and biological factors such as environmental temperature, pH and oxygen content of the environment and overcrowding and handling of the fish. The chemical stressors like heavy metals (in addition to the cortisol-mediated immunosuppression) can also cause immunosuppression through their direct effect on immunocompetent cells and molecules (Fig. 1), (Borges and Wetterham 1989; Costa 1991).

IMMUNOMODULATION BY METALS AND PESTICIDES IN *OREOCHROMIS MOSSAMBICUS* (PETERS)

Although the immunomodulatory effects of heavy metals in fish have been shown in different species (Table 2) a concerted effort

Table 2. Effect of metals on humoral immune responses in fish

Pollutant	Fish	Effect	Reference
Chromium	Carp	–	O'Neill 1981*b*
Chromium	Brown Trout	–	O'Neill 1981*b*
Cadmium	Rainbow trout	–	Viale and Calamari 1984
Cadmium	Cunners	–	Robohm 1986
	Striped bass	+	Robohm 1986
Cadmium	Brown trout	–	O'Neill 1981*a*
Copper	Carp	–	O'Neill 1981*b*
Copper	Brown trout	–	O'Neill 1981
Copper and methyl mercury	Blue gourami	–	Roales and Perlmutter 1977
Lead	Brown trout	–	O'Neill 1981*a*
Nickel	Carp	–	Roales and Perlmutter 1977
Nickel	Brown Trout	–	O'Neill 1981*b*
Zinc	Brown Trout	–	O'Neill 1981*b*

– indicates suppression; + indicates enhancement.

has been made in this laboratory to analyse the immunomodulatory effects of various metals on the antibody response, leucocyte and lymphoid parameters in the tilapia, *Oreochromis mossambicus* (Peters). The effect of pesticides also was studied with reference to antibody response.

Oreochromis mossambicus is a common fresh and brackish water cichlid fish. Fish of 30 g were maintained in uncontrolled ambient temperature (28 ± 1.2°C). To study the effect of metals or pesticides on the immune response, groups of *O. mossambicus* were administered with/exposed to very small concentrations (0.1 to 10%, of the LD/LC_{50} concentrations) of heavy metals or pesticides prior to immunization with optimal dose (5 mg) of the antigen, bovine serum albumin (Bovine albumin, fraction V powder, Sigma Chemical Co, St. Louis, USA). Serial bleeding of fish for antiserum preparation was done following the method described by Michael *et al.* (1994). Anti BSA titres were determined by passive haemagglutination using sheep erythrocytes coupled to BSA by chromic chloride as per Michael (1986). Total peripheral blood and splenocyte were counted in Neubauer's counting chamber and differential counts were made of the blood smears stained with Leishman's. Spleen weight was determined following the method of Michie (1973).

Summary of the results of these studies shown in Tables 3–6 indicates all the metals except selenium caused reduction in the magnitude of antibody response (Table 3), peripheral blood leucocyte lymphocyte counts, spleen weight and splenocyte count (Table 4). Similarly all the pesticides tested (except fenvalerate) induced a reduction in the magnitude of antibody response (Table 5).

Since the duration of exposure and the concentration of metals or pesticides used were very small and they promptly induced reduction in leucocyte, lympocyte and splenocyte counts, it can be predicted that this pollutant mediated immunosuppression was perhaps due to the direct effect of these pollutants on the immunocompetent cells and molecules.

Table 3. Effect of metals on antibody response in *Oreochromis mossambicus*

Sl. No.	Metal compound	Antibody response	Reference
1.	Chromium trioxide	–	Devapriya 1988
2.	Chromium trioxide	–	Rajan 1989
3	Potassium dichromate	–	Rajasekaran 1992
4.	Sodium dichromate	–	Arunkumar
5.	Chromic chloride	–	Rajasekaran 1992
6.	Mercuric chloride	–	Thampidhas 1989
7.	Mercuric chloride	–	Thangam 1989
8.	Mercuric chloride and acetate	–	Agnella 1992
9.	Lead nitrate	Nil	Geetha 1990
10.	Lead acetate	–	Geetha 1990
11.	Zinc sulphate	–	Rajan 1990
12.	Selenium dioxide	+	Ponniah 1991
13.	Copper chloride	–	Varadharaj 1991
14.	Nickel chloride	–	Radhakrishnan 1992
15.	Nickel acetate and sulphate	–	Radhakrishnan 1995
16.	Aluminium sulfate	–	Shajeevana 1993
17.	Aluminium sulfate	–	Rajesh 1996

– indicates suppression; + indicates enhancement.

Table 4. Effect of metals on lymphoid / leucocyte parameters in *Oreochromis mossambicus*

Sl. No.	Metal compound	Lymphoid / Leucocyte parameters affected	Effect
1.	Chromium trioxide	PBL count	–
2.	Potassium dichromate and Chromic chloride	Spleen weight Splenocyte count PBL total count	– –
3.	Mercuric chloride	PBL count and % of lymphocytes	– –
4.	Mercuric chloride and Mercuric acetate	Spleen weight Splenocyte count	– –
5.	Lead nitrate	Spleen weight	–
6.	Zinc sulfate	Spleen weight-Exposure % of lymphocytes	– –
7.	Selenium dioxide	Spleen weight % of lymphocytes	+ +
8.	Nickel acetate and Nickel sulfate	Spleen weight Splenocyte count	– –
9.	Aluminium sulfate	Spleen weight PBL total count	– –

– indicates suppression; + indicates enhancement.

To field-test the effect of environmental pollution/stressors on the antibody response in a large extensive culture system, three groups of tilapia were maintained in fish cages (2 m × 1 m × 1 m) in a highly polluted area (close to the site of continuous pollution), a moderately polluted area and a fairly unpolluted area respectively.

After a couple of weeks of maintenance, the fishes were immunized with bovine serum albumin; and serially bled at 5-day intervals; sera separated and antibody titres were estimated by passive haemagglutination. The results shown in Figure 2 indicate that the pollution/stress-induced immunosuppression and that the degree of immunosuppression was directly proportional to the level of pollution (Jeyakaran 1996).

Table 5. Effect of pesticides on antibody response in *Oreochromis mossambicus*

Sl. No.	Pesticide	Antibody Response	Reference
1.	Dimecron (Organophosphate)	–	Jeyashree 1991
2.	Endosulfan (Organochlorine)	–	Bhooma 1992
3.	Nuvacron (Monocrotophos)	–	Rajavarthini 1993
4.	Sumicidin (Fenvalerate)	+	Retnakumari 1993
5.	Ekalux (Quinolphos)	–	Suresh Kumar 1994

– indicates suppression; + indicates enhancement.

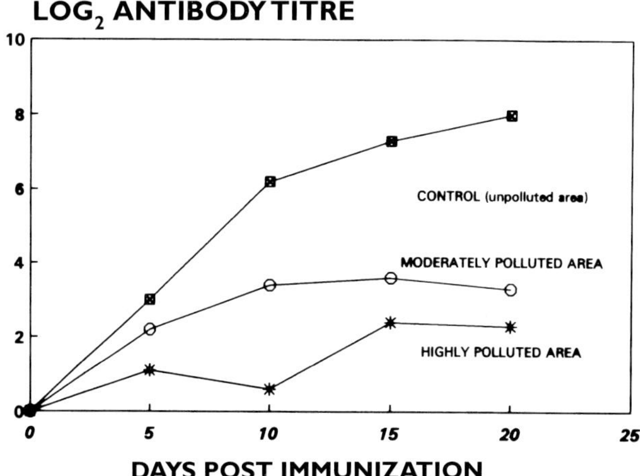

Fig. 2. Effect of environmental stress on Humoral Immune response to Bovine Albumin in fish caged at control and experimental areas.

All these experiments indicate that magnitude of antibody response, leucocyte and lymphoid parameters can be used as indicators of environmental pollution/stress for the prediction of a possible infection by available opportunistic pathogens and perhaps a disease outbreak.

Anderson (1990) has reviewed and listed the immunological assays which can be used as indicators of environmental stress (a slightly modified list is shown in Table 6). The tests range from simple non-specific assays like enumeration of leucocyte and macrophage functional assays to sensitive specific assays like Enzyme linked immunosorbent assay (ELISA) and host resistant assay. The assay can be either controlled field test (monitored cage tests) or laboratory tests (specific or non-specific exposures).

CONCLUSION

'Prevention is better than cure' is an accepted maxim. Effective and timely prevention can be done if a clear indication is available. The immunoindicators are fairly sensitive, relevant, reproducible and easily measurable. The application of immunoindicators in culture systems depends upon the cost effectiveness worked out with reference to equipment, trained personnel and the quantum of the aquaculture production.

REFERENCES

Agnella, L M (1992). Modulatory effect of the organic and the inorganic forms of mercury on the immune system of *O. mossambicus* (Peters). M. Phil dissertation, The American College, Madurai, India.

Anderson, D P (1990). Immunological indicators: Effects of environmental stress on immune protection and disease outbreaks. *American Fish Society Symposium* **8**, 38–50.

Anderson, D P, Roberson, B S, and Dixon, O W (1982). Immunosuppression induced by a corticosteroid or an *alkylating* agent in rainbow trout (*Salmo gairdneri*) administered a *Yersinia ruckerri* Bacterin. *Developments in Comparative Immunology* (Suppl. 2) 197–204.

Anderson, D P, van Muiswinkel, W V, and Roberson, B S (1984). Effects of chemically induced immune modulation on infectious diseases of fish. In 'Chemical regulation of immunity in Veterinary Medicine'. (Eds M Kende, J Gainer and M Chirgos) pp. 182–211 (A L Liss: New York).

Bhooma, R (1992). Modulation of humoral immune response by the organochlorine pesticide endosulfan in *O. mossambicus* (Peters). M.Sc dissertation, The American College, Madurai, India.

Blaxhall, P C (1972). The haematological assessment of the health of fresh water fish. A review of selected literature. *Journal of Fish Biology* **4**, 593–604.

Borges, K M, and Wetterham, K E (1989). Chromium cross-links glutathione and cysteine to DNA. *Carcinogen* **10**, 2165–8.

Costa, M (1991). DNA-protein complexes induced by chromate and other carcinogens. *Environmental Health Perspec* **92**, 45–52.

Table 6. Immune indicator assays (Indicators of IS dysfunction)

Assay	Ease (To Do)	Sensitivity	Specific Immune Impairment	Training Needed	Reference
Non-specific Assays					
1. Total leucocyte count	Easy	Low	None	Little	Pickering and Pottinger 1987
2. Dif. leucocyte count	Easy	Low	None	Moderate	Blaxhall 1972
3. Lymphoid organ weight	Easy	Low	None	Little	Michie 1973
4. Phagocytic index	Fair	Good	Phagocytosis	Moderate	Olivier *et al.* 1986
5. Pinocytic index	Fair	Good	Pinocytosis	Moderate	Weeks *et al.* 1987
Specific Assays					
6. Bacterial agglutination	Easy	Good	Antibody resp.	Moderate	Robohm 1986
7. Passive agglutination	Fair	Excellent	Antibody resp.	Moderate	Stolen *et al.* 1983
8. Immunoelectrophoresis	Fair	Good	Antibody resp.	Intensive	Roitt *et al.* 1989
9. ELISA (Enzyme linked Immunosorbent Assay)	Difficult	Excellent	Antibody resp.	Intensive	Sakai *et al.* 1987
10. RIA (Radio Immuno Assay)	Difficult	Excellent	Antibody resp.	Intensive	Grondel and Boesten 1982
11. Scale Allograft Rejection	Easy	Fair	Allograft Rej.	Moderate	Hildemann 1957
Integrated Immunity Tests					
12. Host Resistance test	Difficult	Excellent	Protection	Moderate	Hetrick *et al.* 1979

Devapriya, N D (1988). Effect of chromium on the humoral immune response and peripheral blood cells in the fish *O. mossambicus* (Peters). M.Sc dissertation, The American College, Madurai, India.

Fevolden, S, Refstie, T, and Roed, K H (1991). Selection for high and low cortisol response in Atlantic salmon (*Salmo salar* L.) and rainbow trout (*Onchorhynchus mykiss*). *Aquaculture* **95**, 53–65.

Fevolden, S, Refstie, T, and Roed, K H (1992). Disease resistance in rainbow trout (*Onchorhynchus mykiss*) selected for stress response. *Aquaculture* **104**, 19–29.

Geetha, R (1990). Studies on the effect of heavy metal, lead on the humoral immune response to the protein antigen Bovine Serum Albumin in *O. mossambicus* (Peters). M. Phil dissertation, The American College, Madurai, India.

Grondel, J L, and Boesten, H J A M (1982). The influence of antibiotic on the immune system. I. Inhibition of the mitogenic leucocyte response *in vitro* by oxytetracycline. *Development in Comparative Immunology* (Suppl. 6) 211–6.

Hetrick, F M, Knittel, M D, and Fryer, J L (1979). Increased susceptibility of rainbow trout to infectious hematopoietic necrosis virus after exposure to copper. *Applied and Environmental Microbiology* **3**, 198–201.

Hildemann, W H (1957). Early onset of homograft rejection. *Transplant Bulletin* **3**, 114. pp.

Jeyashree, U R (1991). Modulation of humoral immune response to a protein antigen Bovine Serum Albumin by an organo-phosphorus pesticide, dimecron in the fish *0. mossambicus* (Peters). M.Sc dissertation, The American College, Madurai, India.

Jeyakaran, J (1996). Immunological indicators: effects of environmental stress on immune response in *Oreochromis mossambicus* (Peters) MSc dissertation. The American College, Madurai, India.

Jhingran, A G (1990). Role of *Micrococcus* sp. in the outbreak of EUS in India. *Fish Health Section Newsletter* **1**(1), 5.

Kumar, D, and, Dey, R K (1988). Fish Diseases in India. In 'Proceedings of the symposium on Aquaculture productivity held in December 1988 under aegis of Hindustan Lever Research Foundation (Eds V R P Sinha and H C Srivastava) pp. 315–43.

Lipton, A P, and Lakshmanan, M (1986). Microbial diseases of the fresh water fishes of India. *Indian Review Life Science* **6**, 141–61.

MacArthur, J I, Fletcher, T C, Pirie, B J S, Davidson, R J L, and Thomson, A W (1984). Peritoneal inflammatory cells in plaice, *Pleuronectes platessa* L.: effects of stress and endotoxin. *Journal of Fish Biology* **25**, 69–81.

Mahoney, J, Midlidge, F, and Deuel, D (1973). A fin rot disease of marine and euryhaline fishes in the New York Bight. *Transactions of the American Fisheries Society* **102**, 596–605.

Manning, M J (1994). Fishes. In 'Immunology — a comparative approach'. (Ed R J Turner). (Jonh Wiley & Sons: New York). 222 pp.

Maule, A G, Schreck, C B, and Kaattari, S L (1987). Changes in the immune system of Coho salmon (*Onchorhynchus kisutch*) during parr-to-smolt transformation. *Canadian Journal of Fisheries and Aquatic Sciences* **44**, 161–6.

Michael, R D (1986). Studies on the immune response to Bovine Serum Albumin in the lizard, *Calotes versicolor*. Ph.D thesis, Madurai University, Madurai, India.

Michael, R D, Srinivas, S D, Sailendri, K, and Muthukkaruppan, VR (1994). A rapid method for repetitive bleeding in fish. *Indian Journal of Experimental Biology* **32**, 838–9.

Michie, D (1973). 'Handbook of experimental immunology.' (Ed D M Weir). (Blackwell Publications: Oxford). 30 pp.

Olivier, G, Eaton, C A, and Campbell, N (1986). Intersction between *Aeromonas salmonicida* and peritoneal macrophages of brook trout (*Salvelinus fontinalis*). *Veterinary Immunology and Immunopathology* **12**, 223–34.

O'Neill, J G (1981*a*). Effects of intraperitoneal lead and cadmium on the humoral immune response of *Salmo trutta*. *Bulletin Environmental Contamination of Toxicology* **27**, 42–8.

O'Neill, J G (1981*b*). The humoral immune response of *Salmo trutta* L and *Cyprinus carpio* L. exposed to heavy metals. *Journal of Fish Biology* **19**, 297–306.

Pickering, A D, and Duston, J (1983). Administration of cortisol to brown trout, *Salmo trutta* L and its effects on the susceptibility to saprolegnia infection and furunculosis. *Journal of Fish Biology* **23**, 163–75.

Pickering, A D, and Pottinger, T G (1985). Cortisol can increase the susceptibility of brown trout, *Salmo trutta* L to disease without reducing the white blood cell count. *Journal of Fish Biology* **27**, 611–9.

Pickering, A D, and Pottinger, T G (1987). Crowding causes prolonged leucopenia in salmonid fish despite interrenal acclimation. *Journal of Fish Biology* **30**, 701–12.

Ponniah, V (1991). Effect of heavy metal selenium on antibody response to cellular and protein antigen in *O. mossambicus* (Peters). M.Phil dissertation, The American College, Madurai, India.

Radhakrishnan, K (1992). The effect of Nickel on the humoral immune response to a protein and a cellular antigen *0. mossambicus* (Peters). M.Sc dissertation, The American College, Madurai, India.

Radhakrishnan, K (1995). Immunosuppressive effects of organic and inorganic forms of Nickel in *O. mossambicus* (Peters). M.Phil dissertation, The American College, Madurai, India.

Rajan, R (1989). Effect of the heavy metal chromium on the antibody response to a protein antigen in the fish *O. mossambicus* (Peters). M.Sc dissertation, The American College, Madurai, India.

Rajan, R (1990). Effect of the heavy metal zinc on the antibody response to a protein and cellular antigen in the fish *0. mossambicus* (Peters). M.Phil dissertation, The American College, Madurai, India.

Rajasekaran, P (1992). Differential effect of trivalent and hexavalent forms of chromium on the immune system of *0. mossambicus* (Peters). M.Phil dissertation, The American College, Madurai, India.

Rajavarthini, P B (1993). The effect of Monocrotophos (Nuvacron) on the humoral immune response to Bovine Serum Albumin in *O. mossambicus*. M.Sc dissertation, The American College, Madurai, India.

Rajesh, M (1996). Immunostimulatory effect of Aluminium in *O. mossambicus* (Peters). M.Phil dissertation, The American College, Madurai, India.

Retnakumari, R N (1993). The effect of Fenvalerate (Sumicidin) on the humoral immune response to Bovine Serum Albumin in *0. mossambicus* (Peters). M.Sc dissertation, The American College, Madurai, India.

Roales, R R, and Perlmutter, A (1977). The effects of sublethal doses of methyl mercury and copper; applied singly and jointly, on the immune response to the blue gourami (*Trichogaster trichopterus*) to viral and bacterial antigens. *Archives Environmental Contamination Toxicology* **5**, 325–31.

Robohm, R A (1986). Paradoxical effects of cadmium exposure in antibacterial antibody responses in two fishes: inhibition in cunners (*Tautosolabrus adspersus*) and enhancement in striped bass (*Morone saxatilis*). *Veterinary Immunology and Immunopathology* **12**, 251–62.

Roitt, I M, Brostoff, J, and Male, D K (1989). Immunological techniques. In 'Immunology'. pp. 25.2–25.3. (Gower Medical Publishing, London).

Sakai, M, Amaaki, N , Atsuta, S, and Kobeyashi, M (1987). Comparative sensitivities of several dot blot methods used to detect bacterial kidney diseases. *Journal of Fish Diseases* **10**, 229–31.

Shajeevana, R V (1993). Effect of Aluminium on the humoral Immune response to Bovine Serum Albumin in *0. mossambicus* (Peters). M.Sc dissertation, The American College, Madurai, India.

Stave, J W, and Roberson, B S (1985). Hydrocortisone suppresses the chemiluminescent response of striped bass phagocytes. *Developments in Comparative Immunolology* **9**, 77–84.

Stolen, J S, Kasper, W, Gahn, T, Lipcon, V, Nagle, J J, and Adams, W N (1983). Monitoring environmental pollution in marine fishes by immunological techniques. The immune response of fish exposed by injection or bath to bacterial isolates from sludge or *in situ* exposure to sludge. *Biotechnology* **1**, 66–8.

Suresh Kumar, J (1994). The effect on humoral immune response of the organophosphate pesticide, Quinolphos (Ekalux) in *0. mossambicus.* M.Sc dissertation, The American College, Madurai, India.

Thampidhas, G D (1989). Effect of the heavy metal mercury on the antibody response and oxygen consumption in the teleost fish *0. mossambicus* (Peters). M.Phil dissertation, The American College, Madurai, India.

Thangam, P V (1989). Modulation of humoral immune response to protein and cellular antigens by the heavy metal mercury in the cichlid fish, *0. mossambicus* (Peters). M.Phil dissertation, The American College, Madurai, India.

Tripathi, S D, Dutta, A, Sengupta, K K, and Patra, S (1978). Effect of parasites and diseases on fish growth. Workshop on Diseases of fish cultured for food in south east Asia. IDRC, Cisaura, Bogor, Indonesia, 28th Nov-lst Dec. 8 pp.

Varadharaj, R (1991). Studies on the effect of the heavy metal copper on the humoral immune response to a protein antigen Bovine Serum Albumin in the teleost *O. mossambicus* (Peters). M.Sc dissertation, The American College, Madurai, India.

Viale, G, and Calamari, D (1984). Immune response in rainbow trout *Salmo gairdneri* after long-term treatment with low levels of Cr, Cd and Cu. *Environmental Pollution* **35**, 247–57.

Wedemeyer, G A, Mcleay, D J. and Goodyear, C P (1984). Assessing the tolerance of fish and populations to environmental stress and the problems and methods of monitoring. In 'Contaminant effects on fisheries'. (Eds V W Cairns, P V Hodson and J O Nriagu) pp.163–95 (John Wiley and Sons: Chichester, New York).

Weeks, B A, Keisler, A S, Myrvik, Q N, and Warinner, J E (1987). Differential uptake of neutral red and macrophages from three species of estuarine fish. *Developments in Comparative Immunology* **11**, 117–24.

Wiik R, Anderson, K, Uglenes, I, and Egidius, E (1989). Cortisol induced increase in susceptibility of Atlantic salmon, *Salmo salar* to *Vibrio salmonicida*, together with effects on the blood cell pattern. *Aquaculture* **83**, 201–15.

Zeeman, M G, and Brindley, W A (1981). Effects of toxic agents upon fish immune system. In 'Immunological considerations in Toxicology.' (Ed R P Sharma Boca Raton) Vol. III pp. 1–60 (CRC Press: Florida).

Towards Sustainability of the Aquaculture Industry in Jervis Bay

Joanne Scarsbrick

Institute for Coastal Resource MAnagement, University of Technology, Sydney, NSW 2000, Australia
Present address: Manly Municipal Council, PO Box 82 Manly, New South Wales 2095, Australia.

Summary

This paper explores the constraints to the development of the New South Wales aquaculture industry, looking specifically at Jervis Bay. Many of the constraints, such as resource use conflict and possible environmental impact, are all too familiar. NSW also has legislative constraints that have been inadvertently imposed on the industry by government. The difficulty lies in a convoluted approvals process that has held up many development approvals in Jervis Bay since the 1970s.

It is suggested that the constraints and forces driving the development of aquaculture at a regional level, and the increasing demand for quality, fresh, fish protein, can be managed through a change mechanism: the Aquaculture Industry Development Plan. The core element of this plan is an Aquaculture Community Impact Agreement, designed to bring about a sustainable aquaculture industry and related sustainable integrated coastal management.

Introduction

Resource use conflict is familiar to any kind of development including aquaculture, that may be perceived to impact on others. An interesting example of this is a study carried out by the author at Wallis Lake, New South Wales (NSW). This coastal region is not only known by many tourists and residents of the area for its scenic qualities and the highly-prized Wallis Lake 'Sydney rock oyster' (*Saccostrea commercialis*), but also for the 'cold war' that waged between two factions of the oyster industry: those growers who represented the respective Sydney rock oyster and the Pacific oyster (*Crassostrea gigas*) causes.

These facts were well known to the author when undertaking a study in 1994 to assess the potential for expanding the aquaculture industry in Wallis Lake. Following numerous studies and wide consultation, including both factions of the oyster growers, it became obvious that the study had become a driving force to make these two groups meet and work together for the good of the estuary in which both industries were established. The oyster growers agreed that they didn't want any other aquaculture in the bay either, as is described in a press release from the *Sydney Sun Herald*, 19/6/94.

The outcome was an Oyster Management Plan for Wallis Lake; the first of its kind in NSW. This Plan was also adopted by the Great Lakes Council into its Development Control Plans for the estuary. The point demonstrated is that by managing driving forces and constraints in a positive way a more sustainable oyster/aquaculture industry was achieved in Wallis Lake and the region.

Another example of such a concept can be found at Jervis Bay, in southern NSW. This is an example of a more complex problem, symptomatic of aquaculture development where there are a number of unresolved issues.

Before exploring the aquaculture issues relating to Jervis Bay,including some of the constraints to achieving sustainability and some of the driving forces that can be utilized to manage the problems the industry is facing in NSW, it is important to achieve a common understanding of what is meant by 'aquaculture'and 'sustainability of the aquaculture industry'. The NSW Fisheries Management Act defines aquaculture to mean:

- 'the cultivation of fish or marine vegetation for the purposes of harvesting the fish or marine vegetation or their progeny with the view for sale; or
- keeping fish or marine vegetation in a confined area for a commercial purpose'.

A 'sustainable aquaculture industry' is one where the resources of society, the economy and the aquaculture environment, are managed with the goal that all three gain in value, both now and into the future.

The achievement of sustainable development which embraces society, economic and natural environment, is a familiar but baffling objective. So it is hardly surprising to see a community react to new developments, such as aquaculture, if the natural environment, which encompasses the physical and biological components of their surroundings and their activities, is perceived to be threatened.

RESOURCE USE CONFLICT

In Jervis Bay there are many conflicting demands on coastal and estuarine resources, including urban development, recreation (sailing, diving and recreational fishing), tourism (whale and dolphin watching), industry (commercial fishing) and mariculture *per se.* Many overseas and interstate examples are often cited as examples of resource use conflicts that must be avoided. The cases of aquaculture expansion in Tasmania, South Australia and New Zealand, for example, conjure up images of what can only be described as 'islands' of pens, or rafts of trays, buoys, etc. Much of the waterways may already be occupied by flotillas of recreational cruisers, fishing and navy vessels, as is the case in Jervis Bay, all of which arguably contribute to environmental degradation and issues of resource use conflict.

If we were to overlay Jervis Bay with some Geographic Information System (GIS) variables, which indicate naval presence, commercial fishing areas and possible areas for mussel farms, we can easily see a resource use conflict without even taking into account the wants, needs or values of other users, such as tourism operators, recreational fishers, real estate investors and residents in the area.

Apart from constraints posed by resource use conflicts on proponents of an aquaculture industry in Jervis Bay, there are socio-economic and environmental factors which indicate that 'a change mechanism' could manage these constraints and the driving forces to development, making it possible for a sustainable aquaculture industry to develop. Some of the fundamental forces that are driving industry development, include a decline in the wild fishery, regional development and a consumer demand for more quality fish protein.

Jervis Bay, located in the Shoalhaven Region of NSW is characterized by the appropriate socio-economic and environmental factors: a well-educated and informed community, suitable to accommodate these driving forces, high levels of retired residents, high unemployment and a Local Government that enthusiastically supports employment opportunities and the establishment of suitable aquaculture sites for extensive farming and offshore intensive farming. During recent conflicts, media support for the initiative of the Shoalhaven Council and local aquaculturalists was strong, as demonstrated by an article in the *Times*, 16/3/94.

PROBLEMATIC CASE OF SUSTAINABLE AQUACULTURE DEVELOPMENT IN JERVIS BAY

Gone are the 'good old days' for developers when the approvals process was simple and developers could mostly do whatever they wished, with little community involvement or consideration for amenity or design of the proposed development. In the past, a developer was free from the complications of: development consents, development control plans, Aquaculture Industry Development Plans, the requirement to prepare a Commercial Farm Plan, requirement for community consultation and participation, and regard for the possible environmental impacts of the development.

The Council and other relevant authorities no longer have to deal with an uncomplicated situation involving a simple relationship with one customer or applicant. There is now a complex relationship with a number of customers who lobby, attempt to influence, politic, and participate in planning groups. Perhaps they have brought the needed balance to *ad hoc* development that was often previously insensitive to aesthetic or environmental issues?

On one side of the fence are those individuals or groups whose economic self-interest is threatened by those who perceive aquaculture as a threat to environmental quality. On the other side are those who view almost any attempt to promote economic development as a potential threat to ecological balance.

In Jervis Bay, tensions arose during organized workshops where a range of issues were highlighted and discussed which:

- involved personal or societal values, as well as scientific and objective data (aesthetics, visual pollution)
- discussed possible habitat destruction or species extinction (escape of exotic fish into the marine habitat, uncontrolled

pollution from the output of fish farms or industrial premises or agriculture farms)

- indicated that aquaculture was not to be sited in areas where naval activities were carried out
- indicated that aquaculture was not to be sited in the proposed Marine Reserve
- determined that the nature, boundaries, participants and costs are uncertain (where and how many fish farms, what size, and with a sense of NIMBY, not in my backyard- i.e. *NOT in my fishing spot!).*

The existing approval process

Aquaculture development throughout NSW, not only in Jervis Bay, is plagued by an approvals process which can involve Local Government, Dept of Urban Affairs and Planning, and Dept of NSW Fisheries and at least seven other agencies as consent authorities in the process of determining the approval of an application for an aquaculture lease. Essentially three Acts (the EPA Act 1979, the Local Government Act 1993 and the Fisheries Management Act 1994) are involved in this convoluted approvals process. It is not a matter of an approval being given within 40 days but maybe a matter of at least ten years, as has been the experience of members of the Jervis Bay Mariculture Association!

With the combination of such a complicated approvals process and an increase in tensions within the broader community (including recreational fishers, recreational sailors, land holders, environmentalists, government agencies and developmental interests), the marine aquaculture industry in NSW is fighting for survival! The flow chart in Figure 1 maps the current approvals process.

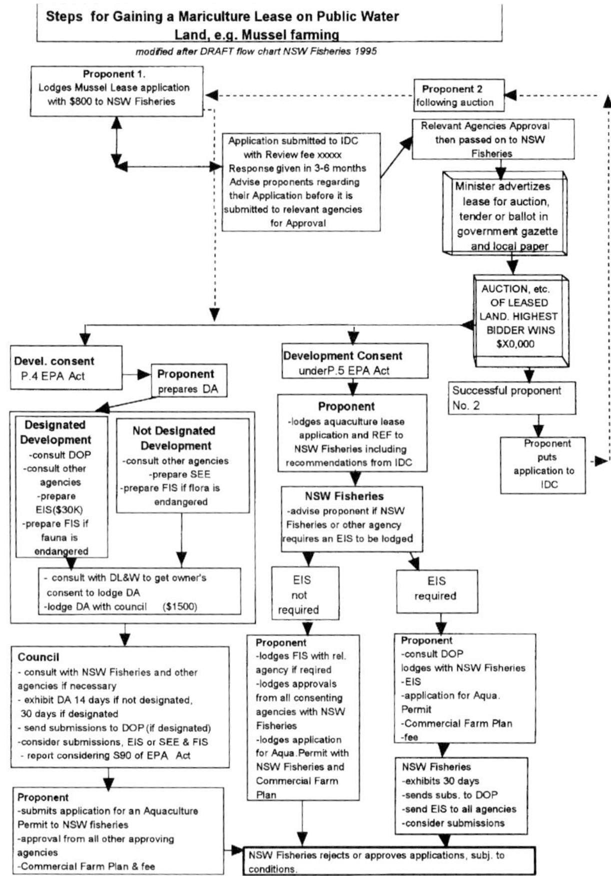

Fig. 1. The current approvals process for aquaculture development in NSW.

Options for sustainability

A glimmer of hope lies in the fact that nowadays there is an increasing recognition that fishery management is as much to do with 'people management' as with biological and economic management. In many countries, such as North America, Canada, and the Australian States of South Australia and Tasmania, fishery experts feel that future management regimes must therefore provide for more extensive participation of stakeholders than is presently the case. This is especially so in NSW, and is particularly the view of Jervis Bay aquaculture industry members. For this trend to influence fisheries planning in NSW, it will require a paradigm shift away from paternalistic fisheries management in NSW with its top-down modes of management.

Pinkerton and Weinstein (1995) agree with the need for participatory fisheries management, by commenting that a fishery is a 'human, socio-cultural and economic system which is articulated with a marine-biological system and effective management cannot be effective without incorporating the two!'

Community participation

Understanding who has structural power *versus* informal influence, and who has neither, assists in recognizing the sources of conflict in the decision-making process. It is well known that there are problems of communication and mistrust between industry and government, but there are also strategies that can bring these groups together to interact in a more facilitative and less combative, consensual decision-making process.

This paper, which describes a process that, given the constraints to the sustainability of aquaculture from:

- resource use conflict and the need for more community participation in planning and decision making; and
- the complex approvals process,

may provide an option to manage these in an integrated way through, what is referred to as, an Aquaculture Industry Management Plan (AIMP)-which is a change mechanism. This plan contains a key process element, the Community Impact Agreement (CIA) (Smith 1994), by which to implement change. Change management with broader community involvement is essential for the development of a sustainable aquaculture industry.

The AIMP manages by integrating higher level bureaucratic planning with State agencies and the regional community, represented by Local Government, and also considers issues such as the Environmental Impact Study (EIS) on a regional basis, not an individual proponent basis.

THE COMMUNITY IMPACT AGREEMENT

The Community Impact Agreement has been defined by Smith (1994) as:

> 'Institutional arrangements that balance business efficiency and equity or fairness. Agreements are mechanisms that both communities and proponents can use to manage the effects of socio-economic (*environmental*) change resulting from development, in a manner consistent with community aspirations and with the principles of sustainable development'.

The effects of *environmental* change have been included into this definition as this is a key element to community acceptance of the industry and its sustainable development. The approach is considered to have universal application, and can be applied to other forms of major developments, especially in coastal regions.

The mechanism for change was first developed in the engineering field, by Ontario Hydro, Quebec, for building electricity generators on the Niagara River, Niagara Falls (Smith 1994).

The mechanism works by industry and the broader community (including local government and state agencies, along with environmentalists, tourism groups and chambers of commerce etc.) being involved in the strategic planning and decision-making process together. Importantly the process is led by industry, and industry is the responsible and accountable partner to an agreement that is drawn up by all the parties involved.

A Community Impact Agreement (Fig. 2) (Smith 1993) would be especially suitable to being incorporated into the strategic planning and development of the NSW aquaculture industry, initially at Jervis Bay and then applied on a regional basis. In achieving this goal a CIA would provide the link between aquaculture and integrated coastal zone management. It demonstrates that aquaculture, or any new development (with coastal and catchment management impacts), can be an essential component in effective coastal zone management.

SOLUTIONS PROVIDED BY AN AQUACULTURE COMMUNITY IMPACT AGREEMENT (ACIA)

Integrated or holistic planning is provided by an ACIA. It adds horizontal and vertical integration processes to strategic management. This process is an essential element for sustainability of the industry in Jervis Bay and for Coastal Zone Management. The outcomes of the process are to:

- provide management of the marine resource through a process that ensures that the Shoalhaven Region stakeholders participate at the community level in decision making, and which is clearly understood by all to be set within the wider context of regional and State processes and perspectives
- equip and build capacity among the regional community stakeholders by working together to realize integrated management; identify roles and responsibilities of stakeholders; provide vehicles for wider participation and conflict resolution; build trust through administrative transparency, i.e. openness with information sharing, etc.
- develop benchmarks not only for community participation, involving local and State government, industry and the community, but also for participatory environment and coastal zone management
- place an emphasis on ecosystem-based management that focusses on sustainable development (this is because it involves local government and their communities, the scientific community, conservation organizations, ecosystem monitoring agencies,resource users and other developers and State agencies)
- recognize municipal decision-making processes as a local management function with legal jurisdiction
- encourage the development of community involvement in environmental monitoring programmes
- encourage industry to voluntarily lead remedial programmes and take responsibility for resolving contamination issues; and
- ensure that developments such as aquaculture are considered in the broader context of municipal, State, federal and, sometimes international, issues (e.g. fresh salmon importation to Australia from Canada and North America) in their plan.

These are some of the goals that came out of the Community Participation workshops held in the Shoalhaven area in 1994 during community consultations relating to proposed aquaculture development. It is suggested that these goals are achievable through an ACIA.

However, in addition to the use of a change mechanism to deal with resource use conflict and the need for greater participation

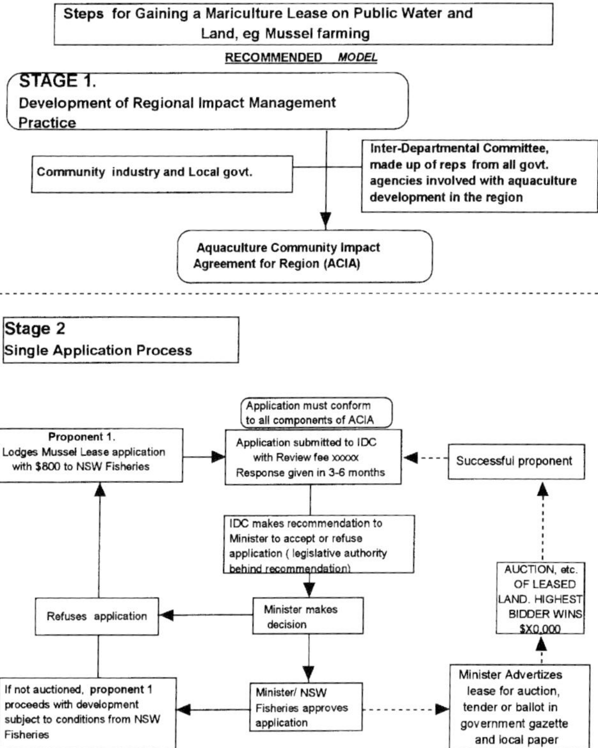

Fig. 2. Suggested Reformed Legislation-Single Approvals Act (Aquaculture).

in the planning processes, it can also be used to assist in implementing change through a reformed approvals process and associated legislative reform of the Fisheries Management Act (primarily). Three recommendations for simplifying and streamlining the approvals process are:

- to reform and streamline the existing approval process which in turn would provide substantial economic benefits to the industry by making the system more effective and efficient;
- to incorporate the ACIA change mechanism into a formal approvals process
- to give legislative standing and authority to the Inter Departmental Committee (IDC) to make recommendations to the Minister for approving applications; the IDC would be represented by the relevant Departmental executives with the appropriate delegations to make decisions.

These recommendations would be achieved by developing a single approval process by the co-ordination of the EPA Act, 1979, the Local Government Act, 1993, and the FMA, 1994. Such an single approval system could possibly be formed by a single Act governing the process, such as a single 'Approvals Act (Aquaculture)' (see Fig. 3).

Conclusion

Society has increasingly given the mandate to planning authorities, to provide their need for more involvement in planning and policy development, and for greater constraints to be placed on development. This is so that society's environmental assets may be protected from degradation resulting from changes brought about by uncontrolled development, including aquaculture.

The proposed ACIA model would play a significant role in establishing a formal, streamlined approval process and could be a mechanism by which to manage the complicated and often volatile issues related to how upward policy integration could be incorporated into management of the industry from a government, industry and community perspective.

The proposed ACIA would provide a cohesive link between externalities such as resource use conflict and pollution etc. The

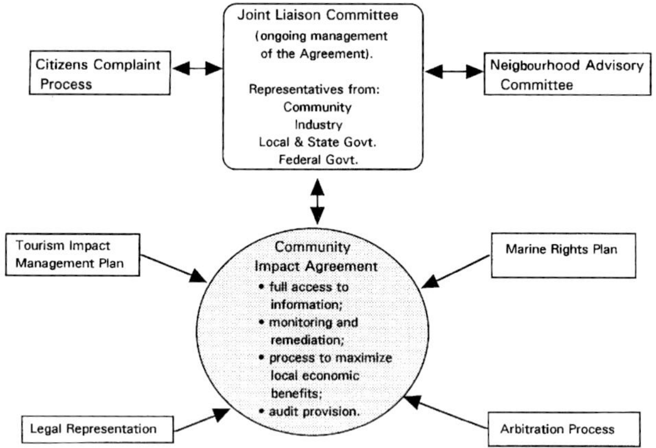

Fig. 3. A Community Impact Agreement (after Smith 1994).

ACIA mechanism could also be the determining factor in facilitating sustainable regional development and reducing regulation of the industry by co-regulatory or self-regulatory practices, and assist industry in bringing about change in a non-threatening and participatory way. Indeed the Agreement could be instrumental in managing the constraints and driving forces of the industry as it contains essential elements that are necessary for the development of an environmentally and economically sustainable industry; a necessary imperative in coastal zone management.

References

Pinkerton, E, and Weinstein, M (1995). Fisheries that Work: Sustainability Through Community Based Management.The David Suzuki Foundation,Vancouver, BC. 109 pp.

Smith, M A (1994). Community Impact Agreements, Mechanisms for Change Management: The Niagara Experience. Proceedings of I.A.I.A., Quebec, Canada.

Sydney Morning Herald (1994). Anger over fish-farm plan for Wallis Lake. 19 June, p. 7.

The Times (1994). Million dollar industry plan for Jervis Bay. 16 March, p. 12.

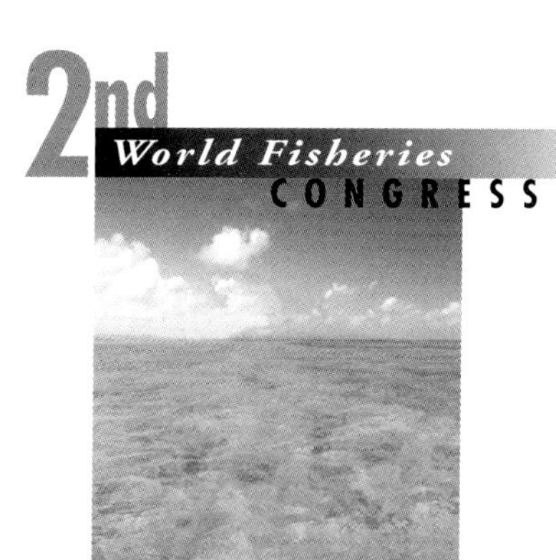

Issues and Outcomes for Theme 4

How Can Aquaculture Help Sustain World Fisheries?

I Chiu Liao

Taiwan Fisheries Research Institute, Keelung, Taiwan 202.

Introduction

Aquaculture is generally regarded as an integral part of fisheries, a fact that is reflected in this Congress. Aquaculture, which may be regarded as the 'artificial' resource of fisheries, is important in today's world because it can help sustain world fisheries. This importance is increasing with time.

For Theme 4, aquaculture experts came from all parts of the world. They brought with them 14 oral papers and 25 posters. After four hectic days of presentations and discussions, we were able to identify the following issues and outcomes. These issues have occurred in every country in which aquaculture is practised, so it is worthwhile to note these experiences as a caveat for countries intending to develop aquaculture and as suggestions for those already undergoing development. Aquaculture is not limitless and carefree, and, although it will prove to be of great benefit, it does carry some risks; these have been identified and discussed by experts at this Congress.

The necessity for aquaculture

The world population is increasing and the demand for fisheries resources is growing. There is evidence that capture fisheries are increasingly being depleted. Therefore, aquaculture is essential to meet the demand for fisheries products and to create job opportunities for people who no longer have a career in capture fisheries. However, from experience in developing our fisheries, we all recognize that we are operating in a competitive environment, characterized by rapid changes in climate, pollution and human activities.

Many reports of successful cases of aquaculture were given at the Congress; for example, the giant clam research in the Solomon Islands reported by Hart *et al.*, the enhancement of the scallop fishery in Tasman Bay, New Zealand, by Hickman, and the dietary experiments and assessment of stock enhancement of barramundi in Australia by McMeniman and Sands and by Russell and Rimmer. Aquaculture can restore the natural resource, supply fisheries products as protein sources and provide important economic benefits for local communities. The paper by Green and Kahn goes even further by saying that subsidization of aquaculture is justified when the net benefits of aquaculture exceed the social cost of the subsidy. This indicates the importance of aquaculture to the community.

The development of aquaculture

In ancient times when few fish were caught from the seas, aquaculture was developed in China because of necessity. In recent years, aquaculture is thriving again because capture fisheries resources are being depleted. It is to be hoped that the gathering of scientists at the Second World Fisheries Congress will have contributed to the development of aquaculture, and that aquaculture will help to create the second spring of fisheries.

As Welcomme notes, aquaculture is expanding world-wide and is becoming an accepted industry in areas outside its traditional confines in Asia and Europe. In fact, papers from Australia, northern and southern America, Africa, and the Pacific Ocean are presented in Theme 4 of this Congress. Because aquaculture is linked with development, many parts of the world are converting from capture fisheries to aquaculture. Therefore, the development of aquaculture has two aspects; one is that many countries are becoming aquaculture-developing countries, and the other is that the aquaculture-developed countries are progressing towards the development of advanced systems, such as recirculation and superintensive systems. The fact that aquaculture production reached 22% of total fisheries production in 1993 shows undoubtedly that aquaculture can directly help sustain world fisheries.

The solutions for aquaculture development

Aquaculture comprises mainly fish farming on land, cage farming in the sea, and stock enhancement. Any one of these needs the resources of water and land, which are also the major elements for other uses such as agriculture, industry, and land development. Because of the location and the operation of aquaculture, it directly conflicts with recreational, navigational and aesthetic interests. It is also criticized by conservationists for its pollution-generating activities. Those problems need to be considered in planning for aquaculture.

However, both Hickman and McLoughlin offer some hints on planning for aquaculture. Their suggestions, such as collaboration instead of competition, and regional planning with legislation, are valuable. Blankenship and Leber go even further and consider the sustainability of stock enhancement, an important topic. Some other important areas of high priority are ecological considerations, environmental stress, control of disease, nutritional requirements, improved diets and economic evaluation, as proposed by Chien and Liao, Green and Kahn, Soares and Hughes, Shiau *et al.*, Williams *et al.*, and Woods and Soares. Further challenges to be considered include cryopreservation of gametes, induced spawning of threatened species, and immunoindicators, as emphasized by Chao, Ingram and Gooley, and Michael.

Gene manipulation

The present development of aquaculture is proceeding in several directions including stock enhancement and transgenetics. Stock enhancement increases the quantity of aquaculture products, but it confronts other users of water and land. Transgeneics improve the quality of aquaculture products, but it presents problems with regard to ecology and ethics.

Technology such as transgenetics is much needed to satisfy the protein demand of the increasing world population. On the other hand, Pullin concludes that the effective conservation of fish genes and genotypes is of great importance to keep options open for the sustainable development of aquaculture and for the generation of a wide diversity of aquatic products.

Concluding remarks

In addition to the above, educating the public and particularly the younger generation is of great importance. This is what the exhibition booths at the Congress strongly suggested. It was also agreed that this action should be taken in as many countries as possible.

How can aquaculture help to sustain world fisheries? In this Congress, we obtained clues to major directions. Through the advancement of aquaculture technology, production from aquaculture can definitely help to supplement the insufficient supply from capture fisheries. Through better management, the adverse impact on the environment from the open types of aquaculture, such as cage culture, can be greatly reduced. Thus, the development of aquaculture can be in harmony with and benefit from the environment. Closed types of aquaculture, such as superintensive eel culture, are independent of the surrounding environment and have little impact on the environment.

Through the release of seed, the stock may be enhanced. The prerequisite is a pollution-free habitat. Where the habitats are not destroyed physically or polluted chemically, released seeds and/or breeding stocks can survive and directly increase populations. The continuation of stock enhancement and its proper management are surely the answers to depleted fisheries caused by overfishing. Stock enhancement can help sustain the world's fisheries.

However, the success or the significance of stock enhancement can be ensured only after considering carefully several key points which are given below:

(1) Can artificially reared seeds stand and survive in the wild?

(2) Are hatchery-reared fry of certain species suitable for being released into the wild?

(3) What quantity of the seeds released can make a significant contribution to the stock?

(4) How should fry be released to maximize their survival?

(5) Do hatchery releases affect the genetic structure of wild conspecifics?

(6) Does enhancement increase the risk of transmitting disease to wild stocks?

Although stock enhancement is one way that aquaculture can directly contribute to capture fisheries, the massive release of seed should not be considered before all critical problems are solved. It is to be hoped that these key points will soon develop into methods and actions to realize the help that aquaculture can contribute to the sustainability of world fisheries. Let this Congress be the frontier to such a realization!

The limits to the exploitation of capture fisheries

Theme 5

Limits to Exploitation of Capture Fisheries

J. R. Beddington, K. Lorenzen and I. Payne

Centre for Environmental Technology, Imperial College of Science, Technology and Medicine, London, UK.

Summary

This paper focuses on three areas of work which are likely to affect the future of capture fisheries: better management of current fisheries, new fisheries and the role of enhancement. The importance of enforcement in ensuring better management of fisheries is explored in the context of simple models and data on the success of enforcement programmes in deterring illegal activities. It is argued that fishery regulations need to not only meet sustainability criteria, but should be framed in such a way that they are enforceable at reasonable cost. The theory involved in estimating the potential yield of new fisheries is explored, and the need for a theory to address the main ecosystem processes which can affect this yield is identified. The importance of enhancement in the development of capture fisheries, particularly in the developing world, is explored, and simple models based on the dynamics of enhanced populations that allow more efficient management are reviewed.

Introduction

This paper addresses one of the conference themes 'What is the scope for the development of wild stock fisheries?'. This theme is somewhat all encompassing so the subject of this paper has been limited to three main factors which are likely to affect the future development of capture fisheries. These are:

1. Better management of current fisheries.
2. The development and sustainable exploitation of new fisheries.
3. The enhancement of capture fisheries.

Without better performance in all of these areas, it seems clear that the future of capture fisheries is likely to be bleak. Many papers, including the Keynote Address to this Congress (Mace 1997), have dealt with the gross levels of overexploitation and overcapitalization that are prevalent in the world's fisheries and it is unnecessary in this paper to review these. Nevertheless, the scale of the problem is an indication of how badly fisheries management has addressed the problems of sustainable exploitation and the avoidance of overcapitalization in the past.

This is not an issue of limited resources and the lack of scientific endeavour. One of the most studied fisheries ecosystems is that of the North Sea with a plethora of scientific work going back

Table 1. Comparison of fishing mortalities for selected North Sea stocks

Species	F_{max}	Current Fishing Mortality
Cod	0.24	0.88
Haddock	0.386	0.85
Whiting	F_{max}?? $F_{0.1}$ = 0.26	0.67
Saithe	0.22	0.50
Plaice	0.24	0.42
Sole	0.24	0.47

Table 2. Estimates of unrecorded landings and discarded fish in waters fished by European Community vessels in the West of Scotland

	1988	1989	1990	1991	1992	1993
weight in '000 tonnes						
unrecorded landings	28	33	192	94	123	30
discarded fish	71	40	100	102	94	143
unrecorded/discarded*	54	32	9	49	36	16
Total	**153**	**105**	**301**	**245**	**253**	**189**

* No separate figures for unrecorded landings and discarded fish given

many decades, including the massive contribution of Beverton and Holt (1957). Despite this enormous level of scientific input, the situation in the North Sea is probably as bad as in any fishery area in the world today. Table 1 shows for North Sea groundfish the levels of fishing mortality currently prevalent compared to the idealized levels of fishing mortality that give maximum yield and the level of $F_{0.1}$.

It is remarkable that only a decade or so ago there was a significant debate about whether indeed a target fishing mortality should ever go as high as that which provides the maximum yield or whether the more conservative alternative of $F_{0.1}$ should be preferred. Clearly, in the case of North Sea groundfish and regrettably many other fisheries, that debate is largely irrelevant today.

The regime operating within the North Sea and indeed elsewhere is one that could be termed sustained overexploitation. The management effectively relies on some degree of resilience in the stocks which prevents complete catastrophic collapse and the disappearance of the stocks concerned. This is an uncomfortable basis for the management of major commercial activities.

Better management: the problem of enforcement

If yields from capture fisheries are to be sustained, then the successful application of management regulations is going to be required. Many of the contributions in this Congress have addressed issues of improved fisheries management and how understanding the dynamics of populations and ecosystems can be used to improve the management process. Others have looked at the way in which economic allocation of property rights can confer benefits both in avoidance of overcapitalization and in generating an environment in which conservation concerns can be directly addressed. Only a few of the contributions to this Congress have looked at the problem of the enforceability of management regulations; however, successful enforcement of regulations is the key to successful fisheries management.

For successful fisheries management, a system will need to have:

1. Regulations that meet sustainability criteria
2. Regulations that are of the type which can be enforced with reasonable levels of cost
3. Regulations which ensure that the information obtained from a fishery is reliable and sufficient to meet the requirements of the management regime.

Many systems have been developed which meet the first criterion, but there are few which meet them all and many fail to meet either of the second two criteria.

The reliability of information is often a key indicator of the enforceability of management regulations. Table 2 illustrates the problem of the reliability of information for some typical North Sea fisheries. The information was highlighted in a recent report of the UK National Audit Office on Scottish Fisheries (Anon. 1995), but the problem is not unique to the N.E. Atlantic. It is well recognized that misreporting of catches by fishing fleets to management authorities is ubiquitous. Such misreporting undermines the scientific basis for assessment of the state of the stocks and hence it can lead to management regulations which are inappropriate. In cases where the catches exceed the total allowable catch, misreporting negates the sustainability of the regime.

The origin of the problem lies in the lack of enforceability of the management regulations. Fishing is an economic process and to understand this process one needs to examine in a straightforward way the motivation of the fishermen and how they take decisions. Fishermen will tend to obey regulations when the expected benefit that can be obtained from ignoring the regulations is less than the expected penalty. This problem was looked at some ten years ago by Sutinen and Anderson (1985) and Sutinen (1987). The simple statement set out above can be formalized in a simple equation that relates the expected profit or loss to the individual fishing operator in terms of the expected benefit that can be obtained from breaking the regulations, which can be in terms of the value of increased catch, lower cost, etc, and the expected penalty. This is the product of the probability of detection of an offence by the authorities, the probability of successful prosecution of the vessel owners for the offence, and the level of penalty once successful prosecution has been achieved, i.e.

$$P = B - p_1 \, p_2 \, F$$

where P = Expected Net Profit/Loss; B = Expected Benefit; p_1 = Probability of Detection of Offence; p_2 = Probability of Successful Prosecution, and F = Level of Fine for Offence.

In a recent study, the probability of detection of vessels fishing illegally by the Scottish Fisheries Protection Service was examined by the UK National Audit Office. Table 3, taken from this study, shows that the probability of detection of an offence is likely to be

Table 3. The probability of Scottish fishing vessels being boarded by fishery protection vessels

	Number of fishery protection vessels	Total boardings of Scottish fishing vessels	Probability of Scottish fishing vessel being boarded on a typical fishing day
Inshore fishery	3	450	0.26%
Offshore fishery	4	1168	0.69%
Royal Navy	1	304	0.18%
Total	**8**	**1922**	**1.30%**

low. Further analysis taken from information contained in the reports of the Scottish Fisheries Protection Service summarized in this report is illustrated in Figure 1. This shows how the probability of prosecution being successfully completed further reduces the overall deterrence of illegal activity.

From Figure 1 it can be seen that different types of regulations aimed at controlling fisheries will have different probabilities of detection and indeed different probabilities of successful prosecution. Such information can be analysed using the statistics published by the regulatory authorities, but it is also known to the fishing community. The authorities in the short term can only control the level of deterrence by the size of the penalty. The level of penalty which will deter potential offenders can be calculated for different types of offence once the probabilities associated with detection and successful prosecution have been estimated.

Figure 2 illustrates this process and indicates the level of penalty that would be required for modest levels of catch above the allocated Total Allowable Catch (TAC). The approximate size for probabilities of detection and successful prosecution are taken from Table 3 and Figure 1. The other statistics are meant to be broadly illustrative of white fish operations in the North Sea. Quite clearly, the level of penalty which will provide an adequate deterrent is unlikely to be acceptable, particularly within judicial systems which require that penalties are in line with the benefits obtained from committing the offence.

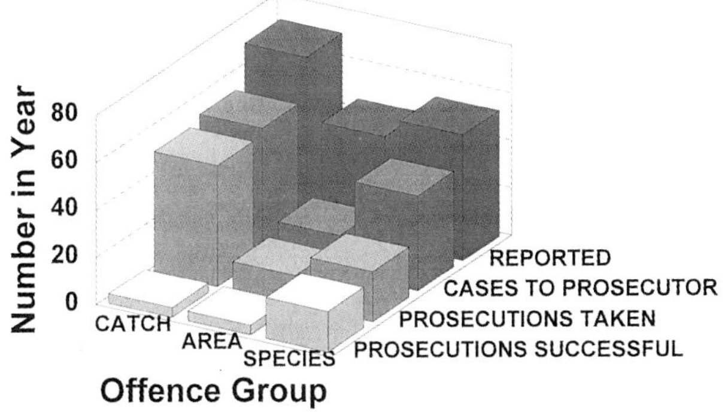

Fig. 1. Offences, prosecutions and convictions.

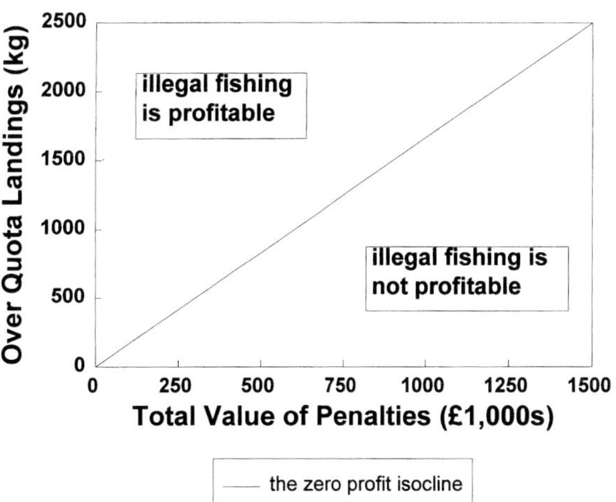

Fig. 2. Profitability depends on combinations of price, the level of over quota landings and penalties.

Regulation of catches by TAC is clearly very difficult to enforce and there is therefore an implicit economic incentive to misreport catch levels: a practice that undermines the scientific basis of stock assessment. It is unlikely that significant increases in fishery enforcement capability will improve this situation in the foreseeable future. Accordingly, some alternative basis for the regulation of fishing activity is needed.

An attractive alternative to the control of fishing by TAC is to control fishing mortality directly by restricting effort. A recent study by the European Union (Larsen *et al.* 1996) has looked at the problem in the context of the need for capacity reduction. The issues are relatively straightforward, but have until recently been little examined. What is required is an understanding of the relationship between some measure of fishing effort, for example days at sea, and the level of fishing mortality that will be imposed on stocks by this level of effort. In the case of mixed fisheries, for example, this is a difficult scientific problem.

Nevertheless, its solution has the potential for significantly improving fisheries management. Effort regulations are vastly easier to enforce than TACs as the probability of detection of an offence is high and thus the deterrent effect is significantly greater. The TAC or effort control issue is perhaps the most important at present, but it is by no means unique.

What is required is work which links the process of regulation intimately into the management measures used to control the fishing mortality. This is not happening at present, but the benefits could be enormous.

THE POTENTIAL OF NEW OR LIGHTLY EXPLOITED FISHERIES

In the last few years, there has been a marked decline in the rate of increase in the world fish catch and there is a well-supported view that some upper limit is being reached to the total catch that can be obtained from current fisheries (Garcia and Newton 1994; Pauly and Christensen 1995). There is at the same time a continuing increase in the world population and an increasing demand for fish products, which imply an increasing demand

Table 4. The potential of new or lightly exploited fisheries

Region	Area	Species	Estimated Potential	Source
Northwest Atlantic	Scotian Shelf Atlantic Ridge	Clam (*Spisula polynyma*) Deep sea ocean perch	Substantial	FAO (1994)
Western Central Atlantic		Cephalopods (Squid and Octopus) Small pelagic species Deep water shrimps Deep water snappers		FAO (1994)
	North coast S. America	Small pelagics (primarily clupeids engraulids and carangids)		
Eastern Central Atlantic	Mauritanian coast Northern shelf of Angola Western Mediterranean	Clams Small pelagic species Deep water slope fisheries — large hake and royal red shrimp	300 000 t	FAO (1992) FAO (1994) FAO (1994)
Southwest Atlantic	South Brazil Uruguay N. Argentina	Anchoita (*Engraulis anchoita*)	estimated 300 000 t	
Northwest Pacific	Korea	Squid		FAO (1994)
East Central Pacific		Squid		FAO (1994)
West Indian Ocean	Sri Lanka	Reef fish (snappers and groupers) Lantern fish		FAO (1994)
East Indian Ocean	Andaman and Nicobar Archipelago	Deep sea fisheries (snappers, groupers) Deep sea prawns Small pelagic species Squid	73 000 t	Sudarsan (1990)
	Continental slope of Myanmar	Deep water lobster (*Puerulus sewelli*) Penaeid shrimp (*Aristeus* spp.)		FAO (1984)
	India	Shrimp (caught as bycatch)	120 000 t	Gordon (1991)
Southern Ocean	Falklands and SW Atlantic	Kingclip (*Genypterus blacodes*)		
	S. Africa and Sub-Antartic regions	Toothfish		

for fish, a demand which is unlikely to be met by conventional fish resources.

As indicated in the earlier section, some of this demand might be met from improved management and the recovery of depleted species. However, there are likely to be quite severe limits to this as many increases in yields of some fisheries have come at the expense of decreases in others and the ecological balances are likely to be a constraint on any significant increase. This point is discussed further later in this section.

The other potential for increased fish catches lies in new or lightly exploited fisheries. In the last few years, FAO has documented a number of fisheries where some potential exists for an expansion of catch. These are relatively modest in size and scope and a tentative compilation is given in Table 4.

The main potential for new fisheries to expand to meet the likely shortfall in supply over the next two decades (estimated in some tens of millions of tonnes) is likely to lie with the fisheries for mesopelagic fish, squid and Antarctic krill.

Clearly, the sustainable management of new fisheries is going to be of critical importance. Fortunately the problems of overcapitalization that pose such difficulties in the management of current fisheries are absent, and there is thus real scope in the management of new fisheries for sustainability.

The key is that the level of fishing capacity needs to be in line with the productivity of the resource. In the case of single species or stocks there is a reasonably well founded theory on which the level of potential yield and hence the fishing capacity that can be allowed to develop can be assessed. Early work in assessing the capacity of new fisheries to sustain yield was based on the work of the late John Gulland. Gulland's rule, which has a number of derivations, is given by the equation:

$$Y = {}^{1}/_{2}MB_0$$

where Y = Maximum Sustainable Yield, M = Natural Mortality rate, and B_0 = Unexploited Biomass.

Recently, work by Kirkwood *et al.* (1994), extending earlier work by Beddington and Cooke (1983), has arrived at results which indicate that the basis for assessing yield is a relatively simple process which relates to the life history characteristics of the species concerned. The idea that life history parameters can act as a close guide to the potential for species to sustain yield is a beguiling one and dates back to Gulland's early work.

The results of Kirkwood *et al.* (1994) indicate that there is an underlying simplicity in the yield equations which is in line with Gulland's proposal. The yield as a proportion of underlying unexploited biomass has a simple relationship with the natural mortality rate and hence life span of the species.

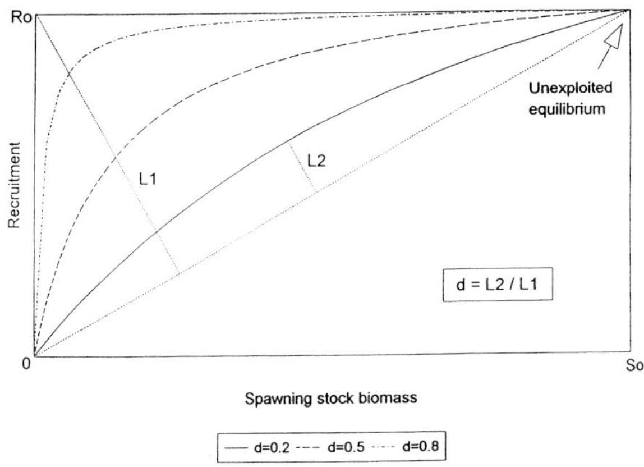

Fig. 3. Beverton-Holt stock recruitment relations with different degrees of density dependence (d).

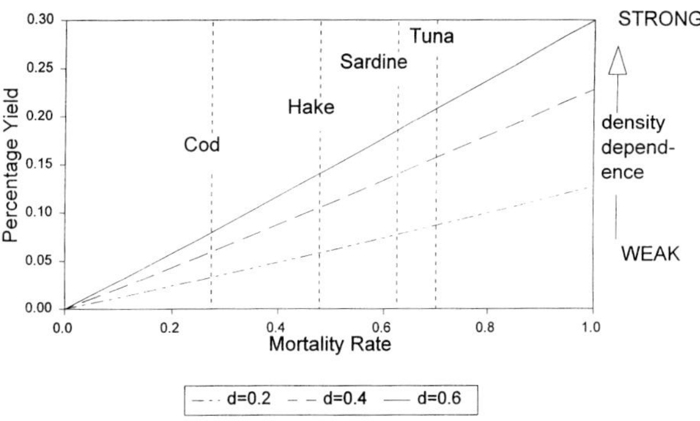

Fig. 5. General effect of differing degrees of density dependence (d) on the maximum sustainable yield of some typical species.

The assumption of constant recruitment produces a strictly linear relationship between yield and mortality rate. For other stock and recruitment relationships, Kirkwood *et al.* (1994) and Beddington and Basson (1994) have shown that the relationship is very close to linear over a very large range of mortality rates corresponding with the range of life spans encountered in exploited populations. The parameterization of one stock and recruitment relationship is illustrated in Figure 3. The relationship for different degrees of the stock and recruitment relationship is shown in Figure 4.

This basic analysis extends from the extremely long-lived, low mortality rate species such as orange roughy, where potential yield as a proportion of standing stock biomass is tiny, through to very short-lived annual species such as squid where the proportion of biomass that can be taken can be extremely high. Indeed, paradoxically, it can appear to be higher than estimates of initial abundance! This latter case proves to be a problem of definition of initial abundance.

The degree of density dependence and the way that this affects the level of yield can be seen in Figure 5. These results are deterministic, but extensions of the results, contained in Kirkwood *et al.* (1994) allow complications of age-dependent mortality, recruitment stochasticity, etc. to be incorporated. These results show that with very minor adjustments the basic formulae derived provide a good basis for calculating the expectation of potential yield.

Given available life history parameters for major fish stocks it is possible to assess what sort of levels of yield that stocks of different types can provide. Figure 6 illustrates this for some of the main types of stock. This allows a good first approximation to be made of the level of yield that an individual fishery can be expected to produce and hence can act as a reasonable indicator of the appropriate level of capacity that can be built up to exploit the fishery sustainably. It thus provides the basic guidelines for regimes which restrict entry to the fishery to the level that is sustainable.

This theory allows estimates to be made directly of the likely level of the capacity that new fisheries can sustain. What it fails to do is to indicate from the perspective of the ecosystem whether the level of the fishery is dependent on the changes that have previously taken place in the ecosystem. For example, the level of biomass may be artificially high due to previous depletion of important predators of the species concerned. In these situations, there is the obvious concern that changes in the ecosystem, recovery of important predators or increases in competition, can undermine the sustainability of a fishery for a

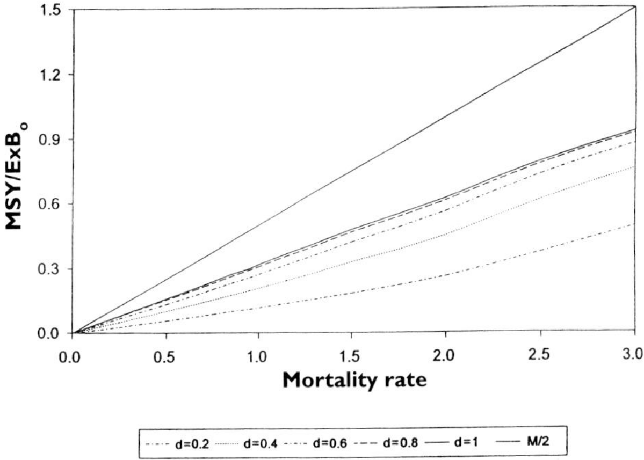

Fig. 4. Effect of differing degrees of density dependence (d) on the yield: exploitable biomass ratios.

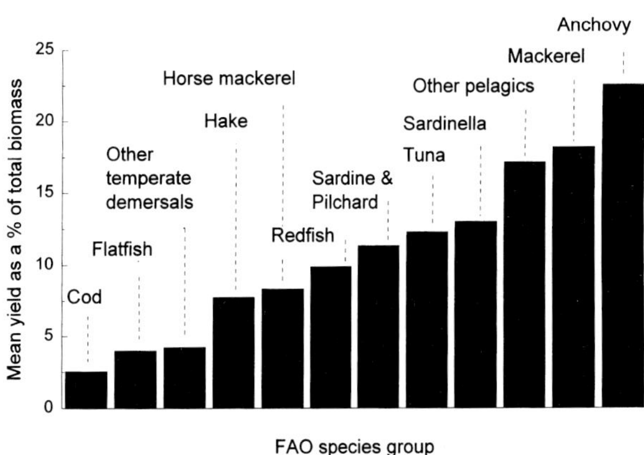

Fig. 6. Potential yield of major FAO species groups.

particular stock. There is little ecological theory to inform these concerns other than the simplest concepts of the effect of predator abundance on yields (Beddington and May 1983).

This problem has been addressed at the global level in the analysis performed by Pauly and Christensen (1995). They examined the level of primary productivity required to sustain fish catches at current levels, and point to very high levels of primary productivity required to sustain fish catches on the heavily fished continental shelves; around 30% of total productivity is required in some ecosystems. Unfortunately, there is no theory to indicate whether such levels of productivity requirements, or indeed levels of exploitation of the different components of the ecosystem, are sustainable. For example, it is obvious that if the catch composition from an ecosystem were to move more towards species lower in the trophic web, then the primary productivity requirements of the same tonnage of catch would be reduced.

This absence of theory is particularly unfortunate in the case of some of the new fisheries with the greatest potential. Krill and a number of the squid species that are likely to provide increased levels of catch in the future may well have increased in abundance due to previous depletion of predator populations, primarily marine mammals. There is thus a distinct possibility that substantial fisheries on some species can either delay recovery of depleted predators or be eroded as predators increase.

This possibility has led certain management bodies to invoke precautionary approaches to harvesting in an attempt to provide guidelines which are sufficiently conservative to preserve ecosystem interactions. Such precautionary guidelines are likely to be overly conservative from the viewpoint of harvest potential, but in the absence of theory they are unlikely to be improved upon. The Commission for the Conservation of Antarctic Marine Living Resources (CCAMLR) has set precautionary limits for krill catches based on very modest proportions of the available abundance, while fisheries for squid (Rodhouse in press) may similarly have to depend on estimates of predator consumption in which catch levels are determined by the constraint that they will not affect predator abundance because they are a sufficiently low proportion of predator consumption.

There is clearly a scientific need for some well developed theory in this area which will relate traditional population dynamics of standard fish stock assessment to models of energy flow and ecosystem dynamics.

Stock enhancement of capture fisheries

The third issue which affects the future of yields from capture fisheries is the possibility of increasing yield by artificial enhancement. When dealing with the issue of enhancement fisheries, a difficulty arises in that there is effectively a continuum between aquaculture and enhancement. An obvious basis for a distinction is that within most aquaculture systems there is an element of ownership. However, there is significant overlap in many of the basic methodologies, in particular those involving rearing young fish and artificially increasing the productivity of water bodies. Here much of the basic research has been done within the field of aquaculture. In his address to this Congress, Liao (1997) has indicated some of the major increases in technological achievements that have come to aquaculture in recent times.

The activities of enhancement can be considered in two ways: those associated with the rehabilitation of a fishery that has for some reason been reduced to an unproductive level, or those involving the improvement of a fishery or indeed the creation of a new fishery.

In marine systems, stock enhancement has been tried for many years and indeed as a technique to attempt to restore marine fisheries, it was extremely popular in the latter half of the 19thC and the early decades of the 20thC. However, there was almost no noticeable effect on the productivity of major stocks, which included Atlantic cod, pollack, winter flounder and Atlantic mackerel. Latterly, the focus of marine stock enhancement programmes has moved to less abundant species and focussed on a relatively modest scale of operations.

Blankenship and Leber (1995) have recently reviewed some of the main issues of marine stock enhancement. These include the problem of evaluating the success of enhancement programmes *via* some form of tagging and the need to successfully rear fish to the juvenile or larger size. Much work in this area has addressed the general issues of environmental responsibility and the need to focus on appropriate species for enhancement activities. There is an exciting future for this research, but it is fair to say that it is at an early stage of development.

By contrast, in freshwater there has been, as in aquaculture generally, more scientific work on enhancement fisheries. Although in terms of global impact these fisheries form a relatively modest proportion of world potential, they are crucial in a number of areas, particularly in Africa and Asia where poverty alleviation is a goal.

The principal mechanisms for stock enhancement are:

1. Stocking natural waters to improve recruitment or to alter the fish community towards a favoured species or to maintain a species which would not naturally breed in the system.
2. Introduction of new species to exploit underutilized parts of the food web or habitat uncolonized by indigenous species.
3. Engineering of the environment to improve or maintain levels of reproduction, shelter or some other resource.
4. Genetic modifications of material to increase growth to provide resistance to disease or improve tolerance to other environmental pressures.

These latter two mechanisms are close to the point on the spectrum where aquaculture enters.

Major enhancement programmes for commercial species in freshwater are a relatively new phenomenon compared to attempts in the marine field. One of the largest was the Third Fisheries Project in Bangladesh. In this floodplain fishery, three major carp species had declined in abundance leaving the potential for enhancement of these underexploited niches. This enhancement experiment involved the introduction of very high numbers of fingerlings in the size range 7–10 cm at a stocking density of around 20 kg per hectare. The programme proved successful, showing a tenfold increase in yield from the stocked fish. Economically, it also appears to be attractive, showing high rates of return.

Despite its success, this programme can only be considered as being at the level of an experiment. What is lacking is an understanding of the two variables that an enhancement fishery can control: the stocking density of young fish and the size at which they can be introduced. Some important insights into this have been made using rather simple models which are very similar in structure to those used for the stock assessment of fisheries. In Figure 7 the relationship between stocking density production and the proportion of the stock recaptured for a particular level of fishing mortality is shown following Lorenzen (1995). This highlights two important issues: that the harvest is optimized at some appropriate stocking density and that this harvest occurs at intermediate levels of stocking biomass and fishing mortality.

There are three control variables in a normal enhancement fishery. These are the level of fishing mortality that is imposed, the stocking density and the size at which young are introduced. In Figure 8, the relationship between yields and the two control variables of stocking density and fishing mortality are illustrated. These yield isopleth diagrams indicate the regions of overstocking and overfishing that can occur when the operation is suboptimal.

If this Figure is reproduced for different sizes of seed fish introduced then an appropriate mix of stocking density, fishing mortality and seed fish size can be determined.

Unfortunately, an assessment of the level at which the yield can be optimized as a function of stocking density and the size of seed fish turns out to depend on the degree of density dependence in growth. Such information is often unavailable and a sensitivity analysis (Lorenzen 1996) indicated the generic result that deviations from the optimal density for stocking are less in their effect on maximal production when the size of the seed fish is relatively small (see Fig. 9). This implies that in the absence of information there is considerable merit in using relatively small sizes of seed fish.

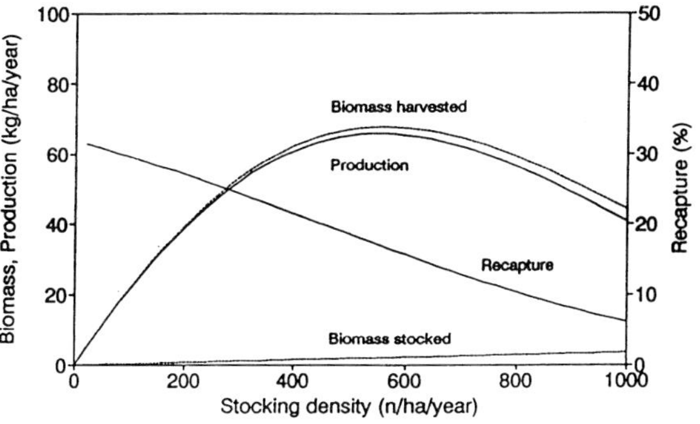

Fig. 7. Stock biomass, harvested biomass and production as a function of stocking density in numbers. Also shown is the fraction of seed fish recaptured in the fishery. Standard parameter values are used, gear selection length l_c of 30 cm, and very high fishing mortality.

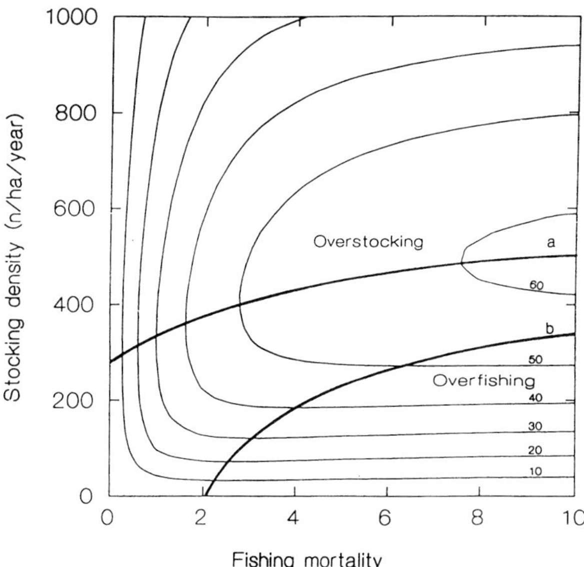

Fig. 8. Production as a function of fishing mortality and stocking density, for a gear selection length l_c of 30cm. Labels on the contour lines indicate production in kg/ha/year. The heavy solid line (a) indicates optimum stocking density in relation to fishing mortality, line (b) indicates optimal fishing mortality in relation to stocking density.

The early results indicate that basic improvements in the efficiency of enhancement fisheries can be achieved with relatively modest scientific input. The methods described, which use simple models of population dynamics, have the opportunity to provide guidelines for what can be expensive capital programmes.

CONCLUDING REMARKS

This paper has singled out three areas where scientific study has the potential to improve the future of capture fisheries. In the case of fisheries management, our understanding of the socio-economic imperatives of fishery operators and the importance and limitations of control, monitoring and surveillance need an integrated approach which can assess the trade off between

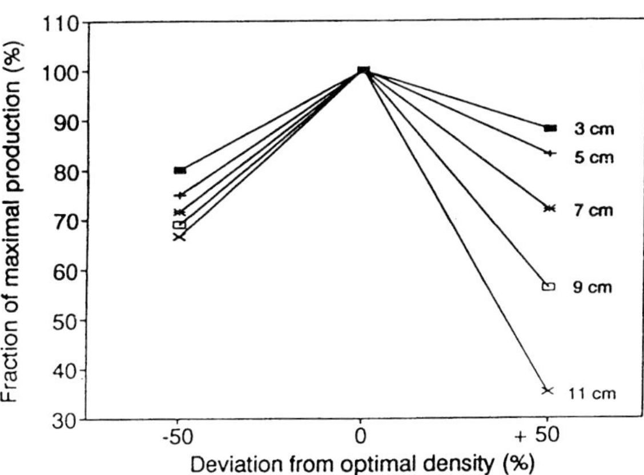

Fig. 9. Sensitivity of production to suboptimal stocking density, for different sizes of seed fish. The effect on production of a 50% deviation from optimal stocking density is shown.

controls on fishing mortality and their likelihood of implementation.

In the case of the sustainable management of new fisheries, there is a clear need to link the traditional methods of stock assessment with an understanding of ecosystem processes. Such work will require new methodologies and have new and substantial data demands.

The immediate future of enhancement fisheries looks promising as simple models of population dynamics and technology developed within aquaculture are applied. Here there is the potential for significantly increased productivity, but this will only be sustainable if capacity (or access) is controlled. Socio-economic investigations that parallel those investigating biological productivity are essential.

The challenges posed by work in the above areas are likely to be met by somewhat different interdisciplinary research groupings than have operated in fishing institutions in the past.

In concluding, it is important to recognize that the majority of people operating in capture fisheries lie outside the commercial fisheries sector and within the artisanal sector. Here the problems of overcapitalization and overexploitation are mirrored, but in somewhat different ways. Population pressure and poverty tend to be the driving mechanism for generating overexploitation, and community agreement on methods of control and regulation are essential. Where fisheries management authorities have a role is mainly in the allocation of rights and the avoidance of conflict, particularly with commercial fisheries.

The difficulty in dealing in any generic way with artisanal fisheries lies in the individual complexity of the fisheries involved and particularly in the social structure of fishing communities. This problem will need to be addressed if the artisanal sector of world fisheries, which is largely unreflected in global statistics, is going to continue to support substantial communities.

Significant progress in this has been made, primarily in the freshwater sector where the issues of overexploitation, ownership and management of the fisheries are rather stark. It is far more complicated in marine and coastal communities. Work on the linkages between the social, environmental and ecological aspects of the resource in the context of alternative or supplementary occupations for the community is needed. This is a central component of the future of capture fisheries.

References

Anon. (1995). The protection of Scottish fisheries. National Audit Office London:HMSO: HC 28 Session 1995–96.

Beddington, J R, and Basson, M (1994). The limits to exploitation on land and sea. *Philosophical Transactions of the Royal Society of London* **B 343**, 87–92.

Beddington, J R, and Cooke, J G (1983). On the potential yield of fish stocks. *FAO Technical Papers* 242, 1–47.

Beddington, J R, and May, R M (1983). Maximum sustainable yields in systems subject to harvesting at more than one trophic level. *Mathematical Biosciences* **43**, 64–85.

Beverton, R J H, and Holt, S J (1957). 'On the dynamics of exploited fish populations'. *MAFF Fisheries Investigations London*, Series 2, **19**, 1–533.

Blankenship, H L, and Leber, K M (1995). A responsible approach to marine stock enhancement. *American Fisheries Society Symposium* **15**, 167–75.

FAO (1984). Marine fisheries resources survey and exploratory fishing, Burma. Project findings and recommendations. FN-DP/BUR/77/003. 40 pp.

FAO (1992). Report of the ninth session of the CECAF working party on resource evaluation. *FAO Fisheries Report* 454. 79 pp.

FAO (1994). Review of the state of the world marine fishery resources. *FAO Fisheries Technical Paper* 335. 136 pp.

Garcia, S M, and Newton, C (1994). Current situation, trends and prospects in world capture fisheries. Presented at Conference on Fisheries Management, Global Trends, Seattle.

Gordon, A (1991). The by-catch from Indian shrimp trawlers in the Bay of Bengal: the potential for its improved utilization. FAO BOBP/WP/68. 1–27.

Kirkwood, G P, Beddington, J R, and Rossouw, J A (1994). Harvesting species of different lifespans. Symposium of British Ecological Society Large-Scale Ecology and Conservation Biology. pp. 199–227 (Blackwells: Oxford).

Larsen *et al.* (1996). The report of the group of independent experts to advise the European Commission on the Fourth Generation Multi-Annual Guidance Programmes.

Liao, I C (1997). How can aquaculture help sustain world fisheries. In 'Developing and Sustaining World Fisheries Resources: The State of Science and Management. Second World Fisheries Congress, Brisbane 1996'. (Eds D A Hancock, D C Smith, A Grant and J P Beumer) pp. 431–6 (CSIRO Publishing: Melbourne).

Lorenzen, K (1995). Population dynamics of culture-based fisheries. *Fisheries Management and Ecology* **2**, 61–73.

Lorenzen, K (1996). Culture-based fisheries and extensive aquaculture in Asian reservoirs: populations dynamics, assessment and management. Ph.D Thesis: University of London.

Mace, P M (1997). Developing and Sustaining World Fisheries Resources: The State of the Science and Management. In 'Developing and Sustaining World Fisheries Resources: The State of Science and Management. Second World Fisheries Congress, Brisbane 1996'. (Eds D A Hancock, D C Smith, A Grant and J P Beumer) pp. 1–20 (CSIRO Publishing: Melbourne).

Pauly, D, and Christensen, V (1995). Primary production required to sustain global fisheries. *Nature* **374**, 255–7.

Rodhouse, P G (in press). Precautionary measures for a new *Martialia hyadesi* (Cephalopoda, Ommastrephidae) fishery in the Scotia Sea: an ecological approach. *Commission for the Conservation of Antarctic Marine Living Resources Science.*

Sudarsan, D (1990). Marine fishery resources potential in the India exclusive economic zone: an update. *Bulletin of Fishery Surveys in India* **20**, 1–37.

Sutinen, J G (1987). Enforcement of the MFCMA: an economist's perspective. *Marine Fisheries Review* **49**(3), 36–44.

Sutinen, J G, and Anderson, P (1985). The economicis of fisheries law enforcement. *Law Economics* **61**, No.4. November 1985.

Recent Advancements in Environmentally Friendly Trawl Gear Research in Australia

D. T. Brewer,[A] *S. J. Eayrs,*[B] *N. J. F. Rawlinson,*[A] *J. P. Salini,*[A] *M. Farmer,*[A] *S. J. M. Blaber,*[A] *D. C. Ramm,*[C] *I. Cartwright*[B] *and I. R. Poiner*[A]

[A] CSIRO Division of Marine Research, Cleveland, Queensland, 4163 Australia.
[B] Australian Maritime College (AMC), Launceston, Tasmania, 7250 Australia.
[C] Northern Territory Department of Primary Industry and Fisheries (NTDPIF), Darwin, Northern Territory, 0800 Australia.

Summary

The effects of demersal trawling on marine communities in Australia are a major concern for the industry, managers, conservation agencies and the Australian public. Australian researchers have recently developed and tested fish trawls that decrease the impact on benthic communities and unwanted bycatch populations in tropical waters; and developed and tested a range of bycatch reduction devices for prawn trawls in New South Wales and North Eastern Australia. Scientific trials of several devices demonstrated significant reduction in the amount of unwanted bycatch, exclusion of turtles and other large animals, and one — the Super Shooter — maintained commercial catches of prawns. Planned commercial trials will improve the performance of these devices and facilitate their implementation into Australia's fisheries. When BRDs are widely adopted the decreased impact of trawling on bycatch populations will address key issues relating to their impact on marine communities.

Introduction

There is a widespread and increasing requirement for using bycatch reduction devices (BRDs hereafter) in trawl fisheries throughout the world. The pressure for these changes to fishing practices has come from different sources including (i) conservation groups that have lobbied government to protect marine bycatch species (e.g. Watson and Seidel 1980), (ii) scientific findings that have reported species community changes due to trawling (e.g. Kennelly 1995; Sainsbury *et al.* in press), (iii) pressure to decrease catches of species that are targets of other fisheries (e.g. Watson *et al.* 1993; Kennelly 1995; Broadhurst and Kennelly 1995*a*) or juveniles of the same fishery (Isaksen *et al.* 1992). In the past seven years, these same pressures have prompted Australian fisheries research funding agencies to support several projects that were specifically designed to develop, test and commercially-trial gears that reduce the amount of unwanted bycatch in trawl fisheries.

Prawn trawl research

Scientists from the New South Wales Fisheries Research Institute have been studying bycatch reduction techniques in New South Wales offshore and inshore prawn trawl fisheries since 1989. These studies have included description of the interaction between shrimp trawling and other commercial and recreational fisheries (Kennelly *et al.* 1992), and the development and testing of several BRDs. They include square-mesh panels (Broadhurst and Kennelly 1994, 1996; Broadhurst *et al.* 1996*b*), Nordmøre grids and other separator panels (Andrew *et al.* 1993; Broadhurst *et al.* 1996*a*). These projects have reported significant reduction in unwanted bycatch without significant loss of prawns. More importantly, they have tested these devices in close collaboration with the industry, which has resulted in widespread voluntary adoption of BRDs in these fisheries (Broadhurst *et al.* 1996*a*).

Fishing Technologists and Scientists from the Northern Territory Department of Primary Industry and Fisheries (NTDPIF) and Queensland Department of Primary Industries have developed and tested a BRD, known as the AusTED (Australian trawl efficiency device), aboard commercial trawlers in subtropical Australian prawn fisheries. This device has a flexible grid to exclude large animals such as turtles, and escape openings and meshes to exclude smaller bycatch (Mounsey *et al.* 1995). Research trials showed no differences in prawn catches between the control and AusTED-equipped nets, but sea turtles and large stingrays were excluded from the AusTED-equipped net and non-commercial bycatch was significantly reduced at most sites trawled (Robins-Troeger *et al.* 1995).

Fish trawl research

In response to growing evidence that demersal fish trawling alters benthic habitats and, as a consequence, demersal fish community structure (Sainsbury 1987; Ramm *et al.* 1993; Sainsbury *et al.* in press), research workers have recently developed and tested fish trawls that operate semi-pelagically (ground rope off the sea bed). Ramm *et al.* (1993) modified a fork-rigged semi-pelagic trawl for use in the tropical Australian fish trawl fishery. They found that this trawl — known as the 'Julie Anne trawl' — had similar catch rates and caught similar sizes of target species (mainly *Lutjanus* spp.) to a standard demersal trawl, but caught less unwanted bycatch and had less impact on the substrate. Brewer *et al.* (1996) tested a semi-pelagic version of a standard demersal wing trawl and obtained similar results. When trawled with the foot rope 0.4–0.5 m above the sea bed, the semi-pelagic trawl caught the same amount of commercially important *Lutjanus* spp. but significantly less fish bycatch and benthic animals than the demersal version of the trawl. The use of semi-pelagic trawls is a requirement in Australia's Northern Fish Trawl Fishery (Newton *et al.* 1994), and the above studies have shown that using a trawl rigged in this manner is more environmentally friendly than previous demersal trawl methods, thus greatly enhancing ecological sustainability of the fishery. There has also been some recent research into fish trawl mesh size-selectivity. Broadhurst and Kennelly (1995*a*) showed that increasing the mesh size in the body of fish trawls from 90 to 100 mm significantly decreased catches of several unwanted fish species in the New South Wales inshore otter-trawl finfish fishery.

Recent trawl research in tropical Australia

In 1993, the Australian Fisheries Research and Development Corporation (FRDC) funded a 3-year collaborative project between the Australian Maritime College (AMC), the Commonwealth Scientific and Industrial Research Organisation (CSIRO) Division of Fisheries, and the Northern Territory Department of Primary Industry and Fisheries (NTDPIF), to develop and test trawls designed to reduce unwanted bycatch and benthos from tropical Australian trawl fisheries. Beside the development of a semi-pelagic fish trawl (Brewer *et al.* 1995), the main thrust of the project was to trial a range of BRDs for potential use in Australia's \$A130 million p.a. Northern Prawn Fishery (NPF hereafter).

This paper summarizes the results from research on several BRDs and describes the first assessment of damage and survival of fishes in tropical fisheries after their escape through square-mesh cod-ends. The impending increased use of environmentally friendly fishing gears in Australian trawl fisheries is also discussed.

Environmentally friendly trawl gear research in the NPF

Environmentally friendly trawl gear research in the NPF was conducted during a series of five cruises in or just outside Albatross Bay (12°30′ S, 142°35′ E) in the Gulf of Carpentaria between November 1993 and November 1995. This research included testing a new semi-pelagic fish trawl (Brewer *et al.* 1995), testing 17 different BRDs or combinations of different BRDs, measuring damage and survival of selected bycatch species after passing through a square-mesh cod-end, and using a multi-level beam trawl to examine the effect of a lower headline height on prawn trawl catches. Video data were also collected to increase the knowledge of the reactions of prawn and bycatch species to trawl nets and to BRDs.

Descriptions of BRDs

The BRDs described below were tested individually and some formed part of a composite BRD made from two devices. Some of the BRDs described below were tested in preliminary trials only. Their catch results can be obtained from the authors. The selection process for the continued testing of a BRD depended on whether the device produced some of the most promising results or whether modification could improve the results already obtained. Only the results from BRDs selected for the final scientific trials are reported here.

Nordmøre grid: This design originated in Norway (Valdemarsen *et al.* 1993). Our version was a 750 mm × 1140 mm inclined aluminium grid angled at about 55° from the vertical, with a bar spacing of 100 mm (Fig. 1a). A panel of netting guided the catch towards the bottom of the grid. Large animals (most commonly turtles, sharks and rays) and some small animals, contact the grid which guides them through a triangular-shaped escape opening in the cod-end at the top of the grid. Smaller species such as prawns can pass between the grid bars into the cod-end. This grid was tested in conjunction with two separate orientations of a fish eye and two different versions of square-mesh windows (0 meshes and 5 meshes behind the grid). These devices were placed in the top of the cod-end behind the grid exit.

Super Shooter: The Super Shooter was first introduced into the United States Gulf of Mexico shrimp fishery in the late 1980s to exclude turtles (Mitchell *et al.* 1995). A funnel of netting guides

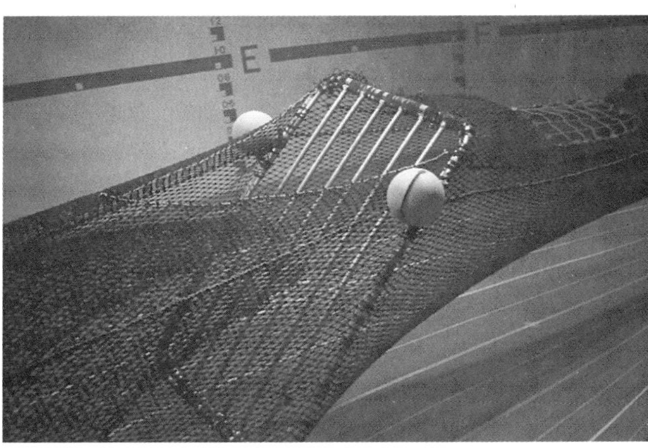

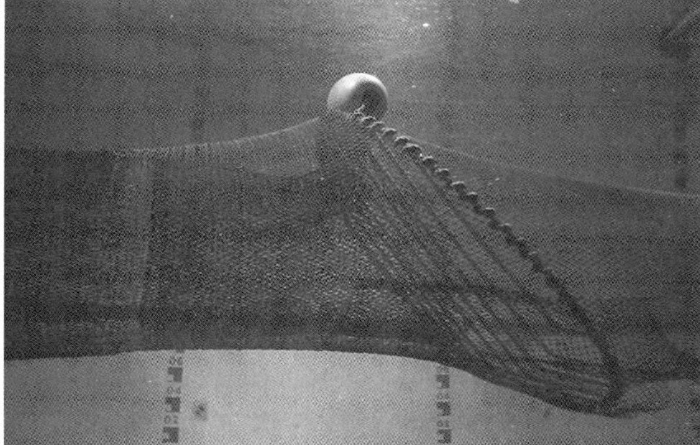

Fig. 1. Photographs of two bycatch reduction devices suspended in the Australian Maritime College flume tank: (a) the Nordmøre grid (escape opening at the top), (b) the Super Shooter (covered escape opening at the bottom)

the catch towards an inclined grid — at about 45° from the vertical with 100 mm bar spacing — where large animals and sea bed debris are guided downwards through an escape opening in the floor of the cod-end (Fig. 1b). A flap of buoyant polyethylene netting covers this escape opening to minimize prawn loss. The Super Shooter was also trialled in conjunction with two versions of a fisheye placed in the top of the cod-end behind the grid.

Square-mesh cod-ends: Prawn trawl cod-ends are usually made of diamond-mesh netting. Diamond meshes close during a tow due to the weight of the catch. Square-meshes maintain their open shape during the trawl, allowing a greater degree of selectivity in the trawl. Square-mesh cod-ends were used as a possible means of reducing the catch of small fish. Two sizes of square-mesh (38 and 45 mm bar length) were trialled.

Square-mesh windows: Square-mesh windows have been investigated as a means of small fish bycatch reduction since the early 1980s (e.g. Robertson 1983; Arkley 1990) and were first studied and used in penaeid fisheries in 1992 (Broadhurst and Kennelly 1994). The device used in our study was made of a 150 mm (6") square-mesh polyethylene netting panel measuring 8 bar lengths wide by 13 bar lengths long, fitted to the top, front section of the cod-end. Small animals with swimming ability can escape upwards out of this window, while most of the poorer swimmers (e.g. prawns) pass into the cod-end. Square-mesh windows were tested in conjunction with a Nordmøre grid. (Fig. 1a).

Three other variations of the square-mesh window were also tested during preliminary trials (results can be obtained from the authors). One with a hummer stimulator device placed 5 meshes behind the window (as used for the Super Shooter); another with a 1.5 m long canvas black cylinder mounted around the cod-end also just behind the window (Glass *et al.* 1995*a*, *b*); and another with the window constructed of square-mesh netting that glows in the dark (developed in Japan by Nichimo Pty Ltd). The 'hummer' and black cylinder were designed to stimulate fish to turn and swim forward, thereby improving their chances of escape through the square-mesh window. The glow netting emits a green glow under dark conditions highlighting the square-mesh window at night.

Fisheye : The fisheye (named from its shape) was developed for the Gulf of Mexico shrimp fishery (Watson *et al.* 1993) and is a steel frame fitted into the top of the cod-end to provide a small elliptical opening to allow fish to escape. The elliptical opening faces forward in the top of the cod-end. Animals pass into the cod-end and must turn to swim forward and upwards to escape through this device. The fisheye breaks the continuum of the panel of netting in the cod-end and the turbulence that results from this may attract swimming animals near the exit hole and so assist their escape.

Trials of BRDs

BRDs were tested on the 66 m CSIRO research vessel, FRV *Southern Surveyor,* by towing dual-rigged prawn trawls and comparing catches between trawls containing different BRDs. For each trawl, cod-ends with different BRDs were switched to give a series of comparisons between all BRDs tested, including a standard cod-end with no BRD. The results summarized here are from a series of half-hour trawls using square-mesh cod-ends (Brewer *and* Eayrs 1994) and two-hour trawls using 8 different BRDs. Estimates of mean catches of prawns and bycatch, for each BRD, were obtained using a linear model (Proc Mixed, SAS 1988) which takes into account variation between hauls and nets. Mean figures were compared against the standard net to obtain percentage differences in prawn retention and bycatch loss. Further details of the methods used can be obtained from the authors.

BRD performance

The performance of BRDs in prawn trawling is measured by a combination of several criteria, such as:

1. weight and numbers of the target species caught;
2. condition (= $ value) of target species caught;
3. hydrodynamic performance of the trawl containing the BRD;
4. ease of use of the BRD;
5. weight and numbers of bycatch excluded;
6. survival rate of excluded bycatch.

The first three criteria relate directly to fishing profitability, by way of the amount and quality of the target species or by the cost of fuel associated with towing a 'light' or 'heavy' trawl rig. Criteria 4 and 5 directly affect the ease and safety of the fishing procedure and criteria 5 and 6 affect the impact of the trawling activity on the marine ecosystem.

Australian fisheries managers have progressed to using a precautionary style of fisheries development that will promote ecological sustainability at an ecosystem level. Hence, the main aim of BRDs is to substantially reduce the amount of unwanted, non-target organisms in trawl catches.

It is also important that BRDs maximize the overall profitability of the trawling procedure; should be able to be used simply and safely; and should significantly increase the amount of bycatch that survives the trawl procedure compared to trawling without the BRD. Maintaining profits can result from maintaining catches of target species, increasing the value of target species, decreasing fuel costs, or combinations of these.

The following results from BRD trials do not report on all of these criteria because not all were measured. However, future research should aim to assess as many different types of performance criteria as possible.

Prawn catches

Australian sea-caught prawns are of high export quality and are worth around $A230 million per year (Battaglene *et al.* 1996). Almost all are exported to overseas markets such as Japan where a high standard of quality is essential. It is predicted that growth in demand for Australian sea-caught prawns will depend on their ability to supply a cleaner, higher quality product such as near-perfect condition prawns for tray-packing (Battaglene *et al.* 1996). BRDs may play an important role in this value-adding process and help to keep the Australian prawn trawl industry in demand in export markets around the world.

Prawn catches in the study area during November 1995 were dominated by the grooved tiger prawn, *Penaeus semisulcatus* (58% by weight), and the red endeavour prawn, *Metapenaeus ensis* (34% by weight). Other commercially important prawns caught included *P. esculentus* (2.2%), *M. endeavouri* (1.2%), *P. merguiensis* (0.2%), *P. latisulcatus* (0.1%) and *P. monodon* (0.1%). Commercially unimportant species made up less than 4% of prawn catches by weight.

There was no significant prawn loss from cod-ends with the Super Shooter + fisheye 1 or the fisheye on its own (Table 1). The Super Shooter + fisheye 2 produced significant losses of prawn weights, but not prawn numbers. Significantly lower catches of prawns were measured for the AusTED, Nordmøre grid + fisheye, and Nordmøre grid + square-mesh windows. However, the AusTED and Nordmøre grid did not lose prawns in other unpublished trials and both have the potential to be effective in the NPF. Moreover, there were less prawns lost from the second versions of both the Nordmøre + fisheye and the Nordmøre + square-mesh window combinations, where the fisheye and square-mesh window were set farther back from the top of the grids and the fisheye was oriented differently.

Square-mesh cod-ends (45 mm mesh) were able to reduce the amount of unwanted bycatch by about one-third while maintaining catches of commercially valuable prawns. Perhaps more importantly, this cod-end allowed the escape of between 58 and 98% of under-commercial-sized tiger prawns (Rawlinson, Brewer and Eayrs unpublished data). The proportion of these that survived to contribute to future stocks is unknown.

Fish catches

In the NPF, small fish make up the bulk of the unwanted bycatch (Pender and Willing 1989). The reduction of small fish in the catch can have an effect on fishing profitability. A decrease in the bulk of small fish bycatch can speed up sorting time, may decrease damage to individual prawns and increase shot time. These could lead to a higher-value product and a more efficient fishing process. However, an important repercussion of decreasing the fish bycatch is the lighter impact on the

Table 1. Bycatch exclusion and prawn retention of bycatch reduction devices during November 1995

Bycatch reduction device	No. trawls	No. prawns caught (%)	Weight prawns caught (%)	Small fish excluded (% weight)	No. large elasmobranchs (>5 kg)	No. turtles caught	No. sea snakes caught
AusTED	15	78.0**	75.2***	26.6***	3	0	3
Super Shooter + fisheye 1	15	95.4	98.2	16.3*	3	0	8
Super Shooter + fisheye 2	21	89.5	89.4*	13.5**	0	0	12
Nordm¿re grid + fisheye 1	15	81.0*	83.3**	30.9***	1	0	8
Nordm¿re grid + fisheye 2	22	85.3**	86.1**	27.5***	1	0	6
Nordm¿re grid+ square-mesh window 1	15	66.0***	62.3***	38.9***	1	0	4
Nordm¿re grid+ square-mesh window 2	22	83.7**	84.4***	28.4***	1	0	4
Fisheye	15	89.7	92.0	10.7	11	4	7
Standard trawl	36				24	7	15

Catch differences that are significantly different from the catches of the standard prawn trawl are shown thus: * $P < 0.05$; ** $P < 0.01$; ***$P < 0.001$. nd, no data.

populations of these species and hence, the marine ecosystem that supports this fishery.

The fish bycatch is made up of over 250 species, most individuals weighing less than 300 g. The species composition is not described here but consisted of a highly diverse range of species similar to those described by Blaber *et al.* (1990), Ramm *et al.* (1990) and Martin *et al.* (1995). Details of the bycatch species composition can be obtained from the authors.

All BRDs except the fisheye significantly reduced catches of small fish. Highest exclusion was achieved by the Nordmøre grid + square-mesh window, but the Nordmøre grid + fisheye and the AusTED, all excluded more than 26% of small fish bycatch. The Super Shooter–fisheye combination excluded about half this amount of small fish bycatch.

Large animal catches

Large animals — known as monsters — are highly undesirable in prawn trawl catches for two main reasons. Firstly, they include species that are vulnerable or endangered, such as turtles; and secondly, large sting rays, sharks and turtles can damage and therefore devalue prawns by crushing them in the cod-end. Other reasons to exclude monsters from catches are to avoid the difficulties, dangers and time delays associated with handling them on the deck, to decrease the damage they cause to fishing gear and to minimize impacts on the marine community (Brewer *et al.* 1995).

The catch of 'monsters', was made up of 11 sea turtles (eight flatback sea turtles, *Natattor depressa*, two loggerhead sea turtles, *Caretta caretta*, and one Olive Ridley sea turtle, *Lepidochelys olivacea*), two sciaenid fishes and 43 elasmobranchs. The elasmobranchs were comprised of 27 rays (mainly 18 *Himantura toshi*, and 4 *Pastinachus sephen*), 13 shovelnose rays (12 *Rhynchobatus djiddensis* and 1 *Rhinobatus typus*), two shark rays, *Rhina ancylostoma*, and one sawfish, *Anoxypristis cuspidata*.

No sea turtles were caught in any of the cod-ends with BRDs containing excluder grids (125 trawls), whereas the two cod-ends without these grids (51 trawls) caught 11 sea turtles (one every 4.6 trawls). Cod-ends with BRDs containing the excluder grids caught 10 large (>5 kg) elasmobranchs (one every 12.5 trawls) as opposed to 35 (one every 1.5 trawls) caught in the two cod-ends without the grids (fisheye and standard trawl).

Sea snake catches

Like turtles, sea snakes are protected by law in Australian waters (Regulation 46, Schedule 1, National Parks and Wildlife Conservation Act 1975) but are caught as bycatch in the NPF. They are of no value to prawn fishers, and in fact, they may be an added danger to crew while sorting the catch. Decreasing catches of sea snakes would avert any future pressure on the prawn trawling industry by conservation groups, improve safety and further minimize impacts on the marine community.

Lowest catch rates of sea snakes were seen in cod-ends containing the AusTED — one every five trawls — and the two Nordmøre grids with a square-mesh windows (one every four or five trawls). The other five devices performed the same or slightly worse than the standard trawl catching about one sea snake every two trawls.

Bycatch damage assessment and survival

An important factor when assessing the effectiveness of BRDs is whether bycatch that is excluded from the trawl survives. Since BRDs usually only exclude some of the unwanted bycatch, high survival rates of the excluded bycatch are essential to fulfil the aim of minimizing the trawl impact on the marine community. Survival of excluded bycatch will probably vary depending on the type of BRD used. This study describes the damage to and survival of fish bycatch after passing through square-mesh cod-ends, and represents the first research of its kind in Australian waters.

Studies of the damage and survival of selected species were made from the *FRV Southern Surveyor*, the *FV Milana J*, and the *FV Island Girl*. For damage assessment, fish were retained in a cod-end cover, then brought on board using a large water-filled scoop. This minimized damage to fish during the period the cod-end was lifted on-board. Fish were then immediately killed with the anaesthetic MS222 (Tricaine methane sulphonate), and examined for damage. Damage to the jaws, preoperculum, operculum, eye, dorsal fin, anal fin, caudal fin and an estimate of scale loss were measured for all fish. Damage, along with other explanatory variables were included in multivariate analysis of variance to assess the differences in damage to fish between the diamond-mesh and 2 square-mesh cod-end types. Details of the methods can be obtained from the authors.

All five of the fish species analysed — *Apogon poecilopterus, Leiognathus splendens, Sardinella albella, Saurida micropectoralis* and *Upeneus sulphureus* — had very low levels of damage after escaping from either the 45 mm square or the diamond-mesh cod-ends. Higher levels of damage were measured from fish that had escaped the 38 mm square-mesh cod-ends, but this damage may not be enough to significantly reduce their survival.

Survival of fish passing through 45 mm square-mesh cod-ends was assessed by retaining cod-end escapees in a cod-end cover and transferring them to either moored sea cages or large, land-based pools for ten days. Their survival rate was compared with fish that had passed through an open cod-end and into the cover, and also fish caught on hook-and-line which were kept under the same conditions. Nine-fathom (16.4 m) and four-fathom (7.3 m) prawn trawls were used to capture the fish in two separate experiments. The Fisher Exact test (SAS 1988) was used to test for differences between the number of fish surviving after passing through the square-mesh and the number of fish surviving that had not passed through the square-mesh. Details of these experiments can be obtained from the authors.

This study shows that survival varies between species and that a significant proportion of the escapees will survive after passing through a square-mesh cod-end. Of the fish that passed through square-mesh cod-ends and were kept in pools, 83% survived (Table 2). The survival rate of fish in cages was lower: 26% (May 1994) and 21% (August 1994), but these results were considered not to accurately reflect true survival rates. These survival rates are conservative; under normal trawling

Table 2. Survival of fish that passed through a 45 mm square-mesh (Treatment) or open cod-end (Control), and including experiments made in sea cages and large land-based pools

Experiment	No. days	No. taxa	Treatment			Control		
			No. fish	No. survived	% Survival	No. fish	No. survived	% Survival
May 1994 (Cage)	10	3	124	32	26	39	22	56
August 1994 (Cage)	8	7	250	53	21	118	27	15
August 1994 (Pool)	8	4	123	102	83	86	79	92

conditions, fish escaping through a cod-end are not subject to the added stress of handling and being held in captivity.

Results for industry

In the last few years there have been some important advancements in the development of environmentally friendly trawl gears in Australia. These include new versions of gears such as the semi-pelagic fish trawls, designed to minimize impacts on the benthos and reduce bycatch (Ramm *et al.* 1993; Brewer *et al.* 1995), and the AusTED, a new design to reduce bycatch in Queensland east coast prawn trawls (Robins-Troeger *et al.* 1995).

Australian researchers have also successfully applied proven BRD technology from overseas work into local prawn trawl fisheries, the most successful being the voluntary adoption of square-mesh panels and Nordmøre grids into the NSW prawn trawl fisheries (Broadhurst and Kennelly 1994, 1995*b*, 1996; Broadhurst *et al.* 1996*a*, *b*). Our study in the NPF has shown that several grid devices can virtually eliminate catches of turtles and other unwanted large animals. One of these devices, the Super Shooter (developed in the Gulf of Mexico prawn trawl fishery), eliminates turtles, large epibenthic animals and other seabed debris, reduces fish bycatch, and maintains prawn catches in NPF waters. Other devices may also have the potential to significantly reduce bycatch while not reducing valuable prawn catches. Their performance will be improved during trials on commercial boats, providing industry with a larger range of effective BRDs for use in the NPF and other Australian prawn trawl fisheries.

Industry adoption

Experience from other fisheries in the world has shown that it is not easy to achieve widespread adoption of environmentally friendly trawl gear into the fishing industry. For example, the introduction of legislation to enforce TEDs in the US Gulf of Mexico fishery caused animosity between fishers and fishery managers as well as an aggressive legal challenge. It has taken researchers years to rebuild the relationship with industry to the point where fishers are now heavily involved in the continuing improvement of TED design and performance (Mitchell *et al.* 1995). Through a process of collaboration and liaison, fishers have learned that TEDs can reduce the amount of labour required to sort the catch as well as improve the quality of their shrimp product. As described earlier, this approach has also been successfully used in NSW prawn trawl fisheries (Broadhurst *et al.* 1996*a*) and should be the next major step towards the introduction of the recently developed environmentally friendly trawl gear to other northern Australian trawl fisheries.

In recent years there have been greatly increased pressures on Australia's trawl fisheries to adopt more environmentally friendly trawl practices. Under recent fisheries Acts, Commonwealth and nearly all State fisheries management agencies are required to manage fisheries under their control in an ecologically sustainable way. There are also impending pressures under Australia's Endangered Species Act to list trawling as a 'key threatening process' based on its alleged impact on endangered or vulnerable species such as certain species of sea turtles and sea snakes. If this submission is upheld the fishers may have to use BRDs that will eliminate catches of vulnerable or endangered species (e.g. sea turtles) in order to protect survival of the industries. Export markets are also under threat following the recent US court order to implement a ban on shrimp imports from nations not using turtle excluder devices (TEDs). Although most of Australia's exported prawns go elsewhere, this ban may indirectly add much greater competition from countries looking for new markets. These mounting pressures on Australia's prawn trawl fisheries are likely to be the catalyst for an escalation in the adoption of more environmentally friendly fishing practices.

Acknowledgments

Brian Taylor, John Watson, Steve Kennelly and Julie Robins kindly provided comments on the manuscript. We also thank the crews of the *FRV Southern Surveyor*, *FV Milana J* and *FV Island Girl.* for their assistance on research cruises. We are grateful to the many scientific staff for their assistance in the field. Brian Macdonald, Marcus Strauss, and Anders Cormie also assisted with BRD construction and gear trials. This project was partly funded by a FRDC grant (number 93/179).

References

Andrew, N L, Kennelly, S J, and Broadhurst, M K (1993). An application of the Morrison Soft TED to the offshore prawn fishery in NSW, Australia. *Fisheries Research* **16**, 101–11.

Arkley, K (1990). Fishing trials to evaluate the use of square mesh selector panels fitted to nephops trawls — MFV Heather Sprig (BCK 181) November/December 1990. Sea Fish Industry Authority, Seafish Report No. 383. 21 pp.

Battaglene, T, Standen, R, and Smith, P (1996). Outlook for the Australian fishing industry in 1996. *Professional Fisherman* March, 26–31.

Blaber, S J M, Brewer, D T, and Harris, A N (1990). Biomass, catch rates and abundances of demersal fishes, particularly predators of prawns, in a tropical bay in the Gulf of Carpentaria, Australia. *Marine Biology* **107**, 397–408.

Brewer, D T, and Eayrs, S (1994). New prawn and fish trawl gear reduces non-target bycatch. *Australian Fisheries* **53**(3), 9–11.

Brewer, D T, Eayrs, S, and Rawlinson, N J F (1995). Bycatch reduction devices show promise in the NPF. *Australian Fisheries* **54**(5), 24-6.

Brewer, D T, Eayrs, S, Mounsey, R P, and Wang, Y (1996). Assessment of an environmentally friendly, semi-pelagic fish trawl. *Fisheries Research* **26**, 225–7.

Broadhurst, M K, and Kennelly, S J (1994). Reducing the by-catch of juvenile fish (mulloway) in the Hawkesbury River prawn-trawl fishery using square-mesh panels in cod-ends. *Fisheries Research* (Amsterdam) **19**, 321–31.

Broadhurst, M K, and Kennelly, S J (1995*a*). Effects of an increase in mesh size on the catches of fish trawls off New South Wales, Australia. *Marine and Freshwater Research* **46**, 745–50.

Broadhurst, M K, and Kennelly, S J (1995*b*). A trouser-trawl experiment to assess cod-ends that exclude juvenile mulloway (*Argyrosomus hololepidotus*) in the Hawkesbury River prawn-trawl fishery. *Marine and Freshwater Research*. **46**(6), 953–58.

Broadhurst, M K, and Kennelly, S J (1996). Effects of the circumference of cod-ends and a new design of square-mesh panel in reducing unwanted bycatch in the New South Wales oceanic prawn-trawl fishery, Australia. *Fisheries Research*. **27**, 203–14.

Broadhurst, M K, Kennelly, S J, and Isaksen, B (1996*a*). Assessment of modified cod-ends that reduce the bycatch of fish in two estuarine prawn-trawl fisheries in New South Wales, Australia. *Fisheries Research*. **27**, 89–112.

Broadhurst, M K, Kennelly, S J, and O'Doherty, G (1996*b*). Effects of square-mesh panels in cod-ends and of haulback delay on bycatch reduction in the oceanic prawn-trawl fishery of New South Wales, Australia. *Fishery Bulletin*. **94**, 412–22.

Glass, C W, Wardle, C S, Gosden, S J, and Racey, D N (1995*a*). Studies on the use of visual stimuli to control fish escape from cod-ends. I. Laboratory studies on the effect of a black tunnel on mesh penetration. *Fisheries Research* **23**, 157–64.

Glass, C W, Wardle, C S, Gosden, S J, and Racey, D N (1995*b*). Studies on the use of visual stimuli to control fish escape from cod-ends. II. The effect of a black tunnel on the reaction behaviour of fish in otter trawl cod-ends. *Fisheries Research* **23**, 165–74.

Isaksen, B, Valdemarsen, J W, Larsen, R B, and Karlsen, L (1992). Reduction of fish by-catch in shrimp trawl using a rigid separator grid in the aft belly. *Fisheries Research* **13**, 335–52.

Kennelly, S J (1995). The issue of bycatch in Australia's demersal trawl fisheries. *Reviews in Fish Biology and Fisheries* **5**, 213–34.

Kennelly, S J, Kearney, R E, Liggins, G W, and Broadhurst, M K (1992). The effect of shrimp trawling bycatch on other commercial and recreational fisheries–An Australian perspective. In 'Proceedings of the International Conference on Shrimp Bycatch, 24–27 May 1992, Lake Buena Vista, FL, Southeastern Fisheries Association, Tallahassee, FL'. pp. 97–113.

Martin, T J, Brewer, D T, and Blaber, S J M (1995). Factors affecting distribution and abundance of small demersal fishes in the Gulf of Carpentaria, Australia. *Marine and Freshwater Research* **46**, 909–20.

Mitchell, J F, Watson J W, Foster, D G, and Bell, R E (1995). The turtle excluder device (TED): A guide to better performance. NOAA Technical Memorandum NMFS-SEFSC- 366. 35 pp.

Mounsey, R P, Baulch, G A, and Buckworth, R C (1995). Development of a trawl efficiency device (TED) for Australian prawn fisheries. I. The AusTED design. *Fisheries Research* **22**, 99–105.

Newton, G, Fowler, J, McLoughlin, K, and Staples, D (Eds) (1994). Northern Fish Trawl Fishery and Northern Shark Fishery 1993, Fisheries Assessment Report compiled by the Fisheries Resource Assessment Group. Australian Fisheries Management Authority, Canberra.

Pender, P, and Willing, R (1989). Trash and treasure? *Australian Fisheries* **48**(1), 35–6.

Ramm, D C, Pender, P J, Willing, R S, and Buckworth, R C (1990). Large scale spatial patterns of abundance within the assemblage of fish caught by prawn trawlers from northern Australian waters. *Australian Journal of Marine and Freshwater Research* **41**, 79–95.

Ramm, D C, Mounsey, R P, Xiao, Y, and Poole, S E, (1993). Use of a semi-pelagic trawl in a tropical demersal trawl fishery. Fisheries Research **15**, 301–13.

Rawlinson, N J F, and Brewer, D T (1995). 'Monsters', 'blubber' and other bycatch. *Australian Fisheries* **54**(5), 27–8.

Robertson, J H B (1983). Square mesh nets help young fish escape. *Fisheries News*, No. 3652, 10–11.

Robins-Troeger, J B, Buckworth, R C, and Dredge, M C L (1995). Development of a trawl efficiency device (TED) for Australian prawn fisheries. II. Field evaluations of the AusTED. *Fisheries Research* **22**, 107–17.

SAS Institute Inc. (1988). 'SAS/STAT Users Guide' 6.03 Edition.

Sainsbury, K J (1987). Assessment and management of the demersal fishery on the continental shelf of northwestern Australia. In 'Tropical Snappers and Groupers; Biology and Fishery Management'. (Eds J J Polovina and S Ralston) pp. 465–503 (Westview Press: Boulder CO).

Sainsbury, K J (1988). The ecological basis of multispecies fisheries and management of a demersal fishery in tropical Australia. In 'Fish Population Dynamics' 2nd edition. (Ed J A Gulland) pp. 349-82 (John Wiley: London).

Sainsbury, K J, Campbell, R A, Lindholm, R, and Whitelaw, A W (1997). Experimental management of an Australian multispecies fishery: examining the possibility of trawl induced habitat modification. In ' Global trends: fisheries management'. (Eds E K Pikitch, D D Huppert and M P Sissenwine). (in press) (John Wiley and Sons: London).

Valdemarsen, J W, Lehmann, K, Riget, F, and Boje, J (1993). Grid devices to select shrimp size in trawls. ICES Statutory Meeting, Fish capture Committee.

Watson, J W, and Seidel, W R (1980). Evaluation of techniques to decrease turtle mortalities in the southeastern United States shrimp fishery. ICES CM 1980/B:31. 8 pp.

Watson, J, Workman, I, Foster, D, Taylor, C, Shah, A, Barbour, J, and Hataway, D (March 1993). Status Report on the Development of Gear Modifications to Reduce Finfish Bycatch in Shrimp Trawls in the Southeastern United States 1990-1992. NOAA Technical Memorandum, NMFS-SEFSC-327. 131 pp.

A Framework for Solving Bycatch Problems: Examples from New South Wales, Australia, the Eastern Pacific and the Northwest Atlantic

Steven J. Kennelly

NSW Fisheries Research Institute, PO Box 21, Cronulla, NSW 2230, Australia.

Summary

A framework for solving bycatch problems is described that involves combining the respective expertises of scientists and fishermen. The first prerequisite for any attempt to ameliorate bycatch problems involves identifying and quantifying bycatches using large-scale observer programmes. These programmes involve scientists collecting information at sea from normal commercial fishing operations and so determine potential problems without relying on anecdotal information or data from research vessels. Once the species-specific distributions and abundances of bycatches are determined, modifications to fishing methods are tested in experiments using commercial fishing vessels. The fishermen's roles in this framework are: (i) to be seen as the driving force in addressing any conflicts that may come from their bycatches; (ii) to provide scientists with their unique practical knowledge of the relevant fishing technology; and (iii) to implement solutions into normal fishing operations efficiently and, in many cases, voluntarily. The scientists' role is to organize, analyse and disseminate the work, provide information on possible solutions through their access to the international literature and to ensure the scientific rigour of the experiments. Finally, both scientists and fishermen are responsible for communicating the solutions (and their adoption) to other fishermen, the public and special interest groups. The success of this approach is described using examples from the prawn trawl fisheries of New South Wales, Australia, the shrimp and fish trawl fisheries off the northeastern United States and the tuna purse-seine fisheries of the eastern Pacific.

Introduction

Earlier this decade, several authors correctly predicted that bycatch would become one of the most important fisheries issues of the 1990s (e.g. Klima 1993; Tillman 1993). Declining fish stocks in many of the world's fisheries, and widespread publicity over the incidental capture of charismatic species like dolphins and turtles, have led to commercial and recreational fishermen, conservationists, environmentalists, politicians, fisheries managers and scientists, all identifying bycatch as a key problem and calling for ways to reduce it (for recent reviews see Andrew and Pepperell 1992; Alverson *et al.* 1994; Kennelly 1995).

In recent years, scientists and fishermen in some countries have successfully solved certain bycatch problems in their respective fisheries. In considering the procedures followed in these cases, it is apparent that a relatively simple and logical framework was adopted that involves fishermen and scientists each applying their respective expertises to the problem. In short, this framework or protocol (see Fig. 1) involves identifying and quantifying the relevant problem (*via* observer programmes), solving the problem through modifications to commercial fishing gears and/or fishing practices, and then 'selling' these solutions throughout the particular fishery and finally to concerned interest groups.

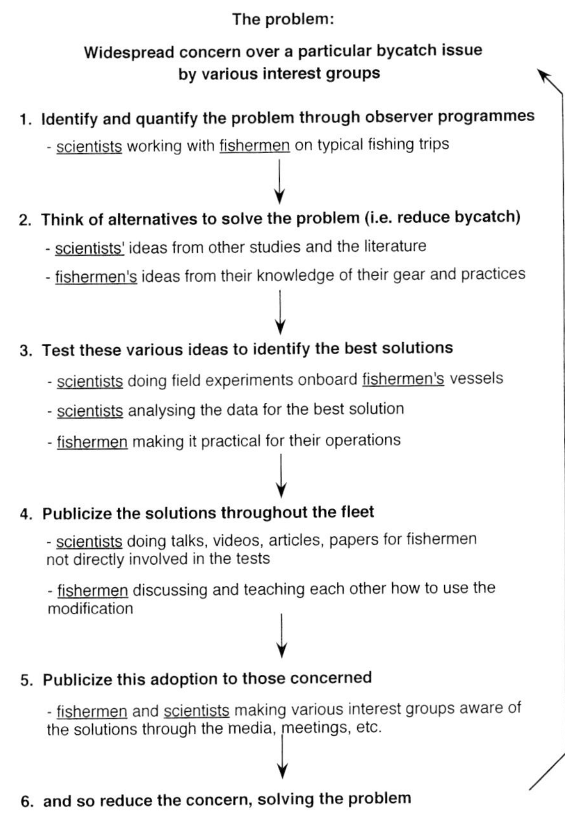

The problem:
Widespread concern over a particular bycatch issue by various interest groups
1. Identify and quantify the problem through observer programmes
- scientists working with fishermen on typical fishing trips
2. Think of alternatives to solve the problem (i.e. reduce bycatch)
- scientists' ideas from other studies and the literature
- fishermen's ideas from their knowledge of their gear and practices
3. Test these various ideas to identify the best solutions
- scientists doing field experiments onboard fishermen's vessels
- scientists analysing the data for the best solution
- fishermen making it practical for their operations
4. Publicize the solutions throughout the fleet
- scientists doing talks, videos, articles, papers for fishermen not directly involved in the tests
- fishermen discussing and teaching each other how to use the modification
5. Publicize this adoption to those concerned
- fishermen and scientists making various interest groups aware of the solutions through the media, meetings, etc.
6. and so reduce the concern, solving the problem

Fig. 1. The framework used to address bycatch problems.

The framework to address bycatch problems

Observer programmes

The first step in this framework is to identify and quantify the particular bycatch issue of concern by determining species-specific spatial and temporal variabilities in bycatches — so removing any reliance on qualitative or anecdotal information. Such data cannot be collected from information on commercial landings, nor can one rely on fishermen to provide accurate data on bycatches (it is often argued, in fact, that it is in fishermen's best interests *not* to provide such information). Further, one cannot rely on information gathered from research vessels because such data may not mimic that gathered with commercial fishing vessels, gears, operating procedures, targeting decisions, etc. It is well established that the best way to obtain bycatch information is for scientists (or scientific observers) to work alongside fishermen on their own vessels and to collect the data *in situ* by sorting, identifying, measuring, counting and weighing retained and discarded catches (see also Howell and Langan 1992; Saila 1983; Alverson *et al.* 1994; Kennelly 1995; Murawski *et al.* 1995). While such observer surveys assume that fishermen do not change their normal operations in the presence of observers, they nevertheless constitute the most accurate form of bycatch information that can be gathered.

It is during this stage of the framework that scientists, observers and fishermen usually forge working relationships that later prove vital in solving identified bycatch problems. It is worth noting that these relationships do not arise out of port meetings, conferences or workshops (these are all important and occur later in the framework), but are developed onboard many different vessels, at sea, in rivers, during long days and nights, working alongside each other sorting catches. Without several years of working alongside fishermen in this way, scientists usually are not in a position to solve bycatch problems for two reasons: (i) they lack the necessary data on bycatches which identify the particular issues that required solving; and (ii) they lack the respect from industry that is needed to work with them on solutions.

Once the observer information is available for a particular bycatch concern, it is necessary to consider whether the concern reflects a real or merely a perceived problem. If biological and stock parameters for the relevant bycatch species are available, can be incorporated with observer data, and show that the concern does not reflect a real problem to populations, then the best way to ameliorate the concern may involve education programmes aimed at particular interest groups. If, however, such analyses prove otherwise (that the concern reflects a real problem to populations) or, as is often the case, the relevant parameters are not available, then alternative ways to reduce bycatches may need to be considered.

Alternative solutions

Once the particular bycatch issue has been adequately identified and quantified, the scientists and fishermen are in a position to develop alternatives that aim to reduce the problem. Developing alternative modifications to fishing gears and practices to reduce unwanted bycatch is best done as a joint exercise: the scientists provide information gleaned from other studies, the scientific literature, conferences, etc. and from liaising directly with other scientists throughout the world; the local fishermen provide their unique practical knowledge of their fishing gears, vessels and grounds and how various modifications may be applied in their operations. In bringing together both types of expertise and experience, the scientists and fishermen eventually identify those modifications that warrant further consideration and field testing.

Testing the alternatives

Once various alternatives have been identified, the scientists and fishermen then test various selected modifications to gears and fishing practices *via* field trials onboard conventional commercial fishing vessels (usually chartered by the scientists). The scientists' role at this stage is to design rigorous field experiments and sampling protocols, and collect, analyse and write up the data as quality papers and reports that can pass critical peer review. The fishermen's role is to ensure that the proposed modifications and their fishing operations can be readily combined. Using commercial vessels to do this research (rather than research vessels) is vital because: (i) it supplies skippers and crews who possess local knowledge of the conventional methods used and local fishing grounds; (ii) it supplies control gears (e.g. conventional nets) and fishing practices against which the modifications can be tested; and

(iii) it ensures the interest and involvement of the rest of the fleet (i.e. those not chartered for the research) because the research is done with their colleagues, alongside them, in their grounds, using similar gears and vessels.

After preliminary trials, refinements to various modifications, and subsequent re-testing and refinement,scientists and fishermen eventually arrive at some solution(s) that they conclude works best in reducing the identified bycatch problem in their particular fishery.

Publicizing the results to other fishermen

While the graphs and analyses of the data from the field trials usually convince scientists and managers of the utility (or otherwise) of the modifications, the next step is to illustrate the success of these modifications to the rest of the fishery (i.e. including those fishermen who were not directly involved in the research). This is done by presenting photographs, videos, data and reports to as many fishermen in the fishery as possible, through meetings, dockside talks, workshops and encouraging the circulation of the information throughout the fishery. The fishermen involved in the trials also discuss the modifications with other fishermen and assist them in making and using the modifications. These new users then inform other users and eventually many members of the various fleets are using the modifications to reduce their unwanted bycatch — in some cases this has even occurred on a voluntary basis without any changes in regulations (see below).

Publicizing the solutions to the public

Unfortunately, reducing bycatch *via* the above protocol is insufficient by itself to resolve the concern from various interest groups over the particular bycatch issue (i.e. the initial problem). This final step can only be done by widespread publicity of the solution, its development, testing and acceptance by fishermen to those most concerned with the issue. This is usually achieved by the fishermen and scientists making presentations to committees that represent other commercial and recreational fisheries, conservationists, and environmentalists, and releasing photographs, videos, interviews, etc. to the print, radio and television media. Armed with such evidence (in addition to verifying the credibility of the work publishing the results in scientific journals), the scientists and fishermen involved are usually able to reduce the perceived problem and so reduce the concern identified as central to the initial problem.

EXAMPLES USING THE FRAMEWORK

Virtually all fisheries in the world have some bycatch associated with them (interestingly, one of the few fisheries that has negligible bycatch is whale harpooning where the only bycatch may include some barnacles). However, some types of fishing are recognized as having greater bycatch problems than others with two of the more infamous being: (i) demersal trawling for fish and prawns (or shrimp), which often results in the capture and discard of many juvenile fish that, when larger, would be targeted in other commerical and/or recreational fisheries; and (ii) purse seining around schools of dolphins for tuna, which has led to mortalities of large numbers of dolphins (for a recent review, see Alverson *et al.* 1994). Of the few cases throughout the world where successful solutions to bycatch problems have been developed, several have dealt with problems in these types of fisheries. The rest of this paper briefly summarizes a few examples of the way in which the above framework was (and is being) used to resolve various bycatch problems in these cases.

The prawn trawl fisheries of New South Wales, Australia

New South Wales, has experienced quite high-profile bycatch problems in its estuarine and oceanic prawn fisheries for many years (going as far back as the late 19th century — see Dannevig 1904; for review see Kennelly 1995). In the late 1980s these concerns reached a maximum and resulted in threats to close certain prawn fisheries to stop the bycatch of juvenile fish. At around this time it was discovered that, despite some anecdotal information, there were very few scientific data concerning this problem and so we began a large-scale observer programme was commenced in 1989 to identify and quantify the issue. This involved censussing catches on replicate, randomly selected vessels doing typical fishing trips in several estuaries and out from several oceanic ports throughout the State.

The data from the observer programme led to quite uncompromising information on the bycatches of juvenile fish by various prawn trawl fleets (for details see Kennelly 1993; Kennelly *et al.* 1993; Liggins and Kennelly 1996). For example, in the Clarence River estuarine fishery in the year 1991–92, it was estimated that in catching 270 t of prawns, this fishery discarded 123 t of bycatch, including ~0.8 million individuals of the recreationally and commercially important yellowfin bream. In the oceanic fishery offshore from this river in the same year, we estimated that in catching 288 t of prawns, 4022 t of bycatch was caught (including ~6 million red spot whiting — an important commercial species). Of this bycatch, an estimated 725 t was landed for sale as 'by-product' (including various species of slipper lobsters, squid, octopus and large fish) while the rest (some 3297 t) was discarded.

Information such as this was given to fishermen throughout NSW at various port meetings and as written reports to all fishermen involved. After some debate, these meetings eventually led to the scientists (and the fishermen) identifying the key bycatch problems in some detail (in terms of species-specific spatial and temporal patterns) and so made it possible to focus on possible solutions. In the example described above, the bycatch and discarding of large numbers of yellowfin bream were clearly seen as a problem for the Clarence River estuarine fishery. For the oceanic fishery, the bycatch of large numbers of small red spot whiting and other finfish was seen as a problem but (unlike the estuarine fishery) any solution in this fishery needed to take account of the fishermen's requirement to keep certain species of bycatch for sale as by-product.

Developing alternative modifications to trawl gears to reduce these unwanted bycatches was a joint exercise undertaken by scientists, fishermen and key net makers. The scientists supplied information from other studies, the scientific literature, international conferences and workshops, and from liaising directly with colleagues throughout the world (especially those

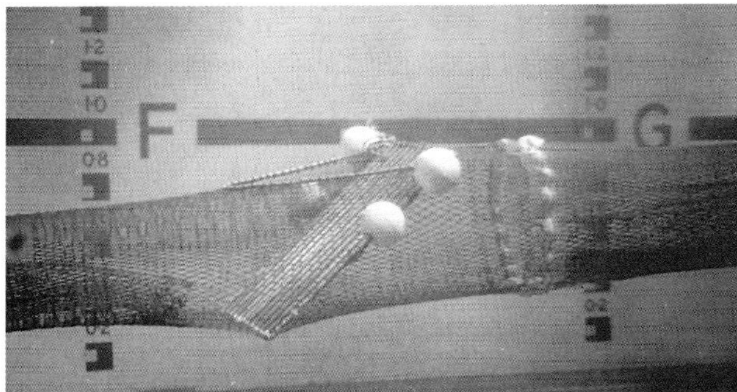

Fig. 2. The Nordmore grid design tested and now used in the Clarence River estuarine prawn trawl fishery

in Norway, Scotland and the United States). The local fishermen and netmakers from the Clarence River supplied their unique practical knowledge of their fishing gears, vessels and grounds and how various modifications could be applied in their operations. In this way, modifications were identified which warranted further consideration and field testing.

These discussions led to the testing of several kinds of Nordmore grids and square-mesh panels in these estuarine and oceanic fisheries *via* manipulative experiments onboard chartered commercial vessels set up to trawl in the conventional way. In general, these experiments took the form of paired comparisons of modified nets with conventional nets and were analysed using paired *t*-tests and analyses of variance (for details, see Broadhurst and Kennelly 1994; Broadhurst *et al.* 1996; and Broadhurst *et al.* in press).

Preliminary trials, refinements to various modifications, further testing, and refining, came up with two modifications for these fisheries that proved successful at reducing bycatch while maintaining and sometimes even enhancing catches of prawns. Because the targeted prawns (eastern school prawns) in the estuarine fishery were smaller than the bycatch to be excluded, a Nordmore grid was found to be most suitable for this fishery (Fig. 2). Figure 3 shows the striking difference in bycatches that came from using this grid in the fishery and the data shown in Figure 4 confirm these results with bycatches (including bream) being greatly reduced while prawn catches were maintained.

For the oceanic fishery, such grids were not appropriate because the targeted prawns (eastern king prawns) were quite large and the grids tended to exclude most of the by-product species which the fishermen wished to retain (slipper lobsters, octopus, squid, larger fish, etc.). For this fishery we developed a unique type of composite square-mesh panel anterior to the cod-end (see Fig. 5) — the theory being that small fish swam out of the cod-end with the water flowing through the panel while the less mobile prawns, slipper lobsters, squid and octopus would go to the back of the cod-end. The sizes of fish excluded in this way could be selected by adjusting the mesh size in the square-mesh panel. Figure 6 shows the effects of this panel on bycatches, and Figure 7 shows the data confirming these reductions in bycatches (including red spot whiting) whilst catches of prawns were maintained and even slightly enhanced. Other data show that this modification also had no effect on most of the by-product species (see Broadhurst *et al.* 1996).

Fig. 3. Example of the catches from paired comparisons in the Clarence River estuarine prawn trawl fishery using a conventional cod-end (on the left) and one with a Nordmore grid (on the right)

Whilst the graphs and analyses of the data from the above trials convinced various anonymous referees of the usefulness of the modifications, it was the photographs (e.g. Figs 3 and 6), videos, and meetings with the chartered fishermen that

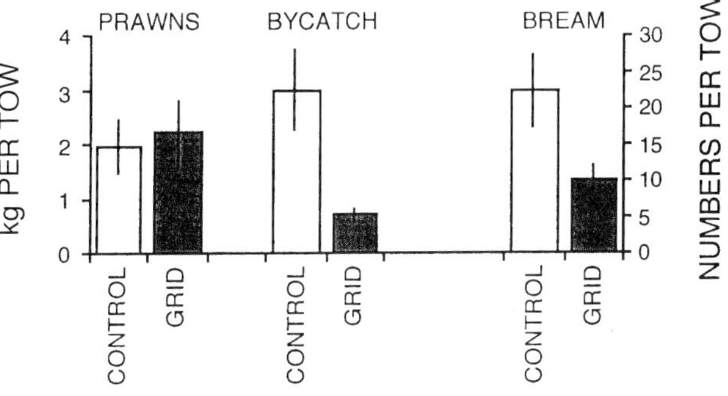

Fig. 4. Summaries of data (for weights of prawns and bycatch and numbers of bream) from comparisons of a cod-end with the Nordmore grid and a conventional cod-end in the Clarence River estuarine prawn trawl fishery

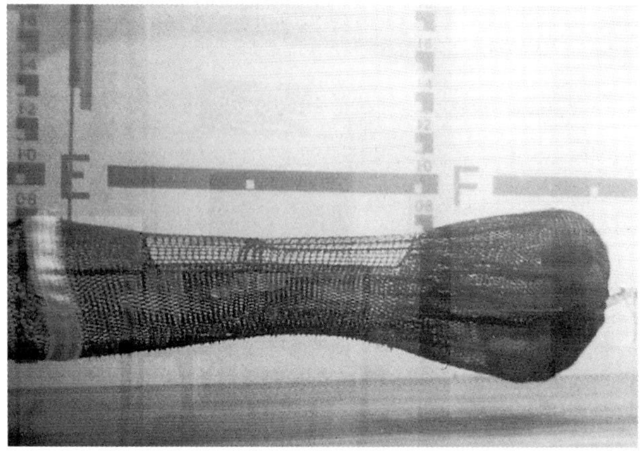

Fig. 5. The composite squaremesh panel modification tested and now used in NSW's oceanic prawn trawl fishery.

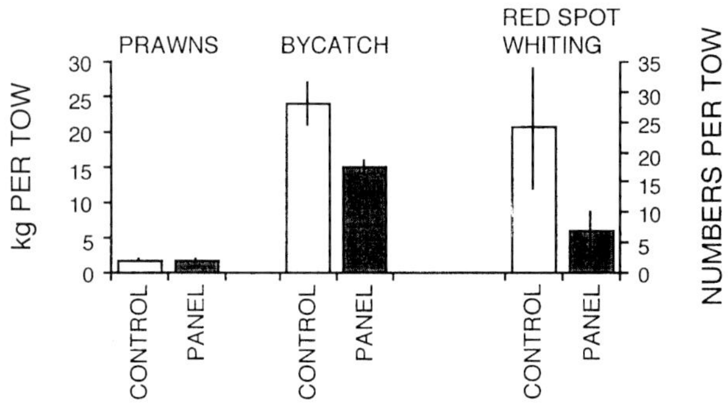

Fig. 7. Summaries of data (for weights of prawns and bycatch and numbers of red spot whiting) from comparisons of a cod-end with the composite square-mesh panel modification and a conventional cod-end in the Clarence River oceanic prawn trawl fishery.

illustrated the success of these modifications to fishermen who were not directly involved in the research. Distributed photographs and videos were distributed to fishermen in the relevant ports and the circulation of the information to other ports was encouraged. The fishermen involved in the trials discussed the modifications with other fishermen and assisted them in making and using the modifications. These new users then informed other users and before long, the majority of fishermen in the Clarence River estuarine and oceanic fisheries were using these gears and reducing their unwanted bycatches — all on a purely voluntary basis without any changes in regulations. News of these modifications has now spread to other ports and estuaries throughout NSW and southern Queensland and many fishermen in these other ports are now also using these gears voluntarily. We are currently in the process of ensuring that most of the fishermen in these other ports know about these modifications before their legislative adoption (which should result in complete compliance across these fisheries). Because of the widespread voluntary acceptance of the new gears, we believe that this last legislative step should be a relatively painless process.

Fig. 6. Example of the catches from paired comparisons in the Clarence River oceanic prawn trawl fishery using a conventional cod-end (on the left) and one with the composite square-mesh panel modification (on the right)

The final stage in this work was to address the widespread concern from various interest groups over the issue. This has been done by widespread publicity of the solution, its development, testing and voluntary acceptance by fishermen to those most concerned with the issue. This is being achieved by the fishermen and scientists making presentations to committees (representing other commercial and recreational fisheries) and releasing photographs, videos, interviews etc. to the print, radio and television media. With such evidence (and the credibility achieved through the publication of the results in scientific journals), perceived problems concerning this issue in these fisheries were reduced.

Trawl fisheries in the northeastern United States

Like the prawn trawl fisheries in New South Wales and many other prawn and shrimp trawl fisheries throughout the world, the shrimp trawl fishery in the Gulf of Maine (in the northeast United States) has attracted substantial concern over its bycatch and discard of small fish (e.g. Howell and Langan 1992). Further, the solution of this issue in this region has followed a very similar path as that described above for NSW. Since 1988, the National Marine Fisheries Service has operated a large-scale observer programme in most of the fisheries in the region, supplying managers and scientists with their chief source of information on discards (e.g. see Murawski *et al.* 1995). The data from this programme identified and quantified the problems associated with bycatch from the shrimp fishery and, after a period of development by scientists and fishermen, a Nordmore grid system was introduced into the fishery to reduce unwanted bycatches (see Kenney *et al.* 1991; Richards and

Hendrickson 1995). While the introduction of this grid into this fishery was not done voluntarily but was mandated, there is now reasonable acceptance of the gear by shrimp trawlers and, because of the publicity surrounding the effectiveness of the gear and its acceptance, there has been a large decrease in concern over shrimp trawl bycatch in this region.

The groundfish trawl fisheries of the northeastern United States have also attracted their share of attention with regard to the discarding of other species and undersized individuals of target species. A summary of the observer data for these trawlers from 1990 to 1994 is seen in Table 1, which shows the average retained and discard rates (per trawl hour) of several important species. These summary figures highlight quite significant discarding rates for those species shown and more detailed analyses (Kennelly 1996) quantified the sorts of spatial and temporal patterns in discarding which, for several years, have caused significant conflicts with various user groups. For example, the discard of lobsters has caused conflict with lobster trappers, the discard of scup has caused problems with various recreational fishing groups and the discard of flatfish has caused problems with various commercial fisheries.

While the solutions to these problems for fish trawl gear may not be quite as simple as using Nordmore grids or square-mesh panels in shrimp trawls, recent developments in sorting devices for finfish and other species in fish trawl gear may provide some possible solutions. Modifications such as downward sorting grids and horizontal panels in nets have been shown to have great potential for reducing the bycatches of unwanted species and unwanted sizes of certain species in Norwegian groundfish trawls (see Isaksen 1993; Larsen and Isaksen 1993; Engas and West 1995) and scientists in the northeastern United States are currently working with local fishermen to test the effectiveness of some of these designs. Because of the existence of the large-scale observer programme, these scientists believe that the most difficult job in solving such bycatch problems is already in hand. That is: (i) they already have good observer data that identifies and quantifies the species-specific spatial and temporal scales of various problems, and (ii) they have established a working environment with fishermen that should enable such solutions to be tested, proven and eventually adopted.

Table 1. Summary of retained and discarded catches in the northeastern United States, 1990–1994: as estimated by the observer programme

	Retained (lbs/hour)	Discarded (lbs/hour)	% Discarded
Total catch (all species)	668.2	611.3	47.8
Lobster	1.4	1.9	57.9
Scup (porgy)	22.9	18.3	44.5
Butterfish	55.9	26.4	32.1
Red Hake	12.4	34.8	73.7
Silver Hake	182.2	33.3	15.4
Yellowtail flounder	9.9	3.9	28.3
Winter flounder	10.9	2.7	20.0
American plaice	4.8	1.5	23.4
Windowpane flounder	4.3	5.0	53.8

Tuna purse-seine fisheries of the eastern Pacific

In recent years, Martin Hall (Head of the Tuna-Dolphin Program at the Inter-American Tropical Tuna Commission) has described in several papers the success of the tuna purse-seine fisheries of the eastern Pacific in reducing the mortality of incidentally-caught dolphins (Hall 1994; Hall in press *a* and *b*). This particular bycatch issue has been one of the most infamous in the world since the 1960s with very dramatic outcries from Greenpeace and other environmental and conservation organizations. Fortunately, this fishery has had very high observer coverage, with most trips by most vessels covered, so information on discarding is quite thorough and has led to an excellent understanding of the dolphin issue and ways to solve it.

The most common way purse seiners fish in this region is to encircle groups of dolphins to catch the tuna that they swim with. In the early years of the fishery, the incidental mortality of dolphins using this method was quite high (an average of ~ 350 000 dolphins/year during the 1960s) which is believed to have caused significant declines in populations of dolphins in the region. In the 1960s and 1970s, fishermen developed ways to reduce these mortalities and dolphin mortalities dropped to 20 000 to 40 000/year in the early 1980s. After this time, however, new entrants into the fishery and more targeting on dolphins led to increases in dolphin mortalities (133 000 in 1986).

However, since that time, dolphin mortalities have been reduced by about 97% (to 3300 in 1995) through the development of a series of modifications to the purse seines used, the release practices of the encircled dolphins, and the education and training of skippers and crews in the techniques and special skills required. These modifications were developed over time by fishermen and scientists and involved a variety of operational and gear modifications, e.g. different mesh sizes were incorporated into certain sections of the purse seines used, a different method was developed for tying the cork line, a manoeuvre termed 'backdown' after dolphins were encircled was developed (Medina 1994), speedboats and dolphin rescue boats were incorporated into operations, and key places containing populations of dolphins particularly prone to discard mortality were avoided. Next, once such modifications were developed, they were shown to participating fishermen in a large-scale education programme that included the training of skippers and crews in the new techniques. Many of the programmes developed by the staff of the IAATC's Tuna–Dolphin programme have been implemented by an international agreement called the Panama Declaration which led to widespread adoption of the various modifications by fishermen. The success involved in reducing this bycatch problem is now being publicized to the various conservation and environmental organizations to the point where the widespread concern over this issue is gradually declining. Currently, Greenpeace, World Wildlife Fund, Environmental Defense Fund and the Center for Marine Conservation are supporting these programmes. In summary, the success of the work done in this fishery by the scientists and fishermen involved has shown that it was possible to save the dolphins without closing a major fishery.

Conclusions

A common point made whenever bycatch issues are discussed is that, while most fisheries have bycatch and many have particular bycatch problems, there are no easy 'quick fixes' to these problems (see also Alverson *et al.* 1994; Pitcher and Chuenpagdee 1994). The number and variety of bycatch problems in the world's fisheries are at least as numerous and complex as the fisheries themselves and no simple solution will work for all problems. (The marked success of the Nordmore Grid in many of the world's prawn and shrimp fisheries suggests that this device may be as close as one gets to a generic solution to prawn and shrimp trawl bycatch problems — although it too has its limitations if the targeted prawns are larger than or of a similar size to the unwanted bycatch). Despite the need for fishery-specific solutions to bycatch issues, however, this paper has demonstrated that in several very disparate cases where bycatch issues have been resolved, the fishermen and scientists involved have tended to adhere to a certain common protocol. These successes would suggest that subsequent attempts to ameliorate bycatch problems in other fisheries should at least consider such a framework as a useful way to begin.

References

Alverson, D L, Freeberg, M H, Murawski, S A, and Pope, J G (1994). A global assessment of fisheries bycatch and discards. *FAO Fisheries Technical Paper* 339 (Rome).

Andrew, N L, and Pepperell, J G (1992). The by-catch of shrimp trawl fisheries. *Oceanography and Marine Biology Annual Review* **30**, 527–65.

Broadhurst, M K, and Kennelly, S J (1994). Reducing the by-catch of juvenile fish (mulloway *Argyrosomus hololepidotus*) using square-mesh panels in codends in the Hawkesbury River prawn-trawl fishery. *Fisheries Research* **19**, 321–31.

Broadhurst, M K, Kennelly, S J, and O'Doherty, G (1996). Effects of square-mesh panels in codends and of haulback-delay on bycatch reduction in the oceanic prawn-trawl fishery of New South Wales, Australia. *Fishery Bulletin* **94**, 412–22.

Broadhurst, M K, Kennelly, S J, and Isaksen, B (in press). Assessments of modified codends that reduce the by-catch of fish in two estuarine prawn-trawl fisheries in New South Wales, Australia. *Fisheries Research.*

Dannevig, H C (1904). Preliminary report upon the prawning industry in Port Jackson. W A Gullick, NSW Government Printer. 17 pp.

Engas, A, and West, C W (1995). Development of a species-selective trawl for demersal gadoid fisheries. ICES CM 1995/B+G+H+J+K:1. 20 pp.

Hall, M A (1994). By-catches in purse-seine fisheries. In 'By-catches in fisheries and their impact on the ecosystem'. (Eds T J Pitcher and R Chuenpagdee) pp. 53–58. Fisheries Centre Research Reports, University of British Columbia, Canada. 2(1) ISSN 1198–6727.

Hall, M A (in press *a*). On bycatches. *Reviews in Fish Biology and Fisheries.*

Hall, M A (in press *b*). Strategies to reduce the incidental capture of marine mammals and other species in fisheries. Proceedings of the Solving By-catch Workshop: Considerations for Today and Tomorrow. Seattle, Washington, USA.

Howell, W H, and Langan, R (1992). Discarding of commercial groundfish species in the Gulf of Maine shrimp fishery. *North American Journal of Fisheries Management* **12**, 568–80.

Isaksen, B (1993). I. Kort oppsummering av forsok med rist i snurrev ad. II. Monteringsbeskrivelse. Havforskningsinstituttet Rapport fra Senter for Marine Ressurser NR. 8-1993. ISSN 0804-2136. 12 pp.

Kennelly, S J (1993). Study of the by-catch of the NSW east coast trawl fishery. Final report to the Fisheries Research and Development Corporation. Project No. 88/108. ISBN 0 7310 2096 0. 520 pp.

Kennelly, S J (1995). The issue of by-catch in Australia's demersal trawl fisheries. *Reviews in Fish Biology and Fisheries* **5**, 213–34.

Kennelly, S J (1996). Summaries of National Marine Fisheries Service Sea Sampling Data for Fish Trawling in the Northeastern United States from July 1990 to June 1994. Manomet Observatory, Massachusetts, US. 44 pp.

Kennelly, S J, Kearney, R E, Liggins, G W, and Broadhurst, M K (1993). The effect of shrimp trawling by-catch on other commercial and recreational fisheries — an Australian perspective. In 'International Conference on Shrimp By-catch', May, 1992, Lake Buena Vista, Florida. (Ed R P Jones) pp. 97–114. Southeastern Fisheries Association, Tallahassee, Florida.

Kenney, J F, Blott, A J, and Nulk, V E (1991). Experiments with a Nordmore grate in the Gulf of Maine shrimp fishery. A report of the New England Fishery Management Council to NOAA, pursuant to NOAA award No. NA87EA-H-00052.

Klima, E F (1993). Shrimp by-catch — Hopes and fears. In 'International Conference on Shrimp By-catch', May, 1992, Lake Buena Vista, Florida. (Ed R P Jones) pp. 5–12. Southeastern Fisheries Association, Tallahassee, Florida.

Larsen, R B, and Isaksen, B (1993). Size selectivity of rigid sorting grids in bottom trawls for Atlantic cod (*Gadus morhua*) and haddock (*Melanogrammus aeglefinus*). *ICES Marine Science Symposium* **196**, 178–82.

Liggins, G W, and Kennelly, S J (1996). By-catch from prawn trawling in the Clarence River estuary, New South Wales, Australia. *Fisheries Research* **25**, 347–67.

Medina, H (1994). Reducing by-catch through gear modifications: the experience of the tuna–dolphin fishery. In 'By-catches in fisheries and their impact on the ecosystem'. (Eds T J Pitcher and R Chuenpagdee) 60 pp. Fisheries Centre Research Reports, University of British Columbia, Canada, 2(1) ISSN 1198-6727.

Murawski, S, Mays, K, and Christensen, D (1995). Fishery observer program. In 'Status of the fishery resources off the Northeastern United States for 1994'. NOAA Technical Memorandum NMFS-NE-108. 35-41.

Pitcher, T J, and Chuenpagdee, R (Eds) (1994). By-catches in fisheries and their impact on the ecosystem. Fisheries Centre Research Reports, University of British Columbia, Canada 2(1) ISSN 1198-6727.

Richards, A, and Hendrickson, L (1995). Effectiveness of the Nordmore grate in reducing by-catch of finfish in shrimp trawls. American Fisheries Society 125th Annual Meeting, Tampa, Florida, Abstracts.

Saila, S B (1983). Importance and assessment of discards in commercial fisheries. *FAO Fisheries Circular* 765.

Tillman, M F (1993). Bycatch — The issue of the 90s. In 'International Conference on Shrimp By-catch', May, 1992, Lake Buena Vista, Florida'. (Ed R P Jones) pp. 13–8. Southeastern Fisheries Association, Tallahassee, Florida.

Assessment by Underwater Infrared Video of the Selectivity of an Experimental Danish Seine for Deep-water Prawn

Tatsuro Matsuoka,[A] *Satoi Ishizuka,*[A] *Kazuhiko Anraku*[A] *and Masaaki Nakano*[B]

[A] Faculty of Fisheries, Kagoshima University, Shimoarata 4–50–20, Kagoshima, 890, Japan.
[B] Kagoshima Prefectural Fisheries Research Station, Kinko-cho 11–40, Kagoshima, 890, Japan.

Summary

Selectivity of a Danish seine net for deep-water prawn was studied to exclude the undersized prawns and bycatch finfishes necessary for resource management and environmental conservation. The behaviour of organisms at depth was observed by an infrared video set, and the exclusion process in catches was studied by pocket/cover nets and by mathematical modelling. Prawns showed an optomotor response, with a repeated rush backwards onto the netting being apparently the only behavioural pattern leading to exclusion. Prawns were primarily excluded through the wings of the net, with few through the side and ceiling. Those excluded through wings were smaller than those through the cod-end. The whole-net selectivity for prawns was steeper than the cod-end selectivity. These observations suggest that cod-end selectivity does not represent selectivity of whole gear and that wing mesh size must be managed because of the major exclusion of prawns through the wings. Net panels parallel to towing are seldom passed through by prawns, and these are consequently the best positions to devise for finfish segregation.

Introduction

Selectivity of the Danish seine nets currently used for a deep-water prawn, *Haliporoides sibogae*, in the East China Sea off Kagoshima, Japan, was studied. Improvement of the fishing gear to exclude (the expression 'excluded' is here used to describe individuals not retained by the net) undersized prawns and bycatch finfishes is desirable for the purpose of management of the prawn resource and for environmental conservation of the fishing ground. In addition to reduction of the time and labour effort expended in sorting catches and clearing enmeshed finfish, the consequent improvement in the quality of the prawn catches is of benefit to the fishermen. Because of the depth of the fishing ground, a newly designed infrared video set was used. By this means, selectivity of the gear and its potential for improvement have been examined in a combination of studies of the behaviour of organisms against the fishing gear, and of catches by pocket and cover nets, and by mathematical modelling of the retention and exclusion process for the prawn. The present paper summarizes part of an extension project by the Kagoshima Prefectural Government and a research project supported by the Scientific Research Foundation of the Japanese Government.

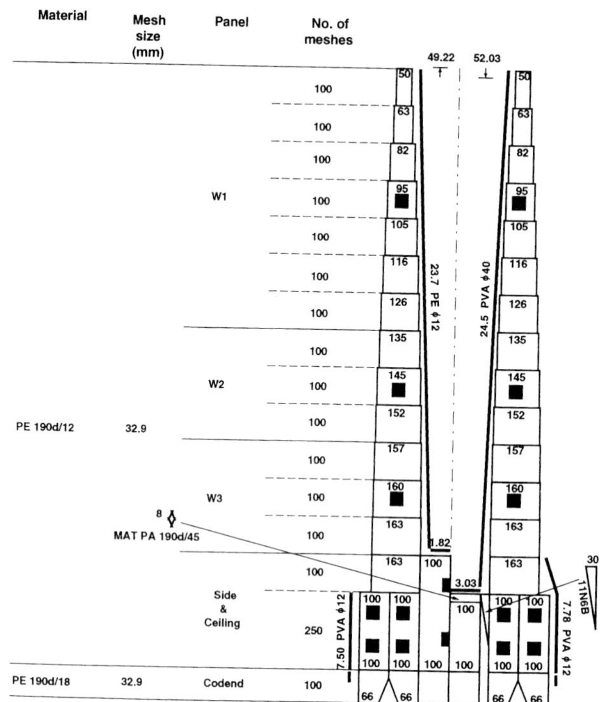

Fig. 1. Tested Danish seine and designation of panels over gear: Dimensions are indicated in mm for mesh size and diameters; in m for rope lengths (following the FAO standard gear illustration); and dark spaces indicate the positions of pocket nets.

(a)

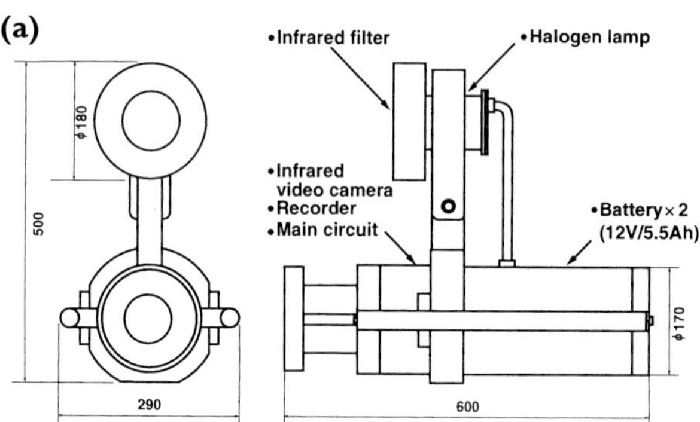

(b)

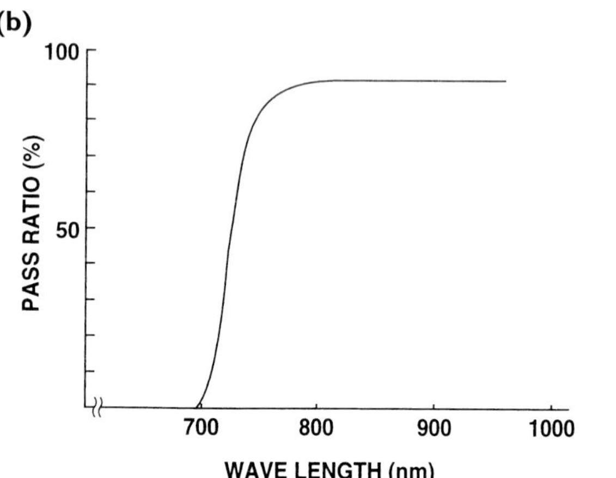

Fig. 2. (a) Underwater infrared video system (dimensions in mm) and (b) the characteristics of the installed infrared filter.

Materials and Methods

The Danish seine net tested was similar to the commercial gear common among fishing boats 7GT to 12GT for the studied fishery (Fig. 1). The major part of the gear consisted of net webbing of PE190d/12, 32.9 mm in measured mesh size, and the cod-end consisted of PE190d/18. The net webbing was rigged at 55% hanging ratio around the wings and headrope, and at 79% along the running rope through the main body. The completed gear had a footrope length of 52 m, seined with a 200-m long sweep and 1200-m long warp on each side.

The net was equipped with an underwater infrared video set newly developed in order to observe the behaviour of deep-water organisms without disturbance by visible light (Fig. 2). The system was equipped with a halogen lamp with an infrared filter which passes light beyond 720 nm in wave length. The video set is pressure-cased to 500 m, to record a black-and-white image for approximately 40 min after a set delay. The system was hung at various positions in the main body or the cod-end during the series of experiments.

The net was divided into six areas of webbing (Fig. 1), which were fitted with pocket nets of 20.3 mm mesh in order to estimate the degree of exclusion occurring through different portions of the gear; these areas comprised three wing panels and one main lateral on each side, the main ceiling and the cod-end. There was one pocket on each wing panel, four on the main lateral and two on the ceiling, and they covered 1.4% to 10.0% of meshes of the allocated panels. The cod-end was entirely covered by a net piece which was approximately 1.5 times longer than the cod-end.

Results

Operation and fishing gear under water

Horizontal encircling by the gear was estimated on the basis of the traces recorded by Global Positioning System (GPS) and angles between the warps measured at the stern of the boat. From an average of three hauls at depths of 370 m to 380 m, the wings were open initially to an angle of 146° which decreased to 27° before hauling. The net was towed for 1312 m in 72.3 min; hence, the average towing speed was approximately 0.30 m s^{-1}.

The shape of the gear underwater was observed by SCUBA-diving while the gear was operated at a depth of 25 m with shortened sweeps and warps. It was established that all the wing panels opened perpendicularly and that the lateral panels of the main body were almost parallel to each other. It was also confirmed that the netting twines were taut and that there was sufficient space between the cod-end and the cover.

Eleven trials were conducted from October 1994 to January 1996. Catches from two hauls were sampled in each trial, except that three hauls were sampled in the first two trials. All specimens from the pockets and cover were taken, whereas one-half to one-eighth of those in the cod-end were sampled, depending on the variation of catch.

The major prawn in the catch was *Haliporoides sibogae*, with a lesser amount of the bycatch *Plesionika martia*. Common finfishes

Table 1. Percentages of exclusion of organisms through different net panels and the numbers of meshes and cover ratio for panels

Species	No. of sample hauls	Net panels					
		W1	W2	W3	Side	Ceiling	Cod-end
H. sibogae	11	8.1	14.7	34.6	19.7	2.2	20.7
P. martia	9	27.5	12.6	33.0	8.1	1.9	16.9
Coelorinchus spp.	11	3.3	6.1	19.5	10.1	0.5	60.6
E. lucifer	11	0	1.3	18.1	28.5	8.0	44.1
V. garmani	9	1.5	13.9	24.0	9.5	0.4	50.6
C. megastomus	9	0	11.6	22.6	5.8	0.2	59.7
C. albatrossis	11	0	9.1	5.0	7.0	1.0	77.9
S. misakia	9	0	20.4	21.7	7.2	0	50.7
Number of meshes (thousands)	127.4	88.4	96.0	115.0	50.0	54.7	
Cover ratio (%)		1.41	5.78	5.21	6.26	10.0	100.0

The listed % were obtained as the averages of % to panels calculated for each trial.

were three species of grenadier, *Coelorinchus jordani*, *C. hubbsi* and *Ventrifossa garmani*, deepwater dogfish, *Eimopterus lucifer*, bigmouth conger, *Congriscus megastomus*, bigeyed greeneye, *Chlorophthalmus albatrossis*, and aglet sculpin, *Stlengis misakia*.

Behaviour of organisms against the net

Detailed analysis of the video image taken around the posterior end of the left wing showed that *H. sibogae* moved freely in the net, including swimming, drifting, bumping and hanging. Swimming against the flow of water in the net appeared to be more frequent than swimming with the flow, but this difference was not significant. A very clear optomotor response was demonstrated by prawns which kept pace with net outside the cod-end. Finfishes swam slowly in general. *E. lucifer*, in particular, swam very slowly and in random directions. The grenadiers *C. jordani*, *C. hubbsi* and *V. garmani* showed a long-lasting optomotor response both inside and outside the net.

Prawns passed through the meshes backwards. When a prawn's body touched the net the abdomen flexed to propel the prawn backwards and then stretched when the prawn bumped onto the net. This was repeated seven times on average until either the prawn passed out through a mesh or the encounter stimuli ceased owing to the passage of the prawn farther into the net. This was the only behaviour that resulted in exclusion of prawns. This reaction was mainly induced by touching the tail piece onto the netting. Other parts of the body such as antennae and chelate legs did not seem to be particular sensors of the stimulus to induce the rushing and bobbing behaviour. No exclusion by forward swimming or crawling was observed. Finfishes occasionally passed through meshes by swimming.

Results from pocket- and cover-net experiments

From the number $c_{k,j}$ of the kth length class retained by pockets/cover on the jth panels on both sides (coded from the wing tip to the ceiling and the cod-end), the number $E_{k,j}$ of the excluded individuals was estimated as:

$$E_{k,j} = K_j \cdot c_{k,j} \quad (1)$$

where K_j is a coefficient which is the inverse of the ratio of meshes screened by pocket/cover nets on each panel (Table 1).

Prawns were excluded primarily through the posterior wing panel adjacent to the main body, whereas the majority of finfishes was excluded through the cod-end (Table 1). Increasing numbers of *H. sibogae* were excluded posteriorly over the wing and, in total, 57.4% were excluded through wing meshes; only 20.7% were excluded through the cod-end. *P. martia* was excluded mainly through the foremost wing panel. In contrast, 44.1% to 77.9% of finfishes were excluded through the cod-end, and few through the anterior wing. Neither prawns nor most finfishes passed through the side or ceiling panels to any great extent.

The average lengths of *H. sibogae* in the pocket/cover nets increased in posterior panels, and for *P. martia* lengths were greater in the cover net than those in the anterior pockets (Table 2); this trend was not seen for finfishes.

Table 2. Lengths of organisms sampled by pockets and cover on different net panels

Species	No. of specimens	Length over net panels (mm)					
		W1	W2	W3	Side	Ceiling	Cod-end
H. sibogae	710	–	53.9	56.6	58.9	59.8	58.4
P. martia	4637	59.8	60.7	58.5	61.5	57.2	64.0
Coelorinchus spp.	2466	–	40.6	39.9	39.2	–	37.3
E. lucifer	487	–	–	146.5	141.1	–	150.3
V. garmani	507	–	28.0	31.9	27.1	–	28.1
C. megastomus	1526		233.3	245.8	231.1	–	237.1
C. albatrossis	586		–	96.7	86.3	–	93.0

Body length was measured for *H. sibogae*, *E. lucifer*, *C. megastomus* and *C. albatrossis*, and anal length for *Coelorinchus* spp. and *V. garmani*. Sample groups of which the specimens were fewer than ten are omitted and shown as –, and spaces represent that no specimens were obtained.

Selectivities of the whole net, S_k, and the cod-end, S'_k, were calculated as

$$S_k = C_k / (C_k + \sum_{j=1}^{6} E_{k,j}) \tag{2}$$

and

$$S'_k = C_k / (C_k + E_{k,6}) \tag{3}$$

where C_k is the number of the *k*th sized catch by the cod-end. As $E_{k,j}$ has been defined above, $E_{k,6}$ is the exclusion through the cod-end. The whole-net and cod-end selectivities were calculated for the two species of prawns, one finfish group (two *Coelorinchus* species were combined because of similar size and form), and one finfish species, and two examples are illustrated in Figure 3. The calculated plots were fitted with respective logistic curves by the least-squares method by iteration.

The respective selectivity curves for prawns had different slopes. The whole-net selectivity for *H. sibogae* was, for example, steeper and more selective against small size than the cod-end selectivity. The two curves were, however, similar at greater lengths. These are attributable to the fact that exclusion of *H. sibogae* occurred mainly through wing panels and that these prawns were smaller than those excluded through the cod-end. For *P. martia* a similar trend was observed. For finfishes, the whole-net selectivity curve was similar to the cod-end selectivity. This reflects the fact that lesser finfish were excluded through panels other than the cod-end and that there was no distinctive difference in size among those excluded through different panels.

The experimental results were summarized in 50%-selective lengths and selectivity indices, I_s, defined as:

$$I_s = (L_{75} - L_{25})/L_{50} \tag{4}$$

where L_{75}, L_{25} and L_{50} are the selective lengths corresponding to those percentages. The results (Table 3) showed that whole-net selectivity possesses larger 50%-selective lengths and smaller selectivity indices, and this difference was distinctive in the case of prawns.

Model of the retention and exclusion process

On the basis of the behavioural observations, it was hypothesized that selectivity for prawns is ruled by the geometrical relationship between the section of the bent abdomen and the mesh opening as well as by the chances for net panels to be bumped by prawns. Prawns in front of the gear are presumed to be distributed at random and to come toward the netting along a direct path (Fig. 4). The number of encounters, $I_{i,i}$, first on the *i*th panel could be, therefore, proportional to the area of the panel transversely projected as:

$$I_{i,i} = \gamma \cdot A_i \cdot \sin\phi_i \tag{5}$$

where γ is a constant related to stock density in a fishing ground and ϕ_i is the horizontal attack angle of the *i*th panel of which the expanded area is A_i. When a prawn encounters a mesh, the probability, P_e, that it will pass through a mesh is represented by:

$$P_e = S'(L) / S(m,r) \tag{6}$$

where $S(m,r)$ is the area of an *m*-size mesh and *r* is the hanging ratio. $S'(L)$ is the area occupied by the trace of the centre of the maximum body section of an *L*-long prawn marked within the mesh, with the body girth touching no strands. This is based on the assumption that prawns can insert their body in a mesh only as above because there is no propulsive force after flexion, the body mass is small and strands of the netting are taut.

Prawns have a weak optomotor response. Presuming that the distances and directions of their bouncing after bumping on to

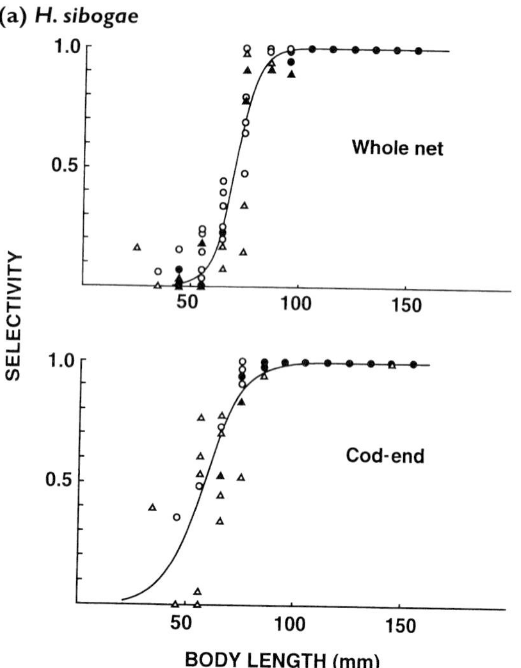

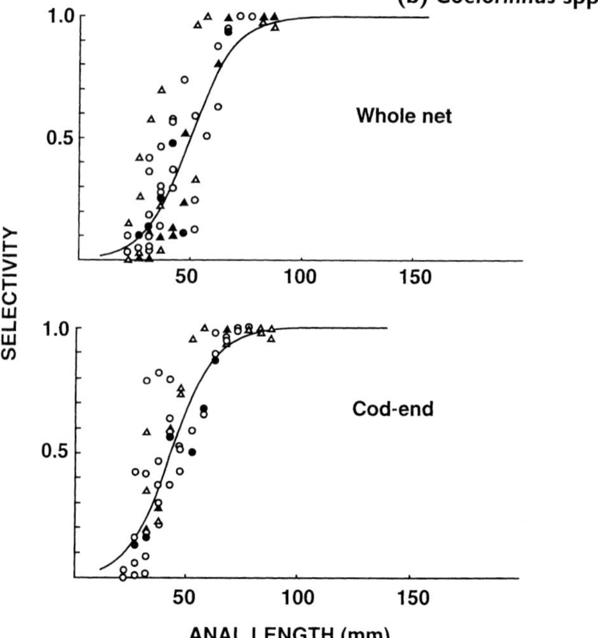

Fig. 3. Whole-net and cod-end selectivity curves for (a) *H. sibogae* and (b) *Coelorinchus* spp.: symbols represent the number of specimens used as denominators to calculate selectivity; △, 20–50; ○, 51–200; ▲, 201–450, and ●, >450. Values obtained from specimens <20 were not used for calculation.

Table 3. 50%-selective length and selectivity index assessed for whole-net selectivity and cod-end selectivity.

Species	50%-selective length (mm)		Selectivity index	
	Whole net	Cod-end	Whole net	Cod-end
H. sibogae	69.7	58.0	0.168	0.335
P. martia	82.9	66.6	0.248	0.504
Coelorinchus spp.	50.6	43.4	0.426	0.482
E. lucifer	182	147	0.443	0.483

the net are randomly distributed (Watson *et al.* 1992; Isaksen and Valdemarsen 1994) and that they drift afterwards, the probable frequency, b_i, of bumping over the ith panel is ruled by the attack angle and the number of meshes, n_i, and was simplified as

$$b_i = \frac{4\phi_i}{\pi} B \cdot n_i \tag{7}$$

where B is the number of bumps against a unit mesh at the standard attack angle, $\pi/4$. The probable number of bumps for prawns first encountering a given panel is half the value of b_i because they travel for half a distance over the panel on average.

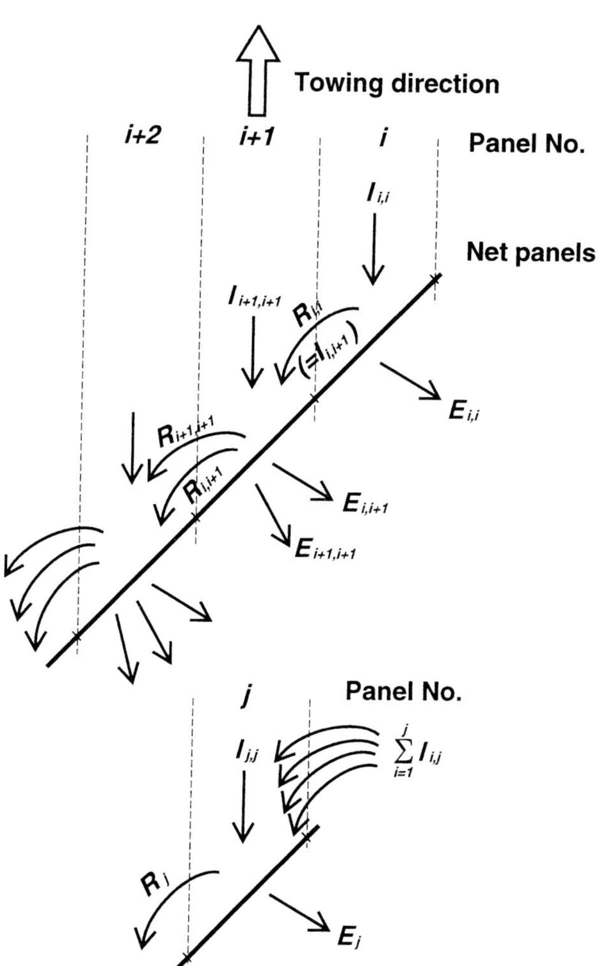

Fig. 4. Schematic diagram representing the process of exclusion and retention over net panels.

Among those encountering the ith panel, the number of retentions, ρ, is

$$\rho = \exp\{\ln(1 - P_e) \,.\, b_i\} \,.\, I_{i,i} \tag{8}$$

Retained prawns are transferred to travel over the next panel for either exclusion or retention again; hence, the numbers of prawns that encountered the ith panel first and end up retained, $R'_{i,j}$, and excluded, $E'_{i,j}$, at the jth panel are

$$R'_{i,j} = \exp\{\ln(1 - P_e) \,.\, b_j\} \,.\, I_{i,j} \tag{9}$$

and

$$E'_{i,j} = I_{i,j} - R'_{i,j} \tag{10}$$

where $I_{i,j}$ is the number of prawns that first encountered the ith panel and then transferred to the jth panel, or is equal to $R_{i,j-1}$ where $i < j$ other than $I_{i,i}$ defined above where $i = j$.

The jth panel receives all the prawns remaining from the panels from first to $(j - 1)$th and also the prawns newly coming to the jth panel. Therefore, the total retention, R_j, and exclusion, E_j, at the jth panel are

$$R_j = \exp\{\ln(1 - P_e) \cdot b_j\} \sum_{i=1}^{j} I_{i,j} \tag{11}$$

and

$$E_j = \sum_{i=1}^{j} I_{i,j} - R_j \tag{12}$$

Simulation on the basis of the above was carried out for *H. sibogae*. For the calculation, the attack angle of the wing was taken to be 43.3°, the average of those found at start and finish of the haul in the experiment analysed above. Those of the main side panels were taken to be 6.7° to 7.7° on the basis of the hanging ratios over the main body. The constant, γ, was not taken into consideration because of its lack of effect under a constant resource condition. The encounter of a mesh by a prawn was simplified as that of a rectangle and a circle, of which the area is the same as an ellipse defined by the maximum breadth, b_m, and height, h_m, of the flexed body section. For this, 26 individuals of critical sizes were measured and the regression of the two parameters against body length were given as

$$b_m = 0.154\ L - 1.90 \tag{13}$$

and

$$h_m = 0.319\ L - 1.33 \tag{14}$$

The constant B was obtained from the observation that prawns in the posterior end of the wing bumped 0.103 times per mesh per individual at $\phi = 43.3°$. The numbers of incoming prawns of different length classes were estimated on the basis of the whole-gear selectivity and average catches from all the trials. The behaviour and exclusion in the cod-end was unknown and could not be modelled. Therefore, the number of bumps on the cod-end panel was adjusted by trial and error so that the ratio of the exclusion by the cod-end was the same as that obtained from the experiment, i.e. 20%.

Table 4 shows the simulated distribution of exclusions over the designated panels. It clearly demonstrates that smaller individuals are mainly excluded from anterior panels, and that larger ones are mainly excluded from posterior panels and the cod-end. Therefore, those excluded from wing panels were smaller in size than those from the cod-end and the average sizes from wing panels increased posteriorly. All these are similar to the trends observed in the experiment. This has also an implication to verify the large number of the observed exclusion of the small shrimp, *P. martia*, through the most anterior panel.

Despite the above, there was a limitation of the present model regarding the distribution of exclusions over wing panels (Table 4). The discrepancy between simulated and observed results may be attributed to over-estimation of the exclusion of small prawns. This is also reflected in the fact that the selective sizes obtained from the simulation did not correspond well with those found in the experiment.

Discussion

Behaviour of organisms and retention/exclusion process

The newly designed infrared video system revealed slow and stable swimming of organisms; this is quite different from the quick movements that have appeared in many non-infrared video images previously available. The optomotor response of prawns is also one of the most important results attained by the method, although it was thought to be very weak. These results suggest that such techniques, which avoid visible light stimuli, are essential for behavioural studies on deep-water fishing.

The primary exclusion of prawns by wings and of finfish by the cod-end is attributable to the difference in optomotor response. This behavioural difference has been emphasized by Watson (1989), Matsuoka and Kan (1991) and Watson *et al.* (1992); however, its implication with regard to the selection process has not been fully delineated. The present model illustrated that the panels to be passed through by organisms are primarily determined by probabilities of bumps or encounters. In a tank experiment (Matsuoka *et al.* unpublished), finfishes such as red seabream, *Pagrus major*, positioned in front of a diagonally set and moving net and dislocated by swimming away from the net when caught up. This suggests that the optomotor response decreases the chances of bumps with the net. Exclusion shifted posteriorly by decreasing values of B in Equation (7), i.e. lower chances of bumping, could be also proved by the model.

A greater number of excluded organisms through posterior panels of towed gear has been reported by Chow *et al.* (1990) and Matsushita *et al.* (1992). This has been attributed to disruption of the optomotor response around there (Watson 1989); however, the pattern of exclusion is now established for prawns, which have only a weak optomotor response. This dislocation is primarily attributable to the fact that the posterior panels are encountered by a greater number of organisms, as discussed above. Hence, the results do not necessarily imply that the panels to be passed through are the rear part of the main body. The panels to be passed must be determined by attack angles among posterior panels.

Implication to selectivity

The differential passage of organisms through different panels has implications for the contribution of wings to the selectivity of towed gear. It can be generalized that organisms of small size and weak optomotor response tend to be excluded anteriorly and, consequently, the cod-end contributes less to selectivity. Cod-end selectivity does not necessarily represent the selectivity of gear as a whole. This is particularly true for weak optomotor-respondents such as prawns.

The sizes at the two extremes of a selectivity range appear to be ruled by different factors according to the present experimental and analytical results. The higher selectivity size is almost spontaneously determined by mesh size, given an appropriate hanging ratio, whereas the lower selectivity size is determined mainly by mechanisms of repetition of bumping against net panels. This is also supported by the selectivity curves for the whole gear and cod-end, of which the upper extremes are similar whereas the lower extremes are largely different. This suggests that the selectivity span or steepness of a selectivity curve is a secondary phenomenon. In order to gain sharp selectivity when a mesh size is given, improvement in exclusion of small individuals is the key issue.

Table 4. Simulation of numbers of *H. sibogae* in length classes excluded from different panels and their average lengths

Length (mm)	No. of encounters	Numbers of exclusions through panels				
		W1	W2	W3	Side	Cod-end
55	35.8	11.42	8.25	9.63	3.14	3.33
65	35.8	9.05	7.51	9.10	3.08	6.36
75	44.1	4.24	4.62	6.28	2.29	9.48
85	81.1	0.09	0.12	0.18	0.73	0.42
Total	196.8	24.80	20.50	25.20	8.59	19.59
Average length (mm)		45.25	46.97	47.64	48.16	60.76

The simulation could not achieve selective sizes equivalent to those experimentally obtained. The main reason is the simplification of the geometric relationship between the prawn's body section and the mesh opening. Fujiishi (1973) modelled the encountering of organisms and a diamond mesh at random directions, from which it is easily conjectured that it is possible to obtain a lower probability of passage for small prawns and less steep selectivity curves than in the present simulation. However, Fujiishi's model was not applicable because the relationship between a prawn body and a mesh at the instant of dislocation has not yet been clarified. The critical body section which determines if a prawn will pass a mesh must be further studied. In line with the above, the behaviour of prawns at the instance of bumping must be quantitatively assessed, whereas the present study could not take their weak optomotor response sufficiently into consideration.

Approaches to selective nets

The present study suggests the following potential approaches for better size- and species-selectivity of trawl and seine nets that have prawns and finfishes as their main catch or bycatch:

1. Mesh sizes of wing panels must be further considered in management because of the high degree of exclusion of prawns by them.
2. Net panels parallel to the towing direction, i.e. around the ceiling or extension tube, are not readily passed by prawns.
3. The best panels to instal such devices as BED (Bycatch Exclusion Device) to dislocate finfishes are those that segregate fewer prawns, as above.

The selectivity curves for prawns and finfishes found in the present study suggest that improvement of size selectivity for prawns by using enlarged meshes could also contribute to species selectivity against bycatch finfishes; this is perhaps applicable in many fisheries to deal with bycatch organisms. This suggests, therefore, that studies on size selectivity and species selectivity, which have been relatively separate in recent years, must be collaborative.

References

Chow, Y S, Chen, C C, and Huang, C S (1990). Study on the escape of fish through different parts of a bottom otter trawl net. In 'Second Asian Fisheries Forum'. (Eds R Hirano and I Hanyu) pp. 817–20 (Asian Fisheries Society: Toyko).

Fujiishi, A (1973). A theoretical approach to selectivity of the net gears-I. A method of estimating selectivity curves of trawl and seine nets. *The Journal of the Shimonosek University of Fisheries* **22**, 1–28.

Isaksen, B, and Valdemarsen, J W (1994). Bycatch reduction in trawls by utilizing behaviour differences. In 'Marine Fish Behaviour in Capture and Abundance Estimation'. (Eds A Ferno and S Olsen) pp. 69–83 (Blackwell: Oxford).

Matsuoka, T, and Kan, T T (1991). Passive exclusion of finfish by trawl efficiency device (TED) in prawn trawling in the Gulf of Papua, Papua New Guinea, *Nippon Suisan Gakkaishi* **57**, 1321–9.

Matsushita, Y, Inoue, Y, Shevchenko, A I, and Norinov, Y G (1992). Selectivity in the cod-end and in the main body of the trawl. In 'Fish Behaviour in Relation to Fishing Operations'. (Eds C S Wardel and C E Hollingworth) pp. 170–99. International Council for the Exploration of the Sea.

Watson, J W (1989). Fish behaviour and trawl design: Potential for selective trawl development. In 'World Symposium on Fishing Gear and Fishing Vessel Design'. pp. 25–9 (Marine Institute: St. John's Newfoundland).

Watson, J W, Workman, I K, and Hataway, B D (1992). The behaviour of fish and shrimp encountering trawls in the southeastern US Penaeid shrimp fishery. In 'Meeting 1992 Global Ocean Partnership'. pp. 336–41 (Marine Technology Society: Washington DC).

Managing the Fishdown of the Australian Orange Roughy Resource

J. A. Koslow, N. J. Bax, C. M. Bulman, R. J. Kloser, A. D. M Smith and A. Williams

CSIRO Marine Laboratories, GPO Box 1538, Hobart, Tasmania 7001, Australia.

Summary

The orange roughy (*Hoplostethus atlanticus*), which lives mainly at depths between 700 and 1200 m, is exceptionally vulnerable to overfishing due to its low productivity and aggregated distribution around seamounts. Large aggregations of orange roughy were first discovered off southeast Australia in 1989. Catches peaked the following year at ~50 000 t and in 1996 are believed to be at sustainable levels (~5000 t). Effective management of the Australian orange roughy resource through rapid reduction in catches from early unsustainable levels depended on several factors. Two independent survey methods (acoustics and the egg production method) were applied the year following discovery of the resource. Early absolute biomass estimates established the approximate sustainable yield, which supported a precautionary Total Allowable Catch (TAC). Consistent results between the two methods and between years enhanced confidence in the assessment. A stock reduction analysis based on both the surveys (acoustic surveys now treated as relative biomass indices) and the catch history gave further support to these results. Involvement of industry, as well as various State and Commonwealth agencies, in the research and assessment process enhanced a sense of broad ownership. Finally, development of probabilistic asessment models enabled industry and managers to select among alternative catch reduction scenarios in order to achieve the management objective for the fishery: that stock biomass be at least 30% of pre-fishery biomass by the year 2004.

Introduction

Orange roughy (*Hoplostethus atlanticus*) is among the deepest living commercially exploited fishes, being generally fished at depths of 700–1200 m. It was first fished commercially in the waters around New Zealand in the late 1970s; exploration for the resource began in the early 1980s in the waters around south-eastern Australia.

In 1988–89, the Tasmanian Department of Sea Fisheries and CSIRO separately conducted random-stratified trawl surveys on trawlable ground at mid-slope depths around south-eastern Australia. The conclusion of both surveys, based upon standard, swept-area methods of biomass estimation, was that the orange roughy resource in these waters was extremely limited: of the order of 10 000–20 000 t (Lyle *et al.* 1989; Bulman *et al.* 1994).

However, in 1989 commercial fishers discovered large aggregations of orange roughy on seamounts off north-eastern and southern Tasmania, where the fish appeared to be spawning in winter and feeding in summer (Fig. 1). Because the seamounts were deemed untrawlable using conventional trawl methods, they had not been sampled in the research vessel surveys. Orange roughy landings from these areas exceeded 30 000 t in 1989.

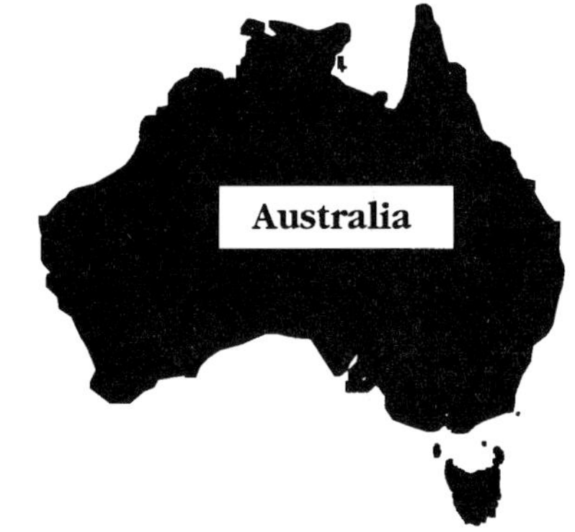

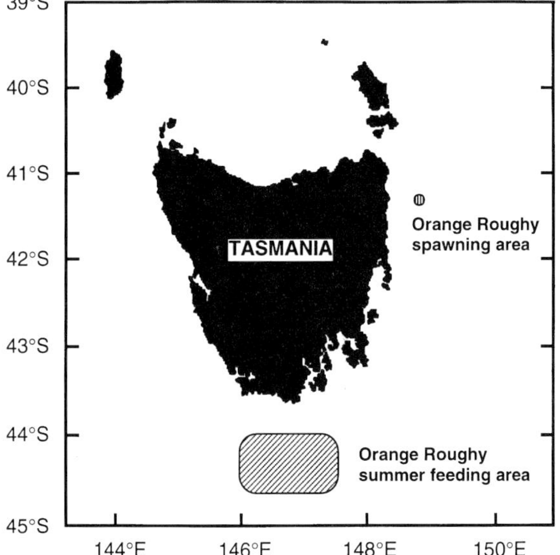

Fig. 1. The location of the primary orange roughy fishing grounds in southeastern Australia: the winter spawning site off northeastern Tasmania and the summer feeding ground off southern Tasmania.

It was apparent that the trawl surveys had sampled only a fraction of the orange roughy resource. There were virtually no data on which a biomass estimate could be based, and estimates of the size of the resource varied from 50 000 to >1 million t. There was considerable urgency to obtain a valid biomass estimate in view of the latent effort in the fishery and the extreme vulnerability of orange roughy to overfishing, especially if the biomass was at the lower end of the estimated range.

Orange roughy are exceptionally vulnerable to overfishing due to the combination of their low productivity and aggregating behaviour that makes them accessible to modern deepwater trawlers. The sustainable yield of an orange roughy stock is only a few percent of its virgin biomass due to the species' longevity (> 100 years) and concomitant low natural mortality (M ~0.05) and recruitment (Francis 1992), slow growth, and late maturity (25–30 years) (Fenton *et al.* 1991; Smith *et al.* 1995). (Although these factors were not fully understood in 1989, there were indications from the bimodal size distribution of the catch that the species was slow-growing and long-lived.) Because orange roughy aggregate on specific topographic features, the fish present virtually fixed targets. This is particularly evident during spawning which, off south-eastern Australia, occurs for the most part on a single seamount. Once the technique of fishing on seamounts was developed, catch rates were very high, ranging from several t to >50 t per shot for bottom times of typically no more than a few minutes.

To assess the sustainable yield of the south-eastern Australian orange roughy resource, the primary issue was to estimate stock size, assuming that ongoing research would resolve the species' age and growth and hence its productivity. Both acoustics and the annual egg production method (Saville 1964) were used to assess orange roughy stock size because of the potential value of the resource and uncertainty regarding its size; the need to assess stock size as quickly and clearly as possible; and the biases and technical problems potentially associated with any single-stock assessment method applied to a new species, particularly one living at mid-slope depths.

In fact, orange roughy proved amenable to both assessment methods, at least on their spawning ground off north-eastern Tasmania. The spawning site consisted of a single seamount ~4 km in diameter and the spawning season was short and highly predictable, extending for approximately one month from early July to early August. The acoustic survey was facilitated by the occurrence of the orange roughy around the spawning seamount in large single-species aggregations, and the annual egg production method by the orange roughy's determinate fecundity and readily identifiable eggs.

On the other hand, the feeding aggregations south of Tasmania were less amenable to fishery-independent survey methods. The fish were found in smaller aggregations around some fifty seamounts, and they occurred with concentrations of several other species, primarily smooth and black oreos (*Pseudocyttus maculatus* and *Allocyttus niger*); the basketwork eel (*Diastobranchus capensis*); and several species of deepwater squalid sharks. The seamounts cannot be trawled by conventional methods and most can only be trawled, even by commercial trawlers, in selected areas. Hence the species composition of acoustic marks often could not be directly sampled. Orange roughy have low acoustic reflectance due to their lack of a gas-filled swimbladder, so acoustic estimates of their abundance are highly sensitive to error in estimation of species composition.

METHODS

To conduct the acoustic surveys, a Simrad EK500 with 38 kHz split-beam transducer was adapted for use with a towed body that could be towed at 6–8 knots at 500–650 m depth or lowered vertically to 1000 m to obtain target-strength measurements. The towed body provided additional stability for the transducer, eliminated the effects of near-surface noise and removed bias caused by beam thresholding and acoustic attenuation due to air bubbles in the near-surface layer. It also reduced by approximately 50% the amount of biomass that had to be extrapolated into the acoustic shadow-zone, caused by the edge of the acoustic beam impinging on the steep slopes of the seamounts (Kloser 1996).

The acoustic surveys of the spawning ground were conducted each year with six north–south and six east–west transects at half-minute intervals of latitude and longitude across the seamount. The acoustic work was supported by demersal and midwater trawls and measurements of acoustic target strength carried out *in situ* or on dead specimens suspended below the transducer and lowered to depth (Kloser *et al.* 1996, in press). The acoustic and trawl data were stratified by 100 m depth intervals based on observed changes in species composition with depth.

The annual egg production method was implemented by means of egg surveys carried out throughout the spawning period. The survey in 1991 was based on a series of east–west transects that extended to 30 nm north and south of the spawning site. However, egg densities were generally too low to be effectively sampled more then 10 nm from the spawning site. In 1992, more intensive sampling was achieved through a random, stratified survey design based on a 'box' that extended 5 nm north and south of the spawning ground and a second stratum that extended a further 5 nm north, south and seaward. This survey design was adequate to sample the eggs fully during their first day after spawning, and sampling was conducted throughout the spawning period (Koslow *et al.* 1995*b*). Samples of adult females were examined over a three year period to assess fecundity (Koslow *et al.* 1995*a*), and egg incubation experiments were carried out to assess hatching time (Bulman and Koslow 1995).

Stock reduction analyses (SRA) for the spawning orange roughy off northeastern Tasmania (the Eastern Zone) and the feeding groups off southern Tasmania (the Southern Zone) were carried out under two scenarios: that the fish from the two zones represent a single stock or that they represent two separate stocks. Although it is considered most likely that there is some mixing of orange roughy between the two areas, stock discrimination studies remain inconclusive to date (Lester *et al.* 1988; Edmonds *et al.* 1991; Elliott and Ward 1992; Smolenski *et al.* 1993; Elliott *et al.* 1994; Elliott *et al.* 1995).

SRA for the spawning stock was based on the acoustic estimates, now treated as relative indices of stock biomass, the egg production estimate, and the record of commercial landings. Two estimates of natural mortality (M) were used, 0.048 and 0.064, derived from alternative estimates of when the fish recruit to the fishery: 31 years, the mean age at which there is a transition from wide to narrow circuli on the otolith; or 52 years, the age of the mode in the catch at age distribution (D C Smith and A D M Smith unpublished). The difference between these two indicators of age of recruitment suggests either that the otolith transition zone is not an indicator of recruitment to the fishery or that recent recruitment has been lower than that earlier on.

SRA for orange roughy south of Tasmania was based primarily upon catch-per-unit effort (CPUE) of selected vessels as an index of abundance, in combination with the catch history, due to the broad confidence limits associated with the acoustic surveys (N. Bax and A. D. M. Smith unpublished). The problems associated with use of CPUE data from an aggregating species for such purposes were recognized, but CPUE appeared to follow closely the trend associated with the decline in stock size based upon SRA for the Eastern Zone data. It was therefore used, albeit cautiously, as the basis for assessment of the Southern Zone stock.

SRA for the single-stock hypothesis was based on the egg and acoustic surveys carried out on the spawning orange roughy and on the catch history for the two areas.

Estimates of virgin and current biomass from SRA were based on the median of the maximum likelihood distributions. Advice to managers was based on the probability of being above or below specified biomass levels over specified time horizons.

Results

Annual acoustic surveys carried out from 1990–93 and an egg survey in 1992 consistently indicated that the virgin biomass of the orange roughy spawning stock (B_0) was approximately 100 000 t. This information, combined with increasing evidence of the extreme longevity of orange roughy (Fenton *et al.* 1991; Smith *et al.* 1995) indicated that the sustainable yield from the spawning stock would be only several thousand t. A precautionary catch limit was therefore first imposed after 1990, when catches peaked at ~50 000 t and the total allowable catch was reduced rapidly thereafter. The acoustic surveys clearly established the rapid fishdown of the virgin orange roughy spawning stock (Fig. 2).

Under the two-stock hypothesis, the SRA for the spawning stock refined previous acoustic and egg survey estimates of its virgin biomass (i.e. B_0 ~100 000 t) (Fig. 3, Table 1) ((N Bax and A D M Smith unpublished). The SRA also indicated that orange roughy biomass in the Southern Zone was comparable to that found in the Eastern Zone (Table 1, Fig. 4). Under the single stock hypothesis, SRA indicated that the virgin biomass from the two areas was approximately 175 000 t (Table 1) (N Bax and A D M Smith unpublished).

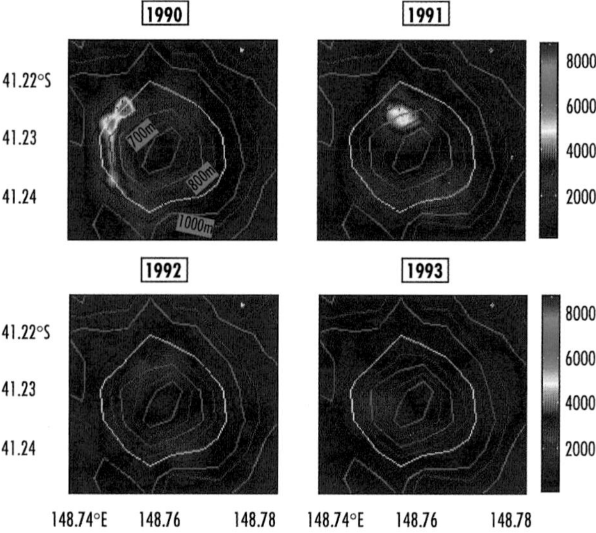

Fig. 2. The distribution and intensity of acoustic backscattering (Sa) on the orange roughy spawning ground off northeastern Tasmania, 1990–93. The latitude and longitude and depth contours (in 100 m depth intervals) of the spawning seamount are shown.

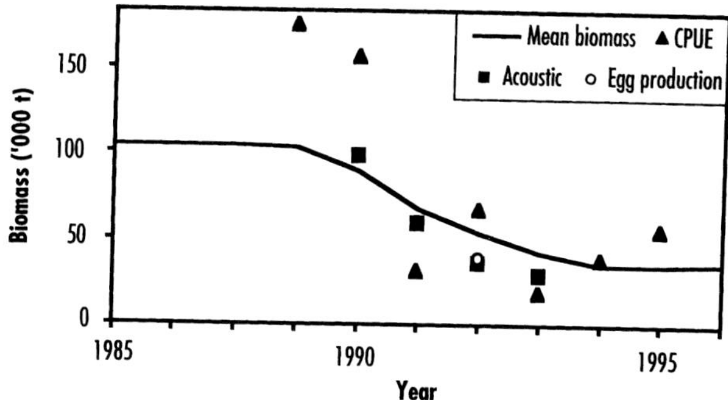

Fig. 3. Stock reduction analysis (SRA) showing the decline in the spawning stock of orange roughy off northeastern Tasmania based on results of acoustic and egg surveys and catch history. Catch-per-unit-effort (CPUE) follows the trend in decline of the stock but was not used in the SRA.

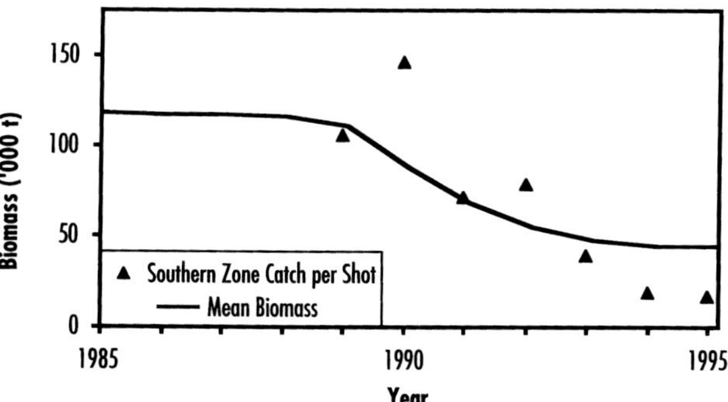

Fig. 4. Stock reduction analysis of orange roughy from the fishing zone south of Tasmania based on a catch-per-unit-effort (CPUE) index of biomass and catch history.

In 1994, the Australian Fisheries Management Authority (AFMA) finalized management objectives to develop the fishery based on the principles of ecologically sustainable development, following extensive discussions between fishers, managers and scientists. The primary objective was to set TACs to maintain the spawning biomass of each stock above 30% of B_0. If below this level, TACs were to be reduced so spawning biomass reached 30% B_0 by the year 2004. The fishery was to close if spawning biomass fell below 20% B_0.

The fishdown of the orange roughy resource around Tasmania now appears complete with the stock(s) between 20% and 30% of pre-fishery biomass, the assessment varying somewhat in relation to assumptions regarding natural mortality and stock structure (Table 1). Under the most optimistic scenario (M = 0.064, two-stock hypothesis), the eastern and southern stocks are close to 0.3 B_0. Under the most pessimistic scenario (M = 0.048, single-stock hypothesis), there is a 61% probability that current biomass is below 0.2 B_0. Probabilistic forecasts of the impacts of future catch scenarios on meeting AFMA's performance objectives have been developed in conjunction with industry and management ((N. Bax and A. D. M. Smith unpublished). These forecasts are used to recommend catch levels in the two zones until the year 2004 that would meet the performance objectives, taking into account associated structural and data uncertainties and the impact of alternative catch scenarios on economic returns to industry.

DISCUSSION

Conditions at the outset of the orange roughy fishery seemed a prescription for disaster: a low-productivity, aggregating species; large latent effort in the fishery; and some highly over-optimistic initial speculation regarding the probable size of the resource (i.e. that stock size was as much as an order of magnitude higher than present estimates).

Several factors contributed to the implementation of a successful quota reduction strategy (Fig. 5). First, the use of independent survey methods that provided absolute estimates of stock biomass quickly clarified the likely sustainable yield. Second, the rapid implementation of a precautionary TAC and its phased reduction, based upon the accumulating scientific evidence, provided sufficient time for the ultimate management objectives for the fishery to be established. Thirdly, overall confidence in the assessment, which was ultimately required to reduce catches ten-fold over a five-year period, was enhanced by the consistency of the scientific results between the two methods and between years, and their further confirmation by the SRA. Fourthly, there was a sense of ownership of the assessment by all sectors. The scientific research was presented publicly at annual meetings, and the fishing industry participated formally in the development and review of the research programme through a Government-Industry Technical Liaison Committee (see Ross and Smith 1997). Aspects of the research were carried out at State fishery agencies and universities, as well as Commonwealth agencies; and the fishing industry was represented in the stock assessment process itself. Another significant factor was an

Table 1. Stock reduction analysis estimates of pre-fishery biomass (B_0) and the probability that current biomass is below 0.3 or 0.2 B_0, based on alternative assumptions of stock structure (single combined stock *v* separate eastern and southern stocks) and natural mortality (*M*)

Stock	*M*	B_0 (t)	Probability $B_{1995} < 0.3{*}B_0$	$B_{1995} < 0.2{*}B_0$
Combined	0.048	178 000	0.86	0.61
	0.064	172 000	0.82	0.49
Eastern	0.048	96 000	0.63	0.21
	0.064	93 000	0.47	0.09
Southern	0.048	93 000	0.68	0.51
	0.064	97 000	0.56	0.37

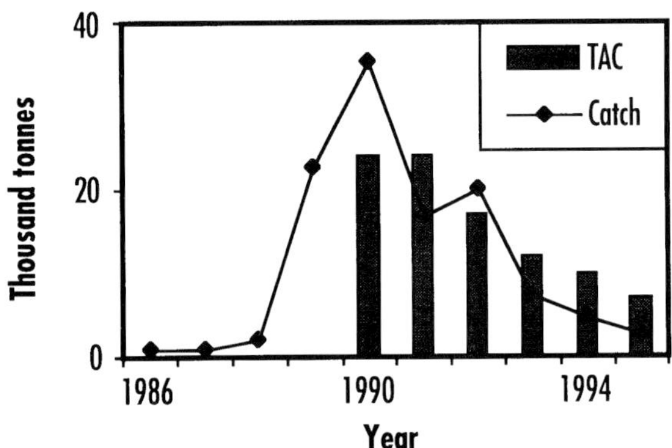

Fig. 5. The history of Total Allowable Catch (TAC) quotas and catch from the orange roughy fishery around Tasmania.

international scientific review of the stock assessment, which confirmed its overall scientific rigour. Finally, the ability of management to develop a long-term management strategy was enhanced by development of stock assessment models that forecast the consequences of alternative catch reduction scenarios. Ultimately, management of the fishdown of Australia's orange roughy required cooperation among scientists, managers, and industry, as well as a strong commitment to sustainable development. This cooperation and commitment will need to be carried into the future if the resource is to rebuild over the next decade to 30% of pre-fishery biomass.

Acknowledgments

We thank M. Lewis and A. Terauds for assistance with field and laboratory work and graphics. K. Haskard provided statistical advice. J. Stevens, T. Davis and J. Young reviewed the manauscript. The work was supported by grants from the Fisheries Research Development Corporation and the Australian Fisheries Management Authority.

References

Bulman, C M, and Koslow, J A (1995). Development and depth distribution of the eggs of orange roughy, *Hoplostethus atlanticus* (Pisces: Trachichthyidae). *Australian Journal of Marine and Freshwater Research* **46**, 697–705.

Bulman, C M, Wayte, S E, and Elliott, N G (1994). Orange roughy surveys, 1988 and 1989; Part A: abundance indices, Part B: biological data. *CSIRO Marine Laboratories Report* 215. 21 pp.

Edmonds, J S, Caputi, N, and Morita, M (1991). Stock discrimination by trace element analysis of otoliths of orange roughy (*Hoplostethus atlanticus*), a deep-water marine teleost. *Australian Journal of Marine and Freshwater Research* **42**, 383–9.

Elliott, N G, and Ward, R D (1992). Enzyme variation in orange roughy, *Hoplostethus atlanticus* (Teleostei: Trachichthyidae), from southern Australian and New Zealand waters. *Australian Journal of Marine and Freshwater Research* **43**, 1561–71.

Elliott, N G, Haskard, K, and Koslow, J A (1995). Morphometric analysis of orange roughy (*Hoplostethus atlanticus*) off the continental slope of southern Australia. *Journal of Fish Biology* **46**, 202–20.

Elliott, N G, Smolenski, A J, and Ward, R D (1994). Allozyme and mitochondrial DNA variation in orange roughy, *Hoplostethus atlanticus* (Teleostei: Trachichthyidae): little differentiation between Australian and North Atlantic populations. *Marine Biology* **119**, 621–7.

Fenton, G E, Short, S A, and Ritz, D A (1991). Age determination of orange roughy, *Hoplostethus atlanticus* (Pisces: Trachichthyidae) using ^{210}Pb:^{226}Ra disequilibria. *Marine Biology* **109**, 197–202.

Francis, R I C C (1992). Use of risk analysis to assess fishery management strategies: a case study using orange roughy (*Hoplostethus atlanticus*) on the Chatham Rise, New Zealand. *Canadian Journal of Fisheries and Aquatic Sciences* **49**, 922–30.

Kloser, R J (1996). Improved precision of acoustic surveys of benthopelagic fish by means of a deep-towed transducer. *ICES Journal of Marine Science* **53**, 407–13.

Kloser, R J, Koslow, J A, and Williams, A (1996). Acoustic assessment of the biomass of a spawning aggregation of orange roughy (*Hoplostethus atlanticus*, Collett) off southeastern Australia 1990–93. *Marine and Freshwater Research* **47**, 1015–24.

Kloser, R J, Williams, A, and Koslow, J A (in press). Problems with acoustic target strength measurements of a deepwater fish, orange roughy (*Hoplostethus atlanticus*). *ICES Journal of Marine Science.*

Koslow, J A, Bell, J, Virtue, P, and Smith, D C (1995*a*). Fecundity and its variability in orange roughy (*Hoplostethus atlanticus*) off southeastern Australia. *Journal of Fish Biology* **47**, 1063–80.

Koslow, J A, Bulman, C M, Lyle, J M, and Haskard, K A (1995*b*). Biomass assessment of a deep-water fish, the orange roughy (*Hoplostethus atlanticus*), based on an egg survey. *Australian Journal of Marine and Freshwater Research* **46**, 819–30.

Lester, R J G, Sewell, K B, Barnes, A, and Evans, K (1988). Stock discrimination of orange roughy, *Hoplostethus atlanticus*, by parasite analysis. *Marine Biology* **99**, 137–43.

Lyle, J M, Kitchener, J A, and Riley, S P (1989). An assessment of the orange roughy resource off Tasmania: Progress report 1989. *Report to DPFRG* 28. Department of Sea Fisheries, Tasmania. 37 pp.

Ross, D M, and Smith, D C (1997). The Australian orange roughy fishery: A process for better resource management. In 'Developing and Sustaining World Fisheries Resources: The State of Science and Management. Second World Fisheries Congress, Brisbane 1996'. (Eds D A Hancock, D C Smith, A Grant and J P Beumer) pp. 256–60 (CSIRO Publishing: Melbourne).

Saville, A (1964) Estimation of the abundance of a fish stock from egg and larval surveys. *Rapports et Proces-Verbaux des Reunions Conseil International pour l'Exploration de la Mer* **153**, 164–70.

Smith, D C, Fenton, G E, Robertson, S G, and Short, S A (1995). Age determination and growth of orange roughy (*Hoplostethus atlanticus*): a comparison of annulus counts with radiometric ageing. *Canadian Journal of Fisheries and Aquatic Sciences* **52**, 391–401.

Smolenski, A J, Ovenden, J R, and White, R W G (1993). Evidence of stock separation in southern hemisphere orange roughy (*Hoplostethus atlanticus*, Trachichthyidae) from restriction enzyme analysis of mitochondrial DNA. *Marine Biology* **116**, 219–30.

Mesh-Size Management in Pelagic Trawl Fisheries — Potential Solutions

Petri Suuronen,[A] *Daniel Erickson*[B] *and Ellen Pikitch*[B]

[A] Finnish Game and Fisheries Research Institute, PO Box, FIN-00151 Helsinki, Finland.
[B] University of Washington, Fisheries Research Institute, Box 357980, Seattle, WA 98195, USA.

Summary

Minimum mesh-size regulations have traditionally been applied in many fisheries to limit the retention and discarding of small fish by fishing gear and to increase long-term yields. However, recent studies show that size selection in some pelagic trawl cod-ends diminishes as catch volume increases, regardless of cod-end mesh-size or mesh type. Underwater observations demonstrate that as the catch begins to accumulate in the cod-end, exhausted fish begin lining (or are pressed against) the cod-end walls. This mechanism of mesh blockage can take place several metres ahead of the eventual catch bulge, effectively preventing fish from reaching open and unobstructed meshes. The blocking problems observed for cod-end meshes may be substantially reduced or eliminated by using special sorting devices in front of the cod-end, such as sorting grids, that permit adequate escapement of undersized fish before they reach the cod-end. A fundamental assumption of most stock-assessment models is that fish managing to escape subsequently survive. However, our recent experiments show that small pelagic fish (such as herring and vendace) that escape through cod-end meshes may be extremely vulnerable to the capture and escape process. Most of the injuries and exhaustion experienced by small pelagic fish take place inside the cod-end and the rear funnelling part of the trawl. Mortality of escapees was strongly dependent on fish size; the smallest escapees suffered the highest mortality. Cod-end mesh-size did not have a significant effect on mortality rates. Clearly, the usefulness of conventional minimum cod-end mesh-size management in certain pelagic fisheries is questionable. Methods that better facilitate escapement of undersized fish without causing damage to escapees, should be tested and evaluated, and alternative means to protect young fish should be explored.

Introduction

Responsible fisheries management requires fishing gear that retains large adult fish while allowing small juveniles to escape. Therefore, minimum mesh-size regulations are mandated for many trawl fisheries to limit the retention of small fish. There is little evidence, however, to support the usefulness of such measures, especially in pelagic trawl fisheries. Large quantities of undersized fish continue to be taken by some commercial fishing fleets in spite of minimum mesh-size regulations (e.g. Dahm 1991; Hildén *et al.* 1991; van Marlen 1991; Casey *et al.* 1992; Suuronen and Millar 1992; Alverson *et al.* 1994). Depending on market conditions and landing restrictions, this bycatch of undersized fish is discarded (as dead or dying) at sea.

Cod-end mesh-size is often the only gear parameter that is regulated in trawl fisheries to prevent the capture of undersized fish, but mesh-size is only one of several gear-design factors determining size-selectivity in the cod-end (Stewart and Galbraith 1989; MacLennan 1992; Reeves *et al.* 1992; Ferro and O'Neill 1994). Furthermore, there are several factors beyond cod-end construction that may significantly affect trawl selectivity. For instance, the potential relation between catch volume and cod-end size-selectivity has been addressed by many investigators but is not yet fully understood and described. We present further evidence that catch volume does affect cod-end selectivity in high volume pelagic trawl fisheries, and we discuss its implications on the mesh-size management of pelagic fish stocks.

The survival of fish that escape through the meshes of nets critically impacts the effectiveness of minimum mesh-size regulations. If escaping fish suffer damage and subsequently die, then minimum mesh-size regulations may not be justified to conserve the stocks. We present evidence, for certain pelagic fish species, that the survival of escapees may be extremely poor.

Effect of catch volume on cod-end size-selectivity

Several researchers have shown that escapement of 'undersized' fish through cod-end meshes may decrease as catch volume increases (e.g. Beverton 1959; Bohl 1961; Hodder and May 1964; Isaksen *et al.* 1990; Dahm 1991). In most cases, however, only a weak relationship has been found, and it is often assumed that the experimental gear (i.e. the cod-end cover), rather than catch, caused the effect. Significant catch-size effects on trawl-cod-end selectivity have recently been reported for pelagic herring (*Clupea harengus*), mackerel (*Scomber scombrus*) and Alaskan pollock (*Theragra chalcogramma*) by Suuronen *et al.* (1991), Casey *et al.* (1992) and Erickson *et al.* (1996), respectively. As catch volume increased, length at 50% retention (i.e. fish length corresponding to a 50% probability of capture = L_{50}) decreased (Fig. 1). Catch volumes in these studies ranged widely enough to clearly demonstrate this effect. These latter studies suggest that, for these fisheries, there likely is not a mesh-size large enough for effective selectivity to be maintained at high catch rates, unless the mesh is so large that eventually all the catch would escape.

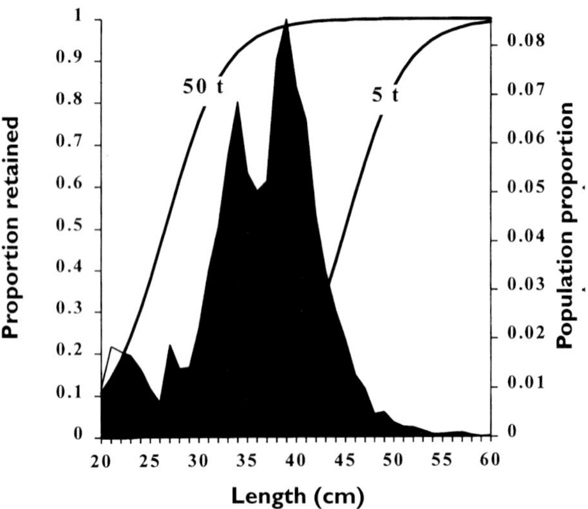

Fig. 1. Alaskan pollock catch size (ln) and cod-end type were regressed against ln(μ), where μ = L_{50} (Erickson *et al.* in press). Cod-end type and the interaction between ln(catch size) and cod-end type were not significant. Catch size, however, significantly affected L_{50} for Alaskan pollock; the relation is described by these selectivity ogives (solid lines) for a given population structure (shaded area).

Numerous underwater observations of pelagic herring trawls clearly demonstrate that decreases in L_{50} with increasing catch volumes are due to the blocking of cod-end meshes with the catch (Suuronen 1995). As the volume of the catch increases, exhausted fish begin lining the cod-end walls, thus forming a hollow cylinder of fish inside the cod-end (Fig. 2). This mechanism of mesh blockage can take place several metres ahead of the eventual catch bulge, effectively preventing fish from reaching open and unobstructed meshes. Other mechanisms of cod-end-mesh blocking have been described. Blocking may occur by extensive meshing or gilling of fish in the cod-end meshes (Claesson 1984; Järvik and Raid 1991; Erickson *et al.* 1996). Also, high densities of fish swimming in the cod-end may preclude other fish from dropping back from the intermediate (the cod-end extension, sometimes made of small mesh) until haul back (Pikitch *et al.* 1996). The rate of mesh-blocking by the catch may be dependent on the size of the cod-end in relation to the size of the catch (Erickson *et al.* 1996), as well as on the rate of arrival and the size composition of the fish. Blocked cod-end meshes may not only impair the selectivity characteristics of a cod-end, but may also reduce water flow through the cod-end. Impaired water flow may affect the catching efficiency of the gear and thus reduce the catch of larger sized fish.

Solution — special sorting devices?

To reduce or eliminate the mesh-blocking problem observed in cod-ends at high catch rates, undersized fish should be able to escape before they reach the blocked area of the cod-end. Our preliminary tests indicate that a sorting grid (Fig. 3) placed in front of the cod-end is an alternative to conventional cod-end mesh selection (Suuronen *et al.* 1993; see also Isaksen *et al.* 1992; Larsen and Isaksen 1993; van Marlen *et al.* 1994). A properly functioning grid is advantageous compared to selective-meshes because (a) the escape area between the bars of the grid are always open; the size and shape of the escape-opening is not dependent on such factors as towing speed, door spread, or catch volume, and (b) grid systems are easier to describe and physically measure (with less variation)

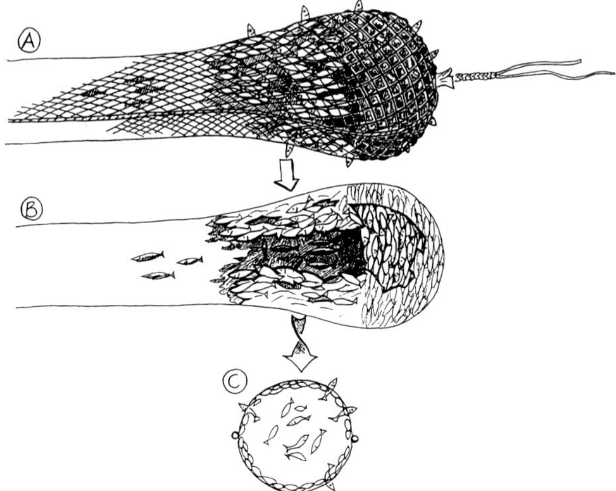

Fig. 2. Schematic presentation of exhausted herring lining the cod-end walls and forming a 'hollow tube'. Views are (A) outside, (B) longitudinal section and (C) cross-section of cod-end. Figure from Suuronen (1995).

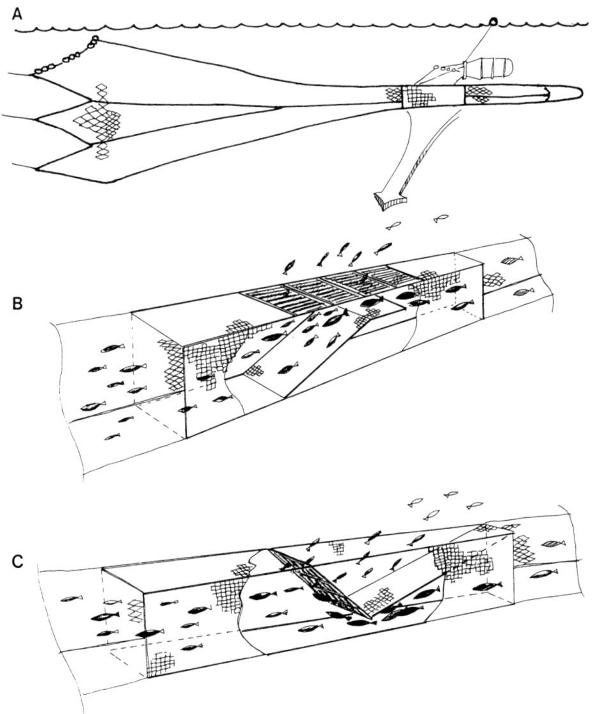

Fig. 3. (A) Overall position of sorting grid tested in a pelagic herring trawl in the Baltic Sea. Design of cover used in selectivity and survival experiments is shown. (B) Grid mounted parallel to netting. Small-meshed lifting netting below the grid is guiding fish closer to the grid. (C) Angled grid modification. Figure from Suuronen (1995).

than meshes, which would benefit enforcement personnel. The development of size-sorting grid systems in pelagic trawl fisheries is, however, at an early stage and more experience and data are required before commercial use is viable. To improve the selectivity performance of the grids, it is necessary to provide conditions which (a) prevent fish from accumulating against the netting in front of the grid, and (b) stimulate and help small fish to pass through the grid.

Numerous practical problems must be solved before grid-systems are introduced into high volume pelagic commercial fisheries. For instance, rigid-grids often result in mesh-stretching and mesh-damage in the vicinity of the grid (P. Suuronen, pers. comm.); this damage is due to forces exerted while lifting the grid and winding it onto the net-reel. Rigid-grids may also become a menace to the crew under certain conditions. Hence, grids should be constructed of a flexible or semi-flexible material so they can be lifted and taken onto the net reel without damaging meshes and to reduce handling difficulties.

The performance of a sorting grid is highly dependent on its correct rigging and position (angle), and requires undisturbed water flow around and through it. To ensure that grids function properly, many fishers have had to purchase expensive grid angle sensors. In addition, grids sometimes cause twisting in the trawl body while setting the gear, another potential high-expense to the fisher if left unnoticed during tows. Finally, no convincing evidence exists to suggest that grids can handle large catches without problems and difficulties. Clearly, further experimentation is needed before the practical problems in sorting grid technology can be solved.

From a fisher's point of view, special sorting devices (e.g. sorting grids) may be attractive, once the technological problems are solved, because only a small part of the existing gear needs modification, and hence, may provide a cheaper and more practical solution than a mandatory increase in the minimum mesh-size applicable to the entire cod-end. Moreover, research shows that the selection performance (e.g. sharpness of selection) is often substantially better for sorting-grids than for meshes of conventional cod-ends (Larsen and Isaksen 1993; Suuronen *et al.* 1993). This means that not only a higher portion of undersized fish are sorted out of the catch, but also a higher portion of large marketable fish is retained inside the cod-end. This latter point would likely enhance fishers' acceptance of this type of gear modification.

Survival of escaping fish?

The justification for traditional cod-end mesh-size management in certain pelagic trawl fisheries is questionable, not only because of poor escapement through cod-end meshes under certain conditions, but also because the survival of escapees may be poor. The mortality of fish escaping from a trawl cod-end is largely dependent on species and size (Soldal *et al.* 1993; Sangster and Lehmann 1994; Suuronen 1995). Small pelagic fish such as herring and vendace (*Coregonus albula*) are extremely vulnerable to damage in the trawl capture and escape process (Suuronen *et al.* 1995*a*, 1996*a*, *b*; Fig. 4). High mortalities have been observed in crowding experiments with mackerel (Lockwood *et al.* 1983), suggesting that this pelagic species is also sensitive to the trawl-capture and escape. On the other hand, demersal gadoid species such as cod (*Gadus morhua*) seem to endure this process without suffering noticeable mortality (Soldal *et al.* 1993; Suuronen *et al.* 1995*b*).

The final act in the capture and escape process — swimming or squeezing though a cod-end mesh — is not the primary reason

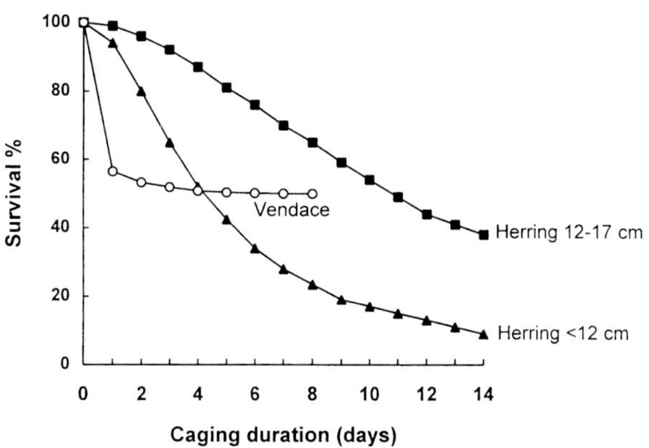

Fig. 4. Many of the small herring (<12 cm) escaping from a cod-end of a pelagic herring trawl died within a few days of their escape. Many larger (12–17 cm) herring escapees were alive after a 7-day observation period, but most were dead after two weeks of caging. For vendace, mortality was highest during the first day after escape. On average, 50% of the escaped vendace died. Modified from Suuronen *et al.* (1995*a*, 1996*a*, *b*).

for the damage and subsequent mortality of escaping herring and vendace (Suuronen *et al.* 1995*a*, 1996*a*, *b*). Sangster and Lehmann (1994) and Soldal *et al.* (1993) also found no evidence that cod-end mesh-size significantly affects the degree of injury and subsequent mortality of gadoid-escapees. These findings are not consistent with those of Treschev *et al.* (1975) and Efanov (1981), who reported a clear mesh-size-dependent mortality for herring escapees.

Skin, mucus and scales act as barriers against infections and help in osmoregulation, and their damage may substantially contribute to fish mortality (e.g. van Oosten 1957; Efanov 1981; Lockwood *et al.* 1983; Hay *et al.* 1986; Main and Sangster 1990, 1991). Skin damage caused by gear contacts is largely responsible for the high mortality rate of herring escapees (Suuronen *et al.* 1996*a*, *b*). Further, muscular exhaustion caused by herding and escape activities within the trawl may strongly contribute to the high mortality, especially for the smallest escapees. Our underwater observations clearly show that small herring and vendace are forced to swim beyond the limits of their endurance during trawl capture, a finding that was supported by liver glycogen depletion of small herring escapees (Suuronen *et al.* 1995*a*, 1996*a*).

In terms of conserving undersized herring and vendace, there is no clear evidence that a rigid sorting grid placed in the cod-end extension is more advantageous than traditional cod-end-mesh-size modifications (Suuronen *et al.* 1996*b*). Fish that escaped between the bars of a sorting grid experienced almost the same mortality as fish that escaped through cod-end meshes. Nevertheless, survival of escapees might improve substantially if fish were selected long before reaching the funnelling rear part of the trawl (i.e. the rear belly), where severe skin damage and exhaustion effects begin to accumulate. Attaining high selectivity performance in this area of the trawl would require substantial changes in trawl and grid design.

Highest mortality (70–80%) of vendace escapees was recorded for hauls conducted in the dark; trawling in daylight caused significantly lower mortality (30–40%; Suuronen *et al.* 1995*a*). The reason for higher escapee-mortality during night-time hauls is probably due to the lower ambient light level. Low ambient light levels at night may cause fish to have more unintended contacts with the net and other fish, and thus more skin abrasion and scale loss than fish captured during the day. Our results suggest that considerable numbers of juvenile vendace could be saved in Finnish lakes during summer months if trawling did not take place in the late evening or at night.

Survival of herring and vendace escapees was highly variable during our experiments. Within tows, some fish escaped the cod-end without apparent damage, whereas other similar-sized escapees experienced considerable damage. This suggests that the likelihood of death for a particular fish can not be explained in terms of a single factor. Hence, escapee-survival is not only dependent on species and size, but is also influenced by other factors such as fish condition and age, water temperature, salinity, towing speed and duration, catch composition, and general gear design (ICES 1994; Sangster and Lehmann 1994; Chopin and Arimoto 1995).

Recommendations and conclusions

The advantages gained by improved gear selectivity are clear for groundfish stocks exhibiting low escape- and natural-mortality rates and high growth rates (e.g. Gulland 1983, 1991; Pikitch 1987, 1991, 1992; Ulltang 1987; Halliday and White 1989). However, the usefulness of conventional minimum cod-end mesh-size management is questionable for pelagic fisheries where commercial catch rates are high and for species that are easily damaged by gear contacts or herding. Based on these results, we recommend that before cod-end mesh-size regulations are mandated on a fishing fleet, the selectivity characteristics of the 'new' gear type be tested, and the survival of escapees be understood. We further recommend that these experiments be conducted under commercial fishing conditions to the extent possible, otherwise conclusions may be inapplicable to actual fishing conditions. Clearly, mandating the use of 'selective' trawl-gear based on erroneous or non-existent selectivity and escapee-survival information can lead to inaccurate stock assessments, and subsequently, less than ideal management of the resource.

Based on the results presented in this paper, we recommend that gear-research efforts, particularly for high catch-volume fisheries, be focussed on attaining selection well ahead of the cod-end rather than attempting to stimulate escapement within the cod-end. Selection devices (meshes or grids) placed far ahead of the cod-end may (a) reduce the dependence of selectivity on catch volume, and (b) improve the survival of individuals that escape. Finally, to effectively protect young fish, alternative methods such as controlling fishing effort in areas important for young fish should be evaluated in combination with the effects of gear-related measures.

References

Alverson, D L, Freeberg, M H, Pope, J G, and Murawski, S A (1994). A global assessment of fisheries bycatch and discards. *FAO Fisheries Technical Paper*. No. 339. FAO, Rome. 233 pp.

Beverton, R J H (1959). The selectivity of a modified form of top-side chafer. *ICES C.M.* 1959/117.

Bohl, H (1961). German mesh experiments on redfish in 1961. *ICES C.M.* 1961/F:88.

Casey, J, Nicholson, M D, and Warnes, S (1992). Selectivity of square mesh cod-ends on pelagic trawls for Atlantic mackerel (*Scomber scombrus* L.). *Fisheries Research* **13**, 267–79.

Chopin, F S, and Arimoto, T (1995). The condition of fish escaping from fishing gears — a review. *Fisheries Research* **21**, 315–27.

Claesson, B (1984). Meshing in pelagic herring trawls. *ICES C.M.* 1984/B:5. 4 pp.

Dahm, E (1991). Doubtful improvement of the selectivity of herring midwater trawls by means of square mesh cod ends and constructional modifications of diamond mesh cod ends. *ICES CM* 1991/B:2. 8 pp.

Efanov, S F (1981). Herring of the Gulf of Riga: the problem of escapement and mechanical impact of the trawl. *ICES C.M.* 1981/J:7. 16 pp.

Erickson, D L, Perez-Comas, J A, Pikitch, E K, and Wallace, J R (1996). Effects of catch size and cod-end type on the escapement of walleye pollock (*Theragra chalcogramma*) from pelagic trawls. *Fisheries Research* **28**, 179–96.

Ferro, R S T, and O'Neill, F G (1994). An overview of the characteristics of twines and netting that may change selectivity. *ICES CM* 1994/B:35. 6 pp.

Gulland, J A (1983). Fish stock assessment. A manual of basic methods. John Wiley and Sons, New York. 223 pp.

Gulland, J A (1991). Under what conditions will multispecies models lead to a better fisheries management? *ICES Marine Science Symposia* **193**, 348–52.

Halliday, R G, and White, G N (1989). The biological/technical implications of an increase in minimum trawl mesh-size for groundfish fisheries in the Scotia-Fundy region. *Canadian Technical Reports Fisheries Aquatic Science* 1691. 153 pp.

Hay, D E, Cooke, K D, and Gissing, C V (1986). Experimental studies of Pacific herring gillnets. *Fisheries Research* **4**, 191–211.

Hildén, M, Mickwitz, P, Paananen, T, Partanen, H, Setälä, J, Söderkultalahti, P, and Vihervuori, A (1991). Capacity of the fishing fleet and processing industry in Finland. (In Finnish). *Kalantutkimuksia* **29**, Helsinki. 72 pp.

Hodder, V M, and May, A W (1964). The effect of catch size on the selectivity of otter trawls. *ICNAF Research Bulletin* **1**, 28-35.

ICES (1994). Report of the sub-group on methodology of fish survival experiments. *ICES C.M.* 1994/B:8. 46 pp.

Isaksen, B, Lisovsky, S, and Sakhno, V A (1990). A comparison of the selectivity in cod-ends used by the Soviet and Norwegian trawler fleet in the Barents Sea. *ICES* 1990/B:51. 14 pp.

Isaksen, B, Valdemarsen, J W, Larsen, R B, and Karlsen, L (1992). Reduction of fish by-catch in shrimp trawl using a rigid separator grid in the aft belly. *Fisheries Research* **13**, 335-52.

Järvik, A, and Raid, T (1991). The problem of meshing in Baltic herring trawl fishery. In 'Proceedings of the International Herring Symposium'. Anchorage, USA, 533-44.

Larsen, R, and Isaksen, B (1993). Size selectivity of rigid sorting grids in bottom trawls for Atlantic cod (*Gadus morhua*) and haddock (*Melanogrammus aeglefinus*). *ICES Marine Science Symposia* **196**, 178–82.

Lockwood, S J, Pawson, M G, and Eaton, D R (1983). The effects of crowding on mackerel (*Scomber scombrus* L.) — physical conditions and mortality. *Fisheries Research* **2**, 129–47.

MacLennan, D N (1992). Fishing gear selectivity: an overview. *Fisheries Research* **13**, 201–04.

Main, J, and Sangster, G I (1990). An assessment of the scale damage to and survival rates of young gadoid fish escaping from the cod-end of a demersal trawl. *Scottish Fisheries Research Reports* **46**. 28 pp.

Main, J, and, Sangster, G I (1991). Do fish escaping from cod-ends survive? *Scottish Fisheries Working Paper* 18/91. 8 pp.

Pikitch, E K (1987). Use of a mixed-species yield-per-recruit model to explore the consequences of various management policies for the Oregon flatfish fishery. *Canadian Journal of Fisheries and Aquatic Sciences* **44** (Suppl. 2), 349–59.

Pikitch, E K (1991). Technological interactions in the US West Coast groundfish trawl fishery and their implications for management. *ICES Marine Science Symposia* **193**, 253–63.

Pikitch, E K (1992). Potential for gear solutions to bycatch problems. Proceedings of the national industry bycatch workshop, February 4-6 1992, Newport, Oregon, Natural Resources Consultants, Inc., Seattle, WA, USA. 128–38.

Pikitch, E K, Erickson, D, Perez-Comas, J A, and Suuronen, P (1996). Cod-end size-selection: good concept, but does it really work? Proceedings of Solving Bycatch — Considerations for today and tomorrow. Alaska Sea Grant Program No. 96-03 pp. 107–14 (University of Alaska: Fairbanks).

Reeves, S A, Armstrong, D W, Fryer, R J and Coull, K A (1992). The effects of mesh-size, cod-end extension length and cod-end diameter on the selectivity of Scottish trawls and seines. *ICES Journal of Marine Science* **49**, 279–88.

Sangster, G, and Lehmann, K (1994). Commercial fishing experiments to assess the scale damage and survival of haddock and whiting after escape from four sizes of diamond mesh cod-ends. *ICES C.M.* 1994/B:38. 24 pp.

Soldal, A V, Isaksen, B, Marteinsson, J E, and Engås, A (1993). Survival of gadoids that escape from a demersal trawl. *ICES marine Science Symposia* **196**, 122–27.

Stewart, P A M, and Galbraith, R D (1989). Cod-end design, selectivity and legal definitions. *ICES C.M.* 1989/B:11. 7 pp.

Suuronen, P (1995). Conservation of young fish by management of trawl selectivity. *Finnish Fisheries Research* **15**, 97–116.

Suuronen, P, and Millar, R B (1992). Size selectivity of diamond and square mesh cod-ends in pelagic herring trawls: only small herring will notice the difference. *Canadian Journal of Fisheries and Aquatic Sciences* **49**, 2104–117.

Suuronen, P, Millar, R B, and Järvik, A (1991). Selectivity of diamond and hexagonal mesh cod-ends in pelagic herring trawls: evidence of a catch size effect. *Finnish Fisheries Research* **12**, 143–56.

Suuronen, P, Lehtonen, E, and Tschernij, V (1993). Possibilities to increase the size-selectivity of a herring trawl by using a rigid sorting grid. *NAFO SCR Doc. 93/119, Serial* No. N2313. 12 pp.

Suuronen, P, Turunen, T, Kiviniemi, M, and Karjalainen, J (1995*a*). Survival of vendace (*Coregonus albula* L) escaping from a trawl cod-end. *Canadian Journal of Fisheries and Aquatic Sciences* **52**(12), 2527–33.

Suuronen, P, Lehtonen, E, Tschernij, V, and Larsen, P O (1995*b*). Skin injury and mortality of Baltic cod escaping from trawl cod-ends equipped with exit windows. *ICES C.M.* 1995/B:8. 13 pp.

Suuronen, P, Erickson, D, and Orrensalo, A (1996*a*). Mortality of herring escaping from pelagic trawl cod-ends. *Fisheries Research* **3–4**, 305–21.

Suuronen, P, Perez-Comas, J A, Lehtonen, E, and Tschernij, V (1996*b*). Size related mortality of Baltic herring (*Clupea harengus* L) escaping through a rigid sorting grid and trawl cod-end meshes. *ICES Journal of Marine Science* **53**(3), 691–700.

Treschev, A I, Efanov, S F, Shevtsov, S E, and Klavsons, U A (1975). Die Verletzung und die öberlebenschancen des Ostseeherings nach dem Durchdringen durch die Maschen des Schleppnetzsteerts. *Fisherei-Forschung. Wissenschaftliche Schriftenreihe* **13**(1), 55–9.

Ulltang, Ö (1987). Potential gains from improved management of the Northeast Arctic cod stock. *Fisheries Research* **5**, 319–30.

van Marlen, B (1991). Selectivity of fishing gears in wider perspective. *ICES C.M.*1991/B:22. 12 pp.

van Marlen, B, Lange, K, Wardle, C S, Glass, C W, and Ashcroft, B (1994). Intermediate results in EC-project TE-3-613 'Improved species and size selectivity of midwater trawls (SELMITRA).' *ICES C.M.* 1994/B.13. 8 pp.

van Oosten, J (1957). The skin and scales. In 'The Physiology of fishes'. Vol 1. (Ed M E Brown). pp. 207–44 (Academic Press: New York).

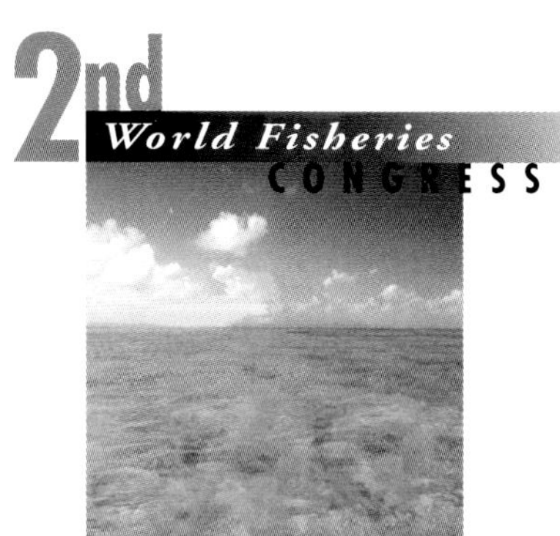

Bycatch Reduction in the Northwest Atlantic Small-mesh Bottom-trawl Fishery for Silver Hake (*Merluccius bilinearis*)

J. DeAlteris, H. Milliken and D. Morse

Department of Fisheries and Aquaculture, University of Rhode Island, Kingston, R.I. USA 02881.

Summary

The management of discards is an important issue in the Northwest Atlantic bottom-trawl fishery. Reduction of flatfish bycatch in the small-mesh bottom-trawl fishery for silver hake (*Merluccius bilinearis*) is critical for the continuation of this fishery. Separation of species in the trawl mouth based on species-specific behaviours was examined using variations of two technical solutions: modifications of the sweep/fishing lines and the use of escape panels in the lower belly of the trawl. Eight modified net designs were evaluated using paired alternate tows. Two proved effective at reducing flatfish bycatch. A small window in the bosum of the sweep formed by elevating the fishing line and providing a discontinuity in the sweep, and a large-mesh, small-diameter, orange-coloured twine panel in the lower belly, resulted in a 96 and 73% reduction in flatfish catch respectively, while not significantly affecting the catch of silver hake.

Introduction

Background

Bycatch and, in particular, the management of discards, is an important issue in the Northwest Atlantic bottom-trawl fishery. Bycatch is defined as the catch of non-target animals whether kept for market or discarded at sea (Alverson *et. al.* 1994). Discarded bycatch is composed of animals for which there is no profitable market or animals for which regulations prohibit landing. In the Northwest Atlantic Ocean, bycatch in the bottom-trawl fisheries has been problematic due to the high rate of discarded juveniles of overexploited regulated groundfish stocks. This discarding has contributed to the stock declines and slowed the stock rebuilding process.

Minimum mesh size regulations for bottom-trawls have minimized juvenile fishing mortality in large-mesh fisheries (Lowry and Sangster 1996). In the small-mesh fisheries, the Nordmøre grate has contributed to a substantial reduction of bycatch in the northern shrimp fishery (Kenney *et al.* 1992). In the small-mesh bottom-trawl fishery for silver hake however, the solution has been more elusive because both the target species and the juvenile groundfish bycatch have similar body sizes.

Therefore, size sorting by mesh or grid is not an option. The only opportunity for species separation is in the net mouth, utilizing species-specific, differential behaviours.

Life history, stock status, fishery and fishery management

The silver hake *(Merluccius bilinearis)* is an abundant and commercially-important finfish occurring throughout the Gulf of Maine, Georges Bank, and Southern New England regions. Two stocks of silver hake are currently recognized based on morphological differences, trawl surveys, and commercial landings: the Gulf of Maine–Northern Georges Bank and the southern Georges Bank– Middle Atlantic stock.

The silver hake fishery occurs in late summer, after their spawning period, when they are found in dense aggregations. Although they are found throughout the water column, the principal capture method is the bottom-trawl (Bigelow and Schroeder 1953). Traditional areas of high concentrations of post-spawners include the coastal regions from Maine to Cape Cod Bay, Cultivator Shoals, and the Southern New England area south and west of Martha's Vineyard in the Northwest Atlantic Ocean.

Both stocks are currently managed under the Northeast Multispecies Fishery Management Plan, and are listed as fully exploited. Bottom-trawl surveys indicate a rising trend for the northern stock biomass, and a declining trend for the biomass of the southern stock. Annual landings have remained relatively constant since 1982 at approximately 6400 t for the northern stock, and 11 940 t for the southern stock. Both stocks co-habit with other regulated and non-regulated species including the flounders, cod and red hake. Because small-mesh trawls are required to harvest silver hake, the fishery is exempted from minimum mesh size regulations required for directed fisheries on the regulated species. Under the fishery management plan, bycatch of regulated species in an 'exempted fishery' must be less than 5% of the total catch and the bycatch must not jeopardize the target fishery mortality rates for the regulated species. Therefore, the bycatch of flatfish in the silver hake fishery is a critical issue.

Previous work

Since the early 1900s, scientists have been concerned about the high rate of discards and bycatch species caught by bottom-trawls. Holt (1895) examined ways to reduce the catch of juvenile fish using square-mesh panels while Herrington (1932) began studying the effects of increasing mesh sizes. More recently, through the use of underwater video data, Wardle (1988, 1993) has demonstrated that groundfish tend to be herded by the net towards the trawl mouth. Carr and Caruso (1992) showed that different species of fish react differently in the mouth of the trawl. Silver hake were found to swim towards the top of the net as they pass through the mouth, whilst most species of flounder passed at the bottom of the net closer to the footrope. In a study on the effectiveness of a separator panel, Arkley and MacMullen (1996) noted that 94.4% of a related species of silver hake (*Merluccius merluccius*) were found in the upper portion of the trawl net and flatfish were found in the lower portion. Rose and Williams (unpublished) successfully used a large-mesh escape panel to reduce the bycatch of juvenile Pacific halibut.

The effect of twine on the behaviour of fish has also been studied. To improve the effectiveness of gill-nets, Wardle (1988) conducted experiments to determine the visibility of certain netting materials and colours. In an attempt to induce certain behaviours of fish within the bottom-trawl net, Wardle (1993) also studied the effect of varying the colours of the twine used in the construction of the net. In order to make a net panel appear invisible when it is near the bottom, Wardle (1993) argues that matching the colour to the sea bed is the important factor.

Rationale and objectives

To harvest silver hake commercially, fisherman must utilize small-mesh (less than 7.6 cm) because the adult fish has a small girth and can easily pass through the 15.2 cm minimum mesh size of the northeast groundfish trawl. Because of the commercial viability of the silver hake fishery and the concern over the take of juveniles of regulated groundfish species, the objective of this project was to develop a technology for the reduction of flatfish bycatch while not affecting the catch of silver hake.

Based on previous research, modifications to the sweep or the installation of a large-mesh panel in the lower belly of the net are techniques that were considered viable for reducing the bycatch of flounder species in the small-mesh silver hake fishery because they utilize the differing behaviour of the species.

METHODS

The experimental design consisted of a paired comparison using alternate tows of experimental and control nets. The areas fished were those normally used by commercial fishermen engaged in the silver hake fishery from Cape Cod Bay to Southern New England. Fishing operation methods were held constant throughout the study. Although time of day, season and area fished changed during the study, they were held constant within a paired tow set. After each tow the catch was sorted and the data, weights and numbers by species for each tow were evaluated using a paired *t*-test and a Wilcoxon non-parametric paired test.

Two trawls, a control and an experimental, were used during all experiments. The trawl design was typical of trawls used in the fishery, and the two trawls were identical except for the construction of the experimental sweeps and lower belly escape panels. The trawls were of a semi-balloon design, and with 12.7 cm stretched-mesh length, nylon webbing in the wing sections, and 6.4 cm nylon webbing throughout the body of the net. Both trawls had a footrope length of 34.1 m (112.0 ft), a headrope length of 25.9 m, and used 34.1 m of 1.0 l cm Trawlex chain for the sweep. Each design is described in detail below and illustrated in Figures 1 and 2.

The variations in the sweep design attempted to provide an escape path for the flounder. Experimental Trawl No. 1s included a sweep that had 47 dropper chains attached, with 0.8 cm proof coil chain connecting the sweep and the traveller. The extended lengths of the droppers, including shackles,

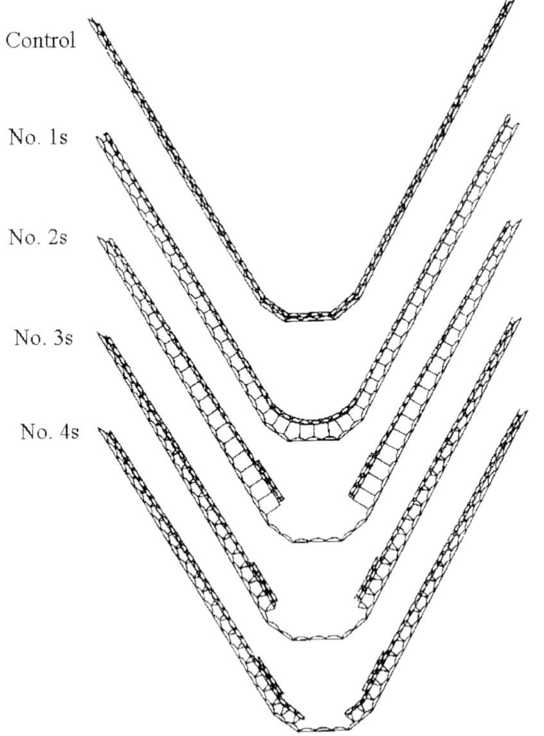

Fig. 1. Sweep arrangements for the control and experimental trawls Nos. 1s, 2s, 3s and 4s. Note the length of the droppers between the traveller and the sweep chain, and the width of the window in the sweep.

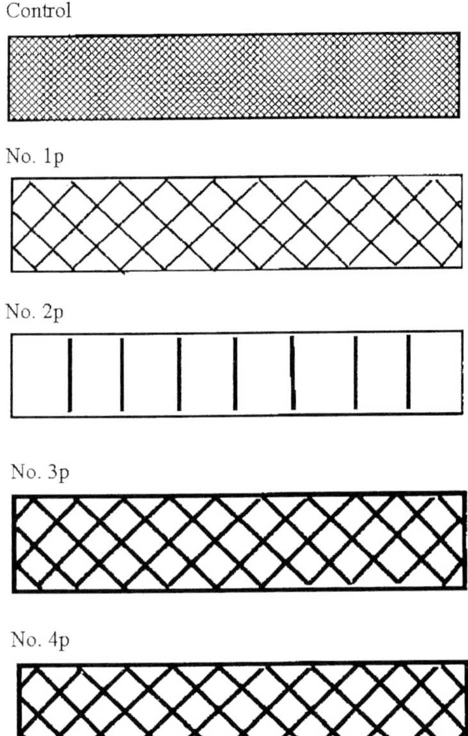

Fig. 2. Escape panel arrangements for the control and experimental trawls Nos. 1p, 2p, 3p and 4p. Note that the colour of the control panel is black, Nos. 2p and 3p are white and Nos. 1p and 4p are orange.

ranged from 33.0 cm at the wing ends to 94.0 cm in the bosum. Experimental Trawl No. 2s incorporated a 'window' in the bosum of the sweep, and maintained the same dropper lengths as Trawl No. 1s. The centre seven droppers were removed, and the sweep chain was cut at the middle link. The resulting unattached sweep chain was brought forward and shackled to the existing sweep, to keep the majority of the sweeps' weight in the bosum. The sweep window was therefore approximately 4.6 m long by 0.6 m high. Kevlar lines replaced all the dropper chains. Experimental Trawl No. 3s had the Kevlar drop lines detached, bent over the traveller, and reattached on the sweep, halving the drop heights throughout the length of the trawl. This effectively reduced the separation between the sweep and the fishing line. In experimental Trawl No. 4s, the width of the window in the sweep was reduced, with the addition of 1.2 m (48 in) of 0.95 cm (3/8 in) chain to the rear section of each half of the sweep, with Kevlar drops of the appropriate height. This reduced the window width to approximately 2.1 m.

The variations in the lower belly escape panel also attempted to provide an escape path for the flounder. Experimental Trawl No.1p included a large-mesh panel that was constructed out of 1.6 mm twisted nylon twine dyed dark orange. The meshes were 40.6 cm hung with a primary hanging ratio of 0.5 and a secondary hanging ratio of 0.87. The webbing was framed in a panel measuring 7.6 m by 2.6 m which equated to 37.5 meshes across by 7.5 meshes deep. Experimental Trawl No. 2p included an escape panel constructed of white polyester ropes with a diameter of 9.5 mm that, stretched from the top of the panel to the bottom of the panel at 38.1 cm intervals, which corresponds to 18 ropes in the 7.6 m wide panel. The depth of the panel was 2.6 m, the same as in the mesh panels. Experimental Trawls No. 3p and No.4p were exactly the same as panel No. 1p except that the twine used to make the webbing in Nos. 3p and 4p was 4.8 mm diameter. In No. 3p the twine was white; in No. 4p, the twine was dyed orange.

Results

Based on 6 alternate paired tows, the experimental net with a continuous sweep and 97 cm height at the centre of the sweep (No. 1s) had a 37% catch of regulated groundfish compared to the control which caught 22.7% (Table 1). There was no significant difference in the catches of silver hake, red hake and flatfish and between the two nets. Based on 13 alternate paired tows the experimental net with a discontinuous sweep and a 4.6 m long by 0.6 m high window (No. 2s) had a 0.5% catch of regulated groundfish as compared to the control net which caught 1.4 % (Table 2). There was a highly significant difference in the catches of silver hake, red hake and flatfish between the two nets. The experimental net lost a high percentage of both hake and flatfish species. Based on 9 alternate paired tows, the experimental net with a discontinuous sweep and a 4.6 m long by 0.3 m high window (No. 3s) had a 2.4 % catch of regulated groundfish compared to the control net which caught 4.6% (Table 3). There was a significant difference in the catch of silver hake and red hake

Table 1. Species composition by weight (kg) and the results of the paired comparison analysis on hake and flatfish for 6 paired tows made during the evaluation of Experimental Trawl No. 1s

Species Group	Weight (kg) Control	Weight (kg) Experimental	Significant Difference
Silver hake	43.3	31.0	No
Red hake	0.0	0.0	No
All flatfish	85.7	90.5	No
Sharks/skates	695.1	647.9	
Other	665.8	698.7	
Total catch	1489.9	1568.1	
Regulated flatfish	85.7	84.7	
Regulated groundfish	339.1	580.7	
% regulated species of total catch:	22.7	37.0	

Table 2. Species composition by weight (kg) and the results of the paired comparison analysis on hake and flatfish for 13 paired tows made during the evaluation of Experimental Trawl No. 2s

Species Group	Weight (kg) Control	Weight (kg) Experimental	Significant Difference
Silver hake	3505.3	875.4	Yes **
Red hake	3501.7	217.3	Yes **
All flatfish	1145.6	25.9	Yes **
Sharks/skates	5213.5	244.2	
Other	3383.1	836.5	
Total catch	16749.2	2199.3	
Regulated flatfish	241.8	4.3	
Regulated groundfish	241.8	4.5	
% regulated species of total catch:	1.4	0.5	

** *P*-value < 0.01.

Table 3. Species composition by weight (kg) and the results of the paired comparison analysis on hake and flatfish for 9 paired tows made during the evaluation of Experimental Trawl No. 3s

Species Group	Weight (kg) Control	Weight (kg) Experimental	Significant Difference
Silver hake	759.0	236.0	Yes *
Red hake	1022.9	64.5	Yes *
All flatfish	829.0	48.2	Yes **
Sharks/skates	3116.3	227.0	
Other	2792.9	544.2	
Total catch	8250.1	1119.9	
Regulated flatfish	381.5	26.0	
Regulated groundfish	381.5	26.5	
% regulated species of total catch:	4.6	2.4	

* *P*-value < 0.05; ** *P*-value < 0.01.

Table 4. Species composition by weight (kg) and the results of the paired comparison analysis on hake and flatfish for 5 paired tows made during the evaluation of Experimental Trawl No. 4s

Species Group	Weight (kg) Control	Weight (kg) Experimental	Significant Difference
Silver hake	98.6	128.5	No
Red hake	68.8	39.5	No
All flatfish	424.3	16.4	Yes *
Sharks/skates	1782.2	83.3	
Other	807.8	218.0	
Total catch	3181.7	485.7	
Regulated flatfish	332.7	13.0	
Regulated groundfish	333.2	22.0	
% regulated species of total catch:	10.4	4.5	

* *P*-value < 0.05).

and a highly significant difference in the catches of flatfish between the two nets. The experimental net again lost a high percentage of both hake and flatfish species. Based on 5 alternate paired tows, the experimental net with a discontinuous sweep and a 2.1 m long by 0.3 m high window (No. 4s) had a 4.5% catch of regulated groundfish compared to the control net which caught 10.4% (Table 4). There was no significant difference in the catch of silver and red hake and a significant difference in the catch of flatfish between the two nets. This design performed well with no loss of hake and a 95% loss of flatfish species.

Based on 28 alternate paired tows, the experimental net with an orange large-mesh panel of 1.6 mm twine (No. 1p) had a 4.7% catch of regulated groundfish compared to the control which caught 12.0% (Table 5). There was no significant difference in the catch of silver hake and a highly significant difference in the catch of flatfish between the two nets. Interestingly, red hake also showed a highly significant difference. This panel design resulted in no significant loss of silver hake, and nearly a 75% reduction in the flatfish catch. Based on 13 alternate paired tows, the experimental net with 9.5 mm ropes spaced 38.1 cm apart (No. 2p) had a 5.7% catch of regulated groundfish compared to the control which caught 16.8% (Table 6). There were highly significant differences in the catches of silver hake, red hake and flatfish between the two nets. Therefore, this experimental design lost a high percentage of both hakes and flatfish species. Based on 9 alternate paired tows, the experimental net with a white large-mesh panel constructed out of 4.8 mm twine (No. 3p) had a 5.1% catch of regulated groundfish compared to the control which caught 3.8 % (Table 7). There was no significant difference in the catches of silver and red hake and flatfish between the two nets. Based on 9 paired tows, the experimental net with an orange large-mesh panel constructed of 4.8 mm twine (No. 4p) had a 4.3% catch of regulated groundfish compared to the control which caught 5.7% (Table 8). There was a significant difference in the catches of silver hake and flatfish between the two nets. Red hake showed no significant difference in catch.

Table 5. Species composition by weight (kg) and the results of the paired comparison analysis on hake and flatfish for 28 paired tows made during the evaluation of Experimental Trawl No. 1p

Species Group	Weight (kg) Control	Weight (kg) Experimental	Significant Difference
Silver hake	10100.7	7549.2	No
Red hake	1463.4	742.7	Yes**
All flatfish	4627.4	1254.6	Yes**
Sharks/skates	3367.2	4429.7	
Other	11376.5	8494.8	
Total catch	30935.3	22471.0	
Regulated flatfish	3706.6	1050.5	
Regulated groundfish	3706.6	1050.5	
% regulated species of total catch:	12.0	4.7	

** *P*-value < 0.01.

Table 6. Species composition by weight (kg) and the results of the paired comparison analysis on hake and flatfish for 13 paired tows made during the evaluation of Experimental Trawl No. 2p

Species Group	Weight (kg) Control	Weight (kg) Experimental	Significant Difference
Silver hake	5940.3	2636.7	Yes**
Red hake	629.3	59.6	Yes*
All flatfish	5056.0	1102.4	Yes**
Sharks/skates	1340.1	1709.9	
Other	1050.5	2314.8	
Total catch	14016.1	7823.3	
Regulated flatfish	2357.9	444.7	
Regulated groundfish	2357.9	444.7	
% regulated species of total catch:	16.8	5.7	

* *P*-value < 0.05; ** *P*-value < 0.01.

Table 7. Species composition by weight (kg) and the results of the paired comparison analysis on hake and flatfish for 9 paired tows made during the evaluation of Experimental Trawl No. 3p

Species Group	Weight (kg) Control	Weight (kg) Experimental	Significant Difference
Silver hake	3111.9	1972.4	No
Red hake	1130.6	732.4	No
All flatfish	557.8	389.5	No
Sharks/skates	481.8	508.6	
Other	2581.9	1548.6	
Total catch	7864.1	5124.7	
Regulated flatfish	297.9	261.7	
Regulated groundfish	297.9	261.7	
% regulated species of total catch:	3.8	5.1	

Table 8. Species composition by weight (kg) and the results of the paired comparison analysis on hake and flatfish for 9 paired tows made during the evaluation of Experimental Trawl No. 4p

Species Group	Weight (kg) Control	Weight (kg) Experimental	Significant Difference
Silver hake	3111.9	1972.4	Yes*
Red hake	1638.6	1596.8	No
All flatfish	1041.2	389.5	Yes*
Sharks/skates	656.3	442.0	
Other	8376.0	4048.7	
Total catch	14824.0	8663.7	
Regulated flatfish	841.3	371.6	
Regulated groundfish	841.3	371.6	
% regulated species of total catch:	5.7	4.3	

* P-value < 0.05.

Discussion

Experiments were conducted to evaluate two different strategies for the reduction of flatfish bycatch in the small-mesh groundfish bottom-trawl. The first set of experiments was devoted to modifications of the fishing line and sweep. Vertical openings of varying height separating the fishing line from the sweep were evaluated, as were discontinuities in the bosum of the sweep. Of the four modifications tested, only the experimental trawl No. 4s with the 2.1 m long by 0.3 m high discontinuity in the sweep resulted in significant reductions in flatfish bycatch (96% on average), while not affecting the silver hake catch. The second set of experiments was directed at modifications that included an escape panel in the lower belly. The escape panels were designed with openings of different sizes and twines (colour and diameter). Of the four modifications tested, only the experimental trawl No. 1p with the 40.6 cm diamond-shaped meshes constructed with 1.6 mm diameter orange-coloured twine proved effective in significantly reducing the flatfish bycatch (73%) while not affecting the catch of the targeted silver hake. Both strategies are simple and could be easily incorporated into the small-mesh whiting nets.

References

Alverson, D G, Freeberg, M H, Murawski, S A, and Pope, S G (1994). A global assessment of fisheries bycatch and discards. *FAO Fisheries Technical Paper* 339. 235 pp.

Arkley, K, and MacMullen, P (1996). Commercial evaluation of a separator trawl in a North Sea fishery. *ICES Marine Science Symposium Working Group Paper.* 12 pp.

Bigelow, H B, and Schroeder, W G (1953). Fishes of the Gulf of Maine. *Fish and Wildlife service Fishery Bulletin* No. 74. US Printing Office, Washington, DC. 577 pp.

Carr, H A, and Caruso, P (1992). Application of a horizontal separator panel to reduce bycatch in the small-mesh whiting fishery. *Proceedings of the Marine Technological Society.* **1**, 401–7.

Herrington, W C (1932). Conservation of immature fish in otter trawling. *Transactions of American Fisheries Society* **62**, 57–63.

Holt, E W L (1895). An examination of the present state of the Grimsby trawl with especial reference to the destruction of immature fish. *Journal of Marine Biological Association., N.S.*, Vol. III, No. 5, 1893–1895, 339–458.

Kenney, J F, Blott, A J, and Nulk, V E (1992). Experiments with a Nordmore grate in the Gulf of Maine shrimp fishery. NOAA Technical Memorandum NOAA-NA87EA-H-00052. 26 pp.

Lowry, N, and Sangster, G (1996) Survival of gadoid fish escaping from the cod-end of trawls. Report of the Working Group on Fishing Technology and Fish Behaviour. ICES CM 1996/6, 2.

Rose, C S, and Williams, G H (unpublished). Behavioural comparisons of Pacific halibut with other groundfish species: A project to reduce halibut bycatch in bottom-trawls.

Wardle, C S (1988). Understanding fish behaviour can lead to more selective fishing gears. World Symposium on Fishing Gear and Fishing Vessel Design, Marine Institute, St. Johns, Newfoundland, pp. 12–18.

Wardle, C S (1993). Fish behaviour and fishing gear. In 'Behaviour of Teleost Fishes'. (Ed J J Pitcher) pp. 609–44 (Chapman and Hall: NY).

A Strategy for Measuring the Relative Abundance of Pueruli of the Spiny Lobster *Jasus verreauxi*

S.S. Montgomery and J.R.Craig

NSW Fisheries Research Institute, PO Box 21, Cronulla, NSW 2230, Australia.

Summary

Attracting devices ('collectors') have been used to collect information on the relative abundance of the post-larvae (pueruli) of shallow-water spiny lobsters. In this paper a comparison is made of the effects on catches of pueruli of *Jasus verreauxi* between the two basic types of collectors. A fully orthogonal experiment was done with factors of Type (seaweed- or crevice-types) and Position (near surface or substratum) of collector to test the null hypothesis that there were no differences in mean catch of pueruli between type of collector. Pueruli were caught only on seaweed-type collectors. It was concluded that seaweed-type collectors positioned near the surface could be used to monitor the relative abundance of pueruli and post-pueruli of *J. verreauxi* and a collector is described that can withstand sea-swells of around 6 m and retain pueruli.

Introduction

Knowledge of the relative abundance of individuals recruiting to a population provides information about the effect of management on recruitment and, together with information on the relative abundance of spawning individuals, about stock-recruitment relationships (see Hilborn and Walters 1992 for a review). Information on the relative abundance of recruits has been used also to predict the level of catch of spiny lobsters in future years. For example, a relationship between the relative abundance of puerulus in one year, and the catch of legal sized lobsters some years later, has been used to predict the level of catch of *Panulirus cygnus* in the Western Australian rock lobster fishery (Phillips 1986).

After hatching, spiny lobsters develop during the larval phase from a nauplius, through a series of phyllosomata and then metamorphose to a post-larva, commonly referred to as a puerulus (see Booth and Phillips 1994 for a review). The relative abundance of post-larvae (pueruli) occurring in nearshore waters has been measured as an index of recruitment to spiny lobster populations.

Attracting devices commonly referred to as collectors usually have been used to capture pueruli (see Phillips and Booth 1994 for a review). Collectors are typically grouped into two categories; those that imitate seagrasses or algae, commonly referred to as seaweed-type collectors, and those that simulate holes and crevices, known as crevice-type collectors. Pueruli of various species of spiny lobster have been captured by seaweed-type collectors set near the surface (Witham *et al.* 1968; Phillips 1972; Serfling and Ford 1975; MacDonald 1986) or near the substratum (Tholasilingam and Rangoranjan 1986). Crevice-type collectors have been set near the substratum to sample *Jasus edwardsii* in waters off New Zealand (Booth and Tarring 1986) and Australia (Kennedy *et al.* 1994).

J. verreauxi occurs off the east coast of Australia from Tweed Heads (28°10' S) southwards, around the coast of Tasmania, and as far west as Port MacDonnell (38°03' S) in South Australia (Fig. 1 inset). The fishery off New South Wales, Australia, presently has an annual total allowable commercial catch of 106 t, worth approximately $A5 million at the point of first sale.

Little is known about the patterns of settlement of pueruli of *J. verreauxi.* Montgomery and Kittaka (1994) showed that pueruli could be caught in seaweed-type collectors of the same design as those described by Phillips (1972). Their experiment compared catches of pueruli in seaweed-type collectors set near the surface and crevice-type collectors set near the substratum. Montgomery and Kittaka (1994) were unable to make conclusions about the optimal type of collector to use to sample pueruli of *J. verreauxi* because collectors were not compared near the surface or the substratum. Further when seas were rough, seaweed-type collectors were lost and crevice-type collectors dislodged.

In this paper we describe an experimental study of some factors affecting catches of pueruli of *Jasus verreauxi.* A modified seaweed-type collector that is capable of withstanding prolonged periods of rough seas is described.

MATERIALS AND METHODS

Study sites

The experiment was carried out between September and November 1992 off Shark Island, and then from November 1992 to January 1993 in an embayment called Curracurrang. Both sites were near the NSW Fisheries Research Institute, close to Port Hacking (Sydney), New South Wales, Australia (Fig. 1). Shark Island is an area of sandstone reef approximately 100 m east of the mainland. It is approximately 0.5 m above sea level at ebb tide and is exposed to ocean swells. Rocky reef surrounding the island extends out to waters of around 10 m depth. Curracurrang is 5 m deep at its deepest point and is sheltered from the wind and swell. It is surrounded by high cliffs, has a rocky shoreline and reef extending out to approximately 5 m depth. Both sites had forests of kelp (*Ecklonia radiata*) present throughout the rocky reef. Encrusting and filamentous algae and sponges, ascidians and bryozoans made up the understorey.

Procedure

The designs of collectors used were the seaweed-type collector described by Phillips (1972) and the crevice-type collector described by Booth and Tarring (1986). In the experiment there were two orthogonal factors: type of collector (seaweed-type or crevice-type), and position of the collector (near the surface or substratum), with four replicates in each treatment.

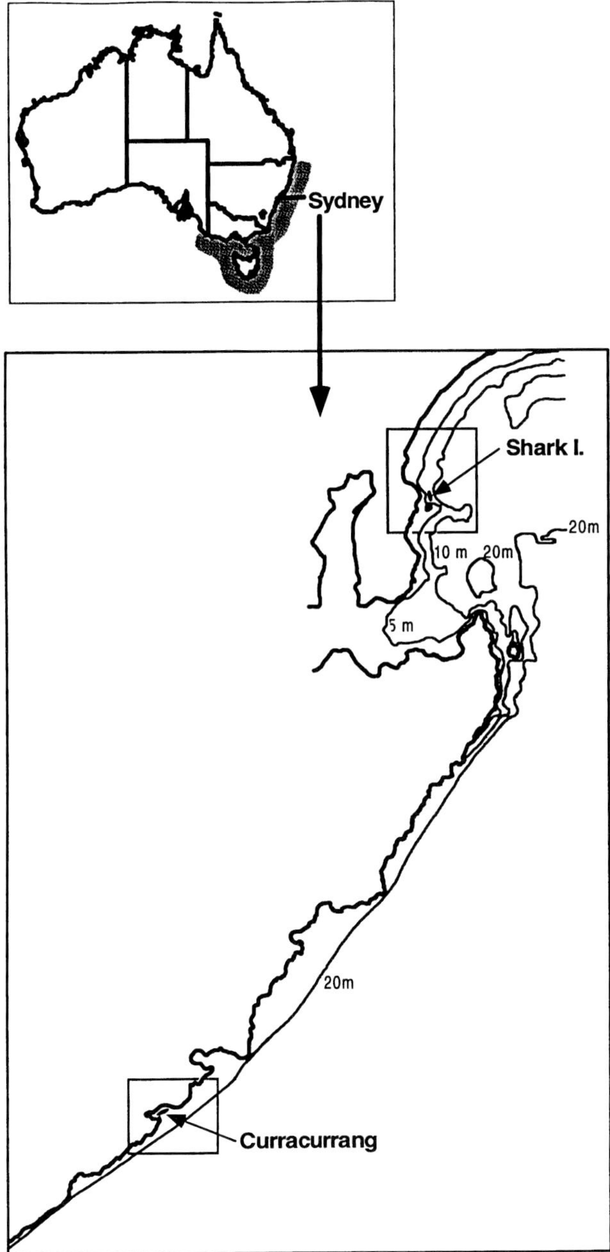

Fig. 1. Location of sites used in the experiment to compare catches of pueruli of *J. verreauxi* between types collector. The inserted map of Australia shows the distribution of *Jasus verreauxi.*

The collectors were sampled weekly off Shark Island and fortnightly off Curracurrang. Collectors had a bag placed over them by a diver before being released from their mooring and taken on board. Those near the substratum were floated to the surface with the aid of an air-bag. Each panel of the seaweed-type collectors was taken from the frame and shaken 30 times into a 320 L container. Crevice-type collectors were placed in the 320 L container and inspected for pueruli.

The number of pueruli (smooth carapace, transparent or brown in colour) or post-pueruli (spines on carapace and brown in colour) were counted. Data were pooled across sites and times, because so few individuals were caught.

RESULTS

A total of 19 pueruli and post-pueruli of *J. verreauxi* were caught in seaweed-type collectors. None were caught in crevice-type collectors. The mean number of pueruli and post-pueruli (combined) caught in seaweed-type collectors positioned near the surface was greater than in seaweed-type collectors near the substratum (ANOVA $P < 0.05$, Fig. 2).

Changes to the design of the collector

Results from the experiment above complemented those of Montgomery and Kittaka (1994) when a total of 28 pueruli and post-pueruli were caught on seaweed-type collectors. Seaweed-type collectors were obviously the best type for collecting information about the relative abundance of pueruli of *J. verreauxi*. There were problems, however, with maintaining the collector described by Phillips (1972) during periods of rough seas — collectors would drag or break away from moorings. We have attempted to address this problem by modifying the Phillip's (1972) design into a flat structure that has less resistance in the water.

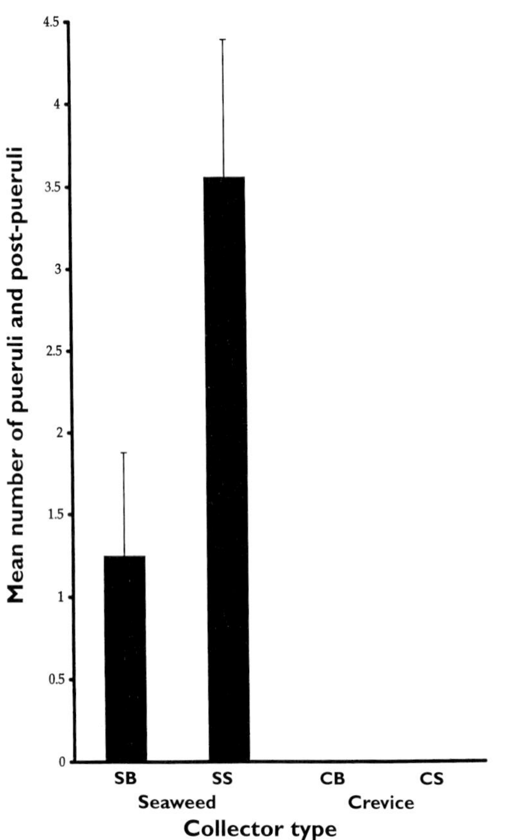

Fig. 2. Mean number (+SE) of pueruli and post-pueruli caught in seaweed-type (S) and crevice-type (C) collectors set near the surface (S) and substratum (B).

The modified collector consisted of two rather than three 61 mm × 35 mm × 0.4 mm PVC panels of 'tufts' with polyethylene split rope fibre (used by Phillips 1972) and was supported by a centre frame with a 300 mm polystyrene buoy on the top rather than a triangular frame supported by 200 mm floats (Fig. 3). A tuft was comprised of 80 g of 500 mm lengths of 125 text fibre. The middle of the tuft was tied with a plastic tie, passed through a hole in the PVC panel and secured by 16 gauge stainless steel wire. The centre frame was made of 1300 × 40 × 6 mm flat stainless steel bar with four 350 × 27 × 16 mm PVC cross members to attach the panels.

Collectors were attached by two 127 mm shark clips to 20 m of 14 mm polyethylene rope and moored to the substratum in waters of 10–12 m depth (Fig. 3). The rope was connected near the substratum by a 10 mm stainless steel shackle to 3 m of 10 mm galvanized chain, which in turn was attached to 3 m of 25 mm long link chain (90 kg) by two 10 mm galvanized shackles. A 25 mm galvanized shackle connected the chain to two 25 kg iron blocks.

DISCUSSION

Results presented in this study supported the conclusion by Montgomery and Kittaka (1994) that seaweed-type collectors were more suited than crevice-type collectors for catching pueruli of *J. verreauxi*. The results provided the additional information necessary to conclude that seaweed-type collectors were best positioned near the surface for sampling pueruli of *J. verreauxi*.

Seaweed-type collectors attracted and retained pueruli of *J. verreauxi* whereas crevice-type collectors did not. The number of pueruli caught was few, but the results were consistent with those of Montgomery and Kittaka (1994). Several studies on different species of spiny lobster have found that seaweed-type collectors were best for sampling pueruli (e.g. *Panulirus argus* by Witham *et al.* 1968, and *Panulirus interruptus* by Serfling and Ford 1975). The greater success of seaweed-type collectors probably indicates that pueruli and juveniles of *J. verreauxi*, like those of many other species of spiny lobster, prefer the complex structure of vegetation as habitat (for a review see Herrnkind *et al.* 1994).

Phillips and Booth (1994) recommended that seaweed-type collectors be placed in sheltered waters. In contrast, we have described in this study a seaweed-type collector capable of withstanding sea-swells of up to around 6 m. The changes made to the seaweed-type collector used by Phillips (1972) and Montgomery and Kittaka (1994) reduced the resistance of the collector to sea-swells, whilst a heavier mooring kept the collector in place. To have restricted the collectors to only sheltered areas would have drastically reduced the number of possible sampling sites along the coast of New South Wales and as a consequence, the information on relative abundance may have been biased. We found that exposure to sea-swell did not affect catches of pueruli of *J. verreauxi* on seaweed-type collectors (unpublished data).

Collectors set near the surface have the advantage that they can be seen between sampling periods, particularly after big sea-swells. Any loss of, or damage to, collectors therefore can be quickly fixed.

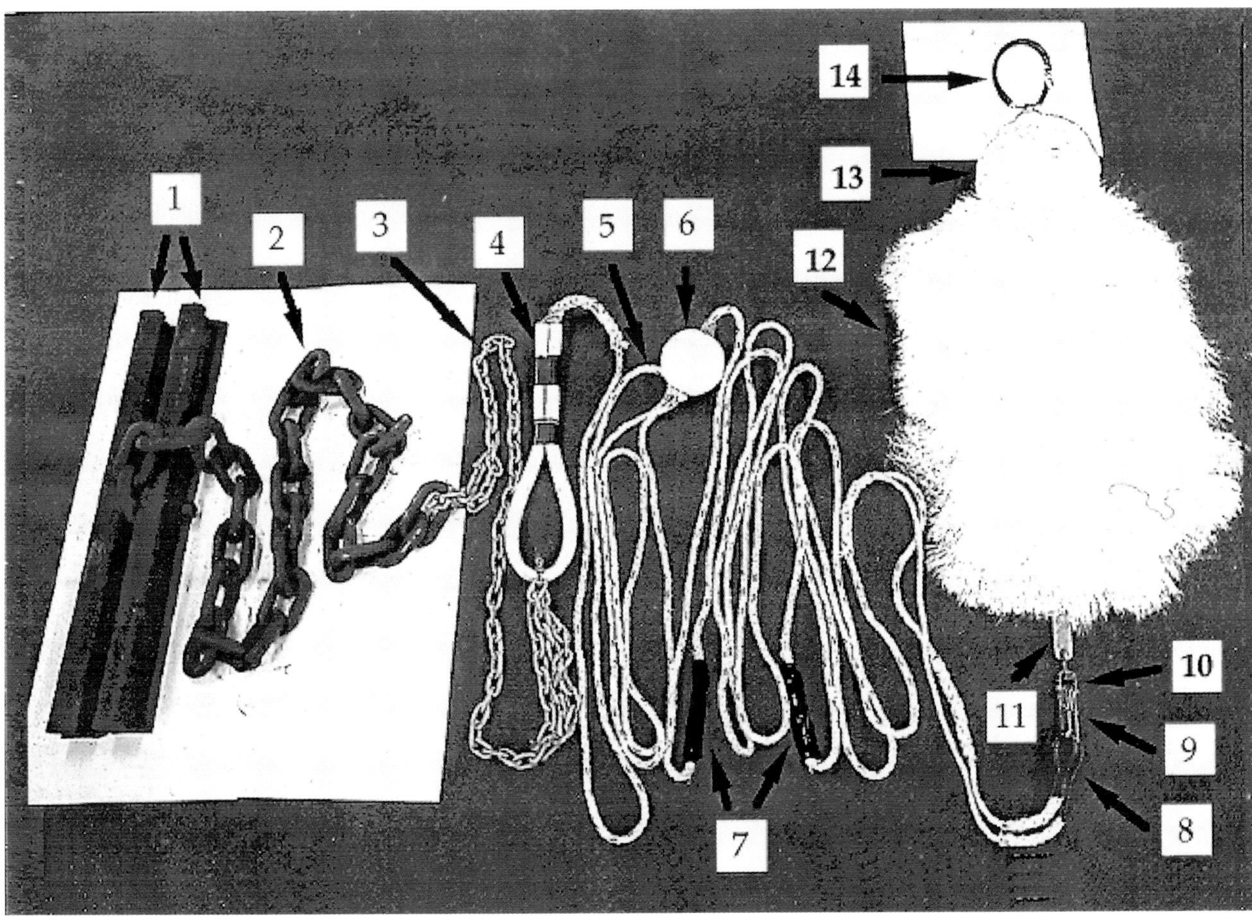

Fig. 3. Design of seaweed-type collector used to sample pueruli of *J. verreauxi*. The mooring for the collector consists of two 25 kg steel weights (1), 25 mm (2) and 10 mm chain (3), protective hosing (4 and 8) around areas of wear on the rope (5), a polystyrene float (6) and weights (7) to hold the rope away from the bottom and surface, respectively and two shark clips (9) to attach the mooring to a swivel (10) at the base of the collector. The collector consists of two panels of tufts of rope fibre (12) supported by a centre frame (11) and a polystyrene float (13). A handle (14) is used to lift the collector.

Other factors that may affect catches of pueruli on collectors other than the type of collector and exposure to sea-swell are the phase of the moon and soak-time of the collector. Some studies on other species of spiny lobster have found an association between the level of catch of pueruli on collectors and the phase of the moon. Peak settlement generally occurs between new moon and first quarter (e.g. *P. cygnus,* Phillips and Sastry 1980). However, this pattern is not typical of all species. Studies on *J. edwardsii* (Booth 1989) and *P. interruptus* (Serfling and Ford 1975) found no association between settlement and the lunar cycle.

Phillips and Booth (1994) recommended that collectors should be checked on the basis of a calendar month when the settlement behaviour of the species being studied is not associated to the phase of the moon. For example, several studies have shown that sampling monthly is adequate for *J. edwardsii* (e.g. Booth and Stewart 1993). When the pattern of settlement of pueruli is associated with the phase of the moon, then collectors need to be checked every four weeks at that time of the lunar month that maximizes catches. For instance, most pueruli of *P. cygnus* settle on seaweed-type collectors around new moon, so collectors are sampled after this phase of the moon to make sure that settlement is completed for that lunar month (Phillips 1986).

We sample collectors around the first quarter of the lunar month and leave collectors to soak for approximately four weeks between sampling times. Although we have found no association between phase of the moon and level of catch of pueruli, the combination of a four week soak-time and sampling around the first quarter of the lunar month provided amongst the greatest catches (unpublished data).

In summary, pueruli and post-pueruli of *J. verreauxi* can be caught using seaweed-type collectors positioned near the surface. A seaweed-type collector designed as a flat structure, such as two panels of rope fibre attached together worked better in rough waters than the triangular structure described by Phillips (1972).

ACKNOWLEDGMENTS

We would like to thank P. Brett, C. Blount, D. Ward and J. Matthews for their assistance with field work during the experiments and for their opinions during the development of the seaweed-type collector. Thanks go also to M. Tanner for preparing final figures and for comments on drafts of this manuscript. The comments by Dr N. Andrew on earlier drafts also have improved this manuscript. Professor A. Underwood

(University of Sydney) provided constructive support throughout this study. This work was funded by the Fisheries Research and Development Corporation of Australia.

References

Booth, J D (1989). Occurrence of the puerulus stage of the rock lobster, *Jasus edwardsii* at the New Plymouth Power Station, New Zealand. *New Zealand Journal of Marine and Freshwater Research* **23**, 43–50.

Booth, J D, and Phillips, B F (1994). Early life history of spiny lobster. *Crustaceana* **66**, 271–94.

Booth, J D, and Stewart, R A (1993). Puerulus settlement in the red rock lobster, *Jasus edwardsii. New Zealand Fisheries. Assessment Reearch Document.* 93/5. 39 pp.

Booth, J D, and Tarring, S C (1986). Settlement of the red rock lobster, *Jasus edwardsii*, near Gisborne, New Zealand. *New Zealand Journal of Marine and Freshwater Research* **20**, 291–7.

Herrnkind, W F, Jernakoff, P, and Butler, M J (1994). Puerulus and post-puerulus ecology. In 'Spiny Lobster Management'. (Eds B F Phillips, J S Cobb and J Kittaka) pp. 213–29 (Fishing News Books: Oxford).

Hilborn, R, and Walters, C J (1992). 'Quantitative fisheries stock assessment; Choice, dynamics and uncertainty.' (Chapman and Hall, New York). 570 pp.

Kennedy, R B, Pearn, R M, Tarbath, D B, and Terry, P (1994). Assessment of spatial and temporal variation in puerulus settlement of the southern rock lobster *Jasus edwardsii. Department of Primary Industires and Fisheries, Sea Fisheries Division Internal Report* **6**, 1 pp.

MacDonald, C D (1986). Recruitment of the puerulus of the spiny lobster, *Panulirus marginatus*, in Hawaii. *Canadian Journal of Fisheries and Aquatic Sciences* **43**, 2118–25.

Montgomery, S S, and Kittaka, J (1994). Occurrence of pueruli of *Jasus verreauxi* (H Milne Edwards 1851) (Decapoda, Palinuridae) in waters off Cronulla, New South Wales, Australia. *Crustaceana* **67**, 65–70.

Phillips, B F (1972). A semi-quantitative collector of the puerulus larvae of the western rock lobster *Panulirus longipes cygnus* George (Decapoda, Palinuridae). *Crustaceana* **22**, 147–54.

Phillips, B F (1986). Prediction of commercial catches of the western rock lobster *Panulirus cygnus. Canadian Journal of Fisheries and Aquatic Sciences* **43**, 2126–30.

Phillips, B F, and Booth, J D (1994). Design, use and effectiveness of collectors for catching the puerulus stage of spiny lobsters. *Reviews in Fisheries Science* **2**, 255–89.

Phillips, B F, and Sastry, AN (1980). Larval Ecology. In 'The Biology and Management of Lobsters'. (Eds J S Cobb and B F Phillips). Vol. I, 11–57 (Academic Press, New York).

Serfling, S A, and Ford, R F (1975). Ecological studies of the puerulus larval stage of the California spiny rock lobster, *Panulirus interruptus. Fishery Bulletin (US)* **73**, 360–77.

Tholasilongam, T, and Rangarajan, K (1986). Prospects of spiny lobster, *Panulirus* spp. culture in the east coast of India. *Proceedings of the Symposium on Coastal Aquaculture* **4**, 1171–5.

Witham, R, Ingle, R M, and Joyce, E A (1968). Physiological and ecological studies of *Panulirus argus* from the St. Lucie Estuary. *State of Florida Board of Conservation Technical Series* **53**. 31 pp.

Relationships Between Different Life History Stages of the Western Rock Lobster, *Panulirus cygnus*, and their Implications for Management

N Caputi,[A] *C Chubb,*[A] *N Hall*[A] *and A Pearce*[B]

[A] Fisheries Dept of Western Australia, PO Box 20, North Beach, WA 6020, Australia.
[B] CSIRO Division of Oceanography, PO Box 20, North Beach, WA 6020, Australia.

Summary

Management of the Western Australian rock lobster fishery is enhanced by an ability to forecast sustainable catch levels and set appropriate catch controls. This fishery has an extensive history of research, management and enforcement of regulations; it was one of the first limited-entry fisheries, having been introduced in 1963. It continues to operate successfully under a system of individually transferable units of effort.

The fishery's database includes information on: abundance of puerulus settlement, juveniles and spawning stock; catch and fishing effort; environmental factors affecting recruitment; and vessels, gear and equipment. Prediction of catches up to four years ahead, based on puerulus and juvenile abundance and on fishing effort, allow management measures to be undertaken before a year-class reaches legal size. The effects of the environment and of the abundance of spawning stock on the level of puerulus settlement are also considered in the management process. In 1993–94, management measures were introduced in response to the decline in spawning stock to below 20% of the virgin biomass. These measures include 18% pot reduction, increase in minimum size, establishment of a maximum size for females, and the return of mature females to the sea. This paper illustrates the array of data that are relevant to stock assessment.

Introduction

The limited-entry fishery for the western rock lobster (*Panulirus cygnus*) in Western Australia is one of the largest and most successful rock lobster fisheries in the world. In 1994–95, 11 000 t of lobsters were landed (valued at $A300 million), slightly above the long-term average catch of 10 500 t. Catches fall into two phases: the 'whites' fishery (15 November to 31 January) targeting lightly-coloured migrating immature lobsters as they disperse into the breeding grounds in the deeper waters, and the 'reds' fishery (February to June) targeting 'resident' lobsters in the shallow to mid depths (Brown and Phillips 1994). Moulting occurs at the start of these two phases. The fishery is managed under input controls over three zones (Fig. 1) and was one of the first limited-entry fisheries in the world.

Knowledge of the relationships between different life-history stages such as the puerulus (first post-larval stage), settlement stage larva, juvenile, recruitment to the fishery, and spawning stock has been of major benefit to the management of this fishery (Brown and Phillips 1994). The ability to forecast catches up to four years in advance has permitted the development of management rules before a year-class has recruited to the fishery. For example, concern for the effect on the spawning stock of high fishing effort on a predicted low catch for the 1986–87 season was the impetus for a 10% pot reduction for that season. This

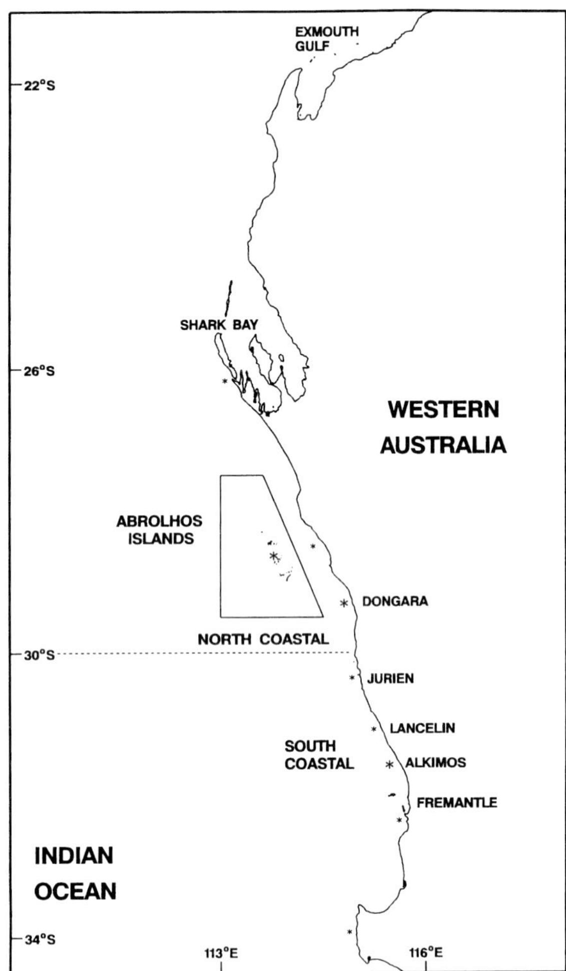

Fig. 1. Map of locations of puerulus collectors (*) and the three management zones (north coastal, south coastal and the Abrolhos Islands) of the western rock lobster fishery of Western Australia. Larger asterisks denote localities described in the text.

contrasts markedly with management in most fisheries which usually reacts to a low recruitment after it has already been fished. If the fishing effort is high on this low recruitment then this may result in a very low spawning stock which may cause recruitment overfishing.

An understanding of the relative importance of the factors (spawning stock and environment) responsible for the variation in recruitment has also proved to be of major benefit in the management of this fishery. For example, the low puerulus settlement in 1982–83 that was used to predict the low catch in 1986–87 was attributed to a strong El Niño in 1982–83 and weak westerly winds (Pearce and Phillips 1988; Caputi and Brown 1993) rather than to low spawning stock. Hence, although it was necessary to reduce fishing on a poor recruitment year-class to avoid a resultant low spawning stock, there was no need for the severe management action that would have been required if the poor year-class had been due to the low level of spawning stock.

This paper outlines the databases and relationships between life-history stages that have been developed for the western rock lobster fishery and describes how they have been used in the recent management plans of the fishery. It also details the current issues facing the managers and the approaches for assessing the likely stock impacts of strategies designed to control the flow of catch both within and between seasons.

Research information and life-history relationships

Long time series of fisheries data are available from a number of sources: commercial fishers' compulsory catch-and-effort returns including an annual vessel, gear and equipment survey; voluntary daily catch-and-effort log-book data from about 30% of the commercial fleet; monthly at-sea monitoring of catches (as well as undersize and mature females returned to the sea) on commercial vessels operating in four representative localities and four depth categories; compulsory processors' returns of total catch and production by product type and grade category; the value of the catch and licences; and an annual mail survey to determine the levels of catch and effort by licensed recreational fishers. In addition to the fisheries data, standardized research surveys are used to monitor puerulus settlement at nine localities within the western rock lobster fishery and to provide a 'fishery independent' index of the breeding stock at five coastal localities and the offshore Abrolhos Islands.

Environmental data are important in explaining the variation in indices of abundance of many of the life history stages. For example, water temperature and swell affect the catchability of lobsters, and puerulus settlement is strongly influenced by the strength of the Leeuwin Current (indexed by coastal sea level and affected by El Niño Southern Oscillation (ENSO) events) and the strength of westerly onshore winds (measured by rainfall) over the continental shelf (Pearce and Phillips 1988, 1994; Caputi and Brown 1993).

Satellite-derived sea-surface temperatures, and wind observations from the Comprehensive Ocean–Atmospheric Data Set (COADS) are being added to the environmental database. Thermal infrared satellite imagery from the NOAA/AVHRR satellite has been used to show mesoscale circulation patterns which may be linked to variations in puerulus settlement (Pearce and Phillips 1988; Phillips *et al.* 1991). The availability of altimeter data from the Topex satellite will enable direct estimates of ocean currents to be made using the topography of the sea surface. Ocean colour imagery from the Ocean Colour and Temperature Scanner (scheduled for launch in 1996 on the Japanese Advanced Earth Observing Satellite) and SeaWiFS will allow monitoring of phytoplankton distributions by measuring chlorophyll concentrations in the upper layers of the global ocean. Both temporal and spatial variability in ocean productivity along our continental shelves may be linked with food resources and hence with recruitment and catches of fish.

An ocean modelling project by the CSIRO Division of Oceanography will assimilate existing oceanographic data collected in Australian waters over the past two decades into a numerical model describing the ocean circulation and water properties, providing detailed information on 3-dimensional ocean dynamics (Craig 1995). Variations in recruitment can then be linked with seasonal and interannual changes in circulation patterns which may play an important role in the oceanic migration of the planktonic phyllosomata and subsequent settlement of the puerulus stage. This modelling project may provide some insight into the spawning regions most likely to contribute to the puerulus settlement at different localities.

The databases have facilitated the investigation of relationships between life history stages of the western rock lobster, the most important of which have been the relationship between catch and indices of puerulus and juvenile abundance (Phillips 1986; Caputi *et al.* 1995*a*, *b*) and relationships between spawning stock and recruitment (Morgan *et al.* 1982; Caputi *et al.* 1995*c*).

Regional catch predictions three and four years ahead based on puerulus abundance and levels of fishing effort (Fig. 2) are used regularly to allow management measures to be discussed fully and introduced before a recruit class reaches legal size. Prediction of catches in the 'white' and 'red' parts of the fishery (Fig. 3) are being examined as part of the assessment of the likely effects of transferring product within and between seasons to improve the economic worth of the catch. The trend in spawning stock abundance (Fig. 4), the effects of spawning stock and environment on levels of puerulus settlement, and the increasing effective effort (Brown *et al.* 1995) are also considered in setting effort levels for the fishery.

MANAGEMENT ISSUES (CURRENT AND FUTURE)

1993–94 management package

Stock assessment in the early 1990s indicated that the northern and southern coastal breeding stocks had been declining at an annual rate of 6% and 2.5% respectively (Fig. 4) (Caputi *et al.* 1995*c*). Modelling of the fishery also indicated that egg production had declined to about 15–20% of its unfished level (using the model of Walters *et al.* 1993). It was considered that the decline had been accelerated in the late 1980s and early 1990s by the introduction of new technology such as sophisticated navigation (e.g. Global Positioning System) and colour echo sounders and the trend to larger more seaworthy vessels (Brown *et al.* 1995). This resulted in an increased exploitation rate, particularly on the deep-water stocks principally comprising the breeding animals. The variation in puerulus settlement at coastal localities was attributed mainly to environmental effects (Fig. 5).

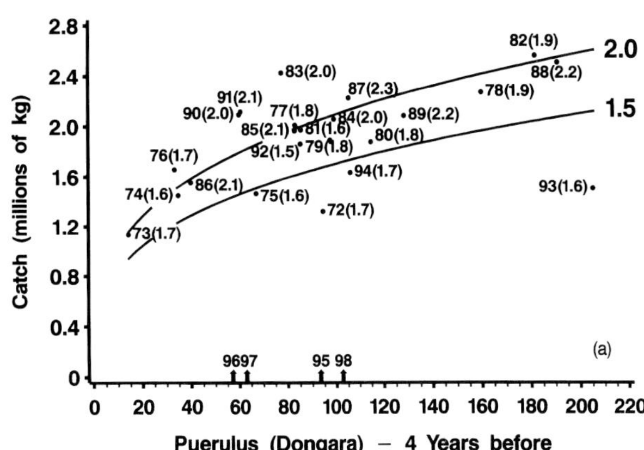

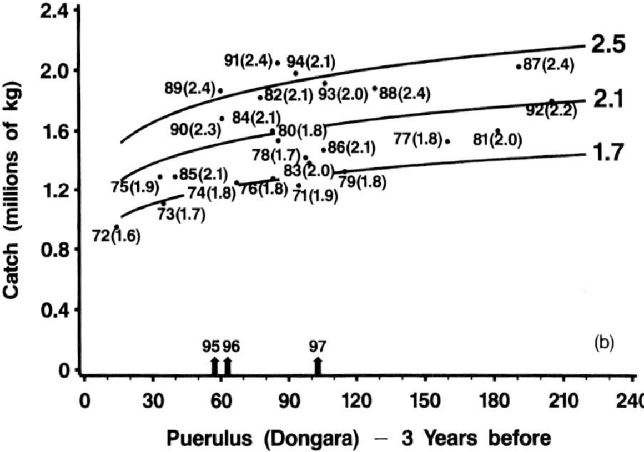

Fig. 3. Catch–puerulus relationship for the north coastal zone for the **(a)** whites and (b) reds fishery. The year of the catch is shown with the nominal effort (million potlifts) in parenthesis. The curves indicate the relationship at different levels of fishing effort (1.5, 2.0, 1.7, 2.1 and 2.5 million potlifts).

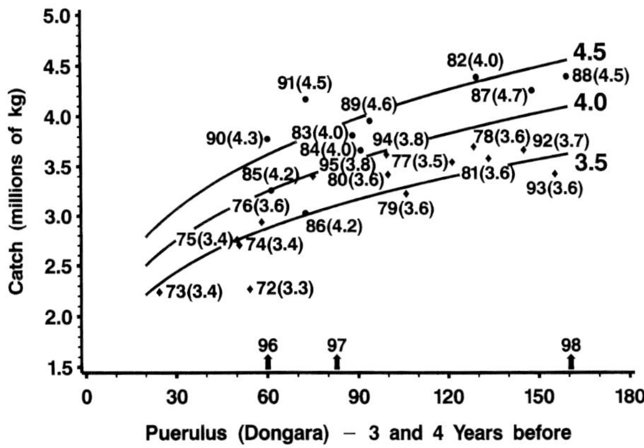

Fig. 2. Catch–puerulus relationship for the north-coastal management zone of Western Australia. The year of the catch is shown with the nominal fishery effort (million potlifts) in parenthesis. Puerulus is number per collector per year. The three curves indicate the relationship at three levels of fishing effort (3.5, 4.0 and 4.5 million potlifts).

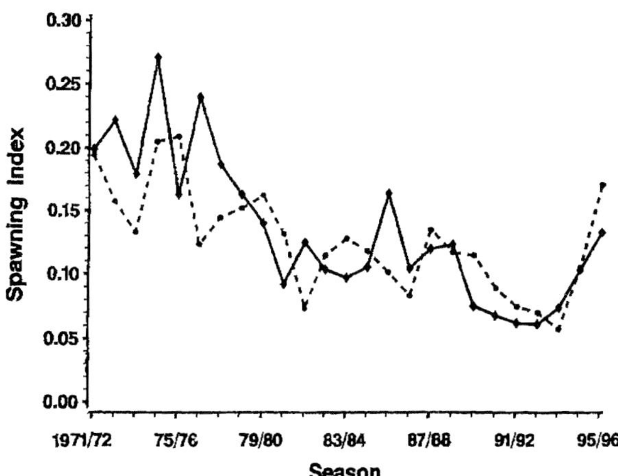

Fig. 4. Time series of spawning stock indices of the western rock lobster fishery for (—◆—) north and (--•--) south coastal regions.

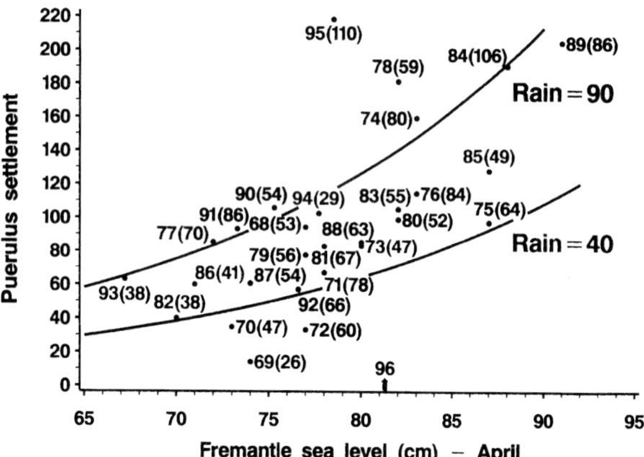

Fig. 5. Relationship between puerulus settlement at Dongara and the strength of the Leeuwin Current (measured by coastal sea level in April) and the westerly winds (measured by rainfall in southern localities during October–November). The year is shown with the rainfall (mm) in parenthesis.

However, there was concern at the average decrease of 50% in puerulus settlement at the Abrolhos Islands from the mid 1980s and 1990s compared with that achieved in the 1970s, and the possible link to the low abundance of coastal spawning stock (Fig. 6) (Caputi *et al.* 1995*c*).

To combat the downturn in egg production, a package of management measures was introduced for the 1993–94 and 1994–95 seasons and later extended to 1995–96. With above-average catch expected in 1993–94 and lower catches predicted for the following seasons until 1996–97, it was considered that a greater increase in the breeding stock and reduced hardship for the fishing industry would be achieved by commencing the effort reduction in the year of good catches. The management measures included a temporary 18% reduction in pots (traps), an increase in the minimum legal size for the 'whites' fishery, total protection for all breeding female rock lobsters (those with ovigerous setae, spermatophores or bearing eggs) and the establishment of a maximum size for large female lobsters. This package aimed at significantly increasing the proportion of female lobsters surviving, to ensure that the spawning potential of the stock returned to levels that historically had provided adequate recruitment each year (Rock Lobster Industry Advisory Committee 1993). Previous measures to reduce effort had been compensated by fishers fishing harder; however, the level of latent fishing effort has been reduced (Fig. 7) so that these management measures represent real reductions in exploitation. At this stage, the upturn in the spawning stock indices (Fig. 4) suggests that the package is starting to achieve its primary aim of increasing the abundance of breeding stock and improving egg production.

Protection of the breeding stock was foremost in importance in the preparation of the 1993–94 package, but it was recognized that economic benefits could flow from elements of the present management arrangements. One such benefit was the transfer of product from the 'whites' fishery to the 'reds' in which the overall value of the lobster on the market is greater. This shift resulted from the 1-mm increase in the minimum legal length during the 'whites' and the reduced exploitation rate from the temporary 18% pot reduction. The effect in the northern sector of the fishery was dramatic, reducing the 'whites' catch by about 1000 t in the first year and increasing the catch in the 'reds' fishery, and evening out the two catch peaks formerly dominated by the 'whites'. This brought significantly higher returns to fishers in this sector. However, owing to the larger sizes of lobsters in the southern sector of the fishery, the impact was not as marked and catches are still dominated by a large peak in catch in December in the 'whites'.

Industry has realized that if the spawning potential of the stock is to be maintained, there is no longer the potential to increase the economic return from the fishery by increasing catch. Further economic improvements will arise only from better use of the available catch. Hence, managers are examining further strategies to control the flow of catch by adjusting the pattern of fishing effort within season, and adjusting the fishing season to take advantage of higher market prices, without adversely affecting the overall exploitation of the rock lobster stock.

Another problem identified by fishery managers is the interannual variability of recruitment leading to highly variable

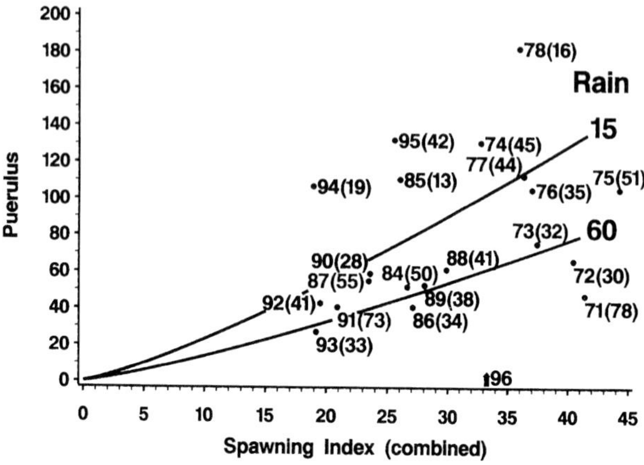

Fig. 6. Spawning stock–recruitment relationship with environmental effects incorporated for the Abrolhos Is. The year is shown with the rainfall during October–November (mm) in parenthesis.

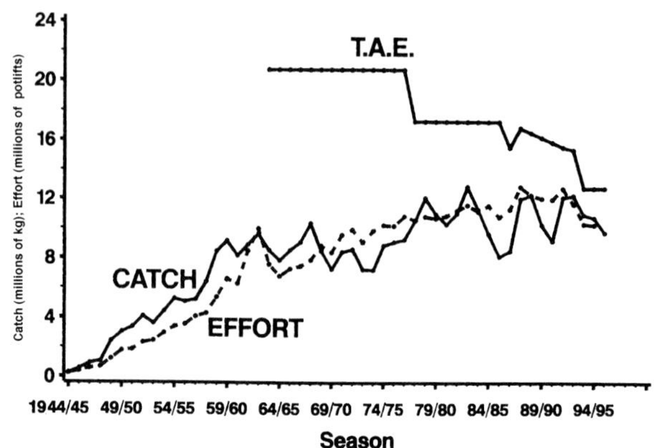

Fig. 7. Trend in nominal effort and an estimate of the total allowable effort (TAE). The difference between the two trends indicates the latent effort in the fishery.

annual catches. The market does not readily adjust to large volumes of product landed in seasons of high recruitment. From a market point of view, it is preferable to reduce the variability and provide a more stable supply of product. For example, the very high catch in 1992–93 resulted in an average price reduction from about \$A25 kg^{-1} in the previous year to about \$A14–18 kg^{-1}. The availability of reliable catch predictions allows managers the opportunity to consider reducing the fishing effort during the season of predicted high catch with an appropriate increase in effort in the years before and/or after the high catch year as a means of levelling out the catch.

The timing of the introduction of the 1993–94 package has provided a trial for industry of the economic benefits of transferring product from a high catch year to subsequent seasons, because the package was introduced in a year of predicted above-average catches (11 800 t) that was to be followed by lower catches in the following two years. The effect of the package was to reduce the overall catch in 1993–94 by about 800 t and to achieve a better price in that season than in the previous season. Hence, the catch in 1993–94 was still slightly above the long-term average, and some of the catch was transferred to the following season which achieved 10 800 t. Until this package was introduced, the overseas markets were using the predicted years of high catches to reduce prices paid as a result of the expected increased supply. This package was the first attempt to take advantage of the catch predictions and regulate catches within and between seasons.

Current management issues

As a result of very good puerulus settlement during 1995–96 and good settlement also in some localities in the previous season, the catch prediction relationships forecast a very good 'reds' catch in 1998–99 followed by a good 'whites' catch in 1999–2000. With the spawning stock currently improving as a result of the 1993–94 management package, management is currently focussed on options to obtain the best economic value from the good recruitment that has been forecast. For example, a reduction in the minimum legal size and increase in fishing effort in the season preceding a high recruitment (with a corresponding decrease in exploitation rate during the season of good recruitment) to allow some of the expected product to be harvested earlier may improve the overall market return. Alternatively, as occurred in 1993–94, some catch could be transferred to the season following the season of good recruitment.

There is also some economic benefit in further reducing the 'whites' catch with an appropriate increase in the 'reds' catch, particularly in the southern sector which achieves about 45% of its catch during the 2.5-month 'whites' fishery. This will be particularly important in view of the very good 'whites' catch forecast for 1999–2000. Reducing the fishing effort on this part of the fishery and increasing the catch in the 'reds' fishery in the previous year would add further product during a period that is already expected to have an above-average catch. However, transferring the product to the following 'reds' season may result in a more even catch distribution. Assessment of the relative benefits of these proposals requires more information on the timing of the recruitment to the fishery, growth, mortality, and relative prices of red and white rock lobsters and of different size categories. That is, a bio-economic model would assist in evaluating the management options.

For such a bio-economic model, it is proposed that model output would allow examination of the economic return from overseas markets of the predicted catch flow resulting from alternative management strategies. The management strategies to be considered would permit the use of a variable level of effort within season, with control extending to the week-by-week level of effort in each management zone, combined with control of the total effort to be applied within future fishing seasons. Controls would also extend to a flexible legal-size range of lobsters of each sex that might be landed. The model would attempt to assess the market response to varying supply of the different size ranges of product, on the basis of historical responses of market price to supply of product. The objective of the modelling would be to use the stream of indices of puerulus settlement and fishery databases in order to determine the strategy of size and effort controls that would result in greatest financial return, yet to ensure that the level of breeding stock was not adversely affected.

Input or output controls

When the 1993–94 management package was being developed and industry was faced with significant pot reductions, industry requested that output controls (catch quotas) be investigated as an alternative harvest strategy in the longer term. The Rock Lobster Industry Advisory Committee, an expert committee comprising mainly industry members advising the Minister for Fisheries on management of this fishery, commissioned an investigation into input and output controls for the future management of the fishery (Bowen 1994; Lindner 1994; Marec Pty Ltd 1994; McLaughlan 1994). Investigation of the relative benefits of the two systems took into account economics, marketing, enforcement, stock assessment and the possible options for implementing catch quotas. Industry has since opted to remain with input controls using the principle of Individual Transferable Effort units (ITEs), and has noted the need to vary the level of fishing effort on the basis of the predicted and desired levels of catch. It accepted that future reductions in effort may be necessary to take into account increases in fishing efficiency. Figure 8 shows how the life-history data and relationships are used in the management of this fishery using ITE units and how this approach could be used in setting catch quotas if necessary.

Puerulus grow-out

Management has also expressed some interest in enhancing the overall production by the possible harvest of puerulus for grow-out. As the catch–puerulus relationship is not a straight-line relationship through the origin, there is a density-dependent mortality occurring between these life-history stages. Thus, the prediction relationship can be used to provide some information on the relative mortality at different levels of abundance of puerulus settlement (Fig. 9). This indicates that much higher mortality must occur between the puerulus and the recruitment-to-the-fishery stages in years of high puerulus settlement than in years of lower settlement. A comparison between the slopes of the

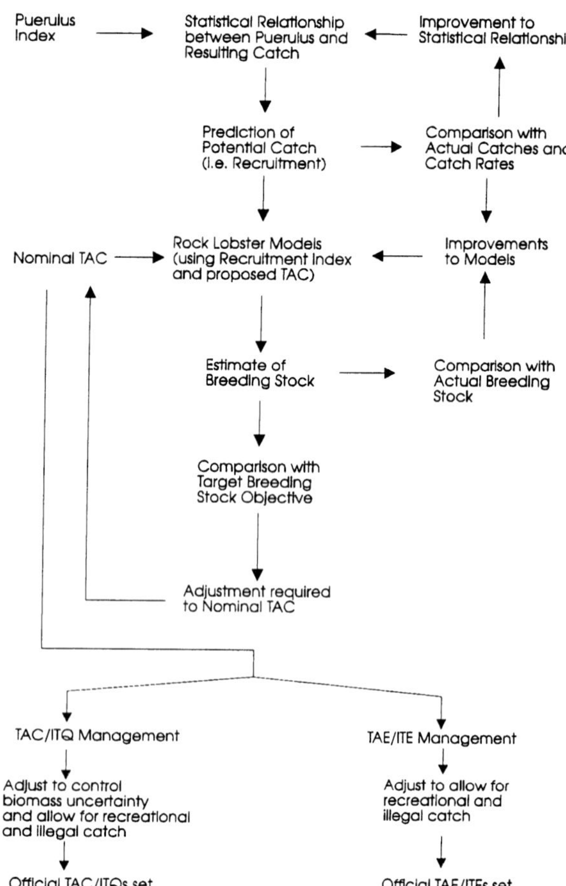

Fig. 8. Summary of the process for assessing the total allowable effort (TAE) or total allowable catch (TAC) of the western rock lobster fishery. Estimates of recruitment and breeding stock are required for each sector of the fishery.

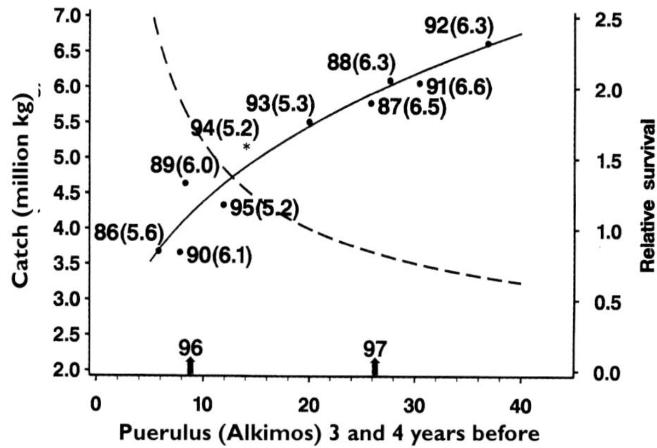

Fig. 9. Catch–puerulus relationship for Alkimos with the corresponding relationship between relative survival (dashed line) and puerulus settlement. The year of the catch is shown with the nominal effort (million potlifts) in parenthesis.

relationships for the different regions indicates that there is a stronger density-dependent mortality relationship at the Abrolhos Islands than in the south coastal region (Caputi *et al.* 1995*a*). For example, a 50% drop in puerulus settlement results in only an 8% drop in catch at the Abrolhos compared with a 20% drop in catch for the south coastal region. This reflects the higher abundance of rock lobsters at the Abrolhos than in the coastal regions. However, these relationships are based on indices of abundance and cannot be used to assess the effect of taking out a certain number of pueruli from a given area or during a given year.

Conclusions

This case study illustrates the array of data inputs that are relevant to the stock assessment process and highlights the value of long time series of data in determining relationships which contribute to the management process. The catch-prediction relationships and the relationship among the breeding stock, the environment and the level of puerulus settlement have been fundamental to the proper management of this fishery

With an improvement in the breeding stock as a result of the current management package, more emphasis is being placed on the use of the catch-prediction relationships to improve the value of the fishery. Hence, there is a greater need for bio-economic modelling with economic information such as variation in monthly prices by grade category and colour (red and white). The impact of possible changes in the monthly catch on prices also needs to be evaluated.

Although the current understanding of the biological processes has proved invaluable for providing advice to managers, a more detailed and broader understanding of the various relationships will be required in future to answer the more complex questions that are being raised. Current management issues, together with the future issues that have been identified, clearly identify future directions for research. Further improvement and refinement of the relationships linking the biology, the fishery, and the environment will continue as new data become available. However, there is also a requirement to address areas where current understanding of the processes is deficient, such as the need for a more detailed description of growth, or the need for improved understanding of the spatial distribution of puerulus settlement and the spawning stock from which it has arisen, through increased knowledge of the oceanographic processes. Research studies must also be initiated on the economic relationships that have, until now, received little attention in models of the rock lobster fishery.

Acknowledgments

Coastal sea levels have been provided by the National Tidal Facility, copyright reserved. Funding for aspects of this research has been provide by the Fisheries Research and Development Corporation. The authors thank research scientists of the Western Australian Marine Research Laboratories and CSIRO Division of Oceanography for reading the manuscript and offering suggestions.

References

Bowen, B K (1994). Long term management strategies for the western rock lobster fishery. Evaluation of management options (volume 1). Fisheries Department of Western Australia, *Fisheries Management Paper* 67. 67 pp.

Brown, R S, and Phillips, B F (1994). The current status of Australia's rock lobster fisheries. In 'Spiny lobster management'. (Eds B F Phillips, J S Cobb and J Kittaka) pp. 31–63 (Fishing News Books. London).

Brown, R S, Caputi, N, and Barker, E H (1995). A preliminary assessment of the effect of increases in fishing power on stock assessment and fishing effort of the western rock lobster (*Panulirus cygnus* George 1962) fishery in Western Australia. *Crustaceana* **68**, 227–37.

Caputi, N, and Brown, R S. (1993). The effect of environment on puerulus settlement of the western rock lobster (*Panulirus cygnus*) in Western Australia. *Fisheries Oceanography* **2**, 1–10.

Caputi, N, Brown, R S, and Chubb, C F (1995*a*). Regional prediction of the western rock lobster, *Panulirus cygnus*, catch in Western Australia. *Crustaceana* **68**, 245–56.

Caputi, N, Brown, R S, and Phillips, B F (1995*b*). Prediction of catches of the western rock lobster (*Panulirus cygnus*) based on indices of puerulus and juvenile abundance. *ICES Marine Science Symposia* **199**, 287–93.

Caputi, N, Chubb, C F, and Brown, R S (1995c). Relationship between spawning stock, environment, recruitment and fishing effort for the western rock lobster, *Panulirus cygnus*, fishery in Western Australia. *Crustaceana* **68**, 213–26.

Craig, P (1995). 'Oceans-EEZ' will describe Australia's oceans. *Australian Fisheries*, April 1995, 10–12.

Lindner, B (1994). Long term management strategies for the western rock lobster fishery. Economic efficiency of alternative input and output based management systems (volume 2). Fisheries Department of Western Australia, *Fisheries Management Paper* 68. 36 pp.

Marec Pty Ltd (1994). Long term management strategies for the western rock lobster fishery. A market-based economic assessment for the western rock lobster industry (volume 3). Fisheries Department of Western Australia, *Fisheries Management Paper* 69. 71 pp.

McLaughlan, N (1994). Long term management strategies for the western rock lobster fishery. Law enforcement considerations (volume 4). Fisheries Department of Western Australia, *Fisheries Management Paper* 70. 12 pp.

Morgan, G R, Phillips, B F, and Joll, L M (1982). Stock and recruitment relationships in *Panulirus cygnus*, the commercial rock (spiny) lobster of Western Australia. *Fishery Bulletin* **80**, 475–86.

Pearce, A F, and Phillips, B F (1988). ENSO events, the Leeuwin Current, and larval recruitment of the western rock lobster. *Journal du Conseil international pour l'Exploration de la Mer* **45**, 13–21.

Pearce, A F, and Phillips, B F (1994). Oceanic processes, puerulus settlement and recruitment of the western rock lobster *Panulirus cygnus*. In 'The bio-physics of marine larval dispersa'. (Eds P W Sammarco and M L Heron). American Geophysical Union, *Coastal and Estuarine Studies* **45**, 279–303.

Phillips, B F (1986). Prediction of commercial catches of western rock lobster *Panulirus cygnus* George. *Canadian Journal of Fisheries and Aquatic Sciences* **43**, 2126–30.

Phillips, B F, Pearce, A F, and Litchfield, R T (1991). The Leeuwin Current and larval recruitment to the rock (spiny) fishery off Western Australia. *Journal of the Royal Society of Western Australia* **74**, 93–100.

Rock Lobster Industry Advisory Committee (1993). Management proposals for 1993/94 and 1994/95 western rock lobster season. Fisheries Department of Western Australia, *Fisheries Management Paper* 54. 8 pp.

Walters, C, Hall, N, Brown, R S, and Chubb, C F (1993). A spatial model for the population dynamics and exploitation of the Western Australia rock lobster, *Panulirus cygnus*. *Canadian Journal of Fisheries and Aquatic Science* **50**, 1650–62.

How to Achieve Sustainable Fisheries Development in a Developing Country: the Case of Mexico

Pablo Arenas Fuentes and Antonio Diaz de Leon Corral

Instituto Nacional de la Pesca, Pitágoras 1,320, Col. Santa Cruz Atoyac, CP 03310 Mexico.

Summary

This paper deals with the special problems of achieving sustainable fisheries development in a developing country. That is, how to implement a complex concept — conceived for the most part in developed countries — in a setting in which the overall lack of financing, organization, research, information flow, legal framework and enforcement is the rule.

We analyse the case of Mexico as an example. Mexico is a developing country with more resources than most such nations, but with acute social problems and overpopulation pressures, and also under extreme international pressure regarding the environment from neighbouring countries. The most important steps the country has taken and some of the results are described.

The key issues are: How to reach a balance between development and conservation? How to reach a workable definition of sustainable development in fisheries? How much and what kind of fisheries research is appropriate to reach this goal? How to take legal and management measures in the face of incomplete information, something that is so prevalent in developing countries? How to make society aware of these problems, and how to manage so that all users have an input into management decisions?

To look at these questions and to assess how the policy changes are shaping fisheries management, several fisheries in Mexico are reviewed. Particular emphasis is given to the most important shrimp fishery in Mexico, as well as analysing other fisheries, ranging from the almost complete open-access (tuna fishery) to the almost private property case (sea urchin concessions), and the very difficult-to-regulate (because of social pressure) coastal shark fishery.

Introduction

How to achieve sustainable fisheries development? This is a central question for most fisheries managers in the developed world. However, for fisheries officials in the rest of the world this is a concept that is still far from being accepted; it is still rather new. Thus, the appropriate question for the developing world could be: How to achieve sustainable fisheries development in a developing country? That is, how to implement a complex concept — developed for the most part in developed countries — in a setting in which the overall lack of financing, organization, research, information flow, legal framework and enforcement is the rule. This is by no means a trivial question, as developing countries now account for more than half of the global fisheries harvest. There has been considerable insight into what is sustainable development, but little into how to actually achieve it; in addition, the examples from the developing world are almost non-existent.

In this paper we analyse the steps taken by Mexico over the past two years as an example of efforts that may lead towards sustainability in fisheries. This is by no means a technical discussion on sustainability; rather, we offer our views candidly and openly, and seek feedback on the search for the right institutions and partnerships required to achieve this goal. Before tackling the central issue, a few words about Mexican Fisheries are necessary.

The Mexican fisheries

The fishing industry in Mexico is the fastest-growing within its food system; since 1992 it has reached almost 1.0% of the Gross National Product and it comprises 1.3% of the total employed population. National fisheries production was under 500 000 t from 1940 to 1976. The maximum historical catch was reached in 1981 with 1.565 million t. In 1989 the total catch was 1.519 million t, and in 1995 1.410 million t were caught.

On a regional basis, the importance of the fishery sector during the last few years has been greater due to increasing standards in both social and economic development in comparison with other sectors. For example, inshore fisheries contribute every year with about 40% of total production. Coastal fisheries are an important source of employment for a number of families and represent in many cases cultural traditions, an anthropological aspect not often shown in statistical tables.

The main fishing activity is located in northwest Mexico where Baja California, Baja California Sur, Sinaloa and Sonora States comprise 53% of total production and 59.1% of the processing trade industries. Catches from the Pacific Ocean contributed 70.31% of the total national production in 1995. In the Gulf of Mexico, there is an important level of exploitation of valuable fisheries, like shrimp, oysters and sharks. The Gulf of Mexico and Caribbean sea represented 26.63% of the total national production (both fishery and aquaculture) in 1995. The production of tilapia is the main activity in inland waters, followed to a lesser extent by carp, trout and crayfish. Inland waters produced 3.06% of the 1995 total production.

Mexico ranked 18th in the world in fish and shellfish production by volume in 1993. Its contribution to world catches in 1993 was 2.8%, and for selected species it was as follows: sharks 5.0%; tuna 3.0%; mollusks 3.0%; shrimp 2.3%; and small pelagics 1.0%.

Total catches seem to have stabilized although, in some cases, a moderate increase in production is considered possible.

The legal framework

The Mexican fishing industry is made up of two major sectors: the privately-owned sector and the social-sector cooperatives. The latter were introduced in 1938. The system gave Mexican fishermen exclusive exploitation of valuable species such as shrimp, abalone, oysters and lobsters. However, after an initial success, the cooperatives developed into marginal economic entities and by 1994 already 90% of the fleet had shifted to private firms. The private sector now controls the offshore fisheries although cooperatives are very important socially and occur throughout in the artisanal fisheries.

The first measures to regulate the fisheries sector in Mexico were implemented in the mid-nineteenth century. These were broad controls until more comprehensive legislation came in 1902 with the Law of Territorial Sea. The first specific fisheries-related Act was enacted in 1923. Amended versions followed in 1925, 1932, 1947, 1986 and 1992.

The latest version of Mexican fishing law, introduced in 1992, includes concepts that had not been considered in earlier versions, such as enhancing preservation of endangered species; deregulation of the fisheries sector; authorization of fishing to private investors allowing them to exploit species formerly exclusive to state-owned cooperatives; and aquaculture promotion.

With respect to non-Mexican participation in fisheries, foreign vessels are allowed to fish within the Exclusive Economic Zone, under very strict guidelines, only when a resource surplus is declared or an international agreement has been established. However, a variety of companies can be formed with foreign capital, in association with Mexican nationals (49% maximum and a Mexican flag).

Although the fishing sector is among the fastest-growing in Mexico, the problems facing the industry are many; they are for the most part common to many developing countries, i.e. social and political pressures are more important than research to manage fisheries; lack of finance; ageing fishing fleet; lack of co-ordination among decision-making institutions; widespread conflicts between artisanal and industrial fishermen, over-capitalization, overexploitation of valuable resources, and underexploitation or waste of many resources that could be used directly for food; and in general no consideration of the roles of the ecology and the environment in fishery matters.

These were some of the problems facing the new administration late in 1994. The goal was stated then, and a strong decision was taken, to incorporate environmental variables within fisheries management. This led to many changes. However, before discussing these, some of the relevant concepts regarding sustainability in fisheries need to be examined.

Sustainable fisheries development

The word 'sustainability' in fisheries has been used for a number of years and yet its meaning remains rather unclear. Since the introduction of the term Maximum Sustainable Yield (MSY) by Schaefer in the 1950s the concept was reduced to the allowable annual harvest of a fish stock that would not drive it to depletion (i.e. the surplus production of the stock).

The simplistic graph proposed to show what MSY and its assumptions mean, has been criticized (Larkin 1977; Caddy and Griffiths 1995). However, MSY has frequently been used, and even recommended, as a mandatory target or as a key component for decision makers in management plans (Pitcher and Hart 1982).

The incorporation of economic factors to the logistic model by Gordon led to the concept of the Maximum Economic Yield (MEY) either Static or Dynamic, which indicates the level of effort where economic rent is maximized. This goal as a theoretical concept is much more accepted as a management tool. However, it has been pointed out that awareness has to be observed for assessing the behaviour of harvesters when discounting makes depleting a stock more profitable than conserving it for the future (Caddy and Mahon 1995).

Discussion on having a suitable definition for Sustainable Development in fisheries is far from over but one of the most referred to definitions in literature is :

> sustainable development is to meet the need of the present without compromising the ability of future generations to meet their own needs.

The whole sustainability conceptualization, and particularly its implementation, has itself been distrusted and proposed to be replaced by policies for management taking into account uncertainty (Ludwig *et al.* 1993). Sustainability, regardless of this point of view, remains a common approach to fisheries management and it has to be said that attain it faces problems like natural variability, scientific uncertainty and conflicting objectives. For example, in the USA about 45% of the assessed resources are considered to be overharvested, in Europe this is 59% (Rosenberg *et al.* 1993).

In general, sustainability has not been achieved in many cases because changes in concepts and organization have not been made nor an integrated decision-making framework developed to include all biological, socio-economic, political and operational elements of the fishery (Stephenson and Lane 1995).

Recognizing this complexity, sustainable development has been defined as a multi-faceted process involving four components within an integrated framework (Charles 1994): ecological, institutional, socio-economic and community, sustainability. This process may be approached from five main points of view: living with uncertainty, coping with complexity, improving local control, establishing appropriate property rights, and combining internal planning with suitable external economic diversification.

At the same time, in recent years, a whole new way of examining fisheries management has been developed in the international arena. Concepts like the precautionary approach, adaptive management and the use of reference points and procedural rules have coalesced into new strategies. However, their links to sustainability have yet to be developed in a clear, operational way for developing countries.

Sustainability concepts have been applied successfully for several fisheries in countries including the USA, Canada, New Zealand and some from Europe, where formal risk assessments are now routinely incorporated into many stock evaluations (Rosenberg *et al.* 1993). Even then, there has been a degree of success only at small-scale community level or through private ownership (Hilborn *et al.* 1996); yet the central concept has not been shown to be effective for developing countries.

One of the reasons is that while management in most developed countries has evolved toward systems that control access to fisheries resources, open-access resource use policy is still a major feature of developing countries (Rosenberg *et al.* 1993). Thus, two major steps need to be taken in such cases, particularly for heavily exploited fisheries: to increase local socio-economic and community fishery benefits within resource limitations and to pursue economic diversification by creating non-fishery employment alternatives in order to lessen the impact of regulations (Charles and Herrera 1994).

Given the lack of an operational definition of sustainability in fisheries, particularly for developing countries, the following general questions come to mind: How to reach a balance between development and conservation? How to reach a workable definition of sustainable development in fisheries? How much and what kind of fisheries research are appropriate to reach this goal? How to take legal and management measures in the face of incomplete information? How to make society aware of these problems, and how to manage so that all users have an input into management decisions? The following offers an account of the steps taken by Mexico, and some of the results up to 1996.

Institutional framework

Late in 1994, the new Mexican administration reorganized all sectors dealing with fisheries, natural resources and the environment into a single agency with a cabinet level. Thus, SEMARNAP was created. This move forced the inclusion of environmental variables into fisheries management and made explicit the consideration of sustainability, by joining ecological thinking and fisheries managers under the same roof. M.Sc. Julia Carabias, a respected scientist, was named by the President to be the head of the new Ministry.

This change brought about a major reorganization of the fisheries administration. The management, legal, promotion and development areas stayed under the Undersecretary for Fisheries, at deputy Minister level. Fisheries research is now the sole responsibility of the National Fisheries Institute (INP), also at Undersecretary level. Finally, the key area of enforcement, is neither within management nor linked with research. It is a separate agency whose only duty is to enforce the legal framework. As the management and political areas cannot operate without the scientific advice of the research Institute, no management action can be taken without the backing of scientific advice.

The separation of these three areas guarantees that the scientific body can do research without the additional hassle and the political risk of enforcement or regulation tasks. This increases confidence among producers and discourages political and social pressures, a situation very different to the one so prevalent before in fisheries management. In a similar way, without enforcement or research duties, management is now also detached to some extent from political pressure. The latter was so overwhelming in previous administrations that the specific input of research was very low. The present system creates a healthy balance of power among regulation, research and enforcement issues for fisheries.

The resolution of merging environmental variables, together with the sustainability approach for managing the fishery resources, led in early 1995 to the assessment of the strong and weakest points of the research body, the National Fisheries Institute. This diagnosis clearly indicated the necessity for improvement by two main approaches: on one side there was a need to transform its bureaucratic and administrative structure, and on the other, there was a compromise to transform it into an institution of research-oriented excellence with strong academic ties.

In this way, the programme for academizing the National Fisheries Institute has become one of the most fundamental goals of the Ministry of the Environment, Natural Resources and Fisheries. The strengthening of fisheries research in Mexico and the transformation of the National Fisheries Institute's duties and framework, are summarized in the National Plan for Fisheries and Aquaculture 1995–2000. This compromise has been openly ratified by both President Zedillo and M.Sc. Julia Carabias.

The National Fisheries Institute administrative restructuring has been carried out since mid-1995. INP is now a research entity which depends directly on the Ministry of the Environment. The Institute is now under a Presidential directorate and has wider responsibilities such as research on assessment and management of fishery resources, research on technological development of fishing gear and food processing, as well as in new areas devoted to research on aquaculture and sustainable development. Parallel to this administrative transformation and the attainment of a higher level within the federal government, there is a specific proposal of flexibility for the management of resources.

The new structure and level of the National Fisheries Institute, with more than 400 research scientists and its 17 regional research centres, has given the Institute a stronger influence and a determinant role in the management of fisheries resources.

The new framework has already started to give results for the planning projects, the creation of specific national regulations, the management of fisheries, the National Program for Fisheries and Aquaculture 1995–2000 and in a more direct way, the shrimp fishery, which is the most important fishery of the whole country.

With respect to research excellence, since 1995 we have been making efforts to incorporate within our institution the most modern techniques and the latest methodology for managing fisheries resources to fulfil the duties of a scientific advisory body. Hence, adaptive management, the need for establishing reference points, the explicit consideration of uncertainty, the development of relational data bases, the use of quantitative models and simulation, the principles of responsible fishing and the precautionary approach, are concepts that we embody in our daily work.

Such a transformation toward academic excellence had not been explicitly suggested before. This acknowledgment forces federal research to have closer ties with universities and other centres of research and, at the same time, makes INP follow, in a very accurate way, the characteristics of the whole fishery sector. In this manner, the Institute becomes a type of mediator between the technically-oriented industry requirements and the mostly scientific academic research, keeping the essential elements and values of both.

In fact, the aim of academization has renewed policies for training, hiring, incentives, promotions, publications and the whole quality of research. For example, the publication of 'Ciencia Pesquera' as a very focussed, peer-reviewed journal with top-level international referees, has been recommended. Regarding promotions, hiring and incentives, the research body is revising the academic statutory principles to have a more realistic blueprint in accord with the present schemes of universities and similar institutions.

Training is an issue that has been of the utmost concern from the beginning. We have implemented an adjustable schedule for our staff in order to allow them to complete their studies and to obtain their degrees accordingly. We aim to have at least 15% of our staff pursuing graduate degrees.

Moreover, we have tightened links with universities and research centres from both Mexico and abroad to share experiences and update the technical knowledge of our staff. For example, five researchers have recently returned from studying for 14 weeks at the School of Fisheries of the University of Washington. Dr. Ray Hilborn, an expert on fisheries matters has been at the National Fisheries Institute twice this year to lead workshops specially designed to solve practical problems involved in the management of the shrimp fishery. These workshops were attended by our best scientists from all the regional research centres. Three further workshops are to be held during 1996.

As well as strengthening of ties with universities and research centres by means of joint programmes, it is planned to start a Master's degree on fishery science within the Institute. The quality of our research projects and the circulation of our results will be increased by establishing an evaluation system formed by a board of external referees.

The results of this academic transformation process are already visible; for example, for the coming National Fisheries Congress to be held in La Paz, Baja California Sur, 80 papers from INP have been accepted. This number is much higher than for papers accepted from other non-governmental institutions also devoted to fisheries research.

The shrimp fishery

Until two years ago, decisions regarding opening and closing of the shrimp fishery were subject to extreme social and political pressure. Decision making was therefore a difficult process where science played a relatively minor role. Thus, this fishery is a good example of how the search for new institutions and partnerships is working.

Firstly over the past two seasons, a very intensive sampling system has been pursued, striving to achieve near real-time information. This sampling programme, together with more conventional schemes of research on processing plants and a logbook information system, have been used to provide a reasonably good tracking of the season as it progresses.

Secondly, a training system, using the best expertise available, in this case Dr Ray Hilborn, was also implemented. This produced a strong quantitative modelling approach, with scientists from all over the Institute participating. The modelling approach, which can be used to cope with uncertainty explicitly, is now used to analyse the risks involved in several management strategies.

Thirdly, the set of options available is discussed openly with scientific advisors from both the artisanal and the industrial sectors. Together with alternatives offered by them, a proposal on specific dates and outcomes is made by all scientists involved, in a very responsible and cooperative way. The system still has some pitfalls, as researchers strive to avoid political issues and try to base recommendations on the best science available, but in general it has produced remarkable agreement on tradeoffs between issues of relevance for two very different fishing sectors.

In a fourth stage, the recommendations are reviewed in an open forum, where all sectors involved in the fishery are represented. Thus, the leaders of the private and social sectors, together with the decision-making managers and the financial advisors listen to talks by researchers and choose a specific date (opening/closure) by consensus.

The system makes all sectors share the responsibility for the management decision and gives very little room for surprises, as the social and political leeway is well defined and based on scientific evidence alone. Results have been generally satisfactory and the yields and returns for both artisanal and industrial sectors have been good.

The system is now in its third season, and decision-tables with specific consideration of management risks, on yields and returns, based on specific reference points have been introduced. With this, decision making will be made more rational, and, as long as there are social and political inputs from resource users, we think it will lead more easily towards sustainability, with shared responsibility and openness.

The shrimp management system is only in its beginnings, but the results so far are encouraging and certainly are better than those of the past, when there was just a closed-door arrangement among a handful of people, with very little or no science input at all.

The giant squid fishery

The giant squid fishery in Mexico started in the 1970s off Baja California and Sonora. As for many squid fisheries, it has been rather unstable and difficult to forecast. Although there have been some studies on stock assessment, squid biology remains poorly understood and the migration patterns are not yet well determined.

The Mexican fleet that harvests giant squid is composed of small boats with outboard motors, called 'pangas'. This artisanal fleet fishes within a range of 10 miles off the coast during the night and using lines called 'poteras'.

During the late 1970s and early 1980s joint ventures between high-technology Japanese boats and Mexican firms started to catch giant squid in adjacent Mexican waters due to a biomass bloom. The joint-venture enterprises stopped their operations as the resource did not reappear in commercial quantities for several years.

For the 1995 fishing season, shrimp trawlers joined the giant squid harvest when the shrimp fishery closure came; these vessels were adapted and it was estimated that one trawler had the equivalent fishing power of six pangas. Once the shrimp fishery season re-opened, the trawlers quit the giant squid harvest; however, their influence and profitability resulted in a phenomenal increase in total catches that reached an historical record of almost 40 000 t.

When the shrimp fishery closure of 1996 was approaching, the shrimp trawlers again joined the giant squid fishery, but this time 160 applications for additional permits were submitted to the Federal authorities. At that time 90 trawlers and a large number of pangas were already fishing for giant squid. As the new permit applications were denied, a blockade was set up in Guaymas Bay, the most important landings port in the Gulf of California.

Representatives from the fishing industry, cooperatives and from the Instituto Nacional de la Pesca agreed in a joint meeting to collect funds for a survey cruise in order to directly estimate whether the stock was healthy enough to support such an increase in effort. The objective was to determine this on a scientific basis and with a precautionary approach for an unstable fishery, and fishermen supported it.

Meanwhile, eight scientists from the Regional Research Centre and from the central offices of Instituto Nacional de la Pesca were assessing the abundance of the stock by means of catch and effort data using several standard methods (e.g. depletion, surplus production) as well as cohort analysis. The best-fitted models were then 'anchored' by the biomass estimation from the survey cruise and validated by assessing the risk and the uncertainty of allowing the entire fleet to fish, subject to specific reference points (escape 40%). For this purpose, a decision-table was calculated, incorporating cost and yield into the management process.

The results of these analyses were well received by users, and freed managers and scientists from most political pressure. It was determined that the biomass levels over the short term were enough to sustain the fishery. The key aspect in this case is that industry and artisanal fishermen agreed and supported specific scientific advice, including the precautionary approach and reference points, and management's role was simply to implement the advice, with a more appropriate role for political negotiation.

Tuna and sea urchins

It is worth reviewing briefly the case of the tuna fishery. Because of the tuna–dolphin issue, international pressures and a commercial embargo, this fishery is perhaps the most highly observed fishery in the world. Mexico is a major player in it, with the largest fleet. There is a 100% trip-coverage by scientific observers, who record fishing activities by the hour. So, we know exactly what a boat is doing, in real time and with a degree of detail unheard of in most fisheries. The approach is working for Mexico and other countries, as the embargo appears to be on the verge of being lifted, but the cost is enormous, in money and human resources. This contrasts highly with the situation in many fisheries, especially artisanal ones of the developing world, where not even the number of fishermen or the overall catch are known. Clearly, a set of standards need to be in place so that priorities can be determined as to how much research effort is going to be devoted to each fishery.

Another example worth commenting upon is the sea urchin fishery of the west coast of Baja California. A system of concessions is in place in Mexico for many resources, especially benthic ones such as this echinoderm. The interesting point here is that the highly valuable sea urchin fishery has driven some users to enforce self-imposed regulations very strictly and to even patrol their concessioned territory so that furtive fishing is diminished. These fishermen set harvest rates, limits on size and effort and forecast production a couple of years ahead, and are very vocal when they are treated in the same way as other fishermen, that do not have the same self discipline. As their territory has a large sea urchin — and other benthic resources — biomass, and they are aware of it, they are confident they can take care of their fishery. There are no private property rights yet, but it has some of the features and the situation could certainly lead towards sustainability more easily.

Concluding remarks

Now, after looking at these changes and some examples, let us consider again the question posed at the beginning of this paper: How to achieve sustainable fisheries development in a developing country?

Of course Mexico does not have all the answers, but we believe we are on the right track, in considering explicitly uncertainty, establishing reference points, improving local control and even considering concessions, quotas and other access-limiting instruments. However, this has taken considerable effort, money and political strength, and these three factors do not come together frequently. We have had to re-engineer the whole National Fisheries Institute, at a considerable cost, both political and in resources. The process is ongoing and still very difficult.

One of the major problems is of course money, and where to allocate it to achieve the required research. The key words here are 'required research'. For example, at INP the resource that takes most of the money is shrimp, but after that, money is devoted to two environmental priority non-fishery issues: research and scientific observers to deal with the tuna-dolphin problem, and the sea turtle conservation programmes. Sharks, and a host of finfish and shellfish fishery resources that could potentially be transformed into food and money, come after that. Clearly, research priorities have been heavily influenced more by international pressure than by local needs. I doubt this will lead us towards sustainability.

These two environmental issues offer insight also into another problem: Mexico has promoted in the international arena the need for multilateral agreements and regional organizations when managing resources. An international agreement for the tuna fishery is in operation, and one for sea turtle conservation has almost been agreed upon. The burden of proof in both cases is rightly placed on scientific advice. This is the reason why the tuna embargo is on the verge to be resolved, and why the threat of a shrimp embargo has reduced.

The question is: Who has the capability to offer the very detailed scientific evidence required for management? Only international organizations or developed countries can, at an enormous cost. In both cases, Mexico is fortunate to have started research programmes a long time ago, but most countries cannot say the same. So, how are developing countries going to show their needs regarding these two resources if they do not have the body of evidence required? What is going to happen when international pressure starts focussing on other resources such as sharks, billfish or even regional or local fisheries? Only a handful of research institutions will have the answers at hand or will be able to devote economic resources to investigate them, and they will certainly not be in developing countries.

The problem is not only a lack of money, but also the lack of adequate research programmes and training, and an appropriate local focus. This is magnified by the general lack of papers, international audience and overall prestige by scientists in developing countries. This is in part due to lack of training, but is also the result of biased peer-reviewed systems that do not favour them, and of the fundamentally different nature of their research, that needs to be devoted to basic inventories and routine technical assessments (Gibbs 1995). These, of course have been carried out for a long time in the countries of Europe or North America and usually within to the gray literature. This is a luxury developing countries can ill afford.

One aspect that is often ignored is the fact that ignorance has an enormous virtue: it is completely free. However, the degree of uncertainty should be acknowledged (Orians 1993). Thus information offers marginal value, and there are analytical tools now that can be used to overcome the problems of incomplete evidence. There are sampling schemes and massive research programmes in some developing countries that are relics of similar programmes in industrial countries that have been abandoned a long time ago. Clearly, a waste of effort will not yield more information or open the road to sustainability, and these resources could go to those strategic areas where small amounts of knowledge will have the largest marginal values.

The strategic importance of serious Non-Government Organizations (NGOs) should be noted, particularly in research matters and for an internal counterbalance of political power. However, governments will do well to offer clear rules of procedure, as grass-roots organizations lend themselves easily towards emotional issues. However, we believe that their responsible participation and that from the academic sector at large, will be a key factor for sustainable fisheries development in non-industrial regions, as they encourage governments to act openly and with clear rules.

As in many instances of relationships between countries, it should be clearly stated here that the problems of sustainability concern everyone — recall that over half the world's fish production comes from this region, but the solutions are the responsibility of the individual governments and the particular social and political makeup of each country, as sovereignty and nationalism issues are very sensitive aspects not to be overlooked.

The problems of overcapitalization are a threat to the achievement of sustainability everywhere, but particularly for non-industrial nations, where investment and employment are usually encouraged. We believe the best counteraction is good science and sensible reference points, but the science must be backed by political muscle so that the advice is taken into account. Otherwise, good advice is not heeded. This is by no means easily accomplished. In Mexico, for example, we are at the present wrestling with the pressures from industry over the need to overhaul the shrimp fishing fleet. The key is how to do it without increasing fishing mortality, currently at the upper limit. We are considering a programme of the actual reduction of the number of vessels, but an increase in their individual fishing power.

A similar problem is that of massive infrastructure programmes to open and clean coastal lagoons. They are very attractive because of the claim of increased production. However, many of these programmes have been unsuccessful in the past because of the lack of scientific checks and appropriate balances. These programmes, and also those for open packing and processing plants are now under the scrutiny of the Research Institute. The idea again is not to increase fishing mortality, but to reduce pressures in the future by not planning for the peak of production.

A few more conceptual questions seem to be pertinent to conclude this paper, with an invitation to reflect upon them. For example, we must think really hard whether there is really a fundamental tradeoff between development and the environment in fisheries, and whether there is really an environmental crisis in this area as seems to be in other areas.

Catch statistics seem to point that way, especially regarding wild populations, so it seems that the developmental model the world has followed has led into a dead end, and is now time to change it. The new way to 'ecological modernization' is this concept of sustainable development, poorly defined in fisheries, difficult to implement, and hampered by the lack of legal instruments and traditions such as private property rights, existent in other fields.

So, we should ask ourselves: is there really a new environmental paradigm in fisheries? Are there any alternatives to sustainable development in fisheries? (perhaps bioregionalization?). Are our developing countries preparing for the challenge of globalization, free trade and new development, by focussing on sustainable development?

We must confess that we are sceptical. We believe that as a target, sustainable development will have the virtue of shifting our thinking, and our organization's research capabilities, towards objectives that are sensible under the present circumstances, such as the precautionary approach. On the other hand, exaggeration of risk assessment, and the virtues of considering uncertainty may hamper processes of development, or may shift even more the scientific capabilities of developing countries towards the needs of developed countries.

Mexico is committed to the objective of sustainability. Results to date are promising, but we must consider how many developing countries will really have the capabilities to change toward sustainable development, with all its promises and its share of uncertainties.

ACKNOWLEDGMENTS

The authors wish to thank M Sc Alonso Aguilar for his insights, which led us to offer a more balanced look at the sustainability problem in fisheries.

REFERENCES

Caddy, J F, and Griffiths, R C (1995). Living marine resources and their sustainable development: some environmental and institutional perspectives. *FAO Fisheries Technical Paper* 353. 167 pp.

Caddy, J F, and Mahon, R (1995). Reference points for fisheries management. *FAO Fisheries Technical Paper* 247. 83 pp.

Charles, A T (1994). Towards sustainability: the fishery experience. *Ecological Economics* **11**, 201–11.

Charles, A T, and Herrera, A (1994). Development and diversification: sustainability strategies for a Costa Rican fishing cooperative. *Proceedings of the 6th Conference of the International Institute for Fisheries Economics and Trade. IFREMER, France,* 1315–24.

Gibbs, W W (1995). Information have-nots. *Scientific American* (May), 8–9.

Hilborn, R, Walters, C J, and Ludwig, D (1996). Sustainable exploitation of renewable resources. *Annual Review Ecological System* **26**, 45–67.

Larkin, P (1977). An epitaph for the concept of maximum sustainable yield. *Transactions American Fisheries Society* **106**, 1–11.

Ludwig, D, Hilborn, R, and Walters, C (1993). Uncertainty, resource, exploitation and conservation: lessons from History. *Science*, **260**, 17–36.

Orians, G H (1993). Ecological concepts of sustainability. *Environment* **32**(9), 10–15, 34–9.

Pitcher, T, and Hart, P B (1982). Fisheries Ecology. Croom Helm, London and Canberra. 414 pp.

Rosenberg, A A, Fogarty, M J, Sissenwine, M P, Beddington, J R, and Shepherd, J G (1993). Achieving sustainable use of renewable resources. *Science* **262**, 828–9.

Stephenson, R L, and Lane, D E (1995). Fisheries management science: a plea for conceptual change. *Canadian Journal of Fisheries and Aquatic Sciences* **52**, 2051–6.

Estimating Fisheries Impacts Using Commercial Fisheries Data: Simulation Models and Time Series Analysis of Hawaii's Yellowfin Tuna Fisheries

Xi He[A] *and Christofer H. Boggs*[B]

[A] Joint Institute of Marine and Atmospheric Research, School of Ocean and Earth Science and Technology, University of Hawaii, Honolulu, Hawaii 96822 USA.
[B] Honolulu Laboratory, Southwest Fisheries Science Center, National Marine Fisheries Service, NOAA, Honolulu, Hawaii 96822 USA.

Summary

This study examines a tool that can be used to estimate fisheries impacts on fish abundance when only limited commercial fisheries data are available. Simulation models were used to produce time series of catch and catch-per-unit-effort (CPUE) under different scenarios of fishing mortality and variability in catchability (VC). The simulated data and real data were then analysed by transfer function models (TFMs). The TFMs had different patterns at low and high fishing mortalities. The TFM results from the simulation models and the real fisheries data were compared, and the power of using TFMs to detect local fishery impacts at different VC levels was evaluated. The type of TFM that fitted the real data most frequently was characteristic of simulated data under low fishing mortality. This study indicated that total catches by the Hawaii fisheries had a low probability of affecting local abundance of yellowfin tuna.

Introduction

Commercial fisheries data are often used to evaluate fisheries impacts on fish populations since they are often the only data available for fisheries scientists and managers. However, fisheries data are often limited to crude catch and catch-per-unit-effort (CPUE) statistics, which do not provide enough information for reliable estimation of the population parameters required to evaluate fisheries impacts using traditional stock assessment methods (Hilborn and Walters 1992). New methods are needed to maximize the usefulness of commercial fisheries data for estimating fisheries impacts on exploited fish populations or on locally exploited fractions of those populations.

Yellowfin tuna (*Thunnus albacares*) is exploited by several of Hawaii's commercial and recreational fisheries. Yellowfin tuna is widely distributed and the locally exploited resource is closely related to stocks across the Pacific Ocean (Boggs and Ito 1993; Skillman *et al.* 1993; Suzuki 1994). Migration patterns of this species between the Hawaii exclusive economic zone (EEZ) and other Pacific regions are poorly understood, as are other population parameters such as natural mortality and catchability (Suzuki 1993). Catches of yellowfin tuna have composed an important fraction of Hawaii's total fish catch since the development of the Hawaii-based longline fishery in the 1920s and the expansion of the small-vessel troll and handline fisheries in the 1970s (Boggs and Ito 1993; Pooley 1993). Recently, rapid expansion of the Hawaii longline fishery has caused concern about whether local catches may cause local depletion of this species (Boggs and Ito 1993; Boggs 1994).

This study had two purposes. First, we examined a tool that can be used to estimate fisheries impacts on relative abundance of fish populations when only limited commercial fisheries data are available. This tool can be applied to other fisheries. Second, we applied this tool to estimate local fisheries impacts on local abundance of the yellowfin tuna resource within the Hawaii EEZ.

We used four steps in our analysis. First, we used simulation models with seasonal and stochastic variation in immigration and fishing effort to produce time series of total catch and CPUE under different scenarios of two key fishery parameters: fishing mortality and catchability. Second, the time series were analysed by transfer function models (TFMs). Third, TFMs were also applied to real commercial fisheries data from Hawaii's yellowfin tuna fisheries. Finally, the TFM results from the simulation models and the real fisheries data were compared, the power of using TFM to detect local fishery impacts was evaluated, and a tentative conclusion was reached regarding the probability of local fisheries impacts on local fish abundance.

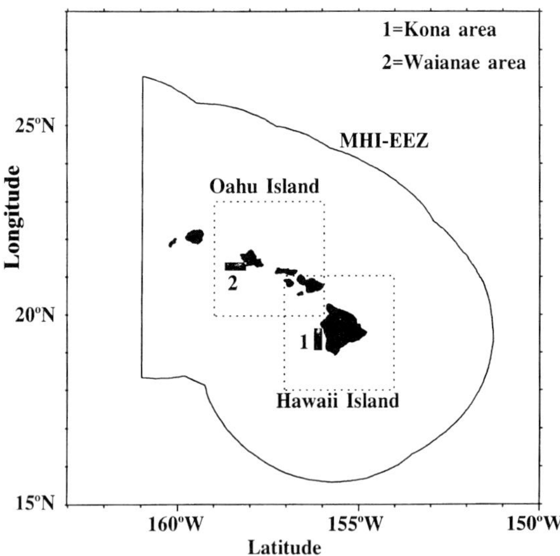

Fig. 1. Map showing three spatial scales and five areas where total catches and catch-per-unit-effort were calculated for the time series analysis. Black areas indicate the Main Hawaiian Islands. MHI-EEZ = Main Hawaiian Islands Exclusive Economic Zone.

METHODS AND MATERIALS

Data sources

Commercial fisheries data used in this paper included (1) domestic total catch of yellowfin tuna from longline, handline, and troll reported to the Hawaii Department of Aquatic Resources (HDAR) from 1962 to 1994 (augmented by data collected by the Honolulu Laboratory of the National Marine Fisheries Service), (2) longline catch by Japanese fishing vessels in the main Hawaiian Islands EEZ (MHI-EEZ) from 1962 to 1980 (Yong and Wetherall 1980), and (3) CPUE for the Hawaii domestic handline and troll fisheries derived from the HDAR data from 1962 to 1994. Total catch was calculated by summing domestic and Japanese catches. There were no Japanese fishing activities within the MHI-EEZ after 1980. Reliable estimates of total effort for all fisheries combined could not be computed owing to lack of information on the relative efficiency of different gears. Details on calculating total catches and CPUEs are presented elsewhere (He and Boggs 1996).

Time series of monthly total catch (all gears) and CPUE (troll and handline) from 1962 to 1994 were used in the analysis. CPUE was defined as biomass caught (kg) per trip (usually a day) for handline and troll. Time series of catch and CPUE were calculated for five areas at three spatial scales (Fig. 1). The largest-scale area was the MHI-EEZ. Two medium-scale areas were 3 × 3 nautical-degree boxes around the island of Hawaii and the island of Oahu. The smallest-scale areas (shaded areas in Fig. 1) were two 20 × 40 nautical-minute boxes off the Kona coast of Hawaii and off the Waianae coast of Oahu, which are among the most intensively fished areas in Hawaii's tuna fisheries. Both handline and troll CPUE were calculated for the MHI-EEZ and the two Hawaii Island areas, but only troll CPUE was calculated for the two Oahu areas where handline fishing is infrequent. During the early years some areas and gears had no data for several months so these years were truncated from the time series.

Simulation models

The simulation model was set up for the MHI-EEZ similar to Boggs' (1994) simulation:

$$\frac{dA}{dt} = I - qFA - MA \tag{1}$$

where A = biomass (t within the local fishery), I = immigration (i.e. 10 000 t/year), q = catchability (i.e. 0.0001/unit of effort), F = fishing effort (arbitrary effort units) and M is natural mortality and emigration (combined, i.e. 0.6/year).

The model assumed that there was no local recruitment, and biomass was supplied only by immigration. The equivalent assumption would be that recruitment and immigration were analogous, combined processes independent of local biomass (A). For each model scenario, annual immigration and fishing effort were modelled as constants subjected to stochastic changes from year to year. Stochastic processes were modelled as standard normal distributions, with standard deviations of 0.1 and 0.25 for annual immigration and fishing effort.

The seasonal patterns of immigration, fishing effort, and natural mortality for the simulation were modelled as monthly proportions of the annual values (Fig. 2). Seasonal patterns of fishing effort were matched to patterns observed in the fishery (i.e. a late summer peak). The model assumed that immigration was low from January to March, was high from April to May, and then decreased from June to December. The model also assumed that natural mortality and emigration (combined) were lowest from January to June, highest from July to September, and at an intermediate level from October to December. These seasonal patterns of immigration and natural mortality (Fig. 2)

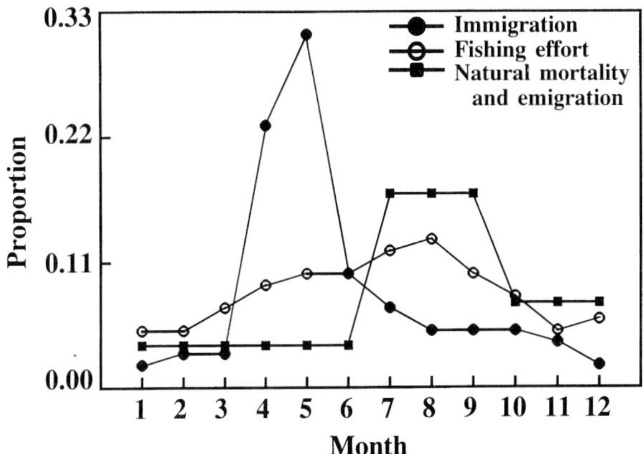

Fig. 2. Monthly immigration, fishing effort, and natural mortality and emigration used in the simulations, modelled as proportions of the annual values.

were chosen so that the simulation model produced seasonal patterns of CPUE which matched seasonal patterns observed in the fishery. Stochastic variation was also applied to the monthly proportions of annual immigration and effort, which were modelled as standard normal distributions with standard deviations of 0.20 and 0.25.

The simulation model was run at different scenarios of annual fishing effort and monthly variability in catchability (VC). Between scenarios, annual fishing effort ranged from 1000 to 19 000 effort units at increments of 3000 units, which was equivalent to an annual local fishing mortality (summed over 12 months) of 0.10 to 0.85. Detectable fishery impacts are expected when fishing mortality rates (F) are relatively high (i.e. high F/M ratio). The range of fishing effort was chosen so that annual F/M had a wide range (0.17 to 1.3). Between scenarios the VC ranged from 0 to 0.6 at increments of 0.1. A VC of 0 represented constant catchability, whereas a VC of 0.6 represented catchability with a mean monthly deviation of 60%.

The simulation models were run at a time step of 1 month for 60 years. Because of the many stochastic processes in the model, 3000 simulations were run for each scenario. For each run, the simulated monthly time series of catch and CPUE from the last 30 years of the model were used for the time series analysis.

Time series transfer function models (TFM)

Transfer function models (TFM) can be viewed as modelling relationships between anomalies of input and output variables (Wei 1990; Bowerman and O'Connell 1993; Carpenter 1993). The anomalies are the remaining time series after serial dependencies and seasonal trends in the original series are removed by ARIMA (Autoregressive Integrated Moving Average) models (Wei 1990). If anomalies in the output variable are not explainable by their own series dependency or seasonality but follow anomalies in the input variable, then there exist relationships between the input and output variables, and a TFM can be fit.

The general form of the simplified linear transfer function can be represented as:

$$y_t = \frac{\omega_s(B)B^b}{\delta_r(B)} x_t + \eta_t \tag{2}$$

where t is time, y is the time series for the output variable, x is the time series for the input variable, b is a delay parameter representing the actual time lag that elapses before the input variable produces an effect on the output variable, η_t is the residual model for y_t (see Equation 5), and B is the backshift operator often used in time series models with the property that $B^bZ_t=Z_{t-b}$ (where Z is any variable). $\omega_s(B)$ and δ_r are parameters of the transfer function that represents linear and non-linear effects of the input variable on the current value of the output variable, and s and r are time units extended to the past. $\omega_s(B)$ can be written as:

$$\omega_s(B) = \omega_o - \omega_1 B^1 - \omega_2 - \ldots - \omega_s B^s \tag{3}$$

and δ_r can be written as:

$$\delta_r(B) = 1 - \delta_1 B^1 - \delta_2 B^2 \ldots - \delta_r B^r \tag{4}$$

The residual model for y_t, ω_t, can be written as:

$$\eta_t = \frac{(1-\phi_1 B - \ldots \phi_p B^p)(1-\phi_{12}B^{12})a_t}{(1-\theta_1 B - \ldots \theta_q B^q)(1-\theta_{12}B^{12})} \tag{5}$$

where the numerator describes a moving average (MA) process and the denominator describes an autoregressive (AR) process, a_t is a zero mean white noise process, and ϕ_p and θ_q are parameters representing an MA process at the order of p and an AR process at the order of q, respectively (Wei 1990). The second factors in both the numerator and denominator represent the seasonality of 12 months in both the MA and AR processes.

In our models, the input variables (x_t) were total catches and the output variables (y_t) were CPUE values, which were assumed to be representative of the relative local abundance of yellowfin tuna. Because the transfer function models require time series to be stationary but both total catch and CPUE were not, all time series data were differenced by 12 months in the model. The differencing not only transforms the time series into a stationary series but also removes the seasonality from the time series (Cohen and Stone 1987; Wei 1990).

The SAS Statistics Program, Version 6.09 (SAS Institute Inc. 1993) was used for the construction of all TFMs and estimation of all parameters. Detailed procedures on construction of models and estimation of parameters can be found in Wei (1990), Bowerman and O'Connell (1993), and SAS Institute Inc. (1993). In fitting TFMs, parameters with significant T-ratios ($|T\text{-ratio}| \geq 1.96$) were obtained, and the Schwartz's Bayesian Criterion (SBC) was provided by the SAS program to measure the goodness-of-fit. In situations where several models with different parameters provided significant fits to the same data, the model with the smallest SBC was chosen.

Although TFMs can be complex, they can be categorized into five types (Table 1). The 'no effect' type, in which all parameters are not significantly different from zero, indicates no effect of the input variable (catch) on the output variable (CPUE). That

Table 1. Five types of transfer function models that were used to categorize results of fitting simulated fisheries data

Type of model	ω_0	$\omega_1, \omega_2 \ldots \omega_s$	$\delta_1, \delta_2, \ldots \delta_r$
No effect	0	0	0
Lag 0 effect	>0	≤0	0
Linear	>0	<0	0
Nonlinear effect	>0	any value	≠0
No fit	Model cannot be fitted		

is, catch has no relationship to CPUE. The 'lag 0 effect' type, in which $\omega_0 > 0$ or $\omega_1, \omega_2 \ldots \omega_s \leq 0$, indicates either that there is a positive correlation between the input (catch) and the output (CPUE) variables at time lag of zero, or that there are positive correlations between the past values of catch and the current values of CPUE. In both cases, catches are considered to have no impact on the fish abundance represented by the CPUEs, because (1) positive correlations between current values of total catch and CPUE can result from a simple lack of independence between the variables (CPUE = catch/effort) in the absence of other effects, and (2) although past or current removals of fish (catches) may have positive correlations with current fish abundance (CPUE), fish removals cannot be interpreted as *causing* high fish abundance.

The 'linear effect' type of TFM, in which only $\omega_s > 0$, indicates that past values of the input variable (catch) are negatively correlated to the current value of the output variable (CPUE). That is, high catches in the past are associated with low current CPUEs, or vice versa. This would be the obvious type of TFM to expect if fishing were having an impact on local fish abundance. The 'nonlinear effect' type, in which $\omega > 0$, $\omega_1, \omega_2 \ldots \omega_s$ equal any value, and $\delta_1, \delta_2, \ldots \delta_r \neq 0$, indicates that there is a nonlinear relationship between past values of the input variable and the current value of the output variable. The 'no fit' type, in which the model cannot be fitted to the data, indicates a failure of statistical procedures or that there might be more complex relations between the input and the output variables.

Time series from both real data and the outputs of the simulation models were analysed using the TFMs and were categorized into the above five types. These results were then compared. The probabilities of different types of TFM fits at known *F/M* levels from the simulation models were used to evaluate the power of using TFMs to detect fisheries impacts using the real data.

Calculations of variability in catchability

The variability in catchability (VC) describes the monthly stochastic changes in catchability (q, Equation 1) in the simulation models. If the VC is large, CPUE will not be representative of the relative abundance of local biomass (A) since CPUE = qA. The simulation models were run under various levels of VC for comparison with real time series, so it was important to try to estimate VC in Hawaii's handline and troll fisheries. For each month and gear, individual VC estimates can be approximated from observed CPUE data by using the equation:

$$\text{VC} = \frac{S_e \text{ of CPUE}}{\textit{Mean}\ \text{CPUE}} \qquad (6)$$

by assuming that all of the within-month variation in CPUE is due to q. The standard error (S_e) is the appropriate statistic to estimate the deviation of the monthly mean for comparison with the simulated stochastic monthly VC.

We used handline data from Kona and troll data from Waianae in the VC estimation since both areas were among the most intensively fished. An individual VC was estimated only if there were ≥30 observations. 275 individual VCs were estimated for the Kona handline fishery and 283 for the Waianae troll fishery. Mean VCs for each fishing gear were calculated by averaging all individual VCs. It should be noted that VC values estimated in this way are certainly overestimates of the true values of VC, which cannot be estimated accurately from the current data because the data contain variability in fishing behaviour, as well as variability in within-month and within-area fish abundance and other kinds of variability.

Results

Hawaii's handline and troll fisheries

All time series of total catch and CPUE showed strong seasonality in all years, and high catches and high CPUE occurred during the summer months (Figs 3 and 4). For the MHI-EEZ (Fig. 3), total catches were relatively low during the 1960s and high from the late 1970s to the 1990s. Handline CPUE exhibited a pattern similar to that observed for the total catch. The troll CPUE fluctuated with highs in the early 1960s, middle to later 1970s, and middle 1980s and with lows in the middle 1960s to the early 1970s, in the early 1980s, and in the late 1980s to early 1990s.

For the Island of Hawaii (not illustrated here) the overall patterns were similar to those for the MHI-EEZ. For troll CPUE, there was a more pronounced decrease from the high seen in the middle 1970s to the low in the late 1980s and early 1990s.

For the Kona area both total catch and handline CPUE were low from the 1960s to the early 1970s, high from the middle 1970s to the middle 1980s, and low from the late 1980s to the early 1990s. A similar pattern was observed for troll CPUE except for a more pronounced decrease from the late 1980s to the early 1990s.

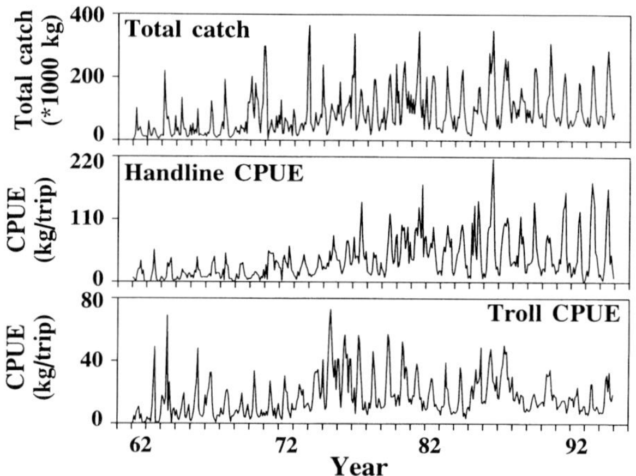

Fig. 3. Time series of monthly total catch, handline catch-per-unit-effort, and troll catch-per-unit-effort of yellowfin tuna for the Main Hawaiian Islands Exclusive Economic Zone from 1962 to 1994.

For the Island of Oahu (Fig. 4), total catch was low throughout the 1960s and higher throughout the 1970s. The highest catches were observed in 1980, 1981, 1989 and 1990, and low catches were observed from 1991 to 1994. Troll CPUE was low from the 1960s to the early 1970s, high from the middle 1970s to the middle 1980s, and low from the middle 1980s to the early 1990s, showing some similar patterns as observed for the MHI-EEZ.

For the Waianae area (Fig. 4), total catches fluctuated, but were particularly low during periods of the early 1980s and the early 1990s. Troll CPUE was low in the late 1980s and early 1990s and showed an exceptional high in November 1985.

Fitting the TFM to the above time series showed that the 'lag 0 effect' type of model best fitted all time series (Table 2), except the troll CPUE for the MHI-EEZ, which was best fitted by the 'linear effect' type of model.

Analysis of the simulation model outputs

After the simulation model was run 3000 times for each scenario the output from each run was analysed using the TFMs and categorized as one of the five types (Table 1). The proportions of each type of TFM out of the 3000 best fits for each scenario were plotted as response surfaces (Fig. 5). The 'no effect' type of fit occurred only in the lowest fishing effort and VC scenario (response surface not illustrated). The response of the 'lag 0 effect' to changes in fishing effort and VC was complex (Fig. 5). In general, when the VC was low, the proportion of the 'lag 0 effect' decreased exponentially to very low values as the fishing effort increased. When the VC was high, the proportion of the 'lag 0 effect' decreased gradually and more linearly as the fishing effort increased. At high effort levels, the proportion increased as VC increased, but at low effort levels the response to increased VC was sinusoid (Fig. 5).

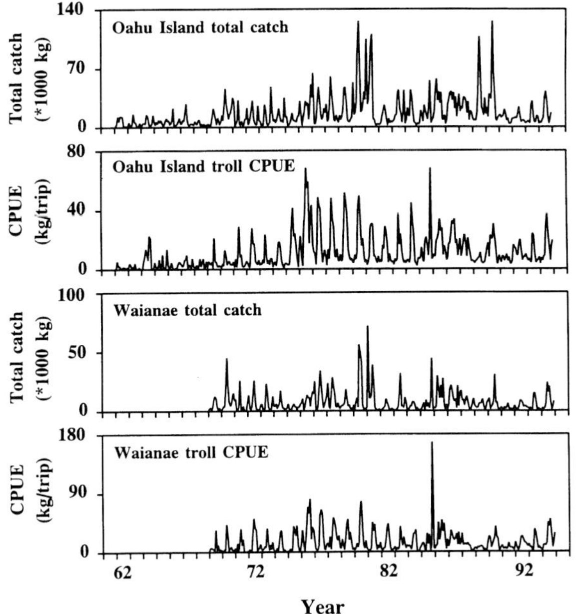

Fig. 4. Time series of monthly total catch and troll catch-per-unit-effort of yellowfin tuna for the Island of Oahu and the Waianae area from 1962 to 1994.

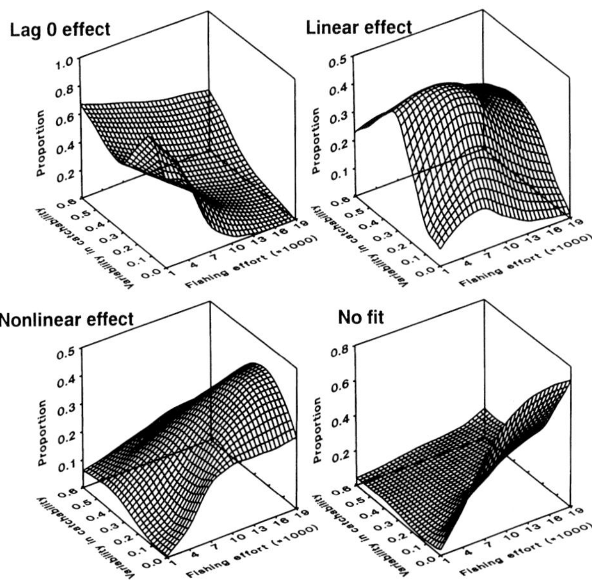

Fig. 5. The proportion of four types of transfer function model fits in relation to fishing effort and variability in catchability, calculated from 3000 simulation runs. Fishing effort (in arbitrary units) was varied to simulate a range in *F* from 0.1 to 0.85.

The proportion of 'linear effect' fits was highest at the intermediate levels of VC, low at high levels of VC, and lowest at low levels of VC (Fig. 5). This same general pattern occurred at all levels of fishing effort, but as effort levels increased the proportion of linear fits first increased and then decreased.

In contrast, the proportion of 'non-linear effect' fits continuously increased as fishing effort increased at all levels of VC (Fig. 5). These results suggest that at high relative fishing mortality (i.e. high *F/M*) the impacts of fishing on subsequent CPUE progressed from linear effects to non-linear effects. The proportion of non-linear fits was highest at intermediate levels of VC. The 'no fit' proportion was highest when VC was low and fishing effort was high (Fig. 5).

Comparisons of TFMs between simulation models and data

The analysis showed that the 'lag 0 effect' type of TFMs best fitted the real data time series in all but one case. In contrast, all five types of TFMs fitted simulated time series with the proportions of each type depending on the scenario. The analysis of simulated data showed that the probability of the 'lag 0 effect' type fits (i.e. the proportion of 'lag 0 effect' fits) under a given scenario depended on both the level of fishing effort and VC. As shown in the summary plot (Fig. 6), when both the fishing effort and VC were low, the 'lag 0 effect' was by far the most likely type of TFM to fit the time series. And when the fishing effort was high and VC was low, the 'non-linear effect' and the 'no fit' types were most likely to result (Fig. 6A). This suggests that the power to detect scenarios with high fishing mortality and substantial fisheries impacts was high when VC was low. However, when VC was high, there was much less contrast between the probabilities of fitting different types of TFMs as the fishing effort increased, and therefore the power to detect fisheries impacts was very low (Fig. 6B).

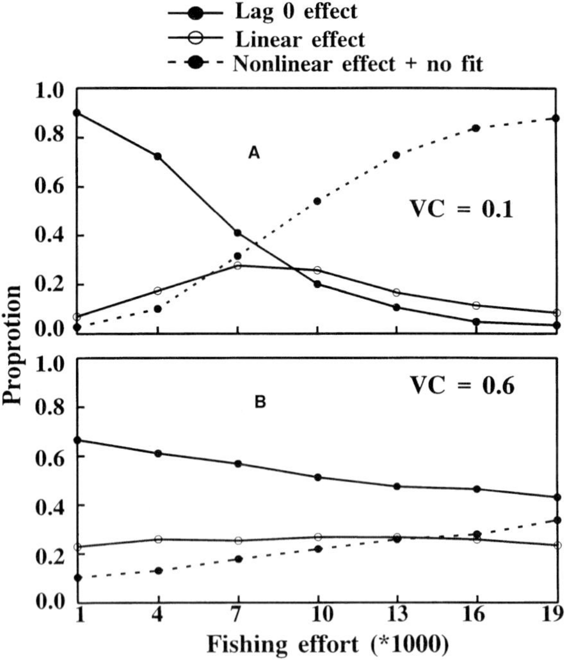

Fig. 6. Summary plot of the proportion of the four types of transfer function model fits in relation to fishing effort at two levels of variability in catchability (VC). The plot is extracted from Figure 5. Fishing effort (in arbitrary units) was varied to simulate a range in *F* from 0.1 to 0.85.

Mean VCs estimated for the Kona handline fishery and the Waianae troll fishery were 0.186 and 0.299, respectively. If it is assumed that the true values of monthly mean VC for the two fisheries are lower than these estimated values, then the power of using TFMs to detect fisheries impacts was high. Moreover, in 7 out of 8 cases (Table 2) the 'lag 0' type TFM best fit the real data, implying that local catch had a low probability of impacting the local abundance of yellowfin tuna.

Discussion

This study found that TFMs can be useful tools for estimating fisheries impacts when only limited commercial fisheries data are available. The basic requirements for using TFMs are that both time series of catch and CPUE are long term, probably at least five life cycles of the species, and that variation in catchability and other variation in measurement of CPUE is not too great. As shown in the analysis of the simulation model output, TFMs can detect fisheries impacts even when there is realistic variation. In real data, variation in q is confounded with error in catch and effort reporting, variation in fishing operations, and variation in availability.

The problem in obtaining long-term and low-error indices of abundance using data from most commercial fisheries is that fishing techniques and fisherman behaviour change over both long and short time scales, often with little documentation. This certainly occurred in Hawaii's yellowfin tuna fisheries. In our analysis, we used several methods to reduce CPUE variation. First, we used a subgroup of CPUE data which excluded data reported from newly found fishing grounds (i.e. seamounts); data from reports with extremely high catches (assumed recording errors); and data from incomplete or inconsistent records, such as monthly catches that appeared to be summed as single trip reports, fishing trips using multiple gears, and inconsistently reported zero catch trips (Curran, Yang and Boggs unpublished; He and Boggs 1996). This subgroup of data exhibits less variation in CPUE than the whole data set. Second, we used the differencing technique in the time series analysis, which removed not only seasonality but also long-term trends in the time series (Cohen and Stone 1987; Wei 1990), such as those that may be caused by improvements in fishing methods over the years.

We used catch as the input variable in the time series analysis instead of the more traditional fishing effort variable because catch data were much more accurate and straightforward to obtain than effort data. This is often true for multigear fisheries, where total effort is difficult, if not impossible, to standardize. In contrast, catch can be easily summed from all fisheries and can often be obtained from or compared with data from fish markets.

The simulation model used in this study had some shortcomings. For example, by assuming all biomass was supplied through seasonal immigration, the model did not include any biomass-dependent stock–recruitment relationship. Inclusion of biomass-dependent recruitment may induce stronger series dependency and higher variations in the time series. Whether the TFMs would have similar power to detect fisheries impacts under such scenarios is under further examination. Also, the model was a single-box fishery model, ignoring spatial interactions within the box. Future analysis certainly needs to include fish movement between boxes, spatial heterogeneities, and environmental factors affecting catchability.

The conclusion of this study, that total catch by Hawaii's fisheries had a low probability of affecting local abundance of yellowfin tuna, is similar to that of the study by He and Boggs (1996), which used a similar TFM analysis but a different subset of data and two temporal scales. Perhaps a more conservative but stronger conclusion would be that fisheries impacts on local abundance of yellowfin tuna at past and current levels of effort cannot be detected.

Table 2. Results of fitting transfer function models (TFMs) to time series of total catch and catch-per-unit-effort (CPUE) for handline and troll fisheries at different spatial scales in Hawaii's yellowfin tuna fisheries. The input/output column indicates the spatial scales (Fig. 1) for the input variables (total catch) and output variables (CPUE). (MHI-EEZ = Main Hawaii Islands Exclusive Economic Zone)

	Fishing gear	
Input/output	**Troll**	**Handline**
MHI-EEZ/MHI-EEZ	Linear effect	Lag 0 effect
Hawaii/Hawaii	Lag 0 effect	Lag 0 effect
Kona/Kona	Lag 0 effect	Lag 0 effect
Oahu/Oahu	Lag 0 effect	No data
Waianae/Waianae	Lag 0 effect	No data

ACKNOWLEDGMENTS

The data were provided by the Hawaii Department of Aquatic Resources and the Honolulu Laboratory of the National Marine Fisheries Service. Many people assisted in this work, especially Jenny Zhou and Dan Curran of the Joint Institute of Marine and Atmospheric Research (JIMAR), University of Hawaii School of Ocean and Earth Science and Technology (SOEST). This work was funded, in part, by a cooperative agreement with the National Oceanic and Atmospheric Administration (NOAA Coop. Agr. No. NA37RJ0199) through the JIMAR Pacific Pelagic Fisheries Research Program. This paper is JIMAR publication No. 97-310 and SOEST contribution No. 4397. The views expressed herein are those of the authors and do not necessarily reflect the views of NOAA or any of its subagencies.

REFERENCES

Boggs, C H (1994). Methods for analyzing interactions of limited range fisheries: Hawaii's pelagic fisheries. In 'Interactions of Pacific tuna fisheries, Volume 1 — Summary report and papers on interaction'. (Eds E S Shomura, J Majkowski and S Langi) pp. 74–91 (*FAO Fisheries Technical Paper* 336/1, Rome, Italy).

Boggs, C H, and Ito, R Y (1993). Hawaii's pelagic fisheries. *Marine Fisheries Review* **55**, 69–82.

Bowerman, B L, and O'Connell, R T (1993). 'Forecasting and time series: An applied approach.' (Duxbury Press: California).

Carpenter, S R (1993). Statistical analysis of the ecosystem experiments. In 'The trophic cascade in lakes'. (Eds S R Carpenter and J F Kitchell) pp. 26–42 (Cambridge University Press: New York).

Cohen, Y, and Stone, J N (1987). Multivariate time series analysis of the Canadian fisheries system in Lake Superior. *Canadian Journal of Fisheries and Aquatic Sciences* **44**(Supplement 2), 171–81.

He, X, and Boggs, C H (1996). Do local catches affect local abundance? Time series analysis on Hawaii's Tuna Fisheries. In 'Proceedings of the Second FAO Expert Consultation on Interactions of Pacific Tuna Fisheries'. (Eds R S Shomura, J Majkowski, and R F Harman) pp. 224–40 (*FAO Fisheries Technical Paper* 365, Rome, Italy).

Hilborn, R, and Walters, C J (1992). 'Quantitative fisheries stock assessment.' (Chapman and Hall: New York).

Pooley, S G (1993). Hawaii's Marine Fisheries: Some history, long-term trends, and recent developments. *Marine Fisheries Review* **55**, 7–19.

SAS Institute Inc. (1993). SAS/ETS users guide, Version 6, 2nd edn (SAS Institute Inc.: North Carolina, USA).

Skillman, R A, Boggs, C H, and Pooley, S G (1993). Fishery interaction between the tuna longline and other pelagic fisheries in Hawaii. *NOAA Technical Memorandum NMFS* **189** (National Marine Fisheries Service, NOAA, Honolulu, Hawaii, USA).

Suzuki, Z (1994). A review of the biology and fisheries for yellowfin tuna (*Thunnus albacares*) in the western and central Pacific Ocean. In 'Interactions of Pacific tuna fisheries, Volume 2 — Papers on biology and fisheries'. (Eds E S Shomura, J Majkowski and S Langi) pp. 108–37 (FAO Fisheries Technical Paper, 336/2, Rome, Italy).

Wei, W W S (1990). 'Time series analysis: Univariate and multivariate methods'. (Addison-Wesley Publishing Company: California).

Yong, M Y Y, and Wetherall, J A (1980). Estimate of the catch and effort by foreign tuna longliners and baitboats in the fishery conservation zone of the central and western Pacific, 1965–77. *NOAA Technical Memorandum NMFS* 2 (National Marine Fisheries Service, Honolulu, Hawaii, USA).

Intrinsic Quality and Fisheries Management: Bio-economic Analysis of the Pacific Whiting Fishery

G. Sylvia,[A] *S. L. Larkin*[B] *and M. Morrissey*[C]

[A] Oregon State University, Coastal Oregon Marine Experiment Station, Hatfield Marine Science Center, Newport, Oregon 97365, USA.
[B] Oregon State University, Agricultural and Resource Economics, Corvallis, Oregon 97331, USA.
[C] Oregon State University, Seafood Laboratory, Astoria, Oregon 97103, USA.

Summary

The management of wild-stock fisheries often overlooks physiological changes caused by maturation, spawning, migration and feeding. For many marine species, such changes impact condition factors, proximal content, and organoleptic properties. In turn, these intrinsic factors affect harvest yields, processor recovery rates, and the market value of harvested and processed products. The Pacific whiting (*Merluccius productus*) fishery serves as a case study to illustrate how intrinsic quality affects seasonal harvest strategies (including the allocation to competing harvest groups) which, ultimately, impact the management goals of resource conservation, maximum economic value, and full utilization. The long-run bio-economic programming model advances the realism of applied analysis by incorporating stochastic recruitment, intra-annual growth, heterogeneous harvest groups, and hedonic equations to capture intra-seasonal (in-season) price effects. Findings demonstrate that by considering the in-season variability in intrinsic quality and its impact on optimal harvest strategies, the management goals are complementary.

Introduction

Many marine species experience considerable physiological changes caused by maturation, spawning, migration, and feeding (Love 1988). Management of wild-stock fisheries often overlooks such changes which can impact condition factors, proximal content, and organoleptic properties. In turn, these intrinsic factors affect harvest yields, processor recovery rates, and the market value of harvested and processed products. In many cases there will be significant management implications if the intrinsic quality of the stock does not remain homogeneous throughout the management period (i.e. during the harvest season).

Existing theoretical and empirical literature concerning the optimal timing of harvests incorporates weight gain as the sole measure of stock growth (Anderson 1989; Christensen and Vestergaard 1991; Blomo *et al.* 1978; Kellogg *et al.* 1988; Clark 1990; Önal *et al.* 1991). This simple representation, however, may not be appropriate for wild-stocks that exhibit considerable in-season change or consist of heterogeneous cohorts. In addition, these studies assume that heavier individuals receive a higher unit price. Although the value of the stock may increase over time due to weight and associated price increases, greater value also may be related to the improvement in flesh composition as stocks are allowed to recover from spawning and/or migration.

The Pacific whiting (*Merluccius productus*) fishery serves as a case study to illustrate the importance of intrinsic quality. Pacific whiting is the largest groundfish stock located off the west coast of North America south of Alaska (Pacific Fisheries Management Council (PFMC) 1993). Contemporary management by the United States (which harvests over 70% of the annual total allowable catch) allocates annual quotas among competing harvest sectors and determines the season opening date (which has been set to immediately follow spawning and migration; PFMC 1993). A licence programme limits entry but the presence of large-capacity harvest sectors generates 'Olympic-style' or 'pulse' fisheries (i.e. annual quotas are captured within weeks). This compressed early-season fishery has raised concerns that the quality and yield of processed products (including surimi, fillets, headed and gutted, and meal) are compromised due to the poor intrinsic quality of the raw product.

A bio-economic programming model is used to evaluate how alternative strategies impact the three primary goals of the Groundfish Management Plan (PFMC 1993), including (1) resource conservation, (2) high economic value, and (3) full resource utilization. The model selects monthly harvest rates and property rights allocations which maximize net present value (NPV) subject to resource conservation. The optimal or 'proposed' solution is compared to a baseline which represents the contemporary management strategy. The objectives of this analysis are to quantify the management effects of seasonal changes in intrinsic quality and to quantify the improvement in meeting each management goal that results from incorporating seasonal quality.

MODEL

The management-level model represents the contemporary annual-based management of the Pacific whiting fishery. The model attempts to maximize the present value of the stream of net benefits that accrue to the fishery over a 10-year period. The maximized value of the objective function and, therefore, optimal values of the variables defining the seasonal harvest pattern, are determined simultaneously by equations defining both population dynamics and the economics of the fishery. Consistent with contemporary management practice, the remaining management goals of conservation and utilization are embedded *via* a lower bound on annual spawning biomass and the specification of product recovery rates (i.e. the conversion ratios of harvested to final product weight), respectively. Due to space limitations only selected equations are presented. This analysis serves as a compilation of individual efforts to model the fishery on an intra-season basis. Those individual efforts focussed on the management implications of using (1) in-season *versus* annual quotas including alternative in-season growth specifications and property rights allocations (Larkin and Sylvia 1996*a*), and (2) hedonic analysis of the surimi market (approximately 90% of total harvests are used to produce surimi; Larkin and Sylvia 1996*b*). These manuscripts are in various stages of development and publication; refer to the authors for details.

Population dynamics

Standard nonlinear population dynamics equations are used to move the stock through time (Ricker 1975). These fish numbers are tracked in the United States and Canada, over all twelve months of each of the ten years, and for each of the fourteen cohorts. The total instantaneous mortality rate consists of natural mortality plus fishing mortality. Fishing mortality is proportionate to (1) the size of the spawning biomass relative to the 'cautionary' spawning biomass level, and (2) the harvest rate which corresponds to a 'low' level of risk that the spawning biomass will fall below the cautionary level. This representation is consistent with the most recent attempts to model this fishery (Sylvia and Enriquez 1994; Methot and Dorn 1995).

Stock size increases each January with recruitment into the fishery. Historically, recruitment has been highly variable and unpredictable and does not correlate with the size of the spawning stock (Francis 1983; Dorn *et al.* 1993). There is, however, an identifiable pattern; 'large' (e.g. 4×10^9 fish) recruiting cohorts occur, at most, every third or fourth year. To avoid mis-specification and incorrect population assessments associated with using averages of previous recruitment levels, we choose a pattern of recruitment from the observed series which reflects poor initial conditions (i.e. first three years have relatively low recruitment, e.g. four million fish). This assumption is examined later.

The data for this biological submodel, and source for model validation, were obtained from the National Marine Fisheries Service (NMFS) which develops the stock assessments used to manage the fishery (Dorn *et al.* 1993).

Fishery economics

This submodel contains the equations which translate the harvest in numbers into quantities and values of products marketed by the industry. Although the fishery is modelled throughout the year, harvest is not allowed during spawning or migration; therefore, the harvest season is restricted to occur between April and October. The fish may be harvested by either small catcher boats which deliver shoreside for processing (the onshore sector), or by relatively large vessels which process at-sea (the offshore sector). In terms of management, it can be very important to distinguish between these alternative sectors (Anderson 1989; Milliman *et al.* 1992). In our model, each sector faces different selectivities, product recovery rates, costs, and, in some cases, prices. In addition, each sector specializes in the production of different products; the offshore sector concentrates on the production of surimi whereas the onshore sector diverts some of its harvest into fillets and headed and gutted products. Most importantly, in terms of contemporary management, each sector differs in its maximum daily capacity and utilization of waste for the production of meal.

The determination of production quantities deserves special attention. First, monthly growth rates are estimated for each cohort. These non-linear, time-dependent, weight gain equations are estimated as a seemingly unrelated regression, or SUR, system (Greene 1990). These equations, together with fish numbers, are used to calculate gross harvest weight. Conversion factors, or product recovery rates, represent the loss associated with processing the catch into final marketable products. These rates vary by product form (surimi, headed and gutted, fillets, and meal) and are multiplied by the gross weight of fish used to produce each type of product to arrive at final production quantities. These rates are also frequently used as indicators of the efficiency of the industry or comparison of the efficiency of the competing harvest sectors.

The specification of product recovery rates is important for the analysis of any processed product, i.e. managers want to maximize the conversion of raw product and minimize waste. For this analysis, instead of using annual averages of historical figures, the product recovery rates were hypothesized to vary throughout the harvest season and were modelled as a function of the intrinsic quality of the fish at the time of harvest. This specification also makes intuitive sense since higher-quality raw materials will likely result in either higher quality final products or larger output volume. This specification is particularly applicable to the Pacific whiting fishery as variable quality has frequently been documented (PFMC 1992; PFMC 1993; Sylvia 1995).

The general form of the intra-season product recovery rates (production yields, *yld*),

$$yld_{m,s,f} = f(X_1, X_2) = f(cf, wl, pro, moi, fat) \tag{1}$$

assumes that yields are determined by both fish size (X_1) and flesh composition (X_2), where fish size is described by either the condition factor (*cf*) or the weight-length ratio (*wl*), and flesh composition consists of the percentage of weight accounted for by protein, moisture, or fat (*pro*, *moi*, and *fat*, respectively). In particular, a larger fish (e.g. heavier or 'plumper') can either increase the recovery rate (processing equipment is generally able to extract more from larger-sized fish) or decrease it (if size is a result of increased gonadal tissue). Similarly, improvements in the composition of the flesh can either increase recovery rates (as protein content and quality are positively related) or decrease rates (as moisture content and quality are inversely related). Moyle and Cech (1988), Busacker *et al.* (1990), Jobling (1993) discuss the seasonal variation in fish quality (i.e. X_1 and X_2 variables) which is associated with biological growth. Partial correlation analysis identified which variables would explain the most variation in each of the production yields. This system of linear equations was estimated using SUR (Larkin and Sylvia 1996a).

Hedonic equations are frequently used to determine the price effects or relative value of the characteristics which collectively define a commodity. In terms of resource management, these models can examine how management tools can be used to alter product characteristics or quality in such a way as to increase the value of the stock (Eleftheriades and Tsalikidis 1990; Englin and Mendelson 1991). Although marine resources are faster-growing and more perishable than the forest resources which have been examined using this approach, both stocks are renewable and thus face similar management concerns. In terms of fisheries management, hedonic models can be used to relate the characteristics of the processed product to its price. In the case of Pacific whiting this representation is applicable to the principal product form, surimi, which is a graded product and one of only a few fisheries-based products which have an established structure for quality determination. The quality characteristics of surimi determine its price and the production formulae for several hundred surimi-based final products. The other product forms are not marketed using a consistent set of quality parameters nor have they experienced significant in-season price variation.

Surimi quality is typically identified by quantifiable levels of certain characteristics; together these characteristics are used to define its grade, and ultimately the price of the product. These characteristics include gel strength, whiteness, brightness, and moisture content. Seasonality is typically incorporated using a dummy variable; however, in our analysis seasonality is implicitly incorporated through changes in the characteristics of surimi which result from variation in intrinsic quality (i.e. flesh composition as described earlier). The equations specifying this relationship were estimated as a SUR system (i.e. the surimi characteristics were regressed against raw product characteristics in order to relate intrinsic quality to surimi quality).

The price hedonic equation, in general functional form, appears as:

$$p_{surimi} = g\,(gel, wat, hw, br) \tag{2}$$

where price (*p*) is expected to be positively related to gel strength (*gel*), inversely related to the moisture content (*wat*), and positively related to measures of whiteness (*hw*) and brightness (*br*). Some of these variables (e.g. *gel* and *wat*) are believed to be correlated and, therefore, significance of the individual coefficients is not necessarily expected; for this analysis we are interested in the overall explanatory power. This equation was estimated by ordinary least squares and appears to adequately capture the observed in-season price variation ($r^2 = 0.86$). Only *gel* was significant at the 5% level; using the coefficient, a reasonable 25% seasonal increase in gel strength would, *ceteris paribus*, produce a $US0.10 per pound, or 16%, price increase. The data utilized were intentionally obtained during a period of relatively stable supply and demand conditions. Obviously supply and demand cannot be expected to remain constant in the long-run; however, our goal for this analysis is only to incorporate price variation to the extent that it is affected by the collective changes in raw quality (for more information see Larkin and Sylvia 1996*b*).

Several data sources were used for this submodel. Weight and length data by age were obtained from 1986 to 1988 (PFMC 1992). The Oregon State University Seafood Laboratory provided information on weights, lengths, and proximal content for 1992–1994 (Morrissey 1995). Data sources for earlier years (mid–1960s) and similar species (cod, pollock) confirm the stability and absolute values of our data (Nelson *et al.* 1985 and Alaska Fisheries Development Foundation 1991). Annual average product recovery rates were obtained from NMFS, and in-season rates were provided by two private firms on a confidential basis. Data for the hedonic estimations were received from four distinct, but confidential, sources which enabled model validation. Remaining price and cost data were obtained from NMFS and correspond to the estimates used by the resource managers (PFMC 1993).

RESULTS

The non-linear, dynamic programming subroutine of the General Algebraic Modeling System (GAMS) was used to optimize the model (Brooke *et al.* 1988). Results are analysed in accordance with the stated objectives. First, we analyse seasonality (i.e. the isolated effects of individual in-season changes), then we evaluate management goals (i.e. how well each goal is met). In addition, we discuss the sensitivity of the model to property rights allocations and the recruitment specification.

Seasonality

The optimized model described in the previous section maximizes the value of the fishery over time by determining when, where, by whom, and how much is harvested. The optimal management plan (i.e. the 'proposed' plan) is compared with the present management practice which ignores seasonality (i.e. the 'contemporary' or base plan). More importantly, we identify the cause of the differences between plans and quantify the individual effects of changes in weight, product recovery rates (*via* intrinsic quality changes, i.e. changes in size and flesh composition), and market prices for the final goods (*via* results of hedonic analysis).

Results show that the in-season timing of harvests is significantly affected by respecifying the model to include seasonality. Incorporating in-season changes delays the optimal harvest and increases net present value NPV by 117%. These results are summarized in Figure 1. Under contemporary management, the offshore sector extracts its quota in April and, on average, the season closes in July. Under the proposed management plan the season would open, on average, in July and the offshore sector would harvest in October.

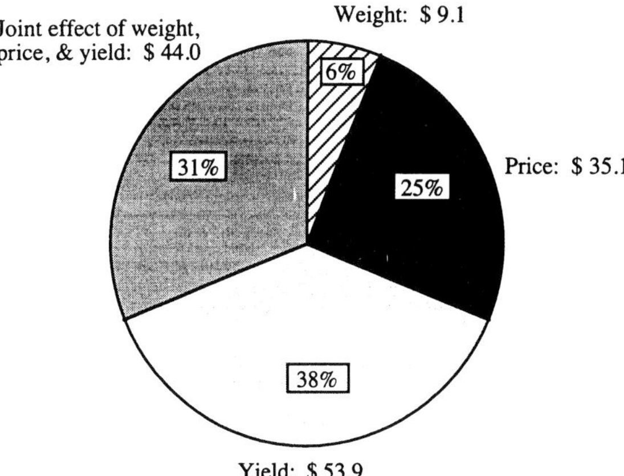

Fig. 2. Relative contribution of weight, price, and production yields to increased NPV resulting from introduction of seasonal variability (in \$US × 10^6).

The individual effects of newly-incorporated in-season changes are disaggregated in Figure 2. This Figure disaggregates the increase in NPV, that is, the difference between the contemporary and the proposed plan. The recovery rate effect dominates by accounting for 38% of the increased NPV. In-season price and weight changes contribute 25% and 6%, respectively. These individual effects were determined by systematically allowing each component to vary while holding the remainder constant and recording the change in NPV. Under the proposed plan, all components (yield, price, and weight) were allowed to vary. By subtracting the individual effects on NPV from the NPV achieved under the proposed plan, the interactive effect of the components is determined. The interactive effect is due to the joint or simultaneous in-season variation in each component.

It is possible to disaggregate the yield and price effects even further by examining the changes in the condition of the fish at the time of capture, i.e. fish size and flesh composition. For example, as the moisture content of the fish declines throughout the season (from approximately 83% to 80%), both production yields (of surimi and fillets) and surimi characteristics (including water content, whiteness, and gel strength) are affected. This is also true for protein and fat content. Seasonal changes in the intrinsic characteristics provide the critical link, in terms of effective management, between biological and economic components.

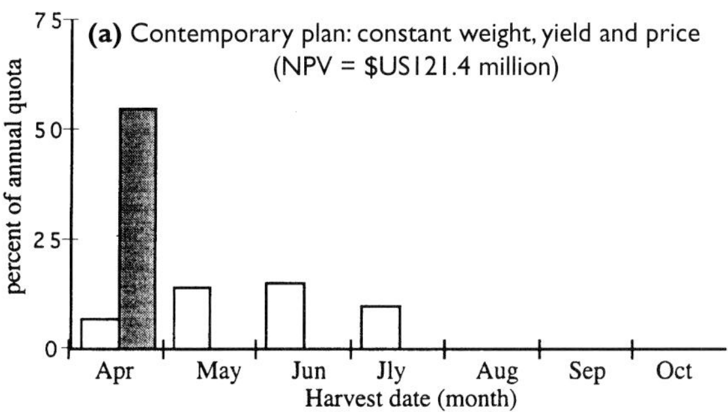

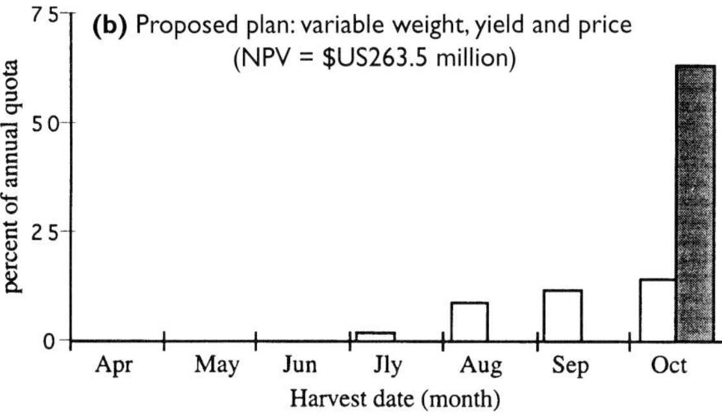

Fig. 1. Net present value (NPV) and the associated average intra-season harvest pattern for (*a*) the contemporary and (*b*) proposed management plans. The harvest pattern by user group (onshore, white; offshore, shaded) is determined by multiplying the optimal property-rights allocation by the optimal monthly distribution of effort (both calculated as averages over the planning horizon).

Management goals

It is important to consider the implications for the remaining management goals, aside from just changes in the pattern of harvest and industry NPV. To this end, we compare the contemporary management plan (base case) with the proposed plan (optimal case) on the basis of the stated goals of the PFMC Groundfish Management plan. Results are summarized in Table 1.

Table 1. Comparison of management goals under the contemporary and proposed plans. Values represent the average of annual statistics unless otherwise noted. If there is a seasonal component, the annual measure is the seasonal average based on when the harvest occurred

U.S. management goals	Management plans Contemporary	Proposed	Relative change
(1) Conservation			
1a) number of fish harvested	513.7 million	461.7 million	−10%
1b) weight of fish harvested	225 100 t	253 000 t	+ 12%
1c) size of spawning biomass (year 10)	1015 million	1028 million	+ 1%
1d) relative size of harvest (harvest/ spawning biomass)	46.6%	41.5%	−11%
(2) Economic value			
2a) net present value (NPV)	$US121.4 million	$US263.5 million	+ 117 %
2b) property rights allocation			
onshore	46%	37%	−19%
offshore	54%	63%	+16%
(3) Utilization			
3a) product recovery rates (*yld*)			
surimi	14%	17.4%	+24%
headed and gutted	56.4%	61.4%	+9%
fillets	23.5%	27.2%	+16%
meal and oil	9.8%	11.0%	+12%
3b) output quantities			
surimi	28 400 t	39 100 t	+37%
headed and gutted	8 000 t	8 000 t	0%
fillets	5 000 t	5 000 t	0%
meal and oil	27 900 t	24 100 t	−16%

No single measure can adequately represent the conservation of a resource. We use four separate measures to understand the possible implications to the ecosystem of the proposed management plan. The proposed plan offers greater conservation since both the absolute and relative number of fish harvested declines (1a and 1d, respectively). In addition, the total annual harvest quota increases, i.e. fewer but heavier fish are harvested. Perhaps more importantly, the size of the spawning biomass remaining at the end of the planning period is not compromised (1c). For the second goal, NPV increases under the proposed plan (NPV is our interpretation of the economic management goal). Alternatively, one could compare changes in employment or other measures of economic value to society (e.g. the relative change in property rights, 2b). In terms of utilization of the resource, average recovery rates increase under the proposed plan (3a). Output of surimi increases as property rights are shifted to the offshore sector (which specializes in surimi production), whereas production of meal declines. In addition, the onshore production of surimi also increases under the proposed plan (resulting from the seasonal price effect). The quantity of fillet and headed and gutted product does not change since both present production capabilities and market opportunities are limited.

Sensitivity analysis

The in-season timing of harvests is significantly affected by the capacity of competing harvest sectors. Under the previous scenarios, the onshore sector was guaranteed 20–50% of the annual harvests, depending on capacity constraints and size of the harvestable biomass, to ensure shoreside community stability. In order to examine the effect of this assumption, the model was allowed to choose the optimal allocation of annual quotas, i.e. the percentage of harvest by the onshore and offshore components which maximizes long-run NPV of the fishery. Comparing the proposed plan with this 'optimal rights' plan, indicates that the property-rights effect (i.e. the transfer of 9% of the allowable catch from the onshore to the offshore sector) is relatively small; the optimal property-rights allocation increased NPV only 4% after accounting for seasonal intrinsic quality.

The level of annual biological recruitment is an extremely important variable for managing the fishery. For many species, recruitment is related to the size of the parent stock; however, for Pacific whiting this relationship has been extremely difficult to predict (Dorn *et al.* 1993; Methot and Dorn 1995). Contemporary short-run management employs the 30-year median value which is assumed to hold across the modelling horizon. Alternatively, using the observed historical pattern (which has only one large recruiting class every three or four years) is not only necessary for a long-run analysis but also increases the realism of the model. Initially, a 'poor' starting point was assumed, i.e. large recruiting class first occurs in the fourth year. Results indicate that the effect of altering the recruitment pattern is rather large; NPV increases 54% if the recruiting class in the first modelling year is large (and followed by three years of low recruitment). In addition, if the median recruitment level is used (which is independent of the pattern and, therefore, the same under both scenarios), NPV increases by 38%. These results demonstrate the significance of initial conditions and recruitment variability to the magnitude of NPV.

IMPLICATIONS

Addressing the question 'does in-season intrinsic quality variability affect optimal management?' may be an important key for successful development and management of many wild-stock fisheries. An interdisciplinary model demonstrated how seasonal changes in the raw product quality (i.e. fish weight at harvest and relative size and composition of the flesh) influence the economics of the fishery and its management. In particular, variations in weight directly affect the harvest quantities, and variations in the relative size and proximate content impact production yields, product quality, and product price.

For many species, management has disregarded the in-season timing of harvest in order to focus on other issues, including allocating the annual quota among the competing harvest sectors. Failure to consider in-season intrinsic variability, however, results in suboptimal management of fast-growing or rapidly changing stocks. The result is foregone benefits to society and potentially the ecosystem. More importantly, management goals may not be mutually exclusive if harvest policies are dictated by the characteristics of the individual fish; that is, goals such as conservation, efficiency, and utilization may often be complementary. Consideration of intrinsic quality does, however, complicate the management of wild-stock resources. Results from this analysis suggest additional benefits of property rights strategies which allow the industry to internalize the opportunity costs of complex biological and economic issues associated with intrinsic quality.

REFERENCES

Alaska Fisheries Development Foundation (1991). Groundfish quality project final report. NA-89-ABH-00008. (National Marine Fisheries Service: Seattle).

Anderson, L G (1989). Optimal intra- and interseasonal harvesting strategies when price varies with individual size. *Marine Resource Economics* **6**, 145–62.

Blomo, V, Stokes, K, Griffin, W L, Grant, W E, and Nichols, J P (1978). Bioeconomic modeling of the Gulf shrimp fishery: an application to Galveston Bay and adjacent offshore areas. *Southern Journal of Agricultural Economics* **10**, 119–25.

Brooke, A, Kendrick, D, and Meeraus, A (1988). 'GAMS: A User's Guide'. (The Scientific Press: San Francisco).

Busacker, G P, Adelman, I R, and Goolish, E M(1990). Growth. In 'Methods for Fish Biology.' (Eds C B Schreck and P B Moyle) pp. 363–87 (American Fisheries Society: Bethesda).

Christensen, S, and Vestergaard, N (1991). A bioeconomic analysis of the Greenland shrimp fishery in the Davis Strait. *Marine Resource Economics* **8**, 345–65.

Clark, C W (1990). 'Mathematical Bioeconomics: The Optimal Management of Renewable Resources.' 2nd edition (John Wiley and Sons: New York).

Dorn, M W, Nunnallee, E P, Wilson, C D, and Wilkins, M (1993). Status of the Coastal Pacific Whiting Resource in 1993. National Marine Fisheries Service (Department of Commerce: Seattle).

Eleftheriadis, N, and Tsalikidis, I (1990). Coastal pine forest landscapes: modelling scenic beauty for forest management. *Journal of Environmental Management* **30**(1), 47–62.

Englin, J, and Mendelson, R (1991). A hedonic travel cost analysis for valuation of multiple components of site quality: the recreation value of site management. *Journal of Environmental Economics and Management* **21**(3), 275–90.

Francis, R C (1983). Population and trophic dynamics of Pacific hake (*Merluccius productus*). *Canadian Journal of Fisheries and Aquatic Sciences* **40**, 1925–43.

Greene, W H (1990). 'Econometric Analysis.' (Macmillan Publishing: New York).

Jobling, M (1993). Bio-energetics: feed intake and energy partitioning. In 'Fish Ecophysiology'. (Eds J C Rankin and F B Jensen) pp. 1–40 (Chapman and Hall: London).

Kellogg, R, Easley, J E, and Johnson, T (1988). Optimal timing of harvest for the North Carolina Bay scallop fishery. *American Journal of Agricultural Economics* **70**(1), 50–61.

Larkin, S, and Sylvia, G (1996*a*). Intrinsic product characteristics that affect fisheries management: A bioeconomic analysis of a whiting fishery. Working paper (Oregon State University: Corvallis).

Larkin, S, and Sylvia, G(1996*b*). Hedonic analysis of the US — Japan surimi market: A historical and current perspective with emphasis on the role of Pacific whiting. Working paper (Oregon State University: Corvallis).

Love, R M (1988). 'The Food Fishes: Their Intrinsic Variation and Practical Implications.' (Farrand Press: London).

Methot, R D, and Dorn, M W (1995). Biology and fisheries of North Pacific hake (*M. productus*). In 'Hake: Fisheries, Ecology and Markets'. (Eds J Alheit and T J Pitcher) pp. 389–414 (Chapman and Hall: New York).

Milliman, S R, Johnson, B L, Bishop, R C, and Boyle, K J (1992). The bioeconomics of resource rehabilitation: a commercial-sport analysis for a Great Lakes fishery. *Land Economics* **69**(2), 191–210.

Morrissey, M (1995). Investigation of the effects of seasonal variation on composition and weight of Pacific whiting. Working paper. (Oregon State University Seafood Laboratory: Astoria).

Moyle, P B, and Cech Jr, J (1988). 'Fishes: An Introduction to Ichthyology.' 2nd edition (Prentice Hall: New Jersey).

Nelson, R W, Barnett, H J, and Kudo, G (1985). Preservation and processing characteristics of Pacific whiting (*Merluccius productus*). *Marine Fisheries Review* **47**(2), 60–7.

Önal, H, McCarl, B A, Griffin, W L, Matlock, G, and Clark, J (1991). A bioeconomic analysis of the Texas shrimp fishery and its optimal management. *American Journal of Agricultural Economics* **73**(4), 1161–70.

PFMC (1992). The effect of altering the US fishery opening date on the yield per recruit of Pacific whiting. In 'Appendices of the Environmental Assessment and Regulatory Impact Review: A Supplement to the Pacific Whiting Allocation'. Appendix 12, attachment G.7.b. (Pacific Fisheries Management Council: Portland).

PFMC (1993). 'Pacific Whiting Allocation: Environmental Assessment and Regulatory Impact Review of the Anticipated Biological, Social, and Economic Impacts of a Proposal to Allocate the Resource in 1994, 1995, and 1996'. (Pacific Fisheries Management Council: Portland).

Ricker, W (1975). Computation and interpretation of biological statistics of fish populations. *Bulletin* 191 (Department of the Environment: Ottawa).

Sylvia, G (1995). Global markets and products of hake, In 'Hake: Fisheries, Ecology and Markets'. (Eds J Alheit and T J Pitcher) pp. 415–36 (Chapman and Hall: New York).

Sylvia, G, and Enriquez, R (1994). A multiobjective bioeconomic analysis: an application to the Pacific whiting fishery. *Marine Resource Economics* **9**, 311–28.

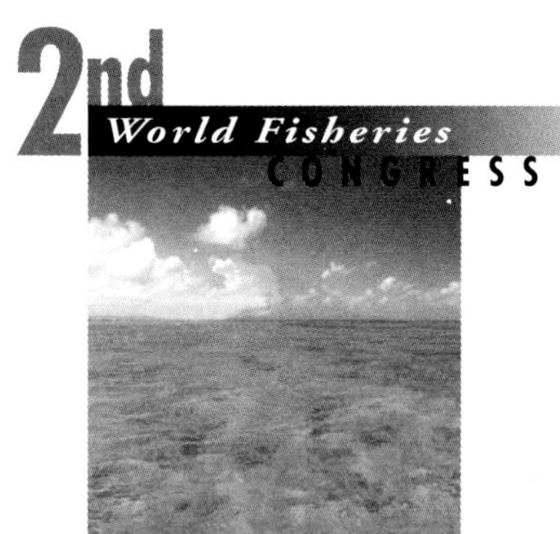

Development of an Ecosystem Model for Managing the Fisheries Resources of Prince William Sound

G. L. Thomas, E.V. Patrick, J. Kirsch and J.R. Allen

Prince William Sound Science Center, PO Box 705, Cordova Alaska 99574. US.

Summary

The Sound Ecosystem Assessment (SEA) program is developing and applying new survey designs to estimate the abundance and distribution of dominant fish stocks (Pacific herring and walleye pollock) in Prince William Sound. The long-term goal is the development of numerical models to improve the prediction of fish population change. However, accurate estimates of biomass are not only needed to initialize and verify model predictions but are essential for implementing conservation practices that will sustain healthy fisheries. The current management of the walleye pollock fishery could be contradictory to the restoration of the collapsed herring stock and production strategies for hatchery salmon. We support the implementation of a biomass monitoring programme for herring and pollock stocks in the Sound and the development of in-season management practices. Furthermore, we urge the adoption of experimental multi-species harvest strategies while the three dominant species (herring, pollock and salmon) are being monitored for changes in biomass. We see this to be an unprecedented opportunity for fisheries science to evaluate a new management paradigm.

Introduction

In 1988-90, a group of scientists working on the Ocean Ecosystem Dynamics Program (GLOBEC) of the National Science Foundation concluded that our inability to predict changes in marine fish populations has prevented us from separating natural and anthropogenic impacts (Cullen 1988). Using the GLOBEC program as a guide (GLOBEC 1991), in 1994, the Sound Ecosystem Assessment program (SEA) was implemented to develop better predictive tools for fish production in the Prince William Sound, Alaska. SEA assumes that the physical (temperature, turbulence, currents, etc.) and biological (predators, prey and competitors) conditions are major contributors to patterns of abundance and production of marine animals.

After failures in recruitment of Prince William Sound pink salmon *Onchorhynchus gorbuscha* and Pacific herring *Clupea harengus pallasi* in 1991–93 (Thomas and Mathisen 1993; ADF&G 1996), these stocks were classified as damaged by the TV *Exxon Valdez* oil spill Trustee Council (Wolfe *et al.* 1993). Efforts to restore these stocks are confounded by the poorly understood effects of climate, food and predators. These factors have been shown to account for more than 99% of the mortality in the early marine period for these fishes (Hjort 1914; Parker 1968; Ricker 1976; Hartt 1980; Bax 1983). During this period, slow growing individuals sustain a higher mortality because they are vulnerable to predators for a longer time (Parker 1971; Ricker 1976; Healey 1982; West and Larkin 1987). The long-term goal of the SEA program is to improve forecasting of fish recruitment in Prince William Sound. SEA assumes that contemporary measurement and modelling technologies are on the threshold of being sufficient to identify, monitor and simulate the primary physical and biological processes that determine juvenile fish survival. However, a more immediate need is to incorporate new knowledge of these species, where possible, into the ongoing management and restoration programmes. Here we focus on the problem of measuring fish populations and the application of contemporary methods to improve accuracy and precision of biomass estimates.

More immediate concerns

The development of new measurement techniques to determine abundance and distribution of nekton and plankton is a prerequisite for testing underlying hypotheses of the SEA models. Although model development is an eight to ten year goal, there is no reason why the new information on stock abundance, distribution and interactions cannot be used to create more effective restoration and management programmes. Since pink salmon, Pacific herring and walleye pollock are all commercially exploited stocks, the establishment of exploitation rates has direct implications on restoration.

Presently, the stocks, pink salmon, herring and walleye pollock, are all managed independently by establishing single-species harvest strategies. Single-species harvest strategies are often based upon spawner-recruit relationships and stock assessment information. Missing are the effects of physics and other populations on the stock. With support that these species are influenced by climate-driven, predator-prey mechanisms and

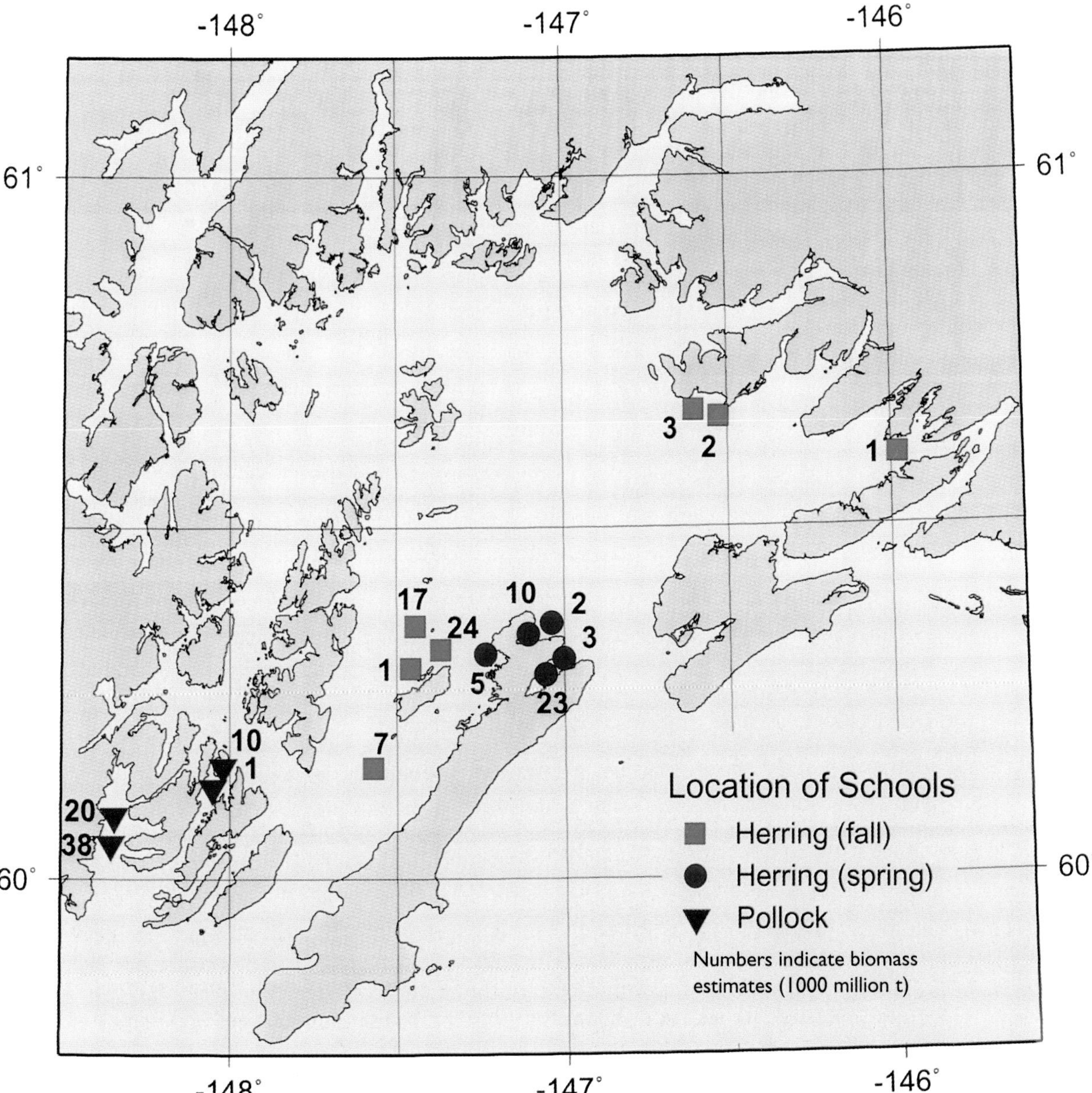

Fig. 1. Map of Prince William Sound, Alaska, showing locations of herring and pollock schools measured on eight pre-spawning surveys, 1993–1996.

the acquisition of reasonably accurate, direct estimates of these stocks, an opportunity exists for testing the efficacy of multi-species management.

This paper addresses the findings after two years of the SEA program. Specifically, we review developments in the ability to measure the abundance and distribution of two dominant species (pollock and herring), and the implications of new information on current fisheries restoration and management.

METHODS

Prince William Sound (Fig. 1) is a complex fiord/estuary located at the northern margin of the Gulf of Alaska. High mountain peaks in excess of 4000 m border the Sound and receive the brunt of the seasonally intense cyclonic storms from the Gulf (Thomas *et al.* 1991). Depths exceeding 400 m occur in the western and central portions of the Sound which support overwintering populations of oceanic copepods.

Sampling design

Acoustic surveys were run from a variety of charter vessels according to standard echo integration practices (MacLennan and Simmonds 1992) using sphere-calibrated BioSonics 101 120 kHz and 10 238 kHz dual beam echosounders and BioSonics ESP/EI and DB software programs. Each sonar system was equipped with a Global Positioning System (GPS) receiver to geo-reference acoustic data. Echo integration, dual-beam target strength and GPS data were stored on hard drives and backed up on magnetic disks/tapes. Unprocessed data were stored on DAT recorders. The parameters of the two acoustic systems were:

120 kHz /SL = +225.075 dB; RG = −159.282dB;
b2 = 0.0010718, PD = 0.4 ms;

38 kHz /SL = +215.784; RG = −144.474, b2 = 0.00219;
PD = 1.0 ms.

Long day-lengths necessitated daylight surveys to monitor predator densities in the spring. Otherwise the acoustic surveys were conducted at night (Burczynski *et al.* 1986). Only red running lights were used on night surveys to minimize boat avoidance and search-light sonars were used to monitor school behaviour relative to the vessel track. The spring predator (walleye pollock) surveys took place mainly in the Wells and Perry Island Passages, in the north-western corner of the Sound. This was to take advantage of the release of over 500 million salmon smolt by the hatcheries in this area. The fall-winter pre-spawning surveys of adult herring and pollock took place mainly in the north Montague Straits and Port Bainbridge areas. Since 1993, we conducted 11 spring predator surveys and eight fall-winter pre-spawner surveys.

On all surveys, measurements were made on systematic grids of parallel transects. On the pre-spawner surveys the grids were conducted only over fish school groups. This was the second of a two-stage survey design, with the first stage a search for school groups. Cochran (1977) described the algorithms for estimating biomass and variance, and discussed the biases associated with single and two-stage, stratified-systematic sampling designs. All acoustic sampling was supported by midwater trawl or purse seine catches to collect biological information.

Data Analysis

Batch processing to transform 2-d arrays of acoustic targets from dB to kg and numbers, estimate and visualize biomass, were conducted after transferring the data to a UNIX workstation. All data are stored in the appropriate format for post processing using Interactive Data Language (IDL), and Advanced Visual Systems (AVS) software.

Measured target strengths of individual fish were compared with length data of fish captured by the nets. To establish a fish size-target strength relationship, we used the relationships advanced by Thorne (1983) for target strength per kg *versus* length, Traynor and Ehrenberg (1979) for walleye pollock target strength *versus* length, and MacLennan and Simmonds (1992) for Pacific herring target strength *versus* length. Weighted mean densities and their variances were computed and extrapolated to biomass and 95% confidence limits using the delta method (Seber 1973).

Visualizations of biomass survey data were made using AVS. Geo-coded volume backscatter values were stored in 1-d scatter arrays and displayed as scatter- dot clouds in 3-d space to view sampled locations. The 1-d scatter array was then converted to 3-d unstructured cell data (UCD) format. In UCD format each point is identified by its spatial relationship to neighbouring points. Interpolation using Delauney triangulation was used to generate the UCD structure. The 3-d UCD volume was converted to a geometry by applying a user-specified colour map to the nodes and then displayed and manipulated in the AVS geometry viewer. Finally, serial slices throughout the UCD volume were taken using horizontal planes to show truncation of school group measurements. Isosurfaces of the school groups were examined at various thresholds, rotated in 3-d space and animated as time sequences to select views that best describe the truncation problem.

RESULTS AND DISCUSSION

Implications of new biomass information

Six surveys of pre-spawning herring and two of pre-spawning pollock were conducted between 1993 and 1996. Both species occur in highly aggregated school groups that are distinct enough in their vertical distribution and density that identification of their acoustic targets is not an issue. Only 12 school groups of herring in excess of 1000 t were observed on the six surveys over three years (Fig. 1). Only four school groups over 1000 t of pollock were observed on two surveys in 1995 (Fig. 1).

Four of the 12 herring concentrations were surveyed twice to determine repeatability of the measurements. In fall 1993, we estimated the same school group on Applegate Rocks to be 12 875 and 16 442 t; in spring 1995, we estimated the same school group in Rocky Bay to be 10 480 and 8050 t; in winter 1996, we estimated the same group of fish in Zaikof Bay at 26 309 and 20 097 t; in the spring of 1996, we estimated the same group of fish in Stockdale Harbor at 3227 and 3791 t.

Visual inspection of the 3-d plots shows that the largest source of error in estimates from the repeated transects over the herring or pollock aggregations was the lack of adequate coverage of the fish concentration. Figures 2–6 show the results of five consecutive surveys on a concentration of herring in Stockdale Harbor. The school isosurfaces show that the measurement of the herring school group was severely truncated on surveys #1, #3 and #5, but not on surveys #2 and #4. The nearshore affinity of walleye pollock suggests that truncation of measurements is also a problem on their surveys. The biomass estimated on #2 and #4 were 3227 and 3791 t, whereas the biomass of the other three surveys never exceeded 1700 t.

Another aspect of repeatability is between fall and winter-spring estimates. The fall 1994 and spring 1995 estimates of adult herring biomass were both about 13 000 t. The fall 1995 and winter 1996 estimates of adult herring biomass were about 24 000 and 23 000 t. The observed increases from spring to fall are due to recruitment and little overwinter mortality has been observed. Finally, in 1995 the biomass of pre-spawning pollock in Port Bainbridge was estimated to be 27 366 ± 7227 t from a survey of nine parallel transects.

The fact that there are so few school groups of pre-spawning herring and pollock in the Sound region, and that they do undergo localized movements which can allow good coverage if the surveys are continually repeated, suggests the use of a two stage survey design: (1) use aerial and sonar searches to locate the school groups, and (2) remain with the school group until repeatable measures are made. These findings suggest that making 'corrections' for fish too close to the shore (Hampton 1996) for Prince William Sound herring and pollock can result in unacceptable inaccuracy in the biomass estimation. Adoption of two stage procedure should result in repeated estimates of school group biomass within 10% of the mean biomass.

The single-species paradigm

Currently, the three dominant pelagic fish populations in the Sound, pink salmon, Pacific herring and walleye pollock are also three of the most important commercial fishes. As is the tradition in fisheries science (Hilborn and Walters 1992), each fishery is managed separately and with sparse data. With new information generated by the SEA program on abundance, distribution and interaction, there is the possibility to test a multi-species approach that minimizes the potential for contradictions between restoration and management activities.

The status of the pink salmon is somewhat independent of the system because 90% are produced by hatcheries (Thomas and Mathisen 1993), and they only spend three months in the Sound as juveniles before migrating to the ocean to feed (Willette 1993), and about two to three weeks in the Sound as adults while migrating to spawning areas. In contrast, Pacific herring and walleye pollock spend much of their life histories in the Sound.

The Sound's Pacific herring stock has been too low to support commercial harvests since 1993 (ADF&G 1996). Concurrently, the Sound's walleye pollock stock is considered to be part of a declining Gulf of Alaska stock (NMFS 1996). Thus, the Sound's walleye pollock are being harvested at a reduced exploitation rate (9% in 1995). Since the status of the Gulf of Alaska pollock stock is determined by a model prediction for the west-central stocks, where the predictions do not agree with NMFS 1995 survey data, this harvest rate is at least controversial. With recent findings that the walleye pollock and herring are the two dominant, competing, pelagic species in the Sound, there is a concern that the current management strategy to build the pollock stock may contradict the restoration of the herring stock.

Walters *et al.* (1986) found herring survival to recruitment to be negatively related to cod (*Gadus macrocephalus*) abundance. Analysis of cod stomach contents supported the estimated mortality rates. We have found that walleye pollock are the primary predators of young pink salmon and herring (Walters *et al.* 1986: Willette 1993). Thus, a negative correlation between walleye pollock abundance and the recruitment of herring is a good hypothesis for testing.

Sainsbury (1988) proposed that managers set up experimental fishing regimes for multi-species analysis. Present fishing regimes in the Sound are established independently for the walleye pollock, Pacific herring and pink salmon by the Alaska Department of Fish and Game and National Marine Fisheries Service, hence they are already subjects of an ongoing experiment. With both of these agencies represented by the EVOS Trustee Council there is unprecedented opportunity to merge information from both management and restoration programmes and evaluate a multi-species management approach.

Hilborn and Walters (1992) identified the data requirements for multi-species biomass dynamics and age structured models as too demanding or expensive for practical utility. However, we anticipate that GLOBEC and SEA research and development programs will produce remote sampling methods that are accurate and cost effective enough to satisfy both the accuracy and cost constraints. To illustrate, one survey is less than $US50 000. Such are well within the amount that costs can be borne by test fishing revenues (Thomas 1992). Such monitoring will also serve our long-term goals by providing the information needed to initialize and verify the next generation numerical models for predicting fish population changes (GLOBEC 1991).

In-season management for escapement

We show that acoustic measurement techniques can make repeatable estimates of the size of herring and pollock school groups in the Prince William Sound region. Repeatable measurements of biomass have been made both within and between fall and spring surveys. The largest source of error in the biomass estimation procedure is truncation of the school groups by the survey volume. Visualization techniques have been developed to provide criteria for accepting repeated measurements as replicates. Use of new criteria for repeating biomass estimates in the field is expected to produce the accuracy needed for implementing in-season management practices. Implementation of in-season management practices is the key to providing the measurements of exploitation rate and biomass needed to evaluate experimental multi-species harvest strategies, while protecting the stocks form overharvest.

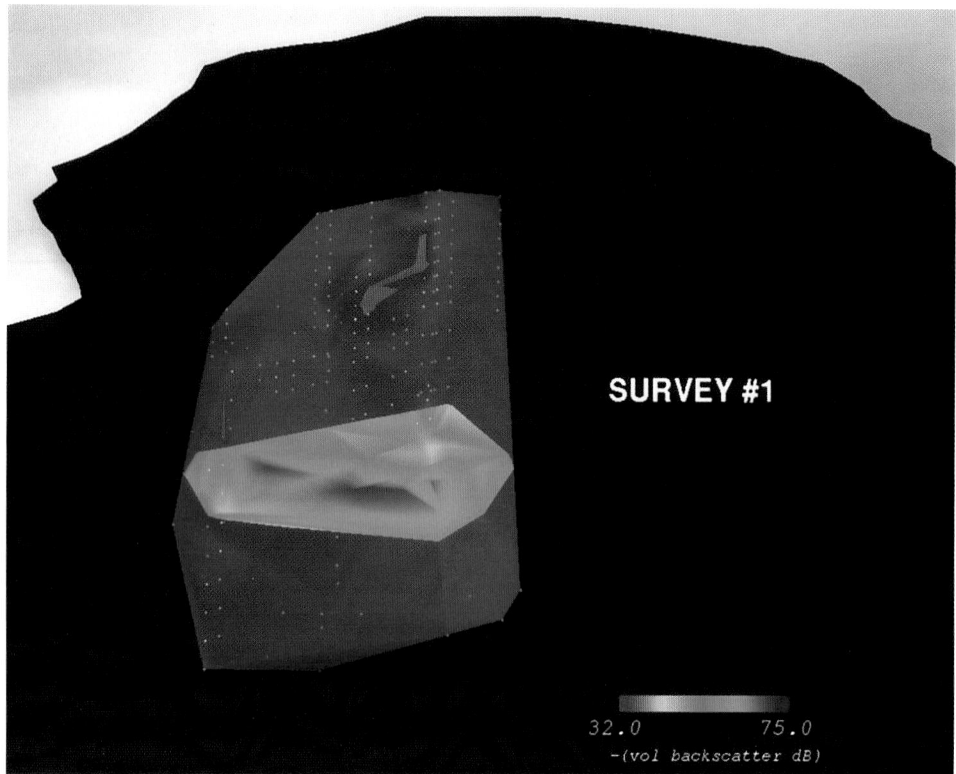

Fig. 2. First of five surveys of a herring school group located in Stockdale Harbor, Prince William Sound, Alaska, on the night of April 17, 1996. Note that the school group was missed by this survey. It was too close to the shoreline.

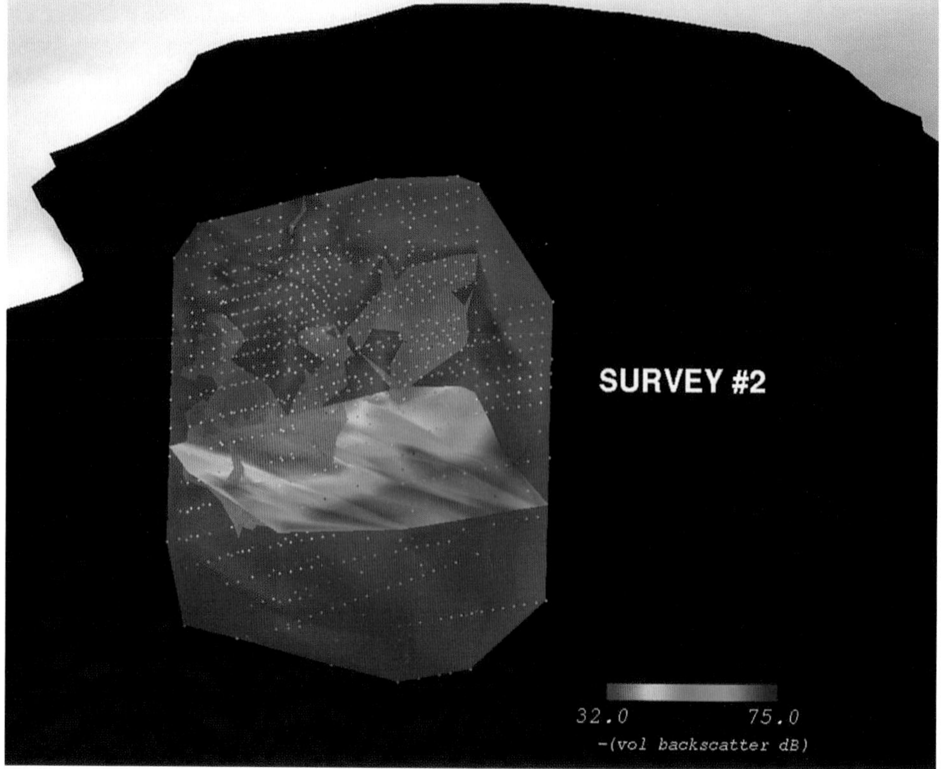

Fig. 3. Second of five surveys of a herring school group located in Stockdale Harbor, Prince William Sound, Alaska, on the night of April 17, 1996. Note that the school group was nearly completely insonified on this survey. Colour density gradient (red = –35 dB, yellow = –40dB) on the parallel plane indicates increasing density toward the shoreline.

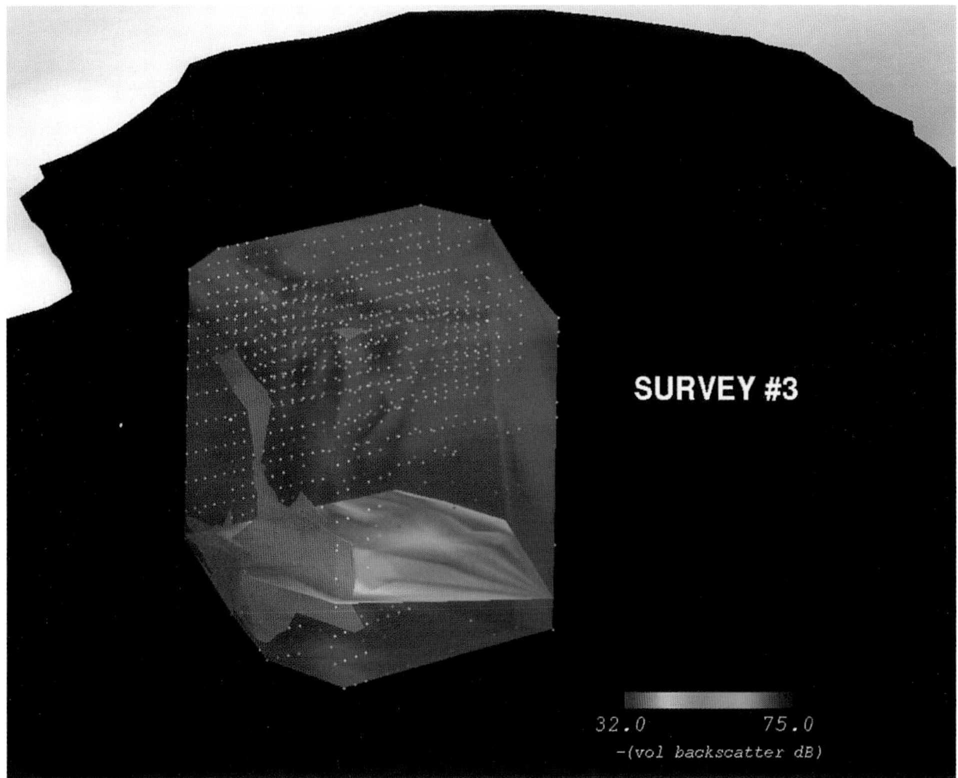

Fig. 4. Third of five surveys of a herring school group located in Stockdale Harbor, Prince William Sound, Alaska, on the night of April 17, 1996. Note that the school group was missed by this survey because it had moved deeper and too close to the shoreline.

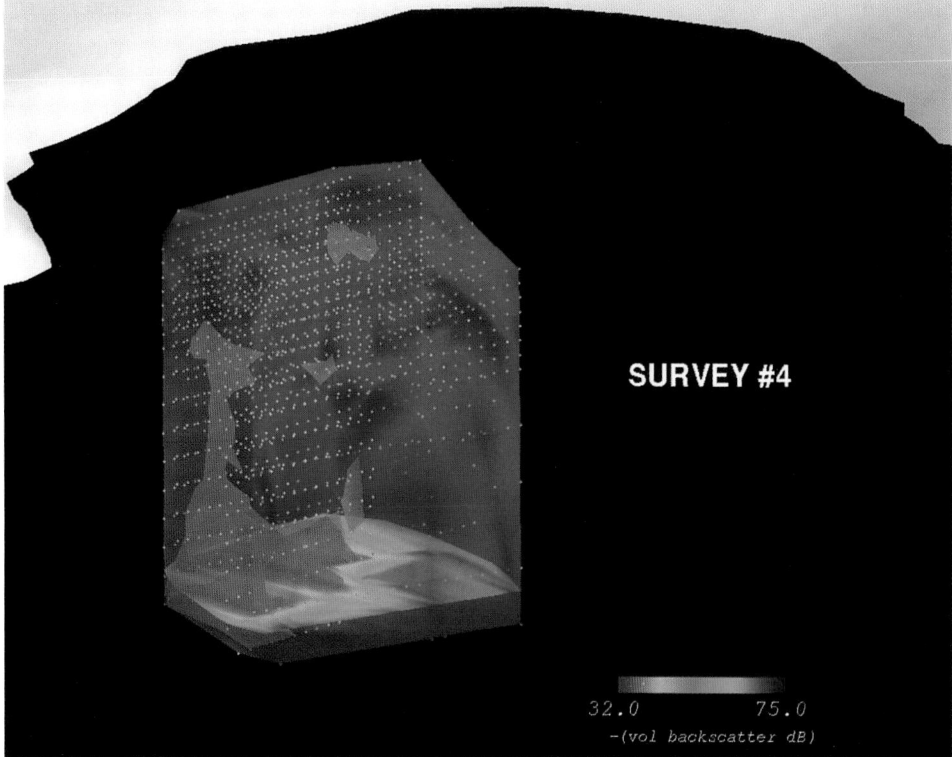

Fig. 5. Fourth of five surveys of a herring school group located in Stockdale Harbor, Prince William Sound, Alaska, on the night of April 17, 1996. Note that the school group had moved away from the shoreline and was nearly completely insonified on this survey.

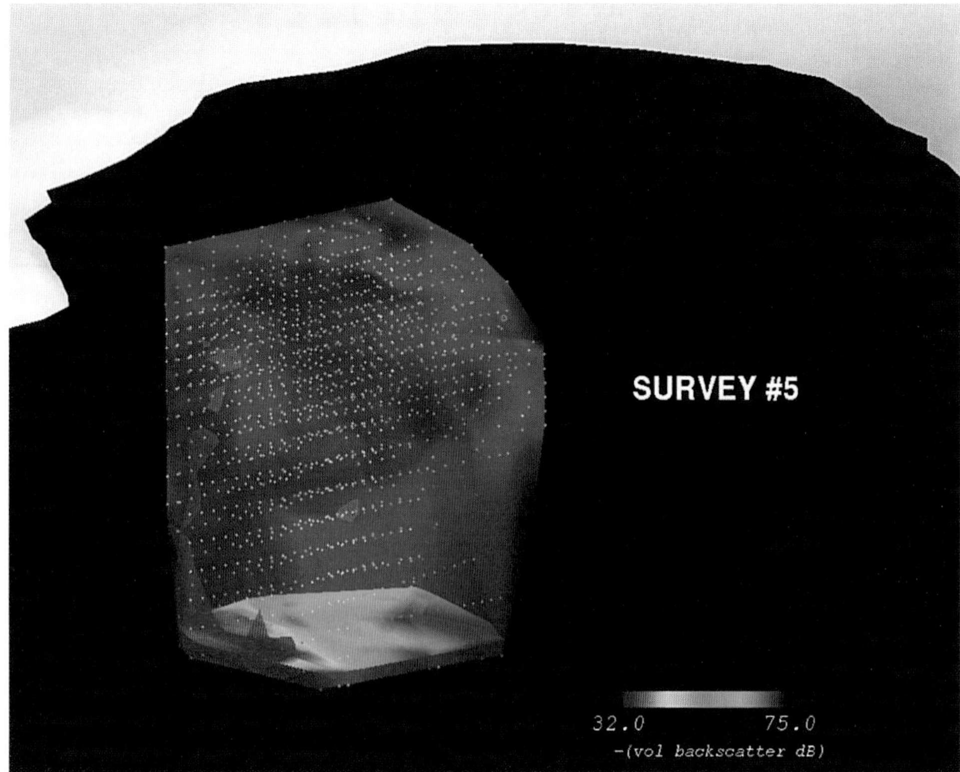

Fig. 6. Fifth of five surveys of a herring school group located in Stockdale Harbour, Prince William Sound, Alaska, on the night of April 17, 1996. Note that the school group had moved deeper and back to the shoreline so was missed by this survey.

Current management of the walleye pollock fishery could be contradictory to the restoration of the collapsed herring stock and production strategies for hatchery salmon. With the ongoing SEA research program, we recommend that management agencies take advantage of this opportunity by experimenting with multi-species harvest strategies and implementing in-season management practices for herring and pollock. Since present management of these stocks establishes harvest strategies on an annual basis, this multi-species experiment can be as simple as a retospective analysis of what is already being done

Predictive models

With short-term concerns over conservation of stocks satisfied by using accurate estimates of fish biomass to implement in-season management practices, research can focus on the development of predictive models. Accurate monitoring of the fish biomass is also essential for initializing and verifying model predictions. Thus, with the development, refinement and implementation of monitoring programmes to accurately assess fish stock biomass, conservation practices and, eventually, the prediction of animal population change is possible. Since the ability to predict animal population change is a prerequisite to the separation of natural and anthropogenic effects, we have a pathway to developing the ability to assess the effects of fishing, hatchery practices, oil spills and other man-induced impacts on natural populations.

Acknowledgments

This research is funded by the EVOS Trustee Council. We thank the Council and its staff, the SEA program researchers, local area managers and Prince William Sound Science Center personnel and associates for support of this research. Special thanks go to the fishermen in Prince William Sound who have contributed valuable insight on the fish distribution and behaviour and are second to none as vessel skippers.

References

ADF&G (1996). Catch statistics and records. Unpublished. Cordova, Alaska.

Bax, N J (1983). Early marine mortality of marked juvenile chum salmon released into Hood Canal, Puget Sound, WA, in 1980. *Canadian Journal of Fisheries and Aquatic Sciences* **40**, 426–35.

Burczynski, J J, Michaletz, P H, and Marrone, G (1986). Hydroacoustic assessment of the abundance and distribution of rainbow smelt in Lake Oahe. *North American Journal of Fisheries Management* **7**, 106–16.

Cochran, W G (1977). 'Sampling Techniques.' (John Wiley and Sons: New York). 428 pp.

Cullen, V (Ed). (1988). Global ecosystem dynamics. Joint Oceanographic Institutions, Incorporated Washington DC. 131 pp.

GLOBEC (1991). Initial science plan. Global Ecosystem Dynamics Report Number 1. Joint Oceanographic Institutions, Incorporated Washington DC. 93 pp.

Hampton, I (1996). Acoustic and egg-production estimates of South African anchovy biomass over a decade: comparisons, accuracy and utility. *ICES Journal of Marine Science* **53**, 493–500.

Hartt, A C (1980). Juvenile salmonids in the oceanic ecosystem—the critical first summer. In 'Salmonid ecosystems of the North Pacific'. (Eds W J McNeil and D C Himsworth) pp. 25–57 (Oregon State University Press: Corvallis, OR).

Healey, M C (1982). Fish behavior by day night and twilight. In 'Behavior of teleost fishes'. (Ed T J Pitcher) pp. 285–305 (Chapman and Hall: New York). 715 pp.

Hilborn, R, and Walters, C J (1992). Quantitative Fisheries Stock Assessment. (Chapman and Hall: New York). 570 pp.

Hjort, J (1914). Fluctuations in the great fisheries of northern Europe viewed in the light of biological research. *Rapports et Proces-Verbaux des Reunions, Conseil International; pour l'Exploration de la Mer* **20**, 1–228.

MacLennan, D N, and Simmonds, E J (1992). 'Fisheries Acoustics.' (Chapman and Hall: London). 527 pp.

NMFS (1996). Catch statistics and records. Unpublished. Seattle. Washington.

Parker, R R (1968). Marine mortality schedules of pink salmon of the Bella Coola River, Central British Columbia. *Journal of Fisheries Research Board Canada* **25**, 757–94.

Parker, R R (1971). Size selective predation among juvenile salmonid fishes in a British Columbia Inlet. *Journal of Fisheries Research Board Canada* **28**, 1503–10.

Ricker, W E (1976). Review of the growth rate and mortality of Pacific salmon in saltwater, and noncatch mortality caused by fishing. *Journal of Fisheries Research Board Canada* **33**, 1483–1532.

Sainsbury, K (1988). The ecological basis of multispecies fisheries and management of a demersal fishery in tropical Australia. In 'Fish Population Dynamics'. (Ed J Gulland) pp. 349–82 (Wiley: Chichester).

Seber, G A F (1973). 'The estimation of animal abundance and related parameters.' (Griffin: London). 506 pp.

Thomas, G L (1992). Successes and failures of fisheries acoustics — an international, national and regional point of view. *Fisheries Research* **14**, 95–104.

Thomas, G L, and Mathisen, O A (1993). Biological interactions of natural and enhanced stocks of salmon in Alaska. *Fisheries Research* **18**, 1–17.

Thomas, G L, Backus, E H, Christensen, H H, and Weigand, J (1991). Prince William Sound/Copper River/Gulf of Alaska Ecosystem. J Dobbins Association WA. DC. 15 pp.

Thorne, R E (1983). Assessment of population abundance by hydroacoustics. *Biological Oceanography* **2**, 254–61.

Traynor, J J, and Ehrenberg, J E (1979). Evaluation of the dual-beam acoustic fish target strength method. *Journal of Fisheries Research Board of Canada* **36**, 1065–71.

Walters, C J, Stocker, M, Tyler, A V, and Westrheim, S J (1986). Interaction between Pacific cod *Gadus macrocephalus* and herring *Clupea harengus pallasi* in Hecate Strait, British Columbia. *Canadian Journal of Fisheries and Aquatic Sciences* **43**, 830–8.

West, C J, and Larkin, P A (1987). Evidence of size selective mortality of juvenile sockeye salmon (*Oncorhynchus nerka*) in Babine Lake, British Columbia. *Canadian Journal of Fisheries and Aquatic Sciences* **44**, 712–21.

Willete, M (1993). Pink Salmon Investigations in Prince William Sound after the *Exxon Valdez* oil spill. In 'Proceedings of the *Exxon Valdez* Oil Spill Symposium'. (Eds D Wolfe, R Spies, D Shaw and P Bergman). February 2–5, 1993. Anchorage Alaska. 355 pp.

Wolfe, D, Spies, R, Shaw D, and Bergman, P (Eds) (1993). Proceedings of the *Exxon Valdez* Oil Spill Symposium. (Eds D Wolfe, R Spies, D Shaw and P Bergman). February 2–5, 1993. Anchorage Alaska. 355 pp.

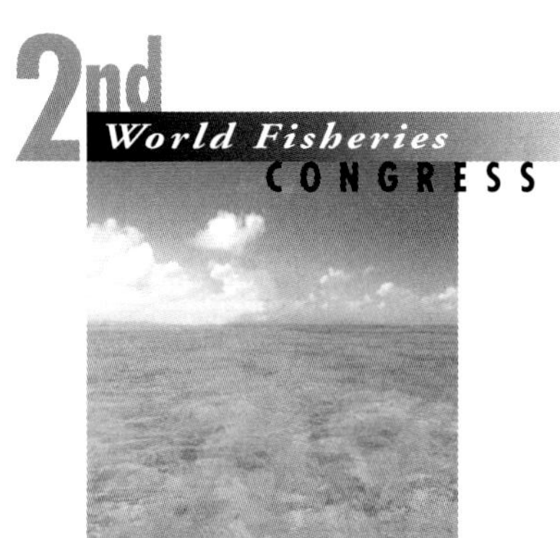

Distribution, Abundance, Reproductive Biology, Age and Growth of *Loligo chinensis* and *Loligo duvauceli* in the Western Gulf of Thailand

C. Chotiyaputta

Marine Fisheries Division, Department of Fisheries, Kasetsart University Campus, Bangkok 10900, Thailand.

Summary

At monthly intervals throughout 1990, trawl surveys were conducted in day-time at 23 stations with depths ranging from 10 to >50 m. Adults of *Loligo chinensis* ranged from 6 to 42 cm mantle length (ML) and were most abundant at depths of >30 m, and adults of *L. duvauceli* ranged from 6 to 30 cm ML and were distributely randomly in all depths and in all areas. Mature females were found throughout the year, but spawning peaked in February–May. Data from the research vessel, from commercial trawlers and from squid light-fishing were used in the estimation of age and growth by length–frequency analysis and by reference to daily increments of the statoliths. Growth was rapid, and mature sizes were attained within about four months for *L. chinensis* and about three months for *L. duvauceli*. For each species, the life span was estimated to be about one year.

Introduction

The western Gulf of Thailand is one of the major fishing grounds of Thailand, especially for squids and other cephalopods. Chotiyaputta (1992) reported that *Loligo duvauceli* is the most abundant species of squid in the western Gulf; it is distributed randomly in all areas from shallow water to depths of more than 50 m. *L. chinensis* occurs in water of >10 m depth and is more abundant in deeper waters. Juveniles of each species are randomly distributed in waters of 10–50 m in depth and are most abundant at depths greater than 20 m, in waters offshore of Chon-Buri Province and from Hua-Hin in Prachuab-Kir-Khan Province to Langsuan of Chumphon Province (Chotiyaputta 1993). In *L. chinensis*, the minimum mature sizes have been reported as 10.5 and 10.0 cm mantle length (ML) and the sizes at 50% maturity are 15.7 and 14.0 for male and female, respectively; in *L. duvauceli* the minimum mature sizes were 8.0 and 7.0 and the sizes at 50% maturity were 13.5 and 9.5 cm ML for male and female respectively (Chotiyaputta 1990). Peaks in abundance of mature females occurred two or three times per year and corresponded with those of the males.

Definitions of sexual maturity differ, but most have been based on the presence of spermatophores and eggs in the gonads

(Fields 1965; Araya and Ishii 1974; Lipinski 1979; Juanicó 1983; Mangold 1983).

Growth rates of Loliginidae vary from 4–5 mm to more than 8–9 cm per month, depending on locality, season, food availability and species (Fields 1965; Natsukari *et al.* 1988; Jackson 1990; Rodhouse and Hatfield 1990; Jackson and Choat 1992; Natsukari and Komine 1992).

Jackson and Choat (1992) determined age and growth of *L. chinensis* in Australian waters from both statolith and length–frequency data. Growth was most rapid during summer, with males showing higher growth rates than females and reaching adult size in <200 days. Length–frequency analysis indicated slow growth, whereas statolith analysis indicated rapid growth and attainment of full size in months rather than years. Growth parameters L_∞ and K estimated from length–frequency analysis were 18.6 cm and 0.8 respectively.

From length–frequency data, Supongpan (1988) estimated L_∞ and K to be 40.9 cm and 0.49 for *L. chinensis*, and 26.6 cm and 0.56 for *L. duvauceli*. Meiyappan and Srinath (1989) determined L_∞ and K of *L. duvauceli* to be 37.2 cm and 1.1 for males, and 23.8 cm and 1.7 for females, respectively.

MATERIALS AND METHODS

RV *Pramong 4*, equipped with standard trawl gear (cod-end 4 cm, cover net 2.5 cm and 1 cm mesh size), was operated at 2.5 knots in day-time at 23 stations of a fixed grid (Fig. 1). About 238 one-hour hauls were carried out over 12 monthly sampling cruises between January and December 1990. Squids from the cod-end and cover net of 2.5 cm mesh were classed as adults and those from the cover net of 1 cm mesh as juveniles. Data were used in analyses of catch composition, distribution, abundance, maturity, recruitment, spawning ground, spawning season, age and growth.

Monthly length-distribution data and some statoliths were collected from commercial trawlers and light-fishing operators; catches were landed at fishing ports in the Prachuap-Kiri-Khan, Chumphon and Surat-Thani provinces. The techniques used for extraction, storage, mounting, grinding and counting of statoliths were as described by Dawe and Natsukari (1991).

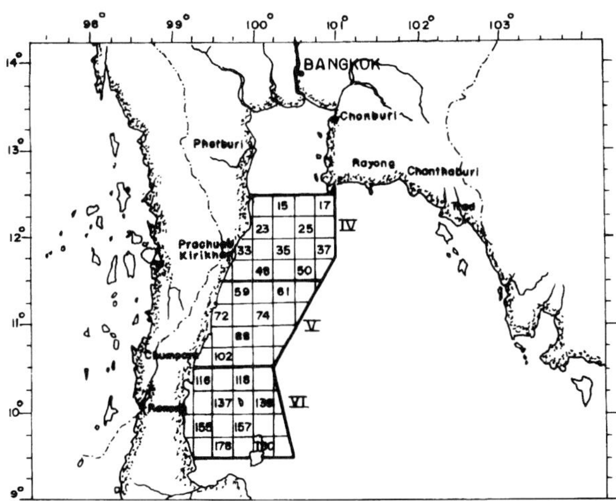

Fig. 1. Survey stations of the western Gulf of Thailand.

Distribution and abundance

Catch data were used to estimate the abundance. The mean catch-per-unit-effort (CPUE) is an index of stock abundance.

Maturity

Five stages of maturity were recognized on the basis of a visual assessment of the gonads and accessory reproductive glands.

Female:

I. Immature: ovary small, transparent, membranous, with no granulate structure; nidamental gland noticeable.

II. Maturing: eggs small, nidamental gland and accessory nidamental gland small to large; ovary granulate, whitish, opaque.

III. Mature: nidamental gland and accessory nidamental gland large; ova of two different stages present: (1) oval or polygonal, whitish opaque; (2) round, reticulate, pale yellowish.

IV. Fully Mature: nidamental gland and accessory nidamental gland very large, accessory nidamental gland bright red; ovary enlarged to fill dorsal portion of body cavity with reticulate pale-yellowish ova; oviduct filled with rounded, transparent, fully mature ova.

V. Spent: gonad small, nidamental gland relatively large and soft; a small number of large ova remaining in the ovary.

Male:

I. Immature: testis membranous, no sperm present in spermatophoric sac.

II. Maturing: testis clearly visible; seminal vesicle and spermatophoric sac well developed; spermatophoric sac with few spermatophores, soft, whitish with structureless particles.

III. Mature: testis compact and voluminous; spermatophores developed in spermatophoric sac, but not full.

IV. Fully Mature: testis rigid; spermatophores densely packed in spermatophoric sac and apparent base and tip of penis.

V. Spent: testis long and thin, few sperm in spermatophoric sac.

Reproductive biology

Fecundity was determined from mature females. The smallest mature sizes of both male and females were observed. The percentages of individuals that were sexually mature were calculated from a logistic equation. The significance of the sex ratio was assessed by χ^2 test. The fecundity index of the population was derived from monthly data on the proportion of mature specimens, fecundity (egg production as a function of size) and abundance (catch rate) of females; these data were used to determine the peak of spawning.

Age and growth

The length–frequency data analyses were based on the von Bertalanffy growth model and used the FiSAT, FAO-ICLARM package for microcomputer as stock assessment tool (Gayanilo *et al.* 1994). L_∞ was derived from the Powell–Wetherall plot, $\bar{L} - L' = a + bL'$, where $\bar{L}$ is mid length, L' is the lower limit of the corresponding length interval, and $L_\infty = -a/b$

The statolith analyses were based on the assumption that the increment is formed daily (Jackson 1990). The relationship between mantle length and statolith daily growth increments could be expressed as a linear regression.

The length of specimens observed (L_t) and daily growth increments (t) from the statolith studied and L_∞ from analyses of length–frequency data were used to determine K and t_0 from the von Bertalanffy equation:

$$-\ln(1 - L(t)/L_\infty) = -Kt_0 + Kt.$$

Length-at-age was derived from the von Bertalanffy growth equation:

$$L_t = L_\infty\left(1 - e^{K(t-t_0)}\right)\cdot e^{-K(t-t_0)}$$

Results

Catch, distribution and abundance

Catches of demersal resources comprised squid 20.9%, other cephalopods 2.9%, demersal fish 36.1%, pelagic fish 5.7%, other invertebrates 3.2% and trash fish 31.2%. Loliginid squids caught were *Loligo duvauceli, L. chinensis, Loliolus sumatrensis* and *Sepioteuthis lessoniana*, with *L. duvauceli* representing 64.22% and *L. chinensis* 16.44% of the total squid catch.

The average CPUE of total demersal resources in the western Gulf of Thailand was 24.43 kg/h, of which squid accounted for 5.11 kg/h. Catches of each species were highest in February–April and October–November.

L. chinensis was abundant in areas IV and V (Fig. 1) at depths of 20 m to >50 m, being most abundant at depths of 30 to >50 m. Catches were highest in January–March and October–December. Adults ranged from 6 to 42 cm ML. *L. duvauceli* was distributed in all survey areas from a depth of 10 m to >50 m and was abundant throughout the year, with peaks around February–April, June–August and October–November. Adults ranged from 6 to 30 cm ML.

Juveniles were found throughout the year but each species was most abundant in January and March–June; *L. chinensis* occurred at depths of 30–50 m and *L. duvauceli* at 20–40 m. Juveniles and adults were found at the same stations (Fig. 2).

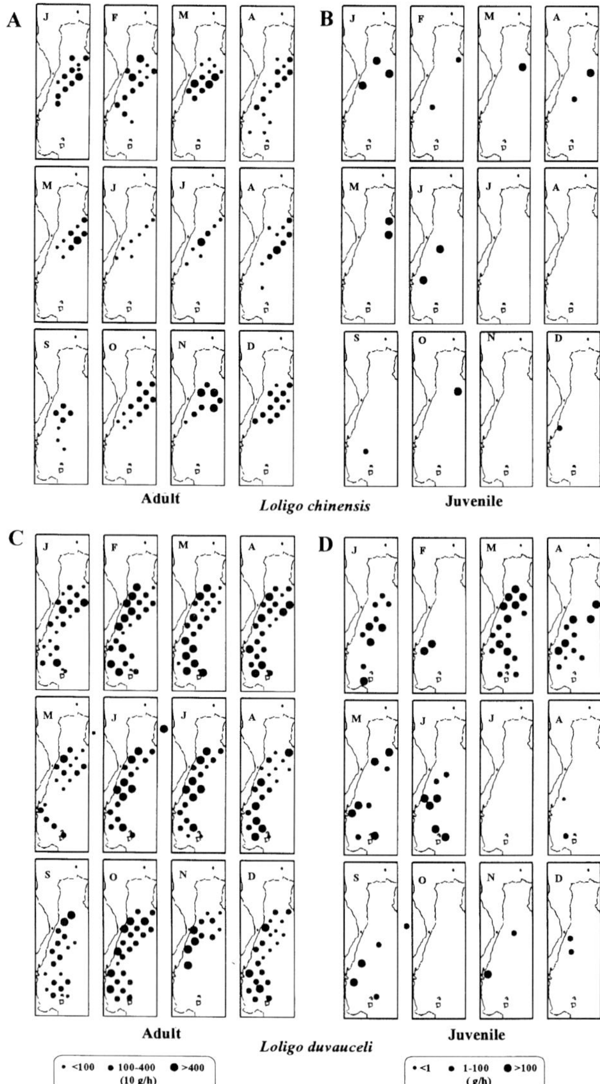

Fig. 2. Distribution and abundance of squid in the western Gulf of Thailand.

Reproductive biology

Loligo chinensis. The minimum mature size for both male and female was about 8.5 cm ML. The sizes of 50% maturity for males and females were 17.3 and 15.4 cm ML respectively. Mature females (maturity stages III to V) were found throughout the year. Peaks in abundance of mature females were in March–April, June–July and, to a lesser extent, November–December. Mature males were also found throughout the year, with peaks in abundance corresponding with those of females. Fecundity (F) was estimated from 10 mature females of 11–20 cm ML, with the number of ova per female ranging from 3000 to 11 000; it could be expressed as $F = 83.44[ML] - 6673.79$ ($r^2 = 0.8957$).

The largest mean mature size (19.5 cm ML) was found during November–December and the smallest (~15.5 cm ML) in June. The number of females caught in any month was usually significantly greater than that of males, although the ratio in a month's catch was 1:1 in February, March, April, June and July. The monthly relative abundance of females in the western Gulf fluctuated, being low in April and June. The population fecundity index in terms of the egg production pattern yielded the peaks in April and June. All reproduction data were combined as an expression of reproductive dynamics that was used to determine spawning season (Fig. 3A). March–April and

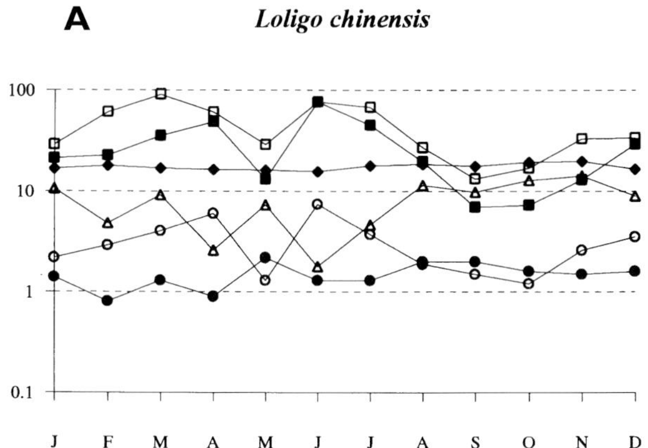

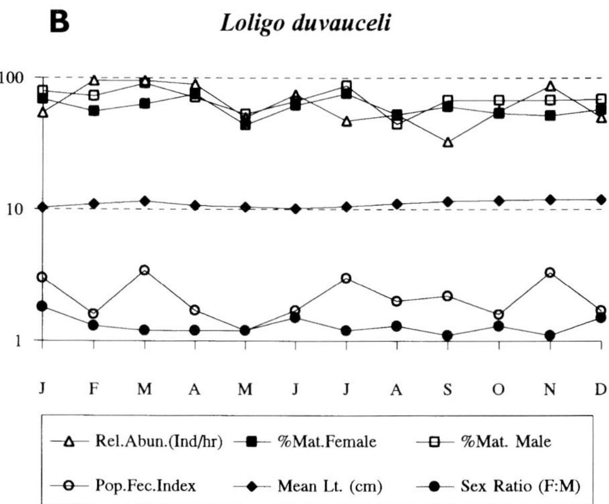

Fig. 3. Reproductive dynamics of squids.

June–July seem to be the most productive spawning peaks, with a high percentage of spawners, high relative abundance and large mean mature size. There was another small peak in November–December as spawners and population fecundity index slightly increased from October. Juveniles of 0.5 to 6 cm ML were abundant at depths of 30–50 m in January and March–June, and there was a small peak in October (Fig. 4); this pattern resulted from corresponding peaks in spawning (Fig. 3A).

Loligo duvauceli. The minimum mature sizes for male and female were 7.0 and 6.5 cm ML, and sizes at 50% maturity were 12.4 and 10.2 cm, respectively. Mature females were found throughout the year. Peaks in abundance of mature females were in January, March–April, June–July and September–December. Maturity of the males coincided with that of the females. Fecundity was estimated from 20 mature females of maturity stages III and IV and 8.0–15.5 cm ML, with the number of ova per female ranging from 1500 to 10 000; it could be expressed as $F = 99.75[\mathrm{ML}] - 6187.29$ ($r^2 = 0.9644$).

The largest mean mature size (11.9 cm ML) occurred during November–December and the smallest size (10.1 cm ML) in June. Females were usually more abundant than males but the ratio was 1:1 in September and November. All data on reproduction were combined to determine spawning season. March, July and November seem to be productive spawning peaks as a consequence of the high percentage of mature females, population fecundity index and relative abundance (Fig. 3B). Juveniles were found throughout the year. They ranged from 0.5 to 5.0 cm ML, and their peaks in abundance (Fig. 4) corresponded to earlier peaks in spawning at depths of 20–40 m.

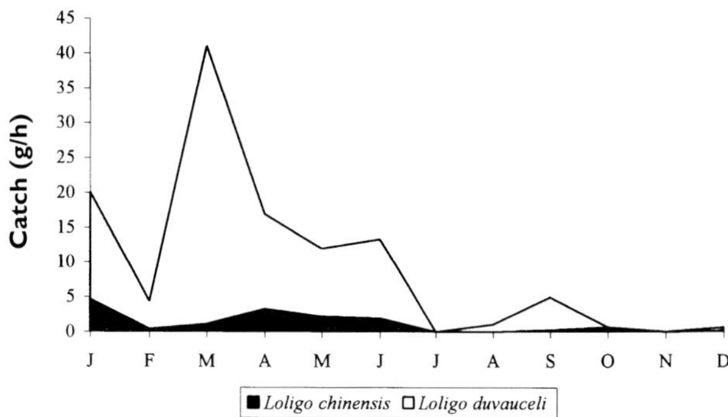

Fig. 4. Abundance of juvenile squids in the western gulf of Thailand.

Juveniles were found throughout the year. They ranged from 0.5 to 5.0 cm ML, and their peaks in abundance (Fig. 4) corresponded to earlier peaks in spawning at depths of 20–40 m.

Age and growth

Loligo chinensis. Length–frequency analysis of 2010 specimens (822 males and 1188 females) ranging from 5 to 41.5 cm ML produced estimates for L_∞ of 42.04 and 36.77 for males and females respectively.

The statoliths studied were 19 from males (8.5–27.1 cm ML) and 18 from females (8.0–23.5 cm ML); they had 76–172 and 74–151 daily increments respectively. The relationship between mantle length (L_t) and daily growth increments (t) can be expressed as:

Male: $L_t\,(\mathrm{cm}) = 0.17386t - 4.6832$ ($r^2 = 0.90526$)

Female: $L_t\,(\mathrm{cm}) = 0.16143t - 4.8541$ ($r^2 = 0.83949$)

Growth parameters K and t_0 were estimated from:

$$-\ln(1 - L_t / L_\infty) = -Kt_0 + Kt$$

Male: $\ln(1 - L_t / L_\infty) = 0.0072t - 0.3504$ ($r^2 = 0.8698$; $n = 19$)

hence, t_0 is 48.91 days (or 0.1340 years); and K is 0.0072 day^{-1} (or 2.62 year^{-1}).

Female: $-\ln(1 - L_t / L_\infty) = 0.0074t - 0.3651$ ($r^2 = 0.7602$; $n = 18$)

hence, t_0 is 49.45 days (or 0.1355 years); and K is 0.0074 day^{-1} (or 2.70 year^{-1}).

Loligo duvauceli. Length–frequency analysis of 28 775 specimens (11 804 males and 16 971 females) ranging from 2.5 to 36.5 cm ML produced estimates for L_∞ of 32.65 and 30.5 for male and female respectively.

The statoliths studied were 16 from males (47–119 mm ML) and 15 from females (42–110 mm ML); they had 66–98, and 50–95 daily increments respectively. The relationship between mantle length and daily growth increments can be expressed as:

Male: $L_t\,(\text{cm}) = 0.19998t - 8.020$ $(r^2 = 0.6799)$

Female: $L_t\,(\text{cm}) = 0.16057t - 4.428$ $(r^2 = 0.9157)$

K and t_0 can be estimated as follows:

Male: $\ln(1 - L_t/L_\infty) = 0.0083t - 0.3819$ $(r^2 = 0.6718;\ n = 14)$

hence, t_0 is 45.86 days (or 0.1256 years), and K is 0.0083 day^{-1} (or 3.03 year^{-1}).

Female: $\ln(1 - L_t/L_\infty) = 0.0069t - 0.2271$ $(r^2 = 0.9182;\ n = 10)$

hence, t_0 is 32.87 days (or 0.0901 years), and K is 0.0069 day^{-1} (or 2.52 year^{-1}).

The fit between length at age-of-squids obtained from analysis and the von Bertalanffy model is shown in Figure 5. It seems that *L. chinensis* grow to first mature size in about four months and *L duvauceli* in about three months. The lifespan is about one year.

Conclusions

The studied areas IV, V and VI of the western Gulf of Thailand yield high CPUE for demersal resources including squids. The most abundant species of squid is *L. duvauceli*, yielding 3.3 kg/h, which represents 56.2% of the cephalopod catch or 13.4% of the total demersal catch. The *L. chinensis* catch is 0.88 kg/h, representing 15.1% of the cephalopod catch or 3.6% of the total demersal catch.

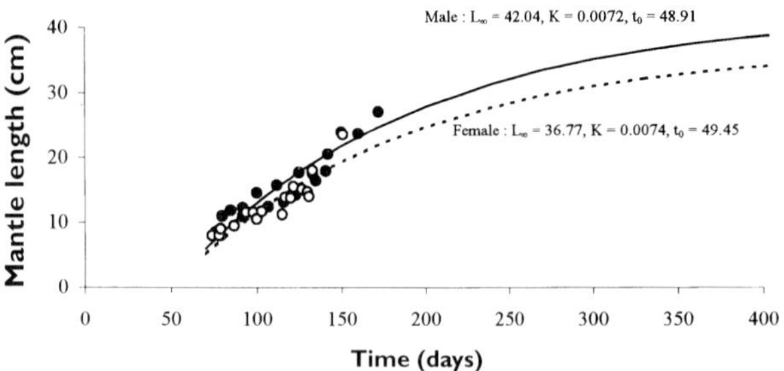

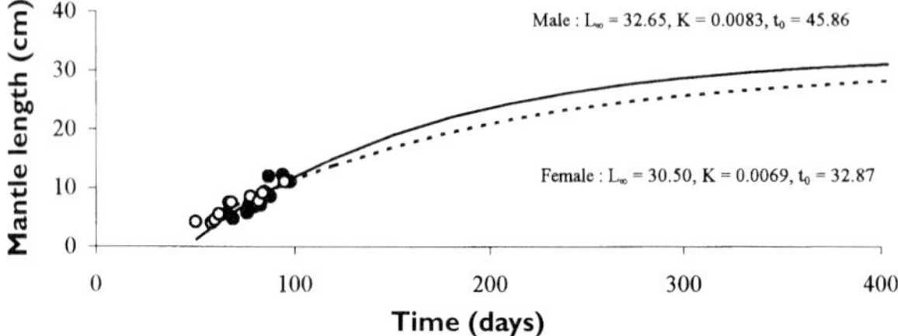

Fig. 5. Length-at-age of squids derived from statoliths (○: female, •: male) fitted with von Bertanlanffy equation.

The data on abundance suggest that spawning activity is highest during the inter-monsoon period (February–May), during the south-west monsoon (June–October) and during the north-east monsoon (November–January), with the most productive peak being during February–May. Some inter-annual variability may occur.

On the assumption that the spawning ground, nursery ground and fishing ground are all the same area, the spawning grounds of *L. chinensis* appear to be off Prachuab– Kiri-Khan Province, and the spawning grounds of *L. duvauceli* appear to be near the shoreline, from Prachuab–Kirikhan Province southward to Chumporn Province, and around Samui and Pha-Ngan Islands.

The results of age determination indicate that these squids grow very rapidly, reaching mature adult size in a few months (four months in *L. chinensis* and three months in *L. duvauceli*); this suggests that squids of the two species are short-lived. They are tropical species, with little seasonal influence on the life cycle, and they spawn throughout the year, although with more productive peaks about twice a year. The life span may be about one year.

Acknowledgments

I am indebted to the staff and crews of the Research Vessel and of the Resources Survey Unit, Marine Fisheries Division, for collecting specimens and data. I am very grateful to Dr Takashi Okutani and Dr Erik Ursin for their valuable advice and comments on the study.

References

Araya, H, and Ishii, M (1974). Information on the fishery and ecology of the squid *Doryteuthis bleekeri* in the waters of Hokkaido. *Bulletin Hokkaido Regional Fisheries Research Laboratory* **40**, 1–13.

Chotiyaputta, C (1990). Maturity and spawning season of squids; *Loligo duvauceli* and *Loligo chinensis* from Prachuab-Kirikhan, Chumporn and Surat-Tani Provinces. Technical Paper Biological Studies Sub-division, Marine Fisheries Division, Department of Fisheries (Thailand) No. 11, 13. (In Thai with English abstract).

Chotiyaputta, C (1992). Trawl survey of marine resources in the Gulf of Thailand from Chonburi to Surat-tani Provinces, 1989. Technical Paper Marine Resources Survey Unit, Marine Fisheries Division, Department of Fisheries (Thailand) No. 2, 48. (In Thai with English abstract).

Chotiyaputta, C (1993). A survey on diversity and distribution of juvenile squids in the inner and western Gulf of Thailand. *Thai Marine Fisheries Research Bulletin* **4**, 19–36.

Dawe, E G, and Natsukari, Y (1991). Light microscopy. In 'Squid age determination using statoliths'. (Eds P Jereb, S Ragonese and SV Boletzky) pp. 83–95. Proceedings of the International Workshop held in the Instituto di Technologia della Pesca e del Pescato. NTR–ITPP.

Fields, W G (1965). The structure, development, food relations, reproduction, and life history of the squid *Loligo opalescens* Berry. California Fish and Game, Fisheries Bulletin No. 131. 108 pp.

Gayanilo, F C, Jr, Sparre, P, and Pauly, D (1994). 'The FAO–ICLARM Stock Assessment Tools (FiSAT) User's Guide.' FAO Computerized Information Series (Fisheries) No. 6. (FAO: Rome). 186 pp.

Jackson, G D (1990). The use of tetracycline staining techniques to determine statolith growth ring periodicity in the tropical loliginid squids *Loliolus noctiluca* and *Loligo chinensis*. *Veliger* **33**, 309–93

Jackson, G D, and Choat, J H (1992). Growth in tropical cephalopods: an analysis based on statolith microstructure. *Canadian Journal of Fisheries and Aquatic Sciences* **49**, 218–28.

Juanicó, M (1983). Squid maturity scales for population analysis. In 'Advances in Assessment of World Cephalopod Resources'. (Ed J F Caddy). *FAO Fisheries Technical Paper* **231**, 341–78.

Lipinski, M (1979). Universal maturity scale for the commercially important squids. The results of maturity classification of the *Illex illecebrosus* population for the years 1973–1977. International Commission for the Northwest Atlantic Fisheries Research Document 79/2/38, Series No. 5364,40.

Mangold, K (1983). Food, feeding and growth in cephalopods. Memoir of the National Museum of Victoria No. 44, 81–93.

Meiyappan, M M, and Srinath, M (1989). Growth and mortality of the Indian squid (*Loligo duvauceli*) off Cochin, India. In 'Contributions to the Tropical Fish Stock. Assessment in India'. (Eds S C Venema and N P van Zalinge) pp. 1–14. FAO/DANIDA/ICAR National Follow-up Training Course in Fish Stock Assessment, Cochin, India, 2–28 November 1987.

Natsukari, Y, and Komine, N (1992). Age and growth estimation of European squid, *Loligo vulgaris*, based on statolith microstructure. *Journal of the Marine Biological Association of the United Kingdom* **72**, 271–80.

Natsukari, Y, Nakasone, T, and Oda, K (1988). Age and growth of loliginid squid *Photololigo edulis* (Hoyle, 1885). *Journal of Experimental Marine Biology and Ecology* **116**, 177–90.

Rodhouse, P G, and Hatfield, E M C (1990). Age determination in squid using statolith growth increments. *Fishery Research* **8**, 323–34.

Supongpan, M (1988). Assessment of Indian squid (*Loligo duvauceli*) and mitre squid (*L. chinensis*) in the Gulf of Thailand. Contributions to Tropical Fisheries Biology. In 'FAO/DANIDA Follow up Training Course'. (Eds S C Venema, M J Christensen and D Pauly.) *FAO Fisheries Report* **389**, 25–41.

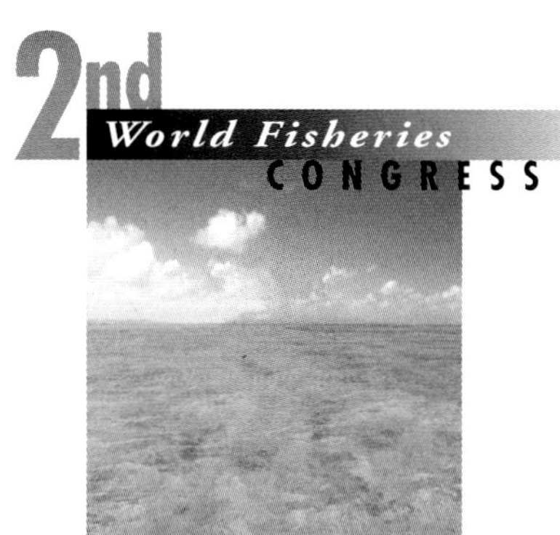

Monitoring the Abundance of Nesting Olive Ridley Sea Turtles (*Lepidochelys olivacea*), in 'La Escobilla', Oaxaca, Mexico

Rene Márquez-M,[A] *Cuauhtemoc Peñaflores*[B] *and Javier Vasconcelos*[B]

[A] INP/CRIP-Manzanillo, Playa Ventanas (AP 591), Manzanillo, Colima CP 28200, Mexico.
[B] INP/Centro Mexicano de la Tortuga-Mazunte, AP 16, Puerto Angel, Oaxaca, Mexico.

Summary

Annual numbers of nests laid in rookeries by marine turtles can be interpreted as an indication of population size. Commonly these numbers have wide variations between years. Reductions in the size and number of 'arrivals' may be interpreted as signs of overexploitation. At 'La Escobilla' beach, the annual number of nests has varied widely. Between 1973 and 1988 a steady decrease was documented, from 401 000 nests in 1975 to only 55 700 nests in 1988. From this minimum in 1988 a rise commenced just after the declaration of a total ban on turtle harvesting in 1990. In 1994 and 1995, there were 718 800 and 679 500 nests respectively. This rapid recovery has resulted from 30 years of beach protection and the end of excessive harvesting. Nowadays, arrivals are larger than historical records, and it will be important to know the maximum carrying capacity for this beach. Because of the rapid recovery, turtle fishermen are pressing the government to re-open the fishery to exploitation, arguing that this population is not now endangered.

Introduction

The annual number of nests laid by marine turtles in certain sandy beaches can be used as an indication of the population size. Usually, numbers of nests have wide variations between nesting seasons.

Some species of turtles, like 'Ridley' turtles, *Lepidochelys olivacea*, nest in 'arrivals' — an arrival being the synchronized event during which large aggregations of female sea turtles approach the beach to nest. Arrivals usually happen around the third quarter of the moon. This event is also erroneously named 'arribada', (a word which is more correctly used to indicate the arrival of a ship to port, than to designate the massive nesting of turtles); the correct name in Spanish is 'arribazón'. Reductions in the numbers of turtles participating in, and the frequency of arrivals, may be a warning sign, which if ignored could affect the stability of the population. On the other hand, the observation of more frequent arrivals with larger numbers of participating turtles can be interpreted as a signal of a recovering breeding population.

Observations and Methods

'La Escobilla', in the State of Oaxaca, is a sandy beach, 7 km long, on the Pacific coast of Mexico. Extreme changes have been recorded at 'La Escobilla', particularly a reduction in the number of mature female turtles that approach the beach to nest. 'La Escobilla' beach has been officially patrolled and protected since 1967 and the numerical abundance of nesting has been recorded since 1973. During the first period of surveys (1973–1988) a steady reduction was observed. In 1975, 401 000 nests were counted in four turtle arrivals (Márquez-M *et al.* 1976), but in 1988, only 55 700 nests were laid in the same number of arrivals. Each female may lay two or three nests per season (Márquez-M 1990). After the minimum abundance of breeding females in 1988, a rise was observed, assisted by the declaration, on 31 May 1990, of a total ban on turtle harvesting (Diario Oficial 1990). In 1990, a remarkable recovery of the turtle population was measured, with over 154 000 nests in four arrivals and four years later the number of nests increased four-fold to 718 800 nests in 13 arrivals between May 1994 and February 1995. In the 1995 breeding season a similar level of turtle nesting was recorded, with 670 500 nests in eight arrivals between May and December. In addition, more than 9000 nests were recorded in a further arrival in March of 1996.

Until now, there has been no other experience in any part of the world that shows this kind of rapid recovery in the turtle populations. Usually the populations do not respond so rapidly to protective measures. This surprising recovery is believed to result from the ending of excessive harvesting, which was practised in front of the nesting beach during the breeding season. This measure could not have been effective without the programme of beach protection instituted at this beach since 1967. In other words, the reproduction and recruitment of the breeding colony was reduced but not halted during the harvesting period and apparently, the production of hatchlings was enough to maintain the population at a healthy level, allowing rapid recovery to the former levels of abundance.

At 'La Escobilla', the largest historical breeding season was in 1975, with more than 401 000 nests, recorded in four arrivals. A new record was obtained 20 years later, in 1995, with almost double the number of nests for 13 arrivals (718 000). In future, it will be interesting to know the maximum carrying capacity of this beach and the optimum size of the breeding population, particularly because turtle breeding success is highly density-dependent, and it is possible that a collapse in hatching and survival can occur when the availability of space is limited by high densities of breeding females. For example, the beach of Nancite, Costa Rica, which is one km compared with 'La Escobilla's' seven km, has a survival rate during incubation and hatchout of less than 5% of the total number of eggs produced (Cornelius 1982); it has a similar numerical abundance of nesting females, but lower density by area, and the survival rate is usually not less than 25%.

During the breeding season some populations of sea turtles assemble in great numbers in front of nesting beaches. In these areas, a kind of sexual segregation occurs, where the sex ratio is highly biased toward females. The olive ridley sea turtles, before the breeding season, assemble several hundred metres offshore, waiting for a signal that triggers the massive, synchronized nesting. Spectacular arrivals of olive ridley sea turtles also occur at other beaches in the world, namely Gahimartha, India (Kar and Bhaskar 1982), Nancite and Ostional in Costa Rica (Cornelius 1982) and Eilanti in Surinam (Pritchard 1969). However on the Pacific coast of Mexico, the beach named 'La Escobilla', Oaxaca (Márquez-M *et al.* 1976), is now the most important breeding site for the species in the western hemisphere.

MONITORING AT 'LA ESCOBILLA'

Seasonal variations in the population sizes of turtles may be of natural origin or induced by man. To understand these variations, detailed monitoring of the abundance of nests must be undertaken every season. Individual female turtles have annual, biennial, triennial or irregular breeding cycles. Therefore, as a result of this non-synchronous breeding behaviour, large seasonal changes in abundance of turtles on nesting beaches can be measured. Such natural fluctuations must be separated from those produced by overharvesting, which are evident as steady declines in the numbers of nests per season.

At 'La Escobilla' nearly all the arrivals have been recorded each year since 1973 (Márquez-M *et al.* 1976) by researchers of the Sea Turtle Program at the Instituto Nacional de la Pesca. During the last ten years, they have been assisted by students of the Oaxaca State University and volunteer workers. Because nesting occurs at night and the number of nesting females is very large, a direct count is not possible. Therefore, a sampling method based on the number of nests was used.

Evaluation of sampling results was based on the model of Márquez-M and van Dissel (1982) and the results in Figures 1 and 2 show the status of the population and its trend. At 'La Escobilla' beach, turtles were harvested just in front of the nesting beach up to 1990. The lowest abundance occurred in 1988, and during this period the most intensive harvesting occurred. The average number of arrivals was usually less than 6 (mean = 4.71) per nesting season. However, from 1990 to 1995, turtle harvesting was banned and there have been more arrivals, between 4 and 13 per nesting season (mean = 8.67). The maximum number of arrivals occurred in 1994 and 1995 (Figs 3A and 3B).

It has been observed that when the olive ridley turtle nesting population is not disturbed greatly, arrivals are highly synchronized, occurring at night around every third quarter of

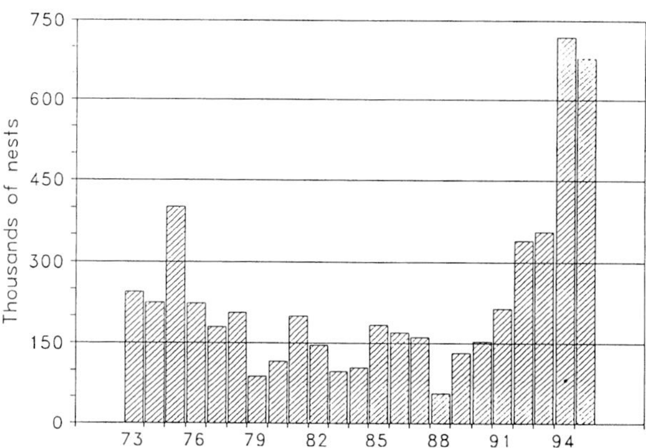

Fig. 1. Annual number of nests in arrivals of olive ridley sea turtle, at 'La Escobilla', Oaxaca, 1973–1995.

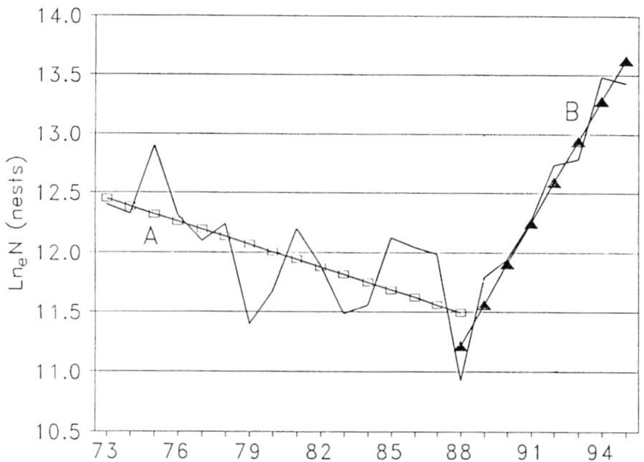

Fig. 2. Change of abundance in adult female population of olive ridley sea turtle, at 'La Escobilla', Oaxaca. Derived from Ln_e of annual number of nests (N) and time (t years).
A: $Ln_e N = -0.0634*t + 12.513$; B: $Ln_e N = 0.345*t + 5.7007$.

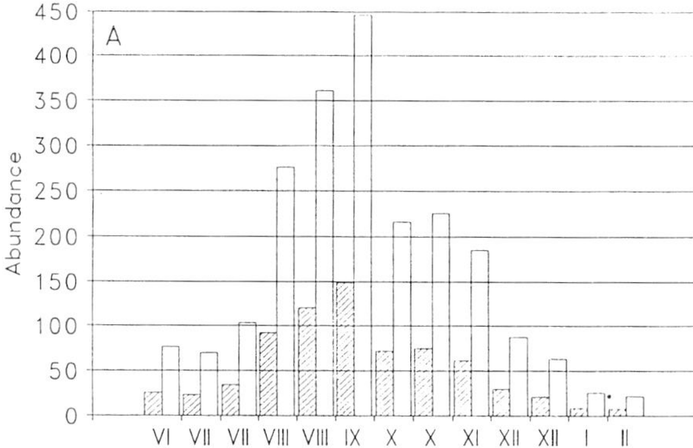

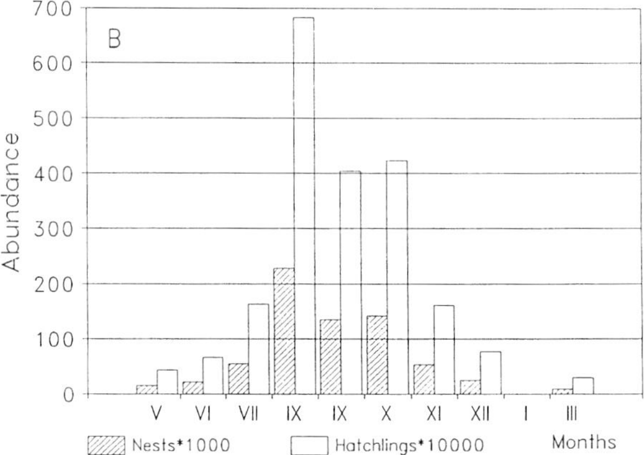

Fig. 3. Arrivals of mature female olive ridley sea turtle (*Lepidochelys olivacea*) and hatchout, at in 'La Escobilla', Oaxaca. A: Breeding season of 1994–1995, B: Breeding season of 1995–1996.

the moon. But during periods of commercial exploitation when overharvesting is obvious, the usual periodicity of arrivals breaks down and the abundance of turtles decreases rapidly. Consequently, when the normal periodicity of arrivals is resumed, it can be assumed that the population is recovering to its optimum level (Figs 1, 2, 3A, 3B).

According to Figure 1, the olive ridley population of 'La Escobilla', has varied greatly in abundance and it was predicted in 1978 that the arrivals (population) would disappear in less than a decade (Cahill 1978). In spite of the prediction, the arrivals were smaller but not completely interrupted, indicating that recruitment to the adult population has never ceased. Conservation work carried out since 1967, has helped produce several million hatchlings per year. During the harvesting period (1955–1989), regulations such as closed seasons, quotas, restricted areas, etc. contributed towards maintaining the population up to 1988. After the lowest number of 57.7 thousand nests laid in 4 arrivals was recorded in 1988, the survival of the species was considered at risk of extinction by governmental researchers and many non-governmental organizations (NGOs) (ecologists and conservationists groups) and it was decided to implement plans to forbid sea turtle exploitation nation-wide on 31 May 1990 (Diario Oficial 1990).

Discussion

The resources of the Mexican Government for the study and conservation of sea turtles during the period of exploitation (1966–1988) were insufficient to cover all nesting beaches, partly because only one federal agency was responsible for the species. But at least the three important breeding grounds — 'Mismaloya' in Jalisco, 'Piedra de Tlalcoyunque' in Guerrero and 'La Escobilla' in Oaxaca — were monitored. The lack of funding in the first years resulted in an incomplete evaluation of the size and number of arrivals, particularly because the field work was discontinued earlier each season (usually before December). Consequently, in years before the 1970s both early and late arrivals in the nesting season were not usually recorded, so the annual numbers of nests were underestimated. Also for some arrivals, turtle working group members were not present, and in these cases rough estimates of numbers of nesting females were supplied by cooperative fishermen and by local fishing inspectors. During the 1980s interest in the turtles increased and Universities and NGOs contributed further to the beach monitoring programme. In the 1990s, due to the great interest in nature conservation and ecology, full-time field workers are located on most beaches during the breeding season.

In the 1980s overexploitation clearly affected arrivals and behaviour of breeding turtles. Also, the dates of arrivals became erratic and some, such as those during October, November and December of 1982, failed. This was probably due to the harvesting of breeding females when they were aggregating in the vicinity of the nesting beach. The disruption caused by so many fishing boats crossing the area probably inhibited the formation and the trigger of some arrivals. As a result of this excessive harvesting of adult female olive ridley turtles, the population declined dramatically to a minimum in 1988. Despite a slight recovery in 1989 (Figs 1 and 2), the government decided, with consensus from NGOs, to declare a total ban on the exploitation of all sea turtle species from 31 May 1990 (Diario Oficial 1990) for an indefinite period.

Surprisingly, in the same year as the total ban, arrivals became more frequent with greater numbers of turtles, and were more precisely synchronized to the third quarter of the moon. The

Table 1. Changes in total mortality and survival rate observed in the population of olive ridley sea turtle, *Lepidochelys olivacea*, at 'La Escobilla', Oaxaca

Period (years)	Total mortality	Survival rate	r^2	Annual decrement (%)	Annual increment (%)
1973–1988	0.0635	0.9385	0.406	6.15	0
1988–1995	-0.3437	1.4115	0.950	0	41.15

numbers of arrivals and nests recovered in less than five years, and now exceed the historical records of abundance and as in good years, there are usually more than eight arrivals per year starting in May or June and continuing up to February or March of the next year..

Figure 2 compares population size trends based on linear regressions of the annual number of nests against time. Two periods were clearly separated, the first one between 1973 when monitoring of arrivals was started and the lowest point of nesting abundance in 1988; and the second between 1988 until 1995. Natural logarithmic values of the total number of nests laid by season were used in the calculation. The total mortality Z is considered the value of the slope in the formula: $S = e^{Zt}$ (Doi 1974), where S is the survival rate of the population and t is the year. The total mortality in the period from 1973–1988 is indicated by a negative slope. If the negative trend, averaging more than 6% per year, had continued, the olive ridley population would have declined to an unrecoverable level in about 16 years. Fortunately after the total ban, in less than five years the trend of the olive ridley population of' 'La Escobilla' beach recovered at an annual rate of 41% (Table 1), and is now almost twice the level of historical records. This is due to the high survival rate (S) from eggs to adults, resulting in recruitment exceeding the total mortality (Z).

The health of the population at 'La Escobilla' now is also demonstrated by the size and frequency of the arrivals during 1994 and 1995. Figures 3A and 3B show the results of monitoring arrivals throughout the year. In 1994, the first arrival started in May and the nestings continued until the last arrival in March 1995. This was the first time that 13 arrivals were recorded (Fig. 3A), with a total of 718 960 nests. The 1995 season again started in May and ended in March 1996. There were nine arrivals, and a total of 679 480 nests (Fig. 3B). However, there were no arrivals during August 1995 and January and February 1996.

Conclusions

The depletion of sea turtle populations is a common consequence of overharvesting and environmental modifications. Recoveries resulting from conservation programmes are not common, in spite of many years of work directed toward management, recovery and conservation. In the case of 'La Escobilla', a nesting beach protected since 1967 and declared as a 'Natural Reserve' in 1986 (Diario Oficial 1986), the population of olive ridley turtles was considered endangered. Perhaps surprisingly, the population has responded well to many years (1967–1995) of patrolling the beach to protect nests, incubating eggs and hatchlings. After only five years of full protection to the arrivals and the enforcement of the Total Ban (1990) on harvesting, the population appears as healthy as it was before the years of high levels of exploitation (1966–1971) (Márquez-M 1976).

The results of this work demonstrate that turtle populations can respond to protection, and recovery is possible if a well-organized programme is developed. However, it is not easy to maintain a good level of population abundance. For success, it is necessary to develop excellent field work and maintain strong enforcement of protective regulations. Despite limitations in Mexico, a good level of work was supported through Government programmes from the start, and lately through the collaboration of University students and voluntary support from NGOs and other institutions. Patience is important for recovery of turtle populations, because recovery usually takes many years and may not have an obvious end point.

Because of the spectacular recovery of the breeding population at 'La Escobilla' in just a few years, the former turtle fishermen are now pressing the government to again open the turtle fishery, arguing that the Oaxacan population cannot be considered as an endangered population any more.

Because the conservation of a natural resource *per se* has ecological value in the biodiversity of nature, it is of interest to know the carrying capacity of this nesting beach and the highest level of healthy population that it can maintain. When this is known it may be reasonable to set new harvesting quotas for eggs and or adult turtles.

Acknowledgments

Many researchers have worked gathering the information, since the Marine Turtle Program started in 1964. Pioneers of Pacific turtle camps were Antonio Montoya, Manuel Grande Vidal, Aristóteles Villanueva, Emmanuel Vargas, Daniel Ríos, etc. Many students assisted the work and, particularly in 'La Escobilla', the University of Oaxaca 'Benito Juarez'. Great assistance has been given by the Mexican Navy and Fisheries Inspectors. All those people must be recognized for the survival of this valuable species. Also the Fisheries Cooperatives of the Oaxaca State, that were harvesting the species for many years, are now a principal factor for its conservation.

References

Cahill, T (1978). The Shame of Escobilla. *Outside* (Feb), 23–7, 62–4.

Cornelius, S (1982). Status of sea turtles along the Pacific coast of Middle America. In 'Biology and Conservation of Sea Turtles'. (Ed K Bjorndall) pp. 211–20 (Smithsonian Institution Press).

Diario Oficial (1986). Decreto de Zonas de Reserva y Sitio de Refugio y Veda para la Protección, Conservación, Repoblación, Desarrollo y Control de las diversas especies de Tortugas Marinas. Diario Oficial de la Federación, México. Octubre 29, 1986, 8–10.

Diario Oficial (1990). Acuerdo que establece veda para todas las especies y subespecies de tortugas marinas en aguas de jurisdicción nacional de los litorales del Océano Pacífico, Golfo de México y Mar Caribe. Diario Oficial de la Federación, México. Mayo 31, 1990, 21–2.

Doi, T (1974). Outline of Mathematical Analysis on Fish Populations for Practical Use in Front. In 'Fisheries Biology and Population Dynamics of Marine Resources — Textbook for Marine Fisheries Research Course'. Japan International Cooperative Agency, 117–210.

Kar, C S, and Bhaskar, S (1982). Status of the sea turtles in the eastern Indian Ocean. In 'Biology and Conservation of Sea Turtles'. (Ed K Bjorndall) pp. 365–72, (Smithsonian Institution Press).

Márquez-M, R (1976). Estado actual de la pesquería de tortugas marinas en México, 1974. *Instituto Nacional de Pesca, México, Serie Información* **46**. 27 pp.

Márquez-M, R, (1990). FAO species catalogue. Vol. 11: Sea turtles of the world. An annotated and illustrated catalogue of sea turtle species known to date. FAO Fisheries Synopsis No. 125. (FAO: Rome). 81 pp.

Márquez-M, R, and van Dissel, H (1982). A method for evaluating the number of massed nesting olive ridley sea turtles, *Lepidochelys olivacea*, during an arribazón, with comments on arribazón behaviour. *Netherlands Journal of Zoology* **32**(3), 419–25.

Márquez-M, R, Villanueva, A, and Peñaflores, C (1976). Sinopsis de datos biológicos sobre la tortuga golfina, *Lepidochelys olivacea* (Eschscholtz 1829). Instituto Nacional de la Pesca, México, *INP Sinopsis de la Pesca* 2, INP/S2. 61 pp.

Pritchard, P C H (1969). Sea turtles of the Guianas. *Bulletin of Florida State Museum, Biological Sciences* **13**, 85–140.

Issues and Outcomes for Theme 5

What is the Scope for Development of Wild Stock Fisheries?

J. R. Beddington

Centre for Environmental Technology, Imperial College of Science, Technology and Medicine, London, UK.

Introduction

A substantial number of papers and posters was presented to the Congress in Theme 5 and a diverse set of topics was covered by these presentations. Any attempt at synthesis of such diversity is therefore going to run the danger of being over-simplistic and also of omitting important aspects of the subject area. Accordingly, this preamble is aimed at providing an apologia for this necessary simplification.

The Basic Issues

In the Theme presentation to this Congress, it was argued that the future of wild stock fisheries and the scope for their development is likely to depend on three key issues. These are: better management, the sustainable management of new fisheries and the enhancement of capture fisheries. To this list, having studied the presentations at the Congress, it seems appropriate to add gear design; this is an area of study where important questions of fisheries management can be resolved in a practical and straightforward way. In the remainder of this summary, some of the contributions made by individual contributors to the Congress will be briefly noted under these four topics.

Better Management

Better management of fisheries demands a better understanding of the biology of individual organisms subject to harvest, a better understanding of the interactions among elements of the ecosystem and the ability to monitor the effects of fishing on stocks. Yet the practice of fisheries management does not work in a vacuum and the other aspects of management include the institutional issues of how management interacts with the activities of the fishing industry and the controlling agencies and how management regulations are monitored and enforced.

Better understanding of biology

Within this theme there were many contributions of importance. Classification is difficult and the simplest is perhaps by species types. Squid, an increasingly important resource, had papers on key biological factors from Chotiyaputta and Wadley. Lobster work included that by Caputi *et al.* on early life-history stages of the western rock lobster, Pitcher *et al.* on the ornate rock lobster, and Treble on southern rock lobster. A key problem in invertebrate fisheries management is the estimation of age, and the paper by Day *et al.* had a novel approach to this issue for abalone.

Work on prawns and shrimps was varied, and the ecological studies by Vance *et al.* on banana prawns and Loneragan *et al.* on the importance of seagrass habitat to juvenile penaeid prawns were good examples of basic ecological work that can lead to an understanding of constraints to management.

Stock separation is a major problem to which the contributions of Lavery and Kline were contrasting approaches, and the review by Ward *et al.* of molecular genetic techniques for assessing yellowfin tuna showed the complexities involved in addressing this important species and its management.

Relatively few studies explored direct interaction between species, an exception being the study by Warburton *et al.* of prawn consumption by barramundi, which indicated the importance of seasonality.

Taken together with other presentations at the Congress which have not been mentioned, it is clear that there is throughout the world active and ingenious research aimed at solving biological problems for management. What is less clear, at least to this author, is that this work is leading to some synthesis even within the various subareas addressed.

Stock Estimation

In contrast to the plethora of different approaches and studies in basic biological research, stock estimation and assessment methods have developed in reasonably well defined areas. Three papers — Bax and Smith on orange roughy, Fuentes *et al.* on reef fish, and He and Boggs on tuna — all address with different emphasis the use of information on abundance as stocks are reduced.

Other methodologies were applied by Pauly and Nicol to krill, and Virgona and Halliday to sea mullet.

Clearly, successful management requires the ability to assess the effect that fishing is having on stock abundance. What seems to this author to be missing is a comprehensive approach that enquires how this information is used, and how important are different degrees of accuracy in the management of species of different life histories.

Institutional Issues

One of the issues that is central to the management of fisheries in both developed and developing countries is the way in which different institutions are involved in the management process. The contribution by Arenas-Fuentes and de Leon Corral demonstrated how Mexico has been addressing these problems by a division of responsibility among different institutional parts. This model of the way in which a developing country has addressed fisheries management issues was interesting and provided an analytical view of the management process from the perspective of the developing country.

Enforcement

No papers within this Theme session addressed the issue of the enforcement of fishing regulations. In another Theme, Hart developed a novel game-theoretical approach to investigate the potential for altering regulations to improve compliance. This intriguing idea appears to have important possibilities for development in the context of fisheries monitoring and surveillance.

As can be seen, this author agrees completely with the conclusion set out in the paper by Hemming and Pierce that enforcement is the most neglected area of fishery management and an area where considerable benefits can be achieved.

New Fisheries

Within this Theme session, only Bax and Smith, focussing on orange roughy, and Pitcher and Bundy had contributions of clear relevance to the sustainable management of new fisheries. The latter work involved estimation of potential yields of introduced species in African lakes using an empirical extrapolation from known yields. This seems a promising new approach.

In other Congress Themes, papers of relevance to this subject included those by Clark and Koslow *et al.* on orange roughy.

Enhancement

Contributions within the general theme of enhancement were primarily those that fell within the aquaculture theme of this Congress. This author would single out the contribution by Welcomme, which emphasized the importance of a poverty focus in the development of enhancement fisheries. Other contributions included Bianchini *et al.* dealing with the slipper lobster and Liao's Theme 4 address.

Gear Design

A significant problem of fisheries management involves the bycatch of non-target species, and a related problem is the environmental impact of fishing. Both problems can be addressed by appropriate gear design as well as by regulations.

Several papers in this Theme session addressed this problem. Indeed, Brewer and co-workers presented some four papers on different aspects. De Alteris *et al.* investigated problems of bycatch in the North Atlantic, and Kennelly examined both the North Atlantic and examples from the Pacific and New South Wales.

A number of contributions focussed on more efficient designs of gear, including Robins *et al.*, Eayrs *et al.* and Matsuoka *et al.*. The paper by Matsuoka *et al.* involved an interesting investigation of prawn behaviour using infrared video technology.

This author believes that clear linkage between scientists studying gear design and those involved in stock assessment offers significant opportunities.

Opportunities for the Future

The clear goal for fisheries management is to have a system that can provide sustainable fisheries that are economically efficient. To achieve this goal, input is needed from a substantial mix of different disciplines. The key is to understand the effect of management actions, not only on the species and the ecosystem, but also on the industry itself. It is in this latter area that the biggest gaps in knowledge occur. The scope for new multidisciplinary work within fisheries management seems to be substantial at present.

For example, to achieve sustainability of a fishery with reasonable enforcement costs, cooperation is needed between those who are involved directly with industry in the activities of monitoring and control and those involved in stock assessment and the setting of management regulations. Similarly, the difficulties of bycatch and environmental damage, which can bedevil the operation of fisheries management, need to be addressed by teams working on gear, on stock assessment and on management regulations including enforcement.

Within the scientific community there is a clear need for work that links the basic population dynamics of stock assessment with the understanding of ecosystem dynamics. A new generation of practical and workable models is required.

These opportunities need to be grasped if fisheries are to have a healthy and viable future.

Fisheries management and sustainability: A new perspective or an old problem?

Theme 6

Fisheries Management and Sustainability: A New Perspective of an Old Problem ?

S. M. Garcia and R. Grainger

FAO Fisheries Department, Via delle Terme di Caracalla, 00100 Rome, Italy.

'Sustainable economy represents nothing less than a higher social order... although it is a fundamentally new endeavour, with many uncertainties, is far less risky than continuing with business as usual.' (Brown et al. *1991).*

Summary

Numerous prescriptions for improved management of fisheries have been given since Michael Graham in 1936 formulated his 'Great law of fishing' and provided his preoccupying diagnosis and prognosis of overfishing. This paper briefly reviews the present situation of fisheries to reiterate the fact that, while fisheries provided livelihood and food to hundreds of millions of people, conventional fisheries management has largely failed to maintain resources and economic efficiency, a failure now widely recognized by governments. It reviews the causes of failure and the consequences and, taking a 'bird's eye' view of the problem, puts fisheries management in the perspective of the sustainable development theory which has become so fashionable, especially after the UNCED Rio Summit. It identifies the critical dimensions of fisheries management as: bio-ecological, technological, economic, socio-cultural, legal and institutional, spatio-temporal, informational, and extra-sectoral (macro-political, macro-economic) and briefly reviews them and their implications. The paper also underlines the potential impact of uncertainty and the need for a precautionary approach, and addresses the potential role of the market in future management strategies. The expected conclusion of the paper is that despite the fact that fisheries management rhetorics recognized sustainability as a central concept decades ago, it failed because it did not properly control extraction rates, and because to do so it would have needed to address as a priority the central political issue of resource allocation, at international and national levels. The sustainable development perspective helps by identifying the critical dimensions of the problem and the prioririties for action. It also helps in stressing the need for a holistic and participative approach. To achieve it, the paper proposes a comprehensive framework taking into account all the traditional and new requirements imposed on fisheries management and on the fishery sector, stresses the need for an integrated approach, through a nested set of solutions, and warns against the temptation to use 'quick-fix' approaches to avoid difficult political decisions.

Introduction

This paper was requested as an introduction and backdrop to Theme 6 of the 2nd World Fisheries Congress on 'What is needed to manage fisheries sustainably'. When considering the invitation, we realized that it was arrogant to pretend to know how to manage future fisheries when so much had been tried and had failed, and so few examples of success were available. However, the field of fisheries management is changing rapidly and the task of summarizing past failures as well as new requirements and initiatives in an attempt to 'map' the possible future of management was both attractive and challenging. Many explanations have already been put forward for the failures, and quick-fix solutions and fashionable approaches have been proposed, such as Individual Transferable Quotas (ITQs) or co-management, often without recognizing the deep complexity of the matter.

This paper attempts, therefore, to avoid unrealistic simplification by broad-brushing the evolution of fishery management concepts, requirements and performance, in the perspective of sustainable development, highlighting the directions of ongoing changes. There is a real risk that the reader will be overwhelmed by the complexity of the issue, but that complexity is real and stems from the accumulation of requirements imposed by only loosely related international agreements on environment, fisheries, biodiversity, trade, etc. The result is a maze of requirements, sometimes conflicting, which will be difficult to meet simultaneously.

The following sections of this paper will therefore: (1) review briefly the global challenge of sustainable development and its evolution in fisheries; (2) summarize the present state of fisheries to underline the management failure and the need for a shift in its operational paradigm; (3) identify the sectoral dimensions of fisheries management (bio-ecological, technological, economic, socio-cultural, institutional, spatio-temporal, and informational) as well as the extra-sectoral ones (macro-economic and macro-political); and (4) propose a general framework for sustainable fisheries management, underscoring the required shifts in objectives and in priority action; the need for a nested set of solutions; the actions required by Government, the sector, and the scientists; with (5) a reflection on the implementation process.

The sustainability challenge

One general deficiency of many textbooks and prescriptions on management (and one source of management failure) is the lack of explicit linkage between development and management. International development philosophies have been characterized by a general faith in growth in the 1960s; a more marked concern for equity in the 1970s; a strong advocacy for economic liberalism in the 1980s; and for sustainable development from the end of the 1980s with the work of the World Commission on Environment and Development and with UNCED and its follow-up. These philosophies have affected fisheries development to a great extent and they will most probably do so in the future. Despite slight differentces in dates, the parallel evolution of fisheries and agriculture developments, from an FAO standpoint, is evident in Table 1.

Despite the difference in geographical scale, the analogy between the redistribution of ocean resources through the establishment of Exclusive Economic Zones (EEZs) and property and use-rights, and the redistribution of land resources through agrarian reforms is evident.

Sustainable development has been defined by the World Commission on Environment and Development (1987) as 'development that meets the needs of the present generation without compromising the ability of future generations to meet their own needs'. It has also been later on defined by the FAO Council in 1991 as: 'the management and conservation of the natural resource base, and the orientation of technological and institutional change in such a manner as to ensure the attainment of continued satisfaction of human needs for present and future generations. Such sustainable development conserves (land) water, plants and (animal) genetic resources, is environmentally non-degrading, technologically appropriate, economically viable and socially acceptable'. This definition implies an objective of optimizing welfare[1] from a limited natural resource, minimizing resource and environmental degradation, and regulating the rate of use of these resources over time.[2]

Development specialists recognize two degrees of sustainability (Goodland 1995). **Weak sustainability** implies that: (a) both natural, man-made, human and social capitals are perfect substitutes for each other; (b) the total capital has to be conserved for future generations; and (c) extinction of a resource could be acceptable provided it generates sufficient man-made capital (e.g. infrastructures, alternative industries, etc.) and other sorts of capital to provide equivalent development opportunities to future generations. **Strong sustainability** assumes that all forms of capital are not equivalent but complementary and should be conserved in their own right. The whole theory of fisheries management is clearly based on strong sustainability concepts and explicitly requires the limitation of annual harvest to the net annual increment of the resource. This requirement, which Goodland considers 'absurdly strong', is central to the concept of Sustainable Yield (and MSY) in fisheries, enshrined in all modern international agreements on fisheries, including the 1982 UN Convention on the Law of the Sea, and the 1995 FAO Code of Conduct for Responsible Fisheries (FAO 1995*c*).

In fisheries, the concept of sustainable fisheries has been reformulated in that of **Responsible Fisheries** which, in an international debate dominated by the issue of sovereign rights, intends to remind States about their related duties and responsibilities. The concept, coined by the FAO Committee on Fisheries, was defined by the International Conference on Responsible Fishing (Cancun, Mexico, May 1992) and 'encompasses the sustainable utilisation of fishery resources in harmony with the environment; the use of capture and aquaculture practices which are not harmful to ecosystems, resources and their quality; the incorporation of added value to such products through transformation processes meeting the required sanitary standards; the conduct of commercial practices so as to provide consumers access to good quality products'. The concept is developed in the International Code of Conduct for Responsible Fisheries adopted by FAO in 1995.

Present situation of fisheries

A brief statement on the present state of fisheries is necessary to illustrate the relative failure of management and provide evidence of the need for immediate action. The present situation of fisheries reflects a number of positive elements which should not be overlooked. Capture fisheries produce ~90 million t of fish annually, ~35% of which is used for animals feeds. They provide livelihood directly and indirectly to ~200 million people. Following years of difficult negotiations, EEZs have been established in most areas allocating resources at international level. The threats to sustainability, from fisheries or

[1] 'Welfare' is defined by Schuh and Archibald (1996) as including '*the value of natural amenities, improvement in environmental quality, reduction of pollution and waste, and value of the inter-generational equity*'.

[2] Pontecorvo and Schranck (1995) refers to '*the welfare goal of maximizing the long-run net economic yield from the resource base*' which would '*eliminate short run profit maximization…and incorporates the biological goal of conservation*'.

Table 1. Comparison between the agriculture and fisheries development processes from an FAO perspective (from Garcia 1994 and Breth1996).

Agricultural Development	Fisheries development
1945–1959: The initial period Countries focussed on developing a world food policy, based on human needs and universal equity. It addressed the challenge of feeding a growing population from a limited resource base and the need for 'good use and good management'.	**1945–58: Post-war reconstruction** Countries focussed on reconstruction of the fishing capacity as a means to re-establish food production. The impact of fishing on resources was recognized as well as the overfishing problem.
1960–1971: Faith in growth and raising productivity The priority was in producing food where it was most needed, i.e. in the progressively independent developing world where a population explosion was ongoing. The decade witnessed major expansion in irrigated agriculture, introduction of high-yield varieties, fertilizers, pest control and disease eradication.	**1958–72: Fisheries expansion** The priority was on discovering new grounds, new resources, new technologies and long-range fleets; fleets of newly independent countries expanded, supported by intensive research programmes aiming at discovery of new resources and new technologies.
1972–1986: Managing natural resources and the human environment Following the Stockholm Conference on the Human Environment (1972), focus was on assessment and management of natural resources, on the need to address equity and participation, and for an agrarian reform. The concept of integrated management was introduced.	**1973–1983: A new economic order of the ocean** Starting the UNCLOS III process countries negotiated a new economic order and a major allocation of resources to coastal countries. UNCLOS was adopted in 1982 and formally established the basis for sustainable fisheries development and management for newly established EEZs.
1987–1992: Sustainable development Building on the work of the WCED (the Brundtland Commission), FAO addressed the need of future generations and adopted its definition on sustainable development, and the concept of Sustainable Agriculture and Rural Development (SARD).	**1984–1992: Towards global concerns** Building on the WCED, the adoption of UNCLOS and the 1984 FAO World Conference on Fisheries, countries experimented with management, discovering resource and environmental degradation, the problems of open access, developing Agenda 21 for the oceans and coastal areas.
1992–2000: The sustainability challenge Following UNCED, attempting to implement Agenda 21, implementing the SARD concept, facing the challenges of population growth, food security, intensification, erosion, desertification, water resources management, integrated pest management, decreasing terms of exchange and other overriding macro-economic forces in a globalized economy.	**1992–2000: The sustainability challenge** Following UNCED, attempting to implement Agenda 21 and developing a Code for Responsible Fisheries, countries face the challenges of population growth, coastal migration, coastal degradation, biodiversity conservation, institutional change, global climate change and overriding macro-economic forces in a globalized economy.

from the development activities on land are identified and Governments have agreed on international instruments, voluntary or legally binding. National policies are already evolving in many countries to more explicitly reflect the FAO Code of Conduct for Responsible Fisheries and other international instruments. In general (and recognizing that their impact on genetic pools may not be insignificant), fisheries have had little impact on species diversity or non-reversible impacts on ecosystems. Following UNCED and the work of the Commission for Sustainable Development (CSD), the level of awareness is very high among both Governments and people who are getting more and more involved (e.g. through Non Government Organizations (NGOs)).

The situation, however, also reflects a number of preoccupations which underscore the need for action. The combined effect of population growth and stagnation of fish supplies, has led to a decline of the *per-capita* availability of fish for human consumption in the 1990s while prices continue to increase due to the widening gap between demand and supply. World-wide, 70% of the stocks for which data are available require urgent intervention to: (a) avoid decline of the fully exploited resources; (b) stop decline of overfished resources; and (c) rebuild depleted resources (Garcia and Newton in press). The proportion of stocks considered overutilized is 45% in the USA and 59% in Europe (Rosenberg *et al.* 1993). In New Zealand, where fishery development is more recent, about 41% of the stocks are fully exploited or overfished (at or below B_{msy}), 9% are above B_{msy} and 50% are in an unknown state (Annala 1995). The world fleet reflects a global overcapacity of about 30–40% and higher figures have to be expected locally. This results in an economic waste of about \$US54 000 million (FAO 1993; Garcia and Newton in press). Discards represent a biological waste of 27 million t or about 25% of the total catch (Alverson *et al.* 1994). The degradation of coastal environments is spreading. Few species are still underutilized (e.g. mesopelagic fish, possibly krill, some whales, and many oceanic squids) and their further development faces difficulties of an economic, ecological, and ethical nature. The question of the potential threat to biodiversity is being raised, particularly in relation to the endangered species. The picture is further complicated by uncertainty about the resource level and resilience, as well as the relative impacts of fisheries, environmental degradation and climate change. The new requirement for holistic management of ecosystems has revealed a lack of understanding of their functioning and of an operational paradigm. Concern exists in relation to sustainable use of high seas resources including straddling and highly migratory stocks. Economic stress is reflected by the spreading of conflicts for resource use at inter-sectoral and international levels, civil disobedience and non-compliance (resulting also in degradation of the information base). As stated by Walters (1995), in reference to Canadian salmon fisheries, 'our historical management approach...has also produced an institutional quagmire, with grossly overcapitalised and bitterly competitive fleets, an allocation system among fishermen that is dominated more by threat of civil disobedience than by reasoned analysis of where rights and privileges ought to lie, and a publicly costly and burdensome apparatus for both biological management and economic support to fisheries'.

While the world-wide recognition of the problem is certainly a positive factor, the risk of over-reaction is shown by the chain-reaction of catastrophic prophecies in the media as reflected in the headlines: 'A decade of devastation! Echoes of a sad song! The fish wars! A way of life on the verge of disappearing! Fisheries on the brink! A failure of duty! Stripping the sea's life! An ecological catastrophe is unfolding! Greed imperils world fisheries! More Atlantic fisheries closed! Fishing fleet's lament: where have the fish gone? Cod fishery facing extinction! UK fishermen fear further cuts in 1993! California salmon fishery only gets worse! US Fisheries in crisis: Stock of fish vanish on barren grounds!', etc. While attracting governments' and peoples' attention world-wide, these headlines convey an image which is only correct for a few fisheries, as shown in the first paragraph of this section of the paper. Fisheries are producing — and have the potential to sustainably produce — food, livelihood, economic returns, development opportunities, and recreation opportunities for billions of people, imposing limited long-term or irreversible damage to the world ecosystem. It would be only fair to realize that, compared to agriculture, as well as chemical and other industrial sectors, with their huge subsidies, and colossal irreversible impacts on biodiversity, environment, food safety, health and climate, fisheries should appear to any objective eye as one of the most (and possibly last?) large-scale way of exploiting sustainably our world ecosystem ... provided some measures are urgently taken.

Sharing responsibilities

The risk and actual occurrence of local overfishing has been known for centuries (Tiphaigne de la Roche 1760; Pauly and Chua 1988; Policansky 1993). The concern is probably as old as human exploitation of aquatic resources but it has conveniently been institutionally 'forgotten' as technology broke the barriers of natural productivity. Overfishing was formally recognized by governments at least half a century ago (FAO 1945; UN 1946), forgotten during the world fisheries expansion phase in the 1960s and 1970s, and disregarded in the 1980s and 1990s despite being repeatedly underlined at the FAO Committee of Fisheries and Conferences on Fisheries as well as through the scientific literature of the last half century.[3]

The fishing industry, managers and scientists should probably share the blame for the present situation. The fishing industry, prime actor in the 'tragedy' had always been aware of the problem but operating under conditions of open access, moved mainly by short-term interests and rising demand and prices, racing for technological progress, and compensated through subsidies, individual fishermen had little possibility and incentive to correct the problem. Adopted management measures, often more lenient than those recommended by scientists, were regularly circumvented through poaching, misreporting, high-grading, and other unlawful practices, allowed by inefficient enforcement.

Fishery managers preferred to use subsidies, promote modern technologies, and long-range fishing instead of facing the more politically difficult solutions required to solve the domestic problem. Instead of addressing the structural causes of overfishing, facing the problems of fishing rights and resource allocation, they used less controversial and less effective technical measures such as gear regulations, closed areas and seasons. The Total Allowable Catch (TAC) approach failed to limit capacity and effort, aggravating the situation in some cases and leading to fisheries lasting from a few weeks to a few minutes per year and about 27 million t of discards (Alverson *et al.* 1994).

The scientific community described the overfishing problem, studied and modelled it, by stock, by area. It developed a sophisticated institutional interface with the national and international management systems, not recognizing sufficiently its limits, underestimating the impact of uncertainty, under-using its social and economic components, and permanently being overtaken by technological developments.

The large-scale awareness-raising process accelerated drastically in the 1990s, in relation to the UNCED process, fostered by a few issues of high symbolic significance such as whaling, large-scale pelagic drift-net fishing, and dolphin mortality in tuna fisheries. It progressively gained momentum, spreading to the depletion of straddling stocks and highly migratory species, environmental degradation, irresponsible fishing techniques, discards, overcapacity and subsidies, putting practically all aspects of fisheries and aquaculture in the limelight of public opinion. Garcia (1992*a*, *b*) described the progressive extension of the overfishing from the North Sea in the 1950s to the entire world ocean by the 1990s. He also described the parallel shift in the dominant paradigm of fishery science from a biological reductionist approach in the 1950s, to a multi-species and economic approach in the 1970s–1980s, and to a holistic approach in the 1990s.

Causes of failure

The causes for management failure are now well known and related to the nature of resource, the dynamics of the fisheries sectors, and the deficiencies in the management system. Resources are limited, fluid, and variable, and subject to environmental degradation. They cross political boundaries, comprise large number of inter-related components having different

[3] A few of these references are: Gordon 1954; Scott 1955; FAO 1967; Gulland and Carroz 1968; Stevenson, 1973; Gulland 1978, 1984, 1989; Pope, 1979; Troadec, 1982, 1991; FAO, 1983, 1993; Alverson and Paulik, 1983; Miles 1989; Saetersdal 1989; Garcia and Majkowski 1990; Glantz 1990; Alverson and Larkin 1992; Johnson 1992; Christy 1993, 1994; Shepherd 1993; Rosenberg *et al.* 1993; Sissenwine and Rosenberg 1993; Walters 1995; Garcia and Newton 1996. These references represent only a small sample of the warnings that the scientific cassandras have given, and the increase in their number with time reflects the reality of a growing world consensus.

requirements. Their increasing scarcity generates a potential rent which attracts economic activity inasmuch as access to the resource is open[4] and free. The fisheries sectors are dynamic and diversified, characterized by intra- and inter-sectoral competition for the resource and related rent, as well as by low mobility of labour and capital (not easily deflected to other activities), antiquated property regimes, and are difficult to monitor and control. The management system is too centralized. It lacks sufficient research and enforcement capacity; has no authority on distribution of wealth, cannot deal with variability of supply or dynamics of demand, fails to integrate long-term objectives, lacks precautionary focus and flexibility, is predominantly based on biological advice, strongly influenced by vested interests, and is often not formally accountable to the nation.

The critical dimensions of fishery management

The outcome of management decisions is affected by the broader political and macro-economic contexts, as well as by micro-economic and local considerations, including interactions with other sectors. As a consequence, conceptual generalizations may be possible at macro-level and in a medium- to long-term perspective, but practical solutions must be custom-designed in a strictly local context with due consideration to short-term consequences.

Seen from a sustainable development perspective, the problem of managing fisheries is mainly a problem with two components: (a) the resource in its environment, and (b) the society, its evolution and system of values — both requiring a holistic approach. When the various dimensions of these two components are considered, it appears clearly that fisheries management is a multi-dimensional problem which can be examined in a number of ways. In the following sections, the dimensions of fishery management have been grouped in the following categories: bio-ecological, technological, economic, socio-cultural, institutional, spatio-temporal, informational and extra-sectoral. In addressing the various dimensions and their facets in the following sections we shall attempt to reflect the trend from traditionally recognized aspects of fishery management to those dimensions and aspects which have been recognized explicitly only recently, highlighting the implications for management action.

Bio-ecological dimension

Achieving sustainability requires an adequate understanding of the biophysical and ecological system. In the sustainable development paradigm, environmental conservation is not considered as an end in itself but as a socially desirable goal and a prerequisite for production of a continuous flow of benefits to successive generations of human beings and the long-term optimal welfare of people. The ultimate goal of fishery management in addressing the above issues is to maintain the activity for fishers' benefits. However, competitive requirements of other users outside the fishery sector constrains the options available to fishery managers and to the sector itself. Such competition often translates into pollution or degradation of the fisheries productive environment related to two main factors: population growth (and concentration in coastal areas) and industrial development. Negative environmental impacts of fisheries include: decrease in biomass, reproductive potential and resilience of resources; increased variability of the bio-system; waste through discards; impact on species dominance and genetic diversity, onshore and at-sea pollution; and habitat degradation (mainly through trawling, prohibited fishing methods, e.g. 'dynamite', and aquaculture). The impact of the various sources of pollution on oceans has been reviewed by Miles (1994).

Biological constraints on target and associated species

The biological limits of a large part of the conventional target fishery resources are being reached or overshot at present levels of capacity and countries have reached broad agreement on main policy goals in relation to the resource and its environment. Some resources such as mesopelagic fish, krill and cephalopods are still underutilized. Krill exploitation goes on mainly in the Antarctic. Mesopelagic fisheries are being developed in Oman and Iran. The ban on large-scale pelagic drift-net fishing has reduced the economic viability of oceanic squid fisheries. If technology improves the economic viability of these fisheries and allows their significant increase, the question of their impact on the food chain and on the food needs of other important species (including cetaceans) may be raised.

In addition, and following on the requirements set forth by the 1982 Law of the Sea Convention, the Convention on Biodiversity has added significant implications for fisheries management, drawing attention to: (a) the impacts of fisheries on species and genetic diversity; (b) the associated and dependent species as well as the target species; (c) the modification of trophodynamics in the ecosystems by fishing; and (d) the modification and degradation of the habitat. Managers should now give more attention to the sustainability of the exploitation and to the conservation of species associated with or dependent on the target species, usually captured as 'bycatch', and sometimes discarded. The lack of precise information and of transparency of this aspect of fisheries has led to renewed expression of concern and to specific principles and provisions in the most modern fisheries texts, such as the UN Agreement on Straddling Stocks and Highly Migratory species (UN 1995) and the FAO Code of Conduct for Responsible Fisheries (FAO 1995*c*). Elements of management to be revisited in light of the new requirements are: multi-species management; area-based management and marine protected areas (including environmental management); ecosystem management; selectivity, discards and bycatch; as well as aspects of management-oriented research, e.g. on indicators of sustainability of ecosystems' use.

Environmental variability and risk

At the time-scales of relevance to fisheries, resources are variable. Ecosystems are in dynamic equilibrium and may shift between

[4] The observation that the fate of common resources was degradation was made at least a century ago, by Lloyd in 1860 (according to Hardin 1968) and then formally stated by Gordon (1954) and Scott (1955) before being coined 'the tragedy of the commons' by Hardin himself (*19*).

two or a few alternative states when subjected to stress, the origin of which could be related to normal natural oscillations of the environment (El Niño, droughts, floods) or anthropogenic stress (fishing, pollution, alteration of river flows, deforestation, siltation, desertification, industrial pollution, and coastal development). Many of these factors are not well understood or predictable, and management operates in a context of uncertainty in which decisions must be made taking into account the probability of error, the potential consequences of such errors, and the possible means to avoid or eventually correct them, preferably by adopting a precautionary approach (*cf.* below, Information gap).

Inter-sectoral linkages

Because of the inter-connectedness of aquatic systems, marine fisheries resources and their environment are more seriously affected by pollution from coastal and inland development sectors than by fisheries.[5] Sources of external impacts on fisheries include: land-based pollution; diversion of fresh water for agriculture; modification of freshwater regimes in estuaries and lagoons (e.g. through damming); modification of coastal ecosystems (e.g. for marinas); dumping of waste and sewage; siltation (from deforestation and bad land-use practices). These interactions must be taken into account and FAO has proposed the concept of Integrated Coastal Fisheries Management (ICFM) (Fallon Scura 1994; Willmann and Insull 1993; Cunningham 1994) which takes into account all relevant impacts on fisheries resources and their critical environment, whether emanating from the fisheries themselves (conventional fishery management) or from other sectoral activities on the coastal zone or catchment basin.

Trade and environment

The relation between trade, environment, and management has been a focus of international debate in recent years, in the Global Agreement on Tariffs and Trade (GATT) as well as in various fora on fisheries or conservation. The debate has two main components: (a) the impact that trade may generate on resources by the incentives it may provide to increase extractive pressure and (b) the ways in which trade regulations (and barriers) could be used to improve environmental and resource conservation. In this respect the Commission for International Trade of Endangered Species (CITES) is trying to play an increased role in species conservation through trade controls for endangered species and related products, with variable results. The criteria used to list species as 'endangered' in the red book annexes have recently been revised, broadening significantly the concept, and attempts have been made to intervene in conventional management mechanisms, as in the case of Atlantic bluefin tuna management in the International Commission for the Conservation of Atlantic Tuna (ICCAT). More interventions of CITES are likely if the situation of some seriously depressed resources is not improved.

GATT has been under great pressure from some governments and NGOs to mitigate the trend towards totally free trade by special controls and barriers on trade of species exploited in a manner considered as non-sustainable. While GATT has been very reluctant to agree, the question is not yet resolved and will have to be addressed in due course by the newly created World Trade Organisation (WTO).

Products certification and eco-labelling

'Ecological' certification schemes for products or exploitations have been promoted (e.g. for forestry) which, in theory, would aim at 'promoting environmentally responsible, socially beneficial and economically viable management..., by establishing a world-wide standard of recognised and respected principles of...management' (Forest Stewardship Council 1995), accrediting and monitoring claims of 'sustainability' and providing consumers with information on products and management systems. While establishing a link between producers and consumers for eventual feedback, certification may also aim, however, at protecting markets or even increasing market shares by convincing consumers that the products being offered are coming from more sustainably managed resource systems than the competing ones (Bourke 1995). In Canada, Sproul (1996) proposes such a system for Pacific commercial fisheries and underlines that: 'the challenge is to establish a mechanism for informing consumers of constructive environmental practice in fisheries that do not create non-tariff barriers and violate international trade conventions'.

The movement is developing in relation to fisheries. In 1996, the World Wide Fund for Nature (WWF) and a major international frozen fish trader (Unilever Plc/Nv) have planned to establish, by 1998, a Marine Stewardship Council (MSC), on the model of the Forestry Stewardship Council. The aim of the MSC is to promote an eco-labelling system aiming at promoting trade of fish and fishery products caught by sustainable fishing practices and certified as such through a mechanism to be still determined and using a set of principles and criteria to be developed.[6] Fisheries meeting these criteria will be eligible for certification by independent accredited certifying firms. As a first step, Unilever has indicated that it will soon stop buying products from industrial European fisheries for fish meals and oils and, by 2005, will be buying all its products from sustainable sources. This extremely important issue has yet to be formally addressed in inter-governmental fisheries fora. It is worth noting that a precedent exists in fisheries with the US Congress Consumer Protection Information Act which provided for the use of a 'Dolphin safe' label on cans of tuna not caught in association with dolphins. This initiative had significant consequences on market flows and fleet operations and distribution, but its exact contribution to the improvement of the situation (in which IATTC played a major role) is not obvious.

[5] Because most fisheries effects (e.g. depletion) are reversible, while coastal degradation of critical habitats is usually not.

[6] The FAO Code of Conduct for Responsible Fisheries, adopted in 1995 by the highest level inter-Governmental forum available for fisheries, contains the basis for such criteria. In addition, criteria for assessing the precautionary nature of fisheries are available in Garcia (1996) and Kirkwood and Smith (1996).

If fisheries were 'privatized' through use rights (just as forests concessions are), the evaluation could be done at national level, by the independent certifying firms. Certification will, however, be technically very complex in the case of multi-species fisheries and would often have to be gear-based,[7] area-based or even country-based. The last two options would amount to certifying a national management set-up or a national policy and, while this approach might increase awareness and transparency, it is doubtful that governments would accept an international system of evaluation of national performance. Another major issue of importance, considering the potential damage to industry, is the legal liability of the certifying firms in case of error either way.

It seems that certification schemes would have no impact on subsistence fisheries and species traded in the local and informal markets. In addition, impacts on resources from non-fishery activities would have to be addressed too if product safety is to be also certified (an issue already addressed by the *Codex Alimentarius*). It is not clear how negative impacts on the fisheries environment from non-fisheries activities will be dealt with considering that the reaction of fishery product consumers is unlikely to have any feedback effect on the remote polluters.

The challenge would also be to determine and agree on: (a) appropriate criteria and indicators for assessing the responsible nature of a fishery (using as a basis the FAO Code of Conduct for Responsible Fisheries); (b) to agree on the independent authority in which the granting authority should be vested and to ensure its independence and international acceptance;[8] and (c) to ensure that the 'responsible nature' of the fishery system is across the entire chain of production, from capture, to processing and retail outlets. While at national level the independent commissions mentioned above could have such authority; at international level, the problem would be more complicated.

Some important questions would also need to be addressed regarding: (a) the relative benefits for exporting (developing) and importing (developed) countries, and for the consumer; (b) the likelihood that the consumer will accept to pay more for certified products, contributing to meeting part of the costs; (c) the risk that certification be used as a non-tariff barrier, incompatible with the fair trade practices recommended in the FAO Code of Conduct.

Technological dimension

About 3 million vessels world-wide, use a large range of technologies to yield about 100 million t of fish. The traditional concern in fishing technology has been to improve efficiency in both fishing and processing, in a cost-effective way, through research and technology transfer, and with the view to fish more, farther offshore, and to deliver better products, under increasingly safe working conditions. A related problem with management has been to allow technological progress while limiting gear conflict (e.g. through zoning). Technological progress in boat and gear design, propulsion, on-board processing and preservation techniques, as well as on navigational aids, has provided access to a very large proportion of the available aquatic resources, contributing substantially to human welfare.

Simultaneously, however, increases in efficiency combined with state-subsidized investments have led to overcapacity with significant negative impacts on fisheries economics and resources, and potential threat to resources conservation and fisheries sustainability. As a consequence, conflicts between gears or for allocation of space (often hiding resource allocation issues) are increasing. Management is being required to adhere to a precautionary approach to technology, including the adoption of Environmental Impact Assessment (EIA), pilot projects, and Prior Consent or Prior Information Procedures to evaluate the consequence of a new technology on the target species, the associated and dependent species and the environment before its full scale deployment (FAO 1995*a*; Garcia 1996).

The problem of fishery technology control and optimization is two-fold. First, new technologies in fisheries are generally borrowed from non-fishery industries including military ones (e.g. propulsion, artificial webbing material, hydraulic power, navigational aids, echo sounding, etc.) and their development and transfer is not under the control of fishery authorities. Second, many essential elements of fishing technology such has the navigation and location devices (e.g. sounders, Global Positioning Systems (GPS), or Argos and Inmarsat automatic tracking systems) have the dual effect of improving fishing capacity and improving safety on board, complicating the issue of their eventual restriction.

Economic dimension

It may be trivial to recall that fisheries are economic activities. Major concerns in traditional fisheries management were related to achieving some economic optima, using technologies which could be cost-effective in the short-term, taking into account the potential rent available and, in the best cases, the cost of research and management, including enforcement. While in many fisheries (and not only international ones) these simple economic considerations have not even been addressed yet, the picture is being complicated by the long-term time horizon now imposed on fisheries management by the requirements for sustainability. The role of governmental subsidies is being seriously questioned as well as the gratuity of access to the resource. The User-Pays Principle (UPP), already applied by coastal countries in relation to access to their resources by foreigners, is being contemplated for nationals too, often in

[7] Certifying a fishing technology or practice as environmentally friendly. This approach would be analogous and complementary to the certification of gears through their insertion in green, orange and red (or white, grey and black) lists depending on their degree of 'friendliness', envisaged in the FAO Guidelines for a Precautionary Approach to Fisheries (FAO 1995*a*). Certifying a fishery would have to be based on internationally agreed criteria (e.g. the FAO Code of Conduct or the *Codex Alimentarius*). Sproul (1996) refers also to the broader ISO system of certification.

[8] The Forest Stewardship Council founded in 1993 as an independent international non-profit NGO to accredit certification organizations is yet to be internationally accepted.

relation to the establishment of exclusive fishing rights. A major difficulty is in determining the appropriate discount rates required to assess the economic implications of fisheries development and management in the long term. Another significant difficulty is in determining the intrinsic economic value of fisheries resources independently of their use by fisheries, in order to fully understand the alternative (non-fisheries) options eventually available, as well as the interests of non-fishery users. A last, but not least, economic issue is that of assessing the actual and potential role of the market on fisheries development and management.

Socio-cultural dimension

An estimated 12.5 million fishermen (of whom 10 million are artisans), operating from more than 3 million vessels, yield around 90 million t of fish per year and provide livelihood directly to about 200 million people. The management of fisheries aims to ensure that the fisheries actors' behaviour is maintained within norms compatible with sustainable use of the resources and the environment. The social dimension of fisheries has always been present in management decisions (particularly the employment problem) but is being broadened and combined with the recognition of the cultural heritage and related rights. The problems traditionally encountered will, in the future, be compounded by the development of the coastal mega-cities of the 21st century, the feeding of which may require a complete revision of the potential role foreseen respectively for artisanal and industrial fisheries. As one of the most often hidden dimensions (if not function) of fishery management is allocation of wealth among subsectors as well as within, all stages of the process require intense participation of stakeholders and are affected by their socio-cultural context.

History and culture

Human societies tend to evolve through adaptive learning, by trial-and-error. Fisheries have existed for centuries and have largely evolved in this manner. Their present state — including people's attitudes and the system of regulations governing them — and level of performance, depend on the history of the people concerned, markets, technology, governance and institutions, including the history of regulations, traditional rights, etc. The consequence is that the reaction of a fishing community to a set of regulations, or the performance of a new institution, will depend on their cultural and historical contexts and cannot be easily modelled and predicted. A related consequence is that apparently similar communities with different historical trajectories and institutional memories may not react in the same way to a given management prescription. Despite this, fishing communities have often largely been left out of the management process, particularly small-scale coastal communities mainly because of a lack of understanding of their historical, cultural, organizational, and functional 'realities' or perceptions.

Employment

In many poor rural communities, overfishing, and environmental degradation in general, is the consequence of poverty and of lack of alternative employment or sources of livelihood. Fisheries have been implicitly considered by governments as an infinite source of employment of last resort, particularly in the case of rural small-scale communities. Demand for manpower has been reduced through mechanization and industrialization and is now further jeopardized by overfishing and overcapacity which call for a reduction in fishing effort. As a result, alternative sources of employment have to be found for fishers. Unfortunately, the situation of other sectors of the national economy is often not very different and fishermen have limited employment possibilities outside the fishery sector, particularly in rural small-scale fisheries. With the present rates of population growth and migration towards the coastal areas, fisheries may, therefore, still be required to continue to provide employment and, around large cities of the developing world, marginal layers of the population will probably continue to attempt to survive in small-scale and subsistence fisheries and the employment issue will remain one of the most serious to be faced by fishery managers and policy makers.

Participation

Fisheries management involves a permanent process of **negotiation** between the administration and the sector, a process which is, however, neither transparent nor accessible to other potential users of the resource. At the interface between the resource and the fishery, the management and development institutions have therefore a major impact on management performance, which depends, *inter alia*, on the decision-making process and to the degree of participation of the users in this process. The **private sector** has a central role to play in forging the short-term objectives of management, contributing to the long-term objectives, ensuring optimal and responsible use of inputs and optimizing outputs (from the individual operator point of view) and participating in planning and management. In order to promote responsible behaviour among users, some authority to make decisions on objectives and strategies, as well as responsibility for implementing them, and accountability, should be vested in those stakeholders who may gain and lose from using the resource. This can be achieved through active people and sector participation with a view to increase commitment of users to long-term conservation, to increase compliance and reduce public costs of implementation, and to improve mutual understanding. Participation can be increased through decentralization of some of the governmental functions.

Since UNCED, **participation** has become a key word in development agencies and fora. To be effective, participation must be enhanced from the present forms of mere consultation and top-down information to participative decision making and interactive management. This change requires appropriate institutional structures and capacity-building programmes. In general, it is proposed to replace or complement the conventional 'top-down' sequence of scientific analysis, advice to managers, negotiation and imposed regulations (often with insufficient consideration of their operational consequences), by a 'bottom-up' negotiation among the parties involved (research, administration, fishermen, traders), particularly among these with conflicting requirements (Rubenstein 1993).

The concept, which is generally applicable, is particularly useful for the management of small-scale commercial or subsistence fisheries where social problems are particularly acute and the cost of command-and-control instruments particularly high because of the geographical dispersion and technological flexibility and diversity of the fishery. On the other hand, existing traditional institutions, instruments, and social cohesion and coercion could be used (when still in existence) to establish more cost-effective management in partnership. Kurien (1988) and others have stressed the role of traditional fishery communities in management in the developing world and their historical progressive weakening and elimination by centralized government systems as a limiting factor to modernization and 'development'. They have advocated their re-establishment and strengthening as a necessary condition to improved management in small-scale fisheries. Modern fishermen's organizations have also played a significant role in the past, either as cooperative services to reduce operational costs or as lobbying groups to collectively defend their interests, and in most internationally agreed fishery management schemes, these associations are called to play an even greater role.

Community-based management, co-management and various other forms of interactive participation have been recently resurrected or proposed to improve management of fisheries, involving both the Government, the managers and the resource users themselves, working jointly, through fishermen associations and other specifically designed institutions, to design management strategies and to implement them, including in monitoring and enforcement (Rettig *et al.* 1989; Pommeroy 1994, 1995). However, with the increased recognition of multiple-use, it may become difficult to devolve all rights of use to a single community, and a co-management approach, capitalizing on local knowledge and long-term self interests, involving shared governance between the State and all the users would be most effective (Feeny *et al.* 1990). These authors indicate also that the concepts of communal property could even be applied at a large scale to global commons (e.g. atmosphere and ozone layer), a suggestion which could be considered for high seas resources and highly migratory species.

Participation, however, has a long history and attempts have been made in the past to involve people in some aspects of planning and implementation with mixed results. In many countries, pressures on management from fishing communities and from the fishery sector in general have aimed more at obtaining subsidies and delaying management action than towards conservation. Paralleling fisheries and forest management in Canada, and while recognizing that 'fishermen have consistently shown strong concern for the future of the resources', Walters (1995) concludes that community members, who are now conscious of modern economic trade-offs, can no longer be relied on for cooperation in assessment and management if they can gain more by acting otherwise (e.g. if financial rates of return remain substantially higher than natural growth rates). A similar view is held by Hannesson (1986) who, after a careful analysis of past experiences concludes that: 'the outcome of giving fishermen associations a say in fisheries management depends crucially on the economic framework and philosophy prevailing in each country'.

Decentralization

Decentralization appears to be an important factor in addressing fisheries overcapacity and, even more so, the integrated sustainable use of the coastal areas, facilitating decision on allocation and cooperative enforcement (see for instance Troadec 1996). However, the often-recommended decentralization (or devolution) of management authority to reduce government intervention and increase people's participation, implies building another layer of bureaucracy at lower level, co-opting local elites to achieve national objectives. Experience has shown that while generating greater local and sectoral involvement and local social peace, decentralization has also often led to increased sectoral lobbying for short-term returns, at the expense of the resource base, and decrease in reliability of statistics. Hence the need to compensate decentralization with the establishment of independent review and control mechanisms.

Institutional dimension

Institutions are usually taken to encompass the rules and organizations established to ensure satisfaction of common interests. Management of fisheries involves managing people, and the social dynamics of human population require and generate institutions to codify people's behaviour and to produce management plans, policies, and strategies to control ecological and socio-economic systems. In the perspective of sustainable development, institutions are needed to ensure the transfer (or conservation) of the present resource endowments for future generations. Their role is particularly essential when established patterns of resource allocation, valuation, use, and consumption have to be changed, and so are the legal instruments put in place for the purpose of determining conservation goals; establishing access rules, property, and use rights; undertaking monitoring and enforcement; dispute resolution; etc. with due regard to the sources of spatial and functional mismatch described below.

Laws and regulations

Specific legislations have been adopted to regulate fisheries development and day-to-day operations and management. However, other legislation, which may have been promulgated without consideration of their impact on fisheries, such as legislation for environmental conservation or economic legislation covering investment, credit, taxes, or trade may have consequences for fisheries. Similarly, a number of international non-fishery instruments may have a significant impact on fisheries' environment (e.g. the Bonn Convention), trade (e.g. GATT or CITES) and fishing operations such as the Regulations for the Prevention of Collision at Sea (COLREGS), and for Safety of Life at Sea (SOLAS), the International Convention for the Prevention of Pollution from Ships (MARPOL), and the London Dumping Convention.

Although regulatory responsibilities may be clearly established, avoiding duplication, conflict, or gaps in organizational jurisdictions, the practical result is that fisheries operate in a very complex regulatory framework, difficult to assimilate totally by operators which most often perceive the regulations as overwhelming and unjust, one obvious source of non-compliance.

The situation is aggravated by the fact that regulations are often inadequately enforced with penalties which are often too low to have any deterrent effect.

The modern necessity to nest fisheries management in the integrated management of coastal areas,[9] leads to a general need to redesign the fisheries management legislative framework and its institutional linkages, and will add a level of complexity. Progress in this area, with due recognition of the rights and duties of fisheries in the coastal zone will require intense interaction and negotiation of the fishery sector within the ICFM institutional context (see below).

User rights

Hardin (1968), in his 'tragedy of the commons' pointed to the major institutional flaw represented by open access[10] and illustrated the divergence between individual and collective rationality in the exploitation of natural renewable resources. He predicted the eventual degradation of all resources used in common (*res nullius)* and recommended that natural resources be under either (a) private property or (b) government property, allocated through use rights. Feeny *et al.* (1990) demonstrated, however, that private, State, and communal[11] properties were all viable management options. They demonstrated also, that none of these options was a guarantee of sustainability, noting that institutional arrangements (to control access and regulate use and to foster participation) as well as cultural factors determined management success.

The problems are in ensuring appropriate price of access to such property, the destination of the rent potentially generated, in allocating the resource at intra- and inter-sectoral levels. Troadec (1988) held the view that the State, on behalf of the Nation, should be the only formal owner of the resource which it should conserve by law and coercion for the present and future generations (Duty of Care). The fishery sector and other competitive users should be given transferable use rights and made responsible, and accountable, for optimizing production and utilization within the constraints established by the State owner. Regardless of the form of right adopted, there is a growing consensus that privileged users should ideally pay a fee in application of the user-pay principle, equivalent to the full long-run marginal cost of resource use and related services, including social opportunity costs and any other costs associated with research, management, surveillance, enforcement, etc. The principle and its potential drawbacks in relation to equity are, however, discussed by Bonus (1993).

There are many forms of rights expressing different levels of 'property' and it is not possible to point to any panacea in this respect. For example, Individual Transferable Quotas (ITQs) are considered by some as a panacea, expected to avoid wasteful economic competition, improve economic performance, increase concern for and effective participation in long-term conservation, and improved contribution to research. They are, however, criticized by Pontecorvo and Shrank (1995) and Walters (1995) for, *inter alia,* creating patterns of concentrated ownership, decreasing equity in income distribution, and failure to eliminate incentives for individual mis-reporting.

Fishery management organizations

At national level, management commissions should be established for each major fishery and for the integrated management of coastal areas, with an effective involvement of the fishery sector and coastal communities. Independent national commissions might also be established to ensure that management systems comply with the conservation requirements and precautionary management strategies (see below). The implementation of Integrated Coastal Fisheries Management (ICFM) to deal effectively with sectoral interactions will require: (a) increased awareness both at sectoral and inter-sectoral level; (b) explicit consideration of resource limitations and conservation through regulations of access and resources allocation at inter-sectoral level; (c) collection of broader information on environment and other sectoral activities; (d) education of fishermen on their responsibilities and rights; (e) improved representation of fishermen in inter-sectoral negotiating meetings; (f) recognition of fishermen's rights; (g) improvement of planning and co-ordination mechanisms (at ministerial level); and (h) more effective participation and improved transparency in decision making.

At international level, regional and sub-regional fishery bodies and arrangements have an important role to play in the implementation of the 1982 Convention, the 1995 UN Agreement on straddling and highly migratory stocks, and the 1995 FAO Code of Conduct. However, many of them have failed to establish sustainable patterns of resource use and need to be strengthened in terms of technical capacity, management mandate and decision processes. In addition, because of the inter-connectedness of ocean environments, the migration of resources, and the mobility of fishing fleets, improved cooperation is needed between management bodies, across space (e.g. across the whole Pacific Ocean for tuna management) and national sectors (e.g. in the coastal zone).[12]

Whether national, regional or international, fishery management organizations and arrangements need to: (a) strengthen their mandate in relation to management; (b) improve their decision-making process (from consensus to voting); (c) ensure full representation of all those involved in the fishery; (d) adopt legally binding agreements; (e) develop efficient monitoring and credible enforcement capacity; and (f) establish mandatory dispute-settlement mechanisms.

More generally, the international cooperation on oceans has been developed since UNCED with the creation of the Sub-Committee on Oceans and Coastal Areas of the UN Committee

[9] An analysis of the legal and institutional aspects of ICAM is given by Boelaert-Suominen and Cullinan (1994).
[10] i.e. in the sense of *res nullius*, in which anyone can decide to extract part of the resource.
[11] i.e. in the sense of *res communes*, in which resources are owned (or exclusively used) jointly by an identified group of interdependent users.
[12] Even the UNEP Regional Seas institutions which have potentially the mandate for such co-ordination, need significant improvement.

of Administrative Coordination (ACC) and its strengthening has been recommended at inter-governmental level. In addition, in 1995, an NGO has established itself as Independent Commission for the Oceans, and the establishment of an inter-governmental forum on the Oceans has been recommended (House of Lords 1996, para. 3.19). The future of all these mechanisms and their potential influence on fisheries development is still far from clear.

Oversight control mechanisms

The interaction of vested interests with research and management is both desirable and dangerous. Past experience shows that management decisions taken under pressure of fishery lobbies have often been at the expense of the resource. The pressures may become particularly difficult to withstand when fisheries have to be managed in the context of Coastal Areas Management or in the context of a precautionary approach to management. The establishment of a system of 'checks and balances' including a Committee on Limits and Standards to monitor management decisions and performance (in relation to management reference points) has been suggested by Caddy (1994). Walters (1995) suggested that final authority for approval of development and management programmes could be vested in quasi-judicial independent systems such as Management Boards, Councils of Elders, Natural Resources Councils, Independent Conservation Commissions, etc. Their members would have to be insulated from political interference, technically competent, deeply committed to long-term sustainability and free from any vested interest in the economic benefits of harvesting. It is interesting to note in this regard that the Australian Government has established a special Appeals Committee to review decisions on allocations of fishing rights having regard to the wider implications of such decisions on the fishery as a whole (Anon. 1989). A similar institution is being established in Morocco[13] composed of eminent national and foreign personalities.

Spatio-temporal dimension

The ecosystem processes leading to biological productivity and turnover, as well as the social and economic processes of industry development cycles have specific spatio-temporal dimensions. When the two systems interact together and with the management system, mismatch of scales leads to dysfunction. Lee (1993*a*) stressed that 'when human responsibility does not match the spatial, temporal and functional scales of a natural phenomena, unsustainable use of resources is likely'. The reason is that mismatches tend to impede feeling by the actors of the consequences of their inadequate action, allowing externalization[14] of impacts and related cost. Mismatches usually stem from reductionist approaches to development policies, sectoral management and decision-support research. They result in artificial (and only **apparent**) barriers between subsystems and give a false sense of control to those who manage. They also block information flows and contribute to uncertainty.

The main issues reviewed below relate to the role of history and culture; the mismatch between short-term and long-term perspectives, and between local and global ones; the functional mismatch between processes; and the trade-offs between present and future generations' welfare or present and future investments and gains.

Temporal mismatch

Resource, vessels, fleets, management systems and fishing communities have different **generation times** (and lifespans) and evolve at different rates. The appropriate time scale for stock recovery assuming reversibility is from the order of several months in the coastal tropics (except coral reef environments) to several years in colder and deeper environments. For modernization, or structural adjustment of fishing fleets, it is in the order of one or two decades. For environmental rehabilitation, it may be in the order of one or more decades. For changes in peoples' values and attitudes it is variable, but deep changes tend to require a generation (two to three decades). Fishermen's decisions tend to be made with the view to maximizing short-term income, taking account of immediate market forces and climatic events. Terms of office of political leaders tend to be shorter than required by fisheries rehabilitation plans, biasing decisions in favour of those which may affect elections. These differences in temporal scale affect the likely response time of the different elements of the system to stress and to management initiatives and are major sources of management inefficiency.

Economic and social **discount rates** reflect the present value of future goods and services as well as the future costs of present mistakes. Management should reconcile these magnitudes which can be rather tangible when they refer to the present (e.g. today's cost of closing down a fishery) and fairly theoretical when referring to the future (e.g. the future value of a coral reef to society). There is not yet agreement on the discount rates to be applied in converting future streams of incomes and costs to present values. There is also a difficulty in determining the acceptable price to be paid by present generations for conservation of the resource for future use (social opportunity cost of conservation).

The controversial issue of **inter-generational equity** is also related to temporal mismatch as it implies inter-temporal allocation of financial and natural capital. At a national level, there is a trade-off between present and future growth and food security, and present compromises with sustainability may be made at the expense of future generations, particularly when non-sustainable strategies result in irreversible negative impacts, obliterating future opportunities. In many instances, however, the impacts of present irresponsible fishery development strategies will be felt, and often are already felt, by present generations, and the issues of inter- and intra-generational equity are confounded.

Matching between the resources and industry's vital rates and between short-term and long-term societal perceptions could be

[13] Conseil Supérieur pour la Sauvegarde et l'Exploitation du Partimoine Halieutique.

[14] i.e. transfer of the impact and cost to another group of people or layer of the society.

improved by changing national accounting rules, (to account for resources rebuilding, conservation or degradation) and establishing user-fees that will compensate for the differences in time-related rates, balancing short-term and long-term opportunities and incentives.

Spatial mismatch

The development of world fisheries (Garcia 1992*b*) has been marked by the formidable geographical expansion of fisheries from the Northern to the Southern Hemisphere, from the 1950s to the 1980s, as well as by progressive extension of coastal States' jurisdiction from the late 1940s to today, with a peak in the mid-1970s when many EEZs were established. The 1982 Convention on the Law of the Sea has formalized the 200-mile limit for EEZs but attempts to extend coastal countries' influence farther offshore are still continuing as illustrated by the concept of 'Mar Presencial' enunciated by the Commander of the Chilean Navy in 1990;[15] the concept of 'Contiguous Zone' of 24 miles beyond the 200-mile zone proposed by Canada (Tobin 1995); the extension of coastal States control on the Donut Hole in the Bering Sea; the unilateral extension of control on the Peanut Hole by Russia in 1993; and the Norway–Iceland conflict on the Barents Sea 'loophole' in 1994 (Miles 1994).

Both people's activities and fishery resources, which management intends to regulate, have a fundamental spatial character. Stocks depend on nursery, feeding, and spawning areas and often migrate across large areas. Fishing vessels can work in isolation or groups, sharing, or not, information. Their potential operational range, favourite target and migrations are determined by the technology used and market signals. The management system's operational range — in an EEZ, a coastal zone, a river basin, or an international management area — depends on political factors, and conflicts for resource use are usually area based. The awareness of the importance of space in management, emphasized by Caddy and Garcia (1986) is reflected in the growing interest for Geographical Information Systems (GIS) in multi-objective decision support systems. A consequential 'mismatch' occurs when the area of competence of the management system does not match the area covered by the stock (and/or the fleet) as it is often the case for shared, straddling, and highly migratory resources; in the high seas, large river basins, and large marine ecosystems (LMEs), etc.

A mismatch occurs also with long-range fleets which, in a very large operational range, often exploit a series of unrelated and often oscillating stocks. This mismatch allows these fleets to escape the consequences of the local overfishing they generate by moving elsewhere, seriously reducing the feed-back control of the resource level on the fleet activity and allowing overfishing to expand through pulse fishing. The progressive extension of overfishing from the North Atlantic and Pacific to the whole world ocean in half a century (Garcia 1992*b*) illustrates the consequences of this mismatch. The logical proposal made by FAO at the UN Conference on Straddling Fish Stocks and Highly Migratory Fish Stocks to ensure compatibility between management schemes (and precautionary measures) across the entire species' range, met, however, with opposition from some coastal countries on the basis of their sovereign rights. Fortunately for the resource and the future of fisheries, after a protracted debate, reason prevailed and this principle is now adopted both in the UN Agreement and in the Code of Conduct for Responsible Fisheries.

Functional mismatch

According to Lee (1993*b*), functional mismatch occurs when the management system does not deal with the processes which ensure sustainability. In fisheries, this could occur when management measures and plans ignore multi-species linkages or ecosystem processes (predator–prey relationships, inter-specific competition, density-dependent factors, multiple states, complex non-linear processes and feed-back responses). As overfishing is a direct consequence of socio-economic processes and technical measures are usually founded on biological processes, conventional management which uses mainly technical measures is a prime example of dangerous functional mismatch.

Functional mismatch may also result from splitting key responsibilities among different authorities, in national and international fishery systems. For instance, the bearing of a national flag on a vessel is usually authorized by the Ministry of Transport based on concerns foreign to fisheries. As a consequence, vessels may change flags to circumvent national legislation (moving under joint-venture agreements to evade fishing fees) or international regulations (moving under a non-party flag of convenience). The mismatch between the authority granting the flag and the authority paying for the consequences of this decision opens a loop-hole and undermines efforts at controlling gear, allowable catches, or fleet size. In many countries, fishing agreements are negotiated by the Ministry of Foreign Affairs, fishing fees are cashed in by the Ministry of Finance, interaction between foreign and national fleets are dealt with by the Ministry of Fisheries, control and surveillance is the Navy's responsibility, and penalties are 'imposed' by the Ministry of Justice! It is easy to understand that such a situation can generate a number of functional mismatches in objectives, competence, jurisdiction, and calendars which can only undermine management efficiency.

Informational dimension

Information gap

Having identified the main dimensions of fisheries sustainable development, a major problem to be faced is the lack or inadequacy of the information available about them and the key biological and economic parameters including the 'assimilative capacity' of the resource and the discount rates to apply in converting future streams of incomes and costs to present values. The potential consequences of changes in these dimensions with time or of mismatch between them are also not fully understood. Despite general awareness of the data problem, there are very few signs of improvement in commitment and the situation is worsening as the number of parameters to be considered by management in its quest for sustainability increases. As stated by

[15] Although it is being made clear that the concept does not represent an extension of limits of national jurisdiction.

Ricker (in Walters 1995) 'with deterioration in monitoring programmes and environmental change... things (are) even more unpredictable than ... a few decades ago'.

Uncertainty, risk and precaution

The potential role of science in improving fisheries sustainability in this uncertain context has been heatedly debated (Levin 1993). Ludwig *et al.* (1993) stressed the multi-dimensionality of fishery systems' scientific understanding and the fact that there will never be enough information to manage the fisheries without risk at all. A major source of uncertainty for managers is the relative extent to which natural fluctuations (compounded by global climate change) and human impacts are responsible for the observed changes in the resource base, and use patterns. Recognition of uncertainty and its potential consequences has led to the adoption of the precautionary approach in UNCED Rio Declaration (Principle 15), the 1995 UN Agreement on Straddling and Highly Migratory Fish Stocks, and the 1995 FAO Code of Conduct for Responsible Fisheries. The implications of the precautionary approach for fisheries operation and management (and related activities) are fundamental. The recognition of uncertainty and of the need for management to limit potentially negative outcomes resulting from it, for the resources and ultimately for the people, has led *inter alia* to recognition of: (a) the risk of using MSY as a target and of the need for a broader and more precautionary range of management targets and limit reference points (Smith *et al.* 1993; Caddy and Mahon 1995; Garcia 1996); (b) the need to quantify better the confidence limits of scientific advice and the robustness of management systems to uncertainty; (c) the need for impact assessment and/or pilot projects as a basis for prior authorization before introducing new fishing gear and methods.[16]

Extra-sectoral dimension

Fisheries development and management are affected by interactions between fisheries and non-fisheries social and economic policies established at national and international levels. At national level, the macro-political context of fisheries is determined by policies established in other sectors, beyond direct control of the fishery authorities. At international level, the evolution of policies (and technology) regarding transportation, information and communication is leading to the development of a global economy and international market, linking all national policies and fishery sectors (and some non-fishery sectors) in a way never experienced before. International trends in macro-economic policies are also accelerating the evolution of management strategies and constraining the options available. Some of the aspects related to: (a) policy interactions with other sectors; (b) policy interactions within the fishery sector; (c) trade liberalization and GATT; (d) delegation of responsibility and privatization; (e) macro-economic context; (f) role of the market; and (g) resource pricing, will be examined below.

Policy interactions with other sectors

At national level, fisheries operate in the broad framework of development, financial, commercial, environmental, etc. policies often elaborated with no or little involvement of the fisheries'actors despite their implications in relation to sharing a common and limited resource base and environment. In the future, fisheries will have to be managed within the framework of Integrated Coastal Areas Management and explicitly taken into account in national development plans and strategies (inter-sectoral planning). Modern trends in national policies towards greater environmental protection, liberalization of trade, reduction of government intervention, and increased role of the private sector will undoubtedly affect the strategy options and performance of fishery management systems. Conversely, some reforms of non-fishery national policies may need to be promoted in order to reduce macro-economic imbalances and liberalize market mechanisms as a means to promote more efficient use of fishery resources.

Fisheries policies may also be affected by international agreements not directly aimed at fisheries. The Convention on Biodiversity has an impact on issues such as discards, or introduction of species for aquaculture. MARPOL would affect fishing operations, disposal of old nets, processing at sea, etc. GATT and trade debates at the newly established World Trade Organisation (WTO) will affect fish trade and possibly even development options and management strategies. This particular aspect is further discussed below because of its importance.

Policy interactions within the fishery sector

Within the fishery sector, interactions occur when two or more countries exploit a common fishery resource. In such cases there is a need for agreements on resources sharing and compatible management schemes across the whole stock range, for shared stocks, straddling, highly migratory, and high seas stocks. The reduction of effort in Europe and the former USSR countries is making available their excess capacity at a relatively low price, aggravating the risk of overcapacity elsewhere and particularly in the developing world. The collapse of the former USSR long-range fleet and the relocation of its activities to closer national waters and adjacent areas have had a strong impact on fishing effort around Europe as well as on food-fish supply and prices in West Africa. The 'dolphin safe' tuna initiative of the USA[17] affected tuna fishing and processing operations in American and Asian countries, modified world-wide trade flows, helped to establish Thailand as a new leader in tuna processing, prompted massive fleet transfers across to the South Pacific, and, ultimately, led the foundation of the international conflict on large-scale pelagic drift-net fishing in that latter region. In addition national fisheries policies will be affected by international agreements on fisheries (e.g. the 1982 UN Convention, the 1995 UN Agreement on Straddling Stocks; the 1995 FAO Code of Conduct for Responsible Fisheries, or the 1993 FAO Compliance Agreement).

[16] A full treatment of the approach and its implications for research, technology transfer and development, fish processing, trade, and management are available in Garcia (1994, 1996) and FAO (1995*a*, 1996).

[17] Apparently the first attempt at 'eco-labelling' in fisheries.

Trade liberalization and GATT

The avowed objective of trade liberalization is to reduce macro-economic imbalances and restore relative prices. The main issues relate to the reflection of environmental costs in the global market, the role of public sector in non-price intervention (research, infrastructure), the reduction of government role and the response of the private sector, and the legal framework for trade regulations.

The value of fish traded internationally, expressed as constant 1994 prices, has increased from about 1200 million US dollars in 1970, to 50 000 million US dollars in 1994. In the mean time, developed countries (e.g. Europe, USA and Japan) became main importers, developing countries increased rapidly their role as exporters with 56.6% of total exports value in 1993 (Garcia and Newton in press) and new importing markets are emerging in Asia. The trade concessions negotiated during the Uruguay Round of Multilateral Trade Negotiations, concluded in December 1993, will positively affect 90% of world fisheries exports, providing increased market access by reducing or eliminating trade barriers, providing legal security for market access through tariff bindings, and lowering tariffs on fishery products by the developed countries more than for other natural resource-based products. If commitments are maintained, protectionism will be practically dismantled if new non-tariff barriers are not created. According to a GATT Secretariat study, fish and fishery products exports could, as a consequence, increase by about 13% in volume terms and the duty-free trade by more than 20% with an impact varying between regions.

As a consequence, in planning fishery development and designing management systems, fishery authorities should, *inter alia*: (a) seek to integrate fisheries planning and management into the context of national development plans and Integrated Coastal Areas Management schemes; (b) take account of the existing international agreements and world-wide trends of relevance to fisheries: (c) study and implement the measures eventually needed to accompany inevitable trade reforms; (d) be aware of the potential consequences of international agreements on environmental conservation and trade, including the impact of an eventual process of certification of fishery products (see below); and (e) evaluate the impact of trade liberalization on rural fisheries development and coastal communities.

Delegation of responsibility and privatization

This issue has been briefly addressed in the section on socio-cultural dimensions. Governments have had a determinant role in shaping today's fisheries. Many essential functions of management require governmental intervention: e.g. decision on ownership and resources conservation; allocation of resources, between uses and between communities; protection of future generations' development opportunities;[18] institutional building; legal frameworks; intra-national co-ordination and dispute resolution; provision of services and infrastructure, including upstream research and market infrastructure; extension and training; interactions with the private sector; enforcement and evaluation of management systems; introduction of macro-economic incentives and disincentives; social welfare, equity, and cultural equilibria; enforcement of national resource boundaries; negotiation and enforcement of international fishing agreements. Under ideal macro-level conditions, some of these roles could be delegated and optimal investment policies still be achieved. These conditions, however, are rarely met in practice and the intervention of governments is required.

Modern trends, partly forced by international development banks and structural adjustment process emphasize the need to privatize some traditional government functions of services such as management-oriented research and monitoring, control and surveillance (MCS). Management-oriented fishery research has already been privatized *de facto* in Australia in some limited entry fisheries where an autonomous research capacity has been acquired by the licence-holders associations. An alternative to government-run systems of monitoring, control, and surveillance (MCS) is to contract a private company acting as an agent of the government, an approach followed by some countries (e.g. in Sierra Leone and in UK Falkland/Malvinas islands) with different degrees of success (FAO 1992, para. 40). In relation to enforcement, it is usually agreed that progress will come from more participation of fishermen and not from more coercion which governments do not have the means to implement in most cases.

Macro-economic context

National macro-economic policies regarding trade, currency fluctuations, investment strategies, interest rates, subsidies, taxes on fuels and oils, etc. are usually as important as sectoral policies in influencing discount rates, people's choices, resource use, income distribution and growth, and thus sustainability. It is therefore essential that, at the national level, the two be made complementary, particularly when management-related responsibilities are split between different Ministries, and the degree of sustainability aimed at in national development policies is important. Finally, the **role of the market** in shaping fisheries is more important than usually recognized and will be carefully examined below.

Discount rates have an important role in sustainability of resources exploitation. Schuh and Archibald (1996) indicate that 'the prevailing real discount rate is an important determinant of the weight policy makers and private citizens place on the future in making their economic decisions'. A perception that discount rates are low will give higher weight to present benefits against future ones, and will promote resource depletion. An important international consideration is that, in a global economic system, with low or non-existent barriers to flows of goods, capital and services, differences in interest rates will create differences in discount rates and, therefore in the perceptions about management. Schuh and Archibald (1996) indicate that interest rates tend to be higher in the developing world, and this

[18] The reference to the need to safeguard options for future generations is often referred to as duty of care or stewardship, concepts not easily accepted by States concerned that it might limit their sovereign rights to decide to an unacceptable extent.

situation, added to the demands in the international fish trade, creates therefore stronger incentives for depletion.

International macro-economic factors may create world-wide distortions which affect the outcome of management or development decisions taken at national level. For example, structural adjustment policies imposed on many developing countries by international financial institutions as a condition for financial assistance, affect dramatically the options available to a fishery manager. Another example is given by Schuh and Archibald (1996) who indicate that 'the immense scale of subsidies offered by Governments of the European Union, the USA, Canada and Japan to their fishery sector provides a grossly unfair competition to similar commodities from developing countries'. The following shows, however, that the situation is not as simple as it may appear. The **concentration of the market** for international fishery products in a few developed countries (USA, Japan and Europe), for example, tends to shape the demand and dictate prices, influencing choices and development strategies in the fish-producing and exporting nations of the developing world. The large **surpluses in fishing and processing capacity** proposed at low price from developed countries and countries in transition to developing countries is likely to lower artificially the cost of fishing in the latter, modifying substantially the options available to them, increasing the risk of overfishing. This is equivalent to subsidizing capital expenditures and may distort the price of production factors in favour of capital-intensive industrial fisheries (as opposed to manpower-intensive artisanal fisheries), possibly pushing the fishery system away from its long-term optimal welfare goal. It could also constrain small-scale communities to irresponsible fishing practices of last resort (e.g. using dynamite and poison or depleting further already endangered species).

The global widening **supply–demand gap** between a limited and already scarce supply and an ever-growing demand will significantly boost world prices for high quality fish, increasing incentives for investment, and even traditionally low-valued fish such as mackerel and horse mackerel may be affected. The problem will be particularly serious if aquaculture does not develop alternative sources of fish as rapidly as required and if the main large stocks are not well managed. In the absence of a national policy to control fishing effort effectively, the resulting increase in unit economic value may compensate decreases in abundance and yields for some time, reducing incentives to manage the resource, increasing incentives for lobby groups to exert pressure on governments in favour of expansionist fisheries investment policies.

Macro-economic factors may have significant effects on costs of fishing, profitability and, ultimately, on affordable levels of investments. A decrease in oil prices would favour development (or delay attrition) in capital-intensive fisheries with medium- to large-scale vessels as well as long range fleets, and facilitate increased pressure on small pelagic fish stocks. On the contrary, an increase in oil prices may favour the development of small-scale local fisheries, particularly for exported products, i.e. on already heavily targeted high value species.

Role of the market

Schuh and Archibald (1996) indicate that the general trend to rely more on markets to determine adequate costs and benefits reflects the need to provide economic agents with incentives to act in their own best interests. The market is not perfect, however, affected as it is by 'natural' imperfections and voluntary distortions. This means that observed market prices do not always reflect the true social costs or value of a given good or service and that Government intervention may be necessary.

International fisheries trade multiplied by three in quantity and by four in value (using constant 1994 prices) between 1970 and 1994 (FAO Yearbook, of Fishery Statistics — Commodities), and the interactions between supply and demand, affected by market imperfections, are leading forces shaping investments and practices and ultimately controlling the rates of use of the resource, at local, national or international level. Despite the recognized needs for government intervention to correct or compensate for market imperfections, modern trends are towards reducing such intervention and relying more on market forces, ensuring through macro-economic, structural and institutional measures, that these forces are indeed conducive to resource use optimization and conservation for future generations.

From a market point of view, management failure results from:

(a) The inability to understand the dynamics of supply and the conditions for its stability and composition (natural variations, multi-species effects, resource erosion);

(b) The inability to understand the dynamics of demand, and its short- and long-term impacts on the sector's dynamics;

(c) The conflicting goals of the different agents of fisheries: the biologists who are expected to determine how to conserve the resource; the economic actors who want to maximize short-term profits; the economists who want to maximize net discounted yield from the assets employed; and the government which wants to ensure social peace and maximize employment.

According to Pontecorvo and Schranck (1995), failure results from the use of **supply-driven management strategies**, aiming at constraining the extracting industry in order to achieve some desired state of the resource, acting mostly on the capture level, with little or no reference to processing and market levels. With the world fishery supply fluctuating close to its upper bound, and the demand from a growing population still increasing, the alternative would be demand-driven management strategies aiming at a stable extractive industry, dynamically responding to the demand and the fluctuations and shifts in supply. These authors propose to consider fishery products as a commodity available on an imperfect world market (with trade barriers, inaccurate reporting, subsidies, etc.) and to deal with fisheries accordingly by: (a) improving biological and economic data; (b) dealing explicitly with variable supplies; (c) improving forecasts of supply and demand (at fishermen, processors and consumers' levels); (d) understanding demand, particularly from the large markets (Europe, Japan, USA, international fish meal);

(e) increasing financial strength and organizational capacity of fishing industries; (f) improving labour mobility to adjust to resources oscillations; and (g) establishing trade monopolies and cartels (but watching public interests).

A **demand-driven management strategy** would indeed imply fundamental changes in the focus of research, the role of the private sector, and the structure and composition of management institutions. It would require a transparent and challenging charter between the private sector and the State to ensure compatibility between, on the one hand, the natural inclination of markets for short-term profits, their volatility and versatility, and on the other hand, the requirement for resource stability and conservation for future generations. In particular, the lack of (or relative indifference to) species differentiation in the market could represent a problem for conservation of biodiversity. For example, the invention, and promotion of demand for non-specific fishery products such as fish protein extracts (*surimi*), fish meals, and oils, justified by the need to find new markets and alternative markets for excess production, may not provide much incentive for conservation of biological diversity, as illustrated for instance by the depletion of whales through a management system based on whale oil quotas.

In considering the advisability of establishing a demand-driven management system, it should be remembered that the steady long-term rise in the price of fish on the world market sustained the overfishing and overcapacity processes. It should also be remembered that the divergence between scientific advice and the TACs and quotas agreed by the policy-makers resulted from the lobbying pressure exerted by the whole industry (including processors and traders). In addition, many issues, such as: (a) the appropriate type of fishery organization; (b) the proper process of transformation towards this type of organization; (c) the way in which inputs (labour and capital) could be made as variable as the natural resource (while minimizing these fluctuations); (d) future supplies and demands, at processing and consumer levels; and (e) the relevance of the approach for integrated management of multiple uses addressing different markets, would have to be clarified. In addition it would be necessary to ensure that the improvement of labour mobility does not merely shift the industry cost of dealing with a variable resource to a governmental (societal) cost of dealing with a variable manpower demand, in contrast with the present stable employment policies of most governments.

Resource pricing

Most fisheries are in a dubious economic state. Many operate under conditions of very low profitability, and require various forms of subsidies to sustain their activities. FAO has calculated that the world fisheries generated a deficit of more than \$US50 000 million, indicating probably the existence of large-scale, partly hidden, subsidies which represent a social cost absorbed by society on top of the private costs incurred by fishing companies. Overcapacity, overfishing and related low yields have been the most commonly reported factor for that situation. Another reason (suggested by Garcia and Newton in press) is that, in many cases, fish might be under-priced at the first production level (ex-vessel price) and while the capture sector often hardly breaks even, the processing and marketing sectors still generate large profits. This would indicate that the end-price to the consumer may be correct but that the share of this price between the various sub-sectors of the fishing industry might not be equitable. This in turn would point to the need to reflect on the present mechanisms of price formation for fish and fishery products.

Getting the prices right is extremely important for promoting efficient use and allocation of a natural resource and it is generally agreed that the market mechanism is more efficient in achieving this result, provided the interests of the individual actors are made compatible with the interests of the Nation. In the case of fisheries, market mechanisms cannot function properly in the absence of transferable property rights and governments must intervene to close the divergence between private and social values through user fees, taxes, and other economic instruments aiming at ensuring that fisheries are made to bear most of their costs.

The purpose of Government action is to bring the cost of fishing close to its social opportunity cost, ensuring that all the costs of fisheries (including the cost of research, monitoring, control and surveillance) are borne by it, accounting for shadow prices, improving the intra-sectoral allocation of the price (with a more equitable share of the consumer price going to the fishermen), in order to ensure that the uncertain returns expected in the medium- to long-term can still justify the required investments. Somers (1994) indicated that, if the '**user pay principle**' was adopted to cover the costs of management, as in Australian shrimp fisheries, the involvement of industry would be boosted in order to reduce these costs.

A FRAMEWORK FOR SUSTAINABLE FISHERIES

In the following sections, the objectives of fisheries management as well as the priorities for action will be briefly examined before presenting the nested set of solutions required and reflecting on the implementation process.

Objectives

A large consensus on generic management objectives appears in the major international agreements on fisheries and more particularly in the comprehensive FAO Code of Conduct for Responsible Fisheries. Management is generally expected to meet the three categories of overall objectives of sustainable economic development:

(a) **techno-economic:** ensuring growth and efficiency; leading to an optimal investment policy; fulfilling the long-term duty of care;

(b) **bio-ecological:** ensuring efficient use of the resources and conservation of ecosystems' integrity, carrying capacity and biodiversity, as well as environmental conservation and rehabilitation; and

(c) **socio-cultural:** ensuring equity, participation and empowerment, social mobility and cohesion, designing proper institutional arrangements, taking into account cultural identities, and ethical requirements.

Modern management will still have to face the traditional and specific objectives of resource conservation and development in relation to supplies, safety of fishery products, employment, and social peace, which are certainly not obsolete. In addition, however, it will have to face new or emphasized objectives made necessary by the relative failure of management and its consequences, in terms of fishing capacity, biological waste (discards), food security, environmental conservation and ecosystem management, equitable allocation of access and use rights, trans-generational equity, and resolution of conflict between fishery and non-fishery uses of the aquatic ecosystems. In many cases, equally meeting all these categories of objectives (and their subsidiaries) will be impossible and some ranking and prioritizing will be essential.

Priorities for action

Referring to the need to constantly and actively improve understanding while institutionalizing flexibility of natural resources management systems, Holling (1993) stressed that 'evolving systems require policies and actions that not only satisfy social objectives but, at the same time, also achieve continuously modified understanding of the evolving conditions and provide flexibility for adaptation to surprises' (see also Walters and Hilborn 1978; Shepherd 1991; Smith *et al* 1993). Referring to the problems of sustainable development of renewable resources in general, Holling (1994) concluded that 'the problems are not amenable to solutions based on knowledge of small parts of the whole or to assumptions of constancy or stability of fundamental ecological, economic, or social relationships. Assumptions that such constancy is the rule might give a comfortable sense of certainty but it is spurious. Such assumptions produce policies and science that contribute to a pathology of rigid and unseeing institutions, increasingly brittle natural systems and public dependencies'.

UNCED, the Rio Declaration, and the beginning of the implementation of Agenda 21 have provided the opportunity to re-examine the approaches to fisheries, to put them into perspective with global efforts towards sustainable development, taking into account new requirements, giving more attention to the environmental impacts of fisheries development as well as to the environmental impacts of general development on fisheries. The **Rome Consensus** signed in FAO by the world Ministers of Fisheries (FAO 1995*e*) recognized that action was urgently required to: (a) eliminate overfishing and prevent further resources decline; (b) reduce overcapacity; (c) rehabilitate productive habitats; (d) minimize wasteful practices and post-harvest losses; (e) develop sustainable aquaculture and stock enhancement; and (f) develop new or alternative sustainable sources of supply compatible with ecosystem conservation. The **1995 FAO Code of Conduct for Responsible Fisheries** as well as the **1995 UN Agreement** on the Implementation of the Provisions of the Convention of the Law of the Sea of 10/12/1982 Relating to the Conservation and Management of Straddling Fish Stocks and Highly Migratory Fish Stocks (1995), reflect a new international consensus on a number of sustainability-related concepts which were not explicitly embedded, or sufficiently highlighted, in the 1982 UN Law Of the Sea Convention such as the need for: (a) recognition that the government has a duty of care;[19] (b) prohibition of all resource uses unless specifically authorized; (c) various forms of the user-pay principle;[20] (d) stricter and more explicit accountability; (e) the precautionary approach; (f) a broader range of management reference points; (g) more active people participation, and communities empowerment; (h) careful consideration of environmental impacts; and (i) compatibility of management measures across the stock's distribution range. Many of these principles have been recently integrated by the 95th Inter-Parliamentary Conference (Istanbul, Turkey, 15–20 April 1996) in its resolution on conservation of world fish stocks (IPU 1996), as well as in the Greenpeace principles for ecologically responsible fisheries (Greenpeace 1996) confirming the consensus.

When examining these prescriptions, however, a manager can only be overwhelmed by the actions required and their implications in terms of information systems, research, decision-making systems, intra- and inter-sectoral interactions, institutional building, etc. As for objectives, it will often be impossible to meet all the requirements together and some prioritization will be necessary and even advisable, provided it is done within a long-term perspective ensuring a step-wise approach to fisheries optimization.

A nested set of solutions

In order to ensure fisheries sustainability, a nested set of inter-linked actions is required, at different levels. The agreed international instruments contain elements from which management packages may be customized for implementation by governments and fishers with the cooperation of NGOs. The choice between measures is, however, not entirely free because many of the elements must be implemented together. Increasing people participation will not be beneficial in the long-term if open access is not suppressed. Limiting fleet size will not prevent overcapacity if allocation is not explicit and quantitative. Resource allocation will fail if property rights are not transferable. Inter-generational equity and long-term conservation are unlikely to be natural outputs of self-management in any exploitation system and require States' vigilance. In addition, sustainable fisheries development cannot be achieved at species or individual fishermen levels and solutions must be found at the ecosystem and community levels, defined on a matching geographical basis, and identifying and

[19] even though the words 'duty of care' do not appear in fisheries international agreements it is largely recognized that Governments should at least aim at transferring to next generations '*a productive capacity that provides a level of welfare (utility) equivalent to that of the present generation*' (Schuh and Archibald 1996).

[20] The user-pays principle has implicitly been accepted by distant-water fishing nations willing to fish, under agreement in some other country, EEZ. Few countries, however, have applied the principle for their nationals in their EEZs.

taking into account major externalities. Action is therefore required at various levels as follows:

- **At the sub-regional, regional and international level:** (a) ratify international agreements; (b) co-ordinate regional resource management bodies at regional or ocean level; (c) improve decision-making procedures for regional mechanisms and arrangements; (d) improve management of shared stocks, straddling stocks, and high seas resources management and; (e) fulfill flag–State responsibility; and (f) improve enforcement.
- **At the national and inter-sectoral levels:** (a) establish macro-economic incentives; (b) sign and ratify the international agreements; (c) establish national overview committees for resource conservation; (d) establish inter-sectoral conflict-resolution mechanisms; (e) address the issue of non-consumptive uses (ethical issues); (f) facilitate integrated management (ICAM and ICFM); (g) adopt environmental protection and rehabilitation programmes; (h) address the population growth issue; (i) reduce land-based sources of pollution; (j) develop contingency plans for global change; (k) control access and establish property and user rights; (l) improve research and monitoring capacity; and (m) promote awareness and social learning.
- **At the sectoral and local levels:** (a) improve decision-support systems and information collection and databases; (b) take account of supply and demand trends; (c) reduce and progressively eliminate subsidies; (d) elaborate development/ management plans for all fisheries and resources; (e) monitor aquatic biodiversity; (f) protect critical habitats (coral reefs, sea-grass, estuaries, nurseries, spawning grounds, etc.); (g) include fisheries in national development planning; (h) improve decision-making procedures and participation in fishery management bodies; (i) establish dispute resolution mechanisms; (j) improve monitoring, control and surveillance; (k) adopt the user-pays principle; (l) introduce the precautionary approach in sectoral planning; (m) master the process of resource enhancement; and (n) protect biodiversity.
- **At the fishery and stock levels:** (a) develop scientific evidence as a basis for management; (b) adopt the precautionary approach including precautionary Target and Limit Reference Points; (c) undertake impact assessment of existing practices; (d) list all authorized gear and practices (environmentally friendly technologies); (e) adopt prior consent and pilot project procedures for new gear and practices; (f) strengthen managerial powers of management authorities and committees; (g) eliminate open access conditions and establish Territorial Use Rights in Fisheries (TURFs; Christy 1993, 1994), property and user rights and user fees; (h) reduce overcapitalization and control fishing effort; (i) elaborate technical measures (command and control); (j) implement management strategies which promote progressive learning; (k) improve selectivity and reduce discards; (l) identify and protect endangered species; (m) reduce pollution, including debris and waste from fishing and processing; (n) reduce accidental (and prohibit voluntary) disposal of fishing gear at sea; and (o) improve participation and forms of management in partnership.

As underlined by Holling (1993, 1994) and explicitly required in many international agreements, a **holistic** or **systemic approach** to fisheries management, addressing all levels and all dimensions, will be necessary to reduce the risk of mismatches. However, fisheries contain a system of systems which, as the *matrioshka* Russian dolls, are contained in one another, and the operational limits of the manageable systems are not easily drawn. A demersal fish stock, for example, is part of a species assemblage, which occupies a particular component of a complex habitat, (sometimes seasonally changing), in a broader coastal environment, which is itself part of a large marine ecosystem (LME), included in or overlapping with a Marine Catchment Basin under the influence of the continental watershed and its land-based sources of pollution. Within a set of possible and successively broader limits, the operational boundaries of the managed system, in which controls will apply, need to be pragmatically selected. The relevance of these limits between fisheries evolves with time, with fishing community structures, and with changes in the nature of the management problem.

The required change towards modern and more effective management frameworks could also be examined in terms of: (a) changes in objectives; (b) changes in policy principles; and (c) the action required from industry, governments and research to effectuate the changes required (Table 2).

Implementation process

The challenge facing governments is in implementation and the task is more difficult because effective action has not been implemented and they now have to face the new requirements as well as the consequences of their past policies. The list of actions required (as listed in Table 2) may look overwhelming even though some elements of this Table have been emphasized in bold, based on a subjective perception of the authors. More objective prioritization of action, however, should be possible case by case, based on an assessment of the local conditions.

However, some of the actions required are milestones without which other actions will fail to produce expected outcomes. The key action for governments, which will condition the success or failure of management, is in the explicit **allocation of wealth** (through decentralization and allocation of access, resource shares, space, property or use rights, etc. as appropriate). The mandate for such action, however, does not usually belong to the fishery management authorities which, however, can promote the establishment of the **institutional framework** which will facilitate the difficult political process required.

Some of the actions have also the potential to reduce the need for others which would more or less automatically follow. For example, if exclusive use rights are given to fishermen, the need for governmental intervention or stronger enforcement will probably be reduced, and the active participation of the fisheries actors more readily ensured.

Once local priorities for action have been determined, it will often be realized that undertaking the action required is not that easy. While there is abundant literature on the theory and principles of fishery and natural resources management with a high degree of generalization (including in this paper), there

Table 2. Main elements of modern fishery management systems (elements subjectively considered as particularly important are in bold)

Changes in objectives		
From:	**To:**	
* Sustaining stocks	* Sustaining assemblages and ecosystems	
* Maximizing annual catches	*** Maximizing long-term welfare**	
* Maximizing employment	* Providing sustainable employment	
* Ensuring full resource use	* Ensuring efficient resource use (no waste)	
*Tending to short-term interests	*Addressing both short- and long-term interests	
*Addressing local considerations	*Addressing both local and global considerations	
Achieved by changes in policy		
From	**To:**	
* Open and free access	*** Limited entry, user rights, and user fees**	
* Sectoral fishery policy	* Coastal zone inter-sectoral policy	
* Command-and-control instruments (C&C)	* C&C and macro-economic instruments	
*Top-down and risk prone approaches	*** Participative and precautionary approaches**	
Translate into action by:		
Industry:	**Government:**	**Scientists:**
***Reduce capacity**	***Allocate resources explicitly**	*** Integrate socio-economics**
*** Improve compliance**	*** Suppress subsidies**	***Account for uncertainty**
* Integrate vertically	*** Reduce overcapacity**	*** Monitor/test managt. performance**
* Reduce gear loss	***Adopt precautionary approach**	*** Design precautionary methods**
* Reduce gear impact	* Ratify international agreements	* Improve resource monitoring
* Reduce pollution	* Co-ordinate regional bodies	*Analyse all possible options
* Organize people (NGOs)	*Promote awareness	* Study ecosystems dynamics
*** Increase participation**	* Improve decision making	* Study fishery system dynamics
* Contribute to research	* Increase participation	* Develop rehabilitation schemes
* Ensure food safety	* Create management committees	* Develop impact assessment
*** Give better information**	* Create overview committees	* Improve technology
	* Improve conflict resolution	* Forecast climate change
	*Establish ICAM plans	
	*** Face duty of care**	
	* Improve enforcement	
	* Rehabilitate environment	
	* Reduce land-based pollution	
	*Protect endangered species	

seem to be very little information and guidelines available on the **dynamic process of implementating change.**

It is suggested that the transition from the present (inadequate) to the future (optimal) situation in most fisheries will have to follow a strategically planned step-wise approach, based on a medium- to long-term fishery rehabilitation scheme, where the rate of implementation will be sufficiently intense to be effective and produce measurable effects but still sufficiently economically bearable and socially acceptable to be implementable in practice. The time horizon adopted for the transition from an inadequate management system to a significantly improved one will determine the **economic and social costs** and stress generated by the process, and, therefore, its degree of acceptability as well as the cost of the eventual enforcement. Too rapid large-scale negative modifications to the livelihood, cultural habits, and wealth distribution among economic agents is unlikely to be accepted and **optimal pathways** may have to be studied to find the best 'trajectory' for a given fishery system, promoting **social learning** in the process (Lee 1993a). In the short-term, it may be necessary to accept less-than-ideal solutions as necessary steps towards optimal ones in a process of **active probing of adaptive management strategies** (Walters and Hilborn 1978).

The **mix of measures and instruments** to be used will have to be determined, taking into account the historical, political, social, economic and environmental contexts in order to minimize resistance and optimize the rate of change. Social and economic sciences will be particularly useful in the process but considering the limitations of the data and the bio-socio-economic and behavioural models, it will be extremely advisable to promote very **active participation of the people** concerned. Even though **Rapid Appraisal** methodologies may help, initially, to compensate for the lack of historical data and understanding where a particular fishery system stands, 'quick fix' solutions are unlikely.

While rationalizing the fishery and ensuring that it bears the costs of as many of its impacts as possible, care will be taken to ensure that fisheries are equitably treated in comparison to other sectors such as agriculture or tourism. This is necessary in order to avoid that non-equitable stringency does not weaken its competitivity *vis-a-vis* other sectors (still heavily subsidized) with which it competes for natural, financial and human resources, leading to undesirable outcomes and reducing global efficiency.

Particular attention will have to be given to small-scale fisheries. The available international legal and guiding instruments are of general application but have not considered fully the specific implications for small-scale fisheries and coastal fishers' communities. In addition management models implicitly found at the foundations of these agreements tend to be based on 'Western' culture. As a consequence, a precautionary and partnership approach will be needed in applying these instruments to these fisheries, with due regard to the socio-economic uncertainty and related risks for the people. Areas of concern are: traditional use rights and resource allocation, generation of incomes, alternative employment, conflict with industrial fisheries and extensive coastal aquaculture, habitat degradation, technology transfer, access to capital and credit, devolution of responsibilities, etc.

Above all, following Holling's (1994) advice, the management strategies and the strategic rehabilitation plans will have to include precautionary devices allowing constant evaluation of progress towards targets, as well as flexibility to adjust sufficiently rapidly to 'surprises'. A key decision in this respect is in the management institutional framework, and in the role of the State in it, including through parastatal agencies, autonomous and decentralized but still under State responsibility in compliance with its stewardship role and duty of care.

Discussion

The present situation of fisheries is characterized both by management failure and by a clear recognition of that failure. There is also a general consensus that, paraphrasing Brown *et al* (1991), 'Sustainable economy... is a fundamentally new endeavour, with many uncertainties' but that it is also 'far less risky than continuing with business as usual'. Fisheries management systems have indeed faced and sometimes addressed the issue of sustainability for decades, but they have done it piecemeal, choosing in the management 'tool box' the less controversial tools and approaches, doing always too little and too late.

Some of the factors which have contributed to the present situation are: (1) the lack of political resolution to undertake the necessary and politically difficult adjustment required; (2) the persistence of heavy direct and indirect subsidies; (3) the lack of control by flag–States of the behaviour of their fleets; (4) the inefficiency of fishery bodies, to which member countries are reluctant to delegate the necessary powers; (5) the lack (or disregard) of scientific advice; (6) the lack of consideration for traditional communities, their rights and potential contribution to management; (7) the power of industrial lobbies in constraining decisions, perpetuating subsidies, and resisting change.; and last but not least, (8) the insufficient implementation capacity in developing countries, particularly small island countries.

Addressing management in a sustainable development perspective should bring more realism into the process, showing the futility of short-term election-oriented approaches or technical 'quick fixes', underscoring instead the need for comprehensive (holistic) approaches. The improvements needed are listed in Table 2 above.

The simple answer to the question contained in the title is that world-wide sustainable development movement offers indeed to fisheries management a *new perspective of an old problem.* The concepts are, however, not new and many of the the causes for failure were identified long ago, as shown in the section on the 'critical dimensions of fishery management' (see also World Bank 1992). What is new is:

(a) the official recognition of the problem;

(b) new concerns for environment and people's participation and empowerment; and

(c) the consensus on the institutional (and political) origin of the failure, all other factors, including scientific uncertainty, being also important but secondary;

(d) the readiness to change expressed at the highest levels of Governance;

(e) the availability of international agreements (first of all, UNCLOS) laying down the path to sustainability (and responsibility).

Conditions are therefore met for a perhaps modest but fundamental shift in the operational paradigm of management and in the context in which it will be applied.

Under these conditions, fisheries management should improve rapidly in the next 10 years. If, for different reasons it does not , and despite expected growth in aquaculture, there could be a shortfall of 10–40 million t of fish for human consumption by year 2010 Westlund (1995). If domestic supplies of major importing countries are not improved, their unsatisfied demand will promote increased international trade. It may in this way promote further depletion of already thinned-out stocks of the exporting (mainly developing) nations as well as a decline of the contribution of fisheries to food security and economic welfare (FAO 1995*d*). With effective management response, however, annual benefits could be very high and of the order of 10–20% of the present landed value of about $US70 000 million.

Fisheries are at a crossroad. The response that will be provided by those in charge of management and the fishery sector will establish whether capture fisheries will emerge as a sustainable activity for the 21st century or are doomed sunset industries. In this respect, the task in front of the fishery sector and its managing authorities (as shown in Table 2) appears daunting. However, the world-wide process has already started. Many countries are considering the task in front of them and preparing for it. Some have already begun practical implementation (*cf.* Commonwealth of Australia 1992) and the high level of awareness in most developed countries and in many developing ones is a good omen for the future.

Despite the deep concerns expressed during the last few years, there is no risk that the global fishery resource will disappear in a foreseeable future as a consequence of fishing even if size and composition change in less desirable directions and the consequence of inadequate action may be more serious for some areas than for others. Even though fisheries have become a global concern, they do not represent a major global societal risk and they contribute much less than other manufacturing and chemical industries or even agriculture to the problem of global change and biodiversity degradation. However, there are numerous 'local risks' in some species (e.g. long-lived ones), in some areas (e.g. high seas but also many EEZs) or habitats (e.g. coral reefs, mangroves). Three types of risk are illustrated below:

1. In the absence of significant improvement of management, the necessary adjustments to human behaviour may be 'forced' by nature in the form of abrupt resource changes (as in the Canadian cod fishery) with very significant socio-economic damage and heavy penalties in terms of resource availability. These dramatic changes have the potential to trigger very significant and rapid corrective action.[21] Catastrophic types of events will be more frequent with small pelagic species linked to upwellings or for species with a very critical and vulnerable life stage (e.g. salmon). The dimension and suddenness of the consequences may lead to lengthy rehabilitation programmes and shifts in the fishery sector's and governments' perceptions and behaviour. In the tropics, where species are generally short-lived, rehabilitation may be obtained in one or a few years. In colder and deeper waters, however, such rehabilitation, if achieved, may take several years, possibly up to decades if unfavourable climatic conditions prevail.

2. Nature may be more forgiving. Through species resilience and slow change in species dominance and trophic relationships and, with time, it may allow some 'adaptation' to excessive fishing pressure, by the biological system and even by people's habits and food preferences. In Mauritania for instance, Nature offered a 'jackpot' when overfished sea-bream resources were progressively replaced by more valuable octopus resources. Surreptitious degradation may continue for some time, unabated, without raising enough concern to trigger action. Some preferred species will become scarcer and may be replaced. Fish formerly discarded may become utilizable as progress will be made in using trash fish for protein pastes and concentrates and other products of substitution. Environmental degradation, from coastal areas and from the hinterland may progressively decrease productivity and increase fish disease. Experience in other natural resources systems predict that this scenario is particularly likely in areas dominated by lack of rights and poverty, where liability cannot be established and alternative livelihoods are non-existent. This is indeed the most likely scenario, and the most preoccupying!

3. A more general and perhaps less obvious risk for the fishing communities is that of being progressively deprived of the traditional rights they have acquired (but which are often not recognized formally) through the pressure exerted by newer and often better organized sectors such as conservation (with which, however, they have so much in common), or tourism, oil industries, and other coastal activities, which seem to be more succesful at getting an allocation of the coastal resources legally recognized despite their often more drastic effect on the resources and the environment.

It may not be possible for fisheries management to totally suppress the first two risks of catastrophic or creeping change, which will in fact co-exist in a given area, but their probability of occurrence could be reduced substantially by effectively complying with the principles of sustainable development in fisheries planning, development, and management including the precautionary approach contained in the comprehensive FAO (1995*c*) Code of Conduct for Responsible Fisheries. The third risk, the probability of which can only be increased by irresponsible fishing behaviour, could be seriously reduced if the fishing communities would: (a) demonstrate a higher level of responsibility; and (b) request recognition of their traditional use rights, opening the way to legal action against those who harm the resource base and its environment.

[21] Fishing effort was reduced by 50% in the Scotia-Fundy bottomfish fishery since 1991 (M. Sinclair pers. comm.)

ACKNOWLEDGMENTS

This paper has greatly benefited from inputs, critical comments and wise suggestions of our colleagues: Helga Josupeit, Jean-Jacques Maguire, Ehrhardt Ruckes, Robin Welcomme, Ulf Wijkstrom, Annick Van Houtte and Paul Vantomme to whom we owe our most grateful thanks.

REFERENCES

Alverson, D L, and Paulik, G J (1983). Objectives and problems in managing aquatic living resources. *Journal of the Fisheries Research Board of Canada*, **30**(12), 1936–1947.

Alverson, D L, and Larkin, A P (1992). Fisheries: Fisheries Science and Management: Century 21. In 'The State of World's fisheries resources'. (Ed C W Voigtlander) pp. 150–67. *Proceedings of the World Fisheries Congress Plenary Sessions* (Oxford and IBH Publishing Co. Pvt. Ltd: New Delhi).

Alverson, D L, Freeberg, M H, Murawsky, S A, and Pope, J G (1994). A global assessment of fishery by-catch and discards. *FAO Fishery Technical Paper* No 339. 239 pp.

Annala, J (1995). New Zealand ITQ system: Have the first eight years been a success or failure? *Reviews in Fish Biology and Fisheries*: 8 pp. (manuscript).

Anon. (1989). New directions for commonwealth fisheries management in the 1990s. A Government policy statement, December 1989. (Australian Government Publishing Service: Canberra). 114 pp.

Boelaert-Suominen, S, and Cullinan, C (1994). Legal and institutional aspects of integrated coastal areas management in national legislation. FAO. 188 pp.

Bonus, H (1993). Implications of the polluter-pay and user-pay principles for developing countries. In 'Fair principles of sustainable development: Essays on environmental policies and developing countries'. (Ed E Dommen) pp. 61–72. UNCTAD, Geneva, New Horizons in Environmental Economics.

Bourke, I J (1995). Forest product: a background note prepared for the Meeting of the Private Forest Industry Sector on UNCED Follow-up. Rome 8 March 1995. 4 pp.

Breth, S A (Ed) (1996). Integration of sustainable agriculture and rural development issues in agricultural policy. (Winrock International: Morrilton, Arkansas).

Brown, L R, Postel, S, and Flavin, C (1991). From growth to sustainable development. In 'Environmentally sustainable economic development'. (Eds R Goodland, H Daly, S El Serafy and B Van Droste) (Building on Brundtland, UNESCO).

Caddy, J F, and Garcia, S M (1986). Fisheries thematic mapping — A prerequisite for intelligent management and development of fisheries. *Oceanographie Tropicale* **1**, 31–52.

Caddy, J F (1994).Checks and balances in the management of marine fish stocks: organizational requirements for a limit reference point approach. ICES CM 1994/T:1.

Caddy, J F, and Mahon, R (1995). Reference points for fishery management. *FAO Fisheries Technical Paper* 347. 82 pp.

Christy, F T (1993). A re-evaluation of approaches to fisheries development: the special characteristics of fisheries and the need for management. *FAO Fisheries Report* **474**(2), 597–608.

Christy, F T (1994). Economic waste in fisheries: impediment to change and conditions for improvement. Paper presented at the Conference on Fisheries Management and Global Trends, University Washington, Seattle, 14–16 June 1994. 33 pp.

Commonwealth of Australia (1992). Fisheries ecosystem management. In 'National strategy for ecologically sustainable development'. pp. 27–29 (Australian Government Publishing Service: Canberra).

Cunningham, S (1994). Proposed FAO guidelines on the integration of agriculture, forestry and fisheries into coastal areas management: fisheries section. FAO Draft. 32 pp.

Fallon Scura, L (1994).Typological framework and strategy elements for integrated coastal fisheries management (ICFM). FAO FI:DP/INT/91/007. 23 pp.

FAO (1945). Report of the Technical Committee on Fisheries, submitted to the United Nations Interim Commission on Food and Agriculture. Washington. 13 pp.

FAO (1967). The Management of Fishery Resources. In: 'The State of Food and Agriculture'. (FAO: Rome). pp. 119–33.

FAO (1983). Report of the Expert Consultation on the Regulation of Fishing Effort (16–26/1/1983). *FAO Fisheries Report* 289. 34 pp.

FAO (1992). Report of a regional workshop on monitoring, control and surveillance for African States bordering the Atlantic Ocean (Accra, Ghana, 2–5 November 1992). FAO/Norway Cooperative Programme. GCP/INT/466/NOR. 94 pp.

FAO (1993). Marine fisheries and the law of the sea: a decade of change. Special chapter (revised) of The State of Food and Agriculture 1992. *FAO Fisheries Circular* 853. 65 pp.

FAO (1995*a*). The precautionary approach to fisheries. Part 1. Guidelines to the precautionary approach to fisheries and species introductions. *FAO Fisheries Technical Paper* 350(1). 52 pp.

FAO (1995*b*). Code of Conduct for Responsible Fisheries. (FAO: Rome). 41 pp.

FAO (1995*c*). Safeguarding future fish supplies: key policy issues and measures. Document KC/FI/95/1. International Conference on the Sustainable Contribution of Fisheries to Food Security, Kyoto, Japan, 4–9 December 1995. 50 pp.

FAO (1995*d*). The Rome consensus on World Fisheries adopted by the FAO Ministerial Conference on Fisheries. Rome, 14–15 March 1995. 4 pp.

FAO (1996). The precautionary approach to fisheries. Part 2. Scientific papers. *FAO Fisheries Technical Paper* 350 (2).

Feeny, D, Berkes, F, McCay, B J, and Acheson, J M (1990). The tragedy of the commons: twenty two years later. *Human Ecology* **18**(1), 1–19.

Forest Stewardship Council (1995). Principles and criteria for natural forest management. Board approved version. June 1994. Oaxaca, Mexico. 4 pp.

Garcia, S M (1992*a*). High seas fisheries management. New concepts and techniques. In: FAO. 'Papers presented at the Technical Consultation on High seas Fishing. Rome, 7–15 September 1992. *FAO Fisheries Report* 484 (suppl.), 37–40.

Garcia, S M (1992*b*). Ocean fisheries management. The FAO programme. (Ed . Fabbri). Ocean management in global change. Elsevier Applied Science. pp. 381–418.

Garcia, S M (1992*c*). Fishery research and management: virtues and constraints of a symbiosis. Paper presented to the First World Fisheries Congress. May 1992, Athens, Greece. 25 pp.

Garcia S M (1994). The precautionary principle: Its implications in capture fisheries management. *Ocean and Coastal Management* **22**, 99–125.

Garcia, S M (1995). The precautionary approach to fisheries with reference to straddling fish stocks and highly migratory fish stocks. *FAO Fisheries Circular* 871, 76 pp. (Trilingual, English, Spanish, French).

Garcia, S M (1996). The precautionary approach to fisheries and its implications for fishery research, technology and management: an updated review. *FAO Fisheries Technical Paper* 350(2).

Garcia, S M, and Majkowski, J (1990). State of high seas resources. In 'The law of the sea in the 1990s: a framework for further international cooperation'. (Eds T Kurobayashi and EL Miles). pp. 175–236 (Law of the Sea Institute: Honolulu, Hawai, USA).

Garcia, S M, and Newton, C (in press). Current situation, trends and prospects in world capture fisheries. In 'Global trends: Fisheries management'. (EDS E K Pikitch, D D Huppert and M D Sissenwine). American Fisheries Society Symposium 20. Bethesda, Maryland, USA.

Glantz, M H (1990). Desertification sur terre et dans l'ocean. In 'L'homme et les ressources halieutiques'. (Ed J P Troadec). pp. 655–679 (IFREMER: Paris).

Goodland, R (1995). The concept of sustainability. *Ecodecision* **15**, 30–2.

Gordon, H S (1954). Economic theory of a common-property resource: The fishery. *Journal of political economy* **LXII** :124–142.

Greenpeace (1996). Greenpeace principles for ecologically responsible fisheries. Preliminary document. February 1996. MS: 6 pp.

Gulland, J A (1978). Fisheries management: New strategies for new conditions. *Transactions of the American Fisheries Society* **107**(1), 1–12.

Gulland, J A (1984). Looking beyond the Golden Age. *Marine Policy* **8**, 137–50.

Gulland, J A (1989). Fishery management: how can we do better? In 'Management of world fisheries: implications of extended coastal States jurisdiction'. (Ed E L Miles) pp. 255–72 (University of Washington: Washington, USA).

Gulland, J A, and Carroz, J E (1968). Management of Fishery Resources. *Advances in Marine Biology* **6**, 1–71.

Hannesson, R (1986). Fishermen's organizations and their role in fisheries management: theoretical considerations and experiences from industrialized countries. In 'FAO: Studies on the role of fishermen organizations in fisheries management'. *FAO Fisheries Technical Paper* 300, pp. 1–23.

Hardin, G (1968). The tragedy of the commons. *Science* **162**, 1243–8.

Holling, C S (1993). Investing in research for sustainability. *Ecological Applications* **3**(4), 555–8.

Holling, C S (1994). New science and new investment for a sustainable biosphere. In 'Investing in natural capital. The ecological economics approach to sustainability. International Society for Ecological Economics'. (Eds A Jansson, M Hammer, C Folke and R Costanza) pp. 57–73 (Island Press: Washington DC).

House of Lords (1996). Fish conservation and management. Report of the House of Lords Select Committee on Science and Technology. Session 1995–96. 2nd report. HMSO, London. 77 pp.

IPU (1996). Conservation of world fish stocks in order to provide an important source of protein and ensure the continued viability and economic stability of fishing around the world. Draft resolution adopted unanimously by the Committee on Education, Science, Culture, and Environment. 95th Inter-Parliamentary Union. CONF/95/6-DR 19 April 1996. 5 pp.

Johnson, R S (1992). Fisheries development, fisheries management, and externalities. *World Bank Discussion Papers, Fisheries Series* 195. 43 pp.

Kirkwood, G P, and Smith, A D M (in press). Assessing the precautionary nature of fishery management strategies. In 'Precautionary approach to fisheries'. Part 2: Scientific papers. *FAO Fisheries Technical Paper* 350(2).

Kurien, J (1988). The role of fishermen in fisheries management in developing countries (with particualr reference to the Indo-Pacific region). In 'FAO: Studies on the role of fishermen organizations in fisheries management'. *FAO Fisheries Technical Paper* 300. 29–48.

Lee, K N (1993*a*). Compass and gyroscope: integrating science and politics for the environment. (Island Press: Washington DC).

Lee, K N (1993*b*). Greed, scale mismatch, and learning. *Ecological Applications* **3**(4), 564–6.

Levin, S A (1993). Forum: Science and technology. *Ecological applications* **3**(4), 546–7.

Ludwig, D, Hilborn, R, and Walters, C (1993). Uncertainty, resource, exploitation and conservation: lessons from History. *Science* **260**, 17–36.

Mangel M. *et al.* (1996). Principles for the conservation of wild living resources. *Ecological Applications* **6**(2), 66.

Miles, E L (1989). Management of world fisheries: implications of coastal Sates extended jurisdiction. (University of Washington Press: Seattle and London).

Miles, E L (1994). Sustainable development and the uses of the world ocean. Paper prepared for Illahee: *Journal for the Northwest Environment*.

Pauly, D, and Chua, T E (1988). The overfishing of marine resources: socio-economic background in Southeast Asia. *Ambio* **17**(3), 200–6.

Policansky, D (1993). Uncertainty, knowledge, and resource management. *Ecological Applications* **3**(4), 583–4.

Pommeroy, R S (Ed) (1994). Community management and common property of coastal fisheries in Asia and the Pacific: Concepts, methods and experiences. *ICLARM Conference Proceedings* **45**, 189.

Pommeroy, R S (1995). Community-based and co-management institutions for sustainable coastal fisheries management in Southeast Asia. *Ocean and Coastal Management* **27**(3), 143–62.

Pontecorvo, G, and Schrank, W E (1995). Commercial fisheries: the results of stochastic supply and economic uncertainty. A paper presented at the Columbia Resource Seminar on 'Sustainable Development and a Managed Resource: the Current Crisis in Commercial Fisheries'. Arden House, 5–6 May 1995. (University British Columbia Press: Vancouver). 61 pp. (MS).

Pope, J G (1979). Population dynamics and management — Current status and future trends. *Investigaciones Pesqueras* **43**(1), 199–221.

Rettig, R.B, Berkes, F, and Pinkerton, E (1989). The future of fisheries co-management: a multidisciplinary assessment. In 'Co-operative management of local fisheries'. (Ed E Pinkerton) pp. 273–89 (University British Columbia Press: Vancouver).

Rosenberg, A A, Fogarty, M J, Sissenwine, M P, Beddington, J R, and Shepherd, J G (1993). Achieving sustainable use of renewable resources. *Science* **262**, 828–9.

Rubenstein, D L (1993). Science and the pursuit of a sustainable world. *Ecological Applications* **3**(4), 585–7.

Saetersdal, G (1989). Mastering resource management. *Samudra* **2**, 3–22.

Schuh, E G, and Archibald, S (1996). A framework for the integration of environmental and sustainable development issues into agricultural planning and policy analysis in developing countries. In 'Integration of sustainable agriculture and rural development issues in agricultural policy'. (Ed S A Breth) pp. 3–44 (Winrock International: Morrilton, Arkansas).

Scott, A D (1955). The fishery: the objectives of sole ownwership. *Journal of Political economy* **LXIII**. 116–124.

Shepherd, J (Convener) (1991). Special session on management under uncertainty. 5–7 September 1990. *NAFO Scientific Council Studies* **16**, 189.

Shepherd, J G (1993). Why fisheries need to be managed and why technical conservation measures, on their own, are not enough. Directorate of Fisheries Research, Lowestoft. Laboratory Leaflet 71, pp. 1–15.

Sissenwine, M P, and Rosenberg, A A (1993). Fisheries: opportunities and concerns. *Oceanus* **36**(2), 48–54.

Smith, S J, Hunt, J H, and Rivard, D (1993). Risk evaluation and biological reference points for fisheries management. *Canadian Special Publication in Fisheries and Aquatic Sciences* **120**, 441.

Somers, I (1994). Through the crystal ball: the next 25 years. In 'Australia's Northern prawn fishery. The first 25 years'. (Ed P C Pownall) pp. 131–40 (NPF25, Cleveland, Australia).

Sproul, J T (1996). Green fisheries? Linking fishery health, practices and prices through eco-certificationn, labelling and crediting. UBC Fisheries Center, Vancouver, (British Columbia, Canada). 12 pp. (Manuscript).

Stevenson, J C (Ed) (1973). Technical Conference on fishery management and development. *Journal of the Fishery Research Board of Canada* **30**(12), 2537.

Tiphaigne de la Roche, G F (1760). Essai sur l'histoire économique sur les mers occidentales de la France. (C.J.B. Bauché: Paris).

Tobin, B (1995). A vision for ocean management. Ministry of Fisheries and Oceans, Canada. 8 pp.

Troadec, J P (1982). Introduction a l'aménagement des pecheries: interet, difficultes, et principales methodes. *FAO DocumentTechnique Pêches* **224**. 65 pp.

Troadec, J P (1988). Institutions et aménagement des pèches au Maroc. Rapport de consultance. Project PNUD/FAO/MOR/86/019/B/01/12. FAO. 61 pp.

Troadec, J P (1991). Fisheries efficiency, resources conservation effectiveness, and institutional innovations. PACEM IN MARIBUS XIX; 1991 Nov 18; Lisbon, Portugal. 15 pp.

Troadec, J P (1996). Produire mieux en pêchant moins. : la rêgulation de l'accès. In 'GREP: Pêches maritimes françaises: Bilan et perspectives'. Revue POUR, (149–150): 89–102. Groupe de recherche pour l'éducation et la prospective. (Diffusion de l'Harmattan: Paris).

United Nations (1946). Final Act and Convention of the International Overfishing Conference. London, 25 March–5 April, 1946. *HMSO Miscellaneous* **7**, 1–12.

United Nations (1995). Agreement for the Implementation of the Provision of the United Nations Convention on the Law of the Sea of 10 December 1982. Relating to the Conservation and Management of Straddling Fish Stocks and Highly Migratory Fish Stocks. UN A/Conf 164/37.

Walters, C J, and Hilborn, R (1978). Ecological optimization and adaptive management. *Annual Review of Ecology and Systematics* **9**, 157–88.

Walters, C (1995). Fish on the line: The future of Pacific Fisheries. A report to the David Suzuki Foundation Fisheries project, Phase 1. Vancouver, British Columbia, Canada. 82 pp.

Westlund, L (1995). Apparent historical consumption and future demand for fish and fishery products — expanding calculations. International Conference on the Sustainable Contribution of Fisheries to Food Security. Kyoto, Japan 4–9 December 1995. 50 pp.

Willmann, R, and Insull, D(1993). Integrated coastal fisheries management. *Ocean and Coastal Management* **21**, 285–302.

World Bank (1992). A study of international Fishery research. (The World Bank:. Washington DC). 103 pp.

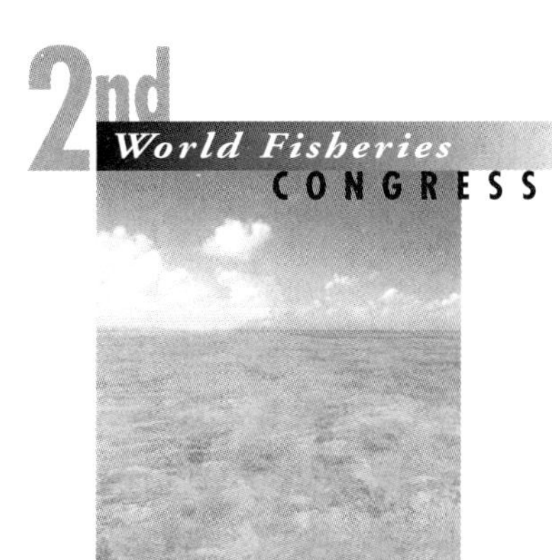

Australian Fisheries Management Authority: Organizational Structure and Management Philosophy

James C. McColl and Richard A. Stevens

Australian Fisheries Management Authority, PO Box 7051, Canberra Mail Centre, ACT 2610, Australia.

Summary

Prior to 1992, responsibility for Commonwealth fisheries management resided with a Division known as the Australian Fisheries Service within the Federal Department of Primary Industries and Energy. In the late 1980s and early 1990s, the Australian Government decided that major and fundamental changes in fisheries policy were required and in 1989, the Government published a policy statement entitled 'New Directions for Commonwealth Fisheries Management in the 1990s'. Subsequently, in 1991, a package of new Commonwealth fisheries legislation was passed by the Australian Parliament which established the administrative and operational structure for the implementation of the Commonwealth's fisheries management objectives. In particular, the legislation established the Australian Fisheries Management Authority (AFMA). The paper assesses the success of the AFMA model and the accompanying 'Management Advisory Committee' framework, together with the management philosophy pursued by AFMA in its dealings with major user groups including recreational and environmental interests.

Introduction

This paper presents an overview of the institutional and policy framework within which fisheries are managed in Australia, and outlines the organizational/institutional framework for fisheries management that existed prior to 1992, and the major changes that have been introduced since that date. It describes the relationship between the Commonwealth government and State governments with respect to fisheries management together with the management philosophy being pursued by the Australian Fisheries Management Authority (AFMA) which was established in early 1992. An assessement is made of the success to date of the AFMA model and the incorporated 'Management Advisory Committee' framework.

The following provides a few basic facts about Australian fisheries:

- the landed value of the Australian harvest is approximately $\$A1.6 \times 10^9$
- most of the harvest is exported, with over 60% as whole fish. Many of the species harvested are of high market value — a single southern bluefin tuna can sell for as much as $A30 000 in Japan.

- the Australian Fishing Zone (AFZ) is the third-largest of any coastal country and is larger than the area of continental Australia.
- in general, State governments are responsible for fisheries management from the low-water mark out to three nautical miles, and the Commonwealth Government has jurisdiction from three miles out to the 200 mile limit of the AFZ.
- institutional arrangements for fisheries management vary between the States and the Commonwealth. Commonwealth fisheries are managed through AFMA, a statutory authority established under specific legislation. One State uses a statutory authority to manage fisheries, another uses management committees that report directly to the Minister, and the remainder employ the government department model.

How the Commonwealth manages fisheries

Prior to 1992

Prior to 1992, the Australian Fisheries Service (AFS), a division within the Department of Primary Industries and Energy (DPIE), had responsibility for Commonwealth fisheries management. Despite its best endeavours, the management of Commonwealth fisheries was subject to a number of criticisms. For example, it was argued that:

- resources were being overfished, and expansion of fishing fleets was continuing;
- there was a lack of trust between industry, management and scientists; and
- industry frequently did not accept stock assessments;
- industry felt that management services were not being provided in a cost-effective manner;
- user-groups in general had little effective say in the setting of management objectives in the development of management arrangements;
- management/user group conflicts were the order of the day; and
- the Commonwealth Minister responsible for fisheries was subject to direct lobbying, and became involved in many no-win decisions. The fisheries portfolio seldom provided political benefits to the Party in power.

In light of these difficulties, and after a number of reviews, it was decided that a major and fundamental change in fisheries policy was needed. In 1989, a major fisheries policy White Paper entitled 'New Directions for Commonwealth Fisheries Management in the 1990s' was developed.

Establishment of AFMA

In 1991 a package of new Commonwealth fisheries legislation was passed by the Australian Parliament establishing AFMA as a statutory authority, and setting in place the administrative and operational structure for the implementation of the Commonwealth's fisheries management objectives. A small fisheries policy unit remained in DPIE.

The government's policy statement at the time stated that:

> The structure of a statutory authority would enable the Government to effect its responsibilities in a flexible, open and less bureaucratic way. It would also allow greater community and industry participation in determining the appropriate management programs for Commonwealth fisheries than has been the case in the past.

AFMA is now responsible for the day-to-day management of Commonwealth fisheries. All management costs, other than those in respect of community service obligations, are recovered through levies on the fishing industry. Of the total costs of AFMA, 52.2 per cent are derived from industry.

AFMA's accountability to both the Parliament, the fishing industry and the community is maintained through the submission of both a longer-term corporate plan and an annual operational plan to the Minister, together with both an annual report to the Parliament and annual audit by the Auditor-General.

The Minister has the power under the new legislation to give directions in writing to the Authority concerning its performance, but these directions must be consistent with AFMA's objectives under legislation and must be tabled in the Parliament. To date, four years after AFMA's establishment, no such directions have been given.

Objectives of AFMA

The objectives under the legislation are:

- implementing efficient and cost-effective fisheries management on behalf of the Commonwealth;
- ensuring that the exploitation of fisheries resources and the carrying on of any related activities are conducted in a manner consistent with the principles of ecologically sustainable development, in particular the need to have regard to the impact of fishing activities on non-target species and the marine environment;
- maximizing economic efficiency in the exploitation of fisheries resources;
- ensuring accountability of AFMA's management of fisheries resources to the fishing industry and the Australian community; and
- achieving government targets in relation to the recovery of costs of AFMA.

The AFMA Board

AFMA is a body corporate consisting of a chairperson, government director, managing director and five nominated directors. Directors, other than the managing director, are appointed by the Minister. Directors are chosen from persons nominated by an AFMA selection committee on the basis of their expertise in fishing operations, processing, natural resource management, fisheries science, marine ecology, and business management The selection committee, which is established

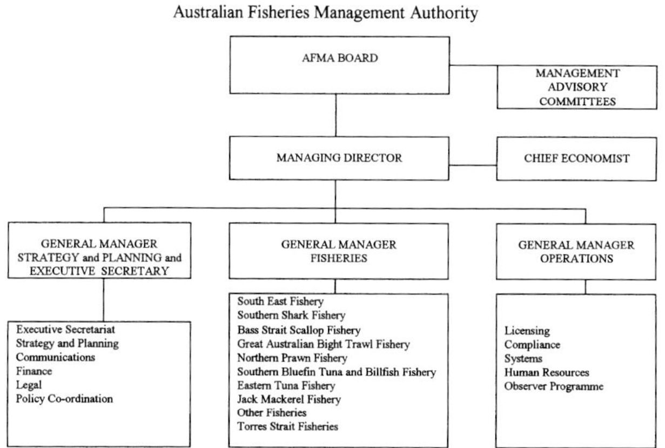

Fig. 1. Structure of AFMA.

under the legislation and appointed by the Minister is comprised of both government and fishing industry representatives.

Management Advisory Committees

AFMA has established Management Advisory Committees (MACs) for each of the major Commonwealth fisheries. MACs are the focal point for joint management/user group participation in fisheries management decision-making.

MACs are normally comprised of an independent chairperson, an AFMA member, a scientific member, a member representing State or Territory governments and five industry members. Basically, MACs are responsible for advising the AFMA Board on management policies for a fishery and for developing cost-efficient management arrangements. Each MAC has an executive officer responsible to the chairperson who generally has fisheries management expertise.

Whilst consultation with industry can be time-consuming and expensive, AFMA believes it is the key to gaining broader industry acceptance and ownership of management decisions. Of course, involving industry in the decision-making process brings with it certain obligations and responsibilities on the part of industry, and AFMA has made a concerted effort to make industry members on MACs aware of their important role. Specifically, industry members must be able to satisfy the following requirements:

- They must be able to put views clearly and concisely and be prepared to negotiate to achieve acceptable compromises where necessary;
- They must have industry's confidence and authority to undertake their functions as members of the MACs;
- They are required to act in the best interests of the fishery as a whole, rather than as an advocate for any particular organization or interest group;
- They must be prepared to observe confidentiality and exercise tact and discretion when dealing with sensitive issues; and
- They must avoid pursuing personal agendas, but participate in discussion in an objective and impartial manner.

The involvement of industry in the decision-making structure through the MACs has been a change which has brought with it significant industry responsibility and accountability. Industry, to its credit, has responded extremely well and the MAC process is now well and truly bedded down. The most important benefit has been far more informed discussion of management arrangements, research priorities and stock assessment.

In its corporate plan for 1995 to 2000, AFMA has indicated its intention to involve a broader range of interest groups, including environmental and recreational fishing groups in the consultation process for developing and refining management arrangements. To this end, a number of the MACs for the major fisheries managed by AFMA now include membership from both the environment/conservation and recreational fishing sectors.

Management plans

Currently, most MACs are focussed on the development of formal management plans. A management plan is a subordinate legislative instrument, that is determined by AFMA, accepted by the Minister and approved by Parliament. Management plans must include: the identification of management objectives and performance criteria; a description of the fishing concessions to be used and determination of how they will be allocated; and the identification of the rules governing those operating in the fishery.

There is an explicit consultation process that must be followed in establishing management plans. AFMA may determine a management plan for a fishery after it has consulted with those persons engaged in fishing, and given public notice of its intention and both invited and considered representations.

AFMA must inform the Minister of plans it has determined, any representations received and consultations conducted prior to the making of the determination. The Minister is required to accept the plan if adequate consultation and consideration were given to representations and the plan is consistent with AFMA's objectives and corporate and operational plan. If the plan is not accepted, the Minister must refer it back to AFMA with reasons for its rejection. AFMA must rectify a rejected plan as soon as practicable and resubmit it to the Minister. This process is continued until an acceptable plan is reached. Accepted plans are gazetted and are subject to disallowance by Parliament.

Cost recovery

Under current Commonwealth policy, industry pays 100% of recoverable management costs in Commonwealth-managed fisheries. The move towards 100% cost recovery is probably the most important factor in setting the dynamics in place for both increased industry participation and the continuing devolution of power to MACs.

Recoverable management activities include: the running costs of MACs, licensing, AFMA's day-to-day management activity, ongoing costs associated with maintaining management plans, log-books, surveillance, and quota monitoring.

Non-recoverable activities which are funded by the Commonwealth Government include: enforcement of domestic fishing, a portion of recoverable costs in exploratory and collapsed fisheries, surveillance and enforcement of illegal foreign fishing activity in the AFZ, and Commonwealth-requested participation of AFMA in international fora (e.g. FAO and OECD).

Costs are recovered on a fishery-by-fishery basis. Therefore AFMA has had to establish a finance system capable of presenting detailed expenditures by fishery. MACs are heavily involved in the preparation of annual budgets for each fishery, and are provided with detailed cost information. Final budgets must be approved by the AFMA board.

Resource allocation

AFMA's legislation provides for user groups to have an influential role in the development and implementation of fisheries management measures, and resource allocation is one of those decisions which requires careful thought. The resource allocation issue not only relates to that between commercial operators but also to that between commercial and recreational user groups. AFMA has established a number of key requirements in making resource allocation decisions, as follows:

(i) *A reliable data base:* Where allocations are to be based on past participation in a fishery, AFMA needs to know who has operated in the fishery and, if it will affect the quantity of rights being allocated, the level of such participation. This is particularly relevant to output-control fisheries, but can also apply to some input-control fisheries. Whilst submission of catch and effort information sometimes gives an incentive for fishermen to be less than truthful, AFMA recognizes that it is more important that there be some alternative means of verifying data.

(ii) *Goals and objectives set*: One of the challenges that fisheries management agencies frequently face is that the goals and objectives are nearly always stated in very general terms in legislation. Almost any outcome could be categorized as being successful or unsuccessful by any particular group. In addition, many fisheries managers often take the approach of getting through the day-to-day 'bushfires' without wishing to be too bothered about general objectives set out in legislation. However, it is important to have well-defined objectives to ensure fisheries management is pro-active, strategic and forward-looking, whilst at the same time recognizing that there will always be uncertainty and risk in the whole process.

In looking at resource allocation in commercial fisheries, it is therefore important to clearly specify the denied objectives you are trying to achieve *prior* to making resource allocation decisions.

(iii) *Impact on user groups assessed:* Decisions on resource allocation inevitably affect the livelihoods of individuals and in some instances whole communities. It is therefore essential that the expected consequences of such decisions are taken into account, and this includes economic, social, legal and political consequences. In particular, user groups are entitled to have as much advance notice as possible of any change to resource allocation which might seriously affect their long-term business prospects. AFMA's legislation has an explicit consultation requirement, thus ensuring that the views of user groups are taken into account.

Adequate consultation is critical. In the case of the fishing industry, it already has to endeavour to accommodate matters over which it has little control or influence, such as changes in broader government policies affecting how it operates, or significant international influences. Accordingly, the industry will always appreciate as much advance warning as possible of any changes in fisheries management arrangements which affect resources.

(iv) *The right climate for decisions:* Whilst resource allocation decisions are best made in a climate of reasonable certainty, this is probably the exception rather than the rule in fisheries. For example, the introduction of major management changes in most fisheries almost always is in response to some kind of crisis such as a significant decline in landings in the face of rising fishing effort, with conservation goals being presently unattainable and costs of management escalating. Resource allocation decisions, including the possibility of not allocating any resource at all, are then forced on the fisheries manager in a climate of adversity, uncertainty and the inevitable debate about the real status of stocks. This in turn makes goal and objective setting essential.

(v) *An Appeals process:* Denying access to a traditional means of livelihood to some fishermen for whatever reason brings with it the distinct prospect of appeals, and in AFMA's case, review of decisions by the Administrative Appeals Tribunal, the Federal Court or even the High Court. The important point to emphasize is the need for fairness and consistency, together with confidence in the legislation and the policy upon which decisions are based. Defective policy in the resource allocation decision-making process is a recipe for real trouble, and it is therefore essential that competent legal input is sought prior to proceeding with a major decision on resource allocation.

Fisheries assessment

AFMA's approach to stock assessment in fisheries involves a cooperative partnership between scientists, industry and fisheries managers through the MAC process. The key elements include:

- Catch and effort data, and gear and fleet dynamics information (e.g. engine power, vessel size, technological accessories) both by fishing method to permit 'standardization' of the effort, and by species for all fisheries;
- Annual 'fishery assessment reports' in an agreed format to describe the state of the stocks that comprise each fishery and the economic performance and status of each fishery;
- A set of research priorities, endorsed by the MAC and formulated into a strategic plan for each major fishery, which identify the major problems in managing the fishery;
- A list of identified projects to solve the major problems identified in the research priorities for the fisheries, which includes both ongoing and new projects, and which identifies ongoing or possible funding sources;
- Where research projects have been identified, suitable arrangements are made for applications to be prepared and examined by the MACs, prior to submission to the funding agency;

- Special studies may be arranged by AFMA to examine bycatch, effects of trawling, environmental questions, and biodiversity aspects.

This process has worked well in the last two years, and has enabled AFMA to establish research priorities within the constraints of available research funding. Importantly, it has provided a sound basis for assessing the performance of fisheries against AFMA's legislated objectives.

Jurisdiction

A very important aspect of fisheries management in Australia relates to the division of responsibilities between the Commonwealth and State and Territory governments. This has led to a very complex network of jurisdictional responsibilities frequently involving several States and the Commonwealth in the management of a specific fishery.

Australia is a federation governed by six State and two Territorial governments and the Commonwealth government. When the six then self-governing British colonies decided to federate at the end of the last century, they agreed that certain specified powers would be given to the Commonwealth government, while the remaining unspecified powers would remain with the colonies (which on federation became the States). The powers ceded to the Commonwealth were mainly broad national powers such as defence, immigration, foreign affairs and customs.

The Commonwealth has two principal powers under the Constitution on which it can regulate sea fisheries. The first is the power to make laws with respect to fisheries in Australian waters beyond 3 nautical miles (nm) measured from the coast of an Australian State. It follows that the Commonwealth can not control fisheries within 3 nautical miles of the coast of an Australian state.

The Constitution also gives the Commonwealth government power with respect to 'external affairs'. There are two aspects of this power relevant to the control of fisheries. The first aspect is that which enables the Commonwealth to give effect to international law and the second which enables the Commonwealth to legislate with respect to matters 'physically external' to Australia. There is a strong body of opinion that this power would probably enable the Commonwealth, if it so wished and was prepared to wear the political consequences, to legislate in respect of fisheries in all waters adjacent to Australia.

There is one significant qualification to State power. That is that any State law with respect to fisheries will be invalid if it is inconsistent with a valid law of the Commonwealth Parliament dealing with fisheries. For example, if the Commonwealth law authorizes the taking of a particular fish, a State law purporting to prohibit the taking of that fish would be invalid.

State fisheries laws therefore applied in the area within 3 nm of the coast and Commonwealth laws applied to the area beyond 3 nm from the coast. Such a rigid geographic demarcation of fisheries jurisdiction inhibited sensible and consistent fisheries management. State management of a particular fishery within 3 nm of the coast could be effectively undermined by a different form of Commonwealth management of the fishery beyond 3 nm. There is enough uncertainty and risk in fisheries management without having say, five different management regimes applying to what is biologically, in respect of some species, one fishery.

Offshore Constitutional Settlement

The Offshore Constitutional Settlement (OCS) legislation which came into effect in 1983, as well as formally transferring the title of the seabed of the then 3 mile territorial sea to the States, established co-operative mechanisms for regulating some marine industries and in particular fisheries.

The problems arising from multiple jurisdiction were intended to be solved by the fisheries elements of the OCS. As part of the settlement the Commonwealth and each of the States and the Northern Territory passed legislation which would form the basis of a co-operative scheme of fisheries management. That legislation authorizes joint arrangements for particular commercial fisheries which span a number of jurisdictions.

Under an arrangement, a particular fishery would be regulated under either Commonwealth or State law, and if necessary, by one of a number of joint State/Commonwealth authorities to be established by legislation. Joint management, where applied, displaces the traditional 3 nm boundary which had previously formed a barrier between State and Commonwealth fisheries laws. In other words the arrangements enable a single fishery which stretches across a number of areas of State and Commonwealth fisheries jurisdiction to be regulated by one set of laws, which could be that of the Commonwealth.

The choice of arrangement made in any particular case will depend on political as well as fisheries management factors. Nevertheless, such a scheme has the capacity to avoid the conflicts in management inherent in what is biologically a single fishery being managed pursuant to a number of different laws. OCS arrangements between the States and the Commonwealth are intended to simplify the lives of fishermen by reducing the number of licences and by removing impediments to compliance with management arrangements.

Although the Commonwealth has succeeded in negotiating simplified jurisdictional arrangements for the northern fisheries, negotiations have been long and protracted for the major southern fisheries and have not yet been finalized.

Concluding remarks

The false impression should not be left that the transition to a statutory authority and the greater involvement of user groups in fisheries management through MACs has been without significant difficulties, or that the process of change is complete.

Active participation in management decision making was generally new to industry, and industry was not always prepared for the new role. The institutional structures needed for effective, broad-based consultations were not always in place within industry. Existing power groups within industry understood how to influence management decisions through the political process, and the institutional change was not in the interests of some industry players. It is important to note here that instances of industry advocates or self-interest groups lobbying the Minister have almost ceased since the establishment of AFMA.

In addition, constructive dialogue amongst managers, scientists and industry on MACs Committees is not always easy to achieve, especially in the beginning. There is often profound distrust and suspicion amongst these parties. Also it has taken considerable time and effort on behalf of individuals from each group to make the new co-operative approach work.

AFMA has been established as an effective fisheries management agency, with a sound financial and operational relationship with industry, governments and the scientific community. The decision-making processes are generally working well, commencing from the MACs and moving through to the AFMA Board and, when required, to the Minister for broad policy direction, possible legislative change and accountability requirements.

Simplified jurisdictional arrangements have not yet been completely achieved, the programme of management plans for the major fisheries has not been fully implemented, and new management arrangements in some fisheries have not yet settled down, as continuing litigation needs to be addressed.

Conservation interests and groups are increasingly focussing on fisheries world-wide and Australia is getting its share of attention. At the same time, the competition between the commercial and recreational user groups for limited fish resources is becoming a major public issue.

Plenty of challenges for fisheries management remain for the future.

International Environmental Instruments and their Impact on the Fishing Industry

Martin Tsamenyi [A] *and Alistair McIlgorm* [B]

[A] University of Wollongong, Northfields Avenue, Wollongong, New South Wales, 2522, Australia.
[B] Australian Maritime College, PO Box 986, Launceston, Tasmania, 7250, Australia.

Summary

International environmental instruments have been a feature of international relations in the past two decades. They address the exploitation of natural resources and seek to protect the environment from degradation or destruction. In the marine fisheries and environment sector, a host of binding and non-binding instruments directly and indirectly address fisheries management, conservation and marine environmental management. These instruments have the potential to impact on fisheries management and fishing practices. Fisheries managers, fishers, and fisheries policy-makers cannot ignore these international environmental instruments. In the long term, the fishing industry has much to gain by complying with the requirements under international environmental instruments that regulate fisheries.

Introduction

Since the recognition of the exclusive economic zone (EEZ) concept under the 1982 United Nations Law of the Sea Convention (LOSC), there has been a significant rise in the fishing pressure on fish stocks under national jurisdiction, and on the high seas in particular. This has led to action at the domestic and international levels to promote a more rational conservation and sustainable utilization of the fisheries resources of the oceans. These 'green' concerns relate to target stocks issues such as species-selectivity (bycatch) and the impact of fishing gear on the marine environment in general. Consequently, a number of international instruments have been developed with the specific purpose of regulating how fishing is carried out. At the same time, other international instruments which have been developed to address wider conservation and environmental issues, have the potential to be applied to fisheries.

Pressure for change

International enviromental instruments that impact on fisheries have developed as a result of pressure from two major sources. In the first category are coastal States in response to the obligations they have assumed under LOSC and in response to distant water fishing nations (DWFNs) which have continued to ignore the new regime, particularly on the high seas. Second, since the Stockholm Conference on the Human Environment in 1972, the United Nations has taken an active involvement in the preservation of the oceans and marine environment and its resources. This has led to negotiation of several international environmental conventions and other instruments under the auspices of the United Nations itself or through one of its subsidiary agencies such as the United Nations Environment Programme (UNEP) and the Food and Agriculture Organization (FAO). The negotiation of Agenda 21, the Convention on Straddling Stocks and Highly Migratory Species (SS/HMS), the ban of drift-net fishing, and the development of a Code of Conduct for Responsible Fishing are examples of such initiatives. In all of these initiatives non-governmental organizations (NGOs) and the fishing industry have been involved to differing extents.

International environmental instruments

The international environmental instruments that directly and indirectly impact on marine fisheries may be divided into two groups in terms of their legal effect. In the first group are those instruments which qualify as treaties or conventions in international law and are therefore binding on the parties to them. In the second category are those instruments which may be described as declarations or resolutions. These are often referred to as 'soft law', that is to say that they are not legally binding but have political and moral force. They

Table 1. An overview of the different international environmental instruments that influence fisheries

A 1) Binding Instruments directly affecting fisheries.

The Law of the Sea Convention (LOSC), 1982 imposes obligations on Parties to conserve the fisheries resources of the sea and to adopt management measures to promote the optimum utilization of the fisheries.

The Convention on the Conservation of Antarctic Marine Living Resources (CCAMLR), 1982 is aimed at the conservation of Antarctic marine living resources. The rate of bycatch on non-targeted species has emerged as significant issue under the Convention.

The Convention for the Prohibition of Fishing with Long Driftnets in the South Pacific Region, 1989 prohibits the use of gill-nets or drift-nets which are more than 2.5 km long in the EEZs of the countries in the South Pacific, including Australia and New Zealand.

The Convention for the Conservation of Southern Bluefin Tuna, 1993 sets quotas for Australia, Japan and New Zealand in respect of southern bluefin tunas to ensure the conservation of such species.

The Agreement to Promote Compliance with International Conservation and Management Measures by fishing Vessels on the High Seas, 1993 empowers parties to it to impose stringent conservation requirements on national fishing vessels fishing on the high seas.

The Agreement for the Implementation of the Provisions of the United Nations Convention on the Law of the Sea, 10 December 1982 relating to the Conservation and Management of Straddling Fish Stocks and Highly Migratory Fish Stocks (SS/HMS), 1995 provides for the conservation and management of straddling fish stocks and highly migratory fish stocks on the high sea; and in limited circumstances, it also applies to fisheries management in the EEZ.

A 2) Binding instruments indirectly influencing fisheries.

The Convention on Wetlands of International Importance Especially as Waterfowl Habitat (RAMSAR Convention), 1971 aims to prevent the loss of habitats through encouraging the wise use of all wetlands. The Convention requires Parties to designate at least one national wetland for inclusion on a List of Wetlands of International Importance which are to be given special protection.

The Convention Concerning the Protection of the World Cultural and Natural Heritage (World Heritage Convention), 1972 seeks the conservation of natural and cultural areas of outstanding universal value through their inclusion on a World Heritage List and a List of World Heritage in Danger.

The Convention on International Trade in Endangered Species of Wild Fauna and Flora (CITES), 1973 aims to prevent overexploitation of endangered species of flora and fauna by means of import and export permits for species identified in the appendices to the Convention. Species listed in appendices are considered to be in danger of extinction and may be subject to trade restrictions.

The Convention for the Conservation of Migratory Species of Wild Animals (Bonn Convention), 1979 aims to conserve species of wild animals that migrate across or outside national boundaries by placing strict conservation obligations on Parties that are range states.

The Convention on Biological Diversity (Biodiversity Convention), 1992 is aimed at the conservation of biological diversity and to promote the sustainable use of its components.

B) Non-binding instruments.

UN Resolutions 44/225 and 46/215, 1989 and 1991. These resolutions called for a complete ban on drift-net fishing in the South Pacific and a world-wide moratorium on all high seas drift-net fishing by December 1992 in all the world's oceans, including enclosed seas and semi-enclosed seas. The Resolutions specifically encouraged all members of the international community to take measures individually and collectively to prevent large-scale pelagic drift-net fishing operations on the high seas.

Agenda 21 (Chapter 17) is the programme of action agreed to by States during the Rio United Nations Conference on Environment and Development (UNCED) 1992. Chapter 17 of Agenda 21 calls on the international community to address environmental issues that affect the marine environment in a comprehensive manner. The adoption of the Precautionary Principle is one of the important aspects of Agenda 21.

FAO Code of Conduct for Responsible Fishing, 1995 was developed by the FAO Committee on Fisheries. The aim of the Code is to provide guidelines for responsible approaches to fishing.

reflect international public opinion and may indicate future trends in the development of binding legal rules. A brief description of relevant international environmental instruments is given in Table 1.

ISSUES FROM ENVIRONMENTAL INSTRUMENTS

The issues emerging from international environmental binding and non-binding instruments are presented in Table 2. The issues can also be analysed from the perspective of the fishing industry. An opportunity threats analysis is given in Table 3. These issues are subjected to further analysis by Tsamenyi and McIlgorm (1995). Another review is given by Bergin and Haward (1995).

IMPLICATIONS FOR FISHERIES MANAGEMENT AND THE INTERNATIONAL FISHING INDUSTRY

From the overview, it is apparent that the objectives of fisheries management such as 'conservation' and 'optimum utilization' of resources are stated in many binding instruments whereas issues in the second wave of non-binding instruments are more problem-specific, e.g. protecting species, restoration, banning of specific gears, minimizing bycatch and specific actions in management plans. The apparent trend identified is from general objectives, in currently binding agreements, to more specific constraints and management methods in subsequent non-binding instruments.

Binding instruments	Conservation of fish & adopt management measures	Optimum utlization	Conservation of areas	Endangered species/potential prohibition of species capture	Adopt measures for rehabilitation of a species	Banning fishing methods or practices	Bycatch	List species	Trade	Precautionary principle
LOSC	X	X	X	X		X	X			
RAMSAR			X	X	X					
BIODIVERSITY	X		X	X	X		X			X
CITES				X				X	X	
DRIFT-NET CONVENTION	X					X	X		X	
MIGRATORY (BONN)	X		X	X	X		X	X		
WORLD HERITAGE			X		X	X				
SBT CONVENTION	X	X								
CCAMLR	X	X	X	X	X	X	X			
HIGH SEAS COMPLIANCE			X			X				X
SS/HMS CONVENTION	X	X		X	X		X			X
Number of features	7	4	7	7	6	5	6	2	2	3
Non-binding instruments										
UN RESOLUTIONS ON DRIFT-NETS	X		X	X		X	X			
AGENDA 21 (Ch 17)	X		X	X	X	X	X			X
FAO CODE OF CONDUCT	X		X	X	X	X	X			X
Number of features	3	0	3	3	2	3	3	0	0	2

Table 2. The main features of binding and non-binding international environmental issues

Table 3. An opportunity threats analysis by area and fishing method

Areas	Threats	Opportunities
Offshore/High Seas	High seas/ straddling stocks/HMS. Code of conduct for responsible fishing. CITES. Unilateral declarations by other countries. UNCLOS Art64.	
In EEZ/DFZ	CITES. UNCLOS III, UNCED (Agenda 21), SS/HMS, Precautionary principle.	
Inshore *Shelf*	UNCED (Agenda 21), UNCLOS RAMSAR (<6 m deep), Biodiversity Convention. Precautionary principle/bycatch issues.	Get environmental agreeement on land-based sources of pollution.
Estuarine	UNCED (Agenda 21), RAMSAR, World Heritage. RAMSAR (<6 m deep), Biodiversity Convention. Precautionary principle/bycatch issues.	Get environmental agreement on land-based sources of pollution.
Fishing Method	**Threats**	**Opportunities**
Trawling		
Demersal	UNCED (Agenda 21), Ch17 – "Minimize bycatch". Bycatch – in Prawn fisheries, particularly estuarine.	Gear modification to reduce bottom contact.
Pelagic	Bycatch.	Pelagic fishing must increase due to reduced bottom contact.
Purse seining	UNCED (Agenda 21), Ch17. Bycatch i.e. on associated species e.g. baitfish.	Adopt industry code of practice for multi-species fisheries management.
Longlining	Endangered species legislation (Albatross). Bonn Convention for migratory animals.	
UNCED (Agenda 21), Ch17.		

The review of international environmental instruments shows that the major issues for fisheries managers are:

- the interpretation of *'conservation and optimum utilization'*— despite these terms having been around for some time their practical implications are not clear;
- *conservation of areas* — ranked highly in both types of instruments. This has implications for vessel access;
- *species protection and restoration* — has implications for closure of areas, banning of fishing methods, and bycatch regulation;
- *greater detail in the management planning process* — for example contingencies in management plans;
- *a shifting in the burden of proof* — for example the precautionary principle.

The tightening of environmental constraints in fisheries management will be gradual, though the diversity of issues makes the time for implementation of policies uncertain. The political and moral power of the non-binding instruments is unpredictable. 'Soft law' instruments tend to become binding given time, though issues such as the precautionary principle and endangered species can be included in national legislation which may be more direct than international developments.

Revisiting fisheries management issues

Fisheries management has been based around maintaining fish stocks, their environment, and the economic well-being of the commercial fishing industry and other user groups. Few of the points identified above are new to fisheries managers, but they are now international obligations and may also come into national law. Many of the issues are central to the application of the precautionary principle and management plans will increasingly reflect this. The binding and non-binding instruments call for better implementation and control of the harvesting strategies by industry, with plans being phased-in during the development of new resources.

Enhanced management plans will require more information from fisheries science than is currently available and will re-open debates on the cost of management information and the funding of research. This raises critical questions about the quality of science, as in some major fisheries we see the 'good science-bad science' debate taking place, for example in the northern bluefin tuna fishery (Drumm 1994). Risk assessment, biological reference points, and the harmonization of science to international standards will be major challenges for scientists.

The declaration of closed areas, endangered species, and endangering fishing practices have been controversial issues in the past. The process used to close an area or declare a species endangered is in need of greater transparency and requires a consistent approach across different government agencies and between countries.

Fisheries management will also have to confront the bycatch issue as part of international obligations, if only due to the now well-established and possibly-increasing public profile on the issue. Trawling may be the first fishing method to be impacted by minimization of bycatch, but other fishing processes are not immune.

Technical fixes for the bycatch problem involve the development of Turtle Excluding Devices (TEDS) or Bycatch Reduction Devices (BRDs) (Crowley 1994). Legislation can be applied to address specific cases, e.g. dolphin kills under Australian legislation lead to large fines for taking protected species. A quantitative limit on bycatch has a similar disincentive and has been used in the United States of America tuna/dolphin management (Warren 1994). The full-cost pricing principle is examined by Campbell *et al.* (in press). These measures may involve 100% levels of observer coverage and are another payment issue for government and industry.

With international environmental instruments it is evident that the complexity of fisheries management will increase. Greater transparency and consultation with the fishing industry, all relevant government agencies and appropriate NGOs, should be part of the policy-making process. Potential sanctions and the impacts of national and international legislation should be considered by Government bodies, which should be aware of their potential impact through established consultation processes involving industry.

The impact on the fishing industry

The fishing industry will be influenced substantially by environmental instruments and will face allocation disputes, closure of areas and reduction or prohibition of catch. The impacts of the environmental instruments on industry can be analysed in several ways. In Table 3 the impacts are compared between the high seas, within the EEZ, inshore, and by fishing method.

From Table 3 it is evident that the impacts of instruments in the EEZs and the high seas are similar, though the SS/HMS convention has more impact on high-seas fishing practices. The issues inshore, in coastal and estuarine fisheries, reflect the visibility of fishing practices in these areas and their inter-relationships with other activities in the coastal zone.

The impacts of fishing methods are varied. Trawling has received much publicity for its apparent ecological destruction, whereas other methods, such as longlining, have been considered as selective fishing methods until recent bycatch problems with birds. There is little explicit international legislation on bycatch other than the requirement of Agenda 21 and the FAO Code of Conduct to 'minimize bycatch'. This term is open to interpretation.

The fishing industry response to international environmental instruments can be seen in the development of the Code of Conduct for Responsible Fishing developed by the FAO. However, at a national level the fishing industry must decide to adopt a code of conduct suited to national requirements. This would usually include a bycatch response strategy, with the development of appropriate exclusion devices as methods of minimizing bycatch.

The issues identified from the analysis of the impacts of international environmental instruments above can be seen as short-term threats to the industry. However, in the long-term the benefits of such policies should flow to the industry from greater sustainable harvests. This may not be the case where other fishing sectors, such as recreationalists, may benefit from changes in the commercial fleet catch. Allocation is an important issue in any of these potential changes.

There are few immediate commercial opportunities for the industry from international environmental instruments. It is recognized that there will be long-term benefits in keeping a clean environment, but higher short-term returns may be forthcoming from conforming to the eco-labelling preferences of discriminating consumers, probably in major foreign markets such as Japan and the USA. Other opportunities for industry may be in getting local authorities to adopt standards in limiting pollution of the coastal area. This will protect the fisher's most fundamental long-term asset — the marine environment.

Conclusions

International environmental instruments have significant implications for the operation of the fishing industry world-wide and for the administration of fisheries by government. The implementation of these international environmental instruments will result in the fishing industry being subject to an increasing number of policies that conserve marine areas by restricting vessel access. Similarly, protection and restoration of endangered fish species will lead to area, fishing method, and bycatch-reducing restrictions on fishers. These may not only come from international instruments but will usually be implemented through national legislation. In fisheries management, international obligations will lead to greater detail in management planning and acceptance of the precautionary principle. There will also be a shifting in the burden of proof required for new fishing ventures, with industry being required to prove that their fishing is not deleterious to the marine environment.

Benefits from international environmental instruments may come from industry implementing policies on the eco-labelling of fishery products. This may help capture any market advantage from conforming to environmental standards.

The fishing industry needs to abide by international environmental instruments such as the FAO code of conduct on responsible fishing and generate national codes of practice on issues such as bycatch. The fishing industry will have compliance problems in meeting any measures adopted and must support their representative organizations to enable communication with government agencies and NGOs. The fishing industry cannot ignore international environmental instruments and their impact on fishing operations and management.

References

Bergin, A, and Haward, M (1995). International Environmental Conventions and actions — implications for the fishing industry. Procedings of the National Agricultural Outlook Conference, Australian Bureau of Agricultural and Resource Economics, Canberra, 281–99.

Campbell, H F, McIlgorm, A, and Tsamenyi, B M (in press). Fishery management, environmental protection and trade liberalization. *International Journal of Social Economics*.

Crowley, M (1994). What's new in TEDs? *National Fisherman*, August, 34–36.

Drumm, R (1994). When scientists become advocates. *National Fisherman*, April, 46–49 and 96.

Tsamenyi, M, and McIlgorm, A (1995). International environmental instruments- their effect on the fishing industry. Final Report to the FRDC (Fisheries Research and Development Corporation). University of Wollongong and the Australian Maritime College. 50. pp.

Warren, B (1994). Dealing with by catch part three: the tuna dolphin solution: A model for other fisheries. *National Fisherman*, July, 14 and 25.

Legal Regimes for Fishery Resource Management

D. E. Fisher

Queensland University of Technology, GPO Box 2434, Brisbane, QLD 4034, Australia.

Summary

In the past, courts have often been reluctant to analyse in detail concepts of the sophistication and generality inherent in notions such as conservation, exploitation, overexploitation, optimum utilization, sustainability and economic efficiency. The nature, terms and language of the legislation have, however, required them to do precisely this. It is in some respects a novel and perhaps an uncomfortable process for the judiciary. The law is seeking precision and objectivity in precisely the same way that science is seeking these same results. Neither discipline has yet achieved this goal satisfactorily. It behoves scientists and lawyers to collaborate by providing their input and expertise not only when problems arise, but in anticipation of problems. This means in practice that a multi-disciplinary approach should be adopted to fishery resource management from initial investigation through assessment and evaluation to policy formulation and implementation leading to operational involvement until termination of the project.

The background

Management deficiencies

Agenda 21 (the preparatory papers for the United Nations Conference on Environment and Development, 1992) identified a number of major deficiencies in the way the fishery resources of the world are managed: for example:

- unregulated fishing in the high seas;
- unauthorized incursions by foreign fleets in waters under national jurisdiction;
- local overfishing in such waters;
- under-evaluation of catch in such waters;
- overcapitalization and excessive fleet size in any waters.

These issues throw up a host of problems. While some are predominantly scientific and others predominantly legal, their solution undoubtedly requires cooperation between the two professions. The problems of unregulated fishing, for example, involve identification of the relevant regulations and a determination whether there has been a contravention or not. The under-evaluation of catch requires a determination of the permissible level of catch, the method for its evaluation, the actual catch and the relationship between the actual catch and the amount of catch claimed. Each of these problems involves a careful investigation of what has happened, an identification of what should have happened and an assessment of the correlation between the two. This requires an inter-disciplinary approach.

Two of the disciplines are science and law. This raises the question of the distinctive contributions of each of these disciplines to the solution of these problems. Science is concerned with the acquisition of knowledge about substances so that human beings can understand their nature, qualities and characteristics. Law is concerned with what human beings should or should not do in certain sets of circumstances. In this sense the function of science is largely descriptive while the function of law is almost totally prescriptive.

Disciplinary problems

The application of prescribed standards involves an understanding of the law which prescribes the standards, the standards themselves and the circumstances to which they are applied. In a complex area such as fishery resources, the lawyer is unlikely to have a sufficiently detailed knowledge of the scientific and technical context of fishery management. Nor is the scientist likely to have a sufficiently detailed knowledge of the legal implications of the procedures for applying the standards and the consequences of applying the standards. The professional involvement of each of the disciplines, however, goes beyond the operational level to include the formulation of plans and strategies for fishery management. This involves the two professions in a much more creative exercise: making predictions for the future based upon information derived from the past, with a view to putting in place systems that will learn from the deficiencies of the past and ensure that future management is more likely to be successful in achieving its objectives.

Current objectives

Agenda 21 has clarified these objectives in relation to the resources of the high seas and those under national jurisdiction. In each case there is a commitment to the conservation and sustainable use of marine living resources. More specific objectives have been identified:

- to develop and increase the potential of marine living resources to meet human nutritional needs, as well as social, economic and development goals;
- to maintain or restore populations of marine species at levels that can produce the maximum sustainable yield as qualified by relevant environmental and economic factors, taking into consideration relationships among species.

These propositions are no more than expectations in legal terms. The obligations of states are found in the United Nations Convention on the Law of the Sea 1982 (the Convention) while the legal rights and obligations of those responsible for managing fisheries are prescribed by the relevant national system of law. Although individual managers and operators are not strictly bound by the international legal regime, it represents the framework within which the national system has effect.

Access to fisheries

The principle of permanent sovereignty

One of the critical questions for resource management both internationally and nationally is the nature and extent of access to the resource. Fundamentally it is a question of whether access is open or closed and, if it is closed, who is responsible for managing the regime and according to what criteria. The doctrine of permanent sovereignty of a state over its natural resources gives to the state total and absolute control of the resources within its territory which for this purpose includes the internal waters of the state and its territorial sea. Although this remains the position, the Rio Declaration on Environment and Development 1992 has qualified it thus:

States have, in accordance with the Charter of the United Nations and the principles of international law, the sovereign right to exploit their own resources pursuant to their own environmental and developmental policies, and the responsibility to ensure that activities within their jurisdiction or control do not cause damage to the environment of other states or of areas beyond the limits of national jurisdiction.

On the other hand, unlike the territorial jurisdiction, access to the resources of the high seas, originally the maritime area beyond the territorial sea, was open and unrestricted. Thus the Geneva Convention on the High Seas 1958 provided that freedom of the high seas comprised among others freedom of fishing. However, this freedom was required to be exercised by all states with reasonable regard to the interests of other states.

Modifications to that principle

This approach has been considerably modified over the last several decades. First, general concern for the quality of the territorial and marine environments has led to an increasingly complex and rigorous international legal regime for preventing, reducing and controlling pollution of the environment. Secondly, the rights of states, particularly coastal states, to secure access to and control of marine resources, have been extended to marine areas beyond the territorial sea. Thirdly, the exercise of this right is controlled in ways designed to protect the resource itself. The question, so far as fishery resources are concerned, is what are the rights and obligations of states, both coastal and non-coastal:

- in internal and territorial waters;
- in the waters of the exclusive economic zone, that is the area of the seas extending from the limits of the territorial sea for a distance not exceeding 200 nautical miles from the baselines from which the breadth of the territorial sea is measured;
- in the waters of the high seas.

The rights of a coastal state

The rights of a coastal state in the exclusive economic zone (EEZ), are set out in Article 56 of the Convention, namely sovereign rights for the purpose of exploring and exploiting, conserving and managing natural resources, whether living or non-living, of the waters superjacent to the seabed and of the seabed and subsoil. The exercise of these rights is constrained in ways that reflect the nature of the fishery resource and the range of interests involved in its management. The Convention recognizes not only the interests of other coastal states but also the interests of non-coastal states, developing countries, landlocked states and geographically disadvantaged states. Also relevant are the nature and characteristics of the fish which comprise the fishery resource itself — for example their migratory habits, their spawning habits, whether they are anadromous or catadromous. The complexity based upon these factors is compounded by the nature of the overriding objective stated by the Rio Declaration of 1992 — the conservation and sustainable use of marine living resources.

Management of the fishery resource

Articles 61 and 62 of the Convention set out the fundamental rules for the management and control of the fishery resource in an exclusive economic zone. These provisions comprise a complicated set of interrelated functions, criteria and objectives. For example:

- determination of the allowable catch;
- the prevention of overexploitation;
- production of the maximum sustainable yield;
- promotion of optimum utilization.

Underlying this management system is the obligation placed upon the coastal state to determine the allowable catch. Is the allowable catch determined strictly in accordance with these articles or is it open to the coastal state to take into account other matters? On the face of it, it should be the former approach. But in practice it is more likely to be the latter approach.

(a) Allowable catch

Paragraph 1 of article 61 simply requires determination of the allowable catch. Paragraph 2 imposes a duty to ensure that maintenance of living resources is not endangered by overexploitation. This is achieved by proper conservation and management measures and these measures are stated by paragraph 3 to include the maintenance or restoration of populations of harvested species at levels which can produce the maximum sustainable yield. This is qualified by relevant environmental and economic factors including the interests of other states and the characteristics of the resource. Paragraph 2 is thus directed towards conservation of the resource by the prevention of its overexploitation while paragraph 3 contemplates the capacity of species to be harvested at maximum sustainable yield levels. The former looks at the fishery as a resource while the latter looks at sustainable yield in terms of species, but with reference to factors external to the species and inclusive of other human and ecological interests.

(b) Harvestable capacity

Then there is article 62. The fundamental obligation in paragraph 2 is for the coastal state to determine its capacity to harvest the fishery resource. Where the allowable catch exceeds the harvestable capacity of the coastal state, access to the surplus must be given to other states in accordance with the criteria set out in paragraph 3. These relate almost entirely to the interests of the coastal state and of other states in the fishery resource in question and are directed towards achieving the objective of optimum utilization of the resource. It should be emphasized that this objective is the best use of the resource and not necessarily the maximum use of the resource. This clearly permits under-utilization of the resource — a conclusion entirely consistent with the statement that this objective is without prejudice to article 61 which is directed towards conservation so as not to overexploit the resource. The coastal state thus determines the allowable catch and its capacity to harvest the fishery resource but it does so to achieve optimum utilization of the resource and the maximum sustainable yield in relation to species but within the overall framework of non-overexploitation.

(c) The state's own interests

The duty in paragraph 2 of article 61 to ensure the prevention of overexploitation is limited by the need to take into account the best scientific evidence. There is no reference to scientific evidence in relation to any of the other decision-making processes. Indeed paragraph 2 of article 62 contemplates the relevance of the interests of all states involved rather than the nature and characteristics of the fishery resource itself. Notwithstanding the need to take into account the best scientific evidence related to the allowable catch, the determination of the allowable catch and the determination of the capacity of the coastal state to harvest its fishery resource are so interlinked that it would be difficult to conclude that non-scientific matters were not relevant to these decision-making processes. Finally, the way in which articles 61 and 62 are framed implies that it is legitimate for the coastal state to make a determination not only in relation to its capacity to harvest the fishery resource, but also in relation to the allowable catch with particular reference to its own interests. There is therefore no cogent reason to suggest that the coastal state in exercising its sovereign rights in its exclusive economic zone cannot make these decisions in accordance with its perception of its own interests. Even if access to the resource is granted to other states or the nationals of other states, it is the legal regime relating to the fishery resource established by the coastal state that applies to the nationals of other states with access rights.

Fisheries with special characteristics

Access to fishery resources of particular kinds is controlled by articles 63 to 67 of the Convention in accordance with their special characteristics, namely:

- fish stocks occurring within two or more exclusive economic zones or other areas (straddling fish stocks);
- highly migratory species;

- marine mammals;
- anadromous stocks;
- catadromous species.

The underlying principles are for the interested states to co-ordinate their management practices, to cooperate in achieving conservation and to take steps to agree upon the appropriate measures necessary. However, in the case of anadromous stocks, it is the state in whose rivers the anadromous stocks originate that has the primary interest in and responsibility for these stocks. In the case of catadromous species, it is the coastal state in whose waters the species spend the greater part of their life cycle that has the responsibility for their management and ensuring the ingress and egress of migrating fish. Fisheries for anadromous stocks may be conducted only in waters landward of the outer limits of exclusive economic zones except in certain circumstances. Harvesting of catadromous species may be conducted only in the same waters but there are no exceptions.

States with special characteristics

Articles 69 and 70 regulate the interests of land-locked states and geographically disadvantaged states. These states have a right to participate on an equitable basis in the exploitation of an appropriate part of the surplus of the fishery resources of the exclusive economic zones of coastal states of the same subregion or region. Consideration is required of the relevant economic and geographical circumstances of all the states concerned. The right to participate, although it is stated to be available on an equitable basis, must be exercised in conformity with articles 61 and 62.

High seas fisheries

Management of the fishery resources of the high seas is subject to a much less detailed regulatory regime. Article 116 confirms the right for the nationals of all states to engage in fishing on the high seas. All states are under a duty to take, or to cooperate with other states in taking, measures necessary for the conservation of the living resources of the high seas by, for example, subregional or regional arrangements. The most restrictive provisions are in article 119. While conservation of the fishery resource appears to be the fundamental objective, it is achieved in terms of determination of the allowable catch. Thus there is an obligation on all states to take measures designed, on the best scientific evidence, to maintain or restore populations of harvested species at levels which can produce the maximum sustainable yield. This is qualified in general terms by reference to relevant environmental and economic factors. These include:

- the special requirements of developing states;
- taking into account fishing patterns;
- taking into account the interdependence of stocks;
- taking into account any generally recommended international minimum standards whether subregional, regional or global.

The means for achieving these conservation measures include not only the creation of these obligations but also the obligation of states to impose relevant obligations upon their nationals.

Conclusion

The regime for management of the fishery resource of the high seas provides for conservation and measures to achieve conservation. This does relatively little to limit freedom of fishing in the high seas. Although the constraints upon the coastal state in the management of the fishery resource in its exclusive economic zone are much more detailed, the practicality is that it is very much a matter for the coastal state to determine precisely how the fishery resource will be managed and whose interests will be recognized and protected. This is not to say that the jurisdiction of a coastal state has been extended to the outer limit of the exclusive economic zone. What the Convention does, however, is leave much of the detail of the management regime to the discretion of the coastal state itself.

The regime in Australia

The constitutional arrangements

Against this background of international law, let us consider how a coastal state such as Australia has attempted to manage its fishery resources. Management of the resource is divided between the Commonwealth and the States. Since the Commonwealth can legislate with respect to activities and events taking place outside Australia and to give effect within Australia to international obligations, the Commonwealth can make laws for the fishery resource of the territorial sea, the exclusive economic zone (known for these purposes as the Australian fishing zone) and the high seas in accordance with the Convention. However, according to an arrangement between them and the Commonwealth, the States and the Northern Territory enjoy legislative capacity in relation to the coastal waters of a State or the Territory. These are in effect Australian waters extending to the three nautical-mile limit. An arrangement may also be made for the Commonwealth and each State or the Territory to undertake the joint management of a particular fishery located in the coastal waters of the State or the Territory or the Australian fishing zone. Such an arrangement is managed in accordance with the law of the Commonwealth or of the State or the Territory. Although the Commonwealth, each State and the Territory have enacted fisheries legislation, let us consider as an example the Commonwealth legislation.

The objectives of the Commonwealth legislation

Section 3 of the *Fisheries Management Act* 1991 (Cth) specifies the objectives to be achieved in the implementation of the legislation. The most important is specified in subsection (1) (b), namely:

> ensuring that the exploitation of fisheries resources and the carrying on of any related activities are conducted in a manner consistent with the principles of ecologically sustainable development, in particular the need to have regard to the impact of fishing activities on non-target species and the marine environment.

This is complemented by the duty specified in subsection (2)(a) and (b) to have regard to additional objectives:

(a) ensuring, through proper conservation and management measures, that the living resources of the Australian fishing zone are not endangered by overexploitation, and

(b) achieving the optimum utilization of the living resources of the Australian fishing zone.

While the objective in subsection (1) does not reflect the terms of the Convention, the objectives referred to in subsection (2) incorporate aspects of articles 61 and 62: in particular, paragraph 2 of article 61 is repeated in subsection (2)(a) and paragraph 1 of article 62 in subsection 2(b). However, section 3 says nothing about allowable catch or harvestable capacity which constitute the substantive duty placed by articles 61 and 62 upon a coastal state. The manner in which these duties are discharged is left to the discretion of the coastal state. This is consistent with the foregoing analysis of the terms of the Convention. In the case of Australia these matters are determined by the plan of management prepared for a specific fishery.

Plans of management

The Australian Fisheries Management Authority determines a plan of management for a fishery under section 17 of the *Fisheries Management Act* 1991 (Cth). A plan sets out objectives, measures for attaining them, and performance criteria. More particularly a plan provides among others for:

- the manner in which the fishing capacity of the fishery is to be measured;
- the fishing capacity, measured in that manner, permitted for the fishery;
- a description of the fishery by reference to area, fish species, fishing methods or any other matter;
- the kind and quantity of equipment that may be used;
- the circumstances in which a statutory fishing right may authorize fishing by or from a foreign boat.

These cover what the Convention contemplates as the allowable catch, harvestable capacity and access by nationals of other states.

A plan of management performs a critical function in the legal regime for managing a fishery resource. It applies to government in exercise of its regulatory functions and to the fishing industry in exercise of its operational functions. A plan has no effect to the extent that it is inconsistent with a provision of the *Fisheries Management Act* 1991 (Cth). While a plan is in force for a fishery, the Australian Fisheries Management Authority as the relevant agent of government must perform its functions in relation to the fishery in accordance with the plan of management. The determination of a plan of management for a fishery is a legislative act and a plan of management has the force of law. It is thus binding upon fishery industry operators. This is consistent with the Convention for the plan is the mechanism by which Australia implements its obligations to determine the allowable catch and to provide access for the nationals of other states.

Conclusion

The foregoing analysis reveals that Australian fisheries are managed in accordance with a complex legal regime that comprises an international component, a Commonwealth component and a State component. Each is related to the others and it is expected that the three components of the system will be consistent with each other. This is an important practical issue. The Convention and the Australian legislation operate within political and cultural systems that are in many respects anthropocentric in nature, and within biological and ecological systems that are simultaneously exploited for anthropocentric reasons and conserved for ecocentric reasons. Decisions on fishery resource management may therefore be based upon a whole range of considerations. It is the distinctive function of the law to indicate as precisely as possible what is the basis for decision making so that those seeking to apply the legislation, including not only governments but also industry operators, can predict the likely outcome of the process. Herein lies the fundamental conundrum with which fishery resource management is confronted. On the one hand there is scientific difficulty in predicting accurately according to objective evidence the likely consequences of certain courses of action. On the other hand the language in which the legal rules are expressed frequently produces similar uncertainty and ambiguity. Although both science and law claim precision, certainty, accuracy and predictability, this is frequently no more than a myth. An examination of plans of management may explain the reason for this.

Plans of management in practice

The context

A plan of management under the Commonwealth legislation is the pivotal link between the objectives of the legislation, reflecting as they do some of the concepts underlying the Convention, and the detailed regulatory and enforcement systems created by the legislation. Notwithstanding that the plan is critical, it demonstrates the uncertainties of the system and the tensions within it. Although the *Fisheries Act* 1952 (Cth) has been superseded by the *Fisheries Management Act* 1991 (Cth), the essential nature of a plan of management remains the same. Particularly important for present purposes are the two objectives to be taken into account, namely the prevention of overexploitation and the achievement of optimum utilization.

The need for flexibility

A plan of management, of course, has the force of law. How therefore can it be designed to reflect the circumstances of a particular fishery and thus provide a degree of flexibility in managing it? The objectives of a plan, for example, may be achieved by mechanisms that complement the plan provided they do not contravene the plan itself. But this may impose additional obligations upon the industry operators. For example, in the case of *Latitude Fisheries Pty Ltd* (1992) it was noted that:

> section 7B of the Act [*Fisheries Act* 1952 (Cth)] leaves open the possibility that fishing capacity may be measured and determined in a variety of ways. It permits the use of provisions relating to the division of fishing capacity into units and their allocation and assignment to persons and boats.

This case was concerned with the Northern Prawn Fisheries Management Plan which:

> has as its objectives the conservation and reduction of pressure upon the stock of prawns in the area of the fishery and the promotion of its economic efficiency. The second objective [economic efficiency] is not independent of the first for overfishing of prawn stock will have obvious consequences for the economic viability of the industry in the area covered by the plan. Consistently with those objectives, the adoption of the system of limiting and dividing notional permitted capacity in the fishery does not prevent the limitation of catches within that capacity by a means which may be introduced *from time to time* by the use of fisheries notices.

The achievement of conservation and the promotion of economic efficiency require flexibility to enable a response to changing circumstances. In this case the court decided that the effect of a plan of management could be modified and at the same time its objectives achieved by the issue of a fisheries notice restricting the number and size of nets that could be used in the area covered by the plan, provided the notices were not inconsistent with the plan.

The need for predictability

There is also a need for certainty and predictability. This requires a plan of management to adopt a prescriptive approach. Section 7B(8) of the *Fisheries Act* 1952 (Cth) provided that, while a plan of management was in force for a fishery, the Minister and the Secretary were required to perform their functions and exercise their powers in relation to the fishery in accordance with the plan of management and not otherwise. According to *Latitude Fisheries Pty Ltd* (1992) this constituted a test of consistency and not of absolute compliance. Thus:

> if the Minister or the Secretary were to purport to perform a function or exercise a power under the Act in relation to the fishery in a way that was in direct conflict with a provision of the plan or in a way which differed from that exhaustively prescribed by the plan for doing that thing, then the purported performance of the function or exercise of power would be *ultra vires*. The test is one of consistency and there is ample authority in a variety of contexts for the way in which it is to be applied.

The court went on to suggest that the terms in which the plan is drafted may reflect its legal effect. Thus:

> the more prescriptive the plan in areas relating to the exercise of statutory powers and functions, the more confining the content of the duty posed by section 7B(8) will be. A plan drawn in more general terms will allow greater freedom of action.

Although the legislation requires government to exercise its powers consistently with a plan of management, the plan affects, as it is intended to do, the private sector fishing industry. A plan has been stated by the courts to have the force of law: it is a legislative instrument. It is therefore binding upon all those affected by it. This includes the operational managers of the fishery. While they are strictly bound by the plan, the government agency responsible for it is obliged merely to act consistently with it. The legal effect of a plan depends therefore upon the function it performs and the persons to whom it applies.

The function of the courts

What is the scope of a court's jurisdiction? Will it determine the merits of a plan of management or simply its legal validity? It is the orthodox view that the function of a court is to determine the validity of a decision. However, in certain cases the criteria for determining the validity of a decision come close to the merits of the decision. This is particularly so in light of the tendency in recent years for legislation to specify what are the objectives of the legislation or what matters must be taken into account in reaching decisions under the legislation. Fisheries legislation is a particularly good example of this. Section 3(1) of the *Fisheries Management Act* 1991 (Cth), of course, sets out its objectives. These include ensuring that the exploitation of fishery resources is conducted in a manner consistent with the principles of ecologically sustainable development, in particular the need to have regard to the impact of fishing activities on non-target species in the marine environment. Subsection (2) requires consideration to be given to two objectives: namely the prevention of overexploitation and the achievement of optimum utilization. Conservation, overexploitation and optimum utilization are concepts of major significance from the scientific point of view. Are these issues approached purely on the basis of the scientific information that is available or are other matters relevant?

(a) Calculating allocations

In the case of *La Macchia* (1992) the court referred to the significance of the language of the legislation. Reference was made in particular to the two objectives in question:

> It is contemplated that these objectives will be served by management plans under section 7B [of the *Fisheries Act* 1952 (Cth)], which contains provisions ... for the allocation of units of fishing capacity. Neither these provisions nor the objectives of the Act suggest that the allocation is to be arbitrary or determined by the capricious consequences of a kind of statistical lottery. Where a statute provides for an allocation of a scarce resource among participants in the relevant industry, in general, and failing some clear indication to the contrary, the statute should be understood as authorizing a method of allocation in accordance with some intelligible principle appropriate to achieve a reasonable division between those participants.

One of the purposes of the legislation might be to indicate what is the intelligible principle according to which the allocation should take place. According to the court, neither the protection of fisheries from overexploitation nor the achievement of the optimum utilization of fisheries was a 'principle' in this sense.

In this case the court was faced with a provision in the management plan which set out, in a series of formulae, a method for the calculation of the number of units of a specified species of fish to be allocated in respect of each boat. One of the components in question enabled account to be taken of an appropriate average share (in respect of each particular boat) of the total catch for a period covering several years, the number of years varying according to the species of fish. This was described as 'inherently fallacious' by one of the expert witnesses. As the court said, 'that is because it involves the simple addition of percentages of annual catches, which percentages are then averaged for the period, without taking account of annual variations of total fleet catches. A procedure of this kind can produce startling results because of the disproportionate effect of small numbers in a year of low total fleet catch, when they may represent significant percentages.' This approach was rejected by the court. As the judge said, 'on the expert evidence before me, the principle upon which the method here in question was selected is not intelligible ... and the result is a demonstrably arbitrary and unequal sharing.' This part of the management plan was therefore void.

(b) Determining the basis for allocations

The issue in the case of *Bienke* (1994) was not so much the validity of the principle according to which fishing capacity should be allocated but the range of considerations relevant to determining the basis for allocation in the first place. This concerned the Northern Prawn Fishery Management Plan 1989. Paragraph 7.1 of the plan said this:

> For the purposes of subsection 7B(2) of the Act [*Fisheries Act* 1952 (Cth)] the objects of this plan are :
>
> (a) to conserve the stocks of prawns in the area of the fishery;
>
> (b) to reduce the fishing pressure on the stocks of prawns in the area of the fishery; and
>
> (c) to promote the economic efficiency of the fishery.

It was argued that the reference to the promotion of the economic efficiency of the fishery went beyond the scope of the objectives of the legislation. Section 5B of the Act, it will be recalled, referred to the protection of the fishery resources from overexploitation and the achievement of their optimum utilization. It was argued that economic efficiency was first irrelevant as a matter of law and secondly undesirable as a matter of principle in the light of the effects that it would produce.

The second argument was based upon a set of bio-economic models which in this case dealt with fleet size and the relationship between catch and the effort involved. It was suggested that seeking economic efficiency in this way would favour those operating large boats and lead to the reduction in the number of small boats. Thus 'the formulation of a plan of management under the influence of such considerations would be to allow the legislative scheme to operate in a way that favoured oligopoly at the expense of the free operation of competitive forces.' In view of these arguments, the court was faced with two issues. The first was the legal validity of the model proposed in this case. The second was the relevance of the objective of economic efficiency.

The court expressed a degree of suspicion about the use of bio-economic models:

> Bio-economic models are simplified and generalised statements which reduce the variety and complexity of actuality to a level that can be understood and clearly specified, thereby providing what is said to be a framework for a policy analysis. Because these models are a simplification of reality, the results must be interpreted with caution. However, the evidence indicates that they are used by government planning authorities in various countries.

The model was nevertheless acceptable — but what about the relevance of economic efficiency? After a careful examination of the legislation, it was concluded:

> Having regard to the subject matter, scope and purpose of the Act, to my mind the statement in section 5B is not exhaustive. Further, the promotion of economic efficiency of the fishery is an objective to which the Minister might properly have regard.

(c) The meaning of optimum utilization

In reaching this conclusion the court was influenced by a number of matters, including the nature of the obligation in section 5B and the notion of optimum utilization. Section 5B, it will be recalled, requires the Minister to have regard to the two stated objectives. This is not prescriptive in the sense that it requires the achievement of these objectives. Its relatively non-prescriptive nature led the court to conclude that the list of objectives did not exclude others that could be argued to be relevant. The meaning of optimum utilization convinced the court on this point.

Reference was made to the use of the same expression in the Convention. The court went on:

> This indicates that the phrase 'optimum utilisation' is not limited to avoidance, by proper conservation and management measures, of the danger of over-exploitation of the living resources of the Australian fishing zone. That the notion of 'optimum utilisation' is broader than what one might call conservation concerns as suggested, in any event, by the form of the two paragraphs in section 5B. Paragraph (b) would, one would expect, cover broader or other grounds than those in paragraph (a). The force of that consideration is not diminished by the final clause as to the particular concerns with conservation and protection of whales.
>
> In my view, 'optimum utilisation' is a phrase broad enough to include economic exploitation for the benefit of the littoral state, namely Australia, by, for example, the raising of revenue from a resources tax. Certainly, in my view, there may be 'optimum utilization' by having regard to the economic interests and prosperity of those, taken as a whole, who exploit the fishery when adopting a regime for the reduction in the number of units.

Economic issues were therefore relevant to a determination of optimum utilization. Although the courts examine the evidence, including the scientific evidence, relevant to the issue, they do not determine the merits of a plan of management or, for that matter, of any other decision. On the other hand the validity of the decision may well depend upon the relationship between the evidence and the rationality of the decision in the light of the evidence. This comes close to, but is not quite the same as, reviewing the decision on its merits. Courts therefore do not manage fishery resources: their role is to ensure that fishery resource managers do so lawfully.

List of Cases

Department of Primary Industries and Energy v Collins (1992) 106 ALR* 351

Azevedo v Department of Primary Industries and Energy (1992) 106 ALR 683

Austral Fisheries Pty Ltd v Minister for Primary Industries and Energy (1992) 37 FCR** 463

Latitude Fisheries Pty Ltd v Minister for Primary Industries and Energy (1992) 110 ALR 209

La Macchia v Minister for Primary Industries and Energy (1992) 110 ALR 201

Minister for Primary Industries and Energy v Austral Fisheries Pty Ltd (1993) 40 FCR 381

Latitude Fisheries Pty Ltd v Minister for Primary Industries and Energy (1993) 41 FCR 536

Minister for Primary Industry and Energy v Davey (1993) 47 FCR 151

Bienke v Minister for Primary Industries and Energy (1994) 125 ALR 151

*Australian Law Reports, **Federal Court Report

Further reading

Geneva Convention on the Law of the Sea 1958

United Nations Convention on the Law of the Sea 1982

Rio Declaration on Environment and Development 1992

Agenda 21 (the preparatory papers for the United Nations Conference on Environment and Development 1992)

Fisheries Act 1952 (Cth)

Fisheries Management Act 1991 (Cth)

Birnie, P W, and Boyle, A E (1992). 'International Law and the Environment.' (Claredon Press: Oxford; Oxford University Press: New York).

Birnie, P W, and Boyle, A E (1995). 'Basic Documents on International Law and the Environment.' (Claredon Press: Oxford).

Burke, W T (1994). 'The New International Law of Fisheries: UNCLOS 1982 and Beyond.' (Claredon Press: Oxford; Oxford University Press: New York).

Fisher, D E (1987). 'Natural Resources Law in Australia.' (Law Book Company: Sydney).

Fisher, D E (1993). 'Environmental Law: Text and Materials'. (Law Book Company, Sydney).

Fisheries Enforcement: Our Last Fisheries Management Frontier

B. Hemming[A] *and B. E. Pierce*[B]

[A] Fisheries Compliance, Primary Industries (South Australia) — Fisheries, GPO Box 1625, Adelaide, South Australia, 5001, Australia.
[B] Inland Waters Research and Development Program, South Australian Research and Development Institute — Aquatic Sciences, PO Box 120, Henley Beach, South Australia, 5022, Australia.

Summary

Fisheries enforcement is universally important because illegal activity is common to all world fisheries, compliance officers are the most visible component of fisheries management, a world resource with an annual commercial value alone worth $US16 × 10^9 needs protection, and enforcement personnel usually represent over one-third of fisheries staff. However, enforcement has often been an afterthought of management. Despite its importance, enforcement/compliance is the focus of less than 0.4% of the recently published literature, rarely has quantitative objectives or quantified outcomes, is often poorly integrated into the management feed-back loop, and may be unable to enforce 50+% of regulations. As world capture-fishery production peaks, major value-added benefits can be obtained for legitimate stakeholders *via* strategic research to optimize enforcement outcomes, by maximizing participation of enforcement staff in crafting policy and legislation, by exploring with sociologists the best ways to meet community desires, and by better sharing of positive results.

Introduction

Ancestral fishers, essentially hunters of the underwater environment, pursued common-property resources upon which their combined demands and capture abilities had an impact secondary to that of habitat and biological factors. Once the combination of human population-increased demand and technologically-increased capture ability resulted in a culturally (perceived) loss of benefits from the resource, efforts to regain benefits were made and fisheries management was born. Cultural mores were the first 'management regulations' and all community members acted as 'enforcement officers' (Ruddle 1994). Recently, with dramatic population growth and technological advancement, mores have largely given way to legislatively-based regulatory structures and paid professionals are employed to enforce measures aimed at providing fishery-related benefits to stakeholders.

The problem is that illegal activity/non-compliance is a fact of life in virtually all modern fisheries world-wide. Activities such as poaching and polluting divert fishery benefits from legitimate 'winners' and thwart management objectives and societal stewardship principles. While some of the motivation for such activity may be explained by economic incentives inherent in unmet demand for specific products, coupled with available and efficient capture methods, some primal 'hunter' mentality also resurfaces to contribute as is obvious to anyone who has ever witnessed a charter boat of average people stuck into a school of feeding fish. Today, a tiny fraction of society is dedicated to enforcement on behalf of the (previously enlisted) entire population, at the same time that incentives to venture outside the legal management constraints are continually increasing. With such expectations of this 'quality control' sector of fisheries management, we would expect it to be well-quantified and contestable, with major ongoing research activity to optimize benefits, deeply and smoothly embedded within the overall fisheries management process, and effectively intertwined with the social fabric of the community. The primary thesis of this paper is that overall, world-wide, nothing could be further from reality — and as such, fisheries stakeholders world-wide are missing out on massive financial, sustainability and stewardship benefits[1].

Our objectives are therefore twofold:

- to demonstrate from available information that fisheries enforcement is:

    critically important to sustaining and increasing *all* stakeholder benefits;

    poorly researched, poorly integrated into the fisheries management process, and sub-optimally implemented sociologically; and

- to point toward areas for value-adding major fisheries benefits in enforcement.

ENFORCEMENT: THE AFTERTHOUGHT OF FISHERIES MANAGEMENT

Enforcement, here taken to include incentive-based compliance activities as well as disincentive-based enforcement actions, typically involves approximately 50% of staff involved in the management process in government fisheries agencies (Sutinen 1994). Although this proportion may decrease to one-third when the contribution of external consultancies is taken into consideration, enforcement still represents one of the top two uses of human and operational resources within the fisheries management process. In South Australia, a typical example, each enforcement officer must service in excess of 10 000 active recreational and commercial fishers operating in diverse fisheries with differing management regulations. An alternative perspective sees each officer responsible for continuously patrolling over 27 500 km^2 of land and water to protect a share of commercial production worth \$A5.4 $\times$ 10^6 (1994–5). World-wide, the enforcement function is responsible for protecting a combined capture fishery value of \$US16 $\times$ 10^9 (1993, FAO 1995), a major component of the steadily-rising culture-fishery value, vast capital investment on behalf of the commercial and recreational sectors, together with the investment ploughed back into the sector through research and, finally, a massive environmental capital value representing society's investment in the basis of production — the habitat.

Such a massive investment and responsibility must surely seem an overwhelming incentive for enforcement to 'get it right' both now and in the future. On the surface, the above evidence seems to imply that enforcement world-wide is a lean, efficient function ably contributing to the fisheries management process and cumulative fishery value. This may be the case, but we have not been able to find conclusive evidence for or against this contention for *any* fishery worldwide. If wrong, the losses to all legitimate players may be both massive and long term.

THE RESEARCH GAP

Review of evidence from the 1988–1995 scientific literature (Cambridge Scientific Abstracts 1996) across broad fisheries investigative areas paints a picture (Fig.1) of a literature relating to enforcement/compliance that is pitiful — constituting less than 0.4% of the total fisheries-related publications for the period. But even these 360 publications are primarily composed of incidental references, the fewer substantive articles commonly being reactive reports on past expenditure and general activity. The literature focusses on high-value pelagic fisheries and particularly those with a historic 'foreign' exploitation component.

While a few seminal publications bring status reports or comparative information to the light of critical review (e.g. Sutinen and Gauvin 1989; Wang and Zhan 1992; OECD 1994), we are forced to conclude either that research/assessment is not being undertaken in the enforcement arena, or that it is virtually never being shared *via* the published literature. This is particularly unfortunate given the extensive literature regarding fisheries management options for which the enforcement outcomes are unavailable for comparison.

THE PROCESS GAP

Assuming that the number of published references by fisheries investigative area (Fig.1) is broadly indicative of the world-wide

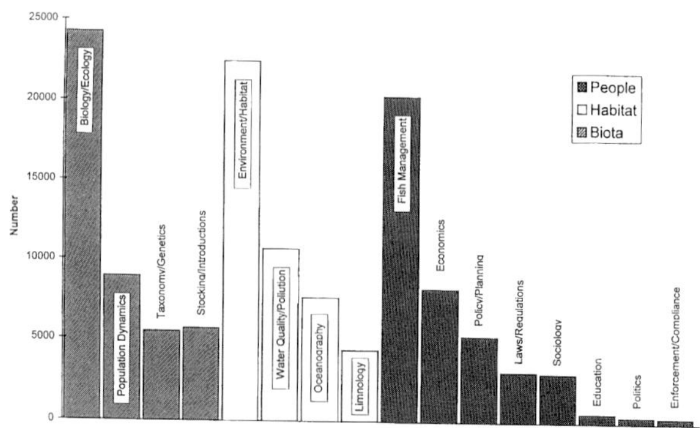

Fig. 1. Fisheries-relevant publications cited in the Aquatic Sciences and Fisheries Abstracts, 1988–1995 inclusive, (Cambridge Scientific Services 1996) by major facet. Category membership is not mutually exclusive.

[1] *Caveat:* The authors do not pretend to fully review world fisheries enforcement; we argue the existing literature is hopelessly insufficient to allow this at present. We acknowledge our western cultural bias. However, we proceed from practical experience and contact with diverse fisheries world-wide, including participation in the commercial and recreational sectors, extensive enforcement expertise, as well as a biological, economic and social research background.

perspective, fisheries management continues to be preoccupied with the input/output components of the system (e.g. biology, stock abundance, management regime) at the expense of the crucial day-to-day management of people. We propose a model of the fisheries management process (Fig. 2) that specifically highlights the concept of biotic and abiotic inputs continually passing through a critical control variable (the people-management process) with the enforcement service the key implementation and quality control factor.

We argue that failure to integrate enforcement adequately into management can be readily seen in most modern fisheries in three ways:

- Pick up virtually any fishery-specific management plan and look for the goals set for enforcement, their objectives, and performance indicators. They are not there, or if they are, they tend to be general to the point of being 'motherhood' statements and lacking teeth. Quantitative objectives, milestones and efficiency-reporting standards all mask accountability, and the ability to 'sell' successes and the full benefits of the work and service. Even within recent fisheries management texts (e.g. Kohler and Hubert 1993), the enforcement function, and priority-setting for it, receives minimal mention.
- Walk into a meeting set to revise the management of a fishery and listen for the input from the field-seasoned enforcement professional — listen hard because it usually isn't there. Far more likely is that the consultation process will produce a revised plan complete with new ideas, input from fishers, allocation compromises, a seasoning of political reality and a dash of biology/environmental concerns. Once the deal is completed and the committee retires to celebrate at the hotel, the package goes to Enforcement/Compliance 'for review' which translates 'make it happen.' In an important historic compilation regarding the United Kingdom experience with inland fishery law (Howarth 1987), enforcement input or feedback is seldom recorded as directing improvements/modification. Finally, the critically important issue of moving from reactive to strategic enforcement is seldom addressed in planning or elsewhere to get enforcement off the back foot (fighting fires) and in a position to head off big problems before they spread beyond control.
- Watch any fisheries compliance officer for a period and you will notice that the same regulations are being enforced, the same incentives being applied. Some of this may be human nature on the part of the officer or errant fishers, but in many fisheries it is equally likely that the other regulations are difficult to enforce, won't hold up in court, or require unavailable resources to implement. There is also a tendency, in order to achieve 'recognizable and tangible' results, for enforcement effort to be targeted at the lower end of the infringement scale (the ignorant and small opportunists) which has no real impact on the big problem of organized fisheries criminals. Better integration of an enforcement feedback function into fisheries management could help improve some of these less workable regulations, target them at the 'meat' of the problem, or else drop them entirely.

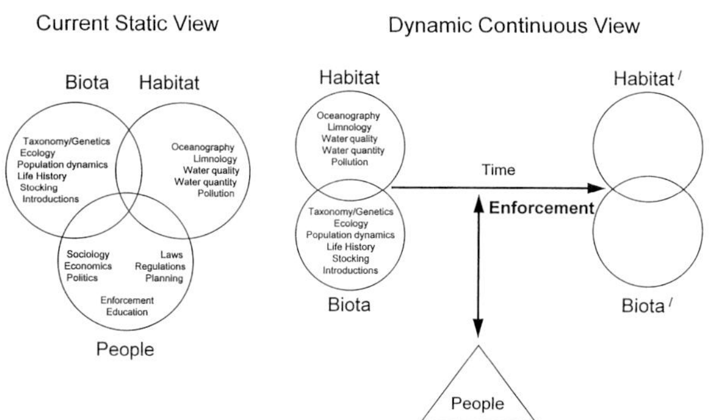

Fig. 2. Two views of fisheries management. The current venn concept on the left (modified from Nielsen 1993) focusses on the inter-relatedness of key factors. The modified model on the right emphasizes the temporal treatment effect and the pervasive human control variable as mediated, to some degree, by the Enforcement function.

This Congress is itself a good practical example of lack of integration of enforcement within the management process. Although each core Congress theme is highly relevant to enforcement functions and staff who will surely have to deal with the implemented outcomes, this major group of professionals is not advertised as being 'beneficiaries' nor explicitly sought as attendees.

THE PEOPLE GAP

The disturbing lack of sociological literature specific to fisheries (Fig. 1) becomes almost an absence when we look for information on the social implementation of fisheries compliance/enforcement functions. If the community doesn't know of, support and accept the actions of the enforcement sector, then in practical terms it is hard to expect the officers or their desired outcomes to survive in that community. Understanding not only what a community wants for its fisheries, but what it understands about them and how it desires the outcomes be achieved, is essential to helping the community to co-manage itself.

HARVESTING GREATER ENFORCEMENT BENEFITS

It is clear that as the values of fisheries production, habitat protection and community education steadily rise in the future, the demand for enforcement services will also rise. Given peaking wild-capture fishery production (FAO 1995), greater emphasis is being placed on value-adding to increase returns to stakeholders. Although some technology transfer elements may be adequately exploited now, we argue that enforcement services generally offer major opportunities for value-adding benefits in areas that have to date gone largely unexploited.

Key areas we would highlight (and invest in) include:

- strategic research into:
  - the degree and adverse effects of non-compliance by fishery and fisheries management.

- optimization techniques for components of enforcement function (e.g. education, recreational compliance, etc.) resulting in predicted resource requirements to gain an agreed level of service.

- the relative merits of outsourced *vs* agency-based enforcement functions.

- the economics of the enforcement function with a focus on success (e.g. the value that current enforcement/ compliance currently retains for the total community).

- increased incorporation of enforcement/offender impacts into existing models of higher management intensity fisheries.

- far greater investigation of the fisheries-related desires of communities and the sociologically-optimal means of achieving management outcomes with emphasis on long-term sustainability and a stable working relationship.

- identifying and comparing differing enforcement paradigms.

- developing fishery-independent and officer-independent indices of compliance as standards for specific fisheries with a view to increasing use of automated procedures.

- investigating relative efficiencies of incentive (e.g. education) *vs* disincentive measures by type of fishery (e.g. recreational, commercial, conservation).

- quantifying the relationship between the value of penalties levied against fisheries criminals, the lost value due to damage inflicted on the resource, and the impact of the penalty as a disincentive to future non-compliance.

• integration of enforcement into the total fisheries management equation, including:

- pro-active setting of performance indicators, goals and objectives for all fisheries (see Baker 1997), not with a view to setting minimum offence records per officer, but focussing on agreed outcomes for an agreed price.

- focus on production of a minimum number of 'compliance tailored regulations' which achieve results.

- attempting to integrate non-fisheries management into the overall fisheries plan. For example, environmental, coastal protection, pollution, water flow regulation, and student education functions are all often 'controlled' by non-fisheries agencies, but success of sustainability or pro-active fishery improvement may rest critically on incorporating/influencing these areas.

- incorporating enforcement 'grassroots' constructive input into management revision and legislative enactment to increase the efficiency, smooth operation and effectiveness of resulting plans/regulations/policy (particularly based on Anderson 1989).

- enhancing feedback loops from all stakeholders into the enforcement/compliance process with emphasis on accountability of all parties.

• integration of enforcement into the total community equation, including:

- greater use of volunteers, students, and other mechanisms for shifting enforcement responsibility back in part to the community beneficiaries.

- greater impact on primary students to instill principles on which future sustainable management can rest.

- quantification of community perceptions regarding enforcement and fisheries in general.

- prioritizing sociological input into formulation of all fisheries management planning.

- reinvesting 'community service' and monetary penalties back into the fisheries resource from where the benefits were originally 'stolen.'

- incorporating 'junior fishing clinics' run through enforcement to help educate the fishers of the future in the right way of recreational fishing, as well as good ethics/stewardship practices, rather than focussing on solving non-compliance problems with huge penalties later in their lives.

• far greater cross-agency transfer of expertise, experiences, and techniques, by:

- regularly bringing together representatives of the world enforcement community to benchmark the current state of play and point toward future trends and directions.

- increase transfer of community policing techniques, research and success to fisheries enforcement. A major literature exists in the broad study of criminology which has largely failed to jump into the 'wet' area.

- facilitate inter-agency and particularly international officer exchange programmes and evaluate their benefits as an integral part of the programmes. Emphasis should be placed on selecting participants at all levels, not just higher management echelons.

- increase the use of as many other professional disciplines as possible to increase the potential for innovation within the enforcement function.

- seek to incorporate at least one enforcement-focussed theme at the Third World Fisheries Congress.

Conclusion

We must push back the enforcement frontier to achieve a positive and sustainable future or watch the world's fisheries continue to degrade.

References

Anderson, L G (1989). Enforcement issues in selecting fishery management policy. *Marine Resource Economics* **6**, 261–77.

Baker, D (1997). Your work is of value, prove it or perish. In 'Developing and Sustaining World Fisheries Resources: The State of Science and Management. Second World Fisheries Congress, Brisbane 1996.' (Eds D A Hancock, D C Smith, A Grant and J P Beumer) pp. 215–19 (CSIRO Publishing: Melbourne).

Cambridge Scientific Abstracts (1996). Aquatic Sciences and Fisheries Abstracts. (Cambridge Scientific Abstracts: London).

FAO (1995). *FAO fishery statistics — Commodities* **77.**

Howarth, W (1987). 'Freshwater fishery law.' (Financial Training Publications: London).

Kohler, C C, and Hubert, W A (Eds) (1993). 'Inland fisheries management in North America.' (American Fisheries Society: Bethesda).

Nielsen, L A (1993). History of inland fisheries management in North America. In 'Inland fisheries management in North America'. (Eds C C Kohler and W A Hubert) pp. 3–31 (American Fisheries Society: Bethesda).

OECD (1994). 'Fisheries enforcement issues.' (OECD: Paris).

Ruddle, K (1994). A guide to the literature on traditional community-based fishery management in the Asia-Pacific tropics. *FAO Fisheries Circular* 869.

Sutinen, J G (1994). Summary and conclusions of the workshop on enforcement measures. In 'Fisheries enforcement issues'. pp. 7–16. (OECD: Paris).

Sutinen, J G, and Gauvin, J R (1989). An econometric study of regulatory enforcement and compliance in the commercial inshore lobster fishery of Massachusetts. In 'Rights based fishing'. (Eds P A Neher, R Arnason and N Mollett) pp. 415–31 (NATO: Brussels).

Wang, S D H, and Zhan, B (1992). Marine fishery resource management in PR China. *Marine Policy* **16**, 197–209.

Seventy-five Years of Halibut Management Success

Donald A. McCaughran

International Pacific Halibut Commission, PO Box 95009 Seattle, Washington 98145-2009, USA.

Summary

The fishery for Pacific halibut is 108 years old, and is still yielding higher-than-average catches. Few other North American groundfish fisheries have been as fortunate. Both fishery-related and environmental factors have played a role in determining stock conditions over the years. The stocks have been maintained at biomass levels larger than that which produces the maximum number of recruits. Maintaining a large spawning biomass has taken precedence over maximizing productivity. The management techniques employed over the years are catch-per-unit-effort (CPUE) driven 'trial and error' production models, cohort analysis and more complete age-structured models, all reflecting the current state of fisheries science of the time. These models have helped reduce uncertainty, but where uncertainty exists the International Pacific Halibut Commission has been very conservative. The Commission uses a great deal of scientifically sampled data from fishermen but does not let their opinions or short-term economic desires influence catch limit determination. Halibut fishermen tend to be conservative and have always shared the conservation goals of the Commission. Science dominates catch limit determination and political interference is absent. In addition to 'state of the art' science and conservative management it must be acknowledged that 'good luck' has played a major role.

Introduction

The 1996 fishery for Pacific halibut marks the 108th year of its existence. The present catch limit is 22 068 t, which is approximately the average catch since the time the fishery became fully exploited. If we add the sport catch, personal use, and the bycatch in other fisheries, the total removals exceed 30 000 t. While there have been 'ups and downs', the fishery has been maintained as an economically-viable industry. This is not the situation with several other north-eastern Pacific fisheries. Major collapses have occurred in the Alaskan king crab and shrimp fisheries, and several finfish fisheries have suffered considerable declines. In other areas of North America, groundfish fisheries have also not fared well. Groundfish stocks off the east coast of the USA and Canada are severely depressed. Why then has the Pacific halibut fishery survived when so many other managed fisheries have not?

The answer to this question is more difficult than might appear. It must be acknowledged that during much of the time the fishery was under formal management the scientific techniques of stock assessment and of setting catch limits were not yet well developed. It has only been during the last 40 years that these methods became useful, and in particular only in the last 20 years have halibut managers been willing to place their faith in modern population assessment procedures for the purpose of setting catch limits. The history of management by the Commission will be discussed and an attempt will be made to shed light on the reasons why the Pacific halibut fishery is still productive and has not joined the long list of fisheries no longer economically viable.

THE FISHERY

The commercial fishery for halibut began in 1888. The average annual removal during the first five years was approximately 900 t. The catches increased, as the fleet expanded, to an annual average of 28 500 t from 1913 to 1916 (Fig. 1). The catches dropped to lower levels by the early 1920s which concerned halibut fishermen and caused them to petition the governments of Canada and the USA to enter into an agreement to study Pacific halibut and to determine if any action was necessary to preserve the stocks. The reduced fishing success around 1920 was understandable since the stock was newly exploited and could not sustain, at that time, catches above 30 000 t. The initial high catch of approximately 30 000 t was an extremely important datum since it estimated a harvest level above which the stocks would decline.

The International Fisheries Commission (later changed to the International Pacific Halibut Commission) was formed in 1923 in response to the demands of fishermen. Initially it was only a research group with no management authority other than to establish a closure of the fishery for three months in the winter. A new treaty in 1930 allowed the Commission to set regulatory areas and to set catch limits for each area. The Commission reduced the allowable removals in 1931 and maintained the annual harvest at less than 26 000 t for the next 20 years. The catches were allowed to increase as stocks grew, which was measured by the success of fishermen (catch-per-unit-effort CPUE), to a high of about 34 000 t in the early 1960s. At that time non-domestic trawl fleets began to operate in the Bering Sea and Gulf of Alaska and their estimated halibut bycatch mortality, while imprecisely measured, exceeded 10 000 t. The bycatch removals, in addition to the high commercial catches, combined to overfish the stocks and they declined. By the early 1970s the Commission had improved estimates of the bycatch and reduced the domestic catch limits to avoid further overfishing. The catch was reduced to a low enough level to allow the stocks to grow, and grow they did, until the biomass reached its largest recorded level by the mid-1980s. Since that time the biomass and harvest have declined. The variation in stock size documented over the past century has no simple explanation. It appears that both the environment and fishing have demonstrable impacts on halibut production.

ENVIRONMENTAL EFFECTS

Considerable discussion has taken place over the years as to whether the environment or fishing removals play the major role in controlling halibut stock size. There was major disagreement between W. F. Thompson and M. D. Burkenroad in the early 1950s about the cause of the decline in catches from 1915 to 1925 (Burkenroad 1948, 1953; Thompson 1950, 1952). This disagreement helped to focus attention on the problem of separating environmental changes from the effects of fishing. The argument seems somewhat less important today since we now have information to show that growth, survival, and year-class strength are probably all affected by environmental conditions. We also have information to indicate that fishing removals may have very large effects as well. The arguments used by Burkenroad were based on mark-recapture estimates from data which violated all the assumptions necessary to obtain statistically reliable stock size estimates. While Burkenroad was

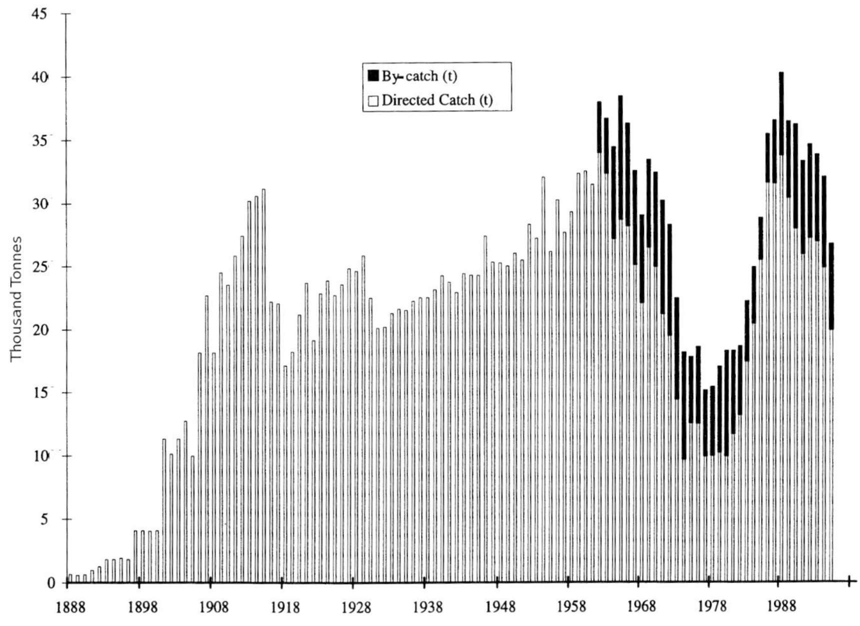

Fig. 1. Historical removals of Pacific halibut.

correct in citing a strong environmental influence on halibut stocks, the arguments he used would not stand present-day review. The main problem of that discussion was that no acceptable estimates of stock size were available at that time, and the confounded effects of fishing and the environment could not be separated. Skud (1975) discussed the Thompson-Burkenroad debate and presented a graph which showed stock size estimates of 90 000 t to 400 000 t. Present analyses show stock size has not exceeded 200 000 t. The debate was conducted using arguments based on poor analysis and should not be taken seriously, except that it did bring into focus the fact that environmental conditions do affect stock abundance.

Beginning in 1975 the Commission began to use age-structure methods (cohort analysis) and for the first time had estimates of historic biomass. Estimates of recruitment became available, and from past survey work, historic estimates of growth were obtained. These estimates produced information that allowed a more critical examination of the effects of the environment on halibut stocks.

Figure 2 shows how growth has changed over the past 60 years. The growth in weight of 12-year-old female halibut from the central Gulf of Alaska doubled from 1930 to 1960, remained fairly constant for about 20 years, and has recently declined to the 1930 levels. No clear explanation for this phenomenon is available, but it is noted that this trend does not appear to be related to the biomass of the stock (Fig. 3) nor to stock size in terms of numbers (Fig. 4). One hypothesis for the present decline, is that since the early 1960s the populations of other flatfish have increased to very high levels. Competition for food by juvenile halibut with the other species of flatfish may reduce growth rates during early years from which they never catch up. The growth rate reduction currently experienced has a major impact on biomass, and even with similar numbers of halibut the present biomass is only 50% of what it was in 1985.

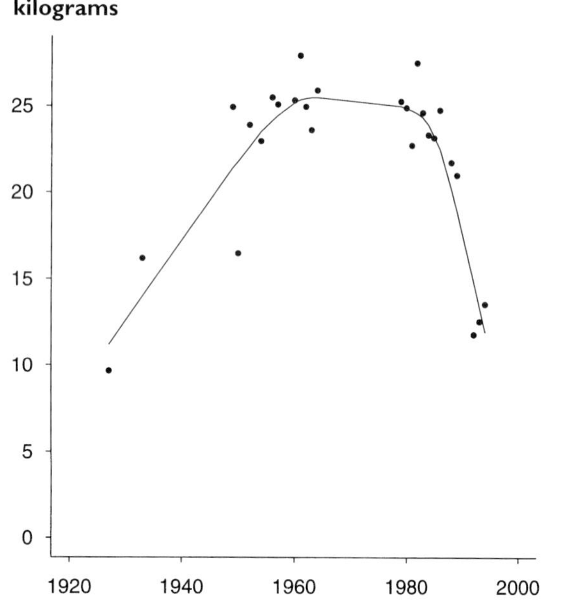

Fig. 2. Mean weight of 12-year-old female halibut in the central Gulf of Alaska.

Figure 5 shows changes in the biomass of spawners and recruits (8-year-olds) over time. The auto-correlated estimates occur in groups indicating an environmental component in the halibut stock-recruitment relationship. The overall form of the relationship shows a degree of dependency on the size of the parent stock in determining recruits. However, recruitment is also a function of similar year-clusters. The specific environmental factors which are responsible for this condition are not known. We believe a combination of both physical and biological factors is responsible. A physical model to explain year-class strength (Parker 1988) proposed that in years when the Alaska coastal current is strong there is also a strong onshore flow. The onshore flow carries halibut larvae into shallow water where they settle in suitable inshore habitat. Parker showed that year-class strength estimates are positively correlated with the strength of the current.

Fishing effects

Changes in the level of exploitable stock size caused by fishing removals are confounded with environmental changes. It can be seen from Figure 5 that fishing-down the spawning stock will change the productivity of the stock. Reducing the Gulf of Alaska female spawning stock from 65 000 t to 25 000 t will, on the average, increase the production of recruits from approximately 3 to 5.5 million. Historically, the spawning stock size has always been maintained at levels at or above the level which produces the highest numbers of recruits. Although not necessarily by intent, productivity has been sacrificed for the maintenance of large spawning biomasses. With a relatively flat spawner-recruit relationship this strategy carries a lower risk of overfishing than attempting to maximize productivity, although over the years this was never a conscious strategy.

Historically, there has been one episode of removals large enough to significantly reduce the spawning biomass. This occurred largely because of bycatch removals of halibut in the foreign trawl fleet. That fleet began fishing in the late 1950s and the size of the halibut removals was not well documented. This occurred at a time when the Commission was setting very high catch limits, close to what was then estimated as maximum sustainable yield (MSY). The combined removal of bycatch plus directed catch, was approximately 38 000 t. Figure 5 shows that these record catches occurred at a time when environmental

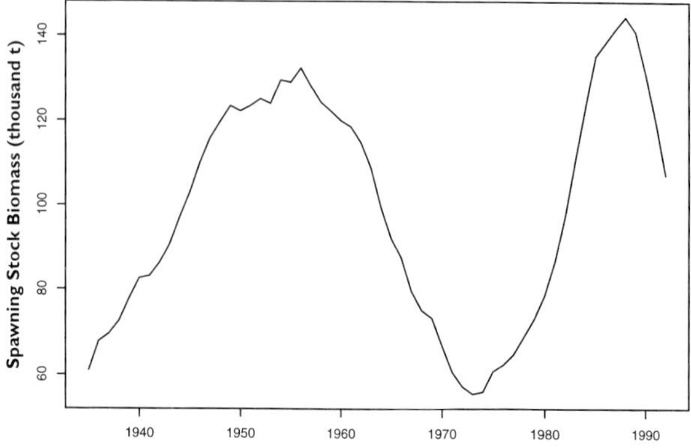

Fig. 3. Historical spawning stock biomass.

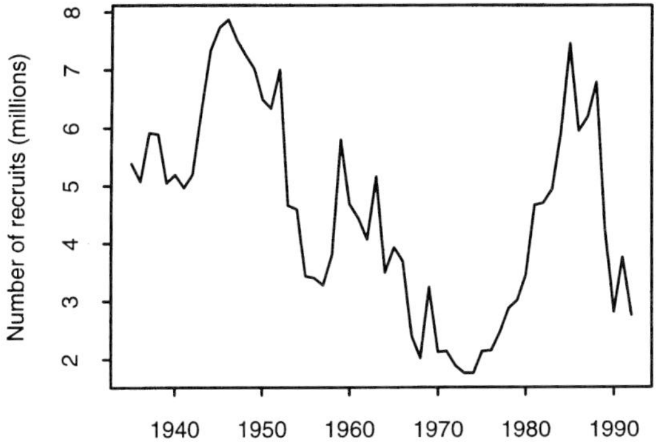

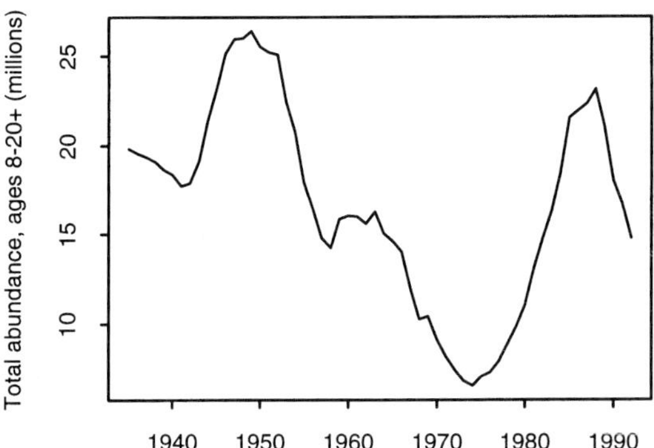

Fig. 4. Historical numbers of halibut recruits and numbers of halibut in the exploitable population (8–20 year-olds).

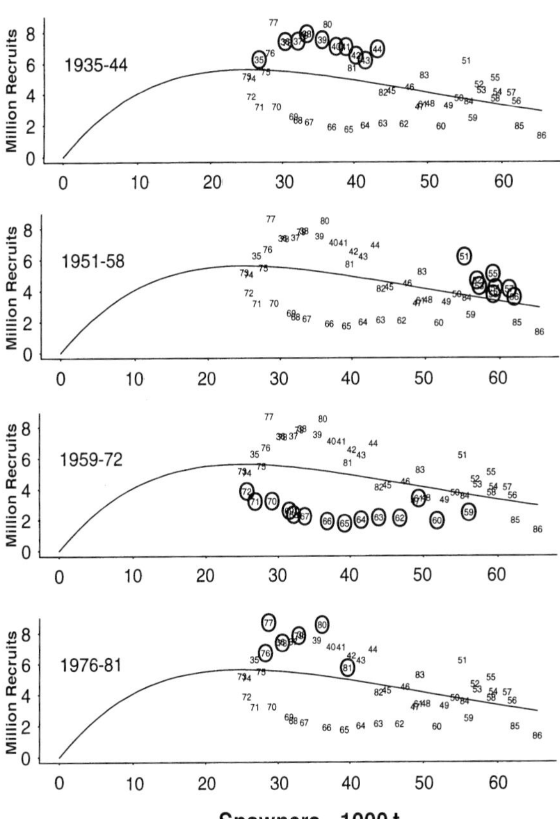

Fig. 5. Central Gulf of Alaska adult female Pacific halibut spawner-recruit relationship showing strong and weak groups of year classes.

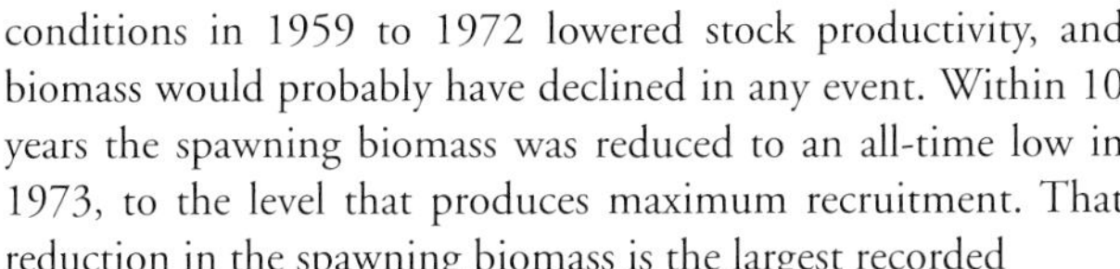

conditions in 1959 to 1972 lowered stock productivity, and biomass would probably have declined in any event. Within 10 years the spawning biomass was reduced to an all-time low in 1973, to the level that produces maximum recruitment. That reduction in the spawning biomass is the largest recorded

Bycatch in the trawl fishery is composed of pre-recruits primarily less than six years of age. Consequently, even under high productivity, bycatch causes decreased juvenile survival. The period 1959 to 1972 had very high juvenile mortality rates due in part to the high uncontrolled bycatch levels. This coincided with a period of unfavourable environmental conditions, and the combined effect lowered the recruitment level.

Juvenile fishing mortality (bycatch), adult fishing mortality (commercial catch), and environmental effects combined during the period 1959 to 1972 to cause a major decline in stock size. The Commission was forced to reduce commercial removals to the lowest in 70 years (9 000 t).

While a quantitative determination of the relative effects of the environment and fishing is not easily obtained, it is clear that both factors affect halibut stock size and productivity.

MANAGEMENT HISTORY

During the first 42 years of halibut fishing there were no restrictions on the commercial fishery except a three-month winter closure that was implemented with the signing of the 1923 treaty between the USA and Canada. A new convention was signed in 1930 which expanded the Commission's management authority. The Commission divided the range of Pacific halibut into four areas, set catch limits for those areas, and closed nursery grounds to fishing. While there were limits placed on the catches there was little science involved; commercial harvest was essentially managed by 'trial and error'. The catches from 1925 to 1929 ranged between 22 720 t and 25 600 t. They were viewed as excessive because from mark-recapture returns it was estimated that 40% of the commercial stock was removed annually in addition to natural mortality. The Commission also had good data on CPUE which showed a steady decline from the initiation of the fishery until the 1920s. Once the treaty allowed the Commission to regulate catches the annual catch was reduced to 20 860 t beginning in 1932. This represented a reduction of ~20% and, if the estimated exploitation rate was approximately correct, this resulted in a reduction from 0.40 to 0.32 in the exploitation rate. It is interesting to note that 0.30 is presently used as the optimal exploitation rate. The 20 860 t catch limit was maintained until 1937 and then gradually increased as the CPUE increased. The period 1935 to 1958 had above-average recruitment and when this was combined with lower catches the stocks grew to very high levels by the late 1950s. The period from 1958 to 1975 was managed mainly by observing the CPUE trend, but by this time a large number of population dynamic models were available. The Commission staff used these new models in an attempt to estimate the maximum sustainable yield

(Chapman *et al.* 1962). The yields were approaching the estimated MSY by 1960 when the foreign bycatch removals began and the combination of a poor environment, a spawning stock which was large and less productive, and overfishing combined to quickly reduce stock size to the low of 1973. The 1973 stock was, however, near the point of maximum productivity (Fig. 5).

In the 1970s, fisheries scientists were becoming disillusioned with the concept of MSY. It was becoming obvious that MSY was not a stable parameter and, depending on environmental conditions, it could take on a variety of values. Attention turned to models for estimating stock size based on the annual catch at each age. Hoag and McNaughton (1978) applied Pope's cohort analysis (Pope 1972) to halibut data. Fortunately, W.F. Thompson, the first director of the Commission, implemented a sampling programme to collect otoliths from the commercial fishery which provided age data date from 1930. These provided Hoag with historical estimated catch-at-age data for his analysis. The estimates of current-year stock size were imprecise and could not be used for setting catch limits, but for the first time CPUE and historical biomass estimates could be related and past exploitation rates were available. This new information was used to set conservative catch limits during the early 1970s. The age model was expanded and improved by Deriso *et al.* (1985). CPUE was used as an auxiliary variable in order to obtain stable estimates of current biomass levels. This allowed the Commission staff to set catch limits as a proportion of exploitable biomass (8-20 year olds). The biomass estimates were, however, initially used to calculate annual surplus production (ASP). The catch limits were set at 75% of ASP from 1979 to 1985.

In 1985 the Commission adopted a constant exploitation yield strategy (CEY). Catch limits were then set at 35% of the exploitable biomass. Considerable research on optimal exploitation rate strategies was undertaken. Simulation studies, which incorporated a variety of stock-recruitment relationships and environmental cycles, caused the staff to reduce the exploitation rate to 30% in 1992. Review of the historical biomass estimates (McCaughran in press) showed that from 1980 to 1990 successive annual estimates of the biomass of a particular year increased, indicating that the first estimate (the one used to set catch limits) underestimated the stock size. Retrospective analyses with the present age structure model has confirmed this problem, and produced estimated exploitation rates between 0.14 and 0.25 from 1978 to 1994, considerably lower than the 0.35 or 0.30 target exploitation rate.

It is believed that the large change in growth rates has also changed the age selectivity to longline gear over that period. Recently, modifications have been made to the age-structured model to incorporate changing growth rates and age selectivity. The retrospective study with this modified model shows a considerable improvement in the stability of successive biomass estimates, and further testing of the model is presently being conducted. Although the Commission has attempted to exploit approximately 30–35% of the adult stock annually over the last decade, history shows that the actual exploitation rates have generally been much lower.

At the present time the latest stock assessment model is being examined carefully and we are re-evaluating the exploitation rate and size limit in an attempt to examine the optimal properties of our annual catch limit estimates. What is needed is an annual biomass estimate that is at least in proportion to the true biomass levels. The chosen exploitation rate will hopefully compensate for the proportional difference between the estimates and the true stock size.

Consultation with industry

Fishermen from the USA and Canada were instrumental in the formation of the Commission. After the first world war, halibut fishermen petitioned the governments to set up a body to study Pacific halibut, which led to the treaty of 1923. After the revision of the treaty in 1930 the Commission set up an advisory body of fishermen from both countries. This group was formed in 1931 and is called the Conference Board.

The processing sector participated only informally until 1955, and they have never had the degree of influence that the Conference Board has enjoyed. In 1995 the processors increased their participation by forming a 'Processors Advisory Group'. They promise to be more active participants in the future.

In recent years it is often heard that the fishing industry should play a bigger role in decision making, and fishery managers should spend more time listening to fishermen. This concept, if applied to stock assessment or quota setting, can be a recipe for overfishing and stock collapses. Apparently, this situation plagued the USA New England Fishery Management Council in its management of groundfish. Scientific information was often ignored, and industry played the major role in management. What is often not understood by advocates for public input is that fishery scientists already use a great deal of input from fishermen. The Halibut Commission uses fishermen's log-books to obtain their actual fishing experiences, but does not use their opinions. To show how inconsistent fishermen's opinions can be, the following anecdote happened in the summer of 1995: Seattle-based Fisherman A claimed the fishing in the Aleutian Islands was better than we thought and we should increase the catch limit. Fisherman B claimed that fishing was average and the catch limits should be left as is. Later that week Fisherman C from Alaska informed a staff member that the fishing in that area was poor and he was convinced that the catch limit would not be taken. These diverse opinions prompted the staff to review the most recent catch-per-stake (1800 ft. of ground line with 100 hooks) log-book data of these fishermen from that area:

Fisherman A	385 pounds per skate
Fisherman B	420 pounds per skate
Fisherman C	415 pounds per skate

It is obvious that these three diverse opinions about the stock condition in this area have little value, whereas their actual experience (CPUE) is an accurate index of stock condition and is incorporated into our stock assessment analysis. The Commission collects ~4000 log-books each year which represent the majority of the fishermen's experience. Using accurately-sampled data from fishermen is necessary for high-quality stock assessment, but allowing fishermen's opinions to become involved in catch limit determination is often counter-productive.

On the other hand economic and sociological information from fishermen is necessary in developing harvesting systems such as individual transferable quota (ITQ) management. The Commission's conference board is given considerable responsibility for setting fishing seasons, and their opinion on fishing methods and juvenile mortality has been very useful. Halibut fishermen generally trust the Commission staff to set conservation-based catch limits. This trust has developed over the years because the staff has no political agenda and encourages review of its stock assessment methods. Stock assessment methodology and all data (except fishermen's catch locations) are open for review by the public. Problems and uncertainties with stock assessment are openly discussed at the Commission's annual meeting. The Commission's consultative process allows for constructive input from industry but does not accommodate self-serving opinion. The Commission's role is stock conservation, while the two national governments have the responsibility for implementing allocative harvesting systems.

Discussion

There are still a lot of halibut in the sea; in fact a larger number than the average over the past hundred years. The reasons for this fortunate condition are many; halibut are very well-suited to their environment and respond strongly to favourable environmental conditions, a single-species management agency, dedicated scientific staff, no political interference, the use of appropriate harvester input and, last but not least, good luck. All these factors have contributed to maintaining the resource.

In the early years there was little science in fisheries management but the Commission did collect data on catch and effort, and age structure, and it knew the maximum catch the stocks could stand, at that time, without decreasing under unlimited fishing. Although the maximum allowable removal changes over time, depending on the environment, it was a piece of baseline data that was extremely useful. In addition, CPUE data were collected over that same period and were related to harvest level. So in 1930 'trial and error' management began. Good fortune played a major role, the reduction in the stocks due to uncontrolled fishing had reduced the spawning biomass to its most productive level and the environment was very favourable (1935–1944). The stocks grew, CPUE got better and catches were cautiously increased. The Commission was very conservative with the allowable catches. W. F. Thompson, the staff director, was a strong-minded conservative individual and he was supported by equally strong minded and conservative halibut fishermen (mainly of Norwegian descent) in both the USA and Canada. Conservation of the resource was the main goal of both the fishermen and the Commission. That principle has remained unchanged to this day.

The Commission staff has dedicated itself to the management of one species and has always applied the latest in fisheries science. Single-species management has been criticized for not accounting for other species in the ecosystem. However, multi-species management, while fashionable, is often an excuse for continuing the use of fishing gear which has less than optimal selective properties toward many of the species involved. While the gear might optimally harvest one species it usually overharvests some species and underharvests others.

The cost of single-species management of Pacific halibut varies from 1 to 3% of the annual ex-vessel value of the harvest. Cost to the two governments has increased with recently-implemented ITQ systems but the benefits far exceed the additional cost.

Even with the most modern population dynamic stock assessment models and high quality input data there is still considerable uncertainty. The Commission has always taken a conservative approach to catch-limit determination in the face of uncertainty. Historical review of the stock-recruitment relationship (Fig. 5) shows the results of this conservative approach. The female spawning biomass in the Gulf of Alaska for example has never been lower than 25 000 t, the optimal level for recruit production, and not much larger than 65 000 t. With fluctuating environmental conditions this is probably a good strategy and is certainly central to the success of maintaining the fishery. Halibut respond very quickly to environmental conditions; review of Figure 5 shows quite clearly that productivity responses are not random but occur in groups of successive years. What is required for stock growth is adequate spawning biomass and the environment to be favourable.

The staff does not make final catch limit decisions; the Commission does. However, the Commission has always accepted the conservation-based recommendations of the staff and has never changed the recommendations by more than a few percent and only when the staff has no objections. Little political pressure is brought to bear on this process, most certainly because of the Commission's international structure. High long-term yields with minimum risk of reducing the spawning biomass below the optimum level is the goal and no consideration is given to the short-term economic desires of the industry. The industry plays a role in setting other regulations such as, season lengths and administrative types of regulations, but catch limits are the staff's concern. Generally, it is a recipe for disaster to have fishermen negotiating catch limits, although there are a number of conservative halibut fishermen who would always argue for lower catch limits, than those recommended by the staff.

The conservative management of the Pacific halibut resource maintained the stock until science produced methods which reduced uncertainty. There is less uncertainty today but it is still present, hence, there is still a need to be conservative. As fishery science progresses it is hoped that uncertainty in management will be further reduced to a level that will allow optimal harvesting while continuing to maintain the stocks over the next hundred years.

References

Burkenroad, M D (1948). Fluctuations in abundance of Pacific halibut. *Bulletin of the Bingham Oceanography Collection* 11(4), 81–129.

Burkenroad, M D (1953). Theory and practice of marine fishery management. *Journal du Conseil International pour l'Exploration de la Mer* 18(3), 300–10.

Chapman, D G, Myhre, R J, and Southward, G M (1962). Estimation of maximum sustained yield, 1960. *International Pacific Halibut Commission, Scientific Report* No. 31.

Deriso, R B, Quinn II, T J, and Neal, P R (1985). Catch-age analysis with auxiliary information. *Canadian Journal of Fisheries and Aquatic Sciences* **42**, 815–24.

Hoag, S H, and McNaughton, R J (1978). Abundance and fishing mortality of Pacific halibut, cohort analysis, 1935–1976. *International Pacific Halibut Commission Scientific Report* No. 65.

McCaughran, D A (in press). The effectiveness of joint US/Canada management of Pacific halibut. Proceedings of the First World Fisheries Congress.

Parker, K S (1988). Influence of oceanographic and meteorological process on the recruitment of Pacific halibut (*Hipppoglossus stenolepis*) in the Gulf of Alaska. University of Washington PhD Thesis.

Pope, J G (1972). An investigation of the accuracy of virtual population analysis using cohort analysis. *International Commission of the Northwest Atlantic Fisheries, Bulletin* No. 9, 65–74.

Skud, B E (1975). Revised estimates of halibut abundance and the Thompson-Burkenroad debate. *International Pacific Halibut Commission Scientific Report* No. 56.

Thompson, W F (1950). The effect of fishing on stocks of halibut in the Pacific. (University of Washington Press).

Thompson, W F (1952). Condition of stocks of halibut in the Pacific. *Journal du Conseil International pour l'Exploration de la Mer* **18**(2), 141–66.

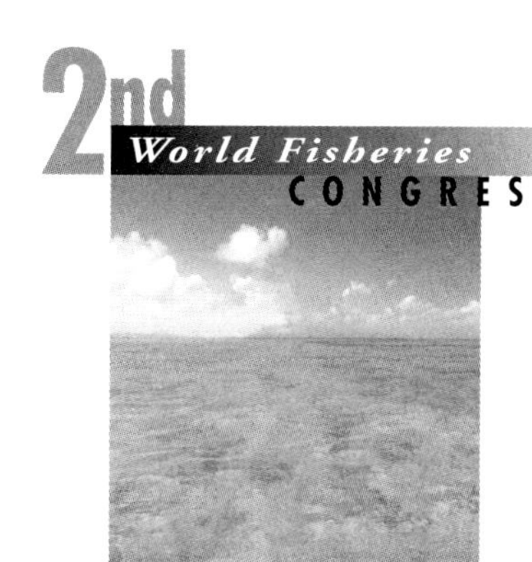

Alaska's Sockeye Salmon Fishery Management: Can We Learn From Success?[1]

Dana C. Schmidt,[A] *Gary B. Kyle,*[A] *Stan R. Carlson,*[A] *Harold J. Geiger*[B] *and Bruce P. Finney*[C]

[A] Alaska Dept. of Fish and Game, Soldotna, Alaska, USA.
[B] Alaska Dept. of Fish and Game, Juneau, Alaska, USA.
[C] Institute of Marine Science, University of Alaska, Fairbanks, Alaska, USA.

Summary

Alaska's sockeye salmon (*Oncorhynchus nerka*) stocks are entering a second decade of historically high production. Despite declines in fisheries globally, this fishery provides an example of how a once-depleted resource can recover and sustain high harvest levels.

A limited entry system, implemented in 1973 when fisheries were depleted, has created a transferable permit system. As a result, individual permits have increased in value and provided a long-term investment in the sustainability of these fisheries. A conservation ethic has developed in many coastal communities because of their collective memory of overfishing and depleted runs in the 1950s and 1960s. This ethic has been maintained through effective communication and conservation efforts of local managers who reside in the fishing communities and who remain neutral on allocation disputes. An uncommon feature of the modern major sockeye salmon fisheries is harvest regulations based on in-season 'spawner escapement' goals, as opposed to harvest quotas or constant harvest rates. Constant harvest rate policy may be deficient when applied to scenarios where escapement affects the productivity of rearing habitat used by subsequent generations of salmon. We present Karluk Lake on Kodiak Island as a case study of how traditional analysis and conservative harvest rates have resulted in declining yield. The evidence provided suggests that violation of the Ricker model assumptions may lead to invalid conclusions concerning the robust nature of fixed harvest rates. Limnology data on the carrying capacity of natal lakes can be used independently or in conjunction with stock-recruit data to determine escapement goals. Escapements goals also result in a public and a political demand for meeting a conservation objective rather than meeting an economic objective such as a catch quota. An escapement goal policy provides incentives for accurate catch reporting and political support for funding stock assessment programmes and improved evaluation of escapement goals.

Introduction

Alaska's sockeye salmon (*Oncorhynchus nerka*) fishery is completing its second decade of prosperity (Fig. 1). Most salmon fisheries in Alaska continue to prosper despite major collapses in fisheries in the western United States. Favourable climatic conditions have contributed to the recent large runs (Beamish and Bouillon 1993) but account for only a part of this success story. Royce (1989) provides a broad description of the successes of Alaskan salmon fisheries and the roles of industry, science, and government in developing the present fisheries management system.

We examine the sockeye salmon fishery in Alaska and how fixed escapement goal policies interact with socio-economic and political entities to maintain viable fisheries. We have not

[1]Contribution PP-132 of the Alaska Department of Fish and Game, Division of Commercial Fisheries, Juneau.

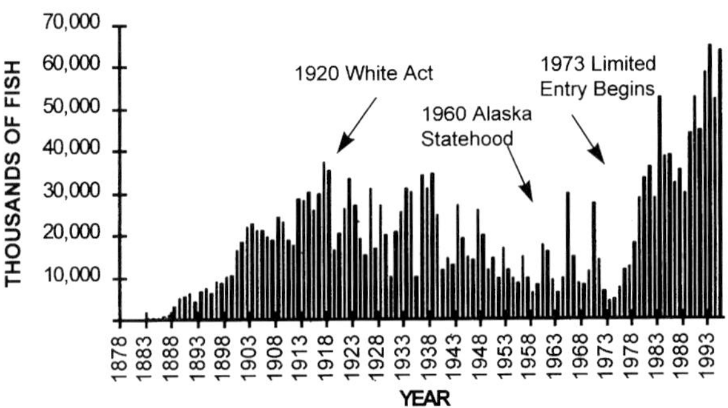

Fig. 1. Historical catch of sockeye salmon in Alaska. Historic events that changed fisheries management direction in Alaska are identified. The White Act resulted in a fixed harvest rate of 50% statewide. The statehood act eliminated fish traps and provided for much more local regulation of fisheries with authority transferred to the Alaska Department of Fish and Game. Licence limitations began in 1973 and provided for transferable permits exclusively granted to prior participants.

attempted to quantify the effects these factors have on salmon production using the techniques of economists or social scientists. Instead, we describe the fisheries management system, in which policy objectives and results are often inconsistent but still result in a successful harvest policy. We also address other practical consequences of the fixed escapement goal policy that often elude investigators relying on incomplete models to make conclusions.

Description of the Alaskan sockeye salmon fisheries management system

The sockeye salmon fishery in Alaska has some distinguishing features that separate it from other salmon fisheries in the Pacific Northwest. The sizes of sockeye salmon populations are usually dependent upon the size and quality of natal rearing lakes. Most of the harvest of these separate stocks occurs in the ocean near the mouth of the outlet stream, allowing reasonable targeting of a single stock. In Alaska all management is *local:* Area Managers reside in fishing communities and have emergency closure authority with law enforcement status. This authority allows managers to close fisheries with limited notice (e.g. one hour) if conservation concerns exist or as needed to meet allocative objectives developed by the Alaska Board of Fisheries (ABOF). These managers spend much of their careers in the same community and become quite familiar with the local fishing industry. They communicate with fishermen throughout the year and represent conservation interests in often heated allocation disputes. Allocation disputes are often debated through local advisory committees that report to a statewide politically-appointed board, comprised of individuals with experience in the fishing industry. Recently, disputes over allocation between sport fishing, subsistence, and commercial users have threatened this process because of federal oversight and legislative and ballot initiatives. Because commercial fishermen have historically dominated the ABOF, other groups have attempted to bypass this process because of perceived inequities. However, the vast majority of allocation disputes and decisions are still completed through the ABOF process.

Enhancement practices, such as lake fertilization, hatchery stocking of barren lakes, or hatchery stocking of spawning-limited lakes are primarily supported by private non-profit commercial fishing organizations. If a conflict exists between enhancement operations and maintaining wild stocks, the state has a legislative policy mandate to provide a priority for wild stock management.

A low human population and modest industrial development have all helped to maintain habitat quality. Logging and related road construction activities remains a highly divisive issue within the state. Because much of the logging is on federal lands, there is little oversight by the State of Alaska. However, private and state lands have been subjected to logging under guidelines that are often less conservative than those used by the federal government. Because land development in the state has been minimal, impacts on overall fisheries production have been minor and localized.

The effectiveness of the Alaska Department of Fish and Game on the social and political front has been paramount to public acceptance of fisheries harvest policies based on science. Without public support, largely developed by area fisheries managers, necessary conservation measures which result in short-term economic hardship, would not have been supported.

Alaskan sockeye salmon fisheries have well-defined conservation objectives or escapement goals. Fisheries are managed in-season to achieve escapement goals based on real-time estimates of fish entering river systems. Weirs, sonar, and aerial counts are used to estimate escapements. Time and area openings, and closures are used by local fisheries managers to ensure that escapement goals are met. Escapements in most major river systems are accurately enumerated, and the current management system has been effective in maintaining escapements close to the goal. Escapement goals are established by a variety of analytical methods, often dependent upon available data. These analyses may establish goals as historical mean escapements that have provided for an apparently sustainable fishery. More detailed analyses project escapements from lake-carrying capacity studies, classical stock-recruitment analyses from Ricker models, and more recently, an integration of both. Although not the sole factor in development of the support for fisheries management in Alaska, the escapement goal harvest policy has been instrumental in maintaining support for conservation of Alaska's fisheries.

Economics of the salmon fisheries have changed dramatically since the early 1970s. A limited entry system developed through licensing in 1973, when fisheries were depleted, has created a transferable permit system. A conservation ethic has developed in many coastal communities because of their collective memory of past periods of overfishing and depleted runs. The development of limited entry with transferable licences, local access to managers, neutrality of managers in allocation disputes, timely data acquisition, and regulatory authority to expand fisheries during strong runs and close fisheries during weak runs, all contribute to both fishermen's and public acceptance. During weak runs, the ability to control fisheries to maintain escapements above some thresholds is the hallmark of the Alaska fishery management system.

A PRACTICAL EVALUATION OF THE FIXED ESCAPEMENT GOAL HARVEST POLICY

A feature of Alaskan sockeye salmon fisheries uncommon in other fisheries is harvest regulations based on in-season spawner escapement goals, as opposed to harvest quotas or constant harvest rates.

Critics of a constant escapement policy argue that it limits variation in stock size for proper stock recruitment analysis (Smith and Walters 1981). Because the constant harvest rate policy produces greater variation in escapement, some view the constant harvest rate policy as more informative (a fisheries argument). The importance of stock recruitment data in current management thinking is stressed by Walters and Collie (1988), who argued that detailed limnological studies are of limited value to fisheries management and that stock size must be based on historical experience.

Critics of the current system also cite Deriso (1985), who examined harvest policies under the logarithmic utility function (an economic argument). Deriso (1985) points out that constant harvest rate policy maximizes logarithm of catch. Supporters of the constant harvest rate policy have gone on to argue that the logarithm of catch better reflects economic goals for fishing fleets that are, in some sense, risk averse.

Beginning in the mid 1970s microcomputers began a revolution in the way fisheries managers directed commercial fisheries. As a result, management became increasingly dependent on statistics and applied mathematics. Under the current thinking, harvest policy discussions for Pacific salmon usually begin with the assumption that a salmon population responds strictly according to the Ricker model:

$$R_t = S_t e^{\alpha - \beta S_t + v_t}$$

with recruitment (R_t) from a stock of a particular size (S_t), an intrinsic productivity parameter (α), and a carrying capacity parameter (β). All random influences and changes in the state of the system, aside from stock size, are assumed to be summarized by v_t, which is a normally distributed random variate, serially uncorrelated, with a mean of zero and a constant variance (σ^2).

Eggers (1993) compared the constant harvest rate policy with the fixed escapement policy by means of simulation using the Ricker model. His results indicated a substantial improvement of escapement goal harvest policies compared to fixed harvest rates when likely scenarios of management error were examined. However, his model assumed static spawner and recruit dynamics, which are inconsistent with autocorrelated residuals; the condition most commonly associated with changing climatic patterns affecting marine survival.

In the following discussion we examine the ramifications of a constant harvest rate policy for sockeye salmon from a different perspective, i.e. how limnological forage base (zooplankton) information can be used to evaluate the carrying capacity of sockeye salmon rearing lakes, the habitat component most critically related to production potential. We examine one of Alaska's major sockeye salmon systems that has a history of fixed harvest rate management policies and, more recently, a fixed escapement goal policy. Finally, we consider the practical aspect of altering the management system for sockeye salmon fisheries and respond to the assertion that constant harvest rate policies for Pacific salmon are more risk averse.

Large time series and the α productivity parameter

Analyses of stock-recruit data usually have required that process error be summarized by a single log-normally distributed random variable (e^{v_t}), with no autocorrelation, and still maintain a two-parameter model (Eggers 1993). Much of the variation in production of Alaskan sockeye salmon is attributed to marine climatic changes (Quinn and Marshall 1989; Beamish and Bouillon 1993). Quinn and Marshall (1989) describe the structure in the catch of odd- and even-year pink salmon *Oncorhynchus gorbuscha*, in Alaska, which they suggest is the effect of climate on salmon production. These trends corresponded to the catch of chum salmon *Oncorhynchus keta*, coho salmon *Oncorhynchus kisutch*, and sockeye salmon in Alaska. These authors' results indicate that the two-parameter model with uncorrelated lognormal error is not adequate to fully describe the actual process of stock and recruitment for Pacific salmon in Alaska. However, if the Ricker model is used with the standard assumptions about uncorrelated process error, these results imply that the fundamental productivity parameter of salmon-producing systems changes in response to relatively low-frequency, large-scale ocean phenomena. In particular, assuming relatively stable management practices since 1960, the 1960s through early 1970s in Alaska were a period of relatively low α parameters. These α parameters began to increase in the mid 1970s. This implies that large data sets will not necessarily lead to better advice from a Ricker analysis. Walters (1985) has previously pointed out that parameter estimates of a system following a Ricker stock-recruit analysis will be seriously biased by residual autocorrelation. Problems such as errors in the measured stock size can further cause large data sets to give a false sense of security from a Ricker analysis (see Schnute 1991).

If the Ricker analysis can fail to precisely give good advice with small sample sizes, and if the stock recruitment situation has completely changed once large sample sizes have been collected, what should be done? How else can the relationship of stock and recruitment be understood except through the summarization of stock and recruitment data?

Limnological responses to escapements

An overlooked assumption of the stock-recruitment analysis is the false acceptance of brood year independence of stock and recruitment observations. This becomes a question of whether management action can interact with stock productivity and whether large escapements or extreme variation in stock size can decrease a system's productivity. To answer these questions we considered the mechanisms that drive stock and recruitment in fresh water. The notion that spawning area and freshwater rearing habitat predominantly control sockeye salmon production have been central to the understanding of the biology of this species for over 20 years (Foerster 1968). Although mechanisms of population regulation in lakes, such as density-dependent growth and production, have been well documented (Burgner 1964; Johnson 1965; Hartman and Burgner 1972; Goodlad *et al.* 1974; Rogers 1980; Kyle *et al.* 1988; Rieman and Meyers 1990), a biological mechanism responsible for negative interaction between runs is not well documented. We have examined limnological data collected in Alaskan lakes over the past two decades to identify the utility of such ancillary information in resolving the shortcomings of relying solely on stock-recruit data in evaluating harvest policies.

Karluk Lake sockeye salmon, located on Kodiak Island in Alaska, provide an example of a fishery that has been monitored intensively with extensive analyses of apparent changes in productivity (Rounsefell 1958; Koenings and Burkett 1987). A fixed harvest rate policy of 50%, based on the 1921 White Act mandate (Roppel 1986), resulted in conservative harvest rates through the 1960s (Fig. 2). An examination of sediment marine nitrogen (Fig. 3A), indicated that marine nitrogen deposition correlates strongly with trends in escapement (Bruce Finney, IMS, University of Alaska, Fairbanks, A K, pers, comm.). This downward trend in escapement, as indexed by marine nitrogen, is strongly correlated with the development of the commercial fisheries and suggests historical escapement several hundred years prior to commercial exploitation was less variable and much higher (Fig. 3B). These data also suggest that nutrient contributions from fish carcasses were broadly integrated into the pelagic food web of Karluk Lake.

Our analysis of the stationarity of Ricker α and β parameters using a time-polynomial regression model (Fig. 4) suggests a fundamental change in the production capacity of the lake for rearing juvenile sockeye salmon as indicated by trends in estimated β. This problem has been identified previously (Walters 1987), but our data suggest that historic escapements are the major determinant in nonstationarity of these parameters. Limnological investigations of nutrients, primary productivity, and the response of the zooplankton community to nutrient changes suggest that salmon carcasses play a major role in the long-term lake productivity and consequently changes in the β parameter (Koenings and Burkett 1987; Stockner 1987). Marine nitrogen trends in the sediment (Fig. 3) provide further support for this hypothesis. The α parameter, in the case of Karluk Lake, did not show significant temporal trends based on the time-polynomial model. This finding is counter to that of Quinn and Marshall (1989) and Beamish and Bouillon (1993) who reported that climate positively influenced productivity of other salmon stocks in the Gulf of Alaska.

Karluk Lake may be exceptional in the influence that carcasses have on productivity. The long-term trend in escapement before harvests occurred — estimated from historical marine nitrogen in the sediment — corresponds with the actual escapements observed by weir counts from 1920 to 1930. These data provide a unique opportunity to independently estimate a replacement abundance (S_r) of 1.2 million ($S_r = \alpha/\beta$, from Ricker 1975). The α parameter is estimated as approximately 1.8 from actual stock recruit data from the past 70 years. Since we have little

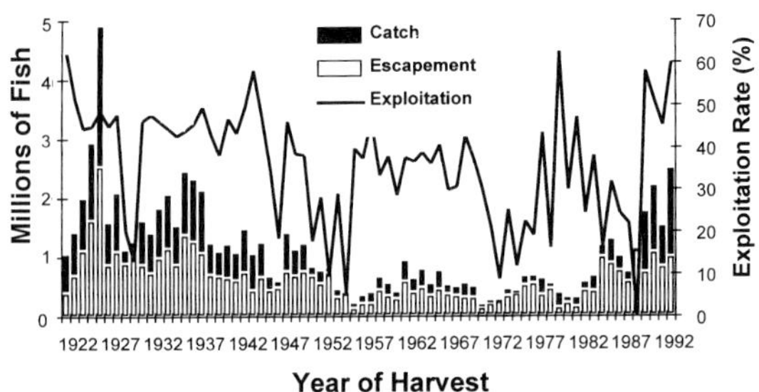

Fig. 2. Catch, escapement, and exploitation rates for Karluk Lake, Alaska from 1922 to 1992.

reason to suspect major changes in this parameter historically, the unfished population would have $\beta = \alpha/S_r$. From Ricker (1975) and Deriso (1985) a fixed escapement goal of about 450000 fish and an exploitation rate of 67% would provide maximum sustained yield. However, even with 400 years of escapement estimates implied from the sediment core data prior to initiating a fishery, this advice is apparently in error. Even with more modest harvest rates a long-term decline was observed, apparently related to oligotrophication from reduced salmon carcasses. These analyses suggest that a negative feedback loop develops when carcass reductions from one generation result in a reduction in the rearing capacity of the lake for subsequent generations. This is supported by limnological data collected from 1980 through 1993 which supports correlation

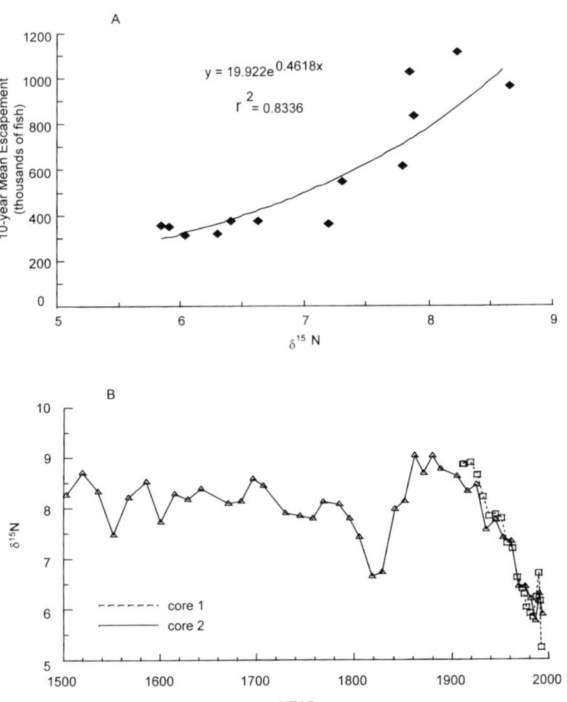

Fig. 3. The relationship of sediment cores to historical escapements for Karluk Lake. where (A) is the regression of Karluk Lake 10-year mean escapement history 1922–83 to δ^{15}N, and (B) shows the temporal trend in δ^{15}N for the last 500 years, including replicate core covering approximately the last 100 years. Age of core layer was approximated by using sediment weight extrapolated from a known ash layer from the 1912 Katmai Volcano.

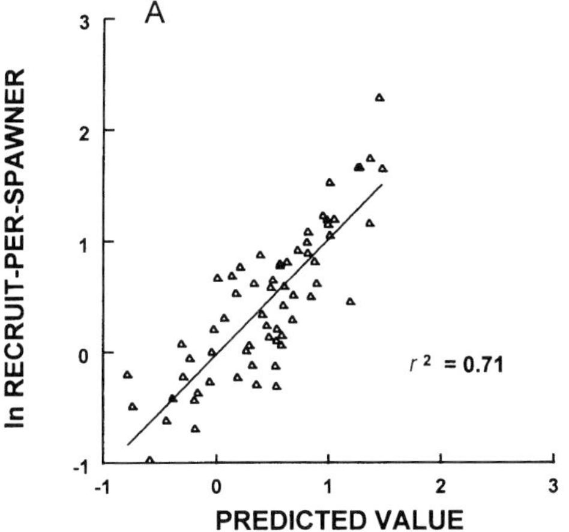

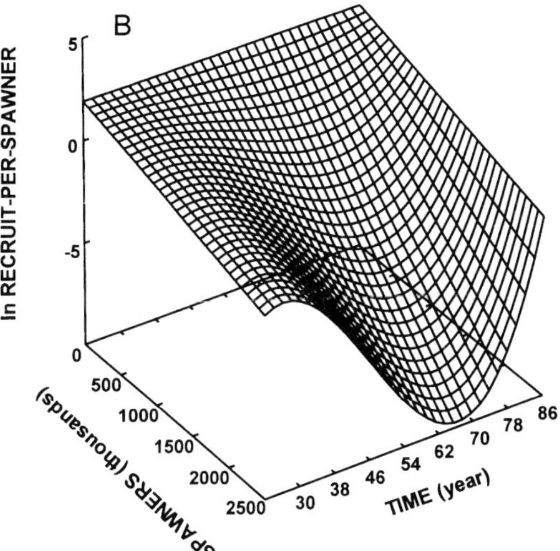

Fig. 4. Results of Ricker time-polynomial analysis of the Karluk Lake sockeye salmon spawner-recruit relationship. Variables were selected using backward elimination (significance level = 0.10) beginning with a full cubic polynomial model. The following regression function was selected: $\ln(R/S) = 1.92 - 3.9\times10^{-3}(S) - 1.13\times10^{-4}(ST) + 2.6\times10^{-6}(ST^2) + 1.1\times10^{-7}(ST^3)$, where S is spawners, R is recruits, and T is time. (A) is the predicted *versus* observed values from the fitted linear regression function, and (B) is the fitted regression surface from the selected model. Note that the intercept (density independent (α) parameter) is constant while the slope (density dependent (β) parameter) is nonconstant, indicating that intrinsic productivity has remained stable while the carrying capacity of the system has exhibited nonstationarity.

of fish carcasses with total phosphorous and chlorophyll *a* production (J Edmundson, ADF and G Limnological Laboratory, Soldotna, Alaska, pers. comm.).

Following a fixed escapement goal policy of 700 000 to 1 000 000 begun in the late 1970s, coupled with a lake fertilization programme in the mid 1980s, Karluk Lake apparently increased in fertility (Fig. 3B) and approached the high productivity observed in the 1920s. Local fisheries managers insisted on meeting high escapement goals, primarily because large production was observed at these escapement levels during the early history of Karluk Lake sockeye salmon fisheries. Any reasonable analysis of the spawner-recruit data from 1940–1980 would have suggested much lower escapement goal objectives and a much higher harvest rate. Such an action would have resulted in a sustainable fishery with about half the harvest that can be achieved under higher escapements.

In systems such as Karluk Lake, management advice based on short-term stock-recruit analyses, whether it be fixed harvest rates or fixed escapement goals, can substantially decrease yield. Walters and Parma's (1996) assumption that nonstationarity of Ricker model parameters is independent of historical escapements may not be valid. If invalid, the robustness of fixed harvest rate strategies will be overestimated. If escapement goals are established based on Ricker parameter estimates from short-term stock-recruit data, evaluation of fixed escapement goal strategies (Eggers 1993) will also overestimate yield and underestimate escapement requirements in systems like Karluk Lake.

Sockeye salmon lakes in which carcasses provide much of the nutrients necessary for sustaining high levels of productivity are most likely rare. However, high escapements affecting the carrying capacity of sockeye salmon lakes in years subsequent to the escapement event is supported by other investigations. Several consecutive year classes of overly-abundant juvenile sockeye salmon may adversely impact subsequent sockeye generations by decreasing zooplankton recruitment or by promoting a predation-resistant community (Kyle 1996; Koenings and Kyle 1997). Contrary to earlier work (Ward and Larkin 1964), experimental lake manipulations provide evidence of persistent adverse effects of high juvenile abundance on food supplies for sockeye salmon fry of subsequent years. A large fry recruitment from a preceding brood year (or stocking year) can affect growth and survival of subsequent rearing fry through carryover effects on the forage base. Correspondingly, the most reasonable interpretation, from the point of view of the Ricker model, is dramatic changes in the parameter values. In the case of top-down effects, α declines as the predominant age class shifts from age-1 to age-2 and freshwater survival drops, and the Ricker β increases (increased density dependence) as the forage base declines. In bottom-up control *via* nutrient enrichment, α increases as the predominant age class shifts from age-2 to age-1 and survival increases (Kyle 1994), and β decreases (less density dependence). The assumption that zooplankton recover quickly after distress, such as from a large escapement or fry stocking, is not supported for barren lakes that have a plankton population that has not adapted to planktivores (Kyle 1996). Similar limnological mechanisms to those observed in barren lakes may also occur in lakes supporting sockeye salmon where a much higher component of Alaska's total salmon production occurs (Schmidt *et al.* 1996). This process may be important in maintaining cyclic dominance of one or two brood years observed in many sockeye salmon systems. These findings reveal problems with the technique of estimating Ricker parameters using simple regression when small, and possibly even large, data sets are used.

Nutrient additions from fertilizer or carcasses can be effective in reducing top-down control, as observed in Karluk Lake. This implies that Ricker parameters, particularly β, can undergo quick shifts in magnitude by single events or may change slowly when nutrient reserves from the hypolimnion and sediment are recirculated during turnover. More importantly, this reveals problems with using stock-recruit data collected before large interventions, whether through degradation caused by logging, development, earthquake, etc. or through enhancement (fertilization), to describe the stock-recruit relation after the intervention.

Limnological study results are not in conflict with the basic tenet of the stock-recruit model: recruitment is controlled by the basic productivity of the stock together with density-dependent limitations. However, the assumptions that recruitment events are independent relative to freshwater productivity and that accumulation of many such 'independent events' results in a clear picture of the basic spawner-recruit relationship is not supported. The examples demonstrate the effect of both top-down and bottom-up processes on brood year interactions. Understanding the underlying cause for the variation in recruitment of fry is essential when examining brood year interactions. In stock-recruit evaluations (Eggers 1993), these effects would require examination of (1) autocorrelated residuals from brood year interactions when fry from different brood years are rearing together, and (2) the compounding effects of long-term changes in the carrying capacity of the system (Ricker's β) from decreasing carcass nutrients and short-term effects of top-down trophic effects. Because these processes are density dependent, it is not surprising that use of spawner-recruit information alone often provides inconsistent direction to harvest managers. However, we agree with Eggers (1993) that a fixed escapement goal strategy will outperform fixed harvest rates, but only if a broader ecological role of escapements is used in establishing escapement goals rather than relying on short-term spawner-recruit responses. Because density-independent events, such as climatic changes, can precipitate management error (Eggers 1993), major escapement changes under fixed harvest rates or under fixed escapement goals, the resulting ecological effects of carcasses or zooplankton predation on subsequent generations of sockeye salmon production will most likely confound explanations based solely on stock-recruit data.

Unfortunately, in actual practice stock-recruit models are used to the intentional exclusion of lake dynamics in estimating management parameters. Contrary to the assertion that bold manipulations of stock size are needed to determine optimal stock sizes (Walters 1986), we assert that trophic level responses and brood year interaction are integral factors in determining stock size and preventing overtaxing the rearing environment. Experimental studies that rely on adult return data for testing alternative regimes (e.g. Walters *et al.* 1993) may result in decreased harvests for extended periods and often provide ambiguous results. Failure to separate non-density-dependent changes in marine survival from density-related effects on stock-recruit model parameters may lead to incorrect interpretations of adult returns over the short duration of such experiments. Juvenile salmon recruitment and growth can be used in conjunction with limnological data to adjust escapement goals.

Management and forecast errors of a fixed harvest rate policy

Management errors, often difficult to quantify, may be amplified by the selection of harvest policies. Fixed escapement goal policies have several management benefits absent from harvest rate policies. For example, an escapement goal policy depends on a relatively simple and observable measure of success — a fixed number of salmon escaping into a refuge near the natal rivers and lakes. With a fixed harvest rate policy, forecasts become a major factor in determining the escapement level. Such forecasts are notorious for error and cause conflicts in the limited number of systems where they are implemented. Eggers and Rogers (1987) reported that management errors during high returns to Bristol Bay were normally distributed, but often deviated from the targeted exploitation rate by 30%. When runs were less than the escapement goals, management error was higher and always exceeded the targeted harvest rate. This suggests that a fixed harvest rate policy would in reality become depensatory, whereas targeted harvest rates would frequently be exceeded when runs are low. When runs are high, harvest rates would be more dependent upon forecasting errors, allocation concerns, or overharvest of weak stocks in mixed stock fisheries. Although, on average, management harvest objectives would be achieved, failure to anticipate large or weak runs would result in errors that correlate with run strength. Fixed escapement goal policies produce these types of management errors; however, fixed harvest rates would be expected to have even greater errors because of more dependence on forecasts and data provided by fishermen and processors.

Under fixed escapement goals, the fishermen and processors that report the catch have no short-term incentive to mis-report harvests, but under a fixed harvest rate policy mis-reporting of catch can create short-term financial gains when managers are led to overestimate stock abundance. For example, over-reporting harvest would provide for an overly optimistic assessment of run strength, thus increasing the harvest rate. In other situations the fleet might under-report harvest to create the impression that harvest rates are low. Regulatory response of management agencies caused by mis-reporting would create a need for large-scale observer programmes, similar to those being developed for regulating fixed harvest rate fisheries on groundfish in the Gulf of Alaska and Bering Sea.

The use of economic objectives to manage the salmon fishery

Is a policy that attempts to maximize average harvest appropriate for the Alaskan salmon industry? In economic and decision theory, analysts often use the notion of utility (e.g. Berger 1980) as the quantity to be maximized. The basic idea is that after a fisherman has made enough money to cover expenses, the relative usefulness or utility of more money is not as high as it was before expenses were made. The utility is very high for the first quantity of money earned; without it, the fisherman will go out of business. In an above-average year, fishermen earn enough money above operational expenses to allow for discretionary expenses. Because the need for those first few dollars is much higher than the need for the final dollars of above-average earnings in an above average year, the fisherman is said to have a risk averse utility function of money.

Deriso (1985) found that a harvest rate policy maximizes average logarithm of catch when compared with constant escapement policies. This finding was based on using the logarithm of catch as a surrogate for risk averse utility functions of harvest. Average log catch is higher with a fixed harvest rate because the constant harvest rate policy increases variability in escapement and reduces variability in catch. By considering the logarithm of catch, years of a large catch are given much less influence in the risk analysis objective function than they are when considering average catch.

We contend that use of the logarithm of catch is not a useful tool for assessing economic risks in salmon fisheries in Alaska. Salmon fishermen in Alaska usually target multiple stocks of sockeye salmon in addition to taking significant harvests of pink, chum, and coho salmon. The season is short, lasting at most a few months, which allows time for many other economic opportunities aside from salmon fishing. Deriso (1985) concluded the 'fixed escapement goal policy is most appropriate when the particular fishery makes up a small portion of the average income, and the price and costs are relatively unaffected by volume of harvest.' Although the proportion of income varies significantly among the many participants in the sockeye salmon fishery, very few are dependent upon a particular stock, and virtually all have alternative species of salmon to exploit. Frequently, Alaskan salmon fishermen also participate in herring, crab, and other groundfish commercial fisheries. Because of the diversity of alternative resources, an economic subject beyond the scope of this paper, the net effect is stabilization of annual income. The risk of a salmon stock failure applies often to only land-based set gill-net sites, which are frequently low-capital operations with minimal fixed costs. Vessel-based fishermen not only have multiple sockeye salmon systems to exploit, but they also exploit other species (e.g. halibut, crab, groundfish). All of these factors reduce the economic risk to fishermen, processors, and communities; concurrently, long-term productivity increases from fixed escapement goals in sockeye salmon fisheries are realized. We contend the real risk of stock failures is biologically-based rather than economically-based. Risks of stock failures are clearly illustrated by many examples of overfishing around the world in which managers argue about the size of surplus production in order to justify conservation actions.

Harvest rate versus escapement goal management policies

The success of Alaskan sockeye salmon fisheries cannot be isolated to a particular harvest policy employed or to a mathematical formula. Any harvest policy will fail if political support is not forthcoming; lack of funds and local support for fisheries managers and law enforcement renders any harvest policy non-functional. In this regard, escapement goal policies have considerable advantages: (1) escapement numbers are collected independent of the fishery, (2) they provide a very rapid and easily understood information source that the public can use to judge the effectiveness of conservation actions, and (3) they are based on an understood concept of 'crop failure'; fisheries closures are understood as necessary to 'seed' the streams for future generations. The logic of harvesting 70% of the salmon runs, regardless of abundance (as is the case with fixed harvest rate policies), defies the common sense of most members of the fishing community.

The fishery arguments for a constant harvest rate policy rest on notions that increased variation in stock size increases the value of stock-recruitment information. The economic arguments rest on the notion that decreased variation in the catch increases value. We have concluded that stock-recruit data have been overvalued, regardless of management policy, and the greatest risk to fishery management is the loss of control of harvest and escapement, which will lead to stock declines.

All of the biological and economic risks associated with variability of stock size are poorly understood; however, in the few cases that have been studied, not only do persistently large (or small) escapements result in adverse trophic level changes, but more importantly, these changes alter the usefulness of stock-recruitment data in the future. While we know of dozens of studies that use Ricker analysis, we know of few critical examinations of this procedure. Koslow (1992) provides a discussion of the intricacy of stock-recruitment relationships and explains why such relationships cannot be defined with historical data in many circumstances, even when the stock size is intentionally varied. The incorporation of environmental variables to improve stock-recruit analysis has been reviewed and discussed by Caputi (1993). However, current models generally assume the environmental variables are independent of spawner biomass and are not affected by them, such as nutrient enrichment by fish carcasses as we have observed in Karluk Lake.

Deriso (1985) found harvest rate policies are preferred when 'we are relatively certain about the intrinsic productivity rate of a fish stock but uncertain about the size of carrying capacity.' In the case of sockeye salmon management we are usually uncertain about the intrinsic productivity, which apparently changes in response to poorly understood large-scale ocean phenomena. Alternatively, a lake's carrying capacity can be approximated from limnological or habitat information (Geiger and Koenings 1991) and is a more reliable source of stock productivity.

CONCLUSIONS

(1) Even if a system strictly follows the classical Ricker model, stock-recruit analysis for Pacific salmon can lead to poor management decisions about optimum stock size.

(2) Harvest policies interact with the fundamental parameters of stock and recruitment, so that information gained under one policy may not be relevant to management under another.

(3) Lake-specific limnological and habitat information are critical to establish escapement goals.

(4) Extremes in escapements can reduce the future productivity of a system.

(5) A constant harvest rate policy would be much harder and expensive to implement, would create incentives for

deception and false reporting, would probably result in much greater management error, and could create depensatory forces on the stock.

(6) The economic risk-averse benefits of constant harvest rate policies are not relevant to salmon fisheries in Alaska.

The late P. A. Larkin (1977) was on the forefront of fisheries professionals to expound on the obligations of fisheries managers to first ensure the stock and stock habitat are conserved, followed closely by economic and social concerns. Alaska's salmon escapement goal management, when properly implemented, clearly aligns stock and stock habitat above economic and social concerns and when integrated into a locally based management system coupled with limited entry, provides a positive model of a sustainable fishery.

References

Beamish, R J, and Bouillon, D R (1993). Pacific salmon production trends in relation to climate. *Canadian Journal of Fisheries and Aquatic Sciences* **50**, 1002–16.

Berger, J O (1980). 'Statistical Decision Theory and Bayesian Analysis.' Springer-Verlag. New York. 617 pp.

Burgner, R L (1964). Factors influencing production of sockeye salmon *(Oncorhynchus nerka)* in lakes of southwestern Alaska. *Verh. Internat. Verein. Limnol.* **15**, 504–13.

Caputi, N (1993). Aspects of spawner-recruit relationships, with particular reference to crustacean stocks; a review. *Australian Journal of Marine Freshwater Research* **44**, 589–607.

Deriso, R (1985). Risk averse harvesting strategies. In 'Resource Management, Proceedings of the Second Ralf Youque Workshop'. (Ed M Mangle). Lecture notes in biomathematics, Number 61. (Springer-Verlag: New York).

Eggers, D M (1993). Robust harvest policies for Pacific salmon fisheries. In 'Proceedings of the International Symposium on Management Strategies for Exploited Fish Populations'. (Eds G Kruse, D M Eggers, R J Marasco, C Pautzke and T J Quinn II) pp. 85–106. *Alaska Sea Grant College Program Report* No. 93-02. (University of Alaska Fairbanks).

Eggers, D M, and Rogers, D E (1987). The cycle of runs of sockeye salmon (*Oncorhynchus nerka*) to the Kvichak River, Bristol Bay, Alaska: Cyclic dominance or depensatory fishing? In 'Sockeye Salmon (*Oncorhynchus nerka*) population biology and future management'. (Eds H D Smith, L Margolis and C C Wood) pp. 343–66. *Canadian Special Publication of Fisheries and Aquatic Sciences* **96**.

Foerster, R (1968). The sockeye salmon *Oncorhynchus nerka*. *Bulletin Fisheries Research Board Canada* **162**, 422 pp.

Geiger, H J, and Koenings, J P (1991). Escapement goals for sockeye salmon with informative prior probabilities based on habitat considerations. *Fisheries Research* **11**, 239–56.

Goodlad, J C, Gjernes, T W, and Brannon, E L (1974). Factors affecting sockeye salmon *(Oncorhynchus nerka)* growth in four lakes of the Fraser River system. *Journal of the Fisheries Research Board Canada* **31**, 871–92.

Hartman, W L, and Burgner, R L (1972). Limnology and fish ecology of sockeye salmon nursery lakes of the world. *Journal of the Fisheries Research Board Canada* **29**, 699–715.

Johnson, W E (1965). On the mechanism of self-regulation of population abundance in sockeye salmon (*Oncorhynchus nerka*). *Mitteilungen der Internationalen Vereinigung für theoretische und angewandte Limnologie* **13**, 66–87.

Koenings, J P, and Burkett, R D (1987). The population characteristics of sockeye salmon (*Oncorhynchus nerka*) smolts relative to temperature regimes, euphotic volume, fry density, and forage base within Alaskan lakes. In 'Sockeye salmon (*Oncorhynchus nerka*) population biology and future management'. (Eds H D Smith, L Margolis and C C Wood) pp. 216–34. *Canadian Special Publication of Fisheries and Aquatic Sciences* 96.

Koenings, J P, and Kyle, G B (1997). Collapsed populations and delayed recovery of zooplankton in response to heavy juvenile sockeye salmon (*Oncorhynchus nerka*) foraging. 'Proceedings of the international symposium on biological interactions of enhanced and wild salmonids'. *Canadian Special Publication of Fisheries and Aquatic Sciences* (in press).

Koslow, J A (1992). Fecundity and the stock-recruitment relationship. *Canadian Journal of Fisheries and Aquatic Sciences* **49**, 210–17.

Kyle, G B (1994). Nutrient treatment of three coastal Alaska lakes: trophic level responses and sockeye salmon production trends. Alaska Department of Fish and Game, *Alaska Fisheries Research Bulletin* 1(2), 153–67.

Kyle, G B (1996). Stocking sockeye salmon (*Oncorhynchus nerka*) in barren lakes of Alaska: effects on the macrozooplankton community. *Fisheries Research*. **28**(1), 29–44.

Kyle, G B, Koenings, J P, Barrett, B M (1988). Density-dependent, trophic level responses to an introduced run of sockeye salmon (*Oncorhynchus nerka*) at Frazer lake, Kodiak Island, Alaska. *Canadian Journal of Fisheries and Aquatic Sciences* **45**(5), 856–67.

Larkin, P A (1977). An epitaph for the concept of maximum sustained yield. *Transaction of the American Fisheries Society* **106**(1), 1–11.

Quinn, T J, and Marshall, R P (1989). Time Series analysis: quantifying variability and correlation in SE Alaska salmon catches and environmental data.. In 'Effects of ocean variability on recruitment and an evaluation of parameters used in stock assessment models'. (Eds R J Beamish and G A McFarlane) pp. 67–80. *Canadian Special Publication of Fisheries and Aquatic Sciences* 108.

Ricker, W E (1975). Computation and interpretation of biological statistics of fish populations. *Bulletin of the Fisheries Research Board of Canada* 191. 382 pp.

Rieman, B E, and Meyers, D (1990). Status and analysis of salmonid fisheries: kokanee population dynamics. Idaho Department of Fish and Game, Job Performance Report, Project Number F–73–R–12, Boise, Idaho.

Rogers, D E (1980). Density-dependent growth of Bristol Bay sockeye salmon. In 'Salmonid ecosystem of the North Pacific'. (Eds W J McNeil and D C Himsworth) pp. 267–83 (Oregon State University Press, Corvallis, Oregon).

Roppel, P (1986). 'Salmon from Kodiak: an history of the salmon fisheries of Kodiak Island, Alaska.' Alaska Historical Commission Studies in History No. 216, Anchorage. 355 pp.

Rounsefell, G A (1958). Factors causing decline in sockeye salmon of Karluk River. *Alaska Fisheries Bulletin* **58**, 83–169.

Royce, W F (1989). Managing Alaska's salmon fisheries for a prosperous future. *Fisheries* **15**(2), 8–13.

Schmidt, D C, Tarbox, D E, King, B E, Brannian, L K, Kyle, G B, and Carlson, S R (1996). Kenai River sockeye salmon: an assessment of overescapements as a cause of the decline. In 'Exxon Valdez Oil Spill Symposium Proceedings'. (Eds S D Rice, R B Spies, D A Wolfe and B A Wright) pp. 604–8. *American Fisheries Society Symposium* Number 18.

Schnute, J T (1991). The importance of noise in fish population models. *Fisheries Research* **11**(3–4), 197–224.

Smith, A D, and Walters, C J (1981). Adaptive Management of Stock Recruitment Systems. *Canadian Journal of Fisheries and Aquatic Sciences* **38**, 690–703.

Stockner, J G (1987). Lake fertilization: the enrichment cycle and lake sockeye salmon (*Oncorhynchus nerka*) production. *Canadian Special Publication of Fisheries and Aquatic Sciences* **96**, 198–215.

Walters, C J (1985). Bias in the estimation of functional relationship from time series data. *Canadian Journal of Fisheries and Aquatic Sciences* **42**, 147–9.

Walters, C J (1986). Adaptive management of renewable resources. Macmillan Publishing: New York. 374 pp.

Walters, C J (1987). Nonstationarity of production relationships in exploited populations. *Canadian Journal of Fisheries and Aquatic Sciences* **44**(2), 156–65.

Walters, C J, and Collie, J S (1988). Is research on environmental factors useful to fisheries management? *Canadian Journal of Fisheries and Aquatic Sciences* **45**, 1848–54.

Walters, C J, and Parma, A M (1996). Fixed exploitation rate strategies for coping with effects of climate change. *Canadian Journal of Fisheries and Aquatic Sciences* **53**, 148–58.

Walters, C J, Goruk, R D, and Radford, D (1993). Rivers Inlet sockeye salmon: An experiment in adaptive management. *International American Journal of Fisheries Management* **13**, 253–62.

Ward, F J, and Larkin, P A (1964). Cyclic dominance in Adams River sockeye salmon. *International Pacific Salmon Fisheries Committee Progress Report* No. **11**. 116 pp.

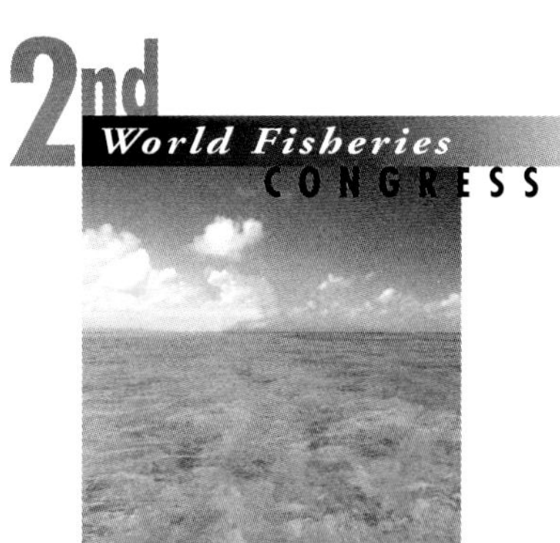

Managing Namibia's Fisheries Sector: Optimal Resource use and National Development Objectives

P. R. Manning

Development Studies Institute, London School of Economics and Political Science, London WC2A 2AE, UK.

Summary

Namibia's fisheries, with their high levels of primary production, have experienced major developments since independence in 1990. The Constitution underpins a policy commitment to the restoration of fish stocks and to their sustainable utilization for the benefit of Namibians. During the last six years a new fisheries regime has been established. In 1993 new rights of exploitation were granted, a requirement for receiving a quota in Namibia's TAC and quota system. New Namibian companies were offered the opportunity of entry into the industry and the fisheries sector has become one of the main contributors to economic growth. The paper reviews these developments.

It then examines the catch histories of the pilchard and hake stocks, the two most important of Namibia's commercial stocks in order to highlight difficulties occurring in the management of Namibia's fisheries. Finally the paper focusses on two issues: the capacity of the economic structure of the industry to accommodate ecosystem demands and the ownership and use of resource rents.

Introduction

Namibia became an independent state in March 1990 and declared its exclusive economic zone (EEZ) three months later. The Benguela ecosystem, on which the Namibian fisheries are based, is one of the world's four major eastern ocean boundary upwelling systems. It has been, and potentially is, very rich in fisheries resources but, at independence, Namibia inherited its most important commercial stocks in a depleted state. Namibia has in recent years won acclaim in the fisheries media for sound fisheries management policies and practice. Yet problems have arisen in the management of the fisheries through recent environmental perturbation.

This paper focusses on the post-independence management of Namibia's fisheries. It will first review the objectives of Government for use of the fisheries and the management system for the sector, then it will examine targets which the Ministry of Fisheries and Marine Resources (MFMR) has set itself so as to meet those objectives. Finally it will focus on two issues essential for long-term sustainability and equity: accommodating the need for maintaining ecosystem resilience and its implications for economic management, and the related issue of the generation, determination and allocation of resource rents.

THE NEW FISHERIES REGIME

Objectives of government policy

The Namibian Constitution *inter alia* requires of the State 'that ecosystems, essential ecological processes and biological diversity are maintained and living natural resources are utilized on a sustainable basis for the benefit of Namibians, both present and future'. (Art. 95 (l)).

A White Paper, entitled 'Towards the Responsible Development of the Fisheries Sector' (Anon. 1991) articulates the policy for the development of the fisheries sector reflecting this constitutional requirement. Two overwhelmingly significant objectives immediately become apparent. The first is effectively to address the serious depletion of several species which took place prior to independence. The second is to maximize benefits for Namibians from the fisheries sector both in the harvesting of fish and in the processing industry.

The new rights of exploitation

When existing rights granted prior to independence terminated in December 1993, the decision was made to grant rights for periods of ten, seven and four years from January 1994, depending on a pre-announced set of criteria. Only companies holding a right of exploitation may be granted a quota, which would not be transferable without permission of the Minister.

The Sea Fisheries Act and the Sea Fisheries Regulations set out criteria to be used in determining who was to be granted rights of exploitation. While the need for technical and financial competence are requirements, the criteria also focus on the degree of Namibian ownership and control and, notably, on 'the advancement of persons in Namibia who have been socially or educationally disadvantaged by discriminatory laws or practices which have been enacted or practised before the independence of Namibia' (Anon. 1993). Fisheries was thus to be used to address the unjust legacy of the apartheid system in Namibia.

A policy document on the granting of rights of exploitation supplementing the White Paper was published in June 1993 and applications were invited for new rights of exploitation in June 1993. During the next three months the Ministry received 565 applications for rights of exploitation from 316 applicants. Of the 165 rights granted to 124 enterprises, around 90 of these were to companies which had entered the industry since independence in 1990. Around 30 of the existing rights were not renewed. Overwhelmingly the companies had a majority Namibian ownership (Konkondi 1994). This statement, however, requires qualification.

With the announcement of the new Fisheries Policy in 1991 there were major changes in the formal ownership of many of the established companies, particularly where the companies were foreign-owned. In some instances, there was a direct sale of interests to Namibian concerns. In other cases a majority of shares was sold to Namibian shareholders but with management contracts which have enabled foreign enterprises to retain control of the 'Namibianized' companies while at the same time being able to claim that they qualify as 51% Namibian owned.

Typically, many new companies in the industry, holding non-transferable quotas and without adequate collateral, have not been able to raise finance from the financial institutions to purchase vessels or establish processing plants. They have subsequently landed in the hands of the larger established companies. In some instances, smaller companies have had to lease their year's quotas to the larger operators as the only means of having their quota caught. In principle, these companies aim to accumulate capital in order to enable them to become operational and the Ministry expects this of them (Angula 1993). Others have found themselves locked into five-year contracts with the larger companies to deliver fish exclusively to them for processing and marketing, at relatively low prices over which they have little choice. In other instances there have been buy-outs of a majority of shares of new companies by the established companies.

Although the Ministry explicitly rejected the introduction of Individual Transferable Quotas (ITQs) for equity reasons (Konkondi 1994), buying a company whose only substantial asset is that it holds a quota is much the same, in effect, as buying a quota. It may be argued that unofficially an ITQ type system is developing although it does not deal with a permanent asset, and it remains to be seen whether the Ministry will use its statutory powers to withdraw quotas from companies which no longer fulfil the criteria for which they were granted quotas in the first instance.

A total allowable catch (TAC) and quota system operates for the most important commercial species, and a system of licences restricting fishing effort operates for the rest. The objective of the Ministry is to move towards vessel quotas with the aim of discouraging the development of over-capacity in vessels. It introduced vessel quotas into the hake fishery in 1994. Once a company accepts a quota it becomes responsible for the payment of a quota fee whether or not the fish are caught. Quota fees are set on a sliding scale to encourage Namibianisation of the industry and the development of land-based processing facilities.

Targets for the Ministry

The MFMR in 1992 established nine targets reflecting Government policy objectives for the sector (Anon. 1994*a*). These can be grouped to reflect the ecosystem objectives and the socio-economic objectives of the Government's policy.

The ecosystem targets are:

- to rebuild the depleted stocks to full potential, specifically 'hake and other demersal species, pilchard and lobster', all of which were considered to be depleted in 1992.
- to ensure that rebuilt stocks are maintained at a biomass level 'that can support maximum sustainable yields'.
- to establish a biodiversity baseline for both marine and freshwater species in Namibia and monitor changes in species composition with the objective of ensuring that no species present in Namibian waters should become endangered.

The socio-economic targets are:

- that fishing and fish processing should contribute 13% of gross domestic product (GDP) at factor cost by the year 2000, increasing the contribution of fisheries from a 1992 level of 3.3%. A preliminary estimate indicates that fisheries contributed 7.6% to GDP in 1995 (MFMR pers. comm. June 1996).
- to treble the value of fish exports, which stood at $US178 million in 1992, to about $US535 million by the year 2000. In 1994 the value of fish exports reached $US360 million (Anon. 1996), which was 27% of the value of total exports.
- to double the contribution to state revenue of quota fees and related charges from the 1992 estimate of $US20.2 million to about $US41 million. In 1993 the quota levy, the research levy and licence fees contributed $US30 million, but this dropped, due to the decrease in catch, to $US27 million in 1994.
- to increase Namibian ownership, with a more specific target to be set in 1994. In the event, no target was set because 'appropriate measures for Namibianization could not be developed at the time' (Anon. 1994*a*).
- to increase employment in onshore processing with the objective of having 15 000 people employed in the sector by the year 2000. By 1994 employment had more than doubled in the sector since 1991, rising to 13 600 employed with the projection that by the year 2000 the number employed would reach 23 000 (MFMR pers. comm. 1996).
- to aim at having Namibians fill 95% of posts for seagoing certified personnel by the year 2004. To this end a training programme for skippers and engineers has been established.

This set of targets at first sight appears to be very ambitious but, given the potential of the resource in relation to the size of Namibia's population of 1.5 million people and the size of the economy, it is not unrealistic, providing that stocks do indeed recover as forecast. It should be noted that the total catch in Namibian waters in 1994 of 639 000 t is still well below the pre-independence estimated total catch for 1989 of 1.3 million t from what are now Namibian waters (Anon. 1991).

A synopsis of the catch histories of pilchard (*Sardinops ocellatus*) and hake (*Merluccius capensis* and *M. paradoxus*, two species treated as one for purposes of catch statistics) puts into perspective the state of the fisheries and the expectations of the Namibian Government. These are two of the most important commercial stocks for Namibia. Together they contribute more than two-thirds of the total value of the catch. Both stocks are at historically low levels.

Figure 1 illustrates the catch history for pilchard. Catches rose to nearly 1.4 million t in 1968, collapsed and then recovered slightly in the early 1970s, only to collapse completely by the end of the 1970s. The biomass has never recovered to a level remotely close to its former size. In 1991, the first full year of post-independence management, the TAC was set low at 60 000 t. Good recruitment during the following two years helped stocks to improve considerably and increases were made to the TAC. Environmental perturbation linked to the cyclical El Niño Southern Oscillation (ENSO) phenomenon began in 1993 and continued through to 1995. The pilchard biomass appeared to drop sharply in late 1994 reaching extremely low levels. Ministry scientists advised setting the TAC for pilchard for 1995 as close to zero as possible; in the event, the TAC was reduced from 125 000 t in 1994 to 40 000 t in 1995 but not lower for short-term economic reasons.

The hake catch off Namibia grew rapidly from the start of the fishery in 1964 to an all-time high in 1972 when some 800 000 t were reported caught (Fig. 2). After reaching a low point in 1980 with a reported catch of 169 000 t (International Commission for the Southeast Atlantic Fisheries, Statistical Bulletins), it increased again in the early 1980s and then declined during the latter part of the decade. The catch in 1989 of 326 000 t was reported to have been 83% juveniles (Anon. 1990). Following the declaration of an exclusive economic zone in 1990, the Namibian Government set a relatively low 1991 TAC of 60 000 t for hake. The objective was to allow the stock to rebuild and to establish the industry in Namibia instead of it being dominated

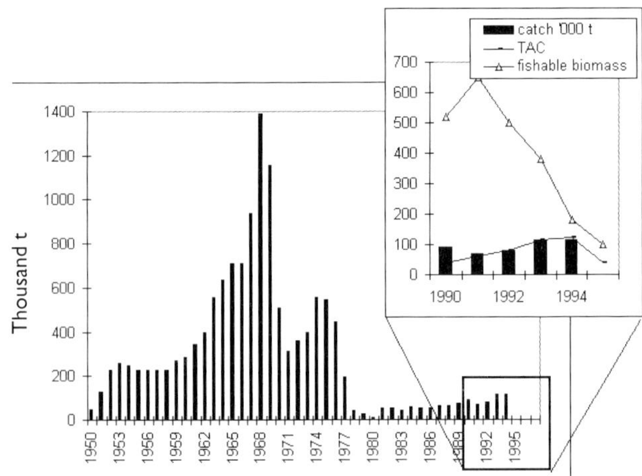

Fig. 1. Pilchard catch; inset post-independence catch fishable biomass and TAC.

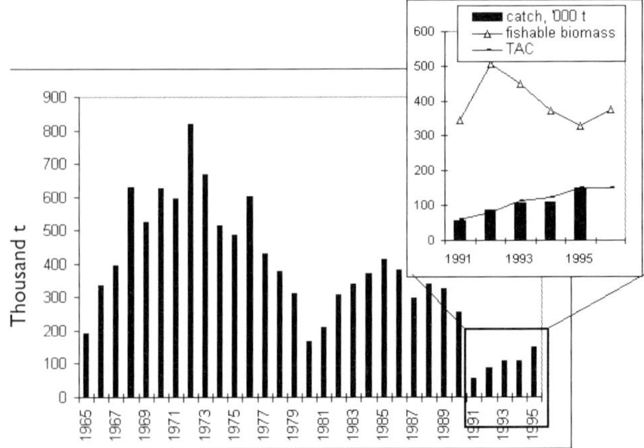

Fig. 2. Hake catch; inset post-independence catch, fishable biomass and TAC.

by foreign freezer trawler fleets which for years had exploited the stock. During the 25 years the fishery had existed before independence, it was effectively unregulated. During this period 10.5 million t of hake, valued at an estimated $US6 × 10^9 at 1992 prices, were reported to have been taken from Namibian waters almost entirely by foreign fleets (Stuttaford 1994). Namibia gained virtually nothing of this vast sum.

The uncertainty that may surround the condition of the biomass of the hake fishery emerged through a debate between MFMR scientists, and consultants employed by the industry. MFMR scientists argued that the fishable biomass in February 1996 was about 380 000 t, a little improved on the previous year. This was based on swept area surveys and analysis of catch and effort data from the commercial fleet. In contrast, consultants for the industry contended that the fishable biomass was in fact between 2 million and 3.5 million t and that 'TAC levels in excess of 200 000 t are therefore advised' (OLRACcc, Namibia Hake Association 1996). The hake TAC was set at 150 000 t for 1996, twice the level of 75 000 t recommended by the MFMR scientists for optimal stock recovery.

The above socio-economic targets assume stock recovery to a point where optimal exploitation can take place by the year 2000, an assumption which is perhaps not realistic. Trebling the export value of fish and fish products, doubling the contribution to state revenue from quota and related fees, and increasing employment by the year 2000, is based on a projection of the recovery of the fish stocks to a significantly higher level by that date and would be feasible only if that were to happen. This, in effect, links the socio-economic targets to the ecosystem targets by setting a timetable for the recovery of the commercial stocks.

While the ecosystem targets are broadly admirable, setting ambitious economic targets generates expectations of ecosystem performance which may be unrealistic. The difficulty is that the fishing industry is entirely reliant on a resource base of which the dynamics are still little understood.

The above raises more questions than are practical to reflect on in this paper. Comment will thus be confined to the ecosystem-related issue of maintaining resilience, and the socio-economic question of abnormal profits or resource rents not collected by the state.

OPTIMAL RESOURCE USE AND NATIONAL DEVELOPMENT OBJECTIVES

Maintaining ecosystem resilience

Survival and growth of a fish stock will vary, *inter alia*, according to the size of the biomass, the salinity and temperature of the water, the prevailing currents, the numbers and feeding habits of other species, both predators and competitors for food, the amount of radiated solar energy, the supply of mineral elements and the rate of photosynthesis. The communities of organisms in a specified environment, depending on and interacting with this environment and with each other, are an ecosystem (Laevastu and Larkins 1981). Namibia's fisheries arise from the Benguela ecosystem and, as with other ecosystems, it is not closed but is part of a vast global system which envelops our planet. Environmental influence on species is complex; the complexity is compounded by influences considered external to the ecosystem. This happens with the Benguela ecosystem. About 12–18 months after the Indo-Pacific El Niño Southern Oscillation (ENSO) phenomenon begins, the south-easterly winds subside off Namibia, the upwelling process consequently weakens, sea surface temperatures rise and the Benguela warm water event is underway (Anon. 1994*b*).

The primary scientific focus has been on the most important commercial species individually. Historically there has been too little recognition of the ecosystem in which they are found. The risk of this approach is not to recognize the importance of ecosystem resilience, 'the ability of a system to maintain its structure and patterns of behaviour in the face of disturbance' (Holling 1986). If the resilience of the ecosystem is lowered by excessive fishing, then the ecosystem could flip to a new equilibrium in which previously abundant species may have only a minor part (Laevastu and Larkins 1981). While caution is needed and more vigorous application of the precautionary principle advisable, evidence does exist that for most major stocks recovery is possible (Myers *et al.* 1995) but it should not be taken for granted. Decisions on the TAC for pilchard and hake have been influenced by wishing to minimize current economic damage to the industry, an entirely legitimate objective. It remains important, however, to ensure that resilience is maintained, as the economist Kenneth Arrow and others asserted in their extensively noted article in Science, 'even though the limits on the nature and scale of economic activities thus required are necessarily uncertain' (Arrow *et al.* 1995).

Accommodating the variability of the resource requires that the economic system also be resilient. A pilchard TAC of 40 000 t instead of zero was set in 1995 because the industry was not considered resilient enough to withstand the shock of a zero TAC. Capacity to adjust to changes in the availability of the resource requires strictly limiting vessel and processing capacity. It may also require increased research into ways in which processing plants may utilize other species, such as the robust horse mackerel stock in the absence of pilchard.

Resource rent

The second focal point of this paper relates to abnormal profits or the resource rent of the fishery, that is the difference between the price of a unit of a good produced using a natural resource and the unit costs of turning that natural resource into the good (Hartwick and Olewiler 1986; Pearce and Turner 1990).

In a world where global total costs of the fishing industry are estimated at $US124 × 10^9, and global revenues only $US70 × 10^9 (FAO 1992), it is widely recognized that potential resource rent is massively wasted on overcapacity. Article 56 (1)(a) of the United Nations Convention on Law of the Sea (1982) grants to the coastal state the right to explore, exploit, conserve and manage its fish stocks. For practical purposes, this has been widely regarded as having granted to the people of coastal states a type of collective ownership of the resources (Hannesson 1993). This resource rent thus belongs to the country as a whole. In the case of Namibia the right is implicitly acknowledged in the socio-economic targets above, the objectives articulated in the White Paper and in the Constitution.

In many fisheries where there has been some measure of success in reducing or eliminating overcapacity, much of the resource rent generated has not been collected by the state. Relative to the performance of many other states, Namibia has been quite successful in capturing resource rent.

Namibia earns considerable net revenue from its fisheries. Revenue from quota fees alone for the financial year 1994/5 amounted to $US30 million while the running costs of the entire Ministry, including the inspectorate, amounted to $US9.6 million for the same financial year (1994/5 Budget, Government of Republic of Namibia (GRN)). (There are both capital costs and operating costs covered by development aid not included in these figures, but these do not invalidate the basic picture). In addition to quota fees, revenues include fuel tax, the research levy, licence fees for vessels and company and income tax generated. Although the Namibian Government is managing to recover a significant portion of the resource rent, generating much needed revenue for purposes of development, there is still considerable resource rent, or abnormal profits, accruing to the industry.

Practical difficulties exist in accurately determining the level of resource rent accruing from any particular species. Many factors, such as the spatial dispersion and size of the biomass which vary over time, will influence the resource rent that may be associated with a particular stock. The state has offered rebates on quota levy as a means of encouraging the establishment of onshore processing in Namibia and rebates to encourage the development of a Namibian-owned and controlled industry. The Government does not collect all the resource rent available but, in effect, offers subsidies in the form of uncollected resource rents for certain purposes. These sums can be quite considerable. Although quotas are not officially transferable in Namibia, companies may lease out quotas on an annual basis. In 1995 hake quota for the year was leased out at up to $US233/t. Larger companies seek out additional quotas, indicating that they are earning at least normal profits. This provides an approximation of the uncollected resource rent for hake which is not being dissipated by overcapacity or other inefficiencies. These subsidies granted to the industry may be justified in terms of the development objectives of increasing employment, Namibianization of the industry and establishing new industries destined to target export markets. But there is no particular reason why fisheries should necessarily receive those subsidies; other economic activity may well be more deserving.

However, accountability by the industry for the use of those subsidies requires that their transparency needs to be improved. If the rent not collected for the fisheries resources could be more explicitly acknowledged as a subsidy, the transparency and rationality of resource allocation would improve. This may require collection of a larger portion of the resource rent through quota levies and the explicit allocation of subsidies to the industry where appropriate.

Conclusion

The focus of this paper has been on significant difficulties that have occurred in the Namibian fisheries sector. It would be disingenuous, however, not to emphasise the context of very considerable achievements by the Namibian MFMR, only some of which have been referred to here. In a few short years it has put an end to a free-for-all taking place off its coast, established a creditable measure of control of the fisheries and now earns net revenue from this resource.

The Ministry has committed to scientific research considerable extra resources, generated largely through a ‘research levy’ on catch, in order better to understand its fisheries resources and their environment. It has made good use of foreign technical assistance and has established institutions that considerably improve the collection and analysis of data, consultation with the industry and its involvement in scientific discussion.

The point remains, however, that ecosystem health is the foundation on which the industry is built. The resilience of the integrated ecological and economic systems is essential to make it possible for the industry and the economy generally to absorb the shocks which the ecosystem might produce; otherwise short-term economic considerations may produce irresistible pressure to risk ecosystem resilience.

References

Angula, H. (1993). Address to the ‘Fisheries Newcomers Conference’, 22 September 1993.

Anon. (1990). Resource Survey 1990. Interim Report Ministry of Fisheries and Marine Resources. Government of the Republic of Namibia.

Anon. (1991). Towards the responsible development of the fisheries sector. White Paper. Government of the Republic of Namibia.

Anon. (1993). Sea Fisheries Regulation, Government of the Republic of Namibia, Government Gazette No 566.

Anon. (1994*a*). Activities and the State of the Fisheries Sector, 1993 and 1994. MFMR Report.

Anon. (1994*b*). Indian Ocean may have El Niño of its own. *Eos, Transactions, American Geophysical Union*, 585–6.

Anon (1996). *Economist Intelligence Unit*, Quarterly Report, 1st Quarter 1996.

Arrow, K, Bolin, B, Costanza, R, Dasgupta, P, Folke, C, Holling, C S, Jansson, B O, Levin, S, Mäler, K G, Perrings, C, and Pimentel, D (1995). Economic Growth, carrying capacity, and the environment. *Science* **268**, 520–21.

Food and Agricultural Organization (1992). ‘State of Food and Agriculture.’ (FAO: Rome).

Hannesson, R (1993). ‘Bioeconomic Analysis of Fisheries.’ (Fishing News Books: Oxford).

Hartwick, J M, and Olewiler, N D (1986). ‘The Economics of Natural Resource Use.’ (Harper and Row: New York).

Holling, C S (1986). The resilience of terrestial ecosystems: local surprise and global change. In ‘Sustainable Development of the Biosphere’. (Eds W C Clark and R E Munn) pp. 292–317 (Cambridge University Press: Cambridge).

Konkondi, R(1994). Nambianised — how development has replaced the last great coastal free-for-all. *Fishing News International* **34**, 24–7.

Laevastu, T, and Larkins, H A (1981). ‘Marine Fisheries Ecosystem.’ (Fishing News Books: Farnham, Surrey, England).

Myers, R A, Barrowman, N J, Hutchings, J A, and Rosenberg A A, (1995). Population Dynamics of Exploited Fish Stocks at Low Population Levels. *Science* **269**, 1106–8.

OLRACcc, Namibia Hake Association (1996). Further Scientific Insight into the season TAC revision in March 1996. 12 pp.

Pearce, D W, and Turner, R K (1990). 'Economics of Natural Resources and the Environment.' (Harvester Wheatsheaf: London).

Stuttaford, M (1994). Recovering from the near collapse of a rich resource. *Namibia Brief* **18**, 12–17.

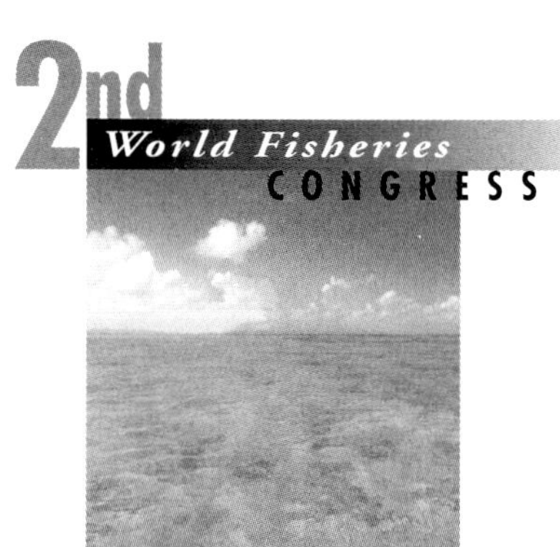

'No-take' Marine Reserve Networks Support Fisheries

W. J. Ballantine

University of Auckland, Leigh Marine Laboratory, Box 349, Warkworth, New Zealand.

Summary

'No-take' marine reserves offer a new and additional form of fisheries management. Over the past 20 years New Zealand has demonstrated that such reserves are practical and provide valuable support for science, education, recreation, conservation and social economics. A representative network is now the official aim. This is an opportunity for fisheries management to upgrade its aims and avoid the traps of data-dependent, stock-specific management. While retaining detailed management for particular fisheries, the provision of a 'no-take' network would inject a new level of decision based on general principles — including ecosystem dynamics, the overall public interest, and the need for insurance against human error and natural disasters. Recent history strongly suggests this move would be actively supported by the general public.

Introduction

'No-take' marine reserves do not immediately appeal to most fisheries scientists and managers. They have to concentrate their efforts at the level of particular stocks. Although managers often have problems, what happens to other stocks is not strictly their business. When making recommendations for action and regulation, they are stock-specific. A detailed analytical approach to management is necessary, for scientific and economic reasons, but it may not be enough. Data-based management is certainly required, but is it sufficient? Could we add other management approaches, with different advantages and problems? In particular, could we stop making all our decisions on precise data, and act on basic principles for a proportion of the business?

This is commonly done. In most human affairs, accidents, error and large variations in natural forces are so frequent as to require active counter-measures. Such measures cannot, by definition, be based on specific knowledge. But practical and efficient systems for this purpose are common and widespread. One of the few areas where they are largely absent is fisheries management.

It is my contention that a network of 'no-take' marine reserves would be a practical and effective addition to current fisheries management. The main barrier to the adoption of this additional approach is the same as its main practical advantage — it is based on general principles, not on specific and detailed knowledge.

THE PRINCIPLES FOR 'NO-TAKE' MARINE RESERVES

Most forms of fisheries regulation have to be justified by detailed data on the likely effects, and require monitoring to see if these occur. Furthermore the justification is concerned with the effects on human activities — which also change. In contrast, an outright ban on any form of extraction can shift the justification to principles, and the basic concern to effects on the natural system. This reverses the burden of proof, changes the type of data regarded as evidence, and enables general experience (i.e. principles) to be used as hard argument.

In standard management the argument is mainly about 'allocation' and 'proof of damage'. For 'no-take' reserves, there is no allocation, and it is assumed that any extraction will have some effect. So while it is still possible to oppose the whole idea, the level of argument is different. The first principle is scientific. The provision of permanent minimally disturbed areas is necessary to act as 'controls'. Without such 'controls', it is much more difficult to produce valid scientific conclusions (e.g. separating and measuring natural changes).

The second principle is both basic biology and social common sense. Areas without deliberate extractions will help the conservation of biodiversity (genetic, specific, habitat and ecosystem). These less-disrupted areas will assist enjoyment of natural heritage (recreational and spiritual) and learning about marine life and its processes (education and advanced training). The overall public interest requires 'no-take' areas.

It should be noted that these principles do not depend on demonstrating that specific extractions have severe or even measurable effects. It is simply a statement that extractions *may* have serious or complex effects, and that it would be helpful to minimize *all* these in *some* places. While it is possible to believe such 'no-take' reserves should be small and rare, it is difficult to argue convincingly that they should not exist at all (which is the case in most countries at present).

Despite a slow start (the first reserve took 12 years!) the above principles are now widely recognized in New Zealand. There are now 13 established 'no-take' marine reserves. They are scientifically useful, valuable in education, popular for recreation, appreciated by tourists, supported by the local communities and clearly helpful in conservation. As a result, the rate of establishment is increasing. At least 25 more are in various stages of active public discussion. All the main political parties are promising 'more marine reserves' and some are advocating '10% by the year 2000'. The official aim is to 'establish a network of marine reserves' including 'representative examples of the full range of habitats and ecosystems' (Department of Conservation 1995)

EXISTING ASSUMPTIONS IN FISHERIES

It is reasonable to assume that the practical management of fisheries should focus on actual stocks — the self-contained breeding units. However, applying this assumption has some odd effects. Managers tend to assume that fish are there to be fished. This is even more basic than the idea of 'fishing rights'. If a potentially-fishable stock exists, there may be discussion about how it should be fished, who should fish it and how much can be taken. But that it should be fished, if possible, is taken for granted. Since this is true for every stock, the question of having 'no-take' areas is not normally considered.

Fisheries scientists, who provide data for management decisions, also focus on particular stocks. They reasonably assume that better data on the stock will help. But scientists also tend to assume that the answers to management problems can be found *within* the stocks. They may admit that the existing data are inadequate or that politics, economics, or natural disasters change the rules of the game. But their job is to get precise and up-to-date stock-specific data. The question of whether 'no-take' areas would help management (or science) in all stocks is not even considered.

EXISTING PROBLEMS IN FISHERIES

Short-term efficiency is promoted by focussed attention to the matter in hand, but only while the assumptions hold true. Stock-specific, data-based fisheries science and management does make sense in the short-term, and will always be the day-to-day focus of activity. Nevertheless uncertainty about the assumptions is a permanent problem and, if ignored, frequently becomes critical.

When a fish stock crashes or has critical problems, we are usually informed that this was due to some combination of:

inadequate data | *changing economic conditions*
inappropriate analysis | *new technology*
political interference | *natural changes*
insufficient enforcement

The main public discussion then becomes a search for scapegoats and compensation. It tends to be assumed that the 'causes' listed above are relatively rare and could be predicted or avoided. A more dispassionate analysis indicates that some levels of ignorance, error and unforeseen changes are to be expected even in the best-regulated systems, and consequently some higher level of control is also required. The types of uncertainty listed above apply to almost every kind of human business, including farming, manufacture trade, government and sports. In most of these, high-level controls based on general principles are regarded as standard management practice.

Problems regarding bycatch and bait fish, vote-catching political decisions, fashions in food and recreation, complex analytical difficulties, natural changes in stocks or currents (and dozens of similar problems) are likely to be with us in the foreseeable future. Fisheries also need some forms of control based on principles, rather than stock-specific data.

Ludwig *et al.* (1993) reviewing the history of fisheries, reject the idea that sustainability can be based on scientifically-agreed stock-specific data. They suggest that we confront uncertainty in fisheries and 'consider a variety of hypotheses and strategies; favour actions that are reversible, informative and robust to uncertainty; probe, monitor, and modify' and above all 'hedge'.

A network of 'no-take' marine reserves would do much of this and would assist with the rest. It would provide an escape from the infinite regression in which we are now trapped. Instead of *always* needing more data to predict precisely the effects on detailed user-group activities, we could also start at the other end. What actions are most robust to uncertainty, most informative, provide a hedge on existing activities, allow better monitoring and are easily reversible? One such action would be to have areas with no fishing at all.

What are the reasonable hypotheses on the level of provision for 'no-take' areas, even if total fishing effort and catches are to be maintained or enhanced? On the basis of general principles, it seems that if 10–30% of the ecosystem area is 'no-take', then overall fish catches would be maintained or enhanced. But, of course, there have been very few trials (and these are mostly small, isolated and poorly monitored). Essentially we have not yet ventured to do the experiment.

The principles for a network of 'no-take' marine reserves

When the scientific, conservation and social values are accepted as principles, the argument turns to the question of quantity. The first principle for this is *representation.* To be representative, 'no-take' marine reserves are needed in each major biogeographic region (simply because these have different biota). Within each region the 'no-take' areas would include some of each of the major habitats/communities (for the same reason). The second principle is *replication.* An example of each habitat in more than one location would be required to provide scientific replication, to give protection against accidents (whether pollution or local natural disasters) and to improve social values (including recreation and tourism).

The principle of *sustainability* needs to be included here. In order to maintain its values, the whole system of 'no-take' reserves must be self-sustaining. Single marine reserves are unlikely to be self-sustaining, since most marine organisms have remote dispersal mechanisms (e.g. planktonic larvae). A 'network' would be needed to provide sustainability. The great value of networks is that they are non-specific in action.

These simple principles add up to a significant total requirement in area. Indeed the total might be a serious problem but for two further points. The first is that the principles are not strictly additive. One reserve can (and should) have more than one habitat. Replication can be included within the network needed for sustainability.

The second point is directly related to fisheries management. A self-sustaining network of 'no-take' marine reserves would automatically act to support all exploited fish stocks. Such reserves are not 'fenced in'. Eggs, larvae, and adults can and will move out. In the absence of exploitation the breeding potential is likely to increase, and such increases are most likely when exploitation outside is high. These points are based on firm biological principles. However, it will be difficult to measure, even after the event, just how much any particular stock has benefited, or which problems had been prevented. It will be impossible to predict stock-specific effects in detail.

Why does that worry us so much? We do not bother about this when promoting 'networks' for transport, education, fire insurance, etc. When the principles are recognized, general action seems wise and practical. Indeed, specific action based on actual data is clearly a later stage. We plan and develop roads, schools and insurance systems without knowing what particular journeys will be made, how many engineers will be needed or which buildings will catch fire.

The important and useful point is that 'no-take' networks will act on known biological principles in predictable directions regardless of whether we have detailed data or not. They are thus valuable additions to management, especially for stocks where the available data are unsatisfactory, the dynamics are complex, the economics and politics are fierce, or large natural changes occur. Since one or more of these points are valid for most stocks, the value of non-specific action is very real.

Most reviews of marine reserves for fisheries management confine themselves to detailed stock-specific points (e.g. Roberts and Polunin 1991; Rowley 1994), but more general principles are beginning to be discussed (e.g. Dugan and Davis 1993; Dayton *et al.* 1995). In 1990 a team of fisheries scientists in Florida, concerned about the multi-species reef fisheries, suggested making 20% of the whole region 'no-take' as the best management option (Plan Development Team 1990). This idea was a shock to many people, but a review panel of independent scientists supported both the arguments and conclusions (Roberts *et al.* 1995). It seems that when we are forced by local or special circumstances to consider many stocks and whole ecosystems, the advantages of 'no-take' areas to fisheries science and management are obvious and robust. The real problem is that such breadth of vision rarely occurs.

Some steps towards ecosystem management

1. The long-term public interest is paramount

   The sea has many uses. Some are extractive, but all ultimately depend on the continued functioning of the ecosystem. The overall public need is for marine ecosystems to continue to operate naturally.

2. Permanent 'no-take' areas in the sea are needed to ensure this

   While our power of disruption is now great, we do not have the knowledge or technique for positive control. The systems must be allowed to 'manage themselves'. 'No-take' areas will assist this and enable us to check it is happening.

3. Non-extractive uses should be encouraged in 'no-take' areas

   The benefits of 'no-take' areas to science, education, recreation, personal experience, etc. are valuable in themselves. These uses also generate an informed public. With a better knowledge of natural conditions and exploitive effects, the public is much more likely to support sensible arrangements for conservation and management.

4. Representation of all marine habitats in every biological region is needed

5. Replication is essential

6. A network design is required (i.e. non-specific to problems, stocks or species)
7. The minimum level of provision is that which is self-sustainable
8. The general pattern of 'no-take' areas is determined by marine topography

   The fractal form of coastlines and depth contours determine the pattern — basically 'stripes' normal to straight coasts with parallel contours, through 'clusters' on indented coasts with islands, to 'geometric networks' in complex (high fractal) regions.
9. The regional arrangement is partly deterministic and partly an optimization

   The actual scale of the general pattern (e.g. the width and spacing of 'stripes') is based on the actual scale, diversity and pattern of the habitats (using topography as proxy data if necessary), but constrained by the optimum retention of replication, etc.
10. The precise location of 'no-take' areas is not deterministic

    Within the constraints of 4–9 above, the precise locations cannot be determined on scientific criteria. There may be several, or many, possibilities. Other factors (e.g. social) are biologically permissible and politically advantageous at this final stage.

    Several of these points, especially the last, are counter-intuitive. The order of consideration is critical. My prime point is that if consideration begins with a particular problem or any sectional interest, the conclusions are radically different. Until recently, solving problems and carefully considering sectional interests did seem to be the name of the game. I suggest that this is no longer the only option, or even a very sensible one.

Upgrading 'fisheries' science and management

Years ago, the public naively assumed that fisheries scientists and managers were the arm of government that looked after the sea for them. New Zealand once had a 'Marine Department' — which was supposed to manage everything from lighthouses and reclamations, to fishing regulations and waste disposal. It was hard to be 'professional' over such a range, so it was broken up. World-wide, over the past few decades, fisheries science and management have become very professional, but the aims have narrowed. This has also happened in 'transport', 'coastal planning' and 'pollution control'. They too are now much more efficient, of course, but they have all become specialized niches. The public no longer knows who is 'looking after the sea' and is beginning to suspect that the answer is no one.

This is starting to worry the public. It is easy to be cynical about movements to save the whales, ban drift-nets, stop ocean dumping, protect coral reefs, etc. It would be more sensible to see these as indicators of a trend. The public no longer accepts as an axiom that existing user-groups should control whole ecosystems — whether they are foresters, miners, or fishermen. The rate of change is very variable — by region and topic, but it is all one way. The trend is predictable and the stages have been labelled (Landner 1994). Fisheries may be late in the game, but they are not immune to the trend.

Can we learn from what is already happening in forestry and mining, or will the same kind of messy confrontations develop in fisheries? Who will promote some intelligent principles to guide the not-so-silent majority as they begin to realize that it is their sea, and that the fish in it belong to their children.

If we can already recognize many values and uses for 'no-take' marine areas, and we can see that a significant network of such areas would be supportive of fisheries science and management, why don't we upgrade the existing professional aims? The public need a lead. They want to believe someone cares about more than details and problem-solving. They would support moves to principles and attempts to prevent more problems in the sea. This process has already started in New Zealand and is accelerating. The politicians were taken by surprise, but are catching up. The professionals are mostly still sceptical. I believe this is unwise. Professionalism is the recognition of relevant principles and an insistence that these control detailed action. Ecosystem management may still be a distant objective in fisheries, but practical moves towards it deserve active professional support. The overall public interest in life in the sea may not yet be the responsibility of 'fisheries managers', but it certainly hasn't been given to anyone else.

Conclusions

'No-take' marine reserves offer a new and additional form of fisheries management. Such reserves avoid detailed arguments about 'allocation' and 'proof of damage', and allow non-specific but strong support for stocks, habitats and ecosystems. The first 'no-take' marine reserve in New Zealand was established in 1977. A further 12 have been added since, and at least 25 more proposals are under active public consideration. A representative network is now the official aim.

To date, the promotion and use of marine reserves has come mainly from marine scientists, divers, conservationists, teachers, tourists, and the general public. Most fisheries scientists and managers have remained sceptical or uninterested. This could change rapidly, due to political pressure or professional recognition of the opportunities. All major political parties in New Zealand now support more marine reserves. Despite the application of many stock-specific management techniques, it is officially recognized that many NZ fisheries have serious problems, and new ideas are needed.

Data-based, stock-specific fisheries management, focussed on the interests of existing user groups, is a political and practical necessity. However, the history of resource management and recent events in fisheries indicate that this approach is not sufficient. Concentrating on the effects of existing activities distracts attention from the intrinsic properties of the ecosystems, and prevents consideration of other values or different uses. Furthermore the standard approach inevitably suffers from 'infinite regression', with more-and-more data being required to predict increasingly precise questions.

Well-understood business systems have many types of insurance, that operate in different ways. Insurance assumes uncertainty, ignores detailed causation, and concentrates on preventing damage without specific prediction. Its success depends on the perception of risks, not on knowledge of causes. The multiplicity of fishery regulations obscures (i) the focus on stock-specific dynamics — that excludes the intrinsic properties of the ecosystem, (ii) the need to demonstrate damage to stocks — which prevents pro-active management and (iii) the assumption of fishing rights — which precludes natural baselines and promotes 'brinkmanship' (e.g. maximum sustainable yields).

Until recently, non-involvement in fishing meant non-interest, except in the price of food fish. However, as has already happened in forestry and mining, the environmental views of the general public are becoming significant. 'No-take' marine reserves appeal to the lay public as a form of insurance, not because they doubt the skill of fisheries scientists and managers but because they believe their aims are too narrow. Such ideas are reinforced by the discovery that fisheries management is not always successful even in sustaining fisheries.

If fishery scientists and managers upgraded their status to 'public guardians of all marine life' (which they earlier held by default), they could escape from the problems of extreme specialization, promote 'no-take' marine reserves as an sensible form of insurance, and earn the active support of an increasingly-concerned general public.

References

Dayton, P K, Thrush, S F, Agardy, M T, and Hofman, R J (1995). Environmental effects of marine fishing. *Aquatic Conservation: Marine and Freshwater Ecosystems* **5**, 205–32.

Department of Conservation (1995). Marine Reserves: a Department of Conservation Information Paper. (Department of Conservation, Wellington, New Zealand). 21 pp.

Dugan, J E, and Davis, G E (1993). Applications of marine refugia to coastal fisheries management. *Canadian Journal of Fisheries and Aquatic Sciences* **50**, 2029–42.

Landner, L (1994). How do we know when we have done enough to protect the environment? *Marine Pollution Bulletin* **29**, 593–8.

Ludwig, D, Hilborn, R, and Walters, C (1993). Uncertainty, resource exploitation and conservation: lessons from history. *Science* **260**, 17 and 36.

Plan Development Team (1990). The potential of marine fishery reserves for reef fish management in the US Southern Atlantic. *NOAA Technical Memorandum NMFS-SEFC-261*. 40 pp.

Roberts, C M, and Polunin, N V C (1991). Are marine reserves effective in management of reef fisheries? *Reviews in Fish Biology and Fisheries* **1**, 65–91.

Roberts, C, Ballantine, W J, Buxton, C D, Dayton, P, Crowder, L B, Milon, W, Orbach, M K, Pauly, D, Texler, J, and Walters, C J (1995). Review of the use of marine fisheries reserves in the US southeastern Atlantic. *NOAA Technical Memorandum NMFS-SEFSC-376*. 31 pp.

Rowley, R J (1994). Marine reserves in fisheries management. *Aquatic Conservation: Marine and Freshwater Ecosystems* **4**, 233–54.

Recent Steps in the Evolution of Fisheries Resource Conservation Institutions in Canada

W. G. Doubleday and H. Powles

Department of Fisheries and Oceans, 200 Kent Street, Ottawa, Ontario, Canada, K1A 0E6.

Summary

Canadian fisheries science and management institutions are undergoing fundamental changes, driven by recent experience with stock dynamics and by the need to reduce government expenditures to sustainable levels.

Canada has adopted the precautionary approach to resource conservation. New programmes are bringing industry and government together to improve the knowledge base and to develop consensus on conservation requirements. Industry contributions to fishery monitoring, to resource assessment, and to enforcement of regulations are increasing. The Fisheries Resource Conservation Council, established in 1993, includes industry and university representation and has developed consensus-based recommendations on conservation requirements for the severely depleted groundfish stocks in Atlantic Canada. Arrangements to provide industry with longer-term security of access to resources are under discussion and may further contribute to industry participation in resource assessment and conservation. An Aboriginal Fisheries Strategy has improved management of constitutionally-guaranteed aboriginal fisheries for food, social and ceremonial purposes.

Research on fishery resource productivity under different climatic regimes and of the impacts of changes in predator and prey dynamics are contributing to setting more realistic expectations for future fishery yields.

Introduction

Under Canada's Constitution, the Minister of Fisheries and Oceans in the federal government has broad powers to ensure conservation of fishery resources and to manage fisheries for the benefit of all Canadians. The Department of Fisheries and Oceans, which supports the Minister in this task, must work to ensure conservation of over 200 stocks of more than 60 marine species in three oceans, and thousands of anadromous salmon stocks at a time when pressure for access to resources is continually increasing. We also have to ensure orderly allocation of access to resources among commercial, recreational and aboriginal sectors and between competing industry sectors, regions and provinces.

Canada has experienced profound changes in its fisheries in recent years. Groundfish landings, after increasing in the 1980s (after Canada declared its 200-mile extended economic zone) have all but ceased since 1992, following closure of fisheries due to low stock levels (Fig. 1). Surprisingly, lobster landings, after fluctuating around a more or less constant level during the 20th century, essentially tripled between 1975 and 1991 (Fig. 1). This enormous increase in landings occurred despite the fact that lobster stocks are very heavily exploited, probably overexploited, through much of their range in Atlantic Canada. As with the

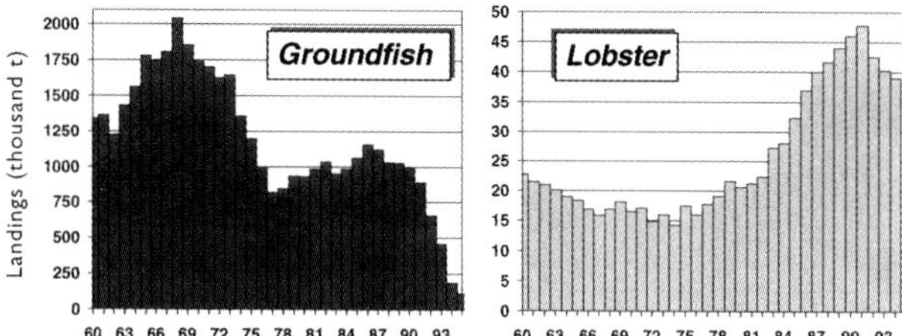

Fig. 1. Landings of groundfish and lobster in Canadian Atlantic region, 1960–94.

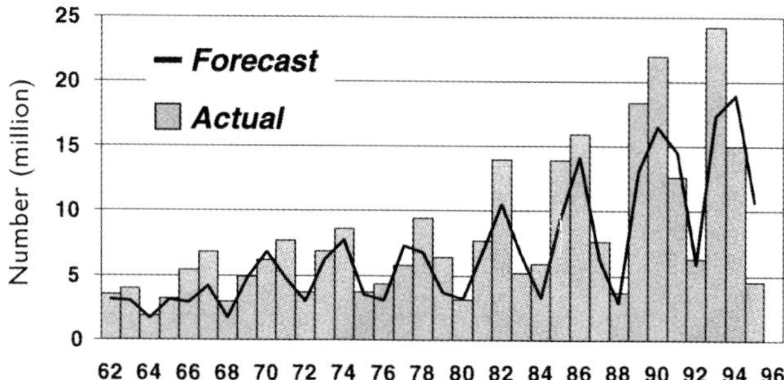

Fig. 2. Forecast and actual returns of sockeye salmon to Fraser River system, 1962–95.

decline and collapse of groundfish, the increase in lobsters was not predicted by assessment models, and it is probably largely due to changes in the marine ecosystem as well as to developments in the fishery. Other invertebrate stocks have also shown increases in abundance in recent years, to the extent that the total landed value of Canadian Atlantic fisheries in 1995 was the highest ever recorded.

Some important Pacific salmon stocks have also shown unexpected declines in recent years. Returns to the Fraser River sockeye stock, one of Canada's most important, were less than half those forecast in 1995 following many years of relatively accurate forecasts (Fig. 2). The unexpectedly low returns follow a period in which abundance and landings of North Pacific salmon were very high, probably owing to favourable production conditions in the North Pacific Ocean as a whole. Returns to other salmon stocks have been much higher than forecast — in the large Skeena/Nass sockeye stock, 1996 returns have been twice the forecast. Again, these increases and decreases in abundance were large, were not predicted, and were due to the interaction between natural environmental factors and the effects of fishing.

Developments in the socio-economic area as well as in the marine environment are forcing sweeping changes to fisheries resource assessment and management in Canada. The declines in some important stocks combined with rising prices due to development of a world market in fish have led to tremendous pressure for access to those resources that are still abundant. The institutional framework for conservation of fisheries resources is coming under pressure as accumulated government debt has led to the need to place restrictions on expenditures throughout the Canadian government including the Department of Fisheries and Oceans. The Department is faced with a 30% reduction in total funding between 1994 and 2000 and an equivalent reduction in personnel.

Precautionary approach to management

Following the unprecedented declines in abundance of some of the most important resources, the Department of Fisheries and Oceans has adopted the precautionary principle as the overall approach to management, under which resource conservation is the first priority. The precautionary approach is engraved in the new UN Straddling and Highly Migratory Stocks Agreement adopted in August 1995. In the 1980s, fishery resources were seen as basically resilient, and understanding of these resources was considered adequate to predict future production. Accordingly, the approach was to attempt to satisfy the requirements of industry for fish to harvest as long as there was no proof that the resource was in danger. We have learned to our cost that resources appear to be more responsive to unpredictable changes in environment than previously thought, and as a consequence our approach has changed to erring on the side of caution when there is uncertainty about resource status and prospects. 'Conservation comes first' has become the guiding principle that overrides all other considerations.

Conservation measures — short-term

In response to unexpected declines in important stocks, and in keeping with the precautionary approach, a number of stringent conservation measures have been implemented. The closure of Atlantic groundfish fisheries was the most visible; the fishery on the northern cod stock was the first to be closed, in 1992, and since then most of our Atlantic cod stocks have been closed to fishing, along with many stocks of other species of groundfish. In 1995, more than 23 groundfish stocks were closed to fishing. The decisions to close these fisheries were very difficult, as some 38 000 fishing and processing jobs have been lost in the Atlantic region as a result of the groundfish closures in an area where there are few other economic alternatives; however, conservation requirements come first. The economic impact of the closures has been mitigated by income support measures instituted in 1992 by the federal government. Currently some $CAN7 million ($US5 million) is being disbursed weekly to those put out of work as a result of this crisis.

We have implemented other measures toward conservation and rebuilding of Atlantic groundfish stocks, including a prohibition on discarding fish (discarding under individual quotas had been shown to be a major problem in the years before the groundfish closures), institution of more comprehensive conservation-oriented annual harvesting plans, improvement of data on catches and landings through expansion of observer programmes and dockside monitoring programmes, increase in mesh size and mandatory use of square mesh, and institution of small-fish protocols. Following these very stringent conservation measures the most recent annual report on the status of groundfish stocks (Anon. 1996*a*) has indicated that the decline of stocks has been stopped. We must now wait for recruitment from these very low stock levels to act to rebuild abundance levels.

Stringent conservation measures have also been taken for Pacific salmon stocks. Harvests from the Fraser River sockeye stock were stopped when only a fraction of the projected allowable catch had been taken in 1995, to allow adequate spawning escapement from a run much smaller than forecast. For 1996, no commercial harvest on this stock will be possible based on forecast run size and only limited catch allocations for aboriginal harvesting will be allowed. Some stocks of chinook salmon, a large species much prized by recreational fishers, are also in serious decline owing to mackerel predation resulting from El Niño events in the early 1990s and no catch will be allowed on this species in extensive areas of the Pacific coast in 1996. A Pacific Fisheries Revitalization Plan is being implemented to reduce fishing capacity in the Pacific salmon fleet by 50%, implement a more effective fisheries management licensing policy and apply the precautionary conservation approach to reduce pressure on stocks. The economic consequences of these stringent measures are very serious for the Pacific coast, both to commercial harvesters and to the recreational fishery operators, but they are absolutely necessary to ensure that conservation requirements are met.

Changes to conservation institutions — commercial fisheries

In addition to these rapid responses to unforeseen declines in key stocks, the Department has implemented fundamental institutional changes to ensure that our fisheries are sustainable for the long-term future.

The fishing industry has become much more centrally involved in resource management over the past five years, in the first instance by substantially increasing its contributions to fishery monitoring. The coverage of at-sea observer programmes has been extended from large offshore vessels to a broader range of fishery sectors including some of the smaller 'midshore' fleets. Dockside catch-monitoring programmes have been introduced in the past several years to many of our fisheries, with a view to verifying landings information previously obtained from sales slips and vessel log-books. For both of these programmes, industry organizations manage funds collected from individual vessels or enterprises and contract directly for the services of observers and dockside monitors, while the Department of Fisheries and Oceans provides advice on sampling design and implementation, and sets standards for data quality and coverage. The information collected is available for public use. These programmes have been very successful in improving the quality of data on fishery removals, bycatch and population characteristics.

Over the past several years, the department has made significant improvements in the strategies it employs to ensure compliance with fisheries legislation and the provisions of the various fishery management plans.

We have moved forward in the development of pro-active or preventative enforcement programmes and strategies which involve both the fishing industry and dependent coastal communities to become involved in delivery of education and information programmes that are designed to promote conservation principles and fishing ethics. These programmes include an Ocean Watch programme being led by the Fishers' Resource Council in western Newfoundland as well as numerous agreements for the provision of conservation and protection with many native communities throughout Canada.

Canada has also made significant improvements in the delivery of its reactive enforcement programme, which provides the capability to ensure compliance with regulations. Canada has moved away from a strict reliance on vessel patrols as the means of ensuring compliance. Although vessels are still an important tool in fisheries surveillance, the department is moving towards more innovative and less costly strategies to ensure compliance. The development and implementation of surveillance data management systems which integrate vessel location and activity data from all surveillance sources, including air, vessels, observers, transponders, fishery Officers and the Fishing Industry has had a positive impact on compliance levels. The interpretation of these data, in real time, provides for the effective and efficient deployment of enforcement resources to correct non-compliance problems. In addition, we have developed state-of-the-art, aerial surveillance low-light and night photography capability to provide a deterrence to potential illegal activities from occurring at night. The Conservation and Protection Program is also developing commercial monitoring and auditing processes to ensure compliance with Individual Quota allocation programmes.

The Enforcement Program is also working with the fishing industry to develop new and cost-effective processes for the out-sourced delivery of a number of enforcement programmes. The intent of the out-sourcing of programmes is to provide a mechanism for Industry to pay for these programmes while ensuring that the programmes maintain a very high level of integrity. The Department is currently developing quality management (ISO) and audit standards for the at-sea and dockside monitoring programmes. It is expected that this approach will reduce the overall cost of the delivery of certain elements of the enforcement programme to government.

A further step toward industry involvement in resource conservation has been the development of comprehensive joint agreements in a few key fisheries to improve scientific assessment of stock status. Examples of this are the plans developed for snow crab in the southern Gulf of St Lawrence, and for the Stimpson surf clam resource in offshore waters of Canada's Atlantic region.

In groundfish stocks, for which commercial fisheries have been closed, information on stock status has been provided by 'sentinel fisheries' organized by scientists and fishing organizations. These are restricted fisheries which emulate commercial practices and which are designed to provide statistically valid information on catch-per-unit-effort (CPUE), distribution and biological characteristics of commercial species. In combination with research-vessel surveys, sentinel fisheries are providing information for scientific monitoring of stock rebuilding and are also allowing fishermen themselves to monitor rebuilding. In 1995, some 500 fishermen were occupied in sentinel fisheries at 114 locations throughout Atlantic Canada.

At the same time, scientists have been seeking much greater input from harvesters into stock assessments, to ensure that fishermen's knowledge as well as scientific knowledge is incorporated. Scientists have also made efforts to ensure that the results of assessments are presented in such a way as to be more accessible to non-specialists.

The next step on the road to bringing industry to the centre of fisheries management will be more fundamental. It should provide industry with longer-term security of access to the resource to encourage a more active role in resource conservation and management. Multi-year security of access represents a fundamental change in the Canadian context. In the past, fishery management plans have been developed annually and the Minister's mandate to manage and conserve resources has been interpreted as requiring the flexibility to make allocation decisions on an annual or even shorter-term basis. Industry is obviously very interested in planning horizons longer than the annual cycle, and may find it advantageous to take a longer-term approach to resource conservation if provided with a longer-term stake in the resource.

Discussions have begun on how a change to longer-term access planning might occur, but the prospect of this fundamental change has raised vigorous resistance in some quarters, notably in some coastal communities whose primary economic activity is small-scale fisheries. The perception has grown that moving toward industry partnership and longer-term security of access will result in 'privatizing' the resource and handing it over to highly capitalized interests. This is certainly not the intention. Indeed a move toward longer-term security of access should contribute to greater stability for coastal communities. The move toward longer-term arrangements is being taken slowly and extensive consultations will be undertaken to ensure that all participants understand the benefits as well as the risks of this approach. Hopefully, this longer-term management approach will gain as wide acceptance as individual quotas which, although hotly contested when first introduced, are generally much appreciated in the wide range of fleets which now account for 40% of landings where they are now part of management.

The process for developing conservation measures for exploited fish stocks is also being fundamentally changed. Prior to 1992, peer-reviewed scientific stock assessments and recommended harvest levels were prepared by government scientists for consideration as the annual fishery management plans were developed. Industry input to resource conservation was through consultation during the process of developing the management plan, but this covered both conservation limits and allocation issues. The fact that conservation and allocation issues were covered simultaneously meant that conservation did not always get the attention it deserved.

In 1992, a new organization, the Fisheries Resource Conservation Council, was established by the Minister of Fisheries and Oceans with a view to clearly separating conservation from allocation decisions and to ensuring that industry had a more direct role in resource conservation (Fig. 3). The FRCC is an independent body whose members are drawn from industry, universities, and (*ex officio*) provincial and federal governments. Based on the annual scientific assessments of stock status and on consultations with industry and the public, the FRCC prepares advice for the Minister on conservation measures. The key difference from the previous process is that the conservation step is no longer a purely government responsibility. Industry and the public have a major say in what conservation measures should be. Further conservation limits are clearly established before decisions on resource allocation are considered.

Overall, fishermen and industry are much more involved in conservation from start to finish now than 3 to 4 years ago. The new process has helped to ensure that the wrenching decisions to close most of Canada's Atlantic groundfish stocks to fishing have been based on a consensus position of industry and government.

Changes to conservation institutions — Aboriginal fisheries

The Government of Canada has also made significant changes in its relationship with aboriginal peoples in fisheries matters. Fisheries are of fundamental importance to aboriginal people throughout Canada. The Department of Fisheries and Oceans has had a long-standing policy of providing access to fish for food as part of the resource allocation process, and a range of innovative co-management agreements had been negotiated and implemented in the 1980s, particularly in northern Canada, covering subsistence fisheries which are an essential part of life for native communities.

In 1990, the Supreme Court of Canada reached a landmark decision that aboriginal peoples retain a fundamental right to fish for food, social and ceremonial purposes, subject only to conservation requirements. The federal government was determined by the Court to have a fiduciary obligation to ensure that aboriginal peoples have access to fish resources as a priority over all other uses.

The Department of Fisheries and Oceans implemented an aboriginal fisheries strategy soon after the Court decision, with the general intent of establishing cooperative mechanisms for fisheries management. Under this strategy, annual communal licences are issued to Native bands for agreed harvest levels, and catch monitoring and enforcement of regulations are conducted by the Native band with assistance and funding from the Department. Funding has been provided to Native bands to carry out fish habitat improvement and fishery enhancement, expand the aboriginal role in fisheries enforcement, and to undertake research and public-awareness activities. In addition,

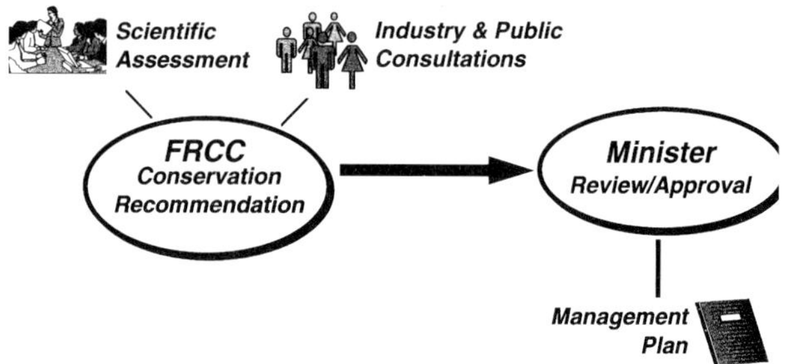

Fig. 3. The Fisheries Resource Conservation Council (FRCC), part of a new process for developing conservation measures for Atlantic fish stocks.

a number of commercial fishing licences have been repurchased (mainly in the Pacific salmon and Atlantic lobster fisheries) for use by Native groups, to ensure more equitable access to commercial fishing operations.

Changes to conservation institutions — International fisheries

Canada has contributed to improvements in the framework for international management of fisheries resources, largely as a response to interest in straddling stocks off the Atlantic coast and the highly-migratory Pacific salmon resource that Canada shares with the United States of America. As a result of the highly publicized dispute between Canada and the European Union over groundfish harvesting on the nose and tail of the Grand Banks of Newfoundland (Fig. 4) early in 1995, the Northwest Atlantic Fisheries Organisation has implemented a series of stringent conservation measures including 100% observer coverage of all vessels fishing in the area and severe reductions in allowable catches. Canada was an active participant in developing the United Nations Agreement on Highly Migratory and Straddling Stocks and was among the initial signatories to this Agreement in December 1995. Ratification of the Agreement is a Canadian government priority. Canada is actively promoting early ratification and implementation of the Agreement.

Considerable progress has been made on conservation issues in bilateral and multilateral fora, but there remain some contentious issues that have not been resolved. Chinook salmon stocks suffering from serious declines due to other causes remain subject to excessive fishing in Alaska in 1996, and Canadian catches have been almost eliminated. The Canada–US Pacific Salmon Treaty has failed to protect these stocks from overexploitation because the USA has adopted a unilateral fishing plan that greatly exceeds the recommendations of a bilateral scientific committee to substantially reduce the harvest rate.

CHANGES TO SCIENCE PROGRAMMES

Constant improvement in research programmes is necessary to ensure that forecasts of potential yields are as accurate as possible. Much of the research effort in recent years has been directed to improving understanding of the dynamics of exploited stocks in an ecosystem context, which is particularly critical in light of the very considerable variations in stock abundance described above.

A recent study has shown a close relationship over a 70-year period between total catches of north Pacific salmon species and a climate index related to the overall biological productivity of the North Pacific (Fig. 5) (Beamish and Bouillon 1993). This suggests that conditions over this broad oceanic area can switch between 'favourable' and 'unfavourable' for salmon production and that changes in salmon catches by a factor of 2–3 times may correspond to these shifts in production regimes. Carrying capacity for harvestable species may fluctuate with environmental conditions over this large area (and over other marine production areas), such that the idea of rebuilding stocks or managing these to some more-or-less constant level has to be reconsidered.

Another type of system-level interaction has been the subject of study as it affects abundance of Canada's Pacific salmon. In warm years, warm water species such as Pacific hake and Pacific mackerel move farther northward than usual. These species are known to be predators on young salmon as they move from rivers to the ocean environment. One hypothesis that may help to explain the recent unexpectedly low returns of Pacific salmon, particularly chinook, is that these fish went to sea in warm years when abundance of predators was presumably higher than normal. The work of understanding these kinds of ecosystem effects and factoring them into our assessments of stock status and forecasts of future abundance will be a major focus of our science programmes for the coming years.

On the Atlantic coast, similar ecosystem-level processes are being studied. The decline and collapse of Atlantic groundfish stocks followed a number of years of very severe climatic conditions, in particular, abnormally cold water temperatures across much of the Northwest Atlantic (Fig. 6) (Anon. 1996*b*). Harsh conditions may have contributed to an unfavourable production regime for groundfish; this, in combination with the very intensive harvesting that is known to have occurred, could have contributed to the stock collapse. With the groundfish

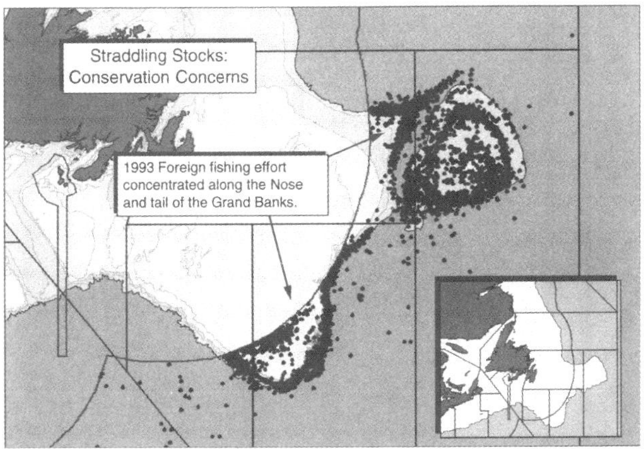

Fig. 4. Foreign fishing effort on the nose and tail of the Grand Banks of Newfoundland, 1993.

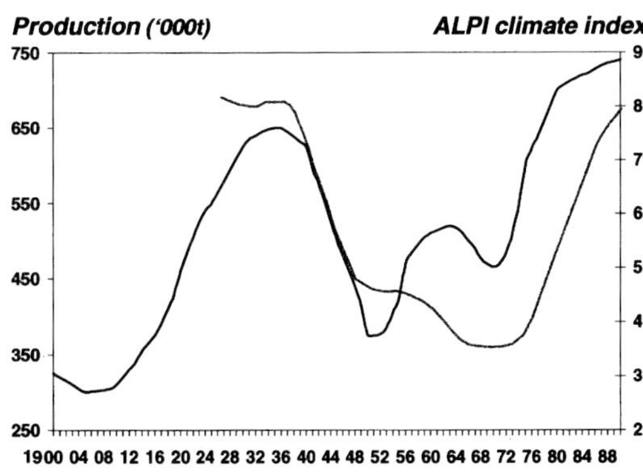

Fig. 5. Long-term climate Aleutian Low Pressure Index (ALPI), and salmon production fluctuations, North Pacific Ocean. After Beamish and Bouillon (1993).

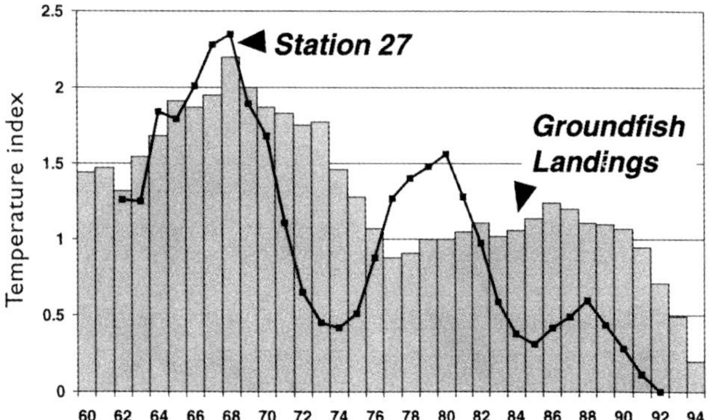

Fig. 6. Ocean temperature at an index station on the Grand Banks of Newfoundland, and groundfish landings.

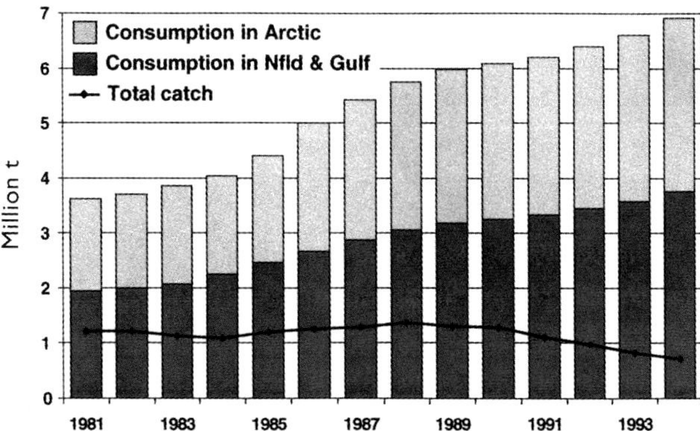

Fig. 7. Estimated consumption of fish by harp seals and total Canadian catches, 1961–95.

stocks so very low now there is considerable concern about the potential impact of Canada's large and expanding seal populations on the rebuilding of stocks. Over the period when groundfish declined and collapsed, and during the period of fishery closures since 1992, populations of the three main Atlantic seal species have been very high and growing rapidly (Anon. 1995). On the base of predation studies and these abundance estimates, seal predation could be accounting for removals of young fish from our major stocks comparable in tonnage to those due to the fishery in the years before closures were imposed (Fig. 7). The impact of seal predation on commercial fish populations has not been precisely quantified as yet, but this is a very significant potential factor in groundfish stock dynamics and it is receiving considerable research effort.

Conclusions

Fisheries resource conservation and management are facing great challenges in the late 1990s, as increasing demands on governmental services must be met with declining public funds. This situation has served to focus governmental attention on its core mandates: resource conservation, as the dominant priority, discharging our fiduciary responsibility to aboriginal peoples, ensuring that conditions are right for industry to be profitable and to create stable employment for communities that rely on the fisheries. The 'Fishery of the Future' as we envision it will be built on several guiding principles:

- conservation comes first
- aboriginal rights must be respected
- government and industry must operate in partnership
- industry harvesting capacity must be balanced with resource availability and supply
- the fishery must be conducted by professionals.

As a more secure, self-reliant, professional industry takes on a greater role in the day-to-day activities in fisheries monitoring and management, government, as part of its core mandate, can continue to lead development of knowledge of the fundamental factors affecting resource productivity, in particular system-level factors. The changes described here should foster rebuilding of those stocks that are presently at very low levels and maintaining good production levels from those which are currently doing well.

Acknowledgment

We are grateful to Kathryn A. Bruce for compilation and analysis of information and to Jennifer Volrath for assistance with the figures.

References

Anon. (1995). Report on the status of harp seals in the northwest Atlantic. *DFO Atlantic Fisheries Stock Status Report* 95/7. 4 pp. Science Branch, Northwest Atlantic Fisheries Center, DFO, PO Box 5667, St. John's, Newfoundland, Canada A1C 5X1.

Anon. (1996*a*). Overview of Canadian managed groundfish stocks in the Gulf of St. Lawrence and the Canadian Atlantic. *DFO Atlantic Fisheries Stock Status Report* 96/40E. 11 pp + Annex. Atlantic Stock Assessment Secretariat, DFO, 200 Kent Street, Ottawa, Ontario, Canada K1A 0E6.

Anon. (1996*b*). State of the Ocean: Northwest Atlantic. *DFO Atlantic Fisheries Stock Status Report* 96/41E: 7 pp. Maritimes Regional Advisory Process, DFO, PO Box 1006, Dartmouth, NS, Canada B2Y 4A2.

Beamish, R J, and. Bouillon, D R (1993). Pacific salmon production trends in relation to climate. *Canadian Journal of Fisheries and Aquatic Sciences* **50**, 1002–16.

The Costs of Fisheries Management Information and Fisheries Research

P. A. Larkin[†]

The North Pacific Universities Marine Mammal Research Consortium, Fisheries Centre, University of British Columbia, Room 18, Hut B-3, 6248 Biological Sciences Road, Vancouver, B.C., Canada V6T 1Z4.

Summary

The cost of a fisheries management organization should not exceed 20% of landed value from fisheries. For groundfish fisheries of temperate latitudes, virtual population analysis (MSVPA) requires information on catch and diet of major species in a fish community, preferably augmented by a fleet interaction model. These requirements are fairly daunting even under ideal conditions. The issue of costs is relevant to enthusiasms for ecosystem management unless adaptive management approaches are adopted. For gauntlet fisheries for anadromous species such as Pacific salmon, information requirements include daily estimates of catch and escapement for several thousands of races, each returning to a stream of origin. Costs associated with management can become greater than the economic return, particularly when the resource has been depleted. To reduce costs, no measure is potentially more effective than reduction of fishing effort. To date that has rarely been accomplished but the institution of various schemes of rights-based fishing are promising.

Introduction

In 1983, Brian Rothschild edited a book entitled *Global Fisheries: Perspectives for the 1980s*. The chapter I wrote was called 'How much is enough? An essay on the structure of fisheries management agencies' (Larkin 1983). Among other gratuitous comments, I ventured some opinions about how the costs might be distributed. Figure 1, adapted from those remarks, illustrates the increasing importance of research, stock assessment and sampling as management becomes more intense. The scientific activities, research, stock assessment, sampling and development take up 50% of the costs of a modest level of management, 65% of intensive management. Percentage costs of administration are cut in half. To round out this brave bit of advice, I suggested that the annual cost of the whole organization should not exceed 20% of the landed value of the fishery with the long term average expenditure on research and information for management in the range of 10–15% of landed values.

Since those words of crackerbarrel wisdom were written, I have received no comments from colleagues either in Canada or abroad, from which it must be concluded that everyone agreed or nobody read the chapter. I prefer the former explanation. Nevertheless, in Canada today, the cost of the Department of Fisheries and Oceans is roughly $Cdn 750 million and the value

[†] 1924–1996 (See p 718).

of the resource is only twice that amount, $\$1.5 \times 10^4$. Admittedly, the fisheries are going through hard times, partly because of natural causes, but the discrepancy between 50% of the value of the resource as the costs of a department of fisheries and the 20% I advocated can hardly be explained as the extra costs of being a department that also has a responsibility for Canada's share of three oceans, Atlantic, Pacific and Arctic.

But perhaps the whole question of the total costs of a fisheries department should not be based on some arbitrary percentage of the economic return from the resource, nor should the research and management costs be reckoned as a certain percentage of the total cost of a department of fisheries. A better approach may be to ask 'on the basis of individual fisheries, what is the relation between expenditures on research and information for management and the landed value of the catch?' This question may frequently be asked in senior administrative circles, often in the context of zero-base budgeting exercises such as have recently been taking place in Canada, but it does not seem to surface in fisheries literature.

There are some obvious reasons. For any approach there are some major difficulties in assessing the adequacy of the return on investment in research and information gathering. First, the return often comes many years after the investment. Second, a great many factors other than research may be responsible for the fate of a fishery. Thus, the research and information advocate will usually be optimistic about the returns and will claim credit for good results. If things go poorly and stocks decline, the causes given will be lack of sufficient data and/or the vagaries of 'Nature and other enemies of the government'. Third, the de-segregation of expenditures into various activities always involves some arbitrariness. In consequence, comparing the performance of research and information-gathering in one region to that in another region (especially another country) is fraught with accounting anomalies.

Rather obviously, doing little or no data collection or research is inexpensive but provides no sound basis for changing management practices. A modicum of activity provides some insight into local problems and enables participation in the professional networks of managers and researchers. With greater investment the risk of management error may be smaller, but the costs may not be commensurate with the increased value of the catch. Going beyond such generalities means close examination of the requirements for various models of fisheries and their population dynamics.

Single-species management

The simplest and most widely-adopted approach to fisheries management has been to make assessments for various species using models that treat one species at a time, as though they were rows of garden vegetables which do not interact with each other. The requirements sound easy: reliable catch and effort statistics, preferably accompanied by research vessel data, and for some models, rates of growth and stock/recruit relationships. The difficulties are well known. Catch statistics are frequently inadequate, effort statistics can be misleading because of technological changes, research vessels are very expensive and stock/recruit models require many years of data. Where these requirements seem onerous, various rules-of-thumb have been devised for rough-and-ready management that errs on the side of sacrificing some catch to take less risk of overfishing (Gulland 1974; Doubleday 1976; Collie and Spencer 1993). But prevailing practice in fisheries has commonly been to press for the maximum sustainable yield (MSY) on a stock- by-stock basis. The concept of optimum sustainable yield (OSY) has not commonly resulted in harvests of lesser amounts than MSY. In consequence, requirements for information and research tend to exceed those for rough and ready management. To push fisheries closer to MSY without having more data and models that consider the interactions among species, implies acceptance of greater risk.

Multi-species management

The most rigorous present assessment technique for dealing with more than one species is multi-species virtual population analysis (MSVPA) which requires, for the major species in the fish community food web, the statistics of catch, including its age composition and data on diet, preferably on a seasonal basis (Sparre 1991). To the extent that some species that are involved in the food web may not be caught or, if caught, discarded, the

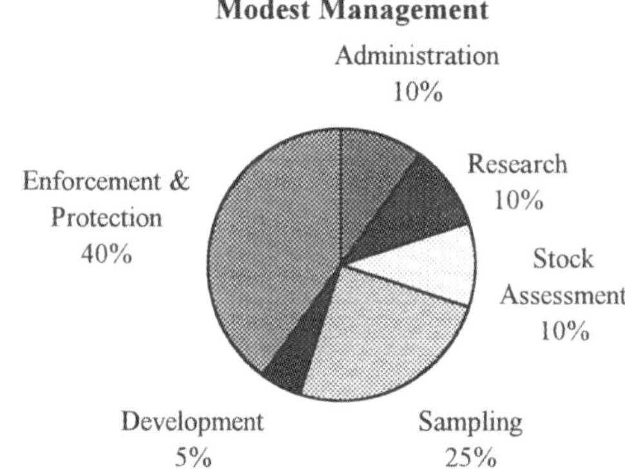

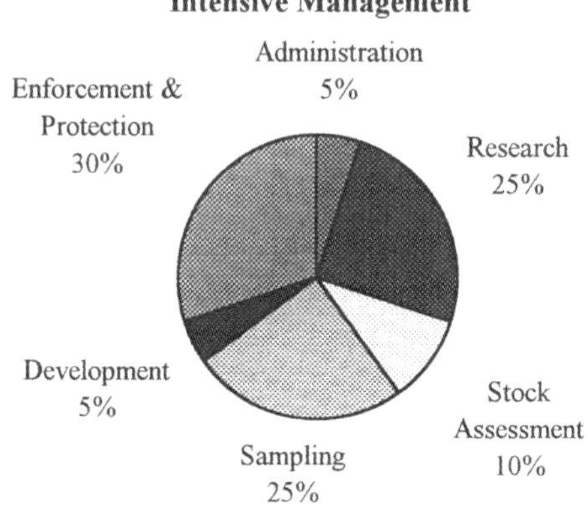

Fig. 1. Ideal proportional expenditures on various activities in fisheries management (adapted from Larkin 1983).

method depends heavily on the assumption that the major commercial species are the major actors in the fish community part of the ecosystem. Independent data from research vessels are virtually essential.

The cost of collecting the necessary data for MSVPA may be formidable. Transponders and satellites can track the movements of vessels but not the makeup of the catch. If there is reason to question the reliability of the catch statistics (and there usually is) it may be necessary to place observers on fishing vessels. Statistics of bycatch have special relevance and most fishermen do not have the time, the staff or the inclination to record bycatches in the necessary detail. The recent review of world bycatch information by Alverson *et al.* 1994 underlines the magnitude of the task of collecting the necessary information. For a comprehensive MSVPA, data on age and diet of the bycatch is also relevant. When the costs of data analysis are added in, MSVPA is not cheap. Some shortcuts may be used to advantage (such as weighing otoliths rather than grinding and counting annuli (Worthington *et al.* 1995)), but in general there is no substitute for spending many more hours at the laboratory bench than in the final synthesis for management. And even with all of this information, MSVPA may still be misleading because of the unseen mortalities caused by fishing gear (Ferno and Olsen 1995; ICES 1995).

MSVPA is probably best assessed as currently the most comprehensive approach to simultaneous assessment of fish community statistics. To fulfil the purposes of management it should be accompanied by a model describing the response of the fishing fleet to changes in relative abundance of species in the catch (so-called fleet interaction models, Brugge and Holden 1991). For these models too, the data requirements are formidable, especially because much of the information needed may be proprietary. Fishermen and fishing companies are intensely competitive and are not enthusiastic about releasing information about their tactics and strategies.

ECOSYSTEM MANAGEMENT

As the decline of world fisheries over the past decade (Garcia and Newton 1994) has become of increasing concern, more attention has been given to what is now being called 'ecosystem management'. Many of the ideas inherent in the concept of ecosystem management have a long history in the literature of fisheries research (May *et al.* 1979). Computers have now made it possible to model the potential implications of the complex interactions among the various components of a community of fishes and the ecosystem in which they are embedded. Now, the key question is 'where are the data to calibrate and validate the models'.

The blithe assumption of enthusiasts of ecosystem management is that ecosystem structure and function can be manipulated to suit human objectives, which presumably at a bare minimum (Larkin 1996) are:

a. a sustainable yield of a desired mix of species;

b. special measures of protection for some species such as marine mammals and birds; and

c. protection from the effects of pollution and habitat degradation.

Unfortunately, (for the manipulator) natural ecosystems are also influenced by changes in the physical environment, to such an extent that stable equilibrium is the exception rather than the rule. In consequence, the relative contributions of the various species to the mix that is harvested may change from time to time. At the same time, the interactions among the species may change.

A comprehensive statement of ecosystem management, such as that of the Ecological Society of America (1995), is breathtaking in its implications for data collection. Ecosystem management is seen to include:

1. Sustainability;
2. Measurable goals that specify future processes and outcomes necessary for sustainability;
3. Sound ecological models and understanding: research at all levels of ecological understanding;
4. Recognition of biological diversity and structural complexity as providing strength against disturbance and greater resources for adaptation to long-term change;
5. Recognition that change and evolution are inherent in sustainability;
6. Multiple scales of time and space for management;
7. Recognition of the values in the active role of humans in achieving management goals;
8. Management approaches viewed as hypotheses to be tested by research and monitoring.

Item 3, calling for sound ecological models and research at *all* levels of ecological understanding alone implies enormous expense. Models such as ECOPATH (Christensen and Pauly 1992) and DYNUMES (Laevestu and Favorite 1981) involve all levels of ecosystems, but at degrees of refinement that depend upon what is available in the literature. For many marine ecosystems, particularly those of tropical latitudes, the degree of detail concerning the benthic community of invertebrates for example, or of the constituents of the zooplankton, is sketchy to say the least. To resolve these problems, greatest detail is given to the major actors of the fish assemblage. In effect, these models are precise at the top but at the bottom are built on progressively darker black boxes.

Item 4, concerning genetic resources, is also enormously demanding of information on the adaptiveness of various component stocks of the major species and also assumes some understanding of what is needed to keep them adapted to each other and to changing circumstances (e.g. see Ryman *et al.* 1995). 'Dream on' is the only possible comment.

Item 8, a plea for adaptive management approaches, is perhaps the most promising suggestion considering the present state of knowledge and what could be done to improve upon it.

Adaptive management

Recognition of the costs of obtaining all of the relevant information either for MSVPA or for holistic ecosystem models (and the follies of acting upon them) has led to enthusiasms for 'adaptive management' as exemplified by the approach to management of the North-West Fisheries of Australia (Sainsbury 1991). The complexity of marine ecosystems defies comprehensive depiction. The sensible and practical approach therefore is to 'try something', observe how the system reacts and learn from the experience, a philosophy developed and advocated by Holling (1978) and his colleagues Walters and Hilborn (1978). The essential message is simple: to understand at the system level, experiment at the system level.

Adaptive management, practised on a system-wide basis can be a scary proposition for fisheries managers. For example, in the Eastern Bering Sea, the relative abundance of many species of marine mammals, marine birds, pelagic and demersal fish species and crabs, has changed remarkably in the past three or four decades. Some of the changes may be attributable to intense fisheries, others to changes in oceanographic conditions (Wooster 1993). What has happened has been a sort of unplanned 'natural experiment' which generates some ideas for large-scale adaptive management to restore conditions to something closer to the historic situation. But some measures to be taken, such as closing fishing for some species and directing major fishing effort to species of lesser commercial value, could be considered as very radical, especially when there would be no great confidence that the results for the ecosystem would resemble the predictions. The argument that much would be learned as a basis for future manipulations is not reassuring to risk-averse managers and industry investors. (You cannot learn much about a watch by trying to repair it with an axe.) In consequence, management is essentially passive and reactive, with much tinkering with regulations and minor adjustments by the fishing industry, and only lip service to what are called 'ecosystem considerations'. There are no major proposals to deliberately perturb the system in a planned way (North Pacific Fisheries Management Council 1994).

Adaptive management on a system-wide scale for major fisheries thus remains more of a theory than a practice which perhaps is as it should be, considering the present state of knowledge of marine ecosystems. The stakes are high and the costs of failure may be great. In the long-term the price may someday have to be paid.

For intense gauntlet fisheries for anadromous species such as Pacific Salmon (*Oncorhynchus*) the requirements of information for management are enormous, reflecting how much is known about the complexities of salmon population structure. In British Columbia alone there are roughly 10 000 individual stocks among the six species of Pacific salmon. The prevailing philosophy calls for management on a stock-by-stock basis because each stock is associated with its particular 'home stream'. For each stock the first provision in the annual run is for escapement, the second to aboriginal peoples, the third to the commercial fishery. The manager is thus faced with the challenge of guessing the size of the run so that what is left after the commercial catch will be enough to meet the remaining provisions. As a further complication, within the commercial fishery there are allocations to three gear types (troll, seine and gill-net). Recreational fishermen also make noises, and as a crowning complication many of the stocks move through United States waters *en route* to their home streams. Because the fishing fleet has at least twice the fishing power necessary to take the permissible catch, there are rarely more than two days fishing in a week over the five-month season. Catches and escapements must be tallied and assigned to stocks on a daily basis, an exercise based on a large accumulated research expenditure, and requiring great expense to maintain as a management information system. Even then, the whole exercise may involve a great deal of fiddling with the numbers at the end of the season when the numbers of fish actually on the spawning grounds can be counted.

Further costs are involved in the understanding of the freshwater ecosystem which the salmon occupy as spawning adults and as juveniles. Substantial research and management costs are involved in estimating escapements, survival and numbers of seaward migrants as well as protection of habitat.

By and large these various costs of salmon research and management have been accepted as necessary and justified on social grounds (salmon as a heritage) but inevitably, questions are now being asked about whether all of the expense is really wise. On the Columbia River in the United States it is said that each chinook salmon now costs tax payers over $US200. In British Columbia the figure may be equally high for at least some stocks of chinook (Pearse 1994). Averaging over all species of salmon in British Columbia, the costs in 1991 of management and research were $Cdn127 million, not including expenditures on environmental protection incurred by the federal department of environment and provincial agencies. The landed value of the catch was $Cdn225 million. Thus, research and management is more than 50% of landed value, about the Canadian average (F.E.A. Wood, pers. comm., DFO Program Planning and Economics Branch, British Columbia). By contrast, the salmon farming industry in British Columbia in 1995 produced $165 million in landed values. The costs of research (largely done by the federal government) and management (by the provincial government) were together of the order of $15 million at most, i.e. less than 10% of the landed value.

Future prospects may be for smaller runs of Pacific salmon in part because of the increasing difficulties of precise management and the resulting decline in salmon abundance. The prospect of climate change is also foreboding (Beamish and Bouillon 1993). Catches will go down and the apprehended needs for more research and better management will go up.

As the complexities of salmon fishing have multiplied it has become increasingly obvious that achievement of the imagined goals of management is virtually impossible. In addition to the ever-expanding requirements for information and analysis, problems of enforcement have come increasingly to the fore, in part reflecting the frustrations of those who go fishing. An endless round of consultations with 'stakeholders' is seen as the 'politically correct' way to go, but achieves little except to make the frustrations a bit better informed.

Rather evidently, salmon management must become much simpler if the costs of information and research are not to exceed the value of the catch, e.g. if fishing were allowed on only one day a week, coastwide, for the whole season, the total catch might be a bit less than it would be otherwise, but the savings in management and enforcement would be substantial.

CONCLUSIONS

Single-species management has the great merit of being simplistic and cheap. Everything other than the species or even the stock of interest is an externality. If the species of interest is ecologically dominant and fishing is not too intense, research and information needs may be slight in relation to value of the catch. Multi-species management is more realistic but more expensive. Maintaining the same relative abundance of species may not be possible in the face of an intense fishery and as stocks of favoured species decline, costs of research and management information increase. Ecosystem management is idealistic and to do it literally would be impossibly expensive. In the present state of knowledge its predictions are far too vague and general to have value for fisheries managers. Adaptive management is inexpensive but highly risky but in the longer term is essential for understanding.

It is tempting to suggest that attempts at management of many fisheries are futile, regardless of what is spent on them. In addition to the difficulties of obtaining the necessary understanding and information, fisheries are notoriously subject to the consequences of public ownership especially where international competition is involved. Politicians are much more likely to subsidize a fishery than to curtail it. The recently prepared FAO (1994), Code of Conduct for Responsible Fisheries is an admirable statement of an ideal, but on the basis of what has happened in fisheries over the past 50 years, the Code is not realistic.

WHAT THEN IS THE BEST PLAN FOR THE FUTURE?

The answer probably lies in extremely rigorous application of the so-called 'precautionary principle' of the Code which in turn implies major reductions in fishing effort. At present, world fishing fleets have three or four times the capacity necessary to take the present catch. With so much excess fishing capacity there is strong competition for fish with many implications for enforcement of regulations and collection of reliable information for management. Research needs have multiplied as fisheries have become more intense and more diversified. Costs of fisheries management and research have continued to increase as progressively more stocks are overexploited and the landed values of catch have declined. The institutional structures have, so far, simply not been strong enough to take the necessary steps for reducing fishing effort. Meanwhile, the costs of research, management and enforcement will continue to grow until it becomes obvious that as a national business, catching wild fish is not a viable enterprise. Fish farming is a much better investment.

Perhaps the most promising trend for wild fisheries is the growing interest in rights-based fishing (Neher *et al.* 1989). Given a guaranteed share in the resource, participants will self-enforce regulations, collect information, support research and more actively advocate the role of fisheries in the context of coastal zone management. There is a danger of concentrating control of the resource in the hands of a few, but that seems a small social price to pay by comparison with the consequences and management costs of open access.

POSTSCRIPT

To prepare for this paper I arranged for a search of four databases: 'Waves', the library catalogue of Fisheries and Oceans Canada; 'Aquaref' (Environment Canada); UBC Library catalogue; and 'Current Contents', a United States service. To my surprise the searches uncovered remarkably few relevant references. There is considerable evidence that Treasury Board officials look at the broad picture of research and development (R&D) expenditures, but have difficulties in making comparisons between various kinds of activities (say for example between electronics, space science and fisheries). Within fisheries departments comparisons are made among various fisheries, the expenditures being very roughly proportional to their landed value when averaged over a reasonable period of years. The disaggregation of all costs by species/fishery is done internally though apparently not systematically. The whole subject of the costs of research and management of fisheries is well worth detailed examination.

REFERENCES

Alverson, D L, Freeberg, M H, Murawski, S A and Pope, J G (1994). A global assessment of fisheries bycatch and discards. *FAO Fisheries Technical Paper* 339. 233 pp.

Beamish, R J, and Bouillon, D R (1993). Pacific Salmon trends in relation to climate. *Canadian Journal of Fisheries and Aquatic Sciences* **50**, 1002–16.

Brugge, W J, and Holden, M J (1991). Multispecies management: a manager's point of view. *ICES Marine Science Symposium* **193**, 353- 8.

Christensen, V, and Pauly, D (1992). ECOPATH II — A system for balancing steady-state ecosystem models and calculating network characteristics. *Ecological Modelling* **61**, 169–85.

Collie, J S, and Spencer, P D (1993). Management strategies for fish populations subject to long-term environmental variability and depensatory predation. In 'Management strategies for exploited fish populations'. (Eds G Kruse, D M Eggers, R J Marasco, C Pautzke and T J Quinn II) pp. 629–50. *Alaska Sea Grant College Program Report* 93-02 (University of Alaska, Fairbanks).

Doubleday, W G (1976). Environmental fluctuations and fisheries management. *International Commission for Northwest Atlantic Fisheries Selected Papers* **1**, 141–50.

Ecological Society of America (1995). Report of the Committee on the scientific basis for ecosystem management. (N L Christensen, Chair). 6 pp.

FAO (1994). Code of conduct for responsible fisheries. FAO Secretariat.

Ferno, A and Olsen, S (Eds) (1995). 'Marine fish behaviour in capture and abundance estimation.' (Fishing News Books: Oxford).

Garcia, S M, and Newton, C (1994). Current trends and prospects in world fisheries. FAO Fisheries Division, Rome.

Gulland, J A (1974). 'The management of marine fisheries.' University of Washington Press, Seattle. 198 pp.

Holling, C S (Ed) (1978). 'Adaptive environmental assessment and management.' (John Wiley and Sons, New York). 363 pp.

ICES (1995). Study group on unaccounted mortality in fisheries. *International Council for the exploration of the Sea*. C.M. 1995/3:1.

Laevestu, T, and Favorite, F (1981). Holistic simulation of marine ecosystems. In 'Analysis of Marine Ecosystems'. (Ed A R Longhurst) pp 702–27 (Academic Press: London).

Larkin, P A (1983). How much is enough? An essay on the structure of fisheries management agencies. Global Fisheries: Perspectives for the 1980s. (Ed B J Rothschild) pp. 229–45 (Springer Verlag New York).

Larkin, P A (1996). Concepts and issues in marine ecosystem management. *Reviews of Fish Biology and Fisheries* **6**, 1-26.

May, R M, Beddington, J R, Clark, C.W, Holt, S J, and Laws, R M (1979). Management of multispecies fisheries. *Science* **205**, 267–77.

Neher, P A, Arnason, R, and Mollett, N (1989). Rights based fishing: Proceedings of the NATO Advanced Research Workshop on Scientific Foundations for Rights Based Fishing. Reykjavik, Iceland, June 27–July 1, 1988. 541 pp.

North Pacific Fisheries Management Council (1994). Ecosystem considerations 1995: The Plan Teams for the Groundfish Fisheries of the Bering Sea, Aleutian Islands, and the Gulf of Alaska. 88 pp.

Pearse, P H (1994). Salmon enhancement: an assessment of the salmon stock development program on Canada's Pacific Coast. Fisheries and Oceans Canada. Vancouver.

Ryman, N, Utter, F, and Laikre, L (1995). Protection of intraspecific biodiversity of exploited fishes. *Reviews in Fish Biology and Fisheries* **5**, 417–46.

Sainsbury, K J (1991). Application of an experimental approach to management of a tropical multispecies fishery with highly uncertain dynamics. *ICES Marine Science Symposium*. **193**, 301–19.

Sparre, P (1991). Introduction to multispecies virtual population analysis. *ICES Marine Science Symposium* **193**, 12–21.

Walters, C J, and Hilborn, R (1978). Ecological optimization and adaptive management. *Annual Reviews in Ecology and Systematics* **9**, 157–88.

Wooster, W S (1993). Is it food? An overview. In 'Is it Food? Addressing marine mammal and seabird declines.' Workshop Summary. pp. 1-3. *Alaska Sea Grant Report* 93-01.

Worthington, D G, Fowler, A J, and Doherty, P J (1995). Determining the most efficient method of age determination for estimating the age structure of a fish population. *Canadian Journal of Fisheries and Aquatic Sciences* **52**, 2320–26.

Obituary

Peter Larkin was one of Canada's premier scholars. It is difficult to find an area of resource management in which his name does not occupy a prominent position. Born in New Zealand in 1924, he moved to Canada as a child. He earned his BA and MA degrees from the University of Saskatchewan. In 1946 he went as a Rhodes Scholar to Oxford University, where he received his DPhil. He began his career in 1948 as the Chief Fisheries Biologist in British Columbia, where he focussed on the management of freshwater fisheries in BC lakes and rivers. His career as a fisheries biologist at the University of BC spanned four decades, during which he held many administrative positions. His influence on national and provincial fisheries, and in environment and science policy, was enormous. He also served on many national and international commissions and committees. Throughout his career he was the recipient of many honours and awards. His legacy of writings, thought, wisdom, special wit and humour will long be remembered and, even more important, will continue to be used in shaping our future actions.

Indicators of Sustainable Fisheries Development

Derek Staples

Fisheries Resources Branch, Bureau of Resource Sciences, Canberra, ACT Australia.

Summary

Despite sustainable development being the catch cry of many international and national governments throughout the 1990s, it is difficult to assess how well these governments are meeting sustainable development objectives. The reasons for this are two-fold. Firstly, the concept of sustainable development is poorly defined and open to many interpretations, often resulting in poorly defined objectives and strategies. Secondly, even if objectives are well specified, there is usually no way to assess progress toward meeting the objectives.

In a fisheries context, adopting sustainable development principles requires broadening the traditional fisheries management concerns of sustainable fishing of target species (with associated concepts such as maximum sustainable yield) to consideration of the total ecosystem health, and both the social and economic well-being of all stakeholders in the fishery.

The development of a framework for evaluating progress towards achieving this sustainable development of fisheries then requires: (i) specifying what is trying to be achieved in terms of the fisheries resource, the environment, as well as the economics and social benefits, (ii) defining what is meant by success in meeting these objectives, (iii) developing performance indicators against these objectives and a basis for comparison, and (iv) implementing the system to include a decision-making framework in which management actions result from a pre-defined change in indicators.

There is an urgent need for this type of development and implementation of fisheries management performance evaluation frameworks to ensure that sustainable development of fisheries becomes a reality not just a poorly defined concept.

Introduction

The concept of sustainable development has been the popular goal of international and national governments throughout the 1990s. However, although the concept of sustainable development has been articulated in a range of fora, including the now famous United Nations Conference on Environment and Development (UNCED), it is difficult to determine whether the principles of sustainable development have in fact been implemented. This is particularly the case in fisheries. This difficulty arises from, firstly, divergent interpretation of what is meant by sustainable development for fisheries and, secondly, the lack of an evaluation framework and performance indicators.

Interpretations of sustainable development of fisheries range from narrow views that state that fisheries management has always been about sustainable development because of the objective of obtaining maximum sustainable yield to those that more broadly embrace the complex concepts underlying sustainable development. These differences in interpretation lead to difficulties in developing appropriate objectives and consequently to difficulties in measuring and assessing progress towards meeting these objectives. In this paper I will examine both the setting and evaluating of sustainable development objectives for the fisheries by attempting to answer the following questions:

- What is sustainable development?
- What is the role of fisheries management in the context of sustainable development?
- How do you evaluate whether sustainable development is being achieved?
- What is the role of indicators in the evaluation?
- How do we use indicators in decision making?

SUSTAINABLE DEVELOPMENT

History of sustainable development

Concerns for the plight of humanity and the earth's ability to cope with the pace of development date back many centuries. It is only recently, however, that people have begun to consider the effects of continuing development on a global scale. During the 1960s and 1970s, local problems of pollution and environmental degradation were seen by some as symptoms of a much larger problem. It was during this time that a group of people from a broad range of professional and cultural backgrounds, the so-called Club of Rome, met to consider 'The Predicament of Mankind'. This resulted in the publication *The Limits to Growth,* which concluded that development could not continue at its 1970 rate because there were absolute limits to the natural resources available and the amount of pollution the world could tolerate. It also predicted that these limits would be met very quickly with the exponential rise in population and consumption, and advocated that the world should move towards a state of equilibrium rather than of continuing growth.

Today, very few people would agree with all the findings and arguments put forward in *The Limits to Growth.* Its major flaw was that it failed to account for advances in technology and efficiency in discovering, producing and consuming resources. It also assumed exponential growth in consumption which proved to be incorrect. Nevertheless, the book did introduce two important concepts. The first was that of sustainable development itself (although it advocated diminishing growth) and the notion of intergenerational equity, which should ensure that acceptable standards of living are handed on from one generation to the next.

Whereas the concerns about the availability of non-renewable resources tended to diminish in the late 1970s and 1980s, awareness of the impact of human activity on the environment (e.g. greenhouse effect, ozone holes) increased dramatically.

The World Commission on Environment and Development report of 1987, *Our Common Future* (the so-called Brundtland Report), made it clear that the world's current pattern of economic growth was not sustainable, and that a new type of development was required. One of the essential ideas of this report was the recognition that there is a strong interdependence between sustaining the environment and sustaining development, and one can not occur without the other. An equally important concept that also came from this report was that sustainable development is a dynamic process that changes and evolves in association with changes in how present and future needs are met.

Definition of sustainable development

Before attempting to define sustainable development, it is important to recognize that different people, groups and societies hold different views about what nature is really like.

The first view is that **nature is extremely robust,** is insensitive to human impacts and is virtually inexhaustible as a resource base. Market forces are seen to be the only necessary regulatory mechanism for the system. The advocates of this view promote managing the environment through market efficiency and the spread of technology and goods. The opposite view is that **nature is extremely fragile,** and the system is vulnerable to irreversible collapse due to ecological degradation or exploration of natural resources. Changes to our current economic, political and social systems are seen to be necessary to preserve the existing ecosystems and return to 'harmony with the natural world'. According to a third view, **nature is robust within limits,** which must not be surpassed. This view is held by many governments, which maintain that the use of natural resources needs to be monitored and managed by a specific agency; and that accurate scientific understanding of the ecological limits and the development of rational management will enable economic growth to be maintained. The fourth view is that **nature is essentially random,** and will continue in some form or other, but is unpredictable. This is the fatalist's view, in which management does not have a place and one must survive as best as one can.

Depending on which view is held, sustainable development will have a different meaning. Because each of the views is probably correct to some degree, there is no single interpretation of sustainable development and no single solution to a given problem.

The above discussion should have convinced the reader that a single, simple definition of sustainable development is not particularly useful. However, the Brundtland Commission suggested 'to meet the needs of the present without compromising the ability of future generations to meet their own needs'. The Australian NESD Strategy suggested 'to promote economic and social development which aims to meet the needs of Australians today while conserving the ecosystem for the benefits of future generations'.

One important point is that the concept is much more than the sum of the two words 'sustainable' and 'development'. On the one hand, 'sustainabilty' relates to questions concerning carrying capacity of the earth and does not consider social issues, particularly those concerning equity and social justice. 'Development', on the other hand, relates to the improvement in the quality of life and ignores ecological constraints and carrying capacity. A very different concept emerges when the two concepts are combined and the result is much more than the sum of the parts.

'Development' is also often equated incorrectly with 'growth'. 'Growth' refers to a change in quantity whereas 'development' refers to a change in quality. It is, therefore, possible to have development without economic growth, growth without development, neither or both. However, in the current decade, it is commonly assumed that development is not possible without economic growth. This obviously depends on one's view of nature as discussed above.

The spatial and temporal scale under consideration will also influence the interpretation of sustainable development. The concept will differ according to whether sustainable development is being considered on a global, regional, national or sectorial scale. Similarly, whether one is considering short-term or long-term implications will also have a profound effect on the view of sustainable development.

Characteristics of sustainable development

As well as providing a definition of sustainable development, the Brundtland Report served as a basis for discussion and implementation of programmes for sustainable development. A global action plan known as Agenda 21 was developed by the United Nations and adopted at the UNCED in Rio de Janeiro in June 1992. Some of the general characteristics that distinguish a sustainable approach to development are:

- consideration of the integrated social, economic and environmental implications of decisions for development,
- consideration of the long-term rather than the short-term view when taking decisions and actions, and
- assessment of the material and non-material well-being of the ecosystem over the longer-term in making development decisions.

In Australia, the Council of Australian Governments endorsed a national strategy for sustainable development in 1992. Australia adopted the term 'ecologically sustainable development' (ESD) to emphasize the importance of considering the well-being of the ecosystem in promoting sustainable development. The strategy suggested the following definition of ESD: To promote economic and social development which aims to meet the needs of Australians today while conserving the ecosystems for the benefit of future generations.

The core objectives of the National strategy are:

- to enhance individual and community well-being by following economic development that safeguards the welfare of future generations,
- to provide for equity within and between generations,
- to protect biological diversity and maintain essential ecological processes and life-support systems.

The guiding principles also adopted the *precautionary principle*, which requires that decisions should be more cautious when information is uncertain, unreliable or inadequate. The absence of adequate scientific information will not be used as a reason for postponing or failing to take conservation and management measures (FAO 1995).

Whatever the definition, it is clear that the concept is intended to be an integrated framework for making decisions about resource use, which involves taking the wider economic, social and environmental implications of decisions and actions, taking into account both short- and long-term implications.

FISHERIES MANAGEMENT IN THE CONTEXT OF SUSTAINABLE DEVELOPMENT

Over the past 50 years, fisheries managers and society in general have changed their view of nature from a belief in a very robust system with inexhaustible resources to one of being either robust within limits or extremely fragile. In association with this change, fisheries management has become concerned with one (albeit important) aspect of sustainable development—sustainability of the target species that provide the economic benefit to the fishing industry and meet the need for food for humans. Concepts such as the maximum sustainable yield (MSY) were introduced to maintain target species populations within the ecological carrying-capacity limits.

Concepts such as MSY became *de facto* objectives for fisheries management during the 1950s and 1960s, were formally adopted in several international fisheries conventions, and still persist in some jurisdictions today. In many countries, the constant catch strategy implied by MSY came to be replaced by a constant effort (or constant exploitation rate) strategy, for example, so-called F zero-point-one ($F_{0.1}$) (Gulland and Boerema 1973) based on biological parameters such as growth and longevity and estimates of current exploitable biomass. This approach came to be widely adopted and applied, particularly in North Atlantic fisheries, in the 1970s and 1980s.

However, concepts such as MSY or $F_{0.1}$ do not capture many of the principles of sustainable development. They tend to imply that if the resource or fishing is sustainable, then sustainable development objectives have been met. However, just because harvesting of a particular species is sustainable in the sense that the fish population is expected to sustain this rate indefinitely without the spawning stock declining any further, it is not necessarily consistent with sustainable development. For example, if the method of harvesting destroys the habitat of other fish species, or threatens an endangered marine mammal, or excludes a large number of people from a desirable recreational activity, the fishing activity may not be acceptable in terms of sustainable development.

To be consistent with sustainable development, the objectives of fisheries management have to adopt a broader perspective that encompasses the broader environmental, as well as the socio-economic, aspects of sustainable development.

Measuring progress towards sustainable development

Performance Evaluation

If the overall objective of fisheries management is to ensure sustainable development through objectives aimed at the environmental, social and economic aspects of the fishery, how do we evaluate whether this is being achieved? Concepts such as biological reference points were introduced to provide a way of evaluating whether the harvesting rate on a particular target species was at an appropriate level in terms of sustaining the resource. The biological reference points are either target points (points or levels for management to aim at, e.g. MSY), or limit points (points or levels below which a given attribute should not fall e.g. minimum spawning biomass).

Although biological reference points are useful in determining whether fisheries management is meeting one of its specific management objectives (sustainable harvesting of target species), the evaluation process, as with the objectives themselves, has to become much broader to encompass other principles of sustainable development.

Conceptually, the evaluation process consists of firstly setting clear objectives (focussing on outcomes), defining what a successful outcome would be, developing performance indicators, and developing a basis for comparing the indicators (Fig. 1).

Setting objectives and outcomes

Multiple objectives

To be consistent with sustainable development, multiple objectives need to be clearly stated for economic, and social, as well as for environmental/resource, components of fisheries. Obviously, the intent and wording of objectives will vary from country to country and fishery to fishery depending on stage of development and overall policies. The important point is that, if we are trying to achieve sustainable development in fisheries, objectives are needed for all components, not just the harvesting of the resource. The objectives need to provide a framework for decision making that will sustain the development in both the short and the longer term.

Hierarchical objectives

Management objectives, strategies for reaching those objectives, and performance criteria, are often confused. This is because there are numerous levels of objectives and strategies to achieve these objectives. Objectives and strategies are linked in a hierarchy, in that the strategy of one level becomes the objective of the next. One classification of objectives is to start at the top level by considering the human needs and strategies to meet these needs (Fig. 2). Policy and legislation are then developed to implement these strategies, which, in turn, form the basis for overarching fisheries management objectives for a given fisheries jurisdiction. More specific objectives are then developed for individual fisheries; they state what management is attempting

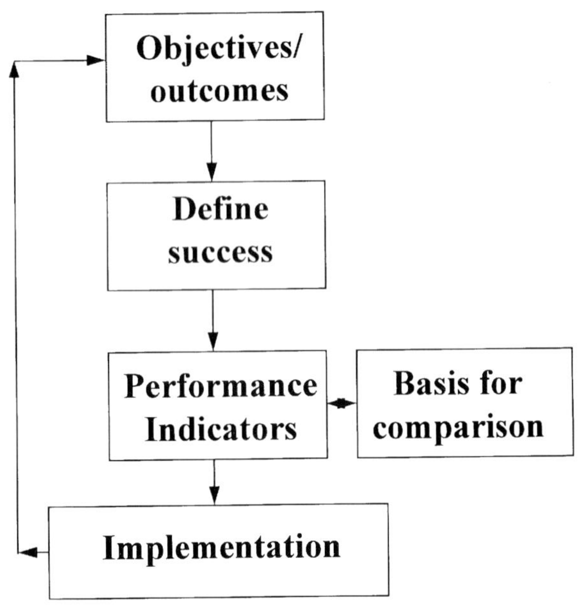

Fig. 1. Flow diagram of steps required to carry out a performance evaluation of fisheries management.

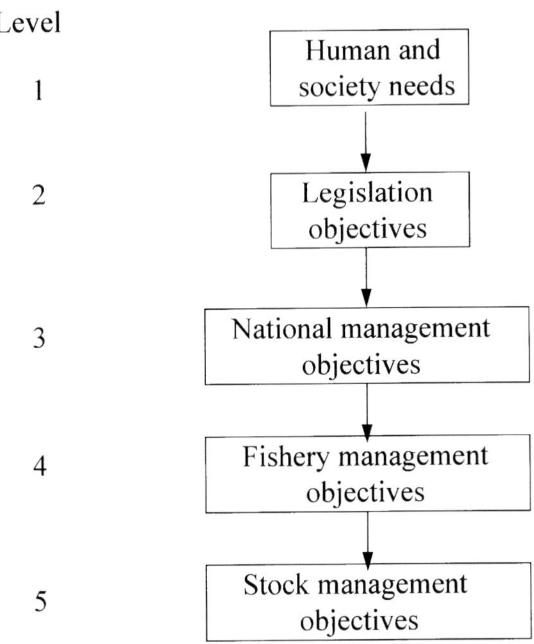

Fig. 2. Hierarchical objectives/outcomes for fisheries management.

to achieve for that fishery. Objectives are often developed for individual stocks based on their biological status and methods of harvesting to make management more achievable and its performance measurable.

A simple example may make this clearer. Two obvious needs are satisfied by fishing: provision of food for other humans and profit for operators (so that they can use the money to supply their needs). In Australia, the legislative objectives to achieve these needs for fisheries under Commonwealth jurisdiction are contained in the *Commonwealth Fisheries Management Act 1991*:

- *ensuring that the exploitation of fisheries resources and the carrying on of any related activities are conducted in a manner consistent with the principles of ecologically sustainable development, in particular the need to have regard to the impact of fishing activities on non-target species and the marine environment.*

[Although not clearly articulated in the legislation, the principles for ecologically sustainable development are contained in the National Strategy for ESD (Anon. 1992).]

- *maximizing economic efficiency in the exploitation of fisheries resources'.*

The Australian Fisheries Management Authority (AFMA), in turn, has developed three specific objectives based on the strategies for achieving ESD principles and economic efficiency:

- *manage Commonwealth fisheries at an ecologically sustainable fishing level,*
- *maximize economic returns to the industry and the broader community while meeting sustainability objectives, and*
- *effectively communicate and consult with government, the fishing industry, other marine resource users and the broader community.*

Several strategies are then given to achieve these, each strategy in turn becoming an objective at a lower level. Even more-specific objectives are then formulated in a Fishery Management Plan for each fishery and, within these fisheries, for particular target species. These lower-level objectives are included in AFMA's annual Fisheries Assessment reports.

Evaluating whether AFMA is managing its fisheries in such a way as to meet sustainable development requires the development of performance indicators for each level of objectives.

Defining success

The next stage is take each objective/outcome and define success. The main reason for this is that the objectives are written in very general terms and, unless success is defined, the outcomes can be interpreted differently by different people, depending on their perspective. Each outcome is taken separately and a perfect outcome described, considering issues such as cost, timeliness, relevance, accuracy, quality and quantity. Success is then defined in terms of the level of performance that can be realistically achieved.

Developing Performance Indicators

A performance indicator is:

- a piece of information that focusses on important and useful information,
- expressed as an index, rate or other ratio in comparison with one or more criteria, and
- monitored at regular intervals.

In developing indicators, consideration should be give to who will be the main users of the indicators; the main users will probably be decision makers in the public and private sectors as well as the general public. As such, indicators for their use should be highly aggregated so that they provide concise information devoid of detail.

Each indicator requires a basis for comparison. Comparisons can be made:

- with a standard
- with a target
- with a threshold or limit value
- before and after a change, and
- with a similar management objective made by another agency (benchmark).

Performance indicators should provide information on appropriateness, effectiveness and efficiency. Appropriateness is a check on how well objectives match government priorities and community needs. Effectiveness measures the extent to which management achieves stated objectives; and efficiency is the extent to which the inputs to management are maximized to provide outputs and outcomes.

Implementing performance indicators

Australia has had limited experience in developing and implementing indicators of sustainable development for fisheries. The main problems encountered include achieving a simple system that is practical and useful, and making rational decisions about the level of investment required to obtain the information needed for a particular indicator. Many of Australia's fisheries are small, low-value and low-investment fisheries. A cost–benefit analysis is needed to determine the level of research and effort that should go into indicators that are expensive to monitor and assess.

Another difficulty is to determine who should be introducing the performance indicators. Ultimately, it is a management responsibility, but in cases where industry is heavily involved in management decision making, use of a 'public watchdog' may be warranted. In Australia, where Commonwealth fisheries are managed under a partnership arrangement with major stakeholders, in particular with the commercial fishing industry, government agencies are responsible for auditing the performance of Commonwealth fisheries management.

Answer: Yes/No/Uncertain.

Fishery:

State of development: Underfished/overfished/fully fished

Fishery ESD Check List	
Fishing pressure	
Is current fishing effort sustainable?	
Is latent effort at an acceptable level?	
Resource/ecosystem status	
Target species	
Is the current catch sustainable?	
Are catch rates stable?	
Is the spawning stock adequate (compared with biological reference points)?	
Non-target species	
Is the level and composition of bycatch acceptable?	
Is the quantity of bycatch declining?	
Is the composition of bycatch stable?	
Impact of fishing on the environment	
Is the impact of fishing on fish habitats and the marine environment acceptable?	
Is the impact declining?	
Impact of other activities on the environment	
Pollution	
Are pollution levels declining?	
Habitat change	
Is the area of critical fisheries habitat stable?	
Socio-economic status	
Is the community obtaining an adequate return?	
Is the fishery profitable or providing net benefits?	
Are the average rates of returns sufficient for economic sustainability?	
Is the rate of new investment commensurate with return?	
Is the capital investment appropriate (not overcapitalized)?	
Is the fishing unit value stable?	
Is employment stable?	
Management	
Is a formal management plan in place?	
Is fisheries management meeting its objectives?	
Are management costs being met by the stakeholders?	
Is habitat management adequate to meet its objectives?	
Is pollution control adequate?	

Fig. 3. Fishery Sustainable Development Indicator Check List.

At a national level, it has been decided to develop and introduce sustainable development indicators in the form of an indicator check list (Fig. 3). In this approach, a series of indicators has been developed and compared against various criteria in the form of a simple yes/no/uncertain list of questions. The indicators cover fishing pressure, status of the resource (both target and non-target species), and health of the marine environment, as well as a range of socio-economic indicators. Considerably more work is required in developing more specific outcomes, definitions of success and performance indicators, but the approach provides a practical entry point for fisheries managers.

Use of indicators in decision making

One of the main uses of indicators should be to guide decisions. As well as the obvious benefit of providing performance criteria against which governments can assess their rate of progress towards meeting UNCED objectives, sustainability indicators can also be used operationally to link future management to agreed actions. These decision rules can provide commitment to remedial action when objectives are not met, providing long-term benefits to both the fishing industry and the public. The process becomes pro-active rather than reactive and is transparent to all stakeholders.

The development of broad policy performance indicators is relatively young and not yet incorporated into the decision-making process. However, a similar process, which considers more specific management objectives at a fishery or stock level, has been developed sufficiently to be incorporated in the decision-making process. This 'management strategy evaluation' uses several different performance indicators to evaluate alternative harvest strategies. The problem is posed within a decision-making framework, and involves the following steps:

- clearly defining a set of management objectives
- specifying a set of quantitative performance indices related to each objective
- identifying various alternative harvest strategy scenarios, and
- evaluating the performance indices for each harvest strategy.

In many cases, the evaluation in the last step involves simulating the future development of the fishery over a specified period under each harvest strategy. Examples of performance indices include total catch (or profit) over the period, variability in catch and the frequency with which threshold stock sizes are transgressed. The results are presented as a decision table to allow trade-offs to be made across conflicting objectives.

The most detailed application of the approach has been in the development of 'management procedures', in the International Whaling Commission and in several fisheries in South Africa. These management plans specify the types of data that will be used to assess the state of the fishery, how they will be collected, the models that will be used to analyse them, and the 'decision rule', which will be used to turn the output of the model (the 'stock assessment') into the annual decision on the management measures to be adopted (such as the annual quota in the case of an output-controlled fishery). The management procedure is developed and evaluated in full consultation between scientists, industry and managers, but, once agreed to, it is set in place for several years and not lightly set aside. One advantage of the approach is to remove annual decision making from the highly uncertain annual stock assessments. Under this type of management procedure, stock assessment becomes almost automatic, although decisions will vary from year to year on the basis of new information. The approach provides for much more certainty in the decision process for all concerned.

This approach appears to have many advantages and should be developed further, incorporating the broader aspects of sustainable development including the impact of fishing on non-target species and the environment, as well as social implications of decisions. In this way, fisheries management will be in a position to lead the world in the application of sustainable development principles, and provide a framework for better decision making in meeting the needs of the current generation without compromising the ability of the next generation to meet its own needs.

References

Anon. (1992). National Strategy for ecologically sustainable development. (Australian Government Publishing Service: Canberra). 128 pp.

FAO (1995). Precautionary approach to fisheries. Part 1. Guidelines on the precautionary approach to capture fisheries and species introductions. *FAO Fisheries Technical Paper* 347. 83 pp.

Gulland J A, and Boerema, L K(1973). Scientific advice on catch levels. *Fishery Bulletin* **71**(2), 325–35.

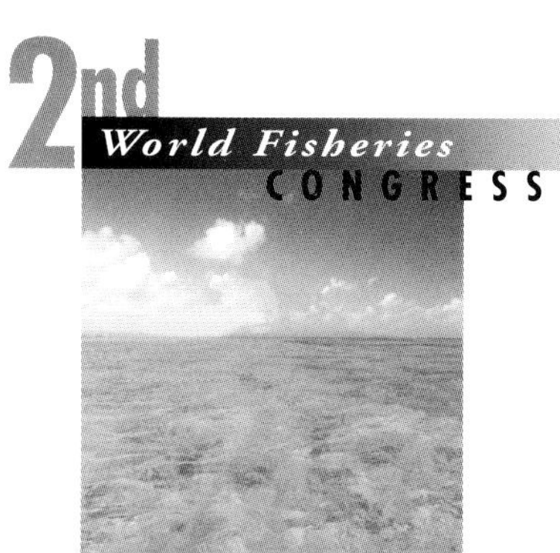

A New Paradigm for Managing Marine Fisheries in the Next Millennium

Michael Sutton

Endangered Seas Campaign, WWF International, Panda House, Weyside Park, Catteshall Lane, Godalming, Surrey GU7 1XR, UK.

Summary

The history of modern fishery management is replete with spectacular failures. In virtually every case, the *real-politik* of fisheries has rendered long-term sustainability of catches a futile management goal. All too often, politicians have compelled fishery managers to ignore the implications of the best available science. Management actions that might have prevented the disastrous collapse of fisheries, but which carried a price unacceptable to industry, have been scrupulously avoided. Society has simply lacked the political will to forestall the fishing industry's tendency to use up its living capital and thereby destroy itself.

Reversing the fisheries crisis will require a major overhaul of contemporary fishery management. Public concern and market forces must be harnessed in support of new international norms. This paper presents a series of necessary reforms and suggests how, by working together, conservationists, scientists, fishers, industry, and governments can help shape the future of world fisheries. World Wide Fund for Nature (WWF) and Unilever NV/plc have taken a first step by launching the Marine Stewardship Council initiative, an innovative plan to bring market forces and consumer power to bear in favour of sustainable, well-managed fisheries.

Introduction

The need for fundamental reform of marine fishery management has become abundantly and painfully clear over the past decade (Earle 1995; Parfit 1995; Safina 1995; Weber and Gradwohl 1995; WWF 1996*a*, 1996*b*, 1996*e*). Fisheries that once sustained coastal communities have suffered catastrophic declines. In some areas, excessive fishing has driven staple species such as northern cod and Atlantic halibut to commercial extinction. The Food and Agriculture Organization of the United Nations (FAO) reports that 70 % of the world's commercially-important marine fish stocks are fully-fished, over-exploited, depleted, or slowly recovering (FAO 1995). Governments pay an estimated \$US54 $\times$ 10^9 each year in fisheries subsidies to catch only \$US70 $\times$ 10^9 worth of fish (FAO 1993). Increasingly volatile 'fish wars,' such as the 1995 dispute between Canada and Spain, have erupted over what remains. Gone forever are the historical estimates that marine catches could top 500 million t per year. Without doubt, modern commercial fisheries have exceeded the limit of the seas.

The essential question is thus not whether the past model of marine fishery management has failed, but why? What lessons can we draw for the future? Throughout modern history, governments have largely managed world-wide marine fisheries for the growth and development of their associated commercial fishing industries. Decision makers have paid scant attention to the sustainability of those fisheries, much less the health of their associated ecosystems or the needs of artisanal fishers exploiting the same species. In virtually every case, the short-term social and economic needs of a region's fishing industry have rendered long-term sustainability of catches a futile management goal. In many parts of the world, subsidized fleets have become grossly overcapitalized (FAO 1995). Unsustainable fishing has literally become an industrial addiction.

This predicament cannot be attributed to a lack of scientific information. Fisheries scientists have for years provided more-or-less accurate models of fish population dynamics and educated estimates of fishery production. But all too often, fishery managers more concerned with political than scientific realities have been compelled to ignore the implications of the best available science. Politicians, often at the highest levels, have frequently intervened in decisions about specific fisheries. Governments facing undeniable disasters have typically devised politically-expedient 'solutions' and then described them as environmentally necessary. But these efforts have mostly been too little, too late. Management actions that might have prevented the disastrous collapse of fisheries, but which carried a price unacceptable to industry, have been scrupulously avoided. Society has simply lacked the political will to forestall the fishing industry's tendency to use up its living capital and thereby destroy itself.

Turning this situation around will require more than merely reinventing contemporary fishery management. Two heretofore overlooked influences will have to be harnessed to help reverse the fisheries crisis and forge a new paradigm of management: public support and market forces (Sutton in press). First, greater public awareness, concern, and involvement in fishery management must be generated. Somehow, the same world-wide public concern that motivated governments to ban the trade in elephant ivory and outlaw commercial whaling must be brought to bear. Second, market-led economic incentives must be created to promote sustainable fishing. 'Sustainable use' means using renewable resources such as marine fisheries 'at rates within their capacity for renewal' (IUCN/UNEP/WWF 1991). Conservationists, working with responsible, progressive seafood companies and other stakeholders, must develop market reforms that will encourage consumers to purchase seafood products that come from sustainable, well-managed fisheries.

This paper will outline ten essential reforms of marine fishery management and suggest how, by working together, conservationists, scientists, fishers, industry and governments can help shape the future of world fisheries and the marine environment.

Restoring abundant seas: ten essential reforms

To reverse the fisheries crisis, long-term solutions that are environmentally necessary must be developed and then made politically feasible through the use of public pressure and economic incentives. The following package of ten reforms will be essential in order to speed the transition to sustainable, well-managed, and ecologically sound fisheries.

Strengthen national, regional, and international capacity to manage marine fishes

Governments must allocate sufficient funds to develop the scientific and technical capabilities necessary to adequately manage their marine fisheries. Those nations suffering from a fishery management system rife with conflicts of interest must reform their laws (WWF 1995*a*). International commissions charged with managing shared fisheries and those on the high seas must immediately implement the provisions of the 1995 UN Agreement on Straddling Fish Stocks and Highly Migratory Fish Stocks. Among other things, this will require regional bodies to open their decision-making procedures to public scrutiny. In addition, fishery management at all levels must be relieved from sweeping political interference aimed at satisfying the short-term economic needs of fishers rather than the long-term requirements of fish populations and the marine ecosystem. To help ensure that these reforms are carried out, the United Nations should create a high-profile Global Oceans Forum to elevate fishery conservation and other ocean issues higher on the international political agenda. The Oceans Forum could report annually to the UN Secretary General on the state of the oceans, especially marine fisheries (WWF 1995*b*).

Focus management programmes on limiting effort and restricting access to fisheries

Past efforts at fishery management have often been characterized by ineffective measures such as mesh-size restrictions and trip limits that simply attempt to mandate inefficiency (WWF 1995*a*). These techniques should be abandoned in favour of management schemes that limit fishing effort, especially in fisheries that are overfished or depleted. Effort should be reduced to levels consistent with sustainable fishing and the recovery of depleted species. Limited-access programmes should prevent new entry into fisheries that are fully subscribed. Such programmes should form a part of comprehensive management plans for each fishery. Private property rights in fisheries, if any, should be created with extreme caution in order to safeguard the public interest in these resources.

Enact and implement recovery plans for depleted species

Many overfished species, even those that have been severely depleted, are not subject to any kind of recovery plans. Fishery managers should as a matter of priority develop and implement effective recovery plans that include target population sizes and timetables for achieving them. The targets and pace of these plans should be driven primarily by the biological requirements of the fish populations involved, not the short-term demands of the local fishing industry. Well-managed fisheries that are allowed to recover from past overfishing would help restore the vitality of the marine ecosystem and concurrently yield far more to fishers.

Reduce and eliminate the subsidies that sustain commercial fisheries

The \$US54 $\times 10^9$ in subsidies that is propping up unsustainable fisheries must be eliminated immediately, including those funds for shipbuilding and construction, refitting of fishing vessels, market research and development, industry bail-outs, low cost industry loans, and development of fisheries for so-called 'underutilized' species. Where subsidies are provided, they should be part of a comprehensive plan for the future of the fishing industry, including decommissioning of fishing vessels and retraining of fishers, where necessary.

Accelerate programmes for decommissioning excess fishing fleet capacity

Overcapacity due to unbridled growth of the world's fishing fleets is one of the most serious problems facing marine fisheries today (WWF 1996*c*). At a minimum, funds available for vessel buy-back and decommissioning programmes should be increased as quickly as possible in order to expand these programmes to achieve an immediate reduction in fishing effort (House of Lords 1996). Funding of future decommissioning programmes should be dictated by capacity-reduction targets specific to each fishery. Governments should make the appropriation of these funds a top priority in drawing up their annual fisheries budgets.

Expand programmes for retraining fishers displaced by overfishing and effort limitation

The legacy of past open-access fisheries is a population of fishers that far exceeds the number required to catch what fish are available. Many of these people have few skills or professional abilities other than fishing or fish processing. Retraining programmes are urgently needed in order to move displaced fishers into productive employment in other sectors as quickly as possible and prevent unnecessary social upheaval. Funds for retraining fishers should be a priority and should go hand-in-hand with those for decommissioning programmes.

Develop social and economic incentives for sustainable, well-managed fisheries

Today's social and economic forces mostly provide an incentive for unsustainable, destructive fishing. Positive incentives for sustainable, well-managed fishing are urgently required. Governments should take the lead in developing and enacting economic incentives for sustainable fishing, thus providing a 'carrot' rather than merely the 'stick' of prescriptive regulations. In addition, voluntary, market-led incentives for sustainable fishing must be created in order to swing market forces and consumer power behind efforts to recover and sustain clean, well-managed fisheries. This will require conservation organizations and progressive seafood companies to work together to educate consumers about the enormous potential effect of their purchasing decisions.

Reduce the 'footprint' of developed countries on third-world fisheries

Northern states pay huge amounts to secure access to the fisheries of other nations, notably in the developing world. In 1993, for example, the European Union spent 36.7 % of its fisheries budget of more than ECU 750 million to subsidize access to foreign fisheries by the EU distant-water fleet (European Commission 1994). International standards for distant-water fishing should be developed as a matter of priority and enforced by UN mandate. No state should be permitted to purchase fishing rights from a foreign government without a full assessment of the impacts of such fishing on the specific fisheries involved, their associated marine ecosystems, and indigenous fishers that have relied on the same resources for generations. The use of fishery development funds by northern countries to coerce political favours from developing nations, such as their votes at the International Whaling Commission, should be actively discouraged.

Eliminate destructive fishing practices such as the use of poisons and explosives

Destructive fishing practices should be phased out immediately in favour of more sustainable, less destructive alternatives. Enforcement of laws that proscribe activities such as cyanide fishing, bleach fishing, and the use of dynamite should be strengthened. Funding should be provided for programmes that train fishers to use techniques other than poisons and explosives (Veitayaki *et al.* 1995). Where practicable, governments and industry should interdict supplies of poisons and explosives used by illegal fishers. Where specific fisheries have traditionally used poisons, such as cyanide in the 'live fish' trade of East Asia, simple tests should be devised that allow inspectors and customers to determine when poisons have been used and to act accordingly.

Reduce and eliminate the bycatch of marine wildlife in commercial fisheries

Commercial fisheries currently kill and waste an estimated 18–40 million t of fishes and other marine wildlife annually (Alverson *et al.* 1994). To add insult to injury, modern fishing practices have a devastating effect on marine biological diversity and the physical environment of the oceans (Dayton *et al.* 1995; Lee and Safina 1995; National Research Council 1995; WWF 1996*d*). This enormous destruction and waste of marine life should be prevented by the use of effective bycatch-reduction devices on fishing gear. Mandatory programmes should be imposed immediately to require the use of such devices wherever they are available. Incentives such as bycatch quotas should be imposed or made available to encourage the use of the least-destructive fishing gear and practices. When implementing programmes to reduce waste by allowing the landing of bycatch, governments should use extreme caution to ensure that these programmes do not impede bycatch-reduction efforts. The reduction of waste should go hand-in-hand with the elimination of bycatch.

Creating incentives for sustainable fisheries

It is becoming increasingly clear that regulation alone, whether at the local, national, regional or international levels, cannot be expected to resolve the fisheries crisis. At best, governments and multilateral organizations, such as treaty bodies, merely instate the lowest acceptable standards of practice. Moreover, recent experience with the International Commission for the Conservation of Atlantic Tunas and other regional fishery management bodies suggests that reliance on governments and international organizations alone to achieve conservation goals would be unwise (Safina 1993; Sutton 1996).

The next century will likely see further world-wide movement towards deregulation, privatization, trade liberalization, globalization, and decentralization of government authority. In fact, according to Elizabeth Dowdeswell, Secretary General of the United Nations Environment Programme, 'The market is replacing our democratic institutions as the dominant force in our society.' (Dowdeswell 1995). In the future, it will be increasingly necessary for conservation organizations to find industry partners in order to build incentives for sustainable fishing. Market forces themselves must be engaged to counter unsustainable fishing and its powerful proponents. History has shown that sustainable use of resources is most likely to occur where conservation and economic goals can be made to coincide (Meadows *et al.* 1992). This is especially true in the case of large-scale extractive industries such as timber and fisheries (Sullivan and Bendall 1996).

In early 1996, as part of its Endangered Seas Campaign, WWF formed a conservation partnership with Unilever NV/plc, a major buyer of frozen fish and manufacturer of the world's best known frozen fish products under such brands as Iglo, Birds Eye and Gorton's (Maitland 1996). Unilever and its subsidiaries control about 20% of the frozen seafood market in Europe and North America. With sales of close to $\$US50 \times 10^9$ in 1995, Anglo-Dutch Unilever is one of the world's largest consumer-products companies. It produces and markets a wide range of foods and beverages, soaps and detergents, and personal care products. Unilever operates through some 500 companies in 90 countries around the globe, and employs more than 304 000 people.

The purpose of the WWF/Unilever partnership is to create economic incentives for sustainable fishing by establishing an independent Marine Stewardship Council (MSC) by early 1997. WWF seeks a new approach to ensure more effective management of marine fisheries. Unilever is interested in long-term fish stock sustainability to guarantee a future for its successful fish business. The motivations are different, but the objective is shared: to ensure the long-term viability of global fish populations and the health of the marine ecosystems on which they depend.

Inspired by the Forest Stewardship Council set up in the early 1990s by conservationists and timber companies, the Marine Stewardship Council will be an independent, non-profit, non-governmental body. The organization will establish a broad set of principles and criteria for sustainable fishing and set standards for individual fisheries. Only fisheries meeting these standards will be eligible for certification by independent, accredited certifying firms. Seafood companies will be encouraged to join sustainable buyers' groups and make commitments to purchase their fish products only from certified sources. Ultimately, products from fisheries certified to MSC standards will be marked with an on-pack logo. This will allow seafood consumers to select fish products with confidence that they come from a well-managed source.

A senior project manager will co-ordinate a team of consultants that will work on the development of the MSC. The project team will combine expertise in certification and labelling schemes with intimate knowledge of the commercial fishing industry. Team members will consult with a broad range of experts representing all stakeholders in marine fisheries. Together, the team will draft the broad set of principles for sustainable fishing that will form the underpinning of the MSC. The team will draw on the standards and guidelines embodied in existing international agreements, such as the FAO Code of Conduct for Responsible Fishing and the UN Agreement on Straddling Fish Stocks and Highly Migratory Fish Stocks. In addition, the team will enlist new information and expertise in the fields of marine conservation biology, economics, seafood marketing, and commercial viability to help move current thinking forward.

The MSC will circulate the draft principles and criteria to a broad spectrum of stakeholders in fisheries: conservationists, fishers, seafood industry officials, fishery managers, lawmakers, etc. The organization will then sponsor series of national and regional consultations and workshops around the world. The purpose of these workshops will be to refine and strengthen the principles and develop a process for international implementation. The MSC, WWF and Unilever are actively seeking the widest-possible involvement of other organizations in this exciting initiative.

If the experience of the Forest Stewardship Council (FSC) is any indication, prospects for the success of the MSC initiative are excellent. By the end of 1995, FSC-accredited companies such as Scientific Certification Systems had certified 20 forests comprising more than four million hectares as conforming to sustainable forestry practices. Perhaps more important, more than 100 timber companies and retailers world-wide had joined buyer's groups and made commitments to purchase only timber certified to FSC standards.

The creation of the Marine Stewardship Council has the potential to significantly alter world-wide fishing practices in favour of more sustainable, less-destructive fisheries. When Unilever and other major seafood companies make commitments to buy their fish products only from well-managed fisheries certified to MSC standards, the fishing industry will be compelled to modify its current practices. Governments, laws, and treaties aside, the market itself will begin to determine the means of fish production.

CONCLUSION

Fisheries are the last major world industry exploiting wild natural resources for food. Only a series of fundamental reforms of fishery management, coupled with heightened public interest and powerful economic incentives, will bring chronic overfishing to a halt and shift the paradigm of fishery management from development and exploitation to conservation and sustainability. If marine fishes are to survive into the next millennium — both as important sources of food and vital components of ocean ecosystems — conservationists must bring to bear the same world-wide public concern that drove the international community to protect the great whales, tigers, and elephants. This increased public support, together with market forces and consumer power, must be used to create social, economic, and political incentives for sustainable, well-managed fishing. That won't be easy: fish neither sing like whales nor look like pandas. But the stakes are high: the future of world fisheries, their associated marine ecosystems, and the millions of people that depend on them for food and employment.

REFERENCES

Alverson, D L, Freeberg, M, Murawski, S A, and Pope, J G (1994). A global assessment of fisheries by-catch and discards. *FAO Fisheries Technical Paper* **339**, vii.

Dayton, P K, Thrush, S F, Agardy, M T, and Hofman, R J (1995). Viewpoint: Environmental effects of marine fishing. *Aquatic Conservation Marine and Freshwater Ecosystems* **5**, 205–32.

Dowdeswell, E (1995). Address to the Annual Conference of the Society of Environmental Journalists. Massachusetts Institute of Technology, Cambridge, Massachusetts, October 27.

Earle, S A (1995). Sea Change: A Message of the Oceans. (G.P. Putnam's Sons: New York).

European Commission (1994). The New Common Fisheries Policy. (European Commission, Directorate-General for Fisheries: Brussels).

Food and Agriculture Organization of the United Nations (1993). Marine Fishes and the Law of the Sea: A Decade of Change. *FAO Fisheries Circular* 853.

Food and Agriculture Organization of the United Nations (1995). The State of World Fisheries and Aquaculture. (FAO Fisheries Department: Rome).

House of Lords Select Committee on Science and Technology (1996). 'Report: Fish Stock Conservation and Management.' (HMSO: London).

IUCN/UNEP/WWF (1991). 'Caring for the Earth: A Strategy for Sustainable Living.' (IUCN-The World Conservation Union, UNEP-United Nations Environment Programme, WWF-World Wide Fund for Nature: Gland, Switzerland). 10pp.

Lee, M and Safina, C (1995). The Effects of Overfishing on Marine Biodiversity. *Current: The Journal of Marine Education* **13**(2), 5–9.

Maitland, A (1996). Unilever in fight to save global fish stocks. *Financial Times* (February 22). 3 pp.

Meadows, D H, Meadows, D L and Randers, J (1992). 'Beyond the Limits: Global Collapse or a Sustainable Future.' pp. 185–8. (Earthscan: London).

National Research Council (1995). 'Understanding Marine Biodiversity.' (National Academy Press: Washington).

Parfit, M (1995). Diminishing Returns: Exploiting the Ocean's Bounty. *National Geographic* **188**(5), 2–37.

Safina, C (1993). Bluefin tuna in the West Atlantic: Negligent management and the making of an endangered species. *Conservation Biology* **7**(2), 229–34.

Safina, C (1995). The World's Imperiled Fish. *Scientific American* **273**(5), 30–7.

Sullivan, F, and Bendall, J (1996). Sleeping with the Enemy? Business-Environmentalist Partnerships for Sustainable Development: The Case of the WWF 1995 Group. In 'Environmentalist and Business Partnerships: A Sustainable Model?' (Eds R Aspinwall and J Smith). (White Horse Press: Cambridge).

Sutton, M (1996). Reversing the Crisis in Marine Fisheries: The Role of Non-Governmental Organizations. Paper presented at the 30th Annual Conference, The Law of the Sea Institute, Al-Ain, United Arab Emirates, May 19–22.

Sutton, M (in press). A New Paradigm for Managing Marine Fisheries for the Next Millennium. In 'Reinventing Fisheries Management'. (Eds T J Pitcher, Paul J B Hart and D Pauly) (Chapman and Hall: London).

Veitayaki, J, Ram-Bidesi, V, Matthews, E, Gibson, L, and Veikila, V (1995). Overview of Destructive Fishing Practices in the Pacific Islands Region. *SPREP Reports and Studies Series No.* 93 (South Pacific Regional Environment Programme: Apia).

Weber, M L and Gradwohl, J A (1995). 'The Wealth of Oceans', Chapter 8 (W. Norton: New York).

WWF (1995*a*). 'Managing U.S. Marine Fisheries: Public Interest or Conflict of Interest?' pp. 25–27 (WWF: USA).

WWF (1995*b*). A Global Framework for the Responsible Management of the Oceans, Panel 2, Paper 2. The London Workshop on Environmental Science, Comprehensiveness and Consistency in Global Decisions on Oceans Issues (November) 64–73.

WWF (1996*a*). 'Current Status of Cod and Haddock Stocks in the North Sea.' (WWF: United Kingdom).

WWF (1996*b*). 'Current Status of Plaice and Herring Stocks in the North Sea'. (WWF: United Kingdom).

WWF (1996*c*). 'The Structure, Overcapitalization and Decommissioning of the UK Fishing Fleet.' (WWF: United Kingdom).

WWF (1996d). 'Impacts of Fishing on the Marine Environment—Bycatch and Discards.' (WWF: United Kingdom).

WWF (1996e). 'Marine Fishes in the Wild: A 1996 WWF Species Status Report.' (WWF: Gland, Switzerland).

Its About Time: Rethinking Fisheries Management

Gary D. Sharp

California State University, Monterey Bay Centre for Climate/Ocean Resources Study, 780 Harrison Road, Salinas, California, USA.

Summary

The recent decades' catastrophes in ocean fisheries are among many signs of lack of societal will in resource management contexts. Although abundant theory, and sometimes adequate information from fisheries activities, exists, continuous surprises and stock failures provide impetus to revise not only the basic theory of resource management, but even the philosophies of conventional fisheries management practice. Gross perturbations of ecosystem structures due to fishing have often been denied. Habitat degradation and losses, along with declining natural biodiversity, define the principal issues of anadromous and estuarine species. Uncertainties of context-free fisheries stock assessments form the bases of legal contentions. Pitting government science against industry lawyers is clearly ineffective. Beyond CPUE, Yield-per-Recruit, VPA, and their associated faulty assumptions, necessary information needs to be defined and integrated into ecosystem-wide monitoring, resource assessments, and management processes. We have a global crisis needing revolution, not consensual fiddling.

Formulation of an Appropriate Fisheries Modelling Context

The 'innocent until proven guilty' legal philosophy has resulted in the systematic reduction of most of the natural resources in the developed world, and the chaotic exploitation of natural resources in the undeveloped world, threatening their elimination. Critical habitat degradations and losses define the principal issues of anadromous and estuarine species. In less sophisticated settings usually associated with but certainly not limited to developing nations, general absence of information along with the underlying ignorance of and about coastal subsistence fishers can only lead to despair for natural resources, and dwindling environmental quality.

Another important issue is that of maintaining biodiversity in natural populations (Anon. 1996). For most exploited fisheries we simply have no idea what levels of diversity ever existed within the exploited populations, particularly the losses of diversity that may have occurred during the recent 'swarming' of fishing fleets over the global oceans. Significant efforts to measure and understand genetic issues have not been adequately funded. Few studies have ever been funded that have permitted researchers the luxury of large enough samples, collected over several years, to provide the statistical power needed to reject any complex null hypothesis. A few tens or a hundred individuals sampled from each area is not a sufficient statistical sample for any rigorous study, for more than inferential or taxonomic purposes. An important related management problem is that the 'inconvenience' of complex stock structures has in most cases been assumed away, as data from ever broader ranging fisheries are merged to produce population biomass estimates, from which most fisheries are managed.

To date, few or no fisheries management plans include environmental contexts. Physical variations that affect all living resources are well defined from basic physiological ecology. Interdependencies of ecosystem components can be resolved from food web structure studies and population energetic dynamics, i.e. fat content and reproductive status at size/age (Parrish *et al.* 1986; Carpenter *et al.* 1994; Schulein *et al.* 1995;).

Only within the last decade and a half has it been possible to openly discuss the consequences of fishery independent causalities such as climate-driven oceanographic processes that affect fluctuations of the major pelagic fish stocks (Csirke and Sharp 1983; Sharp and Csirke 1983). Few resource management success stories will prove capable of coping with dwindling, or highly variable resource bases that do not include understanding of natural environmental patterns and responses of living resources to these variations.

Lastly, there is the issue of where to best apply meaningful resource management regulations. The limited success of direct regulatory manipulation of fishing effort in many cases can be attributed to other, often unregulated sectors of the industry, i.e. markets. In recent years the focus amongst environmental activist groups on market places has been much more effective than conventional negotiated effort limitations, treaties, or state-managed catch prohibitions. Perhaps there is a lesson to be learned.

Background

From at least the mid 1970s to present, I and my colleagues at the Fisheries Department of the Food and Agriculture Organization in Rome set about compiling various arguments that might provide bases for the reorganization of fisheries science in support of the elusive 'rational' management of the world's fisheries resources. We have attempted to reduce the assumptions common to fisheries modelling in support of fisheries management, a result of the equilibrium-based approach. (Sharp 1981; Bakun *et al.* 1982; Caddy 1983; Caddy and Gulland 1983; Csirke and Sharp 1983; Sharp and Csirke 1983; Caddy and Sharp 1986; Sharp 1988; Caddy 1993; Csirke 1995). Fisheries research and related management passed from early empirical observations through the post World War II period of theory development, and on to the mathematicization of basic ecological and population biology concepts that evolved throughout the last century (May *et al.* 1979). Many of the period's steps, and mis-steps, are described in Sharp (1995). What follows is a terse and personal review of the sequences of events that led me (and many colleagues) to make a very strong effort to reset the ideas and thinking of the world's fisheries managers. I make no claim to 'original' thinking in this problem area.

At the 1988 American Fisheries Society symposium in Toronto, Canada, there was a day-long tribute to the late Professor F.E.J. Fry entitled 'From Environment to Fish to Fisheries'. I was invited to speak at a concurrent symposium on fisheries mathematical modelling. During my talk I admonished those attending the modelling session to quit isolating themselves, to take immediate advantage of the symposium, to leave the room and learn from Fry's colleagues about the predictive powers of physiological ecology. I am an advocate of similar cause-response, system-oriented research and interpretations of empirical physiological ecologists, i.e. the late Professors Wrigler and Fry. I cannot condone the guessing or parameterization that have become the hallmarks of context-free ocean fisheries stock assessments (Ricker 1975; Gulland 1983).

Dichotomies abound in fisheries science. On one hand, agency population dynamicists hold their statistical training out as a licence to criticize any and every empirical study made in the attempts to organize a cause-effect, or simple stimulus-response framework for applications in fisheries management. They have, somehow, as a group convinced themselves that anything they agree to accept as an assumption, for the sake of solving messy mathematical equations, is justified, even if these 'principles' do not fit empirical observations (Finlayson 1994). What if they were wrong? Well, then it is always easy enough to 'adjust' the parameters, or restate their conclusions in hind-sight, such that the ranges of possible solutions encompassed any new findings.

Sensitivity analysis has long been in vogue as a 'cheap' substitute for direct experimentation. It remains a puzzle that, once the fisheries modellers construct and adopt another 'convention', contradictory empirical observations can, and likely will, be ignored. That is not science. That is theology (Dayton 1979).

Unvalidated, 'consensus truths' are, of course, 'the enemy' in the real world of scientiflc endeavor. Many empiricists and disbelievers suffer constant lashes from many such truth-by-consensus, pseudo-scientific conventions. This behaviour is rampant in the competition for funds and recognition amongst so-called 'peers'. The problem is not unique to fisheries science, but is rife in natural resource research issues, particularly the 'Global Warming' debate and its associated 'peer-driven' funding grabs.

Dread Factors seem to drive funding for modern research effort. Global Warming and the attendant doom-and-gloom scenarios have never fared well against historical knowledge. Some 'facts' needing to be reconciled are that the peak warming that ended the recent Ice Age was reached about 9000 years ago. We are well advanced into the next Ice Age (Dean *et al.* 1984; Overpeck 1996). That well recognized Warm period also corresponds to the advent of human civilization. The cooling trend since, has been punctuated by several warm and cool epochs, many of which brought civilization to its knees. For example, the Medieval Warm period (from about 600–1180 AD) was a period of relative self-sufficiency over most of Europe, Asia, and the Americas. With the onset of the so-called Little Ice Age (AD 1180 to ~1850) European civilization was in constant flux, and human populations fell due to plagues, famines, and wars, as concentration of resources shifted, newly-blessed regions were identified, and then fought over for those limited resources. In fact, from a comparison of historical records and modern research results, there is abundant information that suggests that Global Warming, is *not* a bad thing. Rapid changes in climate, i.e. local, regional and global regime changes, commonly occur over short periods, less than a decade, or several decades (e.g. Dean *et al.* 1984; Allen and Anderson 1993). Such dynamic shifts are important to the periodic renewals of entire ecosystems, in which production systems change, predators are minimized, nutrients recharged, and the Darwinian Play is acted out. Climate-driven fish population blooms and collapses serve a similar function, as long as adequate genetic plasticity and broad habitat access are retained.

This brings me to my principal point. If we are ever to accomplish Sustainable Resource Management it is imperative that there be full incorporation of the transitory nature of climate, and the cascade of ecological responses (e.g. Allen and Anderson 1993; Carpenter *et al.* 1994) in forecasts. Related observations provide a basis for rehabilitation of our science, as well as reclamation of scientific credibility. While there is always a place for modelling, in defining data needs, it is more important to consider known sequences of events, processes, and consequences, within defined context.

ASSUMPTIONS: THE BASIS OF THE APPARENT CONTEXTUAL CHAOS

Wrong concepts and assumptions are rampant in fisheries models. Production modelling assumes that removal of adult fishes leaves 'space' within the niche for more younger fish of the subject species. An intrinsic compensatory increase in basic population production is assumed. This underlies the fallacy of constant (or varying) 'Carrying Capacity' and related 'Fishing-Up' dualities. To physiological ecologists, the assumptions makes very little sense. Carrying capacity is not a species concept, it is a system concept. Clearly, a truncated, younger population will certainly 'grow' at a greater average rate than an older, age-distributed population, on an aggregated per-unit basis. That is a common thread of living systems, and physiological ecology (as per Ulanowicz 1986). However, decrementation of larger members of an oceanic population does not automatically 'release' either niche space, or resources, to smaller, younger age classes of only that population. The first, immediate options go to competitors, usually of different species (except in the case of Arctic lakes, and similarly constrained systems (Johnson 1981).

Within a production model's assumptions also lies an assumption invoking the conventional fishery stock and recruitment (S/R) concept, disguised as one more blur. Consider also the disconnect within the older and pre-recruit populations that takes place when biomass and fishing effort are the sole contributing information to this approach. For most pelagic species, these are fundamentally wrong assumptions (Sharp 1981, 1995; Koslow 1992).

Coastal, pelagic-spawning fish populations characterize the more productive fishes. Effective positive contributions resulting from decreases in the larger adults to the population mostly derive from primary environmentally determined ecological interactions. The science problem is to identify direct mechanism(s) that result(s) in increased survival of eggs and larvae through stages that share little or nothing with their parents. Once gametes are released into the sea one invokes changes in intra- and inter-specific competition (Carpenter *et al.* 1994), or the consequences of cannibalism (Sharp 1981). Assuming these do not matter seems inadequate to explain regime changes, and often well co-ordinated timing of changes that occur over entire ecosystems, across and amongst oceans (Sharp and Csirke 1983; Crawford *et al.* 1991).

Where is the necessary differentiation and 'understanding' provided by the context-free, biomass management approach? How can such an approach be useful in explaining changes in either population (size–age) structures; or the ecological cascade that results from an expected decrease in the predation rates due to harvesting the larger adults (Sharp *et al.* 1983)? It is difficult to identify many aquatic species that are self-limiting, given naturally heterogeneous habitat (i.e. access to refugia, continuous environmental variation, behavioural plasticity, and subsequent ecological responses), or natural environments that fulfil to any reasonable degree the usual assumptions. It is nearly impossible to find supportive data for many conventional (perhaps convenient is a better term) fishery modelling assumptions, except for fissional, monoclonal closed-cultures of yeast or *Tetrahymena*, that both consume and poison their local environments simultaneously.

ASKING QUESTIONS AND GETTING NO ANSWER

Convergence on the concept of management for near-constant fishable stock biomass during the 1960s to 1980s was misleading, initiating inappropriate reactions once a fishery system began a regime change. Sustainable management practices and adaptability have been subject to much recent rhetoric (Walters and Collie 1988; Kesteven 1996; Pauly 1996; Steele 1996), yet no one is really being heard over the 'chanting priesthood' (Hilborn 1992) to deliver the obvious messages. Few deny that the 'well-managed' Atlantic cod has fallen victim of management by consensus. Hard sought credibility is lost, and 'Truth is Emergent' (Sharp 1995).

One fundamental criterion for validating scientific research is that if a question is asked repeatedly, and no predictive answer appears from the approaches taken, it is probably a wrong question. The quest for a direct S/R (stock and recruitment) relation is a good example of this type of problem. Sharp (1981, 1995) described the details of dichotomies between observed and expected life histories for different fishes.

North Sea plaice (*Pleuronectes platessa*), provide a nearly flat recruitment record, with a few lows and highs over the century-long records. I am amazed that any direct stock and recruitment concept might even be proposed from such information. What is the linking mechanism between adult numbers, or biomass, and recruitment, particularly for nearly constant recruitment — with occasional highs and lows? Well-documented examples arguing against these sorts of processes abound (Koslow 1992;

Sharp *et al* 1983). In studies of the Wadden Sea as a nursery ground for the plaice and other commercial species, Rauck and Zijlstra (1978), and their associates, describe the pelagic plankton egg and pre-feeding stages of plaice. Once feeding begins, larval transformation and settlement out of the plankton onto the bottom is induced. Where they settle out, and what they encounter up until they have settled, has little or nothing to do with the behaviour, numbers, or conditions of their parents, after the gametes are cast into the sea.

Why invoke an S/R relation? What is the source of persistence of this 'conventional wisdom'? Why does this myth persist? Is this fishery science's analogue for the Arthurian quest for the 'Holy Grail'? Neill *et al.* (1994) apply Fry's physiological stimulus-response paradigms in a recent re-evaluation of fish recruitment, including plaice, providing invigorating insights into fishes and their ecologies. Empirical concepts work, at all levels of integration.

What's wrong?

Forgetting to ask 'Why?' is a major flaw in stock assessment paradigms. It is often the case in resource management issues. Recognition of the 'managed' and unmanaged declines of fishable stocks, and the inadequacies of conventional fishery model data, should make a bit more acceptable my earlier arguments against prevailing conventional 'wisdoms' (reviewed in Sharp 1995).

A growing body of evidence indicates that the prime variables that impose the strongest variations, particularly those that enhance population recruitment, are driven by environmental variability, including ecological interactions, rather than inherent population properties. This does not, however, exonerate historically bad resource management. It is critically important to address the damage done by overcapitalized fisheries, and related poor management decisions and practices, in quick order, by creating a new framework for sustainable resource management practices. That must start with a clear definition of what is, and what is not, sustainable. For example: Business as usual, either fishing, or management, is not sustainable.

'Sustainability' does not include any of the following:

- Open access fisheries;
- Constant high levels of catch from season to season or year to year;
- Economic guarantees; or
- Single species management.

Sustainability can be approached by implementing at least:

1. Reduction of total expenditures of non-renewable energy per unit catch, i.e., greater true energetic and economic efficiencies from selection of appropriate technologies, including minimization of redundancies;
2. Creation of better, more inclusive methods for assessing resource status, i.e. monitoring whole contexts, including economics along with ecological and environmental measures;
3. Redefining resource population health in realistic, measurable terms, e.g. monitoring of energy content and reproductive status by size and age, through well-sampled catch aged to monthly birth dates *via* daily increments, and associated measurements of fat content and gonadal status;
4. Real commitment to better methods of regulating fisheries production, e.g. market-based production management, rather than only direct effort based management;
5. Enhancement and co-ordination of catch, processing, and transport into a system that can be controlled at any stage, for optimal economic return from varying resource bases and often changing markets, i.e. adaptive integration of the entire industry;
6. Monitoring and control of the entire global market outlet system, to ensure most effective maintenance of resource bases, through vertical integration of environmental information, including status indicators of local and regional resources, resource demands, and economic indicators for global market places;
7. Return to appropriate, sustainable technologies, (related to 1); and
8. Addressing the issues of cultural support collapses due to recent evolution toward 280+ day per year offshore vessel activities.

New syntheses: empirical thinking: beginning a new awareness

While I will be the first to agree that future states of living resources may remain truly unpredictable, or in many cases only marginally forecastable, there is a very realistic need for understanding of the full set of processes that can and will affect each fishery, within the lifetime of any investment. Empirical evidence shows, depending upon the region of interest, that there are many patterns and quasi-cycles that repeat themselves over decadal to century time scales, and in specific cases, such as the El Niño-Southern Oscillation (ENSO) phenomenon, short-term ecological perturbations are significant.

Therein lies the basic issue. An equilibrium conceptual approach to modelling fisheries cannot provide the necessary insights that will allow both long-term conservation and short-term, efficient exploitation of the major living aquatic resources in the face of continuous environmental change. Typical X–Y plots from fisheries data or models, with 'best fit' lines, *always* underestimate the good and overestimate the bad. Fitting averages through time series has similar logical consequences. Good decision making needs to account for well-documented prognoses about the whole system. The question is not how to project or forecast one population, but for system production changes of meaningful magnitudes? What has not worked is modelling fisheries *via* elegant applications of mean expectations, or worse, through inappropriate Monte Carlo simulations, claiming mathematical rigour, but generating logical chaos. Patterned sequences are the rule for most events and climatic processes, ranging from daily and seasonal patterns, onward through quasi-periodic climate processes (Sharp 1995). This does not imply that each event or process is identical, but only that each follows some unifying pattern. Hence, only certain one-way sequences are likely, as in the case of changes in state or during long-term trends.

Observed regime (or system) state changes occur on either ENSO or longer time scales, and are either continuous, patterned cycles, or abrupt apparent shifts. These system changes are due to crossing of unknown thresholds, that result in immediate shifts in ecological processes. Defining these thresholds, and tracking changes in time are important to revitalization of fisheries management, and regaining scientific credibility. This can be best accomplished through whole-ecosystem monitoring (Loeb and Rojas 1988; Shelton 1992; Sharp and McLain 1993). We've work to do.

REFERENCES

Allen, B D, and Anderson, R Y(1993). Evidence from western North America for rapid shifts in climate during the last glacial maximum. *Science* **260**, 1920–3.

Anon. (1996). Marine Biological Diversity: A special issue commemorating 25 years of science and service by the National Oceanic and Atmospheric Administration. *Oceanography* **9**(1), 110 pp.

Bakun, A., Beyer, J, Pauly, D, Pope, J G, and Sharp, G D (1982). Ocean sciences in relation to living resources: a report. *Canadian Journal of Fisheries and Aquatic Sciences* **39**(7), 1059–70.

Caddy, J F (1983). An alternative to equilibrium theory for management of fisheries. In 'Papers presented at the Expert Consultation on the Regulation of Fishing Effort (Fishing Mortality)'. Rome, January 1983. *FAO Fisheries Report Series* 290(2), 285–327.

Caddy, J F (1993). Towards a comparative evaluation of human impacts on fishery ecosystem of enclosed and semi-enclosed seas. *Review of Fisheries Science* **1**(1), 57–95.

Caddy, J F, and Gulland, J A (1983). Historical patterns of fish stocks. *Marine Policy* **7**(4), 267–78.

Caddy, J F, and Sharp, G D (1986). An ecological framework for marine fishery investigations. *FAO Fisheries Technical Paper* 283. 152 pp.

Carpenter, S, Frost, T M, Ives, A R, Kitchell, J F, and Krantz, T K (1994). Complexity, cascades and compensation in ecosystems. In 'Biodiversity: Its Complexity and Role'. (Eds M Yasumo and M M Watanabe) pp.197–207 (Global Environmental Forum, Tokyo).

Crawford, R J M, Underhill, L G, Shannon, L V, Lluch-Belda, D, Siegfried, W R, and Villacastin-Herero, C A (1991). An empirical investigation of trans-oceanic linkages between areas of high abundance of sardine. In 'Long-Term Variability of pelagic Fish Populations and Their Environment'. (Eds T Kawasaki, S Tanaka, Y Toba and A Taniguchi) pp.319–32 (Pergamon Press, Tokyo).

Csirke, J (1995). Fluctuations in abundance of small and mid-sized pelagics. *Scientia Marina* **59**(3–4), 481–90.

Csirke, J, and Sharp, G D (Eds) (1983). 'Reports of the Expert Consultation to Examine Changes in Abundance and Species Composition of Neritic Fish Resources'. San Jose, Costa Rica, April 1983. *FAO Fisheries Report Series* **291**(3), 1–102.

Dayton, P K (1979). Ecology: a science or a religion. In 'Ecological Processes in Coastal and Marine Systems'. (Ed R J Livingston) pp. 3–20. *Marine Science* 10. (Plenum: New York).

Dean, W E, Bradbury, J P, Anderson, R Y, and Barnowsky, C W (1984). The variability of Holocene climate change: evidence from varved lake sediments. *Science* **226**, 1191–4.

Finlayson, A C (1994). 'Fishing for Truth.' (Institute of Social and Economic Research, St. John's, Newfoundland).

Gulland, J A (1983). 'Fish Stock Assessment.' (Wiley: London).

Hilborn, R (1992). Current and future trends in stock assessment and management. *South African Journal of Marine Science* **12**, 975–88.

Johnson, L (1981). The thermodynamic origin of ecosystems. *Canadian Journal of Fisheries and Aquatic Sciences* **38**(5), 571–90.

Kesteven, G L (1996). A fisheries science approach to problems of world fisheries or: three phases of an industrial revolution. *Fisheries Research* **25**(1), 5–18.

Koslow, J A (1992). Fecundity and the Stock-Recruitment relationship. *Canadian Journal of Fisheries and Aquatic Sciences* **49**, 210–7.

Loeb, V, and Rojas, O (1988). Interannual variation of ichthyoplankton composition and abundance relations off northern Chile. *Fisheries Bulletin, US* **86**, 1–24.

May, R, Beddington, J R, Clark, C W, Holt, S J, and Laws, R M (1979). Management of multi-species. *Fisheries Science* **205**, 267–77.

Neill, W H, Miller, J M, Van Der Veer, H M, and Winemiller, K O (1994). Ecophysiology of marine fish recruitment: a conceptual framework for understanding interannual variability. *Netherlands Journal of Sea Research* **32**(2), 135–52.

Overpeck, J T (1996). Warm climate surprises. *Science* **271**,1820-1.

Parrish, R H, Mallicoate, D L, and Klingbeil, R A (1986). Age dependent fecundity, number of spawnings per year, sex ratio, and maturation stages in northern anchovy, *Engraulis mordax*. *Fishery Bulletin, US* **84**(3), 503–17.

Pauly, D (1996). One hundred million tonnes of fish, and fisheries research. *Fisheries Research* **25**(1), 25–38.

Rauck, G and Zijlstra, J J (1978), On the nursery aspects of the Wadden for some commercial fish species and possible long-term changes. Rapp. p.V. R=E9un. cons. Int. Explor. Mer, **172**, 266–275.

Ricker, W E (1975). Computation and interpretation of biological statistics of fish populations. *Bulletin of the Fisheries Research Board of Canada* 191. 382 pp.

Schülein, F H, Boyd, A J and Underhill, L G (1995), Oil-to-meal ratios of pelagic fish taken from the northern and southern Benguela system: seasonal patterns and temporal trends, 1951–1993. *South African Journal of Marine Science* **15**, 61–82.

Sharp, G D (1981). Report of the workshop on the effects of environmental variation on the survival of larval pelagic fishes. In 'Report and Supporting Documentation of the Workshop on the Effects of Environmental Variation on the Survival of Larval Pelagic Fishes'. Lima, 1980. (Conv./Ed G D Sharp) pp. 1–47. *IOC Workshop Report* 28. (Unesco: Paris).

Sharp, G D (1988). Fish Populations and Fisheries: their perturbations, natural and man induced. In 'Ecosystems of the World 27, Continental Shelves'. (Eds H Postma and J J Zijlstra) pp.155–202 (Amsterdam, Elsevier).

Sharp, G D (1995). Its about time: new beginnings and old good ideas in fisheries science. *Fisheries Oceanography* **4**(4), 324–41.

Sharp, G D, and Csirke, J (Eds) (1983). 'Proceedings of the Expert Consultation to Examine Changes in Abundance and Species Composition of Neritic Fish Resources'. San Jose, Costa Rica, April 1983. *FAO Fisheries Report Series* 291(3) 1294 pp.

Sharp, G D, and McLain, D R (1993). Comments on the global ocean observing capabilities, indicator species as climate proxies, and the need for timely ocean monitoring. *Oceanography* **5**(3), 163–98.

Sharp, G D, Csirke, J, and Garcia, S (1983). Modelling Fisheries: What was the Question? In 'Proceedings of the Expert Consultation to Examine Changes in Abundance and Species Composition of Neritic Fish Resources'. San Jose, Costa Rica, April 1983. (Eds G D Sharp and J Csirke) pp. 1177–224. *FAO Fisheries Report Series* 291(3).

Shelton, P A (1992). Detecting and incorporating multispecies effects into fisheries management in the North-West and South-East Atlantic. *South African Journal of Marine Science* **12**, 723–38

Steele, J H (1996). Regime shifts in fisheries management. *Fisheries Research* **25**(1), 19–24.

Ulanowicz, R E (1986). 'Growth and Development: ecosystem phenomenology'. (Springer-Verlag, New York).

Walters, C J, and Collie, J S (1988). Is research on environmental factors useful to fisheries management?. *Canadian Journal of Fisheries and Aquatic Sciences* **45**, 1848–54.

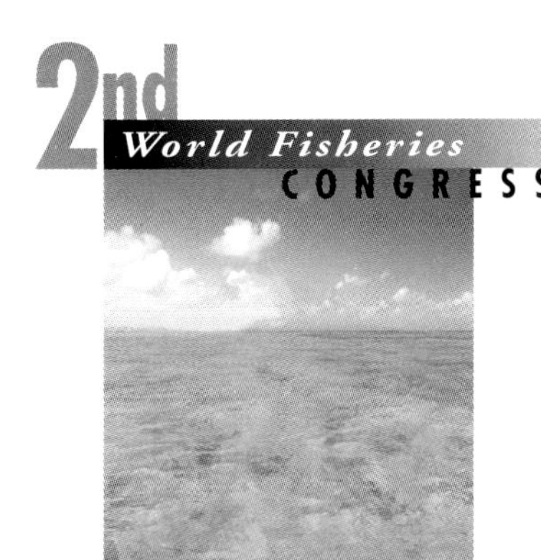

Fisheries Management from a Property Rights Perspective

Hon. Doug Kidd

Minister of Fisheries, Parliament Buildings, Wellington, New Zealand.

Summary

New Zealand's quota management system (QMS), based upon individual transferable quotas (ITQs), has been in place for the major commercial fisheries since 1986 and is widely supported by the New Zealand fishing industry. ITQs are expressed as a proportion of a Total Allowable Commercial Catch, after allowance has been made for other users within an overall TAC. Property rights create incentives for more co- and self-management of fisheries, which means that the role of Government changes from a 'hands on' approach to fisheries management to a role which is focussed on ensuring that the use of the fisheries resources is sustainable (maintaining the productivity of the fisheries resource and managing any adverse effects on the environment). The sustainability and environmental parameters are set by the Government to reflect the values of the wider society and the global community, and the fishers of the resource are given more responsibility for managing the fishery.

Introduction to 'Fisheries Management from a property Right Perspective'

Warwick Tuck

Ministry of Fisheries, PO Box 1020, Wellington, New Zealand.

I am delivering today the address by the Honourable Doug Kidd, Minister of Fisheries of New Zealand. The Minister very much wanted to be here to speak to you of the success of the New Zealand fisheries management regime of individual transferable quotas.

The reason he cannot be here today is that last evening he was required to see our new Fisheries Act through the Parliament in Wellington; he needed to be present to ensure that the Act was finally passed into law. He was simply not prepared to take the risk that delay could be occasioned by his absence, as he believed the changes in the new Act are essential and need to be implemented with urgency. I am happy to tell you that what was until last evening our Fisheries Bill is now law (the Fisheries Act).

The Minister is proud of the provisions of the Act that provide for the simplification of counting catch against quota, or in terms of the new Act ACE, which will ensure a much simplified regime to that which currently exists.

But more importantly, he is particularly proud of what are known as the 'religious provisions' of the Act. These are those parts at the front of the Act dealing with the Act's purposes and principles.

Section 8 of the Act deals with the purpose of the Act which is to provide for the utilization of fisheries resources while ensuring sustainability. Ensuring sustainability means maintaining the potential of fisheries resources to meet the reasonably foreseeable needs of future generations and avoiding, remedying, or mitigating any adverse effects of fishing on the aquatic environment. Utilization means conserving, using, enhancing, and developing fisheries resources to enable people to provide for their social, economic and cultural well-being.

Section 9 of the new Act deals with specific environmental principles to be followed. It reads:

> All persons exercising or performing functions, duties, or powers under this Act, in relation to the utilisation of fisheries resources or ensuring sustainability, shall take into account the following environmental principles:
>
> Associated or dependent species should be maintained above a level that ensures their long term viability.
>
> Biological diversity of the aquatic environment should be maintained.
>
> Habitat of particular significance for fisheries management should be protected.

Section 10 of the new Act deals with the precautionary approach to Fisheries Management and reads:

> All persons exercising or performing functions, duties, or powers under this Act, in relation to the utilisation of fisheries resources or ensuring sustainability, shall take into account the following information principles:
>
> Decisions should be based upon the best available information.
>
> Decision makers should consider any uncertainty in the information available in any case.
>
> Decision makers should be cautious when information is uncertain, unreliable, or inadequate.
>
> The absence of, or uncertainty in, any information should not be used as a reason for postponing or failing to take any action to achieve the purpose of this Act.

The Minister wishes you well in your deliberations here and looks forward to reading the proceedings of the Congress in due course.

The Minister's Address

I would like to develop some themes and ideas which I believe will become important for fisheries decision makers in the future. I will briefly outline what I believe to be a cause of many of the difficulties facing the world's fisheries and how, in the New Zealand context, we have responded to this core problem. Then I will look at some of the implications of this response; a response which has evolved into an approach to fisheries management based on individual property rights.

In many instances, fisheries around the world are characterized by their treatment as common property resources. As is the case with common property resources with open and competitive access, there is conflict with the principle of sustainable utilization of the resource.

There are two ways in which Governments can deal with the problems associated with the common property nature of most fisheries.

The first is to impose input controls which may include limitations on the number of boats, days fished, the size of boats or the type and size of gear used. It should be noted that these limitations are upon inputs and not on the levels of catch of individuals. While this method may meet some resource management objectives, the basic economic problems associated with common property resources remain. A spiral of increased effort and inventiveness to circumvent the intent of input controls occurs, while the core problem of incorrect economic incentives remains.

The second response to the common property problem is to create property rights. Under such a system only those with access to the resource have the right to fish it. A property rights-based system provides those that own access rights a strong incentive, within appropriate overall sustainable catch limits, to maximize profit from their fishing activity. Such an approach provides appropriate economic incentives to fishers as well as providing for effective protection of the sustainability of fisheries.

The most advanced form of fisheries property right to have been developed to date is the individual transferable quota (ITQ). To date only New Zealand, Iceland, Australia and the Netherlands have made widespread use of such rights in various fisheries although they have been implemented in individual fisheries in the United States and Canada.

New Zealand's quota management system (QMS), based upon ITQs, has been in place for the major commercial fisheries since 1986 and is widely supported by the New Zealand fishing industry. In general the QMS has been successful. The value of industry production grew 130% between 1986 and 1996 and employment in the sector grew 25%. The increase in employment is interesting as economic theory often suggests that job losses are associated with the introduction of ITQs.

In New Zealand, ITQs are expressed as a proportion of a catch limit. This catch limit is set, on the basis of scientific information, which will move the fish stock in question towards a biomass size which will support the maximum sustainable yield.

Again, the overall performance in this regard has been good.

A recent study by the Ministry of Fisheries' Chief Scientist, John Annala indicated that of the fish stocks with known status, 85% are above, at, or very close to B_{msy} with rebuilding strategies in place for the remaining 15 %.

I suggest to you that in relative international terms, New Zealand's fish stocks are in very good shape. I am, however, looking forward to the day when New Zealand can report that there are no fish stocks below B_{msy}.

The introduction of property rights in ITQs has led to a set of interesting challenges and opportunities. As an example, and as detailed by Sir Tipene O'Regan (1997), introducing ITQs brought to a head Maori fisheries grievances and at the same time provided one of the important means for obtaining a comprehensive settlement of these grievances. Like other countries we have experienced some management problems in the use of ITQs. Some of these have been addressed in the new Act in terms of the new balancing regime.

One of the opportunities that the QMS has provided is increased incentive for co- and self- management. This is the first theme I would like to note.

In terms of co-management, because fishers now have a clear, strong and valuable interest in the fishery *via* quota, their interest in the management of the fishery increases. As an example, the annual catch limit setting process now involves substantial consultation where fishers fund their own scientists and advisors to represent their interests.

With self-management (the Government sets the broad management principles and has an oversight function leaving stakeholders to manage their own affairs). In one scallop fishery, quota holders have formed their own company (the Southern Scallop Enhancement Company) and are involved in innovative management practices including; co-ordinating their own research and stock assessments according to Ministry of Fisheries' standards and specifications; developing their own management plans including the setting of a Total Allowable Catch and allocating a share of that to the recreational sector; the introduction of an enhancement programme, rotational fishing, and a decrease in the minimum legal size which increased the yield from the fishery. In addition, the Company is working together with the Ministry to develop a joint compliance strategy.

The incentives property rights create for collective activity provide an essential springboard for moves towards more co- and self-management of fisheries. ITQs provide the incentives which support management initiatives for specific fisheries and which can meet the needs of specific local communities.

The second theme I would like to refer to is the implications of this future growth in co- and self-management for Government. With these developments the Government must adapt and change. As fishers exercise their initiative more and more in maximizing, according to their own needs, the value of their property rights, the role of Government needs to change. The same old 'hands on' approach to fisheries management that we are familiar with is no longer necessary when powerful incentives have been created which encourage fishers to organize themselves and their activities to maximize their return from the resource.

In New Zealand we have experienced a paradigm shift as Government has redefined its role in fisheries management. This role is now focussed on ensuring that the use of the fisheries resources is sustainable. Here sustainability, as mentioned before, refers to maintaining the productivity of the fisheries resource and managing any adverse effects on the environment. The precautionary approach to information used in decision making is also embraced. The sustainability and environmental parameters are set by the Government which reflect the values of the wider society and the global community, and the fishers of the resource are given the opportunity to be more involved in managing the fishery.

These trends and changes are exciting and challenging.

The final theme I would like to refer to also relates to property rights; or rather the lack of them. Here I mean the 'tragedy of the commons' which is taking place on the high seas.

In the South West Pacific region, to my dismay, I have seen the mining of high-value fish species on the high seas. To the east of the New Zealand EEZ there are substantive high-seas orange roughy stocks which are being very, very heavily fished. While the activities of the fishers involved is of concern to me, I can understand why they are occurring. It is because of the absurd set of incentives which supports such activity.

At this point in time I acknowledge that international law does not currently provide an adequate framework for vesting property rights for high-seas stocks. However, if we are to be considered responsible stewards of the world's fisheries resources, then management initiatives should be investigated seriously. Economic incentives in an open access regime will drive these fish stocks until they are seriously depleted. I suspect that the answer in the future does lie in property rights, perhaps in the form of country quotas. This is not a problem which I believe we should set aside as 'too hard' and let perhaps irreparable damage be done to the aquatic environment.

The adoption of the new UN fisheries agreement on Straddling Fish Stocks and Highly Migratory Fish Stocks represents a high point in international efforts to stem unregulated and unsustainable fishing practices with respect to these stocks.

The 1982 UN Convention on the Law of the Sea included provisions obliging states to cooperate to conserve high seas fisheries resources, but in recent years it has become apparent that these provisions were not being effectively implemented in certain areas around the world.

One of the most significant steps forward is that the new Agreement provides for the application of the precautionary approach to fisheries management. In essence, this means that when setting up a management regime the best scientific evidence must be taken into account and management should proceed along cautious lines, particularly where information is limited.

Another important aspect of the Agreement is the elaboration it provides of the responsibilities of all States to exercise control over their vessels, and in particular to ensure that appropriate enforcement action is taken if they breach relevant conservation and management measures. The overall objective of this is to provide greater incentives for flag states to comply with their responsibilities under the 1982 Law of the Sea Convention.

While there may be some who fear that their operations on the high seas are going to be unduly constrained, the Agreement puts in place a mechanism by which straddling and highly migratory stocks can be sustainably managed and the long-term interests of all fishers protected.

One minor source of light at the end of the tunnel is southern bluefin tuna (SBT). The efforts here of New Zealand, Australia and Japan to control fishing effort in New Zealand and Australian zones and at the same time on the high sea are, I submit, a model example.

Notwithstanding firm negotiations over a number of years the three countries have established country catch limits agreed on formulas to share the recovery and have established an international treaty with its administrative commission. The next phase is to ensure that the benefits gained by responsible restraint from the founding nations are not lost to others which are not part of the existing arrangement but fish on the high seas, or are nations whose water the migratory route passes through.

New Zealand is committed to persuading other nations who might have an interest in the SBT fishery to join the convention. To this end we will be making initially strong approaches to Indonesia in whose waters the SBT stocks spawn, and Korea, a significant fisher on the high seas, to join the convention on the basis that their interests in the fishery will be enhanced and better protected by being part of the convention rather than remaining outside it.

Gene manipulation

The present development of aquaculture is proceeding in several directions including stock enhancement and transgenetics. Stock enhancement increases the quantity of aquaculture products, but it confronts other users of water and land. Transgenetics improve the quality of aquaculture products, but it presents problems with regard to ecology and ethics.

Reference

O'Regan,T. (1997). Maori Fisheries Rights and the Quota Management System. In 'Developing and Sustaining World Fisheries Resources: The State of Science and Management. Second World Fisheries Congress, Brisbane 1996'. (Eds D A Hancock, D C Smith, A Grant and J P Beumer) pp. 325–8 (CSIRO Publishing: Melbourne).

Great Lakes Fisheries Futures: Using an Ecosystem Approach to Balancing the Demands of a Binational Resource

W. W. Taylor,[A] *C. P. Ferreri,*[B] *L. A. Nielsen,*[B] *and J. M. Robertson,*[C]

[A] Michigan State University, Department of Fisheries and Wildlife, East Lansing, Michigan 48824, USA.
[B] Pennsylvania State University, School of Forest Resources, University Park, PA 16802, USA.
[C] Michigan Department of Natural Resources, Fisheries Division, PO Box 30028, Lansing, Michigan 48909, USA.

Summary

The fishery resources of the Great Lakes have changed dramatically over the past century as a result of an increasing human population and extensive land and water-use changes in the Great Lakes basin. The combined effects of increased exploitation and land and water-use changes resulted in a collapse of the historic fisheries, eutrophication of the lakes, contamination of their biota, and spread of exotics into the basin. Great Lakes fishery managers are struggling to balance a wide variety of demands on the Great Lakes ecosystem which impact their fishery resources. Effective management of production and yield of Great Lakes fisheries depends on understanding and rehabilitating the biological, chemical, physical and social components of the ecosystem. The ultimate success of Great Lakes fisheries management activities is closely tied to our ability to manage the ecosystem as a whole for the long-term optimal production of target fisheries.

Introduction

The Laurentian Great Lakes are a prominent feature of the North American continent. They are located on the border between the United States and Canada and are connected to the Atlantic ocean *via* the St. Lawrence River. Eight states in the US and one province in Canada have jurisdictional responsibilities for this massive aquatic resource. These lakes rank among the fifteen largest freshwater lakes in the world and contain nearly 20% of the world's surface fresh water (Beeton *et al.* in press).

The character of these lakes is intrinsically entwined with the landscape and history of the human population in their watershed. This complex and expansive ecosystem has been directly related to the region's environmental health, economic well-being and general quality of life (Donohue 1991). The diversity of habitat types found in the lakes, ranging from deep oligotrophic basins to shallow eutrophic embayments, supports a wide variety of fish species (Hartman 1988). There are 174 species of native and introduced fish representing 71 genera and 28 families in the Great Lakes (Bailey and Smith 1981).

Before European settlement of the Great Lakes region, the relatively few native peoples who lived along the lake shores fished for subsistence (Hartman 1988). With the influx of European settlers, commercial fisheries for species such as lake trout

(*Salvelinus namaycush*), lake whitefish (*Coregonus clupeaformis*), lake herring (*Coregonus artedii*), yellow perch (*Perca flavescens*), and walleye (*Stizostedion vitreum*) became important. Concurrent with this increase in exploitation were significant changes in the land and water-use patterns in the Great Lakes watershed. These changes accelerated as the human population grew, adversely affecting the fisheries resources of the Great Lakes.

One of the primary reasons for the tremendous population growth in the Great Lakes watershed during the 19th century was the European immigration to the USA, and the development of the canal transportation system from New York. The Erie Canal enabled the transportation of people and materials into the relatively desolate, but natural-resource rich, and agriculturally-productive interior portions of the USA. Once arriving at Lake Erie, access to the other Great Lakes was possible *via* lake transport.

The canal system, which contributed to the expansion of human settlement and supported the well-being of these new settlements, was the beginning of a major ecological disaster for the Great Lakes ecosystems and their fisheries. The canals provided access to these bodies of water to non-indigenous species such as the alewife (*Alosa pseudoharengus*) which would greatly influence the trophic dynamics of the Great Lakes ecosystems (Mills *et al.* 1993). Additionally, introduction of plant materials was reported to have resulted from ballast manipulation by the shipping industry after the Erie Canal was opened. Prior to the Erie canal, falls at Montreal and at Niagara separated the Great Lakes biological fauna from the Atlantic Ocean fauna, although considerable debate by fisheries scientists is still occurring as to the degree of isolation the biological community in Lake Ontario had from the Atlantic Ocean (Smith 1995).

Further assaults on the Great Lakes fisheries occurred as the human population expanded into the upper Great Lakes watershed. Dams were built to harness energy for the industrial revolution, blocking streams for anadromous strains of fishes, thus severely impacting the abundance of these stream-spawning stocks. Concurrent with the construction of these dams was a massive transformation of the landscape in the watershed. Trees were removed rapidly to provide the lumber needed for the increasing urbanization of the human population during this time period and for production of agricultural lands. Additionally, wetlands were drained to provide land for agriculture and other human uses. A better and bigger shipping industry developed to move raw materials from one area of the Great Lakes to another for processing. This allowed for further urbanization and the construction of a bigger lock and dam system, including the Welland canal which directly linked Lake Ontario to Lake Erie. This canal is largely thought to be responsible for the introduction of the sea lamprey (*Petromyzon marinus*) to the upper Great Lakes where it did massive damage to the historic fish populations (Smith and Tibbles 1980).

The consequences of change

The landscape changes associated with the industrialization and urbanization of the Great Lakes watershed, led to changes in the limnological status of the Great Lakes through increased sedimentation and nutrient additions (Beeton *et al.* in press). Sedimentation buried spawning gravels in streams, further diminishing the productivity of Great Lakes tributaries for fish, while increases in nutrients from anthropogenic sources enriched the lakes to the point where Lake Erie was declared dead by 1969. The final insult to the Great Lakes fisheries was the production of contaminants by the industries in the watershed, many of which were directly released into the Great Lakes (Beeton *et al.* in press). The civil engineers of the 19th century viewed the Great Lakes as vast aquatic oceans which could dilute human municipal and industrial waste to levels which would not have an impact on the ecosystem. Unfortunately, they did not foresee the tremendous human growth in the basin nor understand the cumulative impact of human habitation on the Great Lakes ecosystem. All this changed, however, between 1950 and 1970. The first warning that the ecosystem had become seriously out of balance was the demise of the lake trout during the 1950s in all lakes but Superior, where lake trout were severely diminished (Christie 1974). This was followed by the massive die-offs of alewives which littered the beaches (Christie 1974), the significant reduction of mayflies due to poor water quality during the 1960s (Beeton *et al.* in press), and the burning of the Cuyahoga River in Cleveland in 1969. Clearly, humans had dramatically affected a resource which was considered so expansive as to be immune to human insults.

Perception often drives reality, and human perception during the 19th and 20th century in the USA and Canada was that humans control the destiny of the world in their hands (Leopold 1949). If we had a problem, modern science would come up with a technological 'fix' that would make the world an even better place to live in than it was before. The Great Lakes ecosystem from the 1950s through the 1970s told us that was wrong. The old adage 'don't fool with Mother Nature' asserted itself with a vengeance! Clearly, the balance of nature in the Great Lakes had been dramatically altered due to landscape level changes, and the ecosystem was telling us so.

However, had people been observant, they would have seen the warnings that the ecosystem had been giving from almost the beginning of the changes. Fish stocks became locally depleted and waters became foul in our bays and harbours. The response was typical for the time; introduce new technologies to catch fish in different ways and locations, use hatcheries to augment depleting stocks, and introduce new species that may be more adapted to the changed environment (Smith 1994). Fisheries management could best be described during this period as detached from the ecosystem and reactive. Significantly, few understood the revulsion by the ecosystem and what it meant to humans. They were still under the guise of technological fixes and a disbelief that they could impact, at least in the long term, the Great Lakes ecosystem.

Ultimately, however, the successive impacts of human population growth around the Great Lakes basin, including overexploitation, habitat degradation and the invasion/introduction of exotic species caused commercial fisheries to collapse (Christie 1974). Production of high-value species declined from 14 million pounds in 1940, to 12 million in 1955, to seven million in 1965 (Tanner *et al.* 1980). The negative economic impact of the degraded state of the Great Lakes led to the initiation of a fishery rehabilitation programme that aimed to restore valuable fish populations (GLFC 1992). At the time, fisheries managers focussed their attention on single-species management with little understanding of ecosystem dynamics and their relationship to fish production. Thus, during the early 1960s when alewives were dying in droves during the spring and littering beaches with their corpses during the summer, fisheries managers only thought of reducing this 'nuisance' species through the introduction of a new suite of predators; coho (*Oncorhynchus kisutch*) and chinook (*O. tshawytscha*) salmon. Additionally, an aggressive programme of lake trout stocking and sea lamprey control, a primary mortality agent on the lake trout, was begun (GLFC 1992). Interestingly, both the lake trout rehabilitation efforts and the Pacific salmon introduction were undertaken without a clear understanding of the role of these species in ecosystem structure and function. They were reactive measures to failed fisheries and a perturbed ecosystem which had significant social and economic implications.

With the successful introduction of coho and chinook salmon, the priority of the Great Lakes fishery management changed from the historical commercial fishing industry to the newly created sport-fishing industry (Tanner *et al.* 1980). This latter industry tended to fuel an economic renaissance for the Great Lakes and increase their social and economic value to unprecedented levels during the 1970s and 1980s (Talhelm 1988). This interest by the public in the Great Lakes fisheries was coupled with an increasing desire by the citizens to be involved with the priority-setting process of fisheries management. The inclusion of socio-economic considerations into fisheries management led to a shift in management philosophy in the late 1960s, away from the more biologically-based maximum sustainable yield towards one of compatible biological and sociological considerations of optimum yield (Larkin 1977).

The success of the Pacific salmon programme in the Great Lakes reinforced the philosophy that environmental problems could be solved technologically rather than by focussing our attention on remediating ecosystem structure and function. For instance, it is not fully understood why the coho and chinook salmon did so well during the late 1960s through the early 1980s, as the primary focus was on the salmon and their catch rate rather than on understanding the production dynamics of these fishes in relation to the ecosystem in which they reside (Ferreri *et al.* in press). Ignorance of the linkages between salmon production and ecosystem processes became painfully obvious when Great Lakes salmon populations collapsed during the mid-1980s and no obvious single-species solutions were apparent. This has forced the fisheries management community in the Great Lakes to move management strategy away from a single-species focus to more of an ecosystem focus, where ecosystem responses lead to enhanced and sustainable fishery resources (Ferreri *et al.* in press). Additionally, managers also better understood the importance of the social dimension in designing management strategies, as the collapse of the salmon fisheries had a significant effect on the local economies of shoreline communities and the charter-boat industry. Thus, management agencies such as the Michigan Department of Natural Resources set up Lake Advisory Councils to receive input from interested public and other agencies regarding where management of the Great Lakes should go in the future. These actions culminated a shift in fisheries management philosophy away from single-species management towards a more holistic view which included an understanding of the physical, chemical, biological, and social environments which impact the sustainability of our Great Lakes fishery resources.

TOWARDS HOLOSTIC MANAGEMENT

Reinforcing this holistic view of fisheries management was a realization that fish production was a product of ecosystem processes and that these processes were altered by anthropogenic influences on local and global scales (Taylor *et al.* 1995). Examples of impacts at these different spatial scales are illustrated by cultural eutrophication and acid rain. Both processes had a significant impact on ecosystem structure and function which resulted in dramatically different levels and type of fish production. Cultural eutrophication (Beeton *et al.* in press) was a result of watershed level processes, while acidification (Schindler 1988) was both local and global in nature. As fisheries scientists began to better understand aquatic fertility and its relationship to fish and watershed processes, it became clear that fish were integrators of the physical, chemical and biological environments. Alterations in the processes in any of these categories impacted species composition and production as habitat condition changed. Thus, managers came to understand that fish production was a function of the productive capabilities of the environment and that this production was directly related to the quantity and quality of habitat provided by the surrounding watershed and airshed (Taylor *et al.* 1996). Thus, while fisheries managers historically focussed solely on a single aquatic species or population of interest, today's managers have become increasingly aware that fisheries are a by-product of the quantity and quality of the waters in which fish populations live, and that water quantity and quality are dependent on watershed characteristics (see Table 1). The relationship between watershed processes and fish production is dynamic and interactive in nature with water quality and quantity determining the productive base for our fisheries (Fig. 1). A good example of these relationships is provided by an evaluation of the production of sea lampreys in Great Lakes tributaries. Streams where water quality and quantity were degraded by watershed level changes produced fewer sea lampreys than streams that had improved water quality and quantity (Ferreri *et al.* 1995). This study clearly depicted the impact of watershed level modification on fisheries production as critical habitat availability was dramatically altered due to human impacts in the watershed. In this example, spawning and rearing habitat was reduced due to increased sedimentation and organic/chemical pollution, a result of increased human activities in the watershed and lack of appropriate environmental laws. Thus, the quantity and quality of sea lamprey habitat in Great Lakes tributaries were directly impacted by watershed level processes. This finding further

Table 1. Examples of important watershed characteristics related to fish production and fishery demands

Physical	Chemical	Biological	Social
Soil types and characteristics	Soil nutrients	Land use / vegetative cover	Political or jurisdictional bounds
Climate -precipitation -temperature	Contaminants	Avian and mammal predators	Political or social organizations
Topography -slope -land forms -watershed size	Alkalinity / pH of soil and precipitation	Fish species	Resident human population
Groundwater:surface runoff relationship		Prey organisms	Seasonal human use Human demands/goals Human development — dams, roads, drains Water withdrawals Water treatment/discharge Land use

emphasizes the importance of an ecosystem level approach to managing fish production. One must not stop, however, by just evaluating the watershed or a particular stretch of water within the basin because the impact of watershed level changes often reverberates throughout the entire water ways, e.g. Ferreri *et al.* (1995) predicted that the increase in sea lamprey production due to improved water quality and quantity would directly impact the production of lake trout in the Great Lakes. Thus, in this case, changes in water quality and quantity in tributaries to the Great Lakes directly impacted fish production in the Great Lakes; further emphasizing the needs for fisheries managers to be ecosystem-level managers.

Society, in general, has not been very good at doing this type of holistic analysis. From a fisheries management perspective, managers have often been limited by jurisdictional boundaries and their inability to control events beyond those boundaries. Further, the missions of resource agencies personnel have often been split, piecemeal, lacking an ecosystem perspective. For instance, fisheries managers do not directly control the water quality and quantity that provide the productive base of their fisheries in lakes and streams. Their mandate begins and ends at the margin of the lake or river they manage. Other agencies control the use of the landscape in the watershed and these uses are often incompatible with the fisheries production desired by society. Fragmented agency mandates further detract from ecosystem-level management with one agency making decisions which may counteract or impede the desired actions of another agency. The sea lamprey production story described earlier provides an excellent example of this phenomenon.

Sea lamprey control

Sea lamprey control has been a primary mandate of the Great Lakes Fishery Commission (GLFC) since its inception in 1955 by a treaty between the USA and Canada (GLFC 1992). The sea lamprey was a dominant mortality agent on lake trout in the Great Lakes, and the goal of the GLFC was to rehabilitate productive fisheries in the Great Lakes, with lake trout being a primary target. When both nations were attempting to rectify environmental degradation of water quality, they did not evaluate their actions on the fish communities and productivities within the Great Lakes and their tributaries. Rather, they focussed on water quality improvements without understanding the impact of water quality changes on the type and magnitude of Great Lakes fisheries productivities. In fact, it was rare for the environmental regulatory agencies and the resource management agencies to discuss the ramifications of each others' actions on the others' objectives. Thus, as the National Environmental Policy Act of 1969 and the Clean Water Act of 1972 (Grad 1985) led to greatly improved water quality conditions in the Great Lakes, fisheries productivity has likely been reduced (due to a lack of critical nutrients) and sea lamprey productivity has increased. Thus we have two agencies whose mandates, in part, conflict with each other; the GLFC which is attempting to reduce sea lampreys and rehabilitate

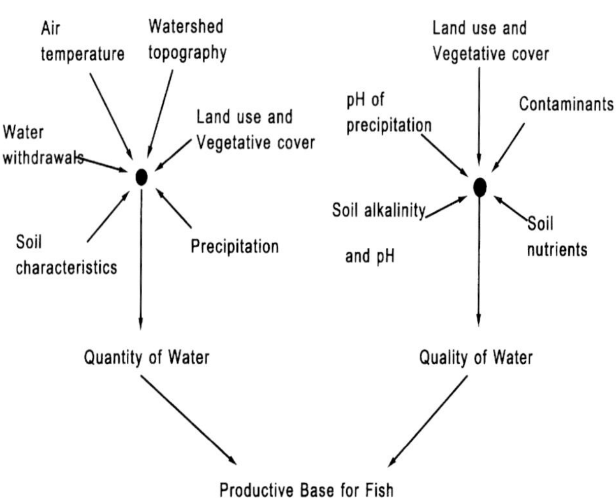

Fig. 1. Integrating factors affecting the quantity and quality of water delivered to streams and lakes. The quantity and quality of water is a main determinant of the productive base for fisheries.

fisheries productivity to historic levels, and the International Joint Commission (IJC) and other environmental agencies which, while attempting to improve water quality, have unintentionally provided a better habitat for sea lampreys in tributary streams to the Great Lakes; thereby increasing the mortality rates of lake trout.

Ecosystem management

While the example described above was the result of ignoring ecosystem processes and narrowly-focussed institutional mandates, the Great Lakes provide a forum for direct conflict between people, cultures and governments over fisheries resources. In our opinion, most of these conflicts are over allocation of scarce resources and could be resolved through formation of a collaborative management framework and an understanding by all involved that fisheries are a limited resource whose production depends on interactive landscape level process (Ferreri *et al.* in press). While it is difficult to overcome what appear to be basic human qualities of greed and self interest, it is by approaching fisheries management as an ecosystem level phenomenon with all stakeholders sharing the same information, that the sustainability of the fisheries of the Great Lakes could be enhanced.

Management of the Great Lakes fisheries at an ecosystem level is, however, complex, not only because of the expanse and diversity of the ecosystem itself, but also because of the intricate governance structure involved. In total, there are over 650 governance units, ranging from municipalities to federal governments, within the Great Lakes watershed (Caldwell 1994) whose activities directly or indirectly affect the fisheries production of the Great Lakes. Such a governance structure imposes further stress on the Great Lakes ecosystem which is best known as 'jurisdictional stress' (Ferreri *et al.* in press). This type of stress arises when the number of specific demands and interests represented by each of the governance structures and public interest groups exceeds the natural ability of the ecosystem to accommodate these demands. Conflicts arise as to whose demand is the most important, often leading to the implementation of management actions that contradict or overlap each other. We believe the solution for this type of stress is to focus on management of the ecosystem rather than on the management of individual jurisdictions. By focussing on the ecosystem, individual interests become subordinate to the collective interest in the welfare of the ecosystem and the sustainability of its natural resources. Thus, jurisdictional boundaries become secondary to the system boundaries, allowing management practices to be developed on a holistic, rather than in a fragmented, jurisdictional basis.

Further, for ecosystem management to be successful, all parties with a vested interest in the Great Lakes ecosystem need to be brought together in a common forum where the interests of each party are considered and opportunities for collaborative management of the resources are sought. The use of consensus in these fora is crucial to success of co-management plans, as this type of decision process creates a sense of ownership, satisfaction and commitment among the participants (Tjosvold and Field 1985; Garcia 1986). We have found that if we start the discussion about what are our long-term goals for the ecosystem, there is general consensus for a healthy Great Lakes ecosystem with a diverse and sustainable fishery (Eshenroder *et al.* 1991; IJC 1991).

Additionally, if we focus on the landscape feature which produces this vision, there is little disagreement by the stakeholders. Where problems occur is when we enter negotiations with each other regarding allocation of the Great Lakes resources. In our opinion, disagreements arise due to a focus on allocation of specific resources without understanding of the production potential of that resource from an ecosystem perspective. Ferreri *et al.* (in press) proposed a management strategy referred to as ACME, the Adaptive Collaborative Management of Ecosystems. This strategy facilitates consensus-building and provides a goal-focussed process for fisheries management in the Great Lakes ecosystem. A primary feature of this system is that the resource is the ultimate customer, and thus the manager's top priority is in maintaining and enhancing the health of the Great Lakes ecosystem, which in turn will enhance the predictability and sustainability of its fishery resources, thereby benefitting all user groups. An aggressive and credible public education programme and an adaptive management approach by management agency personnel, are critical to the success of the ACME approach (Ferreri *et al.* 1996). Through these mechanisms, governance structures participating in the ACME process will more likely arrive at consensus decisions by creating win-win situations for all units involved. A major step toward the ACME approach was the development of the Strategic Great Lakes Fisheries Management Plan (GLFC 1980). However, jurisdictional boundaries and the lack of the IJC participation (and other environmental regulatory organizations) hinders full consensus on the desired state of the future Great Lakes ecosystem.

Great Lakes fisheries professionals have become increasingly aware that management of the fishery resources is an integrative process which requires an understanding of the biology of the species of interest, the productive capacity of the environment and the socio-economic expectations of the public. Thus, fisheries management needs to be holistic in nature, integrating the physical, chemical, biological and social processes of the watershed into the weave of fishery production and allocation. Only by focussing on the ecosystem, and placing fish within the context of the production potential of ecosystem processes, will the health and sustainability of our Great Lakes fisheries be ensured (Taylor *et al.* 1995).

References

Bailey, R M, and Smith, G R (1981). Origin and geography of the fish fauna of the Laurentian Great Lakes basin. *Canadian Journal of Fisheries and Aquatic Sciences* **38**, 1539–61.

Beeton, A M, Sellinger, C E, and Reid, D F (in press). An overview of the physical, chemical, and biological attributes of the Laurentian Great Lakes. In 'Great Lakes Fisheries Policy and Management: A Binational Perspective'. (Eds W W Taylor and C P Ferreri) (Michigan State University Press:Michigan).

Caldwell, L K (1994). Disharmony in the Great Lakes basin: institutional jurisdictions frustrate the ecosystem approach. *Alternatives* **20**, 26–32.

Christie, W J (1974). Changes in the fish species composition of the Great Lakes. *Journal of Fisheries Research Board Canada* **31**, 827–54.

Donahue, M J (1991). Water resources and policy. In 'The Great Lakes Economy: Looking North and South'. (Ed W A Testa) pp. 57–71 (Federal Reserve Bank of Chicago:Illinois).

Eshenroder, R L, Hartig, J H, and Gannon, J E (1991). Lake Michigan: an ecosystem approach for remediation of critical pollutants and management of fish communities. Great Lakes Fishery Commission Special Publication 91-2. (Great Lakes Fishery Commission: Michigan).

Ferreri, C P, Taylor W W, and Koonce, J F (1995). Effects of improved water quality and stream treatment rotation on sea lamprey abundance: implications for lake trout rehabilitation in the Great Lakes. *Journal of Great Lakes Research* **21**, 176–84.

Ferreri, C P, Taylor, W W, and Robertson, J M (in press). Great Lakes fisheries futures: balancing the demands of a multi-jurisdictional resource. *The Environmental Professional*.

Garcia, A (1986). Consensus decision making promotes involvement, ownership, satisfaction. *NASSP Bulletin* **20**, 50–2.

Grad, F P (1985). 'Environmental Law'. (Matthew Bender: California).

Great Lakes Fishery Commission (GLFC) (1980). 'A joint strategic plan for management of Great Lakes fisheries'. (Great Lakes Fishery Commission: Michigan).

Great Lakes Fishery Commission (GLFC) (1992). 'Strategic vision of the Great Lakes Fishery Commission for the decade of the 1990s'. (Great Lakes Fishery Commission: Michigan).

Hartman, W L (1988). Historical changes in the major fish resources of the Great Lakes. In 'Toxic Contaminants and Ecosystem Health: a Great Lakes Perspective'. (Ed M S Evans) pp. 103–31 (John Wiley and Sons:New York).

International Joint Commission (IJC) (1991). 'A proposed framework for developing indicators of ecosystem health for the Great Lakes region'. (International Joint Commission:Ontario).

Larkin, P A (1977). Epitaph on the concept of maximum sustainable yield. *Transactions of the American Fisheries Society* **106**, 1–11.

Leopold, A (1949). 'A Sand County Almanac'. (Oxford University Press: New York).

Mills, E L, Leach, J H, Carlton, J T, and Secor, C L (1993). Exotic species in the Great Lakes: a history of biotic crises and anthropogenic introductions. *Journal of Great Lakes Research* **19**, 1–54.

Schindler, D W (1988). Effects of acid rain on freshwater ecosystems. *Science* **238**, 149–57.

Smith, B R, and Tibbles, J J (1980). Sea lamprey (*Petromyzon marinus*) in Lakes Huron, Michigan, and Superior: history of invasion and control. *Canadian Journal of Fisheries and Aquatic Sciences* **37**, 1780–801.

Smith, S H (1995). Early changes in the fish community of Lake Ontario. Great Lakes Fishery Commission Technical Report # 60. (Great Lakes Fishery Commission: Michigan).

Smith, T D (1994). 'Scaling fisheries: the science of measuring the effects of fishing, 1855–1955'. (Cambridge University Press: New York).

Tanner, H A, Patriarche, M H, and Mullendore, M J (1980). 'Shaping the world's finest freshwater fishery'. (Michigan Department of Natural Resources:Michigan).

Talhelm, D R (1988). Economics of Great Lakes fisheries: a 1985 assessment. Great Lakes Fishery Commission Technical Report No. 54. (Great Lakes Fishery Commission: Michigan).

Taylor, W W, Ferreri, C P, Poston, F L, and Robertson, J M (1995). Educating fisheries professionals using a watershed approach to emphasize the ecosystem paradigm. *Fisheries* **20**, 6–8.

Taylor, W W, Ferreri, C P, Hayes, D B, and Robertson, J M (1996). Using watershed analysis to teach fisheries management: integrating fisheries and the ecosystem. In 'Proceedings of the First Biennial Conference on University Education in Natural Resources'. pp. 190–96. (The Pennsylvania State University: Pennsylvania).

Tjosvold, D, and Field, R H G (1985). Effect of concurrence, controversy, and consensus on group decision making. *The Journal of Social Psychology* **125**, 355–63.

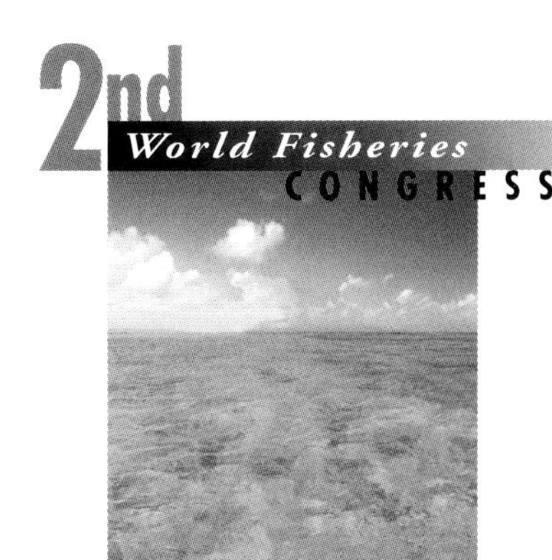

Conservation When Fisheries Collapse: the Canadian Atlantic Experience

Jean-Claude Brêthes

Université du Québec à Rimouski, C.P. 3300, Rimouski (QC) G5L 3A1 Canada.

Summary

Atlantic Canadian groundfish fisheries had been an important industry (landings around 600 000 t/y, employment of 60 000 persons annually). Landings decreased constantly from 1987, and in 1995, 14 of the 32 major stocks had to be closed to direct fishing. In this context, the Fisheries Resource Conservation Council (FRCC) was created as an independent body to advise the Minister on conservation measures and on research priorities, while fostering partnership among the various stakeholders. The FRCC faces a crisis associated with changes in ecological structures (hydro-climate, decrease of every groundfish species, predator-prey system). In the meantime, the socio-economical structure has also been modified. This Council also faces conflicting views in terms of 'conservation' and of 'ecosystemic management'. The global system is evolving and may not be reversible despite drastic actions. While aiming at preparing the fishery of the future, the FRCC only looks at one aspect of the system. Only the future will tell if this structure is part of the solution.

Introduction

The Canadian Atlantic continental shelf is one of the widest in the world and has been recognized as highly productive. In 1990, the Atlantic fisheries, within the Canadian Exclusive Economic Zone (EEZ), provided landings of 1.2 million t (source: Statistics Canada). In 1994, the total landings were less than 500 000 t (Fig. 1). For Atlantic cod, the 'typical species', catches moved from 380 000 t down to 23 000 t during the same period. The decline of the groundfish (Fig. 1) then became a major crisis. The Canadian groundfish fishery is now becoming a 'case study' subject of several analyses (*e.g.* Chantraine 1992; Steele *et al.* 1992; Fortier 1994; Hutchings and Meyers 1994; Rogers 1995).

The Canadian management system is highly centralized, with the global philosophy that a fishing licence is a 'privilege', not a 'property' granted year after year to a specific person for a particular species. The annual groundfish harvesting plan and the resource allocation is approved by the Minister of the Department of Fisheries and Oceans (DFO). Most of the major stocks are managed with an Enterprise Allocation system (EA, transferable or not, the system varies among regions and species according to local characteristics). From the creation of the EEZ in 1977, until 1992, stock assessments and recommendations for a global quota were the result of an internal process within the Science Sector of the DFO, with a peer-review system, through NAFO, for transboundary stocks and those of significant foreign interest, and the Canadian Atlantic Fisheries Scientific Advisory Committee (CAFSAC). For groundfish, the recommendation was also submitted to the Atlantic Groundfish Advisory Committee (AGAC), composed of DFO and industry representatives, which recommended on final quotas and on resource allocation among fishing sectors.

Such a system seems coherent: an internal process of resource assessment 'separating 'public interest' from the private competition' (Rogers 1995), decentralization of research and data gathering and treatment, involvement of the industry in the recommendation on harvesting levels, and centralized final power, which allows rapid decisions (e.g. the closure of the Northern Cod fishery, during mid-season 1992, was decided within a few days). However, the result is disappointing. In 1995, 14 among the 32 major groundfish stocks were closed to direct fishing, and most of the remainder faced reduced quotas from 30% and 90%, compared to 1990 levels. Cod landings represented around 40% of the total Atlantic landings during the period 1980 to 1990 and now represent less than 3%. The fishing industry also raised concerns about that system and criticized the 'confidential' character of the assessment process as well as the apparent contradictions, among years, of scientific recommendations, which induced a great lack in confidence in the whole management system. Some also felt that the AGAC favoured the large companies, to the detriment of small-scale fisheries.

The Fisheries Resource Conservation Council

In 1992, the Minister of the DFO decided to modify the whole process for Atlantic fisheries. From CAFSAC, Science kept the internal peer review process. The AGAC's role of recommending harvesting plans disappeared. In place, he created the Fisheries Resource Conservation Council (FRCC), composed of 14 independent members from universities and industry, along with non-voting representatives from the DFO and from each Province concerned. The Council advises the Minister on conservation measures, including total allowable catches (TACs), and makes recommendations on research priorities. The objectives given to the FRCC are: 'rebuilding stocks to their 'optimum levels' and thereafter maintaining them at or near these levels'. Considering that a more comprehensive approach is needed, the Council has to develop a better understanding of complex fisheries ecosystem. Finally, the Council has to give a more effective role in decision making to those with practical expertise and knowledge in the fishery. The key elements of FRCCs philosophy are then, to develop an 'ecosystemic approach' toward the fishery, to forge a partnership between the various players (fishers, processors, managers and scientists), and to use a 'precautionary approach' toward conservation. It represents a kind of 'nodal point', centralizing the information from all available sources (Fig. 2). In this new system, biologists are asked to provide information, not solutions. The views of the stakeholders on the resource status and on other conservation issues are gathered through public audiences.

The present mandate focusses on groundfishes. After three years, the most 'mediatic' recommendations were the closure of nine stocks and the drastic reduction of TACs for most of the remaining ones even if other less spectacular recommendations may have a large impact (especially on scientific research). The Council also pursues specific activities, especially on the definition of criteria to allow re-opening a closed fishery.

The present context of Atlantic Canadian fisheries

The FRCC was created in a complex and difficult context. The most evident fact is the constant decline of landings, at least since 1986. The high level of science activities and the energy spent on surveillance, were unable to prevent the collapse of a large part of the groundfish stocks.

During that period, effective fishing effort, and the related fishing mortality (F) was very high. The cod stock of the

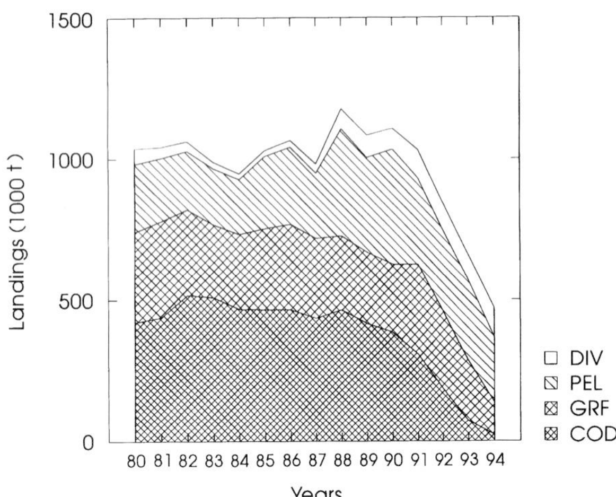

Fig. 1. Fisheries landings in the Canadian Atlantic EEZ: Cod, GRF = other groundfishes, PEL = pelagic fishes, DIV = other species. (Source: Statistics Canada.)

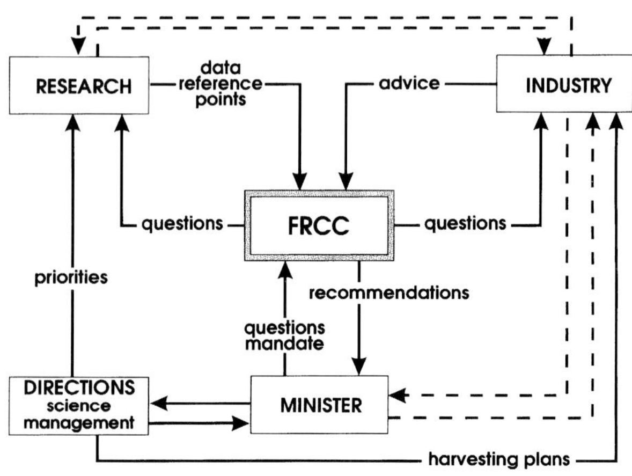

Fig. 2. Place of the Fisheries Resource Conservation Council in the information flow and in the decision process for Canadian Atlantic Fisheries.

southern Gulf of St.Lawrence may be an illustration as F varied between 0.35 and 1.12 during the 1971 to 1992 period (Sinclair *et al.* 1994). It means that, in practice, the TAC level, calculated from a theoretical $F = 0.2$, was set too high, year after year. An independent panel, appointed by the Minister of the DFO to review the status of the Northern Cod, stated that 'there has been a serious underestimating of fishing mortality rates in the years between 1977 and 1989' (Harris 1990).

The Canadian Atlantic also faces harsh environmental conditions. Over the past decade, temperatures have been constantly lower than the long-term average, and the ice coverage has been wider and remained longer than usual (Colbourne 1995). Difficult environmental conditions could have had a detrimental effect on recruitment (Myers *et al.* 1993). In the meantime, the global system seems being changing. On the eastern Newfoundland region, the biomass of all groundfish species is decreasing, for heavily-exploited as well as for lightly-exploited species (Fig. 3). The decline of groundfishes is correlated with increased landings of crustacean species (shrimp, lobster and snow crab; Fig. 4). On the south-east coast of Nova-Scotia and on Georges Bank, an increase of the biomass of elasmobranch fishes ('shark like species') has been noticed (Simon and Comeau 1994). Seal populations, recognized as heavy fish predators, are increasing: the harp seal (*Phoca groenlandica*) population grew from 3 million individuals in 1990 to 4.5 million in 1994 (Stenson *et al.* 1995). However, no clear causal relationships can be drawn, even if fishers strongly believe in them.

The groundfish crisis is also translated into a socio-economical crisis. In several regions of the Canadian Atlantic, human communities were built on fisheries resources. In Newfoundland, the economy is highly dependant on groundfishes: at least since 1985, the provincial Gross Domestic Product appears related to fish landings values, and employment to landings volume (Fig. 5). Generally, one considers that 1000 t of fish landed create 30 person years of employment (Cashin 1993). In a Province like Newfoundland, it can be roughly estimated that 10% of the total labour force has been directly affected (around 23 000 persons).

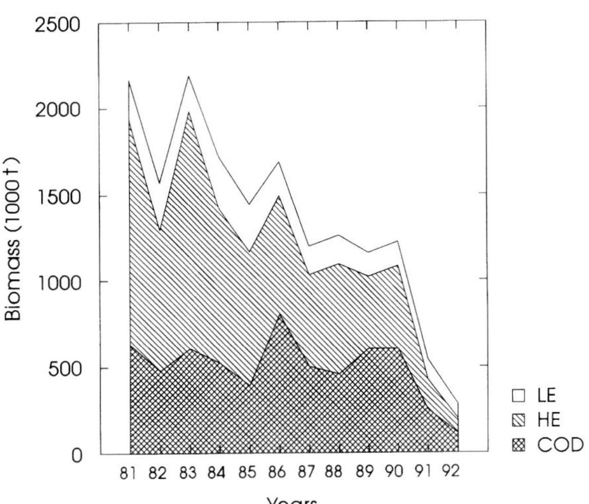

Fig. 3. Global trends of groundfish biomass indices from research surveys in eastern Newfoundland, OPANO area 2J3KL: Cod, HE = heavily exploited species (redfish and flatfishes), LE = weakly exploited species (Original data provided by B. Atkinson, DFO, Atlantic Region, St.John's, Newfoundland, Canada).

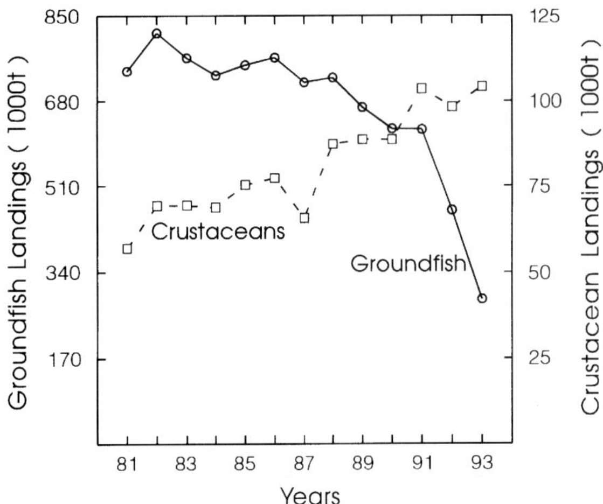

Fig. 4. Trends in landings for all groundfish species (solid line) and for crustaceans (dotted line) in Atlantic Canada (Source: Statistics Canada).

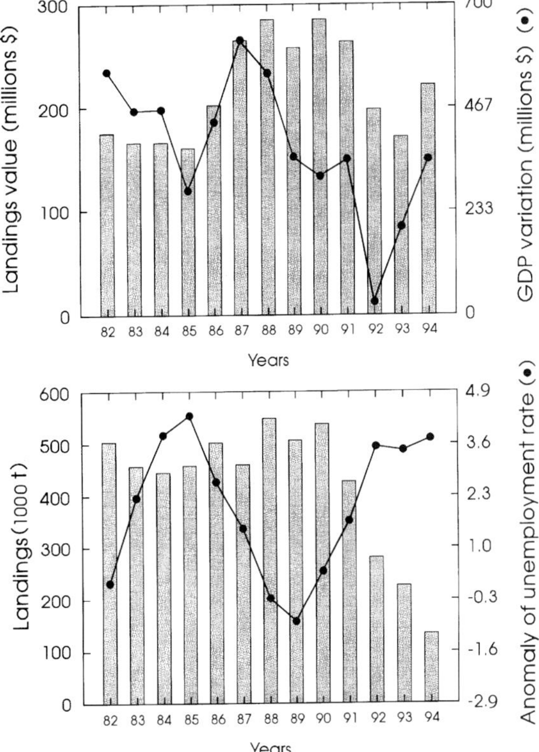

Fig. 5. Province of Newfoundland: above, relationship between the value of landings (bars) and the annual variation of the Gross Domestic Product (line); below, relationship between global landings (bars) and the variation of the unemployment rate (line), related to the 1972–1994 average (Source of economical data: Statistics Canada, Provincial Economic Accounts).

THE FUTURE

After recommending closures, FRCC's approach is to prepare for the future, when the stocks will be in a recovery stage. A first step is to define 're-opening criteria', (FRCC 1995). They are mainly based on basic biological data (such as total biomass, spawning biomass, recruitment indices, age structure). In a closed fishery, those data come from scientific surveys complemented by 'sentinel fisheries' (non-commercial fisheries carried on by fishers under close scientific supervision). In the absence of a widely-accepted definition of a 'healthy' stock and the relative perception of a 'viable' fishery, the threshold to allow the re-opening of a commercial fishery will remain somewhat arbitrary, and the FRCC has organized public debates on that matter.

Another step would be to prepare the 'fishery of the future' under a conservation framework defining a new approach, in order to avoid the errors of the past. The Canadian Atlantic crisis illustrates the failure of the classical reductionist vision of a fishery (the 'Theory of Fishing'): neither the stocks (and the related socio-economical structure) nor the ecosystem have been preserved. This crisis shows the necessity of an approach based on a global understanding of the system: the 'ecosystemic' approach, that the FRCC tries to promote. However, this concept remains ambiguous. An 'ecological' management may signify either exploiting the resource so as not to modify the ecological system, or using knowledge to optimize exploitation and management. In the first case, 'conservation' may signify the preservation of the structure and function of the ecosystem, forbidding any exploitation being the ultimate solution. In the second case, one aims at the maintenance of a resource level allowing an 'optimal' exploitation rate (the goal of the fisheries management system) and the conservation concept refers more to the preservation of an economic activity.

The expected effect of a fishing closure is still related to the Theory of Fishing: setting the fishing mortality at zero means that the biomass will recover, as the surplus of production is no longer exploited. It remains a single-species deterministic approach. We know that the ecosystem is a dynamic entity in which species interact, and those interactions may induce a different result from that expected: the drastic decrease of traditionally-exploited biomasses releases space for competitive species (as may be the case for shark-like species in the Canadian Atlantic) that, from a prey status, become predators on the resource, and competition for food and predation may limit recovery. The resource may attain a new state of equilibrium, lower than the previous one, where the return to the historical stage is hypothetical (Cury 1988). Halting fishing on some species may also signify the disappearance of the predation represented by humans: the biomass of those species, previously limited by fishing, may grow to the point where they limit the growth of other species (as may be the case for seals).

All the various ecological factors, including physical characteristics, move to a new different ecological structure, human activities being an integral part of it. After centuries of exploitation, a close relationship was created between the exploitation system and the natural system. On the one hand, the biological ecosystem has been deeply modified by human activities; on the other hand, societies became dependant on the fisheries resource: permanent economic activities were elaborated and local communities were structured around the resource; they present a 'history' that was shaped by the natural environment. Closing fisheries for a long period may be translated in a loss of 'human expertise' and in a 'way of life'. Overexploitation, and consecutive drastic measures, have disturbed the system. We have to consider that there is no 'pristine state' we can refer to and we cannot imagine what will be the future.

It is recognized that 'there is such a thing as a marine system' and that the 'systems perspective can be the basis for a more powerful management regime' (Platt 1993). We can arbitrarily divide such a system into three large components (Anderson 1984): the resource and its natural environment; the exploitation sub-system; the management system. The biological system is now widely studied, as numerous works deal with the relationships between environment and biological resource, and 'biodiversity' is now a key-word in natural resource management policies. The issue of the relationship between the fishers (with their cultural, social and economical environment: the 'socio-diversity') and the resource is recent in fisheries studies. The management system may be the less known even if 'over-exploitation is more than a moral failing; it is also an institutional failure' (Hardin 1968). Inertia is the rule in the whole system (physical environment → resource; resource → assessment): e.g. in the present system, there is a two-year delay between the gathering of scientific data and the implementation of harvesting plans derived from those data. A re-opening process, involving the gathering of sufficient data, the designing of a new management regime and modifications to fishing practices (as recommended by Cashin 1993) may be slow, if well done. Re-opening will be more difficult than closing, considering the present fishing capacity. The 'classical' management system does not seem to be prepared to deal with the new situation.

In this context, the actual role of a structure such as the FRCC remains conflictual. The Council has to act on conservation on both aspects: resource and fisheries. It means that recommendations may not be 'optimal' from each separate point of view (e.g. on the ecological side: recommending the increase of seal hunting in order to help groundfish stocks recovery; on the industry side: closing fisheries or reducing TACs). Designed as 'a Council for the Fish', it is supposed to only look after the conservation of the resource, while 'erring on the side of caution' and 'acting before scientific consensus is achieved' (in the views of Ludwig *et al.* 1993). The Council, in fact, only looks at one part of the whole system. Its recommendations for a future 'sustainable fishery' (such as compulsory design of conservation plans, reduction of fishing capacity) interact with management and social systems, which are outside its mandate. Council's recommendations, when followed (as is the case for the present), will be effective if the other components are moving at the same pace and in the same direction. Political pressure and increasing social crisis are forces that could jeopardize good intents. Even if the FRCC is an 'independent' body, it remains tied to the political power, which has the liberty to modify its mandate, and, in fact, fishers hardly believe in that 'independence'. The FRCC is new and is still evolving within the general situation. The Canadian Government is preparing a new 'Ocean Act' that will modify the whole landscape of the management of Canadian marine environment, while the DFO would be in charge of marine conservation within the EEZ and would have control of activities in the ocean. The new Act could have a significant impact on FRCC action, especially if fisheries become only one part of its mandate. Only the future will tell if creating such a structure was the right way to go or if it was only another part of the problem in a classical situation.

References

Anderson, L G (1984). Uncertainty in the fisheries management process. *Marine Resource Economics* **1**, 77-87.

Cashin, R (1993). 'Changement de cap: les pêches de l'avenir. Rapport du groupe d'étude sur les revenus et l'adaptation des pêches de l'Atlantique.' (Ministère des Approvisionnements et services du Canada: Ottawa).

Chantraine, P (1992). 'La dernière queue de morue. Comment l'ignorance et la cupidité ont mis en péril l'une des plus vastes ressources alimentaires du onde: les Grands bancs de Terre-Neuve.' (L'étincelle: Montréal).

Colbourne, E (1995). Oceanographic conditions and climate change in the Newfoundland region during 1994. *DFO Atlantic Fisheries Research Document* 95/2. 36 pp.

Cury, P (1988). Pressions sélectives et nouveautés évolutives: une hypothèse pour comprendre certains aspects des fluctuations à long terme des poissons pélagiques côtiers. *Canadian Journal of Fisheries and Aquatic Sciences* **45**, 1077–107.

Fisheries Resource Conservation Council (1995). Considerations on re-opening a closed fishery. *FRCC Discussion Paper*, FRCC95.TD1 (Ottawa), 9+viii pp.

Fortier, L (1994). Des morues et des hommes: enquête sur un désastre. *Interface* **15**, 22–39.

Hardin, G (1968). The tragedy of the commons. *Science* **162**, 1243–8.

Harris, L (1990). 'Independent Review of the State of the Northern Cod.' (Department of Supply and Services: Ottawa).

Hutchings, J A, and Meyers, R A (1994). What can be learned from the collapse of a renewable resource? Atlantic Cod, *Gadus morhua*, of Newfoundland and Labrador. *Canadian Journal of Fisheries and Aquatic Sciences* **51**, 2126–46.

Ludwig, D, Hilborn, R, and Walters, C (1993). Uncertainty, Resource Exploitation, and Conservation: Lessons from History. *Science* **260**, 17–36.

Myers, R A, Drinkwater, K F, Barrowman, N J, and Baird, J W (1993). Salinity and recruitment of Atlantic Cod (*Gadus morhua*) in the Newfoundland region. *Canadian Journal of Fisheries and Aquatic Sciences* **50**, 1599–609.

Platt, D D (1993). 'The System in the Sea: Applying Ecosystems Pinciples to Marine Fisheries.' Volume 1: Conference Summary (Island Institute: Rockland, Maine, USA).

Rogers, R A (1995). 'The Oceans are emptying: Fish Wars and Sustainability.' (Black Rose Books: Montreal).

Simon, J E, and Comeau, P P (1994). Summer distribution and abundance trends of species caught on the Scotian shelf from 1970–92, by research vessels groundfish survey. *Canadian Technical Report of Fisheries and Aquatic Sciences* 1953. 145 pp.

Sinclair, A, Chouinard, G, Swain, D, Hébert, R, Nielsen, G, Hanson, M, Currie, L, and Hurlbut, T (1994). Assessment of the Fishery for Southern Gulf of St.Lawrence Cod: May 1994. *DFO Atlantic Fisheries Research Document* 94/77. 116pp.

Stenson, G B, Hammill, M O, Kingsley, M C S, Sjare, B, Warren, W G and Meyers, R A (1995). Pup production of harp seals, *Phoca groenlandica*, in the northwest Atlantic during 1994. *DFO Atlantic Fisheries Research Document* 95/20.

Steele, D H, Andersen, R and Green, J M (1992). The managed commercial annihilation of northern cod. *Newfoundland Studies* **8**, 34–68.

Managing Fisheries Sustainably: The Australian Solution

Bruce Phillips,[A] *Gerry Morvell,*[B] *Annie Ilett,*[B] *and Neil Hughes*[B]

[A] School of Environmental Biology, Curtin University of Technology, Perth, Western Australia.
[B] Coasts and Marine Branch, Commonwealth Department of the Environment, Sport and Territories, Canberra, ACT, Australia.

Summary

The Australian experience has demonstrated some fundamental principles for successful fisheries management. These include:

- a sound information base with sufficient detail to support informed decision making for the sustainable management and use of fisheries resources;
- co-ordinated, strategic and integrated management plans and decision making across all levels of government and in consultation with key stakeholders;
- a decision-making process that is one step removed from political processes;
- development and implementation of cohesive policy and planning frameworks for Australian oceans.

There is no one solution to the management of a fishery, and each must be dealt with on a case-by-case basis. Each fishery needs to devise its own solution to managing the fishery sustainably, taking into account the social, economic, cultural, political, and biophysical context of that fishery. However, the broader community has a stake in ensuring that its fisheries are managed sustainably, since all coastal activities impact on the fishing resources.

Australia has made advances on all these fronts in the past few years and there have been significant achievements in establishing a framework for sustainable management of Australia's fisheries. Although yet to develop a fully integrated management regime for the oceans, the Australian approach to fisheries management provides a model for many other countries.

Introduction

Australia has a vast coastline of 69 630 km, a continental shelf area of 14.8 million km^2, and over 12 000 islands. However, the population of less than 20 million people is concentrated in a few areas along 30% of the total coastline.

With 80% of Australians living on the coastal fringe and 25% living within 3 km of the sea, Australia's economic, cultural, and social activities are focussed in the coastal zone. This leads to tremendous pressure on coastal land and waters for urban development, recreation, fishing, industry, tourism and mining.

The 200 nautical mile (nm) Australian Fishing Zone (AFZ) (Fig. 1) is the third largest in the world. Despite this, Australia's fish catches are relatively low at about 200 000 t each year, and ranking only around 50th in the world. An additional 15 000 t

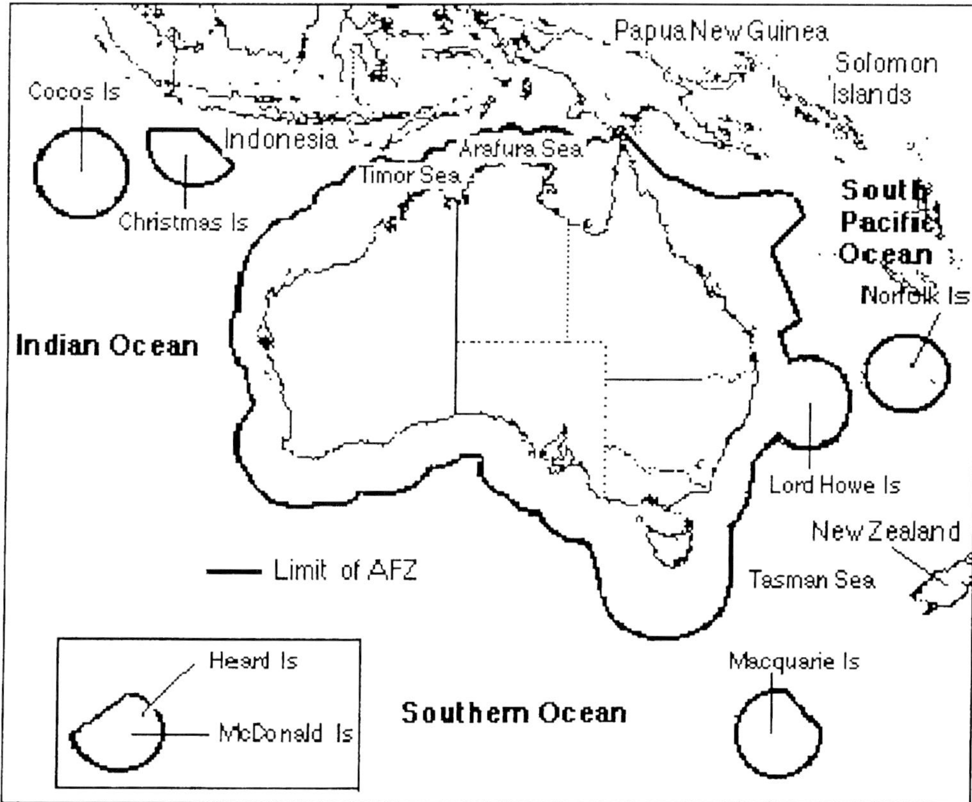

Fig. 1. The Australian Fishing Zone (Figure courtesy of DPIE Fisheries and Aquaculture Branch).

are taken from Australian waters by licensed foreign vessels. The low catch level is because of naturally-limited run-off of nutrients from the land, a relatively small area of continental shelf, and the absence of major upwellings of nutrient-rich deep waters. However, Australia has a number of high value export fisheries such as rock (spiny) lobsters, prawns (shrimp), abalone, and a cultured pearl industry. Annual exports of marine products are valued at $A1.4 $\times$ 10^{12} (ABARE 1995).

At least 4.5 million Australians take part in recreational fishing and this industry is worth at least $A2.2 $\times$ 10^9 annually.

Aquaculture is an emerging industry in Australia, although oysters and pearls have been farmed for a long time. Sixty species, from seaweeds to prawns, and fish to crocodiles, are now farmed and worth $A260 $\times$ 10^6 annually.

Australia's commercial fishing fleet consists of almost 10 000 vessels using more than 40 methods of fishing. Some 200 different species of fish, 60 species of crustaceans and 30 species of molluscs are fished.

Of the 100 main commercial species, nine are considered overfished, 23 are fully or heavily fished, nine are underfished, and 59 are of unknown status (Fig. 2). Catches of many coastal fisheries have declined since the 1980s, and the combined effects of commercial and recreational fishing have probably resulted in a real deterioration of the stocks, but the value of the catch has increased greatly.

The reasons for declines in some fisheries include overfishing, use of non-selective fishing gear, loss of habitat, pollution, and the complexity of Australia's marine jurisdictional, administration and legislative frameworks across Commonwealth, State and Territory governments which hinders management of a fish species or a population.

Fisheries resources in Australia are managed by Commonwealth, State and Territory governments. As a general guide the Commonwealth Government manages the offshore and highly migratory fish stocks while the State and Territory governments manage the inshore fisheries. Some fisheries are managed jointly. Where an agency manages a stock which is both inshore and offshore, such as the western rock lobster *(Panulirus cygnus)*, this is by agreements between the Commonwealth and the State in question.

In 1992 the Commonwealth Government established a new organization, the Australian Fisheries Management Authority (AFMA) to manage Australia's Commonwealth fisheries. AFMA is a statutory authority responsible for ensuring the sustainable use of Commonwealth fishery resources. While accountable to the Minister for Resources, AFMA's governing legislation enables it to operate at a distance from the Minister and the Department of Primary Industries and Energy (DPIE). The Minister for Resources is not involved in AFMA's day-to-day operations, but must approve AFMA's corporate plan and annual operational plan, as well as individual management plans.

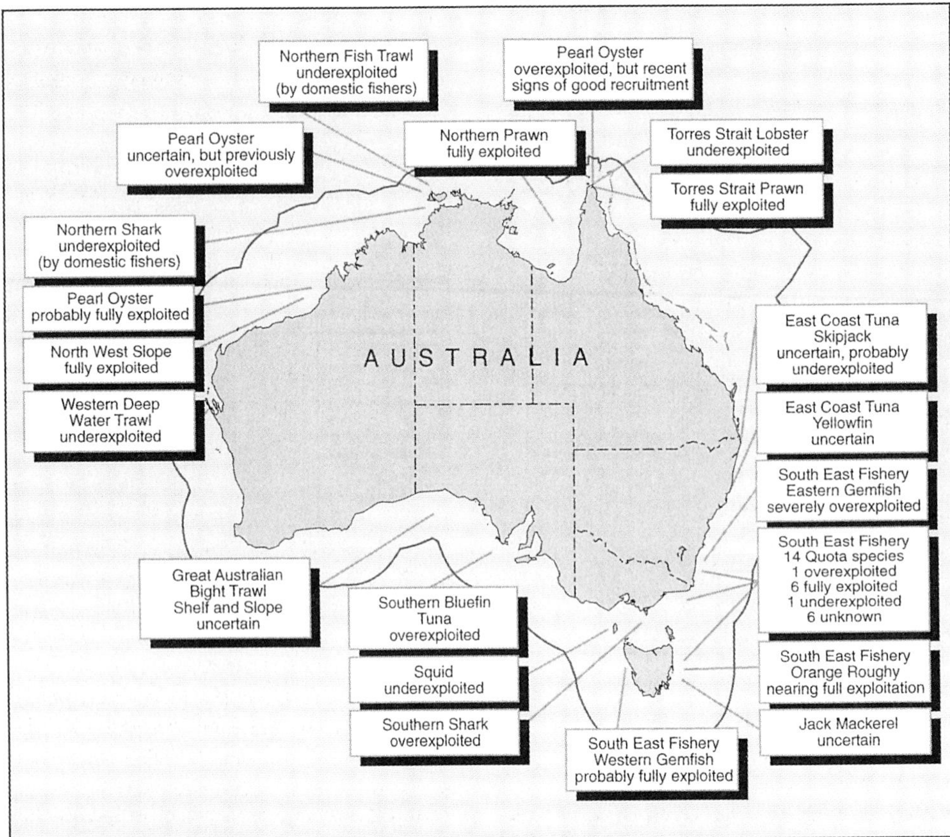

Fig. 2. Status of Commonwealth fisheries (Figure courtesy of DPIE Fisheries and Aquaculture Branch).

A number of Management Advisory Committees (MACs)have been established by AFMA to assist in the management of specific fisheries. The primary role of these committees is to act as liaison bodies between the fishing industry and AFMA and to provide advice on a variety of issues such as management arrangements, research, compliance and finance. AFMA also consults with peak industry groups and is taking steps to ensure that a broader range of interest groups is consulted with and involved in fisheries management issues (AFMA 1995*a*).

While day-to-day fisheries policy is developed and implemented by AFMA, a strategic national policy role remains with the Fisheries Policy Branch of DPIE. In particular DPIE, in consultation with key stakeholders, is responsible for the development of policies in response to international fisheries initiatives.

The Commonwealth Department of the Environment, Sport and Territories (DEST) comprises of a number of conservation and natural resource management agencies, as well as core functional areas dealing with issues such as marine and coastal policy development, impacts of climate change, pollution management and environmental impact assessment. DEST has a variety of interests in fisheries management ranging from the protection of individual fish species, the protection of marine ecosystems, the protection of areas with high heritage value (both natural and anthropogenic) and the consistency of fishing activities within international environmental obligations. The Department promotes the development of an integrated approach to coastal and marine management, including multiple-use regimes for the sustainable management of marine protected areas.

Ecologically sustainable development (ESD) is the widely-accepted basis for natural resource management in Australia. The Australian National Strategy for Ecologically Sustainable Development (NSESD) sets out objectives and guiding principles for the sustainable use of natural resources. The goal is:

> Development that improves the total quality of life, both now and in the future, in a way that maintains the ecological processes on which life depends.

The core objectives of the strategy (Council of Australian Governments 1992) are to:

- enhance individual and community well-being and welfare by following a path of economic development that safeguards the welfare of future generations;
- provide for equity within and between generations; and
- protect biological diversity and maintain essential ecological processes and life support systems.

The NSESD, however, does not prescribe what constitutes sustainable management of fisheries. Ecologically sustainable development of this resource involves a complex web of interactions which ensures that there is no single solution which can be applied universally. Fish species are part of very complex ecosystems which can be adversely impacted by many natural and anthropogenic activities; markets are complex and ever-changing due to a variety of external pressures; community perception of products and industries and social values have a temporal and geographic variability and have a significant impact through political processes.

No management system can be designed to accommodate all of these variables. However, the relative political and economic stability in Australia provides opportunity to focus the attention of fisheries managers on key issues, in the knowledge that many of the external pressures can be accommodated with some certainty into a management regime.

Australian experience has shown there are six key issues in sustainable management of fisheries:

1. Knowledge of the stocks
2. Consistent jurisdictional management
3. Enforceable legislative frameworks and management plans
4. Support of the fishing industry
5. Ecosystem management
6. Consistent policy frameworks.

None of these is new, and each is important in its own right, but their strength is in the synergy created when they mesh together. It is this synergistic value which constitutes sustainable management of the fisheries. The key issues are examined in more detailed below.

1. Knowledge of the stocks

Accurate long-term data on the size and biological condition of the stocks within the fisheries are essential to sustainable management of the fishery. Such information can be used to calculate Total Allowable Catches (TACs), establish levels of effort or other measures designed to set catching levels so as to maintain fish stocks at sustainable levels. These data are expensive and difficult to obtain. They are available for few fisheries.

Stock assessment has played an important role in identifying problems with the eastern gemfish (*Rexea solandri*) stock and has led to the implementation of significant management decisions. The Stock Assessment Report (South East Fishery Assessment Group 1994), reports that eastern gemfish have been caught on shelf fishing grounds off south-eastern Australia since the development of the trawl fishery in the 1920s. Recorded catches from this fishery were low (less than 200 t) and comprised mainly juvenile fish. The TAC was progressively reduced to zero in 1993.

Assessment of the eastern gemfish stock has been based on a cohort analysis of a series of size and age composition samples from the winter catches. The data suggest a sudden and severe decline in recruitment of eastern gemfish; however, there is uncertainty regarding the circumstances under which the recruitment decline occurred. It is likely that gemfish recruitment is influenced by both adult biomass and environmental effects; however it is not possible, with current knowledge of the stocks, to predict recruitment levels in future years. These factors led to a conservative approach to the management of eastern gemfish stock and the resulting closure of the fishery.

According to the Stock Assessment Report (*ibid*), continued updating of the stock assessment is needed based on improved techniques and new data. This requires a co-ordinated data collection programme involving port sampling, log-book records and on-board observers to provide annual information on catch and size distribution, combined with ageing of samples by the Central Ageing Facility at the Marine and Freshwater Research Institute in Victoria. It also includes continuous assessment and evaluation of the stock assessment models and their underlying assumptions.

From the data collected, a problem was found with the eastern gemfish fishery and management steps were taken to deal with it. But there still remains the issue of an estimated 60 t of eastern gemfish being taken each year in the bycatch. It is balancing this bycatch with the controlled take of the fishery which is the next challenge in sustainable management of this fishery.

The Western Rock Lobster (*Panulirus cygnus*) Fishery, which is managed by the State of Western Australia under Offshore Constitutional Settlement (OCS) arrangements, is recognized as one of the most successfully-managed fisheries in the world, with high catch levels and catch value. It has a long data set and there is considerable biological information on the species. The annual catch averaged 10 000 t between 1970 and 1990 (Brown and Phillips 1994). The 668 vessels licensed to participate in the fishery shared a gross income of $\$A250 \times 10^6$ (1992), making it Australia's most valuable single-species fishery. Because of the high prices paid for *P. cygnus* and the good returns a fisher can expect, the stock experienced very high and increasing exploitation since about 1970, balanced by effort reductions (*ibid*).

Databases form the basis of stock assessment and are continually maintained, updated and improved (*ibid*). Information on modelling, recruitment/catch predictions, spawning stock estimates, movements, migration and growth, fishing power, recreational fishing and economic information are all included in a comprehensive database on the western rock lobster. It is this long-term, valuable information which has made it possible to successfully manage the Western Rock Lobster Fishery.

2. Consistent jurisdictional management

An impediment to good fisheries management is the confusing system of political boundaries in place in Australia. Marine ecosystems and the resources they support naturally do not respect these boundaries.

Australia has a federation style of government which operates at three levels: the Commonwealth, the States and Territories, and local governments. Each level has its own areas of responsibility for management and each has its own legislative frameworks in which to operate. Co-ordination of fisheries management has not always been achieved, resulting in inconsistent or incomplete management arrangements in some fisheries.

Historically, fishing within three nautical miles (nm) of the coast was managed by the adjacent State or Territory and the Commonwealth managed fishing from 3 nm to the 200 nm limit. Under the 'Offshore Constitutional Settlement' process the Commonwealth and State/Territory Governments have agreed on management arrangements which place a number of fisheries under a single jurisdiction, either State/Territory or Commonwealth.

In the Torres Strait, Papua New Guinea and Australia have established the Torres Strait Protected Zone under a treaty which provides a basis for single management of fisheries and marine protection. The treaty also provides for the direct involvement of local indigenous communities in this management arrangement.

The lack of consistent jurisdiction has created an unacceptable compliance loophole in the management of some fisheries, e.g. the South Eastern Fishery, a multi-species fishery, has different arrangements applying to trawling and Danish seining in State and Commonwealth waters. In Commonwealth waters quotas have been imposed on the take of a number of species to ensure that the harvesting of these species is sustainable. Of around 120 vessels licensed to fish in Commonwealth waters, ~half of these vessels are also licensed to fish in State waters. Some dual-licensed fishers are able to exploit the lack of compatible management arrangements between State and Commonwealth waters by declaring that their Commonwealth catches of quota species have been taken in State waters where no quotas apply.

While there has been considerable effort made by State and Commonwealth Government fisheries agencies to limit the extent of mis-declarations of catches in Commonwealth waters, there continues to be substantial mis-reporting. In recent years Commonwealth total allowable catch quotas for a number of species have been exceeded and it could be argued that attempts at sustainable management have therefore not been fully effective (AFMA 1995*b*).

For effective sustainable fisheries management to occur there must be consistent management across the full range of target species. In some cases, such as those involving straddling fish stocks or highly migratory species, there is a need to enter into international arangements. The recent *Agreement for the Implementation of the Provisions of the United Nations Convention on the Law of the Sea of 1982 relating to the Conservation and Management of Straddling Fish Stocks and Highly Migratory Fish Stocks* and the FAO *Code of Conduct for Responsible Fisheries* provide an opportunity for fisheries management to be undertaken internationally and for globally-appropriate management measures to be taken to ensure sustainable harvesting.

3. Enforceable legislative frameworks and agreed management plans

To achieve effective management of fisheries, all fisheries must be subject to a legislative framework which allows for the management of individual species, fishing gear and fishing effort. Without this framework fisheries management action cannot be applied and conservation and sustainable use of fisheries resources cannot be guaranteed. Frameworks must be enforceable and sufficient resources must be made available to effectively monitor fishing activity and enforce relevant legislation as necessary.

Commonwealth fisheries legislation provides, for the purpose of protecting the marine environment, wide-ranging powers to regulate activities within the Australian Fishing Zone (AFZ) and Australians fishing outside the AFZ. Regulations made under the legislation include the prohibition or regulation of specified fishing practices or methods; the use of specified fishing equipment; the taking and treatment of bycatch; and littering at sea.

The Fisheries Working Group, which was established as part of the ESD process, recommended (ESD Working Group 1991), that strategic management plans be developed immediately for each fishery. Guiding principles for the development of such plans include incorporation of legally-binding management measures, provision for a clear recognition of the need for risk-averse decisions, extensive consultations with key stakeholders, public input and influence at the draft stage, ongoing and substantive review of plans and assessments of the likely impacts of new and expanding fisheries.

The Working Group specifically recommended that all management plans contain:

- the aims and objectives of the plan for the fishery;
- a clear statement about the fishery's access rights including their specifications and conditions;
- the performance criteria against which the plan will be assessed, including sustainability indicators;
- how timely monitoring and enforcement of regulations will be undertaken;
- how the plans are expected to affect current and future stakeholders;
- the instruments to make the fishery sustainable and efficient, taking cognizance of the likely effect of these measures on the processing and marketing sectors;
- the data set (biological, ecological and economic) needed to manage the fishery on a sustainable basis and identification of gaps in that data set; and
- the key components of the ecosystem on which the fishery depends, and current and potential impacts/ threats to those components.

The Working Group's recommendations have been adopted in part under the Commonwealth fisheries legislation administered by AFMA. To date, management plans for the Southern Bluefin Tuna, Northern Prawn and Great Australian Bight Fisheries have been developed by AFMA in consultation with the fishing industry and the broader community.

To ensure sustainable management of existing fisheries, and the recovery of declining stocks, management plans must also be developed that outline long-term stock-recovery strategies. This type of management was applied to the Orange Roughy

(*Hoplostethus atlanticus*) Fishery in 1994 after serious declines in stock levels were noted. The AFMA board endorsed a ten-year harvesting strategy aimed at achieving a 50% probability of stocks returning to 30% or better of the pre-1986 spawning biomass by the year 2004. This provides for long-term sustainable harvesting consistent with the principles of ESD (Phillips and Rayns 1995).

Environmental impact assessment is an important component in the ecologically sustainable development of fisheries. All levels of Governments in Australia have some form of environmental impact assessment (EIA) process. At the Commonwealth level, EIA is required for all Commonwealth decisions that are 'environmentally significant'. The AFMA has acknowledged that EIA should occur for decisions relating to fisheries management and has identified the assessment of fisheries management plans as the most effective way of assessing the environmental impact of fisheries activities in a fishery. The Environment Protection Agency of the Commonwealth Department of the Environment, Sport and Territories has been assigned responsibility for EIA of fisheries. It is expected that all environmentally significant decisions made by AFMA will undergo comprehensive and independent EIA in the near future.

The actions of AFMA and DPIE are monitored closely by a number of Government departments and community groups. There is a need for capacity to exist outside fisheries management agencies to independently monitor performance against sustainable-use criteria. The right of all groups with an interest in marine conservation to have access to fisheries management information and the opportunity to participate in fisheries management is essential to guaranteeing that fishing is conducted in a sustainable manner.

4. Support of the fishing industry

Legislation alone is not the complete solution. No country can afford, nor is it desirable, to have constant surveillance of fishing activities. The cooperation of the catching sector including the fishers, processors and marketers is essential and can only be obtained by meaningful communication, and long-term involvement of the catching sector in the decision-making process.

The incorporation of Statutory Fishing Rights into fisheries management plans is now being recognized in Australia as an attempt, at least for commercial fishers, to move away from problems caused by the fish being a common-property resource. Fishers identified under a management plan can be recognized as having continuing ownership rights and a right to fish in the fishery. Rights exist even where fishing activity is not occurring. If, for example, for the purposes of stock protection a nil catch is declared, the fishers continue to own the right. When the stock recovers, the fishers are then entitled to take a proportion of the stocks allocated. This ensures that fishers have a long-term interest in stock protection and the sustainability of the fishery.

Industry representatives take an active role in fisheries management in all Australian States and Territories and at the Commonwealth level. This involvement includes formal industry representation on fisheries management advisory committees and consultation during the drafting of management plans or when considering changes to existing regulations. The fishing industry, through National, State or Territory associations, also provides expert representation to a number of committees whose activities deal with fisheries issues. These activities are not restricted to fisheries management matters and can include general environment protection activities, local or regional planning or activities relating to national and international initiatives such as biodiversity, World Heritage protection, climate change and marine pollution.

The Commonwealth Government has recently sought to encourage greater industry involvement and self-regulation in specific fisheries sectors through funding the development of national approaches, such as the National Strategy on Aquaculture in Australia and the Aboriginal and Torres Strait Islander Fisheries Strategy. These are being further elaborated through codes of practice; the first two which were initiated in 1995 relate to aquaculture and recreational fisheries.

In the post-harvesting phase there have been recent developments in product-handling and marketing techniques, leading to a minimization of waste, improved profits and increased economic sustainability of the fishery. Changes in the harvesting and export of southern bluefin tuna (SBT) *Thunnus maccoyii* exemplify these improvements with part of the surface fishery catch transferred to rearing and fattening cages to enhance value. Larger-sized SBT than previously are now exported to supply the high unit value Japanese sashimi market (Caton and Sainsbury 1995).

5. Ecosystem management

In Australia there is now clear recognition of the need for examination of fisheries and the sustainability of the resources within an ecosystem framework. There is a need to integrate the sustainable harvesting of a target species with the need to manage the ecosystem on which the species depends, to ensure that fishing activities or other anthropocentric impacts do not significantly affect the overall ecosystem.

This approach is not yet fully developed in Australia. However, issues of non-targeted catches, including species of mammals, seabirds, turtles, dugongs, whales, sea snakes, and non-retained catches of many other species, are now being addressed in a variety of fora. In addition the effects of aquaculture, benthic habitat damage by fishing, destruction of nursery areas by other activities, declining water quality from land- and ship-based activities, and maintenance of biodiversity, are increasingly important.

The Australian State of the Marine Environment Report (Department of the Environment, Sport and Territories 1995*a*) found that overall the condition of Australia's oceans is quite good. But there are problems including:

- degradation of estuaries and coastal lakes;
- declines in temperate seagrass;
- loss of mangrove and saltmarsh habitats;

- unsustainable coastal development;
- effects of fishing on marine ecosystems and wildlife;
- introductions of foreign species; and
- population increases in non-native species.

The Commonwealth Department of the Environment, Sports and Territories and AFMA are currently engaged in discussions as to how ecosystem management can be achieved as part of fisheries management. The use of management plans to specify how fishing activity can be undertaken in a manner that does not significantly alter the nature of marine ecosystems has been a feature of these discussions.

AFMA has undertaken to pursue additional initiatives relating to the impacts of fishing on marine ecosystems as well as on non-target species. With the assistance of industry, research organizations and other interest groups, AFMA intends to develop the concept of biological reference points to provide a focus for research and specific fisheries management performance criteria against which the organization can report. AFMA also intends to finalize and implement 5-year research plans for each major fishery. This will enable priorities to be set for stock assessment and ecological impact assessment research (AFMA 1995*a*).

In line with new fisheries legislation in the States, Management plans will provide for consideration of ecosystems in the management of their fisheries.

6. Consistent Policy Frameworks

In the past 30 years, over 50 inquiries have found that Australians are concerned that the quality and character of the coastal environment have diminished and that not enough was done to stop further damage. These concerns are reflected in the following reports:

The Resource Assessment Commission's *Coastal Zone Inquiry* (1993); the *State of the Marine Environment Report for Australia* (SOMER) (Department of the Environment, Sport and Territories 1995*a, b, c, d*); the IUCN (1994) paper, *Towards a Strategy for the Conservation of Australia's Marine Environment*; *Ocean Outlook: a blueprint for the oceans* (Ocean Outlook Steering Committee 1995); Prime Minister's Science and Engineering Council Report, *Australia's Ocean Age: Science and Technology for Managing Our Ocean Territory* (Office of the Chief Scientist 1995); *State of the Environment Report* (Department of the Environment, Sport and Territories 1996); *Fisheries Reviewed* (Senate Standing Committee 1993); *Recreational Fishing in Australia: A National Policy* (Standing Committee on Fisheries and Aquaculture 1994*a*); *Commonwealth Fisheries Management Audit Report 32* (Australian National Audit Office 1995/96).

In May 1995, the Commonwealth Government formally responded to these reports and has released the Commonwealth Coastal Policy, '*Living on the Coast*' (Department of the Environment, Sport and Territories 1995*e*). This provides a clear statement of the Australian Government's position on coastal management matters by adoption of common objectives and guiding principles and the initiatives that the Australian Government will take to initiate the objectives to help improve management of the coastal zone.

The Commonwealth Coastal Policy focusses action on the areas of:

- developing integrated solutions to particular management issues — integrated strategic planning is a key to sustained improvements to coastal management. Initiatives focus on integrated area planning and integrated water management planning;
- community participation in coastal management — through a Coastcare programme, and establishment of consultative mechanisms. Aboriginal and Torres Strait Islander peoples' special relationship with and interest in coastal lands and waters and their resources is recognized through a targeted element of the Coastcare programme, development of an indigenous fisheries strategy and establishment of an indigenous coastal advisory group to Commonwealth Ministers;
- increasing the capacity and knowledge of those with coastal management responsibilities to discharge them effectively; and
- developing appropriate links with Australia's regional neighbours.

The Commonwealth Coastal Policy has been strengthened by the adoption of coastal policies and new programme initiatives in the States and Territories. Inter-governmental cooperation underpins this complementary policy regime and is reflected in the signing of Memoranda of Understanding by Commonwealth, State and Local Governments to provide a single, State-based delivery mechanism for programme initiatives.

The Australian Government supports a community-based approach to implement the solutions to marine and coastal environmental problems. The Coastcare programme will link activities on the land and run-off with the declining quality of the coastal waters. Run-off to coastal waters is being reduced through total catchment management practices promoted by the existing Landcare programme. Landcare is one of the most significant developments in land-based natural resource management and conservation in Australia and its community approach is proving to be a successful model. However, implementation of the coastal policy, including management of fisheries resources using this approach will not be easy (Phillips and Ilett in press).

Whilst positive steps have now been taken to improve the management of coastal areas through implementation of Commonwealth and State coastal policies, the oceans in general have been neglected as a policy priority to date. Marine industries have made significant changes over the past few decades to accommodate broad community concerns on protection of the marine environment. However, there are increasing uncertainties over resource allocation in the oceans and no mechanism exists to resolve conflicts. On ratifying the UN Convention on the Law of the Sea (UNCLOS), Australia has custodianship of one of the largest exclusive economic zones (EEZs) in the world. If the resources within this area are not managed sustainably, Australia stands to lose some of these resources to other nations. The attitude of Australian governments to the marine environment is now changing significantly, and, based on wide political support for improved management of the coastal zone, there is now also bipartisan political support, at least at the Commonwealth level, for development of an oceans policy.

The challenge for Australian governments is to provide a rational framework for resource development and environment protection of Australia's EEZ, an area 1.6 times larger than the land mass of Australia. There are numerous pressures on the marine environment which must be addressed now to ensure that the oceans and the marine living resources continue to function as essential ecological systems and support Australia's economic development. Some of these pressures include a lack of essential information on the oceans and the resources they contain; increasing pressure for protection of key ecosystems including breeding and nursery habitats for fish; increasing pollution from land-based activities; the need to address the aspirations of indigenous people reflected in native title claims and calls for increased access to resources; and the need to establish efficient and cost-effective management systems that balance public interest with self-regulation. It is essential that the fishing industry plays an active role in determining the scope and nature of this policy.

The development of an oceans policy is at an embryonic stage, with the input of key stakeholders yet to be sought. It is not possible to predict what form an oceans policy will take at this point. The Commonwealth Coastal Policy provides one model that could be readily translated, including the adoption of clear objectives and principles for management decisions, establishment of an institutional framework for resource allocation and conflict resolution, and a focus of action on gaps in the policy framework. This approach gained wide support from State Governments, non-government organizations and industry, including the Australian Seafood Industry Council.

DISCUSSION

The establishment of AFMA in 1992 marked a new era in fisheries by involving the fishing industry, scientists, governments and recreational fishers directly in the management process. Fisheries management costs are now recoverable from the fishery and this has led to vast improvements in cost-effectiveness of management procedures. AFMA is resolving the jurisdictional issues with the States and Territories Governments which have plagued fisheries research and management in the past. AFMA has adopted the use of biological reference points as the basis for setting Total Allowable Catches (TACs) to achieve ecosystem management of the fisheries. The manner of setting these biological reference points, and consequent TACs, is a fascinating story but unfortunately too long for this presentation.

Successful management of the fisheries resources will require a considerable improvement in the database on the species comprising the fisheries, and better monitoring of the resources and the environment. This is not easily achievable. However, the way ahead in management of reserves is now being considered by a series of cooperative research centres, such as the Cooperative Centre for Coral Reef Research. A task-oriented approach is ensuring a focus on the scientific, sociological and economic research which is needed to answer the problems identified in sustainable management of marine resources. The government has recently allocated \$A13 $\times 10^6$ over seven years to establish a research programme to undertake an integrated programme of applied research and development, training and extension, aimed at enhancing the ecological sustainability of the Great Barrier Reef Marine Park and expanding economic activity with particular emphasis on tourism, and providing an improved scientific basis for management and regulatory decision making for the Great Barrier Reef.

Recreational fishing is a component of sustainable management of fisheries and is being addressed through the *National Policy on Recreational Fishing in Australia* (Standing Committee on Fisheries and Agriculture 1994*a*). AFMA makes allowance in their management plans for the catches of recreational fishers.

Aquaculture is developing rapidly in Australia, and a *National Strategy on Aquaculture in Australia* was completed in 1993 (Standing Committee on Fisheries and Aquaculture 1994*b*). The use of aquaculture for re-enhancement in the marine environment is in its infancy. Some beginnings are being made in Queensland to rear and release some coastal fishes. However, this is to increase possible catches by recreational fishers, and there is as yet no evidence that it will be successful. Artificial habitats have been used in Australia for about 20 years, but mainly for recreational fishing. There is no proper evidence to measure their effectiveness in enhancing the commercial fisheries resources.

Although the SOMER report was a once-only event, the Australian *State of the Environment Report*, released in 1996, marks the first report of a regular reporting system for the total environment. Regular reporting on the state of the marine environment, the pressures upon it and the management responses made to ameliorate these pressures is essential (Department of the Environment, Sport and Territories 1993). Only regular examination of Australia's vision for its marine and coastal environment and reviews of the progress of implementation of policy in an integrated and co-ordinated manner will ensure that sustainable management actions are introduced and are successful in maintaining Australia's fisheries for future generations.

To successfully manage and sustainably use fisheries resources within its EEZ, Australia's challenge is to develop a cohesive policy, planning and management framework to provide a mechanism to resolve conflict in the use of the oceans. With the current political will, this is likely to be achievable within the next few years.

REFERENCES

ABARE (1995). 'Australian Fisheries Statistics.' Australian Bureau of Resource Economics. (Australian Government Publishing Service: Canberra). 48 pp.

AFMA (1995*a*). AFMA Corporate Plan 1995~2000. pp 8–9. (Australian Government Publishing Service: Canberra).

AFMA (1995*b*). Discussion Paper on the Management of the South East Fishery. (Australian Fisheries Management Authority). 2 pp.

Australian National Audit Office (1995/96). 'Commonwealth, Fisheries Management Audit Report 32.' (Australian Government Publishing Service: Canberra). 96 pp.

Brown, R S, and Phillips, B F (1994). The current status of Australia's rock lobster fisheries. In 'Spiny Lobster Management'. (Ed B F Phillips, J S Cobb and J Kittaka) pp. 33–63. (Fishing News Books: London).

Caton, A, and Sainsbury, K (1995). Southern Bluefin Tuna. In 'Fishery Status 1994 — Resource Assessment of Australian Commonwealth Fisheries'. (Ed K McLoughlin, B Wallner and D Staples) pp. 49–55. (Bureau of Resource Sciences: Canberra).

Council of Australian Governments (1992). 'National Strategy for Ecologically Sustainable Development.' (Australian Government Publishing Service: Canberra). 113 pp

Department of the Environment, Sport and Territories (1993). 'The State of the Environment Reporting Framework for Australia'. (Australian Government Publishing Service: Canberra). 42 pp.

Department of the Environment, Sport and Territories (1995*a*). 'State of the Marine Environment Report for Australia. Our Sea, Our Future. Major findings of the State of the Marine Environment Report for Australia.' (Great Barrier Reef Marine Park Authority: Canberra). 112 pp.

Department of the Environment, Sport and Territories (1995*b*). 'State of the Marine Environment Report for Australia. Technical Annex 1. The Marine Environment.' (Great Barrier Reef Marine Park Authority: Canberra). 193 pp.

Department of the Environment, Sport and Territories (1995*c*). 'State of the Marine Environment Report for Australia. Technical Annex 2. Pollution.' (Great Barrier Reef Marine Park Authority: Canberra). 93 pp.

Department of the Environment, Sport and Territories (1995*d*). 'State of the Marine Environment Report for Australia (1995). Summary of the State of the Marine Environment Report for Australia.' (Great Barrier Reef Marine Park Authority: Canberra). 8 pp.

Department of the Environment, Sport and Territories (1995e). 'Living on the Coast — the Commonwealth Coastal Policy.' (Australian Government Publishing Service: Canberra). 100 pp.

Department of the Environment, Sport and Territories (1996). 'State of the Environment Report: Executive Summary.' (Australian Government Publishing Service: Canberra). 47 pp.

Ecologically Sustainable Development Working Groups (1991). 'Final Report-Fisheries.' pp. 115–19. (Australian Government Publishing Service: Canberra).

IUCN Australian Committee (1994). 'Towards a Strategy for the Conservation of Australia's Marine Environment.' *ACIUCN Occasional Paper Number 5* (Australian Committee for IUCN: Sydney). 30 pp.

Ocean Outlook Steering Committee (1995). 'Ocean Outlook: a blueprint for the oceans: a report from the Congress 16–17 November 1994.' (Commonwealth Scientific and Industrial Research Organisation, Division of Oceanograph: Hobart). 18 pp.

Office of the Chief Scientist (1995). 'Australia's Ocean Age: Science and Technology for Managing our Ocean Territory.' (Australian Government Publishing Service: Canberra). 48 pp.

Phillips, B F, and Ilett, A (in press). Coastal zone management.: the Australian experience. *Biologia Marina Mediterranea*.

Phillips, B F, and Rayns, N (1995). AFMA establishes new approach to managing fish stocks. *Australian Fisheries* **52**, 6-8.

Resource Assessment Commission (1993). 'Coastal Zone Inquiry.' (Australian Government Publishing Service: Canberra). 519 pp.

Senate Standing Committee on Industry, Science, Technology, Transport, Communications and Infrastructure (1993). 'Fisheries Reviewed.' (Australian Government Publishing Service: Canberra). 163 pp.

South East Fishery Assessment Group (1994). 'Stock Assessment Report Eastern Gemfish 1994.' pp. 9–18 (Bureau of Resource Sciences: Canberra).

Standing Committee on Fisheries and Aquaculture (1994*a*). 'Recreational Fishing in Australia: A National Policy.' (Department of Primary Industry and Energy, Fisheries Policy Branch: Canberra). 25 pp.

Standing Committee on Fisheries and Aquaculture (1994*b*). 'National Strategy on Aquaculture in Australia.' (Department of Primary Industry and Energy, Fisheries Policy Branch: Canberra). 28 pp.

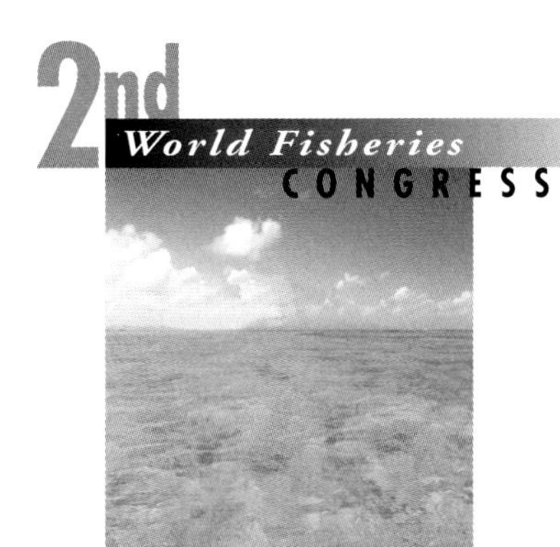

Community-based Management for a Sustainable Seahorse Fishery

A. C. J. Vincent[A] *and M. G. Pajaro*[B]

[A] Department of Zoology, University of Oxford, South Parks Road, Oxford OX1 3PS, UK.
Present address: Department of Biology, McGill University, 1205 Ave, Dr. Penfold, Montreal H3A 1B1, Canada.
[B] Haribon Foundation for the Conservation of Natural Resources, 28 Quezon Ave, Quezon City, 1100 Metro Manila, Philippines.

Summary

The seahorse fishery (for traditional medicines, aquarium fishes and curios) is large, global, and economically important. It also appears to be unsustainable, with seahorse numbers declining markedly where fished. Handumon village (*barangay*) in the central Philippines, where the first seahorse conservation and management project was implemented in January 1995, is particularly dependent on seahorses. Early socio-economic study provides the template for current co-management efforts. These include protective measures, fishery modifications and enhancement efforts: (1) the *barangay* established and now enforces a 33-ha no-exploitation sanctuary and an adjacent traditional fishing zone; (2) fishers place newly caught pregnant males into sea cages whence newborn young can escape before the male is sold; (3) fishers re-seed areas depleted of seahorses and have begun developing seahorse culturing skills. The eventual goal is for the *barangay* to take full responsibility for the project.

Introduction

A broader view of fisheries needs to emerge in which we realize that not all exploited animals are caught by net and trawl to be killed for food. Recent attention has highlighted the threats posed by the expansion of live-food trade in groupers and wrasses (Johannes and Riepen 1995), and the great demand for ornamental fishes for the aquarium trade (Wood 1985), with emphasis on the potentially serious effects of cyanide as a catch method. Virtually no attention has yet been paid to the medicine trade in marine animals, despite the fact that hundreds of marine species are sought in traditional Chinese medicine alone (Tang 1987).

The less-studied trades in marine products commonly arise from small-scale inshore fishing, a sector that has yet to receive the attention it merits. Most such fisheries remain unquantified — often even unmentioned — in global assessments of declining fish resources, probably partly because they are difficult to assess and partly because they are of greatest importance to marginalized fishers who have little political power (White *et al.* 1994). Even when they are considered, relatively few models and management tools are appropriate for the many inshore fishes that have low reproductive potential and highly structured social and spatial patterns.

Seahorses have recently been recognized as a global fishery, caught for medicines, aphrodisiacs, tonic foods, aquarium fishes and curios (Vincent 1996). They may be among the most valuable fisheries commodities in Southeast Asia, with retail prices reaching $US1200 per kg in Hong Kong in 1995 (Vincent 1996). Global consumption of dried seahorses for traditional Chinese medicine exceeds 20 million individuals per annum, with further large trades for Indonesian *Jamu* and other indigenous medicines (Vincent 1996). The aquarium and curio trades also consume hundreds of thousands of seahorses annually (Vincent 1996). Fishers and merchants report that supplies fail to match the limitless demand for dried seahorses. At least 36 countries or territories are already known to trade seahorses and their pipefish relatives, with new seahorse fisheries opening as far afield as Nigeria and the Galapagos Islands (Vincent 1996).

The consensus is that seahorses are declining in both number and size, with serious consequences for the many subsistence fishers who target them directly (Vincent 1996). Heavy fishing pressure will prove particularly problematic for seahorses: most species appear to be characterized by sparse distributions, low mobility, low natural adult mortality, small home ranges, low fecundity, lengthy parental care, and mate fidelity (e.g. Strawn 1953; Vincent 1990, 1994, 1995; Vincent and Sadler 1995). Fishers in certain areas of India, Indonesia, the Philippines, Thailand, and Vietnam estimate catch reductions of up to 50% over the past 5 years and it now takes more seahorses to accumulate 1 kg of dried animals (Vincent 1996). The problem is likely to worsen as demand increases and patent medicines become more common; consumers generally reject small, dark and spiny seahorses when choosing whole animals but all seahorses become acceptable once packaged.

Programmes to ensure the persistence of seahorses will have to consider the needs of dependent fishers, if widespread poaching is to be averted. Community-based co-management is proving useful in enabling stakeholders to conserve local resources, particularly where fished by the local community (White *et al.* 1994). This approach should be particularly applicable to seahorses, for which the labour-intensive nature of the fishery and the need for familiarity with local waters reduces competition from outside the community.

The first project to conserve and manage seahorses was launched in January 1995 in Handumon village (*barangay*) in the central Philippines. This paper begins with an analysis of the problem and then describes initial progress toward co-management.

The project and the problem

The Philippines is a major exporter of seahorses, both dead and live (Vincent 1996). Throughout the Philippines, seahorse collectors cite the municipality of Getafe in Bohol, and *Barangay* Handumon in particular, as a key seahorse area (Fig. 1). A 1993 visit confirmed the importance of this region, and it was decided to implement the Philippines' first seahorse conservation project there. Team members visited Handumon for three months to discuss the planned project with the *barangay* and officials, in an interactive approach that set the tone for all subsequent work together.

Because socio-economic indicators can determine management approaches and act as indices of success (White *et al.* 1994), the community organizer formally interviewed 22 seahorse fishers and 18 non-seahorse fishers for information on their socio-economic status. We also conducted lengthy semi-structured interviews with 25 seahorse collectors and five buyers. They were first given the opportunity to speak freely about their perceived problems and were then posed specific questions about seahorse biology, fishing, trade, marketing and conservation.

Socio-economic base of the fishery

Handumon is one of three *barangays* on Jandayan Island, in the municipality of Getafe, Bohol. Comprising about 800 people in 140 households, it ranks among the poorest *barangays* of a poor municipality in a poor region. Handumon has no running water, no electricity, no roads and no telephones. Declared monthly income is about P800–3500 ($US32–140), lower than the 1991 national poverty threshold of P3676 ($US147) per month for a family of six, set by the Philippines' National Economic Development Authority (Cruz-Trinidad *et al.* in prep).

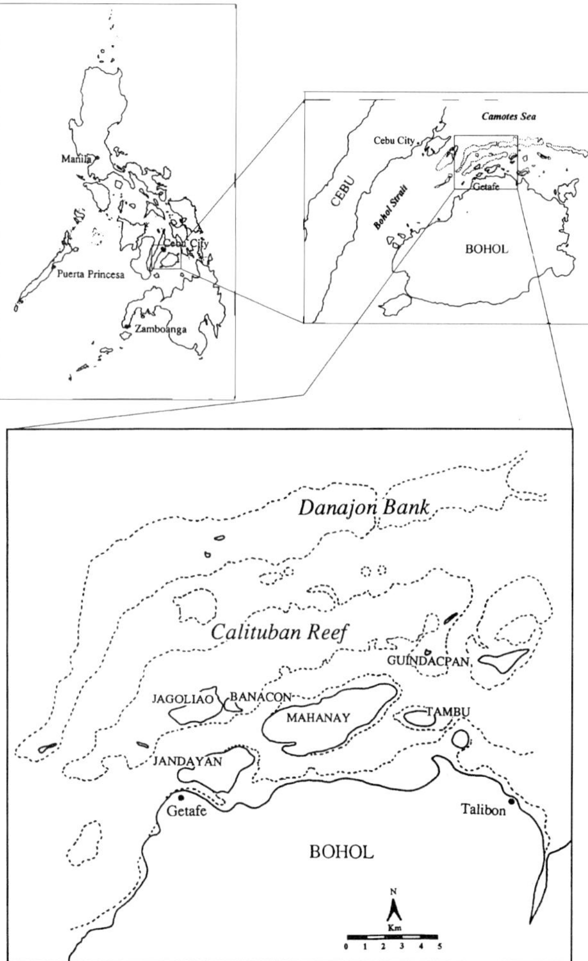

Fig. 1. Maps of the Philippines, Bohol, and north-western Bohol. The community-based seahorse project is located at Barangay Handumon on Jandayan Island in Getafe municipality, and is adjacent to an unusual double barrier-reef formation (Calituban Reef and Danajon Bank).

Households in Handumon include an average of six people (four dependants). Nearly 60% of the inhabitants are under 19 years old, with annual population growth rates of about 3%. Only 32% of people aged 20–80 have completed the six years of elementary schooling (Cruz-Trinidad *et al.* in prep).

Most households derive their income from some combination of fishing, subsistence agriculture and labouring, including water hauling, carpentry, firewood gathering, and shellcraft (Cruz-Trinidad *et al.* in prep). Half of all families rely on fishing as their primary source of income, and fishing is the usual fall-back occupation when other alternatives fail. Handumon is located on the edge of Danajon Bank, an unusual double barrier-reef formation that was previously one of the richest fishing areas in the Philippines. Local waters also include mangroves, seagrasses and sargassum (Fig. 2). Despite this wealth of habitats, overfishing and destructive fishing techniques (e.g. dynamite, fine mesh nets and cyanide) have severely degraded and impoverished local fish stocks and marine resources. Consequently, a night of fishing may yield only enough fish for one family meal, and non-fishers must often compete to buy any excess fish.

The seahorse fishery

About 40% of all fishers (21% of *barangay* households) are involved in seahorse collecting at any one time. Fishers change target species in response to availability, entering and leaving the seahorse fishery at unpredictable intervals, but seahorse collecting requires considerable skill. In early 1996, fishers received a flat $US0.24 for each live seahorse and up to $US0.56 for a dead seahorse, depending on its weight. The greatest number of fishers estimate that seahorses contributed 31–40% of their annual income, and as much as 100% of peak-season income (December to June). Seahorses are sold through one or two levels of buyers, then to exporters in Cebu (dried) or Manila (live). Earnings from seahorses go primarily to buy rice and other food.

Here, as in many other regions, seahorses are a target catch, collected at night in the course of spearfishing (Fig. 3). Each fisher works alone, swimming beside his outrigger boat (*banca*, either paddle or motorized), his way lit by a low-slung lantern. A few fishers work at greater depths, breathing air supplied from the surface by highly unreliable compressors. All fishers are seeking food for their families, taking anything edible. The seahorses are collected by hand and provide most of the cash from the fishing trip. The need to fish for food anyway makes it unlikely that seahorse collecting will become economically non-viable.

Assessing conservation status

Fishers (n = 25) estimated in 1994 that their seahorse catches had decreased by a mean of 69% over the previous ten years, 50% over the previous five years, and 38% over the previous year. Collectors commented that seahorse densities in the wild declined from 5–10 seahorses per m^2 of good habitat in 1985 to less than one seahorse per m^2 ten years later. The number of Handumon seahorse fishers active at any one time had increased from 10–15 in 1985 to 25–30 in 1990, but thereafter had stabilized while catch declines had continued.

Buyers (n = 5) also surmised that catches had decreased by 60–70% over the ten years from 1985 to 1994, although they estimated a decline of only about 10% from 1993 to 1994. Commensurate with changes in seahorse availability, Cebu exporters' prices for dried seahorses increased from P600 ($US24) per kg in 1985 to P2500–3000 ($US100–120) per kg in 1995.

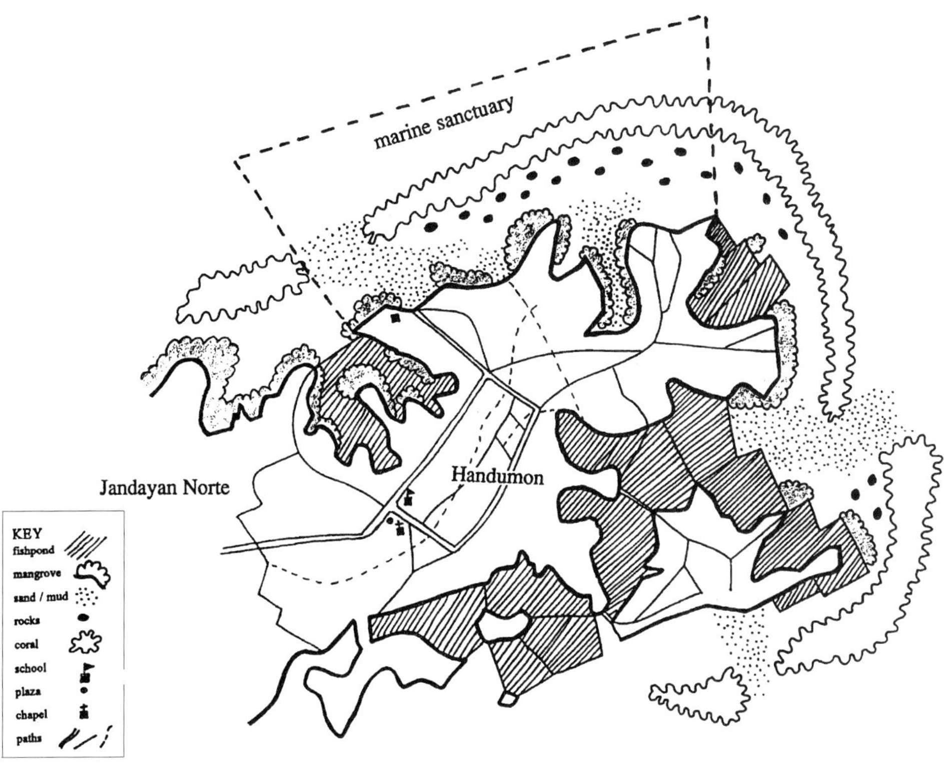

Fig. 2. Map of Barangay Handumon (Jandayan Island), showing key habitat features and the location of the new 33-ha marine sanctuary, in which no exploitation is allowed.

Fig. 3. Drawing of Handumon spearfisher, hand-collecting seahorses at night. Equipped with wood and glass goggles and wooden fins, he tows the outrigger boat along, spearing food fishes and finding seahorses by lantern light.

The minimum saleable size was 100 mm vertical length until the aquarium trade began in the 1970s, but it declined steadily to 50 mm vertical length in 1995, even though most local species mature at about 85 mm vertical length. One kilogram comprised only about 250 seahorses in 1993 but 300–450 seahorses in 1995, reflecting diminishing seahorse size.

All seahorse collectors identified a need to manage the seahorse fishery. Most offered to contribute time, knowledge, skills, materials and money. Only one of 25 felt that efforts should be left to *barangay* leaders, 'who get salaries for such undertakings'. The others were prepared to contribute time for conservation-related activities including patrolling, meetings and lobbying, helping find seahorses for research and assessment, helping build cages and corrals, and helping to re-seed depleted areas with seahorses. This self-help approach facilitates co-management.

PROGRESS IN MANAGEMENT

The lack of alternative livelihood options, in fishing and in other occupations, precluded any attempt to close the seahorse fishery, even had that been thought necessary. Instead, we are carrying out the first field study on the behaviour and ecology of Indo-Pacific seahorses, monitoring fishing yields, and encouraging action to ensure the survival of both seahorses and seahorse fishers. Fishers' own views of what is necessary (see above) influence the project initiatives we propose, and fishers make the final decisions, not always as we expect.

Biological information

The fishers had only limited understanding of seahorse natural history, so biological information had to be sought to facilitate management. Seahorse taxonomy is in chaos but there are probably five real species within the area fished from Handumon; these are segregated by habitat, including seagrass, coral, sargassum and sponge communities. From monitoring tagged animals and gathering fishers' descriptions, it seems that most of these species live as pairs of one male and one female, to the extent that the fishers consider the discovery of a pregnant male to signal the presence of a female within 1–2 m. Pregnant males of most species — the male seahorse provides all parental care — can be found throughout the year, but it is not yet known whether any given individual breeds all year round. The peak breeding season appears to be December and January, leading to a greatly increased number of juvenile seahorses in February and March, although they may also be more visible then because the sargassum thins.

Early observations suggest site fidelity for putative *Hippocampus comes* at least, with some tagged seahorses still resident in the same locality six months after first observation. Twenty-four-hour observations confirmed the fishers' comments that seahorses emerge only at night, disappearing into the corals at dawn. In the past, seahorses were visible by day; fishers speculate that this shift in behaviour may arise from the former heavy fishing pressure by day.

Management data

Fishers participate in collecting management data in two ways. First, they report their seahorse catches as they land each morning. We record species, sex, length, weight and reproductive state of each fish, and we talk to fishers about the costs and yields of the fishing trip, including food fish. Fishers further report the seahorse habitat choice, density, activity patterns, and social behaviour they observe in the field. Second, fishers maintain a diary that indicates days on which they fished and/or sought seahorses, and their catch for each day. This allows an overview of *barangay*-wide catch and effort.

About 85% of the seahorses caught by Handumon fishers are one species (putative *H. comes*) and about 95% are caught on coral-associated habitats, including sponges and sargassum. Most seahorses are caught from January to July. Mean seahorse length in catches is smaller in March than during the rest of the year, because of the many juveniles caught. Diaries show that mean catch per fisher per night's fishing peaked in April at 8.75±2.70 seahorses. Smaller catches later in the year reflected the lower availability of seahorses and the consequent switch to target other species (e.g. cuttlefishes and abalone) as they became more available.

Conservation decisions

Protecting seahorses and other marine resources. Almost immediately after the project began, the *barangay* set aside about 0.25 ha of coral reef for seahorse research. The area was proffered because (a) it was close to the landing area and therefore easily accessible for study and (b) very little life remained on this very degraded reef so its loss would not be detrimental to fishers. Many discussions later, the council expanded this to a 33-ha sanctuary where no fishing of any species is allowed; it encompasses mangroves, sargassum, corals and deeper waters (Fig. 2). The seahorse team gave advice on management, enforcement and legislative protection, ensuring that the sanctuary was approved by a municipal ordinance in July 1995. Simultaneously, the *barangay* also closed adjacent *barangay* waters to fishers using destructive techniques such as dynamite, cyanide and *liba-liba* fine-mesh nets.

Barangay compliance is very high, with volunteer patrols operating several nights a week. The few initial local violations were resolved publicly and effectively by a determined captain and council, with the support of the *barangay.* Patrols have apprehended and fined violators and illegal fishers from other *barangays*, despite threats involving guns and dynamite. A recent donation of a fast outrigger patrol boat, searchlight and two-way radios has enhanced patrols to the point where violators are seldom seen.

The seahorse team helps with liaison between *barangay* and municipality on administrative matters and helps to ensure municipal support for enforcement. Recently, the mayor gave municipal police the mandate to enforce fishing laws, and persisted when a political ally was caught. The police now join the *barangay* patrol on their boat, paying costs, and together they have apprehended violators in other areas of the municipality.

When they can afford it, fishers have been donating (usually small) seahorses to the sanctuary in order to increase the resident population and to create a viable source population from which young may disperse to restock other areas. Every seahorse donated represents an important proportion of a fisher's daily cash income.

Fishers and biologists are cooperating on a series of surveys to detect changes in abundance of seahorses, other fishes and marine resources in general. Although the programme has only begun, fishers are delighted with the greatly increased number of species and fishes in the sanctuary. Selected fishers are now monitoring changes in their neighbours' fishing success (for all species, not just seahorses) on the edge of the sanctuary. There are still insufficient data for formal analysis, but fishers seem generally to feel that catches have improved slightly.

Modifying the fishery. New data on seahorse biology and catches are shared with the *barangay* in feedback sessions. These have been dominated by the issue of how to change fishing practices to reduce pressure on wild populations. There is general agreement that pregnant males and juveniles should not be taken, but two problems arise: (1) fishers usually need money immediately and can sell all but the very smallest seahorses, and (2) any seahorses left behind may be taken by other fishers.

A partial solution has now been found with regards to pregnant males, one that costs fishers nothing. They have built two sea cages in which they place newly caught pregnant males, leaving them there until the young are born, whereupon the young escape to the reef and the male is sold. The seahorse team usually advances the sale price when the male is put in the cage. Hundreds of pregnant males have already been held thus, and local buyers (of dried and live seahorses) are now starting to refuse pregnant males, encouraging compliance. Fishers are concerned about probable high mortality of young when escaping the cage, and are seeking modifications, but they accept that even the lowest realistic estimates of survival mean that many thousands more young settle than would otherwise be the case.

There have been failures. Fishers decided in early 1996 not to catch seahorses smaller than 3 inches (75 mm) vertical length. For the next few days, no catches that were brought to be recorded included such small seahorses. It was then realized that many were being sold on the way home to a buyer on an adjacent island. Compliant fishers and the Handumon buyer were being penalized, so fishers suspended the restriction, while still agreeing it to be a good idea. Further co-ordination with adjacent *barangays* will be necessary before a further attempt to implement this restriction.

Enhancing seahorse numbers. Caging pregnant males (see above) represents a first step toward seahorse aquaculture and allows fishers to develop organizational capability. Fishers took the second step by building a 50 m^2 corral in March 1996, in the zone where only non-destructive fishing is permitted, and donating seahorses to fill it. The goal is a stable breeding community of seahorses, from which young will escape to restock the area. The next step will be to retain and feed the young seahorses for the first month, when they are most vulnerable to predation, before releasing them (sea ranching).

Culturing seahorses on land may be possible but has elsewhere proved problematic (e.g. in China, Indonesia, the Philippines, Thailand and Vietnam) because of difficulties in securing adequate live food for young seahorses and in reducing disease, although these are considered surmountable.

Community involvement and training

The project team holds regular feedback and planning sessions, attends *barangay* council meetings and assemblies every month, and has been allocated a seat on the Municipal Development Council and on the municipal Environment and Natural Resources Council. Educational activities include presentations on topics ranging from bird biology to managing the freshwater resources of the island, slide shows, film showings, training sessions on marine ecology, a tree-planting activity for World Environment Day, and children's activities.

Capacity building is one of the most important objectives. A *barangay* resource management team has been initiated, and is working to define its roles and responsibilities in planning, organization and decision making. Key fishers and leaders were taken to visit the highly successful community-based marine reserve at Apo Island (White 1988), and they returned full of enthusiasm. The project supports a high school scholarship programme, with students receiving full financial support in exchange for spending weekends as apprentices on the conservation project. The first student was elected president of the youth association in May 1996, proposing a programme of conservation activities.

Discussion

The dependency of Handumon on seahorses has encouraged fishers to cooperate in order to ensure a sustainable fishery. Indications of successful community-based management include the following: community participation in planning, implementation and monitoring; increased community awareness and empowerment; sophistication of community organizations; and smoothness of collaboration (White *et al.* 1994). By such standards, the Handumon project is going well. Fishers are beginning to take responsibility for management decisions, whereas they previously flouted externally imposed laws prohibiting certain fishing gear.

We cannot yet judge success in terms of improved socio-economic status, livelihood patterns, and management tools (White *et al.* 1994), but community support for the project has made it possible to embark on conservation and management initiatives quite rapidly. Fishers themselves feel that there is progress. Enthusiasm for caging pregnant male seahorses, corral building, and eventual culturing continues to grow. Seahorse catches were higher in March 1996 than in March 1995, owing to the higher numbers of juveniles; the improvement prompted further efforts from fishers even though it is unlikely that the project can yet be producing an effect.

Fishers are particularly heartened by the evident recovery of the sanctuary. This success suggests that it may sometimes be appropriate to accept poor areas as sanctuaries and reserves: it costs fishers less to cede poor habitat and they can readily observe the success of a sanctuary by an almost-immediate increase in marine life. Enforcement of the Handumon sanctuary has led other *barangays* to evaluate their own resource management; several have begun enforcing existing reserves and others are considering no-exploitation reserves. This is encouraging because the efforts of Handumon alone will not sustain communal marine resources.

Many of the management problems of the seahorse fishery apply also to other fishes. The life histories of inshore fishes commonly have characteristics (low fecundity and structured populations) that would make them vulnerable to heavy fishing pressure, such as that developing in response to the economic growth of China. Traditional Chinese medicine is expanding in both scope and scale: the number of different products used in this medicine now exceeds 10 000, and demand for traditional medicine, aphrodisiacs and tonic foods is soaring. Hundreds of marine species are already in use (Tang 1987), and many may be caught or traded by subsistence fishers who harvest animals past supposed economic extinction, in conjunction with seeking food.

The threat to seahorses is part of the dilemma of subsistence communities confronted with dwindling resources and increasing populations, and it poses the oft-encountered question of whether to ban use of a resource or to try to manage its exploitation. The Handumon project suggests that the latter may be feasible.

Acknowledgments

We are most grateful for the support and involvement of the captain, council and people of *Barangay* Handumon, and the mayor and people of Getafe municipality. The project is funded by the Darwin Initiative for the Survival of Species, United Kingdom Department of the Environment. Early pilot studies were supported by National Geographic and the Whitley Award for Animal Conservation (Royal Geographical Society). We thank British Airways Assisting Conservation for generous support. The British Broadcasting Corporation gave the *barangay* a much-needed outrigger boat and two-way radios for patrols. Sara Lourie and Milo Socias produced the figures.

References

Johannes, R E, and Riepen, M (1995). Environmental, economic, and social implications of the live reef fish trade in Asia and the Western Pacific. Report of The Nature Conservancy and South Pacific Forum Fisheries Agency.

Strawn, K (1953). A study of the dwarf seahorse, *Hippocampus regulus* Ginsburg at Cedar Key, Florida. M.Sc. Thesis, University of Florida.

Tang, W-C (1987). Chinese medicinal materials from the sea. *Abstracts of Chinese Medicine* **1**, 571–600.

Vincent, A C J (1990). Reproductive ecology of seahorses. Ph.D. Thesis, University of Cambridge.

Vincent, A C J (1994). Operational sex ratios in seahorses. *Behaviour* **128**, 153–67.

Vincent, A C J (1995). A role for daily greetings in maintaining seahorse pair bonds. *Animal Behaviour* **49**, 258–60.

Vincent, A C J (1996). 'The International Trade in Seahorses.' (TRAFFIC International: Cambridge, UK.)

Vincent, A C J, and Sadler, L M (1995). Faithful pair bonds in wild seahorses, *Hippocampus whitei. Animal Behaviour* **50**, 1557–69.

White, A T (1988). The effect of community-managed marine reserves in the Philippines on their associated coral reef fish populations. *Asian Fisheries Science* **2**, 27–41.

White, A T, Hale, L Z, Renard, Y, and Cortesi, L (Eds) (1994). 'Collaborative and Community-based Management of Coral Reefs.' (Kumarian Press: West Hartford, Connecticut, USA.)

Wood, E (1985). 'Exploitation of Coral Reef Fishes for the Aquarium Trade.' (Marine Conservation Society: Ross-on-Wye, UK.)

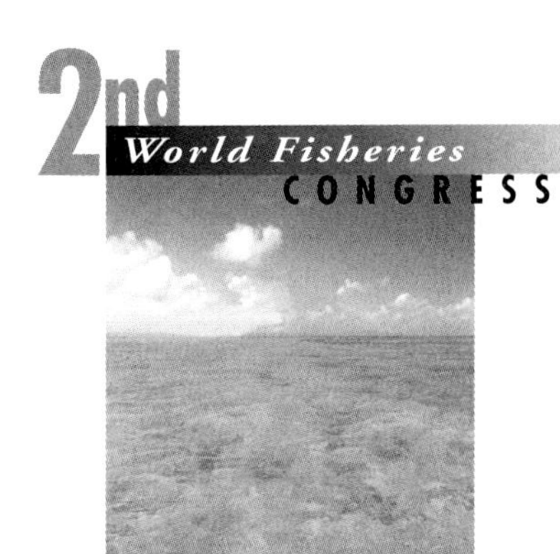

Factors Affecting Management in a Mexican Fishery

Silvia Salas and Ricardo Torres

Fisheries Centre, University of British Columbia, 2204 Main Mall, V6T 1Z4 Vancouver, B.C. Canada.

Summary

In Mexico, fisheries regulation systems mainly involve the control of fishing effort inputs. However, the multi-species nature of most Mexican fisheries, where fishers target different species with the same gears and vessels, fishing effort is difficult to control. In this paper, we describe the fisheries' institutional and legal framework, the status of the main resources, and the participation of the socio-economic sectors in Yucatan fisheries. The likely consequence of recent changes in legislation is also discussed. Information obtained from official documents, research reports, catch statistics, as well as from a questionnaire survey carried out in the fishing communities, was used in this study. Promotion of different kinds of fishing groups in the last decade resulted in an increase in the size of the fleet. Additionally, recent changes in legislation have led to the opportunity to increase fishing effort. This increase was made without consideration of the status of the resources. Even though the number of cooperatives has remained stable in recent years, there is evidence of uncontrolled increases in other kinds of organizations. Thus, the spatial and temporal distribution of the fleet are not well known. Contradictory policies in management, such as fishing inputs control, and simultaneous promotion of more organizations, as well as the lack of knowledge of the dynamic of the resources and their users, affects the success of the management system.

Introduction

Fisheries play an important role in both developed and developing countries, providing directly or indirectly employment and revenue to 200 million people globally. Nevertheless, many fisheries are operating with excess capacity, overcapitalization, economic inefficiency and biological overexploitation (Garcia and Newton 1994).

Mexico is no exception to these global trends, even though fisheries have been an important source of food, employment and foreign earnings. However, gradual adjustments to policies instead of long-term programmes, have led to overexploitation in some of the most valuable fisheries and to underutilization of others.

Uncontrolled access to fisheries resources, limited enforcement, lack of reliable data for planning and stock assessment, and uncertainty in the resource dynamics and market information, have been identified as related problems to the lack of definition in the management process (Arreguin-Sanchez *et al.* 1987; Vazquez-Leon and Mcguire 1993).

The government, in order to generate employment and foreign exchange, has encouraged increases in fleet size in different regions. From 1982 to 1992, the small-scale fleet was expanded by more than 30 000 units.

In Yucatan, the increase in number of organizations has resulted in an increase in the number of boats that have exploited fishing resources in the last 10 years (Pare and Fraga 1994). However, those increases were made without considering recommendations from different scientists who, since 1987, pointed out the need to reduce fishing effort in grouper and octopus fisheries (Arreguin-Sanchez *et al.* 1987; Seijo *et al.* 1987).

Based on the review of the status of the main fishing resources and the development of the fishers' organizations in Yucatan, as well as on a survey carried out in 15 fishing communities in Yucatan, the contradictory policies promoting increases in fleet size and simultaneously attempting to regulate fisheries by input controls, are discussed.

NATIONAL CONTEXT

The national fleet is comprised of nearly 75 000 vessels, of which the small-scale fleet represents over 95%. Consequently, most of the fishing effort is concentrated in the coastal zone (SEMARNAP 1996), exerting a tremendous pressure on it. In the last five years, the large-scale fleet has been reduced in some cases (tuna fishery) or has been stabilized in others (trawlers), whereas the small-scale fleet has increased (SEMARNAP 1996).

Legal access to fish is by fishing concessions (lasting from 5 to 20 years), and fishing permits (lasting four years). Management tactics include control of the number of fishers, size of their boats and kind of gears used, as well as establishing minimum legal sizes, quotas and closed seasons.

Fisheries management is centralized by the SEMARNAP (the Ministry of Environment, Natural Resources and Fisheries). The regional enforcement depends on the Regional Fisheries Delegations. In the case of cooperatives, there is also a self-regulatory system, by which fishers define their own enforcement tactics, according to rules stated by general agreements within the local organizations (Fig. 1).

In the next section, Yucatan fisheries are described as an example of the contradictory policies existing in Mexico, focussing mainly on input controls, but with increasing fishing pressure through socio-economic programmes developed and implemented by the government.

YUCATAN FISHERIES

The fisheries

Around 40 species are captured off the coast of Yucatan, yet 80% of the landings comprises red grouper (*Ephinephelus morio*) and octopus (*Octopus maya*). Other species caught in the region are: demersal fishes (snappers, porgies), sharks, small pelagic fishes and spiny lobster, *Panulirus argus*. The lobster is one of the most profitable resources in the region.

Octopus and grouper are caught by small-scale and commercial fleets. Lobsters are captured only by small-scale boats (Seijo *et al.* 1994; Solis-Ramirez 1994). Groupers are also caught by Cuban fishers, who have a quota established by an international agreement between the Mexican and Cuban governments.

The total fleet in Yucatan includes nearly 4000 vessels, of which 85% are small-scale boats distributed in 15 fishing communities along the coast. Most of them are multi-purpose (Burgos *et al.* 1992), switching gears to catch demersal fishes, lobster or octopus. In general, there are no reliable records of the spatial allocation of the fleet and of the number of fishers' organizations involved in the activity.

As in other parts of Mexico, in the last 20 years, the Yucatan coastal zone has been seen as an alternative employment opportunity for people who depended on agricultural activities. From 1970 to 1990, fisher numbers increased by 300% (Pare and Fraga 1994), making it extremely difficult to control the inputs and outputs of the fisheries.

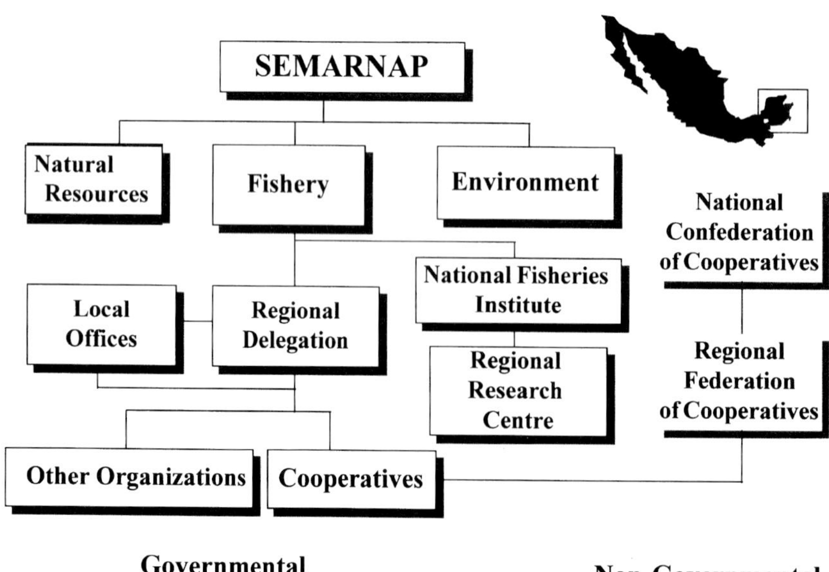

Fig. 1. Regulatory System in Mexico.

The users

The socio-economic sectors involved in Yucatan fisheries are private and social. Both of them own small-scale and large-scale boats (Fig. 2). However, the private sector has the means to process and market fishing products, and to build vessels besides harvesting. The social sector concentrates only on the harvest. The fishers of this sector tend to organize into groups to obtain credits for purchasing their boats and motors.

The private sector consist of 150 members in one organization (Chamber of Fishing Industry). This sector owns 71% of the state small-scale fleet (2407 vessels), and 85% of the commercial fleet (492 vessels). In the last decade, 86% of the catches have been reported by this sector (Fig. 2), and 64% of that comes from only 36 members (Rosado and Rosado 1995).

The social sector in Yucatan include three types of fishers' organizations: cooperatives (36), rural societies (46), and "solidarity societies" (98).

To clarify the context of the organizations' constitution, it is considered important to describe the origins of the different groups and the context in which they were promoted. According to this, three periods can be identified.

First period (1958–1980). In 1958, the Federal government gave the exclusive right to cooperatives to exploit the most profitable species (shrimp, abalone, lobster, conch, oyster), which encouraged fishers to create more groups. During that period, 28 cooperatives were registered, with 36 by 1994.

Second period (1980–1989). At the beginning of the 1980s, due to the market crisis in the henequen (American agave) industry, the government assisted peasants to migrate to the coast and become fishers as an economic alternative (Fraga 1992). Incentives included the creation of a training centre in fishing activities, providing soft credits to purchase boats and motors (Fraga 1993), and 'paying' peasants during the training time. Consequently, in this period, 46 rural societies were created. However, Pare and Fraga (1994) point out that 80% of the people attended the courses as an immediate economic opportunity, rather than as a future employment alternative in the fishery.

Third period (1989–1993). In order to increase production and reduce fishing pressure in the coastal zone, government encouraged switching from small-scale to large-scale fishing, by giving credits to purchase vessels and equipment. It motivated the creation of the so-called *solidarity societies,* which include both traditional fishers and peasants recently incorporated into the fisheries. There was no control over the number of groups created; 98 organizations were established.

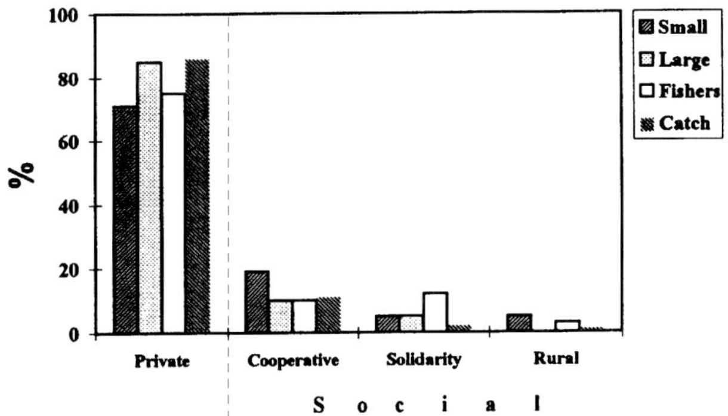

Fig. 2. Participation of private and social sectors in the Yucatan Fishery.

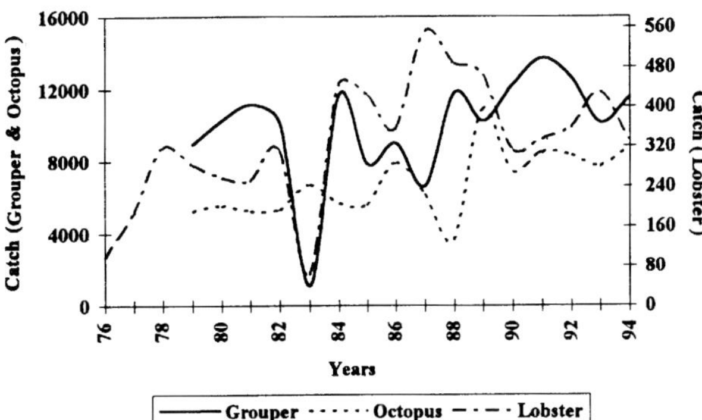

Fig. 3. Trends of catch (t) of the main species captured in Yucatan from 1976 to 1994.

It is interesting to note that from 1958 to 1980, an average of 1.3 cooperatives per year were registered, while from 1980 to 1993, the rate increased to 11 groups per year.

A survey answered by the heads of the organizations in the social sector in 1994 in 15 ports, showed that only 47% of the organizations, which include 3958 fishers, were registered in the National Fisheries Register. As a consequence, catch landings were misreported. Only 72 of 180 organizations reported their catches that year. Of those catches, 80% was recorded by cooperatives, and the remaining 20% by the rural societies and the solidarity societies (Sepesca internal report 1992).

The status of the main resources on the Yucatan coast

Landings of the three main species were highly variable during the last 20 years. Landings fell to their lowest level in 1983 for grouper and lobster, and in 1988 for octopus. At the beginning of the 1990s, the catch generally declined for the three species (Fig. 3).

Even though many studies have been done in the region (Table 1), lack of information on fishing effort makes it difficult to explore the causes of variations in the landings. Most of the studies carried out in the area have focussed on grouper and octopus. Studies on other species are limited mainly to estimation of growth and mortality, and they used mainly commercial data from the small-scale fleet. This situation applies for most of the species; thus, only a fraction of the populations had been considered in stock assessment studies (Arceo *et al.* 1995).

Since 1987 the need to control fishing effort on grouper and octopus has been suggested (Arreguin-Sanchez *et al.* 1987; Seijo *et al.* 1987). At the same time, further development in the lobster fishery has been considered and changes to catching technology were promoted (Table 2). However, the biological capacity of the stock has not been assessed (Arreguin-Sanchez *et al.* 1987; Seijo *et al.* 1994; Torres and Salas 1993).

Even though several scientists (Hilborn and Walters 1992), have pointed out that a harvest strategy sets out how the catch taken

Table 1. Numbers of studies carried out on the main species caught in Yucatan, Mexico

Species	Growth	Mortality	Reproduction	Recruitment	Food habits	Stock assessment
Grouper	14	15	2	11	1	10
Octopus	13	9	3	2		5
Lobster	4	4	1	1		1
Snappers	7	7	2	6	1	1
Shark	5	3		1		1
Mackerel	3	3				
Queen conch			2	1		
Porgy	2	2		1	1	1
Sardine	1	1				
Tuna	1	1				

from a specific stock should be adjusted year to year, depending on various factors, such as biology, ecology, socio-economics, and uncertainty related to the resource, market and environment, few studies in Yucatan have considered the socio-economic framework of the fishery (Seijo *et al.* 1987; Hernandez 1995), and even less have evaluated alternative management objectives (Diaz-DeLeon and Seijo 1992).

Cooperatives and changes in legislation.

In 1994, the following changes in legislation were approved: a) There are no longer any reserved species; b) Any proponent, individually or collectively, is able to request a concession to exploit former reserved species in a defined area. National and foreign capital now has access to exploit the most valuable, formerly reserved, species.

Government has been providing financial aid for a long time to cooperatives and other organizations. However, since the national budget became smaller and subsidies decreased, industrial companies have taken over the government's role as financial providers. In these cases, fishers have to sell their catches to the companies. Fishers don't have control over the pay they receive, and companies may change prices at their convenience.

Under these circumstances, cooperatives cannot compete with the private sector for valuable species. Given these changes in legislation, the expectation of increased incomes might attract private investment in new technology into the lobster fishery. This situation may also contain a risk if it opens access to other resources, once the fleet settles in the region.

Table 2. Status of the main species caught in Yucatan, Mexico

Species	Current status	Current yield (t)	Potential yield (t)	Reference
Red grouper	overexpl	10 000	20 000	Chavez 1994
Octopus	overexpl	7 000	8 000	Solis 1994
Lobster	?	400	6 400	Arceo 1995
Red snapper	increasing	1 000	7 800	Gonzalez 1987
Shark	?	2 100	?	Bonfil 1996
Mackerel	?	1 100	5 000	Arreguin 1987
Queen conch	overexpl	0	?	Solis 1994

Discussion

The history of Mexican fisheries illustrates the results of a heterogeneity of objectives, which attempt to satisfy different users and particular goals in different political contexts. Most of these goals have favoured the short-term yield and revenues above the long-term benefits.

It seems that the implementation of programmes which give alternatives to certain groups, depending on immediate priorities or political goals, have failed. By attempting to solve a problem in a particular area, government has generated conflicts in others.

To promote economic growth and social equity, some programmes have been developed, but at the same time, they have increased pressure on fishery resources through intense immigration to coastal zones. Similarly, recent changes in legislation covering the reserved species go in the same direction, by opening access to those species. Most of the changes have advantaged the private sector, which already has the higher participation in the fishery. This is not exclusive to Yucatan; similar situations have been reported in other states (Ortiz 1993; Vazquez-Leon and Mcguire 1993; Diaz *et al.* 1984; Sada 1984).

Information on harvest rate tendencies, on changes in stock density, and on the spatial allocation of the fleet in Yucatan fisheries is limited. This information is needed in order to make efficient decisions on management.

A new relationship between fishers and government needs to be developed to achieve an efficient fisheries management system. Effective mechanisms should be implemented which allow fishers real participation in the management decision-making process (Stollery 1988; Hilborn and Walters 1992; Hilborn *et al.* 1995).

Government needs to avoid policies that focus on short-term goals. It is necessary to define proper policies in pursuit of sustainability of the fishery system. These approaches include improvement in the local control of fishing effort, establishment of an appropriate property right system to control access, and combination of comprehensive planning with proper knowledge of the natural resources and the dynamics of the users.

ACKNOWLEDGMENTS

The authors acknowladge the helpful comments of Alida Bundy, Daniel Pauly and Vincent Guillett. We are grateful to the fishers who participated in the survey questionnaire and the people who gave us access to official documents.

REFERENCES

Arceo, P, Zetina, C, Riosy, G V, and Mena, R (1995). Evaluacion preliminar de la poblacion de langosta espinosa (*Panulirus argus*) en la costa oriente del estado de Yucatan, Mexico. *Informe Tecnico*. CRIP Yucalpeten, INP, Mexico. 6 pp.

Arreguin-Sanchez, F, Seijo, J C, Fuentes, D, and Solis, M (1987). Estado del conocimiento de los recursos pesqueros de la plataforma continental en Yucatan y region adyacente. Centro Regional de Investigaciones Pesqueras. *Documento Tecnico*. CRIP Yucalpeten, SEPESCA, 1.41 pp.

Bonfil, R (1996). Elasmobranch fisheries: status, assessment and management. PhD. Thesis. Resource Management and Environmental Studies, University of British Columbia.

Burgos, R, Moreuo, V, Contreras, M, and Perez, M (1992). Densidad del mero (*Ephinephelus morio*) en el Banco de Campeche, Mexico. Memorias del Congreso Nacional de Zoologia, Merida, Mexico. 10 pp.

Chavez, E (1994). Los recursos marinos de la Peninsula de Yucatan. 1994). In 'Recursos faunisticos del litoral de la Peninsula de Yucatan'. (Ed A Yañez-Arancibia) *EPOMEX Serie Cientifica: Mexico* **2**, 1–12.

Diaz, M, Iturbide, G, and Garcia, I (1984). Los pescadores de la costa Norte de Chiapas. CIESAS No. 115, Mexico. 101 pp.

Diaz DeLeon, A, and Seijo, J C (1992). A multi-criteria non linear optimization model for the control and management of a tropical fishery. *Marine Resource Economics* **7**, 23–40.

Fraga, J (1992). 'Los efectos de la inmigracion en la region costera. Implicaciones ambientales y socioculturales'. Estudio de caso en dos subregiones. Notas Antropologicas. Escuela de Antropologia de la UAEM **4**, 95–107.

Fraga, J (1993). La immigración y sus principales efectos en la costa Yucateca, estudio de caso en Celeston y Sisal. Tesis de Maestric, U A Y Mexico. 200 pp.

Garcia, S M, and Newton, I (1994). Current situation, trends and prospects in world capture fisheries. *FAO, D/T* 4396. 45 pp.

Gonzales, M E (1982). Analisis de la pesqueria del huachinango, *Lutjanus campechanus*, del banco de Campeche. Tesis de maestria. CINVESTAV Unidad Merida, Mexico.

Gonzales-Cano, J, Arreguin-Sanchez, F, and Contreras, M (1994). Diagnostico del estado de la pesqueria del mero (*Ephinephelus morio*) en el Banco de Campeche. Doc. Interno INP, SEPESCA, Mexico. 26 pp.

Hernandez, A (1995). Analisis bioeconomico, espacial y temporal de la pesqueria del mero (*Ephinephelus morio*) en la plataforma continental de Yucatan. Tesis de maestria. CINVESTAV Unidad Merida. 139 pp.

Hilborn, R, and Walters, C (1992). Quantitative fisheries stock assessment. Choice, dynamics and uncertainty. (Chapman and Hall: New York). 570 pp.

Hilborn, R, Walters, C, and Ludwig, D (1995). Sustainable exploitation of renewable resources. *Annual Review of Ecology and Systematics* **26**, 45–67.

Ortiz, C (1993). Historia de la pesca de tiburon en Puerto Madero, Chiapas. CIESAS, Mexico. 41 pp.

Pare, L, and Fraga, J (1994). La costa de Yucatan: Desarrollo y vulnerabilidad ambiental. Cuadernos de Investigacion. (Instituto de Investigaciones Sociales. UNAM: Mexico) **23**. 120 pp.

Rosado, G, and Rosado, C (1995). El eslabon olvidado: la iniciativa privada y la conformacion y desarrollo de la industria pesquera. In Procesos territoriales de Yucatan.(Ed M T Peraza). Universidad Autonoma de Yucatan, Mexico. 433 pp.

Sada, J (1984). Los pescadores de la laguna de Tamiahua. CIESAS No. 113, Mexico. 143 pp.

Seijo, J C, Solis, M, and Morales, G (1987). Simulacion bioeconomica de la pesqueria del pulpo (*Octopus maya*) en la plataforma continental de Yucatan. Memorias del Simposio de Investigacion en Biologia y Oceanografia Pesquera, La Paz, B C, Mexico, 125–38.

Seijo, J C, Arceo, P, Salas, S, and Arce, A M (1994). La pesqueria de la langosta (*Panulirus argus*) de las costas de Yucatan: Recurso, usuarios y estrategias de manejo. In 'Recursos faunisticos del litoral de la Peninsula de Yucatan'. (Ed Yañez-Arancibia) *EPOMEX Serie Cientifica: Mexico* **2**, 33–41.

SEMARNAP (1996). Programa de Pesca y Acuacultura 1995–2000. Mexico. 96 pp.

Solis-Ramirez, M (1994). Mollusca de la Peninsula de Yucatan, Mexico. In 'Recursos faunisticos del litoral de la Peninsula de Yucatan'. (Ed A Yañez-Arancibia) *EPOMEX Serie Cientifica: Mexico* **2**, 13–31.

Stollery, K (1988). Cooperatives as an alternative to regulation in commercial fisheries. *Marine Resources Economics* **4**, 289–304.

Torres, R, and Salas, S (1993). Tecnificacion de la captura de langosta en Yucatan. In 'La utilizacion de refugios artificiales en las pesquerias de langosta: sus implicaciones en la dinamica y manejo del recurso'. Memorias del Congreso Bi-Nacional Mexico-Cuba. 1993, Mexico. (Eds Jouzales-Cano and Cruz-Izquierdo). pp. 103–12.

Vazquez-Leon, M, and McGuire, T (1993). La inicitiva privada in the Mexican shrimp Industry. *Mast* **6**(1/2), 59–73.

Zetina, C, Rios, G V, and Cervera, K (1992). Algunas relaciones biometricas de la langosta espinosa (*Panulirus argus*) capturada en las costas del Estado de Yucatan. Informe Interno. (CRIP Yucalpeten. I. N. P.: Mexico). 12 pp.

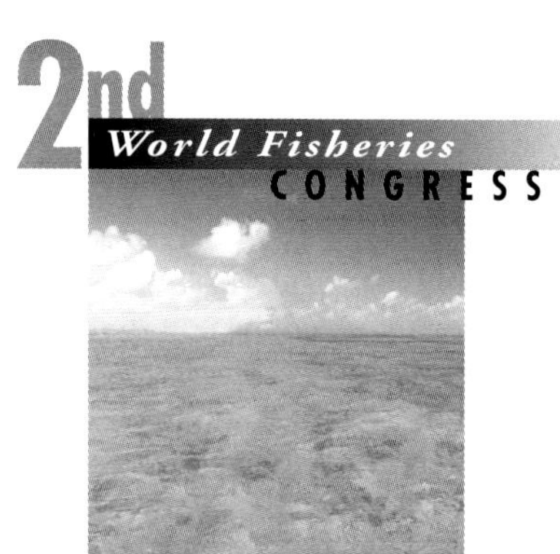

Issues and Outcomes for Theme 6

How to Manage Fisheries in the Future

Serge M. Garcia

FAO Fisheries Department, Via delle Terme di Caracalla, 00100 Rome, Italy.

Introduction

The presentations in Theme 6 stressed that the problems of managing fisheries have essentially two components: (a) the resource in its environment, and (b) the society, its evolution and system of values, both requiring a holistic approach towards sustainable development. For this brief account I did my own selection of the material presented and, in order to avoid a shopping list, I grouped the presentations into three major groups of issues related to: (1) management-related information; (2) management institutions, including legal instruments; and (3) management approaches.

Management-related information

The information required for proper decision making, monitoring, and public information was addressed by a large number of participants, both in terms of relevance, quality and cost. On behalf of Dr Peter Larkin, Tony Pitcher raised the issue of the cost of information for decision making stressing that present systems are not only inefficient but unreasonably expensive, stressing the fact that (a) these costs increase with mismanagement; (b) will increase further to satisfy the new requirements for more complex management systems; and (c) could be decreased by better management methods such as property rights, and effort controls. As an echo to Peter's concerns, Chesson and Phillips described the stock-assessment system operating in the South East Australian fisheries. This system, largely based on fishery-related information, collected in a standardized manner, with data requirements tailored to each species, as part of an integrated scientific monitoring programme, implemented in a cooperative approach, provides a model for an inexpensive assessment system usable in areas where available resources are limited. Linking fishery development with sustainable development, Derek Staples discussed the concept of sustainability indicators in fisheries which have the potential for becoming central to fisheries management during the next few years. Stressing that the concept and the necessity for such indicators stem from UNCED and Agenda 21, and that indicators will be required as part of the process of fisheries accreditation presently developing, he poses that development and use of indicators should be an integral part of the development process and of its evaluation. Echoing his concerns, Campbell examined the potential role of economic performance indicators in user-pays management systems in the Australian context, stressing the need to integrate such indicators in the fishery information system and underlining the trade-offs that may be involved between timeliness, accuracy, level of detail on the one hand and costs on the other.

A number of posters addressed various aspects of the information and assessment problem. For instance, the particular problem of effectively collecting information in recreational fisheries was addressed by Teirney, Sylvester and Cryer, who, in New Zealand, developed and successfully used household telephone surveys and diary schemes yielding remarkably accurate results. Lucy, Ayvazian and Malchoff showed that mortality, in catch-and-release recreational fisheries, can be estimated to provide better understanding of human impact and could help in improving management.

Management institutions

The relative failure of fisheries management is largely due to the deficiencies in institutional structures and, as a consequence, institutional issues received a lot of attention in a number of presentations. These dealt with management systems, organizations, mechanisms and legal regimes. The important institutional change that may be required to implement the principles of ecologically sustainable fisheries, was described by McColl and Stevens based on the example of Australia where a successful shift has been completed from a Government bureaucratic institution to a more responsive parastatal autonomous management agency (Australian Fisheries Management Authority) entrusted with responsibilities related to management plans, resource allocations, appeal procedures, stock assessment, cost-recovery, capacity adjustments, in partnership with industry, and within a complex system of local and federal nested jurisdictions — a change that was not without difficulties despite the assistance of a fairly independent Board and a number of Advisory Management Committees. The issue was also addressed by Bruce Phillips and his colleagues who stressed the emphasis put in Australia on integrated strategic planning, capacity building, community participation, improved research and information, habitat protection, improved marketing, granting of statutory rights, decentralization of decision making, 100% management costs recovery, all of this in a multi-use context. The question of cooperation between the different authorities having jurisdiction on Australian Fisheries has been mentioned in a few papers and Fowler and Brayford described how *ad hoc* cooperative agreements were replaced recently by a formally established mechanism of cooperation between the Commonwealth, the State of Western Australia, and the Northern Territory, based on the concept of ecosystem management. Finally, Gerry Morvell presented the Australian solution to managing fisheries sustainably which consists of: (1) promotion of an integrated strategic planning approach; (2) increasing the capacity of managers; and (3) promotion of community participation. It is estimated that the approach will lead to added benefits to the nation to an amount of \$A5–10 $\times$ 10^9 per year within 25 years.

J.C. Brêthes informed us about the institutional changes implemented in Canada following the demise of Atlantic fisheries and informed us about the difficulty of dealing with new concepts (such as ecosystem management) and new tasks (such as resources rehabilitation) with the limited information available, and in a context of high uncertainty about the future evolution of the resource and of the social system.

Bill Fox gave us his four ingredients of sustainability as: (1) Good science, (2) risk-averse decisions, (3) willing compliance, and (4) independent oversight. He elaborated on the practical experience in the USA, in relation to two major national instruments: the Marine Mammal Protection Act, which aims at ensuring a healthy state of the marine mammal resource, and the Magnusson Fishery Conservation and Management Act which aims at ensuring a certain level of output from the US 200-miles zone. In the USA, as in Australia, Canada or New Zealand, management plans are the important institutional instrument in which, and around which, the assessments, criteria, targets, legal instruments, and institutions come together with the view to achieving the formally adopted objectives. In the USA, all national management plans are judged against a set of seven national standards.

Doubleday highlighted the evolution of fisheries conservation in Canada following accidental depletion of demersal resources and contemporaneous increases of invertebrate resources, resulting in a constant output in economic terms. The system is now aimed at conservation, which is the first priority, and is based on the precautionary approach. Drastic measures are being taken including a 50–60% reduction in fleet capacity and the adoption of an aboriginal fisheries strategy, while the role of the sector in partnership management is enhanced.

Basing his proposition on his conclusion that conventional management systems will not achieve sustainability because of political and socio-economic pressures, Mike Sutton called for a new management paradigm involving greater people involvement and accountability. He looks for a high level of public concern and more effective trade-related measures, referring in particular to the ban on trade of ivory. While I share with him the idea that it is time that all stakeholders get involved, and see the market force as a potentially powerful lever, when I heard the reference to ivory and elephants, I found myself, as an old African, hoping that the system works better for groupers and cods than it did for elephants. Mike Sutton informed us of the developments towards the establishment of a Marine Stewardship Council in 1997 as part of a voluntary scheme of accreditation and eco-certification, based on a set of criteria and principles which are still to be determined. Doug Kidd gave us a first-hand account of the experience he had in managing fisheries through property rights. While there seems to be consensus that property rights are a prerequisite to improvement of management, Salas and Torres have described the difficulties encountered in Mexico, despite the existence of non-transferable land-based exclusive communal fishing rights, because of the absence of effective control on access and on effort levels, illustrating perhaps the fact that, in fisheries management, we are dealing with people and with socio-cultural and political systems, and that generalizations should be avoided.

Altogether, these presentations indicate that the institutional issue is taken very seriously, that important changes are tried and going on, that past deficiencies are identified and being corrected, and that interesting new developments are on the way regarding the interface between trade and management.

The important implications of many fishery- and non-fishery-related international environmental instruments were underscored by Martin Tsamenyi together with the need for their national implementation, stressing that (a) the process of integration in national law had to take account of the sensitive issue of national sovereignty, and (b) that they will make management even more complex than it already is. In response to a question, he expressed doubts as to the likelihood that the fishery sector will ever implement effectively the instruments it had just agreed to but that other global instruments such as the

Convention on Biodiversity might be more effective in forcing fisheries to comply. Considering the commitment demonstrated by fishing and coastal nations in agreeing and adopting UNCLOS, the UN Agreement and the Code of Conduct, and the very active and leading role demonstrated by them in that respect in the UNCED follow-up, I must confess that I personally failed to understand the pessimism and I do hope that, despite the poor past records, this pessimism in undue.

Douglas Fisher reminded us that a number of key fisheries issues had legal implications (e.g. high seas fishing, illegal foreign fishing and poaching, misreporting, overcapitalization, etc.). He provided an analysis of the respective roles of Science (which, according to him, is largely descriptive) and Law (which he sees as essentially prescriptive) stressing the need to increase cooperation between these two fields of expertise. He also noted that states had given, up to now, more attention to their rights than to their obligations. Agreeing totally with that assertion, I would add that this is a bias that the FAO Code of Conduct for Responsible Fisheries is attempting to correct, stressing responsibilities in its very title. The legal regime of fisheries is usually reflected in the management plan and Fisher stressed that it had been established through the Court that *flexibility* in implementation was acceptable, provided plans objectives were not compromised and that *economic efficiency* was indeed an acceptable objective, related to the notion of optimum yield, as enshrined in UNCLOS.

Little was said about enforcement, one of the most essential, expensive, and difficult parts of fishery management as underlined by John Beddington in his introduction to Theme 5. In addressing very effectively the issue in Theme 6, Brian Hemming drew our attention to the fact that fisheries-related crime was often considered as socially acceptable and considered as part of normal business risk. He indicated that enforcement costs were in the order of 50% of total management costs and that in order to reap benefits from enforcement, efforts should be devoted to research on compliance, to better integration between enforcement and management, to better integration in the community life and system of values, and to the exchange among States of information on best enforcement practices. Brian proposed that a major theme of the third World Fisheries Congress be indeed related to this last issue.

Management approaches

Many presentations touched more or less specifically on management approaches.

Gary Sharp criticized the current assessment and management paradigms, provocatively stating that fishermen's cultures were the real endangered species, endangered by faulty assessments and ineffective management. He noted that the present opposition between government scientists and industry lawyers was not very effective. He also reminded us that the premise of stock assessment should be that all stocks vary unless proven otherwise, a different kind of reversal of the burden of proof than envisaged in the precautionary approach he pleads for. He suggested that we should pay greater attention to the ecosystem and biodiversity implications of fisheries, their environmentally driven fluctuations, and the potential role of the market in improving management performance, joining in this Mike Sutton and Derek Staples.

In striking opposition to the present generalized concern for mismanagement, it was refreshing to hear Donald McCaughran telling us about a fishery that lasted more than 100 years and is still around, and thriving: the halibut fishery of the North Pacific. The ingredients of success seem to have been: (1) scrupulous accounting of all removals, (2) reliance on excellent science, (3) a conservative management strategy aiming at the highest yield compatible with low risk, and (4) the ignoring of pressures from states to increase fishing beyond the boundaries established by scientific advice.

Another rather comforting case of fishery management was given by Dana Schmidt in relation to the Alaska sockeye salmon fishery. This fishery shows a good case of decline and effective rebuilding processes and is characterized by a successful decentralization of management authority to local communities and local officials with effective enforcement and decisional powers.

In reporting on fisheries management in Namibia, Peter Manning raised two issues: the allocation and use of the resource rent, and the difficulty of operationalizing ecosystem management. He stressed that catches are still well below historical peaks and that the main species, hake and pilchard, are still in a depressed state following pre-independence overfishing. He suggested that in coping with a highly-fluctuating ecosystem, it was necessary to ensure industry flexibility through a multi-target approach. He also considered that economic rents not extracted by the Government should be considered as a subsidy.

An appeal has been presented by Bill Ballantine for the generalization of networks of no-take, totally protected, marine reserves, basing his rationale on the premise that sustainability requires ecosystem conservation which, in turn, requires the conservation of its essential structures, processes, and patterns.

The Great Lakes Fisheries management system described by William Taylor reflects the complexity of the management of a multi-species resource, in a large ecosystem, severely impacted by human activity, under the authority of a complex governance system including two nations, eight states, one province and a multitude of users, the result of which is overexploitation, collapse of fisheries, eutrophication, contamination and spread of exotic species. The solution proposed is to adopt an ecosystem management approach.

Perez and Batungbacal informed us that, in the Philippines, Sustainable Coastal Areas Development Programmes have been set up aiming at sustainability through empowerment of coastal community, equity, and gender-fairness in an attempt to stop and reverse the present trends in poverty, overfishing and coastal environment degradation.

Boje and Siegstad reported that in Greenland, the shrimp and green halibut fisheries are managed through a system of vessel quotas determined as a percentage of the Total Allowable Catch. These quotas are allocated by geographical areas and may or may not be transferable depending on the fishery (for shrimp) and are combined with technical measures such as mesh size regulations and protection of juveniles through temporary closed areas.

Environmental impact management

A number of presentations and posters referred to the environmental aspects of fisheries management, either originating from fisheries or from other activities.

An interesting case was presented by Clark, Jones and Cole on the management of the urban drain in Cairns (Queensland) where a Plan has been prepared, based on a careful environmental impact assessment, to rehabilitate the drains and maintain them to ensure that both functions, as drains and as fishery habitats, are optimized.

Similarly Harris, Hartley and Bruce reported on efforts made in New South Wales for assessing the health of the State's river system and the species it contains, leading to the development of an Index of Biotic Integrity used to assess and monitor the river and resources conditions.

Inoue, Matsushita and Chopin described an interesting programme of environmental improvement of coastal fisheries in Japan which, despite their high productivity are showing the impact of fishing and of environmental damage. Work is being done, in a cooperative programme between scientists, fishermen, and industry, to (1) reduce biological waste, (2) improve selectivity, (3) reduce discards mortality, (4) study the impact of high speed engine noise, and (5) reduce bottom impact of trawling.

Conclusion

During the course of the meeting I have perceived a general recognition that:

- the situation of the world fisheries is not bright,
- the overcapacity issue must be addressed as a priority,
- together with the issue of environmental protection, and resource rehabilitation, and
- that many Governments seem to be committed to take action.

To succeed in the needed transition towards ecologically sustainable fisheries, there is also agreement that what is needed is, *inter alia*:

- better commitment and political will than in the past,
- a pragmatic approach, as holistic as possible, combining the following:

  fishing rights with some form of property,

  suppression of most subsidies,

  higher level of participation of the people in planning and decision making,

  stronger involvement of the public and the consumers,

  improved controls, in EEZs, in high seas,

  much clearer objectives and management goals,

  improved data and science,

  improved institutions and legislation,

  development of appropriate sustainability indicators,

  development of more sustainable technologies,

  experimentation with market controls.

Constraints are recognized in:

uncertainty about the resource and its resilience,

uncertainty about resources needing rebuilding strategies,

lack of capacity in developing countries,

difficulty addressing ecosystem management for want of adequate theories and methods, and

problems of co-ordination between split responsibilities.

Congress Overview

Fisheries at a Critical Juncture: Understanding the Cause, Correcting the Course

Pamela M. Mace

National Marine Fisheries Service, Northeast Fisheries Science Center, 166 Water Street, Woods Hole, MA 02543, USA.

I would like to begin my overview of the Second World Fisheries Congress by reiterating the sentiments of the theme keynote speakers who reflected on the phenomenal quality and diversity of papers that have been presented at this Congress. I think the professionalism and quality of the presentations, and the thoughtfulness, depth and number of stimulating, innovative ideas has been outstanding, to say the least. Of course, another measure of the success of the Congress is the degree and calibre of participation. As we were informed, almost 1000 delegates representing 66 countries attended, and they provided numerous insightful comments and questions following the presentations.

I also think it is important to note that the overall tenor of the Congress has been remarkably positive. The setting of world fisheries at a 'crossroads' or a 'critical juncture', as stated in the Invitation to Participate in the Congress, is not new News and while it has formed much of the background for the Congress themes, the presentations that we have heard have been mainly about the *progress* that is being made towards resolving the problems most of us here have been aware of for years. Most of us already understand the causes of fisheries failures, at least at a generic level, and there are many indications that we are in the process of correcting the course. I will say more about this at the end of my Overview.

First, I will address the six Congress themes. When I initially thought about the six themes for the Congress, each of which is a challenge in itself, other related questions or subthemes immediately came to mind.

For the first theme (Why do some fisheries survive, while others collapse?), the obvious first question is how do you define survival and collapse. Mike Sinclair suggested a workable definition for 'collapse' based on recruitment overfishing, and most speakers implicitly assumed that survival equates with 'persistence' or 'not-collapsed'. From what we have heard, it is probably much easier to answer the question: Why do stocks collapse?, rather than: Why do stocks survive? As Mike Sinclair said in his keynote presentation, stocks collapse because of two or three factors, usually acting in combination rather than singly: overfishing, adverse environmental conditions, and/or a major change in the abundance of a prey or predator population, seals for example. Other papers identified additional factors such as loss of habitat and pollution. As to why fish stocks survive, the easy or wishful-thinking answer is good management, but often it is as likely to be low demand, low price, long periods of favourable environmental conditions, life history characteristics that facilitate management, good luck, or a combination of all of these factors. The problem is that a sudden change — increased demand, increased price, or a downturn in environmental conditions — can very suddenly turn a success into a failure.

So perhaps a more intriguing question is: are fisheries that have 'survived' to date just in a different stage of evolution? We heard about Atlantic herring, where many of the stocks collapsed in the late 1960s and early 1970s but some survived only to become depleted if not collapsed at a later stage. The suddenness with which a 'success' can become a 'failure' is one of the reasons why we need a buffer: we should not be maximizing yields, we should be minimizing probabilities of stock collapse, taking account of uncertainties in our knowledge of stock status, and taking into account surprises that may occur in the future. We should not be trying to 'manage at the edge'. Unfortunately, it almost seems that we need catastrophes in the form of collapses of large fisheries, causing major economic and social disruption (and to be able to show that these are clearly linked to overfishing more so than environmental factors) before we are willing to diverge from the *status quo*. Why do we need such strong and numerous 'wake-up calls'?

The second theme (What is the role of science, economics, sociology and politics in fisheries management?) was probably one of the best addressed of all by this Congress, particularly if the list is expanded to include other players such as members of the commercial fishing industry, recreationists, environmentalists, lawyers, consumers, communities and the general public. Of course, a bigger challenge than simply defining the roles of each of the professions or players is to determine how to reconcile different perspectives, to the extent that we can at least agree on the objectives of fisheries management. How do we get everyone to think in terms of common currencies? Overall, we have probably got a long way to go on this issue, but in the last few days we heard about numerous individual examples of successful cooperation, or models for successful cooperation. There have been fundamental changes in the management philosophy and structures of management institutions in Canada, Australia, New Zealand, South Africa, Namibia and elsewhere, and fishers' community trusts and other forms of co-management are receiving more attention. A sentiment emphasized by Bernard Bowen and echoed by other speakers was the need for a commitment by all participants to continuous improvement in the fisheries management process. We need mutual goals and mutual understanding, with participants jointly developing management plans and therefore maximizing commitment and minimizing the need for coercion.

The third theme was: How can fisheries resources be allocated? Who owns the fish? As we heard in Rudy van der Elst's keynote address for this theme, and in other presentations including three excellent papers on aboriginal claims to fisheries, three or four very good papers on commercial–recreational conflicts, papers on transferable quota systems, a suggestion for franchising fisheries resources and an excellent comparative analysis of fisheries co-management examples, allocation is a major issue. Papers presented in support of this theme underscored the complexity of the resource allocation issue and the need for socially acceptable distributions of fisheries resources. As Rudy van der Elst stated, in many situations, including the current situation in South Africa, the question becomes 'How to cut up a cake that has already been consumed'. This is particularly true when there is a need to address past inequities.

I think there is general agreement among managers, scientists, economists and others that fishing rights and responsibilities need to be vested with owners — either individually or collectively — but it should not be surprising that those whose livelihoods may be adversely affected may not be particularly willing to subscribe to the need to downscale fishing activities and diversify into other unspecified areas of employment. Until the issue of who is going to get the fish is resolved to everyone's mutual satisfaction, it is probably going to be difficult to get all of the stakeholders to recognize that controlled access (which for many of our fisheries implies some initial downscaling) is even needed at all. We have been presented with many innovative ideas, and one obvious conclusion is that there is no one solution to fit all problems. The issue is multidimensional and the social, economic, cultural and political dimensions will all vary from fishery to fishery. In addition, while fishing rights are essential, fishers are not the only users of the resource.

The fourth theme was: How can aquaculture help sustain world fisheries? I think that the aquaculture papers presented at this Congress — the comprehensive, enlightening and entertaining keynote presentation by I Chiu Liao as well as those providing national or local overviews, and those focussing on particular types of aquaculture — have generally sounded a positive note for the short- and long-term future prospects for aquaculture, provided, of course, that development proceeds in an environmentally responsible way.

We heard a number of suggestions for mechanisms for ensuring aquaculture does proceed in an environmentally or ecologically sustainable manner; for example, the planning process implemented in Tasmania, including requirements for environmental assessment prior to initiating new projects and environmental monitoring afterwards. Suggestions for developing and expanding aquaculture have included stock enhancement, super-intensive systems (e.g. recirculating systems which have much lower water requirements and fewer negative impacts on the environment) and open-ocean cage culture systems. Small-scale integrated agriculture-aquaculture and polyculture will also continue to be important. I Chiu Liao also mentioned another 'commodity' that will ensure aquaculture's future success; namely, resourceful people. The issue that, perhaps, has not been covered so well, at least in the papers I heard, is how we will meet the *social* challenges that have already occurred as a result of expansion of aquaculture, and that will probably intensify in the future, particularly given that most of the environmentally sound technologies require a large initial outlay in capital. The distribution of benefits is an important social problem. Already, there are instances where small-scale subsistence fish farmers have been displaced by entrepreneurs often financed by foreign investors.

In the fifth theme (What are the limits to the exploitation of capture fisheries?), as John Beddington mentioned, we have about three options for expanding catches: better management, proper development and sustainable exploitation of new fisheries, and enhancement activities. A number of papers at this Congress addressed one or more of these issues. Better management means more environmentally friendly gear, reduction of bycatch, more rigorous management objectives, better management strategies, and, in some cases, new management models. Regulations need to be enforceable, at reasonable cost. Information must be sufficient to meet management needs.

Several papers in this and other theme sessions pointed out that there is scope for rehabilitation and expansion of fisheries on a local basis. However, even though the last two years of statistics on global fisheries production do not support FAO's previous assertion that capture fisheries are reaching a plateau, other evidence (e.g. the proportion of fish stocks that are fully-exploited, overexploited or depleted) strongly suggests that a limit is being approached on a global basis, at least for pure capture fisheries (that is, those not involving enhancement). The last frontier has been conquered and subdued. Mankind now has the technology to exploit fisheries resources at the fartherest ends of the earth and is already reaching far into the fartherest depths of the marine abyss.

But perhaps a more substantial challenge than determining 'What are the limits?' is: 'How can we get all players to recognize the limits and act accordingly?' To answer my own question, one part of the solution is obviously communication and education — that is, education of all players, perhaps most especially the politicians. We need rational government policies to provide the framework, the standards and the guidelines, and mechanisms or fora for developing partnerships between stakeholders to solve problems for individual fisheries. On the other hand, perhaps it is unreasonable to expect change to happen too fast. As Serge Garcia noted in his paper, major changes in people's values and attitudes may take at least a generation (2–3 decades). A good example of the type of attitude that simply must change comes from the field of fisheries enforcement. Brian Hemming's statement that fisheries violations are often viewed as socially-accepted crime certainly rang true. But it also reminded me of a telephone call I received a few months ago from a fisher who said 'What is the world coming to; yesterday my son came home from school and called me a criminal just because I catch sharks for a living'. Perhaps this example goes too far, but it does illustrate that fishing 'ethics' are already in the process of a major evolution.

The sixth theme was: What is needed to manage fish stocks sustainably? Serge Garcia's keynote for this theme emphasized the multidimensional nature of the problem and the implementation challenge. There were also other extremely interesting papers addressing this theme, with several innovative ideas including various management structures and philosophies, successful management systems that we can learn from, and ecosystem and other approaches to management in the future. Bill Fox suggested four general principles for sustainability: good science, risk-averse decision making, willing compliance and independent oversight. Mike Sutton outlined the World Wildlife Fund/Unilever agreement to form a Marine Stewardship Council that will certify fisheries products on the basis of the degree to which those fisheries meet conservation objectives.

I definitely agree with Serge Garcia and many other speakers that one of the key missing components in achieving environmentally sustainable and economically viable fisheries is not conceptual shortcomings, but rather *implementation.* How can we translate from theory (words on paper) to reality (practical implementation)? As a member of the audience at a seminar I recently attended remarked: 'We are getting better at writing words and worse at writing cheques'. One would think that we should now be at the point where we have enough experience to know what needs to be done, at least in a very broad sense. Some observers of global trends in fisheries are becoming frustrated with hearing descriptions of the same problems reiterated over and over, when it seems that we are not making very fast progress towards solving them. But, unfortunately, the old saying that recognizing the problem is half the solution does not seem to apply in fisheries. For that matter, even recognizing the solution does not mean that it will be accepted or implemented, as Steven Wright graphically demonstrated in the video clips of Canadian Pacific salmon fishers storming the offices of the Canadian Department of Fisheries and Oceans to protest a plan to reduce fleet capacity by 50%, even though industry representatives had been heavily involved in formulating the plan, and even though Peter Pearse made the same recommendation for Pacific salmon fisheries some 14 years previously. The social implications of reduced employment opportunities in the fishing sector need careful consideration.

In addition to the six key themes of the Congress, a number of other controversial issues also received considerable attention during the course of the meeting. One of these was the issue of scientific peer review of fish stock assessments, a subject addressed during an evening session by a panel of experts. Fisheries science has rapidly progressed from the ivory-tower view of knowledge for knowledge's sake, to science in support of management, to science under fire. The validity of scientific stock assessments is being challenged with increasing frequency, sometimes in an effective and constructive way, but sometimes not. Panellists pointed out that 'peer-review' of stock assessments is fundamentally different from 'peer-review' of research articles submitted for publication in scientific journals, and suggested that the former process could be improved by involving the fishing industry and fisheries managers throughout the assessment process.

Another issue that grew into a lively debate was based on papers presented in themes 1, 2 and 6, although primarily kindled by the paper presented by Sinclair *et al.* and the paper prepared by Peter Larkin namely, the issue of the costs of management relative to benefits such as landed value. Some delegates made the claim that Canada's fisheries are a net loss to the national economy, but other delegates pointed out that the conclusions depend on the factors considered in the calculations. The general issue of the costs and benefits of capture and culture fisheries deserves much more attention. It may well be that the costs of attempting to manage fisheries 'optimally' are too high. As Peter Larkin warned, if we do not solve the problem of excess fishing capacity then 'the costs of research, management and enforcement will continue to grow until it becomes obvious that as a national business, catching wild fish is not a viable enterprise'. In addition, the costs of rectifying fisheries problems need to balanced against the costs of not doing so.

I would like to end this Overview by going back to one of the sections of my Keynote Address: Future Prospects and Challenges. Although I used the word *future* in this title, it is obvious that the challenges I listed do not just apply to the future, they apply to the present, they have already applied for several years and in fact, progress has already been made in all of these areas. The papers presented in this Congress bring home just how much progress has already been made.

With regard to the first challenge (1. Reduce fleet capacity and reduce dependence on wild fish stocks): The recognition of the need to reduce fleet capacity is very widespread and various kinds of fleet reduction programmes are in place all over the world (with notable exceptions, such as China, which is proceeding with major fleet expansion programmes). There were not actually many papers specifically on the topic of reducing fleet capacity but the need to ensure that fleet capacity is commensurate with resource productivity is generally considered to be a fundamental goal for any viable future management plan.

With regard to the second challenge (2. Define and implement workable access rights systems): I would guess that based on the papers presented in the last four days, access rights systems will within a decade or so become the norm that will replace open access. By that time, there will likely be a wide diversity of types of controlled access in place, ranging from community-based systems with little outside intervention, to various forms of co-management (a wide diversity of which already exist; see Sen and Nielsen), to private property rights vested in individuals or corporations with varying amounts of government or public control. It is unlikely that any one system of controlled access will prevail since at some level each fishery is unique and thus requires a tailored approach, a point which several papers emphasized.

With regard to the third challenge (3. Implement the precautionary approach): As we have heard, the precautionary approach is already being embodied in national policy and management objectives, but I suspect there will be a long period where people will claim to be using precautionary approaches which may ultimately prove to be not-precautionary-enough. The precautionary approach needs to be more than just the current 'buzz-word'; it needs to be defined operationally by each fishing nation, with the specifics potentially varying from fishery to fishery.

With regard to the fourth challenge (4. Expand aquaculture without destroying natural environments): As I mentioned earlier, based on the papers presented, authorities reviewing aquaculture development programmes are now becoming acutely aware of the need to consider environmental impacts first and foremost. As a result, the need for integrated coastal zone management to resolve conflicts over land and water use is being increasingly recognized. In addition, more emphasis is being placed on scientific and engineering research to produce environmentally friendly aquaculture technologies.

With regard to the fifth challenge (5. Develop effective policies and institutions to achieve the other objectives): This challenge has been well-addressed in the Congress by the numerous case studies already mentioned, including many from Australia. Many countries are already well on the road to reforming fisheries institutions and further reform is inevitable since, as Tsamenyi and McIlgorm stated, the fishing industry and fisheries administrators will not be able to ignore recent international fishing agreements.

I think it is appropriate to end my Overview with some very positive statements about the state of Science and Management in our host country, Australia. At the Federal level, Australia has had in place a very innovative and progressive management system for about four years now; that is, the Australian Fisheries Management Authority (AFMA). Discussions about AFMA in this Congress have convinced me that it shows great promise in terms of its future likely success, despite the negative press it received just prior to this Congress as a result of a recently-completed audit. Without going into detail, the fundamental positive innovations in AFMA's organization and management philosophy are (i) the active and participatory involvement of all stakeholders including industry and recreational interests, (ii) the requirement for and integration of good science into the whole process, and (iii) a decision-making process that is one step removed from the political process. Coming from the United States, I cannot tell you how important the latter innovation is. Recently, there have been several instances of fishing interests 'end-running' the US fisheries management process by enlisting political support for views contrary to those developed in the established fisheries plan development process, which involves considerable public participation. At present, the United States Congress is considering a ban on ITQs and related management systems, thus removing a useful tool from the already-limited arsenal of fisheries management approaches [and subsequent to the World Fisheries Congress, the US Congress actually passed legislation implementing such a ban for four years].

Of course, AFMA is not the only progressive management institution in Australia. Many of Australia's States are also developing innovative participatory management systems. Australia has also invested well in fisheries science and has a large contingent of exceptional, internationally-renowned scientists. I do believe that Australia's investment in fisheries science and its progressive approach to fisheries management put it at the forefront with respect to the world scene.

But in fairness to the rest of the world, it seems that almost all fishing nations at the moment are experimenting with new approaches to formulating policy and building institutional structures that can succeed. We will all do well to continue to explore each other's attempts to improve the state of our fisheries science and management which, of course, is one of the reasons why international Congresses such as this, the Second World Fisheries Congress, are so invaluable.

Poster Presentations

Poster Presentations

Theme I

Presenter	Title
M. Al-Hossaini, S. Al-Ayoub and J. Dashi	Age validation of nagroor, *Pomadasys kaakan* (Cuvier, 1830) (Family: Pomadasidae) in Kuwait waters
M. Bakes, N.G. Elliott, P. Nichols and C. Strauss	An update on marine oil research in Australia
J.D. Booth	Geographic variation in productivity and a prognosis for the New Zealand rock lobster fishery
S.A. Bortone and R.W. Chapman	Stock identification of red snapper, *Lutjanus campechanus* (Lutjanidae), from the Gulf of Mexico using mtDNA and year-class data
S.F. Bossy	Joint management initiatives in the English Channel
S.B. Brandt	Spatial and temporal complexity in aquatic environments: a key to fluctuating fisheries?
I.W. Brown and R. Clarke	Development and management of the Queensland spanner crab fishery
C.Y. Burridge, C.R. Pitcher and T. Wassenberg	A comparison of interreef benthic communities in zones open and closed to trawling in the northern Great Barrier Reef
I.E. Calderon-Aguilera, M.E. Siu, A.H. Maciel and L. Crisostomo	Assessment of shrimp postlarvae as a proxy variable of stock recruitment in the Gulf of California
M. Cappo, N. Duke, D. Alongi and D. Williams	An overview of Australian fisheries habitat research
C. Chauvet	Existence of a sex-ratio regulation mechanism in Tunisian stocks of bream *Sparus auratus*
K.H. Chu, L.M. Huang, Q.C. Chen and C.K. Wong	Crustacean fisheries in the Zhujiang (Pearl River) Estuary, China
M. Clark	New Zealand orange roughy: steps towards a sustainable fishery
M.R J. Connell, T.J. Wassenberg and N.A. Gribble	Cross-shelf distribution of surficial sediments in the far northern Great Barrier Reef, Australia
D.D. Dauble and D.R. Geist	Rebuilding strategies for endangered stocks of Pacific salmon
D. Del Piero	The striped venus and the razor clam fishery in the Gulf of Trieste: different life strategies and different fishery perspectives
N. Ehrhardt, P. Barria and R. Serra	Sustainability of fisheries supported by dynamic surplus production: the case of the Chilean pelagic fisheries
N.G. Elliott, P.S. Lowry, R.D. Ward, P. M. Grewe, B. Innes and G. Yearsley	Evidence for two depth-separated stocks of the deepwater oreo *Neocyttus rhomboidalis* (Oreosomatidae) off southern Australia
R.A. Englund and R. Filbert	Decline of Oahu's stream fisheries: identifying causes and planning restoration
Lee-Shing Fang	The environmental biology of an endemic freshwater fish, *Varicorhinus alticorpus*, in Taiwan

Poster Presentations

Presenter	Title
E. Garagitsou, C. Papaconstantinou and A. Siapatis	The distribution of larval scombrids in the Aegean Sea (Greece)
P.C. Gehrke, K. Astles, J.H. Harris, D.A. Pollard and L.O. Growns	Effects of urban development on fish and fisheries of the Hawkesbury–Nepean River System, Australia
D.J. Gilbert	A hypothesis to explain fish recruitment
D.V. Gillman	Integrated management in the Great Lakes: the sea lamprey programme
A. Gracia	Brown shrimp sequential fisheries in the Gulf of Mexico
N.A. Gribble, M. Connell and J. Glaister	Spatial variation in the distribution of Penaeidea (shrimp) species on the northern Great Barrier Reef: potential effects on management
S.R. Hare, R.C. Francis, A.B. Hollowed and W.S. Wooster	Effects of interdecadal climate variability on the oceanic ecosystems of the North Pacific
E. Hudson	Marine fish and the IUCN red list of threatened animals
K.D. Hyatt and B. Smiley	Data sets and environmental indicators used to manage Pacific salmon for sustainable fisheries on Canada's west coast
S. Ibrahim	Distribution of benthic organisms around different designed artificial reefs
H. Ishihara, Taniuchi M., Homma K., and T. Itoh	Some interesting evidence concerning the manta ray occurring in the Indo-Pacific
G.K. Jones	The effects of oceanographic variability on the recruitment of two species of juvenile arripid fish to a nursery are hauling net fishery in South Australia
A. Jordan	Recruitment and size structure of jackass morwong (*Nemadactylus macropterus*) and tiger flathead (*Neoplatycephalus richardsoni*) populations in eastern and south-eastern Tasmania, Australia
T Kadiri-Jan	Destruction of the environment of juvenile coconut crabs
Z. Kanamoto	Daily changes in the catch of *Parapristipoma trilineatum* at the hook and line fishing ground of Toshima Fishery Co-op in the Uwa Sea, Japan
S.S. Knight, F.D. Shields Jr and C.M. Cooper	Habitat and fisheries restoration of incised stream channels
M. Kroese and W.H.H. Sauer	Elasmobranch stocks in South Africa – improved utilization or increasing catches?
T. Lee and T.J. Pitcher	An individually based model of fish school dynamics
M.R. Lipinski	Squid fisheries of the world: chaos or pattern?
D. Lluch-Belda	The signal of global interdecadal regime variation on temperate sardine and anchovy populations
J.G. Loesch and D.A. Dixon	Development and application of a juvenile index of abundance for anadromous river herring
E. Lozano-Alvarez and P. Briones-Fourzán	Fluctuations in population parameters of the spiny lobster *Panulirus argus* in a fishery based on artificial shelters

Presenter	Title
J.O. Manyala	Effect of variability in some population parameters of *Rastrineobola argentea* (Pellegrin 1904) on the yield and biomass in Lake Victoria
B.D. Mapstone, C.R. Davies, J.R. Higg and D.J. Welch	The effect of re-opening Bramble Reef to bottom fishing on fishing behaviour and catch rates of commercial and recreational line fishers
O.A. Mathisen	The fall and rise of Pacific salmon in North America
S.C. Mathur	Protozoan parasites of freshwater food fishes of Punjab India, with special reference to gregarines, microsporidians and amoebae
S. McClatchie, R.B. Millar, F. Webster, P.J. Lester, R. Hurst and N. Bagley	Demersal fish community diversity off New Zealand is related to depth, latitude and regional surface phytoplankton
J. McKenzie and N. Davies	Biomass estimation of New Zealand snapper stocks by Petersen Mark recapture using cryptic (coded wire) tags
B.A. Megrey, S.A. Macklin, A. B. Hollowed and P.J. Stabeno	Applied fisheries oceanography: guiding fisheries management by relating environmental and recruitment variability to forecast year-class strength of Alaska walleye pollock
M. Moran, J. Jenke, G. Nowara and P. Stephenson	Effects of fishing gear on benthic habitat
A. Muhlia-Melo, J. Arvizu-Martinz, J. Rodriguez-Romero, D. Guerrero-Tortolero and F. Gutierrez-Sanchez	The robalo fishery in Mexico: management, enhancement and aquaculture
W. Nash	Assessment of pulse-fished abalone stocks by the change-in-ratio method
J.F. Palmisano	Regional abundance variation in Washington's salmonid fisheries
E. Pikitch, D. Erickson, J. Wallace, G. Oddsson and E. Babcock	Mortality estimation of trawl-caught and discarded Pacific halibut (*Hippoglossus stenolepis*)
L.B. Prenski, A. Giussi, O.C. Wohler, S. Garcia de la Rosa, J.E. Hansen, N. Mari and F. Sánchez	Southwest Atlantic long tail hake (*Macruronus magellanicus*): state of stock and management
N.J.F. Rawlinson and D.T. Brewer	Survival of selected fish species after escaping from square-mesh codends
D.D. Reid, J.H. Harris and G.A. White	Decline of freshwater fisheries in a large semi-arid floodplain river system
J.B. Reynolds	Relative length: a comparative approach to fish growth
C. Robins	Factors influencing fishing power in the northern prawn fishery
J.B. Robins	Is it true what they say about trawling and sea turtles?
M. Sanaullah, B. Hjeltnes and K. Pittman	*Myxobolus* spp. (Myxozoa: Myxosporea) as primary causative agents of Epizootic Ucerative Syndrome (EUS) in fishes from Beel Mahmoodpur, Faridpur, Bangladesh
P. Sánchez and M. Demestre	Catch composition of the otter bottom-trawl fishery on the Catalan coast, northwestern Mediterranean

Poster Presentations

Presenter	Title
P.C. Stephenson and M. Moran	Relating fishing mortality to fish trawl effort on the north-west shelf of Western Australia
K. Sunnane	The north-east Euro-Arctic ecosystem – successes and failures of managing multispecies and multifleet fisheries
G. Thorncraft	Assessment of rock-ramp fishways
L. Twagilimana and J. Ntivuguruzwa	Problems of fisheries in developing countries: example of degradation of the aquatic environment by Rwandese war (Africa)
T. Vassiliki	Comparative ecology of the shallow-depth productive lagoons of north and north-east Greece
R.J. Williams, C. Copeland, F. Watford, J. Hannah and V. Balgshov	Estuarine rehabilitation studies in New South Wales, Australia: from intensive to extensive

Theme 2

Presenter	Title
T. Arimoto	Use of artificial stimuli for fish harvesting and controlling purposes
J. Barker	Co-management – fact or fiction?
J. Barrington	"All quiet on the western front" – the application of war and peace strategies in fisheries management
B.L. Brown and L. Kline	Use of ecological and genetic data as a basis for multijurisdictional fisheries management
A. Bundy	Conflicts in fisheries: an integrative approach with management options
A. Coleman and L. West	Fishcount '95: an innovative design for collection of recreational fisheries data
J.C. Cooper	Environmental impact assessment and other regulatory mechanisms: approaches to fisheries protection and mitigation in Senegal and Indonesia
R. Curtotti and D. Brown	Incorporating risk in a bioeconomic analysis of the orange roughy fishery
P. Dayaratne	Bio-socio-economic approach in the assessment and management of the purse-seine fishery in the south-west coast of Sri Lanka
K.J. Derbyshire, S.R. Willoughby, A.L. McColl and D. M. Hocroft	Geographical Fisheries System (GFS): visualising fisheries data
P.G. Du Plessis	The socio-economic effects of longlining versus trawling and of exploiting bycatch in the South African economy and grassroots fishing communities
M.O. Bergh, A. Barkai and S.F. J. Dudley	The effect of netting strategy on shark attack risk: is it predictable?
C. Gomes, H.A. Oxenford and R.B.G. Dales	The use of DNA markers in determination of stock structure of the four-wing flyingfish, *Hirundichthys affinis*, and its implications for fisheries management in the central western Atlantic
D.L. Hammond	The joining of provincial government, sportfishing industry and fishery management to help conserve istiophorid stocks

Presenter	Title
L.R. Higginbottom, T.J. Pauly and M. J. Underwood	The EchoListener: a low-cost and high-resolution acoustic data logger for fishers, marine scientists and hydrographers
J.B. Higgs	Can the recreational fishing community be used to collect information suitable for fisheries management decision making?
M.L. Kangas	Review of the effectiveness of a total closure in the Gulf St Vincent prawn fishery
I.M. Kaplan	Policy, compliance and ecology: a case study of New England fishers
D.P. Karavellas	Fisheries and the conservation of biodiversity: the case of the Mediterranean monk seal *Monachus monachus* in Greece
F. Kirschbaum	Need for further study on environmental control of cyclic reproduction of tropical freshwater fishes
I.A. Knuckey and C.E. Calogeras	Good management and/or good luck? Timely introduction of input controls in the Northem Territory mud crab fishery
A.S. Hurn and A.D. McDonald	An empirical evaluation of important sources of price risk for Tasmania southern rock lobster
D.P. McPhee, G.A. Begg and D.S. Cameron	Cooperative tagging programs: scientists and anglers working together
R.D. Michael and K. Radhakrishnan	Differential immunomodulatory effect of nickel compounds in *Oreochromis mossambicus* (Peters)
M. Mukherjee	White spot disease outbreak in penaeid prawns from the probable cause to control measures
S. Nathanael and E.L.L.Silva	The role of science, economics, sociology and politics in the sustainable management of the fishery at the Victoria Reservoir, Sri Lanka
D.A. Neitzel	Setting objectives: lessons learned from Pacific northwest, United States of America
J.S. Nicolas	Database management: the application of Lotus Macro for on-farm economic analysis of rice-fish data
T.J. Pitcher, T.Hutton, A. Bundy, D. Ramm	Exploration of a multi-disciplinary taxonomy of fisheries
A.E. Punt, A.D.M. Smith	Advances in Bayesian stock assessment and decision analysis – application to orange roughy off Tasmania
K. Peterson	Assessment of the health of the marine environment using otolith microstructure in fish
B.H. Ransom, S.V. Johnson and T.W. Steig	Counting migrating juvenile and adult salmonids (*Oncorhynchus* and *Salmo* spp.) in rivers using split-beam hydroacoustics and target tracking
R. Rawlinson and R. Metzner	Structural adjustment and fisheries in Australia: sustainability and economic efficiency. Policy considerations for a national fisheries adjustment program
N. Rayns	Effective consultation and its impact on the management of eastem Australian gemfish
M. J. Roberts, W. Sauer, W. DeWet and M. van den Berg	Fisheries management: developing an environmentally driven predictive capability (EDPC) for the South African squid fishery
J. Robertson	Fisheries management in the Great Barrier Reef Marine Park

Poster Presentations

Presenter	Title
W.H.H. Sauer and M.J. Roberts	Squid fishers and scientists – creating a symbiotic environment
B. Sawynok	Sportfish tagging data and their role in a recreational fishing database in Queensland
U. R. Sumaila,	Co-operative and non co-operative exploitation of the Arcto-Norwegian cod stock
B.A. Thompson, R.L. Allen, J.H. Render and H. Blancher	Louisiana striped mullet: integration of science, fishery and management
R. J. Trumble	When science isn't enough: improving survival of discarded Pacific halibut
L.E. Williams and R.N. Garrett	Effect of feedback on commercial fisher data collection in remote-area fisheries

Theme 3

Md Islam Ali	"As nets belong to fishers, fisheries belong to them" (Jaal jar jala taar)
C. J. Augustyn.	Management of offshore marine resources in the new South Africa
A. Butcher	Development of the stout whiting fishery in southeast Queensland
D. Cameron	Management considerations for the dusky flathead fishery in Moreton Bay, Queensland, Australia
R. Castro, M. Medellin and E. Rosas	Brown shrimp (*Penaeus aztecus*) fisheries resources in the Mexican waters of the Gulf of Mexico
P. Coutin and S. Conron	The recreational catch in Port Phillip Bay, Victoria
C.R. Davies	Appropriate spatial scales for marine fishery reserves for management of coral trout, *Plectropomus leopardus*, on the Great Barrier Reef
S.T. Fennessy and J. Walsh	South African east coast trawling – what's the (by) catch?
W. Ford	The division of access in a fully exploited fishery: restructuring the Tasmanian rock lobster fishery
M.A. Kinloch and D. McGlennon	Resource allocation in the South Australian marine inshore fishery
B.D. Mapstone, R.A. Campbel, A.D.M. Smith and C.R. Davies	Design of experimental manipulations of line fishing and area closures on the Great Barrier Reef
R. O. Boyle and S. D'Entremont	Strategic planning in fisheries management – a blueprint for sustainable harvesting
R. Page	The Victorian Recreational Fishing Peak-Body
W. Palmer	Property rights in fishery resources
D.J. Passer,	Enforcing quotas in Alaska
T.J. Pitcher, A. Tautz, S. Hemphill and J. Row	The benefits of recreational fisheries: cases from Kenya, South Africa and British Columbia
D.A. Pollard	Marine harvest refugia as a tool for inshore fish population enhancement
E. Slooten	Risk and uncertainty – implications for the sustainable management of bycatch mortality of Hector's dolphin
J.T. Smith	Octopus fishing at Eaglehawk Bay: a case study of conflict and resource management

Presenter	Title
W. Sumpton, S. Jackson, R. Joyce and R. Fioravanti	Problems associated with the assessment of offshore multi-user fisheries
R.D.J. Tilzey	Allocation issues between recreational and commercial fishers: the Australian experience
C. Turnbull and R.A. Watson	Research-directed management or management-directed research? The Torres Strait prawn fishery
N.P. van Zalinge, T.S. Touch, T. Ing, L. Diep, C. Sam, S. Phem, R. Traung, S. Lieng and C. Yim	Fisheries of Cambodia: the Tonle Sap Great Lake and River ecosystem under threat
B.G. Wallner, K.J. McLoughlin and G. Johnson	Sustainable management of northern Australian fisheries resources—access and catch sharing with traditional Indonesian fishers

Theme 4

Presenter	Title
D. Bergero, M. Boccignone, F. Di Natale, G. Forneris, G.B. Palmegiano and I. Zoccarato	Intensive fishculture and its impact on the environment: the role of natural zeolites in the reduction of the ammonium content in the effluents
Nai-Hsien Chao	Can cryopreservation of sperm gametes help sustain aquaculture diversity?
S.H. Cheah and C.L. Lee	Preliminary trials on larval rearing of the Australian eel tail catfish, *Neosilurus ater* (Perugia)
Yew-Hu Chien and I Chiu Liao	Ecological consideration for the planning, implementation and practice of pond mariculture
J. DeAlteris and M. Rice	Transient gear, shellfish aquaculture, an innovative socially acceptable complement to the wild-stock capture fishery
O. Fagbenro	Biopreservation of fish by-products for aquaculture feed in tropical Africa
Chung Huu-Yun and Guang-Hsiung Kou	Review of epizootics in cultured fish in Taiwan
B. Ingram, G. Gooley, B. Larkin, S. De Silva and R. Collins	Progress towards development of culture methods for glass eels of the Australian short-finned eel *Anguilla australis*
B. Ingram and G. Gooley	Hormone-induced spawning of the threatened Macquarie perch (*Macquaria australasica*), an Australian native freshwater fish
B. Ingram, G. Gooley and B. Rae	Evaluation of two electronic juvenile fish counters
G. John	Indian aquaculture: where does it stand?
K. Kikuchi, T. Furuta, N. Iwata, H. Honda and I Sakaguchi	Intensive production of Japanese flounder with a closed recirculating culture system
Chin-Lau Kuo, Chia-Hwei Wan, Chi-Lun Wu, Shih-Tsung Hwang and I Chiu Liao	Survey of leptocephalus of Japanese eel in the western Pacific Ocean

Poster Presentations

Presenter	Title
L.J. McKinnon, G. Gooley and R.J. Gasior	A pilot evaluation of aquaculture integration with irrigated farming systems
N.P. McMeniman and N. Sands	Estimation of digestibility *in vivo* of diets fed to barramundi (*Lates calcarifer*)
P.A.G. Preece, S M. Clarke, J.K. Keesing, K.R. Messner and B.L. Foureur	Feasibility of abalone stock enhancement and rehabilitation by larval reseeding: new developments in South Australia
A. Ramachandran	Strategic planning to tide over the present crisis and sustain the brackish-water aquaculture industry in India
Chyuan-Yuan Shiau, Yu-Jane Pong, Tze-Kuei Chiou andYun-Yuen Tin	Effects of growth and starvation on the concentration of free histidine in the muscle of milkfish (*Chanos chanos*)
J.H. Soares Jr and K.P. Hughes	Improved dietary phosphorus utilization by striped bass fed phytase
Mao-Sen Su and I. Chiu Liao	Status and prospects of small abalone culture in Taiwan
P. Unprasert and H.R. Robinette	Influence of channel catfish, *Ictalurus punctatus*, size-class distribution on protein ulitization
K.C. Williams, C. Barlow and L.J. Rodgers	Improved grow-out diets for farmed barramundi *Lates calcarifer* (Bloch)
L.C. Woods III and J.H. Soares Jr	Nutritional requirements of domestic striped bass broodstock

Theme 5

Presenter	Title
N.J. Bax and A.D.M. Smith	Quantitative assessment and forecasting of Australia's orange roughy stocks
D.T. Brewer, N.J.F. Rawlinson and S.J. Eayrs	A comparison between diamond- and square-mesh codend selectivity in the northern prawn fishery of Australia
D.T. Brewer, S.J. Eayrs, R. Mounsey and Y.G. Wang	Assessment of an environmentally friendly, semi-pelagic fish trawl
D.T. Brewer, N.J.F. Rawlinson, S.J. Eayrs and J.P. Salini	An assessment of eight bycatch reduction devices in Australia's northern prawn fishery
L. Cannizzaro, M.G. Andreoli, G. Garofalo and D. Levi	Catch and abundance in areas with different exploitation rates in the Sicilian Channel
K. Danaher	Documenting fisheries habitat resources – mapping mangroves with remote sensing in Queensland, northeast Australia
R.W. Day, G.P. Hawkes, M.W. Wallace and M.C. Williams	The value of research to age abalone, and how to time-stamp them
D.J. Die, B.J. Hill and B. Taylor	Can biological research influence management decisions in a trawl fishery for tropical prawns? The Australian northern prawn fishery
J. Douglas, G. Gooley and A. Gason	Development and validation of a fish index of biotic integrity as part of an evaluation of aquatic ecosystem condition
S.J. Eayrs, R.C. Buckworth, I. Cartwright and S. Bose	Headline height modifications to improve prawn trawl performance

Presenter	Title
M. Diaz and B. Ely	Evidence for genetic differentiation among striped bass populations in eastern North America
M. Finn and M. Kingsford	Larval supply and replenishment of a coral reef fish population
J.S. Franks, J.M. Lotz, R.M. Overstreet and J.R. Warren	Age, growth and reproduction of cobia, *Rachycentron canadum*, from the northern Gulf of Mexico
H.R. Fuentes, D.J. Die, J.L. Leqata, L. Tuwai and R. Watson	Depletion experiments and estimation of abundance in reef fish stocks of Solomon Islands and Fiji
A. Harris and H. Nona	A cross-check of catch data collected by a traditional fishing community
Shann-Tzong Jiang, Bai-Lin Less and Jai-Jaan Lee	Effects of cathepsin B on the disintegration of minced mackerel
D. Kemp and G. Gooley	Development and validation of an index to evaluate the condition of aquatic habitat
T.C. Kline	Natural stable isotope abundance used for assessment of anadromous and amphidromous migrations in fish ecology: implications for fisheries management and habitat protection
J.D. Koehn	The key criteria to sustaining the wild-stock Murray cod fishery in Lake Mulwala
S. Lavery	Molecular genetic markers in discriminating stocks of Australian penaeid prawns
Hsi-Chang Lui and Chien-Chang Hsu	The management implication for the Indian albacore stock when production model analyses result in discrepancies in estimated parameters
N.R. Loneragan, M.D.E. Haywood, D.S. Heales, R.A. Kenyon, R.C. Pendrey and D.J. Vance	Structure of seagrasses and their carrying capacity for juvenile penaeid prawns: studies of the effects of prawn density and predation
Y. Matsushita, Y. Inoue and K. Nojima	Multi-species separation in the Japanese coastal trawl fishery
L.J. McKinnon, G.J. Gooley and R.J. Gasior	Assessment of the potential of glass eel resources in south-eastern Australia for commercial aquaculture
R. Mohan	Spiny lobster fisheries and management concerns in Sultanate of Oman
J.L. Nielsen, L. Dijkstra, M. Unwin and R. McDowall	Genetic origins of rainbow trout and chinook salmon transferred to New Zealand from California at the turn of the century
G. Nowara and T.I. Walker	Effects of time of night, depth and jogging method on catch rates and size of squid off Victoria
T. Pauly and S. Nicol	Hydroacoustic methods and results of krill stock surveys for three consecutive years in east Antarctica
C.R. Pitcher, T.D. Skewes and D.M. Dennis	Research for management of the ornate rock lobster fishery in Torres Strait
T.J. Pitcher and A. Bundy	A new family of empirical models for the potential yield of lake fisheries

Poster Presentations

Presenter	Title
R.A. Watson and T.J. Quinn II	Performance of transect and point-count underwater visual census methods for reef fish assessments
M.A. Rimmer, P.F. de Guingand, G.M. Meikle, B. Franklin, B.D. Paterson, L.G. Anderson, A. Thomas, J. Handlinger and B.K. Hughes	Development of improved techniques for the transport of live finfish
J.B. Robins, R. Mounsey, J. McGilvray, R. Buckworth and J. McCartie	Trawl industry considers new AusTED design
C.B. Schiller, P. Brown and P.C. Gehrke	Population size-structure and recruitment of freshwater fish species in the Murray–Darling River system, Australia
C.A. Simpfendorfer, R.C.J. Lenanton and K.J. Donohue	Gauntlet fisheries for large, long-lived sharks: approaches to research and management
K. Smith	Importance of physical oceanographic processes in larval recruitment
I. Somers and You-Gan Wang	A bioeconomic analysis of seasonal closures in Australia's multispecies northem prawn fishery
I. Suthers	How do nutrients affect marine food chains?
T. Tanabe and M. Ogura	Study on the early life of skipjack tuna, *Katsuwonus pelamis*, in the tropical western Pacific
R.J. Treble	Stock assessment of southern rock lobster *Jasus edwardsii*) at Apollo Bay, Victoria, Australia
V.S. Troynikov	Estimation and prediction of growth heterogeneity in the context of fish stock assessment problems
D. Vance, M.D.E. Haywood, D.S. Heales, R.A. Kenyon, N.R. Loneragan and R.C. Pendrey	How far do prawns and fish move into mangroves? Distribution of juvenile banana prawns, *Penaeus merguiensis*, and fish in a tropical mangrove forest in northern Australia
J. Virgona and I. Halliday	Assessment of stocks of sea mullet (*Mugil cephalus*) in New South Wales and Queensland waters
V. Wadley	Squid stocks in south-east Australian waters
Yisheng Wang and Ming Qi	Change of lipid in dried mussel during storage
K. Warburton, S.J. Blaber and K. Williams	Applying bioenergetic models to fisheries management: a simulation of prawn consumption by juvenile barramundi
R. D. Ward, N.G. Elliott and P.M. Grewe	Comparison of molecular genetic techniques for assessing the global populations structure of yellowfin tuna
W. Whitelaw, R. Campbell and J. Gunn	Fishing characteristics of tuna longlines – from theory to practice
Y. Yamaguchi	The falling velocity of cuttlefish eggs in water
Y. Yamaguchi, H. Nishinokubi and T. Yamane	The pot fishery for cuttlefish (*Sepia esculenta*) in Nagasaki, Japan
S.Y. Yu	Utilization of *Leiognathus equulus*, a low-value fish species in fishball processing.

THEME 6

Presenter	Title
J. Boje and H. Siegstad	Management regimes for the main Greenland fishery resources
M. Camhi	Fishing in Darwin's paradise: can the Galapagos Islands survive commercial export fisheries?
D. Campbell	Role of economic performance indicators in ensuring fish resource use management objectives
J. Chesson and B. Phillips	Evolution of a fishery assessment process
R. Chuenpagdee and K. Juntarashote	New directions in research, policy and management of Thai fisheries
A. Clarke, L. Johns and R. Coles	The impact of urban drain management on fisheries habitat in north Queensland, Australia
A. Coan	The US purse seine fishery for tropical tunas in the western Pacific Ocean: an example of responsible management
A.J. Conides and S.D. Klaoudatos	Effects of bottom sediment type, salinity and temperature on the dynamics of 0-fry population of European sea bass, *Dicentrarchus labrax* (Linnaeus, 1758) used as seed for aquaculture
M. Cosgrove and A.J. Courtney	Population dynamics and per-recruit analyses of scyllarid lobster (*Thenus* spp.) from coastal waters of Queensland, Australia
G.E. Fenton, D.C. Smith, B.D. Stewart and R. Chisari	Fish ageing: validation by radiometric analysis for oreo dories (Family Oreosomatidae)
N.J. Fowler and H.G. Brayford	Co-operative management: a fisheries ecosystem approach to jurisdiction
W.W. Fox, Jr.	Sustainability and living marine resource management in the United States of America
S.M. Garcia	Chronicles of world fishery landings (1950–1994): trend analysis, fisheries potential and outlook
J.H. Harris, S. Hartley and A. Bruce	Towards sustainable management of NSW riverine fish: developing knowledge on diversity, distribution and biotic integrity
D.K. Hobday, R.A. Officer and G.D. Parry	Changes to fish communities in Port Phillip Bay, Victoria, Australia, over two decades: 1970–1991
Y. Inoue, Y. Matsushita and F. Chopin	Conservation technology research in Japanese coastal fisheries
D.W. Japp	Longlining or trawling – managing change in the demersal fisheries of a new South Africa
G. Johnson	Giving an old fishery a new lease of life
C.P. Keenan, D. Mann, S. Lavery and P. Davie	Genetics and morphology distinguish three species of mud crab, genus *Scylla*
Y. Kuronuma	Key conditions for community-based fisheries management: a case study on self-imposed management in Alfonsino fishing ground off Katsuura, Japan
S.J. Lamberth, B.Q. Mann, S. Brouwer and W. Sauer	An evaluation of fishers' perception, and management, of the South African marine linefishery

Poster Presentations

Presenter	Title
R. Lea	The jurisdictional arrangement applying to internal, inshore and offshore waters of Australia in to the management of renewable living resources and their habitats
J. Lindholm, M. Ruth and L. Kaufman	The biology of yearling groundfish on Georges Bank: a spatially explicit dynamic model and its policy implications
J. Lucy, S. Ayvazian and M. Malchoff	Quantifying hook-release mortality in marine recreational fisheries: working toward more effective fishery management
C.J. Lupton, M.J. Heidenreich and S.G. McKinnon	Fisheries resources assessment of coastal rivers in the Wide Bay–Burnett region of Queensland, Australia – a base-line for fisheries management
K.R. Matthews	To manage fisheries sustainably, new philosophies are required
K.J. McLoughlin and G. Newton	Australian fisheries resources – the status of Commonwealth-managed fisheries
A.D. Tucker, J.B. Robins and D.P. McPhee	Comparing the introductions of trawl excluders and bycatch excluders in Australia and the USA
J. Meehan	Modern fisheries research vessels
M.Z. Mtsambiwa	New approaches to fisheries management in Lake Kariba (southern Africa)
B.R. Murphy, W.D. Harvey and D.W. Fonticiella Garcia	Freshwater fisheries management conflicts in Cuban reservoirs
Thouk Nao and Mahfuzuddin Ahmed	Role of local fisher communities in the sustainable management of the Great Lake fisheries, a case study in Kompong Khleang community, Sot Nikum district, Siem Reap Province, Cambodia
M. Narayanan, W. Thilagavathy, A.J. Thatheyus and A. Jayaseeli	Effect of size and density on the rates of ammonia excretion in freshwater prawn *Caridina weberi*
M. Narayanan, M.E. Poline, A.J. Thatheyus and B. Victor	Acute toxicity of mercuric chloride on respiratory metabolism and tissue pathology of the freshwater prawn, *Caridina weberi* (de Man)
B. Pease	The headaches and benefits of compiling a historial summary of industry-based fisheries statistics
F.R. Perez and E. Batungbacal	Sustainable coastal-area development: a framework for sustainable fisheries
R. Pokrant, S. Rashid, P. Reeves, J. McGuire and A. Pope	Traditional fishers and the "new fisheries management policy", Bangladesh, 1986–96
P. Pownall	Management of Australia's northern prawn fishery: a delicate balancing act
J.D. Prince and R.J. Ruiz-Avila	Spatiality, stock assessment and property rights in Indo-Pacific fisheries
S.G. Robertson, A.K. Morison and D.C. Smith	The Central Ageing Facility: production ageing for stock assessment
C. Safina	Decline of world fisheries
Z. Sary, H.A. Oxenford and J. Woodley	Responses of an over-exploited Caribbean trap fishery to the introduction of a larger mesh size
A.M. Schofield	The Natal Parks Board and small-scale fisheries along the coast of KwaZulu-Natal

Presenter	Title
M.J. Smale and L.J.V. Compagno	Conservation and management of chondrichthyans
L. Teirney, T. Sylvester and M. Cryer	The development of techniques for estimating the recreational harvest of marine fish species in New Zealand
P.N. Trathan, P.G.K. Rodhouse, E.M.C. Hatfield, E.J. Murphy and F.H.J. Daunt	Use of a marine GIS to examine variability in squid fisheries over the Patagonian Shelf in relation to remotely sensed oceanographic variables
T.I. Walker	Can shark stocks be harvested sustainably? – a question revisited
W. Zacharin	Management arrangements for developing fisheries: the precautionary principle and use of adaptive management strategies